AF334047

INDUSTRIAL ELECTRONICS SERIES

THE
INDUSTRIAL INFORMATION TECHNOLOGY
HANDBOOK

INDUSTRIAL ELECTRONICS SERIES

THE
INDUSTRIAL INFORMATION TECHNOLOGY
HANDBOOK

Edited by

Richard Zurawski

ISA Corporation
South San Francisco, CA

CRC PRESS

Boca Raton   London   New York   Washington, D.C.

Library of Congress Cataloging-in-Publication Data

The industrial information technology handbook / edited by Richard Zurawski
      p. cm. — (Industrial electronics series)
Includes bibliographical references and index.
ISBN 0-8493-1985-4 (alk. paper)
1. Information technology—Handbooks, manuals, etc.
2. Automation—Handbooks, manuals, etc. I. Zurawski, Richard. II. Series.

T58.5.I513 2004
670.42'7—dc22   2004051919

Visit the CRC Press Web site at www.crcpress.com

To my wife, Celine

Foreword

Extending the Reach of Automation to Achieve Continuous Productivity Improvements

Automation is largely about productivity improvements, and distributed control systems have been deployed extensively around the world to this end. While the systems were installed to reduce process variability, increase plant availability and automate tasks, the improvements were achieved by focusing primarily on process control. In the 1990s, the open control systems (OCS) were introduced. These had been developed to "open up" the control system environment, making information from the plant floor available to other applications and facilitating interfacing with applications assigned principally to keep the process running smoothly. They were the first such systems to feature major incorporations of "off-the-shelf" technology. However, the focus was still primarily on controlling the process.

All that has now changed. Today, to deliver additional productivity gains, it is necessary to extend the reach of the automation system beyond the world of process control.

Utilizing Data More Intelligently

As mentioned, consistent and predictable process control was achieved in the past with traditional DCS/OCS functionality — operations, engineering, and logic. The operations functions were focused on the needs of the production operator. Interfaces to applications and devices external to the control system with their own data storage typically were collected and duplicated in the control system database to provide accessibility for reporting and operational needs. This approach was found to be difficult to engineer and maintain. The practice of having information stored in multiple locations made it difficult to ensure data integrity as the individual software applications were advanced at asynchronous paces and the customized interfaces between them required frequent maintenance. The increasing deployment of peripheral applications related to productivity improvement, such as quality systems, advanced control, production management, information management, and maintenance management, resulted in different user interfaces for different types of information sources. More recently, the availability of "smart" field devices has vastly increased the amount of data available to improve productivity in a plant. However, each device has its own data collection and communication environment, adding to the complexity of the integration task.

Hence, the data are there, but can an enterprise be sure that they are made available to the right individual, at the appropriate time and in a format that is relevant? The data must be analyzed, and action must be taken to ensure maximum production asset availability, optimal quality, and predictable and appropriate performance — ever-increasing productivity. One of the most consistent demands voiced by end-users is the need to get the right information to the right person to facilitate sound business decisions and appropriate action.

Time for Automation Systems that Can Provide Extended Automation

The systems of today and the future have to have an architecture moving the automation to a new level. From an information access and integration perspective, such systems incorporate functionality far beyond process control.

Specifically, systems have to incorporate process control, production management, safety, discrete logic and sequence control, advanced control, information management, smart instrumentation, smart drives, asset management, and document management capabilities in a singular virtual database environment. Such an integrated system environment allows incorporation of any application solution from any vendor. Consequently, it will be possible to access data directly from its source in the context of the production asset without needing to know where the data come from, and without concern about data integrity and concordance. This elegant solution addresses the engineering and maintenance issues of past solutions mentioned above.

Putting Operations within Reach of Different Stakeholders

From an operations perspective, the modern Human–Machine Interface (HMI) system shall provide a single, consistent HMI intuitive to access and interact with information from all the applications encompassed within an extended automation environment. It allows users of all disciplines who impact production to organize information and navigate throughout the system in the context of their job function. The HMI shall be designed to provide a work environment that helps users to identify events occurring in the process that are relevant to their job function in real-time, while they are happening. This is achieved by customizing the information that is displayed — and, just as importantly, the information that is *not* displayed — as well as the way it is presented for a given process area, according to the job function of the person using the system. As a consequence, the root cause of events can be quickly and intuitively analyzed, and the best course of action can be determined, by virtue of the system's integration and its ability to link to the process object all aspects of information provided by the various system applications.

How Modern Automation Systems Can Improve Productivity

A good example of how a modern automation system will positively affect productivity can be found in the area of plant asset availability and performance. Two methods of instrument maintenance are prevalent today in the industry: *preventive* and *corrective* maintenance. Preventive maintenance is a scheduled activity, based on experience or recommendations designed to reduce or eliminate the possibility of a failure that could cause a production stoppage. Corrective maintenance is carried out "after the fact," that is, when a device has failed in operation and requires immediate repair.

Failures are rare today, since most of the field instruments operating within their design range have a Mean Time Between Failure (MTBF) of 5 years and more. Rearranging and drift require manual intervention, but this need is also greatly reduced when fieldbus and digital transmitters are used. Pressure transmitters, for example, are essentially trouble-free devices. Nevertheless, there is always the possibility of unstable process conditions and unexpected process events significantly affecting device performance. Extended operation in stress conditions will, over time, affect the measurement accuracy. If these events go undetected, the device could continue to send wrong readings to the control system until the next scheduled maintenance, which might be months away.

To prevent this from happening, all pressure transmitters are checked on a regular basis. However, this exercise is wasteful as well as costly; in more than 60% of the scheduled interventions, the result is "negative — no failure found!" That this has an adverse effect on productivity hardly needs mentioning.

Smart Field Devices and their Maintenance

By tying together smart field devices with documentation, asset optimization, and maintenance management functions, a modern automation system ensures that events leading to performance degradation and failures are detected, and that the time between detection and notification, decision, and action is reduced to a minimum.

Today's smart field devices store a wealth of data that can be used to assess their health and performance. The emergence of fieldbus, Fieldbus Device Tools (FDT), and Device Type Manager (DTM) standards has meanwhile made these data available to control systems and other asset management applications.

A modern automation system provides all the capabilities needed for continuous condition monitoring and for streamlining the maintenance process. Continuous monitoring of asset conditions allows preventive maintenance strategies to be replaced by predictive maintenance in response to warnings of degraded asset performance before the asset can fail, and focuses maintenance activity on those devices requiring it. This significantly reduces the number of unnecessary device checks in the field.

Most of these issues and developments are covered in *The Industrial Information Technology Handbook.*

Christer Ramebäck
Vice President
ABB Automation Technologies
Västerås, Sweden

Preface

Aim

The purpose of *The Industrial Information Technology Handbook* is to provide a reference useful for a broad range of professionals and researchers from industry and academia involved or interested in the use of information technology (IT) in industrial applications ranging from industrial automation to industrial enterprise integration. This is the first publication to cover this field in a cohesive and comprehensive way. The focus of this book is on existing technologies used by the industry, as well as newly emerging technologies and trends, evolution of which has been driven by the actual needs and by the industry-led consortia and organizations.

Field of Industrial IT

Over the last decade, IT has had a profound impact on the evolution of industrial enterprises toward fully integrated entities. Nowadays, we witness major efforts led by some of the largest multinationals to harness IT to achieve integration of I/O control, device configuration, and data collection across multiple networks and plant units; seamless integration between automation and business logistic levels to exchange jobs and production data; transparent data interfaces for all stages of the plant life cycle; the Internet- and Web-enabled remote diagnostics and maintenance, as well as electronic orders and transactions. Some of the IT technologies used in the office and enterprise operation have been adopted and/or transformed to suit and advance industrial controls and automation, and industrial enterprise integration. OPC, Real-Time Corba, Real-Time Linux, Windows CE, real-time specifications for Java (RTSJ, RTCE), and Real-Time UML are good examples, to mention some. The Internet technologies extended the boundaries of the operation of industrial enterprises to their customers and suppliers, as well as the management and technical staff through electronic commerce, B2B, remote data access, and monitoring and control. The industrial IT field spans a number of technological areas such as software and hardware technologies, as well as Web and networking technologies. Those technologies, when used in an integrative way, offer a potential for horizontal and vertical integration of functional layers of industrial units and enterprises, thus transforming traditional islands of automation and enterprise operations into more integrated enterprises better adjusted to cope with the demands of competitive markets.

Contributors

The book contains 112 contributions on topics related to industrial IT, written by experts from industry and academia. One third of the contributions are from industry and industrial research establishments; from the leading multinational corporations at the forefront of the development of IT and industrial automation technologies, to mention ABB (Germany, Norway, Sweden, and Switzerland), Alcatel (Belgium), Cadence Systems, Microsoft Corporation, NEC Labs (U.S.), Rockwell Automation, Schneider Electric (France and Germany), Siemens (Austria and Germany), Volvo Truck Corp. (Sweden), and

Yokogawa America. Most of the multinationals mentioned play a leading role in the formulation of long-term policies for technology development, and are key members of the industry–academe consortia implementing those policies.

The contributions from academia and governmental research organizations are represented by some of the most renowned and reputable institutions such as Columbia University, Cornell University, Frauenhofer FOKUS (Germany), Georgia Institute of Technology, Monash University (Australia), National Institute of Standards and Technology (U.S.), Princeton University, Politecnico di Torino (Italy), UC Berkeley, UC Irvine, UC San Diego, University of Texas at Austin and Dallas, University of Tokyo, Stanford University, Technical University of Berlin, Vienna University of Technology, and many others.

How Written

The material is presented in the form of tutorials, surveys, and technology overviews combining fundamentals with advanced issues, making this publication relevant to the beginner as well as seasoned professional from industry and academia. Particular emphasis is on the industrial perspective, illustrated by actual implementations and technology deployments. The contributions are grouped into sections for cohesive and comprehensive presentation of the areas treated. Some of these sections can be used as a reference material for study (including a formal coursework) at the intermediate through advanced levels. The reports on recent technology developments, deployments, and trends frequently cover material released to the profession for the first time.

Audience

The handbook is designed to cover a very wide range of topics that comprise the field of industrial IT. The material covered in this volume will be of interest to a wide spectrum of professionals and researchers from industry and academia, as well as graduate students, from the fields of industrial and mechatronic engineering, production engineering, electrical and computer engineering, computer science, and IT.

Organization

The book is organized into two parts. Part 1, *Fundamentals of Information Technology*, presents material to cover new and fast evolving aspects of IT. Part 2, *Industrial Information Technology*, introduces cutting-edge and newly emerging areas of industrial IT.

The focus of the book is on fast evolving areas with a major impact on the evolution of industrial automation and industrial enterprise integration. Some of those areas have received limited coverage in other publications due to the fast evolution of the technologies involved, or material confidentiality, or limited circulation in the case of industry-driven developments. The areas covered in the book include industrial communication technology, sensor technology, embedded systems, the Internet, Web and IT technologies in industrial automation, and industrial enterprise integration. To complement this material, it was felt appropriate to include background reading material on some of the fastest evolving areas of IT and IP networking. This has been done primarily to assist readers with little or no background in IT and networking technologies, both essential to understand most of the chapters in Part 2 of the book.

Part 1 has two sections: *Computer Software and Web Technologies* and *The Internet and IP Networks*. The section on *Computer Software and Web Technologies* introduces some of the recent software technologies, platforms, and solutions that have had a profound impact on industrial control and automation, and industrial enterprise integration. It covers material in subsections: *Development Platforms and Frameworks, The Unified Modeling Language, Middleware, Web Technologies, Web Programming,* and *Multidimensional Databases.* Specifically, this section introduces J2EE platform and .Net framework in the *Development Platforms and Frameworks* subsection. J2EE is used extensively at the enterprise business level. The .Net is studied and experimented with by the industry for applications ranging from control to enterprise integration. An overview of UML and its extensions, variants, and implementations is provided in the subsection *The Unified Modeling Language.* The UML language has become a *de facto* standard for modeling real-time and embedded systems, industrial control and automation, as well as business systems. The contribution on UML also introduces selected applications of the language in control and automation. The *Middleware* subsection, on network connectivity software, gives a roundup of Microsoft's distributed components technologies, and Object Management Group Corba standard for object-oriented distributed computing, both of central importance to industrial control and automation. The Web servers, clients, and browsers, as well as languages used for Web and the Internet programming are introduced in the *Web Technologies* subsection. The client–server model is dominant in legacy automation systems. Web services, which allow for platform and language independence in the client–server model, are discussed in the *Web Programming* subsection. The *Multidimensional Databases* subsection gives an overview of multidimensional databases, which is the key technology for the analysis of data, and is essential for effective operation of automated industrial enterprises.

The section *The Internet and IP Networks* provides a comprehensive overview of the Internet and IP network technologies. These technologies have had a major impact on the vertical integration of functional levels of industrial enterprises and the operation of the business level through a variety of e-technologies. This section provides a handy reference material useful for the study of the remaining sections.

The topics covered are organized in the following subsections: *Introduction to the Internet, Internet Core Protocols, Quality of Service in IP Networks, Internet Applications and Application Services, Management of IP Networks, Network Security,* and *Ad Hoc Networking.* Over the last decade, the Internet and IP network technologies have become the basis and catalyst for the transformation of industrial enterprises from so-called "islands of automation" into relatively well-integrated, both horizontally and vertically, modern industrial entities. The section offers a comprehensive overview of the Internet technologies. The contributions introduce fundamentals, as well as present new developments, making the material attractive to both novices and advanced users of those technologies.

A general overview of the Internet is given in the *Introduction to the Internet* subsection. Core protocols and IP routing are covered in the *Internet Core Protocols* subsection. This subsection also includes contributions on multicast, congestion control and Quality of Service (QoS), mobile IP routing, and IP-mobility for cellular and wireless systems. The subsection *QoS in IP Networks* gives an overview of QoS in IP networks and introduces the main features of Multiprotocol Label Switching architecture that integrates label swapping forwarding with network layer routing. It also presents material on the Internet Integrated Services (IntServ) architecture and Reservation Protocol (RSVP), and Internet protocols for real-time Applications (RTP, RTCP, RTSP). Mail transfer protocols (SMTP, POP, IMAP), file transfer protocol (FTP), and hypertext transfer protocol (HTTP) are covered in the *Internet Applications and Applications Services* subsection. Selected aspects of IP networks management are presented in the *Management of IP Networks* subsection. This includes Simple Network Management Protocol (SNMP) and Dynamic Host Configuration Protocol (DHCP). Network security and firewalls are introduced in the *Network Security* subsection. The contribution on *Ad hoc* networking concludes the section on *The Internet and IP Networking.*

Part 2 of the book begins with a section on *Industrial Communication Systems.* Industrial data networks play a pivotal role in the automation and integration of the factory floor by collecting process data and distributing control decisions. This is a very fast evolving area of technology with Ethernet becoming a *de facto* industry standard for communication in factories and plants at the fieldbus level. The random and native CSMA/CD arbitration mechanism is being replaced by other solutions allowing for deterministic behavior required in real-time communication to support soft and hard real-time deadlines. The wireless and mobile communication technologies have already been recognized as a viable alternative for data communication on the factory floor, with the number of reported actual deployments on the rise. This necessitates the development of new solutions for the integration of wireline and wireless fieldbuses. The requirement for the process data availability via remote access prompted research and creation of new solutions for linking factory floor with the Internet.

The material in this section is organized in the following subsections: *Introduction to Data Communication Networks, Field Area Networks, Ethernet and Wireless/Mobile Network Technologies, Linking Factory Floor with the Internet* and *Wireless Fieldbusses, Security and Safety Technologies in Industrial Networks,* and *Automotive and Industrial Applications.* The first section begins with an introduction to data communication networks to cover principles of lower-layer protocols for data communication to be followed by material on wireless local area networks (WLAN) and wireless personal area networks (WPAN) is presented. The next subsection, *Field Area Networks,* contains a comprehensive technical overview of some of the most widely used fieldbuses in the industry. The contributions describe and evaluate Profibus, WorldFIP, Fundation Fieldbus, CAN, EIA-709, time-triggered communication networks, and IEEE 1394. The choice of networks and selected issues related to networked control systems are also covered. Ethernet is fast becoming a *de facto* industry standard for communication in factories and plants at the fieldbus level. The random and native CSMA/CD arbitration mechanism is being replaced by other solutions allowing for deterministic behavior required in real-time communication to support soft and hard real-time deadlines. These issues and various solutions, including those proposed and already adopted by the industry, are discussed in the *Ethernet and Wireless/Mobile Network Technologies* subsection. The subsection also provides a comprehensive overview of the issues involved in using wireless communication on the factory floor, ranging from the reliability and timeliness of lower-layer wireless protocols to suggestions for new MAC protocols. Bluetooth is also covered. The need for remote access to factory floor data calls for connecting factory floor with the Internet. The solutions are discussed in the *Linking Factory Floor with the Internet and Wireless Fieldbuses* subsection. The subsection also presents architectures and solutions for linking wireline and wireless fieldbuses, a new trend resulting from the presence on the factory floor of devices using wireless communications such as wireless sensors, for instance. The security and safety issues in industrial communication networks are presented in the *Security and Safety Technologies in Industrial Networks* subsection. An implementation of safety technologies is presented in a case study describing Profisafe technology used with Profibus. The *Automotive and Industrial Applications* subsection presents a case study introducing and evaluating the use of different networks in automotive applications, specifically in the range of Volvo commercial automotive products. This subsection also introduces and illustrates the use of the Manufacturing Message Specification (MMS) protocol, an OSI application layer messaging protocol designed for the remote control and monitoring of devices such as Remote Terminal Units (RTU), Programmable Logic Controllers (PLC), Numerical Controllers (NC), or Robot Controllers (RC).

The use of the Internet, Web, and IT technologies in industrial control, automation, and design is presented in the section *The Internet, Web, and IT Technologies in Industrial Automation and Design.* This is a vast area and reporting has had to be limited to selected topics where the impact of these technologies was most notably felt. This section also includes a contribution on IT security in automation systems.

The contributions are organized into the following subsections: *Internet and Web-Based Technologies in Industrial Automation, Component Technologies in Industrial Automation, Java Technology in Industrial Automation, Standards for Programmable Logic Controllers and System Design, Virtual Reality in Design and Manufacturing,* and *Security for Automation Systems.* The issues, solutions, and technologies involved in remote monitoring and control, as well as maintenance, an aim for industrial enterprises and urban automation, are presented in the *Internet and Web-Based Technologies in Industrial Automation* subsection. This subsection also covers material on telemanipulation over the Internet, and material on the use of IT technologies. OLE for Process Control (OPC), the standard interface for access to Microsoft`s Windows-based applications in automation, is one of the most popular industrial standards among users and developers of Human Machine Interface (HMI), Supervisory Control and Data Acquisition (SCADA), and Distributed Control System (DCS) for PC-based automation, as well as Soft PLCs. OPC is discussed in detail in the subsection on *Component Technologies in Industrial Automation.* (A comprehensive evaluation of Corba technology, and implementations can be found in a contribution on OCEAN in the section on *Integration Technologies.*) An overview of Java technology, real-time extensions, and prospects for applications in controls and industrial automation are given in the *Java Technology in Industrial Automation* subsection. This contribution provides a roundup of two different real-time extensions for the Java language: Real-Time Specification for Java (RTSJ) developed by the Real-Time for Java Expert Group under the auspices of Sun Microsystems and reference implemented by TimeSys Corp, and Real-Time Core Extensions developed by Real-Time Java Working Group operating within J-Consortium. The contribution also introduces Real-Time Data Access (RTDA) specification developed by the Real-Time Data Access Working Group (RTAWG), also operating within J-Consortium. The RTDA specification focuses on API for accessing I/O-data in typical industrial and embedded applications. The subsection on *Standards for Programmable Logic Controllers and System Design* introduces three standards: IEC 60848 Ed. 2, IEC 61131-3, and IEC 61499. The IEC 61499 standard defines a reference architecture for open and distributed control systems, which provides the means for compatibility between automation systems of different vendors. Virtual reality has become an important tool for the design and prototyping of products and complex systems. The subsection *Virtual Reality in Design and Manufacturing* contains two contributions introducing selected aspects of haptic simulation in design and manufacturing, and the use of virtual reality techniques in the design and validation of factory communication systems. The topic of IT security in automation systems is thoroughly explored at the end of the section. The IT security in automation systems poses a particular challenge given the requirement for remote control and manufacturing.

The current trend in the industry for flexible and distributed control and automation has accelerated the migration of the intelligence and control functions to the field devices, particularly sensors and actuators. The increased processing capabilities of those devices were instrumental in the emergence of a trend for networking of field devices around industrial data networks. The benefits are numerous, including increased flexibility, improved system performance, and ease of system installation, upgrade, and maintenance. The increased functional complexity of those devices, however, has pushed their development into the realm of formal design methods and tools characteristic of embedded systems. Another trend in networking of field devices has emerged recently: the wireless sensor networks. A number of deployments on the factory floor have been reported. In this context, the integration of wireline and wireless technologies becomes an important issue. Recent advances in the research on "intelligent space," which initially targeted an office environment, have helped enhance navigation capabilities of industrial autonomous and mobile robots, which are an integral element of automated factories. Some of those issues and trends are presented in the *Intelligent Sensors and Sensor Networks* section.

The material is structured into the following subsections: *Intelligent Sensor and Device Technology, Intelligent Sensors in Robotics, Sensor Systems and Networks,* and *Multisensor Data Fusion.* The *Intelligent Sensor and Device Technology* subsection provides an in-depth overview of the technologies and standards behind intelligent sensors and actuators. It begins with material on the IEEE 1451 standards for connecting sensors and actuators to microprocessors, control and field networks, and instrumentation systems. The standards also defined the Transducer Electronic Data Sheet (TEDS), which allows for the self-identification of sensors. The IEEE 1451 standards facilitate sensor networking, a new trend in industrial automation, which, among other benefits, offers strong economic incentives. The networking issues and supporting integration technologies are discussed in the next contribution, which also outlines evolution of the intelligent device (sensors and actuators) networking concepts. Intelligent sensors, in addition to performing measurements and sensory information processing, embed complex functionalities essential for their operation and communications. The use of semi/formal design techniques and supporting tools, essential for a rapid and cost-effective design, is discussed in the subsection along with the introduction to the CAP modeling language, which allows for rapid prototyping of intelligent devices. The robotic technology has become an integral part of industrial automation. An overview of the fundamental sensor technologies in robotics is presented in the *Intelligent Sensors in Robotics* subsection. The material covers vision, tactile sensing, sense of smell, and ultrasonic sensors. The subsection *Sensor Systems and Networks* presents contributions on intelligent space and sensor networks, and development methodologies. The intelligent space concept provides solutions, among others, for location of mobile robots nowadays frequently operating autonomously on the factory floor. Distributed wireless sensor networks is a relatively new and exciting proposition for collecting sensory data in the industrial environment. The design of this kind of network poses a particular challenge due to limited computational power and memory size, bandwidth restrictions, power consumption restriction if battery powered, communication requirements, and unattended mode of operation in case of industrially hostile environments, to mention some. The subsection provides a comprehensive discussion of the design issues related to, in particular, self-organizing networks, to include Medium Access Control (MAC) layer protocols, routing protocols, security, location determination, and power management. It also discusses existing software solutions and systems supporting the development of the dynamic wireless sensor networks. This presentation includes middleware, operating systems and execution environments, languages, and whole software frameworks. The section concludes with a subsection and material on *Multisensor Data Fusion*.

The advances in the embedded systems particularly in design methodology and supporting tools, allowed for cost-effective migration of the intelligence and control functions into field devices. This, coupled with developments in other areas, promoted a shift toward decentralized architectures of industrial automation systems, and whole enterprises. The embedded systems is one of the fastest growing areas of technology, and crucial for the realization of fully automated plants and factories. The *Real-Time Embedded Systems* section gives a comprehensive overview of the field: basics and emerging trends. It covers design aspects, languages and tools, security, real-time and embedded operating systems, and networked embedded systems.

The contributions are organized in the following subsections: *Embedded Systems, Security in Embedded Systems, System-on-Chip and Network-on-Chip Design,* and *Networked Embedded Systems.* The fundamental and advanced aspects of embedded systems and design are presented in the subsection *Embedded Systems.* This subsection contains material introducing comprehensively real-time systems, design of embedded systems, models of computation, languages for embedded systems to include hardware-level design and verification languages, design of software for embedded systems, real-time operating systems, and power-aware embedded computing. The security in embedded systems is presented in the *Security in Embedded Systems* subsection. The material in this subsection gives a roundup of the security issues and solutions adopted by the industry. The *System-on-Chip and Network-on-Chip Design* subsection provides a comprehensive overview of the issues involved in the design of systems and networks on chip. The material in this subsection covers system-on-chip and network-on-chip design, platform-based and derivative design, hardware/software interfaces design for systems-on-chip, and on-chip networks. Networked embedded systems are introduced in the *Networked Embedded Systems* subsection.

The last decade has witnessed an accelerated evolution of industrial enterprises. This trend has been driven by financial factors, and availability of practically proven and affordable technologies and solutions. The advances in the embedded systems design and falling fabrication cost allowed for deployment on the factory floor of a new generation of smart and relatively inexpensive field devices with increased functionality. The trend for networking smart field devices using industrial data networks has transformed control architecture from centralized into distributed and decentralized. This, in turn, allowed for efficient integration of I/O control, device configuration, and data collection across multiple networks and plant units. In this context, the control network, to date largely used to integrate devices at the fieldbus level, attained a new integrative role. Other equally profound changes of an integrative nature include seamless integration between automation and business logistic levels to exchange jobs and production data; transparent data interfaces for all stages of the plant life cycle; the Internet- and Web-enabled remote diagnostics and maintenance, as well as electronic orders and transactions. This integration has been achieved through an interplay of different technologies and solutions. The IT technologies have played some of the most important roles in this process. This section shows a small cross-section of the integration efforts and results, primarily arising from activities of major control and automation vendors and multinational groupings. These efforts have a potential for lasting solutions and new technologies.

The material is organized in the following subsections: *E-Technologies in Enterprise Integration, IT Technologies in Enterprise Integration, Network-Based Integration Technologies, Agent-Based Technologies in Industrial Automation,* and *Industrial Information Technology Solutions for Energy and Power Systems.* An introduction to e-manufacturing is presented in the *E-Technologies in Enterprise Integration* subsection. This material gives an overview of the e-manufacturing strategies, fundamental elements, and requirements to meet the changing needs of the manufacturing industry in transition to an e-business environment. It covers e-manufacturing, e-maintenance, e-factory, and e-business. The subsection *IT Technologies in Enterprise Integration* contains four contributions discussing the use of XML and Web Services, and other IT technologies in enterprise integration. The first contribution deals with replacing paper flow by electronic technologies, and presents the use of XML as a means for information exchange during the design and development of automation systems. The next contribution introduces the World Batch Forum's (WBF) Business To Manufacturing Markup Language (B2MML), which is a set of XML schemas that are based upon the ISA-95 Enterprise-Control System Integration Standards. The material demonstrates how B2MML can be used to exchange data between the business/enterprise and manufacturing systems. The contribution also gives a roundup of the ISA-95 standard. The Web Services technology provides the means for the implementation of open and platform-independent integrated automation systems. The challenges for using Web Services in automated systems, solutions, and future trends are discussed in the third contribution. The final contribution in this subsection presents IT-based connectivity solutions for interfacing production and business systems. The presented concepts and architectures are a result of an extensive study and prototyping efforts conducted by ABB in the search for cost-effective approaches leveraging existing mainstream technologies such as enterprise application integration (EAI), Web services and XML, as well as emerging industry standards such as ISA 95 and CIM. The subsection *Network-Based Integration Technologies* presents novel industry approaches to the development of distributed and flexible automation systems centered around control networks. Prominent examples of these approaches discussed in this subsection are the PROFInet standard of the PROFIBUS User Organization, and the Interface for Distributed Automation (IDA) standard of the IDA Group e.V. Both the approaches draw extensively from the use of Ethernet TCP/IP, Web, and network security technologies, and allow for horizontal and vertical integration, and interoperability of devices from different vendors to mention some features. Another notable development presented in this subsection is the Open Controller Enabled by an Advanced real-time Network (OCEAN) project with the aim of realizing a real-time-capable platform for distributed control applications for numerical controls based on standardized communication systems and delivered as an open source. The development draws heavily from the results of the Open System Architecture for Controls within Automation (OSACA) Systems project, a precursor of OCEAN. The high degree of complexity of manufacturing systems coupled with the market-dictated requirements for agility lead to a development of new manufacturing architectures and solutions based on distributed, autonomous, and cooperating units, integrated by the plug-and-play approach, called agents. The *Agent-Based Technologies in Industrial Automation* subsection offers a comprehensive treatment of the topic. It begins with an introduction to the concept of agents and technology, cooperation and coordination models, interoperability, and applications to manufacturing systems. Agents dedicated to real-time manufacturing tasks are called holons. The use of holons and related concepts in manufacturing systems is discussed in the next contribution. The material presented in this contribution is largely based on the results of an international project, Holonic Manufacturing Systems, involving some of the leading multinational corporations. FactoryBroker™, a heterogeneous agent-oriented collaborative control system, developed and implemented by Schneider Electric GmbH (Industrial Automation), in cooperation with DaimlerChrysler AG, Research and Technology, Berlin, Germany, is the focus of the next presentation. The last contribution in this subsection introduces PABIS (Plant Automation Based on Distributed Systems), an IMS project under reference IST-1999-60016, which aims at providing a flexible plant automation solution based on the use of agent technologies. The use of IT in energy and power systems is presented in the concluding subsection. The material presented in this subsection introduces the IEC 61850 standard, the JEVis service platform, and the use of ABB IT platform in power network management. The IEC61850, which incorporates technologies such as Ethernet, TCP/IP, MMS (ISO 9506), and XML, aims at seamless information integration across the utility enterprise using off-the-shelf products. The primary target is the electrical substation automation: switchyards and transformers in the medium- and high-voltage transport and distribution. The JEVis system is an Internet-enabled database connected to distributed networks. It collects measurement data and presents it, after processing, to the customers via Web-based graphical front-ends. One of the JEVis application areas is energy management, or Demand Side Management. The use of ABB's Industrial IT Aspect Integrator Platform in power distribution and transmission applications is presented in the last contribution in the section. Some of the key customer benefits provided by Industrial IT include component-based network control system architecture and company-wide migration opportunities with a low investment.

Locating Topics

To assist readers with locating material, a three-tier structure of the table of contents is offered in the book. A complete table of contents is presented at the front of the book. Each of the seven sections is preceded by an individual table of contents. Finally, each chapter begins with its own table of contents.

Two indexes are provided at the end of the book, the index of authors contributing to the book, together with the titles of the contributions, and a detailed subject index.

Acknowledgments

I would like to thank all members of the International Advisory Board for their help with structuring the book, selection of authors, and material evaluation. Thilo Sauter and Andreas Willig provided crucial assistance with Section 2. Luciano Lavango, Grant Martin and Alberto Sangiovanni-Vincentelli provided support for Section 6.

I have received tremendous cooperation from all contributing authors. I would like to thank them for this.

My gratitude goes to David. J. Irwin, Series Editor, The Industrial Electronics Series, to agree to have this book in his series.

I would like to express gratitude to my publisher Nora Konopka, and other CRC Press staff involved in the book production, particularly Jamie Sigal and Jessica Vakili.

My love goes to my wife who tolerated the countless hours I spent preparing this book.

Richard Zurawski

The Editor

Dr. Richard Zurawski received his M.Sc. in Informatics and Automation from University of Mining and Metallurgy, Krakaw, Poland and his Ph.D. in Computer Science from La Trobe University, Melbourne, Australia. He is president and CEO of ISA Corp, South San Francisco and Santa Clara, CA, a company involved in providing solutions for industrial and societal automation. He is also chief scientist with and a partner in Silicon Valley-based start-ups involved in the development of wireless solutions and technology. Dr. Zurawski is a cofounder of The Institute for Societal Automation, Santa Clara, a research and consulting organization.

Dr. Zurawski has over 25 years of academic and industrial experience, including a regular appointment at the Institute of Industrial Sciences, University of Tokyo, and full-time R&D Advisor with Kawasaki Electric, Tokyo, Japan.

Dr. Zurawski provided consulting services to Telecom Research Laboratories, Melbourne, Australia, Kawasaki, Ricoh, and Toshiba Corporations, Japan. He also participated in a number of Japanese Intelligent Manufacturing Systems programs.

His involvement in R&D projects and activities in the past few years includes remote monitoring and control, network-based solutions for factory floor control, network-based demand side management, MEMS (automatic micro-assembly), Java technology, SEMI implementations, development of DSL telco equipment, and wireless applications.

Dr. Zurawski currently serves as an associate editor of the *IEEE Transactions on Industrial Electronics* and *Real-Time Systems — The International Journal of Time-Critical Computing Systems,* Kluwer Academic Publishers.

He was a guest editor of four special sections in *IEEE Transactions on Industrial Electronics:* two sections on Factory Automation, and two on Factory Communication Systems. He was also a Guest Editor of a special issue of the *Proceedings of the IEEE* dedicated to Industrial Communication Systems.

Dr. Zurawski was invited by IEEE Spectrum to contribute material on Java technology to "Technology 1999: Analysis and Forecast Issue."

He is also the series editor for The Industrial Information Technology Series, CRC Press, Florida.

Dr. Zurawski served as vice president of the IEEE Industrial Electronics Society (IES), chairman of the Factory Automation Council, and chairman of the IEEE IES Ad Hoc Committee on IEEE Transactions on Factory Automation. He was an IES representative to the IEEE Neural Network Council and IEEE Intelligent Transportation Systems Council. He was also on a Steering Committee of the *ASME/IEEE Journal of Micromechanical Systems.* In 1996, he received the Anthony J. Hornfeck Service Award from the IEEE Industrial Electronics Society.

Dr. Zurawski established two IEEE events: IEEE Workshop on Factory Communication Systems — the only IEEE event dedicated to industrial communication networks, and IEEE International Conference dedicated to industrial and Emerging Technologies and Factory Automation — the largest

IEEE conference dedicated to industrial and factory automation. He served as a general, program, and track chair for a number of IEEE conferences and workshops.

Dr. Zurawski has published extensively on various aspects of control systems, industrial and factory automation, industrial communication systems, robotics, formal methods in the design of embedded and industrial systems, and parallel and distributed programming and systems. He was Editor of two major handbooks: *The Industrial Communication Technology Handbook* and *The Embedded Systems Handbook* published by CRC Press in 2004 and 2005, respectively.

International Advisory Board

Contributors

Luis Almeida
University of Aveiro
Aveiro, Portugal

Johann Amsenga
Gennan Systems Pty Ltd.
Centurion, South Africa

Jakob Axelsson
Volvo Car Corporation
Goteborg, Sweden

Arvind Balijepalli
State University of New York at
    Buffalo
Buffalo, New York

Pulak Bandyopadhyay
GM R&D Center
Warren

Marcos R. Pereira Barretto
University of Sao Paulo
Sao Paulo, Brazil

João Paulo Barros
Universidade Nova de Lisboa and
    Instituto Politécnico de Beja
Caparica, Portugal

Herbert Barthel
Siemens AG
Berlin, Germany

Kai Uwe Barthel
University of Applied Sciences for
    Technology and Business
Berlin, Germany

Günther Bauer
Vienna University of Technology
Vienna, Austria

Ali Alphan Bayazit
Princeton University
Princeton, New Jersey

Luca Benini
University of Bologna
Bologna, Italy

Fernando De Bernandinis
University of California at Berkeley
Berkeley, California

Ivan Cibrario Bertolotti
IEIIT - CNR
Torino, Italy

Davide Bertozzi
University of Bologna
Bologna, Italy

Carsten Beuthel
ABB Corporate Research Center
Ladenburg, Germany

Paulo A. Blanco
University of Sao Paulo
Sao Paulo, Brazil

Jan Blumenthal
University of Rostock
Rostock, Germany

Peter Bort
ABB Corporate Research Center
Ladenburg, Germany

Bob Brennan
University of Calgary
Calgary, Alberta, Canada

Martin Buchwitz
Jetter AG
Ludwigsburg, Germany

Ralph Büsgen
Siemens AG
Furth, Germany

Luca Carloni
University of California at Berkeley
Berkeley, California

Adriano Carvalho
University of Porto
Porto, Portugal

Salvatore Cavalieri
University of Catania
Catania, Italy

Gianluca Cena
IEIIT-CNR
Torino, Italy

Wander O. Cesario
IMAG
Grenoble, France

Suvendi Chinnapen
University of Pretoria
Johannesburg, South Africa

James H. Christensen
Rockwell Automation
Mayfield Heights, Ohio

Armando Walter Colombo
Schneider Electric
Sophia-Antipolis, France

Anikó Costa
Universidade Nova de Lisboa
Monte de Caparica, Portugal

Mario Crevatin
ABB
Baden-Diitfwil, Switzerland

Mario De Sousa
University of Porto
Porto, Portugal

Jean-Dominique Decotignie
Centre Suisse d'Electronique et de
    Microtechnique - CSEM
Neuchatel, Switzerland

Eric Dekneuvel
University of Nice Sophia Antipolis
    – ESINSA
Biot, France

Christian Diedrich
Institut für Automation und
    Kommunikation eV – IFAK
Barleban, Germany

Arjan Durresi
Louisiana State University
Baton Rouge, Louisiana

Klaus-Peter Eckert
Fraunhofer FOKUS
Berlin, Germany

Stephen A. Edwards
Columbia University
New York, New York

David Emerson
Yokogawa Corporation of America
Denison, Texas

Michael Eyrich
Technical University of Berlin
Berlin, Germany

Jürgen Falb
Vienna University of Technology
Vienna, Austria

Alexander Fay
ABB Corporate Research Center
Ladenburg, Germany

Joachim Feld
Siemens AG
Nurnberg, Germany

Andreas Festag
Technical University of Berlin
Berlin, Germany

Gerhard Fohler
Mälardalen University
Vasteras, Sweden

Jose A. Fonseca
University of Aveiro
Aveiro, Portugal

Joakim Fröberg
Volvo Construction Equipment
    Components
Eskilstuna, Sweden

Josep M Fuertes
Technical University of Catalonia
Barcelona, Spain

Shashidhar Gandham
University of Texas at Dallas
Dallas, Texas

Paul George
Business Services
Columbus, Ohio

Karl M. Goeshka
Vienna University of Technology
Vienna, Austria

Frank Golatowski
University of Rostock
Rostock, Germany

Luis Gomes
Universidad Nova de Lisboa
Monte de Caparica, Portugal

Michael Göschka
Vienna University of Technology
Vienna, Austria

William A. Gruver
Simon Fraser University
Burnaby, British Colombia
Canada

Aarti Gupta
NEC Laboratories America
Princeton, New Jersey

Rajesh Gupta
University of California at San Diego
San Diego, California

Sumit Gupta
University of California at Irvine
Irvine, California

Zygmunt J. Haas
Cornell University
Ithaca, New York

Marc Haase
University of Rostock
Rostock, Germany

Matthias Handy
University of Rostock
Rostock, Germany

Hans-Michael Hanisch
University of Halle-Wittenberg
Halle, Germany

Hans Hansson
Mälardalen University
Vasteras, Sweden

Hideki Hashimoto
University of Tokyo
Tokyo, Japan

Guido Heising
Heinrich-Hertz-Institute
Berlin, Germany

Helmut Hlavacs
Vienna University of Technology
Vienna, Austria

Mai Hoang
University of Potsdam
Potsdam, Germany

Thomas P. von Hoff
ABB
Baden-Diitfwil, Switzerland

Øyvind Holmeide
OnTime Networks
Oslo, Norway

Zaijun Hu
ABB Corporate Research Center
Mannheim, Germany

Bian D. Huggins
Bradley University
Peoria, Illinois

Karin A. Hummel
University of Vienna
Vienna, Austria

Frank Iwanitz
Softing AG
Munchen, Germany

Hans-Arno Jacobsen
University of Toronto
Toronto, Canada

Margarida F. Jacome
University of Texas at Austin
Austin, Texas

Raj Jain
Nayna Networks, Inc.
Milpitas, California

Martin Jandl
Vienna University of Technology
Vienna, Austria

Ray Jarvis
Monash University
Clayton, Victoria, Australia

Thomas Jatschka
Siemens
Vienna, Austria

Ulrich Jecht
UJ Process Analytics
Baden-Baden, Germany

Christian S. Jensen
Aalborg University
Aalborg, Denmark

A.A. Jerraya
TIMA Laboratory
Grenoble, France

Svein Johannessen
ABB
Oslo, Norway

Andreas Kahmen
Werkzeugmaschinenlabor
Aachen, Germany

Wolfgang Kampichler
Frequentis
Vienna, Austria

Holger Karl
Technical University of Berlin
Berlin, Germany

T. Kesavadas
State University of New York at
    Buffalo
Buffalo, New York

Dong-Sung Kim
Kumoh National Institute of
    Technology
Gum-Si, Korea

Lindsay Kleeman
Monash University
Clayton, Victoria, Australia

Eckehardt Klemm
Phoenix Contact GmbH & Co. KG
Blomberg, Germany

Axel Klostermeyer
Otto-von-Guericke-University
Magdeburg, Germany

Muamma Koç
University of Michigan
Ann Arbor, Michigan

Herman Kopetz
Vienna University of Technology
Vienna, Austria

Peter Korondi
Budapest University of Technology
    and Economics
Budapest, Hungary

Dilip B. Kotak
National Research Council Canada
    Innovation Centre
Burnaby, British Colombia, Canada

Christopher Krügel
University of California at Santa
    Barbara
Santa Barbara, California

Eckhard Kruse
ABB Corporate Research Center
Ladenburg, Germany

Michael Kunes
FREQUENTIS GmbH
Guntramsdorf, Austria

Christian Kurz
Vienna University of Technology
Vienna, Austria

Wook Hyun Kwon
Seoul National University
Seoul, Korea

Juergen Lange
Softing AG
Munchen, Germany

Luciano Lavango
Politecnico di Torino
Torino, Italy

Jay Lee
University of Wisconsin-Milwaukee
Milwaukee, Wisconsin

Joo-Ho Lee
University of Tokyo & Ritsumeikan
    University
Tokyo, Japan

Kang Lee
National Institute of Standards and
    Technology
Gaithersburg, Maryland

Edwin H. van Leeuwen
BHP Billiton, Exploration and
    Mining Technologies
Melbourne, Australia

Patrizio Leviti
Applied Logic for Industrial R&D
    and Management
La Spezia, Italy

Jiang Liu
Bradley University
Peoria, Illinois

Lucia LoBello
University of Catania
Catania, Italy

Martin von Löwis
University of Potsdam
Potsdam, Germany

Dietmar Loy
LoyTech
Vienna, Austria

Arndt Lüder
University of Magdeburg
Magdeburg, Germany

Yogesh Mahajan
Princeton University
Princeton, New Jersey

Aleksander Malinowski
Bradley University
Peoria, Illinois

Vladimir Marik
Czech Technical University
Prague, Czech Republic

Pau Marti
Technical University of Catalonia
Barcelona, Spain

Grant Martin
Tensilica, Inc.
Santa Clara, California

Kirsten Matheus
Carmeq GmbH
Berlin, Germany

Fabrizio Meo
FIDIA
San Mauro Torinese, Italy

Giovanni De Michelli
Stanford University
Palo Alto, California

Kazuyuki Morioka
University of Tokyo
Tokyo, Japan

El Mustapha Mouaddib
Universite de Picardie
Amiens, France

Ravi Musunuri
University of Texas at Dallas
Dallas, Texas

Martin Naedele
ABB Corporate Research
Baden-Daettwil, Switzerland

Petra Nauber
Fraunhofer IPMS
Dresden, Germany

Ralf Neubert
Schneider Electric GmbH
Seligenstadt, Germany

Jun Ni
University of Michigan
Ann Arbor, Michigan

Michael Nolin
Mälardalen University
Vasteras, Sweden

Thomas Nolte
Mälardalen University
Vasteras, Sweden

Douglas H. Norrie
University of Calgary
Calgary, Alberta, Canada

Christer Norström
Mälardalen University
Vasteras, Sweden

Peter Palensky
Vienna University of Technology
Vienna, Austria

Enzo Fabrizio Palumbo
STMicroelectronics
Acicastello, Italy

Claudio Passerone
Politecnico di Torino
Torino, Italy

Hiren D. Patel
Virginia Polytechnic Institute and
    State University
Blacksburg, Virginia

Torben Bach Pedersen
Aalborg University
Aalborg, Denmark

Paulo Pedreiras
University of Aveiro
Aveiro, Portugal

Claude Pegard
Universite de Picardie
Amiens, France

Walter Penzhorn
University of Pretoria
Pretoria, South Africa

Jorn Peschke
University of Magdeburg
Magdeburg, Germany

Marco A. Poli
University of Sao Paulo
Sao Paulo, Brazil

Andreas Polze
University of Potsdam
Potsdam, Germany

Manfred Popp
Siemens AG
Furth, Germany

Paulo Portugal
University of Porto
Porto, Portugal

Frank Pospiech
Alcatel
Paris, France

Wolfgang Radinger
Vienna University of Technology
Vienna, Austria

Anand Ramachandran
University of Texas at Austin
Austin, Texas

Praveen Rentala
University of Texas at Dallas
Dallas, Texas

Manuel Ricardo
University of Porto
Porto, Portugal

José Ruela
University of Porto
Porto, Portugal

R. Andrew Russell
Monash University
Clayton, Victoria, Australia

Kristian Sandström
Mälardalen University
Vasteras, Sweden

Alberto Sangiovanni-Vincentelli
University of California
Berkeley, California

Henning Sanneck
Siemens AG
Munich, Germany

Thilo Sauter
Vienna University of Technology
Vienna, Austria

Udit Saxena
Microsoft Corp.
Redmond, Washington

Günter Schäfer
Technical University of Berlin
Berlin, Germany

Uwe Schelinski
Fraunhofer IPMS
Dresden, Germany

Michael Scholles
Fraunhofer IPMS
Dresden, Germany

Ronald Schoop
Schneider Automation SA
France

Karlheinz Schwarz
Schwarz Consulting Company, SCC
Karlsruhe, Germany

Christian Schwaiger
Austria Card GmbH
Vienna, Austria

Michael Seyfarth
ISW-University of Stuttgart
Stuttgart, Germany

Marco Sgroi
University of California at Berkeley
Berkeley, California and
DoCoMo Eurolabs
Munich, Germany

Cartik Sharma
State University of New York at
  Buffalo
Buffalo, New York

Sandeep K. Shukla
Virginia Tech
Blacksburg, Virginia

Dorgham Sisalem
Fraunhofer FOKUS
Berlin, Germany

Tor Skeie
ABB
Billingstad, Norway

Stefan Soucek
LoyTech
Vienna, Austria

Andreas Steffen
Zurich University of Applied
  Sciences
Winterthur, Switzerland

Wolfgang Stripf
Siemens AG
Karlsruhe, Germany

Peter Szemes
University of Tokyo
Tokyo, Japan

Jean-Pierre Thomesse
LORIA-Institut National
  Polytechnique NANCY de
  Lorraine
Vandeuvre les Nancy, France

Kleanthis Thramboulidis
University of Patras
Patras, Greece

Robert Tolksdorf
Freie Universität Berlin
Berlin, Germany

Ulrich Topp
ABB Corporate Research Center
Ladenburg, Germany

Peter Tröger
University of Potsdam
Potsdam, Germany

Robert Tschofen
Siemens
Vienna, Austria

Adriano Valenzano
IEIIT-CNR
Torino, Italy

Pascal Vasseur
Universite de Picardie
Amiens, France

Yauheni Veryha
ABB Corporate Research
Ladenburg, Germany

Claus Vetter
ABB Switzerland Ltd.
Baden, Switzerland

Ricard Villà
Technical University of Catalonia
Barcelona, Spain

Björn Villing
Volvo Truck Corporation
Goteborg, Sweden

Pavel Vrba
Rockwell Automation
Plzen, Czech Republic

Valeriy Vyatkin
University of Halle-Wittenberg
Halle, Germany

Flávio R. Wagner
Federal University of Rio Grande de
  Sul
Porto Alegre, Brazil

Klaus Wehrle
University of California at Berkeley
Berkeley, California

Peter Wenzel
PROFIBUS International
Karlsruhe, Germany

Thomas Werner
ABB Switzerland Ltd.
Baden, Switzerland

Bogdan M. Wilamowski
Auburn University
Auburn, Alabama

Andreas Willig
University of Potsdam
Potsdam, Germany

Hagen Woesner
Technical University of Berlin
Berlin, Germany

Adam Wolisz
Technical University of Berlin
Berlin, Germany

Tanja Zseby
Fraunhofer FOKUS
Berlin, Germany

Table of Contents

PART II   INDUSTRIAL INFORMATION TECHNOLOGY

SECTION 3   Industrial Communication Systems

SECTION 4   The Internet, Web, and IT Technologies in Industrial Automation and Design

SECTION 7   Integration Technologies

Section 1: Computer Software and Web Technologies

1

Web-based Enterprise Computing Development using J2EE

Jiang B. Liu
Bradley University

Enterprise computing has undergone several changes in development strategies in the last decade. Early development strategies were primarily focused on developing client/server computing platforms within the company to empower the distributed computing. The development technology mainly uses procedural language-based distributed computing infrastructures such as an OSF-distributed computing environment (DCE) [1]. Later, the development was redirected using object-oriented technologies such as distributed object computing using OMG common object request broker architecture (CORBA) [2] or Microsoft-distributed component object model (COM+/DCOM) [3]. The switch was obviously to take advantage of the new emerging object-oriented computing platforms. But all these developments have a huge development cost due to the complexity involved in the underlying computing technologies. DCE requires several online corporate-maintained security servers running in order to maintain communication between the client and the application servers. It is also very difficult to port to other computing platforms and an in-depth knowledge of the system is required for the development. DCE development to the distributed computing is analogous to the early generation of programming at the assembly language level. The CORBA and COM+ solved some of the DCE problems by implementing the component model using object-oriented languages. But the development cost did not reduce much. It has achieved some success at the departmental-level and application-level computing, but it never reached the matured enterprise operational level as the early mainframe computing did due to problems of complexity in the development, performance, reliability, and security.

Today, the enterprise computing has undergone other significant changes in its development technologies, with a focus on web-based computing. Two promising development platforms have been developed to address the shortcomings of the previous development strategies. One such strategy is based on Java J2EE [4] using SUN Microsystems Java platform-independent technology, and is therefore widely adapted on the UNIX/LINUX platforms. Another strategy is based on Microsoft .Net [5] using the Microsoft Window technologies, and is therefore widely adapted by applications that rely on the Window platform. In this article, we will introduce the enterprise applications using J2EE development strategy.

1.1 J2EE Java Development Platform

Web-based computing provides the best framework for E-commerce and enterprise computing. It is much easier to develop, deploy, and access the computing components on the web. J2EE provides a complete solution for the development of these components, and comprehensive development tools for a variety of web-based enterprise computations. We will discuss the following Java computing components in this article:

- Java language basics and Java program development
  - Java API Packages
  - Java Virtual Machine (JVM) and Java-enabled browsers
  - Java Client-side programs: Applets; Applications
  - Java Development Tools: Borland Jbuilder, IBM Websphere Studio, SUN ONE Studio
  - Java Security
  - Java Network Protocols: TCP/IP; HTTP; HTTPS
- Java Database Connectivity: JDBC
- Java Server-side programs: Servlets; JavaServer Page (JSP); JavaBeans
- Java-Distributed Object-Computing Technologies
  - Java Naming and Directory: JNDI
  - Remote Method Invocation: RMI
  - Enterprise JavaBeans: EJB
- Java Transaction Service: JTS
- J2EE Web Service

1.2 Java Language Basics and Java Program Development

Java language is a key element in implementing the Java technologies. We will briefly review the Java language features here.

Java is a pure object-oriented language. Its syntax is similar to C++, but the methods of programming are very different. Java program is compiled into a virtual machine (VM) code instead of the machine code as C++ does; therefore, it can be distributed on the Internet irrespective of the platforms. Java VM is a stack machine model, which makes it independent from the real machine register architectures. It also includes a security code verifier to verify the integrity of the distributed Java compiled code. The Java program that is distributed on the Internet and executed inside a web browser is called applet. Java-enabled web browsers such as Microsoft Internet Explorer (IE) and Netscape have a built-in VM interpreter to run

```
Sample Java Applet Program          Sample Java Application Program
Structure                           Structure

/** Java Doc Comments */             /** Java Doc Comments */
import java.applet.*;                import packageName.*;
import ...                           class ApplicationClassName {
public class AppletClassName extends     Type variables;
Applet {                                     public static void main(String[] args) {
    Type variables;                      Statements; //Internal comments
        methodAttribute methodReturnType     }
methodName(method parameters) {          methodAttribute methodReturnType
        Statements; //Internal comments methodName(method parameters) {
        }                                    Statements;
    }                                        }
                                         }
```

FIGURE 1.1 Java applet and application sample programs.

Java applets. Java applets are normally executed inside a so-called sandbox (an enclosed memory address space) so that it cannot do harm to the local recourses. Java applets digitally signed with the public-key encryption can have more access to the local resources, and depends on the security model implemented. Java program can also run as a local applications program, same as the C++ if a Java runtime environment is installed locally. Normally, the JVM is built right into one's Java software download.

Figure 1.1 shows the sample Java applet program and application program.

Java API Packages

Java packages provide a set of units of related Java classes and interfaces for rapid software development. The classes and interfaces in the packages can be imported to the user program by the import statement. Java SDK (Software Development toolKit) provides a core set of packages. The companies and organizations also provide the extended packages for the Java developers to access their software products. Java developers can also create their own packages for their distribution. The following are the Java core packages and some of the commonly used extended packages (see Figure 1.2):

- Language (import java.lang.*)
- Applet (import java.applet.*)
- Utilities (import java.util.*)
- I/O (import java.io.*)
- AWT (import java.awt.*)
- Networking (import java.net.*)
- Security (import java.security.*)
- RMI (import java.rmi.*)
- SQL (import java.sql.*)
- Swing (import com.sun.java.swing.*)
- Java beans (import java.beans.*)
- CORBA (import org.omg.CORBA.*)
- SUN Java servlets (import javax.servlets.*)
- Oracle JDBC (import oracle.jdbc.*)
- IBM DB2 JDBC (import COM.ibm.db2.jdbc.*)

Java Language package provides classes that are fundamental to the design of the Java programming language. Some of the classes are Object class (superclass for all Java classes), Data-type wrapper classes (wrapping Java's fundamental data types into classes), Math class (groups math functions and constants), String classes (handle text strings), System and runtime classes (access system and runtime environment resources), Thread classes (create and manage threads), Class classes (query and load runtime class),

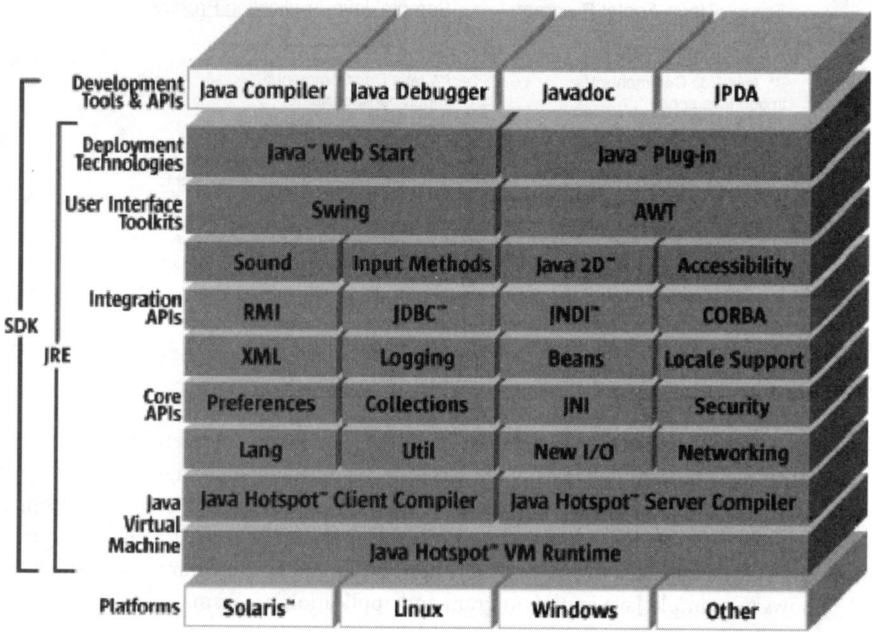

FIGURE 1.2   Java 2 platform. (*Source*: JavaOne conference.)

Exception-handling classes (which deal with runtime error), and Process class (which supports the system process).

Java Applet package provides the classes necessary to create an applet and the classes an applet uses to communicate with its applet context. Some of the classes and interfaces are Applet class (which provides a standard interface between applets and their environment) and AppletContext (which provides the methods for the applet to obtain information about its environment).

Java Utility package provides useful utilities for the Java programmers. Some of the classes are Date class (which represents a calendar date and time based on UTC), Data structure classes and interfaces (which implement popular data structures — BitSet, Dictionary, Hashtable, Properties, Vector, Stack, and Enumeration for storing data), Random class (implements a random-number generator), and StringTokenizer class (which provides a means of converting text string into individual tokens).

Java I/O package provides for system input and output through data streams, serialization, and the file system. Some of the classes are Input stream classes (which handles reading data from an input source such as a file, a string, or memory), Output stream classes (which handles writing data to a output source such as a file, a string, or memory), File classes (which provide methods for file processing), and StreamTokenizer class (which provides methods for converting an input stream of data into a stream of tokens).

Java AWT package provides the simple 2D window implementations. Some of the classes are Graphical classes (which serves as an all-purpose graphical output with all kinds of different drawing functions), Layout manager classes (which provide a framework for controlling the physical layout of GUI elements), Font classes (which provide/specify font information), Dimension class (which represents graphic dimensions such as points, rectangles, and polygons), MediaTracker class (which provides a means to track media when the transmission is completed), and Component/MenuComponent class (an abstract class that defines elements common to all AWT component/menu component classes).

Java Network package provides the classes for implementing TCP/IP networking applications. Some of the classes are InetAddress class (which represents a host name and its IP number), URL classes (which provide access to resources located on the WWW using the HTTP protocol), Socket classes (whcih provide

TCP/IP connection-oriented and connectionless connections), and ContentHandler class (an extension class for new protocols and content types).

Java Security package includes Digital signature classes (which generate a public/private key pair and use them to sign and verifying arbitrary digital data. Digital signatures are used for authentication and integrity assurance of digital data), MessageDigest class (which provides the functionality for a message digest algorithm. Message digests are secure, one-way hash functions that take arbitrary-size data and output a fixed-length hash value. Message digests are useful for producing "digital fingerprints"), and Key management classes (which manage the public and private keys used in the digital signature process).

Java RMI package implements a SUN Microsystems's distributed objects model. Some of the classes are Registry classes (that provide a simple name service that binds a string name to a remote object reference. A server can bind/rebind a name to a remote reference and a client can look up the remote reference for a certain name), RMISecurityManager classes (which manage security issues such as Runtime integrity, Encryption, and Authentication), and UnicastRemoteObject class (which provides support for point-to-point active object references).

Java SQL package provides the JDBC package. JDBC is a standard API for executing SQL statements. It contains classes and interfaces for creating SQL statements and retrieving the results of executing those statements against relational databases. Some of the classes and interfaces are DriverManager (which provides the basic service for managing a set of JDBC drivers), Connection (which makes a connection/session with a specific database), Statement (the object used for executing a static SQL statement and obtaining the results produced by it), PreparedStatement (an object that represents a precompiled SQL statement), CallableStatement (the interface used to execute SQL-stored procedures), and ResultSet (which provides access to a table of returned data).

Java Swing package included a set of "lightweight" (all-Java language) components for all the platforms. It provides a set of 3D graphic components such as JButton, JLabel, JList, JPanel, JTable, and Jtree. Swing has the advantage of a pluggable look and feel and virtual desktop presentation. All the major commercial Java IDEs (Integrated Development Environment) provide virtual programming capabilities for generating the swing component code.

Java Beans package contains classes related to Java Beans development. The graphic components provided by the Java IDEs are an example of Java Beans.

Java CORBA package provides the mapping of the OMG CORBA APIs to the Java programming language and adds CORBA capability to the Java platform. Some of the classes are ORB (providing standards-based interoperability and connectivity), Exception (to throw an exception. Any code using it must have a try/catch block and must handle that exception when it is thrown), and Holder classes (which map the out and inout IDL interface parameters to the Java programming language parameters passing).

Java Servlet package provides the implementation of a server object coded in Java. Server objects coded in Java have the advantage of Java language multithreaded and strong automatic storage management features, which are extremely important for the server executables. Some of the classes and interfaces are ServletRequest (stores information about the client request to the servlet), ServletResponse (sends the server response to the client in the form of Internet MIME-typed message), and ServletContext (which provides information about the server running environment).

Oracle JDBC package provides the JDBC drivers and services to the Oracle databases, and IBM DB2 JDBC package provides the JDBC drivers and services to the IBM DB2 databases.

Java IDE

Java Integrated Development Environment (IDE) provides facilities for constructing high-level objets accessible to HTTP. It usually has extensive editing supports for XML and XSL files and provides facilities for web scripting and lower-level programming, database editing tools, and source code control. It also provides a capable Java and JSP editing environment and EJB template wizards for creating and editing session and entity EJBs.

Some of the popular commercial IDEs are as follows:

- Borland Jbuilder: A leading, cross-platform environment for building industrial-strength enterprise Java applications (EJB, XML), integrated tightly with BEA WebLogic server platform.
- *IBM WebSphere Studio*: An open comprehensive development environment, integrated tightly with IBM Websphere platform. (The IBM VisualAge for Java was withdrawn from the market in 2003. It has a complex visual programming capability. Most of its functionality is incorporated into the Websphere studio.)
- *Sun One Studio (formerly Sun Forte for Java)*: A Java IDE that supports EJB and XML, which paves the way for organization to transform Java objects into Web services.

To illustrate an example, we will briefly describe the IBM Websphere Studio (WS). The IBM WS supports the Java development for the Application Framework for e-business. IBM application framework principles are as follows:

- Server-centric applications
- Managed thin clients
- Scalable architectures
- Java-based techniques
- Platform, Web server independence
- Industry-standard components, services, and interfaces.

IBM WS provides an easy-to-use tool set that helps reduce time and effort when creating, managing, and debugging multiplatform web applications. IBM WS is the first tool in the industry for the visual layout of dynamic Web pages. WS provides tools and wizards support for both individual Web pages development, and large teams advanced Web applications such as servlet, JSP, and EJB development. It has visual layout tools to create dynamic Web sites with Java servlet or JSP components, a built-in XML development environment, and Web services tools for the creation, deployment, and publishing of Web services. Using WS, one can create J2EE applications and make one's application accessible over the Web using the XML Web Services standards such as XML, Simple Object Access Protocol (SOAP), Web Services Description Language (WSDL), and Universal Description Discovery and Integration (UDDI). The major Web development tools provided by WS includes the following:

- *The Web development tools*: The WS web development environment provides the tools necessary to develop applications such as static Web pages with HTML, dynamic Web pages with JSPs or servlets, XML deployment descriptors, and other Web resources, which includes wizards to generate pages driven by databases and Java beans. It automatically updates links when the content changes. It also includes tools for developing images and animated GIFs.
- *Enterprise services toolkit*: The toolkit consists of a set of tools and wizards to facilitate service-oriented development, which provides an additional level of abstraction for developers by providing a consistent view of any service irrespective of implementation.
- *Relational database tools*: The data tools provided allow one to create and manipulate the data design for one's project in terms of relational database schemas. One can create, browse, or import database schemas, and also explore, import, design, and query databases, working with either a local copy of an already deployed design, or by creating an entirely new design to meet one's requirements. The SQL statement wizard and SQL query builder provide a GUI-based interface for creating and executing SQL statements. The relational database tools support connecting to and importing from several database types, such as DB2, Oracle, SQL Server, Sybase, and Informix. The SQL-to-XML wizard can be used to create an XML and related documents that allow one to implement one's query in other applications such as a servlet or JSP.
- *XML tools*: The comprehensive XML tool set includes components for building Document Type Definition (DTDs), XML schemas, and XML files for web service specifications.
- *Java development tools*: The Java tools provide a professional-grade Java development environment.

- *Web services development tools*: The tools help you to discover, create or transform, build, deploy, test, and publish the Web services.
- *Enterprise JavaBeans development tools*: Enterprise JavaBeans (EJB) development tools support the full EJB 1.1 features, includes an updated EJB test client, an enhanced unit test environment for J2EE, and deployment support for Web application archive (WAR) files and enterprise application archive (EAR) files. Entity beans can be mapped to databases, and EJB components can be generated to tie into transaction processing systems. XML provides an extended format for deployment descriptors within EJB.
- *Server tools for testing and deployment*: The Server Tools feature provides a unit-test environment where one can test HTML files, servlets, JSPs, and EJBs. It also provides the capability to configure other local or remote servers for integrated testing and debugging of J2EE applications. The server tools support multiple-server configurations, including WebSphere Application Server and Apache Tomcat web server.

Java Security

Java cryptography architecture uses both symmetric-key and public-key encryption algorithms to generate the encrypted code, and uses the digital signature to authenticate the integrity of the files. The untrusted remote Java code has only a sandbox-restricted access to the local system resources. The trusted signed code can have up to complete access to the local system resources, and depends on the security policy specified. Figure 1.3 shows the Java JDK 1.2 security model.

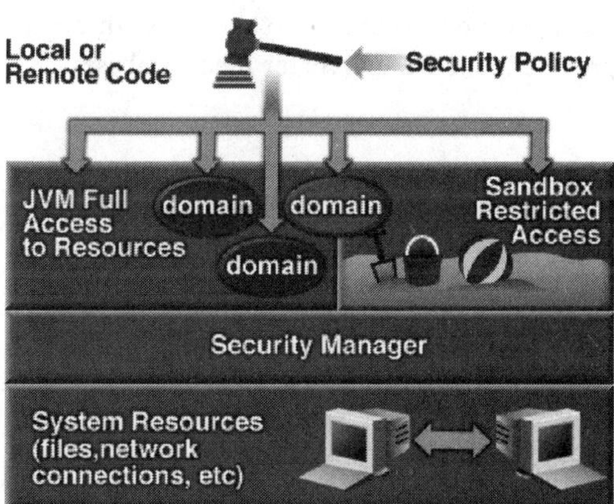

FIGURE 1.3   JDK 1.2 security model. (*Source*: JavaOne conference.)

[X.509v1 certificate, Subject is CN=Jiang B. Liu, OU=CS, O=Bradley University, C=U.S.A. Key: Sun DSA Public Key parameters:
p:fca682ce8e12caba26efccf7110e526db078b05edecbcd1eb4a208f3ae1617ae01f35b91a47e6df 63413c5e12ed0899bcd132acd50d99151bdc43ee737592e17
q: 962eddcc369cba8ebb260ee6b6a126d9346e38c5
g:678471b27a9cf44ee91a49c5147db1a9aaf244f05a434d6486931d2d14271b9e35030b71fd73da 179069b32e2935630e1c2062354d0da20a6c416e50be794ca4
y:5780fc5b73bca1dad3352175a8b17aaf565040c224892968ecd6dbd263462b68df5bb0c966211d 9022f5016563c9343a2d671c2bcbf879f3abe555df687ab262
Validity <Sun July 06 18:00:00 CST 2003> until <Sat Aug 30 18:00:00 CDT 2003> Issuer is CN=Jiang B. Liu, OU=CS, O=Bradley University, C=U.S.A.
Issuer signature used [SHA1withDSA] Serial number =   03ea]

FIGURE 1.4   A Java Certificate generated using the 512-bit DSA.

Java software provides the tools to generate the encryption keys and certificates for signing the Java code. Figure 1.4 shows an example of a certificate generated using the DSA public-key encryption.

Java Network Protocol

Java application programs can directly communicate with the TCP/IP, similar to C/C++ network programs. Java provides three socket classes for TCP/IP connection-oriented and connectionless services.

- Socket class and ServerSocket class are used for connection-oriented service.
- Socket class provides an incoming/outgoing reliable, ordered stream connection service.
- ServerSocket class provides a server-side socket that listens for incoming connections from clients.
- DatagramSocket class is used for connectionless service.

But web-based Java components such as Applet, Servlet, JSP, and EJB use HTTP and HTTPS as their network protocols. HTTPS is a secure HTTP that runs over the SSL (The Secure Sockets Layer). SSL establishes and uses the secure socket connections with encryption and message authentication.

1.3   Java JDBC

JDBC is a Java Database Connectivity specification. The protocol was jointly developed by SUN JavaSoft, Sybase, Informix, IBM, and others to provide a Java object interface to relational databases. It provides a handy set of lower-level classes to manage database connections, operations, and transaction, and supports common database standards. JDBC is a generic way to query and update relational tables and translate the results into Java data types. It is based on the X\Open SQL CLI (Call Level Interface), which is an SQL wrapper that allows one's application to connect to a database through a local driver. All the major relational database vendors provide the JDBC drivers for their databases.

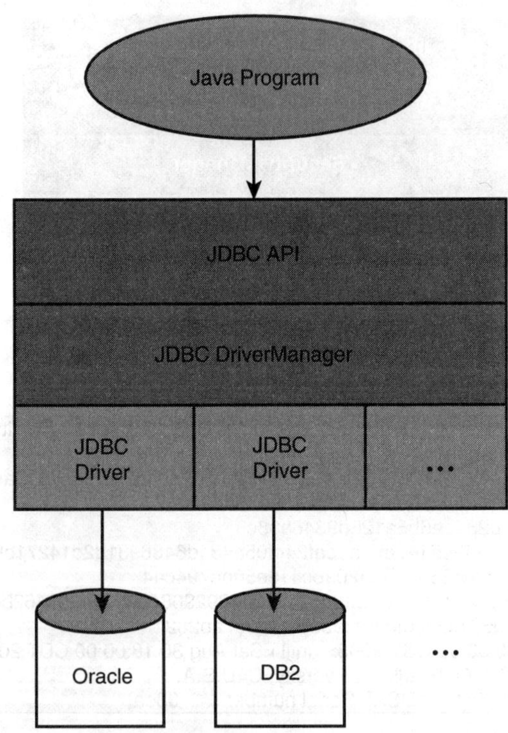

FIGURE 1.5   JDBC architecture.

JDBC consists of a basic set of interfaces (Driver, Connection, Statement, and ResultSet) and a DriverManager class. DriverManager class is a JDBC class that controls the loading of driver-specific classes (Load driver, establish the database connection, etc.). Connection interface is a JDBC interface used by driver developers to create a DBMS-specific implementation of the connection class. Statement and ResultSet classes are JDBC interfaces that are used for executing the SQL statements and processing the return results. The driver provides the functional specifics to a database or database middleware layer.

JDBC Core API locates DBMS drivers, connects to a remote/local database, submits SQL commands, and processes return results. JDBC Java Language Extensions provides new numeric Java data types to meet the precision requirements of SQL. JDBC Java Utilities provide very fine-grained time and date utilities. JDBC Metadata provide the information about the SQL database (see Figure 1.5).

Loading a JDBC Driver

One must select and load a JDBC driver for one's application code. The following show examples of loading three different database drivers.

MS Access: DriverManager.registerDriver(new sun.jdbc.odbc.JdbcOdbcDriver());
Oracle: DriverManager.registerDriver(new oracle.jdbc.driver.OracleDriver());
DB2: DriverManager.registerDriver(new com.ibm.db2.app.DB2Driver()).

Creating a JDBC Connection

A JDBC Connection request can be made on DriverManager class.

```
Connection conn=DriverManager.getConnection(url,userlogin,password)
where URL = jdbc:<subprotocol>:<datasource_name>
```

The following show examples of connecting to a "cs" database on three different database systems.

MS Access:
```
Connection conn = DriverManager.getConnection("jdbc:odbc:cs", user, password);
```
Oracle:
```
Connection conn = DriverManager.getConnection ("jdbc:oracle:thin:@cs1.
bradley.edu: 1521:cs", user, password);
```
DB2:
```
Connection conn = DriverManager.getConnection ("jdbc:db2:cs", user, password).
```

Execute the Query

JDBC has three query statements. Statement is used to execute a static SQL statement. Prepared Statement is used to submit a precompiled SQL command so that it can be efficiently executed multiple times with variable parameters. And Callable Statement is used to submit the stored procedure call. Normally, the Callable Statement executes faster than the prepared statements because a block of sql statements is already stored at the database side (see Figure 1.6).

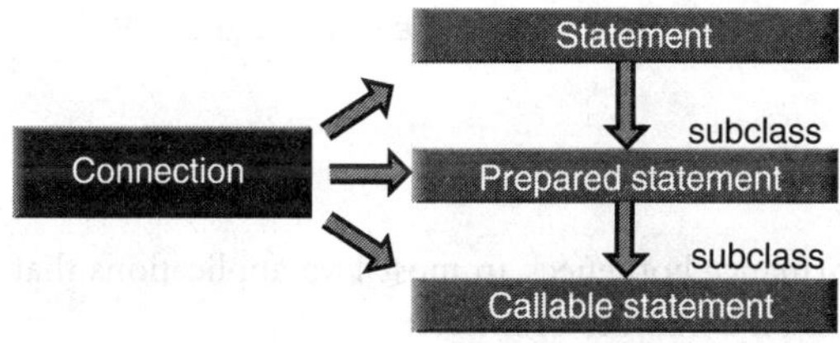

FIGURE 1.6   JDBC query statements. (*Source*: JavaOne conference.)

Example of execution of the SQL SELECT Statement:

```
Statement queryStmt = conn.createStatement();
ResultSet result = queryStmy.executeQuery("SELECT * FROM students").
```

(Similarly, you can execute the delete, insert, and update command using the Statement executeUpdate method.)

Example of execution of the PreparedStatement:

```
PreparedStatement insertStmt = conn.prepareStatement("INSERT INTO students
(firstname, year) VALUES (?,?)");
insertStmt.setString(1, "John"); insertStmt.setInt(2, 3);
insertStmt.executeUpdate().
```

Example of execution of the CallableStatement:

```
CallableStatement cstmt = conn.prepareCall("{call getStudentData(?,?)}");
cstmt.setByte(1, 25); cstmt.setBigDecimal(2, 25.5);
cstmt.registerOutParameter(1, java.sql.Types.TINYINT);
cstmt.registerOutParameter(2, java.sql.Types.NUMERIC);
ResultSet rs = cstmt.executeQuery().
```

Retrieving and Processing Return Data

A ResultSet provides access to a table of data generated by executing a Statement. It maintains a cursor pointing to its current row of data; initially, the cursor is positioned before the first row. The next() method moves the cursor to the next row and it will returns false if there are no more rows left in the table. The table rows are retrieved in sequence and the ResultSet has methods to retrieve column values for the current row.

Example of retrieval and printing of all the first name column of the return records.

```
  while (rs.next()) {
    String fname = rs.getString("firstname");
    System.out.println("Name: " + firstname);
  }
```

Exception Handling

Most of the methods in Statement and Connection can throw SQLException. Checked exception must be caught. SQLException can be handled in one of two ways: catch and recover locally or catch and "convert" the exception by throwing another exception.

Example of catches and printing of the SQL errors.

```
  try {
    Connection conn = DriverManager.getConnection
  ("jdbc:oracle:thin:@cs1.bradley.edu:1521:cs", user, password);
    Statement stmt = conn.createStatement();
  } catch(SQLException e) {
    System.out.println(e+e.getErrorCode()+e.getSQLState())};
  }
```

JDBC Efficiency

Database access is the performance bottleneck in most Java applications that use a relational database. The JDBC efficiently depends on a number of factors:

  1. Use the right driver.

2. Use only those statements that are necessary.
3. Use JDBC's PreparedStatement and CallableStatement when required.

Sun Java JDBC defines four types of drivers:

Type 1: The JDBC–ODBC bridge — slowest, most portable, requires client software.
Example: sun.jdbc.odbc.JdbcOdbcDriver
Type 2: Native-API partly Java driver: Usually fastest, least portable, requires client software.
Example: oracle.jdbc.driver.OracleDriver
com.ibm.db2.app.DB2Driver
Type 3: Net-protocol all-Java: Requires no client software, internet-capable.
Example: com.ibm.db2.net.DB2Driver
Type 4: Native-protocol all-Java: Requires no client software, suitable for Intranets.

All major commercial database vendors are providing the JDBC drivers on their Web sites. Choosing the right driver would be the first step in improving your JDBC efficiency.

Finally, we will list the development steps of accessing a "cs" database using Oracle JDBC Driver from a remote window client.

1. Oracle database on SUN Solaris server development:
   (a) Create and Test the Database tables.
   (b) Create a Database Source ID (SID) for the database.
   (c) Start the Oracle listener on the Solaris server and log the incoming client connection requests for the debugging.
2. Window 98 JDBC client development (using type 2 driver)
   (a) Download the Oracle JDBC driver from Oracle Web site.
   - Choose the thin driver classes for applet/application: classes12.zip
   (b) Develop the JDBC client to test the JDBC connection.
   - The web server and the Oracle database must be in the same machine for testing the Java applet JDBC; otherwise, the sandbox security model will prevent the JDBC connection.

1.4 Java Server-side Programs

Java server-side programming technologies enable developers to create complete, distributed applications in which the clients can access the server programs through the web pages. Java supports server-side servlet Java program and server-side JSP script (JSP will be dynamically compiled into a servlet at the server side). The developer can choose to build either one or both for their applications.

Servlet

Java Servlet provides a Java-based solution to the server-side programming much as the earlier CGI implementation. Servlets are objects that conform to a specific interface that can be plugged into a Java-enabled server, and thus extends the capabilities of a web server. Servlets are to the server-side as what applets are to the client-side. It runs completely on the server; nothing is ever downloaded to the browser. Servlets serve as platform-independent, dynamically loadable, pluggable bytecode objects on the server-side that can be used to dynamically extend server-side functionality.

Servlets have many advantages for the distributed applications. There is a rich set of platform-neutral Java APIs to connect to most backend assets. It is a reusable object. Web server servlet engine plugin provides a consistent servlet environment. Called within the server context can restrict servlet access within the security architecture. Servlets also have high performance in the server object execution. It runs in the same context as the application server and remains in the memory after the execution. Servlets can be preloaded or loaded on demand and sessions can be maintained across HTTP requests. Servlets are multithreaded and can be scaled with multiprocessors; thus, they are the ideal components for the Enterprise-level applications.

Programming the Servlet

HTTP servlets subclass the javax.servlet.http.HttpServlet class and override one of the following methods depending on which HTTP method is used by the client.

```
doGet(HttpServletRequest req, HttpServletResponse res);
doPost(HttpServletRequest req, HttpServletResponse res).
```

HTTP is a request–response protocol. The Get method will attach the request at the end of the URL. Thus, it is only suitable for short request data pairs, but the servlet can then be directly invoked from the URL. The Post method will send the request through an HTML form. Therefore, it is commonly used for the large request data pairs sent from client (see Figure 1.7).

Servlet Invocations

Servlet is usually invoked from a remote web client. The client makes a request to the hosting web server naming the Servlet as part of the URL. Hosting web server forwards the request to the Servlet engine that locates an instance of the Servlet class. The Servlet engine then calls the Servlet's service method. Servlet can also be invoked by another servlet in the servlet chaining, but not all hosting web servers support this (see Figure 1.8).

Example of Servlet using doGet:

```
import javax.servlet.*;
import javax.servlet.http.*; // servlet using the HTTP protocol.
public class HelloServlet extends HttpServlet {
    public void doGet (HttpServletRequest req, HttpServletResponse res)
        throws ServletException, IOException {
        res.setContentType("text/html"); // MIME type: image/jpeg, ...
        PrintWriter out = res.getWriter();
        out.println("<html><head><title>Hello Servlet</title></head><body>");
        out.println("<h1>Hello, Peoria</h1>");
        out.println("</body></html>"); }
}
```

Example of Servlet using doPost:

```
import javax.servlet.*;
import javax.servlet.http.*; // servlet using the HTTP protocol.
public class HelloPerson extends HttpServlet {
    public void doPost (HttpServletRequest req, HttpServletResponse res) throws
        ServletException, IOException {
        res.setContentType("text/html"); // MIME type: image/jpeg, ...
        PrintWriter out = res.getWriter();
        Enumeration enum = req.getParameterNames();
        String fname = (String) enum.nextElement();
        String fvalue = req.getParameter(fname);
        String lname = (String) enum.nextElement();
        String lvalue = req.getParameter(lname);
        out.println("<html><head><title>Hello Person</title></head><body>");
        out.println("<h1>Hello, " + fvalue + ", " + lvalue + "</h1>");
        out.println("</body></html>"); }
}
```

HTML client form to invoke the HelloPerson servlet:

```
<FORM METHOD="POST" ACTION= "... /servlet/HelloPerson">
<P>Please enter your name:
<P>First name: <INPUT NAME="firstname" TYE="TEXT" SIZE="25">
<P>Last name: <INPUT NAME="lastname" TYE="TEXT" SIZE="25">
<INPUT TYPE="SUBMIT"><INPUT TYPE="RESET">
</FORM>
```

FIGURE 1.7 Servlets using doGet and doPost methods.

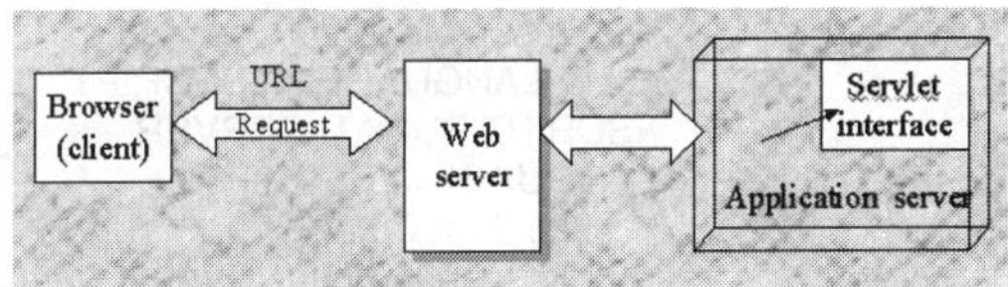

FIGURE 1.8   Servlet Invocation.

Example of a direct invoking of a servlet class name:

```
http:/host_name/servlets/servletClassName  or
<FORM ACTION = http:/host_name/servlets/servletClassName…>
(Servlet class name can also be converted into a user-friendly registered
name.)
```

Example of invoking of a servlet from an applet:

```
<APPLET NAME = "appletClassName" CODE = "appletName">
<PARAM NAME = "server" VALUE = "http://hostname/servletClassName">
<PARAM NAME = "method" VALUE = "GET">
```

Hosting the Servlets

Some free Java web servers, such as Tomcat, have the simple servlet hosting mechanism. After configuring the server.xml, web.xml, and other xml files, the Tomcat web server hosts the servlets in webapps subdirectory with a default path for html files in webapps/applName/servlets/ and default path for classes files in webapps/applName/WEB-INF/classes.

Other commercial web servers such as IBM WebSphere Application Server (WAS) have a set of utilities and tools to facilitate the servlet hosting. WAS starts the AdminServer to start and keep alive all servers. It runs a Name Service based on the JNDI, a Location Service Daemon, and a Security server. An Admin Console Tasks Configuration tool is used for configuring a web application, creating a web application, and adding Servlets and other server objects.

Servlet Design Issues

Servlets are used to generate the dynamic HTML web pages. Do not use the servlet if the contents of the HTML pages are static. Design of the servlet should follow the object component design principle. Each servlet should have well-organized manageable size. It may call the functions in other classes to minimize its size. Each servlet should not allocate too much memory. Check the request and respond data carefully and make sure there are exceptional handling and security checks for all vital client requests.

JavaServer Page

The HTML web page content delivered to client is composed of static or noncustomized content. Customized or dynamic content of a web page is processed programmatically. Server-side scripting is used to process a composite page on Server prior to delivery to the client — the server executes an embedded code to produce dynamic content. JSPs are an HTML extension for server-side scripting in your web pages. They are called from a browser just like HTML pages, but they have special tags that let you insert Java logic directly into the page. Other server-side scripting technologies include the Microsoft Active Server Pages (ASP) on Microsoft web server and PHP (PHP: Hypertext Preprocessor) on Linux Apache web server. Figure 1.9 shows a sample JSP page and a similar ASP page that dynamically display the current date and time.

JSP Execution Model

JSP (HTML page with an embedded Java code) is first converted into a Java servlet within the web server and is then compiled if the source code is new, loaded, and executed (see Figure 1.10).

```
<!-- A sample JSP Page -->        <!-- A sample JSP Page -->
<HTML>                            <%@ LANGUAGE="VBSCRIPT" %>
<HEAD>                            <SCRIPT RUNAT=SERVER
<TITLE>Sample.jsp</TITLE>         LANGUAGE=VBSCRIPT>
</HEAD>                               Sub WriteDate()
<BODY>                                  Response.Write("Current Date and Time: "
    Current Date and Time:        & Date & "");
    <%= new java.util.Date()%>        End Sub
</BODY>                           </SCRIPT>
</HTML>                           <HTML><HEAD>
                                  <TITLE>Sample.asp</TITLE></HEAD>
                                  <BODY>
                                      <% Call WriteDate() %>
                                  </BODY></HTML>
```

FIGURE 1.9   Sample JSP and ASP pages.

FIGURE 1.10   JSP compilation.

Figure 1.11 shows the converted sample$jsp.java servlet code from the previous sample.jsp page.

The JSP servlet is a subclass of HttpJspBase created with an overridden service method. The JSP servlet class created defines the _jspService method where the script content is written. pageContext object is created to hold the output page. Output is written to a JspWriter object that supports buffer management.

JSP coding Technologies

The JSP coding in the above example with an embedded Java code in an HTML file is called scriptlets. To separate the detailed Java code with the HTML code, two other technologies can be used to code the JSP. One is to call the JavaBeans object and another is to use the XML tag like custom tag libraries.

(a) *JSP calls JavaBeans*

A JavaBean is a reusable software component. It presents itself through a single object and uses standard "method" naming conventions that exposes three feature sets: Properties (variables), Methods, and Events. The following example shows a JSP page that calls the CounterBean to count how many times the page has been visited.

```
<h2> JavaBeanCount.jsp </h2>
<jsp:useBean id = "CounterBean" class = "jsp.CounterBean" scope = "session" />
<h3>This page has been visited: <jsp:getProperty name = "CounterBean"
property = "counter"/> time(s)</h3>
```

JSP JaveBeans unifying concepts of JSP objects used to encapsulate dynamic content provide interface for computing dynamic content, and manage complex states through client session. JavaBeans can access the business logic in a parameterized manner and encapsulate the result data (dynamic content). JavaBeans hide the detailed Java code so that the web page designer can concentrate on the presentation logic and leave the business logic coding to the JavaBean designers.

(b) *JSP uses Custom Tag Libraries*

JSP1.2 introduced the Custom Tag Libraries to use the XML-like tags to implement the detailed Java coding. These tags will be processed at the server-side and then generate the output for the client. The tags

```
        package org.apache.jsp;
        import javax.servlet.*;
        import javax.servlet.http.*;
        import javax.servlet.jsp.*;
        import org.apache.jasper.runtime.*;
        public class sample1$jsp extends HttpJspBase {
            static {   }
            public sample1$jsp( ) {   }
            private static boolean _jspx_inited = false;
            public final void _jspx_init() throwsorg.apache.jasper.runtime.JspException {   }
            public void _jspService(HttpServletRequest request, HttpServletResponse  response)
                throws java.io.IOException, ServletException {
                JspFactory _jspxFactory = null;
                PageContext pageContext = null;
                HttpSession session = null;
                ServletContext application = null;
                ServletConfig config = null;
                JspWriter out = null;
                Object page = this;
                String _value = null;
                try {
                    if (_jspx_inited == false) {
                        synchronized (this) {
                            if (_jspx_inited == false) {_jspx_init(); _jspx_inited = true;}
                        }
                    }
                    _jspxFactory = JspFactory.getDefaultFactory();
                    response.setContentType("text/html;charset=ISO-8859-1");
                    pageContext = _jspxFactory.getPageContext(this, request, response,"", true, 8192, true);
                    application = pageContext.getServletContext();
                    config = pageContext.getServletConfig();
                    session = pageContext.getSession();
                    out = pageContext.getOut();

                    // HTML // begin [file="/jsp/Sample.jsp";from=(0,0);to=(7,0)]
                        out.write("<HTML><HEAD><TITLE>Sample.JSP</TITLE></HEAD> \r\n
<BODY>\r\n<H3>Current Date and Time</H3>\r\n");
                    // end
                    // begin [file="/jsp/Sample.jsp";from=(7,3);to=(7,25)]
                        out.print( newjava.util.Date() );
                    // end
                    // HTML // begin [file="/jsp/Sample.jsp";from=(7,27);to=(9,14)]
                        out.write ("\r\n</BODY></HTML>");
                    // end
                } catch (Throwable t) {
                    if (out != null && out.getBufferSize() != 0)
                        out.clearBuffer();
                    if (pageContext!= null) pageContext.handlePageException(t);
                } finally {
                    if (_jspxFactory != null) _jspxFactory.releasePageContext(pageContext);
                }
            }
        }
```

FIGURE 1.11   Generated JSP servlet code.

are defined as Java classes with one class per tag. The tag library is registered with one's web application in the web.xml file. The following is the same example of counting how many times the page has been visited by using tag instead of the Java bean call.

```
<h2> CountTag.jsp </h2>
<%@ taglib uri = "http://brd105nt.bradley.edu:8080/Lib/Jsp" prefix = "ejava" %>
<h3>This page has been visited: <ejava:counter/> time(s)</h3>
```

When the web server JSP container encounters the custom tag, it will create the appropriate tag object. The tag attributes are set using JavaBean style setter methods. The tag object will be executed using doStartTag method. The tag is released at some later time after the run results are returned to the client.

Advantages of using JSP

A JSP processes Java code embedded in an HTML document and returns results to the browser. JSP leverages Java as its language, which is more powerful and flexible than other script languages.

JSP is used to create dynamic contents of the web pages. HTML designers can use a well-defined Java code in JavaBeans or Custom tags inside the HTML pages without necessarily knowing how to write the servlets. Java programmers can use JSP to create dynamic web pages that access server-side business logic and data sources.

JSP vs. Servlets

JSP can generate the dynamic HTML without writing the servlet code (servlet code will be generated by the JSP page compiler.)

Servlet can response to the request from the applet, manage the redirects, and provide a mechanism for more complex dynamic contents creation such as images.

JSP can call the servlet using the normal HTML form action or JSP tags. Servlet can also call the JSP using sendRedirect method in the HttpServletResponse object or forward method in getServletContext().getRequestDispatcher object:

1.5 Java-Distributed Object Computing Technologies

Distributed computing refers to computer applications in which the applications code, the data it works on, and the actual computations performed are spread across multiple computers. Early distributed computing protocols use Remote Procedure Calls (RPC) (such as DCE), which has many disadvantages. The developer is responsible for both the high-level data structures and the tedious data communication details such as data loss and marshaling. The system is difficult to maintain and update.

A distributed object mechanism allows objects implemented on one computer to send messages to objects running in another computer. Objects hide the implementation details, thus providing better solutions than the RPC. Typical distributed object computing system components have a data communication protocol such as TCP/IP for peer-to-peer computing and HTTP/HTTPS for web-based computing, a scheme for locating remote objects, a mechanism for marshaling (packing parameters and return values to be distributed) and unmarshaling the object data over the network, a scheme for providing an interface for a remote object in the local address space, and a reference scheme for keeping track of valid reference to an object. Current major distributed object computing platforms include SUN Microsystems J2EE and Microsoft .Net. We will discuss the SUN J2EE JNDI, RMI, and EJB in this section.

Java Naming and Directory Interface

Java Naming and Directory Interface (JNDI) is an API that describes a standard Java library for accessing naming and directory services. JNDI can provide similar services as provided by Internet Domain Naming Service (DNS), Lightweight Directory Access Protocol (LDAP) directory service, RMI Registry service, and file system services. JNDI can be used to provide a seamless interface for accessing all types of enterprise information resources (see Figure 1.12).

A naming service provides a method for mapping a unique identifier (name) to a local or remote object. JNDI provides a consistent API to access the naming services provided by different service providers. The naming directory contains a slightly more sophisticated API used for filtered searching and modifying the hierarchical value in directory and naming services, such as LDAP. Naming spi provides a set of interfaces and objects used by service provider developers. It is used for mapping JNDI calls to a specific service provider.

Remote Method Invocation

Remote Method Invocation (RMI) is a language-centric distributed computing for the Java platform. It leverages Java ubiquity and enables Java-to-Java distributed objects computing. RMI is a simplified distributed object model as compared with CORBA and is easy to use. It supports seamless remote invocation on objects in different VMs.

Java RMI Classes and Interfaces

The RMI system is based on remote interfaces through which a client accesses the methods of a remote object (server.UnicastRemoteObject). The client stub and server skeleton are generated from the remote interface by the rmi compiler; no interface definition language (IDL) is used as is in the CORBA. RMI naming class provides methods (bind/rebind, lookup) for server classes to make remote objects visible to clients (see Figure 1.13).

Example of development of a "Pizza order" application using RMI (see Figure 1.14).

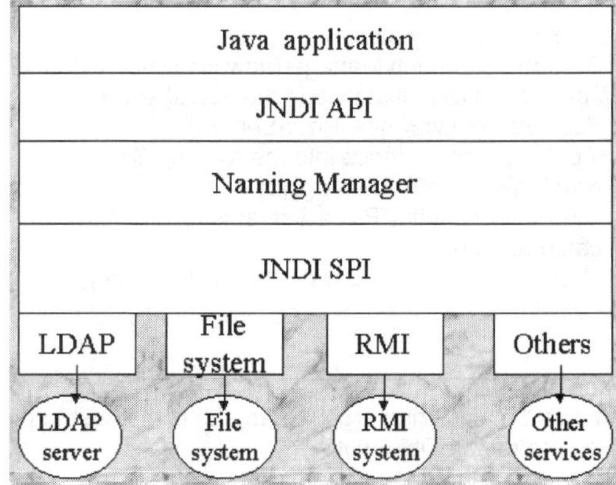

FIGURE 1.12   JNDI architecture.

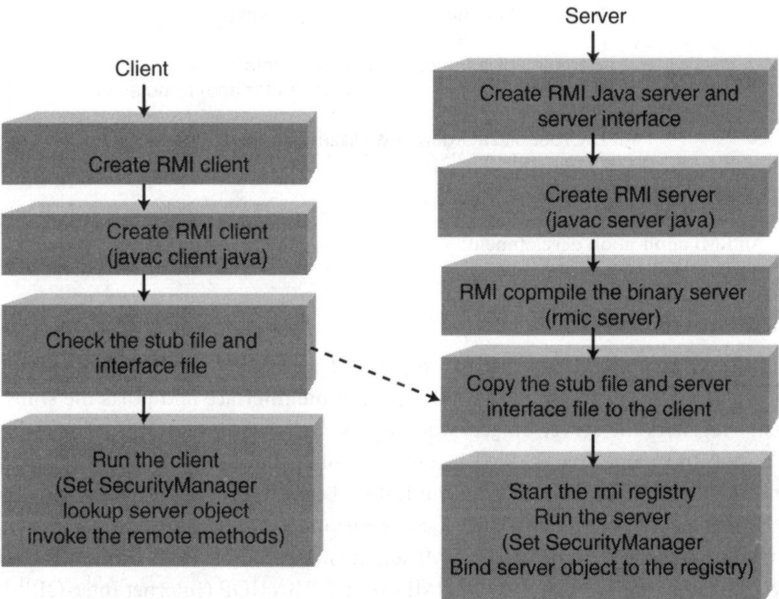

FIGURE 1.13   Client/server development using Java RMI.

```
1.  Develop the Remote Interface
import java.rmi.*;
public interface PizzaServerInterface extends Remote {
      String orderPizza(...) throws RemoteException;
}

2.  Develop the Server Working Object
import java.rmi.*;
import java.rmi.server.*;
public class PizzaServer extends UnicastRemoteObject implements
PizzaServerInterface {
   PizzaServer() throws RemoteException {super();}
   public String orderPizza(...) throws RemoteException {
      // Process pizza order information
      ...
   }
}

3.  Develop the RMI Server
public static void main(String args[]) {//args[0] is the hostname
      try  {
         // set the security manager
         System.setSecurityManager(new RMISecurityManager());
         // create the local instance of the PizzaServer
         PizzaServer svr = new PizzaServer();
         // put the local instance into the Naming Server
         Naming.rebind("rmi://"+args[0]+"/PizzaServer", svr);
         System.out.println("Pizza Server is ready...");
      } catch (Exception exc) {
         System.out.println("Error! - " + exc.toString());
      }
}

4.  Develop the RMI Client
public static void main(String args[]) {//args[0] is the hostname
      Remote remoteObj = null;
      PizzaServerInterface pizzaInterface = null;
      System.setSecurityManage   r(new RMISecurityManager());
      try  {
         remoteObj =Naming.lookup("rmi://"args[0]"/PizzaServer");
      }catch(Exception    exc){
         System.out.println(exc.toString())System.exit(1);}
      try  {
         if(remoteObj instanceof PizzaServerInterface)
             pizzaInterface = (PizzaServerInterface) remoteObj;
      }catch(Exception    exc){...}
      PizzaOrder pizzaOrder=new PizzaOrder(pizzaInterface);
      pizzaOrder.show();
}
```

FIGURE 1.14　An RMI application development.

RMI-over-IIOP

As compared with CORBA, RMI is easy to program. There is no need to learn separate Interface
Definition Language (IDL). RMI simply compiles the remote interface and ports the stubs to the client
site. Rmi registry registers the server objects when the RMI server started. Rmi client will look for the
server object at the Rmi name spaces. RMI is ideal for communication between Java applications. But
CORBA is an OMG standard and allows communication between Java applications and other languages
such as C++. To make it compatible to CORBA, SUN introduced RMI-IIOP API.

RMI-IIOP allows interfaces to be defined in RMI, and then generates the IDL from RMI interfaces. It uses
the IDL Object-By-Value extensions to run the RMI over CORBA IIOP (Internet Inter-ORB Protocol).

RMI is not a web-based computing technology. It is often used as an underlying network protocol
instead of a user-level application protocol. J2EE supports Enterprise Java Bean, which is a much better
choice for the user-level-distributed object computing.

Enterprise JavaBeans

Enterprise applications depend upon a vast array of services that are inherently complex. These services include distribution, multithreading, transaction management, resource management, persistent-state management, and security services. The increasing complexity leads to a desire to have a standardized API to the enterprise service and integrating the utilization of the services within a component-model specification.

Enterprise JavaBeans (EJB) is such a component model developed by SUN Microsystems and its partners for developing and deploying object-oriented, multitier, enterprise Java application. EJB specifications hide the server-side computing complexities, such as transaction processing, database access, and resource management (multithreading, connection pooling) from the programmers. EJB defines a standard for server-side component as the JavaBean defines a standard interface for client-side application components.

EJB was designed to meet the following goals (see Figure 1.15):

- Clearly define the way that applications are partitioned: client vs. EJB vs. Server.
- Create a multitiered model that improves scalability and performance.
- Increase reliability by moving the important code to a well-defined container.
- Centralize resources to increase manageability.
- Promote reuse via well-defined, portable components.
- Provide interoperability with non-Java applications such as legacy systems.
- Be compatible with CORBA.

EJB Architecture

EJB component is created and managed at runtime by a container. It can be customized at deployment time. Client accesses to the EJB are mediated by an EJB container and server. An EJB server is a process that manages EJB containers, which manage the beans. It provides access to system services and Java-related services including a name space accessible via JNDI, a transaction service via JTS, and a security service. EJB container is a program entity in the EJB server hosting the EJBs, similar to that in a browser hosting the applets. Container defines the EJB's interactions with the outside world. It creates a safe environment for the bean and provides services to it, such as database accesses, transaction management, the bean's life cycle, and JNDI. Container provides these services to the EJB programmer through the standard interfaces. It makes the EJB as a reusable, write-once, run-anywhere server components.

EJB is a complex component model that not all the commercial Java web servers support. IBM WebSphere Application Server Enterprise Edition and Sun One Application Server are the two Java web servers that support the full services of EJB (see Figure 1.16).

EJB client can be a servlet, a JSP, an application, or other bean. They find the EJB by locating EJB home interface via JNDI and communicate to the home and remote interfaces of an EJB through the proxies (Home stub and EJBObjet stub).

The beans on the server-side have two interfaces: EJBHome and EJBObject. EJBHome is a container-provided object representing the home interface that allows you to create and find EJBs. It will return references

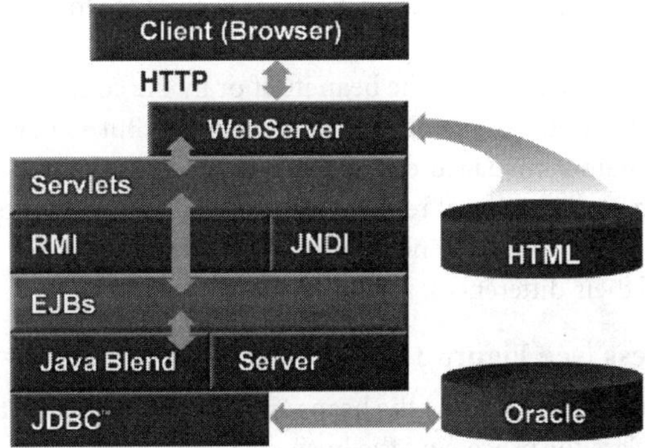

FIGURE 1.15   EJB in web-based computing.

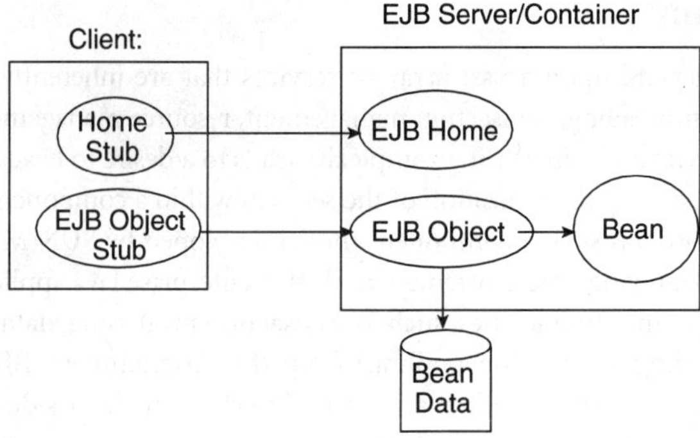

FIGURE 1.16   EJB programming model.

Session Beans	Entity Beans
Single client, Short lived Do not survive a crash.	Multiple clients, Long-lived, Survive a crash.
Stateless Session Beans High Performance & re-usable Single state transaction management.	Bean-Managed Persistence Entity Beans (BMP) Contain code to store and retrieve data
Stateful Session Beans Multi-state transaction management.	Container-Managed Persistence Entity Beans (CMP) Rely on EJB Container to store and retrieve data.

FIGURE 1.17   Differences between the Session Bean and Entity Bean.

to EJBObjects. EJBObject is another container-provided object representing the remote interface that handles tasks such as persistence (in container-managed persistent entity beans), distribution (interface with CORBA and RMI), and support for transactional behavior. The Bean itself implements the business logic.

Type of EJBs

There are two kinds of EJBs: session beans and entity beans. The session beans are normally within a client–web server session to process information and the entity beans are persistent and can reside at the server-side for a long period of time. An example of a session bean can be a login EJB that handles client login functions and an example of an entity bean can be a shopping cart EJB that handles online shopping requests.

The session beans can be stateless or stateful depending on whether some information is stored at the container or not between the client–server interactions. For the stateless session beans, the server can optimize resources by minimizing the number of beans and sharing them. The server does not have to worry about persistence. But clients may come across several beans in one session and the client's remove message may not result in a bean really being removed. For the stateful session beans, clients can reply on a single point of contact on the server. The server manages persistence for long running sessions. But the server has to create one bean per client.

The entity beans can also be managed by the bean itself or by the container. The bean-managed entity beans can optimize resource usage and arbitrary find methods. But it may have to port bean if one changes platforms. The container-managed entity beans have less code in the bean. The container may allow one to change the persistence without recompiling and change containers without recompiling. But it has less control over persistence and may not support all find methods.

Figure 1.17 lists some of their differences.

EJB Development Process (see Figure 1.18)

1. Define Remote Interface for accessing the beans and Home Interface for finding or creating beans, and write the Bean class for processing the business logic.

Example of Create a Stateless Login Session Bean

```
a. Define the remote interface
package loginSessionBean;
import loginSessionBean.*;
/**
 * This is an Enterprise Java Bean Remote Interface
 */
public interface Login extends javax.ejb.EJBObject Remote {

/**
 * @return java.lang.String
 * @param user java.lang.String
 * @param epwd byte []
 * @exception String The exception description.
 */
java.lang.String loginVerify(java.lang.String user, byte [] epwd)
    throws java.rmi.RemoteException;
}


b. Define the home interface
package loginSessionBean;
import loginSessionBean*;
/**
 * This is a Home interface for the Session Bean
 */
public interface LoginHome extends javax.ejb.EJBHome {
/**
 * create method for a session bean
 * @return loginSessionBean.Login
 * @exception javax.ejb.CreateException The exception description.
 * @exception java.rmi.RemoteException The exception description.
 */
Login create() throws javax.ejb.CreateException, java.rmi.RemoteException;
}


c. Develop the Session Bean class
package loginSessionBean;
import java.rmi.RemoteException;
import java.security.Identity;
import java.util.Properties;
import javax.ejb.*;
import javax.crypto.*;
import javax.crypto.spec.*;
import java.io.*;
import loginSessionBean.*;
/**
 * This is a Session Bean Class
 */
public class LoginBean implements SessionBean {
        private javax.ejb.SessionContext mySessionCtx = null;
        final static long serialVersionUID = 3206093459760846163L;
...
/**
 * ejbCreate method comment
 * @exception javax.ejb.CreateException The exception description.
 * @exception java.rmi.RemoteException The exception description.
 */
public void ejbCreate() throws javax.ejb.CreateException, java.rmi.RemoteException {}
/**
 * getSessionContext method comment
 * @return javax.ejb.SessionContext
 */
```

FIGURE 1.18　An EJB Session Bean development.

```java
/**
* @return boolean
 * @param user java.lang.String
 * @param pwd java.lang.String
*/
public String loginVerify(String user, byte[] epwd) {
String decMsg = new String();
String flag = new String();
java.sql.Statement stmt = null;
java.sql.ResultSet rs = null;
java.sql.Connection conn=null;
try{
        byte key[] = "abcdEFGH".getBytes();
        SecretKeySpec secretKey = new SecretKeySpec(key,"DES");
        Cipher cipher = Cipher.getInstance ("DES");
      cipher.init(Cipher.DECRYPT_MODE,secretKey);
        byte[] decryptedMsg = cipher.doFinal(epwd);
        decMsg = new String(decryptedMsg );
        conn = getDBConnection();
        stmt = conn.createStatement();
        rs = stmt.executeQuery("select password from userdetails where userid
='"+user+"'");
        if(rs.next())   {
            if(rs.getString("pa      ssword").equals(decMsg))
                flag = "true";
                }
            else{
                flag = "false";
                }
}catch(Exception e){
        System.out.print  In(e.getMessage());
}
return flag;
}

d. Compile, pack, deploys the Login Session Bean.
e. Develop the Login EJB Client: JSP
<%@ page import = "javax.rmi.*"%>
<%@ page import = "javax.crypto.*"%>
<%@ page import = "javax.crypto.spec.*"%>
<%@ page import = "loginSessionBean.*"%>
<%@ page contentType = "TEXT/HTML" %>

<html><head><title>Login Check</title></head>
<body bgcolor=#a5b56b>
<%String Username = request.getParameter("username");%>
<%String Password = request.getParameter("password");%>
<%String ejbJndiName = "java:comp/env/ejb/LoginBean";%>
<%Login loginbean = null;%>
<%javax.naming.InitialContext ctx=null;%>
<%String pwd = new String();%>
<%String loginverified = new String();%>
<%try {
byte key[] = "abcdEFGH".getBytes();
String msg = Password;
SecretKeySpec secretKey = new SecretKeySpec(key,"DES");
Cipher cipher = Cipher.getInstance ("DES");
cipher.init(Cipher.ENCRYPT_MODE,secretKey);
byte[] encryptedMsg = cipher.doFinal(msg.getBytes());
/////////EJB calling//////////
ctx = new javax.naming.InitialContext();
Object ref = ctx.lookup(ejbJndiName);
if (ref == null) { throw new Exception("ERROR: lookup failed");}
LoginHome home = (LoginHome)PortableRemoteObject.narrow(ref, LoginHome.class);
```

FIGURE 1.18   (continued)

```
loginbean = home.create();
loginverified = loginbean.loginVerify(Username,encryptedMsg);
}catch(Exception ex){ex.printStackTrace();
}
if(loginverified.equals("true")){response.sendRedirect("http://localhost/startpage");
}else{response.sendRedirect("http://localhost/errorpage");}
%>
</body></html>
```

FIGURE 1.18 (continued)

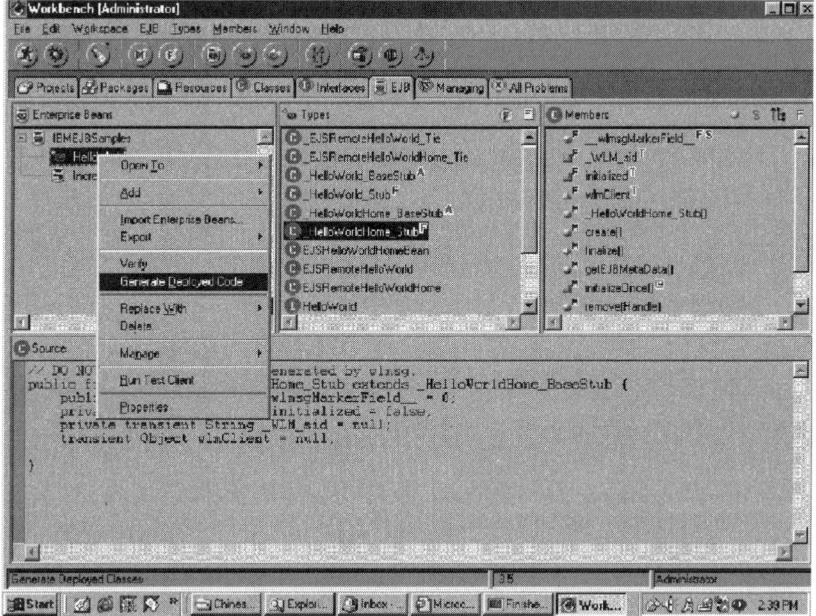

FIGURE 1.19 Generate the EJB deploy code in IBM WS.

2. Deploy Beans to EJB server. (The deployment process generates the client stub and server skeleton files, as well as a series of "helper" and "holder" classes. The exact names, numbers, and contents of these classes will vary with the EJB server and container.)
3. Write EJB clients.
4. Test Beans.

Many Java IDEs such as the IBM WS, SUN One Studio, and Borland JBuilder have the tools built-in to allow the developer to compile, pack, and deploy the EJBs. All the major Java IDEs claim that the EJB compiled and packed in their system can be deployed to any other EJB-supported web servers. But the experiences show that it is much easier to deploy the EJB created from the IBM WS to IBM WAS, deploy the EJB created from SUN One Studio to SUN One Application Server, and deploy the EJB created from Borland JBuilder to BEA WebLogic Server (see Figure 1.19).

Programming rules for session beans (see Figure 1.20):

- Session beans must implement the javax.ejb.SessionBean interface and have at least one method.
- Session beans must implement all of the ejbCreate methods.
- Session beans must implement their remote interfaces.
- The home interface for a session bean must define the create method.
- Stateless beans should only have a create method with no arguments.

<u>Example of Create a Container-Managed Persistence ShoppingCart Entity Bean</u>

```
a. Define the remote interface
package shoppingcartCMP;
/**
 * This is an Enterprise Java Bean Remote Interface
 */
public interface ShoppingCart extends javax.ejb.EJBObject {

/**
 * @return void
 * @param item shoppingcart.PurchasedItem
 * @exception String The exception description.
 */
void addItem(shoppingcart.PurchasedItem item) throws java.rmi.RemoteException;

/**
 * @return boolean
 * @param item shoppingcart.PurchasedItem
 * @exception String The exception description.
 */
boolean purchasedCart() throws java.rmi.RemoteException;
}

b. Define the home interface
package shoppingcartCMP;

/**
 * This is a Home interface for the Entity Bean
 */
public interface ShoppingCartHome extends javax.ejb.EJBHome {

/**
 * create method for a CMP entity bean
 * @return shoppingcart.ShoppingCart
 * @param argCartPK int
 * @exception javax.ejb.CreateException The exception description.
 * @exception java.rmi.RemoteException The exception description.
 */
shoppingcart.ShoppingCart create(String argCartPK)
     throws javax.ejb.CreateException, java.rmi.RemoteException;
/**
 * findByPrimaryKey method comment
 * @return shoppingcart.ShoppingCart
 * @param key shoppingcart.ShoppingCartKey
 * @exception java.rmi.RemoteException The exception description.
 * @exception javax.ejb.FinderException The exception description.
 */
shoppingcart.ShoppingCart findByPrimaryKey(shoppingcart.ShoppingCartKey key)
     throws java.rmi.RemoteException, javax.ejb.FinderException;
}

c. Develop the CMP Bean class
package shoppingcartCMP;
import java.sql.*;
import java.rmi.RemoteException;
import java.security.Identity;
import java.util.Properties;
import javax.ejb.*;
/**
 * This is an Entity Bean class with CMP fields
 */
public class ShoppingCartBean implements EntityBean {
       private    javax.ejb.EntityContext entityContext = null;
```

FIGURE 1.20   An EJB Entity Bean Development.

```java
            final static long serialVersionUID = 3206093459760846163L;
            public final static java.util.Hashtable items = new java.util.Hashtable();
            private java.sql.Connection dbConnection;
            public String CartPK = "";
...
/**
 * ejbCreate method for a CMP entity bean
 * @param argCartPK int
 * @exception javax.ejb.CreateException The exception description.
 * @exception java.rmi.RemoteException The exception description.
 */
public void ejbCreate(String argCartPK) throws javax.ejb.CreateException,
java.rmi.RemoteException {
        _initLinks();
        // All CMP fields should be initialized here.
        CartPK = argCartPK;
}
/**
 * ejbPostCreate method for a CMP entity bean
 * @param argCartPK int
 * @exception java.rmi.RemoteException The exception description.
 */
public void ejbPostCreate(int argCartPK) throws java.rmi.RemoteException {}
/**
 * ejbPostCreate method for a CMP entity bean
 * @param argCartPK java.lang.String
 * @exception java.rmi.RemoteException The exception description.
 */
public void ejbPostCreate(java.lang.String argCartPK) throws java.rmi.RemoteException {}
/**
 * ejbRemove method comment
 * @exception java.rmi.RemoteException The exception description.
 * @exception javax.ejb.RemoveException The exception description.
 */
/**
 * Insert the method's description here.
 * Creation date: (5/2/2003 1:18:52 AM)
 */
public void addItem(PurchasedItem item)
{
        PurchasedItem temp = new PurchasedItem();
        if(item!=null) temp = (PurchasedItem)items.get(item.prodname);
        if(temp!=null){
            item.quantity    +=temp.quantity;
            items.put(item.prodname,item);
            System.out.println    ("Item is Updated");
        }
        else   {
            items.put(item.prodname,item);
            System.out.println("Item    is Added");
        }
}
/**
 * Insert the method's description here.
 * Creation date: (5/2/2003 2:03:55 AM)
 * @return boolean
 */
public boolean purchasedCart() {
        //// Buy Pay and Go concept no exchange
        /// get the hash table, sure with inventory,and update inventory
        java.util.Enumeration   allItems;
        java.sql.ResultSet rs =null;
        java.sql.Statement stmt = null;
        PurchasedItem curItem = new PurchasedItem();
```

FIGURE 1.20   (continued)

```
            allItems = items.elements();
            int curQuantity = 0;
          try  {
            java.sql.Connection conn = getDbConnection();
          while(allItems.hasMoreElements()){
              curItem = (PurchasedItem)allItems.nextElement();
              System.out.println("Current Item name is:"+curItem.prodname);
              rs = stmt.executeQuery("select * from Inventory where
productname='"+curItem.prodname+"'");
              if(rs.next()){
                curQuantity = Integer.parseInt(rs.getString("quantity"));
                System.out.println("Quantity is :"+curQuantity);
                if(curQuantity >= curItem.quantity)
                  stmt.executeUpdate("update inventory set quantity ="+(curQuantity-
curItem.quantity));
              }
          }
          }catch (SQLException s){
                System.out.print     ln(s.getMessage());
          }
          return    false;
}
```

d. Compile, pack, deploys the ShoppingCart CMP Bean.

   Similar to the Login session bean operations, except the container will generate a list of finder, helper, and other management Java files. The following is the list of the files for the deployment of ShoppingCart CMP in Websphere:

```
EJSFinderShoppingCartBean.java
EJSJDBCPersisterShoppingCartBean.java
EJSRemoteShoppingCart.java
EJSRemoteShoppingCartHome.java
EJSShoppingCartHomeBean.java
ShoppingCartBeanFinderHelper.java
ShoppingCartKey.java
```

e. Develop the ShoppingCart EJB client: JSP

```
<%@ page import = "shoppingcartCMP.*"%>
<%@ page contentType = "TEXT/HTML" %>
<html><head><title>Shopping Cart</title></head>
<body bgcolor=cyan>
<form action="http://localhost/tabledetails" method =post name=f1>
<center><img src="shoppingcart.gif"></center><br>
<marquee direction=right><img src="cart.jpg"></marquee><br>
<%String ejbJndiName = "java:comp/env/ejb/ShoppingcartBean";%>
<%Shoppingcart shoppingbean = null;%>
<%javax.naming.InitialContext ctx=null;%>
<%
try{
ctx = new javax.naming.InitialContext();
Object ref = ctx.lookup(ejbJndiName);
if (ref == null) {throw new Exception("ERROR: lookup failed");}
ShoppingcartHome home = (ShoppingcartHome)PortableRemoteObject.narrow(ref,
ShoppingcartHome.class);
shoppingbean = home.create();
Connection con = shoppingbean.getDbConnection();
Statement stmt = con.createStatement();
String query = "SELECT * from tab";
ResultSet rs=stmt.executeQuery(query);
%>
 <h2><font color=red>Please select from the following Tables : </font></h2>
 <center><select MULTIPLE name="SelectList">
 <%
while(rs.next()) {
 %>
<option value="<%=(String)rs.getString("TName")%>" >
```

FIGURE 1.20   (continued)

```
<%=(String)rs.getString("TName")%></option>
<%
}
session.setAttribute("con",con);
session.setAttribute("shoppingbean",shoppingbean);
%>
</select></center><center><br>
<input type=submit value=submit>
<%
}catch(Exception e) {
 out.println(e.getMessage());
}
%>
</form></body></html>
```

FIGURE 1.20   (continued)

- Stateful beans must have at least one create method with the argument.
- The client locates the bean through JNDI. It should remove beans when they are done with them.

Programming rules for entity beans:

- Entity beans must implement the javax.ejb.EntityBean interface and have at least one method.
- Entity beans must define a custom primary key class.
- Entity beans must implement all the ejbCreate and ejbPostCreate methods.
- Entity beans must implement their remote interfaces.
- The home interface for a session bean must define the create method.
- The home interface must define the findByPrimaryKey method and can also define other find methods.

Roles in EJB Development

EJB development is a complex process, which usually requires a team of people working together to build, deploy, and maintain the EJBs. Some of the roles for the EJB development are as follows:

- *EJB Developers*: Create the Enterprise JavaBeans (Home interface, bean's remote interface, and bean).
- *Application Assemblers*: Builds applications from EJBs.
- *EJB Deployers*: Uses tools to configure EJB within the runtime environment.
- *Server Providers*: Create and sell EJB servers.
- *Container Providers*: Create and sell EJB containers (an EJB server can have multiple container types).
- *System Administrator*: Oversees the runtime.

1.6   Java Transaction Service

The enterprise computing involves many business transactions, where a transaction is a unit of work that has the ACID characteristics. The ACID stands for Atomic, Consistent, Isolation, and Durable.

- Atomic (all-or-nothing): If the transaction is interrupted by failure, all effects are undone (rolled back).
- Consistent: Effects of transaction preserve invariant properties.
- Isolated: One transaction will not be allowed to see the working state of another's data; transactions appear to execute serially.
- Durable: Effects of a completed transaction are persistent and never lost.

The Java Transaction Service (JTS) implements the Java mapping of OMG Object Transaction Service (OTS). JTS specifies the implementation of transaction manager (TM) that supports the Java Transaction (JTA) specification.

The Java Transaction API (JTA) specifies local Java interfaces between a Transaction Manager and the parties involved in a distributed transaction system. Two-phase commit protocol is used to allow multiple resource to update within a single transaction with ACID properties.

Phase 1:

1. TM: send "prepare" message to all the Resource Managers (RMs)
2. RMs: return "ready" messages back to the TM.

Phase 2:

3. TM (after received the "ready" message from all the RMs): send "commit" message to all the RMs.
4. RMs: return "OK" messages back to the TM.

Example of Java Code using JTA:

```
try {
  trans.begin();
  ... do work ...
  trans.commit();
} catch (Exception exp) {trans.rollback();}
```

Transactions and EJBs

Currently, the EJB supports the flat transaction (not nested transaction, i.e., transaction inside another transaction). EJB specification supports distributed transactions for both session and entity beans. The EJB container implements the transaction demarcation and management. So the burden of managing transaction is shifted to the container and the EJB server provider and the Java developer can declaratively specify transaction semantics without the implementation details. EJB server and container provide low-level services such as two-phase commit, interaction with databases, and transaction propagation.

1.7   J2EE web services

Web services are independent of any particular operating systems, programming languages, or databases. Therefore, it quickly becomes the preferred technology for the web-based development. J2EE supports the following web service standards developed recently.

- XML (Extensible Markup Language) — Distributed Data Specification.
- SOAP (Simple Object Access Protocol) — Distributed Information Exchange in HTTP.
- WSDL (Web Services Description Language) — Web service description.
- UDDI (Universal Description, Discovery and Integration) — Web service Registration.

Data, services, or systems configured in XML are much easier to process on the Internet. Distributed objects accesses using SOAP simplify the message exchanges than the CORBA-styled ORB protocol. Web services described in WSDL can provide a powerful invocation for the web clients, and UDDI provides a simple web services lookup for the web clients.

J2EE web service is a programmable application component. It provides transparent services through the web to the users. A web service defines a business process, not a technology implementation so that it enables a rapid and simple creation and integration of "business services." J2EE supports the web services via XML and standard web protocols.

J2EE Web Service Description

J2EE describes the web services using WSDL that defines an XML grammar for describing web services as collections of message-enabled endpoints or ports. In WSDL, the abstract definition of endpoints and messages is separated from their concrete deployment or bindings. A binding is a concrete protocol and data format specifications for a particular endpoint type. An endpoint is defined by associating a web address with a binding, and a collection of endpoints defines a service.

J2EE Web Service Implementation

J2EE wraps the server objects using XML-based RPC and exposes them as web services. It complies with the SOAP specification. Once a web service component is implemented, a client can send a message to the component as an XML document and the component can send an XML document back to the client as the response.

J2EE implements web service as a JAX-RPC (Java API for XML-based RPC) service endpoints by wrapping the existing Java classes and applications or as a stateless EJB session bean by wrapping the EJBs. The resulting wrapper is a SOAP-enabled web service that conforms to a WSDL interface based on the original EJB's methods. The J2EE web services architecture is a set of XML-based frameworks, providing infrastructures that allow companies to integrate business-service logic that was previously exposed as proprietary interfaces.

J2EE supports web services via the Java API for XML Parsing (JAXP). This API allows developers to perform any web service operation by manually parsing XML documents. JAXP is for processing XML data using applications written in the Java programming language. It leverages the parser standards SAX (Simple API for XML Parsing) and DOM (Document Object Model) so that one can choose to parse one's data as a stream of events or to build an object representation of it. JAXP also supports the XML Stylesheet Language Transformations (XSLT) standard, giving control over the presentation of the data and enabling one to convert the data in to other XML documents or to other formats, such as HTML. Current J2EE SDK included a Java web services Developer's Pack (JWSDP) for the web service development.

J2EE Web Service Publishing, Discovery, and Binding

J2EE uses the registries to publish, discover, and bind web services. Registries contain the data structures and taxonomies used to describe web services and web service providers. A registry can either be hosted by private organizations or by neutral third parties. Java API for XML Registries (JAXR) is used for interoperating with multiple-registry types. JAXR allows its clients to access the web services provided by a web services implementer, exposing web services built upon an implementation of the JAXR specification.

J2EE Web Service Invocation and Execution

J2EE uses the Java API for XML-based RPC (JAX-RPC) to send SOAP method calls to remote parties and receive the results. JAX-RPC enables Java developers to build Web Services incorporating XML-based RPC functionality according to the SOAP 1.1 specification. The SOAP is a simple, lightweight XML-based protocol that defines a messaging framework for exchanging structured data and type information across the web. Web service recipients operate as SOAP listeners and can notify interested parties (other Web Services, applications, etc.) when a Web Service request is received. The SOAP listener validates a SOAP message against the corresponding XML schemas as defined in a WSDL file. The SOAP listener then unmarshals the SOAP message. Within the SOAP listener, message dispatchers can invoke the corresponding Web Service code implementation. Finally, business logic is invoked to get the reply. The result of the business logic is transformed into a SOAP response and returned to the Web Service caller.

1.8   Conclusion

This article briefly introduced enterprise computing using J2EE technology. Figure 1.21 summarizes the J2EE applications in web-based computing.

J2EE supports both "thin" clients that only need a browser to run and "fat" clients that may need a set of Java packages to execute the applications including the Web services. J2EE Web container supports JSP

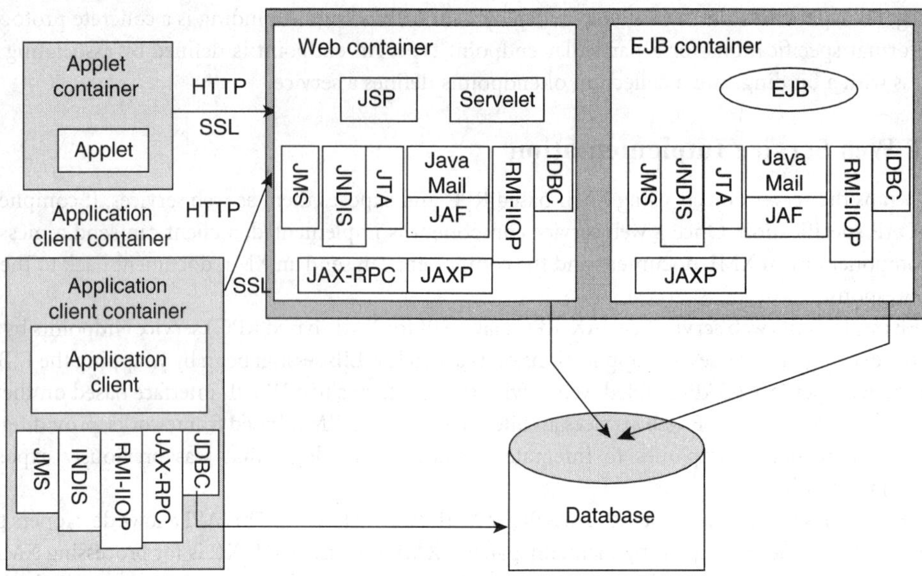

FIGURE 1.21   Web-based enterprise computing using J2EE.

and servlet. Both JSP and servlet can be used to process the business logic and generate the dynamic HTML contents. A common practice is to use the JSPs for the client interactions (similar to the ASPs in Windows .Net) and servlets for the server working objects execution (similar to the VisualBasic, VisualC++, or C# objects in Windows .Net). The EJB container hosts the EJBs for the server components computing. An enterprise Java web server and application server can host many EJB containers, one for each application. J2EE supports a large set of APIs for variety of services. Some of them, such as JMS (Java Message Service), JavaMail, and JAF (JavaBeans Activation Framework), are not discussed in this article.

J2EE is a major web-based enterprise level computing technology. It relies on the Java technologies developed, and adapted many advantages of the distributed object computing component models such as CORBA. J2EE has been chosen as the enterprise computing technology by a wide range of companies and organizations because of its multi-platform capability. (Microsoft .Net works mostly on the Window platforms.) Particularly, J2EE has a dominant influence on the UNIX and LINUX platforms, which are the major enterprise server platforms of today's enterprise computing.

References

[1] Rajan Shah and Jiang B. Liu, "Constructing Client/Server Systems in Heterogeneous Environments," *Proceedings of 10th International Conference on Computer Applications in Industry and Engineering*, San Antonio, TX, November 1997, pp. 131–135.

[2] Jon Siegel, *CORBA — Fundamentals and Programming*, Wiley Computer Publishing, New York, 1996.

[3] Roger Sessions, *COM and DCOM — Microsoft's Vision for Distributed Objects*, Wiley Computer Publishing, New York, 1998.

[4] Stephen Asbury and Scott R. Weiner, *Developing Java Enterprise Applications*, 2nd ed., Wiley, New York, 2001.

[5] Microsoft .Net Developer Training Set, Microsoft Corporation, 2002.

[6] Jiang B. Liu, "Multi-tiered Internet Computing Using Java Technologies," *Proceedings of IECON 27th Annual Conference of the IEEE Industrial Electronics Society*, Denver, CO, December 2001, pp. 1789–1793.

[7] Sun Microsystems J2 Platform, Enterprise Edition (J2EE), http://java.sun.com/j2ee.

[8] Osamu Takagiwa et al., *WebSphere Studio Application Developer Programming Guide*, IBM Redbooks, 2002.

[9] Rufus Credle, *et al.*, *WebSphere Solution Guide: WebSphere Application Server — Express*, Version 5.0, ibm.com/redbooks, April 2003.

2

Microsoft's .NET

Active Data Objects (ADO.NET) • Active Server Pages (ASP.NET) • Interoperability with the Component Object Model • Interoperability with the Win32 API • Web Services

Martin v. Löwis and
Peter Tröger
University of Potsdam

Microsoft has published the .NET framework in 2000 as an integral component of their Windows family of operating systems. It is an execution environment for application code that makes many aspects of the underlying hardware and operating system transparent. In particular, it allows application developers to abstract from the programming language, from the deployment procedures used to get the application to the user, and, to a certain degree, from the operating system, computer hardware, and network infrastructure on which the user runs the application.

Microsoft has accompanied the .NET framework with a number of software tools that support the analysis, design, programming, testing, and deployment of applications of the .NET framework. The Microsoft Visual Studio .NET line of products integrates many of the development activities for a .NET application — in addition to supporting the traditional way of developing software for the Windows operating system (which is now called "native" code).

While .NET is designed to be language independent (using mechanisms that will be presented later), Microsoft has introduced the new programming language C#, which gives full access in a convenient syntax to all aspects of the framework. Consequently, we will use C# throughout this chapter, and only mention other languages supported by .NET in order to demonstrate the language independence.

2.1 Characteristics of the .NET Framework

The .NET framework consists of two major components: the *common language runtime* (CLR) and the *.NET class library*. The CLR is the core of the framework; it includes the execution engine proper for .NET applications. The CLR provides a "virtual" environment for execution, meaning that the application code has no way of directly interacting with the computer hardware. Instead, all application operations are controlled through the CLR. This approach of a virtual machine is shared with a number of other systems, such as the Smalltalk or Java virtual machines. It allows safe and secure execution of code: the virtual machine can detect certain error conditions in the application code, and report the detected errors in a reliable manner (safety). It can also detect attempts of malicious code to gain access to a system that has not been granted to that code, and reject such attempts (security).

Unlike many earlier implementations of virtual machines, the .NET framework is not bound to a single programming language. Instead, an explicit design goal was support for multiple languages. This goal has been achieved through three mechanisms: a *common intermediate language* (CIL, the byte code of the .NET framework), a *common type system* (CTS), and a mechanism to invoke arbitrary "native" code of the underlying platform.

The *common intermediate language* as approach is, in principle, shared with many other virtual machines: all languages implemented in virtual machines translate the high-level constructs of the programming language into lower-level primitive operations, and it turns out that the primitive operations are mostly independent from the programming language. Still, the precise choice of primitive instructions (byte codes) constrains the expressiveness of the virtual machine and the features of programming languages that can be implemented on that machine. Even though all application logic is expressed in terms of byte code instructions, execution is still efficient when run on the Microsoft implementation of the .NET framework. In the Microsoft implementation (and, indeed, most alternative implementations), this is realized by compiling CIL code into machine code just before it is executed (*just-in-time compilation*) (see Figure 2.1).

In addition to the common intermediate language, .NET introduces the notion of a *common type system*. Just like the instruction set defines the building blocks for algorithms, the common type system defines the building blocks for the data structures. The notion of a type system shared across multiple languages is unique in the .NET framework; most virtual machines only target a single language in their type system — even though they could possibly execute different languages with the instruction set. The common type system of .NET covers both built-in types (integers, floating point numbers, pointers etc.), as well as structured types (classes, interfaces). It supports both *value types* and *reference types*. Values of value type get copied in parameter passing and assignment, do not have an identity independent from their state, and they are stored on the stack. Values of reference types do have an identity, assignment or parameter passing only transfers a reference, and they are stored in the heap. Conversion between value types and reference types is supported by means of operations called *boxing* and *unboxing* that are performed implicitly by the runtime when needed, for example, when assigning a value-type variable (such as integer) to a reference-type variable (such as object). Values are stored in typed *locations*. This allows type-safe execution of all CIL code.

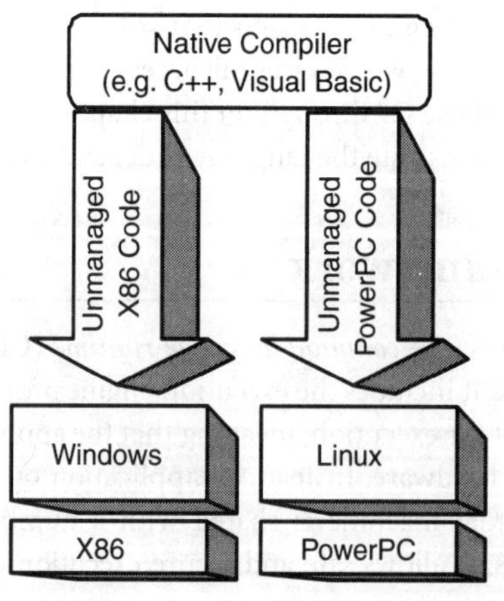

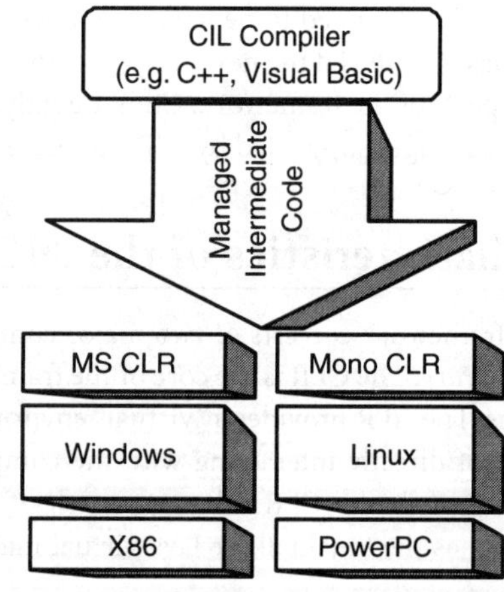

FIGURE 2.1   Managed code concept.

With the CIL, and the common-type system, most features of the widely used programming languages can be represented adequately. However, some language constructs just cannot be expressed using these abstractions. A compiler tool targeting .NET then has the choice of either dropping the feature of the language, or of reverting to native (machine) code. The .NET framework provides explicit hooks to incorporate native code into the runtime system; if these mechanisms are used, the resulting application will lose some of its transparencies (e.g., independence from the microprocessor architecture). An application code that is completely under the control of the runtime system is called *managed code*; a code that runs natively on the processor is called *unmanaged.*

The type system is object-oriented, meaning that references always refer to objects that are instances of classes. Classes support inheritance, polymorphism, and late binding; a class can implement multiple interfaces. Multiple inheritance of classes is not supported in the CTS; languages that support that feature (such as C++) cannot use it in conjunction with managed types. In the case of C++, this leads to the definition of a "managed C++" syntax with several new keywords (e.g., to mark CTS-compliant classes) and slightly changed language semantic. The Microsoft C++ compiler supports a mixture of managed and unmanaged code in a project, mainly to simplify the migration of the existing C++ code to the .NET environment.

The class hierarchy of the type system is single-rooted in the class *System.Object*. All values provide runtime information about their type, including information about the inheritance, implemented interfaces, public and private fields, and public and private methods. This feature is commonly known as *reflection*, and it allows application to dynamically discover and use properties of objects. To implement reflection, .NET stores meta-data about the classes of an application along with the application.

Furthermore, access to fields and operations is not restricted to the current operating system process, or even to the local computer: .NET provides transparent access to objects located remotely — subject to certain semantic and administrative constraints. This functionality is offered in the *System.Remoting* classes of the framework [Rammer02].

As .NET is designed to support multiple programming languages, interoperability across programming languages becomes an issue. Along with the .NET specifications, Microsoft developed the CLS, which describes constraints to which programming languages adhere in order to interoperate with other languages. A CLS-compliant class can be used across programming language, which, for example, allows developers to inherit a class in one language from a base class written in another language. Programming language tools use the meta-data stored in assemblies to discover the available classes, and reverse-map them into the programming language constructs.

Applications of the .NET framework are physically represented in *assemblies*, which are typically files available on a local or remote disk, or downloadable through some network protocol such as HTTP. The assembly contains everything needed to run the application. As a result, deployment of an assembly might not involve anything more than copying the assembly file onto the local computer, without requiring to run a local installation routine to make the software operational. Microsoft has coined the term *xcopy deployment* to describe this approach (named after the Microsoft DOS utility xcopy.exe).

As code is easily downloadable from remote locations, security becomes an important issue: code downloaded over the network might not be trustworthy, and might maliciously attempt to modify or delete data of the computer user, to spy out data and send them to remote locations, or to sabotage the local computer installation. .NET uses a sandboxing approach to execute trusted code, similar to the sandboxing used in Java implementations. However, .NET allows the administrator very fine-grained control over what software does, and it supports the execution of partially trusted code: if a trusted code is invoked from an untrusted code, the trusted code, by default, runs with less privileges than it would normally have. The default settings prevent software downloaded from remote locations to do much harm, and the administrator of a network can easily customize settings networkwide to tighten or loosen control over untrusted code.

Microsoft has submitted major parts of the .NET platform to standardization organizations. In particular, European Computer Manufacturers Association (ECMA) has published the standards ECMA-334 on C#, and ECMA-335 on the CLR. These standards have been forwarded to ISO, and are now also published as ISO/IEC 23270:2003 and ISO/IEC 23271:2003, respectively [ISO03a, ISO03b]. This allows other software vendors to reimplement .NET, and to port it to other computer systems. While the Microsoft .NET framework product (in its version 1.1) is only available on Win32 systems (Windows 98/NT and successors) on Intel x86 processors, ports to other systems are available. Microsoft Research has published the *Rotor* package for noncommercial purposes, which ports .NET to Apple OS X (using PowerPC processors) and FreeBSD (using x86) [Stutz03]. The *GNU Mono* project allows execution of .NET applications on Linux systems, using various processors: x86, PowerPC, IBM S/390, and Intel StrongARM (although only the x86 port uses just-in-time compilation).

Microsoft has also started supporting the .NET platform on embedded devices, targeting primarily hand-held personal information systems. This product is known as *Compact Framework*. At the time of this writing, PocketPC and Windows CE.NET are supported, on ARM, MIPS, and Hitachi SH3 processors. In Compact Framework, tradeoffs have been made between the .NET feature set on the one hand and the available computing power and memory of embedded devices on the other. As a the result, the Compact Framework supports only a subset of the functions available in the .NET framework.

To support development of .NET applications, Microsoft has updated the Visual Studio product to incorporate compilers, debuggers, and other tools specifically designed for .NET; the result product is named *Visual Studio .NET* (VS.NET). New in this tool suite is a C# compiler, which only supports development of .NET application. The C++ compiler available in earlier products is still available, but it got extended with the *managed C++* mode. Furthermore, the *Visual Basic* language including its compiler was redesigned to support .NET only.

2.2 The CLR

The CLR acts as an execution environment in the .NET framework architecture. It is responsible for managing the assembly execution and all runtime-related services. The CLR provides certain "core" functions to the application, which are made available through byte code instructions and library classes. It performs garbage collection on objects that have been allocated and are not used anymore. It also performs several runtime checks such as boundary checks on array indices in order to prevent applications to run over the array end. In case of errors, the runtime system is capable of raising language-independent exceptions. It also supports user-defined exception handling in a language-independent way, allowing those CLR-compliant exceptions to travel across programming language boundaries. The CLR also provides support for multithreading through a set of synchronization mechanisms and an efficient thread-pooling strategy. The realization of all security infrastructure aspects is another major task of the CLR.

In the typical .NET application, the source code written in some programming language is compiled using a language-specific compiler. This compiler generates CIL byte code. The ECMA definition of CIL includes the object model, the distinction of value and reference types, and means to integrate managed (CIL) code and unmanaged (native) code. The exact runtime-semantics of these means is implementation-defined and can differ between vendor-specific variations of the CLR.

For the execution of a CIL binary, a *runtime host* loads a CLR instance. Different runtime hosts for .NET applications allow the customization of the managed code execution regarding the requested CLR/Framework version, CLR build type (workstation vs. server), garbage collection settings, and other specific properties. In case of the Microsoft .NET implementation, there are runtime hosts for Internet Explorer (MIME filter), ASP.NET (ISAPI filter), Microsoft SQL Server (for stored procedures), and Windows Shell. Most developers will get in touch only with the Windows Shell runtime host since it is activated when starting a .NET application in the regular manner from Windows Explorer.

Microsoft .NET compiles the CIL of a called method once into native code before executing it for the first time. Before doing so, a consistency check is performed on the byte code, to detect erroneous or malicious code that violates the type system. This aspect is called *checked execution* (equivalent to the byte code verification in the Java Virtual Machine).

The Microsoft CLR also provides an integrated support for debugging and profiling, allowing the developer convenient access to the runtime state of the .NET virtual machine. Profiling data can be used to detect "expensive" code, and are thus a mechanism to improve the performance of the resulting application.

2.3   Metadata

As one of its most interesting features, .NET allows declarative metadata information in assemblies. Mandatory metadata, which is included by the compiler automatically, describes the assembly itself (i.e., name, version, cultural information, and security settings) and its defined types. In addition, the metadata stored for a type (class, interface) or a member (method, field, property) are extensible, by means of *attributes*. An attribute is an annotation, which is, at runtime, represented by an object. In particular, several predefined attributes in the framework class library are used to represent security attributes, customization of GUI builders, transactional behavior of objects, COM interworking and integration of native code, and declaration of serialization properties of class members:

Custom implementations of attributes are also possible by deriving from the *System.Attribute* class:

These attributes can be used exactly like the predefined ones:

2.4   Application Development and Execution

To develop .NET applications, various development environments are available at the time of this writing. Microsoft itself offers the Visual Studio .NET product. Borland has the "*C#Builder for the Microsoft® .NET Framework*" product. A number of free software projects are enhancing their own development environments; most noteworthy is the *SharpDevelop* tool by Christoph Wille.

Most of these tools support primarily C#, as this language gives access to .NET features in the most direct way. In addition, several tools have been developed that support other languages, either as integrated environments or as stand-alone compilers. In addition to C#, compilers are available for Visual Basic .NET, C++, J# (a Java look-alike defined by Microsoft), Eiffel, COBOL, Oberon, APL, Fortran, Mondrian, Haskell, Mercury, and others.

Independent from the programming language chosen, the compiler will translate the source code into CIL code, and store this code in a file on disk called a *module*. One or more of these modules form an assembly, which is a deployable representation of the program. One file in the assembly will contain the *manifest*, which is an on-disk representation of the metadata of the application (information about classes, fields, methods, and so on). There are two kinds of assemblies: executable applications (.EXE) and libraries (.DLL).

At runtime, an executable and a set of libraries together form the application. In managed code, the operating system's notion of address spaces does not directly translate to the execution of the code — instead, much more fine-grained protection is applied. The notion of a .NET application domain (*AppDomain*) is introduced as an isolated execution context for a set of assemblies and objects created through these assemblies. Within an AppDomain, objects can directly refer to each other, and methods can be called directly. Multiple AppDomains can coexist in a single address space inside the runtime, allowing efficient execution of multiple AppDomains, for example, within a web server. Access to objects in other AppDomains (whether in the same OS process, on the same machine, or a different one) is only possible through network communication, typically done using remoting. Normally, the runtime host assigns a newly loaded assembly to the default AppDomain.

As assemblies are developed, multiple versions of the same assembly might become available that are not be completely compatible. In the past, concurrent deployment of multiple versions of the same "native" library (.DLL) on Windows has caused significant maintenance problems; these problems are colloquially known as "DLL hell" caused by the fact that the identity of a DLL is based only on its file name. With .NET, concurrent installation of multiple versions of the same library is possible, and applications will only use the libraries for which they were tested originally. This is achieved by means of a *strong name* of an assembly, which not only includes the file name but also the version of the library, the culture (locale) for which it was developed, and a cryptographic hash. The latter not only guarantees that no two libraries have the same name; as the hash is cryptographically signed, it also allows detecting malicious replacement of a library with a Trojan horse. The runtime is responsible for checking version information, and it throws an exception in case of a mismatch.

2.5 Security

To allow users to execute code from remote locations without risking the integrity and privacy of their own data, .NET implements access restrictions for "untrusted" code [Freeman03]. Assemblies in an AppDomain are assigned permissions at the time the assembly is loaded, and these permissions are then used to grant or deny operation calls made by the assembly (*Code Access Security*). The permissions granted are based on the *trust* into a specific assembly; trust is established by *evidence* that proves that the assembly is trustworthy. *Evidence* can be based on the zone an assembly originates from (local computer, local network, internet), on the source Uniform Ressource Identifier (URI) of the assembly, on the digital signature, or on the strong name of the assembly.

Using the evidence established, .NET considers certain administrative *policy* settings to compute the permissions that it grants. The policy settings are hierarchical, with the hierarchy levels being enterprise, machine, user, and application domains. Permissions denied on a higher level of the hierarchy cannot be granted on the lower level; however, the lower levels can further restrict the permissions. This is achieved on load time by checking the permissions from the enterprise level down to the application domain level, intersecting the resulting permission sets. If a permission is granted in all policy levels, the assembly gets the permission. This is useful in the context of large company installations where an administrator can override user-specific security settings through the deployment of appropriate enterprise security policies.

On each level of the policy hierarchy, certain *code groups* are defined; the evidence is then used to determine all groups that an assembly belongs to. Each group is assigned a *permission set*, which defines the permissions being granted. The effective permission set for an assembly in the context of one policy level is the union of all applicable code group permission sets. During security check, the runtime traverses the code group tree with "depth search" (see Figure 2.2). For each group, it checks the membership condition with the assembly evidence information and adds the code group permission set to the assembly if the condition evaluates to true. The traversal continues successively to all the subgroups until the leaf group is reached or the condition is not satisfied.

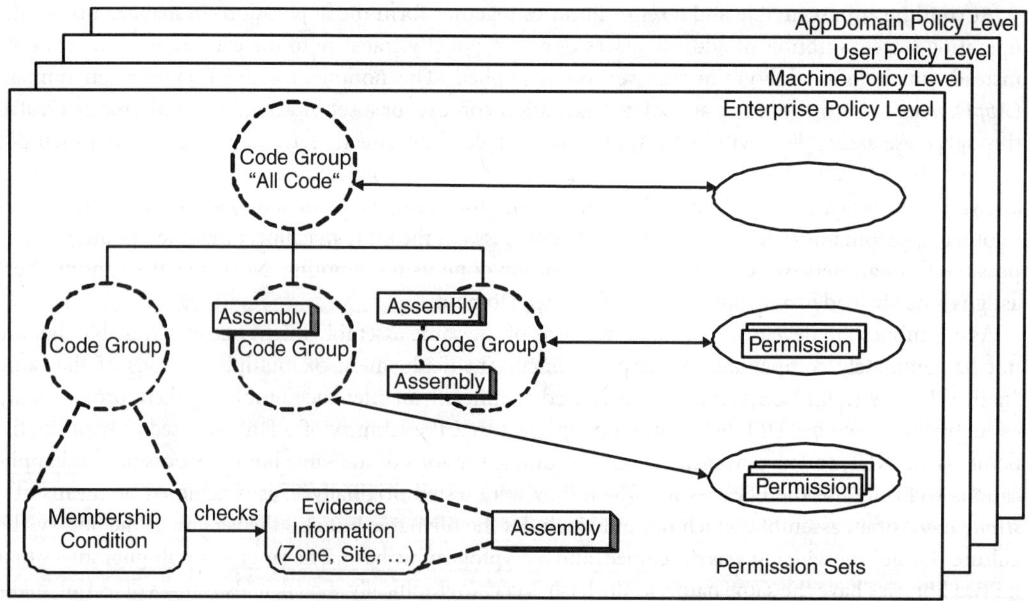

FIGURE 2.2 NET code access security.

The .NET runtime comes with several predefined code groups and named permissions sets (e.g., *FullTrust, Execution, Nothing, Internet, ...*); however, every concept in the .NET security model is extensible through an own implementation.

In addition to the administrative management of security policies, .NET also supports programmatic management in the assembly itself, using both a declarative and an imperative style. Using predefined attributes, an assembly, a class, or method can declare that it requires certain permissions; execution will fail if these permissions have not been granted. An assembly can also declare that certain permissions should not be granted to the application, even if the assembly has the appropriate evidence. This allows users to develop a code that cannot be abused, for example, through malicious application input. As a third possibility, the developer can declare requests for optional permissions.

To determine at runtime whether a permission has been granted, a stack walk is made, so that calls from untrusted code to trusted code can only use the permissions of the untrusted code. A certain mechanism allows trusted code to elevate its permissions beyond those of the caller; this mechanism should be initilized with care.

In cases where the declarative syntax is not powerful enough, an imperative approach (using explicit library calls) can be used. This is more tedious to use, but allows to base decisions on data only available at runtime. Additionally, the security mechanism can be applied to a subset of the execution path, like a part of a method.

Permissions based on evidence of the code only cover execution of an untrusted code. Another common scenario for access control includes restrictions based on the identity of the user. The .NET system provides *role-based security*, where access checks are made based on the role in which the current user acts. To use role-based security, .NET first establishes the *identity* and the *principal* of the user, for example by considering authentication data that the user has provided. At runtime, every thread has one assigned principal, which itself has an assigned identity. Based on the established principal, the application can request assignment of roles. In .NET 1.1, this is done only in an imperative way, without support for declarative or administrative management of principals or roles. This feature interoperates with the role-based security of COM+ 1.0.

2.6　Deployment

In the simplest case, .NET applications can be deployed by just copying them into the target directory, and running them from there. There is no need for registration of applications or libraries with the system, unlike COM. If a library is meant to be used by multiple applications, it must be installed by copying them into the *global assembly cache* (GAC). This can be done programmatically, using command line tools, or interactively by drag-and-drop.

Once an assembly is deployed, management of the assembly on the computer becomes available. The administrator can manipulate the security settings of the assembly (by putting it into an appropriate code group), and she can manipulate the *binding* of the application. By default, an assembly requires the precise versions-dependent assemblies to run. If a new version becomes available, old applications will continue to use the old version. If this is not desired, the administrator can redirect (rebind) the application to use the newer version.

VS.NET supports various additional scenarios for the development of deployable applications. A *setup project* allows the creation of a Microsoft Installer (.msi) file, which provides a user interface for customization and feature selection during the installation of an application. In addition, an installer file allows the automatic deployment of an application to a group of computers, using Windows *group policy*.

A *web setup project* allows automatic deployment to a Web server from within VS.NET, using a proprietary communication protocol. Such a project includes registration of the application in the Web server, and provides means for remote configuration and debugging.

A *merge modules project* allows the creation of *merge modules*, which are reusable building blocks for the creation of installer files.

Finally, a *CAB project* allows the creation of cabinet files, which allow automated installation of ActiveX controls in web browsers.

2.7 Framework Class Library

In addition to the bare execution environment, the .NET framework includes a rich library of classes that support various kinds of applications.
In particular, there are class libraries for:

- Organizing data structures (object container libraries).
- Access to operating system features (file and terminal I/O, threading).
- Reflection and management of code properties.
- Networking libraries (TCP and UDP communication, and libraries on top of TCP, such as HTTP).
- Computing in integral and binary and decimal floating-point numbers.
- Generation and parsing of XML (XPath, XSLT, XSD).
- Serialization of object graphs into a stream of bytes.
- Cryptographic services.
- Security (authentication, authorization, and access control).
- Accessing remote objects over the network (*remoting*).
- Development of graphical applications (GUI: Windows Forms, 2D drawing).
- Access to relational databases (ADO.NET).
- Development of Windows services.
- Development of Web services.
- Development of Web applications (ASP.NET, WebForms).

The *base class library* (BCL) of the .NET framework is part of the ECMA specification, and it contains as a subset the primitive types, string processing (based on Unicode), regular expressions, object collections, custom attributes, reflection, and threading.

The windows forms library provides access to the windowing system, with classes representing various input and output controls. In addition, the drawing library provides access to *GDI+* (graphics device interface) for video and printing device-independent programming.
Some of the other libraries mentioned above will be presented in more detail below.

Active Data Objects (ADO.NET)

The ADO.NET library (in the assembly *System.Data*) provides access to various databases through a consistent programming interface. The databases are typically relational databases that are accessed either locally or over a network. ADO.NET does not directly implement the logic to access a database. Instead, it can incorporate various preexisting generic database adapters, such as ODBC or OLE DB. In addition, database vendors can provide database adapters specifically designed for ADO.NET.

The ADO.NET object model defines a set of interfaces that a database provider must implement. The *DataReader* objects, implementing the *IDataReader* interface, offer access to the result of a query to a particular database type. This result can be generated through the usage of appropriate *Connection* and *Command* objects. The access to the result set is read-only (meaning that data cannot be modified) and with a forward-only cursor (meaning that results are retrieved in a sequential order, processing each result record only once).

The recommended way for working with query results in ADO.NET is the *DataSet* approach. DataSet objects are tightly integrated with the Windows Forms and WebForms library, which allows convenient development of database visualization applications through the usage of a *DataGrid* object. A DataSet object is an in-memory representation of tables, relationships, and constraints. It does not necessarily represent a relational database; instead, the on-disk representation could just as well be a text file. The DataSet is an off-line representation, which means that all cursor operations are local to the cache only. Modifications to the DataSet are possible, and will be written back on application request.

The association between the DataSet and the data source is established through a *DataAdapter* object. Microsoft .NET ships with four data adapters: OleDbAdapater, SqlDataAdapter (for MS SQL Server), OdbcDataAdapter, and OracleDataAdapter. The adaptors rely on the functionality of the underlying DataReader, Command, and Connection object implementation for the particular database. The SQL statements required for retrieving and updating data are customizable for a particular adapter in order to implement specific behavior, that is, the usage of stored procedures.

In addition to the disconnected mode of operation with a DataSet, ADO.NET also supports direct access to data sources by means of database connections. Using the underlying connection and command objects directly, concurrent and transactional modifications of a database become possible.

ADO.NET not only operates on relational database; in particular, it provides access to XML files by means of *XMLReader* objects. The dataset of an XML file can be described through an XML Schema, which the xsd.exe tool compiles into an appropriate application code (in C#, Visual Basic, JScript, or Visual J#). Using this code, a DataSet can then read and write XML files.

Active Server Pages (ASP.NET)

ASP.NET is a framework for the development of dynamic web pages, similar to ColdFusion or PHP. It consists of a .NET library (the assembly *System.Web*), and an Internet Server Application Programming Interface(ISAPI) filter for Internet Information Server (IIS). Alternatively, the Mono project offers an ASP.NET implementation that runs on the Apache webserver. The framework supports various typical aspects of a dynamic web application, such as the state management, the integration of static and dynamic parts of a web page, and the processing of HTML forms.

The framework defines an object model for the structuring of web pages named *WebForms*. A web form is a collection of "server-side controls," which are objects that render themselves as an HTML fragment. ASP.NET combines these fragments together with the static HTML fragments to form the dynamic page. Standard controls are provided for the HTML forms elements; these controls support server-side validation of the input (along with the creation of dynamic HTML displaying the error message in case of a validation failure). User-defined controls give reusable chunks of dynamic HTML, allowing integration of the same logic into multiple pages.

Active server pages are stored in .aspx files, which follow a *code-behind* philosophy: the HTML source contains references to the controls; the actual application logic of the control is stored in a separate C# file. Directly integration of source code into the HTML page is also supported.

Session management is fully transparent in ASP.NET; control objects retain their state across multiple accesses to the same control within the same session. The necessary state information is encoded in the HTML pages, and sent back to the web server by the web client.

An ASP.NET application can be easily extended to support mobile devices through the *mobile web forms* library (*System.Web.Mobile*).

Interoperability with the Component Object Model

In the last decade, Microsoft has favored the Component Object Model (COM) and the ActiveX set of technologies as the primary architecture for developing reusable software components. As a result, huge investments in this technology have been made in the industry, making COM the primary means of programming interface on Microsoft Windows systems.

With the appearance of .NET, developers have a strong interest in continuing the use of such components. In particular, certain kinds of application logic (such as the Internet Explorer) are not directly available in .NET, but would be available through COM interfaces. Consequently, Microsoft has put significant effort into making usage of COM in .NET as seamless as possible.

To use a COM object from a .NET application, .NET creates a *Runtime Callable Wrapper* (*RCW*) as a mediator. This wrapper is transparently derived from a COM-type library in VS.NET, or by means of a command line utility. The RCW interfaces between the managed and the unmanaged code; in particular, it converts the calling conventions, exceptions, and the type system appropriately. The garbage collection in .NET also integrates with the COM memory management: disposal of the RCW causes the COM object to be released. In addition to using the RCW, late binding (through the IDispatch interface) is also possible.

Calling .NET objects through COM interfaces is also possible: .NET transparently generates *COM Callable Wrappers* (*CCW*) as mediators; these primarily offer the *IDispatch* and *IUnknown* interfaces. Assemblies exposing COM interfaces must be signed with a strong name, .NET classes implementing COM classes must have a parameter-less constructor, and classes and methods that are intended to be COM-accessible must have the *COMVisible* attribute. A command line tool takes care of the registration of the assembly as a COM server; the .NET library mscoree.dll then acts as the COM server.

By means of the COM interoperability, WinForms applications can integrate ActiveX controls. Some COM+ features (transactions, object pooling, activation, events) are directly exposed through the assembly *System.EnterpriseServices*; applications must use this library to make use of these features.

Interoperability with the Win32 API

Many of the traditional operating system services (file and terminal I/O, process, thread, and memory management) are available in the .NET base class library. A number of more advanced features of the Win32 API (such as serial ports), or special hardware libraries are not currently accessible through standard .NET API. To give programmers still full access to all platform functions, .NET provides the means to directly bind system functions to .NET functions, by means of the *DLLImport* attribute for function declarations. Code that uses this functionality is, of course, now bound to the Win32 API, and loses its portability.

Web Services

To allow cross-enterprise application integration, the notion of web services and the SOAP protocol has been proposed as an emerging technology. .NET gives easy access to this technology, with support on various levels of the framework [Esposito03]. As the SOAP protocol is based on the exchange of XML documents, the XML processing libraries in .NET (*System.Xml*) provide the foundation for web services support.

In addition to this, two different programming interfaces for Web services are available. On the one hand, the *Remoting* framework (implemented in the assembly *System.Remoting*) provides application developers with the abstraction of remote procedures. While the framework is more general than just web services, a specific configuration allows usage of SOAP as one possible way of communicating remote procedure calls.

In addition, the *Web Services* library (assembly *System.Web.Services*) provides a separate programming interface. It integrates with ASP.NET, and supports the attribution of methods as *WebMethod*. The creation of a web service in VS.NET results in the deployment of an .asmx file on the web server. The ASP.NET runtime framework (in IIS) will then transparently translate SOAP requests into calls to such a method, loading the .asmx page as needed. At the end of the call, the result value or exception is converted back into a SOAP response.

ASP.NET also transparently generates "service help pages," which contain a simple HTML form allowing interactive invocation of the web method; appending "?WSDL" to the service's URL causes dynamic generation of a Web Services Definition Language (WSDL) description of the service. As SOAP is an extensible protocol, the web services library provides the application with means to detect and process additional SOAP headers and extensions.

To consume a Web service, VS.NET supports importing WSDL files, either through a wizard or using a command line utility. The WSDL specification is converted into C# proxy code, which means that the service can be used like a local object. VS.NET can maintain a reference to the WSDL URL in the project, to allow updates of the proxy code if the WSDL specification changes.

2.8   The C# Programming Language

The C# language is in its syntax very similar to Java and C++ [Archer02]. Semantically, it is a "wrapper" around the CLR: it provides syntax for all features of the CLR, and declarations and actions directly map into the corresponding CLR features. Most concepts of C# are similar to Java approaches; however, a number of minor differences become apparent when Java users are first confronted with this language:

- Methods are not automatically *virtual*. Instead, methods that are intended to be overridden must be declared as virtual. Methods that override base class methods must be declared using the *override* keyword. If overriding is not desired, the *new* keyword indicates that the method has no relationship with the base class method of the same name and signature. The *final* keyword indicates that the virtual function cannot be overridden anymore. Applied to a class, it indicates that inheriting from the class is not allowed. On a data member, it indicates that the member is read-only in subclasses.
- Different parameter passing modes are supported, indicated by keywords: *in* indicates call-by-value (this is the default), ref indicates call-by-reference, and *out* indicates call-by-reference without the need to initialize the parameter on input.
- The *internal* keyword declares a class or method to be available only within an assembly; this is similar to the package-private visibility of Java.
- In a *catch* clause, the list of exceptions can be omitted, meaning that all exceptions are caught by the clause.
- Nested classes in C# are always static; inner classes as available in Java are not supported.
- The file names have no significance on the naming of classes and namespaces (unlike Java, where the file name and the class name must match).

A number of features are new in C#, compared to Java:

- C# offers the notion of *properties*, which are used syntactically like class members (i.e., using the syntax "object.property"). Unlike plain members, a property access (read or write) causes the invocation of a property access method.
- C# classes can act like indexable arrays, similar to overloading the operator[] in C++. The index operation method works with an accessor definition similar to properties:
- The *foreach* keyword allows iteration over an indexable collection.
- *Delegates* are a mechanism to support functions/methods as objects. They are similar to function pointers in C, except that they are type-safe, and in that they support *bound methods* (i.e., method pointers incorporating the target instance object). A delegate is declared through the keyword *delegate*:

This declares a delegate taking no arguments, returning no value. Values of the delegate can be created by passing a function or method:

The resulting delegate value can now be called as if it was a function:

- C# supports native, type-safe enumerations.
- C# supports unsigned integer types.
- C# supports structures, which are class-like value types. Structures can have constructors, methods, indexers, properties, operators, can contain nested types, and can implement interfaces. Structures are useful for implementing user-defined values. Unlike classes, they do not support inheritance, and structure constructors must have arguments.
- C# supports multi-dimensional arrays (not just arrays of arrays, like Java).
- Special syntax (using brackets) is provided to declare CLR attributes. These attributes can have parameters and can only be applied to the set of C# language constructs (i.e., methods, members, classes, ...), which are specified by the attribute implementation. By convention, all attributes end with the "Attribute" name; however, C# does not require this suffix:

Normally, it should be relatively easy for an experienced Java programmer to apply his knowledge to the C# programming language. Microsoft pushes C# as the primary implementation language for the framework and included the language specification in the ECMA standard; therefore, C# will maintain the role as most suitable .NET programming language in the future development of the .NET framework.

References

[Archer02] T. Archer and A. Whitechapel, *Inside* C#, 2nd ed., Microsoft Press, Redmond, Washington, 2nd Book and CD-ROM edition, April 24, 2002.

[ISO03a] ISO/IEC 23270:2003, Information Technology — C# Language Specification, ISO: Geneva, Switzerland, 2003.

[ISO03b] ISO/IEC 23271:2003, Information Technology — Common Language Infrastructure, ISO 2003.

[Esposito03] D. Esposito; *Applied XML Programming For Microsoft .NET,* Microsoft Press, ISO: Geneva, Switzerland, 2003.

[Freeman03] A. Freeman and A. Jones, Eds., *Programming .NET Security*, 1st ed., O'Reilly & Associates, June 2003.

[Rammer02] I. Rammer, *Advanced .NET Remoting*, APress, O'Reilly: Sebastopol, California, 2002.

[Stutz03] D. Stutz, T. Neward, and G. Shilling, *Shared Source CLI Essentials*, O'Reilly & Associates, APress: Berkeley, California, 2003.

3

Unified Modeling Language: The Industry Standard for Software Development

Kleanthis Thramboulidis
University of Patras

3.1 Introduction

The Unified Modeling Language (UML) is the new industry standard for modeling software-intensive systems. Modeling, which is as old as engineering, is the designing of software applications before coding. It is an essential part of large software projects, and helpful for medium and even small projects as well (OMG 2003).

UML was conceived as a general-purpose language for modeling object-oriented software applications. The language represents a further abstraction step away from the one provided by high-level programming languages, which are close to the underlying implementation technology. The easiest answer to the question "what is UML?" is, according to Quatrani (2001), the following: "*UML is the standard language for specifying, visualizing, constructing, and documenting all the artifacts of a software system.*" The above terms are explained in Holt (2001) as follows:

To *specify* means to refer to or state specific needs, requirements or instructions concerned with a complex system. To *visualize* means to realize in a visual fashion or, in other words, to represent information using diagrams. To *construct* means to create a system based on specified components of that system. To *document* means to record the knowledge and experience of the software in order to show how the software was conceived, designed, developed verified and validated.

UML is methodology-independent, which means that UML can be used as a notation to represent the deliverables defined by the methodology regardless of the type of methodology used to perform the system's development, that is, whether it is rigorous such as the Unified Process (Jacobson et al. 1999a) or lightweight such as Extreme Programming (Paulk 2001). In any case, the methodology will guide the engineer in deciding what artifacts to produce, what activities and what workers to use to create and manage them, and how to use those artifacts to measure and control the project as a whole (Booch et al. 1999).

The 12 diagram types that can be used to model every aspect of the system make UML extremely expressive, allowing multiple aspects of a system to be modeled at the same time. However, this expressiveness comes at a price. UML is extremely large with many notational possibilities (LeBlanc 2000). In this chapter, a subset of the UML that represents the core of the language is presented. This subset will allow the reader to understand the basic concepts of the language and communicate through UML diagrams, the knowledge of the system that is under development. The evolution of the language and its use in different application domains, especially in the real-time domain and in the control and automation domain, are also considered. A reference is also made to UML CASE tools that support the engineer to analyze the application's requirements and design a solution to meet requirements. Throughout the article, an elevator control system that is used to control an elevator with many cabins is used as a running example.

The remainder of this chapter is organized as follows. The next section provides a brief overview of the evolution of the UML. In Section 3.3, an introduction to the basic object-oriented concepts is given. Then, in Section 3.4, the most important UML diagrams are presented with examples that highlight their basic constructs. The mechanisms for extending UML in different application domains are discussed in Section 3.5. In Section 3.6, the Real-Time UML profile is considered and a brief description of the basic constructs of this profile is given. The current use of UML in control and automation is considered in Section 3.7. In Section 3.8, the problem of selecting a case tool is discussed and a reference to commercially available tools is made. Finally, conclusions are given.

3.2   History of UML

In the last few years, a large number of methods have been proposed, to support the creation of models for software systems. Until 1989, the proposed methods followed the structured approach with the structured analysis and structured design (SA/SD), created by Yourdon and Constantine in 1987, to be considered the most widely used method. An updated version of this method, called Modern Structured Analysis, was published in 1989 by Yourdon. However, this method did not attain wide acceptance since (a) the complexity of today's systems is almost impossible to be handled by the traditional procedural-like approaches and (b) many new proposals following the more promising object-oriented (OO) approach appeared at the same time. The great advantage of the OO paradigm, which has been gradually accepted during the last decade, is the conceptual continuity across all phases of the software development process (Capretz 2003). In the 1990s, at least 20 OO methods have been proposed in books and many more have been proposed in conference and journal papers. A survey and a classification scheme for OO methodologies can be found in Capretz (2003). In the same article, the author states that

after more than thirty years since the first OO programming language was introduced, the debate over the claimed benefits of the OO paradigm still goes on. But there is no doubt that most new software systems will be OO; that no-body disputes.

Wieringa (1998) surveyed state-of-the-art structured and OO development methods. He identifies the underlying composition of structured and OO software specifications and investigates in which respect OO specifications differ essentially from structured ones. The last specification method in his catalog of OO methods is the UML version 1.0.

The UML started out as a collaboration among three outstanding methodologists: Grady Booch, Ivar Jacobson, and James Rumbaugh. As a first step, Booch and Rumbaugh collaborated to combine the best features of their individual OO analysis and design methods (Booch 1994; Rumbaugh et al. 1991) and presented at OOPSLA in 1995 the Unified Method version 0.8. At that time, the Unified Method was both a language and a process. Later, Jacobson joined the group and contributed the best features of the OOSE methodology (Jacobson 1992). The result of this collaboration was the separation of the language from the process, which was later described in Jacobson et al. (1999b). The language was defined and was presented in 1996 as UML 0.9. Later in January 1997, they submitted their initial proposal as UML 1.0. Since then, the standardization odyssey of UML has been in evolution.

Kobryn (1999) placed the end of this odyssey to 2001, but UML 2.0 is still under discussion. Crawford (2002) notes that "proposals of the UML2 are now under consideration, and what's clear is the future is unclear." As Miller (2002) argues, UML1 unified several of the competing schools of modeling; however, some equally important ideas for good OO design did not influence the language. Furthermore, several problems have been cited (e.g., Kobryn 1999; Dori 2002; Kobryn 2002) and are awaiting a solution in UML2. A spirited debate on the future of UML is in evolution. Five groups submitted proposals in response to the OMG RFPs, but hopefully there is a desire for a consensus version (Miller 2002). They all agree that UML2 should be in the context of OMG's Model-Driven Architecture (MDA) initiative (Miller and Mukerji 2001), which considers as primary artifacts of software development not programs, but models created by modeling languages. UML artifacts should be used to create the platform-independent model (PIM), which should then be mapped by a model compiler into a platform-specific model (PSM) as Mellor (2002) states.

The large number of practitioners and researchers who have adopted UML at a rate exceeding even OMG's most optimistic predictions is a strong argument for UML2 to follow an evolutionary rather than revolutionary approach. However, as Selic et al. (2002) state, "the same forces have also created a strong pressure to improve the effectiveness of UML and provide it with a multitude of new features." The one thing that is accepted by all parties is that the language's popularity has confirmed the urgency for a standard communication medium for humans and software tools. It is already widely accepted that UML2 will drive the software development industry for the next decades.

3.3　Basic OO Concepts

Page-Jones and Weiss (1989) state that, "the OO approach is a refinement of some of the best software engineering ideas of the past. " Capretz (2003) claims that the background of the OO paradigm stems from many different research fields. The most important of them are: system simulation with classes and objects, operating systems with monitors, data abstraction with types and encapsulation, and artificial intelligence with frames. Pressman and Ince (2000) in their introduction to OO concepts state:

> We live in a world of objects. These objects exist in nature, in manmade entities, in business, and in the products that we use. They can be categorized, described, organized, combined, manipulated and created. Therefore, it is no surprise that an object-oriented view would be proposed for the creation of computer software- an abstraction that enables us to model the world in ways that help us to better understand and navigate it.

Therefore, the *object* is the basic concept of the new approach instead of the *process*, which was the basic concept in early structured methodologies. An object, as defined by Booch (1994), "represents an individual, identifiable item, unit, or entity, either real or abstract, with a well-defined role in the problem domain." The object is the atomic unit of encapsulation and is used for the decomposition of the system. An object has:

- state,
- behavior, and
- identity.

The structure and behavior of similar objects are defined in their common *class* (Booch 1994). An object contains attributes that represent the object's state, and operations, called methods, which define the behavior of the object. Methods specify the response of objects to received messages. They accept parameters and have access to object's attributes; they usually result in a state change. Each object is composed of two parts: the interface and the implementation. The implementation is the inner workings of the object and is hidden from the outside (information hiding). The interface is visible to other objects that use it to get the services provided by the object.

A system is considered as an aggregation of discrete objects that collaborate in order to perform work that ultimately benefits an outside user. The collaboration between objects that compose the system is obtained through the exchange of messages. *Message passing* is a means of communication among objects within an application.

3.4 UML Diagrams

As was already noted, UML is a language for visualizing, specifying, constructing, and documenting the artifacts of a software-intensive system. A language provides a vocabulary and the rules for combining words in that vocabulary for the purpose of communication. The vocabulary and rules of the UML focus on the conceptual and physical representation of a system. They define how to create and read system models; but they do not define what model should be created and when they should be created. This is why a methodology should be used throughout the development process. Erikson and Penker (1998) discriminated the following three kinds of rules: syntactic, semantic, and pragmatic rules. According to them, (a) the syntax defines how the symbols should look and how they are combined, (b) the semantic rules define what each symbol means and how it should be interpreted by itself and in the context of other symbols, and (c) the pragmatic rules define the intentions of the symbols through which the purpose of the model is achieved and becomes understandable for others.

In UML, a system is represented using multiple models. Each model describes the system from a distinctly different perspective. The following three kinds of views are defined at the top level (Rumbaugh et al. 1999):

Structural classification. Views of this category consider the things of the system and their relationships with each other. The following UML classifiers represent things: class, use case, component, and node. The static view is expressed through class diagrams, while the use case view is expressed through use case diagrams and the implementation view is expressed through component and deployment diagrams that constitute the classification views of this category.

Dynamic behavior. Views of this category describe the different aspects of the application's dynamic behavior, that is, its behavior over time. Views of this category include: (a) the state machine view, which is expressed through statechart diagrams, (b) the activity view, expressed through activity diagrams, and (c) the interaction view, expressed through sequence and collaboration diagrams.

Model management. Views of this category describe the organization of the system models into hierarchical units. The model management view crosses the other views and organizes the application's models during development and configuration control. The generic unit for models is the package. The class diagram is used to express the organization of the system models into hierarchical units.

The remainder of this section presents the basics of the most important diagrams of the above categories.

Use Case Diagram

The use case diagram was introduced by Jacobson (1992) as the basic artifact for requirements modeling. The diagram, which models the functionality of the system as perceived by outside users, usually contains a set of use cases, actors, and their relationships. The construct of actor is used to model what exists outside of the system and the construct of use case to model what should be performed by the system. Figure 3.1 shows the notational elements of the above constructs of the use case diagram. An *actor* characterizes and abstracts an outside user or related set of users that interact with the system. Actors are the only external entities that interact with the system. A *use case* defines a specific way of using the system. It specifies the interaction that takes place between one or more actors and the system. Each use case constitutes a complete course of events initiated by an actor. It defines the behavior of some aspect of the system without revealing its internal structure. However, later, during subsequent analysis modeling, the objects that participate in each use case are determined (Thramboulidis 2003). A *relationship* represents an association between use cases. *Dependency, generalization,* and *association* are the most commonly used relationships.

Figure 3.2 shows part of the elevator control system use case diagram where the "ElevatorCabin at floor," the "Select destination," and the "Request cabin" use cases are shown. Figure 3.3 provides a draft description of the "ElevatorCabin at floor" use case.

Use case diagrams were adopted by the majority of OO methodologies as a means for capturing and documenting requirements. Lee and Xue (1999) note this when they state:

> Use case approaches are increasingly attracting attention in requirements engineering because the user-centered concept is valuable in eliciting, analyzing and documenting requirements. One of the main goals of requirements engineering process is to get agreement on the views of the involved users, and use cases are a good way to elicit requirements from a user's point of view. An important advantage of use case driven analysis is that it helps manage complexity, since it focuses on one specific usage aspect at a time.

Use case models play a key role in the development process. They are used as input to several activities in the software development process. Jacobson, et al. (1999a) note this by capturing the dependencies between the use case model and the other models in the context of the Unified Process. Figure 3.4,

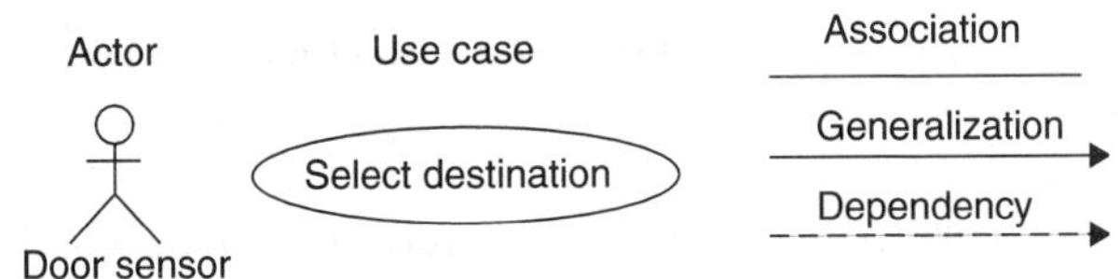

FIGURE 3.1   Basic constructs of use case diagrams.

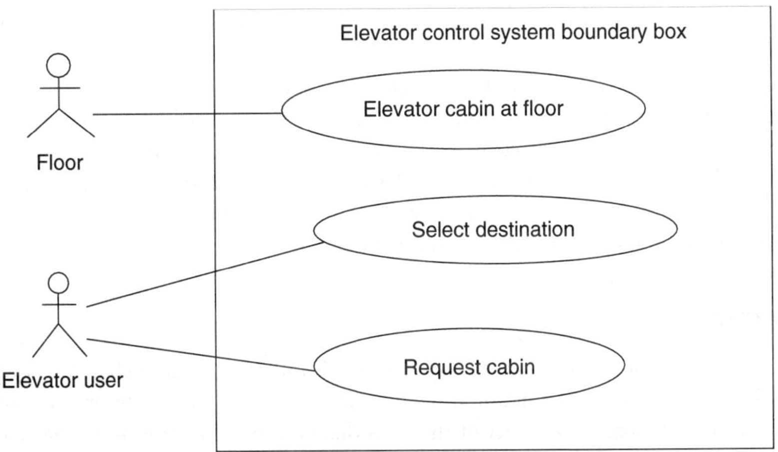

FIGURE 3.2   Example use case diagram of the elevator control system.

> **Use case**: Elevatorcabin at floor
> **Actor**: Floor
> Description: The use case starts when the floor notifies the system through a sensor that a cabin has arrived at its floor. The system checks if the specific cabin has to stop at this floor or there is a pending request from the external button of this floor. If the cabin has to stop, the system stops the cabin and the use case in terminated. If there is a pending request from the specific floor external button, the system checks if the cabin can serve the request. If the cabin can serve the request, the system has to stop the cabin and the use case is terminated. Otherwise the request is postponed to be served later by other cabin and the use case is terminated.
> **Exceptional flow of events**: -

FIGURE 3.3   Description of the "ElevatorCabin at floor" use case.

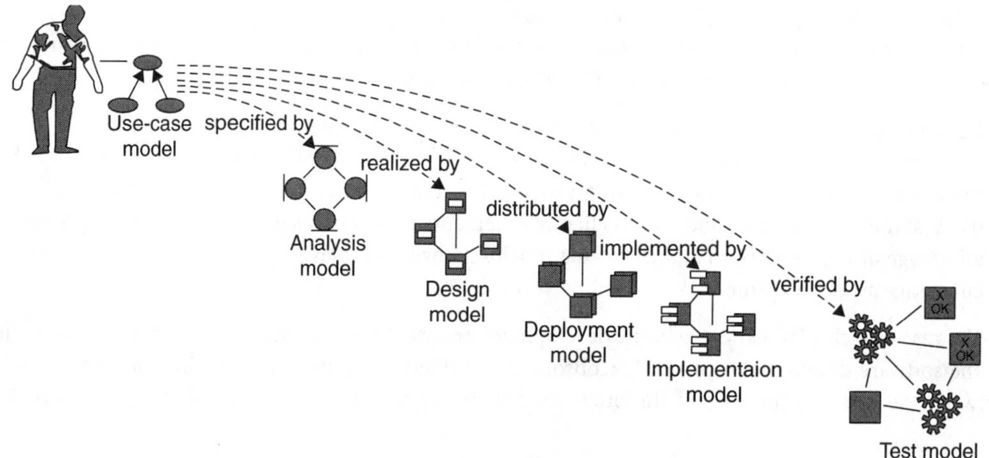

FIGURE 3.4   Use case model dependencies with the other models of the development process (Jacobson et al. 1999a).

taken from Jacobson et al. (1999a), highlights the core role of use case models in the development process.

It is widely accepted that the quality of the use case model has an important impact on the quality of the resulting software product. Software inspection, one of the most efficient methods for verifying software, may be used to ensure the quality of the use case model. Anda and Sjoberg (2002) present a taxonomy of typical defects in use case models and propose a check-list-based inspection technique for detecting such defects.

Although use cases were first introduced as a means to capture functional requirements, there is a trend to expand the whole concept to cover nonfunctional requirements as well. Alexander (2003) presents a way for using use and misuse cases to elicit security requirements. A misuse case is a use case from the point of view of an actor hostile to the system under development. Misuse cases have many interesting applications and interact with use cases in interesting and helpful ways. A use/misuse analysis can complement existing analysis, design, and verification practices.

Class Diagram

The class diagram can be considered as an evolution of the Entity Relationship Diagram (ERD), the main diagram used for many years in information modeling. The large difference between the two diagrams is that the class, which is the basic construct of the class diagram, has, in addition to the structure of the entity, behavior that is described by a set of operations. A class diagram is composed of classes and their relationships.

Berard (1993) defines the class in the following way:

A class is often defined as a template, description, pattern or blueprint for a category of very similar items. Classes are used as templates or "factories" for the creation of specific items that meet the criteria defined in the class.

An implementation-influenced definition is given by Rumbaugh et al. (1991):

A class describes a group of objects with similar properties (attributes), common behavior (operations), common relationships to other objects, and common semantics.

A class is represented in UML by a rectangle that is divided into three compartments as shown in Figure 3.5, which represents the `ElevatorCabin` class. The part in the top contains the name of the class (`ElevatorCabin`), the second part contains the attributes, and the last part contains the operations. `MaxPersons`, `maxLoad`, `load`, and `cabinSize` are the only attributes of the `ElevatorCabin` in this representation. `MoveUp`, `moveDown`, `stop`, `accelerate`, `decelerate`, and `getLoad` are the operations that define the behavior of any instance of this class. The `accelerate` operation instructs the ElevatorCabin instance to increase its speed to the maximum `velocity`. The value of the `ec1.getLoad()` expression is the current load of the specific ElevatorCabin `ec1`.

The main relationships among classes are: associations, composition and aggregation relationships, generalization/specialization relationships, and the various kinds of dependency such as realization and usage.

An association defines a relationship between two or more classes, denoting a static, structural relationship between them. The association `Controls` in Figure 3.6 defines a relationship between `ElevatorControler` and `ElevatorCabin`. An association is inherently bidirectional. Multiplicity must be specified for both sides of the association to define the number of instances of one class that may relate to a single instance of the other class. For the `Controls` association, for example, the multiplicity is *"one"* for the `ElevatorControler` end and *"many"* for the `ElevatorCabin` end. This means that an `ElevatorControler` may control one or more ElevatorCabins; but an `ElevatorCabin` has to be controlled by only one `ElevatorControler`.

Composition and aggregation are used to represent the relationships between the parts and the whole. Composition is stronger than aggregation, which is stronger than association. The composition represents a physical relationship between the whole and its parts. Aggregation is likely to be used to model

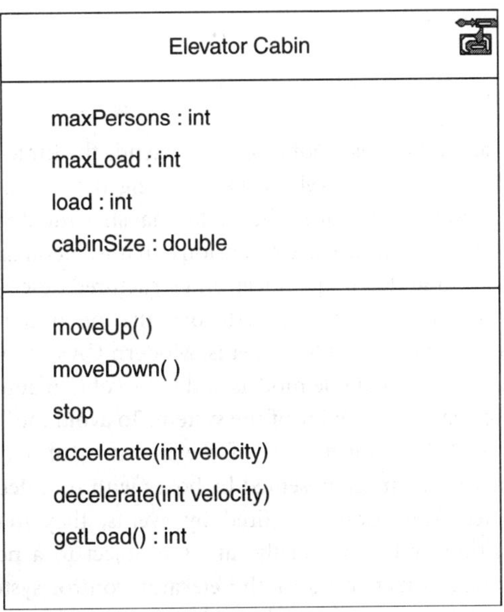

FIGURE 3.5   The `ElevatorCabin` class.

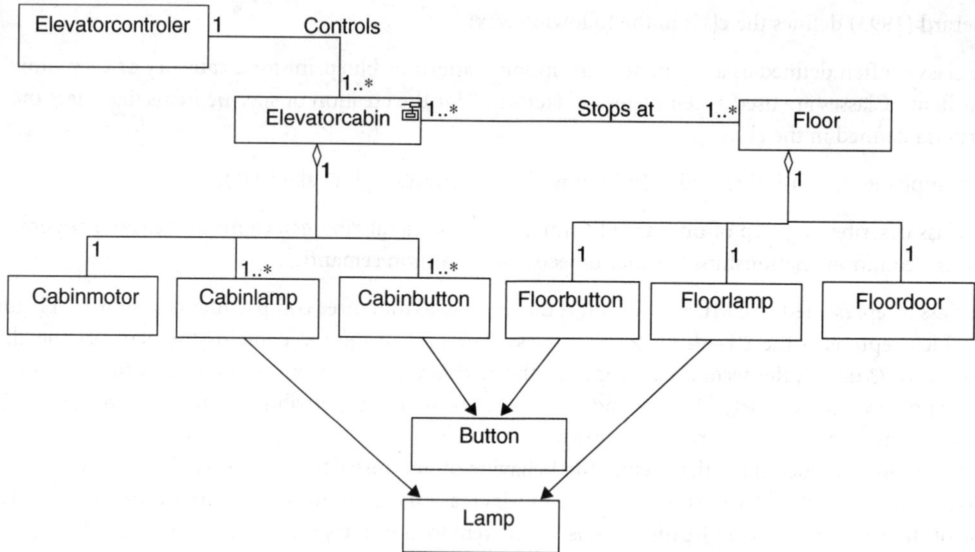

FIGURE 3.6   Class diagram of the elevator control system (part).

conceptual classes rather than physical classes. A part may belong to more than one aggregation. An example of composition relationship is the one connecting the `ElevatorCabin` in Figure 3.6 with its physical components `CabinMotor`, `CabinLamp`, and `CabinButton`.

The generalization/specialization (gen/spec) relationship is used to represent the special kind of association between two classes in which one class represents a general concept and the other a specialization of this general concept. The class `Button` in Figure 3.6, which is referred to as a *superclass* or *ancestor* class, represents the abstract concept Button. It is used to abstract the common structure and behavior of both cabin buttons and floor buttons. The class `CabinButton`, which is referred to as a *subclass or descendent* class, represents the specific type of button used in the ElevatorCabin. Each subclass inherits the properties of the superclass but then extends these properties in different ways. Some authors call this relationship inheritance; but inheritance is the mechanism of the implementation environment used to implement the gen/spec relationship.

Statechart Diagram

Statecharts were defined by Harel (1987) as a notation that extends the finite-state machines. Statecharts overcoming the limitations of traditional FSMs while retaining the benefits of finite state modeling became popular for modeling system behavior in the traditional structured analysis paradigm. Coleman et al. (1992) extended statecharts with three basic extensions, that is, hierarchy, concurrency, and broadcast communication, to produce modular, hierarchical, and structured descriptions in the context of the OO design. Harel and Gery (1997) claim that statecharts form the core of the emerging UML to produce a fully executable language set for modeling OO systems. Modern CASE tools exploit statecharts (a) to allow the developer to produce fully executable models and (b) to obtain automatic code generation.

The statechart diagram describes the behavior of the system. To avoid the "state explosion" problem, a statechart in UML specifies all behavioral aspects of the instances of a class. A statechart contains states connected by transitions. States are represented by boxes with rounded corners, while transitions are indicated by arrowed lines. Transitions are fired by events; they may cause the execution of actions attached to the transition and they usually take the object to a new state. Figure 3.7 shows the statechart of the `ElevatorCabin` class of the elevator control system. When the Elevator is initialized, each `ElevatorCabin` instance is in the `idle` state, which is called the initial state. When an `ElevatorCabin` instance accepts the `MoveUp` message from the `ElevatorController`, a

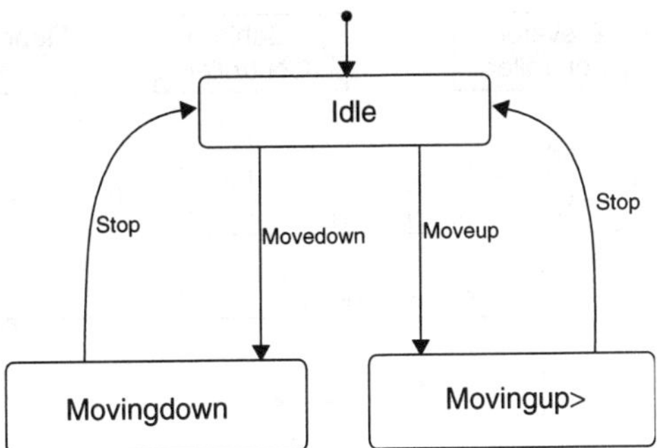

FIGURE 3.7 The ElevatorCabin statechart diagram.

transition is fired and the `ElevatorCabin` instance changes its state to `movingUp`. A `stop` message fires the transition from the `movingUp` state to the `idle` state.

A state may have entry and exit actions, that is, actions that are executed on entering and exiting the state, respectively, as well as internal transitions, that is, transitions that are handled without changing the state. A state may have substates. This feature supports the construction of statecharts of different levels of abstraction and allows concurrency to be handled by using substates (Coleman et al. 1992).

A transition is usually more complex than the one presented in Figure 7. A transition has: (a) a source and a target state, (b) a guard condition (optional), (c) an event trigger that fires the transition if the guard condition is true, and (d) optionally an action, which is an executable atomic computation that may act on the same or other objects.

Sequence Diagram

Class diagrams model the static perspective of an OO system. During the execution of such a system, its objects interact with each other to provide a higher-level behavior. Interaction diagrams in the form of sequence or collaboration diagrams are used to model these interactions, which capture the dynamic aspects of the system.

The sequence diagram shows the sequence of messages, which are exchanged among roles that implement the behavior of the system, arranged in time. It shows the flow of control across many objects that collaborate in the context of a scenario, that is, the individual history of a transaction. The sequence diagram is mainly used to show the behavior sequence of use cases. The sequence diagram of Figure 3.8, for example, shows the sequence diagram of the "ElevatorCabin at Floor" use case. Time is represented vertically in the diagram; hence, the first message is the `cabinAtFloor` that is sent from the `FloorSensor` instance to the `ElevatorControler` object. The `ElevatorController` object then sends a message to the specific `cabinControler` object that is identified by the corresponding parameter of the `cabinAtFloor` message. It then sends a message to the `FloorButton` object that corresponds to the parameter `floorNo` of the `cabinAtFloor` message to check if there is a pending request in this floor.

Sequence diagrams differ from collaboration diagrams in the following features (Booch et al. 1999):

1. The sequence diagram uses the lifeline to highlight the existence of the object for the duration of the interaction. Usually, the objects that appear in an interaction are in existence for the whole duration of the interaction. The lifelines of these objects are drawn from the top of the diagram to the bottom. However, an object may be created or destroyed during the interaction and its lifeline starts after the corresponding message for creation (<<create>> stereotype) and ends with the corresponding message for destruction (<<destroy>> stereotype).

2. The sequence diagram shows the focus of control represented by a thin rectangle that shows the period of time during which the object is active, that is, performs an action as a response to a message that has accepted.

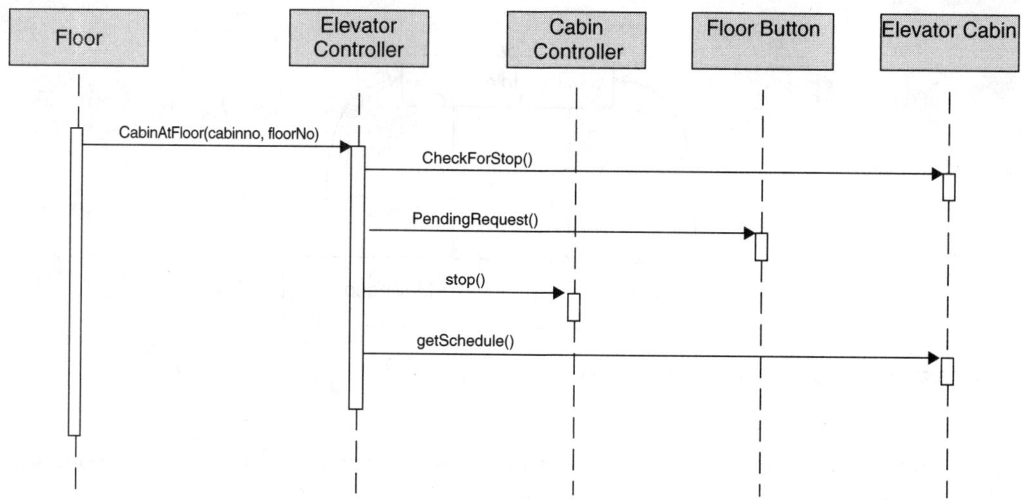

FIGURE 3.8   A simple sequence diagram from the elevator control system.

Collaboration Diagram

A collaboration or Object Interaction diagram (Thramboulidis and Agavanakis 1995) represents the objects that collaborate for the specific behavior to occur and the meaningful links between them. It captures some of the dynamic aspects of message passing required for the realization of a particular behavior. Objects are represented by rectangles, while links representing associations in the context of the particular collaboration are shown by lines connecting rectangles. Sequence numbers preceding the message descriptions indicate the sequence of messages in time. Collaboration diagrams are used during the design phase to model use cases as well as to show the implementation of an operation provided by an object in collaboration with other objects. An example of a collaboration diagram from the elevator control system is given in Figure 3.9.

Collaboration diagrams help in determining the operations of the objects, because the arrival of a message to an object usually invokes an operation. However, during analysis, emphasis is on capturing the information passed between objects rather than the operations invoked (Gomaa 2000). Later during design, it is decided if two different messages arriving at an object invoke different operations or the same.

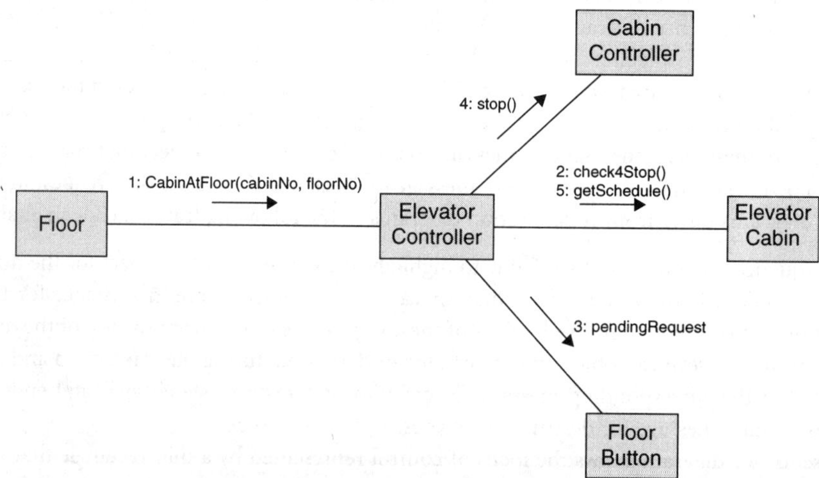

FIGURE 3.9   A simple collaboration diagram from the elevator control system.

In the latter case, the name of the message is handled as a parameter of the operation. The decision for the kind of message, that is, synchronous or asynchronous, is postponed to the design phase.

Collaboration diagrams have two features that distinguish them from sequence diagrams (Booch et al. 1999). The first one can be used to show how an object is linked to another. A path stereotype such as <<local>> is attached to the receiver's object end of a link, to show that the receiving object is local to the sender. The second one is the sequence number that is used to indicate the time order of a message. The sequence number prefixes the message and increases monotonically for each new message in the flow of control. Dewey decimal numbering, for example, 1.1 for the first message nested in message number 1, is used to indicate nesting in messages.

3.5   Extending UML

UML 1.4 is currently used for modeling of just about any type of applications, running on any type and combination of hardware, operating system, programming language, and network. The language's flexibility allows the modeling of distributed applications on any middleware on the market. It is used at different application domains and different levels of abstraction in the system's development process. Examples of application domains where UML has a growing acceptability during the last years are embedded systems (Lee 2000; Svarstad et al. 2001), real-time systems (Selic 2000), and industrial automation and control systems (Thramboulidis 2001; Young et al. 2001; Heck et al. 2003; Thramboulidis and Tranoris 2003).

UML's extensibility mechanisms, namely stereotype, constraints, and required tags, are used to create profiles for the specific application domains. These profiles are primarily intended to tailor UML toward a specific domain by giving to modelers (a) access to common model elements and (b) terminology from the specific domain. A profile is a way of packaging UML specializations. Examples of such profiles are: UML for real-time modeling, UML for data modeling, UML for web modeling, UML for business modeling, UML profile for schedulability, performance and time (OMG 2002), and UML profile for Common Object Request Broker Architecture (CORBA).

A number of other profiles have been proposed. Kobryn (2000), for example, explores the synergies between object modeling and component modeling, by focusing on UML component modeling capabilities to examine the possibilities of using the language to support the leading enterprise component architecture standards EJB and COM+. He claims that standard UML profiles should be defined for these specific component technologies. Moreover, he believes that as components enter more into the mainstream business computing, UML will evolve from OO to component-based modeling language. Yacoub and Ammar (2001) describe an approach that utilizes UML modeling capabilities to construct applications using patterns. They define techniques and composition mechanisms by which patterns can be integrated and deployed in the design of software applications. Design patterns promise early reuse benefits at the design stage (Gamma et al. 1995). As Larsen (1999) states,

Designing systems using components and proven solutions elevates the abstractions at which engineers work. Productivity and quality are the two main drivers this approach brings. *Productivity* will be positively affected by using abstractions for analysis, design and development. *Quality* will be positively affected by reusing known, proven solutions and the components that implement them.

Larsen (1999) discusses and illustrates the use of UML for designing component-based frameworks using patterns. A good interrelationship between patterns, frameworks, and components can also be found in his article. To enhance the UML's usefulness as a graphical programming language, Bjorkander (2000) proposes a merging of UML with the Specification and Description Language (SDL) (Ellsberger et al. 1997). The objective for this combination is to permit the expressive power of UML to coalesce with SDL's strengths of coherence and semantics. Medvidovic et al. (2002) assess the UML's expressive power for modeling software architectures in the manner in which traditional software architectural languages (ADLs) model architectures. They present two strategies for supporting architectural concerns within UML. One of these strategies uses UML without any extension while the other incorporates to UML, using its extension mechanisms, useful features of traditional ADLs.

3.6   Real-Time UML

The use of the OO paradigm in real-time systems has caught on more slowly than in general-purpose applications (Bihary and Gopinath 1992). However, the problems that caused such a delay have already been dealt with and the OO paradigm is becoming the model of choice for the majority of real-time systems. A number of research teams have been exploring real-time OO languages, usually as extensions to existing OO languages. Even more OO analysis and design methods were considered as a means to confront the complexity of the new generation real-time embedded systems. Selic et al. (1994) describe the real-time object-oriented modeling (ROOM) that brings together the power of OO concepts tailored specifically for real-time systems, with an iterative and incremental process that is based on the use of executable models. ROOM is presented as a systems development methodology composed of the following three elements: (a) a modeling language, (b) modeling heuristics, and (c) a framework for organizing and performing development work. Douglas (1999) describes an embedded systems programming methodology for the development of real-time systems. This methodology uses: (a) best practices from object technology and (b) the industry standard UML. Gomaa (2000) presents Concurrent Object Modeling and Architectural Design Method (COMET), an OO methodology that covers requirements modeling and, analysis and design of distributed and real-time applications. A number of other approaches that address the analysis and design of real-time systems using the OO paradigm have also appeared during the last years.

To create a standard in this area, in 1998, the OMG appointed a special working group that is called Real-Time Analysis and Design Work Group, to examine the issue of applying UML in the real-time domain. This group gathered requirements from experts in academia and in industry and from vendors of real-time software, and decided to issue a number of RFPs. Of the proposals, the most important one seems to be the one that combines ROOM with the basic notation of UML. The remainder of this section is mainly based on Selic and Rumbaugh (1998), Selic (1999), and Gullekson (2000), where a detailed presentation of the Real-Time UML constructs and the way they are used for the modeling of real-time systems can be found.

The Real-Time UML utilizes the concepts that have been originally defined in the ROOM methodology and used for years for the development of real-time systems using the OO programming paradigm. It represents them using the extension mechanisms of UML to provide a UML profile for the real-time domain. The Real-Time UML profile focuses on the category of real-time systems that are characterized as complex, event driven, and usually distributed. Selic and Rumbaugh (1998) state that:

> Automatic control applications belong to this category of systems along with telecommunications, aerospace and defense systems. For the development of such applications the architecture plays the primary role. A well-defined architecture must not only support the construction of the initial system but must easily accommodate the system's evolution that is forced by new system requirements. Real-Time UML incorporates among other things the concepts of ROOM that constitute a domain specific architectural definition language. These concepts have proven their effectiveness across hundreds of diverse large-scale industrial projects.

We next describe the constructs of the UML that are divided into two major groups:

1. constructs for modeling structure and
2. constructs for modeling behavior.

Modeling Structure

Class diagrams and collaboration diagrams are used to capture the logical structure of the system. Entities that constitute the system, as well as the relationships between them such as communication and containment relationships, are identified. Class diagrams capture universal relationships among classes, while collaboration diagrams capture relationships that exist only within a particular context, as, for example, within the context of a use case. The following three constructs are defined to model structure:

- capsules,
- ports, and
- connectors.

The *capsule* is defined using the stereotype extension mechanism of UML as a specialization of the general UML concept of class. The capsule is used to represent an entity of the system that has its own thread of control, that is, an active object. Figure 3.10 shows the `DoorController` capsule of the elevator control system. A capsule interacts with the other capsules or the environment through one or more boundary objects called ports. There are no public operations or other public parts to support the interaction of the capsule with the environment. Figure 3.11 shows the `ElevatorController` capsule to interact with the `FloorDoor` capsule.

The *port* is an object that implements a specific interface of the capsule and plays a particular role in the collaboration that the capsule has with other objects. Figure 3.12 shows the ports that the `DoorController` capsule uses to collaborate with its environment. These are `ElevatorPort`, `DoorSensorPort`, and `DoorMechanismPort`. To capture the complex semantics of these interactions, ports are associated with protocols. Each protocol defines the valid flow of information between connected ports, that is, it captures the contractual agreements between communicating capsules. Each port plays a specific role in some protocol, that is, it implements the behavior specified by that protocol role. This is why each port has its own identity and state. Ports do not map directly to UML interfaces that are purely behavioral without implementation structure.

A *connector* is used to represent a communication channel that provides the transmission facilities that are required for the implementation of a connection between capsule ports. Connectors can only be used

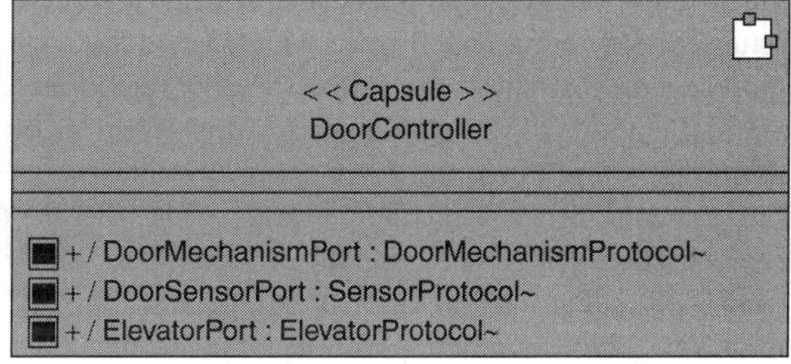

FIGURE 3.10　The `DoorController` capsule.

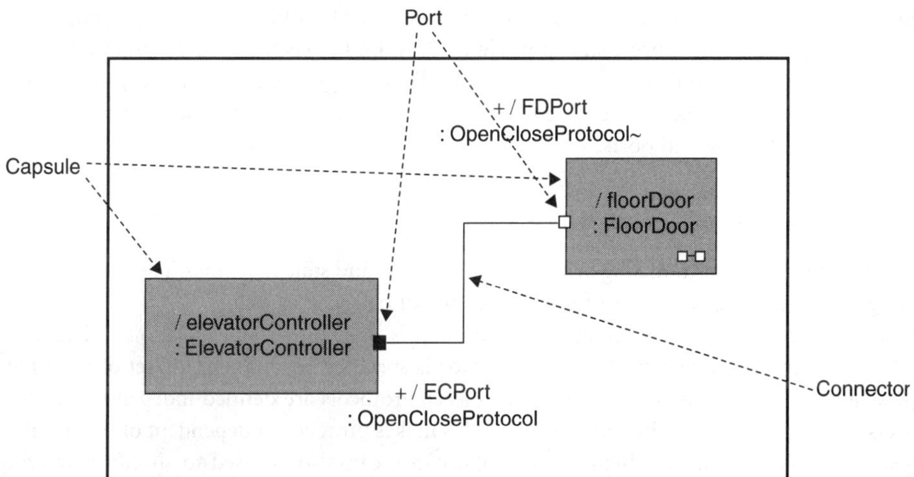

FIGURE 3.11　Capsules interact with each other through one or more boundary objects called ports.

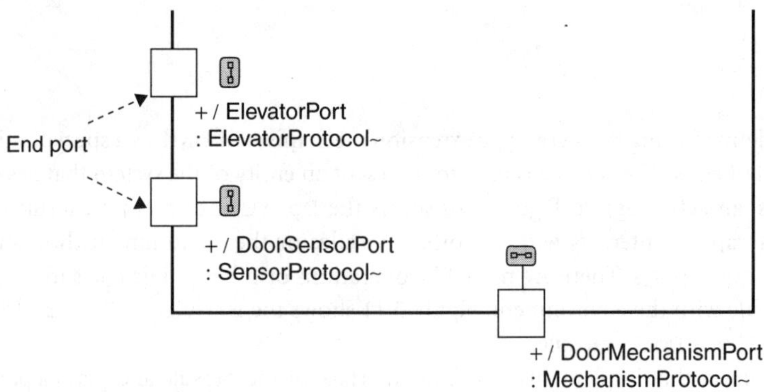

FIGURE 3.12   Ports of the `DoorControler` capsule.

to interconnect ports that have complementary roles in the protocol associated with the connector. The difference between connectors and protocols is that protocols are abstract specifications of desired behavior, while connectors are physical objects whose action is to convey signals between the connected ports. A connector may have physical properties and is usually an object with state and behavior. The actual class that is used to implement a connector is an implementation construct. For the representation of connectors, there is no need for extension mechanisms. A binary protocol connector is represented by a line that is drawn from one port to the complementary one, as the one shown in Figure 11 between the `ElevatorController` and `FloorDoor` capsules.

A capsule may have a state machine that represents its dynamic behavior. The state machine has control of all the internal structures of the capsule and interacts with the capsule's environment, sending and receiving signals through ports. Figure 3.13 shows the state machine of the `DoorController` capsule. According to it, its initial state is `DoorClosed`. The event `openDoor` from the `ElevatorController` fires the transition to the `DoorOpening` state.

The capsule's structure is represented by a collaboration diagram, which specifies the capsule's ports, its subcapsules, and its connectors. The collaboration diagram of Figure 3.14 shows the structure of the `FloorDoor` capsule, which is composed of the `DoorSensor`, `DoorController`, and `DoorMechanism` subcapsules and the `Elevator` port. The same information is represented with the class diagram of Figure 3.15. Ports, subclasses, and connectors are strongly owned by the capsule and cannot exist independent of it. This is why the composition instead of the aggregation relationship was selected.

Ports are of two kinds: relay ports and end ports. Relay ports are used to allow subcapsules to communicate with the outside of the capsule; thus, they are connected to subcapsules as, for example, the port `FDport` of the `FloorDoor` capsule shown in Figure 3.14. End ports, which are connected to the capsule's state machine, are the source or sinks of all messages sent by the capsule. The port `ElevatorPort, DoorSensorPort,` and `DoorMechanismPort` of the `DoorController` capsule in Figure 3.12 are all end ports.

Modeling Behavior

The basic constructs used for modeling behavior are protocols and state machines. The modeling of time, with timing services, is of great interest for the modeling of behavior as well.

A *protocol* is a special type of UML collaboration that is used for the specification of behavior that exists between communicating capsules. This behavior is specified by defining the set of valid message exchange sequences between two or more active objects. Protocols are defined independent of the specific objects that communicate using this protocol. This makes protocols independent of the specific context; thus, their reusability is of a high degree. Inheritance can also be used to specify new protocols as specializations of already defined protocols. Protocols are defined in terms of protocol roles. Each protocol comprises a set of participants, each of which plays a specific role in the protocol. Each protocol

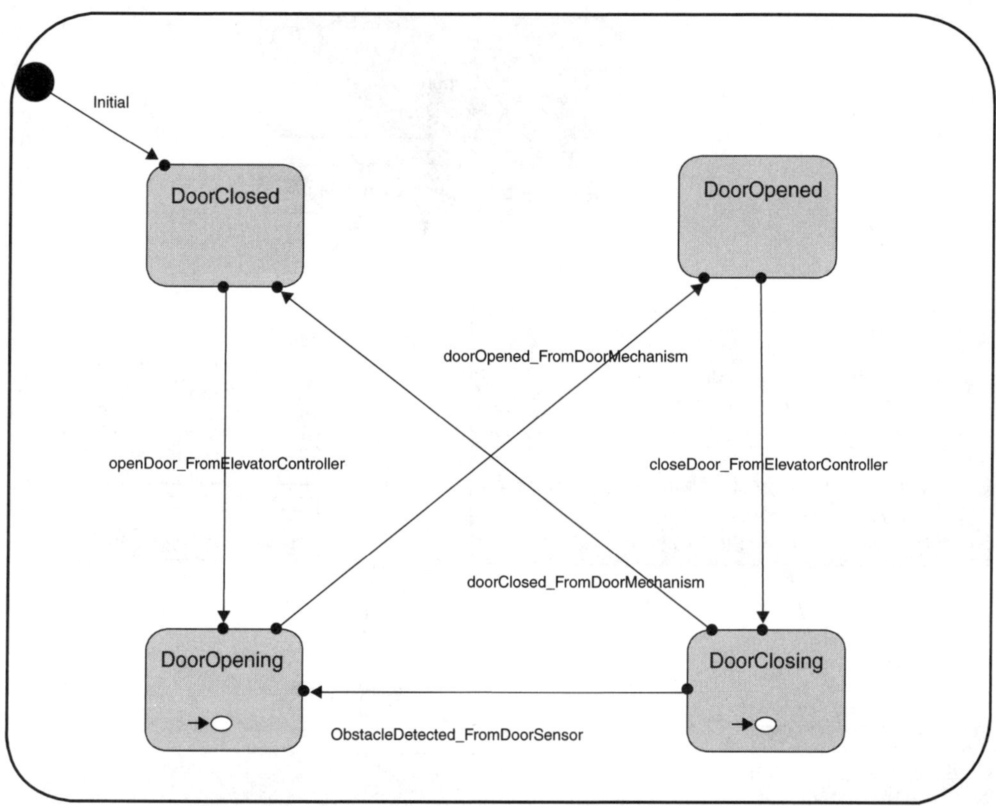

FIGURE 3.13 State machine of the `DoorControler` capsule.

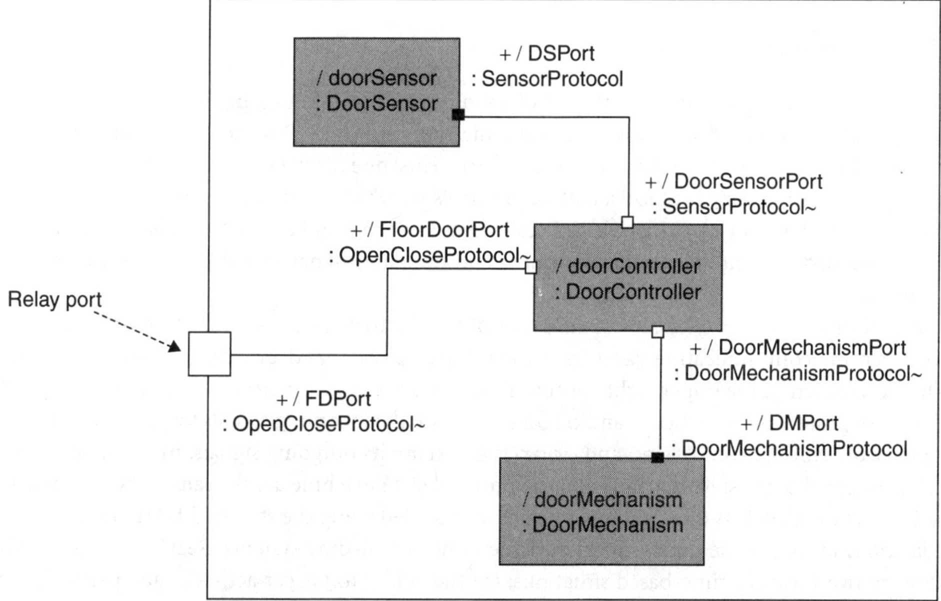

FIGURE 3.14 Collaboration diagram of the `FloorDoor` capsule.

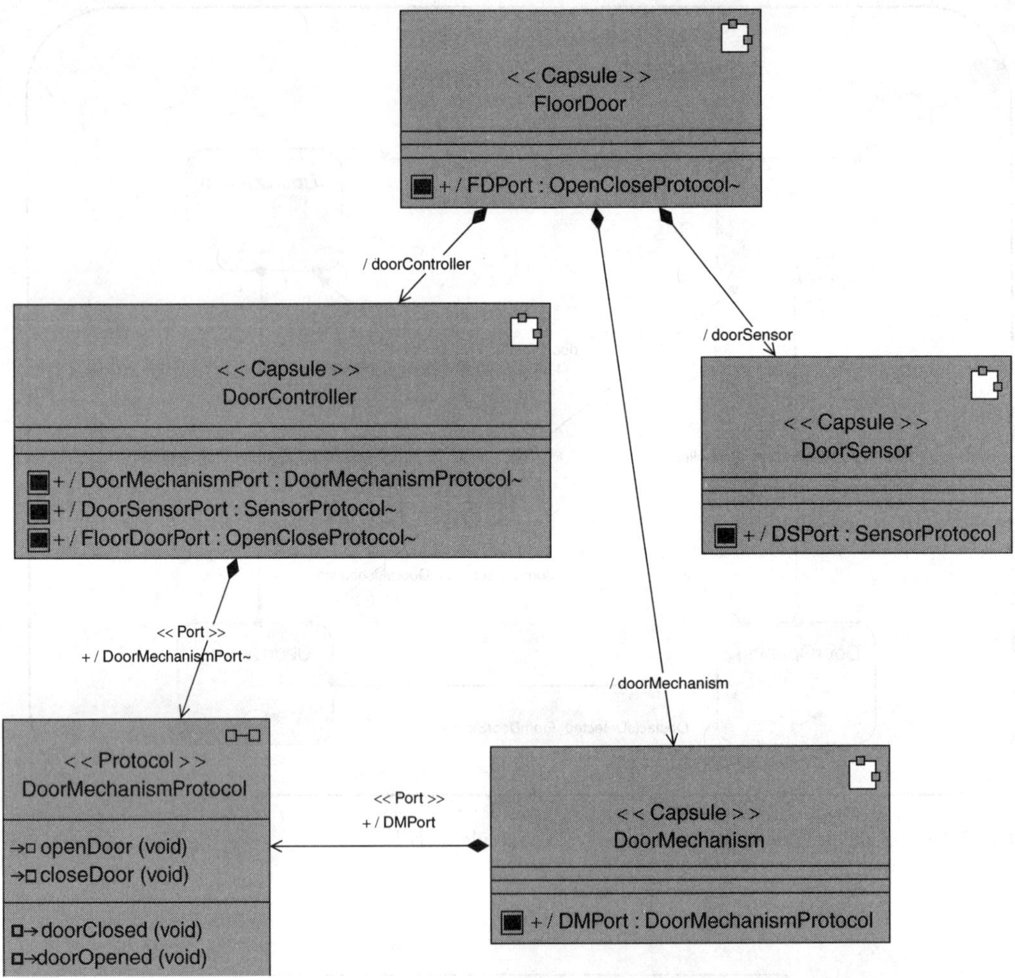

FIGURE 3.15　Class diagram for the `FloorDoor` capsule.

role is specified by a unique name and the set of valid incoming and outgoing signals that are received and sent by that role. It also defines the set of valid message orderings. Protocol roles are represented by the <<protocolRole>> stereotype that has two compartments: one for the incoming signals and the other for the outgoing ones. Binary protocols, that is, protocols with two participants, are the most common. Their advantage is that only one role, which is called base role, has to be specified. The other role called conjugate is defined by simply inverting incoming and outgoing signal sets; this inversion operation is called conjugation.

Figure 3.16 shows the `DoorMechanism` protocol that is used for the specification of behavior that exists between the communicating capsules `DoorController` and `DoorMechanism` of Figure 3.14. The icon shown in the upper right corner of the `DoorMechanismProtocol` capsule is used to identify binary protocols. `OpenDoor` and `closeDoor` are the incoming signals for the base role of this binary protocol, while `doorClosed` and `doorOpened` are its outgoing signals. In this case, since the protocol is binary, the role state machine and the protocol state machine are the same. The state machine and the interaction diagrams of a protocol role are represented using the standard UML notation.

The *handling of time* is the most important issue in most real-time systems. Real-Time UML defines the following two forms of time-based situations: (a) the ability to trigger activities at a particular time of day and (b) the ability to trigger activities after a certain interval has expired from a given point in time. Real-Time UML defines the time service facility that can be accessed through a standard port and

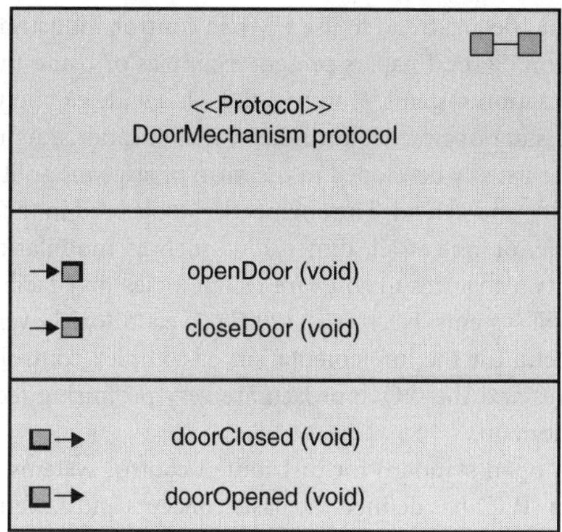

FIGURE 3.16 The binary protocol `DoorMechanismProtocol`.

converts time into events. These events are then handled in the same way as traditional signal-based events. A state machine that uses the time service may be notified with a "timeout" event when a particular time of day has been reached or when a particular interval has expired. The time service is provided by protected end ports, which are of type "TimeServiceSAP," where SAP stands for Service Access Point.

Real-Time UML has wide acceptance in academia and many researchers are working on different aspects related to it. For example, Selic (1999) demonstrates the feasibility of UML to model recourses and thus predict crucial system properties before fully implementing a system. He proposes the incorporation of a generic QoS framework into UML to provide a standard way of producing quantifiable and, hence, precisely analyzable software models. He and Goddard (2000) present a method for capturing the temporal parameters of a real-time application in a Real-Time UML model so that schedulability analysis can be performed early in the design phase. The proposed method enriches Real-Time UML and enhances the modeling of real-time systems by capturing more temporal parameters and making them first-class attributes in the model. It also ensures that a real-time system modeled with Real-Time UML is schedulable and this analysis can begin early in the design phase.

3.7 Applications of UML in Control and Automation

The majority of today's distributed control systems consist of industrial-grade computers or PLCs, which are interconnected by dedicated fieldbus networks. The market is dominated by a small number of very large corporations with proprietary solutions. The most common design and analysis tool for control systems is MATLAB. Unfortunately, the family of MATLAB products tends to generate monolithic rather than component-based C language code, which makes the validation and code update more difficult (Heck et al. 2003).

The OO approach has attracted the interest of researchers in control and automation from the early 1990s (beginning of the 1990s). Aarsten et al. (1996) described the G++ pattern language for the design of distributed control systems, and reported their experience, gained in the field of computer-integrated manufacturing (CIM) and from a project involving cooperative autonomous robots. They claim that although the existence of design patterns is often very valuable for solving isolated design problems, a domain-specific language does the same but to the whole development project. Grady and Liang (1998) describe their OO approach to enhance the ability of companies to respond rapidly to changes in the market environment. They present an OO extension to the design with modules approach, which involves selecting the module combination to best satisfy the given set of requirements. Traub and Scharft (1999) present an OO real-time framework for distributed control systems.

In the last few years, there has been a trend to use UML in control, industrial automation, as well as in industrial enterprise integration. Several papers present examples of using the UML in many different levels of manufacturing information systems. However, there is a wide gap between the state of the art in software engineering and the state of practice in industrial applications. As reported by Lewis (2001), today's control applications are usually developed in the form of large monolithic software packages that are difficult to maintain, modify, and extend. The engineering tools used in the development and deployment processes address a little, or not at all, dimensions such as modularity, flexibility, extensibility, reusability, and interoperability. Advances in software technologies may facilitate the development and deployment of complex control systems. Heck et al. (2003) give a tutorial overview of specific new software technologies that are useful for the implementation of complex control systems. They claim that component-based architectures and the OO approach are very promising technologies for the control and automation application domain.

In an attempt to define an open standard for distributed control systems (DCSs), the International Electro-technical Commission (IEC) has defined the basic concepts and a methodology for the design of modular, reusable DCSs. The IEC61499 standard defines the Function Block (FB) as the main building block and the way that FBs can be used to define robust, reusable software components that constitute the DCS.

So, what is the future in control and automation software development? We discriminate three possible directions:

1. UML-based approach;
2. FB-based approach; and
3. A hybrid approach integrating UML with the FB concept.

We next refer briefly to these approaches.

The UML-Based Approach

According to this approach, the standard UML notation is used to produce the models of the control system. This is the approach proposed by many researchers (Douglas 1999; Gomaa 2000; Young et al. 2001; Mosemann and Wahl 2001; Bonfè and Fantuzzi 2003). Better results will be obtained using the profile for real-time modeling and the profile for schedulability, performance, and time.

The major disadvantage of this approach is that it is a completely new approach for practitioners in control and automation, who are mainly accustomed to IEC languages and concepts.

The FB-Based Approach

This approach, which is the one proposed by the evolving IEC standards 61499 and 61804, is based on the FB construct. Although many researchers are already working on different aspects of the IEC proposal (e.g., Vyatkin and Hanisch 2001; Brennan and Norrie 2001; Fletcher and Norrie 2001; Thramboulidis and Tranoris 2001), the absence of tools and products that are compliant with this approach is evident. The Function Block Development Kit (FBDK) by Rockwell Automation (see at http://www.holobloc.com/fbdk/README.htm for a detailed description of the tool and a free downloaded version) and CORFU-FBDK (Thramboulidis and Tranoris 2003) are the only known tools supporting this approach.

However, it should be mentioned that the *IDA architecture*, which is proposed by the IDA group (http://www.ida-group.org), proposes an application model that is based on the reference architecture defined in IEC61499. Also, the *Profinet Engineering Model*, which is the proposal of the Profibus User Organization to achieve open distributed automation, even though it is non-IEC-compliant, is planning to have an IEC-compliant mapping in the near future.

The major advantage of this approach is that it exploits the experience of industry's practitioners and easily integrates the large number of existing legacy applications and systems. However, it does not exploit all the benefits of object- and component-based technologies.

The Hybrid Approach

This approach integrates the UML notation with the FB concept. However, this integration is done from different perspectives. Brennan et al. (2002) describe two approaches for modeling decentralized Manufacturing Systems with UML capsules. They propose the modeling of the control system at the conceptual level using Real-Time UML and then they propose a mapping of Real-Time UML to IEC 61499 models that are used for the implementation of control application. However, it is quite difficult for control engineers to model in Real-Time UML.

Another approach (Thramboulidis 2001; Thramboulidis and Tranoris 2003; Tranoris and Thramboulidis 2003) considers the use of the UML notation in the requirements specification phase using the concept of use case. It next proposes the transformation of the UML models to FB-based models through a specific tool. The FB design models are next used for the development of the control application. One can consult the CORFU site (http://seg.ee.upatras.gr/corfu) for several white papers that describe this approach and to obtain an evaluation/demonstration version of the engineering tool that is based on this approach.

In any case, a UML profile for control and automation should provide the best alternative for the development of control applications. Such a profile should primarily intend to tailor UML toward the control and automation domain. It should provide to modelers: (a) access to common model elements and (b) terminology from the control and automation domain.

3.8 UML CASE Tools

In the last few years, the number of CASE tools that support the UML notation has increased dramatically. A list of UML CASE tools can be found at http://dmoz.org/ Computers/Programming/ Methodologies/Modeling_Languages/Unified_Modeling_Language/Tools/ as well as at http://www.qucis.queensu.ca/Software-Engineering/_tools.html.

The software developer has several choices; but it is quite difficult to make the right choice. The large number of unsuccessful attempts to introduce CASE tools in the development process forced the software community to give considerable attention to the establishment of processes for the formal assessment of case tools (Mosley 1992; Le Blanc and Korn 1994; Kitchenham and Pickard 1998). Working groups created by authorities such as the Software Engineering Institute (SEI) and the IEEE, which seek to advance the state of practice in software development through the use of good tools for system and software engineering, resulted in the IEEE 1462 standard, which defines guidelines for the evaluation and selection of CASE tools (IEEE Std 1462, 1998). However, a formal assessment following the standards is a very costly and time-consuming task. Even more, some weaknesses have been identified to standards and reported so far (e.g., Lundel and Lhngs 2002).

A less formal assessment is usually adopted. In such an assessment, issues such as ease of use, robustness, functionality, round-trip engineering, model navigation, repository support, support for reusability, ease of insertion, and quality of support must be considered very carefully. A good list of evaluation criteria can be found at http://www.objectsbydesign.com/tools/modeling_tools.html. Several articles report evaluation processes and results that may be used appropriately in the CASE tool selection process (Budgen and Thomson 2002; Frankel 2000; Ypsihe 2000; Eichelberger 2002; Post and Kagan 2000; Artim 2003). We next refer briefly to three commercially available CASE tools:

(A) *The IBM Rational Rose.* A family of model-driven development tools that can be used with many IDEs such as IBM VisualAge, Visual Studio 6.0, Borland JBuilder and other Java, C++, and Ada IDEs. The Rational Rose family includes products that address all types of application needs, including business and enterprise/IT applications, software products and systems, and embedded systems and devices. Information on this family of tools can be found on-line at: http://www.rational.com/products/rose/index.jsp?SMSESSION=NO

(B) *Rhapsody by I-Logix.* Rhapsody provides, according to I-Logix, a collaborative Model-Driven Development (MDD) environment for system and software engineers to work in an iterative fashion, analyze and make rapid system and software design changes, generate the application, and test it quickly and efficiently. The development of real-time embedded applications is supported by a special tool, which is based on executable models (I-Logix 2003). The tool for embedded systems is very flexible. A real-time framework in the source code form is provided with the tool; thus, users can easily adapt the generated code for any other operating system, including proprietary in-house systems. There is a Free Developer's Trial for 30-day provided by Rhapsody. An overview of the tool can be found on-line at: http://www.ilogix.com/products/rhapsody/index.cfm

(C) *Tau Generation2 by Telelogic.* Tau Generation is a family of role-based and model-centric tools for real-time software and systems development. Tau/Developer™ is a stand-alone tool based on UML, for the design, analysis, and development of advanced real-time applications. According to Telelogic, the tool enforces the engineer to make a shift from the traditional labor-intensive code-centric programming to the highly efficient MMD. According to Telelogic, the tool supports executable UML models with behavioral specifications. The tool's philosophy is that the specifications must be validated at an early stage for timely and cost-effective error correction. Information is available on-line at: http://www.taug2.com/

There are also some free CASE tools. Among them, we specify the following:

- *Visual Paradigm for UML.* A visual development platform. http://www.geocities.com/miller-max2001/casetool.html
- *Argo/UML.* A free open source UML tool. http://argouml.tigris.org/
- *Poseidon for UML (Community Edition).* http://www.gentleware.com/adgate.php4?in=odb&banner=poseidonCE&url=/products/poseidonCE.php3

3.9 Conclusions

UML is the *de facto* standard for software development. A large number of practitioners and researchers have adopted UML in many different application domains at a rate exceeding even OMG's most optimistic predictions. Even though UML has emerged as a language for modeling OO systems, it is now evolving to a component-based modeling language. UML2 is expected to be in the context of the OMG's Model Driven Architecture initiative, which considers as primary artifacts of software development not programs, but models created by modeling languages. This ensures that UML will drive the software development industry for the next decades.

However, UML is a very elaborate language. A long time is required to learn the language and efficiently use it in the software development process. In this paper, we have presented a subset of the UML language that includes the most commonly used diagrams to model the most important aspects of software systems. In particular, we have presented the basic concepts regarding use case diagrams, class diagrams, statecharts, and sequence and collaboration diagrams. These diagrams have specific constructs and semantics to enable the software developer to model the most important views of the system under development.

One of the most important features of UML is its ability to be extended and specialized in different application domains. Even though UML's flexibility allows the modeling in different application domains and at different levels of abstraction during the system's development process, the powerful set of extensibility mechanisms, which are provided by the language, allows the definition of profiles for any specific application domain. A profile is primarily intended to tailor UML toward a specific domain by providing modelers access to common model elements and terminology from the specific domain. This considerably enhances productivity and quality in the development process. We have briefly presented the real-time UML profile that focuses on the category of real-time systems that are characterized as complex, event driven, and usually distributed. We have described the most important constructs of the profile, including capsules, ports, connectors, protocols, etc. that are utilized to create executable models for the system under development.

As expected, UML has attracted the interest of researchers from the control and automation application domain. It is believed that component-based architectures and the OO approach are very promising technologies for this application domain and the best way to exploit these technologies is through the use of UML. However, it is not yet clear as to how the UML will be utilized in this domain. We have described three possible directions that are already adopted by researchers in this area. Finally, we have made a reference to the family of CASE tools that support UML. The great number of commercially available and free downloaded CASE tools makes the selection a very costly and time-consuming task. Standards may be used to simplify this task, but a less formal assessment is usually adopted to select the tool that addresses the specific developer's requirements.

References

Aarsten, A., D. Brugali, and G. Menga, Designing concurrent and distributed control systems, *Communications of the ACM*, Vol. 39, Issue 10, Oct. 1996, pp. 50–58.

Alexander, I., Misuse cases: use cases with hostile intent, *IEEE Software*, Vol. 20, Issue 1, Jan.-Feb. 2003, pp. 58–66.

Anda, B. and D. Sjoberg, Towards an Inspection Technique for Use Case Models, SEKE '02, July 15–19, Ischia, Italy, 2002.

Artim, M.J., UML Tools, 2003, Available online at: http://www.primaryview.org/ UML/Tools.html.

Berard, E., *Essays on Object-Oriented Software Engineering,*Vol. I, Prentice-Hall, Englewood Cliffs, NJ, 1993.

Bihari, T.E. and P. Gopinath, Object-oriented real-time systems: concepts and examples, *IEEE Computer*, Vol. 25, Issue 12, Dec. 1992, pp. 25–32.

Bjorkander, M., Graphical programing using UML and SDL, *IEEE Computer*, Vol. 33, Issue 12, Dec. 2000, pp. 30–35.

Bonfè, M. and C. Fantuzzi, Design and Verification of Industrial Logic Controllers with UML and Statecharts, IEEE Conference on Control Applications, CCA 2003, June 23–25, Istanbul, Turkey, 2003.

Booch, G., *Object-Oriented Analysis and Design with Applications*, 2nd ed., Benjamin/Cummings Publishing Company Inc., Menlo Park, CA, 1994.

Booch, G., J. Rumbaugh, , and I. Jacobson. *The Unified Modeling Language User Guide*, Addison-Wesley Longman Inc., Reading, MA, 1999.

Brennan, X. and X. Norrie, Agents, holons and function blocks: distributed intelligent control in manufacturing, *Journal of Applied System Studies, Special Issue on Industrial Applications of Multi-Agent and Holonic Systems*, 2, 1–19, 2001.

Brennan, R., S. Olsen, M. Fletcher, and D. Norrie, Comparing two Approaches to Modelling Decentralized Manufacturing Control Systems with UML Capsules, Proceedings of the 13th IEEE International Workshop on Database and Expert Systems Applications (DEXA'02), 2002.

Budgen, D. and M. Thomson, CASE tool evaluation: experiences from an empirical study, *The Journal of Systems and Software*, Elsevier Science Inc., New York, USA, Vol. 67, Issue 2, August 2003, pp. 55–75.

Capretz, L., 2003 *A Brief History of the Object-Oriented Approach*, Software Engineering Notes, Vol. 28, No. 2, ACM SIGSOFT Software Engineering Notes, Vol. 28, Issue 2, March 2003.

Coleman, D., F. Hayes, and S. Bear, 1992. Introducing objectcharts or how to use statecharts in object-oriented design, *IEEE Transactions on Software Engineering*, Vol. 18, Issue 1, Jan. 1992, pp. 8–18.

Crawford, D., 2002. Editorial pointers, *Communications of the ACM*, Vol. 45, Issue 12, Dec. 2002, pp. 5.

Dori, D., 2002. Why significant UML change is unlikely, *Communications of the ACM*, Vol. 45, Issue 11, Nov. 2002, pp. 82–85.

Douglas, B., *Doing hard time: developing real_time systems with UML, objects, frameworks, and patterns*, Addison-Wesley, Reading, MA, 1999.

Eichelberger, H., Evaluation-Report on the Layout Facilities of UML Tools, Report No. 298, July 2002, Available online at: http://www-info2.informatik.uniwuerzburg.de/mitarbeiter/eichelberger/ reports/eval2002Abstract.html.

Ellsberger, J., D. Hogrefe, and A. Sarma, *SDL: Formal Object-Oriented Languages for Communicating Systems*, Prentice-Hall, Englewood Cliffs, NJ, 1997.

Erikson, E. and M. Penker, *UML Toolkit,* John Wiley, New York, 1998.

Fletcher, M. and D. Norrie, Real-time Reconfiguration using an IEC 61499 Operating System, International Parallel and Distributed Processing Symposium (IPDPS), San Francisco, 2001.

Frankel, D., Alternative UML Modeling Tools, *Software Magazine,* Oct. 2000, Available online at: http://www.softwaremag.com/L.cfm?Doc=archive/2000oct/Lab.html.

Gamma, E., R. Helm, R. Johnson, and J. Vlissides, *Design Patterns: Elements of Reusable Object-Oriented Software,* Addison-Wesley, Reading, MA, 1995.

Gomaa, H., *Designing Concurrent, Distribute, and Real-Time Applications with UML,* Addison-Wesley, Reading, MA, 2000.

Grady, P. and W.-Y. Liang, An object-oriented approach to design with modules, *Computer Integrated Manufacturing Systems,* 11, 1998.

Gullekson, G., Designing for Concurrency and Distribution with rational Rose RealTime, Rational Software White paper, 2000, Available online at: http://www-106.ibm.com/developerworks/rational/library/content/03july/0000/0598/ps-0598.pdf

Harel, D. Statecharts: A Visual Formalism for Complex Systems, *Sci. Comput. Programming,* Vol. 8, 1987, pp. 231–274.

Harel, D. and E. Gery, Executable object modeling with statecharts, *IEEE Computer,* Vol. 30, Issue 7, July 1997, pp. 31–42.

He, W. and S. Goddard, Capturing an Application's Temporal Properties with UML for Real-Time, Fifth IEEE International Symposim on High Assurance Systems Engineering, HASE 2000 , Nov. 15–17, 2000.

Heck, B.S., L.M. Wills, and G.J. Vachtsevanos, Software technology for implementing reusable, distributed control systems, *IEEE Control Systems Magazine,* Vol. 23, Issue 1, Feb. 2003, pp. 21–35.

Holt, J., *UML for Systems Engineering,* The Institution of Electrical Engineers, United Kingdom, 2001.

IEEE Std 1462, Information Technology — Guideline for the Evaluation and Selection of CASE Tools, *IEEE STD,*1998.

I-Logix, Model Driven Development for Real-Time Embedded Applications, 2003, Available on line at: http://www.ilogix.com.

Jacobson, I. *Object-Oriented Software Engineering: A Use Case Driven Approach,* ACM Press, Addison-Wesley, Reading, MA, 1992.

Jacobson, I., G. Booch, and J. Rumbaugh, The unified process, *IEEE Software,* Vol. 16, no. 3, May-June 1999a, pp. 96–102.

Jacobson, I., G. Booch, and J. Rumbaugh, *The Unified Software Development Process,* Addison-Wesley, Reading, MA, 1999b.

Kitchenham, B. and L. Pickard, *Evaluating Software Engineering Methods and Tools: Part 9: Quantitative Case Study Methodology,* ACM SIGSOFT Software Engineering Notes, Vol. 23, Issue 1, Jan. 1998, pp. 24–26.

Kobryn, C. UML 2001: a standardization odyssey, *Communications of the ACM,* Vol. 42, Issue 10, Oct. 1999, pp. 29–37.

Kobryn, C., Modeling components and frameworks with UML, *Communications of the ACM,* Vol. 43, Issue 10, Oct. 2000, pp. 31–38.

Kobryn, C., Will UML 2.0 be agile or awkward?, *Communications of the ACM,* Vol. 45, Issue 1, Jan. 2002, pp. 107–110.

Larsen, G., Designing component-based frameworks using patterns in the UML, *Communications of the ACM,* Vol. 42, Issue 10, Oct. 1999, pp. 38–45.

LeBlanc, C. and L.W. Korn, A phased approach to the evaluation and selection of CASE tools, *Information and Software Technology,* 36, 1994.

LeBlanc, C. UML for Undergraduate Software Engineering, *Journal of Computing Sciences in Colleges,* Ramapo College of New Jersey, Mahwah, New Jersey, USA, Vol. 15, Issue 5, 2000, pp. 8–18.

Lee, E., What's ahead for embedded software? *IEEE Computer,* Vol. 33, Issue 9, Sep. 2000, pp. 18–26.

Lee, J. and N.-L. Xue, Analyzing user requirements by use cases: a goal-driven approach, *IEEE Software,* Vol. 16, Issue 4, July-Aug. 1999, pp. 92–101.

Lewis, R., *Modelling Control Systems Using IEC 61499*, The Institution of Electrical Engineers, United Kingdom, 2001.

Lundel, B. and B. Lhngs, Comments on ISO 14102: the standard for CASE-tool evaluation, *Computer Standards & Interfaces*, 24, 2002.

Medvidovic, N., D. Rosenblum, D. Redmiles, and J. Robbins, Modeling software architectures in the unified modeling language, *ACM Transactions on Software Engineering and Methodology (TOSEM)*, Vol. 11, Issue 1, Jan. 2002, pp. 2–57.

Mellor, S., What UML should be: Make models be assets, *Communications of the ACM*, Vol. 45, Issue 11, Nov. 2002, pp. 76–78.

Miller, J., What UML should be: Introduction, *Communications of the ACM*, Vol. 45, Issue 11, Nov. 2002, pp. 67–69.

Miller, J. and J. Mukerji, Model Driven Architecture (MDA), Document number ormsc/2001-07-01, Architecture Board ORMSC, July 9, 2001, Available online at: http://www.omg.org/docs/ormsc/01-07-01.pdf.

Mosemann, H. and F. Wahl, Automatic decomposition of planned assembly sequences into skill primitives, *IEEE Transactions on Robotics and Automation*, Vol. 17, Issue 5, Oct. 2001, pp. 709–718.

Mosley, V., How to assess tools efficiently and quantitatively, *IEEE Software*, Vol. 9, Issue 3, May 1992, pp. 29–32.

OMG, UML profile for schedulability, performance and time specification, 2002, http://www.omg.org/docs/ptc/02-03-02.pdf.

OMG, Introduction to OMG's unified modeling language, 2003, Available online at: http://www.omg.org/gettingstarted/what_is_uml.htm.

Page-Jones, M. and S. Weiss, Synthesis: an object-oriented analysis and design method, *American Programmer*, Vol. 2, Summer 1989, pp. 7–8.

Paulk, N.C., Extreme programming from a CMM perspective, *IEEE Software*, Vol. 18, Issue 6, Nov.-Dec. 2001, pp. 19–26.

Post, G. and A. Kagan, OO-CASE tools: an evaluation of Rose, *Information and Software Technology*, 42, 2000.

Pressman, R. and D. Ince, *Software Engineering: A Practitioner's Approach, European Adaptation*, 5th ed., McGraw-Hill, New York, 2000.

Quatrani, T., Introduction to the Unified Modeling Language, Rational Rose, White paper, 2001, Available online at: http://www.rational.com/media/uml/intro_rdn.pdf?SMSESSION=NO.

Rumbaugh, J. et al., *Object-Oriented Modeling and Design*, Prentice-Hall International, Englewood cliffs, NJ, 1991.

Rumbaugh, J., I. Jacobson, and G. Booch, *The Unified Modeling Language — Reference Manual*, Addison-Wesley Longman Inc., Reading, MA, 1999.

Selic, B., Turning clockwise: using UML in the real-time domain, *Communications of the ACM*, Vol. 42, Issue 10, Oct. 1999, pp. 46–54.

Selic, B., A generic framework for modeling resources with UML, *IEEE Computer*, Vol. 33, Issue 6, June 2000, pp. 64–69.

Selic, B. and J. Runbaugh, Using UML for Modeling Complex Real-Time Systems, Mar. 1998, Available online at: http://www.rational.com/products/whitepapers/UML-rt.pdf?SMSESSION=NO.

Selic, B., G. Gullekson, and P. Ward, *Real-Time Object-Oriented Modeling*, John Wiley & Sons, Inc., New York, 1994.

Selic, B., G. Ramackers, and C. Kobryn, Evolution, not revolution, *Communications of the ACM*, 45, 2002, pp. 70–72.

Svarstad, K., G. Nicolescu, and A. Jerraya, A Model for Designing Communication between Aggregate Objects in the Specification and Design of Embedded Systems, IEEE Conference on Design, Automation and Test in Europe, 2001.

Thramboulidis, K., Using UML for the Development of Distributed Industrial Process Measurement and Control Systems, IEEE Conference on Control Applications (CCA), Mexico, Sept. 2001.

Thramboulidis, K., A sequence of assignments to teach object-oriented programming: a constructivism design-first approach, *Informatics in Education*, 2, 103–122, 2003.

Thramboulidis, K. and K. Agavanakis, Introducing object interaction diagrams : a technique for A&D, *Journal of Object-Oriented Programming (JOOP)*, June 1995, pp. 25–39.

Thramboulidis, K. and C. Tranoris, An Architecture for the Development of Function Block Oriented Engineering Support Systems, IEEE International Conference on Computational Intelligence in Robotics and Automation, Canada, Aug. 2001.

Thramboulidis K. and C. Tranoris, Developing a CASE Tool for Distributed Control Applications, *The International Journal of Advanced Manufacturing Technology*, Springer-Verlag, Vol. 24, no. 1-2, July 2004, pp. 24–31.

Tranoris, C. and K. Thramboulidis, Integrating UML and the Function Block Concept for the Development of Distributed Control Applications, 9th IEEE International Conference on Emerging Technologies and Factory Automation, Lisbon, Portugal, Sept. 16–19, 2003.

Traub, A. and R. D. Scharft, An Object-Oriented Realtime Framework for Distributed Control Systems, IEEE Inernational Conference on Roborics and Automation, Detroit, U.S.A., 1999.

Vyatkin, V. and H. Hanisch, Formal-Modelling and Verification in the Software Engineering Framework of IEC61499: A Way to Self-Verifying Systems, IEEE Conference on Emerging Technologies in Factory Automation (ETFA'01), Proceedings, Nice, pp. 113–118, Oct. 15-18, 2001.

Wieringa, R., A survey of structured and object-oriented software specification methods and techniques, *ACM Computing Surveys (CSUR)*, Vol. 30, Issue 4, Dec. 1998, pp. 459–527.

Yacoub, S. and H. Ammar, *UML Support for Designing Software Systems as a Composition of Design Patterns*, Lecture Notes in Computer Science, Vol. 2185, Springer-Verlag, Berlin, 2001, pp. 149–165.

Young, K., R. Piggin, and P. Rachitrangsan, An object-oriented approach to an agile manufacturing control system design, *The International Journal of Advanced Manufacturing Technology*, Springer-Verlag, London Ltd., Vol. 17, no. 11, 2001, pp. 840–859.

Yphise, Software evaluation process, Jan. 2000, Available online at: http://www.rational.com/media/products/rose/yphise.pdf?SMSESSION=NO

4

Middleware

Andreas Polze
*Hasso-Plattner-Institute at University
Potsdam*

Middleware is connectivity software that consists of a set of enabling services that allow multiple processes running on one or more machines to interact across a network. Middleware functions as a conversion or translation layer. It is also a consolidator and integrator. Custom-programmed middleware solutions have been developed for decades to enable one application to communicate with another that either runs on a different platform or comes from a different vendor or both. Today, there is a diverse group of products that offer packaged middleware solutions.

Middleware is essential to migrating mainframe applications to client/server applications and providing communication across heterogeneous platforms. This technology evolved during the 1990s to provide for *interoperability* in support of the move to client/server architectures. The most widely publicized middleware initiatives are the Open Software Foundation's *Distributed Computing Environment* (DCE), Object Management Group's *Common Object Request Broker Architecture* (CORBA), and Microsoft's COM/DCOM (*Component Object Model (COM), DCOM*) with its successor Microsoft.NET. The Java programming language and its runtime environment support Remote Method Invocation (RMI) on the language level, which may be seen as an integration of middleware support into the programming language — an idea that can also be found in the .NET languages.

Although middleware systems are often designed to transparently manage the distribution of processes and objects as well as communication across machine boundaries, common programming interfaces between applications are also considered middleware. For example, *Open Database Connectivity* (ODBC) enables applications to make a standard call to all the databases that support the ODBC interface.

4.1   Overview and Classification

As outlined in Figure 4.1, middleware services are sets of distributed software that exist between the application and the operating system and network services on a system node in the network.

Middleware services provide a more functional set of Application Programming Interfaces (API) than the operating system. Network services implemented on the middleware layer allow an application to:

- Locate communication partners transparently across the network, providing interaction with another application or service.
- Be independent of network services.
- Be reliable and available.
- Scale up in capacity without losing function.

Middleware can take on the following different forms.

TP Monitors

The transaction processing monitor (TP monitor) was perhaps the first product to be called middleware. TP monitors provide tools and an environment for developing and deploying distributed applications. Located between the requesting client program and the databases on the server side, a TP monitor ensures that all databases are updated properly.

TP monitors tend to do far more than coordinate and monitor transactions across multiple data resources. They also enhance the performance, reliability, and scalability of server-side systems. To achieve these enhancements, TP monitors establish a framework for creating server-side applications. A TP monitor can reliably and efficiently manage the resources needed by applications that conform to the TP monitor's rules.

Customer Information Control System (CICS) and IMS/TM (a message-based transaction manager) are the transaction processing workhorses of the mainframe environment. On UNIX systems, BEA's TUXEDO, BEA's TOP END, and IBM's Encina are the most widely used TP monitors.

Messaging Middleware

Message-Oriented Middleware (MOM) provides program-to-program data exchange, enabling the creation of distributed applications. MOM is analogous to email, in the sense that it is asynchronous and requires the recipients of messages to interpret their meaning and to take appropriate action. MOM provides a common interface and reliable transport between applications. If the target machine is down or overloaded, it stores the data in a message queue until it becomes available. The messaging system may contain business logic that routes messages to the appropriate destinations and reformats the data as well.

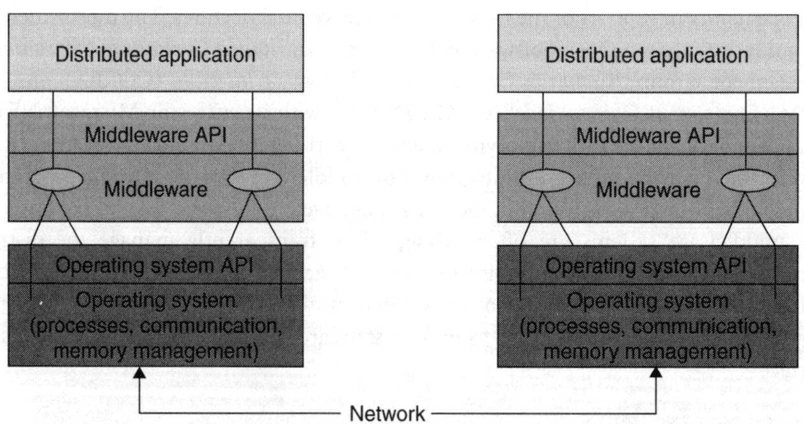

FIGURE 4.1   The middleware layer.

The predominant messaging product for managing asynchronous communication between applications is IBM's MQSeries. MQSeries has been ported to all major server platforms. In conjunction with the Component Object Model (COM), Microsoft introduced its own messaging system, Microsoft Message Queue Server (MSMQ). MSMQ and MQSeries offer much the same functionality.

Distributed Processing

Remote Procedure Calls (RPCs) enable the logic of an application to be distributed across the network. Program logic on remote systems can be executed as simply as calling a local routine. The SUN RPC, introduced with the implementation of the network file system (SUN NFS), as well as the DCE RPC, which also served as the technical foundation of Microsoft's COM, are the most prominent examples of remote procedure call implementations.

Extending the procedural programming model of RPC, *Object Request Brokers* (ORBs) enable the objects that comprise an object-oriented application to be distributed and shared across heterogeneous networks. Distributed object systems such as CORBA, DCOM, .NET, and EJB enable processes to be run anywhere in the network. In contrast to messaging middleware, RPCs and ORBs typically implement synchronous interaction between processes (components/objects).

Database Middleware

Middleware provides a common interface between a query and multiple, distributed databases. Using either a hub and spoke architecture or a distributed architecture, it enables data to be consolidated from a variety of disparate data sources.

Application Server Middleware

A Web-based application server that provides interfaces to a wide variety of applications is used as middleware between browser and legacy systems. Application servers have supported a wide range of server-side processing, with SUN's Java 2 Enterprise Edition (J2EE) as a prominent example. WebServices are a promising route toward standardization of application server interfaces.

4.2   Usage Considerations

The main purpose of middleware services is to help solve many application connectivity and interoperability problems. However, middleware services are not a panacea:

- There is a gap between principles and practice. Many popular middleware services use proprietary implementations (making applications dependent on a single vendor's product).
- The sheer number of middleware services is a barrier to using them. To keep their computing environment manageably simple, developers have to select a small number of services that meet their needs for functionality and platform coverage.
- While middleware services raise the level of abstraction of programming distributed applications, they still leave the application developer with hard design choices. For example, the developer must still decide what functionality to put on the client and server sides of a distributed application (Bernstein 1996).

The key to overcoming these three problems is to fully understand both the application problem and the value of middleware services that can enable the distributed application. To determine the types of middleware services required, the developer must identify the functions required, which fall into one of three classes:

1. Distributed system services, which include critical communications, program-to-program, and data management services. This type of service includes RPCs, MOMs, and ORBs.

2. Application-enabling services, which provide applications access to distributed services and the underlying network. This type of services includes TP monitors and database services such as Structured Query Language (SQL).
3. Middleware management services, which enable applications and system functions to be continuously monitored to ensure optimum performance of the distributed environment.

4.3 Distributed Objects and Distributed Processing

Among the middleware approaches listed above, distributed objects have the biggest potential to solve a wide range of challenges faced by designers of large software systems. Some of these challenges include component packaging, cross-language interoperability, interprocess communication, and intermachine communication. The distributed processing and distributed object architectures, which have been introduced over the last decade, exhibit different strengths and weaknesses based on their ability to meet this challenge.

For our discussion, we divide distributed object architectures into two categories: component architectures and remoting architectures. We define *component architectures* as architectures that focus primarily on component packaging and cross-language interoperability. In contrast, *remoting architectures* focus primarily on support for remote method invocation on distributed objects.

Primary examples for remoting architectures are the Open Software Foundation's (OSF) Distributed Computing Environment (DCE) — which actually is a distributed processing environment based on the RPC paradigm (purely procedural) — and the Object Management Group's (OMG) Common Object Request Broker Architecture (CORBA). The notion of component packaging and deployment has only recently been added to CORBA 3.0.

The predominant component architectures are Microsoft's Component Object Model (COM) — which mainly addresses packaging and deployment of binary component as well as cross-language interoperability — and the JavaBeans and Enterprise Java Beans (EJB) component models introduced by SUN Microsystems. Both COM and EJB address remoting to some extent: the COM model has been extended to Distributed COM (DCOM) using an extended version of DCE RPC as transport. EJB supports client/server communication based on Java Remote Method Invocation (RMI). RMI is special, as it integrates closely with the Java language without requiring a special Interface Definition Language (IDL) to describe component interfaces accessible for remote invocations.

In an evolutionary sense, Microsoft's .NET is the newest and most advanced component architecture available in the market today. Similar to Java, .NET is based on a machine-neutral intermediate language format and just-in-time compilers. .NET adopts a fundamentally new and sound approach to cross-language interoperability as it starts with a common object model and type system applicable to all .NET languages (among them, C#, C++, Visual Basic, Jscript, Eiffel#, Cobol, Scheme, and more to come). .NET extends COM as it makes metadata — runtime-type information, which was optional for COM components — mandatory for each .NET component (so-called assemblies). .NET implements a remoting architecture (.NET Remoting), which is comparable to Java RMI in the sense that it closely integrates with the .NET programming languages and does not require any special IDL.

We will now discuss prominent examples for object-oriented middleware systems in some detail.

4.4 The Distributed Computing Environment (OSF/DCE)

In the mid-1980s, the Open Software Foundation (OSF) was set up by a group of major computer vendors, among them IBM, DEC, and Hewlett-Packard with the initial goal of developing a new version of UNIX, over which they, and not AT&T/SUN, had control. This goal was achieved with the release of the Mach-based OSF/1 operating system. However, many of the OSF consortium's customers wanted to build distributed applications on top of OSF/1 and other UNIX variants. OSF/1 responded to this by issuing a "Request for Technology," which ultimately led to the design of the OSF Distributed Computing Environment (DCE) (DCE 1995).

DCE supports a distributed client/server-programming model. User processes act as clients to access remote services provided by server processes. Some of these services are part of DCE itself, but others belong to the application and are written by application programmers. The DCE RPC is the basis for all communication in DCE. To access a service, a client process issues an RPC to a (possibly) remote server. The server acts on the requests and (optionally) sends back a reply. DCE handles the complete mechanism, including name resolution (locating the server), binding to it, and performing the call. DCE provides an IDL and a compiler, thus allowing a server programmer to describe the remotely callable procedures made available by his application. In order to cope with machine-dependent data representations, the DCE IDL compiler generates code for mapping C data structures onto a network format.

Besides the RPC mechanisms, DCE provides a thread package, allowing multiple threads of control to exist in the same process at the same time. Some versions of UNIX provide threads themselves, but using DCE provides a standard thread interface across systems. The DCE thread implementation is mapped onto native operating system threads whenever possible.

Besides the DCE architecture, Figure 4.2 depicts a number of central services implemented in the DCE environment. Standard services are the time, directory, and security service.

The *distributed time service* provides clock synchronization among machines in a DCE cell. Clock synchronization is a necessary prerequisite for the predictable behavior of many other services, among them the directory service. The DCE *directory service* can be seen as a general naming service, which keeps track of the location of all resources in the system. These resources include machines, printers, servers, data, and much more, and — due to the directory service's hierarchical structuring scheme — they may be geographically distributed across the world. The directory service provides for location-transparent resource access. The *distributed file service* is a worldwide file system that provides a transparent way of accessing any file in the system in a standard manner. It can be either built on top of a host's native file systems or can be used instead of them.

The *security service* implements access control (authentication and authorization) for all kinds of resources in the DCE environment. The DCE authentication procedure uses the Kerberos system developed at M.I.T. Authorization in DCE is handled by associating an Access Control List (ACL) with each resource. The ACL

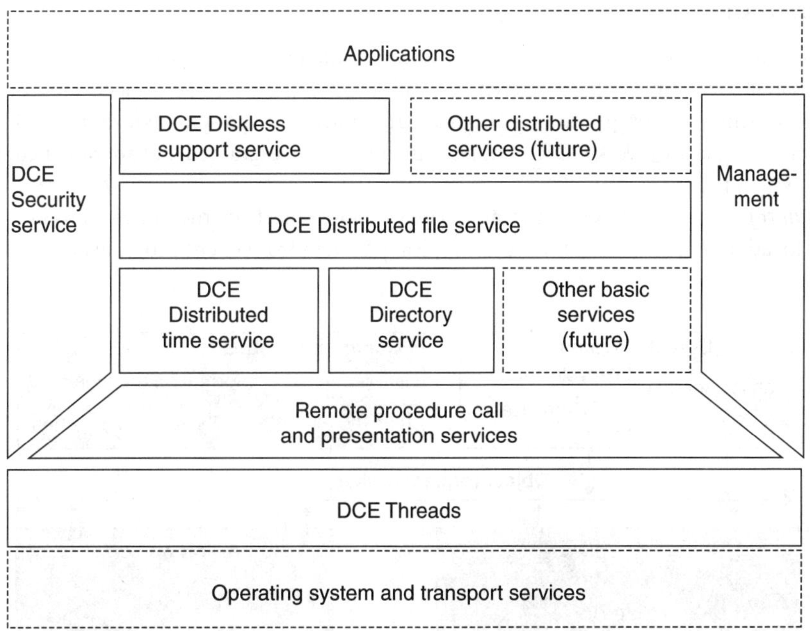

FIGURE 4.2　　DCE architecture and services.

instructs as to which users, groups, and organizations may access the resource and which operations they are entitled to invoke on the resource. Resources may be as coarse as files or as fine as database entries.

Although DCE has pioneered distributed computing in various respects, it has never been complete. DCE provides many facilities and tools; however, it lacks certain runtime management and debugging aids, and did not really keep up with newer software engineering approaches such as object-oriented programming.

4.5   The Common Object Request Broker Architecture (CORBA)

The OMG was founded in 1989 as a consortium of eight companies (3Com, American Airlines, Canon, Data General, Hewlett-Packard, Philips Telecommunications, Sun Microsystems, and Unisys). Today, the OMG exists as an international nonprofit organization of more than 800 software developers, network operators, computer vendors, research institutions, and commercial computer users.

Figure 4.3 depicts the Object Management Architecture (OMA) (Soley 95), as specified by the OMG. OMA describes software architectures for application interoperability, independent of the applications' implementation languages, locations, operating systems, and hardware platforms.

The Object Request Broker (ORB) is the central element of the OMA. Comparable to a software bus, the ORB serves as a universal communication means for various objects in heterogeneous distributed systems. A CORBA object is represented to the outside world by an interface with a set of methods. A particular instance of an object is identified by an object reference. The client of a CORBA object acquires its object reference and uses it as a handle to make method calls, as if the object was located in the client's address space. The ORB is responsible for all the mechanisms required to find the object's implementation, prepare it to receive the request, communicate the request to it, and carry the reply (if any) back to the client. The object implementation interacts with the ORB through either an Object Adapter (OA) or through the ORB interface (see Figure 4.4). In order to implement communication across machine boundaries, multiple ORB instances residing on different machines in the network interact typically using Internet InterORB Protocol (IIOP), which is a specialization of the General InterORB Protocol (GIOP).

The architectural design of the ORB has been described in the CORBA specification, which appeared in version 1.1 in 1991 and exists in version 3.0 today. The CORBA specification emphasizes the following characteristics (CORBA 1995):

Object-orientation: Objects are the basic architectural units in CORBA. An object in the sense of the OMA is characterized by its object reference, a unique identifier. CORBA objects are not necessarily objects in the sense of a programming language. In contrast to fine-grained, short-lived lightweight programming language objects, CORBA objects are often coarse-grained and are rather comparable to operating system processes.

Distribution transparency: CORBA programs access remote objects by the same means as local objects. The exact location of a given server object is typically hidden from client programs.

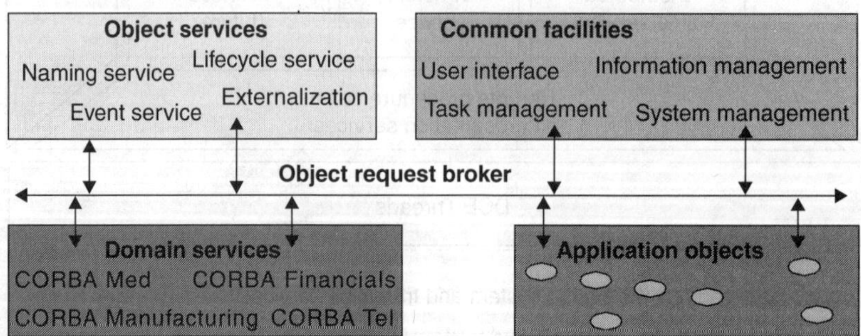

FIGURE 4.3   The object management architecture.

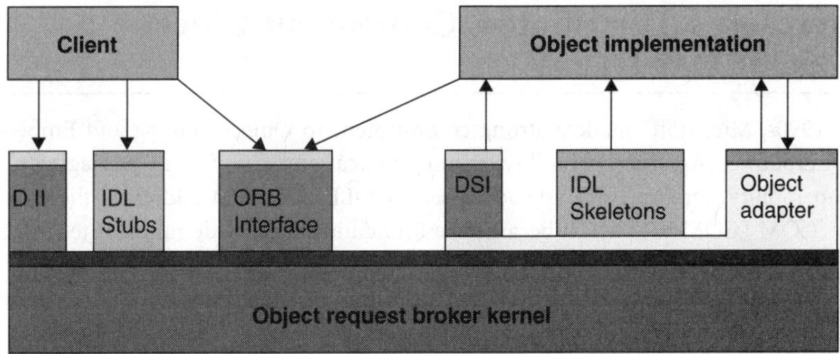

FIGURE 4.4 ORB interfaces.

Efficiency: The ORB architecture specified by OMG is general with respect to actual implementations, allowing for efficient implementations, which may minimize the overhead for local client/server communication (i.e., using shared memory or no interprocess communication at all).

Hardware, operating system, and language transparency: Components of a CORBA program can be implemented using varying operating systems, hardware architectures, and programming languages. This is achieved using a language-neutral interface definition language (CORBA IDL) for components. The OMG has specified language mappings from IDL to a number of implementation languages.

Openess: Using the ORB as a software bus allows for interoperability among components of different vendors. This avoids vendor-lock and opens up the opportunity for domain-specific component markets. However, although the CORBA IDL may serve as a platform-independent description of component interfaces, there is no binary component format standardized by OMG — which makes marketing of binary components for multiple ORB implementations on varying platforms at least difficult.

There exist a variety of commercial and open source CORBA implementations. Compatibility tests and benchmarks (Amar 1998; Gokhale 1997; Vinoski 1997) demonstrate that CORBA is well suited for distributed object-based applications.

The recent CORBA 3.0 standard introduces a number of related new specifications, which specifically address issues in integrating CORBA with Java and the Internet (in contrast to the intranet) as well as quality of service control. Most notable are the following changes (Siegel 2001):

CORBA Component Model (CCM) — introduces the notion of server-side components into CORBA and addresses packaging and deployment for CORBA components. CCM provides for interoperability with EJB.

Objects passable by value (valuetypes) — valuetypes definitely improve integration with Java and also serve as the basis for the XML/Value mapping (released as part of CORBA 2.3).

Java-to-IDL mapping — this mapping allows Java RMI objects to interoperate over the network like CORBA objects using CORBA object references.

XML/Value mapping — standardizes the representation of an XML document as a collection of native CORBA types.

CORBA Firewall Specification — allows firewalls to be configured for CORBA using access rules for IIOP traffic.

CORBA Messaging — encompasses both asynchronous and messaging-mode invocations of CORBA objects as well as Quality of Service Control.

Real-Time CORBA — extends the CORBA architecture with resource control mechanisms for real-time applications running on a real-time operating system in a controlled environment.

Fault-Tolerant CORBA — standardizes redundant software configurations and systems that give CORBA robust and reliable performance (when run on redundant hardware).

Minimum CORBA — defines a small-footprint CORBA configuration that is aimed at embedded and card-based systems.

4.6 Microsoft's (Distributed) Component Object Model (COM/DCOM)

In the early 1990s, Microsoft made a strong commitment to Object Linking and Embedding (OLE). Although OLE allowed for interoperability among applications, component packaging and cross-language interoperability were insufficiently addressed by OLE. Microsoft addressed these issues with the subsequent COM (COM 95), which became the foundation for a wide range of technologies, among them the OLE Control Extension (OCX) — later renamed as *ActiveX* (visual COM controls). COM is the dominant component architecture in use today. This is not very surprising since COM is the component architecture for today's most dominant desktop operating system — Microsoft Windows (98/ME and NT/2000/XP). Recently, Microsoft coined the acronym COM+ to identify the bundling of the COM infrastructure with a number of component services.

DCOM is the distributed extension to COM that builds an object remote procedure call (ORPC) layer on top of DCE RPC to support remote objects. A COM server can create object instances of multiple object classes. A COM object can support multiple interfaces, each representing a different view or behavior of the object. An interface consists of a set of functionally related methods. A COM client interacts with a COM object by acquiring a pointer to one of the object's interfaces and invoking methods through that pointer, as if the object resides in the client's address space. COM specifies that any interface must follow a standard memory layout, which is the same as the C++ virtual function table (Rogerson 1996). Since the specification is at the binary level, it allows integration of binary components possibly written in different programming languages such as C++, Java, and Visual Basic.

COM supports a language-neutral IDL — actually an extension of DCE IDL — that is used to describe the interface of a COM component. Using IDL, the designer of a COM component can describe the interfaces, methods, and properties that are supported by the component. Client applications rely on the IDL definition of the COM component rather than on implementation-specific details such as programming language and implementation platform.

As depicted in Figure 4.5, COM supports the following three styles of servers for implementing components:

1. *In-process server:* An in-process server is implemented as a dynamic linked library (DLL) that executes within the same process space as the application. The performance overhead of invoking an in-process server is small (in the order of an ordinary function call). An in-process server is commonly referred to as an *ActiveX* control.
2. *Local server:* A local server executes in a separate process space on the same computer. Communication between an application and a local server is accomplished by the COM runtime system (implemented in OLE32.DLL and SVCHOST.EXE). The performance overhead of using a local server is typically an order of magnitude greater than that of using an in-process server.
3. *Remote server:* A remote server executes on a remote computer. DCOM provides an RPC-based infrastructure that is used to manage communication between the application and the remote server. The performance overhead of using a remote server is typically an order of magnitude greater than that of using a local server.

An application that uses a COM component may request to be connected to a specific type of server if it wishes to do so — however, it is not required to know what type of server it is using. After a client has obtained a COM object instance handle, client interaction with the COM object is the same regardless of server location. This allows for great flexibility during the design of a component-based system.

COM is a very mature component architecture that has many strengths. Thousands of third-party ActiveX controls (in-process COM components) are available in the market today. Microsoft and other vendors have built many tools that accelerate development of COM-based applications. Microsoft is also providing advanced services such as Microsoft Transaction Server (MTS) and Microsoft Message Queuing Server (MSMQ) to support the development of enterprise multi-tier systems. Microsoft has been using the name COM+ to identify the bundling of the COM runtime with those services.

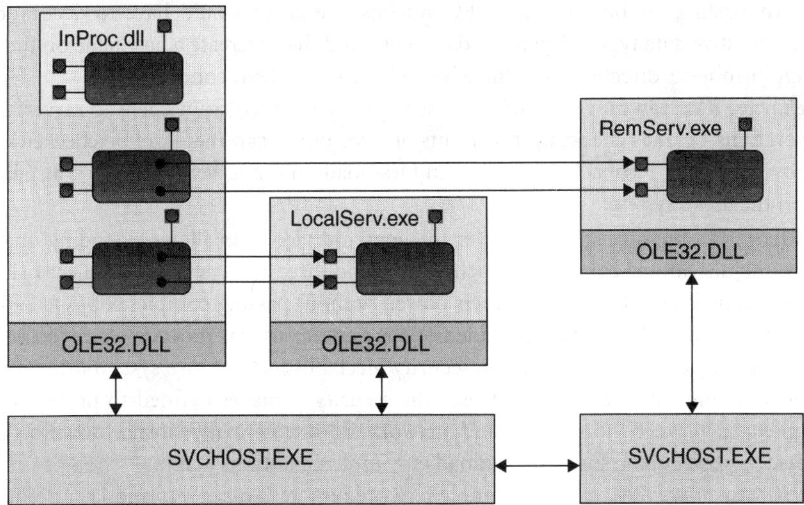

FIGURE 4.5 Styles of COM server components

Especially platform limitations — COM support is extremely limited on UNIX and mainframe platforms — as well as a less than perfect integration of the COM IDL type system with those of the component implementation languages present weaknesses that prevent COM from being an ideal single distributed component-based computing environment. However, Microsoft has carried on many of COM's concepts into the new .NET framework — which has the potential to alleviate most of COM's limitations.

4.7 Java and the Remote Method Invocation Model

The pervasive influence of Java has resulted in several Java-based distributed object architectures. The most ubiquitous is the Java RMI, a remoting architecture that shipped in 1997 with version 1.1 of the Java Development Kit (JDK). Java RMI is a Java-only solution that provides an elegant mechanism for allowing remote invocations on Java objects.

The usage of Java RMI continues to increase because it provides the remoting infrastructure for Enterprise JavaBeans and future systems like SUN Microsystems' JavaSpaces (which actually pick up the concept for de-coupled communication via a tuple space first introduced with the Linda coordination language (Gelernter and Carriero 1992) and Jini. The most significant disadvantage of Java RMI is that it can be used only with the Java programming language. Using both Java RMI and CORBA can alleviate this disadvantage. In fact, the Java-to-IDL mapping recently added to CORBA 3.0 eliminates the need to use Java RMI in many situations.

At the most basic level, RMI is Java's RPC mechanism. RMI has several advantages over traditional RPC systems because it is part of Java's object-oriented approach. Traditional RPC systems are language-neutral, and therefore are essentially least-common-denominator systems — they cannot provide functionality that is not available on all possible target platforms.

RMI is focused on Java, with connectivity to existing systems using native methods. This means RMI can take a natural, direct, and fully powered approach to implement a distributed computing technology that allows to incorporate Java functionality throughout a system in an incremental, yet seamless way.

The primary characteristics of RMI are:

Object orientation: RMI can pass full objects as arguments and return values, not just predefined data types. The programmer can pass complex types, such as a standard Java hashtable object, as a single argument. In existing nonobject-based RPC systems, the client would have to decompose such an object into primitive data types, ship those data types, and than recreate a hashtable on the server. RMI allows shipping objects directly across the wire with no extra client code.

Mobile behavior: RMI can move behavior (class implementations) from client to server and server to client. Operations, such as checking constraints on user input, can therefore be checked on the client side — providing faster feedback to the user and less load on the server — without installing any new software on the user's system.

Design patterns: Passing objects from server to client and *vice versa* allows extending object-oriented technology into distributed computing, such as two- and three-tier systems. All object-oriented design patterns rely on different behaviors for their power; without passing complete objects — both implementations and type — the benefits provided by the design patterns movement are lost.

Safety and securety: RMI uses built-in Java security mechanisms to ensure system integrity when users are downloading implementations. RMI uses the security manager defined to protect systems from hostile applets to protect the systems and network from potentially hostile downloaded code. In extreme cases, a server can refuse to download any implementations at all.

Easy to write/easy to use: RMI makes it simple to write remote Java servers and Java clients that access those servers. A remote interface is an actual Java interface. A server has roughly three lines of code to declare itself a server, and otherwise is like any other Java object.

Connects to existing/legacy systems: RMI interacts with existing systems through Java's native method interface JNI. Using RMI and JNI, Java clients can still use existing server implementations. Similarly, RMI interacts with existing relational databases using JDBC without modifying existing non-Java source that uses the databases.

Distributed garbage collection: RMI uses its distributed garbage collection feature to collect remote server objects that are no longer referenced by any clients in the network.

Because of its use in technologies such as Enterprise JavaBeans, JavaSpaces, and Jini, Java RMI will be found in many large software systems. It currently appears, however, that Java RMI will not dominate as a general remoting architecture because of its reliance on a single programming language.

4.8   The Microsoft .NET Framework

Microsoft announced the .NET initiative in July 2000. The .NET platform is a new development framework with a new programming interface to Windows services and APIs, integrating a number of technologies that emerged from Microsoft during the late 1990s. Incorporated into .NET are COM+ component services, the ASP web development framework, a commitment to XML and object-oriented design, support for new web services protocols such as SOAP, WSDL, and UDDI, and a focus on the Internet (Thai and Lam 2002).

.NET also marks a departure from earlier component systems by requiring each component (assembly in .NET jargon) to carry meta-data. Meta-data are used by the system for integration with legacy technologies, such as COM; however, since component meta-data are extensible from the programming language level through the *attribute* construct, .NET allows the programmer to explicitly express component properties and requirements, such as security, resource usage, timeliness, or fault-tolerance assumptions.

.NET responds to the following trends in the middleware arena today:

Distributed computing: .NET provides a remoting architecture that exploits open Internet standards, including the Hypertext Transfer Protocol (http), Extensible Markup Language (XML), and Simple Object Access Protocol (SOAP).

Componentization: .NET extends the previous COM component model but provides a significantly simpler way to build and deploy components.

Enterprise services: .NET supports the development of scalable enterprise applications without writing code to manage transactions, security, or pooling.

Web paradigm shifts: Over the last few years, web application development has shifted from connectivity (TCP/IP), to presentation (HTML), to programmability (XML and SOAP). A key goal in .NET is to enable software to be sold and distributed as a service.

Maturity factors: Although .NET is a relatively new framework, it builds upon the mature COM+ technology and services.

Comparable to the XML/Value mapping introduced with CORBA 3.0, .NET provides a bidirectional mapping between XML and objects in .NET. For example, a class can be expressed as an XML Schema Definition (XSD), an object can be converted to and from an XML buffer, a method can be specified using the XML-based Web Services Description Language (WSDL), and an invocation (method call) can be expressed using the XML-based SOAP protocol.

The Microsoft .NET platform consists of five main components. At the lowest layer lies the operating system, which can be one of a variety of Windows platforms (CE, ME or 2000/XP) — and, since a substantial part of the .NET framework has been made publicly available as Shared Source Common Language Infrastructure (SSCLI or Rotor) and ported — the UNIX-based FreeBSD and Mac OS X operating systems. On top of the operating system lies the .NET framework (the Common Language Runtime (CLR) plus a collection of library classes), a series of Windows-based .NET Enterprise Server products, and .NET Building Block Services (such as .NET My Services). At the top layer of the .NET architecture resides a development tool — Microsoft Visual Studio.NET.

The COM component framework allowed for component reuse on the binary level. However, in order to use COM components from different languages, the programmer had to obey COM identity, lifetime, and binary layout rules. Also, programmers were required to write plumbing code to create COM components. Being based on a *Common Type System* (CTS), all .NET classes are ready to be reused at the binary level from any of the .NET implementation languages.

Microsoft .NET supports not only *language independence* — as COM did — but allows for *language integration*. This means .NET classes can inherit from each other, catch exceptions, and take advantage of polymorphism across different languages. The CTS specifies that any entity in .NET is an object, derived from a root class System.Object. The CTS supports the general concepts of classes, interfaces, delegates (which support callbacks), reference types, and value types.

Figure 4.6 shows the structure of the .NET framework. The Common Language Runtime (CLR) forms the most important component of the framework. Comparable to the Java Virtual Machine (JVM), the CLR activates objects, performs security checks on them, lays them out in memory, executes them, and garbage collects them.

Conceptually, the CLR and JVM are similar in that they are both runtime infrastructures that abstract the underlying platform differences. However, while the JVM currently only supports the Java language, the CLR supports all languages that can be represented in the CIL. Microsoft has submitted the Common Language Infrastructure (CLI), which is a functional subset of the CLR, to European Computer Manufacturers Association (ECMA) for standardization; thus, a third-party vendor could theoretically implement a CLR for additional operating system platforms.

In Figure 4.6, the layer on top of the CLR is a set of framework base classes, comparable to the Java class library. These classes support input/output functionality, text and string manipulation, security management, network communications, thread management, reflection functionality, as well as other functions. The framework base classes also implement .NET remoting — an object-based interaction scheme that is comparable to Java RMI.

On top of the framework base classes is a set of classes that provide support for data management and XML manipulation. The data classes also support persistent data management. Finally, classes in three different technologies (Web Services, Web Forms, and Windows Forms) extend the framework base classes and the data and XML classes. Web Services support the development of lightweight distributed components, which will work even in the face of firewalls and Network Address Translation (NAT) software. These components support wide-area distributed computing, because Web Services employ standard http and SOAP.

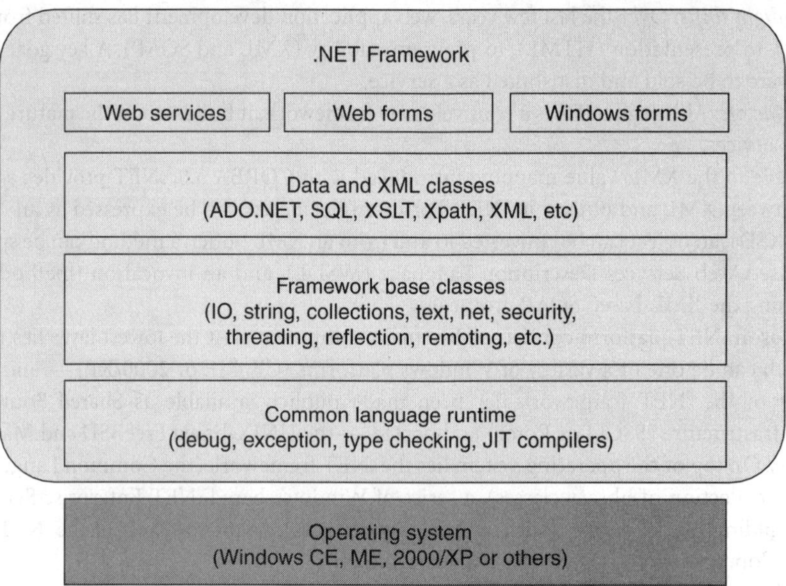

FIGURE 4.6   The .NET framework

4.9   Future Trends: Resource Management and Quality-of-Service

The abstractions offered by various middleware frameworks can be used to provide resource management in a distributed system at a high level. These abstractions can be designed to be rich enough to subsume the three kinds of low-level resources an operating system manages: communication, processing, and storage (memory and disks). Middleware abstractions are also from an end-to-end perspective not just of a single host, which allows for a more global and complete view to a resource management system. Distributed objects are promising, as they not only encapsulate but also cleanly integrate all three kinds of resource into a coherent package. This completeness helps distributed resource management and makes it easier to provide for load balancing, mobility transparency, and overall system reliability.

Distributed systems are inherently very dynamic, which can make them difficult to program. Resource management is a prerequisite for predictable system behavior; but is often not enough for most distributed applications. Starting from the late 1990s, distributed systems research has begun to focus on providing comprehensive Quality-of-Service (QoS), an organizing concept referring to the behavioral properties of an object or system, to help manage the dynamic nature of distributed systems (Zinky *et al.* 1997). The goal of the research is to capture the application's high-level QoS requirements and then translate them down to low-level resource managers. QoS can help runtime adaptation. But it also helps the applications evolve over their lifetime to handle new requirements or to operate in new environments, issues more in the domain of software engineering but also of crucial importance to users and maintainers of distributed systems.

Recent middleware frameworks, like the CORBA 3.0 Component Model (CCM), or the Microsoft .NET framework, allow the expression of nonfunctional component properties, such as resource requirements, timing and security constraints, or fault-tolerance assumptions on the component level using language constructs (like C# and .NET attributes) or component meta-data (like CCM's deployment descriptors). Aspect-Oriented Programming (AOP) is a relatively new discipline that focuses on cross-cutting concerns (targeting many components of a system simultaneously) and nonfunctional component properties. AOP investigates software engineering approaches toward predictable component-based systems. This research opens up new venues for middleware-based architectures on the enterprise level.

References

Amar, V. and Bensoussan, P., CORBA Benchmarks, http://www.beust.com/virginie/Benchmarks/index. html, 1998.

Bernstein, P.A., Middleware: a model for distributed services, *Communications of the ACM*, 39, 86–97, 1996.

Gelernter, D., Carriero, N., Coordination languages and their significance, *Communications of the ACM*, 35, 1992, 97–107.

The Component Object Model Specification, http://www.microsoft.com/oledev/olecom/title.htm.

The Common Object Request Broker: Architecture and Specification, Revision 2.0, July 1995, http://www.omg.org/corba/corbiiop.htm.

(DCE) AES/Distributed Computing — Remote Procedure Call, Revision B, Open Software Foundation, 1995, http://www.osf.org/mall/dce/free_dce.htm.

Gokhale, A.S. and Schmidt, D.C., Evaluating CORBA Latency and Scalability over High-Speed ATM Networks, Proceedings of 17th International Conference on Distributed Computing Systems (ICDCS'97), Baltimore, USA, May 1997, IEEE Computer Society Press, Silver Spring, MD, 1997.

The Object Management Architecture, http://www.omg.org/oma/

Rogerson, D., *Inside COM*, Microsoft Press, Redmond, Washington, 1996.

Siegel, J., *Quick CORBA 3*, Object Management Group/John Wiley & Sons, New York, 2001.

Thai, T. and Lam, H.Q., *.NET Framwork Essentials*, 2nd ed, O'Reilly, Sebastopd, CA, 2002.

Vinoski S., CORBA: integrating diverse applications within distributed heterogeneous environments, *IEEE Communications*, 14, 1997, http://www.iona.com/hyplan/vinoski/ieee.ps.Z.

Zinky, J., Bakken, D., and Schantz, R., Architectural support for quality-of-service for CORBA objects, *Theory and Practice of Object Systems*, 3, 1997, 55–73.

5

Distributed Components in Microsoft Platforms — Technology Overview

Marcos Ribeiro Pereira Barretto
University of Sao Paulo

Paulo Marcelo Porto Alves Blanco
University of Sao Paulo

Marco Antonio Poli
University of Sao Paulo

5.1 Overview

Microsoft component technologies are of central importance for Automation and Control, if for no other reason, because they are the basis for OLE for Process Control (OPC), the standard for SCADA/PLC integration. Another reason is, of course, the predominance of MS Windows as the operating systems for microcomputers, which makes Microsoft's component model the most widely used in the world. In addition to this is the large number of applications developed using Visual Basic, perhaps the most widely used programming language nowadays, which generates components even if the programmer is unaware of them.

The word "component" has different meanings for different people. As defined in this chapter, it refers to the smallest piece of software built to be reused in binary format, packing inside the same file a set of related (according to designer view) functionalities. A component is developed according to a *component model*, which is, simply stated, a set of standards and implementations to allow its use and reuse, including, among others, binary file format, life cycle control (how is it found, how to load/unload), and invocation conventions (how to execute a function inside the component, passing parameters, and getting results back).

It is worth separating the component model from its *runtime environment*, as long as the same component can be run in different environments. For some technologies, there is a one-to-one correspondence (such as in CORBA); but for Microsoft technologies, a component built under one model can be run in different runtime environments. A component runtime environment can provide a large number of services, to allow comfortable development of business applications: transactional control, security, administration are among the most important ones.

Over the recent years, Microsoft provided a number of technology frameworks for components, with similar names but different functionalities. The literature does not provide a uniform view. Given the above definitions, MS component technology includes the following *component models*:

- Component Object Model (COM), which also includes Distributed COM (DCOM) for local (inside only one computer) or remote components.
- .NET[1]: The newest Microsoft achievement in component technology, providing language-independent components.

As *runtime environments*, Microsoft developed:

- MS Windows operating system, providing basic services for locating and component load/unload.
- COM+, providing transactional control, security, administration, and some other services for COM and .NET components.
- .NET, providing about the same services as COM+, but solving many of the problems found in it.

This chapter provides an overview of each of these component models and runtime environments.

5.2 Scope

COM/DCOM/COM+/.NET is a lengthy subject for a single chapter, more so considering different audiences (programmers, architects, end-users, etc.). Therefore, this chapter is devoted to technical audiences, exploiting the subject on its functional aspects. No source code examples are provided; but there are plenty in the references cited.

From all aspects included particularly in component runtime environments, special attention has been devoted to basic mechanisms and Automation and Control requirements, particularly on the shop floor level. Therefore, enterprisewide requirements, such as transactional control and security, more related to the upper levels of CIM pyramid and data-centric systems, are not discussed in this chapter.

Microsoft component-related technologies have a long history. Many concepts were formulated; some have not lasted. As far as possible, the material presented here corresponds to what is available on MS Windows 2000, which closely resembles MS Windows XP. No attempt has been made to provide a broader view, covering the various MS operating systems versions (Windows 95, 98, ME, NT) nor update packages made available by Microsoft. The technologies involved do not have a strict version control numbering (at least, not publicly available), making it impossible to characterize its current status by version numbers.

5.3 Some on the Past and Present

The need for loadable modules to ease application development had been recognized in the late 1960s (Tannenbaum, 1987). In the process of generation of an executable program, after compilation of source code into binary modules, it is necessary to link all compiled modules to libraries of existing modules. This link process can be done:

- Statically, meaning that all object and library modules will be assembled in a single, executable module.
- Dynamically, meaning that some object and library modules will be only loaded in runtime.

---

[1] It shall be read as "dot net."

Microsoft's MS DOS only allowed for static linking; but the possibility of dynamically loaded libraries (known as DLL) has been present since the first version of its Windows operating system, in the mid-1980s. At that time, a DLL was simply a bunch of functions, packaged in a single file, in a predetermined binary format, allowing to determine which functions were included and where they were inside the file.

Object-oriented concepts were spreading very fast and soon Microsoft realized its potential. In the process of creating its client-side (Windows 9x) and server-side operating systems (known as Windows NT and its successors), the need for an object-oriented component model was clear, and the COM model was therefore specified, in 1992 (Microsoft, 1995).

COM components are the basis of the Microsoft's Compound Document Model, perhaps better known as Object Linking and Embedding (OLE) automation. Every MS Office end-user, for instance, uses OLE when he or she inserts an MS Excel worksheet into an MS Word document and edits it from MS Word. In other words, the worksheet is a compound document and every application complying with OLE (as MS Excel) provides a set of functions to allow for in-place (i.e., inside MS Word) activation. OPC, described elsewhere in this book, is also an application based on this standard. As shown in Figure 5.1, OLE is actually a set of higher-level technologies built upon COM, supporting user-interface-oriented features for application integration and task automation.

For programmers, since Visual Basic version 3 back in the very beginning of the 1990s, OLE has become extremely important in the use of visual components. For the first time, appealing visual interfaces could be drawn using drag-and-drop of visual components and with very little programming effort. The acronyms VBX (Visual Basic Custom Control), OCX (OLE Custom Control) and ActiveX correspond to three versions of OLE components, showing an evolution from 16-bit, incomplete object-oriented model to full 32-bit, object-oriented components.

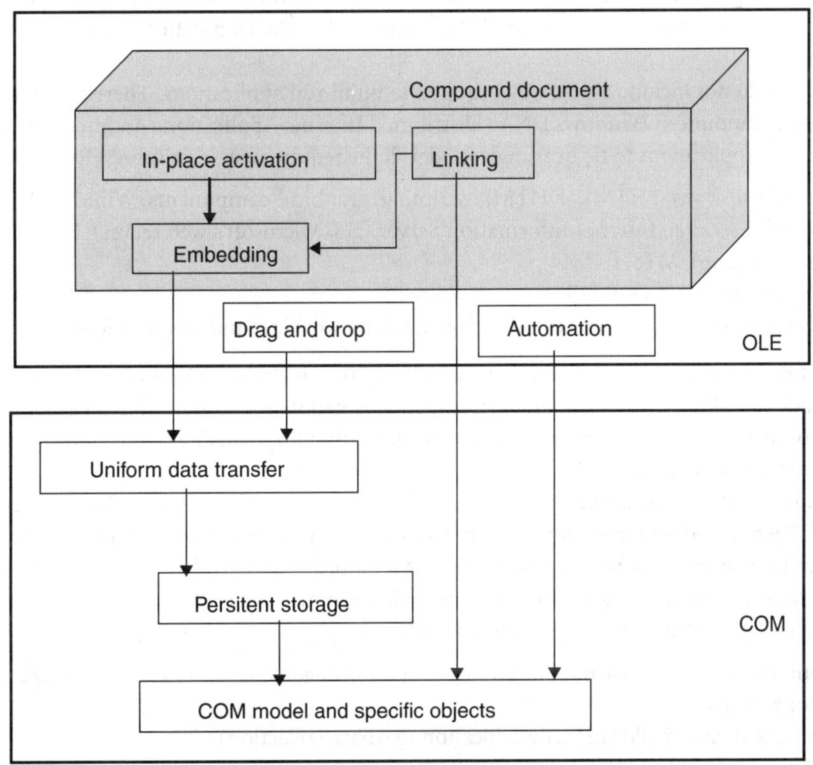

FIGURE 5.1   OLE and COM (based on Microsoft, 1995).

COM specification defines three types of *server*, that is, "some piece of code that structures some object in such a way that COM 'implementor locator' services can run that code and have it create objects" (Microsoft, 1995). They are as follows:

- In-process server, when a server is loaded within client running process. In other words, the component is activated as a local call, as any other function in the same program.
- Local server, when a server runs in a separate process from the client's, but on the same machine.
- Remote server, when a server runs on a machine other than the client's.

The set of specifications for remote servers is known as DCOM. DCOM was previously called "Network OLE" and is based on Open Software Foundation's Distributed Computing Environment, Remote Procedure Call (DCE-RPC) (OSF, 1995), a mid-1980s environment for distributed applications. DCOM appeared in 1995.

But component activation is not sufficient to comfortably develop components for mission critical applications; a comprehensive environment is needed, encompassing transactional control, component administration, security, just-in-time activation, object pooling, synchronization, event, and asynchronous messaging. Microsoft first announced during its 1997 Professional Developers Conference its plans to build such an environment and called it COM+. Today, the acronym COM+ is heavily overloaded and misused, meaning this runtime environment and new versions of COM specification (Lowy, 2001). In this text, COM+ will be used to refer only to the runtime environment, leaving COM as meaning a component model.

In 1998, Microsoft first released its Microsoft Transaction Server (MTS), combining the features of a transaction processing (TP) monitor and an object request broker, the first element to build a scalable environment for application with easy administration by a graphical tool (MMC, Microsoft Management Center). MTS is one of Microsoft's greatest technological achievements, bringing to its platform functionalities only available on mainframes (such as CICS), easy-to-use, configurable, and manageable from graphical interfaces. MTS was then integrated into Internet Information Services (IIS) and Distributed Transaction Coordinator (DTC) remained as the transaction manager for distributed transactions.

Components do not include everything required to build real applications. Therefore, about the same time, Microsoft announced Windows DNA (Distributed Internet Applications Architecture), a set of system services and applications to tie in a single model all the requirements, including (Redmond III, 1999):

- *Presentation services*: HTML, DHTML, scripting, graphical components, Win32 API.
- *Application services*: Internet Information Server (IIS, Microsoft's web server), MSMQ (asynchronous messaging), MTS, COM+.
- *Data services*: ADO, OLE DB.
- *System services*: directory, security, management, networking, and communications.

MS Windows XP, a direct descendent from Windows 2000, includes COM+1.5, the latest COM+ version. It includes SOAP services, configurable transaction isolation level, application partitioning for large-scale environments, pooling and recycling, enabling/disabling/pausing, and improved managing capabilities, among others (Lowy, 2001).

COM+1.5 immediately preceded .NET. Some authors refer to .NET as COM+ (Nilsson, 2002), mainly because .NET was based on ideas (not code) of COM+. COM+ is not dead; as stated before, it can host COM and .NET components, which means they are a natural (perhaps the only) choice to support the addition of new functionalities on legacy systems built upon it.

.NET as a component model brings (Nilsson, 2002):

- A common runtime environment for any programming language, allowing for language interoperability.
- Automatic memory allocation/de-allocation (garbage collection).
- Inheritance, including inter-language inheritance.
- Strong security model, confining execution path and data to allowed portions.

- No need for IDL or interface declarations; methods can be used and declared directly within classes.
- New forms of remote method invocation, especially SOAP.

As a runtime environment, .NET includes:

- Transactional semantics, security, component administration, JITA (just-in-time activation), and other functionalities as in COM+.
- Business-oriented servers, such as BizTalk Server for B2B process automation, Commerce Server for e-commerce, Content Management Server for multimedia content publishing over the Web, Identify Server for single sign-on, and others.

Clearly, .NET builds upon COM/COM+ and the Internet a new application development framework, requiring the revision of well-established concepts and workflows. As Microsoft states, '.NET is a vision and a set of software technologies for connecting information, people, systems and devices.'

In what follows, each of the above-mentioned technologies is briefly described in more technical terms.

5.4   COM/DCOM

COM is an open standard, is publicly documented, and addresses the building of a component-based architecture. It is a distributed, object-oriented model. As stated before, it is the core technology for all Microsoft's component architectures. DCOM is the name by which a part of COM specification is known, related to remote method invocation. Therefore, DCOM is just a part of COM.

To describe a component model, it is necessary to characterize the following:

- The component interface, that is, the methods it will expose to clients.
- An identification mechanism, that is, the "name" or "unique identification" of an object to allow it to be found among hundreds (perhaps thousands) of different objects.
- Runtime environment requirements, that is, what is assumed as runtime responsibilities to allow a component to be created and deleted, or its life cycle.
- Invocation process, that is, how a component is found, its interface returned to client, and how to handle parameters passing to and back from the component, including data type conversion.

It is important to note Microsoft's attempt to conceal all complexities involved in creating COM components. A Visual Basic programmer, for instance, can create COM components without knowing any of the concepts discussed here, just because VB development tools are able to conceal all these details. But they are there, allowing components to communicate.

Client–server Model

Most of the time, a synchronous invocation of a component's method is what is need. Synchronous invocation, also known as the client–server communication model, involves, as shown in Figure 5.2, the following:

1. server object creation,
2. implementation location,
3. interface returned to client,
4. interface method invocation.

COM introduces the idea of location transparency, allowing a server to be:

- In-process, when it runs inside client's process.
- Local, when it runs in a process different from the client's, but on the same machine.
- Remote, when the server runs on a different machine.

These three situations are illustrated in Figure 5.3.

When running in-process, the component becomes a part of client's process, allowing for a direct invocation almost as if they were statically linked. But when running in local- or remote-mode, a

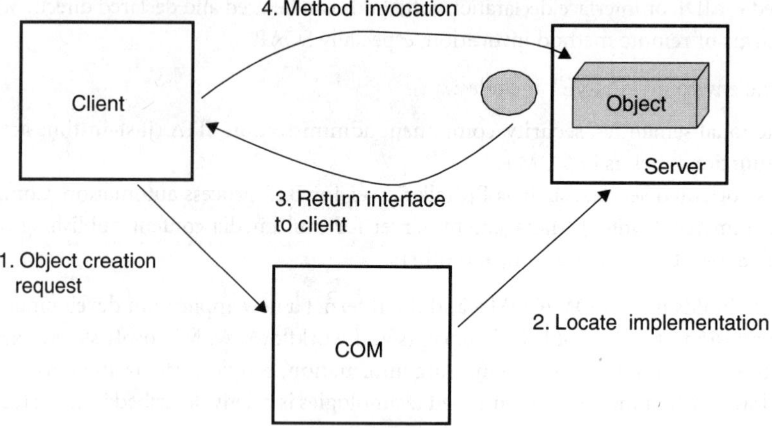

FIGURE 5.2   Client–server model (based on Microsoft, 1995).

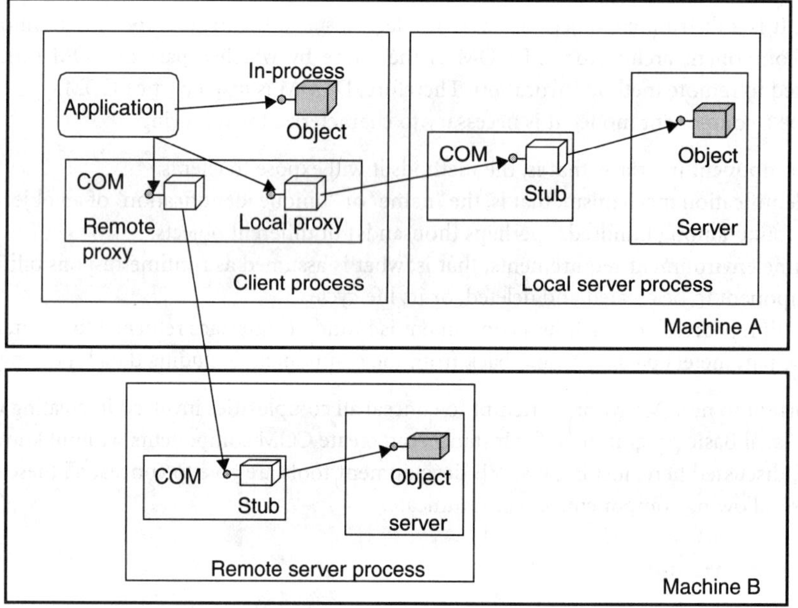

FIGURE 5.3   COM servers (based on Microsoft, 1995).

proxy-stub mechanism (similar to CORBA's stub-skeleton) is generated. A proxy is a local (to client) view of the component; its responsibility is to pass a method invocation request to the corresponding stub, converting parameters into a standard format.[2] On method return, the method output results are passed back to the proxy, which converts it back to the caller's data type and returns control to the client.

Interfaces and Objects

A component is characterized by its interface, that is, the methods it exposes to clients. Often referred as a "contract" between client and object, the interface determines the method's name and input parameters, output results and, in some cases, exceptions thrown.

---

[2]This process of converting parameters in a standard format to/from stubs is known as marshaling.

Interfaces can be specified in Microsoft Interface Definition Language (MIDL). These are discussed here, even bearing in mind that most COM programmers (especially VB) have never heard of them. So, by understanding these interfaces, the "magic" behind them, which would otherwise go unnoticed, is revealed. It is important to note that COM is a binary specification, independent of any source-level tools used to create components. In other words, it is not mandatory to generate an interface definition using MIDL to create COM components; they can be directly generated in binary format.

Code 1 shows an example of interface specification in MIDL.

```
[obiect, uuid (b5483f00-4f6c-101b-alc7-00aa00389acb)]
interface ICalc : IUnknown {
      [id(1),helpstring ("method add")]
      HRESULT add ([in] int x,[in] int y, [out, retval] int* r);
}
```

Code 1 MIDL (simplified) example.

If you have already read the chapter on CORBA, you will immediately notice the similarities between them. It is not a coincidence; both COM and CORBA were inspired by Open Software Foundation's DCE-RPC. But DCE-RPC is function (not object) oriented, while MIDL has been extended to provide object-oriented semantics, especially interface inheritance and method overload, resulting in what is called ORPC (Object RPC).

Code 1 is quite straightforward to understand the following:

- The object attribute determines ORPC coding generation, as opposed to DCE-RPC.
- The uuid attribute (universal unique identifier) provides a universal unique identifier to this interface; more on this in what follows.
- The interface keyword states the beginning of an interface declaration; in this case, it was named as ICalc and inherits (as all COM components) from *IUnknown* interface.
- The attribute id determine a numeric identifier for the add method within ICalc interface, while helpstring gives a human readable name for this method.
- Finally, the add method is declared. All synchronous methods in COM must return HRESULT. The [in] and [out] clauses specify parameters input and output.

An interface must be implemented by some object to be callable. But from COM's point of view, only the interface matters.

Returning to Figure 5.3, the client maintains a pointer to the interface. This pointer is actually a pointer to an array of pointers to interface member functions. Therefore, even on in-process method invocation, there is a double indirect address resolution to actually invoke a method. This situation is illustrated in Figure 5.4.

These data structures are necessary to control an object's life cycle, because COM infrastructure must be aware of which interfaces are in use and when to free those not needed anymore.[3]

Basic Interfaces

COM defines two basic interfaces: *IUnknown* and *IDispatch*. All COM components implement one or the other, which is the same to say that all COM interfaces must inherit from one or another. The bottom line is that all COM components implement IUnknown since IDispatch inherits from it. *IUnknown* has three methods:

- QueryInterface(), which allows a client to request any other interface that the object supports.
- AddRef(), called every time a new client requests access to an interface.
- Release(), a function complementary to AddRef(), called by a client to notify that it is not requiring access to the object anymore.

---

[3]This mechanism is also known as "smart pointers."

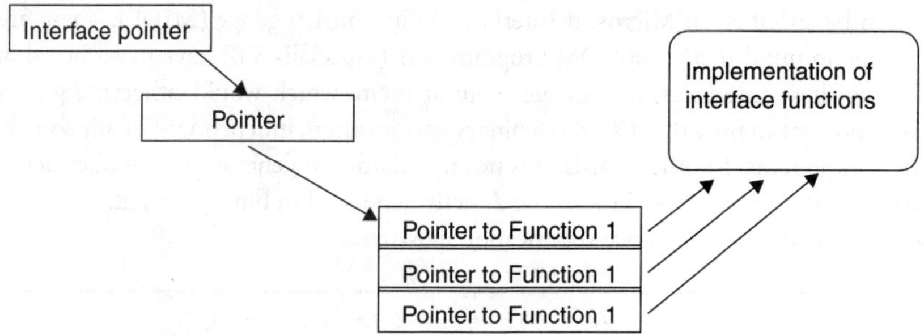

FIGURE 5.4   Interface data structures.

AddRef() and Release() control object life cycle, while QueryInterface() allows a client to gain access to an interface implemented by some object.

IUnknown is central to COM model because it provides the basic mechanism to retrieve an interface. Since all COM components must implement IUnknown, the QueryInterface() method can be called on any component, allowing to retrieve an interface. It reveals another aspect of interface-based component construction: a single object (or component) may support a number of interfaces.

IDispatch inherits from *IUnknown* and has four methods:

- GetIDsOfNames(), which maps a single member and an optional set of argument names to a corresponding set of numeric (integer) method identifier (DISPIDs or Dispatch ID).
- Invoke(), which provides access to properties and methods exposed by an object.
- GetTypeInfo(), which gets the type information for an object.
- GetTypeInfoCount(), which retrieves the amount of type information provided by an object (either zero or one).

Through the use of GetIDsOfNames() and Invoke(), it is possible to dynamically invoke a method, by knowing only its name. Before DCOM, IDispatch was the only way to implement remote method invocation. By using IDispatch, a client can directly invoke a method of an object, instead of retrieving an interface. This is called "late binding" (in contrast to the so-called "early binding" when using IUnknown), and it is the only possibility to access a component when a type library is not available. It requires two calls to the component; one to GetIDsOfNames() to retrieve a DISPID and the second to call the method, eventually passing parameters using the VARIANT data type.

Interface Identification and Registration

In Code 1 above, a universal unique identifier (UUID) was associated with the interface. An interface identifier (IID) is a 128-bit statistically generated number. Microsoft's development environments (such as Visual Studio) have functions to generate these values. They are also known as Globally Unique Identifier (GUID). The object itself (which contains one or more interfaces) is uniquely identified by its Class ID (CLSID), which is a number in the same format as UUID.

When retrieving interfaces through IUnkown, a client requests COM runtime environment to provide access by specifying the UUID. If using IDispatch, a client must first gain access to an object, which is achieved by specifying the component name and retrieving its CLSID (if CLSID was not known in advance). In both situations, the Registration process is central, allowing to maintain, in a kind of database, the UUID/interface/object and CLSID/object association. In some sense, this registration is similar to CORBA's Naming Service; its goal is to allow an interface to be located within an object implementing it.

In the beginning of the 1990s, it was enough to copy the DLL containing the component to the MS Windows installation directory. Nowadays, however, components must be registered in the Windows Registry to become useful. This registration is automatically done when using some development tools (such as Microsoft's) or when installing a new application. To manually register a component, tools such as regsvr32 or MMC (Microsoft Management Console) are available.

Type Libraries

A type library is a description of an interface in a format that can be easily parsed and read by any application or development environment via Windows system calls (Win32 functions). Microsoft decided to include this concept to its component environment to allow for the support of different programming languages, object-orientation, and to automatically find and run server upon client request. In a Visual Basic development environment, for instance, it is the existence of a type library that allows for the browsing of methods and attributes (Properties, in VB terms) of a component, independent of the language the component was written in. It also allows for compile-time syntax checking, because the VB environment (for instance) "knows" what it is inside the component. The "Properties Sheet" on any visual development tool is also an application of type library concept.

Type libraries are also in action during marshaling parameters when calling remote functions (COM invocation other than "in-process" mode). The COM runtime environment reads type information that also contains instructions on parameter data types. Having these instructions, COM is able to convert parameters for remote call.

Sometime ago, type libraries existed in a separate file with .TLB extension. Nowadays, this information is inside the component .DLL file, which eases component distribution because everything is inside a single file.

Strictly speaking, a type library is not required in COM technology, because "in-process" invocation can be done directly in some cases or the IDispatch interface can be used. But with COM+ and modern development environments, it became essential, which makes them, for all practical purposes, mandatory.

DCOM

DCOM is about remote method invocation. Every time a call to an object's method crosses the boundaries of a caller's process, there is a remote invocation. Therefore, it can happen inside a single machine, when the client and component are to run different processes, or when two machines are involved. This section provides some details of the mechanisms involved in a remote call, which is the same to say, details of DCOM.

DCOM general architecture is depicted in Figure 5.5.

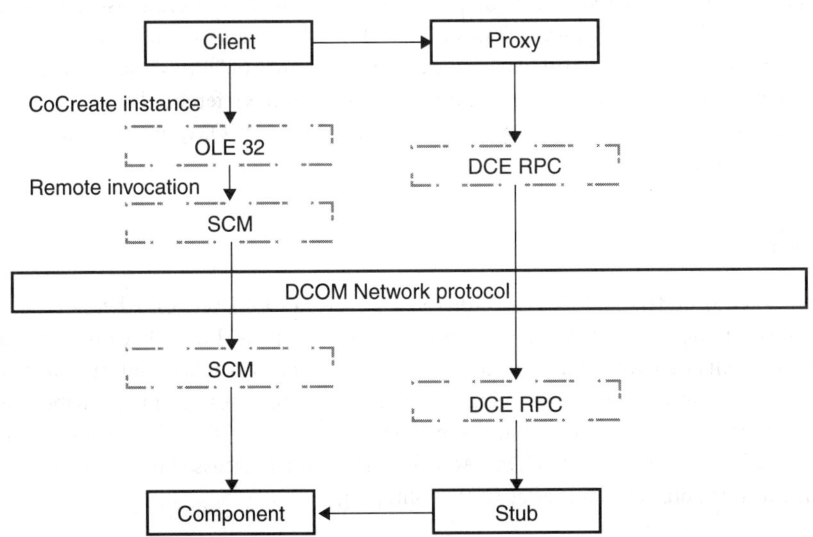

FIGURE 5.5 DCOM architecture (Microsoft, 1996).

The basic mechanism, as mentioned before, is DCE RPC; a proxy object on client's side packs all arguments of a call (which is also known as "marshaling") and sends them (normally, via a TCP/IP socket) to the server. The DCOM network protocol controls this communication. Once at the server, the arguments are unpacked (unmarshaled) and passed to the component. Returned results follow the same path back to the client.

One of the major strengths of the COM architecture is to allow for location transparency. In other words, COM will select the best way a component will run, among the three server modes described above. Whenever possible, "in-process" is selected, because it is the fastest way; in this case, there is no remote method invocation (therefore, no proxies, no RPCs). This transparency is provided by COM's *CoCreateInstanceEx()* method,[4] which will first search the Registry for local DLLs implementing the required interface, then for a local process within client's machine and, finally, on remote machines. Technically, this last case is named as DCOM.

To be able to instantiate an object on a remote machine, COM libraries must know the server's network name and the required interface (CLSID). Once both are known, client's COM library contacts (via its local SCM, Service Control Manager) the remote SCM, through a well-known end-point TCP/IP port (normally, the standard DCE RPC port, number 145). The remote SCM instantiates the object dynamically allocating a port. Then, SCM responds to the client, informing the selected port. From this point, conversation (i.e., method calls) can be made directly from the client to component, without passing through SCM.

CLSID are normally given to the client as fixed data, as long as the interface ids are known in advance. To discover the server's network name, one of the following approaches must be followed:

1 An explicit parameter on CoCreateInstanceEx().
2 As a fixed configuration parameter within the Windows Registry or application's infrastructure (for instance, from a .INI file).
3 In a DCOM class store, which acts as an interface repository.

The first two situations are quite straightforward and require no further comments. The last approach has been possible (or, at least, has been directly supported by MS Windows) since the introduction of Windows 2000, and involves the use of the so-called Active Directory infrastructure to store activation-related information. Active Directory is responsible for holding all user-related information, such as login name, passwords, and others. This way, it behaves in a manner similar to the Naming Services in CORBA Architecture. This simplifies administration, because changes will automatically propagate to all clients.

DCOM also provides a solution for interface reference count, because a remote object must know when an instance has been released. In case of client crash or network failure, an instance must be de-referenced. DCOM provides a "pinging mechanism," which implies the server to "ping" (try contact) the client in a programmable time interval. Each time a certain number of "pings" are not responded to, the server assumes abnormal client termination and removes it from the reference list.

DCOM also defines a mechanism for asynchronous calls (Async DCOM), which is not recommended anymore, as discussed below.

5.5   COM+

There are two ways to understand COM+: either as an evolved COM version (therefore, a component model) or a component's runtime environment. Actually, it is both: because of this, several changes/enhancements were introduced in COM specification and also, it is a runtime environment supporting transactions, asynchronous messaging, and many other features. In this contribution, it will be treated as a runtime environment. The changes in COM specifications, therefore, will be separated from COM+ and considered as belonging to COM. Actually, the material discussed in the previous section also included some features only available after COM+ shipping.

---

[4]Actually, there are other COM functions which can be called; *CoCreateInstanceEx()* is just the more comfortable and used.

As stated by (Brill, 2001),

COM + = COM – older, now unnecessary appendages

    + Event Notification

    + Object Pooling

    + Transactions

    + Queuing

    + Role-based security

Still, following (Brill, 2001), "COM+ is a fusion of traditional COM, MTS and MSMQ," where:

- Microsoft Transaction Server (MTS) enables two or more COM objects to participate in a transaction, adding the "2-phase commit" semantics to transactions.
- Microsoft Message Queueing (MSMQ) enables asynchronous (message-driven) calls between objects (COM only allows for synchronous calls). MSMQ implements guaranteed message delivering, by storing messages in a persistent storage (file, database) prior to sending and only removing after being received.

In the above, IIS, Microsoft's webserver, could also be included since it is a very frequent application server (i.e., the process within which a component runs) in the so-called "web architecture." In fact, since the introduction of Active Server Pages (ASP) with the IIS version 3.0, it became a "de facto" standard for the development of ASP pages activating COM components. An ASP page is a script that contains a mix of HTML tags (to display in a browser) and source code (normally, in a language derived from Visual Basic, called VBScript). In this way, an ASP page can format a screen to be sent to a web browser and do any kind of computational work, including database access. Including IIS in the COM+ definition makes even more sense if it is considered that Microsoft packed MTS within IIS 4.0, making them a single product. Starting with IIS 4.0, a number of IIS components could benefit from this merge, leading to a more component-based product.

Among other requirements, COM+ demands all components to be "in-process" (see the above discussion on COM server modes) and to have a type library. Therefore, a COM component must be "housed" within a process to be executed; normally, they will be within MTS or IIS.

As already mentioned, only topics related to Automation and Control on the shop-floor are discussed here to some extent. COM+ features more related to the upper levels of CIM pyramid and to data-centric systems are not covered, or mentioned.

Threads, Apartment, and Activities

A program is a file containing a piece of code compiled into binary (machine) language. A running program is called "process" and is the basic unit managed by an operating system. Nowadays, almost all modern operating systems (OS), including all Microsoft's OSs, starting with MS Windows 95,[5] offer support for "threads." A process is divided into different threads, each one representing a distinct flow in its computation and having its own memory space, although all threads share the same static memory space and code. Therefore, when an OS schedules a process to run, it selects one of its threads. A process, this way, is just a collective name for all the threads it contains and holds, constraints this collection of threads must deal with (for instance, the total memory available). In the days of single-threaded processes, a cooperative computation required some form of data sharing among processes; on the shop-floor (due to real-time constraints), it was normally accomplished by shared memory or semaphores. In a multi-threaded environment, global variables (static memory allocation) are visible to all threads, simplifying the development of applications. Special system calls allow for a thread start, stop, and the development of critical sections (where scheduling is prevented) for thread synchronization.

---

[5]Some readers may argue that MS Windows 3.11 supported multi-threaded, which is true. But scheduling was on developers' behalf, on a non-preemptive environment. Therefore, for practical (meaning: easy development) purposes, it was not there.

Manual manipulation of threads, and especially of critical sections, is very error-prone and difficult. An error in thread synchronization may lead to race conditions where unexpected results may be obtained (for instance, one thread may alter the value of a variable controlling a loop in another thread, causing this loop to finish before it should do), due to the asynchronous nature of thread scheduling: a thread may be stopped for another one to run, at any time. COM+ incorporated the concept of "apartment" for easy development, although not allowing for a tailored schema. Apartment is a term used to describe groupings of related objects and threads aiming to protect COM objects not prepared to run in multi-threaded mode. The name apartment comes from the idea that a thread and some objects (the initial caller and all its callees) will "live" in one apartment. Apartment is just a concept; it does not map to any well-known operating system construct.

Three threading models are supported in COM+:

- Single-Thread Apartment (STA), when exactly one thread exists in one apartment.
- Multi-Threaded Apartment (MTA), when many different threads coexist in the same apartment. Objects must be thread-safe and, therefore, handle synchronization issues. In this case, invocation is exactly as an ordinary DCE-RPC.
- TNA or NTA (Thread Neutral Apartment), available starting with MS Windows 2000, when threads may enter an apartment momentarily, during a method call.

The desired thread model for a component is set by the component itself and can be chosen between:

- Apartment, which implies running only as STA.
- Free, to be handled as MTA even in cases when it could be handled as STA, being always subject to performance loss due to thread context switching .
- Neutral, which implies TNA (recommended if possible).
- Both, leaving the calling mode to decide among STA, MTA, or TNA. The name "both" came from the days when just two options were available.

The thread model is stored in the Registry beside other object information, because COM needs to know in which mode it initializes.

In STA, since just one thread exists in the apartment, all access is serialized. To accomplish this, a message queue receives all requests directed to the object. Message queues have been present in MS Windows architecture since its early days; at that time, every event (mouse move/click, keyboard hit, etc.) was (as still is) transformed in a Windows Message and queued, to be processed one at a time. This model is also applied to STA; during a call to a remote method, it is converted by COM into a Windows messages and queued. This clearly creates a huge performance penalty, as long as only one invocation can take place at a time. For large-scale (many users) or very fast systems, this will probably be unacceptable. Visual Basic until version 6 was only able to create components to run in the STA mode; C++ programmers do not have such limitations. STA is applicable to graphical interfaces (since all graphical components must run on this mode).

MTA works just like an ordinary DCE-RPC: a running process listens to requests on a determined port, waits for a remote client to connect, and makes a method call. A thread is then retrieved from a thread pool to deal with the request, on the apartment. When the thread pool is exhausted, a new thread is created. In this case, COM does not interfere. With MTA, there is no message loop, only RPC.

TNA is sometimes also called (Brill, 2001) a "rental mode," because during an invocation, assuming client and server are running in different threads, the called thread momentarily enters the caller's apartment. Therefore, the call can be made without context switching. Returning to the image of an apartment inhabited by objects and threads, TNA is just like an empty apartment where an object waits to be acted upon by a thread.

For concurrency control, COM+ introduces the idea of "logical thread" or *activity*. This idea came from a simple observation that there is only one root client. Therefore, from this perspective, there is a single running thread, even if the "physical" thread spans multiple threads, processes, and machines.

Activities are particularly useful for MTA and TNA objects and maintain some resemblance to transactions, in the sense that normally a single transaction is involved when crossing various objects from a root client. An object may:

- DISABLE synchronization support, meaning that COM+ is told to ignore synchronization requirements.
- DOES_NOT_SUPPORT synchronization, preventing an object from participating in an activity.
- SUPPORT synchronization, when COM+ is instructed to share the activation context with the object if the object's creator had one, or will have no synchronization. It is the least applicable mode because of development complexity.
- REQUIRE synchronization, meaning an activity is shared with the creator's, or a new one is created if the creator had none.
- REQUIRE_NEW, when the object must have its own activity, different from the object's creator.

The example in Lowy (2001) makes the situation quite clear. Consider, for instance, the situation in Figure 5.6.

As depicted, Client_1 invokes Object_1 that, in turn, creates a new Object_2. If Object_2 is configured as "Required," a call from Client_2 will wait until the creation process is finished. On the other hand, if Object_2 is configured as "Requires New," Client_2 call is not blocked and Object_2 is capable of serving both requests at the same time.

Object Pooling

To avoid performance penalties due to (very) expensive processes of creating and destroying objects, COM+ provides *object pooling*. Therefore, before activating an object, it searches in the pool for a matching interface and returns it to the client, if found. After use, the object returns to the pool. Object lifetime can be configured using MMC.

To be pooled, an object must support the MTA or TNA threading model; therefore, Visual Basic 6 (and earlier) components cannot be pooled. Since the same object will be used in different activations, it must have no internal state (which is, anyway, the way objects are written using COM, as pure "function-oriented" components). If the object has to refer to external resources (database connections, for instance), it has to retrieve this information from the activation context.

Messaging and Asynchronous Calls

So far, components have been a "tight couple," since only client/server, synchronous calls were discussed. But COM+ also offers the possibility of asynchronous calls between components, using its Queued Components (QC), which rely on MSMQ product. MSMQ allows for:

- Guaranteed message delivery, if requested. In this delivery mode, messages are stored before sending on a persistent media (flat file, database) and are only removed when client receives it.

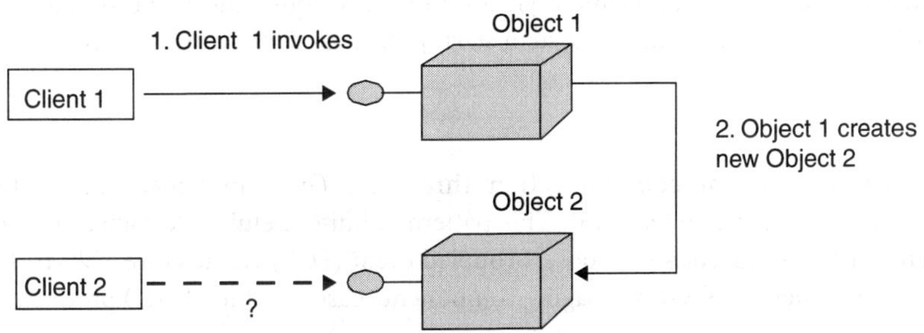

FIGURE 5.6　Synchronization.

- Logging services, allowing messages to be stored for later review.
- Security, allowing permissions to be granted for specific queues to senders/receivers and also including message encryption, avoiding tampering.
- Distributed transactions, associating message sending/receiving with database transactions in a single, two-phase commit, transaction.

QC is just a set of few components, covering the MSMQ C API. Therefore, it is very straightforward to use. Message-oriented communication is based on *queues*. On QC, a queue is neither a fixed connection nor an asynchronous path between two components, or a directional path. Therefore, any component is allowed to send messages through a queue; similarly, any one can read messages from that queue. Queues always follow First-In, First-Out (FIFO) discipline. Queues are completely independent of applications that created them; once created, they will exist forever, unless destroyed by using MMC.

QC is also the basis for asynchronous calls. Any object can be reached via asynchronous call provided:

- Its interface only contains methods with input arguments among OLE automation data types and no output return.
- It is marked as "Queued" when registered via MMC.

When a "queued application" is registered, COM+ creates six queues, which can be seen via MMC, with names beginning with application name. A special queue, "<app name>_deadqueue," is also created and stores messages after a number of unsuccessful delivery retries, probably because of an internal error in the component.

Being able to handle asynchronous calls does not prevent a component from still continuing to be reached by synchronous, normal COM calls. The way a call is processed depends on how it was invoked by the client. If normal COM invocation is used, the call is processed via conventional RPC. To invoke a method asynchronously, a client must first retrieve a queue object pointing to the application queue. This may be accomplished by a single line of code; for instance, when writing a VB6 client, it should be coded as in code 2.

```
' assuming application name is MyAsyncApp implementing SomeInterface
Dim QC As MyAsyncApp.SomeInterface
Set   QC = GetObject   ("queue:ComputerName=somehost/new:MyAsynApp.
SomeInterface")
```

Code 2 Async invocation.

After GetObject() is called, any method can be called, as it occurs normally with COM invocation, as long as GetObject() returns a queue moniker that, for all practical purposes, acts as an interface. The above example also shows how to invoke remote objects asynchronously, by simply specifying *ComputerName* when calling GetObject(), which may be omitted when doing local calls.

Unlike conventional COM components, asynchronous calls require the application to be started from MMC manually or automatically, during system startup; Just-In-Time Activation (JITA) does not apply.

Events

COM+ supports component communication through events, implementing publish/subscribe (observer) design pattern (Gamma, 1995). This pattern is quite useful in Automation and Control, as shown in Blanco (2003). It results in a loosely coupled event (LCE) architecture. COM+ Events benefit from Queued Components, a set of messaging components based on the MSMQ product.[6]

---

[6]COM model also includes the so called Asynchronous COM, which allows about the same kind of functionality. But the limitations of its use are so hard at this time that it can be considered as a legacy feature and only be considered as an alternative for backward compatibility.

In publish/subscribe, a client (subscriber) first calls a server (publisher), requesting to be enlisted to receive information, passing a callback interface (also known as *sink interface*). When new information is available, the server calls the client back through this sink interface.

The COM+ Event Model simplifies publish/subscribe development by placing the connection setup and event firing/callback action beyond the scope of the component. Therefore, a client who wants to receive events registers itself with COM+. Similarly, the publishing object sends events to COM+, which takes care of all callbacks. This is quite similar to the architecture "channel" concept in CORBA's Event Service (OMG, 2001).

An event is fired by the publisher using an *event class*, which is a COM+ implementation of the sink interface. Therefore, a publisher must instantiate the event class, and simply calls one (or more) of its methods. It is COM+'s job to actually make the event reach every subscriber. In its turn, a subscriber must implement this sink interface to be able to be called. Every event class and subscriber must be registered (as any COM component) within MMC. For event class, a special option is available, as shown in Figure 5.7.

Firing can be done synchronously or asynchronously, configurable from MMC. In the synchronous mode, if a subscriber blocks (for instance, because of a modal message box waiting for a user response), event firing also blocks. Therefore, events should be normally fired asynchronously.

Events can be filtered, to avoid waste of bandwidth and CPU cycles. Any event parameter may be used to filter the events a subscriber wants to receive in declarative form in the Properties tab of a subscriber in MMC. Besides subscriber filtering, COM+ also supports publisher filtering, which allows a publisher to decide which subscribers should actually receive an event. This can be accomplished by writing a COM component implementing the IPublisherFilter interface (or a similar interface). When such a filter is created, COM+ invokes the component before publishing. The filter has access to all subscribers and can decide upon which ones to publish to.

Transactions, Security, and additional COM+ features

Perhaps most pages written on COM+ discuss transactions and security. This is so because most systems developed around the world fall into this category. As stated before, these aspects are beyond the scope of this chapter.

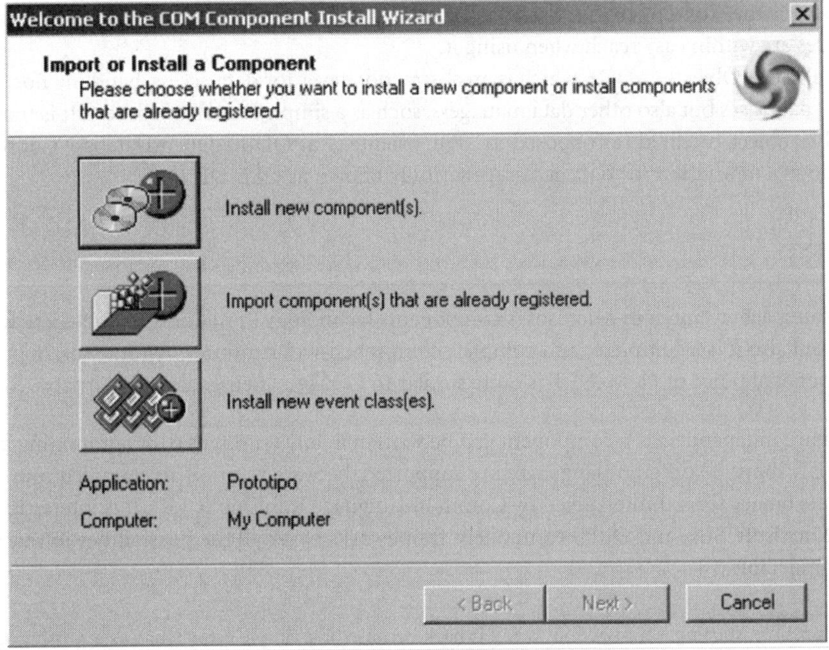

FIGURE 5.7 Registering an event class.

In short, transactional control is about coordinating various Resource Managers (RMs) such as databases and queues to collaborate on a business transaction. The goal is to achieve Atomicity, Consistency, Isolation, Durability (ACID) properties. Transactional control is slightly more related to Atomicity, which implies a transaction is to be fully completed (i.e., all RMs were capable of termination of their work), or all its effects are canceled (i.e., all RMs are requested to roll back all changes already made). This is accomplished via a two-phase commit protocol, which has a:

- Phase 1 (preparation), when all RMs are solicited to do all checks and other work in preparation to complete a transaction. If any RM votes for a rollback, all others are signaled to roll back too.
- Phase 2 (commit), when all RMs are signaled to terminate their work, because all voted not to roll back. There is no going back at this point; changes are made permanent (which is the Durability in ACID).

To coordinate the two-phase commit protocol, Microsoft has been shipping DTC, since 1996.

With COM+, components may also act as RMs, being able to vote for a transaction commit or rollback, as a database does. This may be used if a violation on a business rule is detected: by voting for rollback, a component prevents changes from being committed.[7]

COM+ security services are tightly integrated into the MS Windows 2000 (and later) operating system, allowing for a fine-grained security control. It allows to associate *roles* with users (or groups of them) and then, to control their access to components, interfaces and even at the method level. COM+ security is declarative, meaning that it can be fully configured from the MMC's graphical interface, thus not demanding programming efforts. Also, it is detached from component implementation, passing this responsibility to people in charge of production systems.

Additional important (at least, to data-centric systems) COM+ features involve:

- COM+ Context, holding Transaction, Activity, Security, and Apartment information, shared among objects involved in a processing path started in a root client.
- Compensating Resource Manager (CRM), which facilitates the development of components to act as a Resource Manager (RM), at least as far as the transaction control is concerned.
- COM+ Catalog, which is a graphical interface for deploying, managing, and monitoring components. It is the Component Services part of MMC and is quite important because many COM+ features are within easy reach when using it.
- ActiveX Data Objects (ADO), which is an abstraction layer for data access, handling not only relational databases but also other data managers, such as a simple Excel spreadsheet. It is transaction-capable, object-oriented (as opposed to "SQL oriented," as ODBC-Open Database Connectivity). It follows a new vision for data access, generically known as OLE-DB.

5.6   .NET

.NET is the latest achievement in Microsoft's components technology. It not only includes a new component model but also it is a complete (and complex, comprehensive) runtime environment. In both cases, from the functional point of view, .NET is quite similar to COM+. Their goals are (Barnaby, 2002):

- Language independence: a component can be written in any (supported) programming language. In .NET, about 20 different languages are supported, because they are all compiled into an intermediate binary format interpreted by Common Language Runtime (CLR). It is interesting to note that Microsoft, Sun, and OMG component frameworks cover all the possibilities in this sense, as shown in Table 5.1.

---

[7]Although possible, starting a transaction to roll it back because of a simple error (such as a missing value on a mandatory field) is not a good implementation strategy because votes are only checked at the final of phase 1, when a lot of work on all RMs have been already done, causing avoidable resources comsumption.

TABLE 5.1 Framework's Strategy with Respect to Independence

Framework	Language independence	Operation system independence
.NET	Supports about 20 different languages	Until now, only Microsoft Windows
Java	Just Java	Any OS where a JVM (Java Virtual Machine) is available
CORBA	Any language for which a binding exists	Any OS

As it can be seen from Table 5.1, Microsoft is closer to providing full language- and OS-independence than Java; it only takes a marketing decision.

- *Component interoperability*: since a common set of data types is supported by .NET, complete component interoperability is achieved. Even further, a component written in one language may even inherit from another, written in a different programming language.
- *Location transparency*: local (in-process) and remote (out-of-process) component access are identical, leaving details to a configuration tool.
- *Improved version control*: a new versioning scheme is available, solving a well-known problem in MS platforms (including COM+) when managing DLLs.

If they are similar, why to spend money in rewriting a new component framework? .NET provides a set of improvements over COM+, cleaning up years of technology adaptation and thus simplifying life for Microsoft and developers.

Runtime Environment

What are the basis for a component interoperability between programming languages? The .NET technology answer is a single set of data types, a single set of minimal programming language's features (or restrictions), a uniform runtime environment, and an improved (when compared to COM+) packing scheme. These characteristics are detailed below. Together, they provide a uniform, multi-programming language runtime environment.

Single Set of Data Types

For a component written in one language to call another one, written in a different language, there is a need to convert to/back parameters. Parameters must be in a programming language understandable data type. For instance, if a 16-bit (short) integer is passed from one language to a different one, which only supports 32-bit integers, "someone" must take care of a conversion. Java adopted the simplest solution: just one, single programming language — Java. Therefore, any component uses the same data types because they are Java data types. .NET, because of its goal to support multiple programming languages, decided to define and include a CTS (Common Type System), that is, a set of data types supported in all languages. This also provides a cross-language inheritance; for instance, a VB.NET[8] class may be derived from a C#[9] class.

One major achievement is the .NET Framework, which is a large set of about 3500 classes that can be inherited and therefore, extended. It is available for all .NET programming languages; it takes some time to get to know them, but it does speed up application development.

All classes within .NET Framework inherit from the same class, System.Object, which has the following methods:

- Equals(), allowing for comparison between objects.
- Finalize(), to perform cleanup operations before an object is automatically reclaimed; see discussion on garbage collection below.

---

[8]Visual Basic .NET (VB.NET) is a revision of Visual Basic, making it .NET-compatible. But it includes so many new features that it should be given a new, distinct name.

[9]C# (reads as "C sharp") is a new language, proposed by Microsoft, as a language intermediate between C++ and VB, in terms of simplicity.

- GetHashCode(), which generates a number corresponding to some sort of manipulation of an object's attribute values.
- ToString(), to generate a human-readable string describing an instance of a class.

System.Object plays in .NET the same role as IUnknown and IDispatch in COM: as long as all objects inherit from it and it is a class, it can be dynamically interrogated by using the .NET Reflection API (simply put, a set of functions to determine which methods are inside the component and activate them) and loaded (as JITA, in COM+).

Set of a Minimal Programming Language's Features

To be able to run within .NET framework, a component must be written in a programming language complying with its characteristics. The Common Language Specification (CLS) determines which features must be present. CLS does not mind which keywords are used, just features such as inheritance. That is why Visual Basic (and all other MS programming languages) evolved into VB.NET (or a new .NET programming language).

CLS is actually a subset of CTS. All publicly available methods in a component must comply with CLS. But within a component, non-CLS compliant data types may be used with no risk. For instance, C# supports unsigned types, but VB.NET does not; unsigned types are CTS-compliant but not CLS-compliant.

Uniform Runtime Environment

The last element to achieve full component interoperability is the CLR. As stated by Barnaby (2002), "succinctly put, the CLR is an implementation of CTS," as long as it is an application execution engine and a class library built according to CTS.

All source code in any .NET-programming language is compiled into a platform-neutral code called Common Intermediate Language (CIL or MSIL or simply IL). The CIL code is interpreted by CLR. This approach is quite similar to Java's bytecode and JVM (Java Virtual Machine), but as old as computers.[10] The first time a part of the code is used, it is compiled into native code by a Just-In-Time (JIT) Compiler. Therefore, a user will probably experience a program speed-up after first execution of a sequence of commands.

CLR also introduces the notion of "managed code" and "unmanaged code." In "managed code," CLR runtime takes care of memory allocation/deallocation and security. A garbage collector, again similar to Java's, cleans up unused objects from memory. Modules ("assemblies," in .NET jargon) are loaded only when needed. Managed code frees a developer from many tasks but imposes some barriers. In situations when these barriers are not acceptable, CLR allows for unmanaged code, which allows for full access to memory and Win32 library. Unmanaged code is required for backward compatibility and in some applications where resource control must be in the developer's hands because of resource scarcity and speed.

Assemblies

With COM/DCOM/COM+, components had to be registered in the Registry to be known to an application: the "DLL hell," mainly because of inter-application dependence. Let us assume, for instance, that Application A uses version 1 of a component. As seen before, components are known by their UID. Now, a new application is installed, installing a new version of the same component. If the UID is retained, Application A may be broken. Briefly, component registration is machine-wide, but not exclusive to one application. Component update requires deregistration and reregistration, which, although supported by development tools (such as Visual Studio), is always annoying as it requires special scripts for installation on a production site.

.NET provides a new concept of Assemblies. Assemblies seem to be similar to DLL as long as they are packed into a file with a ".dll" extension. But they are quite different. Packing all types of information and metadata allows for an easy installation.

---

[10]Interpreters are quite frequent in computer's history: Basic and Pascal, for instance, are interpreted languages. But this has been proposed as a solution for platform interoperability since the earliest days in computer's history.

Assemblies may be private or public. Therefore, a component may be packed to be used exclusively by a single application, or available to many different applications.

Private assemblies are stored within the "application tree file structure," that is, in the application's base directory, or some subdirectory. Public assemblies are stored in the Global Assembly Cache (GAC), which is normally the <BASE_WINDOWS_DIRETORY>\Assembly (c:\winnt\assembly). Opening this folder in MS Windows Explorer (over MS Windows 2000 or XP after installing .NET) is quite revealing; each component shows its name, type, version, culture (for internationalization), and public key token (which is part of the security mechanism in Assemblies, see below).

An application declares its dependencies in an XML file, giving a search path for components. Therefore, a much more flexible deployment mechanism may be used, allowing an application to have its own components, or share them among groups of applications, or even with all of them on the same computer. Changing from one option to another may be accomplished by changes in the XML file (or through the visual configuration editor inside Visual Studio .NET). Searching for the correct assembly is one of the key features for solving the "DLL hell"; it is known as "probing" in the .NET jargon.

An integral part of assemblies is its support for security and versioning. An application may be sure of the origin of a component if it requires an assembly with a *strong name* (or *shared name*). A strong name is a combination of a friendly name, culture information, version number, public key, and a digital signature. Because an assembly with a strong name is signed, an application can be sure of the originator. Even more, as long as it is ciphered, an application may be sure of the component's integrity. A "Trojan horse" virus would not be able to impersonate an assembly. Versioning support is also provided. An application may require a specific version (Major:Minor:Build:Revision), or a version range, adapted to a specific culture (for instance, in Portuguese). It is part of the probing mechanism to find a suitable component, that is, security and versioning restrictions are applied.

Application Domains

Operating Systems run processes. But from an application point of view, this abstraction is not adequate. An application may share the same process with another application, or may span across various processes. Therefore, starting and stopping an application (or configuring, resource allocation, and any other management action) require a new concept, which is called in .NET an *application domain*. An application domain is a kind of "logical" process that can be managed as a whole.

An application domain may contain many *contexts*, for sharing data among components, keeping objects with the same synchronization, and threading affinity requirements.

Specific classes within .NET framework allow manipulation of application domains and contexts. Both are used to define the *application boundary*, which plays a central role in the .NET remote invocation. It is somewhat natural that an application domain defines an application boundary. But contexts also may delimit application boundaries; an application may request a new context to be created and to execute one or more components within it. In this case, communication from one component (which is running in the default context) to another (running in this recently created context) also spans an application boundary. In this way, contexts have some correspondence to the Apartment concept in COM+.

Remote Invocation

Every time a component invokes another one crossing an application boundary, a remote invocation occurs. As discussed above, an application boundary exists

- between application domains,
- between objects in different contexts.

Unlike DCOM, to be able to be called from outside of an application boundary, a class must be constructed in a special way. .NET provides support for marshaling by value (MBV), when only data are passed to the remote side, and marshaling by reference (MBR), when a reference is passed to the remote side. When neither MBV nor MBR are involved, objects are called *agile* since the object may be directly accessed. In order to utilize MBV, all parameters must be able to be marked as *Serializable*, a synonym for marshaling. .NET provides means for annotation of *class attributes* such as this. For instance, in VB.NET, a developer will write a code as in code 3.

```
<Serializable>_
Public Class CustomerData
...
End Class
```

Code 3 Declaration of a Serializable class in VB.NET.

Normally, a Serializable class is just a ValueObject, that is, an object containing only attributes.

When invoking a remote method, a communication path is established; this is called *Channel*. The network transport protocol may be chosen between TCP or HTTP. Data are marshaled using Simple Object Access Protocol (SOAP), which is a remote invocation data format based on XML. Within a SOAP message, the following will be found:

- SOAP body, containing all information required to execute a method call, such as method name and parameters.
- SOAP header, specifying nonexecution-related aspects, such as security and transactional information.
- SOAP envelope, encapsulating the body and header.

In Figure 5.8, the Formatter is, normally, a SOAP formatter.

Table 5.2, based on Barnaby (2002), summarizes the existing options:

A server object may be called in Singleton or SingleCall modes. In the Singleton mode, a stateful object is created to service the request. Therefore, a successive call to the same object will reach the same instance. In SingleCall, an instance serves just a single call, that is, to service each call, a new instance is created; this is a stateless call. Instances are destroyed just after finishing a method processing. Of course, this could be a major penalty in performance since instantiation is an expensive process. The .NET framework provides an extensive set of configuration tags (and API calls) to allow a developer to control instance lifetime.

From the client's point of view, all this is not completely transparent. To gain access to a remote component, a client has to know

- hostname hosting the server application,
- type of channel being used (TCP or HTTP),

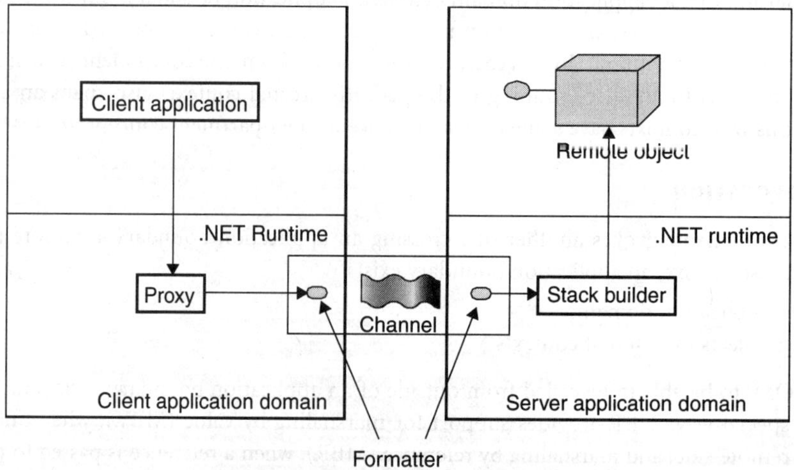

FIGURE 5.8   Remoting in .NET (based on Barnaby, 2002).

TABLE 5.2   Options for Marshaling

Type	Base Class and Class Attributes	Description
Marshal by value	Marked as Serializable	This type is agile within a domain. But when transported across the domain boundary, it is serialized/deserialized
Marshal by Reference	Inherits from MarshalByRef Object; no class attributes needed	A proxy on client side is provided to remote access; therefore, all calls are executed within the application domain where it was originally created
Context bound	Inherits from ContextBound Object; no class attributes needed	A proxy to the object is always involved when crossing boundaries (both application and context); used in situations when a special handling is required (synchronization, for instance)
Not remotable	None of the above base classes or attributes	This type cannot be called outside domain but may be used in different contexts within a domain. Contexts hold a direct reference to them

- port number where the server is listening to incoming requests, and
- remote object URI.

All four values may be specified as an URL, such as
http://localhost:1702/MyURL.soap
In this case, the server object is called a *wellknown* object (or destination), because the client has all information for invocation.

Then, the client can:

- Use Activator.GetObject(), a method within .NET framework that returns a proxy to remote server giving component type and URL or
- Register a well-known end point by using RemotingConfiguration.RegisterWellKnown ClientType() and then simply instantiating objects following language conventions or
- Use its configuration file to declare the URL to a component. During initialization, CLR will register the component. This is probably the preferred method as long as a developer can write a program just following language conventions, providing the same level of location transparency as COM/DCOM.

It is also possible for a client to invoke a remote method without knowing the object's URI in a process called *client-activated object*. In this scenario, a client can either call RemotingConfiguration. RegisterActivatedClientType() or declare protocol/hostname/port values in a configuration file, tying to an object. After that, a developer simply uses the programming language convention to instantiate objects.

Asynchronous Invocation

Within .NET, asynchronous remote calls are completely client-side solution, that is, it has no impact on how the remote object is implemented. Therefore, asynchronous calls are available for all server objects.

Asynchronous calls are accomplished through the use of *delegates*. Delegates are sometimes described as type-safe function pointers. The key programming aspect to implement a delegate is to pass a function address A as an argument of another function B that can "call" A directly. It is a very common construction in C/C++ programs, sometimes known as a "jump table." Asynchronous calls infrastructure uses delegates either to set up a different thread to effectively call the remote method as well as to call back the client (when specified) after the completion of a remote call.

A delegate class has methods to:

- Start a remote invocation asynchronously (BeginInvoke()) eventually also specifying function parameters, callback function for remote call termination, and client-provided context information.
- Synchronize a remote call by waiting for its termination, thus allowing any returned results to be accessible to the client.

The entire remote call occurs asynchronously, including the overhead to marshal data. The major drawback of this approach is that it is a client-side solution; some developers not skilled with the delegates may find it hard to apply. These difficulties are somewhat removed if a callback interface is defined. In this case, the developer has to implement this interface, which is a relatively easy task. The drawbacks of this solution are the need for the client code to explicitly open a communication channel to listen to callbacks, and to make the call synchronous until the remote server is reached.

5.7   Performance

Since there is no standard for middleware performance measurement, a number of partial results may be collected to characterize COM/DCOM/COM+ and .NET performance. It is important to notice that performance is affected in so many ways that it is hard to completely define a test case and environment to conduct analysis.

Since all the technologies come from Microsoft, it seems to be fair to start with MS figures; Table 5.3 in Microsoft (1996) shows some of these.

Even though the results were obtained on relatively slow processors with limited main memory (Pentium@120MHz, 32Mbytes; Alpha@200MHz, 32Mbytes), the results show that the in-process activation is at least 10,000 times faster than a cross-process remote call. Also, cross-platform remote calls are at least 5 times slower than the cross-process activation.

In Adamopoulos (2000), a set of experiments with CORBA and COM/DCOM showed about the same performance for cross-process calls when transferring strings. This experiment also demonstrated that a single call transferring all strings is faster than multiple calls each transferring a single string, suggesting marshaling overhead as being minimal in process, and invocation costs being dominant. The same results were obtained when experimenting with remote calls.

For Automation and Control, Chisholm (1998) demonstrated OPC (which is based on COM) supplying more than 20,000 values per second to four simultaneous clients on a Pentium@233MHz. These figures, obtained on relatively slow processors with limited main memory, are quite important mainly because in Automation and Control this kind of machine is commonplace.

TABLE 5.3   Performance Figures

Parameter	4 bytes (Calls/sec)	50 bytes (Call/sec)
Pentium, in-process	~3.2 M	~3.2 M
Alpha, in-process	~2.8 M	~2.8 M
Pentium, cross-process	~2.300	~2.000
Alpha, cross-process	~1.900	~1.600
Alpha-to-Pentium remote	~400	~300

TABLE 5.4   COM/DCOM — CORBA Comparison (Adapted from Chung, 1997).

	COM/DCOM	CORBA
Basic Programming architecture		
Common base class	IUnknown	CORBA::Object
Object class identifier	CLSID	Interface name
Interface identifier	IID, GUID	Interface name
Invocation (basic)	CoCreateInstance()	bind() (although not really CORBA-compliant)
Object handle	Interface pointer	Object reference
Remoting Architecture		
Name to implementation mapping	Registry	Implementation Repository
Type information for methods	Type library	Interface Repository
Locate implementation	SCM	ORB
Activate implementation	SCM	OA (POA, BOA)
Client- and server-side stubs	Proxy, stub	Stub, skeleton
Wire protocol architecture		
Marshaling data format	NDR	CDR
Object reference	OBJREF	IOR

5.8　COM/DCOM — CORBA Comparison

In Raj (1999), a detailed comparison of CORBA, DCOM, and Java/RMI is presented, by exploring an example application. His conclusion is that "...[both] provide mechanisms for transparent invocation and accessing of remote distributed objects... the approach that each of them take is more or less similar." The same conclusion appears in Chung (1997), which views this comparison in three different layers:

- the basic programming architecture layer,
- the remoting architecture, and
- the wire protocol architecture.

Table 5.4 summarizes their findings.

Acknowledgments

The authors wish to thank Victor Zamora, for his careful revision on the material and suggestions.

References

Adamoupoulos, D., Pavlou, G., and Papandreou, C., Performance Evaluation of Distributed Platforms for Telecommunications Service Engineering Activities, Proceeding of 4th IMACS/IEEE World Multiconference on Circuits, Systems, Communications and Computers (CSCC00), Athens, Greece, July 2000, pp. 131–136.

Barnaby, T., *Distributed .NET Programming in VB.NET*, Apress, Berkeley, CA, 2002.

Blanco, P. et al., OPC and CORBA in Manufacturing Execution Systems : A Review, 9th IEEE International Conference on Emerging Technologies and Factory Automation, Lisbon, Portugal, 2003.

Box, D., *Essential COM*, Addison-Wesley, Reading, MA, 1997.

Brill,G., *Applying-Indianapolis IN COM+*, New Riders Publishing, Berkeley, CA, 2001.

Chisholm, A., DCOM, OPC and Performance Issues, Intellution Inc., 1998, Accessed 31Jul2003 at: http://www.opcfoundation.org/Downloads/White%20Papers/DCOM,%20OPC%20andPerformance%20Issues.pdf.

Chung, P.E. et al., DCOM and CORBA Side by Side, Step by Step and Layer by Layer, Accessed 31Jul2003 at: http://research.microsoft.com/~ymwang/papers/HTML/DCOMnCORBA/S.html.

Gamma, E. et al., *Design Patterns*, Addison-Wesley, Reading, MA, 1995.

Lowy, J. *COM and .NET Component Services*. O'Reilly, Sebastopol, CA, 2001.

Microsoft, The Component Object Model Specification, version 0.9, 1995.

Microsoft, DCOM Technical Overview, Microsoft, 1996, Accessed 31Jul03 at: http://msdn.microsoft.com/library/default.asp?url=/library/en-us/dndcom/html/ msdn_dcomtec.asp.

Nilsson, J., *.NET Enterprise Design with Visual Basic .NET and SQL Server 2000, Indianapolis IN* SAMS, 2002.

OMG., *Event Service Specification*, Version 1.1, OMG, 2001.

Open Software Foundation, Introduction to OSF/DCE, Pearson Education POD, Harlow, U.S.A., 1995.

Raj, G.S., A Detailed Comparison of CORBA, DCOM and Java/RMI, Accessed 31Jul2003, at http://gsraj.tripod.com/misc/compare.html.

Redmond III, F. E., Introduction to Designing and Building Windows DNA Applications, Accessed 31Jul2003 at: http://msdn.microsoft.com/library/default.asp?url=/library/en-us/dndna/html/ windnadesign_intro.asp.

Tannenbaum, A., *Operating Systems: Design and Implementation*, Prentice-Hall, Englewood Cliffs, NJ, 1987.

6

CORBA in Manufacturing — Technology Overview

Marcos Ribeiro Pereira Barretto
University of Sao Paulo

Paulo Marcelo Porto Alves Blanco
University of Sao Paulo

Marco Antonio Poli
University of Sao Paulo

6.1 Overview

As technology evolved toward Object Orientation (OO), distributed computing models also progressed in that direction. By the end of the 1980s, a number of IT companies and individuals, mostly from Universities, founded a new association, the Object Management Group (OMG), devoted to the conception of a new standard for an object-oriented distributed computing framework. OMG is vendor-neutral and vendor-independent. It does not sell any products; its mission is to provide specifications that can be implemented into products.

The first result of this association was the establishment of standards for an object-oriented middleware architecture: the Object Management Architecture (OMA). OMA supplies a group of standards upon which the distributed applications can be developed, and consists of the following:

- definition of standards for remote method invocation (RMI) and object's life cycle control, known as Common Object Request Broker Architecture (CORBA);
- a set of standards defining services to objects, known as CORBAservices; and
- a set of standards defining services to aid application development, denominated CORBAfacilities.

OMG organization also includes a collection of working groups divided according to industry segment (Manufacturing, Medical, etc.), aiming at defining standard services specific for each industry. These groups are called Domain Technology Committee (DTC) and their work is, therefore, a highly valuable source of design ideas for professionals working on each industry segment.

This chapter provides an overview on CORBA, CORBAservices, and the work on DTCs related to manufacturing.

6.2 Scope

OMG's work is lengthy enough to provide material for a full collection of books, more so if distinct audiences (developers, application architects, managers, students, etc.) are considered. Therefore, a narrow focus is necessary due to space limitation.

A conceptual overview approach was considered as being appropriate. This chapter was written keeping in mind a technical audience: professionals and students familiar to programming (but not necessarily developers) and aware of object-orientation concepts. Only concepts are discussed; no source code examples are shown.

Moreover, considering the industrial area, which is the focus of this entire book, a further focus narrowing is necessary, since shop-floor IT demands are quite different from those on the upper levels of Computer Integrated Manufacturing (CIM) pyramid or Engineering departments. In this aspect, it has been decided to focus mainly on shop-floor needs.

Again due to space limitation, it has been decided to describe the technology itself and not its application. A large set of small- and large-scale CORBA applications are in use but they will not be described here.

Finally, OMG work is evolving over time; CORBA standards are up to their 3rd version. Therefore, all materials presented here tend to reflect current technology status, without efforts to show its evolution or trends. Fortunately, most concepts have been present since CORBA's 1st version and might be present in the future, making this decision the least impacting on chapter contents.

6.3 Some on the Past and Future of CORBA

The OMG was founded in Needham, MA, in 1989 as a nonprofit organization, by a group of 11 companies, with the mission of providing "a common architectural framework for object-oriented applications, based on widely available interface specifications." Currently, the OMG consortium includes approximately 800 members.

The work developed by the OMG focuses on establishing industry guidelines and detailed specifications for object management, creating standards that can accelerate the introduction of software components across all major hardware platforms and operating systems. One of the core lines of work inside OMG is the development and maintenance of the OMA, of which CORBA is a part.

The first CORBA version (1.0) was introduced in October 1991, and included the CORBA Object Model, the Interface Definition Language, and a core set of application programming interfaces (APIs) for communication with an ORB. At that time, it only included mapping for the C programming language. It was quickly followed by version 1.1 in February 1992, which clarified some ambiguities and defined the Basic Object Adapter (BOA) interface, and later by version 1.2 in December 1993. More details about the BOA will be given later on this chapter.

The next huge step in CORBA history was taken in 1996, with the introduction of CORBA 2.0, when many important improvements were made, as the ORB interoperability architecture (GIOP, IIOP, and DCE CIOP) and the initial reference resolver for client portability. This version added two new IDL language mappings — C++ and Smalltalk. Version 2.0 was then followed by version 2.1 in 1997 and 2.2 in February 1998, when the JAVA language mapping was introduced.

Since then, there were introductions of version 2.3 in 1999, version 2.4 in 2000 (when the notification service and the real-time CORBA were introduced), version 2.5 (introduction of fault tolerance) in September 2001, and version 2.6 in December 2001.

The next major CORBA release, version 3, was published in 2001. At the time this text was written, version 3.0.2 was approved and version 3.1 was to be voted. The major change introduced was the CORBA Component Model (CCM), which defines a container to transactional, secure, and persistence-aware components, providing a higher level of abstraction than the usage of independent CORBAservices. It is very similar to the container defined in Java's J2EE, which has close integration (Siegel, 2000).

Other efforts toward the development of framework for object interoperability have also been made during this period of CORBA evolution. One of the most relevant is the integration to Microsoft's COM/DCOM.

6.4   Architectural Concept: the OMA Architecture

One of the main results of OMG efforts was the definition of the OMA, of which CORBA is part. Basically, OMA provides a group of standards upon which applications can be developed. It defines the need for:

- an inter-component communication infrastructure (the ORB — Object Request Broker — function);
- a set of components offering basic services to other components (known as CORBAservices);
- a set of components offering basic services to applications (called as CORBAfacilities);
- a set of specialized components for a knowledge domain (such as Medicine or Manufacturing); and
- a set of application objects, forming the applications themselves.

The role of the CORBA architecture in the OMA is implementing the ORB function, providing the programmer with a standard mechanism for the definition of inter-components interfaces, as well as tools to make the implementation of these interfaces easy and language-independent.

In OMG (1996), the OMG presents its technical goals, terminology, and details the conceptual architecture upon which the specifications are defined. This guide also includes the OMG object model that defines the common semantic for the specification of an object's external characteristics in an implementation-independent manner and the OMA reference model.

In order to complement the OMA with detailed specifications of each of the reference model's components, the OMG has issued a series of Requests for Proposals (RFPs) and has thus established specifications for the CORBA architecture, CORBAservices, and CORBAfacilities.

In the reference model presented in Figure 6.1, the OMG identifies and characterizes the components, interfaces, and protocols that form the OMA, including the ORB.

The Object Request Broker, or ORB, the fundamental component of the CORBA architecture, consists of a software element whose purpose is to carry out the communication between objects. This is made possible through the availability of a series of capabilities, among them the capability of locating a remote object through its reference. An ORB is a mechanism through which objects can make requests and receive responses from each other, in a single computer or in a network. With the ORB, the client does not need to be aware of the mechanisms used to communicate with or activate an object, nor of its implementation details or physical location. The ORB is, therefore, the basis for building distributed applications and for the interoperability between applications in homogeneous or heterogeneous environments.

The reference model in Figure 6.1 also presents four categories of object interfaces:

- Object Services (CORBAservices) — interfaces for general services, with potential use in any distributed objects application.
- Common Facilities (CORBAfacilities) — interfaces for horizontal facilities (generic and reusable applications), applicable to most domains, and focused on helping the user with the implementation.
- Domain Interfaces — specific interfaces for each application domain, such as manufacturing, finance, medical, and telecommunications.
- Application Interfaces — nonstandard interfaces, specific for each different application.

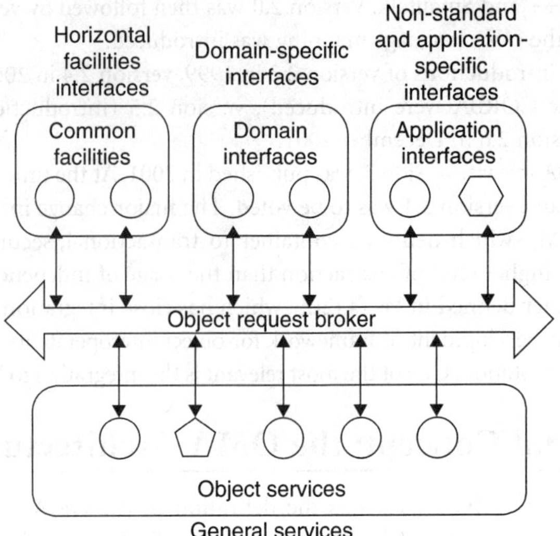

FIGURE 6.1   OMA reference model. (Adapted from OMG, 1997.)

Interfaces are at the heart of all OMA specifications and, in general, any work with components. In other words, to characterize a component's behavior, its interface (i.e., methods, arguments, returned values, and attributes) must be explicit. Semantics are not captured by interface definition. OMA includes the definition of a standard language for interface definition: OMG IDL (Interface Definition Language). IDLs were first introduced by Distributed Computing Environment (DCE) framework (OSF, 1995), proposed and managed by the Open Systems Foundation (OSF). OMG IDL extends OSF's, by adding object-oriented characteristics to it. OMG IDL is explained below (Gorton, 2000).

One of the characteristics of the OMA is that an application does not even have to be developed under the OO paradigm in order to be able to interact with the proposed architecture. It only needs to implement or support the interfaces defined by the OMG, as a way of being encapsulated in OMA-compliant objects.

6.5   Remote Method Invocation

IDL

IDL is a language to specify the interfaces of the CORBA objects, being responsible for the language independence of the CORBA standard. IDL is not a procedural language, that is, it can be used to define only interfaces, not implementations. The IDL can be compared to the header files used in C++ programming. A header file usually does not include classes implementation, being restricted to the interface description.

The IDL specification is responsible for assuring that the data are correctly exchanged between different programming languages. For example, IDL specifies the "long" data type, which is a 32-bit signed integer type, and that can be mapped to the "long" type defined in the C++ language or to an "int" type in JAVA.

Once all the interfaces defined in the CORBA architecture are described through IDL, they can be mapped to any programming language, that is, C++ servers can communicate with JAVA clients, which can communicate with COBOL servers and so on.

The IDL, as well as most programming languages, defines primitive types that include integer numbers, characters, floating point numbers, strings, boolean types, and constants, among others. Besides this, the primitive types can be aggregated in other types, such as "unions" and "structs. "

Input and output parameters representation, methods, modules, attributes, inheritance, and some other language primitives are also defined by the IDL, as a way of creating a complete description of the interfaces defining an application.

Once an IDL server file is compiled by the IDL precompilers included in any CORBA product, a series of classes will be automatically created, forming the basic infrastructure required for the communication of this server with the clients through the CORBA architecture. After this precompilation, the programmer can implement the code portions required for the server to work.

Stubs and Skeletons

In the same way as in other architectures, CORBA has the notion of client and server, being that any CORBA object can work in both ways. When an object is making services available to other objects, it is considered a server, and when an object is accessing other object's services, it is considered a client. Actually, CORBA differentiates a servant (an object implementation that provides the service) from server (a process where one or more servants will "live"). For simplicity, this text will use the term server.

In the client's implementation, the concept of stubs is used. A stub is a code part that allows a client to gain access to a server component. It is linked along with the client part of an application. Similarly, the skeletons are code parts that must be completed when a server is implemented. It is not necessary to manually write the stubs and skeletons codes, once they are automatically generated when the IDL interface is compiled (Figure 6.2).

When a client wishes to invoke a remote method, it issues a method call following its programming language convention. The stub is responsible for converting parameters into CORBA format and to communicate through the network with the server side. This parameter conversion is known as marshaling and is an important part of the language independence provided by CORBA. On the server side, the skeleton unmarshals parameters following the conventions of the language of the object implementation and actually invokes the method. Returned values follow the same way back to the client, being marshaled from an object's implementation language to CORBA language-independent format and, finally, to the client's programming language convention.

CORBA also supports dynamic creation and invocation of requests to objects, using an object's name, which is known as Dynamic Invocation Interface (DII).

Object References

In order to make the communication between objects easier, the CORBA architecture uses the notion of object references or Interoperable Object References (IORs). When a component of a given application needs to access a CORBA object, it initially obtains an IOR for this object, through which it will be able to use methods from this object. IORs are somewhat similar to pointers (or object references) in programming languages such as C++ and Java.

The most basic way of working with IORs is creating small text files with predefined names and writing the IORs in these files. When an object needs to locate a second object, it can open the text file and read the second object's IOR, and thus be able to interact with it. Unfortunately, for this method to work, every object interested in publishing its IOR must do it in a predefined location, in predefined files, which the other objects must be able to access. That is, the sever object must write its IOR in a shared file system known and accessible by the client. Sometimes, in small-to-medium systems, a webserver is used

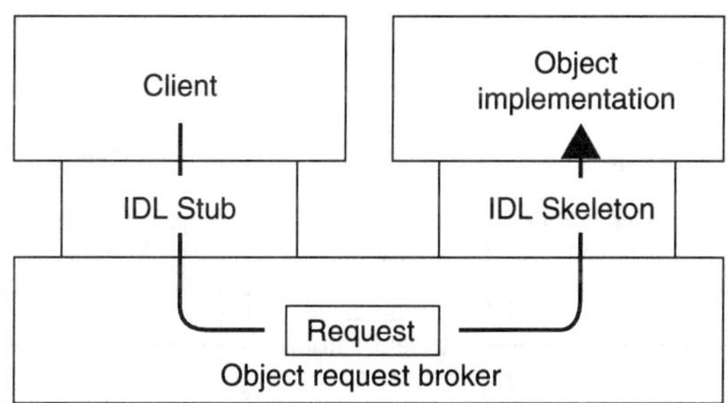

FIGURE 6.2   Request passing from a client to an object implementation. (Adapted from OMG, 2003.)

to store IOR, making them available to clients through conventional HTTP access, as long as they can be stored in text files. This way, the shared file directory may be avoided; a client must only know the object's name and webserver IP address.

For large systems, both alternatives are not applicable. OMG thus defined the Naming Service, whose responsibility is the association of names to objects, through a process called Name Binding. The name bindings are always defined in Name Contexts, which are objects that have a group of name associations, so that each association is unique in a given context. Different names can be associated to the same object, in the same context, or in different contexts; but it is not mandatory that all objects are given names.

There are two basic actions possible in the Name Service. The first one is the association, (or binding, in the Name Service terminology), of a name to an object in a given name context; the second is locating (or resolving) a name, which is the action of identifying the object associated to a name in a context. Each name is always located inside a context, and there cannot be absolute names.

Once the context is an object like any other, it can also be associated to a name in a name context. This association of names to contexts has the effect of creating a name graph, in which contexts are nodes, and that makes possible the creation of composite names, in such a way that a sequence of names can reference an object. This composite names structure works in a manner analogous to the directory tree used by operating systems such as Windows or Unix, a clear analogy between contexts and directories being possible, and also between object names and files.

As all CORBA services defined by OMG, the Name Service is only specified through its interfaces. This means that OMG does not supply source codes to the software developers and vendors; only the interfaces are defined so that compatible solutions may be implemented by various vendors. By supplying IDLs and defining some basic implementation rules to be followed, OMG assures that companies interested in developing products based on the Name Service specification will do so by maintaining the standardization of the basic functions, of the modules' names, and of the methods' names, making the interoperability between different vendors possible.

Communication Protocols

ORBs usually communicate with each other using the Internet Inter-ORB Protocol (IIOP). The term "Internet" on IIOP was a concession to some of CORBA's specifiers, which wanted to reinforce its use with the World Wide Web; it actually means communication over TCP/IP. There are other protocols for inter-ORB communications, but IIOP is the most popular, given the popularity of the TCP/IP protocol. In any case, CORBA is protocol-independent, and could work with any other.

CORBA standard mandates at least the support to IIOP, leaving to implementers the option to implement other communication protocols (such as DCE, for instance, which has also been standardized).

The standard also mandates interoperability among ORBs through the use of IIOP. As shown in Figure 6.3, client and object implementation may be written to different ORBs, whose communication is guaranteed by CORBA inter-operability tests, a set of conformance tests any ORB must pass to be declared CORBA-compliant.

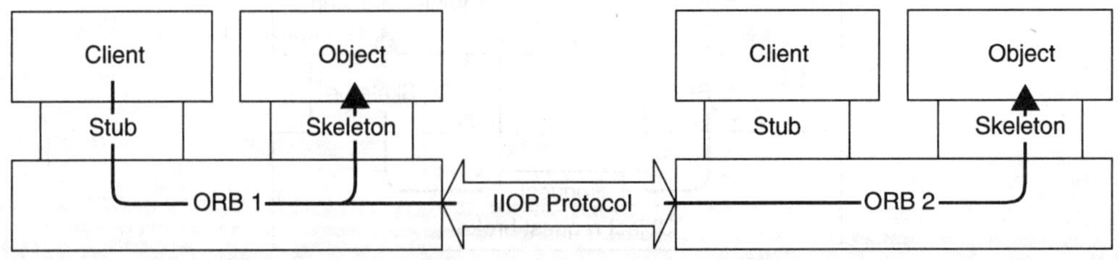

FIGURE 6.3   Remote method invocation using IIOP. (Adapted from OMG, 2003.)

Object Adapters

The CORBA architecture also defines the BOA, which works as the interface between the objects and their ORBs, through the implementation of a series of methods for accessing the ORB functions, such as object activation or persistence.

In the CORBA terminology, the BOA is a pseudo-object, or an object created directly by the ORB, but that can be accessed directly by any other object. The BOA provides operations that the server objects in a system can access and also makes the interface with the ORB and the skeletons. On the other hand, this communication between the BOA, the ORB, and the skeletons is done by a series of private and proprietary interfaces, which means that the BOA is specific to each ORB implementation. This was a problem solved by OMG with the introduction of the Portable Object Adapter(POA), in version 2.2 of the CORBA architecture.

The BOA limitations existed because it had been underspecified when it was introduced in version 1.1. To solve the problem, OMG decided to completely rewrite the BOA specification, once the differences between real CORBA implementations and BOA specification were too numerous to reconcile.

So, instead of risking causing a large migration problem from the original BOA to the new adapter, OMG decided to leave the BOA the way it was and create a portable version: the POA. Nowadays, many CORBA implementations retain both adapters (BOA and POA), in order to maintain backward compatibility with their older systems and also be in conformance with the current CORBA specifications.

Asynchronous Method Invocation

So far, all remote method invocations have been synchronous, in the sense that a client blocks until the server executes. A standard CORBA call causes the execution of an operation by the server object, which is carried out in a synchronous manner and, for this operation to succeed, both the client object and the server object must be on-line and available. There are specific exceptions to be raised when this basic condition is not met.

On the other hand, synchronous method calls do not answer all the needs of a programmer or developer who is working on a distributed system. There are many situations when it is important to decouple the communication between the client and server objects, in order to improve the overall system performance (by making processing asynchronous), or due to particularities of the problem to be solved.

In order to promote the decoupling between client and server objects during data communication, OMG has defined the Event Service, which makes possible the asynchronous data transfer between objects. The Event Service defines two new roles for software objects: the Supplier and the Consumer.

The Supplier objects are those that generate events (that produce data), and the Consumer objects are those that process these events or data, which are passed to them by CORBA requests.

The communication between Suppliers and Consumers can take place in two different ways: through the Push communication model, or through the Pull model (See Figures 6.4 and 6.5). The Push model gives the Supplier the prerogative of initiating the event data transfer to the Consumer objects, which means that the Supplier is in charge of the communication initiative. On the other hand, the Pull model

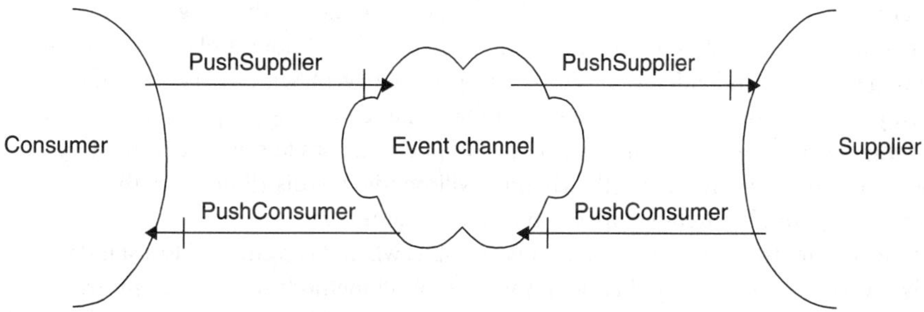

FIGURE 6.4   Push model communication. (Adapted from OMG, 1998.)

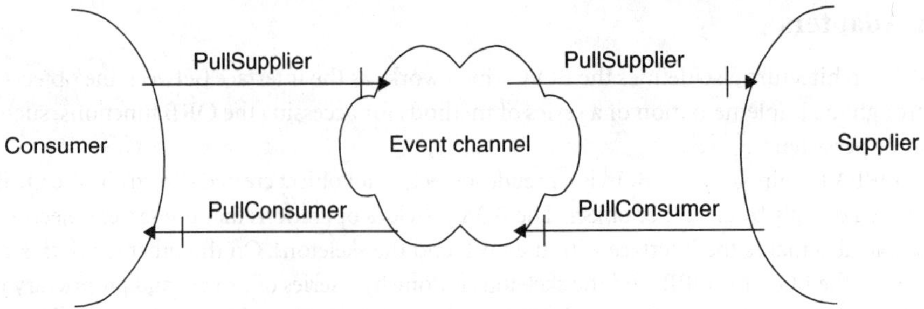

FIGURE 6.5 Pull model communication. (Adapted from OMG, 1998.)

gives the Consumer the ability to request data from the supplier at any given moment, and the consumer is in charge of the communication initiative.

The entire management of the data communication between Suppliers and Consumers as well as the request management are carried out by an object called Event Channel. The Event Channel is at the same time a Consumer and a Supplier, and is what makes the asynchronous data transfers possible. Every data exchange between the Consumer and the Supplier is made through the Event Channel, and never directly between them. The Channels are standard CORBA objects, and the communication between them uses common CORBA calls.

The process of creation of the communication mechanism by the Event Service is started with the instantiation of an Event Channel that supports the standard EventChannel interface, defined in the Event Service specification. This object basically supports three different operations: one operation that returns a Consumer Administrator object (or ConsumerAdmin), one operation that returns a Supplier Administrator (SupplierAdmin), and one operation that destroys the channel.

The administrators' role is to coordinate the addition of Consumers and Suppliers to the Event Channel. This is done through proxies, which are connectors used to link external objects (such as consumers and suppliers) to the channel, without direct contact between them.

This way, the linking process between a supplier application and the Event Channel takes place in two stages: first the event generator application receives a consumer proxy from the channel, through the Supplier Administrator, and then connects itself to this proxy. In the same way, the consumer application must first receive a supplier proxy from the channel and then connect itself to it, in order to gain access to the Event Channel.

As can be seen in Figure 6.6, OMG's specification defines the interfaces used for obtaining the consumer and the supplier proxies, in the Push and Pull communication models, resulting in a total of four different proxy interfaces to be implemented by the Event Service. There is also the event communication process using typed data, which implicates in the definition of a whole new structure, similar to the one shown above, with four more proxy interfaces. Every event communication is carried out through these proxies, so that the consumer and suppliers remain completely disconnected, and do not access each other's methods directly.

One example of applicability for the Event Service is the situation where a given system must provide a sensor reading to several different control panels, spread throughout a plant. Instead of having each panel reading the sensor individually, it is possible, using the Event Service, to create an Event Channel in which each panel is a Consumer, the system that reads the sensor is a Supplier, and the communication is made using the Push model. In this system, the Supplier only has to send the reading data to the channel, through a Consumer Proxy, and the channel will handle the distribution of the information to all Consumers. This example's architecture can be seen in Figure 6.7.

One of the features that can be noticed in this example, when it is compared to a standard CORBA call, is that it is asynchronous. The Supplier does not have to call methods in each Consumer to inform them of the new reading, being enough to send the data to the channel. This prevents potential problems, such as communication failure with the Consumer, which could cause the Supplier to hold until it received a

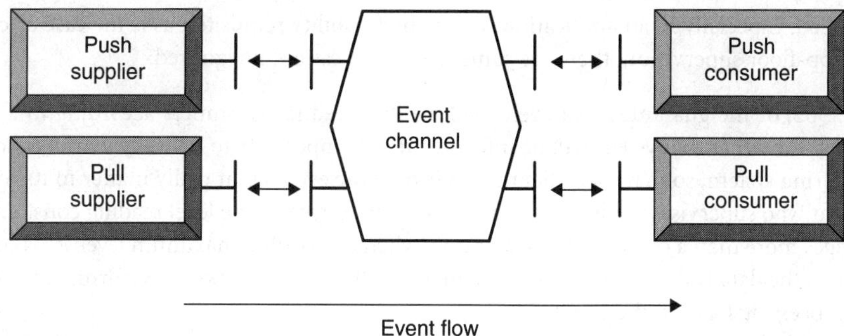

FIGURE 6.6   Untyped Event Channel Architecture, as defined by OMG. (Adapted from OMG, 2000.)

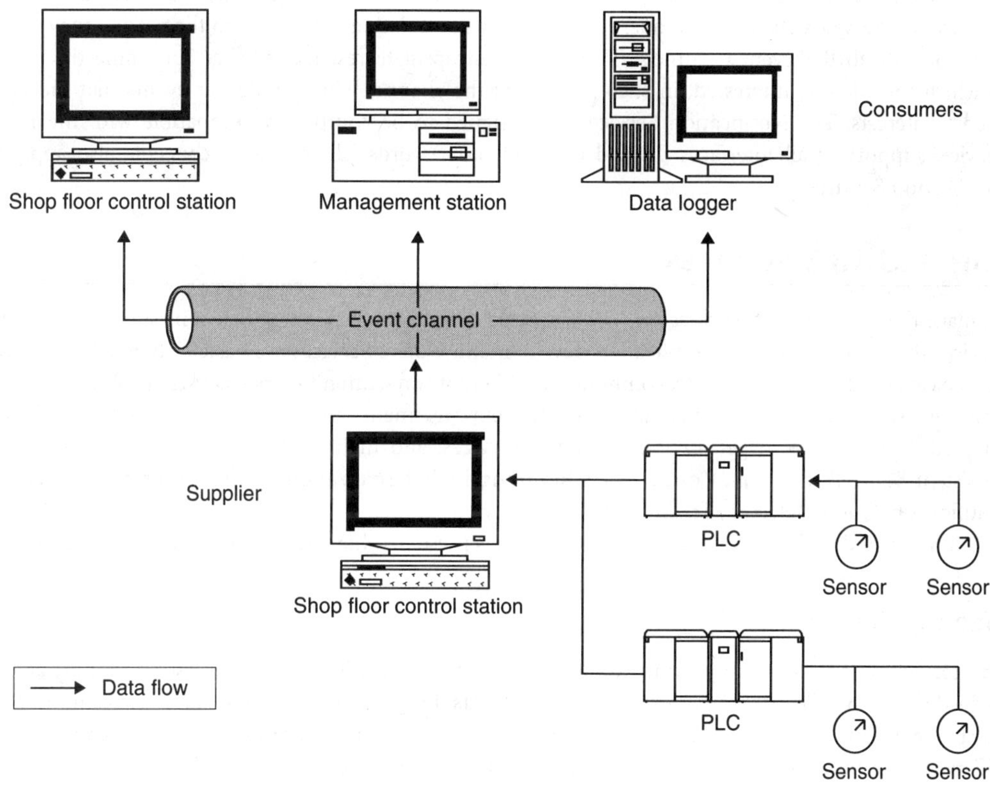

FIGURE 6.7   Example of a control system architecture using the Event Service.

reply (or until a timeout period expired), before it could go on and communicate to the next Consumer. In the original OMG specification for this service, it was not clear which technique should be used to control the event delivery, that is, to control the Quality of Service (QoS). This means that each vendor implementing an Event Service product is free to choose a different method, although all implementations must provide at least the "as quickly as possible" method. This way, it is really up to the vendor to decide what to do when a message cannot be immediately delivered, how to manage the event queue, and other issues inherent to the Event Service.

Although the Event Service is already a good contribution to the agility of the implementation of asynchronous communication between consumers and suppliers, there are situations when this service seems under-specified. Especially in an application with tight reliability requisites, as is the case of control systems and shop-floor supervision, there are some needs that cannot be ignored:

- The QoS, or the guarantee that events will be delivered to consumers according to a predefined priority, and that the system will be informed if it is impossible to deliver them in time.
- A filtering system, so that each client receives only the events that really matter to it. For example, a client who supervises the level of a tank may wish to receive the level reading constantly, when it changes more than a certain quantity, or only when it exceeds a maximum level. This considerably reduces the data traffic in the network, enhancing the overall network performance and reducing the processing load on the clients.
- The capability to transmit structured data, not just simple types, as it happens in the Event Service.

All these needs are addressed by the Notification Service, which is basically an extension to the Event Service. The main goal of the Notification Service is enhancing and refining the Event Service, introducing the concepts of filtering and configurability, according to different QoS requisites. The Notification Service's clients can request specific events associating data filters to the proxies they use to communicate to the Event Channels. These filters encapsulate restrictions that determine the events in which the client is interested, in such a way that the channel will only deliver events that meet the clients' interests. The Notification Service was designed to be completely compatible with the Event Service, supporting all interfaces defined by it. In other words, all the above concepts also apply to Notification Services.

6.6 CORBA Services

To make the development work easier, OMG created the Object Services, which consist of fundamental services that supply a base for application development. Object Services are basic elements for applications based on distributed objects, combining in different ways, attending needs such as object localization, persistency control, concurrency control, event treatment, and others. Object Services already adopted by OMG are in general called CORBAservices, and include the Naming, Event, Life Cycle, Persistent State, Transaction, Concurrency, Relationship, Externalization, Licensing, Property, Security, Notification, Query, Telecom, Trading, and Time services.

A minimum description of main CORBAservices is presented here, since it is a very lengthy subject.

Naming Service

This service aims at associating names to objects, using the name bindings. These associations are always defined in the so-called name contexts, which are objects that contain a group of name associations, each one being unique in a context. Different names can be associated to an object, in one same context or in different ones, but it is not mandatory that all the objects be assigned names.

Event and Notification Services

Event Service was described before, under Asynchronous Method Invocation, together with Notification Service. Both provide CORBA with asynchronous method invocation capabilities.

Collection Service

The Collection Service enables the support for grouping of objects and operations for the manipulation of the objects as a group. Queues, sets, bags, maps, and others can be collection types. Collections are foundation classes used in a broad range of applications.

The purpose of an Object Collection Service is to provide a uniform way to create and manipulate the most common collections generically.

Concurrency Service

The objective of the Concurrency Service is to mediate concurrent access to an object so that it remains consistent when access is needed by more than one concurrently executing computation. The Concurrency Control Service supports both transactional and nontransactional modes of operation through multiple interfaces.

The Concurrency Service serializes clients, transactional and nontransactional. Thus, both kinds of objects may request access to an object at the same time, and the Concurrency Service will make sure that the integrity of both operations is maintained.

Externalization

The Externalization Service specification defines protocols and conventions for externalizing and internalizing objects. Externalizing an object is to record its state in a stream of data. Objects that support the appropriate interfaces and conventions can be externalized to a stream (be it in memory, on a disk, on the network, etc.) and then be internalized into a new object that may be in the same or in a different process.

Externalizing and subsequently internalizing an object is similar to copying the object. The copy operation results in the creation of a new object that is initiated from an existing object.

The Externalization Service is related to the Relationship Service and parallels the Life Cycle Service in defining some kinds of externalization protocols.

Licensing

This Service supports the management of software licenses. Licensing is the most common form used by suppliers to assure that their products are being used according to the contracts signed by the client.

The Licensing Service provides a mechanism for suppliers to control the use of their software and intellectual property as predicted by their customers requests and by their agreement.

Life Cycle

This Service defines services and conventions for actions like creating, deleting, copying, and moving objects. Following one of the main roles of CORBA, Life Cycle Service defines conventions that allow clients to perform life cycle operations on objects in different locations.

Persistent State Service

A datastore is a kind of entity that is responsible for the management of data, a set of files, or a schema in a relational database. This Service presents persistent information as storage objects in a particular storage place called "storage homes". Storage homes are themselves stored in datastores.

Property

Properties, as used in CORBA, are typed, named values dynamically associated with an object, outside of the type system. Examples of cases where the use of properties can become very useful are object classification (version, author, company, etc.) and usage count.

The Property Service implements objects supporting the PropertySet interface or the PropertySetDef interface.

Query

The need to make query operations on collections of objects is fulfilled by the Query Service. Queries are predicate-based and may return collections of objects. They can be specified using object derivatives of

SQL or any other query language. Also, query does not only mean reading but also selection, insertion, updating, and deletion on collections of objects.

The Query Service returns collections of objects that may be selected from source collections given a selection criterion or produced by query evaluations based on the evaluation of a given predicate. The source and result collections may be typed.

Relationship

Objects do not exist in isolation; they are related to each other. The Relationship Service allows entities and relationships to be explicitly represented as CORBA objects. This service defines new objects to deal with a relationship that can be of an arbitrary degree. But the Relationship Services does not define a new type of system. Instead, it uses the IDL type systems to represent relationship and role types. A role represents a CORBA object in a relationship, which is created by passing a set of roles through a relationship factory.

Security

Security is an ever-rising concern of companies of all sizes. This security specification provides means for applications to provide protection against unauthorized access to information, security controls being bypassed, eavesdropping a communication line, and tampering with inter-object communication.

Telecom Log Service

The Telecom Log Service provided by this specification can be used in a pure CORBA environment as well as by TMN systems. This specification defines interfaces that provide logging of any type of event and querying of records based on constraint languages. Capabilities to form log networks for storing and forwarding events are also provided.

Time

The Time Service provides a means for a user to obtain current time together with an error estimate associated with it. Additionally, the specification also suggests that the service provide means to ascertain the order in which events occurred, to generate time-based events, and to compute the interval between two events.

Trading Objects Service

The OMG Trading Object Service aims at facilitating the offer and discovery of instances of services of particular types. A trader is an object through which other objects can advertise their capabilities and match their needs against what is advertised.

Transaction

The Transaction Service provides interfaces that allow multiple objects to cooperate to provide atomicity of a transaction. These interfaces enable the objects to maintain data integrity either by committing all changes or rolling them back in the event of a noncatastrophic failure.

Transaction Service's interfaces combine the transaction and object paradigms, the first being essential to the development of reliable distributed applications and the second providing the productivity and quality to application development.

6.7   Work on Verticals

OMG has established several Technology Committees, divided into three plenaries: the Architecture Board (AB), the Domain Technology Committee (DTC), and the Platform Technology Committee (PTC).

The PTC focuses on infrastructure and modeling specifications, for example, the ORB, object services, protocols, IDL and its language mappings, etc. The DTC focuses on vertical market specifications, for example, finance, manufacturing, etc. The AB's responsibilities are to make sure that specifications developed by the other two plenaries are consistent with the OMG's architecture and existing suite of specifications.

The DTC is divided into many subgroups, usually called "verticals." The most related to industrial IT are Business Enterprise Integration; Consultation, Command, Control, Communications, and Intelligence; Manufacturing Technology and Industrial Systems; Space; Telecommunications; Transportation; Distributed Simulation; Super Distributed Objects; Systems Engineering; GIOP Tunneling over Bluetooth; Historical Data Access from Industrial Systems; SDO for PIM and PSM; XML Telemetric and Command Data Exchange; CAD (Computer Aided Design) Services; Data Access from Industrial Systems; Surveillance Manager; and Telecom Wireless CORBA.

From approved standards produced by these groups, it may be cited as more related to Manufacturing:

- CAD Services, addressing the interoperability between CAD/CAM/CAE systems from different vendors toward a Product Data Management (PDM) system.
- Data Acquisition from Industrial Systems (DAIS), providing support for real-time transfer of large amounts of data and closely related to OPC.
- GIOP tunneling over Bluetooth, specifying how CORBA should be used over wireless links.
- Historical Data Acquisition from Industrial Systems, defining a number of interfaces for a time-series data management facility.
- Laboratory Equipment Control Interface Specification, defining standard interfaces to integrate laboratory equipment.
- Telemetry and Telecommand Data, although devoted to spacecraft telemetry and telecommand, is also somewhat applicable to manufacturing.
- Utility Management Systems (UMS) Data Access Facility, providing interfaces for obtaining analysis data from a UMS.

The items in the list are increasing every day, as new specifications are produced and approved.

6.8   Performance

The study conducted by the Defense Information Systems Agency (DISA) known as "Defense Information Infrastructure Common Operating Environment" (DII COE) (OIS, 1999), conducted by Boeing Phantom Works on real-time CORBA implementations was released in 1999 but it is still considered one of the more important results on CORBA performance. It tested for performance HARDPack (Lockheed Martin), ORBexpress (Objective Interface), and TAO (Washington University). From the publicly available results, it is clear that ORBExpress closely follows raw sockets communication, which can be considered as a lower bound to performance, with TAO being about 3 times slower and HARDPack showing performance degradation with large packets. On a Sun Ultra 167MHz@256Mbytes running SunOS 2.6, ORBExpress has been capable of transferring up to 24 kbytes in less than 1 msec, with client and server processes running on the same computer.

In a more recent study, conducted by Lockheed Martin (Thaker, 2003), TAO and ORBExpress were both capable of delivering 16 kbytes in about 1.1 msec when running on an Ultra 1 170 MHz with Solaris 2.7, with the client and server running on the same computer. In the same study, results for other ORBs are also presented; for instance, Orbix took about 4 msec on the same computer to do an equivalent job.

6.9   Real-time CORBA

What are real-time systems? The answer to this question is not as obvious as it seems. For a citizen of a large city, for instance, real-time information about traffic conditions is that provided on the radio, which reflected the situation on a given avenue a couple of minutes before. On the other hand, for the PLC controlling specific industrial equipment, connected to a critical proximity sensor, real-time information is that which it receives from the sensor just milliseconds after the maximum capacity level of a given reservoir is reached. Therefore, real-time systems are not about speed and efficiency, but about meeting time constraints in applications where timing is as critical for success as the data itself.

Industrial control applications and SCADA systems are, in their very essence, real-time applications, where process-critical information often has to reach its destination in tenths or thousandths of a second.

For example, let us consider a robot manipulator that is required to pick a specific type of part from a conveyor belt. This robot can be connected to a machine vision system, which will provide the specific coordinates of the part on the belt at a given time. If any kind of significant delays are introduced in the process, the manipulator's movement can be compromised, and the operation will fail.

Although the data transfer in this example can be implemented through the use of standard CORBA calls between the machine vision system and the robot controller, the engineer developing this system would not have any way of assuring the system's reliability using only the standard CORBA specifications and services to do the job.

In order to fulfil this kind of need, OMG has released the Real-Time CORBA specification. Real-Time CORBA is an extension of the conventional CORBA specification designed specifically to support the functional aspects of Real-World time-based systems.

On the whole, it can be said that the main goal of Real-Time CORBA is to provide what OMG called "end-to-end predictability," that is, to provide the tools required to build predictable and deterministic CORBA applications. OMG tries to accomplish this goal by:

- Respecting thread priorities between clients and servers during CORBA calls.
- Limiting the duration of thread priorities inversions.
- Limiting the latencies of operation invocations.

It is important to note that the success of real-time data communication in a CORBA application does not depend solely on the application or the ORB capabilities, but in four major aspects of the system:

- The use of a real-time operating system (RTOS).
- The use of a real-time ORB.
- The network communication layer.
- The application itself.

The combination of these four prerequisites is essential for the real-time data transfer, and should not be neglected. If one of these conditions is not met, the system cannot be considered reliable for critical real-time activities.

There are basically two kinds of real-time requirements: soft real-time and hard real-time. Hard real-time applications are those where the resources must be managed to guarantee the constraints are met, all the time, and where missing a constraint is considered a failure to meet the system's requirements.

On the other hand, soft real-time are those applications where the time constraints may be missed infrequently or by a small amount (usually a percentage of the time constraint) or where occasional events may be skipped. In soft real-time applications, the resources must be managed to guarantee that the requirements are met, but not strictly "all the time," which, as a matter of fact, makes soft real-time systems even harder to define and manage than the hard real-time ones. The Real-Time CORBA specification spans these two variations of real-time systems.

An overview of the Real-Time CORBA extensions can be seen in Figure 6.8.

The entire Real-Time CORBA IDL is contained in two modules: *RTCORBA* and *RTPortableServer*, with the exception of only the new service contexts, which have been added to the IIOP module. The *RTCORBA::RTORB* interface is an extension of the ORB interface that handles the operations regarding the Real-Time ORB configuration and manages the instances of the other real-time IDL interfaces.

One of the key concepts necessary for the understanding of the Real-Time CORBA specification is the concept of thread scheduling, which the OMG uses for controlling the scheduling of activities. Basically, a thread can be seen as small parts of an activity, which are conducted at a single node of the distributed

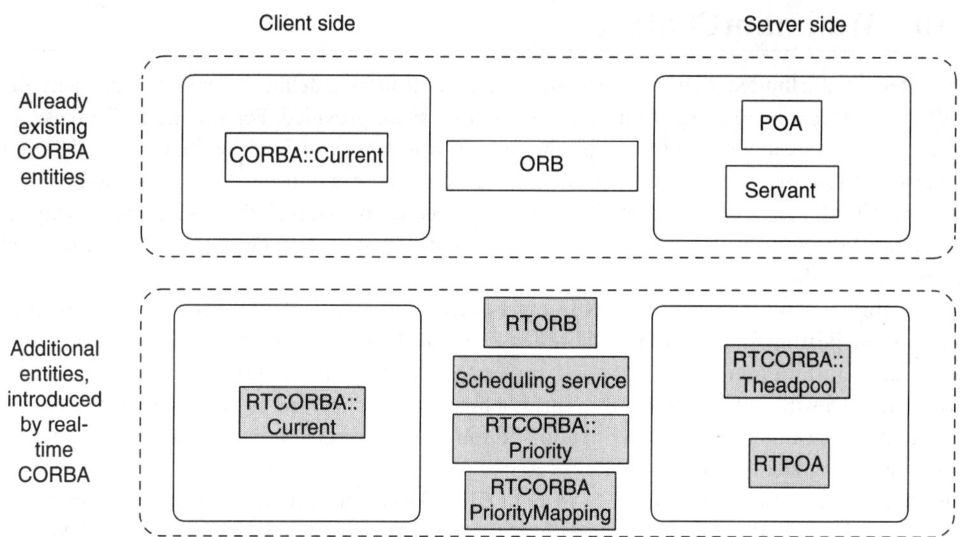

FIGURE 6.8   Real-Time CORBA Extensions. (Adapted from OMG, 2000a.)

objects network. The Real-Time CORBA specification defines interfaces that can be used to control the characteristics of the threads, enabling them to be scheduled, and therefore, allowing a system to control the scheduling of activities. These interfaces are *Threadpool* and *Current*.

Another part of the Real-Time CORBA specification that is quite important is the definition of the Real-Time CORBA *Priority* interface. Every Operating System available on the market has its own priority definition and handling mechanisms, but these mechanisms are not compatible between different OS. Thus, the OMG defined its own thread priority scheme, which overcomes the potential problems created by the utilization of different OSs in a distributed system. Therefore, any Real-Time CORBA application should always use the CORBA priority scheme, even if the same OS is used in all the network nodes. The access to the priority of a specific CORBA thread is achieved through the use of the *Current* interface.

Because of this standard priority scheme use, Real-Time CORBA applications must be able to map thread priorities into the native OS priority scheme. This is done through the *PriorityMapping* interface.

When one works with a shared-resources system, there are basically two different kinds of delays that can cause a thread to have its execution held up:

- Preemption — when a task has its execution delayed by the execution of a higher priority task.
- Blocking, or Priority Inversion — when a task is delayed by the execution of a lower priority task.

One of the goals of the Real-Time CORBA specifications is to minimize the occurrence of priority inversions, and one of the mechanisms used to do so is the propagation of the activity priority from the client to the server. In fact, Real-Time CORBA supports two different priority models:

- Client-Propagated Priority Model — when the server honors the priority of the invocation that had been set by a client.
- Server-Declared Priority Model — when the server handles requests at a priority assigned on the server side.

On the whole, Real-Time CORBA provides tools for the management and configuration of three basic types of resources in a distributed objects system: processor, communication, and memory.

At the processor level, Real-Time CORBA provides the mechanisms of thread pools, priority models, portable priorities, and priority banding. At the communication level, it provides mechanisms for priorities banding, protocol policies, and explicit binding. Finally, at the memory level, the mechanism supplied is the request buffering.

6.10   MinimumCORBA

For systems with limited resources, like embedded controllers, intelligent sensors, and many others, CORBA might just be too large to fit into the memory space provided. For situations like these, where the application possibly does not require the full set of features provided by the CORBA definition, a reduced version of CORBA was specified. This reduced version is called "MinimumCORBA." MinimumCORBA is, in fact, a subset of the CORBA specification, with all the basic features included but with some of the nonessential ones removed. These omitted features can be implemented by the application when needed (OMG, 2002b).

The features cut off in minimumCORBA represent, in fact, a trade-off between usability and resources conservation. Bearing in mind compatibility between CORBA and minimumCORBA, all the IDL features are included in minimumCORBA. So, an application written with CORBA is fully compatible with another written with minimumCORBA. This is a highly desired feature, as there is no need to review or rewrite any application previously written to add some new component like an embedded controller or some new or upgraded sensor.

In short, minimumCORBA is CORBA without DII and its related interfaces, such as Dynamic Skeleton Interface (DSI) and the Interface Repository. It also does not include DCE ESIOP, COM/CORBA Interworking, and Interceptors. As for language mappings, the full mapping must be supported, except for the omitted core objects related to the above items.

6.11   Product Availability

There are a number of suppliers of CORBA ORBs. This section will list some of the software available for CORBA developers. Many of the implementations listed here are freeware, shareware, or free software.

Orbix, by IONA

Orbix is a commercial ORB, supporting Java and C++ mappings, being CORBA 2.5 compliant with some CORBA 3.0 features. It runs on Windows, Linux, and various Unix.

Devoted to the corporate market segment, it includes Naming, Transaction, Security, and Event Services, besides load balancing and fault tolerance capabilities.

Visibroker, by Inprise

This is a CORBA 2.5 compliant product, supporting Java and C++ mapping. Also devoted to the corporate market, it includes clustering and load balancing, portable interceptors, POA, objects-by-value, and RMI-over-IIOP for better Java interoperability. It implements Naming, Transaction, and Notification Services and is available on Windows, Linux, and various Unix.

Inprise also offers Visibroker-RT, a CORBA for real-time, embedded systems. It is offered in two versions, a full CORBA implementation and the Minimum CORBA subset, both with RT CORBA extensions.

The ACE ORB (TAO)

TAO is an open-source, high-performance, real-time implementation of CORBA that provides efficient, predictable, and scalable quality of service (QoS) end-to-end. At the time of this writing, TAO is CORBA 2.3 compliant.

It provides C++ language mapping, in which the generated code closely follows the C++ mapping specified in the latest C++ mapping for CORBA 2.3. TAO supports development in a number of different Operating Systems: many flavors of Unix, Linux, and Windows NT, as well as some real-time operating systems such as Lynx, PSOS, and VxWorks. TAO also supports some embedded operating systems like Windows CE.

TAO also provides the implementation of a number of CORBAservices:

- A/V Streams Service,
- Concurrency Service,
- Event Service,
- Real-time Event Service,
- LifeCycle Service,
- Load Balancing Service,
- Logging Service,
- Naming Service,
- Property Service,
- Scheduling Service,
- Security Service,
- Trading Service,
- Time Service, and
- Notification Service.

It also implements the SSLIOP Pluggable Protocol. TAO's copyright license makes it perpetually and irrevocably an open-source free software, and it can be found on the website: http://www.cs.wustl.edu/~schmidt/TAO.html.

ORBacus

ORBacus is a fully CORBA-compliant ORB that is distributed as a source code. It is a robust, full-featured object request broker with a proven record and enterprise-class features.

Orbacus provides complete C++ and Java language mappings, a simple configuration and bootstrapping, small fooprint, and, at the time of this writing, is CORBA 2.4 compliant. ORBacus also supports some of the features in the CORBA 3 standard, including Portable Interceptors.

It supports the AIX, Tru64, HP-UX, Linux, Solaris, Irix, and Windows NT/2000/XP platforms. ORBacus offers a 30-day free evaluation, and is also free for use, with full support, in application development being offered in a distributed computing course that requires CORBA.

The Naming Service, Event Service, Property Service, and Time Service are included in the package. Orbacus' site is http://www.orbacus.com.

omniORB

The omniORB is a robust high-performance CORBA ORB with C++ and Python language mappings. It is freely available under the terms of the GNU Lesser General Public License (for the libraries) and the GNU General Public License (for the tools). It is one of the only three ORBs to be awarded the Open Group's Open Brand for CORBA.

omniORB version 4.0 adheres to version 2.6 of the CORBA specification. It is fully multithreaded. It has support for Windows NT/XP/95/98, Linux, Solaris, HP-UX, Irix, Digital Unix, AIX, OpenVMS, NextStep Reliant Unix, Phar Lap's Real-time OS, SCO Unixware, Mac OS X, and others.

omniORB supports CORBA 2.6 DynAny interfaces, TypeCode, and type ANY, GIOP, and IIOP 1.0, 1.1, and 1.2, wchar, wstring and code set negotiation, bidirectional GIOP, and others. It also provides included support for the Naming Service and is fully interoperable with other CORBA ORBs.

The omniORB project is hosted at http://omniorb.sourceforge.net/.

jacORB

The jacORB is a high-performance, fully multithreaded ORB with an IDL compiler that supports OMG IDL/Java language mapping revision 2.3. It has native IIOP, GIOP 1.2, and Bidirectional GIOP. CORBA 2.3 specification is supported.

jacORB uses the GNU Lesser General Public License, which means that it is free even if one wishes to sell a product that uses its libraries.

It requires a Java VM with version 1.1 or better, and it should be possible to run jacORB in any operating system with a suitable Virtual Machine. It has been tested on Windows 95/98/NT/2000/XP, Linux, and Solaris; but reports show that jacORB also runs on other Unix flavors and MacOS X.

jacORB supplies CORBA Naming Services, COSS Event Service, Transaction Service Collection, Concurrency, and Trading Services. It should be fully interoperable with other ORB implementations through the IIOP. Portable Object Adapter is supported and some GUI tools like POAMonitor and NameManager are provided, which allow users to inspect their object adapters and browse the Naming Service.

Dynamic Invocation Interface (DII) and Dynamic Skeleton Interface (DSI) and Dynamic Management of Anys (DynAny) are also some features provided by jacORB 1.4.1.

The software and documentation can be found on http://www.jacorb.org.

MICO

The acronym MICO expands as "MICO is Corba." The intention of this project is to provide a freely available and fully compliant implementation of the CORBA standard. MICO has become quite popular as an OpenSource project and is widely used for different purposes. It has recently been branded as CORBA compliant by the Open Group.

MICO is currently tested on Solaris, AIX, Linux, Digital Unix, HP-UX, Ultrix, and Windows 95/NT. Reports indicating that it also works on FreeBSD, Irix, OS/2 DG/UX, and LynxOS have been received by the supplier.

The software is fully CORBA 2.3 compliant, and support Dynamic Invocation Interface (DII), Dynamic Skeleton Interface (DSI), Interface Repository (IR), IIOP as native protocol, IIOP over SSL, and DynAny, among others.

MICO supports IDL to C++ mapping, and supports the following CORBA services:

- Interoperable Naming Service
- Trading Service
- Event Service
- Relationship Service
- Property Service
- Time Service and
- Security Service.

MICO is currently under the GNU General Public License, and can be found at http://www.mico.org.

VBOrb

VBOrb is an Object Request Broker, written entirely in Visual Basic 5. With VBOrb, it is possible to write CORBA Visual Basic clients and servers.

VBOrb can be used instead of a COM-CORBA-Bridge, to access Enterprise Java Beans(EJBs), or to remote control a Microsoft program by using Visual Basic for Applications(VBA). VBOrb comes with an IDL to VB compiler called IDL2VB. The compiler is written in Java.

VBOrb supplies the Naming Service and is distributed under the GNU General Public License. The software can be obtained at: http://home.t-online.de/home/Martin.Both/vborb.html.

ORBit

There are currently two versions of ORBit in use: the original "stable" ORBit and the new ORBit2. ORBit is an implementation of CORBA which is used in the GNOME project, an open-source project aimed at implementing a windowing interface for Linux. Active development is done for ORBit2, being a clean reimplementation from scratch. The old ORBit code is no longer maintained.

ORBit, the original version, is CORBA 2.2-compliant and has a number of language mappings. ORBit2 is a CORBA 2.4-compliant Object Request Broker (ORB) featuring mature C, C++, and Python bindings. Bindings (in various degrees of completeness) are also available for Perl, Lisp, Pascal, Ruby, and TCL; others are under development.

It supports POA, DII, DSI, TypeCode, Any, IR, and IIOP. Optional features including INS and threading are available.

ORBit2 is engineered for the desktop workstation environment, with a focus on performance, low resource usage, and security. The core is written in C and runs under Linux, Windows, and many flavors of Unix.

ORBit and ORBit2 support the following CORBA services:

- Concurrency Service
- Event Service
- Naming Service and
- Transaction Service.

ORBit is distributed under the GNU Public License and GNU Lesser Public License (libraries), and its homepage can be found at http://www.gnome.org/projects/ORBit2; but a number of other sites can be found with consistent documentation about the software.

e*ORB

According to the supplier, OpenFusion, e*ORB (tm) is the smallest and fastest Object Request Broker in the world and has the smallest footprint. It supports C, C++, Java, and Ada mappings. One of e*ORB's most prominent characteristics is its support for Real-time and embedded systems.

e*ORB Java edition was the first Java ORB to be ported to operating systems such as VxWorks, Palm OS, EPOC, and Windows CE. Today, e*ORB empowers embedded systems ranging from driver information systems to battle-tank simulations and optical switches.

It is supported on a wide range of platforms including Lynx, embedded and desktop Linux, Solaris, HP-UX, Windows, VxWorks, Integrity, Stratus VOS, and Palm OS.

Three CORBA services are included with e*ORB:

- Naming Service,
- Notification Service, and
- Logging Service.

The software is licensed and downloads are provided for evaluation free of charge for 30 days. During evaluation, customers are also entitled to 30 days of free support. The e*ORB site can be found at: http://www.prismtechnologies.com

Other Products

A good resource for becoming acquainted with some free CORBA products is the OMG website. OMG maintains a list of links for many free CORBA implementations at: http://www.omg.org/technology/corba/corbadownloads.htm

References

Gorton, I. *Enterprise Transaction Processing Systems: Putting CORBA OTS, Encina++ and Orbix OTM to Work*, Addison-Wesley, Reading, MA, 2000.

OIS. DII COE Real-Time ORB Trade Study Overview, Published by Defense Information Systems Agency (DISA) and available at http://www.ois.com/resources/corb-10-overview.asp, Accessed 01-Sep-2003.

Object Management Group (OMG), Soley, R., Stone, C., *Object Management Architecture Guide*, John Wiley & Sons, New York, 1996.

Object Management Group (OMG), A Discussion of the Object Management Architecture, January, 1997.

Object Management Group (OMG), CORBAservices Common Object Services Specification, December, 1998.

Object Management Group (OMG), Notification Service Specification, June, 2000.

Object Management Group (OMG), Real-Time CORBA Specification, August, 2002a.

Object Management Group (OMG), Minimum CORBA Specification, August, 2002b.

Object Management Group (OMG), CORBA FAQ available at http://www.omg.org/gettingstarted/corbafaq, Accessed 21-Jul-2003.

Open Software Foundation (OSF), *Introduction to OSF/DCE*, Pearson Education POD, Harlow, England, 1995.

Siegel, J., *CORBA 3 Fundamentals and Programming*, John Wiley & Sons, New York, 2000.

Thaker, G., ATL QoS Homepage, Accessed at http://www.atl.lmco.com/projects/QoS on 01-Sep-2003.

7

Web Servers, Clients, and Browsers

Robert Tolksdorf
Freie Universität Berlin

7.1 Introduction

The Web is a network-based information system in which a huge number of components interwork. The main classes of these are Web servers that have and deliver information, Web clients that request information, and intermediate components that help to make the overall system more efficient. In this chapter, we review these components and their functionalities.

7.2 The Architecture of the Web

The Web is a network-based information system where information is delivered by Web servers to Web clients on request. Components involved in this system can be classified into the following three categories:

1. *Web servers* have information and make it available on request. They are accessed by TCP connections on the IP address of the machine they run on and the port they are configured to listen to. On that connection, they communicate with the Hypertext Transfer Protocol HTTP.
2. *Web clients* request information from specific servers by issuing a request following the HTTP. The information received is then further processed according to the kind of client. The most well-known kind of Web clients are Web browsers that display the information transferred from the server to a human user. The most important "invisible" kind of Web clients are *Crawlers* that retrieve information from Web servers to feed it into indexed full-text databases that are the core of Web search engines.
3. *Intermediate components* transform requests or information. One important kind of such components is formed by *proxies* who act on behalf of the actual client, for example, because it is behind a firewall and should not be visible outside. The other important kind of intermediate components are *caches* that also act as proxies but store information retrieved and answer further requests for it without having to contact the actual Web server.

Figure 7.1 shows a configuration of such components. There are two Web servers: one offers some specific information and the other one hosts a Web search engine. There is a traditional Web browser as a client, which is configured to use a proxy to access the corporate Web server and to contact the search engine directly. Close to the search engine, a Web crawler retrieves some information from the corporate Web server that will be fed to the full text index of the search engine on which searches are performed later. All components run on different machines connected via IP. They communicate with HTTP to transfer requests and responses on information.

Figure 7.1 shows a very small excerpt of the real structure of the Web. The Web Server Survey of Netcraft (available at http://www.netcraft.com/Survey) counted 35,424,956 Websites in January 2003. The HTTP service is available on a substantial fraction of the named computers on the Internet — these were counted as 171,638,297 in the January 2003 edition of the Internet Domain Survey of the Internet Software Consortium (available at http://www.isc.org/ds/WWW-200301). The number of clients cannot be counted. However, the number of people having Internet access can give an indication of this. The Global Internet Statistics survey of Global Reach (available at http://www.glreach.com/globstats) estimates the size of the worldwide online population in September 2002 as 619 million people.

7.3   Web Servers

Web servers are the source of information in the Web. They are, in general, passive in that they deliver the information they have on request to clients.

Historically, the first Web page on a server was http://nxoc01. cern.ch/ hypertext /WWW/TheProject. html. The server ran at the European Organization for Nuclear Research (CERN) where the Web project was invented. The page mentioned contained the description of this project. The first server program available to the public was httpd developed at the CERN by Ari Luotonen, Henrik Frystyk Nielsen, and Tim Berners-Lee. It is still available at http://www.w3.org/Daemon. It soon found a more successful rival with the NCSA (National Center for Supercomputing Applications) HTTPd. Today, the most used Web server is Apache (see http://www.apache.org). It ranks first in the Netcraft Web Server Survey (http://www.netcraft.com/survey), with an estimated share of about 62% in January 2003 before the Microsoft product family of Internet-Information-Server with a share of about 27% of all servers.

General Operation

The identification of content on the Web is done by URLs. For HTTP communication, they have the format *http://<Servers IP Address>/<Path to information>*. An example URL is `http://www.w3.org/Protocols/Activity.html`, which identifies the HTML page `Protocols/Activity.html` on the server `www.w3.org`. This Web page is stored in the Web servers file system in the form it is delivered to the client. Therefore, it is called *static content*. Another example is `http://www.google.com/search?q=server`, which addresses the script search on the

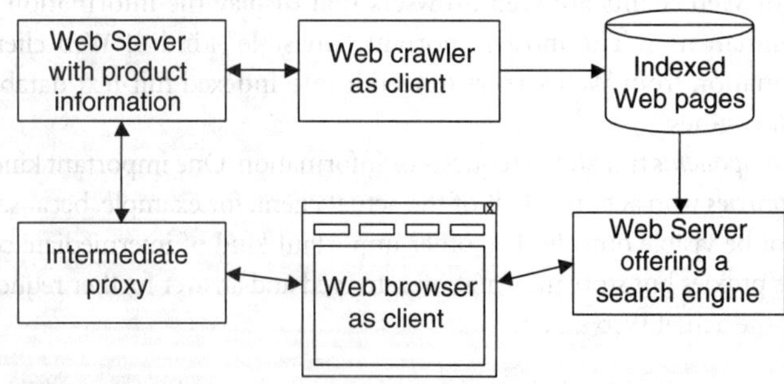

FIGURE 7.1   Components in the Web.

server `www.google.com` and passes a parameter string `q=server` to it. The information returned from the server is generated fresh with each request; therefore, we speak of *dynamic content* in this case.

The resolution of the servers address is part of the standard TCP usage by any client. The resolution of the path information to a specific chunk of data is up to the server. It may map it into some location in a local file system, or to the execution of some programs on the servers' machine.

Web servers in general consist of several subcomponents. Among their duties are the following:

Interfacing with the network: Web servers speak by definition HTTP for communication with clients. HTTP interaction is initiated by contacting the default port 80 on a machine with TCP as the transport protocol. Since Web servers can be configured to accept connections on a different port and since HTTP can be directed to any port, this is actually just a default definition.

A Web server has to set up the standard interface for the respective IP communication, which means to listen on port 80, to accept connections, and to perform the standard HTTP communication over the resulting communication socket.

Mapping of static content to the local file system: Web servers have access to some local file system, which is the persistent storage for the static contents. How the path to information from the URL is mapped to a path in the file system is up to the server and its configuration. `http://www.w3.org/ Protocols/Activity.html` could lead to the delivery of the file `/htdocs/Protocols/ Activity.html`, `/import/htdocs/staticcontent/Protocols/Activity.html`, or `~auser/.public_html/Protocols/Activity.html`. The mapping could also map to different filenames, for example, to `c:\Protocols\Activity.htm`. In the extreme case, it is not necessary to access a file system at all — a server could execute some network access or have static hard-coded HTML pages as constants in its program.

Mapping of dynamic content by execution of programs: The servers' configuration determines whether a URL denotes static or dynamic content. An example configuration could state that —after performing the static mapping — all files in the directory `cgi-bin` and ending in `.cgi` are programs to be executed to generate dynamic content. The server then has to interface the standard input and standard output channels of the programs with the incoming and outgoing network connection and to start the program. A most important side issue of this operation are security aspects, for example, under what user account the program runs and what permissions it has.

Coordination of the components: Web servers must be able to serve multiple requests at a time. Sequential processing is necessary for the acceptance of TCP connections on port 80. All other activities of a server can in principle be executed concurrently in different threads. Managing this degree of concurrency needs coordination activities. These have to take care of shared writable resources, for example, log files, deal with processes started by dynamic resources, or optimize the degree of concurrency.

Dynamic Content

The static content offered by a server is generated by authoring Web pages in advance. This is usually supported by editors for individual pages, by programs that help to handle the complete structure of a site, or by advanced content management systems.

Dynamic content is the basis for interaction on the Web — any forms that accept user input have to be processed on the server. Conceptually, the generation of the page that answers to the form input is the goal of the server. In practice, of course, side effects such as storing inputs in a database or triggering the delivery of some items purchased with the form is the actual goal of this generation of dynamic content.

Figure 7.2 shows a minimal script that generates dynamic content. We can assume that it is accessible by the (imaginary) URL `http://www.foo.ba/cgi-bin/serverdate.cgi`. We also assume that the server runs on a UNIX system that recognizes such a script as executable content. When a request for `/cgi-bin/serverdate.cgi` reaches the server `www.foo.ba`, it executes that script.

The `echo` instructions in this script simply write the following string to the standard output. The line `echo ` `` `/usr/bin/date` `` invokes the program `/usr/bin/date` — which results in the current system date of the server — and forwards its result to the standard output.

```
#!/bin/sh
echo "Content-type: text/html"
echo
echo "<html> <head> <title>Date here</title> </head> <body>"
echo "The current date here:<b>"
echo `/usr/bin/date`
echo "</b>. Thanks for using this service. </body> </html>"
```

FIGURE 7.2   A sample script.

```
Content-type: text/html

<html> <head> <title>Date here</title> </head> <body>
The current date here: <b>
Sun Dec 29 20:54:41 2002
</b>. Thanks for using this service. </body> </html>
```

FIGURE 7.3   The output of the script.

The Web server sends all output of the script to the Web client in its response, resulting in an answer as in Figure 7.3.

The generated content always contains the date of its creation. For interactive Web sites, it is necessary to process input typed by the user in forms with scripts. The *Common Gateway Interface* (CGI) (see http://www.w3.org/CGI) defines a mechanism to pass these data to scripts. The core idea is to use environment variables to communicate this information. The Web server sets a specific set of about 20 variables and then calls the script (which is often also called "CGI program"). An example variable is REMOTE_ADDR, which contains the Internet number from which a client sent the HTTP request. More important is QUERY_STRING, which contains the form-input to be processed.

Figure 7.4 shows the flow of information in the processing of a CGI request. After filling out a form, the client encodes input data in the URL and sends it as part of its request to the server. The server identifies phone.cgi as a script and places the information received from the client into environment variables according to the CGI mechanism. The CGI program processes these data, for example, by querying a database accordingly. The results received are then placed in an HTML page, which is communicated to the server by the standard output of the CGI program. The server passes the generated HTML page to the client, which displays it as the result of the request to the server.

Many Web-Servers offer another mechanism to include dynamic content in Web pages, the so-called *Server Side Includes*. The basic idea is that the server scans the Web-pages to be delivered for a set of special tags. These tags are then replaced by dynamic content by the server. The set of these tags depends on the Web server used.

For most Web servers, the SSI Tags are actually HTML comments of the form

```
<!--#directive attribute₁="value₁" attribut₂="value₂"-->.
```

This makes the approach compatible with all Web servers: if SSI are not parsed and replaced they appear as HTML comments and are ignored by clients.

Figure 7.5 shows an example use of SSI: the regular HTML code includes one SSI tag. It contains the directive flastmod. This tag is replaced by the Web server with the date of the last modification of a file within the Web server's file system. Its name is given in the attribute file. In the case that the attribute has an empty value — as in the example — the modification date of the file that contains the SSI directive is taken.

7.4   Web Clients

Web servers are accessed by programs, which converse in HTTP and interact with the server accordingly. Only at first glance, all clients are Web browsers. There are number of components in the Web that are not intended for user interaction, such as proxies and caches. We provide an overview on them in the following.

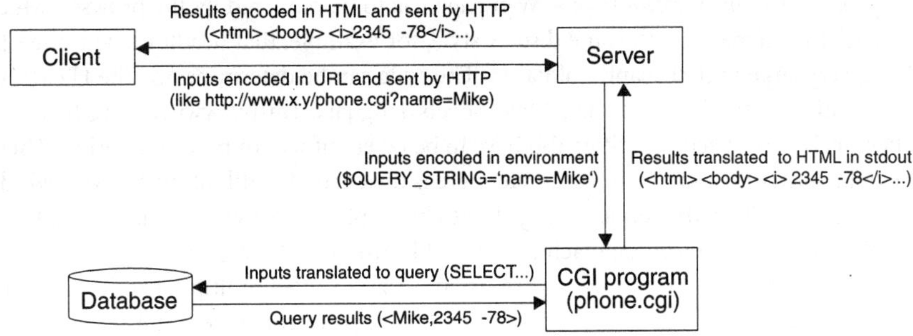

FIGURE 7.4   Processing of a CGI request.

```
With SSI:

<div align="right">
 Last modified:  <! --#flastmod file=" "-->.
</div>

After expansion of SSI:

<div align="right">
 Last modified:  Wednesday, 08-Jan-2003 09:30:15 CET.
</div>
```

FIGURE 7.5   The page with SSIs before and after processing by the server.

Web Browsers

Web browsers are what makes the Web usable — they retrieve information and display it. The success of the Web is due to two mutual reinforcing developments — the availability of more and more information on growing numbers of Web servers and the availability of better browsers as clients.

Historically, the first Web browser was written in 1990 on a NeXT machine by Tim Berners–Lee, who keeps a description including a screen shot at http://www.w3.org/People/Berners-Lee/WorldWideWeb.html. The first widespread client application was the Mosaic browser written by Marc Andreessen and Eric Bina at NCSA. It was available for a number of Unix platforms. Andreessen soon commercialized the application by cofounding Netscape and developing the browser with the same name. The rest is history — after a long time of dominating the browser market, Netscape was outnumbered by the installations of Microsoft's Internet Explorer. Today, the Internet Explorer is the by far the most used Web browser.

A Web browser consists of multiple components that fulfill at least the following functionalities:

Accepting user inputs: A Web browser is a program with a GUI that accepts user inputs like mouse-clicks, menu selection, or keyboard input. The browser has to interpret these inputs, for example, to determine on which link a user has clicked, what bookmark was selected from a menu, or which URL was typed in. Processing this input can affect the current display, affect the browser's state, or lead to a network access.

Retrieving information from servers: When a URL is determined, the respective Web server has to be contacted, which means extracting the server's IP address from the URL and to initiate a HTTP connection with it. Then, the client has to extract the path to information and generate the respective HTTP request. The answer from the Web server has to be read and stored in some internal data format. Then, the network connection has to be closed. The browser has to be able to react properly to the various response-codes as defined in HTTP, for example, by retrieving redirected URL from another server or by asking the user for necessary authentication credentials and retrying the access. The browser also has to scan the retrieved HTML code for further embedded data-like images and retrieve these also from the network.

Rendering and displaying information: A Web page has to be rendered by the browser, which means to associate the HTML elements with some display style, for example, that headlines with the <h1>-Tag are displayed in a very large font-size and boldfaced. The style can be determined by the HTML standard, by a style specification with the Cascading Style Sheets language (http://www.w3.org/Style/ CSS), or be browser-specific. Device specific abilities also have to be taken into account for rendering. Then, the whole display has to be formatted. The main parameter here is the current width of the browser window and any scaling levels selected. Then, the rendered page has to be displayed in a window according to the window-ing and graphic system found on the machine. After this, user inputs are again accepted.

Coordination of activities: Web browsers are usually able to offer multiple windows to the user. This implies that multiple interactions with the network can take place concurrently. Also, embedded elements in a page can be retrieved in parallel. The management of this concurrency needs coordination, which takes care of thread management and optimization of network usage, for example, by limiting the number of concurrent network accesses or ensuring that resources are retrieved only once and then reused.

Management of the users' interaction environment: The activity of browsing the Web is usually supported by some interaction environment that is specific for the user. It includes a set of preferred Web server addresses, the bookmarks or favorites, a user history, the storage of data entered info Web forms and its automatic usage in further forms, account information for Web passwords, functionalities to export Web information to local programs, etc. The variety of functions is large and so is the effort spent on them. Since the basic network access and rendering of information is quite standardized, these user-centered functionalities leave room for browser vendors to come up with unique product features.

Support functionalities: Web information is not limited to HTML and some standard graphic formats. There is a variety of — often proprietary — further formats, for example, for audio. Web browsers usually support a wide variety of such formats and offer an interface for extensions with so-called plug-ins that are able to ren-der further formats. Modern Web browsers also include a set of further functionalities that support the user in other Internet functionalities as well. Usually, a Web browser also includes a client for Netnews following the NNTP protocol and email following POP3, IMAP, and SMTP protocols. Finally, client-side activity is sup-ported by scripting languages like JavaScript and execution environments like Java or ActiveX.

Other Clients

It is not necessary that Web servers are accessed by Web browsers only. Any program that obeys the HTTP can retrieve information from a Web server. Examples are Web site copy programs that retrieve the contents of some Web site by traversing all links starting from some entry site. The content is then stored on a local disk and available for usage even if no network connection is available. More important are, how-ever, so-called *Web crawlers* that are the basis for any search service on the Web like Google or AltaVista.

Historically, one of the first was the WebCrawler (see http://www.ncsa.uiuc.edu/SDG/IT94/ Proceedings/Searching/pinkerton/WebCrawler.html). It used the same principles as today's search engines. However, in 1994, the Web became significantly smaller — WebCrawler indexed about 50000 pages. As of Spring 2003, the largest Web search engine is Google, with more than 3 billion pages indexed.

A Web search engine is usually a full text index of Web pages on which queries can be placed. The tech-nology of these engines differs in the way they determine the similarity between the query and the con-tents of the index and the way they order the results.

But the full text index has to be filled with contents retrieved from the Web. Since the Web is extremely large, this task has to be automated. Web crawlers are clients that automatically traverse the Web and feed the retrieved information into a full text index. They operate according to the following generic steps:

1. Initialize a list of URLs to retrieve.
2. Take a URL from the list and perform a test whether it should be retrieved. This tests whether the URL has already been visited, or the crawler could be configured to focus only on URLs from some specific Internet domains.
3. Retrieve the information referenced by the URL using HTTP.

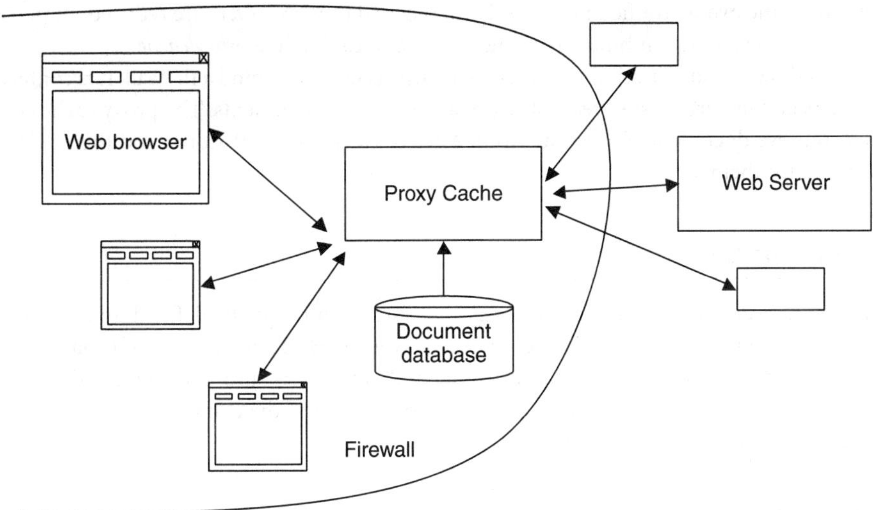

FIGURE 7.6　Browsers connected to servers via a proxy cache.

4. Extract any references from the information (for HTML pages, e.g., the references from the tags `<a>`, `<link>`, `<meta>`, `<img>`, `<object>`, and `<frameset>`) and put it into the URL list.
5. Extract all contents from the retrieved information as text and enter it into the full text index.
6. Extract any metainformation available like authorship, size, date written, etc. and store it in some database.
7. Repeat the process starting with step 2.

This basic process still defines the algorithm used by search engines. However, today, better strategies for selecting the next URL to follow exist; information can be extracted from more file types aside from HTML, and more metadata are retained.

7.5　Intermediate Components

Browsers and crawlers communicate directly with Web servers. However, there are also components that act in between clients and browsers. The most important ones are proxies and caches discussed in this section. Historically, the CERN httpd server was also able to act as a cache. Today, many Web servers offer the cache functionality either integrated or as configurable modules. A popular stand-alone cache program is the Squid proxy cache available at http://www.squid-cache.org.

The direct connection between Web clients and servers is not always possible or advisable. For example, in a corporate network, clients should not be fully exposed to the outside world due to security reasons. All Internet traffic will be routed through a firewall, which is the single and central interface between the — both IP based — LAN and WAN. It could be advisable to hide the internal network usage from the outside world.

A solution to this is to establish a *proxy* that resembles the origin of all Web accesses to servers but actually only forwards requests by clients in the LAN and directs answers to them. A client is configured to access a proxy at a fixed address with its request. The proxy program forwards the request to the actual Web server and retrieves results from there. The results are then forwarded to the original client. This configuration has the consequence that the actual client is never exposed to the outside world in the Internet as a WAN. All traffic is directed through the proxy to whom special security measures can be applied.

Since the proxy is used by all LAN clients, it can be extended to offer an additional functionality. As a *cache,* it can store all information retrieved indexed by the URL used. Any future accesses for the same URL can then be satisfied from that cache without having to contact the actual server. The result is the reduction of WAN usage and lower network latency due to the avoidance of WAN connections to servers.

As a side effect, the overall traffic in the WAN and the load to Web servers are reduced. In practice, proxy and cache functionality is combined in a single program leading to a *proxy cache*.

Figure 7.6 shows a configuration where a proxy cache is located behind a firewall. It is contacted by several browsers and in part satisfies requests from a database of documents. The proxy cache contacts Web servers to retrieve documents not in the store. After retrieval, they are stored and then delivered to the clients asking for them.

7.6 Summary

The Web is conceptually a simple client–server system with clearly defined interfaces. The main components in the Web are clients that access Web servers by HTTP to retrieve information. Mainly that information is displayed to users by a browser, and often it is also used to build search engines. Intermediate components help in optimizing the performance of the components. The most important of these components are proxy caches.

About the Author

Robert Tolksdorf is professor of computer science at the Freie Universität Berlin, heading the group on network-based information systems NBI (http://nbi.inf.fu-berlin.de). His research focuses include the study and development of languages, models, and systems that guide the coordination of activities in networked systems, the study, application, and evaluation of XML-technologies for network-based information systems, among others, and the study and application of Semantic Web technologies, and the use of novel models for the coordination of agent societies on a large scale, especially those models based on Swarm Intelligence. Tolksdorf authored several journal articles, four books on Internet and coordination technologies, and about 90 refereed publications, invited book contributions, and research reports. He is a member of the IEEE Computer Society, the ACM, and the German Informatics Society, where he chairs the special interest group on multimedia.

8

Internet Programming Languages

Wolfgang Radinger
Vienna University of Technology

Martin Jandl
Vienna University of Technology

8.1 Introduction

The development of Internet programming languages was mainly driven by users' needs. Since Tim Berners Lee introduced the first version of the Hyper Text Markup Language (HTML) [9], various Internet programming languages have been developed to meet the demands of the Internet community. At the beginning, the community had to get along with no programming language at all for 20 years. Nowadays, it is hard to find the best and most suitable Internet programming language.

The most popular language on the Web is of course HTML, a markup language. Markup languages are interpreted by the Web browser and are not termed as programming languages. They consist of tags and values, whereby tags declare how to interpret the enclosed value. The main disadvantage of HTML is the miscellaneous content and layout information. This was the main reason for the development of the eXtensible Markup Language (XML). XML [8] has, in contrast to HTML, no predefined tags. Hence, the structure of an XML document can be defined by the developer. The layout of the page will be defined in a separate style file.

A main result of the Web hype was the change of classical software engineering toward a new discipline termed Web engineering. Tools, methods, and data descriptions have been adapted to develop Web Applications more efficiently. Often, Web engineering is described as different from and more complex than traditional software engineering [2]. Web engineering has to deal with the development of Web-based systems and applications. Therefore, it typically has to handle conditions of Web-based systems [4]: rapid growth of the requirements and the continual change of the content. The systems also have to be built as scalable and maintainable because such functionality cannot be added later. In [5], Web engineering is described as the outcome of the convergence process between software engineering and hypertext systems. Until now, Web engineering has been coined, on the one hand, by traditional hypertext-oriented information systems and, on the other, by application-driven e-commerce systems. To overcome these issues, Web engineering and its programming languages attempt to integrate databases and middleware systems, and even use software engineering methods such as the Unified Modeling Language (UML) to design Web systems [1].

The major difficulty in finding a common Internet programming language derives from the diversity of Internet architecture. The traditional client–server structure is heavily influenced by the client type for example, personal computer, personal digital assistant, or mobile phone. The service developer has to keep in mind the devices and the possible bandwidth that the client may use to connect. The business logic often remains the same and only the presentation has to be adapted. Users demand to see their data only in different views. Figure 8.1 shows the Enterprise Application.

The integration (EAI) approach focuses on the integration of different applications by means of connectors[1] [3]. The user integration is defined in a separate layer, which allows to serve several types of user devices and interfaces. At some point, it will be possible to even integrate new user interfaces with as yet unknown technologies. The EAI layer is responsible for the integration of the different applications needed for the service, for examaple, user data or order information. Through EAI, it is possible to combine legacy applications with new technologies and make them accessible via the Internet.

We will now describe the most popular Internet programming languages and their main fields of application. Every description will include a small example to outline the advantages and differences.

8.2 Java

In 1990, Sun Microsystems launched a project under the code name "green project" to develop base technologies for the consumer market. These so-called "set-top-boxes" were planned to be programmable hardware components for households. The goal was to transmit small programs via TV channels for easier update of the functionality of household devices. The result was a portable, network-compatible operating system, and a flexible programming language named Object Application Kernel (OAK). At the same time, the Web began to grow in importance, and with this hype the breakthrough of the language named Java began.

Java [10] is an object-oriented programming language based on the concepts of C and C++. Object-oriented concepts have been implemented more effectively in comparison to C++. The memory management and handling of object references are much simpler, which has the drawback of machine-intimate programming not being possible.

One main advantage of Java is the platform independence. In a first step, the source code is compiled into a platform-independent byte code. This byte code will be interpreted by a Java Virtual Machine (JVM) in a second step. The JVM itself is platform dependent, which means that a JVM has to exist for different target platforms and several operating systems (see Figure 8.2). This portability has the disadvantage of lacking in performance, compared to a compiled program. A conditional corrective is the

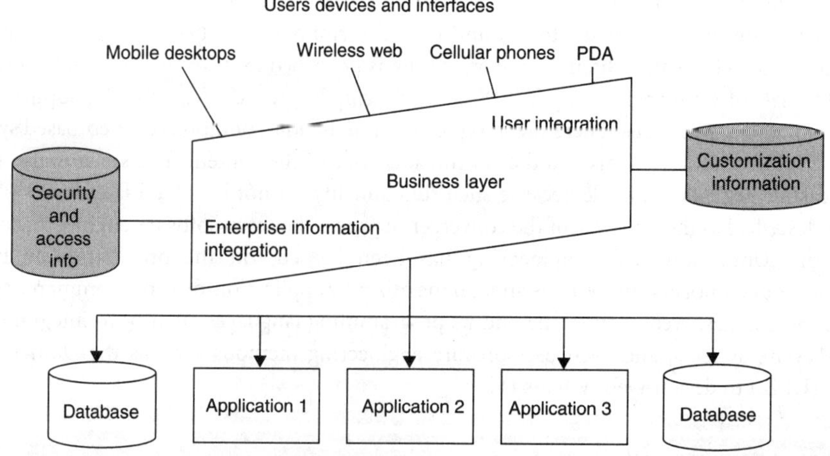

FIGURE 8.1   Enterprise Application Integration Based on [3].

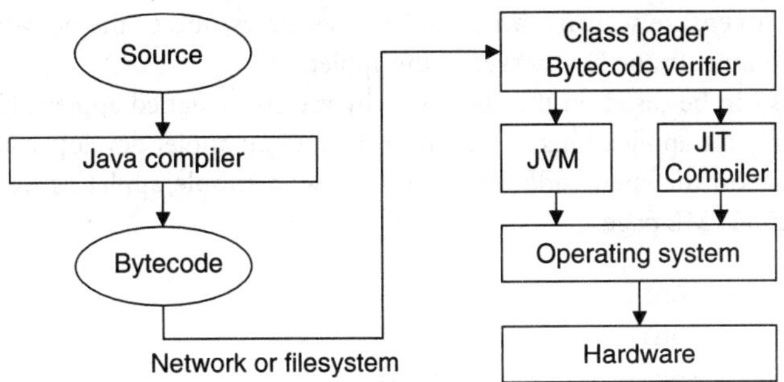

FIGURE 8.2 Design of the Java Compiler.

so-called "Just-in-Time" (JIT) Compiler, which transforms the byte code by the first execution into a platform - and operating system-dependent machine code. During the next processing of this code segment, the machine code can be executed directly by the CPU.

Portable and distributed systems are particularly susceptible to security problems. The security concept of Java is described by using the metaphor of the "Sandbox." The sandbox guarantees that an untrustworthy application cannot gain access to system resources. Therefore, Java implemented a four-level security concept:

- Java itself avoids many problems, for example, no pointers available.
- Verification of byte code during interpretation.
- Dynamic loading of classes is connected with security examinations. Additionally, the runtime environment uses separate name spaces for local classes and classes from the network.
- Access to local resources and network resources is restricted and only possible by means of additional actions. (certification)

The following characteristics of Java — not only as a programming language, but in the sense of the total concept — are to be particularly emphasized:

- Automatic memory management (Garbage Collection).
- Integrated multi-threading allows parallel processing.
- Robustness by explicit declarations, no type casting, and no explicit references (pointers).
- Dynamic loading and binding of classes also over the Internet.

Java offers several options of integration as Internet programming language:

Java-Applications are stand-alone applications without restrictions. In combination with the Java Runtime Environment, they form an operational program.

Java-Applets are special Java programs that can be interpreted directly from the virtual machine of a Web browser. They make Web pages more functional for the client side.

Java-Servlets are Java programs for server-side applications that are using the Common Gateway Interface (CGI). This interface is provided by the Web server and allows to call the Java servlets.

Java Server Pages — JSP are an enhancement of Java servlets with the aim of separating layout information and business logic.

JavaBeans are reusable software components that can be manipulated visually in a builder tool.

Enterprise Java Beans form a framework architecture for component-based and distributed business applications development.

Applets

A Java applet contains classes that can be executed directly by a Web browser. They are embedded in HTML pages and therefore always come with a graphical user interface. Applets are subject to severe restrictions in order to ensure security: access to local resources, and to the network is not permitted. It is possible only to connect to the host server of the applet.

These restrictions can be eased on the client side by means of signed applets. However, this seems to be of importance only for applications in the intranet, since an applet developer has to be in possession of a system structure on the client side. Example 1 shows a simple applet as well as the possibility of embedding it into an HTML page.

```
import java.applet.Applet;
import java.awt.Graphics;
public class HelloWorld extends Applet {
      public void paint(Graphics g) {
            g.drawString("Hello world!", 50, 25);
      }
}
<html>
<head>
<title>A Simple Program</title>
</head>
<body>
      <p>Here is the output of my program:
      <applet code="HelloWorld.class"
            align="baseline"
            width="150"
            height="25">
      </applet>
      </p>
</body>
</html>
```

Example 1. Simple Java Applet — HelloWorld.java

The upper part shows the applet class, which contains a single method to draw the output. Below is the corresponding HTML file to embed the applet by means of the applet tags. Additional parameters to position the applet are provided.

Java Servlets

Java servlets are modules to extend Web servers or application servers that hold appropriate interfaces ready (Java servlet API). They have no user interface and basically serve as alternatives for CGI scripts. While loading, the servlet is initialized, configuration data are loaded, and possible auxiliary threads are started. The servlet is removed from the environment only by the garbage collector; up to then, it operates on the request–response paradigm.

```
public class SimpleServlet extends HttpServlet
{
public  void  doGet  (HttpServletRequest  request,  HttpServletResponse
response) throws ServletException, IOException {
          PrintWriter out;
          String title = "Simple Servlet Output";
          // set content type and other response header fields
```

```
        response.setContentType("text/html");
        // write the data of the response
        out = response.getWriter();
        out.println("<HTML><HEAD><TITLE>" + title );
        out.println("</TITLE></HEAD><BODY>");
        out.println("<H1>" + title + "</H1>");
        out.println("<P>This is output from SimpleServlet.");
        out.println("</BODY></HTML>");
        out.close();
    }
}
```

Example 2. A Simple Java Servlet

As Example 2 shows, the HTML code is directly included in the Java program. Special methods (e.g., doGet) act as entry points to the servlet. The output is written to "standard out" (stdout) and is sent to the client via the Web server.

Java Server Pages — JSP

Java Server Pages (JSP) have developed from Java servlets. They follow the approach of separating the business functionality from the layout information. Normally, the generated layout is pure HTML. Hence, JSP are based on HTML tags and include calls to JavaBeans, which contain the business logic. The JSP are complied into Java servlets with the first call of the page. By means of this approach, it is possible for the Web designer to focus on the layout and for the application developer to focus on the business implementation.

Figure 8.3 shows the JSP architecture. A Web server needs to include an extension to execute the JSP. By the first call, the JSP engine translates the page into a servlet, which will be executed by the servlet engine of the Web server. The HTML output of the servlet will be transferred as an HTTP response to the client.

Example 3 shows the JSP file in extracts in the upper part and below the corresponding Java class. The HTML page consists of wild cards that allow the reference to a Java class.

```
<HTML>
<%@ page language="java" import="Clock" %>
<jsp:useBean id="clock" class="Clock" />
    ...
    Today is the <%= clock.getData() %>
</HTML>
import java.util.*;
public class Clock {
    public String getDate() {
        return new Date();
    }
}
```

Example 3. Simple Java Server Page

8.3   Script Languages

While applications developed with common programming languages have to be compiled before runtime, programs developed with script languages, called scripts, are interpreted at runtime by appropriate

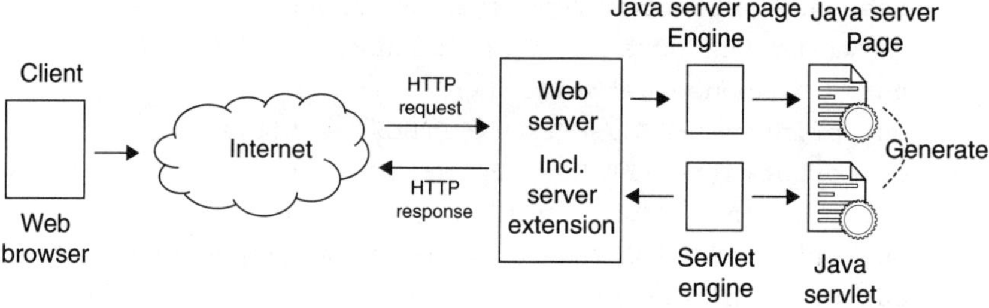

FIGURE 8.3   JSP Architecture.

interpreters. The script itself can contain several lines of commands like a common batch file and may be processed automatically by the interpreter.

Nowadays, many scripting languages are available, and can be divided into server-side scripting (Perl, PHP, etc.) and client-side scripting languages (e.g., JavaScript), depending on where the interpreter processes the script.

In general, scripts can be used to create dynamic Web pages and interactive applications on the World Wide Web. Thereby, scripts may be embedded into HTML pages or may serve as stand-alone Web applications and mostly obtain data from databases or legacy systems via well-defined interfaces. Server-side scripting can be used in the areas of information systems, e-commerce, and for monitoring as well as for loosely coupled steering purposes in the automation systems domain.

Common Gateway Interface — CGI

The Common Gateway Interface (CGI) itself is not an Internet programming language. Referring to [11,12] it is a standard for external gateway programs to interface with information servers such as Hyper Text Transfer Protocol (HTTP) servers. HTTP [6] is used as the protocol between Web clients and Web servers. The disadvantage of the HTTP protocol is its stateless character, which lets the connection between client and server become lost after the connection is closed. Therefore, a separate session handling has to be implemented to overcome these deficiencies. A CGI program, on the other hand, is executed at runtime, and can therefore generate dynamic information. An HTTP server is often used as a gateway to legacy information systems, for example, an existing software system or an existing database application. The Common Gateway Interface is an agreement between HTTP server developers on how to integrate such gateway scripts and programs. It is typically used in conjunction with HTML forms to build dynamic database-based Web applications. The CGI specification is currently maintained by the NCSA Software Development Group [12] and is available in version 1.1, or CGI/1.1. Further revisions of this protocol are guaranteed to be backward compatible.

Figure 8.4 shows the basic architecture of CGI to enhance an existing Web server in order to create a dynamic Web content. The CGI interpreter can vary from a simple shell to more powerful and comfortable interpreters like PHP or Perl, and is chosen in the first lines of the CGI script. Server and interpreter, respective the CGI script, can communicate in four major ways: environment variables, command line, standard input, or standard output.

There are two HTTP methods that can be used to access a certain CGI service: the GET and POST method. Depending on which method is used, the script will receive encoded data of HTML forms from clients:

- *GET method*: The CGI script receives the encoded data in an environment variable.
- *POST method*: The CGI script receives the encoded data directly over "stdin," and can calculate its length.

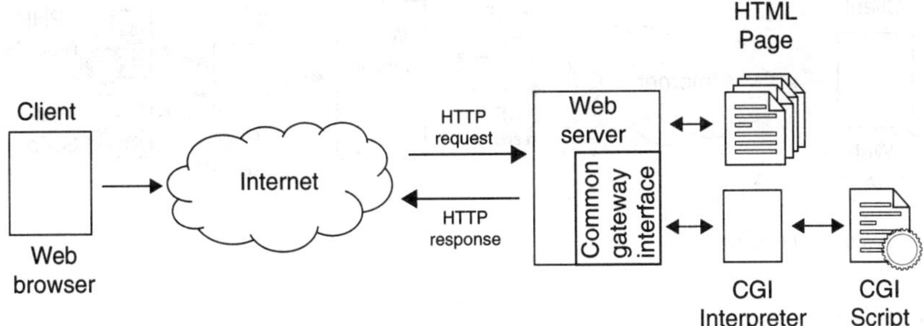

FIGURE 8.4   Web server using Common Gateway Interface.

The results of the CGI script are written to standard output and redirected by the Web server to the inquiring client.

Perl

As mentioned before, the CGI mechanism allows the use of miscellaneous interpreters for parsing and executing scripts. One of these interpreters is Perl, an acronym for "Practical Extraction and Report Language." The corresponding scripts are named Perl-scripts. While Perl was initially developed for UNIX platforms, there are nowadays various interpreters for platforms like Linux, Unix, VMS, OS/2, Macintosh, and all Windows variants in existence. Hence, Perl scripts can be used on all of these systems without making any changes to the scripts.

Perl created, written, developed, and maintained by Larry Wall was initially a language for processing text. In the current version of Perl, features like sophisticated pattern matching capabilities, straightforward I/O, and flexible syntax are included. In fact, by borrowing heavily from C, sed, awk, and the Unix shells, Perl has become the language of choice for many I/O, file processing and management, process management, database administration, client/server programming, and system administration tasks for Web-based information systems [13].

Perl is an ideal choice for small solutions. Example 4 shows a script that sends an email and writes an HTML message to standard output if the script is executed directly in a shell by typing "perl SendMail.pl" on the command line. If the script is called by a Web server, the HTML output is delivered to the Web client. The first line of the script is a comment line and is used to denote the script as Perl script and to define the interpreter application (perl) that should process the script.

Perl is available for free. One of the numerous sites where Perl packages can be downloaded is the Comprehensive Perl Archive Network (CPAN) [14].

PHP

PHP, a project of the Apache Software Foundation, is a widely used general-purpose scripting language especially suited for Web development and can be embedded into HTML documents. This, of course, is a major difference from Perl. Various Web servers can use PHP in the form of the CGI mechanism or by extending the Web server with a PHP module if performance optimization is important. Figure 8.5 shows both configurations.

Most of the following descriptions can be found in [15]. PHP can be used on all major operating systems, including Linux, many Unix variants (including HP-UX, Solaris, and OpenBSD), Microsoft Windows, Mac OS X, RISC OS, and probably others as well. PHP also supports today's most popular Web servers like Apache, Microsoft Internet Information Server, Netscape, and iPlanet. For the majority of these servers, PHP can be integrated as a module or as a normal CGI processor.

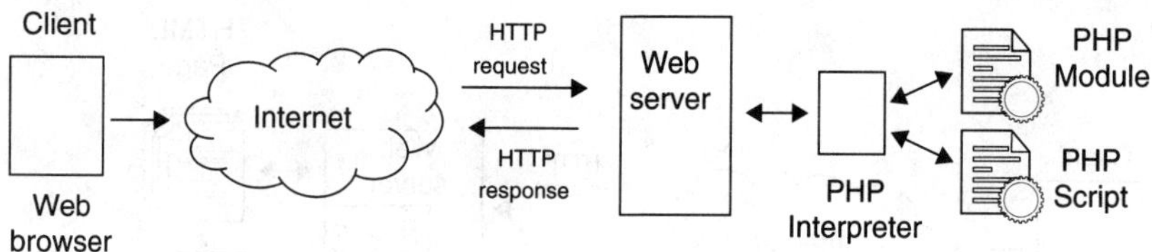

FIGURE 8.5   Web server using PHP.

```perl
#!/usr/bin/perl
use CGI; $cgi=new CGI(); print $cgi->header;
$from=$cgi->param('Sender'); $mailtext=$cgi->param('Mail text'); $sub-
ject=$cgi->param('Subject'); $to='john@doe.com';
send($to, $from, $mailtext, $subject);
print "<HTML><FONT Face=Arial><B>Your Mail was sent.</B><BR>";
print "further text ...<BR></HTML>";
sub send {
my ($to, $from, $mailtext, $subject) = @_;
use Net::SMTP;
my $smtp = new Net::SMTP("smtp.somewhere.com");
$smtp->mail();
$smtp->to($to);                                        #Addressee
$smtp->data();
$smtp->datasend("From: $from\n");                      #Sender
$smtp->datasend("Subject: $subj\n");                   #Subject
$smtp->datasend("\n");                                 #Close heading
$smtp->datasend($mailtext);                            #E-Mail text
$smtp->dataend();
$smtp->quit();
}
```

Example 4. Simple Perl Script — SendMail.pl

PHP is mainly focused on creating dynamic Web pages, also known as server-side scripting. The main fields of operation are: collecting form data, generating dynamic page content, or sending and receiving cookies. There are three main fields where PHP scripts are used:

- *Server-side scripting*: This is the common use for PHP. There have to be three parts to activate PHP scripts: the PHP parser (server module), a Web server, and a Web browser. The browser accesses the script via the Web server and receives the output.
- *Command-line scripting*: In this field, a PHP script can run without any server or browser. Only the PHP parser is needed. This type of usage is suitable for scripts regularly executed automatically, for example, text processing.
- *Writing client-side GUI applications*: A not very widespread approach with PHP is the implementation of window applications. By including the PHP-GTK (Graphical Toolkit) package, PHP allows to include some advanced features even for cross-platform programming.

One of the advantages of PHP is its support for a wide range of databases, for example, Oracle, dBase, mySQL, Microsoft SQL, Sybase, and all ODBC databases. PHP allows to develop database-based Web application at a high speed.

PHP can also communicate with other services using protocols such as Lightweight Directory Access Protocol (LDAP), E-mail protocols, and many others. It is even possible to open raw network sockets and interact using any other protocol.

PHP has support for the WDDX complex data exchange between virtually all Web programming languages. With respect to interconnection, PHP supports instantaneous access of Java objects and use them transparently as PHP objects, in addition to using a CORBA extension to access remote objects. PHP has extremely useful text processing features, for example, for parsing XML documents. It supports the Simple API for XML (SAX) and Document Object Model (DOM) standards.

Example 5 shows a PHP script sending a query to a database and returning the results in the form of an HTML table. The PHP code is embedded in the HTML pages and is processed on the server after calling. It will be replaced by the result of the database query. Thus, the client receives a pure HTML page without PHP code. As the example shows, PHP allows easy integration of a database and thus provides a flexible solution for Web-based database applications.

The file will be normally invoked via a common Web browser.

```
<HTML>
<HEAD>

      ...

</HEAD>
<BODY>
<H1>List of Persons orderd by age:
<TABLE> <TR><TD><b>Lastname</TD><TD><b>Firstname</TD><TD><b>Age</TD><TR>
<?php
// open a connection to a MYSQL database
$link=mysql_connect("localhost","Username","Password");
          or die("Could not connect:" . mysql_error());
// select database
mysql_select_db("DB-Name",$link);
// query persons (ordered by age) from database
$query="select firstname,lastname,age from person order by age,last-
name,firstname asc";
$result = mysql_query($query,$link);
$num = mysql_numrows($result);
// print table of persons
for ($i=0; $i<$num; $i++){
     $row=mysql_fetch_array($result);
     echo
"<TR><TD>$row[lastname]</TD><TD>$row[firstname]</TD><TD>>$row[age]</TD></
TR>;
}
mysql_close($link);
?>
</TABLE>
</BODY>
</HTML>
```

Example 5. Simple PHP Script — Database-Access.php

JavaScript

JavaScript is a client-side script language, which was introduced by Netscape in 1995. It is independent from HTML and has been defined to support HTML developers to optimize their Web pages. JavaScript has nothing in common with the programming language Java.

JavaScript can either be embedded in HTML pages or in a separate file and can only be referenced. The JavaScript code is interpreted during runtime by means of an interpreter software included in the Web browser. It is executed in a "Sandbox" to restrict the possibilities and to ensure security on the client side. It is, for example, not possible to access files out of a JavaScript.

JavaScript does not replace server-side programs, for example, cgi-based programs. It can be used for small calculations, input checks, etc. in HTML files. Example 6 shows a Java Script for the calculation of a square. The function is embedded in the HTML page and executed on the client side after the button "Calculate square" is pressed.

```html
<html>
      <head><title>Simple JavaScript</title>
      <script type="text/javascript">
          <!--
          function square() {
        var Result = document.Form.Input.value *
        document.Form.Input.value;
        alert("The Square of " + document.Form.Input.value + " = " +
        Result);
          }
          //-->
      </script>
      </head>
      <body>
          <form name="Form" action="">
              <input type="text" name="Input" size="3">
            <input type="button" value="Calculate square"
            onClick="square()">
          </form>
      </body>
</html>
```

Example 6. Java Script Example

8.4   Web Services

A new trend throughout the Internet is brought about by Web services. Today's Web is designed for human interaction and is a very successful system in domains like information sharing, distributed content library, Business to Customer (B2C) e-commerce, and nonautomated business-to-business (B2B) interactions. The reasons are the few standards (HTTP and HTML) and a shallow interaction model (only few assumptions made about computing platforms). The result is the ubiquity of the Web where existing applications in various internet programming languages work together.

Open, automated B2B e-commerce, business process integration, resource sharing, or distributed computing are some examples. While an *ad hoc* solution could be used to realize application-to-application (A2A) interaction with HTML forms on top of existing standards, the goal for the future is to enable systematic A2A interaction on the Web.

"Web Services" is an effort to build such a distributed computing platform. According to Gartner Group Inc., Web services will capture substantial attention in 2002, and by 2004 Web services will dominate deployment of new application solutions. According to the W3C, the definition of Web services is [7]:

"A Web service is a software system identified by a URI, whose public interfaces and bindings are defined and described using XML. Its definition can be discovered by other software systems. These systems may then interact with the Web service in a manner prescribed by its definition, using XML based messages conveyed by internet protocols."

A rough outline is shown in Figure 8.6 and in more detail in Figure 8.7.

The goal is to enable universal interoperability, widespread adoption, and ubiquity by enabling an Internetscaled dynamic binding, efficiently supporting both open (Web) and constrained environments, using the following standards to build a Web service:

- Simple Object Access Protocol (SOAP) [16].
- Web Service Description Language (WSDL) [17].
- Universal Description, Discovery, and Integration (UDDI) [18].

The Web service's architecture can be divided into three parts: Service Provider, Service Registry, and Service Customer. The Service Provider publishes services via the Service Registry, and the Service Customer (a person or a program) searches this registry for an appropriate service (Figure 8.8). Web service applications can be seen as encapsulated, loosely coupled Web "components" that can be dynamically connected.

The Web service technology stack, shown in Figure 8.8, is based on the TCP/IP stack and supports, in addition to the network and transport layers, a packaging, a description, and a discovery layer. The discovery layer provides a mechanism to fetch the description of the services (UDDI) provided, the

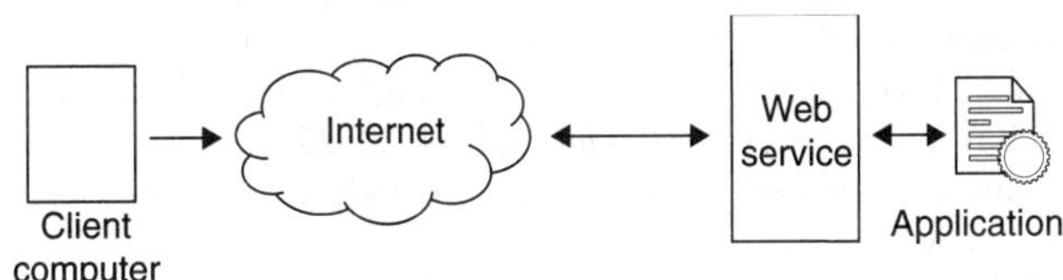

FIGURE 8.6   Web service outline.

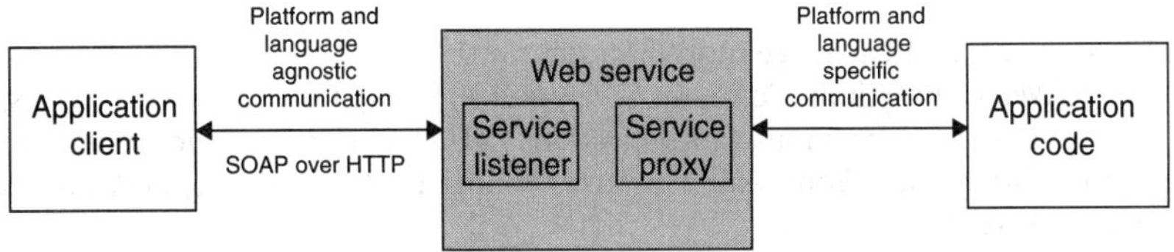

FIGURE 8.7   Web service in more detail.

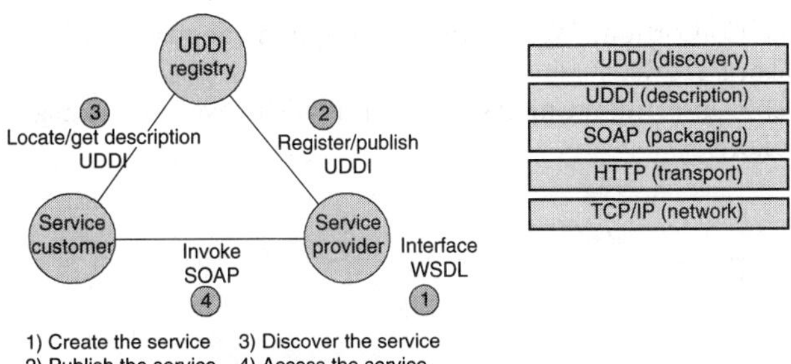

FIGURE 8.8   Web Service Architecture and Technology Stack.

description layer provides a description of services for the customer (WSDL), and the packaging layer provides packaging of the data to serialize it (HTML, SOAP). This layering system helps to guarantee platform neutrality.

Finally, some notes on the technologies involved:

- The SOAP is a lightweight and simple XML-based protocol to allow the exchange of structured and typed information across the Web. It is a standardized Packaging Protocol in the form of an XML application; therefore, XML schemas and XML namespaces can be applied.
- The WSDL is a powerful means for service developers to describe their services. A WSDL description can be used to describe which type of message can be exchanged, how messages are related, and how SOAP messages are exchanged.
- The UDDI standard was designed for building repository-based registry services for the lookup of Web services. It defines how to publish a business information.

The Web Service approach allows either new applications or existing applications (legacy applications) to be accessed via the Internet based on standard protocols. XML is the base technology for the description of the Web services and for the data exchange protocols.

References

[1]   Goeschka, K. and Schranz, M., Client and Legacy Integration in Object Oriented Web Engineering, *IEEE Multimedia.* (ISSN 1070 986X), 8, 32–41, 2001.

[2]   Ginige, A. and Murugesan, S., Web Engineering. An Introduction, *IEEE Multimedia*, 8, 14–18, 2001.

[3]   Moyer, S. and Umar, A., The Impact of Network Convergence on Telecommunication Software, *IEEE Communication Magazine*, 39, 78–84, 2001.

[4]   Ginige, Athula and Murugesan, San, Web Engineering: A Methodology for Developing Scalable, Maintainable Web Applications, *Cutter ITJournal*, 14, 22–25, 2001

[5]   Manninger, M., Goeschka, K., Schwaiger, C., and Dietrich, D., *E-Commerce - Die Technik*, Huethig Verlag, Heidelberg, 2001.

[6]   World Wide Web Consortium, Hypertext Transfer Protocol – HTTP/1.1, RFC 2616

[7]   http://www.w3.org/tr/ws-arch/, Homepage of the Web Service Technology.

[8]   World Wide Web Consortium, eXtensible Markup Language, http://www.w3c.org/XML/

[9]   World Wide Web Consortium, HyperText Markup Language (HTML), http://www.w3.org/MarkUp/

[10] Sun Microsystems, Java Language http://java.sun.com/

[11] World Wide Web Consortium, Common Gateway Interface (CGI), http://www.w3.org/CGI/

[12] National Center for Supercomputing Applications (NCSA), http://www.ncsa.uiuc.edu/

[13] Wall, Larry, Christiansen, Tom, and Schwartz, Randal L., *Programming Perl*, 2nd ed., O'Reilly, Sebastapol, CA, 1996.

[14] Comprehensive Perl Archive Network, Perl Language, http://www.perl.com/CPAN/

[15] The Homepage for PHP, PHP Language, http://www.php.net/

[16] World Wide Web Consortium, Simple Object Access Protocol (SOAP), http://www.w3c.org/TR/SOAP/

[17] World Wide Web Consortium, Web Services Description Language (WSDL), http://www.w3c.org/TR/wsdl/

[18] World Wide Web Consortium, Universal Description, Discovery and Integration (UDDI), http://www.uddi.org/

9

Internet Still Image and Video Formats

Guido Heising
University of Applied Sciences TFH Berlin

Kai Uwe Barthel
University of Applied Sciences for Technology and Business Berlin

Nowadays, it is hard to find a website without images. Even buttons and layout elements of a website consist of small bitmaps. The use of images in the internet has become possible as there are sophisticated compression algorithms to keep the sizes of the images as small as possible. Video clips are still rarely found on typical websites; but the usage of video is increasing. However — opposed to still images — up to now, there is no commonly accepted video coding standard that has been incorporated into web-browsing applications. In addition, the bandwidth that is required for the transmission of high-quality videos—larger than the size of a matchbox — is still too high for most users having access to the internet with moderate bandwidth only (e.g., a modem).

This chapter introduces some basic facts about images and videos. It explains the essential ideas of image and video compression techniques. Standards and quasistandards are described, and their features will be discussed.

9.1 Digital Images

When images like photos or graphics are digitized (e.g., using a scanner), the color and the brightness information of the original image is converted into discrete digital picture elements — the *pixels*. There are two parameters — the *color depth* and the *image resolution* — that control how the analog image information is converted into the digital image data (Figure 9.1).

Due to the fact that our eyes use three kinds of cells for color perception, all colors can be described using three components or channels. Thus, the information that has to be stored for each pixel consists of three values to describe the color components R, G, and B (red, green, and blue). There is a standard sRGB color space that is used by many scanners and other devices. If a different color space is to be used, this information can be described in an *ICC-profile* from the *international color consortium*. This color profile should be joined to the image data. For *grayscale images* (like black and white photos), only one channel for the brightness information is needed. The number of bits that are assigned to a pixel is called the *color depth* and is measured in *bits per pixel* or *bits per channel*. Most computer graphic adapters use a display mode called *True Color*, which uses 8 bits per channel, which corresponds to 24 bits per pixel.

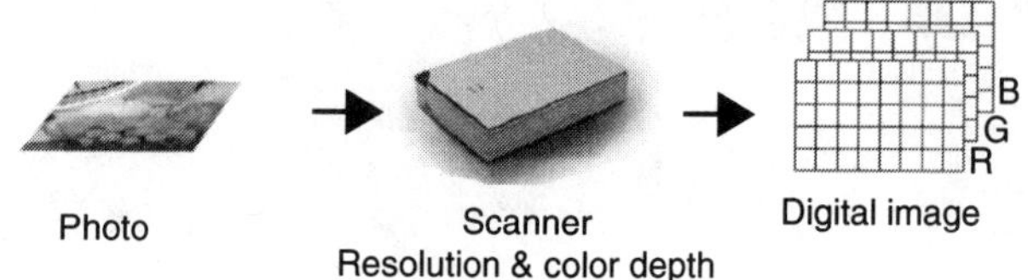

FIGURE 9.1 The resolution and the color depth control the quality and the size of the digitized image.

In this mode, 16.7 million different colors can be described or displayed, which is more than sufficient for typical digital images to be shown on computer monitors. In the prepress domain, four-channel images, Cyan, Magenta, Yellow, and black (CMYK), are used for printing purposes. Other image types like medical images might need a higher color depth such as 12 or 16 bits per channel. A simple fax image uses only 1 bit per pixel to distinguish between black and white. If an image has only a limited number of different colors (up to 256 colors), the RGB-values colors are stored in a lookup table (LUT). In this *palletized image*, every pixel color can be described by an index referring the color from the LUT entry.

By selecting a *resolution*, the scanning device is instructed on how many samples (pixels) are to be generated. The unit for resolution is *dpi*, which stands for *dots per inch*. Typical values for scanning resolutions are in the range between 50 and 1200 dpi. If the resolution is chosen to be a high value, then the quality will be very good as even the finest details of the image can be captured. However, the amount of digital data generated will also be quite high. The pixel number of the image is given by the product of length and width — both multiplied by the dpi number. To obtain the total number of bits for this image, the pixel number has to be multiplied by the color depth.

Websites are viewed by many different users on different computer screens that have dissimilar resolutions. The actual resolution of a monitor depends on the number of pixels that can be displayed and the dimensions of the screen (measured in cm or inch). Typical computer systems have a monitor resolution of approximately 100 dpi. This should be taken into account if images are to be displayed on a website. There is no use displaying an image having more pixels than the area on the screen in which it is displayed. This may seem obvious; however, many web pages can be found that use very large images. Browsers can scale images to any size; however, they do not use sophisticated scaling or interpolation techniques, so that the best visual results are obtained if every pixel will be used only once nonscaled.

Need for Compression

A typical photo (10 cm × 15 cm) scanned at a resolution of 100 dpi in true color (24 bits per pixel) will result in an uncompressed file size of 680 kB (393 rows × 590 columns = 231,870 pixels × 3 Bytes). Using the fastest available modem with a bandwidth of 56,000 bits/sec, this image would need 100 sec to be transmitted. If we were talking about a video clip of this size, even with a reduced frame rate of 10 fps (frames per second), a compression factor of 1000 would be needed for the 56k modem. Typical videos have frame rates of 25 (PAL) or 29.97 (NTSC); the dimension of digitized video in pixels is 720 × 576 pixels for PAL DV, and 720 × 480 for NTSC DV.

These examples show how important it is to use highly efficient image compression techniques. When talking about compression, three different modes should be distinguished: *lossless compression* does a perfect preservation of the image, and every pixel is unchanged in every bit. For typical images (photos), even the best algorithms are not able to achieve a compression factor higher than two to three. "Visual lossless" compression is a term that was introduced by the industry trying to tell their customers that their compression technique is almost perfect. However, there is no exact definition for visual lossless; usually, it means that encoding errors are not noticeable to a typical viewing person. Whether encoding errors may be perceived depends very much on the viewing conditions and how much the viewer is trained. Typical compression factors range between 3 and 10. Then, there is *lossy compression*, mostly used in the Internet due to the need for very small file sizes. A lossy compressed image may look different compared to the original; however, the quality is still acceptable. Needless to say that quality requirements are different in different circumstances. Typical compression factors are in the range from 10 to 100.

Once a compression standard has been established, new improved ideas or algorithms cannot be incorporated. However, as only the decoding part is standardized, there is always room for optimizations on the coder side, as long as the decoder can still decode the compressed data. After a standard has been settled, it usually takes some years until optimal coding settings are determined. This is particularly true for video compression, as there are too many different parameter settings possible. This also means that there are JPEGs and MPEGs of similar size but of different quality. It actually depends on the capability/quality of the coder used, and how well an image or video is compressed.

Image Format Features

If images are to be used in the internet, several aspects have to be considered: there are hundreds of image file formats, compressed and uncompressed; however, standard browsers only support three of them. These supported image types are *Compuserve GIF* [1], *Portable Network Graphics* (PNG) [2], and the *JPEG-standard* from the *Joint Picture Expert Group* [3]. Often, image file formats provide different encoding modes such as support for different color spaces, the choice between lossy and lossless compression, and different transparency modes. However, not all modes of these image file formats are supported by all web browsers. For example, the original JPEG standard knew more than 20 different encoding modes, but only two of them are supported by standard browsers and common image viewing applications. For example, there is a lossless JPEG compressing mode, which cannot be used in the Internet. Image file formats other than the three mentioned above can only be used if appropriate plugins are installed or a special applet image viewer is used.

However, none of the three image formats — GIF, PNG, or JPEG — is suitable for all purposes. They differ in the type of color spaces supported and other modes. Table 9.1 compares the main features of these image file formats.

TABLE 9.1　Comparison of Features of Internet Image File Formats

	Image format		
	GIF	JPEG	PNG 8/24
Color mode			
Bitonal	■		■
Palletized	■		■
RGB		■	■
Lab			
CMYK		□	
Features			
Progressive transmission	■□	■□	■□
Lossless compression	■		
Lossy compression		■	
High compression efficiency		■*)	
Meta data		□	
Animation	■		
Transparency/alpha**)	■		■□
ICC support		□	□
Random access in huge images			

■=supported by image format & web browser.
□=supported by image format, but not supported by some web browsers.
■□=supported by image format, but problems with some web browsers
*) Today JPEG is outperformed by newer & better algorithms like JPEG2000, which has no browser support yet.
**) GIF images allow one out of the possible 256 colors to be fully transparent. PNG has a much higher level of transparency support.

One important distinction is the *support for different color modes. Bitonal images* or palletized images with a reduced number of colors are only supported by GIF and PNG. These images could, however, be converted into RGB images and then stored as JPEG, but this would make these images larger than necessary. JPEG should be used for natural photographic images. The professional device-independent *Lab color space* is not supported by any of the three formats. Four-channel CMYK images can be stored as JPEG, but these images should not be used in the Internet, as browsers cannot display them.

Other distinctions between features are the available *compression mode* and the *support for additional (meta) data* or *special transmission* abilities. JPEG always performs a lossy compression, the desired quality can be controlled by a quality factor. GIF and PNG use lossless compression only. However, some tools can modify the image data to ease subsequent lossless compression.

Compression efficiency has been permanently improved over the last few years. JPEG does a very good compression job; but naturally there are far better schemes by now.

All three formats have support for *progressive* (sometimes called interlaced) *transmission* of the image data, such that a reduced resolution version of the image may be seen before the entire data have been transmitted. Usually, the file size increases by a few percent when using this progressive option. However, some browsers ignore this feature and only display the image if all data have been transmitted.

Meta data like the author, copyright information, etc. become more and more important. Only JPEG supports the *IPTC standard for metadata*; PNG allows some annotations and comments. Usually, browsers cannot display this extra information.

Transparency or an alpha channel describes whether the pixels of an image are fully or partially transparent. GIF images allow one out of the possible 256 colors to be fully transparent. PNG has a much higher level of transparency support, but has very reduced browser support. Transparency support is needed if the background of a website is changed very often. In order to avoid transparency information, the transparent image can be overlaid on the background image/color. The resulting merged image can then be stored with no transparency even using JPEG.

JPEG and PNG have limited *ICC-profile* support, which is ignored by most browsers.

For huge images like maps or aerial views, a *random access* to small fractions of an image would be very desirable. However, none of the described formats has support for this. Usually, many different small images are held on the server, which are then dynamically selected.

The new compression standard *JPEG2000* [4] has support for most of the above-mentioned features and it also outperforms these formats in coding efficiency. To date, there is no browser support for this new powerful standard.

The choice of a particular image file format should be based on the following aspects:

- *For photographic images,* JPEG is the right choice, if a lossy representation can be tolerated. If not, PNG (24 bit mode) should be used for lossless transmission.
- *Nonnatural images* with only a few colors should be compressed to GIF or PNG (8-bit mode). It should be noted that the PNG format does not have any patent restrictions, as opposed to GIF.
- *Animated images* have to be saved as GIF. There is an animated version of PNG called MNG, however, without support from any browser.
- *Images with transparency* need to be stored as GIF or PNG, while PNG has a much higher level of transparency support; but unfortunately is not supported by all browsers. (At *http://entropy-mine.com/jason/testbed/pngtrans/*, a PNG transparency test web page can be found.)

As mentioned before, there are several coding tools that can differ significantly in encoding quality. PNG allows different filter settings that affect the compressed file size. Optimal compression depends on a suitable choice of the filter type for PNG and the usage of a high-quality encoding tool in the case of a JPEG image.

Image Compression Principle

The core idea of image compression is to remove any *redundancy* or correlation present within the image. Redundancy is all the data, that are not important to the information. Consequently, it can be removed by the coder, whereas at the decoder these data are reconstructed and added back to the image.

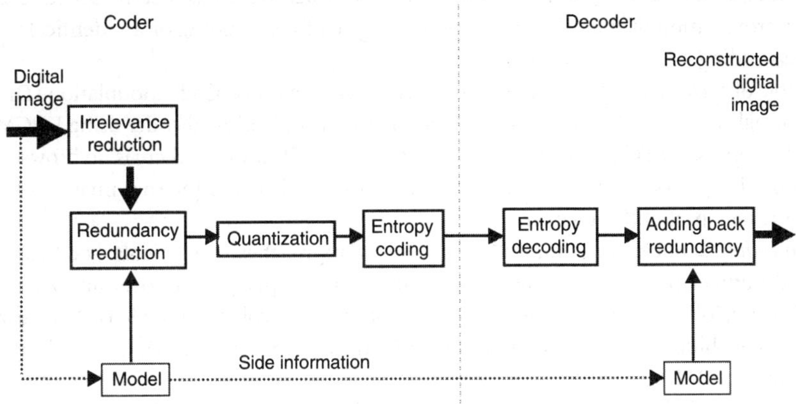

FIGURE 9.2   Block diagram of an (image) compression scheme.

The basic idea of image compression is shown in Figure 9.2. In the first step, all *irrelevant information* is discarded. This might be a very high color depth, which is not needed or a very large image, which is scaled down first before it is compressed and transmitted.

The next step is to remove the redundancy of the image data. The efficiency of a coder depends on how good this redundancy is discovered and removed. Actually, a *model* describes the idea as to how the redundancy can be exploited.

Models have become more and more elaborate over time. A very simple model could use the fact that most pixel values do not change abruptly. The value of the previously transmitted pixel can be used as a guess (a *prediction*) for the next pixel value. Now only the prediction error has to be transmitted or stored. If it is possible to store this prediction error more effectively than the original pixel data (with fewer bits), then there is compression. A more sophisticated model might use a better adaptive predictor taking into account more surrounding pixels. The model for JPEG compression uses the fact that it is more effective to describe an image block through a superposition of suitable basic blocks instead of transmitting it pixel by pixel.

If this model (how to remove the redundancy) is known both to the coder and the decoder, there is no need to transmit the model data. However, if the image data are analyzed first to adapt the model to the particular image, then some extra data (the *side information*) have to be transmitted to the decoder as well in order to guarantee that the same model is used at the coder and decoder sides.

After the redundancy reduction in the next step, the remaining data can be further simplified by using a *quantization* process, which basically performs some sort of rounding. If a quantization is performed, this is the point where the compression gets lossy, since the decoder is not able to reverse the rounding. This is the main parameter for controlling the compression factor.

As an example, let us assume that the pixel values of an 8-bit gray channel are to be coded. Using a quantizer consisting of $2^6=64$ equal distributed values over the input range of $2^8=256$ gray values, the input values can be rounded to the closest value of {2,6,10,…,254}. If only $2^4=16$ output values are allowed, the input values can be rounded to {8,24,40,…,248}. In the first case, 6 bits/pixel are necessary, which results in a compression factor of 8/6=1,33, whereas only 4 bits/pixel are used in the latter case leading to a compression factor of 8/4=2. Needless to say that the losses, the *quantization errors*, introduced in the latter case are larger and the reconstruction quality will be worse.

Finally, the quantized data have to be transmitted or stored in an efficient way: *entropy encoding* tries to assign short codes (bit patterns) to frequent symbols, whereas rare symbols do get longer codes (more bits). Thus, by employing a *variable length code* (VLC), the average code word length will be reduced compared to using a *fixed length code* (FLC) as assumed in the above example.

At the decoder, all the steps are performed in the reverse order to reconstruct the decoded image, which leads to an approximation of the original image in the case of lossy coding or an identical version in the case of lossless coding.

The principle of *predictive coding* — also known as Differential Pulse Code Modulation (**DPCM**) — has already been roughly described in the previous section. A very sophisticated coder using DPCM is the relatively new JPEG lossless (*JPEG-LS*) standard also known as *LOCO*; however, there is no browser support for this. Some other image coding approaches, especially those used by the Internet image formats, shall be explained in the next part of this chapter.

Run length coding is one of the simplest image coding approaches — particularly efficient for images with a limited number of colors. Instead of repeating identical pixel values over and over, the value is reported only once, followed by the quantity of pixels of the same color (Figure 9.3). As an example, "red, red, red, blue, blue, blue, blue, ..." could be transmitted as "(red, 3); (blue,4); ...." This approach is in part in the *Microsoft* **BMP** file format.

Lempel-Ziv-Welch (LZW) is the patented coding algorithm used for the GIF format. It is similar to the techniques used for ZIP or COMPRESS. LZW is a sophisticated lossless compression method, which analyzes the data and searches for repeating patterns within the (image) data. A library is built up during the encoding process. If a new pattern occurs in the data, this pattern will be added to the library at the coder and decoder as well. If the same pattern occurs again, then the library index can be used to describe this pattern. If LZW sees "012012," it is a clever-enough algorithm to spot the trend of alternating symbols and replace each instance with a shorter code for the index, thereby compressing the information.

GIF compression does a line scan of the image data, thus only exploiting the horizontal redundancy. PNG compression can use prediction techniques (called compression filters) to exploit vertical redundancy as well. In addition, it uses a patent-free version for the actual compression.

JPEG coding splits the image into blocks of 8×8 pixels. Every block is then transformed using the *discrete cosine transform* (DCT), which results in energy compaction within the block. This approach corresponds to the idea that instead of describing the block on a pixel-by-pixel basis (Figure 9.4, left), it is more efficient to perform a weighted overlay of only a few well-chosen basis block patterns, which will generate an identical or similar block. The fixed set of basis blocks consists of 64 different cosine-shaped patterns of varying frequency, which is known at the coder and decoder sides. Thus, only the weights (DCT coefficients) have to be coded. Any quality can be achieved from a highly compressed coarse approximation up to a visual lossless representation by selecting the precision (quantizer) of the weights of these blocks. In the example shown in Figure 9.4 on the right, a high precision is chosen, which leads to an almost lossless representation of the small block in the top right position.

The lower the contrast of a basis block, shown in the example, the closer the weight to zero and the lower the importance of that respective basis block. Therefore, one can further reduce the bit rate by coding only the n largest weights of an image block out of 64. The quality of the reconstructed block often is still quite high, as can be derived from Figure 9.5 for $n=16$. There is one exception from the above-stated rule as there is one basis block that does not show a contrast but in most cases does have a large weight. This is the left-most block in Figure 9.5. It represents the mean or DC value of the image block [5].

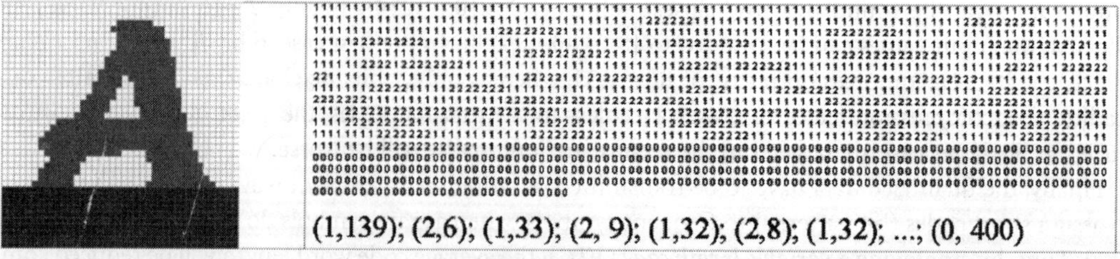

FIGURE 9.3   Run length coding; indices are grouped and described as value followed by the number of occurrences.

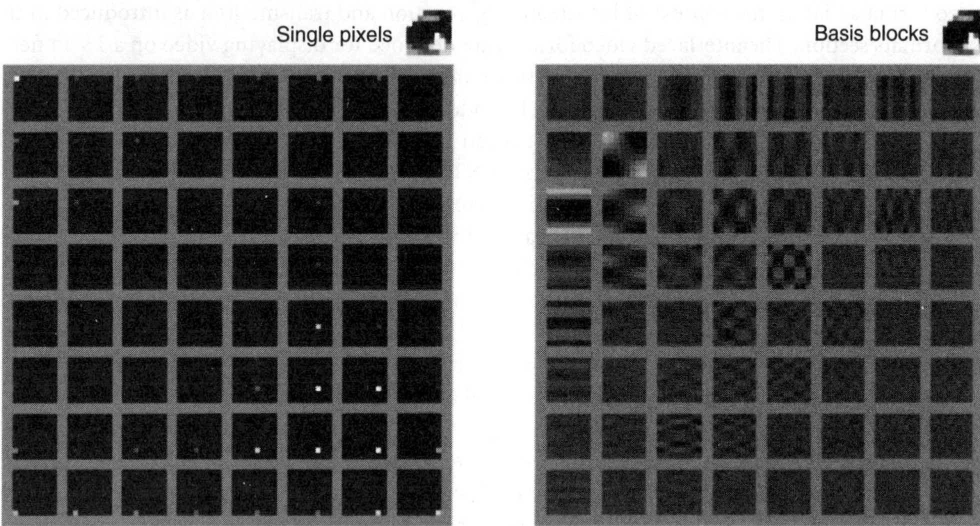

FIGURE 9.4 JPEG; instead of a description by pixels (left) image blocks are approximate by overlaid basis blocks.

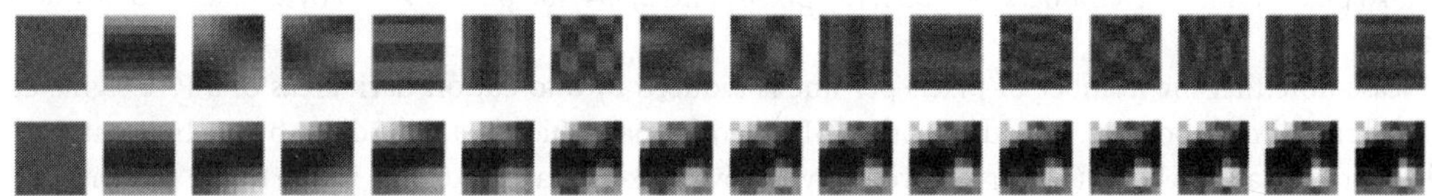

FIGURE 9.5 JPEG; The successive overlay of basis blocks (shown in the upper line) can produce better and better approximations of any image block (shown below).

The DCT approach is very efficient as far as coding of natural images is concerned as in most cases the spatial variation of the image signal is rather smooth. Thus, only a small number of the basis blocks will be required for the approximation, most probably those blocks in the top left corner of Figure 9.4, which represent lower frequencies.

One major disadvantage of the block-based DCT scheme should be mentioned as it is often recognized in decoded images, that is, the so-called *blocking artifacts*. These annoying artificial block structures are related to the choice of a high compression ratio, which led to a poor approximation of the image blocks. This is due to a strong rounding or the omission of DCT coefficients. In JPEG2000, overlapping basis blocks are used to reduce the blocking artifacts. The DCT blocks of size 8×8 are replaced by overlapping blocks of different sizes based on a *wavelet transform*, leading to visually improved decoded images without any blocking artifacts.

9.2 Video Formats

Accessing video via the Internet will become more and more popular as the bandwidths of the Internet connections will increase and the means for acquisition and processing of video data are already highly available on the consumer market. Camcorders, web cams, and graphics cards with a TV tuner or video-in adapter are the devices most commonly used by consumers for bringing video into the computer. The camcorder can be connected to the computer via FireWire (IEEE 1394) for transferring the video and associated audio data. The very popular *DV format* is often used in digital camcorders. Its compression scheme is based on JPEG but with some modifications to better adapt to video signals.

At this stage, one problem may arise, which is associated to *interlaced video formats*, like PAL or NTSC [6]. Please note that the terms *interlaced* and *progressive* have a different meaning in the context of video formats than in the context of bit stream organization and transmission as introduced in the image formats section. The interlaced video formats are designed for displaying video on a TV in fields instead of full frames. There are two alternating fields: *odd* and *even fields*. An odd (even) field consists of all odd (even) numbered rows of a full frame; but odd and even fields are captured and displayed at different time instances. The advantage is that instead of having a full frame rate of 25 Hz for PAL (29.97 Hz for NTSC), a 50 Hz field rate (59.94 Hz for NTSC) is achieved without increasing the bit rate or bandwidth. Thereby, flickering is reduced. Since computer monitors are designed to display *progressive video formats*, that is, full frames, displaying interlaced video can introduce severe artifacts. This is especially the case if two neighboring fields are combined by simply copying the respective lines to a full frame in the presence of motion as the position of a moving object will not coincide in the combined odd and even fields.

One simple way to overcome this problem is to perform an *interlaced to progressive format conversion* also known as *deinterlacing*, which reduces the temporal and spatial video resolutions.

For example, the PAL DV interlaced format with 720×576 full frame pixels consisting of 720×288 pixels each for the odd and the even fields can be converted into a full frame with 360×288 pixels at 25 Hz. This can be easily achieved by omitting one field type and omitting the missing lines in the other fields as well while reducing the number of pixels of every line by a factor of two.

Obviously, the quality of the converted video is lower, but it is good enough for most of the Internet applications. However, the amount of video data is still quite large.

A video sequence in *Common Intermediate Format* (CIF) has $352 \times 288 = 101,376$ pixel/frame and $101,376$ pixel/frame $\times$ 25 frame/s $= 2,534,400$ pixel/s and $2,534,400$ pixel/s $\times$ 1.5 byte/pixel $= 3,801,600$ byte/s.

Please note that the number of pixels per line is reduced by 8 to cut off dark areas of a PAL signal at left and right image borders. In addition, the CIF video format, which is standardized by the International Telecommunication Union (ITU), requires only 1.5 byte/pixel on average. This is due to the fact that the CIF format is not based on the RGB color format but on the *YUV color format*. It consists of a luminance component Y and two chrominance components U and V, each of which can be computed by a linear combination of the RGB components:

$$Y = 0.299R + 0.587G + 0.114\,B, \; U = B - Y \text{ and } V = R - Y$$

This is the common video color format for the transmission and coding of PAL systems; the weighting factors for computing the NTSC or SECAM chrominance components are slightly different [6]. There are two main reasons for utilizing the YUV format: one historical and one related to the human visual perception. The first popular video application was the TV broadcast of monochrome video. Later, when color TV was introduced the analog color TV signal was broadcasted in a *composite video format*, in which the luminance and the two chrominance components were multiplexed in order to make the signal compatible with the monochrome TV systems. In addition, it was realized that the human eye is less sensitive to spatial variations of the chrominance than of the luminance. Thus, the signal bandwidth can be reduced by separating chrominance from luminance and transmitting only a band-limited chrominance, resulting in a hardly visible reduction of the video quality. The CIF format employs the so-called *4:2:0 format*, where the number of chrominance values in each of the two components is reduced by a factor of two in both the vertical and the horizontal directions. This leads on average to one U and one V value for four pixels and to the above-mentioned total of 1.5 byte/pixel when using a color depth of 8 bits per channel.

Since the amount of raw data of the CIF format is still too high for Internet applications, the video resolution can be further reduced and compression should be employed. The *Quarter CIF format* (QCIF) was standardized mainly for videophone-like applications and is well suited for most Internet video applications. It has only half of the resolution of a CIF frame in horizontal and vertical directions (176×144) and a raw data rate of 995,400 byte/s at 25 Hz. For Internet applications, the frame rate is often reduced to 8.33 or even 5 frames/sec (fps). If an advanced video compression scheme like MPEG-4 is additionally applied, the bit rate can be decreased to less than 8 kbyte/s.

Video Compression Principle

In principle, JPEG, like every other image compression scheme, can be used for coding of motion videos as well, simply by coding every video frame separately. This approach is simple but not as efficient as it could be. In addition, some flickering artifacts can be introduced when displaying a moving sequence of images, although individual pictures do not show an annoying artifact. This is due to the coding artifacts related to the quantization process that changes from frame to frame. These artifacts are especially visible at lower bit rates. To achieve better compression, the temporal correlations between consecutive frames have to be exploited.

The most commonly used approach is to perform a *temporal prediction* of the pixels of the current frame to be coded x (Figure 9.6b) out of a previous frame y (Figure 9.6a), subtract the corresponding pixels of the predicted frame from the current frame, and encode the difference frame e (Figure 9.6c) in a JPEG-like manner by transform coding. The difference frame shown in the example of Figure 9.6c is often referred to as the *prediction error frame* in which the mid-gray pixels reflect areas with a low prediction error and the dark and light pixels show areas with large errors. The advantage of coding the prediction error frame instead of the original is that there are normally large areas having only small errors. This will result in only a few, if any, nonzero coefficients after the transform and quantization stages that have to be encoded.

A comparison of frames x and y shows that the image content has moved to the bottom right due to a camera pan. This leads to strong errors in the prediction error frame in textured regions and at object borders, which in turn increases the bit rate required in these areas.

A *motion compensated prediction* is often applied to further reduce the prediction error. This can be achieved by subdividing the current frame into regular, nonoverlapping blocks and finding a corresponding block in the previous frame, which describes each block of the current frame as well as possible. These blocks are then copied for the prediction of the current frame (Figure 9.7). Figure 9.6d shows that motion-compensated prediction leads to less prediction errors. Only those areas of frame x that have new content cannot be predicted properly, for example, at the bottom of the frame.

For every block, a suitable *motion vector* has to be determined during a *motion estimation* process. This vector describes the relative translational motion of the block and has to be coded as additional side information. The quality of an encoder is highly dependent on the way the motion estimation is performed. As this process is generally very time consuming, most encoders allow motion vectors of lower quality,

FIGURE 9.6 Temporal prediction of a video frame with and without motion compensation: (a) previous frame y, (b) current frame x, (c) prediction error frame, $e = x-y$ (d) prediction error frame after motion compensation.

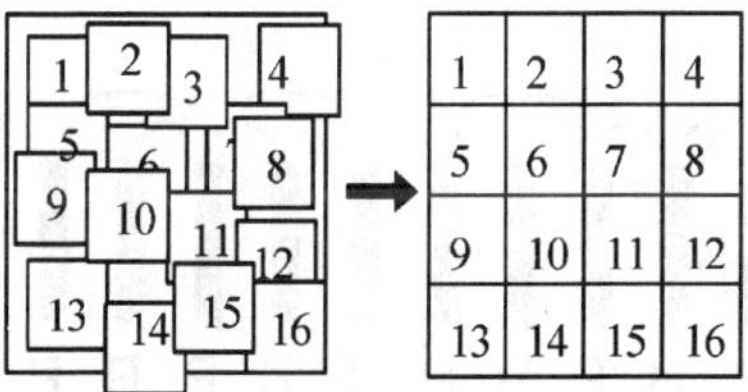

FIGURE 9.7 Motion-compensated temporal prediction of a video frame.

which result in a lower coding efficiency but faster encoding. Fortunately, the decoder does not have to perform the motion search, which reduces the computational complexity considerably.

To enable the encoder to control the video quality at the decoder side, both the encoder and decoder have to perform the prediction equally. Thus, the last decoded frame instead of the original has to be used for prediction, since the decoder does not have the original frame.

This implies that the encoder also consists of a decoder, which feeds back the last decoded frame, including the coding artifacts to predict the current frame. This *coding loop* has a strong impact on the characteristics of the video coding scheme.

- *Coding efficiency*, that is, the visual quality at a given bit rate is much increased. Therefore, almost 100% of the efficient video codecs on the market are based on this scheme.
- Coding artifacts of the previous frame are introduced into the current frame, which is not necessarily a disadvantage. Compared to the separate encoding of the frames, the artifacts are temporarily more correlated, thus reducing the annoying *flickering artifacts*.
- If transmission errors occur, which affect the quality of a frame, these *errors* will be *propagated* to the following frames.
- Since the decoding of a frame depends on the previous frames, there is no *instantaneous access* to the frames in a sequence.

The coding loop introduces a trade-off between coding efficiency and error robustness. The design of a video codec will stress one or the other characteristic depending on the kind of video application.

The frames that are coded by prediction from a previous frame are called *P-frames*. Obviously, the first frame of a sequence has to be coded without prediction. Therefore, it is called intra or *I-frame*, because only information available within this frame is used for coding. A P-frame can be predicted from an I-frame or another P-frame as can be seen in Figure 9.8, where the arrows indicate the prediction direction. There is one other frame type, namely a *B-frame*, which is not predicted from one but from two frames. In most codecs, a *bidirectional prediction* is made from one previous and one following frame, which can further reduce the bit rate. B-frames introduce an additional delay as the following P-frame has to be acquisitioned, coded, and decoded prior to coding the current B-frame coding. Thus, B-frames are not employed in delay-sensitive applications such as video communication.

Further I-frames will be inserted, for example, one per second, to allow for starting the decoding process at more or less arbitrary time instances of the sequence. All frames depending on one I-frame are often grouped into a *group of pictures* GOP structure.

Internet Video Applications

For most of the Internet video applications, the two main problems related to transmission via the Internet are that the available average network bit rate is too low and that some of the video packets do not arrive timely enough to be displayed. The first problem has an impact on all Internet video applications and can be reduced by video codecs with improved coding efficiency. The second problem addresses only those Internet applications where simultaneous transmission and display of video is required, that is, for *live video* reception and *video streaming*. This is a more serious problem, because losing one packet containing, for example, a P-frame means that all following frames of a GOP cannot be decoded. Here, very sophisticated techniques have to be combined to achieve a properly working system. These techniques might fall within the following categories: video buffer management, coder control, feed back,

FIGURE 9.8   Possible frame types in a video codec.

scalable coding, quality of service, streaming and real-time protocols, and error concealment. For more detailed information on this topic, the reader is referred to [6].

A very common technique to improve the error robustness of video a stream for transmission over band-limited channels is to use a *rate-control*. The rate-control controls the encoder coding options in order to achieve at least on average a *constant target bit rate* (CBR). Without rate-control, a constant quality at a *variable bit rate* (VBR) is achieved. Large fluctuations of the amount of bits per frame can occur due to the content of the images, the variations thereof over time and the chosen frame type. An I-frame may need a bit amount 10 times higher than that of a P-frame. The rate-control can influence the bit rate by changing the accuracy of the quantization, which affects the picture quality as well, or by *frame-skipping*. By skipping input frames at the encoder side, the bit rate and temporal resolution will be reduced. The missing frames are substituted at the decoder side by repeating the respective previous frame.

Some encoders provide a *2-pass coding* as an optional feature to better adapt to a target bit rate. In a first pass, the sequence is coded and the required bit amounts per frame for the chosen encoder parameters are evaluated. In the second pass, the actual coding is done with a corrected encoder setting.

In the following, we will concentrate on the more simple application of *video downloading* and successive decoding. For downloading of preencoded video, encoding time and delay are generally not important factors. The main objective is achieving a high coding efficiency in order to allow small video file sizes and a reasonable video quality. Thus, the GOP structure similar to Figure 9.8 including B-frames can be used. An exhaustive but time-consuming motion search can be chosen and also the variable bit rate mode since it often leads to better visual results.

Video Coding Standards, Quasistandards, and File Formats

First of all, a distinction has to be made between the *coding scheme* and the *file format* containing the bit stream. There are file formats that allow compression with a variety of different coding schemes. In addition, a single file may contain more than one video and audio track. Currently, there are more or less four popular internet video formats on the market, that is, *MPEG* (*.mpg), *Windows Media Video* (*.wmv, *.wm), *Real Video* (*.rm), and *QuickTime* (*.mov). The first three have their own video coding scheme and file format, whereas *Apple's QuickTime* format is basically a file format that can be used with several codecs. This is also true for the *Windows AVI* format (*.avi). A tag in the header of the file indicates the used compression scheme. Therefore, it might happen that a *player* cannot decode an AVI file found in the web, since the corresponding decoder is not installed in the player. Thus, the availability of players for the respective format is an important aspect for choosing an appropriate video format. Players for the four formats mentioned above are widely installed and can be downloaded for free at [7–9]. There are several players for the MPEG format, but it can also be played with the *WindowsMediaPlayer*.

In the following, some characteristics of the most popular video formats are given.

Motion JPEG, an extension of the JPEG standard, is a simple but not very efficient scheme for video compression since every frame is stored as an I-frame. The very popular *DV format* is based on motion JPEG but with some slide modifications to better adapt to video signals. It is rarely used for Internet applications.

The different *MPEG formats* have been developed by the ISO Moving Picture Experts Group mainly for video and audio compression, storage, and transmission. There is a file format specified for each version. All MPEG video coding standards are based on motion-compensated prediction and the GOP structure.

MPEG-1 is well suited for compression of progressive video in CIF format at rates of 1–1.5 Mbit/s, including audio. For audio compression, the very popular MPEG-1 layer3 scheme can be used, which is also known as *MP3*. It has been used for storing video on CD-ROM at a quality comparable to an analog consumer VCR.

MPEG-2 has been derived from MPEG-1 but with improved performance and additional features for coding interlaced video at higher bit rates (4–15 Mbit/s) and qualities. It is widely used, for example, for DVD and Digital Video Broadcast (DVB).

Both formats are more appropriate for high bit rates, but can operate at lower bit rates as well; thus, they are used as an Internet video format for downloading.

MPEG-4 is much better suited as an Internet format as it has been especially designed for low bit rates, although it covers a wide range of bit rates (8 kbit/s–1200 Mbit/s), video resolutions, and qualities [10]. Furthermore, it allows to encode arbitrarily shaped video objects, 2D- and 3D-scenes composed of natural and synthetic still image, video, and audio content. Binary format for scenes (**BIFS**) has been specified for the scene description. It was derived from VRML but allows compression of the scene description, user interaction, and streaming. An introduction to the MPEG-4 multimedia standard can be found in [6,11]. Since the MPEG-4 standard is rather new, up to now only rectangular videos compressed by MPEG-4 can be found on the Internet.

A new video coding standard with a much better coding efficiency has been standardized in 2003. It is a joint development of the ITU-T and MPEG, the *H.264 | MPEG-4 AVC* advanced video coding standard. Although being part of MPEG-4 it is not backward compatible with MPEG-4, in comparison to MPEG-4 and MPEG-2, the bit rate at the same visual quality can be further reduced by about 40% and 63%, respectively [12]. Unfortunately, the complexity, especially at the encoder, is largely increased. Due to the good performance, this coding scheme will probably become very popular.

For the *Windows Media Video format*, the *Windows Media Encoder* is available [7]. Since this coding scheme is a proprietary standard of *Microsoft* and the used coding techniques are currently hidden, it can only be decoded by the *Windows Media Player*. But as *Microsoft* participated in the development of MPEG-4 AVC, some of its efficient new features might have gone in the latest Media Encoder. At least its coding efficiency is now better than that of the (older) MPEG-4. The latest *Windows Media* file format can also include video compressed with other codecs such as MPEG-4.

The situation for the *Real Video format* is quite similar [8]. The video coding scheme is hidden; encoding can be performed using *RealSystem Producer* and decoding can only be done by *RealPlayer*. This is the dominant format for streaming applications. It provides slightly better qualities at low bit rates compared to the competitors. In addition, it supports more platforms than *Windows Media* and *QuickTime*, not only Windows and Mac OS but also Linux and Unix. The latest *RealMedia* file format can also include video compressed with other codecs such as MPEG-4.

Apple's QuickTime format is capable of carrying media data other than pure audio and video as well [9]. Animations, graphics, texts, and even 360° virtual reality scenes can be stored in *QuickTime* format. That is the main reason why MPEG has based the MPEG-4 file format on *QuickTime*. *Apple's QuickTime Pro* package provides a variety of different video codecs. Formerly, *Sorenson Video* was often chosen as the preferred video codec, but this may change since the latest version of *QuickTime* supports MPEG-4 video.

Please note that the above-mentioned qualities of the video codecs only reflect the situation at the beginning of the year 2003. Every new release of a codec can lead to large improvements. On the other hand, one should be aware of the fact that standardized coding schemes, like MPEG-4, are public and the format will not change. Proprietary standards will change, and future players may not support older versions.

References

[1] http://www.w3.org/Graphics/GIF/spec-gif89a.txt
[2] http://www.libpng.org/pub/png
[3] http://www.jpeg.org
[4] Taubman, D.S. and M.W. Marcellin, *JPEG2000: Image Compression Fundamentals, Standards and Practice*, Kluwer International Series in Engineering and Computer Science, Dordrecht, 2001.
[5] Clarke, R.J., Transform Coding of Images, *Microelectronics and Signal Processing*, Academic Press, London, 1985.
[6] Wang, Y., J. Ostermann, and Y.-Q. Zhang, *Video Processing and Communications*, Signal Processing Series, Prentice-Hall, Upper Saddle River, NJ, 2002.
[7] http://www.microsoft.com/windows/windowsmedia
[8] http://www.real.com
[9] http://www.apple.com/quicktime/
[10] ISO/IEC JTC1, 2000, Coding of audio-visual objects — Part 2: Visual, ISO/IEC 14496-2 (MPEG-4 Visual version 1), April 1999, Amendment 1 (version 2), February 2000.

[11] ISO/IEC/JTC1/SC29/WG11, MPEG-4 Overview (Version 21), Dokument N4668, Jeju Island, Korea, March 2002. (accessible via the official mpeg web page: http://www.cselt.it/mpeg).

[12] Schwarz, H. and Th. Wiegand, The Emerging JVT/H.26L Video Coding Standard, Proceedings of IBC 2002, Amsterdam, NL, September 2002.

10

The Fundamentals of Web Services

Klaus-Peter Eckert

Fraunhofer Institute for Open Communication Systems

10.1 Introduction to Web Services

Web services are one of the major hypes in today's Internet world. But what is so special about Web services? To answer this question, let us look at two definitions. The first one gives a statement about the problems that Web services promised to solve in their beginning late 2000 [3].

> "A Web service is a collection of functions that are packaged as a single entity and published to the network for use by other programs. Web services are building blocks for creating open distributed systems, and allow companies and individuals to quickly and cheaply make their digital assets available worldwide."

Companies that want to make their "digital assets" available to other companies, to customers, or to roaming employees need a new paradigm for the creation of open distributed systems. They need new kinds of "open interfaces" or standards for specific business domains and they need a better infrastructure for the development, operation, and usage of distributed systems. In short, Web services promised to simplify the syntactical and semantically interworking of programs in a distributed and heterogeneous environment that is operated by different, autonomous companies. Let us now look at the current definition from the W3C [6]:

> "A Web service is a software system identified by a uniform resource identifier URI, whose public interfaces and bindings are defined and described using XML. Its definition can be discovered by other software systems. These systems may then interact with the Web service in a manner prescribed by its definition, using XML based messages conveyed by Internet protocols."

This more technical definition restricts the definition and usage of Web services to the application of XML-based techniques [11]. The definition of the syntax and semantics of a Web service is restricted to the abilities of XML; they are invoked by conveying XML messages by Internet protocols. The most important conclusions from this definition are the following. First, Web services try to solve both the problems mentioned above with the application of XML. Thus, anybody who wants to advertise a new

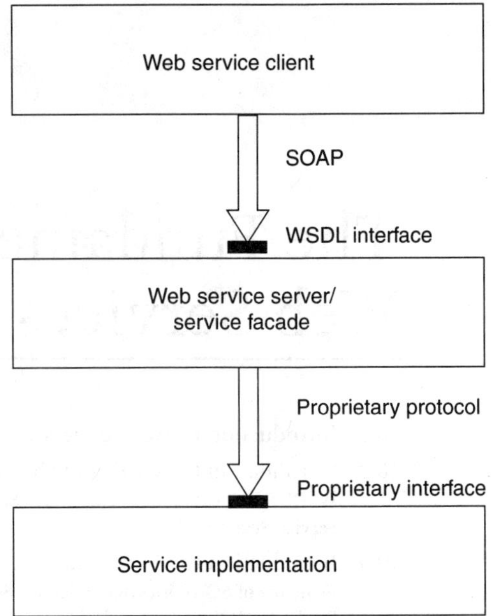

FIGURE 10.1   Constituent parts of a Web service configuration.

Web service or who wants to use an existing Web service has to deal with XML. Second Web service technology provides means for the specification of interfaces and bindings as well as for the exchange of messages. They do not provide any assistance for the implementation of the software system behind the Web service. Web services are an interface technology, not an implementation technology (See Figure 10.1).

Third, the Web service architecture describes three phases

- the provider advertises the existence and contact information in some registry,
- the client looks up the service in a registry, and
- the client contacts and invokes the service.

The means for these three phases, WSDL for the service description, UDDI for the advertisement, and SOAP for the invocation, will be investigated in the following sections.

10.2   Introduction to WSDL and UDDI

The Web service description language WSDL [8] allows the expression of all the information that is necessary to specify the signature of a Web service as well as the binding information to an instance of the service. WSDL defines an XML schema for the description of services. The outer <definitions> element encloses all parts of the service definition. Its attributes specify the XML namespaces used in the specification. The target namespace distinguishes the definitions in this template from definitions in other documents. Additional namespace declarations facilitate the identification of types and elements with the same name, but from different embedded schemas or imported documents. The <definitions> section contains the following information (See Figure 10.2):

- First, the <definitions> element itself has two attributes: the optional *name* is of the XML type *noncolonized name* (NCName), and the optional *targetNamspace* is of the XML type *anyURI*. It is important to consider that the type system for the definition of WSDL is the XML schema Part 2 [11] and that an understanding of the WSDL specification is impossible without knowledge of XML schema.
- <import>ed (associated) XML namespaces that refer to externally defined parts of the Web service definition.

- <types>: the service-specific data types.
  The <types> element encloses data type definitions that are relevant for the exchanged messages. WSDL uses the XML Schema as the canonical type system. The role of this element is to contain an XML <schema> element.
- <message>: the elements of the signatures of the supported messages consist of:
  - the name of the request message (<message name = ...>);
  - the names and types of the input parameters. Each parameter is denoted as a <part> of the <message> :
    <part name = paramName element = paramType> or
    <part name = paramName type = paramType>
    The notion of an *element* refers directly to a similar term in the definition of an XML schema. The notion of a *type* refers to an XSD *simpleType* or *complexType*. The values of both attributes have to be expressed in the form of so-called XML *qualified names* (QName).
  - the name of the response message (<message name = ...>);
  - the names and types of the return parameters:
    <part name = paramName element = paramType> or
    <part name = paramName type = paramType>
- <portType>: the signatures of the provided operations, consisting of:
  - The name of the <operation> and the names of the appropriate request and response <message>s. Operation overloading is possible; thus, operation names do not have to be unique.
  - The order of the parameters.

A port type is a named set of abstract operations and the abstract messages involved. The *name* attribute provides a unique name among all port types defined within the enclosing WSDL document. The WSDL 1.1 specification comprises four different communication patterns called *transmission primitives* that an endpoint can support. Until now it is not clear whether future WSDL specifications will support all these transmission primitives.

- *One-way*: The endpoint *receives* a request message but returns no response.
- *Request–response*: The endpoint *receives* a request message and *sends* a correlated response message. This primitive can be used to model RPC semantics.

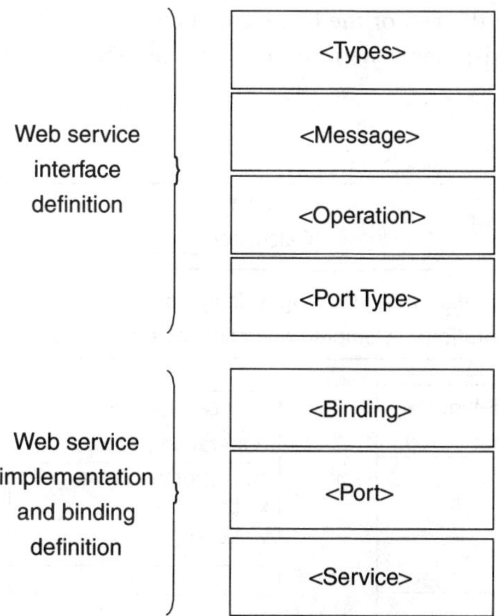

FIGURE 10.2   Sections of a WSDL description.

- *Solicit–response*: The endpoint *sends* a response message, and *receives* a correlated request message. This primitive can be used to model callbacks.
- *Notification*: The endpoint *sends* a response message. This primitive can be used to model notifications in the Web service world.

- <binding>: the specification of:
  - the protocol,
  - the encoding rules, and
  - the supported <operations>

for a particular instance of a <portType> called <port> or *endpoint*. Web services use so-called *extensibility elements* to specify the message format and the used encoding rules. The WSDL specification states

"that it is unreasonable to expect a single type system grammar can be used to describe all abstract types present and future. WSDL allows type systems to be added via extensibility elements."

The binding extensibility elements are used to specify the concrete grammar for the input, output, and fault messages. Peroperation binding information as well as perbinding information may also be specified. For detailed information on the WSDL–SOAP binding, refer to section "SOAP binding in WSDL."

- <port>: the combination of a <binding> with a network address.
- <service>: a collection of <port>s.

WSDL Specification of a Sample Web Service

To clarify this fundamental definition, let us look at a simple calculator service. The service provides two operations. The first one is called *Add*. It returns the sum of two integer arguments. The second one is called *Compute*. It considers a data structure containing an operator and two integer arguments as an input parameter and returns the result of the operation.

It has to be mentioned that the generated WSDL differs from tool to tool. Thus, all the given examples have to be considered as one of several possible syntactical representations of the specification of the same calculator service.(See Fig. 10.3)

The WSDL file contains a top-level <definitions> element. This element defines the imported and exported namespaces used in the rest of the file. Here, the references to XML schema definitions, the WSDL–SOAP binding-related schema definitions, and the WSDL schema definition itself can be found.

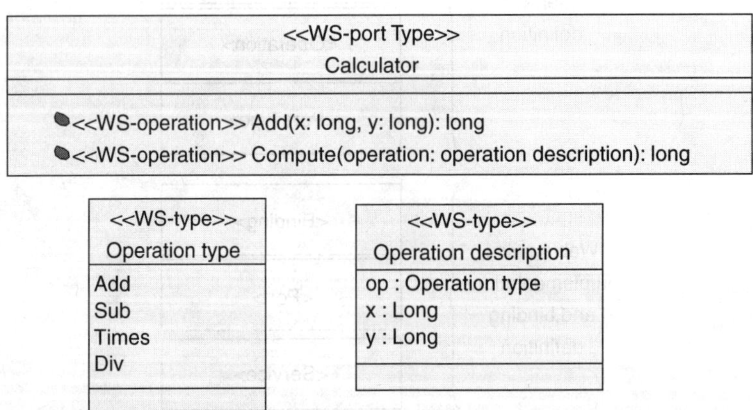

FIGURE 10.3   Specification of a sample Web service.

```
<?xml version="1.0" encoding="UTF-8"?>
<definitions
     name="Calculator"
     targetNamespace="http://www.myorg.org/Calculator.wsdl"
     xmlns="http://schemas.xmlsoap.org/wsdl/"
     xmlns:soapbind="http://schemas.xmlsoap.org/wsdl/soap/"
     xmlns:tns="http://www.myorg.org/Calculator.wsdl"
     xmlns:xsd="http://www.w3.org/2001/XMLSchema"
     xmlns:xsd1="http://www.myorg.org/Calculator.xsd">
...
</definitions>
```

The data type describing the operation is mapped to the <types> section and defined as an XML schema:

```
<types>
     <xsd:schema
               targetNamespace="http://www.myorg.org/Calculator.xsd"
               xmlns:xsd1="http://www.myorg.org/Calculator.xsd"
               xmlns:xsd="http://www.w3.org/2001/XMLSchema">
...
          <xsd:simpleType name="Calculator.OperationType">
               <xsd:restriction base="xsd:string">
                    <xsd:enumeration value="add"/>
                    <xsd:enumeration value="sub"/>
                    <xsd:enumeration value="times"/>
                    <xsd:enumeration value="div"/>
               </xsd:restriction>
          </xsd:simpleType>
          <xsd:complexType name="Calculator.OperationDescription">
               <xsd:sequence>
                    <xsd:element maxOccurs="1" minOccurs="1"
                         name="op" nillable="true"
                         type="xsd1:Calculator.OperationType"/>
                    <xsd:element maxOccurs="1" minOccurs="1" name="x"
                         type="xsd:int"/>
                    <xsd:element maxOccurs="1" minOccurs="1" name="y"
                         type="xsd:int"/>
               </xsd:sequence>
          </xsd:complexType>
     </xsd:schema>
</types>
```

The request and reply messages of the operations are mapped to the <message> section:

```
<message name="Add">
      <part name="x" type="xsd:int"/>
      <part name="y" type="xsd:int"/>
</message>
<message name="AddResponse">
      <part name="AddReturn" type="xsd:int"/>
</message>
<message name="Compute">
      <part name="operation" type="xsd1:Calculator.OperationDescription"/>
</message>
<message name="ComputeResponse">
      <part name="ComputeReturn" type="xsd:int"/>
</message>
```

The signatures of the two operations "Add" and "Compute" are a part of the <portType> section:

```
<portType name="Calculator">
      <operation name="Add" parameterOrder="x y">
            <input message="tns:Add"/>
            <output message="tns:AddResponse"/>
      </operation>
...
</portType>
```

The <binding> section defines that the operations are used in an RPC-like style and that SOAP via HTTP is used as the communication protocol:

```
<binding name="CalculatorBinding" type="tns:Calculator">
      <soapbind:binding style="rpc"
            transport="http://schemas.xmlsoap.org/soap/http"/>
      <operation name="Add">
            <soapbind:operation soapAction="Calculator#Add"/>
            <input>
                  <soapbind:body
                  encodingStyle="http://schemas.xmlsoap.org/soap/
                  encoding/"
                  namespace="Calculator" use="encoded"/>
            </input>
            <output>
                  <soapbind:body
                  encodingStyle="http://schemas.xmlsoap.org/soap/
                  encoding/"
                  namespace="Calculator" use="encoded"/>
            </output>
      </operation>
...
</binding>
```

Finally, the <service> section defines that the calculator service can be accessed at the given URL:

```
<service name="Calculator">
    <port binding="tns:CalculatorBinding" name="Calculator">
        <soapbind:address
               location="http://myHost:8080/serviceDir/Calculator"/>
    </port>
</service>
```

Do not panic after looking at all these XML definitions. Neither for the definition nor for the advertisement or usage of a Web service its WSDL file has to be created by hand. As a programmer, you will specify the "proprietary" interface shown in Figure 10.1 in your preferred language like Java, C#, or CORBA IDL. Afterwards, some tools will take your specification as input and generate the interface-related parts of the WSDL. When one deploys the implementation of the service, that means when one starts it and makes it accessible by other programs, again some tools will generate the binding-related parts of the WSDL (see Figure 10.4).

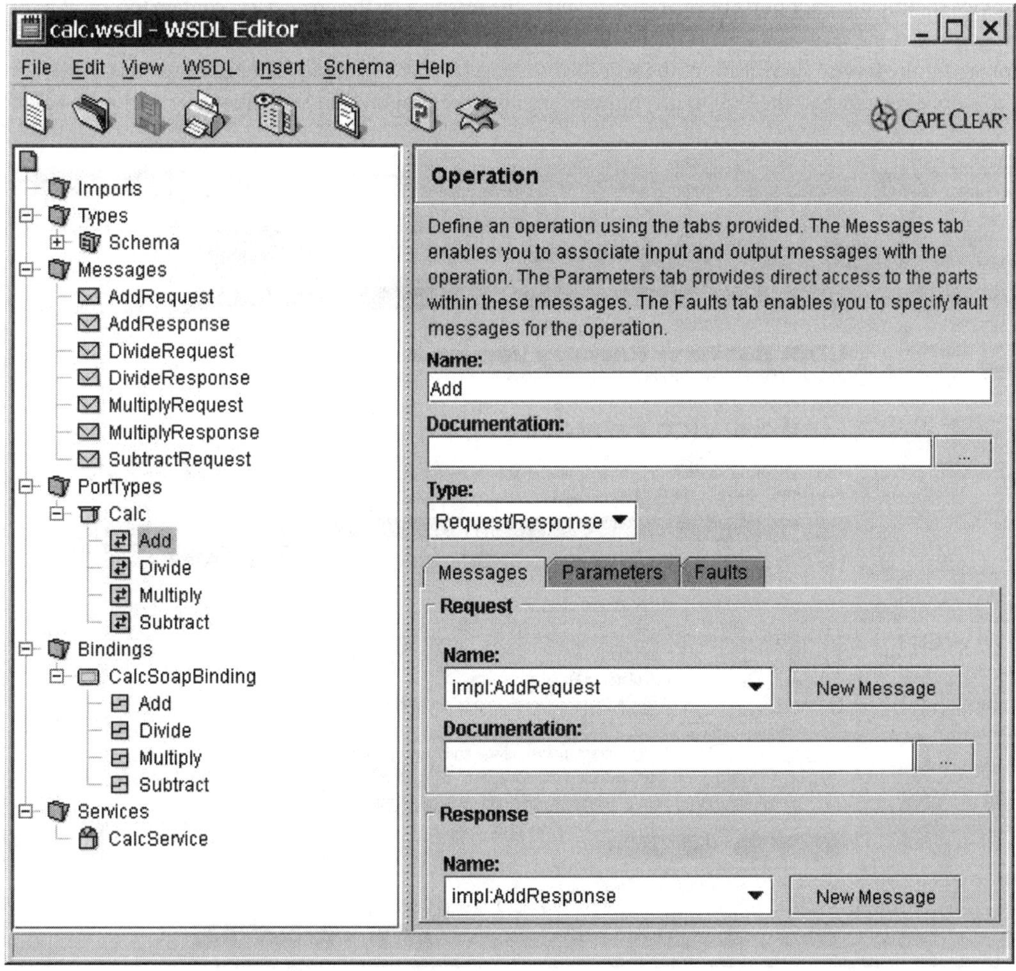

FIGURE 10.4   The sample WSDL shown in a WSDL editor [7].

The UDDI Web Service Registry

Assume now, you have implemented and deployed your Web service and that some tools have generated a WSDL description of your service. The next step is to advertise your service to potential users. All the technical information that is necessary to write a client and access the service is contained in the WSDL file. Consequently, you only have to advertise the location of this file, possibly as a URL, and everybody is able to use the service. But exactly here some major problems occur. Until now you only have specified the syntax of the service and its parameters. No information about the semantics, the orchestration, about you and your organization, about nonfunctional properties like availability, performance, security, charges, etc. is available. Moreover, do you really want to allow everybody to use your service without verifying his identity and authorization?

For the first group of these issues, Web services offer a solution called "universal description, discovery, and integration of Web services" UDDI [4]. The UDDI contains three kinds of information. The so-called *white pages* contain business information like name, address, phone number, or contact of a given organization. This is the right place for service providers to register themselves. The *yellow pages* contain service information that categorizes business domains and services. It should be based on standardized ontologies. This is the right place to advertise the "digital assets" in a standardized way, for example, to store some promotion about our Calculator service.

Although our Calculator is a typical exemplar of an arithmetic processor, some other calculators may provide slightly different ports or operations. This is the reason why the *green pages* contain technical information about the service and its implementation. Here, binding information as well as detailed information about the service specification, the so-called *technical model — tModel*, can be found. The easiest way is to store a reference to the WSDL file; but this excludes the provision of semantic information about the

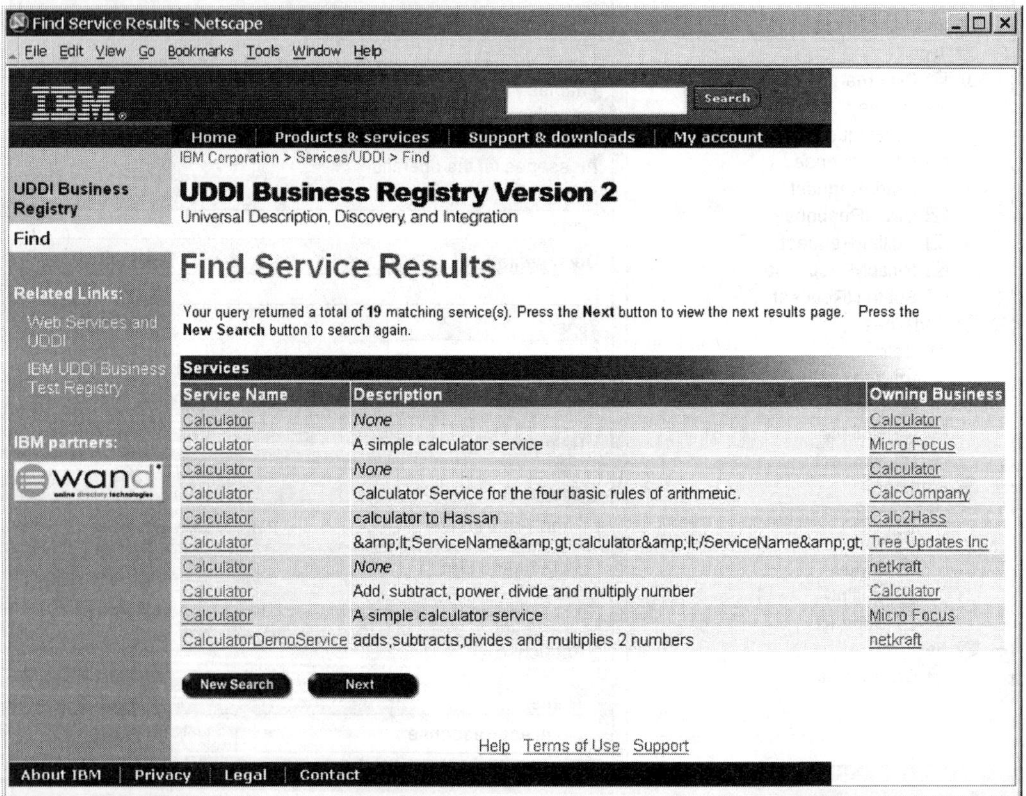

Services		
Service Name	Description	Owning Business
Calculator	*None*	Calculator
Calculator	A simple calculator service	Micro Focus
Calculator	*None*	Calculator
Calculator	Calculator Service for the four basic rules of arithmetic.	CalcCompany
Calculator	calculator to Hassan	Calc2Hass
Calculator	<ServiceName>calculator</ServiceName>	Tree Updates Inc
Calculator	*None*	netkraft
Calculator	Add, subtract, power, divide and multiply number	Calculator
Calculator	A simple calculator service	Micro Focus
CalculatorDemoService	adds,subtracts,divides and multiplies 2 numbers	netkraft

FIGURE 10.5   Searching for a calculator service in a public UDDI.

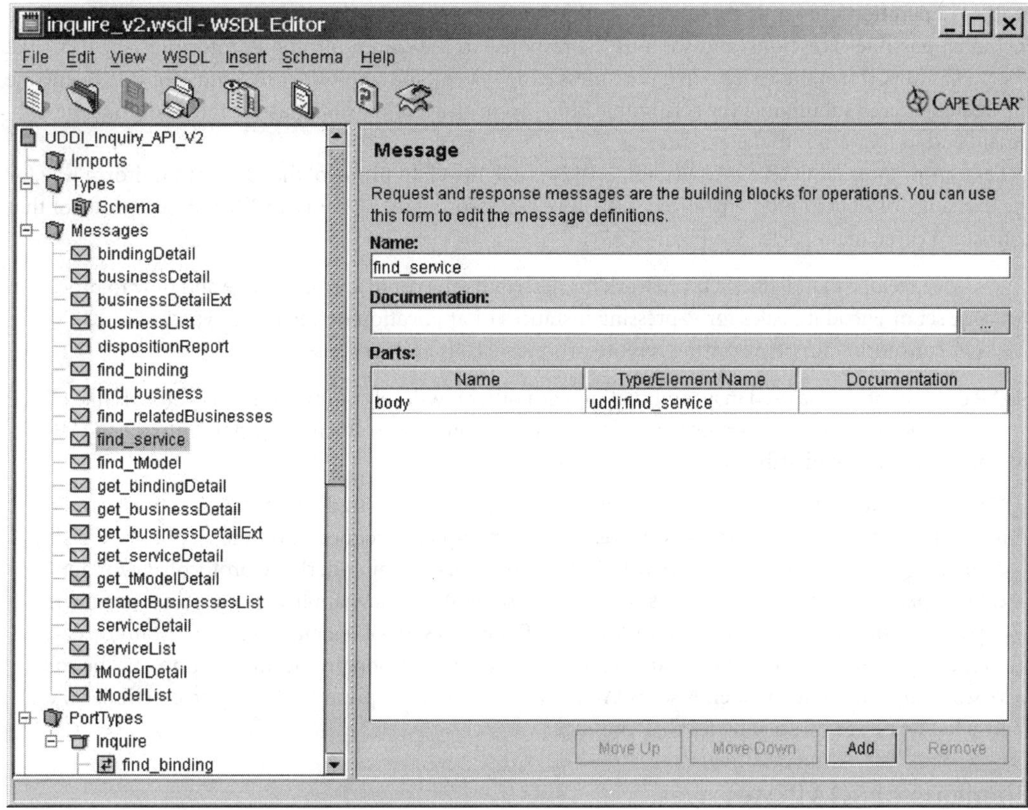

FIGURE 10.6   Access messages of the UDDI Web service.

service, its implementation, and usage. Like everywhere in the Web service's world, UDDI is specified in the form of an XML schema [5] and can itself be accessed via Web service operations. A subset of the appropriate definition is shown in Figure 10.6.

10.3   Introduction to SOAP

Assume that two processes running on different network nodes want to communicate. In such a situation, there exist two fundamental programming models. In a *remote procedure call* (RPC) like style, the client process sends a request to the server process and waits. The server process executes the request and sends a response back to the waiting client, who can continue his execution afterwards. In a *messaging-like* style, the client process sends a document to the server process and continues execution. The server process handles the document and could deliver a response message, containing the response document, later. The SOAP supports both programming models. Before going into more details about SOAP, let us have a look at what happens in an RPC-like communication.

During the implementation of the client, some knowledge about the operations provided by the server is necessary. In general, the client binds to the server and calls one or more of the provided operations. From the client's viewpoint, a local proxy called *client stub* of the server is invoked. The proxy delegates the request to the server, receives the response, and forwards it to its invoker. The request itself consists of the name of the operation and the parameters. The proxy has to encode the request to some "on-the-wire" transferable and convertible representation, to submit the request to the selected communication protocol, to wait for the response, and to decode the response after receiving it via the communication protocol. A similar *server stub* waits for incoming requests. The stub decodes the request, selects an executor for the required operation, forwards the request to this executor, waits for the response, encodes the response including possible exceptions, and submits the response to the communication protocol. Every RPC-like system has to perform these steps. In a pure messaging system, the *transmitter stub* must be able to encode the document and submit it to the communication protocol; a *receiver stub* must be able to decode the document and to forward it to the executor.

Let us now look at SOAP. SOAP Version 1.1 [2] is a lightweight protocol that defines a uniform way for the exchange of data in a decentralized, distributed environment. SOAP is an XML-based protocol that consists of three main parts (See Figure 10.7):

- An envelope that defines a framework for describing the content of a message and how to process it.
- A set of encoding rules for expressing instances of application defined data types.
- A convention for representing remote procedure calls and responses.

SOAP can potentially be used in a combination with other low-level communication protocols like HTTP or SMTP. The current draft specification of SOAP 1.2 [2] provides some more insight into the SOAP paradigm and its role in distributed system protocols:

"SOAP is fundamentally a stateless, one-way message exchange paradigm, but applications can create more complex interaction patterns (e.g., request/response, request/multiple responses, etc.) by combining such one-way exchanges with features provided by an underlying protocol and/or application-specific information. SOAP is silent on the semantics of any application-specific data it conveys, as it is on issues such as the routing of SOAP messages, reliable data transfer, firewall traversal, etc. However, SOAP provides the framework by which application-specific information may be conveyed in an extensible manner. Also, SOAP provides a full description of the required actions taken by a SOAP node on receiving a SOAP message."

Elements of SOAP Messages

As shown in Figure 10.7, a SOAP message consists of an envelope including some optional headers and one mandatory body element. A SOAP header is an extension mechanism that provides a way to include application-independent information in a SOAP message. Headers may contain information about the user, the client process, and their contexts. The SOAP body is the mandatory element within the envelope that carries the end-to-end application-specific information included by the sender and read by the receiver of the message.

Let us look at these elements in some detail. The SOAP *envelope* is the top element of a SOAP message. It contains XML namespace declarations and some additional attributes like the encoding style. The *enve-*

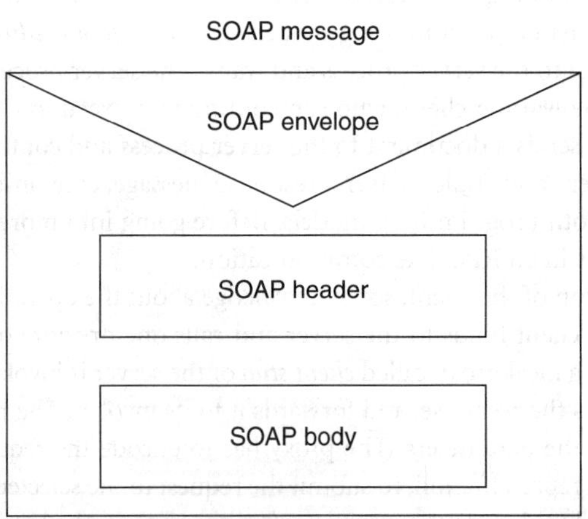

FIGURE 10.7 Elements of a SOAP message.

lope element itself is associated with the namespace http://schemas.xmlsoap.org/soap/envelope. Additional namespaces may refer to XML schema definitions.

The *header* element is the first immediate child element of the *envelope* element. It contains additional information about the message like authentication, payment, context, or transaction-related data. The header contains attributes and application-specific header entries. As a SOAP message may travel from the sender to the receiver by passing several intermediary SOAP nodes, it is necessary to define as to which intermediary is allowed or committed to execute a SOAP header. The *actor* attribute may be used to address the header element to a particular endpoint. The *mustUnderstand* attribute is used to indicate whether a *header entry* is mandatory or optional for the recipient to process. If a *header entry* is tagged with a *mustUnderstand=1* attribute and the receiver is unable to process the entry, a SOAP *fault* element has to be returned.

The mandatory *body* element is used to exchange application-specific information between the initiator and the ultimate recipient of the message. RPC-like request, response, and error messages are typical examples of SOAP bodies. All child elements of the *body* element are called *body entries*. Each *body entry* is encoded as an independent element within the SOAP *body*.

A large part of the SOAP specification deals with SOAP encoding. Because this encoding style is the source of several interoperability problems [10], it is discouraged from using SOAP encoding. Thus, no more information related to SOAP encoding will be given here.

One of the main goals of SOAP is to encapsulate and exchange RPC requests and responses. To represent an RPC request, the following information is needed. First, the URI of the recipient is required. In case of an HTTP-mapping, the URI indicates the target of the invocation.

The RPC request itself is modeled as a *struct*. A *struct* is a compound value in which the unique accessor names are the only distinction among members.

```
<ns:structId>
      <acc_1>val_1</acc_1>
      ...
      <acc_n>val_n</acc_n>
</ns:structId>
```

The struct contains one accessor for each *in* or *in/out* parameter. It is named and typed identical to the method name (see Section 10.2). The accessors are named and typed corresponding to the types of the parameters. They appear in the same order as in the method signature (compare the definition of the WSDL <portType>). An RPC response and a PRC fault are modeled as a *struct* too. The response *struct* contains one accessor for each *out* or *in/out* parameter. It is named and typed identical to the method response. The accessors are named and typed corresponding to the types of the parameters. They appear in the same order as in the method signature. The name of the return value accessor is not significant. A convention is to name it after the method name with the string *Response* appended.

A method fault is encoded using the SOAP *fault* element. The *fault* element defines four different subelements. The *faultcode* subelement is intended for use by software, the *faultstring* subelement is intended to be used by humans, the *faultactor* provides information regarding who caused the fault, and the *detail* subelement is intended for carrying application-specific information related to the body element.

```
<soap:Body>
      <soap:fault xmlns=...>
            <faultcode>soap:VersionMismatch</faultcode>
            <faultstring>Mismatch in used version of WSDL</faultstring>
      </soap:fault>
</soap:Body>
```

Thus, SOAP defines a natural way to map WSDL definitions to RPC-like method invocations and replies. Section "SOAP Binding in WSDL," introduces the relevant parts of this mapping.

Sample SOAP Messages

This is a sample SOAP 1.1 message and the associated HTTP-header for the *Add* request message,

```
<HTTPHeaders>
...
<content-type>text/xml; charset=utf-8</content-type>
<soapaction> HTTP://myHost:8080/serviceDir/Calculator#Add </soapaction>
<content-length>mmm</content-length>
...
</HTTPHeaders>


<?xml version="1.0" encoding="utf-8" ?>
<soap:Envelope
      xmlns:soap="http://schemas.xmlsoap.org/soap/envelope/"
      xmlns:xsi="http://www.w3.org/2001/XMLSchema-instance"
      xmlns:xsd="http://www.w3.org/2001/XMLSchema">
      <soap:Body>
            <Add xmlns="http://www.myorg.org/Calculator">
                  <x>100</x>
                  <y>200</y>
            </Add>
      </soap:Body>
</soap:Envelope>
```

and here follows the same for the *AddResponse* response message. In both examples, the SOAP body contains a reference to the namespace that is associated with the Web service's WSDL definition.

```
<HTTPHeaders>
...
<content-type>text/xml; charset=utf-8</content-type>
<content-length>nnn</content-length>
...
</HTTPHeaders>


<?xml version="1.0" encoding="utf-8" ?>
<soap:Envelope
      xmlns:soap="http://schemas.xmlsoap.org/soap/envelope/"
      xmlns:xsi="http://www.w3.org/2001/XMLSchema-instance"
      xmlns:xsd="http://www.w3.org/2001/XMLSchema">
      <soap:Body>
            <AddResponse xmlns="http://www.myorg.org/Calculator">
                  <AddReturn>300</AddReturn>
            </AddResponse>
      </soap:Body>
</soap:Envelope>
```

The <soapaction> header field in the HTTP header is used to identify the intent of the SOAP request. It is omitted in the SOAP 1.2 specification. The value of the <content-type> header field will change from *text/xml to application/soap+xml.*

SOAP Binding in WSDL

WSDL includes a binding for SOAP 1.1 endpoints (refer to [9], the associated namespace is denoted <soapbind:xxx>) using WSDL extensibility elements. The binding specifies protocol-specific information, such as:

- an indication that a binding refers to the SOAP protocol,
- a way of specifying an address for a SOAP endpoint,
- the URI for the SOAPAction HTTP-header for the HTTP-binding of SOAP, and
- the list of definitions for headers that are transmitted as a part of the SOAP envelope.

The <soapbind:binding> element signifies that the binding is bound to the SOAP protocol format (envelope, header, and body). It makes no claims that the message follows a specific version like SOAP 1.1. The *style* attribute is either set to *rpc* or *document* for each contained operation. The *transport* attribute, which is required, indicates which SOAP transport this binding corresponds to. The value http://schemas.xmlsoap.org/soap/http corresponds to the HTTP binding in the SOAP specification.

The <soapbind:operation> element provides information for the operation as a whole. The *style* attribute indicates whether the operation is RPC-oriented, where messages contain parameters and return values, or document-oriented, and where messages contain one or more documents. The *soapAction* attribute specifies the value of the SOAPAction header for this operation. For the HTTP protocol binding of SOAP 1.1, one must specify a value. For other SOAP protocol bindings, it must not be specified, and the <soapbind:operation> element can be omitted.

The <soapbind:body> element specifies how to assemble the different message parts that appear inside the body element of the SOAP message. Parts of a message can be either abstract-type definitions or physical schema definitions. The optional *parts* attribute indicates which parts appear within the SOAP body portion of the message. If this attribute is omitted, all parts defined by the message are assumed to be included. The *use* attribute, which is required, indicates whether the message parts are encoded using some encoding rules, or if the parts define the physical schema of the message. If *use* is set to *encoded*, each message part references an abstract type using the *type* attribute. If *use* is set to *literal*, each part references a physical schema definition using either the *element* or *type* attribute. The *encodingStyle* attribute is set to a list of URIs, each separated by a single space. These URIs represent encodings used with the message, in order from the most restrictive to the least restrictive. The *namespace* attribute applies to content not explicitly defined by the abstract types.

State of Web Services

As mentioned above, SOAP is a stateless protocol and as a consequence, Web services are inherently stateless too. To implement state full services with Web service technology, an additional state preserving "façade" has to be introduced between the client and the Web service interface implementation.

A pointer identifying the state, for example, a session identification, can be passed at several levels of abstraction. The Apache Axis SOAP engine allows to pass *sessionIds* using cookies or SOAP headers. Both mechanisms are technology (http) or product (Axis) specific. Another possibility is to pass the *sessionId*, or even the whole state description, as an extra parameter. In this case, an appropriate extension of the <portType> definition is required.

10.4 Web Service Interoperability

Web services are an advancing technology. Even though there are only a few basic building blocks used in the Web service specifications, they can be combined in diverse modes, and each specification has its own design cycle. Thus, it is not surprising that some additional rules have to be introduced to define a consistent subset of the specifications. Additionally, guidelines for the usage of these specifications have to be defined to ensure interoperability between the code developed by different programmers or generated by different Web service tools. The Web Service Interoperability organization WS-I [10] addresses this need through the concept of *profiles*.

"The Web Services Interoperability Organization is an open industry effort chartered to promote Web Services interoperability across platforms, applications, and programming languages. The organization brings together a diverse community of Web services leaders to respond to customer needs by providing guidance, recommended practices, and supporting resources for developing interoperable Web services."

A WS-I *profile* is a group of specific versions of Web service specifications, along with conventions about how they work together. The WS-I basic *profile specification* [1] defines the conformance of a Web service instance. The profile consists of the following set of nonproprietary Web services specifications:

- SOAP 1.1,
- WSDL 1.1,
- UDDI 2.0,
- XML 1.0 (Second Edition),
- XML Schema Part 1: Structures,
- XML Schema Part 2: Data types, and
- Diverse transport-oriented RFCs.

The *basic profile* [10] provides constraints and clarifications to those base specifications with the intent to promote interoperability. Where the profile is silent, the base specifications are normative. If the profile prescribes a requirement or constraint, it supersedes the underlying base specification. Some of the constraints imposed by the profile are intended to restrict, or require, optional behavior and functionality, so as to reduce the potential of interoperability problems. Some of the constraints or requirements are provided to clarify language in the base specification that may be the source of frequent misinterpretation. The following key constraints are imposed by the *basic profile*:

- preclude the use of SOAP encoding,
- require the use of HTTP binding for SOAP,
- require the use of HTTP 500 status response for SOAP *fault* messages,
- require the use of HTTP POST method,
- require the use of WSDL1.1 to describe the interface of a Web service,
- require the use of rpc/literal or document/literal forms of the WSDL SOAP binding,
- preclude the use of solicit–response and notification style operations,
- require the use of WSDL SOAP binding extension with HTTP as the required transport, and
- require the use of WSDL1.1 descriptions for UDDI tModel elements representing a Web service.

References

[1]　IBM developer works: First look at the WS-I Basic Profile 1.0
　　　http://www-106.ibm.com/developerworks/library/ws-basicprof.html
[2]　Simple object access protocol SOAP:
　　　Version 1.1 http://www.w3.org/TR/SOAP/ ,
　　　Version 1.2 http://www.w3.org/TR/soap12-part0/
[3]　The Web services revolution: Nov 2000;
　　　http://www-106.ibm.com/developerworks/webservices/library/ws-peer1.html
[4]　UDDI: Universal description, discovery and integration of Web services; http://www.uddi.org/
[5]　UDDI schema definition: http://uddi.org/schema/uddi_v2.xsd
[6]　Web services architecture requirements: Nov 2002; http://www.w3.org/TR/wsa-reqs

[7]   WSDL Editor CapeScience: http://www.capescience.com/
[8]   WSDL schema definition: http://schemas.xmlsoap.org/wsdl/
[9]   WSDL SOAP schema definition: http://schemas.xmlsoap.org/wsdl/soap/
[10] WS-I: Web service interoperability organization; http://www.ws-i.org/
[11] XML schema definition: http://www.w3.org/2001/XMLSchema/

11

Programming Web Services with .net and Java

Klaus-Peter Eckert

Fraunhofer Institute for Open Communication Systems

11.1 Introduction

Most of today's development tools support the generation of Web services. It does not matter if you use Microsoft's Visual Studio.net, the possibly embedded Apache Axis [1] SOAP engine for Java services, or design tools like ArcStyler [2] for the generation of Web service from UML models. Every tool has its own strategy to generate the necessary code fragments for Web service servers and their clients. Every tool uses its own mapping from language-specific signatures to WSDL and *vice versa*.

To explain the necessary steps for the implementation, deployment and usage of a Web service three different models will be explained.

Figure 11.1 shows what happens if the development process starts with the *implementation* of the server code. This can either be a Web service-aware code like in our .net example or a Web service-unaware code like in our Java example. The Web service tool considers the program code as input. It generates the WSDL file and the necessary server and client stubs. In most cases, the server code together with the server stubs will be automatically deployed in the local environment. The user-written client code together with the generated client stubs will be assembled into a client archive that can be deployed in the user's environment. Some exemplary client templates may be generated; but in many cases the developer of the Web service client has to study the tool's manuals for details.

Figure 11.2 shows a similar scenario. Here, the generation process starts with a *specification* of the signature of the Web service. This may be a syntactically correct Java class with empty implementation bodies like the Apache Axis implementation requests. Again, the Web service tool generates the WSDL and stubs and additionally a service template. The service developer has to refine this template and insert the service logic. In case of Apache Axis, the template contains an interface specification that corresponds to the given Java class. Both approaches have the disadvantage that they solely allow the implementation of different clients to a given Web service. They do not support the usage of different implementations of a

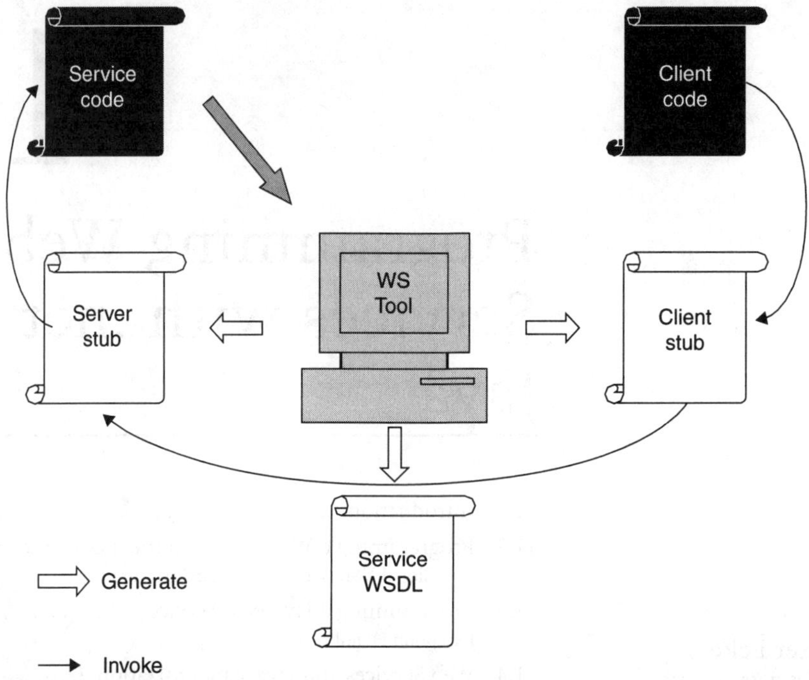

FIGURE 11.1 Starting with the service code.

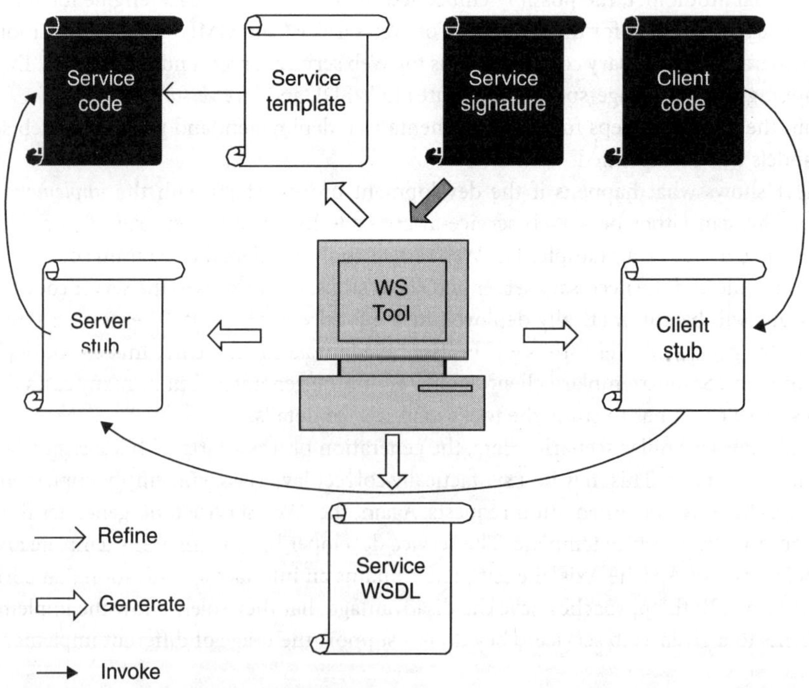

FIGURE 11.2 Starting with the service signature.

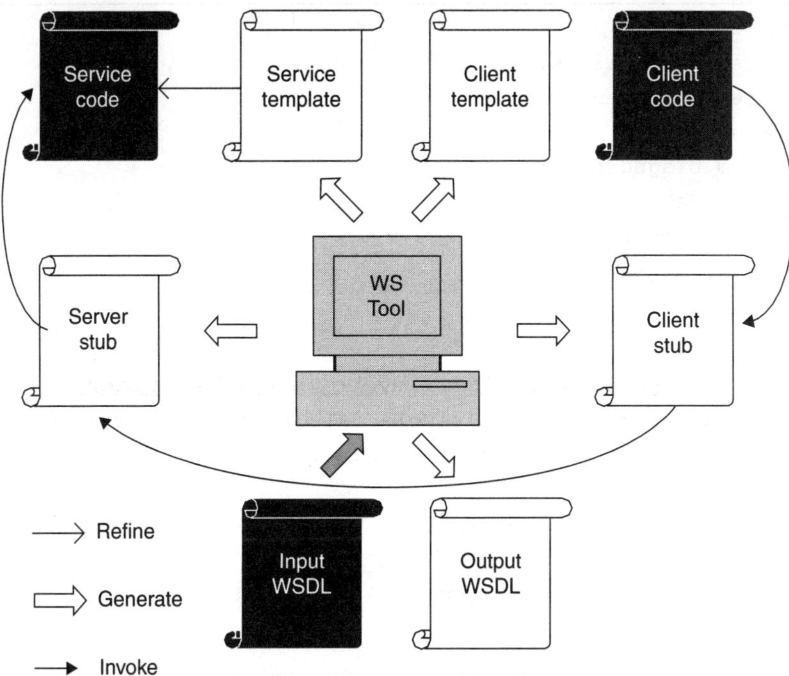

FIGURE 11.3   Starting with the WSDL.

Web service by one client. The latter scenario is important in the case of competing service providers or selective workflows. A solution for this scenario is shown in Figure 11.3.

Here, the Web service tool receives a WSDL file as input. The tool can generate client templates and client stubs that use the entire WSDL information. These clients will bind to the Web service server implementation that is referenced in the *binding* section. More important is the fact that the tool will generate service templates from the *message* and *port type* sections, together with new server stubs and a new WSDL file that contains the binding information to the service that will result from the new implemented and deployed service code. This example shows a problem in the usage of WSDL files. It is not guaranteed that the *message* and *port type* sections of the two WSDL files will remain unchanged in the future. There is neither a logical nor a physical link between both the input and output specifications, and the compliance of the two specifications is the responsibility of the service providers.

To exemplify these abstract models in technical details, let us start with the implementation of a .net Web service, according to the first model.

11.2   Programming a Web Service in the .net Framework

Sometimes, Microsoft Web services and .net are used as synonymous. This is understandable because Web services are one of the core technologies in .net [6]. As a result, the integration of Web services in .net and especially in Visual Studio .net is very high. You only have to say that you want to implement a Web service, and Visual Studio accordingly generates a template and creates several files in the background. The template contains some specific attributes or directives like [WebMethod] or [WebService] that can be used by the compiler to generate the necessary WSDL, stubs, and the other internally used information and files. In the following server implementation, the business code is displayed in *bold*, and the rest of the code is part of the generated template.

```csharp
using System;
using System.Collections;
using System.ComponentModel;
using System.Data;
using System.Diagnostics;
using System.Web;
using System.Web.Services;
namespace CalcWebService
{

    [WebService(
        Namespace="http://www.myorg.org/webservices",
        Description = "The simple calculator Web service" )]
    public class CalcWebService : System.Web.Services.WebService
    {
        public CalcWebService()
        {
            InitializeComponent();
        }
        #region Component Designer generated code
        ...
        #endregion
        [WebMethod(Description = "Add two integers")]
        public int Add(int x, int y){ return x + y; }
        [WebMethod(Description = "Divide two integers")]
        public int Div(int x, int y)
        {
            if(y == 0)
            {
                throw   new   DivideByZeroException("Division   by
zero!");
            }
            return x / y;
        }
    }
}
```

In the implementation of the C# client, the creation of a new *calculator* proxy is required. This can be done in a very natural way. Remember that the binding information is part of the WSDL and that the ws proxy is bound to the server that is specified in the WSDL. The client stubs and the client side SOAP engine are totally hidden to the programmer. The invocation of the Web service's operations on the proxy is done in a very natural way too.

```
using System;
using System.Drawing;
using System.Collections;
using System.ComponentModel;
using System.Windows.Forms;
using System.Data;
namespace CalcClient
{
    public class CalculatorClient : System.Windows.Forms.Form
    {
        private    server.CalcWebService    ws    =    new    server.
CalcWebService ();
        public CalculatorClient()
        {
            InitializeComponent();
        }
        #region Windows Form Designer generated code
        ....
        #endregion
        private void buttonAdd_Click(object sender, System.EventArgs e)
        {
            this.Cursor = Cursors.WaitCursor;
            int ret = ws.Add(
                Convert.ToInt16(textBoxParam1.Text),
                Convert.ToInt16(textBoxParam2.Text));
            textBoxOut.Text = Convert.ToString(ret);
            this.Cursor = Cursors.Default;
        }
    }
}
```

The Global XML Web Services Architecture

The implementation and provision of real Web services, which can be used in a hard business context, involve much more than simply writing C# server and client code. To understand what pieces are missing in the whole puzzle, let us go back to the basics [8]. Every Web service implementation consists of the four layers shown in Figure 11.4.

Web service clients and servers communicate using some application-specific protocols. The used Web service infrastructure is more or less transparent to both the clients and the servers. This basic infrastructure is built on top of the three baseline XML Web service specifications WSDL, SOAP, and UDDI. These specifications provide the foundations for the integration and aggregation of Web applications. Complex Web applications are constructed combining multiple Web services from different vendors running in different technical and organizational domains. But these fundamental specifications leave a gap with respect to the provision of higher-level functionality such as security, reliability, routing, or transactions.

This functionality is often provided in proprietary, vendor-specific, and noninteroperable ways. This is the reason why Microsoft and other companies work on the specification and implementation of the *global XML Web services specifications* as part of the so-called *global XML Web services architecture* GXA [5]. GXA defines a framework for the development of .net Web services that supports additional SOAP modules, a high-level infrastructure, and enhanced description and discovery mechanisms for Web services.

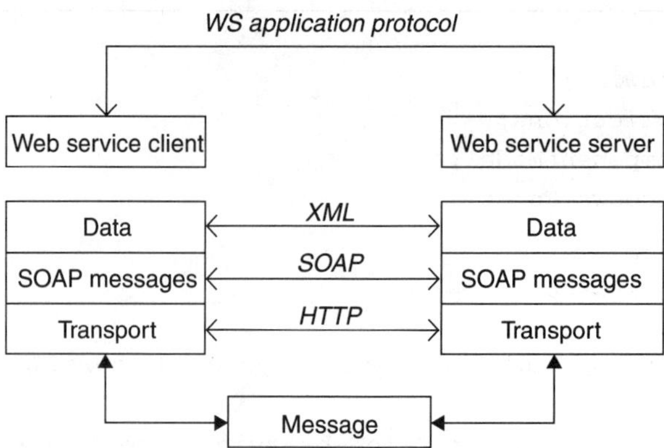

FIGURE 11.4   The Web services protocol stack.

The Web service infrastructure provides two services that support the interworking of Web services. The *WS-Coordination* specification describes an extensible framework for providing protocols that coordinate the actions of distributed Web applications. Such coordination protocols are used to support Web applications that need to reach a consistent agreement on the outcome of distributed transactions. The framework enables a Web service to create a context needed to propagate an activity to other Web services and to register for coordination protocols. It enables existing transaction processing, workflow, and other systems for coordination to conceal their proprietary protocols and to cooperate in a heterogeneous environment.

The *WS-Transaction* specification describes the two coordination types that are used with the extensible coordination framework introduced by the WS-Coordination specification: atomic transaction (AT) and business activity (BA). Developers can use either or both of these coordination types when building Web applications that require consistent agreement on the outcome of distributed activities. WS-Transaction is a building block used with other specifications of Web services like WS-Coordination or WS-Security and application-specific protocols that are able to accommodate a wide variety of coordination protocols related to the coordination actions of distributed Web applications.

UDDI allows to advertise and locate WSDL-based interface and protocol descriptions of Web services in global repositories. The Web service inspection language *WS-Inspection* relies on a distributed model for providing service-related information. It defines both a packaging model described by a schema and a querying model to retrieve a list of service offerings. Requests to retrieve the information are generally made directly to the entities that offer the services. The WS-Inspection specification does not stipulate any particular format for the service information. It relies upon other standards, including WSDL and UDDI, to define the description formats. By providing the ability to disseminate service-related information through existing protocols directly from the point at which the service is being offered, the WS-Inspection mechanism enables focused discovery to be performed on a single target. Due to its decentralized nature, however, the WS-Inspection specification does not provide a good mechanism if the communication partner is unknown. In cases such as this, UDDI would be a better alternative.

The Web services policy framework *WS-Policy* provides a model and a corresponding syntax to describe and communicate the policies of a Web service. At an abstract level, a policy is an expression of a set of conditions on an action that may result in a behavior that reflects these conditions. For Web services, the service provider generally has conditions under which he allows the usage of the service. A requester might use this information to decide whether or not to use the service. *WS-Policy* defines a base set of constructs to describe service requirements, preferences, and capabilities. It consists of two parts. *WS-PolicyAssertions* specify a set of common message policy assertions that can be specified within a policy. *WS-PolicyAttachment* specifies attachment mechanisms for using policy expressions. It defines how to associate policy expressions with WSDL <types> definitions and UDDI entities and how to associate implementation-specific policies with the WSDL <portType>.

The following three specifications define SOAP extensions that define standardized SOAP headers to enable a higher level of security during the usage of Web services. The Web services routing protocol *WS-Routing* defines mechanisms for routing SOAP messages. SOAP does not actually define a mechanism for sending a message from one party to another. WS-Routing defines a new SOAP header and an associated processing model. It is a stateless protocol that extends SOAP by defining a means to specify an ordered route from the originator of the message, through intermediaries, to the ultimate message receiver. The Web services referral protocol *WS-Referral* is a simple SOAP-based protocol for configuring instructions about message routes in intermediary SOAP nodes called *routers*. WS-Referral is orthogonal to *WS-Routing* in that WS-Referral provides a way to configure *how* SOAP routers will build a message route, whereas WS-Routing provides a mechanism for describing an actual path of a message. SOAP routers are used to add nonfunctional properties like load-balancing, mirroring, caching, or client authentication services to a Web service in a manner that is transparent to users of the service.

The *WS-Security* specification defines a set of enhancements to SOAP messaging. Its goal is to provide quality of protection through message integrity, message confidentiality, and single message authentication. The associated mechanisms can be used to accommodate a wide variety of security models and encryption technologies. WS-Security provides an extensible, general-purpose mechanism for associating security tokens with messages. It describes how to encode binary security tokens like X.509 certificates and Kerberos tickets, as well as how to include opaque encrypted keys. It includes extensibility mechanisms that can be used to further describe the characteristics of the credentials that are included within a message. The *Web services security profile for XML-based tokens* describes how to use XML-based tokens such as the Security Assertion Markup Language (SAML) or the eXtensible rights Markup Language (XrML) with the WS-Security specification. *WS-Trust* defines extensions to request and issue security tokens and to manage trust relationships. Finally, *WS-SecureConversation* defines mechanisms for establishing and sharing security contexts, and for deriving session keys from these contexts.

The global XML Web services architecture GXA, as shown in Figure 11.5, consists of these upcoming specifications. It defines a framework for the development of future .net Web services. Some of these specifications result in APIs that will be directly used by the service developers. Others will result in specific deployment or configuration descriptors that have to be mapped by the run time environment to the appropriate specifications.

11.3 Programming a Java Web Service

For the implementation of a Java Web service, let us go into some more technical details. The following description applies to the usage of Apache Axis [1].

As in the C# example, the implementation of the Java server is quite forward. The tool generates the WSDL and server stubs from the implementation and deploys them in the Axis environment, transparent to the developer. The environment consists of an Apache Tomcat Web server and the Axis server-side SOAP engine.

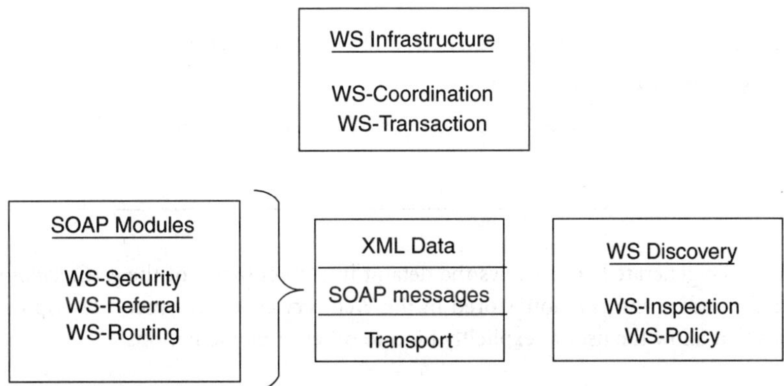

FIGURE 11.5   The global XML Web services architecure GXA.

```java
package calculator;
public class Calculator {
  public int add(int x, int y)
  {
    System.out.println("Add" + x + "+" + y);
    return x + y;
  }
  public int divide(int x, int y) throws DivideByZeroException
  {
    System.out.println("Divide " + x + "/" + y);
    if(y == 0)
    {
      throw new DivideByZeroException("Division by zero!");
    }
    return x / y;
  }
}
```

For the client-side binding and the calculator client stub, two Java interfaces and two Java classes are generated. As shown in Figure 11.6, the class *CalculatorServiceLocator* implements the interface *CalculatorService* and provides an operation *getCalculator* that returns an instance of the client-side proxy called *CalculatorSoapBindingStub*.

```java
/**
 * CalculatorService.java
 *
 * This file was auto-generated from WSDL
 * by the Apache Axis WSDL2Java emitter.
 */
package calculator.generated;
public interface CalculatorService extends javax.xml.rpc.Service {
...
    public calculator.generated.CalculatorPortType
        getCalculator()
            throws javax.xml.rpc.ServiceException;
}
```

This excerpt from the generated code shows the default implementation of the *getCalculator* operation. It uses the default binding information, stored in the Web service's deployment descriptor. There exist more operations that allow the user to explicitly bind to other implementations.

```java
/**
 * CalculatorServiceLocator.java
 *
 * This file was auto-generated from WSDL
 * by the Apache Axis WSDL2Java emitter.
 */
package calculator.generated;
public class CalculatorServiceLocator
      extends org.apache.axis.client.Service
      implements calculator.generated.CalculatorService {
...
    public calculator.generated.CalculatorPortType
          getCalculator()
              throws javax.xml.rpc.ServiceException {
        java.net.URL endpoint;
         try {
             endpoint = new java.net.URL(Calc_address);
         }
         catch (java.net.MalformedURLException e) {
             return null; // unlikely as URL was validated in WSDL2Java
         }
         return getCalculator(endpoint);
    }
    public calculator.generated.CalculatorPortType
          getCalculator(java.net.URL portAddress)
              throws javax.xml.rpc.ServiceException {
        try {
            calculator.generated.CalculatorSoapBindingStub _stub =
              new calculator.generated.CalculatorSoapBindingStub
                    (portAddress, this);
            _stub.setPortName(getCalcWSDDServiceName());
            return _stub;
        }
        catch (org.apache.axis.AxisFault e) {
            return null;
        }
    }
...
}
```

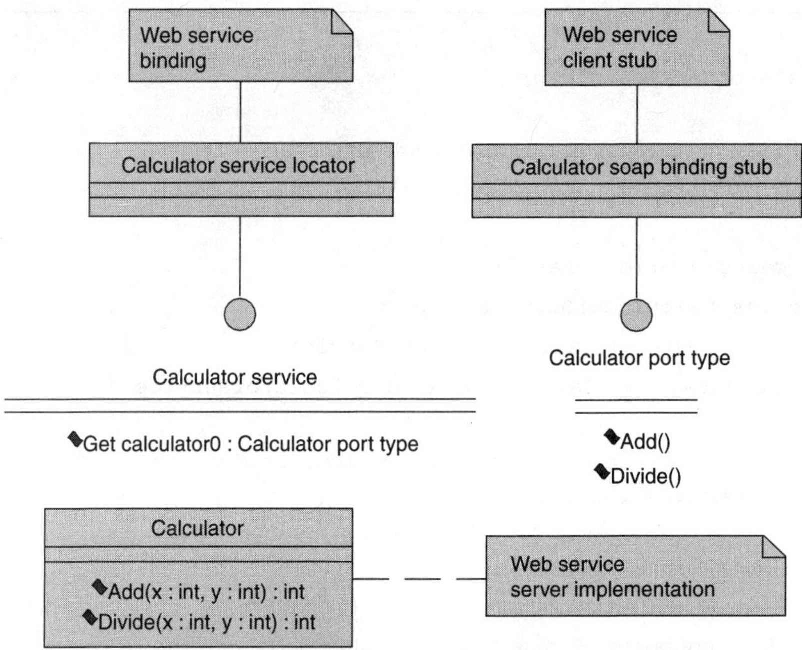

FIGURE 11.6   UML model of the generated Java classes and interfaces.

The generated <portType> proxy interface provides the same operations as the Web service implementation itself.

```
/**
 * CalculatorPortType.java
 *
 * This file was auto-generated from WSDL
 * by the Apache Axis WSDL2Java emitter.
 */
package calculator.generated;
public interface CalculatorPortType extends java.rmi.Remote {
    public int add(int x, int y) throws java.rmi.RemoteException;
    public int divide(int x, int y)
        throws java.rmi.RemoteException, DivideByZeroException;
}
```

The associated class *CalculatorSoapBindingStub* implements these operations. It creates and initializes a dynamic *Call* object and calls its *invoke* operation. This operation passes the object to the SOAP client-side engine that performs all the necessary housekeeping activities: encode the request, submit the request to the selected communication protocol, wait for the response, and decode the response.

```java
/**
 * CalculatorSoapBindingStub.java
 *
 * This file was auto-generated from WSDL
 * by the Apache Axis WSDL2Java emitter.
 */
package calculator.generated;
public class CalculatorSoapBindingStub
        extends org.apache.axis.client.Stub
        implements calculator.generated.CalculatorPortType {
...

    public CalculatorSoapBindingStub(
            java.net.URL endpointURL,
            javax.xml.rpc.Service service)
                throws org.apache.axis.AxisFault {
            this(service);
            super.cachedEndpoint = endpointURL;
    }
...

    public int add(int x, int y) throws java.rmi.RemoteException {
        if (super.cachedEndpoint == null) {
            throw new org.apache.axis.NoEndPointException();
        }
        org.apache.axis.client.Call _call = createCall();
        _call.addParameter(new javax.xml.namespace.QName("", "x"),...
        _call.addParameter(new javax.xml.namespace.QName("", "y"),...
          _call.setReturnType(new javax.xml.namespace.QName(..., int.
class);
        _call.setUseSOAPAction(true);
        _call.setSOAPActionURI("");
        _call.setOperationStyle("rpc");
        _call.setOperationName(
                new    javax.xml.namespace.QName("http://calculator",
"Add"));
        java.lang.Object _resp = _call.invoke(new java.lang.Object[]
          {new java.lang.Integer(x), new java.lang.Integer(y)});
...
}
```

The implementation of the Java test client is quite forward. It creates a *CalculatorServiceLocator* object and calls the *getCalculator* operation to bind itself to the default Web service implementation and to obtain a *CalculatorPortType* interface as a local proxy. This proxy provides the same operations as the Web service itself and can be used in a similar way.

```
package calculator.generated;
public class CalculatorClient {
  public static void main(String [] args) throws Exception {
    calculator.generated.CalculatorServiceLocator locator =
          new calculator.CalculatorServiceLocator();
    calculator.CalculatorPortType obj = locator.getCalculator();
    System.out.println("Test add: " + obj.add (100 , 200));
  }
}
```

As in the .net example, the implementation of a Web service client is quite simple as long as the client binds to the server that is referred in the WSDL file and as long as the Web service tool generates all the necessary client-side classes. It is worth mentioning that the client-side interfaces to the SOAP engine as well as the server side interfaces are totally hidden in this example. If you want to write a Web service client by hand, you have to satisfy these interfaces and you probably have to parse and interpret the WSDL file yourself. More information about the appropriate Java APIs is given in Section 11.4 "The Java community process."

Exception Handling

In distributed environments in many cases, the treatment of exceptions is much more difficult than the treatment of the positive cases. In both the .net and the Java examples, the server implementations throw exceptions on the level of the appropriate programming language. This section shows how WSDL supports the definition of exceptions. It is up to the client-side stub to remap these exceptions to the programming language used by the client stub.

The following example shows an excerpt from the WSDL specification, generated from the source file of the Java implementation of the calculator service. In addition to the previous examples, a new <types> element is included that defines the exception that is thrown by the *Divide* operation in case of an attempt to divide by zero. It has to be mentioned that the generated WSDL differs from tool to tool. Thus, this example has to be considered as one of several possible syntactical representations of the specification of the same calculator Web service too. Different tools may generate different names.

```
<definitions ...
      xmlns:soapbind=http://schemas.xmlsoap.org/wsdl/soap/ ...>
<types>
      <schema ...
            <complexType name="Calculator.DivideByZeroException">
                  <sequence/>
            </complexType>
      </schema>
</types>
```

The exception is expressed as a special return message <part name="fault">.

```
<message name="Div">
     <part name="x" type="xsd:int"/>
     <part name="y" type="xsd:int"/>
</message>
<message name="DivResponse">
     <part name="DivReturn" type="xsd:int"/>
</message>
<message name="Calc$DivideByZeroException">
     <part name="fault"
           type="xsd1:Calculator.DivideByZeroException"/>
</message>
```

The definition of the associated <portType> contains the new message.

```
<portType name="Calculator"> ...
     <operation name="Div" parameterOrder="x y">
          <input name ="Div" message="tns:Div"/>
          <output name ="DivResponse" message="tns:DivResponse"/>
          <fault name="Calc$DivideByZeroException"
              message="tns:Calc$DivideByZeroException"/>
     </operation>
</portType>
```

In the <binding> section, the new *fault* element is included.

```
<binding name="CalculatorBinding" type="tns:Calculator">
. . .
     <fault name="Calc$DivideByZeroException">
          <soapbind:fault
              use="encoded"
              encodingStyle="http://schemas.xmlsoap.org/soap/encoding/"
              namespace="Calculator"/>
     </fault>
</binding>
```

11.4   Web Services and Java standardization

In contrast to .net, the integration of Web service technology into the Java world is still under development. The following two upcoming approaches are of major importance.

Web Services and J2EE

In many cases, the implementation of a Web service is not given by one or more Java classes but by an Enterprise Java Bean. Because Web services are inherently stateless, the only EJB type that is a natural candidate for a Web service is a stateless session bean. Some of today's Web service tools use an EJB archive as input and generate a WSDL file out of a selected remote interface. Additionally, these tools generate a servlet as a Web service proxy that maps the incoming SOAP requests to RMI requests and forwards them to the EJB implementation.

The upcoming EJB specification 2.1 [7] includes an extension for stateless session beans that allows a Web client to access the bean via a *Web service endpoint interface*. The *service–endpoint–interface* element in the Web service client's deployment descriptor defines a fully qualified Java class that represents the service endpoint interface of a WSDL *port*. The Web service itself is directly implemented by the bean and runs in the associated Web container.

The developer is responsible for defining a *service-ref* element for each Web service that the client wants to reference. This includes a logical name for the reference that is used in the client's code, the fully qualified name of the service interface class returned by the JNDI lookup and references to WSDL ports. More technical details can be found in [3].

The Java Community Process

The Java community process JCP [4] is an open organization of international Java developers and licensees, whose charter is to develop and revise Java technology specifications, reference implementations, and technology compatibility kits. JCP publishes a set of so-called Java specification requests JSR that define Java APIs for domain-specific problems. The following JSRs deal with Web service technology and have to be applied by application developers to implement Java Web services by hand or by tool developers to ensure standard conformity and interoperability.

- The XML Parsing JAXP (JSR63) defines a set of implementation-independent portable APIs supporting XML Processing.
- The Java API for WSDL (JSR 110) provides a standard set of APIs for representing and manipulating services described by WSDL documents.
- The Java API for XML-registries JAXR (JSR 93) supports a set of distributed registry services that enables business-to-business integration between business enterprises, using the protocols being defined by ebXML.org, Oasis, ISO 11179.
- The Java API for XML-based PRCs (JSR 101) supports emerging industry XML-based RPC standards.
- The Java API for XML messaging JAXM (JSR 67) supports packaging and transporting business transactions using on-the-wire protocols being defined by ebXML.org, Oasis, W3C and IETF including SOAP with attachment APIs for Java (SAAJ).
- The J2ME Web Service Specification (JSR 172) defines an optional package that provides standard access from mobile J2ME clients to Web services.

References

[1] Apache Axis: http://ws.apache.org/axis/
[2] ArcStyler: http://www.io-software.com/
[3] IBM: Web services for J2EE; Version 1.0; http://www-106.ibm.com/developerworks/webservices/library/ws-jsr109/index.pdf
[4] Java Community Process JCP: http://www.jcp.org/
[5] Microsoft: The global XML Web services architecture GXA; http://msdn.microsoft.com/webservices/understanding/gxa/default.aspx
[6] Microsoft: XML Web Services Developer Centre Home; http://msdn.microsoft.com/webservices/default.aspx
[7] Sun Microsystems: Enterprise JavaBeans Specification; Version 2.1; http://java.sun.com/products/ejb/
[8] W3C: Web services architecture; http://www.w3.org/TR/

12

Multidimensional Databases[1]

Torben Bach Pedersen
Aalborg University

Christian S. Jensen
Aalborg University

12.1   Introduction

The relational data model, which was introduced by Codd in 1970 and earned him the Turing Award a decade later, was the foundation of today's multi-billion-dollar database industry. During the 1990s, a new type of data model, the *multidimensional data model*, has emerged, which has taken over from the relational model where the objective is to *analyze* data, rather than to perform on-line transactions.

Multidimensional data models are designed expressly to support data analyses. A number of such models have been proposed by researchers from academia and industry. In academia, formal mathematical models have been proposed, while the industrial proposals have typically been specified more or less *implicitly* by the concrete software tools that implement them.

Briefly, multidimensional models categorize data as being either *facts* with associated numerical *measures*, or as being *dimensions* that characterize the facts and are mostly textual. For example, in a retail business, *products* are sold to *customers* at certain *times* in certain *amounts* and at certain *prices*. A typical fact would be a *purchase*. Typical measures would be the amount and price of the purchase. Typical dimensions would be the location of the purchase, the type of product being purchased, and the time of

---

[1]Based on "Multidimensional Database Technology" by Torben Bach Pedersen and Christian S. Jensen which appeared in IEEE Computer 34(12):40–46. ©2001 IEEE.

the purchase. Queries then aggregate measure values over ranges of dimension values to produce results such as the total sales per month and product type.

Multidimensional data models have three important application areas within data analysis. First, multidimensional models are used in *data warehousing*. A data warehouse is a large repository of integrated data obtained from several sources in an enterprise for the specific purpose of data analysis. Typically, these data are modeled as being multidimensional, as this best supports data analyses.

Second, multidimensional models lie at the core of *On-Line Analytical Processing* (OLAP) systems. Such systems provide fast answers to queries that aggregate large amounts of so-called detail data to find overall trends, and they present the results in a multidimensional manner. Consequently, a multidimensional data organization is ideal for OLAP.

Third, multidimensional data are increasingly becoming the basis for *data mining*, where the aim is the (semi) automatic discovery of unknown knowledge in large databases, as it turns out that multidimensionally organized databases are particularly well suited for queries posed by data mining tools.

This chapter describes fundamental concepts in multidimensional data models. It aims to communicate the essence of the multitude of proposed models in a clear and easily understood form. In doing so, we hope that the chapter will serve as an introduction of the concepts and benefits of multidimensional databases, which are well known in the database community by now, to the broader community of computer science. In-depth coverage of the features of individual models may be found elsewhere [18, 23]. The chapter also addresses implementation technologies and gives an overview of commercial systems.

12.2 Background

Related Terminology

It is useful to be familiar with a few special terms when studying the literature on issues related to multidimensional databases.

OLAP: OLAP abbreviates *On-Line Analytical Processing*. As opposed to the well-known On-Line *Transaction* Processing (OLTP), the focus is on data analyses rather than transactions. Furthermore, the analyses occur "On-Line," that is, a fast query response is required. OLAP systems always adopt a multidimensional view of data.

Data warehouse: A data warehouse (DW) is a repository of integrated enterprise data that are used specifically for decision support, that is, there is (typically) only one data warehouse in an enterprise. The data in a DW are typically collected from a large number of sources within (and sometimes also outside) the enterprise.

Data mart: A data mart (DM) is a subset of a data warehouse that is specialized to the needs of a special user group, for example, the marketing department.

ETL: Extract-Transform-Load (ETL) is the three-step process that puts data into the DW. First, data are *extracted* from the operational source systems, for example, ERP systems. Second, data are *transformed* from the source system formats into the DW format. This includes combining data from several sources and performing *cleansing* to correct errors such as missing or wrong data. Third, data are *loaded* into the DW. At times, ETL is also referred to as Extract-Transform-Transport (ETT).

Multidimensional History

Multidimensional databases do not have their origin in database technology, but stem from multidimensional matrix algebra, which has been used for (manual) data analyses since the late 19th century.

During the late 1960s, two companies, IRI and Comshare, independently began the development of systems that later evolved into multidimensional database systems. The IRI Express tool became very popular in the marketing analysis area in the late 1970s and early 1980s; it later evolved into a market-leading OLAP tool and was acquired by Oracle. Concurrently, the Comshare system evolved into System W, which was used considerably for financial planning, analysis, and reporting during the 1980s.

In 1991, Arbor was formed with the specific purpose of creating "a multiuser, multidimensional database server," which resulted in the Essbase system. Arbor, now Hyperion, later licensed a basic version of Essbase to IBM for integration into DB2. It was Arbor and Codd who in 1993 coined the term OLAP [2].

Another significant development in the early 1990s was the advent of large *data warehouses* [10], which are typically based on relational *star* or *snowflake* schemas, an approach to implementing multidimensional databases using relational database technology.

In 1998, Microsoft shipped its MS OLAP Server, the first multidimensional system aimed at the mass market. This has led to the current situation, where multidimensional systems are increasingly becoming commodity products that are shipped at no extra cost together with leading relational database management systems.

A more in-depth coverage of the history of multidimensional databases is available in the literature [21].

12.3   Spreadsheets and Relations

Assume that we want to analyze data about sales of products, for which we capture the sales price, the product sold, and the city in which it was sold. A simple example with two dimensions is shown in Table 12.1.

When considering how to analyze such data, *spreadsheets* immediately come to mind as a possibility — Table 12.1 is just a (two-dimensional) spreadsheet.

Our first analysis requirement is that we do not just want to see sales by product and city, but also the two kinds of subtotals, sales by product and sales by city, and the grand total of sales. This means that formulas for producing the (sub)totals must be added to the spreadsheet, each requiring some consideration. It is possible, if rather cumbersome, to add new data to the spreadsheet, for example, if new products are sold. Thus, for two dimensions, we can perhaps somehow manage with spreadsheets.

However, if we go to three dimensions, for example, to include time, we have to consider carefully what to do. The obvious solution is to use separate worksheets to handle the extra dimension, with one worksheet for each dimension value. This will work only when the third dimension has few dimension values and only to some extent. Analyses involving several values of the third dimension are cumbersome, and with many thousands of, say, time dimension values, the solution becomes infeasible. The situation becomes even worse if we need to support four or more dimensions, which in any case will require a very complex setup.

Another problem arises if we want to group, for example, the product into higher-level product types like "food" and "non-food." Then, we must duplicate the grouping information across all worksheets, which results in a system that uses considerable extra space and is very difficult to maintain. The essence of the problem is that spreadsheets tie the storage of data too tightly to the presentation — the *structure* and the *desired views* of the information are not separated. However, spreadsheets are good for *viewing and querying* multidimensional data, for example, using *pivot tables*. A pivot table is a two-dimensional table of data with associated subtotals and totals. For example, if we add subtotals by City and Product and a City/Product grand total to Table 12.1, we obtain an example of a pivot table. To support viewing of more complex data, several dimensions may be nested on the *x*- or *y*-axis, and data may be displayed on multiple pages, for example, one for each product. Pivot tables generally also offer support for interactively selecting subsets of the data and changing the displayed level of detail.

With spreadsheets falling short in meeting our requirements for the management of multidimensional data, we may then consider using an SQL-based, relational system for data management, as the relational

TABLE 12.1   Sales Data

		City		
Product	Aalborg	Copenhagen	Berkeley	New York
Milk	123	555	145	5001
Bread	102	250	54	2010
Jeans	20	89	32	345
Lighbulps	22	213	32	9450

model offers considerable flexibility in the modeling and querying of data. The problem here is that many desirable computations, including cumulative aggregates (sales in year to date), totals and subtotals together, and rankings (top 10 selling products), are difficult or impossible to formulate in standard SQL.

The main underlying issue is that *interrow* computations are difficult to express in SQL — only *intercolumn* computations are easy to specify. Additionally, transpositions of rows and columns are not easily possible, but rather require the manual specification and combination of multiple views. Although extensions of SQL, such as the *data cube operator* [7] and *query windows* [5], advanced by the standards bodies, will remedy some of the problems, the concept of hierarchical dimensions remains to be handled satisfactorily.

To summarize, we have seen that neither spreadsheets nor relational databases fully support the requirements posed by advanced data analyses. To be fair, these technologies may be adequate in more restricted circumstances. For example, if we have only a few dimensions, do not need hierarchical dimensions, and the data volume is small, spreadsheets may provide adequate support. However, the only robust solution to the above problems is to provide data models and database technology that offer inherent support for the full range of multidimensional concepts.

12.4 Cubes

Data cubes provide true multidimensionality. They generalize spreadsheets to any number of dimensions. In addition, hierarchies in dimensions and formulas are first-class, built-in concepts, meaning that these are supported without duplicating their definitions. A collection of related cubes is commonly referred to as a *multidimensional database* or a *multidimensional data warehouse*.

Figure 12.1 shows a cube capturing the product sales from Table 12.1 for the two Danish cities, with the additional dimension Time. The combinations of dimension values define the *cells* of the cube. The actual sales prices are stored within the corresponding cells.

In a cube, dimensions are first-class concepts with associated domains, meaning that the addition of new dimension values is easily handled. Although the term "cube" implies three dimensions, a cube can have any number of dimensions. It turns out that most real-world cubes have 4–12 dimensions [10, 21]. Although there is no theoretical limit to the number of dimensions, current tools often experience performance problems when the number of dimensions is more than 10–15. To better illustrate the high number of dimensions, the term "hypercube" is often used instead of "cube."

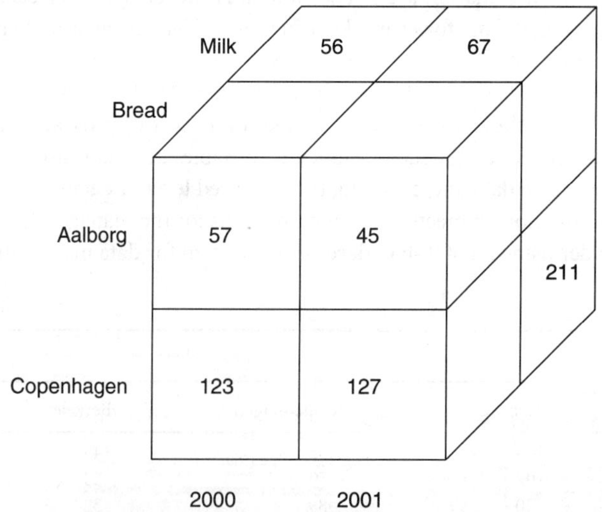

FIGURE 12.1   Sales cube.

Depending on the specific application, a highly varying percentage of the cells in a cube are nonempty, meaning that cubes range from *sparse* to *dense*. Cubes tend to become increasingly sparse with increasing dimensionality and with increasingly finer granularities of the dimension values.

A nonempty cell is called a *fact*. The example has a fact for each combination of time, product, and city where at least one sale was made. A fact has a number of *measures* associated with it. These are numerical values that "live" within the cells. In our case, we have one measure, the sales price.

Generally, only two or three dimensions may be viewed at the same time, although for low-cardinality dimensions, up to four dimensions can be shown by nesting one dimension within another on the axes. Thus, the dimensionality of a cube is reduced at query time by *projecting* it down to two or three dimensions via *aggregation* of the measure values across the projected-out dimensions. For example, if we want to view just sales by City and Time, we aggregate over the entire Product dimension for each combination of City and Time. In our example, we obtain the total sales for Copenhagen in 2001 by adding 127 and 211.

An important goal of multidimensional modeling is to "provide as much context as possible for the facts" [10]. The concept of *dimension* is the central means of providing this context. One consequence of this is a different view on *data redundancy* than in relational databases. In multidimensional databases, controlled redundancy is generally considered appropriate, as long as it considerably increases the information value of the data. One reason to allow redundancy is that multidimensional data are often *derived* from other data sources, for example, data from a transactional relational system, rather than being "born" as multidimensional data, meaning that updates can be handled more easily [10]. However, there is usually no redundancy in the facts, only in the dimensions.

Having introduced the cube, we describe its principal elements, dimensions, facts, and measures, in more detail.

12.5 Dimensions

The notion of a dimension is an essential and distinguishing concept in multidimensional databases. Dimensions are used for two purposes: *selection* of data and *grouping* of data at the desired level of detail.

A dimension is organized into a containment-like hierarchy, composed of a number of *levels* that each represent a level of detail that is of interest to the analyses to be performed. The instances of the dimension are typically called *dimension values*. Each such value belongs to a particular level. In some cases, it is advantageous for a dimension to have *multiple hierarchies* defined on it, for example, a Time dimension that has hierarchies for both *Fiscal Year* and *Calendar Year*. Multiple hierarchies share one or more common lowest level(s), for example, Day and Month, and then group these into multiple levels higher up, for example, Fiscal Quarter and Calendar Quarter to allow for easy reference to several ways of grouping. Most multidimensional models allow multiple hierarchies. A dimension hierarchy is defined in the metadata of the cube, or the metadata of the multidimensional database, if dimensions can be shared. This means that the problem of duplicate hierarchy definitions as discussed in Section 12.3 is avoided.

In Figure 12.2, the schema and instances of a sample *Location* dimension for the data in Table 12.1 are shown.

The Location dimension has three levels, the City level being the lowest. City-level values are grouped into *Country*-level values, that is, countries. For example, Aalborg is in Denmark. The T ("top") level represents *all* of the dimension, that is, every dimension value is part of the T ("top") value.

In some multidimensional models, a level may have associated with it a number of *level properties* that are used to hold simple, nonhierarchical, information. For example, the package size for a given product can be a level property in the Product level of the Product dimension. This information could also be captured using an extra Package Size dimension. Using the level property has the effect of not increasing the dimensionality of the cube.

Unlike the linear spaces used in matrix algebra, there is typically no ordering and/or distance metric on the dimension values in multidimensional models. Rather, the only ordering is the containment of lower-level values in higher-level values. However, for some dimensions, for example, the Time

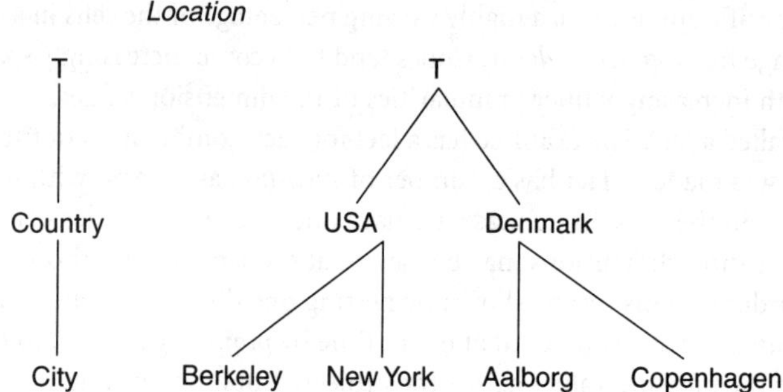

FIGURE 12.2   Schema and instances for the Location dimension.

dimension, an ordering of the dimension values is available and is used for calculating cumulative information such as "total sales in year to date."

Most models require dimension hierarchies to form *balanced trees*. This means that the dimension hierarchy must have a uniform height everywhere, for example, all departments, even small ones, must be subdivided into project groups. Additionally, direct links between dimension values can only go between immediate parent–child levels, and not jump two or more levels. For example, all cities are first grouped into states and then into countries; cities cannot be grouped directly under countries (as is the case in Denmark, which has no states). Finally, each nontop value has precisely one parent, for example, a product must belong to exactly one product group. In Section 12.10, we discuss the relaxation of these constraints.

12.6   Facts

Facts are the objects that represent the *subject* of the desired analyses, that is, the interesting "thing," or event or process, in the enterprise that is to be analyzed to better understand its behavior.

In most multidimensional data models, the facts are *implicitly* defined by their combination of dimension values. If a nonempty cell exists for a particular combination, a fact exists; otherwise, no fact exists. (Some other models treat facts as first-class objects with a separate identity [18].) Next, most multidimensional models require that each fact be mapped to precisely one dimension value at the lowest level in each dimension. Other models relax this requirement [18].

A fact has a certain *granularity*, determined by the levels from which its combination of dimension values are drawn. For example, the fact granularity in our example cube is "Year by Product by City." Granularities consisting of higher- or lower-level dimension levels than a given granularity, for example, "Year by product Type by City" or "Day by Product by City" for our example, are said to be *coarser* or *finer* than the given granularity, respectively.

It is commonplace to distinguish among three kinds of facts: *event* facts, *snapshot* facts, and *cumulative snapshot* facts [10]. Event facts (at least at the finest granularity) typically model *events in the real world*, meaning that a unique instance, for example, a particular sale of a given product in a given store at a given time, of the overall real-world process that is captured, for example, sales for a supermarket chain, is represented by one fact. Examples of event facts include sales, clicks on web pages, and movements of goods in and out of (real) warehouses (flow).

A snapshot fact models the *state* of a given process at a given point in time. Typical examples of snapshot facts include the inventory levels in stores and warehouses, and the number of users using a web site. For snapshot facts, the same object, for example, a specific can of beans on a shelf, with which the captured real-world process, for example, inventory management, is concerned may occur in several facts at different time points.

Cumulative snapshot facts are used to handle information about *a process up to a certain point in time*. For example, we may consider the total sales in year to date as a fact. Then the total sales up to and including the current month this year can be easily compared to the figure for the corresponding month last year.

Often, all three types of facts can be found in a given data warehouse, as they support complementary classes of analyses. Indeed, the same base data, for example, the movement of goods in a (real) warehouse, may often find its way into three cubes of different types, for example, warehouse flow, warehouse inventory, and warehouse flow in year-to-date.

12.7 Measures

A *measure* has two components: a *numerical property* of a fact, for example, the sales price or profit, and a *formula* (most often, a simple aggregation function such as SUM) that can be used to combine several measure values into one. In a multidimensional database, measures generally represent the properties of the chosen facts that the users want to study, for example, with the purpose of optimizing them.

Measures then take on different values for different combinations of dimension values. The property and formula are chosen such that the value of a measure is meaningful for all combinations of aggregation levels. The formula is defined in the metadata and thus not replicated as in the spreadsheet example. Although most multidimensional data models have measures, some do not. In these, dimension values are also used for computations, thus obviating the need for measures, but at the expense of some user-friendliness [18].

It is important to distinguish among three classes of measures, namely *additive, semiadditive*, and *nonadditive* measures, as these behave quite differently in computations.

Additive measure values can be combined meaningfully along any dimension. For example, it makes sense to add the total sales over Product, Location, and Time, as this causes no overlap among the real-world phenomena that caused the individual values. Additive measures occur for any kind of fact.

Semiadditive measure values cannot be combined along one or more of the dimensions, most often the Time dimension. Semiadditive measures generally occur when the fact is of type snapshot or cumulative snapshot. For example, it does not make sense to sum inventory levels across time, as the same inventory item, for example, a specific can of beans, may be counted several times, but it is meaningful to sum inventory levels across products and warehouses.

Nonadditive measure values cannot be combined along any dimension, usually because of the chosen formula. For example, this occurs when averages for lower-level values cannot be combined into averages for higher-level values. Nonadditive measures can occur for any kind of fact.

12.8 Querying

A multidimensional database naturally lends itself toward certain types of queries.

The queries known as *slice and dice* perform selections on a cube, thus reducing the cube. Their effect is similar to that of preparing an onion for cooking. For example, we may slice the cube in Figure 12.1 by considering only those cells that concern "Bread," and we may further slice this resulting slice, for example, by considering only the cells for year "2000." When we select a single value in a dimension, we, in a sense, reduce the dimensionality of the cube (strictly speaking, a degenerate dimension remains), but more general selections are also possible.

The operations known as *drill-down* and *roll-up* are inverses of each other and make use of dimension hierarchies and measures to perform aggregations. Consider the location dimension in Figure 12.2 together with the cube, where the (additive) measure is the count of sales together with the function SUM. We may roll-up from City level to Country level. This results in the measure values for all cities in the same country being combined into one by the associated formula. In our cube, values for "Aalborg" and "Copenhagen" are added to obtain a single "Denmark" value. For the "Bread" slice, this yields two cells, with values 180 (for year 2000) and 172 (for year 2001).

Slicing and dicing may be combined with drill-down and roll-up. Rolling up to the top value in a dimension corresponds in a sense (slightly different from the one above) to omitting the dimension.

When a multidimensional database consists of several cubes that share one or more dimensions, it is possible to combine the cubes via the shared dimensions, via a so-called *drill-across* operation. In terms of relational algebra, this operation performs a join.

Next, operations are also available that enable the user to manipulate the visualization of a cube, for example, by *rotating* it to make a different set of dimensions "face the user," that is, to see the data grouped by other dimensions.

Queries involving order are very important in data analyses and are thus supported by all multidimensional data analysis tools. These may order cells in results, and they may return only those cells that appear at the top or bottom of the specified order. For example, one may wish to retrieve the ten best-selling products in "Copenhagen" in the year 2000. Such queries are often referred to as *ranking* or *TOP N/BOTTOM N* queries [21].

12.9   Implementation Technologies

One may distinguish between two major approaches to implementing multidimensional databases, termed multidimensional vs. relational on-line analytical processing. We first contrast the two and then consider the latter in additional detail.

Multidimensional vs. Relational OLAP

Multidimensional OLAP (MOLAP) systems [21] store data on disk in specialized multidimensional structures. These typically include provisions for handling sparse arrays, and they apply advanced indexing and hashing to locate the data when performing queries [21]. In contrast, *Relational OLAP* (ROLAP) systems [10] use relational database technology for storing the data. In order to achieve good query performance, ROLAP systems employ specialized index structures such as bit-mapped indices [6], as well as materialized views [24]. We offer more information regarding DW performance at the end of this section.

Generally, MOLAP systems provide faster query response and more space-efficient storage, while ROLAP systems scale better in the number of facts, are more flexible with respect to cube redefinitions, and provide better support for frequent updates.

The virtues of the two approached are combined in the *Hybrid OLAP* (HOLAP) approach, which generally stores higher-level summary data using MOLAP technology, while using ROLAP technology to store the detailed data. Some HOLAP systems allow even more flexibility, giving users a choice between MOLAP and ROLAP at the level of individual cubes and/or materialized aggregates.

Relational OLAP Schemas

ROLAP is the most commonly used technology for the detailed data in the DW; hence, we go on to describe the typical schemas found in ROLAP DWs.

ROLAP implementations typically use *star* or *snowflake* schemas [10], both of which store the data in *fact tables* and *dimension tables*. A fact table holds one row for each fact in the cube. It has a column for each measure. In a row, this column then contains the measure value for the fact that the row represents. A fact table also has one column for each dimension. In a row, these columns contain foreign keys that reference dimension tables.

The difference between star and snowflake schemas lies in their handling of dimensions. A star schema has one dimension table for each dimension. The dimension table contains a key column. It also has one column for each level in the dimension. In a row, such a column contains a textual description of a dimension value at its level. Finally, the dimension table contains one column for each level property in the dimension. An example star schema for the Sales cube can be found in Figure 12.3.

The fact table of the star schema holds the sales price for one particular sale. The fact table also has a foreign key column for each of the three dimensions, Product, Location, and Time. The dimension tables have

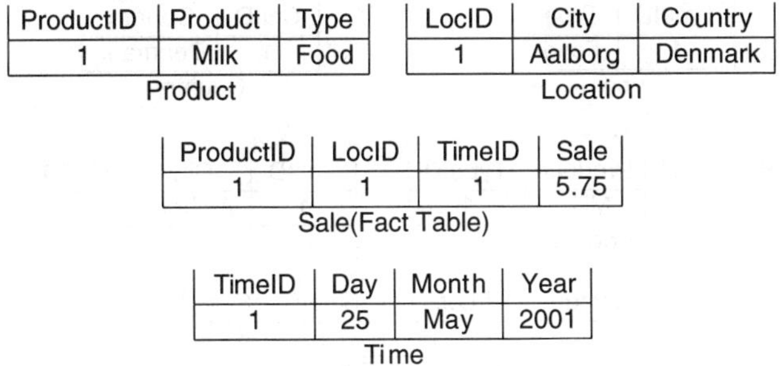

FIGURE 12.3   Star schema for sales cube.

corresponding key columns and one column for each of their levels, for example, LocID, City, and Country for Location. No column is needed for the T level, as that column would always hold the same value. The key column in a dimension table is typically a "dummy" integer key without any semantics. This has several advantages over the option of using information-bearing keys from the source systems, including better storage use, prevention of problems associated with key reuse, and better support for dimension updates [10].

It can be seen that there will be a redundancy in higher-level data. For example, with 31 day values in May 2001, the year value "2001" will be repeated 30 times for this month alone. However, as dimensions typically take up only 1–5% of the total storage required for a cube, redundancy is not a problem space-wise; and since the updates of dimensions are handled centrally, it is also possible to ensure consistency. Thus, it is often a good idea to use redundant dimension tables because this supports a simpler formulation of (better-performing) queries.

Snowflake schemas contain several dimension tables for each dimension, namely one table for each level. This means that redundancy is avoided, which may be advantageous in some situations. The dimension tables contain a key, a column having textual descriptions of the level values, and possibly columns for level properties. Tables for lower levels also contain a foreign key to the containing level.

Figure 12.4 shows a snowflake schema for the Sales cube. For example, the Day table in Figure 12.3 contains an integer key, a date, and a foreign key to the Month table. Note that no year values are replicated.

The choice of star vs. snowflake schemas depends highly on the desired properties of the system being developed. Due to space constraints, we do not provide a full discussion here.

Achieving Fast Query Response Time

The most essential performance-enhancing technique in multidimensional databases is *precomputation* and its more specialized cousin, *preaggregation*. In ROLAP systems, this is referred to as *materialized views*. This technique enables the delivery of response times to queries involving potentially large amounts of data that are fast enough to allow interactive data analysis.

As an example application of preaggregation, we may compute and store (materialize) the total sales of a product by country and month. This enables fast answers to queries that ask for the total sales, for example, by month alone, by country alone, or by quarter and country in combination. These answers may be derived from the precomputed results alone; access to the bulks of detail data in the data warehouse is unnecessary.

Preaggregation has attracted substantial attention in the research community [17], and recent versions of commercial relational database products, as well as the dedicated multidimensional systems, offer query optimization based on precomputed aggregates, as well as automatic maintenance of the stored aggregates when the detailed data on which the aggregates are based are updated [24].

Full preaggregation, where all combinations of aggregates are materialized, is infeasible as it takes too much storage and initial computation time. Instead, modern OLAP systems adopt the *practical*, or partial, preaggregation approach of materializing only select combinations of aggregates and then reuse these

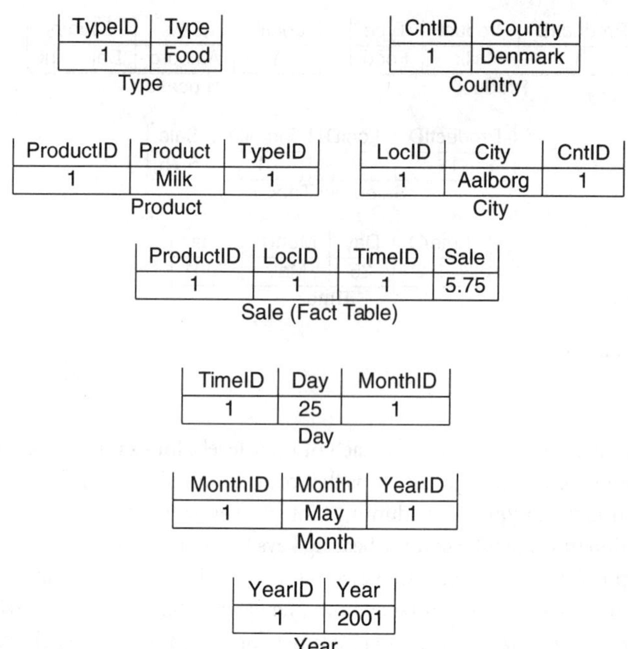

FIGURE 12.4   Snowflake schema for sales cube.

to efficiently compute other aggregates [21, 22]. This reuse of aggregates requires a well-behaved structure of the multidimensional data.

12.10   Complex Multidimensional Data

The traditional multidimensional data models and implementation techniques assume that the data being modeled conform to a quite rigid regime. Specifically, it is typically assumed that all facts map (directly) to dimension values at the lowest levels of the dimensions and only to one value in each dimension. Further, it is assumed that the dimension hierarchies are simply balanced trees. In many cases, this is adequate to support the desired applications satisfactorily. However, situations occur where these assumptions are too rigid for comfort.

In such situations, the support offered by "standard" multidimensional models and systems is inadequate, and more advanced concepts and techniques are called for. A more comprehensive treatment of complex multidimensional data is available in the literature [4]. We proceed to review the impact of irregular hierarchies on practical precomputation.

Complex multidimensional data are problematic as they are not summarizable. Intuitively, data are *summarizable* if the results of higher-level aggregates can be derived from the results of lower-level aggregates. Without summarizability, users will either obtain wrong query results, if they base them on lower-level results, or computation may be prohibitively time consuming because we cannot use precomputed lower-level results to compute higher-level results. When it is no longer possible to precompute, store, and subsequently reuse lower-level results for the computation of higher-level results, aggregates must instead be calculated directly from base data, which is what leads to the increased computational costs.

It has been shown that summarizability requires that aggregate functions be distributive and that the ordering of dimension values be *strict, onto,* and *covering* [11, 18]. Informally, a dimension hierarchy is *strict* if no dimension value has more than one (direct) parent, *onto* if the hierarchy is balanced, and *covering* if no containment path skips a level. Intuitively, this means that dimension hierarchies must be balanced trees. If this is not the case, some lower-level values will be either double-counted or not counted when reusing intermediate query results.

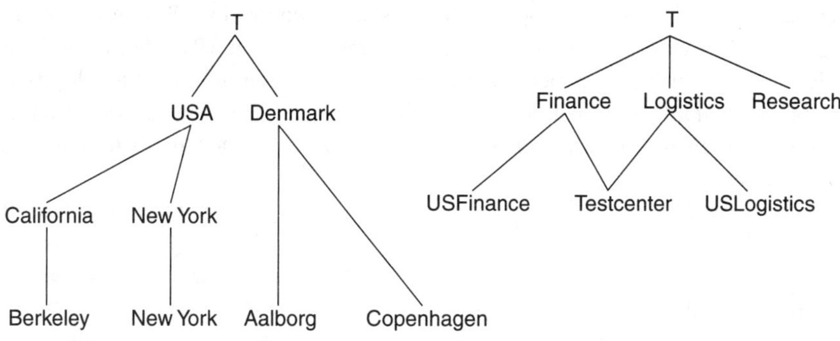

FIGURE 12.5   Irregular dimensions.

Figure 12.5 contains two dimension hierarchies: a Location dimension hierarchy including a State level, and an Organization dimension hierarchy that describes the organization in some company. The hierarchy to the left is *noncovering* because Denmark has no states. If we precompute aggregates at the State level, we will have no values for Aalborg and Copenhagen, meaning that facts mapped to these cities will not be taken into account when computing country aggregates from precomputed State-level aggregates.

The hierarchy to the right in the figure is *nononto* because the Research department has no further subdivision. If we materialize aggregates at the lowest level, facts mapping directly to the Research department will not be counted. The hierarchy is also *nonstrict* because the TestCenter is shared between Finance and Logistics. If we materialize aggregates at the middle level, data for TestCenter will be used twice, for both Finance and Logistics, which is what we want at this level. However, this means that the data will also be used twice if we combine these aggregates into the grand total.

Irregular dimension hierarchies occur in many contexts, including organization hierarchies [26], medical diagnosis hierarchies [14], and concept hierarchies for web portals [25]. A solution to the problems with irregular hierarchies is to *normalize* the hierarchies, a process that pads nononto and noncovering hierarchies with "dummy" dimension values to make them onto and covering, and fuses sets of parents in order to remedy the problems with nonstrict hierarchies. This transformation may be accomplished transparently to the user, rendering preaggregation applicable also to the more generalized types of dimension hierarchies discussed here [17].

12.11   Commercial Systems

As noted in Section 12.2, the development of multidimensional database systems began more than 35 years ago. However, it is only in the last 5–6 years that multidimensional technology has begun to enter the mainstream. Below, we introduce some of the most common commercial systems in the area. We have focused on the offerings from the "big-3" software vendors (IBM, Microsoft, Oracle) as well as the systems having a market share of at least 5% in the 2002 OLAP market, as measured by the acknowledged, independent OLAP market research report "The OLAP Report" [15].

Generally, commercial multidimensional systems fall into two groups: tools that are primarily multidimensional database servers and client-oriented analysis and reporting tools.

The multidimensional database server market is dominated by three products, each of which is backed by one of the "big-3" vendors. The number one server is Microsoft's Analysis Services [12, 22], which comes bundled with the MS SQL Server RDBMS. The number two server is Hyperion Essbase [8], long the market leader. A version of Essbase is sold along with IBM DB2 as DB2 OLAP Services [9]. The number three server is Oracle 9i OLAP R2 [16], which is based on the Oracle RDBMS and the late Express MOLAP server. All three servers offer flexible HOLAP technology, with flexible storage of data using ROLAP and/or MOLAP technology. All three servers generally provide a good query and load performance, as well as advanced query functionality.

The three largest vendors of client-oriented tools are Cognos [3], Business Objects [1], and Microstrategy [13]. They all provide tools for both "real" interactive OLAP analyses as well as more traditional static reporting, including paper-based reports. Another common functionality is support for building web portals that integrate analysis results with other data from the enterprise. They also include embedded server components that are primarily intended as back-ends for the client tools rather than as stand-alone OLAP servers.

The OLAP server vendors also have some client-side offerings for data analysis. Microsoft ships the Data Analyzer OLAP client and an Excel OLAP plug-in for OLAP analysis, and Reporting Services for traditional reports. Hyperion (IBM) has a large suite of client tools and a large market share in customized OLAP applications. Oracle offers both a range of general analysis tools and more specialized systems allowing easy analysis of data from the Oracle Applications ERP system. SAP offers the similar mySAP Business Intelligence and SAP Business Information Warehouse (BIW) products [20], which allow easy DW integration and subsequent analysis of data from the SAP ERP system, as well as from other source systems.

In summary, the most popular server and client tools generally offer sufficient functionality and performance to meet the needs of most customers. The choice of tools should thus be based on criteria such as existing system portfolio, the overall IT strategy, vendor support, and any specialized requirements, to name but a few criteria.

12.12 Summary

The continued advances in key information technology areas as well as the increasing electronic capture of business data have enabled the collection of very large volumes of business data. Advances in software technologies have contributed significantly to the provision of systems with response times that enable interactive analyses of such data.

At the core of these software technologies lies a type of data model that espouses a multidimensional view of data. While this type of model originally evolved from multidimensional matrix algebra, it has in recent years been influenced and enriched by the insights gained in the areas of semantic as well as scientific and statistical data models.

Multidimensional models view data as consisting of business facts, with associated measures, that are characterized by descriptive data values organized in multiple dimensions. Dimension values are organized in containment-type hierarchical structures that enable the computation of aggregate queries at different levels of granularity. This data organization lends itself toward graphical formulation of aggregate queries and graphical display of their results, and it is also conducive to the efficient evaluation of aggregate queries.

A wide variety of commercial systems exist that support multidimensional databases, both on the client and the server side. In recent years, multidimensional database functionality has become available as an integrated part of the leading relational database management systems.

References

[1]   Business Objects Corporation, http://www.businessobjects.com/, current as of July 18th, 2003.
[2]   Codd, E.F., *Providing OLAP (On-Line Analytical Processing) to User-analysts: An IT Mandate*, E.F. Codd and Assoc., Sunnyvale, CA, USA, 1993.
[3]   Cognos Corporation, http://www.cognos.com/, current as of July 18th, 2003.
[4]   Dyreson, C.E., T.B. Pedersen, and C.S. Jensen, Incomplete information in multidimensional databases, in *Multidimensional Databases: Problems and Solutions, Rafanelli, M., Ed.*, Idea Group Publishing, Hershey, PA, USA, 2003.
[5]   Eisenberg, A. and J. Melton, SQL standardization: the next steps, *SIGMOD Record*, 29:63–67, 2000.
[6]   H. Garcia-Molina, J. Ullman, and J. Widom, *Database Systems: The Complete Book*, Prentice-Hall, Englewood Cliffs, NJ, 2002.

[7] J. Gray et al., Data cube: a relational aggregation operator generalizing group-by, cross-tab and sub-totals, *Data Mining and Knowledge Discovery*, 1:29–54, 1997.

[8] Hyperion Corporation, http://www.hyperion.com/, current as of July 18th, 2003.

[9] IBM Corporation, DB2 OLAP Server, http://www.ibm.com/software/data/db2/db2olap/, current as of July 18th, 2003.

[10] R. Kimball, *The Data Warehouse Toolkit*, Wiley Computer Publishing, New York, 1996.

[11] Lenz, H. and A. Shoshani, Summarizability in OLAP and Statistical Data Bases, in *Proceedings of SSDBM*, 1997, pp. 39–48.

[12] Microsoft Corporation, SQL Server Analysis Services, http://www.microsoft.com/sql/evaluation/bi/bianalysis.asp, current as of July 18th, 2003.

[13] Microstrategy Corporation, http://www.microstrategy.com, current as of July 18th, 2003.

[14] National Health Service (NHS), Read Codes version 3, NHS, September 1999.

[15] The OLAP report, 2002 Market Shares, http://www.olapreport.com/Market.htm, current as of July 18th, 2003.

[16] Oracle Corporation, Oracle 9i OLAP R2, http://www.oracle.com/olap/, current as of July 18th, 2003.

[17] Pedersen, T.B., C.S. Jensen, and C. E. Dyreson, Extending Practical Pre-Aggregation in On-Line Analytical Processing, in *Proceedings of VLDB*, 1999, pp. 663–674.

[18] Pedersen, T.B., C.S. Jensen, and C. E. Dyreson, A foundation for capturing and querying complex multidimensional data, *Information Systems*, 26:383–423, 2001 *(Special Issue: Data Warehousing)*.

[19] Pedersen T.B., and C.S. Jensen, Multidimensional database technology, *IEEE Computer Magazine*, 34:40–46, 2001 *(Special Issue: Data Warehouses)*.

[20] SAP Corporation, mySAP Business Intelligence, http://www.sap.com/solutions/bi/, current as of July 18th, 2003.

[21] Thomsen, E., *OLAP Solutions: Building Multidimensional Information Systems*, Wiley, New York, 1997.

[22] Thomsen, E., G. Spofford, and D. Chase, *Microsoft OLAP Solutions*, Wiley, New York, 1999.

[23] Vassiliadis, P. and T. K. Sellis. A survey of logical models for OLAP databases., *SIGMOD Record*, 28:64–69, 1999.

[24] Winter, R., Databases: back in the OLAP game, *Intelligent Enterprise Magazine*, 1:60–64, 1998.

[25] Yahoo! Corporation, www.yahoo.com, current as of July 18th, 2003.

[26] Zurek, T. and M. Sinnwell, Data Warehousing Has More Colours Than Just Black and White, in *Proceedings of VLDB*, 1999, pp. 726–729.

Torben Bach Pedersen is an associate professor of Computer Science at Aalborg University, Denmark. His research interest includes multidimensional databases, OLAP, data warehousing, federated databases, and location-based services. He received his Ph.D. degree in Computer Science from Aalborg University. He is a member of the IEEE, the IEEE Computer Society, and the ACM. He can be contacted at tbp@cs.aau.dk.

Christian S. Jensen is a professor of Computer Science at Aalborg University, Denmark. His research interests include multidimensional databases, data warehousing, temporal and spatio-temporal databases, and location-based services. He received his Ph.D. and Dr.Techn. degrees in Computer Science from Aalborg University. He is a senior member of the IEEE and a member of the IEEE Computer Society and the ACM. He can be contacted at csj@cs.aau.dk.

Section 2: The Internet and IP Networks

13

TCP/IP Architecture, Protocols, and Services

Christian Kurz and
Helmut Hlavacs
University of Vienna

13.1 ISO/OSI Reference Model

The ISO/OSI reference model [ISO7498] was developed by ISO, the International Standard Organization, and completed in 1982. The Open Systems Interconnection (OSI) reference model allows the connection of open systems. This objective is reached by applying a layered approach. The communication system is divided into seven layers (see Figure 13.1) [PET2000]. The lowest three layers are network dependent. They provide support for data communication between and linking of two systems. The upper three layers are application oriented. They allow the end-user application processes

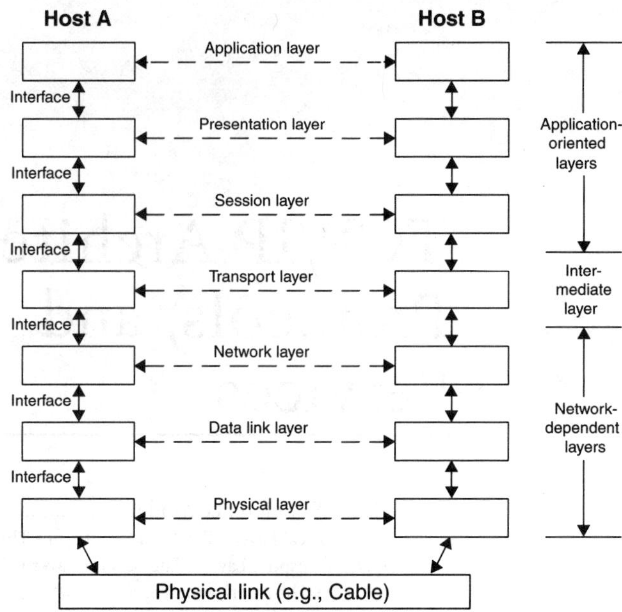

FIGURE 13.1   ISO/OSI reference model.

to interact with each other. The intermediate layer (transport layer) isolates the application-oriented layers from the communication details at the lower layers [HAL1996].

Each layer performs a well-defined function. This allows reducing the level of complexity at each layer and is defined by a protocol. The information flow between the layers is directed through interfaces and should be minimized [TAN1996]. Each layer exchanges messages using services of the layer below. It communicates with the related peer at the same level in a remote system and provides services to the layer above [COL2000]. At each layer, the source host adds a header to the packet, which is read and removed again by the receiver. It is important to note that the implementation of one layer is therefore independent from the implementation of the other layers. In the next section, each of the layers is discussed separately starting from the lowest one.

The Physical Layer

The lowest layer is concerned with the transmission of raw bits from the electrical interface of the user equipment to the communication channel. This can either be an electrical, optical, or wireless medium and transfers a serial stream of data. It has to be assured that a sent 1 bit is seen by the receiver as a 1 bit, and not as a 0 bit. Design issues at this layer are, for example, how long one bit lasts and by which wavelength of light or by which voltage level a 1 and a 0 bit is represented. Additionally, the handling of the initial connection and the closure of the connection are carried out at the physical layer.

Also, mechanical properties of the network equipment such as size and shape of connectors and cables have to be specified. Furthermore, electrical (or optical) parameters must be determined. These are the voltage levels, electrical resistance of the cable, duration of signaling elements, and voltage changes and the coding method. The next issue handled by the physical layer is the functional specification. It concerns the meaning of switched connections distinguishing between data and control wires and specifying the clock rate and ground.

The Data Link Layer

As the physical layer is only concerned with the transmission of raw data, the main function of the data link layer is to recognize and correct transmission errors. For this reason, the sender divides the data stream into frames that are transmitted sequentially. When the frame is received, an acknowledgement may be sent back to the sender. If a frame is destroyed by a burst in the line and is therefore not acknowledged, it is retransmitted by the sender. As the acknowledgement frame could also be lost, care has to be taken that no duplicate frames are inserted into the data stream. The data link layer therefore solves problems arising from lost, damaged, or duplicate frames.

This layer may also offer different service classes, for example, for protected or unprotected services. If the receiver is slower than the sender, frames can be lost because of different processing speeds. To prevent this scenario, a mechanism is implemented to regulate network traffic. Therefore, the sender should know how much buffer space is left at the receiver.

Another task of the data link layer is media access control within broadcasting networks. In these networks, all connected computers perceive all data transferred; they share a common link. Therefore, it has to be ensured that there is only one sender at a time to avoid data collision. If a collision occurs, it has to be detected and retransmission of all affected data has to be initiated. When data can be transmitted in both directions simultaneously, the acknowledgement frame for the sender A sending to receiver B competes with data frames that B is sending to A. A solution for this problem is piggybacking, where the acknowledgement information is added to data packages sent instead of sending additional frames.

The Network Layer

The network layer is responsible for the setup, the handling, and the termination of networkwide connections; it controls the operation of the subnet. There are two possible types of network connections: virtual connections and datagram connections. Virtual connections are set up at the start of a transmission to fix the route for the following data packets; packets are always sent using the same route. Using datagram connection, the route is chosen separately for each package. Sometimes, it has to be assured that packets arrive in the same order that they were sent in. A packet B sent after a packet A may arrive ahead of A using a different route in datagram communication.

Additionally, the network layer is concerned with package routing from source to destination. Routing information can either be stored in a static table or be determined dynamically at the start of each transmission. The chosen route can also depend on the current network load. If too many packets are sent in one subnet, a capacity bottleneck forms. To avoid this situation, the network layer may implement congestion control. To be able to analyze network traffic, an accounting mechanism is incorporated at this layer. This mechanism counts how many packets are sent, also storing information about packet source and destination. The information gathered can be used to produce billing information. Also, there may be problems when a packet is traveling through heterogeneous networks. The network layer handles the issues of different packet sizes, varying addressing schemas or different protocols.

The Transport Layer

This layer is the interface between the higher application-oriented layers and the underlying network-dependent layers. Thus, the session layer can transfer messages independent of the network structure. Seen from layers above, messages can be transferred transparently without having knowledge of the underlying network structure. The transport layer basically cuts messages into smaller packets if needed and passes them to the network layer. At the sender, the messages are assembled again and passed to the session layer.

An important task of the transport layer is the handling of transport connections. Normally, one network connection is created for each transport connection required by the session layer. If the session layer requires an output higher than can be handled by one connection, the network layer might create additional connections. On the other hand, if one wants to save costs, a number of transport connections can be multiplexed onto one network connection. As there is the possibility of setting up multiple connections, a transport header is added to distinguish between them.

The transport layer provides different classes of quality of service (QoS). The lowest service class provides only basic functionality for connection establishment; the highest class allows full error control and flow control. To avoid the situation of a fast sender overrunning a slower receiver with messages, an algorithm for flow control is provided. The most popular type of connection is an error-free point-to-point connection where messages are delivered in the same order that they were sent in. Additionally, messages with no guaranteed order can be sent. It is also possible to send messages not only to one but also to multiple destinations or to send broadcast messages.

The transport layer establishes and terminates connections across the network. Therefore, the need for a naming mechanism arises, allowing processes to choose with whom they converse.

The Session Layer

Layer 6 organizes and synchronizes the data exchange for two application layer processes. It sets up and clears the communication channel for the whole duration of the network transaction between them, and therefore sets up sessions between users on different machines. A session might be used to log into another machine in a remote timesharing environment or to transfer a file. The session layer provides interaction management (also called dialog control). Data can be exchanged using duplex or half-duplex connections. A duplex connection transfers data both ways simultaneously. A half-duplex connection can transfer either one or the other way, where the session layer decides which party is allowed to use the link.

Another task of the session layer is token management. It is useful when both sides are not allowed to perform the same operation at the same time. To schedule these operations, a token is issued only to one process at each given time, allowing only the process that holds the token to perform the critical task. For large data transmissions, synchronization points can be set periodically. If the network connection fails, the transmission is restarted at the last synchronization point set. Thus, retransmission of the whole data can be avoided. Nonrecoverable exceptions during transmissions are reported to the application layer.

The Presentation Layer

The main task of the presentation layer is the representation of data, for example, integers, floating point numbers, or character strings. Therefore, the syntax for these data containers is defined. As different computer may use varying internal data representations, for example, for characters or numbers, a conversion has to be done. The data sent are converted into an appropriate transfer syntax and are transformed back to the receivers internal data format upon receipt. Those converters for the syntax of data do not necessarily have to understand the semantics. This layer may also provide services for data encryption and data compression.

The Application Layer

It provides services not to other layers but directly to application programs. Thus, there is no specific service in this layer but there is a distinct combination of services offered to each application process. As the connected hosts may use different file systems, the application layer handles the differences and avoids incompatibilities. Therefore, this layer provides the means for networkwide distributed information services. This allows the application processes to transfer files, send emails, or perform directory lookups.

Furthermore, the application layer provides services for the identification of intended communication partners to verify the availability of an intended communication partner, verify communication authority, provide privacy services, authenticate communication partners, select the dialog discipline, reach an agreement on the responsibility for error recovery, and identify the constraints in data syntax [HAL1996].

13.2 The TPC/IP Reference Model

TCP/IP was first used when the ARPANET was emerging. This network was developed for use by the US armed forces. Therefore, it was required that even when some parts of the network are destroyed during a battle, it should still provide communication services. As long as two hosts were still functioning and there was a path available between them, communication must be possible. Another important issue was the ability to connect multiple different networks together irrespective of the underlying protocols, physical transport medium, or provided bandwidth. The TCP/IP reference model is structured in a manner similar to the ISO/OSI model introduced in the last chapter; but it consists of only four layers. A comparison of both models is done in the next chapter. The lowest layer is the host-to-network layer, where the internet layer is attached. Above the transport layer, the highest layer is the application layer (Figure 13.2) [SCO1991].

The next section gives an overview of the services provided by the TCP/IP model layers, starting with the lowest one.

The Host-to-Network Layer

The TCP/IP protocol does not specify services or operations at the host-to-network layer. It is only required that the host can somehow connect to the network to enable the internet layer to send packets. As this layer is not defined, the implementation can vary in each system. The network service may be provided by an Ethernet, Token Ring, ATM, WAN technologies, wireless technologies, or any other means of transferring network packets.

The Internet Layer

The main features of the internet layer are addressing, packet routing, and error reporting. Additionally, services for fragmentation and reassembly of packets are provided [HAL1996]. The core protocols at the internet layer are the internet protocol (IP) [RFC791], the address resolution protocol (ARP) [RFC826], the internet control message protocol (ICMP) [RFC792], and the internet group management protocol (IGMP) [RFC2376].

The IP is concerned with packet routing, IP addressing, and the fragmentation and reassembly of packets. It is a packet-switching protocol based on a best-effort connectionless architecture. Packets travel independent of each other from source-to-destination host. Each packet may be routed differently through the network, thus packets may be delivered in a different order they were sent. Packets may also be lost, because delivery is not guaranteed. To be able to route packets across the network, each host has to know the location of a gateway or a router. The gateway decides which path a packet has to travel. For this reason, a routing table is maintained at the internet layer. To send packets across networks that only support small packet sizes, the packets are broken down in size at the source host and are assembled again at the destination host.

ARP [RFC826], the address resolution protocol, translates network-layer addresses to link-layer (hardware) addresses. Thus, an IP-address is translated to, for example, an Ethernet address. The Internet Control Message Protocol (ICMP) [RFC792] is concerned with datagram error reporting and is able to provide certain information about the internet layer. The Internet Group Management Protocol (IGMP) [RFC2376] is used to manage IP multicast groups [RFC1122].

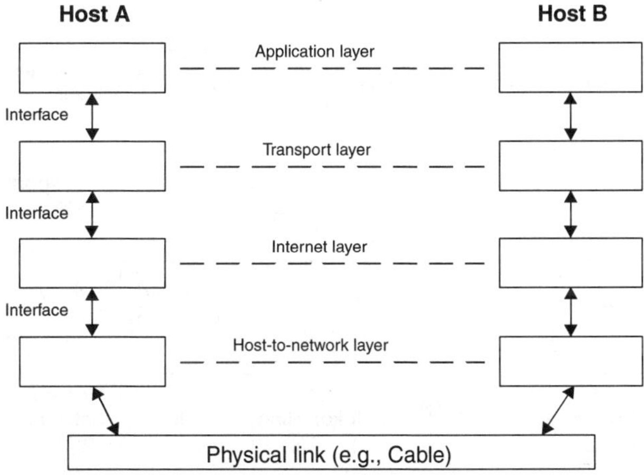

FIGURE 13.2   TCP/IP reference model.

The Transport Layer

The transport layer provides stream and datagram communication services. Protocols specified at this layer are the transmission control protocol (TCP) [RFC793] and the user datagram protocol (UDP) [RFC768]. Both protocols deliver end-to-end communication services (i.e., message transfer).

The transmission control protocol is a connection-oriented and reliable point-to-point communication service [RFC793]. A data stream is sent to any other host in the Internet without errors. This data stream is broken down into messages and handed down to the internet layer. TCP sets up and terminates the connection, and it sequences and acknowledges the packets it sends. It is also responsible for retransmitting packets lost during transmission. Also, a service for flow control is implemented, thus preventing a receiver from being flooded by a faster sender.

The user datagram protocol is a connectionless, unreliable communication protocol [RFC768]. Thus, sequencing or flow control is not provided. It is used when prompt delivery of packets is more important than error-free transmission, as demanded for the transmission of video or audio content. Compared to TCP, there is no connection establishment, no connection state, a smaller packet overhead, and an unregulated send rate [KUR2001].

The Application Layer

This layer provides services to application processes. It accesses services of the transport layer and allows processes at different hosts to communicate with each other using a variety of protocols.

These include the Hypertext Transfer Protocol (HTTP) [RFC2616] to send and receive files that make up Web pages. Also, protocols for sending electronic mail, the Simple Mail Transfer Protocol (SMTP) [RFC821], and interactive file transfer, the File Transfer Protocol (FTP) [RFC959] are implemented at this layer. Another provided and often used service is Telnet, which is a terminal emulation protocol [RFC854]. It enables the user to log on to remote hosts. To access news articles at virtual blackboards, the Network News Transfer Protocol (NNTP) [RFC977] is provided.

Additionally, protocols for the management of TCP/IP networks are available at this layer. The Domain Name Service (DNS) [RFC1034, RFC1035] resolves a host name to an IP address. Network management, including the collection and exchange of management information, is facilitated by the Simple Network Management Protocol (SNMP) [RFC1157].

Besides these basic protocols, a wide variety of other protocols are implemented for use at the TCP/IP application layer. An overview of the assignment of the protocols mentioned in this section to the respective layer is given in Figure 13.3:

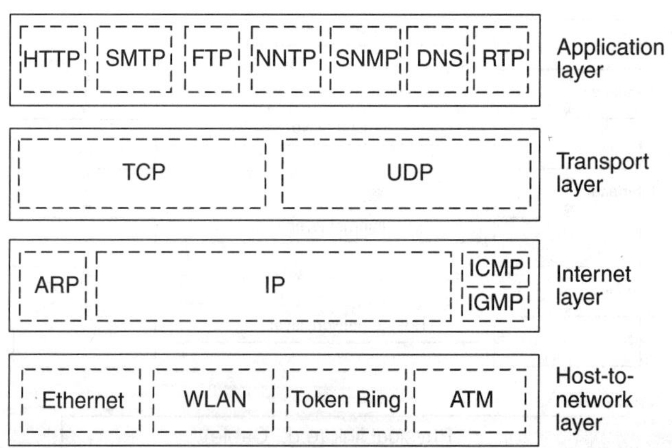

FIGURE 13.3   TCP/IP Protocol Architecture.

13.3   Reference Model Comparison

Both reference models described above are based on a layered approach. Also, the layers provide quite similar services. The application layer of the TCP/IP model corresponds to the application layer of the ISO/OSI reference model. Presentation and session layers are not present in the TCP/IP reference model. Thus, in the TCP/IP model, services provided by these two layers have to be performed by the application process itself. The two transport layers perform similar services. The next layer in the TCP/IP model, the Internet layer, is equivalent to the network layer in ISO/OSI. The data link layer and the physical layer of the OSI reference model are represented by the host-to-network layer in the TCP/IP model (Figure 13.4). In both models, layers above the transport layer are application dependent [TAN1996].

The ISO/OSI model is mainly a conceptual model; it is an example of a universally applicable structured network model. It introduced three main concepts, which were also followed when the TCP/IP reference model was developed. Each layer provides exactly defined *services* to the layer above and uses services of the layer below. These services are accessed using *interfaces* specifying, whose parameters are expected and the results returned. *Protocols* defined in each layer communicate with their peers at the remote host, independent of the underlying network structure. These ideas in both protocols are similar to object-oriented software development [TAN1996].

In the beginning, the TCP/IP model did not strictly separate between services, interfaces, and protocols; these concepts have been introduced later. Thus, protocols in the ISO/OSI model are better encapsulated than in the TCP/IP model. Therefore, it is easier to alter services in the ISO/OSI reference model [TAN1996]. As the ISO/OSI model was developed before the respective protocols and their implementation, it was possible to easily distinguish between services, interfaces, and protocols for each layer. The developers were able to choose the appropriate number of layers, such that each one could perform only a distinct set of matching services. For TCP/IP, the protocols were developed first, and afterwards the abstract model was created. The problem with this approach was that the model did not fit with any existing protocol stacks [TAN1996].

Finally, there are differences in the area of connectionless vs. connection-oriented communication. The ISO/OSI reference model provides services for both kinds of communication at the network layer, but only connection-oriented services at the transport layer. The TCP/IP reference model supports both

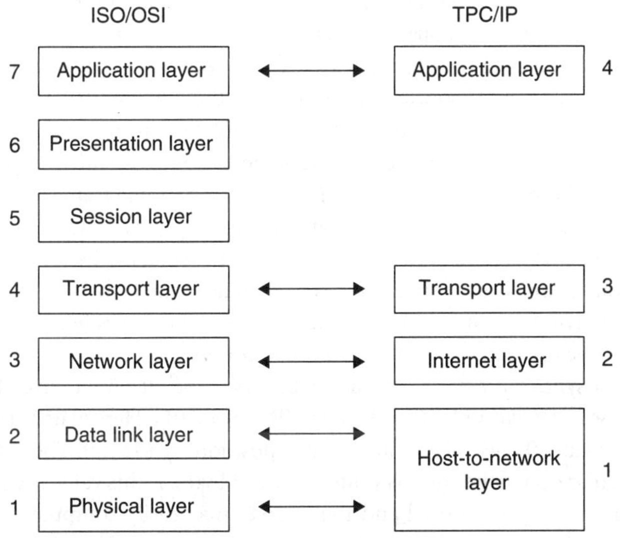

FIGURE 13.4   ISO/OSI vs. TCP/IP.

connectionless and connection-oriented communication at the transport layer, but only connectionless services at the network layer [TAN1996].

In the following, the most important TCP/IP protocols and services, their functionality, and their position in the OSI stack will be described.

13.4   Data Link Layer Protocols and Services

In the OSI model, the *data link layer* is situated at layer 2, and at the host-to-network layer in the TCP/IP reference model. Its purpose is to offer services to OSI layer 3 in such a way that protocols at layer 3 may send data to neighboring computers (i.e., computers directly connected via a network link or via layer 1 or 2 repeaters, bridges, hubs, or switches) in a reliable way. The data link layer may offer one of the following services to layer 3.

- Using an *unacknowledged connectionless* service, no measures are taken to detect lost packets by the sender or the receiver.
- In *acknowledged connectionless* services, the receiver must acknowledge the data it received by sending back an acknowledgment to the sender. If the sender does not receive the acknowledgment after a certain amount of time, it assumes that the data were lost and retransmits them again.
- In *connection-oriented* services, the data link layer must first create a (possibly virtual) path between the sender and receiver before data can be sent. Furthermore, the data link layer adds sequence numbers to the sent data units in order to detect lost or erroneous data units.

The bit-error rate of modern wire-line (electrical or optical) LAN interconnections is too low to justify the additional effort for virtual path creation at this level. In LANs, therefore, usually acknowledged (e.g., Token Ring) or unacknowledged (e.g. Ethernet) connectionless services are used at layer 2. Lost packets or packets delivered out of order can then be often detected at layer 4 or even higher. Wireless networks, however, may severely suffer from lost packets or high bit-error rates. Under these conditions, sophisticated data link layer protocols like IBM's *Synchronous Data Link Control* (SDLC), or the closely related ISO norms *High-Level Data Link Control* (HDLC) and the CCITT recommendation *Link Access Procedure* (LAP), or the IEEE 802.2 norm *Logical Link Control* (LLC) are often used.

Frame Creation

One major task of layer 2 is to pack the data it receives from a higher layer for transfer into so-called *frames*, that is, data packets, which are then modulated onto the physical network medium. This is done in such a way that the desired receivers are able to (i) detect that a frame has been sent, (ii) to decode the frame reliably and retrieve the sender and receiver addresses, and to (iii) identify those frames that are meant for themselves.

Frames are nothing more than a sequence of bits modulated onto a carrier. In order to be able to decode information stored in a frame, a receiver first has to be able to identify the first bit of a frame. This *start bit* is then usually followed by a specific sequence of bits containing frame information like the frame length, the content type, checksums, etc. A simple method for finding the frame start is given by *bit-stuffing*. In protocols like for instance X.25, the start of a frame is signaled by six following 1's. If the data transported in the frame also contain six 1's in a row, then after the fifth 1, a 0 has to be inserted by the data link layer. The receiving data link layer then knows that if it receives five consecutive 1s followed by a 0, the sender must have inserted the 0 and therefore removes it.

Another method for synchronizing senders and receivers at the bit-level is given by sending *sync-bytes*, as, for instance, is done in *Digital Video Broadcast* (DVB) [REI2001]. Here, data frames are 204 bytes long and contain a certain value (0×47) always at the same position (*sync-byte*). The task of a *sync-byte detector* is to detect the regular occurrence of this value every 204 bytes. If this value is detected five times, then the sender and receiver are synchronized and the receiver may easily compute the frame start from it. Other methods include, for instance, *octet counting* or *octet stuffing* and will be described in the context of application protocols.

Error Detection and Correction

Sending data over certain media types is often unreliable and may be severely disturbed by external disruptions causing data to be lost or wrongly received. Thus, another important task of the data link layer is to either correct corrupted frames, or at least to detect the occurrence of bit errors. In order to detect or correct bit errors, the sender must add checksum information additional to the transported headers and user data. The more information is added, the more wrong bits may be detected or even corrected.

A popular method for error detection and correction is given by *Hamming codes*. Here, certain code words are sent, which are different at a specific number of bits, the so-called *Hamming distance*. For instance, a code containing the words 000111, 111000, 000000, and 111111 has the Hamming distance 3, that is, code words differ at at least 3 bits from each other. In order to identify that d bits have been changed during transmission, a Hamming distance of $d+1$ is required. If the receiver has to be able to correct d bits, then a Hamming distance of $2d+1$ must be retained. The code above is thus able to detect 2 wrong bits and correct 1 wrong bit. A simpler way for detecting wrong bits is given by the so-called *parity bit*. Here, only one bit is added to each code word counting the number of 1's in the word. If this number is even, the parity bit is set to 0; otherwise it is set to 1 (or vice versa). Codes with parity bits may detect 1 wrong bit per code word only.

A more sophisticated error detection code is given by the *Cyclic Redundancy Check* (CRC) code. Here, each sequence of bits is treated as a polynomial over the field of binary numbers (modulo 2). The number 101, for instance, is treated as the polynomial x^2+1. Modulo 2 means that each addition of single bits is treated as an exclusive-or (XOR) operation, that is, $0+0=0$, $0+1=1+0=1$, and $1+1=0$. For CRC codes, a fixed polynomial is chosen, called *generator polynomial $G(x)$*. If a code word $W(x)$ is to be sent, it is replaced by another polynomial $R(x)$, which can be divided by $G(x)$ with rest 0, and from which the original code word $W(x)$ can be reconstructed. The polynomial $R(x)$ is then transmitted and received. If the received $R(x)$ can be divided by $G(x)$ without rest, then the transmission has been error-free with high probability. Otherwise, bit errors are detected and the transmitted code word is dropped.

Other error correction techniques include *Reed-Solomon codes* and *Convolutional codes*, but will not be treated here.

Media Access Control

An important part of the data link layer is given by the *media access control* (MAC) sublayer. At this sublayer, access to the physical medium is controlled, which may be shared by several senders concurrently. Depending on the media type, one or several senders may transmit data at the same time. In case of conflicts, several techniques exist in order to grant the right to use the medium.

ALOHA: The *ALOHA* technique, developed at the University of Hawaii, allows all senders to send their data to a commonly shared *broadcast medium* whenever they wish to. In a broadcast medium, the data sent by one host are received by all others listening to the same medium. In case of collisions due to the concurrent sending of two or more senders, the colliding frames are discarded and must be sent anew.

CSMA/CD: For *Carrier Sense Multiple Access/Collision Detection* (CSMA/CD), as, for example, implemented in Ethernet, several network cards share the same broadcast medium, for example, an electrical wire. Each network card listens to the medium (carrier sense); if no signal is detected, then a new sender may use the medium immediately. Due to the limited speed of signals, two or more senders may send simultaneously without noticing each other in time, resulting in collisions. At such an instance, all colliding frames are discarded and each sender waits for a random amount of time until it tries to send again.

TDMA: In *Time Division Multiple Access* (TDMA), time is divided into *time slices*, and each sender is granted one slice where it may send its data into the medium. Here, bandwidth may be wasted, as senders own their time slice whether they have something to send or not.

FDMA: In *Frequency Division Multiple Access* (FDMA), several sending frequencies exist and for each frequency one sender may transmit without fearing interference from other frequencies. For example, GSM uses a mixture of TDMA and FDMA for its calls. Additionally, GSM terminals change their frequency according to a fixed scheme (*frequency hopping*).

CDMA: The concept of *Code Division Multiple Access* (CDMA) is fundamentally different from the previous concepts. Here, each sender is assigned a unique bit sequence of length N bits called *chip*. Each sent bit is then added (modulo 2) to the chip bits, yielding the chip if a 0 is to be sent, or the inverse chip, if a 1 is to be sent. If a terminal wants to transmit R bits per second (bps), then R chips have to be transferred per second, making necessary a much higher bandwidth of $R \times N$ bps in total. Thus, the necessary frequency band is broadened significantly. In essence, the signal is spread over a broad spectrum; thus, the chip is also often called *spreading sequence*. In CDMA, senders with different chips can send concurrently and do not disturb the reception of other signals. This works because different chips are mathematically *orthogonal* to each other with respect to the inner product of chips (which can also be interpreted as bit vectors) and their inverse. Also, due to the use of a broader spectrum, the reconstruction of the signal is more robust with respect to other noise sources.

13.5 Network-Layer Protocols and Services

IPv4

The term *Internet Protocol* (IP) usually denotes IP version 4 (IPv4), which has been specified in [RFC791, RFC1122] and is the established standard protocol for the Internet at layer 3 of the ISO/OSI reference model, in the TCP/IP reference model at the Internet layer. The task of IP is to transport a packet from one source computer to a destination computer, where both computers are interconnected by an Internet. Here, *internet* denotes any (possibly privately managed) heterogeneous network that is interconnected using IP and IP-based routers. In contrast, *the Internet* denotes the well-known worldwide IP-based network interconnecting millions of computers and being managed by *network information centers* (NICs) and *Internet service providers* (ISPs). When traveling through an internet, a packet may pass by several intermediary networks with different network technologies, for instance, Ethernet, Token Ring, ATM, etc., used at layers 1 and 2. At the border between two different networks, the packet's destination network address is examined by a *router*, that is, a computer that is connected to both networks, and that is able to select other routers in the path between sender and receiver, or to find the receiver in its own network. Routing decisions are usually done using predefined and regularly updated *routing tables*. However, the next chosen router is by no means fixed and may depend on runtime situations like congestion or link failures, or it may simply be chosen at random. As a consequence, packets may travel through different paths from the sender to the receiver, and neither the delivery itself nor the delivery in the original order can be guaranteed.

IP packets are called *datagrams*, which may have a total length of 65,535 bytes. Datagrams may be cut into a sequence of smaller datagrams, if the datagram size is larger than the network's *maximum transfer unit* (MTU), that is, the largest OSI layer 2 frame that may be transmitted by the network. For Ethernet, for instance, the MTU is 1500 bytes. This process is called *fragmentation*, and the IP header reflects several fields for reassembling such fragments into the original datagram again.

Each datagram or fragment is led by a 20 bytes header containing the following information:

- The *Version* number of the IP protocol (4).
- The *IP Header Length* (*IHL*), which may be larger than 20.
- The *Total Length* of the datagram including header.
- An *Identification* number for reassembling fragmented datagrams. All fragments with the same ID belong to the same datagram.
- Flags, including the *Don't Fragment* (DF) flag, flagging that the datagram should not be fragmented, and the *More Fragments* (MF) flag, signaling that more fragments are still to come.
- A *Fragment Offset* identifying the offset of the received fragment in the whole datagram.
- The *Time to Live* (TTL) counter, which is decreased by one by each router. A datagram with TTL equal zero is discarded. This prevents faulty datagrams from circling through the Internet forever.
- A number identifying the used transport *Protocol* (6 for TCP, 17 for UDP, ...).

- A header *Checksum*.
- The IP *Source* and *Destination Addresses*.

An important aspect of IPv4 is given by the 32-bit-long IP addresses. The written form follows the *dotted decimal notation* scheme $X_1.X_2.X_3.X_4$, where the X_i are decimals between 0 and 255.

Each address starts with an address class identifier, and is then followed by the network address, and finally by the host address. There are different network classes as shown in Table 13.1.

Each network card attached to the Internet must have a unique IP address. The address assignment scheme is a two-step strategy. First, each site managing a network connected to the Internet is assigned a unique network address by a central authority called *Network Information Center* (NIC). Then, each site may assign the unique host addresses belonging to this network address, which may include $2-2=16,177,214$ addresses (class A), $2^{16}-2=65,534$ (class B), or $2^8-2=254$ (class C) unique host addresses.

IP defines a set of *private addresses* that may be used freely, but whose traffic should not be routed over the Internet without modification [RFC1918]. The three address blocks are:

- 10.0.0.0 − 10.255.255.255 (one class A network).
- 172.16.0.0 − 172.31.255.255 (16 contiguous class B networks).
- 192.168.0.0 − 192.168.255.255 (256 contiguous class C networks).

Multicast addresses are special addresses reserved for groups of hosts receiving the same multimedia program via multicast from a single source [RFC1112]. Multicast addresses may range from 224.0.0.0 to 239.255.255.255.

Two host addresses are reserved in each (sub)network. The host address 0 denotes the network itself; the highest possible host address denotes a broadcast address that is received by all hosts of a given network.

IPv6

The *Internet Protocol version 6* (IPv6) has been designed for replacing the old IPv4 in the next-generation Internet [RFC1883, RFC1887]. It represents a totally new approach and is incompatible with version 4. As most Internet hosts and routers still only support IPv4, IP packets following IPv6 often cannot be transported from the sender to the receiver without further modification. Usually, when leaving the IPv6 subnetwork of the sender, IPv6 packets are tunneled over IPv4, that is, are transported in IPv4 packets, where the whole IPv6 packet is treated as pure IPv4 data.

The header has been simplified and contains only seven fixed fields (the IPv4 header includes 13):

- A *Version* field containing the value 6.
- A *Priority* field distinguishing between data and real-time traffic.
- A *Flow Label* for supporting pseudo end-to-end connections with guaranteed QoS.
- The *Payload Length* specifies the size of the data contained in the packet.
- The *Next Header* points at the next optional header or an ID for the used transport protocol (TCP or UDP).
- The *Hop Limit* is decreased by each passed by router; a packet with zero Hop Limit is discarded. This prevents faulty packets from circling through the network forever.
- Finally, the 16 byte *Source* and *Destination Addresses* are contained.

TABLE 13.1 IP Network Classes

Class	Most significant bits	Network address	Host address
A	0	7 bits	24 bits
B	10	14 bits	16 bits
C	110	21 bits	8 bits
D	1110	28 bits	0 bits
E	11110	Reserved	Reserved

IPv6 offers the following enhancements with respect to IPv4:

- Addresses are 16 bytes long, written in groups of four hexadecimal digits separated by colons (e.g., 8000:0000:1111:2222:3333:4444:ABCD:EFFF). This solves the shortage of IPv4 addresses caused by the exponential growth of the Internet. Even when wasting a lot of such addresses due to the inefficient use of network addresses, thousands of IP addresses could be assigned to each square meter of the earth's surface.
- New address classes exist, including addresses for Internet service providers and geographical regions.
- Due to the simpler header, routing is made more efficient. Additionally, IPv6 supports an arbitrary list of options that may be skipped by routers that do not support them.
- IPv6 supports authentication and encryption.
- IPv6 supports QoS for real-time applications.

Even though IPv6 offers substantial advantages, its implementation is costly and requires to buy new routers and to reconfigure existing hosts. For these reasons, IPv4 still is the Internet protocol today, and IPv6 will not dominate the Internet until the year 2010 or even later.

Address Resolution Protocol

The ARP defined in [RFC826] and its complement, the *Reverse Address Resolution Protocol* (RARP) defined in [RFC903], are a means for connecting OSI layer 2 addresses (MAC addresses) to their corresponding layer 3 IP addresses.

Basically, computers communicate with each other by sending messages on the data link layer (and subsequently the physical layer), for instance, by sending an Ethernet frame over an Ethernet variant. On this level, all network cards are identified by a globally unique 48-bit identifier. In order to successfully send an Ethernet frame, each sending network card must put both its own Ethernet address and the Ethernet address of the receiving card into the Ethernet frame. If too many computers are connected by a single layer 2 network (possibly via hubs, bridges, or switches), senders often know only the IP address of a receiver. However, for Ethernet cards, IP addresses are meaningless. In such situations, ARP can be used to find out the Ethernet address of a network card, which is bound to a given IP address at a higher layer.

If a computer *A* wants to find out the Ethernet address of a network card on computer *B*, which according to its IP address belongs to the same layer 2 subnet, then on computer *A*, ARP is automatically activated. At first, computer *A* looks into a small ARP cache to find out if the desired binding is already stored there. If not, computer *A* generates an ARP request message (*who is B.B.B.B tell A.A.A.A*, where B.B.B.B is the IP address of computer *B* and A.A.A.A the IP address of computer *A*), which is no more than a special Ethernet frame containing the following information:

- Ethernet protocol type is set to 0x806.
- Sender Ethernet address.
- Sender IP address.
- Receiver Ethernet address is set to the Ethernet broadcast address FF:FF:FF:FF:FF:FF.
- Receiver IP address.

As the receiver in this Ethernet frame is the broadcast address, all network cards connected to the same subnet will receive this request, including computer *B*. Upon receiving the ARP request message, computer *B* will then activate its own ARP protocol, which will immediately send an ARP response message (*B.B.B.B is HH:HH:HH:HH:HH:HH*, where B.B.B.B is the IP address and HH:HH:HH:HH:HH:HH is the Ethernet address of the network card of computer *B*).

Once the ARP response message has been received by computer *A*, computer *A* will store this IP-Ethernet address binding for computer *B* in its ARP cache and may start sending Ethernet frames to computer *B*. In order to avoid outdated ARP caches, these caches are periodically emptied.

The purpose of RARP is to let computers find out their IP addresses upon startup, in case they only know their Ethernet address. This can be the case, for example, for diskless workstations, which automatically attach to a server, or for workstations with identical disk images (which do not require a manual setup). RARP works similar to ARP, except that the protocol type value is set to 0x8035. Also, a router is required, which contains a table with the Ethernet-IP bindings. Alternatives to RARP are given by the Bootstrap Protocol (BOOTP) or the Dynamic Host Configuration Protocol (DHCP), which allow to resolve IP addresses in a more flexible way.

Internet Control Message Protocol

The *Internet Control Message Protocol* (ICMP), defined in [RFC792, RFC1122], is used for automatically sending control signals and commands between computers attached to an IP network. Also, ICMP messages can be used for testing connections and measuring interconnection performance. ICMP messages are sent as special IP packets and thus can be handled by routers. As a consequence, ICMP messages can be sent to or received from arbitrary computer connected with each other over an IP network. An ICMP message contains the following data:

- The *Type* defines the purpose of the ICMP packet. There are over thirty different ICMP types.
- The *Code* further defines the packet's purpose.
- A header *Checksum*.
- The rest of the packet may then contain further data depending on the ICMP type.

The most important ICMP packet types are as follows:

- *Echo Request*: When receiving such an ICMP packet, the receiver should answer with an ICMP Echo Reply packet.
- *Echo Reply*: Answer to an ICMP Echo Request packet.
- *Timestamp Request*: The same as Echo Request, except that the receiver answers with a Timestamp Reply packet, which holds additional timestamps.
- *Timestamp Reply*: The answer to an ICMP Timestamp Request, which holds the time points at which the Timestamp Request was received and the Timestamp Reply was sent back.
- *Destination Unreachable*: This message is returned by a router to the source host to inform that the destination of a previously sent packet cannot be reached.
- *Time Exceeded*: Sent from a router to a source host to inform that the lifetime of a previously sent packet has reached zero.
- *Parameter Problem*: Sent to a source host to inform that a previously sent packet contains invalid header data.
- *Source Quench*: Sent to a source host to inform it that due to insufficient bandwidth it should lower its sending bitrate.

13.6 Transport-Layer Protocols and Services

Transmission Control Protocol

The *Transmission Control Protocol* (TCP) operates at OSI layer 4 (in the TCP/IP reference model at the transport layer) on top of IP and is assigned the IP protocol number 6. It constitutes the most important Internet protocol and is defined in [RFC793, RFC1122, RFC1323]. The purpose of TCP is twofold:

- To guarantee the correct delivery of packets sent over an intrinsically unreliable packet-oriented internet and
- To control the output bitrate of each sender in order to minimize packet losses due to congested routers or receivers.

TCP operates connection oriented in full duplex. Applications using TCP may assume that a TCP connection opened from a source host to a receiver host is like a reliable pipeline or byte stream. Data (arbitrary bytes) placed in this pipeline are guaranteed to reach the receiver without losses and in correct order (Figure 13.5).

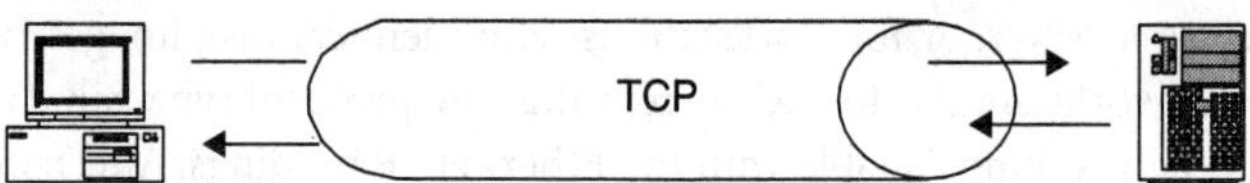

FIGURE 13.5   Full duplex TCP connection.

In order to guarantee this correctness, TCP divides the data to send into so-called segments, which are themselves sent in IP packets. In principle, IP packets may contain up to 65,535 bytes. However, in order to avoid fragmentation, the size of TCP segments is more importantly limited by the network's MTU. Each segment starts with a TCP header, which is at least 20 bytes long, but may hold additional options. The rest of the segment may hold user data, but may also be empty. The TCP header contains the following information:

- *Source Port* and *Destination Port.*
- A *Sequence Number* identifying each sent byte. This is wrapped back to zero, in case the highest number has been used.
- An *Acknowledgement Number* denoting the number of the next expected byte. This field only contains valid data if the ACK bit is set.
- The *Data Offset* holding the size of the TCP header.
- *Explicit congestion notification* (*ECN*) and control bits, including *URG, ACK, PSH, RST, SYN,* and *FIN.*
- Sender receives *Window* size.
- A header *Checksum.*
- An optional pointer to urgent data (URG flag set) and optional TCP headers.

In order to create a TCP connection between two applications X and Y running on computers A and B, both applications first must get a *port number*, an identifier between 0 and 65,535, which can be assigned only once on each computer. The application X initiating the connection must then provide its own port number, the IP address of computer B, and the port number of the partner application Y to TCP. TCP then sends a segment to the given IP address and port number, where the SYN flag is set to 1 and ACK is set to 0, and a random sequence number x is chosen. If application Y correctly waits at the given port, the TCP protocol on computer B answers with a segment, where the SYN and ACK bits are set, the sequence number of side B is set to a random number y, and the acknowledgement number is set to $x+1$. Upon receiving this second segment, the TCP on computer A again sends a segment, where the SYN and ACK flags are set, the sequence number is set to $x+1$, and the acknowledgement number is set to $y+1$. As three segments must be sent for establishing a TCP connection, this process is called *three-way handshake* (see Figure 13.6).

After the establishment of the connection, each side may send arbitrary bytes to the other side. If one side wants to terminate the connection, a segment with set FIN flag must be sent. Otherwise, if, for instance, application X sends data to Y, then the data are placed in one or more TCP segments, which are then sent via IP to computer B. Due to the sequence numbers of each segment, the TCP layer at B is able to realize missing segments or out-of-order delivery of segments. For each correctly received segment, B must send an acknowledgement segment back to A, where the acknowledgement number identifies the number of the next expected byte. The TCP on computer A, on the other hand, starts a so-called *retransmission timer* for each sent segment. If no acknowledgement is received within a certain amount of time, computer A assumes that the segment is lost and has to be sent again.

TCP also maintains two so-called *sliding windows* in order to control the transmission bitrate of each sender (*flow control*). One window simply tells each sender how many bytes the receiver may currently receive without risking a buffer overflow. This information is transmitted in each ACK segment in the receive window size field. The second window is called *congestion window* (CWND). Here, each sender additionally restricts the number of bytes it may send without acknowledgement to the congestion window size. Initially, the window size is set to 1 packet (i.e., the maximum allowed segment size), a strategy that is called *slow start*. For each acknowledged byte, TCP increases the size of its congestion window, at first with exponential speed, but after reaching a certain threshold h, only with linear speed. If a timeout of the retransmission timer

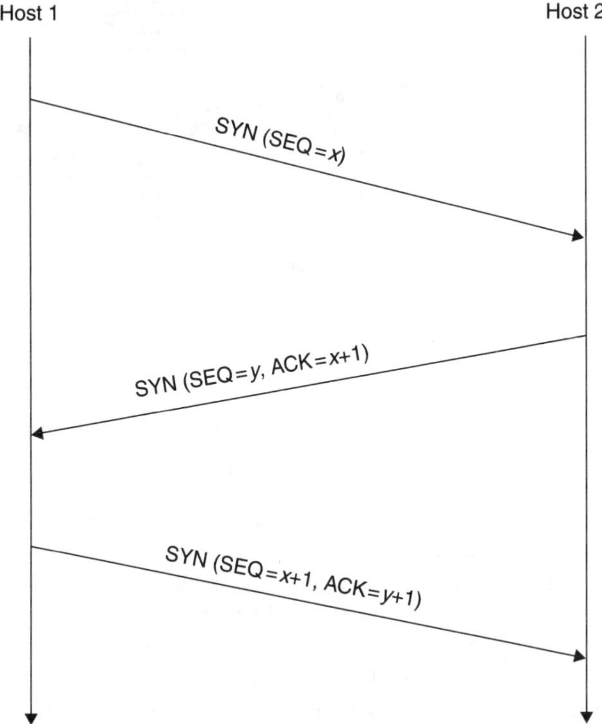

FIGURE 13.6　TCP three-way handshake.

occurs, h is set to $h/2$ and the congestion window is reset to 1 packet. Instead of waiting for the retransmission timer to time out, a strategy called *fast retransmit* enables receivers to send duplicate ACKs to the sender, in case out-of-order segments are received. A sender receiving more than two or three of such duplicate ACKs may deduce that an intermediate segment has been lost rather than that the segments have been just remixed on the way, and may retransmit the missing segment earlier [RFC1122].

User Datagram Protocol

The UDP is the second important IP protocol at OSI layer 4 (in the TCP/IP reference model at the transport layer) [RFC768, RFC1122] and is assigned the IP protocol number 17. It is meant for transporting application data in a message-oriented, unreliable manner from one application to another. As most functionality is already provided by IP, the UDP header only contains the port numbers of the source and receiver applications, the length of the UDP packet, and a checksum.

As UDP does not provide any functionality for detecting lost packets or out-of-order delivery, it is mostly used either in local networks with large bandwidths and reliable layer 2 transport, or for transporting multimedia data like live broadcasts, where a few lost packets will not seriously decrease the perceived quality of the presentation.

In any case, detection of lost packets or out-of-order delivery must be carried out by the receiving applications, usually by including sequence numbers in the UDP application data. The interpretation of these numbers is left solely to the applications.

Resource Reservation Protocol

IPv4 does not contain mechanisms for guaranteeing a minimum quality of service for its traffic, for instance, a minimum sustainable end-to-end bitrate, or a maximum end-to-end delay or jitter (delay

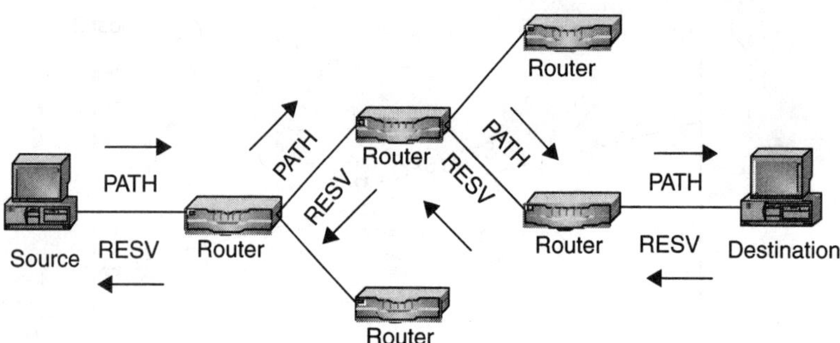

FIGURE 13.7   RESV and PATH messages in a multicast tree.

variation). This may severely affect the presentation quality of real-time transmissions, using, for example, RTP (see application-layer protocols). The *Resource Reservation Protocol* (RSVP) tries to fill this gap by providing means for guaranteeing certain quality of service parameters [RFC2205, RFC2750]. It is an optional add-on for Internet routers and clients using IP (IPv4 and IPv6), and is currently available on a small subset of Internet hosts only. Being at the same level as TCP or UDP, it has its own IP protocol number (47). RSVP is no routing protocol itself but rather a signaling protocol. It cooperates with other routing protocols for controlling efficient unicast and multicast over IP.

RSVP allows two different QoS modes. In the *controlled load service*, RSVP simulates a lightly loaded network for its clients, although the network itself may be overloaded [RFC2211]. Although no hard QoS parameters are met, a lightly loaded network is likely to be sufficient for many load tolerant and adaptive applications like A/V streaming. In contrast, the *guaranteed service* guarantees that the RSVP path will meet the agreed QoS service level at all times [RFC2212].

A client application wishing to receive a multicast multimedia stream passes this request to its local RSVP daemon. This daemon then sends a reservation (*RESV*) request to adjacent RSVP routers toward the multimedia source along the reverse multicast tree path. The RESV request contains a description of the desired quality of service in a so-called *flow descriptor*. Coming from the other side, the multicast source periodically sends *PATH* messages down the multicast tree. PATH messages create and acknowledge valid and active multicasting paths (FIGURE 13.7). Also, they carry information about the quality of service of the path from source to receiver. RSVP routers may merge different QoS requests into one single reservation, here choosing the maximum of each request as the pre-reserved QoS level.

During run time, reservations may be changed to other QoS levels. Also, RSVP paths must be acknowledged periodically by PATH and RESV messages; but RSVP is fault tolerant with respect to a few missing messages. Only if none have been received for a certain time, the whole path is cancelled.

On each RSVP router, an RSVP daemon manages and controls the IP routing process. It consists of the following modules: an incoming QoS reservation request is approved or denied by the *admission control*, depending on whether the QoS request can be satisfied or not. The rights for making reservations are checked in the *policy control* module. Incoming data packets are sorted by the *packet classifier*, which puts them into different queues. Finally, the *packet scheduler* is responsible for granting the agreed QoS to the packets in the routing queues, packets belonging to the same queue are treated identically.

13.7 Presentation-Layer Protocols and Services

Applications may send arbitrary data to others, often embedding complex data structures into their messages. In this process, the data structures have to be transformed (flattened, marshaled) into a sequence of bytes, containing the data itself as well as information about the used data representation. The receiver must be able to understand the structure of the byte sequence and how to interpret the single bytes in order to reconstruct the sent data structures. This is achieved by the *presentation layer* (layer 6 of the ISO/OSI model).

The presentation layer assures that two computer systems may successfully communicate even if they use different data representations. Due to different data representation schemes, the presentation layer is often forced to translate sent or received messages. This, however, should be done in a manner being totally transparent to the OSI application layer above.

Problems may arise, for instance, because of the CPU byte order. In modern 32-bit architectures, CPUs store values and addresses using 32 bits, stored in four consecutive bytes. In Intel processors, for example, the least significant byte is stored first, the most significant byte last. This is called *little endian*. On the other hand, for example, Motorola processors store a 4-byte value in the reverse order, called *big endian*. If an Intel-based computer sends a 32-bit value to a Motorola based computer, without further corrective measures, the receiver totally misinterprets the received value. This may be prevented, for instance, by forcing the sender to convert the data to the receiver's format before sending, or alternatively forcing the receiver to convert the data from the sender's format after receiving. A third approach is to agree to a commonly used format, and to convert to this format before sending or from this format after receiving. TCP/IP, for instance, defines a common *network byte order*. Using, for example, the C programming language, 32-bit values may be converted to and from this format by the macros `hton()` and `ntoh()`. Another system using an external data format is given by the *External Data Representation* (XDR) [RFC1832].

Another problem arising in different computers is the code interpretation. For instance, characters may be stored using one of the following codes: ASCII (common in Intel compatibles, 8 bits/character), EBCDIC (used on IBM mainframes, 8 bits/character), or UNICODE (16 or 20 bits/character). Here, the presentation layer is responsible for automatically translating between the various code schemes.

At the next higher decoding level, received complex data structures should be reconstructed (unmarshalled) from their flattened byte sequence representation. For inhomogeneous data, the data structures must be described by metadata, for instance, defining the data types belonging to each structure, being followed by the data values themselves. This, for instance, can be achieved by using the standardized *Abstract Syntax Notation 1* (ASN.1) [X680].

Other tasks of the presentation layer include the encryption of messages and supporting authentication. Finally, the presentation layer may also be responsible for the compression of data.

13.8 Application-Layer Protocols and Services

In both the ISO/OSI scheme, and the TCP/IP reference model, the *application layer* defines protocols directly to be used by applications for exchanging data with each other. These include, for instance, authentication, distributed databases and files systems, file transport, data syntax restrictions, coordination and agreement procedures, quality-of-service issues, e-mail, or terminal emulation.

Many standard protocols are already specified by the IETF. They define standard data structures that are to be exchanged between applications. Applications following these protocols are guaranteed to be able to successfully interact with other applications over the Internet, even if these applications have been created by different sources. For instance, Web browsers following the HTTP protocol may download Web pages from any Web server connected to the Internet.

The IETF-specified protocols usually use TCP for reliable transport and UDP for the transport of real-time multimedia data (although real-time multimedia data may also be sent over TCP). Usually, both control commands and pure data can be transmitted over the same TCP connection. For signaling the end of a data transmission, one of three approaches is used. In *octet stuffing*, the end of a data transmission is signaled by a certain byte sequence (similar to the bit stuffing used at the data link layer). If the transported data also contain this very sequence, the sequence is changed (escaped) to another sequence. The receiver must detect such a change and undo it. An example for octet stuffing is SMTP. In *octet counting*, transported messages contain special headers that specify the number of data bytes to be transferred. This concept is used, for instance, in HTTP. Finally, in *connection blasting*, the end of a transmission is signaled by closing the TCP connection. This is used, for instance, in FTP.

TELNET

The *TELNET* protocol is meant for providing a general 8-bit interface for the communication between users, hosts, and processes [RFC854]. Generally, a TELNET client running on computer *A* opens a TCP connection to port 23 of a TELNET server on computer *B*. Both sides then emulate a certain simple type of terminal called *network virtual terminal* (NVT), but may negotiate additional services after the connection has been established. An NVT is a bi-directional character device, consisting of a printer that shows the information received from the other side, and a keyboard where keystrokes are produced and sent to the other side.

TELNET defines a set of commands that may be sent in-band with the stream of data. The mechanism used here is the octet stuffing. The byte 255 is called *interpret as command* (IAC) and signals that the following byte specifies a TELNET command, for example, for sending an interrupt to the running process, or for erasing the last character. If a data byte with value 255 is to be sent, then two bytes with value 255 are sent. On receiving two consecutive bytes with value 255, the receiver side must remove one of them automatically.

File Transfer Protocol

The *File Transfer Protocol* (FTP) is used for transporting arbitrary binary data from one Internet host to another [RFC959]. On a computer *A*, an FTP client is started with the IP or DNS address of the Internet computer *B* with which communication is desired. The FTP client then opens a TCP connection to port 21 of computer *B*, representing the *control connection*. The control connection uses the TELNET protocol underneath, and users may send *control commands* to the FTP server on computer *B*, including the request for showing the contents of the current directory at computer *B* (LIST) as well as changing this current directory to another one (CWD), creating new directories (MKD), etc. Additionally, the user may start uploads (STOR) or downloads (RETR) of files to and from the current directory. Upon receiving a control command over the control connection, the server will answer with a reply, sending status or error information to the client. One has to distinguish between the FTP control command that is actually sent over the control channel and commands that are typed in by users into a command line application, which may be different.

Once data are to be sent, a TCP *data connection* is opened by the server on computer *B* from port 20 to the user client on computer *A* listening on port 21. Then, depending on the specified direction, the data are sent either from *A* to *B* or *vice versa*. After transmitting the last byte, the sender must close the data connection, indicating to the other side that the transmission has ended.

It is worth noting that FTP knows different transmission modes. In the *binary* mode, the data are sent without modification. In the *ascii* mode, the FTP protocol automatically changes different character representation codes, for instance, when sending a pure text file from an IBM mainframe (using EBCDIC) to a PC (using ASCII), or when exchanging data between different operating systems like Microsoft Windows and Unix or Unix-like operating systems (having different end-of-line representations in text files).

Hypertext Transfer Protocol

The HTTP is available as version 1.0 [RFC1945] and version 1.1 [RFC2616]. Its purpose is to manage the download of documents being part of the World Wide Web (WWW), usually following the *Hypertext Markup Language* (HTML) [RFC1866].

Most Web browsers and servers nowadays understand HTTP/1.0, although [RFC1945] is not a standard but rather an informational guideline. Newer Web clients and servers also support the standardized HTTP/1.1.

HTTP is a client/server-based protocol following the octet counting approach. A client wishing to download a specific document from a Web server opens a TCP connection to the server port 80 (sometimes 8080). The client then sends a request, containing a request line, various headers, an empty line, and an optional body. The request line specifies what the client wants the server to do. For example, a request line "GET /dir1/dir2/the_document.html HTTP/1.1" informs the server that the client wants to download the document the_document.html, which is situated in the directory /dir1/dir2 by using HTTP/1.1.

Clients may also send data to the server, for example, a form that has been filled out by a user. This can be done, for instance, using the "PUT" command.

The server then answers by sending a status line containing a code for success or an error description, various headers describing the downloaded document, for instance, its size or the time stamp of its last change, followed by an empty line. Finally, in the message body, the HTML document itself is transported to the client.

HTTP/1.1 masters several limitations of HTTP/1.0. For example, an HTML document may contain several other subdocuments, like photos, wall papers, frames, etc. In HTTP/1.0, for each subdocument a new TCP connection has to be created. In HTTP/1.1, all subdocuments can be transported over the same *persistent* TCP connection.

Simple Mail Transfer Protocol

The SMTP defines the exchange and relay of text mails over TCP/IP [RFC821]. If a mail client running on computer *A* wants to send a mail to a receiver on computer *B*, it opens a TCP connection to port 25 of either computer *B* itself, or an intermediate mail server that is able to pass on the mail to the receiver on computer *B*. Then, the sender client sends SMTP commands to the receiver, who replies by sending SMTP responses. Once the sender wants to send an electronic mail, it sends the command MAIL with an identifier for the *sender*. If the receiver is willing to accept mails from the sender, it answers with an OK reply. Now, the sender client sends a sequence of RCPT commands, which identify the receivers of the mail. Each recipient is acknowledged individually by an OK reply. Once all receivers have been specified, the client sends a DATA command followed by the mail data itself. In order to indicate the end of the mail, the client sends a line containing only a period. If such a line is part of the message, the sender will introduce an additional period, which is removed by the receiver automatically (octet stuffing).

In SMTP, a (text) mail must be composed of 7-bit ASCII characters only (byte values 0 – 127), a limitation that was not severe in 1982 when SMTP was designed. Nowadays, electronic mail often contains multimedia attachments like audio or video files, where each byte may contain any value between 0 and 255. In order to be able to transport binary data over SMTP, these data are usually transformed into a sequence of 7-bit ASCII characters by using a byte-to-character mapping like *Base64* or *uuencode*. Upon receiving such a transformed character sequence, the receiver must apply the inverse of the transform in order to retrieve the original binary data.

Resource Location Protocol

Computers connected over an IP network may offer a variety of services to others, including services standardized by the IETF like DNS, SMTP, FTP, etc., as well as self-created services, for instance, for managing personal information. The *Resource Location Protocol* (RLP) has been designed to enable arbitrary computers to automatically find other computers that provide specific services [RFC887].

For this purpose, RLP defines a set of request messages that may be sent by the searching computer. RLP uses UDP as transport protocol. A request message is sent to the UDP port 39 of another host and contains a question and a description of one or more services that are looked for. Depending on the question, hosts that provide the service or know of others that do answer by sending a reply message. RLP defines the following request messages:

- *Who Provides?* is usually broadcast into a LAN. Hosts providing one of the described services may answer, hosts that do not provide any of the specified services may not.
- *Do You Provide?* is directly sent to some specific host. It may not be broadcast. A host receiving this message must answer, regardless of whether it provides any of the specified services or not.
- *Who Anywhere Provides?* also is usually broadcast into a LAN. Hosts either providing any service or knowing other hosts that do so may answer.
- *Does Anyone Provide?* again is sent to a specific host, which must send back an answer, regardless of whether it knows of any host providing any of the services or not.

There are two possible answers. The *I Provide* reply contains a (possibly empty) list of services that are supported by the answering host. The *They Provide* reply contains a (possibly empty) list of supported services, qualified by a list of host-IP addresses supporting them. An RLP message contains the following fields:

- The *Type* field defines the question or reply type.
- The flag *Local Only* specifies, whether only hosts with the same IP network address should answer or be included into the answer list.
- A *Message ID* enables the mapping of received answers to previously sent requests.
- Finally, the *Resource List* contains a description of the looked for or provided services and supporting hosts.

Resources and services may be described by several fields. The first description byte specifies the IP protocol number of the IP transport protocol that the service uses, for instance, 6 for TCP or 17 for UDP. The next byte defines the port that is usually used by the service, for instance, 23 for TELNET or 25 for SMTP. Additional bytes may then define arbitrary self-created services.

Real-Time Protocol

The *Real-Time Protocol* (RTP) has been designed for carrying real-time multimedia data like audio or video information [RFC1889]. Multimedia data usually are produced as a continuous stream of bits. For carrying this stream over the network, it must be packetized and sent as a sequence of packets to one (unicast) or several (multicast) receivers. For real-time traffic, UDP is preferred to TCP, as the delivery of late or lost packets (which is mandatory for TCP) may cause the presentation to stall, which is undesirable, for example, for video conferences. Instead, in case of (a few) lost packets, small artifacts may be visible or audible, which are less annoying than a complete connection breakdown or stall. At the receiver, the original sequence of RTP packets and its content are restored, and lost packets are identified. Pure RTP does not know anything about the payload content. Instead, RTP headers may be altered to fit the needs of specific applications like audio and video conferences. Such changes are then defined in so-called *profile specifications*. Additionally, different RTP payload formats may be defined in payload format specifications, as, for instance, are given by [RFC2190] for H.263. The RTP specification defines the following header fields:

- The RTP *Version* (1 or 2).
- *Padding* and header *Extension* flags.
- Contributing sources (*CSRC*) *count*, that is, length of the CSRC List.
- A marker flag *M* to be used freely by profiles.
- Payload type (*PT*), must be interpreted by the application.
- *Sequence Number*, increased by one for each new RTP packet.
- *Timestamp* of the sampling of the first RTP payload byte.
- RTP *Synchronization Source Identifier*, which must be unique for concurrent RTP sessions.
- An optional contributing sources list (*CSRC List*).

As RTP is transported over the best-effort protocols TCP/UDP/IP, no guarantee can be made that a required bitrate is available for the real-time transport. Instead, RTP provides a means for measuring and controlling the output bitrate and perceived quality of service of a real-time stream. This procedure is provided by the *Real Time Control Protocol* (RTCP). RTCP can carry the following information:

- In a *sender report*, statistics for each active sender are sent to the receivers.
- In a *receiver report*, receivers (which are not senders) send reception statistics to the active senders.
- Sender attributes like e-mail addresses, etc. (*source description*).
- The request for leaving the presentation.
- *Application-specific* control information.

For each real-time session (transporting exactly one medium like audio or video), each participant needs two ports: one for RTP and one for RTCP. RTP is able to multiplex several sessions into one. This is done by a so-called *mixer*. For example, the audio data of several participants of an audio conference may be mixed into one single audio stream and sent over a connection with low bandwidth. Here, the mixer would act as a new synchronization source; the IDs of the original sources, however, may then be stored additionally after the RTP header in the list of *contributing sources*. Another RTP entity is a *translator*, which is able to change payload content or tunnel packets through a firewall.

13.9 Summary

The TCP/IP protocol suite consists of numerous protocols covering several layers of the ISO/OSI stack or alternatively the TCP/IP reference model. Starting at OSI layer 2, protocols are defined for link-level services for secure frame transport, over IP at OSI layer 3 for the unreliable delivery of datagrams from one host connected to the Internet to another. At OSI layer 4, transport protocols regulate the either reliable and controlled or the unreliable transport of data from process to process. Further services may alter the data due to different presentation schemes, or finally offer direct support to applications.

References

[COL2000] Coulouris, G., J. Dollimore, and T. Kindberg, *Distributed Systems: Concepts and Design*, 3rd ed., 1996.

[DAY1983] Day, J.D. and H. Zimmermann, The OSI reference model, *Proceedings of the IEEE*, 71, 1334–1340, 1983.

[HAL1996] Halsall, F., *Data Communications, Computer Networks and Open Systems*, 4th ed., Addison-Wesley, Reading, MA, 1996.

[ISO7498] ISO, *Information Technology — Open Systems Interconnection: Basic Model*, ISO, ISO/IEC 7498-1, 1994.

[KUR2001] Kurose J.F. and K.W. Ross, *Computer Networking: A Top-Down Approach Featuring the Internet*, Addison-Wesley, Reading, MA, 2001.

[PET2000] Peterson, L.L. and B.S. Davie, *Computer Networks: A Systems Approach*, 2nd ed., Morgan Kaufmann, Los Altos, CA, 2000.

[REI2001] Reimers, U., *Digital Video Broadcasting*, Springer, Wien, Berlin, New York, 2001.

[RFC768] RFC 768, User Datagram Protocol, IETF, 1980, http://www.ietf.org/rfc/rfc0768.txt

[RFC791] RFC 791, Internet Protocol. DARPA Internet Program Protocol Specification, DARPA, 1981, http://www.ietf.org/rfc/rfc791.txt

[RFC792] RFC 792, Internet Control Message Protocol, DARPA, 1981, http://www.ietf.org/rfc/rfc792.txt

[RFC793] RFC 793, Transmission Control Protocol, DARPA, 1981, http://www.ietf.org/rfc/rfc793.txt

[RFC821] RFC 821, Simple Mail Transfer Protocol, IETF, 1982, http://www.ietf.org/rfc/rfc821.txt

[RFC826] RFC 826, An Ethernet Address Resolution Protocol, IETF, 1982, http://www.ietf.org/ rfc/rfc826. txt

[RFC854] RFC 854, Telnet Protocol Specification, IETF, 1983, http://www.ietf.org/rfc/rfc854.txt

[RFC887] RFC 887, Resource Location Protocol, IETF, 1983, http://www.ietf.org/rfc/rfc887.txt

[RFC903] RFC 903, A Reverse Address Resolution Protocol, IETF, 1984, http://www.ietf. org/rfc/rfc903.txt

[RFC959] RFC 959, File Transfer Protocol (FTP), IETF, 1985, http://www.ietf.org/rfc/rfc959.txt

[RFC977] RFC 977, Network News Transfer Protocol: A Proposed Standard for the Stream-Based Transmission of News, IETF, 1986, http://www.ietf.org/rfc/rfc977.txt

[RFC1034] RFC 1034, Domain Names — Concepts and Facilities, IETF, 1987, http:// www.ietf.org/ rfc/rfc1034. txt

[RFC1035] RFC 1035, Domain Names — Implementation and Specification, IETF, 1987, http://www. ietf.org/rfc/rfc1035. txt

[RFC1112] RFC 1112, Host Extensions for IP Multicasting, IETF, 1989, http://www.ietf.org/ rfc/ rfc1112.txt

[RFC1122] RFC 1122, Requirements for Internet Hosts — Communication Layers, IETF, 1989, http:// www.ietf.org/rfc/rfc1122.txt

[RFC1157] RFC 1157, A Simple Network Management Protocol (SNMP), IETF, 1990, http:// www.ietf. org/ rfc/ rfc1190.txt

[RFC1323] RFC 1323, TCP Extensions for High Performance, IETF, 1992, http://www.ietf.org/ rfc/ rfc1323.txt

[RFC1832] RFC 1832, XDR: External Data Representation Standard, IETF, 1995, http:// www.ietf.org/rfc/rfc1832.txt

[RFC1866] RFC 1866, Hypertext Markup Language — 2.0, IETF, 1995, http://www.ietf.org/rfc/ rfc1866.txt

[RFC1883] RFC 1883, Internet Protocol, Version 6 (IPv6) Specification, IETF, 1995, http://www.ietf.org/ rfc/rfc1883.txt

[RFC1887] RFC 1887, An Architecture for IPv6 Unicast Address Allocation, IETF, 1995, http://www.ietf. org/rfc/rfc1887.txt

[RFC1889] RFC 1889, RTP: A Transport Protocol for Real-Time Applications, IETF, 1996, http://www.ietf.org/rfc/rfc1889.txt

[RFC1918] RFC 1918, Address Allocation for Private Internets, IETF, 1996, http://www.ietf.org/rfc/ rfc1918.txt

[RFC1945] RFC 1945, Hypertext Transfer Protocol — HTTP/1.0, IETF, 1996, http://www.ietf.org/rfc/ rfc1945.txt

[RFC2190] RFC 2190, RTP Payload Format for H.263 Video Streams, IETF, 1997, http://www.ietf.org/ rfc/rfc2190.txt

[RFC2205] RFC 2205, Resource ReSerVation Protocol (RSVP), IETF, 1997, http://www.ietf.org/rfc/ rfc2205.txt

[RFC2211] RFC 2211, Specification of the Controlled-Load Network Element Service, IETF, 1997, http://www.ietf.org/rfc/rfc2211.txt

[RFC2212] RFC 2212, Specification of Guaranteed Quality of Service, IETF, 1997, http://www.ietf.org/rfc/rfc2212.txt

[RFC2616] RFC 2616, Hypertext Transfer Protocol — HTTP/1.1, IETF, 1999, http://www.ietf.org/ rfc/rfc2616.txt

[RFC2750] RFC 2750, RSVP Extensions for Policy Control, IETF, 2000, http://www.ietf.org/rfc/ rfc2750.txt

[RFC3376] RFC 2376, Internet Group Management Protocol Version 3, IETF, 2002, http://www.ietf.org/ rfc/rfc3376.txt

[SCO1991] Scocolowski, T. and C. Kale, A TCP/IP Tutorial, IETF, Network Working Group, rfc 1180, January 1991, http://www.ietf.org/rfc/rfc1180.txt

[TAN1996] Tanenbaum, A.S., *Computer Networks*, 3rd ed., Prentice-Hall, Englewood Cliffs, NJ, 1996.

[X680] ISO, Information Technology — Specification of Basic Notation, ITU-T Recommendation X.680, Abstract Syntax Notation One (ASN.1), 1997, http://www.itu.int/

14

The Fundamentals of the Quality of Service

Wolfgang Kampichler
Frequentis Nachrichtentechnik Gesellschaft m.b.H.

14.1 Introduction

The Internet offers only very simple Quality of Service (QoS): point-to-point best-effort data delivery. Before IP multicast and real-time applications can be broadly implemented, the Internet infrastructure must be modified to support different levels of services, and to receive secure, predictable, measurable, and guaranteed service.

QoS refers to the ability of a network element to have some level of assurance that its traffic and service requirements can be satisfied. Enabling QoS requires the cooperation of all network layers on every network element from end to end. Thus, such QoS assurances are only as good as the weakest link in the chain between the sender and the receiver.

14.2 What is Quality of Service?

It is difficult to find an adequate definition of what Quality of Service (QoS) actually is. There is a danger that because we wish to use quantitative methods, we might limit the definition of QoS to only those aspects of QoS that can be measured and compared. In fact, there are many subjective and perceptual elements to QoS, and there has been a lot of work done trying to map the perceptual to the quantifiable (particularly in the telephone industry). However, as yet there does not appear to be a standard definition of what QoS actually is in measurable terms.

When considering the definition of QoS, it might be helpful to look at the old story of the three blind men who happen to meet an elephant on their way. The first man touches the elephant's trunk and determines that he has stumbled upon a huge serpent. The second man touches one of the elephant's massive legs and determines that the object is a large tree. The third man touches one of the elephant's ears and determines that he has stumbled upon a huge bird. All three of the men envision different things, because each man examines only a small part of the elephant. In this case, think of the elephant as a concept of QoS. Different people see QoS as different concepts, because various and ambiguous QoS problems exist. Hence, there is more than one way to characterize QoS. Briefly described, QoS is the ability of a network element (e.g., an application, a host, or a router) to provide some level of assurance for consistent and timely network data delivery [3].

By nature, the basic IP service available in most of the network is best effort. For instance, from a router's point of view, this service could be described as follows:

Upon receiving a packet at the router:

- It determines first where to send the incoming packet (the next-hop of the packet). This is usually done by looking up the destination address in the forwarding table.
- Once it is aware of the next-hop, it will send the packet to the interface associated to this next-hop. If the interface is not able to immediately send the packet, it is stored on the interface in an output queue.
- If the queue is full, the arriving packet is dropped. If the queue already contains packets, the new-comer is subjected to extra delay due to the time needed to emit the older packets in the queue.

Best effort allows the complexity to stay in the end-hosts; so the network can remain relatively simple. This scales well, as evidenced by the ability of the Internet to support its growth. As more hosts are connected, network degrades gracefully. Nevertheless, the resulting variability in delivery delay and packet loss does not adversely affect typical Internet applications (e.g., email or file transfer). Considering applications with real-time requirements, delay, delay variation, and packet loss will cause problems. Generally, applications are of two main types:

- Applications that generate elastic traffic — that is, the application would rather wait for reception of traffic in the correct order, without loss, than display incoming information at a constant rate (such as an email) and
- Applications that generate inelastic traffic — that is, timeliness of information is more important to the application than zero loss, and traffic that arrives after a certain delay is essentially useless (such as voice communication).

In an IP-based network, applications run across User Datagram Protocol (UDP) or Transmission Control Protocol (TCP) connections. TCP guarantees delivery, doing so through some overhead and session-layer sequencing of traffic. It also throttles back transmission rates to behave gracefully in the face of network congestion.

By contrast, UDP is connectionless; thus, no guarantee of delivery is made, and sequencing of information is left to the application itself. Most elastic applications use TCP for transmission and, in contrast, many inelastic applications use UDP as a real-time transport. Inelastic applications are often those that demand a preferential class of service or some form of reservation to behave properly. However, many of the mechanisms that network devices use (such as traffic discard or TCP session control) are less effective on UDP-based traffic since it does not offer some of TCP's self-regulation.

Common for all packets is that they are treated equally. There are no guarantees, no differentiation, and no attempt enforcing fairness. However, the network should try to forward as much traffic as possible with reasonable quality. One way to provide a guarantee to some traffic is to treat packets differently from packets of other types of traffic.

Increasing bandwidth is seen as a necessary first step for accommodating real-time applications, but it is still not enough. Even on a relatively unloaded network, delivery delays can vary enough to continue to affect time-sensitive applications adversely. To provide an appropriate service, some level of quantitative or qualitative determinism must be supplemented to network services. This requires adding some "intelligence" to the net, to distinguish traffic with strict timing requirements from others.

Yet, there remains a further challenge: in the real world, the end-to-end communication path consists of different elements utilizing several network layers. Therefore, it is unlikely that QoS protocols will be used independently, and in fact they are designed for use with other QoS technologies to provide top-to-bottom and end-to-end QoS between senders and receivers. What does matter is that each element has to provide QoS control services and the ability to map other QoS technologies in the correct manner. The following gives a brief overview of end-to-end network behavior and some key QoS protocols and architectures. For a detailed description, please let me refer to other articles in this book.

14.3 Factors Affecting the Network Quality

A typical end-to-end communication path might appear as illustrated in Figure 14.1 and consists of two machines, each connected through a Local Area Network (LAN) to an enterprise network. Further, these networks might be connected through a Wide Area Network (WAN). The data exchange can be anything from a short email message to a large file transfer, an application download from a server, or communication data from a time-sensitive application. While networks, especially LANs, have been becoming faster, perceived throughput at the application has not always increased accordingly.

An application is generally running on a host CPU, and its performance is a function of the processing speed, memory availability, and the overall operating system load. In many situations, it is the processing that is the real limiting factor on throughput, rather than the infrastructure that is moving data [17].

Network interface hardware transfers incoming packets from the network to the computer's memory and informs the operating system that a packet has arrived. Usually, the network interface uses the interrupt mechanism to do so. The interrupt causes the CPU to suspend normal processing temporarily and to jump to a code called a device driver. The device driver informs the protocol software that a packet has arrived and must be processed.

Similar operations occur in each intermediate network node. Routing devices pass packets along a chain of hops until the final address is reached. These hops are routing machines of various kinds, which generally maintain a queue (or multiple queues) of outgoing packets on each outgoing physical port [2]. If these queues of outgoing data packets become full, it simply starts discarding packets randomly to ease the buildup of congestion. It is evident that such nodes are customized for forwarding operations, which are mostly processed in hardware.

In recent years, however, the Internet has seen increasing use of applications that rely on the timely, regular delivery of packets, and that cannot tolerate the loss of packets or the delay caused by waiting in queues. In general, the one-way delay is equivalent to the sum of single-hop delays suffered between each pair of consecutive pieces of equipment encountered on the path. Measurable factors [7, 8] that are used to describe network QoS are as follows.

Bandwidth

Bandwidth (better described as data rate in this context) is the transmission capacity of a communications line, which is usually stated in bit/sec. The figure given is a nominal figure. In reality, as data

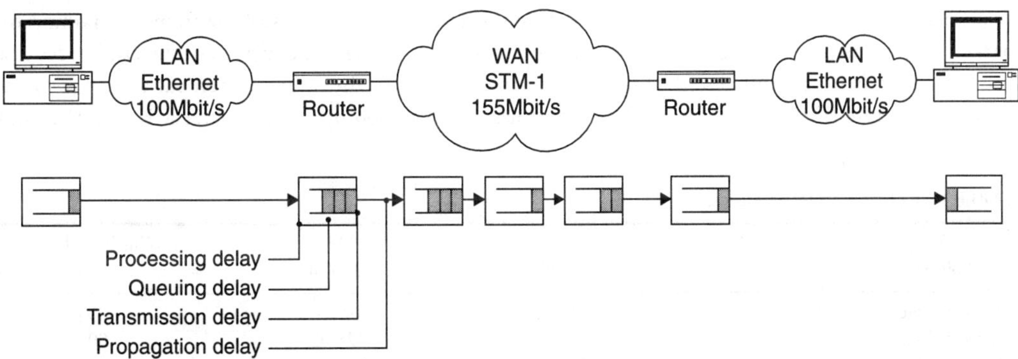

FIGURE 14.1   Network end-to-end communication path.

exchange nears the maximum limit (in a shared environment), delays and collisions might mean a drop in quality. Basically, the bandwidths of all networks utilized in an end-to-end path need to be considered, as the narrowest section provides the maximum speed of data transfer for the entire path. A routing device needs to be capable of transmitting data at a rate commensurate with the potential bandwidth of the network segments that it is servicing. The cost of bandwidth has fallen in recent years; but demand has obviously gone up.

Throughput

Throughput is the average of actual traffic transferred over a given link, in a given time span expressed in bit/sec. It can be seen, for congestion aware transport protocols such as TCP, as *transport capacity* = (*data sent*)/(*elapsed time*) where "*data sent*" represents the unique "data" bits transferred (i.e., not including header bits or emulated header bits). It should also be noted that the amount of data sent should only include the unique number of bits transmitted (i.e., if a particular packet is retransmitted, the data it contains should be counted only once). Hence in such a case, the throughput is also limited by the value of the round-trip time.

Latency

In general, latency is the time taken to transmit a packet from a sending to a receiving node. This encompasses a delay in a transmission path or in a device within the transmission path. The nodes might be end-stations or intermediate routes. Within a single router, latency is the amount of time between the receipt of a data packet and its transmission, which includes processing and queuing delay as described next among other sources of delay.

Queuing Delay

The major random component of delay (i.e., the only source of jitter) for a given end-to-end path consists of queuing delay in the network. Queuing delay depends on the number of hops in the path and the queuing mechanisms used, and it also increases with the offered load, leading to packet loss if the queues are filled up. The last packet in the queue has to wait $(N*8) / X$ seconds before being emitted by the interface, where N is the number of bytes that have to be sent before the last queued packet and X is the sending rate (bit/sec). Typical queuing delay values of state of the art routers are summarized in Table 14.1.

Values are about 0.5 to 1 msec; thus, it can be said that queuing delay in a well-dimensioned backbone network (using priority scheduling mechanisms, as described later) would not dramatically increase latency, even if there are 5 to 8 hops within the path. At this point, it should be mentioned that queuing delay may be impaired by edge-routers connecting high and low bandwidth links and could easily reach tens of milliseconds, thus increasing latency more distinctly.

Transmission Delay

Transmission or serialization delay is the time taken to transmit all the bits of the frame containing the packet, that is, the time between emission of the first bit of the frame and emission of the last bit; see also [4]. It is inversely proportional to the line speed, or in other words, the ratio between packet size (bit) and transmission rate (bit/sec). For example, a transmission of a 1500 byte packet over a 10 Mbit/sec link

TABLE 14.1 Queuing Delays

Number of queued 1000 bit packets	STM-1 (155 Mbit/s)	STM-4 (622 Mbit/s)	Gigabit Ethernet (1 Gbit/s)
40 (80% load)	256 µs	64 µs	40 µs
80 (85% load)	512 µs	128 µs	80 µs
200 (93% load)	1280 µs	320 µs	200 µs
500 (97% load)	3200 µs	800 µs	500 µs

takes 1.2 ms and compared to a 64 kbit/sec link, it takes 187.5 msec (the protocol overhead is not considered in either case). In general, a small packet size and a high transmission rate lower the transmission time.

Propagation Delay

Propagation delay is the time between emission (by the emitting equipment) of the first bit (or the last bit) and the reception of this bit by the receiving equipment. It is mainly a function of the speed of the light and the distance traveled. For local area networks, the propagation delay is almost negligible. For wide area connections, it typically adds 2 msec per 250 mil to the total end-to-end delay. One can assume that a well-designed homogeneous high-speed backbone network (e.g., STM-4) would have a network delay (only propagation and queuing taken into account) of 10 msec when considering 10 hops using priority queuing mechanisms and a network extension of about 625 mil.

Processing Delay

Most networks use a protocol suite that provides connectionless data transfer end-to-end, in our case IP. Link-layer communication is usually implemented in hardware; but IP will usually be implemented in software, executing on the CPU in a communicating end station. Normally, IP performs very few functions. Upon inputting of a packet, it checks the header for correct form, extracts the protocol number, and calls the upper-layer protocol function. The executed path is almost always the same.

Upon outputting, the operation is very similar, as shown in the following IP instruction counts:

- Packet receipt: 57 instructions.
- Packet sending: 61 instructions.

Since input occurs at interrupt time, arbitrary procedures cannot be called to process each packet. Instead, the system uses a queue along with message passing primitives to synchronize communication. When an IP datagram arrives, the interrupt software must en-queue the packet, and invokes a send primitive to notify the IP process that a datagram has arrived. When the IP process has no packets to handle, it calls the receiving primitive to wait for the arrival of another datagram. Once the IP process accepts an incoming datagram, it must decide where to send it for further processing.

If the datagram carries a TCP segment, it must go to the TCP module; if it carries a UDP datagram, it is forwarded to the UDP module. Being complex, most TCP designs use a separate process to handle incoming segments. A consequence of having separate IP and TCP processes is that they must use an inter-process communication mechanism when they interact. Once TCP receives a segment, it uses the protocol port numbers to find the connection to which the segment belongs. If the segment contains data, TCP will add the data to a buffer associated with the connection and return an acknowledgement to the sender. If the incoming segment carries an acknowledgement for outbound data, the input process must also communicate with the TCP timer process to cancel the pending retransmission.

The process structure used to handle an incoming UDP datagram is quite different from that used for TCP. As UDP is much simpler than TCP, the UDP software module does not execute as a separate process. Instead, it consists of conventional procedures that the IP process executes for handling of an incoming UDP datagram. These procedures examine the destination UDP protocol port number, and use it to select an operating system queue for the incoming datagram. The IP process deposits the UDP datagram on the appropriate port, where an application program can extract it [18].

Jitter

Jitter is best described as the variation in end-to-end delay, and has its main source in the random component of the queuing delay. Jitter can be expressed as the distortion of inter-packet arrival times when compared to the inter-packet departure times from the original sending station. For instance, if packets are sent out at regular intervals, they may arrive at varying irregular intervals. Jitter is the variation in interval times. When packets are taking multiple paths to reach their destination, extreme jitter can lead to packets arriving out of order. Jitter is generally measured in milliseconds, or as a percentage of variation from the average latency of particular connection.

Packet Loss

Packets that fail to arrive, or arrive so late that they are useless contribute to packet loss. Lost (or dropped) packets are a product of insufficient bandwidth on at least one routing device on the network path. Some packets may arrive, but have been corrupted in transit and are therefore unusable. Note that loss is relative to the volume of data that is sent, and is usually expressed as a percentage of data being sent. In some contexts, a high loss percentage can mean that the application is trying to send too much information and is overwhelming the available bandwidth. Packet loss becomes a real problem when the percentage of loss exceeds a specific threshold, or when loss occurs in bursts. Thus, it is important to know both the percentages of lost packets and their distribution [5].

14.4 QoS Delivery

As packet-switched networks are operated in a store-and-forward paradigm, a solution for service differentiation in the forwarding process is to give priority to packets requiring, for instance, an upper-bounded delay over other packets. Considering that queuing is the central component in the internal architecture of a forwarding device, it is not difficult to imagine that managing such queuing mechanisms appropriately is crucial for providing the underlying QoS, thus being one of the fundamental parts for differentiating service levels.

The queuing delay can be minimized and kept below a certain value, even in the case of interface congestion. To achieve this, the forwarding device has to support classification, queuing, and scheduling (CQS) techniques to classify packets according to a traffic type and its requirements, to place packets on different queues according to this type. Finally, to schedule outgoing packets by selecting them from the queues in an appropriate manner, see Figure 14.2.

The following descriptions of queuing disciplines focus on output queuing strategies, being the predominant strategic location for store-and-forward traffic management and QoS-related queuing [3], common for all QoS policies.

FIFO Queuing

First-in, first-out (FIFO) queuing is considered to be the standard method for store-and-forward handling of traffic from an incoming interface to an outgoing interface. Many router vendors have highly

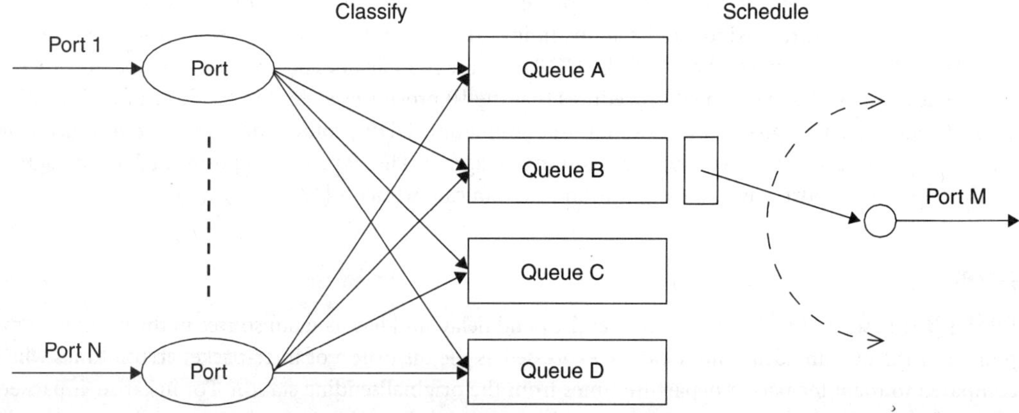

FIGURE 14.2   Classification, queuing, and scheduling.

optimized forwarding performances that make this standard behavior as fast as possible. When a network operates in a mode with a sufficient level of transmission capacity and adequate levels of switching capability, FIFO queuing is highly efficient. This is because, as long as the queue depth remains sufficiently short, the average packet-queuing delay is an insignificant fraction of the end-to-end packet transmission time. Otherwise, when the load on the network increases, the transient bursts raise significant queuing delay, and when the queue is full, all subsequent packets are discarded.

Priority Queuing

One of the first queuing variations to be widely implemented was priority queuing. This is based on the concept that certain types of traffic can be identified and shuffled to the front of the output queue so that some traffic is always transmitted ahead of other types of traffic. Priority queuing may have an adverse effect on forwarding performance because of packet reordering (non-FIFO queuing) in the output queue.

This method offers several levels of priority, and the granularity in identifying traffic to be classified into each queue is very flexible. Although the level of granularity is fairly robust, the more differentiation attempted, the more the impact on computational overhead and packet-forwarding performance. Another possible vulnerability in this queuing approach is that if the volume of high-priority traffic is unusually high, normal traffic to be queued may be dropped because of *buffer starvation*. This usually occurs because of overflow caused by too many packets waiting to be queued and there is not enough room in the queue to accommodate them.

Class-Based Queuing (CBQ)

Another queuing mechanism introduced several years ago is called class-based queuing or custom queuing. Again, this is a well-known mechanism used within operating system design intended to prevent complete resource denial to any particular class of service. CBQ is a variation of priority queuing, where several output queues can be defined. CBQ provides a mechanism to configure how much traffic can be drained off each queue in a servicing rotation. This servicing algorithm is an attempt to provide some semblance of fairness by prioritizing queuing services for certain types of traffic, while not allowing any one class of traffic to monopolize system resources.

CBQ can be considered a primitive method of differentiating traffic into various classes of service, and for several years, it has been considered an efficient method for queue-resource management. However, CBQ simply does not scale to provide the desired performance in some circumstances, primarily because of the computational overhead concerning packet reordering and intensive queue management in networks with very high-speed links.

Weighted Fair Queuing (WFQ)

WFQ is another popular method of queuing that algorithmically attempts to deliver predictable behavior and to ensure that traffic flows do not encounter buffer starvation. It gives low-volume traffic flows preferential treatment and allows higher-volume traffic flows to obtain equity in the remaining amount of queuing capacity. WFQ uses a servicing algorithm that attempts to provide predictable response times and negate inconsistent packet-transmission timing, which is done by sorting and interleaving individual packets by flow, and queuing each flow based on the volume of traffic in each flow [6]. The weighted aspect of WFQ is dependent on the way in which the servicing algorithm is affected by other extraneous criteria. This aspect is usually vendor-specific, and at least one implementation uses the IP precedence bits in the Type of Service (TOS) field to weigh the method of handling individual traffic flows.

WFQ possesses some of the same characteristics as priority and class-based queuing — it simply does not scale to provide the desired performance in some circumstances, primarily because of computational overhead. However, if these methods of queuing (priority, CBQ, and WFQ) could be moved completely into hardware instead of being done in software, the impact on forwarding performance could be reduced greatly.

14.5 Protocols to Improve QoS

Delivering network QoS for a particular application implies minimizing the effects of sharing network resources (bandwidth, routers, etc.) with other applications. This means effective QoS aims to minimize delay, optimize throughput, and minimize jitter and loss. The reality is that network resources are shared with other, competing applications. Some of the competing applications could also be time-dependent services (inelastic traffic); others might be the source of traditional, best-effort traffic. For this reason, QoS has the further goal of minimizing the parameters mentioned for a particular set of applications or users, but without adversely affecting other network users.

In order to regulate network capacity, the network must classify traffic and then handle it in some way. The classification and handling may occur on a single device consisting of both classifiers and queues or routes. In a larger network, however, it is likely that classification will occur at the periphery where devices can recognize application needs, while handling is performed at the core where congestion occurs. The signaling between classifying devices and handling devices can occur in a number of ways, like the ToS of an IP header, or other protocol extensions.

Classification can occur based on a variety of information sources such as protocol content, media identifier, the application that generated the traffic, or extrinsic factors such as time of the day or congestion levels.

Similarly, handling can be performed in a number of ways:

- Through traffic shaping (traffic arrives and is placed in a queue; where its forwarding is regulated, access traffic will be discarded).
- Through various queuing mechanisms (FIFO, priority weighting, and CBQ).
- Through throttling using various flow-control algorithms such as used in TCP.
- Through the selective discard of traffic to notify transmitters of congestion.
- Through packet marking for sending instructions to downstream devices that will shape the traffic.

QoS protocols are designed to act that way, but they never create additional bandwidth; rather, they manage it to be used more effectively. Briefly summarized, QoS is the ability of a network element (e.g., an application, a host, or a router) to provide some level of assurance for consistent and timely network data delivery. The following sections provide a brief overview of some of the key QoS protocols and architectures.

Integrated Services (IntServ)

The IntServ architecture provides a framework for applications to choose between multiple controlled levels of delivery of services for their traffic flows. Two basic requirements exist to support this framework. The first requirement is for the nodes in the traffic path to support the QoS control mechanisms and guaranteed services. The second requirement is for a mechanism by which the applications can communicate their QoS requirements to the nodes along the transit path, as well as for the network nodes to communicate between each other about the requirements that must be provided for the particular traffic flow. All this is provided by a Resource Reservation Set-up Protocol called RSVP [9] that is best described as a QoS signaling protocol. The information presented here is intended to be a qualitative description of the protocol as in [3].

There is a logical separation between the Integrated Services QoS control services and RSVP. RSVP is designed to be used with a variety of QoS control services, and the QoS control services are designed to be used with a variety of setup mechanisms [11]. RSVP does not define the internal format of the protocol objects related to characterizing QoS control services; rather it can be seen as a signaling mechanism transporting the QoS control information. RSVP is analogous to other IP control protocols, such as ICMP, or one of the many IP routing protocols. RSVP itself is not a routing protocol; but it uses the local routing table in routers to determine routes to the appropriate destinations.

In general terms, RSVP is used to provide QoS requests to all router nodes along the transit path of the traffic flows and to maintain the state necessary in the routers required to actually provide the requested services. RSVP requests generally result in resources being reserved in each router in the transit path for each flow.

RSVP requires the receiver to be responsible for requesting specific QoS services instead of the sender. This is an intentional design in the RSVP protocol that attempts to provide for efficient accommodation of large groups (e.g., multicast traffic), dynamic group membership (also for multicast), and diverse receiver requirements.

There are two fundamental RSVP message types, the *Resv message* and the *Path message*, which provide for the basic RSVP operation, illustrated in Figure 14.3.

An RSVP sender transmits Path messages downstream along the traffic path provided by a discrete routing protocol (i.e., Open Shortest Path First, OSPF). The *Resv message* is generated by the receiver and is transported back upstream toward the sender, creating and maintaining a reservation state in each node along the traffic path.

RSVP still can function across intermediate nodes that are not RSVP capable. However, end-to-end resource reservations cannot be made, because non-RSVP capable devices in the traffic path cannot maintain reservation or *Path state* in response to appropriate RSVP messages. Although intermediate nodes that do not run RSVP cannot provide these functions, they may have sufficient capacity to be useful in accommodating tolerant real-time applications.

Since RSVP relies on a discrete routing infrastructure to forward RSVP messages between nodes, the forwarding of *Path messages* by non-RSVP-capable intermediate nodes is unaffected, since the *Path message* is carrying the IP address of the previous RSVP-capable node as it travels toward the receiver.

Summing up, Integrated Services are capable of enhancing the IP network model to support real-time transmissions and guaranteed bandwidth for specific flows. In this case, a flow is defined as a distinguishable stream of related datagrams from a unique sender to a unique receiver that results from a single-user activity and requires the same QoS. The Integrated Services architecture promises precise per-flow service provisioning but never really made it as a commercial end-user product, which was mainly accredited to its lack of scalability [19].

Differentiated Services (DiffServ)

Differentiated Services mechanisms do not use per-flow signaling, and as a result, do not consume per-flow state within the routing infrastructure. Different service levels can be allocated to different groups of users, which means that all traffic is distributed into groups or classes with different QoS parameters. This reduces the maintenance overhead in comparison to Integrated Services. Network traffic is classified and apportioned to network resources according to bandwidth management criteria. To enable QoS, network elements give preferential treatment to classifications identified as having more demanding requirements.

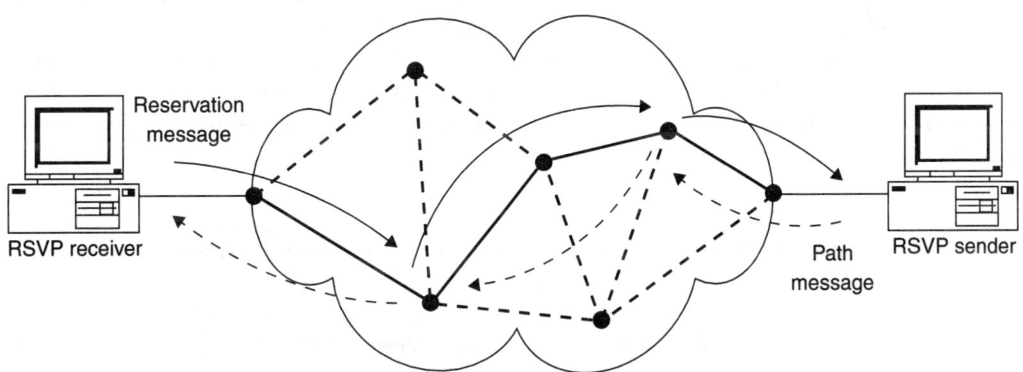

FIGURE 14.3 Traffic flow of the RSVP Path and Resv message.

DiffServ provides a simple and coarse method of classifying services of applications. The main goal of DiffServ is more scalable and manageable architecture for service differentiation in IP networks [16]. The initial premise was that this goal could be achieved by focusing not on individual packet flows, but on traffic aggregates, large sets of flows with similar service requirements.

By carefully aggregating a multitude of QoS-enabled flows into a small number of aggregates, giving a small number of differentiated treatments within the network, DiffServ eliminates the need to recognize and store information about each individual flow in core routers. This basic trick to scalability succeeds by combining a small number of simple packet treatments with a larger number of per-flow policies to provide a broad and flexible range of services.

Each DiffServ flow is policed and marked at the first QoS enabled downstream router according to a contracted service profile, or Service Level Agreement (SLA). Downstream from this router, a DiffServ flow is mingled with similar DiffServ traffic into an aggregate. Then, all further forwarding and policing activities are performed on these aggregates. Current proposals [15] are using a few bits of the IPv4 ToS byte or the IPv6 Traffic Class byte, now called the DiffServ Code Point (DSCP), for marking packets.

There are currently two standard per-hop behaviors defined that effectively represent two service levels (traffic classes):

- *Expedited Forwarding* (EF): It has a single code point (DiffServ value). EF minimizes delay and jitter and provides the highest level of aggregate QoS. This is like a virtual leased line. The EF treatment polices on network ingress and shapes on egress to maintain the service contract to the next provider. Any traffic that exceeds the traffic profile, which is defined by local policy, is discarded.
- *Assured Forwarding* (AF): It defines four priorities (classes) of traffic receiving different bandwidth levels (the "Olympic services" Gold, Silver, Bronze, and best effort). There are three-drop preferences each resulting in 12 different code points. The worse the drop preference, the more chance of getting dropped during congestion. Excess traffic is not delivered with as high a probability as the traffic within profile, which means it may be demoted but not necessarily dropped.

The PHBs are expected to be simple and define forwarding behaviors that may suggest, but do not require a particular implementation or queuing discipline. In general, a classifier selects packets based on one or more predefined sets of header fields. The mapping of the network traffic to the specific behaviors is indicated by the DSCP. The traffic conditioners enforce the rules of each service at the network ingress point. Finally, PHBs are applied to the traffic by the conditioner at a network ingress point according to pre-determined policy criteria. The traffic may be marked at this point, and routed according to the marking, and then unmarked at the network egress.

Each DiffServ-enabled edge router implements traffic conditioning functions, which perform metering, shaping, policing, and marking of packets to ensure that the traffic entering a DiffServ network conforms to the SLA, as illustrated in Figure 14.4.

The simplicity of DiffServ to prioritize traffic belies its extensibility and power. Using RSVP parameters (as described in the next section) or specific application types to identify and classify constant-bit-rate (CBR) traffic might help to establish well-defined aggregate flows that may be directed to fixed bandwidth pipes. DiffServ is more scalable at the cost of coarser service granularity, which may be the reason why it is not yet commercially available to end users either; see also [19].

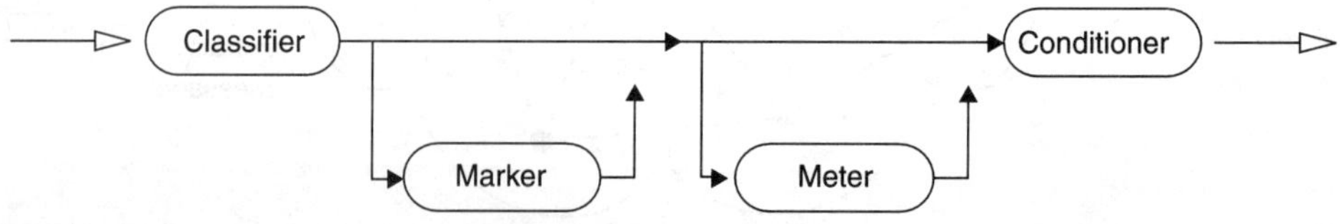

FIGURE 14.4   Edge router: DiffServ classification and conditioning.

Multi-Protocol Label Switching

As stated, we can see that IntServ and DiffServ adopt different approaches to solve the QoS challenge. Meanwhile, another approach exists that is slightly different but already in use: Multi Protocol Label Switching (MPLS). In contrast, it is not primarily a QoS solution, although it can be used to support QoS requirements. MPLS is best viewed as a new switching architecture and is basically a forwarding protocol that simplifies routing in IP-based networks. It specifies a simple and scalable forwarding mechanism, since it uses labels instead of a destination address to make the routing decision. The label value that is placed in an incoming packet header is used as an index to the forwarding table in the router. This lookup requires only one access to the table, in contrast to the traditional routing table access that might require uncountable lookups [1].

One of the most important uses of MPLS is in the area of traffic engineering, which can be summarized as the modeling, characterization, and control of traffic to meet specified performance objectives. Such performance objectives might be traffic oriented or resource oriented. The former deals with QoS and includes aspects such as minimizing delay, jitter, and packet loss. The latter deals with optimum usage of network resources, particularly network bandwidth.

The current situation with IP routing and resource allocation is that the routing protocols are not well equipped to deal with traffic-engineering issues. For example, a protocol such as OSPF can actually promote congestion because it tends to force traffic down the shortest route, although other acceptable routes might be less loaded. With MPLS, a set of flows that share specific attributes can be routed over a given path. This capability has the immediate advantage to steer certain traffic away from the shortest path, which is likely to become congested before other paths.

In conclusion, we may say that label-switching offers scalability to networks by allowing a large number of IP addresses to be associated with one or a few labels. This approach further reduces the size of address (actually label) tables, and allows a router to support more users or to set up fixed paths for different types of traffic. Since the main attributes of label switching are fast relay of the traffic, scalability, simplicity, and route control, label switching can be a valuable tool to reduce latency and jitter for data transmission on packet-switched networks.

Combining QoS Solutions

The QoS solutions described previously adopt different approaches, and each has its advantage and disadvantage. The Integrated Service approach is based on a sophisticated background of research in QoS mechanisms and protocols for packet networks. However, the acceptance of IntServ from network providers and router providers has been quite limited, at least so far, mainly due to scalability and manageability problems [10].

The scalability problems arise because IntServ requires routers to maintain control and forwarding state for all flows passing through them. Maintaining and processing per-flow state for gigabit or terabit links, with several simultaneously active flows, is significantly difficult from an implementation point of view. Hence, the IntServ architecture makes the management and accounting of IP networks significantly more complicated. Additionally, it requires new application–network interfaces and can only provide service guarantees when all elements in the flow's path support IntServ.

MPLS may be used as an alternative intra-domain implementation technology. These architectures in combination can enable end-to-end QoS. End hosts may use RSVP requests with high granularity (e.g., bandwidth, jitter, threshold, etc.). Border routers at backbone ingress points can then map those RSVP "reservations" to a class of service indicated by a DS byte or to a dedicated MPLS path. At the backbone egress point, the RSVP provisioning may be honored again, to the final destination; see Figure 14.5.

Such combinations clearly represent a trade-off between service granularity and scalability: as soon as flows are aggregated, they are not as isolated from each other as they possibly were in the IntServ part of the network. This means that, for instance, unresponsive flows can degrade the quality of responsive

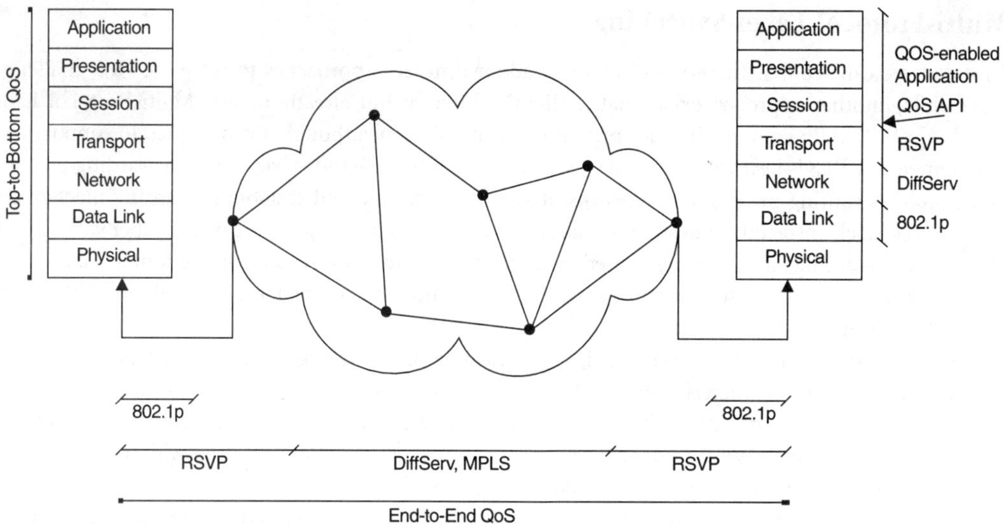

FIGURE 14.5   QoS Architecture.

flows. The strength of a combination is the fact that it gives network operators another opportunity to customize their network and fine-tune it based on QoS and scalability demands, as stated in [19].

Until now, IP has provided a best-effort service in which network resources are shared equitably. Adding QoS support to the Internet raises significant concerns, since it enables differentiated services that represent a significant departure from the fundamental and simple design principles that made the Internet a success. Nonetheless, there is a significant need for IP QoS, and protocols have evolved to address this need. The most viable solution today is a trade-off between protocol complexity and band-width "scarcity" with the following result:

- different QoS levels are used in the core network (e.g., four MPLS levels),
- applications at the user side are distinguished by DiffServ mechanisms, and
- and the marked user traffic is mapped to the appropriate core layers.

Finally, we should always bear in mind that an "application-to-application" guarantee not only depends on network conditions but also on the overall performance of each end-system.

References

[1] Black, Uyless D., *MPLS and Label Switching Networks*, Prentice-Hall, Inc., New Jersey, 2001.

[2] Comer, Douglas E., *Computernetworks and Internets*, 2nd ed., Prentice-Hall, Engleood Cliffs, NJ, 1999.

[3] Ferguson, P. and G. Houston, *Quality of Service: Delivering QoS on the Internet and in Corporate Networks*, John Wiley & Sons, Inc., New York, 1998.

[4] ITU-T Recommendation G.114: One-way Transmission Time, International Telecommunication Union, 1996.

[5] Kalinindi, S., OWDP: A Protocol to Measure One-Way Delay and Packet Loss, Technical report STR-001, Advanced Network & Services, September 1998.

[6] Keshav, S., *An Engineering Approach to Computer Networking*, Addison-Wesley, Reading, MA, January 1997.

[7] Kushida, T., The Traffic and the Empirical Studies for the Internet, in Proceedings of IEEE Globecom 98, pp. 1142–1147, Sydney, IEEE, New York, 1998.

[8] Paxson, V., Towards a Framework for Defining Internet Performance Metrics, Technical report LBNL-38952, Network Research Group, Lawrence Berkeley National Laboratory, June 1996.

[9] RFC2205: Resource ReSerVation Protocol (RSVP) Version 1 Functional Specification, September 1997, http://www.rfc-editor.org/rfc/rfc2205.txt.

[10] RFC2208: Resource ReSerVation Protocol (RSVP) Version 1 Applicability Statement: Some Guidelines on Deployment, September 1997, http://www.rfc-editor.org/rfc/rfc2208.txt.

[11] RFC2210: The Use of RSVP with IETF Integrated Services, September 1997.

[12] RFC2211: Specification of the Controlled-Load Network Element Service, September 1997, http://www.rfc-editor.org/rfc/rfc2211.txt.

[13] RFC2212: Specification of Guaranteed Quality of Service, September 1997.

[14] RFC2215: General Characterization Parameters for Integrated Service Network Elements, September 1997.

[15] RFC2474: Definition of the Differentiated Services Field (DS Field in the IPv4 and IPv6 Headers), September 1997.

[16] RFC2475: An Architecture for Differentiated Services, September 1997.

[17] Seifert, R., *Gigabit Ethernet: Technology and Applications for High-Speed LANs*, Addison-Wesley, Reading, MA, 1998.

[18] Stevens, R.W., *TCP/IP Illustrated: The Protocols*, Vol. 1, Addison-Wesley, New York, 1994.

[19] Welzl, M. and M. Mühlhäuser, Scalability and Quality of Service: A Trade-off?, *IEEE Communications Magazine*, 41, 32–36, 2003.

15

The Internet Protocol

Jürgen Falb
Vienna University of Technology

15.1 Motivation

The Internet Protocol (IP) was designed for use in interconnected packet-switched computer networks. It fulfills two different tasks. First, the IP provides for transmitting blocks of data called IP datagrams from a source host to a destination host. The second task is fragmentation and reassembly of datagrams for packet-switched networks able to handle only small packets. These two functions are named addressing and fragmentation.

The IP is not intended to provide additional services like reliability, flow control, or other common services provided by host-to-host protocols. In the TCP/IP suite, TCP provides these services. The IP only delivers datagrams from one host to another host with the use of interconnected networks.

A comparison of the Internet World with the ISO/OSI model places the IP on layer 3 — the network layer — of the ISO/OSI model. This comparison is often made, but is not sufficient because the IP itself does not define many functionalities of the network layer. For example, the IP does not define mechanisms for routing or congestion control. Therefore, the IP itself only implements a small part of the network layer. The IPs (such as IP, TCP, or UDP) are vendor-independent, and their specification is controlled by open public discussion. The results of the public discussions are written down in so-called "Request For Comments — RFCs." The Internet protocols are widely implemented by a variety of vendors and hence the definition of "open" as used in OSI is fulfilled.

15.2 Basic Operation of The Internet Protocol

The IP delivers datagrams throughout interconnected networks. Its operation makes it necessary that an IP module resides on each host involved in the transmission path as shown in Figure 15.1. Each IP module interfaces with one or more application programs and one or more underlying local networks. Typical application programs are TCP and UDP modules. The most commonly used type of local networks in the IP world is Ethernet. Nevertheless, the setup of IP upon other types of networks like ATM or fieldbus networks is possible.

The basic transmission scenario is as follows: an IP module receives data and a destination address from an application program. Then, the IP module encapsulates the data in IP datagrams, fragmenting

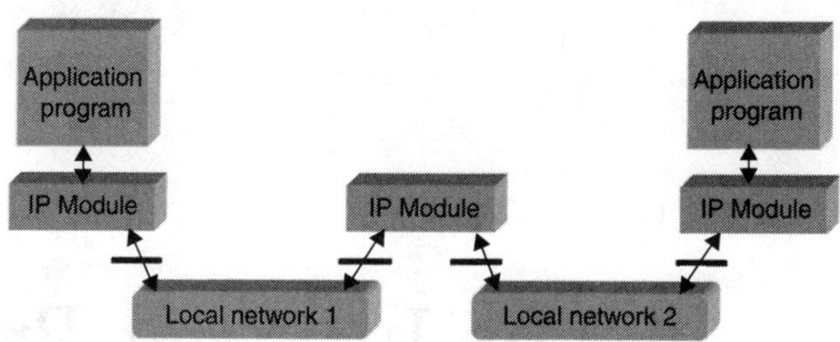

FIGURE 15.1   Interfacing of the IP.

them if necessary, and adding IP headers containing all necessary information to transmit the datagrams. Based on the destination address, the IP module transmits the packets over the local network to the next host. The receiving host decides if the datagrams belong to it by using the same addressing algorithm as the sender. In case the datagrams do not belong to the host, the receiving host forwards them to the next host. If the next local network is only able to transmit smaller datagrams then the prior one, the host fragments the datagrams again. This process is repeated until the datagrams reach their final destination. On the destination host, fragmented datagrams are reassembled. All hosts involved in the transmission make up the transmission path. The selection of a transmission path is called routing. Furthermore, all transmissions are stateless. This means that there is no relationship between subsequent IP datagrams such as a logical connection.

The IP uses four key mechanisms to fulfill its tasks: Type of Service, Time to Live, Options, and Header Checksum.

The *Type of Service* indicates the desired quality of service. Based on the indicated type of service, a host within the transmission path (often named gateway or router) can decide the proper routing and forwarding strategy.

The *Time to Live* mechanism indicates the lifetime of a datagram. The sender of a datagram chooses an initial value for the Time to Live. At each hop the datagram passes, the Time to Live will be decremented. If the Time to Live reaches zero before the datagram reaches its final destination, the datagram destroys itself. This mechanism ensures that a datagram cannot travel though the interconnected networks indefinitely.

Options provide a mechanism for control functions like timestamps, security, or recording of routes. Options are not necessary for normal operation.

The *Header Checksum* provides means for detecting errors. If a datagram was transmitted incorrectly, it will be discarded by the entity detecting the error. Errors are indicated to the sender using the Internet Control Message Protocol (ICMP) as described in section "ICMP".

15.3   IP Addressing

The IP only deals with addressing; finding a path to a destination host has to be done by higher-level protocols. Additionally, the Internet module supports mapping Internet addresses to local network addresses.

An Internet address is an address of fixed-length (32 bits). In the commonly used IP address notation, decimal numbers separated by dots (e.g., 128.131.80.33) represent the four bytes. Each Internet address consists of a network number and a local host address. A common notation for indicating the length of the network number part is the addition of a slash followed by a number specifying the number of bits (e.g., 128.131.80.0/24 refers to 256 IP addresses within the 24-bit network 128.131.80.). The Internet addressing scheme provides three classes of addresses and two classes of special-purpose addresses as shown in Figure 15.2. This original model for distributing IP addresses, based on classful addressing strategies, does not take into account the massive expansion of Internet use and is unable to scale to meet this expansion. Therefore, the authorities responsible for distributing IP address do not use this

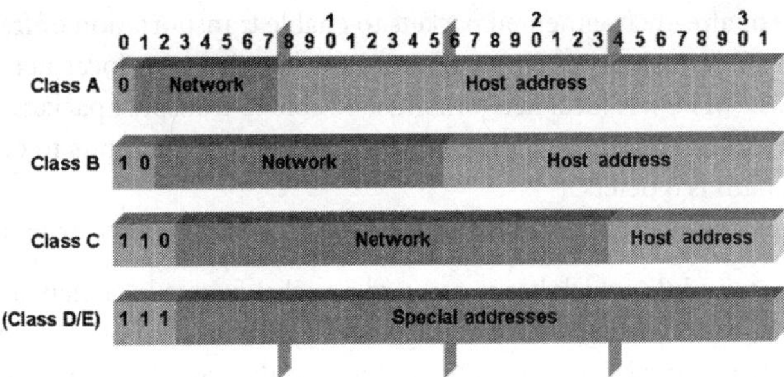

FIGURE 15.2   IP addresses.

class-based terminology anymore. Nowadays, IP addresses are only distributed for requested quantities if there is a documented need for this quantity of IP addresses. Nevertheless, many of the rules applied to the class hierarchy are still valid and discussed in the following paragraphs.

- *Class A*: A Class A address consists of a 7-bit network number and a 24-bit host number. That means there are 128 networks with 16,777,214 hosts. Only 125 of them are available for use, because the other three IP addresses are used for special purposes. The addresses in the network 0.0.0.0/8 refer to hosts on the local network. The special address 0.0.0.0 can be used as a source address for the own host. Addresses in the network 127.0.0.0/8 are used as loopback addresses, but usually in most IP stacks only the address 127.0.0.1 is implemented as a loopback address. Addresses in the network 10.0.0.0/8 can be used for private purposes and should not appear on the public Internet.
- *Class B*: A Class B address consists of a 14-bit network number and a 16-bit host number. Therefore, there are 16,384 networks with 65,532 hosts available. Addresses in the networks 172.16.0.0/12 to 172.31.0.0/12 are used for private purposes and should not appear on the public Internet. The network 169.254.0.0/16 is used as "link local" block. It is used for communication between hosts on a single link. Hosts obtain these addresses by autoconfiguration. This is the case if no DHCP server is available.
- *Class C*: A Class C address consists of a 21-bit network number and an 8-bit host number. Therefore, 2,097,152 Class C networks with 254 hosts exist. The networks 192.168.0.0/24 to 192.168.255.0/24 are reserved for private use and should not appear on the public Internet. The block 192.88.99.0/24 is used for routing between IPv6 and IPv4 networks. The block 192.0.2.0 /24 is called "TEST-NET" and is valid for use in documentation and examples. Addresses in this block should not appear on the public Internet. The last reserved block 198.18.0.0/15 is dedicated to benchmark tests of interconnected network devices.
- *Class D/E*: Classes D and E addresses are used for special purposes. Class D addresses in the range from 224.0.0.0 to 239.255.255.255 are used as IPv4 multicast addresses to address a group of hosts in the Internet for a special purpose. Class E addresses in the range from 240.0.0.0 to 255.255.255.254 are reserved for future use. The special address 255.255.255.255 is used for broadcasts in the local network and should never be forwarded outside the local network.

More on IP addressing can be found in [1].

15.4   IP Fragmenting

If a large packet originates in a local network and it has to traverse another network that only allows smaller packets to get to its destination host, the packet must be fragmented into smaller packets to enable transportation over this particular network. Hereby, each fragment is a valid IP packet again and can be transmitted independent of the other fragments. IP defines an *n*-way fragmentation algorithm to allow further fragmentation of already fragmented packets to enable transportation of fragmented packets that are still too large for some networks. The maximum packet size of a local network is usually called *Maximum Transfer Unit* (MTU). A local network must be able to transport packets with at least 68 octets. This value results from the fact that the maximum IP header length amounts to 60 octets and the minimal fragment block length is 8 octets.

Based on the example in Figure 15.3 the fragmentation algorithm will be explained in this paragraph.

1. First, the Internet module, which has to forward a packet from a local network 1 to a local network 2, has to compare the *total length* (TL) of the packet with the MTU of the local network 2. If the TL is less than or equal to the MTU, it simply can forward the packet without fragmentation. If the TL is greater than the MTU, the packet has to be fragmented. In the example of Figure 15.3, the TL of 124 octets is greater than the next MTU (88 octets). If the packet has to be fragmented, the *don't fragment flag* has to be checked. If this flag is set, the packet should not be fragmented and therefore has to be discarded because forwarding is not possible. In this case, the IP module usually creates an ICMP message to indicate the error to the originator of the packet.

2. In the next step, the Internet module creates the first fragment by copying the original IP header to the new packet. For subsequent fragments, some option headers can be omitted while copying the original IP header to the new packet [2]. Based on the MTU and the *Internet Header Length* (IHL), the Internet module has to compute the *number of fragment blocks* (NFB) it can insert into the new packet: $NFB = (MTU - IHL*4)/8$. The IHL is given in a multiple of 4 octets within the Internet header and each fragment block consists of 8 octets data, except the last one of the last fragment, which can be less than 8 octets.

3. In the third step, the Internet header has to be corrected. If the fragment is not the last one, the *more fragments flag* (MF) has to be set to indicate to the receiver that at least one more fragment will follow. The *fragment offset* (FO) of the Internet header has to be set to the original fragment offset plus the sum of the NFBs of prior fragments. Since the FO field is 13-bit long, the maximum length of the overall packet data (all fragments together) is $2^{13}*8=65,536$ octets. The third field to compute is the total length (TL), which is the sum of the new header length and the fragment data length. Based on the new header values, the *Header Checksum* has to be recomputed for the fragment.

4. The last step is the transmission of the new fragments to the next host in the transmission path. The receiver of the fragments again applies this fragmentation algorithm to the received packets.

The reassembly of the packets occurs only at the final destination host of the original packet. All fragments belonging to one packet are identified by the combination of source address, destination address, identification header field, and protocol header field. The fragments are ordered by the fragment offset header field. Based on the fragment length and the more fragment bit, the IP module on the destination host can decide if all fragments have arrived. If a packet is fully reassembled, it is handed over to the next module for further processing of the data.

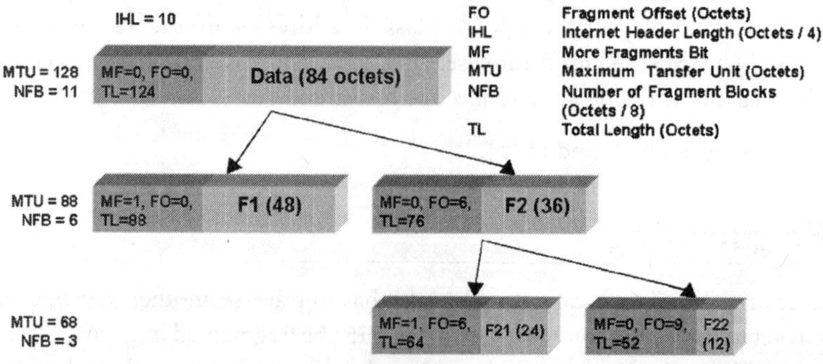

FIGURE 15.3   IP fragmentation example.

15.5 Internet Protocol Header

The IP header contains all information for transmitting a packet from a source host to a destination host and for fragmentation and reassembly of packets. An IP header consists of at least 20 octets up to 60 octets. The structure of the IP header is shown in Figure 15.4. For each field in Figure 15.4 the bit with the lowest number is the most significant bit.

- *Version*: 4 bits. The version field indicates the version of the IP protocol. In this case, the version field contains the value 4.
- *Internet Header Length*: 4 bits. This field denotes the overall length of the IP header in a multiple of 4 octets. Since the IP header consists of at least 20 octets, the minimal value is 5.
- *Type of Service*: 8 bits. The Type of Service field is used to provide different levels of transmission quality. The three most significant bits define the priority of the packet. Packets with higher priority should be processed first. The next three bits define the network parameters to be considered in the packet processing. The three bits represent the following network parameters: delay, throughput, and reliability. Usually, only one of these three bits is set to one. It does not make sense to set more than two bits to one because the improvement of network parameters occurs at the cost of the other network parameters. The last two bits are reserved for future use. More details can be found in [3].
- *Total Length*: 16 bits. The total length field specifies the length of a packet consisting of header and data block in octets. Due to the 16 bits, a maximal length of 65,536 octets is possible. A host must be prepared to accept at least packets with a length of 576 octets. The value is based on a data block of 512 octets and a typical IP header of 20 octets and a reserve for higher-level protocol overhead.
- *Identification*: 16 bits. This field should contain a value unique with respect to the originating host. It is used in the reassembly procedure to identify fragments that belong to a particular packet.
- *Flags*: 3 bits. These bits are used for the fragmentation procedure. Bit 0 is reserved and must be zero. Bit 1 is the Don't Fragment Flag and Bit 2 is the More Fragment Bit indicating that more fragments belonging to this packet will follow.
- *Fragment Offset*: 13 bits. The Fragment Offset field indicates the position of the fragment data within the whole packet. The fragment offset is measured in units of 8 octets.
- *Time to Live*: 8 bits. The Time to Live field indicates the maximum time a packet is allowed to travel through the networks. At each host passed by the packet, this field has to be decremented by at least one. If this field contains the value 0, it has to be discarded. The TTL field represents the maximum time in seconds a packet can live within networks. If a host needs more than a second for processing a packet, the value of the TTL should be decremented by the corresponding value. Since the TTL field is decremented at least by 1 at each host, it just represents an upper bound for the time a packet lives.

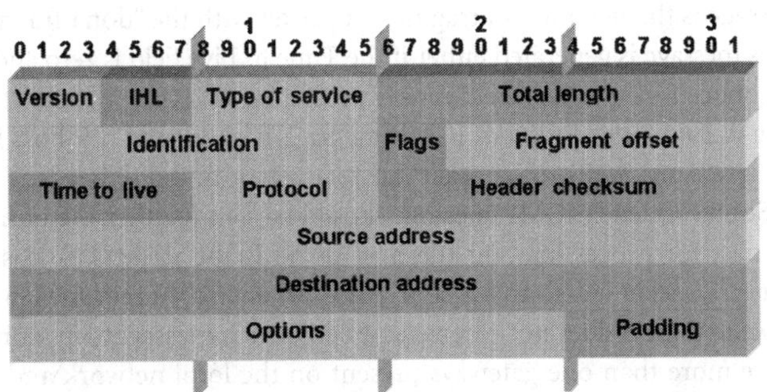

FIGURE 15.4 IP header.

- *Protocol*: 8 bits. The Protocol field identifies the next higher-level protocol used within the data block of the IP packet. The values for this field are maintained by the "Internet Assigned Numbers Authority" [4] and are updated at regular intervals. TCP, for example, has the value 6.
- *Header Checksum*: 16 bits. This field contains a checksum for the header only. Therefore, there is no mechanism in IP for verifying the correctness of the data. Since some header fields change at each host (e.g., TTL), the checksum has to be recomputed at every host in the transmission path.

Source Address: 32 bits. The source address is a structured 32-bit address as defined in section "IP Adressing" identifying the originating source host.

Destination Address: 32 bits. The destination address is also a structured 32- bit address as defined in section "IP Addressing" identifying the final destination host.

Options: Variable. Options are mainly used for control, debugging, and measurement. The IP protocol defines options for recording and supplying route information, transmitting timestamps, and for sending security information. A detailed description of the possible options can be found in [2].

Padding: Variable. The Internet header padding ensures that the Internet header ends at 32-bit boundary. The padding bits are filled with zeros.

One success factor of the Internet is mentioned in the following quote of the originator of the Internet Protocol Jon Postel: "*In general, an implementation must be conservative in its sending behaviour, and liberal in its receiving behaviour. That is, it must be careful to send well-formed datagrams, but must accept any datagram that it can interpret.*" This fact is not as apparent as in the HyperText Transfer Protocol (HTTP), the basis for the World Wide Web, but makes the protocol implementations robust and supports the interoperability of different IP implementations.

15.6   ICMP — Internet Control Message Protocol

The purpose of the ICMP (specified in RFC 792 [5]) is to provide information about problems in the Internet communication environment. To transport the error information to its destination, it uses the IP as a transport mechanism. ICMP is not designed to make the IP reliable; it only provides feedback about problems to the source host. It is not designed for control purposes between routers and hosts.

The ICMP is an integral part of every IP module and therefore must be implemented together with IP. With respect to the datagram structure, it can be seen as a higher-level protocol since it is transported by IP. To avoid too many errors in reporting traffic, there will be no ICMP messages sent about errors in the transmission of other ICMP messages. Also, there is only an error reporting for the first fragment of a fragmented packet (The first fragment is denoted by the fragment offset value 0.)

ICMP is used to provide information about the following network situations:

Destination unreachable: An ICMP message of this type is sent either if a network or a particular host is unreachable or if there is a problem in the next level protocol module. Reasons are the unavailability of the protocol module itself or of the corresponding port. The third case for the generation of a destination unreachable message is the necessity to fragment a packet with the "don't fragment" flag set.

Time Exceeded: This message is generated either if the Time to Live field is zero and the packet has to be dropped or if a timeout occurs during the reassembly procedure of IP fragments at the destination host.

Parameter Problem: A Parameter Problem ICMP message is generated if a host has to drop a packet because it cannot complete processing the packet due to an error in the IP header.

Source Quench: A Source Quench ICMP message should be sent if either the buffer of a host is full or the packets arrive to fast for processing and the host discards the packets. A host can send a Source Quench message for every packet it discards. A source host receiving Source Quench messages should cut back its transmission rate until it does not receive Source Quench messages anymore.

Redirect: If there are more than one gateways present on the local network and a gateway receives a packet that should rather be forwarded by another gateway due to its routing table, it sends a Redirect ICMP message back to the source host containing the address of the other gateway.

Echo or Echo Reply: These messages are primarily used to verify the connectivity between hosts. The data sent in an Echo ICMP message have to be returned in the Echo Reply ICMP message. A common application based on these ICMP message types is the "ping" command.

Timestamp or Timestamp Reply: These two message types are used to record time information. A host sends a Timestamp ICMP message and receives the echoed Timestamp Reply ICMP message. The echoed Timestamp Reply message contains the sent time and received time of the Timestamp message and the sent time of the Timestamp Reply message. With this information, it is possible to calculate the transmission time and the processing time at the destination host.

Information Request or Information Reply: These message types can be used by a host to find out the network number of the network it is on. The sender can set the network part of the source and destination address within the IP header to zero. A network part of zero in an IP addresses identifies "this" network. The destination host, which has to send the Information Reply message, has to specify the full IP addresses within the IP header. By examining the IP addresses in the Information Reply message, the original host can determine its network number.

A typical application based on ICMP messages is the traceroute command. The traceroute command tries to figure out the route a packet will take most probably to get to the destination host. It is not guaranteed that subsequent packets will take the same route as printed out by the traceroute command. Figure 15.5 shows an example traceroute output. The traceroute command works as follows: first, it sends an IP packet to the destination host with a Time to Live of one. The next host in the transmission path has to decrease the TTL header field to zero and therefore has to discard the IP packet. To inform the sender about the error, it sends a Time Exceeded ICMP message (Type 0x0b) back to the sender. The traceroute command recognizes the originator of the ICMP message as the first hop. In the next step, it sends the same IP packet again, but with a TTL field increased by one. This time, the first receiving host decreases the TTL to one and forwards the packet to the second host that decreases the TTL to zero and sends a Time Exceeded ICMP message back to the sender. This procedure is repeated until a packet reaches the final destination host. In this example, the final destination host sends a Destination Unreachable ICMP

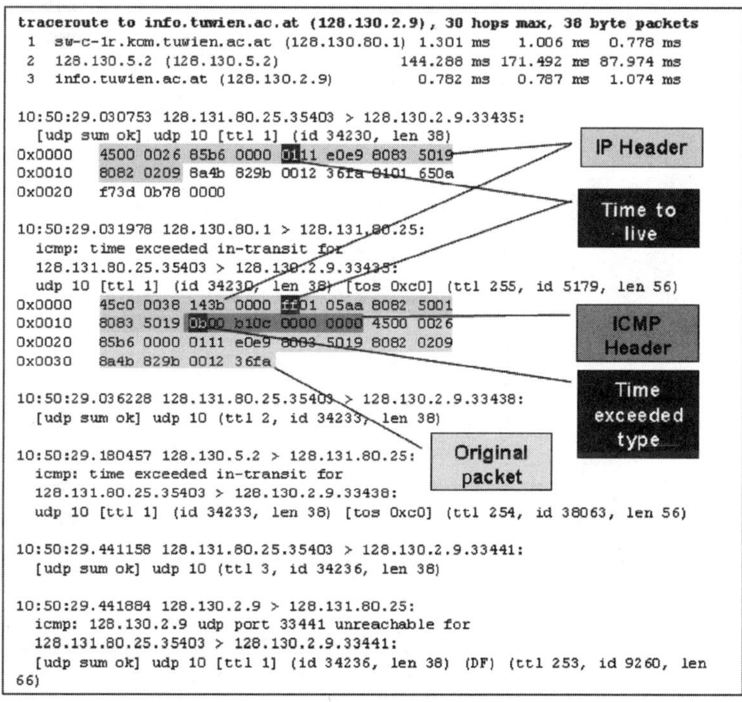

FIGURE 15.5   Example usage of ICMP messages by traceroute.

message, because the IP packet is addressed to port 33,441 unavailable at the final destination host. The ports used by traceroute are usually in the range of 33,434 − 33,600 and are increased by one for each packet. Commonly, three packets are sent with the same TTL value for each hop in case a packet gets dropped or the port is in use.

References

[1] Wegner, J.D., Rockell R., and Blanchet M., *IP Addressing and Subnetting, Including IPv6*, Syngress Media, Rockland, MA, USA, December 1999.
[2] Postel, J., Internet Protocol — DARPA Internet Program Protocol Specification, STD 5, RFC 791, September 1981, http://www.rfc-editor.org/rfc/rfc791.txt
[3] Postel, J., Service Mappings, RFC 795, September 1981, http://www.rfc-editor.org/rfc/rfc795.txt
[4] *IANA*- Internet Assigned Numbers Authority, http://www.iana.org/.
[5] Postel, J., Internet Control Message Protocol — DARPA Internet Program Protocol Specification, STD 5, RFC 792, September 1981, http://www.rfc-editor.org/rfc/rfc792.txt

16

The Transmission Control Protocol

Aleksander Malinowski
Bradley University

Bogdan M. Wilamowski
Auburn University

16.1 Introduction

Transmission Control Protocol (TCP) is a part of the TCP/IP protocol suite. It provides full transport layer services to applications. It belongs to the transport layer in the OSI model as shown in Figure 16.1. TCP is more complicated than User Datagram Protocol (UDP) and provides a reliable connection between both ends of transmission. This connection needs to be set up, maintained, and torn down. The protocol divides a stream of data into units of limited size that are called segments. These segments are reassembled together in a reliable manner at the other end. Reliability is achieved by assembling segments by preserving their original sequence, and arranging for retransmission of lost or corrupted data. While the underlying IP protocol provides logical addressing of segments and is responsible for their transmission through networks, TCP provides additional addressing that allows setting up connections for multiple applications and services on the same end. Each application seeking TCP connection is assigned one or more numbers (one number per connection) that are called port numbers. Figure 16.2 illustrates the concept of port numbers.

TCP provides a logical full duplex connection between two application layer processes. The connection is perceived as a connection-oriented, reliable, in-sequence byte-stream service. The protocol also allows the recipient of the data to control the rate at which the sender transmits data to avoid buffer overflow.

Connection-oriented means that the connection needs to be established and its parameters negotiated between the two processed. This negotiation is performed by sending transmission control blocks (TCB). At the end of transmission, the connection must be torn down. TCP terminates each direction of the connection separately. This allows the data flow to continue even after the other direction has been closed.

Reliable means that either all data are delivered or the connection is broken. TCP is designed to work over the Internet Protocol (IP) and thus it does not assume that underlying network services are reliable. To ensure reliability, each chunk of data is numbered, and a selective automatic repeat request (ARQ) is used to request the retransmission of lost or corrupt data chunk.

In-sequence means that all data are guaranteed to be received by the end receiving process in the same order that they were sent in. Sequencing is implemented by the same mechanism that is used for missing

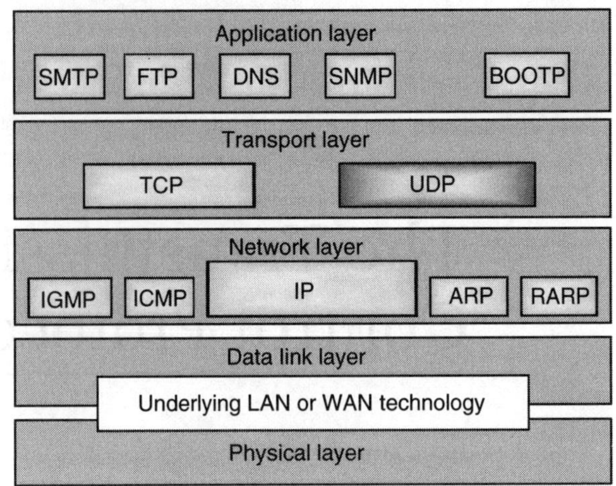

FIGURE 16.1   UDP and TCP/IP in the OSI model.

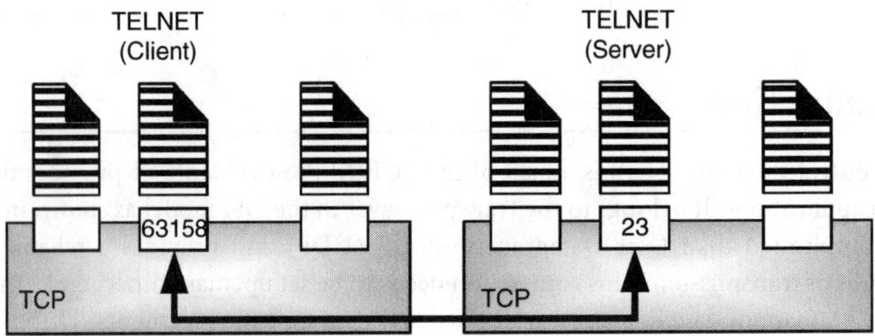

FIGURE 16.2   Process-to-process communication.

data chunk detection. All received data are placed in a receiving buffer in a location corresponding to the data chunk number. The process that receives data reads it from that buffer in the correct sequence.

Byte-stream oriented means that the connection can be perceived as a continuous stream of data regardless of the fact that the underlying network services may transmit data in chunks. These data chunks may be of various sizes and may have a maximum size limit. The sending process, however, may send data either continuously or in bursts. All data to be sent are placed in the sending buffer. Data from the buffer are partitioned into chunks that are called segments according to algorithms of TCP and sent over the network. On the receiving side, the data are placed in the receiving buffer in an appropriate sequence as mentioned earlier. The receiving process may read data in either chunks or byte after byte.

Full duplex means that each party involved in the transmission has control over it. For example, any side can send data at any time or terminate the connection.

16.2 Protocol Operation

Before data transfer begins, TCP establishes a connection between two application processes by setting up the protocol parameters. These parameters are stored in a structure variable called TCB. Once the connection is established, the data transfer may begin. During the transfer, parameters may be modified to improve the efficiency of transmission and prevent congestion. In the end, the connection is closed. Under certain circumstances, connection may be terminated without formal closing.

TCP Segment

Although TCP is a byte-stream-oriented connection and buffering takes care of the disparities in the speed of preparing data for sending, speed of sending data, and the speed of consuming data at the other end, one more step is necessary before data are sent. The underneath IP layer that takes care of sending data through the entire network sends data in packets, and not as a stream of bytes. Sending one byte in a packet would be very inefficient due to the large overhead caused by headers. Therefore, data are grouped into larger chunks that are called segments. Each segment is encapsulated in an IP datagram and then transmitted. Each TCP segment follows the same format as shown in Figure 16.3. Each segment consists of a header that is followed by options and data. The meaning of each header field will be discussed now, and its importance will be illustrated in the following sections, which shows phases of a typical TCP connection.

Source Port Address — this is a 16-bit field that contains a port number of the process that sends options or data in this segment.

Destination Port Address — this is a 16-bit field that contains a port number of the process that is supposed to receive options or data carried by this segment.

Sequence Number — this is a 32-bit field that contains the number assigned to the first byte of data carried by this segment. During the establishment of connection, each side generates its own random initial sequence number (ISN). The sequence number is then increased by one with each one byte of data sent. For example, if a previous segment had a sequence number 700 and had 200 bytes of data, then the next segment has sequence number 900.

Acknowledgement Number — this is a 32-bit field that contains the sequence number for the next segment that is expected to be received from the other party. For example, if the last received segment had a sequence number 700 and had 200 bytes of data, then the acknowledgement number is 900.

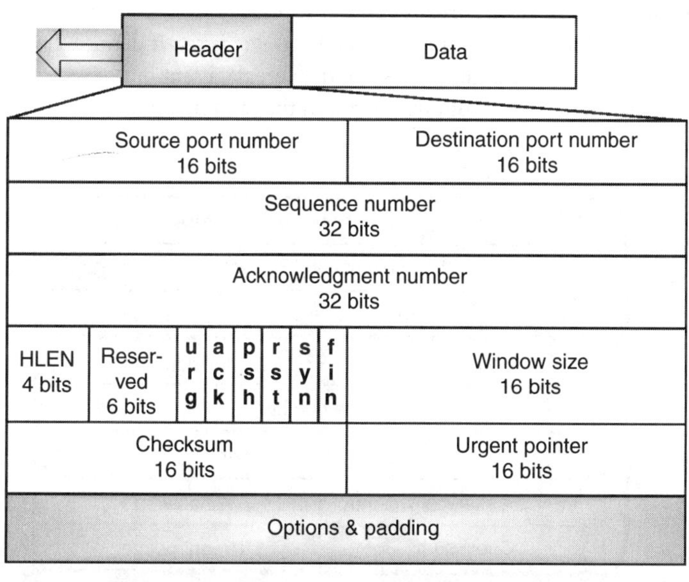

FIGURE 16.3 TCP segment format.

Header Length — this 4-bit field contains the total size of the segment header divided by 4. Since headers may be 20 to 60 bytes long, the value is in the range of 5 – 15.

Reserved — this 6-bit field is reserved for future use.

Control — this 6-bit field contains transmission control flags:

URG — urgent. If the bit is set, then the urgent pointer is valid.

ACK — acknowledgement. If the bit is set, then the acknowledgement number is valid.

PSH — push. If the bit is set, then the other party should not buffer outgoing data but send it immediately, as it becomes available, without buffering.

RST — reset. If the bit is set, then the connection should be aborted immediately due to abnormal condition.

SYN — synchronize. The bit is set in the connection request segment when the connection is being established.

FIN — finalize. If the bit is set it means that the sender does not have anything more to send, and that the connection may be closed. The process will still accept data from the other party until the other party also sends FIN.

Window Size — this 16-bit field contains information about the size of the transmission window that the other party must maintain. Window size will be explained and illustrated later.

Checksum — this 16-bit field contains the checksum. The checksum is calculated by

Initially filling it with 0's.

Adding a pseudoheader with information from IP protocol as illustrated in Figure 16.4.

Treating the whole segment with the pseudoheader prepended as a stream of 16-bit numbers. If the number of bytes is odd, 0 is appended at the end.

Adding all 16-bit numbers using 1's complement binary arithmetic.

Complementing the result. This complemented result is inserted into the checksum field.

Checksum verification — the receiver calculates the new checksum for the received packet, which includes the original checksum, after adding the pseudoheader (see Figure 16.4). If the new checksum is nonzero, then the packet is corrupt and is discarded.

Urgent Pointer — this 16-bit field is used only if the URG flag is set. It defines the location of the last urgent byte in the data section of the segment. Urgent data are data sent out of sequence. For example, a telnet client sends a break character without queuing it in the sending buffer.

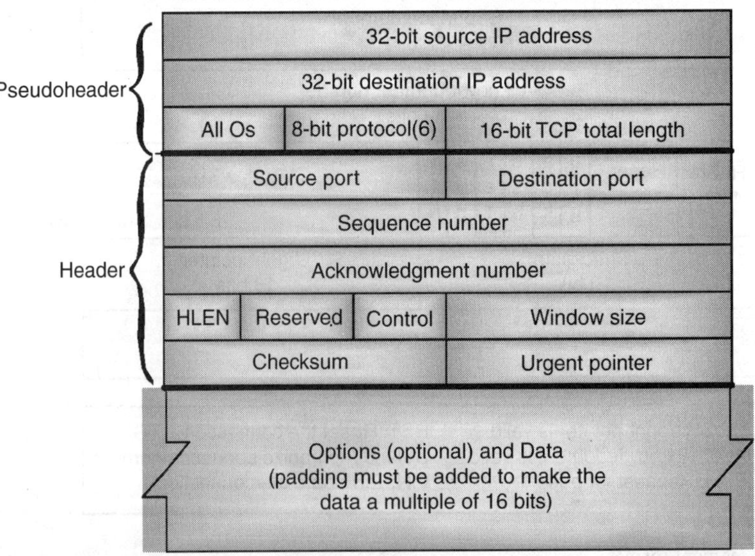

FIGURE 16.4   Pseudoheader added to the TCP datagram.

Options — there can be up to 40 bytes of options appended to the header. The options are padded inside with 0's so that the total number of bytes in the header is divisible by four. The possible options are listed below and are illustrated in Figure 16.5.

No Operation — 1-byte option used as a filler between the options and as padding.

End of Options — 1-byte option used as a padding at the end of options filed. Data portion of the segment follows immediately.

Maximum segment size — option defines the maximum size of a chunk of data that can be received. IT refers to the size of data and not the whole segment. It is used only during the establishment of the connection. The receiving party defines the maximum size of *r* the sender. If not specified, the default maximum is 536.

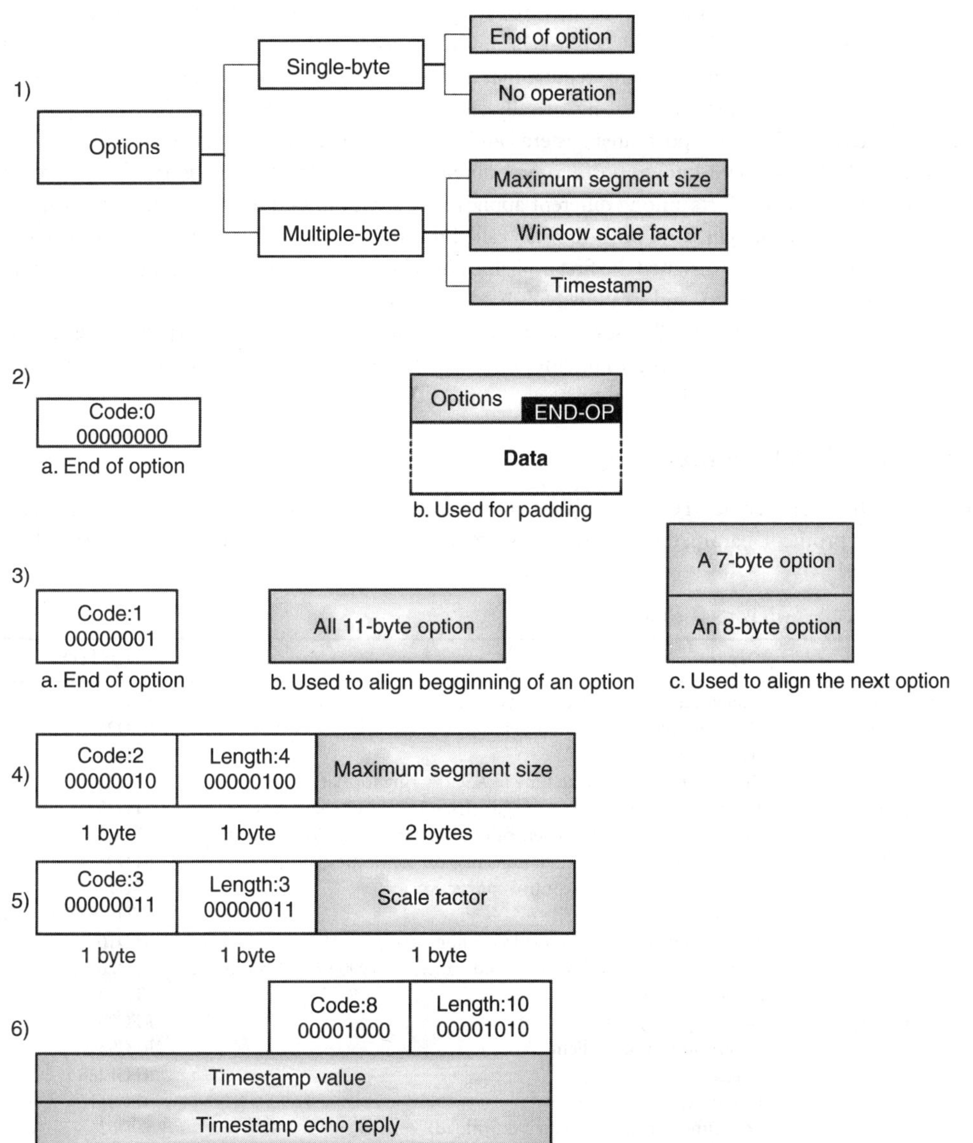

FIGURE 16.5 TCP options: classification — 1, end of options — 2, no operation (used for padding) — 3, setting maximum segment size — 4, setting window scale factor — 5, time stamp — 6.

Window scale factor — for the high throughput channels, the window size from the segment header (16-bit, maximum 65,535) may not be sufficient. This 3-byte option defines the multiplication factor for the window size. The actual window size is computed suing the following formula:

$$ActualWindowSize = WindowSize \cdot 2^{WindowScale\ eFactor}$$

Time stamp — this is a 10-byte option that holds the time stamp value of transmitted segment. When an acknowledgement is sent, the same value is used. This allows the original sender to determine the optimal retransmission time based on the packet round-trip time.

Port Number Assignments

In order to provide platform-independent process addressing on a host, each connection of the process to the network is assigned a 16-bit number. There are three categories of port numbers: well-known (0–1023), registered (1024–49,151) and ephemeral (49,152–6553). *Well-known ports* have been historically assigned to common services. Some operating systems require that the process that utilizes those ports must have administrative privileges. This requirement was historically created to avoid hackers running server imposters on multi-user systems. Well-known ports are registered and controlled by Internet Assigned Numbers Authority (IANA). Table 16.1 shows well-known ports commonly used by TCP. *Registered ports* are also registered by IANA to avoid possible conflicts among different applications attempting to use the same port for listening to incoming connections. *Ephemeral* ports are also called dynamic ports. These ports are used by outgoing connections that are typically assigned the first available port above 49,151. Some operating systems may not follow IANA recommendations, and treat the registered ports range as ephemeral.

For example, a Web Server usually uses well-known port 80, a proxy-server may typically use registered port 8080, a Web browser uses multiple ephemeral ports, one for each new connection open to download a component of a Web page. Ephemeral ports are eventually recycled.

Connection Establishment

Before data transfer begins, TCP establishes a connection between two application processes by setting up the protocol parameters. These parameters are stored in a structure variable called TCB.

TABLE 16.1   Well-known Ports Used with TCP

Port	Protocol	Description	RFC/STD #
7	Echo	Echoes received data back to its sender	STD20
9	Discard	Discards any data that are received	STD21
11	Users	Active users	STD24
13	Daytime	Returns the date and the time in ASCII string format	STD25
17	Quote	Returns a quote of the day, ASCII string up to 512 characters	STD23
19	Chargen	Returns a string of characters of random length 0 to 512 characters	STD22
20	FTP Data	File Transfer Protocol (data transmission connection)	STD9
21	FTP Control	File Transfer protocol (control connection)	STD9
23	Telnet	Telnet	STD8, STD27-STD32
25	SMTP	Simple Mail Transfer Protocol (sending email)	STD10
37	Time	Returns the current time in seconds since 1/1/1900 in a 32-bit format	STD26
53	Nameserver	Domain Name Service	STD13
67	Bootps	Bootstrap Protocol Server	RFC951
68	Bootpc	Bootstrap Protocol Client	RFC951
79	Finger	Finger	RFC1288
80	HTTP	Hypertext Transfer Protocol	RFC2616
110	POP3	Post Office Protocol 3 (email download)	STD53
111	RPC	SUN Remote Procedure Call	RFC1831
143	IMAP	Interim Mail Access Protocol v2 (email access)	RFC2060
443	SHTTP	Secure Hypertext Transfer Protocol	RFC2660

Before the two parties, called here hosts A and B, can start sending data, four actions must be performed:

1. Host A sends a segment announcing that it wants to establish connection, which includes initial information about traffic from A to B.
2. Host B sends a segment acknowledging the request received from A.
3. Host B sends a segment with information about traffic from B to A.
4. Host A sends a segment acknowledging the request received from B.

The second and the third steps can be combined into one segment. Therefore, the connection establishment described is also called *three-way handshaking*. Figure 16.6 illustrates an example of segment exchange during connection establishment. Among the two programs that are involved in the connection, the server program is the one that passively waits for another party to connect. Hereafter, it will be called the server. It informs the underlying TCP library in its system that it is ready to accept a connection. The program that actively seeks connection is called a client program. Hereafter, it will be called the client. The client informs the underlying library in its system that it needs to connect to a particular server that is defined by the computer IP address and the destination port number.

1. The client sends the first segment that includes the source and destination port with SYN bit set. The segment also contains the initial value of Sequence Number that is also called the initialization sequence number (ISN) that will be used by the client in this connection. The client may also send some options as necessary, for example, to alter the window size, or maximum segment size in order to change their values from default.
2. The server sends the second segment that has both ACK and SYN bits set. This is done in order to both acknowledge receipt of the first segment from the client (ACK) and to initialize the sequence number in the direction from the server to the client (SYN). The Acknowledgement number is set to the client ISN plus one. The server ISN is set to a random number. The option to define client window size must also be sent.
3. In reply to the segment from the server, the client sends as segment with ACK bit set and the segment number set to its ISN plus one. The acknowledgement number is set to the server ISN plus one.

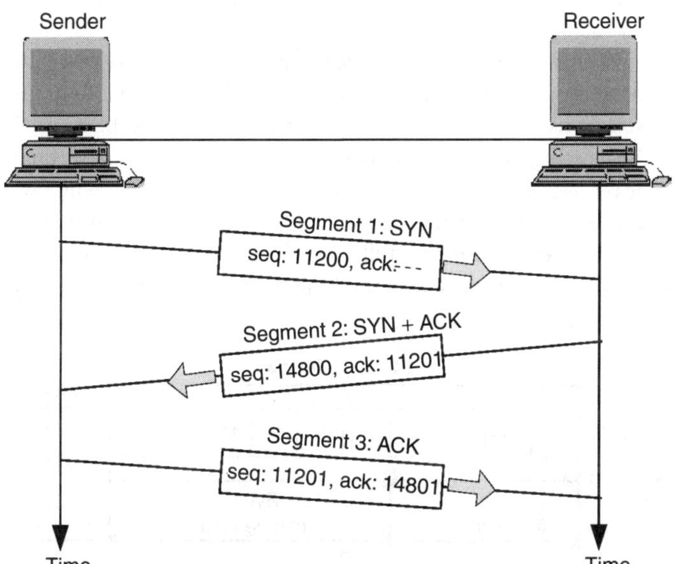

FIGURE 16.6   Three-way handshaking during connection establishment.

4. In very rare situations, the two processes may want to start connection at the same time. In that case, both will issue an active open segment with SYN and ISN, and then both would send the ACK segment. A situation when one of the segments is lost during the transmission and needs to be resent after time out may also be analyzed.

Maintaining the Open Connection

After a connection is established, each process can send data. Each segment has its sequence number increased by the number of bytes sent in the previous segment, with the previous sequence number. The number rolls over at 2^{32}. The process expects to receive segments with ACK bit set and acknowledgement numbers that confirm that data are received. Acknowledgement number is not merely the repetition of the received sequence number. It indicates the next expected sequence number of a segment to be received by the acknowledging process. Acknowledgements must be received within a certain time frame; not every segment that is sent needs to be acknowledged. Sending acknowledgment of a certain segment also acknowledges receiving all previous segments. More segments are usually sent ahead before the sender expects to receive an acknowledgement. Sending an acknowledgement can be combined with sending data in the other direction (*piggybacking*).

Flow Control and Sliding Window Protocol

A certain amount of data are sent before waiting for an acknowledgment from the destination. In extreme situations, the process may wait for an acknowledgement for each segment; but this would make the long-distance transmission very slow due to long latency. On the other hand, the process may send all data without waiting for the confirmation. This is also impractical because a high volume of data may need to be retransmitted in case a segment in the middle of the transmission is lost.

Sliding window protocol is a solution in between the two extreme situations described above that TCP implements in order to speed up the transmission. TCP sends as much data as allowed by a so-called sliding window before waiting for an acknowledgement. After receiving an acknowledgement, it moves (slides) the window ahead in the outgoing data buffer as illustrated in Figure 16.7.

On the receiving side, incoming data are stored in the receiving buffer before data are consumed by the communicating process. Since the size of the buffer is limited, the number of bytes that can be received at a particular time is limited. The size of the unused portion of the buffer is called the *receiver window*. In order to maintain the flow control, the sender must be prevented from sending excessive amount of data and overflowing the incoming data buffer. This is implemented by controlling the timing when the acknowledgements are sent.

If the receiving process consumes data at a faster rate than they arrive, then the receiver window expands. This information is passed to the sender as an option included in a segment so that the sender may synchronize the size of its *sender window*. This allows more data to be sent without receiving an acknowledgement. If the receiving process consumes data slower than they arrive, then the receiver buffer shrinks. In that case, the receiver must inform the sender to shrink the sender window as well. This causes transmission to wait for more frequent acknowledgements and prevents data from being sent without the possibility of storing. If the receiving buffer is full, then the receiver may close (set to zero) the receiver window. The sender cannot send any more data until the receiver reopens the window.

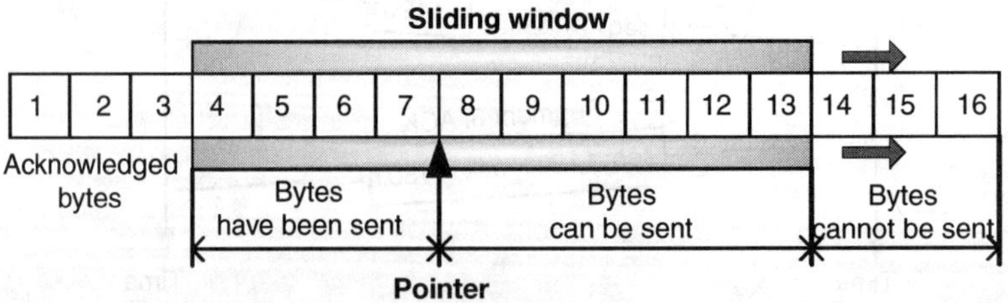

FIGURE 16.7   Sender buffer and sliding window protocol.

Improving Flow Control

In the case when one or both communicating processes process data slowly, a problem leading to significant bandwidth waste may arise. In case of slow transmission caused by the receiving processes, the sliding window may be reduced even as low as to one byte. In case of slow transmission caused by the sender, data may be sent as soon as they are produced even with a large window, one byte at a time as well. This produces segments carrying small data payload, and lowers the ration of carried data size to the total segment size, and to the total packet size in the underlying protocols.

To prevent this situation, data should not be sent as soon as they becomes available; but should be combined in larger chunks provided that this approach is suitable for a particular application. Nagle found a very good and simple solution for the sender that is called *Nagle's Algorithm*. The algorithm can be described by the following three steps:

- The first piece of data is sent immediately even if it is only one byte.
- After sending the first segment, the sender waits for the acknowledgement segment or until enough data are accumulated to fill in the segment of the maximum segment size, whatever happens first. Then, it sends the next segment with data.
- The previous step is repeated until the end of transmission.

On the receiving side, two different solutions are frequently used. *Clark* proposed to send the acknowledgement as soon as the incoming segment arrives but to set the window to zero if the receiving buffer is full (Figure 16.8b), and then to increase the buffer only after a maximum size of a segment can be received or half of the buffer becomes empty. The other solution is to delay the acknowledgment, thus preventing the sender form sending more segments with data. The latter of the two solutions creates less network traffic; but it may cause unnecessary segment retransmission if the delay is too long.

Error Control

TCP as a reliable protocol that uses unreliable underlying IP protocol must implement error control. This error control includes detecting lost segments, out-of-order segments, corrupted segments, and duplicated segments. After detection, the error must be corrected. The error detection is implemented by using three features: segment checksum, segment acknowledgement, and time-out. Segments with *incorrect checksum* (Figure 16.9a) are discarded. Since TCP uses only positive acknowledgement, no negative acknowledgement can be sent, and both the corrupted segment error and the *lost segment* (Figure 16.9b) error are detected by time-out for receiving an acknowledgement by the sender.

The error correction is implemented by establishing a *time-out counter* for each segment sent. If the acknowledgement is received, then the particular and all preceding counters and the segments are disposed. If the counter reaches time-out, then the segment is resent and the timer is reset.

A *duplicate segment* can be created if acknowledgement segment is lost or is delayed too much, since data are reassembled into a byte-stream at the destination data from the duplicate segment is already in the buffer. The duplicate segment is discarded. No acknowledgement is sent.

An *out-of-order segment* is stored but not acknowledged before all segments that precede it are received. This may require resending the segments. The duplicates are discarded.

In case of *lost acknowledgement* (Figure 16.9c), the error is corrected when the acknowledgement for the next segment is sent since an acknowledgement to the next segment also acknowledges all previous segments. Therefore, a lost acknowledgement may not be noticed at all. This could generate a deadlock in case the acknowledgment is sent on the last segment and for a time there are no more segments to send. This deadlock is corrected by a sender who uses a so-called *persistence timer*. If no acknowledgment comes within a certain time frame, then the sender sends a one-byte segment called probe and resets the persistence counter. With each attempt, the persistence timer time-out is doubled until 60 sec time are reached. After receiving an acknowledgement, the persistence timer time-out is reset to the initial value.

Keep-alive timer is used to prevent a long idle connection. If no segments are exchanged for a given period of time, usually 2 h, the server sends a probe. If there is no response after 10 probes, the connection is reset and thus terminated.

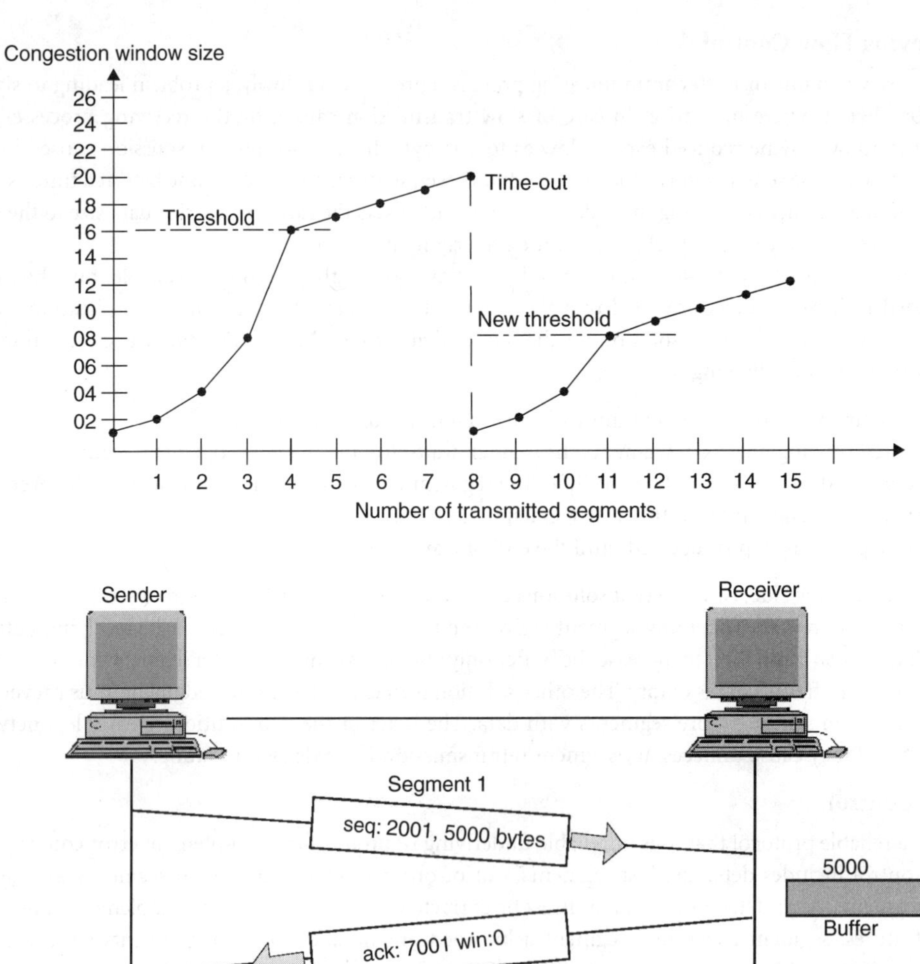

FIGURE 16.8 Congestion Control Window Size. Sender-side A (upper/left), receiver-side control B (lower/right).

Time-waited timer is used during the connection termination procedure that is described later. If there is no acknowledgment to the FIN segment, the FIN segment is resent after time-waited timer expires. Duplicate FIN segments are discarded by the destination.

Congestion Control

In case of congestion on the network, packets could be dropped by routers. TCP assumes that the cause of a lost segment is network congestion. Therefore, an additional mechanism is incorporated into TCP to prevent prompt resending of packets and creating even more congestion. This mechanism is implemented by additional control of the sender window as described below and is illustrated in Figure 16.8a.

- The sender window size is always set to the smaller of the two values: the receiver window size and the *congestion window size.*
- The congestion window size starts at the size of two maximum segments.
- With each received acknowledgement, the congestion window size is doubled until a certain threshold. After reaching the threshold, further increase is additive.
- With each acknowledgement timed out, the threshold is reduced by half and the congestion window size is reset to the initial size of two maximum segment sizes.

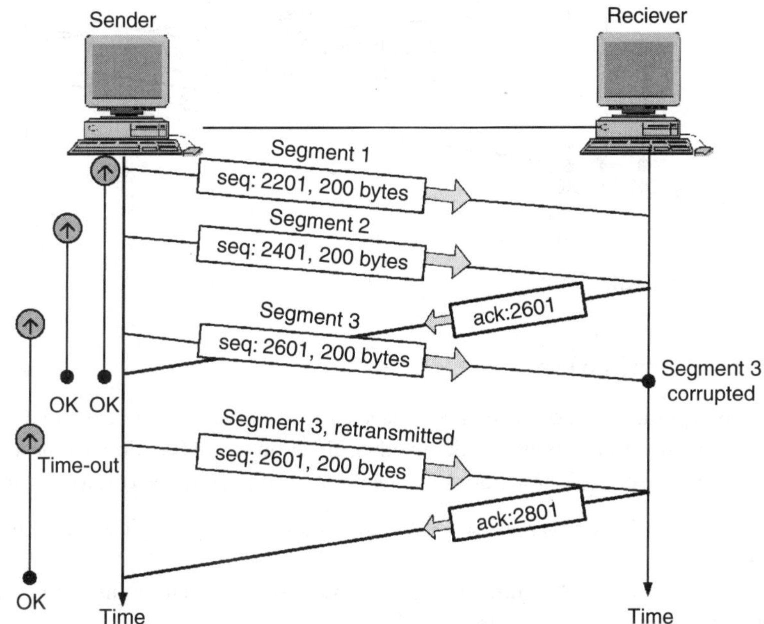

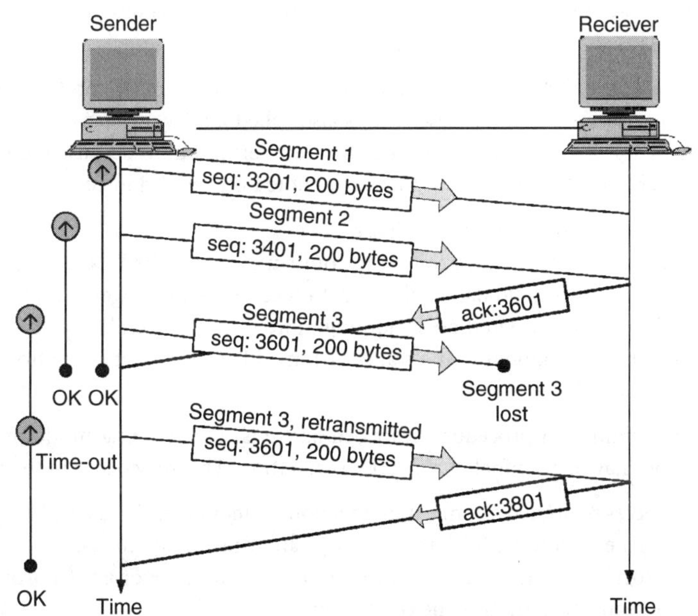

FIGURE 16.9 Corrupted segment (a), Lost segment (b), Lost Acknowledgement (c).

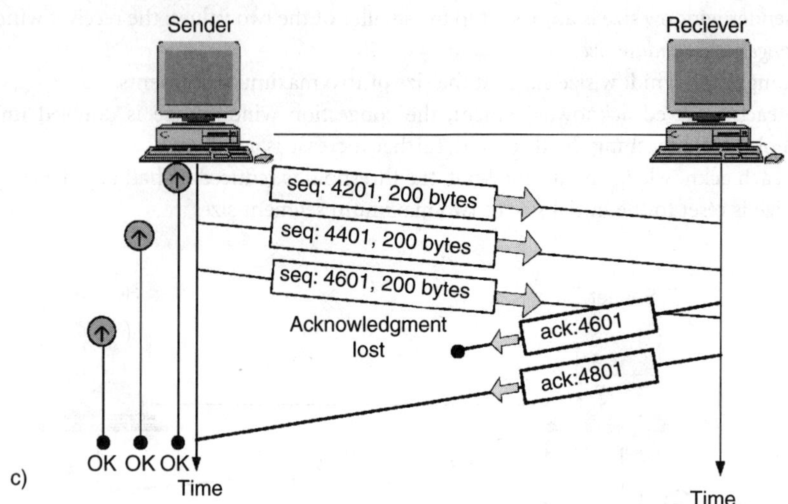

FIGURE 16.9 (Continued).

Connection Termination

Any of the two application processes that established a TCP connection can close that connection. When connection in one direction is terminated, the other direction is still functional, until the second process also terminates it. Therefore, four actions must be performed to close the connection in both directions:

1. Host A (either client or server) sends a segment announcing that it wants to terminate the connection.
2. Host B sends a segment acknowledging the request of A. At that moment, the connection from A to B is closed, but that from B to A is still open.
3. When host B is ready to terminate the connection, for example, after sending remaining data, it sends a segment announcing that it wants to terminate the connection as well.
4. Host A sends a segment acknowledging the request of B. At this moment, the connection in both directions is closed.

Unlike in the case of opening the connection, none of the steps can be combined into one segment. Therefore, the connection termination described is also called *four-way handshaking*. Figure 16.10 illustrates an example of segment exchange during connection termination. Because it does not matter as to which host is a server and which is a client, they are denoted only as A and B.

1. Host A sends a segment with an FIN bit set.
2. Host B sends a segment with ACK bit set to confirm receiving the FIN segment from host A.
3. Host B may continue to send data and expect to receive acknowledgement messages with no data from host A. Eventually, when there is nothing more to send, it sends a segment with FIN bit set.
4. Host A sends the last segment in the transmission with ACK bit set to confirm receiving the FIN segment from host B.

The connection termination procedure described above is so-called graceful termination. In certain cases, the connection may be terminated abruptly by resetting. This may occur in the following cases:

- The client process requested a connection to a nonexistent port. The TCP library on the other side may send a segment with the RST bit set to indicate connection refusal.
- One of the processes wants to abandon the connection because of an abnormal situation. It can send the RST segment to destroy the connection.

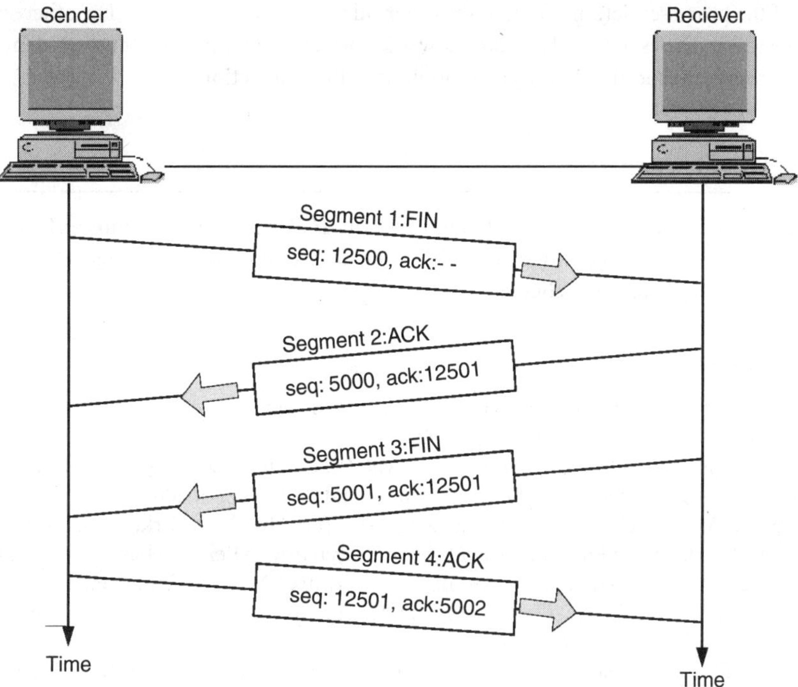

FIGURE 16.10   Four-way handshaking during connection termination.

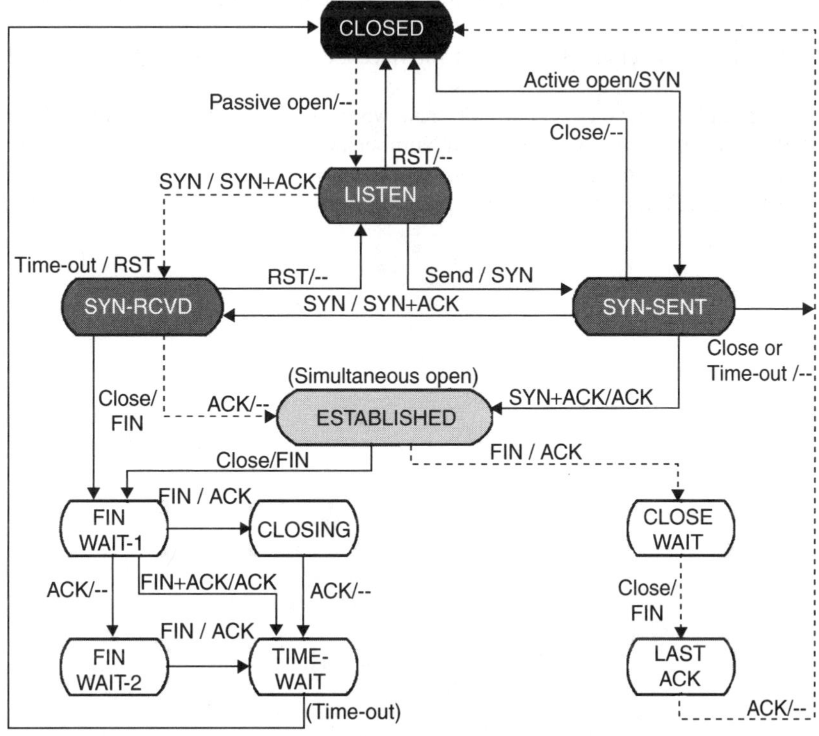

FIGURE 16.11   TCP State Transition Diagram.

- One of the processes determines that the other side is idle for a long time. The other system is considered idle if it does not send any data, any acknowledgements, or keep-alive segments. The awaiting process may send the RST segment to destroy the connection.

16.3   State Diagram

The state transition diagram for software that implements TCP is presented. Figure 16.11 shows the state transition diagram for the server and for the client. The dashed lines indicate transitions in a server. The solid lines indicate transitions in a client.

Additional Reading

[1]   Comer, D., *Internetworking with TCP/IP Vol.1: Principles, Protocols, and Architecture*, 4th ed., Prentice-Hall, Upper Saddle River, NJ, 2000.

[2]   Comer, D. et al., *Internetworking with TCP/IP, Vol. III: Client–Server Programming and Applications, Linux/Posix Sockets Version*, Prentice-Hall, Upper Saddle River, NJ, 2000.

[3]   Forouzan, B.A., *TCP/IP Protocol Suite*, 2nd ed., McGraw-Hill, New York, 2003.

[4]   Hall, E.A., *Internet Core Protocols: The Definitive Reference*, O'Reilly, Sebastopol, CA, 2000.

[5]   Leon-Garcia, A. and I/ Widjaja, *Communication Networks*, McGraw-Hill, New York, 2000.

[6]   Peterson, L.L. and B.S. Davie, *Computer Networks: A System Approach*, 2nd ed., Morgan Kaufmann Publishers, San Francisco, CA, 2000.

[7]   RFC0793/STD0007, Postel, J. Ed., Transmission Control Protocol DARPA Internet Program Protocol Specification, September 1981, http://www.rfc-editor.org/

[8]   RFC1791, Sung, T., TCP And UDP Over IPX Networks With Fixed Path MTU, April 1995, http://www.rfc-editor.org/

[9]   RFC2675, Borman, D. et al., IPv6 Jumbograms, August 1999, http://www.rfc-editor.org/

[10]   RFC2525, Paxson, V. et al., Known TCP Implementation Problems, March 1999, http://www.rfc-editor.org/

[11]   RFC2553, Gilligan, R. et al., Basic Socket Interface Extensions for IPv6, March 1999, http://www.rfc-editor.org/

[12]   RFC2757, Montenegro, G. et al., Long Thin Networks, January 2000, http://www.rfc-editor.org/

[13]   STD2, Reynolds, J., and J. Postel, Assigned Numbers, October 1994, http://www.rfc-editor.org/

[14]   Web Site for *IANA - Internet Assigned Numbers Authority*, http://www.iana.org/

[15]   Web Site for *RFC Editor*, http://www.rfc-editor.org/

17

The User Datagram Protocol

Aleksander Malinowski
Bradley University

Bogdan M. Wilamowski
Auburn University

17.1 Introduction

User Datagram Protocol (UDP) is a part of TCP/IP protocol suite. It provides full transport layer services to applications. It belongs to the transport layer in the OSI model as shown in Figure 17.1. UDP provides a connection between two processes at both ends of the transmission. This connection is provided with minimal overhead, without flow control or acknowledgement of received data. The minimal error control is provided by ignoring (dropping) received packets that fail the checksum test.

17.2 Protocol Operation

The underlying IP protocol facilitates data transport through the network between two hosts. UDP allows addressing particular processes at each host so that incoming data are handled by a particular application. Addressing the processes is implemented by assigning each connection to the network a number that is called a port number, Figure 17.2 illustrates the concept of port numbers and shows the difference between IP and UDP addressing. UDP packets are also called *datagrams*. Each datagram is characterized by four parameters: the IP address and the port number at each end of the connection. Certain classes of destination IP addresses may not describe individual computers but a group of processes on multiple hosts that are to receive the same datagram. This particular case of *multicasting* is described in another article entitled "UDP multicasting routing."

Although the term connection is used here, unlike in the case of Transmission Control Protocol (TCP), there is no persistent connection between the two processes. The data exchange may follow a certain pattern or protocol; but there is no guarantee that all data may reach the destination and that the order of their receiving would be the same as the order they were sent. The client–server approach that is usually used to describe the interaction between two processes can still be used. The process that may be identified as waiting for the data exchange is called the *server*. The process that may be identified as one that initializes that exchange is called the *client*.

UDP Datagram

UDP datagrams have a constant size 8-byte header prepended to the transmitted data as shown in Figure 17.3. The meaning of each header field is described below.

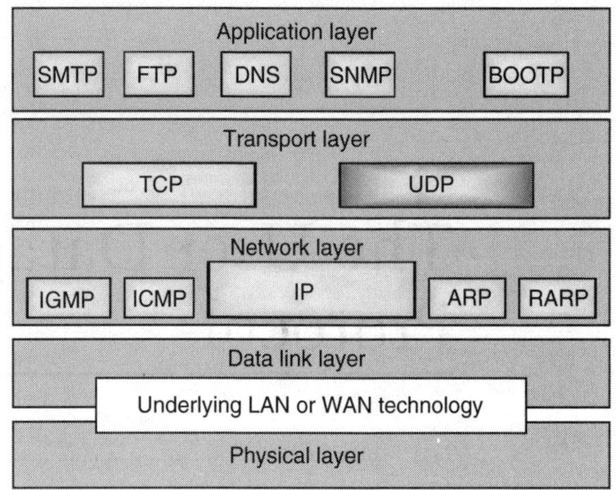

FIGURE 17.1 UDP and TCP/IP in the OSI model.

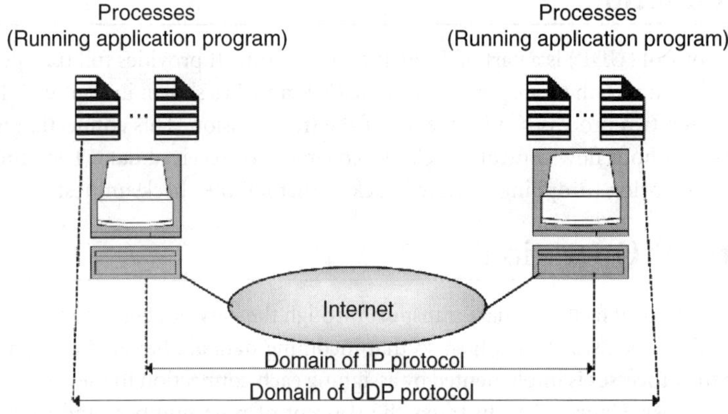

FIGURE 17.2 Comparison of IP and UDP addressing.

Source Port Address — this is a 16-bit field that contains a port number of the process that sends options or data in this segment.

Destination Port Address — this is a 16-bit field that contains a port number of the process that is supposed to receive options or data carried by this segment.

Total Length — this is a 16-bit field that contains the total length of the packet. Although the number could be in the range from 0 to 65,535 the minimum length is 8 bytes, which correspond to the packet with the header and no data. The maximum length is 65,507 because 20 bytes are used by the IP header and 8 bytes are used by the UDP header. Thus, this information is redundant to the packet length stored in the IP header.

Checksum — this 16-bit field contains the checksum. The checksum is calculated by

Initially filling it with 0's.

Adding a pseudoheader with information from the IP protocol as illustrated in Figure 17.4.

Treating the whole segment with the pseudoheader prepended as a stream of 16-bit numbers. If the number of bytes is odd, 0 is appended at the end.

Adding all 16-bit numbers using 1's complements binary arithmetic.

Complementing the result. This complemented result is inserted into the checksum field.

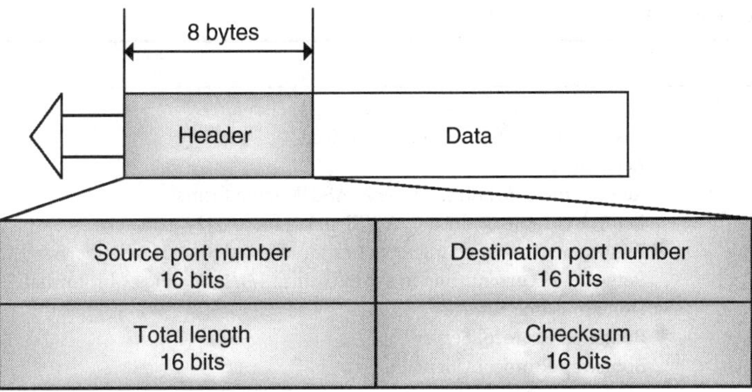

FIGURE 17.3 User datagram format.

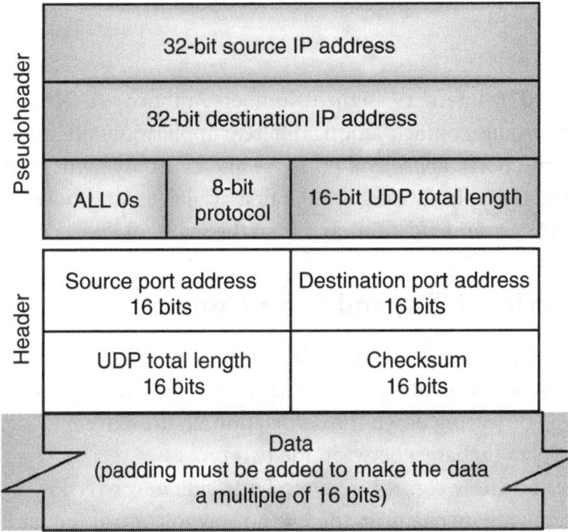

FIGURE 17.4 Pseudoheader added to the UDP datagram.

Checksum verification — the receiver calculates the new checksum for the received packet that includes the original checksum, after adding the so-called pseudoheader (see Figure 17.4). If the new checksum is nonzero, then the datagram is corrupt and is discarded.

In case of UDP, the use of checksum is optional. If it is not calculated, then the field is filled with 0's. The receiver can determine whether checksum was calculated by inspecting the field. Even in case the checksum is 0, the field does not contain 0 as the calculated checksum is complemented at the end of the process, and negative 0 (the field filled with 1's) is stored.

Port Number Assignments

In order to provide platform-independent process addressing on a host, each connection of the process to the network is assigned a 16-bit number. There are three categories of port numbers: well-known (0–1023), registered (1024–49,151), and ephemeral (49,152–6553). *Well-known ports* have been historically assigned to common services. Table 17.1 shows well-known ports commonly used by UDP. Some operating systems require that processes utilizing those ports must have administrative privileges. This requirement was historically created to prevent hackers from running server imposters on multi-user systems. Well-known ports

TABLE 17.1 Well-known Ports Used with UDP

Port	Protocol	Description	RFC/STD #
7	Echo	Echoes a received datagram back to its sender	STD20
9	Discard	Discards any datagram that is received	STD21
11	Users	Active users	STD24
13	Daytime	Returns the date and the time in ASCII string format	STD25
17	Quote	Returns a quote of the day, ASCII string up to 512 characters	STD23
19	Chargen	Returns a string of characters of random length 0 to 512 characters	STD22
37	Time	Returns the current time in seconds since 1/1/1900 in a 32-bit format	STD26
53	Nameserver	Domain Name Service	STD13
67	Bootps	Bootstrap Protocol Server	RFC951
68	Bootpc	Bootstrap Protocol Client	RFC951
69	TFTP	Trivial File Transfer	STD33
111	RPC	SUN Remote Procedure Call	RFC1831
123	NTP	Network Time Protocol	STD12
161	SNMP	Simple Network Management Protocol	STD62
162	SNMP	Simple Network Management Protocol (trap)	STD62

are registered and controlled by Internet Assigned Numbers Authority (IANA). *Registered ports* are also registered by IANA to avoid possible conflicts among different applications attempting to use the same port for listening to incoming connections. *Ephemeral* ports are also called dynamic ports. These ports are used by outgoing connections that are typically assigned the first available port above 49,151. Some operating systems may not follow IANA recommendations and treat the registered ports range as ephemeral.

Connectionless Service, Flow, and Error Control

As already outlined, UDP provides connectionless communication. Each datagram is not related to another datagram coming earlier or later from the same source. The datagrams are not numbered. There is no need for establishing or tearing down the connection. In the extreme case, just one datagram might be sent in the process of data exchange between the two processes.

UDP does not provide any flow control. The receiving side may overflow with incoming data. Unlike in TCP, which has a windowing mechanism, there is no way to control the sender. There is no error control other than the checksum discussed above. The sender cannot be requested to resend any datagram. However, an upper-level protocol that utilizes UDP may implement some kind of control. In that case, unlike in TCP, no data are repeated by resending the same datagram. Instead, the communicating process sends a request to send some information again. Trivial File Transfer Protocol (TFTP) is a very good example of that situation. Since it has its own higher level flow and error control, it can use UDP as a transport-layer protocol instead of TCP. Other examples of UDP utilization are Simple Network Management Protocol (SNMP) or any other protocol that requires only a simple short request-response communication.

Additional Reading

[1] Comer, D., *Internetworking with TCP/IP Vol.1: Principles, Protocols, and Architecture*, 4th ed., Prentice-Hall, Upper Saddle River, NJ, 2000.

[2] Comer, D. et al., *Internetworking with TCP/IP, Vol. III: Client-Server Programming and Applications, Linux/Posix Sockets Version*, Prentice-Hall, Upper Saddle River, NJ, 2000.

[3] Forouzan, B.A., *TCP/IP Protocol Suite*, 2nd ed., McGraw-Hill, New York, 2003.

[4] Hall, E.A., *Internet Core Protocols: The Definitive Reference*, O'Reilly, Sebastopol, CA, 2000.

[5] Leon-Garcia, A. and I/ Widjaja, *Communication Networks*, McGraw-Hill, New York, 2000.

[6] Peterson, L.L. and B.S. Davie, *Computer Networks: A System Approach*, 2nd ed., Morgan Kaufmann Publishers, San Francisco, CA, 2000.

[7]   RFC0768/STD0006, Postel, J., User Datagram Protocol, August 1980, http://www.rfc-editor.org/

[8]   RFC1791, Sung, T., TCP And UDP Over IPX Networks With Fixed Path MTU, April 1995, http://www.rfc-editor.org/

[9]   RFC2675, Borman, D. et al., IPv6 Jumbograms, August 1999, http://www.rfc-editor.org/

[10] STD2, Reynolds, J. and J. Postel, Assigned Numbers, October 1994, http://www.rfc-editor.org/

[11] Web Site for *IANA - Internet Assigned Numbers Authority*, http://www.iana.org/

[12] Web Site for *RFC Editor*, http://www.rfc-editor.org/

18

ARP — Address Resolution Protocol

Jürgen Falb
Vienna University of Technology

In every network, datagrams have to be transmitted over a physical network medium. At this level, network interface cards use MAC addresses (MAC = Media Access Control) for addressing the communication partner. This is necessary because MAC layer protocols support methods of sharing the physical line among a number of computers in contrast to point-to-point protocols like Point-to-Point Protocol (PPP) or Serial Line Interface Protocol (SLIP). The MAC address is used by the MAC sublayer of the Data-Link Layer (layer 2 of the ISO/OSI model responsible for a reliable data transmission over a physical link) to unambiguously identify a host on the physical line. Different physical media use different types of MAC addresses. For example, Ethernet uses 48-bit Ethernet addresses.

By using higher-level protocols like TCP/IP, it is necessary to map protocol addresses (e.g., IP addresses) to MAC addresses (e.g., Ethernet addresses) to be able to transport a datagram to the next host connected to the same physical line. Earlier, a table manually configured on each network node did this mapping. Increasing node numbers and frequent network changes (e.g., notebooks) made it necessary to define a mechanism for automatically configuring the mapping table. The Address Resolution Protocol (ARP) defined in RFC 826 [1, 8] was designed to convert protocol addresses into MAC addresses by solving the prior-mentioned problem. The subsequent sections will describe ARP by concentrating on the most common case: the conversion of IP addresses into Ethernet addresses.

18.1 Operation of The Address Resolution Protocol

If an IP module wants to send a datagram to another host, it looks up the desired IP address in a mapping table called ARP cache. If it finds a corresponding entry in the ARP cache, it uses the associated MAC address to send the datagram to the MAC address. Is there no suitable entry in the ARP cache, the ARP mechanism will be invoked, as explained in the following paragraph.

First, the ARP module builds an ARP datagram containing the required IP address it wishes to resolve. It sends this packet to all hosts connected to the physical line by using a broadcast MAC address (ff:ff:ff:ff:ff:ff in case of 48 bit Ethernet addresses). Each host on the local network reads the datagram and compares the requested IP address with its own IP address. The host identified by the requested IP address then generates an ARP response datagram to convey its MAC address to the sender. The original sender now knows the MAC address of the desired target host on the local network and inserts the mapping information into his ARP cache. In the final step, the host is now able to send IP datagrams to the host with the desired IP address.

The described process relies on the constraint that the underlying local network has to support broadcasts. Ethernet supports broadcast as well as multicast and unicast using special addresses.

To increase efficiency and reduce the network load, the host generating the ARP response caches the address mapping of the originator of the ARP request. Thus, there is no ARP resolution dialog necessary for sending IP datagrams in the opposite direction.

18.2   The Address Resolution Protocol in Detail

The ARP message structure is simple and was designed for use in all kinds of networks. The only constraint of the Address Resolution Protocol is the requirement of broadcast support on the network side.

ARP operates directly on the Data Link Layer, and therefore is directly embedded in the network frame (Ethernet frame). To distinguish ARP from other network protocols, it uses the value 0x0806 for the "Ethernet Type" field within an Ethernet frame. An ARP datagram itself is structured as shown in Figure 18.1.

ARP datagrams consist of the following header fields:

- *Hardware Type*: The hardware type field specifies the network hardware used by the ARP datagram originator. The most important value is "1" for Ethernet. Another typical value is "6" for FDDI (Fiber Distributed Data Interface; for more details, see [3, 4]). This field is most important for switches combining different networks (e.g., Ethernet and FDDI or TokenRing).
- *Protocol Type*: The protocol type field specifies the protocol type of the address to be resolved. A receiving ARP module has to determine if it supports the protocol for the address mapping process.
- *Hardware Address Length*: This header field specifies the length of the hardware address in bytes. For Ethernet addresses, this field has the value "6."
- *Protocol Address Length*: The protocol address length field determines the address length of the higher-level protocol. In case of the Internet Protocol, the value of this field is "4" (32 bit IPv4 addresses).

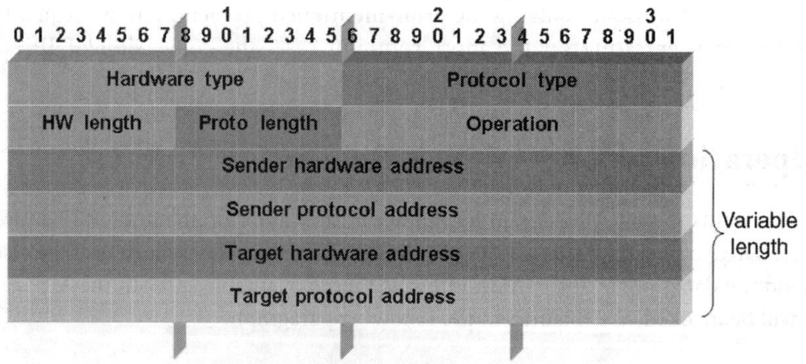

FIGURE 18.1   ARP header.

- *Operation*: The operation field determines if the datagram is a request or a response. There are different values for the different address resolution protocols:

  - 1 — ARP request
  - 2 — ARP response
  - 3 — RARP request (RARP = Reverse Address Resolution Protocol)
  - 4 — RARP response
  - 5 — Dynamic RARP request
  - 6 — Dynamic RARP response
  - 7 — Dynamic RARP error
  - 8 — InARP request (InARP = Inverse Address Resolution Protocol)
  - 9 — InARP response

- *Sender Hardware Address*: This field contains the hardware address of the ARP datagram sender. The length of this field corresponds to the value of the "Hardware Address Length" field. Within an ARP response, this address is the requested hardware address.
- *Sender Protocol Address*: This field contains the protocol address of the ARP datagram sender. This address is typically an IP address. The length of this field corresponds to the value of the "Protocol Address Length" field.
- *Target Hardware Address*: Within an ARP request, this address is unknown and therefore the target address is a broadcast address (Ethernet: ff:ff:ff:ff:ff:ff). An ARP response is always conveyed directly to the originator of the request and contains the hardware address of the ARP request originator.
- *Target Protocol Address*: Within an ARP request, the target protocol address reflects the address that the sender wants to resolve. All hosts on the local network compare this target protocol address with its own protocol address and react by sending an ARP response if they are equal.

An example ARP dialog is shown in Figure 18.2. In this example, the host having the IP address 192.168.1.67 wants to resolve the hardware address belonging to the IP address 192.168.1.142. The sender host broadcasts an ARP request (operation code 0x0001). The hardware type identifies the underlying local network as an Ethernet (0x0001). In the ARP response, the concerned host replies with its MAC address (00:a0:cc:c5:a8:aa).

A problem of ARP lies in the fact that the receiver of the ARP response always assumes that ARP responses are valid. This allows ARP spoofing as explained later.

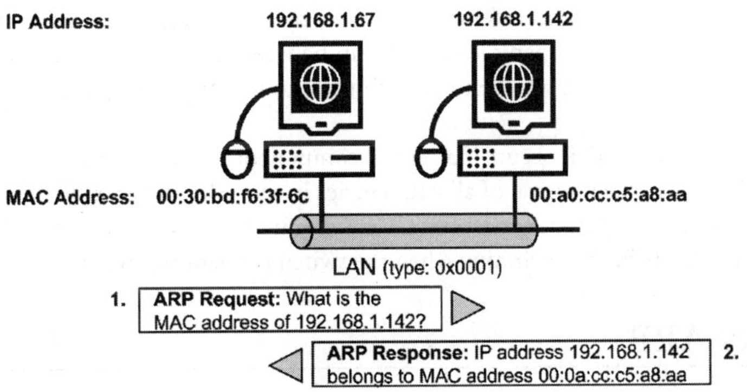

FIGURE 18.2　Example ARP dialog.

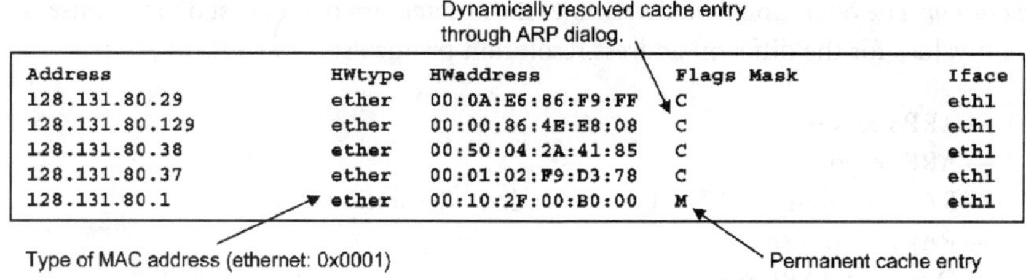

FIGURE 18.3   ARP cache.

18.3   The ARP Cache

The ARP cache contains statically configured mappings and dynamically resolved MAC addresses to reduce the network load and the response time. Figure 18.3 shows a sample output of the "arp" command, which can be used to display and modify the ARP cache of a system. The flag "C" marks an entry as dynamically resolved. Most systems keep the entries in the cache until either the cache is full or the entry is aged out because it was not used for a specified time. Typical aging times are 15 min. Common cache sizes are 128–512 entries. This implies that in usual subnets, entries need not be removed from the cache.

Furthermore, the ARP cache expects a positive feedback from higher-level protocols to mark an ARP entry as valid. For example, a positive feedback can be retrieved from a successful TCP ACK packet. If the cache cannot receive a positive feedback, it marks the entry as stale and removes it after a specified time if it still does not receive a positive feedback.

18.4   ARP Problems

The most common error in conjunction with ARP is based on the situation that two hosts get the same IP address. Let us examine the following scenario:

Hosts A and B get the same IP address. Host A starts a communication with host C by sending an ARP request to resolve the MAC address of host C. Host C sends back an ARP response to host A and inserts the MAC address of host A into its ARP cache. After a while of communication between host A and host C, host B initiates a communication with host C by sending an ARP request to resolve host C's MAC address. Since host C gets an ARP request from host B with the identical IP address, it replaces the MAC address of host A with the MAC address of host B. From now on, host C sends all datagrams intended for host A to host B. Host B discards these unknown datagrams and probably generates error messages. Host A will get a timeout because it does not receive any more datagrams from host C and will restart its communication by sending an ARP request. Now the whole described procedure restarts again. This effect caused by IP address conflicts is often called "flip-flop."

An example "flip-flop" notification message generated by an "arpwatch" daemon can be seen in Figure 18.4. This example shows a time span of about 1 min between the changes of the MAC address in the ARP cache. Network administrators often use such tools to identify changes and problems in the network structure.

The second problem with ARP occurs if a central component like a switch or a physical line fails. The failure of a switch leads to the removal of all ARP entries in the ARP caches of all hosts connected to this switch if there is no communication possible for some minutes. This results in a heavy network load caused by ARP broadcasts in the beginning when the switch is operating again.

18.5   Proxy ARP

Proxy ARP was used for large IP networks (e.g., Class B network) with routers and gateways in between. Nowadays, Proxy ARP has become less important due to the use of subnets and switches.

```
flip flop (pc40.ict.tuwien.ac.at)
                hostname: pc40.ict.tuwien.ac.at
              ip address: 128.131.80.40
        ethernet address: 0:20:af:eb:b3:ea
         ethernet vendor: 3COM Corporation
    old ethernet address: 0:a0:c9:1d:f7:ad
     old ethernet vendor: Intel (PRO100B and PRO100+) [used on Cisco PIX firewall among others]
               timestamp: Thursday, February 27, 2003 10:53:41 +0100
      previous timestamp: Thursday, February 27, 2003 10:52:22 +0100
                   delta: 1 minute
```


FIGURE 18.4   Ethernet address flip-flop.

In a large IP network that has not been split into subnets, it is not possible to connect all hosts to one physical medium since Ethernet, for example, only allows 1024 hosts on one physical line. Therefore, routers are needed, which forward the ARP requests and responses to the other network segments. In this case, the router replaces the sender MAC addresses with its own MAC addresses while forwarding ARP requests and responses and stores the original mapping between protocol address and MAC address together with the network interface in a routing table. A host trying to communicate with a host on the other side of the router now sends all datagrams to the router because it got the MAC address of the router in the ARP request or response. The router forwards datagrams to the other host, which sends IP datagrams back to the router, since it got the router's MAC address in the ARP request or response, too.

Nowadays, using subnets, it is not necessary to forward ARP datagrams since hosts will look up a suitable IP gateway on the local network in their IP routing table and therefore it is not necessary to look up MAC addresses of hosts behind a router. Nevertheless, switches forward ARP datagrams to the involved ports; but they need not rewrite the ARP datagrams if the network interface type is the same on all ports.

18.6   Possible ARP Attacks

The following sections describe the two most common kinds of possible ARP attacks.

Denial of Service Attack by ARP Storms: This attack tries to overload a network component or a whole network segment by broadcasting ARP requests for nonexistent IP addresses. If the system contains Proxy ARP routers, they will start forwarding the broadcasts to all network segments. The effect can be amplified by generating synthetic ARP responses for the ARP requests. Routers forwarding these responses using broadcasts further increase the network load.

ARP Spoofing: An attacker retains his MAC address but tries to acquire the IP address of a trusted host. To achieve this goal, he has to modify the mapping of the spoofed host remotely. This can be done by different methods, for example, being the first answering an ARP request. From this time, all IP traffic belonging to this IP address goes to the attacker. This kind of attack is not very attractive for the attacker. The drawback is that on the one hand, entries in the ARP cache can age quite fast, making it necessary to keep the ARP cache updated with the hackers MAC address and on the other intelligent hardware that can suppress such attacks.

18.7   RARP — Reverse Address Resolution Protocol

The Reverse Address Resolution Protocol was designed for devices that are not able to store their IP addresses. These are, for example, diskless workstations. In contrast to ARP, RARP is designed to look up an IP address for a particular MAC address. RARP is specified in RFC 903 [2].

ARP and RARP are different operations. ARP assumes that every host knows the mapping between its own hardware address and protocol address(es). Information gathered about other hosts is accumulated in a small cache. All hosts are equal in status; there is no distinction between clients and servers.

On the other hand, RARP requires one or more server hosts to maintain a database of mappings from hardware address to protocol address and respond to requests from client hosts.

The datagram format of RARP messages is identical to ARP messages. The differences from ARP are: by using 0x8035 as Ethernet Type Code, it is possible to distinguish RARP from ARP and other higher-level protocols like IP. If the same module handles ARP and RARP messages, the operation code can be used to distinguish between ARP requests, ARP responses, RARP requests, and RARP responses (see Operation Header field on page 2).

By using RARP, one or more RARP servers are needed, maintaining large tables containing the mapping between MAC addresses and network addresses.

If a host needs to know its protocol address (e.g., IP address), it sends an RARP request to an RARP server by unicast or broadcast with its MAC address set in the "Sender Hardware Address" field and the desired protocol type (e.g., IP) set in the "Protocol Type" field. The RARP server sends back an RARP response with all address fields filled out. Therefore, the host knows its IP address and the IP address of the RARP server, which can be used to save an additional ARP request to figure out the RARP server's IP address.

Actually, RARP is not widely used, since it has strong competitors in the Bootstrap Protocol (BOOTP) [5,8] and Dynamic Host Configuration Protocol (DHCP) [6–8] protocols having a richer feature set.

There are some extensions available defined in RFC1931 [9] and called Dynamic RARP (DRARP). The main difference is that a DRARP server always sends an answer back to the originating host. An RARP server, on the other hand, only returns a response if it finds a suitable protocol address for the senders MAC address. In contrast, a DRARP server returns at least an error response if no protocol address was found. Additionally, a DRARP server can return temporary protocol addresses not configured in the mapping table. DRARP was never really used, because it was immediately replaced by DHCP.

18.8 InARP — Inverse Address Resolution Protocol

InARP, specified in RFC 2390 [10], can be seen as an extension to ARP to be able to look up protocol addresses for a given MAC address. InARP originates from Frame Relay [11]. Frame Relay has the following problem: in Frame Relay data are transmitted over so-called "virtual circuits." These virtual circuits will be established by signaling messages and are identified by Data Link Connection Identifier (DLCI). The DLCI can be seen as a MAC address. During communication, the virtual circuits together with their DLCIs can change. Now, the problem is that a host recognizes the reconfigured "MAC addresses," but is not able to associate it with a protocol address again. For this purpose, InARP was developed.

Reverse ARP, for example, seemed like a good candidate; but the response to a request is the protocol address of the requesting host, not the host receiving the request. IP-specific mechanisms were limiting since they would not allow the resolution of protocols other than IP. For this reason, the ARP protocol was expanded.

Basic InARP operates essentially in the same manner as ARP with the exception that InARP does not broadcast requests. This is because the hardware address of the target host is already known. InARP can be distinguished from ARP by the use of different operation codes for InARP requests and InARP responses. After looking up a particular protocol address, the hosts can insert the mapping into their ARP caches.

18.9 Summary

Address resolution protocols are inevitable for using higher-level protocols like IP on physical point-to-multi-point networks like Ethernet. It must be able to associate a protocol address with a hardware address to convey a higher-level protocol datagram to the correct next host.

The most widely used address resolution protocol is ARP in conjunction with Ethernet and IP. It is used for looking up a MAC address belonging to an IP address. RARP is used to look up the requester's own IP address by providing a MAC address; but it is almost substituted by DHCP. InARP is used to look up the protocol address of a destination host and is mainly used in connection-oriented networks transporting IP datagrams.

Although all protocols are mainly used for IP networks, they are specified independent of the protocol and physical network used.

References

[1] Plummer, D., An Ethernet Address Resolution Protocol — or — Converting Network Protocol Addresses to 48.bit Ethernet Address for Transmission on Ethernet Hardware, STD 37, RFC 826, November 1982.

[2] Finlayson, R., Mann, R., Mogul, J., and M. Theimer, A Reverse Address Resolution Protocol, STD 38, RFC 903, June 1984.

[3] ISO, Fiber Distributed Data Interface (FDDI) — Media Access Control, ISO 9314-2, 1989. See also ANSI X3.139-1987.

[4] Katz, D., Transmission of IP and ARP over FDDI Networks, STD 36, RFC 1390, January 1993.

[5] Croft, B., Gilmore, J., Bootstrap Protocol (BOOTP), RFC 951, September 1985.

[6] Droms, R., Dynamic Host Configuration Protocol, RFC 2131, September 1997.

[7] Droms, R., Lemon T., *The DHCP Handbook*, MacMillan Publishing, New York, October 1999.

[8] Wegner, J. D., Rockell R., Blanchet M., *IP Addressing and Subnetting, Including IPv6*, Syngress Media, Rockford, MA, USA, December 1999.

[9] Brownell, D., Dynamic RARP Extensions for Automatic Network Address Acquisition, RFC 1931, April 1996.

[10] Bradley, D., Brown, C., and Malis, A., Inverse Address Resolution Protocol, RFC 2390, September 1998.

[11] Buckwalter, J. T., *Frame Relay: Technology and Practice*, Addison-Wesley Publishing, New York, December 1999.

19

IPv6, IPSec, and VPNs

Walter Penzhorn
University of Pretoria

Johann Amsenga
Gennan Systems (Pty) Ltd

19.1 Introduction

There is growing interest in the use of VPNs as a cost-effective means of building and deploying private networks for multi-site communication over the Internet. In this context, a VPN may be defined as the emulation of a secure, private wide area network over the public Internet. The challenge in developing a VPN for today's global business environment is to utilize the public Internet backbone for both intra- and inter-company communications, while still providing the security of a traditional private, self-administered corporate network.

Within the Internet Engineering Task Force (IETF), the IP Security (IPSec) working group has developed a framework for network layer security over the Internet [11]. IPSec is a flexible framework, enabling a company to configure secure end-to-end solutions that can accommodate both locally attached users and remote-access users, and can support communications both within the company and between different companies [8].

Note that, although VPNs are mostly used over the Internet, they are not restricted to this environment. VPNs can be used over any public network to provide secure communications [1, 10]. The purpose of this article is to give a brief overview of VPNs based on the IPSec protocol, with special emphasis on the forthcoming IP version 6 of the Internet Protocol [4].

19.2 Security Requirements

Types of Security

The security of an organization's electronic information requires a comprehensive plan of action, including the following aspects:

Host security: Controlling access to, and processing on each computer.

Network security: Securing a network, where the network itself acts as an access medium.

Administration: Ensuring that proper functional procedures and operational rules are put in place and adhered to.

Personnel: Making sure that personnel are properly trained, authorized, and alerted to security awareness.

Physical security: Cont\rolling physical access to assets and infrastructure.

Security Framework

Virtual private networks are intended to provide suitable cryptographic services and mechanisms to protect a network. The ITU-T Recommendation X.800, *Security Architecture for OSI*, defines a comprehensive framework in terms of security *threats*, *services*, and *mechanisms* of networks.

Security threat: A threat is anything that poses some form of danger to a network and related resources.

Security service: A service that counters security threats and attacks and enhances the security of a network, by applying a range of available security mechanisms.

Security mechanisms: A mechanism that is intended to detect, prevent, or recover from a security attack.

Security Services

A security service enhances the security of the network and its related components. In general, services function in one of the following categories:

- Prevention
- Detection or
- Recovery

X.800 lists the following five categories of security services:

Authentication: Ensure the authenticity of the identity of all communicating entities, as well as the origin of data or information.

Data Confidentiality: Protect transmitted data or information against eavesdropping or leakage to unauthorized entities.

Data Integrity: Prevent the unauthorized creation, alteration, replay, or destruction of transmitted data.

Access Control: Limit and control access to resources to legitimate users.

Accountability: Ensure the maintenance of a complete record of the actions of every user in the system.

19.3   IP Version 6 (IPv6)

Introduction

The Internet Protocol (IP) is part of the TCP/IP suite, and is the most widely used internetworking protocol [2]. For decades, the keystone of the TCP/IP protocol architecture has been IP version 4 (IPv4). However, IPv4 has several shortcomings and limitations, which prompted the IETF to start working on a next version of IP. In 1995, the specification for a next-generation IP protocol, then known as IPng, was issued. In 1996, this specification was turned into a standard, referred to as IPv6 [4].

IPv6 provides a number of functional enhancements over the existing IPv4, designed to accommodate the higher speeds of today's networks and the mix of data streams, including graphic and video, which are becoming more prevalent. But the prime driving force behind the development of IPv6 was the need for more addresses. IPv4 uses a 32-bit address to specify the source or destination. With the explosive growth of the Internet, and the private networks attached to the Internet, this address length is rapidly becoming insufficient. Ultimately, all installations using TCP/IP are expected to migrate from the current IPv4 to IPv6; but this process may take many years, if not decades.

As shown in Figure 19.1, IPv6 contains 128-bit source and destination address fields (compared to the 32-bit address space of IPv4). Also, instead of the dotted decimal notation used in IPv4, IPv6 uses a different notation to depict an Internet address, that is, hexadecimal numbers separated by colons:

0123:4567:89ab:cdef:0123:4567:89ab:cdef

The IPv6 header has a length of 40 octets, consisting of the following fields:

Version (4 bits): Internet Protocol version number; for IPv6, the value is 6.
Traffic Class (8 bits): To distinguish between different classes or priorities of IPv6 packets.
Flow Label (20 bits): The flow label is intended to assist with resource reservation and real-time traffic processing.
Payload Length (16 bits): Gives the total length of all of the extension headers plus the transport-level payload.
Next Header (8 bits): Identifies the type of header immediately following the IPv6 header.
Hop Limit (8 bits): The hop limit is set to some desired maximum value by the source and decremented by 1 by each node that forwards the packet.
Source Address (128 bits): The address of the originator of the packet.
Destination Address (128 bits): The address of the intended recipient of the packet.

IPv6 Packet Format

An IPv6 packet has the following general form, as shown in Figure 19.2:

The only mandatory header is simply referred to as the IPv6 Header, and has a fixed length of 40 octets. Although the IPv6 Header is longer than the mandatory portion of the IPv4 header (20 octets), the IPv6 packet contains fewer fields (8 vs. 12). Thus, routers have less processing to carry out per packet, which should speed up routing.

The following extension headers are defined for IPv6:

Hop-by-Hop Options Header: Defines special options that require hop-by-hop processing.
Routing Header: Provides extended routing, similar to IPv4 source routing.
Fragment Header: Contains fragmentation and reassembly information.
Authentication Header: Provides packet integrity and authentication (required for IPSec).
Encapsulating Security Payload Header: Provides privacy (required for IPSec).
Destination Options Header: Contains optional information to be examined by the destination node.

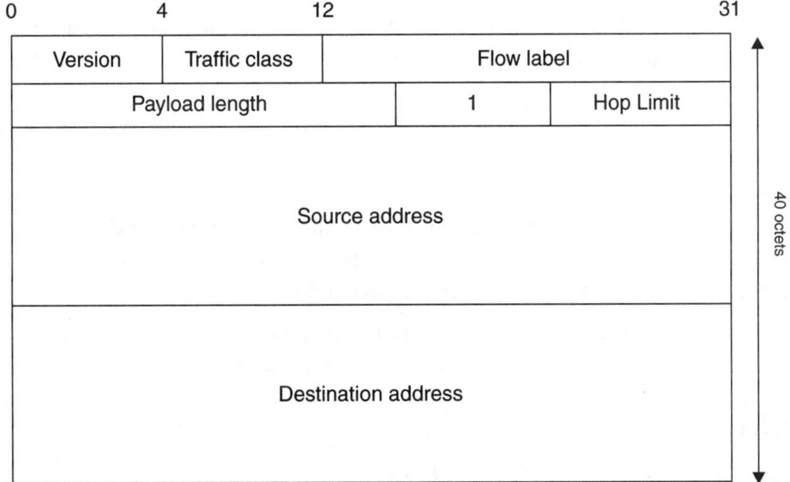

FIGURE 19.1   Illustration of the IPv6 header.

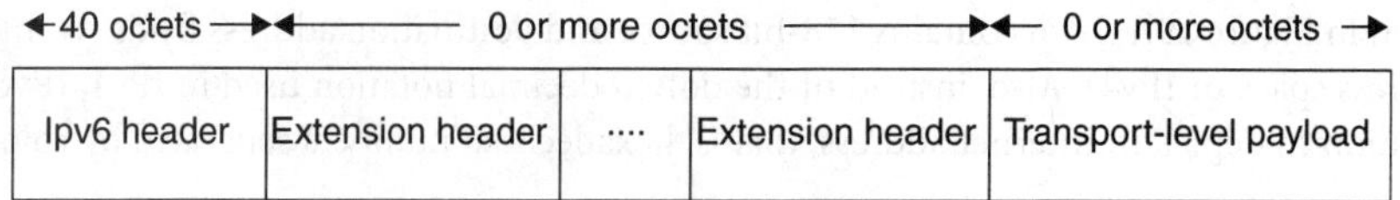

FIGURE 19.2   General form of an IPv6 packet.

Figure 19.3 shows an example of an IPv6 packet that includes an instance of each nonsecurity header. Observe that each extension header includes a Next Header field. This field identifies the type of subsequent header. If the subsequent header is an extension header, then this field contains the type identifier of that header.

19.4 IPSec

Introduction

In 1994, the Internet Architecture Board (IAB) issued a report entitled "Security in the Internet Architecture" (RFC 1636) [11]. The report stated the need for security on the Internet, and identified the following three functional areas for security mechanisms:

Authentication: To ensure that a received packet was transmitted by the claimed source in the packet header, and also that the packet was not altered in transit.

Confidentiality (encryption): Provide encryption of packets to prevent eavesdropping by third parties, as well as the unauthorized monitoring and control of network traffic.

Key management: Establish an infrastructure for the secure management and exchange of cryptographic keys.

In response, the IPSec working group was formed within the IETF, and a framework for network layer security was developed that is usable both with the current IPv4 and the future IPv6.

What is IPSec?

The acronym IPSec, a short form of the term "Internet Protocol Security," is an evolving standard for security at the network or packet layer of network communication. In essence, IPSec is a set of protocols that may be used to implement VPNs that operate over the Internet [5, 8]. The principal feature of IPSec is that it can encrypt and/or authenticate all traffic at the IP level, so that all distributed higher-layer applications, including remote logon, client/server, e-mail, file transfer, Web access, etc. can be secured, as shown in Figure 19.4.

Thus, IPSec is transparent to end users and applications and facilitates a general-purpose solution. Furthermore, IPSec includes a filtering capability so that only selected traffic is affected by the overhead of IPSec processing.

The IPSec specification consists of numerous documents. The most important of these, issued in November 1998, are:

- *RFC 2401*: An overview of a security architecture.
- *RFC 2402*: Description of a packet authentication extension to IPv4 and IPv6.
- *RFC 2406*: Description of a packet encryption extension to IPv4 and IPv6.
- *RFC 2408*: Specification of key management capabilities.

Support for IPSec is mandatory for IPv6 and optional for IPv4. In both cases, the security features are implemented as two extension headers that follow the main IP header.

IPSec Services

The following security services are provided by IPSec:

Data origin authentication: Verifies that each datagram was originated by the claimed sender.

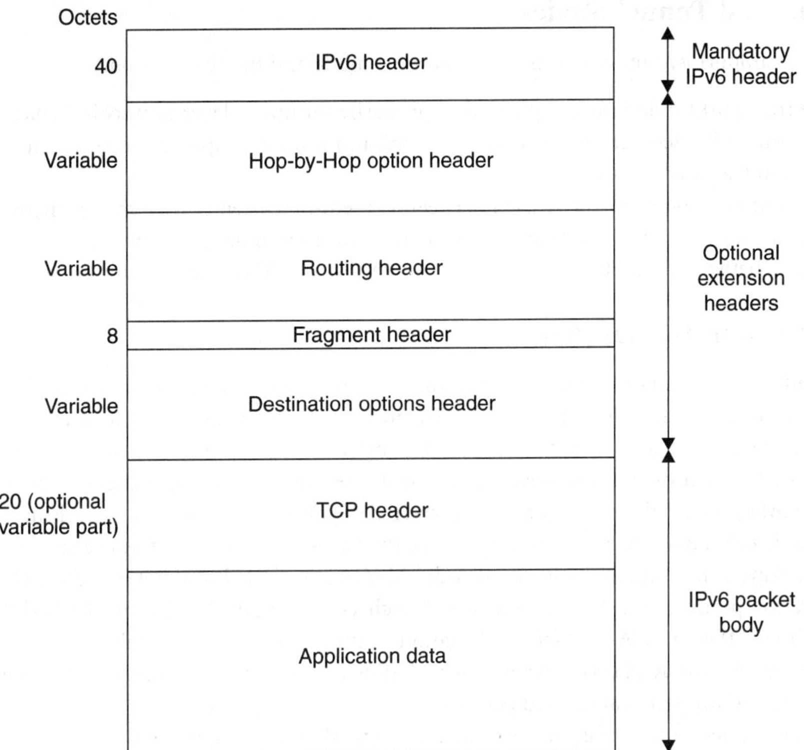

FIGURE 19.3 Illustration of an IPv6 packet with Extension Headers.

FIGURE 19.4 Placement of security in the TCP/IP stack.

Data integrity: Verifies that the contents of the datagram were not changed in transit, either deliberately or due to random errors.

Data confidentiality: Protects the plaintext of a message by means of encryption.

Replay protection: Assures that an attacker cannot intercept a datagram and send it back at some later time.

Key management: Generation and distribution of cryptographic keys across a distributed network environment.

IPSec provides these security services by means of two protocols: an authentication protocol designated by the *header* of the protocol, Authentication Header (AH), and a combined encryption/authentication protocol designated by the *format of the packet* for that protocol, Encapsulating Security Payload (ESP).

Authentication Header (AH): Provides data origin authentication, data integrity, and replay protection.

Encapsulating Security Payload (ESP): Provides data confidentiality, data origin authentication, data integrity, and replay protection.

Internet Security Association and Key Management Protocol (ISAKMP): Provides a framework for the generation, distribution, and management of cryptographic keys.

In the sequel, each of these protocols will be discussed in some detail.

Transport and Tunnel Modes

AH and ESP support two modes of use: the transport mode and the tunnel mode.

- The transport mode provides protection primarily for upper-layer protocols. Typically, the transport mode is used for end-to-end communication, for example, between client and server or between two workstations.
- The tunnel mode is normally used between two entities when at least one of them is not a connection end-point. For example, if secure communication is desired between two gateways that are located between a client and a server, these gateways would utilize IPSec in the tunnel mode.

Authentication Header (AH)

The Authentication Header provides connectionless integrity and data origin authentication for IP datagrams, as well as protection against replay [7]. The format of the AH header is illustrated in Figure 19.5.

The data integrity feature prevents undetected modification of a datagram. Data authentication enables an end system or network device to authenticate the user, and also prevents address spoofing attacks. The AH protects the entire contents of an IP datagram except for certain fields in the IP header, called *mutable fields*, that are being modified while the datagram is in transit. A checksum, generated by means of a (keyed) message authentication code (MAC), is included in the AH header. Hence, the two communication parties must share a secret key. The checksum is calculated using the HMAC algorithm, based on either MD5 or SHA-1. During calculation of the checksum, the mutable fields are treated as if they contained all zeros. Replay is prevented by including a sequence number field within the AH Header, which is increased for each transmitted package.

AH can be applied in the transport or tunnel mode, as shown in Figure 19.6.

In the transport mode, the AH is inserted after the original IP header and before the IP payload. The AH is viewed as an end-to-end payload, that is, it is not examined or processed by intermediate routers. Therefore, the AH appears after the IPv6 base header and the hop-by-hop, routing, and fragment extension headers. Authentication covers the entire packet, excluding mutable fields, which are set to zero for MAC calculation. Note that all information in the datagram is in plaintext form, and therefore is subject to eavesdropping while in transit.

In tunnel mode, a new IP header is generated; the AH is inserted between the new, outer IP header and the original (inner) IP header. The inner IP header carries the ultimate source and destination addresses, while a new IP header may contain intermediate IP addresses (e.g., VPN address). The entire new datagram (new IP Header, AH Header, IP Header, and IP Payload) is protected by the AH protocol. Any change to any field (except the mutable fields) in the tunnel mode datagram can be detected.

0	8	16	31
Next header	Payload length	RESERVED	
Security Parameter Index (SPI)			
Sequence number			
Authentication data (variable)			

FIGURE 19.5 IPSec Authentication Header.

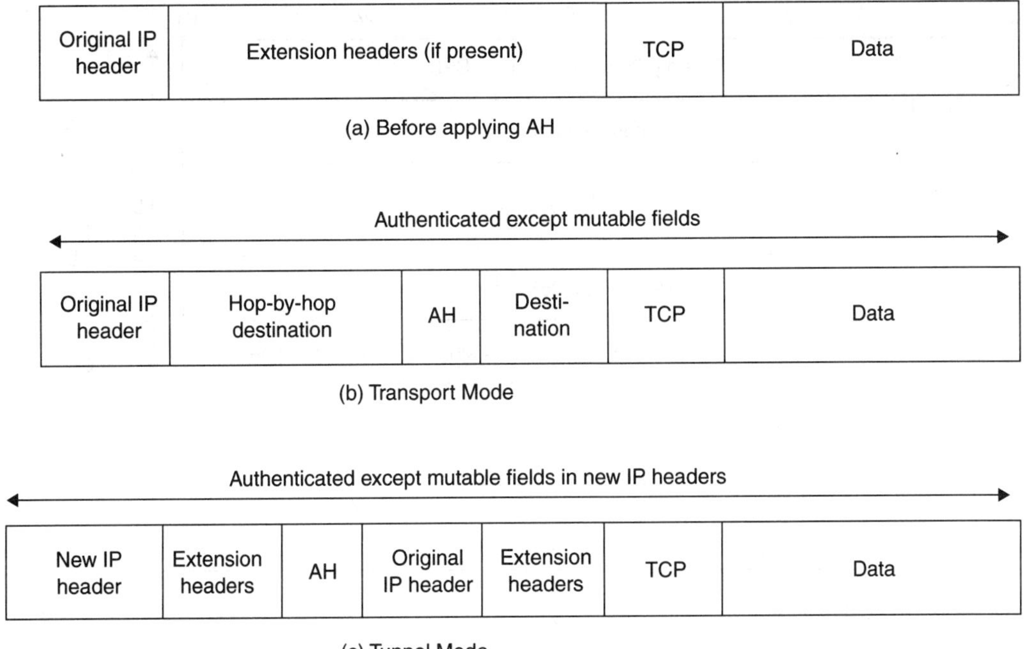

FIGURE 19.6   Illustration of the AH modes.

AH may be applied alone or in combination with the ESP header. The primary difference between the authentication provided by ESP and AH is the extent of the coverage. Specifically, ESP does not protect any IP header fields unless those fields are encapsulated by ESP (tunnel mode).

ESP

The ESP provides data confidentiality, and protection against replay; it can also optionally provide data origin authentication and data integrity checking, similar to AH. The current IPSec specification mandates the use of the Data Encryption Standard (DES) in cipher block chaining (CBC) mode for encryption. Several other algorithms are also accepted, including triple DES, RC5, IDEA, CAST, and Blowfish. When ESP is required to provide authentication as well, it uses the same HMAC algorithms (HMAC-MD5 or HMAC-SHA-1) as the AH protocol. The format of an ESP package is shown in Figure 19.7.

ESP can be applied in either transport or tunnel mode. Transport mode may be used, for example, to provide encryption (and optionally authentication) between two directly connected hosts. On the other hand, a VPN may be established over the Internet between various private, corporate networks by implementing ESP tunnel mode on the security gateways that connect the internal networks to the Internet. In this way, implementation of security on every internal host is avoided. This is illustrated in Figure 19.8.

- In the transport mode, ESP is viewed as an end-to-end payload, that is, it is not examined or processed by intermediate routers. Therefore, the ESP header appears after the IPv6 base header and the hop-by-hop, routing, and fragment extension headers. Encryption covers the entire transport-level segment plus the ESP trailer. Note that the IP Header itself is neither authenticated nor encrypted. Hence, the addressing information in the outer header is visible to an attacker while the datagram is in transit.
- In the tunnel mode, a new IP header is generated. The entire original IP datagram (including both IP Header and IP Payload) and the ESP Trailer are encrypted. Because the original IP Header is encrypted, its contents are not visible to an attacker while it is in transit. Tunneling offers the advantage of hiding original source and destination addresses from users on the Internet, thus diminishing

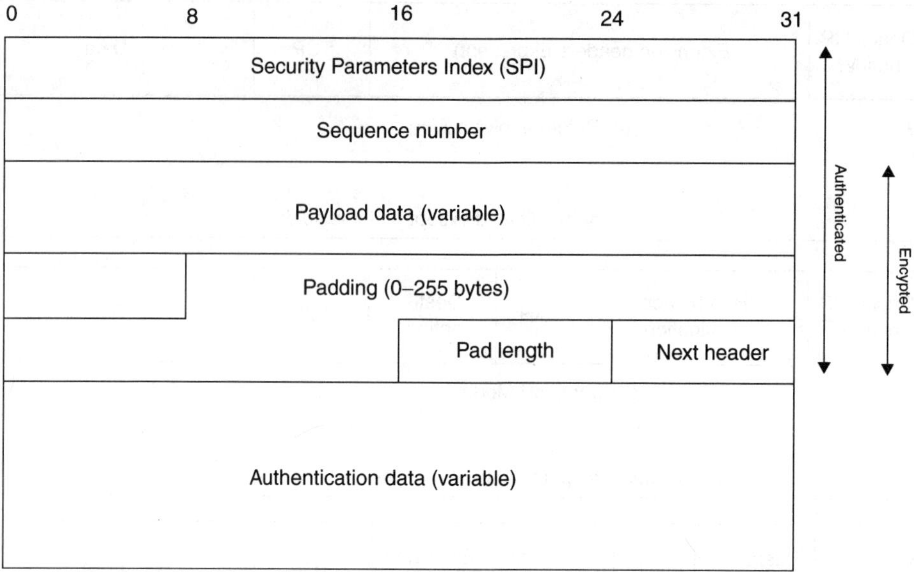

FIGURE 19.7   ESP header format.

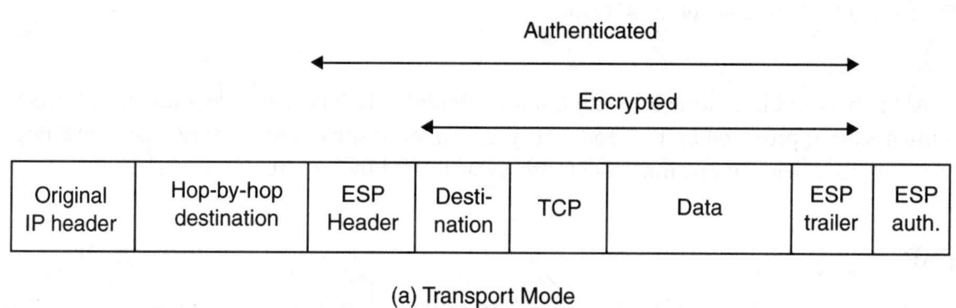

(a) Transport Mode

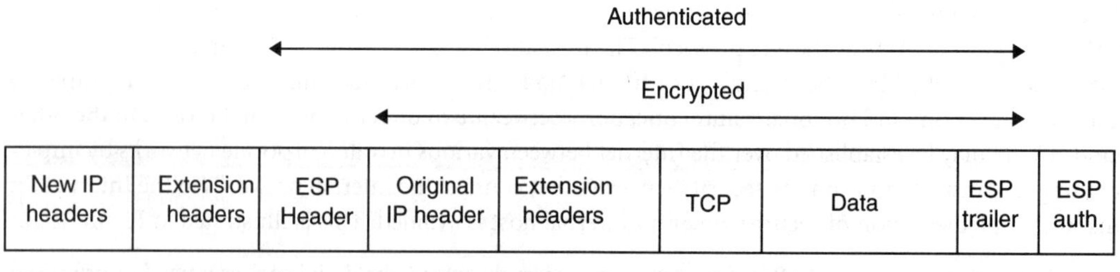

(b) Tunnel Mode

FIGURE 19.8   Illustration of ESP modes for encryption and authentication.

the threat of traffic analysis attacks. A common use of tunnel mode is to protect a datagram that is *tunneled* between two gateways. This relieves hosts on the internal network of the processing burden of encryption and simplifies the key distribution task by reducing the number of required secret keys.

The security services provided by the two IPSec protocols are shown in Table 19.1.

TABLE 19.1   An overview of IPSec Security Services

Security service	AH	ESP(encryption only)	ESP(encryption and authentication)
Access control	Yes	Yes	Yes
Connectionless integrity	Yes		Yes
Data origin authentication	Yes		Yes
Rejection of replayed packets	Yes	Yes	Yes
Confidentiality		Yes	Yes
Limited traffic flow confidentiality		Yes	Yes

Security Association

A Security Association (SA) is a relationship between two or more entities that describes how the entities will use security services to communicate securely. The security association is unidirectional, meaning that for each pair of communicating systems, there are at least two security connections — one from A to B and one from B to A. The security association is uniquely identified by three parameters:

Security Parameters Index (SPI): A randomly chosen unique number that is carried in AH and ESP headers to enable the receiving system to select the SA under which a received packet will be processed.

IP Destination Address: This is the address of the destination endpoint of the SA, which may be an end-user system or a network system such as a fire wall or router.

Security Protocol Identifier: This indicates whether the association is an AH or ESP security association.

In an SA, ESP may be applied alone, in combination with AH, or even nested within another instance of itself. With these combinations, authentication can be provided between a pair of communicating hosts, between a pair of communicating firewalls, or between a host and a firewall. An SA contains all the relevant information that communicating systems need in order to execute the IPSec protocols, such as:

- The mode of the authentication algorithm used in the AH, and the keys for the algorithm.
- The encryption algorithm mode for ESP, and the required keys.
- Authentication algorithm-related information used in the ESP.
- Key management information, such as key lifetime.
- The SA lifetime.

Key Management in IPSec

The operation of IPSec in either AH or ESP mode requires the availability of shared secret keys between communicating entities. Typically, four keys are required for communication between two applications: transmit and receive pairs for both AH and ESP. The IPSec Architecture document mandates support for two types of key management:

Manual: A system administrator manually configures each system with its own keys and with the keys of other communicating systems. This is practical for small, relatively static environments.

Automated: An automated system enables the on-demand creation of keys for SAs and facilitates the use of keys in a large distributed system with an evolving configuration.

The default automated key management protocol for IPSec consists of the following elements:

ISAKMP: ISAKMP defines a standardized framework to support negotiation of SA, initial generation of all cryptographic keys, and subsequent refreshing of these keys [6, 9].

Oakley Key Determination Protocol: This key exchange protocol is based on the Diffie–Hellman algorithm, modified to provide security against known attacks.

ISAKMP by itself does not dictate a specific key exchange algorithm, but consists of a set of message types that enable the use of a variety of key exchange algorithms. Oakley is the specific key exchange algorithm mandated for use with the initial version of ISAKMP. The ISAKMP/Oakley protocol combination was named *Internet Key Exchange (IKE)*, and is specified in RFC 2409.

The ISAKMP methods have been designed with the explicit goal of providing protection against several well-known threats:

Denial of Service: The messages are constructed with unique *cookies* that can be used to quickly identify and reject invalid messages without the need to execute processor-intensive cryptographic operations.

Man-in-the-Middle: Protection is provided against common attacks such as deletion of messages, modification of messages, reflecting messages back to the sender, replaying of old messages, and redirection of messages to unintended recipients.

Perfect Forward Secrecy: Compromise of past keys provides no useful clues for breaking any other key, whether it occurred before or after the compromised key. That is, each refreshed key will be derived without any dependence on predecessor keys.

IKE functions in two phases. Phase one involves two IKE peers establishing a secure channel for performing phase two. Phase two involves the two peers negotiating general-purpose SAs. IKE provides three modes for the exchange of keying information and setting up of IKE SAs. Two modes are for IKE phase-one exchanges, and one mode is for phase-two exchanges.

Main Mode

This mode provides a way to establish the first phase of IKE SA, which is then used to negotiate future communications. The first step, securing an IKE SA, occurs in three two-way exchanges between the sender and the receiver. In the first exchange, the sender and receiver agree on basic algorithms and hashes. In the second exchange, public keys are sent for a Diffie–Hellman exchange. Nonces (random numbers each party must sign and return to prove their identities) are then exchanged. In the third exchange, identities are verified, and each party is assured that the exchange has been completed.

Aggressive Mode

The aggressive mode provides the same services as the main mode. It establishes the phase one SA, and operates in much the same manner as the main mode except that it is completed in two exchanges instead of three. However, the aggressive mode does not provide identity protection for communicating parties. In other words, in the aggressive mode, the sender and recipient exchange identification information before they establish a secure channel where the information is encrypted.

Quick Mode

After two parties have established a secure channel using either aggressive mode or main mode, they can use quick mode. Quick mode has two purposes: to negotiate general IPSec security services and to generate newly keyed material. Quick mode is much simpler than both main and aggressive modes. Quick mode packets are always encrypted under the secure channel (or IKE SA established in phase one) and start with a hash payload that is used to authenticate the rest of the packet.

Perfect Forward Secrecy

For perfect forward secrecy, it is required to generate a new key that does not depend on the current or any previous key. Diffie–Hellman allows the generation of new shared keys that are independent of older keys, thus providing perfect forward secrecy. The derived Diffie–Hellman key can be used either as a session key for subsequent exchanges, or to encrypt another randomly generated key.

19.5   Virtual Private Networks

Introduction

A VPN can best be described as follows:

A Virtual Private Network provides secure, private communications by utilizing an existing public, insecure network infrastructure, such as the Internet.

Within this context, we observe the following:

- A public network is any network that is under the control of another party, or that is accessible by other parties. The term *other parties* is used to refer to other people or organizations that are not part of the relevant organization.
- A *private* network is a network belonging to a specific organization, and is accessible only to authorized members (e.g., employees) of that organization.
- The network is termed *virtual* because it uses a logical connection that is built on the physical connections. Client applications are unaware of the actual physical connection and route traffic securely across the public network in much the same way in which traffic on a private network is securely routed.

In other words, a VPN enables the use of part of a public network for private, confidential purposes [1, 10, 12]. A VPN is intended to give an organization the same capabilities as a private network that is based on leased lines, but at a much lower cost. Privacy is maintained through the use of security services and mechanisms, based on tunneling protocols.

A secure and effective VPN deployment should provide the following functions:

- enforce an overall network security policy,
- ensure that VPN traffic is subject to network access control,
- protect the VPN gateway from security threats,
- provide an overall architecture that optimizes VPN and firewall performance,
- accommodate highly dynamic and growing network environments.

The key technologies that comprise the security component of a VPN are:

- access control to guarantee the security of network connections,
- encryption (confidentiality) to protect the privacy of sensitive data,
- authentication to verify the user's identity as well as the integrity of the data, and
- procedures for exchanging keys and digital certificates (credentials) among different users.

VPN Configurations

User-to-site VPN

A VPN provides an easy mechanism to allow remote access to an organization's private network. In the past, users working at off-site locations such as their homes (telecommuting) or while traveling (mobile or roaming users) had to connect to the company network via dial-up links to a modem pool. The Internet provides a cheaper alternative since the user can connect to the nearest ISP and access the company network via the Internet. Figures 19.9 and 19.10 compare two different scenarios.

Site-to-site VPN

A VPN using a public infrastructure such as the Internet can replace the traditional wide area network (WAN) architecture (using point-to-point leased lines) between offices. WANs are usually privately owned networks that span a large geographical area, typically across a country or even a continent, and interconnect LANs in different locations.

The example in Figure 19.11 shows a company with three offices situated in different locations. Each office requires access to the Internet, as well as being connected to the other offices. Traditionally, connectivity between the offices is obtained by means of a dedicated WAN, based on leased lines. However, if a VPN is used, as shown in Figure 19.12, only one Internet connection is required at each office.

Extranet

An extranet is a network configured to allow specific trusted parties access to selected regions of an organization's internal network. Trusted parties may, for example, include clients and other organizations collaborating on a project. Figure 19.13 shows an example where a company provides access to customers via a dedicated line.

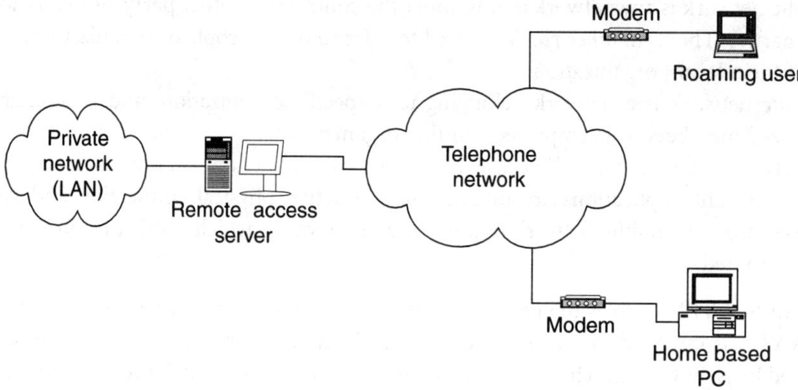

FIGURE 19.9   Remote access using a Remote Access Server.

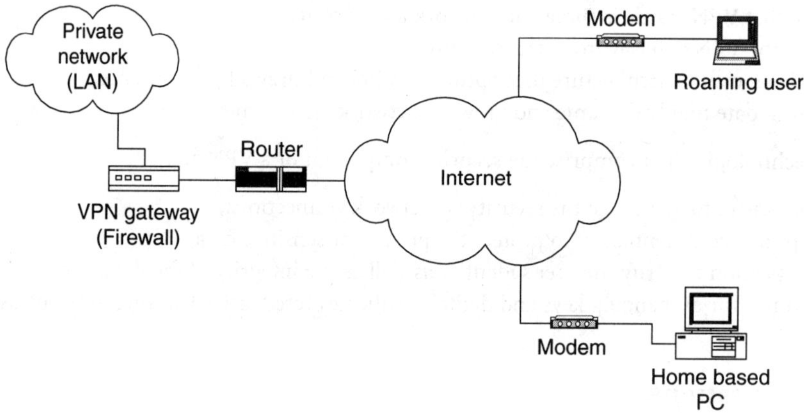

FIGURE 19.10   Remote access via the Internet.

Figure 19.14 shows how an external customer can gain access by means of a VPN connected to the Internet. The access control and authentication services of the VPN are utilized to grant discretionary access to customers and partners to information and resources on the organization's private network.

Internal VPN or Private LANs

A VPN implementation can be used for internal network partitioning. This enables different logical networks on the same network infrastructure. Figure 19.15 shows a LAN configuration with unrestricted access. In the example shown in Figure 19.16, VPN technology can be used to partition the company LAN to restrict Workstation 1 access to Server A and C only, while allowing Workstation 2 access to Server B and C. The partitioning will prevent a user at Workstation 1 from intercepting and reading the traffic between Workstation 2 and Server B.

Methods of VPN Deployment

Some form of VPN gateway is necessary to connect a private network to the Internet. The gateway can be a stand-alone hardware device, or a software component integrated into an existing firewall or router. In the case of remote access, remote users typically run software applications on their own computers.

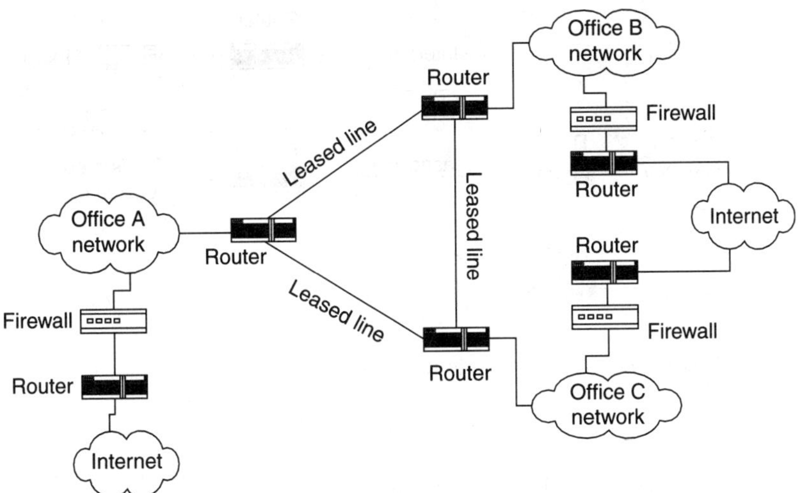

FIGURE 19.11   Traditional site-to-site communication.

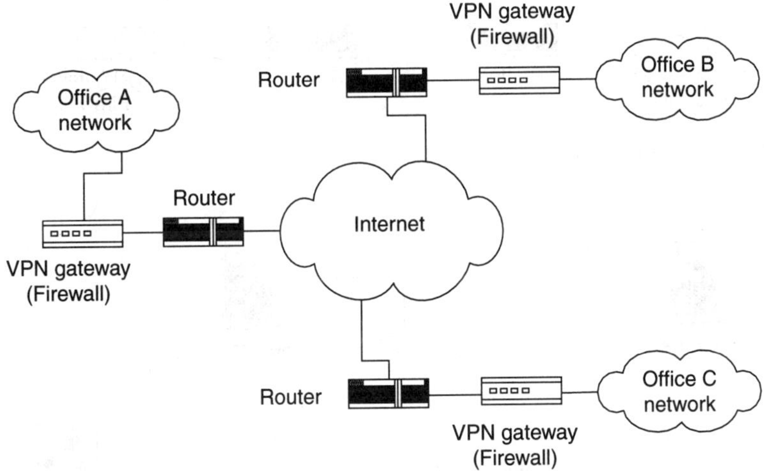

FIGURE 19.12   Site-to-site VPN.

Software Integration

This type of deployment involves the installation of software on existing equipment. In all cases, performance degradation may be a problem.

Routers: Many commercial routers support the addition of VPN software. The processing required for tunneling, etc. may reduce the throughput of the router in cases where the traffic load is high, since all processing is done in software.

Firewalls: Many companies view their firewall as the central component of their network security plan. Firewalls often make provision for the addition of a VPN capability. Again, because of the high processing requirement for both the firewall functionality and VPN service, the throughput may be severely degraded [3].

Servers: Another way to deploy a VPN is to install a software-based VPN on a server. Operating system suppliers and third-party vendors offer VPN applications that are able to link users via a VPN. The software can be installed on existing servers, allowing the existing network configuration to remain intact.

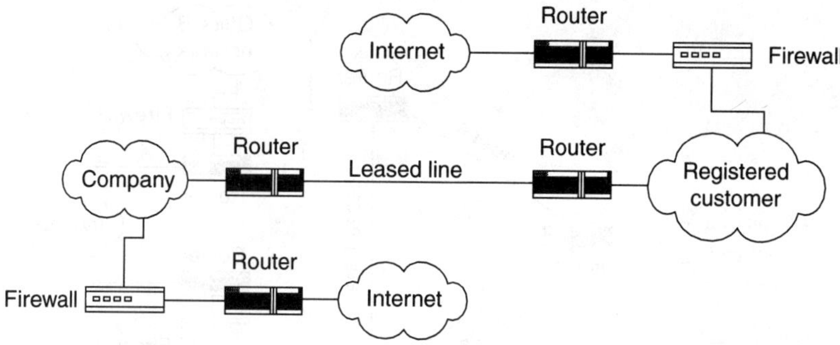

FIGURE 19.13 Enabling customer access to the private network.

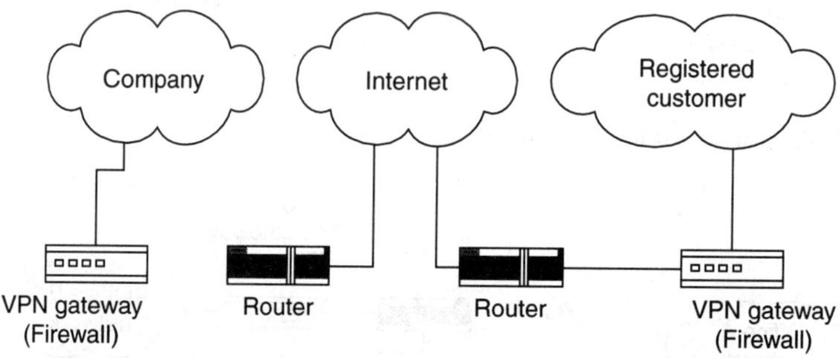

FIGURE 19.14 Enabling customer access via a VPN.

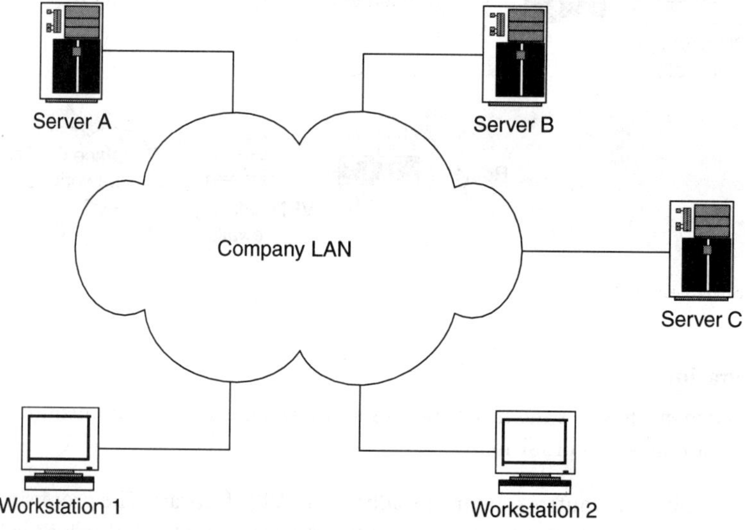

FIGURE 19.15 Company LAN with no access restrictions.

Stand-alone Hardware

As an alternative, VPNs may be implemented on dedicated, stand-alone devices that are specifically designed for tunneling, encryption, and user authentication. These devices are easier to install than to modify existing firewalls or routers, and, in general, they are much more efficient in terms of throughput than software-based solutions.

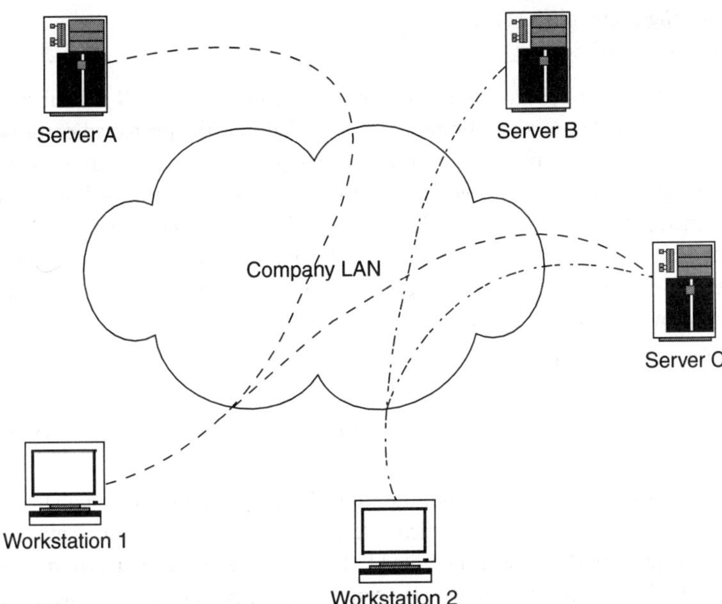

FIGURE 19.16   Virtual private LANs.

Advantages of a VPN

The advantages of using a VPN for connectivity all hinge on one aspect — saving money while employing strong security. A few of the advantages are:

- *Existing infrastructure*: The Internet can be used to provide distributed network services over long distances, and this eliminates the need to create WANs based on expensive leased lines. Using the Internet for remote access leads to substantial savings, since remote users can dial in to a local service provider and gain access to the company network via the Internet.
- *Simpler connectivity*: It is much easier and cheaper to connect offices using the existing Internet infrastructure than to install leased lines.
- *Availability*: In general, the availability of the Internet's communication infrastructure is much higher than with dedicated leased lines. The Internet has a high degree of redundancy — if one link fails, another link is available to carry the traffic.
- *Maintenance*: The cost, time, and resources necessary to maintain and administer a VPN are much less than for a WAN. Furthermore, an organization typically does not need to maintain a dial-in facility to accommodate remote users.
- *Flexibility*: By using the Internet as part of the network infrastructure, companies are no longer bound by long-term contracts as required by leased-line providers. In general, it is easier to change Internet service providers.
- *Extranet applications*: Organizations can allow trusted parties access to certain resources on the private network through an extranet facility. This promotes customer relations and enhances communication with partner organizations, etc.

Competing VPN Deployment Protocols

In recent years, four different protocols have been suggested for creating a VPN over the Internet:

- point-to-point tunneling protocol (PPTP),
- layer-2 forwarding protocol (L2F),

- layer-2 tunneling protocol (L2TP), and
- IP security protocol (IPSec).

One reason for the number of protocols is that, for some companies, a VPN is a substitute for remote-access servers, allowing mobile users and branch offices to dial into the protected corporate network via their local ISP. For others, a VPN may consist of traffic traveling in secure tunnels over the Internet between protected LANs. The protocols that have been developed for VPNs reflect this dichotomy. PPTP, L2F, and L2TP are largely aimed at dial-up VPNs, while IPSec's main focus has been on LAN-to-LAN solutions.

PPTP: Currently, the most commonly used protocol for remote access to the Internet is the point-to-point protocol (PPP). PPP provides a standard method for transmitting IP datagrams over serial point-to-point links, and provides three functions:

- A method for encapsulating IP Datagrams over serial links.
- An extensible Link Control Protocol (LCP).
- A family of Network Control Protocols (NCP) for establishing and configuring different network-layer protocols.

The PPTP is an extension of PPP, to provide remote access that can be tunneled through the Internet to a destination site. PPTP was developed primarily by Microsoft, and at one time enjoyed the status of a *de facto* standard. PPTP creates a secure tunnel in which encrypted PPP packets are sent as IP datagrams over the Internet.

PPTP utilizes the authentication mechanisms within PPP, namely the password authentication protocol (PAP) and Challenge/Handshake Authentication Protocol (CHAP). For encryption, PPTP derives a 40-bit encryption key from the hashed password stored on the client and the server.

Besides the relative simplicity of client support for PPTP, one of the protocol's main advantages is that PPTP is designed to run at open systems interconnection (OSI) Layer 2, or the link layer, as opposed to IPSec, which runs at Layer 3. By supporting data communications at Layer 2, PPTP can transmit protocols other than IP over its tunnels. As currently implemented, PPTP encapsulates PPP packets using a modified version of the generic routing encapsulation (GRE) protocol, which gives PPTP the flexibility of handling protocols other than IP, such as Internet packet exchange (IPX) and network basic input/output system extended user interface (NetBEUI).

PPTP does have some limitations: it does not provide strong encryption for protecting data, nor does it support any token-based methods for authenticating users. Hence, it is not likely that PPTP will become a formal standard endorsed by any of the standard bodies, like the IETF.

L2F: L2F emerged during the early stages of VPN development, and was designed by Cisco Systems. Like PPTP, L2F is intended as a protocol for tunneling traffic from users to their corporate sites. L2F provides tunneling for the encapsulation of non-IP packets, and thus it is able to work directly with other media such as frame relay, X.25, or asynchronous transfer mode (ATM).

Similar to PPTP, L2F uses PPP for authentication of the remote user, but it also includes support for terminal access controller access control system (TACACS)+ and RADIUS for authentication. L2F also differs from PPTP in that it allows tunnels to support more than one connection.

L2TP: In an effort to combine the different functionalities of PPTP and L2F, the IETF has proposed the Layer Two Tunneling Protocol (L2TP). This protocol is able to tunnel PPP traffic over a variety of networks (e.g., IP, X.25, SONET, frame relay, and ATM), using its own tunneling protocol.

Because it uses PPP for dial-up links, L2TP includes the authentication mechanisms within PPP, namely PAP and CHAP. However, PPTP, L2F, and L2TP do not all include encryption, or facilities for management of cryptographic keys. The current L2TP draft standard recommends that IPSec be used for encryption and key management over IP networks.

IPSec: IPSec emerged from efforts to secure IP packets, when the next generation of IP (i.e., IPv6) was being developed; it can now be used with IPv4 protocols as well. IPSec is generally considered the best VPN solution for IP environments, since it provides strong security measures — notably encryption, authentication, and key management.

However, IPSec is mainly intended to handle IP packets, and therefore PPTP and L2TP will probably remain the first choice for a multi-protocol non-IP environment, such as NetBEUI, IPX, and AppleTalk.

Benefits of IPSec

Applying IPSec in a firewall gives the following benefits:

- Strong security can be applied to all traffic crossing a firewall if IPSec is implemented on the firewall.
- If IPSec is implemented on the firewall, traffic on the private network does not incur the overhead of security-related processing.
- IPSec in a firewall is resistant to bypassing if all traffic from the untrusted network must use IP and the firewall is the only access point to the local (trusted) network.
- IPSec operates below the transport layer (e.g., TCP) and is thus transparent to applications. There is no need to change software on a user or server system when IPSec is implemented in the firewall or router. Even if IPSec is implemented in end systems, application-layer software is not affected.

19.6 Conclusion

This article provides a short overview of VPN and the IPsec protocol, one of the technologies available for implementing VPNs. IPSec provides a powerful and flexible security framework for secure communication over the Internet, by means of the following security functions:

- *Authentication:* Ensures that a packet comes from the claimed source and was not altered in transit.
- *Confidentiality:* Prevents unauthorized disclosure of packet contents as well as the monitoring of network traffic.
- *Key management:* Establish an infrastructure for the secure management and exchange of cryptographic keys over insecure links.

IPSec provides two modes of operation, transport and tunnel modes, that are implemented by means of the Authentication Header and Encapsulating Security Payload protocols. The transport mode is generally used for secure end-to-end communication, while the tunnel mode can be used for host-to-host or network-to-network communication.

It is important to note that although VPNs are mostly used for the Internet, they are not restricted to this environment. VPNs can be used over any public network to give secure, private communication. Furthermore, IPSec can provide robust security services for other popular protocols, such as PPTP or L2TP.

References

[1] Brown, S., *Implementing Virtual Private Networks*, McGraw-Hill, New York, 1999.
[2] Comer, D. and D. Stevens, *Internetworking with TCP/IP, Vol. 1: Principle, Protocols and Architectures*, Prentice-Hall, Englewood Cliffs, NJ, 2000.
[3] Cheswick, W.R. and S.M. Bellovin, *Firewalls & Internet Security: Repelling the Wily Hacker*, Addison-Wesley, Reading, MA, 1994.
[4] Deering, S. and R. Hinden, Internet Protocol, Version 6 (IPv6) Specification, RFC 2460, 1998. http://www.ietf.org/rfc/rfc2460.txt
[5] Frankel, S., *Demystifying the IPSec Puzzle*, Artech House, Norwood, MA, 2001.
[6] Harkins, D. and D. Carrel, The Internet Key Exchange (IKE), RFC 2409, 1998. http://www.ietf.org/rfc/rfc2409.txt
[7] Kent, S. and R. Atkinson, IP Authentication Header, RFC 2402, 1998. http://www.ietf.org/rfc/rfc2402.txt
[8] Markham, T., Internet security protocol, *Dr. Dobb's Journal*, pp. 70–77, June 1997.
[9] Maughan, D., M. Schertler, M. Schneider, and J. Turner, Internet Security Association and Key Management Protocol (ISAKMP), RFC 2408, 1998. http://www.ietf.org/rfc/rfc2408.txt
[10] Metz, C., The latest in virtual private networks: part 1, *IEEE Internet Computing*, Vol. 7, no. 1, pp. 87–91, Jan./Feb. 2003.

[11] The Internet Engineering Task Force: http://www.ietf.org
[12] The Virtual Private Network Consortium: http://www.vpnc.org

20

Overview and Classification of IP Routing Protocols — IP Routing: Interior and Exterior Routing Protocols

Lucia Lo Bello
University of Catania

Enzo Fabrizio Palumbo
STMicroelectronics

20.1 Introduction

The Internet Protocol (IP) is the building block of the Internet. Nowadays, IP is also becoming the single technology used to support the different kinds of traffic present in a number of enterprise internetworks. This chapter covers the basics of routing technology and protocols for IP networks. After a review of routing algorithms and design issues, a classification of IP routing protocols is presented, together with details of the different protocols and comparative assessments.

The chapter ends with a case study addressing the routing technologies and design criteria of a global IP network for a large-scale enterprise. The scenario refers to a manufacturing company, which extends over three macro-regions — Europe, North America, and the Asia-Pacific region. The case study provides an overview of the architecture of this global IP network and highlights both technical issues and design approaches toward a large-scale, high-availability infrastructure.

20.2 A Review of Routing

Routers

Routers are an example of Layer 3 Intermediate Systems (ISs), which are generally defined as network devices with the ability to forward packets between portions of a network. More specifically, routers allow IP packets [1] to be switched between interconnected networks. Originally called *gateways* in many early Internet Engineering Task Force (IETF) standards, routers play a crucial role in modern enterprise internetworking, providing transparent connectivity over heterogeneous technology subnetworks. Even routers from different vendors can cooperate to create a holistic view of the network in the form of routing tables.

Routing Protocols

An *internetwork*, also called an internet, consists of two or more networks joined together. In an internetwork, network layer *routing protocols* implement *path determination* and *packet switching*. Path determination consists of choosing which path (or *route*) the packets are to follow across the internetwork from a sending host to a destination, while packet switching refers to transporting them.

Path determination is accomplished by *routing algorithms*. In general terms, the purpose of routing algorithms is to determine, given a set of routers and links connecting them, the best (i.e., least-cost) path from source to destination. In order to support the calculation of the best path for individual packets, routing protocols have to enable routers to communicate with each other exchanging both *topology* information (e.g., about neighbors and routes) and *state* information (e.g., costs). This information is used to build a *routing table* that is consulted whenever a routing decision has to be made.

A routing table consists of a list of *routing entries* indicating which outgoing link should be used to forward packets to a given destination. When a router receives an incoming packet, it checks the routing table in order to find a destination/next hop association for the destination address specified in the packet. Figure 20.1 shows the basic structure of a simplified routing table.

A number of routing algorithms exist, with different characteristics, targets, and metrics. The following subsection presents a classification of the routing protocols used in the IP environment, together with some main design issues.

Destination network	Next Hop	Nhops	Interface
160.4.0.0	160.4.5.2	–	Ethernet0
151.5.0.0	160.4.5.2	5	Ethernet0
192.166.30.0	160.7.3.1	3	Ethernet1

FIGURE 20.1 A simplified routing table.

Routing Design Issues

The first design issue of routing algorithms is *optimality*, as their purpose is to find the optimal (i.e., least-cost) path from source to destination according to a given metric. Paths can be weighted according to a single metric or multiple metrics combined to form a single hybrid one. Metrics give a measure of the overall degree of "goodness" of a path.

The most common routing metric is *path length*. In the simplest case, it is expressed in terms of hop count, that is, the number of hops that a packet must make on its path from a source to a destination. Alternatively, network administrators may assign arbitrary costs (expressed as integer values) to each network link and calculate the path length as the sum of the costs associated with each link traversed.

These costs may be due to several link characteristics, such as:

- *bandwidth;*
- *routing delay,* which gives the length of time required to move a packet from source to destination through the internetwork;
- *load,* which gives the degree of utilization of a router, obtained by monitoring variables such as CPU utilization or packets processed per second;
- *reliability,* which depends on the link's fault probability or recovery time in the event of failure.

Some protocols combine these metrics into a single composite one, which encompasses all these costs according to weighting factors that can be configured by the network administrator.

Different routing protocols generally adopt different metrics and algorithms that are not compatible with each other. In a network where multiple routing protocols are present, it is important to find a way to exchange route information to determine the best path across the multiple protocols.

Administrative distance is the feature used by routers to select the best path when there are two or more different routes to the same destination from two different routing protocols. Each protocol is therefore assigned an administrative distance, that is, an integer value that defines the "trustworthiness" of the protocol. Each routing protocol is prioritized in order of most to least "reliable" using an administrative distance value, and routers will select the route supplied by the protocol with the shortest administrative distance. As an example, Table 20.1 gives the administrative distance assigned to various routing protocols [2]. In some cases, *link monetary cost* also has to be taken into account. For example, companies may prefer to send packets over their own lines, even though slower, rather than through faster but expensive external lines charging money for usage time.

Apart from optimality, another important design issue for routing algorithms is fast *convergence*.

Convergence is the process of agreement, by all routers, on optimal routes. When a router detects a topology change, this information must be propagated through the network and a new routing topology

TABLE 20.1   Administrative Distances for IP Protocols

Protocol	Administrative distance
Unknown	255*
Internal BGP	200
External EIGRP	170
Exterior Gateway Protocol (EGP)	140
Routing Information Protocol (RIP)	120
Intermediate System-to-Intermediate System (IS–IS)	115
OSPF	110
IGRP	100
Internal EIGRP	90
External Border Gateway Protocol (BGP)	20
Enhanced Interior Gateway Routing Protocol (EIGRP) summary route	5
Static route	1
Connected interface	0

*This value indicates that the router does not trust the source of that route and will not insert the route into its routing table.

should be calculated. This is achieved by distributing *routing update* messages to the other routers, thus stimulating recalculation of optimal routes and eventually causing all routers to agree on these routes. The time taken to detect changes in the network topology (such as new routes being added or existing routes changing state) and reconfigure the topology correctly, called *convergence time*, is a very important characteristic of routing algorithms. Slow convergence is undesirable, as it may cause routing loops or network interruption. A routing loop occurs when, due to slow convergence, a packet arriving at a router A or B and destined for a router C bounces back and forth between these two routers until either convergence is reached or the packet has been switched the maximum number of times allowed. Several factors may affect convergence time. Some depend on the network topology and size (e.g., number of routers, link speeds, routing delays), while others are strictly related to the routing protocol adopted and the setting of the relevant timing parameters.

Scalability is also a very important requirement for routing algorithms, especially in large size internetworks. Some routing algorithms behave well in small systems but scale poorly, so they cannot be used in larger internetworks.

Another important issue is *robustness*, that is, the ability of routing algorithms to perform correctly even in the presence of unusual or unforeseen events, such as router failures, misbehavior, or sabotage.

Moreover, routing algorithms should combine *flexibility*, which enables them to adapt to network changes (e.g., in bandwidth, router queue size, network delay) with *stability*, which means that they should not be too sensitive to changes in the network. For example, on a network featuring unstable links, frequent occurrences of changes and subsequent convergences would affect proper network operation.

Last but not the least, routing algorithms should be *simple*, in order to introduce a low overhead on routers (in terms of processing and storage), and *efficient* in resource utilization (especially when running on hosts with limited physical resources).

20.3 Classification of Routing Protocols

Routing protocols may be classified according to several different characteristics [3–5]. Here, we will address the most relevant. As it will be seen in the following, routing protocols may be *static* or *dynamic*, *global* or *decentralized*, *link-state* or *distance-vector*, *single path* or *multipath*, *flat* or *hierarchical*.

Static or Dynamic

Static algorithms are based on table mappings that are fixed by the network administrator. Static routes seldom change, and when changes do occur it is usually a result of human intervention (i.e., editing a router's forwarding table). Thanks to their simplicity, static routing algorithms introduce a low overhead and are suitable for stable environments where network traffic is predictable. Static routing is commonly adopted where there is no need for an alternate path, for example, in permanent point-to-point WAN links to remote sites or dial-up ISDN lines.

Dynamic routing algorithms, on the other hand, automatically generate the routing paths responding to the network traffic or topology changes. When a change occurs, the routing algorithm running on a router recalculates routes, reflects changes in the routing table, and then propagates updates throughout the network, thus stimulating recalculation in the other routers as well.

Dynamic algorithms sometimes have static routes inserted into their routing tables. This is the case, for instance, of *default* routers, to which all traffic should be forwarded when the destination address is unknown (i.e., not explicitly listed in the routing table).

Global or Decentralized

A *global* algorithm makes the routing decision on the basis of complete information about the network, in terms of connectivity and link costs. The calculation of the best path can be up to a single site or replicated over multiple ones.

In a *decentralized* routing algorithm, no site has complete knowledge of the network, and route calculation is iterative and distributed. Each node only knows the status of the links directly connected to it. This information is then distributed to its neighbors, that is, nodes directly connected to it, and this iterative process of route calculations and exchanges enables a node to determine the least-cost path to a destination.

Link-state or Distance-vector

Link-state algorithms (also called Shortest Path First algorithms) compute the least-cost path using complete, global knowledge of the network in terms of connectivity, and link costs. Each router maintains a complete copy of the topology database in its routing table and floods routing information to all the nodes in the internetwork. At the beginning, each router will know only about its neighbors, but it will increase its knowledge through *link state* broadcasts received from all the other routers. The router does not send the entire routing table, but only the portion of the table that describes the state of its own links.

Distance-vector algorithms (also known as minimum hop or Bellman–Ford algorithms) require each router to keep track of its distance (hop count) from all other possible destinations.

Each node receives information from its directly connected neighbors, calculates the routing table, and then distributes the results back to the neighbors. When a change is detected in the link cost from a node to a neighbor, the router first updates its distance table and then, if the change also affects the cost of the least-cost path, it notifies its neighbors.

Distance-vector algorithms are distributed and iterative, as the routing table distribution process goes on until no more exchanges with neighbors occur. Routers can identify new destinations as they come into the network, learn of failures in the network, and calculate distances to all known destinations. Each router on a regular basis advertises all the destinations it is aware of with the relevant distances and sends update messages containing all the information maintained in the routing table to neighboring routers on directly connected segments. Each router can therefore build a detailed picture of the network topology by analyzing routing updates from all other routers. The best route for each destination is determined according to a minimum distance (minimum hop) rule.

Table 20.2 compares link-state and distance-vector routing algorithms.

As in link-state algorithms, all the routers share the same knowledge of the network; they have a consistent view of the best path to a given destination. This would entail a sudden change in the load on the least-cost link, and even congestion, if all the routes decide to send their packets through that link at the same time.

One way of avoiding such *oscillations* would be to ensure that all the routers do not run the algorithm at the same time. However, it has been noted that routers on the Internet can self-synchronize. Even though they initially execute the routing algorithm at the same rate, but at different times, the algorithm execution instance will eventually become synchronized at the routers [6]. To deal with this problem, randomization is introduced into the period between the execution instants of the algorithm at each router.

Single Path or Multipath

Routing protocols may be *single path* or *multipath*. The difference lies in the fact that multipath algorithms support multiple entries for the same destination in the routing table, while single path ones do not. The presence of alternative routes in multipath routing protocols allows traffic to be multiplexed over several circuits (LAN or WAN), thus providing not only greater throughput and topological robustness but also support for load balancing (i.e., splitting traffic between paths that have equal costs). In multipath algorithms, multiplexing may be *packet-based* or *session-based*. In the former case, a round-robin technique is typically used. In the latter case, load sharing is performed on a session basis, typically using a source–destination or destination hash function.

TABLE 20.2   Link-state Vs. distance-vector Routing Algorithms

	Link-state	Distance-vector
Type	Global	Decentralized
Route Advertising	"Tell the world about the neighbors", that is, a router does not sent the entire routing table, but only the portion of the table that describes the state of its own links.	"Tell all neighbors about the world", that is router tend to distribute the entire routing table (or large portions of it) to their directly attached neighbors only
Robustness	More robust, as each router autonomously calculates its routing table	An incorrect node calculation can be spread over the entire network
Convergence	Fast	Slow (routing loops may occur)
Scalability	Good	Poor
Responsiveness to network changes	High	"Good news propagates fast, bad news propogates slowly", that is, a decrease in the cost of the best path propagates fast, while an increase goes slowly.
Message overhead	High: any change in a link cost entails the need to send all nodes the new cost.	Low: when link costs change, the results of the change will be propagatead only if the latter entails a change in the least-cost path for one of the nodes attached to the link.
Implementation complexity	High	Low
Processor and memory requirements	High	Low
Stability	Problematic, due to oscillation in routes	Good

Flat and Hierarchical

Another distinction can be made between *flat* and *hierarchical* routing algorithms. In a *flat* routing algorithm, all routers are peers. Each router is indistinguishable from the other as they all execute the same algorithm to compute routing paths through the entire network. This flat model has two main problems, that is, lack of scalability and poor administrative autonomy. The first derives from the growing computational, storing, and communication overhead that the algorithm introduces when the number of routers becomes large. The second arises from the need to conceal some features of a company's internal network from the outside, a crucial requirement for enterprise networks.

In *hierarchical routing*, logical groups of nodes, called *domains* or *autonomous systems* (ASs'), are defined. A routing domain is a collection of routers that coordinate their routing knowledge using a single routing protocol. An AS is a routing domain that is administered by one authority (person or group). Each AS requires a registered AS number (ASN) to connect to the Internet.

ASs may employ multiple *intradomain* routing protocols internally and interface to other ASs via a common *interdomain* routing protocol.

According to this hierarchy, some routers in an AS can communicate with routers in other ASs, while others can communicate only with routers within their own AS. In very large networks, additional hierarchical levels may exist, with routers at the highest hierarchical level forming a *routing backbone* (as shown in Figure 20.2). Packets from nonbackbone routers are conveyed along the backbone by *backbone routers* until a backbone router connected to the destination AS is found. Then, packets are sent through one or more nonbackbone routers within the AS until the ultimate destination is reached.

As compared to flat routing, a drawback of hierarchical routing is that suboptimal paths may sometimes be found. Nevertheless, hierarchical routing offers several advantages over flat routing. Firstly, the amount of information maintained and exchanged by routers is reduced, and this increases the speed of route calculation, thus allowing faster convergence. Secondly, unlike flat routing, where a single router problem can affect all routers in the network, in a hierarchical algorithm the scope of router misbehavior is limited. This increases the overall network availability. In addition, the existence of boundary

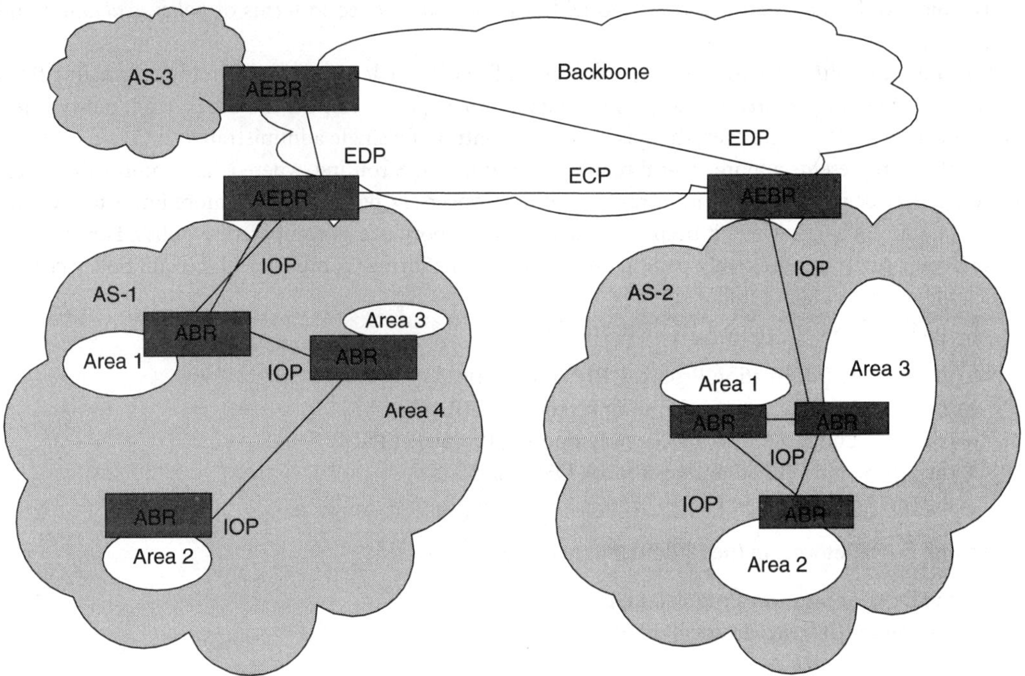

FIGURE 20.2   A Routing Architecture with three ASs connected through backbone routers.

interfaces between different levels in the hierarchy can be exploited to enforce security policies (e.g., access control lists or firewalls) on border routers. Hierarchical routing also improves scalability and protocol upgrades, thus making the task of the network manager easier. Thanks to the above-mentioned advantages, large companies generally adopt hierarchical routing.

20.4   Interior and Exterior Gateway Protocols

A very large internetwork in the IP domain is typically organized as a collection of ASs' (Figure 20.2). An AS is administered by a single entity and has its own routing technology, which may be different for different ASs. An AS can comprise one or more *areas*, composed of contiguous nodes and networks, which may be further split into subnetworks. Within an area, network devices without the capability to forward packets between subnetworks are indicated as *End Systems* (*ESs'*), while network devices equipped with this capability are called *Intermediate Systems* (*ISs'*). ISs' may be further classified into those that can communicate within routing areas only (*intra-area ISs'*) and those that can communicate both within and between routing areas (*inter-area ISs'*).

Figure 20.2 shows a typical structure of a large internetwork, where three ASs (AS-1, AS-2, and AS-3) are connected through a routing backbone to exchange information, while still retaining their own policies and controlling the routing updates received from the other internetworks.

Autonomous System Border Routers (ASBR) on the backbone are entrusted with routing traffic between different ASs, while Area Border Routers (ABR) deal with traffic between different areas within the same AS.

The routing protocol used within an AS is referred to as an *Interior Gateway Protocol* (IGP). A separate protocol is used to interface among the ASs, called the *Exterior Gateway Protocol* (EGP). EGPs are usually referred to as inter-AS routing protocols.

All routers within each AS will run one or more IGPs. Routing information between ASs is exchanged through the routing backbone via an EGP. The use of an EPG limits the amount of routing information exchanged among the three ASs and allows them to be managed differently.

The main differences between IGPs and EGPs can be summarized in terms of *policy, scalability,* and *performance.*

When dealing with inter-AS routing, enforcing policy is crucial. For example, traffic originating from a given AS might be required not to pass through another specific AS. On the other hand, policy is less critical within an AS, where everything is under the control of a single administrative entity.

Scalability represents a more critical requirement in inter-AS routing, where a large number of internetworks may be involved, than in intra-AS routing. Conversely, performance is more important within an AS than in inter-AS, where it becomes of secondary importance as compared to policy. For instance, an EGP may prefer a more costly path to another one if the former complies with certain policy criteria which the second does not fulfil.

Among the IGPs that support IP, there are [7]:

- the Routing Information Protocol, RIPv1 and RIPv2,
- the Cisco Interior Gateway Routing Protocol, IGRP,
- the Cisco Enhanced Interior Gateway Routing Protocol, EIGRP,
- the Open Shortest Path First protocol, OSPF, and
- the Intermediate System-to-Intermediate System protocol, IS–IS.

Among EGP protocols in the IP domain, there are:

- the Exterior Gateway Protocol, EGP and
- the Border Gateway Protocol, BGP.

20.5 Interior Gateway Protocols for IP Networks

In the IP world, interior gateway protocols are used to deal with both intra- and inter-area routings within an AS. RIPv1, RIPv2, and IGRP are distance-vector protocols. OSPF and IS–IS are link-state, while EIGRP is a hybrid. Apart from the two Cisco protocols, all the others are open standards, that is, their specifications are publicly available.

Distance vector IGPs

This section describes and compares two popular distance vector protocols supporting IP, RIP, and IGRP.

RIP

The RIP was one of the first IGPs and has been used for routing computations in computer networks since the early days of the ARPANET. Formally defined in the XNS Internet Transport Protocols publications (1981), its widespread use was favored by its inclusion, as the *routed* process, in the Berkeley Software Distribution (BSD) version of Unix supporting TCP/IP (1982). Two RIP versions exist: Version 1 [8] and Version 2 [9].

In RIP, each router sends a complete copy of its entire routing table to all its neighbors on a regular basis (typical RIP update timer=30 sec). A single RIP routing update contains up to 25 route entries within the AS. Each entry contains the destination address of a host or network, the IP address of the next hop, the distance to the destination (in hops), and the interface. To obtain the cost to a given destination, a router can also send RIP request messages. After receiving an update, a router compares the new information with the information that it already possesses. If the routing update includes a new destination network, it is added to the routing table. If the router receives a route to an existing destination with a lower metric, it replaces the current entry with the new one. If an entry in the update message has the same next hop as the current route entry, but a different metric, the new metric will be used to update the routing table.

If a router does not hear from a neighbor for a given time interval, called a "dead interval" (an *invalid* timer is used for this purpose), it assumes that the neighbor is no longer available (either down or unreachable). As a result, the router modifies its local routing table and then notifies its neighbors of the unavailable route. After another predefined time interval (a *flush* timer is set for this purpose), if nothing is heard from the route, the information is flushed from the router's routing table.

Routers allow for configuration of an *active* or *passive* RIP mode on specific interfaces. The active mode means full routing capability, while the passive mode means a listen-only mode, that is, no RIP updates are sent out.

To speed up convergence, RIP uses *triggered updates*. That is, whenever an RIP router learns of a change, such as a link becoming unavailable, it sends out a triggered update immediately, rather than waiting for the next announcement interval. As it takes time for triggered updates to get to all the other routers in large networks, a gateway that has not yet received the triggered update may issue a regular update at the wrong time, thus causing a bad route to be reinserted in a neighbor that has already received the triggered update.

In order to prevent new routes from reinstating an old link, a *hold-down* period is enforced in the protocol: when a route is removed, no update for that route will be accepted for a given period of time, until the topology becomes stable. Hold-downs have the drawback of slowing convergence.

Besides hold-down, RIP implements other techniques to avoid routing loops between adjacent routers, called *split horizon* with *poison reverse*. The general split horizon algorithm prevents routes from being propagated back to the source, that is, down the interface from which they were learned. As an example, consider the case in Figure 20.3.

During normal operations, router A will notify router B that it has a route to network 1. According to the split horizon algorithm, when B sends updates to A, it will not mention network 1. Now let us assume that router A interface to network 1 goes down. Without split horizon, router B would inform router A that it can get to network 1. Since it no longer has a valid route, router A might select that route. In this case, A and B would both have routes to 1. But this would result in a *circular route*, where A points to B and B points to A.

Using split horizon with poison reverse, instead of not advertising routes to the source, routes are advertised back to the source with a cost of infinity (i.e., 16), which will make the source router ignore the route.

On the whole, RIP is quite robust and very easy to set up. It is the only routing protocol that Unix nodes universally understand and is therefore commonly used in Unix environments. RIP is also commonly used in end system routing as a *dynamic router discovery protocol*. Dynamic router discovery is an alternative to static configurations, which allows hosts to dynamically locate routers when they have to access devices external to the local network.

RIP was designed to work with moderate-sized networks using reasonably homogeneous technology. This makes it suitable for local networks featuring a small number of routers (about a dozen) and links with equal characteristics. RIP has a poor degree of scalability; thus, it is not recommended for use in more complex environments. The RIP protocol cannot be used to build a backbone larger than 15 hops in diameter,[1] as it specifies a maximum hop count of 15. A number of hops equal to 16 correspond to "infinity." This is useful to prevent packets that get stuck in rooting loops from being constantly switched back and forth between routers.

As it uses a simple, fixed metric to compare alternative routes, RIP can generate suboptimal routing tables resulting in packets sent over slow (or costly) links even in the presence of better choices. RIP is therefore not suitable for environments where routes need to be chosen based on dynamically varying parameters such as measured delay, reliability, or load (RIP does not support load balancing).

RIP version 1 (RIPv1) lacks Variable Length Subnet Mask (VLSM) support. The term VLSM refers to a design practice of creating subsubnets in a tree-structured network. In particular, VLSM is the ability to specify a different subnet mask for the same network number on different subnets. More details can be found in [10]. Protocols that do not allow VLSM, because in the update phase they do not propagate the subnet mask and distinguish between address classes, are classified as Classful. Thus, RIPv1 is Classful.

---

[1]RIP is usually configured in such a way that a cost of 1 is used for the outbound link. If a network administrator chooses to use larger costs, the upper bound of 15 can be quickly reached and will become a problem.

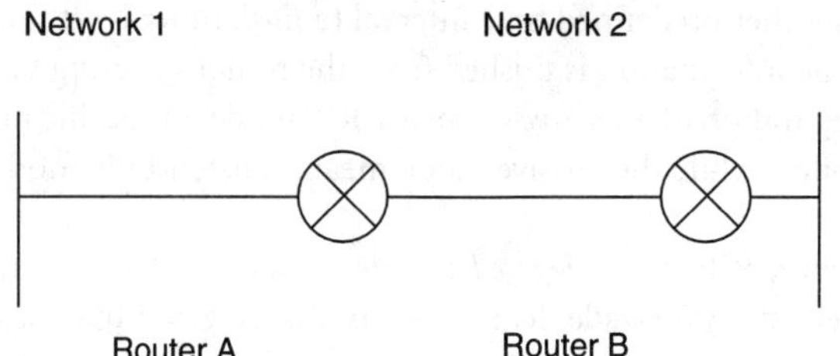

FIGURE 20.3   A simplified internetwork used to explain split horizon with poison reverse.

This may seriously deplete the available address space to the detriment of scalability. On the other hand, protocols that allow VLSM are called Classless. RIP version 2 (RIPv2) is Classless.

RIPv1 uses a broadcast mode to advertise and request routes, while RIPv2 has the ability to send routing updates via multicast addresses. RIPv2 also supports *route aggregation* techniques, that is, the use of a single network prefix to advertise multiple networks [11]. Moreover, RIPv2 provides some support for authorization data link security, implementing authentication support on a per message basis.[2]

RIPv2 also offers some support for ASs and IGP/EGP interaction by means of *external route tags*. External routes are learned from neighbors situated outside the AS, while internal routes lie completely within a given AS. In RIPv2, a Route Tag field is used to separate internal RIP routes from external routes imported from an EGP. Finally, RIPv2 scales better than RIPv1; but as compared to link-state protocols, it suffers from slow convergence and scalability limitations.

Interior Gateway Routing Protocol (IGRP)

The *Interior Gateway Routing Protocol (IGRP)* [12] was created by Cisco in the 1980s. Not being an open standard, it only runs on Cisco routers.

IGRP shares several features with RIP. IGRP sends out updates on a regular basis (every 90 sec) and uses *update, invalid,* and *flush* timers as well (with values different from RIP, depending on the implementation). Like RIP, IGRP uses triggered updates to speed up convergence and hold-down timers to enforce stability. However, while RIP only allows a network diameter of 15 hops, IGRP can support a network diameter of up to 255 hops; thus, it can be used in larger networks.

IGRP differs from RIP with respect to metrics, parallel route support, the reverse poisoning algorithm, and the use of a default gateway. Like RIPv1, IGRP does not support VLSM.

Unlike RIP, IGRP does not use a single metric, but a *vector* of metrics, which takes into account the topological delay time, the bandwidth of the narrowest bandwidth segment of the path, channel occupancy, and the reliability of the path. A single composite metric can be computed from this vector [12], which encompasses the effect of the various components into a single number representing how good a path is. The path featuring the smallest value for the composite metric will be the best path.

The hop count and Maximum Transmission Unit (MTU) that is, the maximum packet size that can be sent along the entire path without fragmentation) of each network are also considered in the best path calculation.

With IGRP, network administrators are supplied with a large range of metrics and they are also allowed to customize them. By giving a higher or lower weight to specific metrics, network administrators can therefore influence IGRP's automatic route selection. For example, suitable constants can be set to several different values to provide different "types of service" (interactive traffic, for instance, would typically yield a higher weight to delay, whereas file transfer would assign a higher weight to bandwidth).

---

[2]A whole RIP entry, denoted by a special Address Family Identifier (AFI) value of 0xFFFF, is used for authentication purposes, thus reducing the total number of routing entries per advertisement to 24.

IGRP is more accurate than RIP in calculating the best path, as a vector of metrics instead of a single metric improves the description of the status of a network.

If, for instance, a single metric is used, several consecutive fast links will appear to be equivalent to a single slow link. While this would be appropriate for delay-sensitive traffic, it would not be so for bulk data transfer, which is more sensitive to bandwidth. This problem may arise with RIP, but not with IGRP, as it considers delay and bandwidth separately.

An interesting feature of IGRP is that it provides multipath routing and can perform load balancing, by splitting traffic equally between paths that have equal values for the composite metrics. To deal with the case of multiple paths featuring unequal values for the composite metrics, a *variance* parameter is defined as a multiplier of the best metrics. Only routes with metrics within a given variance of the best route can be used as multiple paths. In this case, traffic is distributed among multiple paths in inverse proportion to the composite metrics. The variance can be fixed by the network administrator. The default value is 1, which means that when the metric values of two paths are not equal, only the best route will be used. A higher variance value means that any path with a metric up to that value will be considered for routing purposes. However, variance values different from 1 should be carefully managed, as they may cause routing loops.[3]

Unlike RIP, which can only prevent routing loops between adjacent routers, the route poisoning algorithm used in IGRP prevents large routing loops between nonadjacent routers. This is achieved by poisoning routes, whose metrics increase by a factor of 10% or more after an update. The rationale behind this is that routing loops generate continually increasing metrics. It should be noted that if this poisoning rule is used, valid routes may be erroneously deleted from routing tables. However, if the routes are valid, they will be reinstalled by the next update. The advantage of the IGRP poisoning algorithm is that it safely allows a zero hold-down value, which significantly improves network convergence time. IGRP handles default routes differently from RIP. Instead of a single dummy entry for the default route, IGRP allows real networks to be flagged as candidates for being a default. IGRP periodically analyzes all the candidate default routes to choose the one with the lowest metric, which will be the actual default route. This approach is more flexible than the one adopted by typical RIP implementations. The default route can change in response to changes within a network.

A hybrid protocol: the Enhanced Interior Gateway Routing Protocol (EIGRP)

EIGRP was introduced by Cisco in the early 1990s as an evolution of IGRP [13]. As it offers support for multiple network layer protocols (e.g., IP, AppleTalk, and Novell NetWare), EIGRP is commonly used in mixed networking environments.

EIGRP is called a hybrid protocol as it combines a distance-vector routing protocol with the use of a Diffusing Update Algorithm (DUAL) [14], which has some of the features of link-state routing algorithms. The advantages of this combination are fast convergence and lower bandwidth consumption.

DUAL, based on distance information provided by route advertisements from neighbors, finds all loopfree paths to any given destination. Among all the loop-free paths to the destination, the neighbor with the best path is selected as the *successor*, while the others are selected as *feasible successors*, that is, eligible routes to be used if the primary route becomes unavailable. If, after a failure, one feasible successor is found, DUAL promotes it to primary route without performing recalculation, thus reducing the overhead on routers and transmission facilities. If, on the other hand, no feasible successors exist, a recomputation (called a *diffusing computation*) is performed to select a new successor. Unnecessary recomputations affect convergence; thus, they should be avoided.

EIGRP uses the same composite metrics as IGRP. However, it does not send periodic updates, but instead implements neighbor discovery/recovery, based on a hello mechanism, to assess neighbor

---

[3]It should also be pointed out that the load balancing performed by the IGRP may produce out-of-sequence packets. This should be taken into account when neither the Data Link nor Transport layer protocols have the capability of handling out-of-sequence packets.

reachability. Routing updates are only sent in the event of topology changes or a failure in a router or link. Moreover, when the metric for a route changes, only partial updates are sent and they only reach the routers that really need the update. As a result, EIGRP requires less bandwidth than IGRP.

EIGRP relies on the Reliable Transport Protocol (RTP) [15] to achieve guaranteed and ordered delivery of EIGRP packets to all neighbors. It supports VLSM, thus providing more flexibility in internetwork design than RIPv1 or IGRP. This can be particularly useful when dealing with a limited address space.

EIGRP has a modular architecture that makes it possible to add support for new protocols to an existing network. On the whole, EIGRP is robust and easy to configure and use. EIGRP supports both internal and external routes, allowing tags to identify the source of external routes. This feature can be exploited by network administrators to develop their own interdomain routing policies [16,17].

Link-state Protocols

OSPF: The Open Shortest Path First Protocol

OSPF is a link-state IGP designed by the IETF in the late 1980s to overcome the limitations of RIP for large networks. The name of this protocol derives from the fact that it is an Open standard and that it uses the Shortest Path First (also called Dijkstra) algorithm [18,19]. It was specifically designed for the TCP/IP environment and runs directly over IP. OSPF is commonly used in medium-to-large IP networks and is implemented by all major router manufacturers. Several specifications have emerged since the first one [20]. The RFC for OSPF version 2 (OSPF v2) is [21], while OSFP specifications to support IPv6 are given in [22].

One of the most appealing features of OSPF is its support for hierarchical routing design within ASs (although it is purely an IGP). An AS running OSPF can therefore be configured into areas that are interconnected by one *backbone* area (as it is shown in Figure 20.4). There are two types of routing within OSPF, that is, intra- and inter-area routing.

In OSPF, each router monitors the state of its attached interfaces. If a topology change occurs, the router that has detected it distributes information about the change to all the other routers within its area, through broadcast messages called *Link State Advertisements* (LSAs). LSAs include metrics, interface address, and other data. A router uses this information to build a topological database in the form of a direct graph. The costs associated with the various edges (i.e., links) are expressed by a single dimensionless metric configured by the network administrator. A topological database is therefore present in each router.

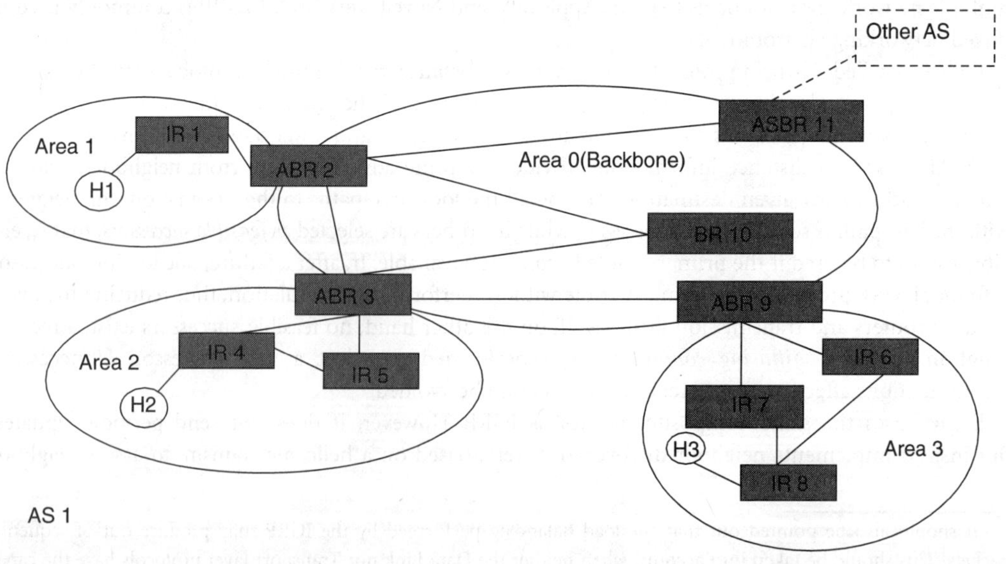

FIGURE 20.4   OSPF Routing Hierarchy.

When two routers have identical topological databases, they become *adjacent*. This is important, as routing information can be exchanged only between adjacent routers. Locally, the router applies a Shortest Path Tree algorithm to all the networks, considering itself as the root node.

As mentioned above, each router sends LSAs only to all the routers in the same area. This prevents intra-area routing from spreading outside the area. Each router within an area will know routes to any destination within the area and to the backbone.

The internal topology of any area is hidden to every other area. Each router within a given area will know how to reach both every other router within its area and the backbone; but it does not have any clue about the number of routers existing or the way they are interconnected for any other area.

Inter-area routing is performed through the backbone (which must be configured as area 0). It should be pointed out that an OSPF backbone is a routing area and not a physical backbone network as the name might suggest. Four different types of router can be configured in OSPF: Internal Routers (IR), ABR, Backbone Routers (BR), and ASBR. All of them are depicted in Figure 20.4.

IRs are only entrusted with intra-area routing. All their interfaces are connected within an area. ABRs route packets between different areas within the AS. They interface to multiple areas (including the backbone); thus, they belong both to an area and the backbone, and maintain topological information about both of them.

BRs belong to the backbone (area 0), but are not ABR; that is, their interfaces are connected only to the backbone (e.g., BR-10). As the backbone operates an area itself, BRs maintain area routing information. ASBRs, also called Boundary Routers, are responsible for exchanging routing information with routers belonging to other ASs (e.g., ASBR-11 in Figure 20.4). They have an external interface to another AS and learn external routes from dynamic EGPs, such as BGP-4 (presented in Section 20.5), or static routes. External routes are passed transparently throughout the AS and kept separate from the OSPF link-state data. External routes can also be tagged by the advertising routes.

A packet to be sent to another area in the AS (inter-area routing) is first routed to an ABR in the source area (intra-area routing) and then routed via the backbone to the ABR that belongs to the destination area. Finally, it will be routed to the ultimate destination.

Route calculation introduces high complexity and, as an OSPF router stores all link states for all the areas, the overhead is high in a large internetwork. On the other hand, OSPF scales well and converges quickly. It also provides several appealing features, such as:

- Security, as all the exchanges between OSPF routers (e.g., LSAs') are authenticated, thus preventing intruders from manipulating routing information.
- Type of Service (TOS) support, as each link may feature different costs according to the traffic TOS requirements.
- Load balancing between multiple equal-cost paths.
- Triggered updates.
- Designated routers: on a LAN, if several routers may be connected, one will be elected as the *designated router* and another one as its backup. The designated router is responsible primarily for generating LSAs for the LAN to all other networks in the OSPF area.
- Explicit support for VLSM and the ability to have discontinuous subnets (i.e., made up of single networks or sets of networks featuring noncontiguous addressing). This feature is very useful in the Internet.
- Tagged external routes; thus, OSPF is capable of receiving routes from and sending routes to different ASs'.
- Route summarization: an area border router can aggregate two subnets belonging to the same area so that only one entry in the routing table of another router will be used to reach both subnets. This minimizes the size of topological database in the routers, thus significantly reducing protocol traffic.

Integrated IS–IS

The *IS–IS* [23] is a link-state protocol. It is an open standard and its name indicates that this protocol is used by routers to talk to each other. Developed for the OSI world, IS–IS was made "Integrated," so that it could route both OSI and IP protocols simultaneously. Similar to OSPF, Integrated IS–IS uses the SPF algorithm and is based on LSAs sent to all routers within a given area and hello packets to check the current state of a router. As it is a link-state protocol, it converges fast, but at the expense of high complexity. Integrated IS–IS supports VLSM, load sharing, and triggered updates. Still not widely deployed at present (it is confined to telco and government networks), Integrated IS–IS has a good potential for medium-to-large IP networks.

20.6 Exterior Gateway Protocols for IP Internetworks

EGPs are responsible for routing between ASs. Possible alternatives in the IP world are static routes, the EGP [24], and the Border Gateway Protocol (BGP) [25].

Static Routes

Static routes can be applied to both intra- and inter-AS routings. However, as they lend particularly appealing features to inter-AS routing, they are included in this subsection. Configuration of static routes is simple, and it is very easy to enforce policy (as no routes equals no access). With static routes, no routing protocol messages travel over the links between ASs. On the other hand, maintenance of static routes in large internetworks may be complex, as they do not scale. For this reason, many network designers adopt the Dynamic Host Configuration Protocol (DHCP) [26], which dynamically allocates a default gateway from a set of candidate gateways. Static routes also lack flexibility (as there is no way to choose a better path that could have been selected if dynamic routing protocols were used); thus, they are not suitable for changing environments. They do not respond to topological changes. To enforce fault tolerance in static routes, a secondary gateway is usually maintained, which could adopt the role of the primary one if the latter becomes unreachable or goes down.

EGP

The EGP [24] was the first EPG to be developed. It runs directly over IP and is a best-effort service. The routing information of EGP is similar to distance-vector protocols; but it does not use metrics. EGP has some design limitations, as it does not support routing loop detection and multiple paths. If more than one path to a destination exists, packets can easily get stuck in routing loops. Nowadays, it has been declared obsolete and has been replaced by the BGP.

BGP

The BGP version 4 (BGP-4), originally specified in RFC 1771 [25], is a very robust and scalable routing protocol, which is becoming *a de facto* standard for inter-AS routing on the current Internet. BGP-4 is an open standard. It somewhat resembles distance-vector protocols, as it is a distributed protocol where information exchange occurs only between directly connected routers. However, it is more appropriate to define BGP-4 as a path-vector protocol. This is because, as stated in [25], "the primary function of a BGP speaking system is to exchange network reachability information with other BGP systems." Network reachability information includes only the list of ASs' that need to be traversed in order to reach other networks. *Path information*, instead of *cost information*, is exchanged between neighboring BGP routers, and BGP-4 does not specify the rule to choose a path from among the advertised ones. Routing mechanism and routing policy are therefore separated. A policy is manually configured in order to allow a BGP router to rate possible routes to other ASs and choose the best path.

Two BGP routers first have to establish a TCP connection. After a negotiation phase, in which the two systems exchange some parameters (such as BGP version number, AS number, etc.), they become *BGP peers* and can start exchanging information. Initially, they exchange the full routing tables. Thereafter, only incremental updates are sent, when some change in the routing tables occurs. No periodic refresh of the entire routing table is needed; thus, a BGP speaker maintains the current version of the entire BGP routing tables of all of its peers for the duration of the connection. In order to maintain the connection, peers regularly exchange *keep-alive* messages. Notification messages are sent in response to errors or special conditions. In the event of error, a notification message is sent and the connection is closed.

Network reachability information is used to construct a graph of AS connectivity, from which routing loops may be pruned and policy decisions at the AS level may be enforced. Each BGP router maintains a routing table with all feasible paths to a given network.

As mentioned above, BGP does not propagate cost information, but path information, and does not specify which path should be selected from those that have been advertised, as this is a policy-dependent decision left to the network administrator. This is because, as was previously stated (in Section 20.3), when dealing with inter-ASs routing, policy is more important than performance.

However, when a BGP speaker receives several updates describing the best paths to the same destination, it should finally choose one of them. The selection criteria are based on rules that take *path attributes* into account. Once the decision has been made, the speaker places the best path in its routing table and then propagates the information to its neighbors. The path attributes generally used in the route selection process are [7,27,28]:

Weight: A Cisco-defined attribute that is local to a router and is not advertised to neighboring routers. If the router learns about more than one route to the same destination, the route with the highest weight will be preferred.

Local preference: Used to prefer an exit point from the local AS. It is propagated throughout the local AS. If there are multiple exit points from the AS, the local preference attribute is used to select the exit point for a specific route.

Multi-exit discriminator, or metric attribute: Used as a suggestion to an external AS regarding the preferred route into the AS that is advertising the metric.

Origin: Indicates how BGP learned about a particular route. The origin attribute can have one of three possible values: *IGP* (i.e., the route is interior to the originating AS), *EGP* (i.e., the route is learned via the Exterior Border Gateway Protocol (eBGP)), *Incomplete* (i.e., the origin of the route is unknown or learned in some other way — this occurs when a route is redistributed in BGP).

AS_path: When a route advertisement passes through an autonomous system, the AS number is added to the ordered list of AS numbers that the advertisement traversed.

Next hop: The IP address that is used to reach the advertising router.

Community: Provides a way of grouping communities, that is, destinations to which routing decisions can be applied. Predefined community attributes are *no-export* (which means that this route must not be advertised to EBGP peers), *no-advertise* (which means that this route must not be advertised to any peer), and *Internet* (which means that this route must not be advertised to the Internet community — all routers in the network belong to it).

Attributes are crucial to achieve scalability, define routing policies, and maintain a stable routing environment. Other BGP features include route filtering, that is, the ability of a BGP speaker to specify which routes to send and receive from any of its peers. Filtering may refer to inbound or outbound links and may be applied as permit or deny.

BGP-4 supports supernetting or Classless Interdomain Routing (CIDR) [29], which enables route aggregation, that is, the combination of several routes within a single-route advertisement. This minimizes the size of routing tables and protocol overhead.

BGP was originally designed to perform inter-AS routing; but it can also be used for intra-AS routing. BGP connections between ASs are called External BGPs (e-BGP), while those within an AS are called Internal BGPs (i-BGP).

Figure 20.5 shows a typical scenario for e-BGP, that is, a multi-homed AS connected to the Internet. All the networks X, Y, Z, 1, 2, and 3 are ASs. In particular, 1 and 3 are stub networks, 2 is a multi-homed stub network, while X, Y, and Z are backbone provider networks. A stub network is such that all traffic

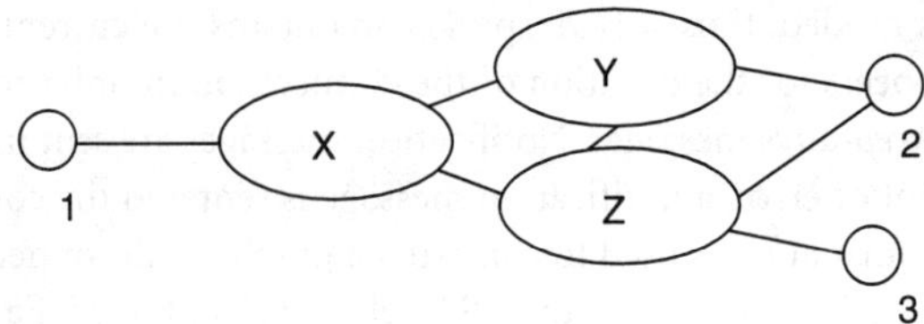

FIGURE 20.5   A scenario for BGP.

entering it should be destined to that network. A multi-homed AS is connected to multiple ASs (e.g., via two different service providers[4]), but does not allow transit traffic.

Stub network 2 will be prevented from forwarding traffic between Y and Z by a selective route advertisement mechanism. BGP routes will be advertised in such a way that network 2 will not advertise to its neighbors (Y and Z) any path to other destinations except itself. For instance, 2 will not advertise path 2 Z 3 to network Y; thus, the latter will not forward traffic to 3 via 2.

The i-BGP protocol can be used to distribute routing information to routers within the AS regarding destinations (networks) outside the AS. Running i-BGP offers several advantages, such as a consistent view of the AS to external neighbors, more control over information exchange within the AS, flexibility, and a slightly shorter convergence time than e-BGP (which is slow).

Table 20.3 (a) and (b) lists the various IGPs and EGPs discussed till now and compares them according to multiple design criteria. Table20.3 can be helpful to guide the designer's choice when dealing with routing protocols in the IP domain.

We have seen that various IGPs and EGPs exist in the IP domain, among them, proprietary protocols and open standards. In a composite, complex internetwork, they may coexist either for historical reasons or due to the presence of multiple vendor solutions. This coexistence may cause incompatibility problems due to both the different metrics adopted and peculiarities of the various protocols. Therefore, a way to overcome incompatibilities and form a holistic view of the network topology is needed, thus enabling interoperability. This is the case for *route redistribution*.

Route redistribution is the process that enables routing information from one protocol to be translated and used by a different one. A router can therefore be configured to run more than one routing protocol and redistribute route information between the two protocols. Redistribution is needed in two cases:

- at the AS boundary, as IGPs and EGPs do not match and
- within an AS, when multiple IGPs are used.

Redistribution has direction. That is, routing information can be redistributed symmetrically (mutual redistribution) or asymmetrically (hierarchical redistribution). Handling redistribution is not an easy task and it is difficult to define a general approach. Solutions are typically vendor-dependent, as the problems that can occur are strictly related to the details of the various protocols.

The following section presents a case study, that is, a global IP network for a large-scale mission-critical enterprise, and gives practical guidelines related to route redistribution in the considered scenario.

20.7   Case Study: A Global IP Network for a Large-scale Mission-Critical Enterprise

This section addresses IP routing technologies and design criteria to be adopted in designing a global IP network for a large-scale mission-critical enterprise. Mission-critical means that high availability (uptime 100%), uninterruptible service, and fault-tolerance (i.e., redundancy, fast, and reliable recovery) must be

---

[4]In this situation, multiple service providers are used to increase availability and to allow load sharing.

TABLE 20.3   (a) and (b) IP routing Protocol Comparison According to Multiple Design Criteria

Protocol		Technology	Type	Metrics	Scalability	VLSM support	Hop Count limit	Load Balance (equal paths)	Load balance (Unequal paths)	Standard	Convergence time	Routing Algorithm
	IGRP		Classful	Band-width, delay, load, reliability, MTU	Medium	No	100 (up to 255)	Yes	Yes	No	Slow	Bellman-Ford
	RIPv1	Distance Vector	Classful	Hop Court	Small	No	15	yes	No	Yes	Slow	Bellman-Ford
IGP	RIPv2		Clasless	Hop court	Small	yes	15	Yes	No	Yes	Slow	Bellman-Ford
	RIGRP	Advanced Distance Vector	Classless	Band-width, delay, load, reliability, MTU	Large	Yes	100 (up to 255)	Yes	Yes	No	Fast	Dual
	OSPF	Link State	Classless	Cost	Large	Yes	200	yes	No	Yes	Fast/Slow	Dijkstra
	IS-IS	Link State	Classless	Clost	Very large	Yes	1024	Yes	No	Yes	Fast/Slow	IS-IS
EGP	BGP-4	Path Vector	Classless	Cost, Hop, Policy	Large	Yes		No	No	Yes	Slow	

provided to guarantee business continuity. This scenario is related to a global-independent manufacturing company, which is a leader in developing and delivering semiconductor solutions across the spectrum of microelectronics applications. The company extends over three macro-regions — Europe, North America, and the Asia-Pacific region. The corporate headquarters are in Europe, with backup from the U.S. headquarters. The main servers are located in both headquarters.

To support the growing demand for data and streaming media, the company chose IP protocols as the basis for voice, video, and data convergence. That is, unlike past solutions, where different protocols were used to support voice and data traffic, respectively, nowadays, IP is the single technology used for all the different kinds of traffic present in the considered scenario.

The architecture of the global IP network system (the infrastructure) has been defined bearing in mind the need for the enterprise to have a supporting network frame capable of providing very high service levels in terms of the available transmission bandwidth, business continuity (for mission-critical environments), and, more generally, the quality of service that users are provided with.

Also, possible new needs that the enterprise may have, with respect to emerging technologies and applications and, above all, in terms of changes in traffic flows, that are anticipated and/or desirable, have been taken into account in the design phase.

The realization expects to implement a high-performance communication architecture at both the backbone and single-area (up to workgroup) level, with a fundamental aspect of redundancy at the single device and system architecture levels.

Infrastructure Requirements for a Large-scale IP Network

From a design viewpoint, analysis of the company network requires an approach oriented toward scalability and high-performance levels that will be adequate for a large number of users (servers and clients) spread over a vast geographical area.

The architecture of this global IP network should be able to provide a large-scale, high-available/reliable, *carrier-grade* infrastructure. The concept of large-scale availability and reliability means that the reachability of an IP packet is guaranteed by improving the reliability and performance of the routing and forwarding functions on the network. Carrier grade means that support of the global IP infrastructure is outsourced to large-scale networks provided by specialized carriers with mechanisms that guarantee Quality of Service (QoS), Service Level Agreement (SLA), redundancy links, etc. [30].

To increase the size of the IP network, it is necessary to improve the switching capacity of the routers and reduce both the time required to search for an IP address in a router and the traffic involved in IP routing between the various routers. In addition, a sophisticated quality monitoring and control mechanism is essential for a real-time infrastructure supporting a "best-effort" approach toward the data and services to be exchanged.

A basic issue in designing this infrastructure is identifying the IP routing technology to support a real multi-service network providing convergence with respect to data, applications, and services.

IP Routing Technologies: Backbone level (Company-wide)

Figure 20.6 shows a scheme of the network infrastructure devised for the large-scale company considered in this case study. It comprises three geographically distributed macro-regions (Europe, America, Asia), each equipped with an ABR gateway and belonging to the same backbone, that is, Area 0. The common layer 2 is Frame Relay [31]. The Company backbone consists of a large set of hosts, routers, and networks under the same administration, which has consistent and complete information about all destinations.

The company infrastructure is an AS Confederation (according to the *divide and conquer* strategy, a large AS is split into sub-ASs), composed of Area 0 (Backbone) and Areas 1,2,..,*n*, where each subarea is a Site of the Company. Sub-AS may use multiple IGPs and metrics, but appear as single AS to other ASs. Internet connections occur only through ASBR routers (Internet is a special kind of AS, which is named "Core AS").

The traffic types throughout the Backbone are divided into:

- *Local,* that is, traffic originating or terminating at each AS and
- *Transit,* that is, nonlocal traffic.

Figure 20.6 also shows a Partner Gateway link toward a partner's extranet and mobileusers' application using Internet Core as the common infrastructure.

The scheme in Figure 20.6 shows the use of dynamic routing protocols that exchange routing information through neighboring routers. The Company Corporate decision is to have an IP infrastructure in which routing policies are transparent to users spread over a wide geographical area with respect to the servers (e.g., mobile users). In general, it will use EIGRP and RIP in a site AS and OSPF as dynamic routing between the various sites of the regions.

The choice of OSPF as the intra-Company and inter-Area Border routing protocol is based on the following considerations:

- Cost of the transmission bandwidth and consequently limited throughput capacity.
- Congestion and delay.
- High network latency.
- Large number of addressable subnetworks.
- Multivendor environments.

OSPF is highly suitable for this context as it converges fast with a low workload as compared with the amount of data transmitted per time unit. It is also standard and therefore does not require homogeneous environments (unlike EIGRP, which requires CISCO in each of the areas indicated).

In addition, OSPF provides fast, efficient addressing of a number of subnetworks as it uses the concept of route summarization (which cannot be implemented in RIPv1).

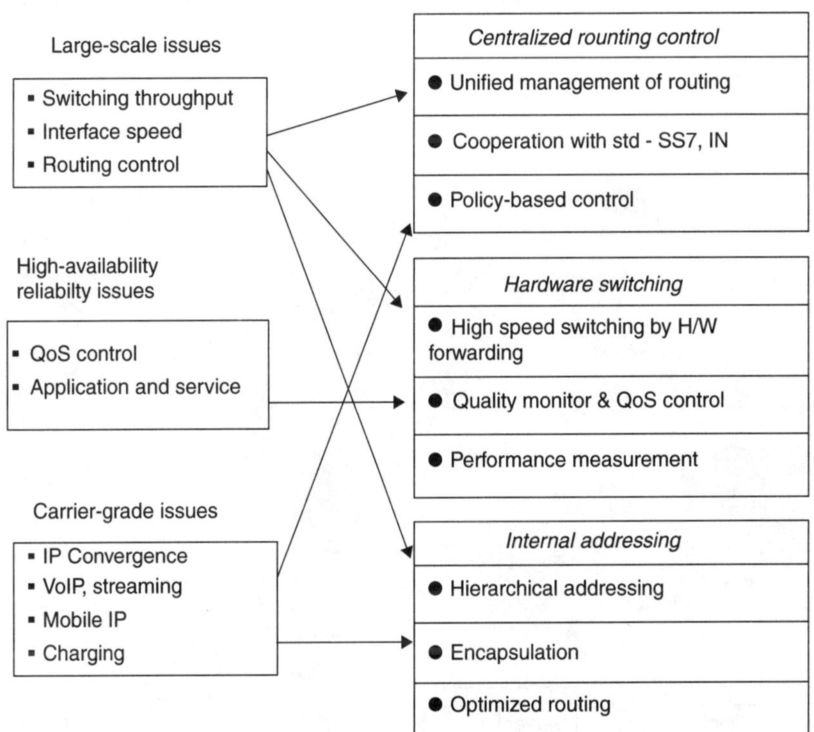

FIGURE 20.6   Large-scale Company IP infrastructure.

Because OSPF supports VLSM, it can be really developed into a true hierarchical addressing scheme. This hierarchical addressing results in very efficient summarization of routes throughout the network. Discontinuous subnets (i.e., composed of single networks or sets of networks featuring noncontiguous addressing) are also supported by OSPF because subnet masks are part of the link-state database [32].

Internet access is only possible through ASBR gateways, by means of a BGP routing, by which the routing tables toward the intranet will be redistributed (see the section on Redistribution of routing tables). Like OSPF and EIGRP, BGP exchanges variations in topology with neighbors (incremental update) in relation to the associated metrics (BGP AS number, path attributes, etc.). A substantial difference lies in the fact that although the ASBR routers of the single ISPs adopt a dual homing mechanism (i.e., redundant links), BGP does not implement load-balancing between the two links, because it always chooses a single path (i.e., the best route) to route data.

Routing Technologies in an AS Domain (Company Site)

Let us consider a single Company Site, such as a campus network model representing a Manufacturing Site, where we distinguish between three large macro-levels related to different application domains that have been identified as Manufacturing, CAD, and Business (Figure 20.7).

We also define three levels of access to the single domain network (Figure 20.8):

- Access layer, hosting only clients (PCs, printers, workstations, etc.).
- Distribution layer, hosting servers connected through an optical fiber connection and implementing the intra-domain routing engine and layer 3 switching (CISCO switch layer 3 [33]).
- Backbone level, which represents the high-throughput core of the infrastructure implementing inter-domain routing.

Access devices have the task of concentrating traffic coming from end users into Ethernet 10/100 technology and then sending it toward the backbone level, which has the task of distributing the traffic toward other networks, both logical VLANs (Virtual LANs) and physical ones, inside and/or outside the campus itself.

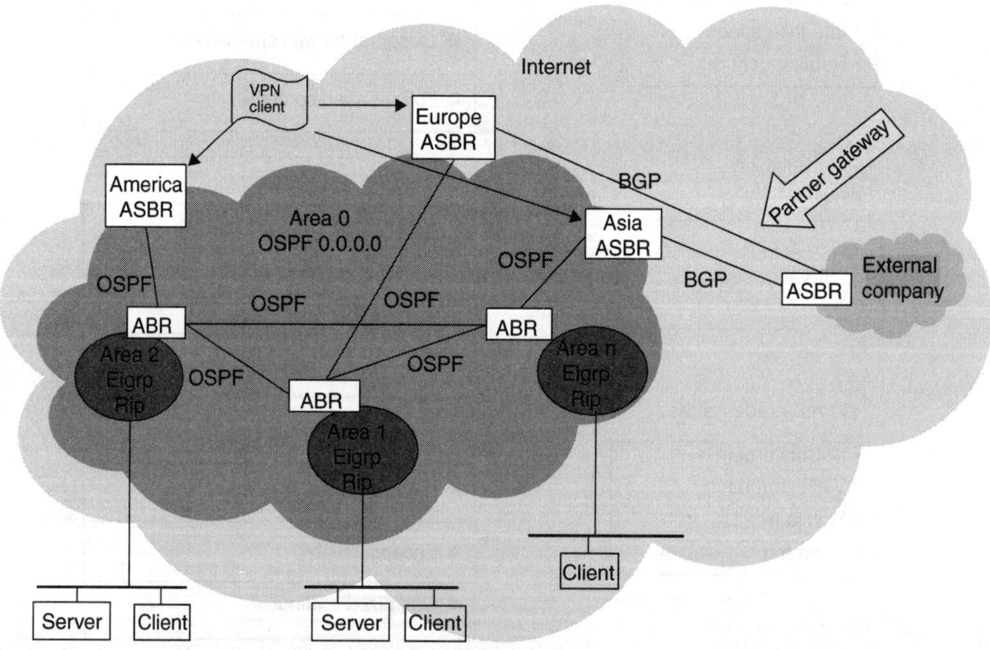

FIGURE 20.7   Large-scale Company IP Infrastructure.

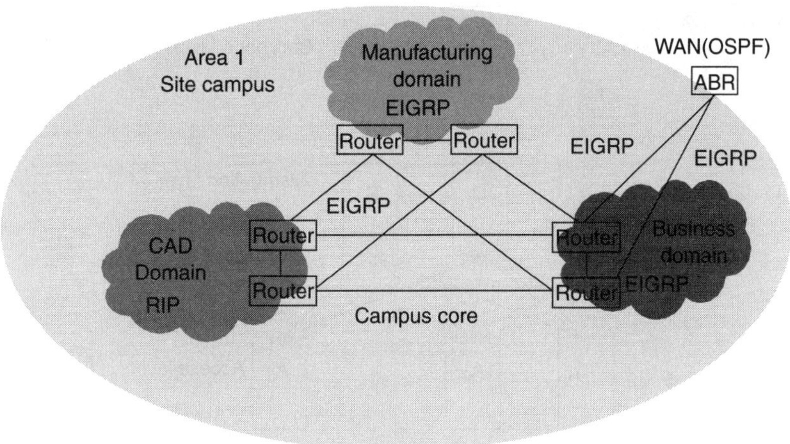

FIGURE 20.8   Company Site IP Infrastructure.

Access devices are planned to be connected in dual homing (with a double LAN connection to two different networks), through an optical fiber connection, a high-throughput (1000 Mbps) Layer 3 Switching device making up the backbone level of the network frame structure. In addition, to ensure maximum continuity of service required for mission-critical environments, they are designed to be completely redundant in power supply and matrix switching.

Both Access level devices and those of the Distribution/Backbone level are endowed with functions allowing completely automatic and simultaneous management of both load sharing on available links and backup functions, with the consequent use of the only link left available in the event of a fault in the other link.

The Distribution/Backbone nodes inside the campus are interconnected by multiple high-throughput links. Together, the nodes will achieve a throughput of more than 4 million packets per second (pps) on the two interconnected devices, which will reduce to two million pps in the event of a complete fault in one of the two devices, without, however, jeopardizing the normal functioning of the network. They have the task of gathering the traffic generated by all the switches present at the access level, concentrating the servers located in the server farm and connecting the campus LAN with the routers needed to guarantee access to a wide area network service, thus allowing interconnection with other sites and/or other organizations. So a high level of performance is available to enable a large amount of data traffic to transit through the campus network (Figure 20.9).

The Backbone devices are therefore capable of forwarding the traffic generated at both level 2 and level 3 (according to needs), using common transmitting/routing protocols. Multi-protocol routing is also allowed by the routers.

The Backbone nodes are required to be completely redundant at the level of both the HW components and the SW functions and at the network architecture level.

Choice of Routing Protocols

In a hybrid environment like a manufacturing site, in terms of the applications and services used, the choice of one routing protocol rather than another may have a considerable impact on performance and efficiency. Thus, the choice of one information routing mechanism rather than another is significant.

In a CAD environment, where the main elements on the network are hosts whose traffic is mostly confined to the environment itself (80% internal traffic and 20% external traffic), it is possible to choose RIP: even if updating due to link faults is not timely, this will not affect normal activity. In some cases of satellite sites, even static routes are used. In environments like Manufacturing and/or Office Automation, on the other hand, where the traffic proportions are 20/80 (20% internal traffic and 80% external traffic) and where convergence times and performance levels have a significant impact on network performance, it is necessary to choose OSPF or, if the hardware architecture allows it, EIGRP.

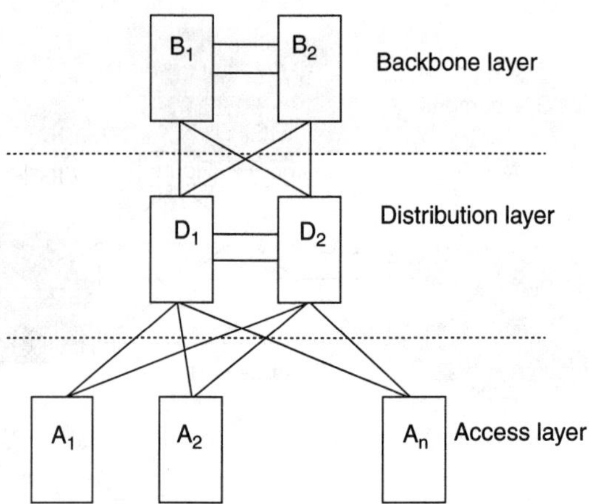

FIGURE 20.9   Manufacturing Domain: Three-level LAN access Infrastructure.

It should, however, be pointed out that choices at the site level are usually homogeneous. If we assume Cisco to be the standard, EIGRP is applicable.

In any case, these three domains can be interconnected via OSPF rather than EIGRP, given the redundancy between the links and the traffic load balancing.

Another difference affecting the choice of one protocol rather than another is the possibility of subdividing networks by using VLSM. In our context, with the exception of the CAD environment, the subdivision into subnets improved the management of addressing and the subdivision of the broadcast domains according to the working environments.

Another decisive factor in choosing a routing protocol is convergence time. This is at the basis of the concept of high availability, which is of great significance in an area like manufacturing, where mission-critical applications require fault tolerance and redundancy. The introduction of QoS policies has to integrate routing policies to support fast convergence of the network topology following a fault or a Spanning Tree event, considering that the convergence time is mainly affected by the update mechanism (hold-down timers, like in the case of DV protocol — RIP), size of topology table, route calculation algorithm, and media type.

In our example, there is also a need to use multiple routing protocols. There are several reasons for this:

- Existing UNIX servers using RIP.
- Multivendor interoperability, as some department may not want to upgrade its non-Cisco routers.
- Groups do not want to work and play nicely with others, etc.

In these cases (as well as in IGP–BGP interaction), *route redistribution* is needed.

Route Redistribution — Implementation Considerations

Redistribution is performed by boundary routers connecting different ASs to exchange and advertise routing information received from one AS to another AS.

Within each AS, the internal routers have complete knowledge of their network. The router interconnecting ASs is the ASBR. The ASBR must have each single routing process as active.

Redistribution is powerful, but increases the complexity of IP routing; key features that may cause complexity are:

- Routing feedback: similar to the routing loop that occurs in distance vector protocols.
- Incompatible routing information (e.g., RIP uses hops and OSPF uses costs): path selection using the redistributed route information may not be optimal.
- Inconsistent convergence time: different routings converge at different rates.

A common global approach to implementing redistribution is how the routers select the best path when more than one IP routing is running. Cisco, for example, uses two parameters:

- Administrative distance, which addresses the trustworthiness of a routing protocol (as seen in Section 20.1).
- Routing metrics, a value representing the path between a local and a remote router.

Once a reliable protocol for a specific area and for each specific destination has been identified, the routing tables are filled with the single routes. This happens for each protocol. The router only distributes the routes from one protocol to another if both have been configured with the respective metrics. Some protocols, for example, IGRP and EIGRP (Cisco proprietary), do this automatically; but this is not the case between RIP and OSPF. It should also be pointed out that only protocols belonging to the same stack can be redistributed (e.g., TCP/IP): if the metrics are configured, RIP and OSPF can redistribute, for example, because they both support TCP/IP.

A significant implementation issue is identification of the core and edge between protocols. In our campus example, EIGRP is the core.

Edges are both the router toward the WAN, which redistributes using OSPF and the routers toward the CAD domain, which redistribute using RIP.

Redistribution: EIGRP vs. RIP

This is the case of the router that in Figure 20.7 connects two networks belonging to CAD and Business domain: the first one uses RIP and the second one uses Enhanced IGRP, respectively. RIP converges slower than EIGRP; thus, if a link goes down, the EIGRP network will learn of it before the RIP network.

The goal of this router is to advertise RIP routes in the Enhanced IGRP network and to advertise Enhanced IGRP routes in the RIP network, while preventing the occurrence of route feedback, (i.e., the router must be configured so that Enhanced IGRP does not send routes learned from RIP back into the RIP network and so that RIP does not send routes learned from Enhanced IGRP back into the Enhanced IGRP network). The redistributed EIGRP router configuration command specifies that routing information derived from Enhanced IGRP be advertised in RIP routing updates. The default-metric router configuration command causes RIP to use the same metric value for all routes obtained from Enhanced IGRP. A default metric helps to solve the problem of redistributing routes that have incompatible metrics. Whenever metrics do not convert, using a default metric provides a reasonable substitute and enables the redistribution to proceed.

Redistribution: IGP vs. BGP

Depending on the mechanism used to propagate BGP information in a given AS, special care must be taken to censure consistency between BGP and IGP. There may, in fact, be time windows between the instant at which a border gateway (A) receives new routing information, originating from another border gateway (B) in the same AS, and the instant at which the IGP in the AS is able to route the routing traffic to BGP (B). During this window, "black holes" may be created. To minimize this problem, BGP (A) should not announce a path to an external network X to all adjacent BGPs in other ASs until all the internal BGPs in the AS are ready to route the traffic destined toward X through the correct border gateway (B). The best choice is OSPF vs. BGP rather than EIGRP vs. BGP, due to the better stability of OSPF at Area 0. EIGRP, in fact, transmits any variation (when it occurs) from every single host across the network due to the lack of areas' definition at the protocol level. It may severely impact the performance. In case of EIGRP at the backbone level, it is mandatory to distribute the only summary toward the BGP. Table 20.4 summarizes qualitative assessments to be taken into account while redistributing from/to AS (IGPs') vs. i-BGP and e-BGP.

TABLE 20.4　IGP and EGP Redistribution

IGP vs. EGP		EGP vs. IGP	
Interior	Exterior	Interior	Exterior
Automatic neighbour discovery	Specifically configured peers	Carries ISP infrastructure addresses only	Carries customer prefixes
Generally trust your IGP routers	Connecting with outside networks	ISPs aim to keep	Carries Internet
Prefixes go to all IGP routers	Set administrative boundries	The IGP small for efficiency and	Prefixes
Binds routers in one AS together	Binds AS's together	scalability	EGPs are independent of ISP network topology

General Guidelines

In the following, useful guidelines to be taken into account in the routing design phase are listed:

- Decisions based on facts, that is, on knowledge of the data flow mechanisms in the company.
- No overlapping of routing protocols in the same domain, if it is Manufacturing or Business. The CAD domain may implement RIP and EIGRP at the same time.
- Redistribution in a single direction (preferable): this avoids loops and convergence problems due to different convergence times. This can be achieved by configuring a default route.
- Redistribution in two directions: if this is necessary, techniques to reduce the risk of loops must be applied, for example, routing table filters (i.e., limits on the path a packet can take), or modifications to the values the metric can take, etc. This prevents changes in routing tables imported from an AS from being reinserted into the same AS as if they were new routing tables. This point is of strategic importance in our example, where the geographical development of the company (WAN) uses lines whose cost is proportional to bit/s transmitted. In this way, one can avoid constant circulation of information regarding routing table alignment, which generates useless and expensive traffic.

References

[1]　Postel, J., Internet Protocol, RFC 791, Sept.1981.
[2]　Cisco Systems Homepage, www.cisco.com.
[3]　Cisco Systems, Inc., *Internetworking Technologies Handbook*, Cisco Press, Indianapolis, 2003.
[4]　Kenyon, T., *Data Networks,*Digital Press, Belford, MA, 2002.
[5]　Kurose, J.F. and K. Ross, Computer Networking, Addison Wesley, Reading, MA, 2001.
[6]　Floyd, S. and V. Jacobson, Synchronization of periodic routing messages, *IEEE/ACM Transactions on Networking*, 2, 122–136, 1997.
[7]　Lewis, C., *Cisco TCP/IP Routing Professional Reference*, The McGraw-Hill Companies, Inc., New York, 1999.
[8]　Hedrick, C.L., RFC 1058- Routing Information Protocol, June 1988.
[9]　Malkin, G., RFC2453/ STD0056- RIP Version 2, Nov. 1998.
[10]　Pummill, T., and B. Manning, RFC 1878 — Variable Length Subnet Table For IPv4, Dec. 1995.
[11]　Chen, E. and J. Stewart, RFC 2519 — A Framework for Inter-Domain Route Aggregation, Feb.1999, ftp://ftp.rfc-editor.org/in-notes/rfc2519.txt
[12]　An Introduction to IGRP, http://www.cisco.com/warp/public/103/5.html
[13]　Cisco Systems Inc., Enhanced IGRP, http://www.cisco.com/univercd/cc/td/doc/cisintwk/ito_doc/en_igrp.htm
[14]　Garcia-Luna-Aceves, J.J., Loop-free routing using diffusing computations, *IEEE/ACM Transactions on Networking,*1,130–141, 1993.
[15]　Schulzrinne, H., S. Casner, and R. Frederick, RFC RFC3550- RTP: A Transport protocol for RealTime Applications, July 2003.
[16]　I. Pepelnjak, *EIGRP Network Design Solutions*, Cisco Press, Indianapolis, 2000.

[17]　Integrating Enhanced IGRP into Existing Networks, http://www.cisco.com.

[18]　Perlman, R., *Interconnections: Bridges and Routers,* Addison-Wesley, Reading, MA, 1992.

[19]　Dijkstra, E.W., A Note on Two Problems in Connexion with Graphs, *Numerische Mathematik,* 1, 269–271, 1959.

[20]　Moy, J., RFC 1131-OSPF specification, Oct. 1989.

[21]　Moy, J., RFC 2328- OSPF Version 2, Apr. 1998.

[22]　Coltun, R., D. Ferguson, and J. Moy, RFC 2740- OSPF for IPv6, Dec. 1999.

[23]　Callon, R.W., RFC 1195 —- Use of OSI IS-IS for routing in TCP/IP and dual environments, Dec.1990.

[24]　Rosen, E.C., RFC 0827-Exterior Gateway Protocol, Oct.1982.

[25]　Rekhter, Y. and T. Li, RFC 1771 — A Border Gateway Protocol 4 (BGP-4), Mar.1995.

[26]　Droms, R., RFC 2131 — Dynamic Host Configuration Protocol, Mar. 1997.

[27]　Halabi, B. and D. McPherson, *Internet Routing Architectures,* Cisco Press, Indianapolis, 2000.

[28]　Huitema, C., *Routing in the Internet,* Prentice Hall, Englewood cliffs, NJ, 1995.

[29]　Fuller, V., T. Li., J. Yu, and K. Varadhan, Classless Inter-Domain Routing (CIDR): an Address Assignment and Aggregation Strategy, RFC 1519, Sept. 1993.

[30]　Hitachi Review, Telecom 99, Vol.48-No.4, Aug. 1999.

[31]　RFC2427/STD0055, Brown, C., and A. Malis, Multiprotocol Interconnect over Frame Relay, Sept.1998.

[32]　Moy, J.T., *OSPF -Anatomy of an Internet Routing Protocol,* Addison-Wesley, Reading, MA, 1998.

[33]　Ballew, S.M. and M.Loukides, *Managing IP Networks with Cisco Routers,* O'Reilly & Associates, Inc., Sebastopol, CA, 1997.

21

Multicast

Tanja Zseby
*Fraunhofer Institute for Open
Communication Systems (FOKUS)*

21.1 Introduction

A wide variety of upcoming Internet applications like audio/video conferencing, multiplayer games, database replication, etc. require group communication. Group communication denotes the communication between more than two end systems. The realization of group communication with multiple point-to-point communication channels is possible but extremely inefficient. Especially, the communication within large and widely distributed groups would lead to overload in network components and end systems. IP multicast provides an efficient way to distribute data to multiple receivers. The basic principle is that the sender sends only one packet, which is then replicated only at junction points on the way to the receivers. With this, it is possible to support applications that rely on group communication without wasting resources in the network and end systems.

21.2 Addressing Principles

Different applications require different kinds of group communication. The following types of group communication can be distinguished (Figure 21.1):

In *one-to-many* communication, a single sender sends data to multiple receivers. One-to-many communication is required, for instance, for audio/video distribution (e.g., TV, radio, remote lectures), file distribution, and caching applications and any kind of distribution or publishing of information to a group of receivers (headlines, network time, stock prices).

In *many-to-many* communication, multiple senders send data to multiple receivers. Many-to-many communication is needed, for instance, for interactive applications like multimedia conferencing, for the synchronization of distributed data bases, for parallel processing, file sharing, interactive simulations, and multi-player games.

In *many-to-one* communication, multiple senders send data to one receiver. This is useful, for instance, for data collection applications (like accounting), for the discovery of resources and for auction and polling applications.

In order to realize the different types of group communication, the following basic addressing principles exist in the Internet:

FIGURE 21.1   Group communication types.

Unicast addresses address one specific receiver. They are used to establish the communication between two endpoints. Most current Internet applications rely on unicast connections.

Broadcast addresses address all reachable stations, for instance, all systems on a local subnet. Broadcast is natively supported in some LAN technologies (e.g., Ethernet). Some protocols use this broadcast functionality to push information to all connected stations.

Multicast addresses address a group of receivers. Multicast is also natively supported in some LAN technologies (e.g., Ethernet). Multicast can be seen as the most generic addressing principle because unicasting can be realized by multicasting to a group with only one member and broadcasting can be realized by multicasting to a group with all stations as members.

Anycast addresses address one station out of a group. In anycast, the first suitable station answers the request. This is a special addressing principle, which is useful, for instance, to find the nearest station that offers a specific service.

21.3   IP Multicast

It is possible to send the same data to a group of receivers using unicast or broadcast. With unicast, the sender can simply send a duplicate of the IP packet to each receiver. This method is very inefficient, especially if the group is large. The sender has to send multiple copies of the same IP packet. Furthermore, equal copies of the same packet travel over the same network links. This leads to unnecessary load on the sender in terms of the network connections (Figure 21.2).

One can also address a group by broadcasting. The IP packet would simply be sent to all reachable receivers. Receivers not interested in the packet just need to discard the packet. This method also leads to unnecessary load on the network and on the receivers who are not members of the group.

IP multicast provides a method that prevents unnecessary resource consumption in the network and in end systems. The sender sends only one IP packet. The packet is then replicated only at the junctions on the way to the receivers. Packet replication takes place in the routers that are the junction points of the distribution tree. This method significantly reduces the required resources in the network and at end stations (Figure 21.2).

IP Multicast communication is unidirectional and runs over UDP, which means there is no flow control available. IP multicast allows a *dynamic membership* [RFC1112], that is, hosts may join and leave groups at any time. Furthermore, there is no restriction on the *location of members* and on the *number of members* in a group. A host can be a member in *multiple groups* simultaneously. A host does not need to be a member of a group to send IP packets to it.

A. IP Multicast Addresses

The address range 224.0.0.0 up to 239.255.255.255 is reserved for IP multicast addresses. Multicast addresses are called *class D addresses*. They consist of the pattern "1110" for the high-order four bits that identify the address as a class D address and the multicast group ID that specifies the group. The Ethernet also provides multicast addresses at the MAC layer. For mapping IP multicast addresses to Ethernet multicast addresses, the last 23 bits of the IP multicast addresses are mapped to the last 23 bits of the Ethernet address (Figure 21.3). Since this mapping is not unambiguous, multiple IP multicast addresses can map to the same Ethernet address. This does not really cause a problem because it is very unlikely that two IP

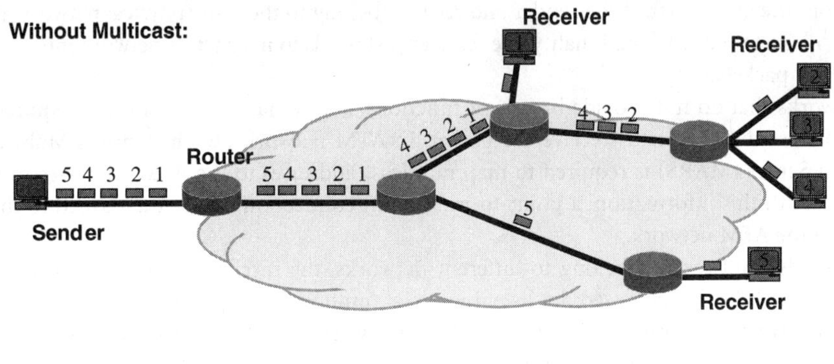

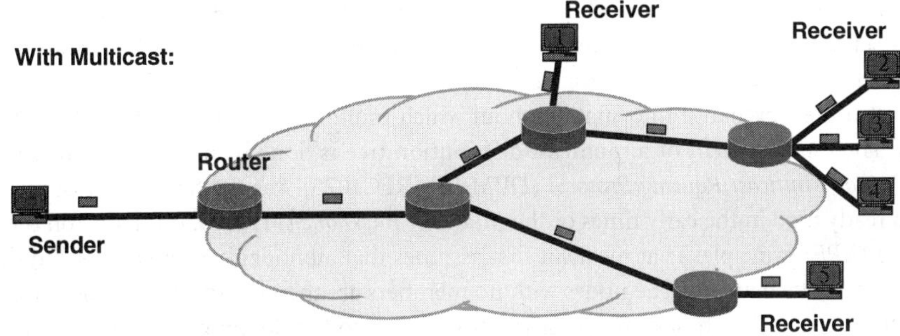

FIGURE 21.2 Group communication with and without IP multicast.

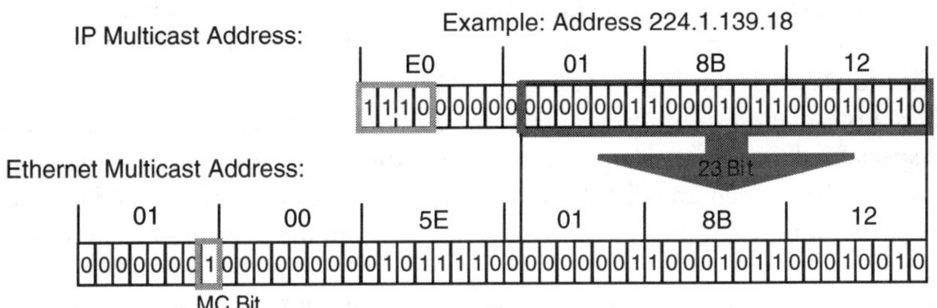

FIGURE 21.3 Mapping of IP multicast address to Ethernet address.

multicast addresses that map to the same Ethernet address are active at the same time in the same network. Furthermore, if a receiver really received a packet that is not intended for it, the IP layer would recognize that no application is listening to the IP multicast address, and may simply discard the packet.

In unicast communication, the distribution of a packet is terminated if it has reached the receiver. For multicast, the scope of the packet is at first undefined. Without any kind of additional *scope control*, multicast packets would be distributed everywhere. A solution to limit the scope is to set the Time To Live (TTL) value in the multicast packet to a specific value and configure thresholds at multicast routers. Routers then discard multicast packets with a TTL value below the threshold. With this, the distribution of packets can be controlled.

B. IP Multicast Protocol Overview

In order to send an IP packet to a multicast group, the sender only needs to specify a multicast address in the destination field. If an end system wants to receive packets that are sent to a specific multicast group, it has to join the group first. If the sender and receiver belong to the same physical network and the network provides a broadcast functionality, the receiver just needs to instruct its network interface to receive the multicast packets.

In networks that do not provide broadcast functions, additional mechanisms are required to ensure that all interested end systems receive the packet. In ATM networks, for instance, a Multicast Address Resolution Server (MARS) is required to map multicast addresses to the ATM addresses that belong to the group. With this information, a point-to-multipoint connection can be established to the multicast receivers in the ATM network.

If the sender and receiver belong to different networks, the receiver needs to inform the router in its network that he is interested in listening to a multicast group. This is done by the *Internet Group Management Protocol (IGMP)*. In IGMPv1 end systems send membership reports to inform the associated router about their wish to join a group. The routers initiate the forwarding of multicast packets [RFC1112]. Routers can also send queries to find out whether there are group members in the network they serve. Newer IGMP versions, IGMPv2 [RFC2236] and IGMPv3 [RFC3376], allow group- and sender-specific queries and membership reports.

Multicast routers exchange information about which multicast packets have to be forwarded to which networks. The establishment of a multicast distribution tree is done by multicast routing protocols. The *Distance Vector Multicast Routing Protocol (DVMRP)* [RFC1075] was the first multicast routing protocol and was already used in the early times of the *Multicast Backbone (MBone)*. It is based on the *Reverse Path Forwarding (RPF)* principle. That means it first assumes that all nodes have members and distributes the data by flooding the network. The nodes with no members are then removed (pruning). It is also possible to explicitly add nodes with new members (grafting). Due to the flooding, it produces a high amount of unnecessary load, especially for sparse multicast groups (few receivers distributed over a wide region). Furthermore, it requires the establishment and maintenance of additional unicast routing tables as helper functions for the establishment of the distribution tree.

The *Protocol Independent Multicast (PIM)* [RFC2117] is a newer improved multicast routing protocol. It operates in two modes: *PIM-dense mode (PIM-DM)* and *PIM-sparse mode (PIM-SM)*. The dense mode is very similar to DVMRP. But in contrast to DVMRP it does not need to maintain additional unicast routing tables. Instead, it uses already existing unicast routing tables at the router (protocol independent). PIM-SM is specialized for sparse groups. Instead of flooding the network, it first assumes that no router has multicast members. Then, the distribution tree is established by explicitly adding nodes with members to the distribution tree. Routers with group members join the tree via some known selected routers (*rendezvous points*).

The sender and receiver may belong to networks with different administrative responsibilities. Different autonomous systems may deploy different multicast routing protocols. Furthermore, certain internal routing information is not relevant for other autonomous systems. Therefore, specific routing protocols for routing between autonomous systems and interoperation among multiple independent multicast routing domains were developed [RFC2715]. The Border *Gateway Multicast Routing Protocol (BGMP)* provides such functions and allows the establishment of a distribution tree over multiple autonomous systems.

Not all routers in the Internet have IP multicast functions enabled. Therefore, it is necessary to tunnel through areas where only unicast forwarding is available. This is done by encapsulating multicast packets in a unicast packet. An IP unicast header is added to the packet at the entry point of the tunnel and removed by the router at the tunnel exit.

21.4 Reliable Multicast

IP multicast is an unreliable protocol, which means that the correct reception of all data at the receivers is not guaranteed. Since many applications (e.g., database replication, file transfer, etc.) require reliable transport, the development of concepts for reliable multicast has become a very active research area (Figure 21.4). One problem with addressing this issue is that the concrete requirements for the different types of applications are

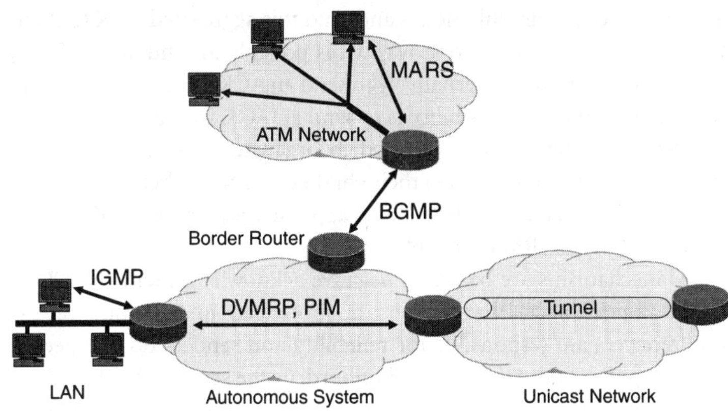

FIGURE 21.4 IP multicast protocol overview.

quite diverse. It seems impossible to find a single solution that serves the needs of all applications. Therefore, a wide variety of methods and protocols have been proposed to support the various application demands.

Within the Internet Engineering Task Force [IETF], an attempt is made to derive a common set of building blocks for reliable multicast support from these approaches. In order to reduce the design space, the Reliable Multicast Transport group (RMT) concentrates on the widely needed but complex problem of bulk data transfer to large groups. [RFC2357] describes primary requirements that need to be met to allow a widespread deployment of a reliable MC protocol in the Internet. *Congestion control* is important to ensure a fair sharing of network resources between applications. *Scalability* is required to allow operation with large groups. Furthermore, *security* issues have to be addressed to deploy reliable multicast in the Internet.

Secondary requirements that allow support for specific applications are summarized in [RFC2887]. Different applications may require different *degrees of reliability*. For applications that require *total reliability*, the loss of one single piece of data renders the whole set of transmitted data useless (e.g., file transfer), whereas applications that require only *semi-reliability* would just experience a quality reduction if data are missing (e.g., audio transmission). Furthermore, it is important to know whether the sender needs some kind of *delivery confirmation* or whether a correct *ordering* or *time-bounded delivery* is required.

Some applications require to *constrain the differences* between receivers, for example, to provide similar delays for the reception of data for all receivers (e.g., stock price distribution, multiplayer gaming). Furthermore, the properties of the network(s) for which the protocol is designed have to be considered. Requirements for a deployment in the Internet differ from requirements in intranets. It is important to know whether a *return path* is available or not (e.g., like in satellite communication). Additionally, there may already be some reliable multicast *support functions* in network that can be utilized.

Packet losses indicate congestion. Furthermore, they need to be defeated to accomplish a good throughput. Therefore, it is essential for reliable multicast protocols to measure packet losses and invoke an appropriate reaction if losses occur. The following basic methods for reliable multicast can be distinguished [RFC2887]:

ACK-based mechanisms are based on positive acknowledgement (ACK) of received data packets. The simplest realization would be that each receiver sends an ACK for every received data packet. This only works for very small groups because the amount of ACKs that arrives at the sender increases with the number of receivers and leads to an *ACK implosion* at the sender. Therefore, this approach is not suitable for most applications. An improvement that allows operation with slightly larger groups can be achieved by sending multiple ACKs in one data packet. To prevent implosion, *ring-based* approaches organize receivers in a ring where an ACK token is passed around (e.g., [WhMK94]). Nevertheless, establishment and maintenance of the ring can be difficult. More robust approaches that are also easier to configure are *tree-based* solutions that arrange receivers in trees. ACKs are only sent to parent nodes. Parent nodes aggregate the ACKs from their child nodes and send this aggregated ACK to their own parent. The aggregation schemes can operate in different ways. It is possible to send an ACK only if all child nodes have confirmed the data reception or alternatively to send an ACK with a list of all children that have acknowledged the reception. A further possibility is to send an ACK with a list of all children that have not confirmed the reception. Trees can also be used to organize retransmissions in a way that lost data are retransmitted by receivers themselves to their child nodes. Nevertheless, the establishment of the tree can be difficult and limit scalability. A tree-based approach is, for instance, used in the *Reliable Multicast Transport Protocol (RMTP)* [RMTP, LiPa96, PaSL97].

NACK-based mechanisms are based on negative acknowledgement (NACK) of missing data packets. Advantages are, for instance, that senders do not need to be aware of the amount of receivers. Furthermore, receivers are responsible for reliability and senders do not need to maintain state for all receivers. This simplifies fault tolerance and unburdens the sender. One NACK is sufficient to inform the sender that a packet is missing by at least one receiver. Therefore, *NACK suppression* mechanisms are used to reduce the number of NACKs. The disadvantages of NACK-based mechanisms are that senders do not know which receivers miss data and that they have no indication at what point of time all receivers have successfully received the data (i.e., when the transmission buffer can be freed). NACK-based protocols mainly differ in the suppression method they use. Examples are the *Scalable Reliable Multicast (SRM)* [SRM, FlJM95] and the *Multicast Dissemination Protocol (MDP)* [MDP, MaAd99].

Redundancy-based methods are based on the sending of redundant information, so that information can be recovered if not all packets are received. The most common method is the *Forward Error Correction (FEC)* [LiCo83]. FEC methods can be grouped in *proactive* methods where the sender directly adds redundancy to the data packets and *reactive* methods where the sender uses feedback from receivers (ACKs' or NACKs') to determine the amount of redundancy that should be added. It is also possible to combine both methods in a way that a certain amount of redundancy is sent proactively; but then the amount is adapted to receiver feedback. FEC-based methods reduce the amount of repair traffic. But due to the processing time needed for adding the redundancy, the end-to-end latency can be increased. This is not a problem for bulk data transfer, but is problematic for interactive applications. A further method is *layered FEC* where the data including redundancy are spread over multiple multicast groups (layers). By joining different amounts of layers, receivers can control the amount of redundancy they receive. With this, it is possible to achieve different reception qualities and data rates for participants of the same application. The method requires no feedback from receivers. It is especially useful in heterogeneous environments where receivers join with different access capabilities or for distinguishing quality levels for commercial data distribution. Furthermore, it provides optimal support if the coding scheme for the application data itself already has a layered structure (layered coding).

It is also possible for reliable multicast protocols to utilize *router assistance* if available. With this constrained NACKs and retransmissions and other network functions can be supported. An example is the *Pragmatic General Multicast (PGM)* [RFC3208]. Furthermore, some *hybrid* approaches exist that are based on a combination of the described methods. A hybrid approach is used, for example, in *RMTP-II* [WhTa00].

21.5 Application Layer Multicast

IP multicast standards have existed for quite some time and router vendors already support IP multicast. Nevertheless, many network operators do not deploy IP multicast. Reasons are, for instance, that native IP multicast does not provide support for error and congestion control. Furthermore, the network management becomes more complex. An additional reason is that there are still no useful accounting methods that appropriately reflect the resource sharing approach that is realized in IP multicast.

In order to support applications that require group communication in networks without native IP multicast support, methods were developed where only end systems handle all multicast functions like group management, multicast routing, and packet duplication. These methods are called *application layer multicast*. Application layer multicast moves the multicast functionality to higher layers and with this unburdens the network from complexity. Since no changes to existing networks are required, application layer multicast pushes the deployment of applications that rely on group communication. Nevertheless, the major advantage of native IP multicast is the resource-sharing concept that prevents sending multiple copies of the same packet over the same paths. Such avoidance of redundant traffic can only be achieved by deploying multicast functions in the network and it is not possible to that extent in application layer multicast [ChRZ00]. Therefore, application layer multicast is more suitable for small groups and interactive applications (audio/video conferencing, multi-player games, etc.) than for the distribution of data from one source to a large number of receivers (e.g., Internet TV).

The provisioning of application layer multicast requires the establishment of an *overlay structure* that consists of unicast paths between end systems. In most cases, a *control topology* is used, which forms a mesh of (unicast) communication channels between members of the group. It is used to exchange control information, like periodic messages to update group membership information. The data are forwarded via the *data topology*. Usually, a tree structure is used for distributing the data. There already exist a variety of application layer multicast protocols. [BaBh03] classifies the protocols according to the sequence of the construction of the control and data topology.

In *mesh-first* approaches, members first establish the control topology. Based on the control topology, a source-specific tree at each member can be established, for example, by using the Reverse Path Forwarding method. An example for a mesh-first approach is the *Narada* protocol [ChRZ00]. Here, designated hosts have the role of the rendezvous points. They maintain the membership information about the group. A host that wants to join a group obtains this information from the rendezvous point and then tries to connect to the mesh by contacting neighbor nodes. Information about the new member is distributed to other members. The data topology is established according to the reverse path forwarding method by using unicast connections between the members. The Narada protocol generates a high control load, because all members of the group have to obtain and retain information about the other members of the group. Therefore, Narada is mainly suitable for small groups. A further mesh-based approach is *Scattercast* [Chaw00], which uses network agents to form the overlay network. Here, small groups of clients are built that connect to the agents. A protocol called *Gossamer* is used to build the client groups.

In *tree-first* approaches, the data delivery tree is established first. Each member is responsible for finding an appropriate parent node to connect to the tree. Again, rendezvous points are used to maintain and provide group membership information. If an end system does not find an appropriate parent, it can become the root of the tree. After establishment of the tree, members periodically try to find better connection points. In some cases, this may lead to a partitioning of the tree. Examples of tree-first approaches are *Your Own Internet Distribution [YOID]* [Fran00] and the *Host Multicast Tree Protocol (HMTP)* [ZhJZ02]. In Yoid additional connections are established with other members that are not direct neighbors on the tree. These connections are used to recover from tree partitions. In HMTP, no explicit control mesh is established but members gather information about some other tree members. This also supports recovery from tree partitions.

Implicit approaches define the data delivery paths implicitly in the control topology, which means that the protocols cover the forwarding of data and control messages. Examples are *NICE* [NICE] ("NICE is the Internet Cooperative Environment") and *Content Addressable Network Multicast (CAN-Multicast)* [RaHK01].

In NICE, members are assigned to different layers of a hierarchical topology. This topology builds the basis for data and control topology. Members within a layer are grouped into clusters. Cluster members periodically exchange control information. The data forwarding is realized by applying specific forwarding rules within the topology.

CAN-Multicast is based on the content addressable network concept proposed in [RaFH01]. A CAN provides an efficient way to map file names to file locations. Such indexing mechanisms are, for instance, required for peer-to-peer file sharing systems. The efficient mapping is achieved by organizing the CAN members in a coordinate space where each member owns a zone. This architecture is used to store key-value pairs. Keys are mapped to points in the coordinate space by a hash function. A key-value pair is then stored at the node that owns the zone to which the hash value belongs. With this, other members can easily locate the data by simply calculating the hash value for the key. In CAN-Multicast, this is used to establish the overlay control topology. For the data distribution messages are simply flooded over the CAN. If all CAN members do want to participate in multicast, mini CANs are established that contain only the multicast members.

21.6 Future Aspects

A. IP Multicast in IPv6

IP version 6 was developed as the successor of the currently used IP version 4. Reasons for developing a new IP standard were, for instance, the limited amount of IPv4 addresses and the inefficient use of the address space, which sooner or later leads to an address shortage. The number of needed addresses increases not just due to the immense increase of computers that are connected to the Internet. Also, the assignment of IP addresses to other devices like mobile phones, personal digital assistants (PDAs), or even household appliances has the effect that one day the number of available addresses will be exhausted. IPv6 addresses consist of 128 bits. Using this, many more addresses can be expressed than by the 32-bit long IPv4 addresses.

IP multicast functions were directly integrated into the IPv6 standard. In order to reduce network load, IPv6 uses multicast for the provisioning of some functions that in IPv4 are realized with broadcast. [RFC2373] describes the IPv6 address format, which is used to express unicast, multicast, and anycast addresses. The first 8 bits of an IPv6 multicast address are set to 1 to identify the address as multicast address. This is followed by a 4-bit field for flags and another 4-bit field for the scope of the address. The remaining bits identify the group ID. The scope field allows an explicit scope control without the need to set specific TTL values. For nonpermanent addresses, the group ID is only valid within the scope. IGMP functions are integrated into the Internet Control Message Protocol for IPv6 (ICMPv6) [RFC2463].

B. Multicast Security

Security concepts for unicast cannot simply be applied to multicast communication. Already the establishment of security associations between multiple participants is much more complex than for a communication with only two participants. In general, three properties of group communication can be identified, which lead to many difficulties for the provisioning of security functions. Due to the *higher number of participants*, security associations have to be established between all members. This severely complicates the distribution and refreshment of cryptographic keys. The *dynamics of groups* leads to the demand for a dynamic establishment of security associations. Furthermore, the *anonymity of receivers* allows arbitrary end systems to join multicast groups without notifying the sender. The main security requirements for multicast communication are the management of *group membership* and *access control*, the *confidentiality* of sent data, and the *authentication* of senders and data. Restricted group membership and access control can be achieved by an authentication of potential group members and a secure distribution and refreshment of keys. Confidentiality of transmitted data can be achieved by encryption.

References

[BaBh03] S. Banerjee and B. Bhattacharjee, A Comparative Study of Application Layer Multicast Protocols, available at http://www.cs.umd.edu/projects/nice/papers/compare.ps.gz, 2003.

[BaBK02] S. Banerjee and B. Bhattacharjee, and C. Kommareddy, Scalable Application Layer Multicast, in Proceedings of ACM SIGCOMM 2002, Pittsburg, USA, Aug., 2002.

[Chaw03] Y. Chawathe: Scattercast: an adaptable broadcast distribution framework, ACM Multimedia Systems Volume 9, Issue 1, Pages: 104–118, July 2003, ISSN:0942-4962, Springer-Verlag New York, Inc., Secaucus, NJ, USA, 2003.

[ChRS01] Yang-hua Chu, S.G. Rao, S. Seshan, and Hui Zhang, Enabling Conferencing Applications on the Internet Using an Overlay Multicast Architecture, in Proceedings of SIGCOMM 2001, San Diego, USA, Aug., 2001.

[ChRZ00] Yang-hua Chu, S.G. Rao, and Hui Zhang, A Case for End System Multicast, in Proceedings of ACM SIGMETRICS 2000, Santa Clara, USA, June, 2000.

[DrCK02] P. Druschel, M. Castro, A.-M. Kermarrec, and A. Rowstron, Scribe: a large-scale and decentralized application-level multicast infrastructure, *IEEE Journal on Selected Areas in Communications*, 20, 2002.

[FlJM95] S. Floyd, V. Jacobson, and S. McCanne, A Reliable Multicast Framework for Light-weight Sessions and Application Level Framing, in Proceedings of ACM SIGCOMM 1995, Cambridge, USA, Aug., 1995.

[Fran00] P. Francis, Yoid: Extending the Internet Multicast Architecture, report available at http://www.icir.org/yoid/docs/yoidArch.ps.gz, April 2000.

[IETF] http://www.ietf.org

[IRTF] http://www.irtf.org

[LiCo83] S. Lin, and D.J. Costello, *Error Correcting Coding: Fundamentals and Applications*, Prentice-Hall, Englewood Cliffs, NJ, 1983.

[LiPa96] J. Lin and S. Paul, RMTP: A Reliable Multicast Transport Protocol, in Proceedings of IEEE Infocom 1996, San Francisco, USA, Mar., 1996.

[MaAd99] J.P. Macker and R.B. Adamson, The Multicast Dissemination Protocol Toolkit, in Proceedings of IEEE MILCOM 1999, Nov., 1999.

[MDP] http://manimac.itd.nrl.navy.mil/MDP/

[NICE] http://www.cs.umd.edu/projects/nice/

[PaSL97] S. Paul, K.K. Sabnani, J.C. Lin, and S. Bhattacharyya, Reliable multicast transport protocol (RMTP), *IEEE Journal on Selected Areas in Communications*, 15, 1997.

[RaFH01] S. Ratnasamy, P. Francis, M. Handley, R. Karp, and S. Shenker, A Scalable Content-Addressable Network, in Proceedings of SIGCOMM, San Diego, USA, Aug., 2001.

[RaHK01] S. Ratnasamy, M. Handley, R. Karp, and S. Shenker, Application-level Multicast using Content-Addressable Networks, in Proceedings of 3rd International Workshop on Networked Group Communication, London, Nov., 2001.

[RFC1075] D. Waitzman, C. Partridge, and S. Deering, Distance Vector Multicast Routing Protocol, Request For Comments 1075, November 1988.

[RFC1112] S. Deering, Host extensions for IP multicasting, Request for Comments 1112, August 1989.

[RFC2117] D. Estrin et. al., Protocol Independent Multicast-Sparse Mode (PIM-SM): Protocol Specification, Request for Comments 2117, June 1997.

[RFC2236] W. Fenner, Internet Group Management Protocol, Version 2, Request for Comments 2236, November 1997.

[RFC2357] A. Mankin, A. Romanow, S. Bradner, and V. Paxson, IETF Criteria for Evaluating Reliable Multicast Transport and Application Protocols, Request for Comments 2357, June 1998.

[RFC2463] A. Conta and S. Deering, Internet Control Message Protocol (ICMPv6) for the Internet Protocol Version 6 (IPv6) Specification, Request for Comments 2463, December 1998.

[RFC2715] D. Thaler, Interoperability Rules for Multicast Routing Protocols, Request for Comments 2715, October 1999.

[RFC2887] M. Handley, S. Floyd, B. Whetten, R. Kermode, L. Vicisano, and M. Luby, The Reliable Multicast Design Space for Bulk Data Transfer, Request for Comments 2887, August 2000.

[RFC3048] B. Whetten, L. Vicisano, R. Kermode, M. Handley, S. Floyd, and M. Luby, Reliable Multicast Transport Building Blocks for One-to-Many Bulk-Data Transfer, Request for Comments 3048, January 2001.

[RFC3208] T. Speakman et. al., PGM Reliable Transport Protocol Specification, Request for Comments 3208, December 2001.

[RFC3376] B. Cain, S. Deering, I. Kouvelas, B. Fenner, and A. Thyagarajan, Internet Group Management Protocol, Version 3, Request for Comments 3376, October 2002.

[RMTP] http://www.bell-labs.com/project/rmtp/

[SRM] http://www.icir.org/floyd/srm.html

[WhMK95] B.Whetten, T. Montgomery, and S. Kaplan, A high performance totally ordered multicast protocol, Theory and Practice in Distributed Systems, Lecture Notes in Computer Science, Vol. 938, Pages 33–57, ISBN 3-540-60042-6, Springer-Verlag, Berlin, 1995.

[WhTa00] B. Whetten and G. Taskale, An overview of reliable multicast transport protocol II, *IEEE Network*, Vol. 14, Issue 1, pp 37–47, January-February 2000.

[YOID] http://www.aciri.org/yoid/

[ZhJZ02] Beichuan Zhang, Sugih Jamin, and Lixia Zhang, Host Multicast: A Framework for Delivering Multicast To End Users, in Proceedings of IEEE Infocom, New York, June, 2002.

22

A Survey of Congestion and QoS Control Mechanisms for the Internet

Dorgham Sisalem
Fraunhofer FOKUS

Adam Wolisz
Technical University of Berlin

22.1 Introduction

The flexible architecture of the Internet and the decline in bandwidth prices have made it possible to use the Internet for various multimedia applications such as telephony and conferencing.

However, transmitting data packets into the network without regard to the actual available resources can easily result in overload situations and data losses. To avoid such a situation, some mechanisms are required to control the amount of data packets entering the network.

In the literature, there have been generally two approaches for traffic management in a network: admission control combined with Quality of Service QoS signaling protocols, on the one hand, and congestion control schemes that adapt the transmission and reception behavior of the end systems in accordance with the available resources in the network on the other.

In Section 22.2 a general survey of congestion control mechanisms as well as some evaluation criteria of congestion control schemes are given. Section 22.3 describes briefly different proposals for QoS control architectures that provide for tighter control on the number of admitted flows to the network and the amount of network resources received by each flow. In Section 22.4, a short overview of the advantages and disadvantages of congestion control schemes compared to QoS architectures is given.

22.2   Congestion Control in the Internet

As the Internet does not provide for any mechanisms for regulating the transmission behavior of sending end systems, a single misbehaving end system sending data at a rate higher than the available bandwidth might lead to severe unfairness problems in the best case and to network congestion collapse in the worst [13].

The unfairness problems arise from the existing architecture of the network elements in the Internet. Currently, most of the network nodes, commonly known as routers, in the Internet employ tail discard queuing systems, mainly due to their simplicity. With such an approach, an end system sending a data stream at a high rate competing with an end system sending at a lower rate will have the same loss probability as the low-rate sender. Hence, the high-rate stream actually receives a larger share of the available bandwidth than the low-rate stream even though it actually caused the congestion [13]. This scenario is even worse for the situation of TCP connections sharing the same bottleneck with UDP flows. In response to the loss of packets, TCP connections reduce their transmission rate. However, without any rate reduction on behalf of the noncongestion-controlled UDP traffic sharing the same bottleneck with the TCP traffic, the congestion situation would prevail and the TCP connections would starve and receive a much smaller bandwidth share than the nonresponsive flows [13].

Severe and consistent congestion states might also lead to congestion collapse. Such a situation occurs when the network is utilized to a very high degree but with only a small amount of useful data reaching the receivers. Congestion collapse in the Internet was first reported by Nagle [22] and was largely due to TCP connections unnecessarily retransmitting data packets that were either in transit or have already arrived at the receivers. This form of collapse has since been avoided by different improvements in the TCP congestion avoidance mechanisms [16].

Other forms of congestion collapse arise when the network transmits fragments of packets that will be discarded at the receiving end systems because they cannot be reassembled into a valid packet. Such a situation might occur from the mismatch between the link level transmission layer and higher transmission layers. For example, when carrying large IP packets over an ATM network, dropping an ATM cell might lead to discarding of the entire IP packet at the receiver. Such a situation might be avoided by mechanisms aimed at providing lower layers with knowledge about the fragmentation of data in higher layers.

The congestion collapse form that we are interested in arises when network resources are wasted by transporting packets through the network only to be dropped before reaching the receiving end system. This situation arises when the sender is transmitting data at a rate higher than that of some bottleneck on the path toward the receiver. Such a situation can be avoided by adjusting the transmission rate of the sender in accordance with the available network resources.

The issue of congestion control in data networks has received considerable attention in the literature. Citing only a fraction of the proposed work is, however, far too voluminous. Therefore, we first give a brief overview of classification possibilities of congestion control schemes and only present some examples for those classes.

- *Control loop*: Depending on whether or not the sender needs to collect information about the network path connecting it to the receiver in order to adjust its transmission behavior, we can distinguish two control possibilities:
  - *Open-loop adaptation*: The sender adjusts the shape of the data to be transmitted in accordance to prior knowledge or estimation of the available resources. For example, when transmitting live video with a variable bit rate over a constant rate link, the sender can adjust the quality or frame rate of the video stream to be suitable for transmission over the available network capacity. This is often realized by using a leaky bucket [21] for enforcing a specific transmission rate. By monitoring the buffer length of the leaky bucket, the application can determine the appropriate quality factor to send the data with. Additionally, instead of altering the transmission behavior, the sender might in a different scenario adapt the priority of its data packets in accordance to their content and the available resources. in [28], Sanneck presents an approach for

improving the quality of audio communication over the Internet. In addition to the audio data, the sender includes information that can be used for the concealment of lost packets. The amount of additional information needed for reconstructing the original content is adapted to the importance of the sent audio data.

- *Closed-loop adaptation*: With open-loop adaptation, the sender can only adjust its behavior based on prior knowledge of its own resources or of the static amount of network resources available to it. Here, the aim is to maintain or improve the quality of the communication in accordance with an amount of fixed resources. To actually avoid network congestion, the sender needs information about the path connecting it to the receiver and back. Such a closed-loop adaptation is a more general case where the sender adjusts its transmission behavior not only based on the locally available resources but also based on feedback information from the receivers or the network. This approach has the major advantage of avoiding network congestion or overwhelming the receiving end-system with data it cannot use.

- *Transmission mode*: Depending on the number of participants in a communication scenario, different issues need to be considered when designing congestion control schemes.
  - *Unicast*: For the case of point-to-point communication, tight control can be realized. Similar to the case of TCP, the receiver of a multimedia flow can send frequent feedback messages to the sender without overloading the sender. This would then allow for fast reactions to changes in the network state.
  - *Multicast*: Multicast communication describes, in general, multi-point to multi-point communication sessions. Having a large number of receivers sending frequent feedback messages to the sender can easily result in overloading the sender or might cause congestion on the link connecting the sender to the network. Hence, for the case of multicast, scalability issues play a crucial role in the design of congestion control schemes. Additionally, due to the heterogeneity of the Internet links and receivers, congestion control schemes need to be carefully designed to achieve a high performance in a variety of situations.
- *Architecture*: Depending on the systems involved in the congestion control, scheme different modes can be distinguished:
  - *End-to-end*: In this case, only the end systems, that is, the senders and receivers, are involved in collecting information about the congestion state in the network and adapting the number of packets entering the network.
  - *Network-supported*: In addition to the end systems, network routers could participate in collecting state information as well as manipulating the amount of data transported in the network.

Note that these classes should only be considered as conceptual ones, which aid us in discussing different features of congestion control schemes. Actually, congestion control schemes usually belong to overlapping classes. Network-supported schemes can, for example, be either open-loop or closed-loop and be used for unicast as well as multicast communication. In this study, we will restrict the scope of investigated congestion control schemes to closed-loop schemes for unicast communication. While we will be mainly concentrating on the end-to-end communication scenario, we will also be briefly looking at network-supported approaches. We first give some criteria for evaluating the efficiency of congestion control schemes. Next, different approaches and schemes for realizing congestion control schemes for unicast are described. A short survey of network-assisted congestion control schemes is then presented.

Evaluation Criteria for Congestion Control Mechanisms

By deploying congestion control mechanisms in the context of multimedia communication, we aim at improving the network performance compared to the case when no control is deployed, thereby increasing the user's satisfaction. In this section, we will describe some of the evaluation criteria for the efficiency of the control.

Network Congestion State

The primary goal of congestion control schemes is to achieve a high network utilization level and at the same time reduce the packet losses at the network level. In estimating the loss performance of a scheme, one should consider the losses as seen at the networking layer and not only at the application layer. Using forward error correction (FEC) mechanisms [9] for reconstructing lost data packets might lead to loss-free data transmission as measured by the application. However, in this case, network resources are wasted transporting redundant data. Hence, while FEC mechanisms improve the quality of single streams, by using congestion avoidance mechanisms, one aims at improving the overall performance of the network and thus all streams traversing the network.

Scalability

Based on the multicast architecture of the Internet, a sender can transmit a media stream to a nearly unlimited number of receivers in an effective way. The performance of an efficient congestion control scheme should therefore not degrade substantially with the increased number of participants in a multicast session. Additionally, the complexity of the schemes used should still be acceptable.

Convergence

Convergence is generally measured by the speed or time with which a system approaches some stable state from any starting state. Efficient adaptation algorithms need to have a short convergence period and lead to a stable state.

Fairness in the Context of Congestion Control for Multimedia Communication

When confronted with the case of connections or users competing for some shared resources such as network bandwidth, we need to consider the aspect of fair resource distribution. However, the term fairness only has a subjective meaning, which differs depending on the situation. Thus, a system can be considered fair if the competing instances receive a share that conforms to some pre-established fairness definition.

In the context of multimedia communication, there have been two major definitions of the values of a fair share a connection can receive: max–min fairness and TCP-friendly bandwidth distribution. Which definition to use depends generally on the control scenario and the information available for the adaptation process.

The Max–Min fairness: The max–min fairness criteria are usually applied when the adaptation involves the distribution of available resources for a known number of instances, for example, bandwidth for a number of streams or processing power for a number of processes.

Taking the example of bandwidth distribution among competing streams, whereas here a stream indicates the flow of data packets from one sender to one or more receivers, the fair share (F_i) of a flow i is minimally

$$F_i = \frac{B}{n} \tag{1}$$

where B is the bandwidth of a network node and n is the number of flows traversing this node. (F_i) can maximally reach

$$F_i = \frac{B - \sum B_{cong}}{n - \sum n_{cong}} \tag{2}$$

where B_{cong} is the bandwidth of the flows bottlenecked elsewhere and n_{cong} is the number of flows bottlenecked elsewhere. A bottlenecked flow indicates that the fair share a flow can receive at some other link is smaller than its minimum fair share at this link. A flow's fair bandwidth share is then set to the minimum fair share calculated at all traversed hops.

This definition can be further extended as follows:

Reservation proportional shares: With each user reserving — or paying for — a guaranteed share (S_i) of some resource R, each user (i) gets a fair share proportional to its S_i. For n flows, the fair share F_i can be calculated as follows:

$$F_i = R * \frac{S_i}{\sum_i^n S_i}$$

Reservation plus equal shares: With each user reserving — or paying for — a guaranteed share S_i of some resource R, each user (i) gets its reserved S_i plus an equal share of the available resources. For n flows, the fair share F_i can be calculated as follows:

$$F_i = S_i + \frac{R - \sum_i^n S_i}{n}$$

Weighted shares: Here, the resource (R) is allocated for each connection proportionally to a pre-determined weight W_i — for example, priority level. For n flows, the fair share F_i can be calculated as follows:

$$F_i = \frac{R * W_i}{\sum_i^n W_i}$$

Note, however, that applying these enhanced definitions requires a mechanism for establishing reservations or setting the allocation weights. As such mechanisms are currently not available in the Internet, these definitions are inappropriate for control mechanisms realized only at the end systems.

TCP-Friendly adaptation: Often, it is the case that the adaptation process is based on an estimation of the performance of a system and the adaptation instance heuristically adjusts its behavior in accordance with the estimated performance without exact knowledge of the available resources or number of other competing instances. Adapting the transmission behavior of a video stream based on the measured loss and delay values in the network provides such an example. In this case, applying the max–min fairness criterium is rather difficult. Instead, we need to consider the effects of using an adaptation scheme on the shares received by other adaptive traffic. As around 95% of the traffic carried over the Internet is TCP traffic [12], there has been a current consensus in various works [13] that adaptation mechanisms and congestion control approaches should be TCP-friendly. TCP-friendliness indicates here that if a TCP and an adaptive flow with similar transmission behavior traverse the same path and thus compete at the same bottleneck, round-trip delays, and face the same loss values, both flows should receive similar bandwidth shares. Due to the different natures of TCP and multimedia traffic, it is expected that a TCP-friendly multimedia flow would acquire the same bandwidth share as a TCP connection only averaged over time intervals of several seconds or even only over the entire lifetime of the flow and not at every time point.

The requirements for fulfilling both the max–min fairness criteria and TCP-friendliness are, however, rather contradictory. With the max–min fairness definition, the resource distribution depends only on the number of flows sharing a bottleneck. Various simulative [32] as well as analytical studies have shown that the bandwidth a TCP connection can utilize depends not only on the number of connections traversing a shared link but also on the round-trip delays and the number of traversed congested hops. That is, connections having large round-trip delays or traversing a larger number of congested hops receive a smaller bandwidth share than a connection with a smaller round-trip delay or less congested hops in its path.

User Satisfaction

Users' satisfaction is in general difficult to measure and quantify. There have already been some proposals for assessing the QoS of multimedia contents using rating schemes as was proposed by the ITU [15], for example.

However, in evaluating the performance of an adaptation scheme, users' satisfaction is just one evaluation criterion out of many. Aspects of network performance or scalability of the scheme are just as important. Due to the complexity of the measurement mechanisms, one could refer to a simple measurement of the loss ratio of the data presented to the user. Studies on the transmission of MPEG video streams over the Internet have shown that losses are a major source of QoS degradation and that losses as low as 3% might render about 30% of a video stream unusable. Hence, an effective adaptation scheme should aim at minimizing the loss ratio observed at the users' end systems.

End-to-End Unicast Congestion Control Schemes

Basically, congestion control schemes for unicast communication can be divided into two major categories of algorithms: window- and rate-based mechanisms. In rate-based adaptation, a limit is placed on the rate at which the source can send packets. In window-based flow control, at any time there is a limit to the number of packets a sender can transmit before having to wait for an acknowledgment from the receiver; but there are no constraints on the rate at which packets can be sent.

The adaptation decision itself can be based on either explicit or implicit indication about the available resources. Explicit information can be as detailed as the actual bandwidth share a sender should be using to avoid network congestion or simply a single bit indicating the congestion state of the network. Explicit indications are created by the network and are either forwarded downstream on the flow's path toward the destination or sent directly back to the senders. With the forward error congestion notification (FECN), the sender transmits special messages that collect information from all traversed routers along the path to the receiver, which echoes them back to the sender. This approach provides the destination with a complete overall view of the congestion state along the flow's path; however, it introduces additional delay as the information needs to traverse all the path up to the destination before being sent back to the sender. FECN also minimizes the needed number of congestion messages compared to the backward error congestion notification (BECN). With BECN each intermediate router creates and sends congestion notification to the source end system when congestion is observed. This can be achieved by creating special messages to be sent to the end systems or by including the congestion notification in the messages that are sent from the receiver to the sender.

On the contrary, the network congestion state might also be implicitly estimated using loss and delay indications measured at the receiving end systems and carried in feedback messages sent by the receivers. In addition to losses and round-trip delays, the end systems might use the delay variations of the received packets as an indication of the buffer lengths in the network and hence as an estimation of the network load.

With the lack of mechanisms for collecting and distributing explicit information about the network congestion state in the Internet, most of the control schemes found in the literature refer to heuristic approaches for adjusting the transmission rate or window, see window-based Adaptation and Rate-based Adaptation sections. This heuristic approach is usually based on some variation of an increase/decrease mechanism (ID). With ID schemes, the bandwidth share of a flow is gradually increased by some amount during underload situations and reduced during overload situations. A popular version of ID schemes is the additive increase and multiple decrease (AIMD) mechanism. With this approach, the sender can increase its resource share by an additive amount during underload periods and reduces it by a multiplicative factor during overload periods. The choice of AIMD can be briefly justified as follows. If the network is operating below the optimal point designated by high utilization and low losses, all users go up equally. If the network is congested, the multiplicative decrease makes users with higher resource shares go down faster than those with smaller shares, making the allocation more fair. In [11], Chiu and Jain show that this approach leads to a stable adaptation behavior and an equal bandwidth distribution.

Window-Based Adaptation

There have been several proposals for adaptation schemes that, similar to the TCP adaptation mechanisms, are based on the so-called sliding window approach.

With the sliding window approach, a window represents the number of packets the sender can transmit before having to wait for an acknowledgment packet from the receiver. In case of losses, TCP reduces the window size by half and starts increasing it again using the slow start and congestion avoidance mechanisms [16]. In addition to these schemes, there have been various other proposals for adapting the window size to the network congestion state. To mention a few out of many:

Slow start and search (Tri-S): The Tri-S scheme [36] uses the changes in throughput as an indication of congestion. The algorithm computes the normalized throughput gradient and compares it to some thresholds to keep the connection at the optimal operating point.

Dual-Window: With this algorithm [37] the sender maintains two windows, one for congestion control and one for flow control. The flow window adapts the transmission rate of the sender in accordance with the available buffers at the receiver and the congestion window adapts the transmission behavior in accordance with the network congestion state.

Delay-based adaptation: In [17], Jain presents an approach for congestion control based on estimating the round trip delay. The sender estimates the round-trip delay as the time passed between sending a packet and receiving an acknowledgment for it. The window size is then adjusted based on the variations in the measured round-trip delay.

Rate-Based Adaptation

With window-based adaptation schemes, the transmission rate of the sender is only controlled indirectly by adjusting the transmission window size. On the contrary, rate-based adaptation schemes control the rate of the sender directly. To mention just a few out of many schemes proposed in this area:

Scalable feedback adaptation: In [5], Bolot et al. present an approach for scalable control of multicast media streams. The senders probe the receivers to solicit feedback information about the losses observed at the receiving end systems. Based on the collected information, the sender either additively increases its transmission rate or decreases it multiplicatively.

Backward congestion notification: In [19], Kanakia et al. present an approach for adapting the transmission of real-time video. The adaptation algorithm uses feedback information sent by the routers about the available buffers to determine the appropriate transmission rate.

RAP: In [27], Rejai et al. present a congestion control scheme called rate adaptation protocol (RAP). Just as with TCP, sent packets are acknowledged by the receivers with losses indicated either by gaps in the sequence numbers of the acknowledged packets or timeouts. Using the acknowledgment packets, the sender can estimate the round-trip delay. If no losses are detected, the sender can periodically increase its transmission rate additively as a function of the estimated round trip delay. After detecting a loss the rate is reduced by half in a similar manner to TCP.

TFRCP: Based on an analytical model of TCP, see Padhye et al. [24], Floyd et al. [14] present a scheme in which the sender estimates the round-trip delay and losses based on the receiver's acknowledgments. In case of losses, the sender restricts its transmission rate to the equivalent TCP rate calculated using the analytical model; otherwise, the transmission rate is increased.

LDA+: LDA+, see [33], is an additive increase and multiplicative decrease algorithm with the addition and reduction values determined dynamically based on the current network situation and the bandwidth share that a flow is already utilizing. During loss situations LDA+ estimates a flow's bandwidth share to be minimally the bandwidth share determined with the theoretical TCP model; see [24]. For the case of no losses, the flow's share can be increased by a value that does not exceed the increase of the bandwidth share of a TCP connection with the same round-trip delay and packet size.

Network-Based Adaptation

Besides various proposals for increasing the fault-tolerance of networks [23] and adapting the transmission of routing information in accordance with network congestion, various studies have investigated network-based mechanisms for adapting multimedia streams in accordance with the network congestion state. Placing the adaptation entities in the network has the clear advantage of solving the congestion problem at the point at which it occurs. However, such an approach requires a network capable of actively reshaping the transferred data, which adds a considerable load on the network nodes and increases the complexity and costs of building and operating such networks.

The proposals for network-based adaptation schemes range form simple buffer management approaches to complicated QoS architectures. To describe just a few examples:

Available Bit Rate (ABR): The ABR service [31] was specified as an ATM service for applications mainly handling data transfer and having the ability to reduce their sending rate if the network requires them to do so. Likewise, such applications may wish to increase their sending rate if there is an extra bandwidth available within the network. Applications using the ABR service can expect the following quality of service commitments from the network:

1. The available bandwidth is fairly distributed among all active ABR connections.
2. A minimum cell rate that is agreed upon during the connection establishment phase.
3. Only a preset fraction of the sent data can be dropped as long as the sending behavior of the application conforms to the negotiated values.

With the ABR service, a source sends a so-called resource management (RM) cell every–sent data cells. These cells indicate the current cell rate (CCR) and the desired one of the senders. The destination end system turns the RM cells around and sends them back to the source. The intermediate switches determine the fair bandwidth shares that the ABR connections should use in order to avoid congestion based on the traffic situation in the network. These values are then written in the backward RM cells in the explicit rate (ER) field. The source should then increase or decrease its rate in accordance with the ER noted in the RM cells. Additionally, the network switches could also just indicate their congestion situation by a single bit indicating an over- or underload situation.

Early congestion notification (ECN): TCP relies primarily on detecting loss events for activating its congestion avoidance mechanisms. That is, a TCP sender only reacts to the congestion situation when it is already too late and packets have been lost. Floyd et al. [26] present an approach in which the network nodes indicate their congestion situation by setting a congestion bit in the TCP packets. As a guideline for the sender's reaction to the congestion bits, Floyd et al. propose the following:

1. TCP's response to ECN should be similar over longer time scales to its response to dropped packets or timeouts.
2. Over smaller time scales, TCP's response to ECN can be less conservative than its response to lost packets.
3. TCP should react to ECN at most once per round-trip time.
4. The response to ECN does not trigger the sending or retransmitting of any data packets.

With ECN, TCP senders can reduce their transmission windows before packets get dropped and thereby avoid the long recovery periods caused by data losses.

Explicit windows: For a better integration of TCP and ATM in [18] Kalampokas et al. propose to adjust the window information carried in the TCP acknowledgment packets based on the available bandwidth share calculated by the network nodes. This approach removes the need for the additive increase and multiplicative decrease used by TCP for setting the transmission window size and, hence, leads to a better overall performance.

PLoP: The predictive loss pattern queue management scheme [29] proposes a mechanism for reducing loss burstiness of audio streams and thus allowing for more efficient usage of forward error correction approaches.

Video gateway: In an effort to reduce the bandwidth of a stream to better suit the capacities of a receiver, in [2] Amir et al. propose using gateways for transcoding high bandwidth video streams into low bandwidth ones. These gateways can be placed at points connecting high- and low-capacity links or where congestion is very likely to happen.

Active networks: Active networks [35] also present a realization possibility of network-based adaptation. With such an approach, a network node can reshape a data stream based on information and procedures provided by an entity with knowledge about the data stream.

Throttling: Another approach is to enhance the routers to identify flows that use more bandwidth than allowed by a theoretical model of TCP and then throttle these flows.

22.3   Quality of Service Control for Multimedia Communication

The packet switching architecture of the Internet provides a flexible infrastructure for the interconnection of networks with different architectures. However, it provides only for a best effort service with little control over the packet delay and loss processes at the network nodes. This best-effort service makes it impossible to provide guarantees for a minimum bandwidth or a maximum delay.

One approach for providing an improved QoS control in the Internet is to augment the current best-effort service to include new services that provide performance guarantees desired by the applications and users. In this section, we briefly describe a few of the most current proposals for such enhanced services.

The Integrated Services Model (IntServ)

The integrated services framework [7] provides the ability for applications to choose among multiple, controlled levels of delivery service for their data packets. To support this capability, two things are required:

1. The individual network elements such as the subnets and IP routers along the path traversed by an application's data packets must support mechanisms to control the QoS delivered to those packets. These mechanisms can be roughly divided into three parts:

   Call admission control: This instance checks if the requested reservation level can be supported by the router.
   Classification: In order to determine the appropriate service to provide for a data packet belonging to some flow, this packet first needs to be classified based on the previously made reservation for this flow and is then inserted into the appropriate scheduling level.
   Scheduling: This instance determines when to forward a data packet based on the reservation made for it.

2. A way to communicate the application's requirements to the network elements along the path and to convey QoS management information between the network elements and the application must be provided.

The integrated services model supports three service types:

Guaranteed load: Maximum delay and bandwidth requested by the application are guaranteed [30].
Controlled load: Only a minimum bandwidth is guaranteed. Traffic sent above this minimal rate is treated as best-effort traffic [38].
Best effort: Data packets are sent without any guarantees.
With the resource reservation protocol (RSVP) [6], Braden et al. have proposed a QoS signaling protocol, in which the sender informs the receivers about the traffic shape of the data flow. In reaction to this, the receivers send reservation requests indicating the amount of required resources to support the flow.

Each network node traversed by the control messages maintains information about the requested resources for each flow. Reservation requests by different receivers for the same data flow are merged together at the network nodes. For each flow that has made a reservation, the network nodes periodically send control messages to their neighboring nodes to indicate the state of this flow. As all the nodes need to maintain state information for each flow and send and receive control messages to and from neighboring nodes to detect the state of each flow, RSVP incurs a significant processing overhead on the network nodes. In addition, various problems resulting from issues like merging of reservations and reservation updates need to be considered with RSVP.

Pang and Schulzrinne propose a sender-based reservation scheme called YESSIR [25] in which the sender issues the reservation request. Each traversed router processes the reservation request, decides to accept or reject it, and establishes some state information for the signaling flow in case of acceptance. After receiving a reservation message from the sender, the receivers issue acknowledgment packets indicating the result of the reservation. After establishing a reservation, the sender periodically transmits control messages to refresh its state at the network nodes. This approach reduces the complexity of RSVP as it does not require the routers to exchange per-flow refresh messages. However, it still requires per-flow state information at the routers.

The Differentiated Services Model (DiffServ)

The need for maintaining per-flow states, periodically refreshing this information as well as the overhead of per-packet classification and processing results in scaling problems for the IntServ model and RSVP when handling a large number of flows. The differentiated services (DiffServ) model [3] was designed to avoid this problem by using service-level agreements (SLA) between the network providers and the users. These SLAs describe the QoS level that the aggregated traffic of a user can expect from the provider. Traffic sent in conformance with the established SLA is marked as belonging to a specific QoS level. At the core routers data packets are serviced differently based on their marked QoS level. As all packets with the same marks are treated equally, the core routers need only to maintain information describing the resources allocated for the supported QoS levels.

As the SLAs are used for traffic aggregates, there is no need for per-flow states. Additionally, with the DiffServ model, only the edge routers need to police incoming traffic and adopt admission control procedures.

Broadly speaking, any traffic management or bandwidth control mechanism that treats different users differently, ranging from simple weighted fair queuing (WFQ) [10] to RSVP and per-session traffic scheduling counts as a DiffServ solution. However, in common Internet usage the term "DiffServ" is beginning to mean any relatively simple, lightweight mechanism that does not depend entirely on per-flow resource reservations.

While such an approach avoids the scalability problems of the integrated services model, it is rather rigid. SLAs are currently mainly thought to be established in a static manner or to be changed only infrequently on the order of days or weeks. Hence, the DiffServ model does not allow the user to increase or decrease the amount of its reserved resources in accordance with its traffic requirements. In addition to the static nature of the SLAs specifying a QoS level for an aggregate of flows can result in unfair distribution of resources among the flows belonging to the same aggregate due to the aggressiveness of some flows, differences in round-trip delays, and the paths taken. Further, the core routers are expected to be dimensioned large enough in a manner as to provide the user with the agreed upon QoS level on any path taken by the user at any time in accordance with the SLA. As the exact path taken by a user's traffic is not known in advance, the network needs to be highly overprovisioned to account for all possible cases. This is even more pronounced for the case of multicast as the user's traffic might take different paths in the providers network and hence sufficient resources need to be available on all paths simultaneously. To allow for dynamical establishment and changes of the resources provided for a user, currently there are different proposals that aim at introducing lightweight signaling protocols as well as different efforts for aggregating state information to reduce the load on the network routers.

QoS Architectures

In addition to the above models, various studies describe complete architectures for reserving network resources and controlling the QoS of the network traffic. While these architectures offer a good basis for the QoS research and first experiences, they are too complicated and could, thus, not gain any significance in the Internet.

While these approaches differ in their scope and complexity, they all have common requirements:

The networking infrastructure needs to be substantially enhanced in order to realize the improved services. This involves improving the network routers by introducing more complicated buffer management schemes than simple FIFO for enforcing the requested QoS levels, admission control mechanisms to effectively utilize the network resources, as well as defining signaling mechanisms for establishing the required QoS by the users.

Provide incentives to encourage applications to request the proper service class for their requirements. In the absence of such incentives, applications could request the highest quality level no matter what their requirements are. One way of providing incentives is via pricing the communication service based on the requested QoS.

In order for the users and applications to benefit from the upgraded networking infrastructure, the applications need to be able to specify their requirements and support the appropriate signaling protocols for doing so.

22.4 Adaptation vs. Reservation

While reservation-based schemes guarantee the required QoS level, deploying such mechanisms leads to a substantial increase in the complexity of the network due to the necessary enhancements of the routers and the introduction of billing mechanisms [4]. Additionally, due to the admission control, the user satisfaction of the network service might be reduced due to the possibility of blocking a service request during overload periods.

On the contrary, end-to-end congestion control schemes require no or only minimal enhancements to the network and rely, mainly, on enhanced end systems. Such adaptive mechanisms aim at improving the quality by adjusting the amount of traffic in the network and hence reducing losses on the one hand and increasing utilization on the other, which is particularly beneficial for multimedia communication. Thus, at the expense of slight degradation in user satisfaction, it is possible to adjust the amount of network resources consumed by a media stream in accordance with the network congestion state. Additionally, it is usually the case that due to the loss of content or the need to use a considerable overhead for forward error correction a low bandwidth video stream, for example, with no or only low losses can have a higher perceived quality than a high bandwidth yet lossy stream. Further, as the adaptive applications are better suited to take advantage of dynamical changes in the network resource availability during a session, which is particularly beneficial during long-lived sessions typical for multimedia communications, higher network utilization levels can be reached. Such a behavior can be achieved by reservation-based schemes only by renegotiating the reservation, which further increases the complexity of these schemes.

In an analytical study, in [8] Breslau et al. show that by overprovisioning the capacity of a network, a best-effort network can provide the same quality levels as a reservation-based network. Thus, for providing a network supporting a high QoS, there is a clear trade-off between the costs of introducing reservation-based schemes and the thereby required network enhancements, billing, and admission control mechanisms on the one hand and the costs of additional network resources required for a best-effort network in order to provide the same QoS level. For both cases, the costs are considerably decreased when deploying adaptive applications. For the reservation case, the applications require a smaller amount of network resources and less stringent control mechanisms. For the best-effort case, users' satisfaction with the QoS can still be high even with a smaller amount of resources, thus allowing for less overprovisioning of the resources.

Whether the future high-quality networks will be based on strict QoS control mechanisms or simply be overprovisioned will be a question of costs. In any case, efficient adaptation mechanisms will be required to reduce the overall costs.

As a combination between the two approaches, one might consider integrating adaptation schemes with a reservation scheme such as YESSIR [25] or RSVP. Through this integration, the receiver can reserve a minimal QoS using YESSIR and try to achieve a better QoS level through utilizing free network resources using adaptation. Alternatively, an approach similar to the scalable resource reservation protocol (SRP) [1] can be integrated with adaptation. That is, the sender keeps informing the network about its desired transmission rate and the routers indicate the actual reserved capacity. Additionally, the senders can still try to use a higher transmission rate using an adaptation scheme.

Another important issue to consider is the aspect of flow protection, that is, protecting adaptive flows from unresponsive ones that do not reduce their transmission rate during congestion states and might thus lead to the starvation of adaptive traffic. Without protection schemes at the routers, a greedy connection can cause the adaptive flows to face losses and thus reduce their bandwidth share. Lakshman et al. [34] present an intelligent buffer management approach that supports the isolation of responsive from unresponsive flows in a simple way.

22.5 Summary

In this article, we have taken a broad look at different congestion and QoS control mechanisms. While both issues have been subject to a great deal of attention in the literature and research, little has found its way to the Internet as we know it today. Data applications such as FTP and WWW use TCP for congestion control. However, applications based on UDP such as voice over IP (VoIP) and some streaming applications still do not deploy any congestion control schemes. The IETF is currently working on proposing another transport protocol more appropriate to multimedia applications, see [20], which will allow the usage of different congestion control schemes. The success of this protocol is, however, still to be seen. With regard to QoS, the only approach used currently for improving the QoS in a network is to overprovision it, that is, to deploy more bandwidth than is expected to be used. This stems mainly from the notion that QoS architectures are considered to be too complicated to introduce and require a considerable change in the infrastructure of the networks as well as in their management, the required billing structures, and inter-domain relations. However, with the introduction of next-generation networks, there is an increased interest in providing QoS based on tighter control by the providers. The exact development in this area is therefore still to be seen.

References

[1] Almesberger, W., T. Ferrari, and J. L. Boudec, Srp:scalable Resource Reservation for the Internet, in IWQoS'98, Napa, California, May 1998.

[2] Amir, E., S. McCanne, and H. Zhang, An Application Level Video Gateway, in Proceedings of the ACM Multimedia, San Francisco, California, Nov., 1995.

[3] Blake, S., D. Black, M. Carlson, E. Davies, Z. Wang, and W. Weiss, An architecture for differentiated service, Request for Comments (Informational) 2475, Internet Engineering Task Force, Dec. 1998.

[4] Bolot, J.-C., Cost-quality tradeoffs in the Internet, *Computer Networks and ISDN Systems*, 645–651, 1996.

[5] Bolot, J.-C., T. Turletti, and I. Wakeman, Scalable Feedback Control for Multicast Video Distribution in the Internet, in SIGCOMM Symposium on Communications Architectures and Protocols, pp. 58–67, London, England, ACM, New York, Aug., 1994.

[6] Braden, B. and L. Zhang, Resource ReSerVation protocol (RSVP) — version 1 message processing rules, Request for Comments (Proposed Standard) 2209, Internet Engineering Task Force, Oct., 1997.

[7] Braden, R., D. Clark, and S. Shenker, Integrated Services in the Internet Architecture: an Overview, Technical report RFC 1633, Internet Engineering Task Force, June 1994.

[8] Breslau, L. and S. Shenker, Best-effort Versus Reservations: A Simple Comparative Analysis, in SIGCOMM'98, Vancouver, Canada, Sept., 1998.

[9] Carle, G. and E.W. Biersack, Survey of error recovery techniques for IP-based audio-visual multicast applications, *IEEE Network Magazine,* Nov./Dec. 1997.

[10] Chipalkatti, R., J. F. Kurose, and D. Towsley, Scheduling Policies for Real-time and Non-real Time Traffic in a Statistical Multiplexer, in Proceedings of the Conference on Computer Communications (IEEE Infocom), pp. 774–783, Ottawa, Canada, IEEE, New York, Apr., 1989.

[11] Chiu, D. and R. Jain, Analysis of the increase/decrease algorithms for congestion avoidance in computer networks, *Journal of Computer Networks and ISDN,* 17:1–14, 1989.

[12] Claffy, K. and G. Miller, The Natureof the Beast: Recent Traffic Measurements From an Internet Backbone, in INET '98, Palexpo Konferenz Zentrum, Genf, Internet Society, July, 1998.

[13] Floyd, S. and K. Fall, Promoting the use of end-to-end congestion control in the internet, *IEEE/ACM Transactions on Networking,* August, 1999.

[14] Floyd, S., M.Handley, J. Padhye, and J. Widmer, Equation-based Congestion Control for Unicast Applications, in SIGCOMM Symposium on Communications Architectures and Protocols, Stockholm, Sweden, Aug., 2000.

[15] ITU-T/CCITT, Interactive test methods for audiovisual communications, Recommendation P.920, International Telecommunication Union.

[16]   Jacobson,V., Congestion avoidance and control, *ACM Computer Communication Review,* 18:314–329, 1988; Proceedings of the Sigcomm '88 Symposium in Stanford, CA, Aug., 1988.

[17]   Jain, R., A delay-based approach for congestion avoidance in interconnected heterogeneous computer networks, *ACM Computer Communication Review,* 19:56–71, 1989; also Digital Equipment Corporation Technical report DEC-TR-566.

[18]   Kalampoukas, L., A. Varma, and K. K. Ramakrishnan, Explicit Window Adaptation: A Method to Enhance TCP Performance, in Proceedings of the Conference on Computer Communications (IEEE Infocom), San Fransisco, Mar., 1998.

[19]   Kanakia, H., P.P. Mishra, and A. R. Reibman, An adaptive congestion control scheme for real time packet video transport, *IEEE/ACM Transactions on Networking,* 3, 1995.

[20]   Kohler, E. et al., Datagram Congestion Control Protocol (DCCP), Internet Draft, Internet Engineering Task Force, Oct. 2002 [Work in progress].

[21]   Lee, D.C., Effects of Leaky Bucket Parameters on the Average Queueing Delay: Worst Case Analysis, in Proceedings of the Conference on Computer Communications (IEEE Infocom), Toronto, Canada, June 1994.

[22]   Nagle, J., Congestion control in IP/TCP internetworks, *ACM Computer Communication Review,* 14:11–17, 1984.

[23]   Nederlof, L., K. Struyve, C. O'Shea, M.H., Y. Du, and B. Tamayo. End-to-end survivable broadband networks. *IEEE Communications,* 33:63–69, 1995.

[24]   Padhye, J., V. Firoiu, D. Towsley, and J. Kurose, Modeling TCP Throughput: A Simple Model and its Empirical Validation, in ACM SIGCOMM '98, Vancouver, Oct 1998.

[25]   Pan, P.P. and H. Schulzrinne, YESSIR: A Simple Reservation Mechanism for the Internet, in Proceedings of International Workshop on Network and Operating System Support for Digital Audio and Video (NOSSDAV), Cambridge, England, July 1998; also IBM Research Technical report TC20967.

[26]   Ramakrishnan, K., S. Floyd, and B. D . The Addition of Explicit Congestion Notification ECN to ip., Technical report RFC 3168, Internet Engineering Task Force, Sept. 2001.

[27]   Rejaie, R., M. Handley, and D. Estrin, An End-to-end Rate-based Congestion Control Mechanism for Realtime Streams in the Internet, in Infocom'99, New York, IEEE, NewYork, Mar., 1999.

[28]   Sanneck, H., Adaptive Loss Concealment for Internet Telephony Applications, in Proceedings INET '98, Geneva, Switzerland, July 1998.

[29]   Sanneck, H. and G. Carle, Predictive loss pattern Queue management for internet routers, in *Internet Routing and Quality of Service,* Civanlar, S., Doolan, P., Luciani, J., and Onvural, R., Ed., Proceedings SPIE, Vol.3529A, Boston, MA, Nov., 1988.

[30]   Shenker, S., C. Partridge, and R. Guerin, Specification of Guaranteed Quality of Service, Technical report RFC 2212, Internet Engineering Task Force, Sept., 1997.

[31]   Shirish, S.S., ATM Forum Traffic Management Specification Version 4.0, Technical report 94-0013R6, ATM Forum, June 1995.

[32]   Sisalem, D. and H. Schulzrinne, Congestion Control in TCP: Performance of Binary Congestion Notification Enhanced TCP Compared to Reno and Tahoe TCP, in International Conference on Network Protocols (ICNP), pp. 268–275, Columbus, OH, Oct. 1996.

[33]   Sisalem, D. and A. Wolisz., LDA+ TCP-friendly Adaptation: A Measurement and Comparison Study, in Proceedings of International Workshop on Network and Operating System Support for Digital Audio and Video (NOSSDAV), pp. 183–192, Chapel Hill, NC, USA, July, 2000.

[34]   Suter, B., T. Lakshman, D. Stiliadis, and A. Choudhury, Design Considerations for Supporting TCP with Per-flow Queueing, in Proceedings of the Conference on Computer Communications (IEEE Infocom), San Francisco, California, Mar., 1998.

[35]   Tennenhouse, D., J.M. Smith, D. Sincoskie, D.J. Wetherall, and J.Gary. A survey of active network research. *IEEE Communications Magazine,* 35:80–86, 1997.

[36]   Wang, Z. and J. Crowcroft, A new congestion control scheme: slow start and search (tri-s), *ACM Computer Communication Review,* 21:32–43, 1991.

[37]   Wang, Z. and J. Crowcroft, A dual-window model for flow and congestion control, *The Distributed Computing Engineering Journal,* 1:162–172, 1994.

[38]   Wroclawski, J., Specification of the Controlled-load Network Element Service, Technical report RFC 2211, Internet Engineering Task Force, September, 1997.

23

Mobile IP Routing

Karin A. Hummel and
Helmut Hlavacs
University of Vienna

23.1 Introduction

Advances in mobile communication technologies such as wireless network standards IEEE 802.Ha/b/g or Bluetooth, and the high availability of portable computing devices like personal digital assistants (PDAs) or notebooks require additional network protocol support for roaming and ubiquitous computing in the Internet. Using IP, which usually means IP version 4 (IPv4 [Pos81]) unless stated otherwise, seamless roaming of mobile devices is only possible up to the physical and link layer (referring to the seven-layer notion of the OSI Reference Model [DZ83]).

23.2 Mobility on the Network Layer

Network layer mobility in the Internet mainly suffers from the IP addressing and routing strategy. When addressing hosts or, more general, network interfaces, the topological structure of the network is explicitly used. An IP address is built up by a prefix referring to the network of the host, followed by a host address. Based on a dynamically changing and best-effort strategy, IP datagrams are routed to the network stated by the receiver host's IP address.

As a consequence, in case a mobile host moves from one network to another, the network prefix of its IP address has to be changed. This change implicates reconfigurations of address parameters in Internet services and applications. However, changing numerous Internet services does not scale and is not desirable from a network protocol design point of view. Figure 23.1 shows a scenario of a PDA with IP address 192.160.32.80 roaming from its home network with a network number or prefix 192.160.32 to a network using 160.80.40 as the network number. The owner of the PDA demands network management by the foreign network (160.80.40) and the redirection of IP packets from the home network (192.160.32) to its mobile device while being reachable using the existing IP addressing and routing scheme.

The Internet Engineering Task Force (IETF) proposed a model [Per96a], which does not require any IP adaptations to support mobile hosts. Thus, the redirection of IP packets requires a managing *home agent* at the home network. It acts as a proxy fetching IP packets addressed to the mobile host and rerouting them to the foreign network currently visited by the mobile host. In the foreign network, the mobile host borrows an IP address, the *care-of address*. This address can be an IP address assigned by a *foreign agent* (*foreign agent care-of address*, usually the IP address of the foreign agent [Tan03]), which receives the rerouted IP packets and delivers them to the mobile host. In case the mobile host should be addressed directly, a local IP address is acquired (*colocated care-of address*), which can be a long-term address or a temporary-address attached using the *Dynamic Host Configuration Protocol* (*DHCP* [Dro97]), Figure 23.2 visualizes the IETF mobility model [Per98]. Home and foreign agents may either be an extended router or a separate system in the managed network.

In such a scenario, several services are required. First, the mobile host demands mechanisms for discovering the *mobility agent*, either home or foreign agent (*agent discovery and advertisement*). To allow the home agent to address the mobile host in the foreign network, the mobile host has to make the care-of address known to the home agent (*registration*). Finally, routing and transferring datagrams to the roaming mobile host have to be provided by the home agent (*tunneling*).

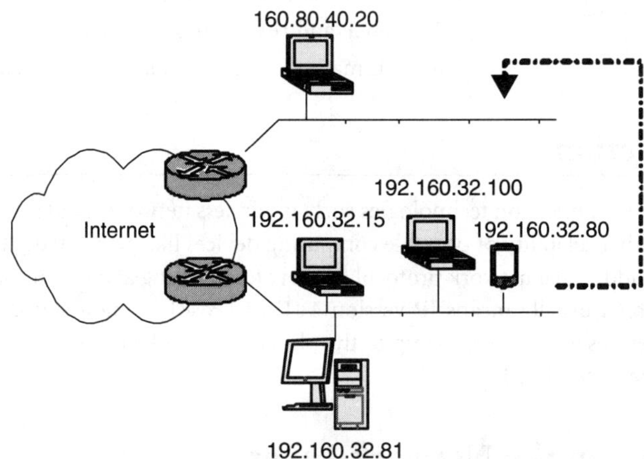

FIGURE 23.1   PDA roaming between different IP networks.

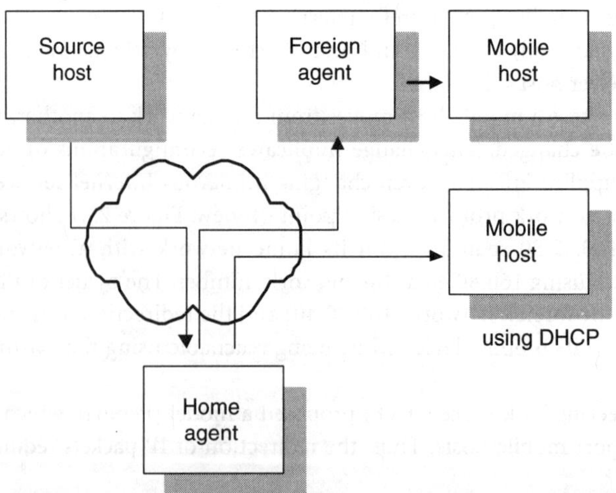

FIGURE 23.2   IETF Mobile IP model.

23.3 Agent Discovery and Advertisement

Whenever a mobile host connects to a network, it needs to find out whether it is linked to the home network. In such a case, it operates without Mobile IP support. In contrast, if a mobile host is connected to a foreign network, it requires a temporary care-of address related to the foreign network. The care-of address is either assigned by a link-layer protocol, or the *agent advertisement* and *agent solicitation* protocol is used.

Agent advertisement messages are sent periodically by mobility agents (home or foreign agents) using broadcasting or multicasting. Using the address information contained in this message, the mobile host is able to contact the mobility agent for further registraion actions. In case the mobile host does not receive an agent advertisement within a reasonable time, it sends an agent solicitation message to force the mobility agent to answer with an advertisement immediately. The interval between two successive solicitation messages has to be iteratively extended using a binary exponential back-off mechanism to reduce message overhead. Both messages are extensions to *Internet Control Message Protocol (ICMP) router discovery messages* [Dee91], used by routers and links to make the routers known to the links. Each ICMP message is encapsulated in an IP packet. Thus, source and destination IP address can be found in the IP header.

To create an agent advertisement message, several ICMP router advertisement message fields are set to selected values and the messages is extended by mobility agent-related fileds. In detail, the IP field *destination address* is required to be either the multicast address for all-systems-on-this-link 224.0.0.1 or the limited-broadcast address 255.255.255.255 and the IP field, TTL (*time to life*) is set to 1, which means that the ICMP message is discarded by the first processing node (e.g., a router or a gateway).

Figure 23.3 describes the ICMP fields in a router advertisement message. As in any router advertisement message, *Type* is set to 9. The field *Code* indicates whether the mobility agent also handles common traffic acting as a router (Code value 0), or not (Code value 16). The *Checksum* is calculated as the 16-bit one's complement sum of the ICMP message and the field *Num addrs* refers to the number of router addresses advertised in this message. The *Addr entry size* field gives the number of 32-bit words for each router address and *Lifetime* controls the maximum number of seconds the advertisement is considered to be valid. Finally, *Router addresses* [*i*] (i=1 … Num addrs) represent possible router's IP addresses followed by a preference value (*Preference level* [*i*]).

Following the router advertisement fields, the agent advertisement extension indicates that the message is an agent advertisement sent by a mobility agent. Figure 23.4 [Per98] shows the fields of the extensions in detail.

<table>
<tr><td colspan="8">0</td><td colspan="8">1</td><td colspan="8">2</td><td colspan="8">3</td></tr>
<tr><td>0</td><td>1</td><td>2</td><td>3</td><td>4</td><td>5</td><td>6</td><td>7</td><td>8</td><td>9</td><td>0</td><td>1</td><td>2</td><td>3</td><td>4</td><td>5</td><td>6</td><td>7</td><td>8</td><td>9</td><td>0</td><td>1</td><td>2</td><td>3</td><td>4</td><td>5</td><td>6</td><td>7</td><td>8</td><td>9</td><td>0</td><td>1</td></tr>
<tr><td colspan="8">Type</td><td colspan="8">Code</td><td colspan="16">Checksum</td></tr>
<tr><td colspan="8">Num addrs</td><td colspan="8">Addr entry size</td><td colspan="16">Lifetime</td></tr>
<tr><td colspan="32">Router address [1]</td></tr>
<tr><td colspan="32">Preference level [1]</td></tr>
<tr><td colspan="32">…</td></tr>
</table>

FIGURE 23.3 ICMP router advertisement message.

<table>
<tr><td colspan="8">0</td><td colspan="8">1</td><td colspan="8">2</td><td colspan="8">3</td></tr>
<tr><td>0</td><td>1</td><td>2</td><td>3</td><td>4</td><td>5</td><td>6</td><td>7</td><td>8</td><td>9</td><td>0</td><td>1</td><td>2</td><td>3</td><td>4</td><td>5</td><td>6</td><td>7</td><td>8</td><td>9</td><td>0</td><td>1</td><td>2</td><td>3</td><td>4</td><td>5</td><td>6</td><td>7</td><td>8</td><td>9</td><td>0</td><td>1</td></tr>
<tr><td colspan="8">Type</td><td colspan="8">Length</td><td colspan="16">Sequence number</td></tr>
<tr><td colspan="16">Registration lifetime</td><td>R</td><td>B</td><td>H</td><td>F</td><td>M</td><td>G</td><td>V</td><td></td><td colspan="8"></td></tr>
<tr><td colspan="32">Zero or more Care-of addresses</td></tr>
</table>

FIGURE 23.4 Mobility agent advertisement extension.

The message *Type* is set to the constant value 16, which identifies the message as an agent advertisement; the *Length* field is calculated by 6+4*N, with *N* being the number of care-of addresses listed at the end of the advertisement *Registration Lifetime* refers to the maximum lifetime value the agent is willing to accept in registration requests (65,535 is used on behalf of infinity). The bits following are used as flags. *R* indicates whether registration with the foreign agent is required. In case of enhancing the reliability of IP datagram redirection, a mobility agent may need to ensure a registration (even if the mobile host has used, e.g., DHCP for care-of address assignment). If the agent is too busy to accept any further registrations from mobile hosts, the flag *B* is set. The flag *H* is set, if the agent offers home agent services on the link, while *F* is set in case foreign agent services are offered. The flags *M* (*Minimal Encapsulation* [Per96b]) and *G* (*Generic Record Encapsulation, GRE* [HLFT94]) give information about the tunneling facilities supported by the agent, while *V* indicates that the agent supports van Jacobson header compression [Jac90].

23.4 Registration

Registration is the method used by mobile hosts to inform the home agent about their care-of addresses. The procedure causes a binding between a care-of address, the home agent, and potentially the foreign agent. Furthermore, it can be used to renew a binding or to deregister in case of returning to the home network. Several care-of addresses may be used simultaneously by one mobile host. After registration, each one is only valid during a *registration lifetime.* Some optional registration steps are the discovery of the home agent in case the mobile host cannot provide information about the home agent's IP address, the selection of tunneling protocols, and the request for van Jacobson header compression.

Registration can either be done via the foreign agent or directly between the mobile host and the home agent. In case the mobile host uses a foreign agent care-of address, it is compulsory to register via the foreign agent. Direct registration is used whenever the mobile host uses a colocated care-of address or it is registering at the home network. Figure 23.5 shows the basic registration message flow where the care-of address is equal to the IP address of the foreign agent (160.80.40.8). Thus, the foreign agent acts as a mediator in the example network structure.

Mobile IP registration uses the *User Datagram Protocol* (*UDP* [Pos80]) on the transport layer. This unreliable protocol without retransmission and ordering facilities is sufficient because retransmission is done via the Mobile IP protocol itself. Besides, an unreliable transport protocol avoids overhead caused by upcoming retransmissions and thus supports wireless links with varying network quality of service far better.

The format of a Mobile IP registration message is a sequence of IP header fields, UDP header fields, the Mobile IP message header, and extensions. The IP fields mainly give information about the source and the destination address, while the UDP fields consist of a destination port with fixed value 434 dedicated to Mobile IP registration and a variable source port.

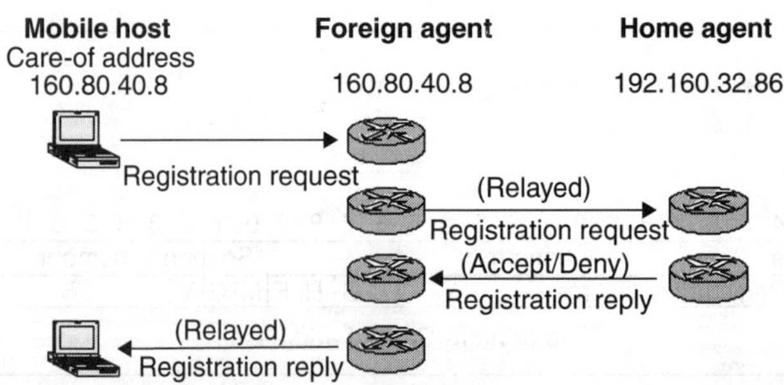

FIGURE 23.5 Mobile IP registration.

Figure 23.6 shows a picture of the Mobile IP fields, starting with a *Type* with fixed value 1 and followed by a sequence of flags. In case the mobile host requests retaining its prior mobility bindings, the *S* flag is set. Setting the *B* flag requests the redirection of any broadcast message received at the home network. The *D* flag is set in case the mobile host itself decapsulates datagrams transferred to the colocated care-of address. The flags *M* or *G* are set, in case Minimal Encapsulation or GRE is provided. *V* is used to enable/disable van Jacobson header compression. After an unused reserved area (*rsv*), the field *Lifetime* states the seconds remaining until the registration is considered to have expired. The following addresses refer to the original IP address of the mobile host (*Home address*), the IP address of the home agent (*Home agent*), and to the datagram tunneling endpoint (*Care-of address*). The *Identification* number is generated by the mobile host and is used for comparing the request with the agent's reply, and for security reasons.

Returning to the home network, the registration request message can be used to deregister all care-of addresses at the home agent by setting the care-of address to the mobile host's home address and setting the lifetime to 0.

To answer the request, a registration reply is sent back from the mobility agent. The most interesting field in the registration reply is an 8-bit *Code* referring to the success of the request [Per98]. In case of acceptance, the Code values 0 or 1 are used (value 1 is used, if simultaneous mobility bindings are not supported). Otherwise, the code indicates the reason for denial, for example, value 66 is used for insufficient resources at the foreign agent, while value 131 is used for a failed mobile host authentication [Per98].

23.5 Tunneling and Route Optimization

To be able to redirect IP datagrams from the home network to the care-of address of the mobile agent, the home agent works as a proxy. First, the home agent fetches all IP datagrams addressed to the mobile host. In the second step, the home agent forwards the datagrams to the registered care-of address.

Since IP addresses are neither used nor understood by the underlying data link layer hardware, a protocol for matching network and data link layer addresses is needed. The *Address Resolution Protocol* (*ARP* [Plu82]) uses broadcast messages to send an IP address to all data link addresses on the network. The host identified by the IP address recognizes its own IP address in the data link broadcast and responds to the sender. Thus, the link between data link address and IP address is found. For a mobile host, ARP is slightly changed. Whenever a mobile host registers at the home agent, the home agent performs *gratuitous ARP* [Ste94] to update ARP cache information of other nodes (routers) in the home network, replacing the home address of the mobile node with its own IP address. Furthermore, the home agent performs *proxy ARP*, that is, sends ARP replies on behalf of the mobile node [Pos84].

After the original datagram has been fetched by the home agent, it is encapsulated into another IP datagram using the care-of address as the new destination address. The encapsulated datagram is received by the owner of the care-of address and is decapsulated. In case a colocated care-of address is used, the decapsulated packet has reached its final destination. Otherwise, if a foreign agent care-of address is used, usually the foreign agent itself is addressed and transfers the decapsulated packet to the mobile host.

```
 0                   1                   2                   3
 0 1 2 3 4 5 6 7 8 9 0 1 2 3 4 5 6 7 8 9 0 1 2 3 4 5 6 7 8 9 0 1
+---------------+-+-+-+-+-+-+-----+-------------------------------+
|     Type      |S|B|D|M|G|V| rsv |           Lifetime            |
+---------------+-+-+-+-+-+-+-----+-------------------------------+
|                         Home address                           |
+----------------------------------------------------------------+
|                          Home agent                            |
+----------------------------------------------------------------+
|                        Care-of address                         |
+----------------------------------------------------------------+
|                                                                |
|                        Identification                          |
+----------------------------------------------------------------+
|                                                                |
|                          Extensions                            |
|                              ...                               |
+----------------------------------------------------------------+
```

FIGURE 23.6 Request for registration: Mobile IP fields.

Using the tunnel metaphor, the home agent and the care-of address owner are the tunnel endpoints. For encapsulation, *IP-in-IP Encapsulation* [Per96c], Minimal Encapsulation, or GRE may be used.

IP-in-IP Encapsulation is used in case no other encapsulation strategy is agreed on. Basically, an outer IP header and some optional additional headers like an authentication header are added. Thus, the original datagram is interpreted as the payload of the new datagram. The original header (now the inner header) is not changed, except the time to life field, which is decremented by 1. Figure 23.7 shows the tunneling in general using IP-in-IP Encapsulation in case of using a foreign agent care-of address.

In case of nonfragmented datagrams, Minimal Encapsulation may be used to save space. A Minimal Encapsulation header is built based on the original destination and the original source address, if this address is available. The outer IP header is built using the original IP header while replacing the source and destination addresses by the tunnel endpoints' IP addresses. Furthermore, the protocol type number is set to the value 55 (for the Minimal Encapsulation protocol).

The most generic encapsulation method is presented by GRE, which fits to a variety of protocols apart from IP. The GRE packet structure consists of a delivery header, where the source and destination addresses (i.e., the endpoints of the tunnel) are included, a GRE header, which can be used for authentication and routing information and the payload [Per98].

The concept of sending IP datagrams to the home agent and tunneling these packets to the care-of address is by its nature inefficient. One topic of route optimization concentrates on overcoming this triangle by using *binding caches* for mobile hosts' care-of addresses at (source) hosts. In case a home agent receives a datagram for a mobile host connected to a remote network, it may send a *binding update* message to the source host, which establishes a direct tunnel to the care-of address as shown in Figure 23.8 via steps 1 and 2. On the other hand, in case a foreign agent receives a datagram for a mobile host that has already left its visited network, it may send a *binding warning* message to the source of the tunnel (e.g., the home agent) to indicate out-of-date cache information [Per98].

In contrast, for example, the handover strategy used in *GSM* networks [MP92], where a handover between a visited *Mobile Switching Center* (MSC) and its successor is specified, the foreign agent is not informed of a new registration of a roaming mobile host (i.e., registration at another network). To make the foreign agents aware of the roaming history, an additional *foreign agent smooth handoff* strategy [Per98] is introduced using an *agent notification extension* in the mobile host's registration request message containing the IP address of the old foreign agent.

Figure 23.8 shows the main messages and actions while performing smooth agent handoff from steps 3 to 6. It is assumed that the mobile host roams between two networks and therefore changes its supporting foreign agent from Foreign agent 1 to Foreign agent 2 (step 3). Step 4 represents the registration procedure, where the IP address of the old foreign agent (Foreign agent 1, 160.80.40.8) is contained in the registration request message. After relaying the request to the home agent (step 5), Foreign agent 2 informs Foreign agent 1 about the new care-of address of the mobile host (step 6). This strategy allows

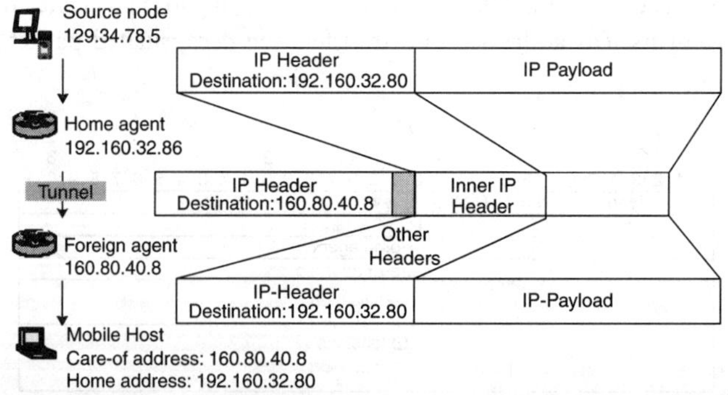

FIGURE 23.7 Tunneling using IP-in-IP Encapsulation.

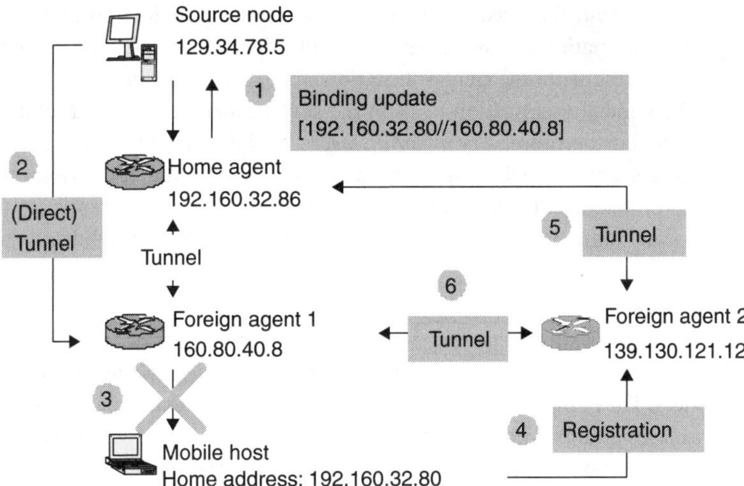

FIGURE 23.8   Route optimization.

the old foreign agent to release any resources assigned to the mobile host accurately and to redirect datagrams still tunneled to the old care-of address.

In case a foreign agent receives a datagram addressed to an unknown mobile host, the foreign agent tunnels the datagram back to the home agent using a *special tunnel* [Per98]. (A mobile host is supposed to be unknown, in case it is neither listed in the visitor list, nor in the binding cache).

23.6   Security Issues

Route optimization uses binding relationships between mobile hosts and several foreign agents. Thus, it facilitates the intrusion of imposters. To secure these relationships, authentication is needed.

During the registration procedure, a *registration key* is used between the anonymous foreign agent and the mobile host [Per98]. In case the mobile host and the foreign host share a security association or the mobile host transmits its public key, the foreign agent can calculate the registration key and send it directly to the mobile host. Otherwise, if security associations both between the mobile host and the home agent, and between the home agent and the foreign agent exist, the home agent calculates the key and transmits it to both partners. The home agent is also used for publishing the registration key in case it knows the public key of the foreign agent. Besides, the registration request/reply messages have to contain a mandatory authentication extension to establish a trusted relationship between the home agent and the mobile host. Although an arbitrary authentication method is possible, at least the *Keyed MD5* (*MD5 Message-Digest Algorithm* [Riv92]) has to be supported, which allows authentication and integrity checking. To avoid replaying of registration messages, the identification field is used for either placing time-stamps or nonces to assure the freshness and originality of the registration message.

23.7   DHCP and IPv6 Issues

DHCP [Dro97] is a mechanism to assign IP addresses dynamically. In case of Mobile IP, the care-of address may be assigned using DHCP services. As a consequence, the mobile host plays the role of the foreign agent itself. Disadvantages can be seen when trying route optimization or agent smooth handoff. If there is no foreign agent in the network (in general), it will not be possible to redirect out-of-date datagrams. Furthermore, the mobile host itself must support decapsulation procedures.

Using *IPv6* (*Internet Protocol version 6* [DH98]), several Mobile IP tasks have to be redesigned because they are much better supported than in IPv4. If it is possible to enable the mobile host to take over all functions of the foreign agent, the need for a supporting agent in the foreign network is eliminated [Per98]. IPv6 supports automatic address configuration, allowing the mobile host to configure its globally routable care-of address without agent solicitation and agent advertisement. Enhanced security options (i.e., encryption and authentication fields in the IPv6 header) render the Mobile IP authentication procedure obsolete. Furthermore, a *neighbor discovery* [NNS98] method may substitute ARP (and therefore proxy ARP and gratuitous ARP). Using this new discovery method, the corresponding nodes are informed about the care-of address by the mobile host itself. The concept of the home agent is still needed with IPv6 to redirect datagrams arriving at the home network using IP6-in-IP6 Encapsulation.

23.8 Summary

To provide mobility support on network level for the Internet, Mobile IP may be used to support and complement IP services. The basic concept uses the model of a home agent, which encapsulates and redirects IPv4 datagrams from a mobile host's home network to the mobile host in the foreign network currently visited (tunneling). To be able to address the mobile host in the foreign network, the mobile host uses a temporarily assigned care-of address, which is usually the IP address of a foreign agent or a colocated care-of address assigned via DHCP. To provide secure and efficient services, route optimization, registration, and authentication techniques, like the MD5, are used. Using the next-generation IP protocol IPv6, some concepts will be changed. Thus, a foreign agent is not necessary any more and enhanced security options may be used.

References

[Pos81] Postel, J., Internet Protocol. Request for Comments 791 (replaces RFC 760) September 1981. http:/www.ietf.org/rfc/rfc791.txt.

[DZ83] Day, J.D. and H. Zimmermann, The OSI reference model, *Proceedings of the IEEE*, 71, 1334–1340, 1983.

[Per96a] Perkins, C.E., Ed. *IP Mobility Support, Request for Comments 2002,* October 1996. http://www.ietf.org/rfc/rfc2002.txt.

[Per98] Perkins, C.E., *Mobile IP Design Principles and Practices,* Prentice-Hall, Englewood Cliffs, NJ, 1998.

[TanO3] Tanenbaum, A.S., *Computer Networks,* 4th ed., Pearson Education/Prentice-Hall, Englewood Cliffs, NJ, 2003.

[Dro97] Droms, R., Dynamic Host Configuration Protocol, Request for Comments 2131, March 1997, http://www.ietf.org/rfc/rfc2131.txt.

[Dee91] Deering, S., ICMP Router Discovery Message, Request for Comments 1256, September 1991, http://www.ietf.org/rfc/rfcl256.txt.

[Per96b] Perkins, C.E., Minimal Encapsulation within IP, Request for Comments 2004, October 1996, http://www.ietf.org/rfc/rfc2004.txt.

[HLFT94] Hanks, S., T. Li, D. Farinacci and P. Traina. Generic Routing Encapsulation (GRE), Request for Comments 1701, October 1994, http://www.ietf.org/rfc/rfcl701.txt.

[Jac90] van Jacobson, Compressing TCP/IP Header for Low-Speed Serial Links, Request for Comments 1144, February 1990, http://www.ietf.org/rfc/rfcl144.txt.

[Pos80] Postel, J., User Datagram Protocol, Request for Comments 768, August 1980, http://www.ietf.org/rfc/rfc768.txt.

[Plu82] Plummer, D.C., An Ethernet Address Resolution Protocol, Request for Comments 826, November 1982, http://www.ietf.org/rfc/rfc826.txt.

[Ste94] Stevens, W.R., *TCP/IP Illustrated,* Volumen 1: The Protocols, Addison-Wesley, Reading, M.A., 1994.

[Pos84] Postel, J., Multi-LAN Address Resolution, Request for Comments 925, October 1984, http://www.ietf.org/rfc/rfc925.txt.

[Per96c] Perkins, C.E., IP Encapsulation within IP, Request for Comments 2003, October 1996, http://www.ietf.org/rfc/rfc2003.txt.

[MP92] Mouly, M., and M.-B. Pautet, The *GSM System for Mobile Communications,* Telecom Publishing, 1992.

[Riv92] Rivest, R., The MD5 Message-Digest Algorithm, Request for Comments 1321, April 1992, http://www.ietf.org/rfc/rfc1321.txt.

[DH98] Deering, S. and R. Hinden, Internet Protocol, Version 6 (IPv6) Specification, Request for Comments 2460, December 1998, http://www.ietf.org/rfc/rfc2460.txt.

[NNS98] Narten, T., E. Nordmark, and W. Simpson. Neighbor Discovery for IP Version 6 (IPv6), Request for Comments 2461, December 1998, http//www.ietf.org/rfc/rfc2461.txt.

24

IP-Mobility for Cellular and Wireless Networks

Andreas Festag
Technical University Berlin

Adam Wolisz
Technical University Berlin

24.1 Introduction

The growing importance of mobile communication deserves support for host mobility in IP-based cellular and wireless networks. Originally, IP networks were designed under the assumption that hosts are stationary. This assumption implies that the IP address does not change and a host is reachable from other hosts by this very address. Data are carried by means of IP packets that contain source and destination addresses. Internet routers inspect the destination address contained in an IP packet. They make a forwarding decision based on the network part of the IP destination address and forward packets to the determined next hop. Consequently, this addressing scheme puts restrictions on the address usage. In particular, an IP address can only be used within the network of its definition. If a mobile host moves to a new network, the old IP address becomes topologically incorrect. Therefore, a new — topologically correct — IP address must be assigned to a mobile host. Such an address assignment allows using standard IP routing; its disadvantage is that existing sessions are interrupted as a session is associated with the old, now invalid IP address. This dichotomy — an IP address represents both host identification and location — is the fundamental problem of mobile IP-based cellular and wireless networks and needs to be overcome by mobility concepts. The requirements for a mobility solution are as follows: (I) Short break in communication during a handover, in particular, for real-time applications. (II) Small protocol overhead, including signaling load, indirect routing, data encapsulation, as well as storage and processing capabilities. (III) Scalability with the number of mobile hosts. (IV) Functional requirements, such as support of heterogeneous end systems and access networks, ease of deployment, as well as location privacy and anonymity.

The mobility concept currently considered to be the most likely candidate for short-term deployment is Mobile IP [RFC3220]. Mobile IP uses additional agents to separate the host identification from the current location and ensures that arbitrary hosts can communicate with a mobile host in an uninterrupted way even while the host moves around. Despite this achievement, Mobile IP has been criticized for its performance problems and for not matching all possible requirements for a mobility concept. Some of these performance problems result from technological developments: higher bandwidth demands necessitate higher frequency bands with a smaller coverage area, resulting in smaller cells with more frequent handovers, thus causing performance degradation. Some requirements are application-dependent: applications might need seamless handover or no packet loss during handover or certain tradeoffs between loss and jitter.

In the following sections, HMIP, the MosquitoNet Mobile IP extensions, RAT, HAWAII, CIP, multicast-based mobility, MPA, ICEBERG, and extended SIP mobility are considered as mobility schemes that attempt to modify, replace, or augment the basic Mobile IP. At first, basic Mobile IP is briefly explained stressing the drawbacks that have triggered the proposal of new schemes for host mobility. Then, the other approaches can be divided into four main categories: (I) extensions of basic Mobile IP preserving the principle of indirect routing and address translation by means of tunneling, (II) host-based routing, (III) multicast-based mobility, and (IV) application-layer proxies (each assigned a separate section). In order to work out the basic assumptions behind the schemes, the motivation to develop a new approach, the required mobility infrastructure, as well as the addressing and routing concept are emphasized.

24.2   Basic Mobile IP and its Weaknesses

With Mobile IP, a mobile host owns an IP home address and in addition gets assigned a temporary care-of-address (CoA) in a visited network. A correspondent host addresses the mobile host via its IP home address. Mobile IP adds two new instances to the network infrastructure: A home agent (HA) and a foreign agent (FA) are executed in IP routers. Routing is performed indirectly using address translation by means of tunneling: a correspondent host directs packets to the mobile host's home network. The HA intercepts and tunnels the packets to the CoA of the mobile host. Tunneling a packet means its encapsulation and forwarding the encapsulated IP packet using the common IP mechanisms. The FA de-capsulates the packets and forwards them via local mechanisms to the mobile host. For the reverse direction from the mobile host to the correspondent host, the mobile host is allowed to send packets directly. This is referred to as triangular routing.

In Mobile IP, the agents periodically send advertisements on the wireless links in order to advertise the offered service and to provide a means for handover detection by the mobile hosts. When the mobile detects that it has moved to a new visited IP network/subnetwork (e.g., it misses a new advertisement from the old FA and receives an advertisement from the new FA) and has obtained a new temporary IP address, it registers its new IP CoA with the new FA. The FA in turn relays the registration to the HA that binds the new CoA to the mobile host's IP home address. Following packets arriving in the home network will be interpreted by the HA and tunneled to the mobile host's new CoA.

In general, the mobility support in IPv6 [PER96] is based on the same main principles as Mobile IP for IP version 4. But in Mobile IPv6, a mobile host is able to create its own CoA using its link-local address and automatic address configuration (combine advertised subnet prefix with own hardware address). Mobile IPv6 introduces two new optional IPv6 header extensions: a Binding Update and a Binding Acknowledgment. These extensions of the header are handled only by the final destination. The mobile host can directly send a Binding Update in the same packets carrying effective traffic to its correspondent host. The correspondent host can then learn and cache the new mobile host's CoA and directly forward data packets to the mobile host without traversing the mobile host's home network.

Mobile IP has a number of problems. Triangular routing causes the first problem: it adds delay to the traffic toward the mobile host. Measurements have shown that Mobile IP increases the delay by 45% in

a campus network [ZHA01]. For two-way real-time communication, this delay is — frequently experienced — not acceptable. Second, encapsulation adds an overhead of about 20 bytes (IP-in-IP encapsulation
[RFC2003]) to each packet potentially resulting in segmentation of IP packets. In comparison to the packet
size of typical real-time applications, such as audio, the overhead is remarkable. For example, the voice
codec G.723.1 [HER00] has a data rate of 5.3 kb/s with a frame size of 20 bytes. With a protocol overhead
for IPv4 (20 bytes), UDP (8 bytes), and RTP (12 bytes), a packet has a length of 60 bytes. Due to the encapsulation the total packet length increases by 33%. Third, triangular routing poses a problem due to ingress
filtering. Ingress filtering is a common security mechanism in IP routers. The mechanism checks for topologically incorrect IP addresses. In Mobile IPv4, the mobile host uses its home address as the source address
to send an IP packet directly to the correspondent host. When a router, such as the access router of the current IP network, examines this packet, the router detects that the packet seems to originate from outside
the network. Therefore, these packets will be discarded. A solution is to use reverse tunneling that again
increases the overhead. Fourth, Mobile IP may cause high handover latency. In Mobile IP, a mobile host
sends a binding update to the home agent. When the home agent receives the binding update, the home
agent starts tunneling IP packets to the new foreign agent. While signaling messages are transported to the
home agent and back, the mobile host is not connected to the network. When the delay between the mobile
host and the home agent is large, the service interruption is unacceptable. Moreover, during the handover
process, packets destined for the mobile host will be misdirected to the old foreign agent. These packets get
lost. The main reason for this performance problem is that in Mobile IP the point for rerouting is located
in the home network that might be distant from the current location of the mobile host. Fifth, in Mobile
IP, the mobile host sends a binding update to the HA each time the mobile host re-registers for refreshing
the binding update in the network. When the current location of the mobile host is distant from the HA,
the aggregate signaling traffic traverses many routers in the network and poses a considerable load.

Some of the weaknesses are solved in Mobile IPv6. Nevertheless, in view of the high number of
installed systems using IPv4 and the unclear evolution of networks from IPv4 to IPv6, a mobility solution for IPv4 is needed. Moreover, some of the problems — such as the performance problems with handover — still exist in IPv6.

24.3 Extensions of Basic Mobile IP

Hierarchical Mobile IP and MosquitoNet extensions modify the basic Mobile IP. RAT simplifies basic
Mobile IP and attempts to replace it. However, all of these proposals are based on indirect addressing and
address translation.

Hierarchical Mobile IP (HMIP)

The Mobile IP extension of hierarchical foreign agents [FOR99] addresses a drawback of Mobile IP: if the
distance between the FA and the HA is large, the signaling delay for the registration may be long, which
then results in long service disruption and high packet losses. Therefore, the FA functionality is distributed to several routers. These FAs can be configured in a tree-like structure. The highest foreign agent
(HFA) is the root of the hierarchy; the lowest foreign agent (LFA) is close to the mobile host. Intermediate
FAs are on the path between the HFAs and the LFAs. The intermediate FA that belongs to the old and the
new path at a handover event is called the switching FA. The classification of the FAs is only conceptual.
The hierarchy may collapse down to a single FA, as in the original Mobile IP approach.

The LFAs send announcements including their own address and the address of the next higher level.
When a mobile host first arrives at a visited domain, it sends a registration request to the LFA that creates
an unacknowledged binding update and forwards the registration request upward to the next higher FA.
This initial registration request creates an address binding in every FA on the path, and finally in the HA.

When a handover occurs, the mobile host generates a registration request that is forwarded by the LFA
(Figure 24.1, the text beside the boxes indicates the routing entries for the mobile host before and after
the handover, respectively). At some point, the switching FA receives the request and detects that a

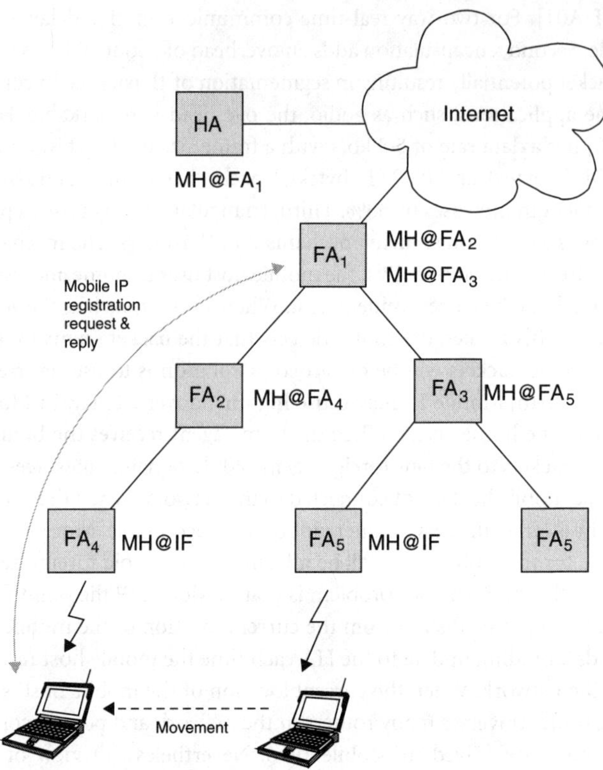

FIGURE 24.1   Handover process in hierarchical Mobile IP.

binding for the mobile host already exists but is coming from a different FA at a lower hierarchical level. This is interpreted as a local handover. The switching FA replies to the mobile host with a registration reply message.

MosquitoNet Extensions of Mobile IP

Other extensions have been proposed by the MosquitoNet group in [ZHA01]. These extensions address some of the problems of Mobile IP, namely the protocol overhead and the growing application of ingress filtering in border routers that prevents the usage of triangular routing in Mobile IP. Instead of modifying the basic Mobile IP principles, the MosquitoNet extensions allow a mobile host to use transparent mobility utilizing the Mobile IP infrastructure or not using it. In the latter case, the mobile host accesses the Internet directly using a temporary IP address in the visited network incorporating the Mobile IP infrastructure. Moreover, the mobile host can decide whether to use triangular routing or bidirectional routing. Mobile IP route optimization (triangular routing) fails when router ingress filtering is used: packets are dropped when they do not carry a topologically correct IP source address. Therefore, a mobile may use the more robust bidirectional tunneling, although it implies an additional overhead.

In the context of this approach, the support of multiple interfaces in a mobile host is essential. Each interface carries a temporary IP address. A flow can be bound to a specific interface. This is done with the help of a socket option. For data transmission, the route lookup has been modified, so that only routes with that specific interface are considered. For reception of data, a flow-to-interface binding (a flow is characterized by IP addresses and port number) is sent to the Mobile IP home agent that forwards datagrams to the appropriate CoA. The handover is similar to Mobile IP, but extends the protocol by an update of the flow-to-interface binding in the Mobile IP home agent. The extension is mainly intended for vertical handover of mobile hosts with multiple network interfaces.

The basic Mobile IP approach is extended by the following functionalities:

A Mobile IP specific routing table extends the regular IP routing table. The Mobile IP-specific routing table contains the information as to whether to use transparent mobility or indirect routing, as well as triangular or bidirectional tunneling on a per-socket basis.

The Mobile IP home agent can manage multiple CoAs for a single mobile host simultaneously and bind a flow to a certain interface.

The protocol supporting registration between the mobile hosts and the home agent is extended.

These functionalities facilitate the flexible usage of Mobile IP in order to cause the protocol overhead only if it is necessary and to obviate the problem with ingress filtering.

Reverse Address Translation (RAT)

The RAT approach [SIN99] is motivated by the limited deployment of Mobile IP. It is intended to simplify mobility support in order to narrow the gap between the lack of applications, which require mobility support, and the poor deployment of Mobile IP. The RAT approach can be considered as a tradeoff: on the one hand, it dispenses with the requirement to maintain TCP connections. On the other, overhead is decreased and most of the traffic can be routed directly. Moreover, the inventors of RAT argue that implementation of Mobile IP functionality is operating system dependent (e.g., registration, tunneling, etc.), whereas RAT aims at a solution that is independent of the operating system.

In the RAT approach, the mobile host owns an IP home address and acquires a temporary IP address in the foreign network. The RAT approach adds new entities to the home network: a registration server and a RAT device. The network infrastructure remains unchanged. In particular, there are no mobility-specific entities required in the foreign network. The RAT approach applies Network Address Translation (NAT) [SRI01]. NAT is an Internet paradigm for extending the IP address space in IP version 4, and also supports the security when used in firewalls.

When a correspondent host sends a packet to the mobile host, it directs the packet to the home address of the mobile host. In the home network, the RAT device intercepts the packet and performs a network address translation. It replaces the destination address with the mobile host's temporary address and the source address with the address of the RAT device. Then, the packet is sent directly to the mobile host without tunneling. In the reverse direction, the mobile host sends a packet to the RAT device, which in turn performs the address translation and sends it to the correspondent host. This scheme is referred to as *reverse address translation (RAT)*. One of the main advantages of this approach is that the indirect routing is deployed for correspondent host initiated sessions only. When the mobile host initiates the session, it will use its temporary address (which is topologically correct) and communicate with the correspondent host directly. The authors argue that the mobile host initiates most of the sessions, and thus no indirect routing via the home network is required. This results in shorter routes and does not increase the packet length by encapsulation, without maintaining mobile host-initiated sessions during handover.

24.4 Host-Based Routing

With host-based routing, an IP address is considered as a unique host identifier without interpreting the network part as a location identifier. The network nodes route IP packets according to tables or caches indexed by host identifiers. The entries in the routing tables/caches are created by means of mobility-related signaling. HAWAII and Cellular IP supplement basic Mobile IP instead of replacing it.

Handoff Aware Wireless Internet Infrastructure (HAWAII)

HAWAII [RAM99] was proposed since Mobile IP results in high control overhead and high latency for local mobility. Also, HAWAII eases the usage of resource reservation protocols (such as RSVP) in a mobile environment, where a mobile host acquiring a new CoA on each handover would trigger the establishment of a new resource reservation.

HAWAII defines a domain as a division of the wireless access network under the administrative control of a single authority. The domain consists of routers and access points. All of them are mobility-enabled by supporting HAWAII-specific signaling in order to optimize routing and forwarding. The router interconnecting the HAWAII domain and the Internet core network is called *foreign domain root router*. Each access point has Mobile IP foreign agent functionality (without decapsulation).

In the HAWAII approach, we distinguish between two types of mobility: *intra-domain* handover and *inter-domain* handover. For both cases different mechanisms are defined. The first case is supported by HAWAII and the second case is supported by Mobile IP. Both cases will be explained below.

In the HAWAII approach, a mobile host has a home domain (similar to the home network in Mobile IP) and a temporary unicast IP address. The home domain may support the HAWAII protocol. When the mobile host is in a foreign HAWAII domain the temporary IP address is assigned once to the mobile host and does not change as long as the mobile host stays in the domain. No address translation mechanism is required, and the Mobile IP home agent is not notified of the mobile host's movement. Instead, connectivity is maintained by using dynamically established paths in the foreign HAWAII domain based on host entries in the routing table of selected routers. Thus, a HAWAII-enabled access network does not rely on IP routing in the sense of routing based on the network's portion of the IP address.

For global mobility support, HAWAII reverts to traditional Mobile IP mechanisms. At first, the case is considered where the mobile host is within the HAWAII home domain. In this case, the mobile host carries a unicast (quasipermanent) IP address. When the mobile host powers up, it sends a Mobile IP registration message to the present access point. The access point then propagates a HAWAII path setup message to the domain root router using a configured default route. Each router in the path between the mobile host and the domain root router adds a forwarding entry for the mobile host. Finally, the domain root router acknowledges to the access point. The access point in turn sends the Mobile IP registration to the mobile host. Packets for the mobile host are sent to the domain root router based on the subnet's portion of the mobile host's IP address. The packets are routed within the domain using the host-based forwarding entries. It is important to note that the entries are in a soft state being kept alive by periodic hop-by-hop messages.

When the mobile host moves within the HAWAII domain (Figure 24.2), the mobile host registers with the new access point by sending a Mobile IP registration request. The new access point then sends a HAWAII path setup update message to the old access point. The old access point performs a routing table lookup for the new access point and adds a forwarding entry for the mobile host's IP address. Then, the message is sent to the upstream router. This router performs similar operations. If the router receiving this message is the crossover router,[1] then this router adds a forwarding entry to the new access point and packets for the mobile host are sent to the new access point. The path via the old access point will time out. This scheme is called *forwarding path setup* scheme since the HAWAII path setup update message is sent from the new to the old access point, and the old access point forwards packets to the new access point only for a limited time. This scheme is optimized for networks where the mobile host listens/transmits to only one access point simultaneously. An alternative scheme is the *nonforwarding* scheme, which is optimized for networks where the mobile host is able to listen/transmit to two or more access points simultaneously. In this path setup scheme, the path setup update message travels from the new access point to the old access point via the cross-over router. Thus, packets are not forwarded from the old access point.

In order to interact with Mobile IP, the mobile host is assigned a co-located CoA from its HAWAII foreign domain. A correspondent host directs the packets to the mobile host's home address. The Mobile IP home agent intercepts the packets and tunnels them to the HAWAII foreign domain root router with the network portion of the outer IP address. This foreign domain root router and the following routers forward the packets according to its host-based routing entries.

The HAWAII approach differentiates between active and idle users as well as appropriate states for the mobile host. For an active user, the network knows the mobile host's current access point, and for an idle

---

[1] This router has a route to the old and the new access point via the same interface.

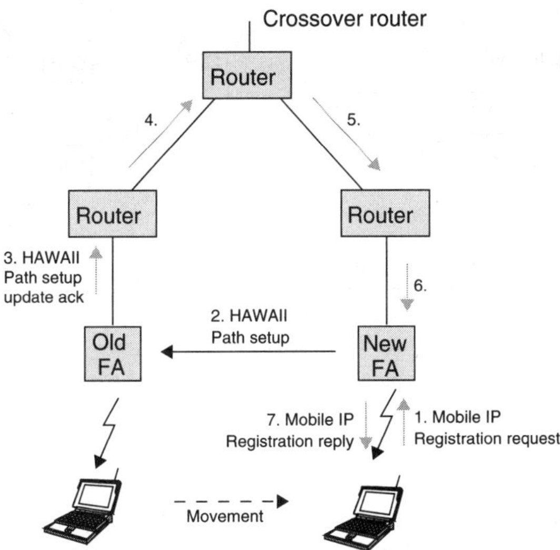

FIGURE 24.2   Forwarding path setup scheme in HAWAII.

user the network only knows the access point approximately, such as a set of access points. When packets for an idle mobile host arrive, the network pages the mobile to determine the mobile's current access point.

Cellular IP (CIP)

Cellular IP [VAL98] envisions a networking environment with ubiquitous computers where highly mobile hosts often migrate during active data transfers and the users expect minimal disturbance to ongoing sessions. The authors argue that Mobile IP is not an optimal solution, because it is optimized for macro-level mobility and relatively slowly moving hosts. Moreover, it is stressed that Mobile IP does not scale for a large number of mobile hosts, since every handover between Mobile IP FAs generates a binding update irrespective of the fact as to whether the mobile host is idle or active.

The Cellular IP approach proposes a hierarchical mobility management that separates global from local mobility. For global mobility, Mobile IP is applied to support handover across the Internet backbone. To support local mobility within the Cellular IP access network, regular IP routing is replaced with routing of packets hop-by-hop via lookup in specific tables (referred to as caches). The entries in the tables have a limited lifetime and time out (soft-state).

In a Cellular IP network, a mobile host is assigned a unique identifier that is used to route packets. It is not required that a mobile host has an IP CoA. For simplicity reasons, the unique identifier is an IP address (e.g., home address) that makes inter-working with Mobile IP easier. However, this is not really required, since within the Cellular IP access network no IP routing is performed.

For mobility support, Cellular IP adds a gateway router (GW) and Cellular IP (CIP) nodes to the network infrastructure. A gateway router interconnects the Internet backbone and the Cellular IP access network. The Cellular IP nodes are located in the Cellular IP access network and can be considered as access points working at the network level. They execute the Cellular IP protocol. It is not required that they be equipped with a wireless interface (if not, they act as a regular network node).

The global mobility support in Cellular IP is provided by basic Mobile IP. The Gateway router is colocated with a Mobile IP FA. The mobile host registers the gateway's IP address with its Mobile IP HA. Packets from a correspondent host are first routed to the Mobile IP HA and then tunneled to the gateway. The gateway detunnels packets and forwards them toward the access points. As long as the host is interconnected to the same access network, local mobility is concealed from the agent in the gateway router.

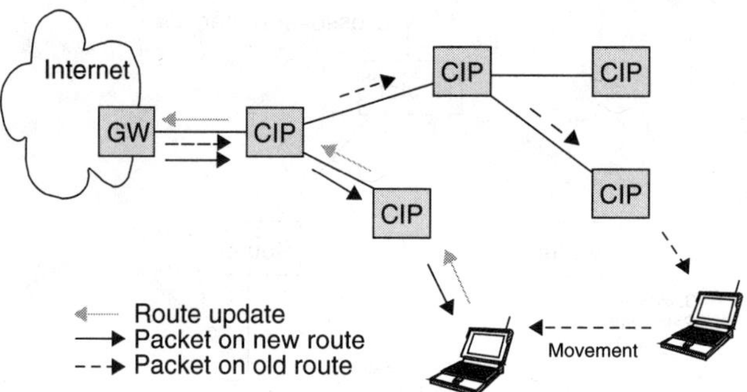

FIGURE 24.3   Hard handover process in Cellular IP.

The local mobility support works as follows (Figure 24.3): inside the Cellular IP access network, nodes are provided with a paging cache and a routing cache. Both contain mappings between mobile host identifiers and node ports (output port similar to a router port) on a soft-state basis. Paging caches are available in a few nodes. A paging cache is updated by data originating from the mobile host (data packets or specific signaling packets). The paging cache is used to locate a mobile host when there is no routing cache entry. In that case, the Gateway Router caches the IP data packets in order to send a paging packet to the mobile host across the Cellular IP nodes. The mobile host replies to that paging packet and creates routing cache entries in every node along the route. Now, the cached IP packets can be sent along this route without address translation and tunneling. Paging cache and routing cache entries are cleared by timers, with different timeout values: the routing cache timeout is on the order of several IP packets, whereas the paging cache timeout is set according to the handover frequency. Thus, an idle and active mobile host can be managed separately with different caches.

When a handover occurs, two cases have to be considered. In the first case, the mobile host generates a route update message when it enters the new cell in order to update the route caches in those nodes where the old and new routes diverge. After the route caches are updated, data packets are sent to the new location of the mobile host via the new route. For a limited time, the old and the new routing cache entry can exist in the routing cache and data packets are sent via the old and the new route. This is used for semisoft handover. In the second case, the routing cache entry in the Gateway was cleared, triggered by a timer, and a new paging packet is generated to locate the mobile host. The explicit search causes a small delay in sending packets; but it allows longer timeouts decreasing the amount of signaling packets.

24.5   Multicast-Based Mobility

The main idea of multicast-based mobility support is to utilize location-independent addressing and routing to support host mobility. In principle, each mobile host is assigned a multicast group address — a special-purpose address that does not contain a network part and, therefore, is independent of the location. The mobile host subscribes to the multicast group through its current access point to the network. Handover is performed by multicast operations, namely subscribe/un-subscribe to/from a multicast group. Data packets are distributed via a multicast tree with branches reaching the current locations of the mobile host on its movement. The branches of the multicast tree can grow and shrink, and hence, follow the mobile host's location. In particular, the branches can be simultaneously set up to multiple locations, including the current as well as the expected locations of the mobile host.

In comparison to the classical solution for IP mobility support, Mobile IP, a multicast approach has the following advantages:

Re-routing for handover is done in a router where the paths to the old and from the new access point diverge (and not in a software agent in the mobile host's home network as in the Mobile IP approach).

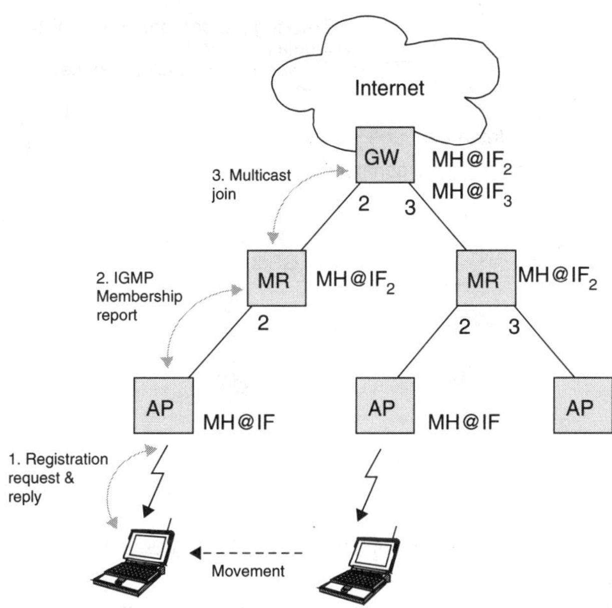

FIGURE 24.4   Handover process in multicast-based mobility support.

A handover-specific signaling and infrastructure is in principle not required; instead, multicast is reused for mobility purposes.

Multicast flexibly offers inherent mechanisms to minimize the service interruption caused by a handover between access points. It can be used to distribute IP packets to a group of access point that buffer the data. When a mobile host registers with one of the access points, data packets are already available and can immediately be forwarded to the mobile host.

Initially, it was proposed to use the classical any-source multicast (ASM) service model of the Internet (using IGMPv2 and DVMRP as multicast protocols) [MYS97]. It was proposed to put the multicast endpoint into the access point and to use an adapted signaling protocol on the wireless link [WU00]. [FES03] proposes to use alternative multicast service models other than ASM for mobility support.

A handover process for multicast-based mobility support with IGMPv2 [RFC2236] and PIM-SM [RFC2362] as multicast protocols is shown in Figure 24.4: a mobile host registers with the new access point (AP). The AP sends an IGMP unsolicited membership report to its designated multicast router (MR). The MR sends a PIM join message, which is forwarded up to the MR already having members of the mobile host's multicast group (rerouting node). In the example in Figure 24.4, the rerouting node is the gateway. The gateway starts forwarding packets on its interface 2, whereas the routing entry for the branch rooted at the gateway's interface 3 will time out.

24.6   Application-Layer Proxies

With application-layer approaches, the support for host mobility is independent of the underlying network-layer components. In particular, these approaches aim at mobility support between different types of access networks (e.g., GSM, WLAN) that are interconnected by the global Internet.

Mobile People Architecture (MPA)

The main goal of the Mobile People Architecture [MAN99] is to maintain a person-to-person reachability while preserving the mobile user's privacy. In the Mobile People Architecture, a *Personal Online ID* identifies

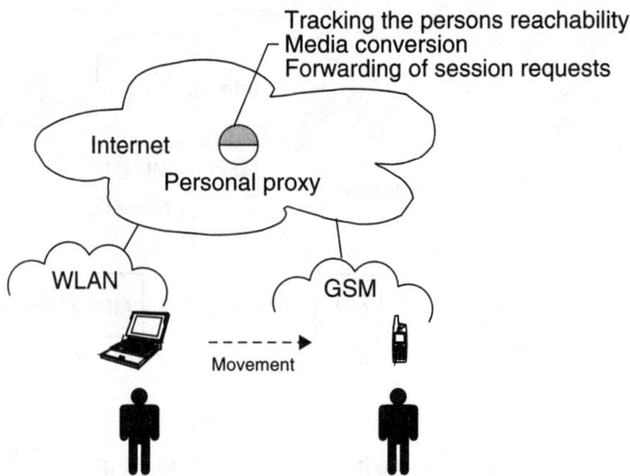

FIGURE 24.5   Mobile People Architecture (MPA).

a user. Additionally, a user is addressed by application-specific addresses (ASAs). Mobility is supported by mapping the Personal Online ID to ASAs.

In the Mobile People Architecture, a new entity is added to the network (Figure 24.5). This entity is called a *personal proxy* and acts as a *person-level router*. (The person level is added to the communication layer model on top of the application level.) The personal proxy tracks the user's current reachability, converts media, and forwards data to a specific end system. It is located in the mobile host's home network (if any), or is offered by a trusted third-party server.

When a user wishes to communicate with the mobile *person,* a call (call is regarded as a kind of session) is directed to the Personal Proxy, and then to the mobile person's preferred end system. When the reachability of the mobile person changes, the tracking agent updates the proxy state. The update can be done in a scheduled manner, manually or automatically.

It is assumed that local mobility is handled within the access network and hidden from the Personal Proxy. The case where a user changes the end system can be regarded as vertical handover. Then, the user updates the Personal Proxy (manually or automatically) and new calls will be directed to the user's new ASA. The case where a user changes the ASA while receiving service is not being considered.

ICEBERG

The motivation of the Internet CorE BEyond thiRd Generation (ICEBERG) [WAN00] is the current diversity of access networks, end systems, and services, in particular, traditional telephony services and data services. Therefore, ICEBERG aims at supporting personal mobility in the sense of seamless access to services independent of the access network and end system. It is intended to give the control of the communication to the callee, and not to the caller. In ICEBERG, a user can be uniquely identified (by means of a *unique-id*). Additionally, the user is associated with one or several *service-ids* (e.g., phone number, email address, and IP address). To achieve mobility, the unique-id is mapped to the service-id.

The ICEBERG network architecture consists of the Internet Core and several different access networks (e.g., GSM, PSTN, WLAN). At the interface between the core network, and an access network, an ICEBERG access point (IAP) transforms services (media converter). Additionally, ICEBERG adds service agents to the core network: preference registries, personal activity tracker (PAT), and extended naming services. The preference registry stores user preference profiles that can be modified by user interaction or by the PAT, which gives inputs about location information.

When a correspondent user wishes to call the mobile user, the call is routed to the IAP. In the access point, a name service lookup is performed, the preference registry of the called user is located, and the preferred end system is determined. After that, the call is established via the correspondent interface. The IAP can perform a service conversion (e.g., fax to jpeg).

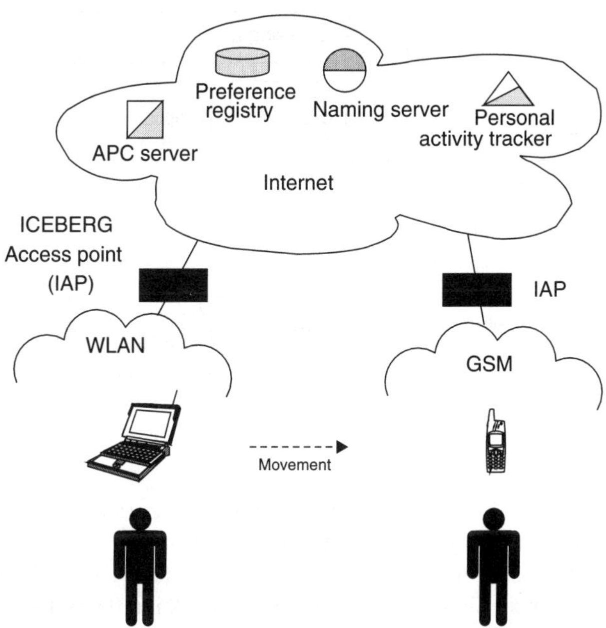

FIGURE 24.6   ICEBERG network architecture.

The ICEBERG approach focuses on user mobility between several access networks (Figure 24.6). It is implicitly assumed that host mobility is transparently supported in the access networks by technology-specific handover schemes (e.g., for GSM, IEEE 802.11).

Extended SIP Mobility

Extended SIP Mobility [WED99] is a mobility approach that utilizes the application-level signaling capabilities of the SIP protocol [RFC2543]. The motivation of the extended SIP mobility can be found in drawbacks of the Mobile IPv4 approach. The authors argue that for real-time traffic over IP, which is mostly RTP [RFC1889] over UDP traffic, there is a need for fast handover, low latency, and high bandwidth utilization. Mobile IPv4 suffers from indirect communication, which increases the delay and causes an overhead due to tunneling, thus decreasing bandwidth utilization.

The extended SIP mobility approach introduces mobility awareness at a layer higher than the network layer. SIP already supports user mobility, and the approach is meant to extend SIP as an application-layer signaling protocol in order to support end-system mobility. The main assumption behind the extended SIP mobility approach is that a mobile user is identified by a unique address (e.g., user@realm). This unique address is mapped to the current IP address of the mobile user's end system. No explicit home IP address is required. SIP introduces a SIP agent on the user's side, and a SIP server (SIP redirect server or proxy server) and location server to the network infrastructure.

User mobility is supported by means of the original SIP protocol: when a user wishes to initiate a session, an invitation is directed to the SIP server, which in turn queries the location server for the current IP address of the mobile user's end system. The SIP server sends the invitation to the called user. The invitation contains the IP address of the callee. If the mobile user moves, the location server is updated, and new sessions will be set up to the new IP address.

Host mobility with this scheme is mainly understood as an increased roaming frequency and as a change of an IP address during an ongoing session. Assuming that a session is already established, the mobile registers the new temporary address with the location server and the mobile reinvites the correspondent host with the same session identifier and the new temporary address (in the contact field of the SIP message). The session can be continued, although the IP address has changed.

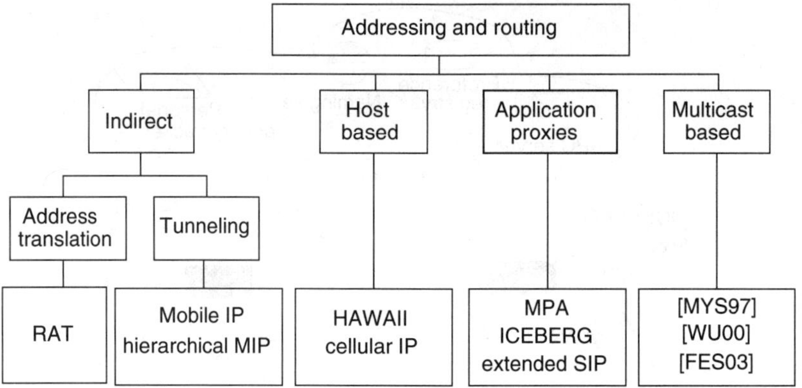

FIGURE 24.7   Classification of IP mobility approaches.

It is important to note that SIP does not support TCP, but UDP traffic only. Therefore, extended SIP mobility supports UDP traffic only. For TCP traffic, it is proposed to use Mobile IP. In fact, both approaches can coexist: for TCP traffic, Mobile IP is applied and for UDP traffic the extended SIP mobility approach is applied. For the simultaneous usage of network interfaces, the MosquitoNet approach of a mobile routing table [ZHA01] is adopted.

24.7   Summary

The basic Mobile IP represents the classical approach to host mobility in IP-based cellular and wireless networks. Despite its achievements, its performance problems have triggered the development of alternative approaches. In particular, it has been recognized that a single solution that attempts to fit all requirements, like Mobile IP, results in high costs for mobility support. The differentiation between global and local mobility, between active and idle hosts, as well as between user and host mobility meet the requirements for host mobility in an economic way. The categorization of existing approaches is shown in Figure 24.7. It can be seen that hierarchical Mobile IP and RAT are based on the same or very similar principles as basic Mobile IP. Hierarchical Mobile IP achieves a shorter service interruption by propagating the handover signaling messages only to a crossover node at the expense of additional infrastructure (hierarchy of foreign agents). RAT simplifies the basic Mobile IP in order to accelerate the hesitant deployment of mobile IP services, but does not support full transparency to the upper protocol layers. Host-based routing facilitates the integration of location management and routing. While in Mobile IP, location management is handled by Mobile IP signaling operations, routing is executed as an overlay on the existing IP routing. With host-based routing, the routes can be updated simultaneously with the location information since the route entries are host specific. Host-based routing, however, replaces standard IP routing. Multicast-based mobility support utilizes the capability of multicast as it potentially exists in future IP networks. It offers great flexibility for handover mechanisms that realize a tradeoff between service interruption/packet losses and protocol overhead. While the former approaches deal with local mobility support within an access network and can work with basic Mobile IP[2] for global mobility, the application-layer proxies attempt to replace (or at least supplement) Mobile IP for global mobility.

References

[FOR99] Forsberg, D., J.T. Malinen, J.K. Malinen, T. Weckström, and M. Tiusanen, Distributing Mobility Agents Hierarchically under Frequent Location Update, in Proceedings of MOMUC'99, pp. 159–168, San Diego, CA, USA, Nov., 1999.

---

[2] or even require Mobile IP as the HMIP approach.

[FES03] Festag, A., H. Karl, and A. Wolisz, Classification and Evaluation of Multicast-Based Mobility Support in All-IP Cellular Netwoks, in Proceedings of KiVS 2003, pp. 233–244, Leipzig, Germany, Mar., 2003.

[HER00] Hersent, H., D. Gurle, and J.-P. Petit, IP Telephony: Packet-Based Multimedia Communications, Addison-Wesley Professional, 2000.

[MAN99] Maniatis, P., M. Roussopoulos, E. Swierk, K. Lai, G. Appenzeller, X. Zhao, and M. Baker, The mobile people architecture, *ACM Mobile Communication Review* (MC2R), 3:36–42, 1999.

[MYS97] Mysore, J. and V. Bharghavan, A New Multicast-Based Architecture for Internet Mobility, in Proceedings of MobiCom'97, pp. 161–172, Budapest, Hungary, Oct., 1997.

[PER96] Perkins, C. and D. Johnson, Mobility Support in IPv6, in Proceedings of MobiCom'96, pp. 27–37, Rye, NY, USA, Nov., 1996.

[RAM99] Ramjee, R., T. LaPorta, S. Thuel, K. Varadhan, and S. Wang, HAWAII: A Domain-Based Approach for Supporting Mobility in Wide-Area Wireless Networks, in Proceedings of ICNP'99, pp. 283–292, Toronto, Canada, Oct./Nov. 1999.

[RFC1889] Schulzrinne, H., S. Casner, R. Frederick, and V. Jacobson, RTP: A Transport Protocol for Real-Time Applications, Internet RFC 1889, January 1996.

[RFC2003] Perkins, C., IP Encapsulation Within IP, Internet RFC 2003, October 1996.

[RFC2236] Fenner, W., Internet Group Management Protocol, Version 2, Internet RFC 2236, November 1997.

[RFC2362] Estrin, D. et al., Protocol-Independent Multicast-Sparse Mode (PIM-SM): Protocol Specification, Internet RFC 2362, June 1998.

[RFC2543] Handley, M., H. Schulzrinne, E. Schooler, and J. Rosenberger, SIP: Session Invitation Protocol, Internet RFC 2543, March 1999.

[RFC3220] Perkins, C., IPv4 Mobility Support, Internet RFC 3220, January 2002.

[SIN99] Singh, R., Y.C. Tay, W.T. Teo, and S.W. Yeow, RAT: A Quick (And Dirty?) Push for Mobility Support, in Proceedings of the Second IEEE Workshop on Mobile Computing Systems and Applications, pp. 32–40, New Orleans, LA, USA, Feb., 1999.

[SRI01] Srisuresh, P. and K. Egevang, Traditional IP Network Address Translator (Traditional NAT), Internet RFC 3022, January 2001.

[VAL98] Valko, A., Cellular IP – A new approach to internet host mobility, *ACM Computer Communication Review*, 29:50–65, 1999.

[WAN00] Wang, H. et al., ICEBERG: An internet-core network architecture for integrated communications, *IEEE Personal Communications*, 7:10–19, 2000.

[WED99] Wedlund, E. and H. Schulzrinne, Mobility Support Using SIP, in Proceedings of WoWMo'99, pp. 76–82, Seattle, WA, USA, Aug., 1999.

[WU00] Wu, J. and G.Q. Maguire, Agent-Based Seamless IP Multicast Receiver Handover, in Proceedings of PWC'2000, pp. 213–225, Gdansk, Poland, Sept., 2000.

[ZHA01] Zhao, X., C. Castelluccia, and M. Baker, Flexible network support for mobile hosts, *MONET Special Issue on Management of Mobility in Distributed Systems*, 6:137–149, 2001.

25

IP-QoS: Scalable and Flexible Quality-of-Service with Differentiated Services

Klaus Wehrle
International Computer Science Institute (ICSI)

25.1 Introduction

In the last decade, the Internet has undergone an exponential growth and currently, there are no signs for an end or attenuation of this trend. In fact, with the convergence of currently separated networks, like telephone, radio, and television, the Internet will even accelerate its growth.

The need for more predictable and quality-based services than today's best-effort delivery is obvious. The Internet evolved from a purely scientific network to an important part of today's business and society. More and more activities are performed on-line, extending the variety of applications in the network constantly. For example, more and more companies use the Internet as a platform to perform their business-to-business (B2B) and business-to-customer (B2C) relationships. However, this increasing importance of the Internet also places strong requirements on its availability and resilience, and even more on the quality of its service. Not just multimedia applications, which are mostly named to motivate the need for value-added services, force the development and deployment of Quality-of-Service (QoS). More and more business applications would benefit from certain assurances. Moreover, the non neglectable market of distributed online games with their special needs also make it attractive for Internet Service Providers (ISP) to think about a deployment of services better than the traditional best-effort delivery.

Based on the Internet Protocol (IP), the Internet is only able to provide a best-effort delivery of data packets. This is implied by its fundamental design principles to keep the network core simple and to move as much actions as possible into end-systems. This allows the network to be scalable and able to continue its exponential growth. Routers in the core only have to perform routing lookups and forward data packets accordingly to the next hop. If a router receives more packets than it is able to forward these will be delayed or even dropped, if the capacity of queues is exceeded. Each packet is treated equally. This is sufficient for typical Internet applications like eMail or World Wide Web. In case of congestion, packets must be dropped. However, when more than best-effort delivery is needed, special actions must be taken, as will be shown in the next section.

This chapter introduces the Differentiated Services (DiffServ) approach and focuses on how a scalable and flexible QoS support can be realized for the Internet. First, Section 25.2 introduces basic elements necessary for an efficient support of quality-based services, and Section 25.3 provides a short insight into the interesting history of QoS approaches for the Internet and why they failed. Section 25.4 presents the Differentiated Services approach describing the main design principles and the overall architecture. After that, Section 25.5 continues by describing the elements of quality-based services in the DiffServ approach. Finally, Section 25.6 gives an introduction on how DS-services can be managed during the introduction phase and how this could be done dynamically in the future. Section 25.7 closes the chapter with a brief summary of the Differentiated Services architecture.

25.2 Elements for Realizing Quality-of-Service

Before taking a closer look at the details of the DiffServ architecture and previous QoS approaches, the following section discusses substantial components of quality-based communication systems. An overview of all functional units necessary for an efficient realization of QoS in packet-based communication systems is given.

The realization of quality-based services is tightly coupled with the management and scheduling of resources. A communication service needs certain resources, for example, CPU time, buffer space in queues, or a portion of a link's bandwidth capacity. Therefore, all QoS-enabled communication systems must consist of special components, whereby their concrete characteristics can be different in each individual case. Nevertheless, the following basic components can be identified and either be arranged in the data path or the control plane (cf. Figure 25.1).

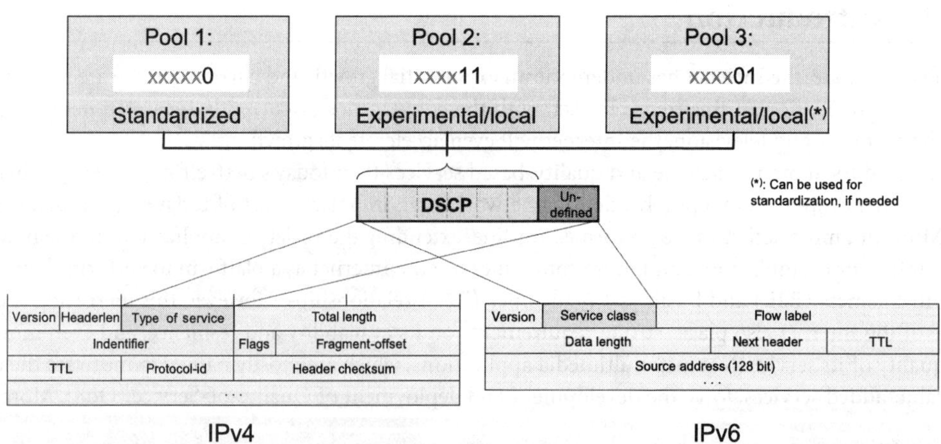

FIGURE 25.1 Elements for realizing Quality-of-Service.

Data Path Components

The data path is defined by the path that data packets take in end-systems and routers, and by the components and mechanisms that deal with the treatment and forwarding of these packets. In the case of packet-based communication systems, the following components can generally be identified in the data path:

The *resource scheduling* is the main component in the data path. It carries out the final differentiation between data streams and thus realizes QoS. Generally, the resource scheduling consists of several components, which apportion the different resources (CPU cycles, buffer space, and bandwidth) to the respective packets, data flows, or aggregates. However, it is necessary that a reservation of the respective resources takes place before they are used. This is usually done via traffic profiles, which contain appropriate information about the exact treatment and the reserved resources of a data stream.

A *usage control* component monitors the data streams in QoS-enabled systems in order to supervise the compliance between resource usage and resource reservations. Therefore, incoming traffic is examined by traffic meters to check its conformity with the associated traffic profile. If necessary, misbehaving traffic can be treated according to the specification in the service-level agreement.

Control Plane

The control plane manages quality-based communication services and their resources. It does not perform any action on data packets.

A *resource manager* administers and assigns the resources mentioned above in a communication system. It has an exact overview of assigned and free resources and stores their assignment to certain data streams in traffic profiles.

The task of the *resource reservation* is the distribution of reservation requests between resource managers. This is usually accomplished over dedicated protocols. By separating resource reservation and resource management, several mechanisms or protocols for the resource reservation can be used simultaneously.

The *resources-based access control* decides on the acceptance of reservation requests. This decision can — next to resource availability — also be based on certain other rules, specified in policies, for example, political interests, agreements with other service providers, etc.

Accounting and billing components perform the task to log and prove concrete resource usage and tariff this to the service user.

The previously described elements are necessary both in data path and in control path to set up an efficient support for quality-based services in the Internet. The following description of failed approaches emphasizes this. Their failure was mostly implied by the lack of one of these elements.

25.3　Evolution of QoS in the Internet

The decision to create the IP as an unreliable and connectionless protocol was made at the beginning of the 1970s. The substantial goal — at that time certainly influenced by the military background of DARPA — was to develop a redundant network in which communication should remain possible, especially after the failure of some systems or links.

The fundamental problem with failing systems is the loss of state information, which normally has to be stored persistently. Therefore, the design decision was taken not to keep or manage status information in intermediate systems and to design all protocol mechanisms in a way that a system can configure itself again after a failure within a short time [14]. Thus, forwarding systems are scalable and can be prevented from carrying an exploding amount of state information. They have only to perform one main task, which is the fast forwarding of packets, and thus can be optimized thereupon.

However, a communication system cannot be designed completely without status information, because at certain points data on connections and their states must be stored. In order to avoid a heavy load on the components within the network, the Internet tries to store this necessary information in the involved end systems as often as possible. In case of a failing end system, only the connections of the failed system — which anyway is no longer operational after a loss — are concerned by the failure (*fate sharing*). This principle of moving the complexity to end systems and to keep the core as simple as possible, is often named as the "*Internet paradigm*" [14].

The following subsection presents a selection of former approaches to realize QoS support in the Internet. Each of them failed due to certain reasons, but mainly either the Internet paradigm was not considered or some of the important elements for QoS support (described in Section 25.2) were missing.

IP Type-of-Service

Since its beginning, the IP offered some support for a different treatment of data packets, which is the fundamental principle behind QoS. But as can be seen, much more is necessary to obtain real quality-based services.

The *Type-of-Service*-field in the IP header allowed to distinguish between different priorities (from 0 as the worst to 7 as the highest priority) and different preferences (low delay, high throughput, high reliability) for a packet [25]. However, there was no architectural framework that determined the concrete usage of the Type-of-Service-field and thus, no explanation of how QoS could be implemented with this approach across the whole Internet. Also, there was no real concept of how network resources should be assigned to ToS-priorities and -preferences and how their usage can be controlled and billed in the end. In practice, although some cases of usage exist, the ToS-field was ignored by applications as well as routers.

Stream Protocol

The Internet Stream Protocol ST (and later ST-II) [5] aimed to add a second network protocol next to IP to support real-time traffic in the Internet. ST was — in contrast to IP — a connection-oriented network protocol that allowed a reliable transport of data packets and, in conjunction with the Stream Control Message Protocol (SCMP), the negotiation of QoS-parameters for a connection.

Although ST-II has been standardized by the IETF, it failed to be deployed in the Internet due to various reasons. The main reason was obviously the establishment of a second network protocol next to IP, which contradicts the traditional hour-glass-model of the Internet with IP as the central protocol. The second reason for ST-II's failure was its connection-oriented nature, which was influenced by the telephone network idea of a connection as the main communication principle [11]. As outlined before, the character of connections requires state information in the network that needs to be maintained persistently and thus breaks with the principles of the Internet paradigm. Furthermore, the expected amount of connection and reservation states would lead to a similar scalability problem as in the case of the failure of the Integrated Services architecture, which is described in the next section.

Integrated Services Architecture

The Integrated Services (IntServ) approach [10] was the IETF's first attempt to develop a comprehensive support for quality-based communication services in the Internet. After the standardization of multicast and the initial experiences with the MBone, the necessity for a QoS-support in the Internet became apparent. Due to bandwidth bottlenecks and fluctuations within the utilization of the network, congestion and packet loss occurred increasingly. Particularly for multimedia applications developed at that time, the prevailing network conditions were insufficient for a satisfying communication.

The substantial goal during the development of the IntServ architecture was the support of multimedia applications by quality-based services without fundamental changes in the Internet architecture. The new services should especially offer support for group communication next to the traditional unicast. Audio and video applications, such as conferencing support were usually mentioned as the main examples. Quality-based services were deployed by a dedicated scheduling of individual data flows in each router. This enabled multiple applications to share one resource reservation, for example, in audio conferences where several participants share one reservation for an audio channel, which is only used by the actual speaker.

In the IntServ architecture network, resources are explicitly assigned to a data flow in each router, for which dedicated reservation and access control mechanisms are needed. To distribute resource reservation requests, the Resource Reservation Protocol (RSVP) [9, 3] was developed. The final decision about accepting a reservation is taken by each router along the path. This implies that every router administers its resources locally and maintains a traffic profile explicitly for each quality-enhanced data stream.

Consequently, flow-specific data must be kept persistently within the core network, thus violating the Internet paradigm. The design decision proved to be problematic since the scalability of the IntServ-architecture cannot be ensured in a constantly growing Internet with an increasing number of data streams [3].

25.4   The DiffServ Approach

The previous examples showed that the realization of quality-based services in the Internet is a goal that has been targeted by numerous approaches so far, but has not been accomplished satisfyingly yet. Various problems led to the failure of the discussed architectures, above all the promising IntServ-approach, which is not suitable for an Internet-wide deployment due to its scalability problems [3]. Nevertheless, the realization of quality-based services in the Internet does not seem to be impossible and a new promising attempt was started, in which the experiences from the failed beginnings were considered

The DiffServ architecture [8] — often abbreviated as DiffServ or DS — was developed from 1998 to 2003 by the Internet Engineering Task Force (IETF). The main goal of the DS working group was to develop a scalable and flexible architecture for the realization of quality-based services in the Internet. According to the definition of scalability in [8], the DiffServ architecture will be scalable if and only if it still works efficiently in an environment with clearly larger characteristics. This means, for example, an environment with a larger number of users or a larger number of flows, even when these parameters grow by several orders of magnitude compared to the situation today.

Further design goals of the DiffServ architecture are considered more detailed in the following section.

Requirements and Goals of DS

The standard RFC 2475 explicitly points out the requirements on the DS architecture regarding the dynamic and unexpected growth of the Internet [8]. Also, it is expected that the number of Internet users, the available bandwidth, and the number of the flows will grow irresistibly, whereas a continuation of the exponential growth is not precluded. Likewise, the variety of applications using the Internet as infrastructure will increase. Consequently, an architecture for quality-based services has to take into account these expected developments and has to support them in an adequate and scalable manner — not only for the foreseeable future, but for a much longer timescale.

As a result, the following demands to the DS architecture have been defined in RFC 2475:

Scalability: As pointed out before, scalability is the main design principle of the DiffServ architecture. Using this, the problems of the IntServ architecture should be avoided, which maintains per flow state information in each router in contrast to the principles of the Internet paradigm [14, 24]. Current router architectures are not capable of performing such a fine-grained classification. Thus, the main design goal of DiffServ is the realization of quality-based services without the need to maintain per-flow-based state information in the core network.

Flexibility and extensibility: DiffServ should also offer many different services to end users, but it should not be the responsibility of the IETF to specify concrete services. Instead, the focus is on defining a basic architecture, on top of which individual services can be realized. The deployment of concrete QoS mechanisms and consequently concrete services is the responsibility of each ISP. This enables them to react faster and more flexibly to their customers' needs and to current market situations.

The IETF will only define basic mechanisms to realize quality-based services without keeping concrete services or certain applications in mind. The architecture should assure the interoperability between different ISPs but should also allow them to define mechanisms and services of their own choice. Concrete services will not be standardized but should evolve very quickly according to the needs of the market. Consequently, the realization of a *flexible* and *extendable* architecture is the second challenge of the DiffServ architecture next to scalability.

Next to these two main challenges, the DiffServ architecture should also fulfill the following demands — which partially result from the need for scalability, flexibility, and extensibility — but are explicitly mentioned in RFC 2475:

Compatibility to legacy applications: Legacy applications should be able to use a network running DiffServ without any need for modification. This implies that no changes on software or user interfaces are required in end systems for taking advantage of the new services.

Separation of traffic conditioning and quality-based forwarding: Mechanisms for traffic monitoring and conditioning should be separated — especially in the core of the network — from mechanisms realizing the differentiation of packets. This can significantly reduce the complexity of the overall architecture and improve scalability and flexibility.

Abdication of a perhop signaling: The DS architecture should not rely on a *hop-by-hop* signaling, as for example, the RSVP protocol in the IntServ architecture. Such a signaling method would finally reinduce the maintenance of states in the core network and thus lead to scalability problems.

The requirements can be summarized by the attributes *scalability*, *flexibility*, and *extensibility*. Based on these three building blocks, the IETF working group *"Differentiated Services"* developed the DiffServ architecture described next.

The Architecture of DiffServ

This section presents the major design principles of the DiffServ architecture. It will mainly be shown how the demands drawn up previously can be fulfilled. The choice of the services finally deployed by an ISP is not the responsibility of the IETF or the DiffServ architecture.

Carrying State Information Instead of its Maintenance

The main goal of the DiffServ architecture is to assure scalability, especially with having a continuously growing Internet in mind. The consideration of scalability in terms of QoS-support focuses mainly on the scalability of maintaining state information in the core network.

As indicated in Section 25.2, the realization of quality-based services in a packet-based network implies some kind of connection, which leads inevitably to maintenance of state information. Thus, it is necessary in a QoS architecture to have a so-called traffic profile for each flow using QoS, in which at least the following information is maintained:

Classification data for identifying a data flow and for assigning its appropriate traffic profile. Normally, for this mapping, a combination of several header fields is used, whereas the following five are the most common: IP source and destination addresses, TCP/UDP source and destination ports, and the protocol identifier in the IP header.

A specification of the *QoS parameters*, for example a specification of the service quality the data stream should experience. These parameters must be negotiated between the service user (customer) and the service provider (ISP) before service usage and should be codified in a Service Level Agreement (SLA).

Operational data arise only at the runtime of a service. Examples are runtime variables, such as the number of tokens in a token bucket meter, and also data about the usage, which can later be used for billing and accounting.

This information is necessary to support quality-based services in the Internet and must therefore be maintained. For example, in the IntServ architecture, this information was stored in each router, and thus led to scalability problems. However, the only way to avoid this problem seems to be by removing the traffic profiles from the inner part of the network, where the expected amount of information cannot be maintained in a scalable way. Consequently, the basis for the classification of individual data streams and their assignment to resources and service classes is missing.

The DiffServ architecture follows exactly this path. It removes the complexity from the core network and shifts it toward the less frequented edges — according to the Internet paradigm (cf. Section 25.3). The state information cannot be completely shifted into end systems since this would induce modifications to system interfaces and applications. Above all, ISPs would lose control over SLAs, since service users can arbitrarily modify them and assign themselves the best services.

For this reason, the traffic profiles are mainly stored in the first router of a communication path, which are typically located in less loaded access networks [3, 8]. The quantity of data streams arising in such low frequented access networks is quite small compared to the Internet's core and therefore the administration of state information can be accomplished efficiently. As will be addressed in Section 25.6, after a certain migration phase the administration of this data can be maintained in dedicated management units and the first nodes will only receive a copy of the data to perform usage control.

The crucial question now is: how can routers in the inner network decide which flows should experience a certain quality-based treatment without the necessary information? The answer to this question is the main design principle of DiffServ, which makes it a scalable and flexible architecture for quality-based services:

After the first router — which still maintains traffic profiles for each data stream — has classified a data packet and retrieved its associated traffic profile, a small tag identifying the service class will be written in the IP packet header. Consequently, each following router can easily determine the Per-Hop Behavior of a packet marked with this little tag called *Differentiated Services Codepoint* (DSCP). A detailed definition of a forwarding behavior will be given in Section 25.5.

To store the DiffServ codepoint in IP, the meaning of the Type-of-Service-field in the IPv4-header and of the Traffic-Class-field in the IPv6-header was changed in RFC 2474 as illustrated in Figure 25.2 [2]. The first six bits in this field store the DiffServ codepoint, which specifies the forwarding behavior (PHB)

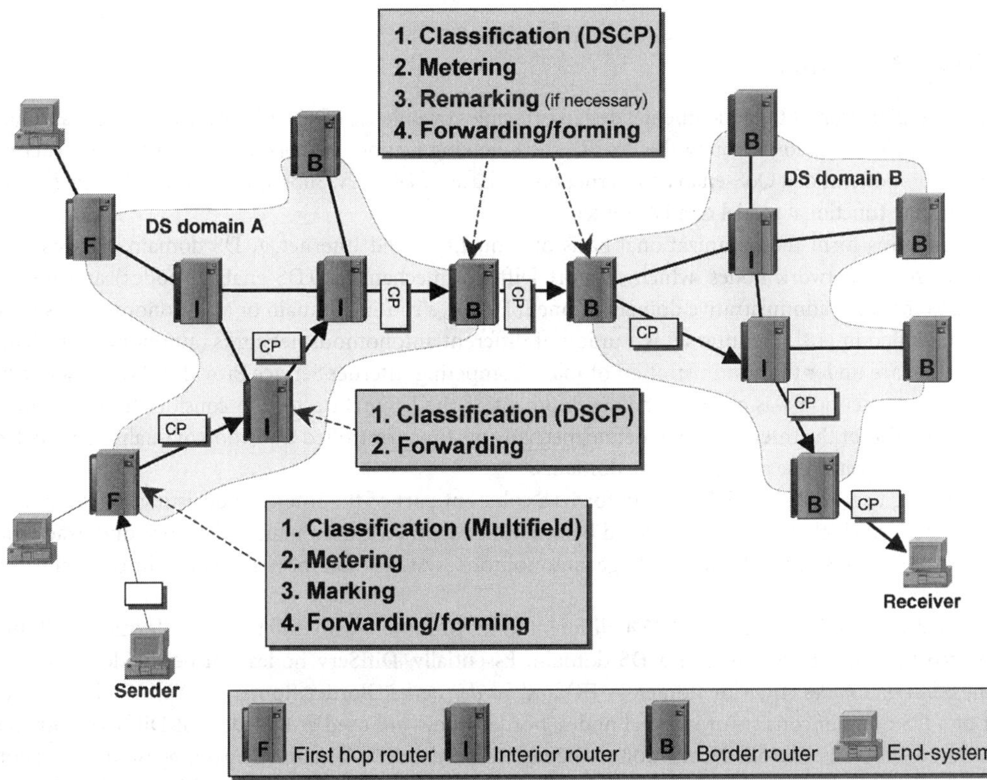

FIGURE 25.2 Format of IPv4 and IPv6 packet headers containing the DSCP-field defined in RFC 2474.

of an IP packet. The DSCP can take values from 0 to 63, whereby only a maximum of 32 is currently released to be used in normal Internet traffic. The remaining 32 values are reserved for test purposes and a potential later standardization.

Consequently, the packet carries the necessary state information in itself; thus, this information does not have to be maintained by the routers, especially by those in the core network. The scalability of the architecture is assured, because no status information per data stream has to be stored anymore. But due to the limited information a DSCP can handle, some difficulties can arise:

Per-Hop Behavior: The DSCP cannot carry further information than the specification of the concrete forwarding behavior. Thus, sometimes special information about individual QoS characteristics can be missing, for example, the reserved bandwidth.

Aggregates instead of individual data flows: With the absence of classification and QoS parameters, it is no longer possible for an internal router to differentiate individual data streams with regard to their different QoS criteria. The only level of differentiation is the assigned Per-Hop Behavior, specified by a DSCP.

Therefore, the set of data streams with the same DSCP will be called an *(behavior) aggregate (BA)* in the following. In the core network, only aggregates can be differentiated. Individual data streams within an aggregate cannot be differentiated inside a DiffServ network due to the missing classification information.

Thus, the differentiation of data flows within the network bases on the DSCP stored in the IP header and all flows carrying the same codepoint will be treated as one aggregate. The isolation of individual data flows, as practiced in the Integrated Services architecture, is given up in favor of scalability.

This raised the question of whether a six bit long codepoint field, which specifies only the forwarding behavior, can be a real replacement for detailed information about individual data flows. This question was the topic of many discussions on the DS working group mailing list, for example, the discussion on aggregation in the Expedited Forwarding PHB and assumptions on their effects [13, 12].

DiffServ Domains

After the explanation of the fundamental design principles (avoidance of states in the network core and carrying service information through the DSCP), the following section presents the administrative structure and the components of a QoS-enabled Internet on the basis of DiffServ. Subsequently, the following section addresses the functional model of a DS-router.

DS domains form the organizational basis of a quality-based Internet. A DS domain consists of a coherent set of network nodes, which support DiffServ mechanisms (DS-enabled nodes) and mostly belong to the same administrative domain of one ISP (e.g., a routing domain or an autonomous system).

As described in [18], the Internet is a union of different autonomous networks (autonomous systems — AS) that are under the administration of many competing Internet Service Providers [16]. Therefore, considerable attention was given to the development of the DS architecture to consider these administrative domains of the Internet and to define mechanisms for a controlled transition of quality-supported data flows between these administrative domains.

According to RFC 2475, a DS domain forms a coherent part of the Internet consisting of several DS-capable nodes, which are all administered by one ISP. Usually, a DS domain will correspond exactly to one autonomous system. However, large autonomous systems can be subdivided into several DS domains.

As illustrated in Figures 25.3 and 25.4, different kinds of routers with different complexities and functions can be distinguished within a DS domain. Essentially, DiffServ nodes can be divided into two groups: *internal nodes* (Interior Router — IR) and *border nodes* (Border Router — BR). While the interior of a DS domain consists of internal nodes, border nodes are used at the edges of DiffServ domains in order to ensure a controlled transition to the next administrative domain. Depending on the direction of a data flow, border nodes can be distinguished either as Ingress Border Routers (IBR) or Egress Border Routers. Also, the first and the last border nodes of an end-to-end communication path receive special

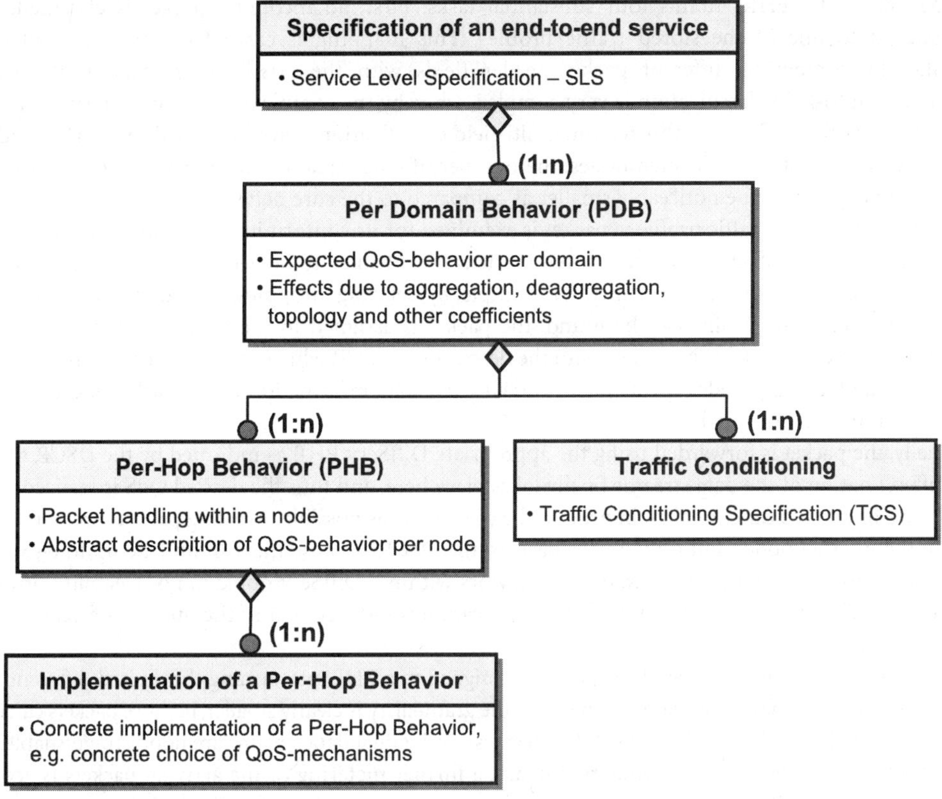

FIGURE 25.3 Components in a DiffServ domain.

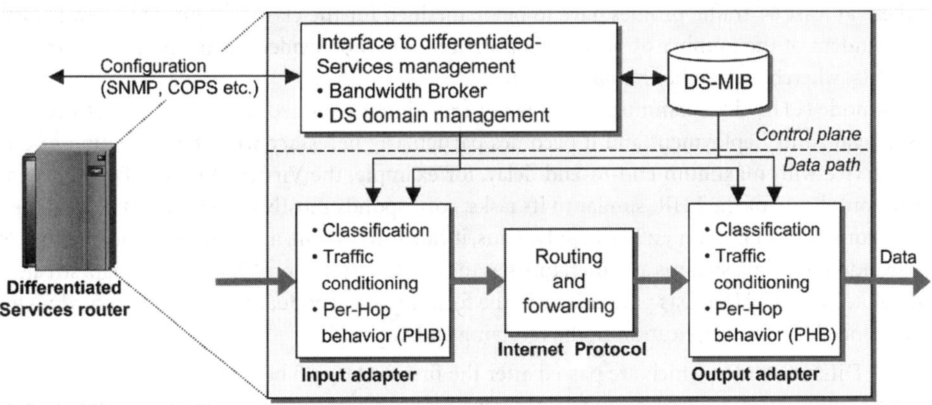

FIGURE 25.4 Hierarchical classification of different DiffServ-router types.

consideration as First Hop Router (FHR) and Last Hop Router (LHR) due to their special duties. It has to be noted that this partitioning is only a logical distinction according to the regarded direction of a data flow. These systems are not physically separated. Thus, for example, a FHR acts as a LHR, when the focus is on the reverse direction. This also emphasizes the unidirectional character of DiffServ.

In the following, the exact functions of the respective DS node types are described, whereby the succession of the explanation corresponds exactly to the path of a packet in a DiffServ-enabled Internet (cf. Figure 25.3):

The first node (FHR) fulfills four substantial tasks. First, an incoming packet is classified, that is, assigned to one of the stored traffic profiles. This mapping is carried out with a multi-field classifier by comparing different packet header fields with the classification data in the traffic profiles. A multi-field classification is very complex, whereby the complexity increases linearly with the number of traffic profiles. For this reason, multi-field classification is accomplished in the DS architecture only in the first node. In such nodes, the number of data streams and thus the number of traffic profiles are expected to be noticeably smaller as compared to the core network.

After mapping to a traffic profile, a packet is examined for its conformity concerning the service class and the negotiated QoS characteristics stored in the traffic profile. In the negative case, appropriate QoS mechanisms can enforce the conformity, for example, by delaying, dropping, degrading the packets, etc. If no matching traffic profile can be found, the packet is assigned to the default traffic profile (Best Effort). In each case, a packet is marked with the determined DSCP, which specifies the forwarding mechanisms in the following nodes (principle of carrying the state information in the packet and not maintaining it in the network nodes).

Finally, the packet is forwarded using the appropriate DiffServ PHB as indicated by the DSCP, that is, the differentiation of the data streams finally takes place here, and thus the desired QoS is realized.

IR have only basic functionality, in order to be as scalable as possible. Arriving packets are also classified in the internal nodes, but only on the basis of the DSCP. Because this can only be up to 64 values, which are statically assigned to forwarding behaviors within a DiffServ domain, no scalability problem occurs here. Therefore, the codepoint-based classification is independent of the number of data streams and traffic profiles.

According to the classified DSCP, the packet is assigned to a PHB, performing the desired differentiated treatment of data streams. Compared to the FHR, the granularity is clearly smaller, because packets can only be differentiated according to their DSCP (aggregates). Individual data streams are nondistinguishable.

In the border node (BR) of a DiffServ domain, a further metering of the arriving packets is accomplished next to the PHB-based forwarding. However, no examination of individual data streams (also called *microflows*) takes place, but only a conformance check on the basis of aggregates is carried out, which consists of all packets with the same DSCP. Therefore, in border nodes, a codepoint classifier is used, where at least 64 traffic profiles have to be maintained for the classification. This way, DS BRs are also independent of the number of passing data streams and independent of the number of traffic profiles in FHRs, whereby scalability remains ensured.

The last node (LHR) in a communication is not explicitly mentioned in RFC 2475. However, there is no reason against its deployment and it becomes particularly necessary with the realization of a deterministic service with maximum end-to-end delay, for example, the Virtual Wire Per-Domain Behavior [20]. The complexity of the LHR, similar to its tasks, corresponds mostly to that of a FHR, and the same scalability considerations as for an FHR apply. Thus, it can also become necessary that the LHR has to differentiate individual data streams and therefore has to maintain traffic profiles for those data streams. The substantial need for a LHR exists particularly in the forming of individual data streams according to traffic patterns needed by an application in the receiving end system.

Thus, all DiffServ nodes, which are passed after the first node (and before the last node) along a way, differentiate only between aggregates with different PHBs. All packets with the same DSCP on a link belong to the same behavior aggregate, and experience the same forwarding behavior.

Functional Model of a DiffServ Node

Quality-based services are realized by a differentiated treatment of packets in routers and end systems. There are numerous algorithms and methods to change the forwarding behavior of a data stream. Since the IETF wanted to develop the DiffServ architecture as flexible as possible, no concrete mechanisms and thus no special services have been standardized. The actual realization of quality-based services is the responsibility of an ISP. Nevertheless, the IETF created a functional model of a DS node [6], allowing manufacturers and network providers to orient themselves on it (cf. Figure 25.5).

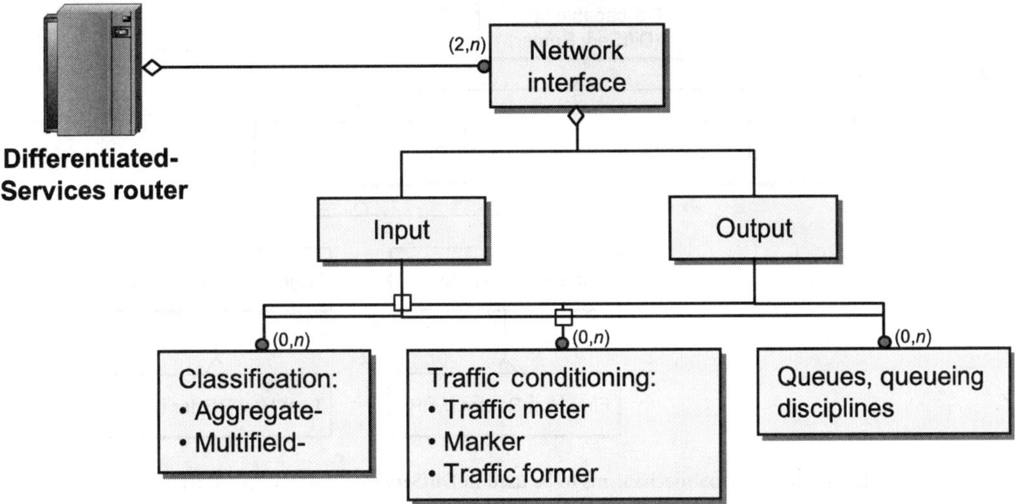

FIGURE 25.5   Components in a DiffServ node.

Thus, according to RFC 2475, the mechanisms used in a DS node can be divided into the following categories, which are illustrated in Figures 25.3 and 25.5:

Classification elements map packets to their appropriate traffic profiles and thus to their PHBs. Classifiers have no direct influence on packets. However, they decide what treatment a packet will experience in the further processing.

Elements for *traffic conditioning* change the actual characteristics of a data stream by performing operations on its packets. This operation does not always have to be active; consider for example, a dropper or a traffic shaper. It can also have an indirect impact, for example, a traffic meter influences the further treatment of a packet.

Queues and *dequeue disciplines* are traditional elements for performing a differentiated treatment on packets, for example, First-In-First-Out, Earliest-Deadline-First, etc.

The actually used algorithms and functions depend mainly on the position and on the kind of node considered. Internal nodes will usually use only DSCP classifiers and appropriate queueing mechanisms in the output interfaces for realizing a PHB. On the other hand, border nodes will have additional functionality, such as multi-field classifiers, metering elements and markers, and droppers or traffic formers.

The concrete implementation of a PHB is not standardized, so that an ISP can select the mechanisms for its implementation. Likewise, in RFC 2475, it is not described as to how these mechanisms have to be combined to form a concrete forwarding behavior (*Traffic Conditioning Block* — TCB), that is, which interfaces exist, which rules have to be considered, etc.

The next section takes a closer look at how in the DS architecture, quality-based services and their components can be described and specified, so that inter-operable end-to-end services can be deployed between different Internet Service Providers and their administrative domains.

25.5   Elements of DS Services

The goal of a QoS architecture for the Internet is the support of quality-based network services for applications. This implies that applications are the real user and therefore these services have to be provided from end system to end system over the entire network (*end-to-end*). Services for individual end-to-end data streams are realized by a differentiated treatment of their packets within the forwarding systems and by traffic-conditioning actions (cf. Figures 25.5 and 25.6). In the case of DS, an end-to-end service is

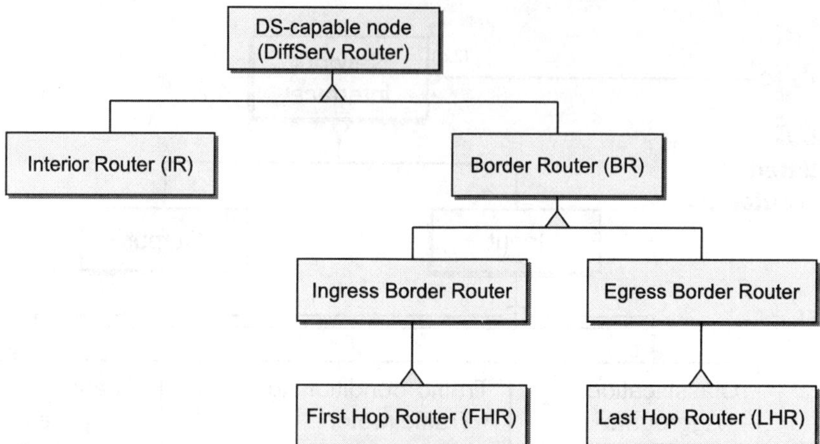

FIGURE 25.6 Classification of QoS mechanisms to be used in DiffServ-nodes according to RFC 2475.

realized gradually by a concatenation of several DS domains along the communication path between the involved end systems. In Section 25.4, it was already mentioned that the DS architecture only specifies a framework for QoS-support and imparts ISPs the greatest possible freedom for the realization of the respective QoS-elements and services. Thus, the question arises regarding how flexible but also how predictable can quality-based services be realized end-to-end over the administrative domains of different, independent, and competing ISPs?

To answer this question, the IETF defined several functional elements, which can be successively combined to deploy end-to-end-services in the DS architecture [8, 23]. The following sections will introduce these functional elements and explain how an end-to-end service can be combined from a set of Per-Domain Behaviors (PDB), which themselves consist of different Per-Hop Behaviors (PHB). The relationship among these functional elements is illustrated in Figure 25.7 using the Object Modelling Technique [26].

Per-Hop Behavior

A PHB represents the basic element for the realization of quality-based services within the DS architecture. It defines the forwarding behavior its packets will experience in a DS node. However, a PHB only defines the behavior as it is visible from the outside and no concrete QoS algorithms are specified. This means that a PHB describes only the basic characteristics that must be fulfilled by a concrete implementation, for example, an upper bound for the delay in the node or the avoidance of reordering within a data stream. Thus, a PHB does not prescribe an implementation, but specifies only the necessary behavior to realize the appropriate PHB to be classified as PHB-compliant [2].

A PHB mainly consists of the description of how the assigned aggregates should be treated differently. Thus, PHBs are usually implemented with certain queues and queueing strategies. However, it can be favorable not to regard several PHBs separately because sometimes packets of one data stream can be assigned to different aggregates. In such a case, the PHBs belonging together will be called a *PHB group* [2].

So far, the DS working group of the IETF has defined three PHBs, in order to conform to the earlier meaning of the Type-of-Service-field in the IP header and to standardize a certain set of PHBs to fulfill the basic demands. Based on these, PHBs PDB were defined, which can be used to realize the two traditional QoS-services for assuring a deterministic and statistic guarantee of a certain bandwidth. The following PHBs were defined so far:

- *Class Selector Compliant PHB Group* (RFC 2474, [2]).
- *Assured Forwarding PHB Group* (RFC 2597,[4]).
- *Expedited Forwarding PHB* (RFC 2598, [21]).

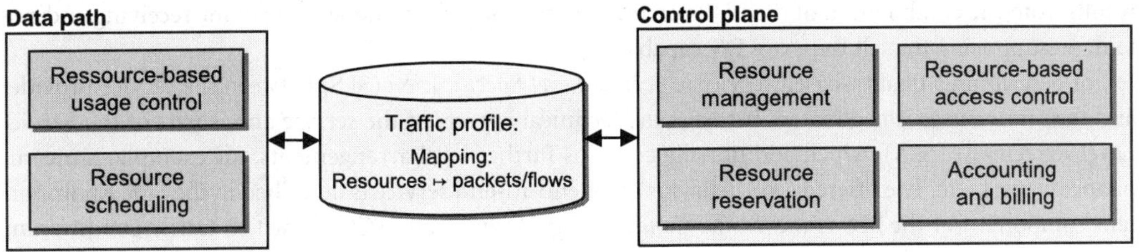

FIGURE 25.7 Relationship between functional elements of services in the DS architecture.

The Class Selector PHB group was defined to guarantee the compatibility with the ToS-field according to RFC 791. It consists of eight values that correspond to the eight relative priorities. The DSCP values were selected in such a way that none of the old ToS-attributes delay, throughput, and reliability were set. Furthermore, each DS-capable node must assign the Best Effort PHB to packets with DSCP 0, as well as packets of the two highest priorities (*network control* and *Internetwork control*) to a PHB that offers a better quality than Best Effort to protect routing traffic from non-DS-enabled systems against the remaining traffic.

Per-Domain Behavior

The differentiated treatment of packets forms the basis for the realization of quality-based communication services. However, as the overview of recent QoS approaches for the Internet in Section 25.3 showed, it requires much more features, like traffic control and traffic conditioning, to offer QoS support. Therefore, based on PHBs RFC 3086 defined the concept of PDB [23]. These describe the expected treatment of packets from edge nodes to edge nodes in a DS domain.

Next to a certain PHB, a PDB-definition also includes requirements for traffic conditioning at domain edges (cf. Figure 25.7). According to RFC 3086, these should be negotiated in a *Traffic Conditioning Agreement* (TCA). The resulting behavior of the PDB that an aggregate experiences when it enters and crosses a domain should be specified by quantifiable and measurable QoS parameters. The behavior will particularly be defined depending on the defining elements of a PDB, that is, dependent on the appropriate PHB and traffic conditioning mechanisms. However, more factors possibly influencing the service quality of an aggregate must be described, for example, the topology of the domain or the load of a node [23].

The description of the PDB can be made with QoS parameters and service classes, as for example, in the following two examples:

Example 1: *The packet loss rate should not be larger than 0.1%, when measured over an interval that is not greater than T.*

Example 2: *The delay can be specified as follows: 50% of the packets should take no longer than d ms, 30% of the packets no more than 2d ms, and 20% of the packets no longer than 3d ms.*

A further difference between a PDB specification and the definitions of a PHB lies in the explicit consideration of aggregation and de-aggregation effects and their impacts on the behavior of an aggregate within a DS domain. One main goal of a PDB definition is the assurance of quantifiable predictions of an aggregate's behavior within a domain. However, if these predictions can be affected by effects of merging or partitioning aggregates within the domain, then these effects should be described as quantifiably as possible. By repeated consideration of these implications for each hop in a domain, it should be possible to predict the domainwide behavior of an aggregate.

End-to-End Behavior

Since the average TCP connection traverses four to five autonomous systems, a quality-based service will reach over several DS domains (*end-to-end service*). The substantial components for an ISP for the realization of an end-to-end service will be PDBs (cf. Figure 25.7). The quality of an end-to-end service thus results from the combination of the PDBs on the resulting path from the sending to the receiving end system, presupposing that all these are DS-capable.

For describing a quality-based service, a *Service Level Specification* (SLS) between the service provider and the service user is negotiated. It covers the technical aspects of the service and is part of the Service Level Agreement (SLA), which additionally contains further legal arrangements, for example, noncompliance clauses, etc. The offered QoS behavior of an end-to-end service is described in the SLS. An important component of the SLS is the *Traffic Conditioning Specification* (TCS), in which traffic-conditioning mechanisms are specified.

The traffic profile in the TCS negotiated between the provider and the user specifies the maximum amount of services and QoS characteristics that the customer can use, for example, the allowable throughput, etc. For the case of violations of these agreements appropriate actions will be taken, which are likewise a part of the TCA. For example, if a sender exceeds the negotiated amount of traffic, the traffic conditioning elements become active and the nonconforming packets are treated accordingly (e.g., rejected, delayed, degraded, etc.).

Beyond this, the SLS contains some more general QoS characteristics that also describe — next to the quality-specific characteristics — operational and security characteristics of the end-to-end service [8]. Some examples are listed below:

- Availability and reliability of the service, with special consideration of the consequences a failure of the service would have, for example, when the traffic has to be bypassed on another path.
- Security, for example, encryption, authentication, etc.
- Special restrictions on certain routing paths, for example, due to operational or political reasons.
- Mechanisms for monitoring and verifying a service.
- Responsibilities as well as the consequences of a breach of contract and customer services.
- Accounting and billing.

A SLS usually differs from a PDB definition in terms of the amount of technical details and parameters. In an SLS, usually not all technical details of the concrete PDB implementation will be revealed to the customer. It is assumed that a service provider decides which service specific characteristics he directly offers to his customers in a contract. These characteristics can sometimes differ clearly from those specified in a PDB definition. QoS-specific attributes are often weakened or removed from contracts in order to avoid recourse payments in case of noncompliance.

Service providers can realize arbitrary services and thus are more flexible in reacting to new applications or market developments. It does not require lengthy standardization processes to deploy a new service. This can be realized in a simple manner, especially if an ISP first offers a new service exclusively in his administrative domains.

However, if the considered end systems are in DS domains that are administered by different providers, arrangements and agreements between the operators, along the communication path become necessary. However, the service user should not notice this fact because he requests a service from the service provider he is directly connected to. It is then the responsibility of his service provider to make the appropriate agreements with the next domain toward the customer's communication partner. This procedure continues gradually until the DS domain of the target system is finally reached. If the service can be provided in each concerned domain, an SLS between the customer and the first service provider will be made.

The implementation of the DS architecture in the Internet of tomorrow will proceed gradually. First, a service provider updates the nodes in his administrative domain with the mechanisms of the DiffServ data path, whereby traffic profiles are installed statically, for example, with SNMP. At this time, only services between directly connected DS domains of the same ISP will be possible.

In a further step, static agreements are made with neighboring domains, in order to be able to offer end-to-end services along several domains. If a customer needs QoS support, this still has to be negotiated between the customer and an authorized person of the ISP. Then, the negotiated traffic profile will be installed manually in the appropriate nodes at the right time. The service can be scheduled at certain times of a day or weekdays — nevertheless, the static character of the service administration remains recognizable, because services cannot be initiated *ad hoc*. Re-negotiations, for example, for necessary adjustments, can be made periodically, in an interval of days, weeks, or months.

The following section surveys the management of services and nodes in DiffServ-enabled networks and presents some efforts for a dynamic establishment of services.

25.6   Management of DS-nodes and -Services

So far, the management of DS domains has not been regarded, that is, the configuration of DS nodes and traffic profiles, as well as the administration of resources, for example, the initiation and processing of reservations, access control, etc. These questions are currently under discussion, and standardization has begun within the IETF [19]. Actually, the DS architecture was developed in such a way that a dynamic management of resources and nodes is not necessary at all. The concept of separating the data path from the control path has already been applied successfully to the Internet routing and was therefore also selected for migration to a quality-based Internet.

The following section illustrates how DS domains will be managed in the beginning, that is, before a dedicated management architecture is introduced. Subsequently, a brief overview regarding promising efforts for a dynamic administration is given.

Static Management

In the DS architecture [8], it is assumed that a service user first contacts the service provider for negotiating an SLA [27]. This SLA should also specify technical parameters, which define the desired service more exactly, for example, which throughput with which kind of service class has to be guaranteed. Before the agreement is given to the traffic contract, the operator examines the validity of the desired QoS support and the availability of the necessary resources. At the beginning, access control and resource administration are performed manually.

If the endpoint of the communication does not lie within the same DS domain, the domain operator must contact the next neighboring domain operator toward the target in order to likewise negotiate a service agreement. This procedure continues until each domain along the whole communication path has reserved the appropriate resources for the desired quality-based service. Before the service is used, the QoS parameters (in terms of a traffic profile) are installed in the appropriate First Hop Router. All arriving packets in the FHR matching this traffic profile will then experience the desired QoS.

All of these mentioned procedures will be carried out manually via appropriate management interfaces in the DS nodes during the introduction phase of DS, for example, with SNMP. In [1], the appropriate administrative data structures — the DS Management Information Base (DS MIB) — have already been specified for this purpose.

However, a manual administration of DS domains is not acceptable for the future, particularly because of inevitably high delays during the installation of a service. In the long term, a static management of QoS services will handicap the acceptance of a quality-supporting Internet. The necessity for a dynamic reservation of resources and an appropriate management of DS domains has already been recognized by the IETF and is currently being discussed in the working group *Next Steps in Signaling (NSIS)* [19].

Next Steps in Signaling

The IETF working group NSIS [19] was formed in 2002 to develop a framework for signaling information about a data flow along its path in the network. This is the current effort of the IETF to develop a framework for dynamic reservation of resources, for example, for DS. But the scope of NSIS is not just limited to QoS reservations. It should also support many other applications that have to signal information, for example, in case of mobility.

So far, the working group has just defined a basic framework for signaling protocols [17], which does not seem to be the final solution but clearly shows the direction of a future signaling approach. One main design decision in this framework was to decompose the overall signaling protocol suite into a generic (lower) layer, with separate upper layers for each specific signaling application. The framework also considers general interactions between signaling and other network-layer functions, for example, routing and mobility.

25.7 Summary

The architecture of DS clearly shows as to what has been learned from the experiences of previous QoS approaches and that the right conclusions were drawn for the support of quality-based services in the Internet. Now with the DS architecture, a promising solution can be found for this challenge. The primary goal consisted of developing a scalable QoS architecture that does not pose problems concerning the feasibility and manageability of services and their necessary administrative information, especially with regard to a further constantly growing Internet. With the principle of carrying the state information in the packets, the management of status information per data stream can be avoided in routers.

Despite the missing state information in the network, resources can still be assigned to the data streams by enforcing a strict access and usage control at network edges, especially at the access network. However, the level of accuracy that can be reached and which problems arise with the aggregated treatment of data streams (e.g., aggregation effects) are still the subject of ongoing research.

The DS architecture was developed consciously with the goals of *flexibility* and *extensibility*, that is, it should be possible to offer various services and to flexibly integrate new services. For this reason, no concrete QoS mechanisms are prescribed by the DS architecture to ISPs. DiffServ is only a framework that regulates the interoperability and the specification of quality-based services and their elements (Per-Hop and Per-Domain Behaviors). The concrete realization of the service elements is left to the respective ISP.

References

[1] F. Baker, K. Chan, and A. Smith, Management information base for the differentiated services architecture, RFC 3289, May 2002.

[2] F. Baker, D. Black, S. Blake, and K. Nichols, Definition of the Differentiated Services Field (DS Field) in the IPv4 and IPv6 Headers, RFC 2474, December 1998.

[3] F. Baker, B. Braden, S. Bradner, M. O'Dell, et al., Resource ReSerVation Protocol (RSVP) Version 1 Applicability Statement, RFC 2208, September 1997.

[4] F. Baker, J. Heinanen, W. Weiss, and J. Wroclawski, Assured Forwarding PHB Group, RFC 2597, June 1999.

[5] L. Berger, L. Delgrossi, D. Duong, S. Jackowski, and S. Schaller, Internet Stream Protocol Version 2 (ST2) Protocol Specification — Version ST2+, RFC 1819, August 1995.

[6] Y. Bernet, S. Blake, D. Grossman, and A. Smith, An informal management model for diffserv routers, RFC 3290, May 2002.

[7] Y. Bernet, R. Yavatkar, P. Ford, F. Baker, et al., A Framework For Integrated Services Operation Over Diffserv Networks, RFC 2998, November 2000.

[8] S. Blake, D. Black, M. Carlson, E. Davies, Zheng Wang, and W. Weiss, An Architecture for Differentiated Services, RFC 2475, December 1998.

[9] R. Braden, S. Berson, S. Herzog, S. Jamin, and L. Zhang, Resource ReSerVation Protocol (RSVP) — Version 1, RFC 2205, September 1997.

[10] R. Braden, D. Clark, and S. Shenker, Integrated Services in the Internet Architecture: an Overview, RFC 1633, June 1994.

[11] B. Carpenter and K. Nichols, Differentiated services in the internet, *Proceedings of the IEEE*, 90,1479–1494, 2002.

[12] A. Charny, J.C.R. Bennett, K. Benson, et al., Supplemental information for the new definition of the ef phb, RFC 3247, March 2002.

[13] A. Charny, B. Davie, K. Benson, et al., Supplemental Information for the New Definition of the EF PHB, RFC 3246, March 2001.

[14] D.D. Clark, The Design Philosophy of the DARPA Internet Protocols, in Proceedings of the ACM Symposium on Communication Architectures & Protocols, ACM, New York, 1988, pp. 106 – 208; SIGCOMM'88, Stanford, CA, U.S.A., Aug., 16–18, 1988.

[15] P. Ferguson and G. Huston, *Quality of Service,* Wiley, New York, 1998.

[16] S. Halabi and D. McPherson, *Internet Routing Architectures,* 2nd ed., Cisco Press, Indianapolis, 2000.

[17] R. Hancock, I. Freytsis, G. Karagiannis, et al., Next steps in signaling: framework, draft-ietf-nsis-fw-02.txt, March 2003.

[18] C. Huitema. *Routing in the Internet, 2nd edition.* Prentice-Hall PTR, Englewood Cliffs, NJ, 2000.

[19] IETF, Homepage of the IETF working group: next steps in signaling, http://www.ietf.org/html.charters/nsis-charter.html.

[20] Van Jacobson, K. Nichols, and K. Poduri, The Virtual Wire Per-Domain Behavior, draft-ietf-diffserv-pdb-vw-00, July 2000.

[21] Van Jacobson, K. Nichols, and K. Poduri, An Expedited Forwarding PHB, RFC 2598, June 1999.

[22] K. Kilkki, *Differentiated Services for the Internet,* Macmillan Technical Publishing, New York, 1999.

[23] K. Nichols and B. Carpenter, Definition of Differentiated Services Per Domain Behavior Aggregates and Rules for their Specification, RFC 3086, June 2001.

[24] Architectural Principles of the Internet, RFC 1958, June 1996.

[25] Internet Protocol, RFC 791, DARPA Internet Program — Protocol Specification, September 1981.

[26] J. Rumbaugh, M. Blaha, W. Premerlani, and F. Eddy, *Object-Oriented Modeling and Design,* Prentice-Hall, Englewood Cliffs, NJ, January 1991.

[27] D. Verma, *Supporting Service Level Agreements on IP Networks,* Macmillan Technical Publishing, New York, 1999.

[28] R. Bless, K. Wehrle: "IP Multicast in Differentiated Services Networks", RFC 3754, IETF, April 2004.

[29] R. Bless, K. Nichols, K. Wehrle: "A Lower Effort Per-Domain Behavior for Differentiated Services", RFC 3662, IETF, January 2004.

26

MPLS — Multiprotocol Label Switching

José Ruela and
Manuel Ricardo
University of Porto
INESC Porto

26.1 Introduction

This article describes the main features of Multiprotocol Label Switching (MPLS), a standard architecture proposed by the Internet Engineering Task Force that integrates label swapping forwarding with network-layer routing. The role of MPLS in overcoming limitations of overlay models, such as IP over Asynchronous Transfer Mode, and of conventional routing in IP networks, is discussed. The main concepts are introduced and the operation of MPLS is explained — classification of packets into Forward Equivalence Classes, label allocation and binding to routes, label distribution, setting up of Label Switched Paths, and route selection. Finally, support of traffic engineering and Quality of Service mechanisms in MPLS networks is analyzed.

26.2 MPLS — Rationale for a New Routing and Forwarding Architecture

The success of the Internet is mainly due to its flexible architecture, based on IP, the ubiquitous internet-working-layer protocol.

IP networks offer unparalleled scalability and flexibility for the deployment of value-added services and are becoming increasingly attractive for carrying services with hard and soft real-time constraints.

This requires extending the traditional best-effort model, designed from the outset for elastic data traffic, with mechanisms that provide differentiated and predictable Quality of Service (QoS) to a wide variety of applications with different requirements [1]. The Internet Engineering Task Force (IETF) has already specified the Integrated Services (IntServ) [2] and the Differentiated Services (DiffServ) [3] models with these goals in mind.

Carriers' Requirements

With the increasing demand and explosive growth of the Internet, service providers require a dependable and controllable network infrastructure that can offer consistent performance. In many cases, the original router-based backbone networks evolved into a two-level structure composed up of a high-speed core network interconnecting edge devices (IP routers) that in turn interface with access networks and provide common services to users, such as security, accounting, Virtual Private Networks (VPNs), web hosting, etc.

Management of such networks requires powerful traffic engineering techniques, that is, the capability of mapping flows into the physical network topology and evenly distributing traffic over the network links, to achieve an efficient utilization of network resources, avoid congestion, and improve network performance.

Integration of IP and ATM

With the deployment of Asynchronous Transfer Mode (ATM) switches in many carriers' backbones, integrating IP and ATM appeared as a natural and attractive architectural choice to fulfill these goals, since it combined simple and robust routing techniques with a scalable, fast switching technology. Moreover, ATM includes provision for traffic management and differentiated transport services [4], which are essential in satisfying carriers' requirements.

However, this integration was not without problems. Initial solutions were based on overlay models, such as Classical IP over ATM (CLIP) [5], which are easy to deploy but have a number of serious shortcomings [6,7], especially in large networks where IP routers are interconnected by a full mesh of ATM connections, to avoid extra hops in the path across the backbone. In fact, in this architecture, two types of nodes (IP routers and ATM switches) run different signaling and routing protocols, on different address spaces, with the IP logical topology segregated from the ATM physical topology. This is inefficient and does not allow joint optimization of resources. In addition, it poses scalability and stability problems, when the logical topology has to be reconfigured, due to the way routing updates are advertised between adjacent nodes by conventional IP routing protocols.

Multilayer Switching and MPLS

To overcome these limitations, alternative architectures, based on the concept of multilayer switching, were investigated and proposed by manufacturers. Two such examples are IP Switching [8] and Tag Switching [9]; but, in general, these solutions share a common principle — the separation of control and forwarding functions. The software-based control component includes layer three routing and additional signaling protocols, while the hardware-based forwarding component uses layer two label switching techniques. A mapping between routes and labels provides the glue between these components, required to build a multilayer switch.

The concept is not restricted to IP and ATM, but when the forwarding component is based on ATM, only the ATM switching fabric is retained (not the ATM-based control protocols).

In this context, the MPLS architecture [10] is the result of the current IETF work to standardize a solution that integrates label swapping forwarding with network layer routing, and that incorporates the basic principles and ideas proposed by manufacturers. Initial efforts are focused on IP, while ATM is a strong but not the unique choice for switching.

MPLS tries to solve most of the critical issues previously identified, with particular emphasis on scalability, fast packet forwarding, and traffic engineering [11]. Although not dealing with specific QoS mechanisms (such as call admission, traffic shaping and policing, packet scheduling, and discard policies), MPLS provides an appropriate framework for supporting a QoS architecture as well.

26.3 Analysis of Traditional Packet Switching Techniques

In order to better understand how MPLS is built upon and extends the capabilities of existing technologies, it is useful to briefly review traditional connectionless and connection-oriented packet switching techniques.

In conventional IP networks, which operate in a connectionless (datagram) mode, packets are forwarded on a hop-by-hop basis, and each router along the path makes an independent forwarding decision. The next hop for a packet is selected based on the IP destination address and on forwarding information updated by means of routing protocols. The routing (forwarding) table is parsed in search of some address prefix that is the longest match for the packet's destination address.

This approach has a number of limitations, especially when we consider the role that IP networks are expected to play in the coming years.

First, forwarding decisions in each node are only based on information that travels with the packet in its header. Conversely, the packet header is not fully utilized, although it contains much more information than required to simply select the next hop.

Second, packets are typically forwarded along a shortest path route to the destination, usually discovered by routing protocols that use an additive link metric (hop count); the consequence is that paths to a common egress router form a tree rooted at the destination. With heavy loads, links on a shortest path tree may become congested, while others remain underutilized due to uneven traffic distribution. This approach does not exploit alternate paths either to reroute traffic around congested nodes, to perform load balancing, or to support dynamic fall over to backup paths. Although it would be possible to redirect traffic by changing link metrics (based, e.g., on traffic characteristics and capacity constraints), this may lead to undesirable effects, such as changing the path of all packets that traverse congested links. In other words, traffic engineering has not been exploited in IP networks [6,7]. Moreover, destination-based forwarding, as used today in IP networks, seriously limits the network services that can be offered, since it does not allow provisioning paths specific to particular sources or services or with a QoS constraint.

Finally, processing the IP header, which includes looking up the routing table, decrementing the Time To Live (TTL) value, and computing a new Cyclic Redundancy Check (CRC), is computationally more intensive and takes more time than processing a label and using it as an index into the forwarding table. Nevertheless, state-of-the-art gigabit routers use optimized algorithms to speed up the IP header processing in hardware.

On the other hand, connection-oriented packet switching is based on the setup of virtual channel connections (virtual circuits) by means of signaling procedures, before data transfer takes place. Packets on a virtual circuit are forwarded along a fixed path, defined at setup; the virtual circuit is identified by a short label in the packet headers. In ATM, the label is structured into two parts and carries a Virtual Path/Virtual Channel Identifier (VPI/VCI), while in Frame Relay the label is a Data Link Connection Identifier (DLCI).

26.4 MPLS Operation

MPLS overcomes the problems found in conventional IP networks as well as the limitations of overlay models by combining traditional IP routing with a label swapping technique, implemented by separate control and forwarding components. Control functions are responsible for building and maintaining routing tables; routing information is used to create the forwarding tables that define how packets are switched and labels are swapped at each node. Figure 26.1 is a simplified representation of this model.

Forward Equivalence Class

An MPLS label is a short, fixed-length identifier, with local significance, used to identify a Forwarding Equivalence Class (FEC). An FEC simply represents a group of IP packets that are forwarded in the same

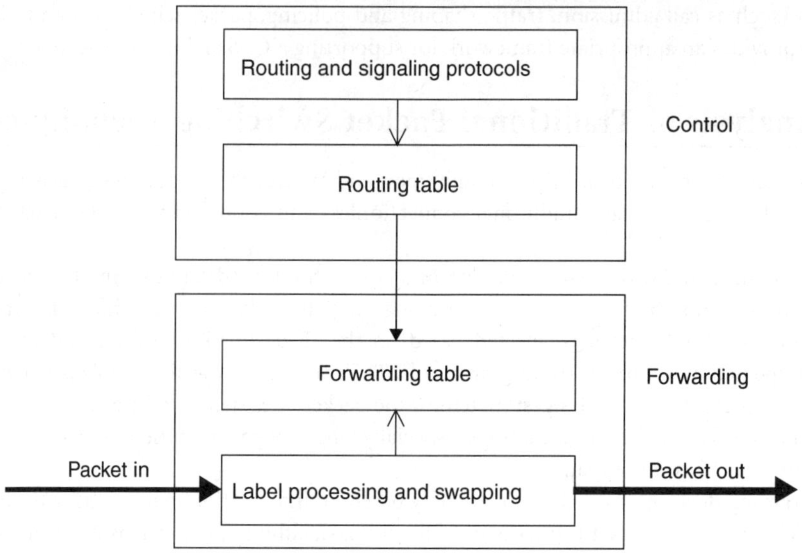

FIGURE 26.1 Mapping between routing and forwarding tables.

manner, that is, over the same path and receiving the same forwarding treatment, since they have similar transport requirements.

The assignment of a packet to an FEC is done just once, at the ingress node of an MPLS domain, where the packet is classified. It is possible to use a variety of forwarding criteria to assign packets to FECs, besides the conventional address prefix used in destination-based routing, such as: Classes of Service (based, for instance, in fields of IPv4 or IPv6 packet headers), Application flows (requiring both source and destination addresses as well as other Network or Transport layer information), IP multicast groups, explicit routing, and VPNs.

An FEC is encoded with a label, which travels with the packet that becomes a "labeled packet." The label may reside in an encapsulation header added for this purpose or may be carried on a layer two header that natively supports labels, such as in ATM or Frame Relay.

Label Switched Paths and Label Switching Routers

Packets in a particular FEC follow a common path through one or more MPLS nodes. This is called a Label Switched Path (LSP), which is defined by the set of labels associated to an FEC at each hop. The forwarding decision taken by each node is simply based on the incoming label, which is used as an index in a table that specifies the next hop and the outgoing label, which is inserted in place of the incoming label (label swapping). An LSP must be set up and labels must be assigned at each hop before traffic can be forwarded.

This is similar to conventional virtual circuit switching, but differs in the way the forwarding tables are created and maintained. We may also say, by analogy, that in hop-by-hop routing an FEC represents a common destination address prefix for a group of packets; but each node makes an independent FEC assignment.

Using LSPs, MPLS can provide many of the advantages of connection-oriented networks, while retaining the simplicity of datagram networks.

MPLS nodes are called Label Switching Routers (LSR); but it is usual to refer to the edge (ingress or egress) nodes as Label Edge Routers (LER). All nodes are aware of MPLS control protocols, run the same layer three routing protocols, and forward packets based on label processing and swapping at wire speed. In addition, an LER performs some specific and more processing-intensive functions, such as interfacing external networks and, in the case of the ingress LER, classifying packets into FECs, assigning the corresponding labels and adding them to the original packet. Therefore, in MPLS, a single family of devices runs the same set of protocols over a common physical and logical topology shared by all nodes, unlike architectures based on the overlay model.

Figure 26.2 shows an LSP and the basic operations performed on a packet. The ingress LER adds a label to the packet, while the egress LER removes it; labels are assigned independently on each hop and are swapped as the packet moves along the LSP.

Label Assignment and Distribution

A key issue in MPLS is the binding between labels (which represent an FEC) and a route. Three steps are required: allocating a label, binding it to a route (LSP), and distributing label binding information among LSRs.

IETF has standardized a new signaling protocol, called Label Distribution Protocol (LDP) [12], for the setup and maintenance of LSPs. It allows the distribution of label binding information between LSRs, thus ensuring that adjacent nodes share a common FEC to label binding, and allowing the creation of LSPs. The forwarding table (label information base) is constructed as a result of label distribution.

Label assignment and route selection are therefore required to set up an LSP. But when should labels be assigned and bound to routes (FECs), and how are routes discovered or selected?

The need for label creation and binding may be driven by data or control traffic. In the data-driven strategy, the arrival of data at a node triggers the process. In the control-driven approach, two methods may be adopted, depending on the route selection scheme: labels are assigned once routes are discovered by conventional routing protocols (topology-based) or in response to request-based control traffic, such as Reservation Protocol (RSVP) [13] messages. Control-driven methods are preferred due to their better scalability properties.

Label Stack and Hierarchical Routing

In the previous description, it was considered that an IP packet is encapsulated with a single label. However, MPLS supports a more general mechanism, in which a labeled packet can carry a number of labels organized in a last-in, first-out manner. This is called a label stack.

The use of a label stack allows hierarchical routing, with different levels of granularity, possibly across various domains. Label swapping is always based on the current top level of the stack and adding or removing a label corresponds to normal push-and-pop operations on the stack. It is also possible to create LSP tunnels that can nest to any depth; one possible application is for the provision of MPLS-based VPNs.

Label Encoding

MPLS does not rely on a specific layer two technology. The only requirement is that LSRs are able to exchange labeled packets across data links.

The IETF defined a general encoding technique that must be supported by LSRs to produce labeled packets before they are transmitted on a data link [14]. It assumes a label stack and is particularly targeted at data links that do not support labels, such as Point-to-Point Protocol (PPP) or LAN media, although it may be used in Frame Relay or ATM-based LSRs.

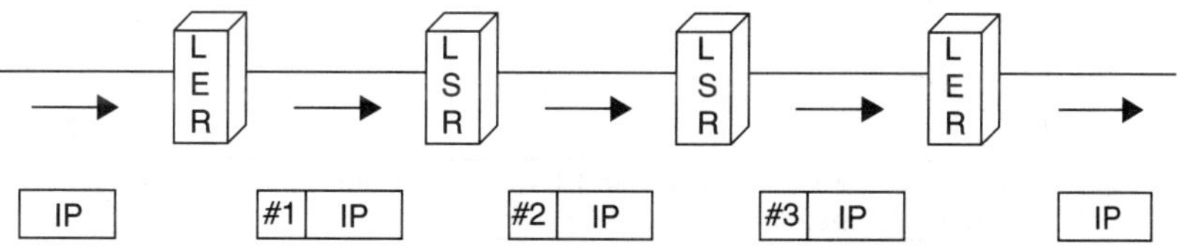

FIGURE 26.2   Label insertion, swapping, and removal.

A label stack is represented as a sequence of label stack entries. An entry is four octets long and includes four fields:

- Label: Label value (20 bits),
- Exp: Experimental use (3 bits),
- S: Bottom of Stack (1 bit),
- TTL: Time To Live (8 bits).

A possible use of the Exp field is as a Class of Service (CoS) identifier.

In PPP and LAN data links, the label stack entries are inserted between layer two and layer three headers, as illustrated in Figure 26.3, and constitute what is usually called a shim header.

In ATM-based LSRs, the top label is directly encoded into the VCI and/or VPI field [15]. In general, when a label stack has to be carried, the label stack entries for a particular packet are carried as a shim header in the ATM Adaptation Layer (AAL 5) frame; the actual value of the top label is encoded in the VPI/VCI field of the ATM cells, and the label value of the top entry in the shim is set to zero.

Route Selection

There are two alternatives to select a route used to set up the LSP for a particular FEC.

The first one is based on hop-by-hop routing; each LSR independently determines the next hop for the LSP based on its IP forwarding table, which is built by traditional IP routing protocols. This is the default, topology-based method and allows the discovery of shortest path routes; a hop-by-hop LSP follows the path that a packet using conventional routing would have used.

The second one is based on explicit routing, which is similar to source routing. In an explicitly routed LSP (ER-LSP), the route for the path is explicitly defined by a single LSR (usually the ingress or egress node) and may include all or only a subset of the LSRs in the path (strict or loose LSP). The route (sequence of LSRs) may be selected by configuration or dynamically and is conveyed in a control message that traverses all nodes along the specified route.

A shortest path tree built by hop-by-hop routing and a single ER-LSP, overlaid on a physical network topology, are represented in Figure 26.4.

26.5   Traffic Engineering and QoS in MPLS Networks

An ER-LSP may be defined and controlled by the network operator or a network management application. Based on administrative or QoS policies or traffic engineering requirements, some traffic can be forced into a path different from the shortest path computed by a routing protocol.

Two approaches have been considered to control (establish, terminate, re-route) ER-LSPs: extending the capabilities of the Label Distribution Protocol to include explicit paths via Constraint-based Routed Label Distribution Protocol (CR-LDP) [16] and extensions to RSVP (MPLS-RSVP) [17]. At an abstract level, the functions of CR-LDP and MPLS-RSVP are rather similar. They both allow an LSR to trigger and control the establishment of an LSP between itself and a remote LER, to strictly or loosely specify the route to be taken by the LSP, and to specify queuing and scheduling parameters to be associated with this LSP at every hop. The relative advantages and disadvantages of these two schemes can be found in [18];

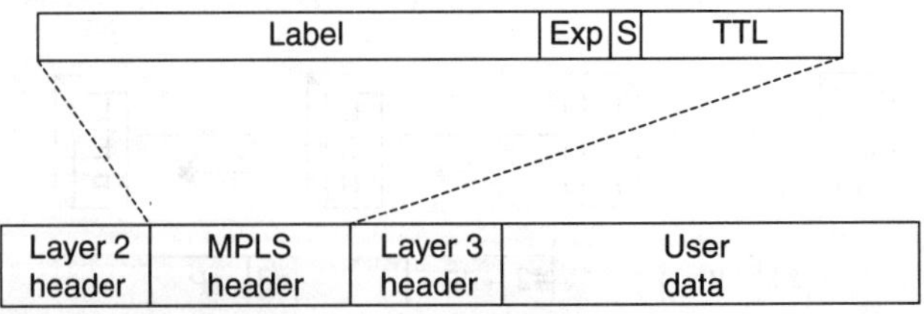

FIGURE 26.3   MPLS header and encapsulation.

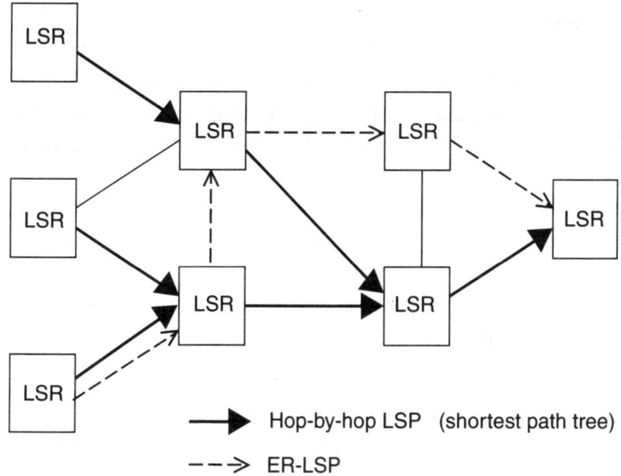

FIGURE 26.4   Hop-by-hop and Explicitly Routed LSPs.

but manageable traffic engineering and QoS control cannot be realized unless one of these protocols is deployed.

By allowing the network to explicitly route an LSP, both traffic engineering techniques [6,7] and provisioning of differentiated services can be supported in an MPLS domain.

Constraint-based Routing and QoS

Explicit routing is a particular case of constraint-based routing, where the constraint is the explicit path. In general, constraint-based routing may take into account link characteristics, such as bandwidth or delay, hop count, or QoS parameters.

MPLS allows setting up explicit paths and forwarding traffic on them; but it does not provide the means to find out paths with constraints. Because MPLS allows traffic engineering and explicit routing, there is keen interest in QoS routing for selecting routes subject to QoS requirements and other policies, instead of the least cost or shortest path route found by traditional routing protocols.

In order to allow the computation of routes with constraints, it is necessary to extend the Interior Gateway Protocols to carry additional information about links that can include, for example, maximum link bandwidth, maximum reservable bandwidth, current bandwidth reservation at different priority levels, default traffic engineering metric, or other attributes used for policy-based routing. Constraint-based routing is just one of the issues being considered by the IETF in the context of Internet Traffic Engineering [19].

MPLS and Differentiated Services

In a DiffServ domain, all packets requiring the same behavior constitute a Behavior Aggregate (BA). At the ingress node of a DiffServ domain, packets are classified and marked with a DiffServ Code Point (DSCP) corresponding to their BA. The DSCP is used in each node visited by the packet to select the Per Hop Behavior (PHB) that determines the treatment it will receive (scheduling, drop precedence, etc.).

MPLS support of Differentiated Services has already been addressed by the IETF [20] and several alternative solutions were considered for mapping BAs onto LSPs. The simplest one consists in using a single LSP to support up to eight BAs of a given FEC. In this case, the Exp field of the MPLS shim header is used by each LSR to determine the PHB to be applied to the packet.

One promising solution for QoS provisioning in IP networks, using currently available IETF standards, consists in combining MPLS with differentiated services and constraint-based routing.

26.6 Conclusions

It is commonly accepted that MPLS offers many advantages over earlier network solutions, as a carrier infrastructure capable of service integration and differentiation. However, some of these advantages are not exclusive of MPLS; on the other hand, MPLS must be combined with other mechanisms, such as QoS, to make use of its features.

Therefore, when evaluating the merits of MPLS, it is useful to adopt a critical view [21], especially when considering competitive solutions, such as gigabit routers, which also promise high throughput and fast switching, as well as traffic and QoS differentiation.

Nevertheless, there are some strong arguments in favor of MPLS. In general, MPLS offers a unique combination of some attractive properties:

- scalability, in terms of number of nodes and traffic flows;
- flexibility, since it is not tied to a single forwarding technology;
- simple and fast label forwarding, which improves network performance; and
- capability to support traffic-engineered paths and service differentiation, essential for QoS provisioning.

Properties like scalability and traffic engineering are especially valuable, when considering short-term deployment of MPLS. This allows offering an efficient transit core network (with high throughput and low latency), improved economy of scale, new services (such as CoS-based forwarding and VPNs), and the ability for fast restoration of data traffic.

Another important issue, in view of the investment many carriers have recently made in ATM equipment, is the possibility of leveraging the installed ATM infrastructure — either using ATM-based LSRs or running MPLS over ATM (overlay model); in the second case, MPLS LSRs communicate over an ATM cloud.

On the other hand, the MPLS paradigm can be extended into the optical domain, by leveraging existing control plane techniques to control optical cross-connects and using wavelengths in place of numerical labels; this is usually called Multiprotocol Lambda Switching (MPλS) [22]. An even more general approach is emerging; the IETF is currently specifying a Generalized Multiprotocol Label Switching (GMPLS) architecture [23] that extends MPLS to include multiple switching dimensions: time (packet and circuit), wavelength (lambda), and spatial (fibre or port) switching. The focus of GMPLS is therefore on a multipurpose control plane architecture that can be applied to a variety of data forwarding techniques, thus requiring further enhancements to existing signaling and routing protocols [24,25].

Although MPLS has been targeted at WAN environments, it can be used in LANs as well. In this case it is an alternative to solutions based on conventional layer two LAN switches and multilayer switches (or router switches) or to ATM LANs based on overlay models, such as LAN Emulation (LANE) and Multiprotocol over ATM (MPOA).

When considering the deployment of MPLS in the Internet, it is necessary to take into account the profound consequences at the architectural level. It changes the basic forwarding model, which has remained essentially unaltered since the early days of the ARPANET. It also impacts the routing architecture, requiring that routing protocols perform new and more complex tasks.

Short-term applications are likely to be within a single network administrative domain; over time, interdomain MPLS is likely to occur, with transit carriers providing services to local or national Internet Service Providers.

The long-term evolution is still unclear. In order to fully exploit the benefits of MPLS (or GMPLS in a broader sense), some open issues still need to be answered and are subjects of intense research. In particular, it is necessary to specify, test, and validate, in operational conditions, the criteria and mechanisms for selecting routes and dynamically establishing paths according to traffic engineering or QoS policies, as well as managing their QoS characteristics.

References

[1] Xipeng Xiao et al., Internet QoS: a big picture, *IEEE Network Magazine*, 13 (2), 1999, 8–18.

[2] R. Braden et al., Integrated Services in the Internet Architecture: an Overview, RFC 1633, June 1994.

[3] S. Blake et al., An Architecture for Differentiated Services, RFC 2475, December 1998.

[4] ATM Forum, *Traffic Management Specification*, Version 4.1, AF-TM-0121.000, March 1999.

[5] M. Laubach et al., Classical IP and ARP over ATM, RFC 2225, April 1998.

[6] D. O. Awduche, MPLS and Traffic Engineering in IP Networks, *IEEE Communications Magazine*, 37 (12), 1999, 42–47.

[7] G. Swallow, MPLS advantages for traffic engineering, *IEEE Communications Magazine*, 37 (12), 1999, 54–57.

[8] P. Newman et al., IP switching and gigabit routers, *IEEE Communications Magazine*, 35 (1), 1997, 64–69.

[9] Y. Rekhter et al., Tag switching architecture overview, *Proceedings of the IEEE*, 85 (12), 1997, 1973–1983.

[10] E. Rosen et al., Multiprotocol Label Switching Architecture, RFC 3031, January 2001.

[11] D. Awduche et al., Requirements for Traffic Engineering over MPLS, RFC 2702, September 1999.

[12] L. Andersson et al., LDP Specification, RFC 3036, January 2001.

[13] R. Braden et al., *Resource ReSerVation Protocol (RSVP) — Version 1 Functional Specification*, RFC 2205, September 1997.

[14] E. Rosen et al., MPLS Label Stack Encoding, RFC 3032, January 2001.

[15] B. Davie et al., MPLS using LDP and ATM VC Switching, RFC 3035, January 2001.

[16] B. Jamoussi et al., Constraint-Based LSP Setup using LDP, RFC 3212, January 2002.

[17] D. Awduche et al., RSVP-TE: Extensions to RSVP for LSP Tunnels, RFC 3209, December 2001.

[18] Anoop Ghanwani et al., Traffic engineering standards in IP networks using MPLS, *IEEE Communications Magazine*, 37 (12), 1999, 49–53.

[19] D. Awduche et al., Overview and principles of Internet Traffic Engineering, RFC 3272, May 2002.

[20] F. Le Faucheur et al., MPLS Support of Differentiated Services, RFC 3270, May 2002.

[21] G. Armitage, MPLS: the magic behind the myths, *IEEE Communications Magazine*, 38 (1), 2000, 124–131.

[22] D. Awduche et al., Multiprotocol lambda switching: combining MPLS traffic engineering control with optical crossconnects, *IEEE Communications Magazine*, 39 (3), 2001, 111–116.

[23] E. Mannie, Ed., Generalized multi-protocol label switching architecture, Internet draft, draft-ietf-ccamp-gmpls-architecture-07.txt, May 2003.

[24] A. Banerjee et al., Generalized multiprotocol label switching: an overview of routing and management enhancements, *IEEE Communications Magazine*, 39 (1), 2001, 144–150.

[25] A. Banerjee et al., Generalized multiprotocol label switching: an overview of signaling enhancements and recovery techniques, *IEEE Communications Magazine*, 39 (7), 2001, 144–151.

27

The Integrated Services Architecture and RSVP

Henning Sanneck
Siemens AG, Information and Communication Mobile–Networks

27.1 Introduction

The Internet is a "best-effort" packet-switched network. It comprises a fundamental trade-off: on the one hand, the bandwidth usage is very efficient using statistical multiplexing of packets on a physical link. Additionally, the network layer protocol (IP) can be easily run on top of several different link layers, because it imposes no requirements for the ability of the link layer to deliver a packet within a certain amount of time or even at all. On the other hand, the properties just pointed out (which were an essential part to make the Internet such a success) make it very difficult to guarantee any type of "Quality of Service" (QoS) to particular packets or flows of packets. Network Quality-of-Service (cf. Chapter 14) means allocating resources (buffers and bandwidth) at the network elements for packets, which match certain criteria resulting in a predictable network performance. This yields a predictable quality at the service level and finally at the level of human perception. Such a predictable quality is particularly needed to meet the requirements of real-time applications like voice and video conferencing.

Several different approaches to enable QoS have been developed in recent years. An approach to QoS always has to be analyzed from two aspects: the achievable quality vs. the overhead (eventually: the costs) imposed by the respective approach. Figure 27.1 shows a taxonomy giving a qualitative classification of the necessary overhead in terms of additional processing and bandwidth consumed at end systems, and network overhead (protocol overhead in terms of operation and deployment and state that has to be maintained).

The schemes designated as "end-to-end" typically do not involve the network at all, but rely on robust end system-based protocols and mechanisms. In contrast, the hop-by-hop approaches require network participation at different levels, thus generally achieving better end-to-end delay properties and lossless service (note the shadings in the figure). Clearly, the associated overhead of both types of approaches

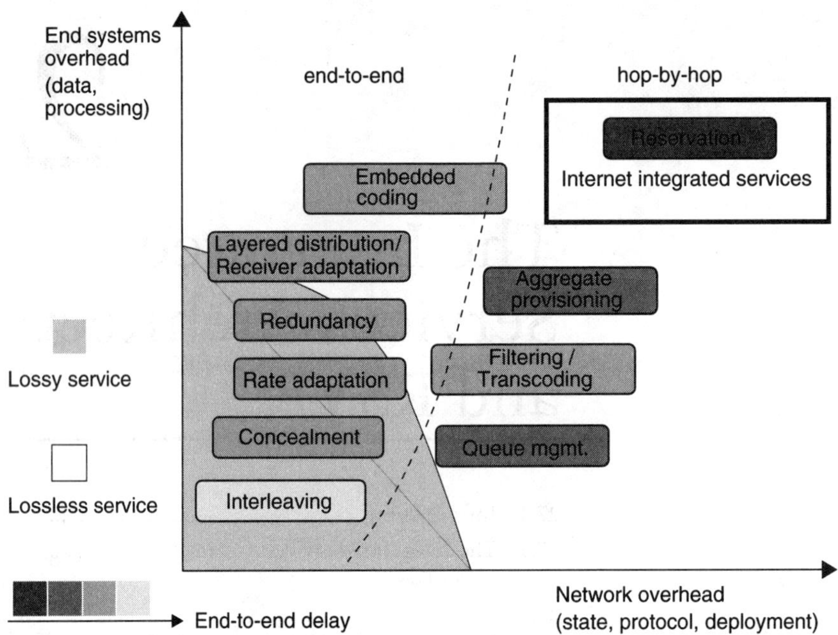

FIGURE 27.1   Qualitative classification of QoS schemes for real-time applications.

influences the scalability and thus the feasibility of deployment. The following categories of QoS-enhancement schemes can be distinguished:

- *Loss alleviation*:
  - interleaving (reordering the sequence of the application data units to spread the potential loss impact),
  - concealment (using audio/video signal processing to conceal the impact of losses at a perceptual level).
- *Loss reconstruction*:
  - redundancy (adding redundancy proactively to the data stream to be able to repair losses at the receiver).
- *Loss avoidance at the application level* (cf. Chapter 22):
  - sender rate adaptation (adapting the transmission rate to the currently achievable throughput in the network),
  - layered distribution/receiver adaptation (splitting the data stream into multiple substreams, which are then selectively subscribed to by a receiver dependent on the (sub-)path properties to this particular receiver; in "embedded coding," the source coding scheme is optimized towards the splitting into substreams),
  - filtering/transcoding (within the network, the transmission rate of the data stream is adapted by exploiting know-how about the particular source coding scheme).
- *Loss avoidance at the network level* (including guarantees on the delay):
  - queue management (optimizing the behavior of individual queues under congestion),
  - aggregate provisioning/per-packet prioritization (cf. Chapter 25) and reservation (explicitly allocating network resources for individual data streams on the entire end-to-end path).

27.2   The Integrated Services Architecture

Integrated Services/RSVP belong to the "reservation" category, which constitutes the "high end" solution to the QoS problem in that real QoS *guarantees* can be provided (Figure 27.1). While reservation is not a new idea in networking, the challenge for the Integrated Services/RSVP is the integration of the necessary mechanisms and protocols into the existing Internet architecture. This implies that the described "best-effort" datagram model should be preserved while enabling resource reservation, particularly for real-time applications. Resource reservation in the Integrated Services model means service differentiation on a per-flow basis, which is accomplished by providing mechanisms to isolate flows from each other, to establish rate and delay guarantees, and to provide controlled sharing of excess bandwidth [14]. Flows are described by their traffic envelope using token and leaky buckets. If flows violate their contracted traffic profile, packets are delayed, discarded, or treated as best effort.

Figure 27.2 shows the Internet protocol architecture needed to support real-time applications. It consists of protocols that provide basic transport functionality (RTP — Real-Time Transport Protocol, cf. Chapter 28), QoS feedback (RTCP — Real-Time Transport Control Protocol, cf. Chapter 28), call-setup signaling (ITU-T H.323 and SIP — Session Initiation Protocol), and QoS signaling (RSVP — Resource Reservation Protocol). In addition to these protocols (a part of which are relevant to QoS), the enforcement of the QoS on the data flows that pass through a router (or which are emitted from a host) is necessary (Figure 27.2: layer 2/3 adaptation). For QoS-passive media, that is, where the link layer does not implement any QoS control mechanisms, the enforcement is realized by a *traffic control* entity that is typically located between IP and the network device driver on an outgoing interface. For complex link layers like ATM, these mechanisms need to either be mapped on or replaced by the respective means available within the link layer (e.g., in [15] this is described for ATM (cell switching) as a link layer using ATM and IP signaling, respectively).

The functional blocks (Figure 27.3), which are required *at every individual hop along the end-to-end path* to establish the QoS guarantees, can be described as follows:

QoS signaling — registration of senders and receivers: The senders advertise the traffic specifications of their traffic flows. Receivers then use this information to request the desired QoS. The signaling in the Integrated Services architecture is realized by the RSVP, which is treated in more detail in the next section.

Policy Admission Control — admission of reservation requests in terms of administrative regulation: The policy admission control interfaces on one side to the other Integrated Services components (QoS signaling, Capacity Admission Control) and on the other side to the policy infrastructure. The policy infrastructure is a set of protocols, information models, and services that allow administrative intentions to be translated into differential treatment of flows (using, e.g., the Integrated Services architecture and RSVP). It gives the network operator the opportunity to regulate which users/applications/hosts should have access to what resources/services under what conditions. The architecture of the policy system (within one domain) consists of the following three tiers:

1. *Policy enforcement points (PEP)*: The network elements where packets are identified and treated and thus policies must be enforced. The PEP is, for example, colocated with the traffic control module.
2. *Policy decision points (PDP)*: The components responsible for determining what actions are applicable to which packets. A PDP interprets policy rules for one or more PEPs based on information

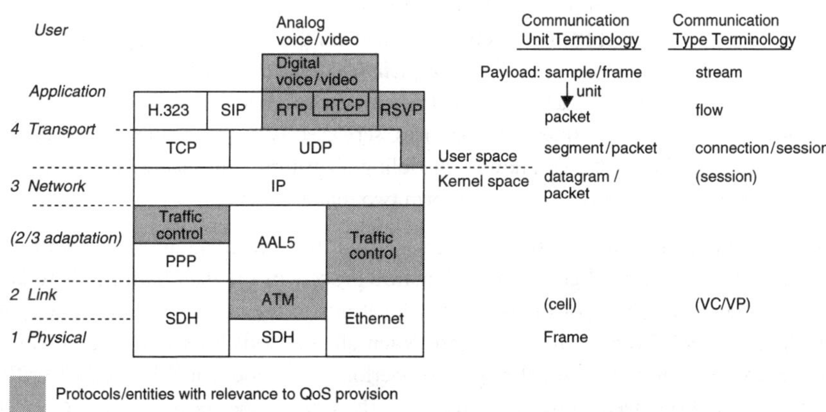

FIGURE 27.2　Integrated Services-enhanced Internet protocol architecture.

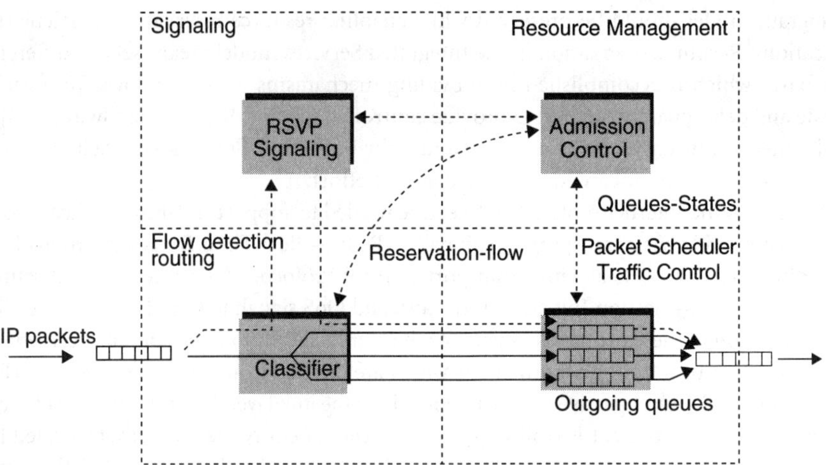

FIGURE 27.3   Functional blocks of a network element in the Integrated Services model.

contained in data and signaling packets, information of current network conditions, account balances, etc.

3. *Policy repository*: The location where the domain policies are stored.

RSVP is able to carry and process "policy objects" together with the conventional reservation requests and responses (see below). These objects may identify and authenticate users, define the relative priorities for reservations, carry accounting and charging information, and much more.

Capacity Admission Control — admission of reservation requests in terms of available resources: Besides global (network domainwide) and local policy constraints, the resource usage at a particular network element needs to be taken into account when admitting reservation requests. A Capacity Admission Control algorithm could use either only the maintained state about the already admitted reservations or it can take into account the actual resource usage (Measurement-Based Admission Control). Another design dimension is if only the current reservation requests/usage is monitored or if future states are taken into account (resource reservation in advance).

Classifier — identification of the association of IP datagrams to flows: The IP layer constitutes a multiplexing layer for packets coming from various network interfaces (at a router) or various local UDP/TCP sockets (at a host). After the routing decision has been taken and the packets are demultiplexed to the correct outgoing interface, it is necessary to associate the packets to the respective flows (or the "best effort" class) to be able to schedule the departure of the packets over the network interface correctly. For IPv4, the classification is done by matching a packet's IPv4 source and destination address, the protocol ID, as well as the transport layer ports against the parameters obtained via the QoS signaling protocol.

Packet Scheduler — schedules the order in which the queues (to which the packets have been associated by the classifier) are served: A "best-effort" packet-switched network implements simple First-Come-First Served (FCFS) scheduling. To support the Integrated Services' service classes (see below), advanced scheduling algorithms are necessary. The most well-known scheduling algorithm is probably Weighted Fair Queuing (WFQ). Packet scheduling algorithms typically perform two major tasks:

1. *Isolation*: A scheduler needs to assure that all QoS requirements that have been communicated for individual flows are met. Of course, this dynamic process in the data path needs to go hand-in-hand with the static control of the resource allocations in the control path (Capacity Admission Control, see above). Even in the worst case when all reserved flows fully exploit their reserved resources and the network element might be overloaded by additional best effort traffic, packets of an individual flow need to be given access to the link by the scheduler to just meet the QoS requirements of that flow.

2. *Sharing*: While providing isolation of flows (to mimic the performance for reserved flows achievable in a circuit-switched network, which provides inherently full isolation), the key advantage of statistical multiplexing in a packet-switched network must be retained. This implies that bandwidth, which has been allocated to reserved flows but which is temporarily not used by those flows, should be available to other reserved and "best-effort" flows. An additional requirement that may be imposed is to distribute this "excess bandwidth" fairly between these competing flows. Note that there is no universally applicable definition of fairness that could be enforced.

Service Classes

To request resource reservations to the network, *invocation information* needs to be supplied to the network. This information consists of a Traffic Specification (TSpec) of a flow and a Reservation Specification (RSpec) to describe the service required by the flow at every hop inside the network. Together, TSpec and RSpec are called Flow Specification (FlowSpec).

The TSpec describes the traffic profile of a flow. The conceptual model underlying the traffic specification (and the corresponding shaping/policing operations at the edge and inside the network) is a sequence of a token bucket and a leaky bucket with the following parameters:

- Token bucket rate r (bytes/sec): Arrival rate of tokens at the bucket.
- Token bucket depth b (bytes): Number of tokens the bucket can contain. The bucket depth translates to the length of data bursts (back-to-back bytes/packets) that can be emitted.
- Leaky bucket rate p (bytes/sec): The maximum rate at which bytes leave the leaky bucket. The leaky bucket constitutes a peak rate limiter, that is, that back-to-back bursts of data (see bucket depth b above) do not leave the shaper at an infinite rate but rather at the rate p.
- Minimum policed unit m (bytes): This variable constitutes the lower bound of the allowed packetization granularity, that is, every packet that is smaller than m bytes will be counted as consuming m bytes.
- Maximum packet size M (bytes): The maximum packet size that can be handled.

The Integrated Services model comprises two service classes: Guaranteed Service and Controlled Load Service.

Guaranteed Service (GS, [2]): This service, which has been standardized in the IETF, is intended for nonadaptive flows that need a strict delay bound (e.g., distributed simulation tools, distributed games, and hard real-time systems). The end-to-end behavior of a path that supports the GS can be seen as a virtual circuit with a guaranteed bandwidth allocation and guaranteed delay properties. GS does not control minimal or average delay, but the *maximum queuing delay*. Additional delays that arise are the propagation delay on the medium and the delay to police/shape the traffic to the profile communicated in the TSpec (the GS expects the traffic to be conforming to the profile; otherwise packets need to be delayed or dropped to make the flow conformant).

The GS RSpec consists of the following parameters:

Service rate R (bytes/sec): The rate at which the network needs to service the flow.
Slack term S (sec): The extra amount of delay that nodes inside the network may add while still meeting the end-to-end delay bound (this is further explained below).

The service provided is similar to those provided on a dedicated circuit with bandwidth R between the source and the destination. Using the T- and RSpec parameters, the worst-case end-to-end queuing delay for a flow can be computed. For $p \rightarrow \infty$ and $R \geq r$ (i.e., that a burst of packets is emitted back-to-back with infinite rate and the reserved rate is larger or equal to the source rate, which needs to be a valid condition to fulfill the reservation at all), we can express the end-to-end worst-case queuing delay as b/R. This means that the queuing delay depends only on the token bucket depth and the service rate. Considering that the data burst of size b is arriving instantaneously to be served, the queue size is b. Therefore, the last byte in the burst is experiencing a delay of b/R until it is received at the sink (note that we have no notion of a packet in this model; therefore, it is called the "fluid flow" model).

If the peak rate is closer to the values of R and r, the worst-case delay is lowered because the arrival of the data burst is spread over a larger time interval where, until the last byte of the burst is serviced, already some data have been transmitted. In this case, the end-to-end worst case queuing delay is (for $p > R \geq r$):

$$\frac{b(p-R)}{R(p-r)}$$

If the peak rate p is smaller than the service rate R, there will be no queue buildup, as the network can deplete the queue faster than the source is transmitting. If $r > R$, however, there is a constant queue buildup and the queuing delay becomes unbounded.

In contrast to the "fluid flow" model, in a real network the service a flow receives is worse than in the idealized model of a dedicated circuit directly connecting the sender to the receiver. Two error terms have been introduced to express the deviation of the GS from the idealized model.

The error term C (measured in bytes) is dependent on the transmission rate and the packet length and represents the deviation due to the packetization and the therefore necessary store-and-forward operation. An intermediate node needs to receive the complete packet until it is able to forward the packet to the next hop.

D (measured in seconds) is the rate-independent error term, which is usually determined at configuration time. Components of D are per-packet processing delays at a node, for example, to perform a route lookup.

The sums of C and D over an end-to-end path are designated as C_{tot} and D_{tot}. These values are exported over the service interface to applications to support a realistic choice of the desired end-to-end delay. Partial sums of C and D for the delay until an interior node on the path is reached are maintained as reservation state.

When the packetized nature and the error terms are reflected in the model, the end-to-end worst-case queuing delay is computed as follows (cf. Figure 27.4):

$$p > R \geq r: \frac{(b-M)(p-R)}{R(p-r)} + \frac{M+C_{tot}}{R} + D_{tot}$$

$$R \geq p \geq r: \frac{M+C_{tot}}{R} + D_{tot}$$

When the delay-bound requested D_{req} is larger than the *maximum* delay D_{max} in the fluid flow system, a delay budget $S = D_{req} - D_{max}$ can be computed, which may be used to reduce the resource allocations at certain nodes with the network. D_{max} can be computed as follows (note that here the error terms are

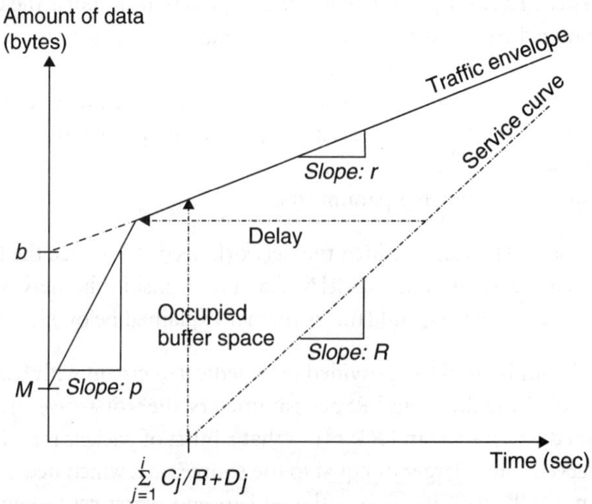

FIGURE 27.4 Delay and buffer calculation for a flow characterized by a (b,r,p,M) traffic envelope serviced with rate R (cf. [14]).

considered and the service rate R is replaced with the flow's rate r): $D_{max} = b/r + C_{tot} / r + D_{tot}$. S is the second parameter in the RSpec (see above) called "slack."

With the GS, it is possible to achieve a mathematically provable bound on the delay and no (congestion-induced) packet losses. A key element of GS is the information exchange at every network element to update the "error terms" for the delay computation, which on the one hand enables to achieve a minimal delay characteristic but also makes it feasible to exploit differences between the desired delay and the minimal delay (slack term) on the other.

Controlled Load Service (CLS, [3]): While the Guaranteed Service comes close to the optimal performance feasible within a packet-switched network, it requires considerable overhead within the network and reserves resources in a conservative way (considering the worst case). Particularly for bursty traffic sources, this will lead to a reduced overall network utilization and high costs for the customer reserving resources. In addition, the exact requirements with regard to the burstiness/delay requirements of a flow are often not known in advance and are therefore difficult to communicate to the network. For many (if not the majority of) applications, a service model with less tight guarantees incurring lower costs is needed. The Controlled Load service, which has also been standardized in the IETF, aims at filling the gap between "best-effort" and a high-end service like the Guaranteed Service by providing a "better-than-best-effort" service. CLS specifies that the behavior of the service should approximate the behavior in a lightly loaded network. This implies that the mean loss seen over time intervals, which are large compared to the flow's burst intervals, should approximate the link error rate, that is, virtually no congestion losses should occur. However, no quantitative commitment about the expected end-to-end delay is made. It is only specified that the delays over time scales significantly larger than the flow's burst intervals should be low on average. These characteristics fit well with adaptive applications, which exhibit some tolerance to losses and delays but need at least some statistical QoS assurance. The CLS is invoked using the TSpec definition introduced above.

27.3 RSVP

If an application wants to receive preferential treatment for its packets, it needs to set up a reservation along the path before it starts to transmit. This requires an additional protocol to be added to the existing TCP/IP protocol suite (Figure 27.2). The protocol developed by the IETF for this purpose is called RSVP [7,10]. The RSVP protocol is employed by hosts to communicate service requirements to the network and to routers to exchange (and subsequently establish) reservation state along the path.

Basic Properties/Design Choices

Uni-directional reservation: RSVP establishes a reservation in only one direction (simplex). Although communicating applications may each act as both sender and receiver, logically the reservation from host A to host B is distinct from the reservation in the opposite direction (B to A). Thus, in this two-way communication scenario, two reservations have to be established.

Receiver orientation: Particularly to support multicast communications (implying large multicast groups, heterogeneous receivers, and dynamic group membership, cf. Chapter 21), the receiver initiates the resource reservation and decides on the amount of resources to be reserved. In a multicast scenario, the reservations travel "upstream" toward the sender, building up a reservation tree that corresponds to (a subset of) the multicast tree.

Soft state: RSVP applies the "soft state" approach to maintain reservations inside the network. "Soft state" means that a timer is associated with the respective state blocks. When this timer expires, the state is automatically deleted. Triggered by the RSVP entity of the host on which the reservation has been requested by the application, the RSVP protocol periodically refreshes the reservation state to allow for continuously maintaining the reservation along the entire path. This allows RSVP to cope with host failures/crashes (where the application cannot communicate that certain resources are not needed any more) and to adapt to changing network topologies (e.g., link outages followed by re-routing) as well changing group membership in the multicast case.

Reservation styles: RSVP provides different styles of reservation, which express how a reservation should be treated. Again, this is of particular importance only for a multicast session. Reservation styles can be used to share a reservation among flows from all senders contributing to a session (Wildcard Filter (WF) style) or to select a specific sender within the shared reservation (Shared Explicit Filter (SE) style). If there is a one-to-one mapping between the sender and the receiver (in either a uni- or multicast session), the reservation style is called Fixed Filter (FF).

Independence of particular policy and traffic control mechanisms: RSVP can be considered to be a general signaling protocol to distribute and maintain state within a network. The data, which are carried in the RSVP messages, are opaque for the RSVP entities. These data are passed over the respective interfaces of the RSVP implementation (cf. Figure 27.3) to other entities for processing. Such an entity is, for example, the Admission Control module, which decides based on the transmitted control data if a reservation request can be granted or not.

RSVP Operation

Figure 27.5 shows the basic mode of operation of RSVP. In RSVP, there are two basic categories of messages: PATH and RESV (reservation) messages. PATH messages are sent from senders toward the receivers ("downstream"). They convey much information: to receivers, they distribute the properties of the sender's flow; they also collect and distribute hop-by-hop characteristics of the path taken. Finally, the PATH messages trigger the establishment of state along the path (and provide the necessary parameters), which can then be used by the RESV messages to actually set up a reservation.

The RESV messages travel "upstream" (from receivers toward the senders). They convey the resource requirements of the receivers (which are based on the state information received with the PATH messages) and actually trigger the installation of reservations on the network interfaces. (Note that the installation/enforcement of the reservation is then again "downstream" — in the direction from the sender toward the receiver.) When the sender has finally received an RESV message, it knows that the reservation has now been established on the end-to-end path and can now transmit data, which are protected within the network.

The path of the PATH messages is determined by the IP unicast and multicast routing protocols (an RSVP implementation has an interface to retrieve the necessary routing information). As paths may be asymmetric (which means that the downstream and upstream paths found by the IP routing may be different), the RESV messages rely on established PATH state information rather than the routing information to follow the exact reverse path taken by the PATH message.

Further RSVP messages are PATHErr and RESVErr to indicate that the establishment of PATH or RESV state has not been successful (e.g., when the admission control has failed). To explicitly tear down the established state inside the network, the PATHTear and RESVTear messages are used (these messages complement the soft state mechanism where outdated state is automatically deleted).

A feature of RSVP with regard to the support for multicast is *merging*, that is, reservations coming from different receivers for the same session are combined (*RESV (E,I)* in Figure 27.5). Merging is required to

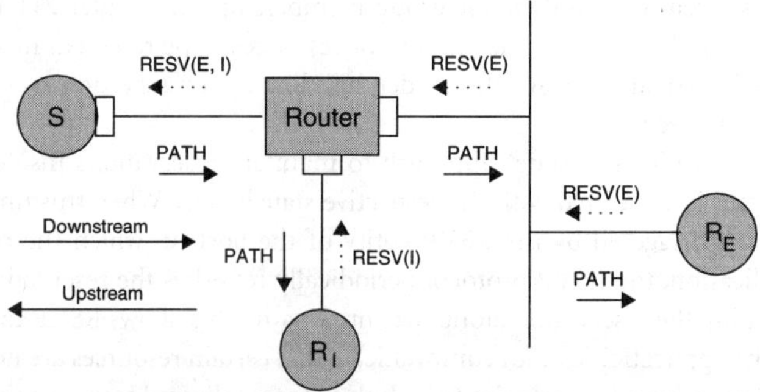

FIGURE 27.5   RSVP operation.

match the RSVP upstream operation with the way multicast forwarding works in the downstream path: replicating packets that have to be delivered to different branches of the multicast tree. Thus, at such replication points, RSVP must merge the reservation requests and compute the maximum of the FlowSpecs of the different receivers. Note that merging FlowSpecs can be very difficult as there are numerous, not necessarily ordered parameters. Different reservation styles cannot be merged as they are fundamentally incompatible. Guidelines for ordering of the parameters and merging are contained in the service class specifications (Controlled Load and Guaranteed Service) rather than the RSVP specification as RSVP itself treats the FlowSpecs as opaque objects.

27.4　Applicability of Integrated Services and RSVP

The Integrated Services Architecture and RSVP have often been described as being very complex and possessing poor scaling properties. To a large extent, the complexity and lack of scalability are due to the approach of dealing with QoS signaling and enforcement at a per-flow granularity inside the network to achieve the "high-end" QoS goals, which have been the original motivation for the Integrated Services Architecture. The resource requirements (processing power, memory) of RSVP (and the corresponding operations and state that need to be maintained in the Integrated Services traffic control) scale with the number of sessions, which may become prohibitive the closer a network element is located to the network core. The development of the IETF Differentiated Services (DiffServ) Architecture (Chapter 25) has followed the design of Integrated Services coming from this conclusion. It aims at treating only flow aggregates inside the network achieving better scalability, which has of course been traded against a less stringent support of QoS guarantees. The operation of Int Serv over DiffServ-enabled network clouds aiming at the combination of per-flow QoS treatment at the network edge and scalable, efficient QoS enforcement in the network core is described in [5]. To reduce the resource requirements of RSVP itself, aggregation of several RSVP reservations into a single one inside the network is proposed [8].

It should also be noted that RSVP started off as a rather simple protocol and became increasingly complex due to inherent difficulties of providing per-hop QoS signaling along the data path (problems that also have to be encountered by alternative signaling protocols). The design of RSVP has been significantly influenced by the predicted importance of IP multicasting: a vision that has not become a reality. The strong support for multicast is today considered rather a burden than an important feature. The merging of reservations coming from different receivers of a multicast session is very complex and can even be counterproductive in that one extremely large reservation request might impair the acceptance of a concurrently issued reservation request by another receiver ("killer reservation problem"). To solve this problem, a complex detection mechanism ("blockade state") had to be integrated into the protocol.

The scalability problem is particularly an issue for Voice-over-IP (which could be seen as the most popular and thus most important application requiring QoS support from the network) as the per-flow bandwidth is low and thus the state overhead to flow bandwidth ratio is high. Additionally, due to the similarity of voice flows, some options of RSVP (which are very useful for other flow types) are again rather a burden than a feature. An example is the initial exchange of sender and receiver TSpecs in PATH and RESV messages, where the sender advertises its traffic properties (which are most probably well known at the receiver anyway through, e.g., the RTP payload type). The receiver then typically reserves exactly with these parameters. The near-immediate setup of a connection in a circuit-switched network should be approximated in a packet-switched network as far as possible. However, the two-way end-to-end reservation setup might take significantly longer (RSVP processing/state update, admission control, traffic control configuration). A sender-based approach to a reservation protocol (see, e.g., [11]) thus seems much better suited to accommodate voice flows.

While the Integrated Service Architecture has not seen significant deployments, RSVP is used as a signaling protocol for Multi Protocol Label Switching (MPLS — cf. Chapter 13) to perform label distribution and support explicit routing (RSVP-TE, [9]). RSVP-TE is intended to be used by label switching routers (LSRs) to establish and maintain label-switched path (LSP) -tunnels and to reserve network resources for such LSP tunnels.

RSVP is also proposed to be used for Third Generation Mobile Networks. In [12], the QoS signaling procedures and policy control within the Universal Mobile Telecommunications System (UMTS) end-to-end QoS architecture are specified. The usage of RSVP is specified as follows: the signaling source and/or destination are the User Equipment (UEs: terminals). The RSVP signaling process can either trigger or be triggered by the Policy Decision Point (PDP) context establishment process. The authorization token and flow identifiers in a policy data object should be included in the RSVP messages sent by the UE. When both RSVP and service-based local policy are used, the Gateway GPRS Support Node (GGSN, the point of interconnection of the 3G network with the Internet) should use the policy information to decide whether to accept and forward RSVP messages.

27.5 Concluding Remarks

Ten years have passed since the basic concepts of RSVP have been devised; meanwhile, the requirements have changed: instead of multicast as an advanced communication model, the issues of node mobility and security are increasingly important for the evolution of the Internet. RSVP has not been designed to accommodate mobility at all; therefore, the following problems may arise: the movement of the mobile nodes may not properly trigger a reservation refresh for the new path and therefore a mobile node may be left without a reservation up to the length of the refresh timer. IntServ/RSVP also does not work properly if the IP address of a mobile node changes or traffic is tunneled toward a mobile node away from its home network, because the classifiers are not able to identify the flow that had a reservation.

From a security point of view, RSVP provides the basic building blocks for deploying the protocol in various environments to protect its messages from forgery and modification, that is, hop-by-hop protection is provided. However, current RSVP security mechanisms do not provide nonrepudiation and protection against message deletion: the two-way peer authentication and key management procedures are missing.

Future QoS architectures and signaling protocols [16] must address the shortcomings of Integrated Services/RSVP that have just been described. Additional requirements are to take inter-domain signaling explicitly into account (where several administrative domains with potentially different per-domain QoS solutions are traversed). Another aspect is a more modular design of the architecture and protocols, which extend the applicability particularly of the signaling protocol well beyond the QoS area. The Next Steps in Signaling (NSIS) Working Group in the IETF is addressing these issues currently.

The margin where QoS mechanisms really lead to a differentiation in the efficiency of how data are carried in the network is relatively small [13] considering the parameters of today's Internet (available bandwidths, typical applications), that is, only when parts of the network are close to being fully loaded, the effect of the QoS mechanisms are visible at all. Then, of course, the differentiation is extremely significant and may provide the distinction between an almost undisturbed operation and complete breakdown of connectivity.

For the development of future Internet QoS schemes, it is important to consider the (predicted) mix of applications used in the network. If there will be a significant portion of applications that are delay-sensitive and at the same time applications will be present that emit very bursty traffic flows, then support for QoS inside the network is almost inevitable. Finally, the economic background for QoS deployment needs to be considered throughout the research and development process, that is, the QoS case where the capacity is maintained and the complexity increased need to be continuously validated against the solution of "overprovisioning," which consists of increasing capacity and maintaining or even reducing the complexity of the network.

27.6 Further Information

RSVP
    Homepage: http://www.isi.edu/rsvp
    Protocol details: http://www.networksorcery.com/enp/protocol/rsvp.htm
    Policy infrastructure
    IETF Resource Allocation Protocol Working Group http://www.ietf.org/html.charters/rap-charter.html

IETF Policy Framework Working Group http://www.ietf.org/html.charters/policy-charter.html

Next-generation (QoS) signaling

Columbia Univ. Lightweight Signaling Project: http://www.cs.columbia.edu/~pingpan/projects/siglite.html

IETF NSIS Working Group http://www.ietf.org/html.charters/nsis-charter.html

References

Integrated Services

1. R. Braden, D. Clark and S. Shenker, Integrated Services in the Internet Architecture: an Overview, IETF, RFC 1633, June 1994.
2. S. Shenker, C. Partridge and R. Guérin, Specification of Guaranteed Quality of Service, IETF, RFC 2212, September 1997.
3. J. Wroclawski, Specification of the Controlled-Load Network Element Service, IETF, RFC 2211, September 1997.
4. J. Wroclawski, The use of RSVP with integrated Services, IETF, RFC 2210, September 1997.
5. Y. Bernet, R. Yavatkar, P. Ford, F. Baker, L. Zhang, M. Speer, R. Braden, B. Davie, J. Wroclawski and E. Felstaine, A framework for integrated services operation over DiffServ Networks, IETF, RFC 2998, November 2000.
6. D. Clark, S. Shenker and L. Zhang, Supporting real-time applications in an integrated services packet network: architecture and mechanism, Proceedings ACM SIGCOMM '92, August 1992.

RSVP

7. R. Braden, L. Zhang, S. Berson, S. Herzog and S. Jamin, RSVP — Version 1 Functional Specification, IETF, RFC 2205, November 1997.
8. F. Baker, C. Iturralde, F. LeFaucheur and B. Davie, Aggregation of RSVP for IPv4 and IPv6 Reservations, IETF, RFC 3175, September 2001.
9. D. Awduche et al., RSVP-TE: Extensions to RSVP for LSP Tunnels, IETF, RFC 3209, December 2001.
10. L. Zhang, S. Deering, D. Estrin, S. Shenker and D. Zappala, RSVP: a new resource reservation protocol, *IEEE Network*, Vol. 7(5), pp. 8–18, September 1993.
11. P. Pan and H. Schulzrinne, YESSIR: a simple reservation mechanism for the Internet, Proceedings NOSSDAV, Cambridge, UK, June 1998, http://www.cs.columbia.edu/~pingpan/projects/yessir.html
12. 3GPP TS 23.207 V5.6.0, End-to-end Quality of Service (QoS) Concept and Architecture, Release 5, December 2002.

Other references

13. P. Ferguson and G. Huston, Quality of service in the Internet: fact, fiction or compromise?, Presentation at INET 98, July 1998, http://www.potaroo.net/prestns/inet98/index.html
14. R. Guérin and V. Peris, Quality-of-service in packet networks: basic mechanisms and directions, IBM Research Report, RC21089, January 1998.
15. H. Sanneck, D. Witaszek, T. Zseby and M. Smirnov, MULTICUBE — IP Multicast over ATM Research, Next Generation Internet in Europe, ACTS Project InfoWin AC 113, pp. 97–103, 1999, ISBN 3-00-004250-4.
16. G. Huston, Next Steps for the IP QoS Architecture, IETF, RFC 2990, November 2000.

28

RTP, RTCP, and RTSP — Internet Protocols for Real-Time Multimedia Communication

Arjan Durresi
Louisiana State University

Raj Jain
Nayna Networks, Inc.

28.1 Multimedia over the Internet

The Internet was originally designed to support data communications. Most of the traffic initially included data such as e-mails and files. Voice traffic was served exclusively by telephone networks. As the Internet grew in terms of number of nodes, applications, and users, the need for multimedia communication over the Internet emerged. Animation, voice, and video clips are now common on the Internet. Multimedia networking products like Internet telephony, Internet TV, and video conferencing are available in the market. In the near future, people will enjoy other multimedia products in distance learning, distributed simulation, distributed work groups, and other areas. Multimedia traffic poses new requirements to the Internet architecture and its protocols. In response to this, researchers have designed a family of protocols, including Real-Time Transmission Protocol (RTP), its control part Real-Time Transmission Control Protocol (RTCP), and Real-Time Streaming Protocol (RTSP), that are the object of this chapter.

For transporting textual data, the best-effort service model of the IP-based Internet was shown to have been an adequate solution. In this service model, IP makes every effort to deliver packets, but does not provide guarantees. Thus, packets can be lost or delivered out of order, and the delay is unpredictable. The advantage of such a model is that it keeps routers as simple as possible and keeps the network architecture very scalable — features that have been crucial for the spectacular success of the Internet. To increase the reliability of the end-to-end services, data packets are transported by TCP, which is one of the most common transport layers used in IP networks. TCP has mechanisms to conceal the unreliability of IP and presents a reliable network channel to applications. For example, TCP retransmits lost packets and keeps them in order. TCP also has mechanisms to keep network congestion under control and to avoid congestion collapse. Hence, TCP is an optimal solution for data traffic. IP, which is implemented in all nodes, is simple and scalable, and TCP, which is implemented only at the end-user nodes, provides the needed reliability for the applications. This solution, however, has problems when attempting to transport multimedia traffic. Many multimedia applications are highly sensitive to end-to-end delay and delay variation (jitter); but can tolerate some data loss. TCP takes care of losses by retransmitting the lost packets, but does not guarantee the delay or its variation. That is why multimedia traffic does not work well with TCP. For example, if some packets carrying part of a video scene are lost, having them retransmitted by TCP could worsen the situation. Such retransmitted video traffic could overlap over a new video scene and may also damage it. To avoid such problems with TCP, most of the multimedia traffic is transported by UDP to make use of its multiplexing and checksum services. UDP is the other transport layer over IP.

Multimedia applications can be divided in two classes: real-time interactive applications and noninteractive streaming applications.

In real-time applications such as voice over the Internet and video conferencing, all communication has to be in real time and the data cannot be buffered for long to smooth out network delay or jitter. There are hundreds of Internet interactive multimedia products currently available such as Microsoft NetMeeting, PC-to-phone and PC-to-PC, H-323 based applications, and MBone tools [1]. Such applications are very delay and jitter sensitive. For example, voice application cannot accept delays more than 250–300 ms and jitter more than 75 ms; otherwise, the conversation becomes almost unintelligible. Some new types of applications are emerging in this category, such as Distributed Virtual Environments (DVE) for collaboration, scientific exploration, instrument operation, and data visualization by supporting continuous media flows, data transfers, and data manipulation tools within virtual environment interfaces. Other specialized equipment, such as haptic devices or head-mounted displays, is used to enable manipulation of instruments or navigation within virtual worlds. These collaborative, distributed, immersive environments pose new requirements to the network in terms of data transfer bit-rates, short one-way delay for real-time operation, and reliable transfer of data. One example of DVE application is a system developed by the National Tele-immersion Initiative, which uses 3D real-time acquisition to capture accurate dynamic models of humans. The system combines these models with a static 3D background (3D model of an office, e.g.), and it also adds synthetic 3D graphic objects that may not exist at all. These objects may be used as a basis for a collaborative design process that users immersed in the system can work on. The system can also be described as a mix of virtual reality and 3D videoconferencing, and can be used in tele-diagnosis and tele-medicine.

Noninteractive applications class can be divided into two subclasses: streaming stored audio/video and streaming live audio/video. Streaming stored applications deliver audio and video streams stored in a server to clients. Examples of such stored files include lectures, songs, movies, and video clips. In streaming stored applications, the user begins to play out of the file, while the file is being received. The main requirement of such applications is that once the playout begins, it should proceed according to the original timing of the recording. This is translated in a requirement for the end-to-end network delay; but such a requirement is less stringent compared to that of interactive applications. Because the playout is not in real time, the file can be buffered to smooth out the effects of random network delay. Examples of streaming stored applications include RealPlayer [2], Apple's Quick Time [3], and Microsoft Windows Media [4]. Streaming live audio and video applications are similar to radio and television broadcasts. Even though the files are not stored in servers, they can be stored locally at clients. In the streaming stored applications, local buffering allows reducing the network jitter and delay compared to interactive applications.

Among the multimedia requirements discussed so far, the requirements for delay and jitter from interactive, real-time applications are the most stringent. The best-effort service of today's Internet cannot satisfy these requirements, and this is one of the reasons why multimedia applications are not yet the new killer applications on the Internet. There is a vivid debate among researchers about how to satisfy such multimedia requirements [5]. Some researchers argue that the Internet's service model should be changed to allow resource reservations, which would enable predictable delays and jitter. A large amount of research work is dedicated to such solutions and protocols, such as RSVP (part of IntServ), that signal the needed resource reservations. The main problem with this approach is that it increases the complexity of protocols and routers, which could lead to scalability issues. Other researchers propose to use more bandwidth (overprovisioning) with the best-effort service. This is still an open debate but for the time being, the proposal to use reservations or other QoS solutions is not being applied and multimedia applications use the best-effort service of the Internet.

Besides the limited delay and jitter, multimedia applications need some transport services with characteristics different from those of TCP, and with more functionality than UDP. The protocol designed to provide services for data with real-time characteristics is called Real-time Transport Protocol (RTP). In the following, we discuss some common needs of real-time applications considered in the design of RTP.

The large variety of multimedia applications of both classes use different coding schemes, each with its own trade-offs between quality, bandwidth, and computational cost. One of the basic requirements such applications have is the ability to interoperate with each other. For example, it should be possible for two independently developed Voice over IP applications to talk to each other. Instead of imposing a given voice-encoding scheme, RTP enables these applications to negotiate and select a scheme that is available to both parties. The same is true for video applications.

In the best-effort service provided by the Internet, the delay in routers is random and the end-to-end delays fluctuate; this phenomenon is called jitter. Jitter can change the original spacing between multimedia packets. If the receiver ignores the presence of jitter and plays the voice or video as soon as they arrive, the result could be unintelligible. Fortunately, jitter can be smoothed out by using a playback buffer, where the arriving packets are stored before being played. In order to play these packets at the right time, the receiver needs to know the timing relationships among the received packets. RTP enables the recipient of the packet stream to determine this information.

Related to the media timing is the need of multimedia applications for synchronization among various media, for example, between voice and video streams in a conference. These streams could have originated at one or more sources. Another important requirement of multimedia applications is to identify packet loss. Then, the specific application can deal in an appropriate manner with such losses.

The output data rate of multimedia applications depends on the information content. Because such applications do not use TCP, they are not congestion sensitive. Therefore, they do not reduce their data rate in response to network congestion, leading to packet loss and more congestion. Yet, many multimedia applications are capable of responding to congestion, for example, by changing parameters in coding schemes to reduce the consumed bandwidth during congestion periods. In order for this to occur, the sender needs the congestion feedback from receivers.

Another requirement for multimedia applications is the concept of frame indication at the application level. For example, it is important to notify a receiver video application that a set of packets belong to the same video scene, so that they can be treated in an appropriate way.

Based on the experience with existing multimedia products over the Internet, researchers have designed RTP.

28.2 RTP

RTP is the Internet-standard protocol for the transport of real-time data, including audio and video [6, 7]. It can be used for media-on-demand as well as interactive services such as Internet telephony. RTP was developed by the Internet Engineering Task Force (IETF) and is in widespread use. The RTP standard actually defines a pair of protocols: RTP and RTCP. RTP is used for the exchange of multimedia data, while RTCP is the control part and is used to periodically obtain feedback control information regarding the quality of transmission associated with the data flows. RTP usually runs over UDP/IP; but efforts are under way to make it transport-independent so that it could be used over other protocols. RTP and the associated RTCP use consecutive transport-layer ports, when used over UDP. Figure 28.1 shows a schematic diagram illustrating the use of RTP. The video/audio is digitized using a particular codec. The blocks formed by such bit streams are encapsulated in RTP packets and then in UDP and IP packets.

The data part of RTP is a thin protocol providing support for applications with real-time properties such as continuous media (e.g., audio and video), including timing reconstruction, loss detection, security, and content identification. RTP does not reserve bandwidth or guarantee Quality of Service (QoS).

The control part of the protocol, RTCP, provides support for real-time conferencing of groups of any size within the Internet. It offers QoS feedback from receivers to the multicast group as well as support for the synchronization of different media streams. RTCP also conveys information about participants in a group session.

Attempts to send voiceover networks began in the early 1970s. Several patents on packet transmission of speech, time stamp, and sequence numbering were granted in the 1970s and 1980s. In 1991, a series of voice experiments were completed on DARTnet. In August 1991, the Network Research Group of Lawrence Berkeley National Laboratory released an audio conference tool **vat** for DARTnet use. The protocol used was referred later as RTP version 0.

In December 1992, Henning Schulzrinne, then at GMD Berlin, published RTP version 1. It underwent several states of Internet Drafts, and was finally approved as a Proposed Standard on November 22, 1995 by the IESG. This version was called RTP version 2 and was published as

- RFC 1889, updated in July 2003 by RFC 3550 RTP: A Transport Protocol for Real-Time Applications.
- RFC 1890, updated in July 2003 by RFC 3551 RTP Profile for Audio and Video Conferences with Minimal Control.

On January 31, 1996, Netscape announced "Netscape LiveMedia" based on RTP and other standards. Microsoft NetMeeting conferencing software also supports RTP.

The design of RTP was made following an architectural principle known as Application Level Framing (ALF). This principle was proposed by Clark and Tennenhouse in 1990 [8] as a new way to design protocols for emerging multimedia applications. ALF is based on the belief that applications understand their own needs better, which means that the intelligence should be placed in applications and the network should be kept simple. For example, an MPEG video application knows better how to recover from lost frames, and how to react to it if an I or a B frame is lost.

RTP is designed to support a wide variety of applications. It provides flexible mechanisms by which new applications can be developed without repeatedly revising RTP itself. For each class of applications, RTP defines a profile and one or more payload formats. The profile provides a range of information that ensures a common understanding of the fields in the RTP header for that application class. A profile can

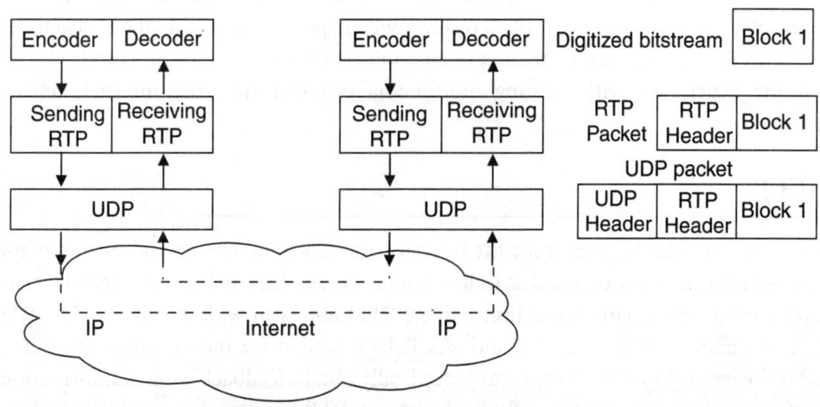

FIGURE 28.1 Real-time Transport Protocol.

also define extensions or modifications to RTP that are specific to a particular class of applications. The payload format specification explains how the data that follow the RTP header are to be interpreted.

Mixers and Translators

Besides the sender and receiver, RTP defines two other roles for systems, those of a translator and a mixer. They reside in between senders and receivers, and process the RTP packets as they pass through. The translators are used to translate from one payload to another. For example, suppose some of the participants in the audio–video conference have lower access bandwidth than that required for the conference. In this case, a translator could convert the stream to audio and video formats that require less bandwidth.

Mixers, on the other hand, are used to combine multiple source streams into one. An example is a video mixer that combines the images of individual people into one video stream to simulate a group scene.

Header Format

Figure 28.2 shows the header format used by RTP. The first 12 bytes are always present, whereas the contributing source identifiers are only used in certain circumstances. After this header, there may be optional header extensions. Finally, the RTP payload, whose format is determined by the application, follows the header. This header contains only the fields that are likely to be used by most of the multimedia applications. Information specific to a single application is carried in the RTP payload.

V — version: 2 bits: The first two bits are a version identifier, which contains the value 2 for the RTP version used at the time of this writing.

P — padding: 1 bit: If the padding bit is set, the packet contains one or more additional padding octets at the end, which are not part of the payload. The last octet of the padding contains a count of how many padding octets should be ignored, including itself. RTP data might be padded to fill up a block of certain size as required by an encryption algorithm.

X — extension: 1 bit: If the extension bit is set, exactly one extension header follows the fixed header.

CC — CSRC count: 4 bits: The CSRC count contains the number of CSRC identifiers that follow the fixed header. This number is more than one if the payload of the RTP packet contains data from several sources.

M — marker: 1 bit: The interpretation of the marker is defined by a profile. The marker is intended to allow significant events such as frame boundaries to be marked in the packet stream.

PT — payload type: 7 bits: PT identifies the format of the RTP payload and determines its interpretation by the application. One possible use of this field would be to enable an application to switch from one coding scheme to another based on the information about resource availability on the network of feedback on application quality. Default payload types are defined in RFC 3551, as shown in Table 28.1.

1	2	3	4	5	6	7	8	9	10					16	17													32
V	P	X	CC			M		Payload type								Sequence number												
Timestamp																												
Synchronization source (SSRC) identifier																												
Contributing source (CSRC) identifier (1)																												
Contributing source (CSRC) identifier (N)																												
Extension header																												
RTP Payload 1-N audio/video frames																												

FIGURE 28.2 RTP header format.

Example specifications include PCM, MPEG1/MPEG2 audio and video, JPEG video, Sun CelB video, H.261 video streams, etc. More payload types can be added by providing a profile and payload format specification. The exact usages of marker bit and the payload type are determined by the application profile.

Sequence number: 16 bits: The sequence number increments by one for each RTP data packet sent, and can be used by the receiver to detect missing and misplaced packets. The initial value is randomly set to make crypto analysis attacks on possible encryption more difficult. RTP does not take any action when it detects a lost packet. It is left to the application to decide what to do when a packet is lost. For example, a video application could replay the previous frame if a frame is lost. Another application, however, could decide to change encoding parameter in order to reduce the needed bandwidth. These decisions need some intelligence, which is left to the applications themselves.

Timestamp: 32 bits: The timestamp indicates the sampling instant of the first octet in the RTP data packet. Its function is to enable the receiver to play back samples at the appropriate intervals and to enable different media streams to be synchronized. It can also be used for calculations in jitter smoothing. The resolution of the clock used should be sufficient for the desired synchronization accuracy and for measuring packet jitter.

The initial value is randomly set. Because different applications may require different granularities of timing, RTP does not specify the units in which time is measured. The timestamp is just a counter of ticks, and the time between ticks is application specific. The clock granularity is specified in the RTP profile or payload format for an application.

SSRC — Synchronization source: 32 bits: This field identifies synchronization sources within the same RTP session. It is chosen randomly to avoid two sources in the same RTP session to have the same SSRC identifier. It indicates where the data were combined, or the source of the data if there is only one source. The sources could be in the same node or in different nodes, such as during a conference.

CSRC — Contributing source list: 0 to 15 items, 32 bits each: The CSRC list identifies the contributing sources for the payload contained in this packet. The CC field gives the number of identifiers. It is used

TABLE 28.1   Payload Types (PT) for Some Audio and Video Encodings

PT	Encoding name	Media type	Clock rate [kHz]
0	PCMU	Audio	8
3	GSM	Audio	8
4	G723	Audio	8
5	DVI4	Audio	8
6	DVI4	Audio	16
7	LPC	Audio	8
8	PCMA	Audio	8
9	G722	Audio	8
10	L16	Audio	44.1
11	L16	Audio	44.1
12	QCELP	Audio	8
13	CN	Audio	8
14	MPA	Audio	90
15	G728	Audio	8
16	DVI4	Audio	11.025
17	DVI4	Audio	22.05
18	G729	Audio	8
25	CelB	Video	90
26	JPEG	Video	90
31	H261	Video	90
32	MPV	Video	90
33	MP2T	Audio/Video	90
34	H263	Video	90

only when a number of RTP streams pass through a mixer. A mixer can be used, for example, in a conference to combine data received from many sources and to send it as a single stream to reduce the needed bandwidth.

To set up an RTP session, the application defines a particular pair of destination transport addresses (one network address plus a pair of ports for RTP and RTCP). In a multimedia session, each medium is carried in a separate RTP session, with its own RTCP packets reporting the reception quality for that session. For example, audio and video would travel on separate RTP sessions, enabling a receiver to select whether or not to receive a particular medium. An audio–video conferencing scenario presented below illustrates the use of RTP.

Example of an Audio and Video Conference

Audio and video media, used in the conference, are transmitted on separate RTP sessions. That is, separate RTP and RTCP packets are transmitted for each medium using two different UDP port pairs. In the pair of ports used for audio, one port is used for data and the other for RTCP. The same is true for other pair of ports used for video pair. The conference can also use Internet multicast service. In this case, two multicast group addresses are needed: one for voice and one for video. There is no direct coupling at the RTP level between the audio and the video sessions, except that a user participating in both sessions should use the same canonical name in the RTCP packets for both so that the sessions can be associated. This separation between audio and video allows users in the conference to receive only audio. The synchronization playback of a source's audio and video can be achieved using timing information carried in RTCP packets for both sessions.

Both audio and video applications used by each participant send chunks of data. Each of these chunks is carried in RTP packets. Each packet has an RTP header that indicates what type of audio or video encoding is contained in the packet. This information is used by the receiver to select the appropriate decoding scheme.

The RTP header contains timing information and a sequence number that allow the receivers to reconstruct the timing produced by the source for both audio and video. The timing reconstruction is done separately for each source of RTP packets in the conference.

Each participants' audio and video application periodically multicasts a reception report plus the identification information name of its user on the RTCP port. The reception report indicates the quality of reception of the current source and could be used to control the adaptive encodings. A participant sends an RTCP BYE packet when it leaves the conference.

28.3　Real-time Transport Control Protocol — RTCP

RTCP is the control protocol designed to work in conjunction with RTP. It is specified in RFC 3550. RTCP's relationship with other layers is shown in Figure 28.3.

In an RTP session, participants periodically send RTCP packets to all the members in the same RTP session using IP multicast. RTCP packets contain sender and/or receiver reports that announce statistics such as the number of packets sent, number of packets lost, and inter-arrival jitter. This function may be useful for adaptive applications that can use this feedback to send high- or low-quality data depending on the network congestion. Such applications will increase the compression ratio when there is little available bandwidth, and will reduce the compression ratio, which will result in higher multimedia quality when there is more available bandwidth. This feedback information can also be used for diagnostic purposes to localize eventual problems.

RTCP provides a way to correlate and synchronize different media streams that have come from the same sender. When collisions of SSRC occur, it is necessary to change the SSCR value of a given stream. This is done using RTCP.

In applications that involve separate multimedia streams, a common system clock is used for their synchronization. The system that initiates the session provides this function, and the RTCP messages enable all systems to use the same clock.

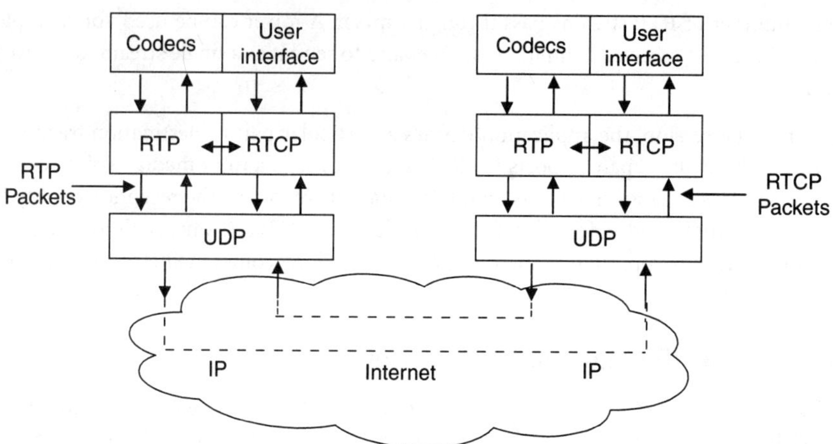

FIGURE 28.3　Real-time Transport Control Protocol.

RTCP is also used to deliver information about the membership in the session.

RFC 3550 defines five RTCP packet types to carry control information. These five types are as follows:

RR (Receiver Report): Receiver reports are generated by participants that are not active senders. They contain reception quality feedback about data delivery, including the highest packets number received, the number of packets lost, interarrival jitter, and timestamps to calculate the round-trip delay between the sender and the receiver.

SR (Sender Report): Sender reports are generated by active senders. In addition to the reception quality feedback as in RR, they contain a sender information section, providing information on inter-media synchronization, cumulative packet counters, and number of bytes sent.

SDES (Source Description Items): They contain information to describe the sources. In RTP data packets, sources are identified by randomly generated 32-bit identifiers. These identifiers are not convenient for human users. RTCP SDES (source description) packets contain textual information called *canonical names* as globally unique identifiers of the session participants. It may include the user's name, telephone number, e-mail address, and other information.

BYE: Indicates end of participation.

APP (Application specific functions): These are intended for experimental use as new applications and new features are developed.

RTCP packets are sent periodically among participants. When the number of participants increases, it is necessary to balance between getting up-to-date control information and limiting the control traffic. In order to scale up to large multicast groups, RTCP has to prevent the control traffic from overwhelming network resources. RTCP limits the control traffic to at most 5% of the overall session traffic. This is enforced by adjusting the RTCP packet generation rate according to the number of participants.

RTP Implementation Resources

RTP is an open protocol that does not provide preimplemented system calls. Implementation is tightly coupled to the application itself. Application developers have to add the complete functionality in the application layer by themselves. It is, however, always more efficient to share and reuse code rather than starting from scratch. The RFC 3550 specification itself contains numerous code segments that can be used directly in the applications. There are some implementations with source code available on the web for evaluation and educational purposes. Many modules in the source code can be used with minor modifications. The following is a list of useful resources:

vat (http://www-nrg.ee.lbl.gov/vat/),
rtptools (ftp://ftp.cs.columbia.edu/pub/schulzrinne/rtptools/),
NeVoT (http://www.cs.columbia.edu/~hgs/rtp/nevot.html).

RTP Library (http://www.bell-labs.com/project/RTPlib/DOCS/rtp_api.html) provides a high-level interface for developing applications that make use of the RTP.

RTP page (http://www/cs.columbia.edu/~hgs/rtp) maintained by Henning Schulzrinne is a very complete reference. Readers might also want to visit the Free Phone site [9], which documents an Internet phone application that uses RTP.

28.4 RTSP

RTSP, defined in RFC 2326, is an application-level protocol that enables control over the delivery of data with real-time properties over IP. Such control includes pausing playback, repositioning playback to future or past point of time, fast forwarding, and rewinding playback. These functionalities are similar to those of a DVD player. RTSP does not typically deliver the continuous media itself, although interleaving of the continuous media stream with the control stream is possible. In the words of the authors: "RTSP acts as a network remote control for multimedia servers"[10]. Sources of data include both live data feeds and stored clips.

RTSP is a client–server multimedia presentation protocol. There is no notion of RTSP connection. Instead, a server maintains a session labeled by an identifier. An RTSP session is in no way tied to a transport-level protocol. During an RTSP session, an RTSP client may open and close many reliable transport connections to the server in order to issue RTSP requests. It may alternatively use a connectionless transport protocol such as UDP.

RTSP is designed to work with lower-level protocols like RTP and RSVP to provide a complete streaming service over the Internet. It provides a means for choosing delivery channels (such as UDP, multicast UDP, and TCP), and delivery mechanisms based on RTP. The RTSP messages are sent out-of-band of the media stream. RTSP works for large-audience multicast as well as single-viewer unicast.

RTSP was jointly developed by RealNetworks, Netscape Communications, and Columbia University. It was developed from the streaming practice and experience of RealNetworks' RealAudio and Netscape's LiveMedia. The first draft of RTSP protocol was submitted to IETF on October 9, 1996 for consideration as an Internet Standard. Since then, it has gone through significant changes, and it was approved by IETF as a Proposed Standard on April 1998.

Numerous products using RTSP are available today. Major online players, such as Netscape, Apple, IBM, Silicon Graphics, VXtreme, Sun, and other companies, have announced their support for RTSP.

RTSP establishes and controls streams of continuous audio and video media between the media servers and the clients. A *media server* provides playback or recording services for the media streams, while a *client* requests continuous media data from the media server. RTSP supports the following operations:

Retrieval of media from the media server: The client can request a presentation description via HTTP or some other method. If the presentation is being multicasted, the presentation description contains the multicast addresses and ports to be used for the continuous media. If the presentation is to be sent only to the client, the client provides the destination for security reasons. The client can also ask the server to set-up a session to send the requested data.

Invitation of a media server to a conference: A media server can be invited to join an existing conference, either to play back media into the presentation or to record all or a subset of the media in a presentation. This method is useful for distributed applications such as distance learning. Several parties in the conference may take turns "pushing the remote control buttons."

Adding media to an existing presentation: The server or the client can notify each other about any additional media becoming available. This is particularly useful for live presentations.

RTSP aims to provide the same services on streamed audio and video just as HTTP does for text and graphics. It is intentionally designed to have similar syntax and operations, so that most extension mechanisms to HTTP can be added to RTSP.

Each presentation and media stream in RTSP is identified by an RTSP URL. The overall presentation and the properties of the media are defined in a presentation description file, which may include the encoding, language, RTSP URLs, destination address, port, and other parameters. The client can also obtain the presentation description file using HTTP, e-mail, or other means.

RTSP, however, differs from HTTP in several aspects. First, while HTTP is a stateless protocol, an RTSP server has to maintain "session states" in order to correlate RTSP requests with a stream. Second, HTTP is an asymmetric protocol where the client issues requests and the server responds; but in RTSP, both the media server and the client can issue requests. For example, the server can issue a request to set playing back parameters of a stream.

The services and operations in the current version are supported through the following methods:

OPTIONS: The client or the server informs the other party about the options it can accept.

DESCRIBE: The client retrieves the description of a presentation or media object identified by the request URL from the server.

ANNOUNCE: When sent from client to server, Announce posts the description of a presentation or media object identified by the request URL to a server. When sent from server to client, Announce updates the session description in real time.

SETUP: The client asks the server to allocate resources for a stream and start an RTSP session.

PLAY: The client asks the server to start sending data on a stream allocated via SETUP.

PAUSE: The client temporarily halts the stream delivery without freeing server resources.

TEARDOWN: The client asks the server to stop delivery of the specified stream and free the resources associated with it.

GET_PARAMETER: Retrieves the value of a parameter of a presentation or a stream specified in the URI.

SET_PARAMETER: Sets the value of a parameter for a presentation or stream specified by the URI.

REDIRECT: The server informs the clients that it must connect to another server location. The mandatory location header indicates the URL that the client should connect to.

RECORD: The client initiates recording a range of media data according to the presentation description.

Note that some of these methods can either be sent from the server to the client or from the client to the server; but others can only be sent in one direction. Not all these methods, however, are necessary in a fully functional server. For example, a media server with live feeds may not support the PAUSE method.

RTSP requests are usually sent on a channel independent of the data channel. They can be transmitted in persistent transport connections, or as a one-connection per request/response transaction, or in connectionless mode.

RTSP Implementation Resources

Several RTSP implementations are available on the web. The following is a collection of useful implementation resources:

RTSP Reference Implementation (http://www.real.com/devzone/library/fireprot/rtsp/reference.html): This is a source code testbed for the standards community to experiment with RTSP compatibility.

RealMedia SDK (http://www.realnetworks.com/devzone/tools/index.html): This is an open, cross-platform, client–server system that allows implementers to create RTSP-based streaming applications. It includes a working RTSP client and server, as well as the components to quickly create RTSP-based applications that stream arbitrary data types and file formats.

W3C's Jigsaw (http://www.w3.org/Jigsaw/): A Java-based web server. The RTSP server in the latest beta version was written in Java.

IBM's RTSP Toolkit (http://www.research.ibm.com/rtsptoolkit/): IBM's toolkits derived from tools developed for ATM/video research and other applications in 1995–1996. Its shell-based implementation illustrates the usefulness of the RTSP protocol for nonmultimedia applications.

28.5 Summary

This chapter discusses the three related protocols for real-time multimedia data over the Internet.

RTP is the transport protocol for real-time data. It provides timestamp, sequence number, and other means to handle the timing issues in real-time data transport.

RTCP is the control part of RTP that helps with quality of service and membership management.

RTSP is a control protocol that initiates and directs delivery of streaming multimedia data from media servers. It is the "Internet VCR remote control protocol." Its role is to provide the remote control; however, the actual data delivery is done separately, most likely by RTP.

References

[1]   Voice On the Net, http://von.com.
[2]   RealNetworks, http://www.realnetworks.com.
[3]   Apple's Quick Time, http://www.apple.com/quicktime.
[4]   Microsoft's Windows Media, http://www.microsoft.com/windows/windowsmedia/.
[5]   J.F. Kurose and K.W. Ross, *Computer Networking — A Top-down Approach Featuring the Internet,* 2nd ed., 2003.
[6]   H. Schulzrinne, S. Casner, R. Frederick, and V. Jacobson, RTP: A Transport Protocol for Real-Time Applications, RFC3550, July 2003.
[7]   H. Schulzrinne, S. Casner, RTP Profile for Audio and Video Conferences with Minimal Control, RFC3551, July 2003.
[8]   D. Clark and D. Tennenhouse, Architectural considerations for a new generation of protocols, Proceedings of the SIGCOMM '90 Symposium, Sept., 1990, pp. 200–208.
[9]   Why use the Plain Old Telephone when you can get so much better on the Internet? 1999, http://www-sop.inria.fr/rodeo/fphone/.
[10] H. Schulzrinne, A. Rao, and R. Lanphier, Real Time Streaming Protocol (RTSP), RFC 2326, April 1998.
[11] L.L. Peterson and B.S. Davie, *Computer Networks — a System Approach,* 3rd ed., Morgan Kaufmann, Los Attos, CA, 2003.
[12] Mark A. Miller, *Voice over IP Technologies,* M&T Books, 2002.
[13] Fred Halsall, *Multimedia Communications,* Addison-Wesley, Reading, MA, 2001.
[14] Alberto Leon-Garcia and Indra Widjaja, *Communication Networks,* 2nd ed., McGraw-Hill, New York, 2003.
[15] RTSP Resource Center, http://www.rtsp.org.

29

Mail Transfer Protocols and File Transfer Protocol

Suvendi Chinnappen
University of Pretoria

29.1 Mail Transfer Protocols

Introduction

It is useful to use the analogy of the old postal system to explain current mail transfer protocols. For example, suppose Tom wanted to send a message to Peter. Tom writes a message on a piece of paper, puts the paper in an envelope, and writes Peter's address details on the front of the envelope and Tom's address details on the back of the envelope (the return address). It is common practice to always put a return address on the envelope, in case the postal service found that Peter's address as specified on the envelope was incorrect; it would need to return the letter to the sender. Tom then puts the letter in a post box. The postal service retrieves the mail from the post box, determines which geographical location the letter is destined to, and *forwards* the letter to Peter's post box where it is *stored* until Peter physically retrieves the letter. A few days later, when Peter checks his post box, he finds the letter addressed to him from Tom.

Current electronic mail transfer protocols use a similar store and forward technology to send e-mail. This means that a person can write a message, which is saved to a file that is addressed to a particular person at a particular computer connected to the Internet. This file is then transferred to a remote computer, where it is stored in the recipient's mailbox, until the person it is addressed to is available to read the file.

As with postage mail, e-mail requires the recipient's address details. An e-mail address is divided into two components that are separated by the @ sign. The left component is the name of the e-mail mailbox and the right component is the name of the computer that contains the mailbox. The name of the host computer is called the domain name. For example, consider a fictional person called John Smith, who works at a fictional company called AlphaWorx. Assume that the AlphaWorx mail server on which each employee is allocated a mailbox is called Alphaworx.com. In addition, assume that there are three other persons by the name of John working at Alphaworx. If Alphaworx uses their surnames to distinguish between the different people called John, and allocates John the mailbox directory "John.Smith," then the e-mail address of John Smith will be John.Smith@Alphaworx.com.

Routers and switches along the Internet use the domain name to send mail to the correct computer (mail server). When the mail server receives the e-mail message, it uses the mailbox name to locate the correct mailbox on the server. If it is unable to find the mailbox, the e-mail is returned to the sender with a message stating that the mailbox is unknown.

Because mail servers are relatively expensive and require more resources, it is not feasible to have a mail server application on each user's computer. Instead, each user in a company is allocated space (the user's mailbox) on a single company mail server. To retrieve mail from the mailbox, a person normally uses a client mail application (which requires fewer resources), to connect to the mail server and retrieve mail from the user's mail box.

A mailbox is actually a directory on the mail server; similar to a folder on a computer's hard disk Host (mail server), computer names are not case sensitive. However, mailbox user names are dependent on the file naming conventions of the underlying operating system and are therefore case sensitive.

Figure 29.1 illustrates the basic mail transfer process.

Sending Mail — How Does Mail Get Transferred from the Sender to the Mailbox?

The Simple Mail Transfer Protocol (SMTP) provides mechanisms for the transmission of mail from the sending user's host to the receiving user's host. SMTP is independent of the particular transport service and requires only a reliable ordered data stream channel. A transport service is an interprocess communication environment that may cover one network, several networks, or a subset of a network.

If the source and destination hosts are not connected to the same transport service, one or more relay SMTP-servers is used. To be able to provide the relay capability, the SMTP-server must be supplied with the name of the ultimate destination host as well as the destination mailbox name [1].

E-mail applications on the Internet typically use a single TCP connection established between the sender process port[1] and the receiver process port, to transfer both control commands and the actual mail data. Since SMTP data are only 7-bit ASCII characters, but TCP is sent in 8-bit bytes, SMTP transmits each character as an 8-bit byte with the high-order bit cleared to zero. There is a predefined standard for sending electronic mail messages over the Internet (RFC 822). An SMTP-application should adhere to this standard when transporting mail messages over the Internet.

How Does SMTP Work?

After a user has typed in a mail message and clicked send, the sender-SMTP application establishes a full duplex TCP (assuming Internet) connection to the receiver-SMTP application. The sender-SMTP application generates and sends commands in sequential order to the receiver-SMTP application that responds to these commands and sends a reply back to the server-SMTP application in response to each command received. The sequence of commands is described in Figure 29.2.

SMTP Command–Response Sequence:

1. SMTP applications establish a TCP connection and send *HELO* greeting and response.
2. Sender-SMTP application sends a *MAIL* command indicating the sender of the mail.
3. If the receiver-SMTP application can accept mail, it returns an *OK* response and a numeric code.
4. Sender-SMTP application sends an *RCPT* command identifying a recipient of the mail.
5. If the receiver-SMTP application can accept mail for that recipient, it returns an *OK* response and a numeric code. If not, it responds with a reply rejecting that recipient (but not the whole mail transaction).
6. If an OK response is received from the receiver-SMTP application, the sender-SMTP application sends the mail *DATA*, terminating with a special sequence. The mail data are text based and may

---

[1]Usually, vendors use the port specified in the RFC821, that is, port 25.

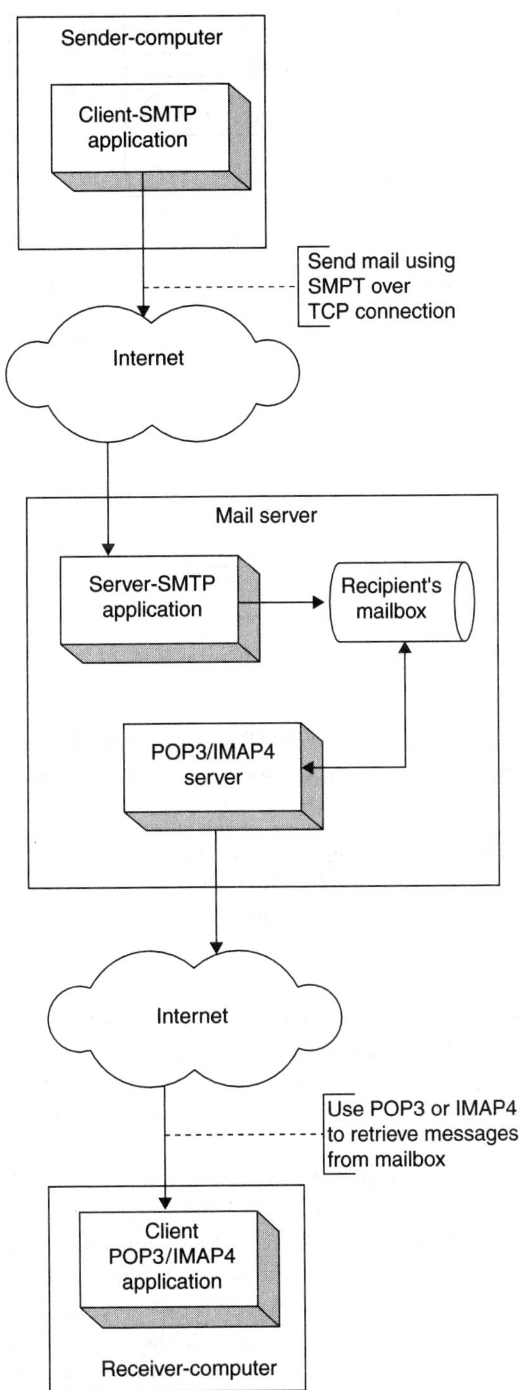

FIGURE 29.1　Overview of mail transfer from sender to receiver.

contain any of the 128 ASCII character codes. A line containing only a period terminates the mail data. This is the end of mail data indication.

7. If the receiver-SMTP application successfully processes the mail data, it responds with an OK reply and a numeric code.

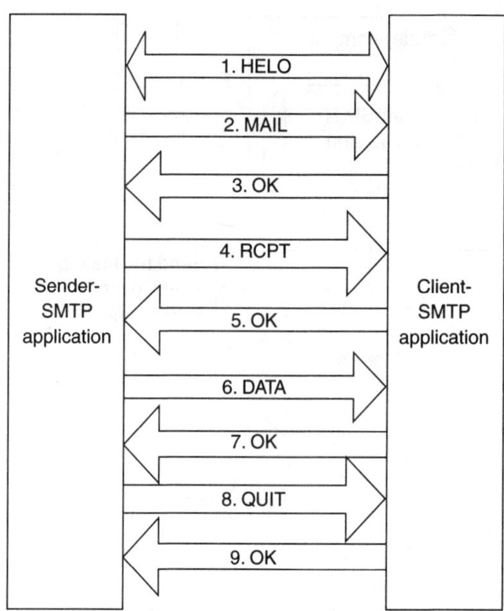

FIGURE 29.2 Basic SMTP Command–Response Sequence.

8. To close the TCP connection, the sender-SMTP application sends a *QUIT* command.
9. The receiver-SMTP application first sends an OK response to the sender-SMTP application, and only then closes the TCP connection.

Note that the sender- and receiver-SMTP applications may negotiate several recipients. When the same message is sent to multiple recipients, the sender-SMTP application tries to transmit only one copy of the data for all the recipients at the same destination host [1].

SMTP Commands

All mail applications developed by vendors have to adhere to the SMTP command and response syntax specified in the standard (RFC821). Commands and replies are not case sensitive. That is, a command or reply word may be upper case, lower case, or any mixture of upper and lower cases. Commands and replies are composed of characters from the ASCII character set. Table 29.1 provides a brief summary of the SMTP commands [1].

There are restrictions on the order in which these command may be used. The first command in a session must be the HELO command. The NOOP, HELP, EXPN, and VRFY commands can be used at any time during a session. The MAIL, SEND, SOML, or SAML commands begin a mail transaction. Once started, a mail transaction consists of one of the transaction beginning commands, one or more RCPT commands, and a DATA command, in that order. A mail transaction may be aborted by the RSET command. There may be zero or more transactions in a session. The last command in a session must be the QUIT command. The QUIT command cannot be used at any other time in a session [1].

Relaying

The *forward-path* consists of the recipient's mailbox identifier and the host name in which the mailbox resides. Sometimes, the forward-path consists of an optional list of hosts and a required destination mailbox. When the list of hosts is present, it is a source route and indicates that the mail must be relayed to the next host on the list. The mailbox is an absolute address, and the route presents information about how to get there.

A receiver-SMTP application checks the host name to determine if the message is intended for itself. If the host name differs from the receiving host's identifier, then the receiver-SMTP can decide to either relay the message or to reject the message.

TABLE 29.1   SMTP Commands

Command	Argument(s)	Description
HELO	The argument field contains the *host name* of the sender-SMTP application.	Identifies the sender-SMTP to the receiver-SMTP.
MAIL	The *reverse-path* is a reverse source routing list of hosts and source mailbox (i.e., a source route from the current location of the message to the originator of the message). This list is used as a source route to report errors and send nondelivery notices to the sender. Refer to section on relaying for more information.	Initiates a mail transaction in which the mail data are delivered to one or more mailboxes. Informs the SMTP-receiver that a new mail transaction is about to begin and to reset all its state tables and buffers, including any recipients or mail data.
RCPT	The *forward-path* consists of an optional list of hosts and a required destination mailbox. When a list of hosts is present, it is a source route and indicates that the mail must be relayed to the next host on the list. Refer to the section on relaying for more information.	Identifies an individual recipient of the mail data. Multiple recipients use multiples of this command.
DATA	The receiver treats the lines following this command as mail data from the sender. This command causes the mail data from this command to be appended to the mail data buffer.	Contains the mail data. A line containing only a period and a carriage-return-line feed indicates the end of mail data.
SEND	Same as for MAIL.	Initiates a mail transaction in which the mail data are delivered to each recipient's terminal. If the recipient is not active (or not accepting terminal messages) on the host, a 450 reply may be returned to an RCPT command. The mail transaction is successful if the message is delivered to the recipient's terminal.
SOML	Same as for MAIL.	SEND or MAIL — Initiates a mail transaction in which the mail data are delivered to each recipient's terminal if the recipient is active (and accepting terminal messages) on the host. If the recipient is not active (or not accepting terminal messages), then the mail data are entered into the recipient's mailbox. The mail transaction is successful if the message is delivered either to the terminal or the mailbox.
SAML	Same as for MAIL.	SEND AND MAIL — Initiates a mail transaction in which the mail data are delivered to the recipient's terminal if the recipient is active (and accepting terminal messages) on the host. The mail data are entered into the recipient's mailbox. The mail transaction is successful if the message is delivered to the mailbox.
RSET	N/A	RESET — specifies that the current mail transaction be aborted.
VRFY	*Destination mail address.*	VERIFY — asks the receiver to confirm that the argument identifies a user.
EXPN	*Destination mail address.*	EXPAND — asks the receiver to confirm that the argument identifies a mailing list, and if so, to return the membership of that list.
HELP	Optional	Causes the receiver to send helpful information to the sender of the HELP command.
NOOP	N/A	Specifies no action other than that the receiver sends an OK reply.
QUIT	N/A	Specifies that the receiver must send an OK reply, and then close the transmission channel.
TURN	N/A	Specifies that the receiver must either (1) send an OK reply and then take on the role of the sender-SMTP or (2) send a refusal reply and retain the role of the receiver-SMTP.

As the message is relayed from one SMTP application to another, the relay host must remove itself from the beginning forward-path and put itself at the beginning of the *reverse-path*. If the relaying host is known by different names in different environments, then the SMTP application uses the name it is known by in the environment it is sending into, not the environment the mail came from. When mail reaches its ultimate destination (the forward-path contains only a destination mailbox), the receiver-SMTP inserts it into the destination mailbox.

If the SMTP application does not implement the relay function, it may use the same reply it would for an unknown local user (550).

If the SMTP application has accepted the task of relaying the mail and later finds that the forward-path is incorrect or that the mail cannot be delivered for whatever reason, then it must construct an "undeliverable mail" notification message and send it to the originator of the undeliverable mail (as indicated by the reverse-path) [1].

Multipurpose Internet Mail Extension (MIME)

SMTP only transmits ASCII text-based messages. It cannot handle non-ASCII messages. To solve this problem, the MIME standard was developed. MIME specifies how attachments and other non-ASCII textual messages are to be associated with e-mail. MIME is a very flexible format, permitting one to include virtually any type of file or document in an e-mail message. To ensure that e-mail messages containing images or other nontext information will be delivered with maximum protection against corruption, MIME provides a way for nontext information to be encoded as text. The encoding method used is known as base64.

How Does MIME Work?

As described previously, an e-mail message consists of a group of bits. The first few bits identify the recipient and sender, followed by the e-mail message and bits that indicate the end of the e-mail. Between the last bit of the e-mail message and the bit that identifies the end of the e-mail are bits that represent the attachment.

An MIME-Version header field uses a version number to declare a message to be conformant with MIME and allows mail-processing agents to distinguish between such messages and those generated by older or nonconformant software, which are presumed to lack such a field. The first few bits of the attachment identify the beginning of it and the type of file that is attached.

A Content-Type header field is used to specify the media type and subtype of data in the body of a message and to fully specify the native representation, (e.g., text, multipart, image, audio, video, or application) of such data. A Content-Transfer-Encoding header field can be used to specify both the encoding transformation that was applied to the body and the domain of the result. Encoding transformations other than the identity transformation are usually applied to data in order to allow it to pass through mail transport mechanisms that may have data or character set limitations. Two additional header fields that can be used to further describe the data in a body are the Content-ID and Content-Description header fields [2].

The five discrete top-level media types are: text, image, audio, video, and application. The two composite top-level media types are: (1) multipart,[2] and (2) message.[3] The five discrete media types provide a standardized mechanism for tagging entities as "audio," "image," or several other kinds of data. The composite "multipart" and "message" media types allow mixing and hierarchical structuring of entities of different types in a single message [3].

When a sender-SMTP application sends a nontext message, it inserts the additional MIME header lines into the message and encodes the binary data in "printable ASCII." The mail message is then transmitted like a text message using SMTP. The receiver-SMTP application interprets the header lines and extracts and decodes the nontext-based part of the message.

---

[2] Multipart — data consisting of multiple entities of independent data types.

[3] Message — an encapsulated message.

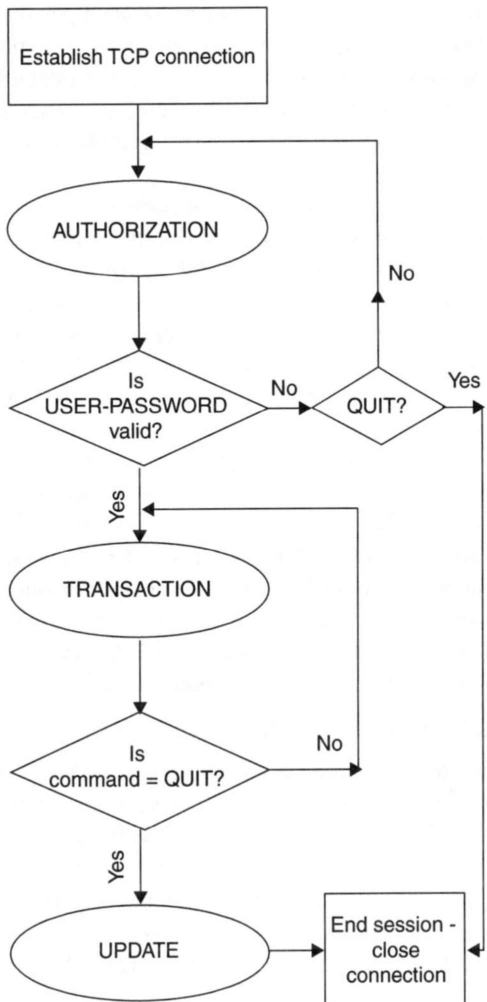

FIGURE 29.3   POP3 session state changes.

Reading Mail — How to Retrieve Mail from the Mailbox?

If every computer required a mail server application running on it in order to receive and send mail, it would turn into a very expensive exercise for most companies and individual user's. A typical user's workstation does not have sufficient resources (cycles, disk space) in order to permit an SMTP server and associated local mail delivery system to be kept resident and continuously running. Also, for individual users, it is expensive to keep a personal computer connected to the Internet for long periods of time.

To solve this problem, a mailbox service was developed, which allows users to use a single dedicated mail server to handle the SMTP send and receive commands. The received mail is stored in individual user's mailboxes. When the user wants to read (or delete) their mail, their client mail application uses either the Post Office Protocol3 (POP3) or Internet Message Access Protocol 4 (IMAP4) to retrieve messages from the user's mailbox on the server system (Figure 29.3).

Post Office Protocol3 (POP3)

POP3 stands for Post Office Protocol Version 3. POP3 allows a client workstation to retrieve mail that the server is holding for it. POP3 is not intended to provide extensive manipulation operations of mail on the server; normally, mail is downloaded and then deleted.

The POP3 architecture requires a POP3 server to be resident on the host computer where the mailboxes are stored. The POP3 server on the host computer runs continuously and listens on TCP port 110. When a client host wishes to make use of the service, it establishes a TCP connection with the server host. When the connection is established, the POP3 server sends a greeting. The client and POP3 server then exchange commands and responses (respectively) until the connection is closed or aborted.

A POP3 session progresses through a number of states during its lifetime. Once the TCP connection has been opened and the POP3 server has sent the greeting, the session enters the AUTHORIZATION state. In this state, the client must identify itself to the POP3 server.

Once the client has been successfully authenticated, the POP3 server acquires an exclusive-access lock on the mailbox. This is necessary to prevent messages from being modified or removed before the session enters the UPDATE state. If the lock is successfully acquired, the POP3 server responds with a positive status indicator. If a lock cannot be acquired or the client is denied access to the appropriate mailbox, the POP3 server returns a negative status indicator. The POP3 server may close the connection. If the server does not close the connection, the client may issue a new authentication command and start again, or the client may issue the QUIT command. After the POP3 server has opened the mailbox, it assigns a message-number to each message, and notes the size of each message in octets.

The POP3 session now enters the TRANSACTION state, with no messages marked as deleted.

The client may now repeatedly issue any of the POP3 commands shown in Table 29.2:

After each command, the POP3 server issues a response. When the client issues the QUIT command, the POP3 session enters the UPDATE state. In this state, the POP3 server releases any resources acquired during the TRANSACTION state and closes the TCP connection [5].

Internet Message Access Protocol 4 (IMAP4)

IMAP4 stands for the Internet Message Access Protocol, Version 4. IMAP4 allows a client to access and manipulate electronic mail messages on a server and supports multiple folders on the server. IMAP4 also provides the capability for an offline client to resynchronize with the server, allows the client to use the mailbox as a semipermanent repository, and can poll an existing connection for newly arrived messages. IMAP4 includes operations for creating, deleting, and renaming mailboxes; checking for new messages; permanently removing messages; setting and clearing flags; RFC 822 and MIME parsing; searching; and selective fetching of message attributes, texts, and portions thereof. Messages in IMAP4 are accessed by the use of numbers [6].

TABLE 29.2 POP3 Commands in the TRANSACTION State

Command	Arguments	Description of POP3 Server response	Restrictions
STAT	None	Server returns information about the mailbox.	Only valid if in TRANSACTION state
LIST [optional]	Message number	Returns information about the specified message, or if no message number specified, returns the total number of messages and a list of the messages.	Only valid if in TRANSACTION state
RETR	Message number	Sends the specified message.	Only valid if in TRANSACTION state
DELE	Message number	Marks specified message as deleted.	Only valid if in TRANSACTION state
NOOP	None	Server does nothing, returns positive response	Only valid if in TRANSACTION state
RSET	None	Unmarks any messages marked as deleted.	Only valid if in TRANSACTION state
QUIT	None	Removes all messages marked as deleted from the mailbox, releases the exclusive lock on the mailbox, and closes the TCP connection.	None

TABLE 29.3 IMAP4 Commands in the AUTHENTICATED States

Command	Argument(s)	Description
SELECT	Mailbox name	Selects a mailbox so that messages in the mailbox can be accessed.
EXAMINE	Mailbox name	Select a mailbox for read-only access.
CREATE	Mailbox name	Creates a mailbox with a given name.
DELETE	Mailbox name	Remove the specified mailbox.
RENAME	Existing mailbox name, new mailbox name	Changes the name of a mailbox.
SUBSCRIBE	Mailbox name	Adds specified mailbox name to server's set of "active" mailboxes to prevent a mailbox from being deleted after its contents expire.
UNSUBSCRIBE	Mailbox name	Remove specified mailbox name from server's set of "active" mailboxes.
LIST	Reference name or mailbox name	Returns a subset of mailbox names from the complete set of all names available to the client.
LSUB	Reference name or mailbox name	Returns a subset of mailbox names from the set of names the user has declared as being "active."
STATUS	Mailbox name and status data items	Requests the status, that is, number of messages, size, etc., of the specified mailbox.
APPEND	Mailbox name and message literal	Appends the literal argument as a new message to the end of the specified destination mailbox.

The IMAP4 architecture requires an IMAP4 server to be resident on the host computer where the mailboxes are stored. The IMAP4 server on the host computer runs continuously and listens on TCP port 143. When a client host wishes to make use of the service, it establishes a TCP connection with the server host. When the connection is established, the IMAP4 server sends a greeting. The client and IMAP4 server then exchange commands and responses (respectively) until the connection is closed or aborted. IMAP4 uses textual commands and responses. Most commands are valid in only certain states.

An IMAP4 session can be in one of four states during the lifetime of the connection. After a connection is established, the IMAP4 server is in a NON-AUTHENTICATED state. In this state, the client can issue either an AUTHENTICATE command, which indicates the authentication mechanism the client would like to use, or the LOGIN command, which identifies the client to the server with a user name and password.

If the client provides valid authentication credentials, the server moves to the AUTHENTICATED state. In this state, the commands shown in Table 29.3 are permitted.

If the client/user sends either the SELECT or EXAMINE command to the server, the server moves to the SELECTED state. A valid mailbox must be selected; if the client/user does not select a valid mailbox, the server stays in the AUTHENTICATED state. In the SELECTED state, the server will accept the commands shown in Table 29.4.

When the client/user has finished, he requests that the session be terminated. The server moves to the LOGOUT state and closes the connection.

Figure 29.4 provides an overview of the IMAP4 session state changes [6].

The state changes in Figure 29.4 are as follows:

1. Connection without pre-authentication (OK greeting).
2. Pre-authenticated connection (PREAUTH greeting).
3. Rejected connection (BYE greeting).
4. Successful LOGIN or AUTHENTICATE command.
5. Successful SELECT or EXAMINE command.
6. CLOSE command, or failed SELECT or EXAMINE command.
7. LOGOUT command, server shutdown, or connection closed [6].

TABLE 29.4   IMAP4 Commands in the SELECTED State

Command	Argument(s)	Description
CHECK	None	Requests a system resource type check of the selected mailbox. This checkpoint refers to any implementation-dependent housekeeping associated with the mailbox (e.g., resolving the server's in-memory state of the mailbox with the state on its disk) that is not normally executed as part of each command.
CLOSE	None	Permanently removes from mailbox all messages that have the deleted flag set and returns to the AUTHENTICATED state.
EXPUNGE	None	Permanently removes from mailbox all messages that have the deleted flag set.
SEARCH	Searching criteria [optional]	Searches a mailbox for messages that match the given search criteria.
FETCH	Message set and message data items	Retrieves either all data or specific data items (e.g., header) associated with a message.
STORE	Message set, message data items, and value for message data item	Alters data associated with a message in the mailbox.
COPY	Mailbox name and message set	Copies the specified message(s) to the end of the specified destination mailbox.

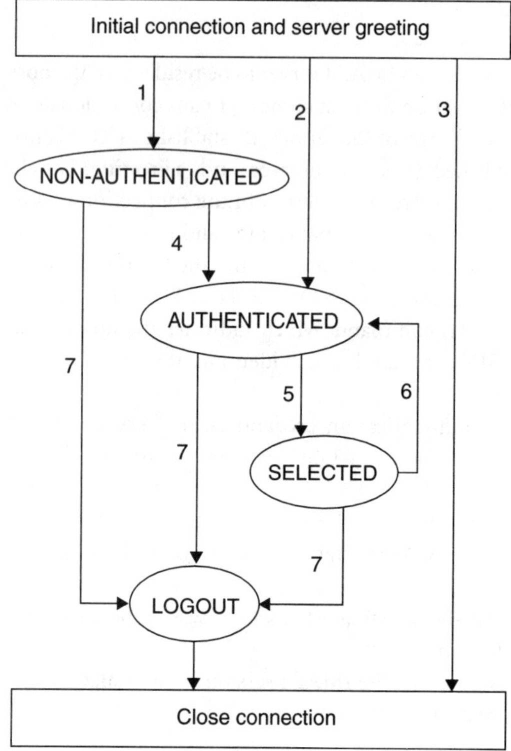

FIGURE 29.4   IMAP4 session state changes.

29.2   File Transfer Protocol

Introduction

The File Transfer Protocol (FTP) is a protocol used for transferring files between interconnected computers. FTP makes use of TCP/IP as the mechanism for communicating between the networked computers.

The main FTP objectives are to [7]

- promote sharing of files (computer programs and/or data),
- encourage indirect or implicit (via programs) use of remote computers,
- shield a user from variations in file storage systems among hosts, and
- transfer data reliably and efficiently.

FTP Model

FTP is a client–server-type application. The FTP server resides on the computer that stores the files. The FTP client can reside on any workstation. The FTP server listens for client connection requests on the specified port number (default is port 21). The FTP client needs to log in to both hosts (e.g., client workstation and remote host) in order to transfer a file from one to the other. To retrieve a specific file, the FTP client needs to know the host the file is located on, and the path name (the directory and possibly subdirectories and the file name) of the file. A user of the FTP client also needs to know if the user's machine uses ASCII, EBCDIC, or other character set to specify the data representation when transferring information.

A typical FTP session includes:

- Initiate a connection to the host, using an FTP client. The host name could either be the domain name or the host's IP address.
- Log into the FTP server, providing a valid user name and password for the server. A valid user name may also include the anonymous user name for servers that provide anonymous access.
- Issue the necessary FTP commands. Some command would be navigation commands and some actual transfer commands.
- When done, quit the client or the close the connection to the host.

Figure 29.5 shows the FTP model.

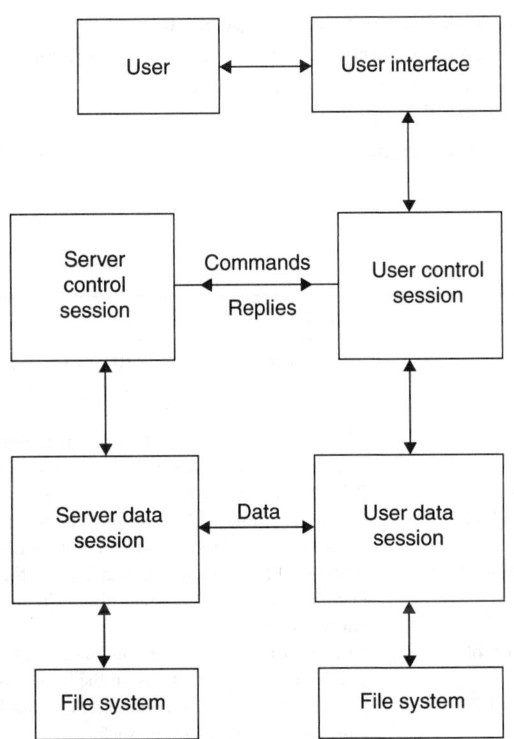

FIGURE 29.5 FTP model [7].

The FTP protocol has two sessions: a control and a data session; both sessions are implemented using TCP/IP connections.

The control session is initiated by the client and is used for client commands and server responses. It is over this session that all user commands are sent and over this session that the client gets the necessary server feedback. The FTP user interacts with this session. The control session's commands and responses are text commands.

The data session is initiated by the server and is used for transferring the actual data from the server to the required destination. A data session does not have to connect to the originating client; in fact, the client can initiate a transfer between two servers.

Data Types

FTP requires the user to specify the representation of data contained in a file when a file is transferred between two computers. The default data type is ASCII; but a user can specify other file types such as EBCDIC, IMAGE, or BINARY data types.

Transmission Mode

The transfer byte size is 8-bit bytes. The completion of all data transfers is indicated either by an end-of-file (EOF) or by closing the data connection.

The following transmission modes are defined in FTP:

Stream Mode: transmits data as a stream of bytes, with little or no processing.

Block Mode: formats data into a series of data blocks and allows for restarts.

Compressed Mode: compresses data for efficient data transfer.

Common FTP Commands

FTP has many commands, and those listed in Table 29.5 are a subset of the most commonly used commands out of all commands available from the FTP specification.

The commands are Telnet character strings transmitted over the control connections. The commands begin with a command code followed by an argument field.

Note that the filenames in brackets are optional. If they are omitted, the receiving file will be created with the same name it has on the originating machine. Other commands can be found by typing *help* at the FTP command prompt.

TABLE 29.5 Common FTP Commands

Command	Argument(s)	Description
Ascii	None	Set the file transfer type to network ASCII (use when transferring text files).
binary	None	Set the file transfer type to support binary image transfer (use for all other files).
Cd	remote_directory	Change the working directory on the remote machine.
Dir	[remote_directory]	Print a local directory listing. Include size and date information.
Ls	[remote_directory]	Print a listing of the directory contents in remote_directory.
Get	remote_file [local_file]	Retrieves remote_file and stores it locally. If local_file is not specified, give it the same name it has on the remote machine.
Mget	remote_file1 remote_file2 ...	Retrieve multiple remote files and stores them locally.
Help	[command]	Print an informative message about the meaning of command or a list of known commands.
Put	local_file [remote_file]	Store a local file on the remote machine. If remote_file is left unspecified, give it the same name it has on the local machine.
Mput	local_file1 local_file2 ...	Store multiple local files on the remote machine.
Quit	None	Quit FTP and return to the shell.

Anonymous FTP

Certain servers provide a service to clients where archives and files are publicly available. The archives are accessible using FTP. Since the data are provided as a public service, general access to the archive sites is allowed. The archive site implements what is known as anonymous FTP.

Anonymous FTP is actually FTP; but the user name and password are publicly available. The common user name for anonymous FTP is "anonymous" and the password may be any string (it is usually the user's e-mail address).

The anonymous user on the archive site is created with limited rights to the host; normally, the only operations allowed are logging in using FTP, listing the contents of a limited set of directories, and retrieving files. A user is not normally allowed to transfer files to the archive site [8].

29.3   Conclusion

The mail transfer protocols, SMTP, POP3, and IMAP4, provide the functionality required to send and receive e-mail. This mail system works in a manner very similar to the postal system except that instead of a physical letter enclosed in a physical envelope and transported via road, rail, or air transport systems, the e-mail message consists of electronic bits enclosed in a digital envelope, which is transported over a digital communications network called the Internet.

Most vendors wrap a user interface around the mail transfer protocols, thus hiding most of the underlying functionality from the user. Typical examples of vendor mail applications are Microsoft Outlook and Netscape Mail. These applications provide additional functionality such as meeting requests, meeting reminders, and planning tools that do not form part of the mail transfer protocol specification.

The FTP is still widely used by users on the Internet to retrieve files from remote archive storage sites. Users also use FTP on local networks to easily move files between computers.

References

[1] Postel, J.B., Simple Mail Transfer Protocol, RFC 821, 1982.
[2] Freed, N. and N. Borenstein, Multipurpose Internet Mail Extensions (MIME) Part One: Format of Internet Message Bodies, RFC 2045, 1996.
[3] Freed, N. and N. Borenstein, Multipurpose Internet Mail Extensions (MIME) Part Two: Media Types, RFC 2046, 1996.
[4] Freed, N. and N. Borenstein, Multipurpose Internet Mail Extensions (MIME) Part Five: Conformance Criteria and Examples, RFC 2049, 1996.
[5] Myers, J. and M. Rose, Post Office Protocol — Version 3, RFC 1939, 1996.
[6] Crispin, M., Internet Message Access Protocol — version 4rev1, RFC 2060, 1996.
[7] Postel, J. and J. Reynolds, File Transfer Protocol (FTP), RFC 959, 1985.
[8] Deutsch, P., A. Emtage, and A. Marine, How to Use Anonymous FTP, RFC 1635, 1994.

30

The Hypertext Transfer Protocol and Uniform Resource Identifier

Karl M. Goeschka
Vienna University of Technology.

30.1 Introduction

The Hypertext Transfer Protocol (HTTP) is an application-level protocol for distributed, collaborative, hypermedia information systems. It is a generic, stateless, protocol that can be used for many tasks beyond its use for hypertext, such as name servers and distributed object management systems, through extension of its request methods, error codes, and headers. A feature of HTTP is the typing and negotiation of data representation, allowing systems to be built independent of the data being transferred. This chapter explains HTTP by example and also considers the Uniform Resource Identifiers (URIs, aka URLs) as short strings that identify resources in the Web.

30.2 The Context of the Hypertext Transfer Protocol

During the early Internet days, information was gathered using File Transfer Protocol (FTP) and Telnet was used for sessions, which proved to be a user-unfriendly way of interaction or searching information. However, since the majority of people on the net were computer literate, this did not prove to be too much of a problem. Things became quite different with the Web: Tim Berners-Lee proposed the first version of Hypertext Markup Language (HTML) in 1989,[1] and the first server and browser prototypes came into existence between 1990 and 1992. From then on, the Web led to an exponential growth of the Internet: in June 1991, about 500.000 nodes were recorded worldwide. As of June 1998, there were about 35 millions nodes recorded, and the monthly growth rate was about 1 million worldwide. Thus, the monthly growth rate far exceeded the total number of nodes only 7 years ago.

The main reasons for the exponential growth rate that the Web caused were the easy-to-use point-and-click graphical user interface and the integration of all relevant services into one tool — the Web browser. As more and more Web servers appeared all over the world, the number of users accessing them also increased.

Mainly, three concepts make up the Web and are responsible for its success: The Hypertext Markup Language (HTML), the Uniform Resource Locator (URL), and the Hypertext Transfer Protocol (HTTP). As HTML is described in a separate chapter, here, we briefly introduce the concept of URLs before focusing on HTTP.

30.3 Uniform Resource Locator and Identifier

Titled "Naming and Adressing," the W3C defines the more general term "Uniform Resource Identifier" (URI) as "the generic set of all names/addresses that are short strings that refer to resources" (http://www.w3.org/Addressing/):

The Web is an information space. Human beings have a lot of mental machinery for manipulating, imagining, and finding their way in spaces. URIs are the points in that space. Unlike web data formats, where HTML is an important one, but not the only one, and web protocols, where HTTP has a similar status, there is only one Web naming/addressing technology: URIs.

Uniform Resource Identifiers (URIs, aka URLs) are short strings that identify resources in the web: documents, images, downloadable files, services, electronic mailboxes, and other resources. They make resources available under a variety of naming schemes and access methods such as HTTP, FTP, and Internet mail addressable in the same simple way. They reduce the tedium of "log in to this server, then issue this magic command ..." down to a single click.

It is an extensible technology: there are a number of existing addressing schemes, and more may be incorporated over time.

Another important term is the "Uniform Resource Name" URN:

1. A URI that has an institutional commitment to persistence, availability, etc. Note that this kind of URI may also be a URL. See, for example, PURLs.
2. A particular scheme, `urn:`, specified by RFC 2141 and related documents, intended to serve as persistent, location-independent, resource identifiers.

Consequently, the URL becomes an informal term (no longer used in technical specifications) associated with popular URI schemes for specifying Internet resources: `http`, `ftp`, `mailto`, etc.

The "generic URI" syntax (RFC 2396) consists of a sequence of four main components:

```
<scheme>://<authority><path>?<query>
```

The *scheme* covers all major Internet protocols, typically `http` or `https` (HTTP over Secure Sockets Layer SSL), but also `file` (Host-specific file names), `ftp` (File Transfer Protocol), `ldap` (Lightweight

---

[1]Although the term hypertext was initially coined in the 1960s, its popularity first grew when the Web was invented.

Directory Access Protocol), `mailto` (Electronic mail address), `news` (USENET news), `nfs` (Netfork File System), `nntp` (Network News Transfer Protocol), `pop` (Post Office Protocol v3), `sip` (Session Initiation Protocol), and `telnet` (Reference to interactive sessions). As of October 2002, the W3C lists 85 schemes, with only some of them being registered (`http://www.w3.org/Addressing/schemes`). RFCs 2717 and 2718 provide registration procedures for scheme names and guidelines for new schemes, respectively.

For schemes that involve the direct use of an IP-based protocol to a specified server on the Internet, the *authority* looks like `<userinfo>@<host>:<port>`, where `<userinfo>` may consist of a user name and optionally, scheme-specific information about how to gain authorization to access the server.[2] The parts `<userinfo>@` and `:<port>` may be omitted. The host is a domain name of a network host (RFCs 1034, 1123), or its IPv4 address as a set of four decimal digit groups separated by dots. The port is the network port number for the server. Most schemes designate protocols that have a default port number. Another port number may optionally be supplied, in decimal, separated from the host by a colon. If the port is omitted, the default port number ("well-known port") is assumed, for example, 80 for HTTP or 443 for HTTPS.

The *path* component contains data, specific to the authority, identifying the resource within the scope of that scheme and authority. In the case of HTTP, it is either a file system path relative to the document root of the respective Web server, or the path has a special, preconfigured meaning for the Web server (e.g., a server-side application).

The *query* component is a string of information to be interpreted by the resource. In the case of HTML and HTTP, this may be a set of name–value pairs resulting from a `FORM` with the `GET` method, a list of names separated by plus signs resulting from an `ISINDEX` query, or a comma-separated pair of x- and y-pixel coordinates resulting from an active server-side imagemap (`ISMAP`).

It is often the case that a group or "tree" of documents has been constructed to serve a common purpose; the vast majority of URI in these documents point to resources within the tree rather than outside of it. Similarly, documents located at a particular site are much more likely to refer to other resources at that site than to resources at remote sites.

Relative addressing of URI allows document trees to be partially independent of their location and access scheme. For instance, it is possible for a single set of hypertext documents to be simultaneously accessible and traversable via each of the `file`, `http`, and `ftp` schemes if the documents refer to each other using relative URI. Furthermore, such document trees can be moved, as a whole, without changing any of the relative references. Experience within the WWW has demonstrated that the ability to perform relative referencing is necessary for the long-term usability of embedded URI.

A URI reference may therefore be absolute or relative, and may have additional information attached in the form of a fragment identifier (following the hash character #). However, "the URI" that results from such a reference includes only the absolute URI after the fragment identifier (if any) is removed and after any relative URI is resolved to its absolute form.

The following examples show typical HTTP URIs:

```
http://some.where.edu/path/mydocument.html#fragment_id
```

This URI accesses the server `some.where.edu` via "well-known" port 80 with the protocol HTTP. Considering a standard Web server configuration, this would result in the delivery of the document `mydocument.html` in the directory path relative to the document root of the Web server. The browser would then open the page and jump to the place in the document, which is marked with the appropriate anchor tag `<a name="fragment_id">`.

```
../stuff/other.html
```

Based on the previous URI, this relative URI would result in the following absolute URI:

```
http://some.where.edu/stuff/other.html
```

---

[2] This practice of using passwords within the URI is *not recommended*, because the passing of authentication information in clear text (such as URI) has proven to be a security risk in almost every case where it has been used.

The following three URIs show possible usages of the query string: the first URI results from pressing the "submit" button of an HTML FORM, which is declared using the GET method. In this case, the name/value pairs of the form inputs are described. The second URI results from an ISINDEX query and enumerates the arguments, and the third URI from a server-side active imagemap (ISMAP), which provides the coordinates in pixels from the upper left corner, where within the image the "click" took place.

```
http://some.where.edu:33333/app/program?a=1&b=2

http://some.where.edu:33333/app/search?arg1+arg2+arg3

http://some.where.edu:33333/app/imagemap?123,45
```

Usually, the resources in these cases are server-side applications prepared to process the incoming query string in order to generate the following HTML document.

30.4 Overall Operation of HTTP

The HTTP is an application-level protocol for distributed, collaborative, hypermedia information systems. It is a generic, stateless protocol that can be used for many tasks beyond its use for hypertext, such as name servers and distributed object management systems, through extension of its request methods, error codes, and headers. A feature of HTTP is the typing and negotiation of data representation, allowing systems to be built independent of the data being transferred. HTTP has been in use by the World Wide Web global information initiative since 1990.

This definition and parts of further explanations are taken from RFC 2616, which defines the protocol referred to as "HTTP/1.1," and is an update to RFC 2068. An overview of the RFCs and other documents relevant for HTTP is provided by the World Wide Web Consortium under `http://www.w3.org/Protocols/`.

The HTTP protocol is a request/response protocol. A client sends a request to the server in the form of a request method, URI, and protocol version, followed by a MIME-like message containing request modifiers, client information, and possible body content over a connection with a server. The server responds with a status line, including the message's protocol version and a success or error code, followed by a MIME-like message containing server information, entity metainformation, and possible entity-body content. An entity is the information transferred as the payload of a request or response. An entity consists of metainformation in the form of entity-header fields and content in the form of an entity-body.

Most HTTP communication is initiated by a user agent and consists of a request to be applied to a resource on some origin server. In the simplest case, this may be accomplished via a single connection between the user agent and the origin server.

A more complicated situation occurs when one or more intermediaries are present in the request/response chain. There are three common forms of intermediary: proxy, gateway, and tunnel. A proxy is a forwarding agent, receiving requests for a URI in its absolute form, rewriting all or part of the message, and forwarding the reformatted request toward the server identified by the URI. A gateway is a receiving agent, acting as a layer above some other server(s) and, if necessary, translating the requests to the underlying server's protocol. A tunnel acts as a relay point between two connections without changing the messages; tunnels are used when the communication needs to pass through an intermediary (such as a firewall) even when the intermediary cannot understand the contents of the messages.

A request or response message that travels the whole chain will pass through several separate connections. This distinction is important because some HTTP communication options may apply only to the connection with the nearest, nontunnel neighbor, only to the end points of the chain, or to all connections along the chain. Moreover, each participant in a certain chain may be engaged in multiple, simultaneous communications.

Any party to the communication that is not acting as a tunnel may employ an internal cache for handling requests. The effect of a cache is that the request/response chain is shortened if one of the participants along the chain has a cached response applicable to that request. Not all responses are usefully cacheable, and some requests may contain modifiers that place special requirements on cache behavior.

In fact, there are a wide variety of architectures and configurations of caches and proxies currently being experimented with or deployed across the World Wide Web. These systems include national hierarchies of proxy caches to save transoceanic bandwidth, systems that broadcast or multicast cache entries, organizations that distribute subsets of cached data via CD-ROM, and so on. HTTP systems are used in corporate intranets over high-bandwidth links, and for access via PDAs with low-power radio links and intermittent connectivity. The goal of HTTP/1.1 is to support the wide diversity of configurations already deployed while introducing protocol constructs that meet the needs of those who build Web applications that require high reliability and, failing that, at least reliable indications of failure.

HTTP communication usually takes place over TCP/IP connections. The default port is TCP 80 (RFC 1700); but other ports can be used. This does not preclude HTTP from being implemented on top of any other protocol on the Internet, or on other networks. HTTP only presumes a reliable transport; any protocol that provides such guarantees can be used; the mapping of the HTTP/1.1 request and response structures onto the transport data units of the protocol in question is beyond the scope of this specification.

In HTTP/1.0, most implementations used a new connection for each request/response exchange. In HTTP/1.1, a connection may be used for one or more request/response exchanges, although connections may be closed for a variety of reasons. Therefore, HTTP should be treated as a connection-oriented but stateless protocol.

30.5 Protocol Parameters

The HTTP uses various parameters; the following section explains the most important ones, and details can be found in RFC 2616.

HTTP Version

HTTP request and response messages include an HTTP protocol version number. The protocol versioning policy is intended to allow the sender to indicate the format of a message and its capacity for understanding further HTTP communication, rather than the features obtained via that communication. No change is made to the version number for the addition of message components that do not affect communication behavior or that only add to extensible field values. The HTTP version of an application is the highest HTTP version for which the application is at least conditionally compliant. Proxy and gateway applications need to be careful when forwarding messages in protocol versions different from that of the application, because converting between versions of HTTP may involve modification of header fields required or forbidden by the versions involved. See RFC 2145 for a more complete explanation.

URI

As far as HTTP is concerned, URI are simply formatted strings which identify — via name, location, or any other characteristic — a resource.

Date/Time

HTTP applications have historically allowed three different formats for the representation of date/time stamps, but a fixed-length subset of that defined by RFC 1123 (an update to RFC 822) is preferred as an Internet standard. All HTTP date/time stamps must be represented in GMT (Greenwich Mean Time), without exception. For the purposes of HTTP, GMT is exactly equal to Universal Time Coordinated (UTC). The format looks like:

```
Sat, 26 Oct 2002 15:58:04 GMT
```

Character Sets

HTTP uses the same definition of the term "character set" as that described for MIME, which is more commonly referred to as a "character encoding." HTTP character sets are identified by case-insensitive tokens. The complete set of tokens is defined by the IANA Character Set registry (RFC 1700), for example: `charset=iso-8859-1`.

Content Codings

Content-coding values indicate an encoding transformation that has been or can be applied to an entity. Content codings are primarily used to allow a document to be compressed or otherwise usefully transformed without losing the identity of its underlying media type and without loss of information. Frequently, the entity is stored in coded form, transmitted directly, and only decoded by the recipient. HTTP/1.1 uses content-coding values in the `Accept-Encoding` and `Content-Encoding` header fields. Although the value describes the content-coding, what is more important is that it indicates what decoding mechanism will be required to remove the encoding. Examples are `gzip` (RFC 1952), `compress` (UNIX file compression program), and `deflate` (RFCs 1950, 1951).

Transfer-Codings

Transfer-coding values are used to indicate an encoding transformation that has been, can be, or may need to be applied to an entity-body in order to ensure "safe transport" through the network. This differs from a content coding in that the transfer-coding is a property of the message, not of the original entity. An example is `Transfer-Encoding: chunked`, which modifies the body of a message in order to transfer it as a series of chunks, each with its own size indicator, followed by an optional trailer containing entity-header fields. This allows dynamically produced content to be transferred along with the information necessary for the recipient to verify that it has received the full message.

Media Types

HTTP uses Internet Media Types in the `Content-Type` and `Accept` header fields in order to provide open and extensible data typing and type negotiation. Media-type values are registered with the Internet Assigned Number Authority (RFC 1700). The media-type registration process is outlined in RFC 1590. The use of nonregistered media types is discouraged. Typical media types for use with HTTP are: `text/html`, `image/gif`, `message/http` (e.g., response on `TRACE` request), `multipart/byteranges` (`HTTP 206 (Partial Content)` response), and `application/x-www-form-urlencoded` (`POST` request resulting from an HTML `FORM`).

Product Tokens

Product tokens are used to allow communicating applications to identify themselves by software name and version. Examples are as follows:

```
User-Agent: Mozilla/4.0 (compatible; MSIE 5.01; Windows NT 5.0)
Server: Apache/1.3.23 (Unix) (Red-Hat/Linux) mod_ssl/2.8.7 OpenSSL/0.9.6b
DAV/1.0.3 PHP/4.1.2 mod_perl/1.26
```

Quality Values

HTTP content negotiation uses short "floating point" numbers to indicate the relative importance ("weight") of various negotiable parameters. A weight is normalized to a real number in the range 0 through 1, where 0 is the minimum and 1 is the maximum value. If a parameter has a quality value of 0, then content with this parameter is 'not acceptable' for the client. "Quality values" is a misnomer, since these values merely represent relative degradation in desired quality.

Language Tags

A language tag identifies a natural language spoken, written, or otherwise conveyed by human beings for communication of information to other human beings. Computer languages are explicitly excluded. HTTP uses language tags within the `Accept-Language` and `Content-Language` fields. The syntax and registry of HTTP language tags are the same as those defined by RFC 1766. Example tags include: `en`, `en-us`, `de-at`, where any two-letter primary-tag is an ISO-639 language abbreviation and any two-letter initial subtag is an ISO-3166 country code.

30.6 HTTP Message

HTTP messages consist of requests from client to server and responses from server to client. Request and response messages use the generic message format of RFC 822 for transferring entities (the payload of the message). Both types of message consist of a start-line, zero, or more header fields (also known as "headers"), an empty line (i.e., a line with nothing preceding the CRLF) indicating the end of the header fields, and possibly a message-body.

HTTP header fields, which include general- , request- , response- , and entity-header fields, follow the same generic format as that given in Section 3.1 of RFC 822. Each header field consists of a name followed by a colon (":") and the field value. General-header fields have general applicability for both request and response messages, but do not apply to the entity being transferred. These header fields apply only to the message being transmitted. Entity-header fields define metainformation about the entity-body or, if no body is present, about the resource identified by the request. Some of this metainformation is *optional*; some might be *required*. Request and response header fields are described in the next section.

The message-body (if any) of an HTTP message is used to carry the entity-body associated with the request or response. The message-body differs from the entity-body only when a transfer coding has been applied, as indicated by the `Transfer-Encoding` header field. The rules for when a message-body is allowed in a message differ for requests and responses. The transfer-length of a message is the length of the message-body as it appears in the message, that is, after any transfer-codings have been applied.

30.7 Request and Response

A request message from a client to a server includes, within the first line of that message, the method to be applied to the resource, the identifier of the resource, and the protocol version in use. Figure 30.1 shows a simple GET request:

The request-line begins with a method token, followed by the request-URI, and the protocol version. The method token indicates the method to be performed on the resource identified by the request-URI. Possible method tokens are `GET`, `HEAD`, `POST`, `OPTIONS`, `TRACE`, `CONNECT`, `PUT`, and `DELETE`.

```
GET /first.html HTTP/1.1
Accept: image/gif, image/x-xbitmap, image/jpeg, image/pjpeg, application/vnd.ms-
powerpoint, application/vnd.ms-excel, application/msword, */*
Accept-Language: de-at
Accept-Encoding: gzip, deflate
User-Agent: Mozilla/4.0 (compatible; MSIE 5.01; Windows NT 5.0)
Host: gutemine.ict.tuwien.ac.at:33333
Connection: Keep-Alive
```

FIGURE 30.1   Simple GET Request.

The `GET` method means retrieve whatever information (in the form of an entity) is identified by the request-URI. If the request-URI refers to a data-producing process, it is the produced data that shall be returned as the entity in the response and not the source text of the process, unless that text happens to be the output of the process.

The request-header fields allow the client to pass additional information about the request, and about the client itself, to the server. These fields act as request modifiers, with semantics equivalent to the parameters on a programming language method invocation.

The `Accept` request-header field can be used to specify certain media types that are acceptable for the response. Accept headers can be used to indicate that the request is specifically limited to a small set of desired types, as in the case of a request for an in-line image. The `Accept-Language` request-header field is similar to `Accept`, but restricts the set of natural languages that are preferred as a response to the request. The `Accept-Encoding` request-header field is similar to `Accept`, but restricts the content-codings that are acceptable in the response.

The `User-Agent` request-header field contains information about the user agent originating the request. This is for statistical purposes, the tracing of protocol violations, and automated recognition of user agents for the sake of tailoring responses to avoid particular user-agent limitations.

The `Host` request-header field specifies the Internet host and port number of the resource being requested, as obtained from the original URI given by the user or referring resource (generally an HTTP URL). The Host field value must represent the naming authority of the origin server or gateway given by the original URL. This allows the origin server or gateway to differentiate between internally ambiguous URLs, such as the root "/" URL of a server for multiple host names on a single IP address (virtual hosting).

The `Connection` general-header field allows the sender to specify options that are desired for that particular connection and must not be communicated by proxies over further connections. Although since HTTP/1.1 persistent connections are the default behavior of any HTTP connection, this header field explicitly demands a persistent connection by using the original HTTP/1.0 form of persistent connections (RFC 2068).

Persistent HTTP connections have a number of advantages:

- By opening and closing fewer TCP connections, CPU time is saved in routers and hosts (clients, servers, proxies, gateways, tunnels, or caches), and memory used for TCP protocol control blocks can be saved in hosts.
- HTTP requests and responses can be pipelined on a connection. Pipelining allows a client to make multiple requests without waiting for each response, allowing a single TCP connection to be used much more efficiently, with much lower elapsed time.
- Network congestion is reduced by reducing the number of packets caused by TCP opens, and by allowing TCP sufficient time to determine the congestion state of the network.
- Latency on subsequent requests is reduced since there is no time spent in TCP's connection opening handshake.
- HTTP can evolve more gracefully, since errors can be reported without the penalty of closing the TCP connection. Clients using future versions of HTTP might optimistically try a new feature, but if communicating with an older server, retry with old semantics after an error is reported.

The response onto the request of Figure 30.1 then may look like Figure 30.2, resulting in a browser view given in Figure 30.3.

After receiving and interpreting a request message, a server responds with an HTTP response message. The first line of a response message is the status-line, consisting of the protocol version followed by a numeric status code and its associated textual phrase. The status-code element is a 3-digit integer result code of the attempt to understand and satisfy the request. These codes are fully defined in RFC 2616. The reason-phrase is intended to give a short textual description of the status-code. The status-code is intended for use by automata and the reason-phrase is intended for the human user. The client is not required to examine or display the reason-phrase. The first digit of the status-code defines the class of response. The last two digits do not have any categorization role. There are 5 values for the first digit:

```
HTTP/1.1 200 OK
Date: Sat, 26 Oct 2002 13:51:17 GMT
Server: Apache/1.3.23 (Unix)  (Red-Hat/Linux) mod_ssl/2.8.7 OpenSSL/0.9.6b
DAV/1.0.3 PHP/4.1.2 mod_perl/1.26 mod_auth_pam/1.0a
Keep-Alive: timeout=15, max=100
Connection: Keep-Alive
Content-Type: text/html

<HTML>
  <HEAD>
    <TITLE>My first document</TITLE>
  </HEAD>
  <BODY>
    <H1>Main Heading of my First Document</H1>
    <P>Some text including a link to the
       <A HREF="second.html">second page</A>
       of my collection.
  </BODY>
</HTML>
```

FIGURE 30.2 Response.

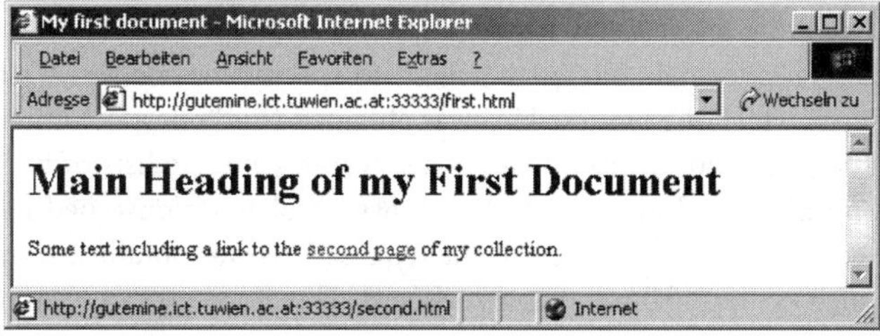

FIGURE 30.3 Resulting browser view.

- 1xx: Informational — Request received, continuing process.
- 2xx: Success — The action was successfully received, understood, and accepted.
- 3xx: Redirection — Further action must be taken in order to complete the request.
- 4xx: Client Error — The request contains bad syntax or cannot be fulfilled.
- 5xx: Server Error — The server failed to fulfill an apparently valid request.

The response-header fields allow the server to pass additional information about the response, which cannot be placed in the status-line. These header fields give information about the server and about further access to the resource identified by the request-URI.

The `Date` general-header field represents the date and time at which the message was originated, having the same semantics as orig-date in RFC 822. The field value is an HTTP-date; it *must* be sent in RFC 1123-date format.

The `Server` response-header field contains information about the software used by the origin server to handle the request. The field can contain multiple product tokens and comments identifying the server and any significant subproducts. The product tokens are listed in order of their significance for identifying the application. *Note*: Revealing the specific software version of the server might allow the server machine to become more vulnerable to attacks against software that is known to contain security holes. Server implementors are therefore encouraged to make this field a configurable option.

The `Keep-Alive` and `Connection` headers are used here only for compatibility reasons with HTTP/1.0 Persistent Connections (see RFC 2068).

The `Content-Type` entity-header field indicates the media type of the entity-body sent to the recipient or, in the case of the `HEAD` method, the media type that would have been sent had the request been a `GET`.

The message body finally contains the HTML page displayed in Figure 30.3.

```
GET /second.html HTTP/1.1
Accept: image/gif, image/x-xbitmap, image/jpeg, image/pjpeg, application/vnd.ms-
powerpoint, application/vnd.ms-excel, application/msword, */*
Referer: http://gutemine.ict.tuwien.ac.at:33333/first.html
Accept-Language: de-at
Accept-Encoding: gzip, deflate
User-Agent: Mozilla/4.0 (compatible; MSIE 5.01; Windows NT 5.0)
Host: gutemine.ict.tuwien.ac.at:33333
Connection: Keep-Alive
```

FIGURE 30.4   Request.

30.8   Authentication: Basic and Digest Access

Authentication is described in RFC 2617. The "basic" authentication scheme is based on the model that the client must authenticate itself with a user-ID and a password for each realm. The realm value should be considered an opaque string, which can only be compared for equality with other realms on that server. The server will service the request only if it can validate the user-ID and password for the protection space of the request-URI. There are no optional authentication parameters. Assume that the link in Figure 30.3 is utilized, resulting in the following request (Figure 30.4).

The `Referer` request-header field allows the client to specify, for the server's benefit, the address (URI) of the resource from which the request-URI was obtained (the "referrer," although the header field is misspelled). The `Referer` request-header allows a server to generate lists of back-links to resources for interest, logging, optimized caching, etc. It also allows obsolete or mistyped links to be traced for maintenance. The Referer field *must not* be sent if the request-URI was obtained from a source that does not have its own URI, such as input from the user keyboard. Therefore, we can see the `Referer` here for the first time.

Now we assume the resource `second.html` to be protected in a protection space called "`Administration  Group`." Upon receipt of an unauthorized request for a URI within the protection space, the origin server may respond with a challenge like the following (see Figure 30.5):

`WWW-Authenticate:  Basic  realm="Administration  Group"`

The `WWW-Authenticate` response-header field must be included in 401 (Unauthorized) response messages. The field value consists of at least one challenge that indicates the authentication scheme(s) and parameters applicable to the request-URI. Usually, the client now has to prompt the user for his username and password (see Figure 30.6). If the client is not capable of promting the user or if the request fails for any other reason, the HTML page is displayed, since that entity might include relevant diagnostic information.

To receive authorization, the client sends the userid and password, separated by a single colon ("`:`") character, within a base64-encoded string in the credentials. If the client wishes to send the userid "myname" and the password "mypasswd," it would use the following header field:

`Authorization:  Basic  bXluYW1lOm15cGFzc3dk`

The following request would be similar to Figure 30.4, with the above header line added. If username and password were correct, the server would respond appropriately. The Basic Access Authentication scheme is not considered to be a secure method of user authentication (unless used in conjunction with some external secure system such as SSL), as the user name and password are passed over the network as uuencoded cleartext. Therefore, a scheme based on cryptographic hashes, referred to as "Digest Access Authentication," is proposed for better security (RFC 2617).

30.9   Caching and Proxies

HTTP is typically used for distributed information systems, where performance can be improved by using response caches. The HTTP/1.1 protocol includes a number of elements intended to make caching

```
HTTP/1.1 401 Authorization Required
Date: Sat, 26 Oct 2002 15:42:08 GMT
Server: Apache/1.3.23 (Unix)  (Red-Hat/Linux) mod_ssl/2.8.7 OpenSSL/0.9.6b
DAV/1.0.3 PHP/4.1.2 mod_perl/1.26 mod_auth_pam/1.0a
WWW-Authenticate: Basic realm="Administration Group"
Keep-Alive: timeout=15, max=100
Connection: Keep-Alive
Content-Type: text/html; charset=iso-8859-1

<HTML>
  <HEAD>
    <TITLE>401 Authorization Required</TITLE>
  </HEAD>
  <BODY>
    <H1>Authorization Required</H1>
    <P>This server could not verify that you
    are authorized to access the document
    requested.  Either you supplied the wrong
    credentials (e.g., bad password), or your
    browser doesn't understand how to supply
    the credentials required.
    <HR>
    <ADDRESS>Apache/1.3.23 Server at
    <A HREF="mailto:webmaster@ict.tuwien.ac.at">gutemine.ict.tuwien.ac.at</A>
    Port 33333
    </ADDRESS>
  </BODY>
</HTML>
```

FIGURE 30.5　Authorization required.

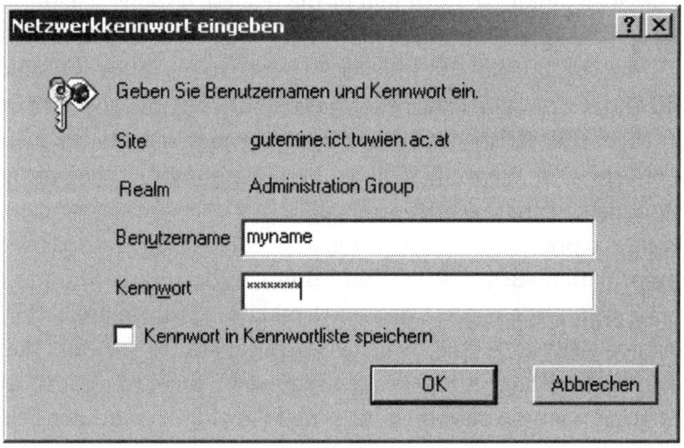

FIGURE 30.6　The client prompts the user for username and password.

work as well as possible. Caching would be useless if it did not significantly improve performance. The goal of caching in HTTP/1.1 is to eliminate the need to send requests in many cases, and to eliminate the need to send full responses in many other cases. The former reduces the number of network round-trips required for many operations; the protocol uses an "expiration" mechanism for this purpose. The latter reduces network bandwidth requirements; the protocol uses a "validation" mechanism for this purpose.

Figure 30.7 shows a GET request being sent to a proxy server instead of being sent to the origin server gutemine.ict.tuwien.ac.at; therefore, the request-URI in the request-line contains not just path and query, but also scheme and authority. In general, a chain of proxy servers is possible. In our example, the proxy server now forwards the request to the origin server, which is shown in Figure 30.8.

We recognize a new header field: the Via general-header field *must* be used by gateways and proxies to indicate the intermediate protocols and recipients between the user agent and the server on requests,

```
GET http://gutemine.ict.tuwien.ac.at:33333/second.html HTTP/1.1
Accept: image/gif, image/x-xbitmap, image/jpeg, image/pjpeg, application/vnd.ms-
powerpoint, application/vnd.ms-excel, application/msword, */*
Accept-Language: de-at
Accept-Encoding: gzip, deflate
User-Agent: Mozilla/4.0 (compatible; MSIE 5.01; Windows NT 5.0)
Host: gutemine.ict.tuwien.ac.at:33333
```

FIGURE 30.7   GET request to a proxy server.

```
GET /second.html HTTP/1.0
Via: 1.1 ADPROSRV01
User-Agent: Mozilla/4.0 (compatible; MSIE 5.01; Windows NT 5.0)
Host: gutemine.ict.tuwien.ac.at:33333
Accept: image/gif, image/x-xbitmap, image/jpeg, image/pjpeg, application/vnd.ms-
powerpoint, application/vnd.ms-excel, application/msword, */*
Accept-Language: de-at
Accept-Encoding: gzip, deflate
Connection: Keep-Alive
```

FIGURE 30.8   GET request forwarded from proxy server to origin server.

and between the origin server and the client on responses. It is analogous to the Received field of RFC 822 and is intended to be used for tracking message forwards, avoiding request loops, and identifying the protocol capabilities of all senders along the request/response chain.

Following the `Via` header field, the received-protocol indicates the protocol version of the message received by the server or client along each segment of the request/response chain. The received-protocol version (1.1 in our example) is appended to the `Via` field value when the message is forwarded so that information about the protocol capabilities of upstream applications remains visible to all recipients.

The received-by field (`ADPROSRV01` in our example) is normally the host and optional port number of a recipient server or client that subsequently forwarded the message. However, if the real host is considered to be sensitive information, it *may* be replaced by a pseudonym. If the port is not given, it *may* be assumed to be the default port of the received-protocol.

Multiple `Via` field values represent each proxy or gateway that has forwarded the message. Each recipient *must* append its information such that the end result is ordered according to the sequence of forwarding applications. For example, a request message could be sent from an HTTP/1.1 user agent to an internal proxy code-named `ADPROSRV01`, which uses HTTP/1.0 to forward the request to a public proxy at `some.where.com`, which completes the request by forwarding it to the origin server. The request received by the origin server would then have the following `Via` header field:

`Via: 1.1 ADPROSRV01, 1.0 some.where.com`

As there is only one proxy in our example, the origin server now responses as shown in Figure 30.9, the proxy stores the response in its cache, and forwards the response to the user agent as shown in Figure 30.10.

If the same request of Figure 30.7 is now sent to the proxy again, the proxy server will verify the validity of its cache regarding the request and if it is valid will answer immediately without first contacting the origin server as shown in Figure 30.11.

The `Age` response-header field conveys the sender's estimate of the amount of time since the response (or its revalidation) was generated at the origin server. A cached response is *fresh* if its age does not exceed its freshness lifetime. Age values are nonnegative decimal integers, representing time in seconds. Age values are calculated as specified in RFC 2616 and are used for cache expiration control.

30.10 Further HTTP Request Methods by Example

This section provides further examples for HTTP request methods.

```
HTTP/1.1 200 OK
Date: Sat, 26 Oct 2002 15:57:32 GMT
Server: Apache/1.3.23 (Unix)  (Red-Hat/Linux) mod_ssl/2.8.7 OpenSSL/0.9.6b
DAV/1.0.3 PHP/4.1.2 mod_perl/1.26 mod_auth_pam/1.0a
Connection: close
Content-Type: text/html

<HTML>
  <HEAD>
    <TITLE>My second document</TITLE>
  </HEAD>
  <BODY>
    <H1>Main Heading of my Second Document</H1>
    <P>More text.
  </BODY>
</HTML>
```

FIGURE 30.9　Response of the origin server received by the proxy.

```
HTTP/1.1 200 OK
Via: 1.1 ADPROSRV01
Connection: close
Content-Type: text/html
Date: Sat, 26 Oct 2002 15:58:04 GMT
Server: Apache/1.3.23 (Unix)  (Red-Hat/Linux) mod_ssl/2.8.7 OpenSSL/0.9.6b
DAV/1.0.3 PHP/4.1.2 mod_perl/1.26 mod_auth_pam/1.0a
Keep-Alive: timeout=15, max=100

<HTML>
  <HEAD>
    <TITLE>My second document</TITLE>
  </HEAD>
  <BODY>
    <H1>Main Heading of my Second Document</H1>
    <P>More text.
  </BODY>
</HTML>
```

FIGURE 30.10　Response of the proxy received by the user agent.

```
HTTP/1.1 200 OK
Via: 1.1 ADPROSRV01
Connection: close
Content-Length: 157
Age: 588
Content-Type: text/html
Date: Sat, 26 Oct 2002 16:07:52 GMT
Server: Apache/1.3.23 (Unix)  (Red-Hat/Linux) mod_ssl/2.8.7 OpenSSL/0.9.6b
DAV/1.0.3 PHP/4.1.2 mod_perl/1.26 mod_auth_pam/1.0a
Keep-Alive: timeout=15, max=100

<HTML>
  <HEAD>
    <TITLE>My second document</TITLE>
  </HEAD>
  <BODY>
    <H1>Main Heading of my Second Document</H1>
    <P>More text.
  </BODY>
</HTML>
```

FIGURE 30.11　Response of the proxy, when the requested information is retrieved from a valid cache.

OPTIONS

The OPTIONS method represents a request for information about the communication options available on the request/response chain identified by the request-URI. This method allows the client to determine the options and/or requirements associated with a resource, or the capabilities of a server, without implying a resource action or initiating a resource retrieval. In Figure 30.12, the server allows the methods GET, HEAD, TRACE, and OPTIONS.

HEAD

The HEAD method is identical to GET except that the server *must not* return a message-body in the response (see Figure 30.13). The metainformation contained in the HTTP headers in response to a HEAD request should be identical to the information sent in response to a GET request. This method can be used for obtaining metainformation about the entity implied by the request without transferring the entity-body itself. This method is often used for testing hypertext links for validity, accessibility, and recent modification.

POST

The POST method is used to request that the origin server accepts the entity enclosed in the request as a new subordinate of the resource identified by the request-URI in the request-line. POST is designed to allow a uniform method to cover the following functions:

- Annotation of existing resources.
- Posting a message to a bulletin board, newsgroup, mailing list, or similar group of articles.
- Providing a block of data, such as the result of submitting a form, to a data-handling process.
- Extending a database through an append operation.

```
OPTIONS / HTTP/1.1
Host: gutemine.ict.tuwien.ac.at:33333
```

```
HTTP/1.1 200 OK
Date: Sat, 26 Oct 2002 17:18:34 GMT
Server: Apache/1.3.23 (Unix)  (Red-Hat/Linux) mod_ssl/2.8.7 OpenSSL/0.9.6b
DAV/1.0.3 PHP/4.1.2 mod_perl/1.26 mod_auth_pam/1.0a
Content-Length: 0
Allow: GET, HEAD, OPTIONS, TRACE
```

FIGURE 30.12   The OPTIONS method request and an appropriate response.

```
HEAD /goeschka/test/first.html HTTP/1.1
Host: gutemine.ict.tuwien.ac.at:33333
```

```
HTTP/1.1 200 OK
Date: Sat, 26 Oct 2002 17:21:32 GMT
Server: Apache/1.3.23 (Unix)  (Red-Hat/Linux) mod_ssl/2.8.7 OpenSSL/0.9.6b
DAV/1.0.3 PHP/4.1.2 mod_perl/1.26 mod_auth_pam/1.0a
Content-Type: text/html
```

FIGURE 30.13   The HEAD method request and an appropriate response.

The actual function performed by the POST method is determined by the server and usually depends on the request-URI. The posted entity is subordinate to that URI in the same way in which a file is subordinate to a directory containing it, a news article is subordinate to a newsgroup to which it is posted, or a record is subordinate to a database.

Consider a Web form, like in Figure 30.14: the FORM Tag is important, describing the request-URI and the request method for submitting the form. Pressing the submit button then causes the user agent to initiate a POST request as shown in Figure 30.15.

The Content-Type application/x-www-form-urlencoded used for submitting Web form data to server-side applications is notable. The Content-Length is also important, because unlike most

```
<html>
  <head>
    <title> Simple HTML FORMS </title>
  </head>
  <body>
    <FORM ACTION="http://gutemine.ict.tuwien.ac.at:33333/cgi-bin/form" METHOD=POST>
    <P> Search string: <INPUT TYPE="text" NAME="srch" VALUE="dogfish">
    <P> Search Type:
      <SELECT NAME="srch_type">
          <OPTION> Insensitive Substring
          <OPTION SELECTED> Exact Match
          <OPTION> Sensitive Substring
          <OPTION> Regular Expression
      </SELECT>
    <P> Search databases in:
      <INPUT TYPE="checkbox" NAME="srvr" VALUE="Canada" CHECKED> Canada
      <INPUT TYPE="checkbox" NAME="srvr" VALUE="Russia"         > Russia
      <INPUT TYPE="checkbox" NAME="srvr" VALUE="Sweden" CHECKED> Sweden
      <INPUT TYPE="checkbox" NAME="srvr" VALUE="U.S.A."         > U.S.A.
      <em>(multiple items can be selected.)</em>
    <P> <INPUT TYPE="submit"> <INPUT TYPE=reset>.
    </FORM>
  </body>
</html>
```


FIGURE 30.14   Web form. HTML and user-agent representation.

```
POST /cgi-bin/form HTTP/1.1
Accept: image/gif, image/x-xbitmap, image/jpeg, image/pjpeg, application/vnd.ms-
powerpoint, application/vnd.ms-excel, application/msword, */*
Accept-Language: de-at
Content-Type: application/x-www-form-urlencoded
Accept-Encoding: gzip, deflate
User-Agent: Mozilla/4.0 (compatible; MSIE 5.01; Windows NT 5.0)
Host: gutemine.ict.tuwien.ac.at:33333
Content-Length: 58
Connection: Keep-Alive

srch=dogfish&srch_type=Exact+Match&srvr=Canada&srvr=Sweden
```

FIGURE 30.15   POST request resulting from the Web form in Figure 30.14.

```
GET /cgi-bin/form?srch=dogfish&srch_type=Exact+Match&srvr=Canada&srvr=Sweden
HTTP/1.1
Accept: image/gif, image/x-xbitmap, image/jpeg, image/pjpeg, application/vnd.ms-
powerpoint, application/vnd.ms-excel, application/msword, */*
Accept-Language: de-at
Accept-Encoding: gzip, deflate
User-Agent: Mozilla/4.0 (compatible; MSIE 5.01; Windows NT 5.0)
Host: gutemine.ict.tuwien.ac.at:33333
Connection: Keep-Alive
```

FIGURE 30.16   GET request resulting from the Web form in Figure 30.14, if GET is used as FORM action instead of POST.

other requests, the POST request usually does have a content. If the FORM action GET would be used instead of POST, the resulting GET request is shown in Figure 30.16. Now, the query string is not delivered within the message content but rather appended to the request-URI. Therefore, no Content-Type or Content-Length is necessary.

TRACE

The TRACE method is used to invoke a remote, application-layer loop-back of the request message. The final recipient of the request *should* reflect the message received back to the client as the entity-body of a 200 (OK) response. A TRACE request *must not* include an entity. TRACE allows the client to see what is being received at the other end of the request chain and use that data for testing or diagnostic information. The value of the Via header field is of particular interest, since it acts as a trace of the request chain. The use of the Max-Forwards header field allows the client to limit the length of the request chain, which is useful for testing a chain of proxies forwarding messages in an infinite loop. If the request is valid, the response *should* contain the entire request message in the entity-body, with a Content-Type of message/http. Responses to this method *must not* be cached. Figure 30.17 shows an example of a TRACE request to a proxy (PAULI) and the respective response.

30.11   Further HTTP Response Status Codes by Example

This section provides further examples for HTTP response status codes. For a complete list of all status codes, please refer to RFC 2616.

Informational 1xx

This class of status code indicates a provisional response, consisting only of the Status-Line and optional headers, and is terminated by an empty line. There are no required headers for this class of status code. A client must be prepared to accept one or more 1xx status responses prior to a regular response, even if

```
TRACE http://www.ict.tuwien.ac.at/goeschka/test/first.html HTTP/1.1
Host: www.ict.tuwien.ac.at
```

```
HTTP/1.0 200 OK
Via: 1.0 PAULI
Content-Type: message/http
Date: Sat, 26 Oct 2002 20:56:12 GMT
Server: Apache/1.3.23 (Unix)  (Red-Hat/Linux) mod_ssl/2.8.7 OpenSSL/0.9.6b
DAV/1.0.3 PHP/4.1.2 mod_perl/1.26 mod_auth_pam/1.0a

TRACE /goeschka/test/first.html HTTP/1.0
Connection: Keep-Alive
Host: www.ict.tuwien.ac.at
Via: 1.0 PAULI
```

FIGURE 30.17　The TRACE method.

the client does not expect a `100 (Continue)` status message. Unexpected 1xx status responses may be ignored by a user-agent.

Successful 2xx

This class of status code indicates that the client's request was successfully received, understood, and accepted.

Redirection 3xx

This class of status code indicates that further action needs to be taken by the user-agent in order to fulfill the request. The action required may be carried out by the user-agent without interaction with the user if and only if the method used in the second request is GET or HEAD. A client should detect infinite redirection loops, since such loops generate network traffic for each redirection.

302 Found

The requested resource resides temporarily under a different URI. Since the redirection might be altered on occasion, the client should continue to use the request-URI for future requests. The temporary URI should be given by the `Location` field in the response. Unless the request method was HEAD, the entity of the response should contain a short hypertext note with a hyperlink to the new URI(s). If the 302 status code is received in response to a request other than GET or HEAD, the user-agent must not automatically redirect the request unless it can be confirmed by the user, since this might change the conditions under which the request was issued. For example, the simple request from Figure 30.4 may result in a response shown in Figure 30.18.

The `Location` response-header field is used to redirect the recipient to a location other than the request-URI for completion of the request or identification of a new resource. For 3xx responses, the location should indicate the server's preferred URI for automatic redirection to the resource. The field value consists of a single absolute URI. If the user-agent understands the response, it will immediately send the following request to the new destination as shown in Figure 30.19. Otherwise, the user-agent will display the HTML content of Figure 30.18 to the user as shown in Figure 30.20.

304 Not Modified

If the client has performed a conditional GET request and access is allowed, but the document has not been modified, the server should respond with this status code. The 304 response must not contain a message-body, and thus is always terminated by the first empty line after the header fields. Figure 30.21 shows a conditional GET request to a proxy server.

```
HTTP/1.1 302 Found
Date: Sat, 26 Oct 2002 22:22:08 GMT
Server: Apache/1.3.23 (Unix)  (Red-Hat/Linux) mod_ssl/2.8.7 OpenSSL/0.9.6b
DAV/1.0.3 PHP/4.1.2 mod_perl/1.26 mod_auth_pam/1.0a
Location: http://some.where.else.ac.at:22222/second.html
Content-Type: text/html; charset=iso-8859-1

<HTML>
  <HEAD>
    <TITLE>302 Found</TITLE>
  </HEAD>
  <BODY>
    <H1>Found</H1>
    <P>The document has moved
    <A HREF="http://some.where.else.ac.at:22222/second.html">here</A>.
    <HR>
    <ADDRESS>Apache/1.3.23 Server at
    <A HREF="mailto:webmaster@ict.tuwien.ac.at">gutemine.ict.tuwien.ac.at</A>
    Port 33333</ADDRESS>
  </BODY>
</HTML>
```

FIGURE 30.18   Found response.

```
GET /second.html HTTP/1.1
Accept: image/gif, image/x-xbitmap, image/jpeg, image/pjpeg, application/vnd.ms-
powerpoint, application/vnd.ms-excel, application/msword, */*
Accept-Language: de-at
Accept-Encoding: gzip, deflate
User-Agent: Mozilla/4.0 (compatible; MSIE 5.01; Windows NT 5.0)
Host: some.where.else.ac.at:22222
```

FIGURE 30.19   Redirected request.

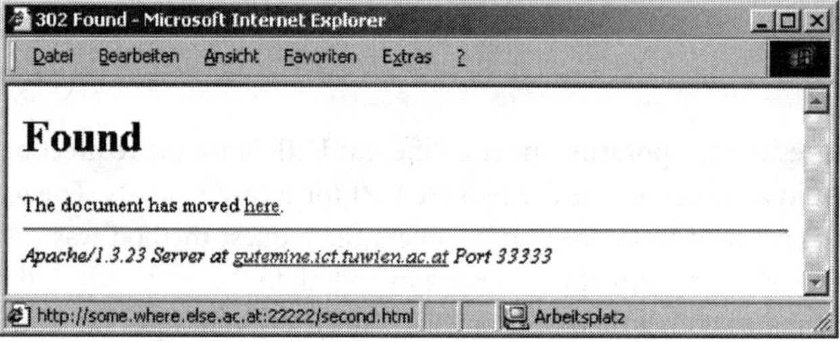

FIGURE 30.20   User information in case of a user-agent not automatically redirecting.

The `If-Modified-Since` request-header field is used with a method to make it conditional: if the requested variant has *not* been modified since the time specified in this field, an entity will not be returned from the server; instead, a `304  (not modified)` response will be returned without any message-body. A `GET` method with an `If-Modified-Since` header and no `Range` header requests that the identified entity be transferred only if it has been modified since the date given by the `If-Modified-Since` header. Assuming the document has not been modified since the given time, Figure 30.21 shows the 304 response from the proxy to the client.

Client Error 4xx

The 4xx class of status code is intended for cases in which the client seems to have erred. Except when responding to a `HEAD` request, the server should include an entity containing an explanation of the error

```
GET http://gutemine.ict.tuwien.ac.at:33333/second.html HTTP/1.1
Accept: image/gif, image/x-xbitmap, image/jpeg, image/pjpeg, application/vnd.ms-
powerpoint, application/vnd.ms-excel, application/msword, */*
Accept-Language: de-at
Accept-Encoding: gzip, deflate
User-Agent: Mozilla/4.0 (compatible; MSIE 5.01; Windows NT 5.0)
Host: gutemine.ict.tuwien.ac.at:33333
If-Modified-Since: Sat, 26 Oct 2002 22:13:35 GMT
```

```
HTTP/1.1 304 Not Modified
Content-Length: 0
Connection: close
Via: 1.1 ADPROSRV01
```

FIGURE 30.21　Conditional GET request and `Not Modified` response.

```
HTTP/1.1 404 Not Found
Date: Mon, 28 Oct 2002 15:30:07 GMT
Server: Apache/1.3.23 (Unix)  (Red-Hat/Linux) mod_ssl/2.8.7 OpenSSL/0.9.6b
DAV/1.0.3 PHP/4.1.2 mod_perl/1.26 mod_auth_pam/1.0a
Content-Type: text/html; charset=iso-8859-1

<HTML>
  <HEAD>
    <TITLE>404 Not Found</TITLE>
  </HEAD>
  <BODY>
    <H1>Not Found</H1>
    <P>The requested URL /third.html was not found on this server.
    <HR>
    <ADDRESS>Apache/1.3.23 Server at
    <A HREF="mailto:webmaster@ict.tuwien.ac.at">gutemine.ict.tuwien.ac.at</A>
    Port 33333</ADDRESS>
  </BODY>
</HTML>
```

FIGURE 30.22　404 — Not found.

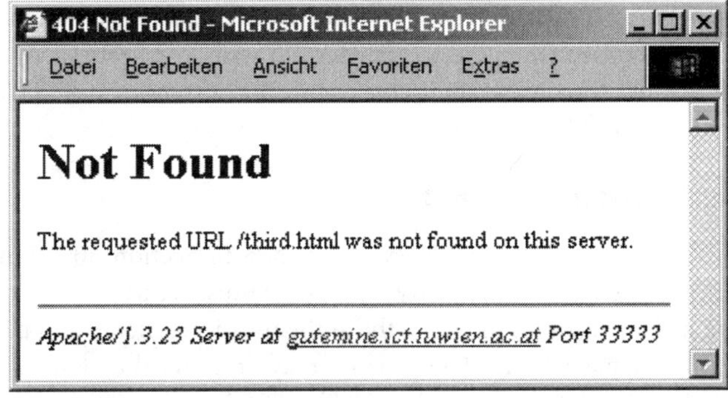

FIGURE 30.23　Not found information displayed to the user.

```
HTTP/1.1 407 Proxy authentication required
Proxy-Authenticate: NTLM
Proxy-Connection: close
Content-Length: 503
Content-Type: text/html

<html>
  <head>
    <title>Error 407</title>
    <meta name="robots" content="noindex">
    <META HTTP-EQUIV="Content-Type" CONTENT="text/html; charset=iso-8859-1">
  </head>
  <body>
    <h2>HTTP Error 407</h2>
    <p><strong>407 Proxy Authentication Required</strong></p>
    <p>You must authenticate with a proxy server before this request can
       be serviced.  Please log on to your proxy server, and then try again.</p>
    <p>Please contact the Web server's administrator if this problem persists.</p>
  </body>
</html>
```

FIGURE 30.24   407 — Proxy authentication required.

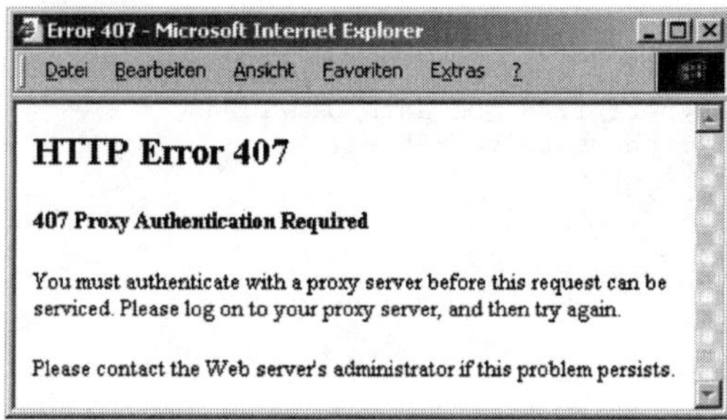

FIGURE 30.25   407 information displayed to the user.

situation, and whether it is a temporary or permanent condition. These status codes are applicable to any request method. User-agents should display any included entity to the user.

404 Not Found

Unfortunately, one of the most common response status codes is as follows: the server has not found anything matching the request-URI. No indication is given of whether the condition is temporary or permanent. The `410  (Gone)` status code should be used if the server knows, through some internally configurable mechanism, that an old resource is permanently unavailable and has no forwarding address. This status code is commonly used when the server does not wish to reveal exactly why the request has been refused, or when no other response is applicable. Figure 30.22 shows the HTTP response and Figure 30.23 shows the respective display for the user.

407 Proxy Authentication Required

This code is similar to `401  (Unauthorized)` explained in Section 30.7, but indicates that the client must first authenticate itself with the proxy. The proxy must return a `Proxy-Authenticate` header field containing a challenge that indicates the authentication scheme and parameters applicable to the proxy for this request-URI. The client may repeat the request with a suitable `Proxy-Authorization` header field (Section 30.8). HTTP access authentication is explained in RFC 2617 (Figures 30.24 and 30.25).

```
GET http://www.ict.tuwien.ac.at/goeschka/test/second.html HTTP/1.1
Accept: image/gif, image/x-xbitmap, image/jpeg, image/pjpeg, application/vnd.ms-
powerpoint, application/vnd.ms-excel, application/msword, */*
Accept-Language: de-at
Accept-Encoding: gzip, deflate
User-Agent: Mozilla/4.0 (compatible; MSIE 5.01; Windows NT 5.0)
Host: www.ict.tuwien.ac.at
Cache-control: only-if-cached
```

FIGURE 30.26   Request with `Cache-control` header.

```
HTTP/1.1 504 Proxy Error ( This operation returned because the timeout period
expired.  )
Via: 1.1 ADPROSRV01
Pragma: no-cache
Cache-Control: no-cache
Content-Type: text/html

<!DOCTYPE HTML PUBLIC "-//IETF//DTD HTML//EN">
<html>
  <head>
    <meta http-equiv="Content-Type" content="text/html; charset=iso-8859-1">
    <meta name="GENERATOR" content="Microsoft FrontPage 2.0">
    <title>Proxy Server Report: Default</title>
  </head>
  <body bgcolor="#BEBEBE">
    <h2>Proxy Reports:</h2>
.............
  </body>
</html>
```

FIGURE 30.27   Proxy Error response (HTML content shortened).

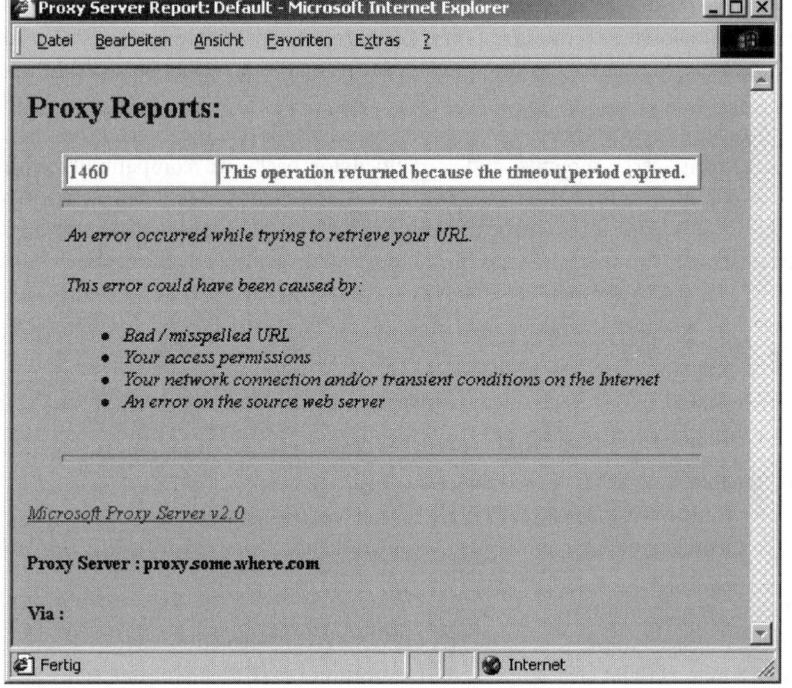

FIGURE 30.28   504 information (proxy error) displayed to the user.

Server Error 5xx

Response status codes beginning with the digit "5" indicate cases in which the server is aware that it has erred or is incapable of performing the request. Except when responding to a `HEAD` request, the server should include an entity containing an explanation of the error situation, and whether it is a temporary or permanent condition. User-agents should display any included entity to the user. These response codes are applicable to any request method.

504 Gateway Timeout

The server, while acting as a gateway or proxy, did not receive a timely response from the upstream server specified by the URI. This status code is also used for cache control. Consider a request to a proxy server using the `Cache-control` header field (see Figure 30.26).

The `Cache-Control` general-header field is used to specify directives that must be obeyed by all caching mechanisms along the request/response chain. The directives specify behavior intended to prevent caches from adversely interfering with the request or response. These directives typically override the default caching algorithms. Cache directives are unidirectional in that the presence of a directive in a request does not imply that the same directive is to be given in the response.

In our example, the cache-request-directive `only-if-cached` is used: in some cases, such as times of extremely poor network connectivity, a client may want a cache to return only those responses that it has currently stored, and not to reload or revalidate with the origin server. To do this, the client may include the `only-if-cached` directive in a request. If it receives this directive, a cache should either respond using a cached entry that is consistent with the other constraints of the request, or respond with a `504 (Gateway Timeout)` status (see Figure 30.27), which is then displayed to the user (see Figure 30.28).

The cache-response-directive `no-cache` has the following meaning in this example: if the `no-cache` directive does not specify a field-name, then a cache must not use the response to satisfy a subsequent request without successful revalidation with the origin server. This allows an origin server to prevent caching even by caches that have been configured to return stale responses to client requests.

The `Pragma` general-header field is used to include implementation-specific directives that might apply to any recipient along the request/response chain. All pragma directives specify optional behavior from the viewpoint of the protocol; however, some systems may require that behavior be consistent with the directives.

When the no-cache directive is present in a request message, this pragma directive has the same semantics as the `no-cache` cache-directive and is defined for backward compatibility with HTTP/1.0. HTTP/1.1 caches should treat `Pragma: no-cache` as if the client had sent `Cache-Control: no-cache`. No new `Pragma` directives will be defined in HTTP. As the meaning of `Pragma: no-cache` as a response header field is not actually specified, it does not provide a reliable replacement for `Cache-Control: no-cache` in a response.

References

There is no other way than to study the Request For Comments (RFCs) if you want to get the information source, for instance, from `ftp://ftp.isi.edu/in-notes/`.

For the HTTP the following RFCs are relevant:
2616   Hypertext Transfer Protocol — HTTP/1.1 (obsoletes 2068) [1999/06]
2617   HTTP Authentication: Basic and Digest Access Authentication (obsoletes 2069) [1999/06]
2145   Use and Interpretation of HTTP Version Numbers [1997/05]
2109   HTTP State Management Mechanism [1997/02]
2068   Hypertext Transfer Protocol — HTTP/1.1 [1997/01] — obsoleted by 2616
2067   An Extension to HTTP : Digest Access Authentication [1997/01] — obsoleted by 2617
1945   Hypertext Transfer Protocol — HTTP/1.0 [1996/05]

With respect to URI/URL, consider the following RFCs:

2717	Registration Procedures for URL Scheme Names [1999/11]
2718	Guidelines for new URL Schemes [1999/11]
2396	Uniform Resource Identifiers (URI): Generic Syntax (obsoletes 1808, 1738) [1998/08]
2141	URN Syntax (updates 1737, 1630) [1997/05]
1808	Relative Uniform Resource Locators [1995/06] — obsoleted by 2396
1738	Uniform Resource Locators (URL) [1994/12] — obsoleted by 2396
1737	Functional Requirements for Uniform Resource Names [1994/12] — updated by 2141
1630	Universal Resource Identifiers in WWW [1994/06] — updated by 2141

The following RFCs are related to HTTP and URN:

1952	GZIP file format specification version 4.3 [1996/05]
1951	DEFLATE Compressed Data Format Specification version 1.3 [1996/05]
1950	ZLIB Compressed Data Format Specification version 3.3 [1996/05]
1766	Tags for the Identification of Languages [1995/03]
1700	Assigned Numbers [1994/10]
1590	Media Type Registration Procedure [1994/03]
1123	Requirements for Internet Hosts — Application and Support [1989/10]
1034	Domain Names — Concepts and Facilities [1987/11]
822 Standard for the format of ARPA Internet Text Messages [1982/08]

Of course, the World Wide Web Consortium provides various ressources on their Web page http:// www.w3.org/, especially http://www.w3.org/Protocols/ and http://www.w3. org/Addressing/, and is an excellent starting point for information search on Web issues in general.

31

Network Management: Basic Notions and Frameworks

Mai Hoang
University of Potsdam

31.1 Introduction

Computer networks and distributed processing systems continue to grow in scale and diversity in business, government, and other organizations. Three facts become evident. First, new networks are added and existing ones are expanded almost as rapidly as new network technologies and products are introduced. The problems associated with network expansion affect day-to-day network operation management. Second, the network and its resources and distributed services become indispensable to organizations. Third, more things can go wrong, which disable the network or degrade the performance to an unacceptable level. Such inhomogeneous large networks cannot be put together and managed by human effort alone. Instead, their complexity dictates the use of a rich set of automated network management tools and applications.

In response, the International Standard Organization (ISO) began work in 1978 to establish a standard for network management, the Open System Interconnection (OSI) Network Management including the management model, functional areas, Common Management Information Services (CMIS), the Common Management Information Protocol (CMIP), and Management Information Base (MIB). The network management model describes the main components of a network management tool for a managed network. For a given managed network, it is necessary to know which problem areas have to be considered. These problem areas are specified as the ISO management functions, which were already contained in the first ISO working draft of the management framework and gradually evolved into what is presently known as the five functional areas of the ISO management framework (performance, faults, configuration, accounting, and security). The important pieces of the ISO management are CMIP and CMIS, which managing and managed devices use for their communication. The CMIP protocol consists of a set of services, the so-called CMIS that define the types of requests and responses and the actions they should invoke. In addition to being able to pass information and backforth, the managing and managed devices need to agree on a set of variables and means to initiate actions. The collection of this information is referred to as the MIB.

Because of the slowness of the ISO standardization process, the complexity of the proposed new standard, and the urgent need for management tools, the IETF has devised the Simple Network Management Protocol (SNMP) [RFC1157], which was originally regarded as a provisional means for network management until the OSI management standards would be completed, but subsequently became a *de facto* standard because of its dissemination and its simplicity. The SNMP consists of three parts: the SNMP protocol, the structure of management information (SMI), and the MIB. The SNMP protocol includes the SNMP operations, the format of messages, and how messages are exchanged between a manager and an agent. The SMI is a set of rules allowing a user to specify the desired management information, for example, by providing a means of naming and declaring the types of variables. Finally, the MIB is a structured collection of all managed objects maintained by a device. The managed objects are structured as a hierarchical tree. In order to address several weaknesses within SNMP, the SNMP version 2 (SNMPv2) was initiated around 1994. SNMPv2 provides more functionality and greater efficiency than the original version of SNMP; but due to various reasons, SNMPv2 did not succeed. Finally, the SNMP version 3 (SNMPv3) was issued in 1998. SNMPv3 describes an overall framework for present and future versions of SNMP and defines security features to SNMP.

This chapter focuses on the basic notions and frameworks of network management standards, which have been developed by ISO and IETF. It aims to serve as a comprehensive survey of conceptual models, protocols, services, and management information bases of these management standards. The chapter is organized as follows. Section 31.2 describes the network management model developed by ISO and IETF. The ISO network management framework is discussed briefly in Section 31.3, while Section 31.4 presents the IETF management framework. These management standards are compared in Section 31.5. Section 31.6 concludes the chapter with an overview of the open problems in network management.

31.2 Network Management Architecture

The network management architecture used in ISO and IETF frameworks is called manager–agent architecture and includes the following key components:

- managed devices,
- management stations,
- management protocols, and
- management Information.

These are shown in Figure 31.1 and described below.

Network management is done from *management stations*, which are computers running special management software. These management stations contain a set of management application processes called *managers* for data analysis, fault recovery, and so on. The manager is the locus of activity for network management; it provides or monitors information to users; issues requests to managed devices in order to ask them to take some action; receives responses to the requests; and receives unsolicited reports from managed devices concerning the status of the devices — these reports are referred to as notification and are frequently used to report problems, anomalies, or changes in the agent environment.

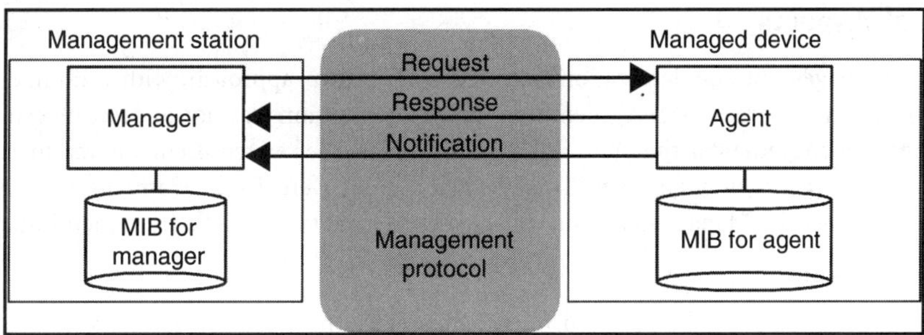

FIGURE 31.1 Manager–Agent architecture.

A *managed device* is a piece of network equipment that resides in a managed network. The managed devices might be hosts, routers, switches, bridges, or printers. To be managed from a management station, a device must be capable of running a management process, called *(management) agent*. These agents communicate with managers running on the management station and take local actions on the managed device under the command and control of the managers. An agent can act upon and respond to request from a manager; furthermore, it can provide unsolicited notifications to a manager.

Each managed device maintains one or more variables (e.g., a network interface card or a set of configuration parameters for a piece of hardware or software) that describe its state. In the ISO and IETF management frameworks, these variables are called *managed objects*. The collection of these managed objects is referred to as a *MIB*. These variables can be viewed and optionally modified by the managers.

The *network management protocol* is needed for communication between managers and agents. This protocol allows the manager to query the status of managed devices and to initiate actions at these devices by triggering the agents. Furthermore, agents can use the network management protocol to report exceptional events to the management stations.

When describing any framework for network management, the following aspects must be addressed:

- *The functional aspect* specifies management functional areas supported by managers and agents. This aspect relates to specific management functions that are carried out by the manager or agent.
- *The information aspect* defines the kind of information that will be exchanged between the manager and agent. The information aspect deals with MIBs and SMI.
- *The communication aspect* addresses the communication protocol between the manager and agent for exchanging this information.
- *The organization aspect* deals with the definition of the principal structural components and the management architecture for a managed network.

The OSI and IETF management frameworks are discussed in the following subsections from the view point of these aspects.

31.3 ISO Systems Management Framework

The first standard for network management was ISO 7498-4, which specifies the network management framework for the OSI model [ISO7498-4]. Although the production of this framework took considerable time, it was not generally accepted as an adequate starting point. It was therefore decided to issue an additional standard, which was called the System Management Overview [ISO10040]. Subsequently, ISO has issued a set of other additional standards for network management. Together, these standards provide the basis for the OSI management framework.

Functional Aspects

OSI Systems Management standardization followed a top-down approach, with a number of system management functional areas (SMFA) identified first. The intention was not to describe exhaustively all relevant types of management activity but rather to investigate the key requirements and to address these through a generic management model. The identified areas were Fault, Configuration, Accounting, Performance and Security Management, and collectively referred to as FCAPS from their initials [Sta93a].

Fault Management

Fault management deals with the mechanisms for the detection, isolation, and correction of abnormal operations. Fault management includes functions to:

- maintain and examine error logs,
- trace and identify faults,
- accept and act upon error notifications, and
- to carry out diagnostic tests and correct faults.

Configuration Management

Configuration management is the set of facilities that allow network managers to exercise control over the configuration of the network components and OSI-layer entities. Configuration management includes functions to:

- record the current configuration,
- record changes in the configuration,
- initialize and close down managed objects,
- identify the network components, and
- to change the configuration of managed objects (e.g., routing table).

Accounting Management

Accounting management deals with the collection and processing of accounting information and for charging and billing purposes. It should enable accounting limits to be set and costs to be combined when multiple resources are used in the context of a service. Accounting management includes functions to:

- inform users of the cost thus far,
- inform users of the expected cost in the future, and
- to set cost limits.

Performance Management

Performance management is the set of facilities that enable the network managers to monitor and evaluate the performance of the system and layer entities. Performance management involves three main steps. First, performance data are gathered on variables of interest to network administrators. Second, the data are analyzed to determine normal (baseline) levels. Finally, appropriate performance thresholds are determined for each important variable so that exceeding these thresholds indicates a network problem worthy of attention. Management entities continually monitor performance variables. When a performance threshold is exceeded, an alert is generated and sent to the network management system. Performance management provides functions to:

- collect and disseminate data concerning the current level of performance of resources and to
- maintain and examine performance logs for planning and analysis purposes.

Security Management

Security management addresses the control of the access to network resources according to local guidelines so that the network cannot be damaged and persons without appropriate authorization cannot access sensitive information. A security management subsystem, for example, can monitor users logging on to a network resource and can refuse access to those who enter inappropriate access codes. Security management provides support for management of:

- authorization facilities,
- access control,
- encryption and key management,
- authentication, and
- security logs.

Soon after the first working drafts of the management framework appeared, ISO started to define protocol standards for each of the five SMFA. After sometime, an interesting observation was made that most of the functional area protocols used a similar set of elementary management functions. ISO decided, therefore, to stop further progression of the five functional area protocols and concentrate on the definition of elementary management functions. Following this, a set of standards, for example, object management, state management, relationships management, alarm reporting, event-report management, and log control, have been issued as the general category systems-management functions (SMF). Each SMF standard defines the functionality to support SMFA requirements. Moreover, these standards provide a mapping between the CMIS services (discussed below) and SMF.

Information Aspects

The information aspects of the OSI system management deal with the resources that are being managed by agents. OSI systems management relies on object-oriented concepts. Therefore, each resource being managed is represented by a managed object. A managed object may represent either a logical resource such as a user account, or a real resource like an ATM switch. Managed objects that refer to resources specific to an individual layer are called (N)-layer-managed objects. Managed objects that refer to resources that encompass more than one layer are called systems-managed objects. According to the OSI Management Information Model [ISO10165-1], a managed object is defined in terms of the *attributes* that it possesses, *operations* that may be performed on it, *notifications* that it may issue, and *its interactions* with other managed objects. The managed objects are defined using two standards: *Abstract Syntax Notation One* (ASN.1) is used to define data types and *Guidelines for definition of managed objects* (GDMO) to define managed objects [ASN90, ISO10165-1].

Under systems management, all the managed objects are represented in the so-called *MIB*. The managed object concept is refined in a number of additional standards that are called the *Structure of Management Information (SMI)* standards [ISO10165-1, ISO10165-2, ISO10165-4, ISO10165-5, and ISO10165-7]. The SMI identifies the data types that can be used in the MIB and how the resources within the MIB are represented and named [Sta93b].

Organization Aspects

The key elements of the OSI architectural model include system-management application process (SMAP), system-management application entity (SMAE), layer-management entity, and management information base. SMAP is the process within a managing device that is responsible for executing the network-management functions; it has access to all parameters of managed devices and can therefore manage all aspects of a managed network and may coordinate with SMAPs on other managed networks. An SMAE is responsible for communication with other devices, especially with devices exercising control functions; CMIP is used as a standardized application-level protocol by SMAE. Layer-management entity is the logic embedded into each layer of OSI architecture to provide network management functions that are specific for this layer.

To provide management of a distributed system, the elements in this architectural model must be implemented in a distributed manner across all the devices in a managed network. The OSI system management is organized in a central manner. According to this scheme, a single manager may control several agents. Each agent contains a number of objects. Each object is a data structure that corresponds to an actual piece of device to be managed. The SMAP is allowed to take on either a manager's or an agent's role. The manager's role for an SMAP occurs in a device that acts as a network control center. The agent's role for an SMAP occurs in managed devices. The manager performs operations upon the agents and the agents forward notifications to the managers. Because of the expansion of the open system, the OSI management environment may be partitioned into a number of management domains. The partition cannot just be based on the required management functions (security, accounting, performance, etc.) but also on other requirements (e.g., geographical).

Communication Aspects

The communication aspect deals with the exchange of SMI between managers and agents within a managed network. With respect to this aspect, ISO has issued two standards; the CMIS and the CMIP [ISO9595, ISO9596].

CMIS provides OSI management services to management applications. It defines a set of management services, specifies types of requests and responses, and defines what each request and response can do. The management processes initiate these services in order to communicate remotely. Seven services used to handle management information have been standardized. Table 31.1 lists the CMIS services with their type and function.

The CMIP provides the information-exchange capability to support CMIS; it defines a set of protocol data units that implement the CMIS services [ISO9596]. In particular, CMIP defines how the requests, responses, and notifications are encoded into messages and specifies which bearer service is used to transport those encoded messages between managers and agents. A CMIP request typically specifies one or more managed objects to which the request is to be sent. The correspondence between CMIS primitives and CMIP data units is described in [Sta99].

31.4   Internet Management Framework

An interesting difference between the IETF and ISO is that the IETF adopts a more pragmatic and result-driven approach than ISO. In the IETF, it is, for instance, unusual to spend much time on architectural discussions; people prefer to use their time for the development of protocols and implementations. This difference explains why no special standards management architecture and function areas have been defined in the first two versions of SNMP; only the communication aspect (as SNMP protocol), the information aspect (as SMI and MIB), and the security aspect have been standardized.

SNMP is an application-layer protocol that facilitates the exchange of management information between network devices. It is a part of the TCP/IP protocol suite and operates over UDP. As described

TABLE 31.1   CMIS Services

CMIS services	Type	Functions
Notification services		
M-EVENT-REPORT	Confirmed/not confirmed	Gives notification of an envent occurring on a managed object
Opertion services		
M-GET	Confirmed	Request for
M-SET	Confirmed/not confirmed	Modification of management data
M-ACTION	Confirmed/not confirmed	Execution of action on a managed object
M-CREATE	Confirmed	Creation of a managed object
M-DELETE	Confirmed	Deletion of a managed object
M-CANCEL-GET	Confirmed	Request to cancel any new responses to a previous request for M-GET services

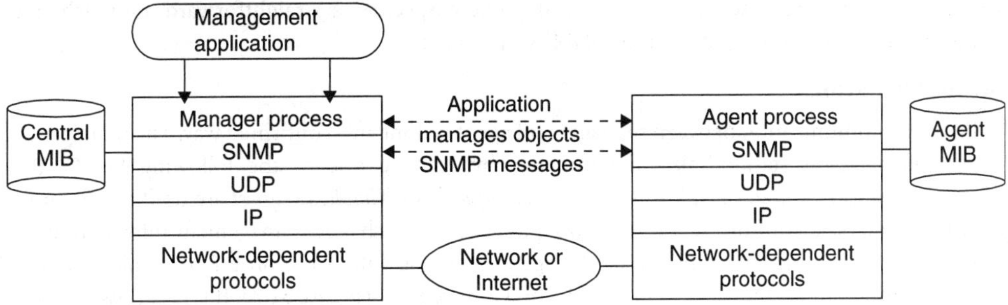

FIGURE 31.2   Internet management architecture.

in Section 31.2, IETF network management is based on the manager–agent architecture. Figure 31.2 shows the architecture of Internet management. In this architecture, a manager process controls access to a central MIB at the management station and provides an interface to the management application. Furthermore, a manager may control many agents, whereby each agent interprets the SNMP messages and controls the agent's MIBs.

In Section 31.2, we have provided an overview of the basic components of a management architecture used by ISO and IETF. The IETF network management framework consists of:

SNMP: SNMP is a management protocol for conveying information and commands between a manager and an agent running in a managed network device [KR01].

MIB: Resources in networks may be managed by representing them as objects. Each object is a data variable that represents one aspect of a managed device. In the IETF network management framework, the representation of a collection of these objects is called MIB [RFC1157, RFC1212]. An MIB object might be a counter such as the number of IP datagrams discarded at a router due to errors, descriptive information such as generic information about the physical interfaces of the entity, or protocol-specific information such as the number of UDP datagrams delivered to UDP users.

SMI: SMI [RFC1155] allows to formally specify the data types that are used in an MIB and specifies how resources within an MIB are named. The SMI is based on the Abstract Syntax Notation One (ASN.1) [ASN90] object-definition language, however, since many SMI-specific data types have been added, SMI should be considered a data-definition language of its own right.

Security and administration is concerned with monitoring and controlling access to managed networks and access to all or part of management information obtained from network nodes.

In the following sections, an overview of several SNMP versions (SNMPv1, SNMPv2, and SNMPv2) with respect to protocol operations, MIB, SMI, and security is given.

SNMPv1

The original network management framework is defined in the following documents:

- RFC 1155 and RFC 1212 define SMI, the mechanisms used for specifying and naming managed objects. RFC 1215 defines a concise description mechanism for defining event notifications that are called traps in SNMPv1.
- RFC 1157 defines SNMPv1, the protocol used for network access to managed objects and event notification.
- RFC 1213 contains definition for a specific MIB (MIB I) covering TCP, UDP, IP, routers, and other inhabitants of the IP world.

SMI

The RFCs 1155, 1212, and 1215 describe the SNMPv1 structure of management information and are often referred to as "SMIv1." Note that the first two SMI documents do not provide the definition of event notifications (traps). Because of this, the last document specifies a straightforward approach toward defining event notifications used with the SNMPv1 protocol.

Protocol Operations

In SNMPv1, communication between manager and agent is done in a confirmed way. The manager at the network management station takes the initiative by sending one of the following SNMP PDUs: *GetRequest, GetNextRequest,* or *SetRequest*. The *GetRequest* and *GetNextRequest* are used to obtain management information from the agent; the SetRequest is used to change management information at the agent. After receiving one of these PDUs, the agent responds with a response PDU, which carries the requested information or indicates failure of the previous request (Figure 31.3). It is also possible that the SNMP agent takes the initiative. This happens when the agent detects some extraordinary event such as a status change at one of its links. As a reaction to this, the agent sends a trap PDU to the manager. The reception of the trap is not confirmed (Figure 31.3(d)).

MIB

As noted above, the MIB can be thought of as a virtual information store, holding managed objects whose values collectively reflect the current "state" of the network. These values may be queried and/or set by a manager by sending SNMP messages to the agent. Managed objects are specified using the SMI discussed above.

The IETF has been standardizing the MIB modules associated with routers, hosts, and other network equipment. This includes basic identification data about a particular piece of hardware, and management information about the device's network interfaces and protocols. With the different SNMP standards, the IETF needed a way to identify and name the standardized MIB modules, as well as the specific managed objects within an MIB module. To do this, the IETF adopted ASN.1 as a standardized object identification (naming) framework. In ASN.1, object identifiers have a hierarchical structure, as shown in Figure 31.4.

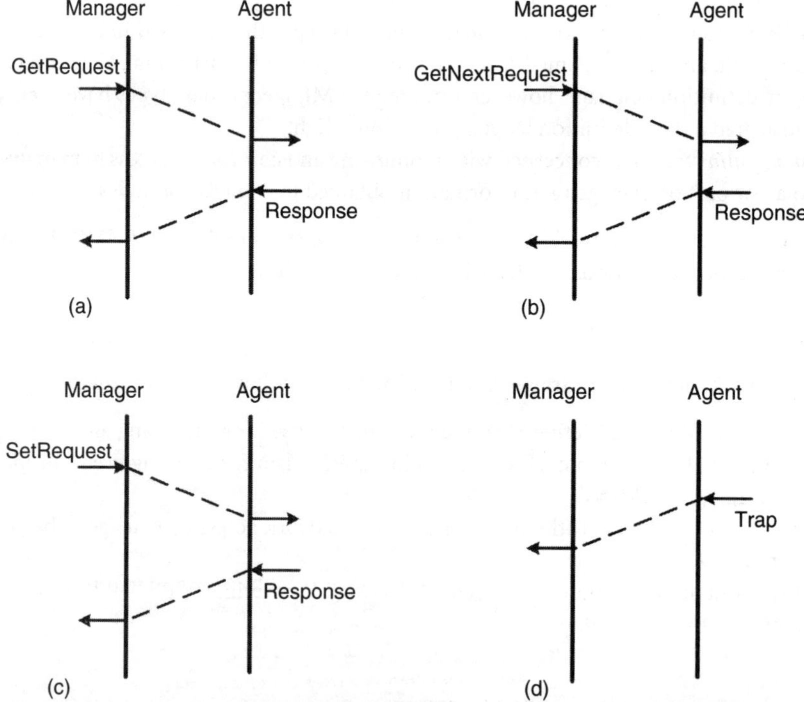

FIGURE 31.3 Initiative from manager (a, b, c) and agent (d).

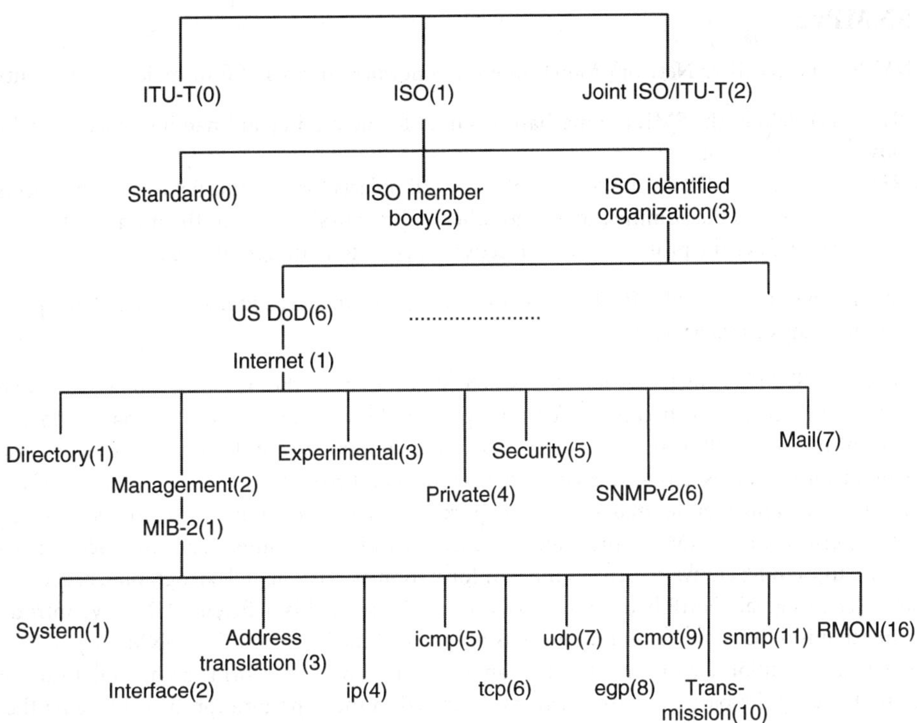

FIGURE 31.4   ASN.1 Object identifier tree.

The global naming tree illustrated in Figure 31.4 allows for unique identification of objects, which correspond to leaf nodes. Describing an object identifier is accomplished by traversing the tree, starting at the root, until the intended object is reached. Several formats can be used to describe an object identifier, with integer values separated by dots being the most common approach.

As shown in Figure 31.4, ISO and the Telecommunication Standardization Sector of the International Telecommunication Union (ITU-T) are at the top of the hierarchy. Under the Internet branch of the tree (1.3.6.1), there are seven categories. Under the management (1.3.6.1.2) and MIB-2 branch (1.3.6.1.2.1) of the object identifier tree, we find the definitions of the standardized MIB modules. The lowest level of the tree shows some of the important hardware-oriented MIB modules (system and interface) as well as modules associated with some of the most important Internet protocols. RFC 2400 lists all standardized MIB modules.

Security

The security capabilities deal with mechanisms to control the access to network resources according to local guidelines so that the network cannot be damaged (intentionally or unintentionally) and persons without appropriate authorization have no access to sensitive information.

SNMPv1 has no security features. For example, it is relatively easy to use the *setRequest* command to corrupt the configuration parameters of a managed device, which in turn could seriously impair network operations. The SNMPv1 framework only allows to assign different access rights to variables (READ-ONLY, READ–WRITE), but performs no authentication. This means that anybody can modify READ–WRITE variables. This is a fundamental weakness in the SNMPv1 framework.

Several proposals have been presented to improve SNMP. In 1992, IETF issued a new standard, the SNMPv2.

The SNMPv2

Like SNMPv1, the SNMPv2 Network Management Framework consists of four major components:

- RFC 1441 defines the SMI, the mechanisms used for describing and naming objects for the purpose of management.
- RFC 1213 defines MIB-II, the core set of managed objects for the Internet suite of protocols.
- RFC 1445 defines the administrative and other architectural aspects of the framework.
- RFC 1448 defines the protocol used for network access to managed objects.

The main achievements of SNMPv2 are the improved performance, better security, and the possibility to build a hierarchy of managers.

Performance: SNMPv1 includes a rule, which states that if the response to a GetRequest or *GetNextRequest* (each of which can ask for multiple variables) would exceed the maximum size of a packet, no information will be returned at all. Because managers cannot determine the size of response packets in advance, they usually take a conservative guess and request just a small amount of data per PDU. To obtain all information, managers are required to issue a large number of consecutive requests. To improve performance, SNMPv2 introduced the *GetBulk* PDU. In comparison with *Get* and *GetNext*, the response to *GetBulk* always returns as much information as possible in lexicographic order.

Security: The original SNMP had no security features. To solve this deficiency, SNMPv2 introduced a security mechanism that is based on the concepts of "parties" and "contexts." The SNMPv2 *party* is a conceptual, virtual execution environment. When an agent or manager performs an action, it does so as a defined party, using the party's environment as described in the configuration files. By using the party concept, an agent can permit one manager to do a certain set of operations (e.g., read, modify), and another manager to do a different set of operations. Each communication session with a different manager can have its own environment. The "context" concept is used to control access to the various parts of an MIB; each context refers to a specific part of an MIB. Contexts may be overlapping and are dynamically configurable, which means that contexts may be created, deleted, or modified during the network's operational phase.

Hierarchy of managers: Practical experience with SNMPv1 protocol showed that in several cases managers are unable to manage more than a few hundred agent systems. The main cause for this restriction is due to the polling nature of SNMPv1. This means that the manager must periodically poll every system under his control, which takes time. To solve this problem, SNMPv2 introduced the so-called "intermediate-level managers" concept, which allows polling to be performed by a number of intermediate-level managers under control of top-level managers via the *InformRequest* command provided by SNMPv2. Figure 31.5 shows an example of hierarchical managers: before the intermediate-level managers start polling, the top-level manager tells the intermediate-level managers which variable must be polled from which agents. Furthermore, the top-level manager informs the intermediate-level manager of the

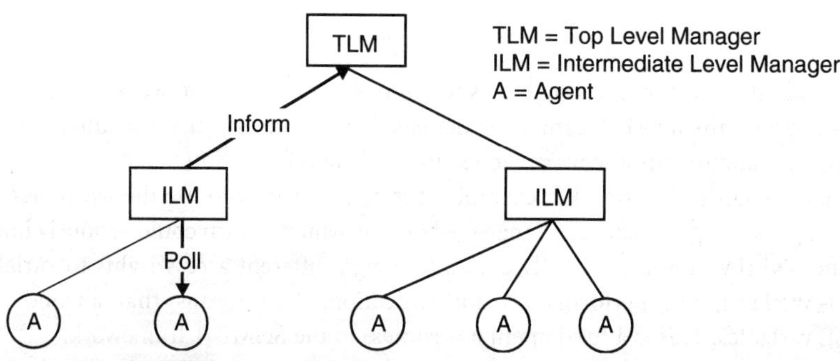

FIGURE 31.5   Hierarchy of managers.

events he wants to be informed about. After the intermediate-level managers are configured, they start polling. If an intermediate-level manager detects an event of interest to the top-level manager, a special *Inform* PDU is generated and sent to the TLM. After receiving this PDU, the TLM directly operates upon the agent, which caused the event.

SNMPv2 dates in 1992, when the IETF formed two working groups to define enhancements to SNMPv1. One of these groups focused on defining security functions, while the other concentrates on defining enhancements to the SNMP protocol. Unfortunately, the group tasked with developing the security enhancements broke into separate camps with diverging views concerning the manner by which security should be implemented. Two proposals (SNMPv2µ and SNMPv2*) for the implementation of encryption and authentication have been issued. Thus, the goal of the SNMPv3 working group was to continue the effort of the disbanded SNMPv2 working group to define a standard for SNMP security and administration.

SNMPv3

The third version of the Simple Network Management Protocol (SNMPv3) was published as proposed standards in RFCs 2271 to 2275, which describe an overall architecture plus specific message structure and security features, but do not define a new SNMP PDU format. This version is based upon the first two versions of SNMP and so it reuses the SNMPv2 standard documents (RFC 1902 to RFC 1908). SNMPv3 can be thought of as SNMPv2 with additional security and administration capabilities [RFC2570]. This section focuses on the management architecture and the security capacities of SNMPv3.

The Management Architecture

The SNMPv3 management architecture is also based on the manager–agent principle. The architecture described in RFC 2271 consists of a distributed, interacting collection of SNMP entities. Each entity implements a part of the SNMP capabilities and may act as an agent, a manager, or a combination of both.

The SNMPv3 working group defines five generic applications for generating and receiving SNMP PDUs: Command generator, command responder, notification originator, notification receiver, and proxy forwarder; a command generator application generates the GetRequest, GetNextRequest, GetBulkRequest and SetRequest PDUs, and handles Response PDUs; a command responder application executes in an agent and receives, processes, and replies to the received getRequest, GetNextRequest, GetBulkRequest, and SetRequest PDUs; a notification originator application also executes within an agent and generates Trap PDUs; a notification receiver accepts and reacts on incoming notifications; and finally a proxy forwarder application forwards request, notification, and response PDUs.

The architecture shown in Figure 31.6 also defines an SNMP engine that consists of four components: dispatcher, Message Processing Subsystem, Security Subsystem, and Access control Subsystem; this SNMP engine is responsible for preparing PDU messages for transmission, extracting PDUs from incoming messages for delivery to the applications, and for carrying out security-related processing of outgoing and incoming messages.

Security

The security capabilities of the SNMPv3 are defined in RFC 2272, RFC 2274, and RFC 2275. These specifications include message processing, user-based security model, and view-based access control model.

The message processing can be used with any security model as follows. For outgoing messages, the message processor is responsible for constructing the message header attached to outgoing PDUs, and for passing the appropriate parameters to the security entity so that it can perform authentication and privacy functions, if required. For incoming messages, the message processor is used for passing the appropriate parameters to the security model for authentication and privacy processing and for processing and removing the message headers of incoming PDUs.

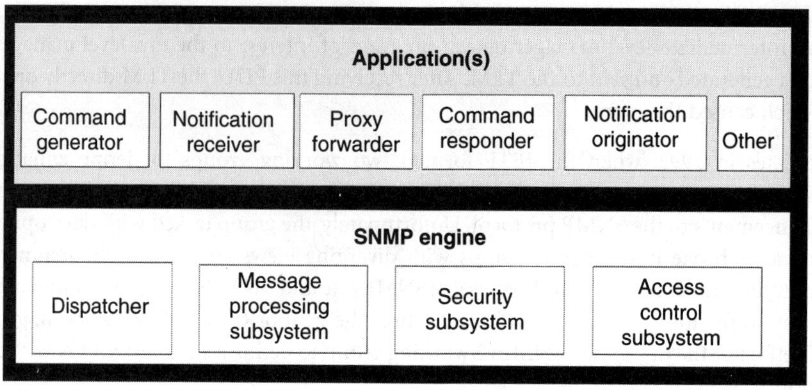

FIGURE 31.6   SNMPv3 entity.

The user-based security model (USM) specified in RFC 2274 uses data encryption standard (DES) for encryption and hashed message authentication codes (HMAC) for authentication [Sch95]. USM includes means for defining procedures by which one SNMP engine obtains information about another SNMP engine, and a key management protocol for defining procedures for key generation, update, and use.

The view-based access control model implements the services required for an access control subsystem [RFC 2275]. It makes an access control decision that is based on the requested resource, the security model, and security level used for communicating the request, the context to which access is requested, the type of access requested, and the actual object for which access is requested.

31.5   ISO and Internet Management Standards: Analysis and Comparison

The purpose of this section is to compare the two different network management frameworks described in the previous sections. This comparison focuses on the four management aspects described above (functional, information, communication, and organization). In particular, the network management protocols (SNMP and CMIP), the MIB and the management functions, management architectures, and security capabilities of these two frameworks are discussed. Possible solutions to some disadvantages are also presented.

SNMP and CMIP

The biggest advantage of SNMP over CMIP is its simple design, which makes it easy to implement and as easy to use on a small network as on a large one. Users can specify the variables to be monitored in a straightforward manner. From a low-level perspective, each variable consists of the following information:

- variable name,
- its data type,
- its access attributes (read-only or read-write), and
- its value.

Another advantage of SNMP is that it is in wide use today around the world. It became very popular; no other network management protocols appeared to replace SNMP. The result of this is that almost all major vendors of network hardware, such as bridges and routers, design their products to support SNMP, making it very easy to implement.

SNMP also has several disadvantages. The first deficiency with SNMP is that it has some large security leaks that can give network intruders access to managed devices. Intruders could also potentially shut down some terminals. To solve this problem, SNMPv2 and SNMPv3 have added some security mechanisms as described above that help combat the security problems.

In comparison with SNMP, CMIP has several advantages. The biggest advantage of CMIP is that an agent not only relays information to and from a terminal as in SNMP, but in CMIP an agent can perform management functions on its own instead of restricting to gather information for remote processing by a manager. Another advantage of the CMIP approach is that it addresses many of the shortcomings of SNMP. For instance, it has built-in security management devices that support authorization, access control, and security logs. The result of this is a safer system from the beginning; no security upgrades are necessary.

CMIP has many advantages, but has not been implemented yet. One problem of the CMIP protocol is that it needs more system resources than SNMP. Furthermore, a full implementation of CMIP requires to add additional processes to network elements. One possible workaround is to decrease the size of the protocol by changing its specifications. Another problem with CMIP is that it is very difficult to program.

MIBs and SMI

MIB and SMI represent the information aspects of a network management framework. In the ISO management framework, managed objects within the MIB are complex and have sophisticated data structures with three attributes: variable attributes that represent the variable characteristics, variable behaviors that define what actions can be triggered on that variable, and notifications that generate an event report whenever a specific event occurs [Sta99]. In contrast, in the SNMP framework, variables are only used to relay information to and from managers.

The SNMP MIB concept has two important disadvantages. The first one is that the user has to know the names and meanings of (thousands of) different variables, which can be a daunting task. The second problem is the lack of variable aggregation: when the user wants to inquire the values contained in an array, he has to ask separately for each element instead of naming the array at once.

The latter problem has been fixed in the newer releases of SNMP, the SNMPv2, and SNMPv3. These versions provide means for aggregating variables, for example, by the new *GetBulkRequest* service. In fact, so many new features have been added that the formal specifications for SNMP MIBs have expanded considerably.

Network Management Functions

One advantage of ISO management framework over the IETF framework is that ISO has issued the five specific management functional areas, which are useful for users to develop management applications. In contrast, the IETF management framework has not defined any specific functions and procedure to do, for example, performance management. These have to be provided entirely by the user. In fact, the IETF management standards explain how individual management operations should be performed; but they do not specify, however, the sequence in which these operations should be carried out to solve particular management problems.

Conclusions

We have surveyed the management nonent architecture, protocols, services, and management information base of the network management frameworks standardized by ISO and IETF. Within each of these frameworks, standards relating to four fundamental aspects of network management (functional aspect, information aspect, communication aspect, and organization aspect) are addressed.

Both management frameworks have their advantages and disadvantages. However, the key decision factor between the two frameworks is in their implementation: it is until now almost impossible to find a system with the necessary resources to support the ISO framework, although it is conceptually superior to SNMP (v1, v2 and v3) in both design and operation.

References

[ASN90] ISO/IEC 8824, Specification of Abstract Syntax Notation One (ASN.1), April 1990.

[BCW90] A. Ben-Artzi, A. Chandna, and U. Warrier, Network management of TCP/IP networks: Present and Future, *IEEE Network Magazine*, Vol. 4, pp. 35–43, July 1990.

[ISO7498-4] ITU-T - ISO/IEC. ITU-T X.700 -ISO/IEC 7498-4, Information Processing Systems — Open Systems Interconnection — Management Framework for Open System Interconnection, 1992.

[ISO9595] ISO 9595, Information Processing Systems — Open Systems Interconnection — Common Management Information Service Definition, Geneva, 1990.

[ISO9596] ISO 9596, Information Processing Systems — Open Systems Interconnection — Common Management Information Protocol, Geneva, 1991.

[ISO10040] ITU-T -ISO/IEC. ITU-T X.701 - ISO/IEC 10040, Information Processing Systems — Open System Interconnection — System Management Overview, 1992.

[ISO10165-1] ITU-T - ISO/IEC. ITU-T X.720 - ISO/IEC 10165-1, Information Processing Systems — Open Systems Interconnection — Structure of Management Information: Management information model, Geneva, 1993.

[ISO10165-2] ISO 10165-2, Information Processing Systems — Open Systems Interconnection — Structure of Management Information: Definition of Management Information, Geneva, 1993.

[ISO10165-4] ISO 10165-4, Information Processing Systems — Open Systems Interconnection — Structure of Management Information — Part 4: Guidelines for the definition of Managed Objects, Geneva, 1993.

[ISO10165-5] ISO 10165-5, Information Processing Systems — Open Systems Interconnection — Structure of Management Information: Generic Management Information, Geneva, 1993.

[ISO10165-7] ISO 10165-7, Information Processing Systems — Open Systems Interconnection — Structure of Management Information: General Relationship Model, Geneva, 1993.

[KR01] J. F. Kurose and K. W. Ross, *Computer Networking — A Top-Down Approach Featuring the Internet*, Addison-Wesley, Reading, MA, 2001.

[RFC1066] K. McCloghrie and M. Rose, Management Information Base for Network Management of TCP/IP-based Internets, RFC 1066, 1998.

[RFC1155] K. McCloghrie and M. Rose, Structure and Identification of Management Information for TCP/IP-based Internets, RFC 1155, 1990.

[RFC1157] J. Case, M. Fedor, M. Schofftall, and C. Davin, The Simple Network Management Protocol, RFC 1157, May 90.

[RFC1212] K. McCloghrie and M. Rose, Concise MIB Definitions, RFC 1212, 1991.

[RFC1213] K. McCloghrie and M. Rose, Management Information Base for Network Management of TCP/IP-based Internets: MIB-II, RFC 1213, 1991.

[RFC1215] M. Rose, A Convention for Defining Traps for use with the SNMP, RFC 1215, 1991.

[RFC1441] K. McCloghrie, M. Rose, J. Case and S. Waldbusser, Introduction to Version 2 of the Internet-standard Network Management Framework, RFC 1441, 1993.

[RFC1445] J. Galvin and K. McCloghrie, Administrative Model for Version 2 of the Simple Network Management Protocol (SNMPv2), RFC 1445, 1993.

[RFC1448] K. McCloghrie, M. Rose, J. Case, and S. Waldbusser, Protocol Operations for version 2 of the Simple Network Management Protocol (SNMPv2), RFC 1448, 1993.

[RFC1902] J. Case, K. McCloghrie, M. Rose, and S. Waldbusser, Structure of Management Information for Version 2 of Simple Network Management Protocol (SNMPv2), RFC 1902, January 1996.

[RFC2271] D. Harrington, R. Presuhn, and B. Wijnen. An Architecture for Describing SNMP Management Frameworks, RFC 2271, 1998.

[RFC2272] J. Case, D. Harrington, R. Presuhn, and B. Wijnen, Message Processing and Dispatching for the Simple Network Management Protocol (SNMP), RFC 2272, 1998.

[RFC2273] D. Levi, P. Meyer, and B. Stewart, SNMPv3 Applications, RFC 2273, 1998.

[RFC2274] U. Blumenthal and B. Wijnen, User-based Security Model (USM) for Version 3 of the Simple Network Management Protocol (SNMPv3), RFC 2274, 1998.

[RFC2275] B. Wijnen, R. Presuhn, and K. McCloghrie, View-based Access Control Model (VACM) for the Simple Network Management Protocol (SNMP), RFC 2275, 1998.

[RFC3411] B. Wijnen, R. Presuhn, and K. McCloghrie, View-based Access Control Model (VACM) for the Simple Network Management Protocol (SNMP), RFC 3411, 2002.

[RFC3415] B. Wijnen, R. Presuhn, and K. McCloghrie, View-based Access Control Model (VACM) for the Simple Network Management Protocol (SNMP), RFC 3415, 2002.

[Ros90] M. T. Rose, *The Open Book. A Practical Perspective on OSI*, Prentice-Hall, Englewood Cliffs, NJ, 1990.

[Sch95] B. Schneider, *Applied Cryptography. Protocols, Algorithms, and Source Code in C*, John Wiley, New York, 1995.

[Sta93a] W. Stallings, *Networking Standards — A Guide to OSI, ISDN, LAN, and MAN Standards*, Addison-Wesley, Reading, MA, 1993.

[Sta93b] W. Stallings, *SNMP, SNMPv2 and CMIP — The Practical Guide to Network Management Standards*, Addison-Wesley, Reading, MA, 1993.

[Sta99] W. Stallings, *SNMP, SNMPv2, SNMPv3 and RMON 1 and 2*, Addison-Wesley, Reading, MA, 1999.

32

Simple Network Management Protocol SNMP

Michael Kunes
FREQUENTIS GmbH

32.1 Introduction

The term "management application" can describe any solution ranging from a cyclically invoked command line tool to a fully integrated, multi-user, and operating system-independent system platform with a graphical user interface. Probably, the most widely used tool for network management is "ping." It provides an easy test if a device is operational and can be reached. But it is limited to exactly that one purpose. Another well-known application to remotely manage network nodes is "telnet." While quite powerful to control a small number of nodes, telnet is highly dependent on the target node. These two samples are standard solutions for network management that are widely used but are not feasible to handle large and diverse networks. Others exist using proprietary solutions or various other open standard protocols.

If a central management system shall gather data about a large and heterogeneous network, a common protocol is needed. At least it should be possible to read data from all the network devices and to trigger actions at these nodes. In addition, the solution should have a method to determine which data is accessible at a remote device. All these requirements have been well addressed by SNMP. The developed standards had evolved through several generations. All these versions of SNMP will be reviewed in this paper.

The remainder of this publication starts with a short historical review followed by a description of the different SNMP versions.

32.2 History

Reviewing the Internet timeline (Figure 32.1) from the first four hosts that formed the core of the ARPA net in the year 1969 [1], one can easily see that the development of SNMP started immediately after data transfer (ftp), remote control of hosts (telnet) and the mailing system protocol (SMTP) were addressed. In 1987, the Simple Gateway Monitoring Protocol SGMP was specified [RFC1028] — it was the predecessor of SNMP. One year later, the first RFCs describing SNMP were made public and shortly afterwards were reviewed and released as SNMPv1.

According to official host counts [1], the number of hosts in the Internet was still quite small in 1990 (some 300.000) but it became clear that a standardized, easy-to-use solution that can be integrated into each and every network device is necessary to manage the fast-growing heterogeneous networks. At this time, only proprietary solutions if any existed for management tasks, mainly based on UNIX. The needs of an approach for a common management solution were filled by SNMPv1, followed by the development of tools having different complexities ranging from simple command-line tools up to integrated management platforms.

In the beginning, SNMP was only intended as a short-term solution until specification of the OSI management framework [2] would be finished and used for all network management tasks. But history has shown that SNMP nowadays not only still exists but can also be found in nearly every network device, while the OSI framework is mostly used for academic purposes. This resembles the situation of OSI and TCP/IP. The phrase "OSI is a beautiful dream and TCP/IP is living it" clearly describes this situation [3].

Only 3 years after SNMPv1, a new version of SNMPv2 was released that should improve security, speed and configuration capabilities. The Internet now had grown to some 1.5 million hosts and mixed-technology environments could be found everywhere. SNMPv2 added new protocol operations, extended the language used for the description of managed objects and addressed the area of security. At this time, the acceptance for the use of SNMP as a common management protocol had reached the

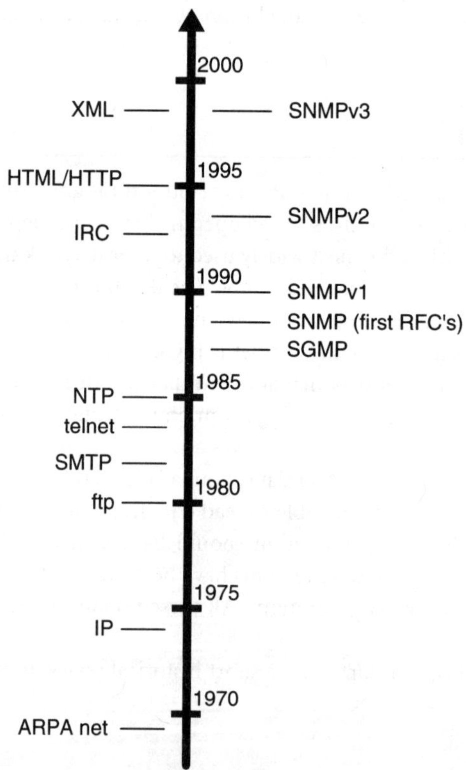

FIGURE 32.1   Internet timeline.

critical mass — standard extensions, integration into different device types, development of management solutions, etc. had broad support.

A complete new set of RFCs was released in 1998, forming SNMPv3. This version redefined security aspects, clarified the structure of the management framework and split the specification into smaller portions that could be replaced or adapted for future needs.

Until now, SNMPv1 still has a large market share while SNMPv2 and SNMPv3 only stepwise begin to take over. It is expected that future networks will consist of a mixture of devices, supporting different types of SNMP. But since the main structure has remained the same, integration is possible and supported by management solutions. Between [RFC1] and [RFC3400], more than 160 RFCs were released for SNMP most of them describing management information bases (about 130 RFCs), while the remainder addresses general parts of the management framework.

32.3   SNMP v1

Many definitions about the components of the SNMP management framework exist. According to a good definition in [RFC 2742], it consists of the following five major components:

- Overall architecture.
- Mechanisms for describing and naming objects and events for the purpose of management.
- Message protocols for transferring management information.
- Protocol operations (PDUs) for accessing management information.
- Set of fundamental applications and the view-based access control mechanism.

The main design goals for SNMPv1 included the following:

- Simplicity (the integration of the protocol in network devices has to be easy).
- Robustness (when everything else fails, network management still has to be possible).
- Low overhead.
- Minimal network resource requirements.

Simplicity also included limiting the complexity of the specification. Thus, the core of SNMPv1 consisted of just three RFCs, ([RFC1155], [RFC1157] and [RFC1213]), refined by some additional definitions ([RFC1212], [RFC1215]). These documents mainly define the *network structure* between managed nodes and managers, a *communication protocol*, a *data encoding method* and a *specification language* for defining information that may be exchanged between systems as well as a *base set of management information* to be supported.

Network Structure

First and foremost, SNMP defines the correlation between network entities that exchange management information. An *agent* provides information about its network node. The *manager* requests data from a number of agents and helps the network administrators accomplish their work. In other words, the agent represents a server and the manager acts as a client. The base relationship is always a 1:n without intermediate elements (Figure 32.2) or a communication between managers. If a network node does not support SNMP, a *proxy* can act as a gateway.

MIB Definition/SMIv1

An SNMP MIB[1] is the database of the agent. All values that can be requested by the manager are accessed through the MIB, which is arranged in a tree-like manner similar to a directory structure of files on a disk.

---

[1]Some explanation about the term "MIB": the name MIB is used either for all data (the whole tree) that is accessible from an agent or some subtree that is specified by some organization or company.

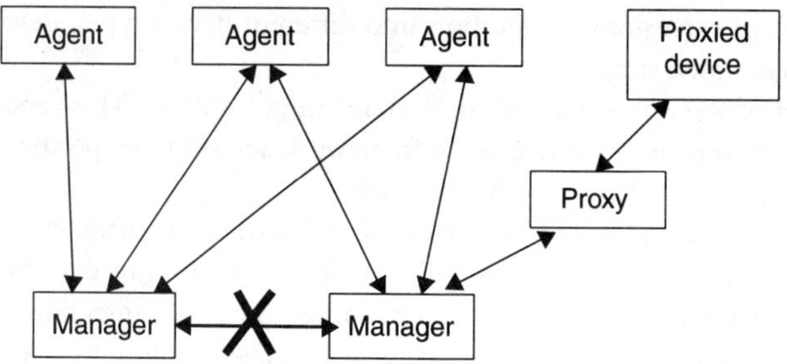

FIGURE 32.2   Manager/agent relationship.

Data is only stored in the leaf elements, never in the intermediated branches. Note that the structure of the MIB is usually fixed — it is constructed by the NW equipment vendor and cannot be added to or modified. The structure and the data provided by a MIB can be seen as a data container; the evaluation of the values depends on the manager application. The huge amount of already defined and supported MIBs can be seen as the most valuable part of SNMP, often defending its existence against alternate network management approaches.

In most cases, MIBs describe simple data containers while complex structures are possible too. One example of a complex MIB structure is the RMON MIB [RFC1271], which defines the interface for a network probe supervising a network on physical and data link layer.

SMI Overview

The Structure of Management Information (SMI, [RFC1155]), is the definition of how managed objects shall be described. By using the SMI, all MIBs are specified in a standardized way. The SMI uses a very small subset of ASN.1 [4] to specify how a MIB has to be defined. Only the following base data types are specified for use with SNMP:

- INTEGER
- OCTET STRING
- OBJECT IDENTIFIER (OID)

Based on these data types, SNMP derives some additional data types like "IpAddress," "Counter," "Gauge," "TimeTicks," etc.

Defining a new MIB module, only the above-mentioned data types can be used. In addition, new types can be derived by defining enumerations and by specification of the value range. Some possibilities are:

- Range for a numeric value: SYNTAX INTEGER (0..65535)
- String of a defined length: SYNTAX OCTET STRING (SIZE (0..255))
- Enumeration of values: SYNTAX INTEGER { enabled(1), disabled(2) }

No complex data types as known by programming languages (structures, objects, unions, etc.) can be defined. Some conventions for the representation of floating-point values, dates, etc., have been established [5].

Using the base data types, only discrete objects and two-dimensional tables can be defined. If the need for more complex structures arises, they have to be mapped to these basic structures. For example, if the MIB designer needs a three-dimensional table, he breaks it apart into several two-dimensional tables thus following the rules of the SMI.

For the specification of the MIB structure and for accessing the values within, a textual representation is used as a human-readable form together with a dotted numeric representation, which is used internally. Both forms are usually combined in the interface of network management tools. Figure 32.3 shows an example.

The ".0" at the end of each line suggests that the accessed object is a discrete object. All discrete objects are accessed by appending a ".0" to the MIB-branch. Appending an INDEX as specified in the table definition accesses the values of a table. This can either be a simple numeric index or even a complex structure like the combination of IP-address and port number appended to the MIB-branch.

Intermediate Objects (OBJECT IDENTIFIER Macro)

For the definition of intermediate MIB branches, an OBJECT IDENTIFIER has to be introduced. This specifies the parent object and the name of a new node in the MIB. Figure 32.4 shows examples for intermediate tree elements taken from MIB-II [RFC1213].

Leaf Objects (OBJECT-TYPE Macro)

The definition of MIB leaf objects is based on the OBJECT-TYPE macro [RFC1212]. Figure 32.5 shows a sample for a discrete object taken from the MIB-II. The object "sysDescr" is a the child of the node "system," under the sub tree "1" and has the following properties:

SYNTAX: The syntax of the object (i.e., the data type) is the type "DisplayString." The value must not be longer than 255 characters.

ACCESS: one of the values *read-only, read-write, write-only* or *not-accessible.*

STATUS: one of the values *mandatory, optional* or *obsolete.* The shown MIB variable is mandatory; thus, the agent must provide it.

DESCRIPTION: a human readable text describing the contents and purpose of the MIB object.

The sample shows the simple structure for the definition of MIB objects. Several conventions have evolved over the years, which increase the complexity of SNMP. These conventions do not have to be followed but make the use more convenient. For more information, see [5]. The sample in Figure 32.5 also reveals that much of the information is in the DESCRIPTION part that is only human readable and not machine-parsable text — one of the criticisms regarding the SMI. The machine-understandable components are very few.

```
Textual form:    Iso.org.dod.internet.mgmt.mib-2.system.sysUptime.0
Dotted decimal:  1.3.6.1.2.1.1.3.0
Combined form:   Iso(1).org(3).dod(6).internet(1).mgmt(2).mib-2(1).system(1).
sysUptime(3).0
```

FIGURE 32.3   SMI sample.

```
mgmt      OBJECT IDENTIFIER ::= {internet 2}
mib-2     OBJECT IDENTIFIER ::= {mgmt 1}
system    OBJECT IDENTIFIER ::= {mib-2 1}
```

FIGURE 32.4   Object identifiers.

```
sysDescr OBJECT-TYPE
    SYNTAX  DisplayString (SIZE (0..255))
    ACCESS  read-only
    STATUS  mandatory
    DESCRIPTION
        "A textual description of the entity. This value
        should include the full name and version
        identification of the system's hardware type,
        software operating-system, and networking
        software. It is mandatory that this only contain
        printable ASCII characters."
    ::= { system 1 }
```

FIGURE 32.5   SMI sample: discrete object.

Table Definitions

The definition of a table is also based on the OBJECT-TYPE macro but is a little more complex than a discrete object. The table columns are defined exactly the same way as a discrete object. To form the table, a SEQUENCE of the columns must be specified to define an entry of the table together with an INDEX clause. This object in turn is used for another SEQUENCE forming the whole table. Figure 32.6 shows these correlations: the *ifTable* object uses a list of *ifEntry* elements to form the table rows. The *ifEntry* element uses a list of discrete objects to define the columns of the table, specifying that the value of *ifIndex* be used to access a row (the index of the table).

The sample also shows some of the mentioned conventions. For example, all table objects start with the same prefix, the table itself uses the suffix "Table" while the entry has the suffix "Entry."

Trap Definitions

The TRAP-TYPE macro as defined in [RFC1215] is used to specify an SNMP trap. Figure 32.7 shows a sample of a trap definition macro with the following fields:

- The value for the enterprise field of the PDU. This part of the message contains an object identifier with the value of sysObjectID to identify the type of hardware that sent the trap.

```
ifTable OBJECT-TYPE                        IfIndex OBJECT-TYPE
    SYNTAX   SEQUENCE OF IfEntry               SYNTAX   INTEGER
    ACCESS   not-accessible                    ACCESS   read-only
    STATUS   mandatory                         STATUS   mandatory
    DESCRIPTION "..."                          DESCRIPTION
  ::= { interfaces 2 }                         "..."
                                             ::= { ifEntry 1 }
ifEntry OBJECT-TYPE
    SYNTAX   IfEntry                        ifSpeed OBJECT-TYPE
    ACCESS   not-accessible                    SYNTAX   Gauge
    STATUS   mandatory                         ACCESS   read-only
    DESCRIPTION "..."                          STATUS   mandatory
    INDEX    { ifIndex }                       DESCRIPTION "..."
  ::= { ifTable 1 }                          ::= { ifEntry 5 }

IfEntry ::=
    SEQUENCE {
    ifIndex
        INTEGER,
    ifSpeed
        Gauge,
    ...
    }
```

FIGURE 32.6   SMI sample: table.

```
linkDown TRAP-TYPE
    ENTERPRISE   snmp
    VARIABLES    { ifIndex }
    DESCRIPTION "A linkDown trap signifies that the sending
        protocol entity recognizes a failure in one of
        the communication links represented in the
        agent's configuration."
    ::= 2
```

FIGURE 32.7   SMI sample: trap.

- Which variables (if any) will be sent as payload of the trap. In the sample, the interface index "ifIndex" will be sent along with the trap.
- A DESCRIPTION clause with human readable text describing the trap.
- The optional REFERENCE clause used for cross-references between traps is not used in the shown definition.

Protocol Operations

SNMPv1 provides just a minimal number of protocol operations. It is a simple request–response scheme between manager and agent. All requests are encoded by the rule of basic encoding rules (BER [6]).

UDP was selected as transport protocol because of low overhead, low bandwidth consumption and high probability to work in faulty networks. Although some of these design issues were revisited and found to be futile later, UDP is still the transport protocol of choice (Figure 32.8).

The simplicity aspect of SNMPv1 was addressed by the low number of different protocol operations supported by the protocol (Figure 32.9). By issuing GET and SET requests, a manager reads or writes variables from/to the agent. Each request can contain either one variable or a collection of variables. The GET-NEXT PDU is similar to the GET request but the agent has to send the next variable that can be found in the MIB tree according to the lexigraphical order. That method is an elegant way to traverse whole MIB subtrees without deeper knowledge about their structure issuing GETNEXT requests in a loop. It must be mentioned that a GETNEXT request can contain any subtree, while a GET/SET operation always has to fully qualify a MIB variable. All the listed requests are answered by a GETRESPONSE PDU.

Finally, the agent can send an asynchronous notification to the manager using the TRAP PDU (e.g., if an error occurred). Six predefined traps like *coldStart* (indication the agent reinitializing itself) or *linkDown* (failure in one of the communication links) were specified by [RFC1157]. In addition, every agent can define new ones.

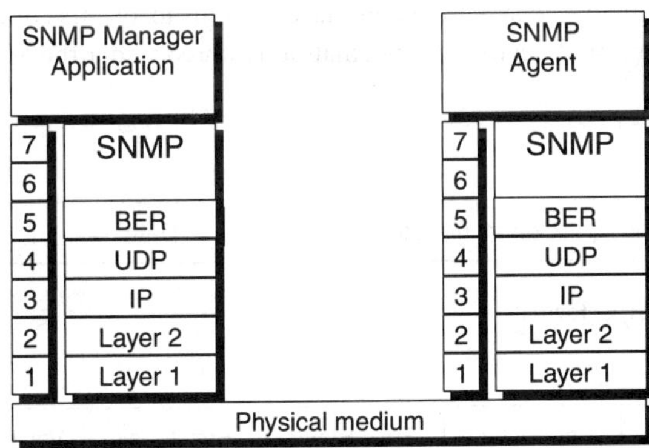

FIGURE 32.8 SNMP protocol stack.

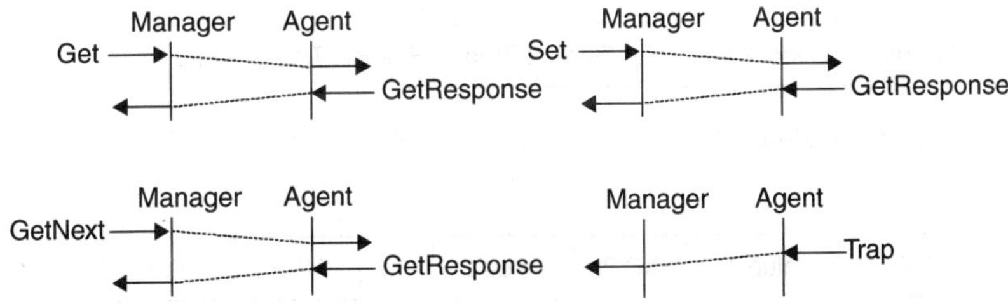

FIGURE 32.9 SNMPv1 protocol operations.

The SNMP message format (Figure 32.10) consists of three parts: the *version information* for SNMPv1 has to be set to the value "0." The *community string* is used for a minimal authentication scheme (See Security Section). The *data* part depends on the PDU type.

All requests except the TRAP PDU share the same structure. The PDU *type* field is used to distinguish the different request types. The *request ID* gives the manager application a possibility for the identification of specific PDUs that he sends. This value is placed inside the PDU by the manager and inserted without change in the same field of the GETRESPONSE PDU by the agent (Figure 32.11). The fields *ErrorStatus* and *ErrorIndex* are filled by the agent to indicate problems answering the request. For the other request types, these values have to be 0. The *VarBind* list of the message is the payload of the PDU and will be described later.

The TRAP message has a special PDU structure (Figure 32.12). It also starts with a *type* field, followed by the *Enterprise* and *Agent Address* fields containing an object ID (OID) identifying the vendor of the equipment that sends the trap and the network address of the device. The *Generic* and *Specific* fields identify why the trap was sent. The *TimeStamp* contains the time since startup of the agent, while the *VarBind* list can contain additional information about the trap and its cause.

One of the intentions at the design time of SNMPv1 was a mechanism called "trap directed polling," where an agent sends a trap to indicate a problem to the manager, which in turn starts to poll the agent for further information. Instead, simple polling approaches are mostly used in reality.

The VarBind fields of all PDU types always contain the payload of the messages. Each VarBind consists of any number of variable bindings. The format is shown in Figure 32.13.

The maximum number of varbinds that can be transported in one request is not specified in SNMP but can be implementation specific for each agent. The only assumption adopted by [RFC1157] was that each agent must be capable of handling a PDU with a size of 484 bytes. It can be expected that most available agents exceed this size by far.

MIB-II

The first standardized set of MIB variables was defined by [RFC1213]. This extensive MIB contains general information about the managed system, its interfaces and protocols. Its content is focused on layers 1–3 information of the monitored device. All information is stored under the MIB subtree

```
Iso(1).org(3).dod(6).internet(1).mgmt.mib-2
```

Version	Community string	Data

FIGURE 32.10 SNMPv1 message format.

PDU Type	Request ID	Error Status	ErrorIndex	VarBind list

FIGURE 32.11 Data part of the Get, GetNext, Set, GetResponse PDUs.

PDU Type	Enterprise	Agent Address	Generic Trap	Specific Trap	Time Stamp	VarBind list

FIGURE 32.12 Data part of the Trap PDU.

OID 1	Value 1	OID 2	Value 2	...	OID n	Value n

FIGURE 32.13 Varbind format.

Group	Mandatory	Description
System	Yes	General information about the network element (e.g. up-time, system location, ...)
interfaces	Yes	Interface specific data (e.g.: operative status, number of bad packets, ...)
At	n.a.	Deprecated part of the MIB
ip, icmp	Both	Network layer information (IP and ICMP protocols)
tcp, udp	If protocol is supported by the device	Transport layer information (UDP and TCP protocols)
Egp	If protocol is supported by the device	Exterior gateway protocol data
transmission	n.a.	Used as a "placeholder" for future MIB about lower network layers (e.g.: Ethernet MIB, Token Ring, FDDI, ...)
Snmp	Yes	Data about the SNMP stack

FIGURE 32.14   MIB-II groups.

The MIB-II is divided into the groups listed in Figure 32.14. Most of these groups are mandatory for devices supporting SNMP — at least if the device supports the mentioned protocols. But the existence of SNMP implies by itself the support of IP, UDP and SNMP. As can be seen by the above list, nearly all groups should be found at any SNMP-enabled device, if the device vendor does not decide to ignore this RFC requirement.

Due to the broad support of the MIB-II by device vendors, each and every management solution should be able to cope with the data provided. The correlation of different values (e.g., correlation of sent packets with discarded packets over a certain amount of time) can help the management staff in tracking down network problems [7, 8].

Security

The community string contained in every PDU is the "security approach" of SNMPv1. But since it is not encrypted and is rather sent as plain text together with the remaining PDU, it can be easily detected by a network sniffer. The only useful function of this field is to give different views of a MIB in a closed environment.

More intrinsic approaches for security have been addressed in SNMPv2 and v3.

Conclusion

SNMPv1 is an easy-to-understand protocol with a broad hardware and software base. Several different solutions either for management applications or for agent implementations do exist. The resource requirements are quite low making the implementation of an SNMP agent feasible even in primitive devices. The SMI provides a formal description of the agent interface that can be used by manager applications. Several standard MIB definitions already exist, together with a countless number of vendor specific MIBs.

SNMPv1 must be criticized for the bad security concept and the waste of bandwidth caused by the PDU format and the polling mechanism that is mostly used.

32.4   SNMP v2

In April 1993, after considerable work by the Network Working Group, the specification of the Simple Network Management Protocol Version 2 (SNMPv2) was released in [RFC1441] to [RFC1452]. Some of these specifications were later marked as historic, while others, like the security model, were reviewed and extended by new specifications [RFC1901] to [RFC1910] released in January 1996.

While SNMPv1 was intended as a quick solution handling a customer's most pressing problems, SNMPv2 should improve its predecessor, especially in the areas of security, capabilities, MIB definition and clarity of the protocol definition. To achieve these goals, new PDUs were defined, the SMI was extended to SMIv2 and several security models were introduced. Unfortunately, indicated by the huge amount of RFCs involved, all these extensions made the specification by far more complex, too.

While the schemes underlying SNMPv1 and SNMPv2 are more or less the same, the two protocols are not compatible. To provide compatibility, agents as well as manager applications have to be bilingual. For easy migration, [RFC1452] defines all steps to convert an SNMPv1 entity to an SNMPv2 entity.

Network Structure

While SNMPv1 defined a clear relationship between a manager and a number of agents, this definition was weakened in SNMPv2, now speaking of "entities." Each entity can take both the roles of a manager and an agent, making even dual role functions possible. Which role is supported by an entity can be distinguished by the supported protocol operations.

An *agent* can generate TRAP and RESPONSE PDUs and must process GET, GETNEXT, GETBULK and SET requests. A *manager* can generate GET, GETNEXT, GETBULK, SET, INFORM and RESPONSE PDUs, while it must process RESPONSE, TRAP and INFORM requests [8, p353].

Protocol Operations

While SNMPv1 is completely based on UDP as a transport protocol, SNMPv2 supports different transport protocol mappings like Novell IPX, AppleTalk, etc [RFC1449].

The overall message format remained the same, but details like the possible states for the field "Error Status" (13 additional error codes providing a better distinction between error conditions) were extended. Also, two new PDUs were introduced: GETBULK for optimized retrieval of larger amounts of data and INFORM for manager-to-manager communications (Figure 32.15).

GETBULK-Request

The GETBULK-Request is a more general form of GETNEXT. It can be used to retrieve table data easier and with less protocol overhead (Figure 32.16).

Using the GETNEXT operation, the lexigraphical successor for each element in the VarBind list is sent back to the requesting application by the RESPONSE message. For each requested element, exactly one element is returned. If the requested data is organized in a table, this method imposes considerable overhead.

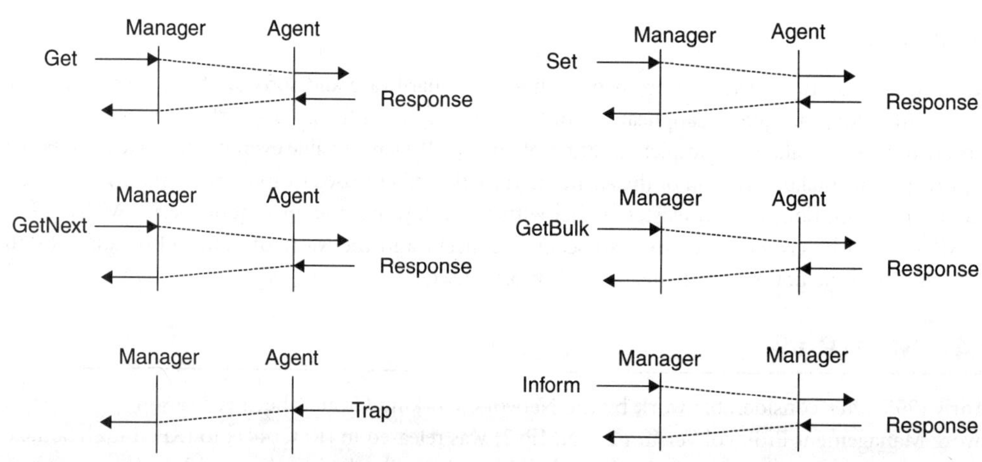

FIGURE 32.15   SNMPv2 PDUs.

PDU Type	Request ID	Nonrepeaters	Max. repetition	VarBind list

FIGURE 32.16 Format of GetBulk PDU.

The GETBULK-Request contains two counters together with the VarBind list: the field *Nonrepeaters* specifies the number of elements in the Varbind list that shall be treated like elements of a GETNEXT request. The first elements up to *Nonrepeaters* each return one field in the RESPONSE PDU. For the remainder of the Varbind list, the *Max. repetition field* specifies the number of GETNEXT requests that shall be issued on each of these elements.

For example, for a Varbind list with six elements, the field *Nonrepeaters* set to 2 and *Max. repetition* set to 3, the response will contain 14 elements: two elements for the *Nonrepeaters* section and three times 4 consecutive elements for the *Max. repetition* part.

Using the GETBULK a manager can retrieve as many variables (e.g., from a large table) as will fit in a RESPONSE without knowing the exact OIDs of all the variables. Tests have shown that the use of this type of request can speed up the performance of an application significantly. In addition, due to the basic similarity to the GETNEXT operation, the extensions for an existing agent implementation are naturally not large.

TRAP PDU

Several improvements were applied to the TRAP PDU, while the number of traps in the management information base for SNMPv2 did not extend the SNMPv1 definitions [RFC1450]. The structure of the TRAP PDU was redefined in SNMPv2. Its format now is identical to all other SNMP PDUs. Furthermore, it is specified that the first elements of the Varbind list have to contain certain values [RFC1448]:

- OID1   ... has to contain the actual value of the sysUpTime MIB variable.
- OID2   ... has to contain the value of snmpTrapOid.0. This is an OID uniquely identifying each trap.

All other elements of the VarBind list depend on the TRAP definition. See that the SNMPv1 method of using two fields (PDU parts "Generic" and "Specific") to identify a trap was replaced by the use of a unique object identifier (OID2). This approach reduces the problem of different agents on different types of hardware using the same values for trap identification.

INFORM Request

The INFORM Request is intended for communications between two entities in the role of a management application. While this PDU is sometimes referred to as a "confirmed trap," it shall be mentioned that it is not intended for use by agents but rather for manager–manager interaction. It is used in situations where one manager has delegated some responsibility to another manager and wishes to be notified about specific conditions. Like the Trap PDU, the first two Varbind fields are predefined [RFC1448]:

- OID1   ... has to contain the actual value of the sysUpTime.0 MIB variable.
- OID2   ... has to contain the value of snmpTrapOid.0 (the authoritative identification of the event type).

SMIv2

The definition of the structure of management information SMI was substantially extended to SMIv2. Again, ASN.1 was used as the base for the formal description notation, but new data types were introduced and a large number of macro definitions were established.

The new data types are Bit String, Counter64, NsapAddress and Uinteger32. Some of the data types of SMIv1 were renamed (e.g., Integer to Integer32).

For the definition of MIB modules, the following macro definitions were introduced:

- MODULE-IDENTITY MACRO
- OBJECT-IDENTITY MACRO

- OBJECT-TYPE MACRO
- NOTIFICATION-TYPE
- TEXTUAL CONVENTION
- OBJECT-GROUP
- MODULE-COMPLIANCE
- AGENT-CAPABILITIES.

All these macros will be reviewed in the remainder of this section.

MODULE-IDENTITY MACRO

This macro has to be provided exactly once in every information module. It is used for the description of the MIB, providing contact information, revision history and a general description of the contents of the information module. The sample in Figure 32.17 is taken from [RFC1450] and shows all elements of the macro except the optional REVISION part.

OBJECT-IDENTITY MACRO

This macro is used to assign an OID value to an identifier in the MIB module. It contains the following fields (Figure 32.18):

- STATUS: one of the values *current* or *obsolete*.
- DESCRIPTION: a human readable text describing the contents and purpose of the MIB object.
- REFERENCE: a textual cross-reference to an object assignment in another information module. The field is optional.

The difference between OBJECT-IDENTIFIER and OBJECT-IDENTITY is that the OBJECT-IDENTITY performs a registration of an OID value, thus making the described element of the MIB unique for all time and space. In contrast, an OBJECT-IDENTIFIER is just a reference to a value of the MIB. As

```
snmpMIB MODULE-IDENTITY
    LAST-UPDATED "9304010000Z"
    ORGANIZATION "IETF SNMPv2 Working Group"
    CONTACT-INFO
        "Marshall T. Rose
        Postal: Dover Beach Consulting, Inc.
        420 Whisman court
        Mountain View, CA 94043-2186
        US

        Tel: +1 415 968 1052
        Fax: +1 415 968 2510

        E-mail: mrose@dbc.mtview.ca.us"
    DESCRIPTION
        "The MIB module for SNMPv2 entities."
    ::= { snmpModules 1 }
```

FIGURE 32.17   MODULE-IDENTITY macro sample.

```
fizbin69 OBJECT-IDENTITY
    STATUS  current
    DESCRIPTION "The authoritative identity of the
        Fizbin 69 chipset."
    ::= { fizbinChipSets 1 }
```

FIGURE 32.18   OBJECT-IDENTITY macro sample.

a consequence, several OBJECT- IDENTIFIERs can have the same assignment, while no other element may be registered with the same OID if an OBJECT-IDENTITY already exists [5]. A MIB compiler must check for and prevent these duplicate registrations while no restrictions are imposed on several OBJECT-IDENTIFIER value assignments to the same OID value.

OBJECT-TYPE MACRO

The OBJECT-TYPE macro was redefined in SMIv2. It now contains the following fields:

- SYNTAX: the objects data type.
- MAX-ACCESS: one of the values *read-only, read-write, read-create* or *not-accessible* (this field replaces the former ACCESS-part).
- STATUS: one of the values *current, deprecated, obsolete.*
- DESCRIPTION: a human readable text describing the contents and purpose of the MIB object.
- REFERENCE: a textual cross-reference to an object in another information module.
- INDEX/AUGMENTS: used for indexing tables.
- DEFVAL: defines a default value for the object.

The fields REFERENCE, INDEX/AUGMENTS and DEFVAL are optional fields and do not have to be present. The usage of the OBJECT-TYPE macro follows the same rules as SMIv1 (see Figure 32.19).

NOTIFICATION-TYPE MACRO

This macro is used for the definition of Traps and confirmed events (INFORM request) (see Figure 32.20). It has the following fields:

- OBJECTS: a list of objects transported together with the notification sent. The field is optional.
- STATUS: one of the values *current, deprecated* or *obsolete.*
- DESCRIPTION: a human readable text describing the contents and purpose of the notification.
- REFERENCE: contains a cross-reference to another notification defined in some other information module. The field is optional.

```
snmpStatsPackets OBJECT-TYPE
    SYNTAX      Counter32
    MAX-ACCESS read-only
    STATUS      current
    DESCRIPTION "The total number of packets received by the
        SNMPv2 entity from the transport service."
    REFERENCE   "Derived from RFC1213-MIB. snmpInPkts."
    ::= { snmpStats 1 }
```

FIGURE 32.19　OBJECT-TYPE macro sample.

```
linkUp NOTIFICATION-TYPE
    OBJECTS { ifIndex }
    STATUS current
    DESCRIPTION
        "A linkUp trap signifies that the SNMPv2 entity,
        acting in an agent role, recognizes that one of
        the communication links represented in its
        configuration has come up."
    ::= { snmpTraps 4 }
```

FIGURE 32.20　NOTIFICATION-TYPE macro sample.

TEXTUAL CONVENTION MACRO

This macro is used to define new data types (See Figure 32.21). It provides more details than the type definition of SMIv1. The macro includes the following fields:

- DISPLAY-HINT: gives a hint for INTEGER or OCTET STRING types and how the value of the corresponding MIB variable shall be displayed. The field is optional.
- STATUS: one of the values *current*, *deprecated* or *obsolete*.
- DESCRIPTION: a human readable text describing the contents and purpose of the notification.
- REFERENCE: contains a cross-reference to a related item in another information module. The field is optional.
- SYNTAX: defines the syntax of the new data type.

Comparing this structure to the definition using SMIv1, the improved description quality becomes obvious (See Figure 32.21 compared to Figure 32.22).

Conformance Statement Macros

To provide a possibility for the definition of the lower limit of an actual implementation to the MIB designers, [RFC1444] introduced several conformance statement macros. The OBJECT-GROUP macro is used to define a collection of related object type definitions. It helps the MIB author to explain the logical grouping of object types showing their correspondence to subsystems. Each MIB object with MAX-ACCESS other than "not-accessible" must be a member of at least one OBJECT-GROUP. The MODULE-COMPLIANCE statement defines requirements for an agent implementation (e.g., which groups are mandatory), while the AGENT-CAPABILITIES can be used by an agent designer to specify implementation characteristics. This macro lists groups, objects and events defined in the imported module that are supported by the implementation. Usually, it is used in a MIB module with no statements other than IMPORT and MODULE-IDENTITY.

MIBs

Together with the protocol specification of SNMPv2, several new MIB modules were released: the management information base for SNMPv2, the party MIB and the manager-to-manager MIB. Other MIBs have later been added to these initial specifications (e.g., MIB modules for token ring, token bus, OSPF, BGP, RIP, Ethernet, FDDI, etc.).

In addition, the Internet-sub tree obtained two new successor elements:

- Security: Internet(5) and
- SnmpV2: Internet(6).

```
DisplayString ::= TEXTUAL-CONVENTION
    DISPLAY-HINT "255a"
    STATUS          current
    DESCRIPTION
        "Represents textual information taken from the NVT
        ASCII character set, as defined in pages 4, 10-11
        of RFC 854. Any object defined using this syntax
        may not exceed 255 characters in length."
    SYNTAX OCTET STRING (SIZE (0..255))
```

FIGURE 32.21   TEXTUAL CONVENTION macro sample.

```
DisplayString ::= OCTET STRING
```

FIGURE 32.22   SMIv1 data type definition.

MIB for SNMPv2

This MIB is defined in [RFC1450] and extends the MIB-II from SNMPv1 by the following groups:

- Statistic groups for SNMPv2: SNMPv2-specific statistics.
- Statistic groups for SNMPv1: SNMPv1-specific statistics.
- Object resources group: contains all objects that can be dynamically configured in an agent.
- Trap group: a table of all objects that can send an SNMPv2 trap PDU.
- Set group: used to coordinate SET requests of different manager applications.

Party MIB (Now Obsolete)

This MIB should help network administrators to define individual access rights to certain users ([RFC1447]). For each party, a certain area of the MIB tree (a MIB view) can be defined. A specific station then only has access to this MIB view (see also the next section). This MIB was later marked as historic.

Manager-to-Manager MIB (Now Obsolete)

[RFC1451] defines the objects of the manager-to-manager MIB. This MIB was introduced to address one of the weak points of SNMPv1, namely the strict distinction of entities acting in the manager or agent role. Now each entity can act either as a manager, as an agent or in a dual role. This MIB should help to coordinate the work of multiple management stations. It was later marked as historic.

Security/SNMPv2 Versions

The main weakness of SNMPv1 was the lack of an appropriate security model. During the development of SNMPv2, different approaches to address security were proposed [9]. Later, all presented approaches except SNMPv2c became obsolete and were changed to "historical" state.

SNMPv2p

This version called "party-based SNMP" was introduced together with SNMPv2. The model is based on the concept of an SNMP party that comprises a protocol entity at a logical network location together with a security context (authentication and/or privacy protocols). Each implementation using SNMPv2p has to implement the Party MIB.

Authentication of entities is achieved by using MD5 algorithm. A 16-byte fingerprint ensures that a message is unchanged and the originator is known. For privacy, the use of DES [10] encryption was proposed. In addition, each PDU has to contain a timestamp to recognize retransmission of a message. The following combinations of authentication an encryption can be used, allowing scaling for different applications:

- No authentication and no privacy.
- MD5 authentication and no privacy.
- MD5 authentication and DES privacy.

One of the problems of this approach was the export restriction of MD5 and DES for outside the USA at the time the model was introduced. Another setback was the complex handling of the parties of this model. Due to these problems, the approach was later set to historic and replaced by SNMPv2u and SNMPv2c.

SNMPv2u

In SNMPv2u, a *user paradigm* is used. Each user has attributes assigned, regarding which type of authentication and/or privacy protocol shall be used if any. In addition, keys for both areas are defined. All these attributes must be known to both the sender and receiver of the message.

Like SNMPv2p, this approach uses MD5 and DES for authentication and privacy with some modifications (e.g., in the party model, the MD5 key is appended after the digest of the message is computed; in SNMPv2u, it is inserted before computing the digest value).

SNMPv2*

This version was never published as an RFC. It combined the best features of SNMPv2p and SNMPv2u. Little documentation exists about this approach.

SNMPv2c

SNMPv2c uses all the enhanced protocol features of SNMPv2 without any of the security models. Instead, it uses the community string paradigm known from SNMPv1. It is defined by [RFC1901]. This version is still experimental but widely used since it is easy to implement and to use.

Conclusion

SNMPv2 was a big step forward. It extended the previous version SNMPv1 in nearly all areas. But these extensions also increased the complexity of each implementation significantly in addition to the base payload required for encryption and authentication if these features are used. The improved SMIv2 provides a good base for a concise definition of MIB modules. While not all macros might be used to the same extent by MIB designers, others will alleviate problems where MIB contents were inaccurate before.

32.5   SNMP V3

The specification of SNMPv2 substantially extended the definitions of SNMPv1 in nearly all areas. The next step was no revolution but a stepwise refinement of SNMPv2. It was released in 1999 by [RFC2570] to [RFC2580]. Citing [RFC2570], SNMPv3 "…can be thought of as SNMPv2 with additional security and administration capabilities."

The base structure again remained similar to the previous SNMP versions. Adaptations only concerned the structure of the documents that were better modularized to allow easier extensions and replacements by future versions and a more detailed description of the architecture. The only areas with larger changes were security and administration.

Architecture

While previous SNMP versions just identified the roles that an entity can accomplish, an important part of the SNMPv3 architecture covers the description of the entities. Each entity consists of an SNMP engine and some associated applications (Figure 32.23).

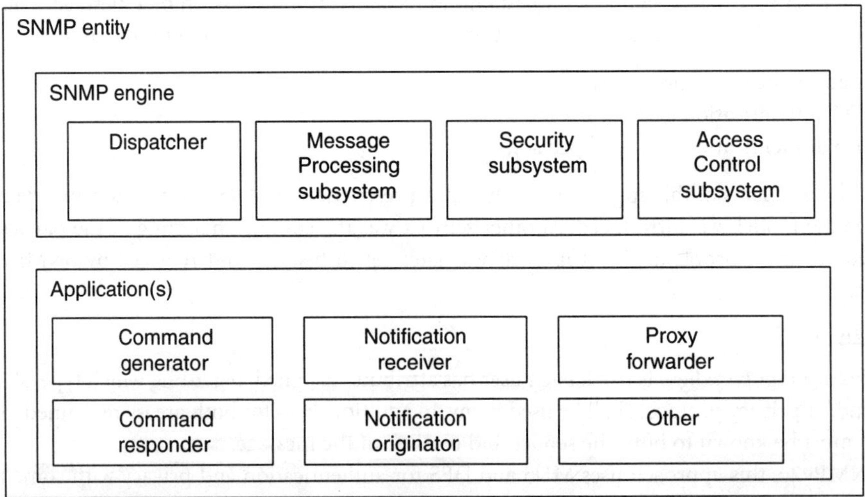

FIGURE 32.23   SNMPv3 entity [RFC2571].

The *Dispatcher* of the SNMP engine is responsible for sending and receiving messages. Depending on the version number in the PDU, each message is passed to the appropriate *Message Processing* subsystem. This subsystem encompasses message handlers for different versions of SNMP. It prepares messages that shall be sent and extracts data from received messages (Figure 32.24).

The *Security subsystem* (Figure 32.25) provides authentication of messages and encrypting/decrypting of messages for privacy. Typical security models would support the community-based security model of SNMPv1 and SNMPv2c or the user-based security model.

The *Access Control subsystem* (Figure 32.25) determines if the access to a managed object shall be allowed or not. At the moment, only one model — the View-based access control model (VACM) — is defined.

In the context of SNMPv3, the term *"application"* is used to describe parts of an SNMP entity. Applications use the functions provided by the SNMP engine to accomplish their tasks. In addition, abstract service interfaces were defined to clarify the interfaces between the various subsystems. They should not constrain the developer but should provide a hint about real-life implementations.

Several types of applications are already defined, including:

- Command generators: generate SNMP commands to collect or set management data.
- Command responders: provide access to management data.
- Notification originators: initiate traps or inform messages.
- Notification receivers: receive and process traps or inform messages.
- Proxy forwarders: forward messages between entities.

An SNMP entity acting in the role of a manager or an agent typically consists of the parts listed by Figure 32.26.

SMI

SMIv2 was updated by [RFC2578] to [RFC2580]. Changes to the former specifications were minimal. Extensions to the older versions only included some new textual conventions. While even the structure of all documents almost remained the same, some effort was spent toward reviewing and clarifying the specifications.

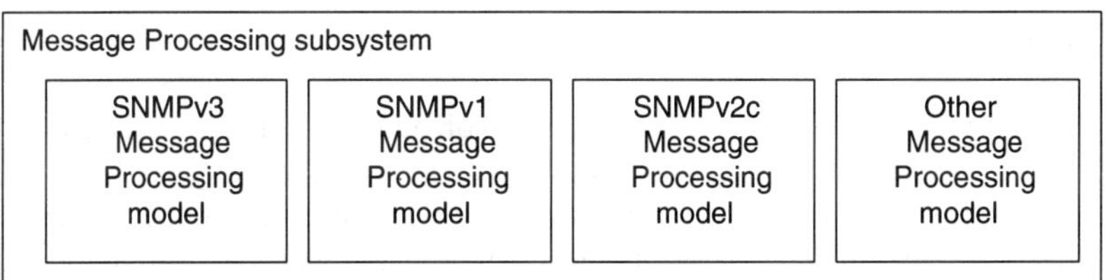

FIGURE 32.24 SNMPv3 Message Processing subsystem [RFC2571].

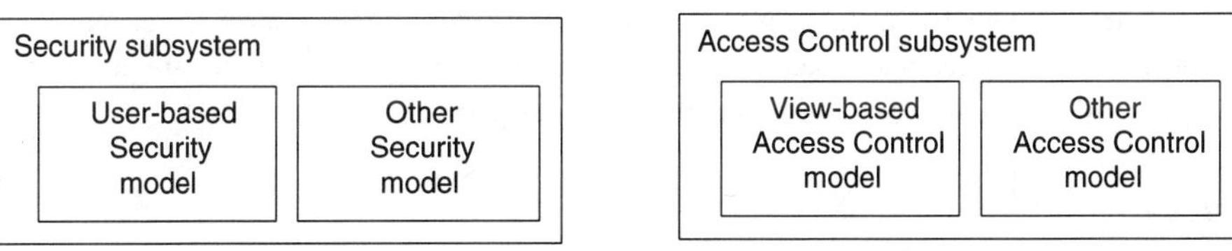

FIGURE 32.25 Security and Access Control subsystems [RFC2571].

	Manager	Agent
Dispatcher	X	X
Message Processing subsystem	X	X
Security subsystem	X	X
Command generator	X	
Notification receiver	X	
Notification originator	X	
Access Control		X
Proxy forwarder		X
Command responder		X

FIGURE 32.26   Elements of SNMP entities.

Protocol changes

No new PDU types were introduced by SNMPv3 but the message format was changed. The format of the PDUs now consists of the following three parts:

- Header: this part contains fields like the version identifier (set to value 3), the maximum message size the sender supports or the identification of the security model used.
- Security parameters: this area depends on the security model used. The data is directly passed to the corresponding Security Model.
- Data: this part is either encrypted or plain text. It contains context information and a valid SNMPv2c PDU.

SNMPv3 messages therefore can be also be described as a message wrapper around SNMPv2 PDUs.

Conclusion

SNMPv3 clarified and extended its predecessors significantly while keeping the overall structure of SNMP the same. Especially, the area of security was consolidated after the different solutions that were proposed and dropped during the evolution of SNMPv2. In addition, the modular approach to the description of an SNMP entity helped in building modular solutions. The future will show if SNMPv3 can completely replace the previous versions.

32.6   Conclusion

All presented SNMP versions provide a good approach for building network management solutions. Significant strengths are found in areas where status information is requested or small amounts of information like switching commands are sent to a network node, while areas like configuration management are still not sufficiently solved. Independent of brand and type, nearly all network equipment on the market today with any form of management interface provides an SNMP agent to access the equipment. Over the years, alternate solutions like WEB-based management were proposed and implemented but none could find the broad acceptance of SNMP.

While the basic structure of SNMPv1 is quite easy to understand, the complexity of the later versions increased considerably. Due to this fact, SNMPv1 solutions are still quite common. Especially, if security is of no concern, for example, in a closed environment and if the new message types are not needed, this version might prove to be an easy solution for network management.

References

[1]   Hobbes Internet Timeline v6.0, http://www.zakon.org/robert/internet/timeline/
[2]   ITU-T Specifications X.700-X.7xx.

[3] M.T. Rose, Industry Comment, *The Simple Times*, vol.1, no.4, 1996, http://www.simple-times.org/

[4] ITU-T: Specification of Abstract Syntax Notation One (ASN.1), X.680, 1993.

[5] D.T. Perkins, *Understanding SNMP MIBs*, Prentice-Hall, Englewood Cliffs, NJ, 1996.

[6] ITU-T: ASN.1 encoding rules, X.690, 1994.

[7] W. Stallings, *SNMP, SNMPv2, SNMPv3, and RMON 1 and 2*, 3rd ed., Addison-Wesley Reading MA, 1999.

[8] D. Zeltserman, *A Practical Guide to SNMPv3 and Network Management*, Prentice-Hall, Englewood Cliffs, NJ, 1999.

[9] D.T. Perkins, SNMP Versions, *The Simple Times*, vol. 5, no. 1, 1997, http://www.simple-times.org/

[10] American National Standards Institute: Data Encryption Algorithm — Modes of Operation, X3.106, 1983.

[RFCxxxx] Request for Comment Nr.xxxx, http://www.ietf.org.

33

The Dynamic Host Configuration Protocol

Michael Eyrich
Technical University of Berlin

33.1 Introduction

Client configuration in large networks requires large effort by administrators if the clients need to be manually configured one by one. The Dynamic Host Configuration Protocol (DHCP) [3] provides a common framework to store network- and client-specific options on a central server and to assign network addresses to clients. It consists of two components: a mechanism for stateful allocation of network addresses to hosts and a protocol for delivering host-specific configuration from a DHCP server to a client. Both functionalities can be used together or separately.

DHCP is based on a client–server architecture where requests are issued by a client and answered by one or more servers (anycast). The servers offer network addresses from a pool of addresses in an automatic, dynamic, or manual allocation scheme depending on the server configuration. While being extremely flexible for configuration, DHCP is nonetheless designed to smoothly interoperate with manually configured, nonparticipating hosts who are configured with an address from the server-managed address pool. The operation of DHCP does not require a client to have any preconfigured network-related data, and furthermore the operation of DHCP is transparent to all other hosts in the same network. Thus, under normal circumstances, a network manager should not have to enter any configuration parameters at the clients. A client that obtained its parameters via DHCP should be able to exchange packets with any other host in the Internet.

As a protocol for basic network configuration, the network address configuration part of DHCP has to fulfill some requirements:

- DHCP has to guarantee that any specific network address is used by at most one client at a time.
- A server should retain DHCP client configuration across a client reboot. Whenever possible, a DHCP client should be assigned the same configuration parameters (e.g., network address) in response to each request.
- DHCP client configuration should be retained across server reboots, and, whenever possible, a DHCP client should be assigned the same configuration parameters despite restarts of the DHCP server.
- It should support the automated assignment of configuration parameters to new clients to avoid manual configuration for new clients.
- It has to support permanent allocation of configuration parameters to specific clients in order to interoperate with older Bootstrap Protocol (BOOTP) [2, 4] clients.
- The DHCP server should not necessarily have to be placed in the same subnet as its clients.

The DHCP evolved from the BOOTP protocol. For this reason, interoperation with BOOTP relay agents and the necessary service for BOOTP clients is required for DHCP. This is achieved by using a compatible message format and the same vendor extension options [1, 12] as defined for BOOTP.

33.2   Protocol Basics

We start with a description of the messages used to exchange DHCP protocol data units (PDU) between server and clients. The set of messages is summarized in Table 33.1. The common message format for BOOTP and DHCP is defined in RFC 951 [2]. A message consists of a header of fixed elements with a total of 236 bytes in length, shown in Figure 33.1. The fields referenced in this figure are explained in Table 33.2. It is followed by a variable length list of options (Figure 33.2) starting with a magic cookie of four octets ('99.130.83.99') followed by the "DHCP message type" and always terminated by the "end" option. The magic cookie helps distinguishing between BOOTP messages and DHCP messages. Messages sent from a client to a server are tagged as a BOOTREQUEST in the op-field, while messages from a server to a client are tagged as

TABLE 33.1   DHCP Messages [3]

Message	Use
DHCPDISCOVER	Client broadcast to locate available servers.
DHCPOFFER	Server to client in response to DHCPDISCOVER with offer of configuration parameters.
DHCPREQUEST	Client message to servers either (a) requesting offered parameters from one server and implicitly declining offers from all others, (b) confirming correctness of previously allocated address after, for example, system reboot, or (c) extending the lease on a particular network address.
DHCPACK	Server to client with configuration parameters, including committed network address.
DHCPNAK	Server to client indicating client's notion of network address is incorrect (e.g., client has moved to new subnet) or client's lease has expired
DHCPDECLINE	Client to server indicating network address is already in use.
DHCPRELEASE	Client to server relinquishing network address and canceling remaining lease.
DHCPINFORM	Client to server, asking only for local configuration parameters; client already has externally configured network address.

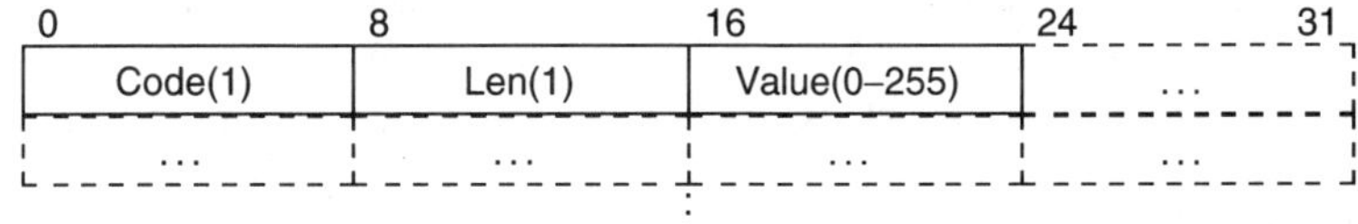

FIGURE 33.1 PDU of DHCP messages.

TABLE 33.2 Description of Fields in a DHCP Message [3]

Field	Octets	Description
op	1	Message op code / message type. 1 = BOOTREQUEST, 2 = BOOTREPLY
htype	1	Hardware address type, see ARP section in "Assigned Numbers" RFC; e.g., "1" = 10mb ethernet.
hlen	1	Hardware address length (e.g., "6" for 10mb ethernet).
hops	1	Client sets to zero, optionally used by relay agents when booting via a relay agent.
xid	4	Transaction ID, a random number chosen by the client, used by the client and server to associate messages and responses between a client and a server.
secs	2	Filled in by client, seconds elapsed since client began address acquisition or renewal process.
flags	2	Flags (e.g., broadcast flag).
ciaddr	4	Client IP address; only filled in if client is in BOUND, RENEW, or REBINDING state and can respond to ARP requests.
yiaddr	4	'your' (client) IP address.
siaddr	4	IP address of next server to use in bootstrap; returned in DHCPOFFER, DHCPACK by server.
giaddr	4	Relay agent IP address, used in booting via a relay agent.
chaddr	16	Client hardware address.
sname	64	Optional server host name, null terminated string.
file	128	Boot file name, null terminated string; "generic" name or null in DHCPDISCOVER, fully qualified directory-path name in DHCPOFFER.
options	var	Optional parameters field. See the options documents for a list of defined options.

FIGURE 33.2 Format of options in the options area (and the respective option file).

a BOOTREPLY message. These flags identify the allowed options within the DHCP message. An option value can be of arbitrary data type, but limited in size to 255 octets. Longer option sizes (e.g., path names, route descriptions) are conveyed by concatenating the values of options with the same code in the same order as they appear in the message. For interoperability reasons, multi-octet quantities are represented in network byte order.

A DHCP packet is further encapsulated as a UDP packet allowing for sizes up to 65,535 bytes. However, since the minimum guaranteed maximum IP datagram size that can be handled by an arbitrary host is 576 octets, larger message sizes have to be negotiated using the "maximum DHCP message size" option. IP fragmentation cannot be used here since the IP protocol stack is not supposed to be completely operational at boot time. Usually, this negotiation is initiated by the client as is all message exchange between client and server.[1]

DHCP servers listen for DHCP and BOOTP messages on port 67. They send replies to the client's port 68. The advantage of using different ports for server and clients is that only servers receive the broadcast messages issued by clients. In the other direction, most messages sent from a server to a client use a unicast address, except in those cases where a server issues a DHCPNACK message informing that a requested IP address could not be allocated to a client (which then clearly has no valid IP address). A second reason is that the client explicitly requests a server broadcast in its request messages (using the `broadcast` bit in the flags field). A reason for this could be the client's inability to process unicast IP messages before having an IP address configured. More specifically, whenever the following conditions hold: (a) the `broadcast` bit is set in a request message, (b) the client is in the same subnet, and (c) the client does not provide a valid IP address (`ciaddr` is zero), then DHCPOFFER and DHCPACK messages are broadcasted to the client using the IP broadcast address 0xffffffff (255.255.255.255) and the MAC broadcast address.

The purpose of the `xid` field is to let the client match its request packets with received response packets. The client should choose this id such that it is not identical to the id chosen by another client, although there is no mechanism to guarantee uniqueness. Nevertheless, no problems should result from such a conflict as a client is further identified either with the "client identifier" or by its hardware address `chaddr`.

Address Discovery

The following description shows the messages exchanged to allow a client to obtain a network address. Due to routing restrictions, broadcast messages are not forwarded over (sub-) network boundaries. Without the support of special BOOTP relay agents (see Section 33.4), DHCP servers need to be placed within the same broadcast domain as their clients. The address discovery process allows for having several servers serving the same subnet with the same set of addresses. However, in the case of multiple servers, it has to be guaranteed by the system administrator that the relevant data (address pool, network configuration information) are consistent across the servers, since the DHCP protocol itself provides no mechanism to ensure this consistency.

The message exchange is usually the following (as shown as a timeline diagram in Figure 33.3 and as a state diagram in Figure 33.4):

1. A client initiates an address discovery by broadcasting a DHCPDISCOVER message on its local subnet. The request is sent using a Medium Access Control (MAC)-layer broadcast with the client's MAC address as the source address. The UDP/IP packet has its IP source address set to 0.0.0.0 and a broadcast destination address (255.255.255.255). Enclosed in the DHCPDISCOVER messages the client provides certain options such as a proposed lease time, a maximum message size, and its previously allocated address.
2. Each of the available servers may respond with a DHCPOFFER message. This message contains an available network address (`yiaddr`) and may contain con-figuration options for the network (such as default router, network mask, name server, Domain Name System (DNS) domain name, etc.). Typically, the offering server should check the availability of the proposed address prior to sending the DHCPOFFER message by probing with an Internet Control Message Protocol (ICMP) echo request.
3. The client can collect several DHCPOFFER messages or already answer the first one. In the latter case, all further DHCPOFFER messages must be discarded. Before accepting an offered network address with a DHCPREQUEST broadcast message, the client should by himself probe for availability of the

---

[1]An exception is the extension of the DHCP protocol with the DHCPFORCERENEW message type shown in point 2 of Section 33.2

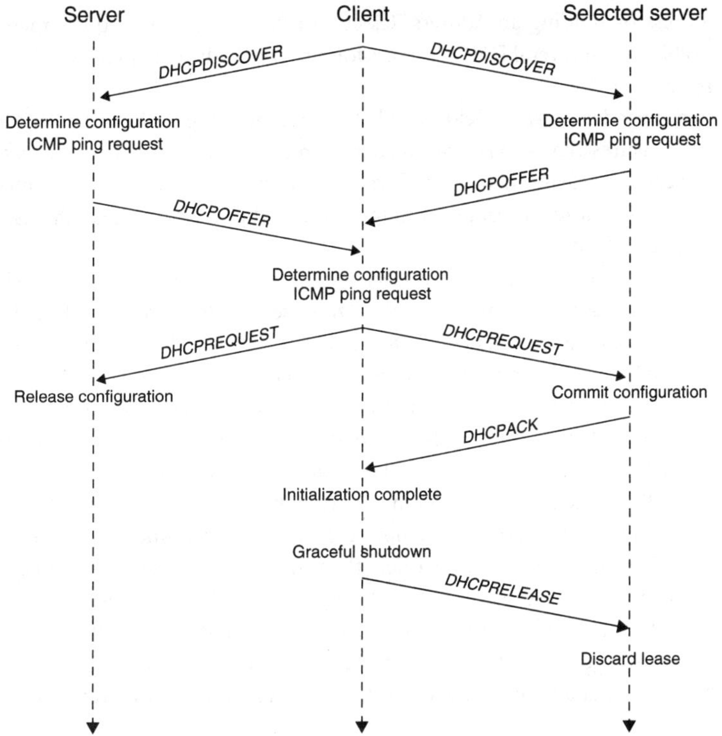

FIGURE 33.3 Timeline diagram of a DHCP address configuration.

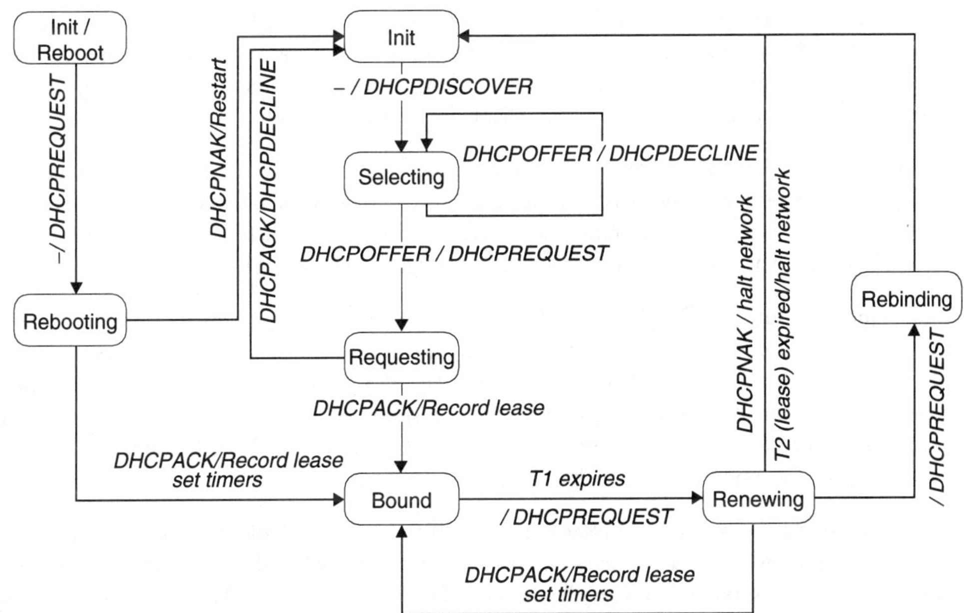

FIGURE 33.4 State chart of the DHCP address configuration.

network address by sending an Address Resolution Protocol (ARP) request message. In order to make it possible to run several DHCP servers for a network, the client's answer has to fulfill a number of constraints:

- the message header's `secs` field must have the same value as the initial DHCPDISCOVER message. This ensures that servers can recognize requests that have been successfully served by other servers and release their own offers to the respective DHCPDISCOVER message.
- the requested IP address must be set to the value of the `yiaddr` field, that is, the selected and available network address.

4. All servers that initially received a DHCPDISCOVER message will likely also receive the DHCPREQUEST message broadcasted from the client. Those servers that are not selected by the DHCPREQUEST command release the offers given to the client. The selected server (as given in the "server identifier option") commits the new binding to persistent storage. A DHCPACK message is used to acknowledge the address to the client and to transmit additional configuration parameters.

   If the server is not able to fulfill the client's request, it sends a DHCPNACK message. There are not many reasons why such a request may fail. Especially, the server should not check an address given in a DHCPREQUEST because the address may already be in use by the requesting client. A request most probably fails if the server does not reserve a network address between the check and DHCPREQUEST message and, therefore, offers the same address to several clients.

5. On receiving a DHCPACK message, the client may again probe the given address, for instance, by issuing an ARP request. If it receives an answer, the client notifies the server by sending an DHCPDECLINE message and restarts the configuration process afterwards.

   If the probe remains unanswered, the client finishes its address configuration by noting the duration of the lease.

   After receiving the DHCPACK message, the client is said to have obtained a lease. This lease is equipped with a permitted lifetime, which the server has given in the DHCPOFFER message (secs field). Before expiry of the lifetime, the client has to renew the lease, as described below.

6. Optionally, the client can relinquish the leased address by sending a DHCPRELEASE message to its server. This is not required, but in networks where addresses are a scarce resource it is useful to relinquish an address, for instance, if the client is shut down. The lease is uniquely identified by the client network address and either the value of the "client identifier option" or the value of the `chaddr` field.

Usually, clients remember their address in a persistent storage. If after a reboot, an old address can be found in the storage or when the client tries to renew its lease, it may use a shortened reservation scheme. The client skips the DHCPDISCOVER message and immediately sends a DHCPREQUEST message containing the remembered network address. There are two reasons why a server might reject a client's request for a given address:

- the requested address is already assigned to another client according to the server's configuration table (this case is different from the case of an initial DHCPDISCOVER message: the server is unable to determine in response to a DHCPREQUEST message if a requested address is already in use by some other system, because the client may already respond to ICMP echo request messages), or
- the requested address is not valid for the network that the client is currently connected to, which is often given for mobile devices such as laptops.

In either case, the server sends a DHCPNACK message. If the server and client are in the same subnet, this message is sent to the broadcast address, as the client may not have a valid address to receive unicast messages. Otherwise, the server has to send the message via unicast to the BOOTP relay agent, as detailed in Section 33.4.

Even if the client receives a DHCPACK message, it has to probe whether the address is already used by another system that does not act as a DHCP client. When receiving an answer to its probe, the client informs the server by sending a DHCPDECLINE message.

In both the cases, a DHCPNACK message or a negative probe result (some answer received), the full DHCP address configuration has to be restarted.

Pushing Reconfiguration Requests

Usually, the DHCP architecture is completely driven by requests initiated from a client. In environments where configuration settings for clients are subject to change at unpredictable times, short lease times can be used to keep the clients up-to-date, at the expense of unnecessarily increased traffic load. In RFC 3203 [5], a new DHCP message type is defined: the DHCPFORCERENEW message. This message is sent to a client as a unicast message, which is possible, as the server keeps track of all clients with valid leases. Upon receipt of the unicast DHCPFORCERENEW message, the client will change its state to the RENEWING state (see Figure 33.4), and will then try to renew its lease according to the shortened DHCP procedures. If the server wants to assign a new IP address to the client, it will reply to the DHCPREQUEST message with a DHCPNACK message. The client will then go back to the init state and broadcast a DHCPDISCOVER message. The server can now assign a new IP address to the client by replying with a DHCPOFFER message. If the DHCPFORCERENEW message is lost, the DHCP server will not receive a DHCPREQUEST message from the client. It retransmits the DHCPFORCERENEW message using a backoff algorithm with an exponentially growing delay as retransmissions fail. The amount of retransmissions is usually limited since the client may have left the network due to power off or movement.

For manually configured devices, a DHCPFORCERENEW message initiates a new DHCPINFORM message to the server.

As with all IP network connections, there are no precautions to avoid session interruptions for clients whose network address is forced to change during a session.

Interpretation of Time

A time data type is needed to specify lease times. Usually, the time is synchronized between computers in a network (e.g., by using the Network Time Protocol [6]). However, since a client enters a network immediately after rebooting, it does not necessarily have a correct absolute time. Therefore, the lease is expressed as a relative time. Later on, the client may update its system time using a time server address provided by the DHCP exchange. The two octet-sized sec-fields in the DHCPOFFER message allow to define lease times up to approximately 18 h, which should be enough for the purpose of DHCP. Still, it is assumed that the drift between both the client's and the server's clock is small. Servers avoid running out their leases database before the clients by sending a lease time to the client, which is less than the one they are storing in their database. The client itself needs to maintain various times. For each request issued, the client has to store its issuing time. Additionally, the client keeps track of the lease times: some time before expiry, the client tries to renew the lease with the granting DHCP server (this time is given by the "Renewal (T1) Time Value" option) . If this is not successful and the lease eventually expires, the client has to start the full configuration procedure with a DHCPDISCOVER message. This procedure is started at a time indicated by the "Rebinding (T2) Time Value" option in the servers DHCPOFFER message.

Failure Recovery

Various situations exist where a server might fail to respond to client requests. DHCP is designed to provide any kind of answer in a reasonable amount of time. Failure recovery is the responsibility of the client. For each request, the client sets a timer. On timer expiry, the request is retransmitted a limited number of times. The timeout value depends on the underlying Internet network type with a lower bound of about 3 sec up to a maximum of 64 sec. The exact timeout values are randomized to avoid unwanted synchronization of several clients. After 60 sec without a valid reply, the client gives up and notifies its user.

To protect the client against server failures for initial configurations or lease renewals, the following rules apply:

1. As long as the client has a valid lease for an address, that is, the lease time for this address has not yet elapsed, and there is no other system using the address, the client can use the address until the lease time has elapsed.

2. If a client does not have a configured address, the system may choose stateless autoconfiguration. Specifically, the client picks one address from the range 169.254.0.0/24 and probes this address

with an ARP request. If nobody answers, the client configures this address; otherwise, it tries another one. The address range 169.254.0.0/24 is reserved for autoconfigured networks. An auto-configured client can only communicate with other autoconfigured clients, since routers do not forward their packets without special measures. For systems supporting the stateless address configuration RFC 2563 obligingly regulates that the "autoconfigure" option has to be included in a DHCPDISCOVER message. The server can indicate in its answer that autoconfiguration addresses are not supported in this subnet (option `DoNotAutoConfigure` set to 0). After receiving this information, the client may notify its user. The client has to respect this option in that it must not configure an autoconfigured address.

Requesting Configuration Information

As we have seen in the beginning, the DHCP protocol is used to fulfill two tasks: network address configuration and provision of additional local configuration information. Both parts may, but do not need to be separated from each other. A manually configured client may request additional configuration information using the DHCP protocol. The special DHCPINFORM broadcast message serves this purpose. It is answered with a unicast DHCPACK message from one or more servers to the client. This answer contains the relevant parameters. In case of multiple servers, the system administrator has to ensure the consistency of the delivered information. The server does not check the client's address for availability and existing bindings of the address (although it should check for consistency!). Furthermore, when formatting the response message, the server fills all address-lease-related fields (e.g., the `yiaddr` field) with zero values.

A client can request specific parameters in its DHCPDISCOVER or DHCPREQUEST message. This is achieved by using the option "Parameter Request List" (code 55) consisting of the lists' length and a list of options the client requests. The response should at least contain all options requested by the client and the order of options should be retained by the server. Additionally, the client should include the maximum DHCP message size it is going to accept. Depending on this size, the server can either send the options completely in a DHCP message, or needs to provide a file, which has to be transferred via the TFTP protocol [9] between the server and client. This file has the same structure as the options part of the DHCP message with the exception that no further indirection to another file is allowed. The use of a file is signaled by setting two flags in the `options` field indicating that the `file` and `sname` field are valid.[2]

The DHCP implementations offer several further options.

Table 33.3 exemplarily lists the basic vendor extensions that are valid for BOOTP and DHCP. The allocation of new options is strictly regulated according to RFC 2489. The rules given there ensure that:

- allocation of new option numbers is coordinated from a single authority,
- new options are reviewed for technical correctness and appropriateness, and
- documentation for new options is complete and published.

33.3   Address Allocation

DHCP supports different modes of address allocation, *automatic, dynamic,* and *manual* allocation. Manual allocation is necessary for BOOTP clients that do not have knowledge about lease times. Instead, BOOTP clients assume that an assigned address remains valid forever. Thus, configuration is done in the server by creating a table of assignments from the client's hardware address (as a Reverse Address Resolution Protocol (RARP) replacement) or a unique identifier (which remains fixed even if the network

---

[2]However, the "option overload" option may alternatively indicate that these two fields are used to transport additional options.

TABLE 33.3 Vendor Extensions [RFC 2132]

Value	Name	Description	Size octets	Octets per element
0	Pad Option	Just padding	1	
255	End Option	Indicates end of options	1	
1	Subnet Mask	Should be before router option	4	
2	Time Offset	positive offset: east of meridian offset, negative offset: west of zero meridian		
3	Router Option	List of IP addresses on the client's subnet given in order of preference	≥ 4	4
4	Time Server Option	List of time servers	≥ 4	4
5	Name Server Option	IEN 116 name servers	≥ 4	4
6	Domain Name Server Option	list of DNS servers	≥ 4	4
7	Log Server Option	list of MIT-LCS UDP log servers	≥ 4	4
8	Cookie Server Option	RFC 865 cookie server	≥ 4	4
9	LPR Server Option	list of RFC 1179 line printer servers	≥ 4	4
10	Impress Server Option	list of Imagen Impress servers	≥ 4	4
11	Resource Location Server Option	RFC 887 Resource Location Servers	≥ 4	4
12	Host Name Option	hostname with or without domain name	≥ 1	
13	Boot File Size Option	length of default boot image for client in 512-octet blocks, specified in unsigned 16-bit integer. Maximum image size is, therefore, 32 Mbytes	2	
14	Merit Dump File	where to dump core image in case of a client crash	≥ 1	
15	Doain Name	the domain name to use when resolving hostnames via DNS	≥ 1	
16	Swap Server	the clients swap server	4	
17	Root Path	pathname to the clients root disk	≥ 1	
18	Extensions Path	A file containing vendor- specific options with unconstrained size and retrievable by Trivial File Transfer Protocol (TFTP); references to tag 18 in this file are ignored	≥ 1	

interface card is changed) to an IP address. In case of automatic allocation, the server assigns addresses from a given pool of addresses with infinite lease times to the clients. It requires that there be as many addresses available as clients may request address. The dynamic allocation mode differs from the automatic mode in that addresses are given to clients for a limited period of time. This is the default mode of DHCP and is used for situations where several systems exist in a network, but only a limited number of them are in operation at the same time.

Multi-homed Servers

When a DHCP server is equipped with interfaces for different networks (multi-homed), it is able to serve all those networks with addresses. It then needs to keep different configurations matching the respective networks and needs to maintain possibly different server identities ("server identifier option") for the different subnetworks in order to be reachable for clients. The server has to accept packets destined to one of its addresses on any interface. Usually, the server selects the network information based on the network interface on which a request was received.

User Groups

Vendor-specific options are a container to provide special information. They are selected by using the "vendor option," which the client can set in a DHCPDISCOVER message. One of these options is the "user class option" as defined in RFC 3004 [10]. It allows the client to select a specific class it belongs to. Based on the class, a selection of options and a particular address range can be chosen by the server. It allows, for instance, to connect certain users to a specific printer or to provide class-specific routes such as routes secured by a Virtual Private Network (VPN). The selection, however, is completely based on the values provided by the client. In a common configuration, there is no way to check if the client is authorized to provide the values or to receive the requested configuration data, although the use of the "authentication option" provides some control.

Behavior for Multiple Servers

DHCP supports that many servers handle a single network. By delivering certain messages as broadcast messages, an explicit synchronization between the servers is not necessary. Still, not all servers have the same rights. This is mainly seen in that only authorized servers are allowed to reject addresses with a DHCPNACK message, which are not in the group of their offered addresses. Apart from this, a client that already has a network address firstly contacts the server that initially assigned the address. Only if this server fails to answer are all other servers contacted by a broadcast message.

33.4 Bypassing Routers: BOOTP Relay Agents

With the help of BOOTP relay agents [2, Section 8], a server can also serve networks to which it is not directly connected.

Figure 33.5 shows a typical application scenario for a BOOTP relay agent. A relay agent can be colocated to the router, respectively, the Network Access Server (NAS), but might as well be located on a separate system.

Mode of Operation

BOOTP relay agents listen on the well-known client port `bootpc` (67) and server port `bootps` (68). Client broadcasts received on the server port are converted to unicast messages and forwarded to the (well-known) DHCP server. Specifically, the relay agent fills the `giaddr` field with its own IP address. This causes the DHCP server to direct its BOOTREPLY answer message to the relay agent's IP address (unicast) instead of the client's MAC address. On receiving a BOOTREPLY message, the relay agent either broadcasts the message (after suitable modification) on the clients network or sends it as a unicast message to the client's MAC address (the client might still lack a valid IP address). With this mechanism, DHCPNACK messages can even be delivered to clients that try to reuse an invalid address, which cannot be

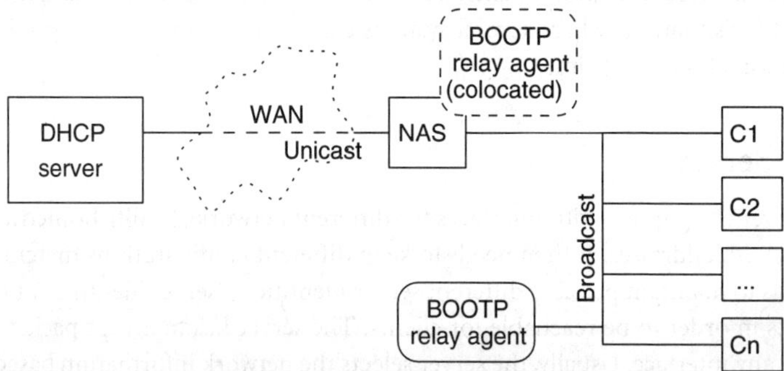

FIGURE 33.5   BOOTP relay agent configuration.

routed within the current network. For the relay agent, it does not make a difference whether it forwards BOOTP or DHCP messages.

Subnet Selection

Usually, a server deduces the network a client belongs to from the interface on which it receives a request. In case of relayed requests, this information is deduced from the `giaddr` field, which usually provides enough information to select the correct network (namely, the IP address of the relay agent within the client's network). However, if this network uses nonroutable private addresses [8], the `giaddr` field should still contain a routable IP address. It is required that the relay agent has a globally routable IP address, which it fills into the `giaddr` field. In this case, the server's automatic selection process must be overridden by using the "subnet selection option" [11]. This option takes precedence over the automatic selection and causes the DHCP server to deduce the network the client is in from the value of the `giaddr` field instead of the incoming interface.

Relay Agent Information

A typical application example of BOOTP relay agents are NASs, which provide public access to the Internet using dial-up lines. The BOOTP agent is often colocated with the NAS and forwards requests coming from the clients (via modem). For proper allocation of IP addresses and configuration information, it is often necessary to let the DHCP server know not only the identity of the client but also the identification of the incoming line. Since there are no standard fields for this in the DHCP packet format, the relay agent is allowed to add a special "Relay agent information option" [7], which contains this information and which allows the (stateless) relay agent to associate the replies of the DHCP server with its request and therefore to the correct incoming line. In addition, this information allows to restrict the BOOTP replies to the single line where the client is reachable instead of broadcasting the packets to all outgoing links. The relay agent removes the "Relay agent information option" before forwarding the BOOTP replies to the client. Since in this situation the relay agent modifies the DHCP packets, it has to be trusted.

33.5　Security Considerations

The DHCP protocol in its basic form is inherently insecure against insider-driven threats even if the network is protected against attacks from outside by a firewall. In a DHCP environment, access rights to a server can only be enforced based on the unique identifier or the hardware address provided by the clients itself. Such values can easily be spoofed, allowing the attacker to gain access to the information provided by the server. Conversely, a rogue DHCP server could spoof replies. The client will not be able to recognize that it is not connected to the regular DHCP server of the network and may be a victim of Denial of Service (DoS) or man-in-the-middle attacks.

The "authentication option" defined in RFC 3118 has been added to prevent these faults. It allows to add replay detection information and authentication information to each DHCP packet. The authentication information is based on the generation of a MAC created, for instance, with an MD5 hash function. This Hashed Message Authentication Code (HMAC) is then signed with a private key of the client or with a shared key. Certain options and fields (such as the `giaddr` field and the "relay agent information") cannot be included in the authenticated part of the packet as they may be purposely inserted or changed by the relay agent during client–server conversation. During message validation, these fields must be set to zero and any further associated options must be ignored. There is currently no mechanism for securing the additional information inserted by BOOTP relay agents. The relay agent is assumed to be trustworthy.

The use of the authentication option necessitates an out-of-band distribution of keys to the clients. In order to make clients distinguishable, each client should have its own unique key. Servers should be able to obtain the necessary client keys in a secure manner. However, scalability problems may arise if the number of clients is high and the client population changes often (nomadic users).

The "authentication option" adds additional overhead to servers and clients, and servers need to commit additional state information to nonvolatile memory such as the information about security associations (which IP address is bound to which key-owner). However, as compared to the threats an open DHCP environment is exposed to, it is well worth the overhead.

33.6 Related Mechanisms

Internet hosts often request informations from servers identified by human readable names instead of IP addresses. If such a server obtains its IP address via DHCP, then this address might change from time to time. In this case, the *dynamic DNS* approach provides a mapping between host names and network addresses. In dynamic DNS, special DNS servers accept frequent updates for "their" hosts and answer queries with very short live times. It makes sense to let the DHCP server perform the address update with the dynamic DNS server automatically. Thus, the DNS table contains only those addresses that are currently assigned and in use, and the clients can be addressed by human-readable names.

33.7 Conclusion

DHCP is a framework that allows to configure clients in large networks without needing to edit configuration data at the client. Instead, all configuration data are stored at a central server and distributed to the clients. This eases network management substantially, since all data are stored in one station and can be kept consistent. By its ability to operate several servers in parallel and in distant networks, DHCP offers a robust solution for the distribution of configuration information. Due to these advantages, the DHCP framework is in widespread use.

Nomenclature

ARP	Address Resolution Protocol
BOOTP	Bootstrap Protocol
DHCP	Dynamic Host Configuration Protocol
DNS	Domain Name System
DoS	Denial of Service
HMAC	Hashed Message Authentication Code
ICMP	Internet Control Message Protocol
MAC	Medium Access Control
NAS	Network Access Server
RARP	Reverse Address Resolution Protocol
TFTP	Trivial File Transfer Protocol
VPN	Virtual Private Network

References

[1] S. Alexander and R. Droms, DHCP Options and BOOTP Vendor Extensions, RFC 1533, October 1993.

[2] W. J. Croft and J. Gilmore, Bootstrap Protocol, RFC 951, September 1985.

[3] R. Droms, Dynamic Host Configuration Protocol, RFC 1541, October 1993.

[4] R. Droms, Interoperation Between DHCP and BOOTP, RFC 1534, October 1993.

[5] C. Hublet and P. De Schrijver, DHCP Reconfigure Extension, RFC 3203, December 2001.

[6] D. L. Mills, Network Time Protocol (Version 3) Specification, Implementation, RFC 1305, March 1992.

[7] M. Patrick, DHCP Relay Agent Information Option, RFC 3046, January 2001.

[8] Y. Rekhter, B. Moskowitz, D. Karrenberg, G.J. de Groot, and E. Lear, Address Allocation for Private Internets, RFC 1918, February 1996.

[9]   K. Sollins, The TFTP Protocol (Revision 2), RFC 1350, July 1992.
[10] G. Stump, R. Droms, Y. Gu, R. Vyaghrapuri, A. Demirtjis, B. Beser, and J. Privat, The User Class Option for DHCP, RFC 3004, November 2000.
[11] G. Waters, The IPv4 Subnet Selection Option for DHCP, RFC 3011, November 2000.
[12] W. Wimer, Clarifications and Extensions for the Bootstrap Protocol, RFC 1532, October 1993.

34

Network Security and Secure Applications

Christopher Kruegel
University of California

34.1 Introduction

In order to provide useful services or to allow people to perform tasks more conveniently, computer systems are attached to networks and get interconnected. This resulted in the worldwide collection of local and wide-area networks known as the Internet. Unfortunately, the extended access possibilities also entail increased security risks as it opens additional avenues for an attacker. For a closed, local system, the attacker was required to be physically present at the network in order to perform unauthorized actions. In the networked case, each host that can send packets to the victim can be potentially utilized. As certain services (such as web or name servers) need to be publicly available, each machine on the Internet might be the originator of malicious activity. This fact makes attacks very likely to happen on a regularly basis.

The following text attempts to give a systematic overview of security requirements of Internet-based systems and potential means to satisfy them. We define properties of a secure system and provide a classification of potential threats to them. We also introduce mechanisms to defend against attacks that attempt to violate desired properties. The most widely used means to secure application data against tampering and eavesdropping, the Secure Sockets Layer (SSL) and its successor, the Transport Layer Security (TLS) protocol, are discussed. Finally, we briefly describe popular application programs that can act as building blocks for securing custom applications.

Before one can evaluate attacks against a system and decide on appropriate mechanisms against them, it is necessary to specify a *security policy* [21]. A security policy defines the desired properties for each part of a secure computer system. It is a decision that has to take into account the value of the assets that should be protected, the expected threats, and the cost of proper protection mechanisms. A security policy that is sufficient for the data of a normal user at home may not be sufficient for bank applications, as these systems are obviously a more likely target and have to protect more valuable resources. Although often neglected, the formulation of an adequate security policy is a prerequisite before one can identify threats and appropriate mechanisms to tackle them.

34.2 Security Attacks and Security Properties

For the following discussion, we assume that the function of a system that is the target of an attack is to provide information. In general, there is a flow of data from a source (e.g., host, file, memory) to a destination (e.g., remote host, other file, user) over a communication channel (e.g., wire, data bus). The task of the security system is to restrict access to this information to only those parties (persons or processes) that are authorized to have access according to the security policy in use. In the case of an automation system that is remotely connected to the Internet, the information flow is from/to a control application that manages sensors and actuators via communication lines of the public Internet and the network of the automation system (e.g., a field-bus).

The normal information flow and several categories of attacks that target it are shown in Figure 34.1 and explained below (according to [20]).

1. *Interruption*: An asset of the system gets destroyed or becomes unavailable. This attack targets the source or the communication channel and prevents information from reaching its intended target (e.g., cut the wire, overload the link so that the information gets dropped because of congestion). Attacks in this category attempt to perform a kind of *denial-of-service(DOS)*.
2. *Interception*: An unauthorized party gains access to the information by eavesdropping into the communication channel (e.g., wiretapping).
3. *Modification*: The information is not only intercepted, but modified by an unauthorized party while in transit from the source to the destination. By tampering with the information, it is actively altered (e.g., modifying message content).
4. *Fabrication*: An attacker inserts counterfeit objects into the system without having the sender doing anything. When a previously intercepted object is inserted, this processes is called *replaying*. When the attacker pretends to be the legitimate source and inserts his desired information, the attack is called *masquerading* (e.g., replay an authentication message, add records to a file).

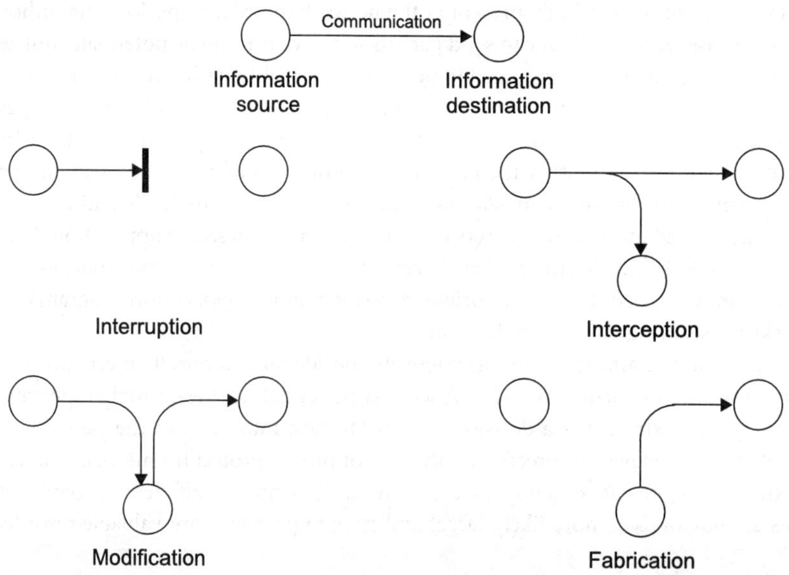

FIGURE 34.1   Security attacks.

The four classes of attacks listed above violate different security properties of the computer system. A security property describes a desired feature of a system with regard to a certain type of attack. A common classification following [5, 11] is listed below.

Confidentiality: This property covers the protection of transmitted data against its release to nonauthorized parties. In addition to the protection of the content itself, the information flow should also be resistant against traffic analysis. Traffic analysis is used to gather information other than the transmitted values themselves from the data flow (e.g., timing data, frequency of messages).

Authentication: Authentication is concerned with making sure that the information is authentic. A system implementing the authentication property assures the recipient that the data are from the source that it claims to be. The system must make sure that no third party can masquerade successfully as another source.

Nonrepudiation: This property describes the feature that prevents either the sender or receiver from denying a transmitted message. When a message has been transferred, the sender can prove that it has been received. Similarly, the receiver can prove that the message has actually been sent.

Availability: Availability characterizes a system whose resources are always ready to be used. Whenever information needs to be transmitted, the communication channel is available and the receiver can cope with the incoming data. This property makes sure that attacks cannot prevent resources from being used for their intended purpose.

Integrity: Integrity protects transmitted information against modifications. This property assures that a single message reaches the receiver as it has left the sender, but integrity also extends to a stream of messages. It means that no messages are lost, duplicated, or reordered and it makes sure that messages cannot be replayed. As destruction is also covered under this property, all data must arrive at the receiver. Integrity is not only important as a security property but also as a property for network protocols. Message integrity must also be ensured in case of random faults, and not just in the case of malicious modifications.

34.3 Security Mechanisms

Different security mechanisms can be used to enforce the security properties defined in a given security policy. Depending on the anticipated attacks, different means have to be applied to satisfy the desired properties. We divide these measures against attacks into three different classes: attack prevention, attack avoidance, and attack detection.

Attack Prevention

Attack prevention is a class of security mechanisms that contains ways of preventing or defending against certain attacks before they can actually reach and affect the target. An important element in this category is access control, a mechanism that can be applied at different levels such as the operating system, the network, or the application layer.

Access control [21] limits and regulates the access to critical resources. This is done by identifying or authenticating the party that requests a resource and checking its permissions against the rights specified for the demanded object. It is assumed that an attacker is not legitimately permitted to use the target object and is therefore denied access to the resource. As access is a prerequisite for an attack, any possible interference is prevented.

The most common form of access control used in multi-user computer systems are access control lists for resources that are based on the user identity of the process that attempts to use them. The identity of a user is determined by an initial authentication process that usually requires a name and a password. The login process retrieves the stored copy of the password corresponding to the user name and compares it with the presented one. When both match, the system grants the user the appropriate user credentials. When a resource should be accessed, the system looks up the user and group in the access control list and grants or denies access as appropriate. An example of this kind of access control is a secure web server. A secure web server delivers certain resources only to clients that have authenticated themselves and that possess sufficient credentials for the desired resource. The authentication process is usually handled by the web client such as the `Microsoft Internet Explorer`, or `Mozilla` by prompting the user for his name and password.

The most important access control system at the network layer is a firewall [4]. The idea of a firewall is based on the separation of a trusted inside network of computers under single administrative control from a potential hostile outside network. The firewall is a central choke point that allows enforcement of access control for services that may run at the inside or outside. The firewall prevents attacks from the outside against the machines in the inside network by denying connection attempts from unauthorized parties located outside. In addition, a firewall may also be utilized to prevent users behind the firewall from using certain services that are outside (e.g., surfing web sites containing pornographic material).

For certain installations, a single firewall is not suitable. Networks that consist of several server machines that need to be publicly accessible and workstations that should be completely protected against connections from the outside would benefit from a separation between these two groups. When an attacker compromises a server machine behind a single firewall, all other machines can be attacked from this new base without restrictions. To prevent this, one can use two firewalls and the concept of a *demilitarized zone (DMZ)* [4] in between as shown in Figure 34.2.

In this setup, one firewall separates the outside network from a segment (DMZ) with the server machines, while a second one separates this area from the rest of the network. The second firewall can be configured in a way that denies all incoming connection attempts. Whenever an intruder compromises a server, he is now unable to immediately attack a workstation located in the inside network.

The following design goals for firewalls are identified in [4]:

1. All traffic from inside to outside, and *vice versa,* must pass through the firewall. This is achieved by physically blocking all access to the internal network except via the firewall.
2. Only authorized traffic, as defined by the local security policy, will be allowed to pass.
3. The firewall itself should be immune to penetration. This implies the use of a trusted system with a secure operating system. A trusted, secure operating system is often purpose-built, has heightened security features, and only provides the minimal functionality necessary to run the desired applications.

These goals can be reached by using a number of general techniques for controlling access. The most common is called *service control* and determines Internet services that can be accessed. Traffic on the Internet is currently filtered on the basis of IP addresses and TCP/UDP port numbers. In addition, there may be proxy software that receives and interprets each service request before passing it on. *Direction control* is a simple mechanism to control the direction in which particular service requests may be initiated and permitted to flow through. *User control* grants access to a service based on user credentials similar to the technique used in a multiuser operating system. Controlling external users requires secure authentication over the network (e.g., such as provided in IPSec [9]). A more declarative approach in contrast to the operational variants mentioned above is *behavior control.* This technique determines how particular services are used. It may be utilized to filter e-mail to eliminate spam or to allow external access to only part of the local web pages.

A summary of capabilities and limitations of firewalls is given in [20]. The following benefits can be expected:

- A firewall defines a single choke point that keeps unauthorized users out of the protected network. The use of such a point also simplifies security management.
- It provides a location for monitoring security-related events. Audits, logs, and alarms can be implemented on the firewall directly. In addition, it forms a convenient platform for some nonsecurity-related functions such as address translation and network management.

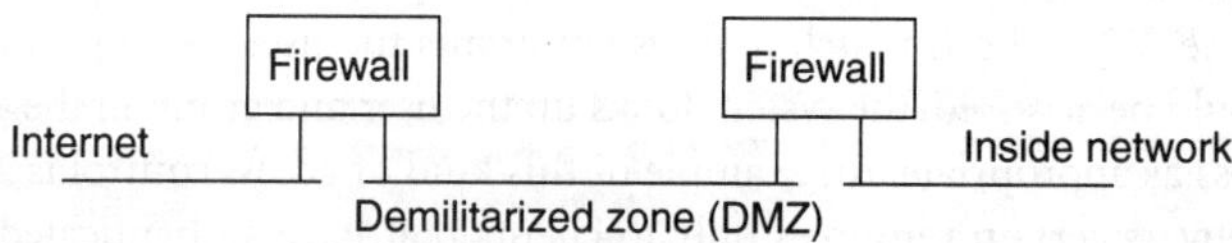

FIGURE 34.2   Demilitarized zone.

- A firewall may serve as a platform to implement a virtual private network (e.g., by using IPSec).

The list below enumerates the limits of the firewall access control mechanism.

- A firewall cannot protect against attacks that bypass it, for example, via a direct dial-up link from the protected network to an Internet Service Provider (ISP). It also does not protect against internal threats from an inside hacker or an insider cooperating with an outside attacker.
- A firewall does not help when attacks are against targets whose access has to be permitted.
- It cannot protect against the transfer of virus-infected programs or files. It would be impossible, in practice, for the firewall to scan all incoming files and e-mails for viruses.

Firewalls can be divided into two main categories. A *Packet-Filtering Router,* or short packet filter, is an extended router that applies certain rules to the packets that are forwarded. Usually, traffic in each direction (in- and outgoing) is checked against a rule set, which determines whether a packet is permitted to continue or should be dropped. The packet filter rules operate on the header fields used by the underlying communication protocols, for the Internet almost always IP, TCP, and UDP. Packet filters have the advantage that they are cheap as they can often be built on existing hardware. In addition, they offer a good performance for high traffic loads. An example for a packet filter is the `iptables` package, which is implemented as part of the `Linux` 2.4 routing software.

A different approach is followed by an *Application-Level Gateway,* also called proxy server. This type of firewall does not forward packets on the network layer but acts as a relay on the application level. The user contacts the gateway, which in turn opens a connection to the intended target (on behalf of the user). A gateway completely separates the inside and outside networks at the network level and only provides a certain set of application services. This allows authentication of the user who requests a connection and session-oriented scanning of the exchanged traffic up to the application-level data. This feature makes application gateways more secure than packet filters and offers a broader range of log facilities. On the downside, the overhead of such a setup may cause performance problems under heavy load.

Attack Avoidance

Security mechanisms in this category assume that an intruder may access the desired resource but the information is modified in a way that makes it unusable for the attacker. The information is preprocessed at the sender before it is transmitted over the communication channel and postprocessed at the receiver. While the information is transported over the communication channel, it resists attacks by being nearly useless for an intruder. One notable exception are attacks against the availability of the information as an attacker could still interrupt the message. During the processing step at the receiver, modifications or errors that might have previously occurred can be detected (usually because the information cannot be correctly reconstructed). When no modification has taken place, the information at the receiver is identical to the one at the sender before the preprocessing step.

The most important member in this category is cryptography, which is defined as the science of keeping messages secure [16]. It allows the sender to transform information into a random data stream from the point of view of an attacker but to have it recovered by an authorized receiver (see Figure 34.3).

The original message is called *plain text* (sometimes clear text). The process of converting it through the application of some transformation rules into a format that hides its substance is called *encryption.* The corresponding disguised message is denoted *cipher text* and the operation of turning it back into clear text is called *decryption.* It is important to notice that the conversion from plain to cipher text has to be loss-less in order to be able to recover the original message at the receiver under all circumstances.

The transformation rules are described by a cryptographic algorithm. The function of this algorithm is based on two main principles: *substitution* and *transposition.* In the case of substitution, each element of the plain text (e.g., bit, block) is mapped into another element of the used alphabet. Transposition describes the process where elements of the plain text are rearranged. Most systems involve multiple steps (called rounds) of transposition and substitution to be more resistant against cryptanalysis. Cryptanalysis is the science of breaking the cipher, that is, discovering the substance of the message behind its disguise.

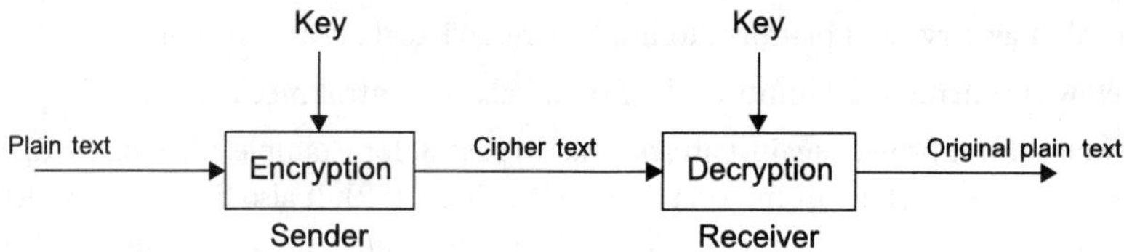

FIGURE 34.3   Encryption and decryption.

When the transformation rules process the input elements one at a time, the mechanism is called a *stream cipher;* in case of operating on fixed-sized input blocks, it is called a *block cipher.*

If the security of an algorithm is based on keeping the way how the algorithm works (i.e., the transformation rules) secret, it is called a restricted algorithm. Those algorithms are no longer of any interest today because they do not allow standardization or public quality control. In addition, when a large group of users is involved, such an approach cannot be used. A single person leaving the group makes it necessary for everyone else to change the algorithm.

Modern cryptosystems solve this problem by basing the ability of the receiver to recover encrypted information on the fact that he possesses a secret piece of information (usually called the *key).* Both encryption and decryption functions have to use a key and they are heavily dependent on it. When the security of the cryptosystem is completely based on the security of the key, the algorithm itself may be revealed. Although the security does not rely on the fact that the algorithm is unknown, the cryptographic function itself and the used key together with its length must be chosen with care. A common assumption is that the attacker has the fastest commercially available hardware at his disposal in his attempt to break the cipher text.

The most common attack, called *known plain text attack,* is executed by obtaining cipher text together with its corresponding plain text. The encryption algorithm must be so complex that even if the code breaker is equipped with plenty of such pairs and powerful machines, it is infeasible for him to retrieve the key. An attack is infeasible when the cost of breaking the cipher exceeds the value of the information or the time it takes to break it exceeds the lifespan of the information.

Given pairs of corresponding cipher and plain text, it is obvious that a simple key guessing algorithm will succeed after some time. The approach of successively trying different key values until the correct one is found is called *brute force* attack because no information about the algorithm is utilized whatsoever. In order to be useful, it is a necessary condition for an encryption algorithm that brute force attacks are infeasible.

Depending on the keys that are used, one can distinguish two major cryptographic approaches — public and secret key cryptosystems.

Secret Key Cryptography

This is the kind of cryptography that has been used for the transmission of secret information for centuries, long before the advent of computers. These algorithms require that the sender and the receiver agree on a key before communication is started.

It is common for this variant (which is also called single key or symmetric encryption) that a single secret key is shared between the sender and the receiver. It needs to be communicated in a secure way before the actual encrypted communication can start and has to remain secret as long as the information is to remain secret. Encryption is achieved by applying an agreed function to the plain text using the secret key. Decryption is performed by applying the inverse function using the same key.

The classic example of a secret key block cipher that is widely deployed today is the Data Encryption Standard (DES) [6]. DES was developed in 1977 by IBM and was adopted as a standard by the US government for administrative and business use. Recently, it has been replaced by the Advanced Encryption Standard (AES — Rijndael) [1]. It is a block cipher that operates on 64-bit plain text blocks and utilizes a key with 56-bits length. The algorithm uses 16 rounds that are key dependent. During each round, 48 key bits are selected and combined with the block that is encrypted. Then, the resulting block is piped through a substitution and a permutation phase (which use known values and are independent of the key) to make cryptanalysis harder.

Although there is no known weakness of the DES algorithm itself, its security has been much debated. The small key length makes brute force attacks possible and several cases have occurred where DES-protected information has been cracked. A suggested improvement called 3DES uses three rounds of the simple DES with three different keys. This extends the key length to 168 bits while still resting on the very secure DES base.

A well-known stream cipher that has been debated recently is RC4 [14], which has been developed by RSA. It is used to secure the transmission in wireless networks that follow the IEEE 802.11 standard and forms the core of the WEP (wired equivalent protection) mechanism. Although the cipher itself has not been broken, current implementations are flawed and reduce the security of RC4 down to a level where the used key can be recovered by statistical analysis within a few hours.

Public Key Cryptography

Since the advent of public key cryptography, the knowledge of the key that is used to encrypt a plain text also allowed the inverse process, the decryption of the cipher text. In 1976, this paradigm of cryptography was changed by Diffie and Hellman [7] when they described their public key approach. Public key cryptography utilizes two different keys, one called the *public key,* and the other called the *private key.* The public key is used to encrypt a message, while the corresponding private key is used to do the opposite. Their innovation was the fact that it is infeasible to retrieve the private key given the public key. This makes it possible to remove the weakness of secure key transmission from the sender to the receiver. The receiver can simply generate his public/private key pair and announce the public key without fear. Anyone can obtain this key and use it to encrypt messages that only the receiver with his private key is able to decrypt.

Mathematically, the process is based on the *trap door* of *one-way* functions. A one-way function is a function that is easy to compute but very hard to inverse. This means that given x, it is easy to determine $f(x)$ but given $f(x)$ it is hard to obtain x. Hard is defined as computationally infeasible in the context of cryptographically strong one-way functions. Although it is obvious that some functions are easier to compute than their inverse (e.g., square of a value in contrast to its square root), there is no mathematical proof or definition of one-way functions. There are a number of problems that are considered difficult enough to act as one-way functions but it is more an agreement among crypto analysts than a rigorously defined set (e.g., factorization of large numbers). A one-way function is not directly usable for cryptography, but it becomes so when a trap door exists. A trap door is a mechanism that allows one to easily calculate x from $f(x)$ when an additional information y is provided.

A common misunderstanding about public key cryptography is thinking that it makes secret key systems obsolete, either because it is more secure or because it does not have the problem of secretly exchanging keys. As the security of a cryptosystem depends on the length of the used key and the utilized transformation rules, there is no automatic advantage of one approach over the other. Although the key exchange problem is elegantly solved with a public key, the process itself is very slow and has its own problems. Secret key systems are usually a factor of 1000 (see [16] for exact numbers) faster than their public key counterparts. Therefore, most communication is stilled secured using secret key systems and public key systems are only utilized for exchanging the secret key for later communication. This hybrid approach is the common design to benefit from the high speed of conventional cryptography (which is often implemented directly in hardware) and from a secure key exchange.

A problem in public key systems is the authenticity of the public key. An attacker may offer the sender his own public key and pretend that it origins from the legitimate receiver. The sender then uses the faked public key to perform his encryption and the attacker can simply decrypt the message using his private key. In order to thwart an attacker that attempts to substitute his public key for the victim's one, *certificates* are used. A certificate combines user information with the user's public key and the digital signature of a trusted third party that guarantees that the key belongs to the mentioned person. The trusted third party is usually called a *certification authority (CA).* The certificate of a CA itself is usually verified by a higher-level CA that confirms that the CA's certificate is genuine and contains its public key. The chain of third parties that verify their respective lower-level CAs has to end at a certain point, which is called the root CA. A user that wants to verify the authenticity of a public key and all involved CAs needs to obtain the self-signed certificate of the root CA via an external channel. Web browsers (e.g., `Netscape Navigator, Internet Explorer`) usually ship with a number of certificates of globally known root CAs. A framework that implements the distribution of certificates is called a public key infrastructure (PKI). An important protocol for key management is X.509 [23]. Another important issue is revocation, the invalidation of a certificate when the key has been compromised.

The best-known public key algorithm and textbook classic is RSA [15], named after its inventors Rivest, Shamir, and Adleman at MIT. It is a block cipher that is still utilized for the majority of current systems, although the key length has been increased over recent years. This has put a heavier processing load on applications, a burden that has ramifications especially for sites doing electronic commerce. A competitive approach that promises similar security as RSA using far smaller key lengths is elliptic curve cryptography. However, as these systems are new and have not been subject to sustained cryptanalysis, the confidence level in them is not yet as high as in RSA.

Authentication and Digital Signatures

An interesting and important feature of public key cryptography is its possible use for authentication. In addition to making the information unusable for attackers, a sender may utilize cryptography to prove his identity to the receiver. This feature is realized by *digital signatures*. A digital signature must have similar properties as a normal handwritten signature. It should be hard to forge and it has to be bound to a certain document. In addition, one has to make sure that a valid signature cannot be used by an attacker to replay the same (or different) messages at a later time.

A way to realize such a digital signature is by using the sender's private key to encrypt a message. When the receiver is capable of successfully decrypting the cipher text with the sender's public key, he can be sure that the message is authentic. This approach obviously requires a cryptosystem that allows encryption with the private key, but many (such as RSA) offer this option. It is easy for a receiver to verify that a message has been successfully decrypted when the plain text is in a human readable format. For binary data, a checksum or similar integrity checking footer can be added to verify a successful decryption. Replay attacks are prevented by adding a time-stamp to the message (e.g., `Kerberos` [10] uses time-stamps to prevent that messages to the ticket granting service are replayed).

Usually, the storage and processing overhead for encrypting a whole document is too high to be practical. This is solved by one-way hash functions. These are functions that map the content of a message onto a short value (called *message digest*). Similar to one-way functions, it is difficult to create a message when given only the hash value itself. Instead of encrypting the whole message, it is enough to simply encrypt the message digest and send it together with the original message. The receiver can then apply the known hash function (e.g., MD5 [13]) to the document and compare it to the decrypted digest. When both values match, the messages is authentic.

Attack and Intrusion Detection

Attack detection assumes that an attacker can obtain access to his desired targets and is successful in violating a given security policy. Mechanisms in this class are based on the optimistic assumption that most of the time the information is transferred without interference. When undesired actions occur, attack detection has the task of reporting that something went wrong and then to react in an appropriate way. In addition, it is often desirable to identify the exact type of attack. An important facet of attack detection is recovery. Often, it is enough to just report that malicious activity has been found, but some systems require that the effect of the attack has to be reverted or that an ongoing and discovered attack is stopped. On the one hand, attack detection has the advantage that it operates under the worst-case assumption that the attacker gains access to the communication channel and is able to use or modify the resource. On the other hand, detection is not effective in providing confidentiality of information. When the security policy specifies that interception of information has a serious security impact, then attack detection is not an applicable mechanism. The most important members of the attack detection class, which have received an increasing amount of attention in the last few years, are intrusion detection systems (aka IDS).

Intrusion Detection [2, 3] is the process of identifying and responding to malicious activities targeted at computing and network resources. This definition introduces the notion of intrusion detection as a process, which involves technology, people, and tools. An IDS basically monitors and collects data from a target system that should be protected, processes and correlates the gathered information, and initiate responses, when evidence for an intrusion is detected. IDS are traditionally classified as anomaly- or signature-based. Signature-based systems act similar to virus scanners and look for known, suspicious patterns in their input data. Anomaly-based systems watch for deviations of actual from expected behavior and classify all "abnormal" activities as malicious.

The advantage of signature-based designs is the fact that they can identify attacks with an acceptable accuracy and tend to produce fewer false alarms (i.e., classifying an action as malicious when in fact it is not) than their anomaly-based cousins. The systems are more intuitive to build and easier to install and configure, especially in large production networks. Because of this, nearly all commercial systems and most deployed installations utilize signature-based detection. Although anomaly-based variants offer the advantage of being able to find prior unknown intrusions, the costs of having to deal with an order of magnitude more false alarms is often prohibitive.

Depending on their source of input data, IDS can be classified as either network- or host-based. Network-based systems collect data from network traffic (e.g., packets by network interfaces in promiscuous mode), while host-based systems monitor events at the operating system level such as system calls or receive input from applications (e.g., via log files). Host-based designs can collect high-quality data directly from the affected system and are not influenced by encrypted network traffic. Nevertheless, they often seriously impact the performance of the machines they are running on. Network-based IDS, on the other hand, can be set up in a nonintrusive manner — often as an appliance box without interfering with the existing infrastructure. In many cases, this makes them the preferred choice.

As many vendors and research centers have developed their own intrusion detection system versions, the IETF has created the intrusion detection working group [8] to coordinate international standardization efforts. The aim is to allow intrusion detection systems to share information and to communicate via well-defined interfaces by proposing a generic architectural description and a message specification and exchange format (IDMEF).

34.4 Secure Network Protocols

After the general concepts and mechanisms of network security have been introduced, the following section concentrates on two actual instances of secure network protocols: the SSL [18] and the TLS [22] protocol.

The idea of secure network protocols is to create an additional layer between the application and the transport/network layer to provide services for a secure end-to-end communication channel.

TCP/IP are almost always used as transport/network layer protocols on the Internet and their task is to provide a reliable end-to-end connection between remote tasks on different machines that intend to communicate. The services on that level are usually directly utilized by application protocols to exchange data, for example, Hypertext Transfer Protocol (HTTP) for web services. Unfortunately, the network layer transmits these data unencrypted, leaving it vulnerable to eavesdropping or tampering attacks. In addition, the authentication mechanisms of TCP/IP are only minimal, thereby allowing a malicious user to hijack connections and redirect traffic to his machine as well as to impersonate legitimate services. These threats are mitigated by secure network protocols that provide privacy and data integrity between two communicating applications by creating an encrypted and authenticated channel.

SSL has emerged as the de-facto standard for secure network protocols. Originally developed by Netscape, its latest version SSL 3.0 is also the base for the standard proposed by the IETF under the name TLS. Both protocols are quite similar and share common ideas, but they unfortunately cannot interoperate. The following discussion will mainly concentrate on SSL and only briefly explain the extensions implemented in TLS.

The SSL protocol [19] usually runs above TCP/IP (although it could use any transport protocol) and below higher-level protocols such as HTTP. It uses TCP/IP on behalf of the higher-level protocols, and in the process allows an SSL-enabled server to authenticate itself to an SSL-enabled client, allows the client to authenticate itself to the server, and allows both machines to establish an encrypted connection. These capabilities address fundamental concerns about communication over the Internet and other TCP/IP networks and give protection against message tampering, eavesdropping, and spoofing.

SSL server authentication allows a user to confirm a server's identity. SSL-enabled client software can use standard techniques of public-key cryptography to check that a server's certificate and public key are valid and have been issued by a CA listed in the client's list of trusted CAs. This confirmation might be important if the user, for example, is sending a credit card number over the network and wants to check the receiving server's identity.

SSL client authentication allows a server to confirm a user's identity. Using the same techniques as those used for server authentication, SSL-enabled server software can check that a client's certificate and public key are valid and have been issued by a CA listed in the server's list of trusted CAs. This confirmation might be important if the server, for example, is a bank sending confidential financial information to a customer and wants to verify the recipient's identity.

An encrypted SSL connection requires all information sent between a client and a server to be encrypted by the sending software and decrypted by the receiving software, thus providing a high degree of confidentiality. Confidentiality is important for both parties to any private transaction. In addition, all data sent over an encrypted SSL connection are protected by a mechanism for detecting tampering — that is, for automatically determining whether the data have been altered in transit.

SSL uses X.509 certificates for authentication, RSA as its public-key cipher, and one of RC4-128, RC2-128, DES, Triple DES, or IDEA as its bulk symmetric cipher. The SSL protocol includes two subprotocols: the SSL Record Protocol and the SSL Handshake Protocol. The SSL Record Protocol simply defines the format used to transmit data. The SSL Handshake Protocol (using the SSL Record Protocol) is utilized to exchange a series of messages between an SSL-enabled server and an SSL-enabled client when they first establish an SSL connection. This exchange of messages is designed to facilitate the following actions.

- Authenticate the server to the client.
- Allow the client and server to select the cryptographic algorithms, or ciphers, that they both support.
- Optionally authenticate the client to the server.
- Use public-key encryption techniques to generate shared secrets.
- Establish an encrypted SSL connection based on the previously exchanged shared secret.

The SSL Handshake Protocol is composed of two phases. Phase 1 deals with the selection of a cipher, the exchange of a secret key, and the authentication of the server. Phase 2 handles client authentication, if requested and finishes the handshaking. After the handshake stage is complete, the data transfer between client and server begins. All messages during handshaking and after, are sent over the SSL Record Protocol layer. Optionally, session identifiers can be used to reestablish a secure connection that has been previously set up.

Figure 34.4 lists in a slightly simplified form the messages that are exchanged between the client C and the server S during a handshake when neither client authentication nor session identifiers are involved. In this figure, {data} key means that data have been encrypted with key.

The message exchanges shows that the client first sends a challenge to the server, which responds with an X.509 certificate containing its public key. The client then creates a secret key and uses RSA with the server's public key to encrypt it, sending the result back to the server. Only the server is capable of decrypting that message with its private key and can retrieve the shared, secret key. In order to prove to the client that the secret key has been successfully decrypted, the server encrypts the client's challenge with the secret key and returns it. When the client is able to decrypt this message and successfully retrieves

Message type	Direction	Data transferred
client-hello	C > S	challenge-data and supported ciphers
server-hello	C < S	server-certificate and ciphers both support
client-master-key	C > S	chosen cipher and
		{secret-key}server-public-key
server-verify	C < S	{challenge-data}secret-key

FIGURE 34.4 SSL handshake message exchange.

the original challenge by using the secret key, it can be certain that the server has access to the private key corresponding to its certificate. From this point on, all communication is encrypted using the chosen cipher and the shared secret key.

TLS uses the same two protocols shown above and a similar handshake mechanism. Nevertheless, the algorithms for calculating message authentication codes (MACs) and secret keys have been modified to make them cryptographically more secure. In addition, the constraints on padding a message up to the next block size have been relaxed for TLS. This leads to an incompatibility between both protocols.

SSL/TLS is widely used to secure web and mail traffic. HTTP as well as the current mail protocols IMAP (Internet Message Access Protocol) and POP3 (post office protocol, version 3) transmit user credential information as well as application data unencrypted. By building them on top of a secure network protocol such as SSL/TLS, they can benefit from secured channels without modifications. The secure communication protocols simply utilize different well-known destination ports (443 for HTTPS, 993 for IMAPS, and 995 for POP3S) than their insecure cousins.

34.5 Secure Applications

A variety of popular tools that allow access to remote hosts (such as telnet, rsh, and rlogin) or that provide means for file transfer (such as rcp or ftp) exchange user credentials and data in plain text. This makes them vulnerable to eavesdropping, tampering, and spoofing attacks. Although the tools mentioned above could have also been built upon SSL/TLS, a different protocol suite called Secure Shell (SSH) [17] has been developed, which follows partial overlapping goals. The SSH Transport and User Authentication protocols have features similar to those of SSL/TLS. However, they are different in the following ways:

- TLS server authentication is optional and the protocol supports fully anonymous operation, in which neither side is authenticated. As such connections are inherently vulnerable to man-in-the-middle attacks, SSH requires server authentication.
- TLS does not provide the range of client authentication options that SSH does — public-key via RSA is the only option.
- Most importantly, TLS does not have the extra features provided by the SSH Connection Protocol.

The SSH Connection Protocol uses the underlying connection, aka secure tunnel, which has been established by the SSH Transport and User Authentication protocols between two hosts. It provides interactive login sessions, remote execution of commands, and forwarded TCP/IP as well as X11 connections. All these terminal sessions and forwarded connections are realized as different logical channels that may be opened by either side on top of the secure tunnel. Channels are flow-controlled, which means that no data may be sent to a channel until a message is received to indicate that window space is available.

The current version of the SSH protocol is SSH 2. It represents a complete rewrite of SSH 1 and improves some of its structural weaknesses. As it encrypts packets in a different way and has abandoned the notion of server and host keys in favor of host keys only, the protocols are incompatible. For applications built from scratch, SSH 2 should always be the preferred choice.

Using the means of logical channels for interactive login sessions and remote execution, a complete replacement for telnet, rsh, and rlogin could be easily implemented. A popular site that lists open-source implementations, which are freely available for many different platforms, can be found under [12]. Recently, a secure file transfer (sftp) application has been developed that makes the use of regular FTP-based programs obsolete.

Notice that it is possible to tunnel arbitrary application traffic over a connection that has been previously set up by the SSH protocols. Similar to SSL/TLS, web and mail traffic could be securely transmitted over an SSH connection before reaching the server port at the destination host. The difference is that SSH requires that a secure tunnel is created in advance, which is bound to a certain port at the destination host. The setup of this secure channel, however, requires that the client that is initiating the connection has to log into the server. Usually, this makes it necessary that the user has an account at the destination host. After the tunnel has been established, all traffic sent into by the client gets forwarded to the desired port at the target machine. Obviously, the connection is encrypted. In contrast to this SSL/TLS connects directly to a certain point without prior logging into the destination host. The encryption is set up directly between the client and the service listening at the destination port without a prior redirection via the SSH server.

The technique of tunneling application traffic is often utilized for mail transactions when the mail server does not support SSL/TLS directly (as users have accounts at the mail server anyway), but it is less common for web traffic.

34.6 Summary

This chapter discusses security threats that systems face when they are connected to the Internet.

In order to achieve the security properties that are required by the security policy in use, three different classes of mechanisms can be adopted. The first is attack prevention, which attempts to stop the attacker before it can reach its desired goals. Such techniques fall into the category of access control and firewalls. The second approach aims to make the data unusable for unauthorized persons by applying cryptographic means. Secret key as well as public keys mechanism can be utilized. The third class of mechanisms contains attack detection approaches. They attempt to detect malicious behavior and recover after undesired activity has been identified.

The text also covers secure network protocols and applications. SSL/TLS as well as SSH are introduced and its most common fields of operations are highlighted. These protocols form the base of securing traffic that is sent over the Internet on behalf of a variety of different applications.

References

[1] Advanced Encryption Standard (AES), National Institute of Standards and Technology, U.S. Department of Commerce, FIPS 197, 2001.

[2] E. Amoroso, *Intrusion Detection — An Introduction to Internet Surveillance, Correlation, Trace Back, and Response,* Intrusion.Net Books, New Jersey, U.S.A., 1999.

[3] R. Bace, *Intrusion Detection,* Macmillan Technical Publishing, Indianapolis, U.S.A., 2000.

[4] W.R. Cheswick and S. M. Bellovin, *Firewalls and Internet Security,* Addison-Wesley, Reading, MA, U.S.A., 1994.

[5] G. Coulouris, J. Dollimore, and T. Kindberg, *Distributed Systems — Concepts and Design,* 2nd ed., Addison-Wesley, Harlow, England, 1996.

[6] Data Encryption Standard (DES), National Bureau of Standards, U.S. Department of Commerce, FIPS 46-3, 1977.

[7] W. Diffie and M. Hellman, New directions in cryptography, *IEEE Transactions on Information Theory,* IT-22, 644–654, 1976.

[8] Intrusion Detection Working Group, http://www.ietf.org/ids.by.wg/idwg.html.

[9] IP Security Protocol, `http://www.ietf.org/html.charters/ipsec-charter.html`, 2002.

[10] J. Kohl, B. Neuman, and T. 'T'so, The evolution of the Kerberos authentication system, *Distributed Open Systems,* 78 – 94, 1994.

[11] S. Northcutt, *Network Intrusion Detection — An Analyst's Handbook,* New Riders, Indianapolis, U.S.A., 1999.

[12] OpenSSH: Free SSH tool suite, `http://www.openssh.org`.

[13] R.L. Rivest, The MD5 message-digest algorithm, Technical report, Internet Request for Comments (RFC) 1321, 1992.

[14] R.L. Rivest, The RC4 encryption algorithm, Technical report, RSA Data Security Inc., 1992.

[15] R.L. Rivest, A. Shamir, and L.A. Adleman, A method for obtaining digital signatures and public-key cryptosystems, *Communications of the ACM,* 21,120–126, 1978.

[16] B. Schneier, *Applied Cryptography,* 2nd ed., John Wiley & Sons, Inc., New York, U.S.A., 1996.

[17] Secure Shell (secsh), `http://www.ietf.org/html.charters/secsh-charter.html`, 2002.

[18] Secure Socket Layer, `http://wp.netscape.com/eng/ss13/`, 1996.

[19] Introduction to Secure Socket Layer, `http://developer.netscape.com/docs/manuals/security/sslin/contents.htm`, 1996.

[20] W. Stallings, *Network Security Essentials — Applications and Standards,* Prentice-Hall, Englewood Cliffs, NJ, U.S.A., 2000.

[21] A.S. Tanenbaum and M. van Steen, *Distributed Systems — Principles and Paradigms,* Prentice-Hall, Englewood Cliffs, NJ, U.S.A., 2002.

[22] Transport Layer Security, `http://www.ietf.org/html.charters/tsl-charter.html`, 2002.

[23] Public-Key Infrastructure X.509, `http://www.ietf.org/html.charters/pkix-charter.html`, 2002.

35

Internet Firewalls

Günter Schäfer

Technical University of Berlin Germany

Internet Firewalls[1] basically realize access control on the subnetwork level, that is, they control which "kind" of IP packets are allowed to flow into a subnetwork and which kind of packets are allowed to flow out of it, thereby protecting certain parts of a network from the intrusion of "undesirable" data units.

In building construction, the word *firewall* has become the established term for describing the task of protecting certain parts of a building from the spread of fire. Based on this term, the components used for protecting specific subnetworks from potential attack from other parts of the Internet are referred to as *Internet firewalls*. This term may not be totally accurate from a technical standpoint since Internet firewalls, unlike normal firewalls, should not be completely impenetrable (disconnecting the physical network connection would suffice in this case). However, it does convey an intuitive understanding of the necessary task.

35.1 Tasks and Basic Principles of Internet Firewalls

Due to the fact that an Internet firewall should not be completely impenetrable, it can easily be compared to the drawbridge of a medieval castle [ZCC00]. A drawbridge typically had the following functions:

- It forced people to enter a castle at one specific location that could be carefully controlled.
- It prevented attackers from getting close to other defense components of the castle.
- It forced people to exit the castle at one specific location.

For obvious reasons, an Internet firewall is normally installed at the network gateway between a protected trusted subnetwork and an untrustworthy network, which allows it to monitor incoming and outgoing data traffic. An example of this is the point where a corporate local area network is connected to the global Internet (compare Figure 35.1). All systems of a protected subnetwork are normally protected in the same way — that is, from the same undesirable data packets — with access control implemented by an Internet firewall at the subnetwork level.

[1]This chapter is an excerpt from the book "Security in Fixed and Wireless Networks", reprinted with kind permission of John Wiley & Sons.

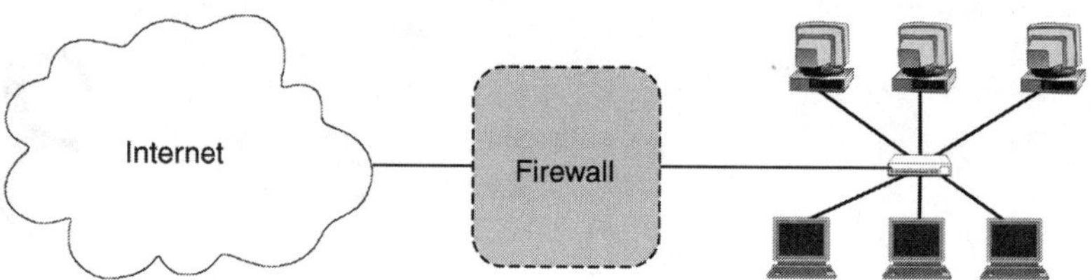

FIGURE 35.1   Firewall placement between a protected network and the Internet.

An Internet firewall therefore provides a central point where access control decisions are made and implemented [Sem96]. It is also an effective means for monitoring the Internet activities of subnetworks and preventing certain security problems from proliferating from one network area to another.

Firewalls can — if at all — provide only limited protection against malicious insiders who are determined to bypass the security strategy of a subnetwork. In particular, a firewall cannot influence any data traffic that is not passing through it. This self-evident fact is often overlooked. An example of this is a modem dial-up access that is set up with a system placed behind the corporate firewall. Furthermore, a firewall cannot protect against totally new attack patterns and also only offers limited protection against the intrusion of software viruses.

A differentiation is made between two diametrically opposed approaches to the basic security strategy that can be implemented by an Internet firewall [WC95]:

The *Default-Deny Strategy* is based on the principle that "Anything that is not explicitly permitted is denied." First an analysis is undertaken to determine which network services users of the protected subnetwork require and how these services can be made available in a secure way. Only those services for which a legitimate need exists and that can be supplied in a secure manner are provided. All other services are blocked through technical means. An example of this approach is permitting the transmission of HTTP traffic to any arbitrary systems and SMTP traffic to an e-mail gateway. All other data traffic would be blocked by the firewall. The advantage of this strategy from the perspective of security is that it also provides limited protection from unknown threats.

The *Default-Permit Strategy* follows an opposite approach, which is, "Anything that is not explicitly forbidden is permitted." With this approach services categorized as high-risk or contravening local security strategies are explicitly blocked through appropriate measures in the firewall. All other services are considered permissible and not blocked by the firewall. For example, such a strategy could mean that traffic from the *Network File System (NFS)* and *X11* protocols are generally blocked and that Telnet connections are only permitted to a specific host. This is the approach most users prefer because it usually gives them more flexibility. However, this strategy can by principle only offer protection from known threats.

As the description of these two approaches indicates, each Internet service blocked or permitted by a firewall always has to be examined specifically. The following section provides a brief overview of popular Internet services and their transport within the framework of the TCP/IP protocol family.

35.2   Firewall-Relevant Internet Services and Protocols

Internet Services: In practice, the following Internet services are of particular importance in the context of Internet firewalls:

- *Electronic mail (e-mail)*, normally implemented with the *Simple Mail Transfer Protocol (SMTP)*.
- *File exchange* on the basis of the *File Transfer Protocol (FTP)*, *Network File System (NFS)*, and *NetBIOS* protocols.
- *Remote terminal access* and *remote command execution* through *Telnet, RLogin,* or *Secure Shell (SSH)* protocols.
- *Usenet News*, which is distributed through the *Network News Transfer Protocol (NNTP)*.

- *World Wide Web* on the basis of the *Hypertext Transfer Protocol (HTTP)*.
- *Information about people* that can be called up using the *Finger* protocol.
- *Real-time conferencing systems*, such as *CUseeMe, Microsoft Netmeeting, Netscape Conference*, and the multicast-based *MBone Tools*.
- *Name services*, which are provided in the Internet through the *Domain Name System (DNS)*.
- *Network management* based on the *Simple Network Management Protocol (SMTP)*.
- *Clock synchronization* using the *Network Time Protocol (NTP)*.
- *Graphical window systems*, with *X11* being the most popular protocol in use.
- *Printing services*, based, for example, on the *Line Printing Protocol (LPR)*.

These services are normally realized through client/server programs and the application protocols used between these programs. The data units of the application protocols are usually transported either in the segments of a TCP connection or in the datagramms of the UDP protocol. TCP segments and UDP datagrams are transported in IP packets that in turn are sent in protocol data units of the layer 2 protocols used on the different communication links between source and destination computer (e.g., Ethernet, FDDI, ATM).

The two tuples *(source-IP address, source port)* and *(destination-IP address, destination port)* are used to address individual application processes. A *port* represents a number comprising two octets that explicitly identifies the Service Access Point (SAP) of an application process within an end system.

Figure 35.2 shows the frame format of an IP packet that is transporting a TCP segment.The following protocol fields are of particular importance for Internet firewalls:

Fields of the access protocol ("layer 2 protocol", not shown): The fields of particular interest are those that identify the contained layer 3 protocol (e.g., IP, Appletalk, IPX) and the source addresses of the access protocol (e.g., Ethernet-MAC address).

Fields of the IP protocol: Of special interest here are the source and destination addresses, the field Flags, which also contain one bit to display IP fragments, the protocol identification of the user data contained (e.g., TCP, UDP), and possible options such as the stipulation of the transmission path by the source. However, IP options are seldom used for anything other than attacks.

Fields of the TCP protocol: The fields of particular interest are the source and destination ports that have a limited influence on determining the sending and receiving application since many popular Internet services use well-defined port numbers. In addition, individual bits of the control field have an effect on connection control. An ACK bit is set in all PDUs except in the first connection establishment PDU, which means that an ACK bit that has not been set identifies a connect PDU. An SYN bit is set in all PDUs except in the first two PDUs of a connection. A packet with a set ACK bit and an SYN bit that is not set

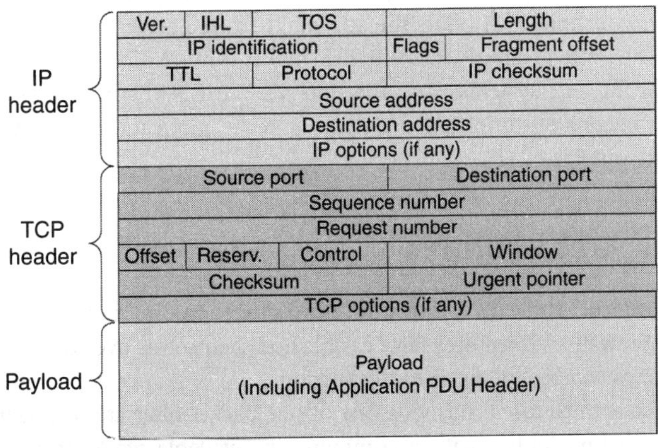

FIGURE 35.2 Frame format of an IP packet with a TCP segment.

therefore identifies the PDU accepting the connection. Lastly, an RST bit permits a connection to be terminated immediately without displaying a further error message and can therefore be used to terminate a connection without providing additional information.

Fields of the application protocol: In some instances, it may also be necessary or meaningful for the Internet firewall to check the protocol fields of an application protocol. However, these vary from application to application and therefore will not be discussed here.

35.3   Terminology and Building Blocks

Internet firewalls are usually made up of certain building blocks that fulfill specific tasks and possess certain characteristics. This section briefly introduces the relevant terminology.

An *Internet firewall* (hereafter also simply referred to as firewall for short) is a component or a group of components that restricts access between a protected network and the Internet or another untrustworthy network. In this context, access refers to the exchange of data packets. Even if the term may appear to be an unusual choice at first glance, it should convey that, from the view of security, access control decisions are made and implemented on the basis of a firewall.

Packet filtering specifies the actions a specific system takes when selectively controlling the flow of data packets from and to a specific network. Packet filtering is an important basic technique for implementing access control at the subnetwork level in packet-oriented networks such as the Internet. The packet filtering process is also referred to as *screening*.

A *Bastion host* is a computer that has special security requirements due to the network configuration that makes it more vulnerable to certain attacks than the other computers in a subnetwork. A bastion host within a firewall often represents the main point of contact between application processes within a protected network and the processes of external hosts.

A computer is identified as a *dual-homed host* if it has at least two network interfaces and is therefore "at home" in two subnetworks.

A *proxy* is a program that deputizes for the client processes of a protected subnetwork and communicates with external server programs. Proxies relay approved requests from internal clients to the actual servers and also forward server responses to the internal clients.

The term *perimeter network* identifies a subnetwork that is inserted between an external and an internal network in order to create an additional security zone. Perimeter networks are often referred to as *demilitarized zones (DMZ)*.

Network Address Translation (NAT) is a technique in which an intermediate system deliberately modifies the address fields in a data packet in order to implement an address conversion between internal and external addresses. For example, with this technique, a large number of systems can be connected to the Internet on the basis of a small number of externally valid addresses (i.e., in the Internet). Although NAT is not actually a security technique, it offers the advantage from a security perspective that the internal addresses will not be known to external attackers. This makes it more difficult for deliberate attacks to be executed to individual internal systems from the outside. With NAT, the "initiative" for specific data streams must basically originate from internal systems, and the application processes of internal systems are not visible from the outside.

35.4   Firewall Architectures

As mentioned above, Internet firewalls can be built from one component or a group of components. As a result, a number of firewall architectures have established themselves over the years. The structures of some of the popular ones will be explained in this section.

The simplest of these architectures consists solely of one packet filter and is illustrated in Figure 35.3. This architecture can be built either with a commercially available workstation computer using two network interfaces or a dedicated router that also incorporates basic packet filter functions. The protection

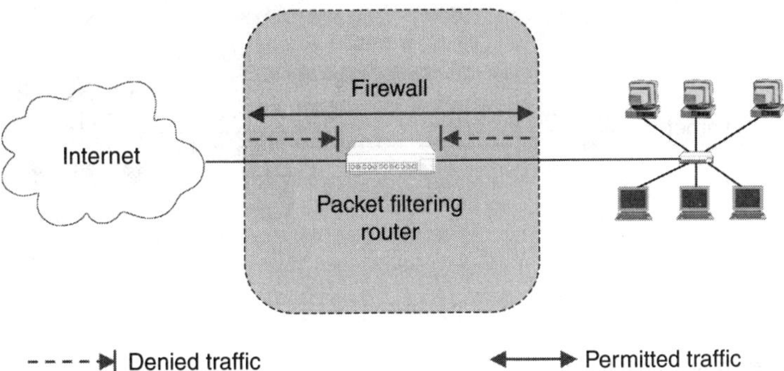

FIGURE 35.3   Architecture of a packet filter firewall.

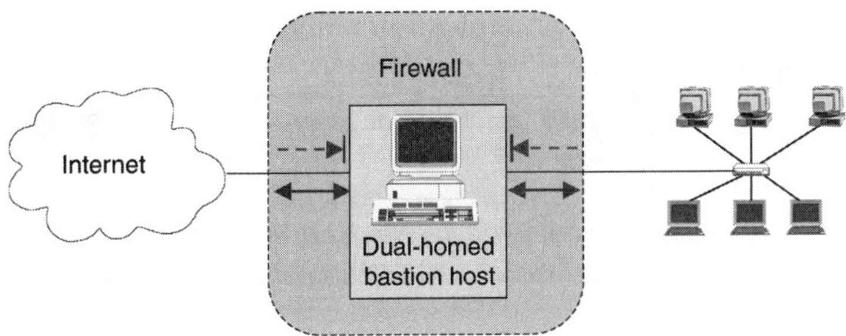

FIGURE 35.4   Dual-homed-host architecture.

function of this architecture is based solely on filtering undesirable data packets between a protected network and the Internet (or other untrustworthy subnetwork). In Figure 35.3 (and in other figures in the rest of this section), a straight arrow shows the "permitted" data traffic and a dotted line shows the "blocked traffic."

Figure 35.4 illustrates the minimally more complex *dual-homed-host architecture*. Although there may be no difference between the hardware configuration of this architecture and a packet filter (if the packet filter is implemented on a workstation computer), the software on the dual-homed host provides additional security functions because its main task is not limited to pure packet filtering.

A dual-homed host provides proxy services to internal and external clients; thus, for security reasons IP packets do not need to be routed directly into and out of the protected subnetwork. The protected network in this case is only attached to the Internet through monitored proxy services. In some cases, the dual-homed host can also handle routing functions for IP packets. However, appropriate packet filter functionality must then also be implemented on the host to prevent the protective effects of the firewall from being lost.

With this architecture, the dual-homed host is the central attack point for potential attackers. Therefore, it should be regarded as a bastion host and protected accordingly (also compare Section 35.6).

The drawback of this architecture is that, depending on the size of the protected subnetwork and the bandwidth of the Internet connection, a dual-homed host can develop into a performance bottleneck.

Figure 35.5 shows a *screened-host architecture*, which spreads the functionality of the packet filter and the proxy server over two components. The packet filter allows permitted traffic to pass between the Internet and the internal bastion host and blocks all direct traffic between other internal computers and the Internet. In some cases, certain traffic streams can also be permitted directly between the internal computers and the Internet; however, this can introduce potential vulnerabilities. The bastion host in turn provides proxy services for communication between internal computers and the Internet.

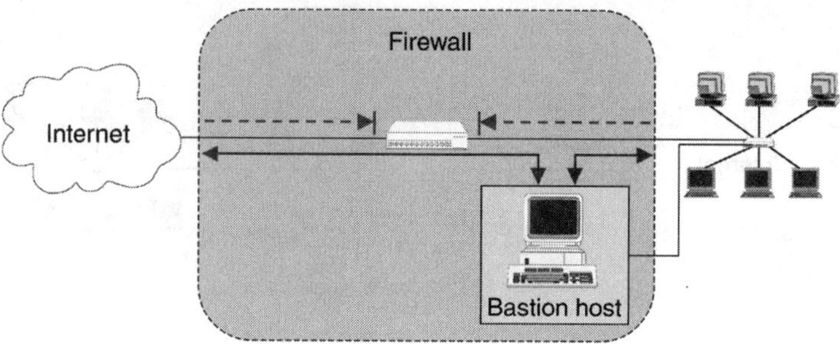

FIGURE 35.5   Screened-host architecture.

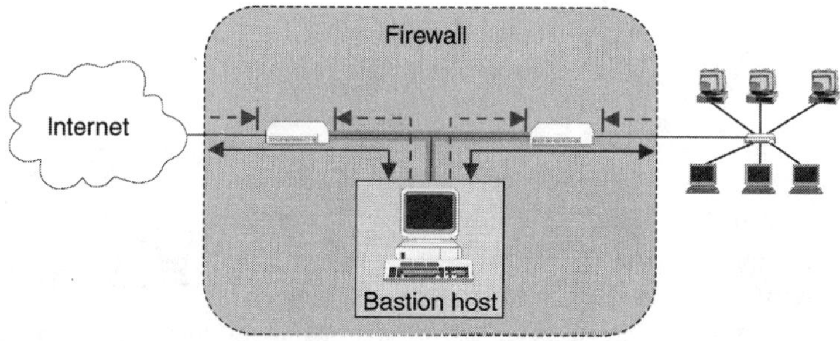

FIGURE 35.6   Screened-subnetwork architecture.

Two main advantages can be gained from separating packet filter functionality and proxy services into two consecutively switched devices:

In terms of security, a bastion host will be better protected from attacks on potentially available services that should not be accessible from outside. Furthermore, compared to proxy services, relatively simple software can be used to implement packet filter functionality, which means that a relatively compact software installation is all that is required on the respective system. This is particularly an advantage because complexity is the "main obstacle" in the construction of secure systems.

Another benefit of this architecture is that it can also provide a performance advantage. When certain data streams (e.g., Internet telephony) are routed directly into an internal network, they do not have to "cross through" the same computer that also provides the proxy services.

In terms of the physical structure of the networking, it should be noted that a bastion host could sometimes also be connected to the same internal subnetwork for which it provides security services. In this case, its relationship with the firewall rather tends to be of a logical nature since it is only conditional on the packet filter rules. It is particularly important to note that a bastion host essentially has special protection needs and therefore should physically be more difficult to access than the systems protected by it (also see Section 35.6). However, this kind of networking presents a security problem even if the bastion host is a physically separate installation: depending on the networking technique used (e.g., Fast-Ethernet-Hub), attackers who are able to compromise a bastion host can then directly sniff the traffic in the internal network.

The *screened-subnetwork architecture* illustrated in Figure 35.6 counters the particular threat of a potentially compromised bastion host by setting up a *demilitarized zone*, also referred to as a *perimeter network*. This zone is created using two packet filters between which one or more bastion hosts is embedded for the purpose of providing proxy services. This zone is also available for the connection of the servers that provide services to external computers, such as WWW and FTP. The main task of the second packet filter is to protect the internal network from the threat of a possibly compromised bastion host.

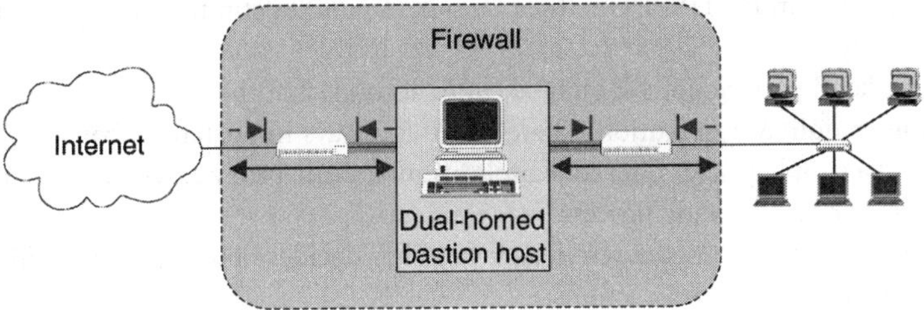

FIGURE 35.7 Split-screened subnetwork architecture.

The *split-screened subnetwork architecture* shown in Figure 35.7 is an even more sophisticated architecture in terms of security. In this architecture, a dual-homed bastion host divides the demilitarized zone into two separate subnetworks. Using proxy services, the bastion host is better able than a simple packet filter to exercise finely granulated control over the data streams. Furthermore, an outer packet filter protects the bastion host from external attackers and an inner packet filter protects the internal subnetwork from the bastion host if it is compromised. The architecture thus provides an "in-depth defense," comparable to that of a knight's castle where attackers normally have to surmount several protective walls before they can penetrate the interior.

35.5 Packet Filtering

As our discussion on the different firewall architectures has shown, the packet filtering function plays a central role in securing individual subnetworks. This section takes a detailed look at the role of packet filtering.

The first question in this context is which protective functions IP packet filtering can provide for access control?

Theoretically, protective measures of any complexity are feasible because all data exchanged between two or more entities in the Internet are ultimately transported in IP packets. In practice, however, the two basic considerations listed below have proven useful in the decision criteria for designing protective measures:

1. Selection operations, which require detailed knowledge about the functioning of higher level protocols or a comprehensive evaluation of the protocol data units transmitted in previous IP packets, are easier to implement as specific proxy services.
2. Simple selection operations that have to be executed quickly and always evaluate individual IP packets can be implemented more efficiently through the use of packet filters.

This kind of *basic packet filtering* uses the following information to make its selection decision:

- Source IP address,
- Destination IP address,
- Transport protocol,
- Source address port,
- Destination address port,
- In some cases, specific protocol fields (e.g., TCPs, ACK-bit, and SYN-bit), and
- Network interface where an IP packet was received.

In addition to these basic functions, *stateful packet filtering* is also sometimes used in selection decisions. This variant, also referred to as *dynamic packet filtering*, ensures that incoming UDP packets are only allowed to pass if they are a response to previously observed outgoing packets.

Lastly, packet filters can also handle basic *protocol checks*. An example is the checking of IP packets that are sent to a DNS port for correct formatting in accordance with the DNS protocol. Another related point is that packets containing an HTTP-PDU in their data part should definitely not be forwarded to certain IP addresses.

The design of packet filter functions should take into account that the more complex they become, the more resources they require for execution. Therefore, it is always important to consider whether a two-step approach using a simple packet filter and application-specific proxy services would produce a more efficient mechanism for the checking process.

Based on the filtering rules configured in it, a packet filter decides how to proceed with each IP packet it looks at. This involves the following actions:

- Forwarding the packet
- Dropping the packet
- In some cases, logging either an entire packet or parts of it
- Possibly sending an error message to the sender, taking into account that error messages could be helpful to any attackers who are preparing further attacks.

Because packet filters normally attempt to separate a protected subnetwork from a less trustworthy network, an implicit understanding of the traffic direction is required for the specification of filter rules. Therefore, IP packets that are received from a network interface located outside the protected network are referred to as *inbound traffic*. Accordingly, packets that are received from the network interface of a protected subnetwork are considered *outbound traffic*. When packet filter rules are specified, the information identifying which network interface received the IP packet is usually based on one of the three alternatives *"inbound," "outbound,"* or *"either."*

Placeholders (also called "wildcards") can be used to specify source and destination IP addresses: for example, "125.26.*" refers to all IP addresses that begin with the prefix "125.26."

The IP addresses in the following examples are often simply indicated as *"internal"* or *"external"* to keep the discussion neutral in terms of any particular network topology. Often, when the source and destination ports are given, entire ranges are specified, for example, ">1023" for all port numbers higher than 1023.

One basic assumption is that packet filter rules are evaluated in the sequence in which they are specified and that the first rule applicable to a packet determines the action to be taken. Over the years, it has been proven that this method is the easiest one for system administrators to understand and the best one for keeping errors in rule specification to a minimum.

The set of packet filter rules developed below illustrate the concepts described so far and are aimed at ensuring that only e-mail traffic is allowed between a protected subnetwork and the Internet (example according to [ZCC00]).

Internet e-mail is being exchanged between two SMTP servers in TCP connections. The initiative for the exchange originates from the server that is delivering the e-mails. This server establishes a TCP connection to destination port 25 of the receiving SMTP server for this purpose, selecting a number higher than 1023 as the source port. Since incoming as well as outgoing e-mails are allowed, packet filter rules have to be designed for both directions.

Table 35.1 shows the first approach that can be taken. The two rules A and B are used to transmit incoming e-mails and rules C and D fulfill the same task for outgoing e-mails. The last one, rule E, is aimed at denying all other data traffic ("Default-Deny Strategy").

TABLE 35.1 Example of a packet filter specification

Rule	Direction	Src. addr.	Dest. addr.	Protocol	Src.port	Dest. port.	ACK	Action
A	Inbound	External	Internal	TCP		25		Permit
B	Outbound	Internal	External	TCP		>1023		Permit
C	Outbound	Internal	External	TCP		25		Permit
D	Inbound	External	Internal	TCP		>1023		Permit
E	Either	Any	Any	Any		Any		Deny

Rule A allows inbound IP packets that are being sent to destination port 25 and are transporting TCP segments as payload. For the TCP connection to be established and maintained between the two SMTP servers, the corresponding response packets of the server receiving the e-mails must be allowed to pass outside. Rule B therefore allows outbound IP packets to port numbers higher than 1023.

The following examples help to explain the protection function of these rules:

Assume that an IP packet with a forged internal source address is being sent to the protected subnetwork from the outside: since inbound packets are only forwarded if they have external source addresses and internal destination addresses (rules A and D), this kind of attack will be successfully defeated. The same applies to outgoing packets from an internal attacker making use of external source addresses (rules B and C).

This filter specification is also effective at blocking incoming Telnet traffic, because Telnet servers normally wait at port 23 for incoming connections and incoming traffic is only allowed by the packet filter if it is being sent either to port 25 or to a port higher than 1023 (rules A and D). On the basis of rules B and C, the same applies to outgoing Telnet traffic.

However, this ruleset is not totally effective. For example, it does not block either incoming or outgoing X11-Protocol traffic. An X11 server usually waits at port 6000 for incoming connections and the corresponding client programs use destination ports from the set of numbers above 1023. Because of rule B, inbound X11-PDUs can therefore pass the packet filter and, because of rule D, outbound X11-PDUs are not blocked either. This provides potential attackers with an attractive opportunity because the X11-Protocol leaves the door open to various different attacks, for example, reading and manipulation of screen information and of keyboard input.

This problem can be dealt with through the inclusion of the source port in the filter specification. Table 35.2 shows the resulting filter rules. With this ruleset, outbound packets are only permitted to ports higher than 1023 if they originate from source port 25 (rule B). Therefore, PDUs from internal X11-client or X11-server programs are blocked. Similarly, rule D blocks incoming traffic to X11-client or X11-server programs.

In practice, however, one cannot assume that an attacker will not use Port 25 as the source port for an attacking X11-client. In this instance, the packet filter would allow the traffic to pass.

Table 35.3 therefore shows an improved version of the filtering specification that also incorporates the ACK bit of the TCP protocol header and includes rules B and D. Since the ACK bit is assumed to be set in rule B, this rule can no longer be used to open an outgoing connection (the TCP connect-request is

TABLE 35.2 Inclusion of source port in a packet filter ruleset

Rule	Direction	Src. addr.	Dest. addr.	Protocol	Src.port	Dest. port.	ACK	Action
A	Inbound	External	Internal	TCP	>1023	25		Permit
B	Outbound	Internal	External	TCP	25	>1023		Permit
C	Outbound	Internal	External	TCP	>1023	25		Permit
D	Inbound	External	Internal	TCP	25	>1023		Permit
E	Either	Any	Any	Any	Any	Any		Deny

TABLE 35.3 Inclusion of an ACK bit in the packet filter ruleset

Rule	Direction	Src. addr.	Dest. addr.	Protocol	Src.port	Dest. port.	ACK	Action
A	Inbound	External	Internal	TCP	>1023	25	Any	Permit
B	Outbound	Internal	External	TCP	25	>1023	Yes	Permit
C	Outbound	Internal	External	TCP	>1023	25	Any	Permit
D	Inbound	External	Internal	TCP	25	>1023	Yes	Permit
E	Either	Any	Any	Any	Any	Any	Any	Deny

TABLE 35.4   Inclusion of bastion host in packet filter ruleset

Rule	Direction	Src. addr.	Dest. addr.	Protocol	Src.port	Dest. port.	ACK	Action
A	Inbound	External	Bastion	TCP	>1023	25	Any	Permit
B	Outbound	Bastion	External	TCP	25	>1023	Yes	Permit
C	Outbound	Bastion	External	TCP	>1023	25	Any	Permit
D	Inbound	External	Bastion	TCP	25	>1023	Yes	Permit
E	Either	Any	Any	Any	Any	Any	Any	Deny

identified by an ACK bit not set). Similarly, the stipulation of a set ACK bit in Rule D prevents the opening of incoming TCP connections sent to ports with a number higher than 1023.

As a basic guideline, it should be noted that each packet filter rule designed to allow TCP-PDUs for outgoing connections (or the reverse) should request a set ACK bit.

If the firewall contains a bastion host, then the SMTP server should preferably be run on this system (or the SMTP host should be regarded as a bastion host). Incorporating the IP address of the bastion host into the packet filter ruleset can increase the security of the protected subnetwork, because in this instance attacks on the SMTP server program will only affect the bastion host. Table 35.4 shows the resulting packet filter ruleset.

With a screened subnetwork firewall, two packet filters have to be equipped with the appropriate rulesets: one for the traffic between the Internet and the bastion host and one for the traffic between the bastion host and internal subnetwork.

35.6   Bastion Hosts and Proxy Servers

As explained in Section 35.3, the bastion host is the main point of contact between application processes within a protected network and the processes of external systems. Depending on the firewall architecture, it is responsible for the task of packet filtering and/or provides dedicated proxy services for specific applications. As the principal point of contact for the protected network to external systems, a bastion host has a higher exposure to threats than the systems to which it provides security functions.

Even though a dedicated packet filter can be used to reduce these risks, special measures should also be taken to secure the proxy server itself. This section begins by presenting some general observations on securing bastion hosts and concludes by briefly reviewing how proxy services are implemented.

The principles applied to building a "secure" bastion host are ultimately merely extensions to the strategies used for securing any key system in a network. The basic rule is that the system configuration of a bastion host should be kept *as simple as possible*. If there are any services that should not be offered on the bastion host, they should not even be installed on it.

One should also basically anticipate the fact that a bastion host might be compromised. Internal systems therefore should not place more trust in the bastion host than is absolutely necessary. For example, no file systems should be exported to the bastion host, no Login should be enabled from the bastion host to internal systems, etc. If possible, the bastion host should be integrated into the network infrastructure in such a way that it is unable to eavesdrop on data traffic in the internal subnetwork. (This can be done through the use of another packet filter to separate the network or through a network interface that does not support "promiscuous mode," which is necessary for eavesdropping.)

In addition, extensive event logging should take place on the bastion host to ensure early detection of any attacks or successful compromising of the host. Note that a successful attacker should not be able to tamper with event logging at a later time. One possibility is to transfer the event entries over the serial interface to a separate log-computer that does not have a network interface itself.

As long as no important reasons exist to the contrary, the bastion host should be a preferably "unattractive" target for potential attackers. For example, slow computers are less attractive targets and also of less use to an attacker than high-speed systems. At the same time, one has to bear in mind that a bastion host should not turn into a performance bottleneck for the internal network and sometimes also has to offer resource-intensive services (e.g., Web proxy services). Nonetheless, the potential usefulness of a bastion host to an attacker should be minimal. Therefore, the only software tools that should be installed on it are those that are essential to its operation. Above all, no user accounts should be available on a bastion host. This will prevent anyone from spying on data or executing password attacks.

Lastly, a bastion host should definitely be installed in a secure location where general physical access is not allowed to ensure that the risk of manipulation is kept to a minimum. Its system configuration should be backed up at regular intervals. A routine reinstallation of the system from a security copy can also be considered for the purpose of correcting any possible manipulation that has occurred. However, it is important to be aware that any existing weakness in a system configuration that has already been exploited by an attacker can be used again for gaining access to the system. A reinstallation of the system, therefore, does not reduce the risk of it being compromised.

A bastion host equipped with proxy services can give users of a protected subnetwork the illusion that all systems in this network are able to access Internet — although in reality, it is only the bastion host that has access.

There are two main different types of proxy services:

If a proxy server analyzes the commands of the application protocol and interprets its semantics, it is referred to as an *application-level proxy.*

If, on the other hand, the proxy server is restricted to forwarding application PDUs between client and server, it is called a *circuit-level proxy.*

Possible "candidates" for proxy services include FTP, Telnet, DNS, SMTP, and HTTP. The following situation usually exists when a proxy server is available: the user of the proxy service thinks that his or her system is exchanging data directly with the actual server, and the server is under the illusion that it is exchanging data with the bastion host.

If a proxy service is used, applications-specific data streams have to be rerouted to the proxy server. In each case, the proxy server must be informed of which application server it is to use for the connection. The following four implementation possibilities exist:

Proxy-aware user procedures: An example of this variant is the use of Telnet proxies. Here, the user first registers on the bastion host and then from there establishes a Telnet session to the actual desired server.

Proxy-aware client software: In this case, the client software is responsible for transferring the corresponding data to the proxy server. This approach is used, for example, with proxy-enabled Web browsers where client software only has to be notified once of the name or address of the Web proxy.

Proxy-aware operating system: With this variant, the operating system (using the appropriate configuration data) is responsible for the rerouting and transfers the required parameters to the proxy server.

Proxy-aware router: This variant transfers the task of activating the proxy to an intermediate system, which reroutes connections initiated by internal client computers to a proxy computer, also sending the necessary address data to the computer.

35.7　Summary

Internet firewalls provide access control at the subnetwork level. For this purpose, Internet firewalls are normally installed between a protected subnetwork and a less trustworthy network, such as the public Internet.

At the same time, it should be noted that an Internet firewall is only capable of monitoring those data streams that pass through it and that it basically cannot provide any protection from internal attackers.

Depending on the level of security being sought and the resources available, an Internet firewall can be constructed using either a single or a group of components. Over time, a number of architectures have evolved as a result, each offering specific security features. In practice, the *packet filter*, the *dual-homed-host*, the *screened-host*, the *screened-subnetwork*, and the *split-screened-subnetwork* architectures are the most popular.

The central elements used by all these architectures are *packet filters*, which make access decisions for each IP packet routed through them, and *proxy servers*, which enable an applications-specific monitoring of data streams. The computing systems with specific functions installed in a firewall are called *bastion hosts*. Because they have a high level of exposure to attackers, bastion hosts must be installed with particular care and should be constantly monitored.

References

[Sch03] Schaefer, G., Security in Fixed and Wireless Networks, John Wiley & Sons, Ltd, UK, 2003.

[Sem96] SEMERIA, C., *Internet Firewalls and Security,* 3Com Technical Paper, 1996.

[WC95] WACK, J.P. and CARNAHAN, L.J., *Keeping Your Site Comfortably Secure: An Introduction to Internet Firewalls,* NIST Special Publication 800-10, 1995.

[ZCC00] ZWICKY, E., COOPER, S., and CHAPMAN, B., *Building Internet Firewalls,* 2nd ed., O'Reilly, Sebastopol, CA, 2000.

36

Ad Hoc Networks

Holger Karl
Technische Universität Berlin

36.1 Introduction

Ad hoc networks are networks that are spontaneously set up, often for a dedicated purpose, and where devices can configure themselves into a connected network. When wireless communication is used, *ad hoc* networks often use multiple wireless communication hops between a message's sender and receiver to overcome the limited range of wireless communication. This relaying of messages is typically done by the participants of the *ad hoc* network themselves, merging the concepts of terminals and network equipment.

Such self-organized *ad hoc* networks make new applications possible, but problems, for example, in the link, routing, and transport layer have to be solved. This chapter gives an overview of these problems and their principal solutions.

36.2 The Case for *Ad Hoc* Networks

Typical communication networks consist of devices connected by wired or wireless communication links where the structure of these links and the communication relationships between these devices are carefully planned and set up: desktop computers are connected to Ethernet switches serving a floor of an office building; these switches are connected to a router, which in turn talks to the Internet backbone. The topology of the entire network is judiciously planned and which device is allowed to communicate how with which other device is carefully preconfigured; integrating new devices in such a network can be cumbersome.

This inflexibility of traditional communication networks can be an obstacle for many types of applications. In any scenario where devices are dynamically deployed or where the communication relationships between devices change frequently, a preplanned network structure is rather onerous. If only devices could spontaneously communicate with each other on an asneeded basis, new application scenarios would be possible. Some dynamic scenarios are:

- Connecting a laptop to either a home, office, or hotel network without having to reconfigure it.
- Automatically connecting the laptops in a meeting room to enable the meeting participants to share documents.
- Electronic price tags in shopping malls update their displayed price and product information via a wireless interface.
- Vehicles in an automated factory floor communicating with each other and with the machinery they pass by as well as accessing a wired backbone network, regardless of the current vehicle position.
- Cars inform each other about driving hazards ahead.
- Wireless communication for rescue teams in disaster relief operations, for example, fire fighters at a wild fire.
- Rapid deployment of numerous tiny, sensor-equipped devices, for example, temperature sensors, over such a disaster area, communicating with each other and with rescue teams.

To support such application types, the concept of *ad hoc networking* has been developed. This concept describes the construction of a communication network out of participating devices on-the-fly in a simple, dynamic manner with as little human intervention as possible. These networks are intended for a dedicated purpose rather than to serve as a general networking infrastructure.

For some of these scenarios, wireline communication would make sense: a laptop can be plugged into a hotel network using a cable; for most scenarios, however, wireless communication is the medium of choice for an appealing solution. Hence, *ad hoc* networks are usually associated with wireless communication. In a wired *ad hoc* network, the questions to solve are more associated with problems of autoconfiguration and seamless integration with existing restrictions, for example, a company laptop might not be allowed to access the company's internal network from any location. These problems are not the topic of this chapter.

Wireless communication, on the other hand, introduces some difficulties, in particular, the limited communication range. While this is no problem for laptops in a meeting room, firefighters will be spread out over a region too large for an individual device to cover with a single wireless transmission. In traditional wireless communication systems, for example, cellular telephony systems like Global System for Mobile Communication (GSM), this problem is overcome by additionally installing infrastructure components — base stations or access points — that both are capable of wireless communication and are connected to the fixed network. These infrastructure components then handle communication between distant devices. But when fighting a wildfire, such infrastructure is typically not available nor would it be cost-efficient to deploy it on a factory floor. The only possible recourse is hence to use other terminal devices to help in communication between distant terminals: these terminals would volunteer to relay data from one terminal toward its destination; data would travel over several wireless communication hops, making such a network a *wireless multi-hop ad hoc network*.

Such a multi-hop *ad hoc* network can be used as a stand-alone solution — communication among firefighters is a prime example. Or they can be combined with infrastructure devices, forming hybrid networks. An evident goal for such hybrid devices is to extend the reach of an access point by relaying data of a terminal too far away for direct communication via intermediate terminals to/from the access point. This would reduce the number of access points required to cover a given area; it is also particularly useful in environments where many obstacles can impede radio communication despite short distances, for

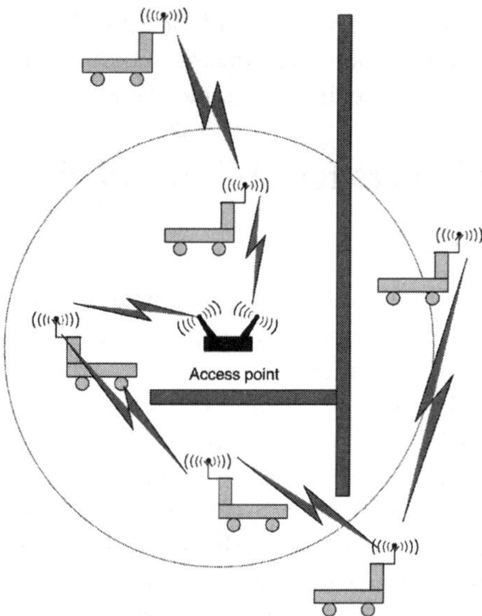

FIGURE 36.1 Typical scenario for hybrid cellular/multi-hop *ad hoc* networks. The dotted circle indicates the maximum range of the access point, and dark rectangles represent obstacles; vehicles are in wireless contact with the access point directly or indirectly.

example, on a factory floor, where an access point would have to communicate "around the corner". One such scenario is illustrated in Figure 36.1.

Using wireless multi-hop *ad hoc* networks in industrial communications carries a number of advantages and disadvantages and a number of challenges that have to be tackled. The biggest advantages are the simple deployment and low cost — no or only little infrastructure is necessary — and the inherent support of mobile devices — partially owing to wireless communication, and also owing to the liberation from coverage restrictions imposed by traditional communication systems.

This ability to support extended mobile scenarios is also the biggest challenge to a Mobile *Ad Hoc* Networks (MANET): when all terminals, including the ones used for relaying data, are free to move about, how to decide where to send data so that it eventually arrives at its destination? Is it possible to give any type of guarantees regarding the connectivity of such an *ad hoc* network as a whole or regarding basic networking metrics like through-put and delay; is it possible to make the communication service rendered by an *ad hoc* network dependable, predictable, and configurable? For some scenarios, additional issues like the energy efficiency of the communication — if the terminals are battery-operated — or processing power required for communication are important as well.

These are the typical problems tackled by wireless mobile *ad hoc* networks research and development. As this is still an active research field, not all of these questions have been completely solved yet. This chapter gives an overview of these problems and some of the currently more popular solutions.

Regarding the physical layer, *ad hoc* networks differ not much from other wireless networks, unless very tight constraints on the processing power are present, for example, in sensor networks where very simple physical layers are used. Other layers differ more substantially. Specifically, the approaches for wireless link layers amenable to support *ad hoc* networks are discussed in Section 36.2. Section 36.3 explains the core problem of mobile *ad hoc* networks, how to find a communication peer despite the changing topology of the network. The following Section 36.4 looks at problems of transport protocols in mobile *ad hoc* networks. Section 36.5 focuses on a specific type of *ad hoc* networks, wireless sensor networks, which are based on tiny devices containing sensors, a simple controller, and a radio modem to form an *ad hoc* network embedded in the physical environment. Finally, Section 36.6 wraps up the chapter by looking at some smaller issues.

This chapter cannot do justice to the entire field of *ad hoc* networks. Some recommendable books on the topic are [8, 12]; but currently many books on wireless communication also include material on *ad hoc* networks that is more detailed than the survey presented here.

36.3   Wireless Link Layers for Mobile *Ad Hoc* Networks

Centralized or Distributed?

The basic tasks of the link layer in general are framing, medium access, local error, and flow control. Not all of them are specific in the wireless case; only the main problems and solutions for the wireless case are briefly summarized here. A link layer for a wireless transmission medium has to contend with additional challenges compared to a wired link layer: the channel quality varies over time, available bandwidth is smaller, the wireless medium usually has broadcast character, and a wireless sender cannot usually sense the medium while it is sending, precluding collision-detection-based mechanisms for medium access.

In an infrastructure-based wireless network where all mobile terminals only communicate with a base stations, a link layer is nevertheless conceptually simple. A straightforward option is to put all medium access and link layer control in such a central base station and use Time Division Multiple Access (TDMA) to assign each terminal a fixed time slot by the base station. This approaches is used, for example, by GSM or HiperLAN/2. A prerequisite for this approach is, however, that every terminal is close enough to the base station to receive synchronization or scheduling commands, for example, when is which terminal allowed to send.

But this proximity requirement to a base station or even the mere need for one is the very property that *ad hoc* networks are intended to overcome; therefore, such centralized link layers are not immediately applicable to *ad hoc* networks. Distributed solutions are required.[1] One such distributed solution is the IEEE 802.11 system, along with its Request To Send (RTS)/Clear To Send (CTS) mechanism to combat the hidden and exposed terminal problems. IEEE 802.11 is based on a Carrier Sense Multiple Access (CSMA) scheme, where a terminal has to check whether the medium is idle before it is allowed to send and has to observe certain times between listening and sending and a given backoff scheme when a busy medium is detected. The goal is to avoid interfering with another ongoing transmission and causing high packet error rates. This turns mere CSMA into a CSMA with Collision Avoidance (CSMA/CA) scheme.

Hidden and Exposed Terminals

While this CSMA/CA scheme is reasonably efficient as long as all terminals are in mutual sending range, it breaks down as soon as this is no longer the case. Consider the scenario in Figure 36.2: terminals A and C both want to send to B but are too far away from each other to receive their respective transmission; they are *hidden* from each other. C would start its own transmission to B when A is already sending and so would destroy A's packet. To solve this hidden terminal problem and avoid the collisions caused by it, a sending terminal A is required to first send, after observing the CSMA regulations, an RTS message to the destination, which answers with a CTS. This CTS is then also observable by C, now knowing that there is an ongoing transmission at B with which C must not interfere.

This solution to the hidden terminal effect is satisfactory but comes at additional overhead for the RTS/CTS messages. Also, it fails in some situations: for example, when B's CTS message collides with the transmission of an additional terminal D's RTS (similar in setup to the one in Figure 36.3). Terminal C would then assume the medium to be still idle and answer the retransmission of D's RTS with a CTS, disturbing the ongoing packet reception at B.

---

[1]There is some research on how to build *ad hoc* networks out of centralized protocols like HiperLAN/2. The core idea is to dynamically elect controlling terminals out of all mobile terminals and use these controlling terminals for coordination. While such an approach has some advantages regarding the possible Quality-of-Service guarantees, it is quite complicated and not the main track of current research.

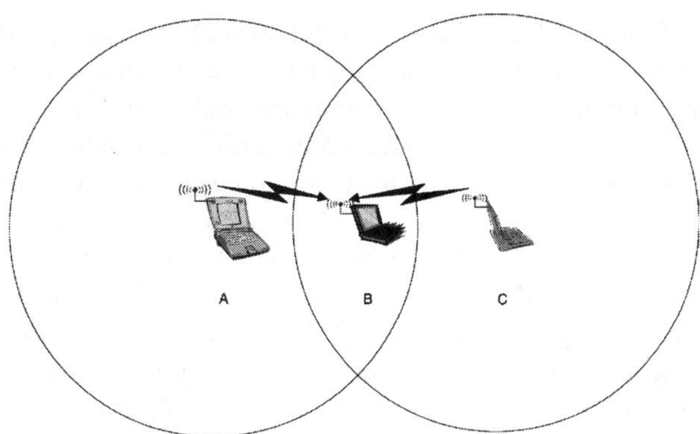

FIGURE 36.2   Hidden terminal scenario.

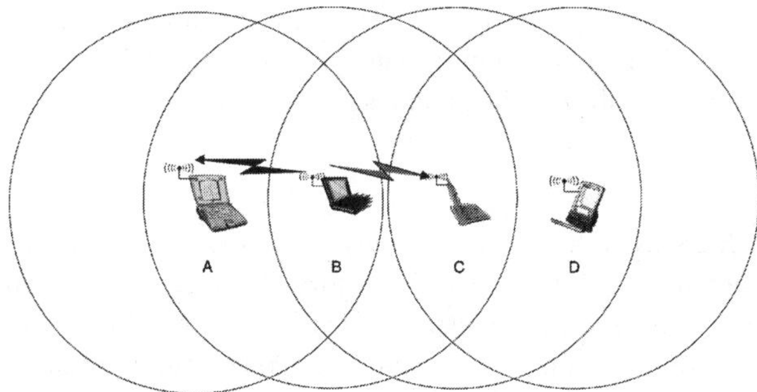

FIGURE 36.3   Exposed terminal scenario.

The *exposed terminal effect* occurs in a similar scenario, shown in Figure 36.3. Suppose terminal B sends to A and C intends to send to D, with C and A again beyond mutual range. As C senses an occupied medium, it defers its transmission according to CSMA/CA rules until the medium is idle again. However, this delay is unnecessary as C's transmission would not interfere with A's reception process. Hence, possible parallelism is wasted.

Shortcomings of Multi-hop *Ad Hoc* Networks

When putting these mechanisms to work in a multi-hop environment, the resulting capacity — measured as the amount of data that reaches its destination per unit time — of the entire network is interesting. Intuitively, one would expect an increase in capacity by the introduction of multi-hopping: an individual transmission covers only shorter distances, allowing multiple transmissions to proceed in parallel, increasing the number of packets delivered per unit time.

As theoretical considerations using a set of idealized assumptions have shown [3], the total capacity of a network does indeed increase with the number of nodes, but only slowly. In fact, the useful throughput per node actually *decreases* as terminals are added to a network. The ultimate limitation is the need to share resources with nearby terminals to communicate over long distances. This result, however, strongly depends on the communication pattern: when terminals only intend to communicate in their vicinity, substantial capacity gains are achievable.

Attempting to apply these theoretical results to a practical multi-hop system reveals some of the major shortcomings of today's protocols. When simulating an *ad hoc* network, a common observation is that the data rate available to each node is indeed only a small fraction of the raw link bandwidth, for example, in a network of 100 terminals each with a 2 Mbit/sec radio transceiver, the available data rate per terminal is on the order of a few kilobits per second. In [6], LI et al. considered several configurations of terminals, in particular, a linear chain of terminals where the first terminal sends data to the last terminal in line, where only neighboring terminals can communicate and interference extends to terminals two hops away. Using such a simple interference model, a Medium Access Control (MAC) protocol should achieve about one-fourth of the raw link bandwidth as end-to-end goodput by pipelining transmissions through the chain. As simulations show, IEEE 802.11 only converges to about one-seventh of the raw link bandwidth as the chain length increases. The capacity is also quite sensitive to the offered load: any load beyond the sustainable throughput will result in congestion losses as IEEE 802.11 cannot effectively arbitrate medium access. Other investigations have attempted to overcome these problems by applying cleverer distributed scheduling schemes where the medium access of a terminal is made contingent upon some partial knowledge of the load situation at other terminals. An alternative is to look at rate adaptation to solve this problem; but no overall solution exists yet. How to solve this problem is still an open research problem.

Apart from medium access control, the link layer has to solve some other problems, in particular, local retransmissions for erroneous packets and flow control. The retransmission schemes are identical for both traditional and multi-hop wireless networks. The flow control, on the other hand, is actually still an open problem as the lack in efficiency of an appropriate flow and congestion control is one of the reasons for the capacity degradations described in the previous paragraph.

Summary

The IEEE 802.11 approach discussed here is not the only possible one, although it is currently by far the most popular one. A reader interested in some of the history of and research on wireless MAC protocols should have a look at Multiple Access with Collision Avoidance (MACA), MACA with Acknowledgement (MACAW), Floor Acquisition Multiple Access (FAMA), Power-Aware Multi-Access Protocol with Signaling (PAMAS), Dual Busy Tone Multiple Access (DBTMA), and Elimination-Yield Non-Preemptive Priority Multiple Access (EY-NPMA), to name but a few protocols.

In summary, some of the currently available link-layer technologies for wireless networks are in principle suitable for multi-hop *ad hoc* networks. As long as the network is expected to work only under low load, the link layer performance is acceptable. In high load situations, however, the sustainable capacity collapses. Moreover, it is difficult to make any guarantees for such distributed link layers. Some work regarding fairness between terminals has been performed, but this problem is also not completely solved. Strict guarantees on the throughput that a particular flow obtains are difficult to make in a distributed multi-hop system and applications therefore should not rely on such guarantees.

36.4   Network Layer for Mobile *Ad Hoc* Networks

Basic Tasks and Main Problems for *Ad Hoc* Routing

The most basic task of a communication network is to move data from its sender to its intended receiver. In most networks, and in *ad hoc* networks as well, senders and receivers are usually not directly connected to each other. Intermediate terminals have to forward an incoming data packet that is not destined for itself toward the packet's destination. However, when a terminal has more than two neighbors, it has to decide to which neighbor to forward a packet. Providing the necessary information to a terminal to make these forwarding decisions is the *routing problem.*

In conventional, fixed networks, the routing problem has to be solved as well; a popular solution is Dijkstra's shortest path algorithm that minimizes the number of hops a packet has to travel. Minimizing the hop count is also desirable in wireless *ad hoc* networks, but the global view of the network that Dijkstra's algorithm requires is not available. While there are other fixed-network routing algorithms that only rely on local information (e.g., distance vector algorithms), they still cannot cope with the idiosyncrasies of *ad hoc* networks. The main problem is the high volatility of the neighborhood relationships: nodes that move around lose contact with their previous neighbors and find new ones. As a consequence, routes from a sender to a receiver can change frequently. Hence, routing protocols are required that can *discover* routes quickly and *maintain* them, if necessary, despite changes in the network's topology. Additionally, energy conservation concerns and relatively high errors rates complicate the design of *ad hoc* routing protocols.

Several families of *ad hoc* routing protocols have emerged during the last few years; the number of individual solutions and enhancements is staggering — in fact, routing has been the most active research area in *ad hoc* networks at all so far. More detailed information can be found in several survey and comparison papers, for example, [9]. An additional complication is presented by asymmetric links, when not all terminals have the same transmission power; this is not covered here.

Most of these routing protocols use an abstract view of the network as a graph with dynamically created and destroyed edges, representing the dynamic nature of the links between terminals. How to detect when a link appears or disappears is actually not a simple problem; common mechanisms are the exchange of "hello" messages to test links or by trying to overhear other terminals' activity, for example, receiving an acknowledgment from a hitherto unknown node. Especially for nodes that are on the borderline of reachability with a highly erroneous link between them, it is not clear whether and when to signal the availability of such a link to the routing protocol. Often, conservative choices are made in that only fairly stable links are used to avoid oscillation of topology information and frequent routing updates.

Besides the routing problem itself, two other important problems also pertain strongly to the networking layer: assigning addresses to nodes that do not previously possess unique identifiers and connecting *ad hoc* networks to larger networks like the Internet. These two topics will also be briefly touched upon here.

Ad Hoc Routing Protocols

Two main types of *ad hoc* routing protocols can be distinguished: "table-driven" and "source-initiated on-demand" protocols. Table-driven protocols are more akin to traditional routing protocols in that they keep a routing table in every participating node. This table contains information via which way any other node in the network can be reached. Handling changes in the network topology — new nodes joining or nodes moving around, forming new neighbors — is the challenge for these protocols, especially since the signaling traffic should be kept at a minimum and routing loops should be prevented. As these protocols try to keep routing information up to date upon topology changes, irrespective of whether this information is actually needed, these protocols are classified as *proactive*. Traditional routing protocols follow similar goals in general, but the requirements in the *ad hoc* case are much tighter. Examples for table-driven protocols are Destination Sequenced Distance Vector (DSDV), Cluster Switch Gateway Routing (CSGR), or Wireless Routing Protocol (WRP).

Source-initiated on-demand protocols adopt a different approach. Instead of attempting to always keep routing tables to all possible destinations up to date, they only become active when a packet is to be sent somewhere and no routing information is available — they are *reactive* rather than proactive. While this economizes on signaling traffic, it introduces delay for determining a route when the first packet to a destination is to be sent; similarly, when a route breaks and a new route has to be found, delay occurs. Additionally, as long as a route exists, it is a challenge for these protocols to capitalize on new, shorter routes becoming available. The classic example is Dynamic Source Routing (DSR); *Ad Hoc* On-Demand Distance Vector Routing (AODV) is currently very popular. Another example is Temporally Ordered Routing Algorithm (TORA), which combines some proactive and reactive components.

A third possibility for an *ad hoc* routing protocol is to not only consider the abstract network topology but to additionally make use of external information. One example for such information would be the location of terminals. As location knowledge is important for several types of context-dependent applications, this information can also be exploited for routing purposes. The basic idea of such *location-aided routing* is to use the sender's and destination's position to choose from the possible next hop terminals one that is closest to the destination. Location information is very useful to easily obtain a packet in the destination's vicinity; however, to actually reach the destination, usually some other *ad hoc* routing protocol is also required (e.g., owing to the limited precision of location information). Also, location-aided routing has to overcome specific problems like dead ends. A popular example is the Location-Aided Routing (LAR) protocol; an overview can be found in [7].

Address Configuration

The dynamic nature of *ad hoc* networks results in some more fundamental problems other than just routing and forwarding: it is not clear as to how to assign addresses to terminals. In the Internet, addresses are assigned either statically or dynamically to reflect the topology of the network, enabling aggregated routing tables. While *ad hoc* routing protocols do not require such a structured address space, addresses still have to be unique. In the simplest case, all nodes have a built-in unique identifier, for example, a MAC id. For some *ad hoc* networks, however, this is not a justifiable assumption (see Section 36.5).

Hence, a mechanism is required that automatically configures the addresses of nodes joining an *ad hoc* network so that no address is assigned twice. A particular challenge here is to handle the case of merging two partitions into one larger *ad hoc* network and to resolve the inevitable address conflicts arising from the independent operation of two sundered partitions.

Several approaches exist in the literature, for example, Duplicate Address Detection (DAD) or weak DAD along with some extensions for large-scale *ad hoc* networks. One approach, Distributed Dynamic Host Configuration (DDHC), is capable of even handling merging partitions: a joining terminal contacts another terminal that is already part of the *ad hoc* network and that negotiates a new address on behalf of the new terminal with the rest of the network. Partitioning is handled by using a partition identifier; when two partitions merge, address tables are exchanged between partitions and for conflicting address, one of the nodes has to newly join the *ad hoc* network.

Gatewaying to the Internet

In many application scenarios, an isolated *ad hoc* network would only be of limited use. Rather, a terminal within an *ad hoc* network usually also desires to communicate with the Internet at large. To do so, some of the terminals in the *ad hoc* network need to be in contact with the fixed network and act as a gateway between *ad hoc* and fixed network; such a gateway could, for example, be an access point in a wireless LAN. *Ad hoc* routing protocols can route packets to the gateway; but the question is how to integrate an *ad hoc*-connected node in the Internet.

The Internet's basic method for supporting mobile terminals with varying points of connection is the Mobile IP protocol. An *ad hoc*-connected terminal should be integrated into Mobile IP; more precisely, the *ad hoc* routing protocol and Mobile IP's agent system have to be coordinated. One proposal for integrating AODV and Mobile IP is describedin [11]. MIPMANET is an alternative approach that uses the *ad hoc* routing to tunnel any Internet-destined data to an Mobile IP foreign agent, which is located on the gateway node.

When connecting *ad hoc* networks to the Internet, a important special case is that the *ad hoc* network moves in its entirety from one gateway to the next — think of a person with a Personal Area Network (PAN) moving around. Here, the amount of signaling traffic should be minimized. This is currently on-going research work.

Summary

Forwarding and routing have long been the research focus in *ad hoc* networks. The dynamic nature of a mobile *ad hoc* network makes routing an interesting challenge, for which a number of stable solutions exist. However, as there is no optimal routing protocol, the scenario requirements do determine the best choice. Additional problems, like address configuration and gatewaying *ad hoc* networks to the Internet, are also solved at least conceptually.

36.5 Transport in Wireless *Ad Hoc* Networks

Reliable Transport over Wireless Links

The task of transport protocols is to abstract away the details of a communication network and to provide a simple-to-use service interface to an application programmer, an interface that does not consider the structure of the network any longer but that allows the programmer to concentrate on the communication on the application level. The service provided by the transport layer is therefore oriented toward the endpoints of communication; it is the first *end-to-end* communication layer.

The simplest end-to-end service is just to take a packet the application wants to send and transport it to the peer endpoint — without attempting to repair errors in the network or worrying about overloading the network or the receiver (congestion and flow control). Such a service is provided by the User Datagram Protocol (UDP) in the Internet. Putting a UDP-like service on top of an *ad hoc* network is no challenge as the promises made are very small.

A more interesting case is reliable service with flow and congestion control, for example, in the manner of the Transport Control Protocol (TCP): a wireless network will produce packet losses, and the service provided by the routing layer is also quite different from a fixed Internet as routes are likely to change, resulting in quite unpredictable round-trip delays. For TCP, both effects challenge its basic assumptions that the network as such is in principle stable and pretty much reliable, and that the only reason for a network to lose packets is congestive overload. Under this assumption, the obvious reaction to a lost packet (e.g., detected by a timeout) is to reduce the load offered to the network in order to reduce congestion.

In a wireless network, however, a packet loss can be due to bad link quality or mobility-induced rerouting and not to congestion. Then, reducing offered load, that is, delaying the retransmission, is counterproductive; the lost packet should rather be repeated quickly. Overcoming this TCP deficiency has been tackled by many research efforts; the challenge is to maintain TCP's end-to-end semantics as best as possible. Some of the resulting approaches work in the end systems (e.g., TCP Westwood), some rely on support from entities within the network, for example, an access point that connects wired and wireless communication to each other; an overview is given in [2].

Some of this research was performed with mostly older radio hardware and pretty poor link-layer performance in mind and for scenarios where a large fixed network is connected via a single wireless last hop to a mobile terminal. With reliable link layers that perform local retransmissions, some of these problems go away; but on the downside, the round-trip time variance is increased. This fools TCP's round-trip time estimation, resulting in somewhat similar problems.

In either case, the case of a single wireless hop combined with a fixed network is not the most relevant case here. What happens when TCP is faced with a multi-hop wireless network?

TCP and Multi-hop Wireless Networks

When running TCP over a wireless multi-hop network, the problems of the error-prone wireless channel combine with additional complications introduced by the multi-hop nature and, possibly, by terminal mobility and ensuing route changes and re-routing delays.

As already discussed in Section 36.2, current wireless link layers cannot efficiently deal with multi-hop wireless networks. When a reliable end-to-end transport protocol like TCP is added, performance degrades further — even if terminals do not move around and routes are stable. One problem is that data packets and TCP acknowledgments conflict with each other, trying to travel over the same route in opposite directions. A second problem can appear by an interaction of TCP with the backoff timers deployed in the MAC protocol. As a consequence, several TCP connections sharing such a multi-hop route are unable to efficiently share the available bandwidth among each other, but one connection can capture the route for seconds or even minutes. Different wireless MAC protocols are more or less susceptible to these problems, but the problem is present for most of them.

TCP's behavior over multi-hop networks becomes even more complicated when terminals additionally become mobile and routes are likely to break. With many *ad hoc* routing protocols, TCP packets or acknowledgments stuck in a now dead end route are likely to be silently discarded; may be the breaking of the route is signaled to the original sender. Two things now occur in parallel: a route repair will take place, searching for a new route between the sender and destination, and TCP will eventually time out and resend packets again. This timing out and resending of packets evidently degrades throughput. If route repair takes a long time, TCP will back off its timers, taking a potentially very long time until even the resending of packets is attempted again — by which time the terminals might already have moved around so much that the just repaired route is already useless. While the details of this process depend on the used routing protocol, mobility will in general decrease the useful throughput that TCP can realize.

Possible Improvements

When attempting to improve TCP performance over multi-hop networks, the basic problem is similar to the wireless last hop scenario: TCP has no information about the actual reason why it observes variations in packet delivery rate and round-trip time; hence, it cannot distinguish between congestion- and mobility-caused problems.

An obvious approach is to provide this information to TCP. This idea is well known even from fixed networks, where routers can provide explicit notifications about imminent or existing congestion to a TCP sender. For an *ad hoc* network, the relevant information is the failure of a (possibly remote) link, which is already provided by many routing protocols. Using such information can result in significant performance improvements. Similarly, such feedback information is used in the TCP-feedback scheme to explicitly freeze TCP timers until a route repair message arrives in order to avoid that TCP mistakenly reduces its congestion window size. The newer TCP-BuS approach uses similar messages, but just stops sending during route repair and extends timer values explicitly (instead of just stopping all timers).

Summary

Using TCP over a wireless multi-hop *ad hoc* network is challenging due to both the nature of the erroneous wireless link itself and the interaction of TCP with wireless MAC and *ad hoc* routing protocols. Errors on the wireless link can, to a certain degree, be handled by reliable link layers; TCP and MAC layer interactions require a careful choice of MAC mechanisms and timer values. Routing protocols and TCP should be modified to explicitly exchange information about route status and on-going route repairs. With these extensions to standard protocols, TCP in *ad hoc* networks becomes a viable concept, but here also considerable research is still required.

36.6   Sensor Networks

Applications for Sensor Networks

In recent years, miniaturization and low-power circuit design have made a new technological vision possible: bring down the cost, size, and power consumption of a device capable of wireless communication, minimal computation, and sensing to a level where such devices can be mass-deployed and embedded into our physical environment. Typical sensing tasks for such a device could be temperature, light, vibration, sound, radiation, etc. The hoped-for size would be a few cubic millimeters, the target price range less than one US$, including radio front end, microcontroller, power supply, and the actual sensor. All these components together in a single device form a so-called *sensor node*.

Joining some or even many of these sensor nodes together forms a new type of an *ad hoc* network, a *wireless sensor network*. Even large-scale sensor networks can be handled by applying *ad hoc* networking technologies. The sensor nodes will configure themselves into a connected network, determining which sensor node is responsible for which task: merely sensing the environment or also forwarding data between different nodes? For a large number of nodes, manual configuration is next to impossible.

Based on such a technological vision, new types of applications become possible. Applications include environmental control such as fire fighting or marine ground floor erosion but also installing sensors on bridges or buildings to understand earthquake vibration patterns; surveillance tasks of many kinds like intruder surveillance in premises; deeply embedding sensing into machinery where wired sensors would not be feasible, for example, because wiring would be too costly, could not reach the deeply embedded points, limits flexibility, represents a maintenance problem, or disallows mobility of devices; tagging mobile items like containers or goods in a factory floor automation system or smart price tags for foods that can communicate with the fridge, etc. The possibilities abound — sensor networks could potentially become a disruptive technology when the basic size and cost problems are solved.

Wireless sensor networks have recently received considerable of attention in the research literature; a good recent survey paper is [1].

Main Challenges

While wireless sensor networks share many similarities with *ad hoc* networks, there are also several important differences. These differences result from both technological structure of sensor nodes and from the intended application scenarios. The most important ones are:

Because sensor nodes have to be small and cheap, the processing and storage capacity of a node is very limited. Also, the radio front ends have to very simple, disallowing high bandwidth or complicated transmission schemes. This fits with intended application scenarios, where only occasionally small amounts of data are to be expected from a simple sensor, on the order of a few kbits/sec at most.

Energy is even more at a premium than in mobile *ad hoc* networks — it is just not acceptable to frequently change batteries of hundreds or thousands of deeply embedded devices; rather, applications demand lifetimes of months or better years on a very small battery. Hence, the entire construction of sensor networks is highly energy efficiency driven, also incorporating technologies like energy scavenging.

Sensor nodes might not necessarily have a unique address (e.g., a MAC identifier). What is more, the individual identity of such a sensor node is not even particularly important; rather, the *data* are.

As data are important, so is the location where data are sensed. For example, in a wildfire detection scenario, it is useless to know that temperature has exceeded a threshold without knowing where this happened.

Given the large numbers of sensor nodes in a sensor network, the simplest possible deployment is crucial — for a wildfire, sensor nodes should just be dropped from an airplane.

Interfacing a sensor network in a meaningful form to an outside network like the Internet is also important.

New Approaches in Sensor Networks

These challenges require some new approaches for communication in sensor networks. Some of the more important issues are outlined here.

Energy-efficient communication protocols: Already in *ad hoc* protocols, the issue of energy-efficient MAC, routing, and transport protocols has come up. In a sensor network, it is possible and useful to take the hardware and physical layer peculiarities as the driving force of protocol and algorithm design. On the other hand, the application characteristics in sensor networks are very different from traditional networks, which also affects the design of communication protocols.

An additional perspective to consider is the low data rates that are generated by a sensor node. For a MAC layer, this means that most of the time should be spent in a sleep mode as even just sensing the channel for an incoming transmission is quite energy-intensive. The most elegant solution would be a so-called wake-up radio, where only a primitive receiver is used that can detect the presence of a signal but is unable to actually decode it. Upon signal detection, the wake-up receiver would power up the more complicated and more power-hungry main receiver to perform the actual reception.

Using more conventional receiver architectures, CSMA schemes can be designed to support such low data rates. Alternatively, sleeping cycles between neighboring nodes could be synchronized (somewhat akin to TDMA schemes) as, for example, proposed in the S-MAC protocol; a combined TDMA/FDMA approach with similar goals is described in [10].

Also, on higher layers, in particular routing, energy efficiency is a driving force. For example, when selecting the route to be taken between source and destination, it stands to reason to choose nodes that currently have a high-energy supply. Work in this area abounds and is summarized in survey papers (e.g., [1]).

Data-centric communication: More specific to sensor networks is a change in paradigm. The interaction patterns are likely to differ from traditional networks: there will be a large amount of short request/reply-type communication, possibly with periodic replies initiated by a single request ("provide sensor reading every one second for the next five minutes"); "file-transfer"-like communications will not take place. Moreover, it should be possible to address the data, not the individual nodes; requests like "Is the temperature anywhere larger than 100° C?" or "Give me the temperature in the living room" are desirable communication primitives. Bearing in mind the required energy efficiency, specific communication protocols are necessary.

These requirements make the communication data-centric rather than node-centric like in traditional networks. One possible programming paradigm to support such data-centric communication is publish/subscribe: a node could publish its *interests* [5] in certain types of events (e.g., a temperature threshold is exceeded). This interest is distributed in the network, and sensor nodes with matching data can subscribe to these interests, providing the relevant information. One option how to implement this programming paradigm is the "directed diffusion" approach, where paths in the network are reinforced implicitly by data traveling toward the interested party. However, setting up such interest structures is a comparatively expensive undertaking; hence, the invested energy must be amortized over many replies.

Data-centric protocols can also be implemented in forms other than directed diffusion, for example, the rumor routing approach can be regarded as one such possibility.

Determining location: For many sensor applications, each sensor needs to know its own location to support correct application semantics or to leverage location information to support protocol operation like routing. This location information must be generated with very simple means: neither is it feasible to explicitly provide the location to a sensor during installation (think of deployment from an aircraft) nor is expensive and bulky location equipment like Global Positioning System (GPS) acceptable. Hence, the location of sensor nodes must be determined by the sensor network itself.

Without any sensor node having access to an absolute location, evidently only location in a relative coordinate system can be determined. Alternatively, a few of the sensor nodes could know their absolute position; these nodes can be used to translate such a relative coordinate system into an absolute one. In either case, the relative location of sensor nodes has to be determined. The basic options here are to use the signal attenuation between a sender and a receiver, which can serve as a rough indicator of distance; the time when signals arrives or the offset in time of two different signal times generated at the same time (e.g., a sender could transmit both a radio and an ultrasonic impulse and the receiver measures the time difference of their arrival), or information about the direction from which signals are received. Each of these options has pros and cons; the differential method using two types of signals tends to have the best accuracy but also requires the largest hardware overhead.

Irrespective of what method to estimate distance and angles between sensor nodes is used, this information will always be inaccurate. Algorithms are required that take this information and compute a most likely estimate for the true locations of all sensor nodes. Several such algorithms exist, but no final solution is available here as all algorithms suffer from different kinds of trade-offs.

A good overview paper on this topic is [4].

Trading off communication and computation — Aggregation: Unlike common sensor systems so far, a wireless sensor network can perform computation inside the network. Compared with wireless communication, computation (on usual hardware platforms) is relatively cheap in power consumption. An evident consideration is therefore how to best trade off communication against some additional computation in the network.

One obvious possibility is to suppress duplicate information to be forwarded to the same receiver. More generally, more or less arbitrary aggregation functions can be usefully computed in sensor networks, for example, averaging or counting functions. Such aggregation is even more useful since the reading of any one individual sensor is usually not of utmost importance or might not even be very reliable as sensor nodes, to be cheap, can only rely on low-quality sensors. More typically, a user would be interested in obtaining an overview of a large area and the "average" situation could be more relevant. How to best choose protocols and aggregation functions, in particular in the presence of node and link failures and the need to be energy-efficient, is still an active research area.

Summary

Wireless sensor networks can be regarded as a specific form of *ad hoc* networks. The stringent miniaturization and cost requirements make economic usage of energy and computational power an even bigger issue than in normal *ad hoc* networks. Moreover, specific applications require a rethinking of some of the basic paradigms with which communication protocols are engineered.

As wireless sensor networks are still a young research field, much activity is still on-going to solve many open issues. As some of the underlying hardware problems, especially with respect to the energy supply and miniaturization, are not yet completely solved, wireless sensor networks are at the time of this writing not yet ready for practical deployment. Nevertheless, these problems could be resolved in the near future.

Perhaps the most pressing conceptual problem is the handling of mobility. In a sensor network, three types of mobility can be distinguished: the sensor nodes themselves can move; an observed phenomenom can move, for example, an intruder in a surveillance application, and the requester of information from a sensor network. None of these mobility types is satisfactorily handled by today's sensor network protocols; the current research results here are still just a first step.

36.7 Further Issues in *Ad Hoc* Networks

Besides the main problems for *ad hoc* networks discussed in the previous sections, there are some additional issues that deserve brief mentioning.

Service discovery: Finding a node that offers a specific service rather than finding a specific node directly can be a handy faculty to solve several problems. For example, when trying to connect an *ad hoc* network to a fixed network, the *ad hoc* routing protocol has to first discover which node(s) offer the service of forwarding a packet onto the fixed network. General service discovery approaches are, in principle, applicable to *ad hoc* networks as well, but are likely to suffer performance degradation. Some initial work on integrating service discovery protocols into *ad hoc* routing protocols exists, but is not yet completed.

Frequency assignment: A key technique in cellular systems is the spatial reuse of spectrum: a given frequency band is used within a cell, neighboring cells use different bands, and a frequency is reused in cells sufficiently far away. This way, terminals/cells do not interfere with each other and multiple transmissions can occur in parallel.

In *ad hoc* networks discussed so far, only a single frequency band is in use, making simultaneous transmissions and hence an efficient forwarding impossible. Using multiple frequencies also in an *ad hoc* network to enable spatial reuse would help to overcome this problem. However, the problem here is more complicated than in cellular networks as terminals would have to listen on one frequency band and to transmit on another one. This requires careful orchestration of switching between frequencies and choosing the times of sending and receiving.

Clearly, a fixed assignment of frequencies to areas is not a good fit with *ad hoc* networks. Rather, a distributed way of choosing frequencies is necessary; the specific challenge is to ensure that a terminal receives at the right moment on the right frequency. There are several "multi-channel" link layers for *ad hoc* networks that try to solve this problem. Alternative approaches do not try to solve the frequency assignment on a per-link basis, but group terminals together in clusters and perform frequency assignments on the basis of such clusters of terminals (of course, membership in such a cluster is dynamic if terminals move around).

In practice, dynamic frequency assignment is not widely deployed. Even for conventional wireless LAN systems, automatic frequency selection is not common practice; the frequency has to be selected manually for each access point.

Multicasting in ad hoc networks: So far, only unicast communication has been considered: a source sending data to a sink. Sometimes, a source intends to deliver data to all reachable terminals (broadcasting) or to an explicitly determined group of terminals (multicasting). An example for broadcasting has already appeared in Section 36.5 on sensor networks, when one terminal distributes its interest in a set of data to all other nodes. Such a broadcast can easily be implemented by flooding: every terminal repeats a received packet (but makes sure that it is only repeated once).

Multicasting is a bigger challenge and several options exist (the classification here follows [12]). Only those terminals repeat a packet that are on a shortest path from the sender to any one of the destinations (for multiple shortest paths, choose the smallest subset of repeaters such that each destination is served). This is equivalent to constructing a tree for this multicast that is rooted in the sender. Constructing such trees can be complicated, depending on the precise requirements; also, maintaining separate trees for each multicast group can result in scalability problems.

An alternative is not to source these trees individually in each sender, but to use only a single tree for all multicast groups. As this tree is rooted at a designated node in the network core (not necessarily in a sender), these approaches are called *core-based tree*. The advantage is the reduced maintenance overhead; the main disadvantages are the introduction of a single point of failure — the core node — and a concentration of traffic on the shared links around the core node. In this sense, these approaches are not ideal for *ad hoc* networks. A popular example is an AODV extension supporting multicast.

Multicasting can also be implemented without trees. An alternative distribution structure is to "overlay" a mesh over the *ad hoc* network and forward data along the mesh lines. The main difference to a tree and both advantage and disadvantage is the existence of multiple paths in such a mesh. It introduces resilience to faults, but also the danger of routing loops.

In sensor networks, also convergecast is relevant: collecting data from multiple senders in a single receiver. Convergecast is usually combined with aggregation of data and is discussed in Section 36.5.

Security: Security for and in *ad hoc* networks is of course faced with the common tasks of security in communication networks. But now, the network infrastructure itself can no longer be trusted and new, additional security problems appear. For example, a node could claim that it forwards all routing signaling messages correctly, but actually acts as a "black hole." Moreover, it becomes necessary to provide and police incentives for nodes to play fair: a node that makes use of other nodes forwarding data on its behalf but never reciprocates should perhaps be excluded from the *ad hoc* network. It becomes necessary to protect against malice and selfishness in the network itself. Zhou and Haas [13] give an overview of some security problems in *ad hoc* networks.

Quality of service: A wireless mobile *ad hoc* network, by its very structure, is a collection of uncertainties: the wireless channel is error-prone, medium access is stochastic or deterministic only with high overhead, routing is constantly endangered by terminals moving around, etc. All these peculiarities make it difficult to provide Quality of Service in its stringent interpretation. Even guarantees on simple connectivity is a difficult problem. Nevertheless, more modest approaches to Quality of Service are, of course, possible even in *ad hoc* networks, such as service differentiation between classes. Moreover, hybrid approaches are conceivable where Quality-of-Service-sensitive data are transported via a cellular infrastructure as far as possible, but best-effort traffic can use *ad hoc* mechanisms, taking advantage of increased capacity and energy efficiency.

These limitations have to be weighted against the advantages of *ad hoc* networks in general. Whether the service rendered by an *ad hoc* network is of sufficient quality largely depends on the application that is to be supported.

Implementations: When trying to build *ad hoc* networks, common communication hardware can be used. IEEE 802.11 systems are by nature of the protocol suited for use in *ad hoc* networks (with the many caveats already mentioned); Bluetooth devices can also be coerced into working as *ad hoc* networks (called "scatternets"), although the master/slave structure of Bluetooth is not really amenable for this purpose.

With respect to practically available protocol implementations, a lot depends on the operating system to be used. For Microsoft Windows, especially Windows CE, only a limited number of implementations are available. The lion's share of academic implementation and experimental work is done using Linux, for which consequently many implementations are available as public domain software. Commercially implemented solutions are not relevant at this time.

For sensor networks, special-purpose operating systems from academic institutions, for example, TinyOS, are currently most relevant.

36.8 Conclusions

Ad hoc and sensor networks open the door to new types of applications. The spontaneous, *ad hoc* deployment of mobile terminals, the self-configuration of the network as a whole are considerable advantages. These networks can be built using readily available communication equipment; implementations for most of the necessary protocols are available at least as a prototype.

These prototypes rest on a large amount of theoretical research. Much understanding regarding the dynamics of *ad hoc* networks in different circumstances has already been gained. This work shows that *ad hoc* networks can be expected to work fine as long as the requirements on them are modest: the load should be low and Quality of Service requirements should not be too tight.

Nonetheless, these investigations also show that there is still quite a bit of counterproductive interaction between protocol layers that is difficult to understand and to forecast. Final solutions for *ad hoc* networking protocols that work perfectly in all conceivable circumstances are not yet available; the application scenario's requirements — the load or mobility to be supported, acceptable delays, and outage ratios — still determine the choice of, for example, link layer, routing, or transport protocols to be used in an *ad hoc* network. Moreover, there is still only limited experience with *ad hoc* networks outside of research laboratories available. In particular, experience with large-scale deployments is still rather limited.

Ad hoc and sensor networking will be a substantial part of networking in the future. It is starting today with limited application fields, and it will expand more and more as the remaining conceptual problems are being solved.

References

[1] Akyildiz, I.F., W. Su, Y. Sankasubramaniam, and E. Cayirci, Wireless sensor networks: a survey, *Computer Networks*, 38:393–422, 2002.

[2] Balakrishnan, H., V.N. Padmanabhan, S. Seshan, and R.H. Katz, A comparison of mechanisms for improving TCP performance over wireless links, *IEEE/ACM Transactions on Networking*, Vol. 5, pp. 756–769, December 1997.

[3] Gupta, P. and P.R. Kumar, The capacity of wireless networks, *IEEE Transactions on Information Theory*, Vol. 46, pp. 388–404, March 2000.

[4] Hightower, J. and G. Borriello, Location systems for ubiquitous computing, *IEEE Computer*, 34(8):57–66, 2001.

[5] Intanagonwiwat, C., R. Govindan, D. Estrin, J. Heidemann, and F. Silva, Directed diffusion for wireless sensor networking, *IEEE Transactions on Networking*, Vol. 11, pp. 2–16, February 2003.

[6] Li, J., D.S.J. De Couto, H.I. Lee, and R. Morris, Capacity of *Ad hoc* Wireless Networks, in Proceedings of the 7th Annual International Conference on Mobile Computing and Networking, Rome, Italy, ACM, New York, July 2001, pp. 61–69.

[7] Mauve, M., J. Widmer, and H. Hartenstein, A survey on position-based routing in mobile ad-hoc networks, *IEEE Network*, Vol. 15, pp. 30–39, November 2001.

[8] Perkins, Charles E., Ed. *Ad Hoc Networking*, Addison-Wesley, Upper Saddle River, NJ, 2001.

[9] Royer, E.M. and C.-K. Toh, A review of current routing protocols for ad-hoc mobile wireless networks, *IEEE Personal Communications*, Vol. 6, pp. 46–55, April 1999.

[10] Sohrabi, K., J. Gao, V. Ailawadhi, and G.J.Pottie, Protocols for self-organization of a wireless sensor network, *IEEE Personal Communications,* 7: 16–27, 2000.
[11] Sun, Y., E.M. Belding-Royer, and C.E. Perkins, Internet connectivity for *ad hoc* mobile networks, *International Journal of Wireless Information Networks (special Issue on Mobile Ad hoc Networks),* 9, 2002.
[12] Toh, C.-K., *Ad Hoc Mobile Wireless Networks,* Prentice-Hall PTR, Upper Saddle River, NJ, 2002.
[13] Zhou, L. and Z.J. Haas, Securing *ad hoc* networks, *IEEE Network,* 13:24–30, 1999.

PART II

INDUSTRIAL INFORMATION TECHNOLOGY

Section 3: Industrial Communication Systems

37

Principles of Lower-Layer Protocols for Data Communications

Andreas Willig
*Hasso-Plattner-Institute of
University of Potsdam*

Hagen Woesner
Technical University of Berlin

37.1 Introduction

In packet-switched networks, the lower layers (data link layer, medium access control layer, and physical layer) have to solve some fundamental tasks in order to facilitate successful communication. The lower layers are concerned with communication between neighbored stations, as compared to the layers above (networking layer, transport layer), which are concerned with end-to-end communications over multiple intermediate stations.

The lower layers therefore communicate over *physical channels* and are thus most strongly influenced by physical channel properties (bandwidth, channel errors, etc.), which in turn affect their design. The importance of the lower layers for industrial communication systems is related to the often found requirement for giving *hard real-time and reliability guarantees*: if the lower layers could not guarantee successful delivery of a packet/frame within a prescribed amount of time, this cannot be compensated by any actions of the upper layers. Therefore, a wide variety of mechanisms has been developed to implement these guarantees and to deal with harmful physical channel properties like transmission errors.

In virtually all data communication networks used in industrial applications, the transmission is *packet-based,* that is, the user data is segmented into a number of distinct *packets* or *frames* and these frames are transmitted over the channel. Therefore, the following fundamental problems have to be solved:

What constitutes packet and how are the bounds of a packet specified? How does the receiver detect packets and the data contained in a packet? To this end, *framing and synchronization schemes* are needed, discussed in Section 37.2.

When to transmit a packet? If multiple stations want to transmit their packets over a common channel, appropriate rules are needed to share the channel and to let each station determine when it may send its packets. This problem is tackled by *medium access control* (MAC) protocols, discussed in Section 37.3.

How to cope with channel errors? The topic of *error control* is briefly touched in Section 37.4.

Which packet to transmit next? This is the problem of *packet scheduling*.

How to protect the receiver against too much data sent by the transmitter? This is the problem of *flow control*.

This chapter surveys framing and synchronization, MAC protocols and error control in some detail in the following sections. Some references discussing packet scheduling algorithms are [57, 59–61, 58, Chapter 9]. Flow control algorithms are explained in [20].

37.2   Framing and Synchronization

The problem of synchronization is related to the transmission of information units (packets, frames) between a sending and a receiving entity. In computer systems, information is usually stored and processed in a binary digital form (bits). A *packet* is formed from a group of bits and shall be transmitted to the receiver. The receiver must be able to uniquely determine the start and end of a packet as well as the bits within the packet.

The transmission of information over short distances, for instance, inside the computer, can be done with parallel transmission. Here, a number (say, 64) of parallel copper wires transport all bits of a 64-bit data word at the same time. In most cases, one additional wire transmits the common reference clock. Whenever the transmitter has applied the correct voltage (representing a "0" or "1" bit) on all wires, it signals this by sending a sampling pulse on the clock wire toward the receiver. Conversely, on receiving a pulse on the clock wire, the receiver samples the voltage levels on all data wires and converts them back to bits by comparing them with a threshold.

This kind of transmission is fast and simple, but cannot span large distances, because the cabling cost becomes prohibitive. Therefore, the data words have to be *serialized* and transmitted bit-by-bit on a single wire.[1]

Bit Synchronization

The spacing of bits generated by the transmitter depends on its local clock. The receiver needs this clock information to sample the incoming signal at appropriate points in time. Unfortunately, the transmitters and receivers clocks are not synchronized, and the synchronization information has to be recovered from the data signal; the receiver has to *synchronize* with the transmitter. This process is called *bit synchronization*. The aim is to let the receiver sample the received signal in the middle of the bit period in order to be robust against the impairments of the physical layer, like bandwidth limitation and signal distortions. Bit synchronization is called *asynchronous* if the clocks are synchronized only for one data word and have to be resynchronized for the next word. One common mechanism used for this uses one *start bit* preceding the data word and one or more *stop bits* concluding it. The Universal Asynchronous Receiver/Transmitter (UART) specification defines one additional parity bit that is added to the 8 data bits, leading to the transmission of 11 bits in total for every 8 data bits [3]. The upper row in Figure 37.2 illustrates this.

For longer streams of information bits, it is necessary to synchronize the receiver clock continuously. The Digital Phase-Locked Loop (DPLL) is an electrical circuit that controls a local clock and adjusts it to the received clock that is extracted from the incoming signal [20]. To recover the clock from the signal, sufficiently frequent changes of signal levels are needed. Otherwise, if the wire shows the same signal level for a long time (as may happen for the *Non-Return to Zero* (NRZ) coding method, where bits are directly

---

[1]The term wire is actually used here as a synonym for a transmission channel. Therefore, it could also be a wireless or ISDN channel.

mapped to voltage levels), the receiver clock could drift away from the transmitter clock. The Manchester encoding that is shown in the second row of Figure 37.1 ensures that there is at least one signal change per bit. Every logical "1" is represented by a signal change from one to zero, while a logical "0" shows the opposite signal change. The internal clock of the DPLL samples the incoming signal with a much higher frequency, for instance, 16 times per bit. For a logical "0" bit that is arriving exactly in time, the DPLL receives a sample pattern of 0000000011111111. If the transition between the "0" and "1" samples is not

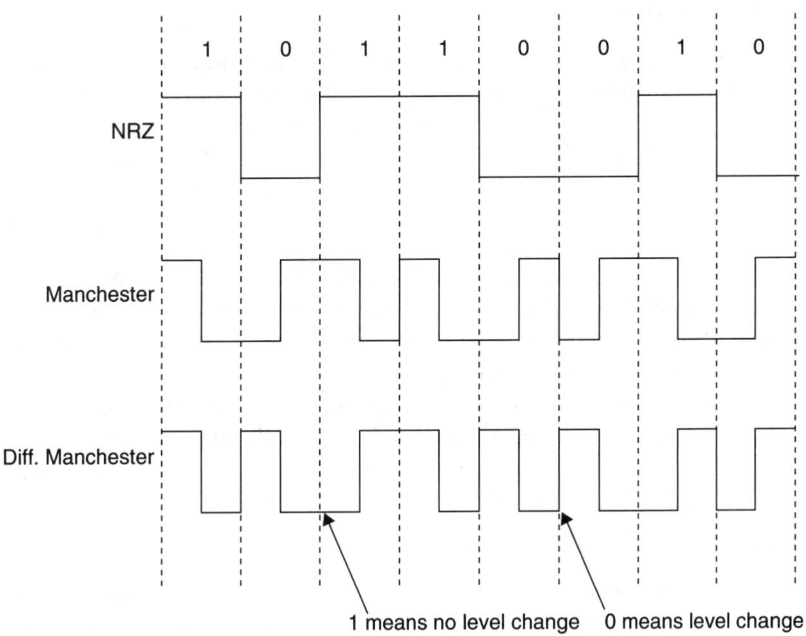

FIGURE 37.1   NRZ, Manchester, and differential Manchester codes.

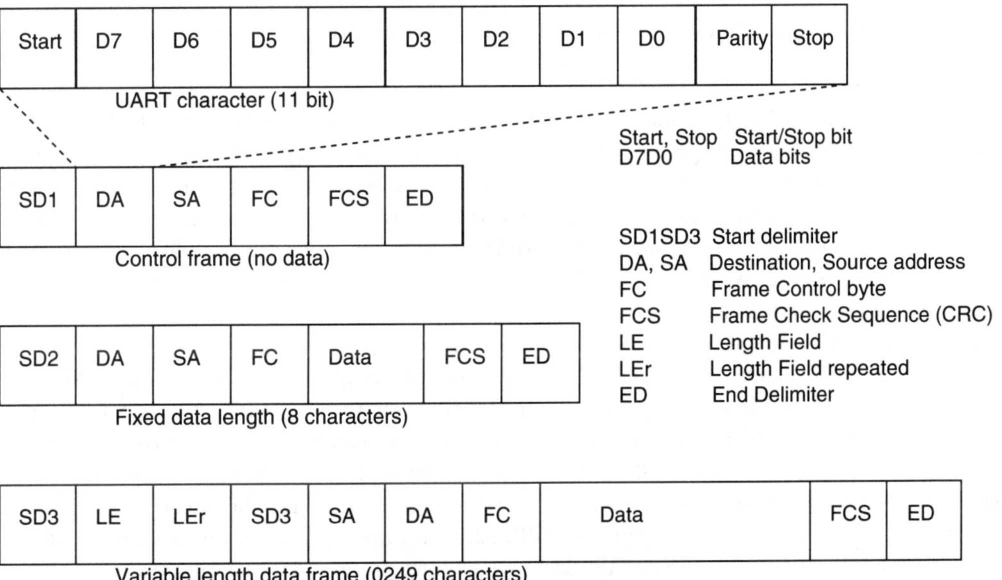

FIGURE 37.2   DIN 19245 Profibus: character and selected frame formats.

exactly in the middle of the bit, but rather left or right of it, the local clock has to be re-adjusted to run faster or slower, respectively. In the classical IEEE 802.3 Ethernet, the bits are Manchester encoded [2]. To allow the DPLL of the receiver to synchronize to the received bit stream, a 64-bit long *preamble* is transmitted ahead of each frame. This preamble consists of alternating "0" and "1" bits that result in a square wave of 5 MHz. A start of frame delimiter of two consecutive "1" bits marks the end of the preamble and the beginning of the data frame.

Frame Synchronization

It is of interest for the receiver to know if the received information was (a) complete and (b) correct. Section 37.4 treats the latter problem in some more detail. To decide about the first problem, the receiver needs to know where a packet starts and ends. The question that arises immediately is that of marking the start and the end of a frame. There are several ways to accomplish this. In real-world protocols, one often finds combinations of them. In the following, the most important will be discussed briefly:

Time Gaps

The most straightforward way to distinguish between frames is to leave certain gaps of silence between them. However, specifically if the medium is shared between several stations, all of them have to obey these time gaps. As it will be seen in CSMA Protocols section, several MAC protocols rely on minimum time gaps to determine if the medium is accessible.

While time gaps are a simple way to detect the start of a frame it should be possible to detect the end of it, too. Using time gaps, the end of the previous packet can be detected only after a period of silence. Even if the receiver detects a silent medium, it cannot be sure if this is the result of a successful transmission or a link or node failure. Therefore, other ways have to be explored.

Code Violations

A bit is usually encoded by a certain signal pattern (e.g., a change in voltage or current levels) that is, of course, different for a "0" and "1" bit. A signal pattern that represents neither of the allowed values can be taken as a marker for the start of a frame. An example for this is the IEEE 802.5 token ring protocol [23], which uses the differential Manchester encoding (see Figure 37.1 and [23]). Here, two special symbols "J" for a so-called positive code violation and "K" for a negative one appear. In contrast to the bit definitions in the encoding, these special symbols do not show a transition in the middle of the bit. Special 8-bit long characters that mark the beginning and end of the frame are constructed from these symbols.

Start/End Flags

Some protocols use special *flags* to indicate the frame boundaries. A sequence of 01111110, that is, six "1" bits in a sequence, marks the beginning and the end of each frame. Of course, since the payload that is being transmitted can be an arbitrary sequence of bits, it is possible that the flag is contained in the payload. To avoid the misinterpretation of a piece of payload data as being the end of a frame, the sender has to make sure that it only transmits the flag pattern if it is meant as a flag. Any flag-like data have to be altered in a consistent way to allow the receiver to recover the original payload. This can be done using *bit* or *byte(character) stuffing* techniques.

Bit stuffing, as it is done in High-level Data Link Control (HDLC) [20], means for the sender to insert a zero bit after each sequence of 5 consecutive "1" bits. The receiver checks if the sixth bit that follows the five ones is a zero or one bit. If it detects a zero, it removes it from the sequence of bits. If it detects a "1," it can be sure that this is a frame boundary. Unfortunately, it might happen that a transmission error modifies the data sequence 01111100 into 01111110 and thus creates a flag prematurily. Therefore, additional mechanisms like time gaps are needed to remove the following bits and detect the actual end of the frame.

Byte stuffing as in Point-to-Point Protocol (PPP) uses the same flag pattern, but relies on a byte-oriented transmission medium [5, 6]. Thus, the flag can be written as a hexadecimal value 0x7E. Every unintentional appearance of the flag pattern is replaced by two characters, 0x7D 0x5E. This way the flag character disappears, but the 0x7D (also called *escape character*) also has to be replaced, if it is found in

the user data. To this end, 0x7D is replaced by 0x7D 0x5D. The receiver, after detecting the escape character in the byte stream, discards this byte and performs an XOR operation of 0x20 with the following
byte to recover the original payload.

In both cases, more data are being transmitted than it would be necessary without bit or byte stuffing
techniques. To make things worse, the amount of overhead that has to be added depends on the contents
of the payload. A malicious user might effectively double the data rate (with byte stuffing) or increase it
by around 20% (with bit stuffing) by transmitting a continuous stream of flags. To avoid this, several
measures can be taken. One is to *scramble* the user data before it is put into data frames [4].

Another possibility is the so-called consistent overhead byte stuffing (COBS), proposed in [1]. Here,
the stream of data bytes is scanned in advance for appearing flags. The sequence of data bytes is then cut
into chunks of at most 254 bytes *not* containing the flag. Thus, every flag that appears in the flow is
replaced by one byte representing the number of nonflag data bytes following it. This way, no additional
data are to be transmitted as long as there is at least one flag every 255 data bytes. Otherwise, one byte is
inserted every 254 bytes, indicating a full-length chunk.

Length Field

To avoid the processing overhead that comes with bit or character stuffing, it is possible to reserve a field
in the header of the frame that indicates the length of the whole frame. Having read this field, the receiver
knows in advance how many characters or bytes will arrive. No end delimiter is needed anymore. Either
a continuous transmission of packets followed by idle symbols or the usual combination of preamble and
start delimiter is needed to correctly determine which of the header fields carries the length information.

Being potentially the best solution concerning transmission overhead, the length field mechanism suffers from erroneous transmission media. If the packet length information is lost or corrupted, then it is
difficult to find it again. Therefore, it has to be protected separately using error-correcting codes or redundant transmission. Additional mechanisms (e.g., time gaps) should be used to find the end of a frame
even when the length field is erroneous.

Example: Bit and Frame Synchronization in the Profibus

As an example to illustrate the mechanisms introduced above, let us look at the lower layers of the DIN
19245 PROcess FIeld BUS (Profibus) [42]. This standard defines a field bus system for industrial applications. The lowest layer, the *physical layer,* is based on the RS-485 electrical interface. A shielded twisted
pair or fiber cable may be used as a transmission medium. The UART converts every byte into 11-bit
transmission characters by adding start, parity, and stop bits. Thus, asynchronous bit synchronization is
used on the lowest layer of the Profibus, no matter which transmission medium is being used.

The second layer is called Fieldbus Data Link (FDL). It defines the frame format as it can be seen in
Figure 37.2. Start and end delimiters are used in every frame, but different SDx (SD1–SD4; the latter is
not shown in the figure) define different frame types. Thus, a receiver knows after reading an SD1 that a
control frame of fixed length will arrive. In addition, time gaps of 33 bit times have to be left between the
frames. After receiving an SD3, the receiver interprets the next byte as LE (length field) and checks this
against the redundant transmission of LE in the third byte, thereby decreasing the probability of undetected errors in the length field. Using the combination of time gaps and the redundant transmission of
the length field, a character stuffing to replace all possible start and end delimiters in the payload becomes
unnecessary.

37.3   Medium Access Control Protocols

All *medium access control* or *MAC* protocols try to solve the same problem: to let a number of stations
share a common resource (namely, the transmission medium) in an efficient manner and such that some
desired performance objectives are met. They are a vital part of local area network (LAN) and metropolitan area network (MAN) technologies, which typically connect a small to moderate number of users in
a small geographical area such that a user can communicate with other users.

With respect to the OSI reference model, the MAC layer does not form a protocol layer of its own, but is a sublayer of either the physical layer or the data link layer [35]. However, in the following, we will consider the MAC as being a layer of its own, since it has a distinguished task. The importance of the MAC layer is reflected by the fact that many MAC protocol standards exist, for example, the IEEE 802.x standards. Its most fundamental task is to determine for each station attached to a common broadcast medium when it is allowed to access the medium, that is, to send data or control frame to it. To this end, each station executes a separate instance of a MAC protocol.

The operation and behavior of a MAC protocol are heavily influenced by the properties of the underlying physical layer and the design goals. Specifically for hard real-time communications, the MAC layer is a key component: if the delays on the MAC layer are not strictly bounded, the upper layers cannot compensate this. A large number of MAC protocols have been developed during the last three decades. The following references are landmark papers or survey articles covering the most important protocols: [7, 8, 26, 27, 37, 43, 38, 19, 32, 33, 39, 40, 15]. Furthermore, MAC protocols are covered in many textbooks on computer networking, for example, [35, 12, 20]. In this survey, we restrict to those classes of protocols that are important for industrial applications and have found some deployment in factory plants, either as stand-alone protocols or as building blocks of more complex protocols.

Requirements and Quality of Service Measures

There are a number of (sometimes conflicting) requirements for MAC protocols, some of which are specific for industrial applications with hard real-time and reliability constraints.

There are two main *delay-oriented measures*: the *medium access delay* and the *transmission delay*. The medium access delay is the time between the arrival of a frame and the time its transmission starts. This time is affected by the operational overhead of the MAC itself, like collisions, MAC control frames, backoff and waiting times, etc. The transmission delay denotes the time between frame arrival and its successful reception at its intended receiver. Clearly, the medium access delay is a fraction of the transmission delay. For industrial applications with hard real-time requirements, both delays must be upper-bounded. In addition, a desirable property is to have low medium access delays in case of low network loads.

A key requirement for industrial applications is the support for *priorities*: important frames (e.g., alarms, periodic process data) should be transmitted before unimportant ones. This requirement can be posed *locally* or *globally*: in the local case, each station decides independently as to which of its frames is transmitted next. There is no guarantee that station A's important frames are not blocked by another station B's unimportant frames. In the global case, the protocol identifies the most important frame of all stations to be transmitted next.

The need to share bandwidth between stations constitutes another important class of desired MAC properties. A frequently posed requirement is *fairness*: stations should get their "fair share" of the bandwidth, which, for example, can be defined as the minimum of a stations load and the overall bandwidth divided by the number of stations. It is also often required that a station receives a *minimum bandwidth*, like for the transmission of periodic process data of fixed size.

With respect to *throughput* clearly, it is important to keep the MAC overhead low. This concerns the frame formats, the number and frequency of MAC control frames, and efficiency losses due to the operation of the MAC protocol. An example for efficiency losses is when the protocol allows collisions: the bandwidth spent for collided packets is lost, since typically the collided frames are useless and must be retransmitted, which is costly in terms of bandwidth. A MAC protocol is said to be *stable* if an increase in the overall load does not lead to a decrease in throughput.

Depending on the application area, other constraints can be important as well. For simple field devices, the MAC implementation should have a low complexity and be simple enough to be implemented in hardware. For mobile stations using wireless media, the energy consumption is a major concern; the MAC should contain different power-saving mechanisms. For wireless transmission media, the MAC should contain additional mechanisms to adapt to the instantaneous error behavior of the wireless channel; possible control knobs are the transmit power, error correcting codes, the bit rate, etc.

Design Factors

The most important factors influencing the design of MAC protocols are the medium properties/ medium topology and the available feedback from the medium.

We can broadly distinguish between *guided media* and *unguided media.* In guided media, the signals originating from frame transmissions propagate within specified geographical bounds, typically within copper or fiber cables. If the medium is properly shielded, then beyond these bounds the communications are invisible, and two cables can be placed close to each other without mutual interference. In contrast, in unguided media (with radio-frequency or infrared wireless media being the prime example), the wave propagation is visible in the whole geographical vicinity of the transmitter, and ongoing transmissions can be received at any point close enough to the transmitter. Therefore, two different networks overlayed within the same geographical region can influence each other.

Guided media networks can show a number of topologies. We discuss a few examples. In a *ring topology* (see Figure 37.3) each station has a point-to-point-link to two neighbors, such that the stations form a ring. In a *bus topology* like the one shown in Figure 37.4, the stations are connected to a common bus, and all stations see the same signals. Hence, the bus is a *broadcast medium.* In the *star topology* illustrated in Figure 37.5, all stations only have a physical connection to a central device, the *star coupler,* which repeats and optionally amplifies the signals coming from one line to all the other lines. A network with a star topology also provides a broadcast medium, where each station can hear all transmissions. When using wireless transmission media, the distance between stations might be too large to allow all stations to receive all transmissions. Therefore, the network is often only partially connected or has a *partial mesh structure,* shown in Figure 37.6. Additional mechanisms have to be employed to implement *multi-hop transmission.*

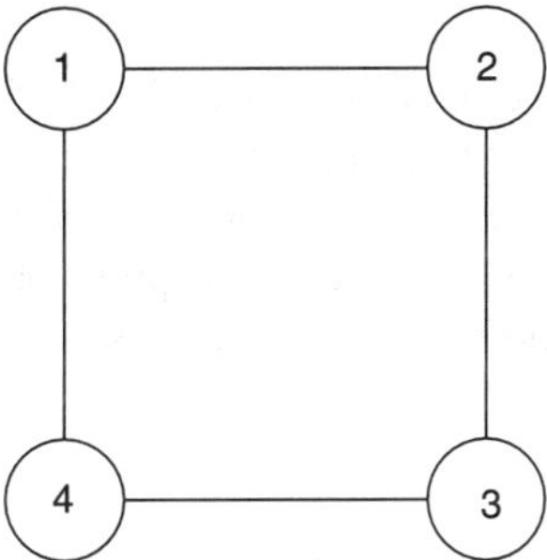

FIGURE 37.3 Ring topology.

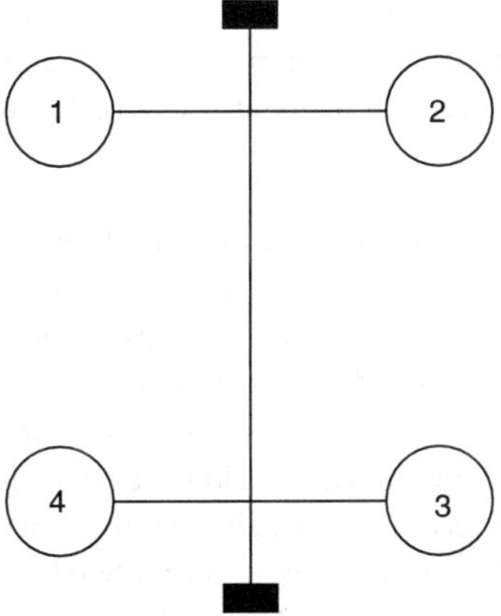

FIGURE 37.4 Bus topology (the black boxes are line terminations).

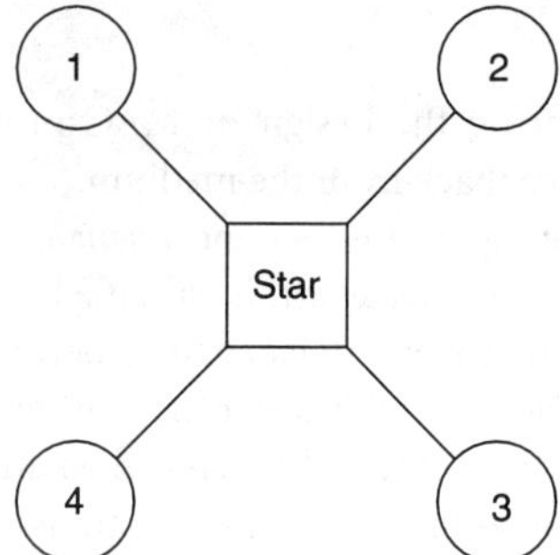

FIGURE 37.5   Star topology.

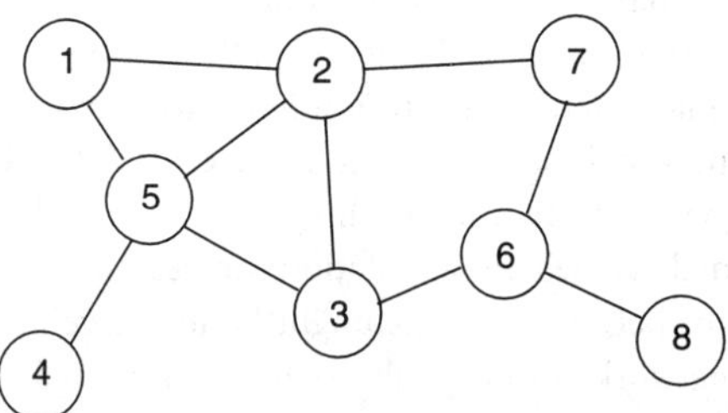

FIGURE 37.6   Partial mesh topology.

An important property of a physical channel is the available feedback. Specifically, some kinds of media allow a station to read back data from the channel while transmitting. This can be done to detect faulty transceivers (like in the PROFIBUS protocol [42]), collisions (like in the Ethernet protocol), or parallel ongoing transmissions of higher priority (like in the CAN protocol). This feature is typically not found in wireless channels. Here, it is not possible to send and receive simultaneously on the same channel.

Random Access Protocols

In random access (RA) protocols, the stations are uncoordinated and the protocols operate in a fully distributed manner. RA protocols often incorporate a random element, for example, by exploiting random packet arrival times, setting timers to random values, etc. The lack of central coordination and of fixed resource assignment allows to share a channel potentially between an infinite number of stations, while fixed assignment and polling protocols support only a finite number of stations. However, the randomness can make it impossible to give deterministic guarantees on medium access delays and transmission delays. There are many RA protocols, which are not only used on their own but also as building blocks of more complex protocols.

ALOHA and Slotted ALOHA

A classical protocol is ALOHA [7], where a station sends a newly arriving data frame immediately without inquiring the status of the transmission medium. Hence, frames from multiple stations can overlap in time (collisions), which may make them unrecognizable. In *slotted ALOHA,* all stations are synchronized to a common time reference and the time is divided into fixed-sized *slots.* In slotted ALOHA, newly arriving frames are sent at the beginning of the next time slot. In both cases, the transmitter starts a timer after frame transmission. The receiver has to send an immediate acknowledgement frame upon successful reception of the data frame. When the transmitter receives the ack, it stops the timer and considers the frame as successfully transmitted. If the timer expires, the transmitter randomly selects a *backoff time,* which has to expire before it attempts to transmit the frame again. The backoff time is chosen randomly to avoid synchronization of colliding stations. This protocol has two advantages: it is extremely simple and it offers short delays in case of a low network load. However, the protocol does not support priorities and with increasing network load the collision rate increases and the transmission delays grow as well. In addition, ALOHA is not *stable:* above a certain threshold load, an increase in the overall load leads to a *decrease* in overall throughput. The maximum normalized throughput of pure ALOHA is $1/2e \approx 18.4\%$ under Poisson arrivals and an infinite number of stations. The maximum throughput can be doubled with slotted ALOHA. A critical parameter in ALOHA is the backoff time, which is typically chosen from a certain time interval *(backoff window)*. A collision can be interpreted as a sign of congestion. If after the backoff time another collision occurs, the next backoff time should be chosen from a larger backoff window to reduce the pressure on the channel. A popular rule for the evolution of the backoff window is the *truncated binary exponential backoff scheme,* where the backoff window size is doubled after each collision, up to a certain threshold, after which the window stays constant. After successful transmission, the backoff window is restored to its original value.

CSMA Protocols

In carrier sense multiple access (CSMA), the stations act more careful than in ALOHA: before transmitting a frame they listen on the medium (carrier sensing) to see whether it is busy or free [39, 36]. If the medium is free (many protocols require it to be contiguously free for some minimum amount of time), the station transmits its frame. If the medium is busy, the station defers transmission. The steps following this distinguish the protocols.

In *nonpersistent CSMA,* the station simply defers for a random time (backoff time) without listening to the medium during this time. All other protocols discussed next wait until the end of the ongoing transmission before starting further activities.

In *p*-persistent CSMA $(0 < p < 1)$, the time after the preceding transmission ends is divided into time slots. A station listens to the medium at the beginning of a slot. If the medium is free, the station starts transmitting its frame with a probability p and with probability $1 - p$ it waits until the next slot. In 1-persistent CSMA, the station transmits immediately without further actions. Both approaches still have the risk of collisions, if two or more stations decide to transmit (1-persistent CSMA) or choose the same slot (*p*-persistent CSMA). The problem is as follows: if one station A senses the medium idle and starts transmission at time τ_0, another station B would notice this earliest at some later time $\tau_0 + \tau$, due to the propagation delay. If B performs carrier sensing at a time between τ_0 and $\tau_0 + \tau$, it senses the medium idle and starts transmission, which results in a collision. Therefore, the collision probability depends on the propagation delay and thus on the maximum geographical distance between stations. As in ALOHA, in pure CSMA protocols the recognition of collisions needs acknowledgement frames.

Although the throughput of CSMA-based protocols is much better than that of ALOHA (ongoing transmissions can be completed without disturbance), the number of collisions and their duration limit the throughput. Collision detection and collision avoidance techniques can be used to relax these problems. These are discussed in the following sections.

Specifically for wireless media, the task of carrier sensing is not without problems, since it is the transmitter who senses the medium ultimately because he wants to know the state of the medium at the intended receiver (collisions are only important at the receiver). However, due to path loss [30, Chapter 4] the signal experiences attenuation with increasing distance. If a minimum signal strength is required, the *hidden terminal problem* occurs (please refer to Figure 37.7): consider three stations A, B, and C with transmission radii as indicated by the circles. Stations A and C are in the range of B, but A is not in the range of C and *vice versa.* If C starts to transmit to B, A cannot detect this by its carrier-sensing mechanism and considers the medium to be free. Hence, A also starts frame transmission and a collision occurs at B.

For wireless media, there is a second scenario where carrier sensing leads to false predictions about the channel state at the receiver: the so-called *exposed terminal scenario,* depicted in Figure 37.8. The four stations A, B, C, and D are placed such that the pairs A/B, B/C, and C/D can hear each other, and all remaining combinations cannot. Consider the situation where B transmits to A, and one short moment later C wants to transmit to D. Station C performs carrier sensing and senses the medium as being busy, due to B's transmission. As a result, C postpones its transmission. However, C could safely transmit its frame to D, without disturbing B's transmission to A. This leads to a loss of efficiency.

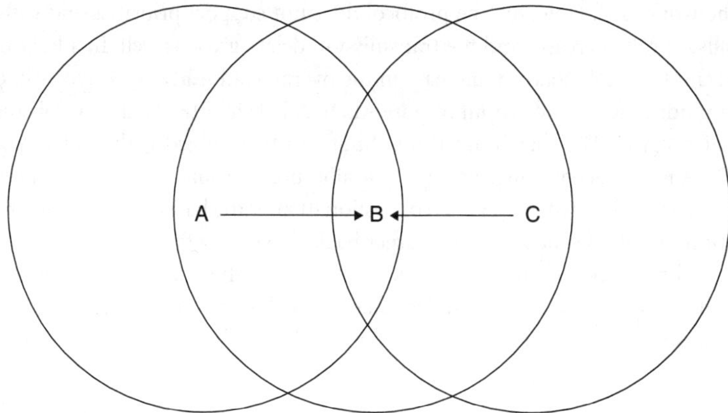

FIGURE 37.7 Hidden terminal scenario.

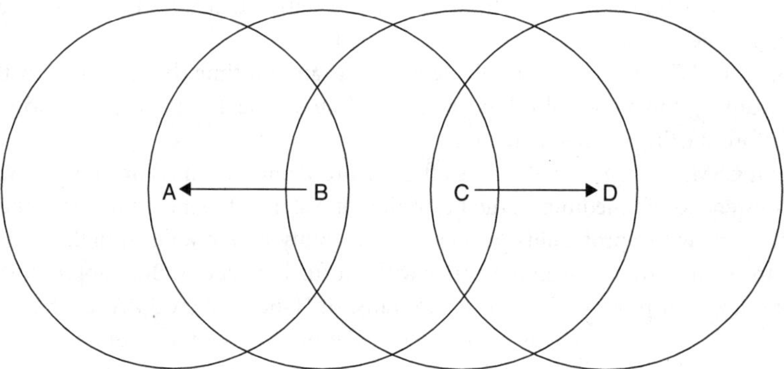

FIGURE 37.8 Exposed terminal scenario.

Two approaches to solve these problems are busy tone solutions [40] and the RTS/CTS protocol, as applied in the IEEE 802.11 wireless LAN medium access control (MAC) protocol [28]. In the busy tone approach, the receiver transmits a busy tone signal on a second channel during frame reception. Carrier sensing is performed on this second channel. This solves the exposed terminal problem. The hidden terminal scenario is also solved, except in the rare cases where A and C start their transmissions simultaneously.

The RTS/CTS protocol attacks the hidden terminal problem using only a single channel. Consider the case that A has a data frame for B. After A has obtained channel access, it sends a short request to send (RTS) frame to B, indicating the time duration needed for the whole frame exchange sequence. If B receives the RTS frame properly, it answers with a clear to send (CTS) frame, which indicates the time needed for the remaining frame exchange sequence. Station A starts transmission after receiving the CTS frame. Any other station C hearing the RTS and/or the CTS frames defers its transmissions for the indicated time, thus not disturbing the ongoing frame exchange. It is a conservative choice to defer on any of these frames, but the exposed terminal problem still exists. If a station C defers only on receiving both frames, the exposed terminal problem is solved. However, there is the risk of bit errors in the CTS frame, which may lead C to start transmissions falsely. The RTS/CTS protocol does not resolve collisions of RTS frames at the receiver; furthermore, it introduces significant overhead.

CSMA Protocols with Collision Detection

If two or more stations collide without recognizing this, they would transmit their entire frames, which, however, are not useful for the receiver. If the stations could detect and abort the collision within a shorter time than needed to transmit the longest frame, less bandwidth is wasted. The class of carrier sense multiple access with collision detection (CSMA/CD) protocols enhances the basic CSMA method with a collision detection facility. The collision detection is performed by hearing back the signal on the cable during transmission and to compare the measured signal with the transmitted one. If the signals differ, a collision has occured [20, Section 6.1.3].

When a station experiences a collision, it can perform a backoff scheme. This scheme works with slotted time, where a time slot is large enough to accommodate the maximum round-trip time to make sure that all stations have the chance to recognize an ongoing transmission reliably. As an example, in the CSMA/CD method of IEEE 802.3 a truncated binary exponential backoff scheme is used: after the first collision a station chooses to wait either 0 or 1 slots. If another station starts transmission during the waiting time, the station defers. After the second collision a station chooses to wait between 0 and 3 slots, and for all subsequent collision the backoff window is doubled. After ten collisions the backoff window is kept fixed to 1024 slots, and after 16 collisions the station gives up and discards the frame.

A stochastic form of priorities can be introduced to CSMA/CD by using different initial sizes or evolution speeds for backoff windows, as is done in [11] for the CSMA variant used in the IEEE 802.11 wireless LAN standard.

In wireless LANs (e.g., in the IEEE 802.11 wireless LAN), frequently acknowledgement frames are used to give the transmitter a feedback, since wireless transceivers cannot transmit and receive simultaneously. The lack of an acknowledgement frame indicates either a collision or a transmission error. Furthermore, two colliding frames do not necessarily result in a total loss of information: when the signal strength of one frame is much stronger than the other one, the receiver may be able to decode the frame (near–far effect).

CSMA Protocols with Collision Resolution

A class of CSMA protocols reacts on collisions not by going into a backoff mode and deferring transmissions; instead, they try to *resolve* the collision, that is, to determine one station among the contenders, which is ultimately allowed to send its frame. Some examples of this class are the adaptive tree walking protocol [13] and protocols with bit-wise priority arbitration like the MAC protocol of controller area network (CAN) [18] and the protocol used for the D-channel of ISDN [31].

The adaptive tree walking protocol works as follows: the time is slotted, just as in the Ethernet CSMA/CD protocol. Furthermore, all stations are arranged in a balanced binary tree $\mathcal{T}$ and know their respective positions in this tree. All stations wishing to transmit a frame (called backlogged stations) wait until the end of the ongoing transmission and start to transmit their frame in the first slot (slot 0). If there is only one backlogged station, then it could transmit its frame without further disturbance. If two or more stations collide, then in slot 1 only the members of the left subtree $\mathcal{T}_L$ are allowed to try transmission again. If another collision occurs, only stations of the left subtree $\mathcal{T}_{LL}$ of $\mathcal{T}_L$ are allowed to transmit in slot 2, and so forth. On the other hand, if only one station from $\mathcal{T}_L$ transmits its frame, then for fairness reasons the next frame transmission is reserved for a station from the right subtree $\mathcal{T}_R$.

The bit-wise arbitration protocols use a similar idea as the tree-walking protocol; however, they do not keep any history information that is needed to achieve fairness. As an example, we present the MAC protocol of CAN. As a prerequisite, the transmission medium guarantees that overlapping signals do not destroy each other, but lead to a valid signal. If two stations transmit the same bit, the medium adapts the common bit value. If one station transmits a zero bit and the other a one bit, the medium adopts a well-defined state, for example, a zero bit. The CAN protocol uses a *priority field* of a certain length at the beginning of a MAC frame. Backlogged stations wait until the end of an ongoing frame and then transmit the first bit of their priority field. In parallel, they read back the state of the medium and compare it with the sent bit. If both agree, the station continues with the second bit of the priority field. If the bits differ, the station has lost contention and defers until the end of the next frame. If it can be guaranteed that all stations use different priority fields, no collisions occur. This protocol supports global frame priorities in a natural way, and the medium access time for the highest priority frame is bounded. However, the assignment of priorities to stations or frames is a nontrivial issue. If the priorities are assigned fixed on a per-station basis, the protocol is inherently unfair. If fairness is desired, the priorities should change over time. Therefore, in CAN applications, the priorities are assigned depending on the type of the message. Another drawback of the protocol is that all stations have to be synchronized with the precision of a bit-time, and the need for all stations to agree on the state of the medium limits either the bit rate or the geographical extension of CAN.

CSMA Protocols with Collision Avoidance

If it is technically not feasible to Immediately detect collisions, one might try to avoid them. Protocols belonging to this class are called carrier sense multiple access with collision avoidance (CSMA/CA) protocols. An important application area are wireless LANs, where: (a) stations cannot transmit and receive simultaneously on the same channel, and (b) the transmitter cannot directly detect collisions at the receiver due to path loss and the need for a minimum signal strength (see the discussion of the hidden terminal scenario in CSMA Protocols section).

One CSMA/CA approach is the RTS/CTS handshake protocol used in the IEEE 802.11 WLAN; see CSMA Protocols section. Other protocols use additional random waiting times. As an example, in the EY/NPMA protocol of HIPERLAN [16], the collision avoidance part consists of three phases: all stations wishing to transmit a frame wait for the end of the ongoing transmission. In the first phase (priority phase), the stations wait for a number of slots corresponding to the frames priority; there are five distinct priorities. If a station A decides to transmit in slot n and another station B starts in slot $m < n$, then A defers, since B has a higher priority frame. In the second phase (elimination phase), the surviving stations transmit a burst of random length before switching to the receive mode. If there is still energy present, the station gives up, since another station sends a longer burst. In the following third phase (yield phase), the surviving stations remain idle for a random amount of time. If another station starts to transmit meanwhile, the station defers. Otherwise, the station starts to transmit its data frame.

Fixed Assignment Protocols

In fixed assignment (FA) protocols, a station is assigned a channel resource (frequency, time, code, space) exclusively, that is, it does not need to contend with other stations for using its share and obtains medium access within a bounded time.

In frequency division multiple access (FDMA) systems, the available spectrum is subdivided into N subchannels, with some guard band between the channels. A channel is assigned exclusively to a station. When a frame arrives, the station can transmit immediately on the assigned channel, the intended receiver has to know the channel in advance. Idle subchannels cannot be used by highly loaded stations.

In code division multiple access (CDMA) systems, the stations spread their frames over a much larger bandwidth than needed while using different codes to separate their transmissions. The receiver has to know the code used by the transmitter; all parallel transmissions using other codes appear as noise. Like in FDMA, stations can transmit incoming frames immediately.

In time division multiple access (TDMA) systems, the time is divided into fixed-length *superframes*, which in turn are divided into *time slots*. Each station is assigned a set of slots in every superframe. During its slots, a station can use the full channel bandwidth. The stations need to be synchronized on slot boundaries; however, the slots contain some *guard times* to compensate for inaccurate synchronization. In a centralized setting where all stations transmit to a central station, such inaccuracies can be introduced by different propagation delays and the problems to determine the exact distance between a station and the central station.

In space division multiple access (SDMA) systems, spatial resources are divided among stations. In wireless systems where a central station is equipped with smart antenna arrays, the central station can form a number of spatially directed spot beams and focuses them to the stations. If a beam covers two or more stations, these have to share the channel by some other protocol; but stations in different beams can transmit in parallel. Another example is the use of sectored antennas in cellular systems.

In all these schemes, the allocation of channel resources to stations can be static or dynamic. In the static case, the allocation may be preconfigured or a station requests the resource once from some resource management facility (which may be part of a central station/access point). The Time-Triggered Protocol is an example of such a scheme [41]. Another example is the cyclic window in the WorldFIP protocol, which offers a preconfigured allocation of time slots. Some dynamics can be introduced by changing the allocation tables at appropriate times. The dynamic approach requires a separate channel, which a station can use to request channel resources from the resource manager. The fixed assignment of channel resources is advantageous specifically for industrial applications for the following reasons:

- It allows to guarantee a minimum bandwidth to a station.
- It allows to guarantee a strictly bounded medium access time as well as strictly isochronous service.

Demand Assignment Protocols

In demand assignment (DA) protocols, channel resources are also assigned exclusively to a station, but on a much shorter timescale than for fixed assignment protocols. In the latter case, assignment happens once and lasts for the lifetime of the session, while in DA protocols resources are assigned for the duration of a *data burst* and have to be requested explicitly using a signaling mechanism. We can broadly distinguish two classes of DA protocols: the distributed protocols use no central authority for resource allocation; instead they often use token-passing schemes. On the other hand, for centralized protocols the stations have to signal their demands to a centralized station, which assigns resources and schedules transmissions. The signaling channel can either be in-band (requests can be piggybacked to transmissions of data or control frames) or a separate logical signaling channel with its own medium access procedure is used.

For industrial applications, DA protocols have two major advantages: they can guarantee a bounded medium access time and they allow to use idle resources. However, for the distributed schemes there is inevitably some jitter in the medium access times, which hinders strictly isochronous services. The centralized schemes introduce a single point of failure, namely the resource manager.

Centralized Schemes: Hub-Polling Protocols and Reservation Protocols

As a very general description [34], a hub-polling system consists of a central station (called *hub)* and a number of stations, with each station conceptually having a queue for requests or frames. The hub carries out two different tasks: first it queries the queue states from the stations, and second, it assigns bandwidth to the stations according to the query results and some polling policy. Typically, it is assumed that a query is less costly than to serve a frame; otherwise, the query overhead could not be justified as compared to a pure TDMA system. To be queried, a station must *register* itself with the hub. Polling schemes differ in the sequence by which stations are polled:

- in *round-robin,* the stations are visited one after another,
- in *table-driven* schemes, the next station to be visited is determined from a prespecified table, and
- in *random polling,* the next station to poll is determined randomly

and in the type of service a polled station receives:

k-Limited service: up to k frames are served per station before proceeding to the next station.

Time-limited service: the station may transmit frames including retransmissions for no longer than a specified time.

Exhaustive service: a queue is serviced until it is empty.

Gated service: the server serves only those frames of station i that were already present when starting service for i.

As an example, the master-/slave protocol of PROFIBUS can be classified as a table-driven and time-limited service (however, with varying masters). In the BITBUS protocol [24], the role of the master does not change over time.

A variation of hub-polling protocols are *probing protocols* [32, 21]. These are based on the observation that polling each station separately is wasteful if the load is low. Instead, it is more effective to poll a group of stations as a whole. For example, the hub may announce that a random access slot follows, which can be used by stations belonging to a certain group to signal their transmission needs. If no station answers, the next group can be polled. If a single station answers, it is granted access to the medium, if two or more stations answer, their requests will collide in the random access slot. Different methods can now be applied to resolve this collision, for example, the tree walking approach discussed in CSMA Protocols with Collision Resolution section, or all stations in the group can be polled separately. In [44], the latter approach is introduced, along with a scheme that adapts the group sizes to the current load.

In reservation protocols, the stations have to send a *reservation message* to a resource management facility (often located in a central station). The reservation message may specify the length of the desired data transmission and its timing constraints. The central station can perform an *admission control test* to decide, whether the request can be satisfied without harming the guarantees given to already admitted requests. After successful admission control, the central station sends some feedback to the station indicating the allocated resources (e.g., the time slots to use). For signaling requests, three approaches are commonly used: (a) in *piggybacking schemes* the reservation requests are sent along with already admitted data or control frames; (b) the stations send request frames on a separate signaling channel under a contention-based MAC protocol (ALOHA or CSMA protocols); and (c) the central station may poll all stations that are currently idle and thus cannot use piggybacking. Many protocols developed in the context of wireless ATM [15] belong to this class, for example, the MASCARA protocol [29] of the European Union Magic Wand project. The FFT-CAN [10] protocol is another example of this class, where stations send reservation requests for periodic transmissions to a central master station.

Distributed Schemes: Token-Passing Protocols

In distributed schemes, there is no central facility controlling resource allocation or medium access. Instead, a special frame, called the *token frame* is sent between stations. The station that currently holds the token is allowed to initiate transmissions. After some time, the current token owner is required to pass the token to another station, that is, to send it a token frame. Token passing schemes can be applied in networks with a ring topology (examples: the IEEE 802.5 Token Ring [23] or FDDI [14, 25]) or with a bus/tree topology (examples: the IEEE 802.4 Token Bus [22] or the PROFIBUS with the FMS profile [42]).

To guarantee an upper bound on medium access delay, the IEEE Token-Bus, FDDI, and the PROFIBUS protocols use variants of the *timed-token protocol* [9]. In this protocol, all stations agree on a common parameter, the *target token rotation time* T_{TTRT}. Furthermore, each station is required to measure the time that passed between the last time it received the token and the actual token reception time. This time is called the *token rotation time*. If the difference $T_{TTRT} - T_{TRT}$ is positive, the arriving token is called an *early token*; otherwise, it is called a *late token*.

Token-passing protocols over broadcast media (bus, tree) construct a *logical token passing ring*. The token frame is passed among all stations in this ring, and each station gets the token once per cycle. The ring members have the additional burden to execute *ring maintenance* algorithms, which include among others: including new stations, excluding leaving or crashed stations, detect and repair lost tokens, etc. These mechanisms use a certain number of control frames and are designed in a way that they do not harm the timing guarantees given by the timed-token protocol.

Meta-MAC Protocols

In [17] "Meta-MAC" protocols are introduced. The basic idea is simple and elegant: a station contains not only a single MAC instance, but several of them, running in parallel. These can be entirely different protocols or the same protocol, but with different parameters. However, only one protocol is really active at a given time in the sense that its decisions (transmit/not transmit) are executed; the other instance decisions are only recorded. From time to time, a new active protocol is selected. This selection is based on history information about transmission outcomes (success, failure). For each candidate protocol, it is evaluated as to how "successful" the protocol would have been given the outcomes in the history. For example, a protocol that produced several "transmit" decisions in "successful" time slots would get a high ranking, while a protocol that "transmits" decisions would have resulted in collisions getting a bad ranking. Based on this ranking, a new protocol is chosen.

37.4 Error Control Techniques

When a packet is transmitted over a physical channel, it might be subject to distortions and channel errors. Potential error sources are noise, interference, loss of signal power, etc. As a result, a packet may either be completely or partially lost (e.g., when the receiver fails to acquire bit synchronization or loses it somewhere), or a number of the bits within a packet are modified. In some types of channels, errors occur quite frequently, for example, in wireless channels [56].

One option to deal with errors is to tolerate them. For example, in Voice-over-IP systems, a loss rate of speech packets of $\approx$ 1% still gives an acceptable speech quality at the receiver, depending on the codec and the influence of error concealment techniques [47, Chapter 7].

However, in safety-critical industrial applications errors are often not tolerable, must be detected, and subsequently be corrected. There are the following fundamental approaches to *error control* [49]:

- In *open-loop approaches*, the transmitter receives no feedback from the receiver about the transmission outcomes. Redundancy is introduced to protect the transmitted data against errors.
- In *closed-loop schemes*, the transmitter receives feedback about erroneously received packets. The receiver requests *retransmission* of these packets from the transmitter.
- In *hybrid schemes*, these two approaches are combined.

The detection of errors is based on *checksums*, which are sent along with the packet, for example, as a *packet trailer*. Well-known kinds of checksums are cyclic redundancy checks (CRC) or parity bits [48]. However, no checksum algorithm is perfect; there are always bit error patterns that cannot be detected by a checksum algorithm. Hence, the *residual error probability* is nonzero. A study of the performance of checksum algorithms over real data is [55].

There is a rich literature about error control. Some standard references are [48–50].

Open-Loop Approaches

In general, open-loop approaches involve *redundant data transmission*. Several kinds of redundancy can be used:

- Send multiple copies of a packet.
- Add redundancy bits to the packet data.
- Diversity techniques.

In the multiple-copies approach, the transmitter sends K identical copies of the same packet [45], each one equipped with a checksum. If the receiver receives at least one copy without checksum errors, this is accepted as the correct packet. If the receiver receives all copies with checksum errors, it might apply a bit-by-bit majority voting scheme [52, Chapter 4] on all received copies and check the result again. A variation of the multiple-copies scheme is to not send multiple copies of the same packet, but to send each bit of the user data multiple times: instead of sending 00110101, the transmitter sends, for example, 000.000.111.111.000.111.000.111. Hence, each user bit is transmitted three times and the receiver applies majority voting to each group of three bits.

In *error correcting codes* or *forward error correction* (FEC) codes to k bits of user data, a number $n - k$ of redundancy bits are appended and the block of n bits is transmitted (the fraction k/n is called *code rate*), such that bit errors can be detected and a limited amount of bit errors can be *corrected* [48–50]. In *block coding schemes*, the user data are divided into blocks of k bits and each block is coded independently. Some well-known block FEC schemes are Reed–Solomon codes, Hamming codes, and Bose–Chaudhuri–Hocquenghem (BCH) codes. In *convolutional coding schemes*, the encoder has some "memory," such that the coding of the current bit affects coding of future bits. Therefore, there are no clear block boundaries. Recently, the class of *turbo codes* has attracted attention [54]. In this class of codes, two convolutional codes are concatenated: first an *inner code* is applied to the user data, and after this an *outer code* is applied to the result of the first coding stage.

Diversity techniques are often applied on wireless channels. In the case of receiver diversity, the receiver is equipped with two or more antennas. If these are appropriately spaced [51], the antennas receive two copies of the transmitted waveform, which, in the best case, are uncorrelated. Hence, it might happen that one antenna receives only a weak signal while the other one experiences good signal quality. The two antenna signals may then be combined in different ways.

Closed-Loop Approaches

In closed-loop approaches, the receiving station B checks the arriving packets sent by station A by means of checksums and provides A with feedback information indicating the transmission outcome (success or failure). Usually, B sends *acknowledgement frames* to provide this feedback to A; but the feedback information may as well be *piggybacked* onto data frames sent from B to A. Automatic Repeat reQuest (ARQ) protocols implement this approach [46, 20]. Four basic ARQ protocols are the Send-and-Wait protocol, the alternating bit protocol, the Goback-N protocol, and the Selective-Repeat protocol.

In the Send-and-Wait protocol, the transmitter sends a packet and starts a timer. The receiver sends an acknowledgement, if the packet is received correctly; otherwise, the receiver keeps quiet. If the transmitter receives the acknowledgment, the timer is resetted and the next packet is transmitted. If the transmitters timer expires without acknowledgment, the transmitter retransmits the packet. This protocol is simple, but does not fill "long fat pipes" (links with a high bandwidth-delay product) efficiently and cannot prevent duplicate data at the receiver. Duplicates can be created if not the data packet but the acknowledgment is lost or erroneous. In the alternating bit protocol, a one-bit sequence number is introduced to detect and remove duplicates. If the receiver receives a duplicate packet, the packet is acknowledged, but the data are not delivered to the user. This works reliably if the delay for data packets or acknowledgments can be upper-bounded. However, the alternating bit protocol is still inefficient on long fat pipes.

Both the Goback-N and the Selective-Repeat protocol are not restricted to a single unacknowledged or *outstanding* frames, but allow for multiple outstanding frames; these protocols are also called *sliding-window protocols*. The frames are identified by sequence numbers. In the Goback-N protocol, there may be N outstanding frames. The transmitter sets a timer for each transmitted frame and resets it as soon as an acknowledgment for this frame is received. If the receiver receives an insequence frame, it delivers the frame to its local user and sends a positive acknowledgment; otherwise the frame, is dropped (even if it is received correctly) and either the receiver sends a negative acknowledgment or keeps quiet. When the transmitter receives a negative ack for an outstanding frame or if the timer for this frame expires, he retransmits this frame and all subsequent outstanding frames. Therefore, it might happen that correctly received frames are retransmitted, which is inefficient.

This drawback is attacked by the Selective-Repeat protocol, which works similar to the Goback-N protocol, but allows the receiver to buffer and acknowledge frames that are not received in-sequence. As soon as the missing frames arrive, the buffered frames are delivered to the user and the buffer is freed.

Hybrid Approaches

Open-loop and closed-loop approaches can be combined, for example, by continuously increasing the amount of FEC applied to subsequent retransmissions of the same packet, or by adapting the number of copies in a multicopy approach to the feedback [45].

Further Countermeasures

The transmitter potentially has further control knobs to reduce the probability of packet errors at the receiver. One control knob is the packet length: in a scenario where a packet is equipped with a checksum but not with redundant FEC data, longer packets have a higher probability to be received in error, but the fixed overhead is only a small portion of a packet. In contrast, shorter packets have a higher probability to arrive successfully, but the fixed overhead requires a larger share of the overall bandwidth. When the transmitter receives feedback about transmission outcomes, it may adapt the packet size accordingly.

It is a fundamental communications law that the bit error rate at the receiver depends on the ratio of the energy expended per bit to the channel noise level [53]. There are two possibilities to use this relationship to increase transmission reliability:

- If the transmit power is increased, the energy per bit is increased and the bit error rate is reduced. However, often, the transmit power is technically or legally restricted.
- If the bits are transmitted at a lower speed, the energy per bit is also increased.

Hence, a transmitter might apply *transmit power control* or *modulation rate control*.

References

Bit and Frame Synchronization

[1] Cheshire, Stuart and Mary Baker, Consistent overhead byte stuffing, *ACM SIGCOMM Computer Communication Review*, 27:209–220, 1997.
[2] IEEE, Carrier Sense Multiple Access with Collision Detection (CSMA/CD) - (ETHERNET), 1985.
[3] International Standardization Organization (ISO), *IS 1177 -1985 Character Structure for Start/Stop and Synchronous Character Oriented Transmission*, 1985.
[4] Manchester, J., J. Anderson, B. Doshi, and S. Dravida, IP over SONET, *IEEE Communications Magazine*, Vol. 36, 136–142, May 1998.
[5] Simpson, W., RFC 1661: The point-to-point protocol (PPP), July 1994.
[6] Simpson, W., RFC 1662: PPP in HDLC-like framing, July 1994, Obsoletes RFC1549, Status: STANDARD.

Medium Access Control Protocols

[7] Abramson, Norman, Development of the ALOHANET, *IEEE Transactions on Information Theory*, 31:119–123, 1985.
[8] Abramson, Norman, Ed., *Multiple Access Communications — Foundations for Emerging Technologies*, IEEE Press, New York, 1993.
[9] Agrawal, G., B. Chen, W. Zhao, and S. Davari, Guaranteeing synchronous message deadlines with the timed token medium access control protocol, *IEEE Transactions on Computers*, 43:327 – 339, 1994.
[10] Almeida, Luis, Paulo Pedreiras, and Jose Alberto G. Fonseca, The FFT-CAN protocol: why and how, *IEEE Transactions on Industrial Electronics*, 49:1189–1201, 2002.
[11] Berry, Michael, Andrew T. Campbell, and Andras Veres, Distributed Control Algorithms for Service Differentiation in Wireless Packet Networks, in Proceedings of INFOCOM 2001, Anchorage, Alaska, IEEE, New York, Apr., 2001.
[12] Bertsekas, D. and R. Gallager, *Data Networks*, Prentice-Hall, Englewood Cliffs, NJ, 1987.
[13] Capetanakis, J. I., Tree Algorithm for Packet Broadcast Channels, *IEEE Transactions on Information Theory*, 25:505–515, 1979.
[14] Chen, Biao, Nicholas Malcolm, and Wei Zhao, Fiber distributed data interface and its use for time-critical applications, in *The Communications Handbook*, Gibson, Jerry D., Ed., CRC Press/IEEE Press, Boca Raton, FL, 1996, pp. 597–610.
[15] Dellaverson, Lou and Wendy Dellaverson, Distributed channel access on wireless atm links, *IEEE Communications Magazine*, 35:110–113, 1997.
[16] ETSI, *High Performance Radio Local Area Network (HIPERLAN) — Draft Standard*, ETSI, March 1996.
[17] Farago, Andras, Andrew D. Myers, Violet R. Syrotiuk, and Gergely V. Zaruba, Meta-MAC protocols: automatic combination of MAC protocols to optimize performance for unknown conditions, *IEEE Journal on Selected Areas in Communications*, 18:1670–1681, 2000.
[18] International Organization for Standardization, *ISO Standard 11898 — Road Vehicle — Interchange of Digital Information — Controller Area Network (CAN) for High-Speed Communication*, ISO-Internation Organization for Standardization, 1993.

[19] Gummalla, Ajay Chandra V., and John O. Limb, Wireless medium access control protocols, *IEEE Communications Surveys and Tutorials*, 3, 2000. http://www.comsoc.org/pubs/surveys.

[20] Fred Halsall, *Data Communications, Computer Networks and Open Systems*, Addison-Wesley, Reading, MA, 1996.

[21] Hayes, J.F., *Modeling and Analysis of Computer Communications Networks*, Plenum Press, New York, 1984.

[22] IEEE, *802.4 Token-passing Bus Access Method*, IEEE, New York, 1985.

[23] IEEE, *802.5 Token Ring Access Method and Physical Layer Specifications*, IEEE, New York, 1985.

[24] IEEE - Institute of Electrical and Electronics Engineers, IEEE Standards Department, 445 Hoes Lane, P.O. Box 1331, Piscataway, NJ 08855-1331, USA, *IEEE Standard 1118 — IEEE Standard Microcontroller System Serial Control Bus*, August 1991.

[25] Jain, Raj, *FDDI Handbook: High-Speed Networking Using Fiber and Other Media*, Addison-Wesley, Reading, MA, 1994.

[26] Kurose, J.F., M. Schwartz, and Y. Yemini, Multiple-access protocols and time-constrained communication, *ACM Computing Surveys*, 16:43–70, 1984.

[27] Lam, S.S., Multiaccess protocols in computer communications, Vol. I: Principles, in *Principles of Communication and Network Protocols*, Chon, W., Ed., Prentice-Hall, Englewood Cliffs, NJ, 1983, pp. 114–155.

[28] The Editors of IEEE 802.11, *IEEE Standard for Wireless LAN Medium Access Control (MAC) and Physical Layer (PHY) Specifications*, IEEE, New York, November 1997.

[29] Passas, Nikos, Sarantis Paskalis, Dimitri Vali, and Lazaros Merakos, Quality-of-service-oriented medium access control for wireless atm networks, *IEEE Communications Magazine*, 35:42–50, 1997.

[30] Rappaport, Theodore S., *Wireless Communications — Principles and Practice*, Prentice-Hall, Upper Saddle River, NJ, USA, 2002.

[31] Rathgeb, Erwin P., Integrated services digital network (isdn) and broadband (b-isdn), in *The Communications Handbook*, Gibson, Jerry D., Ed., CRC Press/IEEE Press, Boca Raton, FL, 1996, pp. 577–590.

[32] Rubin, Izhak, Multiple access methods for communications networks, in *The Communications Handbook*, Gibson, Jerry D., Ed., CRC Press/IEEE Press, Boca Raton, FL, 1996, pp. 622–649.

[33] Sachs, S.R., Alternative local area network access protocols, *IEEE Communications Magazine*, 26:25–45, 1988.

[34] Takagi, Hideaki, *Analysis of Polling Systems*, MIT Press, Cambridge, MA, 1986.

[35] Tanenbaum, Andrew S., *Computer Networks*, 3rd ed., Prentice-Hall, Englewood Cliffs, NJ, 1997.

[36] Tanenbaum, Andrew S., *Computernetzwerke*, 3rd ed., Prentice-Hall, Muenchen, 1997.

[37] Tobagi, Fouad A., Multiaccess protocols in packet communications systems, *IEEE Transactions on Communications*, 28:468–488, 1980.

[38] Tobagi, Fouad A., Multiaccess link control, in *Computer Network Architectures and Protocols*, Green, P.E., Ed., Plenum Press, New York, 1982.

[39] Tobagi, Fouad A. and Leonard Kleinrock, Packet switching in radio channels: Part I carrier sense multiple access models and their throughput-/delay-characteristic, *IEEE Transactions on Communications*, 23:1400–1416, 1975.

[40] Tobagi, Fouad A. and Leonard Kleinrock, Packet switching in radio channels: Part II the hidden terminal problem in csma and busy-tone solutions, *IEEE Transactions on Communications*, 23:1417–1433, 1975.

[41] TTTech Computertechnik GmbH, Vienna, *TTP/C Protocol, Version 0.5*, 1999.

[42] Union Technique de l'Electricité, *General Purpose Field Communication System, EN 50170, Volume 2: PROFIBUS*, 1996.

[43] van As., Harmen R., Media access techniques: the evolution towards terabit/s LANs and MANs, *Computer Networks and ISDN Systems*, 26:603–656, 1994.

[44] Willig, Andreas and Andreas Köpke, The Adaptive-Intervals MAC protocol for a wireless PROFIBUS, in Proceedings of 2002 IEEE International Symposium on Industrial Electronics, L'Aquila, Italy, July 2002.

Error Control

[45] Annamalai, A. and Vijay K. Bhargava, Analysis and optimization of adaptive multicopy transmission ARQ protocols for time-varying channels, *IEEE Transactions on Communications*, 46:1356–1368, 1998.

[46] Haccoun, David and Samuel Pierre, Automatic repeat request. in *The Communications Handbook*, Gibson, Jerry D., Ed., CRC Press/IEEE Press, Boca Raton, FL, 1996, pp. 181–198.

[47] Hersent, Olivier, David Gurle, and Jean-Pierre Petit, *IP Telephony — Packet-based multimedia communications systems*, Addison-Wesley, Harlow/England, London, 2000.

[48] Lin, Shu and Daniel J. Costello, *Error Control Coding — Fundamentals and Applications*, Prentice-Hall, Englewood Cliffs, NJ, 1983.

[49] Liu, Hang, Hairuo Ma, Magda El Zarki, and Sanjay Gupta, Error control schemes for networks: an overview, *MONET — Mobile Networks and Applications*, 2:167–182, 1997.

[50] Michelson, Arnold M. and Allen H. Levesque., *Error-Control Techniques for Digital Communication*, John Wiley and Sons, New York, 1985.

[51] Paulraj, Arogyaswami, Diversity techniques, in *The Communications Handbook*, Gibson, Jerry D., Ed., CRC Press / IEEE Press, Boca Raton, FL, 1996, pp. 213–223.

[52] Shooman, Martin L., *Reliability of computer systems and networks*, John Wiley and Sons, New York, 2002.

[53] Sklar, Bernard, *Digital Communications — Fundamentals and Applications*, Prentice-Hall, Englewood Cliffs, NJ, 1988.

[54] Sklar, Bernard, A primer on turbo code concepts, *IEEE Communications Magazine*, 35, December 1997, pp. 94–102.

[55] Stone, Jonathan, Michael Greenwald, Craig Partridge, and James Hughes, Performance of checksums and crc's over real data, *IEEE/ACM Transactions on Networking*, 6:529–543, 1998.

[56] Willig, Andreas, Martin Kubisch, Christian Hoene, and Adam Wolisz, Measurements of a wireless link in an industrial environment using an IEEE 802.11-compliant physical layer, *IEEE Transactions on Industrial Electronics*, 49:1265–1282, 2002.

Further References

[57] Bennet, Jon C.R. and Hui Zhang, Hierarchical Packet Fair Queueing Algorithms, in Proceedings of ACM SIGCOMM, 1996.

[58] Keshav, Srinivasan, *An Engineering Approach to Computer Networking: ATM Networks, the Internet and the Telephone Network*, Addison-Wesley, Reading, MA, 1997.

[59] Parekh, A. K. and G. Gallager, A Generalized Processor Sharing Approach to Flow Control in Integrated Services Networks - The Single Node Case, in Proceedings of IEEE INFOCOM, Vol. 2, IEEE, New York, 1992, pp. 915–924.

[60] Parekh, A. K. and G. Gallager, A generalized processor sharing approach to flow control in integrated services networks — the multiple node case, in Proceedings of IEEE INFOCOM, Vol. 2, IEEE, New York, 1993, pp. 521–530.

[61] Varma, Anujan and Dimitrios Stiliadis, Hardware implementation of fair queueing algorithms for atm networks, *IEEE Communications Magazine*, 35:54–68, 1997.

38

Wireless Local Area Networks and Wireless Personal Area Networks (WLANs and WPANs)

Kirsten Matheus

Carmeq GmbH

38.1 Introduction

The convenience of true mobility due to wireless connectivity has made wireless technologies gain more and more ground. Next to extensive infrastructure requiring cellular systems like GSM or UMTS, interest is now growing in commercial and industrial deployment of systems that function on a smaller scale and require no costly frequency licensing or infrastructure, systems that are used for wireless PANs and wireless LANs.

As a consequence, Bluetooth (a WPAN representative that costs little, consumes little power, is small in size, as well as flexible concerning the support of voice and data services), IEEE 802.11, and HIPERLAN/2 technologies (WLAN representatives that provide comparably high user data rates and a larger range) all receive a significant amount of public and scientific attention. It is only logical that their original purposes are enhanced. Bluetooth, for example, plays a larger role at the beginning less attended markets like warehousing, retailing, and industrial applications [Bea02b]. IEEE 802.11 is considered for seamless coverage of whole cities [Bea02a].

Whatever the precise application is, the users of these technologies have expectations concerning quality and availability of the systems. How well they can be satisfied largely depends on the environment (i.e., unit density, traffic demand, mobility, changes in the environment, and interference) in which the systems are used. Having a good individual performance while at the same time optimizing the overall radio network capacity is in any case a vital issue. This paper identifies criteria on the network and physical layer level, which should be taken into consideration when deciding on one or the other (or all) technologies to employ. It also describes how well the technologies meet these criteria, especially when taking the specific properties of industrial applications/factory floor environments into account.

This chapter is structured as follows. Section 38.2 first of all defines the terms wireless LAN and wireless PAN and puts them in context to cellular systems, as none of the terms are currently unambiguously used. Section 38.3 introduces the (Quality-of-Service) requirements that are generally placed on a technology. A clear view of these requirements is necessary before starting to look for a technology to use. Sections 38.4–38.6 describe the WPAN technology Bluetooth and the WLAN technologies IEEE 802.11 and HIPERLAN/2. In each section, one part explains the technical background, while the other focuses on the performance of the systems. Section 38.7 shows how Bluetooth and IEEE 802.11b — both are placed in the same frequency band and are possibly used at the same time in the same location — coexist, before Section 38.8 closes this article with a summary and the conclusion.

38.2 Defining WLAN and WPAN

The expressions Wireless Local Area Network (WLAN) and Wireless Personal Area Network (WPAN) are commonly used, often though without consistency and/or explanation of what is exactly meant. Thus, in the following, a clarification of the terminology is given.

WLAN: A wireless LAN has the same functionality as a wired LAN with the difference that the wires are replaced by air links. This means that within a *restricted area* (home, office, hot spot) *intercommunication* between all devices connected to the network is possible and that the focus is on *data communication* as well as *high data rates*. The definition of WLAN says nothing on how the network is organized. The most likely approach is that an infrastructure of mounted access points (APs) enables wireless access to the wired LAN behind the APs. Nevertheless, a wireless LAN functioning on an *ad hoc* basis with the units in range is also imaginable.

WPAN: In a wireless PAN also all devices are *interconnectable*. The difference is that all units are somehow *associated* to someone or something or alike (either because they are yours or because they are shared or public devices you want to use) and are therefore very *near by*. A PAN can consist of a variety of devices and can even comprise different technologies. The applications are therefore not limited to *data transmission;* but *voice communication* can be used in a PAN as well.

While you move *within* the WLAN, you move *with* your WPAN. This means that several independent WPANs can coexist in the same area and that they are self-sufficient without any infrastructure, that is, they generally function on an *ad hoc* basis.

Also note that the expression *ad hoc* network is not unambiguously used. As is illustrated in Figure 38.1, there are several steps that lead to *ad hoc* networking: a "pure" *ad hoc* network employs neither any infrastructure nor a specific unit (like access point or base station) for central control; not for coverage, not for synchronization, and also not for the services. Nevertheless, a network can be "ad hoc" also when it supports only single hops.

It can be seen that WLAN technologies like IEEE 802.11 (infrastructure mode using the point coordination function (PCF)) or HIPERLAN/2 are, when looking at these criteria, in the same classification as typical "cellular" systems like GSM or UMTS/WCDMA; all have infrastructure and a specific unit for central control. A WLAN is equally likely to be laid out in cells and it would thus be correct to call it a "cellular" system. Despite this, common understanding has it that the distinction is made by coverage. This "wide" (instead of "local") area coverage of "cellular" systems like GSM has caused "cellular" systems to be generally associated with complete coverage and access everywhere. The actual differentiating term

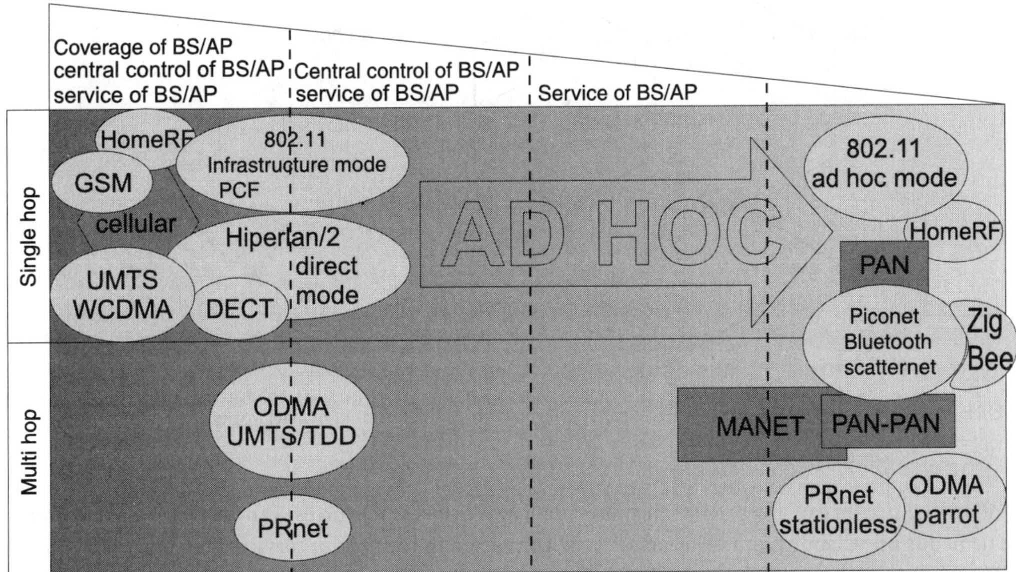

FIGURE 38.1 Classifying wireless technologies between cellular and *ad hoc*; MANET stands for Mobile (or Multihop [JIC99]) *Ad hoc* NETwork (aiming at Internet standards [IET]), PRnet stands for Public Radio network [Mal01], ODMA for Opportunity Driven Multiple Access [3rd99, RBM02]; a Bluetooth scatternet means several interlinked Bluetooth piconets; BS $\triangleq$ base station, AP $\triangleq$ access point.

should be wireless wide area network (WWAN). This chapter will use the common terminology, but wants to point out its inaccuracy.

Naturally, there are more differences: cellular networks focus on voice communication, while WLANs focus on data transmission. Cellular technologies are designed to support user mobility (and roaming) up to very high velocities, while WLANs had more the stationary or portable access in mind. Last but not the least cellular phones represent by far the highest volume product in the wireless market. From the radio network point of view, the most important technical distinction though — with far-reaching consequences to system design, network optimization, etc. — has to be made between systems organized in cells and *ad hoc* systems, for example, regular WLANs and WPANs (and not between cellular systems and WLANs):

As cellular networks are systematically laid out, the minimum distance to the next cochannel interferer (i.e., the next uncoordinated user transmitting at the same time on the same frequency) is generally controllable and known. In contrast, in *ad hoc* networks the next cochannel interferer can be uncontrollably close (see Figure 38.2).

The centralized control in cellular networks allows to distribute the resources effectively in a fair way, because of the accurate knowledge of the overall traffic demand and the possibility of a more global resource management. For *ad hoc* networks or several co-existing WPANs knowledge of overall traffic, demand is generally not available and the systems compete for the resources.

The consequences these differences have for the radio network performance are elucidated in later sections.

38.3 System Requirements and Quality-of-Service

When deciding on a technology to use (or when developing a new one), first of all the performance requirements the technology should fulfill have to be listed. Generally, aspects like throughput (i.e., user data rate), reliability, and delay (or delay variation) come to mind, but also range, cost, power consumption (a long battery life can be especially important for mobile systems) and security (in the sense that no

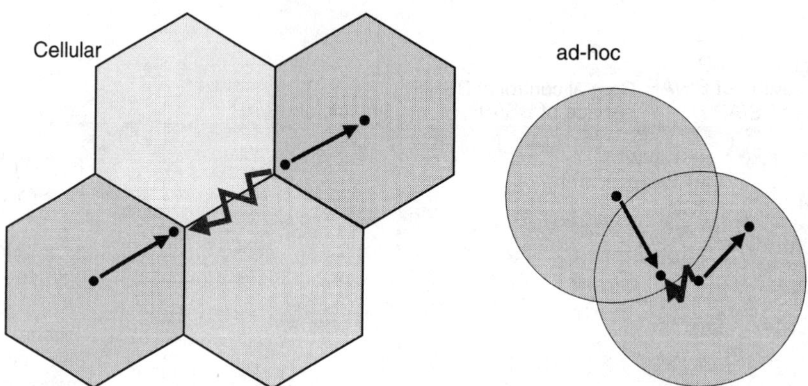

FIGURE 38.2 Possible nearness of next cochannel interferer in cellular and *ad hoc* networks.

unauthorized person can access the transmitted data and network resources) can be decisive parameters. The demands on reliability and delay are directly connected to the type of service intended:

Delay: Delay is especially important for voice transmission, interactive real-time services, and streaming of audio or video in case of small buffer sizes. For example, with a one-way delay larger than ~200 msec, the user in a normal voice communication will feel uncomfortable. At the same time, the user will quite likely still understand the person at the other end of the connection (and not even consider the connection to be inferior) even when 1% of the transmitted data are lost (provided an appropriate speech coding method is used). The delay is first of all caused by the bandwidth available to transmit the required data and the possible need to process the data. Additionally, the transmission itself (depending on the range), the access technology, and the fixed network (due to queuing) can add noteworthily to the overall delay. The main delay causing differences to wired systems are that, due to the air link, wireless systems often provide significantly smaller transmission rates and require additional processing due to extra algorithms needed for coping with the difficult transmission medium.

Reliability: While voice, audio, and video applications can tolerate some errors, typical data applications like web browsing or email require absolute reliability, that is, every bit has to arrive correctly. One mechanism to provide robustness against transmission errors is forward error correction (FEC) encoding (also known as channel coding [CC88, LC83]). To ensure that the bit errors are sufficiently isolated for the FEC to be able to correct them, intra- or interpacket interleaving can additionally be applied. In environments in which every transmitted packet faces different transmission conditions, it might be more effective though to provide an automatic repeat request (ARQ) scheme that initiates a retransmission as soon as the quality of the received packet does not meet the requirements [MZT03].

In case "best-effort" traffic is not sufficient, guarantees of delay, reliability, and throughput values (the latter being in close relation to delay and reliability) are often summarized in the term Quality-of-Service (QoS). There are four categories of QoS:

The *quantitative* QoS gives hard guarantees on the respective delay, throughput, and reliability values [SPG97]. It consequently guarantees the unhindered use of bandwidth and system for the necessary time. *In wireless systems hard quality guarantees cannot be given* due to the possibility of interference, adverse radio conditions,[1] or range limits. Systems requiring a certain amount of data rate within a strict time window because they are, for example, security relevant should not be wireless.

Statistical QoS means that certain requirements can be fulfilled with a certain probability. QoS requirements are thus met in the long run (i.e., in average), while they might not always be reached on short terms. Instead of quantitative QoS, wireless systems that provide QoS generally give statistical QoS guarantees.

---

[1] Note that in industrial environment, they can be especially difficult; metal walls have a significant impact on the transmission conditions. Metal shields radio transmissions while causing, respectively, more reflections.

Subjective QoS gives a certain qualification to the QoS parameters. It naturally leaves more implementation freedom but also insecurity on what quality exactly to expect.

Relative QoS specifies the parameters relative to others. Relative QoS is sensibly used only when there are several traffic classes.

Once the necessary requirements and desired QoS criteria have been agreed upon, the discussion on the technology can commence.

38.4 Bluetooth Technology

Technical Background

Bluetooth (BT) is first of all a cable replacement technology aiming at effortless wireless connectivity in an *ad hoc* fashion, while supporting voice as well as data [Blu99, Blu, Haa98, BS00, Bis01]. Its desired key associations are: easy to use, low in power consumption, and low in cost.

To meet these intentions, the BT Special Interest Group (SIG) placed the technology in the unlicensed ISM-band at 2.4 GHz. This allows close to worldwide deployment without needing to pay fees for frequency licensing. Nevertheless, it requires complying with the respective sharing rules.[2] As a consequence, Bluetooth performs a rather fast frequency hopping (FH) over 79 carriers of 1 MHz bandwidth each such that every Bluetooth packet is transmitted on a newly chosen frequency (which results in a nominal hop rate of 1600 hops/sec). To further reduce cost and support the distribution of the Bluetooth technology, the Bluetooth specification is an open standard that can be used without even needing to pay for the use of its key patents, on the term that the strict qualification procedure is passed. The latter is to ensure the acceptance of the technology, as is the specification of profiles. The profiles describe in detail the implementation of the foreseen applications, thus also enabling units of different manufactures to communicate. The most important characteristics of the physical layer are as follows: The data are GFSK modulated at 1 Mbps and organized in packets consisting of access code, header, and payload. While the use of forward error correction (FEC) for the payload is optional with a 2/3 rate (15,10) shortened Hamming block code, there is no interpacket interleaving (which saves memory).

Bluetooth uses a master–slave concept in which the unit that initiates a connection is temporarily assigned master (for as long as the connection is up). The master organizes the traffic of up to seven other active units, called "slaves" of this "piconet." From the master's device address, the piconet's identity and with it the FH sequence is known. The header of a packet contains the actual addressee, the length of the packet, and other control information. Note though that within one piconet, the slave can only communicate with the master (and not directly with the other slaves) and that only after having been asked (i.e., "polled")[3].

The channel is organized in a TDMA/TDD [GL00] scheme (see Figure 38.3). It is partitioned into 625 μsec time slots. Within this slot grid, the master can only start transmission in the odd numbered slots while the slaves can only respond in even numbered ones. When a unit is not anyway in one of the specific power save modes ("sniff, hold, park"), this supports the "low in power consumption" intention, because every unit has to listen only during the first 10 μsec of its receive slot whether there is a packet arriving (and if not, can "close down" until the next receive slot[4]). This means it needs to listen only 10 μsec/(2.625 μsec)=0.8% of the time. Yet another facet to the long battery life is also the low basic transmit power of 0 dBm (resulting in a nominal range of 10 m)[5].

---

[2]For the U.S. [Fed02, Part 15], for Europe [Eur00b], for Japan [Min02]

[3]Every BT unit can simultaneously be a member of up to four piconets (although it can be master in only one of them). A formation in which several piconets are interlinked in that manner is called "scatternet." Aspects like routing — which are of interest in this constellation — will not be covered in this article. The chapter will thus focus on the properties of a single or multiple independent piconets.

[4]This is quite different from channel access schemes like CSMA as used in IEEE 802.11 (see Section 38.5). Unless asleep, IEEE 802.11 always has to listen into the channel.

[5]Bluetooth can also be used with up to 20 dBm transmit power. This results in a larger range but requires the implementation of power control to fulfill the FCC sharing rules.

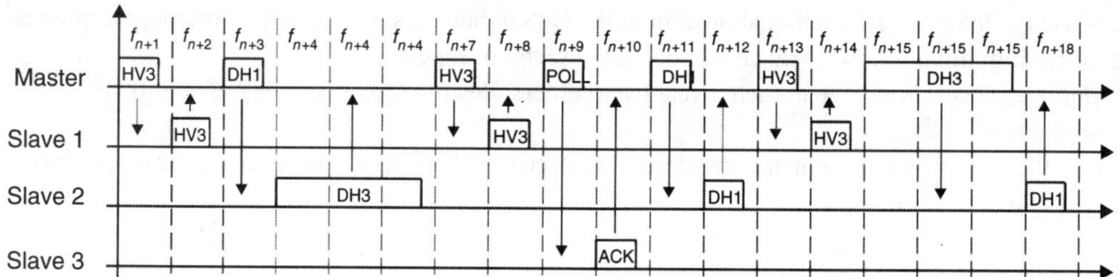

FIGURE 38.3 Example slot occupation within one piconet consisting of master and three slaves; to slave 1 there is an SCO link, to slave 2 (best effort) traffic is transmitted in both directions, slave 3 has currently nothing to send (but has to respond to the POLL-packet with an acknowledgement (ACK)); during the transmission of multislot packets, the frequency is not changed.

Bluetooth provides two different types of connections: an asynchronous connection-less (ACL) link foreseen for data transmission and a synchronous connection-oriented (SCO) link foreseen for speech transmission.

For ACL links, there are six packet types defined. The packets occupy either one, three, or five (625 μsec) time slots and their payloads are either uncoded (called DH1, DH3, DH5) or 2/3-rate FEC protected without any interleaving (called DM1, DM3, DM5). An automatic repeat request (ARQ) scheme reacting on a cyclic redundancy check (CRC, included in each ACL-payload) secures error-free reception of the transmitted information. Table 38.1 gives an overview on the throughput values achievable with ACL connections. The maximum (unidirectional) Bluetooth throughput is 723 kbps.

As speech transmission is delay sensitive, SCO links support three different packet types that are transmitted at fixed intervals. All types were designed to transport CVSD (continuous variable slope delta) encoded speech at 64 kbps. The packet types always occupy just one (625 μsec) time slot only, but they are differentiated by their payload FEC. The packet payloads are either unprotected (called HV3) or 2/3 rate FEC-encoded (HV2) or protected with a 1/3-rate repetition code (HV1). For an HV3 connection a packet is transmitted every 6th slot (see Figure 38.3), for HV2 every 4th slot, and for HV1 every 2nd slot (meaning that with one HV1 connection no other traffic can be transmitted in the piconet). For SCO links, there is currently no ARQ scheme. In case of an erroneous reception of the packet overhead, the SCO packet is replaced by an erasure pattern. In case noncorrectable bit errors occur in the payload only, these errors are forwarded to the speech decoder.

Security is supported in Bluetooth by the specification of authentication and encryption. Changes for future versions of the Bluetooth specifications are:

To further improve coexistence with other systems in the ISM-band, it is intended that upcoming versions of Bluetooth will have the possibility to perform adaptive frequency hopping (AFH), that is, to exclude interfered carriers from the hop sequence. With AFH, the nominal hop rate will be halved, because then the slave will respond on the same frequency as on which it received the packet from the master [Blu02].

A high rate mode is envisioned that allows direct slave-to-slave communication at about a tenfold transmission rate. For this, a 4 MHz channel is chosen at a specifically good location within the bandwidth.

The future SCO link will provide for (some) packet repetitions and negotiation of the transmission rate [Blu03].

Performance

On the factory floor, Bluetooth can be used as a wireless add-on to wired systems or as a replacement of existing cabling. It can cover machine-to-machine communication, wireless/remote monitoring, or tracking and

TABLE 38.1 Throughput values for ACL connections; the reverse link in the unidirectional case consists of DH1 or DM1 packets, depending on whether the forward link uses a DH or DM packet type

name	no. of slots	FEC?	max. no. of user bytes	uni-directional throughput		bi-directional throughput	
				forward	reverse	forward =	reverse
DH1	1	no	27	172.8k	172.8k	172.8k	172.8k
DH3	3	no	183	585.6k	86.4k	390.4k	390.4k
DH5	5	no	339	723.2k	57.6k	433.9k	433.9k
DM1	1	2/3	17	108.8k	108.8k	108.8k	108.8k
DM3	3	2/3	121	387.2k	54.4k	258.1k	258.1k
DM5	5	2/3	224	477.8k	36.3k	286.7k	286.7k

some type of positioning of moving entities [GGH01, Bea02b]. Considering the comparably short range of Bluetooth and the likely association to a specific unit (represented by a machine/person/task), it can happen that several independently active Bluetooth piconets coexist and overlap in space. The use of frequency hopping helps to mitigate the effects of interference among these piconets. When assuming more or less time-synchronized piconets, a worst-case approximation of the loss rate can be made with Equation (1). It calculates the probability $P(x,n)$ that of x other piconets n hop onto the same frequency as the considered piconet

$$P(x,n) = \binom{x}{n}\left(\frac{1}{79}\right)^n\left(\frac{78}{79}\right)^{x-n} \tag{1}$$

The probability that at least one of the other x piconets transmits on the same frequency is then $P(x) = 1 - P(x,0)$. The smaller the number of actually interfering piconets, the better an approximation this approach gives; as for larger numbers, the distances to some of the interferers are likely to be too large to be harmful. In [ZSMH00, Zür00, MZT03], thus, a more sophisticated approach has been chosen and Bluetooth–Bluetooth co-existence results have been obtained with the help of detailed radio network simulations that include traffic, distribution, and fading models as well as adjacent channel effects. All results have been obtained with an office of 10×20 m^2 in mind, assuming an average master–slave distance of 2 m. Naturally, a factory floor is likely to be significantly larger than 10×20 m^2. Nevertheless, the increased delay spread does not really affect Bluetooth due to its small range (which is different for WLAN technologies; see respective performance section). The overall number of piconets can be larger on a factory floor (at a comparable density). At the same time, location and traffic of the units are likely to be more predictable. The results of the aforementioned publications thus give a good idea of what performance is achievable:

A 10×20 m^2 room supports 30 HV3 simultaneous speech connections with an average packet loss rate of 1% (a limit that represents a still acceptable quality).

HV3 packet types are preferable to HV2 and HV1. The subjective quality will not increase with additional payload coding. Using a coded HV packet just increases (unnecessarily) the interference in the network and the power consumption.

100 (!) simultaneous WWW-sessions (bursty traffic with an *average* data rate of 33.2 kbps each) in the 10×20 m^2 size room result in a degradation of the aggregate throughput of only 5%.

The maximum aggregate throughput in the room is 18 Mbps (at 50 fully loaded piconets). These piconets then transmit at a unidirectional data rate of 360 kbps.

Long and uncoded packets are preferable to shorter and/or coded ones. It takes 60 interfering piconets using the same packet type, 10 interfering HV1 connections (worst case), or a link distance of 27 m (which is far beyond the envisioned range of 10 m) before another packet type yields a larger throughput than DH5. It is advisable not to use the optional FEC (DM-packet types). As the coding is not appropriate to handle the almost binary character of the Bluetooth (*ad hoc*) transmission channel,[6] the needed power can additionally be saved.

Bluetooth is cheap and consumes quite little power. The ACL link is reliable with best-effort traffic (with a maximum throughput of 723 kbps). The SCO link has reserved bandwidth (at a relatively small throughput of 64 kbps), although the packets might contain residual bit errors. In principle, Bluetooth is very robust against other Bluetooth interference and a good performance can be achieved even in very dense environments. In case of only one slave per piconet, statistical QoS guarantees can be given; else, the internal scheduler can only with future Bluetooth Specifications give priority to traffic other than SCO traffic. Note that customized implementations, which cannot be based on existing profiles, might be difficult to realize, as the specification of new profiles is quite time-consuming.

38.5   IEEE 802.11

Technical Background

IEEE 802.11 comprises a number of documents that define (only) the lower layers (mainly the physical (PHY) and medium access control (MAC) layers) for a WLAN technology [Ins99b, Ins99a, Ins, OP99]. Being part of the IEEE 802 group means that an interface can be used (IEEE 802.2) to connect to the higher layers, which are then not aware of the — with IEEE 802.11 wireless — network that is actually transporting the data. The key intentions of IEEE 802.11 are thus to provide a high throughput and continuous network connection like available in wired LANs.

To support the wide employment of the technology also the use of IEEE 802.11 does not cause frequency licensing fees. IEEE 802.11 uses either infrared (IR), transmits in the unlicensed ISM-band at 2.4 GHz (like Bluetooth), or (like HIPERLAN/2) in 5 GHz bands that are license exempt in Europe and unlicensed in the U.S. (UNII bands). Other than in case of Bluetooth though, the companies holding key patents within IEEE 802.11 can charge developers of IEEE 802.11 products for using the patents "on reasonable terms" [Ins02, Clause 6b].

In principle, it is possible to have an IEEE 802.11 WLAN consisting of mobile stations (MS) only. It is more likely though that IEEE 802.11 is used as a wireless access technology to a wired LAN to which the connection is made by IEEE 802.11 access points (APs). Should the access point only use the distributed coordination function,[7] the MAC layer supports collision avoidance by using carrier sense multiple access (CSMA). This means that before transmitting a packet, the respective unit has to listen for the availability of the channel.[8] If the channel is sensed free after having been busy, the unit waits a certain period (called "DIFS") and then enters a random back-off period[9] of

$$\underbrace{random\ (0...min(2^{n_{PHY}+n_r}-1, 1023))}_{CW} \cdot t_{slot} \tag{2}$$

---

[6] The reasons are manifold. Without interference, the channel varies already due to hopping over 79 relatively narrowband channels. Additionally, with the wavelength used in Bluetooth, even small changes in position can cause large changes in the received signal strength. When there is interference, the effect becomes more pronounced. The existence or nonexistence of a close cochannel interferer can make the channel change from very good to very bad within the fraction of a moment (see also Figure 38.2).

[7] Which is likely and assumed in the following. In theory, the standard also provides additionally the use of a centralized point coordination function.

[8] The implementor can choose whether the units react (a) on just other IEEE 802.11 traffic, (b) on just other IEEE 802.11 traffic above a certain receive signal strength, or (c) on any signal above a certain receive signal strength [Ins99b, Section 18.4.8.4].

[9] The random back-off period is entered only when the channel was busy before. Else the unit will transmit at once after DIFS.

with $^n_{PHY}$ *being* a parameter depending on the type of physical layer chosen, n_r the index of the retransmission of the packet, t_{slot} presenting the slot duration, and *CW* standing for contention window (with $CW_{min} = 2^{n_{PHY}} - 1$)). Should the cannel by then still not be occupied, the unit transmits its packet (consisting of PHY header, MAC header, and payload). Upon a correct reception, the addressee responds with an ACK packet a short period (called "SIFS") later; see also Figure 38.4. The realized ARQ mechanism secures reliable data.

The IEEE 802.11 WLAN MAC concept is obviously made for best-effort data traffic. Services for which strict delay requirements exist, like speech, cannot well be supported by IEEE 802.11 as it currently is. Any QoS is thus difficult to provide, especially when multiple units coexist in the network. The IEEE 802.11 MAC concept also includes a mechanism to solve the hidden terminal problem. Whether this "ready-to-send/clear-to-send" (RTS/CTS) packet exchange saves more bandwidth (due to avoided retransmissions) than it additionally needs though, depends on the terminal density and payload packet length [Bia00]. As the RTS/CTS mechanism is optional, it will be assumed in the following that the RTS/CTS mechanism is not used.

IEEE 802.11 has a significantly larger power consumption than Bluetooth. Note that this is not only due to the higher transmit power (20 dBm in Europe, 30 dBm in the U.S.) but also due to the CSMA concept. IEEE 802.11 units not specifically in sleep status have to listen into the channel *all* the time (and not like Bluetooth just at the beginning of the receive slot). Simultaneously though, the higher transmit power allows for a larger range of about 50 m (with 20 dBm).

There are four different options for the physical layer implementation of IEEE 802.11:

IR: The infrared mode transmits near visible light at 850–950 nm wavelength. The data are pulse position modulated at 1 or 2 Mbps. In principle, the signal needs line-of-sight (LOS) and cannot go through walls. This and the nonvisibility of IEEE 802.11 IR products are the reasons for not covering the IR-mode further in this chapter.

FHSS: The Frequency Hopping Spread Spectrum mode is placed (like Bluetooth) in the 2.4 GHz ISM-band. The data are GFSK modulated using two levels for 1 Mbps and four for the 2 Mbps modulation rate. The FHSS mode divides the 79 hop frequencies into three distinct sets with 26 different sequences each. The hopping rate can be as slow as 2.5 hops/sec. Despite its comparably good interference robustness [SFEM02], the popularity of the FHSS mode is limited due to its comparably low transmission rates. Note though that its principles have been incorporated in the HomeRF standard [HRF, BS00].

DSSS: The Direct Sequence Spread Spectrum mode is also used in the 2.4 GHz ISM-band. The nominal bandwidth of the main lobe is 22 MHz. The transmit power reduction in the first side lobes and the rest is supposed to be 30 and 50 dB resp. (see also Figure 38.5 for a measured spectrum). In principle, 11/13 (U.S./Europe) center frequencies are available for the DSSS system. Nevertheless, using several systems in parallel requires a spacing of 25/30 MHz (U.S./Europe), which consequently allows to use only three systems in parallel.

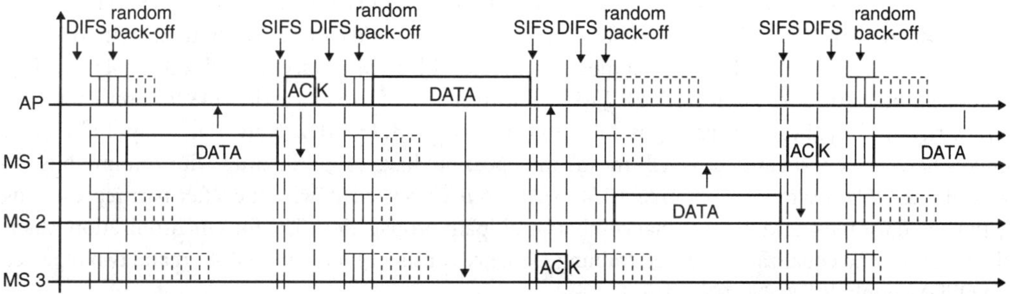

FIGURE 38.4 Principle time behavior of IEEE 802.11 under the distributed coordination function; note that the random back-off timer from units *not* the one to transmit continues after the next DIFS with the remaining number of slots.

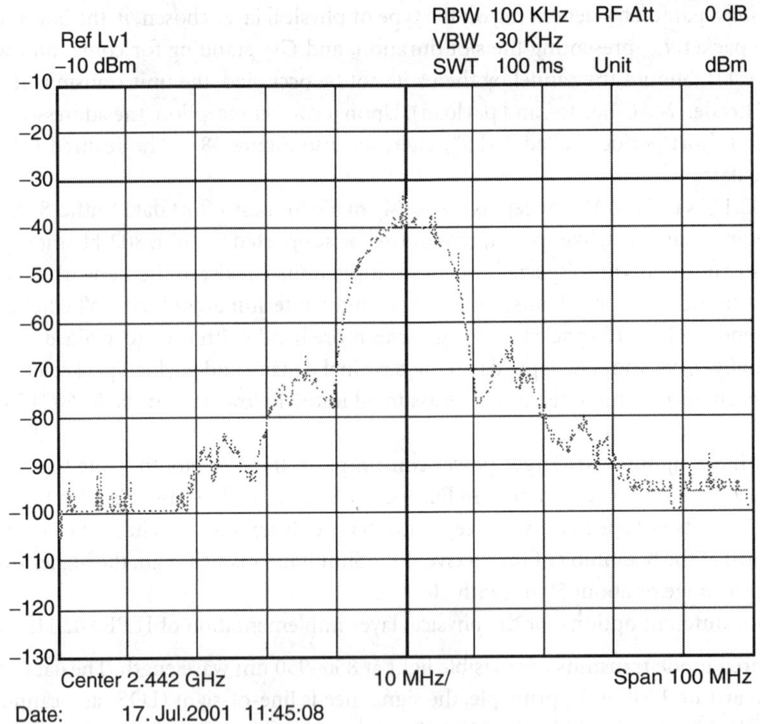

FIGURE 38.5   Measured spectrum of an IEEE 802.11b WLAN PCMCIA card.

The DSSS mode has the original mode (specified in IEEE 802.11) and a high-rate extension (specified in IEEE 802.11b). In the original mode, the chipping of the baseband signal is performed with 11 MHz, using an 11-chip pseudo-random code (Barker sequence). For the 1 Mbps modulation rate, a one-bit DBPSK symbol is spread with the Barker sequence; for the 2 Mbps modulation rate, a two-bit DQPSK symbol is spread with the same sequence.[10] The high-rate extension IEEE 802.11b was at the time of writing the most popular and widespread WLAN technology. For the PHY-header IEEE 802.11b uses the same 1 and 2 Mbps modulations as the plain DSSS mode. Note though that a shortened header of 96 μsec can be used. For the IEEE 802.11b PHY-payload (consisting of the MAC-header and the user data) a 5.5 and 11 Mbps Complementary Code Keying (CCK) modulation is used. The CCK uses a variation of M-ary orthogonal signaling (complex Walsh/Hadamard functions) with 8 complex chips in each spreading code word. For the 5.5 Mbps modulation rate 4 bits are mapped onto 8 chips and for 11 Mbps 8 bits are mapped onto 8 chips.

OFDM: The Orthogonal Frequency Division Multiplexing physical layer (also referred to as IEEE 802.11a) is placed in 5 GHz bands and nearly identical to the one of HIPERLAN/2 (see also Section 38.6). Seven modi are defined ranging from BPSK with rate $R=1/2$ FEC (lowest modulation rate with 6 Mbps) to 64-QAM with rate $R=3/4$ FEC (highest modulation rate with 54 Mbps, see also Table 38.2). The OFDM technique is based on a 64-point IFFT/FFT, while only using 52 of the subcarriers (48 for user data, 4 for pilot carriers). The subcarrier spacing is $\Delta f = 20$ MHz/64 = 0.3125 MHz. Note that always full OFDM symbols have to be transmitted, which means they possibly have to be filled up with dummy bits. To transmit one OFDM symbol $t_{sym} = 1/\Delta f + 1/4 \cdot 1/\Delta f = 4$ μsec is needed, with the latter part representing the time used for the guard interval that combats multipath propagation. The for synchronization, channel estimation, and equalization needed training sequence consists of ten repeated short and two repeated long OFDM symbols [Ins99c, vNAM+99].

---

[10]Note that in contrast to a typical CDMA system like UMTS, *all* users use the same spreading code.

TABLE 38.2 Comparison of different achievable maximum throughput rates in [Mbps] for the different IEEE 802.11 PHY Modes

mode	frq. band	t_{slot}	SIFS	CW_{min}	$t_{PHYh(eader)}$	modulation	ModRate [Mbps]	TP [Mbps] for PayBytes			
								60	576	1500	4061
FHSS	2.4GHz	$50\mu s$	$28\mu s$	15	$128\mu s$	GFSK (2 level)	1	0.29	0.79	0.91	0.96
						GFSK (4 level)	2	0.39	1.40	1.72	1.89
DSS	2.4 GHz	$20\mu s$	$10\mu s$	31	$192\mu s$	DBPSK	1	0.30	0.80	0.91	0.97
						DQPSK	2	0.40	1.42	1.72	1.89
					$96\mu s$	CCK (QPSK)	5.5	0.67	3.14	4.26	4.97
						CCK (QPSK)	11	0.75	4.54	7.11	9.16
OFDM	5 GHz	$9\mu s$	$16\mu s$	15	$20\mu s$ ($t_{sym}=$ $4\mu s$)	BPSK, R1/2	6	1.51	4.57	5.37	5.76
						BPSK, R3/4	9	1.81	6.32	7.79	8.51
						QPSK, R1/2	12	2.02	7.86	10.0	11.2
						QPSK, R3/4	18	2.25	10.4	14.1	16.3
						16-QAM, R1/2	24	2.38	12.3	17.6	21.2
						16-QAM, R3/4	36	2.59	15.2	23.7	30.3
						64-QAM, R1/2	48	2.64	17.1	28.5	38.4
						64-QAM, R3/4	54	2.70	17.9	30.8	42.2

Table 38.2 compares the theoretical maximum throughput values TP of the different IEEE 802.11 PHY versions after the MAC. The maximum payload length is 4095 bytes (which has to include the 34-byte MAC header). 1500 bytes is the common length of an Ethernet packet the values for 1500 bytes payloads (plus 34 bytes MAC header and check sum), 576 a typical length for a web browsing packet, and 60 bytes the length of a TCP acknowledgment. The throughput TP for the FHSS and the DSSS modes are calculated as follows:

$$TP_{FHSS,DSS} = \frac{PayBytes \cdot 8}{\underbrace{DIFS}_{2t_{slot}+SIFS} + \underbrace{\frac{CW_{min}}{2} \cdot t_{slot}}_{\text{average back-off}} + t_{PHYh} + \underbrace{\frac{34 \cdot 8}{ModRate}}_{\text{MACheader}} + \frac{PayBytes \cdot 8}{ModRate} + SIFS + \underbrace{t_{PHYh} + \frac{14 \cdot 8}{ModRate}}_{t_{ACK}}} . \quad (3)$$

$\underbrace{\qquad\qquad\qquad\qquad}_{t_{data\ packet}}$

Due to the specific characteristics of the OFDM symbols, the duration of packet $t_{data\ packet}$ and acknowledgment t_{ACK} are calculated somewhat differently for the throughput of the OFDM mode ("ceil" stands for the next larger integer). The rest of Equation (3) remains the same

$$t_{datapacket} = t_{PHYh} + t_{sym}\, ceil\left(\frac{16 + (34 + PayBytes) \cdot 8 + 6}{ModRate/12\ \text{Mbps} \cdot 48}\right), \quad t_{ACK} = t_{PHYh} + t_{sym}\, ceil\left(\frac{16 + 14 \cdot 8 + 6}{ModRate/12\ \text{Mbps} \cdot 48}\right). \quad (4)$$

For small payload sizes, the throughput values are very small. But when considering Ethernet packets, the highest theoretical throughput rates are 7.11 Mbps for IEEE 802.11b and 30.8 Mbps for IEEE 802.11a. Naturally, these wireless throughput rates are smaller than the wired ones (standard there is today 70–80 Mbps), but at least for the higher modulation rates with 1500 bytes Ethernet packets the throughput values are reasonably good. Note that the throughput rates given further decrease though as higher protocol levels produce additional overhead (or when the channel conditions cause packet losses). For example, measured throughput values for IEEE 802.11b are around 5 Mbps best-effort traffic in uninterfered environments [Mob01, MZ02].

For security IEEE 802.11 supports several authentication processes, which are listed in the specification (none is mandatory).[11] The most important future development of IEEE 802.11 is an extension to

---

[11] Neither Bluetooth nor IEEE 802.11 is renowned for their security concepts, and both have been criticized. Nevertheless, both systems have taken security aspects into consideration.

the standard to enable QoS (IEEE 802.11e). For issues like roaming and hand-over higher protocol layers like mobile IP are responsible. They are not part of the standard.

Performance

Next to aspects like individual link throughput, network capacity, and interference robustness, the transmission environment has to be taken into consideration when intending to use IEEE 802.11 on the factory floor. As the scenarios envisioned for IEEE 802.11 were placed primarily in homes and offices, some differences occur when looking at the delay spread. While in homes and offices the delay spread is assumed to be <50 nsec and < 100 nsec resp., it takes on values of 200–300 nsec on factory floors [OP99].

In case of IEEE 802.11b, a conventional rake receiver supports (only) about 60 nsec delay spread in the 11 Mbps mode and 200 nsec in the 5.5 Mbps version [vNAM$^+$99]. When using an IEEE 802.11b system with such a receiver on a factory floor intersymbol and/or interchip/codeword interference (ISI and/or ICI) are likely to degrade the performance noteworthily. When willing to provide adequate complexity though (in receiver algorithms and therefore in chip space), intelligent algorithms can be found that cope well with delay spreads of $1\,\mu$sec [CLMK02, LMCW02] and even mobility of the user. It is just that the somewhat different factory floor environment has to be taken into account when making the choice. Another option, of course, is to use IEEE 802.11a (or HIPERLAN/2). Due to the guard interval inherent in the OFDM technology, delay spreads of several hundred nanoseconds can be easily supported [vNAM$^+$99].

When considering the overall network performance and not just the individual link performance (or interference performance, as done in Section 38.7), note first of all that the number of publication presenting well-founded results is very limited. It seems to be general thinking that capacity in an IEEE 802.11a or b only network is not an issue. The three publications that do exist do not at all support this. In [Lin01], it is shown that cochannel interference with a carrier-to-interference ratio (CIR) of 5 dB still results in a packet loss rate of 10–20% ($BER=10^{-5}$) and that with a frequency offset of 5 MHz still $CIR=3$ dB is required to achieve the same result. Bianchi [Bia00] and Sadalgi [Sad00] present how the aggregate throughput in a single network decreases with the number of users, either due to hidden or exposed terminal problems or due to additional RTS/CTS overhead. With only 10 stations [Bia00] or a hidden node probability of 5% [Sad00], the system throughput is about halved (!) in case of 1500 byte payloads. Only when there are more than 25 stations is the RTS/CTS implementation justified, while the throughput is still reduced.

Thus, when installing IEEE 802.11a or b in a cellular fashion, some kind of frequency planning should be performed. For IEEE 802.11a or b, a respective mechanism has to be added (refer, e.g., to [Agh00] for frequency allocation algorithms). Note that the most relevant parameter for WLAN frequency planning is the number of mobile terminals that have to be served. From it the optimum number of access points (APs) and distance between APs can be determined. If it is desired that mobile stations can seamlessly change between APs (due to mobile deployment), hand-over algorithms have to be added.

Thus, IEEE 802.11 provides reliable best-effort traffic. The (theoretical) maximum transmission rates for IEEE 802.11a and b are 30.8 and 7.11 Mbps, respectively. To achieve network capacity values anywhere near three times these values (three parallel systems are possible), appropriate receivers and some sophisticated frequency planning should be provided, although a reduction should still be taken into account. In contrast to Bluetooth, IEEE 802.11 systems allow for higher data rates and larger ranges. IEEE 802.11 consumes (significantly) more power than Bluetooth and is not really suitable for speech connections. Currently, IEEE 802.11b chips are larger in size and more expensive than Bluetooth chips (IEEE 802.11a chips are not yet competitive).

38.6   HIPERLAN/2

Technical Background

HIPERLAN/2 is a WLAN standard that defines the physical (PHY), the data link control (DLC), and a convergence layer to enable the wireless access of Ethernet, PPP, UMTS, IEEE 1394, and ATM-based backbone networks with high data rates [Eur01b, Eur00c, Eur01a]. Due to the placement in the 5 GHz range also HIPERLAN/2 can operate globally without expensive frequency licenses. Like for IEEE 802.11 systems (but unlike for Bluetooth), the developers of HIPERLAN/2 products can nevertheless be charged for the use of the key patents on "fair, reasonable and nondiscriminatory terms" [Eur00a].

The PHY layer of HIPERLAN/2 is almost the same as the one of IEEE 802.11a: HIPERLAN/2 uses an OFDM-technique based on a 64-point IFFT/FFT with 48 user and 4 pilot carriers allowing various modulation and error control possibilities using the same carrier spacing and symbol duration as IEEE 802.11a. Depending on the transmit power used — in principle, values between 100 and 1000 mW are possible — the range is 30–60 m indoor and up to several hundred meters outdoor.

The main differences between the systems are found in the control layer. HIPERLAN/2 has a centralized structure that puts all traffic under the control of an access point (AP). Due to the thus smaller overhead but also due to the explicit specification of radio resource control (RRC) functions, larger transmission rates can be achieved with HIPERLAN/2 than with IEEE 802.11a (see also Table 38.3). The RRC functions do not only enable a dynamic frequency selection but additionally support mobility by providing for the required measurements for hand-over algorithms [KYW01]. Furthermore, the provision of radio bearers allows a differentiated treatment of the various traffic flows in the AP and thus enables a HIPERLAN/2 system to give QoS guarantees. Reliability of the data is given by a selective repeat ARQ mechanism [LLM+00]. Note that despite the central control of the AP, traffic can be transmitted directly between two mobile terminals (MTs). In this so-called "direct mode," the AP assigns a certain time window to the mobile units to comply this task.

HIPERLAN/2 organizes its traffic in MAC frames in a TDMA/TDD [GL00] fashion with adaptable time slot sizes. As shown in Figure 38.6, one 2 msec MAC-frame consists of a broadcast channel (BCH) that contains some identifiers and the transmission power, a frame channel (FCH) that defines the structure of the frame, an access feedback channel (ACH) that reports the results of previous access attempts, a period in which downlink or uplink protocol data units (PDUs) are transmitted and random access channels (RCHs) during which the mobile terminals can express their demands for transmit time. The durations of the downlink (**DL**, i.e., from AP to MT) and uplink (**UL**, i.e., from MT to AP) periods are flexible. Each PDU contains a preamble, control-PDUs (with, e.g., acknowledgement information) and user-PDUs. Only one PDU per MT can be transmitted per MAC-frame. Nevertheless, if the AP wants to serve one MT only it can decide to transmit a PDU that takes up the whole DL-period.

As the transmitted data flow are partitioned into 54 bytes long U-PDUs, the actual throughput values for HIPERLAN/2 do not depend so much on the packet sizes that arrive. Nevertheless, each PDU requires its own preamble (of $t_{p,DL} = 8$ μsec and $t_{p,UL, min} = 12$ μsec); hence, the system throughput decreases somewhat with the number of users N_U served within one MAC-frame. Assuming a constant overhead of $t_{OV,min} = 120$ μsec per MAC frame and that for each user served in the DL a 9 bytes long C-PDU has to be transmitted in

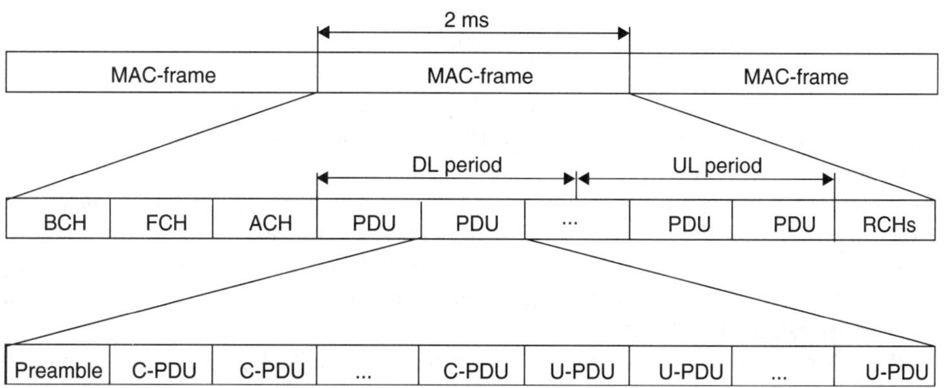

FIGURE 38.6 MAC-frame structure in HIPERLAN/2.

TABLE 38.3   Comparison of different achievable maximum throughput rates in [Mbps] for 1500 bytes Ethernet packets considering different numbers of users that are simultaneously served by the AP

Modulation	CodeRate	ModRate [Mbps]	TP [Mbps] for N_U users			
			1	2	5	10
BPSK	1/2	6	5.36	5.36	4.93	4.5
BPSK	3/4	9	8.14	8.14	7.71	7.07
QPSK	1/2	12	10.9	10.7	10.3	9.43
QPSK	3/4	18	16.5	16.3	15.6	14.6
16-QAM	9/16	27	24.9	24.4	23.6	21.9
16-QAM	3/4	36	33.0	32.6	31.3	29.1
64-QAM	3/4	54	49.7	49.1	47.1	43.9

the UL the maximum system throughput values TP can be calculated with the help of Equations (5) and (6). Like for IEEE 802.11a $t_{sym} = 4\ \mu$ sec.

$$
\begin{aligned}
&2ms - t_{OV,min} \\
&\geq N_U \underbrace{\left(t_{p,UL,min} + ceil\left(\frac{9 \cdot 8}{ModRate/12\ \text{Mbps} \cdot 48}\right) t_{sym}\right)}_{t_{C-PDU}} + \sum_{i=1}^{N_U} \underbrace{\left(t_{p,DL} + ceil\left(\frac{54 \cdot N_{PDU,i} \cdot 8}{ModRate/12\ \text{Mbps} \cdot 48}\right) t_{sym}\right)}_{t_{U-PDU,\,i}}
\end{aligned}
\tag{5}
$$

$$
TP = \frac{54 \cdot 8}{2ms} \sum_{i=1}^{N_U} N_{PDU,i}\, \frac{PayBytes/54}{ceil(PayBytes/54)}
\tag{6}
$$

Naturally, for PayBytes values that just do not fit into one PDU (like PayBytes=60) the throughput values are reduced significantly (still not as badly as small payloads reduce the 802.11a throughput). Nevertheless, for better fitting or larger packet sizes good system throughput values can be achieved (see Table 3 for 1500 bytes Ethernet packets). While in this case the average PHY layer induced reduction is only around 13% for HIPERLAN/2, it is around 25% for IEEE 802.11a. The situation becomes more unfavorable for IEEE 802.11a, the smaller the payload size is.

For security, the HIPERLAN/2 specification foresees encryption (DES and TripleDES) and authentication.

Performance

The investigations of the HIPERLAN/2 performance show first of all that HIPERLAN/2 can well cope with a wide range of channel conditions. In case of delay spreads typical for industrial applications (around 250 nsec, which in outdoor environments corresponds to cell radii of at least 100 m), the system shows no performance degradation, thanks to the robustness of the OFDM PHY [KJSWW99].

The radio network performance in office and exhibition hall environments (the latter scenario being comparable to a factory floor) is promising, too. In office scenarios, system throughput values close to the maximum throughput can easily be reached [MKJST00]. In the exhibition hall, the performance depends somewhat more on the scheduling [LMP+00], link adaptation [TM99], and frequency planning algorithms used [HZ00] as well as on the antenna design [TM99]. With good algorithms though, also in the exhibition hall high throughput values are possible. In [HZ00], it is additionally shown that adaptive frequency planning outperforms a regular fixed frequency plan (which is what would — if at all — be possible for an IEEE 802.11 system), while the frequency reselection rate remains small (less than 1 per hour in the investigated exhibition hall scenario).

HIPERLAN/2 is thus a robust WLAN technology that, due to explicitly considering the obstacles of wireless networking, achieves maximum system throughput values of 50–44 Mbps for 1–10 users simultaneously served. This is the case even for more difficult industrial applications, that is, on factory floors.

As HIPERLAN/2 organizes regular measurements of the channel conditions it can even cope with changing environments and mobility. Due to the centralized structure, HIPERLAN/2 is — compared with IEEE 802.11 systems — very power consumption friendly (the MTs know the periods when to listen) and supports QoS services. HIPERLAN/2 chip sets are not yet competitive.

38.7 Coexistence of WPAN and WLAN (Bluetooth and IEEE 802.11b)

It would be most desirable when every mobile unit can connect effortlessly using whatever technology is most suitable at the time. Multiple wireless technologies will thus coexist in the near future: within an enterprise, for example, WLAN technologies could be used for flexible access to large corporate databases while WPAN technologies handle specific tasks (and cellular systems the voice communication). This is generally not an issue unless the technologies are placed in the same frequency band and/or are linked by the application with each other (example: someone uses a Bluetooth head set with the mobile telephone). The two technologies that are placed in the same frequency band[12] and that are, due to their popularity, quite likely to be used in such a scenario (and possibly even in the same device) are Bluetooth and IEEE 802.11b.

Numerous publications cover the mutual interference and performance impairments of the systems (e.g., [Enn98, Zyr98, GvDS01, Fum01, How01, Mob01, MZ02]). Depending on the investigated scenarios, the assessment of the situation varies from "good reliability even in fairly dense environments" to "the effects of interference can be quite severe". A relative agreement exists in the causes that determine the systems' performances: link distances (BT–BT, 802.11b–802.11b, BT transmitter–802.11b receiver, BT receiver–802.11b transmitter), traffic load, Bluetooth packet type, density of units, local propagation conditions. Some of the results are briefly summarized in the following.

IEEE 802.11b requires about a CIR of 10dB to cope with a (narrow band) Bluetooth hop into its (wide) pass-band. IEEE 802.11b has the disadvantage that its back-off procedure was designed to optimize the IEEE 802.11b WLAN performance but not to handle external interferers: Each loss of a packet due to a collision with Bluetooth will increase the back-off window size by factor two (causing an unnecessary throughput reduction). Furthermore, the protocols overlaying WLAN often incorporate TCP, which includes the risk that packet losses on the air link are mistaken for network congestion, which then might initiate a slow start. In contrast, the main disadvantage of Bluetooth is that its transmit power is 20 dB below that of IEEE 802.11b. Another one is that the BT reverse link packet that contains the acknowledgement is transmitted on a different frequency than the forward link packet. This increases the packet loss probability in case of (frequency static) IEEE 802.11b interference. The packet loss rate PLR_{BT} then yields to

$$PLR_{BT} = PLR_{forward} + \underbrace{(1 - PLR_{forward}) \cdot PLR_{header}}_{PLR_{reverse}} \tag{7}$$

It would otherwise — were the forward and reverse link to use same hop frequency — be $PLR_{BT} \approx PLR_{forward}$ (depending on the IEEE 802.11b system load).

There are, in principle, three different approaches to assist otherwise interfering systems to coexist: Separation in time, separation in frequency, or separation in space.[13]

- *Separation in the frequency domain:*
  1. IEEE 802.11b can be used with an improved transmit filter that reduces the interference power on the side lobe frequencies and thus enhances the separation on those carriers.

---

[12]Next to other technologies like HomeRF networks [HRF], RF-ID systems [Fin00], microwave ovens, etc.

[13]The approaches "separation through code" (keyword "CDMA"), through the channel (keyword "MIMO"), or through the modulation ("I versus Q") allow to unlink several users of *the same* system. It is not obvious though to apply any of these latter methods to improve the coexistence of *different* systems, which is what is needed here.

2. Bluetooth can perform adaptive frequency hopping (AFH) — which is currently being specified [Blu02] —, that is, exclude the most heavily interfered frequencies from its hop sequence. Note that in several realistic situations AFH is sufficient to combat the interference effects.

- *Separation in the time domain*:
  1. The IEEE 802.11b carrier sensing algorithm could also consider Bluetooth signals[14] as well as Bluetooth could be extended with a carrier sensing algorithm [ZKJ00]. The principle problem of carrier sensing though is, that to be really effective, the transmitter has to sense the situation at the receiver correctly, that is, the correlation between transmitter and receiver situation has to be high. In an uncoordinated WLAN–WPAN scenario, the hidden as well as the exposed terminal problems are likely to countermeasure any advantage there might be.
  2. A joint scheduler can allot alternating transmit time shares to both systems in a (to be specified) fair way. This, of course, only works (and with AFH is necessary only) when both systems are in one device, or even on one chip.

- *Separation in space*:
  1. Should IEEE 802.11b and Bluetooth co-exist in the same unit an intelligent antenna design and placement can optimize the isolation between the Bluetooth and IEEE 802.1 1b antennas. This will not prevent collisions but minimizes their impact by maximizing the CIRs for the two systems.[15]
  2. Antenna diversity can help each of the two technologies individually to improve its performance.

38.8   Summary and Conclusions

When deciding on a wireless technology to use, first the characteristics of the foreseen application have to be clarified: do you want to move with or within the network, do you have mobility within a small range or in a larger area, at what speed do the units move, do you need access to large data basis or just locally, is battery life a critical issue, what maximum length can the wireless link have, what length does it have in average, etc.? As a next step, the existing technologies can be viewed for their applicability.

Between the WLAN and WPAN technologies discussed here the main distinction is the range in which connectivity can be established (WPAN small, WLAN larger) and thus whether movement takes place rather with (WPAN) or within the network (WLAN).

Bluetooth is a WPAN technology that is very power efficient. Like any other wireless system, it cannot provide hard throughput guarantees. Bluetooth is nevertheless quite robust for best-effort traffic in coexistence environments. One hundred Bluetooth piconets can transmit at an average data rate of 95%·33.2 kbps in an area of 10×20 m^2. Fifty fully loaded piconets can transmit at an average, unidirectional transmission rate of 360 kbps. Similar results are likely to be achieved on larger factory floors (with comparable piconet density), as the disadvantage of a larger number of units can be outweighed by the more structured and predictable unit location.

IEEE 802.11 is a WLAN technology that is not too power consumption friendly. Additionally, in a relatively dense network scenario, the maximum aggregate throughput of three times 7.11/30.8 Mbps is likely to be seriously impaired. To aid the WLAN performance on a factory floor, it is thus advisable to take the following two measures: apply means to combat the increased delay spread (in case of IEEE 802.11b) and additionally (for any IEEE 802.11 system) carefully plan the frequency layout and access point placement.

---

[14]As has been mentioned before, the IEEE 802.11b specification already provides for the possibility that IEEE 802. 11b senses and not transmits if other than systems IEEE 802.11b systems are active [Ins99b, Section 18.4.8.4]. Next to the fact that most implementations do not seem to support this option, the principle problem is that IEEE 802.11b will refrain from transmission when it senses that Bluetooth *transmits*. To improve coexistence though IEEE 802.11b should refrain from transmission when the nearby Bluetooth units receive.

[15]Helpful to minimize the interference power is of course also power control (when applied by the interfering unit). Nevertheless, even though it can be recommended to implement power control (let alone to save power [MZT03]), in real life situations it cannot be relied upon that the interfering unit can indeed live with less power.

Within HIPERLAN/2, means for dynamic frequency selection are already included and HIPERLAN/2 is certainly the more suitable WLAN technology for industrial applications. Next to allowing for higher throughput rates (three times 44 Mbps should be achievable without problem), HIPERLAN/2 is due to its centralized structure more power efficient and supports different Quality-of-Service classes.

The most efficient measure to aid the coexistence of Bluetooth and IEEE 802.11b is to use Bluetooth with adaptive frequency hopping. If this is not sufficient, hardware-related improvements and/or a common scheduler have to be added.

References

[3rd99]   3rd Generation Partnership Project; Technical Specification Group Acess Network, *Opportunity Driven Multiple Access,* December 1999, 3G TR 25.924 version 1.0.0.

[Agh00]   Aghvami, A. Hamid, *Resource Allocation in Hierarchical Cellular Systems,* Artech House Publishers, Boston, London, 2000.

[Bea02a]   Beaumont, Daniel, Citywide 802.11b Networks Gain Momentum, *Planet Wireless,* 12–14, July 2002.

[Bea02b]   Beaumont, Daniel, More Bluetooth Products but Key Profiles Delayed, *Planet Wireless,* 12–14, July 2002.

[Bia00]   Bianchi, Giuseppe, Performance analysis of the IEEE 802.11 distributed coordination function, *IEEE Journal on Selected Areas in Communications,* 18:535–547, 2000.

[Bis01]   Bisdikian, Chatschik, An overview of the Bluetooth wireless technology, *IEEE Communications Magazine,* 39:86–94, 2001.

[Blu]   http://www.bluetooth.com.

[Blu99]   Bluetooth Special Interest Group, *Specification of the Bluetooth System, Version 1.1,* December 1999.

[Blu02]   Bluetooth SIG Coexistence Working Group, *AFH Specification, Draft Version 0.7,* September 2002.

[Blu03]   Bluetooth Special Interest Group, *Bluetooth 1.2 Core Specification, second draft version,* January 2003.

[BS00]   Bray, Jennifer and Charles F. Sturman, *Bluetooth: Connect Without Cables,* Prentice-Hall, Eaglewood Cliffs, 2000.

[CC88]   Clark, George C. and J. Bibb Cain, *Error-Correction Coding for Digital Communication,* 3rd ed., Plenum Press, New York 1988.

[CLMK02]   Clark, Martin V., Kin K. Leung, Bruce McNair, and Zoran Kostic, Outdoor IEEE 802.11 Cellular Networks: Radio Link Performance, in Proceedings of IEEE International Conference on Communcation (ICC), May, 2002.

[Enn98]   Ennis, Greg, Impact Bluetooth on 802.11 Direct Sequence, Technical report IEEE 802.11-98/319, IEEE, September 1998.

[Eur00a]   ETSI Directives, Annex 6: ETSI Intellectual Property Rights Policy, April, 2000.

[Eur00b]   European Telecommunications Standards Institute (ETSI), ETSI EN 300 328-1 V1.2.2, July 2000.

[Eur00c]   European Telecommunications Standards Institute (ETSI), *HIPERLAN Type 2; Data Link Control Specification; Part 1: Basic Data Transport Functions,* November 2000.

[Eur01a]   European Telecommunications Standards Institute (ETSI), *HIPERLAN Type 2; Data Link Control Specification; Part 2: Radio Link Control (RLC) sub-layer,* April 2001.

[Eur01b]   European Telecommunications Standards Institute (ETSI), *HIPERLAN Type 2; Physical (PHY) Layer,* February 2001.

[Fed02]   Federal Communications Commission (FCC), Code of Federal Regulations, 2002.

[Fin00]   Finkenzeller, Klaus, *RFID-Handbuch,* Hanser, München Wien, 2000.

[Fum01]   Fumolari, David, Link Performance of an Embedded Bluetooth Personal Area Network, in Proceedings of IEEE International Conference on Communication (ICC), Helsinki, June, 2001.

[GGH01]   Green, Jeremy, Rob Gear, and Nick Harman, Bluetooth: Users, Applications and Technologies, *ovum*, June 2001.

[GL00]   Gummalla, Ajay Chandra V. and John O. Limb, Wireless medium access control protocols, *IEEE Communications Surveys and Tutorials*, 3, 2000, http://www.comsoc.org/ pubs/surveys.

[GvDS01]   Golmie, N., R.E. van Dyck, and A. Soltanian, Bluetooth and 802.11b Interference: Simulation Model and System Results, Technical report IEEE802.15-01/195R0, IEEE, April 2001, http://grouper.ieee.org/groups/802/15/pub/2001/May01/01195r0P802-15_TG2-BT-802-11-Model-Results.pdf.

[Haa98]   Haartsen, Jaap, Bluetooth — the universal radio interface for ad hoc, wireless connectivity, *Ericsson Review*, (3):110–117, 1998.

[How01]   Howitt, Ivan, IEEE 802.11 and Bluetooth Coexistence Analysis Methodology, in Proceedings of IEEE Vehicular Technology Conference (VTC), Rhodes, May, 2001.

[HRF]   http://www.homerf.org.

[HZ00]   Huschke, Jörg and Gerd Zimmermann, Impact of Decentralized Adaptive Frequency Allocation on the System Performance of HIPERLAN/2, in Proceedings of IEEE Vehicular Technology Conference (VTC), Tokyo, May 2000.

[IET]   http://www.ietf.org/html.charters/manet-charter.html.

[Ins]   http://grouper.ieee.org/groups/802/11/index.html.

[Ins99a]   Institute of Electrical and Electronic Engineering, Part 11: Wireless LAN Medium Access Control (MAC) and Physical Layer (PHY) specifications, September 1999, ANSI/IEEE Std 802.11.

[Ins99b]   Institute of Electrical and Electronic Engineering, Part 11: Wireless LAN Medium Access Control (MAC) and Physical Layer (PHY) specifications: Higher-Speed Physical Layer Extension in the 2.4 GHz Band, September 1999, IEEE Std 802.11b-1999.

[Ins99c]   Institute of Electrical and Electronic Engineering, Part 11: Wireless LAN Medium Access Control (MAC) and Physical Layer (PHY) specifications: Higher-Speed Physical Layer in the 5GHz Band, September 1999, IEEE Std 802.11a-1999.

[Ins02]   IEEE-SA Standards Board Bylaws, September 2002, Approved by Standards Association Board of Governors.

[JIC99]   Ji, L., M. Ishibashi, and M.S. Corson, An Approach to Mobile Ad Hoc Network Protocol Kernel Design, in Proceecings of IEEE Wireless Communications and Networking Conference (WCNC), New Orleans, Sept., 1999.

[KJSWW99]   Khun-Jush, Jamshid, Peter Schramm, Udo Wachsmann, and Fabian Wenger, Structure and Performance of the HIPERLAN/2 Physical Layer, in Proceedings of IEEE Vehicular Technology Conference (VTC), Amsterdam, Fall 1999, pp. 2667–2671.

[KYW01]   Kadelka, Arndt, Erkan Yidirim, and Bernhard Wegmann, Serving IP Mobility with HIPER-LAN/2, in Proceedings European Mobile Communications Conference (EPMCC), Vienna, Feb., 2001.

[LC83]   Lin, Shu and Daniel J. Costello, *Error-Control Coding*, Prentice-Hall, Eaglewood Cliffs, NJ, 1983.

[Lin01]   Lindgren, Magnus, Physical Layer Simulations of the IEEE 802.11b Wireless LAN-Standard, Master's thesis, Luleå Technical University, 2001.

[LLM+00]   Li, Hui, Jan Lindskog, Göran Malmgren, Gyorgy Miklos, Fredrik Nilsson, and Gunnar Rydnell, Automatic Repeat Request (ARQ) Mechanism in HIPERLAN/2, in Proceedings of IEEE Vehicular Technology Conference (VTC), Tokyo, May, 2000.

[LMCW02]   Leung, Kin K., Bruce McNair, Leonard J. Cimini, and Jack H. Winters, Outdoor IEEE 802.11 Cellular Networks: MAC Protocol Design and Performance, in Proceedings of IEEE International Conference on Communication (ICC), New York, April–May 2002.

[LMP+00]   Li, Hui, Göran Malmgren, Mathias Pauli, Jürgen Rapp, and Gerd Zimmermann, Performance of the Radio Link Protocol of HIPERLAN/2, in Proceedings of IEEE International Symposium on Personal, Indoor and Mobile Radio Communication (PIMRC), London, 2000.

[Mal01]   Maltz, David A., On-Demand Routing in Multi-hop Wireless Mobile Ad Hoc Networks, PhD thesis, School of Computer Science Carnegie Mellon University, Pittsburgh, May 2001, http://reports-archive.adm.cs.cmu.edu/anon/2001/CMU-CS-01-130.pdf.

[Min02]   Ministry of Telecommunications (MKK), Japan, RCR STD-33A, 2002.

[MKJST00]   Malmgren, Göran, Jamshid Khun-Jush, Peter Schramm, and Johan Torsner, 6:3 HIPERLAN Type 2 — An emerging world wide WLAN standard, in Proceedings of International Symposium on Services and Local Access, Stockholm, June, 2000.

[Mob01]   Mobilian, Wi-Fi™ (802.11b) and Bluetooth™; An Examination of Coexistence Approaches, http://www.mobilian.com, 2001

[MZ02]   Matheus, Kirsten and Stefan Zürbes, Co-existence of Bluetooth and IEEE 802.11b WLANs: Results from a Radio Network Testbed, in Proceedings of IEEE International Symposium on Personal, Indoor and Mobile Radio Communication (PIMRC), Lissabon, Sept., 2002.

[MZT03]   Matheus, Kirsten, Stefan Zürbes, and Rakesh Taori, "Fundamental Properties of Ad-Hoc Networks Like Bluetooth: A Radio Network Perspective", in Proceedings of IEEE Vehicular Technology Conference (VTC), Orlando, Sept. 2003.

[OP99]   O'Hara, Bob, and Al Petrick, *IEEE 802.11 Handbook a Designer's Companion*, Standards Information Network IEEE Press, New York, 1999.

[RBM02]   Rouse, T.S., I.W. Band, and S. McLaughlin, Capacity and Power Analysis of Opportunity Driven Multiple Access (ODMA) Networks in CDMA Systems, in Proceedings of IEEE International Conference on Communication (ICC), 2002, pp. 3202–3206.

[Sad00]   Sadalgi, Shreyas, A Performance Analysis of the Basic Access IEEE 802.11 Wireless LAN MAC Protocol (CSMA/CA), http://paul.rutgers.edu/~sadalgi/network.pdf, May 2000.

[SFEM02]   Stranne, André, Fredrik Florén, Ove Edfors, and Bengt-Arne Molin, Throughput of IEEE 802.11 FHSS Networks in the Presence of Strongly Interfering Bluetooth Networks, in Proceedings of IEEE International Symposium on Personal, Indoor and Mobile Radio Communication (PIMRC), Lissabon, Sept., 2002.

[SPG97]   Shenker, S., C. Partridge, and R. Guerin, Specification of Guaranteed Quality of Service, *RFC 2212*, September 1997.

[TM99]   Torsner, Johan and Göran Malmgren, Radio Network Solutions for HIPERLAN/2, in Proceedings of IEEE Vehicular Technology Conference (VTC), pp. 1217–1221, Houston, Spring, 1999.

[vNAM⁺99]   van Nee, Richard, Geert Awater, Masahiro Morikura, Hitoshi Takanashi, Mark Webster, and Karen W. Halford, New High-Rate Wireless LAN Standards, *IEEE Communications Magazine*, 37:82–88, 1999.

[ZKJ00]   Zhen, Bin, Yongsuk Kim, and Kyunghun Jang, The Analysis of Coexistence Mechanisms of Bluetooth, in Proceedings of IEEE Vehicular Technology Conference (VTC), Spring, 2000.

[ZSMH00]   Zürbes, Stefan, Wolfgang Stahl, Kirsten Matheus, and Jaap Haartsen, Radio Network Performance of Bluetooth, in Proceedings of IEEE International Conference on Communication (ICC), New Orleans, June, 2000.

[Zür00]   Zürbes, Stefan, Considerations on Link and System Throughput of Bluetooth Networks, in Proceedings of IEEE International Symposium on Personal, Indoor and Mobile Radio Communication (PIMRC), London, Sept., 2000, pp. 1315–1319.

[Zyr98]   Zyren, Jim, Extension of Bluetooth and 802.11 Direct Sequence Interference Model, Technical report IEEE 802.11-98/378, IEEE, November 1998, http://www.ieee802.org/15/pub/SG.html.

39

PROFIBUS — Open Solutions for the World of Automation

Ulrich Jecht
UJ Process Analytics

Wolfgang Stripf
Siemens AG

Peter Wenzel
PROFIBUS International

39.1 Basics

Fieldbuses are industrial communication systems with bit-serial transmission that use a range of media such as copper cable, fiber optics, or radio transmission to connect distributed field devices (sensors, actuators, drives, transducers, analyzers, etc.) to a central control or management system. Fieldbus technology was developed in the 1980s with the aim to save cabling costs by replacing the commonly used central parallel wiring and dominating analog signal transmission (4–20 mA- or ± 10 V interface) with digital technology. Due to the different industry-specific demands, to sponsored research and development projects or preferred proprietary solutions of large system manufacturers, several bus systems with varying principles and properties were established in the market. The key technologies are now included in the recently adopted standards IEC 61158 and IEC 61784 [1]. *PROFIBUS is an integral part of these standards.* Fieldbuses create the basic prerequisite for distributed automation systems. Over the years, they evolved to instruments for automated processes with high productivity and flexibility compared to conventional technology (Figure 39.1).

PROFIBUS is an open, digital communication system with a wide range of applications, particularly in the fields of factory and process automation, transportation, and power distribution. PROFIBUS is suitable for both fast, time-critical applications, and complex communication tasks.

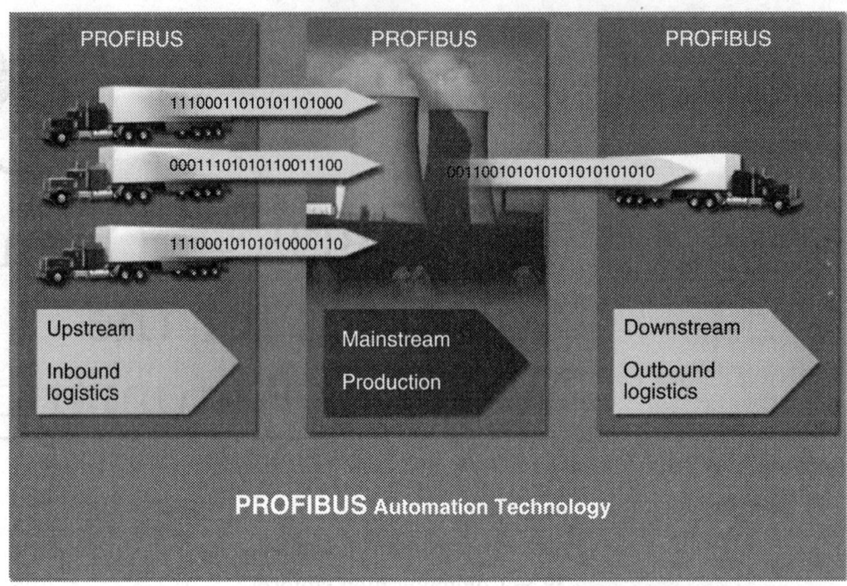

FIGURE 39.1 PROFIBUS suitable for all decentralized applications.

The application and engineering aspects are specified in the generally available guidelines of the PROFIBUS User Organization [2]. This fulfills user demand for standardization, manufacturer independence and openness, and ensures communication between devices of various manufacturers.

Based on a very efficient and extensible communications protocol, combined with the development of numerous application profiles (communication models for device type families) and a fast growing number of devices and systems, PROFIBUS began its record of success, initially in factory automation and, since 1995, in process automation. Today, PROFIBUS is the world market leader for fieldbuses with more than a 20% share of the market, approx. 500,000 equipped plants, and more than 10 million nodes. Today, there are more than 2000 PROFIBUS products available from a wide range of manufacturers.

The success of PROFIBUS stems in equal measures from its progressive technology and the strength of its noncommercial PROFIBUS User Organization e.V. (*PNO*), the trade body of manufacturers and users founded in 1989. Together with the 22 other regional PROFIBUS associations within countries all around the world, and the international umbrella organization PROFIBUS International (*PI*) founded in 1995, this organization now totals more than 1200 members worldwide. Objectives are the continuing further development of PROFIBUS technology and increasing worldwide acceptance.

PROFIBUS has a modular structure (PROFIBUS Tool Box) and offers a range of *transmission and communication technologies,* numerous *application and system profiles* as well as *device management and integration tools.* Thus PROFIBUS covers the various and application-specific demands from the field of factory to process automation, from simple to complex applications, by selecting the adequate set of components out of the toolbox (Figure 39.2).

39.2 Transmission Technologies

PROFIBUS features four different transmission technologies, all of which are based on international standards. They all are assigned to PROFIBUS in both IEC 61158 and IEC 61784: RS485, RS485-IS, and MBP-IS (IS stands for intrinsic safety protection), and Fiber Optics.

RS485 transmission technology is simple and cost-effective and is primarily used for tasks that require high transmission rates. Shielded, twisted pair copper cable with one conductor pair is used. No expert knowledge is required for installation of the cable. The bus structure allows addition or removal of stations or the step-by-step commissioning of the system without interfering in other stations. Subsequent expansions (within defined limits) have no effect on stations already in operation.

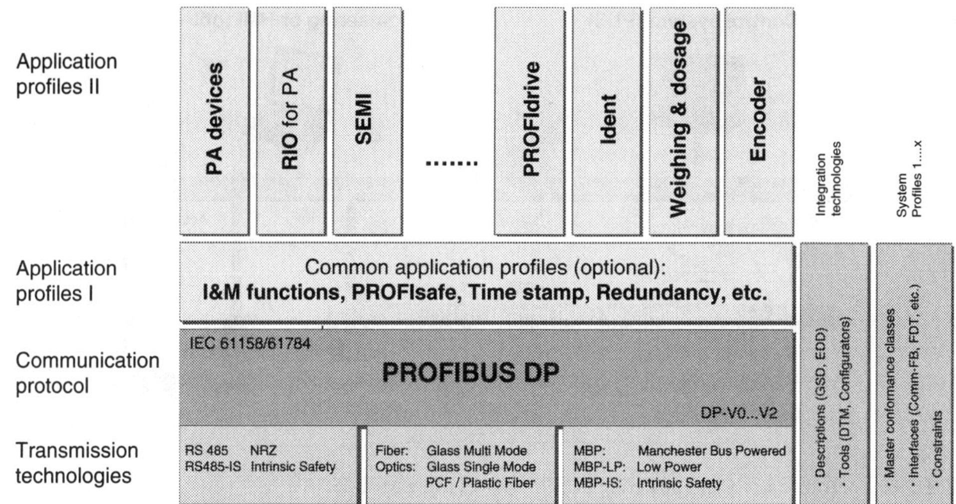

FIGURE 39.2 Structure of PROFIBUS system technology.

Various *transmission rates* can be *selected* from 9.6 Kbit/sec up to 12 Mbit/sec. One uniform speed is selected for all devices on the bus when commissioning the system. Up to *32 stations* (master or slaves) can be connected in a single segment. For connecting more than 32 stations, repeaters can be used. The maximum permissible *line length* depends on the transmission rate.

Different cable types (type designation A – D) for different applications are available on the market for connecting devices either to each other or to network elements (segment couplers, links and repeaters). When using RS485 transmission technology, PI recommends the use of *cable type A.*

RS485-IS transmission technology responds to an increasing market demand to support the use of RS485 with its fast transmission rates within intrinsically safe areas. A PROFIBUS guideline is available for the configuration of intrinsically safe RS485 solutions with simple device interchangeability. The interface specification details the levels for current and voltage that must be adhered to by all stations in order to ensure safe operation during interconnection. An electric circuit limits currents at a specified voltage level. When connecting active sources, the sum of the currents of all stations must not exceed the maximum permissible current. In contrast to the FISCO model (see below), *all* stations represent active sources. Up to 32 stations may be connected to the intrinsically safe bus circuit.

MBP type transmission technology ("Manchester Coding" and "Bus Powered") is a new term that replaces the previously common terms for intrinsically safe transmission such as "Physics in accordance with IEC 61158-2," "1158-2," etc. In the meantime, the current version of the IEC 61158-2 (physical layer) describes *several different* transmission technologies, MBP technology being just one of them. Thus, differentiation in naming was necessary.

MBP is a *synchronous, Manchester-coded transmission with a defined transmission rate of 31.25 Kbit/sec.* In its version **MBP-IS,** it is frequently used in process automation as it satisfies the key demands of the chemical and petrochemical industries for *intrinsic safety* and *bus powering using two-wire technology.*

MBP transmission technology is usually limited to a specific segment (field devices in hazardous areas) of a plant, which is then linked to an RS485 segment via a *segment coupler* or *links* (Figure 39.3).

Segment couplers are signal converters that modulate the RS485 signals to the MBP signal level and *vice versa.* They are transparent from a bus protocol's point of view. In contrast, *links* are providing more computing power. They virtually map the entire field devices connected to the MBP segment into the RS485 segment as a single slave. Tree or line structures (and any combination of the two) are network topologies supported by PROFIBUS with MBP transmission with up to 32 stations per segment and a maximum of 126 per network.

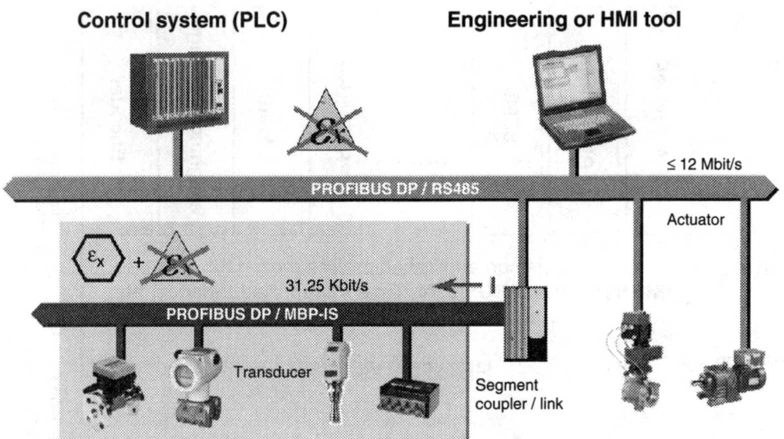

FIGURE 39.3   Intrinsic safety and powering of field devices using MBP-IS.

Fiber Optic transmission technology is used for fieldbus applications with very high electromagnetic interference or that are spread over a large area or distance.

The PROFIBUS guideline for fiber optic transmission [3] specifies the technology available for this purpose including multimode and single-mode glass fiber, plastic fiber, and HCS® fiber. Of course, while developing these specifications, great care was taken to allow problem-free integration of existing PROFIBUS devices in a fiber optic network without the need to change the protocol behavior of PROFIBUS. This ensures *backward compatibility* with existing PROFIBUS installations.

The internationally recognized **FISCO model** considerably simplifies the planning, installation, and expansion of PROFIBUS networks in potentially explosive areas. FISCO stands for *Fieldbus Intrinsically Safe Concept*. It has been developed by the German PTB [4]. The model is based on the specification that a network is intrinsically safe and requires no individual intrinsic safety calculations when the relevant four bus components (field devices, cables, segment couplers, and bus terminators) fall within predefined limits with regard to voltage, current, output, inductance, and capacity. The corresponding proof can be provided by certification of the components through authorized accreditation agencies, such as PTB (Germany) or UL, and FM (USA) and others.

If FISCO-approved devices are used, not only is it possible to operate more devices on a single line, but the devices can be replaced during runtime by devices of other manufacturers or the line can be expanded — all without the need for time-consuming calculations or system certification. So you can simply plug & play, even in hazardous areas!

39.3 Communication Protocol

PROFIBUS DP

At the protocol level, PROFIBUS with Decentralized Peripherals (DP) and its versions DP-V0 to DP-V2 offer a broad spectrum of optional services, which enable optimum communication between different applications.

DP has been designed for *fast data exchange at field level*. Data exchange with the distributed devices is primarily cyclic. The communication functions required for this are specified through the DP basic functions (*version DP-V0*). Geared toward the special demands of the various areas of application, these basic DP functions have been expanded step-by-step with special functions, so that DP is now available in three versions, DP-V0, DP-V1, and DP-V2, whereby each version has its own special key features. All versions of DP are specified in detail in the IEC 61158 and IEC 61784, respectively.

Version DP-V0 provides the basic functionality of DP, including cyclic data exchange as well as station diagnosis, module diagnosis, and channel-specific diagnosis.

Version DP-V1 contains enhancements geared toward process automation, in particular, acyclic data communication for parameter assignment, operation, visualization, and alarm handling of intelligent field devices, in coexistence with cyclic user data communication. This permits online access to stations using engineering tools. In addition, DP-V1 defines alarms. Examples for different types of alarms are status alarm, update alarm, and a manufacturer-specific alarm.

Version DP-V2 contains further enhancements and is geared primarily toward the demands of drive technology. Due to additional functionalities, such as isochronous slave mode and slave-to-slave(s) communication (DXB), etc., the DP-V2 can also be implemented as a drive bus for controlling fast movement sequences in drive axes.

System Configuration and Device Types

DP supports implementation of both mono-master and multi-master systems. This affords a high degree of flexibility during system configuration. A maximum of 126 devices (masters or slaves) can be connected to a bus segment. In *mono-master systems,* only *one* master is active on the bus during operation of the bus system. Figure 39.4 shows the system configuration of a mono-master system. In this case, the master is hosted by a PLC.

The PLC is the central control component. The slaves are connected to the PLC via the transmission medium. This system configuration enables the shortest bus cycle times. In **multi-master systems,** several masters are sharing the same bus. They represent both independent subsystems, comprising masters and their assigned slaves, and/or additional configuration and diagnostic master devices. The masters are coordinating themselves by passing a token from one to the next. Only the master that holds the token can communicate. PROFIBUS DP differentiates three groups of device types on the bus.

DP master class 1 (DPM1) is a central controller that cyclically exchanges information with the distributed stations (slaves) at a specified message cycle. Typical DPM1 devices are programmable logic controllers (PLCs) or PCs. A DPM1 has active bus access with which it can read measurement data (inputs) of the field devices and write the setpoint values (outputs) of the actuators at fixed times. This continuously repeating cycle is the basis of the automation function (Figure 39.4).

DP masters class 2 (DPM2) are engineering, configuration, or operating devices. They are put in operation during commissioning and for maintenance and diagnostics in order to configure connected

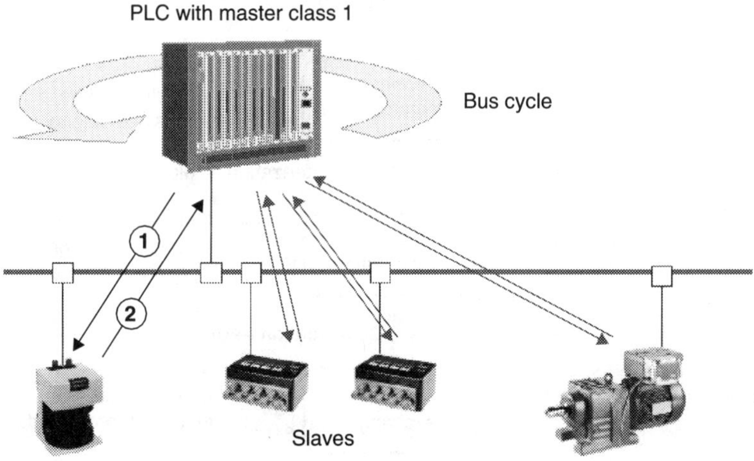

FIGURE 39.4   PROFIBUS DP mono-master system (DP-V0).

devices, evaluate measured values and parameters, and request the device status. A DPM2 does not have to be permanently connected to the bus system. The DPM2 also has active bus access.

Slaves are peripherals (I/O devices, drives, HMIs, valves, transducers, analyzers), which read in process information and/or use output information to intervene in the process. There are also devices that solely provide input information or output information. As far as communication is concerned, slaves are passive devices; they only respond to direct queries (see Figure 39.4, sequence ① and ②). This behavior is simple and cost-effective to implement. In the case of DP-V0, it is already completely included in the Bus-ASIC.

Cyclic and Acyclic Data Communication Protocols

Cyclic data communication between the DPM1 and its assigned slaves is automatically handled by the DPM1 in a defined, recurring sequence (Figure 39.4). The appropriate services are called MS0. The user defines the assignment of the slave(s) to the DPM1 when configuring the bus system. The user also defines which slaves are to be included/excluded in the cyclic user data communication. DPM1 and the slaves are passing three phases during start-up: parameterization, configuration, and cyclic data exchange (Figure 39.5).

Before entering the cyclic data exchange state, the master first sends information about the transmission rate, the data structures within a PDU, and other slave-relevant parameters. In a second step, it checks whether the user-defined configuration matches the actual device configuration. Within any state, the master is enabled to request slave diagnosis in order to indicate faults to the user.

An example for the message structure for the transmission of information between master and slave is shown in Figure 39.6. The message starts with some synchronization bits, the type (SD) and length (LE) of the message, source and destination address, and a function code (FC). The function code indicates the type of message or content of the "load" (Processing Data Unit) and serves as a guard to control the state machine of the master. The PDU, which may carry up to 244 bytes, is followed by a safeguard mechanism Frame Checking Sequence (FCS) and a delimiter.

One example for the usage of the function code (FC) is the indication of a fault situation on the slave side. In this case, the master sends a special diagnosis request instead of the normal process data exchange that the slave replies with a diagnosis message. It comprises 6 bytes of fixed information and additional user definable device, module, or channel-related diagnosis information [1,7].

In addition to the single station-related user data communication, which is automatically handled by the DPM1, the master can also send control commands to all slaves or a group of slaves simultaneously. These control commands are transmitted as multicast messages and enable sync and freeze modes for *event-controlled synchronization* of the slaves [1,7].

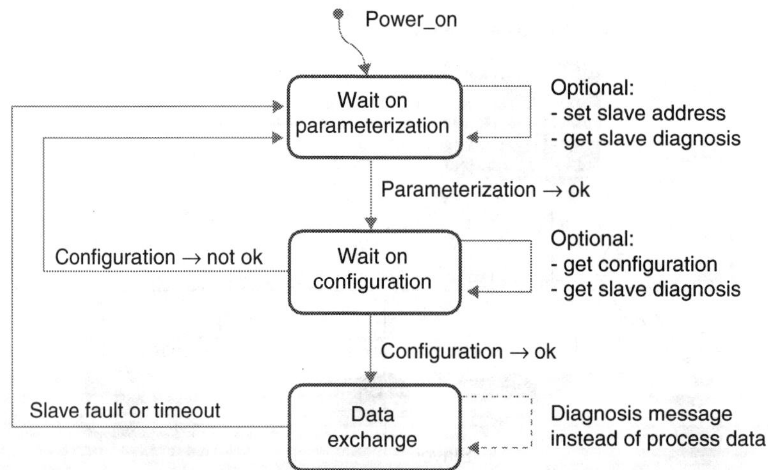

FIGURE 39.5   State machine for slaves.

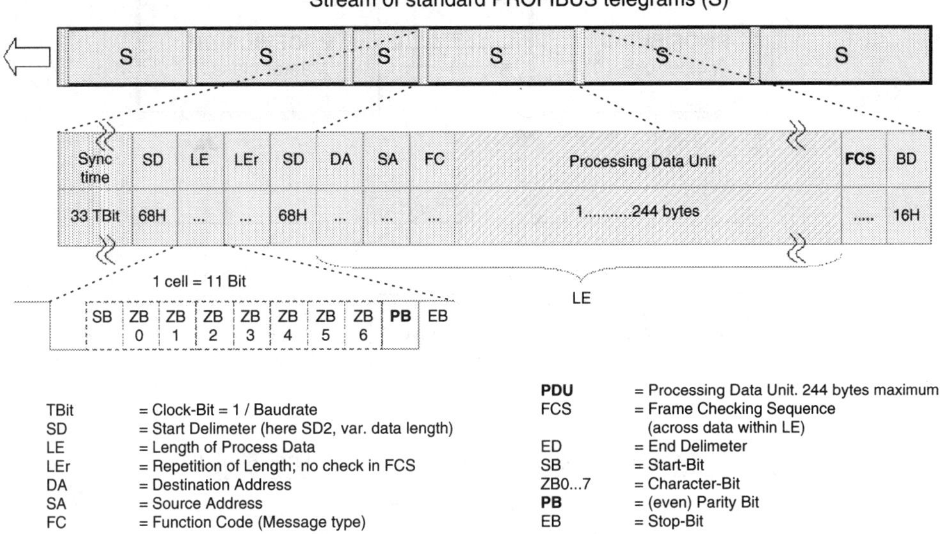

FIGURE 39.6PROFIBUS DP message structure (example).

For safety reasons, it is necessary to ensure that DP has effective *protective functions* against incorrect parameterization or failure of transmission equipment. For this purpose, the DP master and the slaves are fitted with monitoring mechanisms in the form of time monitors. The monitoring interval is defined during configuration.

Acyclic data communication is the key feature of version DP-V1. This forms the requirement for parameterization and calibration of the field devices over the bus during runtime and for the introduction of confirmed alarm messages. Transmission of acyclic data is executed parallel to cyclic data communication, but with a lower priority. Figure 39.7 shows some sample communication sequences for a master class 2, which is using MS2 services. In using MS1 services, a master class 1 is able to execute acyclic communications also.

Slave-to-Slave Communications (DP-V2) enable direct and timesaving communication between slaves using broadcast communication without the detour over a master. In this case, the slaves act as "publisher," that is, the slave response does not go through the coordinating master, but directly to other slaves embedded in the sequence, the so-called "subscribers" (Figure 39.8). This enables slaves to directly read data from other slaves and use them as their own input. This opens up the possibility of completely new applications; it also reduces response times on the bus by up to 90%.

Isochronous mode (DP-V2) enables clock synchronous control in masters and slaves, irrespective of the busload.

The function enables highly precise positioning processes with clock deviations of $<1\mu$sec. All participating device cycles are synchronized to the bus master cycle through a "global control" broadcast message. A special sign of life (consecutive number) allows monitoring of the synchronization.

Clock control (DP-V2) via a new master slave service synchronizes all stations to a system time with a deviation of <1msec. This allows the precise tracking of events. This is particularly useful for the acquisition of timing functions in networks with numerous masters. This facilitates the diagnosis of faults as well as the chronological planning of events.

Upload and download (DP-V2) allows the loading of any size of data area in a field device with the help of a few commands. Within IEC 61158, these services are called *load region*. This enables, for example, programs to be updated or devices replaced without the need for manually loading processes.

Addressing with slot and index is used both for cyclic and acyclic communication services (Figure 39.9) When addressing data, PROFIBUS assumes that the physical structure of the slaves is *modular* or can be

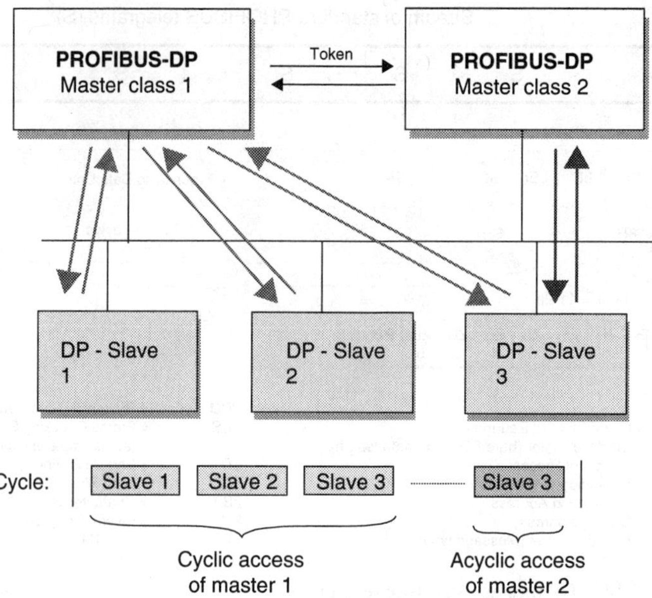

FIGURE 39.7 Cyclic and acyclic data communication with DP-V1.

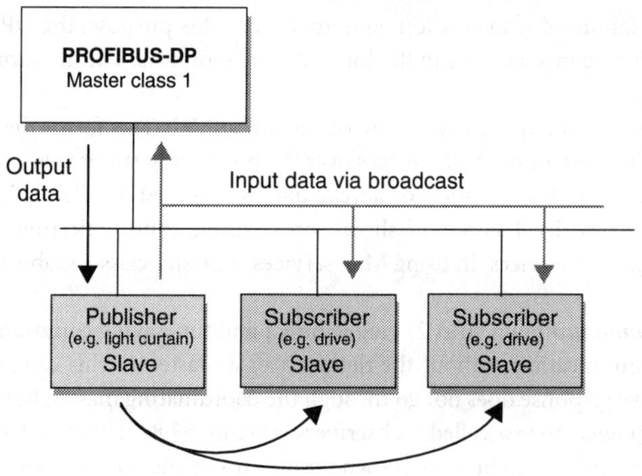

FIGURE 39.8 Slave to slaves data exchange.

structured internally in logical functional units, so-called *modules* (Figure 39.9). The slot number addresses the module and the index addresses the data records assigned to a module. Each data record can be up to 244 bytes. The modules begin at slot 1 and are numbered in ascending contiguous sequence. The slot number 0 is for the device itself. Compact devices are regarded as a unit of virtual modules. These can also be addressed with slot number and index.

39.4 Application Profiles

Profiles are used in automation technology to define specific properties and behavior for devices, device families, or entire systems. Only devices and systems using a vendor-independent profile provide interoperability on a fieldbus, thereby fully exploiting the advantages of a fieldbus. Profiles take into account application and type-specific special features of field devices, controls, and methods of integration (engineering). The

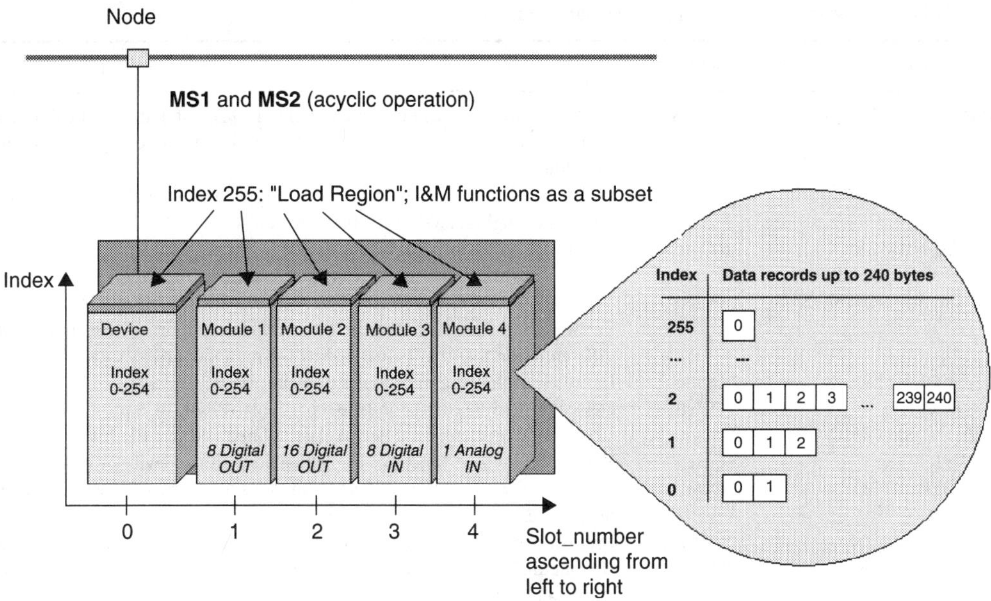

FIGURE 39.9   Slot/index address model of a slave.

term profile ranges from just a few specifications for a specific device class through comprehensive specifications for applications in a specific industry. The generic term for all profiles is *application profiles.*

A distinction is then drawn between *general application profiles* with implementation options for different applications (this includes, e.g., the profiles I&M functions, PROFIsafe, Redundancy, and Time stamp), *specific application profiles*, which are developed for a specific application, such as PROFIdrive, SEMI, or PA Devices, and *system and master profiles*, which describe specific system performance that is available to field devices.

PROFIBUS offers a wide range of such application profiles, which allow application-oriented implementation.

General Application Profiles

Identification and maintenance functions are mandatory for all PROFIBUS devices with MS1 and/or MS2 services. The main purpose of the I&M functions described hereinafter is to support the end user during various scenarios of a device's life cycle be it configuration, commissioning, parameterization, diagnosis, repair, firmware update, asset management, audit trailing, and alike. It is kind of a "type plate" or "boiler plate." The corresponding parameters all are stored at the same address space within the PROFIBUS slot/index address model.

Using the "call" mechanism of the load region services [1] opens up an additional subindex address space of 65,535 data records. The I&M functions are assigned a space between 65,000 and 65,199 for basic, profile-specific, and manufacturer-specific items. Table 39.1 itemizes the individual parameters.

The usage of IDs may not be very helpful for the user if a tool is displaying the information directly out of the device (Figure 39.10). However, nowadays, laptops or engineering tools normally have access to the Internet, at least temporarily. Thus, it is quite easy to reference a central database (e.g., on the PROFIBUS web server) and retrieve comprehensive and always actual information even in a desired language.

PROFIsafe is a comprehensive, open fieldbus solution for safety-relevant applications without the use of a second relay-based layer or proprietary safety buses. PROFIsafe defines how fail-safe devices (emergency stop pushbuttons, light curtains, level switches, etc.) can communicate over PROFIBUS with

TABLE 39.1   Basic Identification and Maintenance Functions

I&M function	Data Format	Notes
MANUFACTURER_ID	2 Octets	I&M functions are using company IDs instead of names. These IDs are harmonized with the list of the HART Foundation and comprise extensions for additional companies.
ORDER_ID	20 Octets	Order number for a particular device type. For virtual modular devices the root or highest level of the basic device.
SERIAL_NUMBER	16 Octets	Unique identifier for a particular device (counter).
HARDWARE_REVISION	2 Octets	The content of this parameter characterizes the edition of the hardware only.
SOFTWARE_REVISION	4 Octets	The content of this parameter characterizes the edition of the software or firmware of a device or module. The structure supports coarse and detailed differentiation that may be defined by the manufacturer: Vx.y.z.
REV_COUNTER	2 Octets	Indicates unplugging of modules or write access.
PROFILE_ID	2 Octets	Device or module corresponds to a particular PROFIBUS profile.
PROFILE_SPECIFIC_TYPE	2 Octets	This identifier references a device class defined within a PROFIBUS profile.
IM_VERSION	2 Octets	Version of the I&M functions implemented within a device or module.
IM_SUPPORTED	2 Octets	Directory for the subset of I&M functions implemented within a device or module.
TAG_FUNCTION	32 Octets	User definable information about the "role" of the device within a plant facility.
TAG_LOCATION	22 Octets	User definable information about the "location coordinates" of the device within a plant facility.
INSTALLATION_DATE	16 Octets	Indicates the data of installation or commissioning of a device or module. E.g. **1995-02-04 16:23**
DESCRIPTOR	54 Octets	User defined comments.
SIGNATURE	54 Octets	Allows parameterization tools to store a "security" code as a reference for a particular parameterization session and audit trail tools to retrieve the code for integrity checks. Used for safety applications according 21 CFR 11 [6] or hazardous machinery (PROFIsafe).

fail-safe controllers in such a manner that they can be used for safety-relevant automation tasks up to category 4 compliant with EN954 (ISO 13849) or SIL3 (Safety Integrity Level) according to IEC 61508. It implements safe communications over a profile, that is, over a special PROFIsafe data frame and a special protocol. PROFIsafe is a single-channel *software solution*, which is implemented in the devices as an additional layer "above" layer 7 (Figure 39.11); the standard PROFIBUS components, such as lines, ASICs, or protocols, remain unchanged. This ensures redundancy mode and retrofit capability. Devices with the PROFIsafe profile can be operated in coexistence with standard devices without restriction on the same bus (cable).

PROFIsafe takes advantage of the acyclic communication (DP-V1) for full maintenance support of the devices and can be used with RS485, fiber optic, or MBP transmission technology. This ensures both fast response times (important for the manufacturing industry) and low power consumption with intrinsically safe operation (important for process automation).

HART on PROFIBUS DP integrates HART devices installed in the field, in existing or new PROFIBUS systems. It includes the benefits of the PROFIBUS communication mechanisms without any changes required to the PROFIBUS protocol and services, the PROFIBUS PDUs (Protocol Data Units) or the state machines, and functional characteristics. This profile is implemented in the master and slave above layer 7, thus enabling mapping of the HART client–master–server model on PROFIBUS. The cooperation of the HART Foundation on the specification work ensures complete conformity with the HART specifications. The HART-client application is integrated in a PROFIBUS master and the HART master in a PROFIBUS slave (Figure 39.12), whereby the latter serves as a multiplexer and handles communication to the HART devices.

The *Time Stamp* application profile describes mechanisms on how to supply certain events and actions with a time stamp, which enables precise time assignment. Precondition is a clock control in the slaves through a clock master via special services. An event can be given a precise system time stamp and

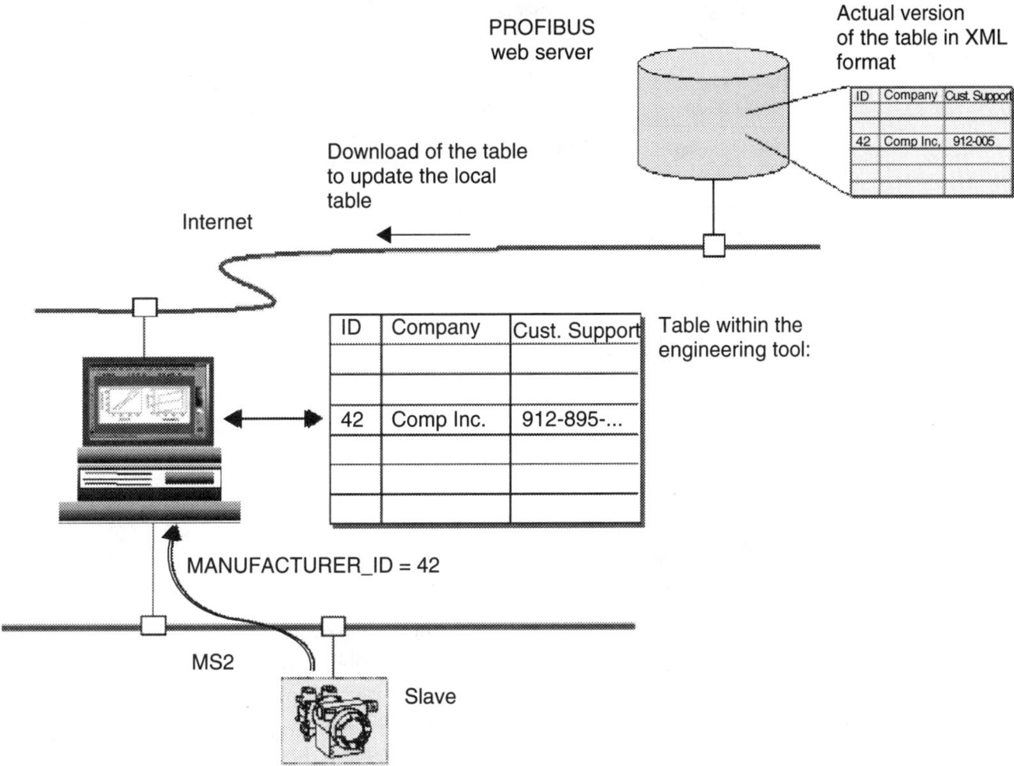

FIGURE 39.10 Referencing IDs via the Internet.

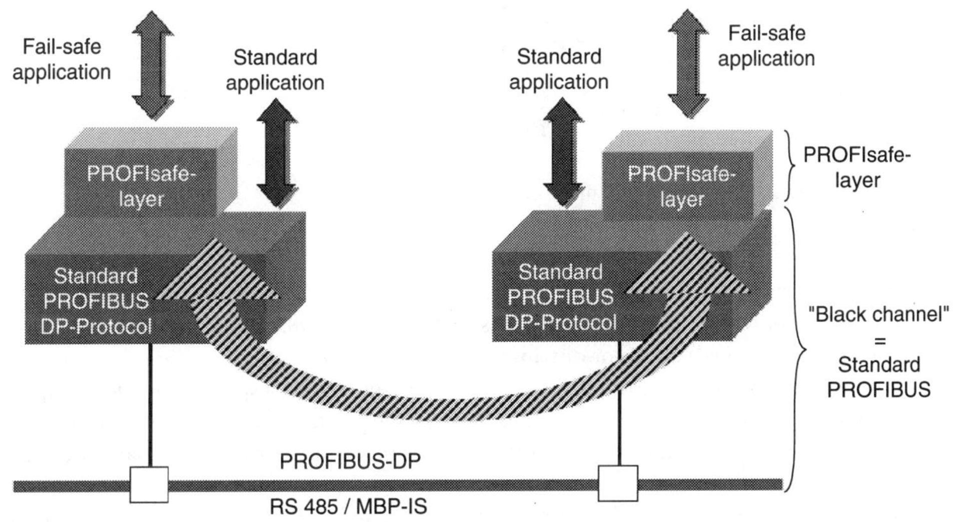

FIGURE 39.11 Safe communication on PROFIBUS DP.

read out accordingly. A concept of graded messages is used. The message types are summarized under the term *"Alerts"* and are divided into high-priority "alarms" (these transmit a diagnostics message) and low-priority "events." In both cases, the master acyclically reads the time-stamped process values and alarm messages from the alarm and event buffer of the field device (Figure 39.13).

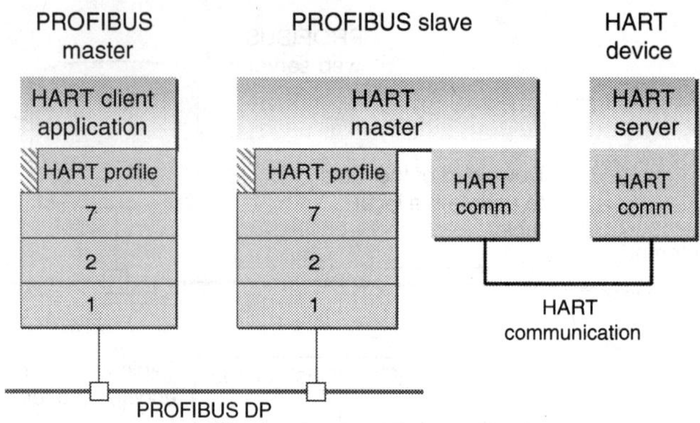

FIGURE 39.12   Operating HART devices over PROFIBUS DP.

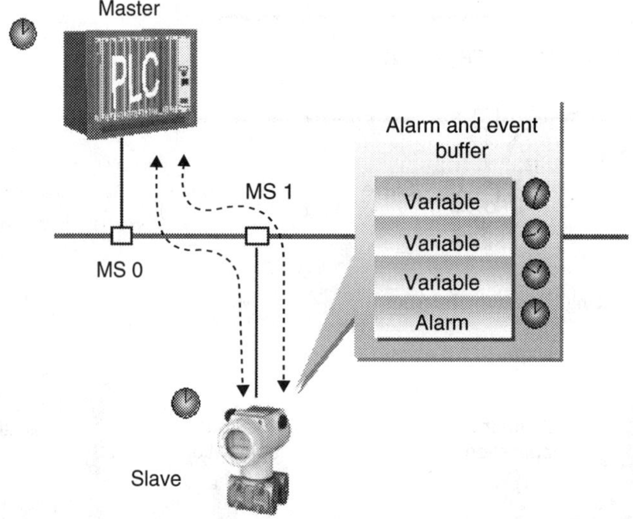

FIGURE 39.13   Time-stamping and alarm messages.

The *Slave redundancy* application profile provides a *slave-redundancy mechanism* (Figure 39.14): Slave devices contain two different PROFIBUS interfaces that are called *primary* and *backup* (slave interface). These may be either in a *single* device or distributed over *two* devices.

The devices are equipped with two independent protocol stacks with a special *redundancy expansion*.

A redundancy communication (RedCom) runs between the protocol stacks, that is, within a device or between two devices. It is independent of PROFIBUS and its performance capability is largely determined by the redundancy reversing times.

Only one device version is required to implement different redundancy structures, and no additional configuration of the backup slave is necessary. The redundancy of PROFIBUS slave devices provides high availability, short reversing times, no data loss, and ensures fault tolerance.

Specific Application Profiles

The **PROFIdrive** application profile defines device behavior and the access procedure to drive data for electric drives on PROFIBUS, from simple frequency converters through to highly dynamic servo-controls.

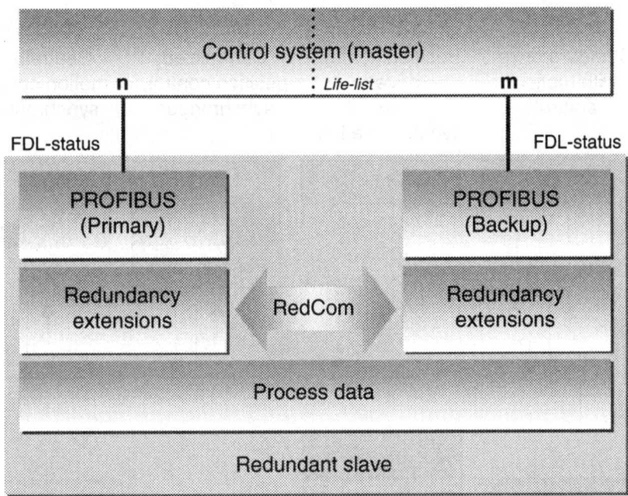

FIGURE 39.14 Slave redundancy in PROFIBUS.

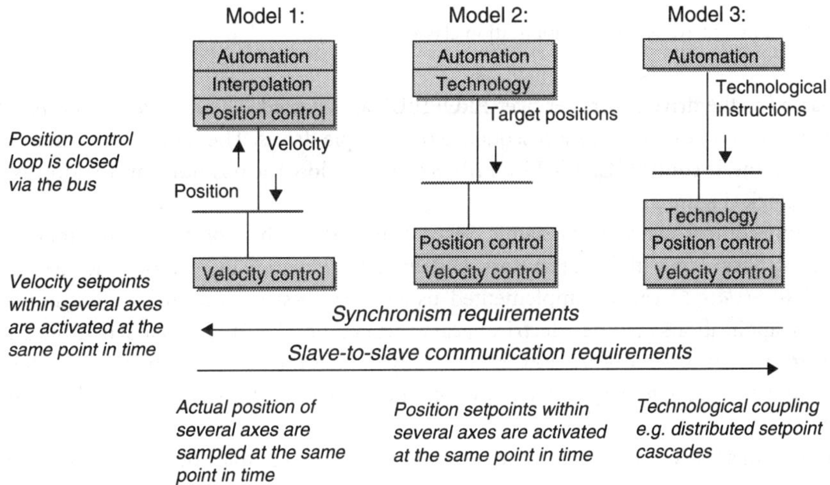

FIGURE 39.15 Different requirements for distributed drive applications.

The method of integrating drives in automation solutions is highly depending on the task of the drives (Figure 39.15). The more the drives act independently from central host controllers, the more they require slave-to-slave communication capabilities. On the other hand, the more the central host controllers are taking over the computing tasks, the more synchronization of the involved drives is required.

For this reason, PROFIdrive defines six classes covering the majority of applications (Figure 39.16).

With *standard drives (class 1)*, the drive is controlled by means of a main setpoint value (e.g., rotational speed), whereby the speed control is carried out in the drive controller. In case of *standard drives with technological function (class 2)*, the automation process is broken down into several subprocesses and some of the automation functions are shifted from the central programmable controller to the drive controllers. PROFIBUS serves as the technology interface in this case. Slave-to-slave communication between the individual drive controls is a requirement for this solution.

The positioning drive (class 3) integrates an additional position controller in the drive, thus covering an extremely broad spectrum of applications (e.g., the twisting on and off of bottle tops). The positioning

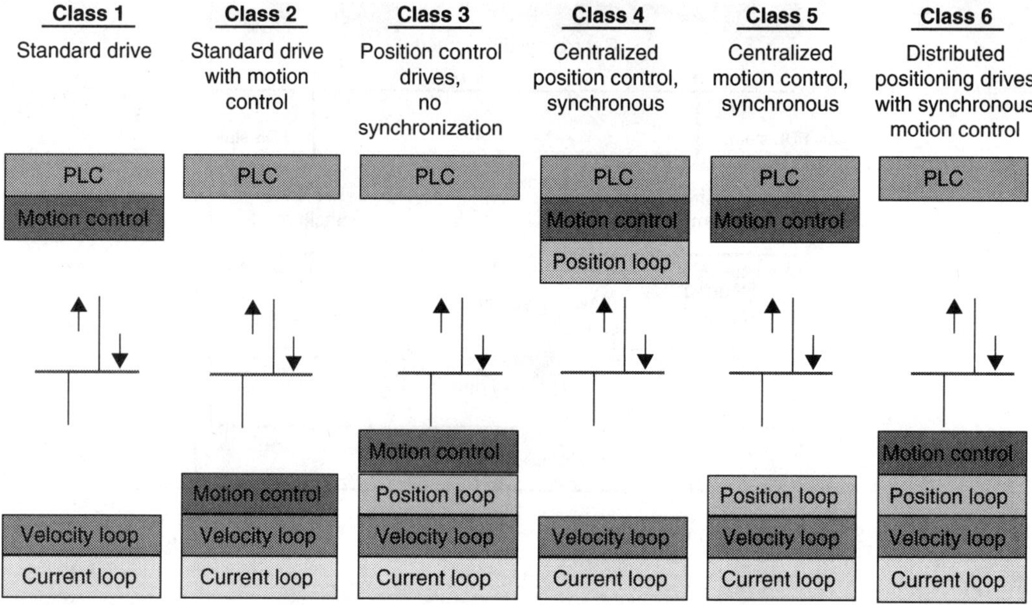

FIGURE 39.16 PROFIdrive defines six application classes.

tasks are passed to the drive controller over PROFIBUS and started. The *central motion control (classes 4 and 5)* enables the coordinated motion sequence of multiple drives. The motion is primarily controlled over a central numeric control (CNC). PROFIBUS serves to close the position control loop as well as synchronize the clock (Figure 39.17).

The position control concept (Dynamic Servo Control) of this solution also supports extremely sophisticated applications with linear motors. *Distributed automation* by means of clocked processes and electronic shafts *(class 6)* can be implemented using slave-to-slave communication and isochronous slaves. Sample applications include "electrical gears," "curve discs," and "angular synchronous processes."

In contrast to other drive profiles, PROFIdrive only defines the access mechanisms to the parameters and a subset of approx. 30 *profile parameters,* which include fault buffers, drive controllers, device identification, etc.

All other parameters (which may number more than 1000 in complex devices) are *manufacturer-specific,* which provide drive manufacturers great flexibility when implementing control functions.

The profile for *PA Devices* defines all functions and parameters for different classes of devices for process automation with local intelligence. They can execute part of the information processing or even take over the overall functionality in automation systems. The profile includes all steps of a typical signal flow – from process sensor signals through the preprocessed process value that is communicated to the control system together with a value qualifier (Figure 39.18).

The profile for PA Devices is documented in a *general model description* containing the currently valid specifications for all device types and in *device data sheets* containing the agreed additional specifications for individual device classes. Version 3.0 of the profile for PA devices includes device data sheets for quantity measurement of pressure and differential pressure, level, temperature, flow rate, and data sheets for valves, actuators, analyzers, analog, and digital inputs and outputs.

In process engineering, it is common to use *blocks* for describing the characteristics and functions of a measuring point or manipulating point at a certain control point and to represent an automation application through a combination of these types of blocks. Therefore, the specification for PA Devices uses a *function block model* according to IEC 61804 to represent functional sequences as shown in Figure 39.19. The blocks are implemented by the manufacturers as software in the field devices and, taken as a whole, represent the functionality of the device.

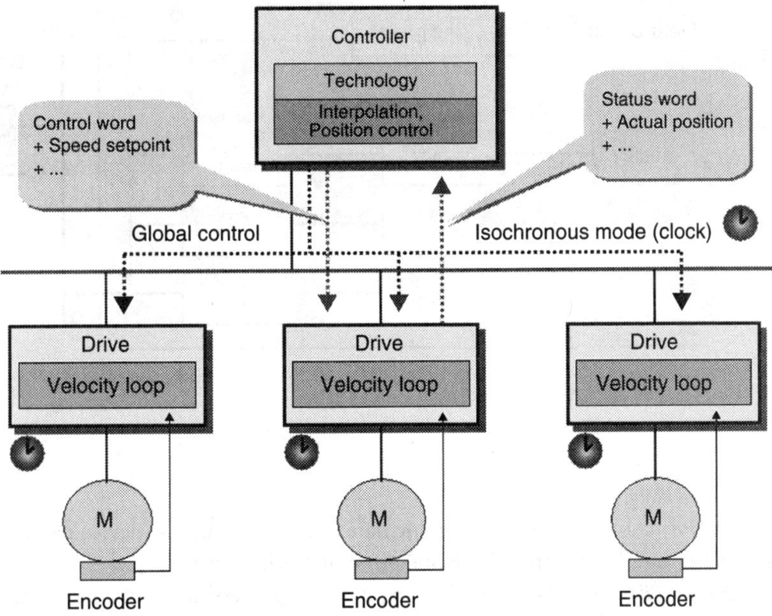

FIGURE 39.17 Positioning with central interpolation and position control.

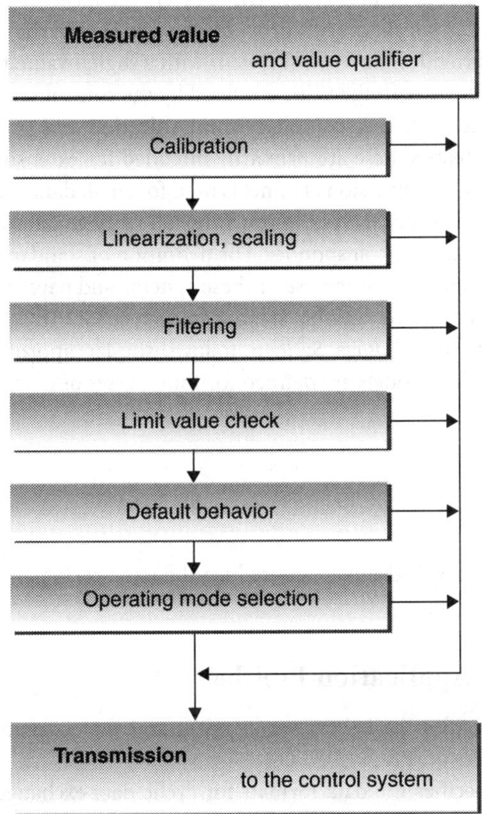

FIGURE 39.18 Signal processing defined in profile PA devices.

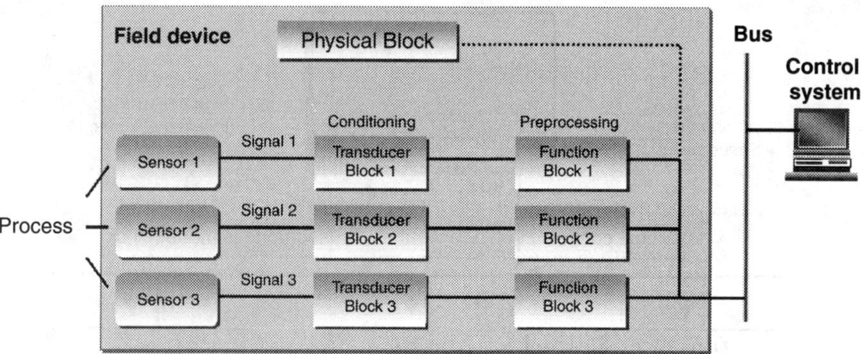

FIGURE 39.19 Block structure of a PA field device.

The following *three block types* are used:

A *Physical Block* (*PB*) contains the characteristic data of a device, such as device name, manufacturer, version, serial number, etc. There can only be one physical block in each device.

A *Transducer Block* (*TB*) contains all the data required for processing an unconditioned signal delivered from a sensor for passing onto a function block. If no processing is required, the TB can be omitted. Multifunctional devices with two or more sensors have a corresponding number of TBs.

A *Function Block* (*FB*) contains all data for final processing of a measured value prior to transmission to the control system, or on the other hand, for processing of a setting before the setting process.

Different FBs are available: *Analog Input Block* (*AI*, delivers the measured value from the sensor/TB to the control system), *Analog Output Block* (*AO*, provides the device with the value specified by the control system), *Digital Input* (*DI*, provides the control system with a digital value from the device), and *Digital Output* (*DO*, provides the device with the value specified by the control system).

The profile for *Ident Systems* defines complete communication and processing models for barcode readers and transponder systems. These are primarily intended for extensive use with the DP-V1 functionality. While the cyclic data transmission channel is used for small data volumes to transfer status/control information, the acyclic channel serves the transmission of large data volumes that result from the information in the barcode reader or transponder. The definition of standard PROXY function blocks [5] according to IEC 61131-3 has facilitated the use of these systems and paves the way for the application of open solutions on completion of international standards, such as ISO/IEC 15962 and ISO/IEC18000.

The profile for *Weighing and Dosage Systems* follows similar approaches as the Ident Systems. Communication and processing models are defined for four classes of devices or systems:

- Simple scale
- Comfort scale
- Continuous scale
- Batch scale.

These new types of profiles will dramatically reduce the engineering costs and improve the bidding process during project execution.

Summary of Specific Application Profiles

PROFIdrive: The profile specifies the behavior of devices and the access procedure to parameters for variable-speed electrical drives on PROFIBUS DP.

PA devices: The profile specifies the data formats for cyclic data exchange and the characteristics for *process engineering* of devices for process automation.

Robots/NC: The profile describes how *handling and assembly robots* are controlled via PROFIBUS.

Panel devices: The profile describes the interfacing of simple *human machine interface (HMI) devices* to control components.

Encoders: This profile describes the interfacing of rotary, angle, and linear *encoders* with single-turn or multi-turn resolution.

Fluid power: The profile describes the control of hydraulic drives via PROFIBUS.

SEMI: The profile defines models of devices for semiconductor production such that they comply with the PA model and the SEMI model.

Low-voltage switchgear: The profile describes data exchange for low-voltage switchgear like circuit breakers, switches, and starters.

Weighing/Dosage: The profile describes the communication and processing models for simple and comfort scales as well as for batch and continuous scales.

Ident systems: The profile describes the communication and processing models for bar code readers and transponders.

Remote I/O for PA devices: This profile takes into account the special conditions of physically modular slaves like limited communication resources and extreme cost sensitivity. The profile follows the model of PA devices as much as possible but has some simplifications.

Master and System Profiles

Master Profiles for PROFIBUS describe classes of controller, each of which support a specific "subset" of all the possible master functionalities, such as cyclic and acyclic communications, diagnostics, alarm handling, clock control, slave-to-slave communication, isochronous mode, and safety.

System Profiles for PROFIBUS go a step further and describe classes of systems including the master functionality, the possible functionality of *Standard Program Interfaces* (FB in accordance with IEC 61131-3, safety layer and FDT), and *integration options* (GSD, EDD and DTM). Figure 39.20 shows the standard platforms available today.

In the PROFIBUS DP system, the master and system profiles provide the much-needed counterpart to the application profiles: master and system profiles describe specific system parameters that are made available to the field devices; application profiles require specific system parameters in order to simplify their defined characteristics.

By using these profiles, the *device manufacturers* can focus on existing or specified system profiles and the *system manufacturers* can provide the platforms required by the existing or specified device application profiles.

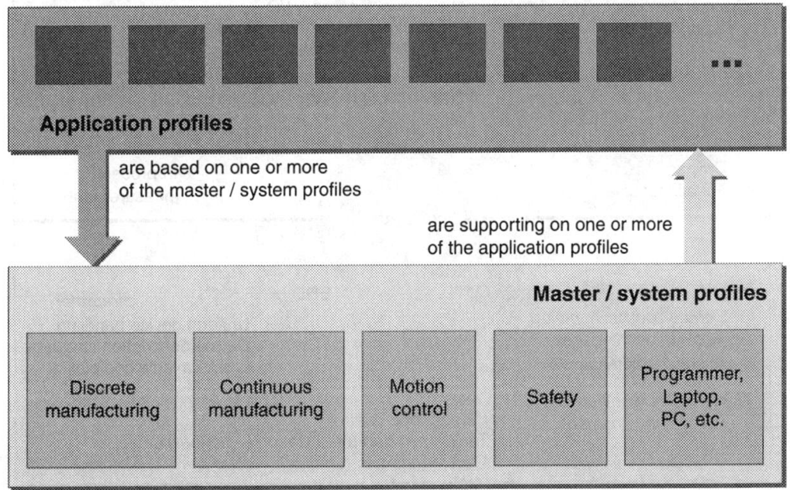

FIGURE 39.20   Master and system profiles for PROFIBUS DP.

39.5 Integration Technologies

Modern field devices in both factory and process automation provide a wide range of information and also execute functions that were previously executed in PLCs and control systems. To execute these tasks, the tools for commissioning, maintenance, engineering, and parameterization of these devices require an exact and complete description of device data and functions, such as the type of application function, configuration parameters, range of values, units of measurement, default values, limit values, identification, etc. The same applies to the controller/control system, whose device-specific parameters and data formats must also be made known (integrated) to ensure error-free data exchange with the field devices.

PROFIBUS has developed a number of methods and tools ("integration technologies") for this type of device description, which enable standardization of device management. The performance range of these tools is optimized to specific tasks (Figure 39.21), which has given rise to the term *scaleable device integration*. GSD and EDD are both a kind of "electronic device data sheets," developed with different languages according to the special scope, whilst a Device Type Manager (DTM) is a software component containing specific field device functions for parameterization, configuration, diagnostics, and maintenance, generated by mapping and to be used together with the universal software interface Field Device Tool (FDT), which is able to implement software components.

A GSD is an electronically readable ASCII text file and contains both general and device-specific specifications for *communication* (General Station Description) and network configuration. Each of the entries describes a feature that is supported by a device. By means of keywords, a configuration tool reads the device identification (ID number), the adjustable parameters, the corresponding data type, and the permitted limit values for the configuration of the device from the GSD. Some of the keywords are *mandatory*, for example, Vendor_Name. Others are *optional*, for example, Sync_Mode_supported. A GSD replaces the previously conventional manuals and supports automatic checks for input errors and data consistency, even during the configuration phase.

Distinction is made between a *device GSD* (for an individual device only) and *profile GSD*, which may be used for devices that comply exactly with a profile such as PROFIdrive version 3 or PA devices version 3.

GSD for *compact* devices, whose block configuration is already known on delivery, can be created completely by the device manufacturer.

GSD for *modular* devices, whose block configuration is not yet conclusively specified on delivery, must be configured by the user in accordance with the actual module configuration using the configuration tool.

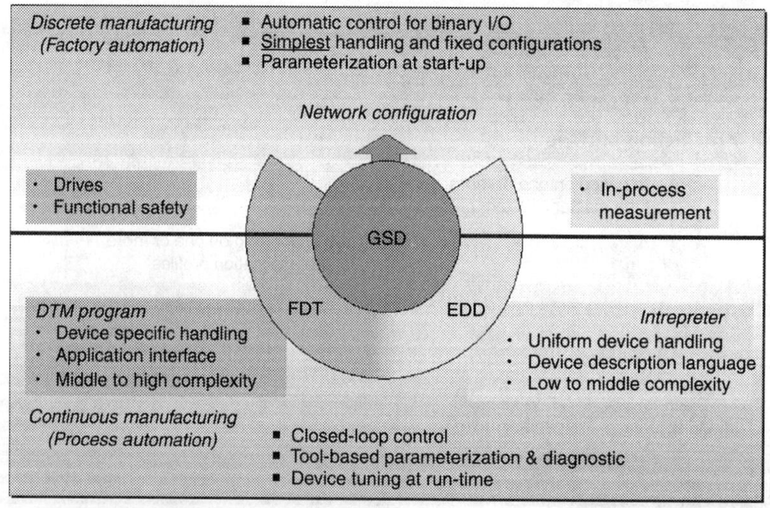

FIGURE 39.21 Integration technologies for PROFIBUS DP.

The device manufacturers are responsible for the scope and quality of the GSD of their devices. Submission of a profile GSD (contains the information from the profile of a device family) or an individual device GSD (device-specific) is essential for certification of a device.

An Electronic Device Description (EDD)

is, like a GSD, an electronic device data sheet, but developed by using a more powerful and universal language, the Electronic Device Description Language (EDDL). An EDD typically describes the *application-*related parameters and functions of a field device such as configuration parameters, ranges of values, units of measurement, default values, etc. An EDD is a versatile source of information for engineering, commissioning, runtime, asset management, and documentation. It also contains support mechanisms to integrate existing profile descriptions in the device description, allow references to existing objects, to access standard dictionaries, and to allow assignment of the device description to a device.

An EDD is independent of operating systems and supports the user by its uniform user and operation interface (only one tool, reliable operation, reduced training, and documentation costs) and also the device manufacturer (no specific knowledge required, existing EDDs and libraries can be used).

The EDD concept is suitable for tasks of low to middle complexity.

A Device Type Manager (DTM)

is a software that is generated by mapping the specific functions and dialogs of a field device for parameterization, configuration, diagnostics, and maintenance, complete with user interface, in a *software component*. This component is called DTM and is integrated in the engineering tool or control system over the FDT interface. A DTM uses the routing function of an engineering system for communicating across the hierarchical levels. It works similar to a printer driver, which the printer supplier includes in delivery and must be installed on the PC by the user. The DTM is generated by the device manufacturer and is included in delivery of the device.

DTM generation may be performed using one of the following options:

- Specific programming in a higher programming language.
- Reuse of existing components or tools through their encapsulation as DTM.
- Generation from an existing device description using a compiler or interpreter.
- Use of the DTM toolkit of MS Visual Basic.

With DTMs, it is possible to obtain direct access to all field devices for planning, diagnostics, and maintenance purposes from a central workstation. A DTM is not a stand-alone tool, but an ActiveX component with defined interfaces. The FDT/DTM concept is protocol-independent and, with its mapping of device functions in software components, opens up interesting new user options. The DTM/FDT concept is very flexible; it resolves interface and navigation needs nowadays, and is suitable for tasks of middle to high complexity.

Quality Assurance

In order for PROFIBUS devices of different types and manufacturers to correctly fulfill tasks in the automation process, it is essential to ensure the error-free exchange of information over the bus. The requirement for this is a standard-compliant implementation of the communications protocol and application profiles by device manufacturers. To ensure that this requirement is fulfilled, the PNO has established a *quality assurance procedure* whereby, on the basis of test reports, certificates are issued to devices that successfully complete the test.

The basis for the certification procedure is the standard EN 45000 (Figure 39.22). The PROFIBUS User Organization has approved manufacturer-independent *test laboratories* in accordance with the specifications of this standard. Only these test laboratories are authorized to carry out device tests, which form the basis for certification.

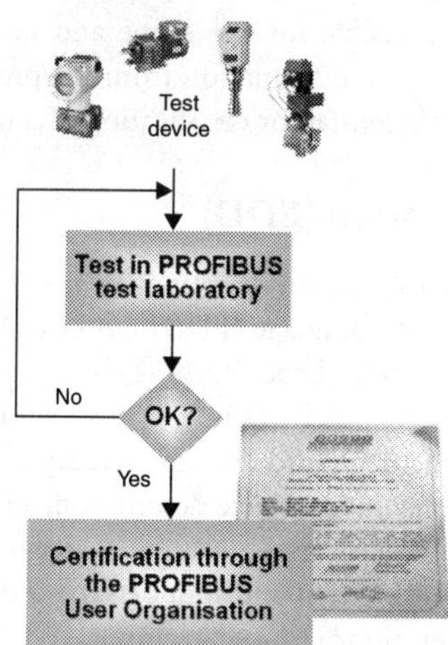

FIGURE 39.22   Certification procedure.

The *test procedure,* which is the same for all test laboratories, is composed of several parts:

- The *GSD/EDD Check* ensures that the device description files comply with the specification.
- The *Hardware Test* tests the electric characteristics of the PROFIBUS interface of the device for compliance with the specifications. This includes terminating resistors, suitability of the implemented drivers and other modules, and the quality of line level.
- The *Function Test* examines the bus access and transmission protocol and the functionality of the test device.
- The *Conformity Test* forms the main part of the test. The objective is to test the conformity of the protocol implementation with the standard.
- The *Interoperability Test* checks the test device for interoperability with the PROFIBUS devices of other manufacturers in a multivendor plant. This checks that the functionality of the plant is maintained when the test device is added. Operation is also tested with different masters.

Once a device has successfully passed all the tests, the manufacturer can apply for a *certificate* from the PROFIBUS User Organization. Each certified device contains a certification number as a reference. The certificate is valid for 3 years but can be extended after undergoing a further test.

Implementation

For the device development or implementation of the PROFIBUS protocol, a broad spectrum of standard components and development tools (PROFIBUS ASICs, PROFIBUS stacks, monitoring, and commissioning tools) as well as services are available, which enable device manufacturers to realize cost-effective development. A corresponding overview is available in the product catalog of the PROFIBUS User Organization [2].

PROFIBUS Interface Modules are ideal for a low/medium volume of devices to be produced. These credit card size modules implement the entire bus protocol. They are fitted on the master board of the device as an additional module.

PROFIBUS Protocol Chips (Single Chips, Communication Chips, Protocol Chips) are recommended for an individual implementation in the case of a high volume of devices.

The implementation of single-chip ASICs is ideal for *simple slaves* (I/O devices). All protocol functions already are integrated on the ASIC. No microprocessor or software is required. Only the bus interface driver, the quartz, and the power electronics are required as external components.

For *intelligent slaves,* parts of the PROFIBUS protocol are implemented on a protocol chip and the remaining protocol parts are implemented as software on a microcontroller. In most of the ASICS available on the market, all cyclic protocol parts have been implemented, which are responsible for transmission of time-critical data.

For *complex masters*, the time-critical parts of the PROFIBUS protocol are also implemented on a protocol chip and the remaining protocol parts are implemented as software on a microcontroller. Various ASICs of different suppliers are currently available for the implementation of complex master devices. They can be operated in combination with many common microprocessors.

An overview for commercially offered PROFIBUS chips and software (PROFIBUS stacks) is available at the PROFIBUS website [2]. For further information, please contact the suppliers directly.

Modem Chips are available to realize the (low) power consumption, which is required when implementing a bus-powered field device with MBP transmission technology. Only a feed current of 10–15 mA over the bus cable is available for these devices, which must supply the overall device, including the bus interface and the measuring electronics. These modems take the required operating energy for the overall device from the MBP bus connection and make it available as feed voltage for the other electronic components of the device. At the same time, the digital signals of the connected protocol chip are converted into the bus signal of the MBP connection modulated to the energy supply.

39.6 Prospects

While the fieldbuses were pioneering the field of distributed automation in discrete and continuous manufacturing facilities within the past 15 years, Ethernet was gaining great success in office automation. The technology matured more and more and evolved a high degree of comfort and flexibility such as high transmission speed, easy-to-handle cables and connectors, efficient control protocols, network devices like switches, and the tremendous success of the Internet. The fieldbus organizations are now eager to provide solutions for a steadily growing demand of the market. The solution from the PROFIBUS organization is PROFINET.

PROFINET Communication

PROFINET is a new automation concept that has emerged as a result of the trend in automation technology toward modular, reusable machines (mechatronic components), and plants with distributed intelligence. With its comprehensive design (uniform model for engineering, communication, and migration architecture to other communication systems, such as PROFIBUS and OPC), PROFINET fulfills all the key demands of automation technology for

- consistent communications from field level to corporate management level such as Enterprise Resource Planning (**ERP**) and Manufacturing Execution Systems (**MES**) using Ethernet,
- a vendor-independent plant-wide engineering model for the entire automation landscape,
- openness to other systems,
- implementation of IT standards, and
- integration capability of PROFIBUS segments without changes.

PROFINET is available as a *specification* and as an operating system-independent *source software* for Ethernet-based communications [9].

PROFINET IO

The PROFINET component model is ideal for intelligent field devices and programmable controllers with data format interfaces that can be standardized. Simple field devices with many IO signals do not fit into the engineering model of PROFINET. Thus, the version PROFINET IO offers an integration methodology based on the PROFINET communication protocols such that a manufacturer of PROFIBUS DP slave devices feels comfortable to switch over. He will find the services described for PROFIBUS DP in PROFINET IO also and more. The most essential feature of this integration is the use of distributed field devices with their input and output data to be processed within the application program of a PLC.

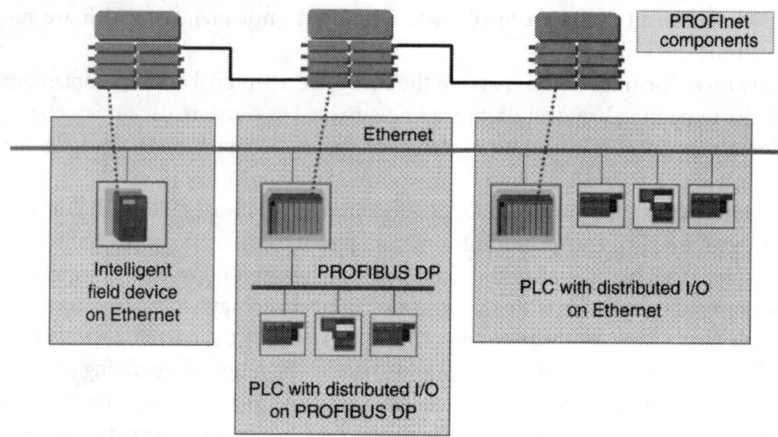

FIGURE 39.23   The migration concept of PROFINET and PROFINET IO.

The PROFINET Migration Model

Allows the integration of PROFIBUS DP segments in PROFINET using *proxies*. These assume a *proxy function* for all the devices connected to PROFIBUS. This means that when rebuilding or expanding plants, the entire spectrum of PROFIBUS devices, including products of PROFIdrive and PROFIsafe can be implemented unchanged, thus providing users with maximum investment protection. Proxy technology also allows integration of other fieldbus systems (Figure 39.23). A second possibility is the usage of PROFINET IO devices directly connected to the host controller (PLC) via PROFINET. Intelligent field devices may be connected directly as a component.

This way the user has all the possibilities to migrate from the current situation to PROFINET at his convenience.

Abbreviations

ASIC	Application-Specific Integrated Circuit
BIA	German Institute of Occupational Safety and Health
CPU	Central Processing Unit
CRC	Cyclic Redundancy Check
DP	De-centralized Peripherals
DPM1	PROFIBUS DP Master Class 1, usually a programmable logic controller
DPM2	PROFIBUS DP Master Class 2, usually a Laptop or PC
DTM	Device Type Manager
DXB	Data Exchange Broadcast (slave to slaves communication)
EDD	Electronic Device Description
EMI	Electromagnetic Interference
EN, prEN	European standard, preliminary ...
FB	Function Block
FDT	Field Device Tool
FISCO	Fieldbus Intrinsically Safe Concept
FM	Factory Mutual Global is a commercial and industrial property insurance company with a unique focus on risk management. www.fmglobal.com
GSD	General Station Description (electronically readable data sheet)
HMI	Human Machine Interface
HW	Hardware
IEC	International Electrotechnical Commission
I/O	Input/Output
ISO/OSI	International Standards Organization / Open Systems Interconnection (Reference Model)
MS0	Cyclic Master Slave communication services of PROFIBUS DP

MS1/MS2	Acyclic Master Slave communication services of PROFIBUS DP
NAMUR	Association of users of process control technology
PA	Process Automation
PC	Personal Computer
PDU	Protocol Data Unit
PLC	Programmable Logic Controller
PTB	Pysikalisch-Technische Bundesanstalt: national institute of natural and engineering sciences and the highest technical authority for metrology and physical safety engineering of the Federal Republic of Germany. www.ptb.de
SW	Software
UL	Underwriters Laboratories Inc. is an independent, not-for-profit product safety testing and certification organization. www.ul.com

References

[1] IEC 61158/61784: Digital data communications for measurement and control — Fieldbus for use in industrial control systems.

[2] PROFIBUS home page: www.profibus.com

[3] Optical Transmission Technology for PROFIBUS, V2.0, 1999, PROFIBUS Order No. 2.021.

[4] IEC/TS 60079-27: Electrical apparatus for explosive gas atmospheres — Part 27: Fieldbus intrinsically safe concept (FISCO), Parts 11, 14, and 25: Constructional and installation requirements.

[5] PROFIBUS Communication and Proxy Function Blocks acc. to IEC 61131-3, V1.2, July 2001, PROFIBUS Order No. 2.182.

[6] Food & Drug Administration: 21 CFR Part 11.

[7] M. Popp, The rapid way to PROFIBUS DP, PROFIBUS Order No. 4.072.

[8] PROFIBUS System Description — Technology and Application, October 2002, free download from www.profibus.com or PNO-Order-No. 4.002.

[9] PROFINET System Description — Technology and Application, November 2002, free download from www.profibus.com or PNO-Order-No. 4.132.

40

The WORLDFIP Fieldbus

Jean-Pierre Thomesse
Institut National Polytechnique de Lorraine

40.1 Introduction

This chapter is dedicated to the study of the WorldFIP[1] Fieldbus. It is one of the first fieldbuses that emerged in the beginning of the 1980s. But it is also at the origin of several main concepts, which are now implemented in other, different fieldbuses. For example, the Producer–Consumer model, the timeliness attributes to qualify the validity of a data, and the time coherence and consistency attributes are some of the most important WorldFIP contributions. A good many of them are coming from research activities (academic and industrial) and from very distributed and real-time requirements analysis. That is why the first sections will briefly relate the origin of this fieldbus (Section 2), the requirements (Section 3), and the choices of WorldFIP specifications (Section 4). The technical aspects will be studied more in detail in the following four sections: the architecture in Section 5, the physical layer in Section 6, the data link layer in Section 7,

---

[1] WorldFIP is the current name of the previous FIP network. FIP stands for "Factory Instrumentation Protocol," but in the French language, the acronym means Flux d'Information (de et vers le) Processus.

and finally the application layer in Section 8. The current state of this fieldbus is given in the last section (Section 9) before conclusion and bibliography. Several theoretical works have been developed for more than 15 years to prove the protocols, to evaluate the performances, and to guarantee the time constraints (Song et al., 1991), or to estimate the performances of distributed applications (Bergé et al., 1995). It is impossible to cite all the authors, but we thank them for their contributions.

40.2 WorldFIP Origin

The first works on the WorldFIP specification started in September 1982, in a working group under the aegis of the French Ministry of Research and Technology. This working group was composed of representatives of end-users, engineering companies, and laboratories. It was important not to include providers and manufacturers of networks at the beginning in order to organize a real end-users needs analysis, without having to consider the possible influence of existing products or existing projects.

The first objective of this work was to analyze the needs for communication in automatic control systems. But it was necessary to take into account in our reflection the following points:

- It was really the starting development of local area networks.
- It was the beginning of the MAP project in the U.S.A. (MAP, 1988).
- And some new ideas appeared on the application architectures, especially the idea of really distributed systems.
- The intelligent devices started their development thanks to the progress of the microelectronics.

The development of WorldFIP started in this context, with essentially two main types of contributions coming from research and from end-users' experiences.

The functional analysis of the communication needs in automatic control systems led to the distinction between two mains flows:

- a flow of information associated to the control rooms in continuous processes, or associated to the plant in discrete part manufacturing applications;
- another flow associated to the field devices called "Flow of Information of the Process," which will be analyzed later and that led to the WorldFIP fieldbus profiles.

To satisfy the former type, different local area networks were already existing when nothing exists for the latter. It was then decided to specify a so-called Instrumentation Network.[2] The first specification of the FIP Fieldbus was then published in May 1984 (Galara and Thomesse, 1984). That is, it was only in the beginning of the 1990s that the name has been transformed in WorldFIP. More information on the origins may be found in Thomesse (1993b, 1998). The first results were presented for sustaining a standardization process at IEC (Gault and Lobert, 1985).

40.3 Requirements

The first (and abstract) requirement was to define a communication system that should take the place of usual connections standards (as 4–20 mA) between the devices and the controllers in an automation system. Another expression was more complete but abstract: the objective was the design of an operating system for instrumentation. It was, in fact, a real need in order to build not only a communication systems but also really distributed systems. It was then important to provide the right and well-suited services for the distribution of the applications (facilities for the management of coherence and consistencies, for the management of the impossible common global state, and of clock synchronization).

The requirements could be specified at different abstraction levels. Starting from the most general (see above), they have led to the following:

The connection between the field devices and the control functions was expensive enough to try the specification of another communication technique.

The access to the data by the network should be standardized.

---

[2] At this moment, the word "Fieldbus" was not yet in use.

The location of data should be transparent for the user.

The system should be built to meet different dependability requirements by using the same basic components.

The competitiveness of companies should be improved by such technologies.

The development should go through the international standardization.

The protocols should be implemented in silicon.

The data flows between the functions and between the set of field and control equipment have then been identified and analyzed, leading to the identification of special needs for a so-called Instrumentation Network.

This led to the identification of the traffics and then to the more technical following requirements:

The exchanged data are coming from sensors or are placed in actuators. Most of them are known and identified (temperature, pressure, speed, position, and so on). But other transmitted data are not identified in the same sense and usual messages must also be transmitted.

The exchanges are periodic or not. They are time constrained, in terms of period and of jitter, as well as in terms of deadline, lifetime, promptness, or refreshment.

Most critical traffic must be periodically managed. But sporadic traffic must also take place.

The timeliness is important for the quality of service and the dependability of the applications.

The distributed decisions must be consistent, that is, the data and the physical process must be seen in a coherent manner by all the application processes. The impossible global state must be approached by a reliable broadcasting of states and events.

40.4 Choices of WorldFIP

According to the previous requirements, the WorldFIP solution is based on a few basic ideas, which give the right quality of service to this fieldbus. These ideas are:

- The distinction of two types of "messages": the notion of identified data vs. the concept of classical messages, associated to the respective cooperation models, producer–consumer vs. client–server.
- The predefined scheduling of periodic traffics, with periods suited to the physical needs, especially the sampling theory.
- The on-line scheduling of sporadic traffic, of messages with priority to the critical traffic.
- The cyclic updating of real-time data at the consumers' sites.
- The timeliness attributes and mechanisms for time-critical systems.

These choices will be presented and analyzed below.

Identified Data vs. Classical Messages

Data provided by sensors, data sent to the actuators, more generally I/O data, and control data are all identified in a given process. They are known within the application. These data are also called identified variables or identified objects. They are often simple objects (temperature, pressure, speed, etc.) and of fixed syntax (integer, real, Boolean, record, list, or other structured data). For instance, a temperature sensor can produce a temperature value coded as an integer, or as real, and the manufacturer identification as a character chain.

The identified data receive a name, which is a global name for the entire application. This name is also used for managing the access to the medium.

Each variable value has a single producer and one or more consumers. Since transferred values correspond to variables in the process, an identifier is attached to each variable whose value is to be transmitted on the network. This identifier is used as the source address to control the medium access. The destination is not indicated. Consumers are responsible for deciding the update of their copy of data on receiving data by recognizing the corresponding identifier. This is the so-called source addressing.

This addressing technique represents several advantages. It allows the communication in a manner "one to many" with broadcast. The communication channel is then not only used efficiently when the same information has to be transmitted to more than one consumer but also, with reliable broadcast, the coherence may be obtained. A new receiver may be added without address modification.

Identifying the variables instead of the sources of the information on the variables offers an additional advantage; the variable is no longer bound to a node of the network. For example, in case of failure of the node providing the variable value, a new source may become active and replace the failed node without any modification of the receivers.

With respect to an identified object, a single active producer is defined and all the other stations may be defined as consumers.

Periodic and Aperiodic Traffics

The control systems are usually based on the system sampling theory; then, the data in inputs and/or in outputs should be transferred periodically.

WorldFIP has chosen to privilege the periodic traffic of identified objects between producer and consumers. Variable values are stored in erasable buffers rather than in queues. There is neither acknowledgement nor retransmission for variable transfers. From this point of view, WorldFIP is a time-triggered system (Kopetz, 1990).

WorldFIP may also be seen as a distributed database updating and management system.

The producer of an identified object updates periodically his own buffer, the WorldFIP protocol updates periodically the buffers at the consumer locations, and then, these consumers may use periodically the copy of the producer value. If a failure occurs during the transmission, the last value is always available for the consumer until it receives a new one.

In WorldFIP, a one-place erasable buffer is associated with each variable at its production and each consumption location. The usual acknowledgements are not necessary and the retransmissions are avoided in case of error.

The question is how to handle some critical data like alarms or rarely occurring events. In WorldFIP, there are two possible ways depending on the criticality. If no real-time reaction is required, the best is to use the usual message transfer. Otherwise, the only good solution is to transform the alarm into a variable whose value reflects the presence of an alarm or not and transfers this value periodically. One may think that this would result in a waste of bandwidth. This is true but is the price to pay to ensure a deterministic response time.

Moreover, multicast transfers are complex when acknowledgements from each receiver are required. In FIP, the choice to suppress acknowledgements drastically simplifies the solution to multicast; for a simple broadcast, appropriate filters in the receivers may be used.

Timeliness attributes and mechanisms for time-critical systems. Due to the periodic transfer of identified data between a producer and their consumers, no acknowledgement has been proposed. No retransmission is basically allowed.

We consider the following three elements: a producer, the consumers, and the bus.

The producer is a process producing a data named X at a given period. Several processes consume X at different periods, and the bus updates at a given period the copy of X at each consumer site from the original of X. The question at each consumption site is: is the value of X fresh, or too old, or obsolete? Therefore, some timeliness attributes have been defined in order to indicate to the consumers if the data are correct or not and in this case the cause of error.

These attributes are called refreshments and promptness. The former type indicates if the production is timely correct or not, and the latter indicates if the reception is correct or not. Based on these elementary attributes, it is then possible to define the time coherence of actions, that is, the fact that different distributed actions take place in a given time interval. This is also the definition of simultaneity of actions.

Other mechanisms have been introduced as synchronization mechanisms between the local operations and the behavior on the network. All these mechanisms will be detailed in Section 8.

40.5 WorldFIP Architecture

The WorldFIP architecture is described by the Figures 40.1 and 40.2, according to the OSI architecture model (Zimmermann, 1980). All the elements were standardized in France in 1992 (AFNOR, 1989). This architecture shows that two main profiles may be used. One is defined to solve the traffic of identified objects; the second one is defined for the usual messaging exchanges. This architecture is directly issued from the need analysis. It is important to remark that the messaging services in the data link layer are related to the point-to-point exchanges of frames, with storage in queues, with or without acknowledgement, and replication detection. The identified traffic services are related to the exchanges of data in a broadcast manner, with storage in erasable buffers, without acknowledgement, except by the space consistency mechanism at the application layer.

Messaging for Periodic and aperiodic Services (MPS) is the service element for the periodic and aperiodic exchanges of identified data. It uses the services of identified traffic at the data link layer. Manufacturing message specification (MMS) is a subset of the well-known MMS standard (ISO, 1990) and uses the messaging services at the data link layer.

We may say that the first profile (left of the figure) is a profile for real-time traffic management, with guaranteed quality-of-service and timeliness properties. The second profile is more used for noncritical exchanges as during commissioning, for maintenance and configuration, or more generally for management. Notice that the messaging services are based on the same Medium Access Control.

Architecture and Standardization

The European Standard EN 50170 (Figure 40.3) contains three national standards in Europe.[3] The volume 3 contains all the WorldFIP specifications according to the following organization: EN 50170 volume 3 (Table 40.1)

Several profiles of WorldFIP have been defined. One of them, the simpler one providing only periodic traffic of identified data, is called DWF (Device WorldFIP) and is standardized (AFNOR, 1996, CEN-ELEC, 1996b).

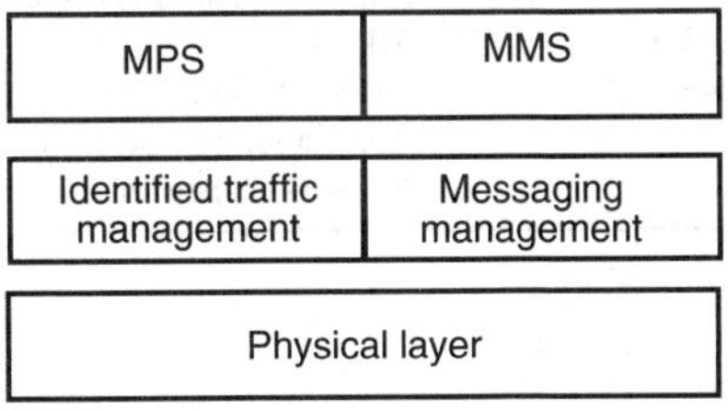

MPS	MMS
Identified traffic management	Messaging management
Physical layer	

FIGURE 40.1 Simplified architecture of WorldFIP.

MPS	MMS
Identified traffic management	Messaging management
(ident, value) transfer	
Physical layer	

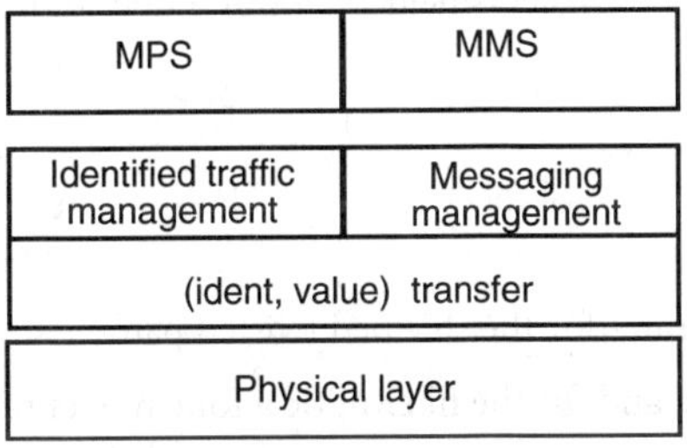

FIGURE 40.2 Architecture of WorldFIP.

---

[3] The other volumes are concerned with P-Net and Profibus.

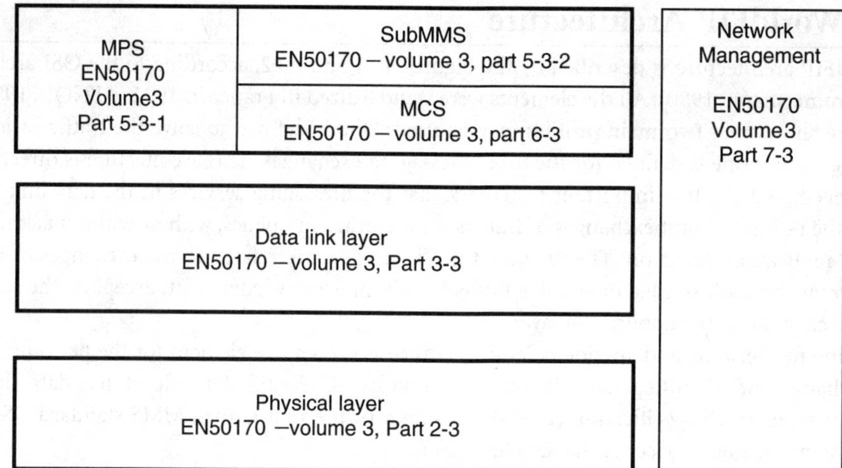

FIGURE 40.3 Architecture and European standard.

TABLE 40.1 Parts of the European 50170-3 Standard

EN 50170 volume 3	Part 1-3	General purpose field communication system
EN 50170 volume 3	Part 2-3	Physical layer
	Subpart 2-3-1	IEC twisted pair (IEC 61158-2)
	Subpart 2-3-2	IEC twisted Pair amendment
	Subpart 2-3-3	IEC fiber optic
EN 50170 volume 3	Part 3-3	Data link layer
	Subpart 3-3-1	Data link layer definitions
	Subpart 3-3-2	FCS definition
	Subpart 3-3-3	Bridge specification
EN 50170 volume 3	Part 5-3	Application layer specification
	Subpart 5-3-1	MPS definition
	Subpart 5-3-2	SubMMS definition
EN 50170 volume 3	Part 6-3	Application protocol specification
EN 50170 volume 3	Part 7-3	Network management

40.6 Physical Layer

The physical layer of WorldFIP was obviously the first to be conformed to the IEC 1158-2 standard,[4] because this standard has been defined starting from the FIP French standard number C46 604.

The medium may be as well be a twisted shielded pair or a fiber optic.

Some Figures

Data Rates

The standard defines three data rates for the shielded twisted pair:

31.25 kbps, 1 Mbps, 2.5 Mbps, and for the fiber optic a fourth data rate, 5 Mbps.

But some experiences have been built with other data rates as, for example, 25 Mbps with the transfer of speed and video.

---

[4] This number was the previous number of the current 61158 standard.

TABLE 40.2   Data Rate and Maximum Possible Lengths

Data Rate	Length Without Repeater	Length With Four Repeaters
31.25 kbps	10 km	50 km
1 Mbps	1 km	5 km
2.5 Mbps	700 m	3.5 km

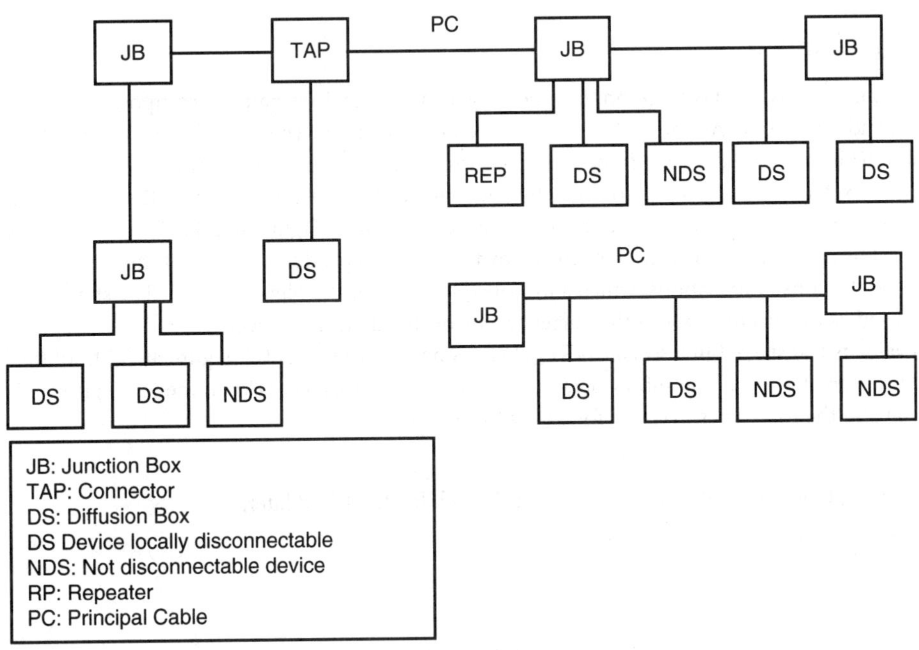

FIGURE 40.4   Example of topology.

Maximum Length

The maximum number of stations is 256 and the maximum number of repeaters is 4. According to the data rate and the number of repeaters, Table 40.2 gives the maximum length.

Topology

The topology for twisted shielded pair may be the following (Figure 40.4).

Coding

The coding is based on a Manchester code. A physical data frame is composed of three parts: a sequence of frame starting composed of a preamble and of a start delimiter (PRE and FSD), the data link information, and the end delimiter (FED). Twenty-four bits are added to each data frame.

40.7   Data Link and Medium Access Control Layers

Introduction

The WorldFIP medium access control is centralized and managed by a so-called bus arbitrator (BA). All the exchanges are under the control of this BA. They are currently scheduled according to the timing requirements and the time constraints (Cardeira and Mammeri, 1995) But the scheduling policy is not relevant of the standard.

The data link layer provides two types of services: for the identified objects and for the messages. But both may or may not take place periodically. Thanks to the MAC protocol, it is obviously easy to manage the periodic traffic, which may be prescheduled before the run time. But it is then necessary to provide the well-suited services for the requirement of sporadic or random traffic, and the associated protocol mechanisms. As is usually known, the random traffic is managed by a periodic server. When a station is polled, it may express a request for an extra polling. Such requests corresponding to the aperiodic traffic are managed dynamically by the bus arbitrator.

Basic Mechanism

The medium access control is based on the following principle: each exchange is composed of two frames: a request and a response. All the exchanges are based on the couple (name of information, value of the information) implemented by two frames: an identification frame and a value frame.

Thus, to exchange a value of an object, the BA sends a frame, which contains the identifier of this object. This frame is denoted ID-DAT (as Identification of Data) (Figure 40.5.1). This frame is received by all the active current stations and recognized by the so-called producer station of the identified object, and also by the consumer stations, which subscribed to this object[5] (Figure 40.5.2). The station that recognizes itself as the producer sends the current value of the identified object.

This value is transferred in a so-called RP_DAT frame (as Response) (cf. Figure 40.5.3). All the interested stations, including all the subscribers and the BA, receive this RP_DAT frame (cf. Figure 40.5.4).

The ID and the RP frames have the following format:

ID Frame:

- a control field (CF) of eight bits, whose roles will be developed later,
- the identifier of the object (16 bits), and
- a CRC (16 bits).

RP Frame:

- a CF of eight bits, whose roles will be developed later,
- the value of the identified object in the previous frame (maximum 256 bytes), and
- a CRC (16 bits).

Now, the idea is to extend this simple mechanism of polling by designation of the data to be sent, in order to solve the problem of messages transfer and of the transfer of aperiodic data or messages.

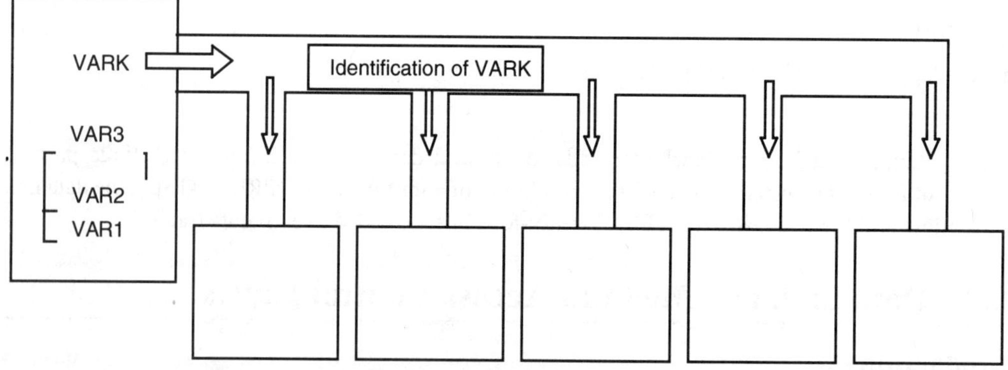

FIGURE 40.5.1   Broadcasting of an identifier by the BA.

---

[5] Notice that the Producer–Consumer model is also called Publisher–Subscriber, following this concept of subscribing to the data by the Consumers.

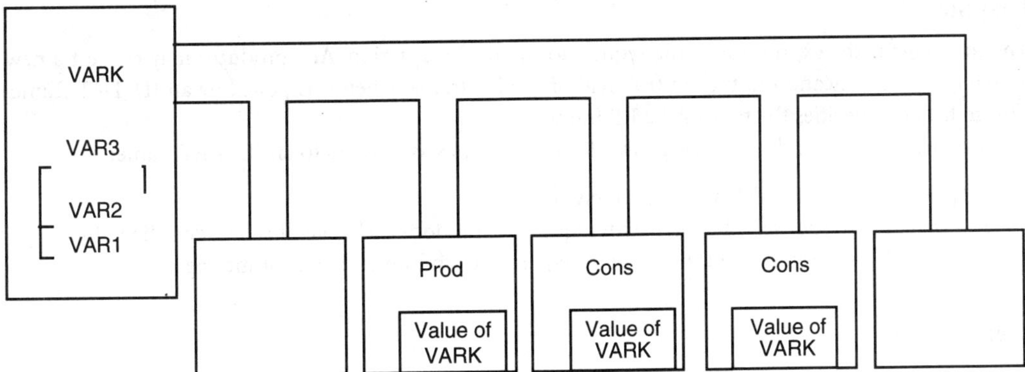

FIGURE 40.5.2 Recognition of the identifier by the producer and by the consumers.

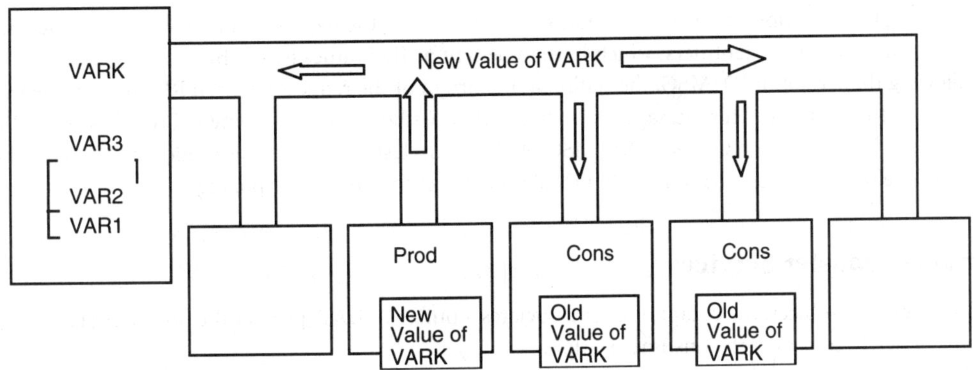

FIGURE 40.5.3 Transmission of the value by the producer.

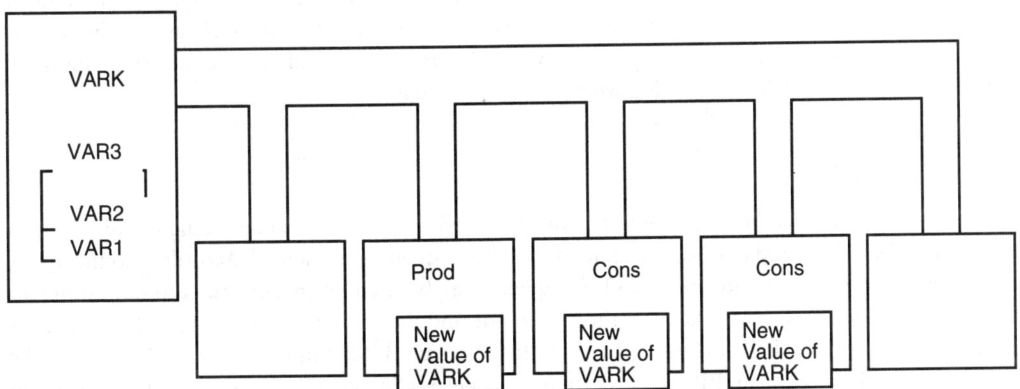

FIGURE 40.5.4 Reception and updating of copies of the value by the consumers.

The Aperiodic Server

The needs for an aperiodic transfer have been identified in Section (40.4). This aperiodic transfer takes place in the free time slots of the periodic one. The aperiodic traffic is dynamically managed by the bus arbitrator.

First Stage

The first stage is the expression of the request to the BA by a station. Any producer may request a new exchange by an indication in the control field of the RP frame, when it is polled by an ID_DAT frame. This indication specifies the type of the RP frame.

We may then observe the following three RP frames types as answers to an ID_DAT frame:

RP_DAT: response to ID DAT without any request.

RP_DAT_RQ: response to ID DAT with request for random exchange of other identified objects.

RP_DAT_MSG: response to ID DAT with request for random exchange of message.

Second Stage

The second stage is to satisfy the request. The BA then has to place the right ID frames in a free time slot of the scanning table, according to its own scheduling policy. The ID frames corresponding to the possible requests are ID_RQ and ID_MSG. The former satisfies the RP_DAT_RQ and the latter satisfies the RP_DAT_MSG.

Following the reception of ID_RQ, the station at the origin of the request sends an RP_RQ frame with, in the data field, a list of identifiers, which have to be sent as ID Frames by the BA.

Following the receipt of ID_MSG, the station at the origin of the request sends an RP_MSG frame with a message in the data field. This message is specified with or without acknowledgment. Both these RP_MSG frames are called RP_MSG NOACK or RP_MSG_ACK. In this last case, the receiver sends an RP_ACK after receipt of the message. A special frame RP_FIN allows the BA to continue its polling.

Variable Transfer Services

To each identified object, the data link layer associates a buffer B_DAT_prod at the producer station and a buffer B_DAT_cons at each consumer station.

Two main services are defined for writing and reading a buffer: L_PUT and L_GET, respectively.

The write service (L_PUT) places the new value in the producer buffer. The previous buffer content is overwritten. The read service (L_GET) gets the value from the consumer buffer. These services do not cause any traffic on the bus (Figure 40.8). The content of each consumer buffer is updated with the value stored in the producer buffer under the control of the Bus Arbitrator. (Figure 40.7) as seen in Section 40.8. A L_SENT.indication informs the producer when the transmission takes place. The consumers are informed by an L_RECEIVED indication when the update occurs.

Message Transfer

For the message as for the identified objects, WorldFIP defines periodic and aperiodic (or on request) message transfers. To periodic messages, an identifier and a queue are assigned. According to the application needs, more than one identifier and one queue may be used when one wants to have different polling periods. Messages are deposited at the source side and the transfer of the content of the queue is periodically triggered by the BA (ID-MSG frame). If the queue is not empty, the source DLL sends the first message in the queue in an RP_MSG_xx frame. The destination data link layer stores the message in the receive queue and, if requested, immediately acknowledges the transfer using an RP_ACK frame. The end of the transaction is signaled by the source to the Bus Arbiter using an RP_FIN frame. If the queue is empty, no transfer takes place, and only the RP_FIN frame is sent. The polling period is a configuration parameter.

For aperiodic message transfer, on the source side, the data link layer defines a single queue F_MSG aper that will hold the pending messages. On the destination side, a receiving queue F_MSG rec is defined. As for an aperiodic variable, transfer requests are signaled to the Bus Arbiter as piggy backs on RP_DAT frames sent in response to ID_DAT frames.

Synthesis on the Data Link Layer

Three types of objects are exchanged according to the basic principle based on the exchange of the couples (name, value) (Thomesse and Rodriguez, 1986).

These objects are: Identified objects or List of identifiers of identified objects, or Messages with or without acknowledgment.

The data link layer protocol may be considered with connection established at the configuration stage. These connections are multicast and the corresponding Services Access Point (SAP) are represented by the identifiers. Associated to each SAP, different Connection End Point (CEP) are represented by the associated objects and necessary resources (Figure 40.6).

A CEP for data exchange represented by the buffer storing the successive values.

A CEP for the list of identifiers exchange represented by another buffer.

A CEP for the sending of messages represented by a queue.

Each of these CEPs is addressed firstly by the identifier addressing the SAP and secondly by the indication in the control field of the ID frame, specifying the corresponding resource.

40.8 Application Layer

Two Application Service Elements compose the WorldFIP Application layer. The first is MPS for real-time data exchange (Thomesse and Delcuvellerie, 1987; Thomesse and Lainé, 1989) and sub-MMS for usual messaging service and compatibility with other networks.

For the real-time data exchange, FIP behaves like a distributed database being refreshed by the network periodically or aperiodically on demand. All the application services related to the periodic and aperiodic data exchange are called MPS.

MPS provides local read/write services (periodic) as well as remote read/write services (aperiodic) of the values of variables or lists of variables. The read services associate indications on the age of the object value. Considering a producer (producer of data named X) at a given period, and the consumers of X consuming X at different periods, the bus itself updating at a given period the copies of X consumers from the original of X, the question at a consumption site is: is the value of X fresh, or too old, or obsolete.

In WorldFIP, this information is based on local mechanisms and provided as two types of status: the *refreshment status* elaborated by the producer and the *promptness status* elaborated by the consumer (Thomesse et al., 1986; Decotignie and Raja, 1993; Lorenz et al., 1994). These statuses are returned by the read services with the value itself. This information may also be used to check whether a set of variables are timely coherent.

As a variable may have several consumers, there is a need to know whether the different copies of the variable value available to the various consumers are identical or not. This information, called spatial coherency status (or spatial consistency) (Saba et al., 1993), is provided by the MPS read services related to lists of variables. The same services also offer a temporal coherence status.

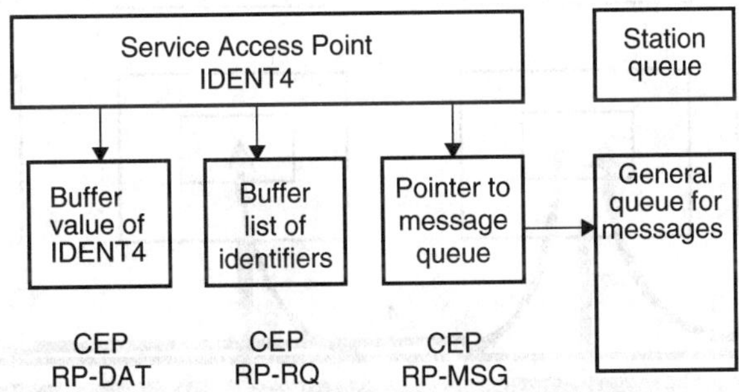

FIGURE 40.6 Service Access Point and Connection End Point.

Services Associated to the Variables

A variable can be of simple type such as integer, floating, Boolean, character, or composite-like array and records. It may have different semantics: variable, synchronization variable, consistency variable, and variable descriptor. Synchronization variables are used to synchronize application processes and also in the elaboration of the temporal and spatial status associated with ordinary variables. Consistency variables are used to elaborate the spatial coherence status. Variable descriptors hold all information concerning variables, type, semantics, periodicity, etc. for configuration, commissioning, maintenance, and management.

Three request services, A_READ, A_WRITE, and A_UPDATE, and two indication services, A_SENT and A_RECEIVED, are available. A_UPDATE is used to request an aperiodic transfer of a variable. A_SENT and A_RECEIVED are optional services. When used, A_RECEIVED informs the consumer at its local communication entity of the reception of new variable value. Similarly, A_SENT notifies the producer of the transmission of the produced value. These indication services can be used by Application Processes (AP) to verify the proper operations of the communication entity and also to synchronize with each other by receiving synchronization variables (see synchronization services later). A_READ and A_WRITE exist in two forms: local and remote.

A local read of a variable (denoted A_READLOC) provides the requesting AP with the current value of the local variable image. A local write (denoted A_WRITELOC), invoked by an AP, updates the local value of the variable to be sent at a next scanning, specified in the request. As already mentioned, these operations do not invoke any communication. The transfer of the value from the producer to its copies in all consumers concerns the distribution function of application layer. The variable value given in a A_WRITELOC will be available for the distributor who is in charge of broadcasting it to all the consumers of this value. The variable value, returned by a A_READLOC at a consumer site, will be the last value updated by the distribution function.

A remote write, A_WRITEFAR, is used by a producer to update the content of the local buffer assigned to the variable specified in the request and to ask for the transfer of this value to the producers. It may been seen as the combination of the invocation of an A_WRITELOC service followed by an A_UPDATE. However, as for the A_WRITELOC, this service may only be invoked by the producer of the variable (Figure 40.7).

Operations:

1. Remote READ request.
2. Ask the distributor for updating.
3. Transfer order from the distributor.
4. Transfer of the producer's variable value to the users.
5. Confirmation of remote READ.

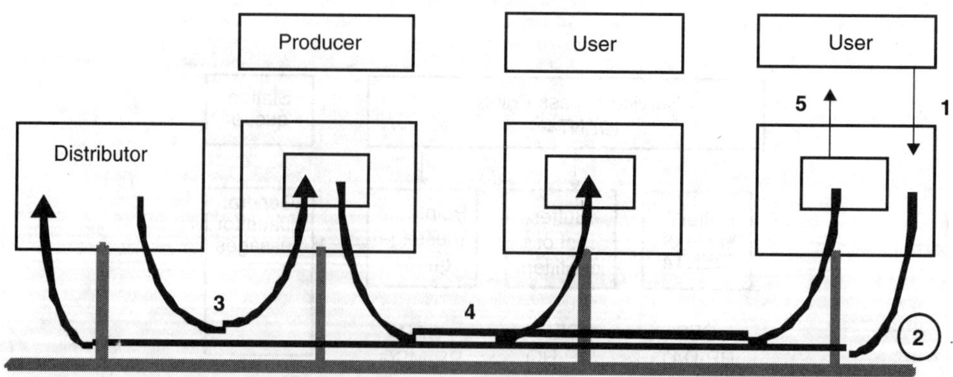

FIGURE 40.7   Remote READ using PDC model.

In a similar way, with a remote read, A_READFAR, a consumer requires the transfer of the variable value from the producer to all consumers. This value is returned as a result of this request. This service is hence a combination of an A_UPDATE service invocation followed by the invocation an A_READLOC service. However, read services may only be invoked on a consumer.

Figure 40.8 shows this service from the Producer–Consumer model point of view. Consider an AP, which is a consumer of variable X. In order to request a transfer explicitly, this AP should also be producer of a variable (Y in the example). Figure 40.8 depicts in detail all necessary primitive calls and frame exchanges to achieve the remote READ of X.

We can see that these remote operations are not symmetrical. The remote write is a service without confirmation, while the remote read is not.

Temporal Validity of Variables

Normally, according to the Producer–Consumer model, operations should proceed in the following order: production, transfer, and consumption. As the receipt of an identifier triggers the transfer, a production and a consumption may be triggered by a production or consumption order. These behaviors may lead to abnormal situations. For example, several productions may occur successively without any transfer. Conversely, a number of transfers may take place between two successive productions of a variable. The same problems may arise on the consumer side. A consumer may read several times the same value of a variable or may not have enough time (or may not be interested) to consume all the received values. It is thus important to detect these deviations from normal behavior. This means that any consumer should be able to know whether the producer has produced on time, the transfer has been handled on time, and the consumer itself has consumed on time. Finally, the consumer should be able to check whether the value is still temporally valid or not (Lorenz et al., 1994).

Refreshment

The refreshment status type for a variable is a Boolean that indicates if a production occurred in the right time window. It is elaborated by the application layer of the producer. The time window is defined by a start event and a duration. From a simplified point of view, the refreshment is correct (true) if the production occurs in the time window, and it remains true during a given delay, which is the normal production period. It is false after this period deadline. The value of the current refreshment status is sent

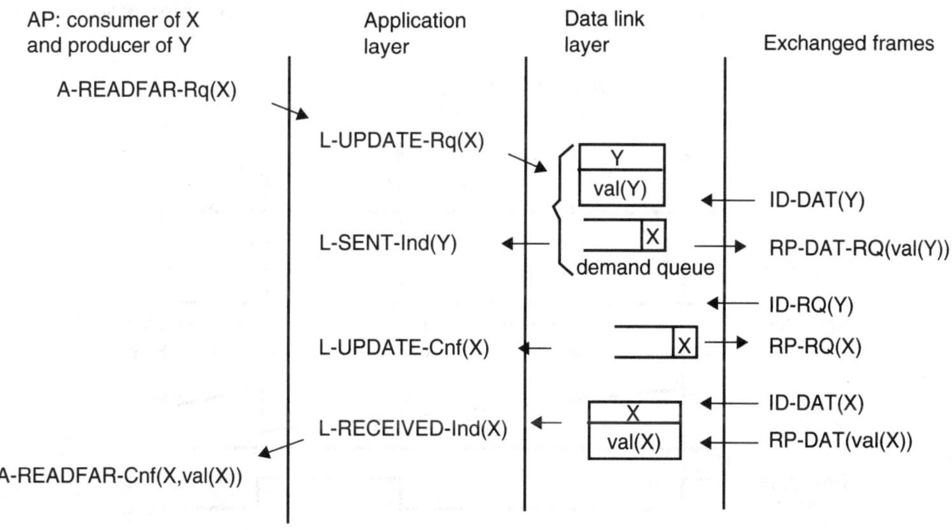

FIGURE 40.8   Primitives and frame exchanges to achieve a remote READ service.

with the value and indicates to the consumer whether from the point of view of the producer and of the transmission the value is valid or not.

In summary, the refreshment status indicates to a consumer that the producer has produced its value by respecting a production delay called production period.

Promptness

The promptness status type for a variable value indicates if the transmission of the data has been done in a right time window. It is elaborated by the communication entities of its consumers. It is returned with the variable value and the refreshment status as a result of an invocation of A_READ.

Synchronous and Asynchronous

Refreshments and Promptness are the timeliness attributes associated to a given variable. They indicate if an event occurs in a time window defined by a starting event and a duration.

The WorldFIP fieldbus has defined two types of timeliness attributes: the synchronous type and asynchronous type.

An attribute is considered synchronous when the starting event of the time window is a Received-indication of a dedicated variable. It is normally periodic, and the BA is, as for other variables, in charge of the respect of the period. The duration is the period of production and is managed by the network, through the BA behavior.

An attribute is considered asynchronous when the starting event is the previous occurrence of this event. The duration is the period of production and is managed here by the device itself.

In fact, in WorldFIP, two attributes are relevant of the synchronous type: the so-called "synchronous" as defined previously and the so-called "ponctual" attribute where the stating event is the same as for the synchronous attributes, but the duration is no more than the period of production. It is a shorter delay.

Further details and the finite state machines of these mechanisms can be found in C46-602 standard and in Lorenz et al. (1994a) and Lorenz et al. (1994b) (Figure 40.9).

Remark

The terms synchronous and asynchronous could be better replaced by "synchronized by the network" and "locally synchronized."

Synchronization Services

The processes of an application can be synchronized or asynchronous. An asynchronous application process is one whose execution is independent of the network behavior. An application process is said to

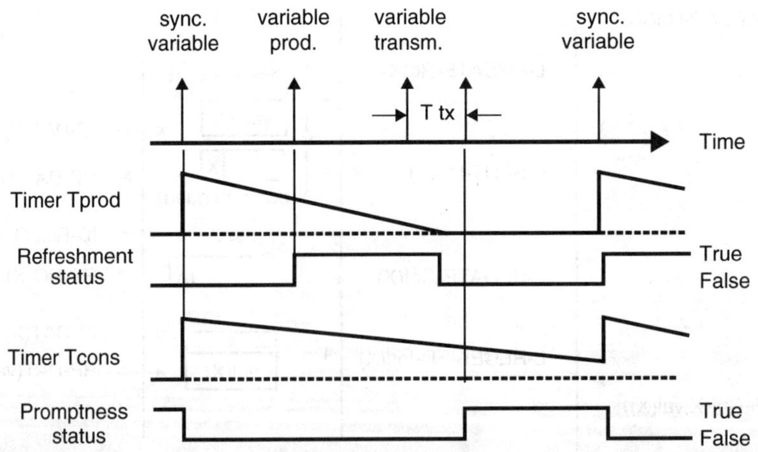

FIGURE 40.9   Refreshment and promptness status elaboration in the synchronous case.

be synchronized when its execution is related to the receipt of some indication of the network. In many cases, the various distributed processes of an application are synchronized. This synchronization may be ensured through the indication of receipt of variables. However, some Application Processes (APS) of an application may not be able to handle synchronization with others. For such asynchronous APs that need to participate in a synchronized distributed application, FIP provides a resynchronization mechanism.

In addition to the existing buffer, called the public buffer, for each variable, the resynchronization mechanism associates a second buffer: the private buffer. The private buffer is only accessible to the corresponding AP. The access can be performed by the local A_READ if the variable is consumed or A_WRITE if the variable is produced. The public buffer is only accessible to the network (Figure 40.10). The resynchronization mechanism consists of copying the content from one buffer to the other according to the synchronization order via the network. Both variable production and variable consumption can be resynchronized.

When a variable consumption has to be resynchronized, its value in the private buffer is kept unchanged until the resynchronization order. If a new value is transferred on the network, it will be kept in the public buffer. This occurs only on receipt of a resynchronization order, so that the value in the public buffer will be copied in the private buffer. It is similar for variable production.

In both cases, the resynchronization order is given by a synchronization variable specified with each variable, produced or consumer, that needs to be resynchronized.

Services Associated with Variables Lists

A variable list is an unordered set of variables that must verify the following property: the time coherence, that is, the fact that they are all produced in a given time window. All the variables of a list are consumed at the same time by at least one consumption process.

In the case of more than one process with respect to the consumption of a list, another property must be verified: the space consistency, that is, the fact that all the copies of all the variables of the list are the same on all the consumption sites.

The variables that compose the list may be produced on different sites. They need not all be produced by the same producer as in the usual application layers such as MMS or MMS-like (FMS or some sub-MMS' e.g.). Usually, the productions are synchronous operations.

The only service defined on lists is A_READLIST, which allows to read all variables of a list in a single invocation. This service returns the last received values for the variables in the list and three optional statuses: a production coherence status, transmission coherence status, and a spatial consistency status. These statuses are provided to account for two important needs in the systems using fieldbuses.

First, a consumer of several variables is interested in knowing whether the corresponding values have been sampled nearly at the same time, which is called time coherence (Kopetz,1990).

Second, when the value of a variable has been distributed to several consumers, it might be useful to know if all the values are identical. This is referred to as spatial consistency. The idea in FIP is not to ensure temporal coherence and spatial consistency but rather to indicate if these properties are present or not. In FIP, temporal coherence indication is given by two statuses: the production coherence status and the transmission coherence status (Saba et al, 1993).

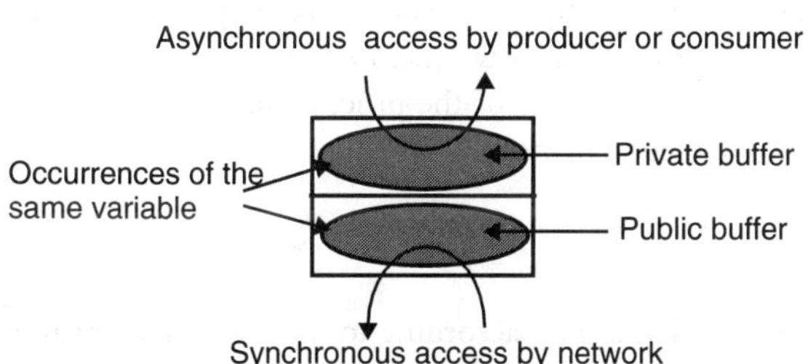

FIGURE 40.10　Resynchronization mechanism.

The production coherence status is a Boolean information elaborated by the consumer application layer. This status corresponds to a logical AND operation of all the corresponding refreshment statuses.

Similarly, a transmission coherence status is calculated as the logical AND of all promptness statuses of the variables in the list. The production coherence status and the transmission coherence status together are an indication of the temporal coherence of the variables in the list.

The space consistency status of a list is elaborated by the application layer of each consumer of the list. The elaboration mechanism relies on the broadcast of a consistency variable by each of the consumers. This variable indicates for each copy of the variable list if it has been received correctly and in a given window or not. After the receipt of all the consistency variables by all the consumers, each of them will have knowledge about the validity of the variables in the list at all consumers. A logical AND operation of all the consistency variables will give the spatial consistency status.

To make the variable list transfer more reliable, FIP defines an error recovery mechanism. If needed, a consumer can trigger a retransmission of its consumed variables when any error is detected. This mechanism performs a retransmission request (using aperiodic variable transfer service of data link layer) for a number of times limited to a maximum defined for each of the instances of this list. The duration of the whole transaction of the list (including retransmission) must be bounded by a time window T smaller than the delay between two consecutive synchronization orders. It has been shown in (Song et al., 1991) that this recovery mechanism (a kind of grouped acknowledgement technique) is very efficient and can be recommended for use for other multicast protocols.

40.9   WorldFIP State and Technology

Technology

Several integrated circuits and of software libraries are available in order to build devices compatible to the standard. They conform to the European standard EN 50170. The circuits cover all the physical-layer protocols, all the profiles of data link layer, and of the application layer. They are referenced as FIPIU2, FULLFIP2 MICROFIP for the communication component. The circuits for the physical layer are essentially FIELDRIVE, CREOL for copper wire, and OPERA-FIPOPTIC for fiber optic. It is important to notice that a redundant channel may be used with the FIELDUAL component. TransoFIP and FIELDTR are possible transformers for connection on a copper wire.

The libraries are used in order to create the link between the user application and the communications controller. Each library is dedicated to a communication component.

FIP: Fieldbus Internet Protocol

FIP may be now be interpreted as a FieldBus Internet Protocol. Indeed, the messaging services are used to transfer HTTP PDU, and then each site on a WorldfIP fieldbus may host a Web server. For remote maintenance applications, for remote configuration, it is then possible to access the stations through a browser.

Two solutions are currently in use.

In the first one, a single station may be seen as the image of all the stations of the fieldbus.

In the second one, each station is directly accessed by tunneling HTTP in the WorldFIP PDUs.

The interest of WorldFIP is that the data flow generated by Internet connections is managed in complete compatibility with the time constraints of the process, and then with its own dependability. The time-critical traffic is always satisfied in priority.

New Development

In 2001, WorldFIP has been certified as safe according to the procedure "proven in use" defined by the IEC 61 508 standard. According to this standard, a fieldbus is considered as a subsystem. It has been certified by Bureau Veritas after considering very reliable applications from detailed records from field users, with a sufficient number of applications, with a high level of confidence in the operational figures. It is the only fieldbus in the world with such a certification.

40.10 Conclusion

The WorldFIP fieldbus is now 20 years old if we considered the beginning of its specification. It is only some years old if we consider the very recent dates of the international standards (Leviti, 2001), knowing that the definition of profiles standard (CENELEC, 2003b) is not yet finished. The current integrated circuits (the third generation for the first IC) are based on the last development of microelectronics. The WorldFIP fieldbus is the only certified SIL3 in the world (Froidevaux et al., 2001). WorldFIP occupies a special place in the international market. It is used in all the types of industry (iron–steel industry, car manufacturing, bottling, power plant, etc.) but also in several time-critical applications, in embedded systems in trains, in autobus, in ships, or in subways and in a very important application : the Large Hadron Collider in CERN (Centre Européen de Recherche Nucléaire) in Switzerland. The main reason for this must be searched in the technical specifications, in the services provided, and in the quality of services, essentially from the timeliness point of view. WorldFIP guarantees that time constraints will be met, periods without jitter, synchronization of distributed actions (productions, consumptions, data acquisitions, controls, etc.). The validation of data is also an important element of the quality of service and the dependability of WorldFIP-based systems. The quality of the physical layer (IEC, 1993) and the redundancy capabilities are also important points for choosing WorldFIP for critical applications. Several of these concepts and mechanisms have been repeated in the TCCA report (ISO, 1991; Grant, 1992) and in the "standard" IEC TS 61158 (CENELEC, 2003a).

With respect to other and newer requirements, WorldFip is able to transport voice and video, without disturbing or influencing the real-time traffic. Some examples of voice transport are already industrial in trains. WorldFIP is also able to transport HTTP protocol data units, allowing the remote access to WorldFIP-based systems for monitoring, maintenance, configuration, and so on.

References

AFNOR (1989), French Standards NF C46601 to C46607, FIP bus for exchange of information between transmitters, actuators and programmable controllers, Published between 1989 and 1992 (in French).

AFNOR, French Standard C46-638, Système de Communication à haute performance pour petits modules de données (WorldFIP Profil 1, DWF), 1996.

Bergé, N., G. Juanole, and M. Samaan, Using Stochastic Petri Nets for Modeling and Analysing an Industrial Application Based on FIP Fieldbus, International Conference on Emerging Technologies and Factory Automation, INRIA, Paris, 1995.

Cardeira, C. and Z. Mammeri, A schedulability analysis of tasks and network traffic in distributed real-time systems, *Measurement*, 15, 71–83, 1995.

CENELEC, European Standard EN 50170. Fieldbus. Vol. 1: P-Net, Vol. 2: PROFIBUS, Vol. 3: WorldFIP, 1996a.

CENELEC, High Efficiency Communications Subsystems for Small Data Packages, CLC TC/65CX, EN 50254, 1996b.

CENELEC, prEN61158-2: Digital Data Communication for Measurement and Control — Fieldbus for Use in Industrial Control Systems, Part 2: Physical Layer Specification, Part 3: Data Link Layer Service Definition, Part 4: Data Link Layer Protocol Specification, Part 5: Application Layer Service Definition, Part 6: Application Layer Protocol Specification, 2003a.

CENELEC, prEN61784-1 (65C/294/FDIS): Digital Data Communications for Measurement and Control — Part 1: Profile Sets for Continuous and Discrete Manufacturing Relative to Fieldbus use in Industrial Control Systems, 2003b.

Decotignie, J.D. and P. Raja, "Fulfilling Temporal Constraints in Fieldbus", Proceedings of IECON'93, Maui, Hawaï, U.S.A., 1993, pp. 519–524.

Froidevaux, J.-P., O. Nick, and M. Suzan, Use of Fieldbus in Safety Related Systems, an Evaluation of WorldFIP According to Proven-in-use Concept of IEC 61508, 4th FET, IFAC Conference, Nancy, France, 2001.

Galara, D. and J.P. Thomesse, Groupe de réflexion FIP, Proposition d'un système de transmission série multiplexée pour les échanges d'informations entre des capteurs, des actionneurs et des automates réflexes, Ministère de l'Industrie et de la Recherche, 1984.

Gault, M. and J.P. Lobert, Contribution for the Fieldbus Standard, presentation to IEC/TC65/SC65C/WG6, 1985.

Grant, K., Users Requirements on Time Critical Communications Architectures, Technical Report ISO TC184/SC5/WG2/TCCA, 1992.

IEC (1993), IEC Standard 1158-2, Fieldbus Standard for Use in Industrial Control Systems — Part 2: Physical Layer Specification and Service Définition + AMD1, 1995.

ISO (1990), International Standard ISO 9506: Manufacturing Message Specification (MMS) — Part 1: Service Definition and Part 2: Protocol Specification, 1991.

ISO, ISO/TC 184/SC 5/WG 2-TCCA-N56, in Draft Technical Report of the TCCA Rapporteurs' Group of ISO/TC 184/SC 5/WG 2 Identifying User Requirements for Systems Supporting Time-critical Communications, Grant, K., Dr, August 1991.

Kopetz, H., *Event Triggered vs Time Triggered Real Time Systems*, Lecture Notes in Computer Science, Vol 563, Springer-Verlag, Berlin, 1990, pp. 87–101.

Leviti, P., IEC 61158, an Offence to Technicians, 4th FET, IFAC Conference, Nancy, France, 2001.

Lorenz, P. and Z. Mammeri, Temporal Mechanisms in Communication Models Applied to Companion Standards, SICICA 94, Budapest, 1994.

Lorenz, P., J.-P. Thomesse, and Z. Mammeri, A State-Machine for Temporal Qualification of Time-Critical Communication, 26th IEEE Southeastern Symposium on System Theory, Ohio, U.S.A., March 20–22, 1994.

MAP, General Motors, Manufacturing Automation Protocol, Version 3.0, 1988.

Pleinevaux, P. and J.-D. Decotignie, (1988) Time critical communications networks: field buses, *IEEE Network Magazine*, 2, 55–63, 1988.

Saba, G., J.P. Thomesse, and Y.Q. Song, Space and Time Consistency Qualification in a Distributed Communication System, Proceedings of IMACS/IFAC International Symposium on Mathematical and Intelligent Models in System Simulation, Vol. 1, Brussels, Belgium, April 12–16, 1993, pp. 383–391.

Simonot, F., Y.Q. Song, and J.P. Thomesse, On message sojourn time in TDM schemes with any buffer capacity, *IEEE Transactions on Communications*, 43, 1013–1021, 1995.

Song, Y.Q., P. Lorenz, F. Simonot, and J.P. Thomesse, Multipeer/multicast protocols for time-critical communication, Multipeer/Multicast Workshop, Orlando, U.S.A., 1991.

Thomesse, J.-P. and J.-L. Delcuvellerie, FIP — A Standard Proposal for Fieldbuses, IEEE–NBS Workshop on Factory Communications, Gaitherburg, U.S.A., March 17–19, 1987.

Thomesse, J.-P. and T. Lainé, The Field Bus Application Services, Proceedings IECON'89 15th Conference IEEE–IES Factory Automation, Philadelphia, USA, November 1989, pp. 526–530.

Thomesse, J.-P. and M. Rodriguez, FIP, A Bus for Instrumentation, Advanced Seminar on Real Time Local Area Networks, Colloque INRIA Bandol, 1986.

Thomesse, J.-P., J.-Y. Dumaine, and J. Brach, "An Industrial Instrumentation Local Area Network", Proceedings of IECON, 1986, pp. 73–78.

Thomesse, J.-P. A Review of the Fieldbuses. Annual Reviews in Control, Pergamon, vol 22, 1998, pp. 35–45.

Thomesse, J.-P. Le réseau de terrain FIP. Revue Réseaux et Informatique Répartie, Ed. Hermés, Vol 3, no. 3, 1993, pp. 287–321.

Zimmermann, H., OSI reference model. The ISO model of architecture for open system interconnection, *IEEE Transactions on Communications*, 28, 425–432, 1980.

41

FOUNDATION Fieldbus:
History and Features

Salvatore Cavalieri
*University of Catania, Faculty of
Engineering*

41.1 Introduction

FOUNDATION Fieldbus is an all-digital, serial, two-way communication system. Its specification has been developed by the for-no-profit Fieldbus Foundation association [1]. Since its very beginning, FOUNDATION Fieldbus has shown two fundamental and unique (at that time, at least) features: an emphasis on the standardization of the description of the devices to be connected to the fieldbus, and the adoption of the main link access mechanisms (i.e., token-based and centralized ones), which the IEC/ISA fieldbus committee (IEC 61158 and ISA SP50) was trying to derive from the existing proposals within a new and complete solution [2–5].

One of the aims of this chapter is to emphasize the value of the choices made by Fieldbus Foundation association as well as their impact on current features of the FOUNDATION Fieldbus communication system. Furthermore, these features will be described in great detail, allowing the reader to clearly understand the key points of the system.

The chapter is organized into two parts: the first part (Section 2) gives an overview of the principles of FOUNDATION Fieldbus; the second part (Section 3) introduces the main features of this communication system.

41.2 Principles of FOUNDATION Fieldbus

Since the first fieldbus communication systems [6,7] have appeared on the market, the need to achieve just one Fieldbus standard was felt immediately. Over 15 years ago, the International Electro-technical Commission (IEC) and Instrument Society of America (ISA) embarked on a joint standardization effort identified by two codes: 61158 on the IEC side and SP50 on the ISA one. The main aim of the standard committee was the definition of a unique communication system able to merge the main features of the Fieldbuses available on the market: FIP [8] and Profibus [9].

Fieldbus Foundation (note that Fieldbus Foundation refers to the name of the association, while FOUNDATION Fieldbus is used for the relevant communication system [1]), established in 1994 as the result of a merger between ISP (Interoperable System Project) and WorldFIP North America, has defined a small set of basic principles.

These basic principles included two main cornerstones:

1. The adoption of both the main Medium Access Control (MAC) mechanisms that the IEC/ISA fieldbus committee was trying to derive from the existing proposals within a truly complete solution
2. Emphasis on the standard description of the devices to be connected on the fieldbus.

Cornerstone (1) made Fieldbus Foundation free from the persistent solution issue: "scheduled access vs. circulated token." IEC 61158 Type 1 DLL stated:

Both paradigms, circulated token and scheduled access were good, but insufficient at the same time; they were complementary, not alternative, and a complete fieldbus solution needs the two together [3].

FOUNDATION Fieldbus, since being established, fully adopted an approach to provide both the "predefined scheduling philosophy" of FIP and the "token rotation philosophy" of Profibus. Section3, provides more details about these two fundamental mechanisms.

Cornerstone (2) allowed Fieldbus Foundation to avoid a situation (which affected most of the previous fieldbus proposals) in which, after defining the communication stack, much more still needed to be done in order to make devices operational after being connected to a fieldbus.

In fact, the previous fieldbus proposals started their developments by focusing on the communication aspects (physical media, access mechanisms, addressing, connections, quality of services, etc.). That mostly motivated by the fact that, when switching from a dedicated low-frequency 4–20 mA signal to a multidata high-frequency serial link, the most evident change was relevant to the communication mechanism itself: Was that cable still good? What's the best encoding method? How can we guarantee noise recovery? Will too many data on the same medium affect their timeliness? And so on.

But once the communication aspects were defined and proven, the data, which such communication was able to transfer between two or more devices, did not make those devices able to interoperate.

FOUNDATION Fieldbus had this concept clear in sight since the beginning, and included the definition of data semantic plus their configuration and use within the first set of specifications.

The only aspect that initially Fieldbus Foundation left intentionally "behind" was the higher speed version H2 of Fieldbus. That was mostly because the market addressed by Fieldbus Foundation was relevant to the replacement of existing 4–20 mA devices with the ones compliant with FOUNDATION Fieldbus, trying to reduce as much as possible the relevant costs. This market strategy, chosen in order to make the adoption of the new fieldbus technology easier, could be realized if the already laid-down twisted-pair cables (connecting the old 4–20 mA devices), which support only the slow-speed H1 version, were maintained. This explains why such a strong effort was put in the development of the H1 technology as opposed to the higher speed H2 version of FOUNDATION Fieldbus.

For H2, Fieldbus Foundation initially planned to adopt the IEC/ISA high -speed standard [2], but ultimately decided to use the High-Speed Ethernet (HSE) instead, mainly due to the wide availability of components and the existence of networks in the plants (at least at backbone level).

41.3 Technical Description of FOUNDATION Fieldbus

FOUNDATION Fieldbus is an all-digital, serial, two-way communication system. Foundation Fieldbus specifications include two different configurations: H1 and HSE [1,10]. H1 (running at 31.25 kbit/sec) interconnects "field" equipment such as sensors, actuators, and Inputs/Outputs (I/Os). HSE (running at 100 Mbit/sec) provides integration of controllers (such as DCSs and PLCs), H1 subsystems (via a linking device), data servers, and workstations. HSE is based on standard Ethernet technology to perform its role. In detail:

- The H1 FOUNDATION Fieldbus communication system is mainly devoted to distributed continuous process control and to replacement of existing 4–20 mA devices. Its communication functionalities, specifically foreseen for time critical applications, are supported by services grouped within levels, as in all other OSI-RM open system architectures. The number of levels is minimal, in order to guarantee maximum speed in data handling. Below the Fieldbus Application Layer, H1 FOUNDATION Fieldbus directly presents the Data Link Layer (DLL) managing access to the communication channel. A Physical Layer deals with the problem of interfacing with the physical medium. A Network and System Management Layer is also present. Figure 41.1 compares the H1 FOUNDATION Fieldbus architecture against the ISO OSI reference model.

- HSE FOUNDATION Fieldbus defines an Application Layer and associated management functions, designed to operate over a standard TCP/UDP/IP stack, over twisted-pair or fiber-optic switched Ethernet. It is mainly foreseen for discrete manufacturing applications, but of course, it can also be used to interconnect H1 segments, as well as foreign protocols through IP/TCP gateways, in order to build complete plant networks. Figure 41.2 compares the HSE FOUNDATION Fieldbus architecture against the ISO OSI reference model.

Looking at Figures 41.1 and 41.2, it becomes evident that the Fieldbus Foundation has specified a User Application Layer, significantly differentiating the solution from the ISO OSI model, which does not define such a layer. The H1 and HSE FOUNDATION Fieldbus User Application Layer is mainly based on Function Blocks, providing a consistent definition of inputs and outputs that allow seamless distribution and integration of functionality from various vendors [11].

H1 and HSE FOUNDATION **Fieldbus User Application Layer**

As above mentioned, the Fieldbus Foundation has defined a standard User Application Layer based on "Blocks." Blocks are representations of different types of application functions. The types of blocks used

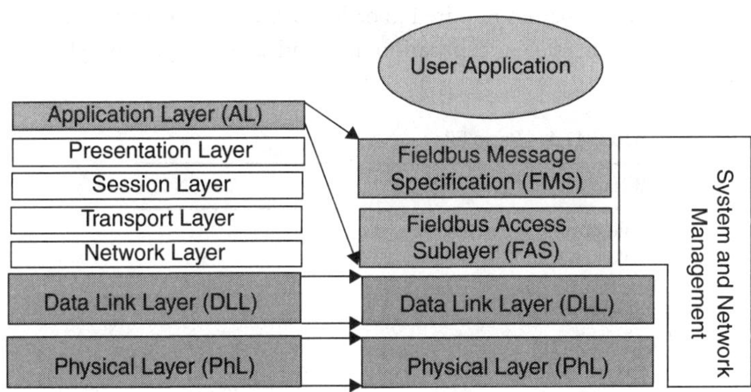

FIGURE 41.1   H1 FOUNDATION Fieldbus vs. ISO OSI Architecture.

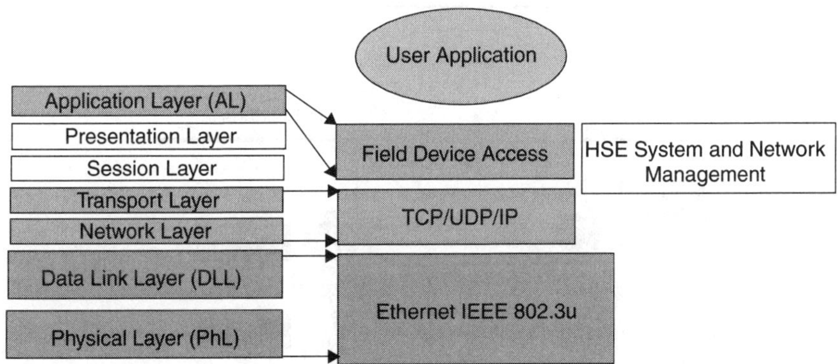

FIGURE 41.2   HSE FOUNDATION Fieldbus vs. ISO OSI Architecture.

in a User Application are Resource, Transducer and Function. Devices are configured by using Resource Blocks and Transducer Blocks. The control strategy is built by using Function Blocks (FBs) [11] instead.

Resource Block

The Resource Block describes characteristics of the fieldbus device such as the device's name, manufacturer, and serial number. There is only one Resource Block in a device.

Function Block

FBs provide the control system behavior. The input and output parameters of FBs running in different devices can be linked over the fieldbus. The execution of each FBs is precisely scheduled. There can be many FBs in a single User Application. The Fieldbus Foundation has defined sets of standard FBs. Ten standard FBs for basic control are defined by the specification in [12]. These blocks are summarized in Table 41.1. Other, more complex, standard FBs are defined by [13, 14].

The Flexible Function Block (FFB) is defined by [15]. An FFB is a user-defined block. FFB allows a manufacturer or user to define block parameters and algorithms to suit an application that interoperates with standard FBs and host systems.

FBs can be built into fieldbus devices as required in order to achieve the desired device functionality. For example, a simple temperature transmitter may contain an AI FB. A control valve might contain a PID FB as well as the expected AO block. Thus, a Complete Control Loop can be built using only a simple transmitter and a control valve (Figure 41.3).

Transducer Blocks

Like the Resource Block, the Transducer Blocks are used to configure devices. Transducer Blocks decouple FBs from the local I/O functionalities required in order to read sensors or to command the actuator's output. They contain information such as calibration date and sensor type [16,17].

TABLE 41.1   Basic FBs

FB Name	Symbol Name
Analog Input	AI
Analog Output	AO
Bias/Gain	BG
Control Selector	CS
Discrete Input	DI
Discrete Output	DO
Manual Loader	ML
Proportional/Derivative	PD
Proportional/Integral/Derivative	PID
Ratio	RA

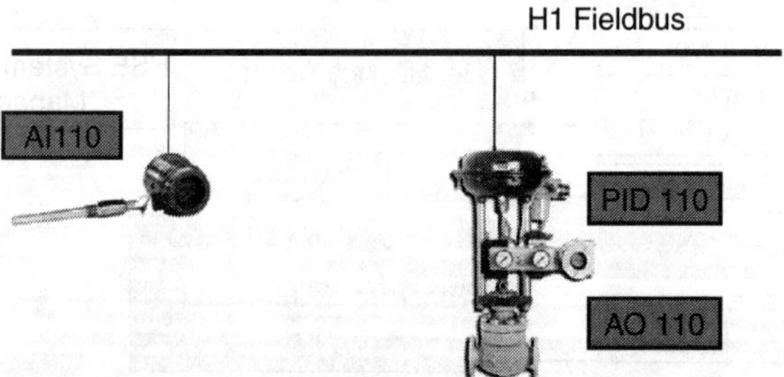

FIGURE 41.3   Example of a Complete Control Loop using FBs in FOUNDATION Fieldbus devices.

H1 Foundation Fieldbus

H1 FOUNDATION Fieldbus is made up of different layers (as shown in Figure 41.1), whose functionalities will be described in the following.

H1 FOUNDATION Fieldbus Physical Layer

The H1 FOUNDATION Fieldbus Physical Layer has been conceived to receive messages from the communication stack in order to convert them into physical signals on the fieldbus transmission medium and vice-versa [18]. Conversion tasks include the adding and removing of preambles, start delimiters, and end delimiters. The preamble is used by the receiver to synchronize its internal clock with the incoming fieldbus signal. The receiver uses the start delimiter to find the beginning of a fieldbus message. After finding the start delimiter, the receiver accepts data until the end delimiter is received.

The Physical Layer is defined by approved standards issued by the IEC and ISA. In particular, the FOUNDATION Fieldbus H1 Physical Layer is the 31.25Kbauds version of IEC(Type 1)/ISA Fieldbus [2,19]. Signals ($\pm$ 10 mA on 50 Ω load) are encoded using the synchronous Manchester Biphase-L technique and can be conveyed on low-cost twisted-pair cables. The signal is called "synchronous serial" because the clock information is embedded in the serial data stream. Data are combined with the clock signal while creating the fieldbus signal. The receiver of the fieldbus signal interprets a positive transition in the middle of a bit time as a logical "0" and a negative transition as a logical "1." Special codes are defined for the preamble, start delimiter, and end delimiter. Special N+ and N − characters are used in the start delimiter and end delimiter. Note that the N+ and N − signals do not have a transition in the middle of a bit time.

Fieldbus Signalling.
The transmitting device delivers ±10 mA at 31.25 kbit/sec into a 50 Ω equivalent load to create a 1.0 V peak-to-peak voltage modulated on top of the direct current (DC) supply voltage. The DC supply voltage can range from 9 to 32 V. The 31.25 kbit/sec fieldbus also supports Intrinsically Safe (IS) fieldbuses for bus-powered devices. To accomplish this, an IS barrier is placed between the power supply in the safe area and the IS device in the hazardous area. In this case (IS applications), the allowed power supply voltage depends on the barrier rating.

Fieldbus Wiring.
H1 FOUNDATION Fieldbus wiring is based on trunk cables featuring terminators installed at each end. H1 FOUNDATION Fieldbus allows for stubs or "spurs" located anywhere along the trunk, and connected to the trunk by a junction box, as shown by Figure 41.4. A single device can be connected by each spur. Existence of spurs allows that a 31.25 kbit/sec device can operate on wiring previously used for 4–20 mA devices [20,21].

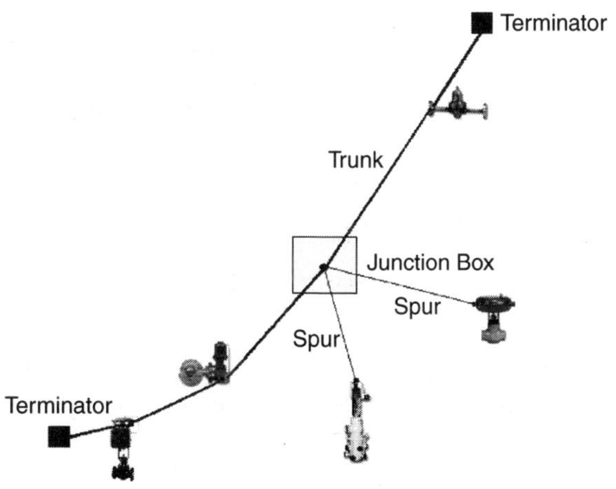

FIGURE 41.4 Trunk, junction box and spurs in H1 FOUNDATION Fieldbus.

More trunks can be connected in a fieldbus link by means of repeaters. Up to five trunks (by means of four repeaters) can be interconnected.

The spur length varies from 1 up to 120 m depending on the number of devices connected to the fieldbus link.

The maximum number of devices on a fieldbus link is 32; the actual number depends on factors such as the power consumption of each device, the type of cable, the use of repeaters, etc. In particular, the maximum number of devices is usually six for IS applications and power delivered through the bus, 12 for non-IS applications but power still delivered through the bus, and 32 for non-IS applications without any power delivered through the bus [22].

The total trunk length (including spurs) is 1900 m and the number of network addresses available for each link is 240.

H1 FOUNDATION Fieldbus DLL

H1 FOUNDATION Fieldbus DLL controls the transmission of messages onto the fieldbus. As already mentioned, FOUNDATION Fieldbus fully adopted the IEC(Type 1)/ISA DLL statement: "Both paradigms, circulated token and scheduled access were good, but insufficient at the same time; they were complementary, not alternative, and a complete fieldbus solution needs the two together" [3,23,24].

Thus, fieldbus needs a good mix of circulated token and scheduled access: well balanced to avoid losing bandwidth for the scheduled access when it is not really needed, but always giving priority to scheduled access over circulated token when a conflict arises.

That is what the IEC(Type 1)/ISA DLL documents propose, and FOUNDATION Fieldbus made real: an overall schedule able to guarantee the needed data at the needed time but also allowing gaps within which a circulated token mechanism can take place while complying with a defined maximum rotation time.

Such a philosophy clearly needs an arbitrator that univocally imposes the transmission of defined data at a defined time by a defined entity, when so required, but also guarantees a defined minimum amount of free time to each entity. This arbitrator is called Link Active Scheduler (LAS) within IEC(Type 1)/ISA and H1 FOUNDATION Fieldbus DLL [3,23,24]. Essentially, LAS performs:

- access to the physical medium on a scheduled basis;
- circulation of the token only when no scheduled traffic is needed; token is passed for a limited amount of time that is always shorter than the interval left before the next scheduled traffic;
- a policy for the management of the token, according to which the token is returned to the LAS instead of passing it onto a new node so that the LAS, depending on the time left, can decide whether to actually pass the token once more or to resume link control to manage the scheduled traffic.

Specific needs of the token mechanism such as

- giving enough contiguous token time to each node and
- guarantee of the most regular, as possible, token rotation time among all the nodes

are granted by the token method of IEC(Type 1)/ISA and H1 FOUNDATION Fieldbus DLL. Other needs such as

- keeping the token cycle short enough and
- satisfying the occurrence of high-priority events

are substituted by the scheduled access method of IEC(Type 1)/ISA and H1 FOUNDATION Fieldbus DLL.

Device Types.
Two types of devices are defined in the DLL specification: Basic Device and Link Master. Link Master devices are capable of becoming the LAS. Basic Devices do not have the capability to become the LAS.

Scheduled Communication.
The way in which the LAS manages the centralized government is based on the following mechanism. The LAS has a list of transmitting times for all data buffers in all devices that need to be periodically

transmitted. When it is time for a device to send the contents of a buffer, the LAS issues a CD message to the device.

Upon receipt of the CD, the device broadcasts or "publishes" the Data Item (DT) in the buffer to all devices on the fieldbus. Any device configured to receive the data is called a "subscriber". Figure 41.5 shows this access mechanism.

Scheduled data transfers are typically used for the regular, periodic transfer of control loop data between devices on the fieldbus.

Unscheduled Communication.

The federal autonomy of each node is given by a bandwidth distribution mechanism based on the use of a circulating token. In unused portions of the bandwidth (i.e., not occupied by the transmission of CDs), the LAS sends a Pass Token (PT) to each node included in a particular list called "Live List" (described in the following). Each token is associated with a maximum utilization interval, during which the receiving node can use the available bandwidth to transmit what it needs. On the expiration of the time interval or when the node completes its transmissions, the token is returned to the LAS by using another frame called Return Token (RT). A Target Token Rotation Time (TTRT) defines the interval time desired for each token rotation. The value to be assigned to this parameter is linked to the maximum admissible delay in the transmission of the asynchronous flow. Figure 41.6 shows the token circulation managed by the LAS.

Live List Maintenance.

The list of all devices that properly respond to PT is called the "Live List."

New devices may be added to the fieldbus at any time. The LAS periodically sends Probe Node (PN) messages to all the addresses not yet present in the Live List. If a new device appears at an address and receives the PN, it immediately returns a Probe Response (PR) message.

When a device returns a PR, the LAS adds the device to the Live List and confirms its addition by sending the device a Node Activation message.

The LAS is required to probe at least one address after it has completed a cycle of sending PTs to all the devices in the Live List.

The device will remain in the Live List as long as it responds properly to the PTs sent from the LAS.

The LAS will remove a device from the Live List if the device does not reply to the relevant PT for three successive tries.

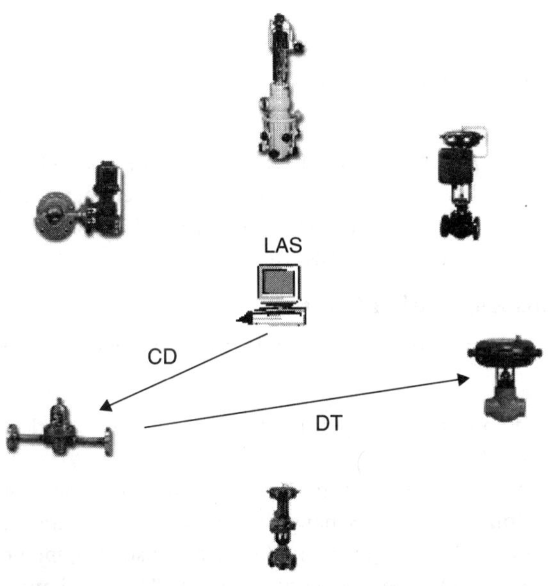

FIGURE 41.5   Centralized access mechanism.

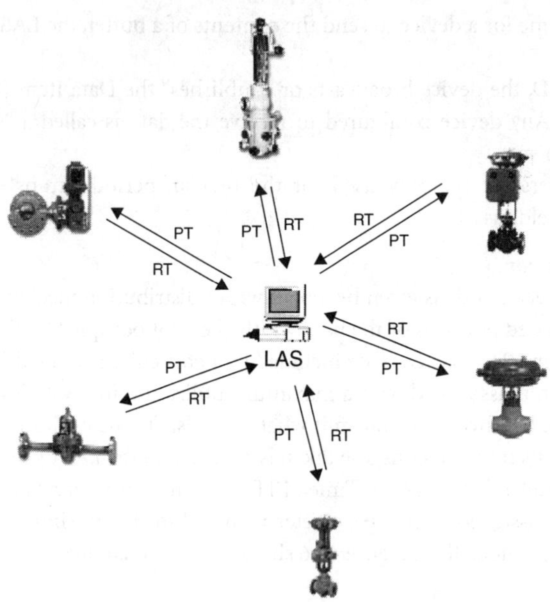

FIGURE 41.6   Token passing mechanism.

Whenever a device is added or removed from the Live List, the LAS broadcasts changes in the Live List to all devices; this allows each Link Master device to maintain a current copy of the Live List in order to be ready to become LAS if needed.

Data Link Time Synchronization.

A DLL time synchronization mechanism is provided so that any node can request the LAS for a scheduled action to be executed at a defined "time" that represents the same absolute instant for all the nodes.

The LAS periodically broadcasts a Time Distribution (TD) message on the fieldbus so that all devices have exactly the same data link time. This is important because scheduled communications on the fieldbus and scheduled FB executions in the User Application Layer are based on timing derived from these messages.

LAS Operation.

The algorithm used by the LAS is shown in Figure 41.7.

LAS Redundancy.

IEC(Type 1)/ISA and H1 FOUNDATION Fieldbus DDL provides the possibility to have more than one potential LAS on each link as well as back-up procedures that are essential for fieldbus availability. In particular, as a fieldbus may have multiple Link Masters, if the current LAS fails, one of the Link Masters will become the LAS and the operation of the fieldbus will continue.

H1 FOUNDATION Fieldbus Application Layer

The H1 FOUNDATION Fieldbus Application Layer includes two sublayers: FAS and FMS [25,26].

Fieldbus Access Sublayer (FAS)

FAS uses both the scheduled and unscheduled features of the DLL to provide services for the Fieldbus Message Specification (FMS) [25].

The type of each FAS service is described by Virtual Communication Relationships (VCR). VCR defines the kind of information (messages) exchanged between two applications. Possible features of the VCR may be the number of receivers (one or many) for each transmitter, memory organization (queue or buffer) used to store the messages to be sent/received, and the DLL mechanism used to send the message (PT or CD).

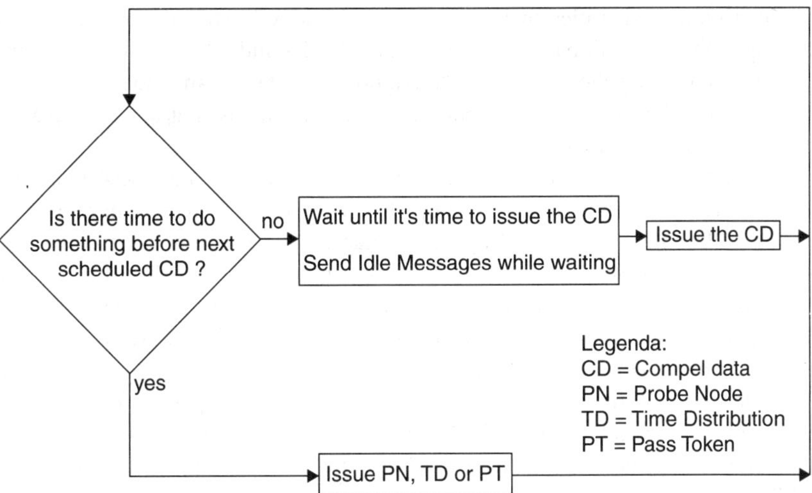

FIGURE 41.7　LAS Algorithm.

The types of VCR defined by Fieldbus Foundation are:

- *Client/Server VCR Type.* The Client/Server VCR Type is used for queued, unscheduled, user-initiated, one-to-one communication between devices on the fieldbus. Queued means that messages are sent and received in the order submitted for transmission, according to their priority, without overwriting previous messages. When a device receives a PT from the LAS, it may send a request message to another device on the Fieldbus. The requester is called the "Client" and the device that received the request is called the "Server." The Server sends the response when it receives a PT from the LAS. The Client/Server VCR Type is used for operator-initiated requests such as setpoint changes, access to and change of a tuning parameter, alarm acknowledgment, and device upload and download.
- *Report Distribution VCR Type.* The Report Distribution VCR Type is used for queued, unscheduled, user-initiated, one-to-many communications. When a device, holding an event or a trend report to send, receives a PT from the LAS, it sends its message to a "group address" defined by its VCR. Devices that are configured to listen for that VCR will receive the report. The Report Distribution VCR Type is normally used by fieldbus devices to send alarm notifications to the operator consoles.
- *Publisher/Subscriber VCR Type.* The Publisher/Subscriber VCR Type is used for buffered, one-to-many communications. Buffered means that only the latest version of the data is maintained within the network. New data completely overwrite previous data. When a device receives Compel Data (CD), the device will "Publish" (broadcast) its message to all devices on the fieldbus. Devices that wish to receive the Published message are called "Subscribers." CD may be scheduled in LAS, or it may be sent by Subscribers on an unscheduled basis. An attribute of the VCR indicates which method is used. The Publisher/Subscriber VCR Type is normally used by the field devices for scheduled publishing of User Application FB inputs and outputs.

Fieldbus Message Specification.

FMS services allow user applications to send messages to each other across the fieldbus by using a standard set of message formats. FMS describes the communication services, message formats, and protocol behavior needed to build messages for the User Application [26].

Data that are communicated over the fieldbus are described by an "object description." Object descriptions are collected together in a structure called an "Object Dictionary" (OD).

The object description is identified by its "index" in OD. Index 0, called the object dictionary header, provides a description of the dictionary itself, and defines the first index for the object descriptions of the User Application. The User Application object descriptions can start at any index above 255. Index 255 and below define standard data types such as boolean, integer, float, bitstring, and data structures that are used to build all other object descriptions.

A "Virtual Field Device" (VFD) is used to remotely view local device data described in the object dictionary. A typical device will have at least two VFDs: the Network and System Management VFD and the User Application VFD. The Network and System Management VFD provides access to the Network Management Information Base (NMIB) and to the System Management Information Base (SMIB). NMIB data include VCR, dynamic variables, statistics, and LAS schedules (if the device is a Link Master). SMIB data include device tag and address information, and schedules for FB execution.

The User Application VFD is used to make the device functions (the function of a fieldbus device is defined by the selection and interconnection of blocks) visible to the fieldbus communication system. The header of the User Application OD points to a Directory which is always the first entry in the FB application. The Directory provides the starting indexes of all of the other entries used in the FB application. The VFD object descriptions and their associated data are accessed remotely over the fieldbus network using VCR.

FMS communication services provide a standard way for user applications, such as FBs, to communicate over the fieldbus. Specific FMS communication services are defined for each object type. Table 41.2 summarizes the communication services available. Detailed descriptions for each service are provided in [26]. All of the FMS services use the Client/Server VCR Type except as noted (see notes 1 and 2 in Table 41.2).

H1 FOUNDATION Fieldbus System Management

Inside H1 FOUNDATION Fieldbus specification, System Management handles important system features [27] such as:

- *Scheduling.* FB must often be executed at precisely defined intervals and in the proper sequence for correct control system operation. System Management synchronizes execution of the FBs to a common time clock shared by all devices. A "macrocycle" is a single iteration of a schedule within a device. According to the type of device, we can have a LAS macrocycle and a device macrocycle. According to the first one, System Management can synchronize execution of the FBs across the entire fieldbus link. On the basis of the device macrocycle, instead, System Management can synchronize execution of FBs inside each device.

- *Application clock distribution.* This function allows publication of the time of day to all devices, including automatic switchover to a redundant time publisher. FOUNDATION Fieldbus supports an application clock distribution function. The application clock is usually set equal to the local time of day or to Universal Coordinated Time. System Management has a time publisher that periodically sends an application clock synchronization message to all fieldbus devices. The data link scheduling time is sampled and sent with the application clock message so that the receiving devices can adjust their local application time. During the intervals between synchronization messages, application clock time is independently maintained within each device relying on its own internal clock.

- *Device address assignment.* Fieldbus devices do not use jumpers or switches to configure addresses. Instead, device addresses are set by configuration tools using System Management services. Every fieldbus device must have a unique network address and physical device tag for the fieldbus to operate properly. To avoid the need for address switches on the devices, assignment of network addresses can be performed by configuration tools using System Management services. The sequence for assigning a network address to a new device is as follows:
  - An unconfigured device will join the network at one of four special temporary default addresses.
  - A configuration tool will assign a physical device tag to the new device using System Management services.

TABLE 41.2 Set of Services in FMS

Group	Service	Meaning	Note
Management and Environment Services	Initiate	Establish a Communication	
	Abort	Abort a Communication	
	Reject	Reject a nonvalid service	
	Status	Gives the status of a service	
	Unsolicited Status	Send a nonrequested status	
	Identify	Read a device specification (vendor, type and version)	
OD Services	GetOD	Read an	
	Initiate Put OD	Start loading an OD	
	PutOD	Load an OD in a device	
	Terminate Put OD	Stop loading an OD	
Variable Access Services	Read	Read a Variable	
	Write	Update the value of a Variable	
	Information Report	Send data	1
	Define Variable List	Define a variable list	
	Delete Variable List	Delete a variable list	
Event Services	Event Notification	Notify an Event	
	Acknowledge Event Notification	Acknowledge an event	2
	Alte rEvent Condition Monitoring	Enable or Disable an Event	
Downloading/Uploading Services	Request Domain Upload	Request of Domain Upload	
	Initiate Upload Sequence	Initiate Upload	
	Upload Segment	Up load data	
	Terminate Upload Sequence	End Upload	
	Request Domain Download	Request of Domain Download	
	Initiate Download Sequence	Initiate Download	
	Download Segment	Download data	
	Terminate Download Sequence	End Download	
Program Handling Services	Create Program Invocation	Create a Program Object	
	Delete Program Invocation	Delete a Program Object	
	Start	Start a Program	
	Stop	Stop a Program	
	Resume	Resume execution of a Program	
	Reset	Reset a Program	
	Kill	Kill a Program	

Note:
(1) Service can only use the Publisher/Subscriber or the Report Distribution VCR.
(2) Service can only use the Report Distribution VCR.

- A configuration tool will choose an unused permanent address and assign it to the device using System Management services.
- The sequence is repeated for all devices that enter the network at a default address.
- Devices store the physical device tag and node address in nonvolatile memory, so they will retain these settings also after a power failure.
- *Find Tag Service.* For the convenience of host systems and portable maintenance devices, System Management supports a service for finding devices or variables by a tag search. The "find tag query" message is broadcasted to all fieldbus devices. Upon receipt of the message, each device searches its VFD for the requested tag and returns complete path information (if the tag is found) including the network address, VFD number, VCR index, and OD index. Once the path is known, the host or maintenance device can access the data by its tag.

All of the configuration information needed by System Management such as the FB schedule is described by object descriptions in the Network and System Management VFD. This VFD provides access to the SMIB and also to the NMIB.

H1 FOUNDATION Fieldbus Network Management

H1 FOUNDATION Fieldbus Network Management mainly provides for the configuration of the communication stack [28].

HSE FOUNDATION Fieldbus

HSE FOUNDATION Fieldbus foresees the architecture depicted in Figure 41.2. As shown, its main feature is the use of Internet Architecture (full TCP/UDP/IP and IEEE 802.3u stack [29–31]) for high-speed discrete control and, more generally, for interconnecting several H1 segments in order to achieve a plant-wide fieldbus network [32].

Before describing the HSE FOUNDATION Fieldbus specifications, a brief overview of its general features will be given, with particular emphasis on the capability to interconnect different H1 FOUNDATION Fieldbus segments.

There are four basic HSE device categories (but several of them are typically combined into a single real device): linking device, Ethernet device, host device, and gateway device. A Linking Device (LD) connects H1 networks to the HSE network. An Ethernet Device (ED) may execute FBs and may have some conventional I/O. A Gateway Device (GD) interfaces other network protocols such as Modbus [33], DeviceNet [34], or Profibus [9]. A Host Device (HD) is a non-HSE device capable of communicating with HSE devices. Examples include configurators, operator workstations, and an OPC server.

The network in Figure 41.8 shows a Host System operating on an HSE bus segment labeled Segment A. Communications to H1 segments (B and C, as shown in the figure) are achieved by means of an Ethernet switch. The same switch is used to connect a second HSE segment (D) and a segment running a foreign protocol (segment E).

Any of the devices connected to the switch may attempt communication to any other device and it is the function of the switch to provide the correct routing and to negotiate transmission without collisions.

The connecting mechanism between HSE and H1 segments is performed by an LD. A typical LD will serve multiple H1 segments, although for simplicity only one segment per LD is shown in Figure 41.8. The connection between HSE and a foreign protocol is made through a GD. The capabilities of the interconnections shown in Figure 41.8 are as follows:

- *HSE Host/H1 segment.* The HSE Host interacts with a standard H1 device through an LD. In this situation, the HSE Host is able to configure, diagnose, and publish and subscribe data to or from the H1 device.
- *HSE Host/HSE segment.* The HSE Host interacts with an HSE device and is able to configure, diagnose, and publish and subscribe data to or from the HSE device.
- H1/H1 segment. In this situation, the interaction is between two H1 devices on two distinct H1 bus segments. The segments are connected to the Ethernet by LDs. Communications between devices on two H1 segments are functionally equivalent to communications between two H1 devices on the same bus segment. But it is clear that real-time communication between devices belonging to different H1 segments cannot be guaranteed due to the lack of a unique scheduler of the communication among different H1 segments.
- HSE Host/foreign protocol. This connection defines the relationship between a foreign device and the FOUNDATION Fieldbus application environment. The connection is made by GD. The foreign device is seen as a publisher to an HSE resident subscriber; the HSE host can handle the data stream from the I/O gateway in the same manner as it treats the data streams from devices on H1/HSE segments.

HSE FOUNDATION Fieldbus Physical, Data Link, Network and Transport Layer

As explained before, a higher speed physical layer specification was always intended for selected process applications and for factory (discrete parts) automation. The original high-speed solution, called H2, was based on H1 protocol and FB application running on different media at either 1 Mbit or 2.5 Mbit/ sec.

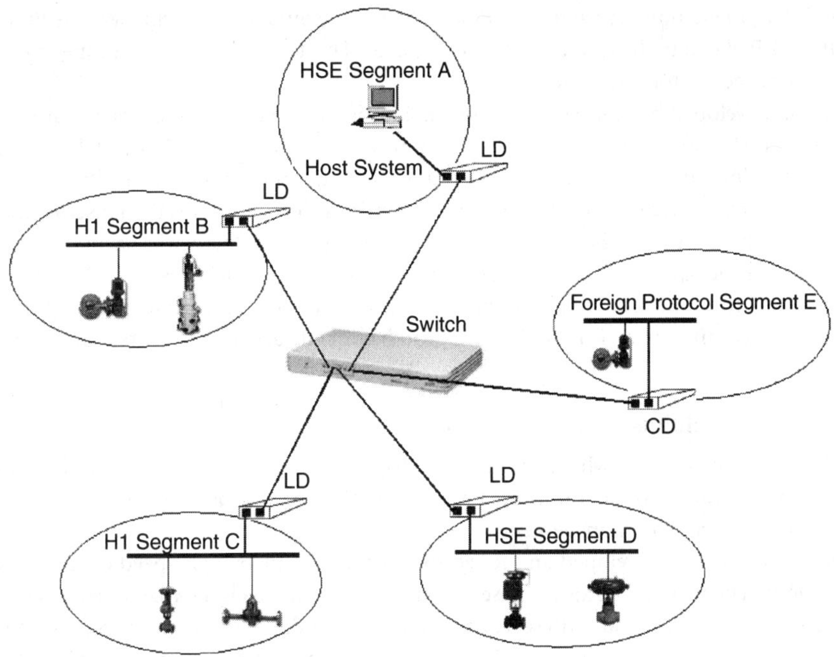

FIGURE 41.8   H1 and HSE FOUNDATION Fieldbus Interconnection.

In March 1998, the Foundation Board of Directors reconsidered the high-speed solution options, and terminated further work on H2. The new approach was based on Ethernet and was intended to make use, as much as possible, of commercially available, off-the-shelf technology (COTS) components and software.

The new solution, HSE, is designed to integrate multiple protocols, including multiple H1 FOUNDATION Fieldbus segments as well as foreign protocols, as described before.

For its high-speed Physical Layer version, Fieldbus Foundation has selected HSE at 100Mbaud. The specifications for the physical layer, as well as for the Ethernet DLL, are maintained by the Institute of Electrical and Electronic Engineers (IEEE) [30,31].

HSE also makes use of well-established internet protocols that are maintained by the Internet Architecture Board. These include Transport Control Protocol (TCP), Unit Datagram Protocol (UDP), and Internet Protocol (IP) [29].

Standard HSE stack components are Distributed Host Configuration Protocol (which assigns addresses), Simple Network Time Protocol (SNTP) and Simple Network Management Protocol (SNMP), which rely on TCP and UDP over IP and the IEEE 802.3 MAC and physical layers.

This has resulted in a practically unlimited number of nodes (IP-addressing) over a star topology network made up of as many links as required, the length of which can be up to 100 m for twisted pair and 2000 m on fiber.

Messages sent on Ethernet are bounded by a series of data fields called frames. The combination of a message and frame is called an Ethernet packet. Typically, a packet encoded according to the TCP/IP protocols will be inserted in the message field of the Ethernet packet.

FOUNDATION Fieldbus uses a similar data structure where messages are bounded by addressing and other data items. What corresponds to a packet in Ethernet is called a Protocol Data Unit (PDU) in FOUNDATION Fieldbus.

Let us consider a communication between two H1 devices over an interposed HSE segment, as illustrated in Figure 41.8. The easiest method for LD might be, upon receiving a communication from an H1 device, to simply insert the entire H1 PDU into the message part of the TCP/IP packet. Then LD on the destination H1 segment, upon receiving the Ethernet packet, would merely strip away the Ethernet frame and send the H1 PDU on to the receiving H1 bus segment. This technique is called tunneling, and is commonly used in mixed protocol networks.

The solution developed by HSE FOUNDATION Fieldbus is somewhat more complex, but more efficient than tunneling. HSE FOUNDATION Fieldbus PDU is inserted into the data field of a TCP/IP message field. However, the Fieldbus address is encoded as a unique TCP/IP address, so the Fieldbus PDU address is used to fill the address field of the TCP/IP packet. The entire TCP/IP packet is then inserted into the message field of the Ethernet packet. Because of the HSE encoding scheme, networks having multiple LDs can locate and transfer messages to the correct destination much more quickly, with far less extraneous bus traffic, as opposed to tunneling. Perhaps even more important, every H1 device (and every HSE device for that matter) has a unique TCP/IP address and can be directly accessed over standard IT and Internet networks.

HSE FOUNDATION Fieldbus Application Layer

Existing Fieldbus specifications, which have been widely tested in H1 applications and had already been maintained by the Fieldbus Foundation, have been used in the HSE standard too, where applicable. These include FMS and SM (System Management).

New specifications were developed and tested to provide complete high-speed communications and control solutions. The new technology is based on the FDA Agent (Field Device Access Agent) [35].

The FDA Agent allows SM and FMS services used by the H1 devices to be conveyed over the Ethernet using standard UDP and TCP. This allows HSE devices to communicate with H1 devices that are connected via an LD.

The FDA Agent is also used by the local FBs in an HSE device. Thus, the FDA Agent enables remote applications to access HSE devices and/or H1 devices through a common interface.

HSE FOUNDATION Fieldbus System Management

The following aspects of management are supported in the HSE System Management Layer [36]:

- Each device has a unique and permanent identity, and a system-specific configured name.
- Devices maintain version control information.
- Devices respond to requests aiming to locate objects, including the device itself.
- Time is distributed to all devices on the network.
- FB schedules are used to execute FBs.
- Devices are added and removed from the network without affecting other devices on the network.

HSE FOUNDATION Fieldbus Network Management

HSE Network Management allows HSE host systems to perform management operations over the HSE network [37]. The following capabilities are provided by the network management:

- Configuring the H1 Bridge, which performs data forwarding and republishing between H1 interfaces.
- Loading the HSE Session List, or single entries in this list. An HSE Session Endpoint represents a logical communication channel between two or more HSE devices.
- Loading the HSE VCR List, or single entries in this list. An HSE VCR is a communication relationship used for accessing VFDs across HSE.
- Performance Monitoring for Session Endpoints, HSE VCRs, and the H1 Bridge.
- Fault Detection Monitoring.

HSE FOUNDATION Fieldbus Redundancy

HSE FOUNDATION Fieldbus specification provides for the management of redundant network interfaces. This capability protects against single and multiple faults in the network. Each device monitors the network and selects the "best" route to destination for each message it has to send. HSE provides for various levels of redundancy up to and including complete device and media redundancy. HSE fault tolerance is achieved by operational transparency, that is, the redundancy operations are not visible to the HSE applications. This is necessary because HSE applications are required to coexist with standard Information Technology applications. The HSE LAN Redundancy Entity (LRE) coordinates the redundancy function. Each HSE device periodically transmits on both of its Ethernet interfaces a diagnostic message (representing its view of the network) to the other HSE devices. Each device uses the diagnostic messages to maintain a Network Status Table (NST), which is used for fault detection and transmission port selection. There is no central "Redundancy Manager." Instead, each device determines its behavior in response to the faults it detects [38].

Open Systems Implementation

One of the main features of FOUNDATION Fieldbus (in both H1 and HSE configurations) is the availability to build open communication systems. It is clear that this represents a key issue in a communication system as building up a perfect communication stack between two devices is completely useless if these two devices are not able to understand the meaning of each other's data, nor each other's behavior.

Implementation of open systems is achieved through the use of FBs and the adoption of a standard way to represent them (the Device Description Language [DDL]). FOUNDATION Fieldbus has defined a set of standard FBs Blocks that can be combined and parameterized to build up a device [11]. Due to the standard format, behavior, and connection of such FBs, their access and use through the bus is immediate, allowing to achieve interoperability and interchangeability.

Further, a manufacturer can improve and innovate on an existing FB, creating again a new standard FB.

The solution adopted inside FOUNDATION Fieldbus to realize this has been the DDL able to provide a formal Device Description (DD), which can then be interpreted by the DD Services Library available through FOUNDATION Fieldbus [39–41]. Such DD acts as a "driver" for each specific device and is supplied together with the device itself.

Within each DD, and for each FB included in the device, a hierarchy of definitions is followed: first, the "universal parameters" of the device itself; second, the common parameters of each FB; third, the common parameters of the Transducers Blocks; and fourth, the parameters specific of the manufacturer.

DD may also include small programs able to interoperate with the device (e.g., for its calibration) as well as download capability for managing manufacturer upgrading.

41.4 Conclusions

Fieldbus Foundation so far appears as the only multisupplier organization able to achieve concrete results in proposing a large-scale fieldbus solution merging the initial FIP/Profibus proposals.

This is mainly due to its features providing true device interoperability and a combination of guaranteed scheduling and token rotation.

References

[1] www.fieldbus.org
[2] IEC 61158, Digital Data Communications for Measurements and Control — Fieldbus for use in Industrial Control Systems — Part 2: Physical Layer Specification, 2001.
[3] IEC 61158, Digital Data Communications for Measurements and Control — Fieldbus for use in Industrial Control Systems — Parts 3 & 4: Data Link Layer Service & Protocol Definition, 2001.
[4] IEC 61158, Digital Data Communications for Measurements and Control — Fieldbus for use in Industrial Control Systems — Parts 5 & 6: Application Layer Service & Protocol Definition, 2001.
[5] IEC 61784, Profile Sets for Continuous and Discrete Manufacturing Relative to Fieldbus use in Industrial Control Systems, 2001.

[6] Decotignie, J.D. and P. Pleinevaux, Time critical communication networks: field buses, *IEEE Network*, 2, 55–63, 1998.

[7] Decotignie, J.D. and P. Pleinevaux, A survey on industrial communication networks, Annales des Telecommunications, 48, 9–10, 1993.

[8] www.worldfip.org.

[9] www.profibus.org.

[10] CENELEC EN50170/A1, General Purpose Field Communication System, Addendum A1: FOUNDATION Fieldbus, 2000.

[11] Fieldbus Foundation FF-890, Function Block Application Process — Part 1.

[12] Fieldbus Foundation FF-891, Function Block Application Process — Part 2.

[13] Fieldbus Foundation FF-892, Function Block Application Process — Part 3.

[14] Fieldbus Foundation FF-893, Function Block Application Process — Part 4.

[15] Fieldbus Foundation FF-894, Function Block Application Process — Part 5.

[16] Fieldbus Foundation FF-902, Transducer Block Application Process — Part 1.

[17] Fieldbus Foundation FF-903, Transducer Block Application Process — Part 2.

[18] Fieldbus Foundation FF-816, 31.25 kbit/s Physical Layer Profile Specification.

[19] ISA S50.02, Physical Layer Standard, 1992.

[20] Fieldbus Foundation AG-140, 31.25 kbit/s Wiring and Installation Guide.

[21] Fieldbus Foundation AG-165, Fieldbus Installation and Planning Guide.

[22] Fieldbus Foundation AG-163, 31.25 kbit/s Intrinsically Safe Systems Application Guide.

[23] Fieldbus Foundation FF-821, Data Link Services Subset.

[24] Fieldbus Foundation FF-822, Data Link Protocol Subset.

[25] Fieldbus Foundation FF-875, Fieldbus Access Sublayer.

[26] Fieldbus Foundation FF-870, Fieldbus Message Specification.

[27] Fieldbus Foundation FF-800, System Management Specification.

[28] Fieldbus Foundation FF-801, Network Management.

[29] Comer, Douglas E., *Internetworking with TCP/IP, Vol. I: Principles, Protocols, and Architecture*, Prentice-Hall International, Englewood chiffs, NJ, 1999.

[30] ANSI/IEEE Std 802.3, IEEE Standards for Local Area Networks: CSMA/CD Access Method and Physical Layer Specifications, 1985.

[31] ANSI/IEEE Std 802.3u, IEEE Standards for Local Area Networks: Supplement to CSMA/CD Access Method and Physical Layer Specifications. MAC Parameters, Physical Layer, MAUs, and Repeater for 100 Mb/s Operation, Type 100BASE-T, 1995.

[32] Fieldbus Foundation FF-581, System Architecture.

[33] www.modbus.org.

[34] CENELEC EN 50325 - 2, DeviceNet.

[35] Fieldbus Foundation FF-588, HSE Field Device Access Agent.

[36] Fieldbus Foundation FF-589, HSE System Management.

[37] Fieldbus Foundation FF-803, HSE Network Management.

[38] Fieldbus Foundation FF-593, HSE Redundancy.

[39] Fieldbus Foundation FD-900, Device Description Language Specification.

[40] Fieldbus Foundation FD-110, DDS User's Guide.

[41] Fieldbus Foundation FD-100, DDL Tokenizer User's Manual.

42

Operating Principles and Features of CAN Networks

Gianluca Cena and Adriano Valenzano

IEIIT-CNR - Istituto di Elettronica e di Ingegneria dell'Informazione e delle Telecomunicazioni

42.1   Introduction

The history of Controller Area Network (CAN) started more than 20 years ago. At the beginning of the 1980s, a group of engineers at Bosch GmbH were looking for a serial bus system suitable for use in passenger cars. The most popular solutions adopted at that time were considered to be inadequate for the needs of most automotive applications. The bus system, in fact, had to provide a number of new features that could hardly be found in the already existing fieldbus architectures. The design of the new proposal also involved several academics and had the support of Intel, as the potential main semiconductor producer.

The new fieldbus was presented officially in 1986 with the name of *"Automotive Serial Controller Area Network"* at the SAE congress held in Detroit. It was based on a multimaster access scheme to the shared medium that resembled the well-known carrier-sense-multiple-access (CSMA) approach. The peculiar aspect, however, was that CAN adopted a new distributed nondestructive arbitration mechanism to solve contentions on the bus by means of priorities implicitly assigned to the colliding messages. Moreover, the fieldbus specifications also included a number of error detection and management mechanisms to enhance the fault-tolerance of the whole system.

In the following years, both Intel and Philips started to produce controller chips for CAN following two different philosophies. The Intel solution (often referred to as FullCAN in the literature) required less host CPU power, since most of the communication and network management functions were carried out directly by the network controller. Instead, the Philips solution (BasicCAN) was simpler but imposed a higher load on the processor used to interface the CAN controller. Since the mid-1990s, more than 15 semiconductor vendors, including Siemens, Motorola, and NEC, have been producing and shipping millions of CAN chips mainly to car manufacturers such as Mercedes-Benz, Volvo, Saab, Volkswagen, BMW, Renault, and Fiat.

The Bosch specification (CAN version 2.0) was submitted for international standardization at the beginning of the 1990s. The proposal was then approved and published as the ISO 11989 standard [IS898] at the end of 1993 and contained the description of the network access protocol and the physical layer architecture. In 1995, an addendum to ISO 11898 was approved to describe the extended format for the message identifiers also used in CAN. The CAN specification is currently in the process of being revised and reorganized, and has been split into four separate parts: [ISO1, ISO2, ISO3, TTC].

The standard documents provided satisfactory specifications for the lower communication layers but did not offer guidelines or recommendations for the upper part of the OSI protocol stack, in general, and for the application layer in particular. This is why the earlier applications of CAN outside the automotive scenario (i.e., textile machines, medical systems, and so on) adopted "*ad hoc*" monolithic solutions. The "CAN in Automation" (CiA) users' group, founded in 1992, was originally concerned with the specification of a "standard" CAN application layer. This effort led to the development of the general-purpose "CAN Application Layer" (CAL) specification. CAL was intended to fill the gap between the distributed application processes and the underlying communication support; but in practice it was not successful, the main reason being that, as CAL is actually application-independent, each user had to develop a suitable profile based on CAL for his specific application field.

Even though it was conceived for vehicle applications, CAN began to be adopted in different scenarios, too. For instance, at the beginning of the 1990s, Allen-Bradley and Honeywell started a joint distributed control project based on CAN. Although the project was abandoned a few years later, Allen Bradley and Honeywell continued their works separately and focused on the higher protocol layers. The results of these activities were the Allen-Bradley DeviceNet solution and the Honeywell Smart Distributed System (SDS). For a number of reasons, SDS remained, in practice, an internal solution to Honeywell Microswitch, while DeviceNet was soon switched to "Open DeviceNet Vendor Association" and was widely adopted in a number of U.S. factory automation areas, thus becoming a serious competitor to widespread solutions such as Profibus-DP and Interbus.

Besides DeviceNet and SDS, other significant initiatives were focused on CAN and its application scenarios. CANopen was conceived in the framework of the European Esprit project ASPIC by a consortium led once again by Bosch GmbH. The purpose of CANopen was to define a profile based on CAL, which could support communications inside production cells. The original CANopen specifications were further refined by CiA and were released in 1995. Later, both CANopen and DeviceNet became European standards and are now widely used, especially in two different areas, that is, factory automation and machine-distributed controls.

42.2 CAN protocol basics

The CAN protocol architecture is structured according to the layered approach of the ISO OSI model. However, as in most of the currently existing networks conceived for use at the field level in the automated manufacturing environments, only a few layers have been considered in its protocol stack. This is to make implementations more efficient and inexpensive. Few protocol layers, in fact, imply reduced processing delays when receiving and transmitting messages and simpler communication software.

The CAN specification [ISO1], in particular, includes only the *physical* and the *data link* layers, as depicted in Figure 42.1. The physical layer is aimed at managing the effective transmission of data over the communication support, and tackles the mechanical, electrical, and functional aspects. Bit timing and synchronization, in particular, belong to this layer.

The data link layer, instead, is split into two separate sublayers: the *medium access control* (MAC) and the *logical link control* (LLC). The purpose of the MAC entity is basically to manage the access to the shared transmission support by providing a mechanism aimed at coordinating the use of the bus, so as to avoid unmanageable collisions. The functions of the MAC sublayer include frame encoding and decoding, arbitration, error checking and signaling, and also fault confinement. The LLC sublayer, instead, offers the user (i.e. the application programs running in the upper layers) a proper interface, which is

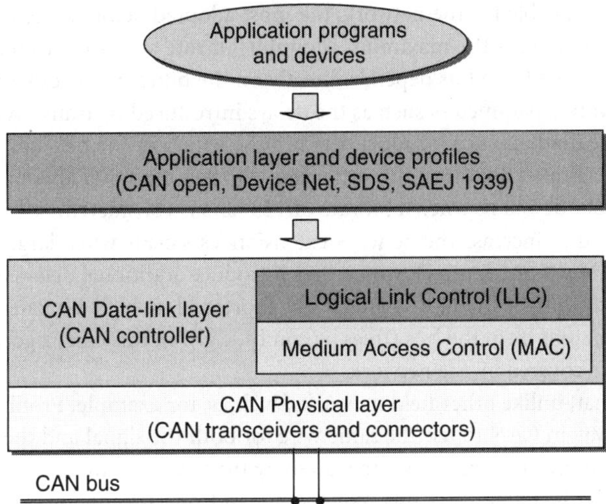

FIGURE 42.1   CAN protocol stack.

characterized by a well-defined set of communication services, in addition to the ability to decide whether an incoming message is relevant or not to the node.

It is worth noting that in CAN, there is considerable flexibility both in terms of the implementation of the LLC services and also in terms of the choice of the physical support, whereas there can be no modifications to the behavior of the MAC sublayer.

Unlike most fieldbus networks, the CAN specification does not include any native application layer. However, a number of such protocols exist that rely on CAN and ease the design and the implementation of complex CAN systems.

Physical layer

The features of the physical layer of CAN that are valid for any system, such as those related to the physical signaling, are described in the ISO 11898-1 document [ISO1]. Instead, the medium access units (i.e. the transceivers) are defined in two separate documents, namely ISO 11898-2 [ISO2] and ISO 11898-3 [ISO3] for high-speed and low-speed communications, respectively. The definition of the medium interface (i.e. the connectors) is usually covered in other documents.

Network topology

CAN networks are based on a shared bus topology. Buses have to be terminated at each end with resistors (the recommended nominal impedance is 120 Ω), so as to suppress signal reflections. For the same reason, the standard documents state that the topology of a CAN network should be as close as possible to a single line. Stubs are permitted for connecting devices to the bus, but their length should be as short as possible. For example, at 1 Mb/s the length of a stub must be shorter than 30 cm.

Several kinds of transmission media can be used:

- Two-wire bus, which enables differential signal transmissions and ensures reliable communications. In this case, a shielded twisted pair can be used to further enhance the immunity to electromagnetic interferences.
- Single-wire bus, a simpler and cheaper solution that features lower immunity to interferences and is mainly suitable for use in automotive applications.
- Optical transmission medium, which ensures complete immunity to electromagnetic noise and can be used in hazardous environments. Fiber optics are often adopted to interconnect (through repeaters) different CAN subnetworks. This is done to cover plants that are spread over a large area.

Several bit rates are available for the network, the most adopted being in the range from 50 kb/s to 1 Mb/s (the latter value represents the maximum allowable bit rate according to the CAN specifications). The maximum length of the CAN bus depends directly on the bit rate. The exact relation between these two quantities also involves parameters such as the delays introduced by transceivers and opto-couplers. Generally speaking, the mathematical product between the length of the bus and the bit rate is approximately constant. For example, the maximum extension allowed for a 500 kb/s network is about 100 m, and it increases up to about 500 m when a bit rate of 125 kb/s is considered.

Repeaters can be used to increase the network extension, especially when large plants have to be covered and the bit rate is low to medium. However, they introduce additional delays on the communication paths; hence, the maximum bus length is effectively shortened at high bit rates. Using repeaters also achieves topologies different from the bus (trees or combs, e.g.). In this case, a good design may increase the effective area that is covered by the network.

It is worth noting that, unlike other field networks, such as, for example, Profibus-PA, there is in general no cost-effective way in CAN to use the same wire for both the signal and the power supply.

Curiously enough, connectors are not standardized by the CAN specification. Instead, several higher-level application standards define their own connectors and pin assignment. CiA DS102 [DS102], for example, foresees the use of a SUB-D9 connector, while DeviceNet and CANopen also suggest the use of either 5-pin mini-style, micro-style, or open-style connectors.

Bit encoding and synchronization

In CAN, the electrical interface of a node to the bus is based on an open-collector-like scheme. As a consequence, the level on the bus can assume two complementary values, which are denoted symbolically as *dominant* and *recessive*. Usually, the dominant level corresponds to the logical value 0, while the recessive level coincides with the logical value 1.

CAN relies on the *non-return to zero* (NRZ) bit encoding, which features very high efficiency in that synchronization information is not encoded separately from the data. Bit synchronization in each node is achieved by means of a *digital-phase locked loop* (DPLL), which extracts the timing information directly from the bit stream received from the bus. In particular, the edges of the signal are used for synchronizing the local clocks, so as to compensate tolerances and drifts of the oscillators.

To provide a satisfactory degree of synchronization among the nodes, the transmitted bit stream should include a sufficient number of edges. To do this, CAN relies on the so-called *bit stuffing* technique. In practice, whenever five consecutive bits at the same value (either dominant or recessive) appear in the transmitted bit stream, the transmitting node inserts one additional stuff bit at the complementary value, as depicted in Figure 42.2. These stuff bits can be easily and safely removed by the receiving nodes, so as to obtain the original stream of bits back.

From a theoretical point of view, the maximum number of stuff bits that may be added is one every 4 bits in the original frame; hence, the encoding efficiency can be as low as 80% (see e.g., the rightmost part of Figure 42.2, where the original bit stream alternates sequences of four consecutive bits at the dominant level followed by four bits at the recessive level). However, the influence of bit stuffing in real operating conditions is noticeably lower than the theoretical value computed above. Simulations show that, on average, only 2 to 4 stuff bits are effectively added to each frame, depending on the size of the identifier and data fields.

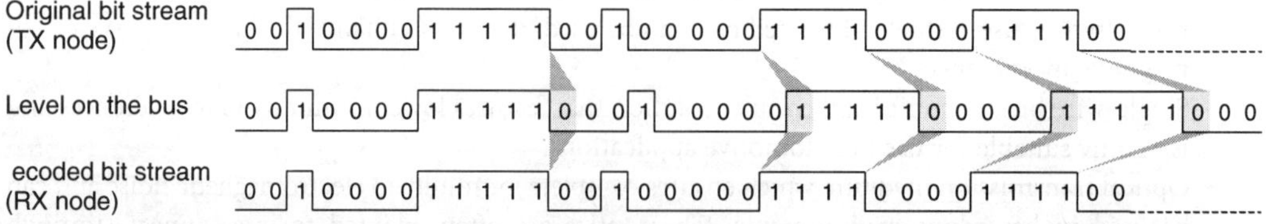

FIGURE 42.2 Bit stuffing technique.

Not all the fields in a CAN frame are encoded according to the bit stuffing mechanism: in particular, it applies only to the parts of the frames from the SOF bit up to the CRC sequence. The remaining fields, instead, are of fixed form and are not stuffed.

Frame format

The CAN specification [ISO1] defines both a standard and an extended frame format. These formats mainly differ in terms of the size of the identifier field and in terms of some other bits in the arbitration field. In particular, the standard frame format (which is also known as CAN 2.0A format) defines an 11-bit identifier field, which means that up to 2048 different identifiers are available to the applications executing in the same network (several older CAN controllers, however, only support identifiers in the range from 0 to 2031). The extended frame format (identified as CAN 2.0B), instead, assigns 29 bits to the identifier, so that up to half a billion different objects could exist (in theory) in the same network. This is a fairly high value, which is virtually sufficient for any kind of application.

Using extended identifiers in a network that also includes 2.0A compliant CAN controllers usually leads to unmanageable transmission errors, which effectively make the network unstable. Thus, a third category of CAN controllers was developed, known as 2.0B passive: they manage in a correct way the transmission and the reception of CAN 2.0A frames, while CAN 2.0B frames are simply ignored.

It is worth noting that, in most practical cases, the number of different objects allowed by the standard frame format is more than adequate. Since standard CAN frames are shorter than the extended ones (because of the shorter arbitration field), they permit a higher communication efficiency (unless part of the payload is moved into the arbitration field). As a consequence, they are adopted in most of the existing CAN systems, and also most of the CAN-based higher layer protocols, such as CANopen and DeviceNet, basically rely on this format.

The CAN protocol foresees only four kinds of frames: *data, remote, error,* and *overload* frames. Their format is described in detail in the following text.

Data frame

Data frames are used to send information over the network. Each data frame in CAN begins with a *start of frame* (SOF) bit at the dominant level, as shown in Figure 42.3. Its role is to mark the beginning of the frame, as in serial transmissions carried out by means of conventional UARTs. The SOF bit is also used to synchronize the receiving nodes.

Immediately after the SOF bit, there is the arbitration field, which includes both the *identifier* and the *remote transmission request* (RTR) bit. As the name suggests, the identifier field identifies-uniquely on the whole network-the content of the frame that is being exchanged. The identifier is also used by the MAC sublayer to detect and manage the priority of the frame, which is used whenever a collision occurs (the lower the numerical value of the identifier, the higher the priority of the frame).

a) Standard frame format

b) Extended frame format

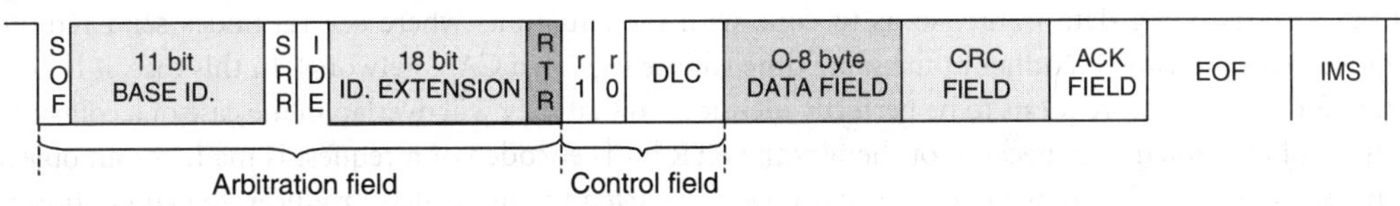

FIGURE 42.3   Format of data frames.

The identifier is sent starting from the most up to the least significant bit. The size of the identifier is different for the standard and the extended frames. In the latter case, the identifier has been split into an 11-bit base identifier and an 18-bit extended identifier, so as to provide compatibility with the standard frame format.

The RTR bit is used to discriminate between data and remote frames. Since a dominant value of RTR denotes a data frame while a recessive value stands for a remote frame, the data frames have a higher priority than remote frames, if the same identifier is considered.

Next to the arbitration field comes the control field. In the case of standard frames, it includes the *identifier extension* (IDE) bit, which discriminates between standard and extended frames, followed by the reserved bit r0. Instead, in the extended frames the IDE bit effectively belongs to the arbitration field, as well as the *substitute remote request* (SRR) bit — a placeholder that is sent at recessive value to preserve the structure of the frames. In this case, the IDE bit is followed by the identifier extension and then by the two reserved bits r1 and r0. After the reserved bits, there is the *data length code* (DLC), which specifies — encoded on 4 bits — the length (in bytes) of the data field. Since the IDE bit is dominant in the standard frames while it is recessive in the extended ones, when the same base identifier is considered standard frames have precedence over extended frames.

Reserved bits r0 and r1 must be sent by the transmitting node at the dominant value. Receivers, however, will ignore the value of these bits. For the DLC field, values ranging from 0 to 8 are allowed. Higher values (from 9 to 15) can be used for application-specific purposes. In this case, however, the length of the data field is meant to be 8. The data field is used to store the effective payload of the frame. In order to ensure a high degree of responsiveness and minimize the priority inversion phenomenon, the size of the data field is limited to 8 bytes at most.

After the data field, there are the CRC and the acknowledgement fields. The former field is composed of a *cyclic redundancy check* sequence encoded on 15 bits, which is followed by a CRC delimiter at the recessive value. The kind of CRC adopted in CAN is particularly suitable to cover short frames (i.e. counting less than 127 bits). The acknowledgement field, instead, is composed of two bits, that is to say the ACK slot followed by the ACK delimiter. Both of them are sent at the recessive level by the transmitter. The ACK slot, however, is overwritten with a dominant value by each node that has received the frame correctly (i.e. no error was detected up to the ACK field). It is worth noting that, in this way, the ACK slot is actually surrounded by two bits at the recessive level, that is to say the CRC and ACK delimiters. By means of the ACK bit, the transmitting node is enabled to discover whether at least one node in the network has received its frame correctly.

At the end of the frame, there is the *end of frame* (EOF) field, composed of seven recessive bits, which notifies the end of an error-free transmission to the nodes. In particular, the transmitting node assumes that the frame has been exchanged correctly if no error is detected until the last bit of the EOF field, while in the case of receivers the frame is valid if there are no errors until the sixth bit of EOF.

Different frames are interleaved by the *intermission* (IMS), which consists of three recessive bits and effectively separates consecutive frames exchanged on the bus.

Remote frames

Remote frames are very similar to data frames. The only difference is that they carry no data (i.e. the data field is not present in this case). They are used to request that a given frame be sent on the network by a remote node. It is worth noting that the requesting node does not know who is the producer of the related information. It is up to the receivers to discover the one that has to reply.

The DLC field in remote frames is not used by the CAN protocol. However, it is set to the same value as the corresponding data frame, so as to cope with the situations where several nodes send remote requests with the same identifier at the same time (this is legal in a CAN network). In this case, it is necessary for the different requests to be perfectly identical, so that they will overlap in the case of a collision.

It should be noted that, because of the way the RTR bit is encoded, if a request is made for an object at the same time as the transmission of that object is started by the related producer, the contention is resolved in favor of the data frame.

Error frames

Error frames are used to notify the nodes in the network that an error has occurred. They consist of two fields: *error flag* and *error delimiter*. There are two kinds of error flag: the active error flag is composed of six dominant bits, while the passive error flag consists of six recessive bits. An active error flag violates the bit stuffing rules or the fixed-format parts of the frame that is currently being exchanged; hence, it enforces an error condition that is detected by all the other stations connected to the network. Each node that detects an error condition transmits an error flag on its own. In this way, following the transmission of an error flag, there can be from 6 to 12 dominant bits on the bus.

The error delimiter, instead, is composed of eight recessive bits. After the transmission of an error flag, each node starts sending recessive bits and, at the same time, it monitors the bus level until a recessive bit is detected. At this point, the node sends seven more recessive bits, thus completing the error delimiter.

Overload frames

Overload frames can be used by the slow receivers to slow down operations on the network. This is done by adding an extra delay between consecutive data and/or remote frames. Their format is very similar to the error frames. In particular, it is composed of an overload flag followed by an overload delimiter. It is worth noting that today's CAN controllers are very fast and so they make the overload frame almost useless.

Access technique

The medium access control mechanism on which CAN relies is basically the *carrier sense multiple access* (CSMA). When no frame is being exchanged, the network is idle and the level on the bus is recessive. Before transmitting a frame, the nodes have to observe the state of the network. If the network is idle, the frame transmission begins immediately; otherwise, the node must wait for the current frame transmission to end. Each frame starts with the SOF bit at the dominant level, which informs all the other nodes that the network has switched to the busy state.

Even though very unlikely, it may happen that two or more nodes start sending their frames exactly at the same time. This is actually possible, in that the propagation delays on the bus — even though very small — are anyway greater than zero. Thus, one node might start its transmission while the SOF bit of another frame is already traveling on the bus. In this case, a collision will occur. In CSMA networks that are based on collision detection, such as, for example, nonswitched Ethernet, this unavoidably leads to the corruption of all the frames involved, which means that they have to be retransmitted. The consequence is a waste of time and a net decrease of the available bandwidth. In high-load conditions, this may lead to congestion: when the number of collisions is so high that the net throughput on the Ethernet network falls below the arrival rate, the network becomes stalled.

Unlike Ethernet, CAN is able to resolve the contentions in a deterministic way, so that neither time nor bandwidth is wasted. Therefore, congestion conditions can no longer occur and the entire theoretical system bandwidth is effectively available for communications.

For the sake of truth, it should be stated that contentions in CAN occur more often than one may think. In fact, when a node that has a frame to transmit finds the bus busy or loses the contention, it waits for the end of the current frame exchange and, immediately after the intermission has elapsed, it starts transmitting. Here, the node may compete with other nodes for which, in the meanwhile, a transmission request has been issued. In this case, the different nodes synchronize on the falling edge of the first SOF bit that is sensed on the network.

This implies that the behavior of a CAN network is effectively that of a network-wide distributed transmission queue where messages are selected for transmission according to a priority order.

Bus arbitration

The most distinctive feature of the medium access technique of CAN is the ability to resolve in a deterministic way any collision that should occur on the bus. In turn, this is made possible by the arbitration mechanism, which effectively finds out the most urgent frame each time there is a contention for the bus.

The CAN arbitration scheme allows the collisions to be resolved by stopping the transmissions of all the frames involved except the one that is characterized by the highest priority (i.e., the lowest identifier). The arbitration technique exploits the peculiarities of the physical layer of CAN, which conceptually provides a wired-and connection scheme among all the nodes. In particular, the level on the bus is dominant if at least one node is sending a dominant bit or, vice versa, the level on the bus is recessive if all the nodes are transmitting recessive bits.

By means of the so-called *binary countdown* technique, each node — immediately following the SOF bit — transmits the message identifier serially on the bus, starting from the most significant bit. When transmitting, each node checks the level observed on the bus against the value of the bit that is being written out. If the node is transmitting a recessive value whereas the level on the bus is dominant, the node understands that it has lost the contention and withdraws immediately. In particular, it ceases transmitting and sets its output port to the recessive level so as not to interfere with the other contending nodes. At the same time, it switches to the receiving state so as to read the incoming (winning) frame.

The binary countdown techniques ensure that, in the case of a collision, all the nodes that are sending lower priority frames will abort their transmissions by the end of the arbitration field, except for the one that is sending the frame characterized by the highest priority (it should be noted that the winning node does not even realize that a collision has occurred). This implies that no two nodes in a CAN network can be transmitting messages related to the same object (that is to say, characterized by the same identifier) at the same time. If this is not the case, in fact, unmanageable collisions could take place, which, in turn, cause transmission errors. Because of the automatic retransmission feature of the CAN controllers, this will lead almost certainly to a burst of errors on the bus, until the stations involved are disconnected by the fault confinement mechanism. This implies that, in general, only one node can be the producer of each object.

One exception to this rule is given by the frames without a data field, such as, for example, the remote frames. In this case, should a collision occur among frames with the same identifier, they overlap perfectly and hence no collision effectively occurs. The same is also true for data frames that have a nonempty data field, provided that the content of this field is the same for all the frames sharing the same identifier. However, it makes no sense in general to send frames with a fixed data field.

All the nodes that lose the contention have to retry the transmission as soon as the exchange of the current (winning) frame ends. They will all try to send their frames again immediately after the intermission is read on the bus. Here, a new collision could take place that also involves the frames sent by the nodes for which a transmission request was issued while the bus was busy.

An example that shows the detailed behavior of the arbitration phase in CAN is outlined in Figure 42.4. Here, three nodes (that have been indicated symbolically as A, B, and C) start transmitting a frame at the same time (maybe at the end of the intermission following the previous frame exchange over the bus). As soon as a node understands it has lost the contention, it switches its output level to the recessive value, so

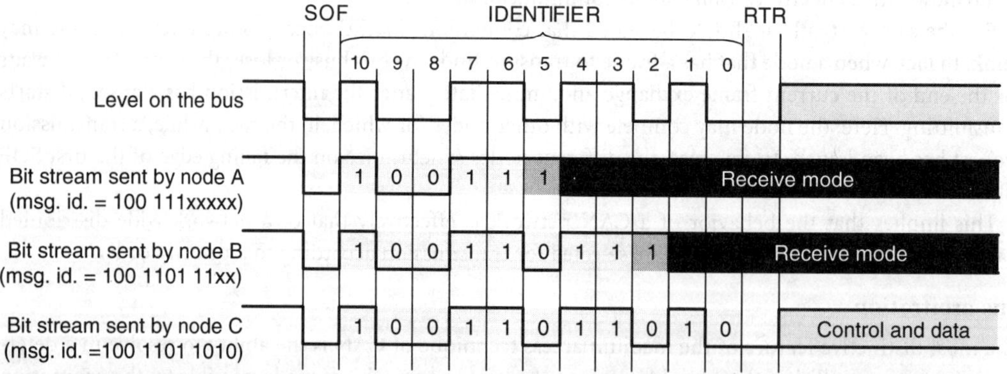

FIGURE 42.4 Arbitration phase in CAN.

that it no longer interferes with the other transmitting nodes. This event takes place when bit ID_5 is being sent for node A, while for node B this happens at bit ID_2. Node C manages to send the entire identifier field, and then it can continue transmitting the remaining part of the frame.

Error management

One of the main requirements that was fundamental in the definition of the CAN protocol was the need to have a communication system characterized by high robustness, that is, a system that is able to detect most of the transmission errors. Hence, particular care has been taken in defining error management.

The CAN specification foresees five different mechanisms to detect transmission errors:

1. *Cyclic redundancy check*: When transmitting a frame, the originating node adds a 15-bit-wide CRC to the end of the frame itself. Receiving nodes reevaluate the CRC to check if it matches the transmitted one. Generally speaking, the CRC used in CAN is able to discover up to five erroneous bits distributed arbitrarily in the frame or errors bursts including up to 15 bits.
2. *Frame check*: The fixed-format fields in the received frames can be easily tested against their expected value. For example, the CRC and ACK delimiter as well as the EOF field have to be at the recessive level. If one or more illegal bits are detected, a form error is generated.
3. *Acknowledgement check*: The transmitting node checks whether the ACK bit has been set to the dominant value in the received frame. On the contrary, an acknowledgement error is issued.
4. *Bit monitoring*: Each transmitting node compares the level on the bus against the value of the bit that is being written. Should a mismatch occur, an error is generated. This does not hold for the arbitration field or the acknowledgement slot. Such an error check is very effective to detect local errors that may occur in the transmitting nodes.
5. *Bit stuffing*: Each node verifies whether the bit stuffing rules have been violated in the portion of the frames from the SOF bit up to the CRC sequence. In the case when 6 bits of identical value are read from the bus, an error is generated.

The residual probability that a corrupted message is not detected in a CAN network — under realistic operating conditions — has been evaluated and is found to be about 4.7×10^{-11} times the frame error rate or less.

Fault confinement

To avoid a node that is not operating properly sending repeatedly corrupted frames, thus blocking the entire network, a fault confinement mechanism has been included in the CAN specification. The fault confinement unit supervises the correct operation of the related MAC sublayer and, should the node become defective, it disconnects that node from the bus.

The fault confinement mechanism has been conceived to discriminate, as long as it is possible, permanent failures from short disturbances that may cause burst of errors on the bus. According to this mechanism, each node can be in one of the three following states:

- Error active
- Error passive
- Bus off

Error-active and error-passive nodes take part in the communication in the same way. However, they react to the error conditions differently. They send active error flags in the former case and passive error flags in the latter. This is because an error-passive node has already experienced several errors, and hence it should avoid interfering with the network operations (a passive error flag, in fact, does not corrupt the ongoing frame exchange).

The fault confinement unit uses two counters to track the behavior of the node with respect to the transmission errors, and they are known as *transmission error count* (TEC) and *receive error count* (REC).

The rules by which TEC and REC are managed are actually quite complex. However, they can be summarized as follows: each time an error is detected, the counters are increased by a given amount, whereas successful exchanges decrease them by one. Furthermore, the amount of the increase for the nodes that first detected the error is higher than the nodes that simply replied to the error flag. In this way, it is very likely that the counters of the faulty nodes reach values higher than the nodes that are operating properly, even when sporadic errors due to electromagnetic noise are considered.

When counters exceed the first threshold (127), the node is switched to the error-passive state, so as to try not to affect the network. When a second threshold (255) is exceeded, instead, the node is switched to the bus-off state. At this point, it can no longer transmit any frame on the network, and it can be switched back to the error-active state only after it has been reset and reconfigured.

Communication services

According to the ISO specification [ISO1], the LLC sublayer of CAN provides two communication services only: L_DATA, which is used to broadcast the value of a specific object over the network, and L_REMOTE, which is used to ask for the value of a specific object to be broadcast by the related remote producer. From a practical point of view, these primitives are implemented directly in the H/W by all the currently available CAN controllers.

Model for information exchanges

Unlike most network protocols conceived for use in automated manufacturing environments (which rely on node addressing), CAN adopts object addressing. In other words, messages are not tagged with the address of the destination and/or originating nodes. Instead, each piece of information that is exchanged over the network (often referred to as an *object*) is assigned a unique identifier, which denotes unambiguously the meaning of the object itself in the whole system.

This fact has important consequences in the way communications are carried out in CAN. In fact, identifying the objects that are exchanged over the network according to their meaning rather than the node they are intended for implicitly allows multicasting and makes it very easy for the control applications to manage interactions among devices according to the producer/consumer paradigm.

The exchange of information in CAN takes place according to the three phases shown in Figure 42.5:

1. the producer of a given piece of information encodes and transmits the related frame on the bus (the arbitration technique will transparently resolve any contention that should occur);
2. because of the intrinsically broadcast nature of the bus, the frame is propagated all over the network, and every node reads its content in a local receive buffer;

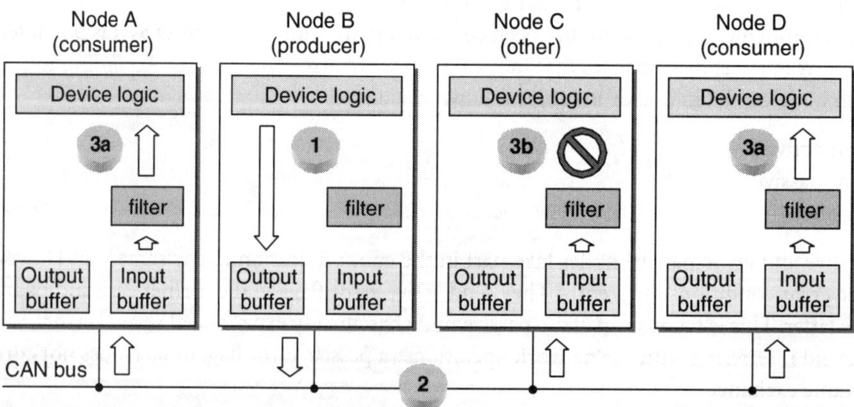

FIGURE 42.5   Producer/consumer model.

3. the *frame acceptance filtering* function (FAF) in each node determines whether or not the information is relevant to the node itself. In the former case, the frame is passed to the upper communication layers (from a practical point of view, this means that the CAN controller raises an interrupt to the local device logic, which will then read the value of the object); on the contrary, the frame is simply ignored and discarded.

In the sample data exchange depicted in Figure 42.5, node B is the producer of some kind of information, which is relevant to (i.e., consumed by) nodes A and D. Node C, instead, is not interested in such data; hence, they are rejected by the filtering function (this is the default behavior of the FAF function).

Model for device interaction

The access technique of CAN makes this kind of network particularly suitable to be used in distributed systems that communicate according to the producer/consumer model. In this case, data frames are used by the producer nodes to broadcast new values over the network, each of which is identified unambiguously by means of its identifier. Unlike the networks based on the producer/consumer/arbiter model such as FIP, information is sent in CAN as soon as it becomes available from either the control applications or the actual system (by means of sensors), without the need for the intervention of a centralized arbiter. This noticeably improves the responsiveness of the whole system.

CAN networks also work equally well when they are used to interconnect devices in systems based on a more conventional master–slave communication model. In this case, the master can use remote frames to ask for some specific information to be remotely sent on the network. The producer of that information, as a consequence of this frame, will reply with a data frame carrying the related object.

It is worth noting that this kind of interaction is implemented in CAN in a fairly more flexible way than in the conventional master–slave networks, such as, for example, Profibus-DP. In CAN, in fact, it is not necessary for the reply (data frame) to follow the request (remote frame) immediately. In other words, the network is not kept busy while the device is trying to send the reply. This allows the entire bandwidth to be theoretically available to the applications. Furthermore, the reply containing the requested value is broadcast on the whole network, and hence it can be read by all the interested nodes, in addition to the one that transmitted the remote request.

42.3 Main features of CAN

The medium access technique on which CAN relies basically implements a nonpreemptive-distributed priority-based communication system, where each node is enabled to compete directly for the bus ownership, so that it can send messages on its own (this means that CAN is a true multimaster system). This can be advantageous for use in event-driven systems.

Advantages

CAN is by far more simple and robust than the token-based access schemes (such as, e.g., Profibus when used in multimaster configurations). In fact, there is no need to build or maintain the logical ring, nor to manage the circulation of the token around the master stations. In the same way, it is noticeably more flexible than the solutions based on the *time division multiple access* (TDMA) or *combined-message* approaches — two techniques adopted by Sercos and Interbus, respectively. This is because message exchanges do not have to be known in advance.

When compared to schemes based on *centralized-polling* such as FIP, finally, it is not necessary to have a node in the network that acts as the bus arbiter, which can become a point of failure of the whole system. Since in CAN all the nodes are masters (at least from the point of view of the MAC mechanism), it is very simple for them to notify asynchronous events, such as, for example, alarms or critical error conditions. In all the cases where this aspect is important, CAN is clearly better than the above-cited solutions.

Thanks to the arbitration scheme, it is certain that no message will be delayed by lower priority exchanges (this phenomenon is known as *priority inversion*). Since the CAN protocol is not preemptive (as in the case for almost all of the existing protocols), a message can still be delayed by a lower priority one whose transmission has already started. This is unavoidable in any nonpreemptive system. However, as the frame size in CAN is very small (standard frames are 135-bit long at most, including stuff bits), the blocking time experienced by the very urgent messages is in general quite low. This makes CAN a very responsive network, which explains why it is used in many real-time control applications despite its relatively low bandwidth.

The above characteristics have to be considered carefully when assigning identifiers to the different objects that have to be exchanged in distributed real-time control applications. From an intuitive point of view, the most urgent messages (i.e., the messages characterized by the tightest deadlines) should be assigned the lowest identifiers (e.g., identifier 0 labels the message that has the highest priority in any CAN network).

If the period of cyclic data exchanges (and the *minimum interarrival time* of the acyclic ones) is known in advance, a number of techniques based on either the rate monotonic or deadline monotonic approaches have appeared in the literature [TIN], which can be used to find (if it exists) a feasible assignment of identifiers to the objects, so that the resulting schedule is feasible (i.e., the deadlines of all the objects are always respected).

Drawbacks

Even though inherently elegant, the arbitration technique of CAN poses serious limitations in terms of the performances that can be obtained by the network. In fact, in order for the arbitration mechanism to operate correctly, it is necessary for the signal to be able to propagate from a node located at one end of the bus up to the farthest node (at the other end) and come back to the first node before this one samples the level on the bus.

Since the sampling point is located roughly after the middle of each bit (the exact position can be programmed by means of suitable registers), the end-to-end propagation delay, also including the hardware delay of transceivers, must be shorter than about one quarter of the bit time (the exact value depending on the bit timing configuration in the CAN controller).

As the propagation speed of signals is fixed (about 200 m/μsec on copper wires), this implies that the maximum length allowed for the bus is necessarily limited, and depends directly on the bit rate chosen for the network. For example, a 250 kb/s CAN network can span at most 200 m. Similarly, the maximum bus length allowed when the bit rate is selected as equal to 1 Mb/s is 40 m only. This, to some degree, explains why the maximum bit rate allowed by the CAN specifications [ISO1] has been limited to 1 Mb/s.

It is worth noting that this limitation depends on physical factors, and hence it cannot be overcome in any way by advances in the technology of transceivers (to make a comparison, it should be mentioned that, at present, several inexpensive communication technologies are available on the market that allow bit rates in the order of tens or hundreds of Mb/s).

Even though this can appear to be a very limiting factor, it will probably not have any relevant impact in the near future for several application areas — including automotive and process control applications — for which a cheap and well-assessed technology is more important than performance. However, there is no doubt that CAN will suffer in a couple of years from the higher bit rates of its competitors, that is, Profibus-DP (up to 12 Mb/s), Sercos (up to 16 Mb/s), Interbus (up to 2 Mb/s), or the networks based on industrial Ethernet (up to 100 Mb/s). Such solutions, in fact, are able to provide a noticeably higher data rate, which is necessary for the systems that have several devices and very short cycle times (1 msec or less).

Among the possible solutions conceived to enhance the performance of CAN, there is the so-called *time triggered CAN* (TTCAN) protocol [TTC], for which the first chips are becoming available. By adopting a common clock and a time-triggered approach, it is possible to reduce jitters and provide a fully deterministic behavior. If asynchronous transmissions are not allowed in the system (which means that the arbitration technique is not actually used), TTCAN effectively behaves like a TDMA system and thus there is no limitation on the bit rate (which could be increased above the theoretical limit of CAN). However, the behavior of the resulting network is noticeably different from CAN.

Other solutions have appeared in the literature, such as, for example, WideCAN [WCAN], which provide higher bit rates and still rely on the conventional CAN arbitration technique. However, at present, their interest is mainly theoretical.

Implementation

According to the internal architecture, CAN controllers can be classified into two different categories: *BasicCAN* and *FullCAN*.

Conceptually, BasicCAN controllers are provided with one transmit and one receive buffer, as in conventional UARTs. The frame filtering function, in this case, is generally left to the application programs (i.e., it is under the control of the host controller), even though some kind of filtering can be done by the controller. To avoid overrun conditions, a double-buffering scheme based on shadow receive buffers is usually available, which permits a new frame to be received from the bus while the previous one is being read by the host controller. An example of a controller based on the BasicCAN scheme is given by the Philips PCA82C200.

FullCAN implementations, instead, foresee a number of internal buffers that can be configured to either receive or transmit some particular messages. In this case, the filtering function is implemented directly in the CAN controller. When a new frame that is of interest for the node is received from the network, it is stored in the related buffer, where it can then be read by the host controller. In general, new values simply overwrite the previous ones, and this does not lead to an overrun condition (the old value of a variable is superseded by a newer one). The Intel 82526 and 82527 CAN controllers are based on the FullCAN architecture.

FullCAN controllers, in general, free the host controller of a number of activities; hence, they are considered to be more powerful than BasicCAN controllers. It is worth noting, however, that the most recent CAN controllers embed the operating principles of both the above architectures; thus, the above classification is actually in the process of being superseded.

42.4 CAN-based application protocols

To reduce the costs of designing and implementing automated systems, a number of higher-level application protocols have been defined in the past few years, which rely on the CAN data link layer to exchange messages among the nodes (it is worth noting that all the functions of the data link layer of CAN are implemented directly in hardware in the current CAN controllers, which increases the efficiency and reliability of the data exchanges). The aim of such protocols is to provide a usable and well-defined set of service primitives, which can be used to interact in a standardized manner with the field devices.

At present, two of the most widely available solutions for the process control and automated manufacturing environments are CANopen [COP] and DeviceNet [DNET]. Both of them define an object model that describes the behavior of devices. This permits interoperability and interchangeability among devices coming from different manufacturers. In fact, as long as a device conforms to a given profile, it can be used in place of any other device (of a different brand) that adheres to the same profile.

CANopen

CANopen was originally conceived to rely on the communication services provided by the CAN Application layer (CAL). However, the latest specifications [DS301] no longer refer explicitly to CAL. Instead, the relevant communication services have been embedded directly in the CANopen documents.

In CANopen, information is exchanged by means of *communication objects* (COBs). A number of different COBs are foreseen, which are aimed at different functions:

- *process data objects* (PDOs), used for real-time exchanges, such as, for example, information used to control the physical system;

- *service data objects* (SDOs), used for non-real-time communications (i.e., parameterization);
- *emergency objects* (EMCY), used by devices to notify the control application that some error condition has occurred;
- *synchronization object* (SYNC), used to achieve synchronized and coordinated operations in the system, and so on.

Even though the CAN access technique foresees that every node is a master, CANopen systems often rely on a master–slave approach, to simplify system configuration and network management. In most cases, in a CANopen network there is only one application master (which is responsible for actually controlling the operations of the automated system) and up to 127 slave devices (sensors and/or actuators). Each device is identified by means of a unique 7-bit address, called the node identifier, in a range from 1 to 127. The node identifier 0 is used in general for broadcast communications.

Object dictionary

The behavior of any CANopen device is described completely by means of a number of objects, each one tackling a particular aspect related to either the communications on the CAN bus or the available functions to interact with the physical controlled system (e.g., there are objects that define the device type, the manufacturer's name, the hardware and software version, and so on).

All the objects relevant to a given node are stored in the so-called *object dictionary* (OD) of that node. Here, each object is addressed by means of a 16-bit index. Each entry in the OD, in turn, may either be represented by a single value or consist of several components, which are accessible through an 8-bit subindex.

The object dictionary is split into four separate parts, according to the index of the entries. Entries below 1000_H are used to specify data types. Entries from 1000_H to $1FFF_H$, instead, are used to describe communication-specific parameters (i.e., the interface of the device as seen by the CAN network).

Entries from 2000_H to $5FFF_H$ are used by manufacturers to extend the basic set of functions of the device. Their use has to be considered carefully, in that they could make devices no longer interoperable. Finally, entries from 6000_H to $9FFF_H$ are used to describe all the aspects related to a specific category of devices (as defined in a device profile).

Process data objects

All the real-time process data involved in controlling a physical system are exchanged in CANopen by means of PDOs. Each PDO is mapped on exactly one CAN frame, so that it can be exchanged quickly and reliably. As a direct consequence, the amount of data that can be exchanged with one PDO is limited to 8 bytes at most. In most cases, this is more than sufficient to encode an item of process data. According to the *predefined connection set* (a standard assignment of identifiers to COBs), each node in CANopen can define up to four receive PDOs (from the application master to the device) and four transmit PDOs (from the device to the application master).

The transmission of PDOs from the slave devices can be triggered by some local event taking place on the node — including the expiration of some timeout — or it can be remotely requested from the master. This gives system designers a very high degree of flexibility in choosing how devices interact in the automated system, and enables the features offered by intelligent devices to be exploited better.

No additional control information is added to PDOs by CANopen, so that communication efficiency is as high as in CAN. This means that the meaning of each PDO is determined directly by the related identifier. As multicasting is allowed on PDOs, their transmission is unconfirmed, that is, the producer has no way to determine whether or not the PDO has been read by all the intended consumers.

Service data objects

SDOs are used in CANopen for parameterization and configuration, which usually take place at a lower priority than process data (hence, they are effectively non-real-time exchanges). In this case, a confirmed transmission service is provided for the applications, which ensures a reliable exchange of information. Furthermore, SDOs are only available on a peer-to-peer communication basis (multicasting is not allowed in this case).

A fragmentation protocol has been adopted for SDOs — which derives directly from the domain transfer services of CAL — so that information of any size can be exchanged. This means that the SDO sender has to split the information into smaller chunks, which are then reassembled at the receiving side. From a practical point of view, SDOs are used to access the entries of the object dictionary directly, so that they can be read or modified by the configuration tools.

Network management

There are two kinds of functions related to network management: *node control* and *error control*. Node control services are used to control the operation of either a single node or the whole network. For example, they can be used to start/stop nodes, to reset their state, or to put a node in configuration (preoperational) mode. Such commands are definitely time-critical and hence they use the highest priority communication object available in CAN.

Error control services, instead, are used to monitor the correct operation of the network. Two mechanisms are available, that is, *node guarding* and *heartbeat*. In both cases, low-priority messages are exchanged periodically in background over the network by the different nodes. Should one device cease sending these messages, the network management layer is aware of the problem and can take appropriate action.

Device profiles

In order to provide interoperability, a number of device profiles have been standardized in CANopen. Each profile describes the common behavior of a particular class of devices, and is usually described in a separate document. Among the available profiles, there are the following:

- I/O devices [DS401], which include both digital and analog input/output devices.
- Drives and motion control, which is used to describe digital motion products, such as, for example, stepper motors and servo-drives.
- Human–machine interfaces, which describes the use of displays and operator interfaces.
- Measuring devices and closed-loop controllers, to measure and control physical quantities.
- IEC 61131-3 programmable device, for describing the behavior of PLCs and intelligent devices.
- Encoders, which define incremental/absolute linear and rotary encoders to measure both position and velocity, and so on.

DeviceNet

DeviceNet [DNET] is a very flexible protocol to be used at the field level in the automated environments. It is worth noting that the implementation of devices that comply to DeviceNet is, in general, slightly more complex than for CANopen devices. However, DeviceNet offers a number of additional features with respect to CANopen, which can be used, for example, in complex multimaster networks.

In addition to the services at the application level, the DeviceNet specification also defines the physical layer in detail, including aspects such as connectors and cables (thin, thick, and flat cables are foreseen). It should be noted that the cable in DeviceNet can be used for both the signal and power supply (by using four wires plus ground).

Each DeviceNet network can include up to 64 different devices, which means that each node is identified by means of a 6-bit MAC ID. The allowable bit rates are limited to 125, 250, and 500 kb/s, which means that a bus length of at least 100 m is allowed.

Object model

The behavior and functions of each device are described in detail in DeviceNet by means of objects. In particular, three kinds of objects are foreseen, that is to say communication, system, and application-specific objects. Two very important objects are, for example, the connection object, which defines all the aspects related to a connection (including the CAN identifier and the triggering mode), and the application object, which defines the standardized behavior of a class of devices.

Data and services within each device are addressed by means of a hierarchical addressing scheme, which is based on the following components: MAC ID (i.e., the device's address), class ID, instance ID, Attribute ID, and service code. The first four identifiers are usually specified on 8 bits, while the service code is composed of a 7-bit integer.

Communication model

Communication among nodes (either point-to-point or multicast) takes place according to a connection-oriented scheme. By using the standard 11-bit CAN identifier, it is possible to provide an addressing scheme based on four message groups, in a decreasing order of priority:

- message group 1 includes the highest priority identifiers, and permits up to 16 different messages per node;
- message group 2 essentially refers to the predefined master/slave connection set;
- message group 3 is similar to message group 1, but it is composed of low-priority frames; and
- message group 4 is primarily used for network management.

Basically, two kinds of communications are possible: *explicit messages* and *I/O messages*. Explicit messages are used for general data exchanges among devices, such as, for example, configuration, management, and diagnostics. This kind of exchange takes place on the network at a low priority. I/O messages, instead, are used to exchange high-priority real-time messages according to the producer/consumer model.

Because the underlying communication system is based on a CAN network, each frame can include at most 8 bytes. Should one item of data exceed this size, a fragmentation protocol has been defined in DeviceNet that manages message splitting and the successive reassembly.

References

[IS898]   International Standard Organisation, Road Vehicles — Interchange of Digital Information — Controller Area Network for High-Speed Communication, ISO IS 11898, November 1993.

[ISO1]   International Standard Organisation, Road Vehicles — Controller Area Network — Part 1: Data Link Layer and Physical Signalling, ISO WD 11898-1, 1999.

[ISO2]   International Standard Organisation, Road Vehicles — Controller Area Network — Part 2: High Speed Medium Access Unit, ISO WD 11898-2, 1999.

[ISO3]   International Standard Organisation, Road Vehicles — Controller Area Network — Part 3: Low Speed Medium Access Unit, ISO WD 11898-3, 1999.

[TTC]   International Standard Organisation, Road Vehicles — Controller Area Network — Part 4: Time-Triggered Communication, ISO WD 11898-4, 2000.

[DS102]   CAN in Automation International Users and Manufacturers Group e.V., CAN Physical Layer for Industrial Applications — Two-Wire Differential Transmission, CiA DS 102, version 2.0, 1994.

[TIN] Tindell, K.W., Burns, A., and Wellings, A.J., Calculating controller area network (CAN) messages response times, *Control Engineering Practice*, 3, 1163–1169, 1995.

[WCAN] Cena, G. and Valenzano, A., A multistage hierarchical distributed arbitration technique for priority-based real-time communication systems, *IEEE Transactions on Industrial Electronics*, 49, 1227–1239, 2002.

[COP]   European Committee for Electrotechnical Standardisation, Industrial Communications Subsystem Based on ISO 11898 (CAN) for Controller–Device Interfaces — Part 4: CANopen, EN 50325-4, 2001.

[DNET] European Committee for Electrotechnical Standardisation, Industrial Communications Subsystem Based on ISO 11898 (CAN) for Controller–Device Interfaces — Part 2: DeviceNet, EN 50325-2, 2000.

[DS301] CAN in Automation International Users and Manufacturers Group e.V., CANopen — Application Layer and Communication Profile, CiA DS 301, version 4.02, 2002.

[DS401] CAN in Automation International Users and Manufacturers Group e.V., CANopen — Device Profile for Generic I/O Modules, CiA DS 401, version 2.0, 1999.

43

LonWorks/EIA-709 Networks
EIA-709 Protocol (LonTalk)

Dietmar Loy
LOYTEC Electronics GmbH

43.1 Introduction

One of the characteristics of LonWorks® is its unique seven-layer protocol implementation called LonTalk® or LonWorks protocol. Unlike other fieldbus protocols, all seven layers of the ISO/OSI reference model are actually defined and implemented in every single network node. One characteristic is its media-independent OSI Layer 2 that supports various communication media like twisted pair cables, power line communication, radio frequency channel, infrared connections, fiber optic channels, as well as IP connections based on the EIA-852 protocol standard. The basic frame layout for a twisted pair media is illustrated in Figure 43.1 The preamble allows for bit synchronization at the receiver and the start bit (SB) indicates the beginning of the byte boundary. The control field carries information about the importance of the data packet and an estimation about the expected network traffic in the near future. The source address holds the subnet/node address of the transmitting node and the destination address holds the address of the receiving node(s) followed by the user data that can be between 1 and 228 bytes long. Finally, a 16-bit CRC protects the frame from bit errors (Hamming Distance 4) and the code violation (CV) indicates the end of the frame. After the CV, a new round of bus arbitration can take place if nodes have data to be sent over the network [1]. On twisted pair cables, a p-persistent CSMA bus arbitration scheme is used. For other communication channels, for example, IP over Ethernet the arbitration scheme defined for this media is used by the EIA-709 protocol stack.

Another characteristic is its elaborated OSI Layer 3, which supports a variety of different addressing schemes and advanced routing capabilities as shown in Figure 43.2. Every node in the network can be identified with a unique 64-bit physical address (Neuron ID) and with a logical address composed of three address elements: the domain identifier, the subnet number, and the node number. Node numbers can be in the range of 1–127, subnet numbers within 1 and 255; hence, a maximum of $127 \times 255 - 32,385$ nodes

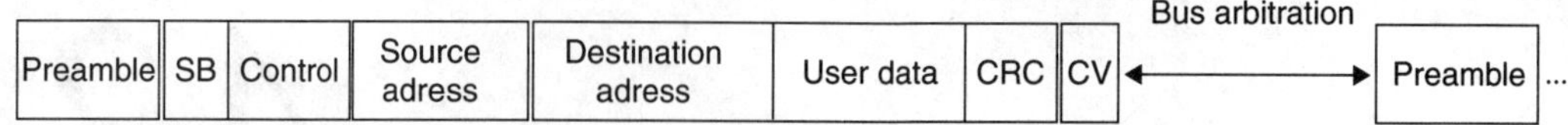

FIGURE 43.1   LonTalk frame layout on a twisted pair cable.

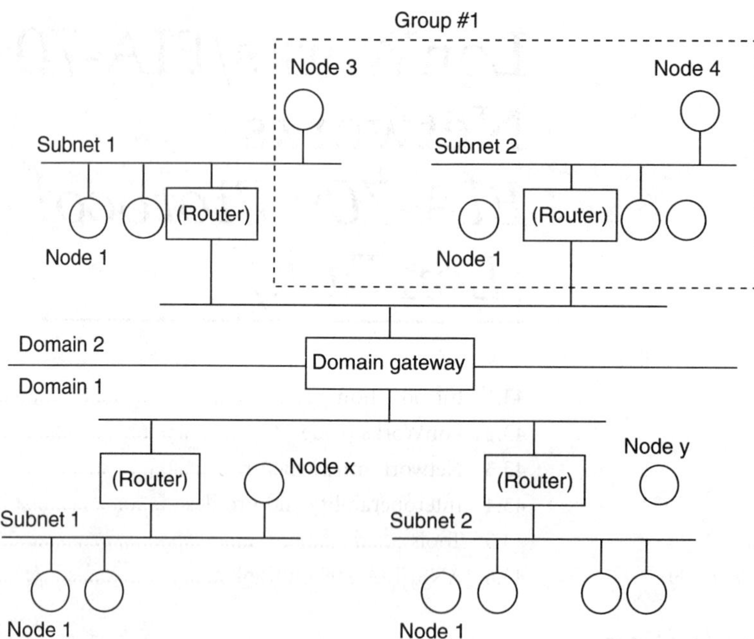

FIGURE 43.2   Addressing elements in EIA-709 networks. Up to 32,385 nodes can be addressed within one domain and up to 2^{48} domains are possible.

can be addressed within a single domain and up to 2^{48} domains can be addressed. Domain gateways can be built between logical domains in order to allow communication across domain boundaries. Groups can be formed to send a single data packet to a group of nodes with a multicast addressed message.

Optionally, routers can be used to keep local traffic within the subnet and only forward packets that are addressed to another subnet as shown in Figure 43.2. An EIA-709 node can send a unicast addressed message to exactly one node using either the physical Neuron ID or the logical subnet/node address or a multicast addressed message to either a group of nodes (group address) or to all nodes in the subnet or all nodes in the entire domain (broadcast address).

OSI Layer 4 supports four service types. The unacknowledged service simply transmits the data packet from the sender to the receiver and hopes that the message does not get lost. The unacknowledged repeated service transmits the same data packet a number of times and assumes that at least one data packet will get through to the receiver. The number of retries is programmable. The acknowledged service transmits the data packet and expects an acknowledgement to come back from the receiver. If the acknowledgement is not received, the transmitter retransmits the same data packet. The number of retries is again programmable. The request response service sends a request message to the receiver and the receiver must respond with a response message. This service type can be used to retrieve, for example, statistics information from a network node. Authentication can be used to authenticate the sender at the receiver side.

The interface to the application program are the so-called network variables (NVs). NVs are variables (like variables in a C program) that can be propagated over the network. NVs have a type, for example, SNVT_temp to represent a temperature in degree Celsius or SNVT_amp to represent an electric current in Amps or mA. SNVT is the acronym for Standard Network Variable Type. Each SNVT has a valid range defined by the upper and lower boundary, a resolution, and an SI unit. Network nodes communicate with each other by exchanging NVs.

43.2 LonWorks Node

For the last 10 years, the Neuron® chip manufactured by Motorola, Toshiba, and Cypress was the only microcontroller that supported the LonTalk protocol. Only in the last few years, with the adoption of the LonTalk protocol as a European and ANSI standard, some platform-independent implementations are available on the market [21]. The traditional LonWorks node is shown in Figure 43.3.

The Neuron chip executes the LonTalk protocol and the application program that interfaces to sensors and actuators through the Input/Output block. Two types of Neuron chips are available. The 3150 provides an external memory bus interface to expand the internal EEPROM and SRAM with external Flash and SRAM. The 3120 is optimized for low node cost applications and has a built-in 10 kbyte ROM, 2 kbyte EEPROM, and 2 kbyte SRAM but has no external memory interface [17].

An ANSI C derivative language called Neuron C is used to program the Neuron chip. Neuron C uses language extensions to schedule application events and to react to incoming data packets (network variables NVs) from the network interface. LonWorks supports a variety of different communication media. The most popular media today are 78 kbps (EIA-709.3) or 1.25 Mbps twisted pair communication, 4 kbps power line communication (EIA-709.2), a 1.25 Mbps fiber optics interface, and RS-485 interfaces at various bit rates between a few hundred bps and 1.25 Mbps. A brand new channel is the EIA-852 IP channel in order to tunnel EIA-709 data packets through IP (Intranet, Internet) networks.

43.3 Network Infrastructure

Great care must be take when designing the network infrastructure for an EIA-709 network. There are a number of requirements that must be met in order to guarantee reliable communication between all the nodes in the network. The maximum cable length must be met, the network cables must be properly terminated, the number of nodes on each physical channel must not exceed the specification, the expected network traffic must be estimated, and a network topology must be found that keeps local traffic local

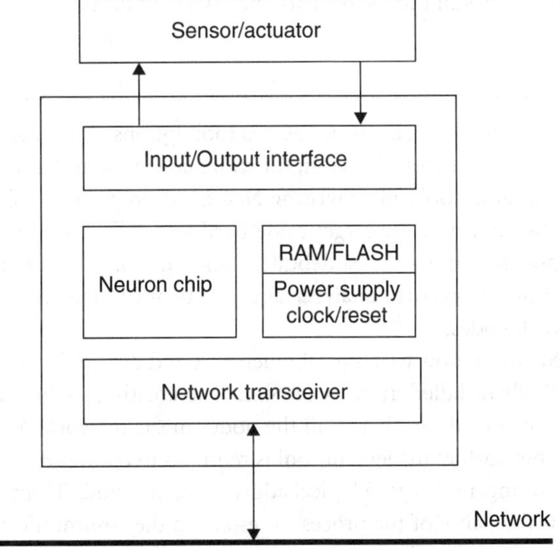

FIGURE 43.3 Typical LonWorks network node architecture.

and only data packets that need to travel across network segments should leave the local segment. The bandwidth utilization must be kept below a certain limit and the number of CRC errors due to collisions and noise on the network media must be measured on each channel and kept within well-defined boundaries. Only a reliable network infrastructure with built-in diagnostics and troubleshooting functionality can guarantee reliable operation of the building or factory.

Depending on the network media and the network transceiver, a variety of network topologies are possible with LonWorks nodes. Traditional bus, ring, star, as well as free topology are supported. Complex networks require networking elements like repeaters, routers, and gateways available in EIA-709 to segment the local traffic from traffic crossing segment boundaries. Such networking elements include Layer 1 repeaters to extend the physical length of a network cable, Layer 2 bridges, and Layer 3 routers to decouple individual segments from a networking backbone, and gateways to bridge between the levels of hierarchy in the automation pyramid (see Figure 43.2).

Modern network infrastructure products [21] have built-in network diagnostics capabilities and can monitor the health state of the network 24 h a day and 7 days a week. They immediately report network malfunctions or deviations from the normal operation to the system operator. All monitoring and reporting can either be done local on site or from a remote through an Intranet or Internet connection.

43.4 Interoperability and Profiles

Open standards allow individual companies to build single pieces of a bigger puzzle and enable system integrators (SI) to design complete systems consisting of those building blocks. Node developers must follow certain rules in order to make the bigger picture work. These rules are defined in interoperability guidelines and in profiles. Compatible does not necessarily mean interoperable and not at all plug and play. More elaborate guidelines need to be established for plug and play, interchangeable, and interworkable products. Different committees have set guidelines to define those terms and to establish the rules. The LonMark organization, for example, has published interoperability guidelines for nodes that use the LonTalk protocol [18]. It is important to note that interoperability must be guaranteed on all seven OSI layers. Interoperability on all seven OSI layers is still not a guarantee to have interworkable products. A typical node only uses a subset of all the possible protocol features. Profiles define these subsets based on the intended use of the node. Task groups within LonMark define functional profiles for analog input, analog output, temperature sensor, humidity sensor, CO2 sensor, VAV controller, fan coil unit, chiller, thermostat, damper actuator, etc. As of today, LonMark hosts different task groups for fire, home/utility, petrol station, HVAC, lighting, sun blinds, security, etc., type applications.

43.5 Tools

In order to program the Neuron chip, the user has two tool options. For node manufacturers, it might be sufficient to buy the NodeBuilder from Echelon; the more advanced and also more expensive tool is the LonBuilder from Echelon. Both tools allow writing Neuron C programs, to compile and link them, and download the final application into the target node hardware. The LonBuilder supports simultaneous debugging of multiple nodes, whereas the Nodebuilder only supports debugging of one node at the time. The LonBuilder has a built-in protocol analyzer and a network binder to create communication relationships between network nodes.

System integrators (SI) need tools to design the network and define the communication relationships before the nodes are actually installed in the field. Similar to creating a schematic of resistors, capacitors, ICs, diodes, etc., the SI creates a floor plan of all the nodes in the network. After physically installing the nodes in the network, a network management tool is required to configure (also called commission) the nodes. During commissioning, nodes get a logical address (subnet/node/Domain) assigned and the binding is created. Binding is the name of the process to establish the communication relationships between sensors, actuators, and controller nodes. Protocol analyzers are used to debug communication problems and to gather traffic and error statistics. During the maintenance phase of the network, the network man-

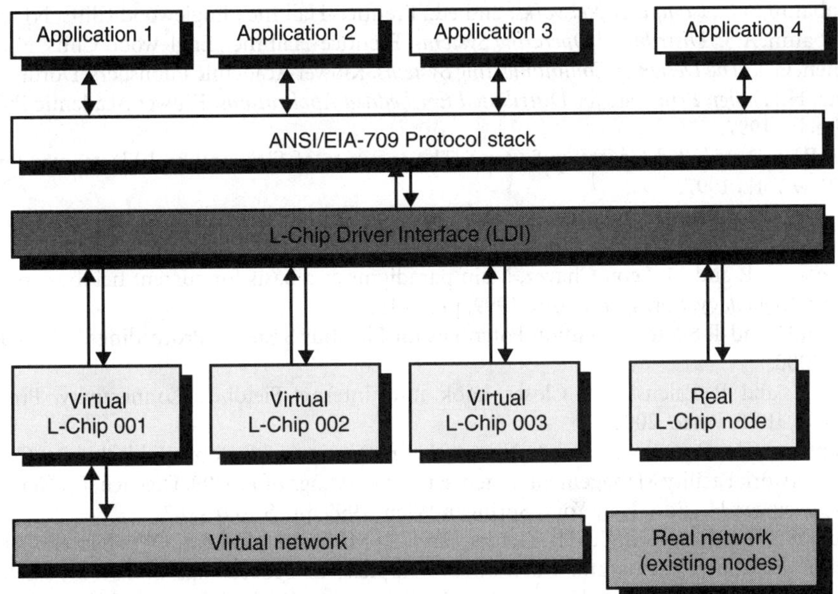

FIGURE 43.4 Platform-independent ANSI/EIA709 implementation. Simulation environment with virtual nodes linked to a physical network.

agement tool also supports replacing faulty nodes and extending the network with additional nodes and changes in the communication relationships.

43.6 ANSI/EIA-709 Outlook

As mentioned earlier, until 1999, the Neuron chip was the only vehicle to execute the LonTalk protocol. Due to the limited resources available on the Neuron chip (CPU performance, memory space, I/O capabilities, etc.), many applications could not be solved with the existing technology. Recent hardware platform-independent implementations of the ANSI/EIA709 protocol allow for good scalability of the CPU performance and memory requirements tailored for the specific application. Furthermore, simulation of the network behavior before actual hardware exists is now possible as shown in Figure 43.4.

Different application programs 1–*n* run on top of the ANSI/EIA709 protocol stack that talks to an interface layer (LDI), which communicates with the virtual network devices. As shown on the right-hand side in Figure 43.4, the LDI can also talk to a real physical network; hence, existing nodes in a network can be part of the system simulation. The application programs on top do not see a difference if the node already exists as a physical device or if it is still a virtual device simulated on a PC or workstation [19, 20]. This architecture is fully portable between different hardware platforms and different operating systems and allows cross development, for example, on a Windows platform for an ARM7-based target hardware. Since the Application Programming Interface (API) on top of the EIA-709 protocol stack is the same on all hardware platforms and under all operating systems, the final application programs can easily be ported between different target systems. This approach allows full hardware software codesign since the application programs can be developed while the hardware development of the network node is still going on.

References

[1] Dietrich, D., D. Loy, and H.-J. Schweinzer, *LON-Technologie, Verteilte Systeme in der Anwendung; 2. Auflage,* Hüthig Verlag, Heidelberg, 1999.

[2] Dietrich, D., P. Neumann, and H. Schweinzer, *Fieldbus Technology — System Integration, Networking, and Engineering,* Springer, Wien, 1999.

[3] Tanenbaum, A.S., *Computer Networks*, 2nd ed., Prentice-Hall Inc., Englewood Cliffs, NJ, 1989.

[4] Tanenbaum, A.S., *Distributed Operating Systems*, Prentice-Hall Inc., Englewood Cliffs, NJ, 1994.

[5] Koomen, C. J., *The Design of Communicating Systems*, Kluwer Academic Publishers, Dordrecht 1991.

[6] Kopetz, H., *Design Principles for Distributed Emebedded Applications*, Kluwer Academic Publishers, Dordrecht, 1997.

[7] Sinha, P.K., *Distributed Operating Systems*, The Institute of Electrical and Electronics Engineers, Inc., New York, 1997.

[8] Gürtler, G.H., Fieldbus standardization, the European approach and experiences, in *Feldbustechnik in Forschung, Entwicklung und Anwendung*, Springer, Berlin, 1997, pp. 2–11.

[9] Thomesse, J.-P. and M. Leon Chavez, Main paradigms as a basis for current fieldbus concepts, in *Fieldbus Technology*, Springer, Berlin, 1999, pp. 2–15.

[10] Dietrich, D. and T. Sauter, Evolution Potentials for Fieldbus Systems, Proceedings WFCS'00, IEEE, Porto, 2000.

[11] Sauter, T. and P. Palensky, A Closer Look into Internet-Fieldbus Connectivity, Proceedings WFCS'00, IEEE, Porto, 2000.

[12] Bangemann, Th., R. Dübner, and A. Neumann, Integration of Fieldbus Objects into Computer-Aided Network Facility Management Systems, in Proceedings of FeT'99, Dietrich, D., Neumann, P., and Schweinzer, H., Eds., New York, Springer, Wien, 1999, pp. S. 180 –187.

[13] Rüping, S., H. Klugmann, K.-H- Gerdes, and S. Mirbach, Modular OPC-Server Connecting Different Fieldbus Systems and Internet Java Applets, in Proceedings of FeT'99, Dietrich, D., Neumann, P., and Schweinzer, H., Eds., New York, Springer, Wien, 1999, pp.S. 240 –246.

[14] Neumann, P. and F. Iwanitz, Integration of Fieldbus Systemns into Distributed Object-Oriented Systems, Proceedings WFCS'97, IEEE, New York, 1997, pp. S. 247 –253.

[15] Döbrich, U. and P. Noury, ESPRIT Project NOAH — introduction, in *Fieldbus Technology*, Doebrich and Noury, Eds., Springer, Berlin, 1999, pp. 414–422.

[16] Palensky, P. The Convergence of Intelligent Software Agents and Field Area Networks, Proceedings of ETFA'99, Barcelona, 1999, pp. 917–922.

[17] Motorola Inc., LonWorks Technology Device Data, DL159, Rev 4, 1997.

[18] LonMark, *Application Layer Interoperability Guidelines*, Version 3, LonMark Interoperability Association, U.S.A.,1996.

[19] Bauer, A., LC-SIM, Loytec electronics GmbH, www.loytec.com, March 2000.

[20] Bauer, A. and S. Soucek, Simulation of ANSI/EIA709 networks, LonWorld, October 2000.

[21] Schweinzer, H., VENUS-Vienna Embedded Networking Utility Suite, Loytec electronics GmbH, www.loytec.com, October 1999.

44

Time-Triggered Communication Networks

Hermann Kopetz and
Günther Bauer
Vienna University of Technology

44.1 Introduction

Computer architectures establish a blueprint and a framework for the design of a class of computing systems that share a common set of characteristics. The time-triggered architecture (TTA) generates such a framework for the domain of large distributed real-time systems in high-dependability environments. It sets up the computing infrastructure for the implementation of applications and provides mechanisms and guidelines to partition a large application into nearly autonomous subsystems along small and well-defined interfaces in order to control the complexity of the evolving artifact. Architecture design is thus interface design. By defining an architectural style that is observed at all component interfaces, the architecture avoids property mismatches at the interfaces and eliminates the need for unproductive "glue" code.

Characteristic of the TTA is the treatment of (physical) time as a first-order quantity. The TTA decomposes a large application into clusters and nodes and provides a fault-tolerant global time base of known precision at every node. The TTA takes advantage of the availability of this global time to precisely specify the interfaces among the nodes, to simplify the communication and agreement protocols, to perform prompt error detection, and to guarantee the timeliness of real-time applications.

This article aims at giving a short yet concise introduction to the TTA in general and time-triggered communication networks in particular. We will discuss the basic concepts of the TTA in Section 44.2. Two representatives of time-triggered network protocols will be presented in Section 44.3. Section 44.4 will discuss fault-tolerance aspects of the TTA, Section 44.5 will elaborate on the design of TTA applications. Finally, the article ends with a conclusion in Section 44.6.

44.2 Fundamental Concepts

The computational model that guides the design of the TTA is the time-triggered (TT) model of computation [Kop98]. The following sections will discuss the fundamental concepts of this architectural model. A more detailed description of the concepts can be found in [KB03].

Model of Time

The model of time of the TTA is based on Newtonian physics. Real-time progresses along a dense time-line, consisting of an infinite set of instants, from the past to the future. A duration (or interval) is a section of the timeline, delimited by two instants. A happening that occurs at an instant (i.e. a cut of the timeline) is called an event. An observation of the state of the world is thus an event. The time stamp of an event is established by assigning the state of the node-local global time to the event immediately after the event occurrence. A fault-tolerant internal clock synchronization algorithm establishes the global time in the TTA. Due to the impossibility of synchronizing clocks perfectly and the denseness property of real time, there is always the possibility of the following sequence of events: clock of node j ticks, event e occurs, clock of node k ticks. In such a situation, the single event e is time-stamped by the two clocks j and k with a difference of one tick. In a distributed system, the finite precision of the global time base and the digitalization of time make it, in general, impossible to consistently order events on the basis of their global time stamps. The TTA solves this problem by the introduction of a sparse-time base [Kop97, p. 55]. In the sparse-time model, the continuum of time is partitioned into an infinite sequence of alternating durations of activity and silence as shown in Figure 44.1. The duration of the activity interval, that is, a granule of the global time, must be larger than the precision of the clock synchronization.

From the point of view of temporal ordering, all events that occur within an interval of activity are considered to happen at the same time. Events that happen in the distributed system at different nodes at the same global clock-tick are thus considered simultaneous. Events that happen during different durations of activity and that are separated by the required interval of silence can be consistently temporally ordered on the basis of their global time stamps. The architecture must make sure that significant events, such as the sending of a message, occur only during an interval of activity. The time stamps of events that are outside the control of the distributed computer system (and therefore happen on a dense timeline) must be assigned to an agreed duration of activity by an agreement protocol.

In the TTA there exists a uniform external representation of time that is modeled according to the Global Positioning System (GPS) time representation. The time stamp of an instant is represented by an eight-byte integer, that is, two words of a 32-bit architecture. The three lower bytes contain the binary fractions of the second, giving a granularity of about 60 nsec. This is the accuracy that can be achieved with a precise GPS receiver. The five upper bytes count the full seconds. The external TTA epoch assigns the value 2^{38} to the start of the GPS epoch, that is, 00:00:00 UTC on January 6, 1980. This offset has been chosen in order that also instants before January 6, 1980 can be represented by positive integers in the TTA. Thus, events that occurred between 8710 years before January 1980 and 26131 years after January 1980 can be time-stamped with an accuracy of 60 nsec. There are different internal time representations

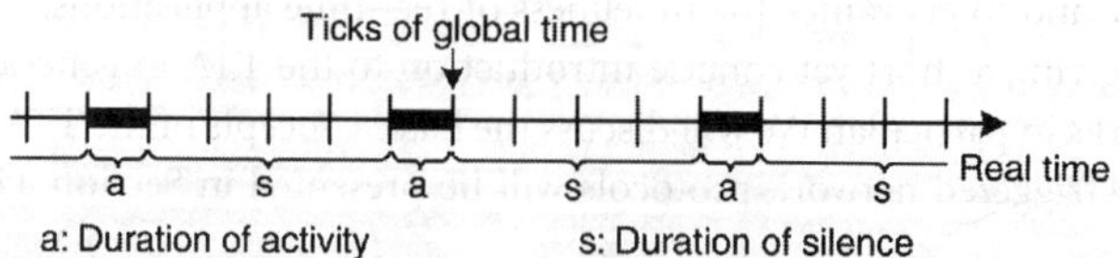

FIGURE 44.1   Sparse time base.

in the TTA that match the time format to the capabilities of the hardware (8 bit, 16 bit, or 32 bit architectures) and the requirements of the application. Since not all time stamps are based on a global time with a precision of 60 nsec, an attribute field is introduced in the external representation indicating the precision of a time stamp.

State Information vs. Event Information

The information that is exchanged across an interface is either state information or event information. Any property of a real-time (RT) entity (i.e. a relevant state variable) that is observed by a node of the distributed RT system at a particular instant, for example, the temperature of a vessel, is called a state attribute and the corresponding information state information. A state observation records the value of a state variable at a particular instant, the point of observation. A state observation can be expressed by the atomic triple

< Name, value, time of observation >

For example, the following is a state observation: "The position of control valve *A* was at 75 degrees at 10:42 a.m." State information is idempotent and requires an at-least-once semantics when transmitted to a client. At the sender, state information is not consumed on sending and at the receiver, state information requires an update-in-place and a nonconsuming read. State information is transmitted in state messages.

A sudden change of state of an RT entity that occurs at an instant is an event. Information that describes an event is called event information. Event information contains the difference between the state before the event and the state after the event. An event observation can be expressed by the atomic triple

< Name, value difference, time of event >

For example, the following is an event observation: "The position of control valve *A* changed by 5 degrees at 10:42 a.m." Event observations require exactly-once semantics when transmitted to a consumer. At the sender, event information is consumed on sending and at the receiver, event information must be queued and consumed on reading. Event information is transmitted in event messages.

Periodic state observations or sporadic event observations are two alternative approaches for the observation of a dynamic environment in order to reconstruct the states and events of the environment at the observer. Periodic state observations produce a sequence of equidistant "snapshots" of the environment that can be used by the observer to reconstruct those events that occur within a minimum temporal distance that is longer than the duration of the sampling period. Starting from an initial state, a complete sequence of (sporadic) event observations can be used by the observer to reconstruct the complete sequence of states of the RT entity that occurred in the environment. However, if there is no minimum duration between events assumed, the observer and the communication system must be infinitely fast.

Temporal Firewalls

An extensible architecture must be based on a small number of orthogonal concepts that are reused in many different situations in order to reduce the mental load required for understanding large systems. In a large distributed system, the characteristics of the interfaces between subsystems determine to a large extent the comprehensibility of the architecture. In the TTA, the communication network interface (CNI, cf. Figure 44.2) between a host computer and the communication network is the most important interface. The CNI appears in every node of the architecture and separates the local processing within a node from the global interactions among the nodes. The CNI consists of two unidirectional data flow interfaces, one from the host computer to the communication system and the other one in the opposite direction.

We call a unidirectional data flow interface elementary, if there is only a unidirectional control flow [Kop99] across this interface. An interface that supports periodic state messages with error detection at the receiver is an example of such an elementary interface. We call a unidirectional data flow interface composite, if even a unidirectional data flow requires a bidirectional control flow. An event message interface with error detection is an example for a composite interface. Composite interfaces are inherently more complex than elementary interfaces, since the correct operation of the sender depends on the control signals from all receivers. This can be a problem in multicast communication where many control messages are generated for every unidirectional data transfer, and each one of the receivers can affect the operation of the sender.

The basic CNI of the TTA as depicted in Figure 44.3 is an elementary interface. The TT transport protocol carries autonomously — driven by its TT schedule — state messages from the sender's CNI to the receiver's CNI. The sender can deposit the information into its local CNI memory according to the information push paradigm, while the receiver will pull the information out of its local CNI memory. From the point of view of temporal predictability, information push into a local memory at the sender and information pull from a local memory at the receiver are optimal, since no unpredictable task delays that extend the worst-case execution occur during reception of messages. A receiver that is working on a time-critical task is never interrupted by a control signal from the communication system. Since no control signals cross the CNI in the TTA (the communication system derives control signals for the fetch and delivery instants from the progress of global time and its local schedule exclusively), propagation of control errors is prohibited by design. We call an interface that prevents propagation of control errors by design a temporal firewall [KN97]. The integrity of the data in the temporal firewall is assured by the nonblocking write (NBW) concurrency control protocol [Kop97, p. 217].

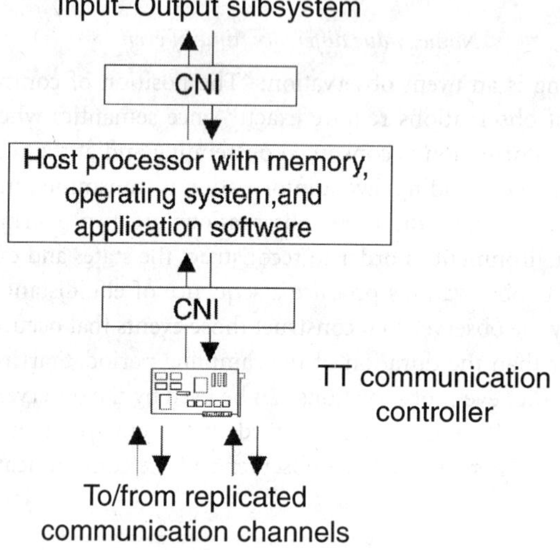

FIGURE 44.2   Node of the TTA.

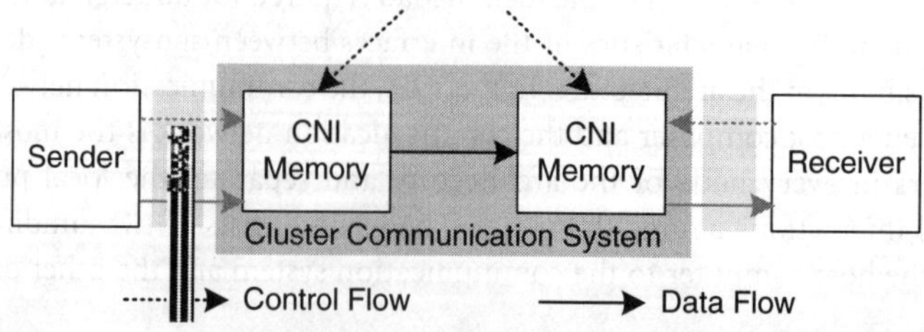

FIGURE 44.3   Data flow and control flow at a TTA interface.

44.3   Time-Triggered Networks

The very basic principle of TT communication is that transmission of messages is triggered by the clock rather than the availability of new information. The upcoming sections will present the view of the application of TT communication.

Principle of Operation

The instants at which information is delivered at or fetched from the CNIs of a node are determined *a priori* and are common knowledge to all nodes of a TTA. *A priori* common knowledge means that these instants are defined before the computation under consideration is started and that the instants are known to all nodes of a cluster beforehand. These instants are the deadlines for the application tasks within a host. Knowing these deadlines, it is in the responsibility of the host to produce the required results before the deadline has passed. Any node-local scheduling strategy that will satisfy these known deadlines is "fit for purpose." It is the responsibility of the TT communication service to transport the information from the sending CNI to the receiving CNI within the interval delimited by these *a priori* known fetch and delivery instants. The TTA contains two communication protocols that provide this communication service, the fault-tolerant TTP/C protocol and the low-cost fieldbus protocol TTP/A.

Communication Interface

From the point of view of complexity management and composability, it is useful to distinguish between three different types of interfaces of a node: the real-time service (RS) interface, the diagnostic and management (DM) interface, and the configuration and planning (CP) interface [Kop00]. These interface types serve different functions and have different characteristics. For the temporal composability, the most important interface is the RS interface.

The RS Interface: The RS interface provides the timely RS to the node environment during the operation of the system. In RT systems, it is a time-critical interface that must meet the temporal specification of the application in all specified load and fault scenarios. The composability of an architecture depends on the proper support of the specified RS interface properties (in the value and in the temporal domain) during operation. From the user's point of view, the internals of the node are not visible at the CNI, since they are hidden behind the RS interface.

The DM Interface: The DM interface opens a communication channel to the internals of a node. It is used for setting node parameters and for retrieving information about the internals of the node, for example, for the purpose of internal fault diagnosis. The maintenance engineer that accesses the internals of a node via the DM interface must have detailed knowledge about the internal objects and behavior of the node. The DM interface does not affect temporal composability. Usually, the DM interface is not time-critical.

The CP Interface: The CP interface is used to connect a node to other nodes of a system. It is used during the integration phase to generate the "glue" between the nearly autonomous nodes. The use of the CP interface does not require detailed knowledge about the internal operation of a node. The CP interface is not time-critical.

The CNI of the TTA can be directly used as the RS interface. On input, the precise interface specifications (in the temporal and value domain) are the preconditions for the correct operation of the host software. On output, the precise interface specifications are the postconditions that must be satisfied by the host, provided the preconditions have been satisfied by the host environment. Since the bandwidth is allocated statically to the host, no starvation of any host can occur due to high-priority message transmission from other hosts. The TTA implements an event-triggered communication service on top of the basic TT service to realize the DM and CP interfaces. Since the event-triggered communication is based on (but not executed in parallel to) the TT communication, it is possible to maintain and to use all predictability properties of the basic TT communication service in event-triggered communication.

The System Protocol TTP/C

The TTP/C protocol [TTT] is a fault-tolerant TT communication protocol that provides the following services:

- Autonomous fault-tolerant message transport with known delay and bounded jitter between the CNIs of the nodes of a cluster by employing a TDMA medium access strategy on replicated communication channels.
- Fault-tolerant clock synchronization that establishes the global time base without relying on a central time server.
- Membership service to inform every node consistently about the "health-state" of every other node of the cluster. This service can be used as an acknowledgement service in multicast communication. The membership service is also used to efficiently implement the fault-tolerant clock synchronization service.
- Clique avoidance to detect and eliminate the formation of cliques in case the fault hypothesis is violated.

In TTP/C the communication is organized into rounds, where every node must send a message in every round. A particular message may carry up to 240 bytes of data. The data are protected by a 24 bits CRC checksum. The message schedule is stored in the message-descriptor list (MEDL) within the communication controller of each node. In order to achieve high data efficiency, the sender name and the message name are derived from the send instant. The clock synchronization of TTP/C exploits the common knowledge of the send schedule: every node measures the difference between the *a priori* known expected and the actually observed arrival time of a correct message to learn about the difference between the clock of the sender and the clock of the receiver. This information is used by a fault-tolerant average algorithm to calculate periodically a correction term for the local clock in order to keep the clock in synchrony with all other clocks of the cluster. The membership service employs a distributed agreement algorithm to determine whether the outgoing link of the sender or the incoming link of the receiver has failed. Nodes that have suffered a transmission fault are excluded from the membership until they restart with a correct protocol state. Before each send operation of a node, the clique avoidance algorithm checks if the node is a member of the majority clique. The detailed specification of the TTP/C protocol can be found at [TTT].

The Fieldbus Protocol TTP/A

The TTP/A protocol is the TT fieldbus protocol of the TTA [OMG02]. It is used to connect low-cost smart transducers to a node of the TTA, which acts as the master of a transducer cluster. In TTP/A, the CNI memory element of Figure 44.3 has been expanded at the transducer side to hold a simple interface file system (IFS). Each interface file contains up to 256 records of 4 bytes each. The IFS forms the uniform name space for the exchange of data between a sensor and its environment (Figure 44.4).

The IFS holds the RT data, calibration data, diagnostic data, and configuration data. The information between the IFS of the smart transducer and the CNI of the TTA node is exchanged by the TT TTP/A protocol, which distinguishes between two types of rounds, the master–slave (MS) round and the

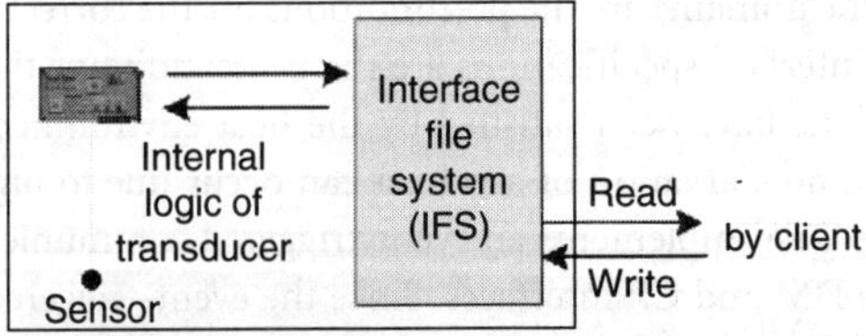

FIGURE 44.4   IFS in a smart transducer.

multi-partner (MP) round. The MS rounds are used to read and write records from the IFS of a particular transducer to implement the DM and CP interfaces. The MP rounds are periodic and transport data from selected IFS records of several transducers across the TTP/A cluster to implement the RS service. MP rounds and MS rounds are interleaved, such that the time-critical RS implemented by means of MP rounds and the event-based MS service can coexist. It is thus possible to diagnose a smart transducer or to reconfigure or install a new smart transducer on-line, without disturbing the time-critical RS of the other nodes. The TTP/A protocol also supports a "plug-and-play" mode where new sensors are detected, configured, and integrated into a running system on-line and dynamically. The detailed specification of the TTP/A protocol can be found at [OMG02].

44.4 Fault Tolerance

In any fault-tolerant architecture, it is important to distinguish clearly between fault containment and error containment. Fault containment is concerned with limiting the immediate impact of a single fault to a defined region, while error containment tries to avoid the propagation of the consequences of a fault, the error. It must be prohibited that an error in one fault-containment region propagates into another fault-containment region that has not been directly affected by the original fault.

Fault Containment

In the field of fault-tolerant computing, the notion of a fault-containment region (FCR) is introduced in order to delimit the impact of a single fault. An FCR is defined as the set of subsystems that share one or more common resources. A fault in any one of these shared resources can thus impact all subsystems of the FCR, that is, the subsystems of an FCR cannot be considered to fail independently of each other. In the context of this article, we consider the following resources that can be impacted by a fault:

- Computing Hardware
- Power Supply
- Timing Source
- Clock Synchronization Service
- Physical Space.

For example, if two subsystems depend on a single timing source, for example, a single oscillator or a single clock synchronization algorithm, then these two subsystems are not considered to be independent and therefore belong to the same FCR. Since this definition of independence allows that two FCRs can share the same design, that is, the same software, software faults are not part of this fault model. In the TTA, a node is considered to form a single FCR.

In the TTA nodes communicate by the exchange of messages across replicated communication channels. Each one of the two channels transports independently its own copy of the message at about the same time from the sending CNI to the receiving CNI. The start of sending a message by the sender is called the message send instant. The termination of receiving a message by the receiver is called the message receive instant. In the TTA, the intended message send instants and the intended message receive instants are *a priori* known to all communicating partners. A message contains an atomic data structure that is protected by a CRC. We make the assumption that a CRC cannot be forged by a fault. A message is called a valid message if it contains a data structure with a correct CRC. A message is called a timely message if it is a valid message and conforms to the temporal specification. A message, which does not conform to the temporal specification, is an untimely message. A timely message is a correct message, if its data structure is in agreement, both at the syntactic and semantic level, with the specification. We call a message with a message length that differs from its specification or with an incorrect CRC an invalid message.

Error Containment in the Temporal Domain

An error that is caused by a fault in the sending FCR can propagate to another FCR via a message failure, that is, the FCR sends a message that deviates from the specification. A message failure can be a message value failure or a message timing failure. A message value failure implies either that a message is invalid or that the data structure contained in a valid message is incorrect. A message timing failure implies that the message send instant or the message receive instant are not in agreement with the specification.

In order to avoid error propagation of a sent message, we need error detection mechanisms that are in different FCRs than the message sender. Otherwise, the error detection mechanism may be impacted by the same fault that caused the message failure. In the TTA, we distinguish between timing failure detection and value failure detection. The timing-failure detection is performed by a guardian (Figure 44.5), which is part of the TTA. Value failure detection is in the responsibility of the host computer.

The guardian is an autonomous unit that has *a priori* knowledge of all intended message send and receive instants. Each one of the two replicated communication channels has its own independent guardian. A receiving node within the TTA judges a sending node as operational, if it has received at least one timely message from the sender around the specified receive instant. It is assumed that a guardian cannot forge a CRC and cannot store messages, that is, it can only output a valid message at one of its output ports if it has received a valid message on one of its input ports within the last d time units. A guardian transforms a message, which it judges untimely into an invalid message by cutting off its tail. Such a truncated message will be recognized as invalid by all correct receivers and will then be discarded. The guardian may truncate a message either because it detected a message timing failure or because the guardian itself is faulty. In the latter case, it is assumed that the sender of the message is correct and thus the correct message will proceed to the receivers via the replicated channel of the TTA.

Error Handling in the Value Domain

Detection of value failures is not in responsibility of the TTA, but in the responsibility of the host computers. For example, detection and correction of value failures can be performed in a single step by triple modular redundancy (TMR). In this case, three replicated senders, placed in three different FCRs, perform the same operations in their host computers. They produce — in the fault-free case — correct messages with the same content that are sent to three replicated receivers that perform a majority vote on these three messages (actually, at the communication-level six messages will be transported, one from each sender on each of its two channels).

Detection of value failures and detection of timing failures are not independent in the TTA. In order to implement a TMR structure at the application level, the integrity of the timing of the architecture must

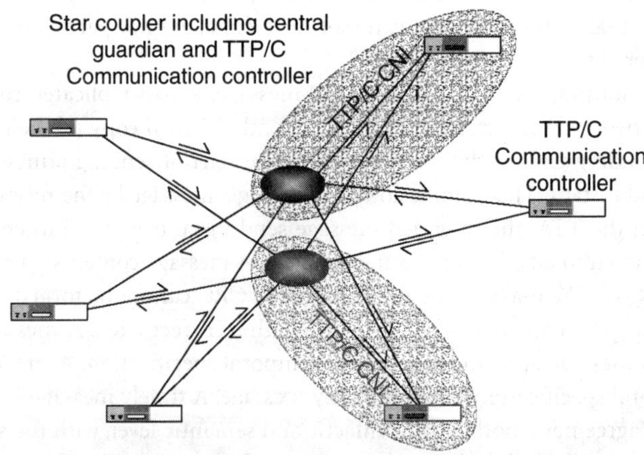

FIGURE 44.5   TTA star topology with central guardian.

be assumed. An intact sparse global time base is a prerequisite for the system-wide definition of the distributed state, which again is a prerequisite for masking value failures by voting. The separation of handling timing failures from handling value failures has beneficial implications for resource requirements. To handle k arbitrary failing nodes, $3k + 1$ nodes are required. All nodes of a cluster, independent of their involvement in a particular application system, can contribute to handling timing failures at the architectural level. Once a proper global time is available, TMR for masking of value failures can be implemented using only $2k + 1$ synchronized nodes in a particular application subsystem. This separation of timing failures and value failures thus reduces the number of components needed for fault tolerance of an application from $3k + 1$ to $2k + 1$.

To summarize, fault containment and error detection is achieved in the TTA in three distinct steps. At first, fault containment is achieved by proper architectural decisions concerning resource sharing in order to provide independent fault containment regions. In a second step, propagation of timing errors is avoided at the architecture level by the guardians. In a third step, handling of value failures is performed at the application level by voting.

44.5   The Design of TTA Applications

Composability and the associated reuse of nodes and software can only be realized if the architecture supports a two-level design methodology. In the TTA such a methodology is supported: the TTA distinguishes between the architecture design (cluster design) and the component design (node design).

Architecture Design

In the cluster design phase, an application is decomposed into clusters and nodes. This decomposition will be guided by engineering insight and the structure inherent in the application, in accordance with the proven architecture principle of "form follows function." For example, in an automotive environment, a "drive-by-wire" system may be decomposed into functional units as depicted in Figure 44.6.

If a system is developed "on the green lawn," then a top-down decomposition will be pursued. After the decomposition has been completed, the CNIs of the nodes must be specified in the temporal and in the value domain. The data elements that are to be exchanged across the CNIs are identified and the precise fetch instants and delivery instants of the data at the CNI must be determined. Given these data, the schedules of the TTP/C communication system can be calculated and verified. At the end of the

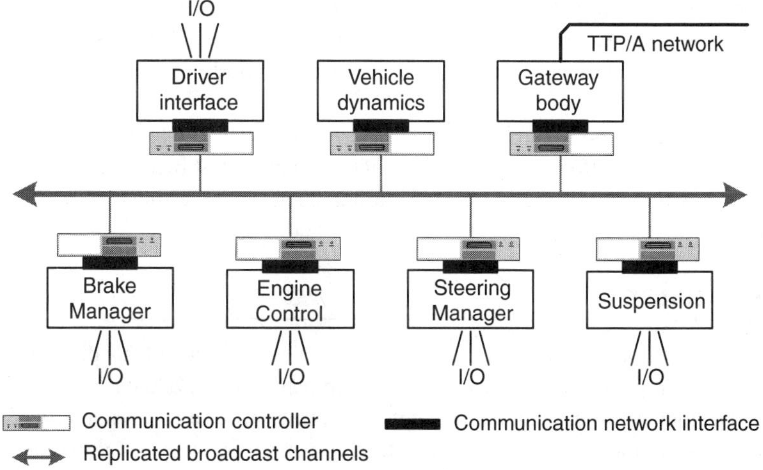

FIGURE 44.6   Decomposition of a "drive-by-wire" application.

architecture design phase, the precise interface specifications of the nodes are available. These interface specifications are the inputs and constraints for the node design.

Given a set of available nodes with their temporal specification (nodes that are available for reuse), a bottom-up design approach must be followed. Given the constraints of the nodes at hand (how much time they need to calculate an output from an input), a TTP/C schedule must be found that meets the application requirements and satisfies the node constraints.

Component Design

During the node design phase, the application software of the host computers is developed. The delivery and fetch instants established during the architecture design phase are the preconditions and postconditions for the temporal validation of the application software. The host operating system can employ any reasonable scheduling strategy, as long as the given deadlines are satisfied and the replica determinism of the host system is maintained.

Node testing proceeds bottom-up. A new node must be tested with respect to the given CNI specifications in all anticipated load and fault conditions. The composability properties of the TTA (stability of prior service achieved by the strict adherence to information pull interfaces) ensure that a property that has been validated at the node level will also hold at the system level. At the system level, testing will focus on validating the emerging services that are result of the integration.

Validation

Today, the integration and validation phases are probably the most expensive phases in the implementation of a large distributed RT system. The TTA has been designed to reduce this integration and validation effort by providing the following mechanisms:

- The architecture provides a consistent distributed computing base to the application and informs the application in case a loss of consistency is caused by a violation of the fault hypothesis. The basic algorithms that provide this consistent distributed computing base (clock synchronization and membership) have been analyzed by formal methods and are implemented once and for all in silicon. The application need not be concerned with the implementation and validation of the complex distributed agreement protocols that are needed to establish consistency in a distributed system.
- The architecture is replica deterministic, which means that any observed deficiency can be reproduced in order to diagnose the cause of the observed problem.
- The interaction pattern between the nodes and the contents of the exchanged messages can be observed by an independent observer without the probe effect. It is thus possible to determine whether a node complies with its preconditions and postconditions without interfering with the operation of the observed node.
- The internal state of a node can be observed and controlled by the DM interface.
- In the TTA, it is straightforward to provide a RT simulation test bench that reproduces the environment to any node in RT. Deterministic automatic regression testing can thus be implemented.

44.6 Conclusions

The TTA is the result of more than 20 years of research in the field of dependable distributed RT systems. During this period, many ideas have been developed, implemented, evaluated, and finally discarded. What survived is a small set of orthogonal concepts that center around the availability of a dependable global time base. The guiding principle during the development of the TTA has always been to take maximum advantage of the availability of this global time, which is part of the world, even if we do not use it. The TTA spans the whole spectrum of dependable distributed RT systems, from the low-cost deeply embedded sensor nodes to high-performance nodes that communicate at gigabits per second speeds, persistently assuming that a global time of appropriate precision is available in every node of the TTA.

At present, the TTA occupies a niche position, since in the experimental as well as in the theoretical realm of main-line computing, time is considered a nuisance that makes life difficult and should be dismissed at the earliest moment [Lee00]. However, as more and more application designers start to realize that RT is an integrated part of the real-world that cannot be abstracted away, the future prospects for the TTA look encouraging.

Acknowledgments

This work has been supported by the European IST project "Next TTA" under project number IST-2001-32111.

References

[Kop97] Kopetz, H., *Real-Time Systems: Design Principles for Distributed Embedded Applications*, Kluwer Academic Publishers, Dordrecht, 1997.

[Kop98] Kopetz, H., The Time-Triggered (TT) Model of Computation, in Proceedings 19th IEEE Real-Time System Symposium, 1998, pp. 168–177.

[Kop99] Kopetz, H., Elementary versus Composite Interfaces in Distributed Real-Time Systems, in Proceedings 4th International Symposium on Autonomous Decentralized Systems, 1999, pp. 26–33.

[Kop00] Kopetz, H., Software Engineering for Real-Time: A Roadmap, in Proceedings 22nd International Conference on Software Engineering, 2000, pp. 201–211.

[KB03] Kopetz, H. and G. Bauer, The time-triggered architecture, *Proceedings of the IEEE*, 91;112–126, 2003.

[KN97] Kopetz, H. and R. Nossal. Temporal Firewalls in Large Distributed Real-Time Systems, in Proceedings of IEEE Workshop on Future Trends in Distributed Computing, 1997, pp. 310–315.

[Lee00] Lee, E., What's ahead for embedded software? *IEEE Computer*, 33;18–26, 2000.

[OMG02] OMG, Smart Transducers Interface, Final Adopted Specification ptc/2002-10-02, Object Management Group, 2002, Available at http://www.omg.org.

[TTT] TTTech Computertechnik AG, Specification of the TTP/C Protocol, available at www.tttech.com.

45

IEEE 1394 for Factory Automation

Michael Scholles, Uwe
Schelinski and Petra Nauber
*Fraunhofer Institute of Photonic
Microsystems*

45.1 Introduction

Almost every modern distributed technical system requires some kind of digital communication infrastructure with high bandwidth, no matter whether it concerns consumer electronics or industrial applications. On the one hand, computers and devices for acquisition and reproduction of digital images and sound converge to so-called multimedia systems. On the other, industrial control systems incorporate multiple sensors like cameras for optical quality inspection, while control and status information, normally distributed via field busses, shall run along the same cable as mass data. All these systems have one common characteristic: they require an efficient peer-to-peer data communication mechanism, since they do not inevitably include a computer that can act as a data hub. Even if such a central node exists, sending data from one node to the computer and then forwarding it to another external device is often not the optimum communication pattern.

Modern consumer electronics and communication systems for industrial and factory automation applications have some more features and requirements in common:

High bandwidth: An industrial black and white camera with VGA resolution, 12-bit color depth per pixel, and a frame rate of 25 Hz, which is often used for automated optical quality inspection, produces a data rate of almost 100 Mbit/sec, which can easily be increased by higher geometrical resolution and use of multicamera systems.

Real-time streaming: If the communication system supports a special type of data transfer that ensures bandwidth for real-time streaming of data or a well-defined latency for message transmission, the definition of higher layer protocols can be significantly simplified. Standard Ethernet does not fulfill this requirement, which makes it difficult to implement hard real-time applications using this bus technology.

One single cabling: Mass data as well as status and control information shall be exchanged via the same cable that is used for all purposes within the system. In order to reduce the costs for cabling and connectors, a serial data transmission is preferred.

Easy reconfiguration: While plug-and-play capability is self-evident for consumer electronics, this feature can also be very helpful in industrial environments, like a complex measurement system with several different signal acquisition and processing devices. If the user just plugs them together without any elaborate setup, the setup time is significantly reduced. This requirement also includes that no specific network topology be enforced.

All these requirements are fulfilled by the bus as defined in the IEEE 1394 standard for a High Performance Serial Bus [1–3], short "IEEE 1394" with commercial implementations known as FireWire and i.LINK.[1] Therefore, it is obvious that this bus standard originally used for consumer electronics and computer peripherals becomes more and more popular for industrial and factory automation applications. The following sections describe the fundamental facts of IEEE 1394, a reference design for industrial IEEE 1394 nodes, and some real-world applications for industrial environments.

45.2   IEEE 1394 Basics

IEEE 1394 is a serial bus connecting nodes with data rates up to 800 Mbit/sec that can be mixed with older devices running at data rates of 100, 200, and 400 Mbit/sec. The standard supports two completely different types of data transfer:

1. Asynchronous transfers are rather short messages that are mainly used for control and setup purposes. Their exchange is controlled by a request and response scheme that guarantees a data delivery for read or write operations, respectively, generating well-defined error codes. This type of communication is used if reliability is more important than timing, as it cannot be exactly determined at which time an asynchronous request of the application is actually sent to the connected node.
2. Isochronous channels are used for mass data that require a fixed, guaranteed bandwidth. The streaming data are divided into packets that are sent every 125 µsec. This 8 kHz clock is distributed across the network by a special packet called Cycle Start Packet. Typical examples of isochronous transfers are real-time video streams where late data are useless. The reception of the data is not secured; the sender does not even know if any other node is listening to the data.

These two transfer types use a common Physical and Link Layers of the IEEE 1394 protocol stack as depicted in Figure 45.1. The Link Layer provides addressing, data checking, and data framing for packet transmission and reception, whereas the Physical Layer transforms the logical symbols used by the Link Layer into electrical signals, which includes bus arbitration in case several nodes want to send data at the same time. The isochronous data are directly fed into the Link Layer; for asynchronous data, there is an additional Transaction Layer as a standard interface for the application on top of the protocol stack. Its main task is to implement the secure request/response protocol for asynchronous transactions. All three layers exchange data with the so-called Serial Bus Management that incorporates some special functions for managing and optimizing the bus. Normally, Physical and Link Layers are realized by dedicated chip sets, whereas Transaction Layer and Serial Bus Management are implemented in firmware. Serial Bus Management is optional and can normally be omitted for design of embedded nodes if a host computer with a common operating system is part of the network.

Until 2002, the physical medium used for IEEE 1394 was restricted to special copper cables and IEEE 1394 specific sockets and connectors. The situation has changed with the passing of the new standard amendment IEEE 1394b-2002 [4]. A matrix with all supported speeds and reach for different kinds of media is shown in Table 45.1. IEEE 1394b offers two promising options for IEEE 1394 in industrial applications. On the one hand, existing CAT5 cabling can be used if a data rate of 100 Mbit/sec is sufficient, for example, if only short status and control messages have to be exchanged. On the other, optical media

---

[1] FireWire is a trademark of Apple Computer, Inc. and the 1394 Trade Association, i.LINK is a trademark of Sony Corporation.

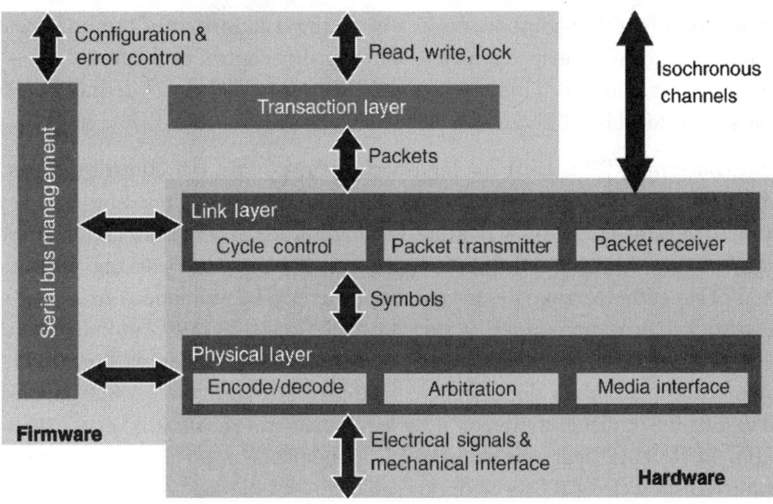

FIGURE 45.1　Layered protocol architecture of IEEE 1394.

TABLE 45.1　Supported media by IEEE 1394

Media	Reach(m)	S100	S200	S400	S800	S1600	S3200
IEEE 1394a STP Copper	4.5	✓	✓	✓			
CAT-5 (Ethernet)	100	✓					
Plastic Optic Fiber	50	✓	✓	(✓)			
Glass Optic Fiber	100			✓	✓	✓	(✓)
IEEE 1394b STP Copper	4.5			✓	✓	✓	(✓)

combine the advantages of high data rates with long distance. If the special Shielded Twisted Pair (STP) IEEE 1394 cables, both IEEE 1394a and IEEE 1394b, are used, not only data but also power can be provided. A maximum current of 1.5 A at a typical voltage of 12 V is sufficient for several applications, so that a separate power supply can be omitted for most nodes. IEEE 1394b only affects the Physical Layer: existing applications that use the original IEEE 1394-1995 standard and the first amendment IEEE 1394a-2000 can easily migrate to IEEE 1394b by just replacing the Physical Layer; no change of software is required.

Every data are transmitted as packets consisting of a header that describes the type and destination address of the packets and the payload. The receiver can check the integrity by means of CRC sums that are also included. The IEEE 1394 standard only specifies how packets are transmitted from one node to another, that is, it covers only the lower layers of the ISO/OSI architecture. The definition of common interpretation of the payload is left to application-specific transport protocols. Currently, more than 60 additional specifications exist, most of them for audio and video consumer electronics. Only a few of them are actually of interest for industrial applications:

1. Serial Bus Protocol 2 (SBP-2) [5] defines a generic method for asynchronous data exchange between two nodes. Principally, it encapsulates an arbitrary command set, but is mainly used for the implementation of the SCSI-2 protocol via IEEE 1394. SBP-2 is a very versatile protocol but has a significant administrative overhead, so that it is not well suited for extremely high-performance applications. This limitation is overcome by a new version of this standard, called SBP-3, which significantly reduces the amount of overhead and also allows isochronous transport of data.
2. The Industrial and Instrumentation Digital Camera (IIDC) 1394-based Digital Camera Specification [6] (short DCAM) describes how industrial cameras delivering uncompressed video

data are accessed by other components as well as the data format of the image data. The main applications are industrial inspection systems for quality control or the like.

3. The Instrument & Industrial Control Protocol IICP Specification [7] defines how to implement the IEEE 488 protocol on IEEE 1394.

The situation concerning IEEE 1394 for high-performance industrial control systems is somewhat complicated. Of course, IEEE 1394 is already used for this application area, but existing solutions are based on proprietary protocols. In the meantime, leading European companies selling products for factory automation, motion control, and the like as well as research institutes working on IEEE 1394, factory automation, and production technologies have formed the "1394 Automation" Association. Their main task was to develop and is to maintain a standard called "1394AP (1394 Automation Protocol)," that allows controllers from different vendors to communicate with each other via IEEE 1394. Additionally, DCAM-compliant commands and video streams can be embedded into the 1394AP packets. More details and issues concerning the implementation of "1394AP" are described below in Section 45.5 However, the exact specification of "1394AP" will only be accessible to members of the "1394 Automation" Association until first products using "1394AP" become available on the market. Afterwards, the responsibility for the protocol will be handed over to the 1394 Trade Association. All this should happen until end of 2004.

45.3 IEEE 1394 System Design

This section describes a generic architecture of an embedded IEEE 1394 device, that is, everything that is not a conventional computer. As stated above, only the lower layers of the IEEE 1394 protocol are realized by dedicated hardware; the upper layers are firmware. Therefore, every IEEE 1394 node must contain some kind of processor that executes the higher layers. This leads to the system architecture as depicted in Figure 45.2. Here, still an IEEE 1394a solution is described. However, if IEEE 1394b is necessary either for speed or because of the necessity of alternative media, only a different hardware for Link and Physical Layers has to be chosen; the overall architecture remains the same.

For the Physical Layer, any standard PHY chip can be used. A three-port PHY gives most flexibility for the network topology. IEEE 1394a forbids closed loops with the system that results in a tree-like network topology, whereas IEEE 1394b automatically breaks up loops and also favors tree-like networks. A variety of Link Layer Controllers (LLC) from different manufacturers exist. For industrial applications, the TSB12LV32 "GP2Lynx" by Texas Instruments is well suited. It provides a memory mapped interface to a microcontroller for asynchronous data and setup of the IEEE 1394 data transmission as well as a 16-Bit high-speed data port that can send or receive data at a full speed of 400 Mbit/sec. Unfortunately, until

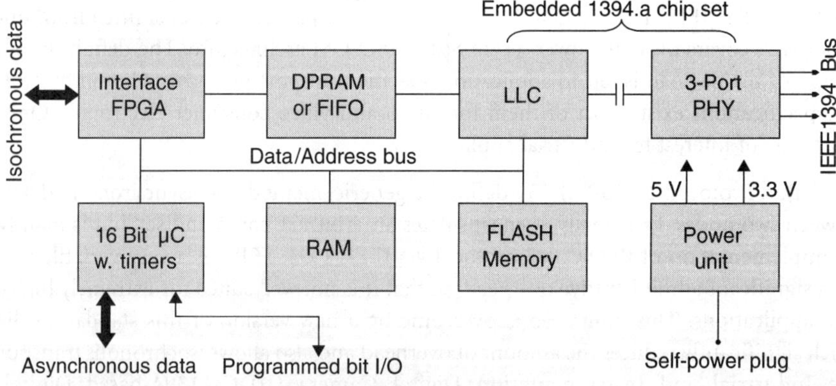

FIGURE 45.2   Hardware architecture of embedded IEEE 1394 device.

now (July 2004) no embedded IEEE 1394b LLC capable of handling 800 Mbit/sec and dedicated to isochronous transmission is available. Oxford Semiconductor produces a chip called OXUF922 that supports 800 Mbit/sec, but is mainly intended for mass storage devices like hard drives as it includes some hardware support for SBP-2 but not for isochronous streaming. One solution might be to use the TSB82AA2 LLC of Texas Instruments that is IEEE 1394b compliant and has a PCI bus interface. If just one application-specific hardware is connected to the PCI interface, it can be treated as a general-purpose 32-bit wide interface. The complexity of PCI only arises if it is actually used as a bus accessed by numerous components.

An important issue, especially for industrial applications, is the electrical interface between PHY and LLC. Since DC voltage levels in IEEE 1394a communicate some information between the PHYs, all PHYs are directly connected with each other. If the nodes are powered from an external supply, they might operate at different ground levels, which may produce a current flow on the IEEE 1394 cables. In order to prevent a faulty operation of the bus or even a damage of the nodes, a galvanic isolation between the LLC and the PHY is strongly recommended. Normally, if the chips include some busholder circuitry, a capacitor inserted into the data and control lines is sufficient. Of course, the problem of additional galvanic isolation disappears if IEEE 1394b and optical media are used.

A crucial point is the choice of the processor for the IEEE 1394 node. Usually, the data handling is split: the processor executes the IEEE 1394 stack, cares about the asynchronous data, and controls additional low bandwidth hardware via programmed bit Input/Output (I/O), whereas the isochronous mass data are processed by some application-specific hardware. In this case, a 16-Bit microcontroller like the SAB-C161PI by Infineon gives sufficient computing power. Another processor that has become very popular for embedded IEEE 1394 devices is the ARM7 type. Of course, the necessary amount of memory depends on the application; for the IEEE 1394 stack, 64 kbyte of both RAM and ROM are sufficient. The use of an FPGA for the hardware processing of mass data provides flexibility, so that the same design can be used for a broad range of applications. Some buffering of data is required at the high-speed port of the LLC in order to transform the streaming data into isochronous packets or *vice versa*. Some LLCs have a sufficient amount of internal FIFO memory; others like the GP2Lynx require an external buffer realized by a Dual Ported RAM or a FIFO. Another solution is to implement the FIFO inside the FPGA.

The architecture of the software implementation of the protocol stack (see Figure 45.3) represents the layer structure as defined in the standard. However, two additional layers are inserted, both of which provide a universal programming interface. The embedded application closely interacts with the Serial Bus Management, the Transaction Layer, and the Link Layer. In order to ease the coding of the application for the user, a common Application Programming Interface (API) must be provided. API calls

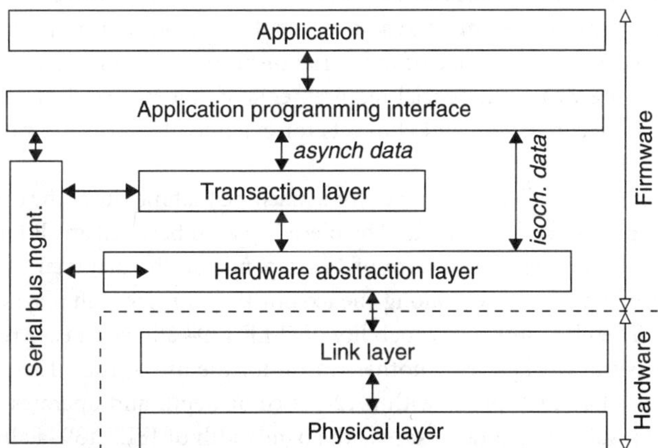

FIGURE 45.3 Embedded IEEE 1394 protocol stack.

include routines for initialization of the bus, the basic asynchronous transactions (Read, Write, and Lock), setup of isochronous transfers, and inquiry of information on the status of the bus and the local node as well as callback functions that are automatically executed in case of external bus events. The latter generate responses to incoming requests without explicit programming in the application. However, the API, the Transaction Layer, and the Serial Bus Management shall be independent of the Link Layer Controller in order to be usable for different embedded systems. Therefore, an additional Hardware Abstraction Layer (HAL) is used that transforms services requested to the Link Layer into corresponding hardware accesses. No embedded real-time operating system is required; time-critical tasks are scheduled via timers of the microcontroller.

For proper operation of IEEE 1394 in industrial environments, a number of requirements must be fulfilled. Firstly, it is important to follow the rules of the PHY manufacturers for PCB design, which include short distance between PHY and LLC, etch traces that match the line impedance of the cable, and avoidance of vias between the PHY and the IEEE 1394 connectors. Secondly, galvanic isolation is self-evident for IEEE 1394a, as described above. Thirdly, EMI protection of the power supply must be applied. Fourthly, shielding and grounding have to be accurate. If these rules are considered, a stable operation of IEEE 1394 nodes can be guaranteed even under harsh industrial conditions. Some examples are described in the following sections.

45.4 Industrial Applications of IEEE 1394

Since the most distinguished feature of IEEE 1394 is its high bandwidth of maximum 50 Mbyte/sec for IEEE 1394a, respectively 100 Mbyte/sec for IEEE 1394b; currently, it is mainly used in industrial systems that comprise some kind of image sensors. They produce a large amount of data that must be transmitted to an image processing system in real time. A typical example is an automated optical quality control of goods. Therefore, the following examples refer to this kind of application. All of them still use IEEE 1394a. The application of IEEE 1394 as field bus replacement will be covered in a separate section.

The first example is an industrial camera that comprises a CMOS image sensor. By using special readout electronics, a dynamic range of 120 dB per pixel is achieved, which allows to capture images with both extremely bright and dark regions, like the welding scene in Figure 45.4. The comparison with an equivalent CCD image (also in Figure 45.4) shows that no artifacts like blooming or smearing occur for the CMOS image. The main applications of this camera will be optical inspection of industrial production like the welding scene in Figure 45.4, but also automotive applications like driver assistance. Here, intensive sun lighting and shadows produce images with enormous contrast that cannot be acquired with conventional CCD cameras.

The IEEE 1394 interface of this camera uses the hardware and software described in the previous section. The camera operates in accordance to the DCAM standard. Since there are already a number of interface cards, image processing systems, and software tools available that support the DCAM standard, this camera can be used for a broad range of industrial inspection systems. Since another application is driver assistance in passenger cars, the special requirements of automotive electronics must be fulfilled. Special IEEE 1394 latching connectors and chip sets for extended temperature range are commercially available.

A second application is a so-called Datalogger for a machine that measures the outer contours of rotation symmetrical workpieces like crankshafts. The piece is placed between an IR lighting unit and some CCD line cameras, whose number (maximum of 10) depends on the diameter of the workpiece. Both the lighting and the cameras are moved along the axis of the workpiece in a synchronous manner, as shown in Figure 45.5. The plug-and-play capability of IEEE 1394 allows an easy reconfiguration of the system for different types of workpieces. Another reason for the use of IEEE 1394 in this application is bandwidth: each camera has 2048 pixels with a 12-Bit color depth and operates at a maximum pixel clock of 20 MHz. The resulting data rate exceeds the bandwidth of IEEE 1394a already for one camera so that areas outside the region of interest are clipped using a lookup table and run length encoded by hardware.

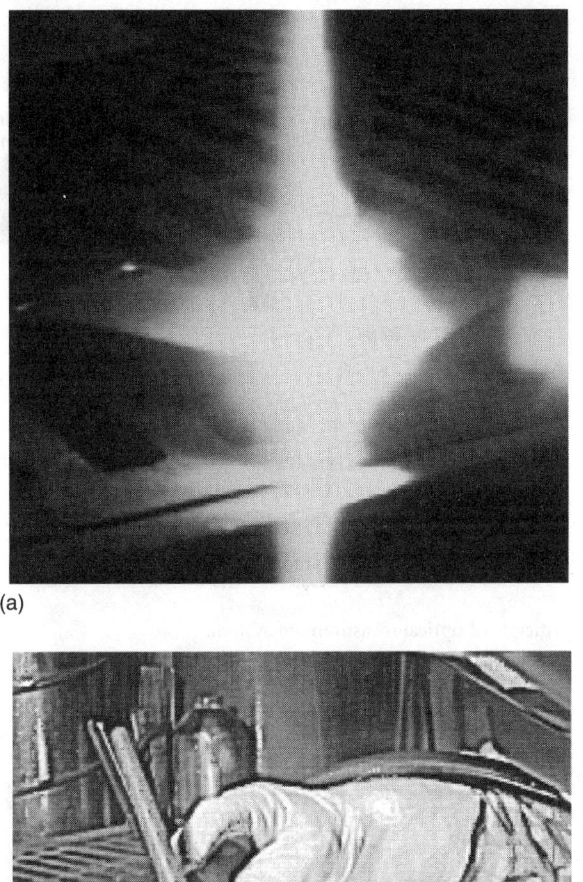

(a)

(b)

FIGURE 45.4 Welding scene acquired with conventional CCD camera (left image) and advanced CMOS camera (right image).

This task, together with the transformation of pixel and additional mechanical data into IEEE 1394 packets, is performed by circuitry implemented in the FPGA. The packets are written into the FIFO and transmitted via the IEEE 1394 subsystem shown in Figure 45.2. With respect to the type of transaction, a trade-off must be made between asynchronous packets that guarantee a secure transmission at the cost of bandwidth for response packets and isochronous transfers that lead to high bus throughput with the lack of feedback upon success or failure. For reasons of security, asynchronous transfers in combination with a proprietary transport protocol have been chosen. However, first experiments have been shown that

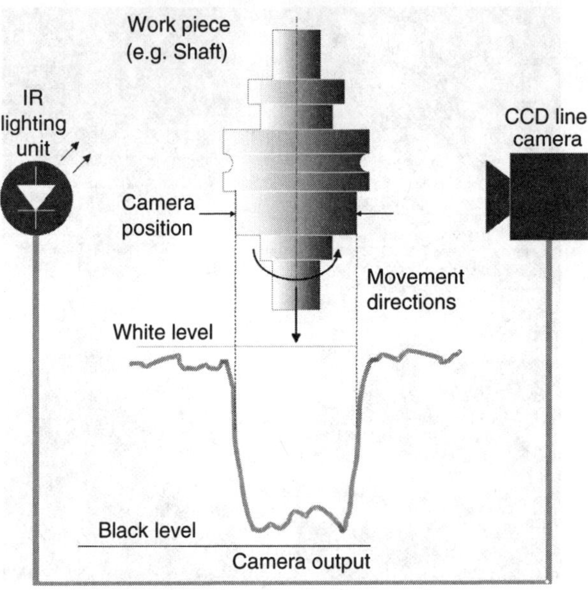

FIGURE 45.5   Working principle of optical measurement system.

FIGURE 45.6   Module for data compression and transmission used in optical measurement machine.

the Datalogger is able to generate data up to 30 MByte/sec; but the connected PC that performs the image analysis cannot handle such amount of data over a longer period of time. Therefore, it is likely that a second version will use isochronous channels for data transmission.

Prototype systems of the Datalogger have been realized (see Figure 45.6) and successfully tested. Due to constructive reasons, only 6 in.^{2} PCB area can be used per camera. The complete system has passed intensive EMC/ESD testing since all rules mentioned in the previous section have been strictly followed. The machine operates in the direct neighborhood of machine tools; but no faulty operation of the IEEE 1394a bus has been observed so far.

The third device is a so-called Color Photo Scanner for extremely high-resolution image acquisition. It is used for applications in which short acquisition time is not the most important issue, but a precise imaging of fine details, like in optical inspection of printed circuit boards. The design goal has been a geometrical resolution of more than 8000 pixels for the shorter edge of the image area and a color depth of 12 bits per elementary color. In order to avoid the enormous costs of a corresponding CCD area sensor, a mixed mechanical–optical principle has been chosen. An arbitrary 3D scene is projected onto a focusing screen via a lens and the resulting 2D image is scanned with a CCD line sensor that is moved by a step

FIGURE 45.7   Color Photo Scanner.

motor. This results in an image size of 8192 times 12,000 pixels per image. In order to reduce the scanning time to the physically minimum duration — integration time per line times number of lines — two hardware tasks must operate in parallel: one must control the CCD sensor, the motor, and the analog, respectively digital image processing circuitry, and the other one the transmission of data via IEEE 1394.

The architecture of the Color Photo Scanner (see photo of the final product in Figure 45.7) is derived from the reference design described in the previous section. The two hardware tasks are implemented inside the FPGA. Instead of the FIFO, a fast SRAM organized as a cyclic buffer is used as an interface between these two tasks. The design makes use of the fact that the high-speed port of the LLC can also handle asynchronous packets at high speed, so that the controller does not care about the mass data. It only executes the IEEE 1394 protocol stack and an implementation of the SBP-2 protocol that is used for the control and setup of the scanner. The microcontroller provides enough computing power for that purpose. The device that has become a commercial product is able to capture a complete image within 90 sec, which can only be achieved by hardware support of the FPGA. All necessary parts fit into the modified housing of an SLR camera.

45.5   IEEE 1394 Automation Protocol

The previous section describes some industrial applications of IEEE 1394 but leaves real industrial communication networks out of consideration. This application field has been dominated by proprietary solutions in the past. Only recently, situation has changed with the passing of the "1394 Automation Protocol," short 1394AP. Its theoretical principles are defined in [8]

In industrial automation, a trend toward decentralized systems can be observed over the last few years. In conjunction with the growing capabilities of embedded hardware, more and more of the system functionality is transferred from the central control unit to distributed control units. A typical example is motion control: the system consists of several I/O modules and decentralized control units to drive the axes. One control node is designated as the host control. This node is responsible for maintaining the network, which includes startup and parameterization, control and supervision of all system components, and network management. Most of the time, the host control is implemented on an industrial computer, but this task may be assigned to an; one of the control units.

Several buses have been established in industrial control in order to connect the decentralized modules, like Ethernet and its derivates, Profibus, CANBus, SerCos, and others. All these technologies lack one important feature: there is no method for cyclic, deterministic distribution of control information with guaranteed delivery at fixed intervals, which is a prerequisite for industrial, especially motion control systems. However, this type of information exchange is one of the characteristics of IEEE 1394: its isochronous mode for data transmission. Therefore, a number of companies selling products for factory automation have integrated IEEE 1394 in their equipment, but are still using proprietary protocols, so that interoperability is not ensured. This limitation shall be overcome by the 1394AP. Its specification as well as the development of a reference implementation have been initiated by the European "1394 Automation" association, which consists of more than a dozen factory automation companies and research institutes.

IEEE 1394 matches the requirements on communication in industrial control systems:

- Communication network has to be fast, robust, and inexpensive.
- IEEE 1394 supports distributed architecture as a tree structure. Isochronous and asynchronous transmission schemes permit unrestricted communication between the system components.
- IEEE 1394 itself already provides the cycle clock synchronization, which has to be expanded to control specific data exchange.
- Low jitter of IEEE 1394 is one reason that makes this bus attractive for industrial control.
- The variety of IEEE 1394 devices guarantee independence from a single supplier and its price policy.

Therefore, IEEE 1394 is an almost ideal choice for industrial communication networks. However, specific features like Network Topology, Application-Specific Control and Status Registers, Packet Payload, Network Management, and Message Synchronization have to be defined, which makes up 1394AP. Because of the special requirements of industrial communication, 1394AP differs somehow from the usual properties of an IEEE 1394-based transport protocol.

One node of the network acts as the so-called Application Master. Most of the times it will be the control PC of the network; but any other node with sufficient computing power may fulfill this function. The capability of becoming Application Master is stated by special entries in the Configuration ROM of the 1394AP node. Since the Application Master is responsible for synchronization of all nodes and the usual IEEE 1394 cycle start packets are used for this purpose, the root of the tree-like network shall become the Application Master. All nonroot nodes are called slaves.

The main task of the Application Master is the cyclic transfer of input data to the slaves. This information is summarized in the Master Data Telegram (MDT), which is the payload of an IEEE 1394 packet. The update rate of the control variables can be adjusted via the 1394AP-specific Application Cycle. Its length is based on the requirements of the application. For high-speed applications, the Application Cycle is identical to the standard IEEE 1394 isochronous cycle of 125 μsec. In this case, the MDT is the payload of isochronous packets, which are sent immediately after the Cycle Start Packet. In case of reduced performance requirements, the Application Cycle is spread over several isochronous cycles. Then, for the MDT, both isochronous and asynchronous packets can be used. In the first case, each packet only carries part of the complete MDT; in the second case, asynchronous broadcast write requests are used. The MDT of 1394AP only defines a data structure for transmission of application-specific variables, but does not specify the meaning of the variables itself. This is left to the application, which gives flexibility for the use of 1394AP. While the slaves receive the MDT and extract the data that are necessary for proper operation, they output their data via the Device Data Telegrams, short DDT. The DDTs transfer the output data both to the master and to the other slaves, so that an actual peer-to-peer data transfer is ensured. The remarks on Application Cycle for the MDTs are also valid for DDTs. Each node provides Network Management services, which include Node activation, suspension, configuration, initialization, and reset as well as Error handling. The Network Master — identical to the Application Master — controls changes in the state of the nodes.

One of the main reasons why IEEE 1394 has been chosen as a control bus for real-time systems is due to its deterministic timing via isochronous data transfer. The clocks running in each individual node are synchronized every 125 μsec by the Cycle Start Packets. Data to be sent either as MDT or DDT are held in local buffers and marked with a time stamp. The time information contained in the Cycle Start Packets is used as a trigger that releases the data for transmission. One problem has to be overcome if isochronous and asynchronous data transfers are mixed. After a cycle starts, first all isochronous packets are sent, and the remaining part of the cycle is used for asynchronous data. However, IEEE 1394 does not check if the transfer of one asynchronous packet can be terminated within 125 μsec since the last cycle start. If not, the next Cycle Start Packet will be delayed. For 1394AP, this results in nondeterministic timing, which is not acceptable. Therefore, the root note also implements some functionality summarized as Asynchronous Resource Manager that prevents the delay by managing the limited asynchronous resources between all nodes of the network.

For first experiments, a communication node that supports features of a preliminary version of the 1394AP protocol by hardware has been designed. It can only be used as a slave node. Principally, it again uses the system design depicted in Section 45.3, but comprises a different Link Layer Controller because of the special communication pattern defined in 1394AP. The "GP2Lynx" is only capable of either sending or receiving isochronous data, but not both simultaneously as required for MDTs and DDTs. Therefore, the Texas Instruments TSB42AB4 "CeLynx" device is used in this design. Originally targeted to consumer electronics, it is also well suited for industrial applications and supports full duplex isochronous streaming.

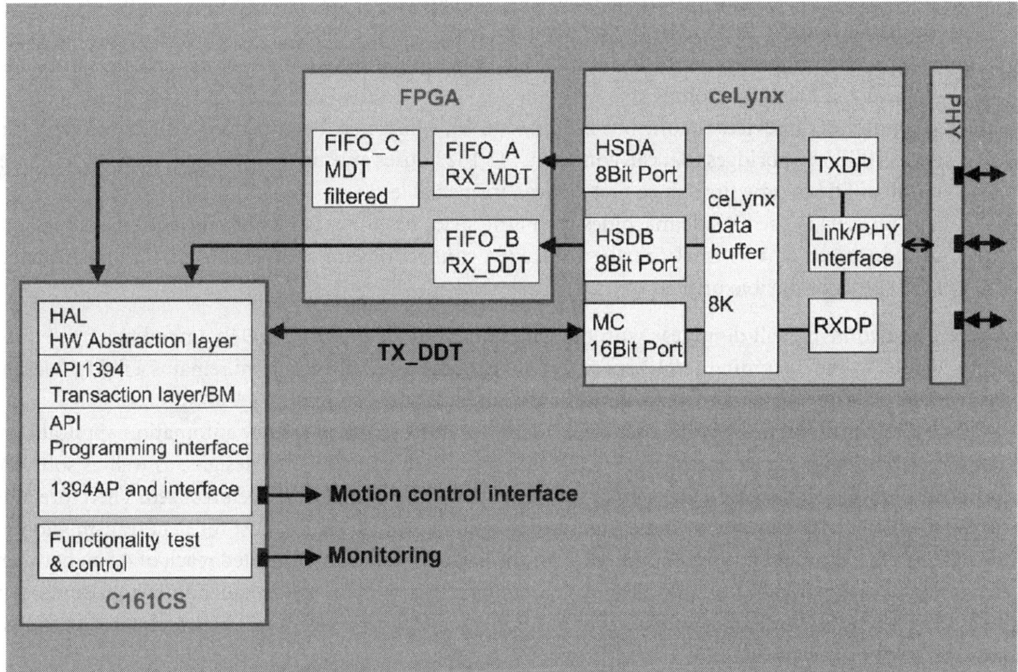

FIGURE 45.8　Architecture of node for synchronous industrial communication via 1394AP.

FIGURE 45.9　Node for synchronous industrial communication via IEEE 1394.

The resulting schematic of the node is shown in Figure 45.8. The FPGA filters the data of incoming MDTs and DDTs, so that the microcontroller of type Infineon C167CS only has to handle relevant data. Outgoing DDTs are sent as asynchronous packets and directly written into the Link Layer Controller by the microcontroller. Figure 45.9 shows the industrial communication node. It is a stacked design using two PCBs; therefore, part of the circuitry is not visible in Figure 45.9. Experiments with this embedded system have shown that 1394AP is suited for industrial real-time communications, especially for motion-control systems.

45.6 Summary

The four examples in the last two sections show that IEEE 1394 is an emergent and well-suited technology for industrial applications. However, some minor problems exist:

1. Currently, almost all IEEE 1394 devices use IEEE 1394a, which limits the distance between two nodes to 4.5 m as defined by the standard, while high-quality cables allow up to 10 m. But for large industrial systems, this might still be not sufficient.
2. The DC coupling of physical layers in IEEE 1394a implies the danger of disturbance problems, especially in harsh environments.
3. The number of devices on one bus is limited to 63, which may be not enough for certain applications. IEEE 1394 bridges that can connect up to 1023 buses are currently in the specification phase, but it will take some time since first devices are available.
4. If IEEE 1394 is used inside machines with little space left for cabling, the conventional cables and connectors are not optimal. Smaller IEEE 1394 connections for short distances, like, for instance, PCB stacks or flexible printed tapes, are desirable.

The first two items will disappear with the widespread use of the IEEE 1394b amendment; the third one is tackled by the upcoming 1394.1 specification. Unfortunately, the last item remains a field of design experiments, because a standardized solution does not exist at the moment.

These items shall not prevent the use of IEEE 1394 for industrial and factory automation applications. The basic technology consisting of IEEE 1394 chip sets, hardware reference designs, as well as software protocol stacks are already available, and first commercial applications have been realized. Therefore, it is very likely that a large number of industrial systems will incorporate IEEE 1394 for data transmission in the near future. The most important limitation of the original standard — limited reach of 4.5 m between two nodes via copper cables — has been overcome by the new IEEE 1394b amendment. In summary, IEEE 1394 is already more than an emergent technology nowadays, since it will be widely used in industrial applications very soon.

References

[1] *IEEE Standard for a High-Performance Serial Bus 1394-1995*, IEEE Press, New York, 1996.
[2] *IEEE Standard for a High-Performance Serial Bus Amendment 1 1394a-2000*, IEEE Press, New York, 2000.
[3] Anderson, D., *Firewire System architecture*, 2nd ed., Addison-Wesley, Reading, MA, 1999.
[4] *IEEE Standard for a High-Performance Serial Bus Amendment 2 1394b-2002*, IEEE Press, New York, 2002.
[5] Serial Bus Protocol 2 (SBP-2), ANSI Standard NCITS 325-1998.
[6] IIDC 1394-based Digital Camera Specification, Version 1.31, Available from 1394 Trade Association (see http://www.1394ta.org/Technology/Specifications/specifications.htm), 2004.
[7] 1394TA IICP Specification for the Instrument & Industrial Control Protocol, Version 1.00, Available from 1394 Trade Association (see http://www.1394ta.org/Technology/Specifications/specifications. htm), 1999.
[8] Beckmann, G., Ein Hochgeschwindigkeits-Kommunikationssystem für die industrielle Automation, Dissertation, Technical University of Braunschweig, 2001 (in German).

46

Which Network for Which Application

Jean-Dominique Decotignie
CSEM (Centre Suisse d'Electronique et de Microtechnique)

46.1 Introduction

Since the advent of industrial communications, dozens of solutions have been designed that all claim to solve the problem. This is true but most of the time in a given context. The reason for this is that there is no single case in industrial communications. Most of the time, industrial processes are organized in a hierarchical manner and, depending on the location in the hierarchy, needs will differ. The manner in which the control software is written also introduces different needs in the communication networks. To select a communication network, it is thus important to understand the different views that the network designers have when they build their solution.

In this chapter, we will first describe the way industrial production is organized. In the second part, the differences of approaches in architecturing control applications will be explained. When introducing a network to support communication between control entities, a number of constraints are added, such as errors and delays. As they have an impact on the system, they will be detailed in the third part. The last section will present selected industrial networks and show which type of application they may support and how well they do the required job.

46.2 Production Hierarchies

Control of a production system is performed by numerous computers organized along several hierarchical levels. Computer networks provide communications between the computers within a level and with some of the computers at adjacent levels.

Before presenting the different levels in more details, let us say that the closer an application is to the process, the higher the temporal constraints (ISO 1994). Conversely, the quantity of transferred data increases with the level in the hierarchy. The position of an application in the production hierarchy also influences the way the application software is built. Applications at the lowest levels often adopt a time-triggered approach (see Section 46.4), while at higher levels an event-driven approach is used in most of the cases.

Numerous factory models, also known as CIM models, have been described in the literature (for a summary, see [Jones et al. 1989]). In terms of hierarchy, all the above proposals do not differ very much. Differences exist mainly with respect to the number and the labeling of the levels. Figure 46.1 depicts a possible summary of these proposals. This figure shows that there exists a more profound difference between process control and manufacturing.

The primary purpose of a plant control system, organized in a hierarchical manner as in Figure 46.1, is to manage the process as seen through the sensors (level 0) by acting upon it through actuators.

Sensors and actuators are linked to the first level of automation (level 1), where each automation device controls a variable of the process in such a way that it stays within limits or follows set point data given by the next level of automation (level 2). Level 1 corresponds to an "axis," control of a single entity, in manufacturing or a local loop in process control. The variable may be discrete, such as a tool changing mechanism in a machine-tool, or continuous such as an axis in a robot or a heater in a distillation column.

Units at level 2 elaborate the set point data or the limits for variables that have to be linked or related to ensure a proper operation. This is typically the purpose of an interpolator in Computer Numerical Controller (CNC) or a tool magazine controller for a machine-tool or, a robot path controller that gives the set point data for joints in order to follow a given path. These units receive their commands from and return their status to the machine or process level (level 3).

Units at level 3 coordinate the actions on, or the operating conditions of, several elements or groups of elements in order to realize or optimize operations or sequences of operations on objects, either solid or fluid. They receive their commands and report to the line or Distributed Control System (DCS) level (level 4).

Units at levels 1 to 3 also have to perform diagnostics on themselves, detect emergency conditions, and take an immediate decision if they have enough information to do so. Otherwise, they report to the next upper level for decisions.

Level 4 units are responsible for optimizing the production, that is, scheduling when and where given operations are performed on objects, and for ensuring that all necessary resources are present.

Network	Level		Manufacturing	Process control
Public network	6	Plant management	Plant management	Plant management
Office LAN	5	Factory controller	Supervisory controller	Supervisory controller
Industrial LAN	4	Cell/line controller	Line	Distributed control system
cell network	3	Workstation controller	Machine	Process controller
Backplane/ cell network	2	Automation module ctrl	Axes	Dedicated controller
Fieldbus	1	Device controller	Axis	Dedicated controller
Fieldbus	0	Sensor or actuator		

FIGURE 46.1 Production hierarchy.

Level 5 normally corresponds to the elaboration of a single product or a family of products on the same devices. That is where process planning for products is performed.

Level 6 represents the top level of a plant. Basic functions performed at this level are product design, resource management, high-level production management with interarea production scheduling, plant policies establishment, etc.

At the top, we may find a corporate management level (not shown in Figure 46.1) that is linked to the different sites or plants via public or corporate networks. At this level, all the nonproduction activities (research, finances, ...) of the enterprise can be linked.

It should be noted that some of these levels might be absent in a given enterprise. Depending on the level of automation and the control structure, level 2 and even level 1 might be included in level 3. Levels 5 and 6 might also be collapsed into a single level. Today, there is a clear tendency to reduce the number of levels. However, the functions described above are always present whatever the number of levels.

46.3 Processes Types

Processes controlled by computers are usually classified according to two paradigms: continuous systems and discrete event systems (Halang and Sacha 1992).

An example of a continuous process is a temperature regulator that keeps the temperature in a reactor within preset limits around a given set point value. Such a system reads the temperature as given by a sensor, compares it with the set point value, computes the necessary correction using an adequate control algorithm, and energizes a heating actuator accordingly. This sequence of operations is repeated at regular intervals, often periodically. The period is set according to the dynamics of the controlled process.

On the other hand, the control of an elevator pertains to the class of discrete systems. Let us describe a part of a typical sequence of operations. While the elevator is approaching its destination floor, the system is expecting that the floor deceleration sensor turns on, in which case, the control system has to react by decreasing the speed of the elevator. In the next step, it has to stop the elevator when the floor sensor indicates that the floor has been reached. During this sequence, other events may occur such as summons requests from external passengers. Such requests may either alter the current sequence of operations or just need to be memorized for later treatment. In addition, a given event may not be interesting at all times. In the above example, the deceleration event is only of interest if the elevator has to stop on the corresponding floor. Event detection needs to be disabled sometimes and enabled again later. An essential characteristic of discrete event systems is that, if the action triggered by the occurrence of an event is perfectly known, the point in time at which this event will occur is *a priori* unknown. Furthermore, the order in which two or more events occur is often very important.

Batch Systems

Several physical systems present both of these aspects, they are called batch, "MIXT" systems.

46.4 Control Systems

Continuous systems and discrete event systems implemented by cyclic or periodic polling are often called time-triggered systems or sampled data systems.

Discrete events systems implemented using interrupts or internal software events are referred as event-triggered systems (Kopetz 1991).

Time-Triggered Systems

From the control system viewpoint, continuous systems will be implemented as looping tasks that wait the beginning of the period, sample their inputs, compute the new output values, and set the actuators according to the new values. Periodicity is not mandatory but is often assumed as it leads to simpler algorithms and more stable and secure systems. Most of the algorithms developed with this assumption are very sensitive to period duration variations, jitters in the starting instant. This is especially the case in motor controllers in precision machines. Simultaneous sampling of inputs is also an important stability factor. Let us assume that the controller implements a state-space control algorithm in which the state vector includes the position, the speed, and the acceleration of a mobile, and that the actual values are obtained by three sensors, a position encoder, a tachometer, and an accelerometer. The control algorithm assumes that measured values were acquired at the same instant. For the control systems, this translates into what is called simultaneous sampling. The acquisitions are indeed not really simultaneous if only one processor is devoted to the three operations. Their instants must, however, be as close as possible and remain apart under a given limit, which may be roughly estimated to two orders of magnitude below the period.

Discrete Event Control Systems

Discrete event systems may be implemented as a set of tasks activated by event occurrences. Let us define an event as a significant change (or a sequence of significant changes) of state of the process (the so-called occurrences). This change, the event, is detected by some input circuitry that continuously monitors the input and is transformed into an interrupt (detection may also been done by software). The event is handled by a task that undertakes the required actions. The time elapsed between the occurrence of the event and the corresponding reaction, often called reaction time, is bounded and given in the requirements. Reaction times may depend on the kind of event. Reactions to events are normally handled according to the order of arrival of the events. However, some events may be more important than others. The emergency stop in an elevator is clearly more important than a summon request. This will translate into priorities in the event handling. In any case, the order in which events occur is important for the application.

The above implementation technique is conventional but is not the only one. It is always possible to implement a discrete event system as a continuous one. In such a case, all the inputs are sampled at regular intervals by the control software that detects the changes and undertakes the necessary reactions. This is the way most Programmable Logic Controllers (PLCs) are implemented. With such an implementation, precedence between events may only be asserted if the events are detected during different polling cycles. If two events are detected during the same cycle or period, they will be considered as simultaneous. The cycle duration or the period should be selected in such a way that all events can be detected. In the elevator example, if the deceleration switch closes during a minimum of 20 msec, the polling period should clearly be lower in order to detect the close event.

In summary, continuous control systems exhibit four important characteristics:

- they are cyclic and often periodic, and period values are set according to the process dynamics,
- jitters in the period should be limited to a few percent of the period,
- the instants of input acquisition and output settings are known in advance and dictated by the control system, and
- all inputs need to be sampled nearly simultaneously.

Discrete event systems may be implemented as continuous ones are. They then exhibit the same characteristics but are not necessarily periodic. However, the cycle time should be kept low enough so that all events may be sensed. These systems may also be implemented using interrupts with the following characteristics:

- event occurrence instants are not known,
- the reaction time to an event is bounded,
- the order of occurrence of events is important,
- reaction to some events may have a higher priority,
- detection of a given event may be temporarily disabled, and
- there is a limit in the density of event occurrence that may be handled by the control system.

From the control point of view, a control system is often composed of a part that is time triggered and of a part that is event triggered.

46.5 Communication Systems

The communication system is there to support the interaction between the control applications. On a given computer, different applications may coexist, some being time triggered some being event triggered. They may communicate with distant applications of the other kind (a time-triggered application may communicate with an event-triggered application).

The communication network itself may be built according to the event-triggered or time-triggered paradigm or a combination of both. Matching a time-triggered application with a time-triggered network is obviously more easily performed than with an event-triggered communication system. The latter requires some additional adaptation. Ideally, the communication system should support both views.

Communications Models

Communications models define how the different application processes may cooperate (Thomesse 1993). The most widely used communication model is the client–server model. In this model, processes interact through requests and responses. The client is the process that requests that an action be performed by another process, the server. The server carries out the work and returns a message with the result. Client and server only refer to the role the processes play during the remote operation. A process that is a client in some operation may become the server for another one. Hence, this model is clearly a point-to-point model.

This model exhibits a number of drawbacks for control applications. First, time is not taken into account. It is impossible to specify a delay between the request and the response and have some means to check that this delay would be respected because the server application is involved in the response. Second, if a client wants to make simultaneous requests to several servers, this may only be performed sequentially, one request after the other. Finally, if two clients make the same request to a server; this one will treat the requests in sequence and the answers may be different. For example, two control nodes may request the value of a sensor attached to a server node. The value returned to the first client may differ from the value returned to the second. The last two problems may be solved by adequate algorithms running on top of the client–server interaction. This results in heavy implementations with often poor performances. However, for a number of remote operations, in particular, during the configuration and setup phases, these problems do not appear and the client–server model is a good solution. For real-time operations, industrial networks need to offer performing and simple solutions that solve the problems of the client–server model. This has led to the "producer–consumer" model (sometimes called publisher–subscriber model), which is a multipoint model. This model is restricted to the transfer of information values and as such well suited to time-triggered systems.

In the producer–consumer model, each piece of information has a producer and one or more consumers. When the producer has some information ready, it gives it to the network. This one transfers the information and makes it available to the consumers. This has some advantages over the client–server model:

- The producer and the consumers need not be synchronized. The consumer need not wait for the response of the producer as in the client–server model. If the information is already transferred, it may use it; otherwise, it considers that no new information is available.
- The same information can be transferred at the same time to all the consumers. The network may thus be used in a more efficient way.
- Two or more consumers will work with the same value at a given time.
- Flow control is no longer necessary as a new information overwrites the previous one. This assumes that the production of a new information outdates the previous one.
- Synchronization between applications can be implemented as production of a synchronization variable.

However, this model comes with a price. As the consumer has no way to relate the information with an explicit request (as in the client–server model), it should be able to know the age of the information. Networks implementing the producer–consumer model should be able to tag the information with some attributes from which this information can be extracted.

The producer–distributor–consumer model is an extension of the producer–consumer model that adds an extra level of separation between the production and the transfer of information. The additional role, the distributor, is in charge of transferring the information from the producer site to the consuming sites. In such a way, transfers are no longer triggered by production but are done according to rules defined for the distributor. This offers more flexibility in the scheduling of transfers in the network and results in an improved efficiency.

Temporal Consistency

Temporal consistency has two facets: absolute temporal consistency and relative temporal consistency.

Absolute time consistency is related to the age of information. A piece of data has been produced at a given time. It is later transmitted over the communication system. It is finally used at some point in time. A piece of data is said to be time consistent in the absolute way as long as the difference between the instant of production and the current instant does not exceed its validity duration. In other words, a piece of data should not be used if it is too old. This behavior is easy to ensure when all takes place on the same computer. When the data are transported over a network, the information is often lost.

In a time-triggered application, the control application expects that all input data have been acquired at the same time. The measurements have been done in a given time window. This property is called "production time coherence" in the Field Instrumentation Protocol (FIP) standard, and more generally, we can speak of "time consistency" each time that some actions must be done in a time interval (ISO 1994). It is called temporal (or relative temporal) consistency (Kopetz and Kim 1990, Kopetz 1988).

It is defined as follows. Let us consider two variables a and b. Let $[a, ta, va]$ and $[b, tb, vb]$ be two observations of a and b, where va and vb are two samples of a and b taken at times ta and tb. The samples are said to be temporally coherent from the production point of view if $| ta - tb | < R$, for a given R, where R is the temporal consistency threshold.

Spatial Consistency

As defined by Le Lann (1990), a distributed system is "A computing system whose behavior is determined by algorithms explicitly designed to work with multiple loci of control." In such systems, there are a number of control nodes that do not work independently but cooperate to perform common tasks. Cooperation may be needed to achieve a certain degree of fault tolerance or because the overall control task is too large to be handled on a single node.

Time-triggered distributed systems work periodically and the various loci of control synchronize their operations according to the passage of time. To do so, they cannot rely on local timers and need to share a common sense of time (Verissimo 1994). This may be achieved through a distributed clock synchronization algorithm (Kopetz and Ochsenreiter 1987) or, with some restrictions, using temporal "events" generated by the networks (He et al. 1990). In such systems, the various loci of control only exchange data, or state information. Obviously, all data necessary for the computations should be available to all loci of control before the synchronization instants; flow control is done statically. A single unit of data may be needed by several control nodes. As these nodes do not share a common memory, the data unit needs to be replicated in each node. This may be done by a broadcast transfer; but we need to have some guarantee that all replicas are identical at the time they are used. This property is called spatial consistency. It may be obtained by a reliable broadcast algorithm (Hadzilacos and Toueg 1993) or, as in FIP, if only an indication of spatial consistency is necessary (Decotignie and Prasad 1994).

Event Ordering

Event-triggered control systems are often very sensitive to the order in which events occur. Networks do not guarantee that requests for information transfers are handled in the order they are submitted. This means that the applications cannot rely on the order in which they receive the events from the networks to establish the order of occurrence of events in time.

Influence of Failures

Failures cannot be avoided but their influence must be minimized. In the value domain, possible failures may be undefined values or defined values considered as correct but incorrect *vis-à-vis* the related input. In the time domain, crash failures, omission failures, and timing, early or late, failures may occur. Coping with failures is the subject of fault-tolerant computing (Le Lann 1992) and, as such, beyond our scope. However, it is worth discussing the impact of the network on the system with regard to failures. With the introduction of a network, faults may occur in a number of additional entities, links, emission and reception circuits, and software. Links may be cut intermittently or permanently and may be subject to perturbations that mutilate the transmitted information. Emission and reception parts in nodes may stop, respond too quickly or too early, and emit when they are not allowed to or even constantly. Networks have been designed to resist faults by detecting and correcting errors using three types of redundancy: time redundancy, physical resource redundancy, and information redundancy. For example, redundant information is added to each message transferred on the network in the form of an error detection code, parity, or Cyclic Redundancy Check (CRC). Each code exhibits a given detection capacity, which means that there are a number of errors that may not be detected at the network level. Crash, omission, and timing failures are normally detected through the use of timers. In time-triggered systems, this detection is easy because each node has to transmit periodically. The absence of message arrival in the period indicates a possible failure, and appropriate countermeasures may be taken in the application. Furthermore, error correction may be based on temporal redundancy by retaining the previous value and just waiting for a new value at the next period. Event-triggered systems are more difficult to handle. Mutilated frames may be detected as previously and signaled to the sender by a negative acknowledgment from the receiver. The sender waits for the acknowledgment, either positive or negative. Omission failures are detected by the absence of acknowledgment in a given delay. In case of negative or absent acknowledgment, the emitter retransmits the same message. This process is repeated until success or a maximum number of retries have been reached. This means that, to cope with possible faults, acknowledgments are necessary. Furthermore, the time necessary to transfer a message from a sender to a receiver may greatly vary with a corresponding degradation of the application response time. It also introduces a delay in the transmission of other messages. A second potential problem in event-triggered systems is the difficulty to distinguish a node crash or an omission failure from an absence of event occurrence. A receiver node, for instance, an actuator node, that does not receive any message may either assume that the control node has failed or that there is no new command. In the first case, the actuator should be put in some safe mode, while in the latter no special treatment is to be done. This means that liveness messages should be sent at regular intervals in addition to normal messages.

Some failures may prevent the network from functioning if the network has not been designed properly. For instance, if all traffic is ruled by a single node, the case of centralized medium access control, a crash, or an omission failure of this node causes the network to stop unless this node is duplicated and some recovery protocol is implemented. Some node may also start transmitting messages constantly or at point in time where they are not allowed to do so. This may be the case with some deterministic Carrier Sense Multiple Access (CSMA) protocols where collision resolution is based on a priority.

46.6 Parameters to Consider in a Choice

When selecting a network for a given application, there are a number of parameters to consider:

Communication models: As described above, two main models are used in building an application layer: the client–server model and the producer–distributed–consumer model (sometimes also called the publishe–subscribe model). Distributed applications are likely to use the latter, while hierarchically organized applications may use the former.

Traffic types: Information transfers may be sporadic (event-triggered), cyclic, or periodic. In the last case, the application may impose strict limits in the jitter between consecutive transmissions of the same data.

Topology and transmission medium: If most solutions permit a tree-like topology, there might be restrictions in the branch lengths and the number of nodes in the tree. Some solutions also offer transmission media other than copper-based twisted pairs. Examples are optical fibers or radio transmissions.

Immunity to noise, environment: Networks can be used in difficult environments. This is often the case in transportation and petrochemical applications. The physical layer may be more or less resistant to such environments. Selecting an inadequate solution may lead to an increase in error rate in the transmission or even a complete fading in some cases.

Errors: As described above (Section 46.5), transmission errors cannot be avoided even if, with proper cabling and an adequate physical layer, their level can be kept very low. In many applications, temporary errors can be tolerated even if the application is not informed of the occurrence of the error. However, in some cases, this is not acceptable and the network should provide ways to inform the application that an error has occurred.

Throughput: The raw bit rate of a given network is a poor measure of the actual throughput as seen from the application. Overheads in the protocols, response delay in the network stack, medium access control schemes, traffic scheduling algorithms, and delays in the application reduce the actual throughput. As an example, transferring a 16-bit value requires the actual transfer of around 1000 bits with Ethernet, 450 bits with Profibus FMS, 200 bits with CAN, and 90 bits with FIP. Due to the other effects, the actual throughput may be one or two orders of magnitude below the raw bit rate.

Guarantees: Considerable research work has been devoted to calculating the guarantees offered by industrial networks as this is of prime importance to the applications. Guarantees give answers to different questions such as:

- will the network be able to withstand a given traffic load?
- what will be the maximum transfer time of a sporadic event?
- what happens in case of network overload?
- what will be the maximum jitter of a periodic transmission?

If most solutions offer some guarantees, few give answers to all of the questions listed above. Furthermore, most of the time, guarantees are given under the assumption that no transmission error will occur, which cannot be taken for granted.

Consistency: As mentioned above, temporal consistency is assumed in most control applications. Unfortunately, most solutions do not offer any support for temporal consistency and spatial consistency. This implies that this aspect should be added on top of the selected network.

Horizontal vs. vertical traffic: As described above, a network may be used to support traffic between computers of adjacent levels (vertical traffic). It may also be used to ensure communication between control devices of the same level to ensure coordination. Traffic patterns and communication relationships are often different in each case, and the network solution should be able to sustain both.

Services: The application layer of a network provides a number of services to the applications. Services that are not available will need to be implemented in the application. Examples of such services are:

- read and writing typed variables (or objects) with possible indication of freshness,
- read sets of data with possible indication of temporal consistency,
- download of data and programs,
- remote invocation (start, stop, pause, resume) of programs,
- synchronization between applications, and
- receive, send, or subscribe to events with possible indication of time of occurrence.

Configuration ease: For most networks, vendors offer tools that allow configuring a network. This may include setting addresses, initial values to the timers used in the network, periods, priorities, etc. Tools are also available to monitor the network. The availability of such tools may greatly reduce commissioning and management efforts.

Connection with the Internet: With the wide spread of web browsers, it is tempting to put a web server inside each industrial device. This server is used to access non-real-time information such as configuration, user manual, etc. To implement such a capability, the transport protocol should be able to carry the HTTP protocol messages. However, Transport Control Protocol (TCP) is not mandatory.

46.7   An Overview of Some Solutions

There are dozens of networks used in the industrial domain. Here, we present briefly some of the most well-known solutions. The objective is to put in light the features that make them fit a type of application rather than another.

Actuator Sensor Interface

Actuator Sensor Interface (ASi) (IEC 62026-2) (EN 50295) is a communication bus targeted at simple remote inputs and outputs for a single industrial computer. It is based on a low-cost electromechanical multidrop connection system designed to operate over a two-wire cable, over a distance of up to 100 m, or more if repeaters are used. Data and power are transmitted on the same two-wire cable. It is especially suitable for lower levels of plant automation where simple, often binary, field devices such as switches need to interoperate in a stand-alone local area automation network controlled by PLC or PC.

The master polls the network by issuing commands and receiving and processing replies from the slaves. Connected slaves are polled cyclically by the master with a data rate of 1666 kbit/sec, which gives a maximum latency of 5148 μsec on a fully loaded network (31 slaves). Sixty two slaves can be connected in the extended addressing mode. Each slave may receive or transmit four data bits. There is provision for automatic slave detection.

ASi does not define any application layer but provides management functions to configure or reset slaves and detect new slaves.

Controller Area Network

Controller Area Network (CAN) results from an effort from Robert BOSCH to provide a serial communication multiplexer for transmission of information inside a vehicle. CAN layer 1 and 2 are normalized by ISO (ISO 1993). Strictly speaking, there are a few application layers for CAN, among which CAN Open is the most well known.

CAN is a bus without repeaters that restricts the possible topologies. Up to 30 stations can be connected. Every message to be transmitted is uniquely identified and there is no means to address directly a given station unless a unique identifier is attached to this station. Every station may access the bus when no other station is transmitting. If two stations start transmitting at the same time, there is an arbitration, and the message with the lowest identifier will win. This means that the transmission of a message bearing a high value as an identifier may be deferred for a long period of time and time constraints may only be fulfilled statistically unless an additional protocol (time slots or central access controller) is added.

It should be noted that by using an "open collector" connection to the cable, CAN implements an immediate and collective acknowledge to any transmitted frame. Hence, the protocol is efficient and spatial consistency may be easily implemented.

At the application layer, CAN Open offers adequate services for sporadic (event) transfers using the producer–consumer model with minimal support of consistency. CAN is not adequate for periodic transfers. As a response, TT-CAN has been developed.

Through its deterministic collision resolution mechanism, CAN offers ways to determine the guarantees offered by the network. Its behavior in the presence of overload is known: low-priority messages cannot access the network any longer. This implies a careful configuration of the network.

DeviceNet (IEC 62026-3 2000) and SDS (Smart Distributed System) (IEC 62026-5 2000) are partly based on CAN.

HART

HART (Rosemount 1991) has been designed by ROSEMOUNT to interconnect transmitters in process control applications. HART may run over existing 4–20 mA lines on a point-to-point mode or on twisted pairs in a bus configuration. In the first case, process values are usually transmitted by analog means and HART is used for configuration and tests.

In the multipoint (bus) mode, up to 15 devices may be remotely powered from the master station while fulfilling intrinsic safety requirements. If they are powered locally, much larger number of devices may be connected. All traffic either comes from or goes to the master station. An additional master station (Hand Held Terminal) may be connected to the bus mainly for management purposes.

HART defines an application layer suited to process control applications and is well established in the field. Beyond this, HART is restricted to very slow master–slave applications.

INTERBUS

Interbus has been developed by Phoenix Contact to implement distributed inputs and outputs on a PLC. It complies with the collapsed 3-layer OSI model and has been standardized by CENELEC (EN50254).

Interbus uses a ring topology with a master station (PLC) and up to 256 slave stations. Twisted pairs are used; but fiber optic cable may be used easily. Topologies are slightly restricted by the ring cabling.

In Interbus, most of the traffic is cyclic or periodic with a single period for all traffic. The traffic is from the master to slaves and *vice versa*. As such, Interbus is targeted to remote inputs and outputs for time-triggered applications.

Simultaneous sampling is available and network guarantees can be easily determined.

LON

LON has been designed by ECHELON especially with building control in mind. It complies with the full seven-layer OSI model and provides a variety of transmission media, including wireless at limited speed (5 kbit/sec). It has been standardized by the Electronic Industries Alliance (EIA-709.1).

On twisted pairs, LON uses the RS485 standard. Repeaters are permitted and a rather general topology may be implemented.

The medium access control is decentralized and contention based (predictive CSMA). Time constraints may hence only be fulfilled statistically. LON has a network layer, and routers are available to interconnect up to 255 subnetworks.

The application layer provides services for sporadic variable and message exchanges without temporal relationship (no support for time stamping). Variables and messages may have different priorities and may be authenticated. The authentication key is assigned to the node. LON has no support for cyclic transfers and cannot ensure periodicity.

Variable may be typed by using one of the predefined types or a C declaration. In the former case, a remote station will be able to read the variable type. Each node may define a variable as an input from or output to the network. In the latter case, when the value of the variable is updated, the new value may be automatically sent to all nodes that have declared the variable as input. Such transmissions seem to take place in a series of point-to-point transfers and not in a broadcast mode as in FIP. A second difference is that the same variable may be updated (declared as output) by several nodes.

Messages may be sent with or without acknowledge in point-to-point or multipoint mode. The content of a message is user defined. Messages provide the means to extend the LON functionality. Loss of interoperability is the price to pay.

MIL-STD-1553

MIL-STD-1553 has been developed for the U.S. air force and the U.S. navy (Haverty 1986). It may be considered as the precursor of fieldbuses, and most of the concepts have been used in other proposals.

Information is communicated over a bus in Manchester II bi-phase encoding at a rate of 1 Mbits/sec. A maximum of 31 Remote Terminals (RTs) can be connected to the bus. Each RT can have up to 30 sub-addresses. They are coupled to the Bus Controller (BC) either directly or via transformers, over a screened twisted pair.

The BC initiates all message transfers. It issues a command frame to a given RT to transmit or receive data. The RT responds by sending a data frame if necessary and terminates the transfer with a status frame. The BC may simultaneously issue a command to receive data to a given RT and a command to transmit data to another RT. This mechanism allows cross-communication between RTs. Command, data, and status frames consist of 16 data bits, 1 parity bit, and 3 bits of Manchester code violation for synchronization at the beginning of the frame.

Very strong requirements are set for cables, coupling transformers, and isolation resistors to ensure very good noise immunity.

MIL-STD-1553 does not define any application layer. There are, however, some attempts to use it in the industry. Due to the high interface cost, its success is very limited.

PROFIBUS-FMS

PROFIBUS is a European standard (EN50170) that adopts a three-layer architecture and includes a full network management part. PROFIBUS has strong analogies with MiniMAP (CCE-CNMA 1995) and may be considered as a rather cheap implementation of this network. It makes the difference between master stations that may transmit on their own and slave stations that may only respond to inquiries from master stations. This allows simple devices to be implemented in an economic manner.

A rather general topology is possible with up to 127 devices connected. Intrinsically safe devices are available from some vendors. Medium redundancy is supported.

PROFIBUS relies on a simplified token passing mechanism that does not offer any guaranteed transfer time. Elaboration of spatial and temporal consistency indications as well as age of data is not standardized and is left to the users.

The application layer may be considered as a subset MMS (MAP application layer) (CCE-CNMA 1995a) even if the syntax in somehow different. There is room for profiles that are more or less equivalent to Companion Standards in MMS. In addition, data access protection is defined and applies to all objects (variables, programs, domains) manipulated by the application services.

PROFIBUS FMS introduces the idea of cyclic exchanges of data. However, this has very little impact on the application services. In fact, cyclic data exchanges are added to handle data transfers from slave stations and not for cyclic user data transfers.

As a cell network, PROFIBUS fulfills most of the requirements. One may regret the absence of application services for event management and application synchronizations (semaphores).

As a fieldbus, i.e., a network placed between sensors and actuators, whether intelligent or not, and the first level of automation, PROFIBUS suffers from five main drawbacks:

- it does not offer periodic services,
- it does not offer any services related to temporal and spatial consistency,
- it does not offer any services for evaluation of the age of a data sample,
- it only fulfills time constraints statistically, in other words there no guaranteed response time, and
- it forces an equal traffic volume on all masters.

Other problems are the absence of standardization of the maximum response delay, the absence of segmentation and the exclusive use of a client-server model (point-to-point) although the Information Report service on a broadcast or multicast connection may be used to implement a *producer-distributor-consumer* model. To overcome these problems, the easiest solution is to design a single master PROFIBUS network which is far from a good use of the standard.

However, PROFIBUS is a very serious replacement for MiniMAP. Efficient implementations exist at a reasonable price. It should be carefully considered when looking for a cell-like network implementation.

PROFIBUS-DP

PROFIBUS-DP is also covered by European standard EN50170. It shares its physical and data link layers with PROFIBUS-FMS. DP is targeted at remote inputs and outputs and has normally one master station and a number of slave stations.

Traffic is essentially cyclic from the master to the slaves and vice versa. In the absence of other master stations, some real-time guarantees can be calculated.

SERCOS

SERCOS results from a common effort of the German Machine-Tool (VDW) and Drive Manufacturers (ZVEI) to design a communication means between CNC and motor drives. It has been standardized by IEC (IEC 61491 2002) and CENELEC (CENELEC 1998a).

SERCOS is a ring where the link between each node is an optical fiber. High noise immunity is thus provided. There is a central station in the CNC and all traffic either comes from or goes to this master station. However, medium access control is not under the control of the master, which would have rendered the network much more efficient. Rather, every slave (drive) has a time slot in which it may transmit information. The selection of the slot is done at startup and takes into account the response time of each device (drive).

Each transmitted frame may contain real-time data as well as messages for configuration. As a maximum of 2 bytes may be sent per frame, messages must be segmented in a large number of segments and reassembled at reception.

Even if SERCOS is not really structured according to the OSI model, a large number of application-layer services are defined. These services mainly apply to drives.

SERCOS provides simultaneous sampling of the inputs and handles periodic traffic at a minimum of 62.5 msec. Real-time traffic is guaranteed. Up to 254 drives can be connected.

WorldFIP

FIP is a European standard (CENELEC 1996) that complies with the ISO collapsed three-layer model. Its physical layer conforms with the Fieldbus international standard (IEC61158) and provides IEC level 3 EMC noise immunity. It includes a full network management part.

FIP assumes that most of the traffic is cyclic or periodic. In this case, each transfer is from a producer to a number of consumers. For real-time data, the producer–consumer model is used. Different periods or cycle durations may coexist on the same network.

The deterministic nature of the traffic justifies a central medium access controller, the *distributor,* or *Bus Arbiter.* For reliability purposes, redundant Bus Arbiters can be added. In FIP, cyclic traffic scheduling is performed by the Bus Arbiter based on the needs expressed during the initialization phase by each communicating entity. Sporadic traffic is scheduled based on the needs expressed at run time.

FIP supports two transmission media: shielded twisted pair and fiber optics. The topology is very flexible. Medium and line-driver redundancy are supported. Up to 64 stations can be connected without repeaters.

As mentioned above, FIP medium access control is centralized. All transfers are under the control of the Bus Arbiter that schedules transfers to comply with timing requirements.

The data link layer provides two types of transmission services: those for variable exchange and those for message transfer. Transfers of variables and messages may take place under request from any station or cyclically (or periodically) according to system configuration.

Variables are exchanged according to the *producer–distributor–consumer* model, and identified by a unique 16-bit identifier known from the producer and the consumer. The identifier is not related to any physical address.

Messages are transferred from a source station to a single or all destination stations according to a *client–server* model. Each message holds its source and its destination address. These addresses are 24 bit long and identify the segment number and the address of the station on the segment (in fact, the link service access point). Messages are optionally acknowledged.

For real-time data exchange, FIP behaves like a distributed database being refreshed by the network periodically or upon demand. All the application services related to periodic and sporadic data exchange are called Manufacturing Periodical/Aperiodical Services (MPS). MPS provides local read/write services (periodic) as well as remote read/write services (sporadic). Accessed objects are variables or lists of variables. For variables, information on freshness is available. For lists of variables, FIP provides information on temporal and spatial consistency status.

In addition, FIP provides a conventional client–server model for messages and events with a subset of MMS as the application layer. The available services are defined according to classes: sensor, actuator, I/O concentrator, PLC, Operator Console, and Programming Console. The MMS subset covers Domain Management, Program Invocation, Variable Access, Semaphore and Journal Management, and the basics of Event Management. Syntax and encoding rules conform to ASN-1.

Ethernet-Based Solutions

Ethernet has been used since the beginning of the 1980s in industrial communications. Even if a few solutions (Le Lann and Rivierre 1993) introducing some degree of predictability in this technology have been designed and produced, users in the industrial domain were reluctant to use Ethernet due to its lack of guarantees. More recently, due to its low cost and wide spread, there is a revival in the interest in using Ethernet (IEEE 802.3) as a communication network in factories.

Ethernet (IEEE 2000) defines the lowest two protocol layers of the OSI model. Its topology is a tree in which each node is either a switch or a hub. This decreases cabling flexibility and increases the cost of the solution. The main question is: which protocol to use over Ethernet? Layer 3 (network layer) is often Internet Protocol (IP) (IETF 1981), which is adequate for real time. Selecting TCP (IETF 1981a) as a transport protocol is not really adequate for real-time use (Decotignie 2001). Other solutions such as XTP (Sanders and Weaver 1990) do a much better job. TCP may still be used for non-real-time traffic. If adequate application layers can be found easily, the missing part is what we could call a real-time layer that can provide an indication of consistency and event ordering. For the time being, a number of organizations are pushing for Ethernet-based solutions including a Profibus version. Unfortunately, there is no compatibility between these solutions. Let us just state that it is certainly possible to build adequate solutions on top of Ethernet but this is still to come.

Solutions from Nonindustrial Markets

It is tempting to use, in the industrial domain, solutions that were designed for the consumer market. They are often cheaper and several tools exist to support them. Universal Serial Bus (USB) and Firewire or IEEE 1394 are examples of such technologies. They were not designed for computer-to-computer networking but as remote input and output media for a single computer. The main questions are:

- is the solution able to withstand industrial use?
- is there a clear advantage over other solutions?

The answer to the first question is case dependent. Successful experiences were obtained with Firewire (Ruiz et al. 1999). However, the physical layer of both USB and Firewire was not designed for industrial use.

The second question has several facets. In terms of throughput, there is no definitive advantage over other high-speed solutions (Dallemagne and Decotignie 2001). The main problem is that most of the functionality is missing. These networks only define the lowest two layers of the OSI model and completely lack the other layers.

46.8 Conclusion

The selection of a communication system to interconnect industrial applications on different computers depends on the type of application. We have presented taxonomy of control applications and have shown that a related taxonomy exists for the industrial communication networks. However, distributing industrial application over several sites connected by a network introduces a number of problems that differ depending on the type of application. Industrial networks should support the resolution of these problems. Finally, selected examples of networks were briefly described and the ability to support the different types of applications and problems related to distribution were outlined.

References

CCE-CNMA, CCE: An Integration Platform for Distributed Manufacturing Applications, Research Report ESPRIT, Springer-Verlag, Berlin, 1995.

CCE-CNMA, MMS: A Communication Language for Manufacturing, Research Report ESPRIT, Springer-Verlag, Berlin, 1995a.

CENELEC EN 50170, *General Purpose Field Communication System*, Vol. 3/3 (WorldFIP), 1996.

CENELEC EN 50170, *General Purpose Field Communication System*, Vol. 2/3 (Profibus), 1996a.

CENELEC EN 50170, *General Purpose Field Communication System*, Vol. 2/3 (P-NET), 1996b.

CENELEC EN 50254:1998 (E), High efficiency communication subsystem for small data packages, CENELEC, Dec. 1998.

CENELEC EN 50295:1999, Low-voltage switchgear and controlgear — controller and device interface systems — actuator sensor interface (AS-i), 1999.

CENELEC EN 61491:1998, Electrical equipment of industrial machines — serial data link for real-time communication between controls and drives (SERCOS), 1998a.

Dallemagne, P. and Decotignie, J.-D., A Comparison of USB 2.0 and Firewire in Industrial Applications, Proceedings of the FeT 2001 International Conference on Fieldbus Systems and their Applications, Nancy, Nov. 15–16, 2001, pp. 16–23.

Decotignie, J.-D., A Perspective on Ethernet-TCP/IP as a Fieldbus, Proceedings of the FeT 2001 International Conference on Fieldbus Systems and their Applications, Nancy, Nov. 15–16, 2001, pp. 138–143.

Decotignie, J.-D. and Prasad, P., Spatio-Temporal Constraints in Fieldbus: Requirements and Current Solutions, 19th IFAC/IFIP Workshop on Real-Time Programming, Isle of Reichnau, June 22–24, 1994, pp. 9–14.

EIA-709.1, *Control Network Specification*, Electronic Industries Alliance, Arlington, VA, U.S.A., March 1998.

Hadzilacos, V. and Toueg, T., Fault-tolerant boradcasts and related problems, in *Distributed Systems*, Mullender, S., Ed., ACM Press, New York, 1993.

Halang, W. and Sacha, K., *Real-Time Systems*, World Scientific, Singapore, 1992.

Haverty, M., MIL-STD 1553 — a standard for data communications, *Communication & Broadcasting*, 10, 29–33, 1986.

He, J., Mammeri, Z., and Thomesse, J.-P., Clock Synchronization in Real-Time Distributed Systems based on FIP Field Bus, *2nd IEEE Workshop on Future Trends of Distributed Computing Systems*, Cairo, Egypt, Sept.30 – Oct. 2, 1990, pp. 135–141.

IEC 61158-2 (2000-08), Fieldbus standard for use in industrial control systems — Part 2: physical layer specification and service definition.

IEC 61491, Electrical equipment of industrial machines — serial data link for real-time communication between controls and drives, 2002.

IEC 62026-2, Low-voltage switchgear and controlgear — controller-device interfaces (CDIs) — Part 2: Actuator sensor interface (AS-i), 2000.

IEC 62026-3, Low-voltage switchgear and controlgear — controller-device interfaces (CDIs) — Part 3: DeviceNet, 2000.

IEC 62026-5, Low-voltage switchgear and controlgear — controller-device interfaces (CDIs) — Part 5: Smart distributed system (SDS), 2000.

IEEE Std 802.3, Part 3: Carrier sense multiple access with collision detect on (CSMA/CD) access method and physical layer specifications, 2000 ed., pp. i –1515.

IETF RFC 791, Internet Protocol, Sept. 1, 1981.

IETF RFC 793, Transport Control Protocol, Sept. 1, 1981, 1981a.

ISO 11898: 1993, Road Vehicles — exchange of digital information — Controller Area Network (CAN) for high-speed communication, 1993.

ISO, Time Critical Communication Architectures — User Requirements, ISO TR 12178, Geneva, 1994.

Jones, A. et al., Issues in the design and implementation of a system architecture for computer integrated manufacturing, *International Journal of Computer Integrated Manufacturing, 2*, 65–76, 1989.

Kopetz, H., Consistency Constraints in Distributed Real Time Systems, Proceedings of the 8th IFAC Workshop on Distributed Computer Control Systems, Vitznau, Sept. 13–15, 1988, pp. 29–34.

Kopetz, H., Event-triggered versus time-triggered real-time systems, in *Operating Systems of the 90s and Beyond*, Lecture Notes in Computer Science, A. Karshmer, J. Nehmer, Eds., Vol. 563, Springer, Berlin, 1991.

Kopetz, H. and Kim, K., Temporal Uncertainties in Interactions among Real-Time Objects, Proceedings of the 9th Symposium on Reliable Distributed Systems, Huntsville, AL, Oct. 9–10, 1990, pp. 165–174.

Kopetz, H. and Ochsenreiter, W., Clock synchronization in distributed real-time systems, *IEEE Computer, 36*, 933–940, 1987.

Le Lann, G., Critical Issues in the Development of Distributed Real-Time Computing Systems, *2nd IEEE Workshop on Future Trends of Distributed Computing Systems*, Cairo, Sept. 30 – Oct. 2, 1990, pp. 96–105.

Le Lann, G., Designing real-time dependable distributed systems, *Computer Communications, 14*, 225–234, 1992.

Le Lann, G. and Rivierre, N., Real-Time Communications over Broadcast networks: the CSMA-DCR and the DOD-CSMA-CD Protocols, Research Report INRIA 1863, 1993.

ROSEMOUNT Inc (1991), HART Smart Communications Protocol Specifications, rev.5.1.4, Jan. 1991, 1993.

Ruiz, L. et al., Using Firewire as an Industrial Network, SCCC'99, Talca, Chile, 1999, pp. 201–208.

Sanders, R. and Weaver, A., The xpress transfer protocol (XTP) — a tutorial, *Computer Communications Review, 20*, 65–80, 1990.

Thomesse, J.-P., Time and Industrial Local Area Networks, in The *7th Annual European Computer Conference on Computer Design, Manufacturing and Production* (COMPEURO'93), Paris-Evry (France), May 1993, pp. 365–374.

Verissimo, P., Ordering and timeliness requirements of dependable real-time programs, *Real-Time Systems, 7*, 104–128, 1994.

47

Networked Control Systems Overview

Pau Martí, Ricard Villà and
Josep M. Fuertes
Technical University of Catalonia

Gerhard Fohler
Mälardalen University

47.1 Introduction

Networked Control Systems (NCS) can contain a large number of interconnected control devices that exchange data through communication networks; examples of application areas include industrial and building automation, office and home automation, intelligent vehicle and transportation systems, and advanced aircraft and spacecraft, among other automated systems. NCS provide several advantages such as modular and flexible system design (e.g., distributed processing and interoperability), simple and fast implementation (e.g., small volume of wiring and powerful easy-to-use configuration tools), and powerful system diagnosis and maintenance utilities (e.g., alarm handling and supervisory packets). However, the combination of sensors, controllers, and actuators with communication networks also makes the analysis and design of NCS a complex task [WIT95] because it requires the integration and good understanding of several disciplines including control systems, communications systems, and real-time systems.

First of all, by distributing the three main functions of the closed-loop operation (sampling, control computation, and actuation) across a communication network, the conventional control theory that supports the control strategy must be reconsidered. Traditionally, the control community has seen the computing platform as providing the determinism that discrete-time control theory requires [AST97]. Control theory has considered implementation other than dedicated processors systems only to a very small extent. Consequently, when computing resources (processor time and communication bandwidth) are limited, control theory rarely advises on how to design controllers to take these limitations into account [ARZ00]. In order to ensure system performance, communication-induced time delays have to be considered at the controller design stage [MAR01].

In addition, the choice of the communication network will be subject to the timing requirements imposed by the closed-loop control application. The use of communication networks such as field buses [LIA01] and more recently TCP/IP-based networks, typically using Ethernet [DEC01], is an increasing tendency in all automation fields. However, each type of network has properties that may preclude its utilization in an NCS or may affect the performance of the closed-loop operation if not properly analyzed.

Moreover, the performance of NCS depends not only on the control strategy and reliable communication network but also on the message scheduling algorithm [WAL01]. When the communication resources are limited, sensors, controllers, and actuators, along with other nodes, have to compete for accessing the network. The criterion to assign the network to nodes (or messages), that is, message scheduling, is aimed to solve this problem. However, different scheduling strategies will derive in shortening or increasing the message latencies, thus influencing system performance [NIL98].

The contribution of this paper is to review fundamental aspects of control systems, communications systems, and real-time systems in order to make NCS analysis and design easier. In Section 47.2, we review the basics of computer-based control systems and characterize NCS with other computer-based control systems. In Section 47.3, we review control aspects of NCS, focusing on the timing required by well-known control strategies that are typically used when designing discrete-time controllers. In Section 47.4, we analyze communication networks with respect to NCS as well as message scheduling strategies. Finally, Section 47.5 concludes and points out open research topics.

47.2 Fundamental Issues in Networked Control Systems

The objective of *computer-based control* is to use computers (and associated technologies, e.g., networks) to manipulate the available inputs of a dynamic system in order to cause this system to behave in a manner more desirable than it otherwise would [LUE79]. The general functionality of a computer-based closed-loop control system can be split into three main activities: *sampling, control algorithm computation* (*calculation* for short), and *actuation*.

The basic concepts of control systems, like the idea of feedback, are reviewed in the next section. In the subsequent section, we describe the closed-loop operation of a computer-based control system and further we analyze different types of architectures (based on processing on devices and networks) in order to characterize the role that a communication network may play in control applications.

The Idea of Feedback for Control Systems

A *control system* is an interconnection of components forming a system configuration that will provide a desired system response [DOR95]. The basis for analysis of a system is the foundation provided by linear system theory, which assumes a cause–effect relationship for the components of the systems. A component or process to be controlled (also called a physical system or plant) can be represented by a block (as shown in Figure 47.1, top) where the input–output relationship represents the cause–effect relationship.

In control terms, the *controlled* variable is the quantity or condition that is measured and controlled (i.e., process output). The *manipulated* variable is the quantity or condition (i.e., process input) that is varied by the controller so as to affect the value of the controlled variable. Depending on the type of information that the controller uses to vary the manipulated variable, we distinguish between *open-loop* control and *closed-loop* control. The defining feature of an open-loop control is that the controller function that varies the manipulated variable is determined completely by an external process that it alone accounts only for the desired output response (Figure 47.1, middle).

A *closed-loop* control system (also called a closed-loop system) utilizes an additional measure of the controlled variable (the actual outputs or other significant variables) to compare it with the desired output. Consequently, the controller function is determined on a continuing basis by the behavior of the

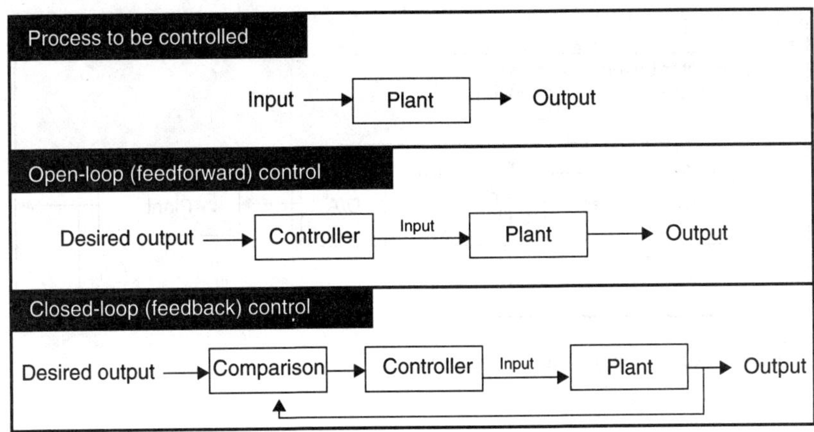

FIGURE 47.1　Feedback in control systems.

system itself (as expressed by the behavior of the outputs). The measure of the output is called the *feedback signal*, because it is fed back (possibly in a modified form due to the controller action) to the process. In summary, in closed-loop systems, *to control* means measuring the value of the controlled variable of the system and applying the manipulated variable to the system to correct or limit deviation of the measured value from a desired value. A simple closed-loop control system is shown in Figure 47.1, bottom.

There are many reasons why closed-loop control is often preferable to open-loop control. Feedback is often superior to open-loop from a performance standpoint and it can automatically adjust to unforeseen system changes or to unanticipated disturbance inputs. A *disturbance* is a signal that tends to adversely affect the value of the output of the system.

The *feedback* concept has been the foundation for control systems analysis and design, which enables us to control the desired output and improve *accuracy* while maintaining *stability*. Traditionally, closed-loop systems were *analog-control systems* (also called *continuous-time systems*); practically all the control systems implemented today are based on computer control, that is, they are *computer-controlled systems* (also called *discrete-time systems*).

An Overview to Computer-Based Control

A computer-based control system (see, e.g., [AST97]), containing both continuous-time signals and discrete-time signals, can be described schematically as in Figure 47.2.

The functionality of such a system can be described as follows:

Sampling: The *controlled* variable, $y(t)$ (the quantity or condition that is measured and controlled, i.e., process output) is usually a continuous-time signal. It is converted to digital form by the Analog-to-Digital (A/D) converter that provides a sequence $\{y(t_k)\}$ at the sampling instants, t_k. The elapsed time between successive sampling instants is called the sampling period and is denoted by h. The A/D converter acts as a sampler that returns a digital snapshot value of a continuous-time signal in the form of an n-bit word.

Calculation: The computer interprets the converted signal, $\{y(t_k)\}$, as a sequence of numbers, processes the input sequence using the control algorithm (implementing a previously designed control law), and gives a new sequence of numbers, $\{u(t_k)\}$.

Actuation: This sequence is converted to an analog signal, $u(t)$, by a Digital-to-Analog (D/A) converter in order to be applied to the process. The D/A converter acts as a hold circuit that takes a discrete-time signal and converts it into a continuous-time signal. Usually, a zero-order hold is used, in which case the resulting continuous-time signal is piecewise constant between successive D/A conversions. The *manipulated* variable $u(t)$ is the quantity or condition (i.e., process input) that is varied by the controller so as to affect the value of the controlled variable.

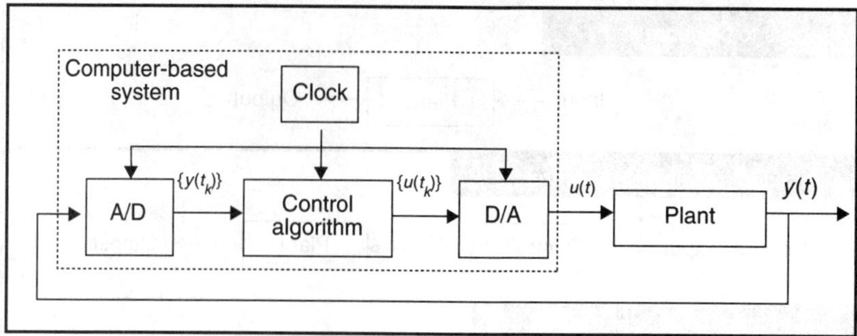

FIGURE 47.2   Diagram of a computer-based control system.

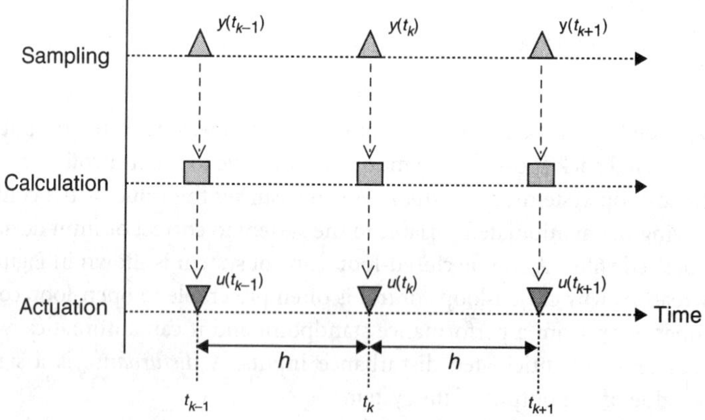

FIGURE 47.3   Diagram of the three main operations in a closed-loop system.

Note that a real-time clock in the computer-based system synchronizes the events.

In summary, we can describe the functionality of a computer-based closed-loop system as the sequence of the three main operations (sampling, calculation, and actuation) that have to be repeatedly executed at every sampling period h, as illustrated in Figure 47.3. The sampling period h, beyond conforming to the Shannon sampling theorem [SHA48], can be chosen following one of various rules-of-thumb, depending on the desired performance of the closed-loop system and the dynamics of the plant to be controlled. An accepted rule-of-thumb is that the sampling frequency should be 4 to 20 times the system's cut-off frequency (which is usually approximated by the system natural frequency ω_n). See [AST97] for further discussion on the selection of the sampling period.

Architectures for Control Systems

Computer-based control is used in many application areas, such as factory automation, process control, robotics, automotive systems, and others. In such applications, different types of computer-based platforms are used to control processes, and are expected to react within precise time constraints to external events according to the application requirements. They are expected to behave correctly both in the value and timing domains: inputs are processed and the adequate outputs are provided with an accurate timing. Therefore, the computer-based architecture that supports the control application, and that will provide the different timings, is of main concern in terms of system performance. Basic architectures for computer-control are discussed next.

The traditional communication architecture for control systems in all fields of automation has been point-to-point (see [BEN00] for a short review of the history of digital computers for control systems). In such an architecture, systems components (sensors and actuators) that interact with the physical process are directly wired to controllers using the traditional 4 to 20 mA current loop. Systems of this type are also referred to as direct digital control systems or local closed-loop systems (see Figure 47.4, top). In point-to-point architectures, each sensor or actuator exchanges data with the controller using a dedicated communication link. Therefore, control data (sensor data and command data) do not suffer delays other than the one expected when transmitting in a point-to-point link. Nowadays, for digital data, this architecture is widely used for Programmable Logic Controller (PLC)-based control architectures. For analog data, point-to-point architectures based on Proportional, Integral and Derivative (PID) controllers are commonly used in the automation field (as stressed in [AST00], PID controller is the first solution that should be tried when feedback is used).

However, such isolated point-to-point architectures do not suit modern automation systems [SCH97] where the goal is to design flexible systems that can accomplish various tasks with small reconfiguration cost. The reconfiguration of a point-to-point architecture to expand physical setups and functionality is not an easy task and it is often easier to build a completely new system than to rewire all the existing system components. In addition, for such systems, diagnosis and maintenance have to be performed locally as no remote operation (control and supervision) is allowed. In the end, this results in inefficient systems and increased costs.

The common bus network architecture can solve the problems of the point-to-point architecture for control systems [LEN93]. Depending on how the control tasks are distributed across the network, we can distinguish two architectures: Distributed Control Systems (DCS) (Figure 47.4, middle) and NCS (Figure 47.4, bottom). Such architectures, main goal is to reduce maintenance, lower cost, reduce weight and power, simplify installation, and improve reliability.

In DCS, most of the real-time control tasks (sensing, calculation, and actuation) are carried out within the individual control nodes. The interconnected nodes mainly use the network to transmit alarm signals, monitoring information and local control actions. The network is also used for system configuration and setup. Therefore, a DCS can be viewed as a networked collection of point-to-point control systems. Note that for such architectures, the control data suffer the same delays as for point-to-point architectures, because it does not travel in the network. DCS are widely spread in all fields of automation.

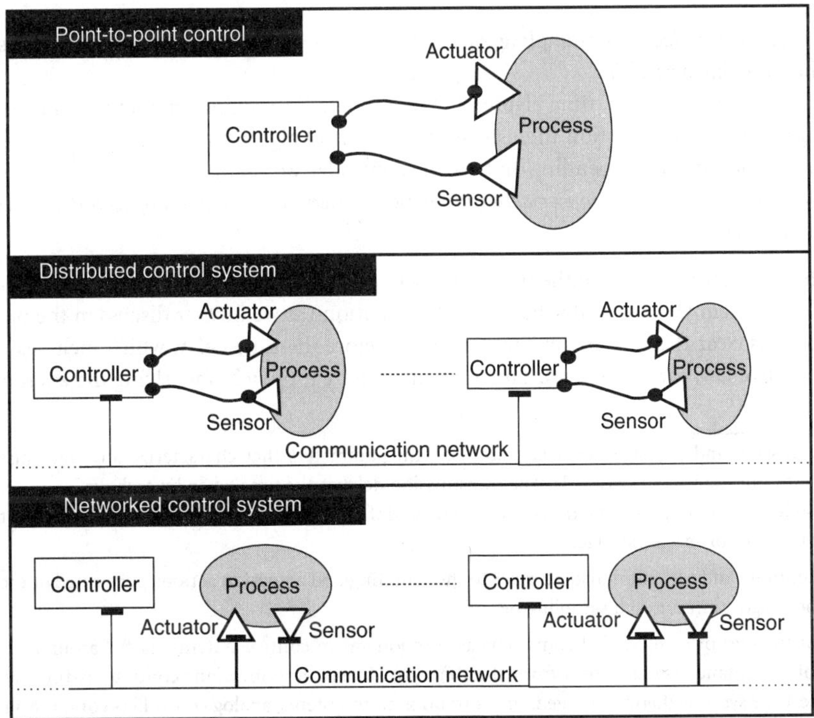

FIGURE 47.4 Architectures for control systems.

One step further in distributing the processing intelligence are NCS. NCS are the application of the intelligent network concept (where all nodes have processing power, allowing for the modularization of the functionality and standard interfaces for interchangeability and interoperability) to control systems [RAY89]. An NCS is a completely distributed and networked feedback control system, where plant sensor–controller–actuator nodes in closed-loop operation drive the network principal traffic. In such architectures, all processing control nodes (sensors, controllers, and actuators) are interconnected through the communication network and, apart from the data typically exchanged in DCS (mainly supervisory data), they transmit sensory and command messages. Note that sensory and command messages (which we also call control data) carry the information required to accomplish the closed-loop operation: sensory messages carry the sampled data, which are used by the controller node to compute the control signal, and command messages carry the command signals that are applied to the process by the actuator node. Therefore, for such architectures, control data are subject to delays caused by the communication network (apart from necessary transmission and propagation delays).[1] Depending on the network and the message scheduling policy, the induced delays will take different values, thus affecting the quality of the closed-loop control operation at different levels.

47.3 Timing Analysis of Control Systems

In this section, we summarize the timing assumptions considered by discrete-time control theory in the closed-loop operation and compare them with the timing analysis that different architectures for computer control provide for closed-loop systems.

Timing Assumptions in the Closed-Loop Operation

From classical discrete-time control operation (which we briefly described in the section An overview to computer-based control), we can derive the following timing assumptions about the three main parts of a closed loop (recall Figure 47.2):

- The *input data collection or sampling* is performed by the sampler at equidistant time instants given by the *sampling period, h.*
- The *time delay*, τ, that is, time elapsed between related[2] sampling and actuation instants (which should include the execution time spent by the control algorithm calculation) is *constant* (either *instantaneous*[3] or not, depending on how the controller was designed).
- The *output data transmission or actuation* is performed at the *actuation instants*, which occurs at the time delay completion.

Although these assumptions are the basis for discrete-time control theory, in practice, it is not possible to keep all of them in a computer-based implementation (as we further discuss in the next section). For example, it is clear that calculations take time, therefore the time delay, which includes the calculation of the control law, cannot be *zero*. The implementation approach underlying the assumption of an

---

[1]The transmission and propagation delays are constant parameters that characterize any transmission of data through a communication media [STA03]. The transmission delay is the time taken by each device — sensor or controller — to send each message given a transmission rate and the propagation delay is the time taken by the communication link to transport each message.

[2]Related sampling and actuation instants refer to the sampling and actuation actions performed at each execution of a closed-loop algorithm for each sampling period.

[3]Recall that the idea of using digital computers as components in control systems started around 1950 [AST97]. The notion of instantaneousness comes from the theory of linear time-invariant continuous-time systems, from which discrete-time systems theory emerged. In continuous-time systems, analog controllers compute almost instantaneously. As a consequence, some of the models and methods used in discrete-time theory, coming from continuous-time theory, maintain the instantaneous assumption.

instantaneous time delay is to build the system in such a way that the time delay is minimized in order to ignore it in the controller design. However, this does not mean that the control computation execution time is zero. It means that the control computation execution time is not relevant to the controlling purposes if compared to the closed-loop system dynamics.

Note that *sampling period* and *time delay* are the timing parameters of interest in the analysis and design of discrete-time controllers. In fact, the well-known control models and methods used in discrete-time control theory were developed for systems characterized with a constant sampling period (h) and a constant time delay (τ).

Timing Analysis of Different Architectures that can Support the Closed-Loop Operation

In the previous section, we have explained the timing assumptions that discrete-time control theory expects in the closed-loop operation. In this section, we give an overview of the real expected timing behavior of closed-loop systems that are either implemented using the point-to-point architecture or the networked architecture. Note that we do not discuss distributed control systems architectures because as we argued in the section Architectures for control systems, in this architecture, control data (sensor and command messages) suffer the same delays as for point-to-point architectures.

Looking at the closed-loop operation for any of the two architectures (Figure 47.5), the following functionality is expected: the sensor samples the process to be controlled with a given sampling period h. Each time the data have been collected, the sensor forwards it to the controller using the corresponding communication link, introducing a communication delay τ_{sc} (sampler to controller delay). The kth control computation execution start time is given by the sampling time t_k plus the τ_{sc} delay. The control computation introduces a delay (τ_c) used to calculate the command signal(s). The controller forwards the command signal(s) to the actuator, introducing another communication delay τ_{ca} (controller to actuator delay). Finally, the actuator performs the actuation at the time given by $t_k + \tau_{sc} + \tau_c + \tau_{ca}$.

In both architectures, the time delay τ (time elapsed between related sampling and actuation instants) is given by $\tau = \tau_{sc} + \tau_c + \tau_{ca}$. However, depending on the architecture, this delay will be either constant or time varying. Note that in most of the actual closed-loop implementations, controllers are implemented in dedicated processors. Therefore, we will also consider the time delay τ_c introduced by the controller to

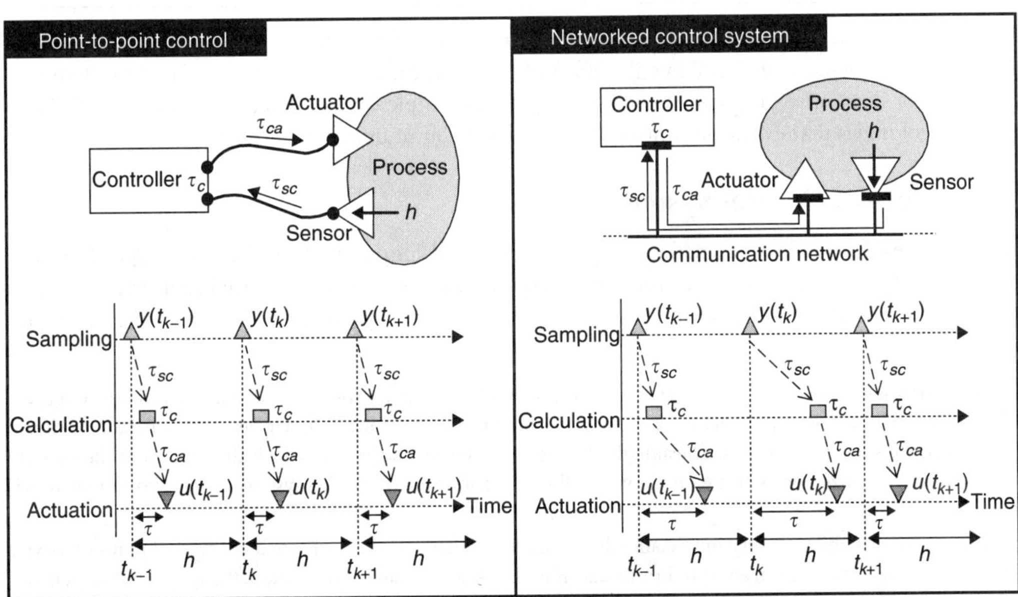

FIGURE 47.5 Timing analysis of the closed-loop operation: point-to-point (left) vs. networked (right) architecture.

be almost constant. We regard the conversion operation delay for A/D conversion in the sensor and D/A conversion in the actuator to be negligible. If these operations take too long a time, they could be included for the timing analysis in the time delay τ_c. For implementations other than dedicated processors, the τ_c delay will depend on the concurrency of the other executing tasks (see [MAR02] for further reading).

Note also that the sampling period is constant for both architectures (Figure 47.5). Samples, $y(t_k)$, are taken at equidistant times, given by h. Therefore, the sampling period does not raise challenges other than the usual ones that are found in the analysis and design of traditional computer-based closed-loop control systems. Consequently, from now on, we will focus on the time delay as a critical timing parameter for ensuring system performance. Specifically, we are interested in the communication-induced time delays, τ_{sc} and τ_{ca}.

If we focus on the point-to-point architecture (Figure 47.5, left), we can model the time delays (τ) introduced by the point-to-point communication links as constant delays. Both τ_{sc} and τ_{ca} will depend on constant parameters such as the transmission time and the propagation delay of the link. Therefore, the control signal $u(t_k)$ is applied to the process at equidistant times, given by the sampling time plus the time delay.

However, looking at the networked architecture (Figure 47.5, right), communication-induced time delays may vary at each closed-loop instantiation. Such delays, apart from being characterized by constant parameters such as the transmission time and propagation delay (as for the previous point-to-point architecture), will depend on new parameters because control nodes exchange data (sampling and controlling messages) through a shared communication link. Issues like the network medium access protocol and the message scheduling will also determine τ_{sc} and τ_{ca} (see the section Network type and message scheduling vs. system performance for further discussion). Therefore, the actuation signals $u(t_k)$ are sent at nonequidistant time instants, given by the sampling time plus a time delay that may vary at each execution, as illustrated in Figure 47.5, right.

After this timing analysis, we can conclude that the theoretical expected timing, constant time delays, assumed by discrete-time control theory is not realistic if compared to the timing derived from networked architectures, varying time delays (see [TÖR98] for further reading in timing analysis). This mismatch will degrade system performance, as we illustrate in the next section.

47.4   Effect of Time Delays in the Performance of Control Systems

In classic control theory, several properties are used to evaluate the performance of closed-loop systems. After discussing them, we will evaluate the effect of communication-induced time delays in system performance. For illustrative purposes, we will use a specific example: an inverted pendulum controlled by a set of control nodes that exchange control data through a communication network.

Performance of Control Systems

The primary evaluation is concerned with meeting the closed-loop response performance specifications[4] and stability[5] (see, e.g., [DOR95]). Beyond these requirements, looking at the closed-loop response, since controller designs attempt to minimize the *system error* to certain anticipated inputs or perturbations,

---

[4]The desired controlled system characteristics, in a formal specification of the problem, are given through the design parameters, which can be met by specifying the closed-loop poles location. However, rather than specifying design parameters, usually it is more meaningful to specify quantities, such as at which time the controlled system recovers from a perturbation (settling time), or the allowed error of the controlled system response to certain anticipated inputs.

[5]The concept of stability is very important when analyzing dynamic systems. From a conceptual point of view, a controlled system is said to be in an equilibrium state if in the absence of any perturbation, the system output remains in the same state. And stability is related to the ability to return to an equilibrium state from a perturbation. For more reading on stability, see [AST97].

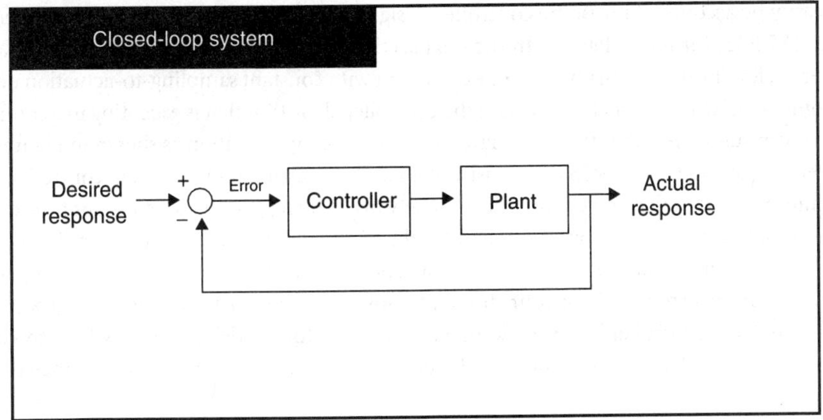

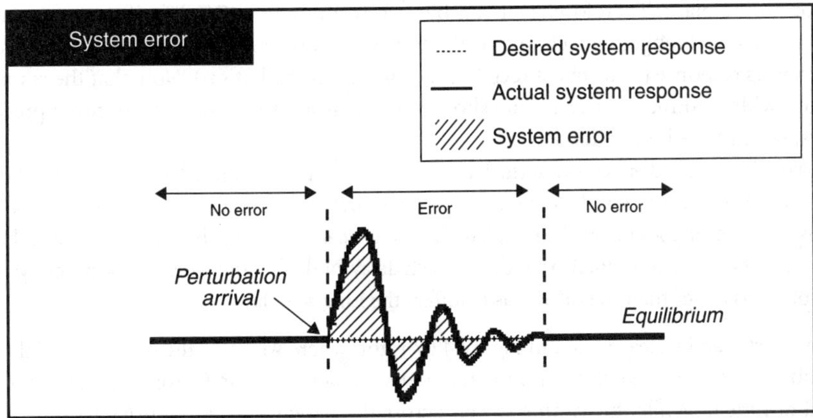

FIGURE 47.6   Performance of closed-loop control systems.

traditional performance criteria focus on the system error. The system error is defined as the difference between the desired response of the system and the actual response of the system. The smaller the difference, the better the performance. Figure 47.6 illustrates these concepts. In Figure 47.6, right, we represent the response of a process that is being controlled. Before the perturbation arrival, the system is in equilibrium. Upon perturbation arrival, the system loses its equilibrium state. Therefore, the actual controlled system response is different from the desired system response, causing the system error. From this moment on, the controller action tries to correct the deviation in order to bring the controlled system back to the equilibrium state again (this is illustrated by the oscillatory curve with decreasing amplitude). The time it takes to reach again the equilibrium state and the maximum amplitude of the deviation determine the performance of the controller.

Constant Communication-Induced Time Delays vs. System Performance

In the section Timing assumptions in the closed-loop operation, we have discussed that controllers designed using classical discrete time control theory are dependent on the sampling period and time delay and that for NCS, as we saw in the section Timing analysis of different architectures that can support the closed-loop operation, time delays are the timing parameters of main concern. Therefore, it is interesting to outline which are the effects of communication-induced time delays on the performance of closed-loop control systems. Recall that we are evaluating time delays in the sense of the time elapsed from sampling to actuation. To do so, in this section, we will focus on constant time delays, because such type of delays can easily be accounted for in the controller design and included in the closed-loop operation (see [MAR01] or [YEP02] for more details). In the next section, we will focus on varying time delays.

Consequently, for illustrative purposes, if we experiment with constant sampling-to-actuation delays (multiples of a nominal delay τ_n) and include them in the controller algorithm that is executing in a controller node of an NCS, we can summarize the effect they have in the closed-loop operation as shown in Figure 47.7. Note that the system we consider is an NCS that consist of a set of processing nodes (sensor, controller and actuator) that communicate data across a communication network, which introduces a constant time delay for the sampling message (form sensor to controller) and for the command message (from controller to actuator).

In Figure 47.7, we show four responses ($y(t)$) of a generic controlled system affected by a perturbation (of magnitude 1 in the error axis) for four different constant sampling-to-actuation delays (τ_n, $2\tau_n$, $3\tau_n$, and $4\tau_n$). Although the controller has been designed to account for the delays, there will always be degrading effects on the controlled system response. The degrading effects on system performance can be summarized as follows:

Delayed response: The first effect is that a sampling-to-actuation delay delays the application of each control action to the plant. Therefore, as it can be seen from Figure 47.7, left, if the time elapsed from sampling to actuation is longer at each one of the four system responses, the more delayed each closed-loop system has its response (in terms of recovering from the perturbation). Note that the response of the system labeled with a nominal delay τ_n (the shortest one) always reacts before (its curve precedes) than all the others, which have longer delays.

Increased error: The second effect is that the later the control action is applied to the process (the longer the delay), the larger the error the closed-loop system response has. As it can also be seen in Figure 47.7, left, the initial error is 1. However, looking at the detailed view (Figure 47.7, right), the error that has to be corrected is not 1 but $1+E_i$, where E_i is the added error due to each delay. Note that the added error of the system labeled with a delay of $4\tau_n$ is larger than the others, as it suffers the longest delay.

These two effects can be clearly seen in Figure 47.7 right, where we show the detailed initial response of each system. In each system, the first sampling that detects a perturbation (error equal to 1) on the system is performed at time 0. While the sample travels across the network in the sensory message to the controller, the controller calculates the control action, and sends it to the actuator, the error of the controlled system increases. It is when the actuator applies the command action to the system (at the delay completion) that the error starts being corrected. The later this action is performed, the larger the system error that must be corrected. Compare, for example, E_4 with E_2. E_4 is the error of the fourth system (with a $4\tau_n$ delay) at the first actuation instant. E_2 is the error of the second system (with a $2\tau_n$ delay) at the first

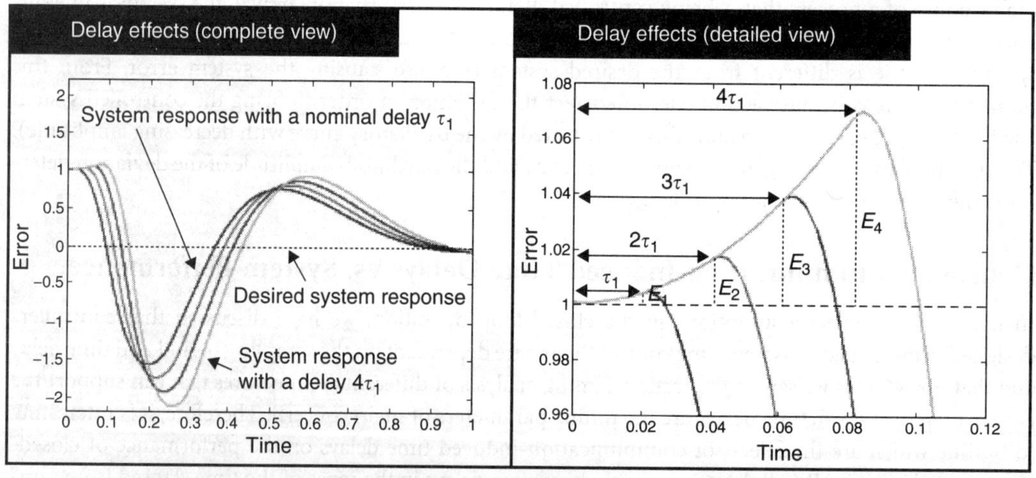

FIGURE 47.7   Performance of a generic controlled system for different time delays.

actuation instant. As it can be seen, the error E_4 is more than twice E_2. However, $4\tau_n$ is exactly $2 \times 2\tau_n$. The error increases faster than the delay increases. For further reading on this topic, see, for example, [YOO00].

From this generic study, it has been shown that the system error generated by a perturbation (which has to be corrected by a controller) may dramatically increase if, from the corresponding sampling, the actuation occurs too late. And this will depend on the network-induced time delays. However, it has to be pointed out that time delays can be assumed in closed-loop systems as far as they are constant, known at design stage, and that the degradation they introduce can be made to conform to the performance specifications. Alternatively, if the communication delays are not constant but exhibit fluctuations, care must be taken because the system performance may drastically decrease as we show in the next section. As we will discuss in the subsequent section, the type of network that supports the networked application and the message scheduling policy will determine the delay pattern that should be accounted for in the analysis and design of networked controllers.

Varying Communication-Induced Time Delays vs. System Performance

In this section, we illustrate, using an example, the problems that varying communication time delays can introduce in control loops closed over communication networks. First of all, we introduce the setup. Afterwards, we analyze the effect of varying time delays in the system performance.

The system we consider is an NCS that consist of a set of control and noncontrol nodes that communicate data across a communication network. The control nodes (sensor, controller, and actuator) are in charge of controlling an *inverted pendulum* mounted on a motor-driven cart (see Figure 47.8, left). Each control node executes the corresponding control law code shown in Figure 47.8, right. The noncontrol nodes do not participate in the closed-loop operation but use the network whenever needed, increasing the traffic load and introducing unexpected longer message latencies. Therefore, the network introduces varying time delays for the sampling message (from sensor to controller) and for the command message (from controller to actuator) due to the interferences of the other noncontrol nodes.

The inverted pendulum control problem can be stated as follows: the inverted pendulum (of length l and mass m) can only swing in a vertical plane parallel to the direction of the cart (of mass M). In the

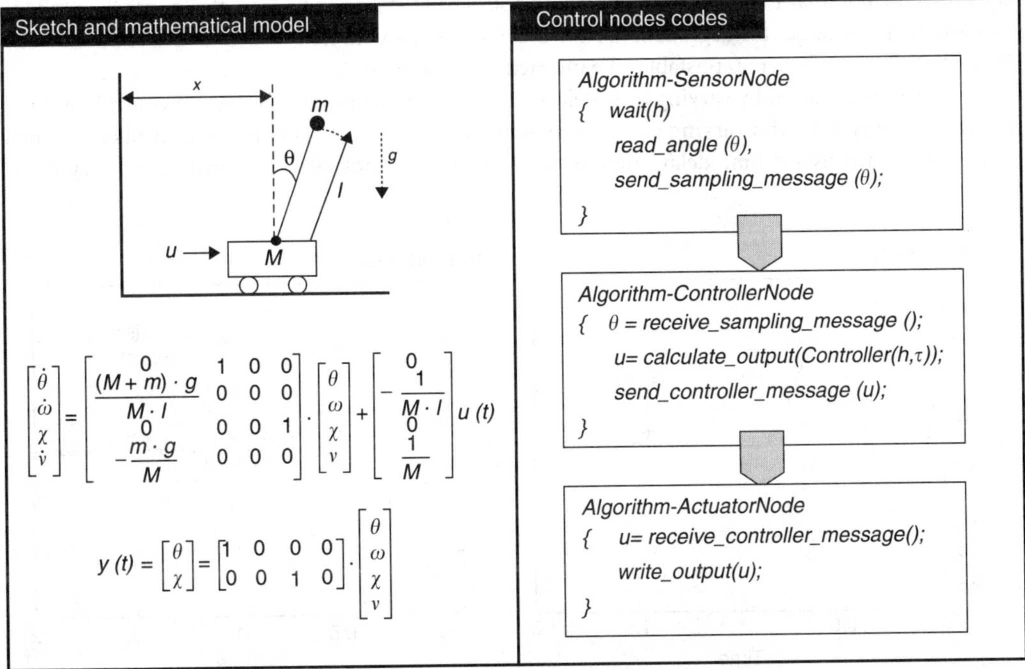

$$\begin{bmatrix} \dot{\theta} \\ \dot{\omega} \\ \dot{\chi} \\ \dot{v} \end{bmatrix} = \begin{bmatrix} 0 & 1 & 0 & 0 \\ \dfrac{(M+m)\cdot g}{M\cdot l} & 0 & 0 & 0 \\ 0 & 0 & 0 & 1 \\ -\dfrac{m\cdot g}{M} & 0 & 0 & 0 \end{bmatrix} \cdot \begin{bmatrix} \theta \\ \omega \\ \chi \\ v \end{bmatrix} + \begin{bmatrix} 0 \\ -\dfrac{1}{M\cdot l} \\ 0 \\ \dfrac{1}{M} \end{bmatrix} u\,(t)$$

$$y\,(t) = \begin{bmatrix} \theta \\ \chi \end{bmatrix} = \begin{bmatrix} 1 & 0 & 0 & 0 \\ 0 & 0 & 1 & 0 \end{bmatrix} \cdot \begin{bmatrix} \theta \\ \omega \\ \chi \\ v \end{bmatrix}$$

FIGURE 47.8 Inverted pendulum model (left) and algorithms for the distributed operation (right).

presence of a perturbation, to balance the pendulum, the cart is pushed back and forth on a track of limited length. Balancing fails when the inclination of the pendulum exceeds preset limits, or when the cart hits the stops at the end of the track. The aim is to find a controller to balance the inverted pendulum, preventing it from failing, and to bring the cart to the center of the track.

The state of the inverted pendulum on a cart is described by the cart position (x), its velocity (v), the pendulum angle (θ), and its angular velocity (ω). The force applied to the cart, (u), is the manipulated controlling command calculated by the controller according to the actual angle and position (controlled variables). A linear time-invariant state-space model of the inverted pendulum, used for the control design can be viewed in Figure 47.8 (where, e.g., we take M=2 kg, m=0.1 kg, l=0.5 m, and g=9.81 m/sec^2).

For the sake of simplicity, we will focus only on the angle (θ). The goal of our controller is to maintain the desired vertical position of the inverted pendulum at all times. Specifically, the performance requirement is to recover from a perturbation in less than 1 sec. To do so, we close the networked loop by using a state-feedback controller designed using classical methods, with strictly periodic sampling h, and accounting for a constant time delay τ.

The inverted pendulum response obtained by the distributed controller if it would execute in isolation (on a networked architecture but with no competitors, thus, with a constant time delay) in the presence of a perturbation (modeled as an initial condition) can be seen in Figure 47.9, left. Since it is beyond the scope of this work to specifically discuss state-feedback controller design, we assume that our design is good enough for illustrative purposes (we call it ideal). Note that the performance requirement is met; that is, the pendulum recovers from the perturbation in less than 1 sec.

The inverted pendulum response obtained by the distributed controller executing in the networked architecture and suffering varying time delays due to the competitor's interferences is quite different. As it can be seen in Figure 47.9, right, the inverted pendulum response suffers different degrees of degradation depending on the randomly induced communication delays of different magnitudes that we have generated (simulating the interference produced by the messaging from the noncontrol nodes). Concretely, in Figure 47.9, right, we show four curves: the dotted line corresponds to the ideal response of the inverted pendulum, with a constant time delay of τ. The other three curves correspond to three responses obtained by three controllers, each one suffering random delays of different magnitudes (delays longer than the nominal τ but bounded by 10, 25 or 50% of τ). As it can be seen, the longer the random generated delay, the larger the degradation. Note that for the case with longest delays (delays bounded by 50% of τ), the system becomes unstable (the inverted pendulum falls).

The degradation caused by varying time delays can be explained as follows. From a control perspective, the control system with varying delays is no longer time-invariant. That is, the controller designed to account for a constant time delay (time invariant system) is not effective when delays vary (time

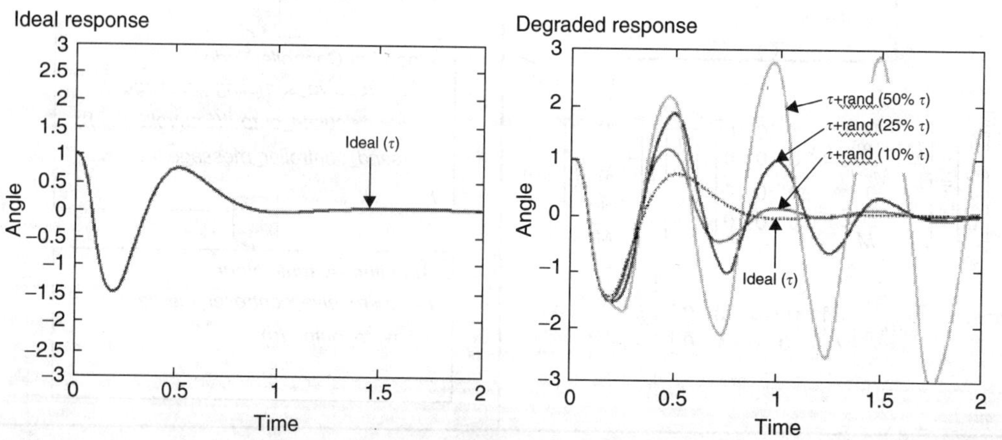

FIGURE 47.9 Inverted pendulum system response: ideal (left) vs. degradation (right) due to varying time delays.

varying system). Therefore, the standard computer control theory (targeted for time-invariant systems) cannot be used in the analysis and design of NCS with varying time delays. The problem of analysis and design of control systems when the communication delays are varying in a random manner is complex and it is still an open research field (see [CHO01] for an introductory tutorial).

Following the seminal work on communication and control [HAL90], an approach to overcome these problems is to model the communication delays as probabilistic distributions. The design of the controller has to account for those delays or alternatively has to view the temporal nondeterminism of network-induced delays as an uncertainty, similar to a plant uncertainty, or a disturbance, and then design the control systems to be robust against such uncertainty.

For example, [NIL98] presents various probabilistic communication delays models and accordingly, solves an LQG optimal control problem for them. Using similar techniques, in [XIA01] communication resources allocation and linear systems design are jointly optimized.

Another approach is to assume a constant sampling-actuation delay, forcing a synchronous actuation (synchronous with respect to the sampling) [WIT98]. This can be achieved by forcing the actuation to occur at equidistant times, given by the worst-case-message-latency. However, this may unnecessarily impose a longer time delay than the actual ones that appear during system runtime, thus forcing a pessimistic timing behavior in the closed-loop system that may result in a graceful but unnecessarily system performance degradation [YEP02].

Up to now, we have not discussed stability issues. As can be seen in Figure 47.9, right, the degradation that the controlled system suffers drives the system to almost instability. It is well known that delays can cause instability. Therefore, in the analysis of control loops closed over communication networks, delays have to be included for the stability analysis. See [ZHA01] for a starting reading point.

Note, however, that the randomness of the communication-induced delays will depend on the type of communication network and message scheduling that supports the closed-loop operation, as introduced in [LIA02]. This is discussed next.

Network Type and Message Scheduling vs. System Performance

Varying communication time delays can seriously degrade the performance of the controlled system, as outlined in the previous section (see Figure 47.9, right). Moreover, when delays appear in a random manner, the lack of well-established control techniques for time-varying systems makes the analysis and design of NCS a complex task. Taking into account that the type of delays determine system performance, the type of network and more concretely, the rules that allow the exchange of messages, that is, message scheduling policy, is a key issue in terms of system performance. By choosing the adequate message scheduling (or protocol), the impact of communication-induced time delays can be drastically reduced because delays can be somehow controlled.

It is well known that random-based scheduling policies (e.g., Ethernet [DEC01]) conflict with the determinism that control theory imposes on an implementation. Standard Ethernet communication introduces random delays and no guarantee can be given on the maximum message latency. Therefore, to design NCS with Ethernet as a communication backbone is not appropriate from the control tractability point of view. However, there is a strong interest in using the cheap and simple Ethernet technology for industrial and embedded systems. This far, however, the lack of real-time services has prevented this change in the used network technology. Nevertheless, in recent literature, works can be found addressing the lack of determinism by using switched Ethernet techniques (see [LOB01] for further reading on Ethernet and control issues in the automation field).

To overcome this problem of the lack of determinism in Ethernet, we can think of well-known communication networks with deterministic scheduling policies such as table-driven (e.g., WorldFIP [ALM02]), master/slave (e.g., Profibus [TOV99]), or nonpreemptive fixed priority-based (e.g., Controller Area Network, CAN [TIN94]) scheduling policies. Such scheduling policies limit message latencies, thus facilitating the analysis and design of closed loops over communication networks and also ensuring average or priority-based levels of performance.

For example, table-driven or master/slave scheduling policies can guarantee constant message latencies. Moreover, latencies (delays) are known before runtime. For example, in a master/slave configuration, the order of the polling determines the order in which nodes have access to the network. And when a node is accessing the network, it uses all the available bandwidth. Therefore, the network-induced delays depend on known parameters (transmission time and propagation time). If delays are known before run time, the analysis and design of the distributed controller taking into account the delay induced by the networked architecture are straightforward [AST97]. Note that such protocols, although providing constant time delays (which makes the control design job easy), may not provide the optimum performance. This is because they may be enforcing delays longer than the ones that could be obtained using other protocols based on more flexible priority-based mechanisms such as CAN. Recall that a delay is always a source of degradation, as we outlined in the section Constant communication-induced time delays vs. system performance.

For nonpreemptive fixed priority-based scheduling policies such as CAN, although message latencies are bounded, they are not known before system runtime. This is because messages may suffer collisions. In this case, the deterministic collision resolution mechanism that CAN incorporate ensures that the message with highest priority will continue being transmitted. Moreover, using [TIN94], the maximum latencies for all messages can be precalculated. Therefore, although message delays are not exactly known, the known upper bounds can be used for the analysis and design of the networked controller. Moreover, such scheduling techniques can be used to shorten the message latencies by adequately assigning the priorities.

47.5 Conclusions

NCS constitute a particular type of applications over communication networks where the principal traffic is driven by plant sensor–controller–actuator nodes in closed-loop systems. As we have reviewed in this chapter, the analysis and design of such systems are complex tasks involving the integration and good understanding of several disciplines: control systems, communication systems, and real-time systems. In particular, when a network is used dedicated just for one control loop, without any external or other node interferences, the resulting networked system becomes trivial. But when the network is shared among different nodes and there is even just one sensor–controller–actuator closed-loop control subnetwork, care must be taken on how the expected or unexpected delays can affect the controlled system performance. This chapter has shown the way to characterize the different architecture issues in a networked control system, mainly in what refers to the different delays that occur in the messaging related to the control actions. In general, if those delays are negligible with respect to the dominant dynamics of the controlled plant, they can be overcomed; but when those time delays are significant, a more precise analysis has to be conducted, both in the design or redesign of the control algorithm, on the network characteristics, and on the system's scheduling. These last topics are still an open issue that is currently being investigated by several research groups.

References

[ALM02] Almeida, L., Tovar, E., Fonseca, J.A., and Vasques, F., Schedulability analysis of real-time traffic in WorldFIP networks: an integrated approach, *IEEE Transactions on Industrial Electronics*, 49, 1165–1174, 2002.

[ÅRZ00] Årzen, K.-E., Cervin, A., Eker, J., and Sha, L., An Introduction to Control and Scheduling Co-Design, Proceedings of the 39th IEEE Conference on Decision and Control, Vol.5, Sydney, Australia, Dec. 2002, pp. 4865–4870.

[AST97] Åström, K.J. and Wittenmark, B., *Computer-Controlled Systems. Theory and Design,* 3rd ed., Prentice-Hall, Englewood Cliffs, NJ, 1997.

[AST00] Åström, K.J. and Hägglund, T., The Future of PID Control, Preprints of IFAC Workshop on Digital Control. Past, Present and Future of PID Control, Terrassa, Spain, 2000, pp. 19–30.

[BEN00] Bennett, S., The Past of PID Controllers, Preprints of IFAC Workshop on Digital Control. Past, Present and Future of PID Control, Terrassa, Spain, 2000, p. 13.

[CHO01] Chow, M.-Y. and Tipsuwan, Y., Network-Based Control Systems: A Tutorial, Proceedings of the 27th Annual Conference of the IEEE Industrial Electronics Society, Vol.3, Denver, CO., U.S.A., 2001, pp. 1593–1602.

[DEC01] Decotignie, J.-D., A Perspective on Ethernet as a Fieldbus, Proceedings of the 4th IFAC International Conference on Fieldbus Systems and their Applications, Nancy, France, November 2001.

[DOR95] Dorf, R.C. and Bishop, R.H. (1995). Modern Control Systems, Seventh Edition, Addison-Wesley

[HAL90] Halevi, Y. and Ray, A., Analysis of integrated communication and control system networks, *ASME Journal of Dynamic Systems, Measurement and Control*, 112, 365–372, 1990.

[LEN93] Lenhart, G., A fieldbus approach to local control networks, *Advances in Instrumentation and Control*, 48, 357–365, 1993.

[LIA01] Lian, F., Moyne, J. and Tillbury, D., Performance evaluation of control networks: ethernet, controlnet, and devicenet, *IEEE Control Systems Magazine*, 21, 66–83, 2001.

[LIA02] Lian, F., Moyne, J. and Tillbury, D., Networked design consideration for distributed control systems, *IEEE Transactions on Control Systems Technology*, 10, 297–307, 2002.

[LOB01] Lo Bello, L., and Mirabella, O., Design Issues for Ethernet in Automation, Proceedings of the 8th IEEE International Conference on Emerging Technologies and Factory Automation, Vol.1, Antibes Juan-les-pins, France, Oct. 2001, pp. 213–221.

[LUE79] Luenberger, D.G., *Introduction to Dynamic Systems. Theory, Models and Applications,* John Wiley & Sons, New York, 1979.

[MAR01] Martí, P., Fuertes, J.M., and Fohler, G., An Integrated Approach to Real-Time Distributed Control Systems Over Fieldbuses, Proceedings of the 8th IEEE International Conference on Emerging Technologies and Factory Automation, Vol.1, Antibes Juan-les-pins, France, Oct. 2001, pp. 23rd 177–182.

[MAR02] Martí, P., Fuertes, J.M., Fohler, G., and Ramamritham, K., Improving Quality-of-Control using Flexible Timing Constraints: Metric and Scheduling Issues, Proceedings of the 23rd IEEE Real-Time Systems Symposium, Austin, TX, USA, Dec. 2002, pp. 91–100.

[NIL98] Nilsson, J., Bernhardsson, B., and Wittenmark, B., Stochastic analysis and control of real-time systems with random time delays, *Automatica*, 34, 57–64, 1998.

[RAY89] Ray, A., Introduction to networking for integrated control systems, *IEEE Control System Magazine*, 9, 76–79, 1989.

[SCH97] Schickhuber, G. and McCarthy, O. Distributed fieldbus and control network systems, *Computing & Control Engineering Journal*, 8, 21–32, 1997.

[SHA48] Shannon, C.E., A mathematical theory of communication, *Bell System Technical Journal*, July and October, 27, 379–423 and 623–656, 1948.

[STA03] Stallings, W., *Data and Computer Communications,* 7th ed., Prentice-Hall, Englewood Cliffs, NJ, 2003.

[TIN94] Tindell, K. and Burns, A., Guaranteeing Message Latencies on Controller Area Network (CAN), Proceedings 1st International CAN Conference, 1994, pp. 1.2–1.11.

[TÖR98] Törngren, M., Fundamentals of implementing real-time control applications in distributed computer systems, *Journal of Real-Time Systems*, 14, 219–250, 1998.

[TOV99] Tovar, E. and Vasques, F., Real-time fieldbus communications using profibus networks, *IEEE Transactions on Industrial Electronics*, 46, 1241–1251, 1999.

[WAL01] Walsh, G.C. and Ye, H., Scheduling of networked control systems, *IEEE Control Systems Magazine*, 21, 57–65, 2001.

[WIT95] Wittenmark, B., Nilsson, J., and Törngren, M., Timing Problems in Real-Time Control Systems: Problem Formulation, *Proceedings of the American Control Conference,* Vol.3, Seattle, WA, U.S.A., 1995, pp. 2000–2004.

[WIT98] Wittenmark, B., Bastian, B., and Nilsson, J., Analysis of Time Delays in Synchronous and Asynchronous Control Loops, Proceedings of the 37th IEEE Conference on Decision and Control, Vol.1, Tampa, FL, U.S.A., 1998, pp. 283–288.

[XIA01] Xiao, L., Johansson, M., Hindi, H., Boyd, S., and Goldsmith, A., Joint Optimization of Communication Rates and Linear Systems, Proceedings of the 40th IEEE Conference on Decision and Control, Vol.3, December 2001, pp. 2321–2326.

[YEP02] Yepez, J., Marti, P., and Fuertes, J.M., Control Loop Performance Analysis over Networked Control Systems, Proceedings of the IEEE 28th Annual Conference of the Industrial Electronics Society, Vol.4, Sevilla, Spain, November 2002, pp. 2880–2885.

[YOO00] Yook, J.K., Tilbury, D.M., and Soparkar, N.R., A Design Methodology for Distributed Control Systems to Optimize Performance in the Presence of Time Delays, Proceedings of the American Control Conference, Vol.3, June 2000, pp. 1959–1964.

[ZHA01] Zhang, W., Branicky, M.S., and Phillips, S.M., Stability of networked control systems, *IEEE Control Systems Magazine*, 21, 84–89, 2001.

48

The Quest for Real-Time Behavior in Ethernet[1]

P. Pedreiras, L. Almeida and
J. A. Fonseca
Universidade de Aveiro

48.1 Introduction

Nowadays, intelligent nodes, that is, microprocessor-based with communication capabilities, are extensively used in the lower layers of both process control and manufacturing industries [8]. In these environments, applications range from embedded command and control systems, to image processing, monitoring, human–machine interface, etc. Moreover, the communication between the different nodes has specific requirements [9] which are quite different from, and sometimes opposed to, those found in office environments. For instance, predictability is favored against average throughput, and message transmission is typically time and precedence constrained. Furthermore, the nonrespect of such constraints can have a significant negative impact on the quality of the control action in Distributed Computer Control Systems (DCCS), or on the quality of the observation of the system state in Distributed Monitoring Systems (DMS). Therefore, to deliver the adequate quality of service, special-purpose networks have been developed, essentially during the last two decades, which are generically called fieldbuses and are particularly adapted to support frequent exchanges of small amounts of data under time, precedence, and dependability constraints [8]. Probably, the most well-known examples existing today are Controller Area Network (CAN), DeviceNet, ProfiBus, WorldFip, P-Net, and Foundation Fieldbus.

In the early days of DCCS, network nodes presented simple interfaces, and supported limited sets of actions. However, the quantity, complexity, and functionality of the nodes on a DCCS have been increasing steadily. As a consequence of this evolution, the amount of information that must be exchanged over the network has also increased, either for configuration or for operational purposes.

[1]This work was partially supported by the Portuguese Government through project CIDER-POSI/ 1999/CHS/33139.

The increase in the amount of data exchanged between DCCS nodes is reaching the limits that are achievable using traditional fieldbuses due to limited bandwidth, typically between 1 and 5 Mbps [9]. Machine vision, and its growing use, is just one example of a killer application for those systems. Therefore, other alternatives are required to support higher bandwidth demands while retaining the main requirements of a Real-Time (RT) communication system: predictability, timeliness, bounded delays, and jitter.

Starting in the 1980s, several general-purpose networks, exhibiting bandwidth higher than the traditional fieldbus protocols, have also been proposed for use at the field level. For example, two prominent networks, FDDI and ATM, have been extensively analyzed, for both hard RT and soft RT communication systems. However, due to high complexity, high cost, lack of flexibility, and interconnection capacity, these protocols have not gained general acceptance [10].

Another communication protocol that has been evaluated for use at the field level is Ethernet. The main factors that favor the use of this protocol are [9]: cheap silicon available, easy integration with Internet, clear path for future expandability, and compatibility with networks used at higher layers on the factory structure.

However, the nondeterministic arbitration mechanism used by Ethernet prevents its direct use at field level, at least for hard RT communications. Therefore, in the past, several attempts have been made to allow Ethernet to support time-constrained communications. The methods that have been used to achieve deterministic message transmission over Ethernet range from modifications to the Medium Access Control (MAC) layer (e.g., [11]), to the addition of sublayers over the Ethernet layer to control the instants of message transmission (e.g., [12]) and therefore avoid collisions. More recently, with the advent of switched Ethernet, and therefore the intrinsic absence of collisions, a new set of works with respect to the ability of this topology to carry time-constrained communications has appeared (e.g., [10]).

This chapter presents a brief description of the Ethernet protocol, followed by a discussion of several techniques that have been proposed or used to enforce RT communication capabilities over Ethernet in the last two decades. The techniques referred to include those that support either probabilistic or deterministic analysis of the network access delay. The paper concludes with a reference to recent work that concerns the use of the Ethernet protocol, highlighting current trends.

48.2 Ethernet Roots

Ethernet originated about 30 years ago, invented by Bob Metcalfe at the Xerox's Palo Alto Research Center. Its initial purpose was to connect two products developed by Xerox: a personal computer and a brand new laser printer. Since then, this protocol has evolved in many ways. For instance, concerning the transmission speed, it has grown from the original 2.94 to 10 Mbps [1–3, 5], then to 100 Mbps [6] and more recently to 1 Gbps [7] and 10 Gbps [26]. With respect to physical medium and network topology, Ethernet also has evolved: it started with a bus topology based firstly on thick coaxial cable [2] and afterwards on thin coaxial cable [3]. In the mid-1980s, a more structured and fault-tolerant approach, based on a star topology, was standardized [4], running, however, only at 1 Mbps. In the beginning of the 1990s, an improvement on this latter technology was also standardized [5], running at 10 Mbps over category 5 unshielded twisted pair cable.

Along this way, two fundamental properties have been kept unchanged:

1. Single collision domain, that is, frames are broadcast on the physical medium and all the Network Interface Cards (NIC) connected to it receive them.
2. The arbitration mechanism, which is called Carrier Sense Multiple Access with Collision Detection (CSMA/CD).

According to the CSMA/CD mechanism (Figure 48.1), an NIC with a message to be transmitted must wait for the bus to become idle and, only then it starts transmitting. However, several NICs may have sensed the bus during the current transmission and then attempted to transmit simultaneously thereafter, causing a collision. In this case, all the stations abort the transmission of the current message, wait for a random time interval, and try again. The number of retries is limited to sixteen.

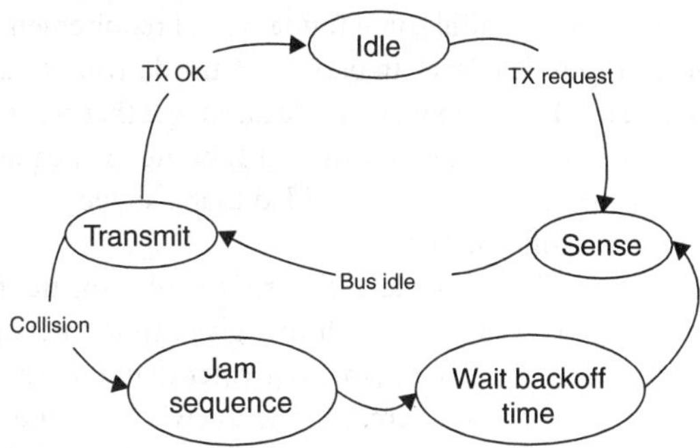

FIGURE 48.1 Ethernet CSMA/CD simplified state diagram.

The use of a single broadcast domain and the CSMA/CD arbitration mechanism has created a bottleneck when encountering highly loaded networks: above a certain threshold, as the submitted load increases, the throughput of the bus decreases, a phenomenon referred to as *thrashing*. In the beginning of the 1990s, the use of switches in place of hubs has been proposed as an effective way to deal with thrashing. A switch creates a single collision domain for each of its ports. If a single node is connected to each port, collisions never actually occur unless they are created on purpose, for example, for flow control. Switches also keep track of the addresses of the NICs connected at each port by inspecting the source address in the incoming messages. This allows forwarding incoming messages directly to the respective outgoing ports according to the respective destination address, a mechanism generally known as *forwarding*. When a match between a destination address and a port cannot be established, the switch forwards the respective message to all ports, a process commonly referred to as *flooding*. The former mechanism, forwarding, allows a higher degree of traffic isolation so that each NIC receives the traffic addressed to it, only. Moreover, since each forwarding action uses a single output port, several of these actions can be carried out in parallel, resulting in multiple simultaneous transmission paths across the switch and, consequently, in a significant increase in the global throughput.

48.3 Why Use Ethernet at Fieldbus Level

The first question that should be answered concerns the reasons that make Ethernet appealing for use at the field level, to convey time-constrained traffic. In fact, its designer has not envisaged this kind of applications, and therefore some properties of this protocol, such as the nondeterministic arbitration mechanism, pose serious challenges concerning its use for this purpose.

Several works address this subject (e.g., [9, 12, 17]). In [9], Decotignie presents a thorough reasoning on the pros and cons on this issue, highlighting two concise sets of arguments, one in favor and the other against the adoption of Ethernet as a fieldbus. Commonly referred arguments in favor can be summarized as follows:

- It is cheap, due to mass production.
- Integration with Internet is easy (TCP/IP stacks over Ethernet are widely available, allowing the use of application layer protocols such as FTP, HTTP, and so on).
- Steady increases in the transmission speed have occured in the past, and are expected to occur in the near future.
- Due to its inherent compatibility with the communication protocols used at higher levels, the information exchange with plant level becomes easier.
- The bandwidth made available by existing fieldbuses is insufficient to support some recent developments, like the use of multimedia (e.g., machine vision) at the field level.
- Availability of technicians familiar with this protocol.
- Wide availability of test equipment from different sources.
- Mature technology, well specified, and with equipment available from many sources, without incompatibility issues.

On the other hand, Ethernet does not fulfill some fundamental requirements that are expected from a communication protocol operating at field level. In particular, the destructive and nondeterministic arbitration mechanism has been regarded as the main obstacle faced by Ethernet concerning this applications domain. The answer to this concern is the use of switched Ethernet, which allows bypassing the native CSMA/CD arbitration mechanism. In these cases, provided that a single NIC is connected to each port, and the operation is full duplex, no collisions occur.

However, just avoiding collisions does not make Ethernet deterministic: for example, if a burst of messages destined to a single port arrive at the switch in a given time interval, they must be serialized and transmitted one after the other. If the arriving rate is greater than the transmission rate, buffers will be exhausted and messages will be lost. Therefore, even with switched Ethernet, full RT behavior still requires some kind of higher-level transmission control. Moreover, bounded transmission delay is not the only requirement of a fieldbus; some other important factors commonly referred to in the literature are: temporal consistency indication, precedence constraints, efficient handling of periodic, and sporadic traffic. Clearly, Ethernet, even with switches, does not provide answers to all these demands.

48.4 Making Ethernet RT

In the previous years, a considerably large amount of work has been carried out, leading to many different proposals for adaptations and add-ons to Ethernet in order to support RT communication. In the remainder of this section, some paradigmatic efforts are presented, with a brief discussion of the strong and weak points of each one.

Modification of the Medium Access Control Sublayer

One of the possible approaches consists in modifying the Ethernet MAC sublayer so that a bounded access to the bus may be achieved (e.g., [11, 18]). For instance, the solution presented in [11] (CSMA/DCR) consists of a binary tree search of colliding messages, that is, there is a hierarchy of priorities in the retry.

During normal operation, the CSMA/DCR follows the standard IEEE 802.3 protocol (*Random Access mode*). However, whenever a collision is detected, the protocol switches to the *Epoch mode*. In this mode, lower priority message sources voluntarily cease contending for the bus, and higher priority ones try again. This process in repeated until a successful transmission occurs (Figure 48.2). After all frames involved in the collision are transmitted, the protocol switches back to random access mode.

Figure 48.2, together with Table 48.1, depicts the CSMA/DCR operation in a situation where six messages collide. Considering that lower indexes correspond to higher priorities, after the initial collision the right branch of the tree (messages 12, 14, and 15) ceases contending for the bus. Since there are still three messages on the left branch, a new collision appears, between messages 2, 3, and 5. Thus, the left subbranch is selected again, leaving message 5 out. In the following slot, messages 2 and 3 will collide again. The subbranch selected after this collision has no active message sources, and thus in the following time slot the bus will be idle (step 4). This causes a move to the right subbranch, where messages 3 and 5 reside, resulting in a new collision. Finally, in step 6 the branch containing only the message with index 5 is selected, resulting in a successful transmission. The algorithm continues this way until all messages are successfully transmitted.

Despite assuring a bounded access time to the transmission medium, this approach exhibits two main drawbacks:

- In some cases (e.g., [11]), the firmware must be modified; therefore, the economy of scale obtained when using standard Ethernet hardware is lost.
- The worst-case transmission time, which is the main factor considered when designing RT systems, can be orders of magnitude greater than the average transmission time. This forces any kind of analysis to be very pessimistic, and therefore leads to low bandwidth utilization, at least with respect to RT traffic.

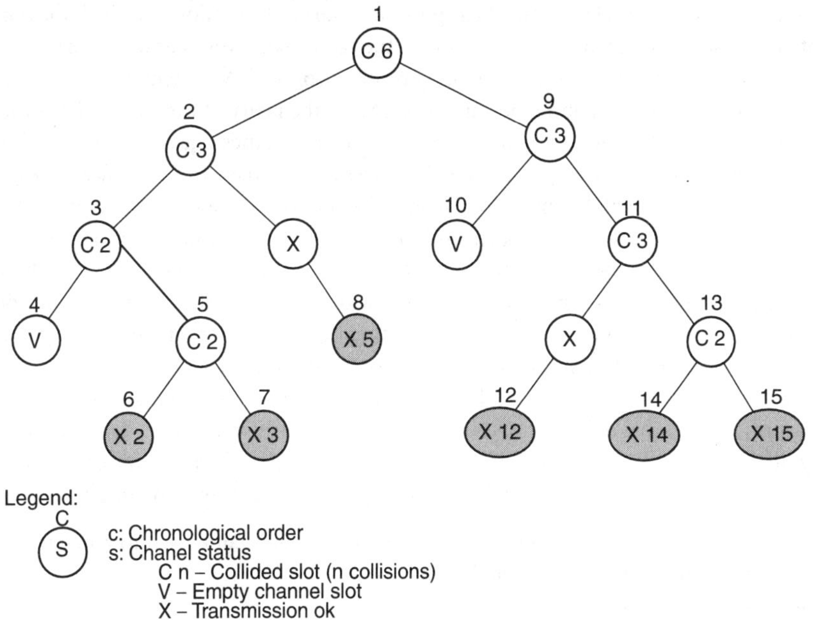

FIGURE 48.2 Example of tree search with CSMA-DCR.

TABLE 48.1 Tree Search Example (Contending Sequence)

Search Order	1	2	3	4	5	6	7	8	9	10	11	12	13	14	15
Channel Status	C	C	C	V	C	X	X	X	C	V	C	X	C	X	X
Source index	2	2	2		2	2	3	5	12		12	12	14	14	15
	3	3	3		3				14		14		15		
	5	5							15		15				
	12														
	14														
	15														

Addition of a Transmission Control Layer Over Ethernet

Another way to achieve time-constrained communication over Ethernet consists of adding an interface layer above the native medium access control, intended to control the instants of message transmissions, eliminating collisions or, at least, bounding their number. The major advantage of this kind of approach when compared to the modification of the MAC layer is that standard Ethernet hardware can be used.

Virtual Time Protocol

One of the approaches present in the literature is known as Virtual Time Protocol and addresses CSMA/CD networks [19, 20]. The protocol allows implementing different scheduling policies (e.g., Minimum-Laxity First [MLF]), and bases its decisions on the assessment of the communication channel status. When the bus becomes idle and a node has a message to transmit, it waits for a given amount of time, related to the scheduling policy implemented. For example, if MLF scheduling is used, the waiting time is derived directly from the laxity using a proportional constant. When this amount of time expires, and if the bus is still idle, the node tries to transmit the message. If a collision occurs, then the scheduler outcome resulted in more than one message having permission to be transmitted at the same time (e.g., when two messages have the same laxity in MLF). In this case, the protocol can either recalculate the waiting time using the same rule or use a probabilistic approach according to which the messages involved in a collision are retransmitted with probability p (p-persistent). This last option is important to sort out situations in which the scheduler cannot differentiate messages, for example, messages with the same laxity would always collide.

Figure 48.3 shows the operation of the Virtual-Time protocol, with MLF scheduling. During the transmission of message m, messages a and b become ready. Since the laxity of message a (i.e., deadline minus message transmission time) is shorter than the laxity of message b, message a is transmitted first. During the transmission of message a, message c arrives. Messages b and c have the same deadline and the same laxity. Therefore, an attempt will be made to transmit them at the same time, causing a collision. Then, the algorithm uses the probabilistic approach, with message b having a lower waiting time than message c, and thus being transmitted next. Finally, message c is transmitted on the bus. Since the only global information is the channel status, there is no way to know that there is only a single message pending. For this reason, after the transmission of message b the waiting time corresponding to message c is computed, and only after the expiration of this interval message is c finally transmitted.

Beyond the advantage of using standard Ethernet hardware, this approach also has the advantage of not requiring any other global information but the channel status, which is readily available at all NICs. Thus, a fully distributed and symmetric implementation is possible, which, in this case, also incurs a relatively low computational overhead. Nevertheless, this approach presents some important drawbacks:

- Performance highly dependent on the proportional constant value used to relate the waiting time with the scheduling policy in use, leading to:
  - Collisions if it is too short.
  - Large amount of idle time if it is too long.
- Proportional constant depends on the properties of the message set; therefore, on-line changes to that set can lead to poor performance.
- The waiting times are computed locally using relative parameters only. There is no global time base and thus, relative phasing is harder to implement.
- Due to possible collisions, worst-case transmission time is much higher than the average transmission time and only probabilistic timeliness guarantees can be given (soft RT systems).

Windows Protocols

Another possible approach is to use the *Windows protocols*. These protocols have been proposed both for CSMA/CD and token ring networks [19]. With respect to the CSMA/CD implementation, the operation is as follows. The nodes on a network agree on a common time interval (referred to as *window*). All nodes synchronize upon a successful transmission, restarting the respective window. The bus state is used to assess the number of nodes with messages to be transmitted within the window:

- if the bus remains idle, there are no messages to be transmitted in the window;
- if only one message is in the window, it will be transmitted; and
- if two or more messages are within the window, a collision occurs.

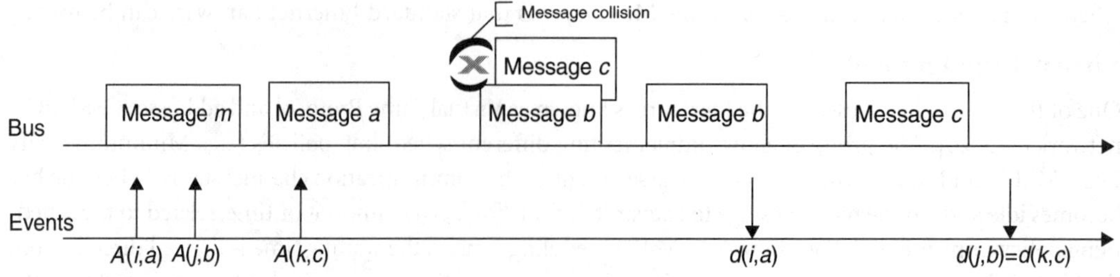

FIGURE 48.3 Example of Virtual-Time protocol (with MLF).

Depending on the bus state, several actions can be performed:

- if the bus remains idle, the window duration is increased in all nodes;
- in the case of a collision, the time window is shortened in all nodes; and
- in case of a successful transmission, the window is restarted and its duration is kept as it is.

In the first two cases, the window duration is changed but the window is not restarted. Moreover, the window duration varies between a maximum (initial) and minimum value. Whenever there is a sufficiently long idle period in the bus, the window will return to its original maximum length. If a new node enters dynamically in the system, it may have an instantaneous window duration different from the remaining nodes. This may cause some perturbation during an initial period, with more collisions than expected. However, as soon as an idle period occurs, all windows will converge to the initial length. A probabilistic retry mechanism may also be necessary when the windows are shrunk to their minimum and collisions still occur (e.g., when two messages have the same transmission time).

Figure 48.4 shows an example of the operation of the windows protocol used to implement MLF message scheduling. The top axis represents the latest send times (*lst*) of messages A, B, and C. The *lst* of a message is the latest time instant by which the message transmission must start so that the respective deadline is met. The first window (Step 1) includes the *lst* of three messages, thus leading to a collision. The intervenient nodes feel the collision, and the window is shrunk (Step 2). However, the *lst* of messages A and B are still inside the window, causing another collision. In response to this event, the window size is shrunk again (Step 3). In this case, only message A has its *lst* within the window, leading to a successful transmission.

This method exhibits properties that are very similar to those of the previous method (virtual time protocol). However, it is somewhat more efficient due to its adaptive behavior. In general, it also aims at soft RT systems and uses a fully distributed symmetrical approach with a relatively low computational overhead. Notice that all message parameters are relative and that there is no global time base again. Moreover, the protocol efficiency is substantially influenced by the magnitude of variations in the window duration, either when increasing or decreasing it.

Traffic Shaping

It is a known fact that, if the bus utilization is kept low, the probability of collisions is also low (although not zero). Therefore, if the network average load is kept below a given threshold and bursts of traffic are avoided, a given probability of collisions can be obtained, as an estimation of the network-induced delay.

An implementation of this paradigm is presented in [21]. An interface layer, called traffic smoother (Figure 48.5), is placed between the IP network layer and the Ethernet data link layer (for this reason, traffic shaping

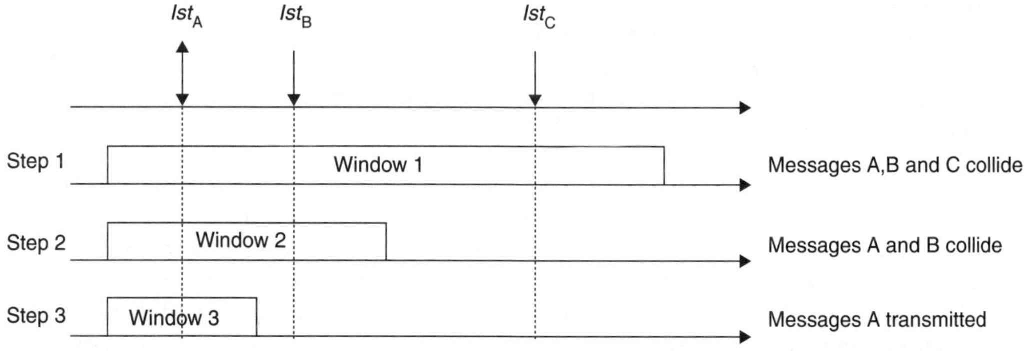

FIGURE 48.4   Windows protocol example.

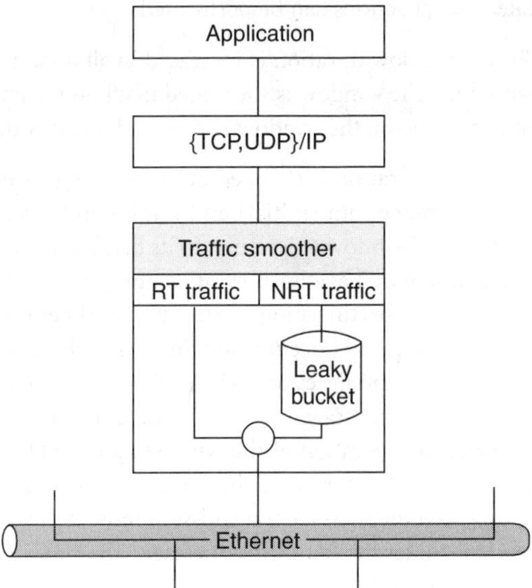

FIGURE 48.5   Software architecture of traffic smoothing.

is also commonly referred to as traffic smoothing). To avoid additional delays, RT traffic is transmitted on demand and it is not handled by the traffic smoother. Notice that real-time (RT) traffic is normally composed of periodic or sporadic message streams with a single packet per message instance, and thus the probability of collisions it induces is low. On the other hand, the Non-Real-Time (NRT) traffic is captured by the traffic smoother, which keeps track of previous message transmissions performed by the node. According to this historical record, it releases NRT messages in a controlled manner to achieve a desired node's traffic generation rate commonly referred to as *station input limit*. Moreover, inside each node, RT messages get higher priority than NRT ones. At the network level, the interference due to NRT traffic is kept under a probabilistic bound.

The station input limit, that is, the parameters of the leaky bucket, can either be static or they can be dynamic, being adjusted to the current network load in an adaptive way. A recent evolution of this latter approach has been presented in [13] in which the network load is estimated using two parameters, the number of collisions and the throughput, both observed in a given interval. These parameters are then used by a fuzzy controller that sets the instantaneous station input limit. The resulting efficiency is substantially higher than with other approaches to traffic shaping, allowing stations (nodes) to take advantage of periods of low throughput to transmit more data over the bus.

In any case, traffic shaping supports probabilistic timing guarantees only being suited to soft RT systems. Typical applications can be either multimedia applications or even some robust control applications, which can cope with a few sample losses.

Token Passing

Another approach to achieve RT behavior in Ethernet consists in explicitly assigning time slots for exclusive use of a given node. This way, collisions are completely avoided. One possible way of implementing this concept in any shared bus network is by means of a timed-token mechanism [14].

In [12], Venkatramani and Chiueh present an implementation of this concept called RETHER. This protocol operates in normal CSMA/CD mode until the arrival of RT requests, upon which it switches to token-bus mode.

In the token-bus mode, RT data are considered to be periodic and the time is divided into cycles. During the cycle duration, access to the bus is regulated by a token, both for RT and NRT traffic. First, the token visits all nodes that are sources of RT messages. Afterwards, if there is enough time until the end of the cycle, the token visits the sources of NRT data. An on-line admission control policy assures that all accepted RT requests can always be served and that new RT requests cannot jeopardize the guarantees of existing RT messages. Therefore, in each cycle, all RT nodes can send their RT messages. However, with respect to the NRT traffic, no timeliness guarantees are granted.

Figure 48.6 illustrates a possible network configuration with six nodes. Nodes 1 and 4 are sources of RT messages, forming the RT set. The remaining nodes have no such RT requirements and constitute the NRT set. The token first visits all the members of the RT set and after, if possible, the members of the NRT set. A possible token visit sequence could be: cycle i {$1 - 4 - 1 - 2 - 3 - 4 - 5 - 6$}, cycle i+1 {$1 - 4 - 1 - 2$}, cycle i+2 {$1 - 4 - 1 - 2 - 3 - 4$}.... . In the ith cycle, the load is low enough so that the token has time to visit the RT set plus all nodes in the NRT set, too. In the following cycle, besides the RT set, the token only visits nodes 1 and 2 of the NRT set and, in the next cycle, only nodes 1 through 4 of the NRT set are visited.

Due to the complete elimination of collisions, this approach supports deterministic analysis of the worst-case network access delay, particularly for the RT traffic. Furthermore, if the NRT traffic is known *a priori*, it is also possible to bound the respective network access delay, which can be important, for example, for sporadic RT messages. However, since the bandwidth available for NRT messages is distributed according to the nodes order established in the token circulation list, the first nodes always get precedence over the following ones, which end up with very long worst-case network delays. Moreover, this method involves a considerable communication overhead caused by the circulation of the token.

Master/Slave Techniques

One of the simplest ways of enforcing RT communication over Ethernet consists in using a master/slave approach in which a special node, the master, controls the access to the medium of all other nodes, the slaves. The traffic timeliness is then reduced to a problem of scheduling that is local to the master. However, this approach typically leads to a considerable underexploitation of the network bandwidth because every data message must be preceded by a control message issued by the master resulting in a substantial communication overhead. Moreover, there is some extra overhead related to the turnaround time, that is, the time that must elapse between consecutive messages, since every node must fully receive and decode the control message before transmitting the respective data message. Nevertheless, it is a rugged transmission control strategy that has been used in many protocols. It is also the basis for the current Ethernet Powerlink protocol [15], a recently proposed industrial solution for RT communication over Ethernet.

A variant mechanism has also been proposed recently, which allows reducing the referred communication overhead. It is called the master/multislave approach [30] according to which the bus time is split into cycles and the master issues one control message for the whole cycle only indicating which data messages must be produced therein. This mechanism has been developed within the FTT framework, Flexible Time-Triggered communication, which has been implemented over different network protocols. One such implementation has been over Ethernet, leading to the FTT-Ethernet protocol [23]. This protocol combines the master/multislave transmission control technique with centralized scheduling, maintaining both the communication requirements and the message scheduling policy localized in one single node, the Master, and facilitating on-line changes to both, thus supporting a high level of operational flexibility.

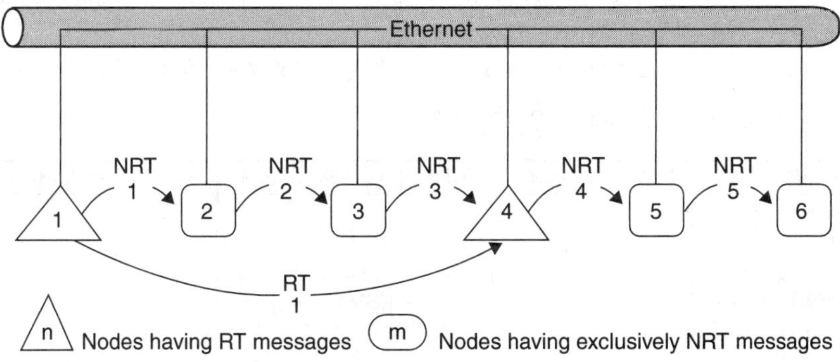

FIGURE 48.6 Sample network configuration for RETHER.

The bus time is divided into fixed-duration time-slots called Elementary Cycles (ECs) that are further decomposed into two phases, the synchronous and asynchronous windows (Figure 48.7), which have very different characteristics. The synchronous window carries the periodic time-triggered traffic that is also scheduled by the master node. The expression time-triggered implies that this traffic is synchronized to a common time reference, which in this case is imposed by the master. The asynchronous window carries the sporadic traffic either related to protocol control messages, such as those conveying change requests for the time-triggered traffic, event-triggered messages, such as those related to alarms, and other NRT traffic. There is a strict temporal isolation between both phase so that the sporadic traffic does not interfere with the time-triggered one.

Despite allowing on-line changes to the attributes of the time-triggered traffic, global timeliness is enforced by the FTT–Ethernet protocol by means of on-line admission control. Due to the global knowledge and centralized control of the time-triggered traffic, the protocol may easily support dynamic QoS management as a complement to admission control.

Beyond the flexibility and timeliness properties that this protocol exhibits, there are also some drawbacks that concern the computational overhead required in the master to execute both the message scheduling and the schedulability analysis on-line. This is, however, confined to one node. Nevertheless, slaves also need some computational power to decode the trigger message in time and start the due transmissions at the right moments. Finally, in safety-critical applications, the master must be replicated, requiring further mechanisms to ensure coherency between their internal databases holding the system communication requirements.

Switched Ethernet

Since roughly one decade ago, the interest in using Ethernet switches has been growing as a means to improve global throughput, traffic isolation, and reduce the impact of the nondeterministic features of the original CSMA/CD arbitration mechanism. Switches, unlike hubs, provide a private collision domain for each of its ports, that is, there is no "direct" connection between its ports. When a message arrives at a switch port, it is buffered, analyzed with respect to its destination, and moved to the buffer of the destination port (Figure 48.8). The "packet handling" block in the figure includes the switch fabrics, which transfers messages from input to output ports, the switch controller or CPU, which controls the switch fabric and manages the internal traffic, and the switch memory, which is used to hold the port queues.

When the arrival rate of messages at each port, either input or output, is greater than the rate of departure, the messages are queued. Currently, most switches are fast enough in handling message arrivals so that queues do not build up at the input ports (these are commonly referred to as nonblocking switches). However, queues may always build up at the output ports whenever several messages arrive in a short interval and are routed to the same port. In such case, queued messages are transmitted sequentially, normally in FCFS order. This queue handling policy may, however, lead to substantial network-induced delays because higher priority or more important messages may be blocked in the queue while lower priority or less important ones are being transmitted. Therefore, the use of several parallel queues for different priority levels has been proposed (IEEE 802.1p). The number of distinct priority levels is limited to eight, but many current

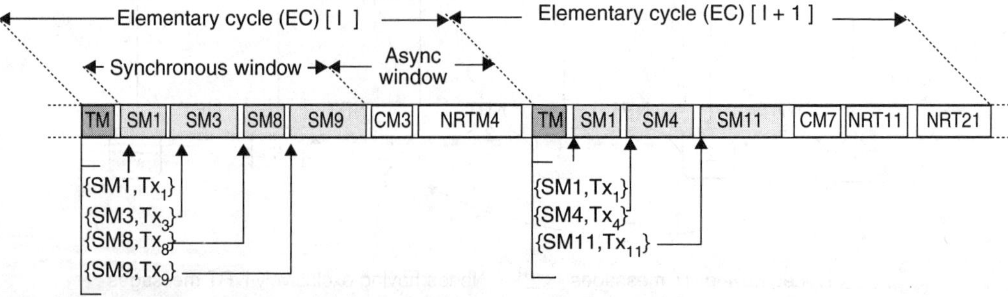

FIGURE 48.7 FTT–Ethernet traffic structure.

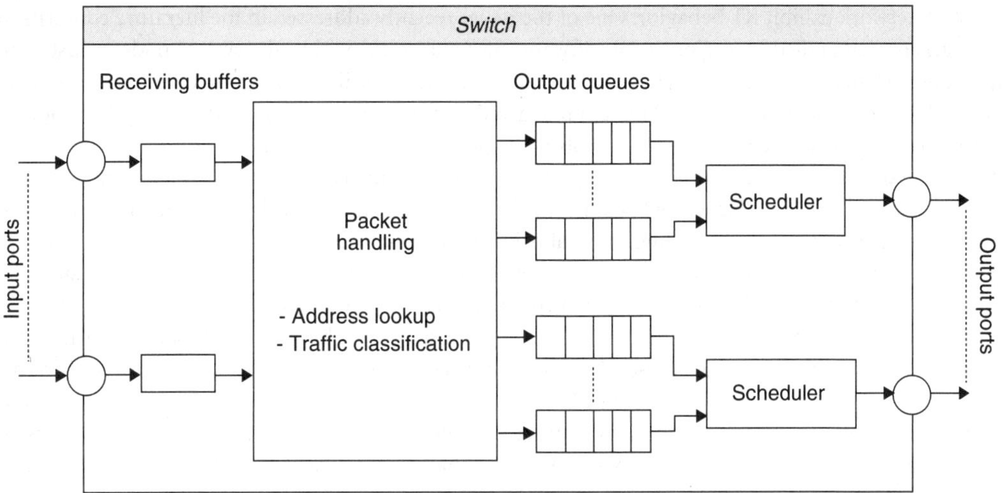

FIGURE 48.8 Switch internal architecture.

switches that support traffic prioritization offer even a further limited number. The scheduling policy used to handle the messages queued at each port also has a strong impact on the network timing behavior [22].

A common misconception is that the use of switches, due to the elimination of collisions, is enough to enforce RT behavior in Ethernet. However, this is not true in the general case. For instance, if a burst of messages destined to the same port arrives at the switch, output queues can overflow, thus leading to a loss of messages. This situation, despite seeming somewhat unrealistic, can occur with a nonnegligible probability in certain communication protocols based on the producer–consumer model, for example, Control Information Protocol (CIP) and its lower level protocols such as Ethernet/Industrial Protocol (IP) [24], or based on the publisher–subscriber model such as RTPS [25] used within Interface for Distributed Automation (IDA). In fact, according to these models, each node that produces a given datum (producer or publisher) transmits it to potentially several nodes (consumers or subscribers) that need it. This model is efficiently supported in Ethernet by means of special addresses, called multicast addresses. Each network interface card can define the multicast addresses related to the information that it should receive. However, the switch has no knowledge about such addresses and thus treats all the multicast traffic as broadcasts, that is, messages with multicast destination addresses are transmitted to all ports (*flooding*).

Therefore, when the predominant type of traffic is multicast/broadcast instead of unicast, one can expect a substantial increase of peak traffic at each output port that increases the probability of queue overflow, causing degradation in network performance. Furthermore, in these circumstances, one of the main benefits of using switched Ethernet, that is, multiple simultaneous transmission paths, can be compromised. A possible way to limit the impact of multicasts is using virtual LANs (VLANs) so that flooding affects only the ports of the respective VLAN [16].

Other problems concerning the use of switched Ethernet are referred in [9] such as the additional latency introduced by the switch in the absence of collisions as well as the low number of available priority levels that hardly supports the implementation of efficient priority-based scheduling.

These problems are, however, essentially technological and are expected to be attenuated in the near future. Moreover, switched Ethernet does alleviate the nondeterminism inherent to CSMA/CD medium access control and paves the way for efficient implementations of RT communication over Ethernet.

48.5 Recent Advances

Most of the recent work performed on RT Ethernet targets switch-based implementations. However, as discussed in Section Switched Ethernet, just replacing a hub by a switch is not enough to make an

Ethernet network exhibit RT behavior. One of the issues recently addressed in the literature concerns the way packets are handled by the protocol software (protocol stack) within the system nodes. Most of the operating systems implement a single queue, usually working according to a First-Come First-Served policy, for both RT and NRT traffic. This approach induces important delays and priority inversions. A methodology that has been proposed to solve this problem is the implementation of multiple transmit/receive queues [27]. With this approach, the RT traffic is intrinsically separated from the NRT traffic. NRT traffic is only sent/processed when the RT queues are empty. It is also possible to build separate queues for each traffic class, providing internal priority-aware scheduling.

An other important issue concerns the degree of freedom in the network topology (e.g., bus, star). The topology impacts the number of switches that messages have to cross before reaching the target, which impacts the temporal properties of the traffic [28, 29]. For instance, the bus (or line) topology, in which each device integrates a simplified switch eases the cabling, but is the most unfavorable topology for RT behavior.

Another aspect concerning switch-based Ethernet networks respects the scheduling policy within the switch itself. Switches support up to eight distinct statically prioritized traffic classes. Different message scheduling strategies have a strong impact on the RT behavior of the switch [28]. Particularly, strategies oriented toward average performance and fairness, which are relevant for general-purpose networks, may negatively impact the switch RT performance.

Finally, the interest in shared Ethernet is not over yet, either for applications requiring frequent multicasting, in which case the benefits of using switches are substantially reduced, as well as for applications requiring precise control of transmission timing, such as high-speed servoing. In fact, switches tend to induce a higher delay and jitter in message forwarding than hubs, caused by internal mechanisms such as MAC address to port translation in forwarding and spanning-tree management protocol. In the previous sections, several examples of this interest were discussed, such as the recent work around adaptive traffic smoothing [13] and master/slave techniques including both the Ethernet Powerlink [15] as well as FTT–Ethernet [23] protocols. This last protocol is also being analyzed for implementation on switched Ethernet, taking advantage of the message queuing in the switch ports and thus simplifying the transmission control. This has the potential to ease the implementation of slave nodes since then it would not be necessary to enforce fine control of the transmission instants of both synchronous and asynchronous messages, strongly reducing the computational overhead. Several existing Ethernet-based industrial protocols, such as Ethernet/IP, are also taking advantage of switches to improve their RT capabilities [16]. Particularly, this protocol is now receiving unprecedented support from major international associations of industrial automation suppliers, such as Open DeviceNet Vendor Association (ODVA), ControlNet International (CNI), Industrial Ethernet Association (IEA), and Industrial Automation Open Networking Alliance (IAONA).

48.6 Conclusion

Due to several reasons, Ethernet became the most popular technology for LANs today. This makes it very attractive even in application domains for which it was not originally designed, in order to benefit from its low cost, high availability, and easy integration with other networks, just to name a few arguments. Some of such application domains, for example, industrial automation, impose RT constraints on the communication services that must be delivered to the applications. This conflicts with the original medium access control technique embedded in the protocol, CSMA/CD, which is nondeterministic and behaves very poorly with medium to high network loads. Therefore, in its nearly 30 years of existence, many adaptations and technologies for Ethernet have been proposed in order to support the desired real-time behavior.

This chapter has presented an overview of those techniques, ranging from changes to the bus arbitration, to the addition of transmission control layers, and also to the use of special networking equipment, such as switches. Such techniques have been described and briefly analyzed in what concerned their pros and cons for different types of application. This chapter concludes with a reference to recent trends, where the growing impact of switches is clear, despite a few specific situations in which shared Ethernet might still be preferable, such as when the traffic is mainly of a multicast nature or a precise transmission timing control is required.

With the current high pressure to bring Ethernet more and more into the world of distributed automation systems, it is likely that such technology will end up taking the place of existing fieldbuses and establishing itself as the *de facto* communication standard for this area. Although its efficiency in terms of bandwidth utilization is still low when considering short messages, particularly lower than with several fieldbuses, its high and still growing bandwidth seems more than enough to supplant such an aspect. Ethernet will then become the long-awaited single networking technology within automation systems, which will support the integration of all levels, from the plant floor to the management, maintenance, supply-chain, etc.

References

1. IEEE, DIX Ethernet V2.0 specification, 1982.
2. IEEE, IEEE 802.3 10BASE5 standard.
3. IEEE, IEEE 802.3 10BASE3 standard.
4. IEEE, IEEE 802.3c 1BASE5 StarLan standard.
5. IEEE, IEEE 802.3i 10BASE-T standard.
6. IEEE, IEEE 802.3c 100BASE-T standard.
7. IEEE, Supplement to IEEE Std 802.3, 1998 Edition: Physical Layer Parameters and Specifications for 1000 Mb/s Operation over 4-Pair of Category 5 Balanced Copper Cabling, Type 1000BASE-T, 1999.
8. Thomesse, J.-P., Fieldbus and interoperability, *Control Engineering Practice*, 7, 81–94, 1999.
9. Decotignie, J.-D., A Perspective on Ethernet as a Fieldbus, in Proceedings of the 4th IFAC International Conference on Fieldbus Systems and their Applications (FeT'2001), IFAC Publications, Elsevier Science, Oxford, England, Nov. 2002, pp. 138–143.
10. Song, Y., Time Constrained Communication Over Switched Ethernet, in Proceedings of the 4th IFAC International Conference on Fieldbus Systems and their Applications (FeT'2001), IFAC Publications, Elsevier Science, Oxford, England, 2002, pp. 152–159.
11. LeLann, G. and N. Rivierre, Real-time communications over broadcast networks: the csma-dcr and the dod-csma-cd protocols, Technical report, INRIA Report RR1863, 1993.
12. Venkatramani, C. and T. Chiueh, Supporting Real-Time Traffic on Ethernet, in Proceedings of IEEE Real-Time Systems Symposium, IEEE Press, New York, Dec. 1994, pp. 282–286.
13. Carpenzano, A., R. Caponetto, L.L. Bello, and O. Mirabella, Fuzzy Traffic Smoothing: an Approach for Real-Time Communication Over Ethernet Networks, in Proceedings of the 4th IEEE International Workshop on Factory Communication Systems (WFCS'2002), IEEE Press, New York, Aug. 2002, pp. 241–248.
14. Malcolm, N. and W. Zhao, The timed-token protocol for real-time communications, *IEEE Computer*, 27, 35–41, 1994.
15. Ethernet Powerlink Standardization Group, Ethernet powerlink protocol, www.ethernet-powerlink.org.
16. Moldovansky, A., Utilization of Modern Switching Technology in Ethernet/IP Networks, in Proceedings of the 1st International Workshop on Real-Time LANs in the Internet Age, Edições Politeama, Porto, Portugal, 2002, pp. 25–27.
17. Bello, L.L. and O. Mirabella, Design Issues for Ethernet in Automation, in Proceedings of 8th IEEE International Conference on Emerging Technologies and Factory Automation (ETFA'2001), Vol. 1, IEEE Press, New York, Oct. 2001, pp. 213–221.
18. Court, R., Real-time ethernet, *Computer Communications*, 15, 198–201, 1992.
19. Malcom, N. and W. Zhao, Hard real-time communication in multiple-access networks, *Real Time Systems*, Vol. 9, Kluwer Academic Publishers, Boston, U.S.A., 1995, pp. 75–107.
20. Molle, M. and L. Kleinrock, Virtual time csma: Why two clocks are better than one, *IEEE Transactions on Communications*, 33, 919–933, 1985.
21. Kweon S.-K. and K. G. Shin, Achieving Real-Time Communication Over Ethernet With Adaptive Traffic Smoothing, in Proceedings of IEEE Real-Time Technology and Applications Symposium, RTAS'00, IEEE Press, Washington, DC, U.S.A., June 2000, pp. 90–100.
22. Jasperneit, J. and P. Neumann, Switched Ethernet for Factory Communication, in Proceedings of 8th IEEE International Conference on Emerging Technologies and Factory Automation (ETFA'2001), Vol. 1, IEEE Press, New York, Oct. 2001, pp. 205–212.

23. Pedreiras, P., L. Almeida, and P. Gai, The FTT-Ethernet Protocol: Merging Flexibility, Timeliness and Efficiency, in Proceedings of 14th IEEE Euromicro Conference on Real-Time Systems, IEEE Press, New York, June 2002, pp. 134–142.

24. Open DeviceNet Vendor Association, Ethernet/IP (industrial protocol) specification, www.odva.org.

25. IDA, Interface for Distributed Automation group, Rtps, real-time publisher/subscriber protocol, www.ida-group.org.

26. IEEE, IEEE Standard for Carrier Sense Multiple Access with Collision Detection (CSMA/CD) Access Method and Physical Layer Specifications-Media Access Control (MAC) Parameters, Physical Layer and Management Parameters for 10 Gb/s Operation, 2002.

27. Skeie, T., S. Johannessen, and O. Holmeide, The Road to an End-to-End Deterministic Ethernet, in Proceedings of the 4th IEEE International Workshop on Factory Communication Systems (WFCS'2002), IEEE Press, New York, Aug. 2002, pp. 3–9.

28. Jasperneite, J., P. Neumann, M. Theis, and K. Watson, Deterministic Real-Time Communication with Switched Ethernet, in Proceedings of the 4th IEEE International Workshop on Factory Communication Systems (WFCS'2002), IEEE Press, New York, Aug. 2002, pp. 11–18.

29. Rondeau, E., T. Divoux, and H. Adoud, Study and Method of Ethernet Architecture Segmentation for Industrial Applications, in Proceedings of the 4th IFAC International Conference on Fieldbus Systems and their Applications (FeT'2001), IFAC Publications, Elsevier Science, Oxford, England, Nov. 2002, pp. 165–172.

30. L. Almeida, P. Pedreiras, and J.A. Fonseca, "The ftt-can protocol: why and how", *IEEE Transactions on Industrial Electronics*, 49, 2002, pp. 1189–1201.

49

Switched Ethernet in Automation Networking

Tor Skeie
*ABB Corporate Research Center and
Simula Research Laboratory*

Svein Johannessen
ABB Corporate Research Center

Øyvind Holmeide
OnTime Networks

49.1 Introduction

In recent years, the Ethernet technology has taken several giant evolution steps. From the 10 Mbit/sec shared cable in 1990, the state of the art is now a switch-based communication technology running at 100 or 1000 Mbit/sec. The network switches are also becoming more sophisticated in that they have started including support for packet priority and virtual networks. Since all this power comes at a steadily decreasing cost, it is steadily invading other cost-sensitive areas, most notably automation networks. This chapter will look at some critical aspects of switched Fast Ethernet as an automation network.

49.2 The Switches are Not the Complete Network

Exchanging the classic Ethernet infrastructure for a switched infrastructure does wonders for the infrastructure but not necessarily for our automation network. The reason for this fact is illustrated in Figure 49.1, which shows all the stages in an information transfer between an automation controller and an I/O node.

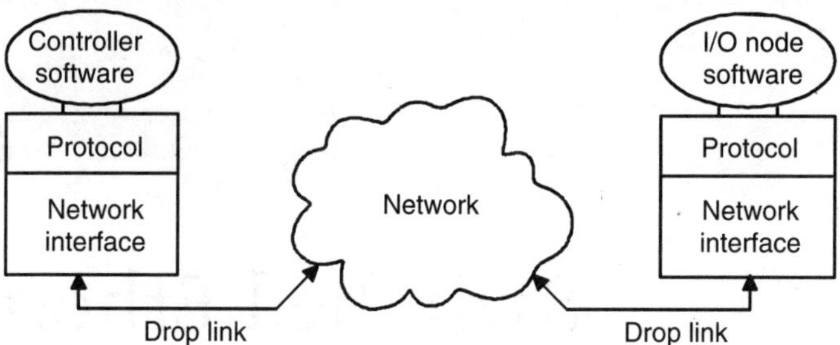

FIGURE 49.1　The complete end-to-end network.

The purpose of this chapter is to

- analyze the advantages, disadvantages, and peculiarities of a switched Ethernet automation network,
- point out the remaining bottlenecks between the controller software and the I/O node software, and
- discuss how the characteristics of the switched network affects some specialized automation network tasks.

49.3　Analyzing Switched Fast Ethernet

Let us start by recapitulating some basic facts about switched Ethernet automation networks.

1. Just as in a hub, an Ethernet switch contains a number of *ports*. Each port is either active (connected to an active Ethernet node) or passive (disconnected or connected to an inactive Ethernet node).
2. The connection between an active port and its associated Ethernet node is point-to-point. The connection may be full duplex (send and receive simultaneously) if the associated node supports it; otherwise, it is half duplex.
3. Each port in an Ethernet switch has a table of network addresses associated with that port. These tables are created by inspecting network packets sent from the node and extracting the source address from the packets.
4. Ethernet switches use a store-and-forward approach, which means that the complete network packet is received and verified by the switch before it is transferred to the output port.
5. The transfer of a network packet from one port to another inside the Ethernet switch is done by memory-to-memory copy at a very high speed.
6. An Ethernet switch does not use the collision mechanism of classic Ethernet.
7. Several transfers between different ports may take place more or less simultaneously. If an Ethernet switch has N ports, it should also be capable of handling N simultaneous connections running at full link speed (assuming that none of them request the same output port). This means that an Ethernet switch has a potentially much greater data transfer capability than a hub.

The Learning Process Inside the Switch

Item (3) above gave the algorithm the switch uses for the internal transfer of network packets. What happens if the switch does not have the destination node address in any port table? In this case, the switch plays it safe and transfers the packet to every output port in the switch. It reduces the switch performance to that of a hub; but it ensures that if the destination node is anywhere on the network, the packet will get through. In the meantime, the switch has learnt the location of the source node and has stored it in the correct table.

This learning process is usually very fast, since each node only has to answer once in order for it to be associated with the correct port. There are, however, some cases when the learning process fails, the most important being:

- Ethernet broadcast and
- Ethernet multicast.

Broadcast: Use with Care

Ethernet inherited the concept of *broadcast* (an address that is accepted by every node on the network) from HDLC. This functionality turned out to be very useful in several network protocols:

- Name announcing in *NetBIOS* ("Hi, everybody, my name is…").
- The Service Advertising Protocol (SAP) in *IPX/SPX*.
- Network address inquiry in *ARP* ("Hello everybody, who has got this IP address?").
- IP address deployment in *BOOTP* and *DHCP* ("I just woke up and I want an IP address!").

The drawbacks inherent in excessive usage of broadcast messages are:

- Broadcast messages cannot be filtered by hardware and must perforce be handled by software, thus consuming CPU resources.
- If broadcast messages are allowed to pass through bridges and routers unchecked, they may create a *broadcast storm* (broadcast messages propagated in circles and thereby clogging the whole network).
- If a broadcast address is used as a *source address*, it may create several network problems (this is a well-known way of creating *network meltdown*).
- Of course, it reduces the performance of a switched network to that of a hub-based network.

Multicast: A Useful, but Dangerous Concept

A huge number of Ethernet addresses are reserved for *multicast* usage. The Multicast concept addresses the need for creating network *groups* of some kind. Traditionally, the multicast concept has been popular in some automation contexts, usually associated with a philosophy called publish-and-subscribe.

This philosophy allows independently developed distributed applications to be able to exchange information in an event-driven manner without needing to know the source of the data or the network topology. Information producers publish information anonymously. Subscribers anonymously receive messages without requesting them.

Like other broadcast-based models, publish-and-subscribe is efficient when used on a classic network. For example, if the cost of energy changes in a distribution system, only a single transmission is required to update all of the devices dependent on the energy price (this is, of course, in the best or most optimistic case).

On a switched Ethernet network, the situation changes drastically. For a standard (unmanaged) switch, multicasting actually uses *more* bandwidth than sending the same message to one recipient after another. This surprising fact is best illustrated with an example.

If we have a 16-port switch and send the same message to four nodes, the message will occupy the sending port 4 times and the four receiving ports one time each, for a total of eight message times. Sending it as a multicast means that the message will occupy all 16 port for one message time — a total of 16 message times.

The moral in this case is: know your protocols and your infrastructure — and make sure they work together, not against one another.

49.4 There are Always Bottlenecks

Referring again to Figure 49.1, we observe that the network infrastructure is just one part of the communication path from the controller application to the I/O node application. The drop links are unchanged from the hub-based network and the end nodes (the controller and the I/O station) may well be unchanged from the days of the 10 Mbit network, or even older. We shall now proceed to look at the most important bottlenecks in our switched network.

Even Highways have Queues

Queues in the Switches

The nondeterministic behavior of traditional switched Ethernet is caused by unpredictable traffic patterns. At times, packets from several input ports will be destined for the same output port and some of them must perforce be queued up waiting for the output port to be free. The reason is that at times, there will be considerable low-priority traffic in the system (node status reports, node firmware update, etc.). If we do not introduce some sort of traffic rules, such a situation will give an unpredictable buffering delay depending on the number of involved packets and their length. In the worst case, packets can be lost when the amount of packets sent to an output port exceeds the bandwidth of this port for a time period that is longer than the output buffer is able to handle.

An automation network will have such coexisting real-time traffic (raw data, time sync data, commands, etc.) and noncritical data (TCP/IP — file transfer, etc.). A maximum size packet (1518 bytes) represents an extra delay of 122 μsec for any following queued packets in case of a 100 Mbps network.

Queues on the Drop Links

Even if we manage to introduce some sort of priority mechanism in our system, we have one queue mechanism left. Since some standards (IEC 61850-9 springs to mind) propose to transfer real-time data by using a "publish-and-subscribe" approach (which is again based on the use of multicast groups), a standard unmanaged Ethernet switch has to route these data packets on to every drop link in the system adhering to the broadcast paradigm. Since all those real-time packets might have the same priority, they will be put in the same switch queue on all output ports. The processing rate for these queues is equal to the bandwidth of the drop link between the switch and the nodes, and is therefore effectively a drop link queue. This way, important data packets destined for one node may be delayed by multicast traffic destined for another node. Obviously, we need to introduce some traffic rules/mechanisms for multicast data packets as well in order to reduce the worst-case transfer time.

Queues in the Nodes

Introducing traffic rules in the Ethernet switches will improve the worst-case latency across the network (from the Ethernet controller in the source node to the Ethernet controller in the destination node). Inside the node, however, there is usually only one single network task, and just a single hardware queue associated with the Ethernet controller. Since a network packet spends a large percentage of its total end-to-end time inside a node, internal packet prioritization is needed in order to have maximal control over the total packet transfer time (80–90% of the end-to-end message latency is spent within the end nodes, at least when adhering to the IETF protocols UDP/TCP/IP).

- For the transmit operation, one typically could have a situation where multiple max size Ethernet packets, for example, fragments of an FTP transfer, queued up at the Ethernet driver level. In a standard implementation, real-time packets will be added to the end of the queue. Such behavior will cause a nonpredictable delay in transmission.
- For the receive operation, the protocol stack implementation represents a possible bottleneck in the system. This is mainly due to the first-come first-served queue at the protocol multiplexer level (the level where the network packets are routed to the appropriate protocol handler software).

Introducing a Standard for Priority and Delivery

High-Priority Packets Jump Ahead in Switch Queues

IEEE 802.1D (see [3]) has been introduced to alleviate the switch queue problem; moreover, the standard specifies a layer 2 mechanism for giving mission-critical data preferential treatment over noncritical data. The concept has primarily been driven by the multimedia industry and is based on priority tagging of packets and implementation of multiple queues within the network elements in order to discriminate packets [1]. For tagging purposes, IEEE 802.1Q [4] defines an extra field for the Ethernet MAC header. This field is called *Tag Control Info* (TCI) field and is inserted as indicated by Figure 49.2. This field contains three priority bits; thus the standard defines eight different levels of priority.

Multicast Distribution Rules Reduce Drop Link Traffic

Figure 49.2 defines more than some priority bits. If you observe closely, you can see something called a "12 bit 802.1Q VLAN identifier." This Virtual Local Area Network (VLAN) identifier is a mandatory part of

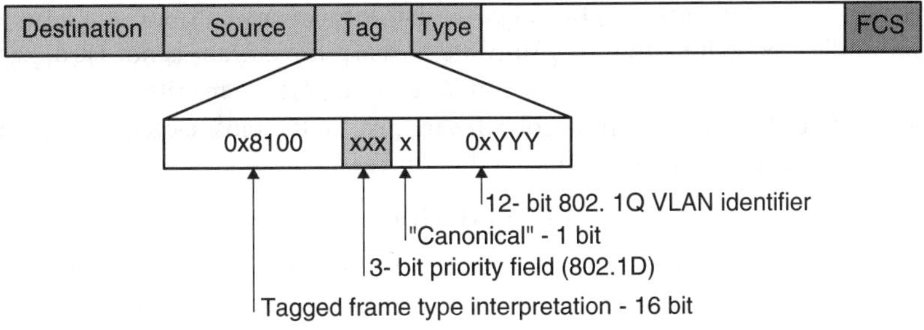

FIGURE 49.2 MAC header (layer 2) with tag.

each TCI field and can, when properly used, remove all unnecessary drop link traffic in an automation network based on publish-and-subscribe. We will discuss the handling and usage of this field in the section on Virtual LANs.

High-Priority Packets Get High-Priority Treatment

A network data packet spends a relatively small percentage of its application-to-application transfer time on the physical network. The actual percentage varies with the speed of the network and the performance of the node, but for a 100 Mbit/sec Fast Ethernet, the average percentage is between 20 and 0.1%. For this reason, implementing network priority will have little influence on the average application-to-application transfer time (it will, however, have a large influence on the worst-case transfer time). In order to improve the average application-to-application transfer time, the concept of priority must be extended to include the protocol layers in both the sending and receiving end. In order to accomplish this, we must consider:

- Adjustable process priority for the protocol stack software.
- Several instances of the protocol stack software running at different priority levels.
- Multiple transmit queues at the Ethernet driver level.

Matching Network Process Priority to Packet Priority

In most real-time operating systems, the protocol stack runs as a single thread in the context of some networking task. If we want the task priority to depend on the packet priority, we can implement it in a very simple way:

1. At compile time, decide on the task priority that should correspond to each packet priority and to an untagged packet. Put those priorities in a table.
2. Set a high basic networking task priority (the priority it uses when no packet is being processed).
3. When the networking task starts to process a packet, it should extract the packet priority and use it with the priority table described above to find the appropriate task priority.
4. Change the task priority by executing a system call.

Multiple Instances of the Protocol Stack Software

There is, however, a problem with the previous solution. Adjusting to correspond to the packet priority ensures that the processing of high-priority network packets does not get interrupted by less important administrative tasks. Still, networking tasks process incoming messages one at a time in a linear manner. What we really want is to process high-priority messages before low-priority ones. In fact, if we could suspend the processing of low-priority network packets when a high-priority message arrives, we would have the perfect solution.

At first glance, there exist an ideal solution. It goes like this:

1. At compile time, decide on the task priority that should correspond to each packet priority and to an untagged packet. Create one task for each priority and put the task IDs in a table.
2. When a packet arrives, extract the packet priority, use it with the task table described above, send the packet to that task, and send a signal to the task in order for it to start processing.

The problem with this solution is that it supposes that the network software is *reentrant*, a condition that is seldom fulfilled. Rewriting the stack software to make it reentrant is not hard, just tedious and time-consuming. Of course, it also means that you have to support the rewritten software in the future.

Running multiple instances of the protocol software may be the most elegant and efficient solution, but one must be prepared to spend some time and resources on it.

Multiple Receive Queues and Adjustable Priority

If we do not want to spend time and money on making the network software reentrant; there is an alternative solution available. This solution does not suspend the processing of lower priority packets, but selects the next packet to be processed from a set of criteria based on the packet priority. One possible implementation goes like this:

5. At compile time, decide on the task priority that should correspond to each packet priority and to an untagged packet. Create one queue for each priority and put a pointer to the queue in a table.
6. Whenever the network software is ready to process the next packet, it pulls all packets from the input queue and distributes them to the priority queues.
7. When the input queue is empty, the priority selection algorithm is run. This algorithm may be implemented in several different ways, for example:
   - Always pick the packet from the highest priority nonempty queue.
   - Pick the packet from the highest priority nonempty queue a certain number of times. Then, move the first packet from each nonempty queue to the next higher priority queue before picking the packet from the highest priority nonempty queue again.
   - Introduce a LOW flag. Wait for a packet to appear in the highest priority queue or for the LOW flag to be set. If the packet appeared in the highest priority queue, set the LOW flag and process that packet. If the LOW flag was set, reset it and process the packet from the highest priority nonempty queue.

Implementing Multiple Transmit Queues

On a transmit request, a standard Ethernet driver takes a buffer, does some housekeeping, and transfers it to the Ethernet controller hardware. Such a driver is simple and "fair" (first come, first served), but may be unsuitable for an efficient priority implementation. One or more large low-priority packets, once they are scheduled for transmission, will delay high-priority packets for the time it takes to transfer the low-priority ones. If we want high-priority packets to go to the head of the transmission queue, and the hardware does not support multiple queues, we must do some priority handling at the driver level.

The simplest solution is to use two queues: the hardware queue and a software queue. Low-priority packets go into the software queue; high-priority packets go straight into the hardware queue. From this point onward, there exist several algorithms addressing different needs.

- If the real-time requirements are moderate, move the first packet in the software queue to the hardware queue whenever the hardware queue is empty.
- If the real-time requirements are strict, move the first packet in the software queue to the hardware queue whenever a high-priority packet has been placed in the hardware queue.

Bottleneck Conclusions

The conclusion, based on a large set of real-world measurements (see [2]), is that:

1. At 100 Mbit/sec, the Ethernet switch and the drop links do not constitute a bottleneck under FTP load.
2. If several switches are interconnected by standard drop link cables, these cables might represent bottlenecks. Most switches have provisions for Gigabit Ethernet (1000 Mbit/sec) interconnections, however, thereby removing this possibility.
3. The main communication delays are inside the nodes. A suitable queue strategy for high-priority packets assures that those will be processed before low-priority packets.

4. For a high-performance node processor, implementing prioritized transmit queues has a greater overall impact on high-priority packet delays than implementing prioritized receive queues.
5. If you want to implement internal receive priority queuing, ensure that it is possible to configure the chosen switch not to remove the priority tagging information.

49.5   Time Synchronization Across Switched Ethernet

Now and then you come across measurement problems that are tightly associated with the notion of *synchronicity*, meaning that things need to happen simultaneously. The usual things that need such *synchronicity* are data sampling and motion control. In the case of data sampling, you need to know the value of two different quantities measured at the same time (within a narrow tolerance). If the measurement sources are close together, this is fairly easy to accomplish, but if they are far apart and connected to different measurement nodes, it suddenly becomes harder. The usual choices are:

1. Use a *special hardware signal* on a separate cable between the controller and all nodes that need synchronization. If the nodes are far apart and the tolerances are tight, make sure that all cables that carry the synchronization signal have the same length.
2. Add a local clock to each node and use the present automation network to keep them in synchronization. Tell each node how often the measurement sources should be sampled and require the node to *time stamp* each measurement.

We shall now take a look at the hardest synchronization requirements for automation purposes and discuss the possibility of implementing class T5 (1-μsec) and class T3 (25-μsec) synchronization in a multitraffic switched Ethernet environment. Common for both solutions is that they adhere to the same standardized time protocol. Such a step would significantly reduce the cabling and transceiver cost, since costly dedicated (separate) links are used for this purpose today.

The Concept of Time Stamping

Let us start at the very beginning — the concept of time stamping:

Time stamping is the association of a data set with a time value. In this context, "time" may also include "date."

Why would anybody want to time stamp anything? The closest example may be on your own PC — whenever you create a document and save it, the document is automatically assigned a data-and-time value. This value enables you to look for

- Documents created on a certain date (e.g., last Monday).
- Documents created within a certain time span (e.g., last half of 1998).
- The order in which a set of documents were created (e.g., the e-mails in your inbox).

If we just look at the examples above, we see that the accuracy we need for the time stamping is about the same as we expect from our trusty old wristwatch. This again means "within a couple of minutes," but as long as the clock does not stop it does not really matter much how precise it is.

Let Us Synchronize Our Watches!

Now we know about time stamping on our own PC. The next step is to connect the PC to a network, maybe even to the Internet and start exchanging documents and e-mails. What happens if the clock in your PC (the clock that is used for time stamping) is wrong by a significant amount?

- If you have an e-mail correspondence with someone, a reply (which is time stamped at the other end) might appear to be written before the question (which is time stamped at your end).
- If you collaborate on some documents, getting the latest version might be problematic.

Therefore, when several PCs are connected together in any kind of network, the PC clocks are still accurate enough, but a new requirement is that they should be *synchronized* (show the same time at one point in time). Now, we could go around to each PC, look at our wristwatch, and set the PC clock to agree with it. The trouble is that this is a boring and time-consuming job and we should look for a better solution.

One solution is to elect one PC to be the "time reference," which means that every other PC should get the current time from it at least once a day and set its own clock to agree with that time. This solution works satisfactorily on a Local Area Network (LAN), but all PC clocks will lag the time reference by the time it takes a clock value to travel from the time reference to the synchronizing PC. Except for very unusual cases, this lag is less than 1 sec and thus good enough for office purposes.

Enter the Internet. Suddenly the synchronization problem escalates, since two collaborating PCs may be located in different time zones (remember to compensate for that) and a synchronization message may take a long time from one PC to the other. Fortunately, the Internet Network Time Protocol has a solution to both problems. This protocol involves sending a time-stamped time request message to a "timeserver." This timeserver adds an arrival time stamp and a retransmit time stamp before returning the request message to the requesting PC. The requesting PC time stamps the message when it returns and uses all the time stamps in calculating the correct time. This protocol and it little brother, the Simple Network Time Protocol, are able to synchronize computers across the Internet with a precision in the low milliseconds.

Synchronization Requirements in Substation Automation

In the energy distribution world, a *substation* is an installation where the energy is combined, split, or transformed. A Substation Automation (SA) system refers to tasks that must be performed in order to control, monitor, and protect the primary equipment of such a substation and its associated feeders. In addition, the SA system has administrative duties such as configuration, communication management, and software management.

The communication within SA systems is crucial from the point that the functionality demands for very time-critical data exchange. These requirements are substantially harder than the corresponding requirements in general automation. This is also true for the required synchronization accuracy of the Intelligent Electrical Device's (IED's) internal clock in order to guarantee precise time stamping of current and voltage samples. Various SA protection functions require different levels of synchronization accuracy. IEC has provisionally defined five levels — IEC classes T1 – T5 (IEC 61850-5, Secs 12.6.6.1 and 12.6.6.2):

- IEC class T1: 1 msec
- IEC class T2: 0.1 msec
- IEC class T3: ±25 μsec
- IEC class T4: ±4 μsec
- IEC class T5: ±1 μsec

Since these definitions and classes are not frozen as yet, we will refer to them as class T1, class T2, etc. (without IEC) here.

At this point in time, the SA field is also on the edge of migrating toward the usage of Switched Fast Ethernet as the automation network infrastructure. The ultimate vision is to achieve interoperability between products from different vendors on all levels within the substation automation field. A proof of this new trend is the upcoming IEC 61850 standard on *Communication networks and systems in substations* issued by Technical Committee. Inventions of de facto standard concepts, and adoption of off-the-shelf technologies are the key instruments to reach the interoperability goal. Figure 49.3 illustrates the communication structure of a future substation adhering to switched Ethernet as a common network concept holding multiple coexisting traffic types [5].

We already know that Switched Fast Ethernet has sufficient real-time characteristics to meet very demanding automation requirements. What is left to show is that it is possible to implement the various IEC classes of synchronization accuracy over such a network. This is considered to be the final obstacle of fully migrating to Ethernet in SA.

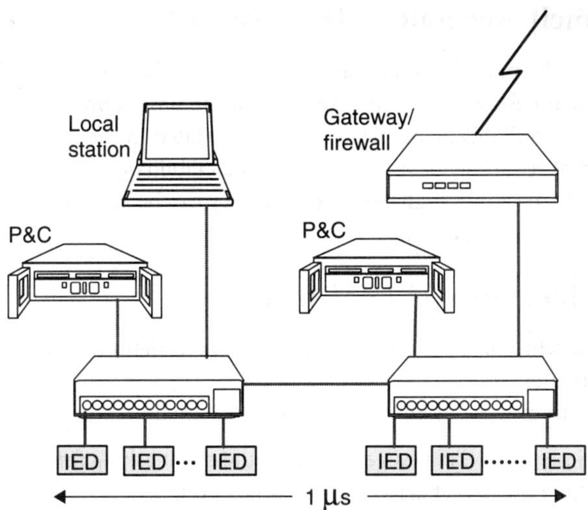

FIGURE 49.3   The communication network in a future SA system using Ethernet as a common network infrastructure.

Time Synchronization Status

There is a plethora of proposed theory and methods for synchronizing clocks in distributed systems. The most prominent public domain synchronization method is the Network Time Protocol (NTP), which is standardized by the Internet Engineering Task Force in RFC1305. A subset of NTP (SNTP — Simple Network Time Protocol) is also defined, and is protocol-compatible with NTP.

The intended use of NTP is to synchronize computer clocks in the global Internet. For this purpose, it relies on sophisticated mechanisms to access national time, organize time synchronization subnets possibly implemented over various media, and adjustment of local clock in each participating peer. SNTP, on the other hand, does not implement the full set of NTP algorithms and is targeting simpler synchronization purposes. Common for this body of work is that it does not present solutions for low microsecond accuracy, instead targeting synchronization of LANs and WANs in the general sense where a precision of some milliseconds is sufficient.

Looking at the automation field in general and especially at the SA world, we find a diversity of proprietary and patented solutions in order to achieve highly accurate time synchronization over Ethernet. Interoperability concerns are, however, present.

Stating the Problem — Why Network Synchronization is Difficult

The delays from the time stamping of a time synchronization message in the message source node until it is time stamped in the message destination node are:

- Message preparation delay
- Communication stack traversal delay (transmission)
- Network access delay
- Network traversal delay
- Communication stack traversal delay (reception)
- Message handling delay

Variations in the delays are due to:

- Real-Time Operating System (RTOS) scheduling unpredictability
- Network access unpredictability
- Network transversal time variations

Time stamping at the lowest stack level helps eliminate the stack delay variations and real-time OS scheduling unpredictability but introduces some complications in the implementation.

How to be Extremely Accurate — IEC Class T3

We have mentioned that the precision that may be achieved by the traditional NTP/SNTP implementations is one millisecond at best. Basically, this stems from the time stamping of incoming and outgoing NTP/SNTP packets at the NTP/SNTP application layer. As stated in the previous paragraph, this makes time stamping a victim of real-time OS scheduling unpredictability.

In this section, we describe how a high degree of accuracy can be achieved using a tuned SNTP implementation and standard Ethernet switches.

A Tuned SNTP Time Protocol Implementation

The NTP/SNTP time synchronization algorithm presented has definite limits to the level of attainable accuracy. Let us recapitulate:

The NTP/SNTP algorithm is based on a network packet containing three time stamps:

- t_1: The (client) time the packet was generated in the client asking for the current time.
- t_2: The (server) time the packet arrived at the timeserver.
- t_3: The (server) time the packet was updated and put into the transmission queue at the server. In addition, the calculations require:
- t_4: The (client) time the packet arrived back at the client.

Now, t_2 and t_4 can easily be determined down to microseconds (and perhaps even better) using hardware or software time stamps based on packet arrival interrupts. The other two have definite problems, however.

For full accuracy, t_1 and t_3 should have been *the time when the packet left* the time client or timeserver, respectively. The problem is that this time stamp is not available until the packet has really left the timeserver or client and then it is, of course, too late to incorporate it into the packet. Therefore, the time synchronization inaccuracy for an NTP/SNTP setup is the variation in the delay between t_1 and the time the packet leaves the time client plus is the variation in the delay between t_3 and the time the packet leaves the timeserver.

It turns out that there are several ways of determining the time when the synchronization packet actually leaves the time client. One algorithm runs like this:

1. Create the NTP packet and fill in the t_1 (Originate Timestamp) as usual. Transmit the package to the timeserver.
2. Get hold of the time stamp when the packet leaves the time client, using any appropriate mechanism. Store this time stamp (t_{11}) in a data structure together with t_1.
3. When the packet returns from the timeserver, extract t_2 from the packet and use t_1 to look up the corresponding value of t_{11}. Since t_{11} and t_1 are stored together, there is no chance of confusion here.
4. Substitute t_{11} for t_1 in the packet.

Achieving Precise Time Stamping Within a Node

The nature of real-time operating systems (they guarantee a maximum response time for an event but allow for a wide variation below that) introduces a substantial variation in the time spent in the communication stacks. This fact has necessitated an interrupt-level time stamping both in the time client and timeserver. The IEC class T3 solution described here adheres to the principle of *interrupt-level time stamping of the SNTP request packet when sent from the time client and when received at the timeserver*. Moreover, we propose that the synchronization is based on the (corrected) *transmit time stamp* set by the client (referred to as t_1 in SNTP terminology) and the *receive time stamp* set by the server (referred to as t_2). Usage of a possible low-level transmit time stamping of the corresponding SNTP reply packet (referred to as t_3) necessitates novel techniques for controlling the nondeterministic access of an Ethernet packet to an Ethernet bus.

A side effect of using $t_1 - t_2$ only is that no mechanism for automatic calibration of the network latency will be available and therefore a manual calibration of the propagation delays of the drop links and the minimum switch latency[1] must be performed.

---

[1]The minimum Ethernet switch delay is usually given in the switch data sheet.

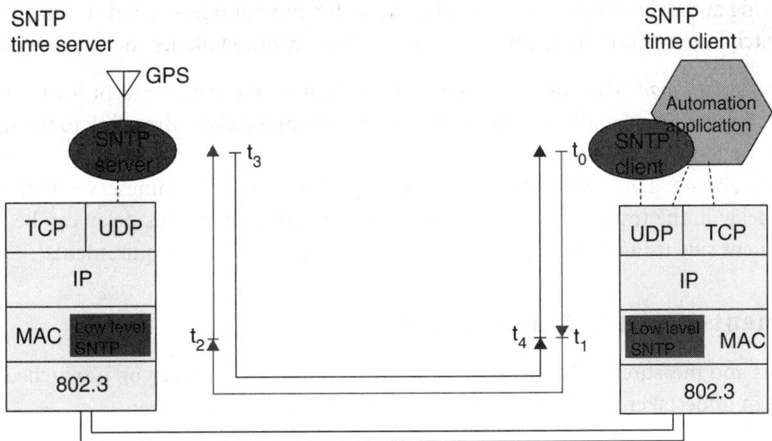

FIGURE 49. 4   The SNTP time client – server relation using low-level time stamping.

Figure 49.4 illustrates the setup of an SNTP time client and timeserver adhering to interrupt-level time stamping.

Time Client Implementation Issues

There are several ways of time stamping a network packet. We shall look at three of them and show that only the first two are suitable for accurate time synchronization:

1. Hardware time stamping in the Ethernet controller.
2. Software time stamping in an Interrupt Service Routine (ISR) outide the RTOS. This ISR should be connected to the Ethernet Interrupt Request signal and have a top hardware priority.
3. Software time stamping in an ISR controlled by the RTOS (Ethernet driver). This ISR is connected to the Ethernet Interrupt Request signal with a normal hardware priority.

Use of any of this low-level time stamping methods is considered an implementation issue and will not cause any incompatibly between a low-level time stamping client and a standard high-level time stamping server. In addition to low-level time stamping, the time client must consider the following aspects:

- The interval between time updates.
- The specifications of the local time-of-day clock with respect to resolution, accuracy/stability, and the availability of drift and offset correction mechanisms.
- The usage of adaptive filtering and time stamp validation methods in order to remove network delay variations.

Timeserver Implementation Issues

In order to achieve class T3 accuracy, the timeserver should be able to time stamp an incoming message with an accuracy of better than 2 μsec independent of network load. The exact time should be taken from a GPS receiver, and the time parameters distributed from the timeserver should be based on Global Positioning System (GPS) time representation instead of absolute time (i.e., UTC timing) in order to cope with the leap second problem. It is also convenient if the timeserver supports full duplex connectivity in order to avoid a situation where upstream data introduce extra switch latency in downstream data (i.e., time requests).

Ethernet Infrastructure Implementation Issues

Only one switch should preferably be allowed between a time client and a timeserver. Having multiple switch levels will impose increased jitter[2] through the infrastructure, which again might call for more

---

[2]Jitter: variations in the delay.

complex filtering at the time client side. The Ethernet switch must also have good switch latency characteristics. The switch latency from the client drop link to the server drop link depends on several parameters:

- *General switch load.* This means all the network load on the switch except for the packets sent to the timeserver. The variations in the switch latency from the client drop link to the server drop link should be less than 2 μsec.
- *Timeserver load.* This parameter means other packet sent to the timeserver that may introduce extra delay in the transmission of a given SNTP request packet. This delay can be handled at the time client side using various filtering techniques (see time client requirements).

Measurements on an Actual Network

Extensive tests and measurements regarding time synchronization accuracy on a switched Ethernet network have been undertaken [6]. The conclusions from this body of work are:

- Traffic not destined for the timeserver does not interfere with traffic to the timeserver.
- The switch latency for Ethernet packets to the timeserver depends to a great extent on other traffic to the timeserver.

The full network measurements conclusions are:

- Software time stamping using a sufficiently high-priority interrupt (preferably nonmaskable) is for all practical purposes indistinguishable from time stamping using special-purpose hardware.
- Software time stamping using an interrupt under RTOS control needs sophisticated filtering and statistical techniques before it can be used for time synchronization purposes. In this respect, this time stamping method is not suitable for IEC class T3 synchronization.
- IEC class T3 time synchronization using tuned SNTP over a switched Ethernet has shown to be eminently feasible.

Beyond the Speed of Light — Class T5

It is now possible to procure industrial-class Fast Ethernet switches fulfilling an extensive list of environment requirements relevant for substation automation applications. Some of these switches can even be delivered with an integrated SNTP timeserver.

Since the internal switch logic has full control over input and output ports, time stamping an SNTP request packet on arrival is no problem. In addition, the switch logic can insert the transmit time stamp whenever the output port is ready for the reply packet (and even adjust the time stamp for the delay between time stamping and actual transmission). Thus, the traditional problem related to the nondeterministic access to the Ethernet is not a problem here due to the tight interaction between the SNTP timeserver and the switch architecture.

This time synchronization scheme provides the following:

- Timing synchronization accuracy better than one microsecond if time stamping in the time client is performed in hardware; see Time client implementation issues.
- Both server time stamps — t_2 (receive) and t_3 (transmit) — may be used at the time client for synchronization purposes, and the drop link propagation delay can easily be removed based on the calculated round-trip delay.
- The timing accuracy is independent of the network load.
- No clever filtering/erasure techniques are needed in the time client.

Summary and conclusions

We have presented general solutions for achieving class T5 (1 μsec) and class T3 (25 μsec) time synchronization over switched Ethernet. The former is based on a dedicated Ethernet switch/Time Server combination, while the latter relies on standard switches. Common for both solutions is that they adhere to the low-level time stamp implementation of the SNTP.

Hardware time stamping or low-level software time stamping outide the real-time operating system eliminates the client inaccuracy from the error budget of the SNTP time synchronization loop. If the SNTP timeserver relies on the same time stamping techniques, the only remaining factor to be handled in the error budget is possible time delay variations within the infrastructure. In these settings, class T3 synchronization is possible over switched Ethernet.

49.6 Introducing Virtual Subnetworks

What are VLANs? A concept made possible by Ethernet switches, the multitude of vendor-specific VLAN solutions and implementation strategies has made it very difficult indeed to define precisely what a VLAN is. Nevertheless, most people would agree that a VLAN might be roughly equated to a broadcast domain[3]. More specifically, VLANs can be seen as analogous to a group of end-stations, perhaps on multiple physical LAN segments, which are not constrained by their physical location and can communicate as if they were on a common LAN.

Currently, three types of VLANs are of interest:

- Port Group VLANs
- Layer 2 — Multicast Group-based VLANs
- Layer 3-Based VLANs

Here, we will only give a brief overview of these different types of VLANs. An important aspect to note in this context is the fact that it is now possible to define dynamic VLANs that correspond exactly to a multicast group. This means that multicast frames will only propagate to members of the indicated VLAN and not to anyone else. In particular, the frames will not occupy drop links or CPU resources in nodes that do not belong to the multicast group.

Port Group VLANs

Port-based VLAN is a concept relying on a pure manual configuration of Ethernet switches (bridges) to set up VLAN membership. Many initial VLAN implementations defined VLAN membership by groups of switch ports (e.g., ports 1, 2, 3, 7, and 8 on a switch make up VLAN A, while ports 4, 5, and 6 make up VLAN B). Furthermore, in most initial implementations, VLANs could only be supported on a single switch.

Second-generation implementations support VLANs that span multiple switches (e.g., ports 1 and 2 of switch #1 and ports 4, 5, 6, and 7 of switch #2 make up VLAN A, while ports 3, 4, 5, 6, 7, and 8 of switch #1 combined with ports 1, 2, 3, and 8 of switch #2 make up VLAN B).

Port grouping is still the most common method of defining VLAN membership, and configuration is fairly straightforward. Defining VLANs purely by port group does not allow multiple VLANs to include the same physical segment (or switch port). However, the primary limitation of defining VLANs by port is that the network manager must reconfigure the VLAN membership when a node is moved from one port to another.

Group-Based VLANs: GARP VLAN Registration Protocol

In the 802.1D standard (see [3]), the Generic Attribute Registration Protocol (GARP) is introduced. This is a general protocol that allows network nodes to control selected switch properties. The first two implementations of this protocol are the GARP Multicast Registration Protocol and the GARP VLAN Registration Protocol (GVRP).

GVRP provides a mechanism that allows switches and end stations to dynamically register (and subsequently, deregister) VLAN membership information with the Ethernet switches attached to the same LAN segment, and for that information to be disseminated across all switched in the LAN that support Extended Filtering Services. Moreover, no manual configuration of the switches is required opposed to

---

[3]Broadcast Domain: A collection of all nodes that can be reached by a broadcast message from one of them.

the port-based solution — on the other hand, the switches must implement extra software as specified by the IEEE 802.1D standard [3].

The operation of GVRP relies upon the services provided by GARP. The information registered, de-registered, and disseminated via GVRP is of the following forms:

1. VLAN membership information. This indicates that one or more GVRP participants that are members of a particular VLAN (or VLANs) exist, and the Ethernet frames carry a 12-bit VLAN Identifier (VID) (see Figure 49.2) that state the membership. The act of registering/deregistering a VID affects the contents of Dynamic VLAN Registration Entries to indicate the Port(s) on which members of the VLAN(s) have been registered.
2. Registration of VLAN membership information allows the Ethernet switches in a LAN to be made aware that frames associated with a particular VID should only be forwarded in the direction of the registered members of that VLAN. In this way, the VLAN membership is propagated through the Ethernet infrastructure, and forwarding of frames associated with the VID therefore occurs only on Ports on which such membership registration has been received.

GVRP is a very new protocol concept and is not yet widely supported by Industrial Ethernet switches, but is foreseen to be one of the future technologies for handling the publish-and-subscribe automation network philosophy. One such handling algorithm consist of three steps:

1. Map every multicast group to one specific VLAN identifier. This is, of course, an off-line activity.
2. At system start-up time, all nodes send out one small packet for each VLAN (and thereby multi-cast group) it wants to join. This packet, which is not a part of any complicated protocol stack, gives the network infrastructure the information it needs in order to map VLAN identifiers to physical drop links. For simple data source nodes, these packets can be precompiled.
3. Whenever a multicast packet is to be transmitted, it should be tagged with "real-time priority" and the VLAN identifier corresponding to the multicast group.

These simple steps ensure that multicast packets do not block any unnecessary drop links. Below, we discuss the possible solutions for taking control of the latency within the end-nodes.

Layer-3 Based VLANs

VLANs based on layer 3 information take into account protocol type (if multiple protocols are sup-ported) or network-layer address (e.g., subnet address for TCP/IP networks) in determining VLAN mem-bership. Although these VLANs are based on layer-3 information, this does not constitute a "routing" function and should not be confused with network-layer routing.

Even though a switch inspects a packet's IP address to determine VLAN membership, no route calcu-lation is undertaken, and frames traversing the switch are usually bridged according to implementation of the Spanning Tree Algorithm (the purpose of this algorithm is to make sure that no network packet is caught in an endless loop between switches).

There are several advantages to defining VLANs at layer 3. First, it enables partitioning by protocol type. This may be an attractive option for network managers who are dedicated to a service-or applica-tion-based VLAN strategy. Second, nodes can physically move their workstations without having to reconfigure each workstation's network address — a benefit primarily for TCP/IP nodes. Third, defining VLANs at layer 3 can eliminate the need for frame tagging in order to communicate VLAN membership between switches, reducing transport overhead.

One of the disadvantages of defining VLANs at layer 3 (vs. MAC- or port-based VLANs) can be per-formance. Inspecting layer 3 addresses in packets is more time consuming than looking at MAC addresses in frames. For this reason, switches that use layer 3 information for VLAN definition are generally slower than those that use layer 2 information. It should be noted that this performance difference is true for most, but not all, vendor implementations.

References

[1] Holmeide, Ø. and T. Skeie, *VoIP Drives Realtime Ethernet,* Industrial Ethernet Book, GGH Marketing Communications Ltd., Vol. 5, March 2001 (http://ethernet.industrial-networking.com/default.asp).

[2] Skeie, T., S. Johannessen, and Ø. Holmeide, "The Road to an End-to-End Deterministic Ethernet", in Proceedings of 4th IEEE International Workshop on Factory Communication Systems (WFCS), Sept., 2002.

[3] IEEE 802.1D, Information Technology — Telecommunications and Information exchange between systems — Local and Metropolitan Area Networks — Communication Specification — Part 3: Media Access Control Bridges, 1998.

[4] IEEE Std 802.1Q-1998, IEEE Standards for Local and Metropolitan Area Networks: *Virtual Bridged Local Area Networks.*

[5] Skeie,T., S. Johannessen, and C. Brunner; Ethernet in substation automation, *IEEE Control Systems Magazine,* 22, 43–51, 2002.

[6] Skeie, T., S. Johannessen, and Ø. Holmeide, Highly Accurate Time Synchronization over Switched Ethernet, in Proceedings of 8th IEEE Conference on Emerging Technologies and Factory Automation (ETFA), pp. 195–204, 2001.

50

Wireless LAN Technology for the Factory Floor

Andreas Willig
University of Potsdam

50.1 Introduction

Wireless communication systems diffuse into an ever-increasing number of application areas and achieved wide popularity. Wireles telephony and cellular systems have entered our daily lives, and wireless local area network (WLAN) technologies serve more and more as a primary means to access business and personal data. Two important benefits of wireless technology are key to this success: the need for cabling is greatly reduced, and computers as well as users can be truly mobile. These properties not only save costs but also enable new applications. This also holds true for factory plants, where wireless technology can be used in many different ways [18, chap. 2]:

- Applications with mobile subsystems, for example, autonomous transport vehicles, robots, and turntables.
- Implementation of distributed control systems in explosible areas, in the presence of aggressive chemicals or heat.

- Rapid prototyping of industrial plants without putting much effort into cabling.
- Mobile plant diagnose systems, and wireless stations for programming and on-site configuration.

However, adopting WLAN technologies for the factory floor must take two things into account: on the one hand, the special requirements of industrial applications, most notably the reliability and hard real-time constraints for communications, and on the other the properties of wireless transmission channels, which often exhibit a substantial and time-varying amount of transmission errors, potentially harming safety-critical data packets. Therefore, an important challenge for wireless industrial LANs is to find transmission schemes and protocols that give optimum performance in terms of real-time and reliability despite channel errors and certain other specific properties of wireless channels. Protocols should be specifically tailored for wireless media and thus they may well be different from the protocols used in wired industrial communication systems.

In this article, we survey some issues in designing and investigating protocols for wireless industrial LANs and provide an overview on the state of the art. The focus is on aspects influencing the time and reliability behavior of wireless transmission, and how these are addressed on the physical layer, the medium access control (MAC), and the link layer. These layers play a key role in delivering reliable and timely service to upper layers and to applications, and they are exposed most to the wireless link characteristics on the other hand. Furthermore, we provide an overview on current wireless field-bus systems and on the IEEE 802.11 WLAN standard, which might play an important role in industrial applications as well.

This chapter is structured as follows: in Section 50.2 important general considerations for wireless industrial communications and wireless field-bus systems are presented. In Section 50.3, we discuss some basic properties of wireless LAN technology and wireless wave propagation effects. The transmission impairments resulting from these effects and some approaches to target them on the physical layer are presented in Section 50.4. For the MAC and link layers, some specific problems as well as solution approaches are discussed in Section 50.5. The following two sections 50.6 and 50.7 adopt a more technology-oriented perspective. Specifically, in Section 50.6, we survey the state of the art regarding wireless industrial communication systems and wireless field-bus systems, and in Section 50.7 we present the important aspects of the IEEE 802.11 WLAN standard w.r.t. transmission of real-time data. Finally, in Section 50.8, we provide a brief summary.

This chapter is restricted to protocol-related aspects of wireless transmission, other aspects like signal processing, analog and digital circuitry, energy aspects, management, etc. are not considered. There are many introductory and advanced books on wireless networking, for example, [48, 1, 9, 26, 45, 50,–52] and several separate topics in wireless communications are treated in [19].

50.2 Wireless Industrial Communications and Wireless Field-Bus

In industrial applications, often *hard real-time* requirements play a key role. In accordance with [46], we assume the following important characteristics of hard real-time communications: (a) safety-critical messages must be transmitted reliably within an application-dependent deadline; (b) there should be support for message priorities; and (c) messages with stringent timing constraints typically have a small size. The qualifier "hard" stems from the fact that losses or deadline misses of safety-critical packets can lead to damage for life or equipment. Often, so-called *field-bus systems* are used in industrial communications. These systems are explicitly designed to meet hard timing and reliability requirements in harsh environments.

To date, the error properties of wireless links as experienced in different measurements (see the section Effects on transmission) suggest that wireless technology is not a candidate for systems where lost messages or deadline misses lead to catastrophes. As an example, the measurements presented in [64] have shown that in a certain industrial environment for several seconds no packet could be transmitted. Therefore, instead of seeking absolute guarantees, it is appropriate to formulate another optimization criterion: the percentage of safety-critical messages that can be transmitted reliably within a prespecified time-bound should be as high as possible (e.g., $> 99.x\%$), even at the cost of other performance measures like throughput or mean delay. We call this approach *stochastic hard real-time*. Of course, this limits the application areas of a wireless industrial LAN: when deterministic guarantees in the range of 10 to 100 msec are essential, wireless transmission is ruled out. However, in applications, where occasionally emergency stop conditions due to message loss or missing deadlines are tolerable, wireless technology can offer its potential. The goal is then to reduce the frequency of losses and deadline misses.

It depends on the communication model how reliability can be implemented. Many field-bus systems (e.g., PROFIBUS), address *stations*, and packets are transmitted from a sender to a receiver station without involving other stations. Reliability can be ensured using different mechanisms, for example, retransmissions, packet duplications, or error-correcting codes, which can all be implemented on wireless media. On the other hand, systems like factory instrumentation protocol (FIP) and controller area network (CAN) implement the model of a real-time database, where not stations but data are identified. A piece of data has one one producer and potentially many consumers. The producer broadcasts the data and all interested consumers copy the packet into an internal buffer. Typically, in these systems, there are no retransmissions, but reliability can be increased by other mechanisms, for example, by using error-correcting codes.

A *wireless field-bus system* aims at providing as stringent time and reliability properties as could be achieved on wireless media [10]. The vast number of already deployed wired field-bus installations and the desire to continue operation of these running systems is an incentive to focus research on *wireless extensions* of already existing field-bus systems instead of creating a new system from scratch. Having a wireless extension allows to integrate wireless and wired stations into a single field-bus LAN.

From a protocol architecture perspective, the different approaches to integrate wireless stations into wired field-bus LANs can be classified according to the layer of the OSI reference model, where the integration actually occurs [63, 10]. This classification reflects the generic architecture of many field-bus systems, where only the layers 1, 2, and 7 of the OSI reference model are used [11]. We use the term *wired station* for a station with a wired transceiver; consequently, a *wireless station* has a wireless transceiver.

Wireless repeater approach: All stations are attached to a cable (hence are wired stations), and just a piece of cable is replaced by a wireless link, using bridge-like devices translating the framing rules of the wireless and wired stations but without modifying the MAC- and link-layer-related fields of the packet. We call this type of bridge a *MAC-unaware bridge*. In this approach, no station is aware of the wireless link. A typical application scenario is the wireless interconnection of two field-bus segments.

Wireless MAC-unaware bridging approach: The LAN comprises both wired and wireless stations, and integration occurs solely at the physical layer. A MAC-unaware bridge translates the different framing rules used by wired stations and wireless stations. The wireless stations use "only" another physical layer (PHY), while MAC and link layer protocols are not modified. Hence, wired and wireless stations run the same set of protocol.

Wireless MAC-Aware bridging approach: The LAN comprises both wired and wireless stations, and integration occurs at the MAC- and data link-layers. There are two different MAC- and link-layer protocol stacks for wired and wireless stations, but both offer the same link-layer interface. The wireless MAC- and link-layer protocol should be: (a) specifically tailored to the wireless medium and (b) easily integrable with the wired MAC and link-layer protocol. An intelligent bridge-like device is responsible for both translating the different framing rules as well as interoperation of the different MAC protocols.

Wireless gateway approach: Integration at the application layer. In these systems, the PHY, MAC-, and link-layer protocol as well as application support services can be different for wireless and wired stations.

Clearly, these scenarios can be mixed.

50.3 Wireless LAN Technology and Wave Propagation

In this section, we discuss some basic notions of WLANs and present fundamental wave propagation effects. The wave propagation effects result in certain transmission problems, which are discussed in Section 50.4.

Wireless LANs

Wireless LANs are targeted for packet-switched communications over short distances (up to a few hundred meters) and with moderate bit rates. Typical values for the bit rate range between 1 and 54 Mb/sec in IEEE 802.11a [43]. They work with infrared or radio frequencies; in the latter case often license-free bands are used, for example, the 2.4 GHz Industrial, Scientific, and Medical band (ISM band). In these bands, the transmit power is legally restricted. Radio waves below 6 GHz propagate through walls and can be reflected on several types of surfaces, depending on both frequency and material. Thus, with radio frequency non-line of sight (NLOS) communications are possible. In contrast, systems based on infrared only allow for line of sight (LOS) communications over a short distance (e.g., IrDA [62]).

Wireless LANs can be roughly subdivided into *ad hoc networks* [56] and *infrastructure-based networks*. In the latter case, some centralized facility, often administered by the network provider, is responsible for tasks like radio resource management, data forwarding to distant stations, mobility management, etc. In general, stations cannot communicate without using the infrastructure. Infrastructure-based WLANs offer some advantages for industrial applications. Many industrial communication systems already have an asymmetric structure, which can be naturally accommodated in infrastructure-based systems. The often used master–slave communication scheme serves as an example. Furthermore, the opportunity to offload certain protocol processing tasks to the infrastructure allows to keep mobile stations more simple and to make efficient centralized decisions.

As compared to other wireless technologies like cellular systems and cordless telephony, WLAN technologies seem most suitable for industrial applications due to the offered compromise between data rate, geographical coverage, and license-free and independent operation.

Wave Propagation Effects

In the wireless channel, waves propagate through the air, which is an unguided medium. The wireless channel characteristics are significantly different from those of guided media like cables and fibers, and create unique challenges for communication protocols.

A transmitted waveform experiences several phenomena: path loss, attenuation, reflection, diffraction, scattering, adjacent- and co-channel interference, thermal or man-made noise, and finally imperfections in the transmitter and the receiver [48, 7].

The path loss/attenuation characterizes the loss in signal power when increasing the distance between a transmitter T and a receiver R. In general, the mean received power level $\widehat{P_{Rx}}$ can be represented as the product of the transmit power P_{Tx} and the mean path loss $\widehat{PL}$:

$$\widehat{P_{Rx}} = P_{Tx} \cdot \widehat{PL}$$

A typical path loss model for omnidirectional antennas is given by [48, chap. 4.9]

$$\widehat{PL}(d) = C \cdot \left(\frac{d}{d_0}\right)^n \; (d \geq d_0)$$

where d is the distance between T and R, $\widehat{PL}(d)$ is the mean path loss, d_0 is a reference distance that depends on the antenna technology, C is a technology- and frequency-dependent scaling factor, and n is the so-called path-loss exponent. Some typical values for n are given in Table 50.1. A second source of signal attenuation are obstacles in the LOS path.

A waveform is reflected on smooth surfaces having a much larger dimension than the actual wavelength. Not all signal energy is reflected; but a fraction of it is absorbed and the amount of absorption depends on the material. The diffraction mechanism can be explained by Huygens principle and allows a wave to propagate into a shadowed region, provided that some sharp edge exists. Scattering is produced when a wavefront hits a rough surface having structures smaller than the wavelength. Then, the signal diffuses.

The most important types of interference are cochannel interference and adjacent-channel interference. In cochannel interference, a signal transmitted from T to R on channel c_1 is distorted by another

TABLE 50.1 Path-Loss Exponents for Different Environments (taken from [48, chap. 4.9])

Environment	n
Free-space	2
Urban area cellular radio	2.7–3.5
Shadowed urban cellular radio	3–5
In-building LOS	1.6–1.8
Obstructed in-building	4–6
Obstructed in factories	2–3

transmission on the same channel c_1. In the case of adjacent channel interference, the interferer I transmits on an adjacent channel c_2, but due to imperfect filters R captures frequency components from c_2, or the interferer I leaks some signal energy into channel c_1 due to nonperfect transmit circuitry (amplifier).

The notion of noise refers to any unwanted signal energy that occurs in the interesting frequency band and that is not generated by any communication system. Thermal noise is present in almost any communications channel. Man-made noise in industrial environments can have several sources, for example, remote controls, motors, or microwave ovens.

50.4 Physical Layer: Transmission Problems and Solution Approaches

The previously discussed wave propagation effects can lead to channel errors. In general, their impact depends on a multitude of factors, for example, frequency, modulation scheme, and the current environment, which comprises the distance between stations, interferers, number of different paths, and their respective loss, etc. These factors can change with time, for example, the number of paths and their attenuation change if the transmitter, receiver, or something in the environment moves. Consequently, the transmission quality is time-variable.

Effects on Transmission

The notion of *slow fading* refers to significant variations in the mean path loss, as they occur due to changes in distance between transmitter T and receiver R, or by moving through tunnels or beyond large obstacles. Slow fading phenomena usually occur on longer timescales, and they often coincide with human activity (e.g., mobility). For short durations in the range of a few seconds, the channel can often be assumed to have constant path loss.

An immediate result of reflection, diffraction, and scattering is that multiple copies of a signal may travel on different paths from T to R (see Figure 50.1). Since these paths usually have different lengths, the copies arrive at different times (delay spread) and with different phase angles at the receiver and overlap. This has two consequences:

- The resulting signal can be amplified or attenuated (constructive or destructive interference), depending on the relative phase shift. This may lead to loss of received power of up to 40 dB.
- The delay spread leads to intersymbol interference, since signals belonging to neighbored information symbols may arrive at the same time.

If the stations move relative to each other or to parts of the environment, the number of paths and their phase shifts vary in time, thus giving a fast fluctuating signal strength at the receiver (*fast fading* or *multipath fading*), however, with nearly constant mean value on short timescales. If the delay spread is small relative to the duration of a channel symbol, the channel is called *nonfrequency selective* or *flat*; otherwise, it is called *frequency-selective*.

These problems translate into bit errors or packet losses. Packet losses occur when the receiver fails to acquire bit synchronization [64], while for bit errors synchronization is successfully acquired, but a number

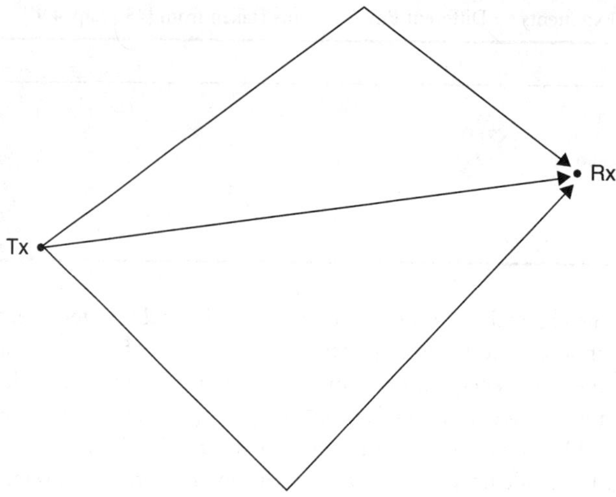

FIGURE 50.1　Multipath fading.

of channel symbols is decoded incorrectly. Bit errors can be combatted with forward error correction (FEC) techniques, packet losses cannot. The characteristics of bit errors and packet losses were investigated in a number of studies [13, 40, 64]. While the results are not immediately comparable, certain trends show up in almost every study:

- Both bit errors and packet losses are "bursty," that is, they occur in clusters with error-free periods between the clusters. The distributions of the cluster lengths and the lengths of error-free periods often have a large coefficient of variation or even seem to be heavy-tailed. Bit errors and packet losses appear to be instationary and time-variable on several time scales.
- The bit error rates depend on the modulation scheme; typically, schemes with higher bit rates exhibit higher error rates.
- Often, bit error rates of 10^{-3}–10^{-6} were observed; hence, the wireless channel is much worse than wired channels. Furthermore, the bit error rate can vary over several orders of magnitude within minutes.

The knowledge about error generation patterns and error characteristics can be helpful in designing more robust protocols.

Wireless Transmission Techniques

A number of different transmission techniques have been developed to combat the impairments of the wireless channel and to increase the reliability of data transmission.

Many types of WLANs (including IEEE 802.11) rely on *spread spectrum* techniques [20], where a narrowband information signal is spread to a wideband signal at the transmitter and despread back to a narrowband signal at the receiver. By using a wideband signal, the effects of narrowband noise or narrowband interference are reduced, since by the despreading operation their bandwidth shrinks, too. The two most important spread spectrum techniques are *direct sequence spread spectrum (DSSS)* and *frequency hopping spread spectrum (FHSS)*. In DSSS systems, an information bit is multiplied (XORed) with a finite digital *chip seqence* such that transmission takes place at the *chip rate* instead of the information bit rate. The chip rate is much higher than the information rate and consequently requires a much larger bandwidth. The channel is assumed to be frequency-selective with respect to the chip rate. This can be explored in two ways. To explain the first one, let us assume that the receiver receives a signal S from an LOS path and a delayed signal S_d from another path, such that the lag between S and S_d is

k chips. The chip sequences are designed such that the autocorrelation between the sequence and a shifted version of it is low for all nonzero lags. If the receiver is synchronized on the direct signal S, the delayed signal S_d appears as white noise and produces only a minor distortion for the receiver. The IEEE 802.11 WLAN specifies a DSSS physical layer with a receiver of this type. In the second approach, the information contained in delayed copies of the signal is explicitly explored by using *RAKE receivers* [51, Sec. 10.4]. Put briefly, a RAKE receiver tries to acquire the direct signal and the strongest time-delayed copies of it in order to combine them coherently. However, RAKE receivers are much more complex than simple DSSS receivers.

In FHSS, the available spectrum is divided into a number of subchannels. The transmitter hops through the subchannels according to a predetermined schedule, which is also known to the receiver. The advantage of this scheme is that a subchannel currently subjected to transmission errors is used only for a short time before the transmitter hops to the next channel. The hopping frequency is an important parameter of FHSS systems, since high frequencies require fast and accurate synchronization. As an example, the FHSS version of IEEE 802.11 hops with 2.5 Hz and many packets can be transmitted before the next hop. In Bluetooth, the hopping frequency is 1.6 kHz and at most one packet can be transmitted before the next hop. Typically, there is no hopping during transmission of a packet.

Recently, there has been considerable interest in orthogonal frequency division multiplexing (OFDM) techniques [59]. OFDM is a multicarrier technique, where blocks of N different symbols are transmitted in parallel over a number of N subcarriers. Hence, a single symbol has an increased symbol duration $N \cdot \tau$ as compared to full-rate transmission with symbol duration τ. The symbol duration $N \cdot \tau$ is usually much larger than the delay spread of the channel, thus combating intersymbol-interference and increasing channel quality. The IEEE 802.11a standard [43] as well as HIPERLAN/2 [15, 16] use an OFDM physical layer.

50.5 Problems and Solution Approaches on the MAC- and Link-Layer

It is the MAC- and the link-layers, that are exposed most to the error characteristics of wireless channels and that should do most of the work to improve the quality and reliability of the channel. Specifically for hard real-time communications, the MAC layer is a key component: if the delays on the MAC layer are not bounded, the upper layers cannot compensate this. In general, the operation of the MAC protocol is largely influenced by the properties of the physical layer. For the case of wireless transmission, there are some unique problems and solutions to these, discussed next. For a general discussion of MAC- and link-ayer protocols, we refer to references [21, 29, 58, 54, 12].

Problems for Wireless MAC Protocols

Several problems arise due to path loss in conjunction with a *threshold property*: wireless receivers require the signal to have a minimum strength, before it is recognized. This translates immediately into an upper bound on the distance between two stations wishing to communicate.[1] Therefore, if the distance between two stations exceeds some threshold distance, they cannot receive each other's transmissions. For protocols using carrier sensing (carrier sense multiple access [CSMA]), this gives rise to the *hidden terminal scenario* [55], depicted in Figure 50.2. Consider three stations A, B, and C with transmission radii as indicated by the circles. Stations A and C are in the range of B, but A is not in the range of C, and vice versa. If C starts to transmit to B, A cannot detect this by its carrier sensing mechanism and considers the medium to be free. Hence, A also starts packet transmission and a collision occurs at B.

---

[1]To complicate things, it is not necessarily always the case that station B hears station A given that station A hears station B. This can be the result of both stations controlling their transmit power independently.

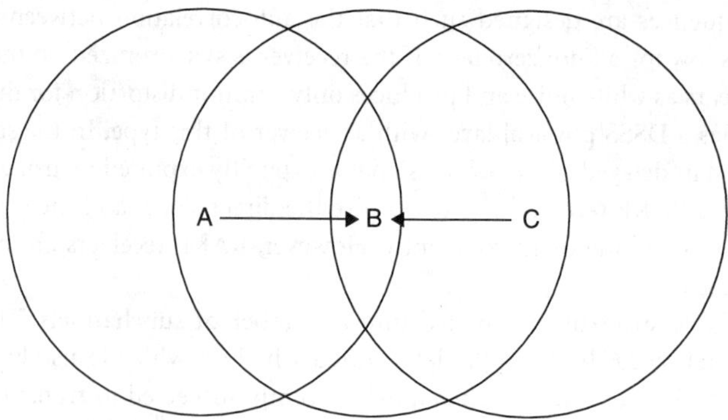

FIGURE 50.2   Hidden terminal scenario.

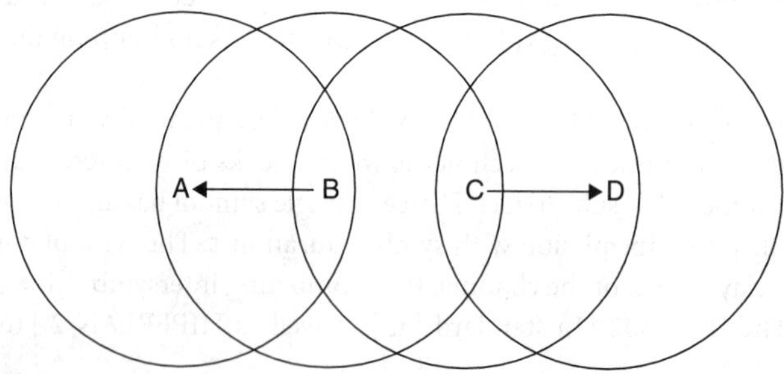

FIGURE 50.3   Exposed terminal scenario.

There is a second scenario where carrier sensing leads to false predictions about the channel state at the receiver: the so-called *exposed terminal scenario,* depicted in Figure 50.3. The four stations A, B, C, and D are placed such that the pairs A/B, B/C, and C/D can hear each other, and all remaining combinations cannot. Consider the situation where B transmits to A, and one short moment later C wants to transmit to D. Station C performs carrier sensing and senses the medium busy, due to B's transmission. As a result, C postpones its transmission. However, C could safely transmit its packet to D, without disturbing B's transmission to A. This leads to a loss of efficiency.

Two approaches to solve these problems are busy tone solutions [55] and the request to send/clear to send (RTS/CTS) protocol, as applied, for example, in the IEEE 802.11 WLAN MAC protocol. In the busy tone approach, the receiver transmits a busy tone signal on a second channel during packet reception. Carrier sensing is performed on this second channel. This solves the exposed terminal problem. The hidden terminal scenario is also solved except in the rare cases where A and C start their transmissions simultaneously.

The RTS/CTS protocol attacks the hidden terminal problem using only a single channel. Consider the case that A has a data packet for B. After A has obtained channel access, it sends a short RTS packet to B, indicating the time duration needed for the whole packet exchange sequence. If B receives the RTS packet properly, it answers with a CTS packet, which indicates the time needed for the remaining packet exchange sequence. Station A starts transmission after receiving the CTS packet. Any other station C hearing the RTS and/or the CTS packets defers its transmissions for the indicated time, thus not disturbing the ongoing packet exchange. It is a conservative choice to defer on any of these packets; but the exposed terminal problem still exists. If a station C defers only on receiving both packets, the exposed terminal problem is solved. However, there is the risk of bit errors in the CTS packet; which may lead C to start transmissions falsely. The RTS/CTS protocol does not resolve collisions of RTS packets at the receiver; furthermore, it introduces significant overhead.

It is not possible to transmit and receive simultaneously on the same frequency band, due to over-charge of receive filters. Hence, the transmitter cannot detect collisions by itself, as is required by, for example, the collision detection part of a CSMA/CD MAC protocol. A possible solution would be to use feedback given by the receiver on another channel, as is done in the busy tone approach [55]. In addition, this property makes implementation of CAN-like protocols impossible.

Even when the collision detection is done at the receiver, it may not work reliably due to the *near–far effect*: consider two stations A and B transmitting packets in parallel to a station C. For simplicity, let us assume that both stations use the same transmit power. Station A is very close to C, station B is far away but still within reach of C. This means that A's signal at C is much stronger than B's. In this case, it may happen that C decodes a packet sent by A despite B's parallel transmission. This situation is advantageous for the system throughput, but disadvantageous for MAC protocols relying on collision detection or collision resolution.

Methods for Combatting Channel Errors and Channel Variation

A challenging problem is the error-prone and time-variable channel, caused by slow fading, multipath fading (delay spread, intersymbol-interference), and interference phenomena. There are many control knobs for improving the channel quality, for example, transmit power, bit rate/modulation, coding scheme/redundancy scheme, packet length, choice of retransmission scheme (automatic repeat request (ARQ)), postponing schemes/timing of transmissions, using spatial diversity, and adaptation as a meta method [17]. In general, adaptation at the transmitter requires some kind of feedback from the receiver, for example, by using immediate acknowledgment packets after every data packet.

The variation of transmit power and of the bit rate/modulation scheme are both equivalent to varying the energy per bit, which in turn influences the bit error rate [47, 48]. In order to increase the probability of successful packet reception, a higher transmit energy or a modulation scheme with a lower bit rate can be chosen [23].

A suitable way to protect data bits against bit errors is to use redundancy, either in the form of error-detecting and correcting codes [33] (forward error correction (FEC)), redundancy packets, or by transmitting multiple copies of a packet [2], which is also helpful for combating packet losses. It has been proven beneficial for the overall throughput to control the amount of redundancy according to the current channel state, such that none or only a little redundancy is added, if the channel currently shows only a few errors [8, 13].

A second standard way to deal with transmission errors are retransmissions and suitable ARQ schemes. For channels with bursty errors immediate packet retransmissions are not clever, since, if the channel is now in a bad state, likely it will still be in the bad state one packet transmission time later. As an example, consider the case of a centralized scenario with one access point (AP) and several mobile stations (MS), channel errors, and downlink traffic. It has been shown that system throughput can be increased significantly for the case of bursty errors (as opposed to "smooth" independent errors) when retransmissions from the AP to an MS are postponed for a while, and other MS are served in the meantime [3, 4, 5]. Furthermore, ARQ schemes can be integrated with FEC schemes (hybrid error control). An overview on ARQ- and FEC-based error control schemes is given in [34].

For bursty channels, the need to retransmit a packet is often taken as a sign of bad channel conditions. However, this approach is costly, since a (possibly) long data packet is wasted. To circumvent this, *channel probing* can be used. Here, the transmitter sends a short probe packet to the receiver, which the latter has to acknowledge. If the transmitter receives no ack packet, it determines the channel to be in a bad state and defers transmission.

Another possibility to reduce the probability of erroneous packet transmissions is to make packets smaller, at the expense of increased overhead. If the transmitter has estimates on the current channel conditions, it can choose the "appropriate" packet size giving the desired tradeoff between reliability and efficiency [35].

If the error behavior is dominated by multipath fading, it may happen that a signal transmitted by station A is subject to destructive interference at geographical position p_1 and strong enough to be properly received at another position p_2. This property can be exploited using transmitter or receiver diversity techniques [48]. With receiver diversity, the receiver has two or more antennas, spatially separated by at least half a wavelength. For example, a receiver can choose the strongest signal.

A second way to explore this property is the following: assume that station A transmits a packet to station B. The channel from A to B is currently in a deep fade, but another station C has successfully received the packet. If the channel from C to B is currently in a good state, the packet can be successfully transmitted from C to B. In [63], this approach is applied to retransmitting data and poll packets in a polling-based MAC protocol.

50.6 Wireless Field-Bus Systems: State of the Art

Field-bus systems are explicitly designed to deliver hard real-time communications under harsh environmental conditions. Correspondingly, a *wireless field-bus* [10] is designed to provide as stringent timing and reliability guarantees as possible over wireless links, that is, it should strive to optimize its stochastic hard real-time performance. However, in most of the literature surveyed here, this issue is not covered. Nonetheless, we discuss existing approaches for different popular field-bus systems.

CAN

A group at the university of Sussex investigated the topic of wireless CAN. Since the CAN MAC protocol cannot be implemented directly on top of a wireless medium, two different approaches were developed [31].

The distributed WMAC protocol uses a CSMA/CA scheme, where a station waits for some backoff time after the medium become idle. The backoff time is chosen according to the packet's priority. If another station starts transmission during the backoff time, the station defers. In the centralized RFMAC protocol, the base station broadcasts the identifiers of the variables to be polled, and their producers transmit them according to the given schedule. In another distributed scheme, the CAN message priority value is mapped onto the channel using an on-off-keying scheme [30]: a station transmits a short burst, if the current priority bit is a logical one, or it switches to receive mode if it is a zero. If the station receives something in the receive mode, it resumes from contention. The priority bits are handled from the most significant bit to the least significant bit. This approach requires tight synchronization and fast switching between transmit and receive mode for the radio modem, which is a problem for certain WLAN technologies. In all references neither station failures nor transmission errors were considered, and thus no statements about their stochastic hard real-time behavior can be made.

FIP/WorldFIP

A group at EPFL Lausanne worked on transparent integration of wireless stations into FIP [37]. FIP uses a polling table to implement a real-time database [57]. A main element of the approach presented in [37] is a wireless-to-wired gateway, which serves as central station for the wireless part. The wireless MAC protocol is built upon a time division multiple access (TDMA)-scheme. In [38], it is investigated how the MAP/MMS application layer protocol can be enhanced with mobility. In the proposed system, the IEEE 802.11 WLAN with distributed coordination function (DCF) is used, time-critical transmissions and channel errors are not considered. In [36], the same question was investigated with digital European cordless telephone (DECT) as underlying technology.

The goal of the European community project OLCHFA (June 92 until September 94) was to add wireless stations to FIP. A prototype system worked in the 2.4 GHz ISM band using a DSSS physical layer [25]. The available publications put emphasis on the management of configuration data and on distributed algorithms for clock synchronization. Within the project, a communications controller was developed, which can transmit over either wireless or wired media. The MAC and data link protocol of FIP was not modified. The system thus uses wireless MAC-unaware bridging. Since FIP extensively uses periodic broadcasting of data, the protocol contains no retransmission scheme for the time-critical data. Instead, the OLCHFA approach is to enhance the FIP process variable model with so-called time-critical variables, which also provide freshness information to the applications. It is thus up to the application to handle cases of repeated losses.

IEC FieldBus

For the IEC FieldBus [24] (which uses a centralized, polling-based access protocol for supporting periodic data and a token passing protocol for asynchronous data) in reference [6] an architecture that allows coupling of several fieldbus segments using a wireless backbone based on IEEE 802.11 with point coordination function (PCF) is proposed.

PROFIBUS

The R-FIELDBUS project (*www.rfieldbus.de*) within the European Union Information Society Technologies (IST) program evaluates how IEEE 802.11 with DSSS can be used in a PROFIBUS fieldbus system, and how such a system can be used for transmission of IP-based multimedia data [27, 49]. The system uses the given PROFIBUS MAC- and link-layer protocol on top of a wireless PHY, and hence is an incarnation of the wireless MAC-unaware bridging approach. A vital part of the project is to set up a prototype and to perform field trials. In [32], a scheme for integration of wireless nodes into a PROFIBUS-DP network (single master, many slaves, no token passing) is described. An application layer gateway is integrated with a "virtual master" station. The virtual master acts as a proxy for the wireless stations, it polls them using standard IP and IEEE 802.11 DCF protocols.

In [63], a system following the wireless MAC-aware bridging approach is proposed. More specifically, on the wireless side specifically tailored polling-based protocols are used, while wired stations run the unmodified PROFIBUS protocol stack. Special care is taken to avoid token passing on wireless segments, since the error behavior of wireless links disturbs the token passing process seriously [65] (the same is true for the similar IEEE 802.4 Token Bus protocol [28]). In fact, it is shown that for bursty channel errors, the polling-based protocols achieve a substantially better performance in terms of stochastic hard real-time behavior than the PROFIBUS token-passing protocol; for certain kind of channels, the 99% quantile of the delay needed to successfully transmit a high-priority packet is up to an order of magnitude smaller than for the PROFIBUS protocol. To combine both protocols, a special coupling device is introduced, which provides a *virtual ring extension*: the coupling element acts on the wired side like a set of PROFIBUS stations on behalf of the wireless stations. Specifically, it creates token frames and ring management frames on behalf of wireless stations.

50.7 Wireless Ethernet/IEEE 802.11

Instead of developing WLAN technology for the factory floor from scratch, existing technologies might serve as a starting point. The prime example is the IEEE 802.11 WLAN standard [41, 43, 42], since it is the most widely used WLAN and was also used in different projects investigating wireless industrial LANs (see Section 50.6). Some alternative systems are HIPERLAN [14, 15], Bluetooth [22], and HomeRF [39].

Brief Description of IEEE 802.11

IEEE 802.11 belongs to the IEEE 802.x family of LAN standards. Basically, the standard describes an architecture, services, and protocols for an Ethernet-like wireless LAN, using a CSMA/CA-based MAC protocol with enhancements for time-bounded services. The protocols run on top of several PHY's: a FHSS PHY, a DSSS PHY offering 1 and 2 Mb/sec [41], a 5.5 and 11 Mb/sec extension of the DSSS PHY [42], and an OFDM PHY with 54 Mb/sec [43]. As for the architecture, the standard describes both an *ad hoc* mode and an infrastructure mode. In the latter, all communications is relayed through APs. An access point constitutes a cell, and mobile stations have to register with the closest AP. The APs are connected through a *distribution system*, which allows to forward data packets to mobile stations in different cells. A detailed description of IEEE 802.11 can be found in [44].

The basic MAC protocol of 802.11 is called the DCF. It is a CSMA/CA protocol using the RTS/CTS scheme and different interframe gaps to provide priorities for certain kinds of frames, for example, acknowledgement frames; but data frames cannot be differentiated according to priorities. The IEEE 802.11 MAC provides a connectionless semireliable best-effort service to its users by performing a bounded number of retransmissions. The user cannot specify any quality-of-service requirements for his packets; he can only choose between contention-based and contention-less transmission (see below). Furthermore, it is not possible to specify attributes like transmit power, modulation scheme, or the number of retransmissions on a per-packet basis. This control would be desirable for treating different packet types differently, for example, for transmitting high priority packets with a high transmit power and small bit rate to increase their reliability.

The enhancement for time-bounded services is called *PCF* and works only in infrastructure mode. The PCF defines a *superframe structure,* with variable- but maximum-length superframes. A superframe consists of a *superframe header,* followed by a *contention free period* (*CFP*) of variable length and a *contention period* (*CP*) of variable length. During the CP all stations, including the AP operate in DCF mode. To initiate start of the CFP, the AP (also called *point coordinator,* PC) has to acquire the wireless medium before it can transmit its *beacon packet.* This implies that the beacon packets and therefore the start times of the CFPs show some jitter, preventing strictly isochronous services. This is called *superframe stretching.* The AP broadcasts the beacon packet to all stations. The beacon indicates the maximum time duration of the CFP. The AP has a poll list of station addresses. Members of this list are polled during the CFP and can transmit their data in a contention-free manner. When polled by the AP, a station A is allowed to transmit only one data packet or an ack packet in case of empty queues. The polling scheme is not fully specified. A station that wants to be polled has to signal this to the AP during the association process. The poll list membership ends upon disassociation or when the station reassociates itself without requesting contention-free service.

Real-Time Transmission over IEEE 802.11

The PCF is designed for providing time-bounded services, for example, to transmit interactive speech data. Many studies [61, 60, 53] confirm that indeed packets transmitted during the CFP receive substantially smaller delays than those transmitted during the CP, however, at the cost of substantial overhead: [61] show a scenario where eight voice calls of each 8 kbit/sec data rate require 50% bandwidth of a 1 Mbit/sec transmission medium (without channel errors). There seem to be no studies assessing the (stochastic) hard real-time capabilities of IEEE 802.11; furthermore, all cited studies do not take transmission errors into account.

When transmission has to be both timely and reliable despite channel errors, retransmissions are needed. When a time-critical packet transmitted during the CFP fails, the station can choose to try the retransmission during the following CP or during the next CFP one superframe later (except the case where multiple entries in the polling list are allocated to the mobile and thus it receives multiple polls during a single CFP). Hence, retransmissions of important packets receive no priority in 802.11.

50.8 Summary

This article presented some problems and solution approaches to bring WLAN technology to the factory plant and to benefit from reduced cabling and mobility. The basic problem is the tension between the hard requirements of industrial applications on the one hand and the serious error rates and time-varying error behavior of wireless channels on the other. Many techniques have been developed to improve the reliability and timeliness behavior of lower-layer wireless protocols, but up to now wireless fieldbus systems have not been deployed at a large scale as the problem of reliable transmission despite channel errors is not solved satisfactorily. It is not clear as to which combination of mechanisms and technologies has the potential to bound the number of deadline misses under all channel conditions.

It seems to be an open question whether just more and better engineering is needed to make wireless transmission suitable for fulfilling hard real-time and reliability requirements, or whether there is a fundamental technological limit or a limit in user acceptance, which prevents a more widespread use.

Fortunately, WLAN technology is a very active field of research, and development. New technologies are created and existing technologies are enhanced, for example, the IEEE 802.11g and IEEE 802.11e working groups are working on delivering higher bit rates and better quality of service to users. It will be exciting to see how industrial applications can benefit from this.

References

[1] Ahlin, Lars and Jens Zander, *Principles of Wireless Communications*, Studentlitteratur, Lund, Sweden, 1998.

[2] Annamalai, A. and Vijay K. Bhargava, Analysis and optimization of adaptive multicopy transmission ARQ protocols for time-varying channels, *IEEE Transactions on Communications*, 46: 1356–1368, 1998.

[3] Bhagwat, Pravin, Partha Bhattacharya, Arvind Krishna, and Satish K. Tripathi, Using channel state dependent packet scheduling to improve TCP throughput over wireless LANs, *Wireless Networks*, 3: 91-102, 1997.

[4] Cam, Richard and Cyril Leung, Multiplexed ARQ for time-varying channels - part i: System model and throughput analysis, *IEEE Transactions on Communications*, 46: 41–51, 1998.

[5] Cam, Richard and Cyril Leung, Multiplexed ARQ for time-varying channels — part ii: postponed retransmission modification and numerical results, *IEEE Transactions on Communications*, 46: 314–326, 1998.

[6] Cavalieri, S. and D. Panno, On the Integration of Fieldbus Traffic Within IEEE 802.11 Wireless LAN, in Proceeding of the 1997 IEEE International Workshop on Factory Communication Systems (WFCS'97), Barcelona, Spain, 1997.

[7] Cavers, James K., *Mobile Channel Characteristics*, Kluwer Academic Publishers, Boston, Dordrecht, 2000.

[8] Chen, R., K.C. Chua, B.T. Tan, and C.S. Ng, Adaptive error coding using channel prediction, *Wireless Networks*, 5: 23–32, 1999.

[9] David, Klaus and Thorsten Benkner, *Digitale Mobilfunksysteme*, Informationstechnik. B.G. Teubner, Stuttgart, 1996.

[10] Decotignie, Jean-Dominique, Wireless Fieldbusses — A Survey of Issues and Solutions, in Proceedings of the 15th IFAC World Congress on Automatic Control (IFAC 2002), Barcelona, Spain, 2002.

[11] Decotignie, Jean-Dominique and Patrick Pleineveaux, A survey on industrial communication networks, *Annals of Telecommunication*, 48: 435ff, 1993.

[12] Dellaverson, Lou and Wendy Dellaverson, Distributed channel access on wireless atm links, *IEEE Communications Magazine*, 35: 110–113, 1997.

[13] Eckhardt, David A. and Peter Steenkiste, A trace-based evaluation of adaptive error correction for a wireless local area network, *MONET — Mobile Networks and Applications*, 4:273–287, 1999.

[14] ETSI, High Performance Radio Local Area Network (HIPERLAN) — Draft Standard, ETSI, March 1996.

[15] ETSI, TR 101 683, HIPERLAN Type 2: System Overview, ETSI, February 2000.

[16] ETSI, TS 101 475, BRAN, HIPERLAN Type 2: Physical (PHY) Layer, ETSI, March 2000.

[17] Farago, Andras, Andrew D. Myers, Violet R. Syrotiuk, and Gergely V. Zaruba, Meta-MAC protocols: automatic combination of MAC protocols to optimize performance for unknown conditions, *IEEE Journal on Selected Areas in Communications*, 18: 1670–1681, 2000.

[18] Funbus-Projektkonsortium, Das Verbundprojekt Drahtlose Feldbusse im Produktionsumfeld (Funbus) — Abschlußbericht, INTERBUS Club Deutschland e.V., Postf. 1108, 32817 Blomberg, Bestell-Nr: TNR 5121324, October 2000, http://www.softing.de/d/NEWS/Funbusbericht.pdf.

[19] Gibson, Jerry D., Ed., *The Communications Handbook*, CRC Press/IEEE Press, Boca Raton, FL, 1996.

[20] Glisic, Savo and Branka Vucetic, *Spread Spectrum CDMA Systems for Wireless Communications*, Artech House, Boston, 1997.

[21] Gummalla, Ajay Chandra V. and John O. Limb, Wireless medium access control protocols, *IEEE Communications Surveys and Tutorials*, 3: 2000, http://www.comsoc.org/pubs/surveys.

[22] Haartsen, Jaap C., The Bluetooth radio system, *IEEE Personal Communications*, 7:28–36, 2000.

[23] Holland, Gavin, Nitin Vaidya, and Paramvir Bahl, A Rate-Adaptive MAC Protocol for Wireless Networks, in Proceedings of the Seventh Annual International Conference on Mobile Computing and Networking 2002 (MobiCom), Rome, Italy, July 2001.

[24] IEC — International Electrotechnical Commission, *IEC-1158-1, FieldBus Specification, Part 1, FieldBus Standard for Use in Industrial Control: Functional Requirements.*

[25] Izikowitz, Ivan and Michael Solvie, Industrial Needs for Time-critical Wireless Communication & Wireless Data Transmission and Application Layer Support for Time Critical Communication, in Proceedings of Euro-Arch'93, München, Springer-Verlag, Berlin, 1993.

[26] Jakes, W.C., Ed., *Microwave Mobile Communications,* IEEE Press, New Jersey, 1993.

[27] Lutz Rauchhaupt Jörg Hähniche, Radio Communication in Automation Systems: The r-fieldbus Approach, in Proceedings of 2000 IEEE International Workshop on Factory Communication Systems (WFCS 2000), Porto, Portugal, 2000, pp. 319–326.

[28] Hong ju Moon, Hong Seong Park, Sang Chul Ahn, and Wook Hyun Kwon, Performance degradation of the IEEE 802.4 token bus network in a noisy environment, *Computer Communications,* 21:547–557, 1998.

[29] Kurose, J.F., M. Schwartz, and Y. Yemini, Multiple-access protocols and time-constrained communication, *ACM Computing Surveys,* 16:43–70, 1984.

[30] Kutlu, A., H. Ekiz, M.D. Baba, and E.T. Powner, Implementation of "Comb" Based Wireless Access Method for Control Area Network, in Proceedings of the 11th International Symposium on Computer and Information Science, Antalaya, Turkey, Nov. 1996, pp. 565–573.

[31] Kutlu, A., H. Ekiz, and E.T. Powner, Performance analysis of MAC protocols for wireless control area network, in Proceedings of the International Symposium on Parallel Architectuers, Algorithms and Networks, Beijing, China, June 1996, pp. 494–499.

[32] Kyung Chang Lee and Suk Lee, Integrated Network of PROFIBUS-DP and IEEE 802.11 Wireless LAN with Hard Real-time Requirement, in Proceedings of IEEE 2001 International Symposium on Industrial Electronics, Korea, IEEE, New York, 2001.

[33] Shu Lin and Daniel J. Costello, Error Control Coding — Fundamentals and Applications, Prentice-Hall, Englewood Cliffs, NJ, 1983.

[34] Hang Liu, Hairuo Ma, Magda El Zarki, and Sanjay Gupta, Error control schemes for networks: an overview, *MONET — Mobile Networks and Applications,* 2:167–182, 1997.

[35] Modiano, Eytan, An adaptive algorithm for optimizing the packet size used in wireless ARQ protocols, *Wireless Networks,* 5:279–286, 1999.

[36] Morel, Philip, Mobility in MAP Networks Using the DECT Wireless Protocols, in Procedings of the 1995 IEEE Workshop on Factory Communication Systems, WFCS'95, Leysin, Switzerland, 1995.

[37] Morel, Philip and Alain Croisier, A Wireless Gateway for Fieldbus, in Proceedings of Sixth International Symposium on Personal, Indoor and Mobile Radio Communications (PIMRC 95), 1995.

[38] Morel, Philip and Jean-Dominique Decotignie, Integration of Wireless Mobile Nodes in MAP/MMS, in Proceedings of 13th IFAC Workshop on Distributed Computer Contorl Systems DCCS95, 1995.

[39] Negus, Kevin J., Adrian P. Stephens, and Jim Lansford, HomeRF: wireless networking for the connected home, *IEEE Personal Communications,* 7:20–27, 2000.

[40] Nguyen, Giao T., Randy H. Katz, Brian Noble, and Mahadev Satyanarayanan, A Trace-based Approach for Modeling Wireless Channel Behavior, in Proceedirags of the Winter Simulation Conference, Coronado, CA, Dec. 1996.

[41] IEEE LAN/MAN Standards Committee of the IEEE Computer Society, *IEEE Standard for Wireless LAN Medium Access Control (MAC) and Physical Layer (PHY) Specifications,* IEEE Press, New York, November 1997.

[42] IEEE LAN/MAN Standards Committee of the IEEE Computer Society, *IEEE Standard for Information Technology — Telecommunications and Information Exchange between Systems — Local and Metropolitan Networks — Specific Requirements — Part 11: Wireless LAN Medium Access Control (MAC) and Physical Layer (PHY) Specifications: Higher Speed Physical Layer (PHY) Extension in the 2.4 GHz band,* IEEE Press, New York, 1999.

[43] IEEE LAN/MAN Standards Committee of the IEEE Computer Society, *IEEE Standard for Telecommunications and Information Exchange between Systems —LAN/MAN Specific Requirements*

— *Part 11: Wireless Medium Access Control (MAC) and Physical Layer (PHY) Specifications: High Speed Physical Layer in the 5 GHz band,* IEEE Press, New York, 1999.

[44] O'Hara, Bob and Al Petrick, *IEEE 802.11 Handbook — A Designer's Companion,* IEEE Press, New York, 1999.

[45] Pehlavan, K. and A.H. Levesque, *Wireless Information Networks,* John Wiley and Sons, New York, 1995.

[46] Pimentel, Juan R., *Communication Networks for Manufacturing,* Prentice-Hall International, Englewood Cliffs, NJ, 1990.

[47] Proakis, John G., Digital Communications, 3rd ed., McGraw-Hill, New York, 1995.

[48] Rappaport, Theodore S., *Wireless Communications — Principles and Practice,* Prentice-Hall, Upper Saddle River, NJ, U.S.A., 2002.

[49] Lutz Rauchhaupt, System and Device Architecture of a Radio-based Fieldbus — the Rfieldbus System, in Proceedings of Fourth IEEE Workshop on Factory Communication Systems 2002 (WFCS 20002), Vasteras, Sweden, 2002.

[50] Santamaria, Asuncion and Francisco J. Lopez-Hernandez, Eds., *Wireless LAN — Standards and Applications,* Mobile Communication Series, Artech House, Boston, London, 2001.

[51] Stallings, William, *Wireless Communications and Networks,* Prentice-Hall, Upper Saddle River, NJ, 2001.

[52] Stojmenovic, Ivan, Ed., *Handbook of Wireless Networks and Mobile Computing,* John Wiley & Sons, New York, 2002.

[53] Takahiro Suzuki and Shuji Tasaka, Performance evaluation of video transmission with the pcf of the IEEE 802.11 standard mac protocol, *IEICE Transactions on Communications,* E83-B(9):2068–2076, 2000.

[54] Tanenbaum, Andrew S., *Computernetzwerke,* 3rd ed., Prentice-Hall, Muenchen, 1997.

[55] Tobagi, Fouad A. and Leonard Kleinrock, Packet switching in radio channels: Part II the hidden terminal problem in csma and busy-tone solutions, IEEE Transactions on Communications, 23:1417–1433, 1975.

[56] Chai-Keong Toh, *Ad Hoc Mobile Wireless Networks — Protocols and Systems,* Prentice-Hall, Upper Saddle River, NJ, 2002.

[57] Union Technique de l'Electricité, *General Purpose Field Communication System, EN 50170, Volume 3: WordFIP,* 1996.

[58] van As, Harmen R., Media access techniques: the evolution towards terabit/s LANs and MANs, *Computer Networks and ISDN Systems,* 26:603–656,1994.

[59] van Nee, Richard and Ramjee Prasad, *OFDM for Wireless Multimedia Communications,* Artech House Publisher, Boston, 2000.

[60] Veeraraghavan, Malathi, Nabeel Cocker, and Tim Moors, Support of Voice Services in IEEE 802.11 Wireless LANS, in Proceedings of INFOCOM 2001, Anchorage, Alaska, IEEE, New York, Apr. 2001.

[61] Visser, Matthijs A. and Magda El Zarki, Voice and Data Transmission over an 802.11 Wireless Network, in Proceedings of IEEE Personal, Indoor and Mobile Radio Conference (PIMRC) 95, pp. 648–652, Toronto, Canada, Sept. 1995.

[62] Williams, Stuart, IrDA: past, present and future, *IEEE Personal Communications,* 7(1):11–19, 2000.

[63] Willig, Andreas, Polling-based MAC protocols for improving realtime performance in a wireless PROFIBUS, *IEEE Transactions on Industrial Electronics,* 50(4):806–817, August 2003.

[64] Willig, Andras, Martin Kubisch, Christian Hoene, and Adam Wolisz, Measurements of a wireless link in an industrial environment using an IEEE 802.11-compliant physical layer, *IEEE Transactions on Industrial Electronics,* 49:1265–1282, 2002.

[65] Willig, Andreas and Adam Wolisz, Ring stability of the PROFIBUS token passing protocol over error prone links, *IEEE Transactions on Industrial Electronics,* 48:1025–1033, 2001.

51

Bluetooth

Thomas Jatschka
Siemens Austria AG

Robert Tschofen
Siemens Austria AG

51.1 Introduction

The Basic Idea

The keynotes for Bluetooth include the following points:

- Universal radio interface for *ad hoc*, wireless connectivity
- Data connection via short-range (10 m) *ad hoc* networks using the license-free 2.45 GHz (ISM) band
- Low-cost RF modules for mass market
- Low power consumption
- Voice and data transmission
- Applications mainly for all portable devices

The major technological difference between Bluetooth and other communication standards is that Bluetooth is a "lower layer" in communication model handling. Bluetooth's radio communication includes sophisticated security. Higher layers of the communication models include well-known application protocols such as RFCOMM, TCP/IP, and PPP. This means that two Bluetooth devices can communicate regardless of the application protocol.

History

In February 1998, Ericsson, Nokia, IBM, Toshiba, and Intel (promoter group) founded the Bluetooth Special Interest Group (SIG). In December 1999, the promoter group was extended by Microsoft, Lucent, 3Com, and Motorola.

Several other companies and institutions joined the SIG as adopters of the technology.

The aims of the project were as follows:

Common Bluetooth specification (*de facto* standard)
Certification of Bluetooth products ("Bluetooth inside")
Zero cost license agreement for all interested companies.

Today's Common Bluetooth Solution in Factories

The most widely spread Bluetooth application in industrial environment may be Bluetooth to serial adapters. These are usually internal modules or external devices directly attached to RS232 or RS485 serial port.

Some examples of actual used Bluetooth solutions are [Cox_John] as follows:

- Temperature and moisture sensors in a cement factory
- Water-pumping stations in Oslo
- Bluetooth-enabled sensors in metal industry
- Wireless physiological monitoring system

51.2   The Physical Layer

Bluetooth operates in the 2.4 GHz license-free ISM band, which is nearly worldwide harmonized (there are still restrictions in some countries. For example, reduced max. power for outdoor applications in France and Italy. Please contact the local authorities for details). The maximum transmitting power is limited to 20 dBm (100 mW), which leads to communication ranges from a few meters up to hundreds of meters, depending on power class, sensitivity, and the environment.

Channel Arrangement

Spectrum spreading is accomplished by Frequency Hopping (FH) with 1600 hops/sec with a pseudorandom sequence over 79 channels with 1 MHz spacing. During one hop, a bandwidth of less than 1 MHz is occupied. The hopping sequence is determined by the master of a piconet (Figure 51.5) and can be calculated by the masters Bluetooth address and a "phase offset," which is given by the master during call setup.

Frequency Hopping

The FH of Bluetooth solves several problems that appear in a license-free frequency band:

Collisions with other Bluetooth piconets: Different nets need not be synchronized in any way; there is only a statistical probability of collision that leads to single lost packets.

Coexistence with other systems in the 2.4 GHz Industrial Scientific and Medical band (ISM band): Packets can get lost if there are strong RF-Transmission systems in the same ISM band as Bluetooth; but due to the fast changing of the Bluetooth frequencies, the link remains stable.

Fading in usual environments, especially if there are several metal objects, single frequencies are transmitted perceptibly worser.

Security especially with point-to-point connections, FH is an additional security feature, because it is nearly impossible to track the hopping sequence if you do not obtain the needed information from the master.

Adaptive Frequency Hopping

In the current Bluetooth core specification 1.1 [BT_1.1], each Bluetooth channel is used; but the subsequent Bluetooth core specification 1.2 [BT_1.2] includes a feature called Adaptive Frequency Hopping (AFH). This AFH allows to avoid (mark the channel as bad) usage of channels, which are used by other communication systems such as WLAN, microwave oven, or analog video transmission.

The "bad" channels are placed in a table at the master device. The entries in this table can be set by the host; this will be the usual case for a combined Bluetooth/WLAN device or may be derived during the communication by the master or by a slave.

Power Classes

Bluetooth features three different power classes:

Class I with 0 to 20 dBm, usually defined as 100 m device
Class II with $-$ 6 to 4 dBm, "10 m device"
Class III with up to 0 dBm

The minimum sensitivity is defined with -70 dBm. It should be noticed that the max. Equivalent Isotropic Radiated Power (EIRP) output power is limited by ETSI and FCC; hence, a 20 dBm device must not be used with an antenna gain >0 dBi.

Range Estimation

An exact definition of the range is not possible, because die propagation of the electromagnetic wave depends strongly on the environment.

A rough estimation can be made using the following formula:

$$Range(m)=10 \wedge ((P_{OUT}+Antenna\ Gain\text{-}Sensitivity-41)/K)\ [dB,dBm]$$

where $K=25$ is for line-of-sight and 35 is for non-line-of sight connections.

Table 51.1 shows some calculations for typical Bluetooth Modules and Antennas with 0 dBi gain.

As can be seen, a Class I device does not necessarily provide a higher Range as a Class II device.

Power Control

A Class I device must feature Power Control; for Classes II and III, this is an optional feature. Power Control works with a feedback message from the remote device, which gives information about the received signal strength.

Power consumption and interference with other devices are reduced. The Power Control must be able to reduce the output power at least down to 0 dBm in not less than 8 steps. If the remote device does not support information about the received signal strength, the transmitting power should be reduced to Class II or III limits.

TABLE 51.1   Range Estimation for Different Bluetooth Modules

Description	Output Power (dBm)	Receiver Sensitivity (dBm)	Line-of-sight range (m)	Non-line-of-sight range (m)
Excellent Class I device	20	−90	575	94
Average Class I device	15	−77	110	29
Bad Class I device	5	−72	28	11
Excellent Class 2 device	4	−90	132	33
Average Class II device	1	−80	40	14
Bad Class II device	−4	−71	11	6

Modulation

The Bluetooth radio module uses Gaussian Frequency Shift Keying (GFSK), where a binary one is represented by a positive frequency deviation and a binary zero is represented by a negative frequency deviation. $B \times T$ (*Bandwidth* $\times$ *Time*) is set to 0.5, and the modulation index must be between 0.28 and 0.35.

The raw data rate of Bluetooth is 1 Mbit/sec. In future specifications [BT_2_RF], Enhanced Data Rate with up to 3 Mbit/sec will be introduced. In this case, the modulation scheme of the payload will be switched to 8 DPSK (Differential Phase Shift Keying) or $\pi/4$-DQPSK (Differential Quadratur Phase Shift Keying) for 2 Mbit/sec. The header and the access code of each packet remain unchanged; thus, the system remains downward compatible and does not change the occupied bandwidth. Also, an operating mode called "high data rate" is planned; this mode will change the modulation and channel allocation completely.

Physical Links

There are two types of links in the Bluetooth specification:

SCO   Synchronous Connection-Oriented for speech transmission and
ACL   Asynchronous Connection-Less for data transmission.

An SCO channel is always a symmetric point-to-point link. Its packets are transferred in fixed timeslots and never retransmitted in Bluetooth core specification 1.1; Bluetooth core specification 1.2 [BT_1.2] defines new packet types that allow retransmission of speech packets. One master can handle up to 3 SCO channels in parallel.

ACL channels can be both symmetric or asymmetric; usually, lost or damaged ACL packets are retransmitted.

Packets

Basic structure:

ACCESS CODE 72 bit	HEADER 54 bit	PAYLOAD 0-2745 bit

There are three different types of access Codes:

Channel Access Code (CAC)
Device Access Code (DAC)
Inquiry Access Code (IAC)

The Header consists of the following parts:

AM_ADDR: 3- bit active member address,
TYPE: 4-bit type code
FLOW: 1-bit flow control
ARQN: 1-bit acknowledge indication
SEQN: 1-bit sequence number

Packet Types

In Bluetooth core specification 1.1, 13 different packet types for data, voice, and signaling are defined (Table 51.2):

For speech transmission, usually HVx packets are used. These packets do not contain a CRC and are not retransmitted. An HV1 packet carries 10 bytes of information with 1/3 FEC (Forward Error

TABLE 52.2   Bluetooth 1.1 Packet Types

	Slot occupancy	SCO	ACL
0	1	NULL	NULL
1	1	POLL	POLL
2	1	FHS	FHS
3	1	DM1	DM1
4	1	Undefined	DH1
5	1	HV1	Undefined
6	1	HV2	Undefined
7	1	HV3	Undefined
8	1	DV	Undefined
9	1	Undefined	AUX1
10	3	Undefined	DM3
11	3	Undefined	DH3
12	3	Undefined	Undefined
13	3	Undefined	Undefined
14	5	Undefined	DM5
15	5	Undefined	DH5

Correction). For a 64 kbit/sec PCM (Pulse Code Modulation) signal, every second timeslot must be used for one packet. HV2 carry 20 bits with 2/3 FEC and HV3 30 bytes without FEC; hence, HV3 packets have to be sent every sixth timeslot.

The Bluetooth core specification 1.2 introduces new packets for speech transmission, called extended SCO (eSCO), which allow packet retransmission and/or reduced bandwidth and power consumption while sacrificing the very low latency of the SCO packets. This packet type will lead to strongly improved audio quality, because up to 2 packet repetitions are possible. With HVx packets, a Packet Error Rate (PER) of 1% means a perceptible reduced quality, while a PER of up to 10% does not seriously affect a data link.

For data transmission DH1, DH3, DH5, DM1, DM3, and DM5 packets are used. DH are without FEC; DM packets use 2/3 FEC. These packets occupy 1 to 5 timeslots and carry a maximum number of data bytes between 18 (DM1) and 339 (DH5). Depending on the quality of the channel, DH5 packets do not necessarily feature the highest data rate. The maximum possible data rate is 723/56 kbit/sec for an asymmetric operation and 433 kbit/sec for a symmetrical operation.

51.3   Power Consumption

The power consumption of a Bluetooth device depends on four main parameters:

- Operation mode (duty cycles)
- Power consumption during transmitting,
- Power consumption during receiving, and
- Power consumption during standby/deep sleep.

The operation mode may have the strongest impact on the current consumption. Depending on the amount and frequency of the data throughput, the optimal configuration of packet size and sleep periods can save more power than any chip technology.

A typical Class II Bluetooth device needs about 50 mA peak current during transmitting and receiving and declines to 25 uA during deep sleep.

Some examples:

For a voice connection with HV3 packets it will need: 1 out of 6 timeslots for receiving, 1 out of 6 timeslots for transmitting, this results in $(1\times50+1\times50)/6=17$ mA with HV1 Packets: $(3\times50+3\times50)/6=50$ mA.

An RS232 Cable replacement with 115 200 Bit/sec:

With DH5 Packets (less than 1/6 of the possible bandwidth used)	: 8 mA (as Master)
With DH1 Packets (~2/3 of the possible bandwidth used)	: 30 mA

If a device is not permanently connected but shall be detectable and/or connectable for other Bluetooth devices, the inquiry and/or page scan must be enabled. This means that the device activates its receiver in specified intervals for a specified period. In this case, you have to trade between the speed of detection and current consumption. The default settings are a 10 ms window every 1.28 sec for both page and inquiry scans. For the Bluetooth device described above this results in 0.85 mA.

If the window size is increased or the interval is decreased for faster detection or connection setup, the power consumption increases.

The Bluetooth core specification 1.2 features "fast connection setup," which allows to set up a connection in less than 2 sec by only increasing the current consumption on the active device.

51.4   Bluetooth in Practical Experience

Integrating Bluetooth HW

Today, there are different kinds of Bluetooth chip(sets)on the market:

- Single Chip solution with ROM or Flash ROM
- Single Chip with external Flash
- Dual Chip solution

Most of the chipsets are also available as modules; this solution is usually preferable for small and medium production quantities (Table 51.3).

Bluetooth Software Efforts

Standard Bluetooth Modules are so called "HCI-modules." This means that the lower layers up to HCI of the Bluetooth-Stack (the Protocol Stack) are running on the chip, and all layers above are running on the host processor. These Bluetooth stacks have a demand on processing power, which is too much for many common microcontrollers, and they are not available for every processor.

For simple applications, which do not need the full Bluetooth data rate, it is more effective to focus on a solution where the higher layers up to the application (Profile) are running on the Bluetooth chip. These solutions are available for widely used Bluetooth Profiles such as HID [BT_HIDP], serial cable, and Headset [BT_HSP]; some chipset manufacturers also provide SDKs to develop on-chip applications.

Cable Replacement

An obvious application for Bluetooth is the replacement of an RS232 cable in industrial environments to avoid, for example, service-connectors, which are sensitive to dirt and hard to reach.

TABLE 51.3   Comparison Module Solution

	Module	Chipset on main PCB
Advantages	Dramatically reduced effort in qualification and approval	Smaller solution, especially in height, possible
	No trimming in production necessary	Reduced cost
	Less RF know-how necessary	
Disadvantages	Many modules do not utilize the rf performance of the chipset	High-quality main PCB, with micro-vias needed
	One additional manufacturer in the production chain	Extra shielding might be necessary
	Not all Pins of the chipset accessible	More extensive production testing

There are different ways to realize this cable replacement:

- The simplest way may be to replace both ends of the cable by two paired (*Pairing*) Bluetooth "dongles" with SPP [BT_SPP] running on it.
- A more sophisticated way would be to integrate a Bluetooth module(s) with SPP [BT_SPP] in the industrial device(s) and connect it to a laptop computer with a USB Bluetooth device via a virtual COM-port. The application running on the laptop computer can remain completely the same; it just has to use a virtual COM-port instead of the "physical" one.

Wireless Mouse or Keyboard

Most wireless keyboards or mice are based on infrared; the new Bluetooth HID profile enables this device to have Bluetooth as a wireless interface. The Bluetooth interface is much more powerful than the infrared one. It has more distance, application, and wireless interface in a single chip solution. Both are connected via a Bluetooth USB dongle on the backside or integrated in to the PC or laptop. Therefore, a simple mouse can be used as a long-distance mouse or as a powerpoint presenter with several additional functionalities as well. Instead of infrared interface, it is not necessary to have a Line of Sight (LOS) connection. For example, in an industrial application, it is very simple to connect a key panel or a keyboard on a long-distance position from the engine or the switchboard. This makes it much more easier to service or control the engine as well as to improve the security!

Sensors

In addition to all the mainstream applications, it is useful to have a powerful wireless technology with an interference robust Radio interface.

Most single chip solutions on the market have programmable pins, whereas the application has the possibility to measure voltage, temperature, and so on. The architecture tendency of the Bluetooth single chip is going toward ARM7. Therefore, a powerful core with several MIPS is available for intelligent applications for a new Sensor technology. Built-in and powered with a special Bluetooth battery package, it gives a long-term intelligent sensor technology with strong capability.

For example, in a medical center or at home, a patient gets an on-body unit for daily heart diagnosis; if a heart attack occurs, the Bluetooth wireless technology establishes a connection with an access point and sends the status of the patient in the hospital or at home.

Identification Plus Positioning

Bluetooth seems to be an overkill for an identification or positioning system, but it might be cheaper and easier than a proprietary system, especially if there is already a Bluetooth system.

For example, an automatic stockkeeping system with Bluetooth equipped trolleys could also have simple access points with reduced range, which detect every trolley passing by and read out its identity and charging. In this case, the identification can be done by the unique Bluetooth address, or in more safety-related tasks, by an authentication procedure.

Remote Control

Bluetooth has the potential to be used for a wide range of remote control tasks. On the one hand, there is the field of home entertainment and domestic work.

A variant for a very all-purpose remote control could be based on a simple webserver running on the Bluetooth chip integrated in every device to be controlled. Some of today's Bluetooth chips already have the potential to run a web-page and do controlling tasks via several Programmable Input–Output (PIO) pins. The speciality of this method is that any Bluetooth-enabled device with a web-browser can be used as a remote control and no special software at all is needed on this device.

On the other hand, industrial remote control scenarios with special access points, which control up to seven remote devices at once, are also possible. The Bluetooth bandwith is sufficient for control and monitoring of most tasks in the environment of a production line and stock keeping; but it must be carefully checked if the latency and connection setup times are good enough.

51.5   The Protocol Stack

The Bluetooth core specification does not only define a radio, it also defines a software protocol stack to enable devices to work with one another.

This Bluetooth Protocol Stack is defined as a series of layers; beyond this, the Bluetooth core specification encompasses several of different profiles. These profiles enable the devices to have different capabilities like establishment of a voice connection between a mobile phone and a headset or establishment of a dial-up connection like a modem to serve in the World Wide Web, etc.

Figure 51.1 shows the fundamental part of the Bluetooth Protocol Stack, whereas the Radio, Base Band (BB) and Link Controller (LC), Link Manager Protocol (LMP), Host Controller Interface (HCI), Logical Link Control and Adaptation (L2CAP), and the Service Discovery Protocol (SDP) are used from all Applications.

In the *Radio*, the data are modulated and demodulated for the transmission and the reception.

Baseband [BT_BB] *and Link Controller* has to control the asymmetric data connections (max. 7) and/or the symmetric voice connections (max. 3) via the radio. This controlling includes all necessary encroachments like packets reassembling and controlling of the FH. The new Bluetooth core specification [BT_1.2] provides better techniques to control and adapt the FH Kernel to improve the coexistence with other wireless connectivity techniques like WLAN (802.11b) (Figure 51.2).

The *Link Manager* [BT_LM] is used to control and negotiate all aspects of the operation of the Bluetooth between two devices. For this, a separate Protocol (LMP) is provided in the Bluetooth Core.

The *Host Controller Interface* [BT_HCl_Func] provides a few different hardware interfaces to handle the communication between the host controller and the host. This is necessary if the host and the host controller are separated (see Figure 51.3) and connected via one of the possible interfaces like the Universal Asynchronous Receiver Transmitter (UART) [BT_HCl_UART], USB, RS232, and the new 3 Wire UART Interface [BT_HCl_3 Wire]. In case of an embedded solution where the host controller and host are not separated, the host controller interface is not required.

The *Logical Link Control and Adaptation* [BT_L2CAP] provides connection-oriented and connectionless data services to upper-layer protocols with protocol multiplexing capability, segmentation and reassembly operation, and group abstractions. L2CAP also permits per-channel flow control and retransmission via the Flow and Error Control modes.

The *Service Discovery Protocol (SDP)* [BT_SDP] provides a means for applications to discover as to which services are available. It also gives the possibility to characterize these available Services (Profiles) more in detail, for example, how to connect.

The *RFCOMM* can emulate the typical nine RS 232 lines and the signaling on air. This gives Bluetooth the capability to be an RS232 cable replacement. Several Applications (Profiles) are based on the RFCOMM. RFCOMM also provides the capability to have multiple concurrent connections by relying on L2CAP to handle multiplexing over single connections, and to provide connections to multiple devices.

Wireless Application Protocol *(WAP)*, Object Exchange *(OBEX)*, Bluetooth Network Encapsulation Protocol *(BNEP)*, Human Interface Devices *(HID) class driver* .., provide interfaces to the higher layer parts of other Communications Protocols.

The *Telephony Control Protocol Specification (TCS)* defines the way to send telephone calls across a Bluetooth link; it also gives guidelines regarding how to establish a point-to-point or a point-to-multipoint (max. 3) call.

In mobile phones, for example, different Profiles like Handsfree [BT_HFP], Headset [BT_HSP], or Phone Access Profiles [BT_PAP] are used to phone via a headset or a car kit unit. The source for the Synchronous (SCO Audio Signals) is plugged in at PCM or I^2S Interface.

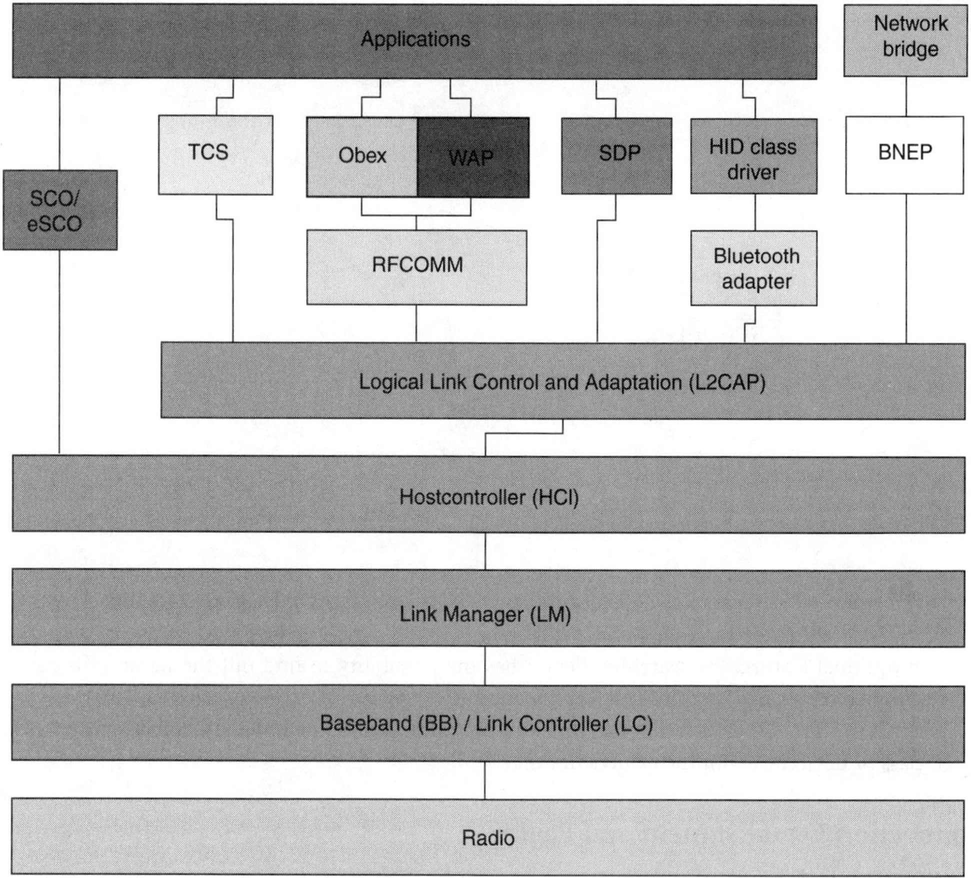

FIGURE 51.1 The Bluetooth protocol stack.

The *Applications* (Bluetooth Profiles) enable the devices to have access to different Services and therefore these Profiles give guidelines regarding how to serve the Bluetooth Protocol Stack to guarantee the interoperability with Bluetooth devices with the same Application from different Vendors. This capability should improve the success and the quality as well as make it more fun to use Bluetooth in the modern way of life.

51.6 Using Bluetooth

The normal way to use Bluetooth enforces a few steps within the Bluetooth connection establishment process that has finished successfully.

Discovering Bluetooth Devices

Discovering means that the device sends a series of inquiry packets (ID) and eventually all the devices in the range max. 100 m reply with a Frequency Hop Synchronization (FHS) packet. This FHS includes all required information that the device needs to create a connection to this device. Also, it is required that the discovered devices in the area are in discoverability mode, which means that the device activates the inquiry scan mode. The inquiry scan mode enables the device every 1.28 sec to catch the ID packets; after the first ID is received the scanning devices use a random back-off to reduce the interference with other responses; hence, it answers with the FHS after it receives the ID the second time. The ID packets contain

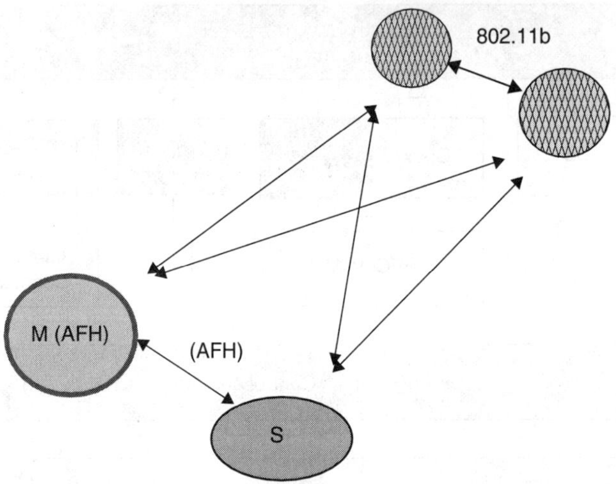

FIGURE 51.2   Coexistence improvement between Bluetooth and 802.11b with AFH.

search information, for example, General Inquiry Access Code (GIAC), Dedicated Inquiry Access Code (DIAC),…; therefore not all devices reply with these FHS.

As an optional feature, the searching device has the possibility to find out the name of the remote device without requiring an explicit ACL connection.

After this process, the application has considerable information like Bluetooth address, Name, Clock offset, class of device,.., about the devices in range.

Connection Establishment and Paging

To understand the paging procedure (connection establishment is called paging), it is useful to understand the Bluetooth behavior in its roles like the master and the slave. To establish a connection, one device (master role) must initiate the connection by addressing the other device (slave role), the request of a connection. The slave is listening for such a request, and this is called page scanning. The page scanning is activated in the page scanning mode, which means that every 1.28 sec, the receiver is listening for the ID packets [BT_CWC].

The master sets the frequency hopping sequence and the slave synchronizes to the master's frequency and clock by following the master's hopping sequence. In earlier versions of Bluetooth devices, the slave was only able to follow one master in frequency and time. The newer Bluetooth devices as well as the Bluetooth core specification allow the devices to follow more than one master in frequency and time. This means that one device has the possibility to handle more than one connection as a slave in a so-called scatternet slave/slave scenario (Figure 51.4). Another scatternet scenario is called slave/master, where the master can serve more than one slave.

The normal point-to-point connection is called piconet. A master in a piconet has the ability to serve up to seven slaves (Figure 51.5).

The pager transmits ID packets with the page scanner's address. If the page scanner receives its ID, it replies with another ID, again using its own address. Bluetooth gives the guarantee that every device around the globe has its unique ID. The pager receives the requested ID and sends the FHS to the page scanner. The FHS includes all the required information for the scanner and replies with another ID packet. The scanner now has the ability to extract all information to calculate the pager's hop sequence; it stops its own and swap with this. Both devices move into connection state, the pager as master and the scanner as slave. Further, both devices have the ability to switch their roles.

In the new Bluetooth core specification, the inquiry and paging processes are modified to obtain connections faster and more successfully. This is very important in industrial applications to guarantee the connection within a defined time window.

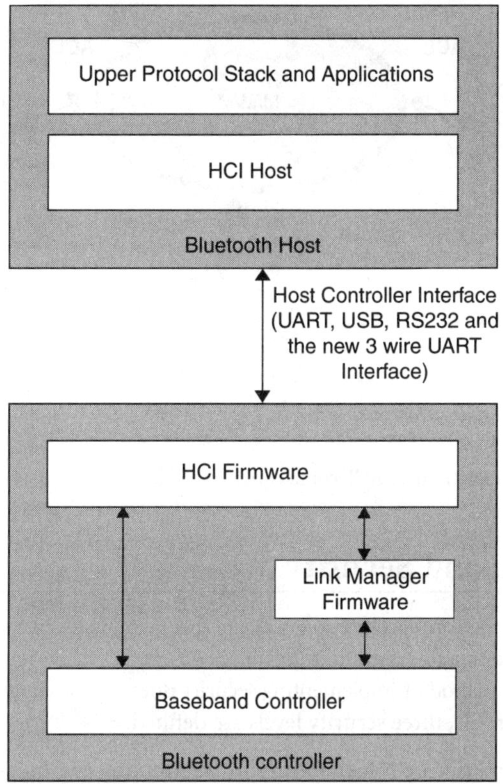

FIGURE.51.3   HCI interface.

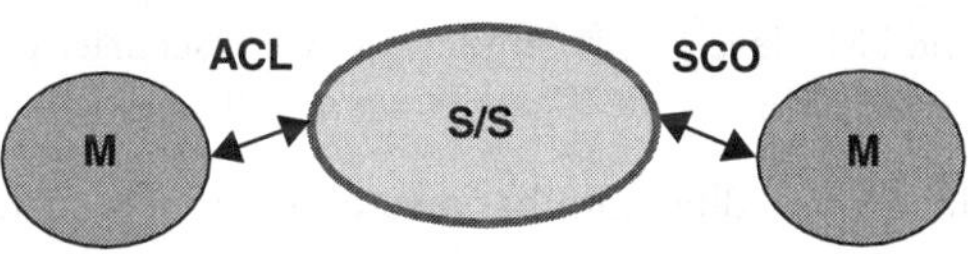

FIGURE 51.4   Typical slave/slave scatternet scenario.

One modification in the inquiry process is the first FHS; this means the scanning device answers without any random back-off after the first hit. This provides more success to find all devices in the range.

There are two other modifications within the setup to speed up the process even more. The first includes using special DIACs to limit the inquiry responses to devices of the correct type and second, using RSSI to help filter the responses. Both limit the number of SDP records that need to be returned.

More complicated changes are made in the scanning behavior of the page or inquiry scanning device. Paging can also be slow when the estimated clock starts the procedure on the wrong train. Depending on the paging configuration, paging may take 1.28 sec (R1) or 2.56 sec (R2) longer than the nominal time. It is desirable to speed up connections when the clock estimate is wrong or out of date. If a scan is unsuccessful, the scanning device will switch to another frequency. The other frequency is chosen such that it belongs to the complementary train. This single mechanism serves to speed up inquiry, paging, as well as to improve reliability in setting up connections. The improvement is called an interlace scan.

In this chapter, it is too complicated to explain the complete scanning process. The need for faster connections is a trade-off between speed on the one hand and bandwidth, power consumption, latency, time to market, and cost on the other.

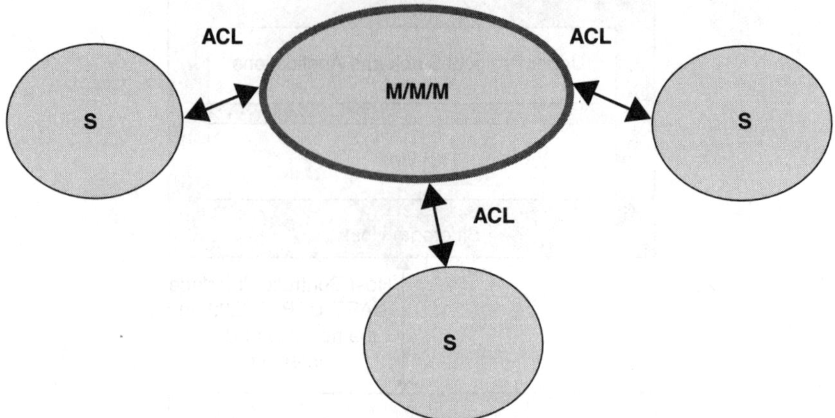

FIGURE 51.5   Typical piconet with three ACL connections.

51.7 Encryption and Security

In many levels of the Bluetooth core specification, security [Sec] is an issue. In Bluetooth security, white paper is a security architecture that may be used to implement Security Mode 2 (Service Level-enforced security) on Bluetooth devices. This method of implementing security does not have any effect on interoperability.

In the Generic Access Profile, three security levels are defined:

1. Security Mode 1 is nonsecure, which means no security process will never be initiated.
2. Security Mode 2 gives Service Level-enforced security; after an L2CAP [BT_L2CAP] channel has been established, it is decided whether or not it needs authorization, authentication, and encryption and its processes.
3. Security Mode 3 is Link Level-enforced security; the security process is initiated before the link is fully established on the LMP level. The completion of the security process in this level is mandatory to complete the connection.

In addition to the Security Modes, Bluetooth has several different security Levels that can be defined for devices and services. In the security white paper [Sec], the devices are split into three categories and two trusted levels [Sec_Arch]:

Trusted devices: paired or bonded device; the device has an entry in a database as trusted; therefore, it gains access without any restriction to all services.

Untrusted or unknown devices: it has no security information stored in the database to this device. The access to services is restricted.

The security white paper suggests that the security requirements for authentication, authorization, and encryption of services could be set separately:

No additional security for the services: any device may access this service.

Authentication only service: any device that can go through the authentication process may have access.

Authentication and authorization services: only trusted devices, after the security process in mode 2 a device is trusted, may access these.

Authentication and authorization are defined as follows [Sec_Arch]:

Authentication is the process of verifying "who" is at the other end of the link. Authentication is performed for devices (BD_ADDR). In Bluetooth, this is achieved by the authentication procedure based on the stored Link Key or by pairing (entering a PIN).

Authorization is the process of deciding if device X is allowed to have access to service Y. This is where the concept of "*trusted*" exists. Trusted devices (authenticated and indicated as "trusted") are allowed access to services. Untrusted or unknown devices may require authorization based on user interaction before access to services is granted. This does not principally exclude that the authorization might be given by an application automatically. Authorization always includes authentication.

Pairing is the name of the combination of Link Key generation on Link Level and the authentication procedure. If this process is done and the known Link Key between two devices is stored, these devices are *bonded*.

In industrial applications the pairing and bonding are a major part, for example, two serial dongles as serial cable replacement have to be paired at production level. Later, it is also necessary to give the guarantee to pair these devices again, for example, if one of the dongels is damaged. On the other hand, a headset device is produced in a nonpaired state. The first action to use a headset with a mobile phone is the pairing process; therefore, an initial PIN Key has to be delivered with this headset device. The PIN Key is required to generate the Link Key.

Connection Setup Procedure General Concept

The following procedures are performed:

1. Connect request to L2CAP
2. L2CAP requests access from the security manager
3. Security manager: lookup in service database
4. Security manager: lookup in device database
5. If necessary, security manager enforces authentication and encryption
6. Security manager grants access
7. L2CAP continues to set up the connection.

51.8 Low-Power Operation

Bluetooth specifies three different ways to remain connected and reduce the power consumption at the same time. This is especially important for battery-driven devices to maximize their lifetime. The Bluetooth core specification solved the problem with the three low-power modes. Within these modes, the device remains connected, but switch off the receiver for as long as possible.

(a) *Hold mode*: The hold mode is used to stop the ACL traffic for a specified period of time. During this period, the slave shall not support ACL packets on the channel; apart from this, this process does not have any effect on the synchronous SCO or eSCO traffic. The hold mode gives the devices the chance to enable major process like scanning, paging, or attending another piconet, for example, in a scatternet scenario.

(b) *Sniff Mode*: The sniff mode may reduce the duty cycle of the slave's activity in the piconet. The slave in the active mode on an ACL logical transport shall listen in every ACL slot to the master traffic. The only exception is if the device is being treated as a scatternet link or is absent due to the hold mode or an absence mask. In some cases, the normal link operating mode can be sniff, since with a short sniff interval, for example, the HID profile may still be able to transmit data fast enough to meet the latency and data reporting requirements of the application. Instead of the Park mode, the Sniff mode maintains its active member address, and data reports may be sent without exiting this mode.

(c) *Park Mode*: As explained in the sniff mode, in the park mode, the device loses its active member state (AM_ADDR), it enters into the park state, and gets a parked member address (PM_ADDR). During the parked state, the parked member uses the masters beacon to be synchronized. One of the major problems of the parked mode in the number of possible parked members as well as a completly used piconet (e.g., master with 7 slaves), it is impossible to get an active member. Therefore, the reaction time is very long. First the master has to park another device or the parked member has to wait until a connection is released.

51.9 Profiles

The Bluetooth Special Interest Group (SIG) has a series of different Working Groups. These Groups develop and specify various new Bluetooth Profiles to expand the usability of Bluetooth. The purpose of

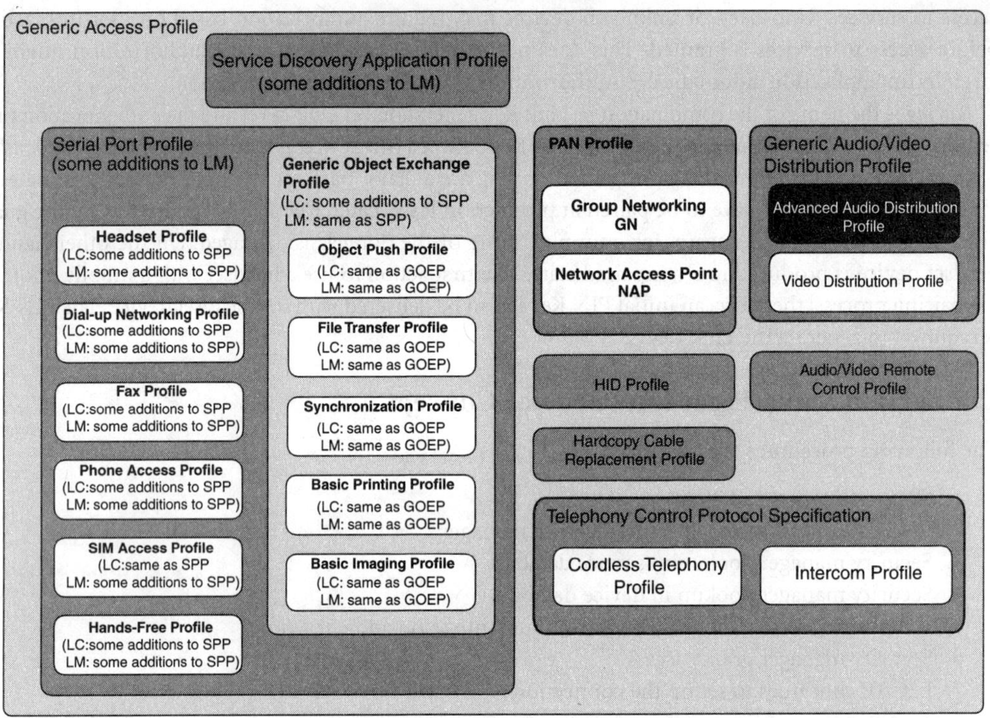

FIGURE 51.6 Profile dependencies and their effects on requirements on LC and LM protocol layers (with each profile, it is indicated whether requirements different from the ones defined for the parent profile or underlying protocol layers are valid).

Bluetooth Profiles is to provide a clear description of how to use and fulfill the specification of a system to impart a special function to a device. All these Profiles go through a complicated review process; after a series of draft versions, the Profile is adopted to the core specification.

Figure 51.6 gives an overview of several profiles; all these profiles provide a way for the Bluetooth technology to slot into different devices [BT_CWC]. The profiles in this overview are organized into groups, whereas the profiles are built upon the one beneath and also includes these features. The Bluetooth profiles should avoid the implementation of proprietary applications and profiles by different vendors to reduce the problem of interoperability; therefore, it gives Bluetooth more chance to be successful on the market.

FIGURE 51.6 Profile dependencies and their effects on requirements on LC and LM protocol layers (with each profile, it is indicated whether requirements different from the ones defined for the parent profile or underlying protocol layers are valid).

51.10 Bluetooth Qualification

The Bluetooth qualification process gives the guarantee that products comply with the Bluetooth core specification. To use the Bluetooth brand on the device and the package as well as the free license to all patents required to implement Bluetooth wireless technology, it is necessary to pass the Bluetooth qualification process. This qualification process also improves the interoperability of Bluetooth devices. It is also necessary to pass all government-specific radio tests where it will be sold.

Four categories are defined:

- Bluetooth radio link requirements
- Bluetooth protocol requirements

- Bluetooth profile requirements
- Bluetooth information requirements.

Abbreviations

ACL	Asynchronous Connection Less
AFH	Adaptive Frequency Hopping
BB	Base Band
BD_ADDR	Bluetooth Device Address
BNEP	Bluetooth Network Encapsulation Protocol
BT	Bluetooth
CRC	Cyclic Redundancy Check
DM	Data-Medium Rate
DQPSK	Differential Quadratur Phase Shift Keying
EIRP	Equivalent Isotropic Radiated Power
ETSI	European Telecommunications Standards Institute
FCC	Federal Communications Commission
FEC	Forward Error Correction
FH	Frequency Hopping
FW	Firmware
GFSK	Gaussian Frequency Shift Keying
HCI	Host Controller Interface
HW	Hardware
ISM	Industrial Scientific and Medical
L2CAP	Logical Link Control and Adaptation Protocol
LC	Link Controller
LM	Link Manager
LMP	Link Manager Protocol
OBEX	Object Exchange
PA	Power Amplifier
PCM	Pulse Code Modulation
PIO	Programmable Input–Output
PSK	Phase Shift Keying
QoS	Quality of Service
RF	Radio Frequency
SCO	Synchronous Connection Oriented
eSCO	extended Synchronous Connection Oriented
SDP	Service Discovery Profile
SIG	Special Interest Group
SW	Software
TCS	Telephony control Protocol Specification
UART	Universal Asynchronous Receiver Transmitter
USB	Universal Serial Bus
WAP	Wireless Application Protocol

References

[BT_1.2] SIG Specification of the Bluetooth System: Core Specification, Version 1.2 5th Draft (2003-07-24).

[BT_1.1] Specification of the Bluetooth System: Core Specification 1.1, (2001-02-22).

[BT_RF] SIG Specification of the Bluetooth System: Volume 2 Part A, Radio Specification, Version 1.2 5th Draft (2003-07-24).

[BT_BB] SIG Specification of the Bluetooth System: Volume 2 Part B, Baseband Specification, Version 1.2 5th Draft (2003-07-24).

[BT_LM] SIG Specification of the Bluetooth System: Volume 2 Part C, Link Manager Protocol, Version 1.2 5th Draft (2003-07-24).

[BT_L2CAP] SIG Specification of the Bluetooth System: Volume 3 Part A, Logical Link Control and Adaptation Protocol Specification, Version 1.2 5th Draft (2003-07-24).

[BT_SDP] SIG Specification of the Bluetooth System: Volume 3 Part A, Service Discovery Protocol, Version 1.2 5th Draft (2003-07-24).

[BT_IEEE] SIG Specification of the Bluetooth System: Volume 0, Part C, Chapter 2, IEEE Language Update, Version 1.2 5th Draft (2003-07-24).

[BT_HCI_Func] SIG Specification of the Bluetooth System: Volume 2 Part E, Host Controller Interface Functional Specification, Version 1.2 5th Draft (2003-07-24).

[BT_HCI_UART] SIG Specification of the Bluetooth System: Vol. 4, HCI UART Transport Layer, Version 1.2 5th Draft (2003-07-24).

[BT_HCI_3 Wire] SIG Specification of the Bluetooth System: Vol. 4, HCI THREE-WIRE UART TRANS-PORT LAYER ,Revision 0.90 (2003-04-14).

[Sec] Bluetooth Security White Paper, Bluetooth SIG Security Expert Group, Revision 1.00 (2002-04-19).

[Sec_Arch] Bluetooth Security Architecture White Paper: Special Interest Group (SIG); (1999-07-15).

[BT_2_RF] SIG Specification of the Bluetooth System, Medium Rate RF Specification, Proposal for Version 0.70 (2003-04-03).

[BT_CWC] Bray Jennifer, and Charles F Sturman, Bluetooth 1.1, *Connect Without Cables,* 2nd ed., Bernard Goodwin, Prentice-Hall, Inc.

[BT_HFP] SIG Specification of the Bluetooth System, Hands-Free Profile, Version 1.00VD (Voting Draft) (2002-07).

[BT_HSP] SIG Specification of the Bluetooth System: Vol. 7, Part G, Headset Profile, Version 1.2 5th Draft (2003-07-24).

[BT_PAP] SIG Specification of the Bluetooth System, Phone Access Profile, Version 0.70 (Draft) (2003-05-08).

[BT_SPP] SIG Specification of the Bluetooth System: Vol. 7, Part B: Serial Port Profile, Version 1.2 5th Draft (2003-07-24).

[BT_HIDP] SIG Specification of the Bluetooth System, Vol. 10, Part A, Human Interface Device Profile, Version 1.2 5th Draft (2003-07-24).

[Cox_John] Bluetooth Gets Greasy and Gritty, Network World, 07/28/03, http://www.networkworld.com/print/default.htm.

52

Linking Factory Floor and the Internet

Thilo Sauter
*Vienna University of Technology,
Institute of Computer Technology*

52.1 Introduction and Historical Background

If one believes contemporary advertisements and marketing articles in various automation domains, the interconnection of fieldbus systems and the Internet is *the* topic in automation. More specifically, but less appropriate for marketing purposes, we are talking about a linkage between arbitrary automation systems — which may also involve dedicated automation networks — and networks based on the Internet Protocol (IP). For the sake of simplicity, we will stick to the familiar terms fieldbus and Internet, however, keeping in mind that the fieldbus is not a must and that the Internet may as well be an intranet or any other network governed by IP and the protocol suite on top of it. What is being advertised as the major benefit of such an interconnection essentially boils down to two major, nonetheless interwoven aspects: remote access to automation systems and the promise of easy integration of automation data in a user-friendly environment. In other words, there are two reasons why a fieldbus/Internet connection can be worthwhile:

To extend the physical dimensions of an automation system. Mostly for historical reasons, the extension of a typical automation network is rather limited due to length restrictions in fieldbus segments and the lack of routing capabilities. If an Internet infrastructure is available, it can be used as a kind of backbone to connect distant segments of the installation.

To provide vertical integration. In the automation domain, this widely used term essentially means bringing information from an automation system into a framework used in the office domain, where it can be used not only for data acquisition but also for strategic operations such as system management or resource planning.

Especially, the second aspect is currently the focus of substantial marketing efforts connected to the growing use of Ethernet in industrial automation. However, it must be stated that the idea of vertical integration is nothing entirely new. The roots date back to the 1980s when the Computer Integrated

Manufacturing (CIM) concept was developed. The hierarchical model with typically five levels (Figure 52.1) was an early attempt to structure the information flow within a company. There are many ways to draw this pyramid, and the names of the individual levels differ according to the application area. But the ultimate goal has always been the same: to provide transparent data exchange between the various levels.

The first attempts to implement the CIM concept were more or less futile, for a number of mostly technological reasons. On the one hand, the protocols designed for the communication inside the automation system — essentially the Manufacturing Automation Protocol (MAP) as a full-fledged implementation of the ISO/OSI model — were too complex; on the other, the progress of microelectronics as a technological backbone had not yet been far enough to provide sufficient computing resources at a reasonable price. Therefore, integration of CIM in the field level turned out to be nearly impossible. In addition, fieldbus systems as low-level automation networks were only at a very early stage of their evolution; hence, there was, from a networking point of view, a missing link between the actual source of the process data and the already existing company networks.

Figure 52.2 shows a few milestones in the evolution of both IT and automation networks and demonstrates how the situation has changed over the years [1]. Partly driven by the insight that MAP was too clumsy to use (but still following its basic ideas), partly driven by application-specific needs, fieldbus systems as dedicated automation networks emerged. They filled the networking gap in the lowest levels of the automation pyramid, providing an anchor point for subsequent integration efforts. The great leap forward for integration itself emerged with the success of the Internet and more precisely, with the invention of the World Wide Web (WWW). While in the fieldbus domain, a large variety of approaches still exist even after a lengthy and cumbersome standardization process [2, 3], the office world is dominated by the Internet Protocol suite and a small number of well-known, widely used applications.

The reason why the old idea of vertical integration was revived in recent years was merely a psychological one. As the timeline in Figure 52.2 shows, the Internet and its basic technologies have been available (and used) for a long time. However, it took the sheer simplicity of the web browser as a ubiquitous tool to boost the acceptance of the Internet and to secure the dominant role of the IP suite. From the user's point of view, a web browser allows access to distant data in a nearly trivial manner. Hence, it is no wonder that the easy navigation through hypertext documents was soon adopted as a model also for the remote access to automation data. Consequently, many solutions of fieldbus/Internet connectivity rely on WWW technology and a web-browser interface. Still, the impression that "the Internet" naturally entails a uniform way of remote access to automation networks is deceptive. The user interface is only one aspect, the underlying mechanisms and data structures are another. In fact, when it comes to implementing an interconnection between IP-based networks and an automation system fieldbus, there is a surprising variety of possibilities even in the considerably "standardized" Internet environment. In particular, the web-based approach is by no means the only one.

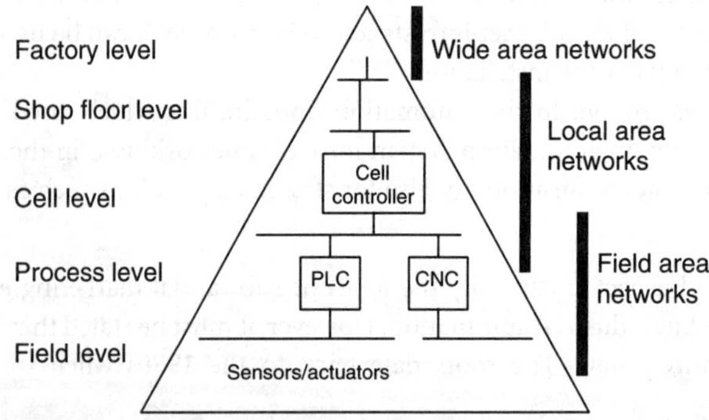

FIGURE 52.1   CIM pyramid and network types used in the various levels.

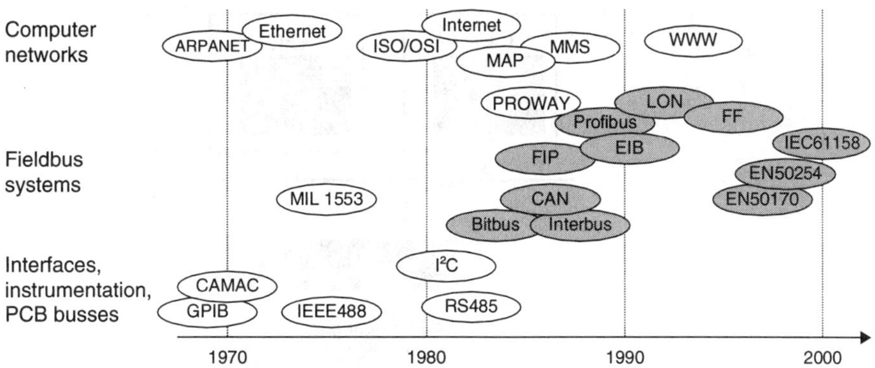

FIGURE 52.2   Evolution of networks in automation and information technology.

52.2   Interconnection Approaches

From an architectural point of view, there are two main possibilities to achieve a fieldbus/Internet interconnection, both of which are being used in practice:

- Tunneling of one protocol over the other
- Providing protocol and service translation via a gateway.

As far as the network topology is concerned, both approaches are very similar. In both cases, a central
access point between the networks is either required or at least reasonable. What differs is the way the
information is processed by this link or more accurately, how the processing is distributed between the
devices involved in the communication.

Protocol Tunneling

In communication networks, tunneling essentially means that data frames of one protocol are simply
wrapped into the payload data of another protocol without any modification or translation. The tunneling approach in the automation context falls into two categories. The currently more common solution
is to encapsulate IP packets in fieldbus messages, which also opens a channel for the upper-layer protocols required, for example, for direct web access to the field devices [4]. While in most cases not foreseen
in the beginning, this possibility has recently been included in several fieldbus systems, a move that has
to be seen mainly in connection with the discussion about Ethernet in automation. At first glance, IP tunneling does provide an easy way to achieve vertical integration, as Figure 52.3 shows. The field devices run
an IP-based service such as a web server providing data for a respective application in the Internet, which
can directly access the device or process data.

A second look, however, reveals that the integration is not at all that easy. There are a few critical points
to be considered to properly design such a system.

Computing resources at the field device: First of all, it is evident that the field device as an endpoint of
the tunnel must be able to run the complete IP stack with all additional protocols required for the particular application. In addition, memory space is required for the actual application program handling
the data that are to be accessed via IP. With today's embedded systems, all this may not be overly problematic, as there are very lightweight protocol implementations available (typically at the expense of
detailed error handling). At any rate, this often requires additional hardware at the device and might be
a cost factor, especially for simple devices like sensors or actuators.

Traffic handling at the access point: The node that connects the IP-based network and the fieldbus has
to encapsulate the IP packets and forward them to the appropriate fieldbus node, where they are
unwrapped and handed over to the IP stack. To this end, the access point has to act as an IP router. More

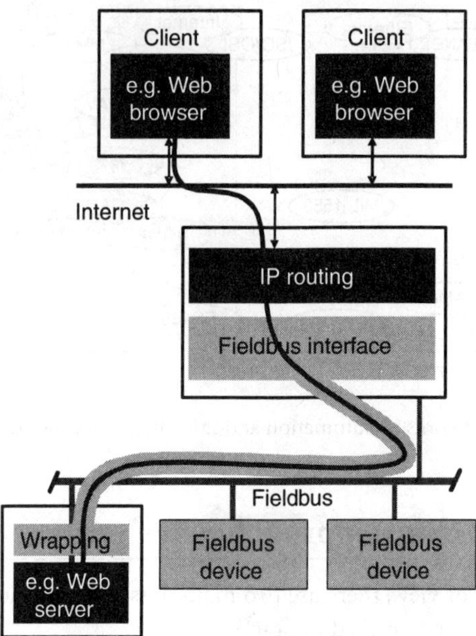

FIGURE 52.3 Structure of IP tunneling over a fieldbus.

than that, it has to map the IP connections to the respective fieldbus communication channels. This requires an address translation on the fieldbus side of the access point. On the Internet side of the node, an additional translation can become necessary if the IP addresses in the fieldbus are not public ones, but taken from a private address range, which is the usual network configuration. The access point thus acts as a simple masquerading firewall with network address translation (NAT). Unlike the case where the IP addresses in the fieldbus are public, the fieldbus as a whole is accessible only via one single IP address, and the distinction between the individual devices (e.g., web servers) hidden behind the firewall can only be done via port forwarding. This means that the services are not reachable through their well-known standard ports, but via dedicated ones which, of course, have to be configured appropriately at the client side in the Internet.

Performance issues: IP packets tend to be rather large, whereas fieldbus data frames are usually optimized for the transmission of only small pieces of data at a time. Squeezing IP packets in the small payload data fields of a fieldbus message therefore requires segmentation, which in turn increases the transmission time. An extreme example is Interbus, where IP traffic can be transported in the parameter channel [5]. This channel has only a small capacity of, say, eight bytes per cycle in order not to interfere with the real-time process data, and one byte is needed for control purposes. Depending on the size of the network and the selected baud rate, one cycle may take a few milliseconds. As an IP packet usually is of the size of several hundred bytes, it is clear that the performance is rather limited. In other fieldbus systems, the data frames are large enough to avoid excessive segmentation. At any rate, the efficiency of IP over fieldbus tunneling depends strongly on the fieldbus.

The alternative way of tunneling works the other way round. Fieldbus messages are wrapped into IP packets (or those of other protocols such as TCP, depending on how the tunnel is set up) and sent to a distant network node (Figure 52.4). From the viewpoint of vertical integration, the disadvantage of this approach is immediately visible. The fieldbus data must be interpreted at the client side, which requires a fieldbus-specific application or at least an experienced user with detailed knowledge about the particular automation system.

This approach does not fulfill the requirements of user-friendliness posed by the idea of vertical integration. In fact, fieldbus tunnels over the Internet are rather used to connect remote segments of an

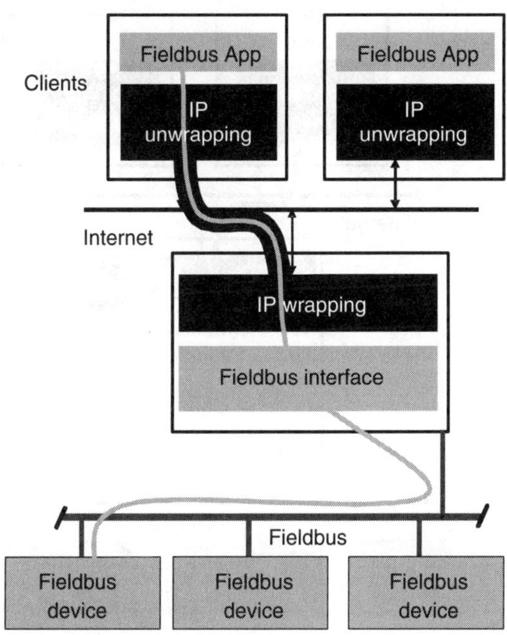

FIGURE 52.4   Structure of a fieldbus tunnel over the Internet.

installation as mentioned in the introduction. A wide field of application for this concept are network-based control systems, for instance, in building automation [6]. It has to be noted, though, that the interconnection of two fieldbus segments over an Internet tunnel is not fully transparent, but has to cope with timing problems introduced by the processing of the data packets and the transmission delays. Therefore, this concept is useful only if the fieldbus protocol provides some routing functionality, where timing requirements can be relaxed.

Gateways

The alternative approach to achieve fieldbus/Internet connectivity is to design the central access point as a gateway. As with the tunneling approach, the gateway is a full member of the fieldbus on the one side and can be accessed through IP-based mechanisms via the Internet (Figure 52.5). What is different, though, is the way the information exchange is handled. In the IP tunneling approach, the client connects directly to the server running on the field device. The access point just forwards the IP traffic, but is not further involved in the data transfer. In the gateway approach, the access point takes the role of a proxy representing the fieldbus and its data to the outside world. It fetches the data from the field devices using the usual fieldbus communication methods and is the communication partner addressed by the client.

Unlike the tunneling approach, where each field device has to provide and maintain its data on an individual basis, information processing for the client in the Internet here is centralized. This enables the gateway to set up a comprehensive and consistent view of the automation system, which is undoubtedly a benefit. In addition, the field devices need no special equipment to process IP-based data and services.

For the sake of completeness, it should be noted that a gateway would in principle also allow for the reversal of the communication direction. Therefore, field devices could as well access resources in the Internet, for example, different fieldbus systems connected to the LAN via compatible gateways. Although there have been efforts to foresee this possibility by providing harmonized interfaces (as in the NOAH project [7]), it is irrelevant in practice.

There are implementations for both the tunneling and the gateway approach available. The gateway is the more pragmatic and also more flexible solution. Furthermore, it is compatible with existing

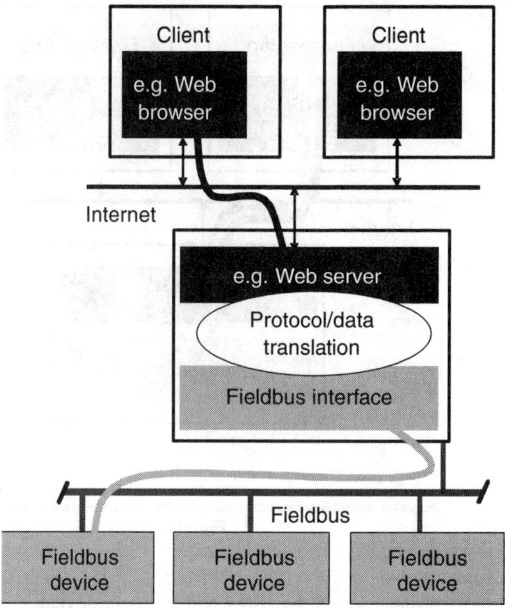

FIGURE 52.5   Network topology for a gateway-based interconnection.

installations, whereas the tunneling approach requires changes on the fieldbus level, both inside the field devices and in the protocol. Gateways are therefore dominant, and the remainder of this paper will focus on the gateway approach. Still, many considerations are also valid for the tunneling concept.

52.3   Design Considerations for a Gateway

There are no general guidelines regarding how to implement a gateway between a fieldbus and the Internet. From the user's perspective, however, there exists a clear-cut requirement for the integration: as the automation data are brought into the office world, the ways and means employed must comply with the standards used there. In particular, this requires some sort of abstraction for both data and functionality. There is a good reason for this. Within the scope of vertical integration, the automation data are accessed from a "strategic" level (rather than control level); thus, the user wants to have a clear picture of the most relevant data. Too many fieldbus-specific details can obscure this picture and be distractive. After all, the user is typically not a fieldbus expert, let alone the one who set up the automation system.

Another argument in favor of an abstract view to the automation system is the desired user-friendliness of the interface. Typically, the user wants to retrieve data from different systems with only one tool. Therefore, the user interface should be as independent of the underlying system as possible. It is important to notice that the way data and functionality abstraction can actually be achieved also depends on the protocols and mechanisms used on the Internet side of the gateway. In the sequel, we list and discuss a few design issues. They may vary with the concrete target application. The model for the present discussion is a simple and cost-effective residential gateway for linking a home network to the Internet [8].

Services

The basic function of a gateway evidently is to provide read and write access to process and device data on the fieldbus for monitoring and control operations, but also for network management purposes. However, as the gateway has to accomplish protocol (and maybe data) conversion and therefore needs a certain amount of computational intelligence anyway, additional services can be included. With the gateway acting as a proxy server for the automation data, there are a number of autonomous functions it can (or should) provide. Some of these add-on services can be described as *events* in an abstract way. Events are triggered by predefined criteria: if a criterion is fulfilled, an action is taken. Criteria can be the expiration of timers (time-based events) or the crossing of thresholds by the value of a data point (value-based events). The actions to be performed upon the occurrence of an event can be manifold.

Data logging: The storage of data point values based on either time or the recognition of certain trigger conditions enables the gateway to perform stand-alone operation independent of the availability of an on-line connection to the high-level network. Especially in remote access via unreliable channels (e.g., wireless or heavily congested service providers), the possibility of creating log files (and retrieving them at a later time) may be an essential requirement of the application. A management station can then retrieve logs (fully or incrementally) to perform subsequent analysis or feed specialized software with control system data.

Asynchronous notifications: Contrary to the predominant client/server communication model in the Internet, where the client has to continuously poll the server (the gateway) to obtain information, the gateway should become active as a resonse to certain events. The availability of such a mechanism is an important criterion for the choice of a suitable protocol on the Internet side of the gateway. In practical implementations, the lack of native asynchronous notification mechanisms is usually circumvented by the use of services based on some kind of e-mail.

Two other types of services are independent of the fieldbus data and cannot be modeled as events.

Security: Fieldbus systems are traditionally security-unaware. When a connection to the Internet is provided, however, the security of protocols that are able to communicate with the gateway is vital for both the field area networks that the gateway is connected to and the gateway itself [9]. The gateway has to take care of access control, the minimum level of security being authentication.

Network structure updates: A last point that might be relevant for the gateway depending on the desired application area is some kind of plug and play functionality. By this, we understand that the gateway is capable of recognizing changes in the network structure autonomously and adapt its internal data point structure. This does not necessarily eliminate the need for preconfiguration, but it may help to keep the gateway operational. A typical application for this feature are residential gateways that — once installed — have to be as autonomous as possible and must cope on their own with devices joining or leaving the field area network. In this case, the underlying fieldbus must of course also support basic plug and play features. A key issue here is the ability of the high-level protocol on the Internet side to reflect structural changes in the field area network: nodes in a field area network can be removed, added, or replaced. This can influence the mapping between protocol addresses and fieldbus addresses. It has to be taken into account as to how a protocol deals with such changes in the network structure and how objects that existed before a change can be consistently addressed after a change.

Gateway Structure

There is no best architecture for a gateway. Its primary function is to act as an abstract interface to a automation system, independent of both the fieldbus protocol (because it operates on application level) and the fieldbus-specific coding of the data. Therefore, the gateway provides unified access to maybe more than one fieldbus simultaneously. The software structure of the gateway implementation can reflect this demand for versatility and exhibit a modular structure [8, 10]. Alternatively, the architecture can as well be monolithic and tailored to one specific case. A modular approach will have to provide abstraction as a matter of course, the three-level approach shown in Figure 52.6 is just one possibility.

What has to be considered separately is the low-level access to the fieldbus. One degree of freedom is the strategy used to fetch the data. This can be based on a caching mechanism, where the gateway autonomously retrieves the data with a predefined refresh rate. Upon request from the client, the response is taken from the cache and returned without further delay. This approach provides immediate answers and can reduce the fieldbus traffic, especially in situations where multiple access from different clients needs to be handled. For proper operation, a time stamping should be foreseen, so that the client can keep track of the age of the data values.

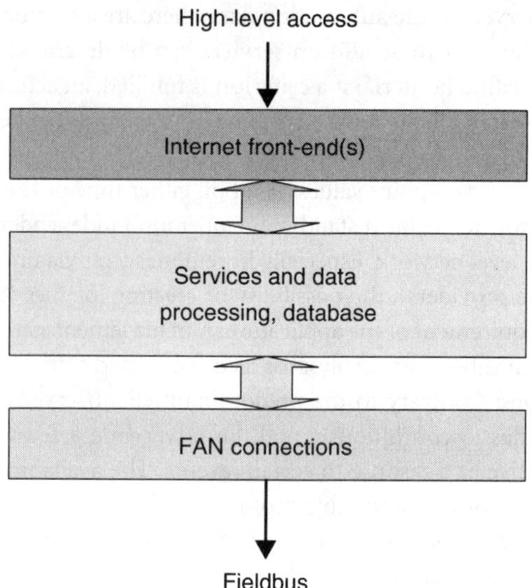

FIGURE 52.6   Modular three-level gateway structure.

The alternative is to obtain the data from the fieldbus only when requested to do so by the client. This avoids cyclic updates as well as time stamps; but the gateway may have to cope with substantial delays of the response from the field device. The connection to the client has to be kept alive in the meantime.

Which strategy can be used depends both on the properties of the fieldbus and the capabilities of the interface used. To avoid excessive bus load in the caching scheme, the gateway can passively listen to the network traffic. Provided the fieldbus interface supports such a monitoring mode, this approach is convenient to track cyclic (in systems like Interbus or Profibus-DP) or event-driven data transfer (in fieldbusses based on the producer/consumer model like CAN and EIB). Unfortunately, many of today's integrated fieldbus controllers perform a low-level data filtering inhibiting straightforward monitoring. In these cases, the gateway must actively retrieve the data from the bus. The same applies to data that are not normally transferred on a regular basis, such as diagnostic data (e.g., in Profibus-DP). Even worse, the transmission of such specialized information often requires the execution of dedicated commands that cannot be mapped to simple read/write commands. Such fieldbus-specific communication sequences (but also other peculiarities like remote program invocation or special network management functions) must be handled differently. As the high-level access via the Internet most likely will not provide methods for direct command execution, such special functions can be triggered by an ordinary read/write access to specially assigned data points.

Another important restriction may be the medium access control scheme employed by the fieldbus. There are no problems with multi-master systems; here, any node can host the gateway. Single-master systems, on the other hand, imply that the gateway must either be the fieldbus master or cooperate with it, in the sense that there is a dedicated and typically nonstandard communication channel through which the gateway can tell the bus master as to which data values are to be read or written. Otherwise, writing data points will inevitably lead to inconsistencies, when the gateway sets a value and the bus master overwrites it during a regular update cycle.

Data Representation

A crucial point in the design of a gateway is the way the fieldbus data are represented. A *data point* is the smallest information unit available via the gateway. To achieve fieldbus independence, the data points have to be arranged in a rather general, high-level way. The question regarding how to find a unified view and coding for the sometimes extremely heterogeneous fieldbus objects is an interesting topic on its own and will be left aside here. More relevant for our purpose is the way the complete set of data points is structured inside the gateway. Basically, there are two different possibilities to do so. The first is a function-oriented approach. In this case, emphasis is put on the content of the individual data point, and the underlying network structure is completely irrelevant. The gateway thus offers a simple flat list of data points [10].

The second approach is a structure-oriented one. Here, the data points are arranged according to the structure of the field area network. The gateway offers a list of nodes, and each node can contain one or several data points [8]. In addition to the simple form of a data point, a data point can also contain a set of values instead of only a single value. This type of data point is referred to as an *aggregate* data point. Data points can also contain multiple values of different types (structures). The advantage of the structure-oriented solution is that properties of nodes (and not only properties of data points) can also be modeled on the gateway. These properties are, for example, a self-identification of the node or a location information that describes the position of a node in a field area network.

Which of the two possibilities is actually used is a matter of taste and also depends on the target application. Furthermore, the choice might be limited by the high-level access protocol and its capabilities of describing and managing data. Every object that is available to a client needs an address that uniquely identifies the object. Protocols typically define their own addressing scheme; the gateway, on the other hand, also defines an addressing scheme that may be similar to or completely abstract from the fieldbus network addressing. These two schemes need to be mapped onto each other to make data points available.

Another point to be observed is how the information of a data point is encoded. This again might be limited by the capabilities of the way the data are transferred from the gateway to the client. Not every protocol may support special data types or encodings used in a particular fieldbus, so that there must be some form of data translation and — necessarily — abstraction. This might not be too problematic for scalar data points like temperatures (although there are obviously many possibilities to encode a temperature value), but it will become increasingly difficult for such common things as date and time, not to speak of aggregate data points available only as a structure on the fieldbus level. For all these more complex data points, no one-to-one mapping is possible, and there are basically only two options to cope with the situation: one is to find a compromise in the form of abstract data representations that omit excess information, and the other is to collapse the complete data point into a flexible and unspecific data format such as a string — with the disadvantage that the client needs to take the string apart and to interpret it, which again requires fieldbus-specific knowledge.

In addition to the actual value, a data point may also need some attributes to facilitate interaction with the client. Such attributes can be a description of the data point type and encoding (to tell the client how the data are to be handled) and maybe a string providing a textual description of the data point. In some fieldbus systems, such a self-description is foreseen for data objects anyway. Last but not the least, there can be a specification of the access mode, that is, what the client is allowed to do with the data point. It should be noted that although the restrictions concerning data representation have been discussed for the gateway approach, they also apply to the IP tunneling solution. In this case, the list of data points has to be maintained by every field device on its own. Typically, the implementation is up to the vendor, and unless there are very stringent rules, this may lead to a substantial inhomogeneity stemming from the many degrees of freedom mentioned before.

52.4 Gateway Access From the Internet

The primary goal of vertical integration is to bring automation data into an IT context, and consequently there are many ways to incorporate data provided by a gateway into office-level applications. To this end, it is not necessary to employ full-fledged application-level protocols on top of IP/UDP/TCP. A very common approach used in contemporary SCADA systems is OPC, which relies on the Windows-specific mechanism OLE and COM/DCOM [11]. With the extension OPC DX, it is in principle even possible to interconnect different fieldbus systems because OPC DX also permits server–server communication over an Ethernet network [12]. A more unconventional idea is to treat the gateway as a database and use SQL to retrieve the data [13].

The protocols discussed in the sequel have been evaluated with respect to their feasibility for home network access, specifically for remote meter reading [9]. The criteria for the evaluation were described in the previous section: a reasonable way of structuring and handling data, a possibility to send asynchronous notifications, and of course security provisions. In addition, the protocol (together with its data representation) should be able to cope with changes in the network structure. This requirement is a direct consequence of the target application in an intended mass-market with a high number of installations, where updates calling for specially trained personnel are not affordable. Likewise, the overall costs of such a gateway should be low, which typically means limited computing power and memory space. Compact implementations on possibly embedded devices are therefore a must.

SNMP

SNMP is a comparatively old protocol and has been designed for remote administration of networked devices [14]. Data are represented in a tree structure called *Management Information Base* (MIB) where each branch and each leaf have a unique identifier, called *Object Identifier* (OID). SNMP *managers* (i.e., the client that connects to an SNMP-enabled device) use this MIB to get extended information about the managed *agent* (i.e., the device that allows SNMP access; in the present case, the gateway). The MIB is a static representation of all the objects available at the managed agent. As SNMP does not provide mechanisms to download the MIB from the agent, the mapping of OIDs and data points needs to be transferred to the manager over mechanisms, which are beyond the scope of SNMP (usually this is done by the administrator of the system during setup). SNMP is based on the connectionless User Datagram Protocol (UDP) running on top of IP. This is rather efficient because of the low overhead; but unlike TCP, UDP does not guarantee delivery of packets, so packets can be lost on the network without being noticed by the sender.

The major commands provided by SNMP are GET and GETNEXT for reading data (GETNEXT allows the manager to traverse an unknown MIB tree and to retrieve the OIDs) and SET for writing to data points. In contrast to the standard manager-driven communication (in typical client/server fashion), SNMP also defines the unacknowledged TRAP command (from version 2 on also the acknowledged INFORM command), which can be sent by an agent to asynchronously inform a manager about special conditions in the agent. This possibility to reverse the client/server communication is a rare exception among the investigated protocols and together with the good protocol efficiency is a strong argument in favor of SNMP.

One of the major problems of SNMP is the static structure of the MIB together with the lack of native transfer mechanisms between the agent and manager, which does not allow dynamic updates of the network (or data point) structure. Therefore, the list of nodes and data points needs to be stored in an SNMP table rather than as a set of separate objects in the MIB [15]. To access individual objects, the manager needs the OID of the table, which is fixed once and for all, and the number of the row containing the object. If the structure of the field area network is changed, the table can be updated by adding or removing the rows that represent the according nodes. This is a critical issue in that SNMP requires the rows in an SNMP table to be numbered consecutively; therefore, the insertion or removal of a row changes the numbering of all subsequent rows.

The workaround of using tables to add a somehow dynamic behavior also raises problems if the gateway is accessed by more than one client simultaneously. It is possible that data in a table are modified by one client (e.g., rows are deleted or added), while another client is in the process of reading the table. This results in an inconsistent set of data. A possible solution used in the case study was to introduce a sequence number for every table. This sequence number changes every time data in the table change. The workaround for achieving data consistency is to read the sequence number, read the table, and afterwards read the sequence number again to check if it changed. In case it remained unchanged, the table data are consistent; otherwise, the whole process has to be repeated. To this end, for write access the protocol had to be slightly changed in a way that every SNMP SET command contains a sequence number that uniquely identifies the version of the table to which a data point is written.

Another problem of SNMP is security. The commonly used version of SNMP is version 1, which does not contain any considerable security features. In the most recent version 3, security mechanisms have been introduced that are sufficient (although not widely used at present) [16]. Still, difficulties arise from the connectionless UDP used as a transport protocol. UDP traffic is usually not permitted to pass through a typical firewall. But even if communication is possible, the underlying UDP does not guarantee packet delivery. This means that an SNMP manager communicating with an agent cannot rely on the last action he has taken on an agent to be successfully finished. Neither can an agent be sure about the status of a communication with a manager. Hence, error monitoring must be implemented within the manager application.

LDAP

LDAP [17] is the lightweight counterpart to X.500 directory services. It is oriented toward directories that are organized in a tree, the *Directory Information Tree* (DIT). The DIT is a hierarchically structured organization of data items, the directory entries. Every entry consists of a collection of attributes. Every attribute has an attribute name, an attribute type, and a value associated with it. One special attribute that is mandatory for every tree entry determines the position of the entry in the tree — the *distinguished name* (DN), which is used as a unique "address" for the entry. It specifies the path by listing all parent entries up to the root entry. The directories contain mostly (but not necessarily) small pieces of data, which are mainly read and searched for. Writing of data is intended to occur only rarely, as well as changes in the tree structure.

LDAP offers a set of data querying and manipulating commands for accessing the directory service. The SEARCH command is very powerful and allows to specify search filters that are applied to each entry. It is possible to request just a subset of the available attributes and in order to narrow a search even more, the scope can be limited to a certain level of the directory tree (search only the current level, one level deeper or the whole subtree). With the commands ADD, DELETE, and MODIFY, single entries can be added, deleted, or modified, respectively. The COMPARE command is used for comparison of two entries.

Since LDAP uses TCP as a transport protocol, there are no problems with firewalls. Also, all security extensions designed for TCP-based protocols such as the *Transport Layer Security* (TLS) [18] can be used. LDAP version 3 also provides appropriate authentication mechanisms in combination with those already defined in TLS [19].

Dynamic update of the directory information tree is also not particularly problematic, although the changes are not automatically taken over by the client. The DIT as such is not confined to a static structure like the SNMP MIB, which facilitates the adding and deleting of entries even at runtime. Consequently, there is also no problem with the consistency of data, because the address of a data point can be added as an attribute to the directory entries to unambiguously identify the data points. As operations on a DIT entry are always atomic, simultaneous and consistent access to one single data point from multiple clients is inherently handled by the built-in mechanisms of the protocol.

A general difficulty of LDAP is that its original application idea does not exactly fit the purpose of fieldbus gateway access. The main application of LDAP is the access to data that are related to humans, for example, telephone numbers, employee data, or classical white pages services. The client support reflects this field of application, and although clients exist, they are not widely used and very different in their performance with respect to network load. A second point is that the access to white pages directories is typically read-only. Therefore, the powerful search command can provide a convenient means to retrieve fieldbus data. Writing, on the other hand, is more difficult and likely becomes a bottleneck. Hence, LDAP is not so well suited for control tasks where intensive write access is required. Another problem is the lack of an asynchronous notification mechanism, so that workarounds (like e-mails or the like) have to be used.

Web-Based Approaches

For many solution providers in the area of fieldbus/Internet connectivity, web-based approaches seem to be the only conceivable solution; at least they are strongly favored. However, unlike the protocols described so far, methods based on the WWW are very heterogeneous. They include communication and data representation using web browsers as clients. The common transport protocol is HTTP, which is typically used to transmit HTML documents. HTML documents can contain embedded objects and script languages (e.g., JavaScript) and can be created by multiple technologies on the server side (such as CGI, ASP, JSP). They all have in common that the client receives a predefined view of a set of data. This view can be modified by the user (e.g., by selecting specific objects in a page and filtering out the others), which requires the creation of a new document on the server side. This process is usually not very responsive because of the necessary communication and computation overhead.

The two main commands that the HTTP protocol supports are GET (for sending requests to a server to retrieve data) and POST (so send data to the server). This, however, does not cover the whole spectrum of possibilities that are based on this mechanism. In many cases, however, the user is presented with a view based on a template, which is filled with data by some server mechanism.

A definite strength of web technologies is the availability of many different clients. Several Internet browsers for multiple platforms exist, leaving hardly any white space on the map of supported systems. The main problem of pure HTML is the response behavior that is imposed by the necessity to refresh (i.e., completely reload) a page in order to receive current data (this can be circumvented by advanced mechanisms, which only retrieve, e.g., the data point value). There is also another performance problem with this approach: As the view on the data depends on HTML documents, typically several data points will be collected on one page to have a better overview. Therefore, even for very small amounts of data (e.g., the value of just one data point), a complete document needs to be created and transmitted. This causes a considerable amount of both computational overhead and network traffic. Likewise, it is not possible to collect data from multiple servers in a single document, which considerably limits the use of this approach. In an IP tunneling scheme, for example, every device providing data needs to maintain a set of HTML pages, and the client needs an instance of the browser for each device. An integrative view of the entire installation, let alone an automatic data acquisition, is not feasible and has to be done manually.

A positive point about the web approach is the comparatively unproblematic security issue. As HTTP is based on TCP, TLS can be used. Furthermore, HTTP is conveniently routable through firewalls, which is an important point in an environment becoming increasingly aware of security — although it has to be noted that the simple routing possibility actually opens a way to circumvent security. A drawback is that there is no way for the server to send data to the client without request. Like with all other protocols except for SNMP, other ways have to be found to accomplish such notifications (e.g., special applets or e-mail-based services). The support of different data types is also very poor. In fact, the data have to be sent as strings and processed appropriately by the client. This lack of ample data type support ideally supports (and also demands) pragmatic, proprietary solutions, but is a severe obstacle to standardized, interoperable implementations. A possible way out of this dilemma is the increasing use of XML. If the data are based on (standardized) XML descriptions, they are not only readable for a human operator (a major drawback of HTML) but also machine parsable, which makes gateway access based on web technologies finally also suited for automatic data acquisition.

A further step in this direction could be the XML-based simple object access protocol (SOAP), which defines methods for data exchange typically used over HTTP. A tremendous benefit is an utmost platform independence as both HTTP and XML are available everywhere. However, as SOAP is a protocol on its own, client applications are still needed; today, a web browser is not sufficient. A potential weak point is, as with the somehow related technology CORBA, the relatively complex structure, which does not lend itself to a lightweight implementation on those resource-limited devices that make up a significant portion of a typical automation system today.

52.5   The Role of Industrial Ethernet

A discussion of fieldbus/Internet connectivity is incomplete without consideration of the growing use of Ethernet in automation. After all, Industrial Ethernet is often praised as *the* solution to vertical integration overcoming all problems encountered in the design of heterogeneous network interconnections. Beyond all well-sounding marketing slogans, it is a matter of fact that the dominant use of Ethernet in local area networks has renewed efforts to use Ethernet also in the automation domain. These efforts date back to the early days of Ethernet development, but only recent technological advances in the development of full duplex and switching techniques made high data rates possible without the old drawbacks of the CSMA/CD medium access control. Although it is still disputed whether or not Ethernet really fits the requirements of a field-level network, the main benefit is clear: if Ethernet was used as a replacement for the fieldbus, and if IP plus the common transport protocols were used on top of it (which is not necessarily the case in all Industrial Ethernet approaches), Internet technologies would be the natural way of accessing automation data. It is thus not astonishing that many contemporary concepts of vertical integration rely strongly on Ethernet as an automation network. For the old CIM pyramid, the penetration of LAN technologies into the classical fieldbus domain has the ultimate consequence that intermediate levels become obsolete. The hierarchy thus gradually turns into a flatter structure with only two, maybe three levels.

How do the facts of Industrial Ethernet compare with the characteristics of vertical integration discussed so far? It is of course short-sighted to postulate that the use of Ethernet alone entails an interoperability with "the office world" — although it is often done. In fact, Ethernet is but a transport medium. Using Ethernet for both office and automation applications may allow for a coexistence on a shared medium (within certain limits posed by, e.g., bandwidth considerations), but does not mean integration. What is much more important to this end is the protocol stack used on top of Ethernet. Indeed, most solutions rely on the Internet and LAN standards TCP/UDP/IP as a transport layer; however, this is already where compatibility ends. The higher layers of all serious approaches are in fact *fieldbus* protocols:

- Ethernet/IP (IP standing for Industrial Protocol) uses the Control and Information Protocol (CIP) already known from ControlNet and DeviceNet [20]. This application layer protocol is simply sent over TCP or UDP, depending on whether configuration or process data have to be transmitted.
- The High Speed Ethernet (HSE) variant of Foundation Fieldbus is in fact an application of the existing Fieldbus Foundation's H1 protocol wrapped in UDP/IP packets [21].
- Modbus/TCP [22] is based on standard Modbus frames encapsulated in TCP frames.
- PROFInet is rather a new protocol development [23]. Its integrability into the IT environment relies on the COM/DCOM mechanism used for communication between devices. Legacy fieldbus systems can be included via so-called proxies (which are essentially gateways in our sense).
- The IDA (Interface for Distributed Automation) proposal is an entirely new protocol on top of UDP for real-time data communication plus a web-based human–machine interface [24].

In the light of our taxonomy sketched in Section 52.2, all these approaches (with maybe the exception of PROFInet and IDA) essentially implement some sort of fieldbus protocol tunneling approach. Of course, they have the benefit of a — thanks to the possibility of using the same medium for office and automation tasks — tight integration from a network point of view. Unless a traffic separation seems reasonable for performance or security reasons, no dedicated connection node is necessary. On the application level, however, integration is by no means so tight. Even though the inclusion of automation data into office applications is simplified by, for example, object-oriented techniques borrowed from the IT world, there is still a substantial amount of fieldbus-specific information to be handled. Consequently, an abstract view on the automation data — one of the requirements for vertical integration — is not a matter of course. Nor is an "automatic" inclusion of Industrial Ethernet into the office world easy to achieve. In fact, the effort that needs to be spent on a gateway design has in this case to be put into the development of appropriate non-standard application-level tools. The only exception where Ethernet together with standard protocols definitely brings an advantage is a web-based human–machine interface used, for example, for configuration purposes. But this is only one small aspect of automation.

52.6 Summary

The classical fieldbus systems have been devised as highly specialized communication systems for the factory floor. Their development, although innovative at that time, was aimed at conventional self-contained automation applications. Today's automation tasks, however, are usually no longer stand-alone systems. At least supervisory control and data acquisition is required in many cases, with the respective SCADA software running in a typical office environment as regards hard and software platform or network environment. More advanced demands are remote data access for monitoring or maintenance, or the inclusion of automation data in company-wide management systems. All these functions are in fact not a part of the factory floor or the process control level; they belong to a higher, more strategic level of information processing, which is governed by office networks. This circumstance calls for an interconnection between automation and office networks.

As different as the goals of interconnection are the possible realizations. Although the systems on either side are standardized, the link is not — not even in a coarse architectural sense. One reason for this is the still existing plethora of fieldbus systems. On the Internet side as well, the IP suite is just a least common multiple of an equally large set of possible applications serving as targets for the interconnection. The preceding sections have highlighted the basic system architectures and the many design options to be considered. Evidently, the properties of the networks together with the applications have an important influence on these options and in fact limit the designer's flexibility with respect to the implementation.

Contrary to an initial guess, the architectural approach (i.e., gateway or tunneling) does not seem to affect the overall system complexity very much. As the discussion showed, it rather influences the distribution of the implementation efforts among the involved entities: field devices, clients on the Internet side, and a possible central access node. The particular choice must largely depend on the intended application; there is no golden rule for it. In fact, all approaches have their own strengths and shortcomings. Even though the major part of the article discussed the subject from the viewpoint of a gateway solution, this does not imply that the tunneling approach is inferior. It is simply suited for slightly different types of application, and most of the points described for a gateway are equally valid.

The overall complexity of the interconnection is — according to practical experience — chiefly determined by the desired level of data and functionality abstraction and the degree of vertical integration to be achieved. A direct forwarding of fieldbus data packets "as is," with the interpretation left up to the client (and, ultimately, the user), is easier to achieve than a reasonable data and protocol translation with the aim of a smooth integration into a non-fieldbus-specific application. Not always is true vertical integration and abstraction really an issue. But if it is, the price to be paid is necessarily a loss of precision, mostly from a functional point of view. The higher the abstraction level, the fewer functions and specific pieces of information can be mapped to the "other side" without problems. Timing precision or real-time behavior should not be a requirement at all because timing relations are typically lost in the node linking the two different networks. Timestamps can help, but only to a certain extent.

The great flexibility of fieldbus/Internet connections sketched in this article is a bit deceptive. In practical implementations, one has to live with a number of inconveniences. Timing is just one. Especially if one goes for vertical integration, the high-level target application and the underlying protocol may not support services provided by the automation system. This frequent lack of a one-to-one relation between the two networks requires some creativity from the designer, especially on the Internet side. A good example are the basic gateway services discussed in Section 52.3. Vital mechanisms such as asynchronous notifications rarely exist as native protocol features and have to be replaced by workarounds outside the actual application. In a number of case studies, it was found that none of the investigated high-level protocols really satisfies all needs. In this respect, the large number of different interconnection approaches is not astonishing, but only reflects the fact that the one ideal solution for all occasions does not exist.

References

[1] Dietrich, D, and T. Sauter, Evolution Potentials for Fieldbus Systems, IEEE International Workshop on Factory Communication Systems, Porto, 6–8 Sept. 2000, pp. 343–350.

[2] Felser, M. and T. Sauter, The Fieldbus War: History or Short Break between Battles?, IEEE International Workshop on Factory Communication Systems (WFCS), Västerås, 28–30 Aug. 2002, pp. 73–80.

[3] Thomesse, J.P., Fieldbuses and interoperability, *Control Engineering Practice*, 7, 81–94, 1999.

[4] Wollschlaeger, M., Framework for Web Integration of Factory Communication Systems, IEEE International Conference on Emerging Technologies and Factory Automation (ETFA), Antibes Juan-les-Pins, 15–18 Oct. 2001, pp. 261–265.

[5] Volz, M., Quo vadis Layer 7?, *The Industrial Ethernet Book*, GGH Marketing Communications Ltd., Titchfield, Issue 5, spring 2001, http://ethernet.industrial-networking.com/articles/i05quovadis-layer7.asp.

[6] Soucek S., T. Sauter, and T. Rauscher, A Scheme to Determine QoS Requirements for Control Network Data over IP, 27th Annual Conference of the IEEE Industrial Electronics Society (IECON), Denver, Colorado, 29 Nov. – 2 Dec. 2001, pp. 153–158.

[7] Döbrich, U. and P. Noury, ESPRIT Project NOAH — Introduction, in *Fieldbus Technology*, D. Dietrich, P. Neumann, Eds., Springer, Berlin, 1999, pp. 414–422.

[8] Pratl, G., M. Lobachov, and T. Sauter, "Highly Modular Gateway Architecture for Fieldbus/Internet Connections, IFAC International Conference on Fieldbus Systems and their Applications, FeT 2001, Nancy, France, Nov. 15–16, 2001, pp. 267–273.

[9] Palensky, P. and T. Sauter, Security Considerations for FAN-Internet Connections, *IEEE International Workshop on Factory Communication Systems*, Porto, 6–8 Sept. 2000, pp. 27–35.

[10] Kunes, M. and T. Sauter, Fieldbus-Internet connectivity: the SNMP approach, *IEEE Transactions on Industrial. Electronics* 48, 1248–1256, 2001.

[11] http://www.opcfoundation.org.

[12] Iwanitz, F. and J. Lange, *OPC — Fundamentals, Implementation and Application*, 2nd ed., Hüthig, Heidelberg, 2002.

[13] Lobashov, M. G. Pratl, and T. Sauter, Applicability of Internet Protocols for Fieldbus Access, IEEE International Workshop on Factory Communication Systems (WFCS), Västerås, 28–30 Aug. 2002, pp. 205–213.

[14] Stallings, W., *SNMP, SNMPv2, SNMPv3, and RMON 1 and 2*, 3rd ed., Addison-Wesley, Reading, MA, 1999.

[15] Perkins, D. and E. McGinnis, and *Understanding SNMP MIBs*, Prentice-Hall, Upper Saddle River, NJ, 1997

[16] Blumenthal, U. and B. Wijnen, *RFC 2574: User-based Security Model (USM) for version 3 of the Simple Network Management Protocol (SNMPv3)*, The Internet Society, Reston, VA, April 1999, http://www.rfc-editor.org.

[17] Kille, S., W. Yeong, and T. Howes, *RFC1777: Lightweight Directory Access Protocol*, The Internet Society, Reston, VA, March 1995, http://www.rfc-editor.org.

[18] Dierks, T. and C. Allen, *RFC 2246: The TLS Protocol Version 1.0*, The Internet Society, Reston, VA, January 1999, http://www.rfc-editor.org.

[19] Hodges, J., R. Morgan, and M. Wahl, *RFC 2830: Lightweight Directory Access Protocol (v3): Extension for Transport Layer Security*, The Internet Society, Reston, VA, May 2000, http://www.rfc-editor.org.

[20] P. Brooks, "Ethernet/IP — Industrial Protocol", IEEE International Conference on Emerging Technologies and Factory Automation (ETFA), Antibes Juan-les-Pins, 15–18 Oct. 2001, pp. 505–514.

[21] http://www.fieldbus.org/

[22] Schneider Automation, Modbus messaging on TCP/IP implementation guide, May 2002, http://www.modbus.org/

[23] http://www.profibus.com/

[24] http://www.ida-group.org/

53

LonWorks™ over IP

Dietmar Loy
LOYTEC electronics GmbH

Stefan Soucek
LOYTEC electronics GmbH

53.1 Introduction and Overview

Control networks also known as fieldbus systems started to emerge rapidly in the early 1990s. Back then, control networks were designed to be used in closed systems exchanging data packets on dedicated network channels. These channels included twisted pair media, power-line networks, wireless, and infrared communication. Dedicated network channels have more or less well-defined properties for bit error rates, propagation delays, average and maximum response times, and they guarantee to maintain the packet order on the network.

With the upcoming hype of the Internet in the late 1990s, requests were made to extend control networks to not only span local communication within relatively small areas but to cover networks that are spread out over cities, countries, and even continents. In building automation, facility managers wanted to connect facilities to a central location for monitoring, logging, alarming, trending, and remote maintenance.

This meant that the engineers had to either go back to the drawing board and redesign the control network protocol to make it work in large networks or use the existing technology like the IP service available on the Internet to transport control network data packets across continents. Of course, there are still restrictions regarding defined maximum response times, proper packet ordering, defined packet loss — all properties that are easily manageable on dedicated network channels but can be more or less unknown or undefined in IP (Internet Protocol) networks.

For the engineers, there was the challenge to live with the restrictions of the control network protocol and add a software layer in between the IP service and the control network protocol to conceal these limitations of the IP transport service from the upper layers of the control network protocol. Figure 53.1 shows the software architecture for a typical fieldbus node connected to a dedicated network channel on the left and if it is using the IP transport service on the right side. One can see the software module that is called ANSI/EIA-852, which is sitting between the control network protocol stack and the TCP/IP and UDP socket interface, and tries to conceal the limitations of the IP service from the control network protocol stack but also packs the control network data packet into an IP packet when transmitting a packet and

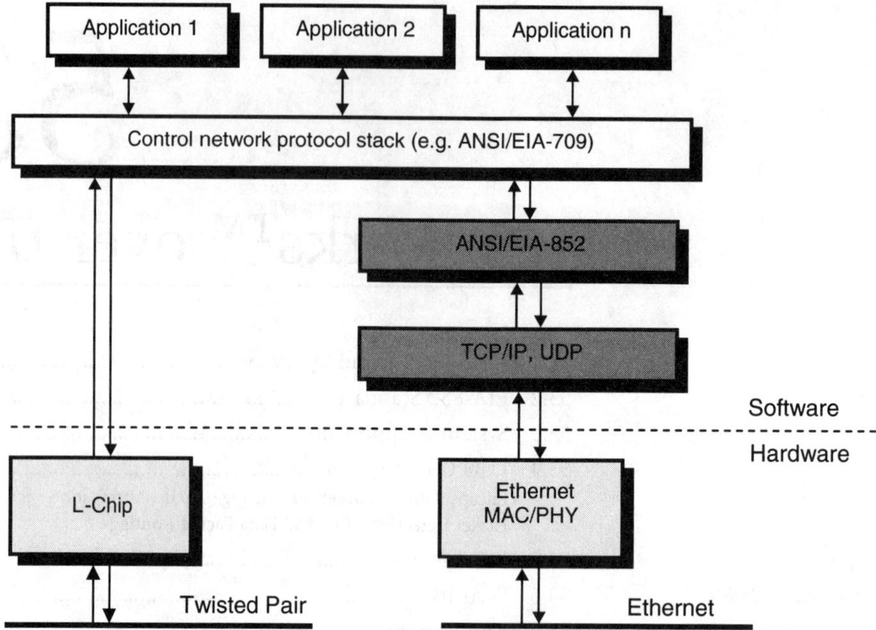

FIGURE 53.1 A dedicated software layer named ANSI/EIA-852 conceals limitations of the IP transport service from the control network protocol stack and sticks the control network data packet into an IP packet.

unpacks the control network data packet from the IP packet when receiving a packet. In this chapter, we concentrate on the ANSI/EIA-709 control network protocol but the technique described here will also be used for EIA-600 and other control network protocols.

The challenge in tunneling control network data packets through IP networks is to minimize the effect of the unpredictable IP timing parameters on the control network protocol stack.

53.2 EIA-852 Standard

The EIA-852 standard [8] was created to ensure interoperability of control network (CN) devices, which communicate over an IP network using a native CN protocol. The basic idea is to exchange CN packets over an IP network by embedding them in IP packets. This technique is known as tunneling. In EIA-852, there are no application-layer transformations involved as in other approaches, which use gateways. The tunneling approach is beneficial in systems where a number of CN devices communicate and use the IP network purely as another medium.

One major design criteria of the EIA-852 standard is to be generic enough to apply it to different CN networks. Currently, the usage for EIA-709 (LonWorks) and EIA-600 (CEBus) is defined. Most EIA-852 implementations on the market today are LonWorks-based. Therefore, the description of the standard is focused on its application to EIA-709. The network elements defined in the context of the EIA-852 standard are CN devices. These CN devices are computing systems (e.g., PCs, embedded systems) equipped with a TCP/IP stack, the EIA-852 software, and a LonWorks protocol stack. A number of CN devices are connected over the IP network where they form a logical network known as the *IP channel*. The IP channel functions as a LonWorks communication channel and CN devices exchange data over the IP channel. The CN devices part of the IP channel are referred to as *channel members*.

It shall be noted that there may exist more than one IP channel on the same IP network. Although the CN devices are connected to the same physical network in this case (e.g., Ethernet), the IP channels are still isolated in the LonWorks domain.

CN devices can have different functions. Depending on their primary function, the EIA-852 standard defines the following CN device types:

CN nodes: They operate solely as nodes in a distributed control application and are members of one IP channel. They have the same functionality as LonWorks nodes on their native communication channels and are connected to the IP channel only.

CN/IP routers: These CN devices are LonWorks routers that connect one native LonWorks channel with one IP channel. These devices are available on the market as stand-alone LonWorks/IP routers or EIA-852 routers, for example, the L-IP by LOYTEC or the i.LON 1000 by Echelon Corp.

IP/IP routers: These CN devices are LonWorks routers that connect two IP channels. These devices are less common on the market today.

CN proxies: These are CN devices, which function as LonWorks proxy nodes to establish cross-domain communication, for example.

The EIA-852 addressing scheme defines to address types. First, there are IP addresses associated with each CN device on the channel. Second, there are logical CN addresses based on the LonWorks address space. The Unique ID is equivalent to the LonWorks node ID or Neuron ID and functions as a unique hardware address. Each CN device is accessible by its unique ID. Logical network addresses include a subnet/node (S/N) scheme, where individual channels are assigned subnet numbers and CN devices on that channels' node addresses. As a convention, subnets must not span over multiple channels, if the CN routers are used as configured LonWorks routers. The group address identifies a number of nodes without looking at subnet addresses. Consequently, groups can span different channels. Finally, those logical addresses are local to a domain.

The mechanisms in the EIA-852 standard are designed to take care of typical CN protocol properties, which differentiate them from common IP application protocols such as Email, Web access, or multimedia streaming. The key characteristics of control network traffic are

- low throughput,
- small packet sizes, and
- higher sensitivity to packet loss and latency.

While low throughput allows CN devices to perform more lengthy per-packet calculations, small packet sizes will lead to significant overhead in the encapsulation process. The implications of packet loss and latency on the IP network are two-fold in EIA-852 systems:

- The control application has to be designed to cope with timing parameters, which differ from native CN channels. This has an impact on the soft real-time properties of some applications. It is important to know the time constants in the application and determine whether the application could break if run over an IP channel [15,16].
- The EIA-852 standard has to ensure the functionality of the CN protocol itself. LonWorks, for instance, makes strict assumptions on maximum end-to-end delay and packet ordering.

The resulting functional domains defined in the EIA-852 standard to implement a CN protocol tunnel over an IP-based network can be summarized as follows:

Assuring correct and efficient data communication: CN packets must be encapsulated and decapsulated, packet ordering must be ensured, message overhead must be minimized, timeliness must be checked, and optional security measures can be used.

Routing CN payload to their correct destination on the IP network: The EIA-852 standard defines two ways to choose destination CN devices, either by using IP multicasting, or by performing selective forwarding.

Managing the IP channel members: This includes configuring the CN devices, distributing information between the CN devices, performing access control on the IP channel, and retrieving statistic information.

53.3   System Components

In addition to assigning logical addresses to the network node as specified in the control network protocol, network nodes must get an IP address assigned. Hence, IP-based network nodes now have a logical CN address and an IP address. In order for a node to transmit data packets to another node on the network, it must not only know the CN address but also the IP address of the target node. Hence, an instance is required that manages the relationship between the CN address and the IP address for a logical IP channel. This logical IP channel is called CN over IP or short IP channel and the instance that is managing the address information for the IP channel is called configuration server or short CS. Figure 53.2 outlines the systems components required to set up and manage a IP channel. The network in Figure 53.2 is split in half and consists of two traditional EIA-709 networks with subnet numbers 2 and 3 and a brand new IP channel consisting of 4 IP client devices and a configuration server CS node. Two out of the four IP client devices act as routers between the traditional EIA-709 network channels and the IP-852 channel. Let us assume that network node with S/N address 1/1 on the IP-852 channel wants to send a data packet to the network node with S/N address 1/2. Both nodes reside on the IP (IP-852) channel. In order for node 1/1 to send out a packet to node 1/2, it must know its IP address 192.168.1.102. This information is maintained in the configuration server CS hosted on IP node 192.168.1.105 and distributed to all nodes on the IP-852 channel at system startup. From now on, node 1/1 knows the relationships between S/N address and IP addresses for each node on the IP-852 channel. In our example, node 1/1 will generate a UDP packet, which is sent to node 1/2 at IP address 192.168.1.102. If this packet was sent using acknowledged service, then node 1/2 would send an Acknowledgement packet to node 1/1 at IP address 192.168.1.101 like it would do on a traditional EIA-709 channel except that node 1/2 must now know that the packet must be addressed to the IP address of node 1/1, which is 192.168.1.101. To summarize the above, every node that joins the IP channel must send out its so-called channel routing packet to the configuration server CS, which defines its relationship between CN protocol addresses and IP addresses. Whenever this information changes for a client node on the IP channel, this client node must create a new channel routing packet and send it to the CS, which in turn distributes the new information to all other members on the IP channel.

CN router devices can bridge the gap between traditional EIA-709 networks and IP channels as shown in Figure 53.2. If, for example, node 1 on Subnet 2 wants to send a message to node 1 on Subnet 3, then the router with IP address 192.168.1.103 forwards the packet to the router with IP address 192.168.1.104, which

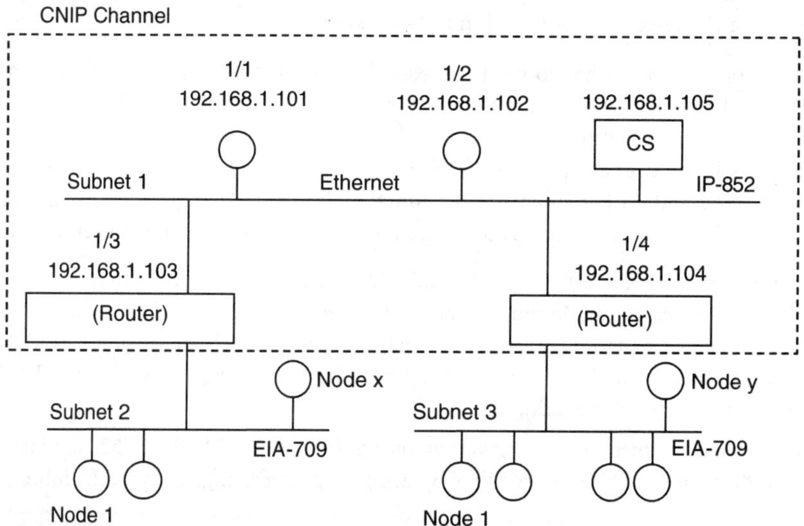

FIGURE 53.2   System components for an IP channel including the IP channel, the configuration server CS, configuration clients called nodes.

forwards the packet to the destination node 1 on Subnet 3. One can see that there is also an IP channel on the communication path between node 1 on Subnet 2 and node 1 on Subnet 3. The source node 1 on Subnet 2 does not even notice this IP channel in between the source and the destination node. As a result, it is now possible to connect different local control networks with IP channels that can either be Intranets or the Internet.

The IP channel can be built- up using different network media that are offering the IP service. Most of the time, 10BaseT or 100BaseT Ethernet networks are used for the IP channel but emerging wireless technologies like 802.11b or 802.11a and Bluetooth, POTS and ISDN lines, cable networks, and fiber optic cables are used as well. This gives existing CN protocols new application areas that could not be satisfied with the traditional technology.

53.4 Data Communication

CN devices exchange CN data packets over the IP channel. This is referred to as data communication to distinguish the traffic from management communication between the CS and the CN devices. There is always a source of a CN data packet and a sink. In the following, we shall refer to a CN device which acts as a source as the sender, to the CN device, which acts as the sink as the receiver. It shall be noted that the sender–receiver relation is only valid in the context of a single data packet or a series of data packets in the same direction. Generally, each CN device acts as a sender and a receiver on the IP channel at the same time.

CN nodes are true sources and sinks of CN data. CN/IP routers are more intermediary devices between the actual sources and sinks of the communication. This detail shall be neglected as the CN/IP routers appear on the IP channel as senders and receivers as well. The diagram in Figure 53.3 shows the functional blocks in the sender and the receiver. When a data packet is generated in the CN node or has to be passed on to the IP channel by a CN/IP router, the sender packs the CN packet in an EIA-852 data packet, adds a sequence number and a timestamp, routes the EIA-852 packet to the appropriate channel members, and finally may collect a bunch of packets before actually sending them on the wire (bunching). The receiver performs the reverse operations: it separates the aggregated packets (debunch), runs them through a sequencer, a stale packet detector, and unpacks the original CN frames. These details are explained in more detail in the following sections.

Encapsulation

The function of packing EIA-709 packets in EIA-852 data packets is referred to as encapsulating. The native EIA-709 packets are encapsulated in UDP frames and routed to the respective CN devices on the IP channel. EIA-852 uses the reserved port number 1628 for EIA-852 communication on CN devices. The configuration server uses port 1629. This makes it possible to combine a configuration server and a CN device on the same system.

The choice of UDP transport over TCP has several advantages: first, because of the connectionless nature of LonWorks communication, it would incur unnecessary overhead to manage connection setup in TCP. Second, TCP ensures delivery through retransmissions. Because of the (soft) real-time requirements

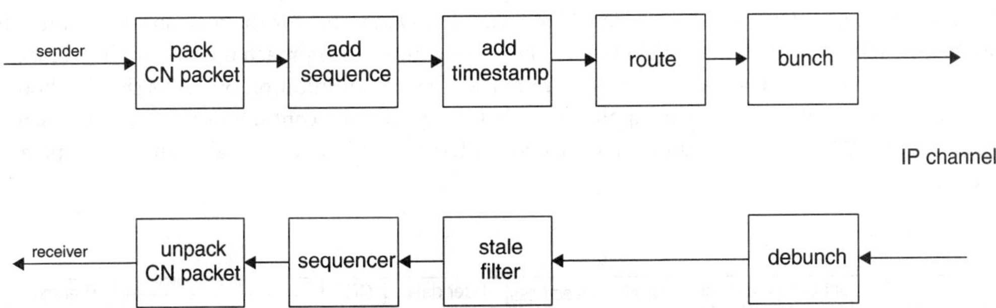

FIGURE 53.3 Functional blocks in EIA-852 data communication: sender and receiver.

of control applications, retransmission will most likely miss the deadlines. Third, the LonWorks protocol implements its own retransmission scheme where necessary.

Figure 53.4 shows the format of the native LonWorks packet format. The preamble bits and the code violation after the CRC are used for bus arbitration and end-of-frame marking on native EIA709 channels. The frame data beginning with the control field up to and including the CRC information are subject to encapsulation. Figure 53.5 shows the EIA-852 common header encapsulation format, which is the payload of a UDP frame. Data fields in EIA-852 packets are defined in network-order, transmitting the most significant byte first. The data packet length specifies the complete length of the EIA-852 packet including the two length bytes. The version number is currently set to 1. The packet type specifies how the EIA-852 payload after the common header is to be interpreted.

An EIA-852 data packet, which encapsulates the CN frame, is of type 0x01. For more packet types, refer to Table 53.1. The extended header size can be used to add more fields to the common header. It specifies them in quantities of 4 bytes. This ensures that the common header is always a multiple of four bytes. The protocol flags define as to, which CN protocol is to be tunneled, for example, 0x00 for LonWorks. The vendor code allows implementors to include their own vendor-specific extensions, which can be transported along with standard EIA-852 messages. The session ID is a 32-bit value that a CN device keeps at random. After a CN device reboots or resets, the session ID must be different from the previous one. The sequence number field is valid for EIA-852 data packets and is used to detect out-of-sequence packets. The timestamp is also used for EIA-852 data messages and is a millisecond value. Therefore, it wraps around every 49 days.

Packet Sequencing

Packet sequencing in EIA-852 is based on the pair session ID and sequence number, which are both part of the EIA-852 common header. An EIA-852 receiver remembers the session ID/sequence number pair for each sending CN device (i.e., for each source IP address). If EIA-852 data packets from a given source are received with the same session ID they are in-sequence, if their sequence numbers are increasing. If data packets get reordered or are dropped on the network, there appear gaps in the sequence numbers. A receiver basically has two choices: (i) continue and drop the missing packets later, or (ii) wait for the missing packets.

If the receiver waits, it has to keep data packets after such a gap in escrow until a defined escrow timeout (ETO). If the missing packets arrive before the timeout, the packets are passed on in their correct sequence. If they do not arrive in time, they will be dropped later. Without escrowing, all out-of-sequence packets are dropped immediately. This may cause unnecessary packet loss on IP channels where packets may get reordered in a small time window.

If the session ID changes, the receiver has to assume that the source has been reset and it accepts the first packet with that new session ID as in-sequence, clearing the sequence numbers from the previous session. This is a design choice of EIA-852 and can cause problems with replay attacks. See the Section on security for more details on this issue.

The ETO is a CN device parameter in most implementations. This time constant has to be considered when designing a control application. It is always a trade-off between unnecessarily losing reordered packets and delaying escrowed packets for too long when packets actually are lost on the IP link. If the ETO is large, more packets can be reordered, because the waiting time for them is longer. For the same reason, if a packet is actually lost on the IP link, the waiting time can approach the ETO in the worst case.

Applications that primarily rely on request/response style communication on LANs should choose a small ETO (e.g., 5 ms) or disable it. Applications that are based on a continuous unidirectional sample stream over a WAN link should choose a large enough ETO. A typical default value on most implementations is 64 ms.

| Preamble | SB | Control | Source address | Destination address | User data | CRC | CV | ◄— Bus arbitration —► | Preamble | ... |

FIGURE 53.4   Native EIA-709 frame format.

Byte 0	Byte 1	Byte 2	Byte 3
Data packet length		Version	Packet type
Extended header size	Protocol flags	Vendor code	
Session ID			
Sequence number			
Timestamp			
: Packet-specific data :			

FIGURE 53.5   EIA-852 Common header format.

TABLE 53.1   ETA-852 Packet Types

Packet name	Packet type	Function
Data Packet	0x01	Encapsulates the the CN frame.
Device Configuration Request(DC-REQ)	0x63	Request a DC from the CN device.
Device Registration (DEVREG)	0x03	Device registers with the CS.
Device Configuration (DC)	0x71	DC requested firm the device or unsolicited DC sent by the CS.
Channel List Request (REQ-CM)	0x64	Device requests CM from the CS.
Channel List (CM)	0x04	The CS responds with the CM.
Send List Request (REQ-SL)	0x66	Request the send list from the CS.
Send List (SL)	0x06	The CS responds with the send list.
Channel Routing Request (REQ-CR)	0x68	The CN device requests a CR from the CS for a specific member.
Chancel Routing Packet (CR)	0x08	The CS responds to a CR request.
Acknowledge (ACK)	0x07	Acknowledgment message from device or CS.
Segment (SEG)	0x7F	Segment of a large data structure.
Statistic Request (REQ-STAT)	0x60	Request statistics from a CN device.
Statistic Response (STAT)	0x70	CN device responds with statistic data.

Packet Aggregation

The native message sizes of typical CN traffic are small, around 10 to 20 bytes in LonWorks. Encapsulating each CN packet into a UDP frame adds the UDP, IP, and Ethernet headers (or other media specific headers), which account for more than 40 bytes. This incurs unnecessary overhead, which may lead to noticeable performance problems on embedded implementations. Those systems typically suffer from an IP transmission bottleneck.

A technique to reduce this overhead is to aggregate more than one EIA-852 data packet in a single UDP frame (packet bunching). Clearly, the more packets are bunched, the less per-CN packet overhead is added by the encapsulation. When aggregated, EIA-852 data messages are concatenated in the UDP frame. The UDP packet is then transmitted to the destination IP address. Therefore, messages can only be aggregated when they share the same destination. The sender may thus be configured to keep data packets and aggregate them until an aggregation timeout (ATO) before sending the UDP frame on the IP link. The ATO value is a CN device parameter in most implementations.

The effect of packet bunching has to be considered in the control application design. There is a trade-off between overhead reduction and extra delay, In the worst case an EIA-852 data packet is delayed for the full ATO until it finally gets on the wire. For highly responsive applications with low packet rates on LANs packet bunching should be disabled. For applications with high packet rates over WANs packet bunching should be enabled. A typical default value in most implementations is 16 ms.

Stale Packet Detection

One of the major issues on EIA-852 channels are packet delays on the IP link. They can be highly random with a large standard deviation on wide area IP links [17,18]. There are two problems with this: (i) the control application may miss some of its deadlines, and (ii) the CN protocol may break if packets are received after they are delayed too long. For example, the LonWorks duplicate detection mechanism is sensitive to high delay variance on the network. While the first problem must be solved by carefully designing the control application [10], the second problem can be eliminated by discarding packets that are delayed beyond a certain time limit (stale packets). The mechanism is called stale packet detection and is based on one-way packet delay measurements. In order to make such measurements, CN devices must synchronize their clocks. Clock synchronization in EIA-852 is performed via SNTP [21]. The sender generates a millisecond timestamp and puts it into the EIA-852 common header. If its measured delay exceeds the channel timeout (CTO), the packet is called a stale packet and the receiver discards it in the stale packet filter.

The CTO is a channel parameter and distributed among the CN devices. Applications on LANs can disable stale packet detection because the delay variance it typically very low and comparable with native LonWorks channels. On WAN links, it is recommended to enable the CTO. Typical values are the average ping delay on the IP channel plus aggregation timeout. The lower bound of the CTO is restricted by the resolution of the clock synchronization. On certain systems (e.g., Windows), the resolution may drop to 50 ms. Practical CTO values start at 200 ms. If certain CN devices are Windows implementations, it must be ensured that the Windows system clock is synchronized. Dropped packets due to stale packet detection can be observed in the ETA-852 statistics of stale packets; see Table 53.2. Stale packet filtering also plays an important role in the context of security. Refer to the Section on security for more details.

EIA-852 Data Packet Routing

One of the EIA-852 CN device tasks is to route EIA-852 data packets to the appropriate IP channel members. In this context, it is important to notice that each channel member is associated with an IP address. If a given CN data packet needs to be routed to multiple IP addresses, it is copied into several EIA-852 data packets, each sent to one of the selected CN devices on the channel. It shall be noted that sequence numbers are associated per destination address. If CN data packets are routed to different IP addresses, the sequence numbers need not necessarily be the same among all destinations. This stems from the fact that some messages are routed to a specific destination and some are not.

Depending on the available channel data, three variants of routing the CN data packets are possible:

1. *Send List Routing:* A send list (SL) is an optimized list of IP addresses to reach all channel members. Typically, SLs are used when IP multicast addresses are assigned to groups of channel members. In

TABLE 53.2 EIA-852 Statistics Data

Data	Explanation
Seconds since cleared	Number of seconds since the statistics have been reset.
Date/Time of clear (GMT)	Date and time when the statistics have been reset.
No. of members	Current number of channel members.
LT Packets received	CN packets received from LonWorks.
LT packets sent	CN packets sent to LonWorks.
IP Packets sent	CN packets sent onto the IP channel.
IP Bytes sent	Aggregate number of bytes in CN packets sent onto the IP channel.
IP Packets received	Number of CN packets received from the IP channel.
IP Bytes received	Aggregate number of bytes in CN packets received from the IP channel.
LT Stale Packets	Number of packets dropped because they are stale.
RFC Packets sent	Number of management data packets sent to the CS.
RFC Packets received	Number of management data packets received from the CS.
Avg. aggregation to IP	Average aggregation of transmitted CN packets on the IP channel.
Avg. aggregation from IP	Average aggregation of received CN packets from the IP channel.
UDP Packets sent	Total number of UDP packets sent (including both data and management).
Multicast Packets sent	Total number of multi-cast packets sent (only for send lists).

the best case, all channel members are in the same IP multicast group. In this case, the SL contains this one multicast address. CN devices send EIA-852 data packets to all destinations in the SL.

2. *Channel Routing*: Channel routing is used when only unicast IP addresses are available. The basic idea is to select those destination IP addresses that will accept the packet based on the LonWorks address information. To advertise which LonWorks addresses a CN device accepts, it publishes channel routing (CR) information to the other channel members. Each channel member collects all the available CR and can route outgoing CN data packets in an optimized way. It shall be noted that with CR, all data packets have to be copied multiple times if there are multiple advertised recipients. Especially for large groups and broadcast addresses, send lists appear much more efficient. For S/N or NID) addressed packets, CR is beneficial because the packets are routed to a single unicast address only, not burdening other CN devices with packets that they do not want.

3. *Brute Force*: Neither SL nor CR information is available. The CN device resorts to the channel list and transmits every data packet to all channel members using unicast addresses. This mode is the least efficient.

The CR algorithm is based on the CR information provided by the CN devices on the channel. The CR information for each channel member contains the following fields:

- The CN device IP address and port
- The CN broadcast flag
- The CN device type (router, node, IPIP router, proxy)
- The CN router mode (configured, bridge, repeater)
- A list of node IDs
- A list of S/N addresses
- A list of domain items (including subnet and group forwarding flags).

The domain items deserve a more thorough discussion. First, they contain a domain number. Then, they specify forwarding flags for both subnets and groups. If the nth subnet flag is set, the CN device accepts all packets destined for subnet n. If the mth group flag is set, the CN device accepts all group packets for group m. CN routers need to specify the subnet and group forwarding flags, while CN nodes only specify the group flags. This is because CN nodes do not forward any packets, but they can be members of multiple groups.

Figure 53.7 summarizes the CR algorithm implemented in most CN/IP router implementations. The CR algorithm iterates over all channel member CR information for each EIA-852 data packet. § I checks the CN broadcast flag and routes all packets if the router type is a repeater and it has at least one domain item. The CN broadcast flag is set when the channel member is unconfigured and does not yet have a valid domain. Otherwise, those devices could not be reached over the IP channel by broadcasts.

Leaving the special case of unconfigured CN devices aside, there are two types of devices on a channel: (i) CN nodes accept only CN packets addressed to them (e.g., their subnet/node address, their NID, their group), (ii) CN router, which forward packets to other channels behind them. The CR algorithm therefore distinguished those two cases (Figure 53.7).

In § II, the packet is routed by looking at the channel member as a CN router (e.g., configured router, bridge). This step only regards the domain items in the CR information. Therefore, a forwarding device must define at least one domain item in its CR information. § II(i) skips the channel routing for a certain domain item if the domain does not match. § II(ii) is responsible for forwarding group addressed messages if the corresponding group forwarding flag is set in the domain item. § II(iii) checks the destination subnet part for S/N, NID, and broadcast-addressed messages. If they are domain-wide (i.e., subnet equals "0"), the message is forwarded. Otherwise, those messages are only forwarded in § II(iv) if the corresponding subnet forwarding flag is set in the domain. item. As a consequence, CN routers must supply domain items for all domains they need to receive packets in.

In § III, the message is routed by looking at the channel member as a CN node on the IP channel. The IP channel is assigned a separate individual subnet number, and all nodes on that channel get their individual node numbers. Therefore, devices that represent nodes on the IP channel must supply at least one

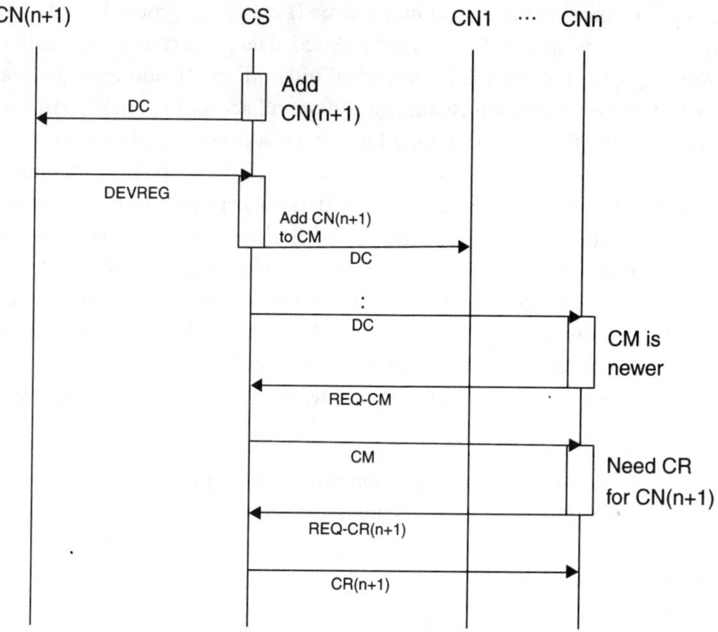

FIGURE 53.6   CS configuration data distribution.

I.   If (CN broadcast flag is set and address == FMT_BCAST) or (router type == REPEATER and there is a domain item) then route to member.

II.   Route to CN router

   (i)   If domain is incorrect skip to III.

   (ii)   If address == FMT_GROUP and group forward flag is set for GROUP route to member.

   (iii)   If SUBNET == 0 route to member.

   (iv)   If subnet forward flag is set for SUBNET route to member.

III.   Route to CN device

   (i)   If address == NID and NID matches route to member.

   (ii)   If DOMAIN does not match skip to IV.

   (iii)   If address == FMT_SN and S/N item matches route to member.

   (iv)   If address == FMT_BCAST and SUBNET == 0 route to member

   (v)   If address == FMT_BCAST and SUBNET matches S/N item route to member.

IV.   Go to next member CR information.

FIGURE 53.7   Channel routing algorithm.

S/N, one NID, and one domain entry in their CR information. § III(i) checks if an NID addressed message matches with one of the NID items in the member CR information. If the NID matches, the message is forwarded to that member not regarding the domain. Therefore, unconfigured nodes on the IP channel can be reached via NID addressed messages. For all other messages, § III(ii) checks if the domain matches with one domain item. If it does not, the algorithm skips to § IV. In § III(iii) S/N addressed messages are routed to the member if they match one of the S/N entries. § III(iv) routes all domain-wide broadcasts to a node on the IP channel. § III(v) routes all subnet broadcasts to the channel member if one of its S/N items has a matching subnet number. §IV iterates to the next member in the channel until the CR algorithm has gone through all channel members.

The CR mechanism has the following consequences for CN devices regarding the CR they must provide:

- Unconfigured channel members must set the CN broadcast flag in order to get configured by LonWorks network management tools such as LonMaker.
- CN routers (configured router, bridges, etc.) must provide corresponding domain items in the CR for all domains they want to exchange packets in.
- CN nodes must provide corresponding S/N, NID, and domain items.

53.5 Management

Apart from data communication, a large part of the EIA-852 functionality is used for the management of IP channels. Management functions include access control on which CN devices can be a member of a specific IP channel, configuration of individual CN devices, distributing information between CN devices, and retrieving statistics information about CN devices.

The EIA-852 management concept is based on a client/server model. A configuration server (CS) is the central part in the system, which is responsible for managing an IP channel. CN devices act as configuration clients (CCs) and request data structures from the CS. The CS only actively sends the device configuration (DC) to CN devices, if the IP channel has been updated and it needs to announce that change. Therefore, the DC is an unsolicited message from the CS. If no channel changes are pending, the CN devices can operate without a CS. As soon as the IP channel needs to be updated, for example, a CN device is added or recommissioned, the CS must be available again to distribute the change.

The management data structures are the per-device DC, a channel membership list (CM) specifying the members on the IP channel, an SL for SL routing that is optional, and a per-device CR information . All management data in EIA-852 are versioned. The version is a 32-bit value, which increases for newer versions. The version is usually date and time information in NTP second format, which may or may not be synchronized to UTC. Hence, the version is called datetime and is nonzero for valid data structures. Local data structures are checked against published datetime versions and requested if newer data are available. The CM, SL, and per-device DC are versioned by the CS, whereas the CR data are versioned by the individual CN devices.

The CN devices, which constitute the IP channel, are defined in the channel list on the CS. Each CN device is uniquely identified by its IP address in the channel list. An important restriction in EIA-852 is that there cannot be two CN devices on a channel, which share the same IP address but have different IP ports.

Each CN device in the channel list is configured by a DC. The DC is entered on the CS, for example, through a serial console or a web interface. The following information needs to be supplied, in order to add a CN device to an IP channel:

- IP address and port of the CN device
- Name of the CN device (typically 15 characters)

This message flow when adding or configuring a CN device is depicted in Figure 53.6. The CS adds a CN device to the channel by sending an updated and unsolicited DC message to the device. Once a CN device is added or has accepted a new configuration, it registers with the CS by sending a device registration (DEVREG). Knowing the channel is updated, the CS sends a new unsolicited DC to all the other members to announce the channel update. The other CN devices in turn request the updated data structures from the CS, starting to request the channel list (REQ-CM). If the CN device finds out that the CM advertises new CR information for certain members, it requests them from the CS (REQ-CR) one by one.

If a CN device is not available when it is added on the CS, it is usually marked as "unregistered" in the channel list. In this case, the CS tries to recontact the CN device using an exponential back-off for the retry interval. It starts at one second, doubles each time, and is limited by 32 sec. As a consequence, it may take up to 32 sec for a device to become functional on the IP channel when it was not reachable at the time of adding it. CS implementations usually display CN devices, which have registered successfully, as "registered." If a device has been registered once, but is not responding to a channel update, a CS may display it as "not responding." The operator can then take appropriate action for those devices.

Access control on the CS is also defined through the channel list. If a CN device tries to register with a CS that does not include this particular device in its channel list, the CN device is rejected from the IP channel. Rejected CN devices cannot send or receive CN packets over the IP channel and usually indicate this condition as an error, for example, a red status LED. Some CS implementations collect information about those CN devices in a so-called orphan list. The operator can then select an orphan device and add it manually to a given IP channel.

An important part of EIA-852 is the routing of CN packets to the correct CN devices over the IP network. As described in the previous section, there are two possibilities: (i) creating an SL, or (ii) distributing CR information. The CS generates the SL by looking at the IP address information in the channel list. By appropriately grouping IP multicast addresses and unicast addresses, the SL typically becomes much shorter than the channel list. The SL is also versioned and distributed to the CN devices by sending an updated DC to the members.

The other possibility is to distribute CR information. The CR, however, does not originate at the CS but at the individual CN devices because they are commissioned and only they have knowledge of their CN address information. Each CN device sends a CR update to the CS if its CR changes. The CS in turn generates a new channel list, which reflects the updated CR datetime version of the device. The other devices will request the new channel list and discover that a specific device has a new CR. The other devices then request this CR from the CS as described before. If more than one CR information has changed, the CN devices request all newer CR data from the CS. This is an important detail, because CN devices never request the new CR from the original device directly. As an implication, the CS must be available, when any CN device is recommissioned.

The management part of EIA-852 also defines a statistics information structure. This structure can be requested outside of the client/server model. Any CN device or the CS can request the statistic information from a CN device. The statistic data supplied are listed in Table 53.2.

53.6 Security

Security is a growing area of concern on IP networks. This also includes EIA-852 traffic. While native LonWorks channels are local (e.g., to a building) and can be accessed only locally, IP channels are established on an open media that may even span across WAN links. Typical security issues are privacy, authenticity, and integrity [19,20]. Security measures defined in EIA-852 refer to the problem of authenticity only. They assure that a received data packet is actually sent by an authentic channel member and not by an attacker.

The principle is based on secure message authentication codes and a shared secret. The secure hash function algorithm MD5 [22] is used to create a unique fingerprint of an EIA-852 message. A Secret key is included in the fingerprint calculation, which is not transmitted over the wire. The receiver can perform the same calculation, supplying its own copy of the secret key and compare the hash codes. If they match, the message is authentic.

The MD5 hash is transmitted right after the EIA-852 message. Its fixed length is 128 bits and is not included in the EIA-852 packet size field. Instead, channel members are configured for secure operation (MD5 mode) and expect the extra 128 bits after each EIA-852 packet. This is especially important for bunched UDP frames. If packets are bunched, the next EIA-852 message follows the MD5 hash.

The calculation principle is depicted in Figure 53.8. The MD5 hash following the EIA-852 message it set to zero (MBZ). The MD5 algorithm is run over the EIA-852 message, the zero hash placeholder, and the 128-bit shared secret. The calculated hash value is copied into the placeholder and the message is

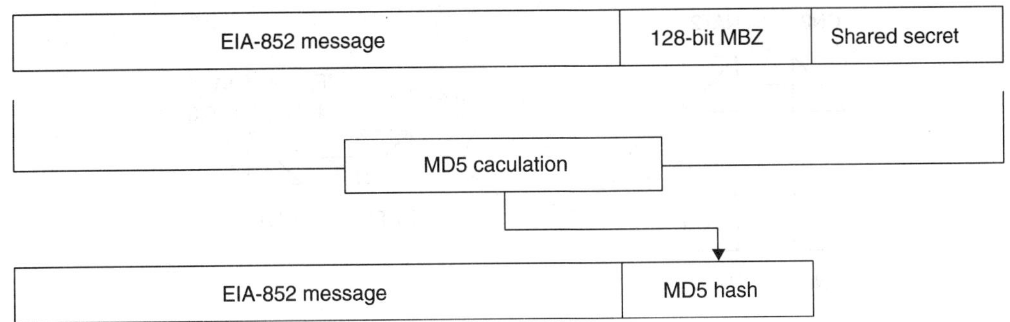

FIGURE 53.8   MD5 authentication of EIA-852 packets.

transmitted excluding the shared secret. The receiver performs the same calculation. It just saves the MD5 hash and then sets it to zero before calculating the MD5 hash locally. If the saved and the locally computed hash values match, the message contents are authentic.

To prevent replay attacks, stale packet detection must be enabled. The reason for this is the sequencing algorithm in EIA-852. An attacker may record a sequence of authenticated messages known to trigger a specific action. These messages will share a common session ID. At a later point in time, the CN device may have chosen a different session ID (e.g., after a power outage). In this moment, the recorded sequence can be replayed successfully because the CN device would accept the old session ID as a new one and start over with counting the sequence numbers. Stale packet detection solves this problem because the EIA-852 messages also include timestamps. Since the CN devices are synchronized to UTC, a playback at a later time is not possible. The CN devices would drop those packets as stale packets.

Another security aspect in modern networks are firewalls and network address translators (NAT). Firewalls in principle only filter traffic. Therefore, CN devices can be operated behind a firewall if the filter rules are set accordingly, namely allowing UDP traffic on ports 1628 and 1629 to pass.

The case with NAT routers is different. NATs actually alter IP addresses and port numbers. This is a problem for EIA-852 because channel members on the IP channel are uniquely identified by their IP address. If that address is changed in the public realm, they could not be part of the IP channel. Some implementations of CN/IP routers therefore allow for a special NAT mode (e.g., the L-IP by LOYTEC). In this mode, a CN device is configured with the public address of its NAT router. It represents this public address in its DC and other members will see its public address in the channel list and CR information. The NAT itself must be configured with a port-forwarding rule to route all EIA-852 packets from the public interface to the CN device in the private realm. Figure 53.9 illustrates this setup. NAT1 contains a port-forwarding rule to route all packets it receives at IP address 80.41.6.3 on ports 1628 and 1629 to the private IP address 192.168.1.250 on the respective ports.

The drawback of the current solution is that only one CN device per IP channel can be operated behind a NAT. This is because there would be only one public IP address visible for multiple private CN devices. This violates the rule that each CN device must have a unique IP address in the channel list. System designers have to split the IP channels if they want to operate more CN devices in the private realm. Figure 53.10 shows a possible solution where one IP channel is established over the public realm having one CN device behind each NAT. The CN device can be either an IP/IP router or two back-to-back CN/IP routers to transfer the CN packets to the private IP channel. In any case, NAT configurations require a careful and handcrafted configuration on the NAT routers.

53.7   Applications

So far, we have learned what the new IP tunneling technology following the ANSI/EIA-852 standard can do in theory. Here, we show the application scenarios that utilize this new technology. We have seen that it is possible to connect CNs via Intranets and this is exactly what is starting to emerge in larger buildings.

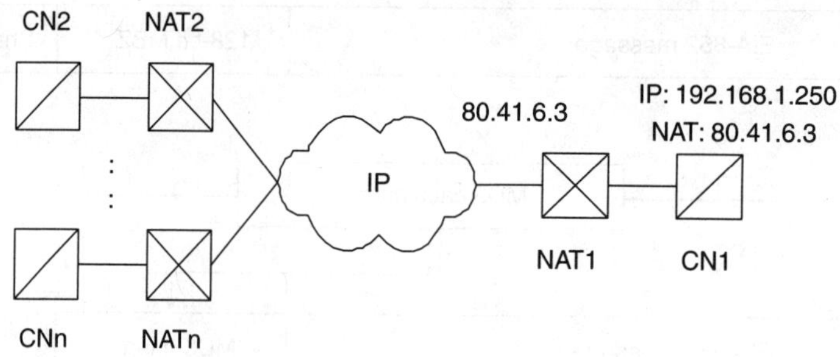

FIGURE 53.9 Simple NAT configuration.

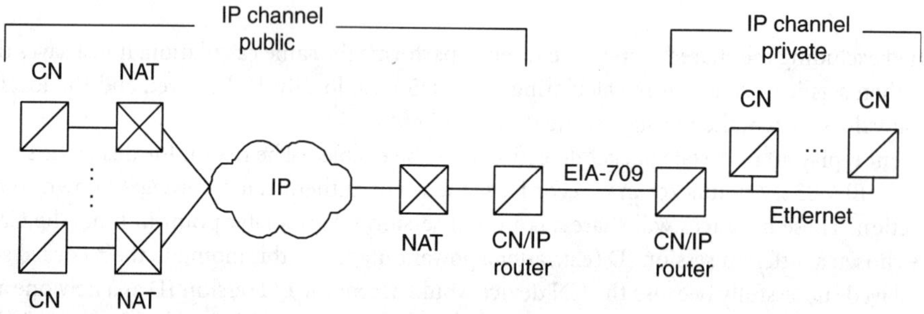

FIGURE 53.10 Complex NAT configuration.

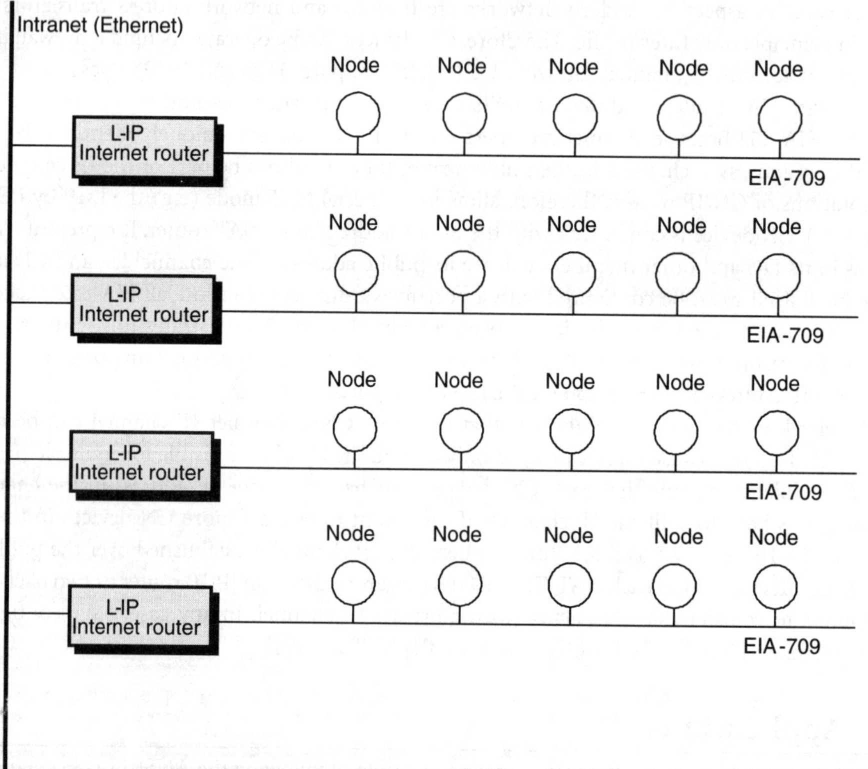

FIGURE 53.11 The IP channel can utilize the existing IP infrastructure in a building to connect several CNs to a large building automation network.

Typically, office buildings do have a sophisticated IP infrastructure; therefore, it is beneficial to utilize this IP infrastructure for services other than PC networking. As shown in Figure 53.11 CN to IP routers like the L-IP from www.loytec.com can utilize the existing IP infrastructure to form a high-speed backbone network in order to connect the control networks on the different floors. This setup allows transparent communication between offices on different floors and to a central SCADA system using the IP backbone.

Another steadily growing application area is internetworking between different buildings that are either connected via an Intranet or the Internet. Once the building infrastructure is "on-line," it is only a small step to remote facility management and remote maintenance. Figure 53.12 shows a typical scenario regarding how building complexes are networked to form one large CN. A management PC connected anywhere to the network allows remote network maintenance, remote trending, alarming, logging, and preventive system repairs.

Another advantage, which was initially not quite obvious, is the fact that network management tools and network-troubleshooting tools can be used together with the new IP technology. This means that thousands of installers and system integrators that have been trained over the past 10 years will use this knowledge with a little bit of additional training to install networks that span Intranets and the Internet.

Once the CN is able to use the IP service as a transport media, modern wireless technologies can be utilized to create new application areas. One can, for example, walk around a building with a wireless PDA and control the building infrastructure such as lighting, heating, AC, blinds, and even carry out network maintenance tasks on the spot by himself. It is now possible to monitor and control remote wastewater pump stations that are connected to the control center via an RF link.

53.8 Conclusion

The world is getting more and more connected and one of the driving factors is the Internet Protocol (IP). Connecting PCs to the Internet was the hype of the 1990s, connecting everyday devices to the

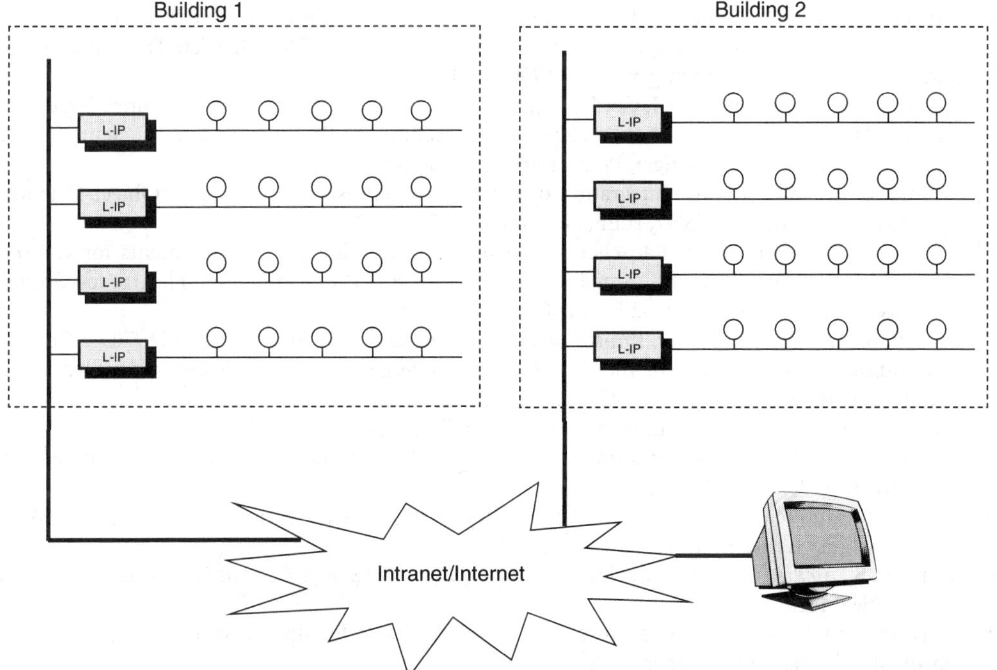

FIGURE 53.12 Connecting different buildings over an Intranet or the Internet allows transparent communication of CN data between nodes on either network.

Internet is the challenge of the new millennium. The groundwork has been done to provide technologies that will satisfy those demands. It is now a matter of time until your bathroom fan can be controlled from your cell phone. In the meantime, facility managers are concentrating on applications that are saving money, protecting environmental resources, and improving the reliability of the network.

We have shown that existing CN protocols have by far not reached the end of their lifecycle, but experience a revival by the push of the IP technology.

References

[1] Dietrich, D., D. Loy, and H.-J. Schweinzer, LON-Technologie, Verteilte Systeme in der Anwendung; 2. Auflage; Hüthig Verlag, Heidelberg, 1999.

[2] Dietrich, D., P. Neumann, and H. Schweinzer, *Fieldbus Technology — System Integration, Networking, and Engineering*, Springer, Wien, 1999.

[3] Tanenbaum, A.S., *Computer Networks*, 2nd ed., Prentice-Hall Inc., Englewood Cliffs, NJ, 1989.

[4] Tanenbaum, A.S., Distributed Operating Systems, Prentice-Hall Inc., Englewood Cliffs, NJ, 1994.

[5] Koomen, C.J., *The Design of Communicating Systems*, Kluwer Academic Publishers, Dordeecht, 1991.

[6] Kopetz, H., *Design Principles for Distributed Emebedded Applications*, Kluwer Academic Publishers, Dordeecht, 1997.

[7] Sinha, P.K., *Distributed Operating Systems*, The Institute of Electrical and Electronics Engineers, Inc., New York, 1997.

[8] EIA Standard: Tunneling Component Network Protocols Over Internet Protocol Channels. EIA/CEA-852, November 2001.

[9] Thomesse, J.-P. and M. Leon Chavez, Main Paradigms as a Basis for Current Fieldbus Concepts, *Fieldbus Technology*, Springer, Berlin, pp. 2–15,1999.

[10] Soucek, S., T. Sauter, and G. Koller, Effect of Delay Jitter on Quality of Control in EIA-852-based Networks, in Proceedings of the 29th Annual Conference of the IEEE Industrial Electronics Society, Virginia, U.S.A., Nov. 2003, pp. 1431–1436.

[11] Sauter, T. and P. Palensky, A Closer Look into Internet-Fieldbus Connectivity; Proceedings WFCS'00, IEEE, Porto, 2000.

[12] Bangemann, Th., R Dübner, and A. Neumann, Integration of Fieldbus Objects into Computer-Aided Network Facility Management Systems, Proceedings of FeT'99, Dietrich, D., Neumann, P., and Schweinzer, H. Eds., Springer, Wien, 1999, pp. 180–187.

[13] Rüping, S., H. Klugmann, K.-H. Gerdes, and S. Mirbach, Modular OPC-Server Connecting different Fieldbus systems and Internet Java Applets, Proceedings of FeT'99, Dietrich, D., Neumann, P., and Schweinzer H. Eds., Springer, Wien, 1999, pp. 240–246

[14] Neumann, P. and F. Iwanitz, Integration of Fieldbus Systemns into Distributed Object-Oriented Systems, Proceedings WFCS'97, IEEE, New York 1997, pp. S.247–S.253

[15] Soucek, S., T. Sauter, and T. Rauscher, A Scheme to Determine QoS Requirements for Control Network Data over IP, 27th Annual Conference of the IEEE Industrial Electronics Society (IECON), Denver, CO, 29 Nov.–2 Dec. 2001, pp. 153–158.

[16] Soucek, S., T. Sauter, G. Koller. Impact of QoS Parameters on Internet-Based EIA-709.1 Control Applications, In Proceedings of the 28th Annual Conference of the IEEE Industrial Electronics Society, Seville, Spain, 5–8 Nov. 2002, pp. 3176–3181.

[17] Motorola Inc., LonWorks Technology Device Data, DL159, Rev 4, Q4/97.

[18] Paxon, V. and S. Floyd, Wide-area traffic: the failure of Poisson modeling, *ACM/IEEE Transactions on Networking*, 3, 226–224, 1995.

[19] LonMark, Application Layer Interoperability Guidelines, Version 3 U.S.A., LonMark Interoperability Association, 1996.

[20] Schneier, B., *Applied Cryptography: Protocols, Algorithms, and Source Code in C*, 2nd ed., John Wiley & Sons, New York, 1995.

[21] Mills. D., Simple Network Time Protocol (SNTP) Version 4 for lPv4, IPv6 and OSI, RFC 2030, University of Delaware, October 1996.

[22] Rivest, R., The MD5 Message Digest Algorithm, RFC 1321, April 1992.

[23] Schweinzer, H., VENUS-Vienna Embedded Networking Utility Suite, Loytec electronics GmbH, www.loytec.com, October 1999.

54

Interconnection of Wireline and Wireless Fieldbusses

Jean-Dominique Decotignie
CSEM (Centre Suisse d'Electronique et de Microtechnique)

54.1 Introduction

Networking at the factory floor is now commonplace. Solutions are available at different levels: sensory, cell, plant, etc. In this chapter, we focus on the sensory level where networks usually called fieldbuses (Pleinevaux and Decotignie, 1988) link sensors and actuators to the first level of automation. This kind of network is no longer exclusive of factories and finds its way into cars, planes, and buildings.

Since the first solutions were designed in the early 1980s, a number of proposals have flourished and the field is well established both in terms of research and industrial use. Most solutions use wired transmission (twisted pairs, coaxial cables, optical fibers). Very early, the need for mobile nodes and the difficulty in installing cables have pushed for wireless solutions based on radio or light transmission. Today, most solutions use radio transmission. Due to the special properties of wireless transmission, it is not desirable to have all nodes of the fieldbus as wireless. On the other hand, the wired devices need to communicate with the wireless nodes thus prompting the necessity for interconnection means. However, the differences between the properties of wired and wireless transmissions introduce constraints in this interconnection. This is even exacerbated by the special requirements put on fieldbuses.

The chapter is organized as follows. Section 54.2 gives some definitions that will be used in the sequel of the paper. It also explains the relevant properties of wireless transmission. Section 54.3 details the different interconnecting mechanisms that can be used. Section 54.4 presents the different architectural options that are used when designing an interconnection architecture. Section 54.5 details how the various options are used in the different proposals and the associated difficulties. Section 54.6 uses the model to derive solutions that were not yet proposed, and discusses their applicability. Some conclusions and possible future work are given in the last section.

54.2 Context and definitions

As mentioned above, this chapter concentrates on field-level devices that communicate through a mix of wireless and wireline transmission. Although this is a very important topic, in this chapter, we do not consider the issue of power consumption in the wireless nodes.

Fieldbus Requirements

Requirements for fieldbuses are numerous and cover all the lifecycle activities. We restrict ourselves here to those properties that are particularly relevant for the interconnection (for a detailed explanation, the reader is referred to the chapter entitled "Which network for which application" in the present book):

- handle periodic traffic with different period durations. In many fieldbus solutions, this requirement is translated into some cyclic traffic. The real need it to be able to transport the information well before the end of the period at which the data have been sampled;
- handle sporadic traffic with bounded latency;
- allow for quasisimultaneous sampling of a number of inputs on different network nodes;
- provide an indication for temporal consistency. The fact is that control or acquisition systems expect that different sensed values correspond to sampling instants that should be within a few percent of the sampling period. The network should hence provide ways to know if a set of values exhibits this property, named relative temporal consistency. Sometimes the age (time elapsed since sampling) of a data, also called absolute temporal consistency (Kopetz, 1988), is also important to its users;
- for sporadic traffic, provide ways to know the order in which events have occurred. An application will take different decisions depending on the order in which events have occurred. As the events are potentially detected on different nodes of the network, there should be a way to find out the order;
- transfer data from one node to another or one node to a number of others; and
- rugged solutions in terms of resistance to interference, vibrations, etc.

It is important to notice the special temporal requirement, which will have a direct effect on the interconnection.

Important Radio Transmission Properties

Wireless transmissions come in two main categories: radio- and light-based systems. They both exhibit characteristics that make them very different from cable-based transmission.

For radio transmission, the main properties are as follows:

Property 1: Compared to cables, radio transmissions suffer from bit error rates (BER) that are some orders of magnitude higher. BER of 10^{-3} to 10^{-4} are usual, whereas in cables one may expect BER ranging from 10^{-7} to 10^{-9}. Error detection schemes should hence be enhanced accordingly. This is especially a concern for token-based systems because the token recovery takes quite a long time compared to the temporal constraints.

Property 2: Spatial reuse is low as spectrum is limited. This means that coexistence of several systems in the same area should be either planned (code or frequency allocation) or the medium access control should be designed in a way that takes care of the interference between systems.

Property 3: Perturbing systems can easily jam radio transmission. This is especially true in the ISM (instrument, scientific, and medical) bands. For instance, in the 2.4 GHz band, high-power medical devices are allowed. They may completely suppress all communications for long periods of time.

Property 4: Transmission distances are smaller. Typical ranges are a few tens of meters indoors and up to 300 m outdoors. Obstacles may further limit this distance.

Property 5: Collisions cannot be detected while emitting. The power of remote emitters is much lower than the power of the transmitter emission that masks the others.

Property 6: A transceiver needs a longer time (up to a few milliseconds) to switch from emission to reception and vice versa. This has to be taken into account when designing the protocol. In particular, protocols that require immediate answers to incoming requests cannot be used.

Property 7: Radio transmissions suffer from frequency-selective multipath fading. Waves may follow different paths that interfere destructively at the receiver site. Communication may be impossible at some points.

Optical waves may be used as an alternative to radio transmission. Here, we will restrict the discussion to infrared transmission as it is the most commonly used. Its special properties are as follows:

Property 8: Transmission operates only in the line of sight. An emitter and a receiver should have direct visibility. This constraint is often relaxed by using satellite-like techniques. A special device that acts as a repeater is located in a place where it "sees" all the other devices. This location is often on the ceiling of the room, thus named satellite. Communication may, however, be impossible at some points.

Property 9: Sources of heat (sun, machines, heaters, etc.) interfere with transmission and induce errors.

Property 10: Spectrum reuse is limited as all systems share the same wavelength.

As a summary, wireless transmission means exhibit properties that are significantly different from those of wires. This will have an impact on the solutions that can be used to interconnect wired and wireless nodes.

Definitions

In order to clarify the different architectural options, it is useful to define some terms:

Data circuit (ISO/IEC 7498:1: 1996) — "a common path in the physical media for OSI among two or more physical entities together with the facilities necessary in the physical layer for the transmission of bits on it."

Subnetwork (ISO/IEC 7498:1: 1996) — "an abstraction of a real subnetwork."

A real subnetwork (ISO/IEC 7498:1: 1996) — "a collection of equipments and physical media which form an autonomous whole and which can be used to interconnect real systems for the purpose of data transfers."

Data link (ISO/IEC 8802:2: 1998) — "an assembly of two or more terminal installations and the interconnecting communication channel operating according to a particular method that permit information to be exchanged. In this context, the term terminal installation does not include the data source and the data sink."

LAN — a data link using the same physical layer and medium access control protocols.

Segment — synonymous to data circuit when the nodes are connected through wires.

Cell — synonymous to data circuit but in the case of wireless medium.

54.3 Interconnection means

Generally speaking, networks may be interconnected in various ways, repeaters, bridges, routers, and gateways (Perlman, 2000).

Repeaters

Repeaters operate above the physical layer. Conventionally, they work bit-by-bit receiving the input signal, regenerating the signal, and emitting it on the other side. In the last decade, we have seen the blossoming of repeaters with multiple ports that are called hubs (Ethernet, USB). In the context of wireline to wireless (or vice versa) repeaters, this may imply changing the encoding scheme (i.e., NRZI to Manchester). However, theoretically, repeaters operate transparently to the protocols above the physical layer.

As wireless transmission is more error-prone than transmission on cables, a different kind of repeater may be designed, (Morel, 1996) namely a word repeater. Instead of repeating the incoming signal bit by bit (bit repeater), the word repeater waits until a number of bits have been received from the wired side, calculates a forward error correction code (FEC), and transmits the bits together with the error correction code. A different modulation scheme (for instance, spread spectrum) may be used to improve the spectral efficiency or the robustness. When receiving from the wireless side, the word repeater receives uses the FEC code to correct possible errors and retransmits the (possibly) corrected information bits to the wire side. This kind of repeater introduces a longer delay, although shorter than the delay of a bridge, a router, or a gateway. It reduces the error rate on the wireless side and brings it closer to the wired side bit error rate (Morel, 1996).

Some practical considerations limit the use of repeaters. Repeaters should, in principle, forward what comes from on port to the other port and vice versa. This would be possible with full duplex lines with on line for each direction of transfer. However, all popular wireline fieldbuses use a single line for both directions (half duplex). The repeater must hence switch from one direction to the other one according to the flow of data. This may be performed using additional lines but is not practical. In practice, wireline repeaters use some mechanism to sense the direction of travel (Murdock and Goldie, 1989) and autonomously switch accordingly. When an incoming signal is sensed on one side, the repeater is switched so that the signal is regenerated and emitted on the other side. The opposite happens when the signal is sensed on the other side. When signals are sensed on both sides, the repeater is put in isolation mode. This behavior is adequate as long as the medium access control does not rely on collisions. In such a case, when a signal is sensed on both sides, the repeater must emit a jamming signal on both sides so that the collision can be detected on both sides. It must also send this jamming signal on the input port when, after switching to one direction of travel, it senses a collision on the output port. A repeater is thus not completely independent of the medium access control scheme used in the overlying network.

Using radio communications on one side of the repeater does not change this picture much unless the medium access control uses contention. Remember that collisions cannot be detected directly on the wireless side (property 5). The repeater thus has no means to detect the collision and propagate it on the other side (assumed to be wired). This precludes the use of protocols exclusively based on collisions. Property 6 will also lengthen the delay incurred in the repeater.

It is important to notice that, if the bit rate is different on both sides of a repeater, the repeater must buffer the information. As repeaters are usually bi-directional, the buffering policy will differ from one direction to the other. Let us assume that the bit rate is higher on side A than on side B. As soon as information is received from side A, emission may start on side B. However, bits are emitted more slowly on side B and the repeater must buffer the incoming bits before they can be emitted on side B. Conversely, when something is received from side B, the repeater cannot start relaying the information on side A immediately. It must wait until a sufficient number of bits has been received to emit the complete packet at the A side bit rate. This means that the repeater must know the maximum size of a packet.

Bridges

A bridge is a data link layer relay. A bridge receives a complete MAC or LLC frame, checks it, and possibly forwards it on the other side. Contrary to repeaters, bridges filter the information that is relayed from one LAN to another. The different types of interconnected LANs lead to various categories of bridges (Varghese and Perlman, 1990):

- passthrough bridges can be used when the LANs on both sides offer identical data link layer functions and addressing. The frames can then be passed unchanged. Ethernet Switches are a recent example of this category of bridges;
- translation bridges are used when both LANs have data link-layer function and addressing that are sufficiently similar to allow a direct translation of data link-layer PDUs;
- encapsulation (or tunneling) bridges may be used when the translation is not possible. An incoming frame is encapsulated in the data link layer format on the other link before being forwarded. This is the kind of bridge used when two, or more, LANs of identical technology need to be interconnected through another one of a different kind.

A bridge participates as a node in each of the LANs it interconnects. It receives a copy of all frames transmitted on each one. Obviously, not all frames need to be relayed. For instance, a frame received on side A, whose destination is also on side A, does not need to be relayed on side B. The bridge thus has to learn which frames should be relayed and which ones should not. In particular, special care should be taken to avoid frame looping when several paths exist between two nodes that communicate (ISO/IEC 7498:1: 1996).

Routers

Routers operate at the network layer level. The main difference between bridges and routers is that the latter are not transparent. Routers modify the packets they forward, in particular, their address fields. Routers exchange information between themselves in order to find a route on which the packet is conveyed. They can thus find an optimum path between two nodes, whereas bridges only use a subset of the available topology. As the network layer is most of the time absent in fieldbuses, we will not deal with this case further.

Gateways

Gateways relay the information on top of the application layer. When a gateway node receives some application service indication, it converts it in a serviced request on the other LAN on which it is connected. When the corresponding confirmation is received, it transforms it in a reply on the other side. Depending on the available services, an invocation may correspond to more than one request. Similarly, the reply may be constructed from a number of confirmations.

In order to overcome the additional delay in the reply to an indication on the LAN on one side, some gateway may make requests on the other side in advance. This is especially true in fieldbuses where the gateway may keep an image of all inputs on one side so that, when a read indication is received on the other side, the cached value is returned in the response. We shall refer this kind of gateways as proxy gateways.

54.4 Major Design Alternatives

If the interconnection mean is the prime choice, there are other degrees of freedom in the architecture of a mixed wired/wireless network.

Single vs. multiple wireline segments: The wireline portion of the system may be composed of a single fieldbus segment on which all the devices are hooked. It may be made of several segments that need to be interconnected.

Single vs. multiple wireless cells: As in the previous case, the nodes that are connected through wireless means may be organized in a single cell or multiple cells.

Separate or integrated wireless and wireline subnetworks: In the first case, the various nodes connected through wireless means form a single subnetwork that is interconnected to the wireline network at some points. This is very similar to the integration of the GSM network and the telephone system. The other solution is to have one or more subnetworks in which wireless and wired nodes are mixed.

Single point or multiple points of interconnection: The interconnection between wireless and wireline transmission can be done at a single point or at a number of points.

Ad-hoc, single base station or multiple base stations for the wireless subnetwork: In the first case, all nodes cooperate without any one playing a role different from the others. In the second option, a single node acts as a coordinator for the traffic between the various wireless nodes. A third one is to have a number of these stations. Another option is whether the base station is the node that serves as an interconnection point.

Wireless networks vary along other parameters such as:

Total or partial interconnection: All wireless nodes can see each other when total interconnection is achieved.

Multiple hops or single hop: In case of multiple hops, traffic from a node addressed to a node that cannot be seen (beyond transmission range) is routed by other nodes to the destination.

Satellite: The absence of full visibility can be compensated by a satellite node that retransmits all the emissions from the nodes at a different frequency. This is equivalent to a network of mobile nodes using a satellite to relay the communications.

However, we believe that these additional parameters are not prime parameters in an architectural model. We shall not consider them further.

Although a given interconnection solution is a combination of these options, not all combinations lead to a feasible system. For instance, a wireless cell (data circuit) and a wired segment cannot be interconnected through more than a single repeater per cell. In the next paragraph, we will detail and discuss a number of proposals. When available, relevant references will be mentioned and commented.

54.5   Solutions for the Interconnection

In this paragraph, we explore the design space and look at the feasibility of various solutions. The practicality of each solution is discussed and reference to existing work is given when available. Solutions based on routers will not be explored as most of the fieldbuses do not offer any network layer, thus excluding this solution. The only noticeable exception is LON (EIA-709.1 1998), which has a 7-layer stack and provides routing between wired segments and wireless nodes.

Repeater-Based Solutions

A solution based on a repeater gives the impression that all nodes share the same medium. The same medium access control scheme must be used on the wireless part and on the wired part. This kind of interconnection has a number of advantages. There is no need for a base station in the wireless cells. It is easier to ensure periodicity because a single MAC scheme is used. The same applies to the guarantees for the latency bounds. However, the latency may increase as an effect of a higher number of nodes sharing the same medium. Furthermore, the higher BER on the wireless part has a direct impact on the frame error rate on the network. This causes higher latency bounds because of the necessary retransmissions. Finally, repeaters introduce some restrictions in the design freedom. Between a segment and another or a wireless, only a single repeater can be used. Furthermore, loops are excluded from the topology (for instance, S6 could not be a repeater in Figure 54.2). This precludes using multiple interconnection points.

The properties of wireless communications impact all the possible solutions in the same way. Assuming that the same bit rate is used on both sides of the repeaters, properties 1, 3, and 7 will limit the throughput and increase the latency. Property 5 imposes restrictions on the type of MAC protocol. All protocols that need collision detection cannot be used directly. Finally, property 6 will increase the latency and the response time for request response protocols.

According to the degrees of freedom explained above, several solutions are possible:

RPS1: single wireless cell and single wired segment: This is an interesting solution to interconnect a number of isolated wireless nodes to a single segment. An example of this solution is given in Morel et al. (1996) (Figure 54.1). The solution can support multiple wireless cells as long as the wireless nodes in two different cells do not receive the radio signals of each other.

RPS2: single wireless cell and multiple wired segments: When there is no isolated wireless node, this solution can be used to interconnect two or more wired distant segments together when no cable can be installed in between.

RPS3: multiple wireless cells and single wired segment: The wired segment may be considered as a backbone for the wireless cells. Alternately, the cells may form a single subnetwork (using wireless to wireless repeaters) connected to the wired segment through a single repeater. For this, the wireless cells must overlap and it is necessary to use a different channel for each cell. When wireless nodes may move and traverse different cells, they have to change the frequency of emission and reception. To trigger this process, special messages are sent regularly by a given node of the network. A mobile station that does not receive such a message starts to listen on other channels until it receives one. More sophisticated procedures are also used, in particular, to reduce the time necessary to find the right channel (Rauchhaupt, 2002).

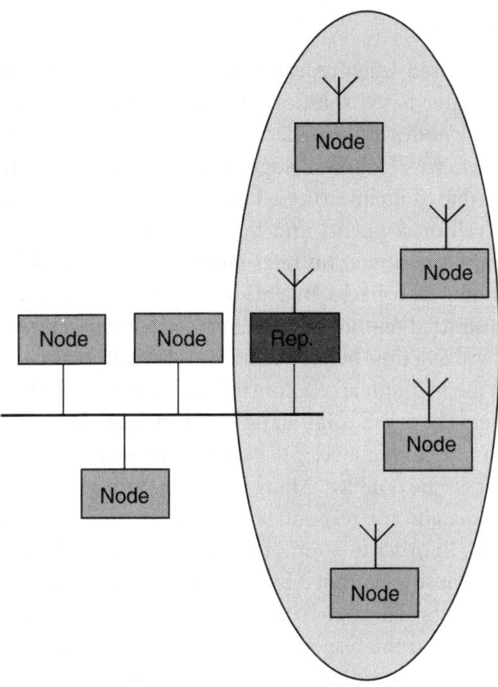

FIGURE 54.1 Repeater-based single segment multiple cell architecture.

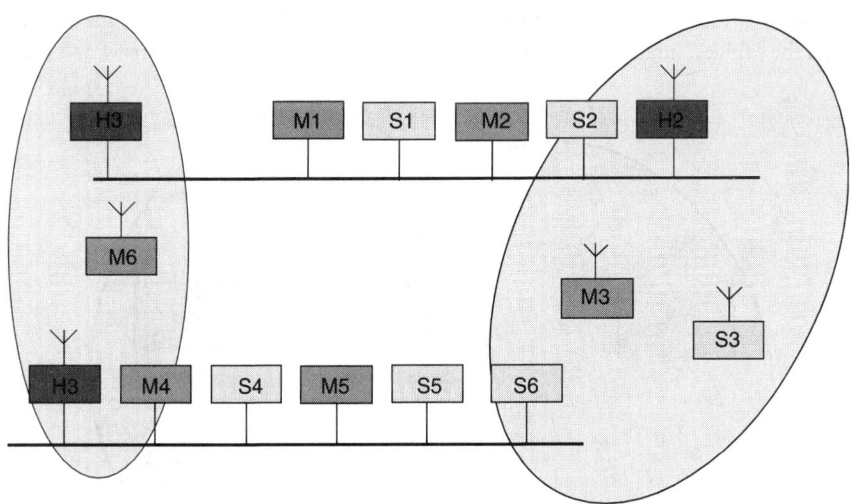

FIGURE 54.2 R-Fieldbus architecture (M: master station, S: slave station, H: repeater).

RPS4: multiple wireless cells and multiple wired segments: This solution is basically a generalization of the previous one. The integrated wireless fiend wireline case is the basis of the R-Fieldbus approach (Rauchhaupt, 2002)(Figure 54.2). R-Fieldbus uses a token bus medium access control and is thus very sensitive to token losses. As tokens go through the wireless cells, it has been necessary to reduce the BER on the wireless part. This has been done using a different encoding, direct sequence spread spectrum, and special countermeasures on the receiver side (Rake receiver). As the bit rate is not the same on the wired part (1.5 Mbit/sec) and on the wireless part (2 Mbit/sec), buffering is necessary as explained above.

Bridge-Based Solutions

Bridge-based solutions may be used when repeaters cannot be used. For instance, in the presence of a large number of nodes, a repeater-based solution may lead to unacceptably long periods or latencies. Using a bridge will partition the nodes into two data links. The number of nodes than will share each link will hence be reduced. The latency will follow. Bridge-based solutions may offer a significant reduction in latency compared to repeater-based architectures. This is particularly the case when most of the traffic remains on each data link (cell or segment) and the traffic that goes through the bridge is minimal. Bridges also remove the need to use similar bit rates on the wired part and on the wireless one.

They are, however, not without drawbacks. In a bridge-based solution, a frame emitted on the segment (or cell) on one side is retransmitted on the other using the medium access control protocol of the latter. There is hence an additional waiting time before the frame is actually emitted on the second segment. The maximum delay depends on the medium access control protocol used. With a token passing protocol, the maximum delay is at minimum the token rotation time. If a master–slave configuration is used, the maximum delay is the maximum time that may elapse between any two successive polls.

In many cases, this delay has to be doubled. Many fieldbuses proceed by transactions in which one station, the initiator, sends a frame and the responder replies with another frame that can be an acknowledgment or contain some data. In order to bound the transfer latency, the delay that may elapse between the initiator frame and the response if bounded by a upper limit. This is also used to set a timer that will trigger a retry if it times out. The bridge introduces an additional delay in the initiator frame as well as in the response when the initiator and the responder are on different segments or cells. These two delays must be added to the timeout limit, accordingly increasing the maximum latency of the fieldbus. In other words, bridges have a negative impact on the real-time guarantees.

The properties of wireless communication have a lower impact than on repeater-based solutions. Property 1 may be mitigated using forward error correction on the wireless cell. Using error detection and consecutive retries would not be a good solution because this would further increase the maximum additional delay in the bridges. Imagine, for instance, the architecture depicted in Figure 54.3. A station on the

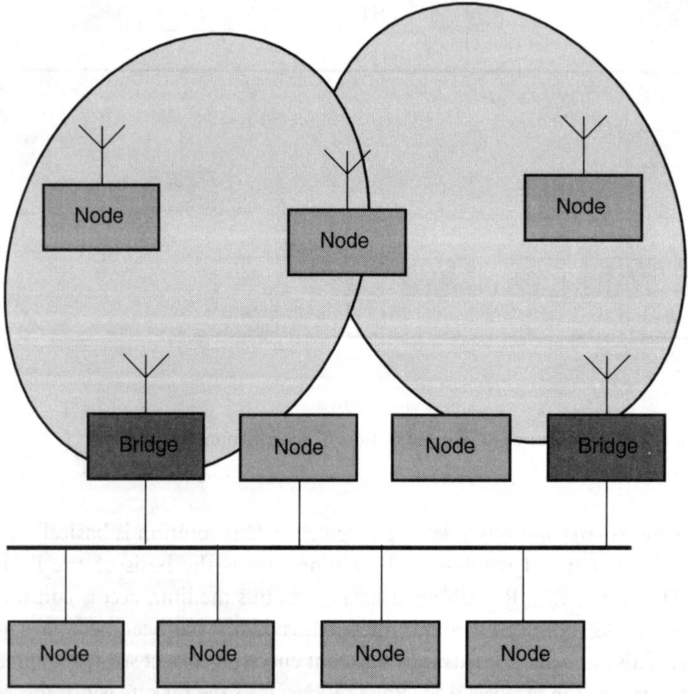

FIGURE 54.3   Isolated node approach.

wired segment initiates a transaction. The responder is a wireless node. One of the bridges retransmits the request over the corresponding wireless cell after some medium access delay. The responder replies before the maximum response delay. The response is not received correctly by the bridge that retries, thus extending the maximum delay before the response is sent to the initiator. The same happens if the reply is lost as a result of properties 7 or 9.

According to the degrees of freedom explained above, several solutions are possible:

BRS1: single wireless cell and single wired segment: The bridge may be used as the base station for the wireless cell. In the office world, this is typically what is obtained using an 802.11 cell coordinated by an access point connected to an 802.3 segment. In the fieldbus domain, this architecture would be a alternative to RPS1 that relaxes the constraints of the repeater but is likely to increase the achievable latencies and periods.

BRS2: single wireless cell and multiple wired segments: This solution has been suggested to interconnect two fieldbuses that cannot be linked through wires (Cavalieri and Panno, 1998). This is depicted in Figure 54.4. In the proposed solutions, both segments use the same protocols and the connection is made through tunneling bridges. The solution offers a better reliability than solution RPS2 because the wireless part may be rugged. Buchholz (Buchholz et al., 1991) reports a similar approach in the in-building context. The system is composed of a number of Ethernet segments that are interconnected through a number of wireless cells. The options are identical, with two exceptions. There may be more than a single wireless cell. Due to the higher bit error rate, each Ethernet frame is segmented before it is encapsulated in an LLC frame of the wireless protocol.

BRS3: multiple wireless cells and single wired segment: This architecture is of two types. Each cell may use a different channel as presented by Leung (1992). In this approach, a number of wireless nodes are connected to a wired segment through a number of bridges. The set of wired nodes (including the bridges) use a protocol different from the wireless protocol but the functions of the data link layer are identical and the addressing is common to all nodes whether wireless or wired. The bridges are thus of the transparent or translating categories. To avoid duplicate transmission, bridges have to learn the topology so as to decide whether to forward a received frame or not.

A second option is to use the same channel for all cells and rely on time or code to separate the traffic.

BRS4: multiple wireless cells and multiple wired segments: Contrary to RPS4, in this solution, there may exist more than a single path between any two nodes. Bridges protocols will ensure that no message looping will occur.

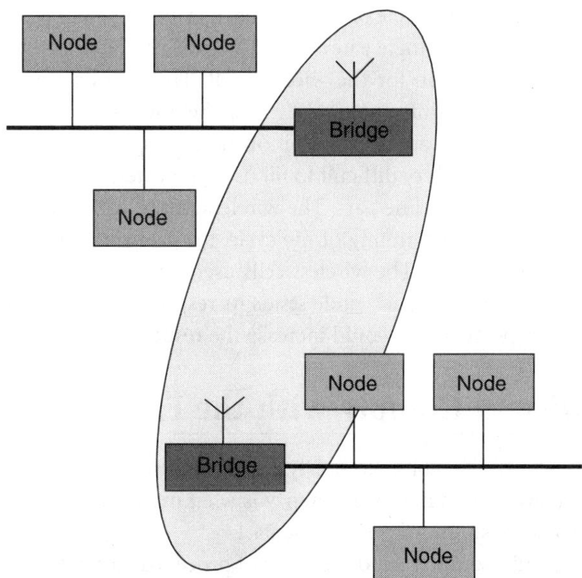

FIGURE 54.4   Interconnection of two wired fieldbuses through a wireless cell using tunneling bridges.

Bridges are sometimes interesting solutions either when repeaters cannot be used or to partition the network into smaller domains. Note that, in general, a bridge requires that the upper layers (network to application) be identical on both sides.

Gateway-Based Solutions

With gateways, operation of each segment or cell may be governed by completely different protocols at all OSI (Open System Interconnection) levels. This is useful when the wireline segment and the wireless cell are built around protocols that are not compatible at the data link layer precluding the deployment of bridges. For instance, a protocol based on the client server model such as Profibus (CENELEC EN 50170, 1996a) will not include the same information at the data link layer than a protocol based on producer consumer on top of broadcast source addressing such as worldFIP (CENELEC EN 50170, 1996b). The first one will include the address of the source node and the address of the destination node. The second will only include the identification of the data. It is clearly impossible to use a bridge in such a case. It is also necessary when the application layers differ on each side. A wireless worldFIP (Roberts, 1993) (Solvie, 1994) on one side and a wired CAN Open (CENELEC EN 50325-4, 2002) on the other, despite their compatible medium access control frames, cannot be bridged.

As both networks operate asynchronously, in terms of latency, gateway-based solutions suffer from the same drawbacks as the bridge-based solutions. The latency is even slightly higher because of the additional overhead introduced by higher protocols layers. Except in special cases (see below), the worst-case latencies (worst case response time) may be estimated as the sum of the values for the individual networks traversed increased by the penalty introduced by the gateways themselves. This may or may not be worse than using bridges, although as a rule of thumb, worse is more likely. Let us give a case in which the worst-case latency is better. In a bridge-based solution, the timeout length for transactions will be increased to take into account the longer response times of nodes that are on another data link (segment or cell). If we take into account these new values for timeout and the higher probability of loss due to the wireless part, the worst-case latency may be increased significantly. Basing the approach on a gateway means that the original values for timeouts are preserved. The higher probability is accounted on the wireless part (not on all transactions). The result is that the bridge-based solution will give a higher worst-case latency than the solution using a gateway.

An interesting solution to improve the worst-case latency or decrease the sampling period for process values is to have a gateway that acts as a proxy. The gateway responds on one of its side as it was a node of the network on its other side. An example of this approach is illustrated in (Morel and Croisier, 1995). A wireless cell is connected through a single gateway to a WorldFIP (CENELEC EN 50170, 1996b) fieldbus. The gateway acts as a base station for the wireless cell. The gateway approach has been selected because of the strict temporal constraints in WorldFIP. This fieldbus requires that a maximum of 70 bit times separate the last bit of a request and the first bit of the corresponding response. This requirement cannot be fulfilled using bridges. It is very difficult to fill using repeaters. The gateway acts as a proxy representing the wireless nodes to the wireline part. The wireless cell is exploited in a round-robin manner according to a constant cycle. At the beginning of the cycle, the gateway broadcasts to all wireless nodes their update values in a single message. The wireless cells use this message as a sampling order, capture the values, and prepare their responses. Each node sends its response after the previous one. No explicit polling is done because of property 6 that would increase the response time (Figure 54.5).

54.6 Amenability to Comply with the Fieldbus Requirements

Table 54.1 gives a summary of the capability to satisfy the fieldbus requirements (see beginning) with the different architectures. For events and messages, latency is what matters. This issue has been discussed at length above, and the results are summarized in Table 54.1.

The objective to have periodic sampling and exchange of process data can be easily satisfied using some broadcast mechanism to sample and an adequate control of the medium access control (Fonseca and Almeida, 1999) when repeaters are used. With bridges and gateways, the same level of control is possible

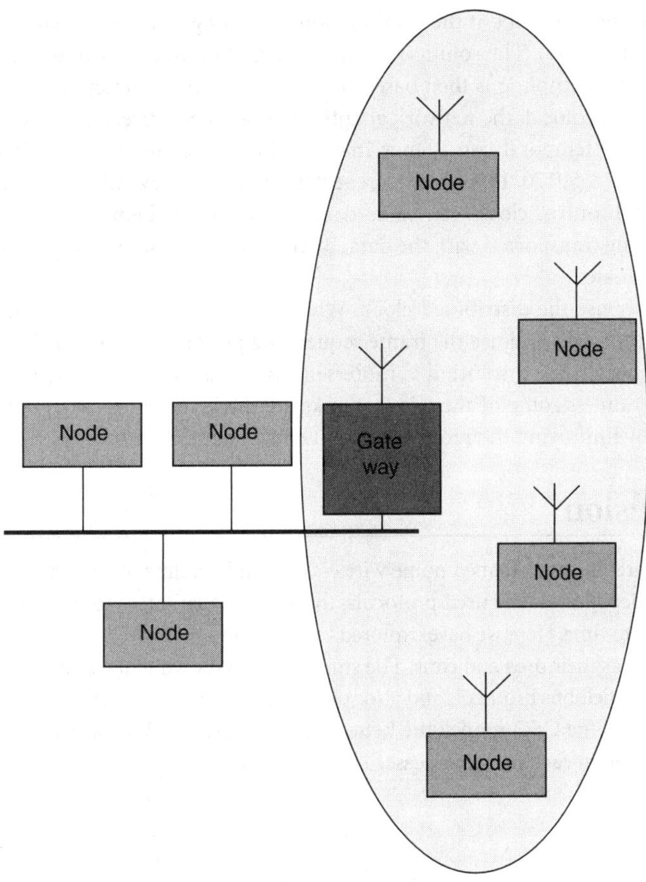

FIGURE 54.5 Interconnection using a proxy gateway.

TABLE 54.1 Repeater Ability to Help in Fulfilling the Fieldbus Requirements

Field bus requirement	Degree of fulfilment using		
	Repeaters	Bridges	Gateways
Bounded latency	Easier to prove if MAC is adequate. Longer delay due to higher number of station in the same link	Possible but more complex to calculate than with repeaters	Possible, sum of figures for each individual network
Periodicity	Easier as the same MAC scheme is used	Difficult to achieve unless synchronized clocks are used	Difficult to achieve unless synchronized clocks are used
Simultaneous sampling	Easily implemented through broadcast	Requires synchronized clocks	Requires synchronized clocks
Consistency indication	Simple to implement (i.e. CENELEC EN 50170 1996)	May be obtained using time stamps	May be obtained using time stamps
Event ordering	May be based on frame ordering	Requires synchronized clocks	Requires synchronized clocks
Rugged	Higher BER in wireless part will increase overall frame error rate and latency	Better than repeater because wireless part may use a different protocol	Similar to bridge based case

in each individual segment or cell but the combination does not give the required periodicity and simultaneous sampling (Saba, 1995). The solution lies in the definition of a distributed clock synchronization algorithm. Simultaneous sampling is then based on this clock. The accuracy is highly dependent on the medium access control protocol, the network adapter interface hardware, and software implementation.

Absolute and relative temporal consistency may be obtained easily using mechanisms like those of WorldFIP (CENELEC EN 50170, 1996b) when repeaters are used. They fail when bridges and routers are used. Here again, synchronized clocks can be used, each information being stamped with its production instant, the stamp being transported with the data. By comparing the stamp values, the two kinds of consistencies may be evaluated.

Event ordering may use the distributed clock. When repeaters are used, one can save the burden of implementing such protocols and use the frame sequencing provided the required granularity is not too small. Finally, the repeater-based solution is more sensitive to the higher bit error and frame drop rate (properties 1, 3, 7-9) unless some of the wireless links are made more robust by some form of forward error correction or by improving the receiver design (Rauchhaupt, 2002).

54.7 Conclusion

There are numerous architectures that combine wireless cells and wireline segments. The choice of a solution for a given problem depends on the used protocols, the kind of guarantees to offer, and the constraints, in particular with regard to time. Here,we have explored the solution space and discussed the applicability of the various options as well as their pros and cons. The study has also shown that some architectural choices may not be offered for some fieldbus protocols and also some combinations of protocols. In general, the lower the interconnection mean in the OSO model, the better the performances. We have seen that this general rule must be carefully verified in each particular case. This can be verified in a few cited examples.

References

Buchholz, D. et al., Wireless in-building network architecture and protocols, *IEEE Network Magazine* 5, 31–38, 1991.

Cavalieri, S. and Panno, D., On the Integration of FieldBus Traffic within IEEE 802.11 Wireless LAN, Proceedings of the 1997 IEEE International Workshop on Factory Communication Systems, pp. 131–138, 1997.

Cavalieri S. and Panno D., A novel solution to interconnect FieldBus systems using IEEE wireless LAN technology, *Computer Standards and Interfaces* 20, 9–23, 1998.

CENELEC EN 50170, General Purpose Field Communication System, Vol.2/3 (Profibus), 1996a.

CENELEC EN 50170, General Purpose Field Communication System, Vol.3/3 (WorldFIP), 1996b.

CENELEC EN 50325-4, Industrial Communications Subsystem based on ISO 11898 (CAN) for Controller-Device Interfaces — Part 4: CANopen, CENELEC, 2002.

EIA-709.1, Control Network Specification, March 1998.

Fonseca, J.A. and Almeida, L.M, Using a Planning Scheduler in the CAN Network, 7th IEEE International Conference on Emerging Technologies and Factory Automation, Vol.2, 18–21 Oct. 1999, pp. 815–821, 1999.

IEC 61158-2 (2000-08) Fieldbus Standard for Use in Industrial Control Systems - Part 2: Physical Layer Specification and Service Definition.

ISO/IEC 7498:1: 1996, Information Processing Systems — Open Systems Interconnection — Basic Reference Model: the Basic Model, 1996.

ISO/IEC 8802.2: 1998, Information Technology — Telecommunications and Information Exchange Between Systems — Local and Metropolitan Area Networks — Specific Requirements. Part 2: Logical Link Control, 1998.

Kopetz, H. Consistency Constraints in Distributed Real Time Systems, Proceedings of the 8th IFAC Workshop on Distributed Computer Control Systems, Vitznau, Sept. 13–15, pp. 29–34, 1988.

Leung, V., Diversity interconnection of wireless terminals to local area networks via radio bridges, *Electronics Letters* 28, 489–490, 1992.

Morel, Ph., Intégration d'une liaison radio dans un réseau industriel, PhD thesis, 1571, Swiss Federal Institute of Technology (EPFL), Lausanne, 1996.

Morel, Ph. and Croisier, A., A Wireless Gateway for Fieldbus, Sixth IEEE International Symposium on Personal, Indoor and Mobile Radio Communications PIMRC'95, Vol.1, pp. 105–109, 1995.

Morel, Ph., Croisier, A., and Decotignie, J.D., Requirements for Wireless Extensions of a FIP Fieldbus, Proceedings of 1996 IEEE Conference on Emerging Technologies and Factory Automation EFTA'96, Vol.1, pp. 116–122, 1996.

Murdock,-G. and Goldie,-J., Build a direction-sensing bidirectional repeater, *Electronic-Design*, 11 May 1989; 37: 105–8, 110, 1989.

Perlman R., *Interconnections — Bridges and Routers*, 2nd ed., Addison-Wesley, Reading, MA, 2000.

Peter M., The Use of Radio Technology in the Fieldbus Area – Using Interbus as an Example, Proceedings of the FeT'99, Magdeburg, pp. 55–60, September 23–24, 1999.

Pleinevaux, P. and Decotignie, J.D., Time critical communication networks: field buses, *IEEE Network Magazine*, 2, 55–63, 1988.

Rauchhaupt, L., System and Device Architecture of a Radio Based Fieldbus — the R-fieldbus System, Proceeding of the 2002 IEEE International Workshop on Factory Communication Systems, pp. 185–192, 2002.

Rauchhaupt, L. and Hähniche, J., Opportunities and Problems of Wireless Fieldbus Extensions, Proceedings of the FeT'99, Magdeburg, pp. 48–54, September 23–24, 1999.

Roberts, D., 'OLCHFA' a Distributed Time-critical Fieldbus, IEE Colloquium on Safety Critical Distributed Systems, pp. 6/1–6/3, 1993.

Saba, G., Mammeri, Z., and Thomesse, J.P., Some solutions for FIP Networks Interconnection, Proceedings of the WFCS'95, Leysin, pp. 13–20, Oct. 1995.

Solvie, M., Configuration of Distributed Time-critical Fieldbus Systems, Proceedings of 2nd International Workshop on Configurable Distributed Systems, p. 211, 1994.

Varghese, G. and Perlman, R., Transparent interconnection of incompatible local area networks using bridges, IEEE Journal on Selected Areas in Communications 8, 42–48, 1990.

55

Security in Automation Networks

Christian Schwaiger
Austria Card Gmbh, Research &
Development

55.1   Introduction

During the last 20 years, the main focus in the development of field area networks (FANs or fieldbus systems) was on meeting the technical requirements of different application areas, which led to a wide variety of different systems. The accompanying standardization efforts (that still continue) led to the —widely accepted today — conclusion that in order to successfully cope with the requirements found in the application areas of interest, different co-existing FAN solutions are needed [1].

At the same time, the technologies that make up today's Internet were conceived, which ultimately led to the enormous surge in the use of the Internet in the 1990s together with the wide adoption of the IP protocol family for use in LANs. The ubiquitousness of the Internet sparked new interest in vertical communication flows according to the CIM pyramid and led to different solutions to connect FANs to the Internet either via a tunneling approach or through a gateway where different solutions based on either Web technologies such as HTTP, Java, and XML [2,3] or on higher-level protocols such as SNMP or LDAP [4,5] can be found.

While it is not clear today as to which is the best means (if one exists) to connect a FAN to the Internet, the notoriously bad security reputation of the Internet, which in its original form provided no security at all, mandates an awareness in FAN security issues. In this respect, it is necessary not to limit oneself to the study of security issues that arise in the connection of a FAN to the Internet. Rather, a more general approach has to be chosen, which includes security measures on the FAN level itself, where appropriate. Clearly, the final goal has to be the integration of security characteristics to the list of properties (like topology, structure, bus access, safety features, and others) that is ultimately used to decide which FAN to choose to realize a task at hand.

The rest of the article is structured in the following way: Section 55.2 introduces the most basic security measures that are applicable to a wide range of systems. Section 55.3 gives an overview of where security measures can be applied at the FAN level. Section 55.4 covers remote security for connections to a FAN and presents an exemplary solution for a FAN–Internet gateway, which might today be the security topic that is of the most interest to the FAN community. Finally, section 55.5 points to several topics that are of immediate interest if secure FAN technology should become a reality.

55.2   Basic Security Measures

Security measures for IT systems in any organization aim at the achievement of three basic security goals, namely *confidentiality, integrity,* and *availability* (CIA), which protect data from unauthorized entities and unauthorized manipulation and ensure that data are accessible when needed, respectively. Another security goal that is often desired is *nonrepudiation,* which binds an entity to a commitment it made earlier. Solutions that are implemented to achieve these goals have to be based on a *security policy* that typically:

- explicitly states the security objectives and the scope of the policy and
- states who is responsible for implementing the policy.

Any security policy has to be backed by the management and communicated to the users of the IT system in question. A good informal starting point for developing a security policy is the "Site Security Handbook" [6], which is geared toward systems that are connected to the Internet. Another good introduction is the "IT Baseline Protection Manual" [7], which aids in the establishment of a security policy and additionally offers common best practices for the implementation of security measures similar to the British Standard 7799, which was adopted as ISO/IEC 17799 [8]. Unfortunately, no guidelines for developing a security policy for FANs exist.

Below, we discuss security in context with the life cycle of IT systems and give an overview of the most common technical means to enforce security goals that are also applicable to FANs.

Security System Life Cycle

While security activities can be started at any time in the life cycle of a system, it is usually desirable (although often not possible) to integrate them in the whole life cycle as described in [9], where five stages are distinguished[1] as shown in Figure 55.1.

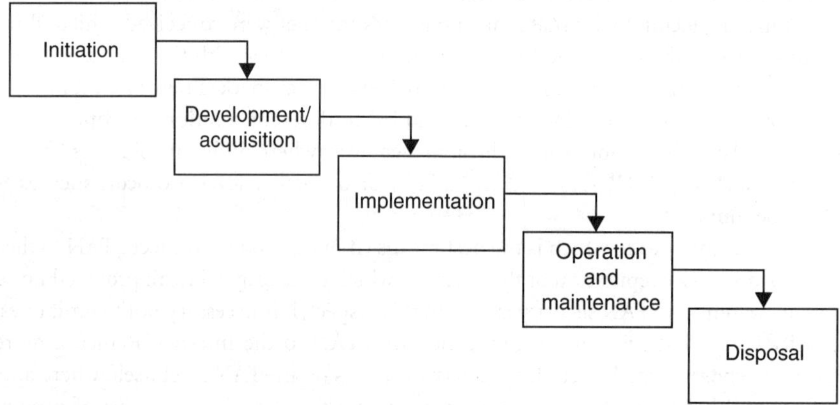

FIGURE 55.1   Basic waterfall life cycle model.

---

[1]The activities that follow can also be applied (in a slightly modified way) to other life cycle models like, for example, the spiral model by Boehm [34].

During the *initiation* phase where the system is designed, a basic security design is also established by means of a sensitivity assessment where the data to be handled by the system as well as its impact on security are estimated. During the *development* stage, the security requirements as well as methods to implement them are selected and included in the overall system specification. During *implementation,* security features of the system are activated and tested. In the *operation and maintenance* phase, security-related operations such as key changes and others take place. Also, in this stage, system audits and monitoring may lead to system changes if new security risks are found that need to be addressed. Finally, the disposition of information hard- and software needs to be handled where the long-time storage of cryptographic keys is a major task.

Common Security Measures

Cryptography

Cryptography is often used to implement services where confidentiality and integrity are needed such as, for example, access control, described below. The major building blocks of cryptography are as follows:

- Cryptographically secure pseudorandom number generators (CSPRNG) to generate seemingly random entities, for example, for secret keys or as random nonces in protocols,
- Cryptographic hash functions, which are often used in conjunction with electronic signatures or to generate message authenticate codes (MAC) to protect data integrity,
- Symmetric and asymmetric encryption functions used to protect the confidentiality of data or may be used to construct MACs or electronic signatures.

An in-depth discussion of cryptography can be found in [10], which covers the above topics (except for randomness) and in [11], which covers the above topics and some more.

Authentication and Access Control

Authentication measures are used to establish the identity of an entity (which might be an actual user, a process that acts on behalf of a user or a role a user assumes during a session). A major distinction is made between *weak authentication* as found in password-based systems like Unix or Windows and *strong authentication,* which is usually based on *challenge–response* protocols that either rely on symmetric or asymmetric cryptography like Kerberos [12] and X.509 [13], respectively.

Authenticated entities can be subjected to access control mechanisms that allow to decide if an entity's desired action is allowed. Apart from simple models like access control lists and other discretionary models, mandatory models exist that are mainly found in military systems. Both models can be implemented by the newer *role-based access control* (RBAC) [14] schemes.

Firewalls and Intrusion Detection

Firewalls (see, e.g., [15–17]) and Intrusion Detection Systems (see, e.g., [18, 19]) are somehow complementary technologies, which are usually associated with LAN/FAN connections to the Internet. While the former try to prevent illegal access to inbound and outbound connections by, for example, packet filtering, the latter monitors and analyzes network access patterns and tries to recognize security breaches and/or attacks, which are referred to a system administrator.

Apart from protecting LANs from intruders from the Internet, both technologies can also be used in a LAN-only environment or where a LAN is connected to a FAN.

Security Evaluations

The construction of a secure system is a complicated task involving a variety of different skills and knowledge and might still not succeed even if the necessary care is taken [20]. A prominent example for a failed security system is *wireless equivalent privacy* (WEP), which was designed to protect 802.11 wireless LANs and miserably failed in doing so [21].

A method to build confidence in a new system is a security evaluation by a third party according to international security standards. The *Common Criteria 2.0* (CC) [22] is the latest in an evolving series of international security standards. A positive evaluation of a security target that is evaluated against the CC according to a security standard chosen by the system manufacturer leads to an internationally acknowledged certificate. To help with the preparation for a certification, so-called *protection profiles* (PP) for different security areas such as firewalls already exist, which can be built upon.

55.3 FAN Security

A natural way to bring security to a FAN would be the adoption of well-known security principles and protocols from the Internet world. Unfortunately, it is not so easy. One reason for this is that security protocols for the Internet domain such as SSL/TLS [23] or the newer IPSec protocol suite [24] operate on top of or at network layers (according to ISO/OSI) that are not defined in most FANs, where — for reasons of speed and simplicity — we usually find only layers 1 and 2, as well as sometimes specifications of layer 7 and above (sometimes called layer 8 or user layer, but outside the OSI model). Another problem in applying Internet security standards is that no provisions for any real-time support are made, which could lead to problems if applied to a FAN. Furthermore, because of the assumption that communicating parties do not know each other in advance, many Internet security protocols heavily rely on asymmetric cryptography, including the ones listed above, with the notable exception of the Kerberos authentication protocol, which was designed with an institutional scale in mind. Finally, in consideration of FAN security, we may face a situation where one or all of the communicating nodes are not under the control of their rightful user(s), which is an unusual situation in conventional LANs.

Security of a FAN system itself has seldom been discussed in the literature. In [25], a secure FAN node is discussed and [26] gives an overview of possible security classes for FANs that have different security implications, which will be detailed below. The major problems that have to be faced when FANs should be made secure are:

- FANs have no built-in provisions for security and are usually designed with a specific application area in mind, which will contribute to the problems that are inevitably encountered when security services should be added to an existing system.
- Cryptographic operations such as encryption needed to secure FAN communication might be prohibitive because of the limited processing power of current FAN nodes, and even more so if real-time needs have to be addressed.
- The FAN nodes as well as the communication media might be under total control of the adversary.
- The number of FAN systems is overwhelming and securing all of them is an infeasible task.

On the other hand, some traits of FANs make them more easily accessible to security engineering than other systems, most notably:

- FANs have a very limited complexity. In comparison to the general-purpose machines found on the Internet, FAN nodes are designed to solve one special problem, which makes them easier to understand and analyze.
- The protocols used in FAN systems are also less complex than those found in other network systems. This can be attributed to either a restriction to only the lower layers of the protocol stack and the application layer, or in cases where this does not hold, to the limitation to a single protocol, contrary to the protocol proliferation that can be found on the Internet. Additionally, the easier a protocol is, the more amenable it is for efficient implementation, which is desired if real-time behavior is of concern.
- As a consequence of the above, the operating systems (OS) found in FAN nodes need not be as complicated as a typical computer OS.

The following discussion will use the data listed in Table 55.1, which shows basic security services and the layers in the OSI model [27] where they may possibly be implemented.

According to [26], the following subsections try to classify typical FAN application scenarios with respect to their impact on security. Essentially, we can distinguish between controlled and uncontrolled FANs.

Security for Controlled FANs

A controlled FAN (CFAN) environment is a FAN installation where the physical access to FAN nodes as well as may be the access to the underlying communication media is controlled by the owner of the FAN system. This situation is comparable to the situation found in the usual Internet environment where one computer or a LAN of a company is connected to the Internet.

For a CFAN, we can further distinguish three different scenarios, which are mutually inclusive as shown in Figure 55.2. The stand-alone FAN (SAF) consists of a FAN-only installation that is not connected to the outside. The LAN-integrated FAN (LIF) and Internet-integrated FAN (IIF) correspond to a FAN installation, which can be accessed via a connected LAN or via the Internet, respectively. The inclusion relations in Figure 55.2 correspond to the actual security threats and solutions (STS) that are applicable to the different scenarios. While the inclusion relationship always holds, it has to be noted that the set STS is highly dependent on the underlying security policy, which shows the importance of deriving a security policy for a specific system at hand.

Considering the CIA security goals as mentioned in Section 55.2, we focus on possible solutions for confidentiality and integrity and mention that provisions to achieve availability in a FAN environment (apart from provisions that are already designed into the FAN) will be hard to come by. In the following discussion, the placement of security functions according to ISO/OSI (see Table 55.1) will be helpful. Moreover, the fact that most FAN systems only implement OSI layers 1, 2, and 7 is of importance. The

TABLE 55.1 OSI Security Services in the Different Layers.

Security Service	Possible Layer(s)
Peer entity authentication	3, 4, 7
Data origin authentication	3, 4, 7
Access control service	3, 4, 7
Connection confidentiality	1, 2, 3, 4, 6, 7
Connectionless confidentiality	2, 3, 4, 6, 7
Selective field confidentiality	6, 7
Traffic flow confidentiality	1, 3, 7
Connection integrity with recovery	4, 7
Connection integrity without recovery	3, 4, 7
Selective field connection integrity	7
Connectionless integrity	3, 4, 7
Selective field connectionless integrity	7
Nonrepudiation origin	7
Nonrepudiation delivery	7

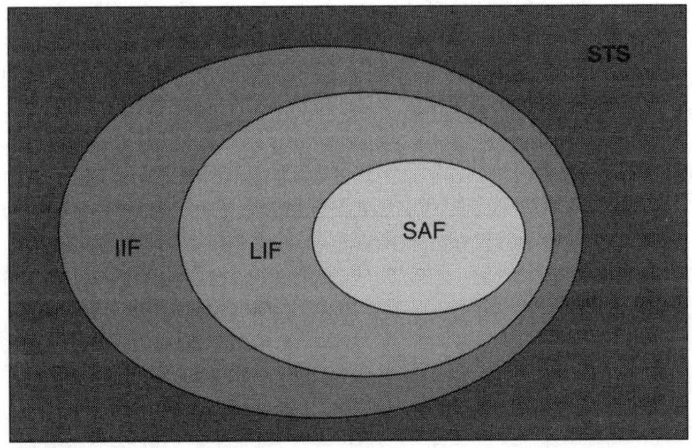

FIGURE 55.2   Three controlled FAN scenarios.

reduced OSI stack of modern FANs is usually a result of efficiency considerations. In application domains such as in building automation where time requirements are more relaxed and extensive networking capabilities are needed an implementation of the full OSI stack may also be found.

With respect to the placement of confidentiality services in the network stack, we find that the majority of the services can be placed at one of the layers 1, 2, or 7. Additionally, placement of such services above layer 7 is possible.

Integrity services can be placed at layers 3, 4, and 7 of the OSI stack, or — as with confidentiality services — above layer 7. Restriction to layers 1, 2, and 7, which will often be necessary in a FAN environment, limits the application of integrity services to layer 7 or above.

Taking into account that in a FAN environment we often find dedicated microcontrollers that are designed to handle a specific FAN interface and protocol, a security solution that is targeted above layer 7 appears to be a sensible choice in all cases where we try to secure an established FAN system. In FAN systems where we find an abstraction layer above layer 7, an inclusion of security services in this layer would be of special interest. Such a solution would have the special benefit that the user layer of a FAN system is typically targeted at the standardization of interoperable services or application profiles and the process of standardizing such services is carried out by a user group (usually comprising the companies that have an interest in the FAN), which leads to a broad acceptance in industry. Apart from the expected acceptance, this solution is also of interest as it provides end-to-end security in its narrowest sense as it secures all communication between two nodes of a FAN at the application level. In the absence of a (widely accepted) user layer into which security services could be integrated or where a specific user group does not see a necessity to add such services (as will presumably be often the case), a generic solution above layer 7 can still be devised with the drawback of losing interoperability with other solutions, which results in increased development efforts and consequently increased costs.

In both approaches, a prudent approach to providing security services could be to use a wrapping mechanism that takes an engineering value (which is a measured input value as it was sampled with a sensor or first sampled and then preprocessed at the node) and apply this value to either an integrity mechanism or to both an integrity and a confidentiality mechanism. Figure 55.3 shows an example of the second case where the field *Header* contains information about the security transformations applied. The *MAC* field contains integrity check information.

The integration of both integrity and confidentiality mechanisms could be prohibitive in most of today's FANs as they require processing power that might be out of range for the typical low-cost 8-bit microprocessors that can be found in typical FAN environments. Another drawback of solutions such as depicted in Figure 55.3 is that the (maximum) payload of FAN messages might not allow the message expansion that would be necessary to add confidentiality and/or encryption. This is a problem especially for integrity mechanisms as they always add additional data to the transmitted message, while for confidentiality only, we could think of just encrypting the engineering value and omitting the header of the message. In this case, two communicating nodes need to know in advance if the data to be sent/received will be in encrypted form or not.

When missing processing power in the nodes is a concern, one could also think of applying the necessary security services to the lower layers in the OSI stack. Although a first look at Table 55.1 shows us that only confidentiality services can be added at the lower level, it is still possible to add the required services at layer 2 if the necessity arises, as has been the case in Ethernet where [28] augments the services found

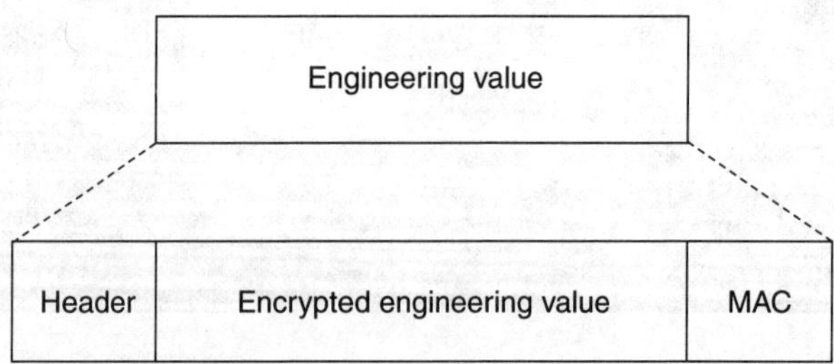

FIGURE 55.3   Wrapping an engineering value.

in the OSI security model to also include integrity, authentication, and access control. While this could be a feasible solution, one has to keep in mind that the resulting protocol will also need to support non-layer 2 management functionalities that handle, for example, key exchange.

A solution for providing confidentiality and/or integrity would still leave a major security threat unanswered: *network management*. In today's FANs, this functionality is typically achieved in an unauthenticated manner; such an approach is synonymous to an invitation to a knowing attacker who has gained access to the FAN. Management functionality might even include the possibility to totally reprogram a node, for example, by directly overwriting parts of the memory of a node as well as to gain complete knowledge about the programming of the node by reading out configuration and application data from the memory. It is easy to imagine situations where such behavior is not acceptable. The least functionality that is needed to achieve security in this area is the provision of an *access control* mechanism. This task can be supported by integrity mechanisms that allow to authenticate configuration communication messages. The implementation of access control for network management should pose little problems. While the data transmitted might be larger than usual FAN messages and even multipart messages might be needed, such operations are usually not time-critical and the restrictions to the acceptable maximum length of a message and the maximum time a transmission may take are usually very relaxed.

Security for Uncontrolled FANs

An *uncontrolled FAN* (*UFAN*) environment is a FAN installation where physical access to communication nodes that comprise the FAN as well as access to the underlying communication media is controlled by a user (or a group of users) who is not the owner of the system and has to be considered as an adversary. Such a situation is found, for example, if a remote metering device is deployed at the site of a customer by a utility company or if a FAN is deployed in a car. A comparable situation that has been well studied (but where neither security mechanism nor attack results are published in the open) is the transmission of secured pay-per-view programs via cable or satellite, which only entitled subscribers should be able to descramble. The examples make it clear that for a UFAN environment, we also need to distinguish between the kind of system access that the owner/manufacturer has after the FAN has been deployed.

In a *remotely maintainable FAN* (RMF), there exists a rightful system owner, which has the possibility to *regularly* communicate with the system. The communication may take place on a periodic basis or when the need arises. In the remote metering example, we could imagine regular communication between the utility company and the meter every month in the case of a contract between the user and the company. On the other hand, if the user has to prepay his/her energy communication could be on a per-payment basis.

Nonmaintainable FANs (NMF) are those systems where the rightful system owner has no *regular* possibility to communicate with the system after its deployment (although some communication could become possible in the unforeseeable future, e.g., in the case of a system failure). This is, for example, the case if a FAN is deployed in a car.

If we compare the RMF and the NMF scenarios as above by looking at the set of possible STSs, we easily see that both scenarios are different in that they overlap in some parts (e.g., the adversary has access to the physical parts of the FAN), whereas neither is included in the other (e.g., the RMF is exposed to outside threats, which are adversarial third parties being neither the system owner nor the entity where the system is deployed). They are also not proper sub- or supersets of the SAF, LIF, and IIF scenarios described in the Security for controlled FANs section, mainly because the whole FAN infrastructure is accessible by the adversary. But we can see again that the sets are highly dependent on the underlying security policy and security assumptions. On the whole, UFAN systems seem to be more at risk than CFAN systems, because the adversary has more possibilities to circumvent possible security measures as well as because of the fact that analysis and results of security issues for such configurations have not been extensively dealt with yet.

As far as ways and means to secure UFANs are concerned, one natural train of thought leads to the investigation of physical measures to secure the different nodes of a FAN as well as the sensors and actuators connected to them. Depending on the security policy, this can lead to simple solutions where plumbs are attached to the devices (today, this is often done to prevent manipulation of power meters) in which case it is the legal implications associated with the removing of the seal that act as a deterrent. A (technologically) stronger solution are attempts to construct tamper-proof devices that either notify the system owner if they are tampered with in the case of an RMF system or that (permanently) deactivate themselves in the case of an NMF system. In the latter case — apart from trying to prevent (physical) access to the nodes — we usually find special security modules such as smart cards that are useful to protect sensitive data as well as to recognize attempted tampering. If it is not possible or feasible to physically protect the nodes, their programs, and their memory; security considerations regarding lower layers of the network stack become in vain as the protection possibly achievable at those layers does nothing to prevent the possible attacks at higher levels. Together with physical node protection, lower layer security is of course of interest again and might — as in the CFAN scenario — help to remedy problems that could arise with the use of integrity services.

Continuing considerations regarding physical security and automatic deactivation of nodes in case of tampering, the possibility of including intrusion detection mechanisms in a node seems to be an interesting topic of research for the future. Such a mechanism could become feasible in UFANs that — after deployment — remain unchanged, meaning that no nodes are added or removed as well as that the communication patterns between the nodes do not change or change in a way that can be anticipated. If such an analysis is possible, a node that detects an intrusion can either contact the system owner or deactivate itself, depending on the actual system configuration (in case of an RMF, a failure to contact the system owner after detection of an intrusion must also result in deactivation).

Regarding the different reasons that might lead to the deactivation of a system, it is also necessary to determine the right point at the time of deactivation. To give an example, it would be quite inconvenient and adverse to any safety considerations if a UFAN in a car would stop working during car operation.

Finally, a critical point in UFAN security is again the network management. In an RMF scenario, it is easy to see that a point-to-point security solution is indispensable. Here, it is conceivable that the same mechanisms that may be utilized in a CFAN are helpful. Conversely, in the NMF scenario we would like to prevent any management activities that concern the configuration of the nodes as well as the programs that are executed on the nodes. While in this case configuration data that reside in a ROM (or maybe in a once writeable WORM) and are therefore not changeable at all would be the easiest solution, such an approach might be prohibitive in the case where due to maintenance reasons single nodes might need to be exchanged or updated.

55.4   Security for External FAN Connections — An Example

If a FAN is connected to either an intranet or to the Internet (or any other public internet for that matter) or to the Internet via a connection to an intranet, we find ourselves in a situation as shown in Figure 55.4.

Access to the FAN is typically achieved via a central access point, in many cases, a gateway translating between the FAN protocol and the protocol used to access the gateway, which we subsequently assume to be part of the TCP/IP protocol suite. A common implementation is a web-based solution, that is, the gateway translates FAN data to accommodate a transport format such as, for example, XML over HTTP (see [29] for a thorough description of XML). Yet this structure also allows for implementations other than the pure gateway approach. Actually, other solutions are possible where FAN data are tunneled over the external network or where the IP protocol is tunneled over the FAN and handled in the node that is the

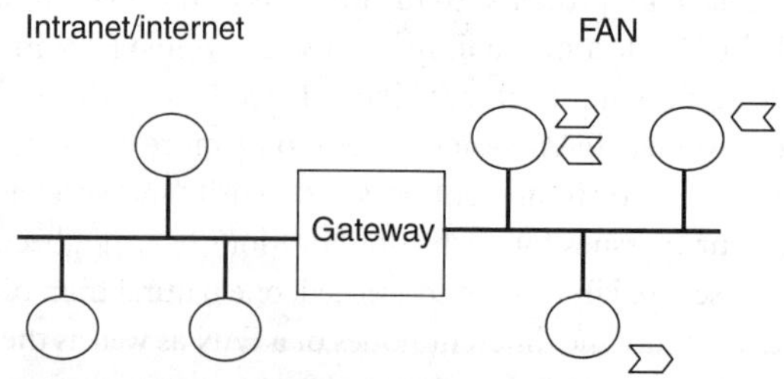

FIGURE 55.4   External connection to a FAN.

sender/recipient of a message (although this last possibility seems to be very awkward if we compare the size of IP messages to that of typical FAN messages).

Access to the gateway will usually be achieved either through a firewall (which might be incorporated into the gateway itself) or via a dial-up connection. For more information about the different possibilities of where to place a firewall (and accompanying IDS system) and how to possibly configure it, which heavily depends on the IT environment at hand, see the references of Section 55.2.

It is important to note (and to recognize) that the secure connection of a FAN to an external IP network is more than just the application of any IP security protocol that we find on the Internet, as can be seen in the example presented below. Contrary to common conceptions, there is no need to rely on any of the very time-consuming asymmetric security primitives that are often assumed when Internet–FAN connections are considered. The reason for this lies in the fact that all parties involved in the communication are previously known to each other. This is a very typical property of Internet–FAN connections, often alleviating system design.

The rest of this section gives an example of a security architecture for a gateway with a need for a very high level of security and a limited demand for high bandwidth. The target application is a residential gateway primarily used for remote access to energy meters and consequently for simplified and automatic billing, may be including the application of flexible (time- and consumption-dependent) tariff schemes. For add-on services, the gateway could be connected to a home automation network. A detailed description of such a gateway can be found in [30]. In the light of the different scenarios described before, such an application is a UFAN, specifically the RMF subtype. From the security point of view, the configuration is a kind of worst case. The owner of the home has some interest to pay as little as possible for the energy consumed; therefore, he has to be considered as an adversary. Even worse, he has unlimited physical access to the entire installation, unless gateway and energy meter are locked in a sealed cabinet and there is no in-home network at all. Therefore, security demands are high.

Overview of the Security Architecture

An important point to note in this example is that the relationship between gateway and a client in the Internet is not just one-to-one. In fact, the utility company desiring remote access to the energy meters has numerous customers as well as more than one data acquisition client. In addition, the whole concept must take into account that the communication infrastructure can be used by third-party service providers who need to be granted access to the gateway as well, albeit to different sets of data. The security architecture therefore becomes rather complicated and consists of an off-line and an on-line part as shown in Figure 55.5. As can be seen, the key distribution center that operates off-line for security reasons generates the necessary keys that securely reside on tamper-proof smart cards [31], which are also used for the execution of cryptographic algorithms.

The on-line participants are the security clients and administrators (which can also be one entity) that access a remote FAN over the gateway using the smart cards to provide the necessary keys and security algorithms. While the clients are only entitled to access the FAN for monitoring and configuration purposes, administrators may additionally change access control settings or the configuration of the security server. On the gateway itself, the security server accepts and handles connections from the Internet using an access control component to enforce access restrictions to the FAN. For the actual FAN access, a FAN-specific module translates between the different protocols.

Implementation

The secure communication between clients/administrators and the gateway requires a minimal set of information to be carried in a message. In the example, the message format shown in Figure 55.6 was chosen providing integrity as well as confidentiality. The particular choice is arbitrary but illustrates the typical elements found in secure communication. Apart from the UID that indicates the sender of the message, the MODE field allows to implement and use different security primitives, should the need arise,

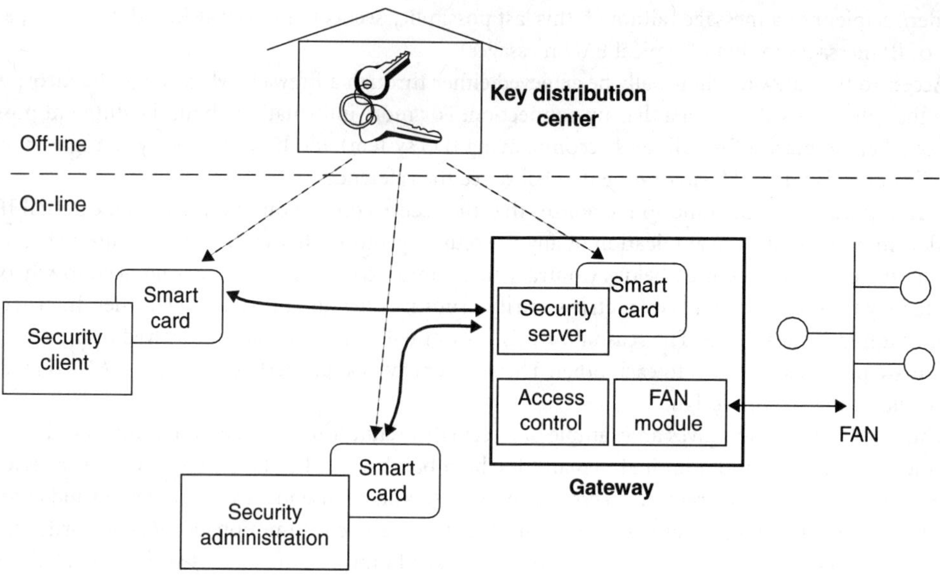

FIGURE 55.5 Architecture of the secure FAN–Internet gateway.

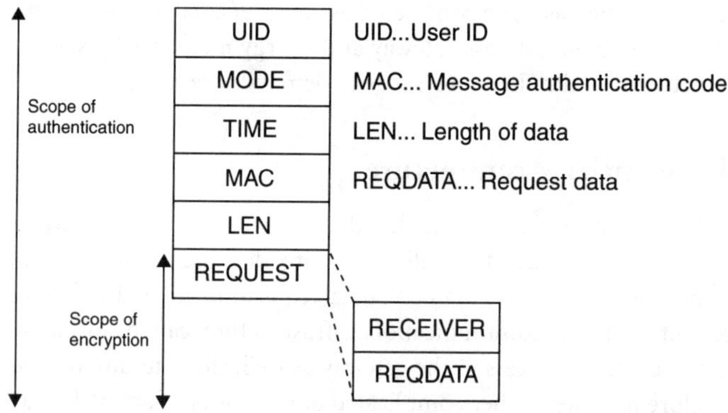

FIGURE 55.6 The communication message format.

because, for example, the security of one of the used primitives is broken. The inclusion of a time field (TIME) that contains the (notion of) time at the gateway involved in the communication allows to define a time window of validity for a message and prevents replay attacks of previously sent messages. In messages from the gateway, the field is used to update the local notion of the gateway's time. The MAC protects the integrity of the message and is calculated over the whole message. Finally, the LEN field indicates the length of the following encrypted REQUEST that is split into the actual request data and the receiver of the data, which is either the security server in case of configuration messages or the underlying FAN. The RECEIVER field also facilitates the support of more than one FAN connected to the gateway, which could be realized by incorporating more than one FAN module.

The security mechanism used to provide data encryption is the *remotely keyed encryption scheme* (RKES) described in [32], which utilizes the secret keys of the smart card to initialize the encryption, while the bulk work is done afterwards by the gateway processor, which calculates faster than the smart card and is also able to cope with larger messages. The authentication mechanism uses secure hash function *SHA-1* [33] and sends the resulting data to the smart card where they are encrypted using 3-DES. The result of this operation is then used as the MAC of the message.

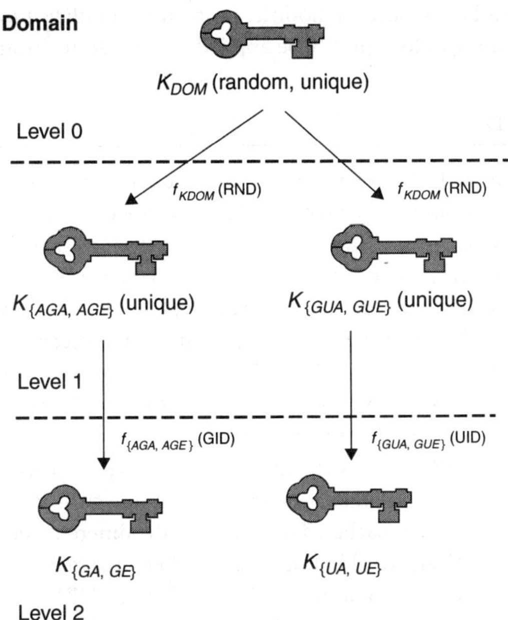

FIGURE 55.7　Key hierarchy.

Figure 55.7 shows the key hierarchy needed in the system. As we foresee that more than one gateway will be needed, we use a domain system where users and administrators of a domain can exchange messages with all gateways of the domain.

To achieve this functionality, a random and unique key K_{DOM} is generated in the key distribution center for each domain. This key is subsequently used to derive all other keys of that domain. Looking at the left portion of Figure 55.7, we see that we derive the keys K_{AGA} and K_{AGE} from the master key using a randomly chosen input at level 1 of the key hierarchy. Those keys are the master keys for authentication and encryption of administrator-to-gateway messages, respectively. They are used to derive the keys K_{GA} and K_{GE} of the gateways of the domain from the unique gateway identifiers (GID) of a gateway. This scheme allows an administrator to communicate with every gateway because she/he can derive every gateway key of level 2 by applying the key derivation function f to the GID of the requested gateway using the master keys of level one. Additionally, if the secret keys of a gateway are compromised, the system can continue to work because every gateway has its unique set of keys. The right portion of Figure 55.7 shows the same procedure for the generation of keys that are used in the communication between a gateway and a user.

Function f, which is used to derive the keys of levels 1 and 2 in Figure 55.7, is the application of SHA-1 to the argument followed by an application of 3-DES in EDE (encrypt–decrypt–encrypt) mode,

$$f_{KEY}(\cdot) = 3\text{-}DES - EDE_{KEY}(SHA\text{-}1(\cdot)) \tag{55.1}$$

which finally yields the key. The index *KEY* in (1) designates the master key used as a parameter in the DES algorithm. To derive, for example, the secret encryption key of a user with the UID X at a gateway, we need to calculate

$$K_{UE} = f_{KGUE}(X) = 3\text{-}DES - EDE_{KGUE}(SHA\text{-}1(X)) \tag{55.2}$$

The final functionality needed in the system is an access control mechanism. The easiest way to realize this functionality is the implementation of an access matrix, which operates on the FAN node level, and allows either *read access*, *write access*, or *create access* to data. Depending on the access rights of the user, and in the case where memory is a limited resource, it is possible to assign a Boolean flag to each entry that indicates if the following access modifiers constitute an *allowance* or *forbiddance* list for the specific user, where the latter would typically be applied for an administrator to keep the list short.

55.5 Conclusion

The necessity for security at the FAN level still has to be brought to the attention of the larger part of the FAN community. While the awareness for security issues in the connection of FANs to the Internet is starting to grow — mainly because of the frightening security record of the Internet — it will be of utmost importance to incorporate security considerations into the next generation of FAN systems themselves. The ultimate goal is to alert the community to the fact that security provisions should be an important factor in the decision about which FAN system to use for a specific task. It is to be hoped that contrary to the Internet, where it took more than 20 years (and the commercialization of the Internet) until people realized this fact, this process will be faster — especially if we consider safety-critical applications where FAN systems are deployed in increasing numbers.

While the pure realization of FAN–Internet connections profits immensely from the experience gained in attempts to secure the Internet — thanks to the gateway approach — we must realize that FAN-level security needs different and new approaches. This is due to the different specialized protocols as well as the very low-end processors typically used in this area. Another large field for further research and development where approaches are not very mature can be found in the UFAN scenarios.

We have discussed some potential security solutions (and pitfalls) for today's FAN systems that are worth a try where the need for security arises. However, the discussion also shows that this can only be an intermediate solution. Applying reasonable security strategies to nonsecurity-aware systems has to overcome significant obstacles, up to an extent where a meaningful operation is no longer practical. It seems that there is no way around the design of a *really* secure FAN infrastructure including all necessary security services as well as the necessary administration framework. The ultimate aim of such efforts has to be a standardized framework that is widely accepted.

References

[1] Kriesel, W., T. Heimbold, and D. Telschow, *Bustechnologien für die Automation*, 2nd ed., Hüthig Verlag, Heidelberg, 2000.

[2] Wollschlaeger, M., Framework for Web Integration of Factory Communication Systems, IEEE International Conference on Emerging Technologies and Factory Automation (ETFA), Antibes Juan-les-Pins, France, Oct. 15–18, 2001, pp. 261–265.

[3] Weaver, A.C., Internet-based Factory Monitoring, 27th Annual Conference of the IEEE Industrial Electronics Society (IECON), Denver, CO, U.S.A, Nov. 29 –Dec. 2, 2001, pp. 1784–1788.

[4] Kunes, M. and T. Sauter, Fieldbus-Internet connectivity: the SNMP approach, *IEEE Transactions on Industrial Electronics*, 48, 1248–1256, 2001.

[5] Sauter, T., M. Lobashov, and G. Pratl, Lessons Learnt From Internet Access to Fieldbus Gateways, 28th Annual Conference of the IEEE Industrial Electronics Society (IECON), Sevilla, Spain, Nov. 5–8, 2002, pp. 2909–2914.

[6] Fraser, B. Ed., *RFC2196 Site Security Handbook*, 1997, http://www.ietf.org/rfc/rfc2196.txt?number = 2196.

[7] Bundesamt für Sicherheit in der Informationstechnik, *IT Baseline Protection Manual*, Bundesanzeiger-Verlag, http://www.bsi.de/gshb/english/menue.htm.

[8] International Standards Organisation, *ISO/IEC 17799:2000, Information Technology — Code of Practice for Information Security Management*, 2000.

[9] National Institute of Standards and Technology, *An Introduction to Computer Security: The NIST Handbook*, NIST Special Publication 800-12, U.S. Government Printing Office, 1995, http://csrc.nist.gov/publications/nistpubs/800-12/handbook.pdf

[10] Douglas R. Stinson, *Cryptography, Theory and Practice*, 2nd ed., Chapman & Hall/CRC, London, 2002.

[11] Alfred J. Menezes, Paul C. van Oorschot, and Scott A. Vanstone, *Handbook of Applied Cryptography*, CRC Press, Boca Raton, FL, 1996.

[12] Steiner, J.G. B. Clifford Neuman, and J.I. Schiller, *Kerberos: An Authentication Service for Open Network Systems*, in Proceedings of the Winter 1988 Usenix Conference, Dallas, TX, U.S.A., February 1988, pp. 191–202.

[13] International Telecommunication Union, *ITU-T Recommendation X.509 (1997 E): Information Technology — Open Systems Interconnection — The Directory: Authentication Framework*, June 1997.

[14] Sandhu, R. E.J. Coyne, H.L. Feinstein, and C.E. Youman, Role based access control models, *IEEE Computers*, 29, 38–47, 1996.

[15] Cheswick, B. The Design of a Secure Internet Gateway, in Proceedings of the Usenix Summer 1990 Technical Conference, Anaheim, CA, U.S.A., June 1990, pp. 233–238.

[16] Wack, J.K. Cutler, and J. Pole, *Recommendations of the National Institute of Standards and Technology Guidelines on Firewalls and Firewall Policy*, NIST Special Publication 800-41, U.S. Government Printing Office, 2002, http://csrc.nist.gov/publications/nistpubs/800-41/sp800-41.pdf

[17] Goncalves, M. *Firewalls Complete*, McGraw-Hill, New York, 1997.

[18] von J. Helden and S. Karsch, *BSI-Studie: Intrusion Detection Systeme Grundlagen, Forderungen und Marktübersicht für Intrusion Detection Systeme (IDS) und Intrusion Response Systeme (IRS)*, Bonn, 1998 http://www.bsi.de/literat/studien/ids/ids-stud.htm.

[19] Bace R. and P. Mell, *Intrusion Detection Systems*, NIST Special Publication 800-31, 2001.

[20] Anderson, R. *Security Engineering*, Wiley, New York, 2001.

[21] Borisov, N., I. Goldberg, and D. Wagner, Intercepting Mobile Communications: The Insecurity of 802.11, in Proceedings of the Seventh ACM SIGMOBILE Annual International Conference on Mobile Computing And Networking, Rome, Italy, July 16—21, 2001, pp. 180–189.

[22] International Standards Organisation, ISO/IEC 15408-1:1999 Information Technology — Security Techniques — Evaluation Criteria for IT security, 1999.

[23] Dierks, T. and C. Allen, *RFC 2246 The TLS Protocol Version 1.0*, January 1999, http://www.ietf.org/rfc/rfc2246.txt.

[24] Loshin P. (Compiler), *Big Book of IPsec RFCs: Internet Security Architecture*, Morgan Kaufmann, Los Altos, CA, 1999.

[25] Palensky P. and T. Sauter, Security Considerations for FAN-Internet Connections, IEEE International Workshop on Factory Communication Systems, Porto, Portugal, Sept. 6–8, 2000, pp. 27–35.

[26] Schwaiger C. and T. Sauter, Security Strategies for Field Area Networks, 28th Annual Conference of the IEEE Industrial Electronics Society (IECON), Sevilla, Spain, Nov. 5–8, 2002, pp. 2915–2920.

[27] International Organization for Standardization, *Basic Reference Model for Open System Interconnection — Part 2: Security Architecture*, International Organization for Standardizations 1989.

[28] *IEEE Std 802.10-1998, IEEE Standards for Local and Metropolitan Area Networks: Standard for Interoperable LAN/MAN Security (SILS)*, The Institute of Electrical and Electronics Engineers, Inc., 1998.

[29] *The XML CD Bookshelf*, O'Reilly & Associates Inc., Sebastopol, CA, 2002.

[30] Schwaiger C. and T. Sauter, A Secure Architecture for Fieldbus/Internet Gateways, IEEE International Conference on Emerging Technologies and Factory Automation (ETFA), Antibes Juan-les-Pins, France, Oct. 15–18, 2001, pp. 279–286.

[31] Rankl W. and W. Effing, *Smart Card Handbook,* 2nd ed., John Wiley & Sons, New York, 2000.

[32] Blaze, M., J. Feigenbaum, and M. Naor, A Formal Treatment of Remotely Keyed Encryption, Advances in Cryptology — EUROCRYPT '98, International Conference on the Theory and Application of Cryptographic Techniques, Espoo, Finland, May 31 – June 4, 1998, Lecture Notes in Computer Science, Vol. 1403, Springer, Berlin, 1998, pp. 251–265.

[33] National Institute of Standards and Technology, *Secure Hash Standard*, Federal Information Processing Standards Publication 180-1, U.S. Government Printing Office, 1995.

[34] Boehm, B. A spiral model of software development and enhancement, *IEEE Computers*, 21, 61–72, 1988.

56

PROFIsafe — Safety Technology with PROFIBUS

Wolfgang Stripf
Siemens AG

Herbert Barthel
Siemens AG

56.1 Why do we Need Safety in Automation?

Any active industrial process is more or less associated with the risk

- To injure or kill people
- To destroy nature
- To damage investments.

With most of the processes, it is quite easy to avoid risk without special requirements imposed on automation systems. However, there are typical applications associated with high risk, for example presses,

saws, tooling machines, robots, conveying and packing systems, chemical processes, high-pressure operations, off-shore technology, fire and gas sensing, burners, cable cars, etc. These applications need special care
and technology.

Over time, the market balances out the reliability and availability of standard automation technology
to a certain economic cost level. This means the failure or error rate of standard automation technology
under normal circumstances is just acceptable for normal operations but not sufficient for the above-
mentioned applications.

The situation may be compared with a public mail system. While normal letter expedition is expected
to be as affordable as possible at certain reliability, everybody will use special mail for important messages.

56.2 Dichotomy of Standard and Safety Automation

In the past, microcontrollers, software, personal computers, and communication networks dramatically
influenced the standard automation means and thus led to cost reduction, higher flexibility, and availability. With respect to safety, existing standards and regulations prohibited any usage of those means. Safety
automation had to be "hard-wired" and based on "relay" technology. This dichotomy or gap is quite natural due to the fact that safety relies on trusted technology or material, trust on experience, and experience
on time. But adding "classical" safety to modern automation solutions always led to inadequate cost due to
additional wiring and engineering, due to less flexibility and availability than expected, and due to other
disadvantages. In the meantime, the situation has changed dramatically. Microcontrollers and software
have proved useful in millions of applications. And the preconditions for their usage in safety applications
have been given since the introduction of the international standard IEC 61508 [1].

56.3 Motivation and Objectives for PROFIBUS

During its lifetime of more than 10 years, PROFIBUS has been emerging as one of the most important
fieldbus systems in the world and just got standardized within IEC 61158 and IEC 61784 [2], respectively.
It is enabling decentralized applications in factory and process automation with its variety of appropriate transmission technologies like RS485, MBP-IS,[1] and fiber optics. Thus, it was merely a matter of time
to integrate the necessary means for safety applications into PROFIBUS DP in a seamless manner and
provide a similar flexibility and availability also for powerful safety devices like remote I/O, laser scanners,
light curtains, level switches, shutdown valves, drives, robots, and the like.

Back in 1998, when the PROFIBUS organization started its project "safe communication across
PROFIBUS DP," more than 25 renowned companies on safety decided to follow the vision of connecting the above safety devices to the same transmission line as the standard devices and let them communicate with an additional programmable safety-related controller (F-Host in Figure 56.1). No cables,
ASICs, layer stack software, or other communication devices like repeaters, links, and couplers should
have been changed. Configuration, parameterization, programming, and diagnostic means should be as
familiar to the user as possible in order to simplify the usage of safety: "what he/she is using, he/she will
not be losing." Fortunately, the working group was able to base its design and development efforts on
the new IEC 61508 and on the spadework of the EN 50159-1, now IEC 62280-1 [3], entitled "Railway
applications — Communication, signalling and processing systems — Part 1: Safety related communication in closed transmission systems." PROFIBUS is such a closed transmission system as it only allows
communication between configured and well-known participants. Open transmission systems in contrast are, for example, the public phone system or the Internet.

Already during the first sessions of the working group in 1998, it became apparent that the pure definition
of the safe transmission of messages via the standard PROFIBUS cable would not prove sufficient for the new
generation of safety devices that had to be connected to it. What sense would it make to a sensor

---

[1] MBP-IS = "Manchester Coded — Bus Powered and Intrinsically Safe" replaces the previous PROFIBUS name
"IEC 1158-2." Further development has listed additional procedures in the corresponding IEC standard, so that creating an unambiguous designation has become necessary.

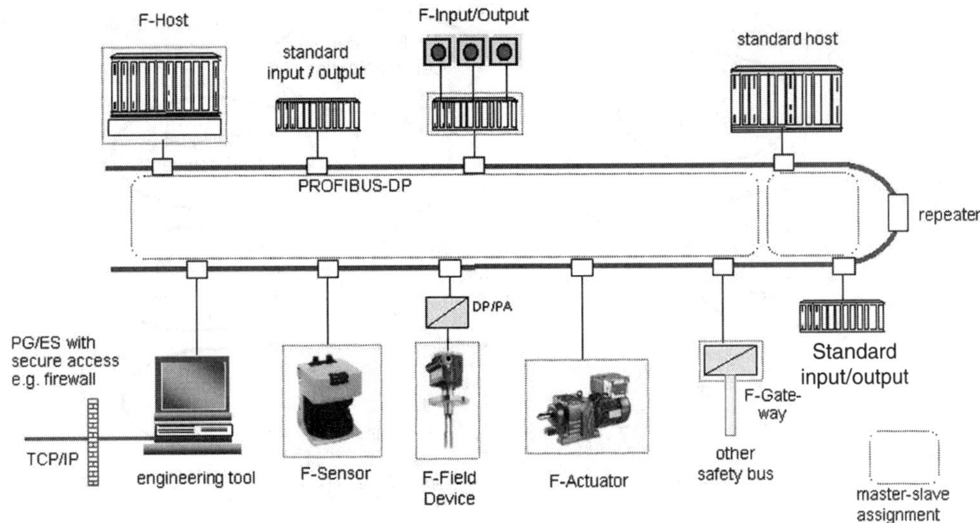

FIGURE 56.1   The PROFIsafe vision.

manufacturer, for example, if merely the shutdown signals could be transported via the bus, while parameter value assignment or diagnosis in the event of a failure still requires a time-consuming PC connection on site via the RS232 interface? The request for the system manufacturers to support the scenario of the rapid device replacement and the integration of the commissioning and diagnosis software of the field device into the engineering software of the system manufacturer for a joint utilization of the communication paths and the project storage was pending (Figure 56.2). Only such integration permits the requests of customers for more production flexibility (e.g., program-controlled parameter value assignment) and for increased availability (predicting and/or faster diagnosis and preventive maintenance) to be satisfied. However, the working group agreed to strive for solutions that keep the device manufacturers as independent as possible from the system manufacturers.

From the very beginning, it was planned to add safety gateways to other safety bus systems like ASi-Safety-at-Work.

Safe communication could be achieved by using redundant transmission lines. However, the working group decided to look for a "single channel" solution for safety applications such that redundancy can still be added to the system as an option to provide additional higher availability/reliability(Table 56.1). Eventually, the safety solution for PROFIBUS was called PROFIsafe and a sign was created.

56.4   PROFIsafe, The Solution

As we know, PROFIBUS DP, like most fieldbus systems, only uses layers 1, 2, and 7 of the ISO/OSI model [4]. No changes to any of these layers means to add the safety measures as a safety layer on top of the PROFIBUS layer 7, thus increasing the size of the OSI application layer. Since this safety layer is merely responsible for the transport of safety-relevant "user" or process data, it takes the rest of the application layer to look after the acquisition and processing of this data. These higher layers can be provided, for example, in a safe field device (e.g., light curtain) by the technology firmware of this device. Usually, parts of this firmware are of a safety-oriented design anyway — typically a redundant hardware/software structure, in which the PROFIsafe functionality can be embedded (Figure 56.1). Like in standard mode, the process data (signals and/or process values) are packed in the processing data unit (PDU) of a PROFIBUS message frame. In case of PROFIsafe, the raw process data are just supplemented by additional information, as we will see in the following sections. The so-completed safety-relevant process data are called a "PROFIsafe frame." Ideally, a "PROFIsafe frame" shall be passed completely unmodified from a (safety)

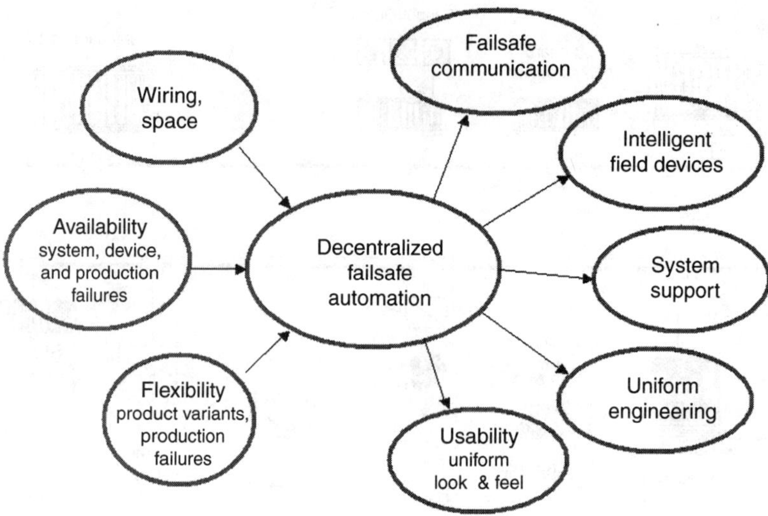

FIGURE 56.2   Requirements and tasks for distributed safety technology.

sender to a (safety) receiver no matter what kind of transmission system both had been using. Thus, the safety measures are *encapsulated* in the communicating end devices.

56.5 Black Channel

Such a communication system is called a "Black Channel" analogical to a "black box." This means first, that the chosen communication technology does not matter except for a few basic constraints to be defined later on in this document. Second, none of the error detection mechanisms of the chosen communication technology are taken into account to guarantee the integrity of the transferred process data. Basically, there are no restrictions with respect to transmission rate, number of bus devices, or transmission technology — as long as its parameters are tolerated by the required reaction times of a given safety application. The following example (Figure 56.3) shows that PROFIsafe also uses the "Black Channel" principle for complex PROFIBUS structures.

Safety-related sensor signals from a modular slave (F-Slave), that is, a PROFIBUS device that can be equipped with several safety-related modules with input/output (I/O) channels, are routed via the

TABLE 56.1   Safety and redundancy options

	PROFIBUS DP	PROFIsafe	Redundancy	PROFIsafe and Redundancy
Application	Suitable for all kinds of distributed automation	Factory and process automation. Presses, robots, level switches, shutdown valves, as well as burner control and cable cars	Process automation. Chemical or pharmaceutical productions, refineries, offshore	Process automation. Chemical or pharmaceutical productions, refineries, offshore
(high) Availability	-	-	No downtimes at best (fault tolerence)	No downtimes at best (fault tolerence)
Safety	-	Avoid hazards (required by laws or insurances)	Redundancy by itself does not provide safety	Avoid hazards (required by laws or insurances)

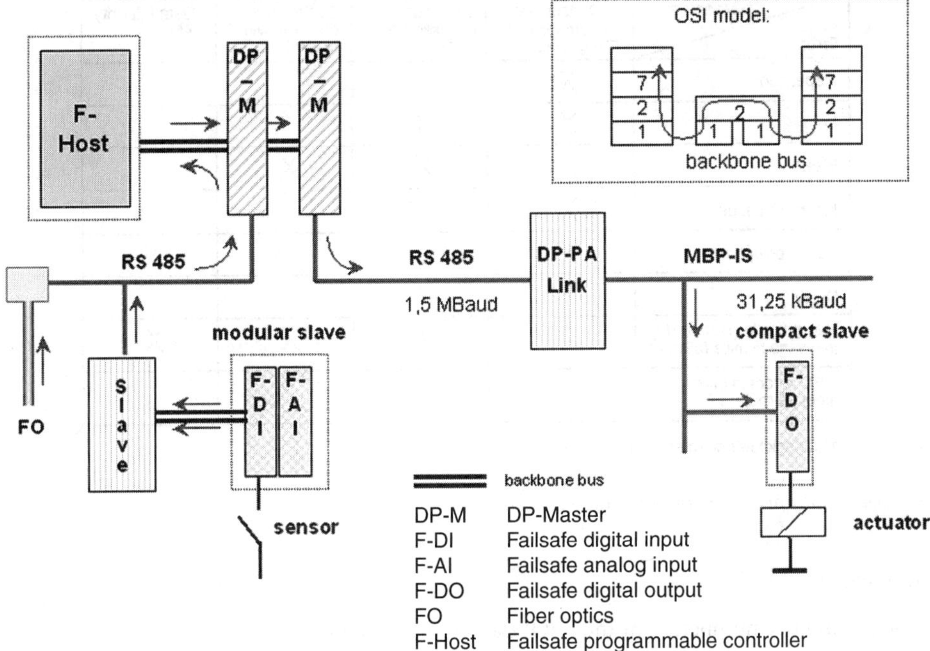

FIGURE 56.3　Complete communication paths for PROFIsafe frames.

PROFIBUS slave node to one of the two DP master nodes of a controller. From here, they proceed via a local backbone bus to the F-Host, the programmable safety-related system (e.g., Safety-PLC or F-Host). After the safe logic operation, a corresponding output signal is routed via the local bus to a second DP master node and to a second PROFIBUS segment. The transmission rate is reduced there in a DP– PA link, and the intrinsically safe transmission technology (MBP-IS) is used for routing the signal to the safety-related PA (process automation) slave. At no point on its communication path has the signal used a redundant communication path. In other words, this is a *single-channel* transfer.

Up to now, we simply dealt with the communication paths for safety-related message frames. Thus, it is still open who is responsible for the transfer, and when the transfer will take place. Here too, a PROFIBUS standard mechanism is used — the master–slave operation. A master, which may be assigned here to an F-Host, *cyclically* exchanges PROFIBUS messages with all its configured slaves one after the other. This means that there is always a *1:1 relationship* between master and slave. The polling operation has the advantage that any failed device will be detected immediately — one of the basic principles of safety technology. Now we are able to identify the basic constraints for the "Black Channel" for PROFIsafe operations: the polling principle and the 1:1 relationship.

56.6　Possible Transmission Errors and Their Remedies

Various errors may occur when messages are transferred in network topologies of the described complexity, be it due to hardware failures, or due to extraordinary electromagnetic interference, or other influences. A message can be lost, occur repeatedly, be inserted from somewhere else, appear delayed or in an incorrect sequence, and/or show corrupted data. In the case of safety-related communication, there may also be incorrect addressing — that means, a standard message erroneously appears at a safety device and pretends to be a safety message. Different transmission rates may additionally cause storage effects to occur. Out of the numerous remedies known from the literature, PROFIsafe concentrates on those presented in the matrix shown in Figure 56.4.

Measure: Error:	Consecutive number (sign of life)	Time-out (with acknowledgement)	Codename (for sender and receiver)	Data integrity (CRC)
Repetition	X			
Loss	X	X		
Insertion	X	X	X	
Incorrect sequence	X			
Data corruption				X
Delay		X		
Masquerade (standard message mimics failsafe)		X	X	X
FIFO errors in intermediate routers		X[1]		

1) no acknowledgement from routers

FIGURE 56.4 Transmission error types and remedies.

These include:

- the consecutive numbering of the PROFIsafe frames ("sign-of-life"),
- a time expectation with acknowledgement ("watch-dog"),
- a codename between sender and receiver ("password"), and
- data integrity checks (CRC = cyclic redundancy check).

Using the consecutive number, a receiver can see whether or not it received the PROFIsafe frames completely and within the correct sequence. When it returns a PROFIsafe frame with the consecutive number only as an acknowledgement to the sender, the sender, too, will be assured. Basically, a simple "toggle bit" would have proven sufficient. Due to the storing bus elements (e.g., routers), however, a counter from 0 to 255 has been selected for PROFIsafe. Zero is an exception in this procedure reserved for the startup transitions.

In safety technology, it not only matters that a message transfers the correct process signals or values. In addition, updated actual values must arrive within a fault tolerance time, thus enabling the respective device to automatically initiate safety reactions on-site, if necessary (e.g., stop of movement). For this purpose, the devices are using a watch-dog timer that is restarted whenever new PROFIsafe frames with incremented consecutive numbers arrive.

The 1:1 relationship between the master and a slave facilitates the detection of misdirected message frames. Master and slave must simply have an identification (password) that is unique in the network, and can be used for verifying the authenticity of a PROFIsafe frame.

Detecting corrupted data bits through an additional cyclic redundancy check (CRC) plays a key role. The necessary probabilistic examination can benefit from the definitions within the IEC 61508 [1] that considers the probability of failure of entire safety functions. PROFIsafe is following this approach (Figure 56.5).

Accordingly, a safety circuit includes all sensors, actuators, transfer elements, and logic processes that are involved in a safety function. IEC 61508 defines overall values for the probability of failures for different safety integrity levels. For SIL3, for example, this is 10^{-7}/h. For the transmission, PROFIsafe merely takes up 1%. This means that the permissible probability of failure is 10^{-9}/h. This permits suitable CRC polynomials to be determined for the intended PROFIsafe frame lengths. The resulting residual error probabilities of undetected corrupted PROFIsafe frames guarantee the required order of magnitude (see the following sections for details). The quality of the chosen CRC polynomials are such that in the case of PROFIsafe, we no longer depend on the basic error detection of standard PROFIBUS DP using frame checking sequence (FCS) and parity check. Thus, proof is not required that the error detection probability of the basic PROFIBUS DP mechanisms is independent of one of the additional PROFIsafe CRC mechanisms.

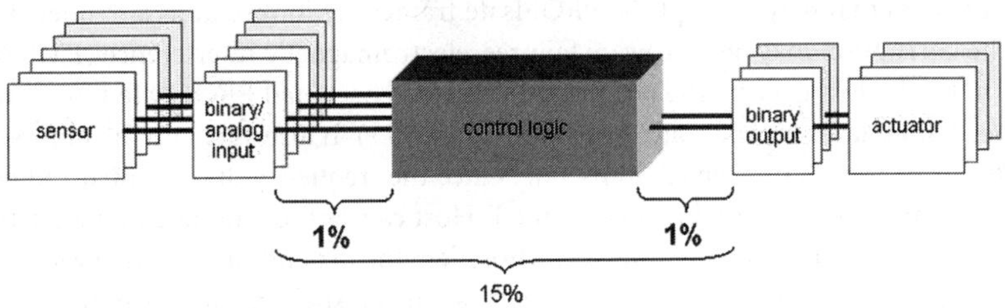

FIGURE 56.5   Safety function acc. IEC 61508.

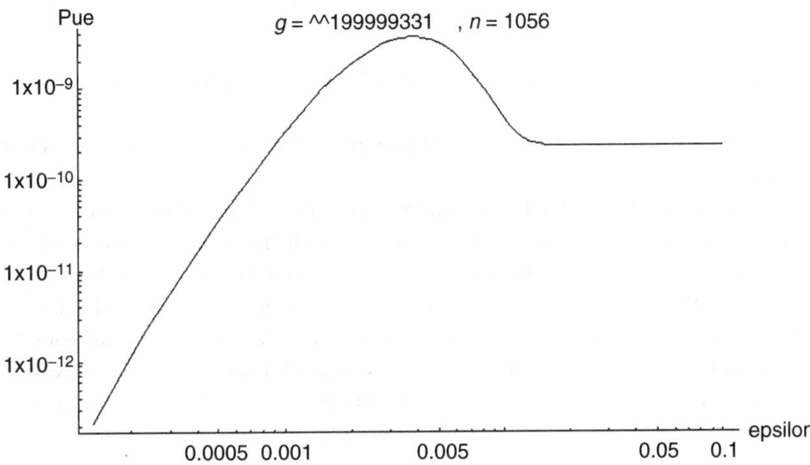

FIGURE 56.6   Example of an improper polynomial.

Besides other things, we need so-called "proper" polynomials. A polynomial is called "proper" when the residual error rate over an increasing bit error rate does not show a curve with pronounced peaks (i.e., when the curve rises continuously monotonic). This has been proven mathematically for the polynomials used for PROFIsafe [5].

16 Bit: 14EABh   32 Bit: 1F4ACFB13h

Figure 56.6 shows an example of an *improper* polynomial. Epsilon on the *x*-axis stands here for the bit error probability.

The residual error probability (*y*-axis) becomes problematic with a high bit error probability (i.e., with a very large number of corrupted bits in a PROFIsafe frame). In order to avoid any insecurity, PROFIsafe uses a patented procedure — the SIL Monitor.

56.7   The SIL Monitor

As mentioned before, PROFIsafe is not relying on the basic data integrity checks of PROFIBUS DP. The entire error detection that is necessary for attaining the required category and/or SIL is advantageously implemented in the additional PROFIsafe protocol. Being dependent on the basic safety structure would require complex verifications for all possible bus configurations. Thus, a mechanism was created that guarantees the compliance with SIL levels over the service life of a distributed safety-related automation solution, irrespective of the components and the configuration used.

Here, all influences that may corrupt the PROFIsafe frames are looked at as a frequency f_w, whatever the cause of the corruption may be: hardware failures, electromagnetic interference (EMI), etc. For each safety function, the F-Host collects the number of detected corrupted PROFIsafe frames reported to it from the associated F-Slaves via the status byte (see Figure 56.7). It also sums up the number of detected corrupted PROFIsafe frames it received on its side. Once the frequency of corrupted PROFIsafe frames exceeds a certain limit (e.g., high bit error rate), the F-Host causes the safe state in the safety function to be activated. This requires monitoring intervals whose lengths depend on the SIL level. For example: $T = 5$ h for SIL3 and $T = 0.5$ h for SIL2. The values are calculated such that a maximum of *one* corrupted PROFIsafe frame is tolerated inside the monitoring interval T. The basis for the calculation is the following formula:

$$\Lambda_{HW+EMI+other} = f_W \{P_{UB(typ)} P_{US} + P_{US(typ)}\} = 1/T \{P_{UB(typ)} P_{US} + P_{US(typ)}\} < 10^{-9}/\text{h}$$

P_{US} = maximum residual error probability for 16/32-bit CRC, at a bit error rate of 0 ...0.5.

The SIL monitor is no separate component. It is implemented as a part of the PROFIsafe driver software within the F-Host.

There are two groups of safety-related parameters. The "F" parameters are associated with the PROFIsafe layer and the "i" (= individual) parameters with the individual safety-related technology firmware of a safety slave. The above-mentioned CRC protection is not only used for cyclically ensuring the data integrity of the process signals and process values but also for ensuring the data integrity of the safety-related parameters stored in the individual slaves (such as codename, watch-dog time, etc.). For this purpose, a separate CRC value of these parameters is generated at longer time intervals (≤ 8 h). This CRC value is then the new start value of the cyclic CRC generation. The same procedure is executed in the F-Host, where the parameters of the slaves are stored.

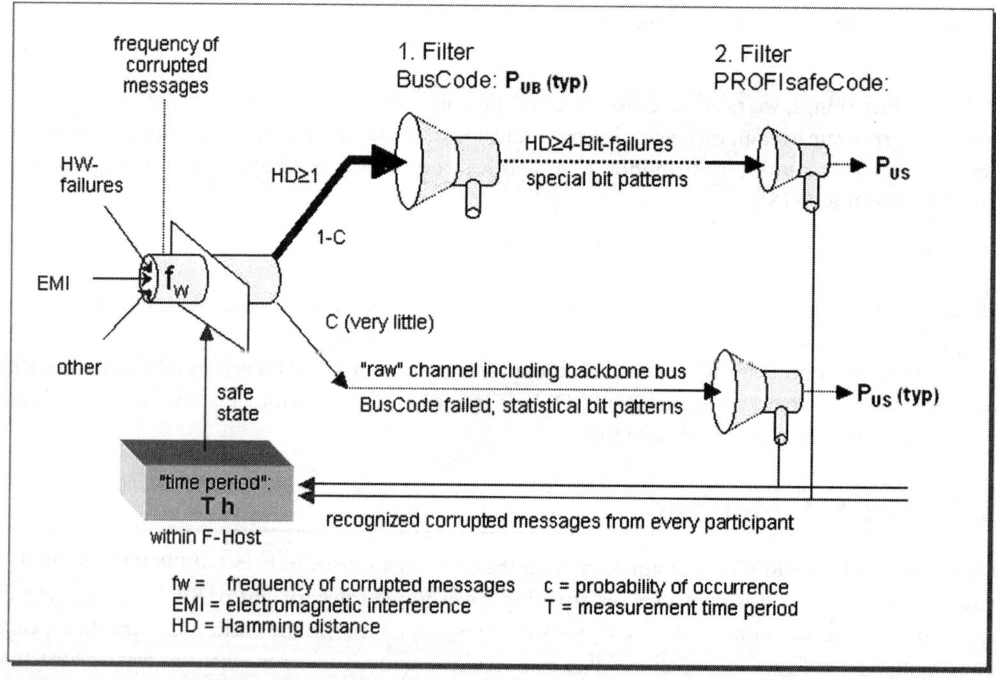

FIGURE 56.7   The SIL monitor.

56.8   PROFIBUS Messages with PROFIsafe Frames

Up to now, we have discussed the processes that PROFIsafe uses for safely transporting the PROFIsafe frames. This section now deals with the concrete mapping onto the PROFIBUS DP communication. Figure 56.8 shows the structure of a PROFIBUS DP message frame.

In this case, only the data unit, the parity bit (PB), and the frame checking sequence (FCS) are of interest. The configuration part of an engineering tool uses the electronic Generic Station Description (GSD) of a particular slave device to arrange for the format of the net process data to be transferred within the data unit of a message between a PROFIBUS DP master and its slave device. The same occurs in the safety case.

Due to the encapsulation via PROFIsafe layers, control and status information must be exchanged between an F-Host and an F-Slave to synchronize the sequences of the state machines. This is necessary, for example, in case an F-Slave requires i-Parameter sets or if the user program within the F-Host wants to change i-Parameter values. Furthermore, the F-Slave can report the detection of an incorrect PROFIsafe frame to the F-Host in order to support the SIL monitor function (Figure 56.8). For the described information exchange, PROFIsafe features a byte that follows the F-process data (Figure 56.9). Another additional byte contains the above-mentioned consecutive number. This number, which is entered by the sender of a PROFIsafe frame ("source-based counter"), is verified by the receiver and returned to the sender in the acknowledgement PROFIsafe frame. Counting is performed in a cycle from 1 to 255. 0 is reserved for the system start.

We have already discussed the fact that factory automation and process automation place different requirements upon a safety system. The former deals with short ("bit") signals that must be processed at a very high speed; the latter involves longer ("floating point") process values that may take slightly more time. PROFIsafe therefore offers two different "F" process data lengths that require CRC protection of different complexity. One length is limited to a maximum of 12 bytes; it requires a 2-byte CRC following the consecutive number. The other length is limited to a maximum of 122 bytes; it requires a 4-byte CRC. The remaining space in the PDU of the message may be used for standard data or other PROFIsafe frames. This results in a very efficient communication with modular F-Slaves or F-Gateways to other safety busses, for example (Figure 56.9).

The consecutive number enables the receiver to monitor the vitality of sender and communication link. Using the acknowledgement mechanism, it is also used for monitoring the propagation times between sender and receiver (Figure 56.10).

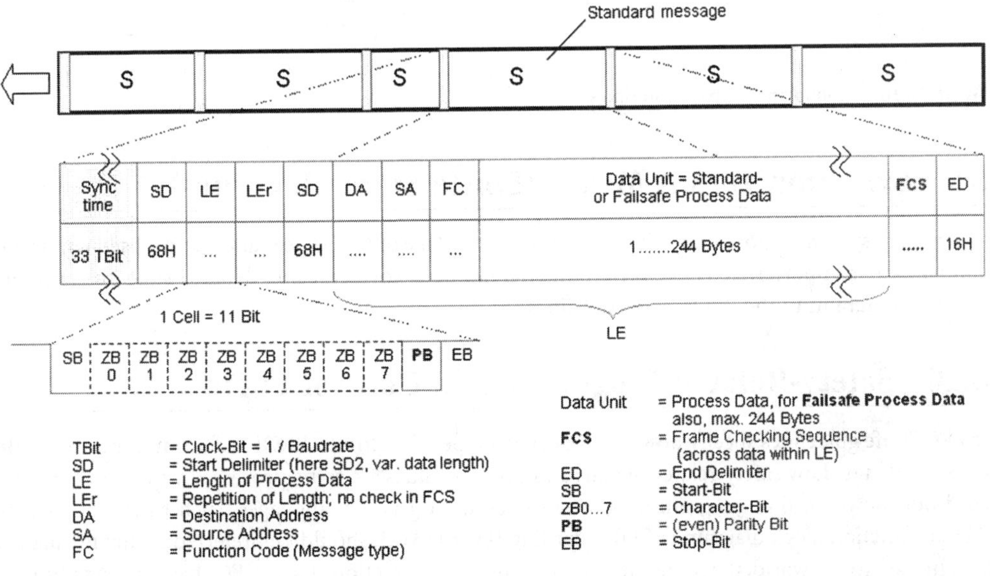

FIGURE 56.8   PROFIBUS DP message structure.

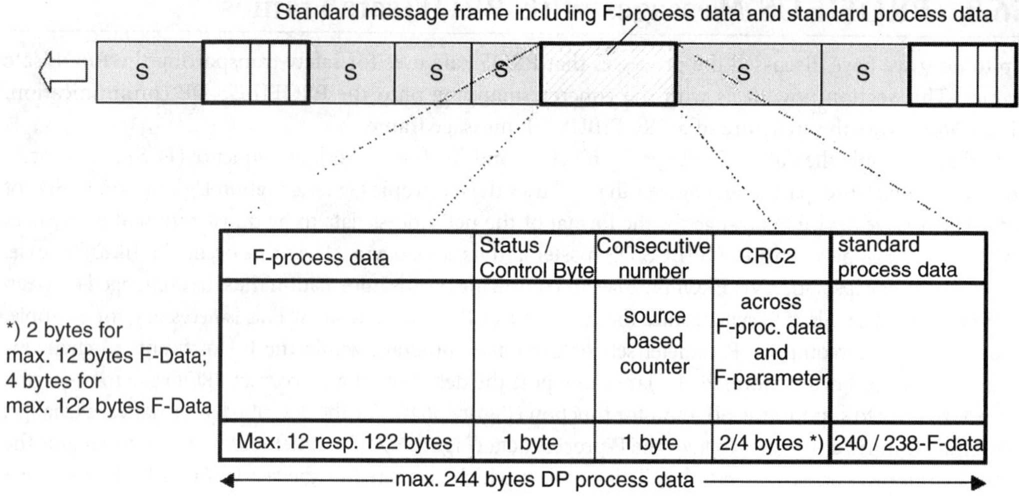

F-process data	Status / Control Byte	Consecutive number	CRC2	standard process data
*) 2 bytes for max. 12 bytes F-Data; 4 bytes for max. 122 bytes F-Data		source based counter	across F-proc. data and F-parameter	
Max. 12 resp. 122 bytes	1 byte	1 byte	2/4 bytes *)	240 / 238-F-data

FIGURE 56.9 PROFIsafe frame structure.

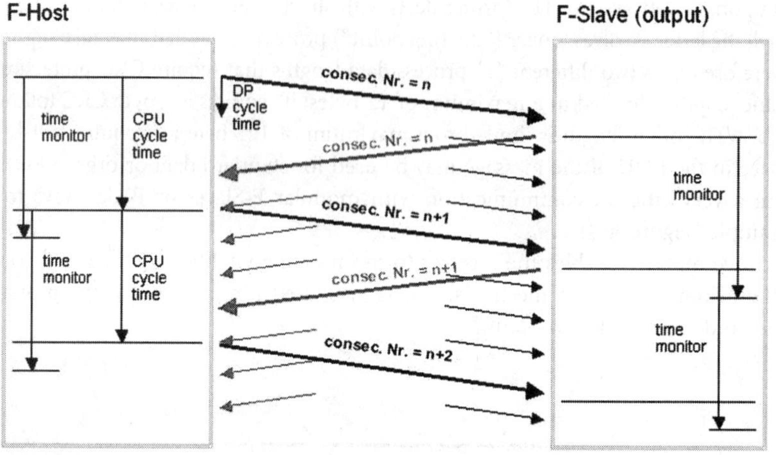

FIGURE 56.10 Monitoring propagation times.

56.9 Synchronization Means (Finite State Machines)

In distributed systems, a potential failure of the supply voltage of the devices and their subsequent restart are of particular importance. This hazard potential has been covered using detailed state diagrams and interaction diagrams in the PROFIsafe guidelines [5].

56.10 Safety-Related Programmable Control Logic

The PROFIsafe guidelines specify how to connect any type of F-Host to PROFIsafe communications. But it does not dictate how safety-related signals and process values are to be processed within an F-Host. It turned out, however, that programming safety circuits in graphic languages, such as ladder diagram (LAD) or function block diagram (FBD) according IEC 61131-3, should be preferred to any textual language. In the process world, the representation in continuous function chart (CFC) has proved to be helpful. Approved relay-based safety circuits now can be offered as a standard certified software library [6].

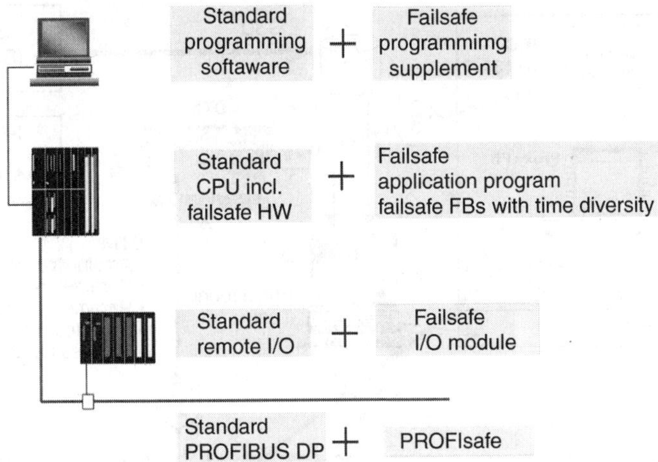

FIGURE 56.11   Combined standard and safety controls.

There are manufacturers offering the option of retrofitting safety-related user programs for some of their standard programmable logic controllers, which fulfill the necessary requirements for safety-related applications. The safety software is of a time diversity structure; it is executed in protected areas where it cannot be influenced by the standard application program. Special F I/O modules can supplement the associated standard remote I/O field devices (see Figure 56.11). CNC control systems, too, provide the connection for PROFIsafe field devices.

56.11   Commissioning and Repair

In order to be able to "address" the individual signals and/or process values during programming, the fieldbus must be *configured* using an engineering tool. In this configuration, the fieldbus participants arrange for their network address, the input signals, the output signals, the data formats, the transmission rate, etc. In case of F-Slaves, additional steps are required that supply parameters to the PROFIsafe layer: Codename, watch-dog time, etc. For the computer-aided calculation of total reaction times, in particular, and, consequently, the determination of watch-dog times, the device manufacturers specify the processing time of sensors and actuators in their electronic data sheets (GSD). During startup of the fieldbus network, the host/master sequentially provides all slaves with the necessary parameters including F-Parameters before it starts cyclic operation [7].

Due to the wide acceptance of microcontrollers, various new devices such as laser scanners and light curtains have turned up in the past even for safety applications that feature a certain complexity and require system support to be able to develop their full functionality. These are, for example, teach-in, rapid device replacement in the event of a malfunction, predictive diagnostics, self-testing and trend analysis, and more.

In its profile guidelines, PROFIsafe has shown ways for mastering these tasks — the so-called 3-component model that separates the responsibilities of device manufacturers and system manufacturers by using suitable interfaces. In the simplest case of an F-Slave that does not require any individual ("*i*") parameter, providing the safety device with PROFIsafe connection and a corresponding electronic data sheet (GSD file) proves sufficient. A possible example is a level switch. Slightly more effort is required in the case of a safety device. It can be connected to a laptop via a separate RS-232 interface. In order to perform diagnostics, for example, it uses a special parameterization and diagnostics program. In this case, the laptop or PC communication can also be routed via the PROFIBUS, using acyclic services that are accessible via the new PNO standard *FDT/DTM* interface technology [8]. Typical representatives of such devices are ESD (electronic shutdown) valves. A device manufacturer would have to provide the safety

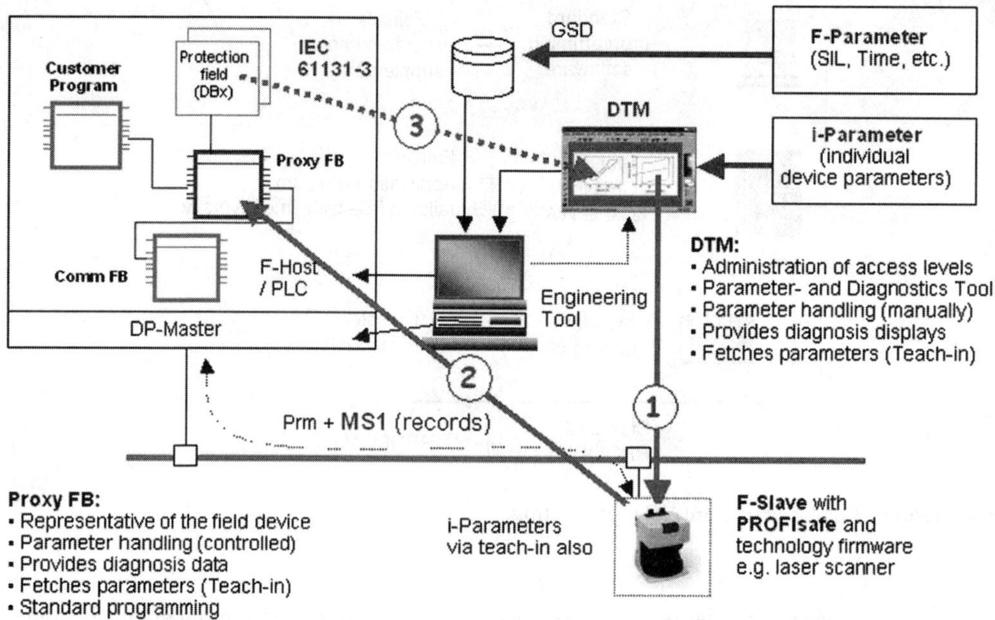

FIGURE 56.12 Parameter handling with powerful safety devices.

device, a GSD file, and a DTM program. The DTM program corresponds to the special parameterization and diagnostics program but it uses the standard FDT interface instead.

The most demanding devices are the ones that require user-program-controlled parameter value assignment in addition to the above-mentioned features. This requires an additional so-called proxy function block (FB) in the controller (e.g., F-Host). Typical representatives of this device class are the laser scanners for which we want to "play" a user scenario (Figure 56.12). DTM represents the parameterization and diagnostic program of the laser scanner manufacturer.

The FDT framework of the engineering tool arranges for the acyclic connection between the DTM and the safety field device. A user can carry out the parameter value assignment of the laser scanner (configuration of the protection field, e.g.) as he/she used to do it via the previously used RS-232 interface. The safety measures in the software, such as divers representations and various authenticity mechanisms, can remain the same. During each upload or download, the "i" (individual safety-related)-Parameters are CRC-protected block wise.

There are new steps now. With the help of the DTM, the user triggers the upload of the i-Parameters from the device into the F-Host. To do this, the proxy FB generates instant data for accommodating the i-Parameters. To ensure correct addressing and correct i-Parameters, the DTM — with the help of the engineering tool — reads the instance data of the proxy FB, and compares it with the i-Parameters directly read from the safety device. From here on, the proxy FB would be able to load the i-Parameters automatically into the laser scanner in the event of a device replacement. By checking its device type and its production number, the proxy FB finds out whether the device is correct or is a different one.

56.12 Availability

The availability plays a significant role in safety technology. It is simply a precondition. There is too much of a risk that a frequent — and apparently groundless — trip of the safety devices causes the system to shut down ("Bhopal" effect) [9]. There are mainly five areas that influence the availability of a system:

- Design/construction of the components
- Facility layout
- Operating conditions
- MTBF of the components (bathtub curve)
- Requirements for operator actions.

In the meantime, most standard automation components and systems on the market have reached a design and construction quality standard that ensures high reliability (high meantime between failures). Among the causes that led to this situation are standards, test institutes, and certifications. However, the availability (up-time in relation to down-time plus up-time) may deteriorate in safety systems that are based on standard systems. Redundant standard components (e.g., two or three controllers) that require synchronization can cause this. The safety device trips if one component takes longer than the other one due to an external event. This risk is also in the redundant implementation of sensors whose measured values diverge briefly (discrepancy). PROFIsafe therefore endeavors to support certified safe integrated F-Slaves that take the risk away from the user as much as possible. Standard fieldbus systems, even proven-in-use, without additional mechanisms like PROFIsafe or special overall risk assessments of an application may not be considered safe.

In order to prevent electromagnetic interference from affecting the operation of a system, the PROFIBUS User Organization has published installation guidelines [10] whose utilization in bidding procedures is recommended, thus creating a clear situation right from the outset. They specify aspects like grounding, shielding, lightning protection, etc. PROFIsafe makes the satisfaction of these installation guidelines a prerequisite for the operation.

An additional redundant structure of the entire system provides a significant increase of the availability. PROFIsafe has taken this mode of operation into account right from the outset.

The operating conditions of a system have a strong influence on the reliability/availability. As we all know, a temperature rise of 10 K reduces the service life of an electronic device by 50 percent. Here, PROFIsafe can merely give recommendations. Devices should be designed for the usage in one of the following zones: Office environment, "enclosed-type" operation within racks, process participation inside and outside a building.

The means of the *predictive diagnosis* is used in order to avoid unexpected late failures toward the end of the service life of a field device. The proxy FBs are provided in the PROFIsafe concept. These proxy FBs can execute, automatically and periodically, test programs in the field device in order to determine its states and the trend. This enables risky units to be replaced at a time when the process is stopped anyway.

The availability of a system is also reduced if the user is unable to handle it properly due to complexity or variety. PROFIsafe therefore currently strives in the PROFIBUS User Organization to support all activities of standardizing the device performance characteristics and the software user interfaces [11].

56.13 Status of Profile Guidelines

At Hanover Fair 1999, the PROFIBUS User Organization had published version 1.0 of the PROFIsafe guidelines. It has been confirmed by positive reports from the German Institute for Occupational Safety and Health (BIA) and the worldwide operating Safety-Assessment Organization (TÜV). In the meantime, these guidelines have been enhanced by the requirements for the integration of any given safety-related control system (e.g., computerized numerical controls). Since most F-Host manufacturers currently use diversity hardware, the PROFIsafe working group also dealt with possible connections to these systems. The most common data types and rules for its arrangement have been defined. Via the information within a GSD file of a field device, the corresponding PROFIsafe layer driver in the F-Host can now be set up. Version 1.20 of the PROFIsafe guidelines was published in the fall of 2002. The guidelines and current notes can be found in the Internet under http://www.profibus.com.

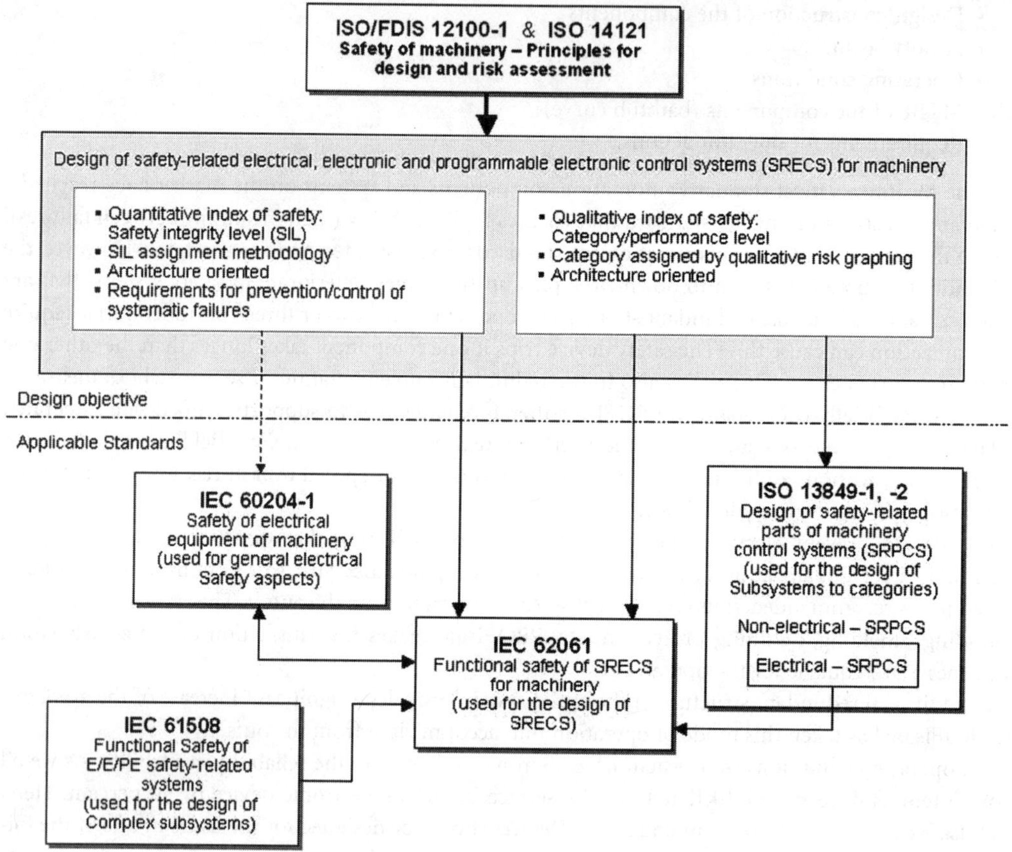

FIGURE 56.13   Relationship between relevant standards for machinery.

56.14   Standards Catching Up

The development of PROFIsafe was based on the new IEC 61508, which recently became European standard and mainly deals with the software development procedures. Up to now, various standards do not go yet into the new possible solutions and are therefore revised and/or newly created. This includes IEC 61511, which deals with the particular problems of process safety on the basis of IEC 61508.

For safety-relevant machine controls with programmable electronics, IEC 62061 also refers to IEC 61508. Figure 56.13 shows how this standard is embedded within other safety-relevant standards. In the U.S.A., existing restrictions against programmable electronics for emergency stop functions disappeared with the new NFPA 79 version that was published by the end of 2002.

56.15   Peculiarities for Different Industries

Within factory automation, the discussion about the need of using a separate safety bus in addition to the standard fieldbus for the implementation of distributed safety technology comes up time and again. For PROFIBUS and PROFIsafe, this vexatious second bus is not a problem. The single-cable solution with a combined standard and safety-related control in one CPU and the solution with separate transmission lines and separate CPUs can both be implemented using the same means. However, compared to a heterogeneous solution with two different bus systems, this "homogenous" solution

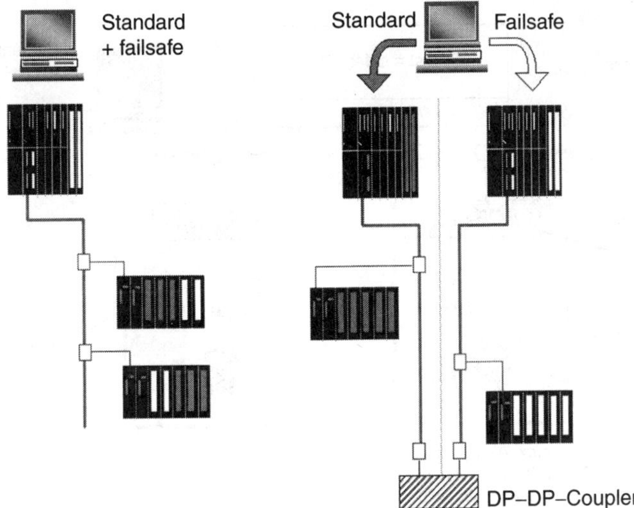

FIGURE 56.14 Safety on one transmission line or separately.

provides advantages through the same technique and engineering tools of uniform handling (Figure 56.14). Many users are following a migration strategy. First, within existing facilities, the new PROFIsafe technology will replace the relay-based safety technology via a second fieldbus segment. At the next opportunity and after some experience, it is planned to go with the combined solution due to the many obvious advantages.

In the meantime, several activities have taken place at the process automation side. As we know, safety applications in process engineering require a consideration that exceeds the functional safety. Pressure and temperature in a process, for example, cannot always be "monitored" independent of each other. A high availability of sensor functions requires a large degree of design skill and experience. To a large degree, the topic of "proven-in-use" plays a large role.

Up to now, the part of the protection equipment — which mostly represents cases similar to SIL2 — has been solved with standard field devices and 4–20-mA transmission technique. Communication faults, such as line interruptions or short circuit, were recognizable, the devices were proven-in-use, and safety monitoring was performed in the F-Host system by interpreting redundant signals and voters. In the case of microprocessor-equipped equipment, guidelines like the NAMUR recommendation NE79 defined the necessary prerequisites. These include the proof that the software/firmware has been developed according to generally recognized quality standards. Furthermore, the user expects facilities such as watch-dog timer, memory test, and protection against inadvertent parameter changes. The new recommendation NE97, which NAMUR published at the end of 2002, deals with the subject "Fieldbus for Safety Tasks." Among other things, this recommendation shows two possible solutions that can easily be implemented with PROFIsafe.

A proven-in-use field device, developed according to NE79 that is connected to PROFIBUS, could actually be used for applications up to SIL2 if the possible communication faults that are known in bus operation: Address corruption, loss, delay, etc., could be eliminated. As we already know, this is possible if we use a PROFIsafe protocol that is implemented in a single channel manner. In the future, this solution permits proven-in-use standard PROFIBUS field devices to be operated in standard operation mode or optionally in safety mode, just by changing a single parameter. The user is at liberty to use certified products (see Figure 56.15).

In many systems, the installed proven-in-use field devices are retained, and an additional fieldbus wiring is installed via a so-called remote I/O. For this operating mode, too, NE97 provides for a solution.

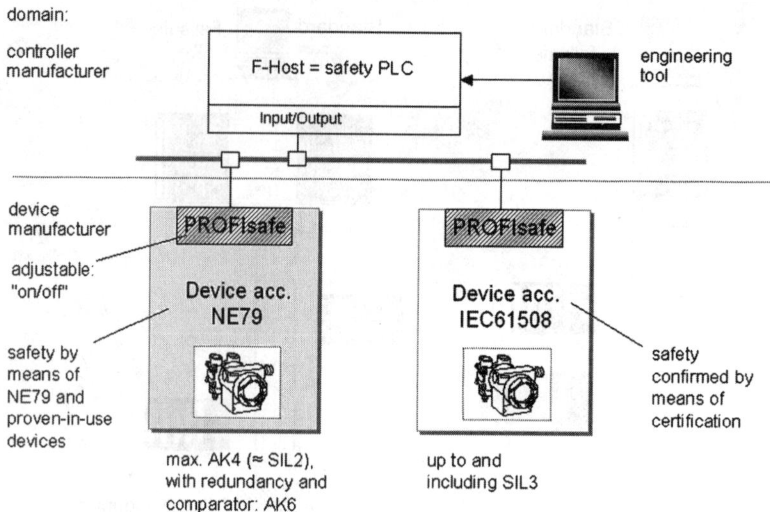

FIGURE 56.15   PROFIsafe in field devices for process automation.

The process automation field devices are usually subject to a strong standardization of device model and device parameters. Parameter value assignment, too, is expected to be as uniform and consistent as possible. To take this into account, the PROFIBUS User Organization has started to integrate PROFIsafe communication compatibly into the existing "PA device model" [12]. The related specification [13] especially deals with the following topics:

- switching between standard/safety mode,
- preparation and commissioning phases,
- new parameters in the "Physical Block" of the PA device model,
- safety-related data structures for the cyclic data exchange, and
- configuration data.

As already mentioned, the field devices should be usable in standard mode and in safety mode. Thus, it is planned to achieve this switch by corresponding parameter settings in the related GSD file, and thus, during the start-up of the field device. The required safety integrity level (SIL) is then communicated via an F-Parameter (F_SIL).

Due to the contingency that parameters have to be edited during operation, a state machine has been designed for the individual commissioning phases. The three states are: "None Safety Mode — S1," Safe Commissioning — S2," and "Safe Operation — S3."

This extended state machine provides the option of using additional states that permit existing parameter assignment tools based on Electronic Device Descriptions (EDD) to be used [14]. In this case, the parameter assignment usually is performed offline and the device is sealed by soldering or similar procedures.

The use of safe parameter value assignment procedures also permits i-Parameter value assignment according to the PROFIsafe Profile Guidelines [5] to be performed at runtime directly in the safe state S3 using FDT/DTM.

Additional parameters are required in such a safety-related field device for mastering the new tasks. Currently, there is one parameter for setting the SIL level, one parameter for displaying the operating states S1 – S3, and a device-individual password for write protection.

Additional data formats for the PROFIsafe frame have been supplemented with the standardized data formats for the PA device model for cyclic data exchanges. The corresponding configuration identifiers have been defined.

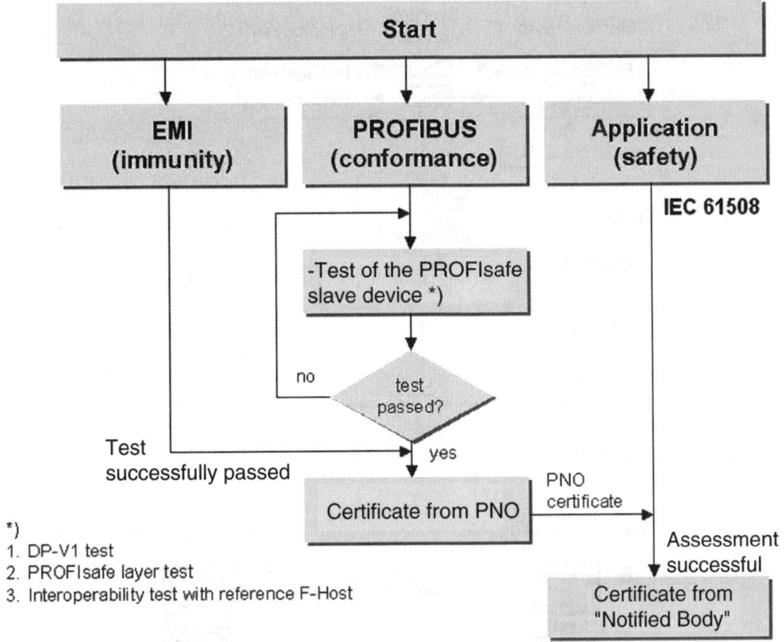

FIGURE 56.16   The PROFIsafe certification procedure.

56.16   Environmental Conditions

In the meantime, the PROFIBUS User Organization has published further specifications since most safety standards ignore bus operations. One of these is defining the boundary conditions for safe communication [15]. It deals with the topics of installation guidelines, electrical safety (overvoltage shock protection — SELV/PELV), power supply units, and immunity to electromagnetic interference. The other is describing a policy for all PROFIBUS member companies and providers of safety-related devices dealing with responsibilities and the quality of products and services [16].

56.17   Test and Certification

The PROFIsafe mechanisms are based on finite state machines. Thus, it was possible via a validation tool for finite state machines to mathematically prove that PROFIsafe is working correctly even in cases where more than two independent errors or failures may occur. This was achieved systematically by generating all possible cases for "test-to-pass" and "test-to-fail" situations. They have been extracted as test cases for a fully automated PROFIsafe layer tester, which is used to check the PROFIsafe conformance of safety-related devices. It is part of a three-step procedure within the overall certification process (Figure 56.16) [17].

56.18   Development Tools and Support

Using PROFIBUS/PROFIsafe in factory automation with short reaction times, and — on the other hand — in hazardous areas in process automation with lowest power dissipation places contradictory tasks. Since most modern devices use microprocessors, implementing the PROFIsafe procedure in software is an obvious solution. The operating conditions determine the selection of the microprocessor with respect to high performance or low power dissipation. PROFIsafe automatically adapts ideally; it requires merely some Kbytes of free memory: no additional power supply and no additional space in the confined enclosures. PROFIsafe becomes a part of the device-specific safety software.

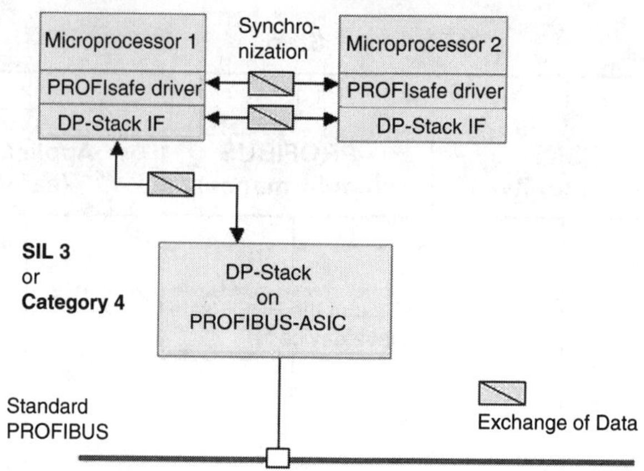

FIGURE 56.17   Generic PROFIsafe driver for slaves.

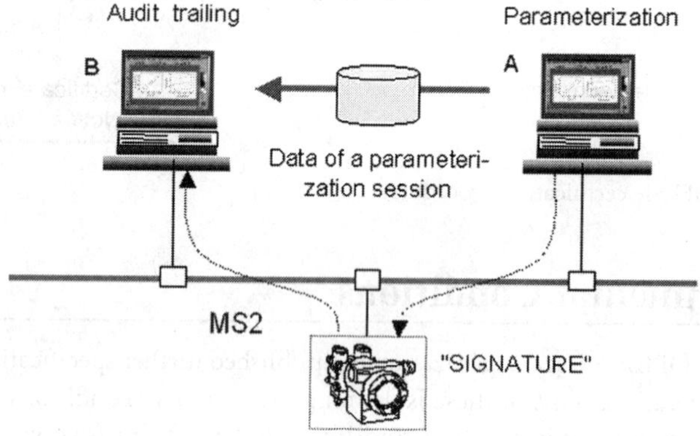

FIGURE 56.18   Audit trailing for secured safety parameters.

Immediately after completion of the specification, a company consortium started a joint development of a "generic PROFIsafe driver software" for slaves. Generic in this context means that the driver has been implemented in ANSI-C, and according to the encoding rules for safety technology for different microprocessors and C compilers. Although it should preferably be used in dual-channel mode (see Figure 56.17), it may also be used in single-channel mode (see Figure 56.15).

For the specifications, the approval authorities merely issue so-called "positive technical reports." With the acceptance of the software in a concrete hardware, we now have a TÜV certificate with the number: Z2 01 12 20411 008.

Without the help of simplifying step-by-step development equipment, the development of safety-related fieldbus devices turned out to be difficult due to the safety measures (consecutive numbering, watch-dog timers, etc.) that have to be fulfilled at any time in a real environment. The members of the PROFIBUS User Organization are now providing development boards and PC Software (PROFIsafe Monitor) to develop step by step from PROFIBUS DP to the PROFIsafe communication level.

56.19 Products

Currently available products are safety-related programmable controllers and numerical controllers, remote-I/O for IP20 and IP67, laser scanners, light curtains, motor starters, frequency converters, drives, gas and fire sensors, etc.

For further actual details, see www.profibus.com.

56.20 Prospects

The sophisticated safety-related devices mentioned in this chapter with its many possibilities to use i-Parameters to adapt to all kinds of safety applications in a flexible manner are leading to the requirement for continuous automatic monitoring of security, safety, and data integrity. Thus, regulations like 21 CFR 11 [18] are defining

> Rules for the use of secure, computer-generated, time-stamped audit trails to independently record the date and time of operator entries and actions that create, modify, or delete electronic records. Record changes shall not obscure previously recorded information. Such audit trail documentation shall be retained for a period at least as long as that required for the subject electronic records and shall be available for agency review and copying.

PROFIBUS suggests in its device-independent Profile Guideline, Part 1 "Identification & Maintenance Functions" [19] the use of asset identification parameters and the parameter SIGNATURE to allow parameterization tools to store a "security" code as a reference for a particular parameterization session. Audit trailing tools may retrieve the code for integrity checks at any time. Together with the "asset identification parameters" consisting of MANUFACTURER_ID, ORDER_ID, and SERIAL_NUMBER, the SIGNATURE parameter allows unambiguity of the session data.

PROFINET: Due to its "Black Channel" approach, the PROFIsafe principle will be easily portable to PROFINET once this new technology is well established on the market. In the meantime, safety device manufacturers are well advised to start with a PROFIBUS DP/PROFIsafe version and find an already existing, fast-growing market that is receptive to the new safety technologies of PROFIsafe.

Abbreviations

ASi	Actuator Sensor Interface
ASIC	Application-Specific Integrated Circuit
BIA	German Institute of Occupational Safety and Health
CPU	Central Processing Unit
CRC	Cyclic Redundancy Check
DP	Decentralized Peripherals
DTM	Device Type Manager
EMI	Electromagnetic Interference
EN, prEN	European standard, preliminary ...
F	Failsafe
FB	Function Block
FDT	Field Device Tool
GSD	Generic Station Description (electronically readable data sheet)
HW	Hardware
IEC	International Electrotechnical Commission
I/O	Input/Output
IP	Ingress Protection, for example IP20
ISO/OSI	International Standards Organization/Open Systems Interconnection (Reference Model)
MS1/MS2	Acyclic Master–Slave communication services of PROFIBUS DP
NAMUR	Association of users of process control technology
PA	Process Automation
PLC	Programmable Logic Controller
SW	Software
TÜV	Safety Assessment Organization

References

1. IEC 61508-SER-Ed.1.0-Bilingual: Functional Safety of Electrical/Electronic/Programmable Electronic Safety-Related Systems; all parts, 2004; www.iec.ch.
2. IEC 61158/61784: Digital Data Communications for Measurement and Control — Fieldbus for Use in Industrial Control Systems-part 1-6; 2003; www.iec.ch.

3. IEC 62280-1/EN50159-1: Railway Applications — Communication, Signalling and Processing Systems — Part 1: Safety-Related Communication in Closed Transmission Systems; 2002; www.iec.ch.
4. ISO/IEC 7498-1: Information technology — Open Systems Interconnection — Basic Reference Model; 1994; www.iso.org.
5. PROFIsafe — Profile for Safety Technology, V1.30, June 2004, PROFIBUS Order No. 3.092.
6. www.plcopen.org.
7. Popp, M., The Rapid Way to PROFIBUS DP, PROFIBUS Order No. 4.072.
8. Specification for PROFIBUS Device Description and Device Integration, Vol. 3: FDT Interface Specification, V1.2, PROFIBUS Order No. 2.162.
9. Gruhn, P. and H. Cheddie; *Safety Shutdown Systems: Design, Analysis and Justification*, ISA, 1998; ISBN 1-55617-665-1.
10. Installation Guideline for PROFIBUS DP/FMS, V1.0, September 1998, PROFIBUS Order No. 2.112.
11. DTM Style Guide, V1.1, PROFIBUS Order No. 2.172.
12. PROFIBUS PA Profile for Process Control Devices, V 3.0, October 1999, PROFIBUS Order No. 3.042.
13. PROFIsafe for PA Devices, Working Draft 0.6; www.profibus.com.
14. Specification for PROFIBUS Device Description and Device Integration, Vol. 2: Electronic Device Description, V1.1, PROFIBUS Order No. 2.152, January 2001.
15. PROFIsafe — Requirements for Installation, Immunity, and Electrical Safety, V1.0, PROFIBUS Order No. 2.232.
16. PROFIsafe Policy, V1.3, PROFIBUS Order No. 2.282.
17. PROFIsafe — Test Specification for Safety-related PROFIBUS DP Slaves, V1.1, June 2004, PROFIBUS Order No. 2.242.
18. Title 21 Code of Federal Regulations, U.S. Food and Drug Administration, www.fda.gov.
19. PROFIBUS Profile Guidelines, Part 1: Identification & Maintenance Functions, V1.1, PROFIBUS Order No. 3.502.

57

A Comparative Case Study of Distributed Network Architectures for Different Automotive Applications

Jakob Axelsson
Volvo Car Corporation

Joakim Fröberg
*Volvo Construction Equipment
Components and Mälardalen
University*

Hans Hansson
Mälardalen University

Christer Norström
Mälardalen University

Kristian Sandström
Mälardalen University

Björn Villing
Volvo Truck Corporation

57.1 Introduction

One of the initial driving forces for the introduction of communication networks in automotive vehicles was to replace the numerous cables and harnesses and thereby reduce the number of connection points, cost, and weight. Moreover, *multiplexing*, as vehicle networking is traditionally referred to in the automotive industry, is an important enabler of new and increasingly complex functions. Today, using software and networking, it is possible to create new functionality, such as an antiskid system, that was considered unfeasible, both with respect to cost and functionality, some ten years ago.

The vehicle industry strives to enable cost-effective implementation of new functionality. A network enables reuse of sensor data and other calculated values. Moreover, distributed functions can be facilitated for truly distributed problems like coordination and synchronization of brakes.

Another driving force is the demand for increasingly efficient diagnostics, service, and production functionality. The network in a vehicle should provide functionality not only during normal operation of the vehicle but also in the after market and during production. It should also provide communication for diagnostic functions in control units, and provide a single point interface to service tools to meet goals on more efficient service.

There is a wide span of requirements on the communication infrastructure in today's vehicles. The vehicle industry works with demands on functionality, reliability, and safety, but also with demands related to product variation, extendability and maintenance of delivered products, and integration of supplier components. This implies high requirements on network flexibility in terms of adding or removing nodes or other components. Moreover, part of the functionality has stringent requirements on real-time performance and safety, for example, safety-critical control applications. Other parts of the functionality, such as the infotainment applications, have high demands on network throughput. Yet, other parts require only lightweight networks, as, for example, locally interconnected lights and switches. All of these varying requirements in vehicle networks are reflected in the architecture, implementation, and operation of a modern in-vehicle network.

In this chapter, we try to describe the context in which today's vehicle industry work and develop distributed systems. This is uncovered in three case studies covering passenger cars, trucks, and construction equipment. Although the three different vehicle domains have much in common, there are also distinct differences, for example, in production volume and the number of different models. In fact, these differences are so important that they result in three quite different network architectures.

The contributions of this chapter are (1) the presentation of three case studies and (2) an analysis of the relation between business needs and architectural design issues of networking in automotive vehicles. The three case studies cover much of automotive industry and describe the business context and demands for automotive networking.

The rest of this chapter is organized as follows: Section 57.2 provides a structuring and presentation of important issues and challenges in automotive networking. An introduction to automotive network technologies is provided in Section 57.3. Section 57.4 presents the three case studies and in Section 57.5, we analyze the architectures with respect to the different business needs. Finally, in Section 57.6, we draw some conclusions; in Section 57.7, links to relevant web sites are provided.

57.2 Challenges

In this section, we present some current trends in vehicle networking, together with descriptions of some of the context in which vehicle networks are developed today. The presentation is structured into sections on the main aspects of vehicle networks, namely functionality, cost, standards, and architecture.

Functionality

Functionality in a vehicle is not limited to end-user functions, but also includes functions to support, for instance, production and service. In this section, we will outline some important groups of functionality, both supportive and end-user functions that are often addressed in vehicle development.

Feedback control includes functions that control mechanical systems in the vehicle, for example, engine control and anti-lock-brake-systems. Feedback control systems can be combined to achieve advanced control functions for vehicle dynamics. Examples are electronic stabilizer programs, ESP, and other chassis control systems like antiroll systems. Furthermore, the vehicle manufacturers strive to achieve cheaper and more flexible functionality by going toward x-by-wire solutions. X-by-wire solutions, such as steer-by-wire, are achieved by replacing, for example, mechanical or hydraulic solutions by computer-controlled electrical systems.

Discrete control, in this context, includes simple functions to switch on or off devices, for example, control of lamps or wipers. The challenges for this group of functions often relate to the sheer number of such simple devices and thereby the amount of traffic on the network. Another problem is that these functions often interact logically in nontrivial ways.

Diagnostics and service: Functions in this group are used in vehicles to support the maintenance of delivered products. Diagnostic functions provide means to diagnose physical components, such as sensors, as well as software properties, such as version number and network load. Service functions provide means for updating the electronic system by downloading new software and to test vehicle operation. Because of the large amount of retrievable information, solutions are needed for automatic, or at least tool-supported, diagnoses, and service.

Infotainment refers to in-vehicle systems not related to driving the vehicle, but more related to information and entertainment. Examples are Internet connection and video consoles. This leads to requirements on high bandwidth for vehicle networks. Components like network controllers and software are often purchased off-the-shelf, and must be integrated in a harsh physical environment. Components must also be integrated without impacting safety-critical functionality in the vehicle.

Telematics [1] is a name of the set of functions that uses wireless communication networks outside the vehicle to perform their task. This includes functions in the vehicle and outside the vehicle. There is a strong trend in the vehicle industry to increase the use of telematics. Examples include fleet management systems for commercial vehicles, maintenance systems, and antitheft systems.

Cost

Providing cost-efficient network architectures is a challenge in several respects. The architecture should exhibit properties that support a variety of business needs. Business needs in the context of vehicle network systems often include lifecycle aspects of development, production, maintenance, and service.

Fixed and variable cost: Building a vehicle is a process of finding the best compromise between conflicting aspects, and one of the most important trade-offs is to find the balance between cost, performance, and functionality that provides the best business case. The cost can be divided into variable cost (the cost of purchasing the physical components that go into the vehicle and the resources consumed during production) and fixed cost (the investments made in development, production facilities, tooling, after market support, etc.). There is always a trade-off to be made between the two, which depends heavily on the production volumes, and the relations between various cost factors[2].

Maintenance and service: To reduce the lifecycle cost of the product, it is often important to consider various aspects of maintenance. To facilitate maintenance, it is desirable to develop architectures that allow future upgrades and extensions to the delivered product. Servicing delivered products requires the ability to upgrade both software and physical components. Configuration management and distribution is then a crucial issue, for example, to determine the compatibility of new components in an existing configuration.

Standards

The use of standards is motivated by the need to meet goals related to:

- Cost reduction,
- Integration of supplier components,
- Increased reliability of components, and
- Commonality in tools used in, for example, development, diagnostics, and service.

One example is the Controller Area Network (CAN) standard that has provided the vehicle industry with cheap network controllers. Due to the large volumes of these controllers, they are tested to a great extent (in the field) under diverse conditions, which increases reliability. A challenge is to standardize properties of software in similar ways, to accommodate reuse of software components, and to allow for common development tools. Although there are great benefits in using standards, they should also allow for proprietary solutions in key areas of a domain in order not to hinder competitive and brand-specific solutions.

Architecture

The IEEE has the following definition of architecture[3]:

Architecture: the fundamental organization of a system embodied in its components, their relationships to each other and to the environment and the principles guiding its design and evolution.

For automotive electronic architectures, the components are mainly the electronic control units (ECU), and the relationships between them are the communication networks. The environment is the vehicle itself, as well as the lifecycle processes that must support it. An important aspect when designing the architecture is the physical packing of the ECUs. A guideline is to place them as close as possible to their inputs and outputs, to reduce the wiring length (and thereby the cost and weight). At the same time, the architecture must be easy to assemble on the manufacturing line, and the overall architecture of the car puts restrictions on the possible locations, since space is generally very limited. Aspects of maintenance must also be considered when designing the architecture in order to facilitate aftermarket service of ECUs and software. Also, environmental factors like temperature and EMC influence the location of the ECUs. As the product development is increasingly focused on platforms and commonality, the ability to create many variants from the same overall structure is important.

57.3 Network Technologies

In this section, we will introduce present and emergent network technologies in the automotive industry. The network technologies in focus are different types of field busses that meet the requirements of automotive applications and communication software and high-level protocols. Throughout the section, we will give an overview of automotive network technologies in general and provide more details about technologies covered in the case studies of Section 57.4.

Automotive Field Busses

Control Busses

The dominant bus technology for power train and body electronics in vehicles is the CAN standard[4]. The protocol has gained widespread acceptance because of the relatively high bandwidth (at the time of introduction) in conjunction with its predictability and reliability. Due to the extensive use of the protocol, dedicated controllers for CAN are cheap and, moreover, integrated in many microcontrollers. The CAN specification is divided into standard CAN (CAN 2.0A) and extended CAN (CAN 2.0B), where the length of the message frame differs. The wide use has made processors available with integrated CAN controllers. CAN is a broadcast bus designed to operate at speeds of up to 1 Mbps. Data are transmitted in messages (frames) containing between 0 and 8 bytes of data. An identifier is associated with each message. The identifier serves two purposes: (1) assigning a priority to the message and (2) enabling receivers to filter messages.

CAN is a collision-detect broadcast bus, which uses deterministic collision resolution to control access to the bus. During arbitration, competing stations are simultaneously putting their identifiers, one bit at the time, on the bus. By monitoring the resulting bus value, a station detects if there is a competing higher priority message and stops transmission if this is the case. Because identifiers are unique within the system, a station transmitting the last bit of the identifier without detecting a higher priority message must be transmitting the highest priority queued message, and hence can start transmitting the body of the message.

Low-Cost Busses

Low-cost busses have been introduced in automotive applications in order to facilitate cost-effective integration of components such as smart sensors and actuators into the vehicle network. Smart sensors and actuators have some ability to process (typically filter, or translate) measurements and send signals on the network, whereas nonintelligent ones are wired to the I/O of an ECU that handles processing. The introduction of low-cost controllers and single-wire networks is made at the expense of bandwidth, which is relatively low for these busses. Besides being a low-cost alternative to adding a full featured ECU, the low-cost busses present a way of reducing the complexity of the master node and facilitate differently equipped product variants with only one ECU configuration.

One bus in this category is the Local Interconnect Network (LIN)[5], which is a serial bus based on the Universal Asynchronous Receiver Transmitter (UART) present in most microcontrollers. LIN implements a time-triggered protocol, where a master node handles synchronization on the bus. By letting a master node handle all synchronization on the network, temporal predictability is achieved without the need for arbitration by slave nodes. LIN provides low bandwidth (up to about 20 kbit/sec) and data packages can be of the size 2, 4, and 8 bytes. The performance, in terms of bandwidth and network services, of a LIN network is limited compared to CAN; but the LIN concept provides for simple inexpensive slave nodes making it possible to also connect very simple components to a network. The LIN standard provides not only the protocol specification but also an application interface and a network configuration description that conforms to the LIN configuration language description. The application interface provides a software-level abstraction for interfacing the network, whereas the configuration description holds information on the composition of the complete LIN network.

Other busses included in this category are the TTP/A bus[6], which is also a time-triggered bus based on UART, and the SAE 1850[7], an asynchronous serial protocol.

Infotainment Busses

Since vehicles are becoming equipped with more and more multimedia and telematics applications, the need for dedicated infotainment busses has arisen. A network in this category is MOST (Media Oriented Systems Transport)[8], which is based on optical fiber technology, and provides bandwidth up to about 20 Mbit/sec. MOST have services optimized for infotainment applications. The topology supported is ring, star, and daisy chain. The network is synchronous with a timing master, making it possible for devices to synchronize their operation; this is suitable for streaming of audio and video. MOST also has defined mechanisms for asynchronous, packet-based data, which can be found in, for example, Internet traffic.

Other infotainment busses relevant to automotive applications are the IEEE1394 [9] and the wireless Bluetooth [10], and IEEE 802.11b [11] protocols. (Bluetooth and IEEE 802.11b are used mainly to connect external devices.)

Time-triggered Safety-Critical Busses

Emergent safety-critical functions, such as x-by-wire applications, where x may be, for example, steer or brake, have forced the development of bus technologies for use in automotive vehicles that meet demands of very high reliability and timeliness. A group of busses that meet this demand are evaluated by the automotive industry. These busses are all based on the time-triggered paradigm where the progression of time initiates data transfers greater than asynchronous events. The time-triggered busses provide synchronous communication without the need for arbitration. Moreover, they offer mechanisms for redundant networks and have built-in support for a global time base. Therefore, the time -triggered protocols are suitable for implementing safety-critical control functions with stringent demands on low latency and low jitter. Three time-triggered protocols developed for automotive use are FlexRay, TTP/C, and TTCAN. All these protocols offer services, such as global time and time-triggered communication enabling pre-run-time scheduling of communication. Moreover, all three protocols also make it possible for event-triggered traffic to coexist with time-triggered communication.

The FlexRay communication protocol [12] supports bandwidth up to 10 Mbit/sec with the possible topologies bus, star, and multiple star. Available communication controllers for the TTP/C [13] protocol support 25 Mbit/sec for time-triggered transmission and 5 Mbit/sec for event-triggered transmission. TTP networks can contain up to 64 nodes and the cabling topology can be bus, star, or any combination of the two. Finally, TTCAN [14] is a further development of Extended CAN (version 2.0B), which like Extended CAN is limited to 1 Mbit/sec.

Summary

Table 57.1 shows the typical use of some different busses in automotive industry. As indicated, CAN is used in many different application areas. However, for safety-critical control, CAN does not have the same abilities as Flexray, TTP/C, and TTCAN, which offers global time base and redundant networks. Furthermore, for simple nodes the CAN solution is more expensive than LIN.

High-level Protocols and Communication Software

High-level protocols and communication software have been introduced in automotive industry to address problems related to integration, configuration, and timing properties. The high-level protocols span from proprietary representation of a given type of engineering data by a specific CAN identifier to powerful methods and tool sets automating large parts of the network configuration. Depending on application area, both proprietary and standardized protocols are in use.

In this section, we will describe three different high-level protocols; Volcano, SAE J1939, and MOST. These protocols are intended for different areas of application within the automotive domain.

Volcano [15] is a communication concept used throughout the Volvo Car Corporation for managing network traffic. Currently, Volcano supports the CAN and LIN busses. The basic concept in Volcano for communication between software components is signals, where a signal typically represents some engineering data. Through the Volcano API, the underlying network technology is hidden from the application engineer. Moreover, the engineer is not concerned with issues regarding assignment of signals to network frames. Instead, this is done automatically ensuring signal timing requirements.

The Volcano concept also addresses vehicle manufacturer-controlled integration of components developed by suppliers. This is done through the use of the Volcano API and by separate specification of the signals used by a component and the network configuration. The network configuration is provided by the integrator and specifies how signals are to be transferred over the network.

In the domain of heavy vehicles, the J1939 standard [16] for communication between components has been created by a number of vehicle manufacturers and subsuppliers. SAE J1939 uses standard CAN 2.0B for communication and communicates with 250 kbit/sec. J1939 defines a transport service in the network layer. In the application layer, the protocol defines data (signals, e.g., vehicle speed) and the packaging of signals in frames. Moreover, J1939 also defines the interaction between components, for example, the interaction between engine and transmission during gear shifting. To provide for vehicle manufacturer-specific functionality, the protocol also allows some proprietary messages.

Both Volcano and J1939 address integration of supplier components; but the approaches differ. Volcano does not, as J1939, explicitly define assignment of specific signals to given frames. Instead, the assignment is resolved in the implementation phase. This makes is possible to do application specific network configurations that are optimal with respect to, for example, timing and bandwidth usage.

For vehicle infotainment systems, MOST, in addition to defining the low-level communication protocol, also defines architecture at the application level of a component connected to a MOST network. At

TABLE 57.1　Primary Use and Key Characteristics of Different Busses in Automotive Applications

	Primary Use					Key Characteristics	
	Feedback Control	Fault-tolerant Control	Discrete Control	Diagnostics and Service	Infotainment and Telematics	Node Cost	Maximum Speed (Mbits/sec)
CAN	x		x	x		Medium	1
LIN			x			Low	20
MOST					x	High	20
FLEXRAY		x				High	10
TTP/C		x				High	25
TTCAN		x				Medium–High	1

the application level, a MOST application such as, for example, an audio system is represented by a *device*. Devices are supported by the system services, which provide a standard interface to network management functions and services for sending and receiving data. A MOST device on the top level can logically be described by a set of function blocks, where a function block represents an application in the device, for example, a CD player or tuner. Moreover, each function block contains a set of functions, for example, play and stop, which are visible to the rest of the MOST system. Function blocks interface other function blocks on the same device or devices connected to the network through services constructed from the system services.

By the definition of the application architecture, devices connected to a MOST system are logically constructed in the same manner and can be interfaced in a consistent way.

57.4   Case Studies

Volvo Car Corporation

The Car Industry

Volvo Car Corporation (VCC) is a subsidiary of the Ford Motor Company, where it is one of the premium brands aimed at the upper end of the car market. The European premium car brands are driving the development of vehicle electronics, having both the demand for advanced functionality and the production volumes to support the costs associated with the introduction of new technology. The components developed for cars are often later used in other parts of the automotive industry. Cars are consumer products, and the customers tend to be sensitive not only to the functionality of the car but also to how it feels and its visual appearance.

Cars are typically manufactured in volumes in the order of millions per year. To achieve these volumes, and still offer the customer a wide range of choices, the products are built on platforms that contain common technology that has the flexibility to adapt to different kinds of cars. As an example, the Volvo XC90, which appeared in 2002, is based on the same platform as four previous Volvos launched since 1998. This reduces the development cost, but also makes it possible to reuse the same manufacturing facilities and strengthens the brand image through an increased similarity between the models.

The component technology is, to a large extent, provided by external suppliers, who work with many different car companies (or OEMs, original equipment manufacturers), providing similar parts. The role of the OEM is thus to provide specifications for the suppliers, so that the component will fit a particular car, and to integrate the components into a product. Traditionally, suppliers have developed physical parts; but in modern cars, they also provide software. As the computational power of the electronic control units (ECUs) increases, it will be more common to include software from several suppliers in the same nodes, which increases the complexity of integration.

Functionality

The driving factor behind the development is increasing demands on functionality. There are several classes of functionality, and in the following paragraphs we provide examples in each of them, which represent some of the largest future challenges for VCC.

Feedback control systems were one of the earliest uses for electronics in cars, and the early applications were engine control, ABS brakes, and vehicle dynamics. These areas are still developing, and one of the main challenges is to cope with new environmental constraints, in particular, related to the reduction of the level of CO_2 emissions. The systems are refined in the sense that more and more sensors are added, and new modes of interaction are included, thereby increasing the overall complexity of the functionality.

Discrete control systems are also common in current cars, in particular, in body electronics. Applications include driver information, security (locks and alarms), and lights. Due to the fact that the overall functionality increases, as well as the abilities of the owner to configure the car through various parameter settings, the complexity of the driver interface becomes a bottleneck. This is caused both by the physical space around the driver seat, and the ability of the driver to process information while driving the car.

Novel ways of interaction are thus needed, in addition to more intelligent systems that, in most cases, can make correct decisions without driver intervention.

Safety-critical control systems become more common as traditional mechanical solutions are replaced by electronics. For the functions currently implemented, there is always a natural fall-back solution if the electronics fail, but future by-wire systems may not have that possibility, which increases the need for fault-tolerance in the electronics and communication. The first such application is likely to be brake-by-wire, and later steer-by-wire will follow. The driving factors behind this development are that the cost and weight could be reduced, and also that it enables new control systems to support the driver, for example to enhance safety. Again, this means a considerable increase in system complexity.

Diagnostic systems provide information about the status of the vehicle. Initially, this was driven by legal requirements that mandated monitoring of emission-related components, but it is also an important factor in increasing the perceived quality of the system. The diagnostic system consists of an in-car part and a workshop tool. The former is usually distributed to all the nodes of the on-board network, and consists of fault detection routines and diagnostic kernels that interact with the workshop tool. As the number of sensors increase, so does the need for diagnostics, and there is also a wish to increase the intelligence of the on-board system, to, for example, detect the need for preventive maintenance so that the customer never experiences critical problems, and thereby gains a perception of high quality.

Infotainment systems implement entertainment functionality, extending from traditional radios to multimedia applications such as TV, video, and gaming, and also contains information functions such as navigation systems. As the number of devices for audio and video data increase in cars, sharing of input and output is essential to bring down cost and conserve space, and this means that complexity moves from hardware to software and communication.

Telematic systems are used in cars for wireless data communication with the world outside via a built-in mobile phone. The applications range from automatic emergency calls in case of an accident to Internet access, and many ideas exist for services that the car owner could be interested in. The area is still in its infancy, and the business cases are currently unclear; but the underlying technology is being developed rapidly.

As can be seen above, complexity is a keyword that must be handled in the development. (For an introduction to the nature of complexity in technical systems, see [17].)

Cost

In the car industry, the development cost is huge in absolute numbers, but still comparatively small in relation to the total variable cost of the production, or the investment in tooling. This means that it is usually profitable to invest in development cost to optimize the components, or to increase commonality between car models on the same platform. Since the cost of development is closely related to the complexity (i.e., the information that needs to be processed to describe the product), it is thus profitable to increase complexity to obtain more flexible components that can be used in many different cars.

For software, the cost relations are somewhat different. There is an indirect variable cost, in that the characteristics of the software influence the resource needs in terms of memory size and processing capability. This cost can, to some extent, be manipulated by optimizing the performance, but this makes the software more complex and thus the development more expensive. In the long term, Moore's law will make hardware increasingly cheap and thereby making it more profitable to optimize software development cost rather than hardware resource needs.

One way to decrease software development cost is to raise the level of abstraction when describing the functionality. At VCC, model-based system development is being introduced [18], where the system is described using a tool chain based on UML, Statecharts, and data flow models. Code generation is then used to reduce the cost of producing the final software. In a way, the complexity is moved from the specific applications to general tools that can be used over and over again. Another example is in network communication over the CAN buses. The Volcano system [15] provides tools for packing data into network frames, and for verifying the end-to-end communication timing to ensure the control performance.

Standards

As indicated above, the OEM's role is to integrate systems from suppliers into a product. This means that it is important to have well-defined interfaces so that the various systems fit together. Also, standards are important as a means to reduce cost of components by sharing development, and allow competition between suppliers. One area where standardization has been particularly vivid is in communication protocols. The following protocols are now used or planned by VCC:

- *CAN* is used for backbone control-oriented peer-to-peer communication.
- *LIN* is a low-cost alternative for control-oriented master–slave communication. Originally developed at VCC, it is now an international standard.
- *MOST* is used for multimedia communication in the infotainment system.
- *Flexray* is expected to be used instead of CAN for safety-critical applications where fault-tolerance is needed.

Selecting a bus protocol is thus a trade-off mainly between cost, bandwidth, predictability, and fault-tolerance. Another area where standards are important is in diagnostics, where authorities mandate it so that they may check that a vehicle fulfills emission regulations, using a single tool.

The current development trends in automotive software also call for increasing standardization of the software structure in the nodes. In particular, the use of code generation requires a clear interface between the support software and the application, and the need to integrate software from different suppliers in the same node also calls for a well-defined structure. The node architecture (see Figure 57.1) includes several important components:

- *Operating systems* (RTOS) provide services for task scheduling and synchronization. Traditional real-time operating systems are usually too resource-consuming to be suitable for automotive applications, and do not provide the predictable timing that is needed. Therefore, the new standard OSEK has been developed. There are several suppliers of OSEK compliant operating systems.
- *Network communication software* provides a layer between the hardware and the application software, so that communication can be described at a high level of abstraction in the application, regardless of the low-level mechanisms used to send data between the nodes. At VCC, the Volcano concept [15] is used for both CAN and LIN communication.
- *Diagnostic kernels* provide an implementation of the diagnostic services that each node must implement to act as a client toward the off-board diagnostic tool. It relies on the communication software to access the networks and on the operating system to schedule diagnostic activities so that it does not interfere with the application functionality.

All these components interact with each other and with the application, and must therefore have standardized interfaces, and at the same time provide the required flexibility. To conserve hardware resources, the components are configurable to only include the parts that are really necessary in each particular instantiation.

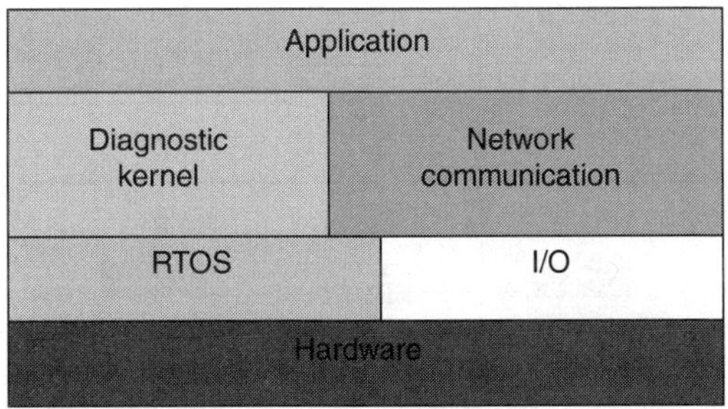

FIGURE 57.1   The node architecture.

Architecture

An example of a contemporary car electronic architecture is that of the Volvo XC90 (see Figure 57.2). The boxes in Figure 57.2 represent ECUs and the lines represent communication buses. The intent of the figure is to show the complexity rather than the details. The maximum configuration contains about 40 ECUs. They are connected mainly by two CAN networks: one for powertrain and one for body functionality. From some of the nodes, LIN subnetworks are used to connect slave nodes into a subsystem. The other main structure is the MOST ring, connecting the infotainment nodes together, with a gateway to the CAN network for limited data exchange. Through this separation, the critical powertrain functions on the CAN network are protected from possible disturbances from the infotainment system. The diagnostics access to the entire car is via a single connection to one ECU. The figure shows approximately how the ECUs are placed in various locations in the car. The partitioning of functionality is decided by the location of the sensors and actuators used, but also by the combinations of optional variants that are possible. If a car is sold with only a subset of the full functionality, the amount of physical hardware installed should be limited to the minimum necessary.

In the future, when safety-critical functionality is introduced, the architecture will be extended with a Flexray-based network, and this will again be isolated from the less critical parts using a gateway. Another important aspect is to create a more flexible partitioning. The main use for this is probably not to find the optimal partitioning for each car on a given platform, since that would create too much work on the verification side, but to allow parts of the software to be reused from one platform to the next. This places even higher demands on the node architecture, since the application must be totally independent of the hardware, through a standardized interface that is stable over time. Therefore, further standardization work is needed, in particular, for sensor and actuator interfaces.

Volvo Trucks

The Trucking Industry

The functionality in trucks has grown dramatically during the past 10 years. Earlier, there was a separate stand-alone system for handling each function; today, all these subsystems are integrated into a single complete system. The development time has decreased and the vehicles are much more complex.

There are different demands depending on market in the truck industry. One example is the system voltage, which is 12 V in the North American market and 24 V in the rest of the world. Moreover, it is

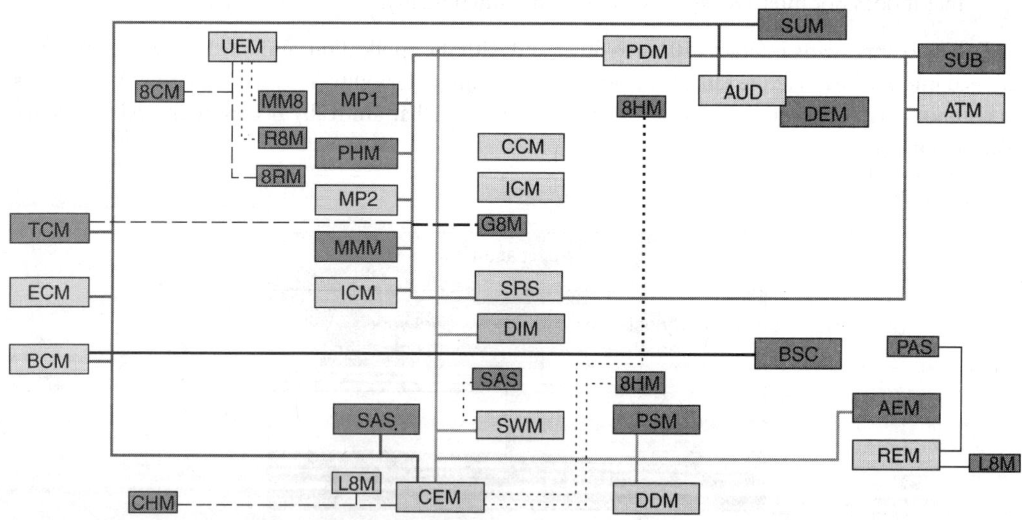

FIGURE 57.2   The electronic architecture of Volvo XC90.

very common for the customers to choose their own driveline in the U.S., which puts special demands on OEMs to integrate engines and transmissions from various suppliers.

One way to obtain cost-effective solutions is to use a common platform covering both mechanical, electrical, and software systems. The challenge is to have a shared platform and yet maintain the unique truck brands.

The truck market has changed from delivering vehicles, to providing a complete transport solution, which may include several different types of services, for example, "around the clock" support and on-line fleet management, etc. A complete transport solution could mean providing the full logistic routing system for a town.

Functionality

Trucks have many areas of use. The requirement on functionality can be split into three different segments.

- Goods transportation and logistics,
- Building and construction, and
- City distribution and waste handling.

"Goods transportation and logistics" means transportation of goods over long distances, for example, food from southern Europe to Sweden. "Building and construction," for example, concrete trucks, crane trucks, or gravel trucks operate under rougher conditions such as on a construction site, in mines or in roadworks. "City distribution and waste" refers to local transport, for instance, a garbage truck.

Feedback control systems were among the first electronic systems introduced in the beginning of the 1980s (e.g., electronic motor control and antilock braking system, ABS). These systems were complete stand-alone systems. Over time, the systems became more complex and integrated. For example, the ABS system can command the engine not to apply the exhaust brake when ABS is activated. Furthermore, some sensors are shared between the systems and data can be exchanged through the vehicle network.

Discrete control systems include functions such as driver information, but also systems like climate control, exterior light, central locking, tachograph, etc. With increasing amount of functionality and information on the network, the requirements on the Human Machine Interface (HMI) are getting increasingly complex. It becomes a challenge to support the driver in deploying the functionality the right way.

Superstructures: Trucks can be supplemented with superstructures, such as concrete aggregates, tipping devices, refrigerator units, etc. One way to decrease the total cost of the vehicle is to have a well-defined interface between the electrical system and the superstructure, to allow, for example, the crane equipment to control the engine speed to facilitate the right flow in the hydraulic pump.

Safety-critical control systems: The increase in safety-critical systems has been striking in the last few years. One driving factor for this is to prevent personal and property damage in case of accident. Earlier mechanical systems are being replaced or supplemented. Many systems are common with the car industry, for instance, ABS and airbags. One difference is that the gross combination weight (weight of vehicle and trailer) is much greater. Trucks also have many variants and it is common that they have more than four wheels.

Recently, the Electronically Controlled Brake System (EBS) and Electronic Stability Program (ESP) were introduced. In the EBS, an electronically controlled valve located close to each wheel applies individual braking force to each wheel. There is also a centrally positioned ECU that controls the vehicle's braking effect, both in the disk brakes and the engine brake. There is a back-up system using pressurized air.

The ESP system is a supplement to EBS. By means of a yaw rate sensor, a lateral accelerator sensor and a steering wheel sensor the new functionally can be added. ESP is an active safety-enhancement system whose task is to stabilize the vehicle, for example, prevent jack-knifing when the driver makes a rapid avoidance maneuver.

Diagnostic system: An increasing part of the vehicle electronics has demands on efficient built-in diagnostics not only for the aftermarket but also to check the mounting of components in production. The goal is to be able to check all components, not only ECUs but also very simple components like switches and bulbs.[1]

---

[1] Compare to the semiconductor industry where diagnostics is included in the chips for production tests only.

There is an aftermarket tool that communicates with the control units. Through this tool, it is possible to read out fault codes, sensor values, etc. The tool is also able to run tests in the control units and download new software and parameters. Since the North American market requires support for third-party components, it must also be possible for the suppliers' aftermarket tools to coexist.

Infotainment has not been very common in trucks, but because many drivers that are transporting goods far away are living in their trucks, the need has increased. Because of the low product volumes compared to the car industry, it is likely that the truck industry for cost reasons will inherit from the car industry.

Telematics in trucks is mainly used for traffic information and tracking of goods. Volvo has its own telematics system called Dynafleet. Tracking of goods gives the possibility to enhance logistics. This type of system becomes more and more common due to the demand for just-in-time transports. Because of the growth of the Internet, the telematics systems will become more integrated in all business systems. By using the information on the truck location and data about the goods (e.g., the temperature in a cold transport), the transports can be planned with a positive effect both on economy and the environment.

Cost

Reducing costs is becoming increasingly important. The vehicles offer more and more functionality but the cost per function is decreasing. Because the truck industry has lower volumes compared to passenger cars, it is important to get the right trade-off between development cost and product cost. Systems that are fitted in all truck models, like the Vehicle Control Unit (VECU), must have a low product cost. But the systems that are only available in some variants produced in a couple of 1000s/year are less sensitive for product cost. Instead, here, a gain is to be made by using "general-purpose" components that realize more than one variant.

To get the best trade-off between development cost and product cost, the architecture design is crucial. The architecture should identify which components are convenient to share with respect to cost. During the life cycle of a platform, the development cost for software is higher than the development cost for hardware; but hardware costs have a more direct effect on the product cost.

Standards

A truck is essentially built by integrating systems from many suppliers. Volvo Trucks has some in-house development of core control units that is important to give the truck the characteristic "feeling." Still, there are several systems that are not profitable to develop in-house. One such system is the ABS brakes. In this area, there are a few big suppliers that have key-knowledge and their own development of systems. But still, Volvo Trucks is deeply involved in formulating requirements on the system and adding unique Volvo functionality. It is, however, important to use as much standard as possible to make the integration in different brands as simple as possible.

For heavy vehicle, there are two standard protocols that are used in today's production:

- SAE J1939 [16] and
- SAE J1708/J1587 [19], an older protocol. It uses RS485 as base and communicates with 9600 bit/sec, and is mainly used in the vehicles for diagnostics and for some fall-back for J1939.

Other protocols that will be used or are under investigation at Volvo trucks include:

- LIN [5] will be used for sensor networks but also for sub-busses.
- MOST [8] is the optical ring bus that is used for infotainment by some manufacturers in the car industry.
- Flexray [12] is under evaluation for safety-critical systems.
- TTCAN [14]. Because TTCAN is built on the well used CAN protocol, TTCAN might be very interesting for control-intensive vehicle systems.

Architecture

The market pull in the U.S. for selecting truck components from different vendors introduces additional complexity, when more and more electronics is integrated in the trucks. To handle the electrical integration, Volvo Trucks uses the standardized protocol J1939. Since J1939 not only defines network and application layer services but also the interaction between vehicle components it partly acts as architecture; see Figure 57.3. Volvo trucks have also added some in-house strategies and guidelines to J1939 and for other electrical installations. Together with the standard, this defines the architecture.

The advantage of using a standard is the relatively straightforward integration of components from different vendor suppliers. On the other hand, there are some disadvantages, for example, development or changes to the protocol can only be made through standardization work.

J1708/J1587 is also used. It is used in parallel with J1939 for diagnostic, information sharing, and as a backup for J1939.

These standards, strategies, and guidelines have been a good support when all stand-alone systems have been integrated to a distributed system. With software and communication, it is now possible to create functionality that was not possible 20 years ago. Nonetheless, it also introduces some new problems. One example is startup and shutdown of the systems, where all systems must synchronize through the network.

Despite the added complexity, the demand for reduced product development time has increased, and at the same time the demands on decreased product cost have also increased. To be able to reduce the development time, it is important to find new solutions. One way is to use a reference architecture that covers all products and include both software and hardware. This will also set demand on new tools that can handle complex system models in real time, real time because there will be a lot of persons working in parallel with the model. Because Volvo Trucks is a global organization with product development on four sites, it is also important that the tools support this way of work. A common misunderstanding is that the high product development cost is related to HW. This is usually wrong. It is therefore very important to have an architecture that supports both hardware and software. This gives the possibility to have a high degree of reuse between the products. This saves both product development time and cost. Of course, this will also give a profit in higher quality.

To enable reduction of the number of physical control units, it is important to have a system view instead of considering each control unit separately. Thereby, software functions can be deployed in the control unit that is closest to the necessary sensors and actuators.

The superstructure interface must also be included in the architecture. A well-defined interface is believed to decrease the cost and time for adding superstructures. This is believed to increase customer satisfaction and thereby profit for the company (Figure 57.4).

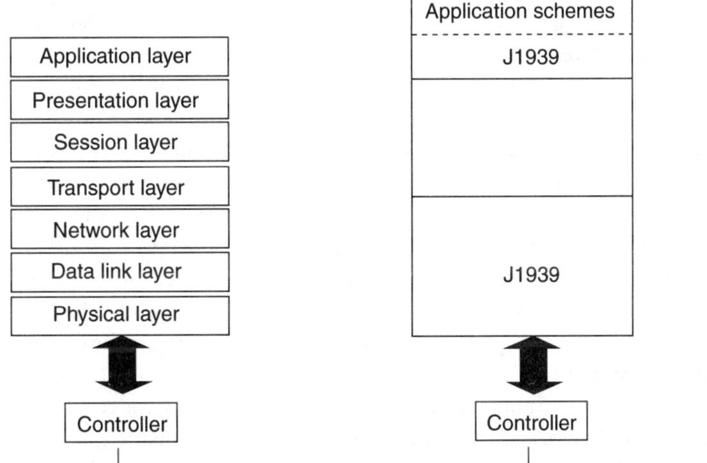

FIGURE 57.3   SAE J1939.

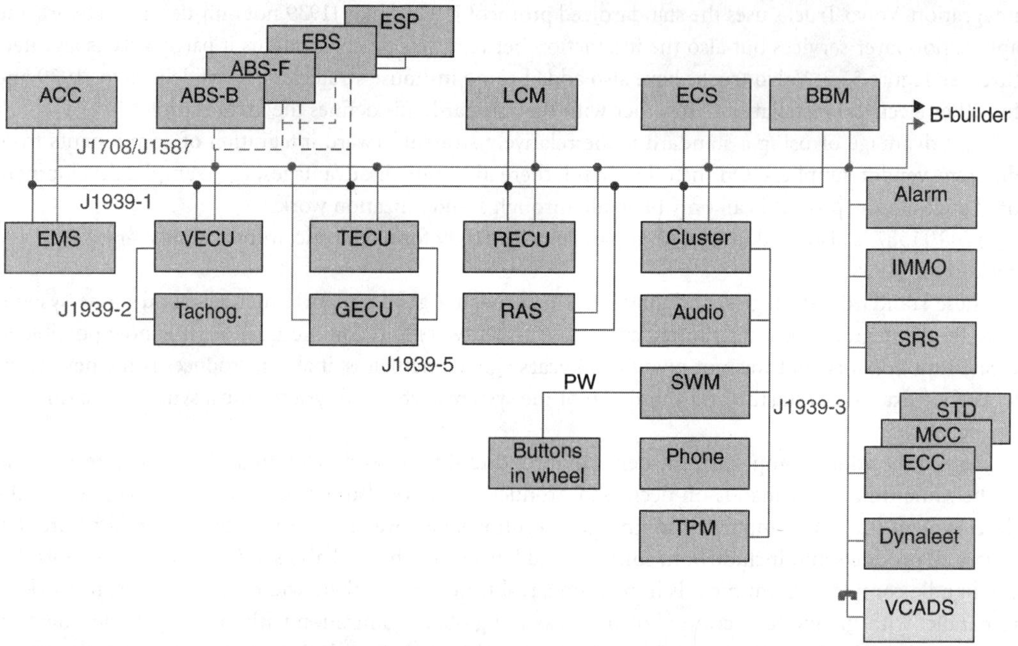

FIGURE 57.4   Volvo FH electronic architecture.

Conclusion

The progress in the last ten years has moved VTC vehicles from centralized computer systems to distributed systems, and much functionality has been added. This has of course increased the complexity and together with demands on a shortened product development time and increased quality, there is a need not to reinvent the wheel. One way to handle this situation is to use reference architectures supported by tools and better methods for product development.

Volvo Construction Equipment

Volvo Construction Equipment (VCE) develops and manufactures a wide variety of construction equipment vehicles, such as articulated haulers, excavators, graders, backhoe loaders, and wheel loaders. The products range from relatively small compact equipment (1.4 t), to large construction equipment (52 t). VCE is divided into product companies with focus on one type of equipment, and that typically manufactures product lines of similar products, for example, excavators or wheel loaders.

Compared with passenger cars and trucks, construction equipment vehicles are equipped with less complex electronic systems and networks. Also, the focus in product development is somewhat different. The products are to be used in construction sites, and the most important aspect of the vehicle is to provide a reliable machine to increase production. A customer of passenger cars, on the other hand, may be interested in a number of styling attributes related to the look and feel of the car, whereas the dominant requirements for customers of construction equipment are related to production goals. The requirements for reliability and robustness are equally high in both cases, but styling requirements are typically low for construction site machinery.

In-vehicle electronic systems and networks are an important part of the construction equipment product and are crucial to provide end-user functionality, such as automatic gearbox, as well as providing diagnostic and service functions. The use of a distributed electronic system also accommodates functions that reduce the cost of the vehicle, for example, reuse of sensors and displays, and adaptive solutions to accommodate cheaper mechanical parts.

Functionality

Here, we describe today's situation in functionality requirements for construction equipment at VCE. To each class of functionality, we present some major drivers and examples.

Feedback Control: The main feedback control functionality in VCE products includes control of engine, gearbox, retarder brake, differential lock, and engine cooling fan. Besides engine control, the automatic gearbox is a complex feedback control system that can include control of clutches and brakes for the gear pinions as well as converter, lockup, and drop box. The functionality for an automatic gearbox includes logics for when to shift gears, minimization of slip, avoidance of hunting, various efficiency optimizations, and self-adapting solutions to accommodate a variance of mechanical properties. There are also distributed control functions that synchronize engine speed for good comfort in shifting gears. Feedback control systems also include the cooperation of brakes to minimize wear, for xample, the cutin of a retarder brake, or exhaust brake.

Discrete Control: Like in the truck case, discrete control systems include control of driver information, wipers, lamps, and other on—off-type devices. The challenges arise due to configuration issues rather than constructing a functional system.

Diagnostics and service: Diagnostic functions are used to determine status and operational history of electronic components. Some diagnostic functions reside in the on-board software and some in service tools. On-board diagnostic functions implement criteria for faults, and can send fault codes via the network. Diagnostic functions also include logging functions to store operational data, for example, fault history or general operation statistics on buffers, network load, or sensor input. A service tool (a PC with service applications) can be connected to the vehicle network via a service connector. The service tool can run tests to diagnose faults either by invoking on-board functions, or by running test programs to verify functionality.

The service tool can, thus, extract diagnostic data for a service technician, but also download new software or parameters into the vehicle control units. The tool can be connected to a central configuration database that holds information on compatibility between versions and configurations. When downloading software or parameters, the central system is updated to reflect the current configuration. Also, the operational data can give valuable feedback to the development department.

Infotainment: As mentioned, customers of construction equipment purchase products with the intent of increasing production at construction sites. There is little incentive to pay for entertainment systems and therefore VCE does not provide any such systems today. There are demands for information systems; but this does not usually include general systems such as Internet connection and video. Instead, tailored applications to increase production are requested.

Telematics: In the field of construction equipment, telematics can be used to achieve the increased production in several ways. The challenge is to accommodate telematic functions in a cost-efficient way. Construction sites could be located in remote regions and wireless technology like satellite or radio communication must be considered as opposed to the car industry where mobile phone communication seems to be sufficient. Fitted equipment and cost for communicating over commercial networks are expensive and may not be crucial to every customer.

Applications running on an office desktop computer can be developed to access information in a fleet of vehicles and to present and analyze data far away from the actual vehicles. Examples are fleet management systems, maintenance systems, and antitheft systems. A fleet management system can provide information to increase the efficiency of a fleet of vehicles. A maintenance system could, for instance, report on status of mechanically worn components like brakes and thereby reduce maintenance cost.

The trend toward telematic systems is very strong and a variety of systems are expected to be available in a few years.

Cost

The electronic systems in VCE products are, in total, less complex and are sold in smaller volumes compared to cars or trucks. The final products are also much more expensive than passenger cars; but the electronic content is a much lower percentage of the total product cost. Therefore, the cost for developing electronic systems in VCE is relatively large compared to the variable costs for production, at least compared to VCC and VTC. This implies that it is usually not profitable to optimize hardware to a large extent since it would generate increased complexity of the system and increased development cost, for example, the need for configuration handling.

Accommodating commonality is considered very beneficial because VCE has a large number of products (although sold in smaller volumes). Reuse is beneficial for both hardware and software as it directly affects development cost. This results in VCE focusing heavily on commonality, even though it may mean that a lower-end product is produced with some spare resources in terms of electronics.

Compared to both cars and trucks, VCE builds on-board electronic systems that have a lower development cost, smaller volumes, and lower overall complexity.

Trends indicate that the electronic content (and complexity) will increase quite rapidly in construction equipment over the coming years. The situation for construction equipment is likely to resemble the situation of trucks today and later may be the situation for cars. However, the volumes will not equal those of trucks or cars. VCE have fewer products with a higher price and will thereby not focus as much on optimizations, but rather at handling complexity and commonality. Commonality can also help in reducing development time.

To customers of construction equipment, providing efficient maintenance and service is probably as important as providing a functional product. This is because of the long-term service contracts where 80% of the vehicles are in use after 20 years.

VCE, like VCC and VTC, also anticipate a trend of increasingly cheap hardware and will focus on optimizing software development cost rather than optimizing hardware content in products, that is, designing electronic systems for a minimum slack of hardware such as memory and processing power.

Standards

VCE uses the same standards for communication protocols as VTC, that is, SAE J1939 / CAN and SAE J1587 / J1708 (see the High-level protocols and communication software, and standards sections for details).

Architecture

The focus of VCE's electronic architecture effort is mainly concerned with assuring system properties that are judged essential to provide the "right" product (a good business case) and to have an architecture that supports working with platforms. System properties include scalability to support product variation, reusability and partitioning of SW components to lower development cost, as well as safety and reliability issues.

This means that methods for designing SW in control units and handling communication are also considered architectural issues. The goal is to have an architecture that helps in designing on-board electronic systems with respect to the wanted system properties that are identified as the most important for the VCE business case.

As temporal behavior is considered very important, VCE uses a design process [20] that focuses on high-level design and temporal attributes of the system. VCE uses the operating system Rubus, which provides a configuration tool allowing specification of temporal constraints, communication, and synchronization. This method separates the design of timing characteristics from the design of functionality, and enables early verification of temporal behavior.

Partitioning, scalability, and commonality: VCE has a setting different from VCC and VTC, in that there is almost no use of externally developed vendor control units. This leads to the possibility of using the same software component model, operating system, and reusing software components. By doing this, the partitioning of functionality is likely to be easier than in the case of a network with many differently developed control units. Easy partitioning provides the possibility to scale the system with respect to hardware and optimize hardware content in a specific product. For instance, a low-end product with fewer features requires less hardware resources, and can be realized by placing software components on

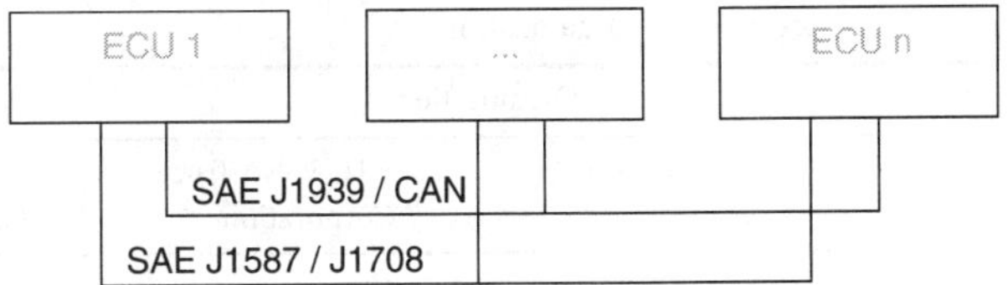

FIGURE 57.5   VCE Network Architecture.

other nodes and thereby reduce the number of control units. This means that the number of control units can be chosen according to resource needs for a given configuration of a product. Thus, the architecture is reused; but the numbers of ECUs differ between products (see Figure 57.5).

To have a situation with a large degree of in-house developed control units also gives benefits in terms of commonality. The use of common design methods for control units in itself leads to reduced cost, but also helps in decreasing the complexity of the system. (The overall information that must be processed during development is decreased.) Reusing infrastructure like drivers, communication components, and other software components is a goal in commonality that can be met as long as common design methods are used.

On the other hand, VCE platforms are used in a wider variety of different products. Facilitating commonality in very different products, such as an excavator and an articulated hauler, presents a different situation compared to VTC or VCC.

The different situation in VCE is caused by lower demands for many variants. While trucks are delivered in numerous configurations, construction equipment often provides quite few. Examples are an optional tachograph, or an optional enhanced suspension system.

In the car and truck industry, there are several vendors that specialize in certain systems, deliver in large numbers, and can provide competitive prices. VCE is likely to move toward the VTC situation with products including external vendor nodes. This will likely generate a need for new methods for accommodating scalability partitioning commonality.

Furthermore, the trend toward supplying services rather than vehicles could be relevant to VCE. Currently, the trend of increasing rental products can be seen as a step toward providing services, as VCE accepts responsibility for up time of products.

57.5   Analysis

The different business demands on cars, trucks, and construction equipment led to different focus in the design effort of the respective vehicles. In this section, we present analysis of the correlation between different business and product characteristics and key properties of the resulting network architectures. In Table 57.2 below, the business and product characteristics are given for the three different organizations.

The case study has shown that the product volumes are different for the three organizations, and thereby also the focus on fixed cost and hardware optimization. The willingness to reduce fixed cost at the expense of variable cost increases with the product volume. One way of achieving a reduced fixed cost is to optimize vehicle hardware content to include a minimum of resources. Software components are not subject to the optimization profit, due to an increase in variable cost but almost no gain in fixed cost. VCC that produces vehicles in the range of 10^6 can benefit to a larger extent by reducing fixed cost, and therefore an increased cost for the design of optimal hardware is more profitable than for VCE that has volumes in the range of 10^4.

The number of vehicle models sold for the respective organizations is indicated in the table by "Number of products." The number of products and the product volume directly affect the profitability of reusing components, that is, commonality, and this also includes software components. VCE has a high number of products, but smaller volumes, while VCC and VTC have high volumes. Thus, the effort to achieve commonality is emphasized in all the three organizations.

                                          The Industrial Information Technology Handbook

TABLE 57.2 Business Characteristics for Each Organization

Characteristics	Organization		
	VCC Volvo Car Corporation	VTC Volvo Truck Corporation	VCE Volvo Construction Equipment
Annual production volume (Order of magnitude)	10^6	10^5	10^4
Products	~8 products	~8 products	>50 products
Platforms	3	3	8
Number of physical configurations per product	Many	Very many	Few
Amount of information	Huge	Very large	Moderate
Standards application level	Proprietary (Volcano)	J1587–J1939	J1587–J1939
Number of network Technologies	~4	2	2
Hardware optimization	High	Medium	Low
Openness	None	High	Some
Safety-critical	Yes	Yes	Yes
Advanced control	Yes	Yes	Yes
Infotainment	Much	Some	None

The "Number of physical configurations" means the different options of network topologies that can be fitted in a certain product. VTC products, which may be configured in many ways, achieve a high extendability and can facilitate change of configuration in the aftermarket or adding superstructures by other vehicle developing organizations. The large amount of data and the many configurations place high requirements on the management of different components, for example, ECUs, connected to the network in different variants.

The large amount of information together with the requirements for optimization in the VCC case imply that using several tailored networks for specific needs can be profitable. The use of LIN networks provides a cost-effective network for handling locally interconnected lights and switches, and a high bandwidth MOST network serves the needs of infotainment applications. Especially for VCE, the increase in development cost for designing tailored networks for a certain purpose is deemed unprofitable and this is reflected in the small number of network technologies.

The use of in-vehicle networks opens up the possibility for efficient diagnostics and service, by the ability to extract information via the bus. Although the amount of information varies in the three cases, the needs for diagnose and service are emphasized in all three organizations. The reason for this is that there is sufficient information in all systems to substantially ease analysis of the distributed system.

As mentioned, VTC needs to facilitate superstructures, and this is reflected in the large number of physical configurations. In order to support extensions to the network by other parties, standardized communication interfaces like SAE J1939 and J1708/J1587 are used. VCE uses the same standards, since both VTC and VCE belong to the Volvo group and there is commonality in service tools.

Safety aspects on networking require that messages be transferred with correct timing and without being corrupted. One step toward guaranteeing the real-time properties and integrity of messages related to safety-critical functionality is to use communication protocols with support for deterministic and analyzable timing behavior. Examples are CAN and LIN and the coming protocols Flexray and TTCAN, which are all used or evaluated by the three considered organizations. Another step is to use several networks that are interconnected through gateways. Safety-critical communication can thereby be separated from communication that is not trusted to the same degree.

57.6 Conclusion

In recent years, networking issues have become more and more important in the design of vehicle control systems. One of the initial driving forces for introduction of communication networks in automotive vehicles was to replace the numerous cables and harnesses and thereby reduce cost and weight. Today, it is possible using software and networking to create new functionality that was considered unfeasible some ten years ago.

We have presented case studies of the context, use, and requirements of networking in construction equipments, trucks, and passenger cars. Furthermore, we have identified challenges with respect to functionality, cost, standards, and architecture for development of vehicles. Based on these case studies with different business and functionality demands, we have provided analysis of the design principles used for the communication architectures in these domains. Despite a common base of similar vehicle functionality, the resulting network architectures used by the three organizations are quite different. The reason for this becomes apparent when looking at different business and product characteristics and their effect on the network architecture. An important lesson from this is that one should be very careful to uncritically apply technical solutions from one industry to another, even when they are as closely related as the applications described in this paper.

57.7   Further information

Listed below are links to sites where detailed information about the presented technologies can be found:

OSEK (www.osek-vdx.org)
CAN (www.can-cia.org)
LIN (www.lin-subbus.org)
MOST (www.mostnet.de)
Flexray (www.flexray-group.com)
Rubus (www.arcticus.se)

References

[1]   Yilin Zhao, Telematics: safe and fun driving, IEEE Intelligent Systems (see also IEEE Expert), Vol. 17, Issue 1, pp. 10–14, Jan.-Feb. 2002.

[2]   J. Axelsson, Cost Models for Electronic Architecture Trade Studies, in Proceedings of the 6th International Conference on Engineering of Complex Computer Systems, Tokyo, 2000, pp. 229–239.

[3]   IEEE Recommended Practice for Architectural Description of Software-Intensive Systems, IEEE Std 1471, 2000.

[4]   Road Vehicles — Interchange of Digital Information — Controller Area Network (CAN) for High-Speed Communication, International Standards Organisation (ISO), ISO Standard-11898, November 1993.

[5]   LIN — Protocol, Development Tools, and Software Interfaces for Local Interconnect Networks in Vehicles, 9th International Conference on Electronic Systems for Vehicles, Baden-Baden, Oct. 2000.

[6]   H. Kopetz, W. Elmenreich, and C. Mack, A comparison of LIN and TTP/A, Proceedings, 2000 IEEE International Workshop on Factory Communication Systems, 2000, pp. 99 –107.

[7]   SAE Standard, Committee: Vehicle Architecture For Data Communications Standards, Document number: J1850, Class B Data Communications Network Interface, May 2001.

[8]   MOST Specifcation Framework Rev 1.1, MOST Cooperation, November 1999, www.mostnet.de

[9]   IEEE Standard for a High Performance Serial Bus, IEEE Std 1394–1995, August 1996.

[10] K.V.S.S.S.S. Sairam, N. Gunasekaran, and S.R. Redd, Bluetooth in wireless communication, *IEEE Communications Magazine*, 40, 90 –96, 2002.

[11] IEEE Standard for Information Technology — Telecommunications and Information Exchange between Systems — Local and Metropolitan Area Networks — specific requirement, Part 11: Wireless LAN Medium Access Control (MAC) and Physical Layer (PHY) Specifications, IEEE Std 802.11b-1999/Cor 1-2001, 2001.

[12] FlexRay Requirements Specification, Version 2.0.2 / April 2002, www.flexray-group.com

[13] Time-Triggered Protocol TTP/C, High-Level Specification Document, Specification edition 1.0.0 of 4-July-2002, TTTech Computertechnik AG.

[14] Road Vehicles — Controller Area Network (CAN) — Part 4: Time-Triggered Communication, International Standards Organisation (ISO), ISO Standard-11898-4, December 2000.

[15] L. Casparsson, A. Rajnak, K. Tindell, and P. Malmberg, Volcano a revolution in on-board communications, Volvo Technology Report, December 1998.

[16] SAE Standard, SAE J1939 Standards Collection, www.sae.org

[17] J. Axelsson, Towards an Improved Understanding of Humans as the Components that Implement Systems Engineering, in Proceedings of the 12th Symposium of the International Council on Systems Engineering (INCOSE), Las Vegas, 2002.

[18] M. Rhodin, L. Ljungberg, and U. Eklund, A Method for Model Based Automotive Software Development, Proceedings of the Work in Progress and Industrial Experience Sessions, 12th Euromicro Conference on Real-Time Systems, Stockholm, 2002, pp. 15–18.

[19] SAE Standard, SAE J1587, Joint SAE/TMC Electronic Data Interchange Between Microcomputer Systems in Heavy-Duty Vehicle Applications, www.sae.org.

[20] C. Norström, K. Sandström, M. Gustafsson, J. Mäki-Turja, and N.-E. Bånkestad, Experiences from Introducing State-of-the-art Real-Time Techniques in the Automotive Industry, in Proceedings of 8th Annual IEEE International Conference and Workshop on the Engineering of Computer Based Systems (ECBS01), Washington, U.S., April 2001, IEEE Computer Society. Press, Silver Spring, MD, 2001.

58

The Standard Message Specification for Industrial Automation Systems — ISO 9506 (MMS)

Karlheinz Schwarz
Schwarz Consulting Company (SCC)

58.1 Introduction

The international standard Manufacturing Message Specification (MMS) is an OSI application layer messaging protocol designed for the remote control and monitoring of devices such as Remote Terminal Units (RTU), Programmable Logic Controllers (PLC), Numerical Controllers (NC), or Robot Controllers (RC). It provides a set of services allowing the remote manipulation of variables, programs, semaphores, events, journals, etc. MMS offers a wide range of services satisfying both simple and complex applications.

For years, the automation of technical processes had been marked by increasing requirements with regard to flexible functionalities for the transparent control and visualization of any kind of processes. The mere cyclic data exchange will be increasingly replaced by systems that join together independent yet coordinated systems — like communication, processing, open- and closed-loop control, quality protection, monitoring, configuring, and archiving systems — to a whole. These individual systems are interconnected and work together. As a common component, they require a suitable real-time communication system with adequate functions.

The MMS standard defines common functions for distributed automation systems. The expression "manufacturing," which stands for the first M in MMS, has been inappropriately chosen. The MMS standard does not contain any manufacturing-specific definitions. The application of MMS is as general as the application of a personal computer. MMS offers a platform for a variety of applications.

The first versions of MMS documents were published in 1990 by ISO TC 184 (Industrial Automation) as an outcome of the GM initiative *Manufacturing Application Protocols* (MAP). The current version was published in 2003:

Part 1: ISO 9506-1 Service definition (2003) [1]: Describes the services that are provided to remotely manipulate the MMS objects. For each service, a description of the parameters carried by the service primitives is given. The services are described in an abstract way, which does not imply any particular implementation.

Part 2: ISO 9506-2 Protocol specification (2003) [2]: Specifies the MMS protocol in terms of messages. The messages are described with a notation ASN.1, which gives the syntax.

Today, MMS is being implemented — unlike the practice 15 years ago and unlike the supposition still partly found today — on all common communications networks that support a safe transport of data. These can be networks like TCP/IP or ISO/OSI on Ethernet, a fieldbus, or simple point-to-point connections like HDLC, RS 485, or RS 232. MMS is independent of a seven-layer stack. Since MMS was originally developed in the MAP environment, it was generally believed earlier that MMS could be used only in connection with MAP.

MMS is the basis of the international project *Utility Communication Architecture* (UCA™, IEEE TR 1550) [13], IEC 60870-6-TASE.2 [14,15] *(Inter control center communication)*, IEC 61850 [17–21] *(Communication networks and systems in substations)*, and IEC 61400-25 [22] *(Communications for monitoring and control of wind power plants)*.

This chapter introduces the basic concepts of MMS applied in the above-mentioned standards.

58.2   The MMS Client/Server Model

MMS describes the behavior of two communicating devices by the client/server model (see Figure 58.1). The client can, for example, be an operating and monitoring system, a control center, or another intelligent device. The server represents one or several real devices or whole systems. MMS uses an object-oriented modeling

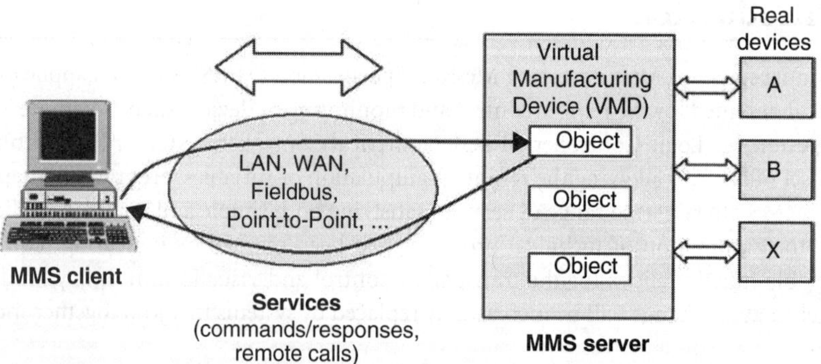

FIGURE 58.1   MMS client/server model.

method with object classes (Named Variable, Domain, Named Variable List, Journal, etc.) and instances from the object classes and methods (services like read, write, information report, download, read Journal, etc.).

The standard is comprehensive. This does not mean that an MMS implementation must be complex or complicated. If only a simple subset is used, then the implementation can also be simple. Meanwhile, MMS implementations are available in the third generation. They allow the use of MMS both on PC platforms and embedded controllers.

The MMS server represents the objects that the MMS client can access. The Virtual Manufacturing Device (VMD) object represents the outermost "container" in which all other objects are contained.

Real devices can play both roles (client and server) simultaneously. A server in a control center for its part can be a client with respect to a substation. MMS basically describes the behavior of the server. The server contains the MMS objects and it also executes services. MMS can be regarded as "server centric." In principle, in a system, more devices are installed that function as server (e.g., controllers and field devices) than devices that perform the client role (e.g., PC and workstation).

The "calls" that the client sends to the server are described in the part ISO 9506-1 (services). These "calls" are processed and answered by the server. The services can also be referred to as remote calls, commands, or methods. Using these services, the client can access the objects in the server. It can, for example, browse through the server, that is, making visible all available objects and their definitions (configurations). The client can define, delete, change, or access objects via reading and writing.

An MMS server models real data (e.g., temperature measurement, counted measurand, or other data of a device). These real data and their implementation are concealed or hidden by the server. MMS does not define any implementation details of the servers. It is only defined as to how the objects behave and represent themselves to the outside (from the point of view of the wire) and how a client can access them.

MMS provides the very common classes. The Named Variable, for example, allows to structure any information provided for access by an application. The content (the semantic of the exchanged information) of the Named Variables is outside the MMS standard. Several other standards define common and domain-specific information models.

IEC 61850 defines the semantic of many "points" in electric substations. For example, "Atlanta26/XCBR3.Pos.stVal" is the position of the third (3rd) circuit breaker in substation "Atlanta26." The names "XCBR," "Pos," and "stVal" are standardized names.

The forth coming standard IEC 61400-25 (communication for wind power plants) defines a comprehensive list of named "points" specific for wind turbines. For example, "Tower24/WROT.RotSpd.mag" is the (deadbanded) measured value of the rotor position of Tower24. "RotSpd.avgVal" is the average value (calculated based on a configuration attribute "avgPer." These information models are based on common data classes like Measured value, 3-phase value (delta and Y), and single point status.

58.3 The VMD

According to Figure 58.2, the real data and devices are represented — in the direction of a client — by the VMD. In this regard, the server represents a "standard driver" that maps the real world to a virtual one. The following definition helps to clarify the modeling in the form of a virtual device:

If it's there and you can see it	It's REAL
If it's there and you can't see it	It's TRANSPARENT
If it's not there and you can see it	It's VIRTUAL
If it's not there and you can't see it	It's GONE

Roy Wills

The VMD can represent, for example, a variable "Measurement3," whose value may not permanently exist in reality; only when the variable is being read will a measurand transducer get started to determinate the value. All objects in a server can already be contained in a device before the delivery of a device. The objects are predefined in this case.

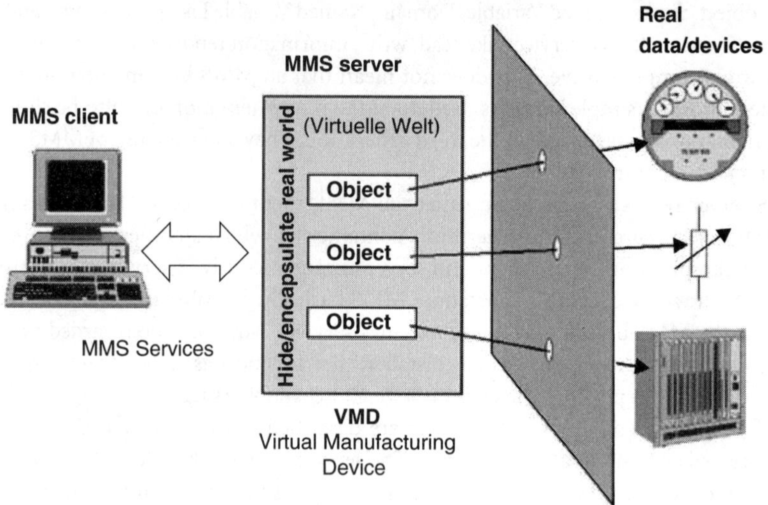

FIGURE 58.2   Hiding real devices in the VMD.

Independent of the implementation of a VMD, data and the access to data are always treated in the same way. This is completely independent of the operating system, the programming language, and memory management. Similar to how printer drivers for a standard operating system hide the various real printers, a VMD also hides real devices. The server can be understood as a communication driver that hides the specifics of real devices. From the point of view of the client, only the server with its objects and its behavior is visible. The real device is not visible directly.

MMS merely describes the server side of the communication (objects and services) and the messages that are exchanged between client and server.

The VMD describes a virtual manufacturing device completely. This virtual device represents the behavior of a real device as far as it is visible "over the wire." It contains, for example, an identification of manufacturer, device type, and version. The virtual device contains objects like variables, lists, programs and data areas, semaphores, events, journals, etc.

The client can read the attributes of the VMD (see Figure 58.3), that is, it can browse through a device. If the client does not have any information about the device, it can view all the objects of the VMD and their attributes by means of the different Get services. With this, the client can perform a first plausibility check on a just installed device by means of a Get(Object-Attribute)-Service. It learns whether the installed device is the ordered device with the right model number (Model Name) and the expected issue number (Revision). All other attributes can also be verified (e.g., variable names and types).

The attributes of all objects represent a self-description of the device. Since they are stored in the device itself, a VMD always has the currently valid and thus consistent configuration information of the respective device. This information can be requested on-line directly from the device. In this way, the client always receives up-to-date information.

MMS defines some 80 functions:

- browsing functions about the "contents" of the virtual device: "Which objects are available?",
- functions for reading, reporting, and writing of arbitrarily structured variable values, and
- functions for the transmission of data and programs, for the control of programs, and many other functions.

The individual groups of the MMS services and objects are shown in Figure 58.4. MMS describes such aspects of the real device that shall be open, that is, standardized. An open device must behave as described by the virtual device. How this behavior is achieved is not visible and also not relevant to the user who accesses the device externally. MMS does not define any local, specific interfaces in the real systems. The

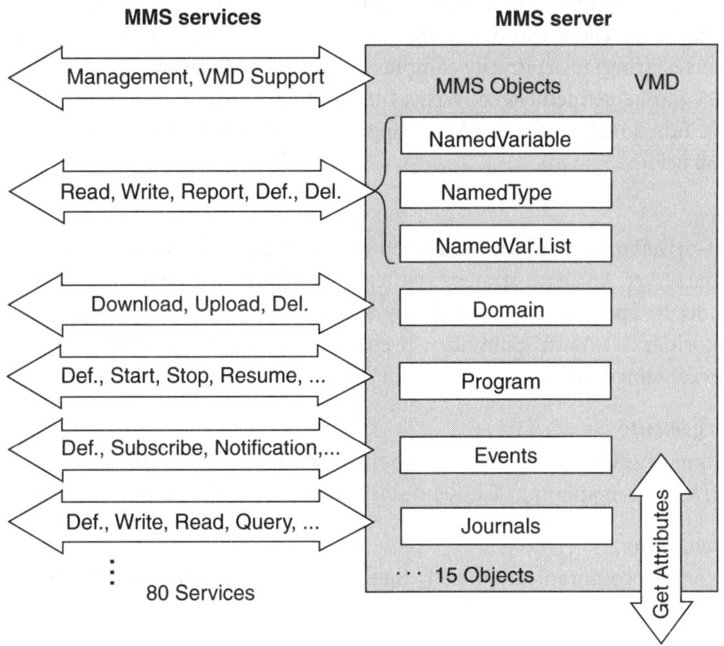

FIGURE 58.3　VMD attributes.

FIGURE 58.4　MMS objects and services.

interfaces are independent of the functions that shall be used remotely. Interfaces in connection with MMS are always understood in the sense that MMS quasi-represents an interface between the devices and not within the devices. This interface could be described as an external interface. Of course, interfaces are also needed for implementations of MMS functions in the individual real devices. These shall not and cannot be defined by a single standard. They are basically dependent on the real systems — and these vary to a great extent.

MMS Models and Services

ISO 9506-1 (part 1) — Service Specification

Environment and General Management Services.
Two applications that wish to communicate with each other can set up, maintain, and close a logical connection (initiate, conclude, abort).

VMD Support.
The client can thereby query the status of a VMD or the status is reported (Unsolicited Status); the client can query the different lists of the objects (Get Name List), query the attributes of the VMD (Identify), or change the names of objects (Rename).

Domain Management.
Using a simple flow control (Download, Upload, Delete Domains, ...), programs and data of arbitrary length can be transmitted between the client and server and also a third station (and vice versa). In the case of simple devices, the receiver of the data stream determines the speed of the transmission.

Program Invocation Management.
Services to create, start, stop, and delete modularly structured programs by the client (Start, Stop, Resume, Kill, Delete, ...).

Variable Access.
This service allows the client to read and write variables that are defined in the server or a server is enabled to report the contents to a client without being requested (Information Report). The structures of these data are simple (octet string) to arbitrarily complex (Structure of Array of ...). In addition, data types and arbitrary variables can be defined (Read, Write, Information Report, Define Variable, ...). The variables consitute the core functionality of every MMS application; therefore, the variable access model will be explained in detail below.

Event Management.
It allows an event-driven operation; that is, a given service (e.g., Read) is only carried out if a given event has occurred in the server. An alarm strategy is integrated. Alarms will be reported to one or more clients if certain events occur. These have the possibility to acknowledge the alarms later (Define, Alter Event Condition Monitoring, Get Alarm Summary, Event Notification, Acknowledge Event Notification, ...). This model is not explained further.

Semaphore Management.
The synchronization of several clients and the coordinated access to the resources of real devices is carried out hereby (Define Semaphore, Take/Relinquish Control, ...). This model is not explained further.

Operator Communication.
Simple services for the communication with operating consoles integrated in the VMD (Input and Output). This model is not explained further.

Journal Management.
Several clients can enter data into journals (archives, logbooks) that are defined in the server. Then these data can be selectively retrieved through filters (Write Journal, Read Journal, ...). This model is not explained further.

ISO9506-2 (Part 2) — Protocol Specification

If a client invokes a service, then the server must be informed about the requested type of the service. For a Read service, for example, the name of the variables must be sent to the server. This information that the server needs for the execution is exchanged in so-called Protocol Data Units (PDU). The sets of all the PDU that can be exchanged between client and server constitute the MMS protocol.

In other words, the protocol specification — using the standards ISO 8824 (Abstract Syntax Notation One, ASN.1) and 8825 (ASN.1 Basic Encoding Rules, BER) — describes the abstract and concrete syntax of the functions defined in part 1. The syntax is explained below exemplarily.

58.4 Locality of the VMD

VMDs are virtual descriptions of real data and devices (e.g., protection devices, measurand transducers, wind turbine, and any other automation device or system). With respect to the implementation of a VMD, there are three very different possibilities where a VMD can be located (see Figure 58.5):

1. *In the end device*: One or several VMDs are in the real device that is represented by the VMD. The implementations of the VMD have direct access to the data in the device. The modeling can be carried out in such a way that each application field in the device is assigned to its own VMD. The individual VMDs are independent of each other.
2. *In the gateway*: One or several VMDs are implemented in a separate computer (a so-called gateway or agent). In this case, all MMS objects that describe the access to real data in the devices are at a central location. While being accessed, the data of a VMD can be in the memory of the gateway — or they must be retrieved from the end-device only after the request. The modeling can be carried out in such a way that for each device or application, a VMD of its own will be implemented. The VMD are independent of each other.
3. *In a file*: One or several VMDs are implemented in a database on a computer, on a FTP server, or on a CD ROM (the possibilities under 1. and 2. are also valid here). Thus, all VMDs and all included objects with all their configuration information can be entered directly in engineering systems. Such a CD ROM, which represents the device description, could also be used, for example, to provide a monitoring system with the configuration information: names, data types, object attributes, etc. Before devices are delivered, the engineering tools can already process the accompanying device configuration information (Electronic Data Sheet)! The configuration information can also be read later on-line from the respective VMD via corresponding MMS requests.

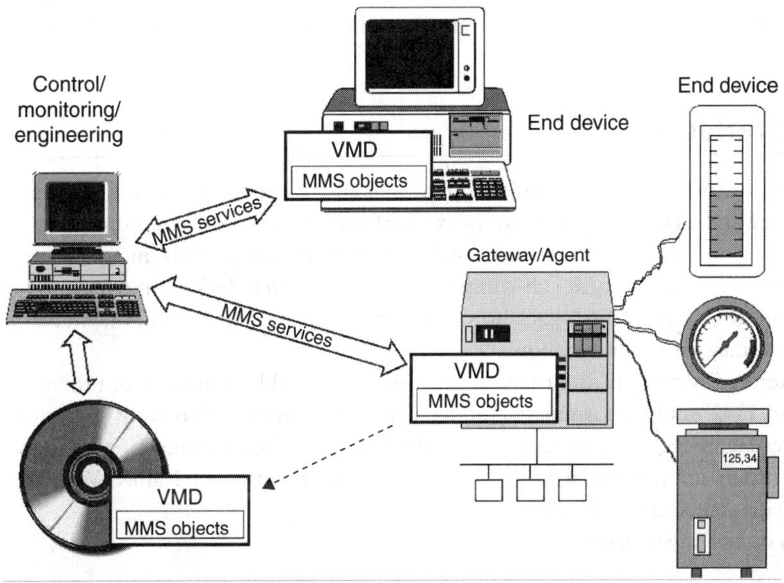

FIGURE 58.5 Location of VMDs.

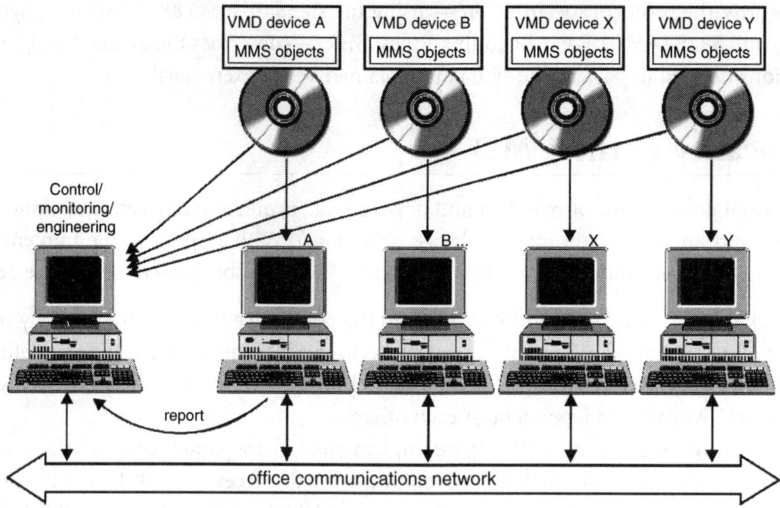

FIGURE 58.6 VMD testing using PC in an office environment.

The VMD is independent of the location. This also allows, for example — besides the support during configuration — that several VMD can be installed for testing purposes on a computer other than the final system (see Figure 58.6). Thus, the VMD of several large robots can be tested in the laboratory or office. The VMD will be installed on one or several computers (the computers emulate the real robots). Using a suitable communication (e.g., intranet or also a simple RS 232 connection — available on every PC), the original client (a control system that controls and supervises the robots) can now access and test the VMD in the laboratory. In this way, whole systems can be tested beforehand regarding the interaction of individual devices (e.g., monitoring and control system).

If the Internet is used instead of the intranet, global access is possible to any VMD that is connected to the Internet. The author himself tested the access from Germany to a VMD that was implemented on a PLC in the U.S.A. That is to say, through standards like MMS and open transmission systems, it has become possible to set up global communications networks for the real-time process data exchange.

The previous statements about the VMD are also valid in full extent for all standards that are based on MMS.

58.5 Interfaces

The increasing distribution of automation applications and the exploding amount of information require more and more and increasingly more complex interfaces for operation and monitoring. Complex interfaces turn into complicated interfaces very fast. Interfaces "cut" components into two pieces; through this, interactions between the emerged sub-components — which were hidden in one component before — become visible. An interface discloses as to which functions are carried out in the individual subcomponents and how they act in combination.

Transmitter and receiver of information must likewise be able to understand these definitions. The request "Read T142" must be formulated "understandably," transmitted correctly, and understood unambiguously (see Figure 58.7). The semantic (named terms that represent something) of the services and the service parameters are defined in MMS. The content, for example, of Named Variables is defined in domain-specific standards like IE 61850.

Interfaces occur in two forms:

- internal program–program interfaces or APIs and
- external interfaces over a network (WAN, LAN, fieldbus, …).

Both interfaces affect each other. MMS defines an external interface. The necessity of complex interfaces (complex because of the necessary functionality, not because of an end in itself) is generally known and accepted. To keep the number of complex interfaces as small as possible, they are defined in standards or industry standards — mostly as open interfaces. Open interfaces are, in the meantime, integral components of every modern automation. In mid-1997, it was explained in Ref. [22] that the trend in automation engineering obviously leads away from the proprietary solutions to open, standardized interfaces — that is, to open systems.

The reason why open interfaces are complicated is not because they were standardized. Proprietary interfaces tend to be more complicated or even very complicated. The major reasons for the latter observation are found in the permanent "improvement" of the interfaces that express themselves in the quick changes of version and in the permanent development of new, apparently better, interfaces. Automation systems of one manufacturer often offer, for identical functions, a variety of complicated interfaces that are incompatible with each other.

At first, interfaces can be divided into two classes (see Figure 58.8): internal interfaces (e.g., in a computer) and external interfaces (over a communication network). The following consideration is strongly simplified because, for example, in reality both internally and externally several interfaces can lie one above the other. However, it nevertheless shows the differences in principle that require attention. MMS defines an external

FIGURE 58.7　Sender and receiver of information.

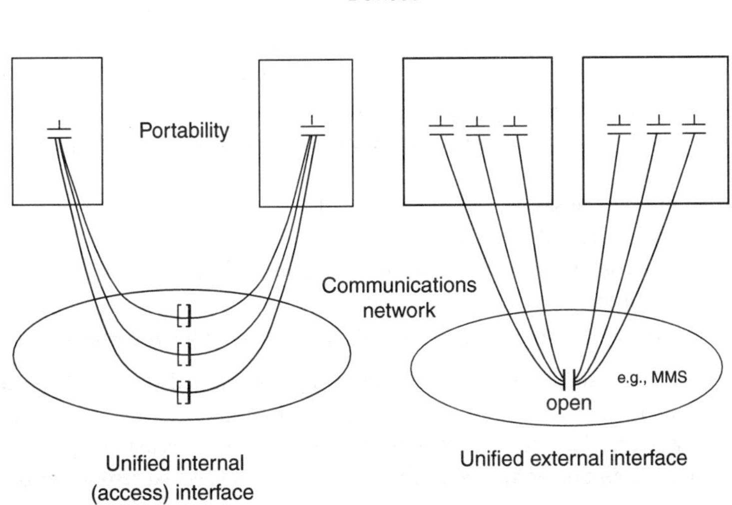

FIGURE 58.8　Internal and external interfaces.

interface. Many understand MMS in such a way that it offers — or at least also offers — an internal interface. This notion results in completely false ideas. Therefore, the following consideration is very helpful.

The left-hand side of the figure shows the case with a uniform internal interface and varying external interfaces. This uniform internal interface allows many applications to access the same functions with the same parameters and perhaps the same programming language — independent of the external interface. Uniform internal interfaces basically allow the portability of the application programs over different external interfaces.

The right-hand side of the figure shows the case where the external interface is uniform. The internal interfaces are various (since the programming languages or the operating systems, e.g., are various). The uniform external interface is independent of the internal interface. The consequence of this is, for example, that devices whose local interfaces differ and are implemented in divers environments can communicate together. Differences can result, for example, from an interface being integrated into the application in a certain device, but being available explicitly in another device. The essential feature of this uniform external interface is the interoperability of different devices. The ISO/OSI reference model is aimed at exactly this feature.

The (internal) MMS interface, for example, in a client (perhaps: $READ (Par. 1, Par. 2, ... Par. N)) depends on the manufacturer, operating system, and programming language. MMS implementations are, for example, available for UNIX or Windows NT. On the one hand, this is a disadvantage because applications that wish to access an MMS server would have to support, depending on the environment, various real program interfaces. On the other hand, the MMS protocol is completely independent of the, fast-changing, operating system platforms. Standardized external interfaces like MMS offer a high degree on stability, because in the first place the communication can hardly be changed arbitrarily by a manufacturer and in the second place several design cycles of devices can survive.

Precisely the stability of the communication, as defined in MMS, also offers a stable basis for the development of internal interfaces on the various platforms such as under Windows 95, NT, or in UNIX environments.

In the ISO/OSI world, openness describes the interface on the "wire." The protocol of this external interface is executed according to defined standardized rules. For an interaction of two components, these rules have to be taken into account on the two sides; otherwise, the two will not understand each other.

58.6 Environment and General Management Services

MMS uses a connection-oriented mode of operation. That is to say, before, for example, a computer can read a value from a PLC for the first time, a connection must be set up between the two.

MMS connections have particular quality features such as:

- Exclusive allocation of computer and memory resources to a connection. This is necessary to guarantee that all services (e.g., five Read, ...) that are allowed to be carried out simultaneously find sufficient resources on both sides of the connection.
- Flow control in order to avoid blockages and vain transmissions, if, for example, the receive buffers are full.
- Segmentation of long messages.
- Routing of messages over different networks.
- Supervision of the connection if no communication takes place.
- Acknowledgment of the transmitted data, and
- Authentication, access protection (password), and encoding of the messages.

Connections are generally established once and then remain established as long as a device is connected (at least during permanently necessary communication). If, for example, a device is only seldom accessed by a diagnostics system, a connection then does not need to be established permanently (waste of resources). It suffices to establish a connection and, later, to close it to release the needed resources again. The connection can remain established for rare but time-critical transmissions. The subordinate layers supervise the connection permanently. Through this, the interruption of a connection is quickly recognized.

The MMS services for the connection management are:

- *initiate* (connection setup),
- *conclude* (orderly connection tear-down; waiting requests are still being answered), and
- *abort* (abrupt connection tear-down; waiting requests are deleted).

Besides these services that are all mapped to the subordinate layers, there are two further services:

- *cancel* and
- *reject*.

After the MMS client has sent a Read request to the MMS server, for example, it may happen that the server leaves the service in its request queue and — for whatever reason — does not process it. Using the Service Cancel, the client can now delete the request in the server. On the other hand, it may occur that the server shall carry out a service with "forbidden" parameters. Using Reject, it rejects the faulty request and reports this back to the client.

Although MMS was originally developed for ISO/OSI networks, a number of implementations are available in the meantime that also use other networks such as the known TCP/IP network. From the point of view of MMS, this is insignificant as long as the necessary quality of the connection is guaranteed.

58.7 VMD Support

The VMD object consists of 12 attributes. The Key Attribute identifies the "Executive Function." The Executive function corresponds directly with the entity of a VMD. A VMD is identified by a presentation address:

```
Object:  VMD
    Key Attribute:  Executive Function
    Attribute:      Vendor Name
    Attribute:      Model Name
    Attribute:      Revision
    Attribute:      Logical Status (STATE-CHANGES-ALLOWED,
                    NO-STATE- CHANGES-ALLOWED,
                    LIMITED-SERVICES-SUPPORTED)
    Attribute:      List of Capabilities
    Attribute:      Physical Status  (OPERATIONAL,  PARTIALLY-OPERATIONAL,
                    INOPERABLE NEEDS-COMMISSIONING)
    Attribute:      List of Program  Invocations
    Attribute:      List of Domains
    Attribute:      List of Transaction Objects
    Attribute:      List of Upload State Machines (ULSM)
    Attribute:      List of Other VMD-specific Objects
```

The attributes Vendor Name, Model Name, and Revision provide information about the manufacturer and the device.

The Logical status defines as to which services may be carried out. The status Limited-Services-Supported allows that only such services that have read access to the VMD be executed.

The Physical Status indicates whether or not the device works in principle.

The following two services:

- *Unsolicited Status* and
- *Status*

are used to get the status unsolicited (Unsolicited status) or explicitly requested (Status). Thus, a client can recognize whether a given server — from the point of view of the communication — works at all.

The List of Capabilities offers clients and servers a possibility to define application-specific agreements in the form of features. The available memory of a device, for example, could be a capability. Through the Service Get Capability List, the current value can be queried. The remaining attributes contain the lists of all the MMS objects available in a VMD. The VMD therefore contains an Object Dictionary in which all objects of a VMD are recorded.

The following three services complete the services of the VMD:

- *Identify* supplies the VMD attributes Vendor Name, Model Name, and Revision. With this, a plausibility check can be carried out from the side of the client.
- *Get Name List* returns the names of all MMS objects. It can be selectively determined from which classes of objects (e.g., Named Variable or Event Condition) the names of the stored objects shall be queried. Let us assume a VMD is not yet known to the client till now (because it is, e.g., a maintenance device), the client can then browse through the VMD and systematically query all names of the objects. Using the Get services that are defined for every object class (e.g., Get Variable Access Attributes), the client can gain detailed knowledge about a given object (e.g., the Named Variable "T142").
- *Rename* allows a client to rename the name of an object.

58.8 Domain Management

Domains are to be viewed as containers that represent memory areas. Domain contents can be interchanged between different devices. The object type "domain" with its 12 attributes and 12 direct operations, which create, manipulate, ..., delete a domain, are part of the model.

The abstract structure of the domain object consists of the following attributes:

```
Object:   Domain

    Key Attribute:   Domain Name
    Attribute:       List of Capabilities
    Attribute:       State (LOADING, COMPLETE, INCOMPLETE,  READY, IN-USE)
    Constraint:      State = (LOADING, COMPLETE, INCOMPLETE)
      Attribute:        Assigned Application Association
      Attribute:        MMS Deletable
      Attribute:        Sharable (TRUE, FALSE)
      Attribute:        Domain Content
      Attribute :       List of Subordinate Objects
    Constraint:      State = (IN-USE)
      Attribute:        List of Program Invocation References
      Attribute:        Upload In Progress
      Attribute:        Additional Detail
```

The domain name is an identifier of a domain within a VMD.

Domain Content is a dummy for the information that is within a domain. The contents of the data to be transmitted can be coded transparently or according to certain rules agreed upon before. Using the MMS version (2003) [1, 2], the data stream can be coded per default in such a way that a VMD can be transmitted completely — including all MMS object definitions that it contains. This means, on the one hand, that the contents of a VMD can be loaded from a configuration tool into a device (or saved from a device), and on the other, that the contents can be stored on a disk per default.

Using a visible string, List of Capabilities describes which resources are to be provided — by the real device — for the domain of a VMD.

MMS Deletable indicates whether or not this domain may be deleted by means of a MMS operation.

Sharable indicates whether or not a domain may be used by more than one program invocation.

List of Program Invocation lists those Program Invocation Objects that use this domain.

List of Subordinate Objects lists those MMS objects (no domains or program invocations) that are defined within this domain: objects that were (a) created by the domain loading, (b) created dynamically by a Program Invocation, (c) created dynamically by MMS operations, or (d) created locally.

State describes one of the ten states in which a domain can be.

Upload in Progress indicates whether or not the Domain Content of this domain is being copied to the client at the moment.

MMS defines loading in two directions:

- data transmission from the client to the server (download) and
- data transmission from the server to the client (Upload).

Three phases can be distinguished during loading:

- open transmission,
- segmented transmission, controlled by the data sink, and
- close transmission.

Transmission during download and upload is initiated by the client, respectively. If the server initiates, then it has the possibility to initiate the transmission indirectly (see Figure 58.9). For this purpose, the server informs the client that it (the client) shall initiate the loading. Even a third station can initiate the transmission by informing the server, which then informs the client.

What is the Domain Scope?

Further MMS objects can be defined within a domain: variable objects, event objects, and semaphore objects. That is to say, a domain forms a scope (validity range) in which named MMS objects are reversibly unambiguous.

MMS objects can be defined in three different scopes, as shown in Figure 58.10. Objects with VMD-specific scope (e.g., the variable "Status_125") can be addressed directly through this name by all clients. If an object has a domain-specific scope such as the object "Status_155," then it is identified by two identifiers:

Domain Identifier "Motor_2" and Object Identifier "Status_155."

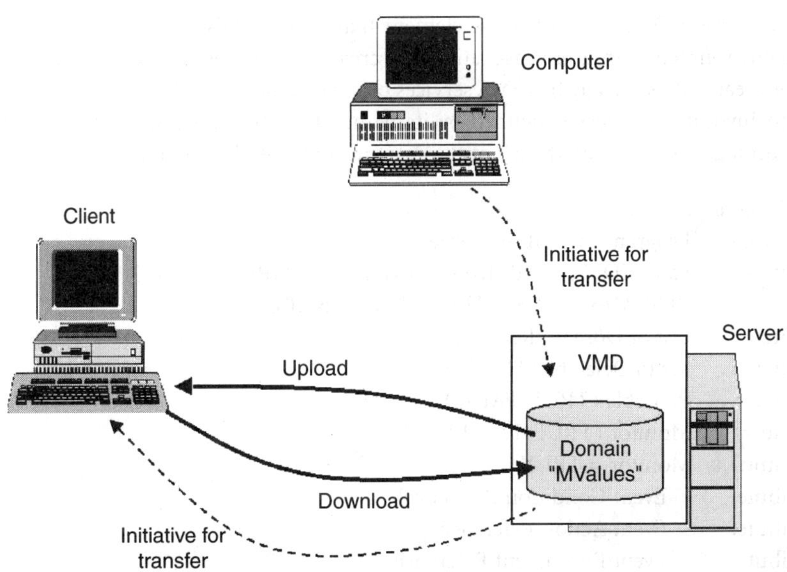

FIGURE 58.9   MMS domain transfer.

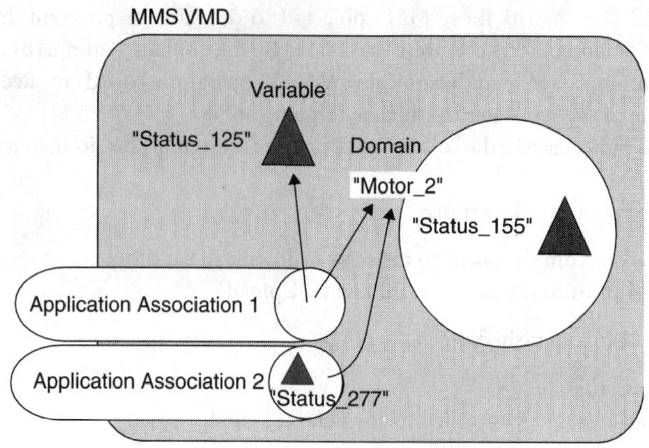

FIGURE 58.10   VMD- and Domain-scope.

A third scope is defined by the Application Association. The object "Status_277" is part of the corresponding connection. This object can only be accessed through this connection. When the connection is closed, all objects are deleted in this scope.

MMS objects can be organized using the different scopes. The object names (with or without domain scope) can be compounded from the following character set:

The identifiers can contain 1 to 32 characters and they must not start with a number.

The object names can be structured by agreement in a further standard or other specification. Many standards that reference MMS make considerable use of this possibility. This way, all named variables with the Prefix "RWE_" and similar prefixes, for example, could describe the membership of the data (in a trans-European information network) to a specific utility of an interconnected operation.

58.9   Program-Invocation-Management

A Program Invocation Object is a dynamic element that corresponds with the program executions in multitasking environments. Program Invocations are created by linking several domains. They are either predefined or created dynamically by MMS services or created locally.

A Program Invocation Object is defined by its name, its status (Idle, Starting, Running, Stopping, Stopped, Resuming, Unrunnable), the list of the domains to be used, and nine operations.

```
Object:   Program Invocation
   Key Attribute:   Program Invocation Name
   Attribute:       State (IDLE, STARTING, RUNNING, STOPPING, STOPPED,
                    RESUMING, RESETTING, UNRUNNABLE)
   Attribute:       List of Domain References
   Attribute:       MMS Deletable(TRUE, FALSE)
   Attribute:       Reusable (TRUE, FALSE)
   Attribute:       Monitor (TRUE, FALSE)
   Constraint:      Monitor = TRUE
      Attribute:        Event Condition Reference
      Attribute:        Event Action Reference
      Attribute:        Event Enrollment Reference
      Attribute:        Execution Argument
      Attribute:        Additional Detail
```

Program Invocations are structured flatly — although several Program Invocations can reference the same domains (Shared Domains). The contents of the individual domains are absolutely transparent from the point of view of both the domain and the Program Invocations. What is semantically connected with the Program Invocations is beyond the scope of MMS. The user of the MMS objects must therefore define the contents; the semantics result from this context. If a Program Invocation connects two domains, then the domain contents must define what these domains shall do together — MMS actually only provides a wrapper.

The Program Invocation Name is a clear identifier of a Program Invocation within a VMD.

State describes the status in which a Program Invocation can be. Altogether, seven states are defined.

List of Domains contains the names of the domains that are combined to a Program Invocation. This list also includes such domains that are created by the Program Invocation itself (this can be a domain in which some output is written).

MMS Deletable indicates whether or not this Program Invocation may be deleted by means of an MMS operation.

Reusable indicates whether or not a Program Invocation can be started again after the program execution. If it cannot be started again, then the program Invocation can only be deleted.

Monitor indicates whether or not the Program Invocation reports a transition to the client when exiting the Running status.

Start Argument contains an application-specific character string that was transferred to a Program Invocation during the last start operation; this string, for example, could indicate which function has started the program last.

Additional Detail allows the companion standards to give application-specific definitions.

Program Invocation Services

Create Program Invocations: This service arranges an operational program, which consists of the indicated domains, in the server. After installation, the Program Invocation is in the status Idle from where it can be started. The monitor and the monitor type indicate as to whether or not, and how the program invocation shall be monitored.

Delete Program Invocation: Deletable Program Invocations are deleted through this service. Primarily, that is to say that the resources bound to a Program Invocation are released again.

Start: The start service causes the server to transfer the specified Program Invocation from the Idle into the Running state. Further information can be transferred to the VMD through a character string in the start argument. A further parameter (Start Detail) once again contains additional information that can be defined by companion standards.

Stop: The stop service changes a specified Program Invocation from the Running into the Stopped state.

Resume: The Resume service changes a specified Program Invocation from the Stopped into the Running state.

Reset: The Reset service changes a specified Program Invocation from the Running or Stopped into the Idle state.

Kill: The Kill service changes a specified Program Invocation from arbitrary states into the Unrunnable state.

Get Program Invocation Attribute: Through this service, the client can read all attributes of a certain Program Invocation.

58.10 The MMS Variable Model

MMS Named Variables are addressed using identifiers composed of the domain name and the Named Variable name within the domain. Components of an MMS Named Variable may also be individually addressed using a scheme called Alternate Access. The Alternate Access address of a component consists of the domain name and the Named Variable name, along with a sequence of enclosing component names of the path down to the target component.

The Variable Access Services contain an extensive variable model that offers the user a variety of advanced services for the description and offers access to arbitrary data of a distributed system.

A wide variety of process data are processed by automation systems. The data, their definition, and representation are usually oriented toward the technological requirements and the available automation equipment. The methods that the components use for the representation of their data and the access to them correspond to the way of thinking of their implementers. This has resulted in a wide variety of data representations and access procedures for one and the same technological datum in different components. If, for example, a certain temperature measurement shall be accessed to in different devices, then a huge quantity of internal details must generally be taken into account for every device (request, parameter, coding of the data, ...).

As shown in Figure 58.11, the number of the protocols for the access of a client (on the left in the figure) to the data from n servers (S1 – Sn) can be reduced to a single protocol (on the right in the figure). Through this, the data rate required for the communication primarily in central devices can be reduced drastically.

In programs, variables are declared, that is, they get a name, a type, and a value. Described in a simplified way, both the name and the type are converted by the compiler into a memory location and into a reference that is only accessible to the compiled program. Without any further measures, the data of the variable are not identifiable outside the program. It is concealed for the user of the program how a compiler carries out the translation into the representation of a certain real machine.

The data are stored in different ways depending on the processor; that is, primarily that the data are stored in various memory locations. During the runtime of the program, only this representation is available.

These data are not visible from the outside. They must first be made visible for the access from the outside. To enable this, an entity must be provided in the implementation of the application. It is insignificant here whether or not this entity is separated from or integrated into the program. This entity is acting for all data that shall be accessible from the outside.

This consideration is helpful for the explanation of the MMS variable model: what do protocols through which process data are accessed have in common? Figure 58.12 shows the characteristics in principle. On the right is the memory with the real process data that shall be read. The client must be able to identify the data to be read (gray shade). For this purpose, pointer (start address) and length of the data must be known. By means of this information, the data can be identified in the memory.

Yet how can a client know the pointer of the data and what the length of the data is? It could have this information somehow and indicate it when reading. Yet if the data should move some time, then the pointer of the data is not correct any more. There can also be the case that the data do not exist at all at the time of reading but must be calculated first. In this case, there is no pointer. To avoid this, references

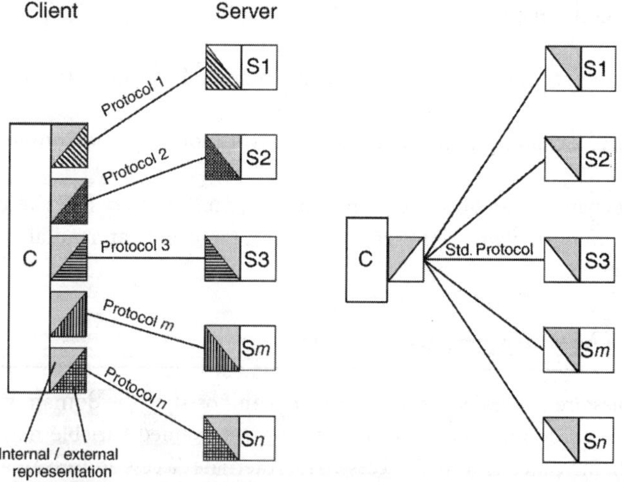

FIGURE 58.11   Unified protocols.

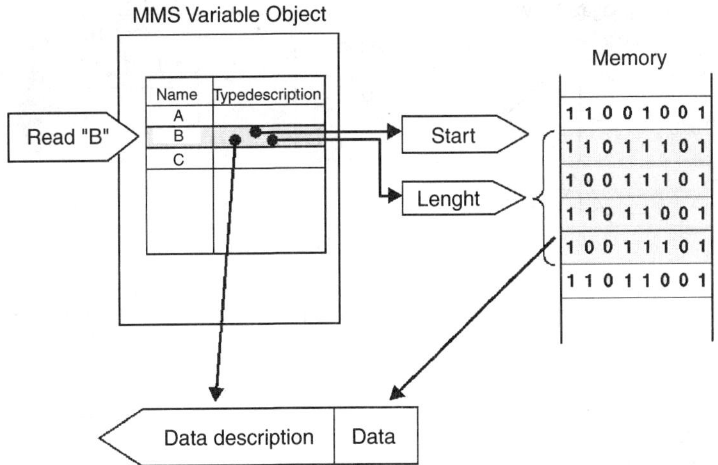

FIGURE 58.12   Data access principle.

to the data, which are mapped to the actual pointers (table or algorithm), are used in most cases. In our case, the reference "B" is mapped by table to the corresponding pointer and the length.

The pointer and the length are stored in the type description of the table. The pointer is a system-specific value, which is generally not visible on the outside. The length is dependent on the internal representation of the memory and on the type. An individual bit can, for example, be stored as an individual bit in the memory or also as a whole octet. However, this is not relevant from the point of view of the communication.

The data themselves and their description are important for the message response of the read service. The question of the external representation (is, e.g., an individual bit encoded as a bit or as an octet?) is — unlike the internal representation — of special importance here! The various receivers of the data must be able to interpret the data unambiguously. For this purpose, they need the representation that is a substantial component of the MMS variable model. The data description is therefore derived from the type description.

For a deeper understanding of the variable model, three aspects have to be explained more exactly:

- objects and their functions,
- services (read, write, …) that access the MMS variable objects, and
- data description for the transmission of the data.

The object model of an MMS variable object is conceptually different from a variable according to a programming language. The MMS objects describe the access path to structured data. In this sense, they do not have any variable value.

Access Paths

The access path represents an essential feature of the MMS variable model. Starting from a more complex hierarchical structure, we will consider the concept. The abstract and extremely simplified example are deliberately chosen. Here, we are merely concerned about the principle.

Certain data of a machine shall be modeled using MMS methods. The machine has a tool magazine with n similar tools. A tool is represented — according to Figure 58.13 — by three components (tool type, number of the blades, and remaining use time).

The machine with its tool magazine "M" is outlined in the figure on the right. The magazine contains three tools "A," "B," and "C." The appropriate data structure of the magazine is shown on the left. The structure is tree-like; the root "M" is drawn as the topmost small circle (node). "M" has three components (branches) "A," "B," and "C," which are also represented as circles. These components in turn also have three components (branches). In this case, the branches end in a leaf (represented in the form of a

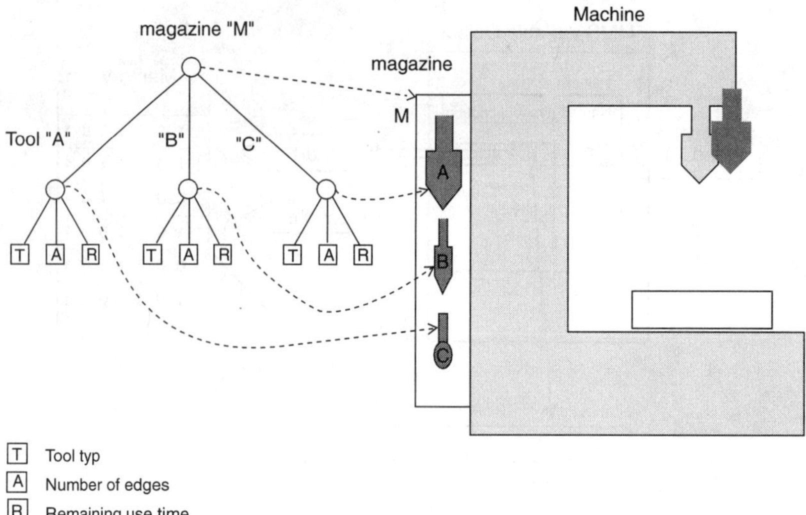

FIGURE 58.13   Data of a machine.

square). Leaves represent the end-points of the branches. For each tool, three leaves are shown: "T" (tool type), "A" (number of blades), and "R" (remaining use time). The leaves represent the real data. The nodes ordered hierarchically are merely introduced for reasons of clustering. Leaves can occur at all nodes. A leaf with the information "magazine full/not full" could, for example, be attached at the topmost node.

With this structure, the MMS features are explained in more detail. The most essential aspect is the definition of the access path. The access to the data (and their use) can be carried out according to various task definitions:

1. selecting all leaves for reading, writing …,
2. selecting certain leaves for reading, writing …, and,
3. selecting all leaves as component of a higher-level structure, for example, "machine" with the components magazine "M" and drilling machine, and
4. selecting certain leaves as component of a higher-level structure.

Examples of cases 1. and 2. are shown in Figure 58.14. The case that the complete structure is read is shown in the left top corner (the selected nodes, branches, and leaves are represented in bold lines or squares). All nine data are transmitted as response (3 × (T+A+R)). At first, the representation during transmission is deliberately refrained from here and also in the following examples.

Only a part of the data is read in the top right corner of the figure: respectively only the leaf "R" of all three components "A," "B," and "C." The notation for the description of the subset M.A.R/.B.R/.C.R is chosen arbitrarily.

The subset M.A.R represents a (access) path that leads from a root to a leaf. It can also be said that one or several paths represent a part of a tree. The read message contains three paths that must be described in the request completely. A path, for example, can also end at "A." All three components of "A" will be transmitted in this case.

Besides the possibility to describe every conceivable subset, MMS also supports the possibility to read several objects "M1" and "M5" in a read request simultaneously (see the lower half of the figure). Of course, every object can only be partly read too (which is, however, not represented).

An example of case 4 is shown in Figure 58.15 (case 4 has to be understood as the generalization of case 3). Here, a new structure was defined using two substructures. The object "machine" contains only the "R" component of all six tools of the two magazines "M1" and "M5." The object "machine" shall not be mixed up with a list (Named Variable List). Read "machine" supplies the six individual "R" values.

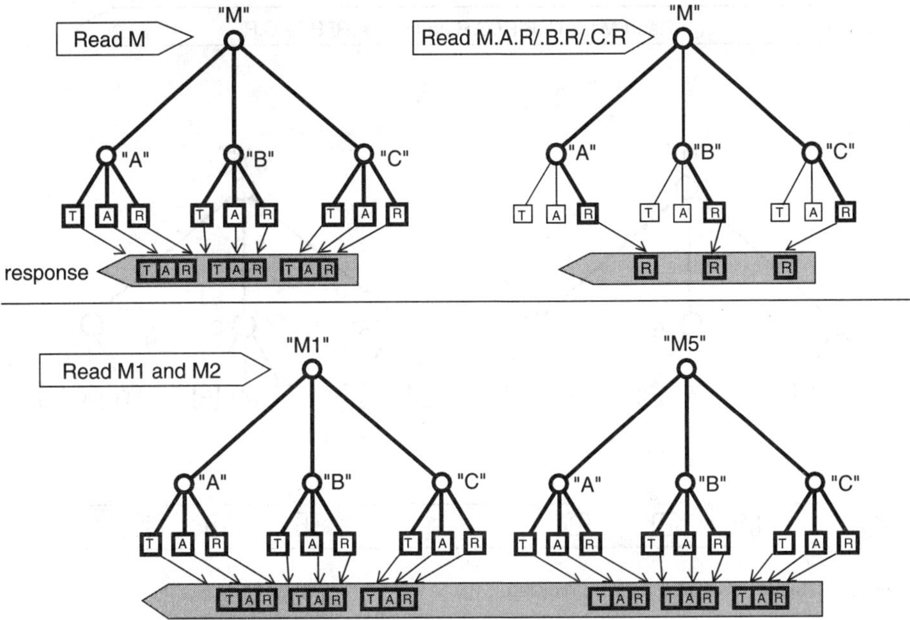

FIGURE 58.14　Access and partial access.

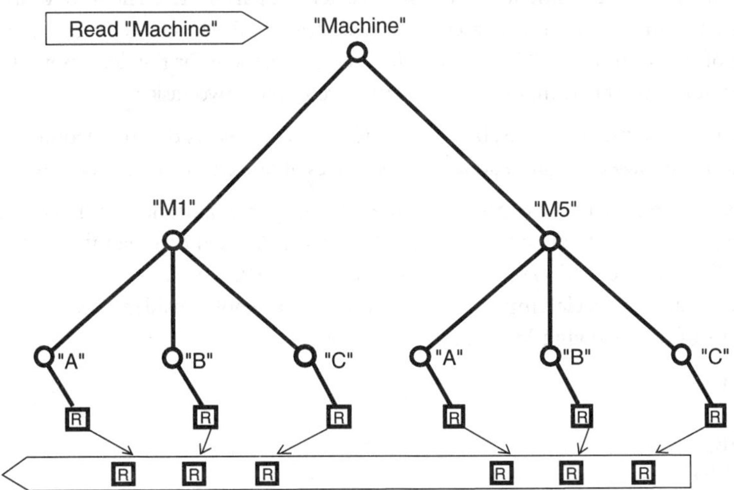

FIGURE 58.15　"Partial" trees.

The component names, such as "A," "B," or "C," need to be unambiguous only below a node — and only at the next lower level. Thus, "R" can always stand for the remaining lifetime. The position in the tree indicates the remaining lifetime of a particular tool. The new structure "machine" has all features that were described in the previous examples also for "M1" and "M5."

These features of the MMS variable objects can be applied to (1) the definition of new variable objects and (2) the access to existing variable objects. The second case is also interesting. As shown in Figure 58.16, the same result as in Figure 58.15 can be achieved by enclosing the description (only the "R" components of the tools shall be read from the two objects "M1" and "M5") in the read request every time. The results (read answer) are absolutely identical in the two cases.

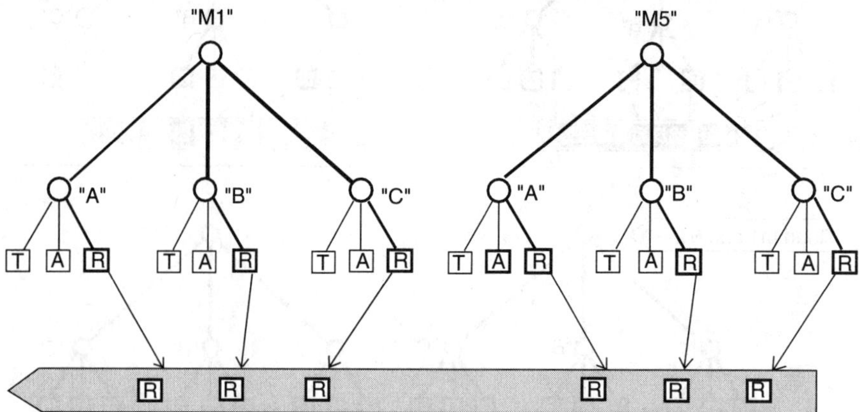

FIGURE 58.16 "Partial" trees used for read requests.

Every possibility, to assemble hierarchies from other hierarchies (1), or to, for example, read parts of a tree during the access (2), has its useful application. The first case is important in order to avoid enclosing the complete path description every time when reading an extensive part of tree. The second case offers the possibility to construct complex structures based on standardized basic structures (e.g., the structure of "tool data" consisting of the components "T," "A," and "R") and to use them for the definition of new objects.

Summarizing, it can be stated that the access paths accomplish two tasks:

- description of a subset of nodes, branches, and leaves of objects during reading, writing, …, and
- description of a subset of nodes, branches, and leaves of objects during the definition of new objects.

In conclusion, this may again be expressed in the following way: path descriptions describe the "way" or the "ways" to a single data (leaf) or to several data (leaves). A client can read the structure description (complete tree) through the MMS service Get Variable Access Attributes.

Another aspect is also of special importance. Till now, we have not considered the description of leaves. Every leaf has one of the following MMS basic data types:

- BOOLEAN,
- BIT,
- INTEGER,
- UNSIGNED,
- FLOATING POINT,
- REAL,
- OCTET STRING,
- VISIBLE,
- GENERALIZED TIME,
- BINARY, and
- BCD.

Every node in a tree is either an array or a structure. Arrays have n elements (0 to $n-1$) of the same type (this can be a basic type, an array or a structure). When describing a part of a tree, any number of array elements can be selected (e.g., one element, two adjacent elements, two arbitrary elements, …).

Structures consist of one or several components. Every component can be marked by its position and if necessary by an "Identifier" — the component name. This component name (e.g., "A") is used for the access to a component. Parts of trees can describe every subset of the structure.

The path description contains the following three elements:

- All of the possibilities of the description of the structures (individual and composite paths in the type description of the MMS variables) are defined in the form of an extensive abstract syntax.
- For every leaf of a structure (these are MMS basic data types), the value range (size) is also defined besides the class. The value range of the class INTEGER can contain one, two, or more octets. The value range four (4) octets (often represented as Int32) for example, indicates that the value cannot exceed these four octets. On the other hand, with ASN.1 BER coding (explained later) and the value range Int32, the decimal value "5" will be transmitted in only one octet (not in 4 octets!). That is to say, only the length needed for the current value will be transmitted.
- The aspect of the representation of the data and their structuring during transmission on the line (communication in the original meaning) is dealt with below in the context of the encoding of messages.

The path description is used in a quintuple way (see Figure 58.17):

- during the access to and during the definition of variable objects,
- in the type description of variable objects,
- in the description of data during reading, for example,
- during the definition of Named Type Objects (an object name of its own is assigned to the type description – i.e., to one or several paths – of those objects), and
- when reading the attributes of variables and Named Type Objects.

Objects of the MMS Variable Model

The five objects of the MMS variable model are:
Description of simple or complex values:

- Unnamed Variable,
- Named Variable.

List of several unnamed variables or named variables:

- Named Variable List,
- Scattered Access (is not explained here).

Description of the structure by means of a user-defined name:

- Named Type.

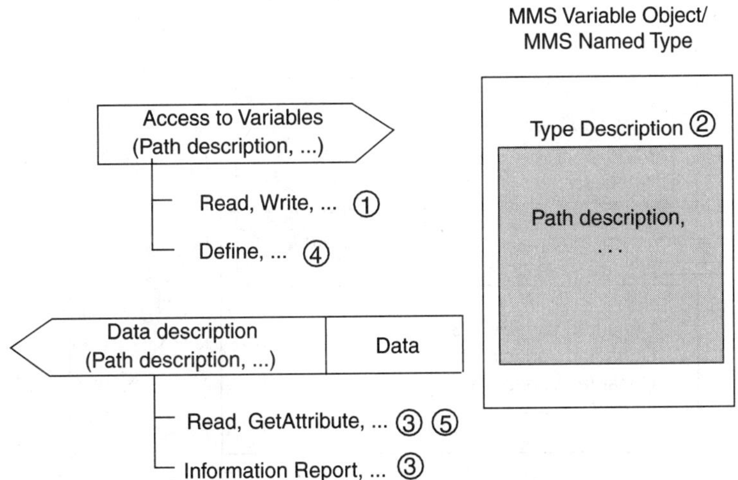

FIGURE 58.17 Application of path descriptions.

The Unnamed Variable

The Unnamed Variable Object describes the assignment of an individual MMS variable to a real variable that is located at a definite address in the device. An Unnamed Variable Object can never be created or deleted. The Unnamed Variable has the attributes:

    Object: Unnamed Variable
      Key Attribute: Address
      Attribute: MMS Deletable (FALSE)
      Attribute: Access Method (PUBLIC)
      Attribute: Type Description

Address: Address is used to reference the object. There are three different kinds:

1. Numeric Address (nonnegative integer values),
2. Symbolic Address (character string), and
3. Unconstrained Address (implementation specific format).

Even though in (2) the address is represented by character strings, this kind of addressing has absolutely to be distinguished from the Object Name of a Named Variable (see domain scope and explanations below).

MMS Deletable: The attribute is always FALSE here.
Access method: The attribute is always PUBLIC here.

Type Description: The attribute points to the inherent abstract type of the subordinate real variable as it is seen by MMS. It specifies the class (bitstring, integer, floating-point, etc.), the value range of the values, and the group formation of the real variable (arrays of structures). The attribute Type Description is completely independent of the addressing.

Figure 58.18 represents the Unnamed Variable roughly sketched. The Unnamed Variable with the address "62" (MMSString) has three components with the names: value, quality, and time. These component names are only required if individual components (specifying the path, for e.g., "62"/"Value") shall be accessed.

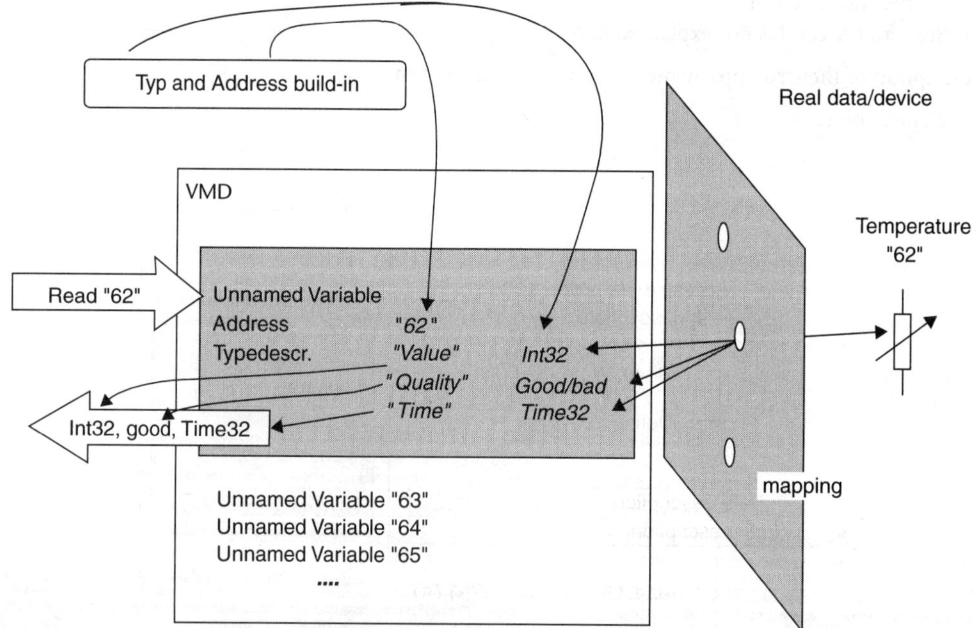

FIGURE 58.18   Unnamed Variable Object.

MMS Address of the Unnamed Variable

The MMS Address is a system-specific reference that is used by the system for the internal addressing — it is quasi-released for the access via MMS. There, the address can assume one of three forms (here, the ASN.1 notation is deliberately used for the first time):

```
Address ::= CHOICE {
    numericAddress              [0] IMPLICIT Unsigned32,
    symbolicAddress             [1] MMSString,
    unconstrainedAddress        [2] IMPLICIT OCTET STRING
    }
```

The above definition has to be read as follows: address defines as (::=) a selection (keyword CHOICE) of three possibilities. The possibilities are numbered here from [0] to [2] to be able to distinguish them. The keyword IMPLICIT is discussed later.

The numeric address is defined as Unsigned32 (4 octet). Thus, the addresses can be defined as an index with a value range of up to 2**32. Since only the actual length (e.g., only one octet for the value 65) will be transmitted for an Unsigned32, the minimal length of the index, which can thus be used, is merely 1 octet. Already, 255 objects (of arbitrary complexity!) can be addressed with one octet.

The symbolic address can transmit an arbitrarily long MMSString (e.g., "DB5_DW6").

The Unconstrained Address represents an arbitrarily long octet string (e.g., 24FE23F2A1hex). The meaning and the structure of these addresses are beyond the scope of the standard.

These addresses can be used in MMS Unnamed Variable and Named Variable Objects and in the corresponding services. MMS can neither define nor change these addresses. The address offers a possibility to reference objects by short indexes.

The addresses can be structured arbitrarily. Unnamed Variables could, for example, contain measurements in the address range (1000 to 1999), status information in the address range (3000 to 3999), limit values in the address range (7000 to 7999), etc.

Services for the Unnamed Variable Object

Read: This service uses the V-Get function to transmit the current value of the real variable, which is described by the Unnamed Variable Object, from a server to a client. Variable Get represents the internal, system-specific function through which an implementation obtains the actual data and provides them for the communication.

Write: This service uses the V-Put function to replace the current value of the real variable, which is described by the Unnamed Variable Object, by the enclosed value.

Information Report: As Read, although without prior request by the client. Only the Read.response is sent by the server to the client without being asked. The Information Report corresponds to a spontaneous message. The application itself determines when the transmission is to be activated.

Get Variable Access Attributes: Through this operation, a client can query the attributes of an Unnamed Variable Object.

Explanation of the TypeDescription

Features of the structure description of MMS variable objects were explained in principle above. For those interested in the details, the formal definition of the MMS Type Specification is explained according to Figure 58.19.

The description in ASN.1 was deliberately also selected here. The type specification is a CHOICE (selection) of 15 possibilities (Tags [0] to [13] and [15]). Tags are qualifications of the selected possibility. The first possibility is the specification of an ObjectName, a Named Type Object. If we remember that one Named Type Object describes one or several paths, then the use is obvious. The path description referenced by the name can be used to define a Named Variable Object. Or, if during reading the path must be specified, it can be referenced by a Named Type Object in the server.

```
TypeSpecification ::= CHOICE {
    typeName                    [0] ObjectName,
    array                       [1] IMPLICIT SEQUENCE {
        packed                      [0] IMPLICIT BOOLEAN DEFAULT FALSE,
        numberOfElements            [1] IMPLICIT Unsigned32,
        elementType                 [2] TypeSpecification },
    structure                   [2] IMPLICIT SEQUENCE {
        packed                      [0] IMPLICIT BOOLEAN DEFAULT FALSE,
        components                  [1] IMPLICIT SEQUENCE OF SEQUENCE {
            componentName [0] IMPLICIT Identifier OPTIONAL,
            componentType           [1] TypeSpecification } },

    Simple                  Size                          Class
    boolean                 [3] IMPLICIT NULL,            - BOOLEAN
    bit-string              [4] IMPLICIT Integer32,       - BIT-STRING
    integer                 [5] IMPLICIT Unsigned8,       - INTEGER
    unsigned                [6] IMPLICIT Unsigned8,       - UNSIGNED
    floating-point          [7] IMPLICIT SEQUENCE {
        format-width            Unsigned8,     - # of bits in fraction plus sign
        exponent-width          Unsigned8      - size of exponent in bits },
    real                    [8] IMPLICIT SEQUENCE {
        base                    [0] IMPLICIT INTEGER(2|10),
        exponent                [1] IMPLICIT INTEGER,   - max number of octets
        mantissa                [2] IMPLICIT INTEGER    - max number of octets },
    octet-string            [9] IMPLICIT Integer32,       - OCTET-STRING
    visible-string          [10] IMPLICIT Integer32,      - VISIBLE-STRING
    generalized-time        [11] IMPLICIT NULL,           - GENERALIZED-TIME
    binary-time             [12] IMPLICIT BOOLEAN,        - BINARY-TIME
    bcd                     [13] IMPLICIT Unsigned8        - BCD
    objId                   [15] IMPLICIT NULL}
```

FIUGRE 58.19   MMS Typespecification.

Note that the ASN.1 definitions in MMS are comparable with XML schema. ASN.1 BER (Basic encoding rules) provide very efficient message encoding compared to XML documents. The forth coming standard IEC 61400-25 provides applies ASN.1 as well as an XML schema for the specification of messages.

The next two possibilities (array and structure) have a common feature. Both refer — through their element type or component type — back to the beginning of the complete definition (type specification). This recursive definition allows the definition of arbitrarily complex structures. Thus, an element of a structure can in turn be a structure or an array.

Arrays are defined by three features. Packed defines whether or not the data are stored optimized. Number-Of-Elements indicates the number of the elements of the array of equal type (Element Type).

The data of structures can also be saved as packed. Structures consist of a series of components (components [1] IMPLICIT SEQUENCE OF SEQUENCE). This series is marked by the keyword SEQUENCE OF ..., which describes a repetition of the following definition. Next in the list is SEQUENCE {component Name und component Type}, which describes the individual component. Since the SEQUENCE OF ... (repetition) can be arbitrarily long, the number of the components at a node is also arbitrary.

Then follow the "simple data types." They start at the tag [3]. The length of the types is typical for the simple data types. For example, integers of different lengths can be defined. The length (size) is defined as Unsigned8, which allows for an integer with the length of 255 octets.

It should be mentioned here that — in the ASN.1 description of the MMS syntax — expressions like integer (written in small letters) show that they are replaced by another definition (in this case by the Tag [5] with the IMPLICIT-Unsigned8 definition). Capital letters at the beginning indicate that the definition is terminated here — it is not replaced any more. Here, it is a basic definition.

Figure 58.20 shows an example of an object defined in IEC 61850-7-4 [20]. The circuit breaker class is instantiated as "XCBR1." The hierarchical components of the object are mapped to MMS (according to IEC 61850-8-1 [21]). The circuit breaker is defined as a comprehensive MMS Named Variable.

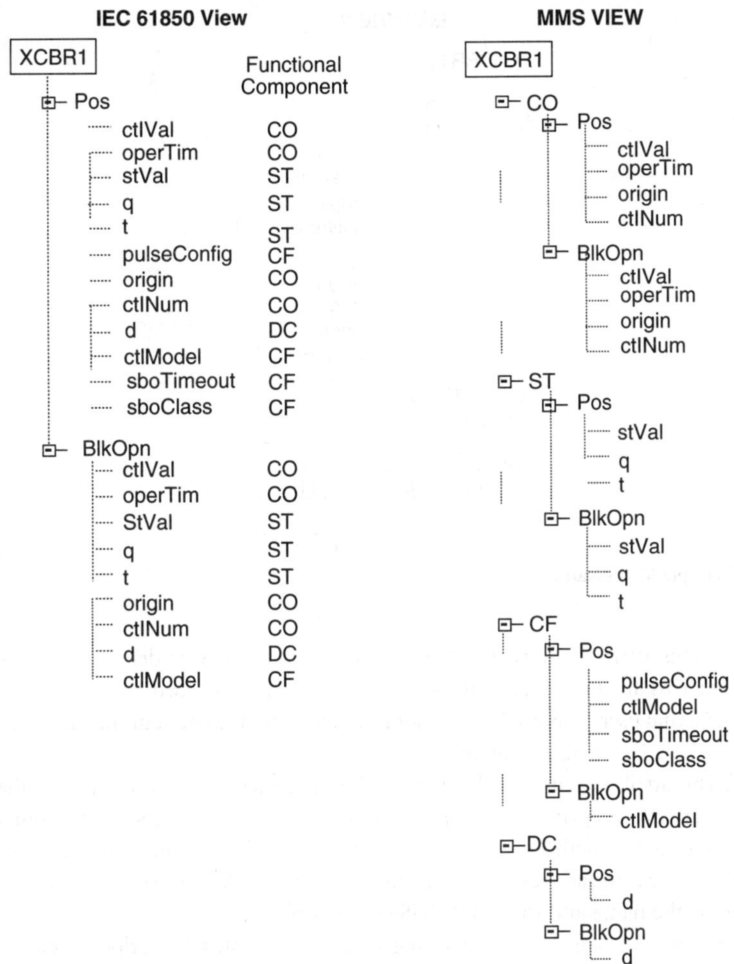

FIGURE 58.20　Example MMS Named Variable.

The components of the hirarchical model can be accessed by the description of the Alternate Access: "XCBR1" → component "ST" → component "Pos" → component "stVal." Another possibility of mapping the hierarchy to a flat name is depicted in Figure 58.21. Each path is defined as a character string.

The Named Variable

The Named Variable Object describes the assignment of a named MMS variable to a real variable. Only one Named Variable Object should be assigned to a real variable. The attributes of the object are as follows:

Object:　Named Variable
　　Key Attribute:　Variable Name
　　Attribute:　MMS Deletable
　　Attribute:　Type Description
　　Attribute:　Access Method (PUBLIC, ...)
　　　Constraint:　Access Method = PUBLIC
　　Attribute:　Address

Variable Name: The Variable Name unambiguously defines the Named Variable Object in a given scope (VMD specific, domain specific, or Application Association specific). The Variable Name can be 32 characters long (plus 32 characters if the object has a domain scope).

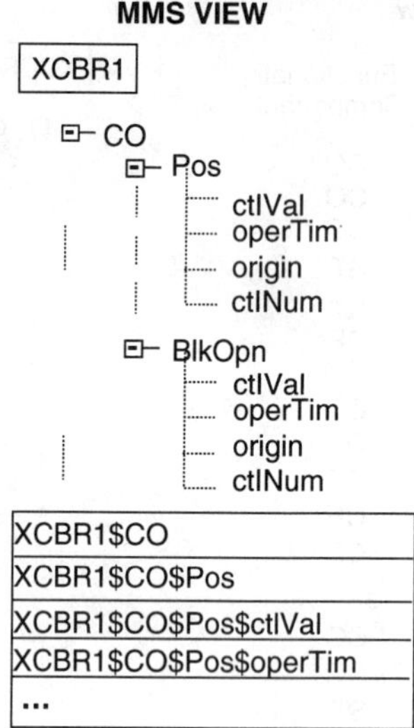

FIGURE 58.21   Example MMS Named Variable.

MMS Deletable: This attribute shows whether or not the object may be deleted using a service.

Type Description: This attribute describes the abstract type of the subordinate real variable as it represents itself to the external user. This attribute is not inherently in the system unlike the Unnamed Variable Object, that is, this Type Description can be defined from the outside.

Access method: This attribute contains the information that a device needs to identify the real variable. It contains values that are necessary and adequate to find the memory location. The contents lie outside MMS. A special method, the method PUBLIC, is standardized. The attribute Address is also available in the case of PUBLIC. This is the address that identifies an Unnamed Variable Object. Named Variables can thus be addressed by the name and the ADDRESS (see Figure 58.22).

Address: See Unnamed Variable Object. Defining a Named Variable Object does not allocate any memory either, because the real variable must already exist; it is assigned to the Named Variable Object with the corresponding name.

Altogether, six operations are defined on the object:

Read: The service uses the V-Get function to retrieve the current value of the real data, which is described by the Object.

Write: The service uses the V-Put function to replace the current value of the real data, which is described by the Object, by the enclosed value.

Define Named Variable: This service creates a new Named Variable Object, which is assigned to real data.

Get Variable Access Attribute: Through this operation, a client can query the attributes of a Named Variable Object.

Delete Variable Access: This service deletes a Named Variable Object if Attribute Deletable is (=TRUE).

Figure 58.23 shows the possibilities to reference a Named Variable Object by names, and if it is required, also by Address (optimal access reference of a given system). For a given name, a client can query the Address by means of the service Get Variable Access Attribute.

This possibility allows the access through technological names (MeasurementTIC13) or with the optimal (index-) address 23 24 hex.

As shown in Figure 58.23, an essential feature of the VMD is the possibility for the client application to define by request Named Variable Objects in the server via the communication. This includes the definition of the name, the type, and the structure. The name by which the client would like to reference the Named Variable later is "TIC42" here. The first component "Value" is of the type Integer32,

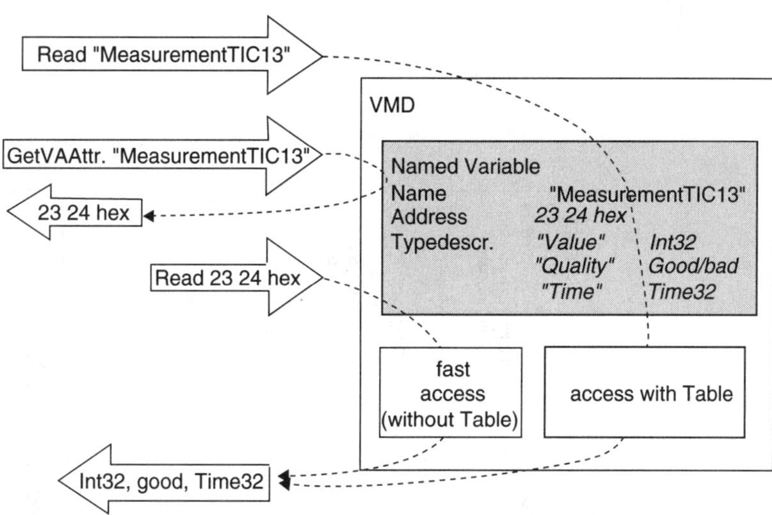

FIGURE 58.22   Address and Variable Name of Named Variable Objects.

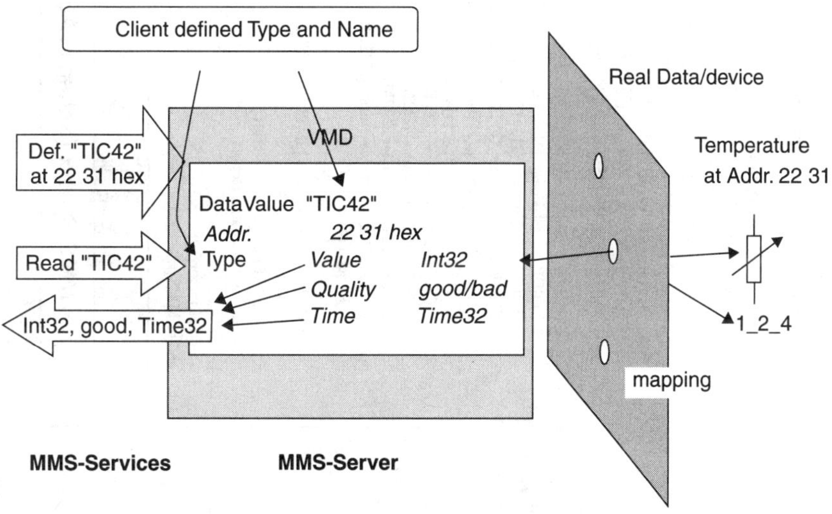

FIGURE 58.23   Client-defined Named Variable Object.

the second is "Quality" with the values "good" or "bad," and the third is "Time" of the type "Time32." The type of the Data Value Object can be arbitrarily simple (flat) or complex (hierarchical). As a rule, the Data Value Objects are implicitly created by the local configuring or programming of the server (they are predefined).

The internal assignment of the variable to the real temperature measurement is made by a system-specific, optimal reference. This reference whose structure and contents are transparent must be known when defining the Named Variable, though. The reference can, for example, be a relative memory address (e.g., DB5 DW15 of a PLC). So a quick access to the data is allowed.

The Named Variable Object describes how data for the communication are modeled, accessed, encoded, and transmitted. What is transmitted is described independent of the function. From the point of view of communication, it is not relevant where the data in the server actually come from or where in the client they actually go to and how they are managed — this is deliberately concealed.

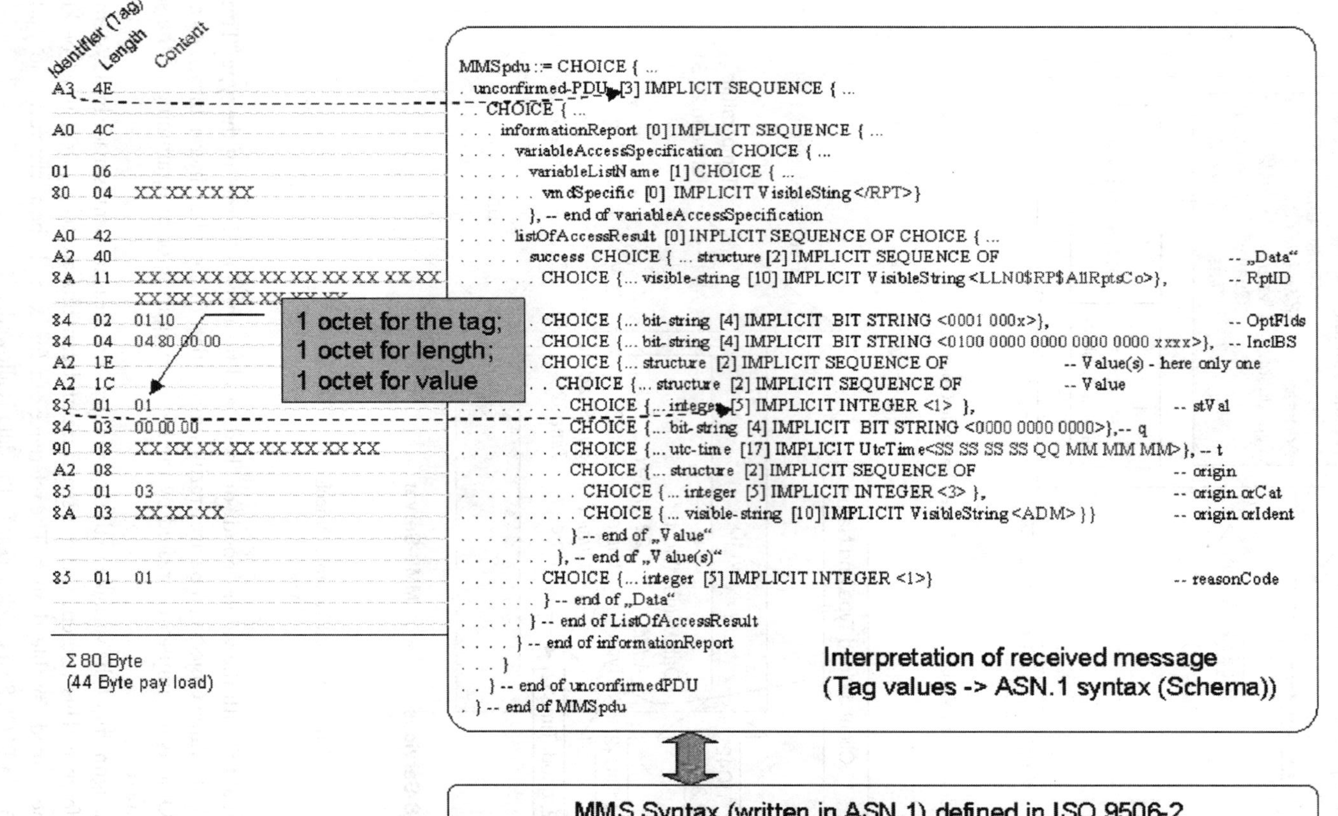

FIGURE 58.24   MMS Information Report (spontaneous message).

Figure 58.24 shows the concrete encoding of the Information Report message. The message is encoded according to ASN.1 BER (the basic encoding rule for abstract syntax notation number one — ISO 8825).

The encoding using XML would be several times longer than when using ASN.1 BER. These Octets are packed into further messages that add lower layer-specific control and address information, for example, the TCP header, IP header, and Ethernet frames.

The receiving IED is able to interpret the Report message according to the identifier, lengths, names, and other values. The interpretation of the message requires the same stack, that is, knowledge of all layers involved — including the definitions of IEC 61850-7-4 [20], IEC 61850-7-3 [19], IEC 61850-7-2 [18], and IEC 61850-7-1 [17].

The Access to Several Variables

Named Variable List

The Named Variable List allows to define a single name for a group of references to arbitrary MMS Unnamed Variables and Named Variables. Thus, the Named Variable List offers a grouping for the frequently repeated access to several variables (see Figure 58.25). Although the simultaneous access to several MMS variables can also be carried out in a single service (Read or Write), the Named Variable List offers a substantial advantage.

When reading several variables in a Read request, the individual names and the internal access parameters (pointers and lengths), corresponding to the names in the request, must be searched for in a server. This search can last for some time in the case of many names or a low processor performance. By using the Named Variable List Object, the search is not required — except for the search of a single name (the name of the Named Variable List Object) — if the references, for example, have been entered into the Named Variables on the list system specifically and thus optimally. Once the name of the list has been found, the appropriate data can be provided quickly.

Thus, the Named Variable List Object provides optimal access features for the applications. This object class is used in the known applications of MMS very intensively.

The structure of the Named Variable List Object is as follows:

```
Object:   Named Variable List
   Key Attribute:   Variable List Name
   Attribute:       MMS Deletable (TRUE, FALSE)
```

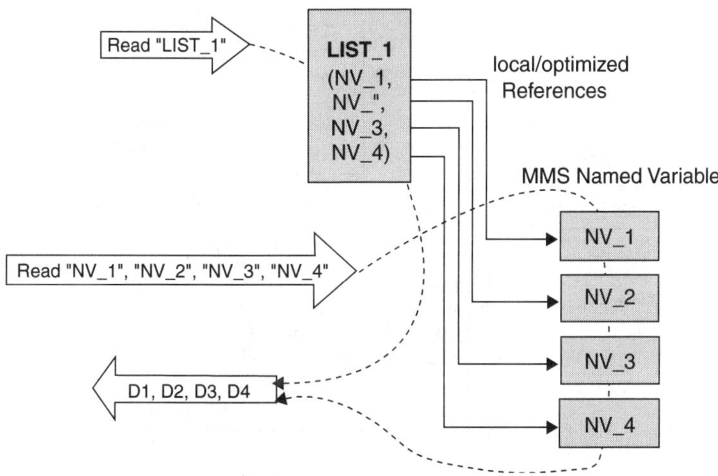

FIGURE 58.25   MMS Named Type and Named Variable.

Attribute:	List of Variable
Attribute:	Kind of Reference (NAMED, UNNAMED, SCATTERED)
Attribute:	Reference
Attribute:	Access Description

Variable List Name: The Variable List Name unambiguously identifies the Named Variable List Object in a given scope (VMD specific, domain specific, or Application Association specific). See also MMS Object Names above.

MMS Deletable: This attribute shows whether or not the object may be deleted.

List of Variable: A list can contain an arbitrary number of objects (Unnamed Variable, Named Variable, or Scattered-Access Object).

Kind of Reference: Lists can refer to three object classes: Named Variables, Unnamed Variables, and Scattered Access; no Named Variable Lists can be included.

Reference: An optimal internal reference to the actual data is assigned to every element of the list. If a referenced object is not (any more) available, the entry into the list will indicate it. When accessing the list, for example, by Read, no data other an error indication will be transmitted to the client for this element.

Access Description: In the same way as for the access of a variable — for example in the request during Read — could be defined that only parts (part of a tree) of a variable shall be read, can this also be applied to every element of the Named Variable List.

Services

Read: This service reads the data of all objects being part of the list (Unnamed Variable, Named Variable, and Scattered Access Object). For objects that are not defined an error is reported in the corresponding place of the list of the returned values.

Write: This service writes the data from the write request into the objects being part of the list (Unnamed Variable, Named Variable, and Scattered Access Object). For objects that are not defined an error is reported in the corresponding place of the list of the returned values.

Information Report: As the Read service, as if the read data were sent by the server to the client without prior request (Read Request) by the client, that is, as if only a Read Response would be transmitted.

Define Named Variable List: Using this service, a client can create a Named Variable List Object.

Get Named Variable List Attributes: This service queries the attributes of a Named Variable List Object.

Delete Named Variable List: This service deletes the specified Named Variable List Object.

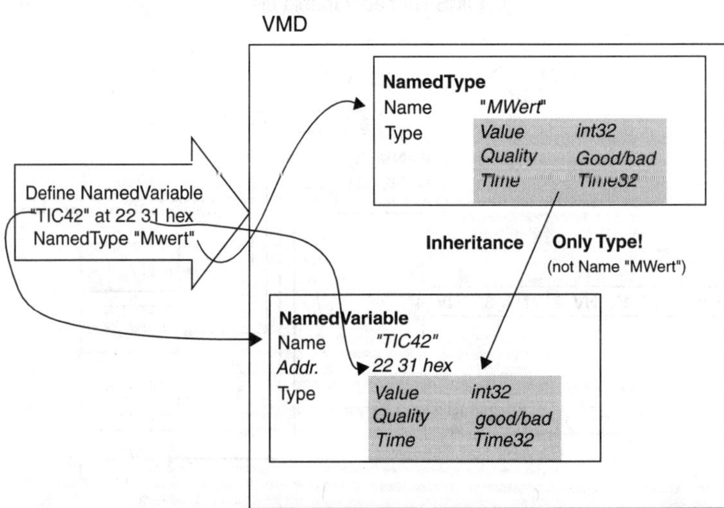

FIGURE 58.26   Inheritance of Type of the MMS NamedType Objects.

The Named Type Object

The Named Type Object merely describes structures. The object model is very simple:

    Object: Named Type
       Key Attribute: Type Name
       Attribute: MMS Deletable (TRUE, FALSE)
       Attribute: Type Description

The essential attribute is the Type Description, which was already discussed before for Named and Unnamed Variables. On the one hand, TASE.2 standard data structures can be specified by means of Named Types. This is the most frequent application of the Named Type Objects. On the other hand, Named Types can be used for the access to the server. The read request can refer to a Named Type Object. Or the Named Type Object will be used to define Named Variables.

Figure 58.26 describes the application of the Named Type Objects for the definition of a Named Variable. A variable will be created by the request Define Named Variable.

It shall have the name "TIC42," the address 22 31 hex, and the type that is defined in the Named Type Object "Mwert." The variable inherits the type from the Named Type Object. The inheritance has the consequence that the variable will only have the type but not the name of the Named Type Object. This inheritance was defined so strictly in order to avoid that through deleting the Named Type the type of the variable would get undefined or that by subsequent definition of a differently structured Named Type with the old type name "MWert" the type of the variable would be changed (the Named Type Object and with that the new Type Description would be referenced by the old name).

Perhaps it is objected now that this strict inheritance has the consequence that the type would also have to be saved for each variable (even though many variables have the same Type Description). Since these variables can internally be implemented in a system in whatever way the programmer likes it, they can refer through an internal index to a single (!) type description. He must only make sure that this type description is not deleted. If the accompanying Named Type Object gets deleted, then the referenced type description must remain preserved for these many variables.

The disadvantage that the name of the "structure mother," that is, the Named Type, is not known any more as an attribute of the variables has been eliminated in the MMS Revision.

Services

Define Named Type: This service creates a Named Type Object.

Get Named Type Attribute: This service delivers all attributes of a Named Type Object.

Read, Write, Define Named Variable, Define Scattered Access, Define Named Variable List, and Define Named Type use the Type Description of the Named Type Object when carrying out their tasks.

58.11   Resume

MMS is a standard messaging specification (comparable to web services), widely implemented by industrial device manufacturers like ABB, Alstom, General Electric, and Siemens. It solves problems of heterogeneity so often found in automation applications. MMS is the lingua franca of industrial devices.

MMS provides much more than TCP/IP, which essentially offers a transfer stream of bytes. MMS transfers commands with parameters between machines.

MMS allows a user to concentrate on the applications, application data to be accessible — and not on communication problems, which are already solved. It provides a basis for the definition of common and domain-specific semantic. Examples are the standards IEC 60870-6 TASE.2 [14,15], IEC 61850 [17–21], and IEC 61400-25 [22].

References

[1] Manufacturing Message Specification (MMS) — Part 1: Service Definition ISO International Standard ISO 9506-1, 2003.

[2] Manufacturing Message Specification (MMS) — Part 2: Protocol Definition ISO International Standard ISO 9506-2, 2003.

[3] Manufacturing Message Specification (MMS) — Part 3: Companion Standard for Robitics ISO/IEC 9506-3, 1991.

[4] Manufacturing Message Specification (MMS) — Part 4: Companion Standard for Numerical Control ISO/IEC 9506-4, 1992.

[5] Manufacturing Message Specification (MMS) — Part 5: Companion Standard for Programmable Controllers ISO/IEC CD 9506-5, 1993.

[6] Manufacturing Message Specification (MMS) — Part 6: Companion Standard for Process Control ISO/IEC 9506-6, 1993.

[7] ESPRIT Consortium CCE-CNMA, Preston, U.K. (Editoren). *MMS: A Communication Language of Manufacturing,* Springer-Verlag, Berlin, 1995.

[8] ESPRIT Consortium CCE-CNMA, Preston, U.K. (Editoren). *CCE: An Integration Platform for Distributed Manufacturing Applications,* Springer-Verlag, Berlin, 1995.

[9] Inter-Control Center Communication, *IEEE Transactions on Power Delivery,* 12, 607–615, 1997.

[10] Telecontrol equipment and systems — Part 6: Telecontrol Protocols Compatible with ISO Standards and ITU-T Recommendations — Section 503: Services and Protocol (ICCP Part 1) IEC 60870-6-503, 1997.

[11] Telecontrol equipment and systems — Part 6: Telecontrol Protocols Compatible with ISO Standards and ITU-T Recommendations — Section 802: Object Models (ICCP Part 4) IEC 60870-6-802, 1997.

[12] März, W. and Schwarz, K., Powerful and Open Communication Platforms for the Operation of Interconnected Networks, ETG-Tage/IEEE PES Summer Meeting 1997, Proceedings, Berlin, 1997.

[13] IEEE Technical Report 1550, Utility Communications Architecture, UCA, 1999, http://www.nettedautomation.com/standardization/IEEE_SCC36_UCA

[14] IEC 60870-6-TASE.2: Telecontrol Application Service Element 2.

[15] Becker, G., Gärtner, W., Kimpel, T., Link, V., März, W., Schmitz, W., and Schwarz, K., Open Communication Platforms for Telecontrol Applications — Benefits from the New Standard IEC 60870-6 TASE.2 (ICCP), etz-Report 32, VDE-Verlag, Berlin, 1999.

[16] Wind Power Communication — Verification Report and Recommendation, Elforsk rapport 02:14; Stockholm, April 2002, www.nettedautomation.com/download/02_14_rapport.pdf

[17] IEC 61850-7-1, Communication Networks and Systems in Substations — Part 7-1: Basic Communication Structure for Substation and Feeder Equipment — Principles and models.

[18] IEC 61850-7-2, Communication Networks and Systems in Substations — Part 7-2: Basic Communication Structure for Substation and Feeder Equipment — Abstract Communication Service Interface (ACSI).

[19] IEC 61850-7-3, Communication Networks and Systems in Substations — Part 7-3: Basic Communication Structure for Substation and Feeder Equipment — Common Data Classes.

[20] IEC 61850-7-4, Communication Networks and Systems in Substations — Part 7-4: Basic Communication Structure for Substation and Feeder Equipment — Compatible Logical Node Classes and Data Classes.

[21] IEC 61850-8-1, Communication Networks and Systems in Substations — Part 8-1: Specific Communication Service Mapping (SCSM) — Mappings to MMS (ISO/IEC 9506-1 and ISO/IEC 9506-2) and to ISO/IEC 8802-3.

[22] IEC 61400-25, Wind turbines — Part 25: Communications for Monitoring and Control of Wind Power Plants.

Usefull web pages that provide additional information:

http://www.livedata.com

http://www.sisconet.com

http://www.tamarack.com

http://www.scc-online.de

http://www.nettedautomation.com

IEC 61850 circuit breaker model:

http://www.nettedautomation.com/qanda/iec61850/information-service.html#

Section 4: The Internet, Web, and IT Technologies in Industrial Automation and Design

59

Remote Monitoring and Control over the Internet

Hans-Arno Jacobsen
University of Toronto

59.1 Introduction

Remote monitoring and control refers to the observation and collection of data in remote locations, the processing of these data, and the emission of control information as a function of the observations to influence the behavior of the distant environment and the monitored entities. Applications range from environmental monitoring (without control) to security and surveillance applications and a high degree of accuracy requiring control tasks (e.g., reactor control, energy control, and industrial process control.)

The remote data are collected through sensors embedded in the environment, which either periodically emit measured values or transmit measurements upon request. Sensors are connected with the monitoring and control system through a communication network. Network connections may be wireless or wired; network protocols may be standard based, such as Internet protocols, or proprietary. The control loop is closed through actuators that influence the controlled process. The monitoring and control system may be distributed to increase the availability of the system and protect against failure of parts of the system. The measured data may be processed at the monitoring and control system or inside the network of sensors. Processing at the monitoring system can be as simple as storing the data in a database or monitoring the data for unexpected trends or events. Processing of data inside the network can deal with preaggregation of the measured values at the source or at nodes inside the network of sensors.

As can be seen from the above discussion, the design space of remote monitoring and control systems is quite large and will depend highly on the particular application context. In this chapter, we will choose one particular application case study, discuss and motivate the application, and present a detailed account of the design and architecture of the application's remote monitoring and control system. The application case study is a Remote Building Monitoring and Operations (RBMO) system for the management of energy resources in buildings over the Internet. The case study is fully implemented and was experimentally deployed on several large buildings.

The remainder of this chapter is organized as follows. Section 59.2 motivates the application context and reviews the technologies underlying the remote building monitoring and control system described. In Section 59.3, the RBMO system architecture is presented and major design decisions are discussed. The application software is described in Section 59.4. Section 59.5 summarizes our experience in developing, deploying, and operating the system. Sections 59.6 and 59.7 present related work and point out directions that require further exploration.

59.2 Case Study: Building Energy Management

Energy Management

Buildings consume one third of all energy in the United States at a cost of $200 billion per year, with $85 billion per year for commercial buildings. A large amount, perhaps half of this energy, is wasted compared to what is cost-effectively achievable. Much of this waste is related to our inability to optimally control and maintain today's complex building systems. Recent building performance case studies suggest that typical savings of about 15%, and as much as 40% of annual energy use can be gained by compiling, analyzing, and acting upon energy end-use data. The largest source of savings arises from turning off equipment when not needed. Other sources of savings include better equipment scheduling and optimization of temperature set point controls [18, 7]. For an example of the building energy efficiency analysis, which detailed building monitoring enables, see the work by Piette, et al. [26].

The applications under development for the RBMO project are based on utilizing the capabilities of, and expanding upon information available through legacy Energy Management Control Systems (EMCSs), which control the Heating Ventilating and Air Conditioning Equipment (HVAC) in commercial buildings (Figure 59.1).

Several studies have explored EMCS monitoring capabilities [17, 32]. EMCSs for commercial buildings are readily available. There are over 150 EMCS manufacturers [14]. About 50% of all buildings over 45,000 m^2 (500,000 ft^2.) have an EMCS, and they are present in nearly all large office buildings [12]. For buildings built since 1992, almost 50% of the floor area is in buildings with EMCSs.

The RBMO system supports remote monitoring and control of multiple buildings, that is, HVAC, lighting, etc., across the Internet, using CORBA – Common Object Request Broker Architecture protocols. It is designed to work with heterogeneous EMCS and HVAC systems. The RBMO system is also designed for integration with a remote building monitoring and control center, which provides data visualization, database management, building energy simulation, and energy usage analysis tools (cf. Section 59.4 for a more detailed presentation of the application.) The remote monitoring facility is composed of two components:

1. A data acquisition and database management server, which is built on a Unix system.
2. Several analysis (Unix & Windows) workstations, which are used for data visualization, statistical analysis of the data, and simulation.

The ultimate users of the system will be owner/operators of multiple buildings, for example, U.S. GSA (federal buildings), school districts, universities, retail chains, banks, property management firms, and utilities.

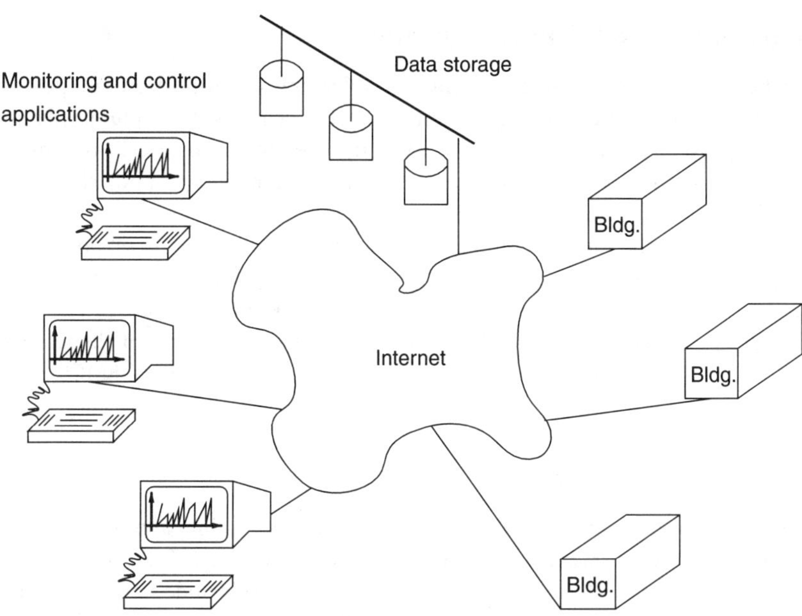

FIGURE 59.1 The RBMO system setup.

The RBMO system pursues three primary design goals: demonstration of technical feasibility of the proposed design, evaluation and experimentation with component software technology and their inter-operability across heterogeneous environments, and demonstration of the utility of the system.

Close attention and careful analysis of the data from building EMCSs typically afford many opportunities to improve building operations, resulting in improvements to occupant comfort and reductions in energy use. However, for most buildings, such close attention and analysis are not undertaken due to lack of appropriate software, operator training, staff, and inaccurate or malfunctioning sensors. Remote building monitoring offers the possibility to reduce the labor costs involved in monitoring and analyzing EMCSs by spreading such labor costs over a large number of buildings. This will also make it economically feasible to use expert HVAC engineers.

The use of the Internet as an underlying communication infrastructure has two aspects: use of the Internet communications protocols and use of the actual Internet network.

The use of Internet *networking protocols* permits access to a wide variety of affordable interoperable hardware and software from many different vendors over many different media. Internet markets are huge, in terms of millions of users; hence, vendors can readily amortize development costs. In contrast, proprietary protocols typically entail more expense, because of lack of competition and because development costs are being amortized over smaller markets. Industry-specific protocols, such as BACnet, fall in between these two extremes. Internet protocols are clearly the most popular for wide area networking applications.

Use of the Internet *network* permits users to share the costs of a common communications infrastructure. When building automation applications can share existing Internet infrastructure, substantial savings compared to tele-phone dial-up systems are possible [16]. The Internet also offers the prospect of inexpensive access to high bandwidths when needed, for example, for remote video surveillance for security, fire, or remote diagnostics. Finally, the Internet offers a reliable, redundantly connected, backbone. The downside of using the Internet is the necessity of more stringent security precautions, discussed further below.

A more detailed discussion of the work presented in this chapter has been published in [23, 25, 24]. In [23], we present an overview of the RBMO application and motivate its deployment. In [25], we present a technical account of the system's architecture and design. Based on this investigation, we synthesize in [24] a set of requirements that future middleware platforms for control applications should exhibit.

Energy Management and Control Systems

Today's EMCS have been built primarily for control. EMCSs are special-purpose computerized control systems, programmed to operate building equipment such as chillers, fans, pumps, dampers, valves, motors, and lighting systems. EMCSs are installed for many different purposes such as: controlling equipment, alerting operators to possible equipment malfunction, maintaining comfortable conditions, and permitting modification of control specifications.

EMCSs typically monitor building operational data such as damper positions, set points, state variables of the working fluids, and equipment status. They are not generally used to monitor electrical demand. In some cases, more accurate sensors may be needed for diagnostics than are typically included in an EMCS. For example, to track chiller efficiency, a very accurate measurement of the temperature drop across and water flow rate through the chiller is needed. The sensors in a typical EMCS may not be sufficient for this calculation.

Building Data Model Standards

In 1995, ASHRAE approved a new communications standard, for building automation systems — The Building Automation and Control Network (BAC-net) Standard [1, 3]. This standard views the HVAC system as a collection of objects. However, the standard does not (yet) include standard applications-level objects such as chillers or cooling towers. We expect that BACnet will shortly be used to connect building gateways to EMCSs.

The International Alliance for Interoperability [20, 6] is a consortium of architectural, engineering, and construction software vendors (e.g., Autodesk and Bentley Systems), who are working on developing a standard data model for description of buildings. The data model is intended to be used by architectural and engineering design tools, construction cost estimation software, construction scheduling software, etc. We are tracking these efforts for use in our naming conventions for monitoring points.

59.3   RBMO System Design

In this section, we describe the architecture of the RBMO system in further detail, motivate our primary design decisions, and describe the following components of the system:

- Gateway systems in each building.
- Data acquisition subsystem.
- Application-level object specifications and unit conversion.
- Time series database.

System Architecture

The RBMO system architecture is shown in Figure 59.2. It constitutes a three-tier architecture with the individual components being: the applications, the database management system, and the building gateway. The individual tiers are entire subsystems interconnected through the Internet via CORBA adaptors.

The system has been designed in a modular manner to ease the evaluation of alternative technologies for database, user interfaces, and visualization components (cf. Time Series Database section 3.5). The following subsection will describe the individual components and their design in further detail.

The RBMO system constitutes a highly heterogeneous system incorporating several different hardware platforms and operating systems, ranging from extremely proprietary systems (building EMCSs) to common platforms (Sun workstations/UNIX and PC/NT). This large heterogeneity was one primary motivation for using CORBA instead of alternate protocols, such as OLE, Active X, DCOM, SNMP, MMS, and netDDE protocols. See Table 59.1 for a brief comparision of these along several dimensions.

For additional information on MMS, see [4] (including a bibliography) and [9], or [13] for utility applications. For information on SNMP, see [8, 31, 21, 30, or 34]. For information on netDDE, see [33].

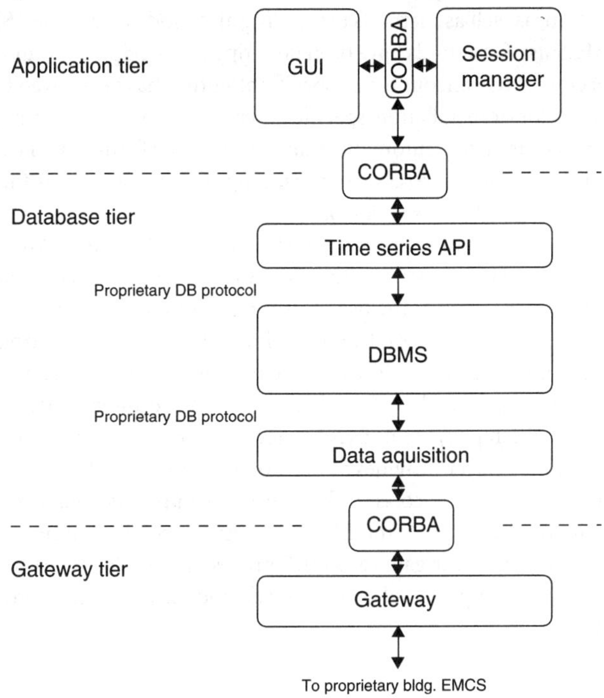

FIGURE 59.2   RBMO system architecture.

TABLE 59.1   Comparison of Interoperability Protocols

Protocol	Application Domain	Characteristics	Platforms	Comments
COBRA	Distributed enterprise applications	OO RPC	All major	Our choice
DCOM	Desktop	OO RPC	PC	Mictosoft only
MMS	Manufacturing, electric utilities,	Messaging	Many	
SNMP	Network management	No methods	Many	
netDDE	Process control	Text messages	PC's	Microsoft only

Building Gateway

The system architecture requires that each participating building have an Internet point-of-presence, also referred to as a building "gateway." These building gateways are addressed in the usual Internet manner (e.g., `north-TowerBuilding.myCompany.com`). Within the gateway itself, further addressing of building-specific objects (more below) is accomplished by a CORBA-compliant Object Request Broker (ORB). The gateway ORB acts as a mediator between incoming CORBA object references and the actual objects resident in the gateway process. Currently, building gateways are Pentium-class PCs running the Windows NT operating system.

One of our goals was to present a network view of a building's monitoring and control data that was consistent across multiple buildings — regardless of EMCS system differences. In order to provide this homogeneous view at the building gateway level, it was necessary to have building-to-building (i.e., EMCS to EMCS) differences resolved within the gateway process itself. This has led to a three-part gateway design consisting of an EMCS adaptation layer, an internal building data point database, and an external CORBA interface layer.

The building data point database portion of the gateway contains entries for all building monitoring and control data points accessible through the building's EMCS. Each entry contains attributes for units, point name, and description, as well as methods for reading and modifying values. Since this database has to mirror that of the EMCS itself, methods have been developed for deciphering and recreating data point lists for each vendor-specific EMCS encountered. Specific attention has been given to mapping data point units information from vendor or installation-specific descriptions to standard representations. Standard unit representations are required for automatic unit system transformation facilities located further downstream. Since the logical place to resolve any mapping discrepancies is at the building level, these issues are dealt with by the internal gateway database.

The EMCS adaptation layer is responsible for interacting with the building EMCS via the specific communications formats and protocols required by the EMCS vendor. In some cases, for example, the interfacing to the Barring-ton EMCS, involves the use of proprietary message protocols using standard serial communications media. In others, for example, interfacing to the Johnson Controls EMCS, the netDDE protocol over LAN-based media is used. In all cases, specific message formatting and handling algorithms are encapsulated in methods associated with each element of the internal data points database (see above), and vendor-specific differences are not visible above the EMCS adaptation layer.

The external CORBA interface layer provides a common layer of C++ objects that are accessible to the gateway's ORB. These objects have full access to the internal data points database and are able, through appropriate overloaded methods, to query the EMCS system for current monitor and set point values. Since these objects execute as part of the gateway's ORB process and not as a part of the gateway program itself, synchronization mutexes are provided to assure safe and exclusive access to building data object methods and attributes.

Data Acquisition System

Our telemetry data on the building state are initially available as cross-sectional data, that is, we sample the entire building state periodically, typically at 1–2 min intervals. Such high-resolution data permit us to study oscillations in the control system. However, typical analysis uses only a few state variables, but will typically wish to look at long time series of each such state variable. Thus, it is expedient for analysis purposes to have the stored data in a separate time series for each state variable.

The question then arises as to when and where the data are to be transposed from cross-sectional to time series storage formats. This can be done at collection time, analysis time, or some intermediate time.

Transposition at collection time slows the collection process, but speeds the analysis phase. Transposition at analysis time permits very fast data collection, but slows the analysis phase. Transposition at intermediate times permits both very fast data collection and fast analysis. However, the software architecture becomes more complex and costly to implement and maintain. Transposition can also occur either at the gateway machines, the database server, or the analysis workstations.

We chose to transpose the data at the database server as they are collected. This will generate at least one disk I/O operation per data point monitored for each sampling period. This appears to be the simplest architecture to implement, and should have adequate performance for our prototype system.

The Data Acquisition System (DAQ) presently is a polling system residing on the central server together with the time series Database Management System (DBMS). Placing the DAQ on the TS DB server is useful for performance reasons as the DAQ imparts considerable I/O to the DB server.

Polling (pull architecture) was chosen initially over a push architecture from the gateways due to perceived simplicity of implementation. We anticipate moving to a push architecture, which we expect will offer better reliability against single point failures of the central system (assuming it has been replicated), and easier support for replicated or partitioned central DAQ/DBMS systems.

Our data acquisition system is multithreaded, with one DAQ process serving multiple buildings. This offers performance (and software licensing) advantages over a single threaded, one process per building design, at the expense of reliability. Any failure in the DAQ process then causes data acquisition from all buildings to fail. Replication or partitioning of the DAQ would be desirable for large-scale applications.

Measurement Units and Time Zones

In dealing with multiple buildings, we were forced to confront the diversity of the measurement units used for various monitoring points and EMCSs.

There are three approaches to dealing with this issue:

1. record all measurements as pairs (value, units),
2. record the units for each monitoring point as part of the metadata for the monitoring point, and
3. record all measurements in a canonical set of units.

Conventional practice in EMCS systems is method 2. The IAI data model adopts method 1. However, the variation in measurement units across monitoring points and EMCSs renders comparisons and computations in the DBMS extremely awkward. We have adopted the convention of storing all measurements in standard SI (metric) units in the DBMS (the conversion is done by the data acquisition system). This expedites the construction of indices and comparisons or computations in query processing. We plan to permit users to specify a preferred systems of units for display of query results.

Similar difficulties arise with time zones. There are two problems — a collection of monitored buildings may span multiple time zones (e.g., a large retail chain) and the use of daylight savings time introduces anomalies. We address both problems by converting all timestamps to (Universal Coordinated Time (UTC)) (a.k.a. Greenwich Mean Time (GMT)). This is done in the data acquisition system prior to storing data.

Time Series Database

The bulk of the data we are collecting will be time series, for example, of temperatures, power usage, etc. Although the raw data may be irregular time series, we will transform these to regular time series for analysis. Regular time series assumes periodic temporal sampling. Hence, our most important criterion for selection of the DBMS is its suitability for time series data and queries.

It is possible to support time series data on any of several DBMS: relational, object-oriented (OO), or hybrid object–relational. The use of relational DBMS for regular time series data typically offers mediocre performance, and weak support for operations such as smoothing and calendars. Object-oriented DBMSs permit the implementation of time series data collections with better support for calendars, smoothing, etc. To minimize development costs, we initially chose an early object–relational DBMS, Illustra (since merged with Informix) with built-in time series support. When this DBMS proved unreliable we switched to a relational DBMS, Oracle. See discussion below. Recently, relational DBMS vendors (e.g., Oracle) have announced built-in time series support.

Most of the data that are not time series are building description data — that is, Computer Aided Design (CAD) data describing the physical configuration of the building, HVAC system, and the EMCS system. Object-Oriented DBMSs (OODBMSs) are generally much better suited to CAD data than relational DBMSs (RDBMSs).

CORBA interfaces, for at least some of the commercial CORBA implementations, are available for several object-oriented DBMSs, but not yet for most of the relational or object–relational DBMSs. However, Oracle has announced that it intends to support CORBA interfaces. Third-party CORBA interfaces are available for relational DBMSs [19].

Support for client–server DBMS configurations is typically available for all of the DBMSs. This will allow us to move the analysis and visualization codes to the PC-workstations while maintaining the database on a robust database server.

System Partitioning

We have introduced CORBA interfaces at several points in the system. This partitioning serves two key purposes:

- distribution of components and
- decoupling for reengineering.

The CORBA protocols between the time series database, the graphical user interface, and the session management facility (cf. Figure 59.2) permit us to vary the distribution of these components depending on the communications bandwidth and client machine performance. Moreover, they allow us to better decouple the individual components for reengineering purposes, for example, Internet browser integration of GUI.

59.4   The RBMO Applications Software

The RBMO applications are designed around providing the following three capabilities:

- Archiving historical time-series data in a database.
- Providing visualization means for building energy performance analysis.
- Performing a series of regressions and statistical analysis techniques to (1) define "baseline" conditions and (2) evaluate energy performance after a retrofit, operations and maintenance changes, occupancy changes, or for historical tracking.

The archiving and visualization applications presently run on Sun workstations in C++. We envision, however, that a future deployment of the visualization and statistical analysis applications will primarily be PC-based, running Java under Windows NT or a web browser.

The whole building and cooling plant monitoring at one installation of the system consists of a few dozen sensor points (out of 1500 total sensor points in the building) that cover whole-building electricity use, weather variables, plus chiller power and cooling tower status, and several temperatures and water flow rates. We added several power and flow meters, and temperature sensors to the cooling plant to compare them with the original EMCS sensors and augment information that was not available from the EMCS. At another installation, we collected 120 monitoring points concerned with the chiller plant, from a total of 8000 points monitored by the EMCS.

We are developing a standard set of performance tracking plots to allow users to view archived data with a user-friendly menu. This includes daily (24 h) load shapes of 5-min operating data for all values; week-day, weekend, and weekly load shapes (hourly data); and monthly average load shapes. The chiller and cooling plant analysis will consist of plots such as power vs. cooling load (kW vs. tons) and efficiency versus load (kW/ton vs. tons).

The applications are intended for use by a remote expert. Capabilities exist for the user to look at detailed data such as time series of cooling plant water temperatures or cooling tower fan start–stop intervals. The concept is to provide a tool for a remote user who might be responsible for tracking the performance of building systems at multiple locations. The menu of standard graphs are designed to allow the user to easily view the most critical, macrolevel performance data, while providing additional means for more detailed analysis of microlevel data. In other words, the scripted graphs should allow the user to answer the question, "is the building and cooling plant operating well?"

To keep the application extensible, we developed a scripting language that drives the visualization and GUI software. Certain extensions and additions may, however, either be performed on-line, or by preparing a script. The scripting language developed is simple; we therefore omit a detailed discussion. In retrospect, we should have adopted a conventional scripting language.

59.5   Discussion

This section presents a detailed account of our experiences in the design, development, and operation of a prototype of the RBMO system.

Interoperability Platform Concerns

Event Notification Capabilities

Central to any remote monitoring and control system is some mechanism for the remote sites to notify the monitoring sites of "interesting" events. Such event notifications need to be asynchronous, persistent, and (often) multicast. Ideally, events should be typed (so that certain processes can listen for specific types of events).

Event notification services address the need to propagate notification of significant occurrences to relevant system components. Depending on system functional requirements, this notification may be considered unreliable (e.g., periodic monitoring value updates) or may be required to be reliable (e.g., equipment status changes). In a distributed environment where system components can fail or become unreachable, the technical requirements for implementing these services can be substantial.

Although low-level event delivery protocols are defined in CORBA, the semantics of describing and specifying higher level event service behavior remains (thus far) vendor-specific. Event service reliability and delivery persistence capabilities are not yet consistently defined and implemented. It is therefore important to identify and describe a significant subset of high-level event service functionality. These event service functions should be sufficiently well described to allow multiple implementations that exhibit interoperable behavior. Some of these issues are being addressed in recent OMG proposals on improved Event Notification Services.

Security and Privacy

The principal security requirement for RBMO is to preclude unauthorized persons from altering the programming of the HVAC equipment. Besides disrupting the activities of building occupants, it is possible to damage HVAC equipment with inappropriate (or malicious) commands. Hence, some scheme for secure authentication and access control is required. The absence of such facilities in CORBA implementations at the beginning of the project led us to defer control applications.

In addition, some sites have indicated a desire for privacy of HVAC utilization data (e.g., occupancy data), and hence asked for encryption of all communications.

One of our major concerns has been the interoperability of the security services. We have been unable to identify interoperable CORBA security services software as yet. Furthermore, we would have liked to be able to construct interoperable software composed of different services (e.g., event services, security services, naming services, etc.) from different vendors. We were unable to do so.

Performance

Our experience with CORBA was that the RPC facilities were too slow to permit the modeling of individual measurement points as separate objects. It proved necessary to aggregate many of these measurement points together into named collections (called "scan lists") and to query the groups of points. In database terms, these constituted precompiled queries. This dramatically reduced the number of CORBA invocations (messages) and improved performance. Such aggregation of messages appears to be a commonplace strategy for improving the performance of distributed systems. However, it can be tedious and error prone to implement and maintain. It forces implementors to mix logical and physical design issues. While faster ORBs would certainly help, we believe that network latency in wide area networks will ultimately constrain the performance of fine-grained monitoring systems. Many distributed systems have found it efficient to aggregate messages.

Automatic (or convenient semiautomatic) tools for aggregating individual monitoring points into larger scan objects are very desirable. Such tools are also needed to support multicast publish/subscribe communication protocols. Such tools would permit a clean separation of logical and physical design considerations in the design and implementation of distributed systems.

For latency and throughput measurements of several CORBA ORB implementations (albeit over much higher speed networks), see the work of Schmidt et al. in [15, 28]. Brose [2] also reports recent performance numbers across a high-speed local area network, indicating that commercial ORBs require about 2 msec for a round-trip no-op. Note that many remote building monitoring applications require operation across wide area networks (with latency and lower bandwidth longer than local area networks).

Part of our performance problems arose from a desire to permit run-time binding of the schema of the monitoring points. Such late binding was perceived as necessary to accommodate changes in the HVAC configuration (e.g., due to maintenance, renovation, etc.) without requiring recompilation of the RBMO code. We could have used CORBA's dynamic invocation capabilities — however, these were perceived as too slow for large-scale RBMO performance requirements.

Finally, in this application and other related applications such as electric power grid monitoring, there is often a need to deal with many objects/points being monitored. It is our understanding that current ORBs maintain in-memory object tables (several tens of bytes per object) for all objects in the system. Such an approach has obvious problems when scaling to very large collections of objects. It is our view that some kind of automatic migration to secondary storage (and related pointer swizzling) similar to that in object-oriented databases will be needed to properly support very large applications.

Legacy System Integration Issues

In the integration of heterogeneous building EMCS into a seamless global distributed system, we have made the following observations.

Wrapping a legacy system with objects can hide essential system features that are critical to both robustness and operation. Even if one is able to hide most of the system's distributed nature using distributed object technologies, the underlying system may still consist of separate components that were designed as stand-alone systems with no expectation of cooperating with, or being controlled from, outside processes.

Centralized input and maintenance of configuration and description data are critical. This feature was mostly discovered by its absence. Although EMCS configuration information may be naturally maintained by that subsystem, these systems are designed as single, stand-alone entities. While it may be possible to create automatic mechanisms for incorporating configuration change information from one component (e.g., EMCS) into other systems (e.g., building simulation), there are typically no mechanisms to effectively signal participating components that such changes have been made. The result is that, although the collection of individual systems into a single larger one has been accomplished via distributed OO techniques (see above), the new, "larger" system depends on a higher, heuristically driven mechanism (e.g., system manager) for both coordination and correctness of operation.

The ability to seamlessly interact with objects within this new "larger" system can lead one to believe that the underlying components are cooperating in a manner that yields synchronized, correct operation. Since each of the respective components (EMCS, automated data loggers, etc.) is typically designed as stand-alone systems, they often lack any mechanism for synchronization or exclusive control path lockout. Therefore, although it may be possible for our system to have a component that tracks chiller performance, factors in near-term weather effects, and changes cooling plant operating setpoints, there is no way to ensure that such a component has exclusive control of all relevant system components. A control layer can be inserted into the architecture; but, its connection to all components may prove problematic.

We believe that part of these issues may be addressed by having stronger support for event specification, event notification, and a system design based on this. We envision a loose coupling of stand-alone components sending and receiving events. For a more complete discussion on event service support, see our discussion above.

Semantic Heterogeneity and Data Management

Naming and Directory Services

Effective configuration description and control is the bane of distributed systems. Distributed systems are constructed, in part, by binding together objects distributed over many platforms. Without a mechanism for describing these bindings, the resulting system lacks flexibility and its complexity remains hidden to all but the original implementors. Naming services provide a way to bind meaningful human names to object instances.

These problems take on greater importance in real-time control environments where systems must accommodate equipment replacement and/or upgrade cycles as well as "live" system reconfiguration. Stable and ubiquitous naming and directory services are a key tool in the design of systems that will accommodate these changes. Examples include the CORBA naming services and LDAP Directory services.

The function of these services goes well beyond the nuts and bolts issues of publishing specific object names and attributes. An effective distributed naming and directory service can expose, through the use of multiple hierarchical views, objects aggregated by function, location, control subsystem, etc. In addition, these services can provide a single, ubiquitous repository for object metainformation. In the case of control systems, this can include information about sensors and instrumentation used in generating object method values, resolution of object name aliases due to differing local and supervisory object namespaces, and provide automated mechanisms for locating physical and/or logical control system components.

Mediation Services

While standard efforts offer some hope in the long run for reconciling semantic heterogeneity among various network devices and controllers (HVAC or electric power grid), the existence of many incompatible legacy systems requires that some facilities for mapping between the global schema and local schemas be provided. At present, for example, in RBMO (and other systems) this is often done in an *ad hoc* manner with handcrafted mapping tables. Such facilities are variously referred to as mediators (partial mappings, late binding) or schema integration tools (nearly complete mappings, early binding). Such tools will require access to metadata (e.g., DB or object schemas). Examples of related work include OMG Object Interface Repository, and Meta Object Facilities.

Static vs. Dynamic Schemas

Traditional database designs specify complete detailed schemas statically. Such static schemas facilitate systematic design, query optimization, etc. However, for many monitoring and control applications (shipped as shrink wrap software), it is desirable to be able to alter the schema (e.g., to accommodate new types of equipment) without rebuilding the DB or recompiling the code. Examples of such approaches include frame-based knowledge representation systems, and the dynamic invocation interface of CORBA, together with interface repositories. Because such dynamic schema capabilities are typically significantly slower than static interfaces, some groups have built hybrid schemas — examples include the International Alliance for Interoperability (building data model), which provide both static schemas (e.g., for geometry) and dynamic schemas (frames) for extensibility. Conceivably, on-the-fly compilation techniques might provide a reasonable performance.

Units Conversion Services

We found it necessary to provide measurement unit conversion capabilities for our RBMO project.

Unit conversions have been a chronic source of errors in data acquisition/management systems. The ubiquity of these issues (in both industrial measurement and commercial trading) suggests their provision as standard services. Ideally, one would like an extensible-type system that incorporated dimensionality and automatic units conversions. However, impure unit conversions (e.g., between mass and volume) are commonplace and difficult to deal with systematically.

Our choice was to store all measurements in canonical units in the DBMS. While this facilitated indexing and querying, it made it very difficult to diagnose errors in the data acquisition system.

Time Zones and Synchronization

For large-scale monitoring applications, we believe that recording all times in Universal Coordinated Time (UTC) is essential. Conversion routines to local time zones, including daylight savings time, are essential.

Some protocol for synchronizing the clocks is needed for monitoring systems. Such protocols, for example, NTP [5], have been developed for other distributed applications. We note that synchronization requirements for monitoring of electric power grids are sufficiently stringent (to measure phase differences among geographically distributed generators) as to lead to the use of GPS timing signals for synchronization.

Time Series DBMS

Our first implementation used the early release of the Illustra DBMS product, chosen because it offered facilities for storing and querying time series data, including calendars. We encountered reliability problems with the DB server. We therefore switched to storing our time series data in a more mature conventional relational DBMS, that is, Oracle Version 7.3 (at present). This has solved our server reliability problems.

Whereas Illustra provided direct support for storing and querying time-series data, in Oracle we have had to construct these facilities atop a relational system. We have not (yet) replicated the calendar facilities, etc. Note that Oracle has announced time series facilities similar to Illustra/Informix.

Whereas Illustra storage usage was asymptotically 4 bytes/data value (for real numbers), Oracle is presently consuming approximately 25 bytes/data value, because each time series data value has become a full tuple in the DBMS.

For additional research work concerned with time series database management systems and data models, see the works [10, 29, 27, 11].

Data Quality Assurance

We greatly underestimated the difficulty and effort required to assure proper data quality. We have encountered a variety of problems with respect to data quality — some due to errors in units conversion, some due to missing or erroneous documentation of system configuration files, and some due to overflow problems with small (16 bit) counters.

We should have paid much more attention to the development of filtering software in the data acquisition system. The presence of dirty data in the time series database has been a barrier to the use of conventional database query facilities for aggregation. Dirty data are a ubiquitous problem in all kinds of data warehousing applications. Obviously, the preferred approach is to clean up the data at the source. If this is not feasible, concise, powerful methods are needed to specify constraints on data validity.

System Operation

A prototype of the system was operational for over one year. Data were acquired from two remote sites. Approximately 1 year of data (plus one year of historical data) were collected. The data collection involved several dozen monitoring points and sampling rates for both buildings are once per minute. The basic data collection volume is approximately 1 KB per minute.

Soda Hall

Soda Hall, a 109,000 ft^2 Computer Science building, was selected as our case study site for several reasons. The building was equipped with an Energy Management Control System that the research team had worked with before. Second, the building was located near LBNL, permitting easy access. Finally, the building has two, 220 ton screw chillers, for a total of 440 tons of cooling plant capacity (or 248 ft^2/t). Early in the project, we decided to focus on chillers, which are the largest single energy-using component in buildings with central cooling plants. The building has served as the subject of several research tasks that involved commissioning and performance.

Results from this work are reported in [26, 22]. A total of 46 points were monitored for about two years, starting in September, 1995 during the first few months of initial occupancy.

Oakland Federal Building

A total of 107 points were monitored at the Oakland Federal Building. (Note, not all of these points are sampled; many are computationally derived from sampled points.) This building is a 1.1 M ft^2. twintower, 18-story office building complex. The building was constructed in 1989 and houses Federal agencies such as the IRS, U.S. District Courts, the U.S. DOE, and Department of Veteran Affairs. The building is served by a central plant with five chillers, three at 1000 t and two at 450 t totaling 3900 t of capacity. This is 280 ft^2/t.

The 107 points cover systems similar to those at Soda Hall, including whole-building power, chiller power and tons, plus cooling tower, and pump data. Additional points include heating loop flow, a series of outside temperatures from five air handlers, and several internal zone temperatures.

One of the interesting questions we are examining at both sites is chiller capacities and oversizing. The operators report that the last 1000 t of cooling are not needed, even in the warmest weather.

59.6   Related Work

Honeywell recently announced a new building monitoring system, Atrium (Honeywell, 1998), quite similar to our project. Atrium uses a different operating system and window system (Windows NT) and a different relational database management system (SQLserver). However, it has a very similar overall architecture. They have also chosen to use a (popular) proprietary distributed object management system, Active X/DCOM, in contrast to our use of the standard CORBA distributed object system. Atrium also uses a standard building automation communications protocol, BACnet, to communicate with the building EMCS systems. BACnet is a recently deployed standard. BACnet equipped EMCSs were not available to us when we developed RBMO.

Other major differences between RBMO and Atrium lie in the time series database. Honeywell has chosen to only store changed data values (discarded repeated measures within dead bands). This results in a smaller, but irregular time series. Honeywell has also implemented a simple time series query language and temporal join operator to query and extract align multiple (irregular) time series.

The Enflex System from CTI Ltd. (CTI, 1998) uses proprietary applications protocols atop TCP/IP.

Finally, readers should be aware that remote building management presents many of the same issues addressed in the management of communications networks. Network management systems have used SNMP and RMON as the principal (standard) communications protocols. See the discussion above under alternative protocols and citations there.

59.7   Future Work

We plan to investigate the use of the CORBA Security Service and Secure HOP for remote control of the HVAC operations, (e.g., authentication).

Remote monitoring of large heterogeneous collections of buildings poses serious problems of software distribution to the buildings. We plan to study static and dynamic software distribution over the Internet onto heterogeneous target platforms.

59.8   Summary

We have described the design and development of the remote building monitoring and operation system, which uses the CORBA protocol across the Internet for communications between remote buildings and a time series database, and to distribute individual *application* components *locally* across different machines. We have discussed design tradeoffs, our development and operational experiences, and related work.

We have argued for the use of CORBA within our project context. We have discussed the design of a distributed system that has to monitor events across large distances, across multiple time zones, heterogeneous sources, and diverse measurement units. The development of the time series database atop different commercial DBMSs has been described.

We found it necessary (for performance reasons) to aggregate individual monitoring points into larger objects for CORBA-based retrieval. Such aggregation is typical of many distributed applications.

None of the commercial DBMS available at the time of the project proved entirely satisfactory for storing and retrieving time series. Recently announced products may remedy this.

The most serious issues included dealing with the heterogeneity of the various building EMCSs, that is, units and naming, and problems with dirty and missing data. Developers of distributed monitoring systems for legacy systems would do well to address such issues thoroughly early in their projects.

This CORBA-based approach is both feasible and competitive with other systems based on proprietary communications protocols. The attraction of our approach lies in the extensive availability of commercial software products, which support distributed computing applications based on the CORBA model and underlying Internet communications protocols.

Acknowledgment

This chapter draws from research carried out in the context of the Remote Building Monitoring and Operations (RBMO) project at Lawrence Berkeley National Laboratories in 1997. The author would like to thank Frank Olken and Chuck McParland. The RBMO system described in this chapter was operational and has been experimentally deployed on two buildings: Soda Hall and the Oakland Federal Building. The central data storage and management site was located at Lawrence Berkeley National Laboratory.

References

[1] ASHRAE, BACnet — A Data Communication Protocol for Building Automation and Control Networks, Technical report ANSI/ASHRAE Standard 135-1995, American Society of Heating, Refrigerating and Air-conditioning Engineers, 1995, URL: `http://www.ashrae.org`.

[2] Brose, G., JacORB Performance, Technical report, Free Unviversity, 1998, URL: `http://www.jacorb.org/performance/`.

[3] Butler, J.F., BACnet: An Object-oriented Network Protocol for Distributed Control and Monitoring, Technical report, Cimetrics Technology, 1996, URL: `http://www.cimetrics.com`.

[4] Castori, P., Manufacturing Messaging Specification, Technical report, Ecole polytechnique federale de Lausanne (EPFL), Institute for Computer Communications and Applications, 1997.

[5] Chang, N. et al., Time Synch — Time Synchronization Server, Technical report, University of Delaware, 1997, URL: `http://www.eecis.udel.edu/`.

[6] IAI UK Chapter, United Kingdom Chaper of Industry Alliance for Interoperability home page, URL: `http://cig.bre.co.uk/iai_uk/new/index.jsp`, 1996.

[7] Claridge, D., J. Haberl, M. Liu, J. Houcek, and A. Athar, Can You Achieve 150% of Predicted Retrofit Savings? Is Time for Recommission-ing?, in Proceedings from the ACEEE 1994 Summer Study on Energy Efficiency in Buildings, Vol. 5, American Council for an Energy-Efficient Economy, 1994.

[8] Cohen, Y., SNMP - The Simple Network Management Protocol, Technical report, RAD Data Communications, 1995, URL: `http://www2.rad.com/networks/1995/snmp/snmp.htm`.

[9] ESPRIT CCE-CNMA Consortium, *MMS: A Communication Language for Manufacturing*, Springer-Verlag, Berlin, 1995.

[10] Dreyer, W., A. Kotz Dittrich, and D. Schmidt, Research perspectives for time series management systems, *SIGMOD Record (ACM Special Interest Group on Management of Data)*, 23, 10–15, 1994.

[11] Dreyer, W., A. Kotz Dittrich, and D. Schmidt, Using the CALANDA time series management system, in Proceedings of the 1995 ACM SIGMOD International Conference on the Management of Data, Carey, Michael J. and Schneider, Donovan A., Eds., SIGMOD Record, 24, 489, 1995.

[12] EIA, *Annual Energy Review for 1991*, Energy Information Administration, U.S. DOE., 1992.

[13] EPRI, Utility Communications Architecture, URL: `http://www.epri.com/`, 1997.

[14] EUN, EUN product guides: Building automation systems, *Energy User News Magazine*, Oct. 1994, pp. 50–51.

[15] Gokhale, A. and D.C. Schmidt, Evaluating the Performance of Demultiplexing Strategies for Real-time CORBA, in Proceedings of the GLOBE-COM '97 Conference, Nov. 1997.

[16] Haberl, J., 1996. personal communication,

[17] Heinemeier, K.E. and H. Akbari, Proposed Guidelines for Using Energy Management and Control Systems for Performance Monitoring, in Proceedings from the ACEEE 1992 Summer Study on Energy Efficiency in Buildings, Vol 3, American Council for an Energy-Efficient Economy, 1992.

[18] Herzog, P. and L. LaVine, Identification and Quantification of the Impact of Improper Operation of Midsize Minnesota Office Buildings on Energy Use, a Seven Building Case Study, in Proceedings from the ACEEE 1992 Summer Study on Energy Efficiency in Buildings, Vol. 3, American Council for an Energy-Efficient Economy, 1992.

[19] I-Kinetics, I-Kinetics home page, URL: `http://www.i-kinetics.com/`, 1997.

[20] IAI, Industry Alliance for Interoperability home page, URL: `http://www.iai-international.org/iai_international/`, 1996.

[21] IETF SNMPv3 Working Group, SNMP Version 3 Working Group Home Page, URL: http: //www. ietf . org/html. charters/snmpv3-charter. html, 1997.

[22] Meyers, S.R., Chiller modeling error analysis: Implications for Energy-Saving Retrofits and Control Strategies, Technical report, Mechanical Engineering Department, UC. Berkeley, 1996. Master's thesis.

[23] Olken, F., H.-A. Jacobsen, and C. McParland, Development of a Remote Building Monitoring System, in Proceedings of the 1998 ACEEE Summer Study on Energy Efficiency in Buildings, American Council for an Energy Efficient Economy, August 1998.

[24] Olken, F., H.-A. Jacobsen, and C. McParland, Middleware Requirements for Remote Monitoring and Control, in Proceedings of the OMG-DARPA-MCC Workshop on Compositional Software Architectures, Jan. 6–8 1998, URL: http://www.objs.com/workshops/ws9801/index.html.

[25] Olken, F., H.-A. Jacobsen, and C. McParland, Object lessons learned from a distributed system for building monitoring and operation, in *Object Oriented Programming Systems, Languages and Applications (OOPSLA)*, Vancouver, ACM, 1998, pp. 284–295.

[26] Piette, M.A., G. Carter, S. Meyers, O. Sezgen, and S. Selkowitz, Model-Based Chiller Energy Tracking for Performance Assurance at a University Building, in Proceedings of the Cool Sense National Forum on Integrated Chiller Retrofits, Sept. 23 – 24, 1997, LBNL Report-40781.

[27] Schmidt, D., W. Dreyer, A. Kotz Dittrich, and R. Marti, Time series, a Neglected Topic in Temporal Database Research?, in Recent Advances in Temporal Databases, Proceedings of the International Workshop on Temporal Databases, Workshops in Computing, Clifford, James and Tuzhilin, Alexander, Eds., Springer-Verlag, Berlin, September 1995, pp. 214–232.

[28] Schmidt, D.C., A. Gokhale, T.H. Harrison, and G. Parulkar, A high-performance endsystem architecture for real-time CORBA, *IEEE Communications Magazine*, Vol. 35, Issue 2, pp. 72–77, Feb. 1997.

[29] Segev, A. and R. Chandra, A data model for time-series analysis, in *Advanced Database Systems*, Adam, Nabil R. and Bhargava, Bharat K., Eds., *Lecture Notes in Computer Science*, Vol. 759, Springer-Verlag, Berlin, 1993, pp. 191–212.

[30] SimpleGroup, The SimpleWeb, Technical report, Centre for Telematics and Information Technology (CTIT) of the University of Twente (UT), 1997, URL: http://www.simpleweb.org/.

[31] SNMP Research International, Inc, SNMP Research International Web Page, URL: http: //www. int. snmp.com/, 1997.

[32] Tseng, P.C., D.R. Stanton-Hoyle, and W. Withers, Use of a Supplemented EMCS in Commissioning, in Proceedings of the ACEEE 1994 Summer Study on Energy Efficiency in Buildings, American Council for an Energy-Efficient Economy, 1994.

[33] Wonderware, Common questions about NetDDE, URL: http://www. wonderware.com/products, 1996.

[34] Yahoo, SNMP (simple network management protocol), URL: `http: //www.yahoo.com/ see Computers and Internet >...> Protocols > SNMP, 1997.`

60

Internet-based Telemanipulation

P. Korondi
Budapest University of Technology and Economics

P. Szemes
University of Tokyo

H. Hashimoto
University of Tokyo

60.1 Introduction

Brief History of Telemanipulation

In about 1945, the first modern master–slave system teleoperator was developed by Goertz at Argonne National Laboratory (ANL) [1]. This system, which was a mechanical pantograph mechanic system, allowed a human operator to manipulate radioactive materials in a "hot cell" from outside. By using a master handle, the human operator could move the slave tong located inside the hot cell and receive force reflection. Soon (in 1954), electrical servomechanisms replaced the direct mechanical master–slave system and able linkages [2]. Closed circuit television was introduced so that the operator could stay at an arbitrary distant place. In addition, the contact force between the slave and its environment was returned to the master arm. In 1964, Mosher developed an impressive work, which was a handy-man containing electrohydraulic arms. Each arm had ten DOF in each arm. The problem of remote control of robotic systems has been the subject of much research in recent years. Remote control of robotic systems has been applied in manufacturing, underwater manipulation, storage tank inspection, nuclear power plant maintenance, space exploration, etc. See an excellent review of relevant historical development in telemanipulation (telerobotics) in [1].

What is Telemanipulation

Telemanipulation is a process where the operator has some task done at the remote hazardous or inaccessible environment as space, underwater, nuclear plants, where he/she cannot physically be. A teleoperator system extends the operator's capability to be able to work at the Remote Workplace. This extension is achieved by a master–slave system. Telemanipulation is divided into two strongly coupled processes. One process is the interaction between the operator and the master device; the other is the interaction between the slave device and the remote environment contact. The master device represents the distant environment at the operator site, and the slave device represents the operator at the remote site. When humans use this device only to move objects and the reaction force from the distant environment has no significant effect on the performance, then measurement of the operator's position and visual feed back may be enough. However, if such tasks are to be performed when the reaction of the environment is important and the salve device can do damage in the remote environment, for example, when screwing a bolt or assembling something, then a force feedback is needed to improve efficiency. Force feedback increases the feeling of being there. The information flow between the operator and remote site can be seen in Figure 60.1 (adapted from [3]), where only three types of information are fed back: visual, audio, and sense of touch. A human being receives five types of sensing from his/her surrounding environment, but only some of these senses are used during telemanipulation.

The structure of this chapter is as follows. The next section overviews the general approach of telemanipulation. It introduces three main layers of a telemanipulation system. Section 60.3 summarizes the basic type of master devices. Section 60.4 provides a short introduction to control modes used for telemanipulation. Section 60.5 presents an application example of a handshaking via Internet.

60.2 General Approach of Telemanipulation

Basic Definitions

Figure 60.2 shows the general concept of telemanipulation. The "world" is divided into two sets: the Master Site and the Slave Site. The master–slave system is realized as two information channels: Action and Reaction channel. The Action channel transfers the information from the operator to the Remote Workplace. The Reaction channel transfers the information to the opposite direction: from the Remote Workplace to the Operator. In the early days of telemanipulation, the human operator's hands and the remote environment were mechanically coupled. The information was transferred mechanically, which limited the type of the tasks that they could perform. Nowadays, the information is transferred electrically, which opened a new dimension at the slave side. A Human–Robot System may be considered as a type of telemanipulation,

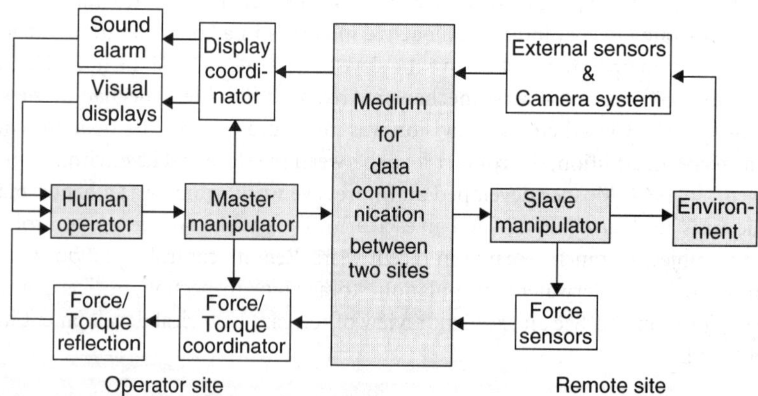

FIGURE 60.1   Information streams of the Telemanipulation (adapted from [3]).

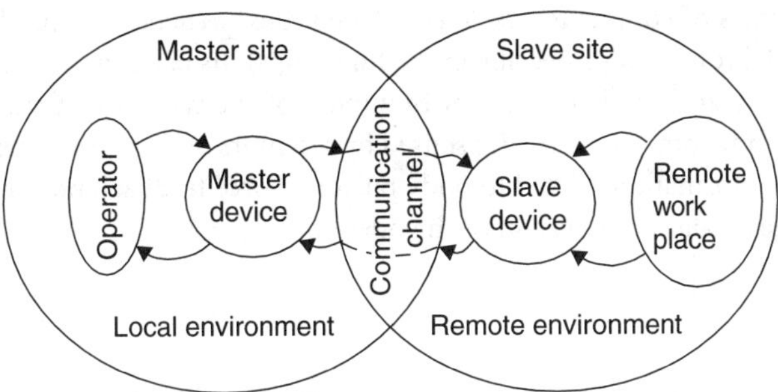

FIGURE 60.2   General concept of the telemanipulation.

where a human and a robot work simultaneously together to achieve a task. For example, in the case of remotely deactivating of a bomb, the robot must reach the bomb and then the timer must be disassembled with the help of a pair of slave manipulators. In this example, the navigation and manipulation tasks may be separated. When a group of mobile robots move a large object cooperatively, the operator must control the overall action. In general, the Slave Device can be a Multi Robot System as well, where more than one robots are working together to achieve a task. Computer Networked Control enables systems to be distributed. The concept of an operator and robot can be replaced with some combination of multiple operators, multiple robots, and maybe even multiple assistant agents that work together in a cooperative environment [4]. The Human–Robot type of telemanipulation can be classified into the following:

- Singlemaster–single slave.
- Multimaster–single slave.
- Singlemaster–multislave.
- Multimaster–multislave.

We can conclude that the meaning of telemanipulation has extended, and new terminologies have appeared. Some of them are presented here.

Teleoperator is the machine itself, which can perform the task in a distant place. Using a general sense of teleoperator, *Telerobot* is a subclass of it. Originally, telemanipulation means the action in which a slave manipulator tracks the motion of the master manipulator. In this case, the salve and master manipulator have the same structure. The extension of this concept means a manipulation in the remote environment with a slave device controlled by a master device moved by a human operator. Using this extended meaning, the structure of master and slave device can be different, but the whole manipulation process is controlled by the human operator, that is, the slave side has no intelligence of its own. The next step is the distribution of intelligence. The slave device may learn the behavior of a human operator and may follow the learned behavior during the period when the command is delayed because of the latency caused by Internet communication [5]. Some authors use the word of teleoperation for the extended meaning of telemanipulaton, and some authors use "telemanipulation" and "teleoperation" in the same sense.

A human operator needs the illusion of direct contact with the remote environment to perform fine tasks such as telesurgery or remote assembling (disassembling) fine mechanism (timer bomb). A human operator receives five types of sensing (sight, hearing, touch, smell, taste) but the combinations of the first three are used during a manipulation. The psychological feeling of "being there" in an environment based on a technologically founded immersion environment is called as *Telepresence* [6,7], which should provide the ideal sensation, that is, we obtain the necessary information fed back from the slave to the master side in a natural way to have the feeling of being physically at the distant location. *Telepresence systems* are a subclass of teleoperators. The technology of audio–visual feedback is well developed and available nowadays. The main challenge is the *haptic feedback*, which is necessary to feel the physical contact with the environment. It is an extension of force feedback. A human operator has different types of receptors for sensing different types of contact forces. *Tactile (cutaneous) feedback* is sensed by mechanoreceptive nerve endings in the glabrous skin (especially in the finger tips). Its bandwidth is about 50–350 Hz [8]. It provides information on small-scale details, for examples of the type of surface. *The force (kinesthetic) feedback* has a lower bandwidth (up to 10 Hz) sensed by receptors on the muscle tendon and in the skeletal system. It provides information related to body posture. Force feedback not only increases efficiency, but it helps to filter the smaller imprecision of the operator.

Ideal Telepresence

In a typical telepresence, information about task at the remote site is required to help a human operator to feel as if he/she physically is present at the remote place. Accordingly, a teleoperator must provide complete *transparency,* which means the positions, velocities, and forces of the master and slave device are equal. An example of simple remote device is remote pliers used to handle objects in a nuclear plant [5]. If this plier is stiff and easy to move, the operator simply performs the task. Images of the ideal connection between the master and slave for revolute motion and linear motion are shown in Figure 60.3(a) and (b). If the operator rotates (move) the master stick by q_m (x_m), the motion of the slave stick must be the same ($q_m=q_s$ or $x_m=x_s$) and the operator must sense the reaction force of the environment ($F_o=-F_e$), in an ideal case. From this example, the aimed dynamic behavior of the teleoperator is therefore close to a rigid rod, whose mechanical properties are minimal inertia and maximal stiffness. The connection of the master and slave arms should have 0 mass with ∞ stiffness.

However, our work deals with a teleoperator with a time delay, whose master and slave are assumed to be physically equivalent. The ideal responses for the Telemanipulation with time delay are definite and as follows:

- The teleoperator must remain stable.
- The force that the human operator applies to the master arm is equal to the force reflected from the environment in the steady state. This can help operators to realize force sensation.
- The master position is equal to the slave position in the steady state.

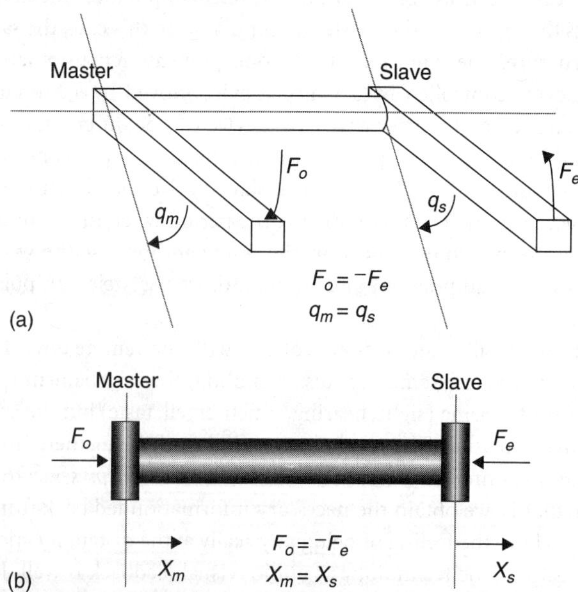

FIGURE 60.3　Ideal Telepresence systems: (a) revolute motion manipulation, (b) linear motion manipulation.

Layer Definitions

The Master Site and the Slave Site are partitioned into three main layers (see Figure 60.4). The layers can communicate vertically with the local layers as well as horizontally with the remote layers via the Internet. The Sensor Layer contains the necessary sensors to observe the Remote Environment, and the Remote Workplace. In general, these sensors are noncontact sensors. The Manipulator Layer makes the effective work in the Remote Workplace. This layer involves the Slave Device typically. This layer may contain contact sensors. The Transporter Layer includes a Transporter (mobile unit), which can carry the slave device. The Monitor Layer at the Master Site shows the processed data of the Sensor Layer. These monitors are usually noncontact displays or indicators. The Manipulator Layer has real contact with the Operator. This layer involves the Master Device typically. This layer, included with the Master Device, may contain contact sensors and contact actuators. The contact actuators relay the sensed data of the contact sensors of the Manipulator Layer. The Transporter Control Layer controls the Transporter at the Slave Site. This layer may contain services that can help in the navigation of the Transporter Layer at the Slave Site.

Sensor Layer

The configuration of the Sensor Layer can be seen in Figure 60.5. The noncontact sensors on the Slave Site are observing the Far Environment. The Intelligent Sensed Data Computing System at the Slave Site is a preprocessor for the sensed data. The purpose of this device is filtering, correction, and compression. The audio and video compression formats are widely used in the Internet to decrease the necessary bandwidth. The Computing System may contain a pattern recognition module to improve the efficiency and intelligence of the Slave Site. The purpose of the Intelligent Sensed Data Computing System at the Master Side is data decompression and recovery of lost data.

Manipulator Layer

There is another approach for the telemanipulation. The process of the telemanipulation can be partitioned into two coupled subprocesses. One subprocess is running in the Local Environment: Local

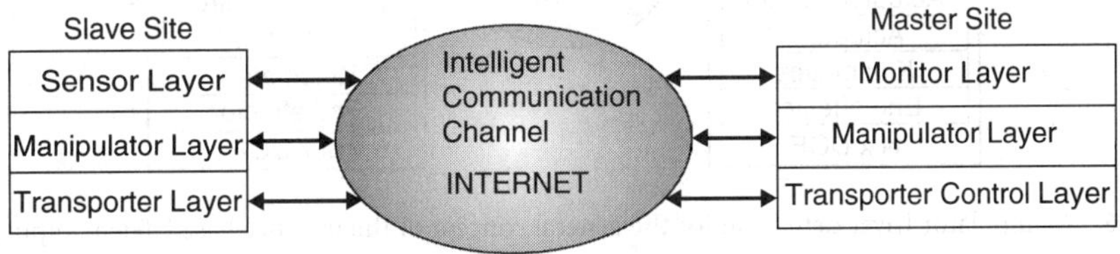

FIGURE 60.4 Layer definition for the general concept of the Internet-based Telemanipulation.

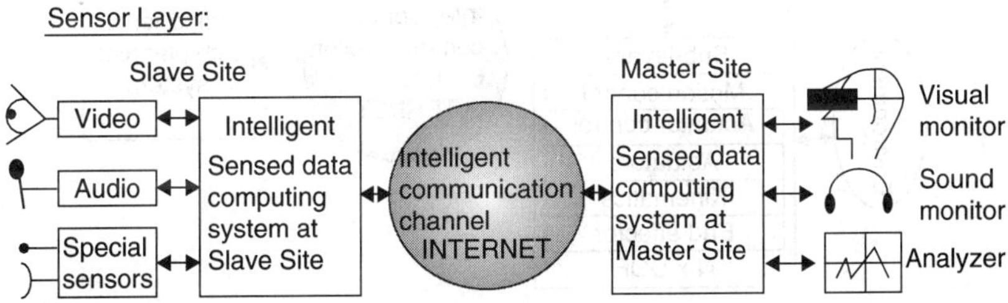

FIGURE 60.5 Sensor Layer definition for the general concept of the Internet-based Telemanipulation.

Process. The other subprocess is running in the Remote Environment: Remote Process. The Local Process is the information circulation between the Operator and the Master Device. The Distant Process is the information circulation between the Slave Device and the Remote Environment. If telemanipulation is ideal, the two processes are the same. In case of ideal telemanipulation, the Master Device represents the Remote Workplace in the Local Environment and the Slave Device represents the Operator in the Far Environment. The Communication Channel is the coupler between the two subprocesses. The main task of the Communication Channel is to ensure total transparency in case of ideal telemanipulation.

The configuration of the Manipulator Layer can be seen in Figure 60.6. Typically, this layer includes the Slave and the Master device, as mentioned above. The main purpose of this layer is the position control and the force-feedback. The force-feedback is an information flow between the layers and between the Master and Slave Site. The force- feedback is a typical action–reaction process. The exact realization of the force-feedback depends on the specific approach. The configuration of the Master Device is not equal to the configuration of the Slave Device in a general case. The main purpose of the Intelligent Path Planning is to couple the master and the slave in position and force.

Transporter Layer

The configuration of the Transporter Layer can be seen in Figure 60.7. The configuration of the Slave Site of Transporter Layer is very similar to one side of the Manipulator layer. The Transporter Layer has a connection with the Far Environment. The Transporter Layer carries the Slave Device and the necessary sensors.

The Task-Oriented Path Planning System controls the transporter on the Slave Side. This device may use advanced services for navigation such as Global Positioning System (GPS), Geographic Information System (GIS), and Intelligent Transport System (ITS).

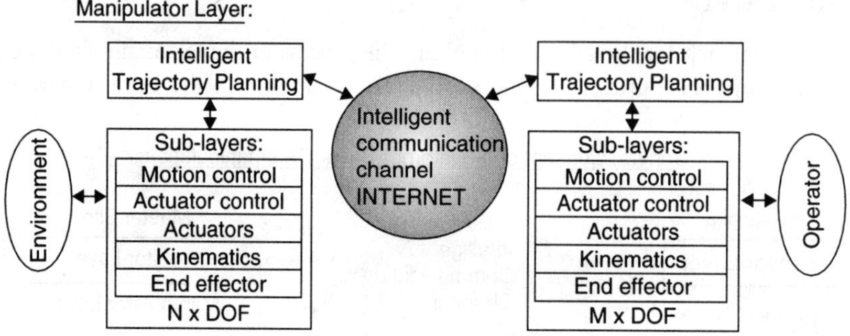

FIGURE 60.6 Manipulator Layer definition for the general concept of the Internet-based Telemanipulation.

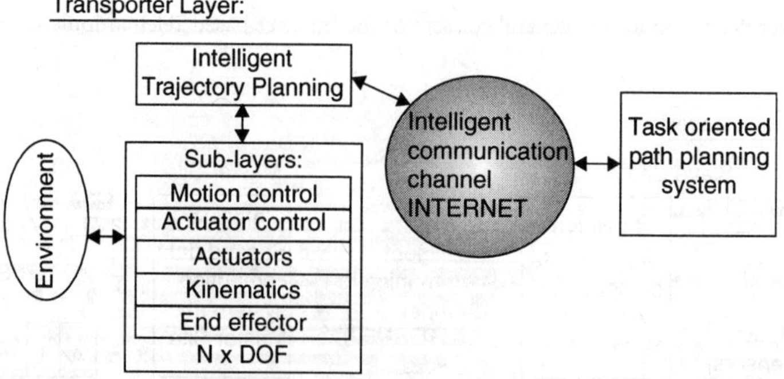

FIGURE 60.7 Layer definition for the general concept of the Internet-based Telemanipulation.

Special Types of Telemanipulation

Telemanipulation in virtual reality is beneficial when an operator is learning to perform dangerous or difficult tasks such as working with radioactive materials or disassembling explosives, or driving an airplane, helicopter, or submarine in a real-time simulator. In addition, "Flight and other vehicles simulators" are very popular in the field of entertainment. It can be a very attractive target in both main fields: professional and commercial. Telemanipulation in a virtual reality application is shown in Figure 60.8. The operator wears a *head-mounted device* that provides visual feedback and *an arm and glove type master device* giving tactical sensation. The operator can see the visual objects and can touch and move them [10].

Micro- and nanotechnology opened a relatively new, emerging dimension in the field of telemanipulation. Going from macro to micro/nanoworld, the main phenomenon is the reduction of the size of objects where the effect of length change is defined as scaling. It is obvious that *Force and position must be scaled* (see Figure 60.9). There is another problem called as *bandwidth effect*, which means that bandwidths of the master device including human operator's reaction and slave device including nano/micro environment are different. The micro/nano actuator has a smaller time constant than the macro force feedback device. Thus, an impedance scaling is necessary to avoid instabilities and unreliable force feeling. It can be considered as a kind of time scaling. In other words, the micro/nanophenomena might be too fast for a human sensation and it must be slowed down by virtual coupling impedance [11].

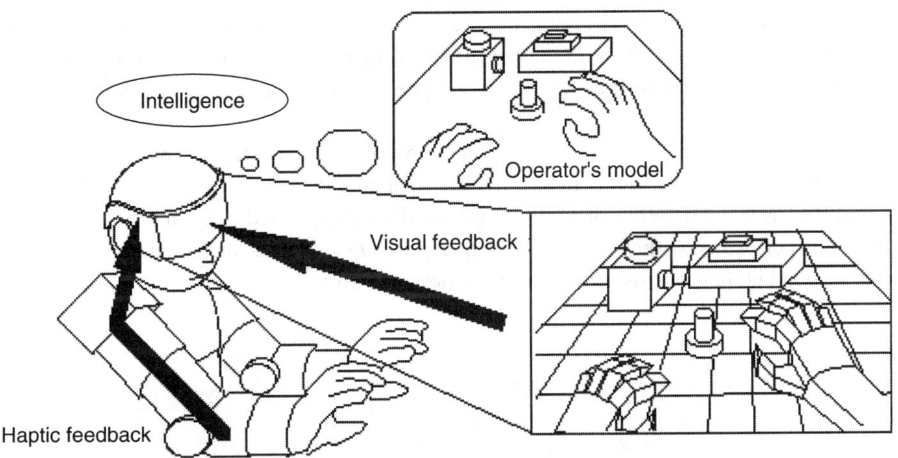

FIGURE 60.8 Telemanipulation in the virtual reality.

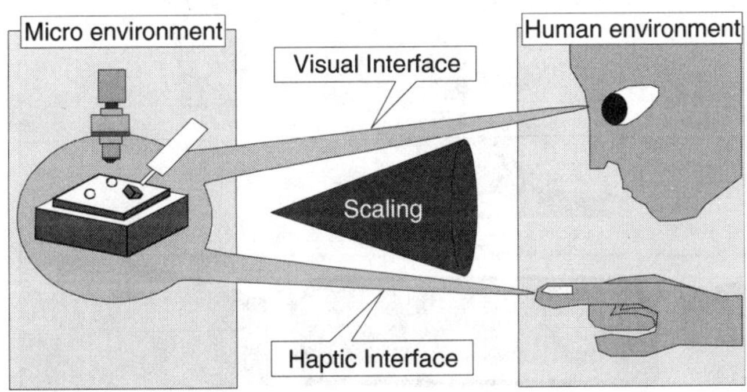

FIGURE 60.9 Micro/nano teleoperation system.

60.3 Master Devices as Haptic Interfaces

Human operators perform their tasks mostly by hand, and the master device must fit the operator's hand. In the field of telemanipulation, many haptic devices have been developed. The applications of the haptic devices cover various fields, such as medical application, working in nuclear power plants, and entertainment. In this section, four important master device concepts are presented.

- *Joystick with force feedback* was developed in the early stage of telemanipulation. Nowadays, several commercial joysticks with force feedback are available. This device is optimal for low-quality telemanipulation application. This device is optimal for "pick and place" operation. Joysticks with force feedback are very popular in recent video game systems. Another important application area is the real-time simulators of vehicles, such as airplanes, helicopters, and submarines. The main problem with these devices is the dynamics change "on the fly." An additional device is needed for balancing the dynamics of the joystick. This feature is necessary in applications where different precision operations are required. Another disadvantage of the device is the two controlled degrees of freedom.
- *The point type master device* is similar to a six DOF manipulator. The operator can point at the remote site. The Phantom [12] is a point-type device (see Figure 60.10) using direct drive and is noted for its high resolution. However, it can cover only fingertips. The operator can control a tip-type slave device with this device. Tip-type slave systems are widely used in surface manipulation applications, such as telemanipulation in the nanospace [11].
- *The arm type master devices* in Figure 60.11 cover the human arm from the shoulder to wrist. It has 6 degrees of freedom. The arm-type haptic device is useful in applications where the operator uses his arm to do some task in the remote environment, for example, some mining workplace. In several applications, the arm-type master device is a subsystem of a complex master device.
- *The glove type master device* allows the most complex and dexterous manipulation [13]. The human hand is the most widely used tool. The final goal is to make such a device that the operator can feel as if the function of whole of his hand were expanded. Several types of manipulation cannot be performed without force feedback. A 20 DOF sensor glove-type device with force feedback (see in Figure 60.12) is discussed in the next section.

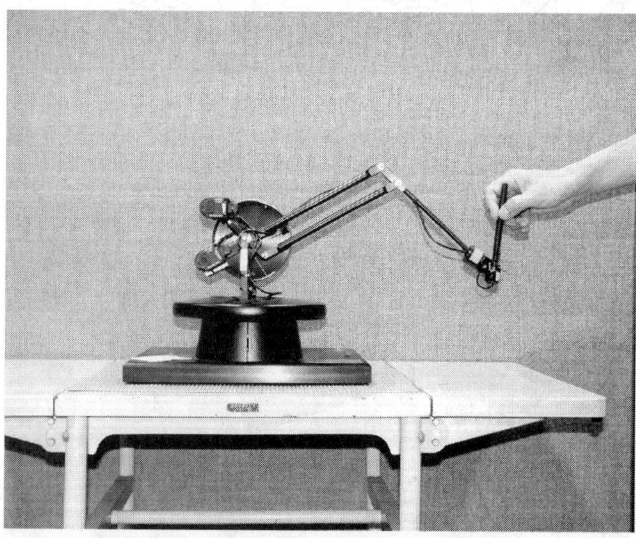

FIGURE 60.10   A point-type master device.

A Glove-type Haptic Interface

Mechanical Structure

The wrist location is measured by an inductive sensor. The sensor glove is equipped with force feedback capability to each joint of the operator fingers; hence, the operator can feel the remote environment as he/she is there. The mechanical structure of a finger is shown in Figure 60.13. The device is designed with consideration of space for sensors and actuators. The device is attached to a human hand by bands. A fixed base for all movements of the glove is constructed on a metallic plate fixed on the back of a human palm. Five yawing drive units are fixed in the plate. Three pitching drive components are placed on the yawing drive units.

Figure 60.14 shows the structure of one DOF of the device. The torques of the finger joints are measured by strain gauges. Rotary encoders in the motors are used to measure the angles of the joints. Ideally, the motors move the joints directly; however, there was no motor available that was small and light enough to be attached on the hand. The pulleys and wires in tubes are used to transmit forces to the motors. Single motors can control all the 20 joints of a human hand independently. One joint of the finger corresponds to two links of the device. The main problem is the friction. Because of this additional force, the operator may not distinguish which force has emerged in a remote environment. The friction compensation is discussed in Section 60.4.

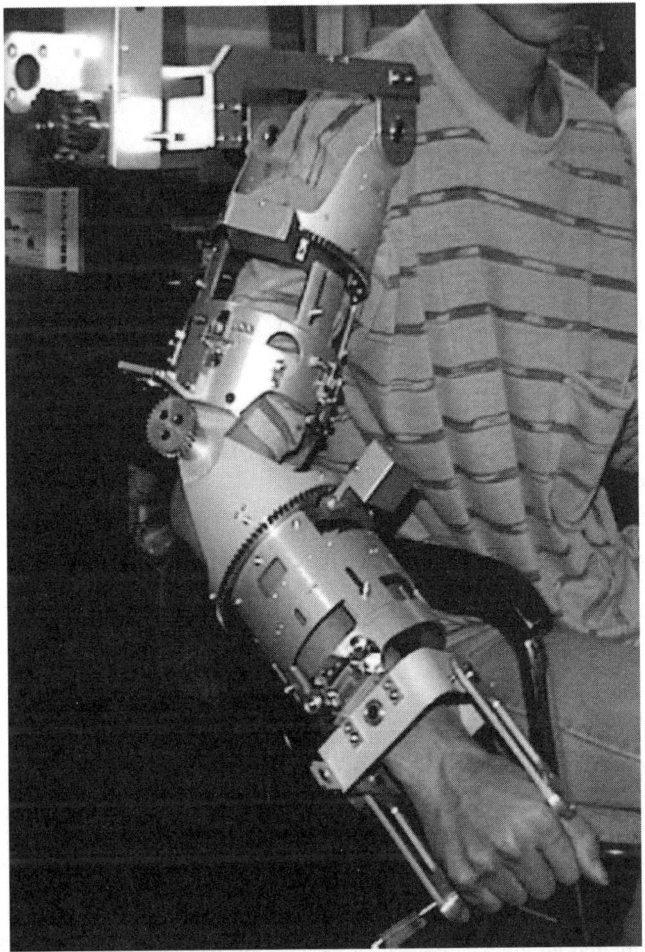

FIGURE 60.11   An arm-type master device.

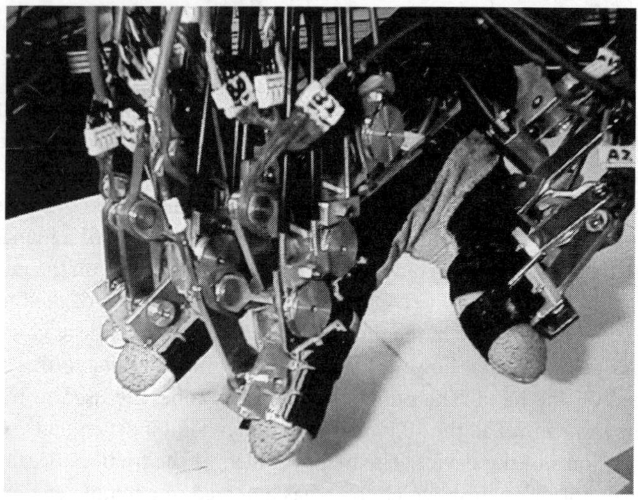

FIGURE 60.12   A glove-type master device.

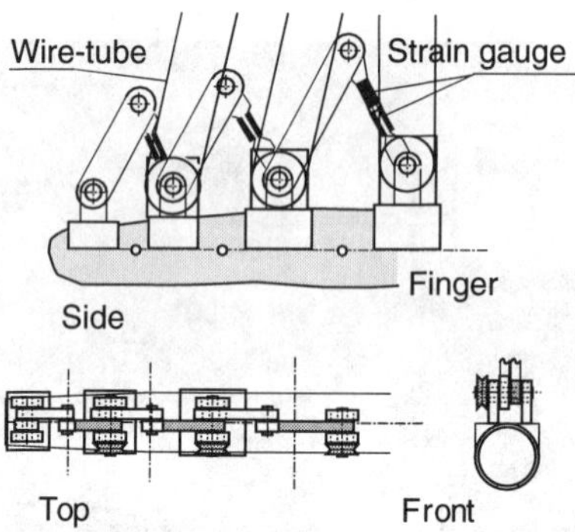

FIGURE 60.13   Mechanical structure of the sensor glove.

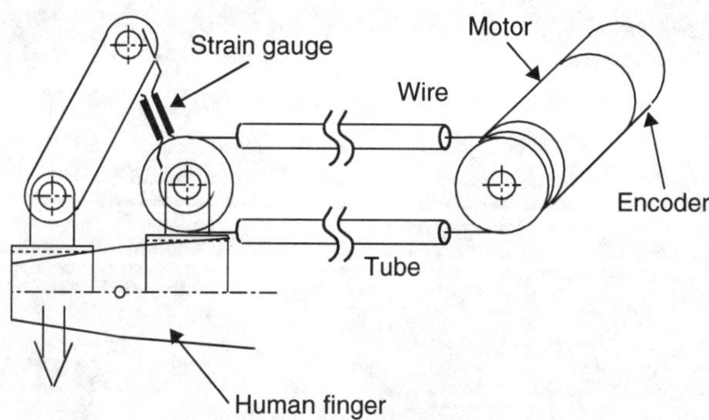

FIGURE 60.14   Structure of one DOF of the Sensor Glove.

Animation of the Operator's Hand Wearing the Sensor Glove

Visual feedback is aimed at giving a quite real representation of the operator's hand, and the program enables the user to wander it in full three-dimensional space (Figure 60.15).

The program was designed to run either locally in full simulation mode or over TCP/IP connection, such as Internet or Local Area Network. In simulation mode, the program can be used to represent all the motions of the human hand. To accomplish this, both anatomical and mathematical models were built, and these models were implemented in OpenGL. The mathematical deduction uses the *Denavit–Hartenberg notation*, because it can be quite well applied both for hands and in OpenGL. The model uses several basic assumptions as follows:

- The base coordinate frame is the same for all fingers. This is used for having a completely general case, where the fingertip coordinates are calculated with respect to a common frame of the wrist.
- This common frame is considered to be at joint 0. This joint is used in order to have the possibility to extend the model at a later time if needed.
- One finger contains three joints: the first has two degrees of freedom, the other two have only one, which means planar rotation. The position of the first joint is determined by a variable l_0, which is the distance between the first, common coordinate frame and the first joint.
- Along one finger, the value of d_i is 0, which means that the common normal line on the same line.
- Along one finger, the value of a_i is the length of the link.
- The twist angle is taken into account only at the base coordinate frame and at the first joint, since lateral movement of the finger is also considered.

It is important that coordinate frame 0 be the same for all fingers, and is a common reference to the whole hand. The individual θ_i angles and l_0 variables determine the exact position and orientation of the next frame of the next joint in each finger. Between the two frames d_i is just equal to the length of the link, l_0. Since joint 0 is of the revolute type, q_0 is θ_1, the relationship between the two joints is as follows:

$$X = A_1^0 X^1 \tag{60.1}$$

In Equation (60.1) A_1^0 is the 4×4 *homogeneous transformation matrix* between the two adjacent joints and X^s are position vectors in the respective coordinate frames.

$$A_1^0 = \begin{bmatrix} \cos\theta_1 & -\sin\theta_1\cos\alpha_1 & \sin\theta_1\sin\alpha_1 & a_0\cos\theta_1 \\ \sin\theta_1 & \cos\theta_1\cos\alpha_1 & -\cos\theta_1\sin\alpha_1 & a_0\sin\theta_1 \\ 0 & \sin\alpha_1 & \cos\alpha_0 & d_0 \\ 0 & 0 & 0 & 1 \end{bmatrix} \tag{60.2}$$

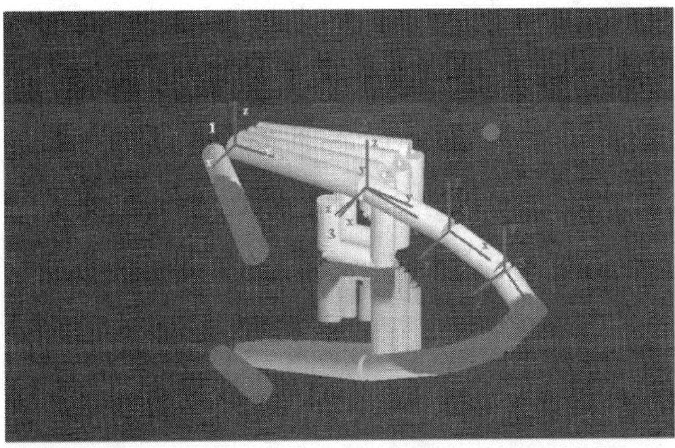

FIGURE 60.15   Computer animation of the robot hand with the coordinate systems.

$$\begin{bmatrix} x_0 \\ y_0 \\ z_0 \\ 1 \end{bmatrix} = \begin{bmatrix} \cos\theta_1 & -\sin\theta_1\cos\alpha_1 & \sin\theta_1\sin\alpha_1 & a_0\cos\theta_1 \\ \sin\theta_1 & \cos\theta_1\cos\alpha_1 & -\cos\theta_1\sin\alpha_1 & a_0\sin\theta_1 \\ 0 & \sin\alpha_1 & \cos\alpha_0 & d_0 \\ 0 & 0 & 0 & 1 \end{bmatrix} \begin{bmatrix} x_1 \\ y_1 \\ z_1 \\ 1 \end{bmatrix} \qquad (60.3)$$

As stated, in this case α is $0°$, which means that its cosine is 1 and its sine is 0. For the two joints, d is considered to be 0 and a is the length of the link. Substituting these values simplifies the previous equation and yields.

$$\begin{bmatrix} x_0 \\ y_0 \\ z_0 \\ 1 \end{bmatrix} = \begin{bmatrix} \cos\theta_1 & \sin\theta_1 & 0 & l_0\cos\theta_1 \\ \sin\theta_1 & \cos\theta_1 & 0 & l_0\sin\theta_1 \\ 0 & 0 & 1 & 0 \\ 0 & 0 & 0 & 1 \end{bmatrix} \begin{bmatrix} x_1 \\ y_1 \\ z_1 \\ 1 \end{bmatrix} \qquad (60.4)$$

Now, the position and the orientation of the coordinate frame at joint 1 are received; further transformations can be carried out.

Now there is a general formula for calculating the position of the finger tip with reference to the common base coordinate frame. Equation (60.3) shows this, and it is very similar to (60.1). Matrix T contains the position and orientation of the tip in the base coordinate system.

$$T = A_1^0(\theta_1)A_2^1(\theta_2)A_3^2(\theta_3)A_4^3(\theta_4)A_5^4(\theta_5) \qquad (60.5)$$

It is useful to state that joint 2 has two transformation matrices and frames because of the two possible axes of rotation.

$$P_1 = T_1 X^{15} \qquad (60.6)$$

where X^{15} is the position vector of the fingertip in its local coordinate frame.

The position and orientation of all five fingertips with respect to the common base frame (the wrist), are given by (60.7) as a function of joint angles.

$$P = [f(\theta_{11},\theta_{12},\theta_{13},\theta_{14})\ f(\theta_{21},\theta_{22},\theta_{23},\theta_{24})\ f(\theta_{31},\theta_{32},\theta_{33},\theta_{34})\ f(\theta_{41},\theta_{42},\theta_{43},\theta_{44})\ f(\theta_{51},\theta_{52},\theta_{53},\theta_{54})] \qquad (60.7)$$

Grasping

Figure 60.16 illustrates the transformations between finger-tip, contact, and object coordinate frames [10]. Let us define homogeneous transformation matrices $U_i \in R^{4\times4}$ and $V_i \in R^{4\times4}$, where U_i transforms

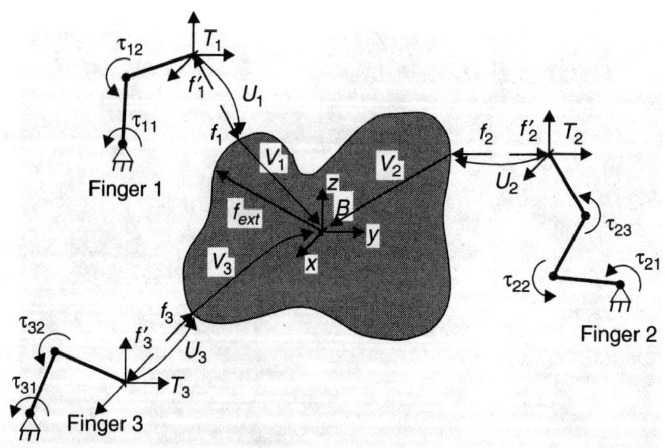

FIGURE 60.16 Object grasped by three Fingers.

from the finger-tip coordinate frame $T_i \in \mathrm{R}^{4\times4}$ to the contact frame $C_i \in \mathrm{R}^{4\times4}$ and $V_i \in \mathrm{R}^{4\times4}$ from the contact frame to the object frame $B_i \in \mathrm{R}^{4\times4}$. The x- and y-axis of the contact coordinate frame C_i are tangents and the z-axis is perpendicular to the object surface curvature. (Figure 60.17). The transformation matrices $\Gamma_i \in \mathrm{R}^{6\times6}$ and $\Lambda_i \in \mathrm{R}^{6\times6}$ are defined in such a way that Γ_i transforms finger-tip forces $^{T_i}f_i$ into forces $^{C_i}f_i$ in the contact coordinate frame and Λ_i transforms these to forces $^{B}f_i$ in the object reference frame B.

The relationship of coordinate transformations can be written as

$$
\begin{aligned}
C_i &= T_i\,U_i \\
B &= C_i\,V_i = T_i\,U_i\,V_i \\
^{C_i}f_i &= \Gamma_i\;^{T_i}f_i \\
^{B}f_i &= \Lambda_i\,^{C_i}f_i = \Lambda_i\,\Gamma_i\;^{T_i}f_i
\end{aligned}
\tag{60.8}
$$

where U_i, V_i are homogeneous transformations and Γ_i, Λ_i are force transformations [10]. The sensor glove has two types of sensor outputs, the torques actuated by the human operator $\tau \in \mathrm{R}^{5\times4}$ and the finger joint angle values $\theta \in \mathrm{R}^{5\times4}$. T_i are calculated from the forward kinematical equations using θ.

60.4 Overview of Control Modes

Basic Architectures

As can be seen in Figure 60.2, the communication takes place in two directions in a telemanipulation system. The controller with two directions for the control commands collectively referred as *bilateral controller* [9]. Before the electrical master–slave system was developed, in order to realize force reflection, the traditional bilateral controlled master–slave system had been proposed since in 1952 (see Figure 60.18) [1]. Figure 60.19 shows another traditional control approach for the master–slave system, which was introduced in 1954 [2]. The two position controllers control both the master and slave positions.

These basic architectures are often used as a starting point to design new controllers. Three main problems and some possible solutions are discussed here.

- Nonlinear scaling (Virtual coupling impedance).
- Time delay compensation of internet-based telemanipulation.
- Friction compensation of master device.

Nonlinear Scaling (Virtual Coupling Impedance)

There are basically two approaches: linear and nonlinear scaling. In the case of linear scaling, a single corresponding constant is used between the slave force/position and master force/position:

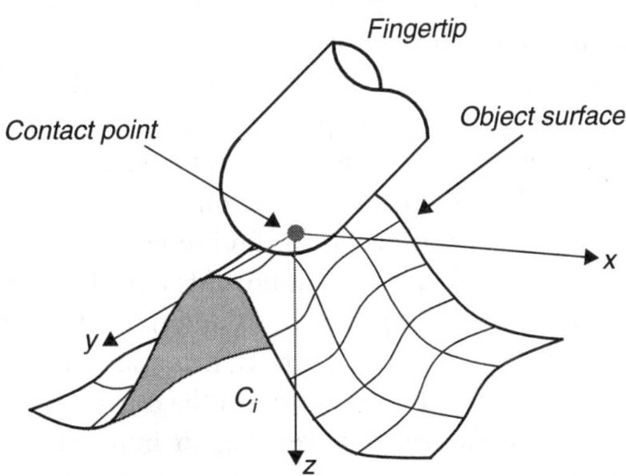

FIGURE 60.17 Contact point and contact frame.

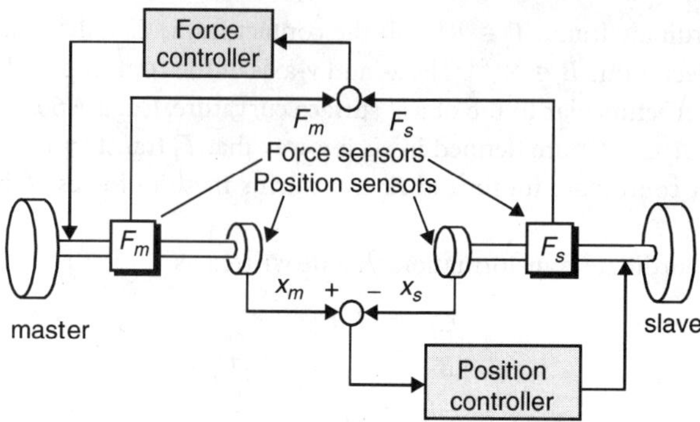

FIGURE 60.18   Conventional bilateral control schema with force and position feedback.

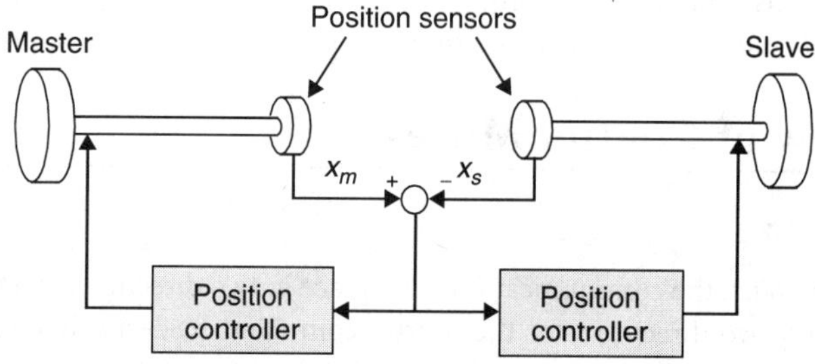

FIGURE 60.19   Conventional bilateral control schema with two position control loops.

$$x_s^*(t) = \alpha_p x_s(t)$$
$$F_s^*(t) = \alpha_f F_s(t) \tag{60.9}$$

where x_s^* and F_s^* are the scaled values, α_p and α_p are the constants of position and force scale, respectively. In case of nonlinear scaling, which is also called impedance scaling [14], forces are scaled independently with respect to their length scaling power relation as follows:

$$x_s^* = \alpha_p x_s(t)$$
$$F_s^* = \alpha_p^3 M_v \ddot{x}_s + \alpha_p^2 D_v \dot{x}_s + \alpha_p K_v x_s \tag{60.10}$$

where M_V is the virtual mass, D_V is the virtual viscosity constant, and K_V is the virtual stiffness. Note that this scaling method is similar to the concept of virtual impedance, which is widely used in the field of robot manipulators. Impedance scaling improves the performance, especially when the sizes or structures of the master and slave devices are different. The mechanical size has some correlation to its bandwidth. There are certain frequency and amplitude ranges, that are necessary for comfortable manipulation. The reaction force must be transformed into these ranges by the virtual coupling impedance. For example, in case of micro/nanomanipulation, the reaction times of the micro/nanoactuator and the micro/nanoenvironment may be too fast for human tactile receptors. The reaction force must be slowed down, that is, the high-frequency components must be shifted to the lower frequency range with the help of virtual coupling impedance. In addition, the virtual coupling impedance behaves as a low-pass filter in the communication channel. It can suppress the high-frequency components, which can destabilize the system with time delay. On the other hand, it contradicts the requirements of ideal telepresence and transparency. In general, an impedance scaling may help prevent instabilities and unreliable force feeling but the price that must be paid for this is the loss of the degree of telepresence and transparency.

Time Delay Compensation of Internet-Based Telemanipulation

The efficiency of telemanipulation has grown in the last few years. The application of the Internet gives new possibilities to the researcher [15]. The medium for data communication is the Internet in Figure 60.1. It allows connecting places electronically that are thousands of miles apart. By combining network technologies with the capabilities of mobile robots, Internet users can discover and physically interact with places far away from them, creating opportunities in resources sharing, remote experimentation, and long-distance learning. Burgard et al. [16] proposed an interactive museum tour guide robot for guiding the user to visit a museum. Simmons [17] presented an autonomous tour guide robot at Carnegie Mellon University. These tour guide robots have the capabilities of obstacle avoidance, self-referring, and path planning. Web users can control the autonomous robot through a web browser such as Netscape and Microsoft Internet Explorer, and click mouse on a map to assign the desired visiting position. Farzin and Goldberg [18] developed an interactive on-line lightbox system to provide users with the function of controlling the camera view for captured images. Stein [19] provided web users with a robot arm to paint a picture through a Java™-based user interface. Kosuge et al. [20] proposed the VISIT system using predictive strategy. Hirukawa and Hara [21] constructed a visual model robot arm to assist the teleoperation.

The main problem of Internet-based Telemanipulation is the variable time delay. Transport Control Protocol/Internet Protocol (TCP/IP) is mainly used to stream data via Internet. It is a packet-switched protocol. This means that the driver divides the data into packets, and sends the packets individually. The receiver tries to collect all the packets and rebuild the original data from the packets. A time delay always occurs when the signal extend from point A to point B. In addition, variable time delay is an immanent characteristic of a packet switched network. Variable time delay has two main reasons. One, the routes of the packets might be different on the network. Second, some packets might be lost. In this case, the receiver asks the driver to send the lost packet again and again. This time delay can destabilize a bilateral teleoperator system. The instability caused by time delay has been revealed by Ferrell [22] as early as in 1965. Since then, various approaches for handling the problem of time delay [23] have appeared in the literature, like delay modeling and control system design [24], predictive displays and advanced operator interfaces [25], shared compliance control [26], impedance control [27], or supervisory control [28]. One possible solution for elimination of the effect of time delay, namely virtual coupling impedance with position error correction method [29], is discussed in detail. This control approach is an extension of the control approach by Otsuka et al. [30]. Virtual coupling impedance is applied both at the operator and the remote site. An additional position correction loop is used to synchronize the two distant sites. This control approach is useful to compensate the effect of variable time delay, also appearing on the pocket-switched computer network, like TCP/IP.

The Smith Predictor

The Smith Predictor is a classical configuration for time delay compensation [31]. Let the $P(s)$ be a first-order filter with transportation leg T_d. The $P_0(s)$ is the model of the plant $P(s)$ in which the time delay is eliminated

$$P(s) = \frac{e^{-T_d s}}{T_f s + 1}, \quad -P_0(s) = \frac{1}{T_f s + 1} \tag{60.11}$$

There are two loops in Figure 60.20. In the inner loop, a compensation signal $v(t)$ contains a prediction of $y(t)$. $C_0(s)$ is a PID controller.

The closed-loop transfer function is

$$W(s) = \frac{y(s)}{r(s)} = \frac{C(s)\,P(s)}{1 + C(s)\,P(s)} = \frac{C_0(s)P(s)}{1 + C_0(s)P_0(s)} \tag{60.12}$$

Note that the transcendental term is eliminated in the characteristic equation of the closed-loop transfer function. In practice, the perfect model $P(s)$ is not known, that is, perfect elimination is impossible, but the Smith predictor reduces the effect of time delay.

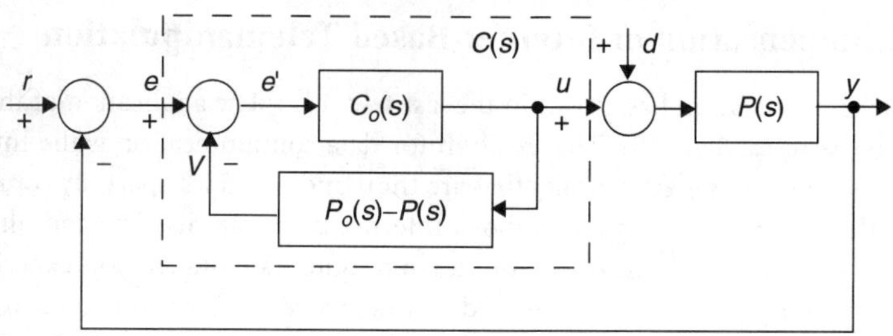

FIGURE 60.20　The configuration of the Smith Predictor.

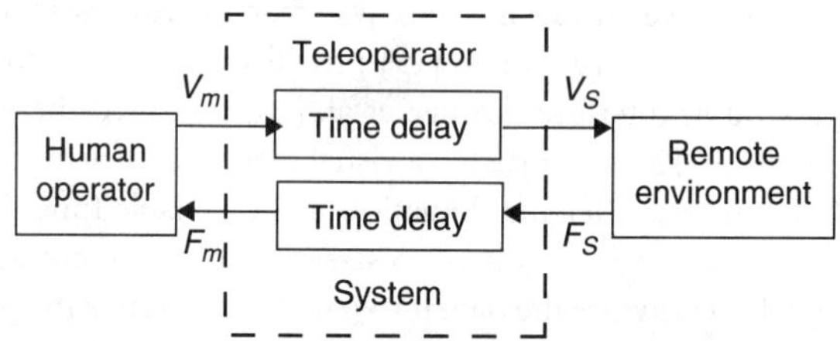

FIGURE 60.21　A simple teleoperator with time delay T_d.

Wave Variable Approach

This approach is based on the *passivity theory* [32], which has a close relationship with the Lyapunov stability. The passivity formalism represents a mathematical description of the intuitive physical concepts of power and energy. The basic idea behind passivity is to limit the input power flow such that it does not exceed the output power flow. If the power dissipation is zero, the system is called lossless. In case of a simplified telemanipulation, the operator moves a master device and its velocity v_m is transmitted to the slave device, which is forced to follow the master movement; ideally, $v_s = v_m$. The slave device contacts the remote environment and the reaction force F_s is transmitted back to the operator, who senses $F_m = F_s$. Figure 60.21 shows a simple telemanipulation with a constant time delay T_d in the communication channel, where

$$v_s(t) = v_m(t - T_d)$$
$$F_m(t) = F_s(t - T_d)$$

(60.13)

To handle the stability problem caused by the time delay, the bidirectional communication channel is considered as a transmission line with a pressure-type wave $w_V(t)$, a flow rate-type wave $w_I(t)$, and a power $P(x,t) = w_V(x,t) w_I(x,t)$, where x indicates that the wave variables must be measured at the same position. For stable operation, the transmission line must be passive, and the dissipated energy must be nonnegative.

$$\int_0^t P_{diss}(\tau)\, d\tau \geq 0$$

(60.14)

where P_{diss} is the power dissipation.

The trivial selection of wave variables ($w_I = v_m$ and $w_V = F_s$) does not lead to a passive transmission line. It highlights the source of the original problem. If the master velocity signal is sent and the slave force signal is received via the Internet, the time delay may destabilize the operation of the telemanipulation system. To achieve a passive, that is, stable system, the "damping" or "leakage" of the transmission line must be increased. Introducing the following normalized wave transformation proposed in [33]:

$$w_I = \frac{bv_m + F_m}{\sqrt{2b}}, \quad w_V = \frac{bv_s - F_s}{\sqrt{2b}}$$

(60.15)

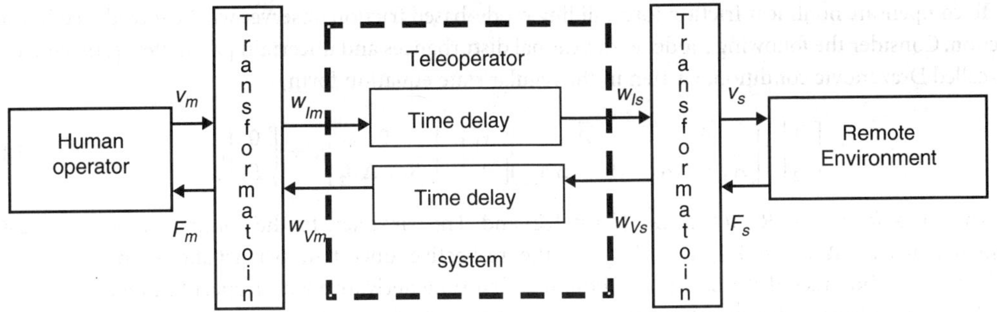

FIGURE 60.22 Telemanipulation with wave variables.

the transmission line becomes passive and lossless, which means the dissipated energy is zero. The velocity and force signal is transferred into wave variables before sending and they are re-transferred after arriving (see Figure 60.22). A possible interpretation of (60.15) is that there are direct feedbacks from the velocity to the force at the master side and from the force to the velocity at the slave side.

Mechanical impedance can be defined in the frequency domain:

$$Z(s) = \frac{w_V(s)}{w_I(s)} \tag{60.16}$$

According to [36], perfect transparency can be ensured by perfect impedance matching. It can be interpreted as the reflected wave deteriorates the transparency. The terminating impedance of this transmission line includes the characteristics of the remote environment. For ideal transparency, the frequency response of the remote environment must be known, and the transmission line have to be modified to match it in the whole frequency domain. There is no such ideal solution but several papers are published to propose an optimal solution.

Friction Compensation for Master Devices

The main problem is that the mechanism applied as a human interface device usually has a reasonable immanent friction. When the operator moves the master device, the friction might be too large to move it smoothly. Because of this additional force, the operator cannot feel which force is exerted in remote environment and which force moves the master mechanism. The whole effect of the friction and the whole dynamics of the mechanic construction cannot be eliminated (since it would need infinite input power) but it can be significantly reduced. The aim of the compensation is to make the system follow an ideal model with small friction. In other words, because of the compensation, the operator feels that it is very easy to move the master device. Usually, the following simplified, linearized reference model is selected for each degree of freedom of the master device:

$$G_m(s) = \frac{1/J_m}{s + D_m/J_m} \tag{60.17}$$

where torque is the input and velocity is the output signal, J_m is the artificial inertia, and D_m is the artificial viscous friction constant. The subscription m refers to the model. Parameters J_m and D_m must be small enough for comfortable manipulation but they must be large enough to avoid the undesirable motion (jitter).

To compensate nonlinear friction force, sliding mode-based friction observer will be introduced in this section. Consider the following model with external disturbances and uncertain parameters satisfying the so-called Drazenovic condition, written in the regular state equation form,

$$\begin{bmatrix} \dot{x}_1 \\ \dot{x}_2 \end{bmatrix} = \begin{bmatrix} \bar{A}_{11} & \bar{A}_{12} \\ \bar{A}_{21} + \Delta A_{21} & \bar{A}_{22} + \Delta A_{22} \end{bmatrix} \begin{bmatrix} x_1 \\ x_2 \end{bmatrix} + \begin{bmatrix} 0 \\ \bar{B}_2 + \Delta B_2 \end{bmatrix} u^0 + \begin{bmatrix} 0 \\ E_2 \end{bmatrix} f(t) \tag{60.18}$$

where $x_1 \in R^{n-m}, x_2 \in R^m, u \in R^m, \bar{A}_{ij}\ (i, j = 1,2)$ and The bar refers to the nominal or desired (ideal) system matrices. $\Delta A_{2j}(j = 1,2)$ and ΔB_2 are the respective uncertain perturbations and $f(t)$ is an unknown, but bounded disturbance with bounded first time derivative with respect to time.

Now, η is defined as the uncertainties and the disturbance of the system.

$$\eta = \Delta \bar{A}_{21} x_1 + \Delta \bar{A}_{22} x_2 + \Delta \bar{B}_2 u^0 + E_2 f(t) \tag{60.19}$$

The second line of (60.18) can be rewritten by η.

$$\dot{x}_2 = \bar{A}_{21} x_1 + \bar{A}_{22} x_2 + \bar{B}_2 u^0 + \eta \tag{60.20}$$

According to [34], $\hat{x}_2$ is estimated by a discontinuous observer:

$$\dot{\hat{x}}_2 = \bar{A}_{21} x_1 + \bar{A}_{22} \hat{x}_2 + \bar{B}_2 (u^0 + v) \tag{60.21}$$

where v is the discontinuous feedback. The goal of the design to find a feedback signal, denoted v, such that the motion of the system (60.18) is restricted to belong to the manifold S.

$$S = \{x_2 | \sigma = x_2 - \hat{x}_2 = 0\} \quad \text{where } \sigma \in R^m. \tag{60.22}$$

In sliding mode ($\sigma = x_2 - \hat{x}_2 = 0$), v contains information on the system's parametric uncertainties and the external disturbance, which can be used for feedback compensation (compare (60.20) and (60.21).

The simplest control law, which can lead to sliding mode, is the relay:

$$v_i = M_i\ sign(\sigma_i). \tag{60.23}$$

If η is in the range of B_2 ($\eta \subset$ range (B_2)), the ideal sliding mode occurs [35].

In the practice, there is no way to calculate the equivalent control v_{eq} precisely, but it can be estimated by a low-pass filter for v as shown in Figure 60.23. There are two loops in Figure 60.23. The observer-sliding mode controller loop should be as fast as possible to achieve an ideal sliding mode. This is granted because

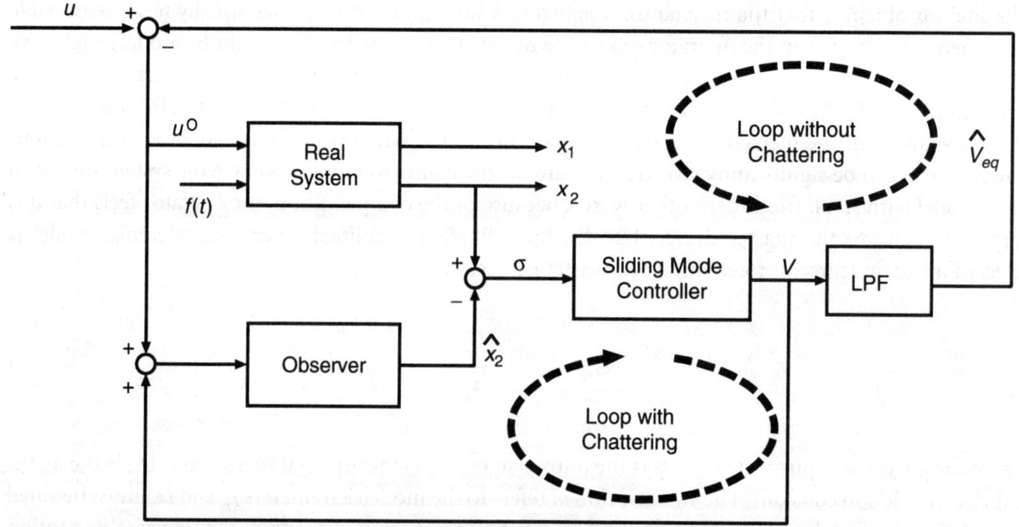

FIGURE 60.23 Sliding mode-based feedback compensation.

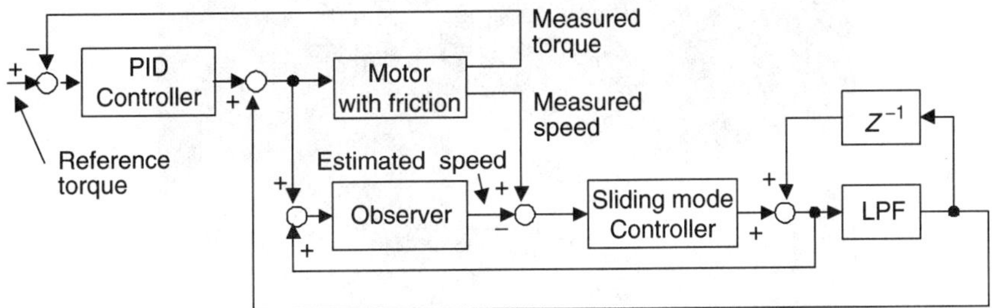

FIGURE 60.24 Overall control scheme for force control.

this loop is realized in a computation engine in the recent application. The real system-compensator (consisting of Sliding Mode Controller and LPF) loop should be faster than the change of the disturbance. On the other hand, the smallest un-modeled resonant frequency of real system should be out of the bandwidth of that loop to avoid chattering.

Applying the above method for friction compensation of a single joint, the parameters are

$$\overline{A}_{21} = 0, \ \overline{A}_{22} = \frac{D_m}{J_m}, \ \overline{B}_2 = \frac{1}{J_m} \tag{60.24}$$

The friction compensator provides an additional compensation signal to any conventional closed-loop controller and it improves its performance. A possible overall control scheme for force control of a single joint is shown in Figure 60.24. There is a simple torque control loop, with a PID controller. Because of the nonlinear characteristics of the friction, it cannot be tuned well. The additional sliding mode-based friction compensation may improve the performance significantly [13].

60.5 A Complete Application Example: A Handshake Via Internet

A handshake via Internet is an interesting type of telemanipulation [29]. The task is to realize a handshake between two people at two different corners of the world. Let one be the operator and the other be the far environment reacting to the operator (hand) movement. The master device will be a hand-shaped device in the operator's hand. The slave device is identical. When the operator grasps the master device (see in Figure 60.25 a and b), the master device transforms the movement and force to the slave device. The slave device gives this movement to the far environment (the other person's hand). This person gives a reaction force and a movement to the slave device. This information is transmitted to the operator via a master device. The handshake will be realized between two people far from each other. The main challenge is compensation of the time delay.

Virtual Impedance with Position Error Correction

Virtual Impedance with Position Error Correction (VIPEC) is based on the concept of the impedance control. VIPEC is composed of two major parts: a virtual impedance model (VI) and a position error correction part. The virtual impedance model generates a motion trajectory for a manipulator based on input forces and its dynamics. The position error correction part is added to decrease the position difference of the master and slave.

Virtual Impedance

In the case of a 1 DOF linear motion manipulator, it is assumed that a virtual mass, M_v, is attached to a manipulator arm. This virtual mass is supported by a virtual damper, D_v, and a virtual spring, K_v. The structure of the VI is shown in Figure 60.26.

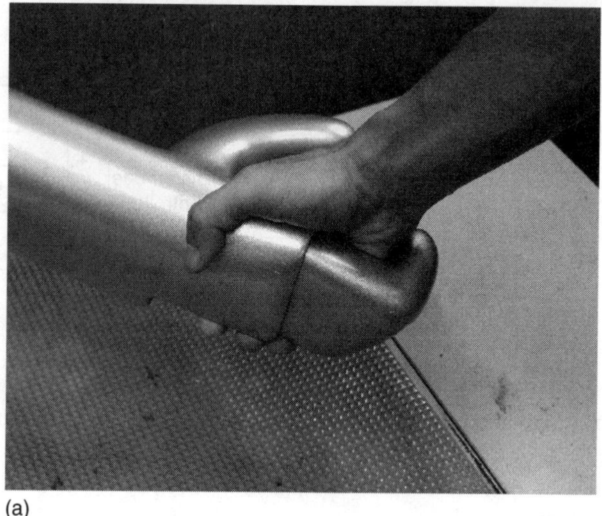

(a)

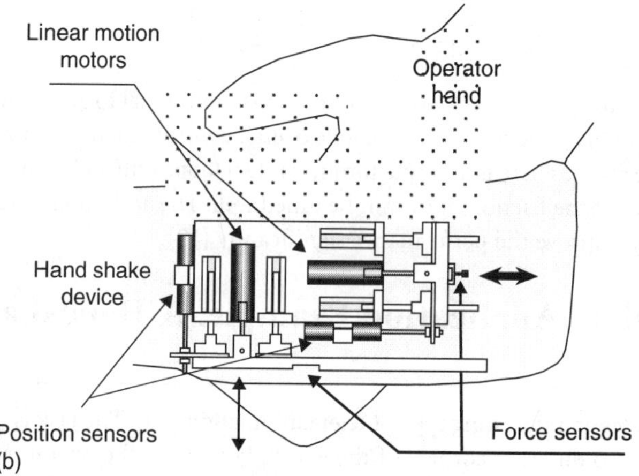

FIGURE 60.25   Tele Handshaking Device. (a) photo, (b) structure.

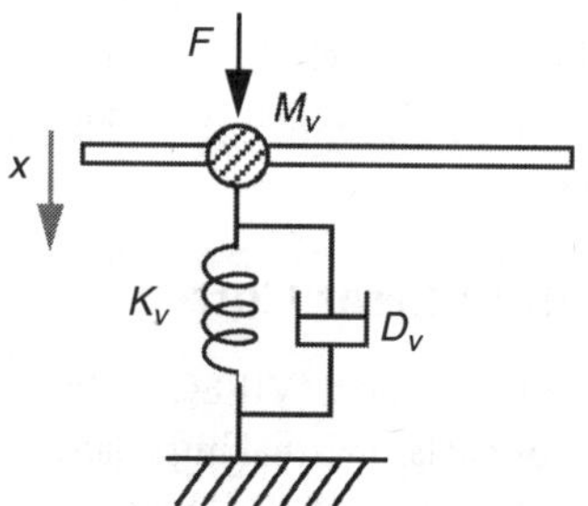

FIGURE 60.26   One DOF linear motion manipulator with virtual impedance.

When a force is applied to the virtual mass, the position of the virtual mass is modified according to the dynamics of VI. The dynamic equation of VI is as follows:

$$F(t) = M_v\, \ddot{x}(t)\, D_v\, \dot{x}(t) + K_v\, x(t) \tag{60.25}$$

where $F(t)$ is the force detected at the virtual mass and $x(t)$ is the displacement of the virtual mass. The transfer function of the VI acts as a second-order low-pass filter to convert force into displacement. Thus, the displacement is always smooth even if the force contains high-frequency components.

Position Error Correction

The structure of a teleoperator with time delay utilizing the VIPEC is illustrated in Figure 60.27. The PEC part (the shaded part) is added to the original approach [30] in order to improve the response of the teleoperator by reducing the position error between the master and slave positions.

In Figure 60.27, F_o denotes the force that an operator applies to the master arm, and F_e denotes the reflected force when the slave arm contacts the environment. x_m and x_{mr} denote the position and the reference position of the master arm. x_s and x_{sr} denote the position and the reference position of the slave arm. T_d represents a delay time in data transmission between the two sites. A_f is a scalar unit, which can be varied depending on the size ratio of the slave and the master.

Each VI generates the reference motion trajectory for each manipulator arm based on the sum of two forces: the force from its own site and the force received from the other site. The data transmission from one site to the other site is delayed by T_d. The reference motion trajectory from VI is sent to a servo controller to control the position of the manipulator arm.

The control approach can be used for the master and the slave arms, which are different in size by appropriately adjusting A_f and the parameters of each VI, for example, when the slave is a micro robot, which is smaller than the master.

The dynamic equation of VI at the operator site is

$$F_o(t) - \frac{1}{A_f} F_e(t-T_d) = M_v \ddot{x}_{mr}(t) + D_v \dot{x}_{mr}(t) + K_v x_{mr}(t) \tag{60.26}$$

The dynamic equation of VI at the remote site is

$$F_e(t) - A_f F_o(t-T_d) = M_v \ddot{x}_{ms}(t) + D_v \dot{x}_{ms}(t) + K_v x_{ms}(t) \tag{60.27}$$

Without loss of generality, $A_f = 1$ and both VIs have the same parameters. When the parameters of VI are appropriately determined, the system should remain stable. In the steady state, F_o is approximately equal to F_e when K_v is set to a low value, and the master and the slave positions should be identical. However, when a time delay is present in the teleoperator, the control approach VI without PEC part causes the position error between the master and slave positions, since there is no comparison between these two positions. To reduce this position difference, the PEC part is added in order to make a closed-loop system. Therefore, the position error can be reduced. The slave position, delayed by T_d, is transmitted to the master site and compared to the master position. The difference between the two positions is sent through the PEC gain, and then modifies the reference motion trajectory of the master arm.

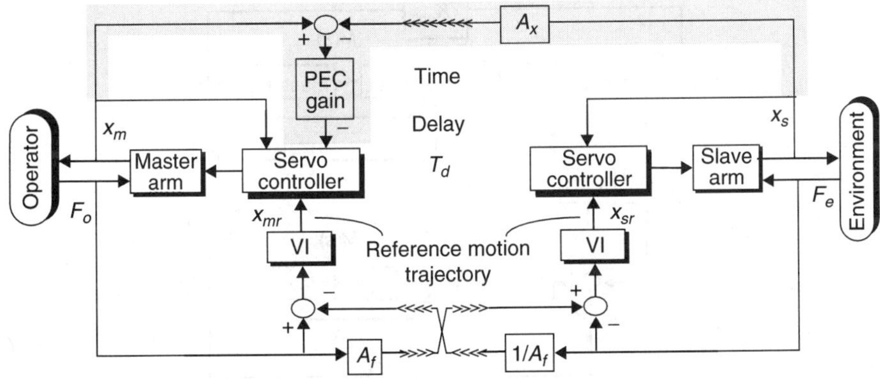

FIGURE 60.27 Virtual Impedance with Position Error Correction for a teleoperator system with time delay.

Control Model of The Handshaking Device

Figure 60.28 shows the control system diagram of the master–slave system inside the telehandshaking system with VIPEC approach. The armature inductance, L_a of each motor is neglected. The two linear motion motors represented by the master block and the slave block can be modeled as simple DC-motors with the transfer function.

$$C(s) = \frac{X(s)}{E_a(s)} = \frac{K_m}{s(T_m s + 1)} \; K_p \tag{60.28}$$

where X and E_a are the Laplace transformed of motor shaft displacement and the applied armature voltage, K_m is the motor gain constant, T_m is the motor time constant, and K_p is a constant to convert the angular displacement into displacement.

The servo controller is a proportional controller whose gain is G. The closed-loop transfer function of the manipulator arm and the controller is

$$M(s) = \frac{C(s)G}{1 + C(s)G} = \frac{\dfrac{GK_pK_p}{T_m}}{s^2 + \dfrac{1}{T_m}s + \dfrac{GK_pK_p}{T_m}} \tag{60.29}$$

Gain P represents PEC gain. For simplicity, the environment is modeled by a spring with stiffness constant K_e. Therefore, the dynamics of the spring interacting with the slave arm is modeled as follows:

$$F_e(t) = K_e x_s(t) \tag{60.30}$$

where x_s is the displacement of the spring. The dynamics of the operator including the dynamic interaction between the operator and the master arm is approximated by

$$F_{op}(t) - F_o(t) = M_{op}\ddot{x}_m(t) + D_{op}\dot{x}_m(t) + K_{op}x_m(t) \tag{60.31}$$

where F_{op} is the force generated by the operator's muscles. M_{op}, $B_{op,}$ and K_{op} denote mass, damper, and stiffness of the operator and the master arm, respectively. Similar to Equation (60.30), the displacement of the operator in Equation (60.31) is represented by x_m since it is also assumed that the operator is firmly grasping the master arm and never releases the master arm during the operation.

Experiment

Using the telehandshake system, two human operators, operator A and operator B, shake hands with each other. When operator A tries to shake the hand of operator B by moving the master arm A, the slave arm

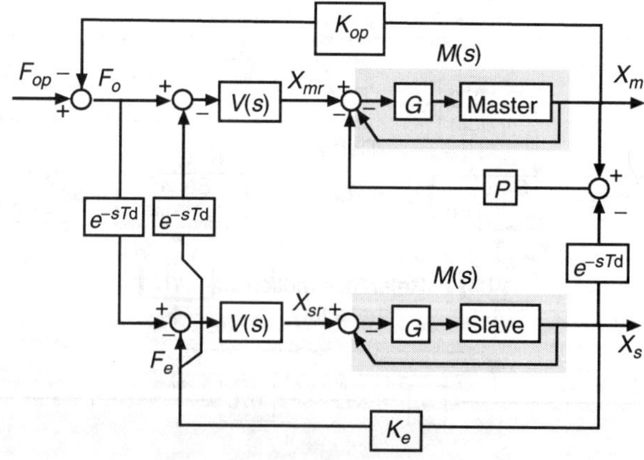

FIGURE 60.28 Control diagram of the handshaking device.

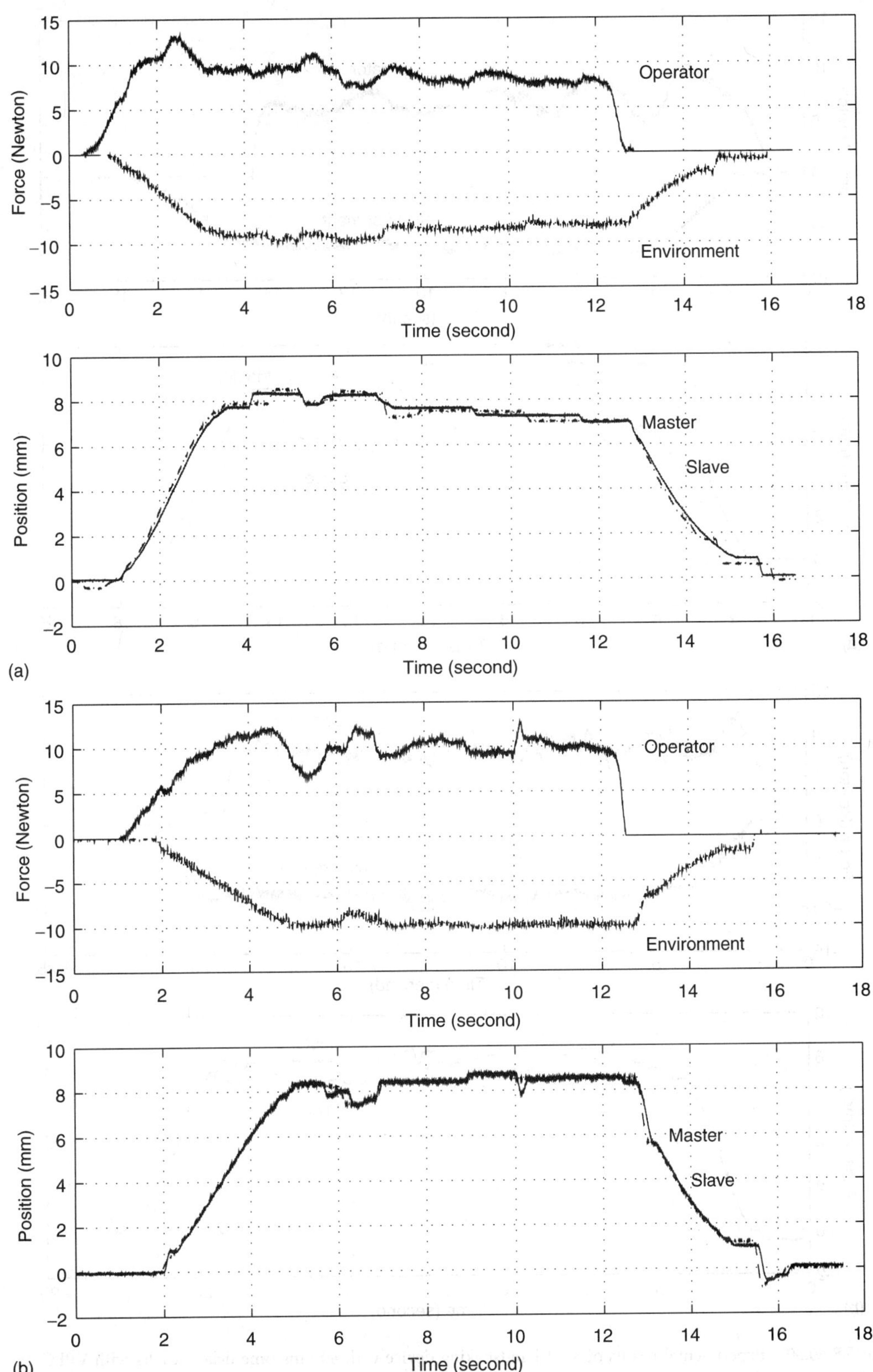

FIGURE 60.29 Experimental results of a telehandshaking device without time delay: results with VIPEC. (a) results with VI and without PEC, and (b) results with VIPEC.

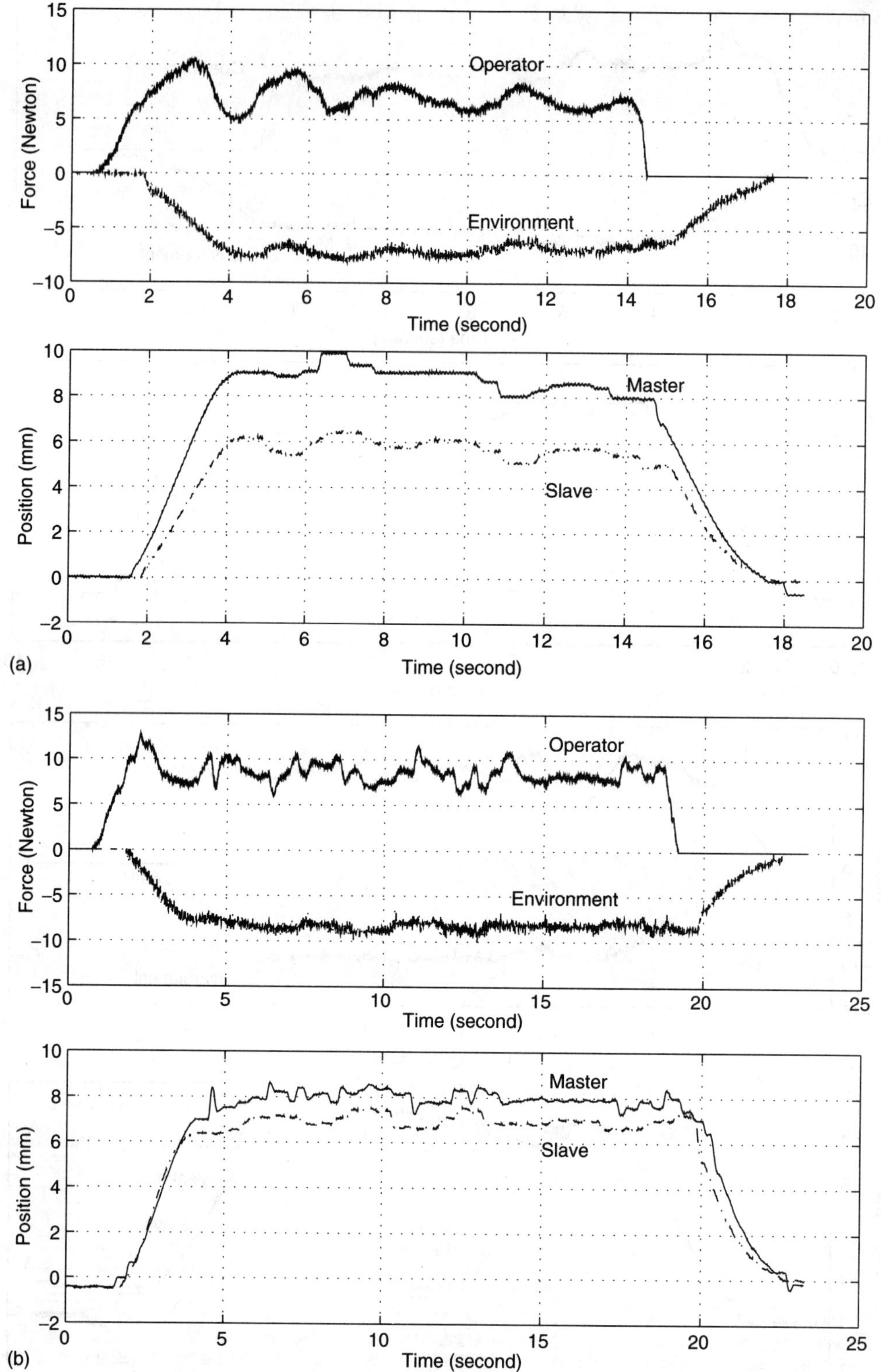

FIGURE 60.30 Experimental results of a telehandshaking device with 400 ms time delay: results with VPEC. (a) results with VI and without PEC, and (b) results with VIPEC.

A should move the same as the master arm A does. When the slave arm A contacts the palm of operator B, operator B can feel the grasping force of operator A. Operator B can also grasp the virtual hand of operator A in the same way. Two operators succeeded in shaking hands with each other and achieved force sensation through the tele-handshaking Device. Experimental results of simple VI and VIPEC methods are compared in Figures 60.29 and 60.30. In Figure 60.29, the time delay is negligible. The difference between the two methods is not significant. In both cases, the positions of the master and slave device are approximately the same. Figure 60.30 illustrates the experimental results with 400 ms time delay. In both cases, the system is stable and the operators can feel the grasping force. However, the position difference between the master and slave device is smaller in case of the VIPEC method.

60.6 Conclusions

The Internet-based telemanipulation is a relatively new and attractive research field. It is a fruitful ground for applying new and sophisticated control methods to eliminate the time delay and nonlinear friction force, which are the main challenges in this field. Of course, total elimination is impossible, but a stable operation may be achieved. This chapter presented a promising method, namely Virtual Impedance method with Position Error Correction. The effectiveness of this method has been proved by measurement results of a telehandshaking device.

Acknowledgments

The authors wish to thank the National Science Research Fund (OTKA T034654) and Control Research Group of Hungarian Academy of Science for their financial support.

References

1. Sheridan, T.B., Telerobotics, *Automatica*, 25,487–507, 1989.
2. Goertz, R.C. and W. C. Thompson, Electroically controlled manipulator, *Nucleonics*, 12, 46–47, 1954.
3. Lee, S., B. Bekey and A.K. Bejczy, Computer Control of Space-borne Teleoperators with Sensory Feedback, Proceedings of IEEE International Conference on Robotics and Automation,1985, pp. 204–214.
4. Dalton, Barney and Ken Taylor, A Framework for Internet Robotics, Proceedings of IEEE/RSJ International Conference on Intelligent Robots and Systems, WS2, 1998, pp. 15–22.
5. Mizik, S., P. Baranyi, P. Korondi, and M. Sugiyama, Virtual training of vector function based guiding styles, *Transactions on Automatic Control and Computer Science* 46(60), 81–86, 2001.
6. Handlykken, M. and T. Turner, Control Systems Analysis and Synthesis for a Six Degree-of-freedom Universal Force-reflecting Hand Controller, in proceedings of the IEEE Conference on Decision and Control, 80, Vol. 2, 1980, pp 1197–1205.
7. Sheridan, T.B., *Telerobotic, Automation, and Human Supervisory Control,* The MIT Press, Cambridge, MA,1992.
8. Burdea, G. and P. Coiffet, *Virtual Reality Technology,* John Wiley & Sons, Inc., New York, 1994.
9. Dudragne, J., C. Andriot, R. Fournier, and J. Vuillemey, A Generalized Bilateral Control Applied to Master–slave Manipulators, Proceedings of 20th Conference of International Society for Intelligent Research, 1989, pp. 435–442.
10. Kunii, Y. and H. Hashimoto, Dynamic force simulator for multifinger force display, *IEEE Transactions on IE*, Vol. 43, 1996, pp. 74–80.
11. Sitti, Metin and Hideki Hashimoto, Two-Dimensional Fine Particle Positioning Using a Piezoresistive Cantilever as a Micro/Nano-Manipulator, IEEE International Conference on Robotics and Automation, 1999, pp. 2729–2735.
12. Massie, Thomas H. and J. Kenneth Salisburg, The PhantomHaptic Interface: A Device for Probing Virtual Objects, Proceedings of the 1994 ASME International Mechanical EngineeringCongress and Exhibition, Chicago, Vol. 55-1, 1994, pp. 295–302.

13. Korondi, Péter, Péter T. Szemes and Hideki Hashimoto, Sliding mode friction compensation for a 20 DOF sensor glove, *Transaction of ASME, Journal of Dynamic Systems, Measurement and Control,* 122, 611–615, 2000.

14. Colgate, J.E., Robust Impedance shaping telemanipulation, *IEEE Transactions on Robotics and Automation,* 9:373–384, 1993.

15. Ren C. Luo, Tse Min Chen, and Chih-Chen Yin, Intelligent Autonomous Mobile Robot Control through the Internet, ISIE 2000, W. Cholula, Puebla, Mexico, 2000.

16. Burgard, A. Cremens, D. Fox, D. Hahnel, G. Lakemeyer, D. Schulz, W. Steiner, and S. Thrun, The Interactive Museum Tour-Guide Robot, Proceedings of National Conference on Artificial Intelligence, 1998, pp. 11–18.

17. Simmons, Reid, Xavier: An Autonomous Mobile Robot on the Web, Proceedings of IEEE/RSJ International Conference on Intelligent Robots and Systems, WS2, 1998, pp. 43–48.

18. Farzin, B.R. and K. Goldberg, A Minimalist Telerobotic Installation on the Internet, Proceedings of IEEE/RSJ International Conference on Intelligent Robots and Systems, WS2, 1998, pp. 7–13.

19. Stein, M.R., Painting on the World Wide Web: The PumaPaint Project, Proceedings of IEEE/RSJ International Conference on Intelligent Robots and Systems, WS2, 1998, pp. 37–42.

20. Kosuge, K., J. Kikuchi, and K. Takeo, VISIT: A Teleoperation System via Computer Network, Proceedings of IEEE/RSJ International Conference on Intelligent Robots and Systems, WS2, 1998, pp. 61–66.

21. Hirukawa, H. and I. Hara, The Web Top Robotics, Proceedings of IEEE/RSJ International Conference on Intelligent Robots and Systems, WS2, 1998, pp. 49–54.

22. Ferrell, R., Remote manipulation with transmission delay, *IEEE Transactions on Human Factors in Electronics,* Vol. 6, 1965, pp. 24–32.

23. Sukhan Lee, Hahk Sung Lee: Advanced Telemanipulation Under Time-delay, SPIE OE/Technology '92, Telemanipulator Technology, 1992.

24. Bejczy, A.K., W.S. Kim, and S.C. Venema, The Phantom Robot: Predictive Displays for Teleoperations with Time-delays, IEEE International Conference on Robotics and Automation, 1990, pp. 546–551.

25. Won S. Kim, Blake Hannaford, Antal K. Bejczy, Force-reflection and Shared Compliant Control in Operating Telemanipulators with Time-delay, IEEE Transactions on Robotics and Automation, 1992, pp. 176–185.

26. Backes, Paul G., Multi-sensor Based Impedance Control for Task Execution, IEEE International Conference on Robotics and Automation, 1992, pp. 1245–1250.

27. Ferrell, W.R. and T.B. Sharidan, Supervisory control of remote manipulation, *IEEE Spectrum,* Vol. 4, 1967, pp. 81–88.

28. Dalton, Barney and Ken Taylor, A Framework for Internet Robotics, Proceedings of IEEE/RSJ International Conference on Intelligent Robots and Systems, WS2, 1998, pp. 15–22.

29. Manorotkul, S. and H. Hashimoto, Virtual Impedance with Position Error Correction for Teleoperation with Time Delay, Proceedings of the 4th International Workshop on Advanced Motion Control, Vol. 2, 1996, pp. 524–528.

30. Otsuka, M., N. Matsumoto, T. Idogaki, K. Kosuge and T. Itoh, Bilateral Telemanipulator System with Communication Time Delay Based on Forcesum-Driven Virtual Internal Model, Proceedings of the IEEE International Conference on Robotics and Automation, 1995, pp. 344–350.

31. Smith, O.J. M., A controller to overcome dead time, Chemical Engineering Progress 53, 217–219, 1959.

32. R.J. Anderson and M.W. Spong, Bilateral control of teleoperators with time delay, *IEEE Transactions on Automatic Control,* 34, 494–501, 1989.

33. Niemeyer, G. and J-J.E. Slotine, Stable adaptive teleoperation, *IEEE transactions on Automatic Control,* 16. 152–162, 1991.

34. Young, K.D., *Manifold Based Feedback Design, Advances in Control and its Applications,* Lecture Notes in Control and Information Sciences, Springer-Verlag, London, 1995.

35. Utkin, V.I., *Sliding Modes in Control and Optimization,* Springer-Verlag, London, 1992.

36. Lawrence, D., Stability and transpareny in bilateral teleoperation, *IEEE Transactions on Robotics and Automation,* 9, 624–637, 1993.

61

Knowledge Connect–
An Approach for an
Industrial^{IT} Service
Tool

Carsten Beuthel

ABB Corporate Research

Paul George

ABB US Business Services

Ulrich Topp

ABB Corporate Research

61.1 Introduction

A service engineer's task of maintaining a huge installation is an immense task. He relies on information that is available about his plant. Previously, information about installed components or the process itself was provided by paper documents delivered by the system manufacturer at the time the system was installed. This documentation was then updated and maintained by its users, normally service engineers, machine operators, or service personal. After a period of time and numerous minor changes, this documentation became outdated and incomplete because it had not been updated adequately. Another problem with paper documentation is that it is hard to search for specific information on equipment in a timely manner.

The shift from paper to electronic documentation was an easy decision — it made it possible to search an entire library for information at one workplace. Secondly, it occupies much less space than the paper documents. The drawback, however, is to know what you were looking for, for example, which component does the process include, how is it connected, and what are the parameters and its current status. In large installations, this can typically be found only by walking to the physical component, writing down the information off of the label, and comparing the data to some connected information system to gain information about the current status. With this information, the electronic documentation can be searched or a help desk can be called. In cases where the information is showing white spaces, it becomes more difficult to find a solution for a specific problem. Modern search engines, like the ABB SolutionsBank [1], use wizards to lead the user to the requested information. Even in this case, the user needs to know specifically what he is searching for.

This fits very well with recent studies, which clearly show that critical problems in the service business arise due to lack of information. This begins with faulty failure diagnosis followed by calling the wrong expert and asking irrelevant questions often not related to the hardware but to the process, and ending with difficulty finding spare part information [2].

ABB offers several different web applications for helping with this problem including SolutionsBank, the e-mmediate Agent, and the ABB University. For most service engineers, these tools are helpful but it is difficult to find the right information in a very limited time frame. The KnowledgeConnect project was initiated to overcome this problem by automatic filling in the white spaces of a request through the use of plant information.

The remainder of this chapter will describe the scope of the project. Section 2 will provide an overview of the architecture based on the aspect object model [6]. Section 3 will summarize the experience of using KnowledgeConnect on a real paper mill control system. Finally, in Section 4, some constraints, advantages, and prerequisites will be discussed together with an outlook.

61.2 Architecture

The basic concept of KnowledgeConnect is based on ABB Industrial^{IT} [1] and consists of the connection from the control system and its engineering with several external databases such as SolutionsBank[3] and the e-mmediate agent. This concept will be made available to different user groups, although the service engineers will mainly use it. The initial project used process graphics available on any Process Portal application for accessing the implemented services. Figure 61.1 shows an overview of the project and the connected external services.

The procedure used for all three Aspect Systems is similar:

- Collect information about the object that the user wishes to search information for
- Request the information from an external application
- Display the results to the user.

While this sounds fairly simple, it is a tricky task, which requires considerable engineering effort to create such a system and keep it up to date while it is used. The main task of using this architecture is to find a solution that needs very little engineering and that is usable straight out of the box.

KnowledgeConnect makes use of the Aspect Directory [5] of the ABB's 800xA System. Any real object is represented in the process portal by an aspect object. Each aspect object has properties, which describe

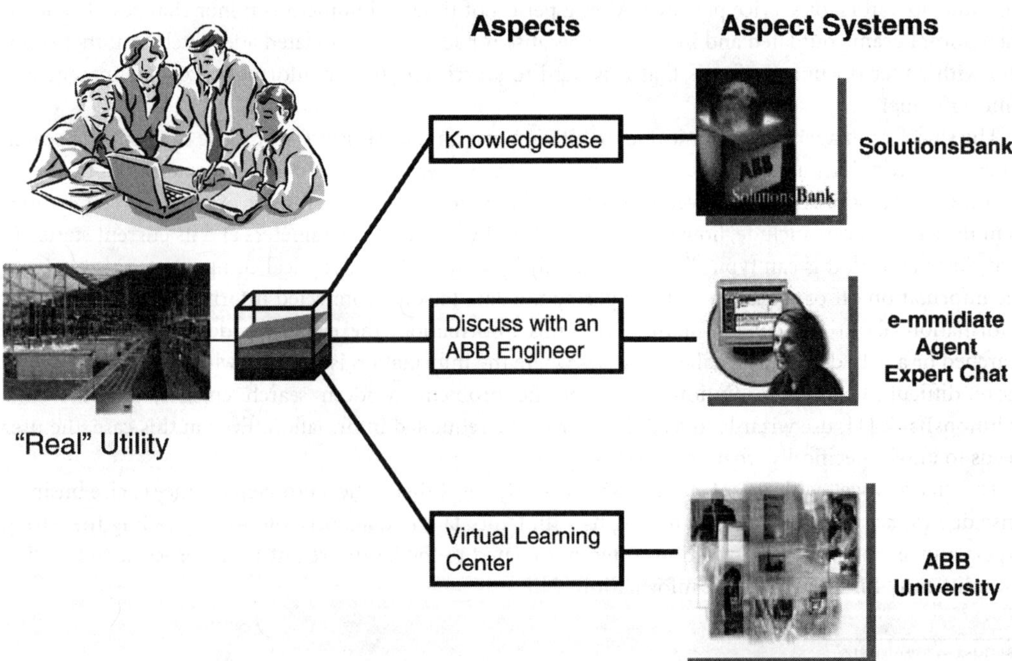

FIGURE 61.1   Architectural overview of KnowledgeConnect.

the object such as the name, the type of component, for example a thermo sensor, the configuration in the engineering tool, or the actual status of it. All this information is available as OPC DA data items for any application. Special information, for example, related to dependencies with other objects might need special applications to calculate those and can be accessed by standardized interfaces through the aspect directory. This information source is used in KnowledgeConnect to describe a very specific search string, which can be sent to the external application, that is capable of returning precise information to the user. This information does not need to be engineered for KnowledgeConnect. This information can be in any form including MS Word, MS Excel, or Adobe PDF. All KnowledgeConnect has to do is to make intelligent links to this data source.

In the aspect directory, the aspect objects are hierarchically sorted [7] and each instance refers to a specific object type, which again belongs to an object category. Each object type owns a special set of aspects, for example configuration data, faceplates, symbol information, and control configuration [8]. These aspects are inherited when a new object is instantiated in a hierarchical structure, for example, Functional Structure (see Figure 61.2).

KnowledgeConnect is formulated as a set of aspect systems, which are added to the desired object types, and categories, which will later refer to service information. When an object is instantiated, all KnowledgeConnect aspects automatically come ready to be used.

When a user selects a KnowledgeConnect aspect, the aspect starts building up a query from the information specified in the object's relevant aspects. Additionally, a Global Configuration aspect is used to provide general information about the customer's site, the database to refer to, and the required log-on information for this database (see Figure 61.3).

The advantage of this procedure is that the system only needs a minimal amount of preconfiguration effort. The object-type definition only has to be defined once, which is typically done by the solutions provider of the SolutionsBank. They are knowledgeable of the object types used in their specific industry systems. Their task is to add the KnowledgeConnect aspects to the appropriate object categories and types and fill them with minimum information, typically the object name. They can also point to other aspect system information, which does not come with the standard package or add specific keywords to these aspects. These preconfiguration data are stored in an extra aspect also available to any other application, which might want to make use of the KnowledgeConnect features (see Figure 61.4). The object types can then be released and distributed to customer sites.

Currently, the following KnowledgeConnect aspects have been developed and tested:

- Technical Information such as release notes, feature changes, etc. (TechInfo).
- Equipment manuals and user documentation (ManualsOnline).
- "Fix a problem" with a component (KnowledgeBank).
- Enter an on-line Chat-Session with an Service Engineer (e-mmediate Agent).
- e-Learning courses available over the web (ABB University).
- On-line Part ordering (Parts Online).

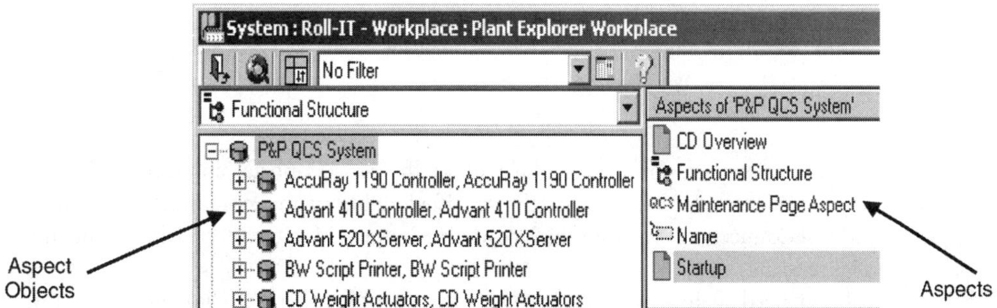

FIGURE 61.2 An object instantiated in the Functional Structure.

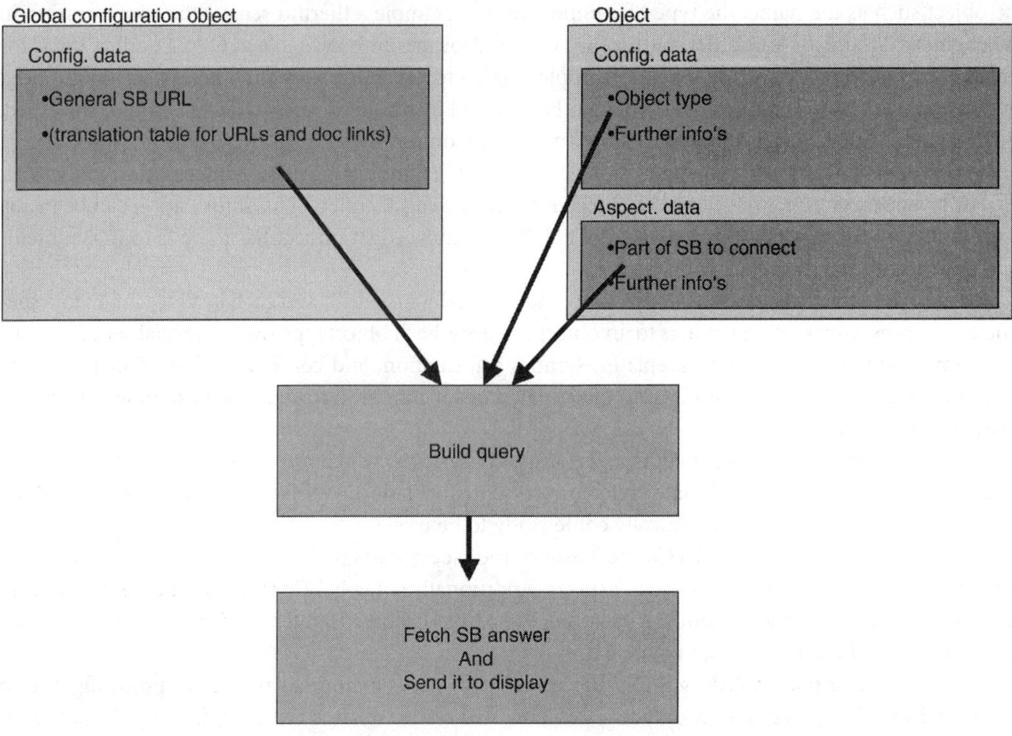

FIGURE 61.3   How a query is built up.

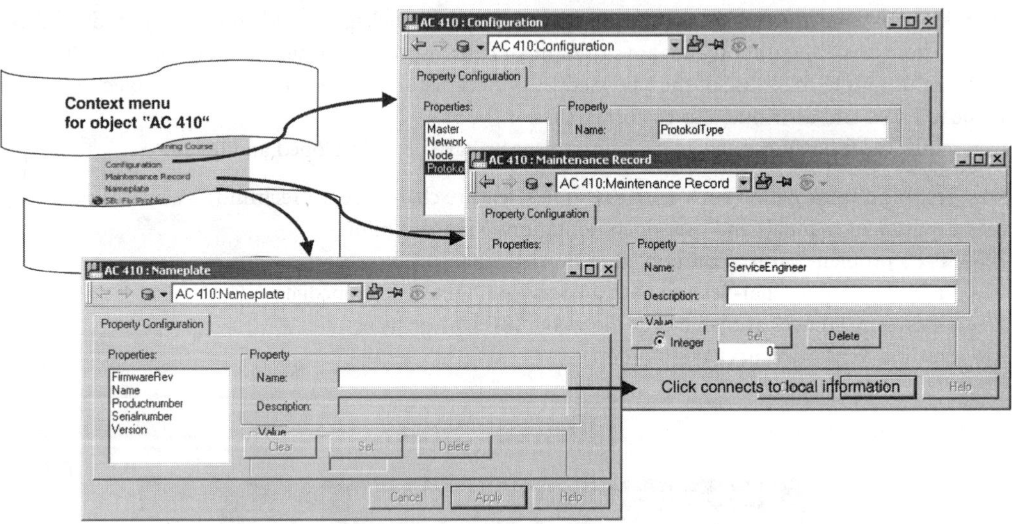

FIGURE 61.4   Standard aspect configuration.

The KnowledgeConnect aspects are an integral part of the process portal application and can be accessed from any user's workplace that owns the specified rights in the system. Typically, the user accesses the KnowledgeConnect aspects from the KnowledgeConnect Maintenance Display as mentioned previously (see Figure 61.5).

When the local user needs help with a question or with a problem, a right click on the asset opens the aspect context menu, which leads to all functionality available to this specific object. The process

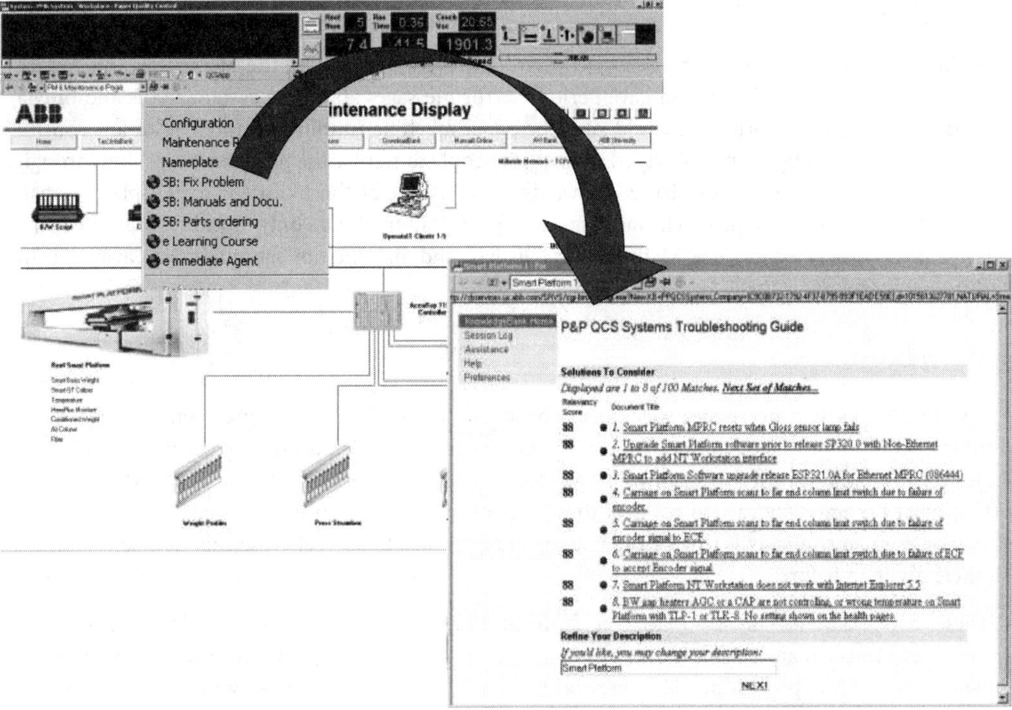

FIGURE 61.5   Maintenance Display with troubleshooting example.

portal application opens the aspect view showing the solutions available for the selected asset, and is ready for further analysis to pinpoint the desired solution using the mechanism described above.

61.3   Experience with a Pilot Installation

KnowledgeConnect was successfully installed on one of ABB's customer sites, a paper mill near Jackson, MS. This site had a new Pulp and Paper quality control system based on Industrial^{IT} installed.

The installation was simple, and the adoptions to the process could be managed within half a day. After the KnowledgeConnect installation and introduction, the service technicians were left alone with testing out the system. Two months later, a follow-up interview was conducted with the service responsible person as well as the service technicians.

The first finding was that all users stated that the system was of great value for them. The system considerably helps to find the appropriate information with a single mouse click for components, that are actually installed. The value is that it normally takes the service engineer an enormous amount of time to figure out which component, version, etc., is actually installed in his system and how it is configured and related to other system components. This information is made accessible on the Maintenance Display and the relationship between the components is visualized as a process display. As a graphical representation, the Maintenance Display acts like a plant explorer for the service engineer. From this one location, information regarding all components can be retrieved. The process portal locates the necessary search criteria and displays the results to the user. Due to this visualization, the Maintenance Display can also be used to look up information, which might be required at a later date.

Two Examples of the Maintenance Display Usage include the following.

Example one

A failure in the control network was discovered and even after calling the Help Line, this failure could not be resolved. The system gave periodic error messages about the system status and only a bypass with some network cable could keep the paper machine running, although with less control and quality. KnowledgeConnect helped the service people to resolve the problem. Searching through the control net-object entries of the KnowledgeBank aspect of the Maintenance Display, a small component was discovered, which was suspected for the failure. It was only found because the component was indicated on the Maintenance Display and information about it was available in SolutionsBank.

Example two

A second example concerned a specific sensor at the frame. The service engineers had very little information about this component and they did not find it in SolutionsBank because the naming and versioning were different from what was used in the plant. Through the use of the Maintenance Display, it became very easy to connect to all the information relevant to this component and the service engineers quickly learned from the on-line manuals to correctly use this new sensor more efficiently than before.

These two examples illustrate that it is important to have an on-line help system like the KnowledgeConnect maintenance display available to retrieve information. Since these capabilities are embedded into ABB's process portal framework, access is not limited to the service staff. Even more, the KnowledgeConnect aspects are naturally coupled with all other functionalities in the system, for example alarm and event lists, process graphics, etc.

More comments were made on the following issues: while the operators, the controllers, and the optimization people have their own views of the control system and the production process, the maintenance people still do not have their own user interface to the process. Service engineers are using the views of the other user groups but they wish to have their own displays to collect important information for their work. The Maintenance Display is exactly what they are looking for. Additionally, they wish to be able to extend and use it as part of their daily work. To do so, it must be easy to add new functionality and details of their system to this display. New links to information sources should be easily added to the display along with objects containing even more details.

Another important task is the illustration of the components. Having only the descriptive name of an object does not help in a complex system to identify it in the process. This means that it is important that a good illustration is available for each component of the system stored as an image icon with the object type. It should then be possible to extend the Maintenance Display with the new information and illustration.

61.4 Discussion and Conclusion

This chapter presented a new aspect system out of the world of Industrial[IT]. KnowledgeConnect reuses information, which it collects from other sources, for example applications running on the same software platform, the ABB Process portal [4]. From this information, the user can easily find relevant information to solve typical service engineer problems. The key mechanisms used are aspect objects, which lead to an efficient and low-effort engineering of such a support system.

There are some constraints that have to be mentioned here. To use such a support system, it is necessary that the service engineer's workplace be connected to both the Internet and the control system at the same time. Only in this configuration can the optimum performance of the system be gained. Unfortunately, connecting a control system directly to the Internet is normally not accepted by most customers due to network security problems. On the other hand, very often, access to control data is already available by links to the company's office network. This makes it essential that network administrators install the required network protection so that it remains possible to use such services that will have a very important role in the future of automation.

KnowledgeConnect demonstrates a very small fraction of the possibilities of communicating applications in the Industrial[IT] world, but the results of this project are very promising for future services to our customers.

References

1. Industrial[IT] — the next way of thinking, URL: http://www.abb.com.
2. ABB SolutionsBank ,URL http://solutionsbank.abb.com.
3. O'Brian, L., Hill, Dick, and Chatha, Andy, *PAS Worldwide Outlook — Market Analysis and Forecast till 2006*, ARC Advisory Group, Dedham, USA, 2002.
4. Rytoft, C. and Normark, B., Industrial[IT] and the utility industry, *ABB Review: Focus for the Utility Industry*, 3, 23–28, 2002.
5. Halásek, Tomáš, Using ABB Aspect Integration Platform for Implementation of Engineering Design Work, ASR '2003 Seminar, Instruments and Control, Vol.XXVIII, Ostrava, May 6, 2003.
6. Ragaller, Klaus, An Inside Look at Industrial[IT] Commitment, ABB Technology Day, 14 November 2001.
7. Krantz, Lars, Industrial[IT] — The Next Way of Thinking, *ABB Review* 1, p. 4–10, 2000.
8. Bratthall, L.G., van der Geest, R., Hofmann, H., Jellum, E., Korendo, Z., Martinez, R., Orkisz, M., Zeidler, C., and Andersson, J.S., *Integrating Hundred's of Products Through one Architecture — The Industrial[IT] Architecture*, ICSE, Orlando, USA, 2002.

62

OPC — Openness, Productivity, and Connectivity

Frank Iwanitz
Softing AG

Jürgen Lange
Softing AG

62.1 Introduction

This chapter provides an introduction in the Openness, Productivity, and Connectivity (OPC) technology. After explaining the history of OPC, the structure of the OPC Foundation, use, cases, and advantages of OPC, the introduction of the specifications follows. The chapter closes with an outlook into the future.

62.2 Open Standards — Automation Technology in Flux

The pace of change in industrial control and automation technology is accelerating. Demands on machines and systems concerning flexible retrofit, production speed, and fail safety are increasing, as are cost pressures. Software is becoming more and more the essential factor with products, systems, and complete plants.

At the same time, changes in the field of automation brought about by the use of the PC as an automation component, by the Internet, and by the tendency to more open standards, can be clearly seen — to the benefit of the user and the manufacturer.

The PC is more and more used for visualization, data acquisition, process control, and the solution of further tasks in automation. It complements or replaces the traditional PLC and the operator terminal. The reasons for this are the continual decrease in the price of the mass-produced PC, the permanent multiplication of the computing capacity of the CPU, the availability of even more efficient and comfortable software components, and the ease of integration with Office products.

Efficiency and cost savings are achieved through the reuse of software components and the flexible compilation of such components into distributed automation solutions.

Horizontal integration of the automation solutions through communication between the distributed components also plays an important role. Immense additional savings are achieved by means of vertical integration by optimizing the process of product planning, development, manufacturing, and sales. This optimization is realized through a consistent data flow, permanent data consistency, and the availability of data on the field level, control level, and office level.

The use of standardized interfaces by several manufacturers is a prerequisite for the flexible compilation and integration of software components.

OPC is now generally accepted as one of the most popular industrial standards among users and also among developers. Most of the Human Machine Interface (HMI), Supervisory Control and Data Acquisition (SCADA), and Distributed Control System (DCS) manufacturers in the field of PC-based automation technology, as well as the manufacturers of Soft PLCs, are offering OPC client and/or OPC server interfaces with their products. The same is true for suppliers of devices and interface cards. In the last few years, OPC servers widely replaced Dynamic Data Exchange (DDE) servers and product-specific drivers in this field.

Today, OPC is the standard interface for access to Windows-based applications in automation technology. Most of the OPC Specifications are based on the Distributed Component Object Model (DCOM), Microsoft's technology for the implementation of distributed systems. Besides the DCOM-based communication in future, in the context of new OPC concepts, more and more data will be exchanged via Web Services.

OPC specifications define an interface between clients and servers as well as servers and servers for different fields of application — access to real time data, monitoring of events, access to historical data, and others. Just as any modern PC can send a print task to any printer, thanks to the integration of printer drivers, software applications can have access to devices of different manufacturers without having to deal with the distinct device specifications. OPC clients and servers can be combined and linked like building blocks using OPC technology.

At present, OPC clients and servers are mainly available on PC systems with Windows 9X/Me/NT/2000/XP and x86 processors. Due to the availability of Web Services for multiple operating systems, the use of OPC components in different environments and in embedded systems will become more important in future.

Why is OPC so successful? The approach of the OPC Foundation has always been to avoid unnecessarily detailed discussions and political disputes and to create practical facts within a very short time.

The development of the Data eXchange specification can serve as an example. More than 30 companies joined the effort and delivered the specification 18 months after starting.

OPC has succeeded in defining a uniform standard worldwide, which has been adopted by manufacturers, system integrators, and users.

62.3 History of OPC

Since reusable software components made their entry into automation technology and replaced mono-lithic, customized software applications, the question of standardized interfaces between components has increased in significance. If such interfaces are missing, every integration is connected with cost-intensive and time-consuming programming supporting the respective interface. If a system consists of several software components, these adaptations have to be carried out several times.

Following the immense distribution of Windows operating systems and their coherent Win32-API in the PC area, different technologies were created to enable communication between software modules by means of standardized interfaces. A first milestone was DDE, which was complemented later on by the more efficient technology Object Linking and Embedding (OLE). With the introduction of the first HMI and SCADA programs based on PC technology between 1989 and 1991, DDE was used for the first time as an interface for software drivers to access the process periphery.

During the development of Windows NT, the DCOM was developed as a continuation of the OLE technology. Windows NT was rapidly accepted by industry. In particular, the highly expanding HMI, SCADA, and DCS systems were made available for NT.

With the increased distribution of their products and the growing number of communication proto-cols and bus systems, software manufacturers faced more and more pressure to develop and maintain hundreds of drivers. A large part of the resources of these enterprises had to be set aside for the develop-ment and maintenance of communication drivers.

In 1995, the companies Fisher-Rosemount, Intellution, Intuitive Technology, Opto22, Rockwell, and Siemens AG decided to work out a solution for this growing problem, and they formed the OPC Task Force. Members of Microsoft staff were also involved, and supplied technical assistance.

The OPC Task Force assigned itself the task to work out a standard for accessing real-time data under Windows operating systems, based on Microsoft's (OLE/)DCOM technology OLE for Process Control, or OPC. The members of the OPC Task Force worked intensively, so that already in August 1996 the OPC Specification Version 1.0 [3] was available. In September 1996, during the ISA Show in Chicago, the OPC Foundation was established; it has been coordinating all specification and marketing work since then.

An important task of the OPC Foundation is to respond to the requirements of the industry and to consider adding them as functional extensions of existing or newly created OPC specifications. The strat-egy is to extend existing specifications, to define basic additions in new specifications, and to carry out modifications with the aim of maximum possible compatibility with existing versions. In September 1997, a first update of the OPC Specification was published in the form of version 1.0A [4]. This specifi-cation was no longer named "OPC Specification" but, more precisely, Data Access Specification. It defined the fundamental mechanisms and functionality of reading and writing process data. This version also served as the basis for the first OPC products, which were displayed at the ISA Show 1997. Consideration of further developments in Microsoft DCOM and industry requirements led to the creation of the Data Access Specification version 2.0 [5] in October 1998.

Already, rather early, after the release of version 1.0A, it could be seen that there was a need for the specification of an interface for monitoring and processing events and alarms. A working group formed to solve this problem worked out the Alarms and Events Specification, which was published in January 1999 as version 1.01. Version 1.10 of the Alarms and Events Specification is available since October 2002 [9].

In addition to the acquisition of real-time data and the monitoring of events, the use of historical data offers another large field of application in automation. The work on the Historical Data Access Specification already began in 1997 and was completed in September 2000 [11].

Defining and implementing security policies for use with OPC components is also of great impor-tance. A corresponding specification has been available since September 2000 and is titled OPC Security Specification [15].

In particular, from the field of industrial batch processing, additional requirements have been for-warded to the OPC Foundation, which have led to the OPC Batch Specification [13].

During work on version 2.0 of the Data Access Specification and the other specifications, it emerged that there are elements common to all specifications. These elements have been combined in two specifications. The document OPC Overview [1] contains explanatory aspects only, while the OPC Common Definitions and Interfaces Specification [2] contain normative definitions.

With the increasing implementation of the OPC specifications in products, and their application in multiple environments, further requirements arose. New working groups have been created by the OPC Foundation. The Data Access 3.0 working group extended the existing Data Access Specification with further functionality. The specification [7] has been available since March 2003. The OPC and XML group defined a way to read and write data by using Web Services, and enables the use of OPC components via the Internet and on operating system platforms without DCOM. The specification Version 1.0 [16] is available since July 2003. The OPC DX working group defined a specification for server-to-server communication without using a client. It is available since March 2003 [8].

The rapid growth in the number of OPC products, from only a few in 1997 to some thousands in 2003, shows the enormous acceptance of this technology. OPC has succeeded in developing from a concept to an industrial standard within only three years.

62.4 OPC — An Overview

OPC is the technological basis for the convenient and efficient link of automation components with control hardware and field devices. Furthermore, it provides the condition for integration of Office products and information systems on the company level such as Enterprise Resource Planning (ERP) and Manufacturing Execution Systems (MES).

Today, OPC, on the one hand, is based on Microsoft's DCOM; on the other OPC XML-DA uses the concept of Web Services. The DCOM describes an object model for the implementation of distributed applications as per the Client–Server Paradigm. A client can use several servers at the same time and a server can provide its functionality to several clients at the same time.

At the core of DCOM is the term interface. DCOM objects provide their services through interfaces. An interface describes a group of related methods (functions).

The most diverse OPC components of different manufacturers can work together. No additional programming for adaptation of the interfaces between the components is necessary. Complex correlation, for example, dependencies of the software component on hardware components, remain concealed behind this abstract interface. Complete components (hardware and software) can be exchanged, provided the interface described in the specifications is supported.

The OPC standards are freely accessible technical specifications that define sets of standard interfaces for different fields of application in automation technology. These interfaces allow a highly efficient data exchange between software components of different manufacturers.

Figure 62.1 shows the currently available specifications or those under work and their relations to each other. These specifications concern different fields of application and are thus largely independent of one another. However, it is possible to combine them in one application.

Areas of OPC Use

What is the industrial environment for using OPC products?

Today, and especially in the future, information drives more and more production. This information exists in different forms in various devices at different levels of production. And this information must be available in different forms at different places. OPC provides a way to access and deliver data at the different levels. Figure 62.2 shows an example for the use of OPC in real applications.

OPC technology is well established in a number of industries (energy, building automation, chemical engineering, ...).

OPC technology can be used to immediately monitor and influence the production process. Process information is visualized, and control information is sent to the devices.

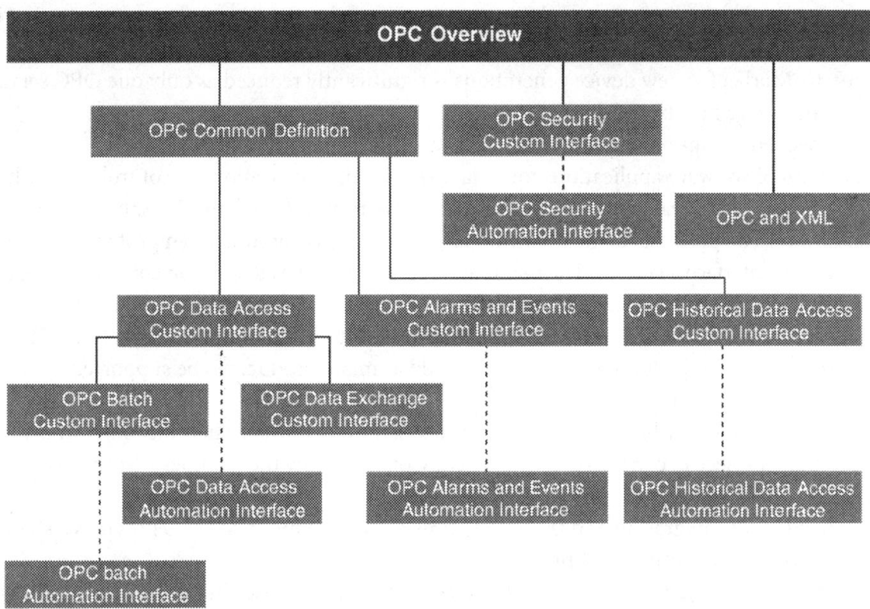

FIGURE 62.1   Available and in-progress specifications.

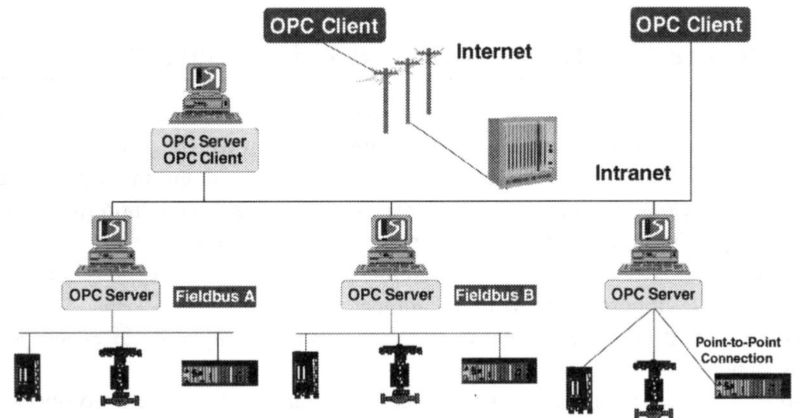

FIGURE 62.2   Examples for the use of OPC in real applications.

In other areas (administration, planning), there is more interest in aggregated information (machine use per hour, ...).

There are a number of OPC specifications and products that can be used in various areas (production, administration, planning).

62.5   OPC — Advantages for Manufacturers and Users

For hardware manufacturers, for example, manufacturers of devices (PLC, barcode reader, measurement devices, embedded devices ...) or PC interface boards (fieldbus interfaces, data acquisition systems, etc.), usage of OPC technology provides a number of advantages:

The product can be used by all OPC-compatible systems in the market and is not limited to an individual system for which a corresponding solution (i.e., specific drivers) must be developed. Due to the existence of standardized interfaces and the interoperability related to them, there is no need to become familiar with the specific requirements of other systems.

The Time-to-Market for new device generations is significantly reduced as only one OPC server has to be updated, not a large number of drivers.

The effort needed for support is also reduced as less products have to be supported.

Manufacturers of software applications for data acquisition, visualization, or control benefit like hardware manufacturers by a clear encapsulation of the software interface from the specific features of the accessed hardware. The product can be used with all devices and communication protocols on the market that make an OPC interface available. The manufacturer no longer has to develop corresponding solutions (specific drivers). Due to the existence of standardized interfaces and the interoperability related to them, there is no need to become familiar with the specifications of other devices and communication protocols.

The time needed for support is considerably reduced as many products to be supported (product-specific drivers up to now) no longer exist.

Using the OPC technology brings much benefit to system integrators. Their flexibility in the choice of products for their project is considerably increased. Consequently, the number of projects that can be processed increases considerably.

The time needed for integration and training is considerably reduced, as OPC provides a standardized interface that remains the same for all products.

Last but not the least, many advantages are the result for the end user by the usage and the huge distribution of OPC: OPC provides additional flexibility (distribution of components, use of new technologies, choice between products, etc.) during the design of the overall system as products of various manufacturers can be combined.

62.6 Structure and Tasks of the OPC Foundation

An important prerequisite for the success of a standardization initiative is an authority coordinating the interests of the members involved. The task of this authority is to protect the common objective from the political interests of individuals. Specification work has to be initiated and marked with clear mission statements in order to avoid a proliferation of variants and derivatives. Furthermore, public relations have to supply the market with information and support the common standard.

The OPC Foundation was founded in 1996 as an independent nonprofit organization with the aim of further developing and supporting the new OPC standard.

Besides specification development, other tasks are distributed to different offices and persons within the OPC Foundation.

The Board of Directors is the Foundation's decision-making body. It is elected once a year at the General Assembly by the members entitled to vote.

Once a year, the Board of Directors appoints the persons who safeguard the interests of the Foundation between the general meetings.

The Technical Steering Committee (TSC) establishes working groups for several target projects. It consists of representatives of the same companies as the Board of Directors plus the chairmen of the working groups of the OPC Foundation.

There are two kinds of OPC Foundation members: the OPC technology users and the OPC technology providers. The technology provider companies are further categorized as profit and nonprofit organizations. The latter do not have voting rights.

The annual membership fees for technology users and nonprofit organizations are independent of their size. The annual membership fees for the technology providers depend on their annual turnover.

In March 2003, the OPC Foundation had over 300 members worldwide, from North America, Europe, and the Far East.

The OPC Foundation has an Internet Web site. At www.opcfoundation.org, visitors can find information regarding the organization, its members, the working groups, and current events. Furthermore, visitors can download released specifications and technical reports in the form of "white papers." An electronic product catalog allows visitors to search for OPC subjects under several headings, such as

FIGURE 62.3   Logo to be used by OPC Foundation Members.

manufacturer, client, server, development tool, and training. In addition, it is possible to exchange questions and opinions in discussion forums.

From the central Web site of the OPC Foundation, the links www.opceurope.org, www.opcjapan.org, and www.opcchina.org lead to the sites of the European, Japanese, and Chinese subcommittees.

The OPC Foundation provides a membership application form on its Web site. Members of the OPC Foundation may use the logo shown in Figure 62.3 for public relation activities.

Specification work counts among the tasks of the OPC Foundation with the greatest importance. The specification process has to be clearly defined, progress has to be monitored, and the results have to be released.

First of all, the Board of Directors defines the specification issue in the form of a mission statement and appoints the chairman of the working group. The chairman addresses the member companies of the OPC Foundation and asks them to cooperate. Interested companies then appoint members. The working group meets several times and prepares the specification. Creation of a sample code, which proves the principal implementation possibility and use of the new specification, is effected in parallel. The specification and the sample code are then passed to the Technical Steering Committee for approval. If approved, the specification and code are submitted to the Board of Directors for release. Otherwise, they are returned to the working group for further processing.

62.7   Technological Basis of OPC

As it has already been mentioned, today's (and future) OPC specifications are based on two technologies — DCOM and Web Services. That is why, a short introduction into these technologies is provided, before explaining the specifications.

DCOM

The DCOM describes Microsoft's solution for the implementation of distributed, object-oriented applications in heterogeneous environments. A component object is the basic component of such applications. It has one or more interfaces providing methods that permit access to data and functionality of the object (Reading and writing of data, accessing properties, adding and deleting of objects).

One or more component objects belong to one server component providing a large number of services. To make use of these services, a client accesses methods at interfaces of the server's component objects. The individual services to be provided are described in specifications (e.g., OPC specifications). The structure of the component objects, of their interfaces, methods, and parameters is defined in an IDL file, which describes the contract between a client and server. DCOM ensures binary interoperability at runtime. A client can query at runtime whether the server supports a certain interface.

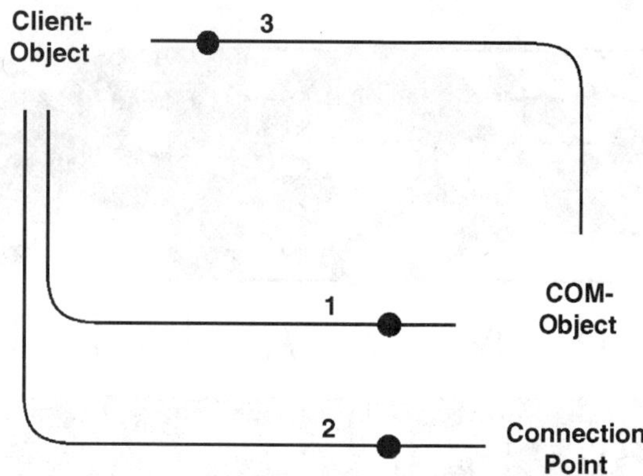

FIGURE 62.4   Creation and use of a callback connection between server and client.

A number of identifiers are used to designate server and interfaces uniquely, for example, CLSID (ClassID) and IID (InterfaceId). The life cycle of an individual component object and of the server is managed via reference counters.

DCOM permits the implementation of interactions between client and server as well as between server and client (callback). This possibility is provided by using connection points. Figure 62.4 shows the relations. The client queries the server whether it supports a specific connection point (1). In this case, the client transmits a reference to an interface on its side to the server (2). The server can later call methods at this interface (3).

62.8   XML, SOAP, and Web Services

There are a large number of products that implement DCOM-based OPC Specifications. But there are also some restrictions that have to be considered during the development and use of this kind of products.

- DCOM does not pass firewalls, that is, direct addressing of computers through the firewall is not possible. However, this is precisely what DCOM needs to perform an internal check.
- There are some devices and applications that provide or require data and that do not run on Microsoft systems. They include, for example, applications in the ERP or MDS areas as data consumers or Embedded Devices as data sources.

These restrictions have been the reason why OPC foundation has started the OPC XML-DA specification effort. This specification is no longer based on DCOM, but on a technology independent of a specific operating system. Since this specification is explained later, the relevant components of this technology will be introduced below.

The eXtensible Markup Language (XML) is a flexible data description language, which is easy to comprehend and learn. Information is exchanged by means of readable XML documents. An XML document is called well-formed if it corresponds to the XML syntax; it is called valid if, in addition, it corresponds to a default schema. The creation of XML documents and schemas as well as validation and processing of the files are supported by a variety of tools.

Today, the support of XML is guaranteed by practically all systems. Thus, even heterogeneous systems can easily interact by exchanging XML documents.

The Simple Object Access Protocol (SOAP) is an interaction protocol that links two technologies: XML and HTTP. HTTP is used as the transport protocol. The parameters of the interactions are described with XML. SOAP is thus predestined specifically for the Internet. SOAP is a protocol independent of object architectures (DCOM, CORBA). A SOAP telegram consists of a part describing the structure of the HTTP call (request/response, host, content type, and content length). This part is included in all the HTTP telegrams. A UniversalResourceIdentifier (URI) was added, which defines the end point and the method to be called. The method parameters are transferred as XML. The programmer is responsible for mapping the SOAP protocol to a concrete implementation. In the meantime, SOAP has been submitted to the World Wide Web Consortium (W3C) for standardization. In this context, the name has changed to XML protocol. Version 1.2 has been available since December 2001.

Based on the technologies introduced above, it is already possible to implement distributed applications that interact via SOAP and are independent of the operating system and the hardware. However, something is still missing: a way of describing an application's interface and of generating program components from this description that are, on the one hand, compliant with the existing infrastructure (HTTP etc.) and that, on the other, can be integrated in existing programs. This is where Web Services come into play.

The World Wide Web is used increasingly for application-to-application communication. The programmatic interfaces made available are referred to as Web Services. SOAP is used as the interaction protocol between components.

Web Services are described using XML. The language used is Web Services Description Language (WSDL), which is standardized in the W3C. An application interacting with the Web Service will deliver a valid XML message that is compliant with the schema. The function call is sent as an XML message. This also happens with the response and a possible error information. Components that support or use Web Services can be implemented on any platform supporting XML and HTTP.

The introduced technologies were not defined by individual companies or company groups, but by the W3C.

This fact is also of importance for future OPC specifications. In the past, the fact that OPC is only based on DCOM has been criticized. This point should disappear if specifications are based on XML and Web Services.

Table 62.2 shows the current status of the different specifications. Recommendation stands for agreed standard and draft stands for a standard that is still not agreed. Note stands for a rather detailed working paper.

TABLE 62.1   OPC Specifications — Contents and Release Status (Status July, 2003)

Specification	Contents	Release Status
OPC Overview [1]	General description of the application fields of OPC specifications	Release 1.00
OPC Common Definitions and Interfaces [2]	Definition of issues concerning a number of specifications	Release 1.00
OPC Data Access Specification [7]	Definition of an interface for reading and writing real-time data	Release 3.0
OPC Alarms and Events Specification[9]	Definition of an interface for monitoring events	Release 1.1
OPC Historical Data Access Specification [11]	Definition of an interface for access to historical data	Release 1.1
OPC Batch Specification [13]	Definition of an interface for access to data required for batch processing. This specification is based on the OPC Data Access Specification and extends it	Release 2.0
OPC Security Specification [15]	Definition of an interface for setting and utilization of security policies	Release 1.0
OPC XML-DA Specification [16]	Integration of OPC and XML for the building of Web applications	Release 1.0
OPC Data eXchange (DX) Specification [8]	Communication between server and server in process	Release 1.0
OPC Complex Data [17]	Definition of possibilities to describe the structure of Complex Data and of ways to access this type of data.	Release 1.0

TABLE 62.2   XML Specifications — Release State (July 2003)

XML 1.0	W3C Recommendation
XML Schema Part 1 and 2 1.0	W3C Recommendation
SOAP/XMLP 1.2	W3C Recommendation
WSDL 1.1	W3C Draft

62.9 OPC Specifications

OPC Overview [1]

As it has already been mentioned, there are several OPC specifications for different applications in automation technology. All specifications describe software interfaces. The existing specifications and their relationships are shown in Figure 62.1.

The "OPC Overview" contains general, nonnormative guidelines for OPC. It contains, for example, facts about OPC applications and OPC basic technology.

OPC Common Definitions and Interfaces Specification [2]

Before the specifications for data accessing are explained, some remarks related to the content of the OPC Common Definitions and Interfaces Specification will be made.

In the preceding part of the chapter, some information about the history of OPC was provided. The specification discussed here came into being when the Data Access Specification was being prepared together with other specifications. During this process, the OPC Foundation members realized that there are some definitions relevant to all specifications. They are summarized in this specification and comprise:

- Functionality to be provided by all servers. This includes the possibility of adapting the server to the geographical area of application (setting the language for the textual messages from the server to the client).
- The procedure of server recognition. Entries in the registry database contain information necessary for starting the server. The entries are shown in Figure 62.5 and explained below. A client will search the database to obtain this information, which is simple in the local registry, but difficult on remote computers. A component offering this functionality was specified and made available.
- The procedure of installation. There are several components (proxy/stub) used together by all servers and clients of one specification. These components must be available on the computer as long as OPC products are used.

An OPC Server is characterized by the following registry entries:

- *ProgId*: Every DCOM Server is characterized by a Program Identifier (ProgId). Rules exist regarding how to generate this Identifier. But these rules do not guarantee that this identifier is unique. Below the ProgId key, a specific key OPC exists. This entry is used to differentiate between OPC Servers and other DCOM servers.
- *CLSID*: This 128-bit-long numerical identifier uniquely describes a DCOM Server, that is, also an OPC Server. There is a way of generating this number that makes it unique. The implementers will use the generated number for exactly one server. The LocalServer32 key contains a reference to the location where the server executable can be found. The Implemented Categories keys contain information regarding which specification is implemented by a server. This is used by a client or the Server Enumerator.
- *AppId*: The Application Identifier contains more information about the server. This includes security settings. The AppId can, but must not be the same as the ClassId.

Data Access Specification [4, 5]

Data Access Servers permit transparent read and write access to any kind of values. These values can be made available by field devices and fetched via different communication systems. Also, servers permitting access to hardware and software (plug-in cards, other programs) in the PC are possible.

OPC specifications are supposed to support interoperability and plug'n'play. One prerequisite is that a client can obtain information on the available values in the server very conveniently. For this purpose, a namespace and functionality for browsing the namespace were defined. This functionality is implemented in the server and used by the client.

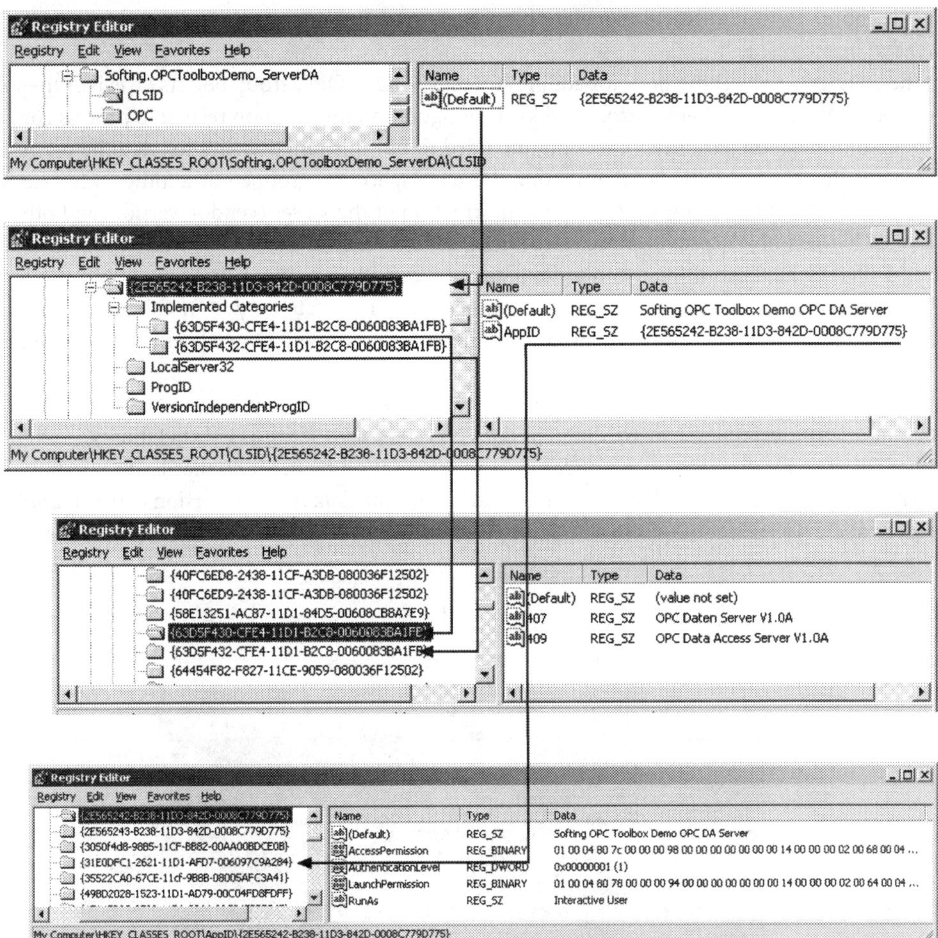

FIGURE 62.5   Registry keys for OPC Servers.

A client is not necessarily interested in all values. Different clients may want to register values under different aspects, for example, all temperature values. Therefore, the specification defines different COM objects organized in a hierarchy. By creating the specific objects, the client can adapt the server to its requirements.

Figure 62.6 shows the components that form part of a Data Access application — OPC Data Access Client and Data Access Server with namespace and object hierarchy.

A namespace can be hierarchical or flat. It can be a result of the commissioning process for the server. The specification does not define how the namespace has to be created, how many hierarchical levels the namespace may have, or how the nodes and leaves are to be designated. Only methods at an interface are defined, which enable the client to read out the information.

The namespace is identical for all clients having access to the same server. The object hierarchy, in contrast, is specific to the client. After the server is started, the client has access to an interface of the OPCServer object. It can create OPCGroup objects and, in this way, determine how access to values is structured. The value that a specific client is interested in is represented by an OPCItem object. For the OPCGroup objects, the client can have OPCItem objects created by the server, which permit access to data. OPCItem objects have no interfaces. This is due to the requirement that several values have to be efficiently read and written at the same time. The necessary functionality is available at interfaces of the OPCGroup object.

A server can also call methods at interfaces of the client. It can, for instance, inform the client of process value changes.

Furthermore, methods with parameters have been specified for the various objects. Methods are grouped at interfaces.

At the OPCServer object methods for adding and removing of OPCGroup objects do exist. Other methods can be invoked by the client to store and load configuration information related to process communication (communication parameter). It is not intended to use this functionality to store the recent object hierarchy of the server. Methods for browsing the namespace are grouped at another interface of this object. The client can furthermore access state information of the server (vendor, version, and others).

The OPCGroup object provides access to methods that can be used to add and remove OPCItem objects. Different methods to read and write data are grouped at different interfaces. This also includes methods supporting the creation of callback connections. The state of the OPCGroup object can be obtained and influenced by various methods. Parameter values can be changed, which influence the data are acquired from the process.

One standardized format for exchanging data between server and client is necessary for interoperability. In automation technology, many different data types are used (IEC 61131, fieldbuses, visualization, ...). The OPC specification creates a standardized representation by defining that DCOM data types are used when values are exchanged between server and client. The data type conversion between application data types and DCOM data types by the server and the client is application-specific. The data format

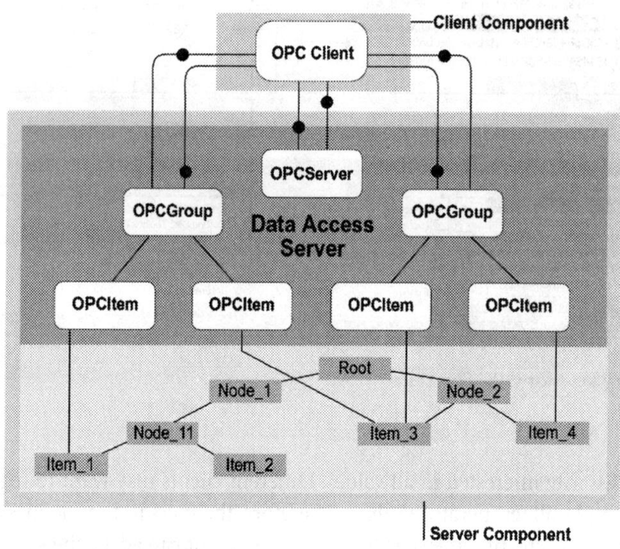

FIGURE 62.6 Components of a Data Access application.

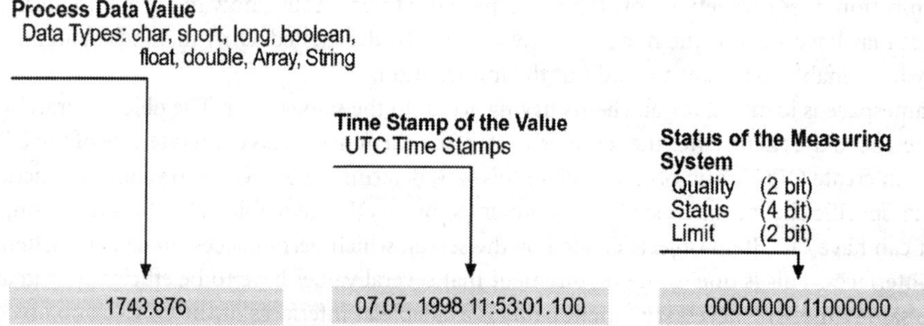

FIGURE 62.7 Data format.

shown in Figure 62.7 contains in addition to the process value also a time stamp and quality information. The time stamp can be either generated in the server or, if it already exists, it can be transferred from the device. The quality information contains a description of the value validity (good/bad/uncertain). The quality is described in more detail by the status value (e.g., bad — not connected).

For the data exchange between client and server, there are a number of requirements that the specification takes into account in defining different types of data exchange.

A client is supposed to read or write values. For this purpose, it can use synchronous and asynchronous read and write requests. Reading can take place for the server cache or for the device (place where the process value is created, e.g., the device connected by a communications link). With synchronous calls, the request is processed completely before information is sent to the client. Under certain circumstances, this may not be efficient. Therefore, there is also a possibility of asynchronous reading and writing. Here, the client calls the specific method at the corresponding interface of the OPCGroup object and passes some information. The server then processes the request and provides the client with the remaining information at a later point in time.

Both types of request processing are also available for writing. However, a client can only write to the device and not to the cache.

For synchronous and asynchronous read requests, the client always has to address the corresponding OPCItem objects in the OPCGroup object. With a refresh, values of all OPCItem objects of an OPCGroup object can be requested either from the cache or from the device. Here, addressing is implicit.

In the procedures mentioned above, the client is always active — it polls for data. However, a type of data exchange where the server automatically transmits values to the client is required as well. The transmission is based on the evaluation of relevant default values. This procedure is also supported in the specification.

After this short overview about the Data Access Specification some functionality will be explained in more detail in the next paragraphs.

As already mentioned, the namespace contains all data points provided by the server. From these points, the client selects those in which it is interested and requests the server to create OPCItem objects for them. As a criterion for the assignment of points in the namespace to OPCItem objects in the server, fully qualified ItemIds are used. They normally consist of sections of the namespace that uniquely identify a data point. In a hierarchical namespace, a fully qualified ItemId will therefore contain the identifier of the leaf representing the data point as well as one or more identifiers of nodes in the namespace. The different identifiers are separated by server-specific delimiters.

The client will first query the structure of the namespace and, in the next step, identifiers for leaves and nodes. By setting method parameters, the client can influence the return values. Different values for one parameter determine whether only node identifiers or leaf identifiers as well are to be passed by the server. Another parameter contains filters. They are used as a criterion, which leaf identifiers must be passed. A limitation in terms of data type, access rights and character string is possible. The client can navigate in the namespace by indicating node identifiers and thus informing the server that this is the current view of the namespace. From a specific node, the client will finally query all identifiers for the items. In a real application, many different types of information may be relevant for clients. This concerns all values that change or are to be changed by a client. However, some rather static values are also of interest to clients (manufacturer, revision number of devices, description of measurement methods, telephone number of maintenance staff, etc.). If all this information were mapped to leaves in the namespace, the latter might become very large. Also, this information on devices and values (manufacturer, version...) might occur repeatedly. For efficient access to this information, properties were introduced with version 2.0 of the Data Access Specification.

As already mentioned, the client has access to an interface of the OPCServer object right after the server component was started. After browsing the namespace, the client will start creating a corresponding object hierarchy in the server. To do this, it can generate one or more OPCGroup objects for structuring data access. Later, the client can assign all values to be read or written with a request to OPCGroup objects (i.e., create OPCItem objects). By defining three parameter values, the client determines how values are to be acquired automatically by the server. With an update rate (in msec), the client defines at which rate values are to be read and written into the cache. The value of PercentDeadband determines the conditions under which values are automatically sent to the client. An OPCGroup object can have the state "active" or "inactive." If the latter is the case, the values of the OPCItem objects are not obtained automatically.

In the last step, the client creates the OPCItem objects for the different OPCGroup objects, which can also be "active" or "inactive." This determines whether or not they are included in automatic data acquisition.

The object hierarchy and object properties can be changed at any time.

The client is able to read data from the server by invoking synchronous or asynchronous calls. A more efficient way to obtain data from the server will be explained in the remaining part of this section.

First, the client has to create the desired object hierarchy and to build the callback connection. This is used to pass values from the server to the client. Besides the hierarchy of the object their behavior also influences the method of data exchange explained below. It works only if both OPCGroup and OPCItem objects have "Active" state.

Timeliness and sensitivity of the data access are influenced by the parameter values "UpdateRate" and "PercentDeadband" of the appropriate OPCGroup object. The "UpdateRate" (msec) value defines how often values of (active) OPCItem objects are automatically read. The "PercentDeadband" value (in%/100, e.g., 0.01) influences the sensitivity of the data exchange.

Besides changes of the value, state changes for the values are also reasons for passing data automatically from the server to the client.

A number of OPCGroup objects can be created of course at the same time, which support this way of data exchange.

The diagram in Figure 62.8 explains the automatic data exchange in detail.

The namespace was configured for the server. For different variables, the EngineeringUnit Type and EngineeringUnit information were defined. The following explanations will only be valid if "analog" is indicated as EU type and the value is of a simple data type.

The client has created an active OPCGroup object and assigned values for UpdateRate (1000 ms in the example) and PercentDeadband (0.1 in the example). Furthermore, the client has created an active OPCItem object for a temperature value. After all objects were generated and the callback connection exists, a value is immediately returned to the client via the callback connection. This value (44) is used as a reference value for the following calculations in the server. According to the given update rate, the temperature value is read from the process every 1000 ms. For each value that is read, a calculation takes place according to the following algorithm: first, the absolute difference between the values of the EU information is calculated (30 in the example). After creation of the OPCGroup object, this value is multiplied by the percent deadband (in the example, $0.1 \times 30 = 3$). For every scanning process, the absolute difference between the value last transmitted to the client and the current value is calculated and compared

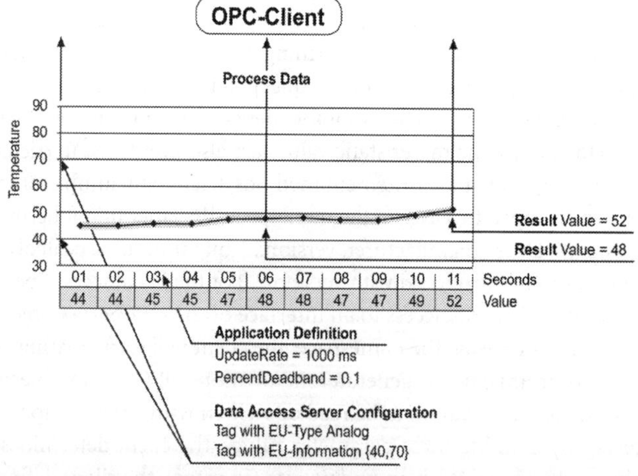

FIGURE 62.8 Automatic data exchange between server and client.

with the other calculated value. If the former is larger than the latter (after 6 sec, the result for the example is 4>3), the scanned value is sent to the client and used as a new reference value.

As has already been mentioned, the procedure in this form works only for values with simple data types and EU type "analog." For other EU types (e.g., "discrete") and structured variables, the value is transferred with each change that takes place. The sensitivity of the data transmission can be influenced by the EU information for every value or by the percent deadband for all OPCItem objects of an OPCGroup object. If the criterion applies to several OPCItem objects, all values are transferred with one call.

OPC Data Access 3.0 [7]

In Version 3.0 of the Data Access Specification, some new functionality and enhancements are defined. Products implementing Data Access 3.0 have to be able to interact with products implementing Data Access 2.0 or Data Access 1.0A, of course. The server is characterized by a CategoryId (refer to chapter Common Definitions and Interfaces). That is why the client already has a hint, as to which functionality can be used.

The following specification issues are new for Data Access 3.0:

- Data can be read and written without the need to create OPCGroup and OPCItem objects. This concerns applications, where Data Access Server are used as an IO layer together with PC-based Control Systems.
- Deadband and SamplingRate can be set at OPCItem object level in addition to the settings for the OPCGroup object. This provides for a more precise setting as the way to set the value only at the OPCGroup level.
- Browser interface, with a functionality comparable to OPC XML-DA. It makes it easier to implement browsing at the client side.
- Connection monitoring functionality has been added to the specification.

OPC XML-DA [16]

The use in the Internet is a case of application for OPC XML-DA products. Therefore, the following conditions had to be taken into account when the specification was written:

- The interaction parameters are coded using XML, which leads to an overhead.
- Interactions take place by HTTP, which is a stateless protocol. Due to the desired scalability of the servers and the line costs, interactions in the Web mainly take place on a short-term basis. A client fetches information from the server; after this, the server "forgets" about the client. Some solutions implement state but this is a specific approach.

Therefore, the Data Access Specification model with an object hierarchy for each client and with callbacks cannot be applied to a Web Service.

An OPC XML-DA Service is stateless. There is no functionality for the creation of objects as defined in the Data Access Specification.

During browsing as well, no information about the position of the client in the namespace is stored in the OPC XML-DA Service, but all information about the namespace (or a defined part of it) is transferred to the client at the same time. The client can poll for values at the server, but it should also be possible, to receive changed values automatically.

Nevertheless, the definition of subscriptions does provide the service with state information. The service must know which data the client is interested in, at what rate they have to be recorded (UpdateRate) and when they have to be transferred (deadband). Since, however, a server cannot call the client on its own initiative with HTTP (it does not know about the client), the specification defines a query of the subscription values, which is initiated by the client. If nothing has changed, the client will not receive any values.

Another important point is monitoring of the connection between client and server and, in case of a subscription, monitoring of the client's availability. Therefore, most function calls contain time values indicating the maximum time that the client will wait for a response from the server or the minimum number of times it will call the server (subscription).

Table 62.3 contains the defined methods.

How does the subscription work in OPC XML-DA?

The sequence of function calls is shown in Figure 62.9 and explained below.

To set up a subscription, the client sends a request indicating the variables it is interested in as well as the values for RequestedSamplingRate and Deadband. These two values can be defined both for a number of variables or single variables in one call. The SubscriptionPingRate defines with what frequency the service is supposed to check whether the client still exists.

The service responds with a subscribe response that contains the handle for the callback and the supported SubscriptionPingRate. Based on RequestedSamplingRate, the server acquires the values and decides based on Deadband regarding whether to send it to the client or not.

By sending a SubscriptionPolledRefreshRequest message to the server, the client requests for the data for a subscription. It receives the data with SubscriptionPolledRefreshResponse. In the request, the client can tell the service how long it will wait for the response and, in this way, determine how long the connection will remain open. The server will delay the sending of the response accordingly. The service will at least wait the Holdtime to send the SubscriptionPolledRefreshResponse to the client. It will continue to acquire data if no data have changed for another period — the WaitTime. If a value has changed, the service will immediately send the SubscriptionPolledRefreshResponse. If WaitTime also expires and

TABLE 62.3   Methods of the OPC XML-DA Specification

GetStatus/GetStatusResponse	Client obtains server status
Browse/ BrowseResponse	Client obtains name space information
GetProperties/GetPropertiesResponse	Client obtains property information
ReadRequest/ ReadResponse	Client reads data
WriteRequest/ WriteResponse	Client writes data
Subscribe (Client)/ SubscribeResponse (Server)	Client establishes subscription
SubscriptionCancel/Response	Client cancels subscription
SubscriptionPolledRefresh(Client)/	Client initiates requests for the values that are provided
SubscriptionPolledRefreshResponse(Server)	in the subscription

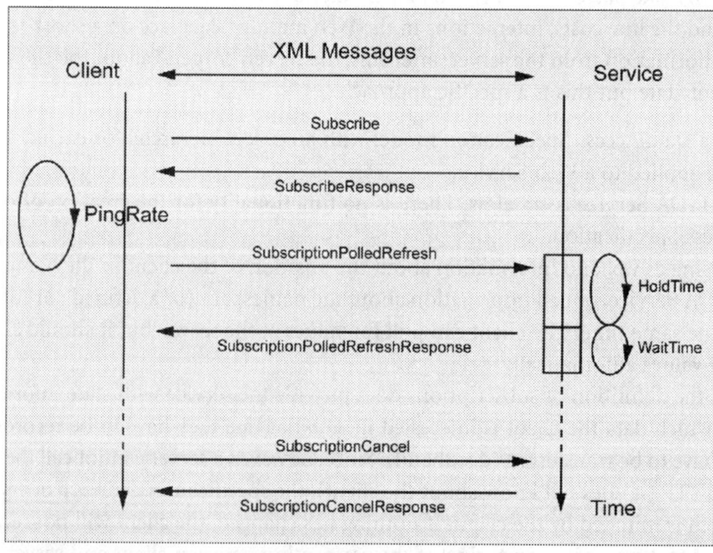

FIGURE 62.9   Subscription based on OPC XML-DA specification issues.

nothing has changed, the service will send an "empty" SubscriptionPolledRefreshResponse, that is, a message without data.

The client will cancel a subscription by passing a SubscriptionCancelRequest message. The service responds with SubscriptionCancelResponse message.

OPC Data eXchange Specification [8]

Why do we need the OPC DX Specification?

As already mentioned, OPC has a huge installed base. There are cases when OPC Servers have to directly interchange data, for example, if information must be transferred from one area of production to another area. Currently, transferring data between servers is only possible using clients or proprietary bridges, which slows down the exchange and excludes overall solutions.

There are a number of fieldbus systems that use TCP/IP on Ethernet as their transport and network protocol. All of them use the same medium, but cannot directly intercommunicate. Here also, OPC DX is supposed to offer a solution.

Figure 62.10 shows the structure of an OPC DX application. A DX Server can receive data from one (or more) DA Server(s) or from one (or more) DX Server(s). For this purpose, a DX Server has implemented DA Client functionality. By means of this client, the DX Server creates OPCGroup and OPCItem objects. The data point from which the data are recorded is called Sourceitem; the data point where they are written is called Targetitem. This linkage and their property are called "connection." All connections taken together are called "configuration." Connections are defined by a configuration client and mapped in the namespace of the DX Server. A monitoring client monitors the connections. Data Access functionality is used for this purpose. Other clients (visualization) can access the existing items in a "normal" way.

The specification distinguishes between the configuration model and the run-time model.

The configuration model defines the semantics of the connections as well as possibilities of creating, deleting, and changing connections and transferring events.

The run-time model defines the data transfer between DX and DA/DX Servers. Here, the manner in which the data are recorded as well as error and status monitoring are important.

The information on connections is stored in a defined branch of the namespace The attributes of a connection and, consequently, the structure of this branch are specified. A Data Access Client is able to browse the name space and create OPCItem objects for values of interest. A DX server therefore also implements Data Access Server functionality besides Data Access client functionality.

Connections are created, modified, and deleted by different commands issued by a configuration client.

The specification not only defines the configuration but also the behavior of a DX Server at runtime, that is, during horizontal data exchange.

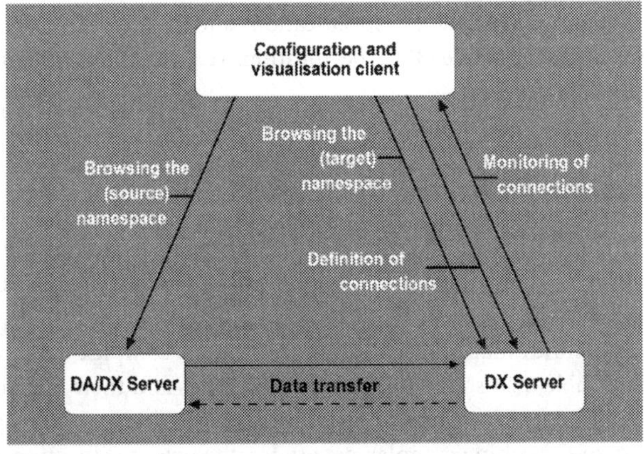

FIGURE 62.10   Structure of an OPC DX application.

The DX Server can subscribe to values from a Data Access Server that supports the specification versions 2.04, 2.05, and 3.0. Another possibility is for the DX Server to support the OPC XML-DA Specification and to receive data from OPC XML-DA Servers.

The target item and the source item can have different data types. During creation of the OPCItem object, the DX Server can request the DA/DX Server to perform the conversion. If this is not supported, the DX Server must perform the conversion itself.

The specification also defines the run-time behavior of a DX Server if the connection to the source server is interrupted. In this case, the DX Server tries to reestablish the connection with the frequency of a ping rate, which can be set as desired. If this is not possible, this information is mapped to the corresponding attributes of the target item. Depending on the properties of the connection, the DX Server can set a substitute value instead of the actual value.

A monitoring client can sign up for the corresponding items in the configuration branch of the namespace and thus obtain information on the state of the connections.

The DX Server stores the entire current configuration at runtime. This means that all results of the calls for adding, modifying, and deleting connections are stored. The specification does not define how and where this is done. During starting of the DX Server, the current configuration is reloaded and available again.

Complex Data Specification [17]

The Data Access Server supports access to simple and complex data. A temperature value is an example of simple data, whereas records of diagnostic information serve as examples of complex data. Both data types (simple and complex) can be read or written in OPC Data Access applications. Yet, the existing OPC specifications do not define the means for a client and a server to exchange structural information about complex data. Therefore, the client can only pass complex data as octet strings to other applications (e.g., database and visualization applications).

The following behavior is requested:

- The client understands the structure of the data. In this case, the client may forward individual elements of the complex data item to other applications.
- The client understands the structure of the data and knows its semantics. In this case, the client is not only able to distinguish between elements but also knows about their type and relations with each other.

The description of complex data type is a prerequisite for this behavior.
The new specification proposes two approaches for the description of complex data:

- type descriptions defined within OPC specifications and
- type descriptions defined outside OPC specifications (e.g., Fieldbus organizations)

OPC-type descriptions are concerned with the definition of the structure of complex data. This allows the client to know the elements of complex data items. The type description is effected using XML. This description system is not very flexible, but it provides for easy implementation.

The type description systems of other organizations can also contain semantic definitions and allow a client to know the semantics. These descriptions can be more flexible but they increase the requirements for implementation, as client and server should be able to understand a number of type description systems.

The new specification defines that information on and about complex data items must be provided using defined properties. Table 62.4 shows the defined properties and their meanings.

Only items with complex data types must support these properties.

Besides the possibility of describing complex data types, the new specification also defines the behavior during writing of complex data.

TABLE 62.4 Properties Used to Describe Complex Data

PropertyId	Name	Meaning
109	Complex Data Type Description System	Identifies the type description system used (e.g., OPC, Fieldbus Consortium)
110	Complex Data Type Description ID	Identifies the type description of a complex data item (e.g., reference to an XML file, reference number related to a type description system of consortia)
111	Complex Data Type Description Version	Identifies the version of the type description
112	Complex Data Type Description	A BLOB that contains the information necessary for clients to interpret the value of the complex data item

Not all elements of the complex data item, once written, may be writable. In this case, the client supplies the entire buffer in a write request and the server ignores the values of nonwritable elements.

In a case where the value for one or more writable elements is invalid, or cannot be applied, the server rejects the entire write request.

OPC Alarms and Events [9]

By implementing Data Access Specification definitions, values can be automatically transmitted from the server to the client if these values have changed within a certain time period or if the values' states have changed. This is not sufficient or not efficient regarding the following requirements:

- It is not only important to be informed about changes of the value, it is also important to be informed that a parameter has exceeded a certain limit.
- The value changes sometimes during the UpdateRate.
- Alarms have to be acknowledged.

The Alarms and Events specification has been written to fulfill these (and other) requirements. The next chapter explains the model specification issues and their practical use.

The specification model differentiates between things that happen (simple events and tracking-related events) and things that exist (condition-related events, i.e., alarms).

The occurrence of an alarm can be acknowledged. An object hierarchy has been defined, where a client can adapt the server to its requirements.

The client normally will not be interested in all events that can be monitored by a server. It can select the interesting events by applying filter. The filter consists of parts from the event area and the filter space. The event area provides a topological structuring of events. The filter space provides event structuring based on event attributes (event type, event category, event severity, ...). Both can be configured during adapting the server to the real environment.

Figure 62.11 shows the components that are part of an Alarms and Events application — OPC Alarms and Events Client and Server. Events can be structured in an event area that is always hierarchical. In the server itself, different DCOM objects with interfaces and methods must be implemented. After launching the server, the client has access to an interface of the OPCEventServer object. Then, the client can either create the OPCEventAreaBrowser object to obtain information on the content of the event area, or it can create OPCEventSubscription objects. These objects are used to monitor event sources and, in case of events, transmit relevant messages to the client. Events are assigned to the OPCEventSubscription objects via filters.

This specification also permits both directions of interaction (client → server and server → client).

Filters are used for assigning events to OPCEventSubscription objects. In this way, a client determines which events are to be monitored by the server. Values of filter parameters are derived from information that concerns the event area (fully qualified AreaId, fully qualified SourceId) or properties of events (type, category...).

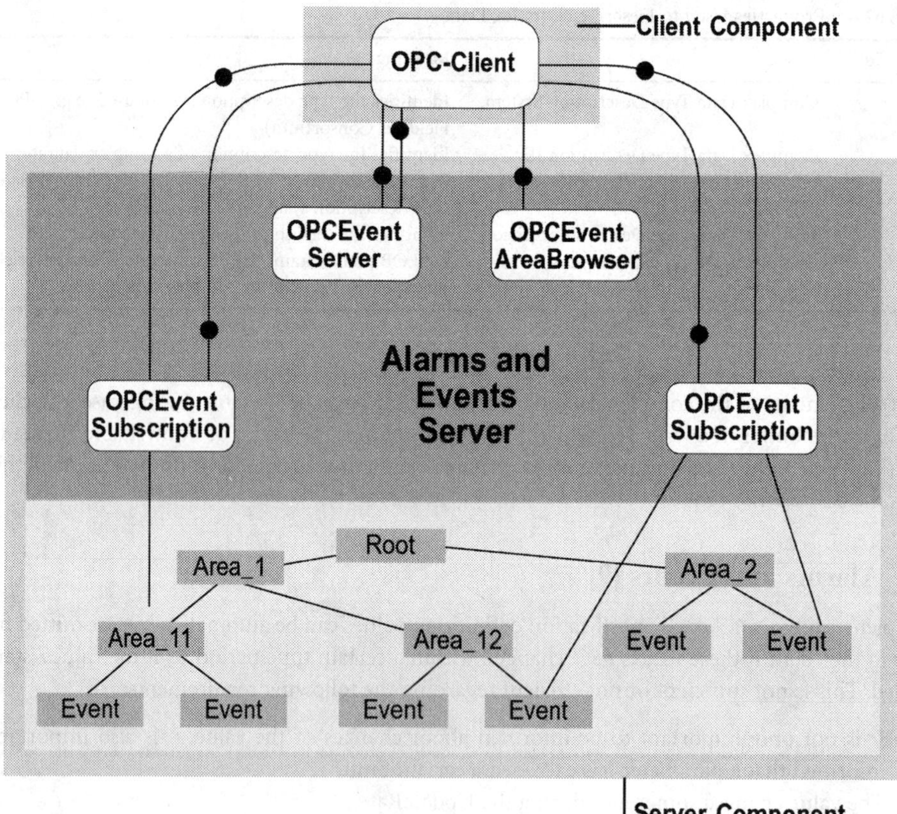

FIGURE 62.11 Components of an Alarms and Events application.

There are three types of events already defined in the specification. An example of a simple event would be a device failure, and that of a tracking-related event, the changing of a setpoint. An example of a condition-related event would be a positive or negative deviation from temperature limit values. For all these three types, there are different categories. The specification proposes only some settings for this value. More categories can be defined based on real application scenarios. For condition-related events, the different categories have conditions and subconditions. The specification already contains some proposals for categories, conditions, and subconditions. Events also have a priority. If necessary, application-specific priorities must be converted into OPC priorities.

A large variety of filters can be defined, for example, <type=condition-related events, category=level, priority=500–700, area=area_1, and source=source_A>.

In case of event occurrence, the server sends an event notification to the client, which has assigned the event to an OPCEventSubscription object via a filter. Depending on the type of event, the notification can have different numbers of parameters. Figure 62.12 shows the mandatory parameters for the various event types. Examples in the figure are related to an event notification informing the client about the occurrence of a condition-related event.

During the notification about a simple event, values for the parameters Source (information from the event space), Time (when did the event occur?), Type, Event Category, Severity, and Message are transferred. The language of the message can be set by means of a method at the general interface mentioned in Part "Common Interfaces and Definitions."

For notification about a tracking-related event, there is the additional parameter ActorId. It contains a numerical identifier to indicate the cause of the event. The creation of the identifier and the meaning of the value are not described in the specification.

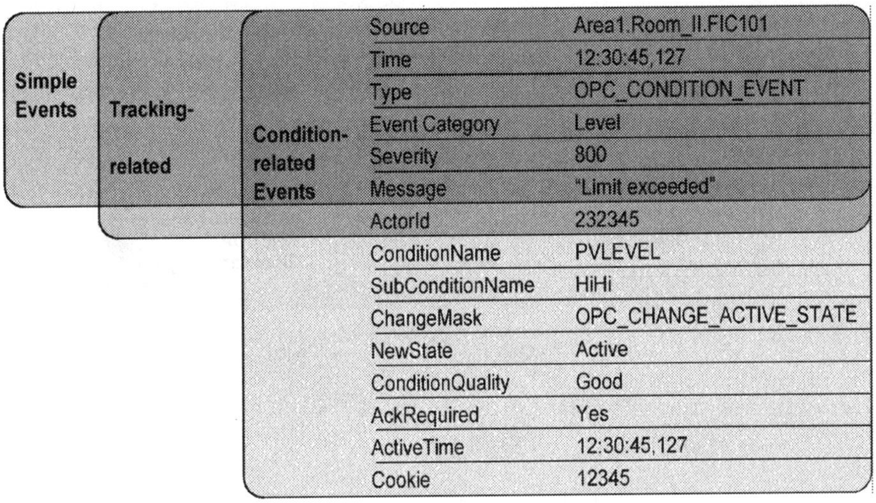

Source	Area1.Room_II.FIC101
Time	12:30:45,127
Type	OPC_CONDITION_EVENT
Event Category	Level
Severity	800
Message	"Limit exceeded"
ActorId	232345
ConditionName	PVLEVEL
SubConditionName	HiHi
ChangeMask	OPC_CHANGE_ACTIVE_STATE
NewState	Active
ConditionQuality	Good
AckRequired	Yes
ActiveTime	12:30:45,127
Cookie	12345

Simple Events — Tracking-related — Condition-related Events

FIGURE 62.12 Structure of an event notification.

The largest number of parameters must be transferred for notification about a condition-related event. The following parameters are added to the ones already mentioned:

- *ConditionName*: Name of the condition from the event area.
- *SubConditionName*: Name of the subcondition, if any.
- *ChangeMask*: Indicates in what way the state of the condition has changed, for example, inactive → active.
- *State*: State of the condition, for example, active, acknowledged, ….
- *ConditionQuality*: The parameter can be compared with the quality of a value from the Data Access Specification. A notification is also sent if the state changes.
- *AckRequired*: The event must be acknowledged.
- *ActiveTime*: Indicates the time at which the state became active. This value is not identical to "Time" since the latter indicates the time of event occurrence. The receipt of an acknowledgment also leads to an event. After the receipt, the values of the parameters differ.
- *Cookie*: Used by the client in the acknowledge and by the server to relate the acknowledge to the event.

There may be other attributes in addition to these mandatory ones. They have to be supported for all events of a category, that is, they are defined for a category.

OPC Historical Data Access [11]

The Historical Data Access Specification defines an API to access historical data. Historical data have been collected over some time frame and is now available. The specification does not define how the data are collected and stored. The recent specification only considers parameter values; access to stored event information is not specified.

Accessing data by using Data Access clients and storing it in a database could be one way to collect data later accessible for Historical Data Access Server. Other possibilities could include other ways of accessing data. Another use of case of interest would be access to data that has been stored over some time in a measuring device and is accessed only once a week, for example, environmental parameter. Figure 62.13 shows the object hierarchy and the relation to the namespace.

Different from Data Access Server, Historical data Access Server will provide access to larger amounts of data. This is caused by the larger number of variables (all variables in a process) and by the amount of

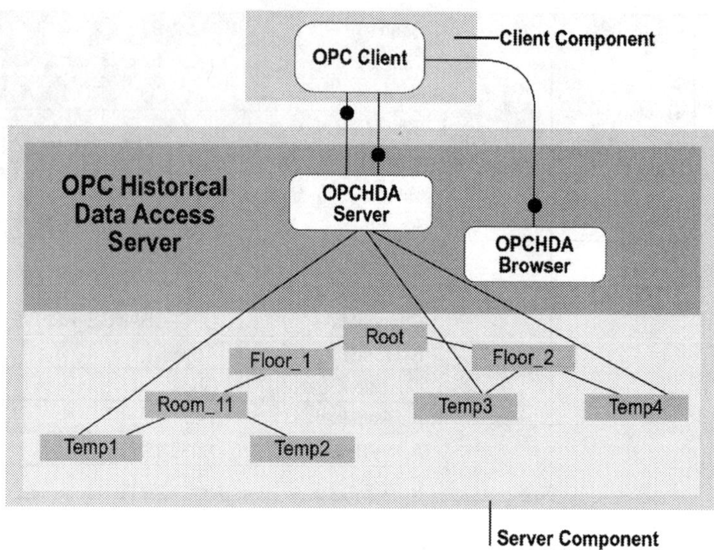

FIGURE 62.13   Components of an HDA application.

values per variable. Stored values do not change. But they can be deleted and new values or attributes can be added. Different from Data Access Server, the access to aggregated values plays an important role (average over a time frame, smallest value over a time frame, ...).

Only two DCOM objects were specified for the server. Immediately after launching of the component, the client has access to the OPCHDAServer object where the entire functionality is available. The OPCHDABrowser object is used for searching the namespace of an HDA Server. This namespace contains all data points for which values are available. Other than with the Data Access Server, there is no explicit object for the access to data points; this is not necessary since access takes place very rarely. For the client, methods for synchronous and asynchronous reading, recording, and changing entries in the database of historical data are available.

The tables in Figure 62.14 show a few aggregates and attributes for stored raw data. For both parameters, the specification already contains default values. The server manufacturer can also provide other possibilities.

An HDA Client can use four different ways to access historical data:

- *Read*: Using this approach, the client can read raw values, processed values, values at a specific point in time (AtTime), modified values, or value attributes. The client can invoke synchronous or asynchronous calls.
- *Update*: The client can insert, replace, or insert and replace values. This can be done for a time frame or for specific values. Synchronous and asynchronous calls are also available.
- *Annotation*: The client can read and insert annotations. Again, this can be performed synchronously and asynchronously.
- *Playback*: The client can request the server to send values with a defined frequency for a defined time frame (every 15 sec values stored over 15 min). This can be applied both for raw and aggregated data.

OPC Batch [13]

The OPC Batch Specification defines an interface between clients and servers for a certain type of application called batch processing. With this procedure, recipes are processed by resources and reports are generated. Functionality that is made available by Data Access Specification is required. For batch

Raw Data

	1	2	3	4	5	6	7	8	9	10
Temp1	18	18	19	20	21	20	19	18	17	18
Temp2	17	17	18	19	20	17	17	17	18	18
Temp3	16	17	18	18	18	17	18	19	19	18
Temp4	17	18	18	19	20	20	20	20	19	19

Attributes

	Temp1	Temp2	Temp3	Temp4
Datatype	VT_I2	VT_I2	VT_I2	VT_I2
Description	Temperature in Room11	Temperature in Room12	Temperature in Room13	Temperature in Room14
NodeName	128.7.16.123	128.7.16.123	128.7.16.124	128.7.16.124

Aggregates

	Temp1	Temp2	Temp3	Temp4
Average	18,8	17,8	17,8	19
Delta	0	1	2	2
Minimum	17	17	16	17
Maximum	21	20	19	20
Range	4	3	3	3

FIGURE 62.14 Relations between Raw Data, Attributes, and Aggregated Data.

processing, there is the international standard IEC 61512-1 defining a model for this kind of data exchange. Many products in this area have been implemented accordingly. Therefore, the idea was to combine the existing functionality of the Data Access Servers with the specifications in this standard. The consequence was an extension of the Data Access Specification, which defines a special namespace and some additional methods adapted optimally to the conditions during batch processing.

Figure 62.15 shows a part of the namespace that is specific to a server used in batch processing. The namespace is defined to be always hierarchical. Specific nodes exist on the highest hierarchy level (OPCBPhysicalModel, OPCBBatchModel and OPCBBatchList). These nodes represent models of IEC 61512-1. Batches contain parameters and results. For each batch, there are corresponding OPCBParameters and OPCBResults nodes. The parameters and results are represented by specific properties.

OPC Security [15]

OPC Clients and Server can run on different PCs. Security aspects have to be considered in such applications. Security settings can be applied in two different ways:

- by using the utility program "dcomcnfg" and/or
- by using the security API functionality of the operating system.

By using "dcomcnfg," security settings are applied to the components, that is, the server or to the overall system. If smaller granularity is necessary (settings for objects, methods, data, ...), the security API must be used. It is the goal of the Security specification to foster interoperability between security-aware applications, that is, applications that use the security API.

The Security Specification defines a model for security, different levels of security, and possibilities of how the client and the server can exchange security information.

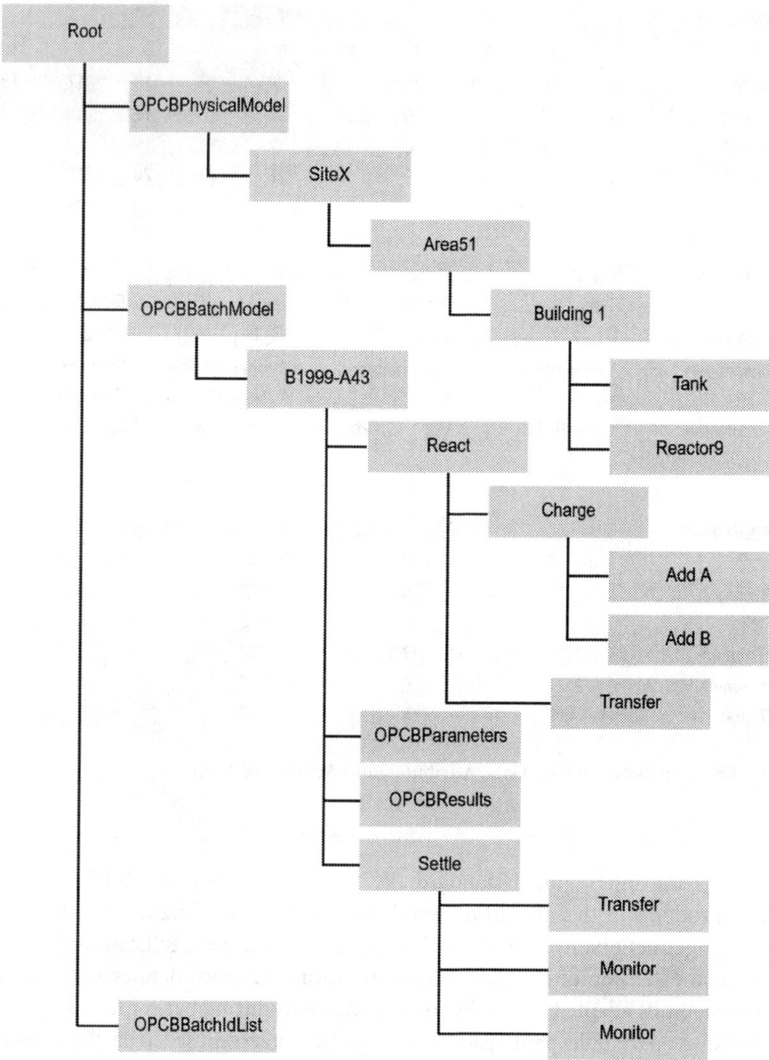

FIGURE 62.15   OPC Batch Server — part of a namespace.

The specification does not define which objects are to be secured and how.

Figure 62.16 shows the Windows NT/2000 security model on which the specification is based.

The model differentiates between principals described by access certificates and security objects. All processes in Windows NT/2000 are principals. If a user logs in, a process is started. Depending on which group a user belongs to (administrators, guests, etc.) and on the corresponding access rights, an access certificate is assigned to the process. This is a kind of "ticket" describing the properties of the principal. On the other hand, there are security objects. These are objects to which access is monitored. If a principal tries to access a security object, the reference monitor decides, using the access control list, whether the principal may do that. If "Dcomcnfg" is used, the ACL is edited by means of this tool, and the reference monitor is part of the DCOM run-time environment. If COM Security API methods are used, the reference monitor and creation of the ACL must be implemented as parts of the server. In this case, security objects of any granularity can be implemented.

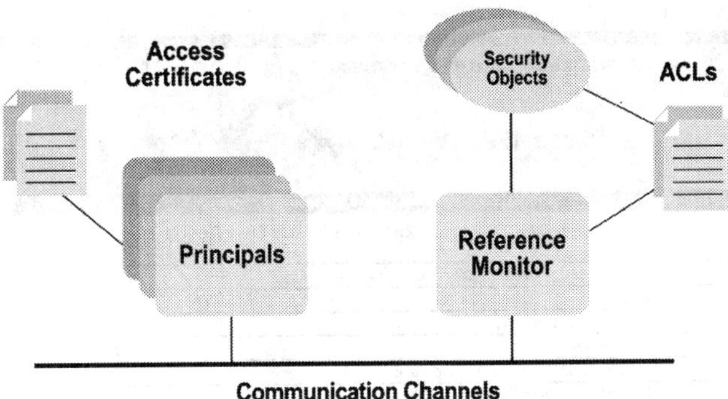

FIGURE 62.16   Windows Security model.

Furthermore, a differentiation is made in the specification regarding the way to characterize principals. Security decisions can be based on access certificates for NT user (NT Access Token) or specific OPC user. NT users are already known in the system. For each of these different certificates, one interface has been specified.

Compliance Test

The OPC Foundation defines compliance testing as a way to verify whether a server implementation conforms to the specification. There are no test tools to check clients!

Compliance testing is based on the following prerequisites:

- *Test cases*: A huge number of test cases have been defined, by generating a number of different parameter values for the method calls. These parameter values can be used to verify the server's behavior in response to valid, invalid, and impossible method calls. Default and extreme values have been defined for the various parameter.
- *Test system*: A Compliance Test Client has been developed by the OPC Foundation.
- Test procedure: During running the Compliance Test Client against a server, test result files are generated. These have to be submitted to the OPC Foundation. It will derive the result. The information is added to a page on the OPC Foundation Web site.

The Compliance Test Client supports the following test possibilities:

- *Stress tests*: Here, it can be determined that a selectable number of objects can be added and removed in the server component to be tested.
- *Logical tests*: Here, it is tested whether the Proxy/Stub-DLLs are installed correctly.
- *Interface tests*: Here, methods are tested at the interfaces of the objects. Valid, invalid, and meaningless parameter values are used for all the methods.

Configuration of the Compliance Test Client has to be carried out prior to starting the test run.

Now, the test run can be started. The results of the proceeded test cases are displayed. If test cases have not been passed successfully, the implementation of the server has to be changed. At the end of the test, after passing all test cases successful, a binary result file will be generated by the test client. This file must be sent to the OPC Foundation. The result is then published on the OPC Foundation Web site. Figure 62.17 shows an example.

Table 62.5 shows the current availability of compliance tests for the different specifications.

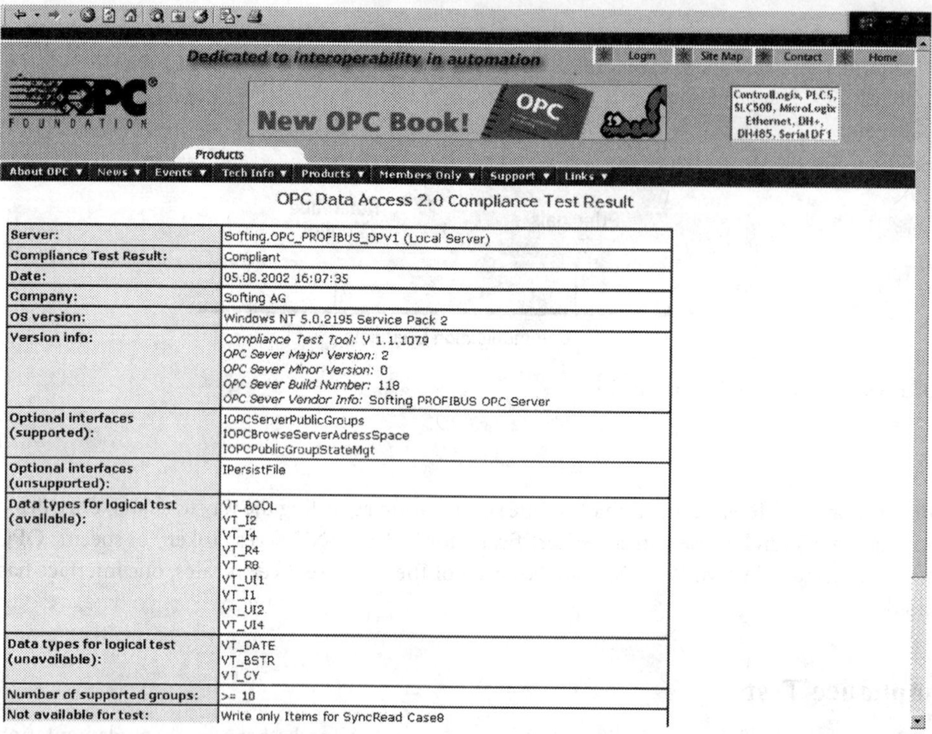

FIGURE 62.17   Test result as published on the OPC Foundation Web site.

TABLE 62.5   OPC Compliance Test — Release States

Specification	Compliance Test Availability
DA 2.05	V2.0.2.1105
DA 3.00	Beta
AE 1.1	V1.0.2.1105
HDA 1.1	Beta
DX	Test Specification Release Candidate
XML-DA	Test Specification Release Candidate

62.10   Implementation of OPC Products

With the release of the OPC XML-DA and OPC DX Specifications, not only the implementation of DCOM based Client and Servers have to be considered, it also becomes important to think about implementation based on Web Services.

There are some major differences between these two approaches:

- DCOM is available mainly on Microsoft systems. Portation exists for Unix systems, but only a few OPC products for this area are available.
- OPC XML-DA products can be easily made available everywhere where HTTP and XML are supported. Even if XML support does not exist, products can be implemented. Coding and decoding of XML messages can be implemented without a parser.

In the following paragraphs, first implementation of DCOM-based OPC products is considered and then Web Service-based solutions are considered.

OPC DCOM Server Implementation

OPC Servers can be implemented in three types:

- *InProc Servers*: The server is implemented as a DLL and runs in the process space of the client. In this manner, high-speed data access is possible since no process or computer boundaries have to be overcome. There will only be restrictions if the server needs to access a protected kernel resource.
- *OutProc Servers*: The server is implemented as a stand-alone executable. It can run on the same computer as the client or on a different one. The advantage is that several clients have access if protected resources are used. Since there are process and maybe even computer boundaries between client and server, data traffic is slower than with the InProc solution.
- *Service*: This is a special kind of OutProc Server. Services can be configured in such a way that they are immediately started during booting without the necessity of user log-in. Thus, the initialization of communications links is shifted to the booting process of the computer. The advantage of this solution is that the server is available immediately after booting. The disadvantage is that services are only available for Windows NT/2000/XP. On other operating systems (Windows 98/ME), they run as a "normal" OutProc server.

OPC DCOM Client Implementation

An OPC Client must provide functionality in two phases of the life cycle of an automation system (during commissioning/configuration and on-line operation). It must be possible to:

- Find and start available OPC Servers during the configuration phase. The namespace/eventspace of the servers must be browsed to obtain fully qualified item identifier necessary to add OPCItem objects and to receive filter criteria. The same is true for properties. ValidateItem checks whether the passed parameters would result in a valid OPCItem object. To accomplish this, OPCGroup objects must also be able to be added. The client must query for supported filter criteria. The result of the configuration process can be stored somewhere. Access to other methods also needs to be implemented.
- At runtime, the OPC Client requests that all objects have to be created in the OPC Server. The OPC Client supplies the application with the requested data and monitors communication. During data exchange, it must also be possible to start additional Servers, add and manipulate OPCGroup/OPCEventSubscription objects, as well as add and manipulate OPCItem objects.

Creating OPC DCOM Components by Means of Tools

As far as DCOM knowledge is seen as core competence, the development of OPC clients and servers can be done by means of different developing environments. Detailed knowledge of DCOM and the OPC specifications is necessary for this.

A simpler way to create OPC components is to use toolkits. Toolkits encapsulate the DCOM and OPC functionality. They exchange information through an API with an application using the toolkits. Toolkits are available for client and server components. Most toolkits support Windows 9X/NT/2000/XP. Versions for Windows CE and Linux also exist.

The toolkits can be distinguished by their possibilities of optimization. There are compact solutions intended for a simple server type and more complex toolkits that permit extremely flexible solutions.

Characteristic of complex toolkits is that they can be combined with each other resp. are based on each other. With the OPC Toolbox shown in Figure 62.18 components can be developed that contain both Data Access and Alarms and Events functionality, which can be used on standard Windows operating systems but also on Windows CE.

FIGURE 62.18   OPC Toolkit sample.

There are toolkits for different use cases that are performed as a C++ class library, so-called C++ toolkits and those that provide a more or less complete OPC component in which only some few function calls must be integrated. ActiveX Controls is a special kind of OPC client implementation: the complete OPC Client functionality is provided as OPC Client Controls and can be used in Visual Basic, Excel, or other ActiveX container applications without any programming.

Implementation of OPC XML Servers and Clients

Three approaches can be distinguished for the implementation. They are described below.

- Starting from scratch — the most demanding approach.
  This approach mainly has to be used for impl ementing OPC products on embedded devices. If HTTP and XML support is available, the infrastructure components for receiving and sending of SOAP telegrams and coding of XML messages have to be implemented. These components must be combined with the remaining parts of the client and server application. If this is not available, the necessary parts have to be ported. There are a number of solutions available in the public domain.
- Using toolkits supporting WSDL.
  Toolkits exist in the market that generate the necessary infrastructure components and templates to integrate with parts of client and server application based on the WSDL file.
- Using OPC toolkits — the easiest approach.
  These toolkits not only implement infrastructure components but also internal logic of client and server application.

62.11   Outlook into Future

The huge acceptance and fast penetration of the OPC technology in many areas of automation technology are unique. Many standards, for example, fieldbuses in industrial communication or in the area of PLC programming have established in one decade or more or not at all. OPC has been much more successful. With the defined specifications, multiple areas of application from fast and equidistant data transfer (Data Access) via processing of large amounts of historical data (Historical Data Access) up to the collection and acknowledgement of volatile and critical events (Alarms and Events) can be covered.

The availability of Data eXchange and OPC XML-DA opens up further areas of application in vertical integration and peer-to-peer interaction. On the basis of the specifications, clients and servers can run in the same process on one PC (InProc Server) or in different processes (OutProc Server) on PCs that can be close together or far apart. One or several clients can access a server. A client, in turn, can communicate with one or more servers. With the availability of Web Services-based products, OPC can be used in Embedded Devices or any other kind of devices supporting this technology.

There are a variety of products that translate the OPC specifications into reality. The scope ranges from sample clients and servers to tools for creating OPC components and servers for a wide variety of communication systems and devices. A multitude of visualization systems, systems for data acquisition, and diagnosis have been equipped with OPC Client interfaces. These products are used in a vast number of actual applications. A large variety of programming languages and development environments are available for the creation of OPC components. OPC products can be used on a wide range of hardware and software platforms.

Solutions for the compliance test are available. The use of successfully tested products raises the confidence of end users in product quality and interoperability.

A nonprofit organization, the OPC Foundation is in charge of marketing and developing the technology.

These points characterize the current actual state described in this handbook. One question still remains to be answered — what does OPC's future look like?

62.12 The Future of OPC

There are two issues that drive the future of OPC:

- OPC XML-DA is available. The market will request to also provide other specifications on this technological basis, for example, OPC XML-AE.
- The market will request for migration paths between technologies and solutions.

Microsoft pushes .NET. More and more potential OPC client providers will go this way. A request for an interoperable .NET interface for accessing the legacy DCOM OPC server arises. OPC XML-DA servers and

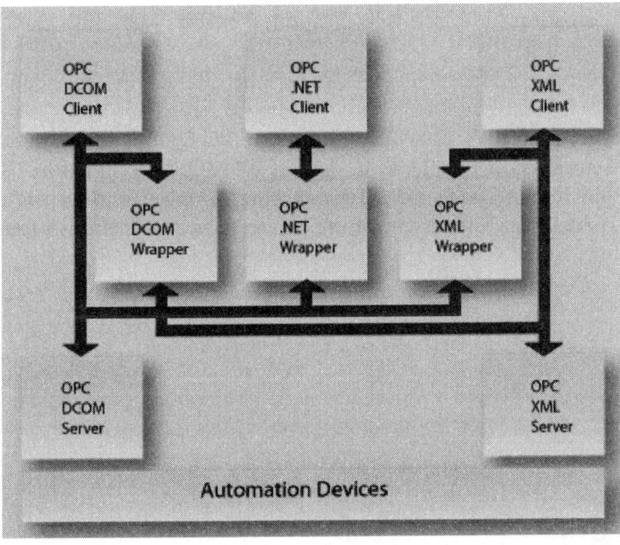

FIGURE 62.19 Migration paths for technologies and specifications.

clients on different platforms will be available in the future. Legacy DCOM clients and servers should be used in mixed applications. Wrappers between these components are necessary. Figure 62.19 shows the different constellations and the necessary wrapper. The OPC Foundation provides a skeleton for a .NET to DCOM wrapper. It implements the DCOM object view and makes it accessible from .NET languages. Wrapper between DCOM and Web Services and vice versa will be provided by different vendors.

It can be foreseen that many more OPC products will be developed in the future and the success story of OPC will continue.

References

1. OPC Overview, Version 1.0, October 27, 1998, OPC Foundation, www.opcfoundation.org
2. OPC Common Definitions and Interfaces Version 1.0, October 27, 1998, OPC Foundation, www.opcfoundation.org
3. OPC Specification Version, Version 1.0, September 1996, OPC Foundation, www.opcfoundation.org
4. OPC Data Access Specification Version 1.0A, September 1997, OPC Foundation, www.opcfoundation.org
5. OPC Data Access Custom Interface Standard Version 2.05a, December 2001, OPC Foundation, www.opcfoundation.org
6. OPC Data Access Automation Interface Standard Version 2.02, February 1999, OPC Foundation, www.opcfoundation.org
7. OPC Data Access Custom Interface Standard Version 3.00, March 2003, OPC Foundation, www.opcfoundation.org
8. OPC Data eXchange Specification Version 1.00, March 2003, OPC Foundation, www.opcfoundation.org
9. OPC Alarms and Events Custom Interface Standard Version 1.10, October 10 2002, OPC Foundation, www.opcfoundation.org
10. OPC Alarms and Events Automation Interface Standard Version 1.01, (Draft) December 15, 1999, OPC Foundation, www.opcfoundation.org
11. OPC Historical Data Acces Custom Interface Standard Version 1.10, January 2001, OPC Foundation, www.opcfoundation.org
12. OPC Historical Data Acces Automation Interface Standard Version 1.0, January 2001, OPC Foundation, www.opcfoundation.org
13. OPC Batch Custom Interface Specification, Version 2.00, July 2001, OPC Foundation, www.opcfoundation.org
14. OPC Batch Automation Interface Specification, Version 1.00, July 2001, OPC Foundation, www.opcfoundation.org
15. OPC Security Custom Interface Version 1.00, October 2000, OPC Foundation, www.opcfoundation.org
16. OPC XMLDA Specification Version 1.00, July 2003, OPC Foundation, www.opcfoundation.org
17. OPC Complex Data Version 1.00, Draft 02 August 2002, OPC Foundation.
18. Iwanitz Frank and Jürgen Lange, *OPC Fundamentals, Implementation and Application*, 2nd ed., Hüthig Verlag, Heidelberg, 2002. www.softing.com/en/communications/products/opc/book.htm

63

Java Technology and Industrial Applications

Jörn Peschke and Arndt Lüder
Otto-von-Guericke University Magdeburg

63.1 Introduction New Programming Paradigms in Industrial Automation

Industrial automation is currently characterized by a number of trends induced by the current market situation. The main trends are the pursuit of high flexibility, good scalability, and high robustness of automation systems, the integration of new technologies, and the harmonization of used technologies in all fields and levels of automation. Of special interest is the integration of technologies, which were originally developed for the office-world into the control area. This trend is characterized by the emergence of industrial PCs, operating systems for embedded devices like WindowsCE, embedded Linux, and RTLinux, data presentation technologies like XML, and communication technologies like Ethernet and Internet technologies.

In this context, object-oriented languages like Java have gained importance. Particularly Java, which has evolved with the Internet and related technologies, fits very well in different areas of industrial automation.

In the following sections, the boundary conditions, advantages, and problems of using Java in the area of industrial automation will be described in more detail. A special focus will be on the lower levels of the automation pyramid, where real-time requirements are of significance.

63.2 Requirements in Automation and Typical Application Areas of Java

Java can be characterized as a high-level, consequently object-oriented programming language with a strong type-system. One important feature is the idea of a Virtual Machine abstracting the concrete underlying hardware by translating Java programs to an intermediate language called Bytecode, which is executed on the Java Virtual Machine (JVM). Thereby, the concept "Write Once, Run Anywhere" (WORA) is realized, enabling a platform-independent application design. This means that Java applications are (of course, under consideration of different Java versions and editions) executable on every platform, which can run a JVM. Together with typical concepts of object-orientation, like encapsulation of functionalities in classes, inheritance, and polymorphism, Java opens many possibilities for a reusability of the code. Java also provides high stability of applications. This is realized by extensive checks (e.g., regarding type or array boundaries) during compile-, load- and runtime. Due to the fact that error-prone concepts like direct pointer manipulations are beyond the scope of the language and memory allocation (and deallocation) is realized by an automatic memory management, the so-called Garbage Collection (GC), the efficiency of software development with Java is very high. Although it is not easy to specify this in figures, some sources state at least 20% improvement compared to C/C++ [9]. In contrast to the most used PLC programming languages specified in the IEC 61131, the efficiency increases by more than 50%. The extensive, easy-to-use networking abilities of Java can help to reduce the difficulty of programming distributed systems.

Besides general advantages, the typical potential application areas for Java in industrial automation and their specific requirements have to be considered. On the upper levels of the automation pyramid (ERP and SCADA/MES-Systems), Java is already used as one alternative. The requirements in this area are very similar to typical IT-applications and characterized by powerful hardware. Here, Java2SE (Standard Edition) and Java2EE (Enterprise Edition) with a wide variety of APIs supporting different technologies for communication, visualization, and database access are the proper Java platforms. Examples for Java applications can mainly be found for different interface realizations of ERPs and the implementation of advanced technologies on the MES level [11].

At the field device level, Java is still not a very common language as the requirements here are a problematic issue for standard Java versions. Many features of Java, normally responsible for typical advantages, cause problems at the field device level. The hardware on this level is very heterogeneous and most devices have only limited resources with respect to memory and computing power. For communication purposes, several different field bus protocols are used, although the increasing relevance of Ethernet-based protocols shows potentials for a common communication medium [15].

For indicator and control elements, Java provides powerful APIs for User-Interface programming (ATW and Swing) and there are no special requirements regarding computing power. In contrast to this, for control devices like PLC, SoftPLC, or IPC, there are constraints that do not fit in every case to the features of standard Java. These devices are characterized by:

- real-time requirements for application parts that realize control functionalities (cyclic control program execution with a defined cycle time),
- direct hardware access to the I/O level,
- today's usage of PLC typical programming concepts (e.g., IEC 61131).

A special case are smart I/O-devices where the computing resources are even more limited than in normal PLCs.

63.3 Problems of Using Java at the Field Level Under Real-Time Conditions

It can be stated that for the use of standard Java in non-real-time conditions with limited resources, several problems have to be solved, even if the advantages mentioned above make Java an interesting programming language for control engineering.

Before describing the existing problems, the usual necessities for the design of applications in the field control area will be given. A (distributed) control application requires, of course, real-time behavior as well as the realization of communication with other applications, field I/O and/or remote I/O. Normally, hardware access regarding memory allocation, access to local I/O, and similar things are also necessary. Figure 63.1 gives an overview of the strengths and problems of Java against the background of these requirements. In general, Java provides advantages for tasks like user interaction over an HMI or communication with other applications/remote I/Os over different protocols. The close relation of Java to the Internet world makes it easy to support communication via protocols like http, ftp, or SMTP.

Unfortunately, other features of Java make it difficult to use it in control engineering without modifications or enhancements. This concerns the resource consumption of Java as well as the real-time capabilities and direct access to the hardware [6]. In the following sections, the reasons for these problems shall be described more in detail.

Resource Consumption

With respect to resource consumption, for the use of Java, two requirements have to be taken into account: first, the necessity to run a JVM on the target system (depending on the system architecture) and second, the need for large standard libraries, if the full features of Java are to be used.

To ensure the applicability of Java on different systems in a common way, three Java2 Editions were introduced. One of them, Java2ME (Micro Edition), was specially developed with respect to limited and embedded devices. With Java2ME consisting of two different configurations and offering the possibility to add profiles and optional packages, it is possible to tailor a Java platform to the special needs of devices and application areas. While the Connected Device Configuration (CDC) is used for devices with a memory of 2 MB and more, the Connected Limited Device Configuration (CLDC) was designed for smaller devices. With enhancements like personal profile and RMI-optional package, the CDC provides nearly the same functionality as J2SE. The CLDC-based K(kilo)-VM is much more limited, but runs on devices with memory down to 128 KB.

The more serious problems are the insufficient real-time capabilities of Java. Besides others, especially some features normally providing basic advantages of Java are problematic: the concept of JVM and GC.

Execution Speed and Predictability

Initially, the interpretation of Bytecode by the JVM was the rule. To improve the performance, later techniques such as just-in-time-compiler (JIT) were introduced. In a JIT, methods are translated into

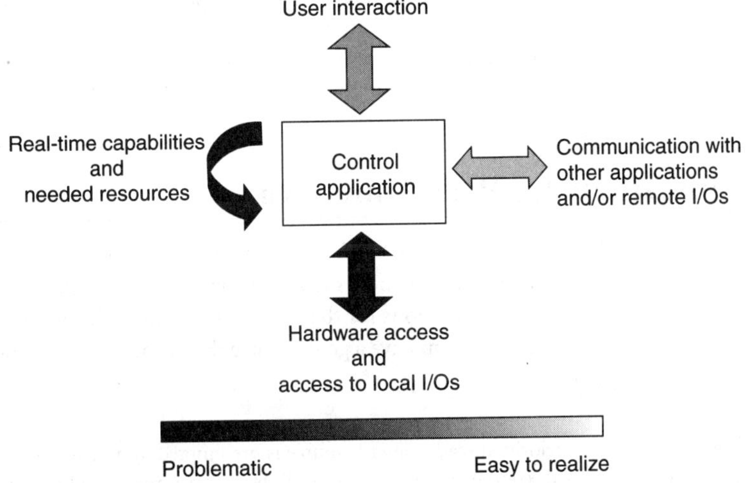

FIGURE 63.1 Requirements around a control application.

machine-code of the current platform during the execution. Hence, at their next invocation they can be executed much faster. For real-time systems, both variants are not sufficient. The interpretation may be too slow (depending on the speed requirements) and the JIT is highly nondeterministic as the execution time of a method can depend on how many times it was performed before! For a real-time system, the worst case is always important and here a JIT compiler is normally not better than an interpreter. Hence, the solution for such systems is to use Ahead-Of-Time compiling (AOT). Here, the application is either translated completely *before* the execution or *during* the class-loading. Both methods enable a considerable optimization and especially the first one reduces the resource consumption on the target system. Unfortunately, with a complete compilation in advance, an important Java feature — the dynamic class loading — is no longer possible. The compilation during class loading retains this possibility, but needs more resources on the target system (class loader).

Garbage Collection

Although the Java Language Specification (JLS) does not require a GC, the most JVM implementations use one for automatic memory management. A GC scans the memory for objects with no references and frees the memory, mostly also followed by a defragmentation of memory.

The GC is normally not interruptible (except incremental GC, which is rarely used) and so it stops any execution of application threads at an unpredictable point of time for an unknown time span. This is an extremely nondeterministic behavior and, of course, not acceptable under real-time constraints.

Synchronization/Priorities

The possibilities for thread scheduling and prioritization offered by Java are insufficient. Real-time systems require a reasonable number of priorities to enable the use of different scheduling strategies. Against this requirement, the number of priorities (the JLS defines 10) is too small. Usually, it is not guaranteed that the Java threads are mapped to threads at the operating system level. However, this is a prerequisite for a correct scheduling in a Real-time operating system (RTOS). Although there is a concept for synchronization of threads (keyword `synchronized`) with monitors, no mechanisms for avoidance of priority inversion[1] are specified by the JLS.

Hardware Access

Finally, as a result of the concept of platform independence realized by the JVM (abstracting the underlining hardware), direct hardware access is not possible in Java. This applies to direct memory access, as well as for specific device functionalities. Although this is no direct requirement for a real-time system, it can be necessary for the I/O-access in control engineering.

All these problems regarding different aspects of standard Java implementations result in the necessity for special modifications or enhancements, if Java is to be used with field-level devices under real-time constraints.

63.4 Specifications for Real-Time Java

Although the creation of a uniform specification for real-time behavior in Java is vitally important right against the background of platform independence, it was not possible to find a common approach of all parties thereto. At the moment, there are suggestions for the enhancement/modification of the standard-Java-specification by two consortia. Both of them have a goal to solve the problems stated above and make Java applicable to real-time systems.

---

[1] Priority inversion occurs, if a low-priority thread using a resource is pre-empted by a medium-priority thread. If now a high-priority thread also needs the resource, it is blocked until the low-priority thread is executed again and can free the resource.

Real-Time Specification for Java

The first consortium is the "Real-Time for Java Expert Group," under the leadership of Sun Microsystems, which has developed the "Real-Time Specification for Java" (RTSJ) [2]. The formal scope for the development of this specification is the Java Community Process (JCP), where the RTSJ runs as Java Specification Request (JSR)-000001. The JCP defines a procedure that requires an internal and public review of the draft specifications as well as a reference implementation and a test suite called "Technology Compatibility Kit" (TCK).

The JSR-000001 has reached the state of Final Release on 07 January, 2002, and a reference implementation has been developed by TimeSys Corporation [13].

General principles for the development of the RTSJ were to guarantee the temporally predictable execution of Java programs and to support the current real-time application development practice. For the realization of real-time features, the boundary conditions were set so that no syntactic extension of the Language needed to be introduced, and the backward compatibility was to be maintained by mapping Java Language Semantics (JLS) to appropriate entities providing the required behavior under real-time conditions. Furthermore, the RTSJ should be appropriate for any Java Platform.

The more general WORA concept of standard Java is replaced by the so-called Write Once Carefully Run Anywhere Conditionally (WOCRAC) principle. This is necessary because the influence of the platform has to be taken into account. For instance, if the performance of the platform is not fast enough to meet some deadlines (this should normally not be the case, but may be, if the original platform was much faster), this is a serious problem that cannot be ignored.

In this context, the goal of the RTSJ is not to optimize the general performance of a JVM, but to improve the features causing problems for real-time systems like the GC, synchronization mechanisms, and the handling of asynchronous events.

To achieve this, the RTSJ enhances the Java specification in seven areas. All these additions improve critical aspects in the behavior of Java, or add typical programming features for real-time system development [10]. In the following, these enhancements will be explained more in detail.

Thread Scheduling and Dispatching

The RTSJ introduces real-time threads with scheduling attributes to improve the scheduling possibilities compared to Standard-Java. Every real-time thread (implementing the interface `schedulable`) has a reference to a scheduler. Although the RTSJ is open for implementations of various scheduling principles, the only required version is (as default scheduler) a strict fixed-priority pre-emptive scheduling for real-time threads. For these threads, at least 28 unique priority levels are required (plus the ten priorities of the JLS). The assignment of thread-priorities is, as usual in most real-time systems, left to the programmer. The algorithm for the determination of the next thread to run (feasibility algorithm) requires an assignment of priorities following the Rate-Monotonic Analysis.

In the RTSJ the interface `scheduleable` is implemented by `RealTimeThreads` and `AsyncEventHandlers`. It contains several parameters to describe relevant data for the scheduling. For instance, the abstract class `SchedulingParameters` is used as the base class for the `PriorityParameters` describing the priority of a thread. Besides the basic RealTimeThread, which may use memory areas other than the normal heap, the `NoHeapRealtimeThread` must be created with a scoped memory area (see next paragraph) and is able to preempt the GC. Hence, there are strong restrictions regarding the interaction with objects on the heap (no reading or writing, nor manipulations of references except objects in the `ImmortalMemory`).

Memory Management

The RTSJ provides different kinds of memory to allocate objects outside the memory area controlled by the garbage collector. As the JLS does not require a GC (although nearly every JVM-implementation has one), the RTSJ also does not define the necessity for a GC. ScopedMemory is used to define the lifetime of objects depending on the syntactical scope. If such a scoped memory area is entered, every use of a new-statement results in an allocation of memory exclusively within this area. These objects are not garbage collected, but live until the control flows out of this scope. Then the whole memory area will be reclaimed. A scope (real-time thread or closure) can be associated with more than one memory area and, also, a scoped memory area can be associated with one or more scopes. There are two types of this kind of memory distinguishable by the relation of allocation time to object size: linear for LTMemory and variable for VTMemory.

Objects created within the third kind of memory, the ImmortalMemory, will never be affected by any GC and will exist until the Java runtime terminates. For each JVM, there is only one ImmortalMemory, which is used by all real-time threads together. The RTSJ provides a restricted support for memory allocation budgets, which can be used for defining the maximal memory consumption of a thread.

Synchronization and Resource Sharing

The RTSJ specifies an implementation of the synchronized primitive, which avoids an unbound priority inversion by using the priority inheritance protocol.[2] Using this method, problems can occur when a RealTimeThread is synchronized with a normal thread. A `NoHeapRealTimeThread` generally has a higher priority than the garbage collector, and a normal thread always has a priority below. Thus, a Java thread cannot have the same priority as a RealTimeThread. To solve this problem, the RTSJ introduces special wait-free queue classes with unidirectional data flow and nonblocking read/write methods.

Asynchronous Event Handling

The RTSJ introduces a mechanism to react on events that occur asynchronously to the program execution. Therefore, a schedulable object (`AsyncEventHandler`) is bound to an event (represented by an instance of the class `AsyncEvent`). If an event occurs, the event handler changes its state to "ready" and the event will then be scheduled like any other schedulable object (implementing the interface `Schedulable`).

Asynchronous Transfer of Control

Similar to the normal exception handling in Java, the RTSJ defines the possibility to transfer the flow of control to a predetermined point in the program. It is important to note that the possibility of an interruption has to be declared explicitly before the execution (interface `interruptible`).

Asynchronous Thread Termination

As the current JLS does not provide a mechanism to terminate a thread (the existing method `thread.stop` is marked as deprecated as it can cause inconsistencies), the RTSJ provides such a capability. It is realized by using asynchronous event handling and asynchronous transfer of control, and typically is deployed to terminate a thread when external events occur.

Physical Memory Access

The RTSJ defines two low-level mechanisms for a direct memory access. `RawMemoryAccess` represents a direct addressable physical memory. The content of this memory can be interpreted, for example, as byte, integer, or short (`RawMemoryAccessFloat` can be used for floating point numbers). The classes `ImmortalPhysicalMemory` and `ScopedPhysicalMemory` provide the possibility to allocate Java objects in the physical memory.

For practical use, the RTSJ has several powerful easy-to-use mechanisms. Other features like the immortal memory are potentially dangerous and have to be used very carefully. As objects in the immortal memory never free their allocated memory (until the runtime is shutdown), continued or periodical object creation can easily lead to a situation where the system runs out of memory.

Combining the RTSJ with a real-time GC [5], as explained in [7], can overcome some limitations. In doing so, it is possible to access the heap from the real-time part without limitations and to directly synchronize real-time and non-real-time parts of an application. Unfortunately, the resulting application will not be executable on each RTSJ-compliant JVM.

Also, some conceptual problems of the RTSJ have to be noticed. While the RTSJ provides a defined API, some aspects of the behavior can depend on the underlying RTOS. The API is the same, but the semantics

---

[2] If a low-priority task holds a resource and therefore blocks a high-priority task, its priority is increased to the same as the high-priority task.

may change! This particularly applies to scheduling and the possibility of implementing scheduling strategies other than the primary scheduler.

The reference implementation (RI) of the RTSJ is provided by TimeSys Corporation. The RI runs on all Linux versions, but the priority inheritance protocol is only supported by TimeSys Linux — a special real-time-capable Linux version. The JVM is based on the JTime-system VM (corresponding to the J2ME-CDC). As the JVM works in an interpreting mode, it is a good system for testing and getting experiences. For product development, one of the available commercial implementations of the RTSJ would be a better choice. Meanwhile, several companies have developed RTSJ-compliant platforms or implemented at least some of the basic concepts of the RTSJ [7, 13, 14].

Real-Time Core Extensions

The second consortium working on real-time specifications for Java is the JConsortium. One group within the JConsortium is the Real-Time Java Working Group (RTJWG). This working group has created the Real-Time Core Extensions (RTCE), finished in September 2000 [2].

The general goal of this specification can be characterized by reaching real-time behavior for Java applications with a performance regarding throughput and memory footprint comparable to compiled C++ code on conventional RTOS. Thus, this specification aims at providing a direct alternative to the existing real-time technologies. It was assumed that for typical applications in this field, the cooperation between the real-time part and the non-real-time part is limited.

As a result of these requirements, the general idea of the RTJWG was to define a set of standard real-time operating system services wrapped by a Java API (the "Core"). In doing so, the standard JVM was not changed, but extended by a Real-Time Core Execution Engine. The components of the core are portable, can be dynamically loaded, and may be extended by profiles providing specialized functionalities. Consequently, all objects implementing real-time behavior are derived from `org.rtjwg.CoreObject`, which is similar to `Java.lang.object`. This means the separation between real-time ("Core") and non-real-time ("baseline Java") is type based.

Following the concept of limited cooperation between core components and baseline Java, the core objects have some special characteristics regarding memory management, synchronization, or scheduling. These will be described in the following subchapters in more detail.

Memory Management

All real-time objects are allocated the core-memory area and are not affected by the GC until the appropriate task[3] terminates. The RTCE also specifies a stack allocation of objects, which provides a more efficient allocation and better reliability and has a performance that is easier to predict. To allocate an object on the stack, it has to be declared as "stackable" and some restrictions have to be observed, for example, a value of a stackable variable cannot be copied to a nonstackable variable. If a task terminates, all objects of this allocation context may be eligible for reclamation, but as there might still be references from the baseline objects, the core engine has to verify if this is the case or not. If not, the objects can be reclaimed.

Synchronization

To avoid priority inversion, the RTCE supports two protocols: the typical priority inheritance protocol and the priority ceiling for "protected objects." The problem of synchronization between baseline and core components (as described for real-time tasks in the RTSJ section) occurs here in the form that the synchronization protocols must not cause unpredictable delays in the real-time tasks. The mechanism to avoid this is the usage of a BufferPair (write/read buffer) in combination with the priority ceiling protocol.

---

[3]In the RTCE, mainly the term "task" is used instead of "thread," which is not unusual for real-time systems.

Scheduling

The RTCE defines 128 priorities, which are all above the normal Java priorities, so that the GC cannot preempt core tasks. The assignment of priorities is done by the programmer and typically a preemptive priority-based scheduling is used.

Cooperation with Baseline Java Objects

As core objects are not garbage collected before release, they must not access baseline Java objects. In contrast, objects in the baseline heap may invoke some special methods to access core objects (although the most fields and methods of core objects are not visible for baseline objects).

Besides this basic behavior, the RTCE provides a set of further features, which are useful for the programming of real-time systems like signaling and counting semaphores as well as interrupts and I/O ports with integrity. For the practical handling of the RTCE, it has to be noticed that the specification was developed under the assumption that core programmers are "trusted experts" (in contrast to baseline Java programmers), who are aware of the typical problems and pitfalls of real-time systems and can benefit from the functionality provided by the RTCE.

Real-Time Data Access

Another working group within the JConsortium is the RTAWG. The focus of this group is the application field of industrial automation. Hence, the resulting "Real-Time Data Access" (RTDA)-specification focuses on an API for accessing I/O-data in typical industrial and embedded applications rather than on features supporting hard real-time requirements.

Analyzing the typical requirements for the usage of Java in industrial applications, the general idea behind the RTDA is that the support of real time is a basic requirement, but that hard real time with sophisticated features is only needed in a few cases. More important is a concept for a common access to I/O data for real time and non-real-time parts of an application as well as the support of typical traditional procedures regarding configuration and event handling in this domain.

Following these conditions, the RTDA considers real-time capabilities as a prerequisite, assuming that the real-time and non-real-time applications run on a real-time-capable JVM.

Real-Time Aspects

The RTDA supports real time by using permanent objects and creating an execution context for these objects. Permanent objects are not garbage collected and have to implement the interface `PermantentMemoryInterface`, or the "permanent memory creation" has to be enabled (by invoking the method `enable()` of the class `PermantentMemory`). As the main intention is to create all these objects in a setup phase, the permanent objects can also be created by a non-real-time thread. This principle simplifies real-time programming, as the problems of memory allocation for object creation under time constraints and availability of memory are not critical for such a procedure.

The RTDA requires 32 priorities and a system priority for Interrupt Service Routines (ISR) and non-Java threads. The ranges are defined by the constants MIN_PRIORITY – MAX_PRIORITY (typically 1–10) for the non-real-time threads and MIN_RT_PRIORITY – MAX_RT_PRIORITY for real-time threads (12–32); the GC runs typically with priority 11.

The existing `synchronized` keyword will be used, but with different behavior depending on the context: usage of the priority inheritance protocol for real-time threads and permanent objects and raising the priority for permanent objects in normal threads to block out the GC.

Event Handling and I/O-Data Access

As the event handling and the I/O-access are the core components of the RTDA, it defines a dynamic execution model supporting asynchronous as well as synchronous access to I/O-data. The main instances responsible for creation of appropriate objects are the Event-Managers (asynchronous) and the Channel-Managers (synchronous). Every Event-Manager is created out of a `DataAccessThreadGroup`, which controls the priority of this Event-Manager. Depending on the type of event, there are different Event-Managers like:

- IOInterruptEventManager   (InteruptEvent)
- IOTimerEventManager   (PeriodicEvent or OneShotEvent)
- IOSporadicEventManager   (SporadicEvent)
- IOGenericEventManager   (all types of events, more than one event)

For the realization of a hardware-independent I/O-data access, three components are important: "Device-Descriptions," the classes `IONode`/`IONodeLeaf`, and the I/O-Proxy classes.

A DeviceDescription specifies a hardware component (static in a native DLL or as Java class). As this description does not represent a concrete instance of a hardware, but describes the type of a device, the concrete instantiation is realized by an `IONodeLeaf`-object.

The complete overall (hierarchical) structure of a given I/O-system is described by a tree of `IONodes`. It represents the concrete configuration of a system where the leafs of this tree are objects of the type `IONodeLeaf`. The configuration realizes a mapping on memory addresses as well as on I/O-space. The access path of an `IONodeLeaf` can be described in a way starting from the root node down to the `IONodeLeaf`. During the configuration, which is executed in the setup phase of the Java system, the instances of the `IONodes` (describing a device node) are created and a mapping regarding the addresses is executed.

The entity to access the I/Os in the application is the I/O-proxy, representing physical or software entities. These proxies are generated out of a `IOChannelManager` and can be identified by a physical or a symbolic name. The mapping between both these names takes place in a name map table of the appropriate `IOChannelManager`. Thereby, the design of hardware-independent applications is enabled if only symbolic names are used. Hence, changes in the hardware require only changes within the map table and not within the application.

`IOChannel` is the superclass for all types of I/O-proxies. Every I/O-proxy has a cache for holding the appropriate I/O-data and provides methods for updating the cache from the input channel as well as flushing the cache to the output channel. Furthermore, there is a common error handling.

As the concept of using I/O-proxies in the control application is an important part of the RTDA, there are different types of proxies. Hardware proxies represent a concrete hardware-I/O-device and therefore have a physical name corresponding to the respective entry in an `IODeviceDescription`. Furthermore, there are two different types of software proxies: Empty I/O-channels and generic Software I/O-channels. Empty I/O-channels give a generic possibility for I/O-data access and can be used by manifold non-RT Java components (e.g., HTTP-server). The generic Software I/O-channels provide the possibility to extend the existing I/O-classes by new functionalities without changing the RTDA implementation. Examples for such possible extensions, given in the specification [3], are, for instance, I/O-proxies for remote access, simulation, or accessing legacy software.

As the superclass `IONode` is the base for all I/O-channels and events, the basic concept for the handling of events is similar to the concept of I/O-handling; a similar mechanism for the handling of events are specified.

Although there is no direct reference implementation for the RTDA, the Java package for SICOMP industrial-PCs by Siemens closely follows this specification of the J-Consortium. This particularly applies to the JFPC system (Java for process control) in relation to the I/O concept of the RTDA [3,8] (The resulting structure of RTDA is given in Figure 63.2).

Comparison

Comparing all three specifications, it can be stated that all of them follow different approaches, but also focus on different application areas. The RTSJ and RTCE have the goal to provide a more general way of enabling Java to become usable in the real-time domain, while the RTDA focuses on the special application field of industrial automation and understands real-time capabilities only as one of several requirements, which are typical for this area.

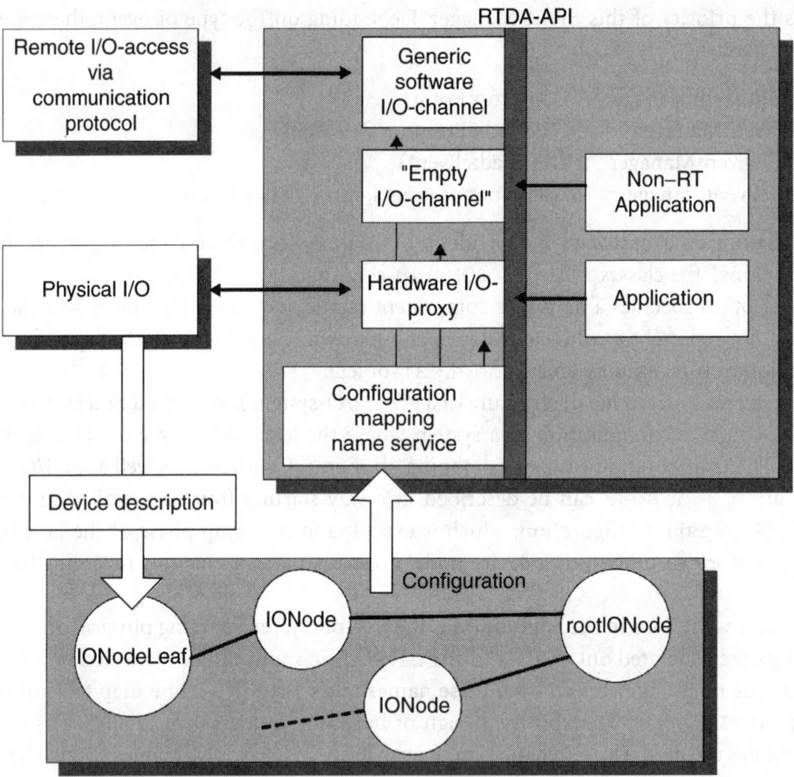

FIGURE 63.2   RTDA — overall archirecture.

The RTSJ is the most general approach, following the idea of "making Java more real-time." Therefore, the RTSJ can be used well by experienced Java programmers, as all enhancements are close to common Java concepts. Considering the number of products implementing the specification, currently, the RTSJ has the highest acceptance.

In contrast, the RTCE follows other premises, focusing on providing a performance comparable to the state of the art (e.g., C++) solutions on commercial RTOS for hard-real-time requirements. The JVM was not changed, but extended by a separated "core," which realizes the real-time features. The RTCE provides real-time functionality following the typical concepts of today's real-time application development and therefore it is assumed that it will be used by experienced real-time programmers. Hence, the RTCE is the first choice for problems dealing with hard-real-time constraints and high-performance requirements, although, at the moment, the lack of a sufficient number of available products reduces the applicability.

The RTDA provides a complete concept for the application development in industrial control engineering. Real-time capabilities are seen only as a needed prerequisite for such applications, but the main part deals with the handling of I/O-data and interrupts. The concept is very similar to typical concepts used in conventional control systems and is therefore easy to understand for programmers working in that domain. Nevertheless, the RTDA is a very "closed" world and therefore it is not easy to adapt all ideas of using reusable components when programming control applications in Java according to the RTDA. Besides the JFPC system, which is available only for very powerful and expensive hardware (a Linux implementation will be developed and opens here new possibilities [12]), the RTDA also lacks a wide range of products.

Currently, it is still not clear if there ever will be *the* one specification, or if each of these specifications will find acceptance in a certain area.

63.5 Java Real-Time Systems

To ensure the real-time capability of a Java application, independent from the used JVM, the underlying operating system (OS) has to be real-time capable as well. Therefore, in particular in the area of embedded systems, different possibilities of combining JVM and real-time-OS exist. In general, there are four different categories:

- *Real-time JVM on conventional RTOS*
  In this case, a real-time JVM is running on a "normal" RTOS. This is typical for systems where the possibility to execute Java is an additional feature (of course, it can be the only feature that is used) and where the resources allow to run a JVM with sufficient execution time. Normally, this is the case for systems like IPCs and SoftPLCs, less for conventional PLCs because the operating systems and processors of conventional PLCs make the integration of a real-time JVM difficult. Examples of such systems are the SICOMP-IPC with JavaPackage [8] or the Jamaica-VM [7].
- *JVM and RTOS as one integrated system*
  If the only function of the RTOS is to run a JVM (this means the whole RT-application is written in Java), it can be useful to integrate both, JVM and RTOS, as one product and tailor the RTOS exactly to the needs of the JVM. Such systems can be a basis for full Java-PLCs.
- *Conventional RTOS with Java-co-processor*
  If the resources of the system are limited with respect to the requirements of a JVM, it can be useful to have a special Java-co-processor for the execution of the Java-Bytecode. This results in an increased execution speed, and also in an increased complexity of the system hardware and software.
- *Java processors*
  Finally, there are systems with a processor the executes Java-Bytecode directly as native processor code. Here the JVM is realized in "hardware," so that the conventional OS level can be dropped. This results in a very efficient (and therefore fast) Bytecode execution, but the flexibility of such systems is reduced due to the fact that non-Java applications cannot be executed. The typical application for the last two combinations are intelligent I/O devices.

63.6 Control Programming in Java

Requirements of Control Applications and New Possibilities in Java

Today, IEC 61131-compliant languages are typical for programming control applications. Although there are concepts for encapsulation of functionalities and data (e.g., function blocks provided in IEC61131-5), the main advantage of Java, compared to these languages, is the possibility to use object-oriented features like encapsulation and inheritance. Of course, networking abilities and stability also play an important role, but object orientation enables completely new ways for code reusability and increases the efficiency of application developments in control programming.

As it is not efficient to implement the whole application from scratch for every new project, it is important to encapsulate functionalities in classes for reasons of reuse. Depending on the concrete device, by means of these classes (or interfaces), generic functions like specific communication protocols or easy access to specific devices can be realized. These existing classes can (if necessary) be modified or extended and then be integrated into the application.

Hence, notable potentials result for industrial automation. Based on the platform independence of Java, parts of applications can be easily reused. Manufacturers of components like I/O-devices or intelligent actors/sensors can deliver classes (representing their components) together with the hardware. Instances of these classes can then be used in the concrete application and provide methods for using the device functionality, concealing the concrete hardware access. In doing so, the manufacturer can provide a clear simplification for the programmer and reduce the danger of improper use of his own hardware.

For complex applications, the concrete hardware access (I/O), communication functions, or other basic functions can be abstracted on different levels. Here, special tailored classes are imaginable, for instance, the usage of IEC 61499 function blocks. A simple example for such a concept shall be explained now more in detail.

Structure of a Control Application in Java — An Example

The goal of the following is to give recommendations for the structure of an application to support efficient programming. This can be ensured on the one hand by as much as possible device-independency, and on the other, by using a consistent procedure for the handling of I/O-variables, independently if these variables represent local I/Os or remote I/O-devices, which are accessible over different communication protocols. The communication can be realized by conventional bus-systems (e.g., CANBus), or via Ethernet (e.g., Modbus/TCP or EtherNet/IP). How should an application for these requirements be structured? In general, it is advantageous to use several levels for the encapsulation of functionalities in Java classes (see Figure 63.4). To explain the structure, a simple example shall be used. The hardware structure of this example is shown in figure 63.3. A conveyer system consisting of several belts and turntables is controlled by a PC-based Java-PLC. As I/Os, two Modbus/TCP-Ethernet-coupler are used.

Objects of the lower level, called here communication level, realize the access to the concrete hardware. For local I/O access (e.g., memory access or GPIOs of a processor), these classes are device-specific. In contrast to this, for the communication with remote I/O-devices or other controls, generic classes (availability of appropriate hardware assumed) can be used on different platforms. Today, such Java classes are available for the Ethernet-based protocols Modbus/TCP and EtherNet/IP. In our example, the Ethernet

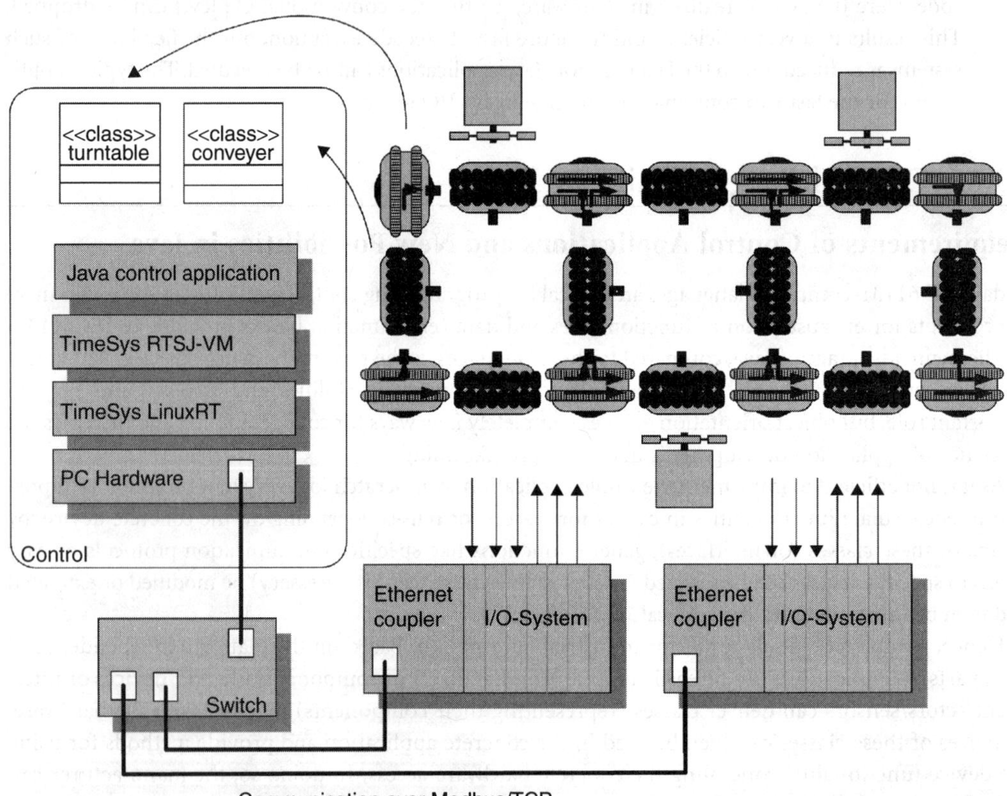

FIGURE 63.3   Hardware structrure of the example.

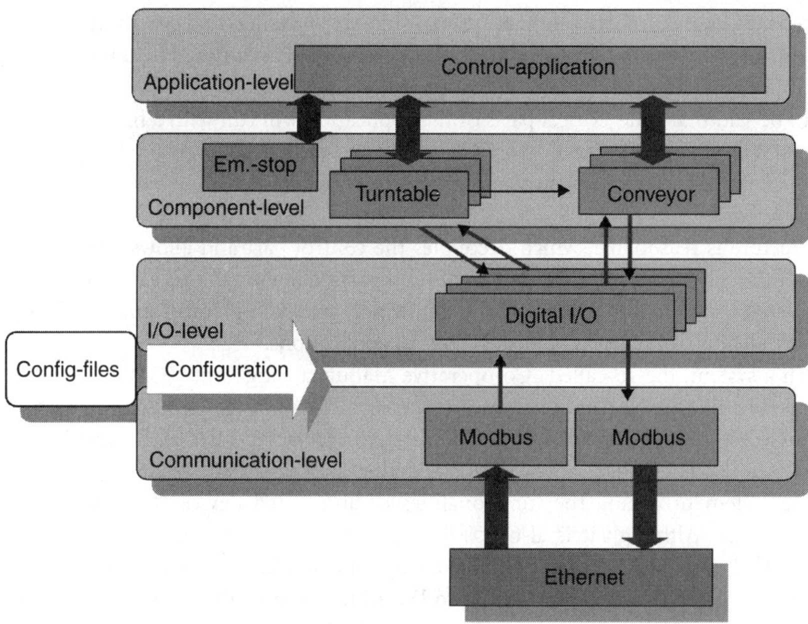

FIGURE 63.4   Structure for control applications in Java.

coupler (and its I/Os) are represented by two instances of a class `Modbus`. These objects ensure the read- and write-access to all I/Os of the appropriate coupler and make it possible to read and write the I/Os. These objects, basically realizing a communication, can be used by classes of the next level (I/O-level), which represents the logical I/O-variables (e.g., digital or analog inputs or outputs). Here, a mapping from physical variables (representing physical I/Os) to logical variables is realized. If this mapping is flexible by using configuration information, for example, stored in a configuration file or as an input from an IDE, the control application can be implemented device-independently. In the example, every input or output is represented by an object of the class `DigitalIn` or `DigitalOut`, which is linked with the help of a configuration file to the appropriate Modbus-objects.

On the third level, complex control functionalities (e.g., emergency-stop) as well as representations of parts of the plant control system (e.g., components like conveyors, drives, or generically usable actors) can be encapsulated in classes. Objects on this level, as well as groups of I/O-variables on the I/O-level, typically run in their own real-time thread. In the example, there are objects for each element of the transport system as instances of the classes `Conveyor` and `Turntable`. These objects provide methods for controlling these elements (e.g., `start()`, `stop()`, `turnleft()`, `turnright()`,...).

This structure simplifies the reusability of huge parts of the control application. Thus, on the highest level (application level), an application can be implemented in a more abstracted, function-related way.

Integration of Advanced Technologies

The use of Java for control programming at the field level shows advantages especially in systems where Java is also used for non-real-time applications. This applies for visualization or remote access for specific functions like maintenance and code download/distribution. In this context, new technologies, also in the area of industrial automation such as agent-systems or plug-and-participate technologies, play an important role.

For implementing such systems, special attention has to be paid to restrictions regarding the mutual interactions of the control application running under real-time conditions and the non-real-time application. If the real-time-Java platform provides support in this respect (like the software I/O-proxy of the

RTDA), this can be realized very easily. Unfortunately, most real-time-Java products and specifications do not follow the requirements of industrial control applications; hence, special attention has to be paid to these aspects.

As a general rule, it can be said that time-critical parts of the application will run with a higher priority, while non-real-time parts have priorities below the GC and will be executed as normal Java applications and, of course, restrictions like synchronous calls from real-time to non-real-time part have to be observed.

For this case, an architecture is reasonable, which decouples the control part and allows a synchronization at certain states of the system. The loose coupling of both parts, allowing the access to the control only in exact defined states, can be implemented by a connection layer using, for example, a finite state machine. This allows to load, parameterize, and start control applications.

As an example for such a system, the so-called "Co-operative Manufacturing Unit" (CMU) shall be mentioned. It was developed within the international research project PABADIS [11]. This project aims at creating a highly flexible structure for automated production systems, replacing parts of the traditional MES-layer by concepts using technologies like mobile agents and plug-and-participate technologies. The CMU is the entity in the system providing the functionalities of automated devices (e.g., welding, drilling) to the PABADIS-system. Although it is also possible to connect conventional controls (e.g., PLCs) to the system, the fully Java-based CMU is the most advanced concept. It avoids the additional communication effort from the object-oriented Java-world to IEC 61131-compliant languages and provides all the advantages of Java stated before. The outcomes of the PABADIS-project give an idea of the possibilities that technologies like Java can bring to future automation systems.

Migration Path for the Step from Conventional Programming to Java Programming

As mentioned before, it is necessary to provide a possibility to migrate in a certain way from the conventional IEC 61131-based programming to Java programming. This way can be opened by applying the ideas of IEC 61499 and the definition of special function blocks for the application parts given in the component level, IO level, and communication level of Figure 63.4. Based on these predefined structures and self-designed control application blocks programmed in IEC 61131-compliant dialects (which can be automatically translated to Java), a new way of programming can be established.

63.7 Conclusion

Summarizing, it has to be stated that Java technology has reached a status where the technical prerequisites regarding use in industrial automation even in the area of control applications on limited devices are fulfilled. There are Java2 Editions allowing an adaptation to the requirements of the platform and several products using different approaches provided real-time capabilities in Java are available. This allows for the use of Java on nearly all types of devices at the field level. A common standard for real-time-Java is important against the background of retaining platform independence of Java also for real-time devices. Now the development of concepts regarding how the advantages of object-oriented high-level languages can best be used for increasing the efficiency of application development (by supporting reusability and providing abstract views on the Java application) is momentous. Besides this, Java opens new possibilities for an easy integration of technologies like XML, Web-services, (mobile) agents, and Plug-and-participate technologies [11].

References

[1] Bollella, G., B. Brosgol, P. Dibble, S. Furr, J. Gosling, D. Hardin, M. Turnbull, and R. Belliardi, *The Real-Time Specification for Java™* (First Public Release), Addison-Wesley, Reading, MA, 2001.

[2] Real-Time Core Extentions, International JConsortium Specification, 1999, 2000.

[3]   Real-Time Data Access, International JConsortium Specification 1.0, November 2001.

[4]   Dibble, P., *Real-Time Java Platform Programming,* Sun Microsystems Press, Prentice-Hall, March 2002.

[5]   Siebert, F., *Hard Realtime Garbage Collection in Modern Object Oriented Programming Languages,* BoD GmbH, Norderstedt, 2002.

[6]   Pilsan, H. and R. Amann, *Echtzeit-Java in der Fertigungsautomation Tagungsband SPS/IPC/Drives 2002,* Hüthig Verlag, Heidelberg, 2002.

[7]   Siebert, F., Bringing the Full Power of Java Technology to Embedded Realtime Applications, Proceedings of the 2nd Mechatronic Systems International Conference, Witherthur, Switzerland, Oct. 2002.

[8]   Hartmann, W., Java-Echtzeit-Datenverarbeitung mit Real-Time Data Access, Java™ SPEKTRUM, 3, 2001.

[9]   Brich, P., G. Hinsken, and K.-H. Krause, *Echtzeitprogrammierung in JAVA,* Publicis MCD Verlag, München und Erlangen, 2001.

[10] Shipkowitz, V., D. Hardin and G. Borella, The Future of Developing Applications with the Real-Time Specification for Java APIs, JavaOne Session, San Francisco, June 2001.

[11] The PABADIS project homepage, www.pabadis.org, 2003.

[12] Kleines, H., P. Wüstner, K. Settke, and K. Zwoll, Using Java for the access to industrial process periphery — a case study with JFPC (Java For Process Control), *IEEE Transactions on Nuclear Science,* 49, pp. 465–469, 2002.

[13] TimeSys, Real-Time Specification for Java reference Implementation, www.timesys.com, 2003.

[14] Hardin, D., aJ-100: A Low-Power Java Processor, Presentation at the Embedded Processor Forum, June 2000, www.ajile.com/Documents/ajile-epf2000.pdf.

[15] Schwab, C. and K. Lorentz, *Ethernet & Factory, PRAXIS Profiline — Visions of Automation,* Vogel-Verlag, Wuerzburg, 2002.

64

The GRAFCET Specification Language

Paulo Portugal and Adriano Carvalho
Faculdade de Engenharia da Universidade do Porto

64.1 Introduction

Due to the complexity of automated systems and the increasing application requirements, the use of appropriate modeling tools becomes essential to achieve a correct system design. These tools must be capable of simultaneously specifying the system behavior in a concise and unambiguous manner, and of validating this specification. Support of a method to easily map the formal model into the final implementation, on possibly varying hardware, is also required.

For several years, the absence of a common modeling tool led system designers to use different methodologies, each supported by a specific tool. This approach resulted in a large number of problems, such as higher development costs, incompatible specifications, absence of formal validation techniques, and difficulties in translating specifications into implementations.

GRAFCET was initially proposed by several academic and industrial research groups, and later standardized by International Electrotechnical Commission (IEC), in the hope that it would be accepted as the common modeling tool. Its aim is to allow the designers to describe and specify a control system's sequential behavior.

This chapter presents a description of the GRAFCET language, from an automated systems' perspective. All aspects of the language are described, although the number of application examples was reduced to a minimum due to space constraints. Interested readers are encouraged to refer to [3][5][6] for more detailed examples.

64.2 The GRAFCET Standards

In 1975, a working group called Logical Systems from AFCET (*Association Française pour la Cybernétique Economique et Technique*) decided to create a committee in charge of defining the requirements an automated system's modeling tool should fulfill. Having evaluated the state-of-the-art modeling tools of that time — flowcharts, petri nets, and state graphs — they concluded that a new more appropriate modeling language was required. In 1977, GRAFCET (*Graphe Fonctionnel de Commande Etapes-Transitions*), which is translatable to Functional Graph of a Step-Transition Command, was proposed to fill the gap [4].

In 1982, a French initiative created the first GRAFCET standard: NF C 03-190 — *Diagramme fonctionnel "GRAFCET" pour la description des systèmes logiques de commande*. Later, the widespread use of GRAFCET led IEC to promote a standardization process, which resulted in 1988 in the first international GRAFCET standard: IEC 848 Ed. 1 — *Preparation of function charts for control systems*. This document was later revised, and a second edition was published in 2001: IEC 60848 (2001) Ed. 2 — *Specification Language GRAFCET for Sequential Function Charts* [1].

Since the IEC 60848 Ed. 2 does not define how to implement GRAFCET in control systems, other initiatives were adopted to implement it as a programming language for programmable logic controllers (PLCs). These efforts resulted in the definition of the language SFC — *Sequential Function Chart*, which is part of the IEC 61131-3 (2001) *Programmable Controllers — Programming Languages* standard [2].

64.3 The GRAFCET Context

A system can be defined as a set of organized interacting components that work together to achieve some defined objective, which is to confer an added value — physical modification, particular arrangement, position change, data processing — to a group of raw materials — products, energy, data, people — in a specific environment. When the operations necessary to achieve the system objective are executed automatically under the control of computational resources, the system is defined as an *automated system*.

From a conceptual viewpoint, the structure of an automated system is usually considered as being composed of two cooperating parts: *operative* and *control* (Figure 64.1). The *operative part*[1] represents the physical process that creates the added value. It is composed of a set of elements such as sensors, actuators, machines, robots, transportation systems, etc. The *control part* is composed of computational resources that coordinate the sequence of operations performed by the operative part. Additionally, it interacts with other external systems in its environment, such as human–machine interfaces and communication networks. All interactions may be modeled by the sending of commands and the reading of data.

The implementation of an automated system requires a formal, accurate, and complete description of its behavior. Without this understanding, the system designer could introduce faults during the development and implementation phases, which could lead to serious problems during system operation. It is important to notice that automated systems are commonly found operating in very sensitive areas — nuclear power plants, railways, industrial facilities — where a design fault could have a catastrophic impact on people and the environment.

The description of any system component's behavior is performed by first defining a virtual interface that isolates it from the external elements with which it interacts, followed by characterizing the interactions that occur across that interface. Therefore, interactions with the control part, which control engineers will typically be focusing on, can be functionally represented by isolating it from other system elements. Data and commands exchanged with both the operative part and the environment are, respectively, represented by *input* and *output variables* (Figure 64.2).

---

[1]Also commonly referred to as the controlled part.

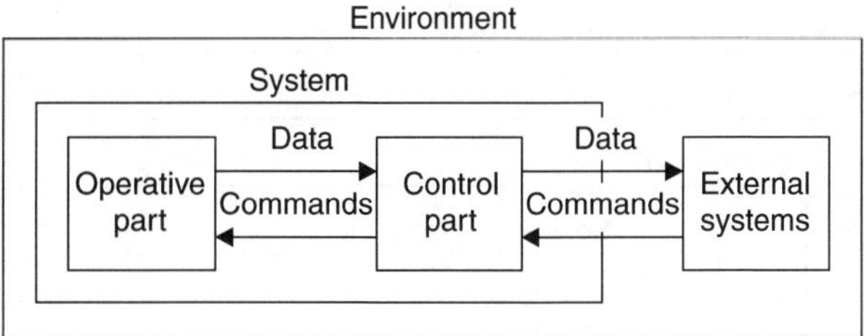

FIGURE 64.1 Automated system structure.

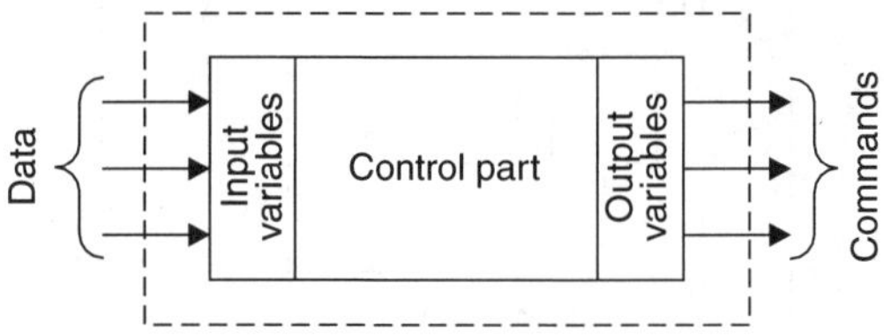

FIGURE 64.2 Control part interactions.

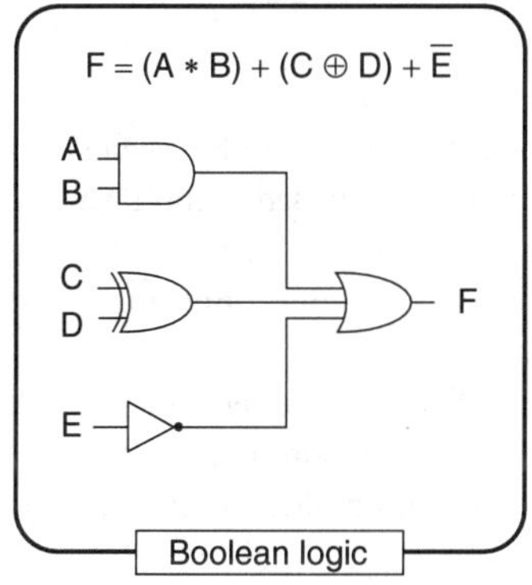

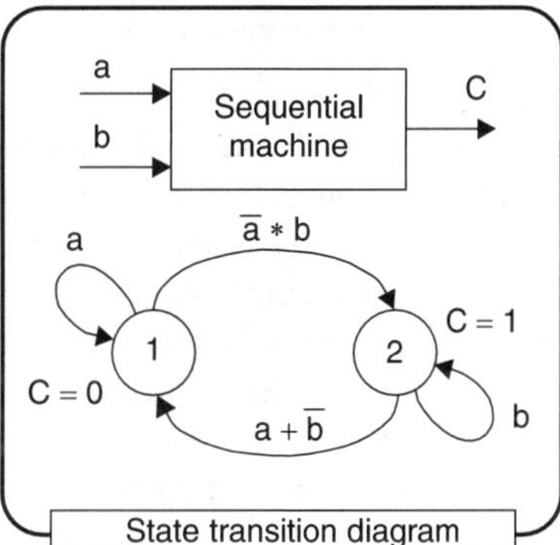

FIGURE 64.3 System behavior modeling.

The control part behavior may be described by the way in which the output variables depend on the input variables, which is necessarily different among distinct system types. Nevertheless, it is possible to find behavioral patterns common to every system and to use appropriate modeling tools to describe it. Among these patterns, two of them are distinguished by being present in any automated system and by representing the majority of interactions related to the control part — *combinational* and *sequential behaviors*. Both of them assume that system inputs and outputs are boolean variables, but other variable types can also be considered within specific contexts.

The *combinational behavior* is characterized by system outputs being a logical combination of their inputs at any given instant. Several modeling tools are available to describe this behavior — boolean logic, Karnaugh maps — which can be easily implemented (Figure 64.3).

The *sequential behavior* introduces the notion of state to describe the system evolution. At any given instant, the system outputs are logical functions of the inputs and the internal system state. This behavior can be modeled as a *sequential machine* by means of a *state transition diagram*.

The diagram is a simple directed graph consisting of circles and arcs, where circles represent the system states (Figure 64.3). At any given instant, only one state is active, representing the current system state, referred to as *situation*[2]. For each system state, a group of actions operating on the output variables

---

[2]Adopting the IEC 60848 Ed. 2 terminology.

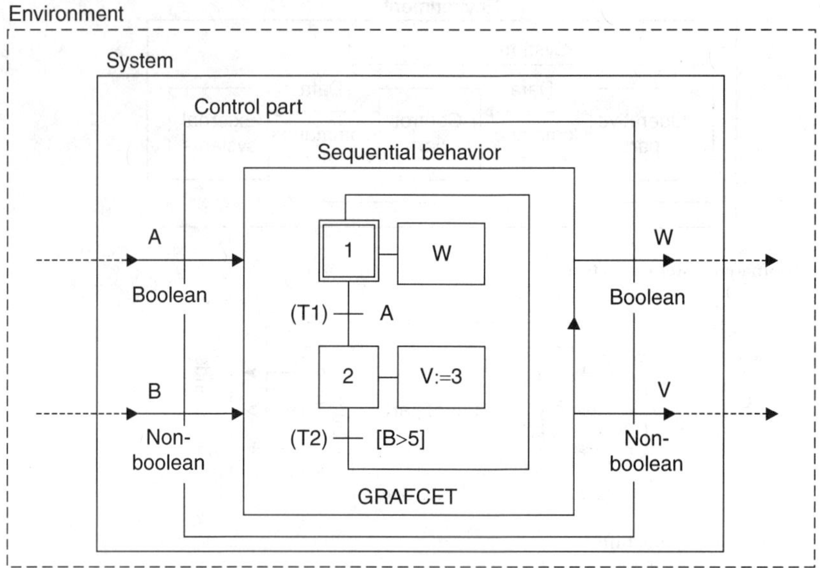

FIGURE 64.4　GRAFCET and system behavior.

may be defined, which are only executed while the state is active. The evolution between system states is modeled by directed arcs connecting the states. A boolean expression composed of input variables is associated to each arc, representing an evolution condition. When this condition is true, a transition from one state to another occurs. The definition of output variables and evolution conditions imply that the combinational behavior is embedded in the sequential behavior.

Although a state transition diagram could describe any type of sequential behavior, they are difficult to use when it becomes necessary to model behaviors commonly found in automated systems, which usually involve a certain degree of complexity, such as synchronization, parallelism, hierarchical relationships, and different detail levels.

In order to overcome these limitations, a new formalism is required, with the same descriptive power as sequential machines, but also capable of representing complex behaviors in a simple, concise, and intuitive way. It should also be well-adapted to represent the sequential evolutions of a system, which will potentially ease the mapping to the final hardware/software implementation. With all these properties, it becomes the appropriate tool for both automated system designers and end users.

These requirements resulted in the specification of the GRAFCET standard — IEC 60848 Ed. 2 [1].

GRAFCET's aim can be synthesized as to describe and specify, in a consistent and unambiguous manner, the sequential behavior of the system control part (Figure 64.4). The description is performed by using a graphical language composed of charts, which includes graphical elements to represent the system states and possible evolutions, associated with an alphanumerical representation of the system input and output variables. Therefore, it can be said that GRAFCET is a *behavior specification language* for sequential systems.

Although any kind of system that exhibits a sequential behavior can be modeled by using GRAFCET, until now its main application domains have been in the specification of control applications running on PLCs.

64.4　Basic Elements

Introduction

In a sequential system at any given instant the output variables are established from the system state and the input variables[3]. Since GRAFCET models the behavior of these systems, it becomes necessary that it includes

---

[3]Both variables are assumed to be of the boolean type, unless otherwise stated.

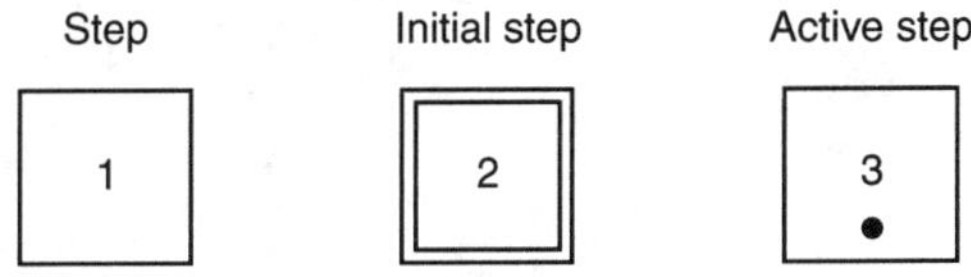

FIGURE 64.5   Step representation.

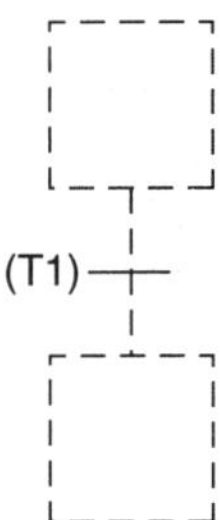

FIGURE 64.6   Transition representation.

all system data in its representation. This is accomplished by producing a chart (or group of charts), referred as *Grafcets*[4], using a set of graphical elements that represent different, but complementary, viewpoints of the system behavior. This representation distinguishes two perspectives: *structure* and *representation*.

Structure

The structure models the system evolution between situations. It is represented as a graph composed of *steps* and *transitions*, disposed in alternating mode, which are interconnected by *directed links*.

Steps

A *step* represents part of the system state. It may have two states: *active* or *inactive*. The set of active steps at any given instant represents the system, and *Grafcet*, situation at that instant.

A step is graphically represented by a square box, which is identified by means of a distinct label (Figure 64.5). The initial situation, *initial step*, represents the state of the system at the initial time ($t = 0$). The corresponding steps are represented by a double box. During the *Grafcet* evolution, the active states, at any given instant, are represented with a small dot inside the box.

Transitions

A *transition* represents a possible system *evolution* from one step to another, which is a change in the *Grafcet* situation. The evolution is performed by *clearing*[5] the transition. It is graphically represented by a thin horizontal bar (Figure 64.6), and it is identified by means of a distinct label, between brackets, placed to the left. This is known as its *designation*.

Directed Links

A *directed link* is used to connect steps to transitions and vice versa. Therefore, two steps cannot be directly connected, unless a transition exists between them. This property is known as the *alternation step-transition* rule.

A directed link is graphically represented by a line, with the direction of evolution path, assumed from top to bottom (Figure 64.7). An arrow can be used to avoid misunderstandings.

---

[4]We have adopted "GRAFCET" to refer to the language in general, and "*Grafcet*" to refer to a particular chart model.
[5]Also commonly referred to as *firing*.

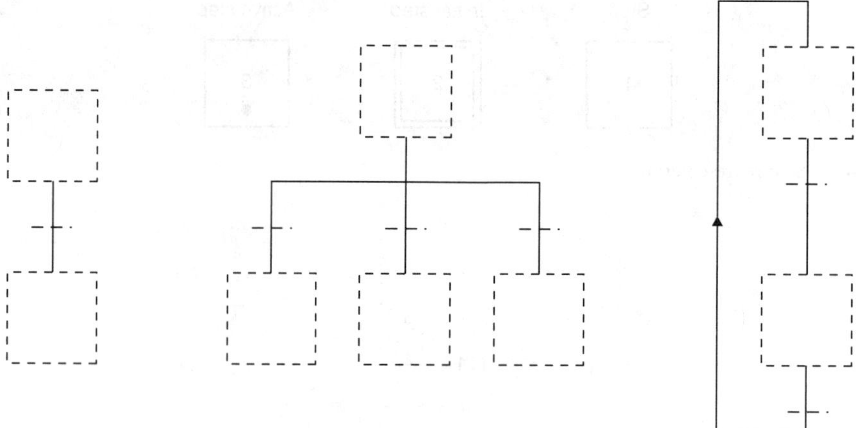

FIGURE 64.7   Directed link representation.

Interpretation

The interpretation establishes the relationships between the input and output variables and the *Grafcet* structure. It is composed of alphanumerical elements — *transition-conditions* and *actions* — representing the functional aspect of GRAFCET.

Transition-Conditions

A *transition-condition*[6] is a logical expression composed of input variables and/or *internal variables* (Section 64.5) representing an *evolution condition*. This condition is associated with a transition and establishes the necessary conditions to its clearing. Input variables are provided by both the operative part and the environment, while internal variables represent data internal to the system control part.

Transition-conditions are graphically represented by a logical expression placed at the right of the transition (Figure 64.8). A condition that is always true is represented as '$\underline{1}$'.

Actions

An *action* represents an operation performed on an output variable, corresponding to a command that is sent either to the operative part or to the environment. The action is executed only while the associated step is active. Several actions can be associated to the same step.

Actions are graphically represented by a rectangle connected to the associated step (Figure 64.9). A literal or symbolic label inside it designates the output variable that takes the true value.

Evolution Rules

Since GRAFCET specifies the behavior of a dynamic system, it is necessary to define a set of rules that establish, without any ambiguity, the system evolution from one situation to another. The following five rules apply:[7]

- *Rule 1 — Initial Situation.* This situation is defined by the system designer and represents the state of the system at initial time ($t = 0$).
- *Rule 2 — Clearing of a Transition.* A transition is cleared if and only if both the following conditions occur:
  - All immediately preceding steps, linked to the transition, are active. The transition is said to be *enabled*.
  - The condition associated with the transition is true.

---

[6]Also commonly referred to as *receptivity*.
[7]According to IEC 60848 Ed. 2.

- *Rule 3 — Evolution of Active Steps.* The clearing of a transition results in the simultaneous deactivation of all immediately preceeding steps and the activation of all immediately succeeding steps.
- *Rule 4 — Simultaneous Evolutions.* All transitions that can be cleared simultaneously are simultaneously cleared.
- *Rule 5 — Simultaneous Activation and Deactivation of a Step.* If during rule 4 phase, an active step is simultaneously deactivated and activated, it remains active.

The systematic application of the previous rules to a *Grafcet* describes all possible evolution situations, which helps the system designer to verify if the model's behavior diverges from the initial specification. Application examples are presented in Figure 64.10.

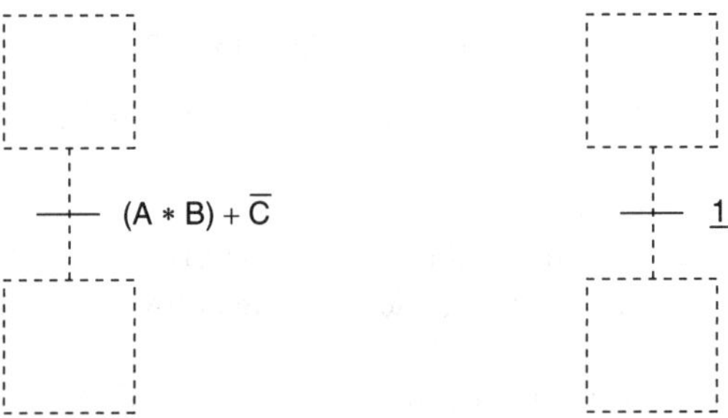

FIGURE 64.8　Transition-condition representation.

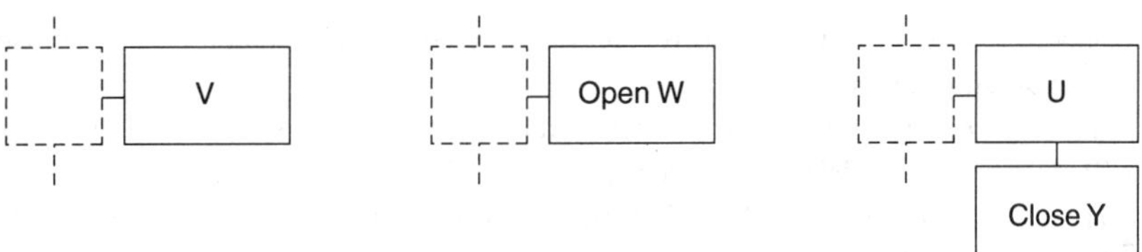

FIGURE 64.9　Action representation.

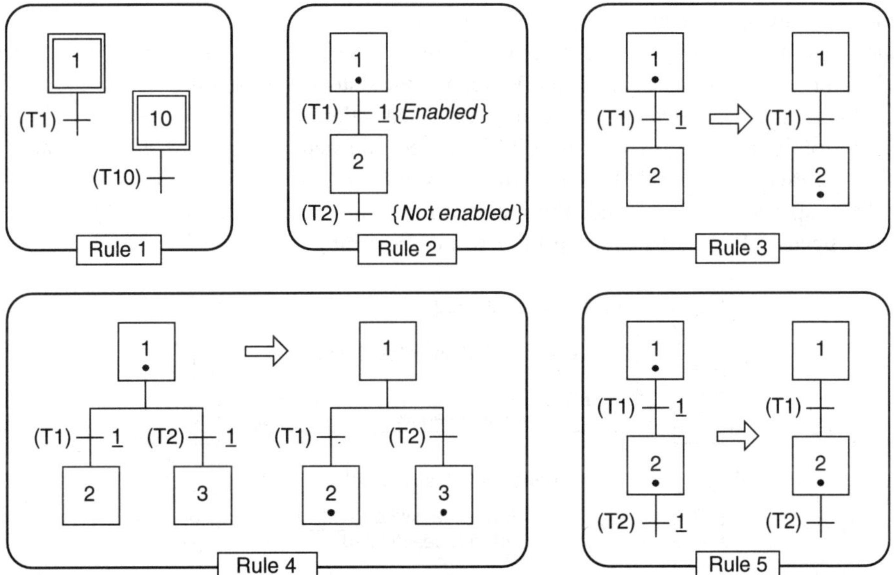

FIGURE 64.10　Evolution rules.

Unstable Situations

The application of the evolution rules, combined with the existence of particular situations, may lead the *Grafcet* to a scenario where an *unstable situation* will occur.

As an example, consider the *Grafcet* presented in Figure 64.11, where at a given instant the situation is defined by: **step 1** is active, A=0 and B=1. Now assume that A changes to 1, initiating the *Grafcet* evolution (Figure 64.11). From the instant where A changes to 1, the *Grafcet* situation instantaneously changes from **step 1** active to **step 3** active. The simultaneous activation and deactivation of **step 2** is referred to as an *unstable situation*. Actions associated to steps in unstable situations are not executed, unless they are *stored actions* (Section 64.5).

Evolution Algorithm

The evolution rules establish a simple algorithm that defines the GRAFCET evolution:

1. Initialization (assumed stable). All initial steps are activated and the actions associated are executed.
2. Define the set of cleared transitions.
3. Clear all cleared transitions, until a stable situation occurs. During this phase, execute eventual stored actions associated with transitions clearings, step activations, and deactivations, including unstable situations.
4. Execute the actions associated with the active steps.
5. Goto to step 2.

Tank Filling Example (I)

The following example was chosen to demonstrate the basic GRAFCET capabilities, when used to specify the behavior of an automated system.

The system represented in Figure 64.12 is composed of a tank that must be filled with a liquid stored in a reservoir. Two valves (V and W), respectively, control the input and output of liquid in the tank. Two level sensors (H and L) monitor the tank level. A button (S) is used to start the system operation.

The tank is empty when the level is less than L (i.e., L=0) and is full when the level is greater than H (i.e., H=1). Initially, the tank is empty. When button S is pressed (i.e., S=1), the tank is filled by opening V and closing W. The filling operation stops when the tank is full, by closing V, and its contents are drained by opening W. When the tank is empty, W is closed. Filling may only restart when S is pressed again. The valves are open when the respective outputs are true, otherwise are closed.

A *Grafcet* that meets these specifications is also presented in Figure 64.12. It has three input variables: L, H, and S, and two output variables: V and W. Initially, the system waits for button S to be pressed. Only **step 1** is active — the initial step — without any associated action. Therefore, both valves will be closed. When the button is pressed, S becomes true, T1 is cleared, and **step 2** is activated. The associated action — open V — is executed. When H becomes true, transition T2 is cleared and **step 3** is activated. In this situation, W is opened and V is closed. The system is maintained in this situation until L becomes true. When this happens, T3 is cleared and **step 1** is activated, returning to the initial situation.

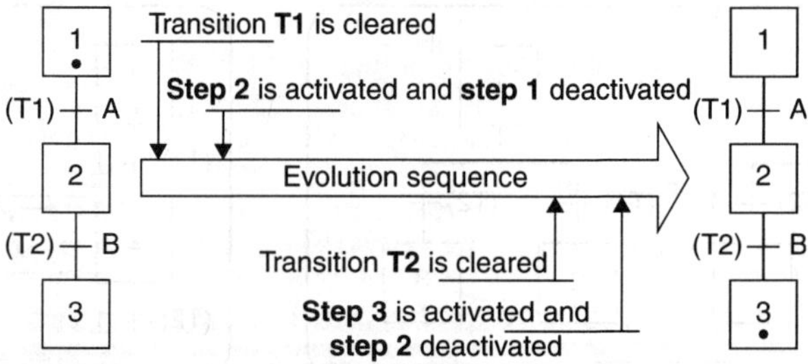

FIGURE 64.11   Unstable situation.

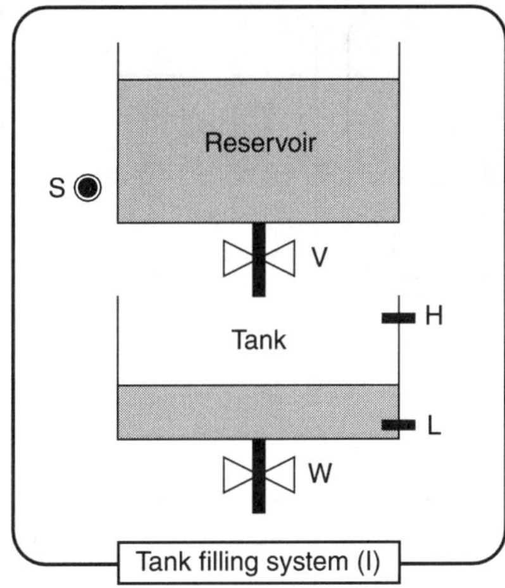

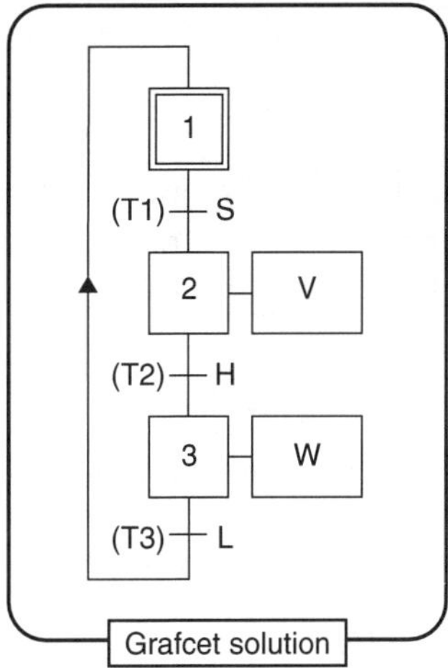

FIGURE 64.12 Tank filling example (I).

64.5 Advanced Elements

The GRAFCET elements already described are not a significant improvement to state transition diagrams. This may be easily understood considering that the GRAFCET basic elements are in fact based on these diagrams; therefore, the modeling capabilities must be similar. The introduction of a new set of advanced elements becomes necessary to enhance the GRAFCET modeling power.

Sequence Structures

A sequence consists of a succession of steps, where each step has only one preceding and succeeding transition, and represents the simplest sequential behavior that can be found. However, the behavior of an automated system is frequently more complex, requiring the use of advanced modeling structures. GRAFCET provides this by defining extensions to the graphical notation that are able to represent those behaviors:

- *Selection.* This structure is intended to represent a choice between alternative sequences. Example (Figure 64.13): transitions **T1, T2,** and **T3** are simultaneously enabled. One or more can be cleared simultaneously. Mutually exclusive transition clearing can only be achieved if their transition-conditions behave accordingly.
- *Parallelism.* This structure is intended to describe the behavior of parallel sequences. It is graphically represented as a double bar after a transition. Example (Figure 64.13): when transition **T4** is cleared, step **5** is deactivated and steps **6, 7,** and **8** are simultaneously activated. After their simultaneous activation, the evolution of the active steps in each parallel sequences becomes independent.
- *Synchronization.* This structure can be interpreted as the inverse of parallelism, since its intention is to join parallel sequences. It is graphically represented as a double bar before a transition. Example (Figure 64.13): transition **T5** is only enabled when all previous steps (**9, 10,** and **11**) are simultaneously active. When transition **T5** is cleared, steps **9, 10,** and **11** are simultaneously deactivated and step **12** is activated.

Variable Type Extensions

The interactions between the control part and the remaining system elements have been formalized by means of boolean input and output variables. Nevertheless, in a real system not all the variables are of the

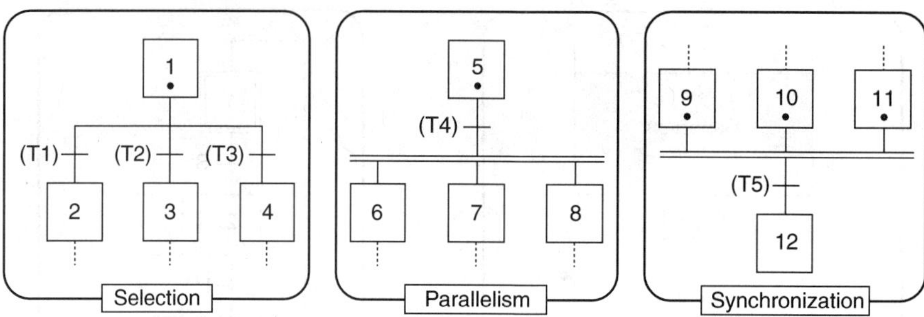

FIGURE 64.13   Sequence representation.

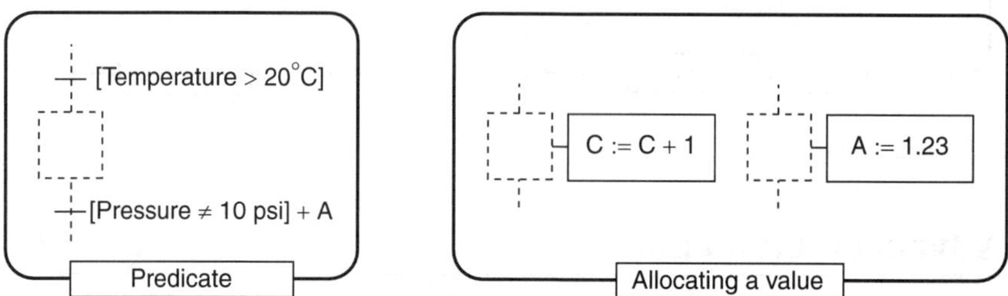

FIGURE 64.14   Predicate and allocation representation.

boolean type (e.g., temperature, pressure, counters, etc.). The system specification must include references to these non-boolean variables; otherwise, it would become inconsistent.

To overcome this limitation, GRAFCET provides an extension to the types of variables supported, without, however, establishing a formal definition of their type (e.g., integer, float, string) as classic programming languages do. Support for the extended types was achieved by naturally extending the use of boolean variables.

Since input variables are associated with transition-conditions, which are logical expressions, supporting these non-boolean input variables only requires that they be included as boolean *predicates,* graphically represented as a logical condition between square brackets (Figure 64.14).

Output variables are associated with actions. When an action is executed, a value is assigned to the variable. In the case of a boolean variable, the inclusion of its label is sufficient to establish its true value. Nonboolean variables are processed in a straightforward way, by *allocating* them a value (Figure 64.14).

Internal Variables

Another extension related with variables reflects the possibility of accessing data that are internal to the *Grafcet* structure. This is performed by distinguishing between *external* and *internal* variables:

- *External variables* are those that are external to the control part — the previously defined input and output variables.
- *Internal variables* are associated with the GRAFCET structure. For each step, a boolean variable is defined with the following syntax: X*step_label*. When the corresponding step is active the variable takes a true value; otherwise, it is false (Figure 64.15). These *step variables* have the same application scope as boolean input variables and they are used to access the step state. They become useful when synchronizing several *Grafcets* (Section 64.6).

The logical value of a boolean expression that comprises one or several boolean variables — input variables, predicate, step variables, partial *Grafcet* variables or macrostep variables (Section 64.6) — can be represented by a *logical variable.*

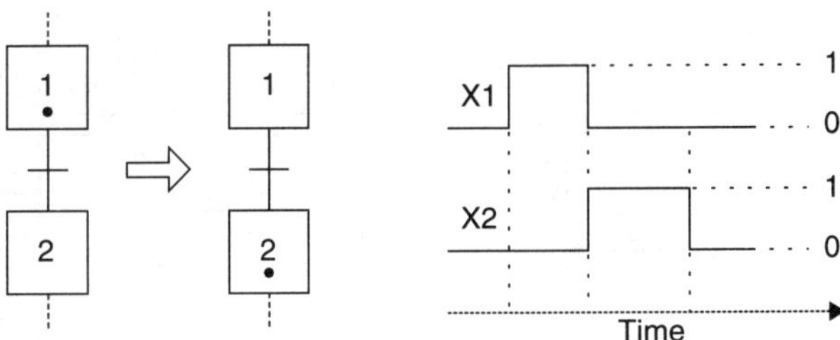

FIGURE 64.15　Step variable behavior.

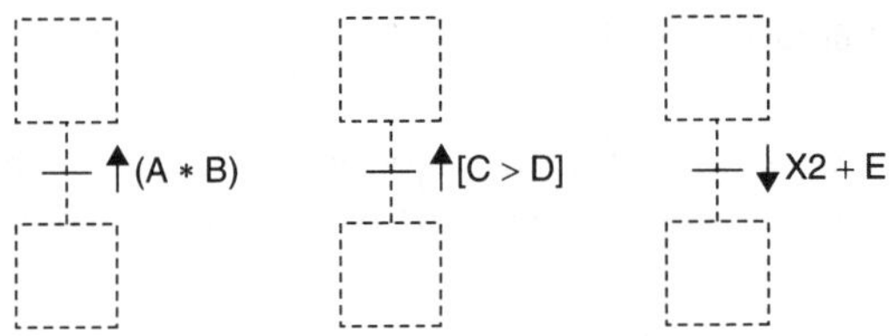

FIGURE 64.16　Input event representation.

Events

At first, it may seem that the input variable definition is sufficient to represent all the conditions that enable the *Grafcet* evolution. Nevertheless, there are scenarios where this representation is insufficient — for example, how would it be possible to express the rising edge of a variable? To overcome this, GRAFCET introduces the notion of *event*. An event is characterized by the falling or rising edge of a variable and takes a boolean value.

It is possible to distinguish between two types of events:

- *Input events*, associated with the falling or rising edge of a logical variable. When the respective edge occurs, this event takes a true value; otherwise, it is false. Since this event is related to logical variables, it has the same application scope. It is graphically represented as a ↑ (rising edge) or ↓ (falling edge) symbol, followed by the associated logical variable (Figure 64.16).
- *Internal events*, associated with a step activation, deactivation, and transition clearing. These events are closely related with *stored actions*. They are interpreted as follows: when the internal event occurs, the associated action is executed. A detailed explanation is presented in the Stored Actions Section.

Time Representation

Time plays a central role in the operation of automated systems, intervening in many aspects of their evolution. In order to track time, GRAFCET defines a *time-dependent condition*. This is a boolean condition related with an input or internal event by the notation: T1/*variable*/T2. The semantics is at follows: the condition is true after a time T1 of the occurrence of the rising edge of the logical variable, and becomes false after a time T2 from the occurrence of the falling edge (Figure 64.17). T2 may be omitted. The logical value of this condition can be represented by a *timed variable*.

Time-dependent conditions have the same application scope as standard boolean conditions — transition-conditions and timed *continuous actions*.

Action Types

Actions represent the connection between the GRAFCET evolution and the system outputs. Although the action types presented until now are capable of describing the behavior of the most common output

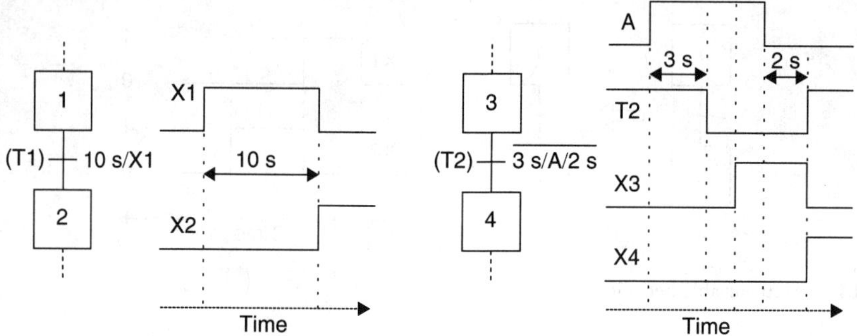

FIGURE 64.17   Time-dependent condition behavior.

types — on, off, assigning a value — they cannot represent more complex behaviors, such as timed or event actions. GRAFCET approaches this issue by extending the basic action definition into *continuous* and *stored actions*.

Continuous Actions

Continuous actions are continuously executed while an associated *assignation condition* is true. This boolean assignation condition has the same characteristics of a logical or timed variable. When the assignation condition is true, and the step associated to the action is active, a true value is assigned to the boolean output variable referenced in the action. Otherwise, the output variable value is considered false. If the assignation condition is omitted, it is considered as always true, which corresponds to the basic action definition.

A continuous action is graphically represented as a basic action with the assignation condition placed at the right of a vertical line, which is connected to the action. Some application examples are presented in Figure 64.18.

Stored Actions

Stored actions are only executed when an event occurs. A value is allocated to the output variable indicated in the action whenever the event occurs. The output variable remains unchanged until a new allocation modifies its value. Typically, this kind of action is used to process non-boolean data such as incrementing/decrementing counters, computing expressions, starting of internal tasks, etc.

A stored action is graphically represented as basic action with symbols referring to the event type. Some application examples are presented in Figure 64.19.

Output Values

From the previous discussions, the output variables behavior results as follows:

> *Continuous actions.* For a given situation, the value of the outputs variables is assigned:
> - to the true value, for each output related to an action associated with an active step for which the assignation condition is true.
> - to the false value, for the remaining outputs.
>
> *Stored actions.* The output variable is allocated to a value when the associated event occurs, and

remains unchanged until a new allocation is performed.

When an output variable is referred in more than one action, a conflict could occur if there are actions being simultaneously executed and if a contradictory assignment or allocation is performed. This problem can only be circumvented if the system designer takes the necessary measures to avoid it; otherwise, an inconsistent behavior results.

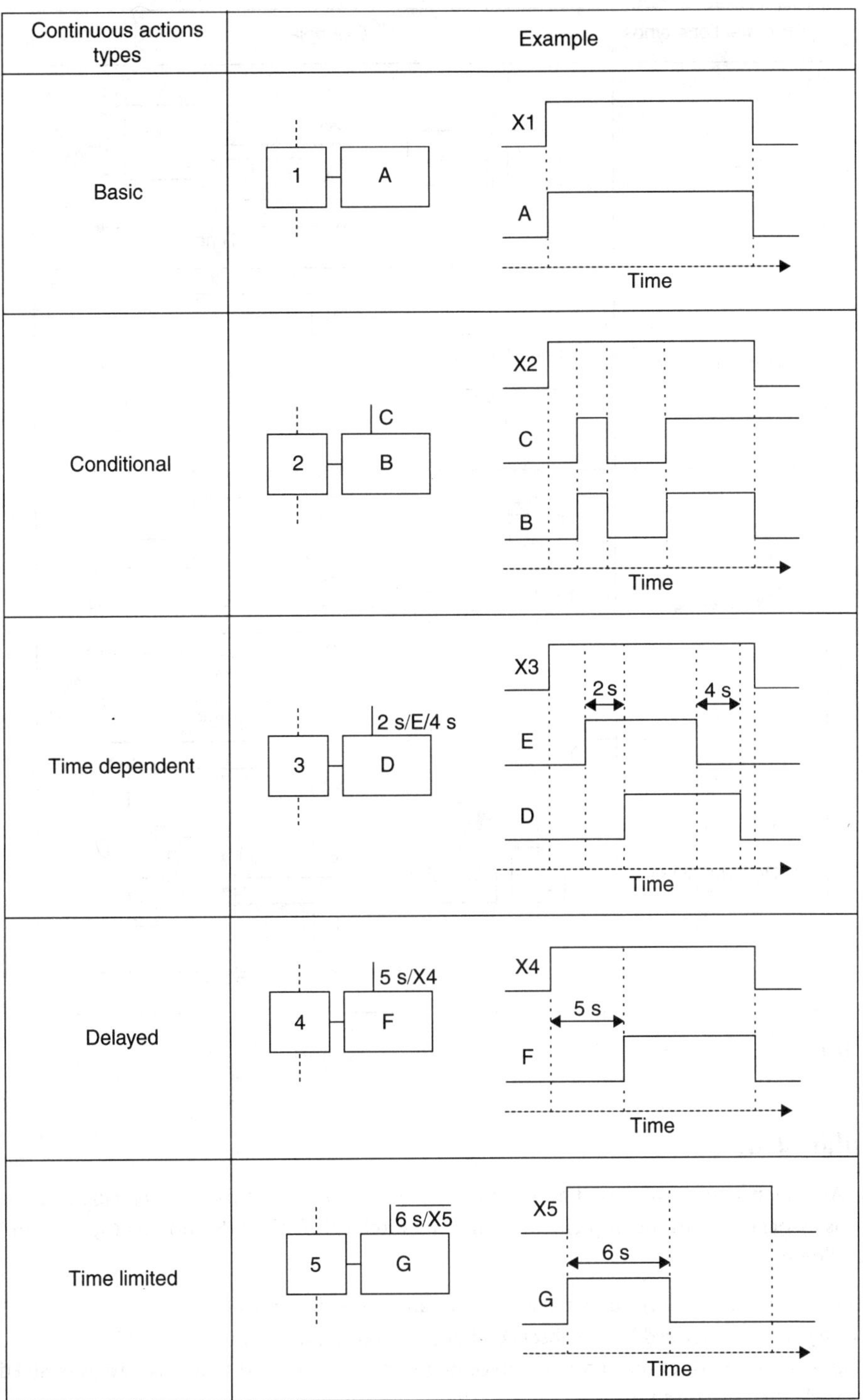

FIGURE 64.18 Continuous actions.

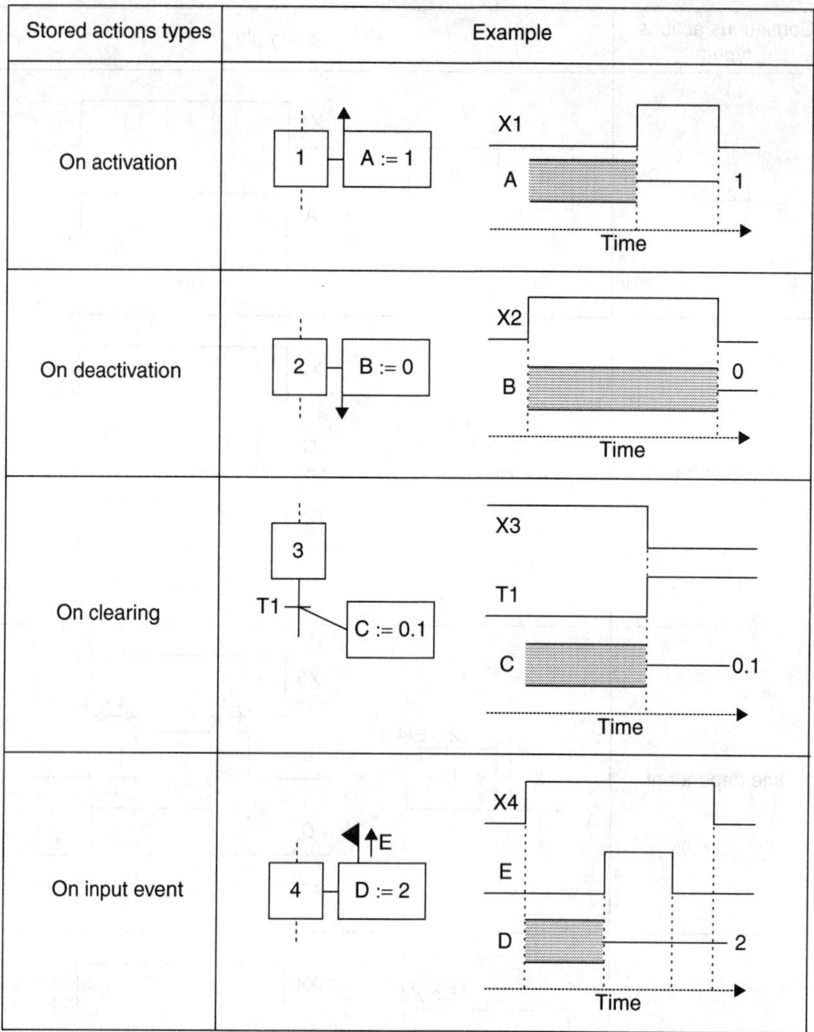

Stored actions types	Example
On activation	
On deactivation	
On clearing	
On input event	

FIGURE 64.19   Stored actions.

Particular Structures

Some GRAFCET structures represent an exception to the *alternation step-transition* rule (Figure 64.20). Their use is associated with the implementation of hierarchical *Grafcets* (Section 64.6), and can be classified as follows:

- *Source step*, is characterized by the absence of any preceding transition.
- *Pit step*, is characterized by the absence of any succeeding transition.
- *Source transition* does not have any preceding step. It is assumed that it is always enabled, being cleared when its condition-transition is true.
- *Pit transition* does not have any succeeding step.

Graphical Composition

Like any other language, GRAFCET includes some elements that help the reader to have a clear description of the charts either textually, using *comments*, or graphically – using *page references* (Figure 64.21).

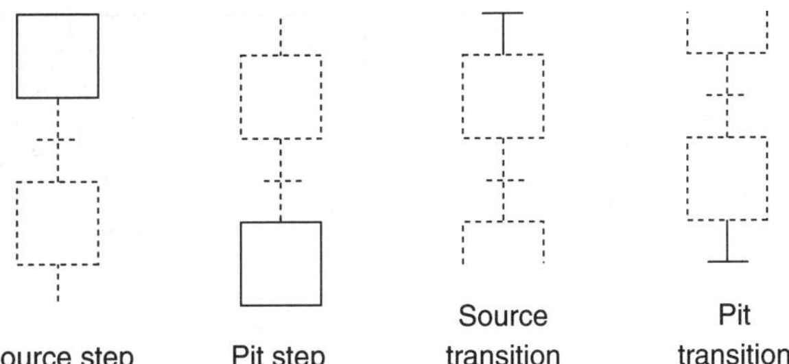

FIGURE 64.20　Particular structures.

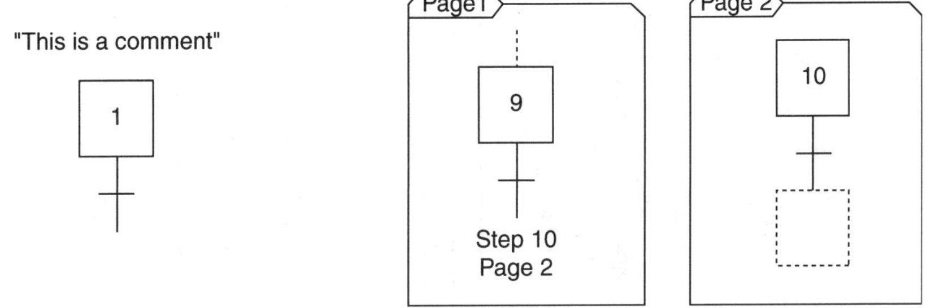

FIGURE 64.21　Graphical composition.

Tank Filling Example (II)

Let us now consider the following extensions to the specifications of the tank system presented previously. The system is now composed of two tanks used in a similar way (Figure 64.22), with the following changes:

1. The filling process is triggered by the rising edge of S.
2. Both tanks are filled simultaneously.
3. Filling may only start up again when both tanks are empty.
4. V1 is closed only 2 seconds after H1 becomes active.
5. W2 is opened only 10 sec after H2 becomes active.

A *Grafcet* that meets these specifications[8] is presented in Figure 64.22. Initially, with **step 1** active, the system waits for button **S** to be pressed. When this occurs, the event ↑S becomes true, and **steps 2** and **6** are simultaneously activated. This initiates two parallel sequences where both tanks are filled.

On the left sequence, **step 2** is activated and a stored action is used to open **V1**. When **H1** becomes true, **V1** is closed 2 seconds after that. This delay is represented by **step 3** and the time-dependent condition-transition **T3**. When **T3** is cleared, **V1** is closed by a stored action. Immediately, **step 4** is activated and the associated action, open **W1**, is executed. The tank 1 is maintained in this situation until **L1** becomes true. When this happens, **step 5** is activated and **W1** is closed. This is a waiting step required to allow synchronization between tanks.

In the sequence on the right, **step 6** is activated and the associated action, open **V2**, is executed. When **H2** becomes true, **step 7** is activated and **V2** is closed. This step has a delayed action, open **W2**, which is only executed 10 seconds after its activation. The tank 2 is maintained in this situation until **L2** becomes true. When this happens, **step 8** is activated and **W2** is closed. Just like with the sequence on the left, a waiting step is required to synchronize behaviors.

---

[8] Our aim is to present several alternative solutions, and not necessarily the simplest one.

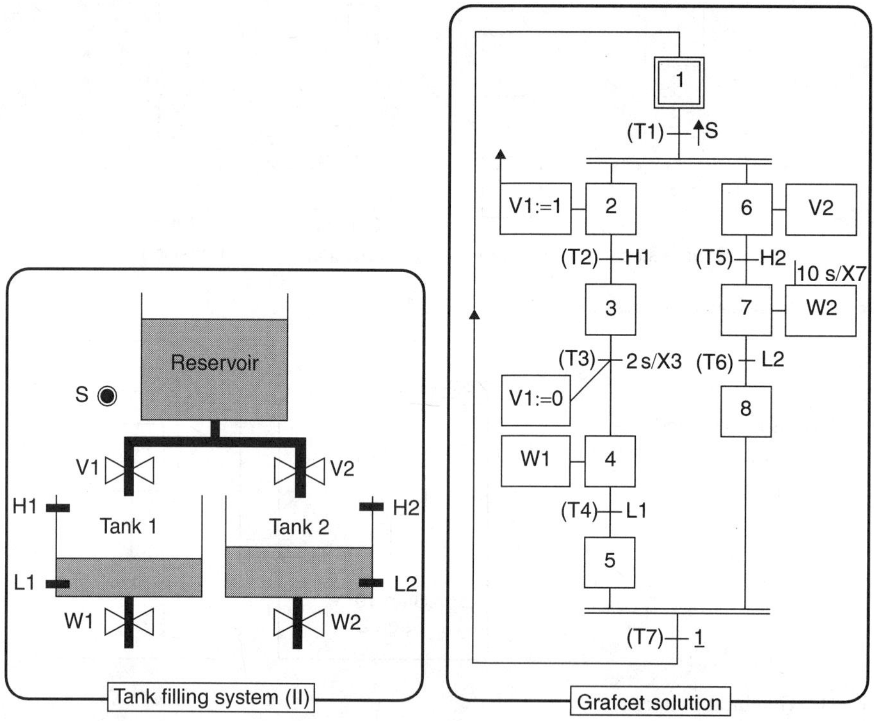

FIGURE 64.22   Tank filling example (II).

When both tanks are empty — **steps 5** and **8** are both activated — the system immediately returns to the initial situation. This is achieved by using a synchronization structure associated to a transition, **T7**, with a condition that is always true.

64.6   Hierarchical Grafcets

Some automated systems present very complex behavior. In these cases, an attempt to model the behavior using a single abstraction level results in a model almost impossible to understand. To overcome these scenarios, system designers usually use a methodology based on a top-down approach, which decomposes a single complex system in several less complex subsystems. This results in several advantages, such as: (i) better specifications, since the behavior of any subsystem is easier to understand; (ii) definition of different detail levels, which helps to identify the relationships between subsystems; and (iii) reuse of specifications — if a subsystem is replicated, a single specification suffices.

In order to support this methodology, GRAFCET defines *hierarchical Grafcets*. Using this approach, a complex *Grafcet* is decomposed into smaller ones, *partial Grafcets*, each representing part of the system behavior. Since a hierarchical relationship exists between *Grafcets*, it is naturally asymmetrical, with high-level *Grafcets* sending commands to the lower ones, and accessing their internal situation. These mechanisms allow the *Grafcets* to synchronize their evolutions (Figure 64.23).

A partial *Grafcet* is formed by one or several *connected Grafcets*. A connected *Grafcet* is one where there is always a continuity of links between any pair of elements (steps or transitions). It is identified by means of a name, with the following syntax: **G***name*. Figure 64.24 presents two partial *Grafcets*, G1 with a single connected *Grafcet* and G2 with two connected *Grafcets*.

The global state of a partial *Grafcet* is represented by a boolean variable, **XG***name*, which is true if at least one of their steps is active. Their internal situation is represented by G*name*{*steps*}, where *steps* is a list of the currently active steps. Some application examples are presented in Figure 64.24.

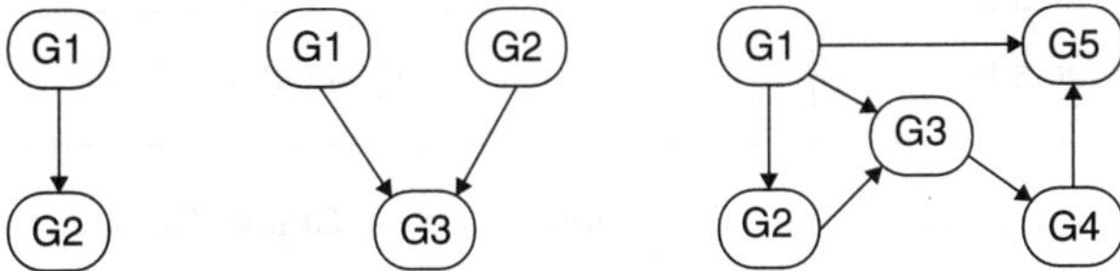

FIGURE 64.23 Hierarchical *Grafcets*.

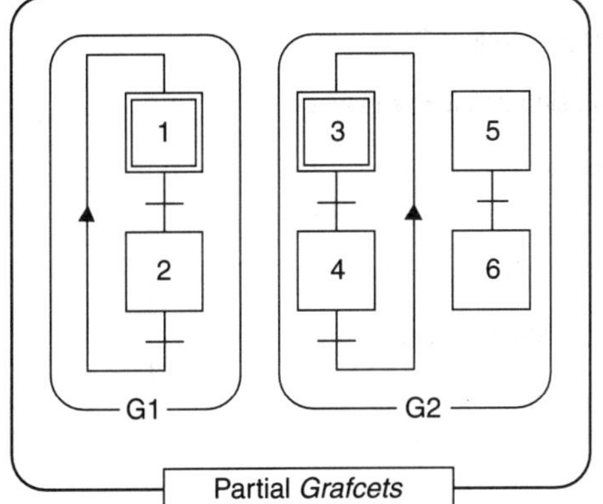

Notation	Example
G2{3,5}	Steps 3 and 5 are active
G1{*}	Current situation
G2{ }	Empty situation, i.e., deactivated
G3{INIT}	Initial situation

Internal situation

FIGURE 64.24 Partial *Grafcet* representation.

As stated previously, structured *Grafcets* need to synchronize their evolution. The GRAFCET notation was therefore extended in order to allow a partial *Grafcet* to modify the current situation of another partial *Grafcet*. The following subsections present the three distinct mechanisms that were defined to support this feature.

Forcing

With this approach, a partial *Grafcet* forces another to a specific situation. This is accomplished by associating a *forcing order* to a step. While the step is active the order is continually executed. It has precedence over the evolution rules and during their execution the forced *Grafcet* cannot evolve. This is referred as a *frozen Grafcet*.

A forcing order is graphically represented as a basic action, but with a double rectangle to distinguish it. Some application examples are presented in Figure 64.25.

Enclosure

An enclosure is an alternative to forcing. It is more powerful, although a bit more complex. The concept is based on the definition of an *enclosure* — a set of steps that are enclosed in an *enclosing step*. The set of enclosed steps can be considered a *partial Grafcet*.

The enclosing step has the same properties of a basic step and a similar graphical representation, but with an octagon inside it (Figure 64.26). The enclosure is graphically represented as partial *Grafcet* inside a rectangle, which has two labels referring to the enclosing step and the enclosure name.

When the enclosing step is activated, the steps in the enclosure identified by an *activation link* are also activated. This activation initiates the evolution of the partial *Grafcet*. When the enclosing step is deactivated, all steps in the enclosure are also deactivated. The activation link is graphically represented by an asterisk "*" at the left of the enclosed steps (Figure 64.26). This behavior is analogous to the release of a task in a programming language.

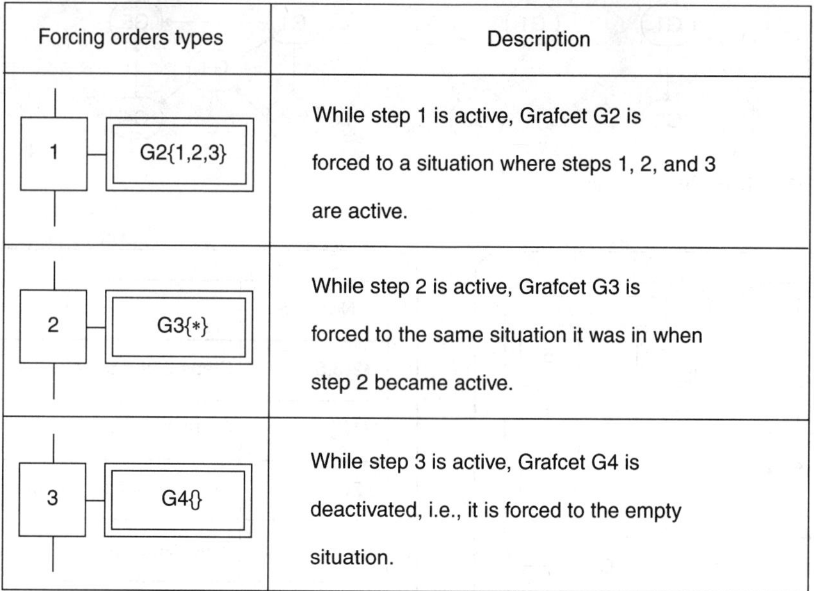

Forcing orders types	Description
1 — G2{1,2,3}	While step 1 is active, Grafcet G2 is forced to a situation where steps 1, 2, and 3 are active.
2 — G3{*}	While step 2 is active, Grafcet G3 is forced to the same situation it was in when step 2 became active.
3 — G4{}	While step 3 is active, Grafcet G4 is deactivated, i.e., it is forced to the empty situation.

FIGURE 64.25 Forcing orders.

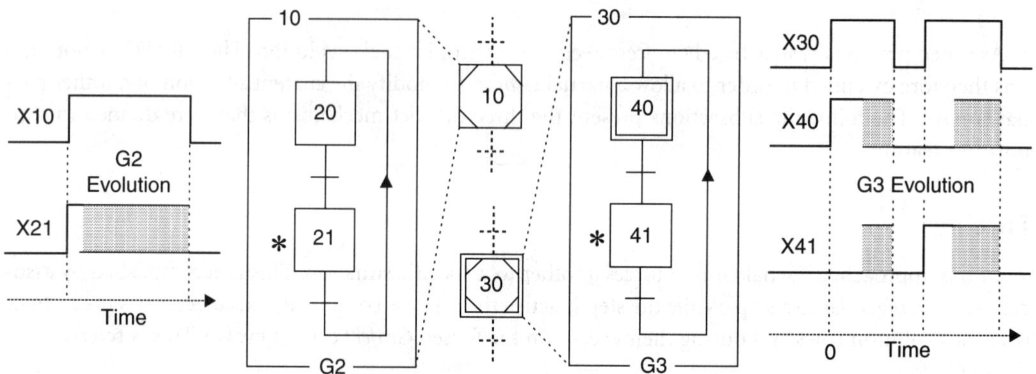

FIGURE 64.26 Enclosure representation.

An enclosure participates in the initial situation if, at least, one of the enclosed steps in each of its enclosures is also an initial step. In this scenario ($t = 0$), only the initial steps are activated (and not the activation links), the enclosure evolution being defined by the standard *Grafcet* rules. However, if the initial enclosing step is deactivated, the enclosure is also deactivated. Posterior reactivations of the enclosing step follow the defined enclosure activation rules (Figure 64.26). Its graphical representation results from merging both the initial and enclosure step representations.

An enclosing step can contain several enclosures simultaneously, which enables the parallel execution of several partial *Grafcets*. An enclosure can itself contain other enclosures, enabling the modeling of hierarchical abstraction layers.

MacroSteps

A partial *Grafcet* may be encapsulated by a special step, referred as a *macrostep*, which when activated, is expanded and substituted by the partial *Grafcet* (Figure 64.27). This encapsulation allows a gradual

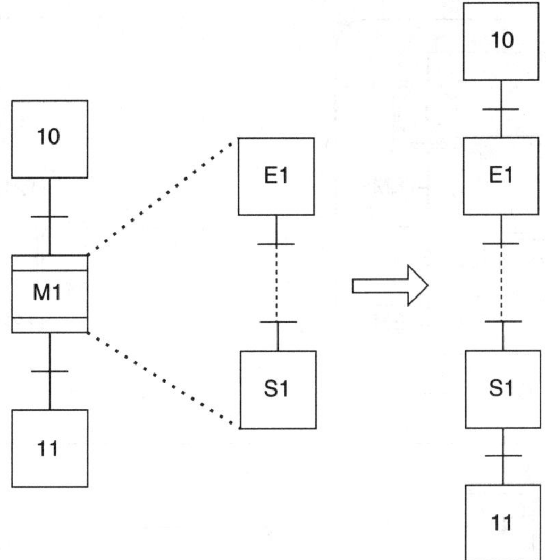

FIGURE 64.27 Macro-step representation.

description from general to particular, with a behavior analogous to the execution of a subroutine in a classical programming language.

To perform this expansion, the partial *Grafcet* must have a particular structure, being initiated by an *entry step*, and terminated by an *exit step*. The entry step becomes active when one of the preceding transitions of the macrostep is cleared. The succeeding transitions of the macrostep are enabled when the exit step becomes active. The expansion of a macrostep can have initial steps, and can recursively include other macrosteps.

A macrostep is graphically represented as a basic step with two horizontal bars inside, and is identified by means of a label with the following syntax: M*label*. Their global state is represented by a boolean variable, XM*label*, which is true if at least one of the encapsulated steps is active.

Tank Filling Example (III)

Consider the following extensions to the system specified in Tank filling example (II):

(i) A suspend–resume button (SR) that when activated suspends the filling operation. If SR is once again activated, the operation is resumed.

(ii) An emergency button (EM) that when activated forces the system to the initial situation. This button has priority over all others.

The *Grafcet* that meets these specifications is presented[9] in Figure 64.28. The system behavior is structured through the use of three *Grafcets* in a hierarchical relationship — G1, G2, and G3.

The *Grafcet* G1 is at the highest level, and related to the operation of the emergency button (EM). On the second level, *Grafcet* G2 describes the operation of the suspend-resume button (SR). Finally, *Grafcet* G3 describes the tank filling operation. Initially **steps 1, 21,** and **31** are active. This initiates the independent evolution of *Grafcets* G1, G2, and G3.

When the emergency button (EM) is pressed, **step 2** is activated, which disables the enclosure G2 and executes a forcing order on *Grafcet* G3, forcing it to its initial situation. During this period, any activations of the start button (S) or the suspend–resume button (SR) are ignored, since G3 is frozen and G2

---

[9] Our aim is to present several alternative solutions, and not necessarily the simplest one.

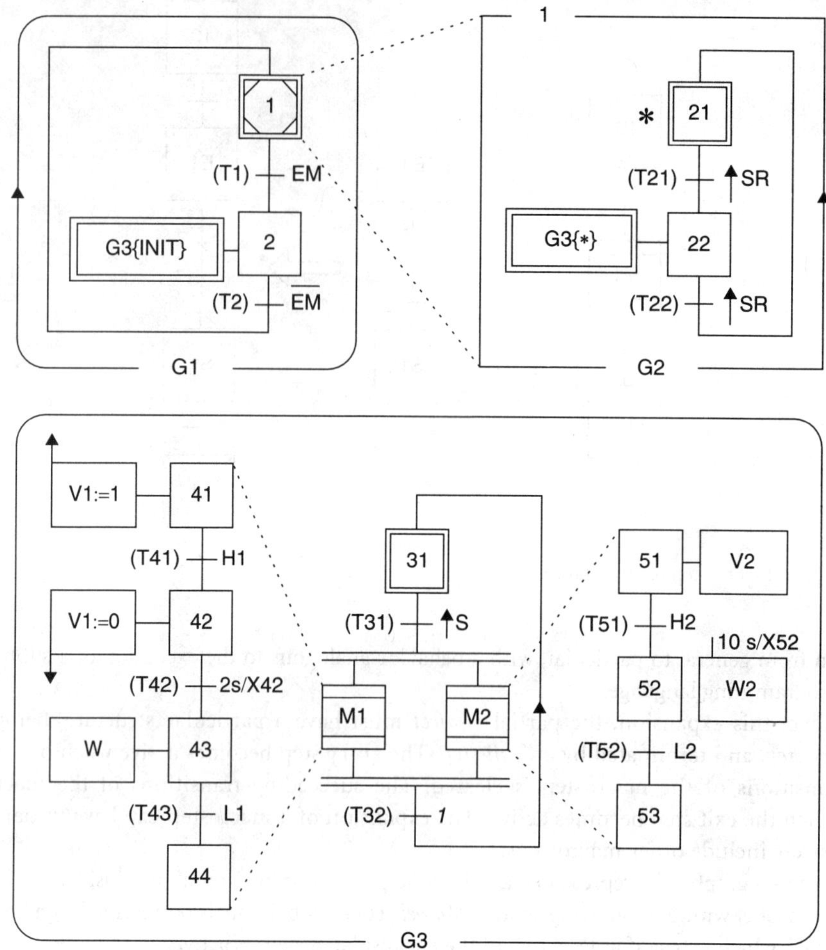

FIGURE 64.28   Example (III): *Grafcet* solution.

is deactivated. When the emergency button (**EM**) is released, the system returns to the initial situation. Since **step 1** is also an enclosing step, its activation initiates the evolution of enclosure **G2** by activating **step 21**.

If during normal system operation the suspend–resume button (**SR**) is pressed, **step 22** is activated. Since its associated action is a freezing order, *Grafcet* **G3** cannot evolve and maintains the current situation. If during this order it is necessary to disable all output variables, then the use of *conditional actions* must be employed. Therefore, the internal variable XZZ should be used as an *assigmation condition* of the actions in steps 41, 42, 51 and 52. For clearance reasons this is ommited. When the suspend–resume button (**SR**) is pressed once again, *Grafcet* **G3** is allowed to resume its normal evolution.

Grafcet **G3** has a structure similar to the example presented in the Tank filling example (II) section, with the exception that the parallel sequences were replaced by macrosteps.

Acknowledgments

The authors wish to thank Mário de Sousa for his contribution, including discussions and revision of the document. He also has contributed a chapter to this book — Programming with the IEC 61131-3 Languages and the MatPLC.

References

[1] IEC 60848 Ed. 2, Specification language GRAFCET for sequential function charts, International Electrotechnical Commission, 2001.

[2] IEC 61131-3, Ed. 2, Programmable Controllers — Programming Languages, Final Draft, International Electrotechnical Commission, 2001.

[3] David, R. and Alla, H., *Petri Nets and Grafcet — Tools for Modelling Discrete Event Systems,* Prentice-Hall, Englewood Cliffs, NJ, 1992.

[4] Home of Grafcet, http://www.lurpa.ens-cachan.fr/grafcet.html

[5] Reeb, B., *Développement des grafcets,* Éditions Ellipse, Paris, 1999 (in French).

[6] Grepa, *Le Grafcet, de nouveux concepts,* Cepadues Éditions, Toulouse, 1985 (in French).

References

[1] IEC 60848 Ed. 2 Specification language GRAFCET for sequential function charts, International Electrotechnical Commission 2002.

[2] IEC 61131-3 Ed 2 Programmable controllers — Programming languages, Final Draft International Electrotechnical Commission 2001

[3] David, R. and Alla, H., Petri Nets and Grafcet — Tools for Modelling Discrete Event Systems, Prentice Hall, Englewood Cliffs, NJ, 1992.

[4] Home of Grafcet, http://www.lurpa.ens-cachan.fr/grafcet.html

[5] Blanchard, M., Du Grafcet aux réseaux de Petri, Editions Hermès, Paris, 1992 (in French).

[6] Grafcet, Le grafcet de nouveau concept, Cepadues Editions, Toulouse, 1988 (in French).

65

Programming with the IEC 61131-3 Languages and the MatPLC

Mário de Sousa and Adriano
Carvalho
*Faculdade de Engenharia da
Universidade do Porto*

65.1 Introduction

The proliferation of Programmable Logic Controllers (PLCs) used in an industrial setting is indicative of their usefulness. These have evolved with the times, to the point that many modern top-of-the-range PLCs are actually full-fledged computers in disguise, executing modern operating systems. The programming languages with which the PLCs are programmed have also evolved with time, simultaneously becoming more powerful and complex. Nevertheless, and unlike the hardware, the programming languages used between different vendors were diverging, leading to longer learning times for the programmer when switching between PLCs.

To this end, the International Electrotechnical Commission (IEC), an international standards body, has approved a collection of standards with the intention of creating a common user experience when configuring and programming industrial controllers. One of the components of this standard, namely the IEC 61131-3 [1], refers to the programming languages.

This chapter is an introduction to the IEC 61131-3 programming languages, but assumes that the reader has previous knowledge of other procedural programming languages. Although the main aspects of the languages are described, a complete description of the languages' syntax is beyond the scope of this chapter. Readers wishing to program using these languages may use the examples in this chapter as a starting point, but are nevertheless advised to read a manual or book [2, 3] with a complete description of the IEC 61131-3 languages.

The IEC 61131 Standards

The IEC 61131 standard has its origins in 1979, when an international committee was formed with the intention of generating a standard for a common user interface for PLCs. In 1982, this committee came up with a draft that was so complex that it was then decided to divide it into five parts. The IEC 61131 standard may therefore be better described as a general framework that tries to establish the rules to which all PLCs should adhere to, encompassing mechanical, electrical, and logical aspects, and consisting of five parts: part 1 — general information; part 2 — equipment and testing requirements; part 3 — programming languages; part 4 — user guidelines; and part 5 — communications.

The third part (IEC 61131-3) deals with the programming aspect of the industrial controllers and defines the programming model, composed of three program organization units and five programming languages. The first version of part 3 was approved in March 1993. This was later revised and extended, resulting in a second edition of the standard being published in early 2003.

65.2 IEC 61131-3 Programming Basics

The IEC 61131-3 standard, which shall henceforth be referred to as simply the standard for brevity, specifies four distinct programming languages. Additionally, it also defines a state machine specification language, which is considered by many as a fifth programming language.

More importantly though, the standard also specifies basic building blocks (referred to as Program Organization Units in the standard), which may be composed into other more complex building blocks, and ultimately, the intended program. The building blocks fall into one of three types: functions, function blocks, or programs. An IEC 61131-3 programming environment comes with predefined functions and function blocks, commonly referred to as basic building blocks. A user may also program new building blocks, of any type, using any of the programming languages.

All languages support the same data types. This allows the programming language used to program the building blocks to be decoupled from the programming language used to access the building block. In other words, a building block written in one language may be called from any other programming language, without having to deal with issues of converting data types between the two languages.

It should be noted, however, that the languages are not completely interchangeable. Although a one-to-one mapping between any two languages may be defined for most of the syntax of these languages, each programming language has some features that do not map directly to the syntax of the others. This, of course, implies that some languages are better suited to specific tasks than others.

Building Blocks

As already mentioned, there are three variations of building blocks: functions, function block types, and program types. All three consist of a variable declaration part, and a code part. Variables may have local scope, that is, they are only accessible from within the building block in which they are declared, or may be input and output parameters of the building block. The code part is specified using the programming languages defined in the standard. The code may contain calls to other building blocks, with the restriction that functions may only call other functions, while programs and function blocks may call other functions or function blocks (Figure 65.1). In any of the three types of building blocks, the code part may not include recursive calling of the building block itself, either directly or indirectly through the call of another building block.

Functions have semantics similar to those in traditional procedural languages, and directly return a single output value. However, besides one or more input values (equivalent to variables passed as values),

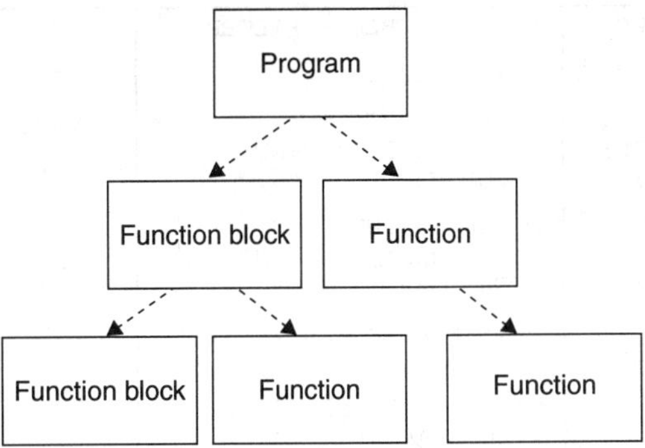

FIGURE 65.1   Calling of building blocks from within other building blocks.

the function may also have parameters used as outputs (equivalent to passing variables as references), or as input and output simultaneously. Functions may also have local private variables. This is a change from the first version of the standard, which only allows a function to have input parameters and local private variables.

Function semantics state that they are idem potent, that is, all invocations of the same function with the same input values should always yield the exact same result, whatever the state of the rest of the system, including the time at which the function is executed. When calling a function, it is not necessary to specify all the function parameters. Unspecified parameters are automatically set to their default values.

Function block types are similar to classes in object-oriented languages, with the limitation of not supporting inheritance, and having a single default public member function and no constructors or destructors. Function blocks are instantiated as variables, each with their own copy of the function block state. The default function of a function block does not directly return any value, but, like functions, may have parameters to pass data as input, output, or bidirectional. Since the function block has only a single function, calling this function is commonly referred to as "calling the function block." Likewise, this function's parameters are often referred to as the function block parameters.

When calling a function block, it is once again not necessary to specify all the parameters. Nevertheless, unlike functions, unspecified parameters are not reinitialized to their default parameters but rather retain their values of the previous call (unless it is the first time the function block is being called, in which case the default values are used). The function block parameters may therefore be said to be persistent between calls. The output and bidirectional (simultaneously input and output) parameters may be read directly from outside the function block, but may not be written to. On the other hand, input parameters may only be given new values during a function block call.

Internal variables may be persistent or temporary, but may only be accessed from within the function block itself. Temporary variables are initialized to their default values every time the function block is called, unlike persistent variables whose values remain unchanged between calls.

Since a function must be idem-potent, it can neither instantiate nor call a function block instance. It may, however, access the current values of the output or bidirectional parameters of a function block. Note that in this case the function block instance must be passed to the function as an input parameter, since a function may not instantiate a function block itself.

Program types are very similar to function blocks, with the exception that these may only be instantiated inside a configuration, and not inside other functions, function block types, or program types. Configurations are described in more detail in Section 65.7.

Figure 65.2 shows an example of interface definitions of a function, function block, and program. The text delimited by "(*' and '*)" are comments, and are ignored by the compilers. Variable declarations may be appear in any order, with local, temporary, and interface variables intermingled. Nevertheless, these appear separated in the examples to make it easier to identify the building block parameters.

```
FUNCTION WEIGH : WORD       FUNCTION_BLOCK FILTER       PROGRAM AGV

   (* Interface *)             (* Interface *)             (* Interface *)

   VAR_INPUT                   VAR_INPUT                   VAR_INPUT
      w_command: BOOL;            signal_in: INT;
      gross_w: WORD;           END_VAR                       fut_dest: loc_type;
      tare_w: INT;
   END_VAR                     VAR_OUTPUT                     add_dest: BOOL;
                                  signal_out: INT;         END_VAR
   (* Local Vars. *)           END_VAR
                                                           VAR_OUTPUT
   VAR                         (* Persistent Vars. *)         cur_dest: loc_type;
      w1, w2, w3: INT;
   END_VAR                     VAR                            dest_reach: BOOL;
   (* Body *)                     s1, s2, s3: INT;         END_VAR
   (*                          END_VAR
                                                           VAR_IN_OUT
                               (* Temporary Vars. *)
      ...                                                     emg_stop: BOOL;
                               VAR_TEMP
   *)                                                      END_VAR
   END_FUNCTION                   t1: REAL;
                                                           (* Local Vars. *)
                               END_VAR
                               (* Body *)                  VAR
                               (*                             (* ... *)
                                                           END_VAR

                                  ...                      (* Body *)

                               *)                          (*
                               END_FUNCTION_BLOCK
                                                              ...

                                                           *)
                                                           END_PROGRAM
```

FIGURE 65.2   Examples of building block declarations.

Language Overview

The five programming languages are

ST — Structured Text
IL — Instruction List
LD — Ladder
FBD — Function Block Diagram
SFC — Sequential Function Chart

Of these, LD and FBD are graphical-based languages, while ST and IL are text based. Programming with SFC may be either graphical or text based, although the text version is hardly ever used. It should be noted that the IL language is also commonly referred to as STL (Statement List). This latter nomenclature, however, is not defined in the standard, and its use should therefore be deprecated.

Programming in LD is similar to designing a relay-based electrical circuit. In fact, this language exists specifically to accommodate electrical technicians with little to no programming knowledge and skills. It can be said that LD is a historical artifact. The first PLCs were competing with existing control equipment based on hardwired relay circuits, and therefore adopted a language similar to the electrical circuits in order to ease platform acceptance by the existing technicians.

An LD program is composed of horizontal rungs connecting two vertical power rails. A rung consists of one or more relay contacts connected in series or parallel, in essence expressing a logic function, and at least one relay coil (see Figures 65.3 and 65.4), representing the variable in which the result of the logic function will be stored. The standard has also extended the language to allow calling of other building blocks (functions or function block instances), and to handle referencing data in variables more powerful than the simple Boolean type, such as analog values, counters, and timers.

```
FUNCTION_BLOCK EDGE
  (* Interface *)
  VAR_INPUT
     fall, rise: BOOL;
     signal: BOOL;
  END_VAR
  VAR_OUTPUT
    edge: BOOL;
  END_VAR  (* Local Vars. *)
  VAR
     old_signal: BOOL;
     (* ... *)
  END_VAR
  (* Body *)
  (*

        ...

  *)
  END_FUNCTION_BLOCK
```

FIGURE 65.3   Interface declaration for language examples in Figures 65.4–65.7.

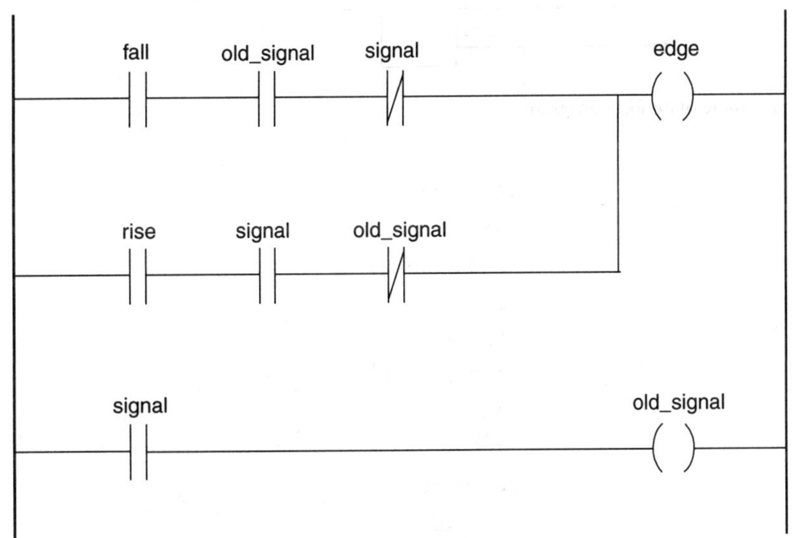

FIGURE 65.4   Example of an LD program.

The graphical FBD sits at the other extreme; a very high-level language where the user simply places boxes representing the building blocks (functions or function blocks), and defines data exchange between these blocks by drawing lines connecting the outputs and inputs of the boxes (see Figures 65.3 and 65.5). Inputs are typically drawn on the left-hand side of the boxes, while outputs are drawn on the right-hand side. Bidirectional parameters (input and output) appear on both sides, and are connected by a line within the box. The standard specifies basic building blocks that implement the basic logic operations, counters, and timers, which makes this programming language somewhat similar to designing a digital electrical circuit based on logic gates, counters, and timers.

The IL language is a low-level programming language similar to assembly. It consists of a list of simple instructions, with an optional parameter. The instruction is an operation on the value stored in a default register, and using the optional parameter, with the result being stored back in the same default register (see Figures 65.3 and 65.6). Like LD, this language has also been extended to handle calling of functions and function blocks, and deal with complex data variables.

ST caters to the user with more knowledge in Computer Science. It is a procedural language consisting of a list of statements (see Figures 65.3 and 65.7). Assignment statements may contain expressions on the right-hand side. In turn, expressions may contain calls to functions, while function block calls, which do not return a value, constitute a single statement. Iteration statements such as FOR, WHILE, and REPEAT as well as flow control statements such as IF THEN, ELSE, and CASE are supported.

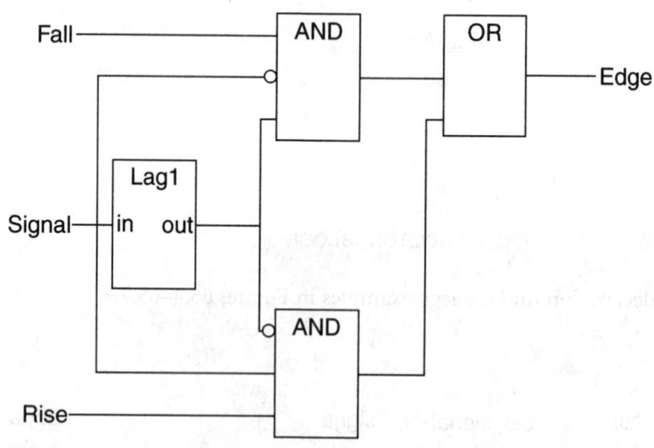

FIGURE 65.5   Example of an FBD program.

```
LD fall
AND old_signal
ANDN signal
OR (
LD rise
AND signal
ANDN old_signal
)
ST edge
LD signal
ST old_signal
```

FIGURE 65.6   Example of an IL program.

```
fedge := fall AND (old_signal AND NOT signal);
redge := rise AND (signal AND NOT old_signal);
edge := fedge OR redge;
old_signal := signal;
```

FIGURE 65.7   Example of an ST program.

Both text-based languages (ST and IL) share the same syntax for declaring the interfaces to programming building blocks, as well as for the declaration of derived data types.

The SFC is a graphical language for specifying state machines, mostly based on Grafcet. Grafcet originated in France in the 1970s and was later standardized by the IEC itself in IEC 60848 (See Chapter 64). In SFC, states of the state machine are called steps. Any number of steps may be active or deactivated at any one time. Directed links between the steps convey the step activation sequence. Between any two linked steps, exactly one transition must be placed. Likewise, between any two linked transitions, exactly one step must be found. A transition is enabled when all the preceding steps (connected to the transition by directed links) are simultaneously active. A transition may fire only when it is enabled, and when a condition associated with it is true. Firing of a transition has the effect of deactivating all preceding steps, and simultaneously activating all successor steps. One or more actions may be associated with each step. Actions are only executed while the step is active. The conditions associated with the transitions, as well as the actions associated with the steps may be specified in any one of the remaining IEC 61131-3 languages, that is, in IL, ST, FBD, or LD. Additionally, actions may also be specified using another SFC. It should be noted that without reverting to the other languages it is not possible to write a complete program using only an SFC chart.

Data Types

As stated previously, all programming languages support the same data types. Supported elementary data types, along with their size, are listed in Figure 65.8.

Note how the standard supports data types to handle time and the passing of time. The TIME_OF_DAY data type is used for absolute times in the day, such as the time to sound the siren marking the end of the shift. The TIME data type is used for relative times, such as periods, offsets, and the difference between two TIME_OF_DAYs.

All data types have a default value. This default value is used when variables are created (unless otherwise specified), and when calling a building block for parameters for which no explicit value is specified.

Derived data types are also supported. The most simple of these is a data type based on a previously defined data type and with a new default value. Besides these, the user may define subranges, enumerations, arrays, and structures. The standard does not include syntax for declaring derived data types in

Data Type	Size (bytes)	Default Value
Boolean		FALSE
Byte	1	0
Word	2	0
DWord	4	0
LWord	8	0
SINT / USINT	1	0
INT / UINT	2	0
DINT / UDINT	4	0
LINT / ULINT	8	0
Real	4	0
LReal	8	0
String	1 (per character)	`' '` (empty string)
WString	2 (per character)	`" "` (empty string)
Time	(implementation dependent)	`T#0S`
Time_of_Day	(implementation dependent)	`TOD#00:00:00`
Date	(implementation dependent)	`D#0001-01-01`
Date_and_Time	(implementation dependent)	`DT#0001-01-01-00:00:00`

FIGURE 65.8   Data types.

graphical languages, from which one may surmise that these must therefore be defined using the syntax defined for the text-based languages. In practice, however, existing programming environments allow the programmer to define the new data types through the use of graphical interfaces such as pull-down menus, lists, etc. This, of course, is not standardized.

Subranges may only be defined based on elementary integer types. Subranges based on derived integer types are not supported, including subranges of subranges. Their default value is the lowest value in the subrange, unless otherwise specified.

Data types consisting of an enumeration of possible values use the first value for the default, unless otherwise specified. The names of the values in an enumeration may be reused for other constructs such as variable names, function names, etc. In these cases, and when being used in an expression, the enumeration value may be qualified by the enumeration type in order to resolve any ambiguities.

Structures are aggregates of variables of possibly different types, and arrays sequences of variables of the same type, both similar to constructs available in other languages such as C and Java. Default values for arrays may contain different values for each element of the array. Arrays may be defined of any data type, which excludes arrays of a function block type, as function blocks are not considered a data type.

Variable Location

Before the IEC 61131-3 standard, it was common for the PLC memory to be divided into blocks, and the physical inputs and outputs to be mapped onto specific memory blocks. When using variables while programming these PLCs, the variables had to be explicitly mapped onto a memory offset in a specific memory block by the programmer.

The IEC 61131-3 standard replaces this with a programming model where the programmer merely declares the variables required, and memory allocation is done automatically by the underlying operating system, or the compiler/interpreter. Nevertheless, in order to be able to access the physical inputs and outputs of the PLC (which are still mapped to memory locations), the standard allows the explicit declaration of the memory location of variables that are declared in function blocks, programs, and configurations. Functions may not access these directly mapped variables as their value may change between calls, destroying the idem-potency property of functions. Directly mapped variables may nevertheless be passed as parameters to functions.

Variables may also be explicitly declared to be retainable or nonretainable. Retainable variables will not lose their values after a power-off/power-on cycle, while nonretainable variables are guaranteed to be reset to their default values after a PLC reset.

65.3   Programming in ST

An ST program consists of a sequence of statements, each ended by a semicolon. The most basic statements are the function block call, and the assignment statement (i.e., assigning the value of an expression to a variable).

Variables may be replaced by expressions in most locations except on the left-hand side of an assignment statement. Expressions may contain logic operators (including exclusive or), comparison, and arithmetic operators (including an exponentiation operator), besides function calls. They are evaluated from left to right, and Boolean expressions may be evaluated only to the extent necessary to determine the resultant value.

A function or function block call may follow either the formal or the nonformal method of specifying parameters. In the nonformal method, the parameters are attributed values in a comma-separated list, in the same order in which they were declared in the function or function block type declaration.

In the formal method, the parameters are assigned values in a comma-separated list of assignment statements, where the assignment symbol changes from ":=" for input and in_out parameters, to "=>" for output parameters. In this case, the parameters may be given values in any order, and a full parameter list is not required. In function calls, the remaining parameters are given their default values, while in function block calls the parameters retain their previous values. Additionally, calls to functions require at least one parameter, while calls to function blocks may be made with an empty parameter list.

The language includes two selection statements, if and case, whose syntax may be found in Figure 65.9, expressed in BNF.

Three iteration statements are also supported: for, while, and repeat. Their syntax may be found in Figure 65.10. All selection and iteration statements have similar semantics as found in other languages such as Pascal, C, and Java.

The last two statements of the ST language allow early exiting from a building block (RETURN) and early exit from the innermost iteration loop (EXIT).

An important consideration is that PLC programs are executed cyclically, and therefore should run to completion as soon as possible. For this reason, the while and repeat iteration loops should not be used for synchronization purposes, for example, by waiting on a physical input until it changes state. Synchronization should only be implemented using SFC (with conditions associated to transitions), this because an SFC program is typically compiled to test the conditions on each execution cycle, and not block on that condition.

A further issue is that, unlike the first version of the standard, the second version does not define the IL operators as reserved words. This means that it should be possible to declare variables, functions, and function block types with names that clash with the IL operators such as LD, LDN, ADD, etc. Current IEC 61131-3 compilers were based on the first version of the standard and do not accept such use of these keywords, even though the generated error messages may be somewhat cryptic and difficult to understand. We strongly recommend not using IL operators as described above, even in compilers based on the standard's second version, since using it may lead to ambiguities.

```
IF <expression> THEN <statement_list>
{ELSEIF <expression> THEN <statement_list>}
[ELSE <statement_list>]
END_IF


CASE <expression> OF
<case_list> : <statement_list>
{<case_list> : <statement_list>}
[ELSE <statement_list>]
END_CASE


case_list: <value> {, <value>}
```

FIGURE 65.9   Syntax of IF and CASE statements in ST.

```
FOR <control_variable> := <expression> TO <expression> [BY <expression>]
DO <statement_list>
END_FOR


WHILE <expression> DO <statement_list> END_WHILE


REPEAT <statement_list> UNTIL <expression> END_REPEAT
```

FIGURE 65.10   Syntax of iteration statements in ST.

65.4 Programming in IL

An IL program consists of a sequence of instructions, one per line. Each instruction may be preceded by an optional label, and followed by an optional instruction modifier and an optional parameter. Each instruction typically executes an operation on the value stored within a default register, and possibly the optional operand, and stores the result back in the same default register. The operand may take the form of a variable name or a literal value.

The most basic instructions copy the operand to the register (LD), or the value in the register to the operand (ST). Other simple instructions may be executed or not, depending on the current value in the default register. This is the case of the set (S) and reset (R) instructions that will set or reset the operand if the register contains the Boolean true value, or simply not do anything otherwise.

Binary logic instructions (AND, OR, XOR, ANDN, ORN, XORN) operate on the register and the operand, and store the result back in the register. The same occurs with the arithmetic instructions (ADD, SUB, MUL, DIV, MOD) and comparison instructions (EQ, NE, GT, GE, LT, LE). For these instructions, the type of value currently stored in the default register has to match the type of value of the operand, for example, it is an error to try and execute a logical instruction (AND, OR, ...) when the default register does not currently contain a boolean value.

The above instructions may be followed by the "(" modifier, in which case the operand is not required. This will result in the current value of the register being pushed onto a last-in-first-out stack, and a new sequence of instructions executed. Only when the corresponding ")" is found by itself in a line is the instruction behind the "(" modifier executed, using as operands the value at the top of the stack and the value in the default register.

The sequence of instruction execution may be changed with the JMP <label> instruction, which jumps to the instruction that is preceded by the specified label. Similarly, the RET instruction exits from the current building block. Both these instructions may be modified by the C (e.g., JMPC) or CN (e.g., RETCN) modifiers, in which case they will only be executed if the default register currently contains the boolean true value in the case of C, or false value if CN is used.

In IL, a function is called by placing the function name in the operator field, followed by a parameter list (see Figure 65.11). Similar to ST, a function may be called using either the nonformal or formal

```
(* Calling function LIMIT *)
(* assume LIMIT requires 3
 * parameters...
 *)


(* Using formal argument list *)
LIMIT (
    in  := 1,
    min := 2,
    max := 3
  )
ST result


(* Using non-formal argument list *)
LD 1
LIMIT 2, 3
ST result
```

FIGURE 65.11   Calling functions from IL.

method. However, and unlike ST, in the nonformal method the first parameter is taken from the value stored in the default register. In either case, the value returned by the function call will be stored back in the default register.

Function blocks may be called using one of four options (see Figure 65.12). Three of these options require the use of the CAL operator, which may be followed by the C or CN modifiers. Using CAL, the function block parameters are passed using: (1) a formal argument list, (2) a nonformal argument list, and (3) storing of the parameters in the input variables, followed by reading of the output variables. The third option is an exception to the previously mentioned rule stating that values to the input parameters may only be attributed during a function block call.

The fourth option only supports calling any of the following function blocks defined in the standard library: R_TRIG and F_TRIG (rising and falling edge detector, respectively), CTU, CTD, CTUD (up, down, and up-down counter, respectively), TP, TON, TOFF (pulse, on-delay, and off-delay timers, respectively), SR and RS (bi-stable flip-flops, set and reset dominant, respectively). This fourth option uses the following special input operators that replace the CAL operator: S, S1, R, R1 (set and reset for SR and RS), CLK (clock for R_TRIG and F_TRIG), CU, CD, R, PV (count up/down, reset, and present value for counters), IN, PT (input and pulse time for timers). In this case, the value in the default register is given to the function block parameter with the same name as the input operator, all other function block parameters remain unchanged, and the function block is called. Using this method, only a single parameter of the function block may change its value between consecutive calls.

```
(* Calling function block *)
(* assume function block COUNT
 * is an instance of the CTU
 * function block type.
 * CTU has 3 input parameters: CU, R, PV
 * and 2 output parameters: Q, CV
 *)


(* Using formal argument list *)
CAL COUNT (CU:=%IX3, Q=>OUT)


(* Using non-formal argument list *)
CAL COUNT (%IX3, FALSE, A, OUT, B)


(* Using pre-loading of values *)
LD %IX3
ST COUNT.CU
CAL COUNT
LD COUNT.Q
ST OUT


(* Use of special input operators *)
LD %IX3
CU COUNT
LD COUNT.Q
ST OUT
```

FIGURE 65.12 Calling function blocks from IL.

65.5 Programming in FBD

As stated previously, an FBD is composed of boxes representing functions and function block instances, with lines connecting data inputs with data outputs. Data are transferred from the outputs to the inputs without any associated data retention.

The body of a function block type may contain an FBD consisting of several networks, where each network is self-contained and does not exchange data with the other networks. In this case, the order by which these networks are evaluated is either implementation defined, or may be explicitly set by the user using an implementation-defined method.

Each network is evaluated according to the following rules:

1. A function or function block is only evaluated once all the values of its inputs have already been evaluated.
2. The evaluation of a network is not complete until all its outputs have been evaluated.

Networks containing loops are allowed, that is, networks where a function or function block output is connected to one of its inputs, or the input of another function or function block that precedes it in the network. In this case, the sequence of evaluation may be defined using an implementation-dependent method. Another alternative is to explicitly specify a feedback variable in one of the output to input connections in the loop. In this case, the evaluation of the network will use the previous value of the feedback variable to determine a new value.

A name that terminates in a colon ":" may be associated to each network. This allows the transfer of program control from one network to another. Transfer of control (conditional or unconditional jump) is indicated by a Boolean signal line terminated in a double arrowhead pointing to the name of the network to which execution shall be transferred. The jump is only executed if the Boolean value of the line is true.

Return from a building block is similarly indicated by a Boolean signal line terminated by the symbol "<RETURN>." The return statement is also only executed if the value of the signal line is true.

65.6 Programming in LD

An LD diagram is composed of two vertical "power" rails, connected by horizontal rungs. A rung consists of contacts and coils connected in series or parallel (resembling an electrical circuit). The left vertical power rail is considered having "power," that is, at logical state 1. Like an electrical circuit, "power" flows from this rail to each rung.

The semantics of a contact resembles the workings of a contact in an electrical relay. It receives power from one side and, depending on the logical state of the variable associated with the contact, it may or may not allow the power to flow onwards to the other side on the right.

Similarly, the semantics of a coil resembles the workings of a coil in an electrical relay. If it receives power from the left, it activates the variable associated with the relay, and allows the power to flow onwards to the right. By connecting the contacts appropriately, and choosing the variables associated with each contact and coil, it is possible to define logical expressions.

Several variations of contacts and coils are accepted, as described in the table of Figure 65.13.

Just like FBD, LD diagrams may also be given a name terminated by a colon ":", and transfer of program control is also indicated using the same symbols (with the same semantics) as those used in FBD.

65.7 Configurations

With the building blocks described thus far, it is already possible to write complete PLC programs. While this may be sufficient for many simple applications that are controlled by a single program, applications that require a PLC with multiple CPUs also require multiple programs executing concurrently.

The standard takes this into account by defining a new modeling layer at a higher abstraction level (see Figure 65.14). This new modeling layer allows the programmer/user to specify a configuration for the

Symbol	Semantics
--| |--	**Normally open contact** Power is allowed to flow if the associated boolean variable is True.
--|/|--	**Normally closed contact** Power is allowed to flow if the associated boolean variable is False.
--|P|--	**Positive transition-sensing contact** Power is allowed to flow during a single evaluation of this element if the associated boolean variable changes from False to True.
--|N|--	**Negative transition-sensing contact** Power is allowed to flow during a single evaluation of this element if the associated boolean variable changes from True to False.
--( )--	**Coil** If power reaches the left link, it flows onwards to the right link and the associated variable is set to True. Otherwise the variable is set to False.
--(/)--	**Negated coil** If power reaches the left link, it flows onwards to the right link and the associated variable is set to False. Otherwise the variable is set to True.
--(S)--	**Set (latch) coil** If power reaches the left link, it flows onwards to the right link and the associated variable is set to True. Otherwise the variable is left unchanged.
--(R)--	**Reset (unlatch) coil** If power reaches the left link, it flows onwards to the right link and the associated variable is set to False. Otherwise the variable is left unchanged.
--(P)--	**Positive transition-sensing coil** If power reaches the left link, it flows onwards to the right link. The associated variable is set to True during a single evaluation of this element whenever the power on the left link transitions from off to on.
--(N)--	**Negative transition-sensing coil** If power reaches the left link, it flows onwards to the right link. The associated variable is set to True during a single evaluation of this element whenever the power on the left link transitions from on to off.

FIGURE 65.13 Symbols and semantics of relays and contacts in LD.

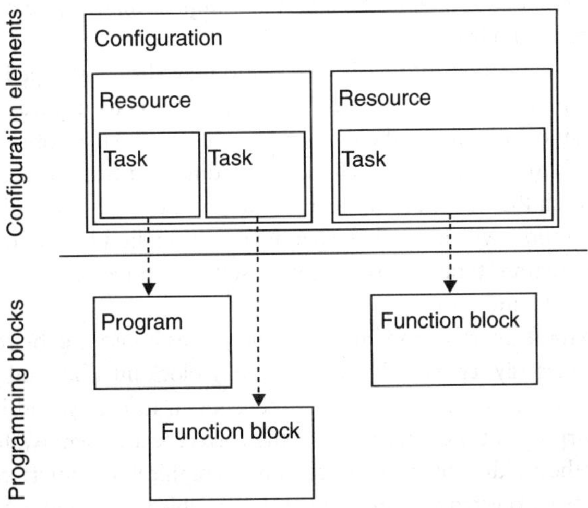

FIGURE 65.14 The components of a configuration.

PLC with possibly multiple CPUs. A configuration contains enough information to completely specify which programs should execute in which CPUs, as well as new variables with global scope. It does not contain executable code, but rather instantiates programs and/or function blocks, creates tasks, and assigns these programs and/or function blocks to tasks.

More specifically, a configuration contains the definition of all the resources it requires, and the type of each resource. In reality, a resource is nothing other than a CPU on the PLC rack, while a resource type defines the CPU model. Each vendor is expected to provide a library of resource types that it supports.

Each resource may be attributed one or more tasks. Tasks are similar to processes in common operating systems, but are intentionally very poorly defined by the standard, which limits itself to defining how tasks should behave, but not how they should be implemented by each PLC vendor. In reality, these may be implemented as processes running on an operating system, as threads in a single process, or ultimately not at all. Tasks may be configured to execute periodically, or to execute upon the occurrence of the rising edge of a specified Boolean variable.

Each task is configured to execute one, or more, program or function block instantiations. The number of tasks a resource type supports, as well as the properties that may be applied to it (such as the period), are implementation dependent. Most simple CPUs support a single fixed task, while some may support a fixed number of two or three tasks. Currently, few CPUs permit the programmer/user to define large numbers of new tasks, which the standard allows.

Besides programs and function blocks, variables may also be instantiated inside a configuration, or inside a resource within the same configuration. The scope of variables, programs, or functions blocks instantiated within a resource is limited to that same resource.

The standard refers to variables created within a configuration or a resource as global variables. These may be accessed by the programs and function blocks created in the same configuration. Nevertheless, from the point of view of the programs or function blocks, these variables are external to them, and must be declared in the program or function block type declarations as external variables before they may be accessed. The mapping between external and global variables must be explicitly specified when instantiating a program or function block in a configuration. External variables may therefore be better viewed as references to variables instantiated outside the program or function block type.

65.8 Programming Example

The example we have chosen is a small part of a real industrial application, and consists of a conveyor that crosses three ovens. Each oven is composed of a heat exchanger, where the temperature is controlled by opening or closing a valve that regulates the amount of hot oil entering the heat exchanger. The oil is heated in a single gas burner and pumped to the ovens. To simplify matters, we consider that the valves have an analog input to control their state.

The system may work in manual or automatic mode, and may also be stopped. Manual mode allows the operator to control the position of each valve directly. This is done by using a pair of up/down pushbuttons for each valve. In automatic mode, the temperature of each oven is controlled by a PID loop, as is the temperature of the oil leaving the heater. The same up/down pushbuttons are now used to change the target temperature of each PID loop. From the above description, it is fairly obvious that the system has two main modes of operation, which should be handled at the highest level of our program. This shall be a building block of the program type, and is basically a state machine; hence, we use an SFC to define its algorithm (see Figures 65.15 and 65.16).

Notice how we have given the interface and body separately. This is because the standard itself does not specify a method to simultaneously represent both the building block interface and graphical body, which means that the method of doing so is implementation-dependent. In reality, and in most programming environments, instead of writing out the interface with the correct textual syntax, the environment allows the programmer to specify the building block interface using a graphical user interface, for e.g. a list of variables. The programmer is then expected to simply insert the variable names and use drop-down menus to choose the data types, data direction (input, output, or in_out), etc. Defining the body usually involves

```
PROGRAM OVENLINE
   (* Interface *)
   VAR
      run  AT %IX1.2.3 : BOOL;   (* run/stop mode_switch *)
      auto AT %IX1.2.4 : BOOL;   (* auto/manual mode_switch *)
      gas_valve AT %QW3.0 : INT;
      oven_valve1 AT %QW3.1 : INT;
      oven_valve2 AT %QW3.2 : INT;
      oven_valve3 AT %QW3.3 : INT;
   END_VAR
   VAR
      ovenline_auto: OVENLINE_AUTO_T;
      ovenline_man: OVENLINE_MAN_T;
   END_VAR
   (* Body -see Figure 65.16 *)
END_ PROGRAM
```

FIGURE 65.15 Interface of the main program.

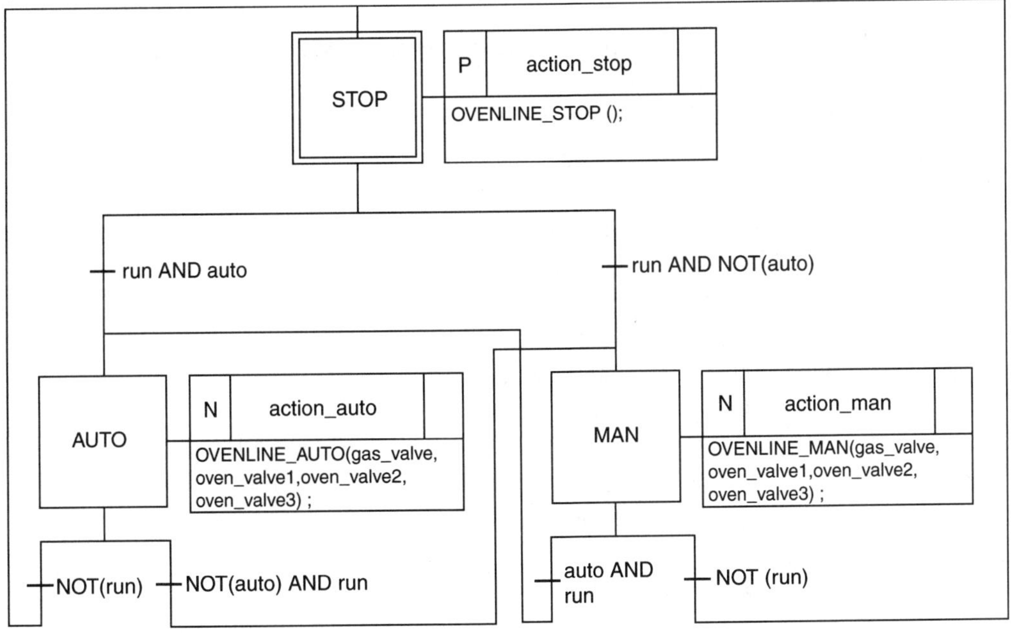

FIGURE 65.16 Body of the main program.

double-clicking the icon representing the building block (or some other similar action on the part of the programmer), choosing the language (SFC, FBD, LD, IL, or ST), and actually writing or drawing the body.

The SFC of Figure 65.16 contains three steps or states (STOP, AUTO and MAN), of which only STOP is active upon startup. From this step, the SFC will evolve to AUTO if the "run" button is activated while the "auto" switch is on, or to the MAN mode if the "run" button is pressed while the "auto" switch is off. In this case, "on" and "auto" refer to the variables declared in Figure 65.15, which are themselves mapped onto digital inputs of the PLC.

From the AUTO mode, the SFC may evolve to either the STOP or MAN modes. Notice how the condition associated with the transition from AUTO to MAN mode contains a test to verify if "run" is still True. If this were not the case, a situation could occur whereupon both the transitions from AUTO (to STOP and MAN) could be simultaneously fired. If this were allowed to happen, then the SFC would evolve to a state where both the STOP and MAN steps would be simultaneously active, which is not desirable in our example. The same may be said of the transitions from AUTO to MAN and STOP.

The action "action_stop" is associated with the STOP step, which consists of calling the ovenline_stop building block. This building block will be called once, and once only, whenever the STOP step is activated, due to the pulse (P) association of the action with the step. The ovenline_stop building block is expected to shut down the production line, such as stopping the oil pump and the conveyer, and closing all the valves if required.

Similarly, associated with the AUTO step is the action "action_auto," which consists of calling the ovenline_auto building block. Unlike the ovenline_stop building block, the ovenline_auto building block will be called continuously, as long as the AUTO step is active, due to the nonstored (N) association of the action with the step.

We now have to write the building blocks that are called from the main SFC. The ovenline_man building block (for the manual mode) will consist of reading up/down pushbuttons for several valves, and changing the state of those valves accordingly. Although the standard up/down counter function blocks could be used for this purpose, we will write our own for demonstration value (see Figure 65.17). We will then instantiate one counter function block for each valve.

```
1          FUNCTION_BLOCK UPDOWNCOUNTER
2            (* Interface *)
3            VAR_INPUT
4              up    : BOOL;
5              down  : BOOL;
6            END_VAR
7            VAR_OUTPUT
8              counter : INT := 0;
9            END_VAR
10           (* Body in IL *)
11           LD up
12           JMPC inc
13           LD down
14           JMPC dec
15           RET
16      inc: LD 1
17           ADD counter
18           ST counter
19           RET
20      dec: LD -1
21           ADD counter
22           ST counter
23           RET
24           END_FUNCTION_BLOCK
```

FIGURE 65.17 Up/down counter building block.

The up/down counter function block maintains a counter (line 8) that starts off with the value 0. Since this building block will be called once every iteration, it will continually increment (decrement) its value while the up (down) Boolean input is True. Line 11 loads the default register with the value stored in the up variable. The next line will then jump to line 16 (labelled inc) if the Boolean value currently in the default variable is True. On line 16, the default register is loaded with the value 1. This is then added to the value stored in counter (line 17), and the result stored back in the default register. Line 18 then copies the result in the default register into the counter variable, and the function returns (line 19). The remaining code is similar to the lines previously explained.

The continuous increment or decrement of the counter value will allow the user controlling the factory floor to control the position of a valve by simply keeping a pushbutton pressed. This function could be further augmented to allow for an increasing rate of change, starting off slowly at first, and increasing as time goes by.

The building block that will control the ovens while these are in manual mode (see Figure 65.18) simply instantiates an up/down counter for each valve, and calls these counters once every iteration with the appropriate pushbuttons as parameters.

The automatic mode requires that we implement several PID loops. Although it is common for vendors to supply functions that implement these loops, we implement them ourselves for demonstration purposes, starting off by writing an open-loop PID controller. Due to the mathematical operations that need to be programmed, it is best to use the ST language (see Figure 65.19). The KP, KI, and KD are the proportional, integral, and derivative parameters of the PID controller, respectively. The controller is expected to be called continually every delta_time time units. It ends by storing the current value of the input (in the prev_in variable); hence, the rate of change may be determined in the next iteration (which is stored in D). The I variable contains the integral of the input, which is constantly updated every iteration.

We are now prepared to define the function block to handle the automatic mode. Since it will have to close the PID loops, using the PID function block, it is probably easiest to program it using FBD (see Figures 65.20 and 65.21).

```
FUNCTION_BLOCK OVENLINE_MANUAL_T
  (* Interface *)
  VAR_IN_OUT
    gas_valve : INT;
    oven_valve1, oven_valve2, oven_valve3 : INT;
  END_VAR
  VAR
    gas_valve_c    : UPDOWNCOUNTER;
    oven_valve1_c, oven_valve2_c, oven_valve3_c: UPDOWNCOUNTER;
  END_VAR
  VAR
  gv_up_pushbutton AT %IX1.5.4 : BOOL;
    ...
  END_VAR
  (* Body in ST *)
  gas_valve_c(gv_up_pushbutton, gv_down_pushbutton, gas_valve);
  oven_valve1_c(ov1_up_pushbutton, ov1_down_pushbutton, oven_valve1);
    ...
  END_FUNCTION_BLOCK
```

FIGURE 65.18 Manual mode building block.

```
                    FUNCTION_BLOCK PID
                    (* Interface *)
                    VAR_IN
                        KP, KI, KD : REAL;
                        in : REAL;
                        delta_time : REAL;
                    END_VAR
                    VAR_OUTPUT
                        out : REAL;
                    END_VAR
                    VAR
                        D, I, prev_in : REAL = 0;
                    END_VAR
                    (* Body in ST *)
                    I := I + in*delta_time;
                    D := (in - prev_in) / delta_time;
                    out:=  KP*in + KD*D + KI*I;
                    prev_in := in;
                    END_FUNCTION_BLOCK
```

FIGURE 65.19   PID building block.

```
FUNCTION_BLOCK OVENLINE_AUTO_T
  (* Interface *)
  VAR_IN_OUT
    gas_valve : INT;
    oven_valve1, oven_valve2, oven_valve3 : INT;
  END_VAR
  VAR
    desired_oiltemp    : UPDOWNCOUNTER;
    desired_oventemp1, desired_oventemp2, desired_oventemp3: UPDOWNCOUNTER;
    oil_pid, O1_pid, O2_pid, O3_pid: PID
  END_VAR
    VAR
    measured_oventemp1 AT %IW5.1: INT;
    measured_oventemp2 AT %IW5.2: INT;
    measured_oventemp3 AT %IW5.3: INT;
    oil_up_pushbutton AT %IX1.5.4 : BOOL;
    ...
  END_VAR
  (* Body -see Figure 65.21 *)
END_FUNCTION_BLOCK
```

FIGURE 65.20   Interface to the auto building block.

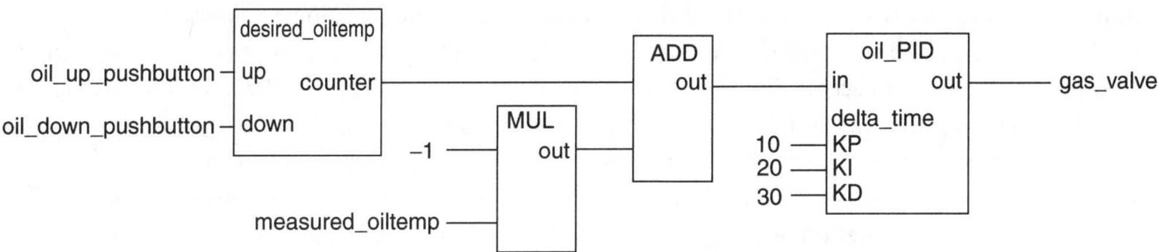

FIGURE 65.21   Body of the ovenline-auto-t building block.

The "desired_oil_temp" function block in Figure 65.21 is declared in Figure 65.20 as being an "updowncounter" (see Figure 65.17). The "up" and "down" inputs of this function block are connected to variables declared in Figure 65.20. These variables are mapped onto the digital inputs of the PLC to which the relevant pushbuttons must be physically connected. The user can then change the target temperature for the oil by simply pressing these up and down pushbuttons.

The difference between the desired temperature and the current oil temperature consists of the error, which the PID loop must eliminate. The error value, computed with the help of the MUL and ADD standard functions, will be the input to the PID block, along with the P, I, and D constants. The output of the PID block is sent to the gas_valve variable, once again declared in Figure 65.20, and mapped to an output.

65.9 The Future

The IEC 61131-3 standard, along with its programming languages, is already well understood and widely used in the industrial world. Nevertheless, this standard relates only to PLCs working in isolation. The companion standard IEC 61131-5 specifies the mechanism of reading and writing of specific values remotely over a network, but it falls short of defining a framework in which PLCs may cooperate in a networked environment.

The IEC soon recognized that the future belongs to networked equipment, and therefore started the IEC 61499 standard [4], which defines a framework for configuring cooperating devices connected by a data network. In this framework, each PLC is considered a device. The framework allows each device to contain zero or more resources (i.e., CPUs), which is in accordance with the IEC 61131-3 standard. Nevertheless, the framework now considers applications that span multiple resources and multiple devices, which means that an application may be distributed among several PLCs, and the same PLC may be executing several applications or subapplications.

Applications are composed of a group of cooperating subapplications or function blocks, with inputs and outputs connected like a network (similar to an FBD). Subapplications are a network of function blocks. Function blocks themselves may be composed of a network of other function blocks. Unlike IEC 61131-3, the framework does not contemplate other building blocks such as functions and programs.

The IEC 61499 function blocks have substantial extensions in relation to the IEC 61131-3 function blocks. Besides normal data passing through input and output variables, information may also be transferred through events. Each function block defines input ports through which it may receive events, and output ports through which it may generate events. A network of function blocks is comprised of several function blocks with output event ports connected to input event ports, and output data variables connected to input data variables.

Basic function blocks (i.e., those that are not composed of a network of other function blocks) may also have several execution states. An execution state chart defines the state transitions and when certain actions or algorithms will be performed. The algorithms are specified using one of several programming languages, including all the languages specified in IEC 61131-3, such as ST and IL.

Applications interact with the real world through specialized function blocks known as service interface function blocks. The standard defines service interface function blocks to access the physical inputs and outputs of the PLC, to access the data network connecting the distributed PLCs, and to manage the state of the user's applications and the PLC itself. This means that access to the PLC's physical inputs and outputs is no longer achieved through the use of located variables, as is done in IEC 61131-3.

The standard merely specifies the interface and the semantics of each service interface function. It does not specify their implementation, leaving this to the PLC vendors. This allows, for example, the coexistence of several network service interface function blocks, each implementing a specific network protocol, but all furnishing the exact same interface to the application. The network service interfaces defined in the standard support the publish/subscribe and the client/server models. As long as the application only accesses the network through these service interface function blocks, the application itself does not have to change in order to use a new network protocol.

65.10 Advantages of the IEC 61131-3 Languages

The main advantage of the standardization of the PLC programming languages is the portability between PLC vendors of both the software itself, as well as the programmers writing the software. Nevertheless, PLC vendors have still been able to lock users into their product family by simply saving the software into a nonstandard file format. This is allowed since the standard merely specifies the language syntax, and not the format of saving the programs in a file. This is currently being worked on by the IEC, which will shortly release a standard specifying a common file format based on XML.

Apart from the standardization advantage, the languages themselves include features that greatly ease the development process, such as automatic memory allocation for data variables, strong consistency of data types, complex data structures, and the definition of the programming blocks, which allow better code structuring and code reuse.

65.11 The MatPLC Project

The MatPLC is the result of a project that is creating a PLC to run on any POSIX-compliant operating system (OS). The entire source code is distributed under the GPL license (GNU Public License), and may be obtained from http://mat.sf.net.

The MatPLC [5] is composed of several modules, each of which is executed by its own OS process. The modules all access the same shared PLC state, which stores the internal PLC coils, or points. Modules may be configured to run periodically, and synchronized among themselves using a Petri Net model. Examples of currently implemented modules are the modbus master, the hilscher module (that interfaces with industrial network interface cards, such as Devicenet, Profibus, ASI, etc.), digital signal processor (including PID controllers and digital filtering), IEC ST and IL compiler, and tcl- or gtk-based graphical interfaces.

The authors have implemented the IEC 61131-3 ST and IL compiler for the MatPLC [6]. This compiler executes in four plus one stages: lexical analyzer, syntax parser, semantics analyzer, code generator, and binary code generator. At the time of writing, the semantic analyzer had not yet been coded, and only a C++ code generator existed for the fourth stage. The last stage of the overall compiler generates the final executable from the C++ code. In our case, we are currently using the GNU C/C++ compiler (gcc) to generate the final executable.

A few conflicts were found in the syntax definition given in the standard. These were mostly due to the existence of more than one route for reducing several constructs. All of these conflicts were easily resolved, as the expected semantics of either route were identical. Other more thorny issues were related to the fact that the language requires more than one look-ahead token to be correctly parsed, while the bison utility we used generates a parser that uses a single look-ahead token.

We also found that the syntax does not contain sufficient redundancy to allow it to be parsed successfully without the help of a symbol table that keeps track of the type of construct to which an identifier (name) refers to. Before any parsing commences, the global table is initialized with the names of all the default functions and functions blocks defined in the standard. The use by the standard of the identifiers "NOT" and "MOD" as either keywords or standard functions, added considerable complexity to the parser, as these identifiers may be correctly interpreted as one of two different tokens by the lexical parser.

References

[1] IEC 61131-3, 2nd ed. Programmable Controllers — Programming Languages, Final Draft, International Electrotechnical Commission, 10th December, 2001.

[2] John, Karl-Heinz and Tiegelkamp, Michael, *IEC 61131-3: Programming Industrial Automation Systems*, Springer-Verlag, Berlin, 2001, ISBN: 3540677526.

[3] Lewis, R.W., *Programming Industrial Control Systems Using IEC 1131-3*, revised edition, Inspec/IEE, 1998, ISBN: 085 296 9503.

[4] Function Blocks for Industrial-Process Measurement and Control Systems, Part 1 — Architecture, Voting Draft, International Electrotechnical Commission, 28th March, 2000.

[5] de Sousa, Mário, MatPLC-The Truly Open Automation Controller, Proceedings of the 28th Annual Conference of IEEE Industrial Electronics Society (IEEE Catalog number 02CH37363), 2002, pp. 2278–2283.

[6] de Sousa, Mário and Adriano de Carvalho, An IEC 61131-3 Compiler for the MatPLC, Proceedings of the 9th IEEE International Conference on Emerging Technologies and Factory Automation, IEEE Product No: TH8696-TBR, 2003, ISBN: 0-7803-7937-3.

66

Achieving Reconfigurability of Automation Systems by Using the New International Standard IEC 61499: A Developer's View

Hans-Michael Hanisch and
Valeriy Vyatkin
*Martin Luther University of
Halle–Wittenberg.*

66.1   Reasons for a New Standard

The development of the IEC 61499 has been stimulated by new requirements coming mainly from the manufacturing industry and by new concepts and capabilities of control software and hardware engineering.

To survive growing competition in more and more internationalized global markets, the production systems need to be more flexible and reconfigurable. This demand is especially strong in highly developed countries with high labor costs and individual customer demands. Acting successfully in such markets means decreased lot sizes and therefore frequent changes of manufacturing orders that may require changes of the manufacturing system itself. This means either a change of parts of the machinery or a change of the interconnection of subsystems by flow of material.

Each change corresponds to a partial or complete redesign of the corresponding control system. The more frequent the changes are, the more time and effort has to be spent on the redesign. As a consequence, there is a growing economic need to minimize these costs by application of appropriate methodologies of control system engineering. The desire of control engineers is to reach so-called "plug and play" integration and reconfiguration.

When the production systems are built from automated units, the control engineers naturally try to reuse the software components of the units. This is especially attractive in case of a reconfiguration, when the changes of units functionality may seem to be minor. However, the obstacles come from the software side.

The currently dominating International Standard IEC 61131 for programming of Programmable Logic Controllers (PLCs) [2] is reaching the end of its technological lifecycle, and its execution semantics does not fit well into the new requirements for distributed, flexible automation systems. The IEC 61331 systems rely on the centralized programmable control model with cyclically scanned program execution. Integration of this kind of systems via communication networks may require quite complex synchronization procedures. Thus, overheads of the integration may be as complex as the startover system development. Furthermore, even different implementations of the same programming language of IEC 61131 may have different execution semantics. Moreover, PLCs of different vendors are not interoperable and require vendor-dependent configuration tools.

For dealing with distributed nonhomogeneous systems, it would be convenient to define control applications in a way that is independent from a particular hardware architecture. However, the architectural concept for such a definition is missing.

These are severe obstacles toward the plug-and-play integration and reconfiguration of flexible automation systems.

The newly emerging International Standard IEC 61499 [1] is an attempt to provide the missing architectural concept. The standard defines a reference architecture for open, distributed control systems. This provides the means for real compatibility between automation systems of different vendors. The standard incorporates advanced software technologies, such as encapsulation of functionality, component-based design, event-driven execution, and distribution. As a result, specific implementations of different providers of field devices, controller hardware, human–machine interfaces, communication networks, etc. can be integrated in component-based, heterogeneous systems. The IEC 61499 standard stimulates the development of new engineering technologies that are intended to reduce the design efforts and to enable fast and easy reconfiguration.

This chapter provides an overview of the concepts and design principles of the standard. For the internal details of the standard, the reader is referred to [1], for a more systematic introduction into the subject to [4], and to [7] for the evolution of the idea. Since the design principles differ considerably from those of the well-known IEC 61131, the general issues are introduced in a rather intuitive way. To illustrate the design principles and new features, an example is presented in this chapter. Real applications — although sparse — do exist but are too complex to be explained here in detail. The interested reader is referred to [9]. This chapter is based on the Publicly Available Specification (PAS) of the IEC TC65 from March 2000. As of July 2004, the parts 61499-1 "Architecture" and 61499-2 "Software Tools Requirements" were voted and approved as IEC Standards. The texts are expected to be published by the end of 2004.

66.2 Basic Concepts of IEC 61499

The standard defines several basic *functional* and *structural* entities. They can be used for specification, modeling, and implementation of distributed control applications. Some of them stand for pure software components. Others represent logical abstractions of a mixed software/hardware nature. The main functional entities are function block types, resource types, device types, and applications that are composed of function block instances. System configurations include instances of device types and applications that are placed (mapped) into the devices. The following subsections will briefly discuss all these entities.

The specification of functionality of a control application and the specification of a system architecture can be performed independently. An implementation of an application on a given system architecture is done by mapping of its function blocks onto the devices and their resources. Each function block must be mapped to a single resource and cannot be distributed over different resources.

Thus, the next section discusses the means to define the functionality of a distributed control application, and the section "Specification of the system architecture" shows how a particular system configuration can be defined.

Describing the Functionality of Control Applications

Overview of the Function Block Concept

The basic entity of a portable software component in IEC 61499 is a so-called *function block*. The new standard defines several types of blocks: basic function blocks, service interface function blocks, and composite function blocks. Although the term function block is known from IEC 61131, a function block of IEC 61499 is different from IEC 61131 function blocks. Figure 66.1 shows a *function block interface* following the standard (note that the graphical appearance is not normative). The upper part is often referred to as the "head" and the lower as the "body" of the function block. A block may have inputs and outputs. There is a clear distinction between data and event input/output signals. Events serve for synchronization and interactions among the execution control mechanisms in interconnected networks of function blocks. The concept of data types is adopted as in any programming language. In particular, the standard refers to the data types of IEC 61131. In the graphical representation in Figure 66.1, the event inputs and outputs are associated with the head of the function block, and the data inputs and outputs are associated with the body.

The IEC 61499 defines a number of standard function blocks for manipulations with events, such as splitting or merging events, generation of events with delays or cyclically, etc.

The definition of the function block external interface also includes an association between the events and relevant data. It is denoted by vertical lines connecting an event and its associated data inputs/outputs in the graphical representation. The association literally means that the values of the variables associated with a particular event (and only these!) will be updated when the event occurs.

The IEC 61499 standard uses the typing concept of function blocks that is similar to that in IEC 61131 and in the object-oriented programming. Once a function block type is defined, it can be stored in the library of function block types and later instantiated in applications over and over again.

The standard does not determine in detail the languages to define the internal functionality of a function block. However, for each type of blocks, the ways of structuring the functionality are identified. These will be discussed in the following sections.

Basic Function Blocks

Basic Function Blocks are software structures intended to implement basic functions of distributed control applications. The standard says that in addition to inputs and outputs, the internal structure of a basic function block may include *internal variables*, one or several *algorithms*, and an *execution control chart*.

An algorithm is a structure of finest granularity. It represents a piece of software code operating on the common input, output, and internal data of the function block. The algorithms can be specified, for

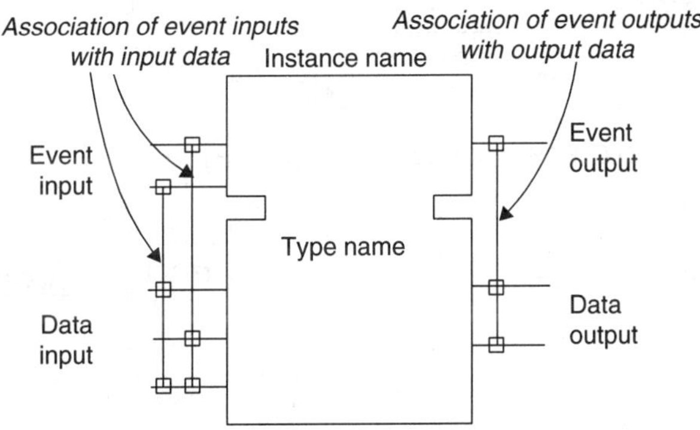

FIGURE 66.1   External interface of a function block.

example, in languages defined in IEC 61131. But, in general, they can be given in any form supported by the implementation platform. It is important to note that an algorithm has to be implemented in one programming language. The internal data of a function block cannot be accessed from outside and can only be used by the internal algorithms of the particular function block (Figure 66.2).

The execution control function specifies the algorithm that must be invoked after a certain input event in a certain state of the execution control function. It is specified by means of Execution Control Charts (ECC for short). Figure 66.3 shows an example. An ECC is a finite state machine with a designated initial state. An ECC consists of states with associated *actions* (designated by ovals in the Figure) and of state transitions (designated by arrows). The actions contain algorithms to invoke and output events to issue upon the completion of the algorithms' execution.

Each state transition is labeled with a BOOLEAN condition that is a logic expression utilizing one or more event input variables, output variables, or internal variables of the function block. The event inputs are represented in the conditions as BOOLEAN variables that are set to TRUE upon an event and cleared after all possible state transitions (initiated by a single input event) are exhausted.

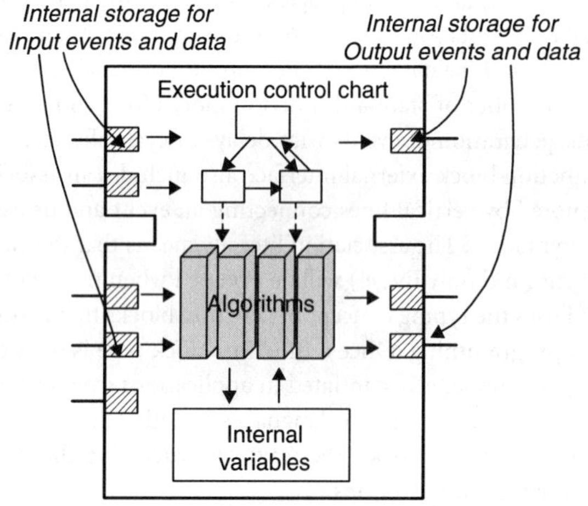

FIGURE 66.2   A basic function block.

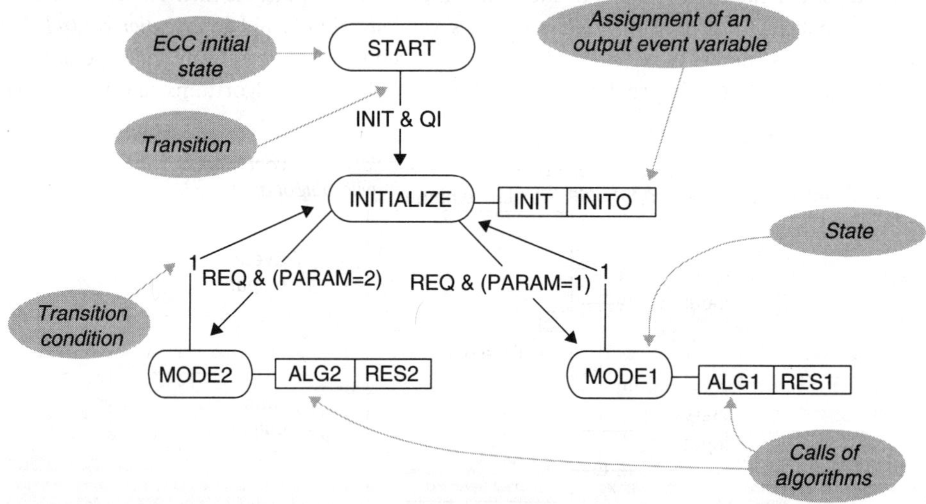

FIGURE 66.3   An example of the ECC [1].

An input event causes the *invocation of the execution control function* that in more detail is as follows (see Figure 66.4):

Step 1: The input variable values relevant to the input event are made available.

Step 2: The input event occurs, the corresponding BOOLEAN variable is set, and the execution control of the function block is triggered.

Step 3: The execution control function evaluates the ECC as follows. All the transition conditions going out of the current ECC state are evaluated. If no transition is enabled, then the procedure goes to the Step 8. Otherwise, if one or several state transitions are enabled (i.e., if the corresponding conditions are evaluated to TRUE), a single state transition takes place.[1] The current state is substituted by the following one. The algorithms associated with the new current state will be scheduled for execution. The execution control function notifies the resource scheduling function to schedule an algorithm for execution.

Step 4: Algorithm execution begins.

Step 5: The algorithm completes the establishment of values for the output variables.

Step 6: The resource scheduling function is notified that algorithm execution has ended.

Step 7: The scheduling function invokes the execution control function. The procedure resumes from Step 3.

Step 8: When some of the output event variables were set during this invocation of the execution control function, then the execution control function signals the corresponding event output and clears the BOOLEAN variables corresponding to the triggered input and output events.

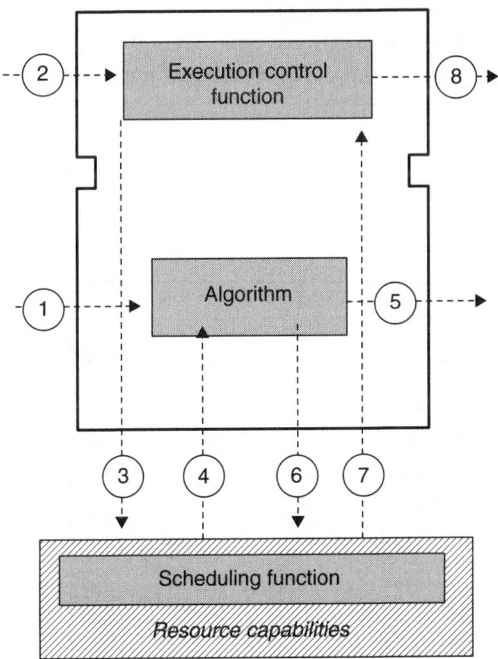

FIGURE 66.4 Execution of a function block.

---

[1]The standard does not determine a rule regarding how to choose a state transition if several are simultaneously enabled.

First conclusions can be drawn at this point:

- A Basic Function Block is an abstraction of a software component adjusted for the needs of measurement and control systems. Its execution semantics is event-driven and platform independent. Basic Function Blocks are intended to be main instruments of an application developer.
- The standard implies separation of the functions, implemented by algorithms, from the execution control. The algorithms encapsulated in a function block can be programmed in different programming languages.
- The execution of function blocks is event-driven. This means that algorithms are executed only if there is a need to execute them in contrast to the cyclically scanned execution in IEC 61131. The need has to be indicated by events. The source of events can be other function blocks. Some of them may encapsulate interfaces to the environment (controlled process, communication networks, hardware of a particular computational device).
- The execution function of a Basic Function Block is defined in the form of a state machine that is available for documentation and specification purposes even if the algorithms are hidden.
- The function block abstracts from a physical platform (the resource) on which it is located. This means that the specification of the function block can be made without any knowledge of the particular hardware on which it will be later executed.

Composite Function Blocks

The standard also defines Composite Function Blocks, the functionality of which, in contrast to Basic Function Blocks, is determined by a network of interconnected function blocks inside. Figure 66.5 shows the principle.

More precisely, members of the network are instances of function block types. These can be either Basic Function Blocks or other Composite Function Blocks. Therefore, hierarchical applications can be built.

It is important to note that Composite Function Blocks have no internal variables, except for those storing the values of input and output events and data. Thus, the functionality of Composite Function Blocks completely depends on the behavior of the constituent function blocks and their interconnections by events and data.

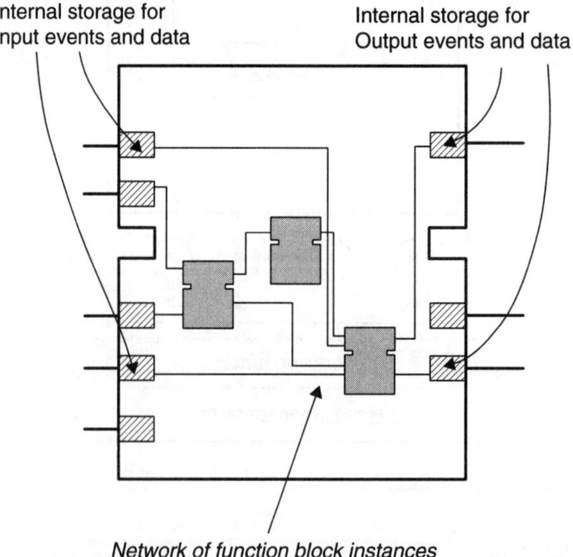

FIGURE 66.5 Composite Function Block.

Along with Basic Function Blocks, the Composite Function Blocks are intended to be main instruments of an application developer.

Service Interface Function Blocks

In contrast to Basic and Composite Function Blocks, Service Interface Function Blocks are not intended to be developed by an application developer. These have to be provided by vendors of the corresponding equipment, for example, controllers, field buses, remote input/output modules, intelligent sensors, etc. The application scope determines the differences of this kind of function blocks from the previously considered ones.

To conceal the implementation-specific details of the system from the application, the IEC 61499 defines the concept of services that the system provides to an application. A service is a functional capability of a resource that is made available to an application. A service is specified by a sequence of service primitives that define properties of the interaction between an application and a resource during the service. The service primitives are specified in a graphical form of the time-sequence diagrams described in Technical Report 8509 of the International Standard Organization (ISO) [3]. This is rather a qualitative specification form as it does not specify exact timing requirements to the services. An example of time-sequence diagrams is presented in Figure 66.7 and will be briefly discussed later in this section.

A Service Interface Function Block is an abstraction of a software component implementing the services. Figure 66.6 shows an example of a Service Interface Function Block REQUESTER that provides some service upon request to an application (examples of possible services: read the values of sensors, increase the memory used by a resource, shut down a resource, send a message, access remote database, etc.). The standard predefines some names for input/output parameters of Service Interface Function Blocks such as INIT for initialization, INITO for confirmation of the initialization, QI for input qualifier, etc.

Some of the services provided by this block are specified in the form of time-sequence diagrams in Figure 66.7. These are "normal establishment" of the service, "normal service," and "application initiated termination" of the service.

The input event INIT serves for initialization/termination of the service, depending on whether the BOOLEAN input QI is true or false. The notation INIT+ means the occurrence of the event INIT with the qualifier value QI=true, and INIT-correspondingly with QI=false.

The input parameter PARAMS stands for the service parameters that have to be taken into account during the service initialization. At the end of the initialization/termination procedure, the Service Interface Function Block responds by event INITO its completion and indicates by the BOOLEAN data output QO whether initialization/termination was successful (QO=true) or not (QO=false).

The input data needed to perform the service are denoted as inputs SD_1 ... SD_m. Note that these data are associated with the event input REQ. The data outputs RD_1 ... RD_n stand for the data computed as a result of the service. These data are associated with event CNF that represents confirmation of the service. The output STATUS provides information about the status of the service upon the occurrence of an event output.

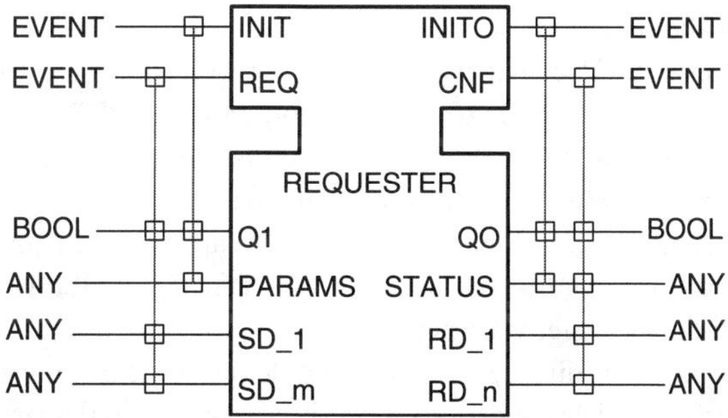

FIGURE 66.6 Generic REQUESTER [1].

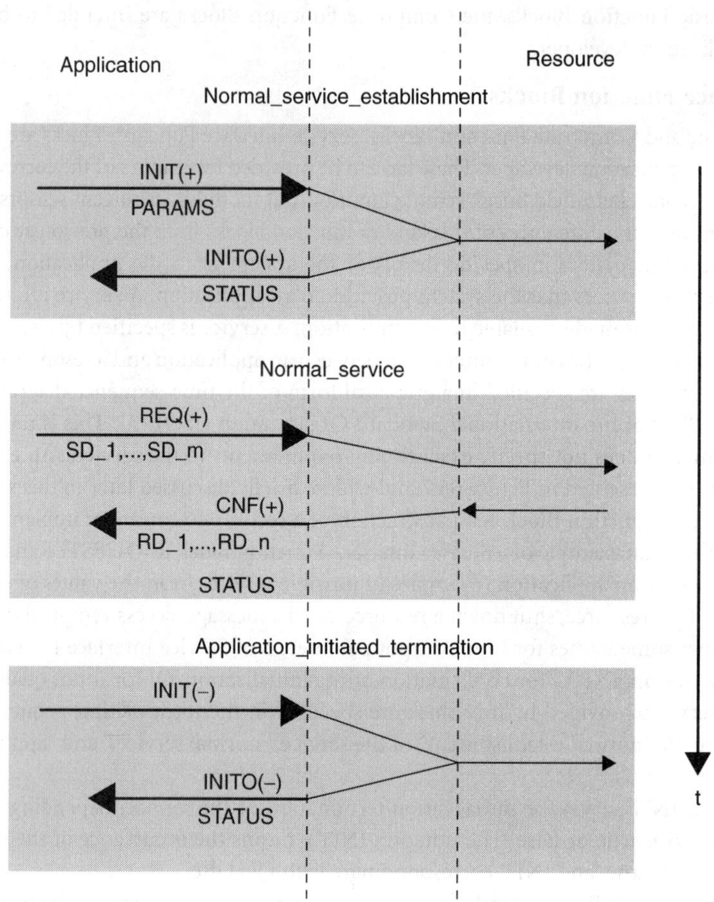

FIGURE 66.7   Diagrams of service sequences for application-initiated interactions.

The execution of Service Interface Function Blocks is initiated by input events. The internal structure of Service Interface Function Blocks is not specified as firmly as for Basic Function Blocks. For example, a programming implementation of a Service Interface Function Block can be done in the form of several encapsulated algorithms (methods, procedures) that are invoked upon a particular event (say algorithm **init** stands for the event INIT). The algorithms may check the value of the qualifier QI and then call either the subroutine responsible for the "normal service establishment" (if QI=true) or the one responsible for the "application initiated termination" (if QI=false). Note that the concept of Service Interface Function Blocks does not presume the need for internal variables — the conditions for initiating services are described by input events and data (qualifiers).

A particular case of Service Interface Function Blocks are Communication Interface Function Blocks. The standard explicitly defines two generic communication patterns: PUBLISH/SUBSCRIBE for unidirectional transactions and CLIENT/SERVER for bidirectional communication. These patterns can be adjusted to a particular networking or communication mechanism of a particular implementation. Otherwise, a provider of communication hardware/software can specify his own patterns if they differ from the above-mentioned ones. Figure 66.8 illustrates the generic PUBLISH and SUBSCRIBE blocks performing unidirectional data transfer via a network.

The PUBLISHER serves for publishing data SD_1 ... SD_m that come from one or more function blocks in the application. It is therefore initialized/terminated by the application in the same way as described above.

Upon the request-event REQ from the application, the data that need to be published are sent by the PUBLISHER via an implementation-dependent network. When this is done, the PUBLISHER informs the publishing application via event output CNF.

The SUBSCRIBER function block is initialized by the application that is supposed to read the data RD_1 ... RD_m. Normal data transfer is initiated by the sending application via the REQ input event to the PUBLISHER. This is illustrated in Figure 66.9 by means of time-sequence diagrams. The PUBLISHER sends the data and triggers the IND-event at the outputs of the SUBSCRIBER to notify the reading applications that new values of data are available at RD_1 ... RD_m outputs of the SUBSCRIBER. The reading application notifies the SUBSCRIBER by the RSP-event that the data are read.

In summarizing this subsection, one can see that Service Interface Function Blocks implement the interface between an application and the specific functionality that is provided by control hardware or system software. The content of Service Interface Function Blocks can be concealed, but the means are reserved to specify their functionality in a visual form.

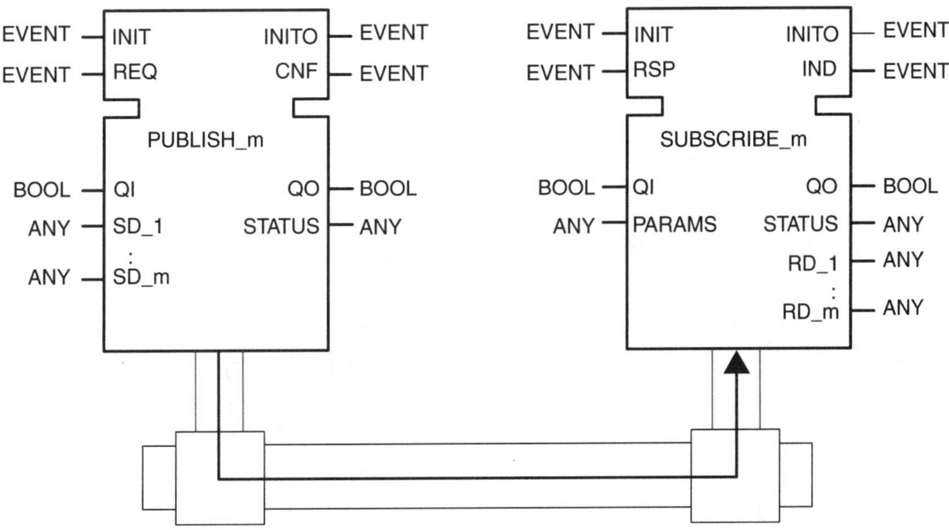

FIGURE 66.8 PUBLISH and SUBSCRIBE communication interface blocks.

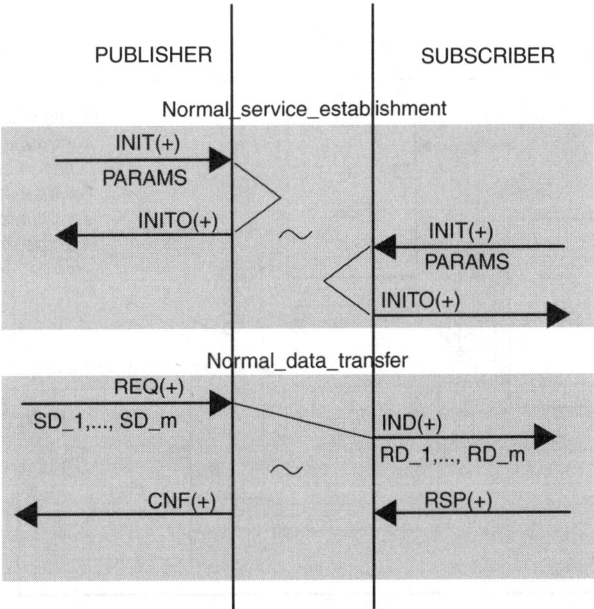

FIGURE 66.9 Communication establishment and normal data transfer sequence.

Application

An *application* following IEC 61499 is a network of function block instances whose data inputs and outputs and event inputs and outputs are interconnected (see Figure 66.10).

An application can be considered as an intermediate step in the system development. It already defines the desired functionality of the system completely, but it does not specify the system's structure in terms of computational devices where the function blocks can be executed. The next step in the engineering process is to define a particular set of devices and to "cut" the application, assigning the blocks to the devices as illustrated in Figure 66.11.

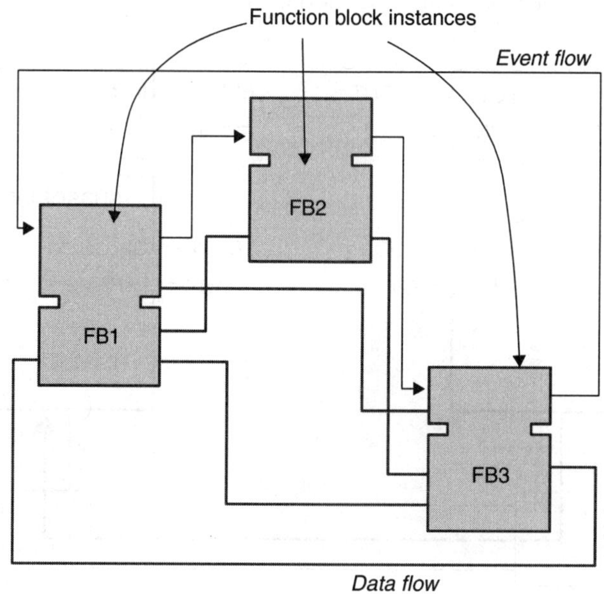

FIGURE 66.10   An application.

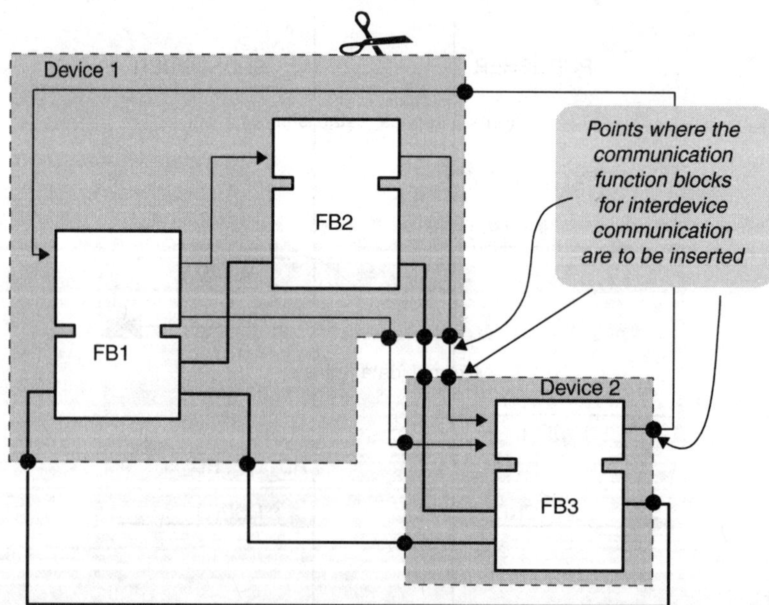

FIGURE 66.11   The application distributed onto two devices.

The way in which the separated parts of the distributed applications communicate with each other has to be explicitly defined. This can be done by adding Communication Function Blocks in the places where the "cut" took place (see Figure 66.12).

A network of function blocks (forming the application) can be encapsulated in a Composite Function Block if needed. In this case, however, it could not be distributed across several devices or resources as a function block can be executed only in a single resource.

In fact, the standard provides a structure (called *subapplication*) that combines features of Composite Function Blocks and of applications. The content of a subapplication can be distributed across several devices. However, the practical applicability of this structure is quite questionable.

The following subsection will show the concepts and specifications of a system following IEC 61499 as a platform for implementation and execution of an application.

Specification of the System Architecture

Resources and Devices

A device in IEC 61499 is an atomic element of a system configuration. The standard provides architectural frames for creating models of devices, including their subdivision in computationally independent resources.

A device type (Figure 66.13) is specified by its process interface and communication interfaces. A device can contain zero or more resources (see the description below) and function block networks (this option is reserved for the devices having no resources).

A "process interface" provides a mapping between the physical process (analog measurements, discrete I/O, etc.) and the resources. Information exchanged with the physical process is presented to the resource as data or events, or both.

Communication interfaces provide a mapping between resources and the information exchanged via a communication network. In particular, services provided by communication interfaces may include presentation of communicated information to the resource and additional services to support programming, configuration, diagnostics, etc.

The interfaces are implemented by the libraries of corresponding Service Interface Function Blocks. The libraries of these blocks form the "identity" of a device. A device that contains no resources is considered to be functionally equivalent to a resource.

A resource (Figure 66.14) is considered to be a functional unit, contained in a device, that has independent control of its operation. It may be created, configured, parameterized, started up, deleted, etc., without affecting other resources within a device. The functions of a resource are to accept data and/or

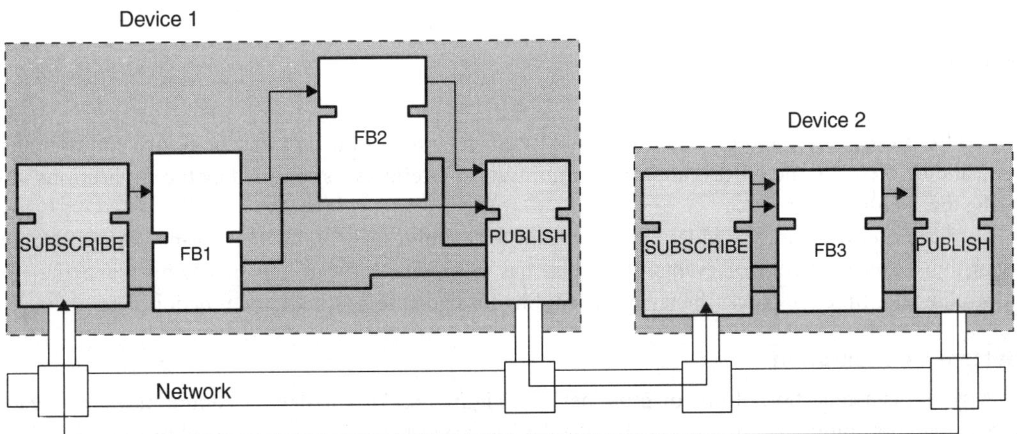

FIGURE 66.12   Communication Function Blocks explicitly connecting parts of the distributed application.

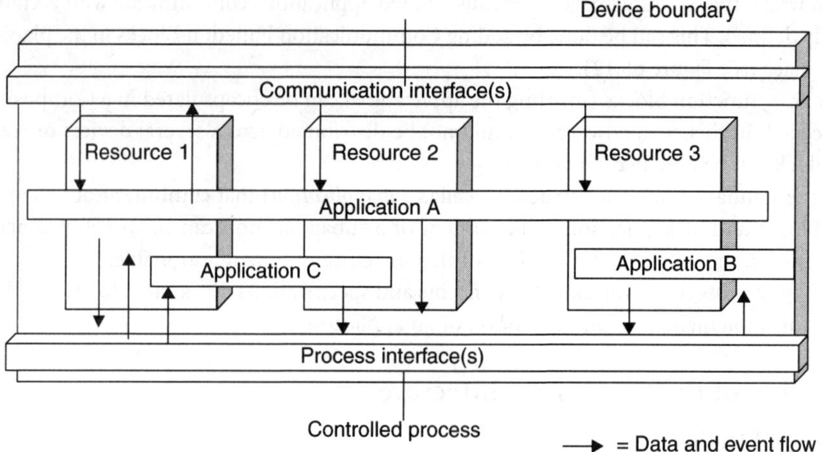

FIGURE 66.13   A device model.

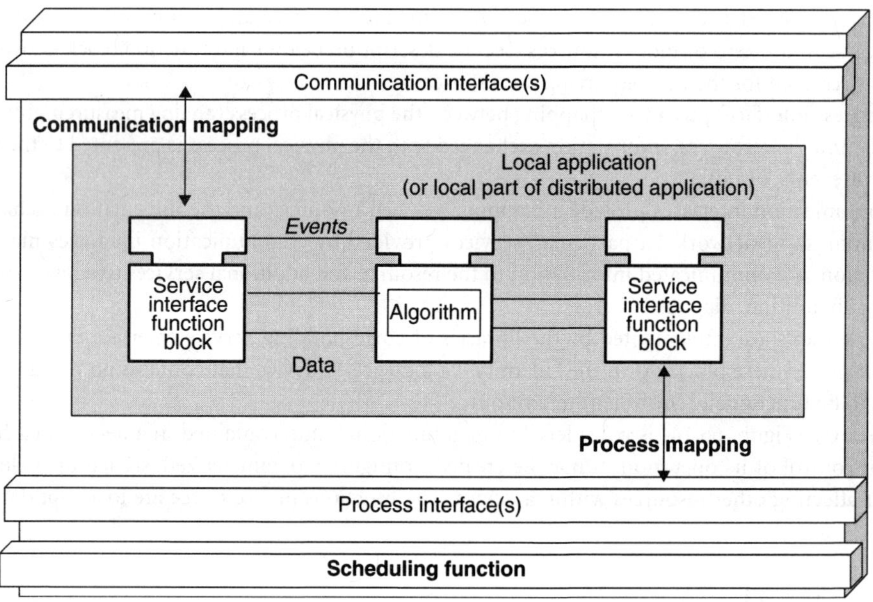

FIGURE 66.14   A model of resource.

events from the process and/or communication interfaces, process the data and/or events, and to return data and/or events to the process and/or communication interfaces, as specified by the applications utilizing the resource.

Furthermore, a resource provides physical means for running algorithms. This means storage for data, algorithms, execution control, events, etc. It also has to provide software capabilities for managing and controlling the function blocks' behavior, scheduling its algorithms (scheduling function), etc.

System Configuration

A system is a collection of one or more devices (Figure 66.15). The devices communicate with each other over communication networks. Also, the devices are linked with the controlled process via sensor and actuator signals.

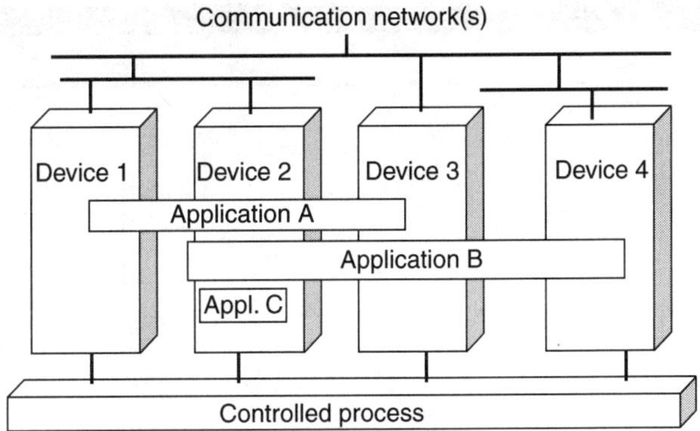

FIGURE 66.15   A system configuration.

Applications are mapped onto the devices. This means that their function blocks are assigned to the resources of the corresponding devices. In this way, a system configuration is formed. A system configuration is feasible if each device in it supports the function block types that are mapped on it. Otherwise, the block would not be instantiated, and the system will not run.

66.3   Illustrative example

Desired Application Functionality

The example "FLASHER" that is used in this section was borrowed from the set of samples provided with the Function Block Development Kit (FBDK). This is the first software tool supporting IEC 61499 system development. It was developed by Rockwell Automation, U.S.A. The toolset can be downloaded from [5]. The example represents an abstraction of an automation system. It is supposed to make four lamps blinking according to a preprogrammed mode of operation.

The system consists of human–machine interface components and of a core functional component. The core component generates the output signals determined by input parameters. The output values are then delivered to the visualization device (lamps). In the form presented here, the system is completely simulated in a computer. All human–machine interface components and lamps exist only on the computer screen. The example visually shows how easily each of the software components that interact with a simulated object can be replaced by components that would interact with the real physical equipment (e.g., buttons, switchers, knobs). After such a reconfiguration, the whole system will then show the same functionality without changing its structure and without redesigning the other components.

Figure 66.16 shows a screenshot of the FBDK containing a system configuration. As a result of its execution, the output frame is produced. The output frame is located in the figure below the FBDK screen. The arrows connect the function blocks with the screen objects created by them.

The system configuration includes one application. It is placed in one device with only one resource. An instance of the device type FRAME_DEVICE is used in this example. This type of device creates a windows frame on the computer screen. The resource of type PANEL_RESOURCE creates a rectangular panel within this frame. If a function block creates an output to the screen, it will be placed in the corresponding panel.

In this particular case, the system configuration implements the application in a centralized manner. The application includes all the blocks shown in Figure 66.16 in the shaded rectangular area. The only block that falls out of the application is the block START of type E_RESTART. It belongs to the resource type PANEL_RESOURCE.

The application creates the following models of human–machine interface primitives. Buttons START and STOP are created by the blocks START_PB and STOP_PB. Both are instances of the type IN_EVENT.

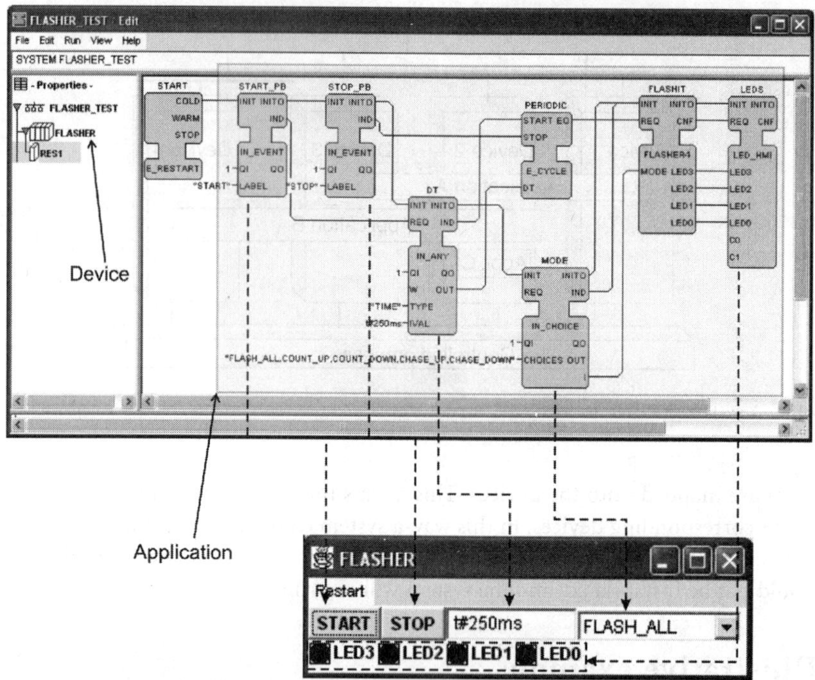

FIGURE 66.16   FLASHER in centralized system configuration.

Block DT creates the input field for the TIME parameter. It is an instance of type IN_ANY. Block MODE creates the pull-down menu to select a desired mode of operation.

The block that produces the output combination is FLASHIT. It is an instance of type FLASHER4. The output values are then visualized by the block LEDS of type LED_HMI. The FLASHER4 generates the output values at every pulse of the event input REQ. The pulses are generated by the block PERIODIC with the frequency determined by the value in the field DT that is received from the block DT.

The operation of the system configuration is started when the resource is initialized, for example, when the device is created (or switched on for more realistic devices). At that moment, the Service Interface Function Block START produces the event COLD (cold restart). It is connected to the input event INIT of the block START_PB. This input event causes the block of type IN_EVENT to place a button image in the resource's panel. The caption of the button is given by the input parameter "LABEL" of the block, that is, "START" in our case. After that, an output event INITO is generated that is connected to the input INIT of the block STOP_PB. It leads to the creation of the button "STOP" and so forth in the chain until the whole picture is created on the resource panel as shown in Figure 66.16.

Once the button START is pressed (e.g., by a mouse click), it ignites the event output IND. This event propagates through the chain of blocks DT and PERIODIC. It enables the latter to generate the output event EO with the desired frequency. Every moment the event comes to the input REQ of the FLASHIT, the output combination of LED0..LED3 is created, and the output event CNF notifies the block LEDS, which updates the picture.

If the operating mode is changed during the operation, the corresponding event IND of the block MODE would notify the FLASHIT. Then, FLASHIT will change the pattern according to which its outputs are generated.

A more detailed description of the function block type FLASHER4 is given now. Figure 66.17 shows the Execution Control Chart of this block. The state machine either switches to the desired algorithm corresponding to the selected MODE of operation (number in the interval 1..5) at input event REQ, or to the initialization algorithm INIT at the input INIT.

   Note that one half of the algorithms encapsulated in the FLASHER4 is programmed in Structured Text (ST), while the other half is programmed in Ladder Logic Diagrams (LD). This is shown in Figures 66.18 and 66.19. This illustrates the opportunities the IEC 61499 function blocks provide to reuse the legacy code and even to combine different programming languages in a single software component.

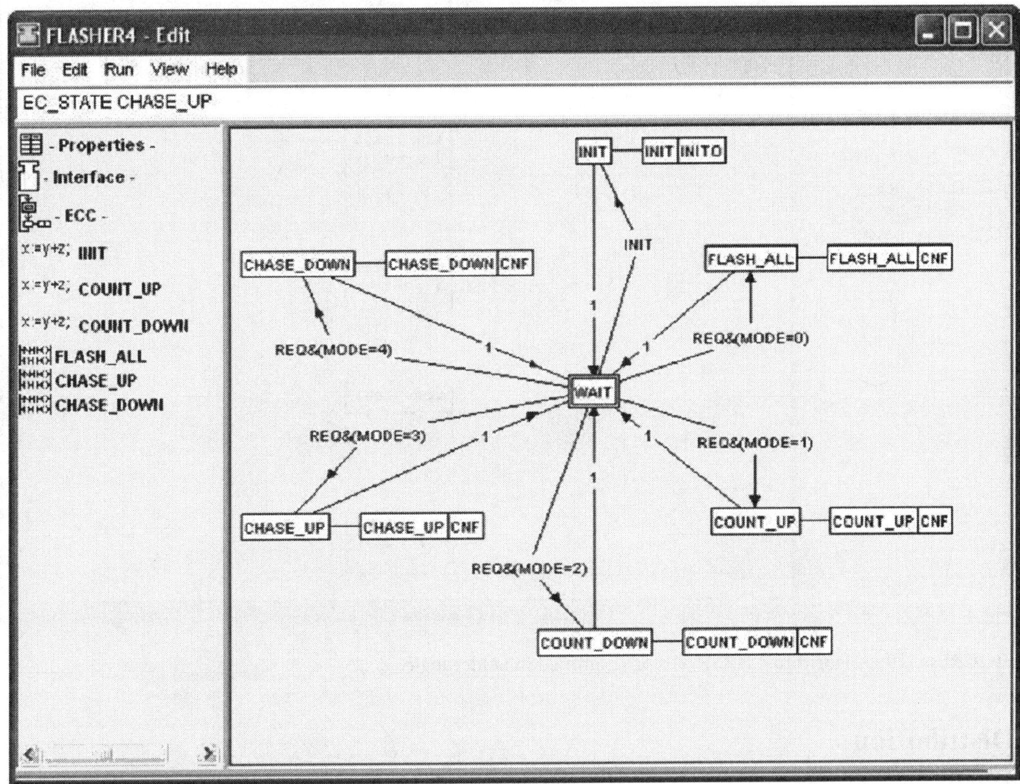

FIGURE 66.17   FLASHER4: Execution Control Chart.

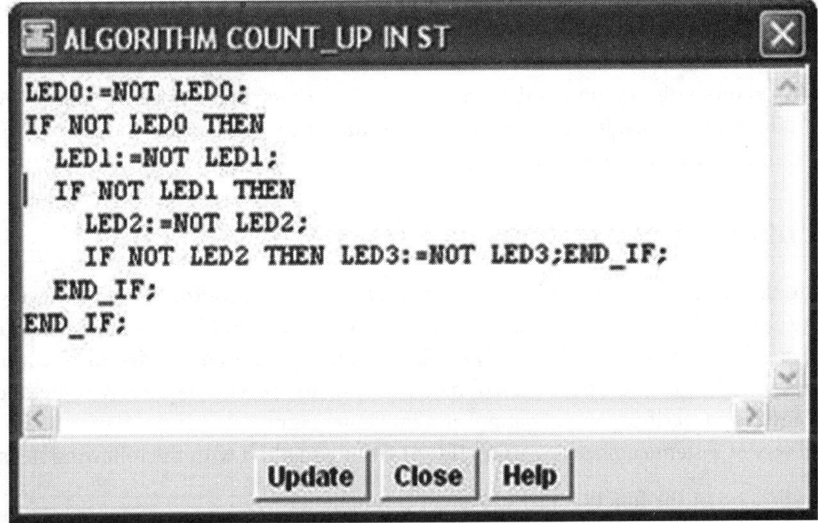

FIGURE 66.18   Algorithm COUNT_UP programmed in structured text.

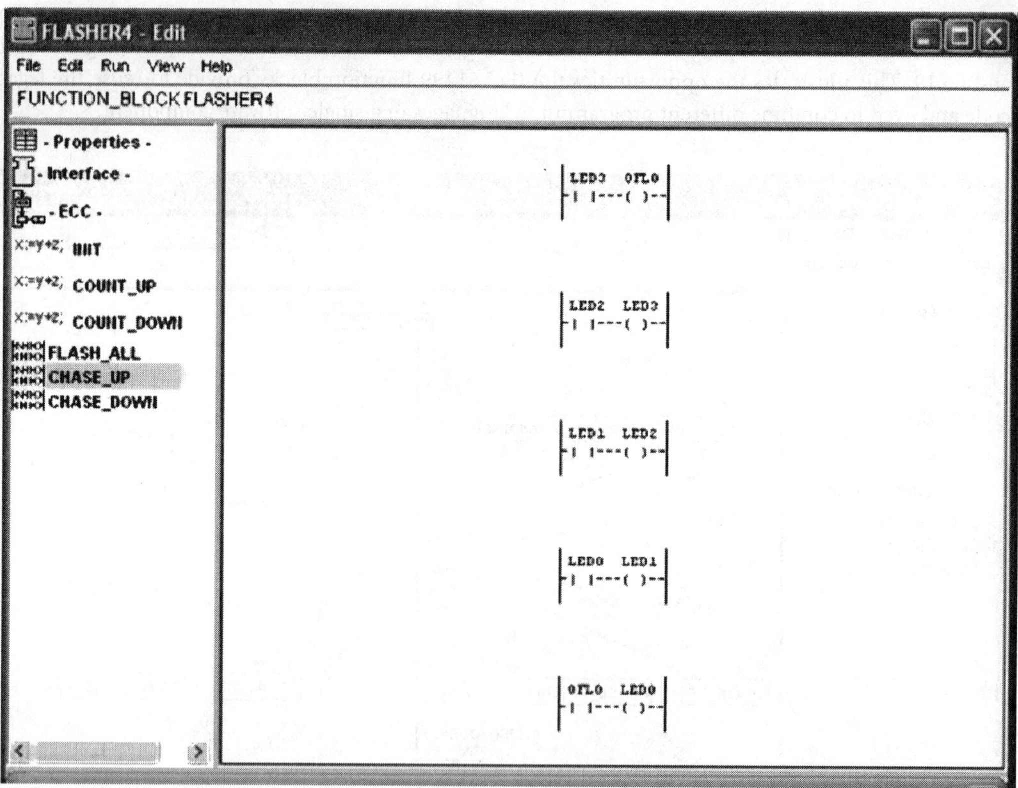

FIGURE 66.19   Algorithm CHASE_UP programmed in ladder logic.

Distribution

The distributed version of the same application is shown in Figure 66.20. The system configuration includes two devices: CTL_PANEL and FLASHER. The function blocks of the application are mapped to either of these devices. In addition, the Communication Function Blocks PUBLISH and SUBSCRIBE are added to connect the parts of the application. The devices may be started or shut down completely independently from each other. As soon as the part located in CTL_PANEL produces any changes in the operation parameters, it will notify the FLASHER's part via the communication channel.

Figure 66.21 shows a distribution of the same application across three devices. Device "FLASHER" (on the right) produces no picture. It produces the output values and sends them to the device DISPLAY given the input parameters received from the CTL_PANEL.

66.4   Engineering Methods and Further Development

In contrast to the present practice of IEC 61131, the specification of a control system following IEC 61499 is a complete change of paradigms. Execution of control code in IEC 61131 is sequential and time-driven. The control engineer who uses the IEC 61499 needs to think in terms of event-driven execution of encapsulated pieces of code. The execution is distributed and concurrent. This requires new methodologies for control system design, verification/validation, and implementation.

A natural way of system engineering using IEC 61499 is a method with the following steps:

1. Identification of the functionality of system components.
2. Encapsulation of basic functionality in these components. This gives the Basic Function Blocks or even Composite Function Blocks.

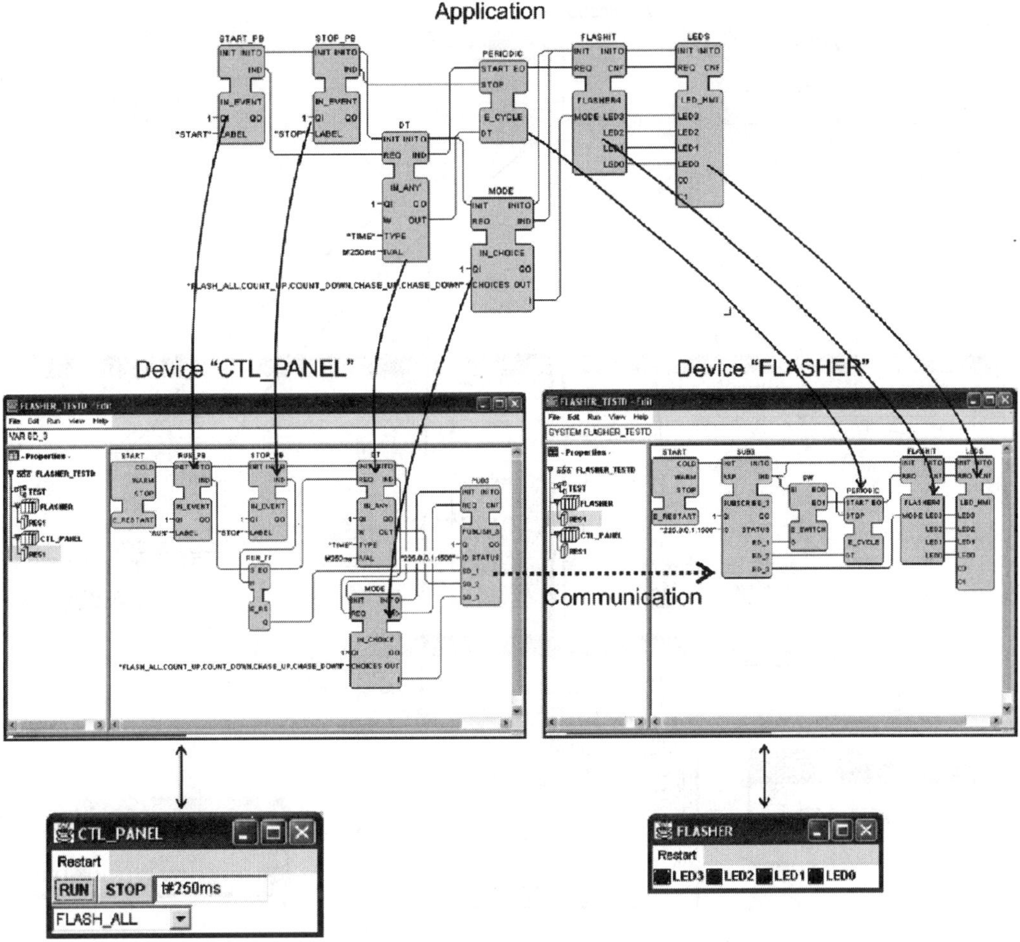

FIGURE 66.20   Application FLASHER distributed across two devices.

3. Interconnection of the function blocks to build an application. External coordination of the network of function blocks may be needed and added.
4. Mapping of the application into a control system architecture.

At least the following questions must be answered to make such a methodology applicable.

- How to encapsulate the existing controllers (sometimes implementing very sophisticated *ad hoc* algorithms) into new event-driven capsules?
- How to specify the component controllers in a way allowing (semi-)automatic integration into distributed systems?
- Conversely, given a desired behavior of the integrated system, how to decompose it to the control actions of the component controllers?

This requires further research and development, in particular, revolving around the concept of Automation Objects [11]. The concept of Automation Objects is understood as a framework for encapsulation and handling of the diverse knowledge relevant to automation systems. This includes operation semantics as well as layouts, CAD data, circuitry, etc. As the scope of IEC 61499 cannot cover all these issues, it probably needs to be combined with ideas arising from other developments, in particular, with IEC 68104 [12] for a more detailed description of device types.

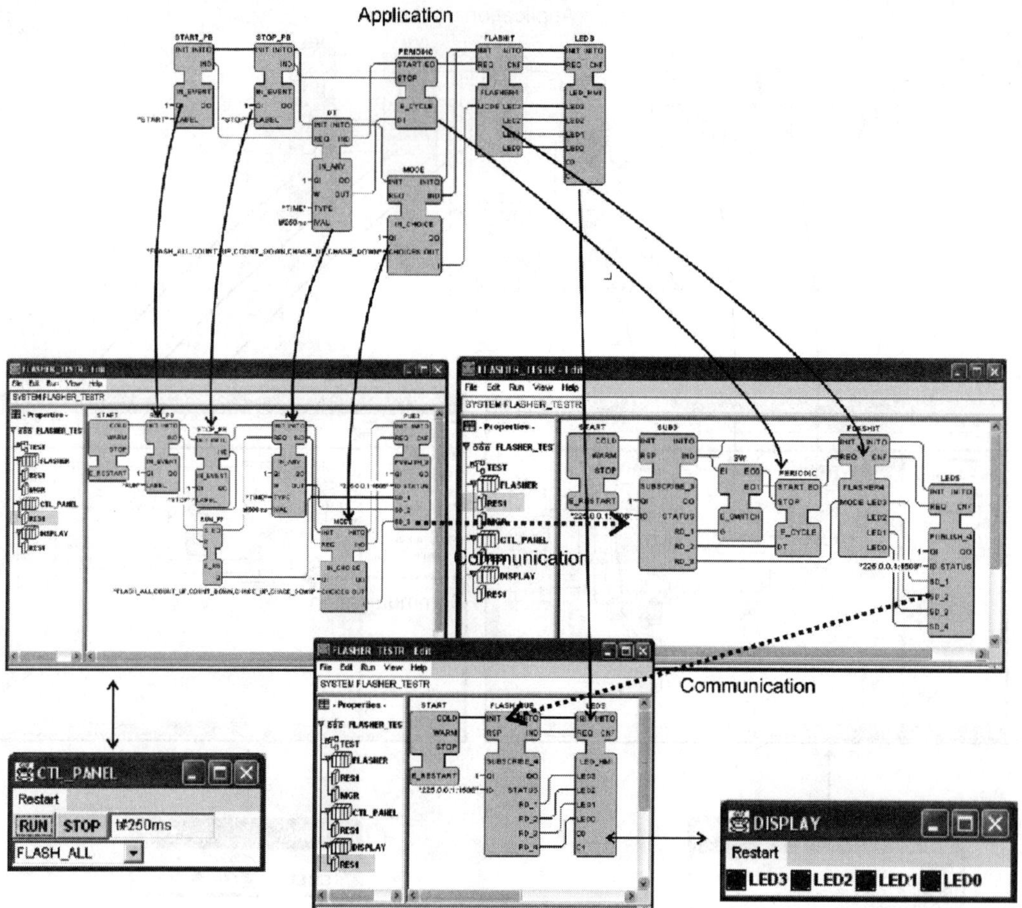

FIGURE 66.21   Distribution of the FLASHER application across three devices.

Conducting such a work requires integration of development activities among leading vendors and users of automation technologies on a global scale. Currently, such an activity is being organized under the support of the Intelligent Manufacturing Systems Research and Development program (www.ims.org) in the form of OOONEIDA project [13].

Currently available engineering methodologies can be found in [6,9]. Also, the idea of combining Unified Modeling Language (UML) with the IEC 61499 promises to provide a consistent engineering methodology for control system engineering. Some details on this can be found in [10].

Another major future issue is how to bring the ideas of reconfiguration to practical applications. The standard itself provides the following means.

1. Basic Function Blocks do not depend on a particular execution platform. They will show equivalent execution semantics on different platforms, regardless of, for example, in which sequence they are listed. This is not the case in the current implementations based on IEC 61131.

2. The functionality of whole applications that are represented as hierarchical networks of function blocks also does not depend on a particular number and topology of computational resources. Thus, system engineering can be done in an implementation-independent way.

3. Models of various devices are represented as device types. This will allow anticipation of the system's behavior after reconfiguration. On the one hand, the way of modeling of devices and resources is very modular. On the other, it uses a very limited number of constituent instruments. It allows modeling of a great variety of system configurations without going into unnecessary details.

4. There are also some other, more technical means provided in IEC61499 to support reconfiguration. One of these is the use of adapter interfaces to minimize interblock connections by means of predefined patterns. Another one includes standard function blocks for manipulations with events.

Another open question for future research and development is how the correctness of the system can be validated. This requires more formal models for simulation or even for formal analysis of the behavior. The concept of encapsulation is extremely useful for embedding such models in the design. In principle, it allows modeling of the whole measurement and control systems with all their diverse components, such as sensors and actuators, networks, computational resources, etc. The models could reproduce the execution semantics of the real system also taking into account such factors as communication delays of particular networks. Moreover, models of the controlled objects can be encapsulated in function blocks as well. This allows to study the behavior of the controller and of the controlled object in the closed loop. The desired properties could be validated by simulation or could even be formally verified. For the first time this approach was applied for formal modeling and verification of IEC 61499-compliant designs in [8]. This approach can also be used for prototyping of the distributed control systems based on their formal models.

It is obvious that all these future developments must be supported by appropriate tools, compliant devices, runtime systems, etc.

Last but not the least, there is a considerable need for training and education of engineering staff to understand and to apply the new concepts of the standard.

Nonetheless, despite the amount of development that still lies ahead, the potential benefits of using IEC 61499 are very clear, and the ideas and concepts define cornerstones of future development and design of distributed control systems.

Acknowledgments

The authors thank Sirko Karras for help in managing the graphic material of this contribution. The authors furthermore express their gratitude to Rockwell Automation and personally to Dr. James Christensen for providing the FLASHER example, for the permission of using Figures 8 and 9, and for the fruitful ideas expressed in the personal communication.

References

[1] Function Blocks for Industrial Process Measurement and Control Systems. Publicly Available Specification, International Electrotechnical Commission, Part 1: Architecture, Technical Committee 65, Working group 6, Geneva, 2000.

[2] International Standard IEC 1131-3, Programmable Controllers — Part 3, International Electrotechnical Commission, Geneva, Switzerland, 1993.

[3] ISO TR 8509-1987, Information processing systems — Open Systems Interconnection - Service conventions, 1987.

[4] Lewis, R., *Modeling Distributed Control Systems using IEC 61499*, Institution of Electrical Engineers, London, 2001.

[5] www.holobloc.com — web site devoted to IEC 61499.

[6] Christensen, J.H., IEC 61499 Architecture, Engineering, Methodologies and Software Tools, 5th IFIP International Conference BASYS'02, Proceedings, Cancun, Mexico, Sep. 2002.

[7] Schoop, R. and H.-D. Ferling, Function Blocks for Distributed Control Systems, DCCS'97, Proceedings, 1997, pp.145–150.

[8] Vyatkin, V. and H.-M. Hanisch, Verification of distributed control systems in intelligent manufacturing, *Journal of Intelligent Manufacturing*, 14, pp. 123–136, 2003.

[9] Vyatkin, V., Intelligent Mechatronic Components: Control System Engineering Using an Open Distributed Architecture, IEEE Conference on Emerging Technologies and Factory Automation (ETFA'03), Proceedings, Vol. II, Lisbon, Sept. 2003, pp. 277–284.

[10] Tranoris, C. and K. Thramboulidis, Integrating UML and the Function Block Concept for the Development of Distributed Applications, IEEE Conference on Emerging Technologies and Factory Automation (ETFA'03), Proceedings, Vol.II, Lisbon, Sept.2003, pp. 87–94.

[11] IEC 61804 Function Blocks for Process Control, Part 1 — General Requirement; Part 2 — Specification, Publicly Available Specification, International Electrotechnical Commission, Working group 6, Geneva, 2002.

[12] Automation Objects for Industrial-Process Measurement and Control Systems, Working Draft , International Electrotechnical Commission, Technical Committee No. 65, 2002.

[13] OOONEIDA: open object-oriented knowledge economy in intelligent industrial automation; Official web-site: http://www.oooneida.info

67

Applications of Haptics in Design and Manufacturing

T. Kesavadas, Arvind Balijepalli, and Cartik Sharma
University at Buffalo

67.1 Introduction

Design problems have increased in complexity over the past few years. The use of Virtual Environments (VE) to simulate and review real-world design problems is being accepted as the norm in most industries today. One factor that impacts the realism of such virtual simulations is the sense of touch and force feedback, which is a very important part of our daily lives [1]. Integration of force and tactile feedback have hence received attention as a new mode of interaction in simulated design and manufacturing environments in recent years. Studies have also been conducted to determine at what point there is a sensory overload, wherein the metric critical to the experiment is being overrepresented, causing confusion and thereby a doubt in the confidence of the person using the system.

67.2 Background on Haptics

Force feedback in virtual reality is often referred to as *haptics*. The word *haptic* is derived from the Greek word *haptikos* or *haptesthai*, which means to grasp or to touch. The principal difference between tactile and haptic feedback is that while sensory input using tactile devices can only be used to impart a superficial sensation of touch, haptic feedback is actively capable of sustaining and returning force inputs to the user. To illustrate this difference, we can consider the example of a simple task of picking up an object in a virtual environment. Using purely tactile feedback, the user will be able to feel this object when he/she comes in contact with it but will not be able to grasp it as his/her hand will pass right through the object due to a lack of resistance, which can only be provided with a haptic "display." Surface textures like surface roughness can be best simulated using a combination of haptic and tactile devices. However, present-day haptic devices that have a fine resolution and sufficient bandwidth are able to simulate surface roughness independently.

A number of haptic devices have been developed in recent years. In the simplest form, haptic devices resemble robotic arms with joints and linkages. With the use of highly accurate encoders, these devices are capable of precisely recording the position and force inputs in three to six degrees of freedom and providing force feedback.

67.3 Computer Interface Technology

The traditional windows and mouse interface that most people are familiar with serves adequately for two-dimensional applications like word processors and spreadsheets. In the realm of the three-dimensional world, these devices have been found to be lacking in a number of ways. Tactile feel of an object in a VE is considered of paramount importance in physical simulations of any kind. The recent appearance of cheap 3-D force feedback joystick for gaming has expanded the use of these devices; but their functionality is still primitive in comparison to more sophisticated haptic hardware.

The sense of touch is a two-way process. Environmental stimuli constantly evoke conscious and subconscious tactile responses. Humans respond naturally in 3-D environments. Computer interface technology has struggled to keep up with the constant demand for information and the explosive growth in program complexity.

Limited capabilities of the computer interface technology severely limit the potential of the user and hamper the potential effectiveness of applications. To demonstrate the importance of physical feedback in a simulation, we can take the case of tool design for aerospace applications that looks at an astronaut training for tool manipulation in outer space. In the absence of a reliable physical simulation, NASA resorts to full-scale mockups of the space shuttle [2]. If this were translated to a virtual environment without force feedback, the user would be able to point to a bolt and it would loosen itself. This does not give the user a good feel for the task as those tasks in the real world; especially in space applications, they are far more complicated.

67.4 Haptic Framework

Haptic applications rely on an accurate representation of force and positions. Most serious simulations have a force model capable of quickly and accurately rendering force and stiffness in correlation with the position of the stylus. Many a time, the virtual object is composed of more than one material having different properties. An example is a simulation involving a dental examination. In order to accurately represent a dental model, one has to account for not only teeth but also surrounding tissue like the gums and tongue. While it may be possible to precompute stiffness and resultant force at every point and interpolate at run-time, using a highly optimized search method or an intelligent data storage algorithm, the application may not lend itself well to changes at runtime. The preferred solution is therefore to create a model map that relies on real-time force calculations for maximum versatility.

Haptic force calculations are very similar to free body calculations based on Newton's third law as represented in Figure 67.1.

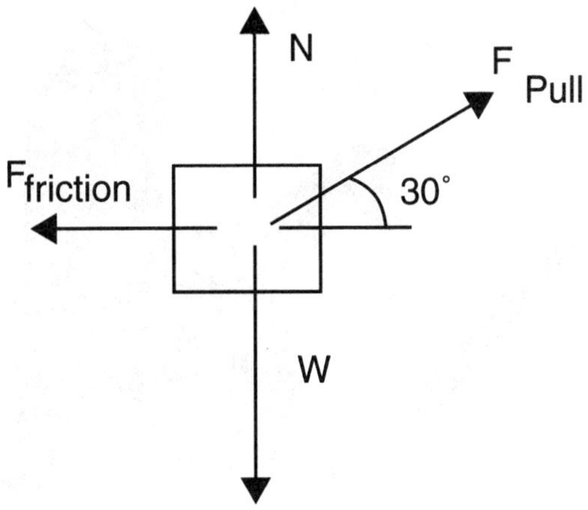

FIGURE 67.1 Free Body Diagram.

67.5 Haptic Hardware

The first haptic devices were derived from the telerobotic arms used in controlling remote robots. This section looks at some of the common variants of such force feedback devices.

Force-Feedback Arms

The simulation of the weight of an object, its inertia, and contact requires force feedback. Robotic arms are common in telerobotics using a master–slave configuration to perform remote manipulation tasks. The master arms employ force-feedback using electric actuators at the joints. Replacing the slave with a computer in the control loop created the earliest force-feedback arms.

Tanie and Kotoku [3] created a force feedback arm specifically for VR applications. Their system consisted of a 4 degree of freedom (DOF) link mechanism with direct drive electrical actuators. A 6-DOF wrist force sensor was placed on the hand grip.

The arm is gravity and inertia compensated, so that the user feels the arm to be weightless when it is not in contact with an object in the virtual environment. However, on contact, the forces felt by the user are computed as:

$$F = [T][B]\Delta r + [K]\Delta r \tag{67.1}$$

where

$\quad [T]$ = Cartesian coordinate transformation matrix
$\quad [B]$ = damping matrix introduced to assure stable contacts
$\quad [K]$ = stiffness matrix which simulates a virtual spring law
$\quad \Delta r$ = deformation vector of the virtual surface in contact.

The torque vector necessary at the joints in order to replicate the above force is given by:

$$\tau = J(\theta)^{\mathrm{T}} F + \tau_C(\theta) \tag{67.2}$$

where

$\quad J(\theta)^{\mathrm{T}}$ = transpose of the master arm Jacobian matrix
$\quad \tau_C(\theta)$ = compensation torque for arm inertia and weight.

The PHANToM™ Desktop haptic device from Sensable Technologies Inc. is similar in construction but is capable of sensing force in 6-DOF and returning force in 3-DOF. This device is shown in Figure 67.2.

FIGURE 67.2   The PHANToM Desktop (Courtesy Sensable Technologies).

Magnetic Levitation Haptic Devices

The Human–Computer Interaction Institute (HCII) at Carnegie Mellon University developed a 6-DOF haptic device based on the Lorentz force magnetic levitation principle. The magnetic levitation device allows noncontact sensing and feedback, high control bandwidth, and good position resolution. The device is specifically developed for fine haptic motion using the fingertips. It consists of an interactive handle suspended in a "flotor" bowl. Forces are exerted on the interactive handle through six magnets in the flotor bowl. Berkelman and Hollis [4] gave the forces on the flotor to be

$$F = AI \qquad\qquad (67.3)$$

where

F = vector of length 6, containing the forces and torques at the center of the flotor
I = vector of six coil currents
A = transformation matrix that maps coil currents to forces and is calculated by combining forces and torques.

Utah Hand Master

The Utah Hand Master consists of a master–slave configuration where a four bar link mechanism-based hand master is used to control the slave arm. The pads on each finger rotate in relation to the fixed base and these angular motions are measured via Hall Effect sensors in the pads. Finger movements along the axis of the base pad translate to linear motion and are measured by a separate set of Hall Effect sensors. Similar systems with different actuators are implemented by EXOS Inc. and Bouzit et al. [5] called Rutgers Master II.

67.6   Control of Haptic Hardware

Haptic interfaces have two basic functions. The first is to measure position information from the haptic device and the second is to render the feedback forces so that the user feels them. Typically, haptic applications have force feedback control, which means that given a certain position input from the user, haptic forces are returned. However, some applications use position feedback control wherein force input by the user returns position feedback. Position feedback controllers require the use of force and torque sensors from which the position is calculated. Position feedback control is not the preferred control method as true position feedback is very hard to achieve and may require the use of extremely stiff actuators. The advantage of using position feedback control is that it reduces the computational load on the model since forces are sensed rather than computed.

Force feedback control systems are open-loop systems. Thus, while a reaction force is returned to the user, there is no force sensing to ensure that the level of force produced in the actuators is accurate. The amount of force actually produced by the actuators depends on physical factors like friction and inertia of the mechanism. To compensate for the above effects and improve control accuracy, the interface needs to incorporate both position and force sensors. This essentially amounts to closed-loop control.

An example of such a system is the one developed by Hannaford and Venema [6]. By expanding network theory, they presented a unified model of the two modes of haptic interaction. Using an analogy of mechanical forces with voltage and position or velocity with current, they reduced haptic interaction to a unified model.

An adaptation of Kirchoff's voltage law to describe force feedback is shown as follows:

$$f_1 - z_1 (v_p) = f_2 - z_2 (v_p) = f_p \tag{67.4}$$

where

$$
\begin{aligned}
f_1 &= \text{forced sensed} \\
f_2 &= \text{force returned} \\
z_1 \text{ and } z_2 &= \text{are mechanical impedances} \\
v_p &= \text{velocity of the motion.}
\end{aligned}
$$

67.7 System Performance and Bandwidth Requirements

The performance of haptic devices is influenced by components like the sensors and dampers. Most haptic devices resemble robotic arms and are incapable of being perfectly stiff. The system's overall performance is summarized by its bandwidth. Sensations that require a crisp feeling need higher sampling rates, while soft sensations do not require very high sampling rates. Low sampling rates can therefore have a detrimental effect on the stiffness of the sensation being transmitted. It has been verified experimentally that the PHANToM™ haptic device maintains a uniform bandwidth of 1 KHz during the simulation. Scenegraph or graphics complexity needs to be adjusted to ensure that this minimum bandwidth is maintained. This can be achieved by limiting the number of polygons being rendered. Often, the entire model does not fit in the cone of vision of the user. Culling these polygons from the scenegraph can save valuable computational time.

67.8 Haptic Rendering

Haptic rendering is a computationally intensive process. As described earlier, in order to maintain a continuous sense of touch, it is essential to maintain a minimum update rate of 1000 Hz. Some real-time applications using haptic arms in medical simulations or as tools in surgery may require higher refresh rates. Despite the steady improvement of processing power in the last few years, true real-time rendering of complex haptic models is still difficult to achieve. Many haptic rendering algorithms have been developed over the years to address the problem of displaying complex haptic models within the constraints of available memory and processing power. Some of these techniques have been reviewed in this section.

Most link mechanisms used in robotic arms are inherently compliant systems. Hence, there is a limitation on the maximum stiffness generated by a PHANToM-like haptic device. Most haptic rendering algorithms rely on a vector-based approach to calculate and render a feedback force. It can be assumed that the resultant force vector should be normal to the surface of the object. However, compliance in the mechanism as discussed above can lead to a change in the direction of these vectors and thereby result in an incorrect force rendering. The problem is compounded when the object in question is thin and there is a merging of the boundaries on either side of the object. Other methods like volume subdivision work well for simple models but produce force discontinuities under certain conditions. Zilles and Salisbury [7] presented the concept of the God-object that addressed the above issues and allows a more realistic force rendering of complex arbitrarily shaped models. In this method, a virtual God-object is superimposed over the point of contact with the object. Unlike the stylus, the virtual God-object does not penetrate the surface and is actually used to compute the resultant forces based on the laws of physics. In their work, they achieved a simulation speed of 1 kHz for an object with 600 triangular polygons on 66 MHz Pentium computer. They claim that significant speedups will be possible by using an improved collision detection scheme as opposed to the simple collision detection used in their test case. Most simulations use a parent–child hierarchical structure to execute geometric and haptic rendering. This makes program execution faster as access to the various nodes is easier. To explain the scenegraph structure, we can consider the example of a simple haptic cube. The parent node would be the transformation node that will have the geometric, lighting, texture, and haptic nodes as its children. Keeping with the top-down hierarchy that governs the scenegraph, any change made to that parent node will affect all the children. The same is true of any of the children who may have children of their own. The real potential of scenegraph representations is apparent in complex VE's where the scenegraph extends several levels deep. Figure 67.3 shows a graphical representation of the scenegraph structure applied to the simulation of a robot.

67.9 Potential Applications of Haptics in Engineering

The information transmitted to the user through touch finds potential use in a variety of engineering applications. A brief review of the applications of haptics indicated its applications in several interesting engineering domains. Complex probabilistic systems are known to contain a factor of risk or uncertainty. Basapur et al. [8] have combined graphic and haptic environment to model uncertainty environment, with the haptic feedback loop accounting for the factor of risk. Hollerbach [9] has employed a haptic interface to probe heterogeneous finite element mesh fitting errors. The amount of error at a particular mesh element was played back as a vibration. This is used as an aide in the pre-processing of finite element meshes. Balijepalli and Kesavadas [10] have explored haptics as a simulation tool in the manufacturing environment (described as case study in this paper), while Joshi and Kesavadas [11] have explored a sympathetic haptic frame work for Product Life Cycle Management (PLM). A more direct force field effect was used by Sharma and Kesavadas [12] to represent vector fields for design problems. An initial point and search vector for exploring design space optimization was implemented in their approach. Costa and Cutkosky [13] have explored perception issues in haptically rendering fractal surfaces. Volkov and Vance [14] studied how a haptic device affects the ability of a person to make design decisions based on virtual prototyping concepts. The specific task involved

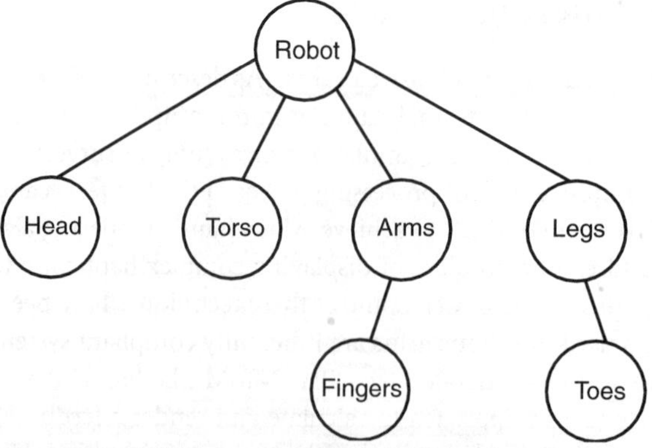

FIGURE 67.3   Scenegraph representation applied to a robot.

evaluation of position and range of motion of a virtual hand brake in a virtual automotive cockpit. This study provides some interesting and informative results regarding preference of haptic treatment. More significantly, it was shown that the users prefer the haptic treatment and this requires lesser time than the nonhaptic treatment for the handbrake design.

In the following sections, two applications of haptics are provided in some detail. The first application involves the use of haptics in simulating complex grinding and polishing task, while the second application explores the use of haptics in design decision-making process.

Operator training of complex grinding tasks

The aim of grinding is to remove asperities on the surface of the workpiece. The theory governing grinding with robots is different from that in conventional machines and needs to be modeled in detail to handle compliance factors of the link mechanism. The force pattern generated during the grinding of scallops depends on the topography of the workpiece.

The contact conditions between the tool and scallops on the surface can be seen in Figure 67.4.

In order to derive the forces at the end effector arm, a model of forces between the grinding wheel and scallops is used. A grinding wheel with a stiffness K_m is assumed and the normal force produced is F_N. However, certain weaknesses were found in this model. They are

1. Low stiffness causes deflection of the end effector arm. As a result, scallops are not removed completely.
2. The high compliance due to the low stiffness can also cause geometrical error.

Prior knowledge of the grinding direction can eliminate the above problems. The normal force F_N is accurately calculated using the Gatlin et al. [15] derivation. If δr is the depth of cut, for complete contact along the width of the grinding tool,

$$F_N = K[C_1]\,^\epsilon[V_w/V_s]^{2\epsilon-1}[D]^{1-\epsilon}[\delta r]^\epsilon[D_w] \tag{67.5}$$

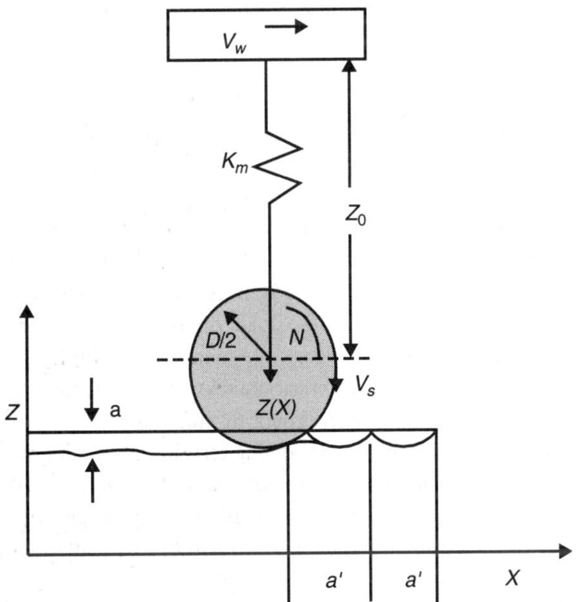

Grinding across the scallops

FIGURE 67.4 Contact conditions on a workpiece.

where

D_w = wheel width
C_1 = empirical constant accounting for static cutting edge
D = wheel diameter (mm)
F_N = normal grinding force
K = machine stiffness normal to the part
V_s = wheel surface speed
V_w = relative workpiece velocity
δ_r = wheel depth of cut
ϵ = empirical constant accounting for kinematics of contact
γ = empirical constant accounting for static cutting edge.

The above equation is valid for complete contact of the grinding wheel. However, during a grinding operation, complete contact is not achieved. The length of contact is given as

$$l_k = \sqrt{D_w\, h} \tag{67.6}$$

The force model is now adapted to account for the contact ratio, which is given as

$$w = (nW/D_w h) \tag{67.7}$$

where

h = contact depth
n = number of scallops in contact
W = width of one scallop from root to root.

Substituting the contact ratio into the normal force equation, we obtain

$$F_N = [\frac{n\,W}{h}]K[C_1]^\epsilon[V_w/V_s]^{2\epsilon-1}[D]^{1-\epsilon}[\delta r]^\epsilon[D_w] \tag{67.8}$$

where

$$K' = K[C_1]^\epsilon \tag{67.9}$$

Image Processing to Obtain Surface Roughness

Traditionally, computer graphics are modeled as a collection of smooth surfaces stitched together to form a composite structure. It is assumed that any or all of the components that make this model can be defined deterministically using mathematical functions. While some complicated structures can be defined by Bezier or B-spline surface patches that use higher-order polynomial functions, this process can be time consuming and very memory intensive. The problem is compounded with complex virtual environments consisting of millions of polygons. With advances in graphics hardware and the ability to texture map digital photographs onto polygons, the load on the graphics engine is reduced. Greater approximations are effectively masked by realistic texture maps. For example, smart billboards consisting of only a single polygon and mapped with an appropriate texture can be made to constantly face the user, thereby eliminating the need to model the environment in multiple views.

These methods are, however, not suitable for modeling terrains from data sources like satellite or aerial photographs. Naturally occurring phenomena do not necessarily conform to well-defined mathematical functions and hence may not be effectively represented using the stochastic methods. The particular method selected for generating a terrain, however, will depend on the end application.

Height Maps

Height maps can hold detailed information about the surface of an object. They represent an effective means of displaying terrain data extracted from photographs. Since the generation of a height map is dependent on the lighting conditions present at the time, it is not the most accurate method of generating a terrain. However, in computer games like flight simulations where the margin for error in the accuracy of the representation is high, these methods are an effective way of representing background terrain formations. In our application, we are interested in using height maps to represent surface roughness of a workpiece based on the image of that workpiece. For very precise models where accuracy is of paramount importance, precise methods like laser scanning or a roughness meter can be used to extract features of the surface of the workpiece. In our case, by specifying the light conditions in advance, it is possible to reduce the variations in image quality.

Height information can be added to a 2-D image by analyzing the light intensities of its pixels, assuming that the light conditions at the time of taking the picture are constant. Shadows formed on an object are directly related to the height of a particular point, respective to its surroundings. While the lighter areas are elevated, the darker areas of an image are depressed. An example of this principle can be found in photographs of a human face. Incident light causes the elevated areas of the nose and cheekbones to appear lighter than the depressed orbs of the eyes, which usually are darker as shadows tend to form on them. Using this fundamental principle, it is possible to extract some basic height information from a simple image.

Height information is extracted as a function of the light intensity values. The z-coordinate at each pixel location is then calculated choosing a suitable amplification or scale factor. Height is calculated with respect to the minimum intensity value in the image using the relation:

$$height = (intensity - min)/(max - min) \tag{67.10}$$

where

$height$ = calculated height(z-coordinate) from pixel (x,y)
$intensity$ = intensity at pixel (x,y) of image
max, min = maximum/minimum intensity value in image.

Creating the Height Map Surface

3-D height map information can be extracted from a 2-D image source, such as a photograph. An uncompressed grayscale image in Targa (TGA) format is used. An OpenGL TGA loader was developed to load the image header and data into dynamic arrays. The loader is capable of loading uncompressed grayscale and RGB(A) images. In the case of RGB(A) images, the image is converted to grayscale before any further processing is performed. Conversion takes place using the empirical relation:

$$grayscale = 0.30R + 0.59G + 0.11B \tag{67.11}$$

Intensity values are stored for every pixel of the image. Height information for the image is generated using these intensity values. Grayscale intensity values range from 0 to 255, 0 representing the darkest (black) and 255 representing the brightest (white).

The final height values are obtained by multiplying the height values by the scale factor. Finally, the height map generated is saved into the PNG format.

Figure 67.5 shows an image of a rough surface with a high number of surface asperities. The conversion into height map is shown in Figure 67.6.

The Haptic Model

Two threads are allocated: one for updating the graphics at 60 Hz for mono or 100 Hz for stereo and the second, a haptic servo loop running at 1 kHz. The refresh rate of 1 kHz in the haptic loop has been found to be necessary to maintain a sense of continuity. All developmental work was carried out on a high-end Dual Pentium III computer.

The model was loaded as a VRML file. By reading the field values of the VRML, into the program, a scene graph was constructed giving control over the various parameters of the model. The VRML file

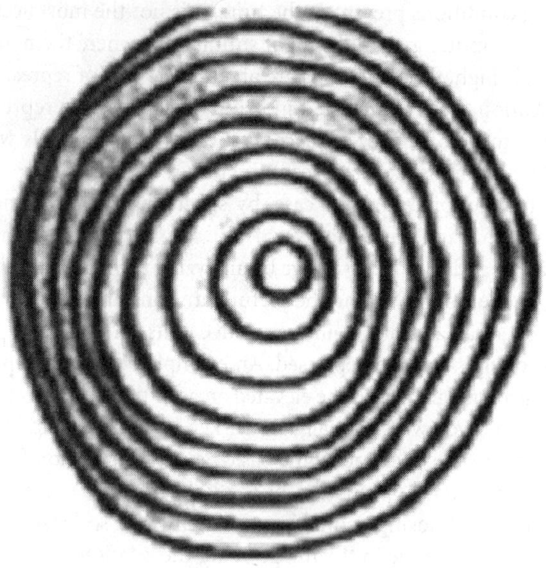

FIGURE 67.5   Image of a patterned rough surface.

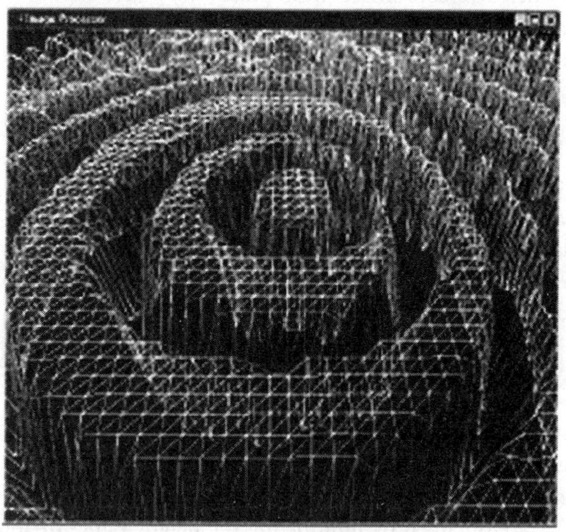

FIGURE 67.6   Surface after extracting height data from image in Figure 67.5.

was used only to load the initial configuration. A simple geometric object, a box in this case, was created. The height map texture was applied on the geometry. All the data extracted from the image (Figure 67.5) were stored in the allocated memory and the necessary lookup functions are precomputed. The rough surface generated by the height maps in Figure 67.6 could then be felt using the PHANToM. Positional information of the stylus was stored sequentially and correlated with the force information at that point. The recorded data are exported for analysis. Using this system, the operator could interactively choose the machining path and the specific surface of the workpiece to be machined. The haptic interface is shown in Figure 67.7.

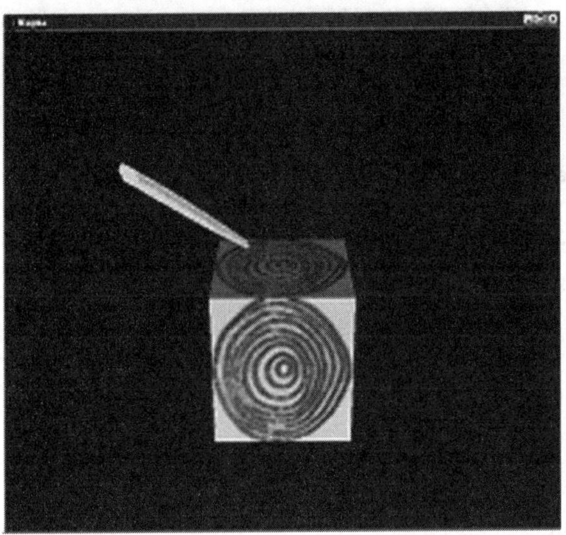

FIGURE 67.7 The haptic interface.

Dynamic Texture Modification for Enhanced Realism

The effectiveness of this simulation is dependent on both the physical characteristics of the height map and the accuracy of the predicted forces. A height map texture that remains unchanged when the applied force is defined as a *static texture*. A static texture greatly limits the practicality of the application. While performing a grinding task, one of the factors that help determine the user's progress is a perceivable change in the feel of the surface being machined. It is extremely difficult for the user to gauge his/her progress without this tactile feedback. In the real world, as the operation progresses, material is continuously removed from the surface. This causes the surface to smoothen out. At the end of the task, if performed correctly, the entire surface should feel and visually appear to be smooth. Hence, a good simulation must take this into account and make suitable adjustments so that the user can perceive changes in the surface texture both haptically and visually.

The variables affecting material removal are speed of the tool, machining velocity, depth of cut, tools stiffness coefficient, and the force applied. Keeping all other factors constant, the amount of material removed is directly a function of the force applied. This information is stored in a table within the program along with the value of the color of the surface texture observed for a particular texture height value. In our case, shades of gray represent different levels of imperfections with white (no z-value) to represent the smoothest surface. The maximum range of height of roughness is small enough to allow a linear approximation. Moreover, the haptic sensors in our hands do not have the resolution to perceive extremely fine changes in roughness and hence the overall error due to the linear approximation is negligible.

A suitable data structure is devised to store this information. At runtime, the haptic device samples the force applied at a particular point and the corresponding color interpolated from the lookup table. The color of the pixel at that point is changed based on the result from the table. The height map is recalculated during the next update, which takes place 1 msec later and the new point is automatically calibrated to reflect the changed height. When the height at a particular point reaches the optimum value, the lookup function is disabled in order to prevent a reverse effect that might signify addition of material.

Another common phenomenon observed in a grinding process arises from the application of excessive force on the workpiece. This leads to an irreversible scarring (gouging of the material from the surface) of the surface. A visual and haptic representation of this phenomenon has been developed, which occurs when the force applied at a particular point exceeds a preset threshold force. Figure 67.8 shows the effects of texture modification as seen by the user.

FIGURE 67.8 Dynamic texture modification.

The User Interface

A user interface has been created using OpenGL to display position and force. When training human subjects, in order for them to get a better feel for the force, the new data are displayed once every 5000 cycles of the haptic loop. A 3-D wire frame model of the work piece is displayed in one window (Figure 67.9[a]). The position values of every point, sampled at 1 kHz, are superimposed on this model. The points are color keyed to highlight force distribution on the model, red being the highest force and green being optimum. Shades of orange represent a higher than optimum force. The second window (Figure 67.9[b]) is a plot of the forces at each point, combined with other relevant data like the target force and the moving average. The target force represents the optimum force that the user should apply in order to effectively machine the surface. The moving average is the grouped average of 100 samples of force applied by the user. This plot is also useful in observing the learning curve of the person using the haptic simulator. The third window (Figure 67.9[c]) quantifies the data in the plots and also displays the total time taken. This user interface can read in VRML files containing the indexed face sets, thus allowing virtually any model that is loaded in the haptic framework to be displayed in our GUI.

Experimental Validation

Sets of experiments were performed to bring out the value of using haptics in operator training simulations. Experiments were broadly split into two groups [10]. One group could feel the roughness of the height map texture, while the second could feel only a smooth surface. Both groups were provided with visual updates of the textured surface. The main groups were then subdivided into two subgroups. One was given a specific instruction to follow the direction of the surface pattern and the other was asked to use his/her intuition to select a method in performing the task — that is, to remove the texture and make the surface smooth. Each subject performed the test with two different patterns. The time taken for each subject was recorded. The average force was computed based on samples collected at 1 msec intervals. A questionnaire to assess the user's perception of the haptic sensations fed to him/her was administered. The questions contained a scale from 1 to 5 that gauged the confidence, realism, and security they associated with the force feedback.

Inferences

When an operator performs this task, he/she subconsciously relies on the vibro-tactile feedback communicated through the tool to the hands. The amount of force applied and subsequent force corrections are adjusted automatically because of the perception of this surface roughness. Data clearly show that subjects who could feel the surface roughness could better relate to the actual physical process than subjects who did not.

(a)   (b)   (c)

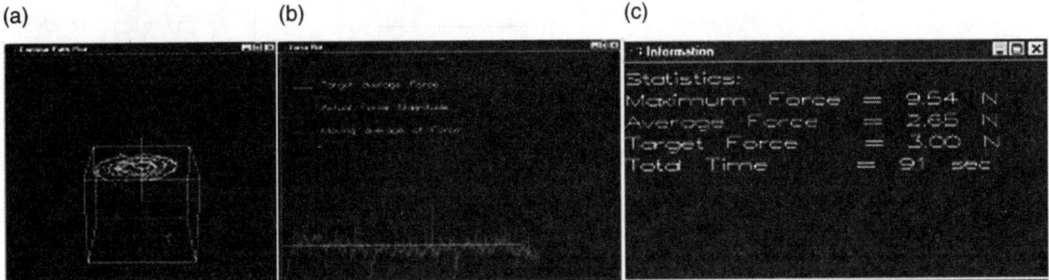

FIGURE 67.9   Graphical user interface.

Without a real-time interface that gives the current force being applied against the desired target, sometimes the only option left to the user is to make an estimate by guessing. This induces a learning effect that in time lets the user predict the force fairly accurately, without referring to any numerical feedback information being displayed. Participant feedback shows that the users with force feedback are in general more confident about their actions than users without force feedback.

67.10 Haptics in Design Optimization

The effect of improved cognition on algorithm performance and evaluation is an exciting area of research. Virtual reality is becoming increasingly popular as a tool for interacting with data and steering the solution, rather than simply displaying results. Constantly evolving haptic technology allows people to directly interact with digital objects and data as they do in the real world using their sense of touch. Haptics is revolutionizing the way engineers interact with computers. Hitherto, designers and artists have used haptics to intuitively sculpt models using digital clay and doctors have trained on haptics-based virtual surgical simulators. Most applications, however, merely seek to make up for the lack of sense of touch and its effects. The current work investigates the application of the haptics loop for decision support activities in core engineering areas like design optimization and shop floor layout. An architecture combining large-scale visualization environment with haptics is proposed. The Simulated Annealing (SA) algorithm, a popular heuristic for solving the Quadratic Assignment facility layout problem, is used as an example to showcase the concept. The use of combined haptics and graphics for sensitivity analysis provides interesting results.

Conceptual Framework for Haptics in Engineering Analysis

In complex design processes, often the user is faced with the task of analyzing very large number of variables. The effect of design decisions on quality, safety, and other important feedback may escape the designer since some of the design or mathematical elements may lack an intuitive feel to it. The ability to present some of these elements in forms of haptic feedback such as force, vibration, resistance, etc., may add a new dimension to the process of design. This work explores a framework for such a system for solving a facilities layout problem using the SA technique. A what-if analysis for shop floor decision making, involving a single choice over several scenarios, can include haptic effects for decision making. Facilities Planning (FP) is the process of locating and laying out of industrial and service facilities to best support the purpose of the facility while respecting constraints on resources such as space and budget. The FP function involves strategic, tactical, and operational decisions depending on the nature of the facility [16].

The multiple FP layout scenarios are visualized using the Virtual Factory (VRFACT) software developed at the University at Buffalo [17]. The PHANToM desktop developed by Sensable Technologies Inc. is used to move the facilities (and machines) around inside the shop floor. The PHANToM desktop produces a buzz effect when the facilities are located at positions corresponding to an optimal configuration. The intensity of this effect is proportional to the solution quality measure. The backbone architecture to provide an integrated GHUI (graphical haptic user interface) is shown in Figure 67.10.

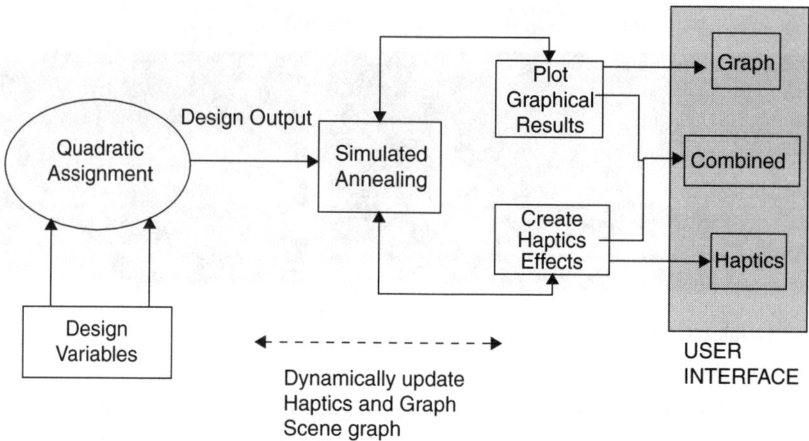

FIGURE 67.10 Haptics-based integrated visualization environment.

The quadratic assignment problem [18] for facility location used to determine the optimum solution is described in the next section. The locations of facilities on the shop floor are obtained and stored using the SA algorithm, described in the following section.

Quadratic Assignment Problem (QAP) Formulation

The layout problem consists of determining the relative positions of the k entities in the set (m_1, m_2, m_k), which may represent either the set of machines, belong to a cell or the set of manufacturing cells within the shop. This analysis seeks to minimize the total distance traveled by the manufactured parts, or pallets of parts, between these entities.

The objective function is defined as

Minimize:

$$E = \sum_{i=1}^{k-1} \sum_{j=i+1}^{k} T_{ij} d_{ij} \tag{67.12}$$

where

Tij = number of pallet transfers between entities m_i and m_j

d_{ij} = distance between entites m_i and m_j

x_{ij} = 1 if the department i is assigned to site k

       0 otherwise

where T_{ij} is the number of pallet transfers between machines m_I and m_j, and distance d_{ij} is the corresponding distance. The value of T_{ij} is the number of pallet trips required for transferring all parts, which use the two entities consecutively. Note that the quantity T_{ij} may also be weighted by any combination of parameters to consider special material handling requirements, part weights, actual material handling costs, etc. To quantify the distance d_{ij} between the entities m_i and m_j, the "Manhattan" distance measure is used

$$d_{ij} = |x_i - x_j| + |y_i - y_j| \tag{67.13}$$

where (x_i, y_i) and (x_j, y_j) are the coordinates of the geometric centers of entities mi and mj, respectively. Since the coordinate space is discrete and finite, each point in the space is assigned a unique "position number" in order to simplify the analysis. The problem thus consists in determining the position number corresponding to each of the k entities corresponding to the objective function of Equation (67.13). It is assumed that entities do not overlap when they are assigned to adjacent positions. Furthermore, each position cannot be occupied by more than one entity. The quadratic assignment problem is a difficult combinatorial optimization problem. An exact solution for problems larger than 14 departments is generally very difficult. Accordingly, the problem is usually solved using heuristic methods.

Proposed Solution Methodology

SA [19] is an approach to solving difficult combinatorial problems that has gained considerable attention in recent years. The approach is based on a Monte Carlo model used to study the relationship between atomic structure, entropy, and temperature during the annealing of a sample of material. In the context of QAP, simulated annealing resembles the pairwise heuristic, with one very important difference: if an interchange increases the total cost, it may still be accepted with some predetermined probability. Thus, there is at least some chance that an unlucky choice for an intermediate solution will not cause the search to be trapped at a suboptimal solution. The sample plot obtained for the SA algorithm is as shown in Figure 67.11 with a snapshot of the layout corresponding to one of the multiple optima.

The SA algorithm produces several optimal layouts making the choice of one particular layout configuration difficult. As shown in Figure 67.11, a minimum material flow may correspond to different combinations of machine locations.

Haptics Process for Design Evaluation

The goal of this research is to develop an interactive analysis tool for helping a designer to try various design options and study its effectiveness using haptic-based multimodal interface. Figure 67.10 shows the sequence of the evaluation process. Thus, the process would require the optimization to be carried out first and then an interactive session would ensue where strategic alternatives would be studied in further detail. An example of such an analysis in the domain of facilities layout would be to study the human factor or ergonomic issue in conjunction with the optimization results. Thus, even though a set of optimization results are in place, the need for two similar machines to be adjacent to each other may override the original goal of the minimizing the distance that a part/pallet moves. However, when altering the machine layout, the effect in terms of the original goal is also ascertained using the proposed interface.

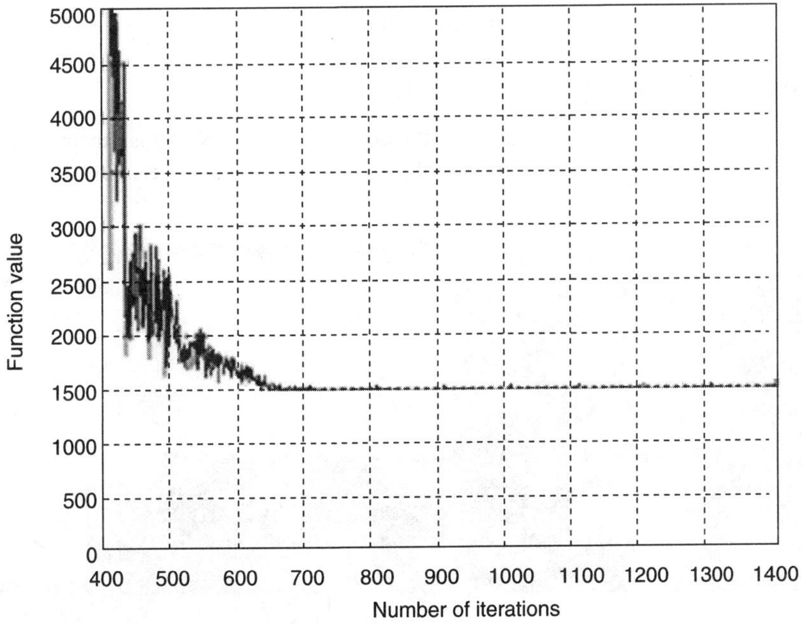

FIGURE 67.11 Sample plot for variation of objective.

The present work provides three modes of this interactive design analysis: (i) using only force effects to represent the optimization results, (ii) graphical plots showing the results, and (iii) a graphical + haptics interface where the user is provided with both graphical and haptic effects simultaneously.

The machines/facilities in the problem are shown as cubes within a shop floor grid as shown in Figure 67.12 (middle). The user can click and drag the machines to different grid locations.

In the haptics interface mode, the cube is moved using the PHANToM haptic device. While a move is being made, based on the efficiency as described in the next section, an appropriate force will resist the movement of the block. The viscose effect creates a feel of moving the block in a resistive environment, such as on a high-friction surface. The resistance to movement of the block is inversely proportional to the efficiency of the solution based on the optimization results already obtained.

In the graphical plot mode, the blocks are moved with the mouse and the results are presented as a 2-D plot. In this case, there is no haptic feedback to the user. In the third mode, both the haptic and graphical results are presented simultaneously to the user. The goal of the above design process is to prompt the user to arrive at the best solution based on the three modes of feedback.

The proposed steps of operations for obtaining a haptics-based integrated decision-making and computational environment is as follows (Figure 67.10):

1. With the proper choice of input parameters, the SA algorithm is run, to obtain layouts corresponding to multiple optima.
2. These objective function values corresponding to these optimal layouts are stored in a file.
3. Choosing a good layout is a combination of the layout corresponding to time as well as a good solution quality measure and are represented as spring stiffness 'k' and viscosity 'c' for viscous force effects in a haptic scene graph. These values are normalized to lie between 0 and 1 to represent spring stiffness and 0.002 and 0.009 for viscosity. By moving facilities around to different locations, a viscous effect corresponding to the solution quality measure and CPU time is represented. The resultant force expression is of the form:

$$f = kx + cx \tag{67.14}$$

where "x" is the velocity of the probe at location "x."

The designer makes his/her decision using one of the three modes as described earlier.

Results of a Testing of the Environment

A PHANToM desktop haptic device was used for the experiments. PHANToM is capable of simulation of 3-D force feedback actions such as viscous effects, softness, hardness, vibrations, etc. In order to appropriately

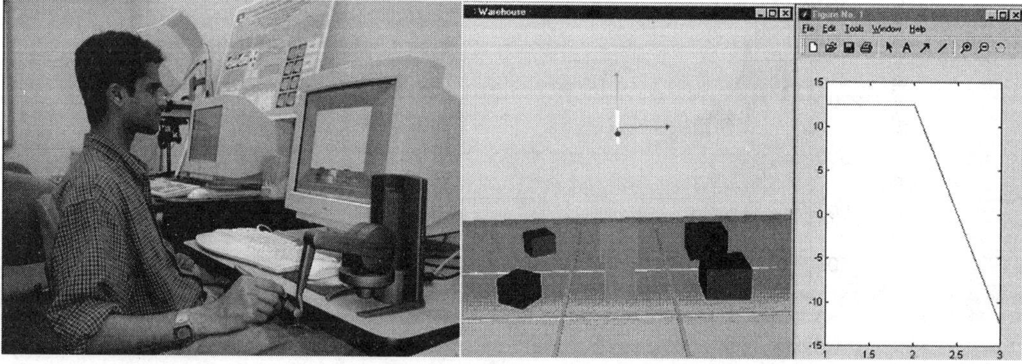

FIGURE 67.12   Experimental setup (left) used in the experiment. The 3-D blocks represent machines (middle) and graph interface shows the efficiency as a plot (right).

represent the machine layout problem, the scene graph consisted of six blocks, in red, green, and blue colors. For each color, two different shapes were created. The placement of the blocks in the grid corresponds to one of three costs: optimal (-12.5), satisfactory (0), or unwanted (12.5). The "goodness" of the layout can be evaluated by plotting the cost corresponding to a particular layout arrangement graphically through a 2-D plot or by getting a resistive force corresponding to the costs associated with the solutions or as a combination of both graphical and haptic effects. Guided by these visual and haptic cues, the layout designer is asked to arrive at a solution. The time taken to arrive at a solution for each set of experiments is recorded and later compared using ANOVA and the "t"-test to decide whether there is a significant difference in the results obtained. Twelve subjects carried out the experiments in a completely randomized order. Each subject also answered a questionnaire for a subjective evaluation on using the Phantom desktop and the experiment in general. The designers were encouraged to give their views during the trial run.

Several interesting results were observed in this work. In all, 66% of the subjects preferred a combination of haptics and graphics to pure haptics or pure graphics interface. It was also observed that the time to arrive at a decision was fastest with the haptic interface. Haptics interface took on an average 35–45% less time than the graphics and the graphics+haptics interface, suggesting that haptics is indeed a powerful new way to interact with complex design scenarios.

References

[1] Burdea, G.C., *Force and Touch Feedback for Virtual Reality*, John Wiley and Sons Inc., New York, 1996.

[2] Ellis, R.E., Planning tactile recognition paths in two and three dimensions, *The International Journal of Robotics Research*, Vol. 11, 87–111, 1992.

[3] Tanie, K. and Kotoku, T., Force display algorithms for virtual environments, *Transactions of The Society of Instrument and Control Engineers*, 29, 347–355, 1993.

[4] Berkelman, P.J. and Hollis, R.L., Dynamic Performance of a Hemispherical Magnetic Levitation Haptic Interface device, Proceedings of the SPIE International Symposium on Intelligent Systems and Intelligent Manufacturing, Vol. 3602, Greensburgh, PA, 1997.

[5] Bouzit, M., Burdea, G., Popescu, G., and Boian, R., The Rutgers Master II — New design force-feedback glove, *IEEE Transactions on Mechatronics*, 7, 256–263, 2002.

[6] Hannaford, B. and Venema, S., *Virtual Environments and Advance Interface Design*, Oxford University Press, Oxford, 1995.

[7] Zilles, C.B. and Salisbury, J.K., A Constraint-Based God-Object Method for Haptic Display, Proceedings of International Conference on Intelligent Robots and Systems, Vol. 3, Pittsburgh, PA, 1995, pp. 3146–3151.

[8] Basapur, S., Bisantz, A.M., and Kesavadas, T., The Effect of Display Modality on Decision-making in Uncertainty, Proceedings of the Human Factors and Ergonomics Society 47th Annual Meeting, Denver, CO, 2003.

[9] Hollerbach, J.M., Some Current Issues in Haptics Research, Proceedings of IEEE International Conference on Robotics and Automation, San Francisco, CA, 2000, pp. 757–762.

[10] Balijepalli, A. and Kesavadas, T., An Exploratory Haptic Based Path Planning and Training Tool for Polishing, ASME International Mechanical Engineering Congress and Exposition, New Orleans, LA, 2002.

[11] Joshi, D. and Kesavadas, T., A Framework for Haptics Based Collaborative Design Environment, CD-ROM Proceedings of the International Symposium on Product Lifecycle Management, Bangalore, India, 2003.

[12] Sharma, C. and Kesavadas, T., A Haptics-Based Virtual Environment for Engineering Design and Manufacturing Applications, CD-ROM Proceedings of the 2001 ASME Design Engineering Technical Conferences, Pittsburgh, PA, 2001.

[13] Costa, M.A. and Cutkosky, M.R., Roughness Perception of Haptically Displayed Fractal Surfaces, Proceedings of the ASME International Mechanical Engineering Congress and Exposition, Vol. 69(2), Orlando, FL, 2000, pp. 1073–1079.

[14] Volkov, S. and Vance, J.M., Effectiveness of haptic sensation for the evaluation of virtual prototypes, *Journal of Computing and Information Science in Engineering*, 1, 123–128, 2001.

[15] Gatlin, D.S., DeVries, W.R., Lauderbaugh, L.K., and Reiss, J.S., A model for cusped surface grinding used in designing a robotic die finishing system, *Computer-Aided Design and Manufacture of Molds PED*, 32, 1–17, 1988.

[16] Kusiak, A. and Heragu, S.S., The facility layout problem, *European Journal of Operations Research*, 29, 229–251, 1987.

[17] Ernzer, M. and Kesavadas, T., Interactive Design of a Virtual Factory Using Cellular Manufacturing System, Proceedings of the IEEE International Conference on Robotics and Automation (ICRA), Vol. 3, Detroit, MI, 1999, pp. 2428–2433.

[18] Koopmans, T.C. and Beckmann, M.J., Assignment problems and the location of economic activities, *Econometrica*, 25, 53–76, 1957.

[19] Kirkpatrick, S., Gelatt, C.D., and Vecchi, M.P., Optimization by simulated annealing, *Science*, 220, 671–680, 1983.

68

Implementation of a Virtual Factory Communication System using the Manufacturing Message Specification Standard

Dong-Sung Kim
Kumoh National Institute of Technology

Zygmunt J. Haas
Cornell University

Wook Hyun Kwon
Seoul National University

68.1 Introduction

For interconnection purposes, a factory automation (FA) system can be combined with various sensors, controllers, and heterogeneous machines using a common message specification. In particular, interconnection of heterogenous machines through a common message specification promotes flexibility and interoperability.

For this reason, the *manufacturing message specification (MMS)* standard has been developed. The standard specifies a sets of communication primitives and communication protocols for the factory communication environment. In particular, the MMS standard specifies various functionalities of the different FA devices in a compatible way. Thus, users of MMS applications have to use functions from only one unique set of functionalities to operate various kinds of automation machines. Moreover, the different automation machines can communicate among themselves through the standard automation language. This enables transition from the traditional centralized control to the distributed control systems.

The MMS standard is composed of the common standards [1, 2] and the *MMS companion standard (MMS-CS)*. The MMS-CS includes device-dependent specifications for a robot [3], numerical controller (NC) [4], programmable logic controller (PLC) [5], and process control (PC) [6]. In its initial stage, the MMS standard was developed for an application layer of the manufacturing automation protocol (MAP) [7]. Nowadays, MMS has been implemented on top of the TCP/IP protocol suite and is used as a reference model for industrial network or other message protocols, such as the home network protocol [8]. In particular, MMS is implemented in a minimized form as the application layer for Profibus (Process Fieldbus) [9] and field instrumentation protocol (FIP) [10].

A *virtual factory* environment can be used to examine the correctness of an implementation and as a test solution of an FA system prior to installing the system in a real factory communication environment [11, 12]. Furthermore, by using the virtual factory communication system, developing time and costs can be minimized. In addition, it can be used as a training tool of MMS users. In this article, the virtual factory communication system was designed and implemented with the use of the MMS standards and its CS part.

Researches in the MMS technology include application of factory devices, middleware, and multimedia communication. In [13], a real NC machine was implemented using the MMS-enabled application program. In [14], an MMS with common object request broker architecture (CoRBA) was studied. A modified architecture was proposed for multimedia communication in an FA system [15, 16]. The performance evaluation and an analysis methodology of the MMS standard have been studied as well [17–19]. However, to the best of the authors' Knowledge, implementation of a virtual factory communication system using the MMS standard and its CS part has not yet been reported in the technical literature.

In this chapter, we describe the implementation of a virtual factory communication system using the MMS and the MMS-CS standards. In addition, we discuss here the implementation of an MMS Internet monitoring system (MIMS) for on-line monitoring of the developed virtual factory communication system.

This chapter is organized as follows. In Section 68.2, the design of the MMS communication on top of TCP/IP is described. The developed virtual factory communication system is presented in Section 68.3. In Section 68.4, MIMS, the monitoring program of virtual factory, is introduced. Finally, summary and conclusions are presented in Section 68.5.

68.2 MMS on Top of TCP/IP

Two types of protocol structures for the MMS communication on top of TCP/IP, referred to here as "MOTIP," have been studied [3]. One method, which is based on requests on comments (RFC) 1006, uses the open systems interconnection (OSI) 7-layer model [20]. The other method is based on N578 that defines the direct mapping relation between the M-services and the TCP functions [21]. In particular, we chose to use the N578 method in the implemented system because of its simplicity. The differences between the structures of a Full-Map, Mini-Map, and MOTIP are presented in Figure 68.1. In the case of Mini-MAP, IEEE 802.4 token-bus network is used instead of the Ethernet (IEEE 802.3) subnetwork. From

1	MMS	ACSE	MMS	MMS
2	Presentation			
3	Session			
4	Transport			
5	Network			TCP
6	Data Link		Data Link	IP
7	Ethernet		Token Bus	Ethernet

FIGURE 68.1   The various types of the MMS Protocol Structures (Full-Map, Mini-Map, and MOTIP).

Figure 68.1, a Full-Map structure has a scheme similar to MOTIP based on the RFC 1006 method, and the Mini-Map structure is similar to MOTIP based on the N578 method.

The implemented virtual factory communication system is based on the description of a practical test plant shown in Figures 68.2 and 68.3. These figures show an overview (Figure 68.2) and the layout (Figure 68.3) of a practical test plant using the MOTIP-based program. MOTIP-based communication programs have been tested with the Mini-Map and the Full-Map system using a developed gateway, and the developed virtual factory communication system is a miniature of the test plant in Figure 68.3.

FIGURE 68.2   Overview photo of the referenced practical test plant.

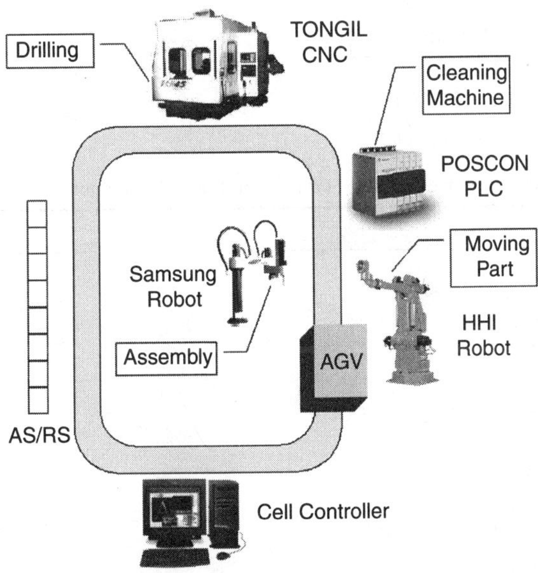

FIGURE 68.3   Reference layout of the virtual factory.

M-Services and MMS Interface in MOTIP

As depicted in Figure 68.4, the MMS functions convert a data control block (DCB) to the corresponding parameters of the M services.

Encoding and decoding can be used for converting DCB to PDU directly. The encoding and the decoding rules are based on abstract syntax notation 1 (ASN.l) [25, 26]. The protocol data unit (PDU) is then forwarded to the TCP layer. Table 68.1 presents the mapping relations between M-services and the TCP functions in the MOTIP implementation [14].

The protocol stack of MOTIP using N578 rule is shown in Figure 68.4. Once an MMS function is requested at the MMS client side, the synchronous service or the asynchronous service can be used for communication. Using the synchronous service, the MMS client receives a confirmation through a response from the MMS server. In the asynchronous service, the functions can be executed according to a received event through the response of the MMS server. Most of the implemented MMS services are executed by synchronous mode in the developed system.

68.3   Virtual Factory Communication System

The developed virtual factory communication system was implemented with the Microsoft Visual C++ and OpenGL library [27]. The virtual factory is embodied by integrating each virtual real machine (VRM), the MMS-CS program, and the MOTIP programs. Each VRM can be communicated with through the developed MMS-CS program based on the SNU MMS library [23].

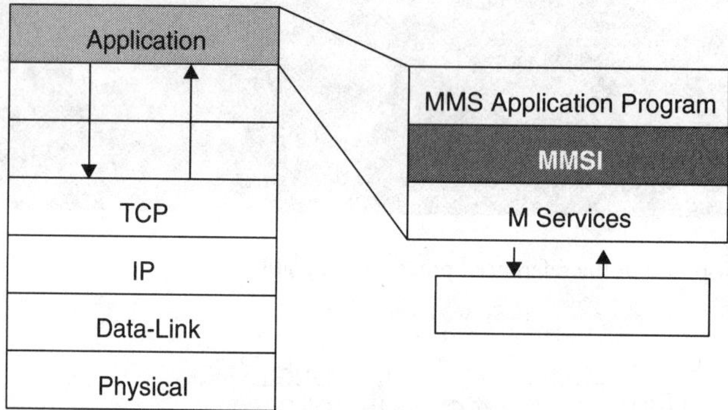

FIGURE 68.4   The protocol stack of MOTIP.

TABLE 68.1   Mapping Functions between the M-Services and the TCP Functions

M-Service	TCP
M_ASSOCIATED.request	OPEN, SEND
M_ASSOCIATED.indication	OPEN, RECEIVE
M_ASSOCIATED.response	SEND, CLOSE
M_ASSOCIATED.confirm	RECEIVE, CLOSE
M_RELEASE.request	SEND
M_RELEASE.indication	RECEIVE
M_RELEASE.response	SEND, CLOSE
M_RELEASE.confirm	RECEIVE, CLOSE
M_DATA.request	SEND
M_DATA.indication	RECEIVE
M_U_ABORT.request	ABORT
M_U_ABORT.indication	TCP error signal
M_U_ABORT.response	TCP error signal

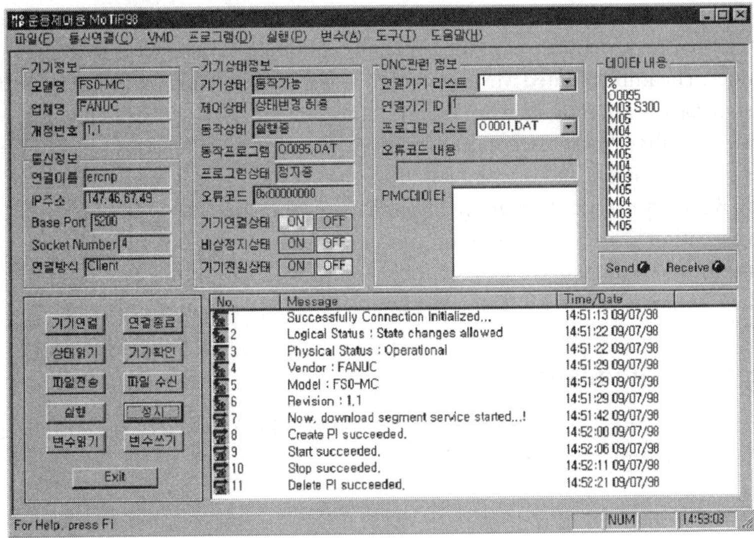

FIGURE 68.5 Screenshot of the MOTIP-based central control program.

FIGURE 68.6 The practical model of a 6-axis robot VRM.

The MOTIP central server program (see Figure 68.5) can be located in a control room of a factory plant. The MOTIP central server can monitor and command the MMS services to the networked manufacturing machines. The implemented MMS services are as follows: virtual manufacturing device (VMD) service for the status acquisition of a networked machine, domain service for file transfer, program invocation (PI) service for the remote execution of a program, and miscellaneous services for reading and writing services of device status and operations. PI services includes creation, initialization, termination, and deletion of a remotely executed program. PI service can be carried out with other services. For example, the remote execution of a specific program can be performed after downloading the status and segment service using the VMD and Domain service. Figure 68.5 shows the execution of the VMD, the Domain, and the PI services in the MOTIP program.

MMS Companion Standard

The implementations of the MMS-CS VRMs are based on the MMS services specification in [1–6]. Specifically, the services of the implemented MMS-CS VRMS consist of: VMD, Domain, PI, and miscellaneous services.

The implemented services are as follows:

- VMD services (3): *Status, Identify,* and *Unsolicited Status.*
- Domain service (5): *Initiate Download, Download Segment, Terminated Download, Initiated Upload Segment,* and *Terminate Upload.*

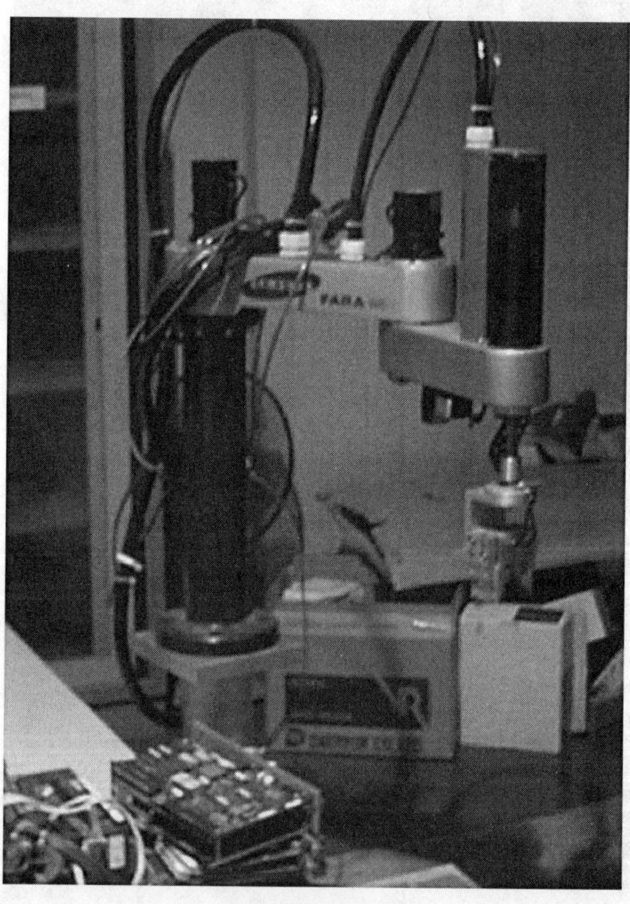

FIGURE 68.7   The practical model of a 4-axis robot VRM.

- PI services (4): *Create*, *Start*, *Stop*, and *Delete*.
- Variable service (2): *Read* and *Write*.

Semaphore, *Operator*, *Event*, and *Journal* services have been implemented partially only.

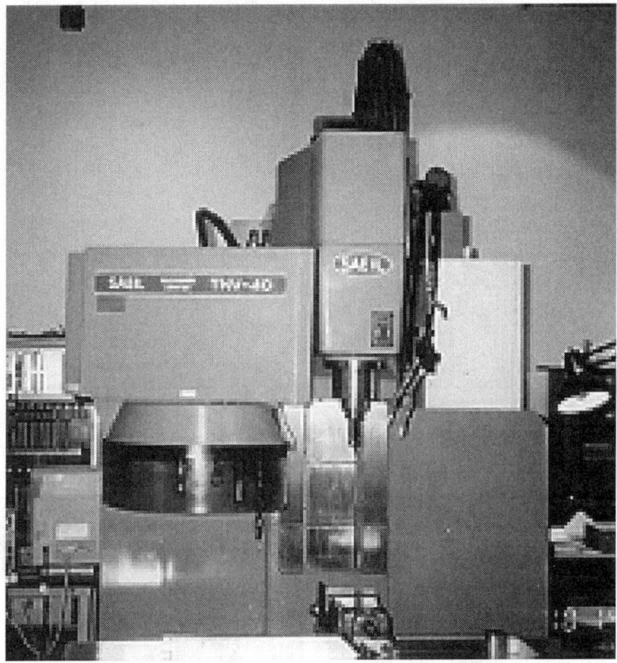

FIGURE 68.8   The practical model of an NC VRM.

FIGURE 68.9   The practical model of a process control VRM: AGV approaches ASRS (Automated Storage and Retrieval System) to load a pallete.

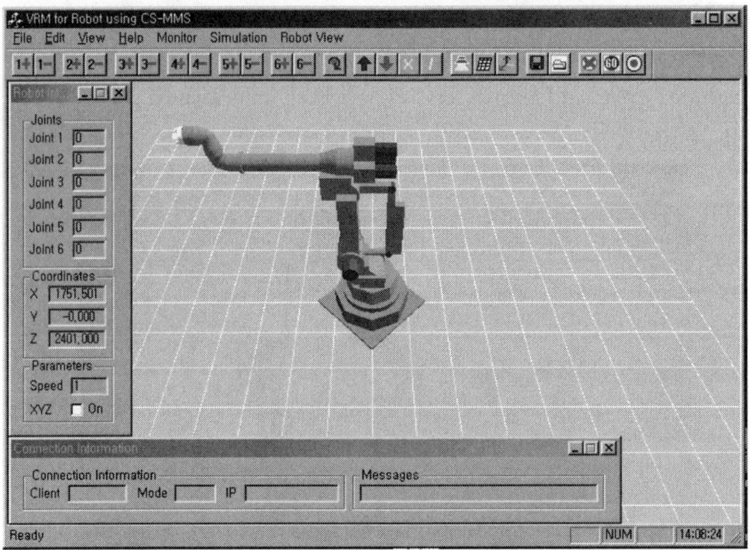

FIGURE 68.10	Screenshot of the 6-axis Robot VRM program.

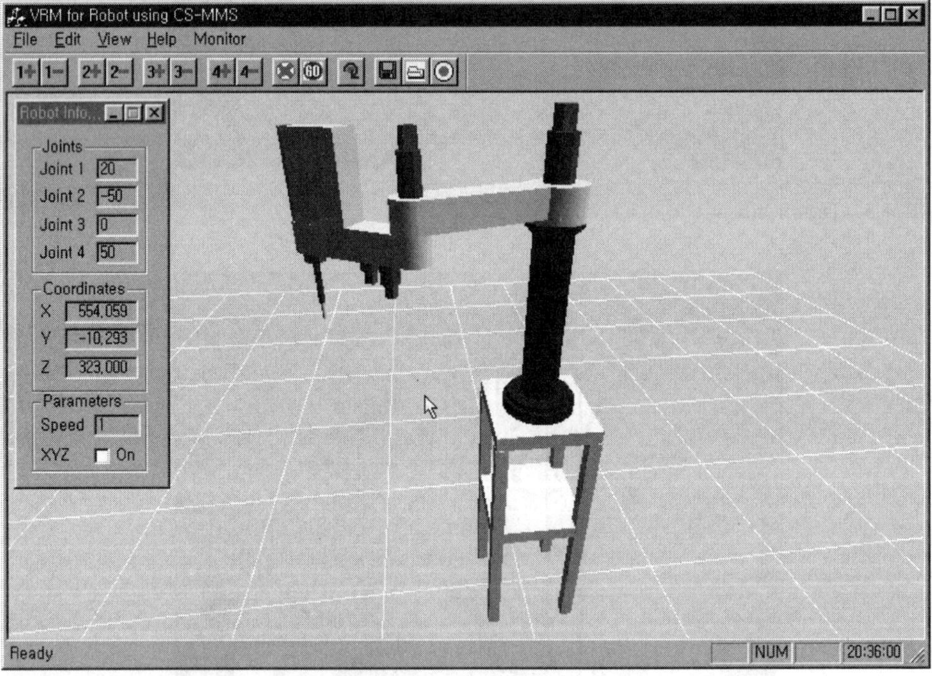

FIGURE 68.11	Screenshot of the 4-axis Robot VRM program.

VRM Using MMS-CS

The virtual factory communication system is operated by controlling each VRM with MOTIP based communication programs. Figures 68.6 – 68.9 show reference models of practical machines. Figures 68.10 – 68.14 show virtual miniatures of the practical machines. The application programs of Robot CS VRMs are modeled after a 6-axis robot made by HYUNDAI Heavy Inc. (Figure 68.6) and the 4-axis scalar robot made by SAMSUNG Heavy Inc. (Figure 68.7). The 6-axis robot is used for conveying the material with pallet throughout the implemented virtual system. The 4-axis robot takes part in a bolt assembling task of the overall process.

NC-CS VRM is modeled after an NC machine (TNV-40) made by TONGIL HEAVY Inc. (Figure 68.8). NC-CS VRM is responsible for drilling a hole in the sample material.

The application program of PLC-CS VRM is modeled after a cleaning machine with an air compressor and clamps. It is controlled by PLC, which is installed under the cleaning machine. PLC-CS VRM takes part in the cleaning works of assembled parts after the drilling process performed by the NC-CS VRM (Figure 68.12).

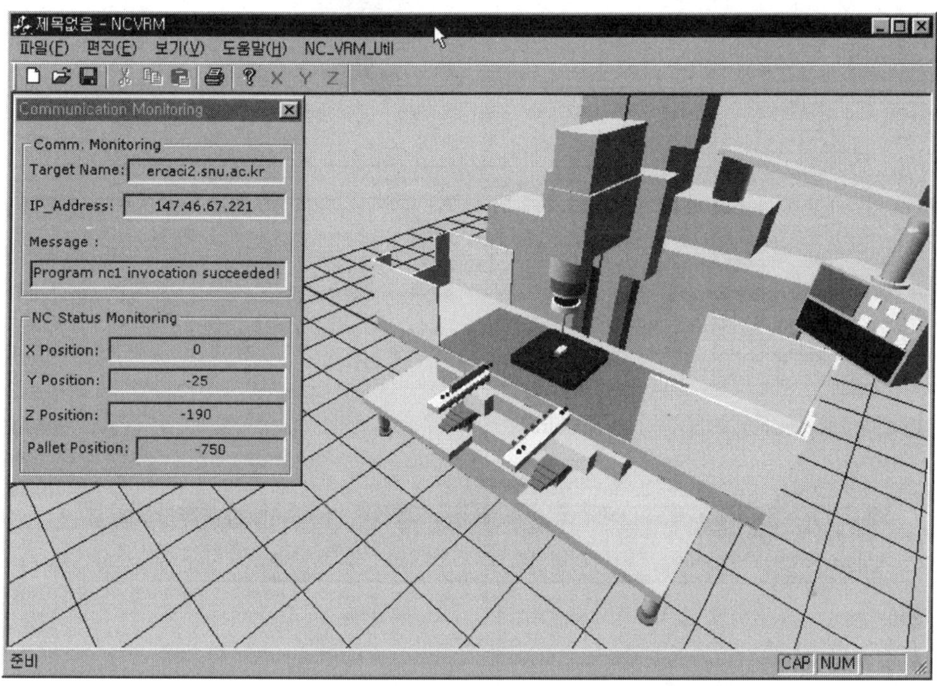

FIGURE 68.12   Screenshot of the NC CS VRM program.

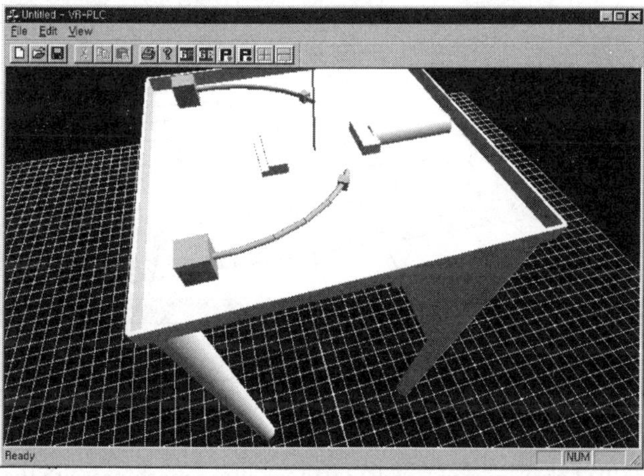

FIGURE 68.13   Screenshot of the PLC-CS VRM program: PLC VRM is operated by a PLC machine, which is installed under the cleaning machine.

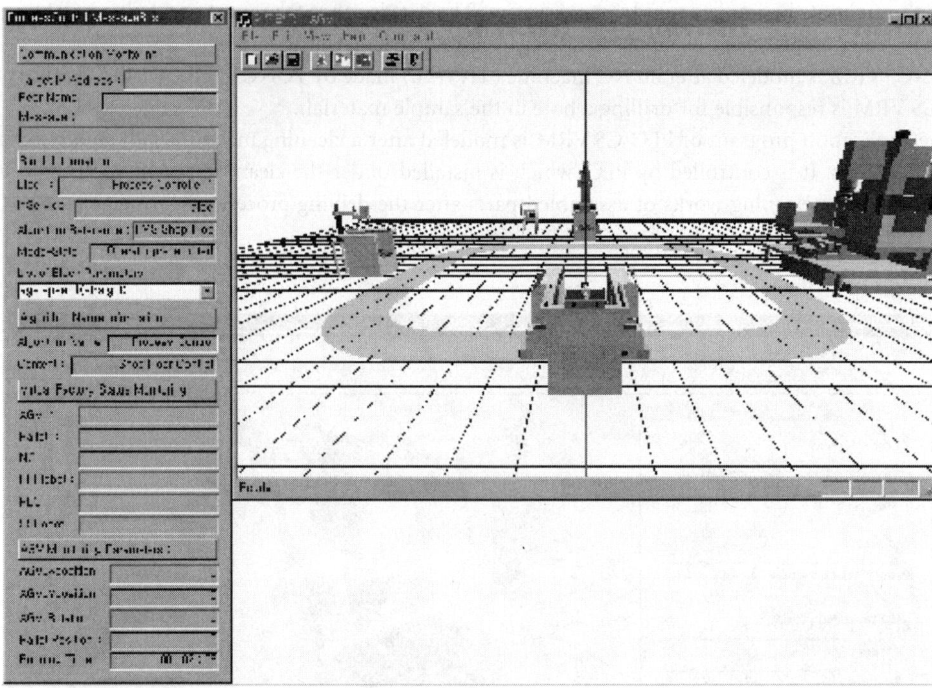

FIGURE 68.14   Screenshot of the PC-CS program.

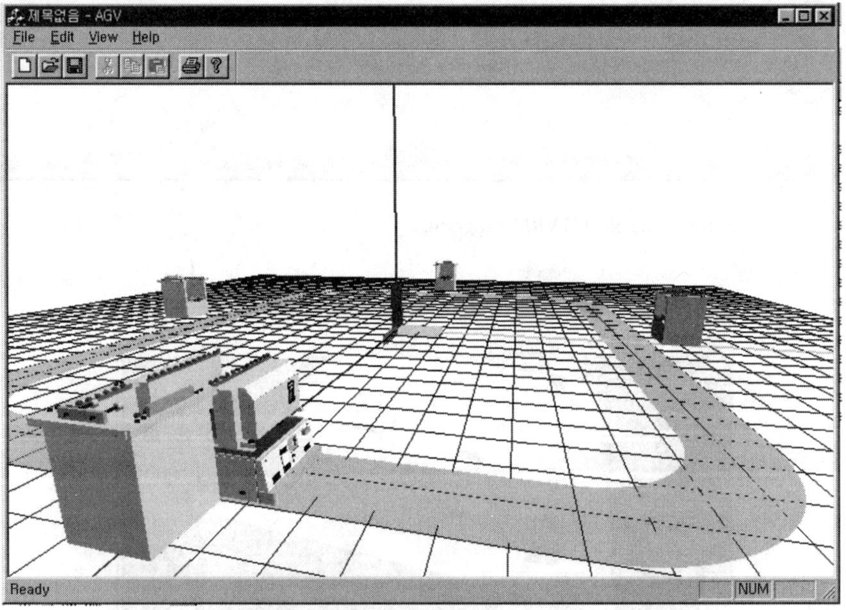

FIGURE 68.15   Screenshot of the AGV movement controlled by the schedule.

The process control VRM program can be scheduled for all the machines with the autonomous guided vehicle (AGV) movement (Figure 68.15). The CS application program includes all the VRMs, the common MMS, and MMS-CS services. The simple test scenario of the application program is presented in

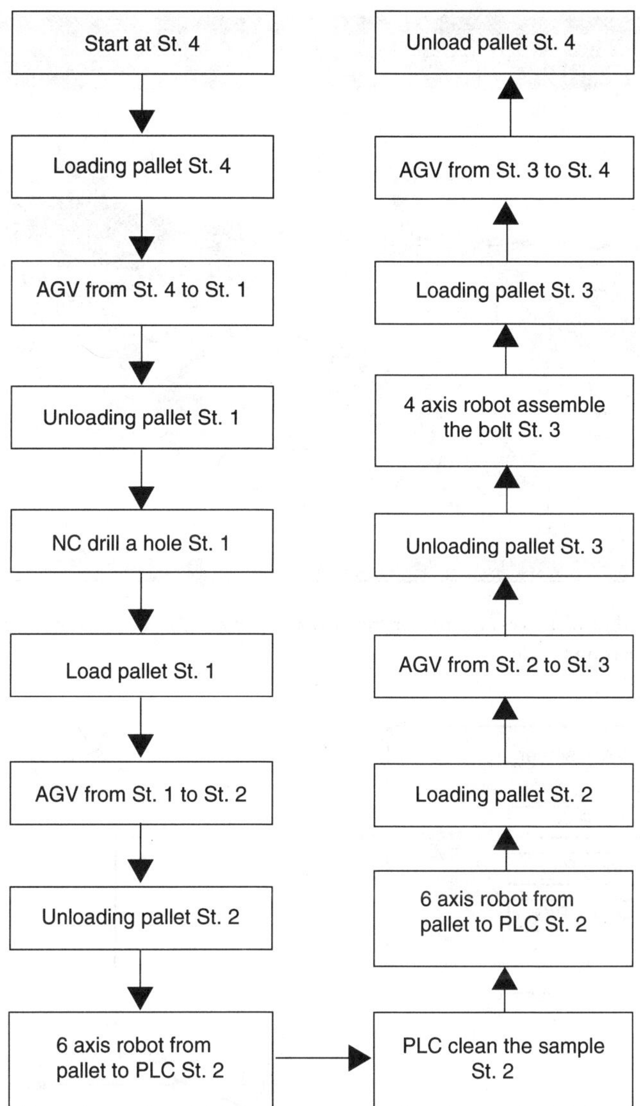

FIGURE 68.16 Block diagram of the demo program.

Figure 68.16. All machines are operated in accordance with their scheduled operation scenarios and they can be modified by the MMS user.

The test scenario begins with the AGV VRM idling status at station 4. The station 4 is the starting and ending station in the developed virtual factory. AGV VRM loads a palette from station 4 (assuming that the AS/RS is aside) and moves to station 1 for unloading the palette. In station 1, the NC VRM executes a hole drilling process in the material. After finishing the drilling process, AGV VRM loads the palette and moves it from station 1 to station 2. In station 2, the 6-axis robot VRM transfers the sample material from the palette to the clamps of the cleaning machine. The PLC CS VRM starts cleaning the sample material using an aircompressor. Then after loading the palette, AGV moves from station 2 to station 3. The 4-axis robot assembles the bolt into a hole of the sample material at station 3. After finishing the assembling process, the AGV moves from station 3 to station 4 and unloads the palette at station 4 (Figure 68.17).

The overall scenario can be scheduled by the MMS user. For reference, the source of the VRMs and the MOTIP program can be downloaded from the site given in [22].

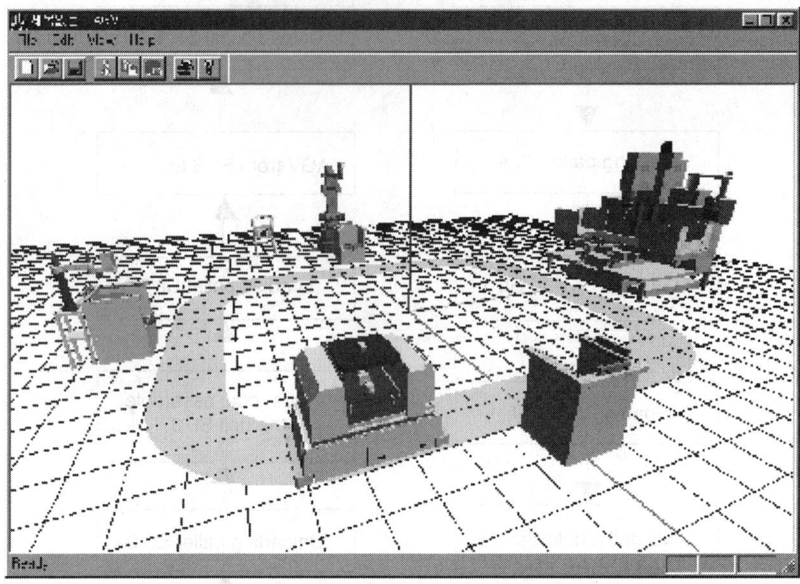

FIGURE 68.17   Screenshot of Process control of the virtual factory: AGV moves from station 1 to station 4 for unloading a processed part on pallete.

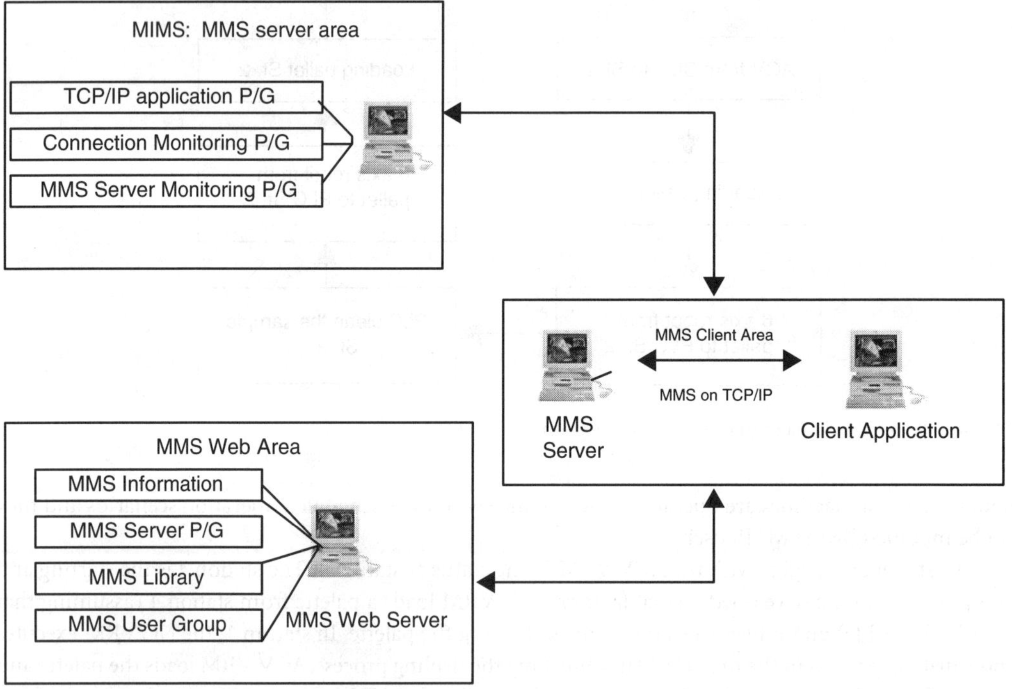

FIGURE 68.18   Structure of the MIMS system.

68.4   MIMS

MIMS is designed for simultaneous monitoring and controlling the status of multiple clients. MIMS transfers the operational information of the virtual factory to an MMS administrator and external MMS users. Therefore, MIMS can concurrently monitor the status of each networked VRM and several MMS-CS services.

The structure of MIMS is shown in Figure 68.18. The MIMS is composed of an MOTIP application program, the application program using MMS-CS, and each networked VRM.

MIMS is designed for up to 100 multiple users. MMS users can download the MMS-CS and the client-specific program from the site at [22]. Whenever MMS clients use the downloaded program based on the MMS-CS, MIMS can monitor the operation status through the user information and the communication data between the MMS client and server.

The administrator can monitor all the messages from the multiple clients and check the problems in the networked system. The MMS user can then fix their technical problem through the on-line advisories from the MMS server administrator. Therefore, the developed system can serve as a training tool for MMS developers and users.

68.5   Conclusions

A virtual factory communication system using the general MMS, its CS part, and the robot VRMs, NC, PLC, and PC were designed and implemented. The applicability of the virtual factory using the MMS and its CS part described here can be extended to the MMS implementation of a practical plant. In particular, the implemented system can be used as a design guideline of an MMS system for a practical plant. Furthermore, the developed virtual factory communication system and the VRMs can also be used for the verification of the operation of a real factory automation environment.

Acknowledgment

The authors acknowledge the partial financial support for this work by the Post-doctoral Fellowship Program of Korean Science and Engineering Foundation (KOSEF). The authors also thank the ERC-OTT members for their comments.

References

[1] ISO 9506-1: Industrial Automation Systems — Manufacturing Message Specification — Part1: Service Definition, 2002.

[2] ISO 9506-2: Industrial Automation Systems — Manufacturing Message Specification — Part2: Protocol Specification, 2002.

[3] ISO 9506-3: Manufacturing Message Specification — Companion Standard: Protocol Specification of Robot, 2000.

[4] ISO 9506-4: Industrial Automation Systems — Manufacturing Message Specification — Companion Standard: Protocol Specification of NC, 2000.

[5] ISO 9506-5: Industrial Automation Systems — Manufacturing Message Specification — Companion Standard: Protocol Specification of PLC, 2000.

[6] ISO 9506-6: Industrial Automation Systems — Manufacturing Message Specification — Companion Standard: Protocol Specification of Process Control, 2000.

[7] MAP 3.0 Specification 1993 Release, 1993.

[8] Kim, D.S., K.Y.Cho, W.H.Kwon, Y.I.Kwan, and Y.H.Kim Home network message specification of white goods and its application, *IEEE Transaction on Consumer Electronics*, 48, 1–10, 2002.

[9] DIN 19 245, Profibus Standard, Profibus Trade Organization, 1993.

[10] WorldFIP, General Purpose Field Communication System, prEN, 50170, 1995.

[11] Brugali, D., D. Dragomirescu et al. A Customizable Software Infrastructure for Virtual Factories Development, Proceedings of the IEEE International Conference on Robotics and Automation, Vol. 3, 1999, pp. 2440–2445.

[12] Florent, F., Experiences with a Control Architecture for the Virtual Factory, IEEE International Workshop on Robot and Human, 1999, pp. 387–393.

[13] Maref, B., Communication System for Industrial Automation, IEEE International Symposium on Industrial Electronics, 1997, pp.1286–1291.

[14] Glavinic, V. and B. Motik, Object-oriented Interface to MMS Services, Mediterranean Electrotechnical Conference, Vol. 2, 1998, pp. 1375–1379.

[15] Matraix V. and J. Tordera E., Video Transmission on Industrial Processes over MAP Networks using MMS Sepere, V., Matraix, J., Tordera, E., IEEE International Conference on Emerging Technologies and Factory Automation, Vol. 1, 1999, pp. 423–433.

[16] Wu, Q., T. Divoux, and F. Lepage, Integrating multimedia communications into an MMS environment, *Computer Communications*, 22, 907–918, 1999.

[17] Maaref, B., S. Nasri, and P. Sicard, Performance Evaluation of an MMS/ATM Implementation, Factory Communication Systems Proceedings, IEEE International Workshop, 1997, pp. 239–244.

[18] Akazan, J. and Z. Mammeri, On tasks synchronizations with the MMS protocol, *Journal of Real Time Systems*, 2, 213–234, 1995.

[19] Watanabe, T., Y. Maeda, et al. Total Simulation System for Steel Plate Manufacturing Based on the Virtual Factory Concept IEEE Conference on Industrial Electronics, Vol. 2, 2000, pp. 1280–1285.

[20] RFC 1006, ISO/IEC Transport Service on Top of the TCP, 1987.

[21] ISO/TC184/SC5/WG2 N578, Communications and Interconnections, 1996.

[22] Virtual Demo System Using MOTIP, Engineering Reserch Center for Advanced Control Instrumentation, Seoul National University, Available from: http://icat.snu.ac.kr:4444/mms/demo_frame.html.

[23] SNU MMS Library, Engineering Reserch Center for Advanced Control Instrumentation, Seoul National University, Available from: http://icat.snu.ac.kr:4444/mms/demo_frame.html.

[24] ISO/IEC8824-l:1995, Information Technology — Abstract Syntax Notation One (ASN.l): Specification of Basic Notation, 1995.

[25] ISO/IEC8825-1:1995, Information technology-ASN.l encoding rules: Specification of Basic Encoding Rules(BER), Canonical Encoding Rules(CER) and Distinguished Encoding Rules(DER), 1995.

[26] Carr, M., Visualization with OpenGL: 3D made easy, *IEEE Antennas and Propagation Magazine*, 39, 116–120, 1997.

69

IT Security for Automation Systems

Martin Naedele
ABB Corporate Research

69.1   Introduction

These days, more and more automation systems, both systems for automating manufacturing processes and for controlling critical infrastructure installations, for example, in power and water utilities, are directly or indirectly connected to public communication networks like the Internet. While this leads to productivity improvements and faster reaction on market demand, it also creates the risk of attacks via the communication network. This chapter surveys how network-connected plants and automation systems can be secured against information system and network-based attacks by state-of-the-art defensive means, and it will provide an outlook on future research.

69.2   Motivation

The influence of automation systems pervades many aspects of everyday life in most parts of the world. In the form of factory and process control systems, they enable high productivity in industrial production, and in the form of electric power, gas, and water utility systems they provide the backbone of technical civilization.

Up to now, most of these systems were isolated, but for the last couple of years, due to market pressures and novel technology capabilities, a new trend has emerged to interconnect automation systems to achieve faster reaction times, to optimize decisions, and to collaborate between plants, enterprises, and industry sectors. Initially, such interconnections were based on obscure, specialized, and proprietary communication means and protocols. Now more and more open and standardized Internet technologies are used for that purpose.

In security terminology, a risk exists if there is a vulnerability, that is, an opportunity to cause damage, together with a threat, that is, the fact that someone will try to find and exploit a vulnerability in order to inflict damage. The importance of automation systems for the functioning of modern society together with market pressure and competition on the one hand and geopolitical tensions on the other make the existence of security threats from terrorism, business competitor sabotage, and other criminal activity appear likely. The pervasiveness of automation systems that are nowadays accessible from anywhere in the world via communications and information technologies for which there are thousands of experts worldwide and that have a large number of well-known security issues creates many IT security vulnerabilities. As a consequence, there are good reasons to investigate and invest regarding how to reduce the IT security vulnerabilities of automation systems, and thus the resulting risks of large financial damage, deteriorated quality of life, and potentially physical harm to humans. This chapter presents an overview of state-of-the-art best practices in that respect, and an outlook on future opportunities.

69.3 Scope

The scope of automation systems considered in this chapter ranges from embedded devices, potentially in isolated locations, via plant control systems to plant- and enterprise-level supervisory control and coordination system, both in the distributed control system (DCS) flavor, more common in factory automation, and the supervisory control and data acquisition (SCADA) flavor, widespread in utility systems [5].

In the associated types of applications, in contrast to commercial and administrative data processing, often not typical data security issues (e.g., confidentiality, integrity) as such are the most important goal; but IT security is one component of the safety and fault-tolerance strategy and architecture for the plant.

69.4 Security Objectives

IT security has a number of different facets that are, to some extent, independent of each other. When defining the security requirements for a system, these facets, on which risk analysis and in turn design of countermeasures are based, can be expressed in terms of the eight security objectives explained in the following.

Confidentiality

The confidentiality objective refers to preventing disclosure of information to unauthorized persons or systems. For automation systems this is relevant both with respect to domain-specific information, such as product recipes or plant performance and planning data, and to the secrets specific to the security mechanisms themselves, such as passwords and encryption keys.

Integrity

The integrity objective refers to preventing modification of information by unauthorized persons or systems. For automation systems, this applies to information coming from and going to the plant, such as product recipes, sensor values, or control commands, and information exchanged inside the plant control network. This objective includes defense against information modification via message injection, message replay, and message delay on the network. Violation of integrity may lead to safety issues, that is, equipment or people may be harmed.

Availability

Availability refers to ensuring that unauthorized persons or systems cannot deny access/use to authorized users. For automation systems this refers to all the IT elements of the plant, like control systems, safety systems, operator workstations, engineering workstations, manufacturing execution systems, as well as the communications systems between these elements and to the outside world. Violation of availability may lead to safety issues, as operators may lose the ability to monitor and control the process.

Authorization

The authorization objective, also known as access control, is concerned with preventing access to or use of the system or parts by persons or systems without permission to do so. In the wider sense, authorization refers to the mechanism that distinguishes between legitimate and illegitimate users for all other security objectives, for example, confidentiality, integrity, etc. In the narrower sense of access control, it refers to restricting the ability to issue commands to the plant control system. Violation of authorization may lead to safety issues.

Authentication

Authentication is concerned with determination of the true identity of a system user (e.g., by means of user-supplied credentials such as username/password combination) and mapping of this identity to a system-internal principle (e.g., a valid user account) under which this user is known to the system. Authentication is the process of determining who the person trying to interact with the system is, and whether he really is who he claims to be. Most other security objectives, most notably authorization, distinguish between authorized and unauthorized users. The base for making this distinction is to associate the interacting user by means of authentication with an internal representation of his permissions used for access control.

Nonrepudiability

The nonrepudiability objective refers to being able to provide irrefutable proof to a third party of who initiated a certain action in the system. This security objective is mostly relevant to establish accountability and liability with respect to fulfillment of contractual obligations or compensation for damages caused. In the context of automation systems, this is most important with regard to regulatory requirements, for example, FDA approval. Violation of this security objective has typically legal/commercial consequences, but no safety implications.

Auditability

Auditability is concerned with being able to reconstruct the complete behavioral history of the system from historical records of all (relevant) actions executed on it. While in this case it might very well be of interest to also record who initiated an action, the difference between the auditability security objective and nonrepudiability is the ability of proving the actor identity to a third party, even if the actor concerned is not cooperating. This security objective is mostly relevant to discover and find reasons for malfunctions in the system after the fact, and to establish the scope of the malfunction or the consequences of a security incident. In the context of automation systems, this is most important in the context of regulatory requirements, for example, FDA approval. Note that auditability without authentication may serve diagnostic purposes but does not provide accountability.

Third-Party Protection

The third-party protection objective refers to averting damage done to third parties directly via the IT system, that is, damage that does not involve safety hazards of the controlled plant. The risk to third parties through possible safety-relevant failures of the plant arising out of attacks against the plant automation system is covered by other security objectives, most notably the authorization/access control objective. However, there is a different kind of damage only involving IT systems: the successfully attacked and subverted automation system could be used for various attacks on the IT systems or data or users of external third parties, for example, via distributed-denial-of-service (DDOS) or worm attacks. Consequences could range from damaged reputation of the automation system owner up to legal liability for the damages of the third party. There is also a certain probability that the attacked third party may retaliate against the subverted automation system causing access control and availability issues. This type of counter attack may even be legal in certain jurisdictions.

69.5   Differences to Conventional IT Security

As the security objectives in the previous section are generally valid, many security issues are the same for automation systems and conventional, office-type IT systems, and many tools can be used successfully in both domains. However,there are also major differences between these two domains with respect to requirements and operating environment, characteristics, and constraints, which make some security issues easier and others more difficult to address. In the following, some of these differences are explained.

Requirements

While office IT security requirements center around confidentiality and privacy issues, for any automation and process control system the foremost operational requirement is safety, the avoidance of injury to humans. Second after that is availability: the plant and the automation system have to be up and running continuously over extended periods of time, with hard real-time response requirements in the millisecond range. In many cases, this precludes standard IT system administration practices of system rebooting for fixing problems, and makes the installation of up-to-date software (SW) patches, for example, addressing security problems in the running application or the underlying operating system difficult if not impossible.

On the other hand, in contrast to e-commerce applications, connectivity to outside networks including the company intranet is normally not mandatory for the automation system, and although extended periods of disconnection are inconvenient, they will not have severe consequences — after all, many automation systems nowadays still run completely isolated.

Operational Environment

The configuration, both of HW and SW, of the automation system part, which contains the safety critical automation and control devices, is comparatively static. Therefore, all involved devices and their normal, legitimate communication patterns (regarding communication partners, frequency, message size, message interaction patterns, etc.) are known at configuration time, so that protection and detection mechanisms can be tailored to the system. Modifications of the system are rare enough to tolerate a certain additional engineering effort for reconfiguring the security settings and thus being able to trade in the convenience of dynamic, administration-free protocols like DHCP against higher determinism and, as a consequence, security of, for example, statically setting up tables with communication partners/addresses in all devices. Static structure and behavioral patterns also make the process of anomaly discovery for intrusion detection easier.

The hosts and devices in the automation system zone are not used for general-purpose computing, preventing the risks created by mainstream applications like e-mail, instant messaging, office application macroviruses, etc. Often, they are even specialized embedded devices dedicated to the automation functionality, such as power line protection in substation automation. All appropriate technical and administrative means are taken to ensure that only authorized and trustworthy personnel have physical access to the automation equipment. Automation system personnel are accustomed to a higher level of care and inconvenience when operating computer systems, than office staff. This increases the acceptance and likelihood of correct execution of security-relevant operating procedures even if they are not absolutely straightforward and convenient. In many plants, additional nonnetworked (out-of-band) safety and fault-tolerance mechanisms are available to mitigate the consequences of failure of one or multiple components of the automation system.

One problem with security mechanisms is that they cannot prevent all attacks directly themselves, but produce output, like alerts and log entries, which humans need to review to decide on the criticality of an event and to initiate appropriate responses. This is often neglected as the expert IT staff is not available around the clock to monitor the system. Automated plants, in contrast, are usually continuously monitored by dedicated staff. A defense architecture could make use of this fact, even though these plant operators do not have IT or even IT security expertise.

Challenges

On the other hand, the characteristics of automation systems and devices create some additional security challenges.

Automation devices often have a lower processing performance compared to desktop computers, which limits the applicability of mainstream cryptographic protocols.

The operating systems of such devices in many cases do not provide authentication, access control, fine-granular file system protection, and memory isolation between processes — or these features are optional and are not used due to the above-mentioned limited processing power.

Especially in telemonitoring applications (e.g., SCADA), communication channels with small to very small bandwidth like telephone, mobile phone, or even satellite phone lines are used, which makes it imperative to reduce communication overhead and thus collides with certain security protocols.

Automation systems tend to have very long lifetimes. This has consequences both for the currently operative systems and for newly implemented systems: those currently operative "legacy" automation systems, as far as IT security was given a thought at all, were designed based on a philosophy of "security by obscurity," assuming that the system would be isolated and only operatable by a small, very trustworthy group of people. This kind of thinking persists even today, as can be seen from the 2002 IEEE 1588 standard on precision time synchronization for automation systems for which it is explicitly stated as a design rationale that security functionality is neglected as all relevant systems can be assumed to be secure. Another consequence of longevity is that automation system installations tend to be very heterogeneous with respect to both subsystem vendors and subsystem technology generations.

For newly built automation systems, the long expected lifetime means that the data communication and authentication/access control functionality must be designed so that it will be able to interoperate with reasonable effort with systems and protocols to appear on the market 10 or 20 years later.

Last but not the least, automation systems are operated by plant technicians and process engineers. Due to their training background, they have a very different attitude toward IT system operation and security than corporate IT staff, and frequently a mutual lack of trust has to be overcome to implement an effective security architecture.

69.6 Building Secure Automation Systems

Building secure systems is difficult, as it is necessary to spread effort and budget so that a wide variety of attacks are efficiently and effectively prevented. For automation systems, the challenges mentioned in the section Challenges create additional difficulties. In the following, two common approaches to secure systems are explained and their effectivity is assessed.

Hard Perimeter

A popular doctrine for defense, be it of cities or IT systems, is the notion of the hard perimeter. The idea is to have one impenetrable wall around the system and to neglect all security issues within. In general, however, this approach does not work, for a variety of reasons. The hard perimeter approach does not make use of reaction capabilities: at the time of detection of a successful attack, the attacker has already broken through the single wall and the whole system is open to him. As a consequence, this means that the wall would need to be infinitely strong, because it needs to resist infinitely long [15]. Also, monoculture is dangerous: the wall is based on one principle or product. If that principle or product fails for some reason to resist the attack, the whole defense is ineffective. The wall must have doors to be usable, which opens it to both technical and nontechnical (social engineering) security risks. Once the attacker has managed to sneak inside, the system is without defense — the risk of the proverbial Trojan horse. A hard perimeter is also, by definition, ineffective against insider attacks. Progressing technology gives the attacker continuously better wall-penetration capabilities. Last but not the least, humans make mistakes: it is illusory to assume that we can design a wall that is without weak spots either in design, implementation, or operation — various border walls in history serve as example.

Defense-In-Depth

The alternative approach is defense-in-depth. Here, several zones/shells are placed around the object that is to be protected. Different types of mechanisms are used concurrently around and inside each zone to defend it. The outer zones contain less valuable targets; the most precious goods, in this case the (safety critical) automation system, are in the innermost zone. In addition to defense mechanisms, there are also detection mechanisms, which allow the automation system operators to detect attacks, and reactive mechanisms and processes to actively defend against them. Each zone also buys time to detect and fend off the attacker. In the spirit of Schwartau's time-based security [15], this allows to live with the fact of imperfect protection mechanisms, as only a security architecture strength of $P \geq D + R$ has to be achieved, where P is the time during which the protection offered by the security system resists the attacker, D is the delay until the ongoing attack is detected, and R is the time until a defensive reaction on the attack has been completed.

Conclusion: There are two basic approaches for securing systems commonly used today, but only one, defense-in-depth, will result in a secure system, provided it is properly implemented.

69.7   Elements of a Security Architecture

In this section, the most important technical elements of a security architecture for automation systems are surveyed. Note, however, that a system cannot be secured purely using technical means. Appropriate user behavior is essential to ensure the effectiveness of any technical means. Acceptable and required user behavior should be clearly documented in a set of policies that are strictly and visibly enforced by plant management. Such policies should address, among other things, user account provisioning, password selection, virus checking, private use, logging and auditing, etc.

The topic of policies and user behavior will not be further discussed here. Example documents are available from various government agencies, IT security organizations, as well as in a number of books.

According to their physical and logical location in the architecture, in the spirit of a defense-in-depth, security mechanisms can be classified as belonging to one or multiple of the following categories. These categories are orthogonal to the security objectives of Section 69.3.

Deterrence. Means of pointing out to the potential attacker that his personal pain in case of getting caught does not make the attack worthwhile. However, in most threat scenarios for safety-critical and infrastructure systems, the deterrence component, especially the threat of legal action, is ineffective.

Connection authorization. Means to decide whether the host trying to initiate a communication is at all permitted to talk to the protected system, and to prevent such connections in case of a negative decision.

User authorization. Means to decide whether and with which level of privileges a user or application is permitted to interact with the protected system and to prevent such interaction in case of a negative decision.

Action authorization. Means to decide whether a user or application is permitted to initiate specific actions and action sequences on the protected system or application, and to prevent such interaction in case of a negative decision. Action authorization is an additional barrier assuming a preceding positive user authorization decision.

Intrusion detection. Means to detect whether an attacker has managed to get past the authorization mechanisms. Most intrusion detection systems are based on monitoring and detecting whether anything "unusual" is going on in the system.

Response. Means to remove an attacker and the damage done by him from the system. Means to lessen the negative impact of the attack on the system and its environment. Means to prevent a future recurrence of the same type of attack.

Mechanism protection. Means to protect the mechanisms for the above-listed categories against subversion. This refers, for example, to not sending passwords in clear text over public networks and hardening operating systems and applications by fixing well-known bugs and vulnerabilities, etc.

Deterrence

The technical mechanisms for deterrence are warning banners at all locations, for example, log in screen, where an attacker — not necessarily an outsider — could access the system. These warning banners should state clearly that unauthorized access is prohibited and legal action may be the consequence. This type of statement could be required to be able to prosecute intruders in certain legislations.

Connection Authorization

The following sections list and explain connection authorization mechanisms for dial-in remote access, WAN, and LAN situations.

Firewall

A firewall or filtering router passes or blocks network connection requests based on parameters such as source/destination IP addresses, source/destination, TCP/UDP ports (services), protocol flags, etc. Depending on the criteria, the specific firewall product uses and how clearly the permitted traffic on the network is defined, a firewall (or, more often, dual-firewall architecture) can be anything from a highly effective filter to a fig leaf.

Intelligent Connection Switch/Monitor

As the information and message flows between the individual devices and applications in an automation system are often deterministic and well defined at system configuration time, this information can be used by a special, intelligent connection monitoring device to determine the legitimacy of a certain message. This goes beyond the parameters that conventional firewalls use, as it also takes into account criteria like message size, frequency, and correctness of complex interaction sequences. Nonconforming messages can be suppressed in collaboration with a switch. In this case, the device is acting as a connection authorization device. On the other hand, an alert can be raised, in which case the device is acting as an automation system-level intrusion detection systems.

Personal Firewall

A personal firewall is an application running on a host as an additional protection for the main functionality of this host, for example, as a firewall, proxy server, or workstation. It monitors and controls as to which applications on the host are allowed to initiate and accept connection requests from the network. In contrast to a (network) firewall, a personal firewall does not control a network segment, but only its own host.

Switched Ethernet

With standard Ethernet, all hosts on a network segment can see all traffic, even that not addressed to them, and multiple hosts sending at the same time can cause nondeterministic delays. The first issue is directly security relevant, and the second one is indirectly security relevant (denial-of-service attacks). Using switches to give each host its own network segment and to directly forward each message only to the intended receiver remedies both issues. Note that there exist attacks on switches to force them to operate in hub mode (broadcast).

Dial-Back

After authentication, a dial-back modem interrupts the telephone connection and dials back to the telephone preconfigured for the authenticated user. This prevents an attacker from masquerading as an authorized user from anywhere in the world, even if he managed to obtain this user's credentials, because the attacker would also need to obtain physical access the authorized user's phone line. Note that in a number of countries, the technical behavior of the public telephone network is such that dial-back modems can be defeated by an attacker.

Access Time Windows

For automation systems, it is often true that the remote connection is not necessary for operation, but only to upload configuration changes and download measurement values, which are not urgent and irregular in timing. Therefore, if a continuous remote connection is not required, the remote access can be restricted to certain time windows only known to authorized users in order to reduce exposure to attacks. Outside these time windows access is disabled, for example, by electrically switching off the modem or router device.

Mutual Device Authentication

As all communication partners in the automation system (devices and applications) and message flows are known at configuration time, it is possible to require mutual authentication of each communication relationship at run-time, provided the available computing power of the automation devices tolerates the execution of the necessary protocols. This renders the installation of rogue devices ineffective.

User Authorization

Login Mechanisms

Most access control schemes rely on passwords ("something you know") as the underlying authentication principle to establish that the user is who he claims to be. This makes the selection of suitable passwords, as well as their management and storage one of the most important aspects, and potential weaknesses, of each system. Alternatively, instead of a fixed password, a one-time password, generated just-in-time by individualized devices (tokens, smartcards) that are given to all authorized users, is used. It replaces or augments the "something you know" principle by "something you have." The third option for proving identity are biometrics, for example, finger prints. This is the most direct mechanism, but expensive, inconvenient, only applicable to humans (not other applications), and there have been incidents where biometrics products have been fooled by an attacker. In any case, if authentication fails, the user has to initiate a new session to continue communication with the protected system, for example, for another login attempt.

Action Authorization

Role-Based Access Control in Applications

In many situations, not all authorized users are in the same way permitted to execute all types of activities on the system; therefore, the system log-on alone does not offer a sufficient granularity of access control. With role-based access control, each authorized user has one or multiple roles, which correspond to sets of actions he can execute in the application. Role-based access control needs to be designed into each application, as it cannot be normally provided by external add-on devices or applications. Defining the rights in the application in terms of roles instead of actual users improves scalability and makes maintenance of the security configuration easier.

System Architecture for Data Exchange

If the remote access functionality is not necessary for interactive operation and examination of the automation system, but only for upload and download of preconfigured parameters and data, the system can be architected so that direct remote access to the automation device, which is the source or target of the data, is not required. Instead, the remote access occurs only to a less valuable FTP or web server, which acts as a cache and communicates over a series of time-shifted, content-screening data-forwarding operations with the actual data source or target. This strongly restricts the type of interactions that the outside system can have with the automation device.

Managed Application Installation

Through centralized administration and monitoring, it is enforced that only the authorized and necessary applications and services are running on the automation system workstations and servers, and that configurations are not changed by the users. The hosts do not carry user-specific data or configurations and thus can be reinstalled from a known-good source at regular, frequent intervals to remove any unauthorized modification in the system, even if the modification was never detected.

Application-Level Gateway

The process-level network to which the automation equipment is connected interfaces with the higher level LAN, for example, the one to which the operator workstations are connected, only via a small number of gateway servers for application-level communications. Direct interactive access (log-in) to the automation device is disabled in this scenario. As the purpose, the software applications, and the topology/configuration of the automation system are well known, these gateways can be customized for the authorized applications and communications to screen messages passing the interface for validity based on domain-specific criteria, such as predefined interaction sequences. Illegitimate or invalid messages can be suppressed and alerts can be raised.

Dual Authorization

Certain commands issued by a user in an automation system can have drastic consequences. If there are no situations in which these commands need to be entered extremely quickly in an emergency, they can be protected by requiring confirmation from a second authorized user. This mechanism secures the system against intruders, malicious insiders, and serves as a protection against unintentional operator errors.

Code Access Security

Code access security is concerned with restricting what an application is allowed to do on a host (e.g., reading or writing files, creating new user accounts, initiating network connections, etc.), even if it executes in the context of an authorized, highly privileged user. This type of "sandboxing" is available for several computing platforms and can serve as additional protection against applications that are not completely trusted, for example, because they have been developed externally.

Intrusion Detection

Network-Based Intrusion Detection System

A network-based intrusion detection system (NIDS) tries to discover attacks based on known attack profiles and/or unusual system behavior from communication traffic seen on a network segment (type, content, frequency, path of the transmitted messages).

Host-Based Intrusion Detection System

A host-based IDS (HIDS) tries to discover attacks based on known attack profiles and/or unusual system behavior from information seen locally on the host on which it is running. A host-based IDS obtains its information, for example, from file system integrity checkers, which monitor whether important system files change without operational reason, from personal firewall logs, or from application logs, for example, for application-level role-based access control.

Honeypot

A honeypot (single host) or honeynet (subnet) [16] is a subsystem that appears particularly attractive to an attacker, for example, from the naming of the host or files on it, or by simulating certain weaknesses in the installation. It is, however, a dedicated and isolated system without importance for the functioning of the automation system, which is especially instrumented with intrusion detection systems. The idea is that an attacker who breaches the first line of defense successfully will be attracted to the honeypot host first, and thus is at the same time delayed and kept away from the really sensitive areas, as well as detected by the intrusion detection systems.

Authentication/Authorization Failure Alerts

Alerts are sent to the operators or IT staff whenever one of the authentication mechanisms fails, which could indicate an unsuccessful attempt to attack the system. While this would probably not be useful in an office system, due to the high false alarm rate of legitimate users mistyping their passwords, it may be a feasible security mechanism in an automation system with a very small number of concurrent legitimate users and few login actions per time period.

Log Analysis

All actions of the authentications devices are logged, and these logs are manually or automatically screened for unusual occurrences or patterns, for example, that an authorized user is suddenly accessing the system outside his normal work hours.

Malicious Activity Detection/Suppression Protocol

As is shown in [9], network-based electronic attacks originating from malicious devices in an automation system, for example, in a power substation can be categorized as either message injection, message modification, or message suppression. Using a suitable communication protocol for detection of invalid messages, one can reduce these three categories to message suppression, which can in many cases be regarded as a system failure that conventional fault-tolerance and fault-response mechanisms such as redundant devices and emergency shutdown sequences can handle.

Response

System Isolation

The connections between the compromised subsystem, for example, the outer part of the security zone at the interface between automation system and other networks, and other, more important parts of the automation system, are closed to avoid further spreading of the attack. Depending on system/remote access functionality and importance, and on whether delaying the attacker and collection of further evidence, or quick restoration of operation is of higher importance, it is an option to shut down all remote connections, both for the affected and the not-yet affected systems, until the effects of the attack are removed. Like electric power grids, the automation system should already be architected and designed such that it can be partitioned into zones, which can be isolated with minimum disturbance of the whole system.

Collect and Secure Evidence

All logs and media (e.g., hard disks) that contain evidence of the malicious activity are gathered, copied, and stored at a secure location. This evidence can be used to identify the attack and thus the system weaknesses in detail, to locate and assess the amount of damage done by the attack, and also to support legal prosecution of the attacker. In this case, the legally correct handling of evidence is especially important.

Trace Back

This refers to activities that aim at discovering the source of the attack, both as part of the evidence collection process (see above) and to enable stronger defense mechanisms for identified sources of attacks (e.g., blocking). Due to various possibilities of faking packet data, such as the originating address, this is technically not easy. Lee and Shields [6] discuss the various technical options for trace back and their obstacles.

Active Counter-Attack

An active counter-attack has as aim to selectively disable the attacker's computer to prevent further attacks and to "punish" the attacker. Due to the fact that an unambiguous trace back is difficult (see above), that often "innocent" systems are used as intermediary stages for staging an attack, and that the legality of a counter-attack is dubious in most situations and localities, a counterattack response is normally not recommended.

Information Sharing

Early sharing of information about on-going attacks, especially novel types of attacks, with the IT community represents good "Internet citizenship" because it gives more defenders a chance to increase alertness and to remove weaknesses. Many types of attacks, such as distributed denial of service and viruses, rely on the fact that the same weakness can be exploited on a large number of systems, which are then used to launch further attacks. Therefore, reducing the number of systems vulnerable to an attack is in every defender's interest. On the other hand, many companies might be concerned about the effects of making the facts and circumstances of attacks known to competitors and the public. For this purpose, several institutions exist, which receive and distribute information about attacks without disclosing the sources of the information. Examples are the SEI CERT (http://www.cert.org) and various industry branch-specific Information Sharing and Analysis Centers (ISACs).

Selective Blocking

If the origin of the attack and the location of entrance into the system can be identified, blocking rules of firewalls, routers, and access servers, perhaps already at the Internet Service Provider, can be temporarily or permanently modified to close the in-roads of the attack.

Switching to Backup IT Infrastructure

If a system is compromised and its unavailability during evidence collection and restauration is unacceptable, a backup automation system, perhaps with minimum functionality, should be available for immediate switch-over. This backup system consisting of automation system workstations and servers, access servers, firewalls, IDS host, etc., should be preconfigured with different passwords, different network addresses, and, even better, also use software different from the primary system, to avoid that it can be immediately subverted by the attacker with the knowledge gained from the primary system.

Automated, Periodic Reinstallation of Applications, Operational Data, and Configurations

The SW applications of the automation system, as well as static operational data and configurations, especially of the security components, are periodically reinstalled from a known-good read-only storage device (e.g., CD). This removes applications with attacker-installed backdoors (Trojans) or modified system files, for example, password files even if the attack was never detected. Of course, this provides only a weak defense if the security vulnerabilities that allowed the system compromise in the first place are allowed to persist, but it may prevent follow-up exploits by other attackers and frustrate the original attacker sufficiently to turn to easier targets.

New Passwords

When automatically or manually reinstalling system configurations in case of an attack, the passwords should also be changed, in particular those, for the security components, to prevent the attacker from immediately reentering the system with a previously compromised set of credentials.

Activation of Safety Mechanisms

If the attack on the automation system endangers the safety of the plant, standard safety mechanisms such as reverting to manual operation or emergency shut-down are a last resort to protect plant equipment and human life. As a consequence, such safety mechanisms should be decoupled from the networked automation system.

Mechanism Protection

Dedicated Lines

Instead of connecting the dial-in modems to telephone lines, which are accessible to anybody from anywhere in the world, dedicated telephone cables are used, which connect only secure, authorized systems. Among other things, this protects the information between the protected system and the remote user, in particular, his credentials such as passwords during the authentication process, from eavesdropping. However, this scheme offers no protection against the telecommunication company and (government) organizations that can force access to dedicated lines. Also, there have been incidents where attackers were able to subvert the telephone switches of telecommunication providers to access dedicated lines.

Virtual Private Network

A virtual private network (VPN) uses encryption and digital signatures to achieve the effect of a physical dedicated line over a shared medium such as a normal telephone connection or the Internet. A VPN is both cheaper and more secure than a line leased from a telecommunication provider; but the computational overhead for strong cryptography may be unacceptable for certain communicating automation devices.

Disable Remote Reprogramming of Dial-In/Dial-Back Modems

The dial-back mechanism described in the section Dial-Back is only effective if the remote attacker cannot redirect the dial-back call to his own telephone.

Network Address Translation

With network address translation (NAT), the system uses IP addresses internally different from those shown in the externally visible messages. A border device, for example, the firewall, is responsible for on-the-fly translation of addresses in both directions. NAT makes remote probing of the internal network topology for interesting or vulnerable targets much more difficult and prevents certain attacks that bypass the firewall.

Diversity

System diversity, for example, by using different operating systems like Windows and Unix/Linux for the production systems and the intrusion detection systems or other security mechanisms, or by selecting different brands of firewalls for different zones and subzones increases security, as an attacker cannot rely on a single vulnerability in one product to break through all defenses. Thus, it also offers a bit more resilience in the time between publication of an exploit for a vulnerability and the design and installation of corresponding patches in the system.

Role-Based Access Control

Role-based access control is also important for the security functionality itself, to ensure that only security administrators and not all inside users can change security settings and read/edit logs. This is the basis of all security precautions against insider attacks and also creates an additional hurdle for attackers that have broken into the account of one authorized user against taking complete control over the system.

Hardened Host

Mechanisms like role-based access control, logging, and intrusion detection rely on the basic functions of the operating system on the host not having been corrupted by the attacker. Today's applications, operating systems, and system configurations are often so complex and complicated that they have many security vulnerabilities caused by misconfiguration or by applications with security relevant bugs, which are installed as part of the operating system, but that are not really necessary for the automation system functionality. Hardening a system means to remove all unnecessary applications and services, to fix known bugs, to replace critical applications with more trustworthy ones of the same functionality, and to set all system configuration parameters to secure values. Guidelines for the hardening of various common operating systems and applications are available from multiple sources.

69.8   Further Reading

A large number of technical and research publications exist on the issue of IT security for home and office information systems. Schneier [14] gives a good general introduction to the topic, and [13] is the reference on cryptographic algorithms and protocols. Reference [10] is a comprehensive resource on the practical issues of securing a computer network, while [1] and [17] address the issue of engineering secure (software) systems from a larger perspective.

On the other hand, apart from some vendor whitepapers, there is almost no literature on the specific security needs and capabilities of industrial automation systems. Palensky [12] investigates remote access to automation systems, specifically home automation systems, with potentially malicious devices, and proposes the use of smartcards as trusted processors to achieve end-to-end security between each device and its legitimate communication partners. In [8] IT security mechanisms applicable for automation systems are presented, according to which conceptual zone they defend — remote access, operator workstations, or automation devices. Byres [3] investigates networking and network-level security issues for Ethernet networks on the plant floor. Falco [5] motivates and describes efforts to create a protection profile for process control systems according to the Common Criteria security evaluation standard. Moore [7] reports on a survey about IT security conducted among automation system users. Dafelmair [4] promotes the use of a Public key Infrastructure (PKI) for process control systems, and [2] suggests a lightweight PKI for power-utility SCADA systems. Oman [11] applies standard IT security mechanisms to power utility control systems, with special emphasis on password management.

69.9 Research Issues

As can be seen from the scarcity of published work, automation system IT security is a comparatively new field. Much more research will be necessary until both security requirements and opportunities specific for automation systems have been explored to the current level of home and business system IT security. The following are just some of the topics to be addressed in the future:

> Considering that plant control systems have a lifetime of 20 to 30 years, how can effective defense-in-depth security mechanisms cost-efficiently be retrofitted onto them?
>
> What security mechanisms can and should be required for automation systems? This topic is being urrently addressed by various industry standard organizations, such as the ISA SP99 working group.
>
> What criteria should be used for assessing and auditing the security mechanisms in automation devices?
>
> How to cope with the vulnerabilities inherent in networking protocols like SNTP for time synchronization or XML web services/OPC-XML by modifying the protocols or adapting the system architecture?
>
> How to ensure appropriate levels of access control, data integrity, and data confidentiality for automation devices with low computing power [18]?
>
> How can the auditability of every interaction and individual accountability of every member of the plant staff who interacts with the automation system be ensured according to regulatory requirements, while at the same time achieving both ease of use in normal operation and fast response times for emergency interventions?
>
> How can a plant operator, who is not an IT security expert, effectively contribute to plant security?

69.10 Summary

Nowadays, realistic scenarios for network-based attacks on infrastructure/utility automation systems and manufacturing/plant automation systems with respect to both motivation as well as technical feasibility exist. In contrast to business systems, which need to be available for business and protect their confidential data, for automation system IT security, the most important security objective is the integrity of the control system to prevent physical damage and human injury.

This chapter has presented arguments regarding why a defense-in-depths approach with layered mechanisms is a better strategy for securing systems than the often-used approach of placing a security "wall" around the system and leaving the inside unchanged. It has also given an overview of available security mechanisms, pointing out which are specifically applicable to automation systems due to the specific operational characteristics of automation systems.

It remains to recall that, as the saying goes, security is not a destination, but a journey — both the notifications generated every day by an appropriate security system, and the whole architecture of the security system need to be reviewed regularly to detect and adapt to new vulnerabilities and threats.

References

1. Anderson, Ross, *Security Engineering*, Wiley, New York, 2001.
2. Beaver, Cheryl, Donald Gallup, William Neumann, and Mark Torgerson, Key Management for SCADA, Technical report SAND2001-3252, Cryptography and Information Systems Security Department, Sandia National Laboratories, March 2002.
3. Byres, Eric, Designing secure networks for process control, *IEEE Industry Applications Magazine*, 6: 33–39, 2000.
4. Dafelmair, Ferdinand J., Improvements in Process Control Dependability Through Internet Security Technology, in Proceedings Safecomp 2000, Lecture Notes in Computer Science, Vol. 1943, Springer, Berlin, 2000, pp. 321–332.
5. Falco, Joe, Keith Stouffer, AlbertWavering, and Frederick Proctor, IT Security for Industrial Control Systems, Technical report, Intelligent Systems Division, (U.S.) National Institute of Standards and Technology (NIST), 2002.
6. Lee, Susan C. and Clay Shields, Technical, legal and societal challenges to automated attack traceback, *IEEE IT Professional*, Vol. 4, no. 3, May/June:12-18, 2002.
7. Moore, Bill, Dick Slansky, and Dick Hill, Security Strategies for Plant Automation Networks, Technical report, ARC Advisory Group, July 2002.
8. Naedele, Martin, IT Security for Automation Systems — Motivations and Mechanisms, *atp-Automatisierungstechnische Praxis*, 45: 2003.
9. Naedele, Martin, Dacfey Dzung, and Michael Stanimirov, Network security for substation automation systems, in Computer Safety, Reliability and Security (Proceedings Safecomp 2001), Voges, Udo, Ed., Lecture Notes in Computer Science, Vol. 2187, Springer, Berlin, 2001.
10. Northcutt, Stephen, Lenny Zeltser, Scott Winters, Karen Fredrick, and Ronald W. Ritchey, *Inside Network Perimeter Security: The Definitive Guide to Firewalls, Virtual Private Networks (VPNs), Routers, and Intrusion Detection Systems*, New Riders, Indianapolis, 2003.
11. Oman, Paul, Edmund Schweitzer, and Deborah Frincke, Concerns About Intrusions into Remotely Accessible Substation Controllers and SCADA Systems, Technical report, Schweitzer Engineering Laboratories, 2000.
12. Palensky, Peter and Thilo Sauter, Security Considerations for FAN-Internet Connections, in Proceedings 2000 IEEE International Workshop on Factory Communication Systems, 2000.
13. Schneier, Bruce, *Applied Cryptography*, 2nd ed., Wiley, New York, 1996.
14. Schneier, Bruce, *Secrets and Lies — Digital Security in a Networked World*, Wiley, New York, 2000.
15. Schwartau, Winn, *Time Based Security*, Interpact Press, 1999.
16. Spitzner, Lance, The Honeynet project: trapping the hackers, *IEEE Security and Privacy*, 1:15–23, 2003.
17. Viega, John and Gary McGraw, *Building Secure Software*, Addison-Wesley, Reading, MA, 2001.
18. von Hoff, Thomas P. and Mario Crevatin, HTTP Digest Authentication in Embedded Automation Systems, 9th IEEE International Conference on Emerging Technologies and Factory Automation, Lisbon, Portugal, 2003.

Section 5: Intelligent Sensors and Sensor Networks

70

A Smart Transducer Interface Standard for Sensors and Actuators

Kang Lee
National Institute of Standards and Technology

70.1 Introduction

Sensors are used in many devices and systems to provide information on the parameters being measured or to identify the states of control. They are good candidates for increased built-in intelligence. Microprocessors can make smart sensors or devices a reality. With this added capability, it is possible for a smart sensor to directly communicate measurements to an instrument or a system. In recent years, the concept of computer networking has gradually migrated into the sensor community. Networking of transducers (sensors or actuators) in a system and communicating transducer information via digital means vs. analog cabling facilitates easy distributed measurements and control. In other words, intelligence and control, which were traditionally centralized, are gradually migrating to the sensor level. They can provide flexibility, improve system performance, and ease system installation, upgrade, and maintenance. Thus, the trend in industry is moving toward distributed control with intelligent sensing architecture. These enabling technologies can be applied to aerospace, automotive, industrial automation, military and homeland defenses, manufacturing process control, smart buildings and homes, and smart toys and appliances for consumers. As examples: (1) in order to reduce the number of personnel to run a naval ship from 400 to less than 100 as required by the reduced-manning program, the U.S. Navy needs tens of thousands of networked sensors per vessel to enhance automation, and (2) Boeing needs to network hundreds of sensors for monitoring and characterizing airplane performance.

Sensors are used across industries and are going global [1]. The sensor market is extremely diverse, and it is expected to grow to $43 billion by 2008. The rapid development and emergence of smart sensor and field network technologies have made the networking of smart transducers a very economical and attractive solution for a broad range of measurement and control applications. However, with the existence of a multitude of incompatible networks and protocols, the number of sensor interfaces and amount of hardware and software development efforts required to support this variety of networks are enormous for both sensor producers and users alike. The reason is that a sensor interface customized for a particular network will not necessarily work with another network. It seems that a variety of networks will coexist to serve their specific industries. The sensor manufacturers are uncertain of which network(s) to support and are restrained from full-scale smart sensor product development. Hence, this condition has impeded the widespread adoption of the smart sensor and networking technologies despite a great desire to build and use them. Clearly, a sensor interface standard is needed to help alleviate this problem [2].

70.2 A Smart Transducer Model

In order to develop a sensor interface standard, a smart transducer model should first be defined. As defined in the IEEE Std 1451.2-1997 [3]

> a smart transducer is a transducer that provides functions beyond those necessary for generating a correct representation of a sensed or controlled quantity. This functionality typically simplifies the integration of the transducer into applications in a networked environment.

Thus, let us consider the functional capability of a smart transducer. A smart transducer should have:

- integrated intelligence closer to the point of measurement and control,
- basic computation capability, and
- capability to communicate data and information in a standardized digital format.

Based on this premise, a smart transducer model is shown in Figure 70.1. It applies to both sensors and actuators. The output of a sensor is conditioned and scaled, then converted to a digital format through an analog-to-digital (A/D) converter. The digitized sensor signal can then be easily processed by a microprocessor using a digital application control algorithm. The output, after being converted to an analog signal via a digital-to-analog (D/A) converter, can then be used to control an actuator. Any of the measured or calculated parameters can be passed on to any device or host in a network by means of network communication protocol.

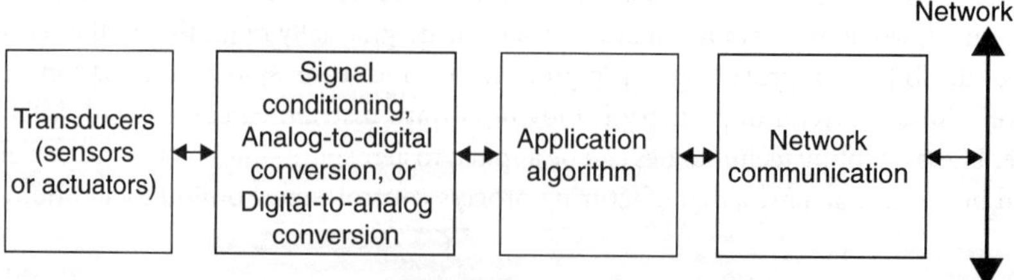

FIGURE 70.1   A smart transducer model.

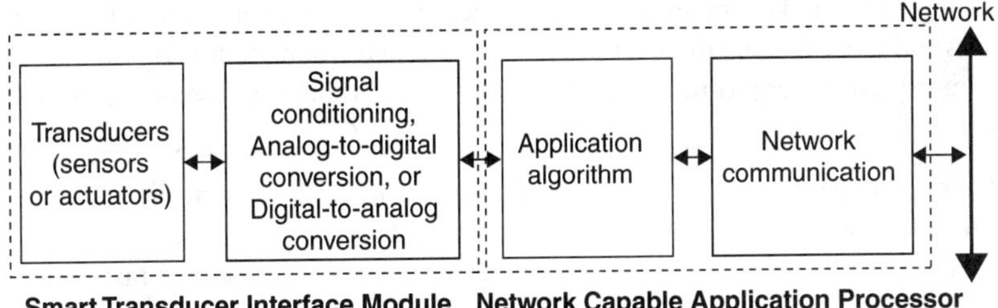

FIGURE 70.2   Functional partitioning.

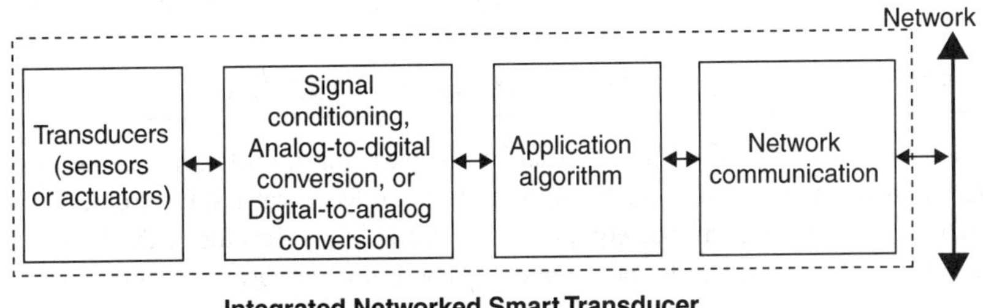

FIGURE 70.3   An integrated networked smart transducer.

The different modules of the smart transducer model can be grouped into functional units as shown in Figure 70.2. The transducers and signal conditioning and conversion modules can be grouped into a building block called a Smart Transducer Interface Module (STIM). Likewise, the application algorithm and network communication modules can be combined into a single entity called a Network-Capable Application Processor (NCAP). With this functional partitioning, transducer to network interoperability can be achieved in the following manner:

1. STIMs from different sensor manufacturers can "plug and play" with NCAPs from a particular sensor network supplier,
2. STIMs from a sensor manufacturer can "plug and play" with NCAPs supplied by different sensor or field network vendors, and
3. STIMs from different manufacturers can be interoperable with NCAPs from different field network suppliers.

Using this partitioning approach, a migration path is provided to those sensor manufacturers who want to build STIMs with their sensors, but do not intend to become field network providers. Similarly, it applies to those sensor network builders who do not want to become sensor manufacturers.

As technology becomes more advanced and microcontrollers become smaller relative to the size of the transducer, integrated networked smart transducers that are economically feasible to implement will emerge in the marketplace. In this case, all the modules are incorporated into a single unit as shown in Figure 70.3. Thus, the interface between the STIM and NCAP is not exposed for external access and separation. The only connection to the integrated transducer is through the network connector. The integrated smart transducer approach simplifies the use of transducers by merely plugging the device into a sensor network.

70.3 Networking Smart Transducers

Not until recently have sensors been connected to instruments or computer systems by means of a point-to-point or multiplexing scheme. These techniques involve a large amount of cabling, which is very bulky and costly to implement and maintain. With the emergence of computer networking technology, transducer manufacturers and users alike are finding ways to apply this networking technology to their transducers for monitoring, measurement, and control applications [4]. Networking smart sensors provides the following features and benefits:

- enable peer-to-peer communication and distributed sensing and control,
- significantly lower the total system cost by simplified wiring,
- use prefabricated cables instead of custom laying of cables for ease of installation and maintenance,
- facilitate expansion and reconfiguration,
- allow time-stamping of sensor data,
- enable sharing of sensor measurement and control data, and
- provide Internet connectivity, meaning *global* or *anywhere,* access of sensor information.

70.4 Establishment of the IEEE 1451 Standards

As discussed earlier, a smart sensor interface standard is needed in industry. In view of this situation, the Technical Committee on Sensor Technology of the Institute of Electrical and Electronics Engineer (IEEE)'s Instrumentation and Measurement Society sponsored a series of projects for establishing a family of IEEE 1451 Standards [5]. These standards specify a set of common interfaces for connecting transducers to instruments, microprocessors, or field networks. They cover digital, mixed-mode, distributed multidrop, and wireless interfaces to address the needs of different sectors of industry. A key concept in the IEEE 1451 standards is the Transducer Electronic Data Sheets (TEDS), which contain manufacture-related information about the sensor such as manufacturer name, sensor types, serial number, and calibration data and standardized data format for the TEDS. The TEDS has many benefits:

- Enable self-identification of sensors or actuators — A sensor or actuator equipped with the IEEE 1451 TEDS can identify and describe itself to the host or network via the sending of the TEDS.
- Provide long-term self-documentation — the TEDS in the sensor can be updated and stored with information such as location of the sensor, recalibration date, repair record, and many maintenance-related data.
- Reduce human error — automatic transfer of TEDS data to the network or system eliminates the entering of sensor parameters by hands, which could induce errors due to various conditions.
- Ease field installation, upgrade, and maintenance of sensors — this helps to reduce life cycle costs because only a less skilled person is needed to perform the task by simply using "plug and play."

IEEE 1451, designated as Standard for a Smart Transducer Interface for Sensors and Actuators, consists of six document standards. The current status of their development are as follows:

1. IEEE P1451.0, [1]Common Functions, Communication Protocols, and TEDS Formats — *In progress.*
2. IEEE Std 1451.1-1999, NCAP Information Model for Smart Transducers [6] — *Published standard.*
3. IEEE Std 1451.2-1997, Transducer to Microprocessor Communication Protocols and TEDS Formats — *Published standard.*
4. IEEE std 1451.3-2003, Digital Communication and TEDS Formats for Distributed Multidrop Systems — *Published standard.*
5. IEEE std 1451.4-2004, Mixed-mode Communication Protocols and TEDS Formats — *Published standard.*
6. IEEE P1451.5, Wireless Communication and TEDS Formats — *In progress.*

---

[1]P1451.0 — the "P" designation means that P1451.0 is a draft standard development project. Once the draft document is approved as a standard, "P" will be dropped.

70.5 Goals of IEEE 1451

The goals of the IEEE 1451 standards are to:

- develop network- and vendor-independent transducer interfaces,
- define TEDS and standardized data formats,
- support general transducer data, control, timing, configuration, and calibration models,
- allow transducers to be installed, upgraded, replaced, and moved with minimum effort by simple "plug and play,"
- eliminate error prone, manual entering of data, and system configuration steps, and
- ease the connection of sensors and actuators by wireline or wireless means.

70.6 The IEEE 1451 Standards

The IEEE 1451 Smart Transducer Model

The IEEE 1451 smart transducer model parallels the smart transducer model discussed in Figure 70.2. In addition, the IEEE 1451 model includes the TEDS. The model for each of the IEEE 1451.X standards is discussed in the following.

IEEE P1451.0 Common Functionality

Several standards in the IEEE 1451 family share certain characteristics, but there is no common set of functions, communications protocols, and TEDS formats that facilitate interoperability among these standards. The IEEE P1451.0 standard provides that commonality and simplifies the creation of future standards with different physical layers that will facilitate interoperability in the family.

This project defines a set of common functionalities for the family of IEEE P1451 smart transducer interface standards. This functionality is independent of the physical communications media. It includes the basic functions required to control and manage smart transducers, common communications protocols, and media-independent TEDS formats. The block diagram for IEEE P1451.0 is shown in Figure 70.4. P1451.0 defines functional characteristics, but it does not define any physical interface.

IEEE 1451.1 Smart Transducer Information Model

The IEEE 1451.1 Standard defines a common object model for the components of a networked smart transducer and the software interface specifications to these components [7]. Some of the components are the NCAP block, function block, and transducer block.

The networked smart transducer object model provides two interfaces.

1. The interface to the transducer block, which encapsulates the details of the transducer hardware implementation within a simple programming model. This makes the sensor or actuator hardware interface resemble an input/output (I/O)-driver.
2. The interface to the NCAP block and ports encapsulate the details of the different network protocol implementations behind a small set of communications methods.

Application-specific behavior is modeled by function blocks. To produce the desired behavior, the function blocks communicate with other blocks both on and off the smart transducer. This common network-independent application model has the following two advantages:

1. Establishment of a high degree of interoperability between sensors/actuators and networks, thus enabling "plug-and-play" capability.
2. Simplification of the support of multiple sensor/actuator control network protocols.

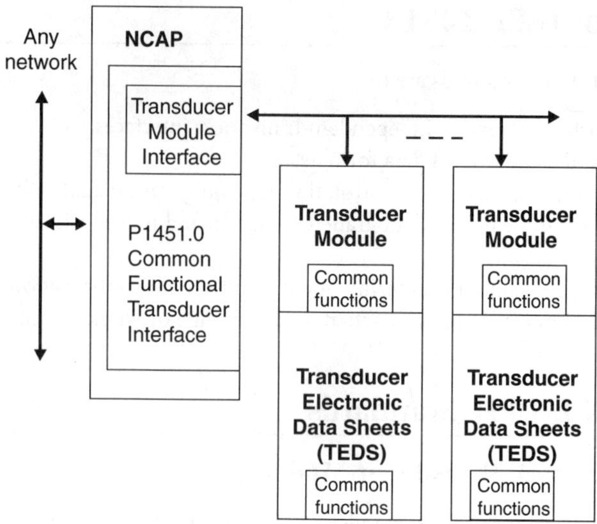

FIGURE 70.4   The block diagram for IEEE P1451.0.

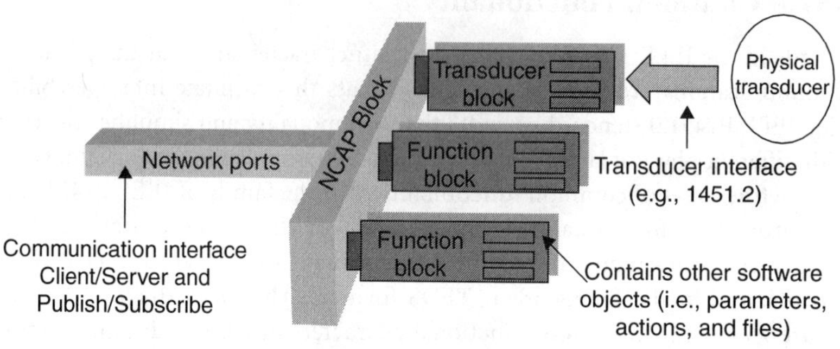

FIGURE 70.5   Conceptual view of IEEE 1451.1.

A conceptual view of IEEE 1451.1 NCAP is shown in Figure 70.5, which uses the idea of a "backplane" or "card cage" to explain the functionality of the NCAP. The NCAP centralizes all system and communications facilities. Network communication can be viewed as a port through the NCAP, and communication interfaces support both client–server and publish–subscribe communication models. Client–server is a tightly coupled, point-to-point communication model, where a specific object, the client, communicates in a one-to-one manner with a specific server object, the server. On the other hand, the publish–subscribe communication model provides a loosely coupled mechanism for network communications between objects, where the sending object, the publisher object, does not need to be aware of the receiving objects, the subscriber objects. The loosely coupled, publish–subscribe model is used for one-to-many and many-to-many communications. A function block containing application code or control algorithm is "plugged" in as needed. Physical transducers are mapped into the NCAP using transducer block objects via the hardware interface, for example, the IEEE 1451.2 interface.

The IEEE 1451 logical interfaces are illustrated in Figure 70.6. The transducer logical interface specification defines how the transducers communicate with the NCAP block object via the transducer block. The network protocol logical interface specification defines how the NCAP block object communicates with any network protocol via the ports.

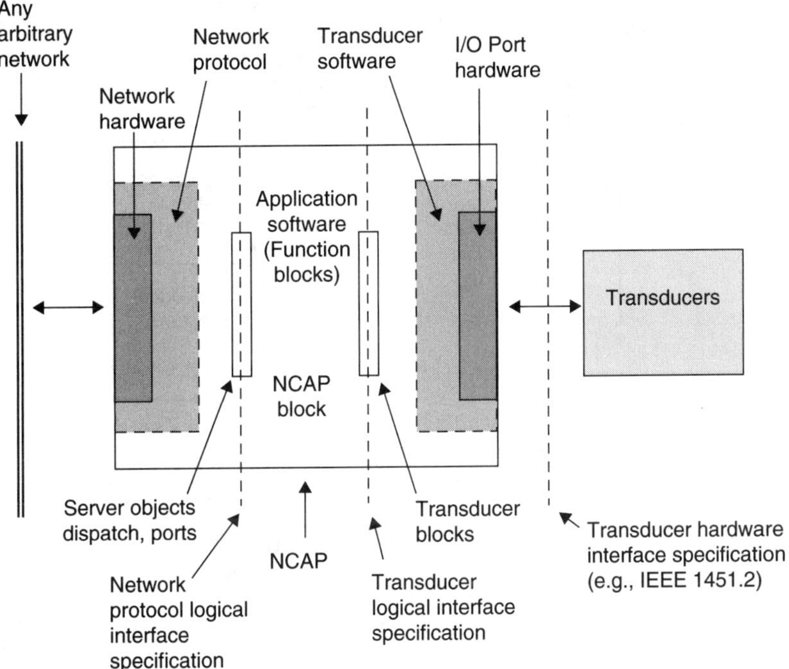

FIGURE 70.6　IEEE 1451 logical interfaces.

IEEE 1451.2 Transducer-to-Microprocessor Interface

The IEEE 1451.2 standard defines a TEDS, its data format, and the digital interface and communication protocols between the STIM and NCAP [8]. A block diagram and detailed system diagram of IEEE 1451 are shown in Figures 70.7 and 70.8, respectively. The STIM contains the transducer(s) and the TEDS, which is stored in a nonvolatile memory attached to a transducer. The TEDS contains fields that describe the type, attributes, operation, and calibration of the transducer. The mandatory requirement for the TEDS is only 179 bytes. The rest of the TEDS specification is optional. A transducer integrated with the TEDS provides a very unique feature that makes possible the self-description of transducers to the system or network. Since the manufacture-related data in the TEDS always go with the transducer, and this information is electronically transferred to an NCAP or host, human errors associated with manual entering of sensor parameters into the host are eliminated. Because of this distinctive feature of the TEDS, upgrading transducers with a higher accuracy and enhanced capability or replacing transducers for maintenance purpose is simply considered "plug and play."

Eight different types of TEDS are defined in the standard. Two of them are mandatory and six are optional. They are listed in Table 70.1. The TEDS are divided into two categories. The first category of TEDS contains data in a machine-readable form, which is intended for use by the NCAP. The second category of TEDS contains data in a human-readable form. The human-readable TEDS may be represented in multiple languages using different encoding for each language.

The Meta TEDS contains the data that describe the whole STIM. It contains the revision of the standard, the version number of the TEDS, the number of channels in the STIM, and the worst-case timing required to access these channels. This information will allow the NCAP to access the channel information. In addition, the Meta TEDS includes the channel groupings that describe the relationships between channels. Each transducer is represented by a channel.

Each channel in the STIM contains a Channel TEDS. The Channel TEDS lists the actual timing parameters for each individual channel. It also lists the type of transducer, the format of the data word being

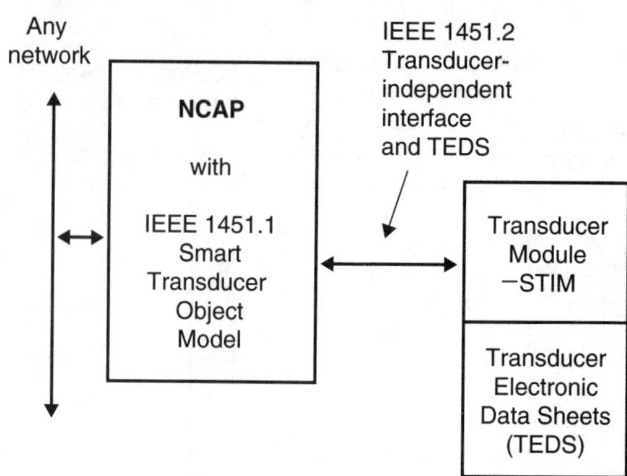

FIGUR.E 70.7   Block diagram of IEEE 1451.

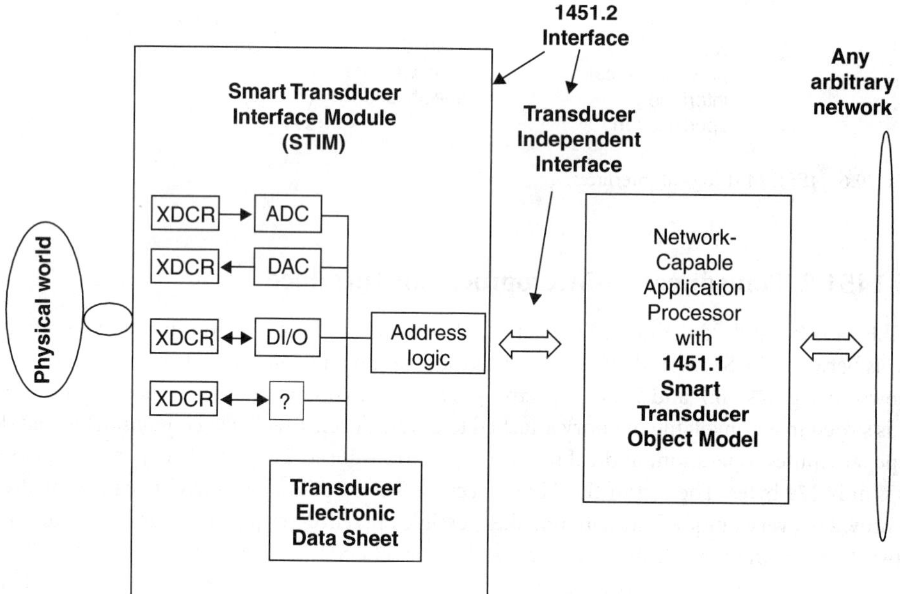

FIGURE 70.8   Detailed system block diagram of an IEEE 1451 smart transducer interface.

TABLE 70.1   Different Types of TEDS

TEDS Name	Type	Optional/Mandatory
Meta TEDS	Machine readable	Mandatory
Channel TEDS	Machine readable	Mandatory
Calibration TEDS	Machine readable	Optional
Generic extension TEDS	Machine readable	Optional
Meta-identification TEDS	Human readable	Optional
Channel identification TEDS	Human readable	Optional
Calibration identification TEDS	Human readable	Optional
End user application-Specific TEDS	Human readable	Optional

output by the channel, the physical units, the upper and lower range limits, the uncertainty or accuracy, whether or not a calibration TEDS is provided, and where the calibration is to be performed.

The Calibration TEDS contains all the necessary information for the sensor data to be converted from the analog-to-digital converter raw output into the physical units specified in the Channel TEDS. If actuators are included in the STIM, it also contains the parameters that convert data in the physical units into the proper output format to drive the actuators. It also contains the calibration interval and last calibration date and time. This allows the system to determine when a calibration is needed. A general calibration algorithm is specified in the standard.

The Generic Extension TEDS is provided to allow industry groups to provide additional TEDS in a machine-readable format.

The Meta Identification TEDS is human-readable data that the system can retrieve from the STIM for display purposes. This TEDS contains fields for the manufacturer's name, the model number and serial number of the STIM, and a date code.

The Channel Identification TEDS is similar to the Meta Identification TEDS. When transducers from different manufacturers are built into an STIM, this information will be very useful for the identification of channels. The Channel Identification TEDS provides information about each channel, whereas the Meta Identification TEDS provides information for the STIM.

The Calibration Identification TEDS provides details of the calibration in the STIM. This information includes who performed the calibration and what standards were used.

The End-User Application-Specific TEDS is not defined in detail by the standard. It allows the user to insert information such as installation location, the time it was installed, or any other desired text.

The STIM module can contain a combination of sensors and actuators of up to 255 channels, signal conditioning/processing, A/D converter, D/A converter, and digital logics to support the transducer-independent interface (TII). Currently, the P1451.2 working group is considering an update to the standard to include a popular serial interface, such as RS232, in addition to the TII for connecting sensors and actuators.

IEEE 1451.3 Distributed Multidrop Systems

The EEE 1451.3 defines a transducer bus for connecting transducer modules to an NCAP in a distributed multidrop manner. A block diagram is shown in Figure 70.9. The physical interface for the transducer bus is based on Home Phoneline Networking Alliance (HomePNA) specification. Both power and data run on a twisted pair of wires. Multiple transducer modules, called Transducer Bus Interface Modules (TBIM), can be connected to an NCAP via the bus. Each TBIM contains transducers, signal conditioning/processing, A/D, D/A, and digital logics to support the bus and can accommodate large

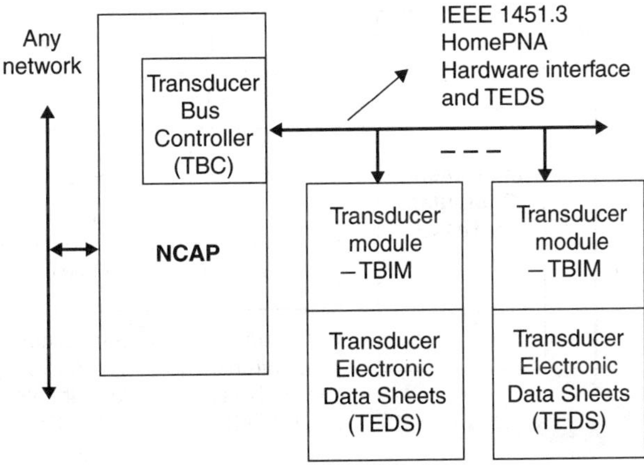

FIGURE 70.9　Block diagram of IEEE 1451.3.

arrays of transducers for synchronized access at up to 128 Mbps with HomePNA 3.0 and up to 240 Mbps with extensions. The TEDS is defined in the eXtensible Markup Language (XML).

IEEE 1451.4 Mixed-Mode Transducer Interface

The IEEE 1451.4 defines a mixed-mode transducer interface (MMI), which is used for connecting transducer modules, Mixed-mode Transducers (MMT), to an instrument, a computer, or an NCAP. The block diagram of the system is shown in Figure 70.10. The physical transducer interface is based on the Maxim/Dallas Semiconductor's one-wire protocol, but it also supports up to 4 wires for bridge-type sensors. It is a simple, low-cost connectivity for analog sensors with a very small TEDS - 64 bits mandatory and 256 bits optional. The mixed-mode interface supports a digital interface for reading and writing the TEDS by the instrument or NCAP. After the TEDS transaction is completed, the interface switches into analog mode, where the analog sensor signal is sent straight to the instrument and NCAP, which is equipped with A/D to read the sensor data.

IEEE P1451.5 Wireless Transducer Interface

Wireless communication is emerging, and low-cost wireless technology is on the horizon. Wireless communication links could replace the costly cabling for sensor connectivity. It could also greatly reduce sensor installation cost. Industry would like to apply the wireless technology for sensors; however, there is a need to solve the interoperability problem among wireless sensors, equipment, and data. In response to this need, the IEEE P1451.5 working group is working to define a wireless sensor communication interface standard that will leverage existing wireless communication technologies and protocols [9]. A block diagram of IEEE P1451.5 is shown in Figure 70.11. The working group seeks to define the wireless message formats, data/control model, security model, and TEDS that are scalable to meet the needs of low cost to sophisticated sensor or device manufacturers. It allows for a minimum of 64 sensors per access point. Intrinsic safety is not required but the standard would allow for it. The physical communication protocol(s) being considered by the working group are: (1) IEEE 802.11 (WiFi), (2) IEEE 802.15.1 (Bluetooth), and (3) IEEE 802.15.4 (ZigBee).

IEEE 1451 Family

Figure 70.12 summarizes the family of IEEE 1451 Standards. Each of the IEEE P1451.X is designed to work with the other. However, they can also stand on their own. For example, IEEE 1451.1 can work without any IEEE 1451.X hardware interface. Likewise, IEEE1451.X can also be used without IEEE 1451.1, but software with similar functionality should provide sensor data/information to each network.

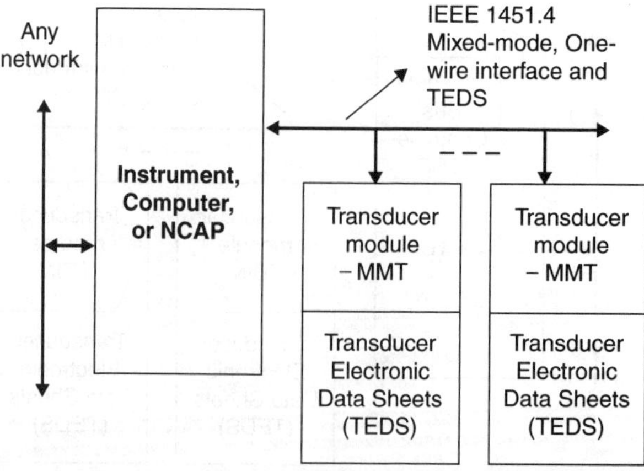

FIGURE 70.10   Block diagram of IEEE 1451.4.

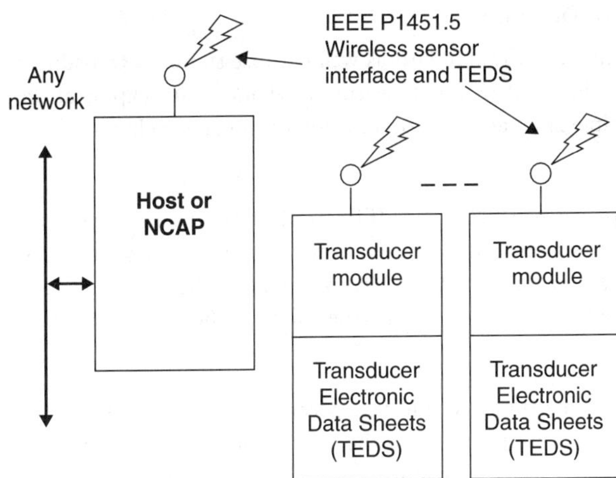

FIGURE 70.11 Block diagram of IEEE P1451.5 wireless transducer.

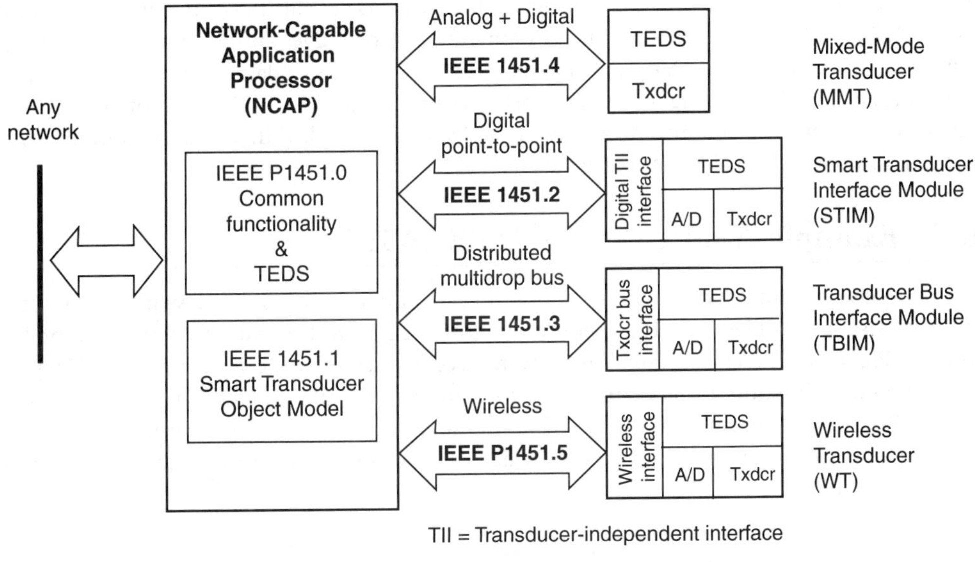

FIGURE 70.12 Family of IEEE 1451 Standards.

Benefits of IEEE 1451

IEEE 1451 defines a set of common transducer interfaces, which will help to lower the cost of designing smart sensors and actuators because designers would only have to design to a single set of standardized digital interfaces. Thus, the overall cost to make networked sensors will decrease.

Incorporating the TEDS with the sensors will enable self-description of sensors and actuators, eliminating error-prone, manual configuration.

Sensor Manufacturers

Sensor manufacturers can benefit from the standard because they only have to design a single standard physical interface. Standard calibration specification and data format can help to design and develop multilevel products based on TEDS with a minimum effort.

Application Software Developers

Applications can benefit from the standard as well because standard transducer models for control and data can support and facilitate distributed measurement and control applications. The standard also provides support for multiple languages — good for international developers.

System Integrators

Sensor system integrators can benefit from IEEE 1451 because sensor systems become easier to install, maintain, modify, and upgrade. Quick and efficient transducer replacement results by simple "plug and play." It can also provide a means to store installation details in the TEDS. Self-documentation of hardware and software is done via the TEDS. Best of all is the ability to choose sensors and networks based on merit.

End Users

End users can benefit from a standard interface because sensors will be easy to use by simple "plug and play." Based on the information provided in the TEDS, software can automatically provide the physical units, readings with significant digits as defined in the TEDS, installation details such as instruction, identification, and location of the sensor.

"Plug and Play" of sensors

IEEE 1451 enables "plug and play" of transducers to a network as illustrated in Figure 70.13. In this example, IEEE 1451.4-compatible transducers from different companies are shown to work with a sensor network. IEEE 1451 also enables "plug and play" of transducers to a data acquisition system/instrumentation system as shown in Figure 70.14. In this example, various IEEE 1451.4-compatible transducers such as an accelerometer, a thermistor, a load cell, and a linear variable differential transformer (LVDT) are shown to work with a LabVIEW-based system.[2]

70.7   Example Application of IEEE 1451.2

IEEE 1451-based sensor network consisting of sensors, STIM, and NCAP are designed and built into a cabinet as shown in Figure 70.15. There were a total of four STIM and NCAP network nodes as shown in Figure 70.15. Thermistor sensors were used for temperature measurements. They were calibrated in the laboratory to generate IEEE 1451.2-compliance calibration TEDS for all four STIMs and NCAPs. The thermistors were mounted on the spindle motor housing, bearing, and axis drive motors of a 3-axis vertical machining center, which is shown in Figure 70.16. Since each NCAP has a

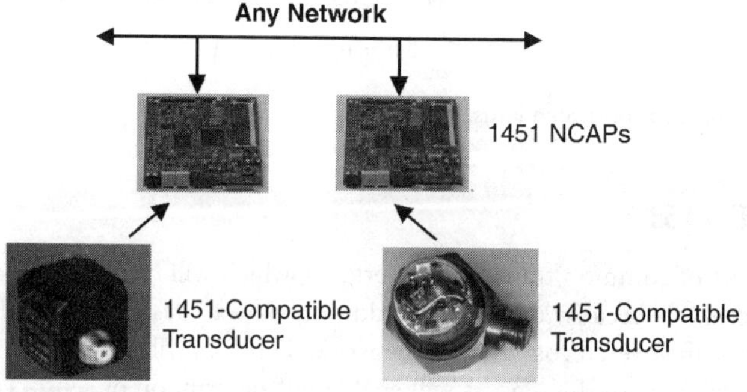

FIGURE 70.13   IEEE 1451 enables "plug and play" of transducers to a network.

---

[2]Certain commercial products are identified in this paper in order to describe the system. Such an identification does not imply recommendation or endorsement by the National Institute of Standards and Technology, nor does it imply that the products identified are necessarily the best or the only ones available for the purpose.

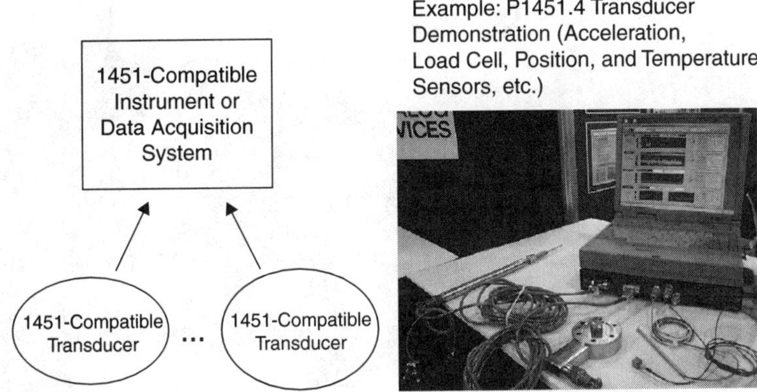

FIGURE 70.14 IEEE 1451 enables "plug and play" of transducers to data acquisition/instrumentation system.

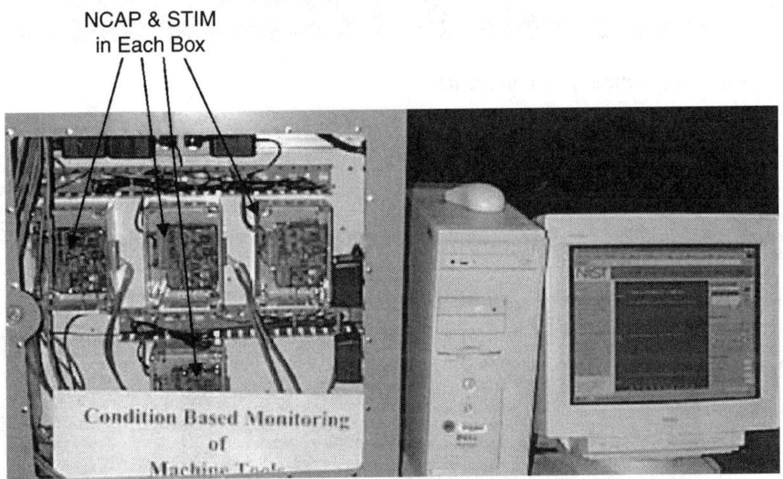

FIGURE 70.15 NCAP-based condition monitoring system.

built-in micro web server, a custom web page was constructed using the web tool provided with the NCAP. Thus, remote monitoring of the machine thermal condition was easily achieved via the Ethernet network and the Internet using a readily available common web browser. The daily trend chart of the temperature of the spindle motor (top trace) and the temperature of the Z-axis drive motor (bottom trace) in the machine is shown in Figure 70.17. The temperature rise tracks the working of the machine during the day and the temperature fall indicates that the machine is cooling off after the machine shop is closed.

70.8 Application of IEEE 1451-Based Sensor Network

A distributed measurement and control system can be easily implemented based on the IEEE 1451 standards [10]. An application model of IEEE 1451 is shown in Figure 70.18. Three NCAP/STIMs are used to illustrate the distributed control, remote sensing or monitoring, and remote actuating. In the first scenario, a sensor and actuator are connected to the STIM of NCAP #1, and an application software running in the NCAP can perform a locally distributed control function, such as maintaining a constant temperature for

FIGURE 70.16 Three-axis vertical machining center.

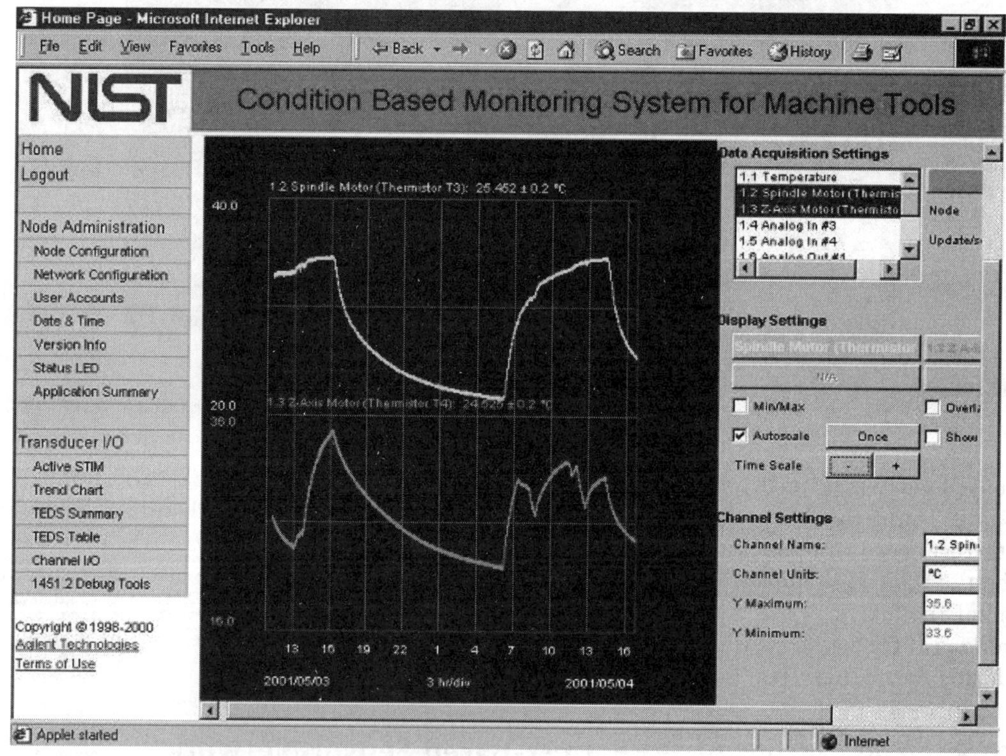

FIGURE 70.17 Temperature trend chart.

a bath. The NCAP reports measurement data, process information, and control status to a remote monitoring station or host. It frees the host from the processor-intensive, closed-loop control operation. In the second scenario, only sensors are connected to NCAP #2, which can perform remote process or condition

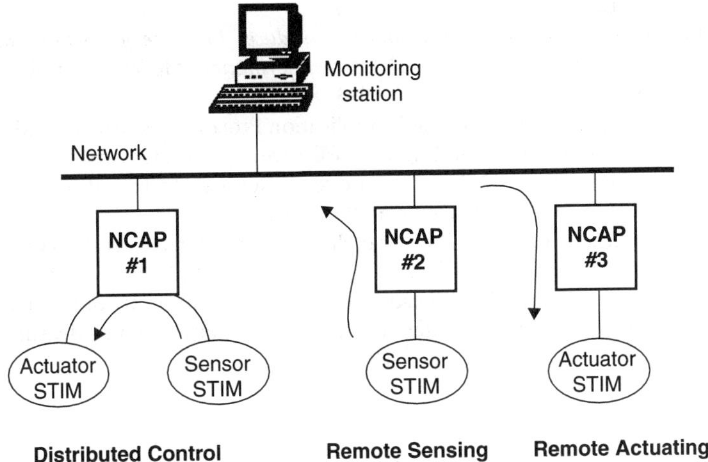

FIGURE 70.18　Application model of IEEE 1451.

monitoring functions, such as monitoring the vibration level of a set of bearings in a turbine. In the third scenario, based on the broadcast data received from NCAP #2, NCAP #3 activates an alarm when the vibration level of the bearings exceeds a critical set point. As illustrated in these examples, an IEEE 1451-based sensor network can easily facilitate peer-to-peer communications and distributed control functions.

70.9　Summary

The IEEE 1451 smart transducer interface standards are defined to allow a transducer manufacturer to build transducers of various performance capabilities that are interoperable within a networking system. The IEEE 1451 family of standards has provided the common interface and enabling technology for the connectivity of transducers to microprocessors, field networks, and instrumentation systems using wired and wireless means. The standardized TEDS allows the self-description of sensors, which turns out to be a very valuable tool for condition-based maintenance. The expanding Internet market has created a good opportunity for sensor and network manufacturers to exploit web-based and smart sensor technologies. As a result, users will greatly benefit from many innovations and new applications.

Acknowledgments

The author sincerely thanks the IEEE 1451 working groups for the use of the materials in this chapter. Through its program in Smart Machine Tools, the Manufacturing Engineering Laboratory of the National Institute of Standards and Technology has contributed to the development of the IEEE 1451 standards.

References

[1]　Amos, Kenna, Sensor market goes global, *InTech — The International Journal for Measurement and Control*, June, 40–43, 1999.

[2]　Bryzek, Janusz, Summary Report, Proceedings of the IEEE/NIST First Smart Sensor Interface Standard Workshop, NIST, Gaithersburg, Maryland, March 31, 1994, pp. 5–12.

[3]　IEEE Std 1451.2-1997, *Standard for a Smart Transducer Interface for Sensors and Actuators — Transducer to Microprocessor Communication Protocols and Transducer Electronic Data Sheet (TEDS) Formats*, Institute of Electrical and Electronics Engineers, Inc., Piscataway, NJ, 1997.

[4]　Eidson, J. and Woods, S., A Research Prototype of a Networked Smart Sensor System, Proceedings Sensors Expo Boston, Helmers Publishing, May 1995.

[5] URL http://ieee1451.nist.gov.

[6] IEEE Std 1451.1-1999, *Standard for a Smart Transducer Interface for Sensors and Actuators — Network Capable Application Processor (NCAP) Information Model*, Institute of Electrical and Electronics Engineers, Inc., Piscataway, NJ, 1999.

[7] Warrior, Jay. IEEE-P1451 Network Capable Application Processor Information Model, Proceedings Sensors Expo Anaheim, Helmers Publishing, April 1996, pp. 15–21.

[8] Woods, Stan et al., IEEE-P1451.2 Smart Transducer Interface Module, Proceedings Sensors Expo Philadelphia, , Helmers Publishing, October 1996, pp. 25–38.

[9] Lee, K.B., Gilsinn, J.D., Schneeman, R.D., and Huang, H. M., First Workshop on Wireless Sensing, National Institute of Standards and Technology, NISTIR 02-*6823*, February 2002.

[10] Lee, Kang and Schneeman, Richard, Distributed Measurement and Control Based on the IEEE 1451 Smart Transducer Interface Standards, Instrumentation and Measurement Technical Conference 1999, Venice, Italy, May 24–26, 1999.

71

Integration Technologies of Field Devices in Distributed Control and Engineering Systems

Christian Diedrich
ifak Magdeburg, Germany

71.1 Introduction

Control system engineering and the instrumentation of industrial automation belong to a very innovative industrial area with a scope for cost reduction. That is why the costs for equipment and devices are already at a relatively low level. This trend is accompanied by a paradigm shift to digital processing in devices and digital communication among them.

This chapter describes the historical development steps from analog electronic devices connected via 4–20 mA, or 24 V technology, to digital devices with industrial communication connections such as fieldbus and Ethernet/TCP/IP. From these steps, the changes of the commissioning and system integration requirements are derived and the arising technologies to support device manufacturers and system integrators are introduced. All these integration technologies support the instrumentation tasks of field devices. Examples of these technologies are the Device Description Languages (provided, e.g., by

PROFIBUS, Device Net, Fieldbus Foundation), standardized interfaces such as OPC and Field Device Architecture (FDT), control application integration using proxy function blocks written in PLC languages (e.g., in IEC 61131-3 languages), and vertical communication from field devices to SCADA, Decentral Control Systems (DCS), and Manufacturing Execution Systems (MES) by means of XML. The reader will become familiar with up-to-date technologies used in field device engineering and instrumentation tools. The chapter concludes with a formal modeling approach, specifically with a device model. This model provides an abstract view of the described technologies and makes the relations among them visible.

71.2 History of smart devices

Digital information computation in field devices and digital fieldbus communication leads to a change in the handling of automation systems in manufacturing and process control. The field devices now contain much more information than the 4–20 mA signal. In addition, they carry out some functions that are/were originally programmed within the PLC or DCS. These field devices are also known as smart devices.

The consequence is a distributed system. The tools for the design and programming of control applications, commissioning, as well as maintenance need both access to the data of the field device and/or an exact machine-readable product description of the field devices including their data and functions. These device descriptions, from the design of the description to its use, are called device description technology. Using these device product descriptions, it is necessary to integrate standard interface specifications and standardized industrial communication systems.

The following example of a transmitter shows the transformation from an analog 4–20 mA device to a smart fieldbus device. Other types of automation devices were or are moving in a similar direction. Even digital, discrete, and I/O field devices are much simpler, for example, the 24 V technology will also be replaced by fieldbus connected devices.

The transmitter in Figure 71.1 is composed of electronics, which detects the specific measurement value (e.g., mV, mA) and which transforms the detected signal into the standardized 4–20 mA. The adjustment to the specific sensor and wiring is done by trim resistors. Each transmitter is connected to the PLC by its own wires.

Digital signal computation provides a higher accuracy. Therefore, the signal processing is carried out by microprocessors (Figure 71.2). An analog/digital and a digital/analog unit transform the signals two times. The signal processing may be influenced by several parameters, which make the transmitter more flexible. These parameters have to be accessed by the operator. For this purpose, the manufacturer provides a local operator panel at the transmitter, which consists of a display and very few buttons. PC

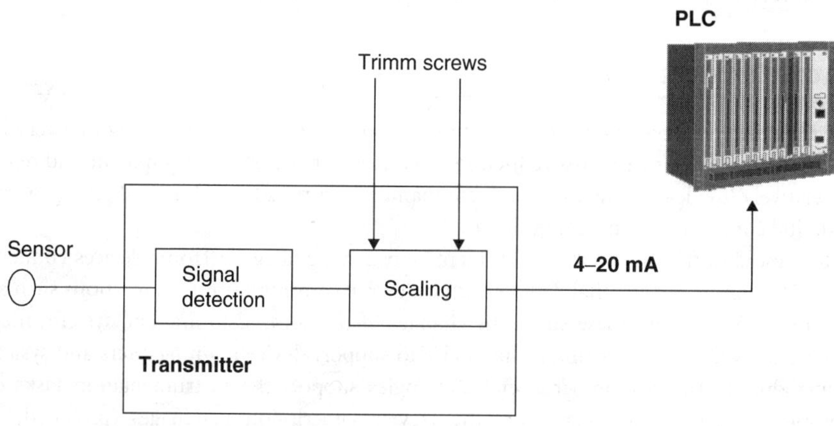

FIGURE 71.1 Structure of analogue transmitter.

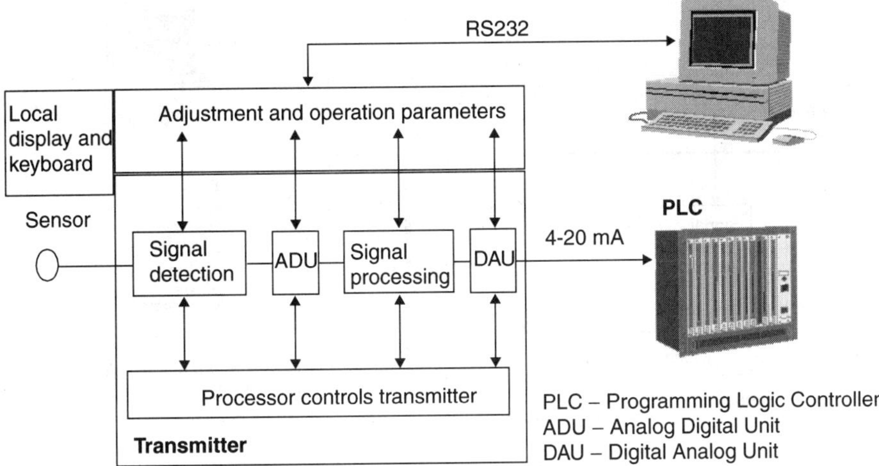

FIGURE 71.2 Structure of smart 4–20 mA transmitters.

tools provide more ergonomic solutions for the commissioning of such smart transmitters, which is necessary if they are more complex.

In principle, in smart fieldbus transmitters, the digital/analog unit at the end of the signal chain in the transmitter is replaced by the fieldbus controller. This increases the accuracy of the devices again. In addition, devices according to Figure 71.3 need specific communication commissioning using additional fieldbus configuration tools.

The smart field device parameterization has moved from manual screwing over local device terminals to PC application software. Additionally, the communication system configuration has to be managed during the commissioning of the instrumentation.

The field device is typically integrated as a component in an industrial automation system. The automation system performs the automation-related part of the complete application. The components of an industrial automation system may be arranged in multiple hierarchical levels connected by communication systems, as illustrated in Figure 71.4.

The field devices are components in the process level connected via inputs and outputs to the process or the physical or logical subnetworks (IEC Guideline, 2002). This also includes programmable devices and router or gateways. A communication system (e.g., a fieldbus) connects the field devices to the upper-level controllers, which are typically programmable controllers or DCS or even MES. Since the engineering tools and the commissioning tools should have access to the field devices and to the controllers, these tools are also located in the controller level. The field devices may directly communicate via the fieldbus or the controller (Programmable Controller). In larger automation systems, another higher level may exist connected via a communication system such as LAN or Ethernet. In these higher-level visualization systems (HMI), DCS, central engineering tools, and SCADA are located. Multiple clusters of field devices with or without a controller as described above may be connected over the LAN with each other or to the higher-level systems. MES, Enterprise Resource Planning (ERP), and other Information Technology (IT) systems can have access indirectly to field devices via the LAN and the controllers or direct via routers.

The result of this development is a large variety of PC-based tools for planning, design, commissioning configuration, and maintenance (Figure 71.4). This problem is also known in the home and consumer electronics. TV, CD player, video recorder, and record player are bought gradually by a family, and may be produced by different manufacturers. Each device is purchased with remote controllers. The user has to deal with different styles, programming approaches, and outlines. This is unacceptable both in consumer electronics and in the industrial environment. Of course, manufacturers desire to be unique regarding functions and features of their products and provide special approaches to interact with the device. The end user gets products very suitable to specific requirements, but he is faced with a broad variety of device interfaces.

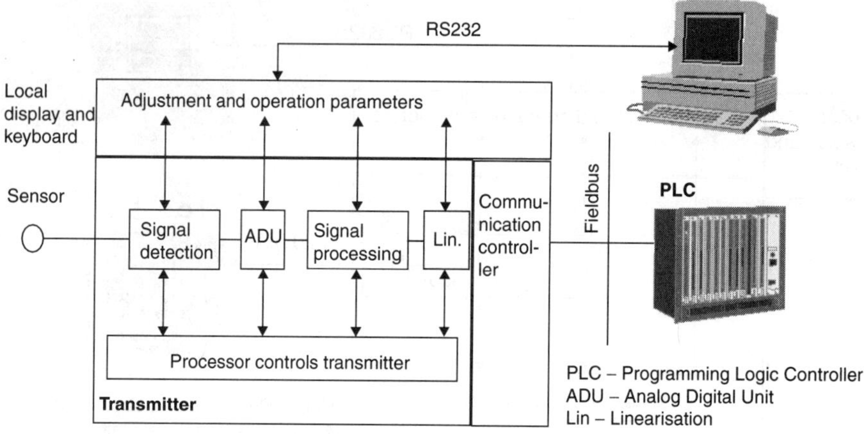

FIGURE 71.3   Structure of a smart fieldbus transmitter.

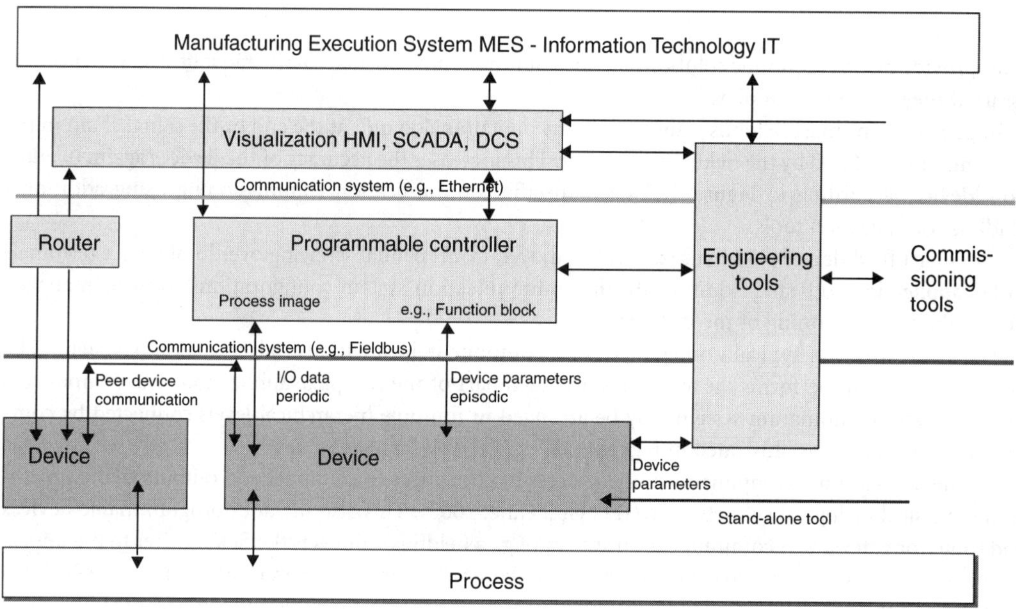

FIGURE 71.4   Typical automation configuration (IEC Guideline, 2002).

The field devices are an integrated part of the entire life cycle of control systems starting with planning/design, over purchase, system integration/commissioning, operation and ending with maintenance. During each phase of the life cycle, specific information represented in different formats is used from the tools. Planning and design define the requirements coming from the process that defines the type and the properties of the field device. The results are the pipe and instrumentation diagram (P&ID) in process control and electrotechnical drawings (E-CAD) in facturing automation combined with device lists. Purchase is made by requests of offers or by web market places directly. Commissioning carries out the parameterization and configuration of field devices in connection with the programming of the programmable controllers. During operation, there are mostly interactions between the controllers and the field devices. The field devices interact with special tools during maintenance. The view to field devices, that is, the used subset of the entire information range, the format of the information (text in manuals, files, data base entries, html pages), and the source of the information (paper, machine readable in the system, on-line from the device) differ between the lifecycle phases. Therefore, integration technologies deal with multiple technologies. B2B and MES are not within the scope of these collections of technologies. The main reason is that this paper concentrates on the functional aspects of the field device integration. Even planning and design are only mentioned in certain sections because product data management does not yet have tight connections to the functional design of the control system.

71.3 Field Device Instrumentation

Today, the large number of different device types and suppliers within a control system project makes the field device parameterization and configuration task difficult and time-consuming. Different tools must be mastered and data must be exchanged between these tools. The data exchange is not standardized; therefore, data conversions are often necessary, requiring detailed specialist knowledge. In the end, the consistency of data, documentation, and configurations can only be guaranteed by an intensive system test.

The central workplace for service and diagnostic tasks in the control system does not fully cover the functional capabilities of the field devices. Furthermore, the different device-specific tools cannot be integrated into the system's software tools. Typically, device-specific tools can only be connected directly to a fieldbus line or directly to the field device (Figure 71.5).

In order to maintain the continuity and operational reliability of process control technology, it is necessary to fully integrate field devices as a subcomponent of process automation.

Field devices often have to be adjusted to their concrete application purpose; therefore, additional software components are necessary for parameterization reasons. These components are necessary because local operator keyboards, with only a few bottoms and small displays, are not suitable to provide the complex parameterization issues to the operator. In principle, the following cases of application can be found (Diedrich et al., 2001):

- Devices with fixed functionality and without parameterization of its device application. These devices have to provide their communication properties, which is only done according to descriptions such as PROFIBUS GSD and CAN EDS. A communication configuration tool verifies all properties of all devices that are connected at one communication segment and generate the communication configuration in terms of the communication parameters such as baud rate and device addresses.
- Devices with only little parameter data have to be fixed only once during the commissioning phase (e.g., minimum and maximum speed of a drive, calibration for a transmitter). This device should receive the data via fieldbus from the control station. Therefore, many fieldbus systems have added parameterization keywords in their communication-related descriptions, such as PROFIBUS GSD and CAN EDS. Fieldbus configuration tools provide the possibility to edit the parameter values; the controller (e.g., PLC) ensures the persistence of the parameter data, for example, for a restart. This case of application does not need any additional parameterization tools. The replacement of a device after failure can be made easily.
- Devices with many parameters and complex parameterization means (e.g., transmitters, actuators in the process control field) have, for a long time, had local terminals and/or their own commissioning tools. Tools that are able to parameterize many different types of devices from different manufacturers have been established in the market. Well-known languages are HART DDL (Device Description Language [HART, 1995]), Fieldbus Foundation DDL (FF, 1996), and the PROFIBUS EDD (Electronic Device Description [PNO, 2001b]), which belong to the same language family (IEC61804, 2003). These languages are characterized by off-line parameterization, device-specific guidance by the operators, and extensive consistency checks. The management of the persistence data is directly related to parameterization tools; therefore, a replacement of these devices without these tools is not possible.
- Devices with complex parameterization sequences and complex data types (e.g., graphical information representation or video representation) cannot be described by the languages mentioned. Therefore, these devices need device-specific commissioning tools (e.g., Laser scanner).

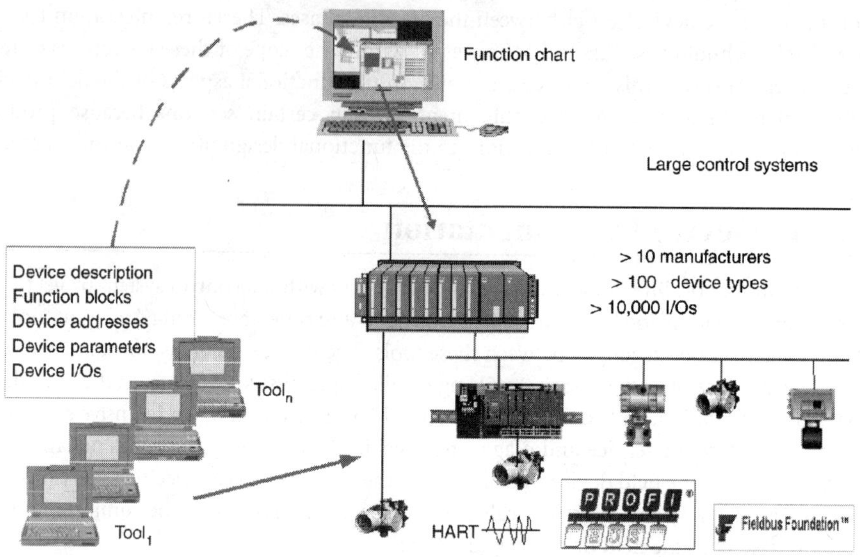

FIGURE 71.5   Different tools and multiple data input have determined field device integration to date (Bruns et al., 1999).

Fieldbus Communication Configuration

GSD Language

A communication segment has to be configured for a certain device configuration. The baud rate, device addresses, and special communication timing parameters have to be adjusted according to the collection of devices and their properties. Therefore, fieldbuses provide communication feature lists in a machine-readable format. Each device is purchased with such a list, which is used by a communication configuration tool to generate the possible or optimal communication configuration.

One example is the PROFIBUS GSD. The abbreviation GSD stands for German Device Data Base (GeräteStammDaten). There is no translation of the abbreviation at all. The GSD describes the communication features and the cyclic data of simple devices such as binary and analog I/O. Communication features are potential baud rates, communication protocol time parameters, and the support of PROFIBUS services and functions (e.g., remote address assignment).

Simple I/O devices (also known as Remote I/O or intelligent clip) are mostly modular. A basic device with power consumption and a central processing unit has several slots, offered to plug in modules with different signal qualities (analog, binary) and quantities (e.g., 2, 4, 8, 16 channels). According to the chosen modules, a different set of data has to be transferred in the cyclic transfer between the PROFIBUS master and slave devices. In other words, every module contributes with a specific set of data to the cyclic telegram. If we consider that the configuration tools of the PROFIBUS master devices (PLC mostly) have to be used device manufacturer-independent, an unambiguous and manufacturer-independent description of the modules is necessary. This is the second main task of the GSD description.

The GSD language is a list of keywords accompanied by their value assignment for each feature of the PROFIBUS DP protocol, which is configurable, and the module specification. This list is accomplished by a device and GSD file identification. The following example shows what a GSD file looks like (Figure 71.6).

Modules are represented in the GSD by the keyword "Module." Between the Module and the End_Module keyword, the declaration of the contribution of the module to the cyclic data telegram is specified in terms of binary codes, the so-called Identifier Bytes. Each module has a name that is printed at the configuration tool screen. If the module is chosen, the specified Identifier Bytes will be concatenated to the configuration string, which is transferred from the master to the slave during the startup of the cyclic data transfer (CFG service of PROFIBUS-DP [2]). In addition, the range and the default value of variables can be configured, which is supported by the variable declaration possibility of GSD. These data specified in the module GSD construct are transferred with the so-called PRM service (PRM from parameterization) of PROFIBUS DP ([2]).

GSD Tool Set

All PROFIBUS DP manufacturers are obliged to sell these devices with a GSD file, which is checked by the PROFIBUS User Organisation (PNO). The PNO offers a special GSD editor, which guides the engineer during the GSD development (Figure 71.7). The engineer chooses the features of the device with a mouse click from the superset of all existing PROFIBUS-DP features. The editor generates the syntactic right GSD ASCII file. This file is not compiled.

The configuration tools read these GSD files and present the specified device features for communication configuration and adjustment of some application parameters. The result of this configuration process are the configuration and parameterization strings for the PROFIBUS DP services (Figure 71.8).

The handling of GSD is very simple and needs no specific training. It matches exactly the needs of simple I/O with no or less application parameterization.

Field Device Application Parameterization — Device Description Languages

Two kinds of users can be distinguished for the task of parameterization:

- the end user or operator of a plant or machine and
- the system integrator.

For the end user of a distributed control system, the most important thing concerning device description is its transparency. The end user wishes to see only the graphical user interface of a device represented in the SCADA-software or other Human Machine Interfaces (HMI). Therefore, the electronic device description has to be constructed following the plug and play concept. Furthermore, exchanging a device in an application is required not to lead to a big engineering task, but is required to be done simply and securely by a technician. The device description has to support this.

In contrast to the end user, the system integrator, who carries out the engineering, instrumentation, and documenting, has different targets to meet. His goal is to reduce the engineering time for solving

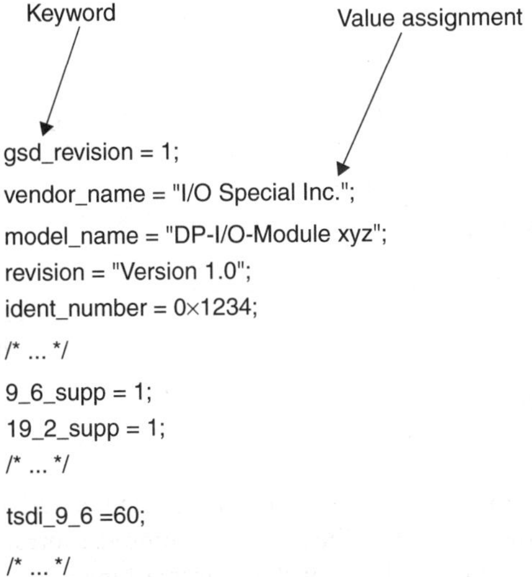

FIGURE 71.6 Example GSD file subset.

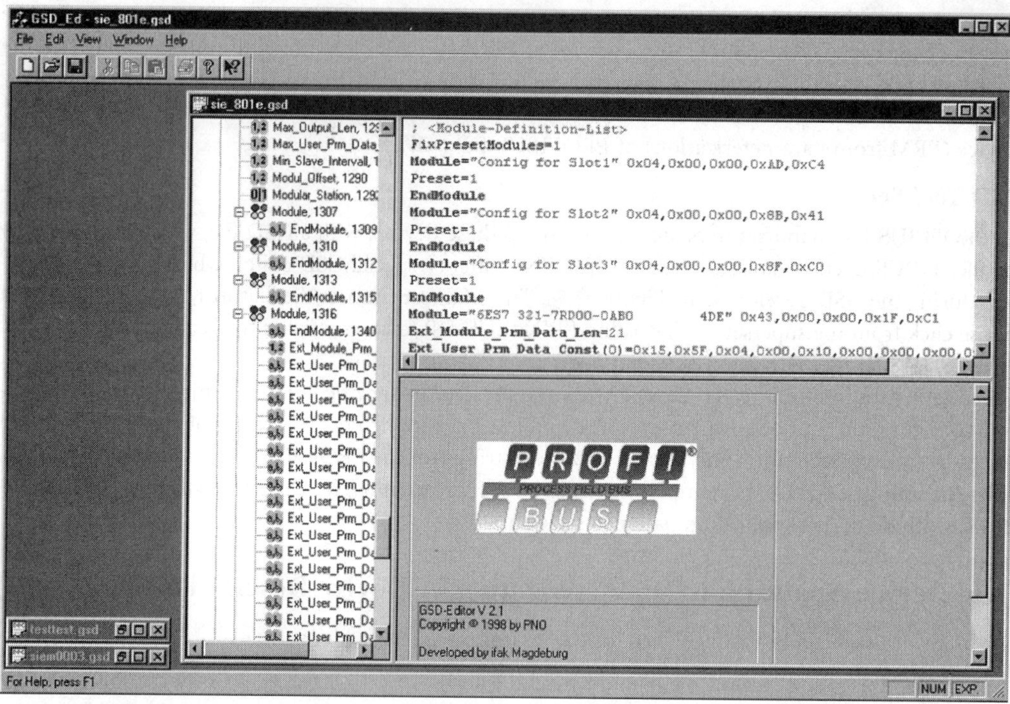

FIGURE 71.7   GSD editor snapshot.

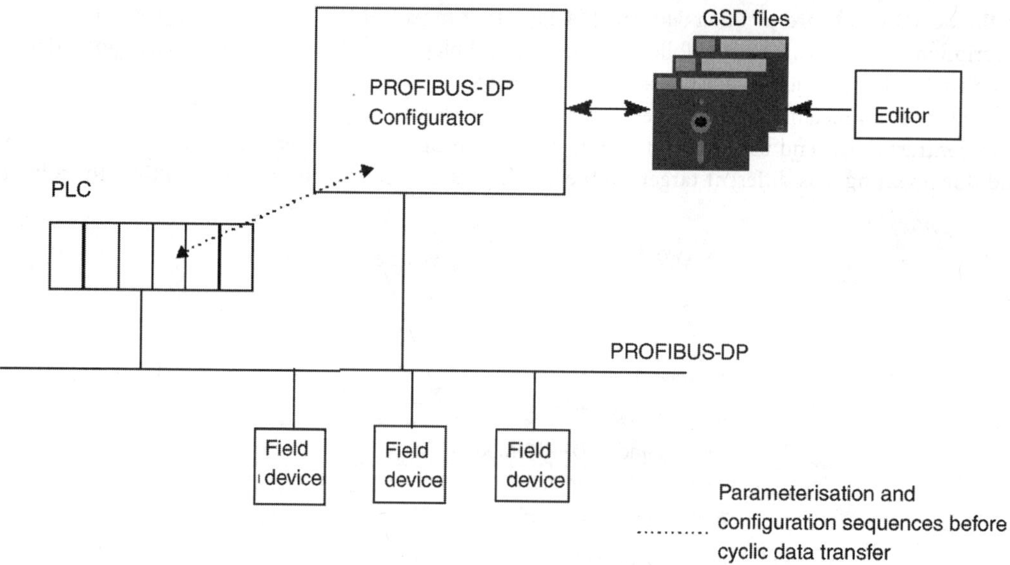

FIGURE 71.8   Tools and handling of GSD.

interoperability problems but to enhance the engineering effort to design the optimal application with regard to the quality of the product to be produced with the process control system. The effectiveness of his work, comprising all preproducing phases of the plant/machine life cycle, depends directly on the support of a well-defined, standardized and a complete machine-readable description of the devices he has to deal with.

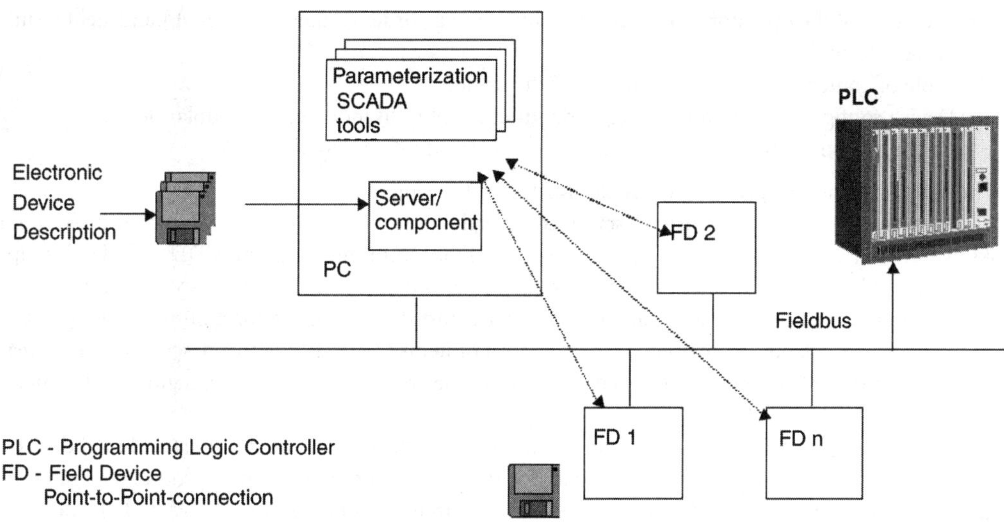

FIGURE 71.9 Using EDD in parameterization tools.

The conclusions from this analysis are that the Device Description Language has to be designed mainly from the system integrator's point of view, because he has to deal directly with the electronic device description. The fieldbus community developed some approaches to exchange data by electronic means. This includes DDL (e.g., for HART, FF, and Electronic Device Description for PROFIBUS).

The main features of this description technology, including the according language, can be summarized as follows:

- An EDD (in terms of a file) is delivered by the device vendor together with the device.
- EDD is used in the engineering process of the distributed control system, supporting planning, commissioning, operation, diagnostics, and maintenance.
- There are different EDD presentations possible, source in a human-readable way and binary format.
- The EDD is mostly stored on disk and can additionally be stored within the device (transport via fieldbus).
- The EDD is nearly independent from the underlying fieldbus system.
- The EDD is used to describe information identifying each item and defining relationships between them (hierarchical, relational).
- The EDD offers language elements for presentation within an HMI and for communication access.
- An EDD-file represents a static description, that is, the declarative part of the device; therefore, only the external interface/behavior of the device is described (no interest in internal code).

A parameterization tool needs special adaption to functions and parameters of each device, which should be commissioned or visualized (Figure 71.9). The EDD contains all device functions and parameters. The functions and parameters differ between the devices, but they are described with a defined language. The tools understand this language and adapt themselves to the described functionality. Adaptation means modification of the screen outline (e.g., content of the menus and bars), and the interactions with the devices. Generally speaking, the devices will be purchased with disks containing their own device description, and the tools get their specific configuration by reading and interpreting this device description. The main parts of the EDD are as follows:

- Description of variables and parameters including their attributes (Process Variable — PV with label for screen prints, e.g., Level hydrostatic, data type floating point, access right read/only, default value = 0).
- Presentation of variables in HMI tools (e.g., PV with label "Temperature").

- Guidance of the operator during commissioning (e.g., order of menu entries at hand-held terminals or PC tools).
- Table of content of all visible elements of the device.
- Device configuration parameter (e.g., identifier for plug-in modules of modular devices).
- Communication configuration parameters (e.g., baud rate = 500 kBaud).

The following example gives a short look-and-feel of the language:

EDD is a formal language used to describe completely and unambiguously what a field instrument looks like when it is seen through the "window" of its digital communication link. EDD includes descriptions of accessible variables, the instrument's command set, and operating procedures such as calibration. It also includes a description of a menu structure that a host device can use for a human operator. The EDD, written in a readable text format, consists of a list of items ("objects") with a description of the features ("attributes" or "properties") of each. Some example fragments from an (imaginary) flowmeter EDD are shown in Figure 71.10.

The major benefit of DDL for suppliers is that it decouples the development of host and field devices. Each designer can complete product development with the assurance that the new product will interoperate correctly with current and older devices, as well as with future devices not yet invented. In addition, a simulation program can be used to "test" the user interface of the EDD, allowing iterative evaluation and improvement, even before the device is built.

For the user, the major benefit is the ability to mix products from different suppliers, with the confidence that each can be used to its full capacity. Easy field upgrades allow host devices to accept new field devices. Innovation in new field devices is encouraged. The EDD is restricted for the description of a single device and use in a mostly stand-alone tool.

Software tools for automation are very complex and implements a lot of know-how. The number of sold products is relatively small in comparison with office packets. The definition of standardized DDL increases the potential users of the tools and speeds up the use of fieldbus-based automation.

Programming of the Control Applications with Integrated Field Device Functions

The Proxy-Concept organizes the functional and data relations between field devices and programmable controllers (e.g., PLCs; Figure 71.11). Field devices implement some signal processing, such as scaling or limit check, which was a part of the programmable controller libraries using the 4–20 mA technology. The consequence of this change is an interruption of the sequence of functions, which have already been carried out in one compact PLC program. This program has been developed with one software tool. Now different devices with different resources, which are scheduled in an asynchroneus manner, carry out the functions and it is unclear if the sequence order works in the necessary way. Additionally, PLC and field devices are programmed with different tools. The IEC 61499 (IEC 61499, 2001), PROFInet (PROFInet, 2002), and IDA (IDA, 2001) standards provide means to address the problems; however, these standards are not yet available on the market.

It would be favorable if the PLC programmer can use the decentralized functions in its normal programming environment with all the characteristic features. This would be possible if there is a device function block in the PLC library. This device function block represents the field device as its data input and output interface. The internals of this device function block are manufacturer specific. This concept of device function blocks of remote functions is known as the proxy concept. The communication between the proxy and the field device is carried out by cyclic (for process variables) and acyclic (for device parameterization, e.g., batch) means, that is, communication function blocks of the PLC. If these communication function blocks are standardized and the PLC IEC 61131-3 [11] compliant, the device function blocks are portable in an easy manner. Portable device function blocks mean that the communication function blocks are PLC manufacturer independent at its interface and main behavior. The specification of IEC 61131-5 that specifies such communication function blocks does not fulfill the requirements for an interoperable data transfer and is not well accepted in the market. Therefore, PROFIBUS has built its own specification, which is based on the IEC [xx].

```
VARIABLE v_low_flow_cutoff
{
LABEL LowFlowCutoff,
HELP „The value below which the process variable will indicate zero,
to prevent noise or a small zero error begin interpreted as a real flow rate".
TYPE FLOAT
{
IF v_precision = high
DISPLAY_FORMAT „4.61",
ELSE
DISPLAY_FORMAT „4.21",
}
CONSTANT_UNIT „%",
HANDLING READ & WRITE;
}

MENU m_configuration
{
LABEL „configuration"
ITEMS
{
v-flow_units,                 /*variable*/
range,  /*edit-display*/
v_low_flow_cutoff,            /*variable*/
m_flow_tube_config,           /*menu*/
m_pulse_output_config         /*menu*/
}
}

COMMAND write_low_flow_cutoff
{
NUMBER 137;
OPERATION WRITE;
TRANSACTION
{
REQUEST
{
low_flow_cutoff
}
REPLY
{
response_code,
device_status,
low_flow_cutoff
}
}
RESPONSE_CODES
{
0. SUCESS,              [no_command_specific_errors];
3. DATA_ENTRY_ERROR,    [passed_parameter_too_large];
4. DATA_ENTRY_ERROR,    [passed_parameter_too_small];
6. MSC_ERROR,           [too_few_data_bytes_received];
7. MODE_ERROR,          [in_write_protect_mode];
}
}
```

FIGURE 71.10 Device Description example.

Proxies, that is, device function blocks, are additional components that have to be provided by a field device manufacturer. This is a consequence of the increase of device functionality, the decentralization of the system and its engineering, and of the resulting distributed system. Proxies are a very good basis for manufacturer- and user-specific functionalities, which can be implemented as PLC programs

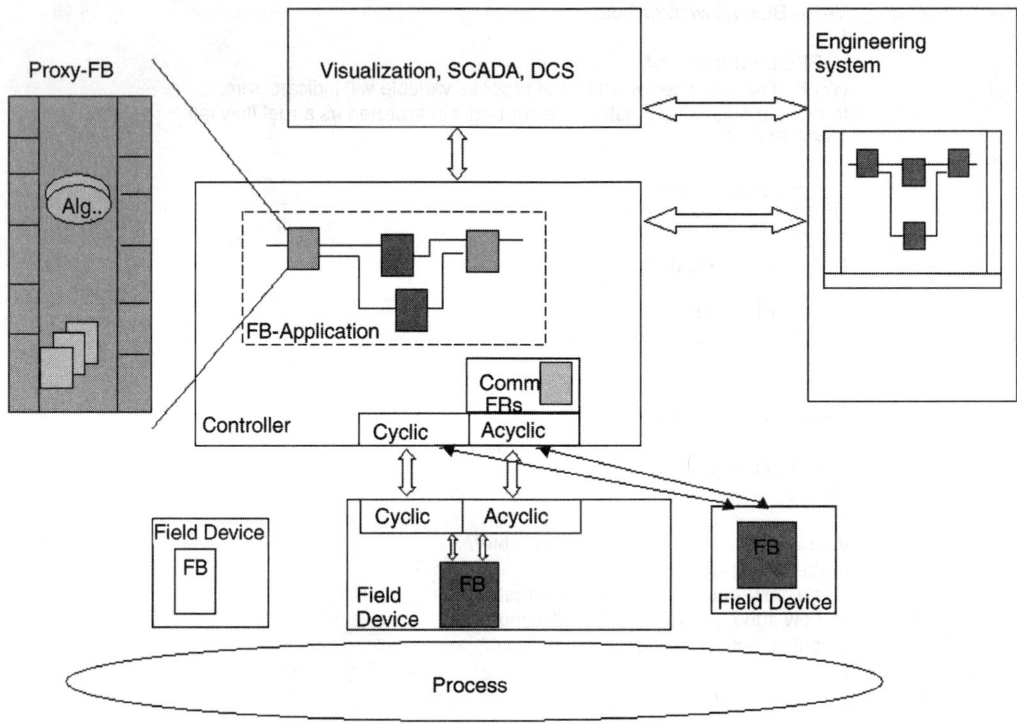

FIGURE 71.11   Proxy-concept based on a function block.

instead of embedded software in the field device. Therefore, more flexibility of the manufacturer satisfying user requests is possible. This flexibility can be implemented using different platforms of different PLC manufacturers.

Field Device System Integration

Common for the technologies described in the previous subsections is the limited scope of data in so-called "stand-alone" tools. The engineering and supervisory system (Figure 71.4) needs data from all devices and components in order to provide a noninterrupted information flow between the tools.

In order to maintain the continuity and operational reliability of process control technology, it is therefore necessary to fully integrate field devices as a subcomponent of process automation [2]. To resolve the situation, the German Electrical and Electronic Manufacturers' Association (ZVEI) initiated a working group in 1998 to define a vendor-independent FDT architecture. This FDT concept defines interfaces between device-specific software components (DTM, Device Type Manager) supplied by device manufacturers, and engineering systems supplied by control system manufacturers. The device manufacturers are responsible for the functionality and quality of the DTMs, which are integrated into engineering systems via the FDT interface. With DTMs integrated into engineering systems in a unified way, the connection between engineering systems (e.g., PLC applications) and inconsistent field devices becomes available. The FDT specification specifies what the interfaces are, not the implementation of these interfaces [2]. Figure 71.12 shows the FDT architecture. From the figure, we can see that DTMs act as bridges between the frame-application and field devices.

FDT Frame-Application

As one component of the FDT structure, the frame-application is supposed to manage data and communication with the device and embedded DTMs. According to the environment a DTM is running in, the frame-application can be an engineering tool, or even a web page. But here, we consider frame-application as an engineering environment that integrates field devices and their configuration tools and controls DTMs within a project. The project is a logical object to describe the management and controls, at least in the lifetime of device instances in terms of the DTM within a frame-application (PNO 2001a). In view of a DTM, interfaces such as IPersistPropertyBag (Connection Storage in Figure 71.12), IfdtCommunication (Connection Communication in Figure 71.12) specified in PNO (2001a), determine what the frame-application can do.

DTM

The DTM is the key concept of field device integration, with each DTM representing one field device. In addition, more than one device of a common type can also be represented by a single DTM object. Moreover, a device can be subdivided into modules and submodules, for example., a device with its remote I/O modules. If a device supports remote I/O channels (in the FDT concept, the I/O of a device is called a channel), it also acts as a gateway. For example, a gateway from Profibus to HART is logically represented by a gateway DTM that implements the operations for channel access. Each time a user wants to communicate with a certain device connected via an interface card between the communication section of the frame-application and the device, an instance of the corresponding DTM is asked to be built. During the user's operation on the device, the whole lifetime of this DTM object is controlled by the frame-application. This is the concept of COM technology. In fact, DTM is implemented as a Microsoft COM object, which will be dynamically loaded when the user intends to obtain information of the device or set parameters to the corresponding field device, and then released after the operations. All these interactions are carried out via interfaces specified in the FDT specification (Figure 71.13).

The interfaces IDtm and IDtmInformation specified in PNO (2001a) provide basic functions used by the frame-application to obtain information for the DTMs and execute certain operations on the corresponding devices. If the device represented by a DTM has its process values, another interface specified in PNO (2001a), IFdtChannel, acts as the gateway to deal with the connection to substructure, such as remote devices connected via Remote I/O channel to the master device. For some special tasks, such as documentation and communication, a DTM must implement the task-related interfaces if it supports the tasks.

The device manufacturer has multiple choices for the development of these DTMs. The DTM can be written manually in a high-level language, for example, in C++ or VB. It is also possible to generate DTMs out of

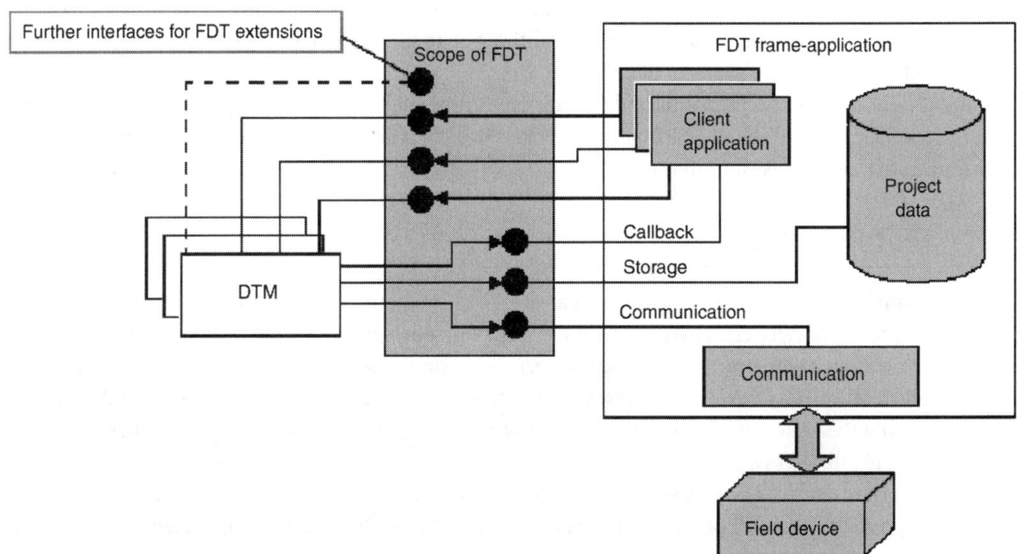

FIGURE 71.12   The architecture of Field Device Tools /PNO 2001a/.

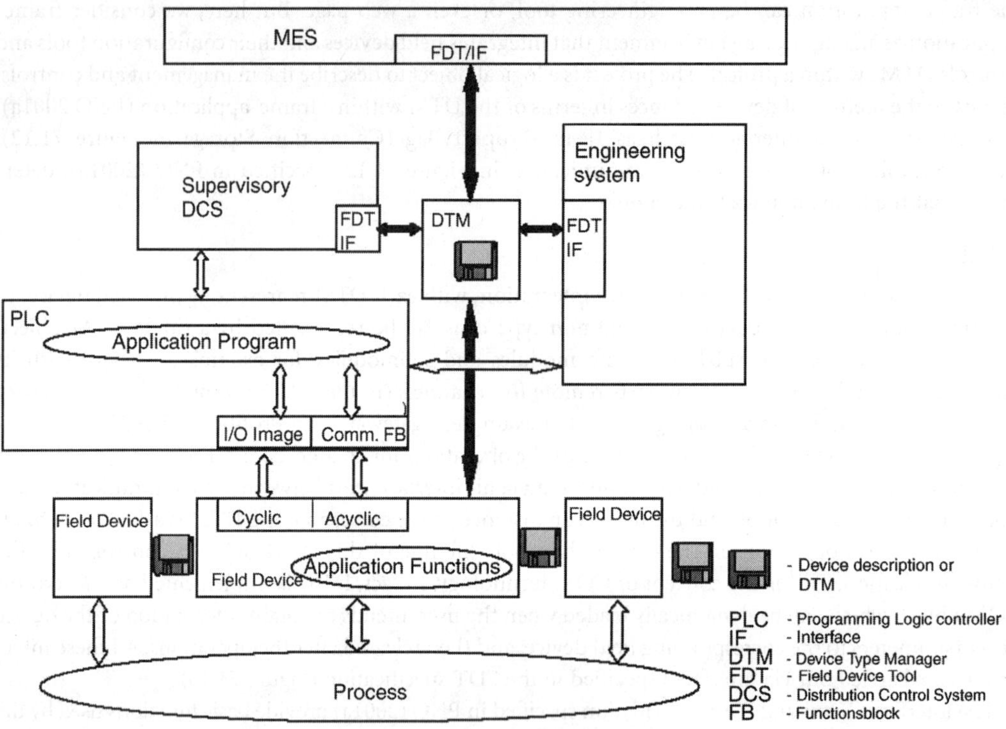

FIGURE 71.13 Integration of DTMs in the automation architecture using FDT interfaces.

existing device descriptions or provide DTM interpreting existing DTMs. Parameterization tools can be transformed to DTMs developing a wrapper around the tools that are conformed with the FDT interfaces.

FDT Interfaces

From the above parts, interfaces of frame-application and DTMs are mentioned and shown as the bridges between frame-application and DTMs. From the outside of an object, we can see only these interfaces that are specified according to their special functionalities but their implementation remains invisible and encapsulated.

With the help of FDT specifications and the corresponding interface technology, the user (engineering system manufacturer) will be able to handle devices and their integration into engineering tools and other frameworks in a consistent manner.

71.4 Fieldbus Profiles

Profiles are functional agreements between field device manufacturers to provide interoperable device functions of a certain device class to other field devices, controllers or DCS, SCADA, and engineering systems. Device classes have common functional kernels accompanied by the variables and parameters. These common set of variables, parameters, and functions are called profiles. Profiles reduce the degrees of freedom using the variety of the communication system and the choice of application functions of the device classes (Figure 71.14).

There are certain degrees of compatibility and accordingly degrees of cooperation between profile-based devices (compatibility levels). Compatibility levels are applicable for various roles of a device, for example, control, diagnosis, parameterization/configuration, and even applicable to subsets of its functionality. This means that one device can have different compatibility levels regarding different interfaces to the system. The levels are dependent on well-defined communication and application device features (Figure 71.15).

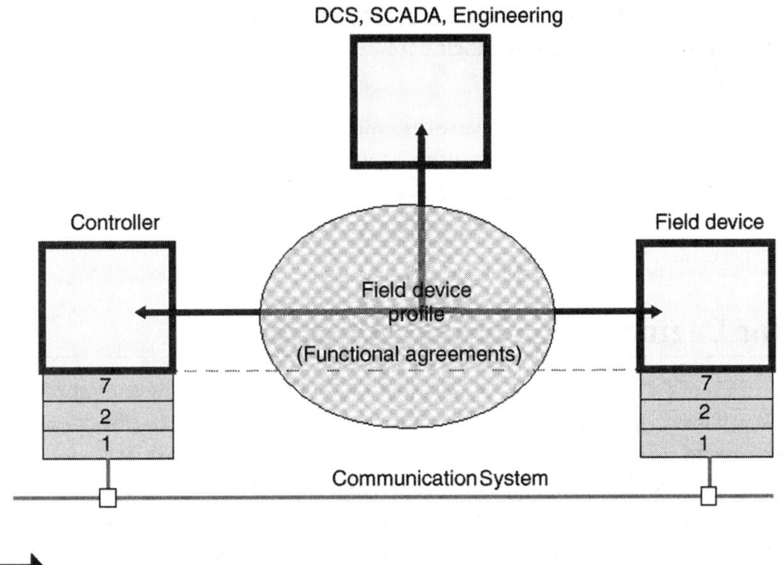

FIGURE 71.14 Profile in the automation architecture.

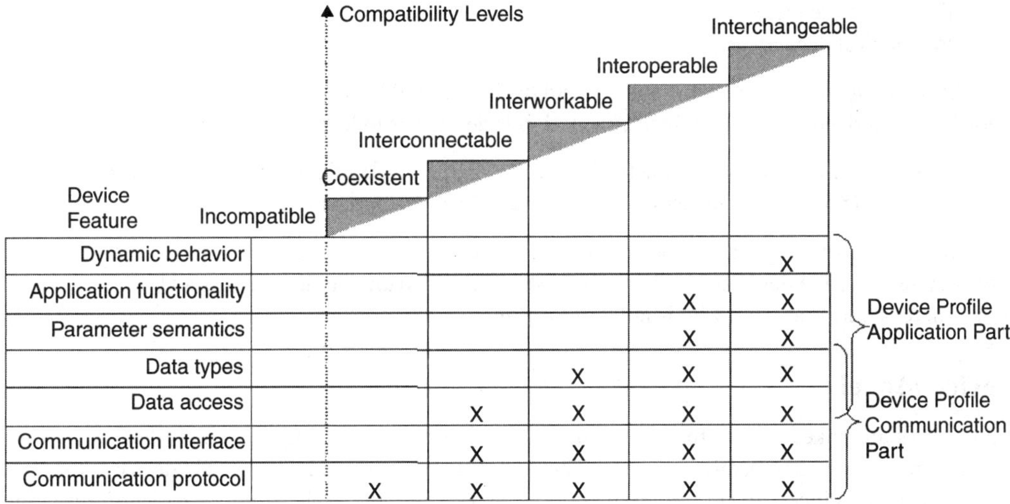

Device Feature	Incompatible	Coexistent	Interconnectable	Interworkable	Interoperable	Interchangeable	
Dynamic behavior						X	Device Profile Application Part
Application functionality					X	X	Device Profile Application Part
Parameter semantics					X	X	Device Profile Application Part
Data types				X	X	X	Device Profile Communication Part
Data access			X	X	X	X	Device Profile Communication Part
Communication interface			X	X	X	X	Device Profile Communication Part
Communication protocol		X	X	X	X	X	Device Profile Communication Part

FIGURE 71.15 Levels of functional compatibility (IEC 61804, 2003).

The device features are either related to the communication system as it is specified in the standard (e.g., protocols, service interfaces, data access, data types specified in IEC 61185, and IEC 61784) or related to the device application such as data types and semantic of the parameters, application functions, and dynamic behavior of the application. Profiles usually provide a mixture of compatibility levels regarding parts of the profile and their different users. For example, the main measurement value of a device is defined very precisely regarding data type, semantic including dynamic behavior; hence, devices are interchangeable for this measurement value. The same profile may skip to specify parameters that are used in the function chain from the electrical signal at the process attachment to the measurement value. Then, devices are not fully interoperable regarding their parameterization.

The main benefit of profiles are:

- The state-of-the-art functionality of device classes are specified including their parameter semantic. This makes it possible that human device users as well as tools find the same functions and parameters with the same names and behavior in devices from different manufacturers.
- It is possible to provide communication feature lists (eg., GSD), EDDs, DTMs, and proxies for device classes with profiles.

71.5 Model for Engineering and Instrumentation

The handling of the life cycle of DCS is a complex process that can only be done using sophisticated tools (hardware, software). Here, it is very important to design a noninterrupted life cycle, that is, to achieve a information transfer from one step to another step without losing information, and to ensure a single-source principle while putting information into the system. This information is used in each step to create a special view for the user. This cannot be done using paper documents for storing and transporting information. However, the usage of database management systems is not sufficient as long as open technologies are not applied. It is necessary to use databases and communication systems that follow a commonly agreed transfer syntax and standardized information models that ensure the meaning of the information (semantics). Basically, a connection between all tools in a DCS must be created (Diedrich, Ch. and Neumann, P., 1998a).

There are several possibilities to classify the life cycle/the engineering of DCS (Alznauer, 1998). Such classification can be made using:

- the hierarchy of the control functions,
- their timing sequence, and
- their logical dependencies.

All life cycle phases that are connected to field devices should be united under the rubric "instrumentation" and described as use cases. Instrumentation is defined as follows:

Instrumentation comprises of all activities within the life cycle of the distributed control system where handling of the field devices (logically or physically) is necessary.

Therefore, instrumentation can be considered as the intersection of the life cycle of the distributed control system and the field device. Figure 71.16 shows the instrumentation steps as a UML use case diagram. There is only one actor, who is not described in detail.

Device Model

Field devices are linked both with process, via I/O hardware/software, and with other devices via communication controllers/transmission media. The center of our attention is on the field device as the computational power is increasing rapidly as mentioned above. Thus, the applications are run more and more on these devices, and the application processes are becoming more and more distributed. We have to solve the problem of configuring and parameterization of these field devices during the operation for real-time data processing purposes, diagnosis, parameter tuning, etc. Therefore, there is a need to model such field devices (Diedrich and Neumann, 1998a). A field device can be characterized by

- internal data management (process I/O image, communication parameters, application parameters),
- process interface,
- information processing (e.g., Function Blocks),
- communication interface (Fieldbus, Ethernet-TCP/IP, etc.),
- (optional) man/machine interface (local display, buttons, switches, LEDs), and
- (optional) persistent memory and others.

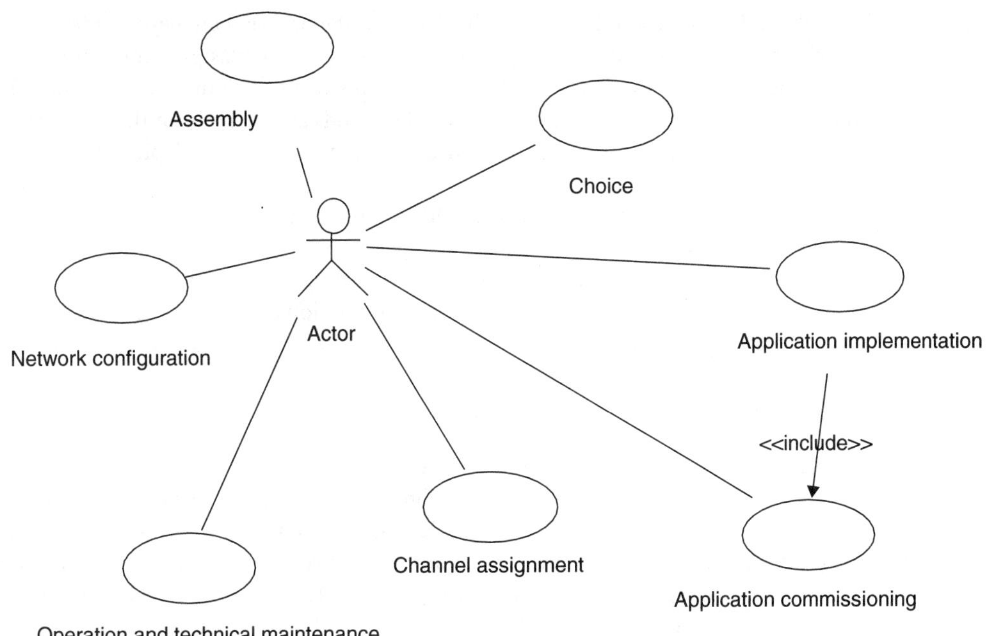

FIGURE 71.16 Use case Diagram Instrumentation (Simon, 2001).

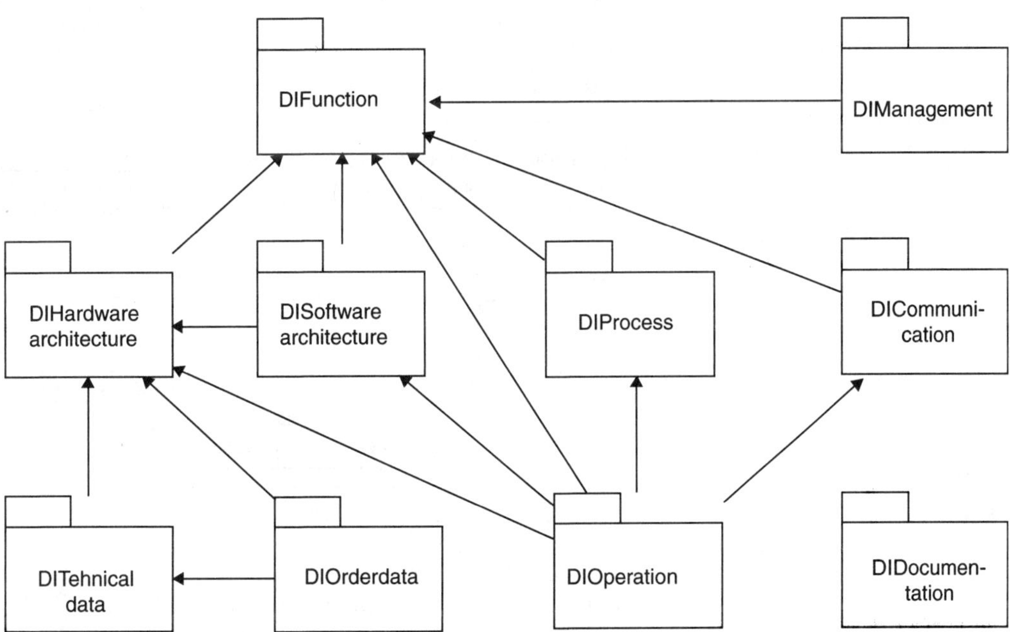

FIGURE 71.17 Device model related to the instrumentation (Simon, 2001).

We can define a device model as shown in Figure 71.17 (represented by UML packages), which supports the data exchange between instrumentation steps. This is a very abstract presentation of a device model.

The packages DIFunction, DIHardwareArchitecture, DISoftwareArchitecture, DIProcess, DICommunication, DIManagement, and DIOperation (DI stands for Device Instrumentation) are important. The packages depicted in Figure 71.17 contain the detailed model represented by UML class diagrams modeling the different views on a device. The details of the packages are beyond the scope of this chapter. However, the figures show the internal structure of elements (i.e, classes), for example., variable, function, and function block are described in detail in terms of its attributes and methods. The described relations between the classes and the attributes and methods are the basis for the development of the tools for the noninterrupted engineering and instrumentation. Figure 71.18 depicts the class diagram of the package DIFunction.

Figure 71.19 depicts the class diagram of the package DICommunication.

Figure 71.20 depicts the class diagram of the package DIOperation.

Simon, (2001) contains all other class diagrams needed for modeling the semantics of field devices as well as further explanations. This model has to be described by description languages to generate the basic information for an uninterrupted tool chain.

Description and Realization Opportunities

The device model can be implemented (realized) in several ways. It is possible to derive the FDT/DTM structure and interfaces, EDD, field device proxies, function blocks, and other technologies from this device model. For this paper, we chose EDD and an Extended Markup Language (XML)-based language. For the computable description of device parameters for automation systems components, the so-called Electronic Device Description Language (EDDL) has been specified (NOAH, 1999); (PNO, 2001b; Simon and Demartini, 1999). EDD is used to describe the configuration and operational behavior of a device and covers the following aspects (Neumann et al., 2001):

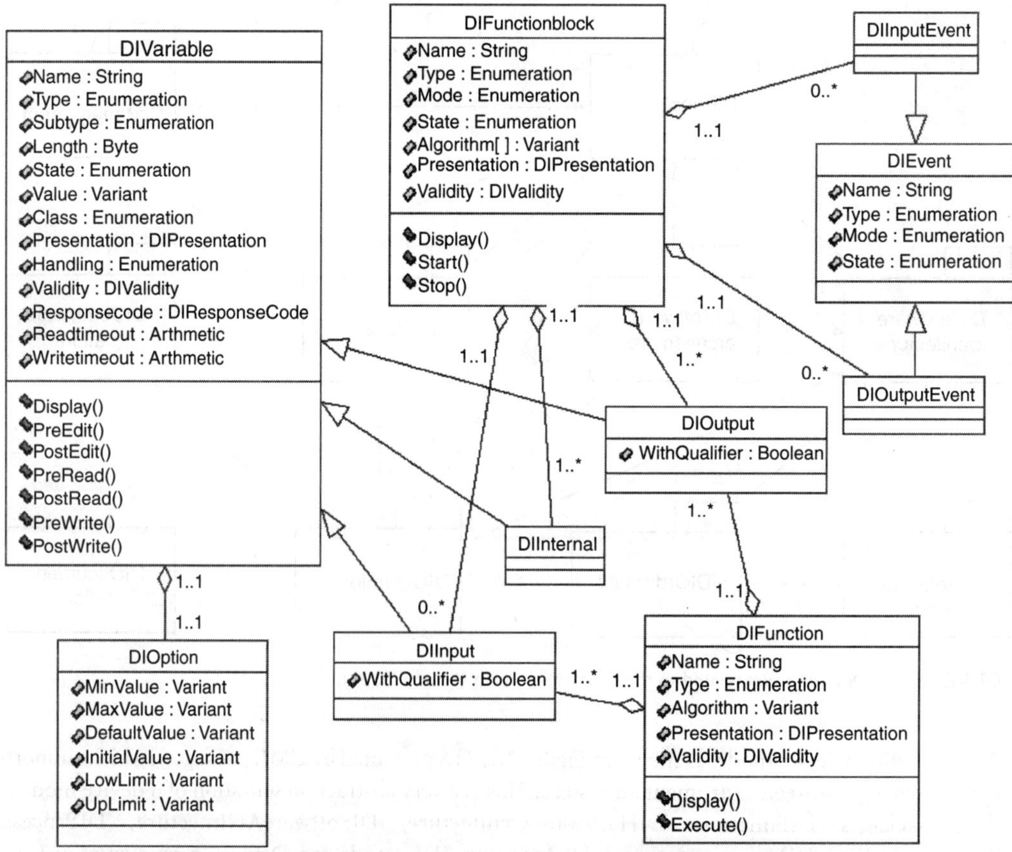

FIGURE 71.18   Package DIFunction (Simon, 2001).

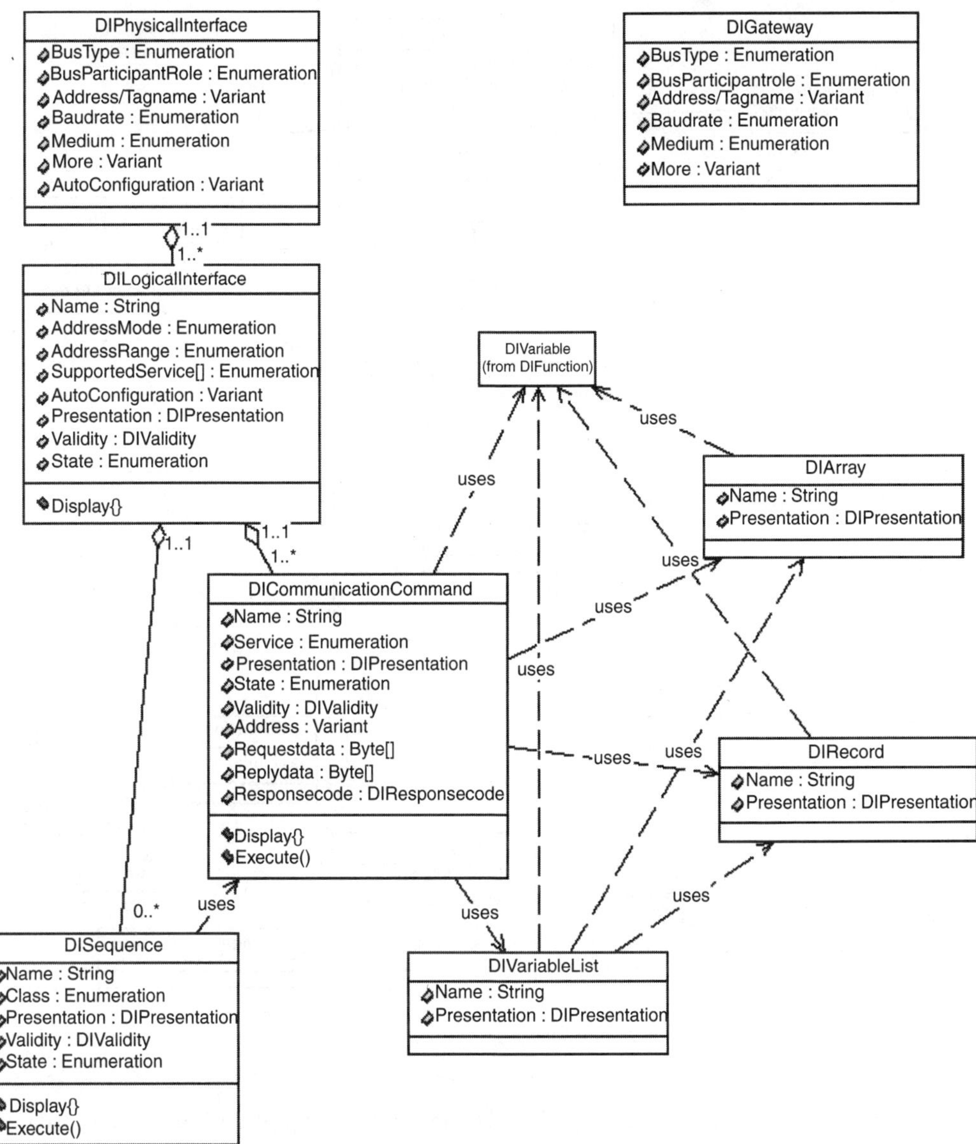

FIGURE 71.19 Package DICommunication (Simon, 2001).

- description of the device parameters, semantically defined by the field device model mentioned above,
- support of parameter dependencies,
- logical grouping of the device parameters,
- selection and execution of supported device functions, and
- description of the device parameter access method.

Overall Example Using EDDL

EDD is based on the ASCII standard. XML could be a promising approach for the future, especially because of its use in other areas. Both approaches contain definitions for the exchange of device descriptions using files. These definitions are not given here; however, a small example is used to show that different realizations can and must be based on the same solid foundation — the device model.

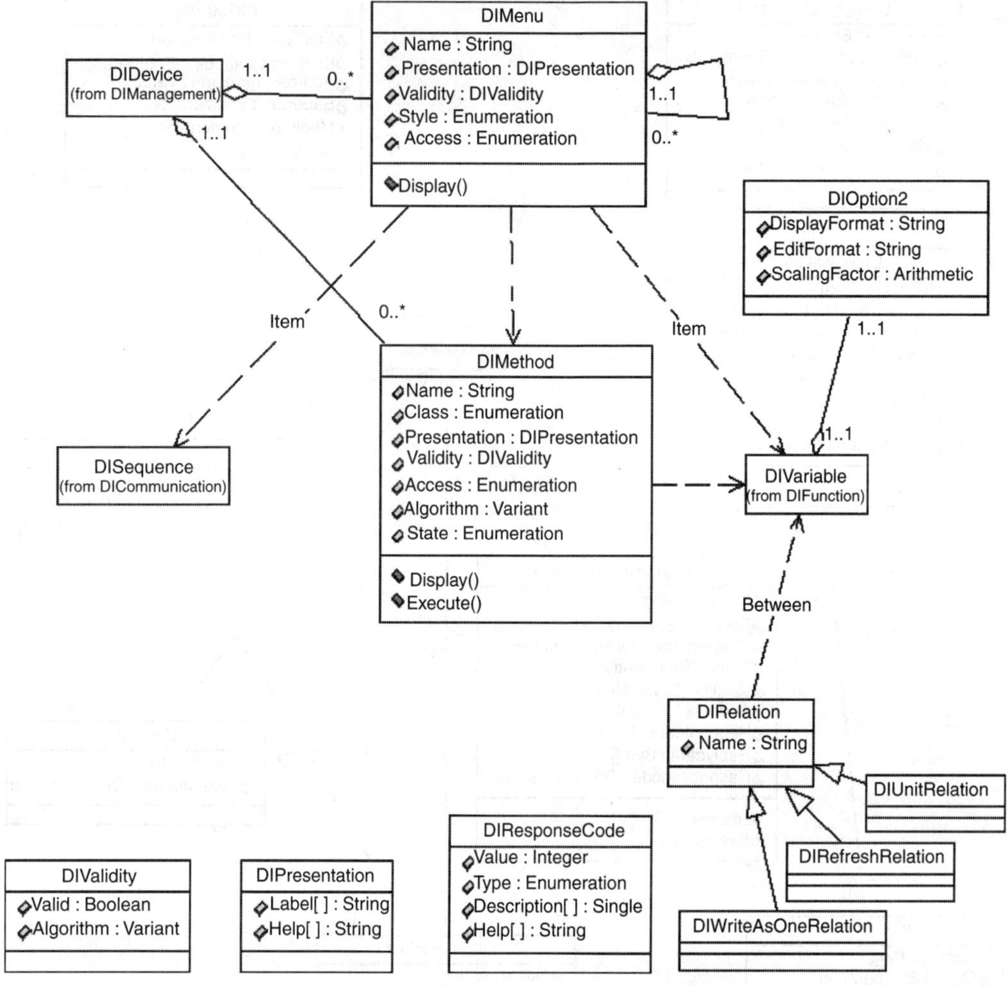

FIGURE 71.20   Package DIOperation (Simon, 2001).

The example comprises of a variable (package DIFunction), which is described by name, data type, label, and help.

The class DIVariableExample as shown in Figure 71.21 is the starting point.

The realization as an EDDL is described using language production rules shown in Figure 71.22.

A sentence created according to these rules may resemble Figure 71.23.

Using this definition of a variable, a commissioning tool provides the human–machine interface (Figure 71.24):

The XML Approach

The XML (Bray et al., 1998) expands the description language HTML with user-defined tags, data types, and structures. In addition, a clear separation between the data descriptions, the data, and their representation in a browser have been introduced. Furthermore, declaring syntactical and semantical information in a separate file (Document Type Definition, DTD) allows reusing the description structure in different contexts. This provides a number of benefits when using the same XML description file for different tasks. Different views can be implemented on top of the same data. The description can be hierarchically organized. Depending on the functions to be performed, the XML data can be filtered and

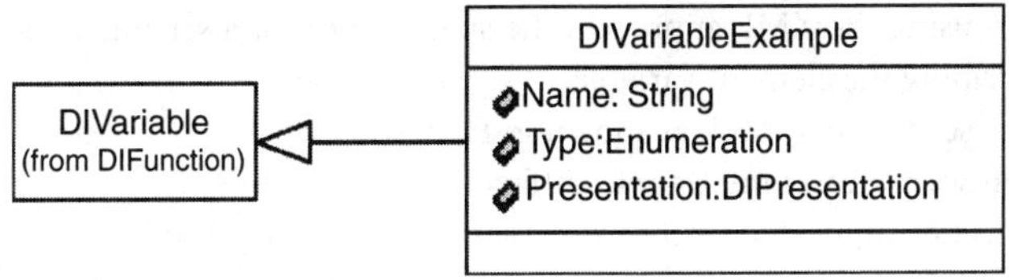

FIGURE 71.21 Class DIVariableExample.

variable
  =   *'VARIABLE'*   *Identifier*   '{'
variable_attribute_list '}'

variable_attribute_list
  = *variable_attribute_listR*

variable_attribute_listR
  = *variable_attribute*
  = *variable_attribute_listR variable_attribute*

variable_attribute
  = *help*
  = *label*
  = *type*

FIGURE 71.22 Language production rules (EDD).

```
VARIABLE     Temperature
{
    LABEL        "Temperatur";
    TYPE         DOUBLE;
    HELP         "Temperatur";
}
```

FIGURE 71.23 A variable definition (EDD).

FIGURE 71.24 Human–machine interface showing a variable.

associated to software components (controls, Java beans, etc.). The selection of the necessary information and the definition of their presentation details can be performed by means of scripts and style sheets. The style sheets are a part of the development of XML (Boumphrey, 1998). In most cases, they are implemented using the extensible Style Language (XSL). The XML file, the scripts, and the different style sheets can be used to generate HTML pages, special text files, and binary files (components, applets) necessary to build the certain functions of the software tools. The distribution of the generated HTML pages and associated software components is done following the concepts used in an Internet environment. The major benefit of this solution is a unique, reusable description with an excellent consistency and reduced efforts of the description process.

For the realization using the XML approach, the specification of a schema is necessary. Figure 71.25 shows part of it describing the element variable.

An instance of this schema may look as shown in Figure 71.26.

A standard web browser creates interface as in Figure 71.27.

The unit (Kelvin) is not supported by the example model. The help text is not visible, and the name of the variable is not used. Based on the data type, the value provided by the device is shown.

The presentation of the small example underlines the objectives targeted by the modeling approach. If the internal structures of different realizations are similar, that is, they follow the same field device model, then it is possible to build translators from one realization to another and to secure investments already done. Similar presentations with the same contents provide the opportunity to simplify training and education through previous recognition.

Device Descriptions are necessary for the integration of intelligent field devices in commissioning tools, maintenance tools, engineering systems, or MES/ERP systems. Device Descriptions comprise device models and presentations based on the models (e.g., ASCII files). The XML concept may help to extend the application scope of Device Description because it is becoming one of the basic technologies of the fast-growing Internet and various e-engineering activities. At present, we can observe a transition to XML-based Device Descriptions, which is characterized by the following issues:

- XML- and ASCII-based Device Description have to use the same device model to retain the semantics already developed by automation industry,
- new application functions supporting different phases of the life cycle of an automation system have to be defined using different views on the Device Description, and

```
<ElementType name="DIVariable" content="mixed' model="closed">
        <attribute type ="Name" required="yes"/>
        <element type="operation:DIPresentation" minOccurs="1" maxOccurs="1"/>
        <element type="Type" minOccurs="1" maxOccurs="1"/>
    </Element Type=>
```

FIGURE 71.25   Schema specification.

```
<DIVariable
        Name="Temperature"
        <operation:DIPresentation>
            <operation:Label
                    String="Temperatur">
            </operation:Label>
            </operation:Help
                    String="Temperature">
            </operation:Help>
        </operation:DIPresentation>
        <Type>
            <Arithmetic>
                <Integer
                </Integer>
            </Arithmetic>
        </Type>
    </DIVariable
```

FIGURE 71.26   A variable definition (XML).

Temperatur   42.0   K

FIGURE 71.27   Web browser showing a variable.

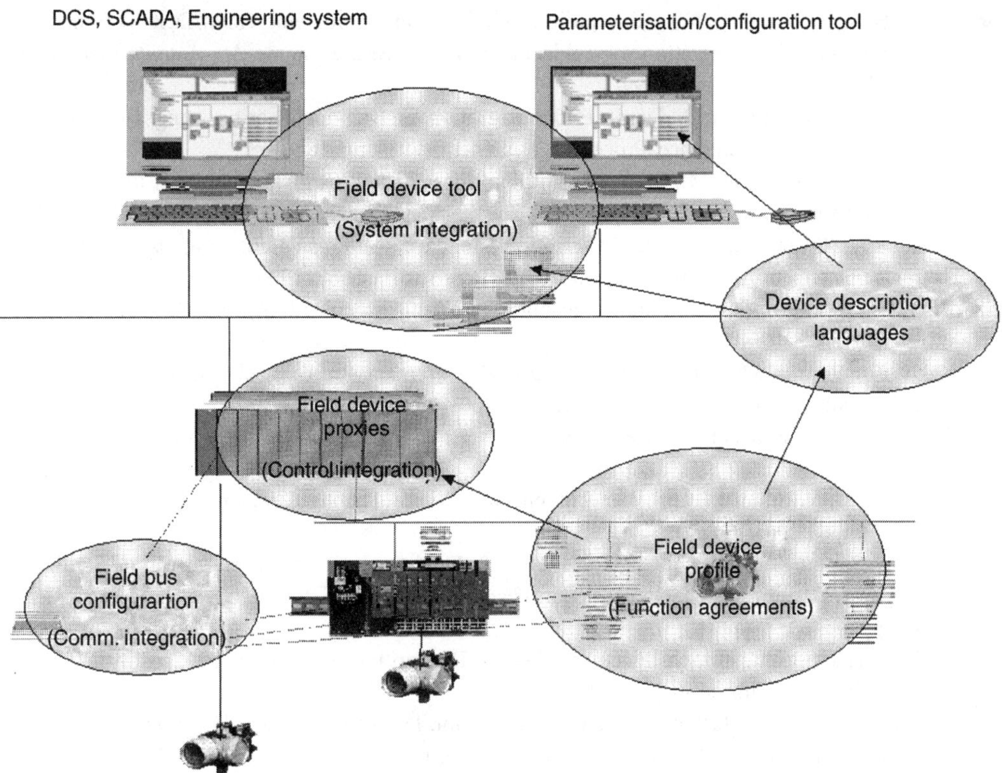

FIGURE 71.28 Integration technologies in automation systems.

- consequently, a new chain of tools for creating and using Device Descriptions has to be developed.

A necessary precondition is the international standardization in this area, which is currently done in the European standardization CENELEC and the IEC.

71.6 Summary

Field devices are integrated parts of the entire automation system. Therefore, new technologies arise to provide means for the device manufacturer, system integrators, and system suppliers to design, commissioning, operate, and maintain systems with a large variety of field device classes (Figure 71.28). These technologies are named integration/instrumentation technologies.

Field device profiles define common sets of functionality that are provided by devices of different manufacturers in an interoperable way. This is the semantic basis for the other technologies. The Bases for all integration are configured communication systems, which use feature list languages. The EDDL the field device proxies, and FDT/DTMs provide means to integrate device-specific and innovative features in the system. All these are integration/instrumentation technologies of field devices in distributed control and engineering systems.

References

Boumphrey, F., *Professional Style Sheets for HTML and XML*, Wrox Press, Birmingham, UK, 1998.
Bray, T., Paoli, J., and Sperberg-McQueen, C.M., Extensible Markup Language (XML) 1.0., http://www.w3.org/TR/REC-xml, 1998.

Bruns, H., Hempen, U., Ott, W., and Vahldieck, R., Ein Konzept für eine COM/DCOM-basierte herstellerunabhängige Integration von Feldgeräten in Engineeringsysteme, *Atp-Automatisier ungstechnische Praxis,* 41, 10, S.34ff, 1999.

Diedrich, Ch. and Neumann, P., Field device integration in DCS engineering using a device model, IECON'98, IEEE Conference, Proceedings, Aachen, 1998a, pp. 164–168.

Diedrich, Ch., Wollschlaeger, M., Riedl, M., and Simon, R., Three Component Model for Field Device Integration in Control Systems, IFAC/FET 2001, Nancy, France, Proceedings, 15–16, Nov., 2001.

FF, Foundation™ Specification, Device Description Language, Fieldbus Foundation, Austin, TX, 1996.

HART, Device Description Language Specification: HCF, Austin, Texas, 1995.

IDA, IDA — the Internet of Automation Technology, White paper V1.0, April 2001, www.ida-group. org.

IEC 61499, Function Blocks for Industrial-Process Measurement and Control Systems — Part 1 and 2, Public Available Specification (PAS), Geneva, 2001.

IEC 61804, Function Blocks for Process Control — Part 2, Committee Draft for Vote (CDV), Geneva, 2003.

IEC Guideline, Preliminary Joint Working Group: Device Profile Guideline, IEC Document for Comment, 65/290/DC, 2002.

Neumann, P., Simon R., Diedrich, Ch., and Riedl, M., Field Device Integration, 8th IEEE International Conference on Emerging Technologies and Factory Automation, ETFA 2001, Proceedings, Antibes, 2001, pp. 63–68.

NOAH, Language Specification of Electronic Device Description, Deliverable 321, Network Oriented Application Harmonisation (NOAH), 1999.

PNO, Specification for PROFIBUS Device Description and Device Integration, Vol. 3, Field Device Tool, Version 1.2, PROFIBUS Guideline, PROFIBUS User Organisation, 2001a.

PNO, Specification for PROFIBUS Device Description and Device Integration, Vol. 2, Electronic Device Description, Version 1.0., PROFIBUS Guideline, PROFIBUS User Organisation, 2001b.

PROFInet, PROFInet — more than just Ethernet, Broschure www.profibus.com, 2002.

Simon, R., Methods for Field Instrumentation of Distributed Computer Control Systems, Ph.D. thesis, Otto-von-Guericke University Magdeburg, 2001 (in German).

Simon, R. and Demartini, C., Electronic Device Description, Proceedings of FET'99, Magdeburg, 23–24 Sept. 1999, Springer-Verlag, Wien, 1999, pp. 429–436.

72

Intelligent Sensors: Analysis and Design

E. Dekneuvel

University of Nice Sophia Antipolis

72.1 Introduction

Today, thanks to the advances in numerical processing and communications, more and more functionalities are embedded into distributed components with the charge for them of providing the right access to these services. Complex systems are then seen like a collection of interacting subsystems embedding control and estimation algorithms. The inherent "modularity" concept underlying this approach is the key answer to the increasing complexity of the systems and this has led to the definition of new models and languages for the formal specification of the components [Medvodovic-00]. In this chapter, we are more particularly interested in "intelligent sensors," components associating computing and communication devices with sensing functions [Staroswiecki-96]. In order to reduce the complexity, the design of an intelligent sensor requires the necessity to provide a model of the sensor at a high level of abstraction of the implementation. The disparity of the knowledge encapsulated within the instrument renders the modeling process very sensitive to the modeling strategy adopted and to the models used. A real-life component like the intelligent instrument usually involves the cooperation of three kinds of programs [Halbwachs-93]:

- A level of data management to perform transformational tasks.
- One or more reactive kernels to compute the outputs from the logical inputs, selecting the suitable reaction (computations and output emissions) to incoming inputs.
- Some interfaces with the environment to acquire the inputs and process the outputs. This level includes interrupt management, input reading from sensors, and conversion between logical, and physical inputs/outputs. Communication with the other components of the system will also be managed at this level.

Data management covers research fields like the probability theory, the possibility theory, the measurement theory, and uncertainty management. Unlike a numeric sensor that provides an objective quantitative description of objects, a symbolic sensor provides a subjective qualitative description of objects [Benoit-2001]. This qualitative description, adapted to the sensor measurement, can be used in

Knowledge-Based Systems (KBS), verifying the validity of a measurement or improving the relevance of a result [Dekneuvel-92]. The reactive part is probably the most difficult part of the design of the intelligent sensor. Like all reactive systems, the intelligent sensor must continuously react to its environment at a speed determined by this environment. This often involves the ability of exhibiting a deterministic behavior, allowing concurrency, and satisfying strict real-time requirements.

A generic intelligent sensor model has been developed to help during the specification step of the sensor functionalities [Riviere-96]. The purpose of the intelligent sensor generic model is to provide a high level of abstraction of the implementation of the sensor, focusing on the fundamental characteristics that the sensor must exhibit. For this, the generic model uses the point of view of the user to describe the services and the operating modes in which the services are available [Staroswiecki-96]. Then, by using a language to compute the formal description, we are in a position to evaluate the component, from a static and/or dynamic point of view. The availability of a language to compute the formal model allows the evaluation of the component. Once the component is validated, a prototyping step can be launched in order to obtain an operational system prototype. This prototyping step being usually expensive in time and resources, the final implementation should be made as much as possible using automatic synthesis from the high-level description to ensure implementation that are "correct by construction" [Edwards-97].

In this chapter, after reviewing the main characteristics of the generic intelligent sensor formal model, we discuss an implementation of the model provided by the CAP language, a language specifically developed for the design of intelligent sensors.

72.2   Designing an Intelligent Sensor

Analysis

As stated earlier, the diversity of the embedded functions, the flexibility, and the reuse argue for a distribution of the functionalities inside a complex system into areas of responsibility [Harel-90]. From an external point of view, an intelligent sensor will be considered as a modular unit behaving as a server. As such, it will be designed to offer its customers (the operator, others instruments, or other modules) an access to the various functionalities encapsulated inside the sensor.

Let us consider the following simple example. A navigation system has to be designed using a closed loop on a surface like a wall to control the locomotion. As can be seen in Figure 72.1, the environment of the system to be designed exhibits various entities or *actors*, such as the axes, the operator, and the obstacles. Every occurrence of a *start_moving* request, a closed loop is activated until a new request like *stop_moving* is emitted by the operator. Once the links between the system and the environment are defined, the functional specifications can be established. For this, a dataflow diagram (see Figure 72.2) can be easily defined by identifying data necessary for the navigation goals: a position measurement value useful for the closed loop to compute the values of the speed to be applied on the various axes.

Suppose we now decide to include another activity that will enable the system to follow a predefined trajectory. In this way, the operator (or a high-level decisional system) is provided with a possibility to

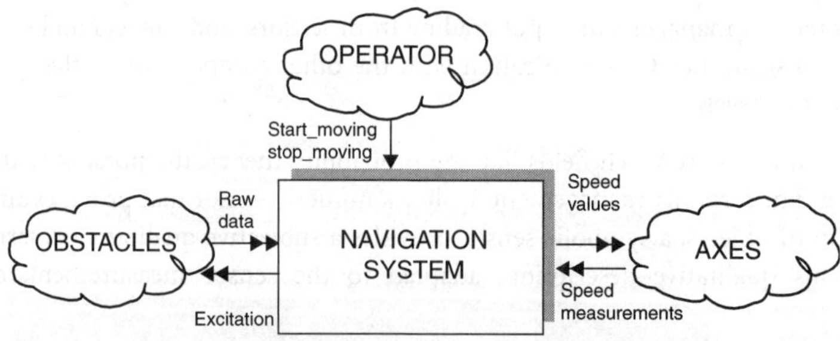

FIGURE 72.1   Context diagram of the application.

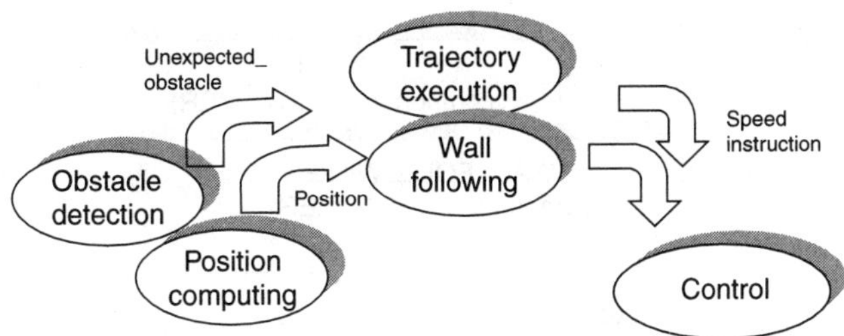

FIGURE 72.2 A global dataflow diagram.

choose between two methods according to the current context of the execution. This trajectory execution activity can be an interesting way of knowing if unexpected obstacles are encountered along this trajectory in order to stop the system before striking one of the obstacles. If both functionalities, the obstacle detection and the computing of the position, use the same physical resource (a set of ultrasonic sensors, e.g.), it is better to encapsulate them for homogeneity inside a subsystem in charge of the physical resource. The intelligent sensor module interacts with the environment through several messages. This is the interface of the module. The structure of a message (its *signature*) is generally limited at this level to a list of parameters such as the sender identity, the communication medium used, and the contents. We usually differentiate between messages for data communication and messages for control. Control messages enable the customers to communicate with the sensor in a bidirectional way, using a client–server protocol (see Figure 72.3). The customer requests the launching of an activity through the request link. The customer receives an identification number for its demand and can be informed about the status of the request (activity launched, terminated, etc.), thanks to another message of control (reply).

To be effective, the intelligent sensor interface description must be complemented by the *behavioral* description of the module. While the structural viewpoint describes the internal organization of a complex system, the behavioral viewpoint will express all the information that characterizes the module to be designed from an external viewpoint [Calvez-93]. A generic model of an intelligent instrument has been developed for this purpose, using the concept of *external services* to qualify the set of operations offered to the outer entities. Bouras [Bouras-98] gives the following definition:

Definition I: From an external point of view, a service is the result of the execution of a treatment, or a set of treatments, for which one can provide a functional interpretation.

In other words, the execution of a service typically results in the production of output values according to input values and consumed by the execution of a processing. The services are not limited to measurement aspects. The set of the services covers a large spectrum of functionalities that we can expect from intelligent sensors. Intelligent sensors must be configured, calibrated, and enabled, so that they can provide their measurements to the rest of the system. Selection of a particular sensor, an alimentation, a reference voltage and a sampling frequency are common examples of configuration services that can be used to set the value of these parameters. Processing embedded inside services can be as simple as the acquisition of a value but usually, it involves more complex treatments like signal processing (to improve, e.g., the resolution of a given value), the data processing, the validation of the measurement, and so on.

In the generic model of an intelligent sensor, a service will consequently be modeled by two sets of parameters:

External: that is, how the service communicates with other services. The services are gathered into User Operating Modes.

Internal: that is, how the external service is decomposed into internal basic processing units.

Let us examine in details both aspects successively.

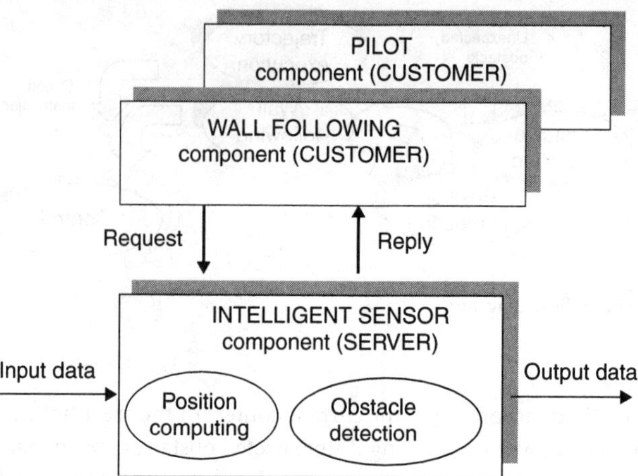

FIGURE 72.3   The intelligent sensor interface.

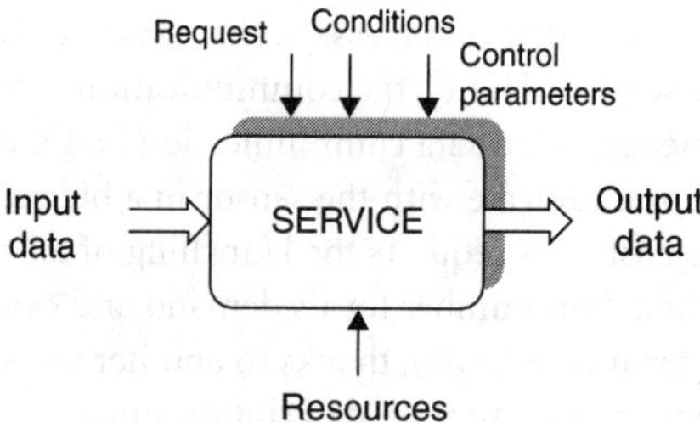

FIGURE 72.4   Graphic model for representing a service.

The External Model

Figure 72.4 depicts the external model of a service in use. A service is mainly described by the *input/output data* and is triggered by an external event. The used data and events can be organized into classes, with the description of a set of characteristics such as the format, the accuracy, the refresh period, etc., for each data class. The description of the input and output behaviors exhibits the possible interconnection between the service and those that precede or follow this particular service. This is then a data-driven representation of the service relationships equivalent to an explicit representation with the advantage of being more efficient and general (new services can be added without being obliged to physically interconnect the entire system).

Control parameters are received from the "parent" activity that requests the service. The control parameters affect the modalities of processing and the modalities of the underlying sensor that the service might encapsulate. The control parameters are usually passed in conjunction with the service activation request. For example, one can easily imagine the obstacle detection service running in a cyclic shooting modality or in a single shooting modality, depending on the nature of the activity that requests the service.

The launching of a service can be conditioned by the verification of its *activation conditions*. The distinction between a request and a condition is that the request for a service is emitted by the user, while the condition is processed by the system. These conditions are often related to the access rights or the security aspects and induce the verification of the origin of the request, the mode of transmission used, etc.

Resources include both hardware (sensor, CPU, memory, etc.) as well as software (extended Kalman filter, etc.) Input and output data can also be considered as resources with the problem of the data obsolescence.

Other properties can be added like *time and complexity measures,* which can help to select between different methods.

The set of services can be easily assimilated to a set of instructions we find in a regular computer. To cope with various states of the sensor that can occur during its life (out of order, in configuration, in manual exploitation, in automatic exploitation, etc.), the different external services are organized into coherent subsets of user operating modes (USOMs). In the model, a sensor service can be requested, and thus accepted, only if the current active USOM includes this service. This prevents the request of services when they cannot be available. According to Bouras [Bouras98]:

Definition II: An external mode is a subset of the set of external services included in the intelligent instrument. An external mode includes at least one external service and each service is included in an external mode at least.

The operating modes can be easily described thanks to a labeled transition system where the label is the external event matching the request of commuting the current mode (see Figure 72.5). Moreover, in each USOM, a notion of context may exist. The context is the subset of the services that are implicitly requested as long as the system remains in the given USOM. The external services included inside an USOM are supposed to behave independently. We say that they belong to orthogonal regions of a state, sometimes termed as *constellations*. As an example, the external services inside the intelligent sensor of the navigation system could be structured into a *wall following* or an *execution trajectory* mode. The *position computation* and *the obstacle detection* services would be implicitly executed when entering the corresponding mode. If the sensor reveals a complex state space, it can be decomposed into nonoverlapping substates to reduce the complexity. For example, the *wall following* and *execution trajectory* states can belong to a more general *measuring* state, often called a macrostate or a superstate, itself belonging to an *active* macrostate, etc. Some properties that the design must satisfy and that can be checked against the functional specifications have been elaborated. They complement the formal model. For example, properties may express some axioms such as:

1. An external mode is a nonempty set of external modes.
2. Each external service belongs to one external mode, at least.
3. In an intelligent instrument, the set of disconnected vertexes in the state-transition diagram is empty, that is, there is no external disconnected mode.
4. A transition between two modes is unique in the graph.
5. Each external mode must be reachable and each external mode can be left, etc.

These properties must be verified to guarantee a safety production of the intelligent instrument. For example, the verification of the property 4 can be easily done by checking if the fan-in and fan-out degrees of every vertex of the graph are not equal to zero. An incidence matrix can help in doing this.

The external point of view of the intelligent instrument will be usually complemented with a second level of description, to capture the algorithmic flow. This level must exhibit the treatments that concur to the global functionality, which is usually called *internal services.*

Functional Decomposition of a Service

Complex operations often need to be decomposed into multiple primitive operations in order to produce the overall behavior. For example, an external measurement service can induce a very complex treatment,

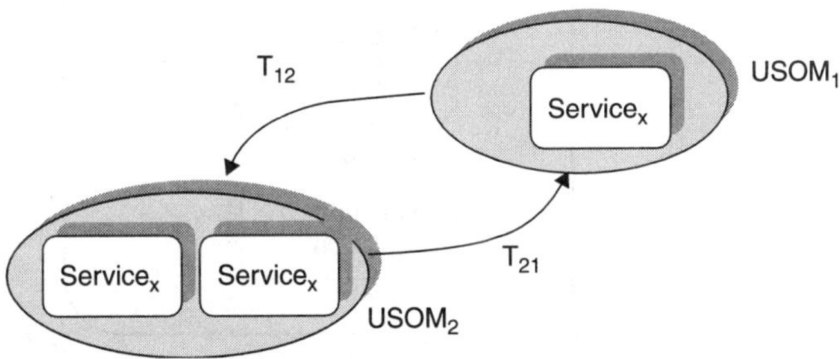

FIGURE 72.5 The Concept of USOMs.

following probably a step of initialization and, for self-terminated services, followed by a step of termination. Thus, Definition I is usually complemented with the following definition [Tailland-00]:

Definition III: An external service is the result of the execution of internal services.

From the point of view of the designer, an internal service is an elementary operation, possibly extracted from a library of components, for which no further decomposition is needed. Its I/O behavior can be easily described through an algorithm. Depending on the area of applications, such a conceptual unit can be known under various appellations such as a module, a codel [Fleury-94], etc.

The functional decomposition of a complex external service into internal services clearly has the following advantages:

- A structured programming helps the designer to describe the different steps of the treatment, without using internal state variables.
- Transitions from one step to an other explicitly define the possible interruption points of the service. Between these points, the operation is considered to be an atomic transaction. This preserves the functionality of coherency problems (loss of data and so on).
- Reuse of common units of programming: they are common to different services or are a result of previous developments. For example, an obstacle detection service and a position computing service can share several common portions of code: the signal emission, the signal acquisition, and so on. These units can be part of a library of IP modules.

As the reader can see, there is no mention at this level of the nature of the realization. Design units can be implemented on customized or software processors. The hardware units can also be freely implemented in the discrete domain using FPGA or DSP components, or by using analog components in the continuous domain. In order to reduce the complexity of the design, the definition of the executive architecture (a problem known as the *partitioning* problem) must be postponed during a detailed design step, taking into account various design constraints like economic and real-time constraints. The internal description of a complex service can be expressed using an activity diagram. The activity diagrams are well suited to show algorithmic details or procedural flow inside the service. They are often compared to flowcharts but they are more expressive. Such a diagram describes the internal operations to be achieved on the incoming flow and their temporal dependencies. Depending on the complexity of the service, the detailed refinement of an activity can be performed using several hierarchical successive levels. The processing step in the Figure 72.6 can, for example, be refined into a feature extraction step followed by a data classification step and a final decision step in sequential ordering. Elementary operations, those for which no refinement is needed, will be described by their internal behavior, usually through an algorithm. The activation of the internal services is controlled by *internal events*.

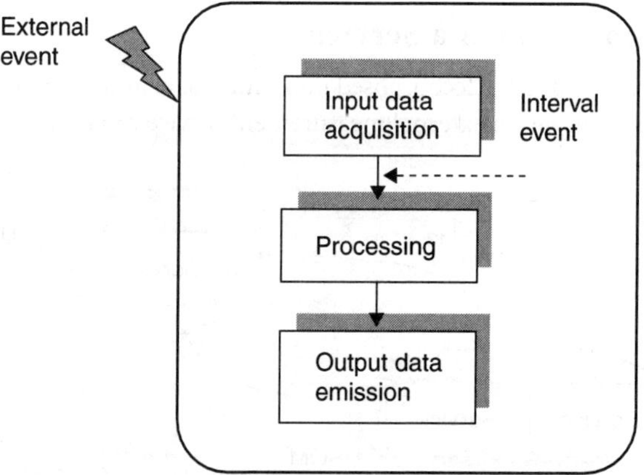

FIGURE 72.6   Example of a functional decomposition.

Definition IV: An internal event in an intelligent instrument is an event that is produced and consumed by the instrument itself.

The producer gives rise to the event. Consumers can react to this event in order to start their processing. The activation of an operation will often depend on the completion of the previous operation; but more complex temporal dependencies will also frequently occur like the activation of a signal processing operation conditioned to the end of an external conversion operation. In this way, an external service can itself be in the position of a client of another component, dynamically starting an external activity to request the data necessary to the achievement of its mission. While an external event is associated with a unique external service, an internal event can be associated with several internal services, leading to an *n producer–m consumer* relationship. In this, activity diagrams differ from conventional flowcharts relative to their capacity to represent concurrency. In Figure 72.7, the execution of the internal service V is followed by the simultaneous execution of the internal services W and Y.

Expressing sequential and parallel compositions of treatments is not always sufficient. The execution of a service can be affected by the state of the resources. The concept of version has been created with the aim of providing alternative versions of treatments that will enable the service to operate under nonnominal conditions. This is a means to take the fault-tolerance problem into account. All the versions of a given service will share the same request and produce the same output; but the inputs, procedures and/or resources will differ from one version to another. For example, a measurement service uses two transducers in a *nominal* mode of the service, to compute a data value using a sophisticated data analysis method. If a defect is detected in one of the transducers, the measurement service can continue to operate using a subset of the features extracted from the input data. Of course, the quality of the result will decrease. The versions are typically ranked and classified into internal modes like the nominal mode and the degraded mode. The management of the versions of a service can be straightforward: at time *t* when the request for service is emitted, the version to be carried out will be the one with the lowest rank whose resources are all nonfaulty [Staroswiecki-99]. As for the USOMs, the description of the internal modes can be made using a state diagram.

Having reviewed the generic formal model of the intelligent sensor, let us now turn to some validation aspects of the sensor.

Sensor Architectural Design

We have seen the mathematical properties underlying the intelligent sensor generic model of computation. These properties can be efficiently used to answer questions about system behavior without carrying out expensive verification tasks. The formal validation generally uses an automata-theoretic approach, modeling

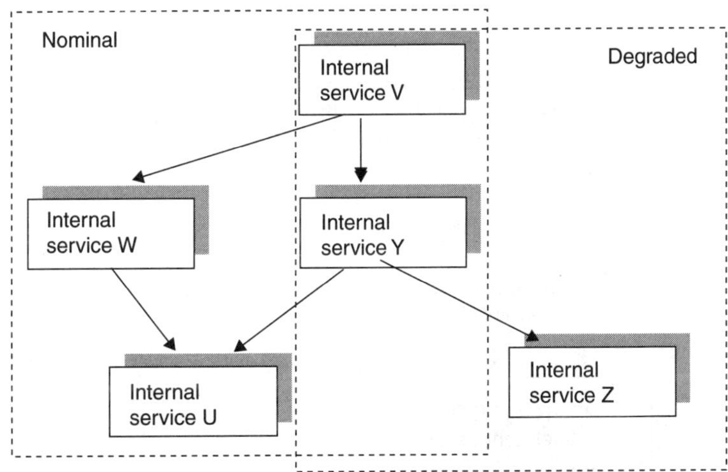

FIGURE 72.7　Version and internal mode concept.

the formal description by FSM and the language of the automaton [Kurschan-94]. As stated earlier, the final implementation of the intelligent sensor should be made as much as possible, using automatic generation from the generic model, to ensure implementations that are "correct by construction." For example, the protocol defined at the high level of abstraction uses the concept of message passing, where a message is an abstraction of data and/or control information passed from one component to another. Various mechanisms (the message signature) can be envisioned at a lower level of definition, including a function call, an interruption, an event using a Real-Time Operating System (RTOS), an ADA rendezvous, or a Remote Procedure Call (RPC) in a distributed implementation. Consequently, a prototype of the sensor is also a useful means to validate the specification in the presence of the real-time inputs, with physical characteristics similar to those of the final implementation and that will be produced by the synthesis stage. Rapid prototyping aims at analyzing the performance of an implementation, to validate its capability of satisfying hard real-time constraints, etc. To do so, the key technologies are the use of software synthesis, hardware synthesis, and the synthesis of interfaces between software and hardware using programmable components.

The prototype to be generated will be highly dependent on the physical architecture selected and the physical communication links. The targeted architecture is itself strongly dependent on the cost of components and production. Consequently, as shown in Figure 72.8, there are some nontrivial aspects to be analyzed, to be in a position to produce a prototype:

- The definition of the hardware/software architecture (partitioning, mapping) and
- the sequencing of the software on each software processor (scheduling).

Figure 72.9 shows a typical hardware/software architecture for an intelligent sensor. We can observe the hybrid character of the intelligent sensors, combining analog and numerical components. Each component description can be refined to exhibit the detailed architecture of a component. In the figure, we can see an architecture organized around a microprocessor that processes the functionalities for which it is in charge. This processor can be a Digital Signal Processor, a processor that has a CPU customized for data-intensive operations such as digital filtering. Bidirectional communication is ensured through various means, using a serial link, a Controller Area Network, (CAN) an Ethernet link, etc. [Warrior-97]. Finally, memories (ROM, RAM, etc.) ensure the memorization of the information located inside the sensor. In the future, the hardware architecture will tend to combine more and more customized hardware with embedded software. The definition of a hardware/software architecture involves checking if the sensor can be schedulable, that is, if all the performance requirements can be guaranteed. A deadline (a point in the time or a delta-interval by which a system action must occur) is an example of a requirement that, when missed, constitutes an erroneous computation. Consequently, the definition of hardware/software architectures generally requires more complex modeling, by defining the external timing requirements of the messages. The requirement

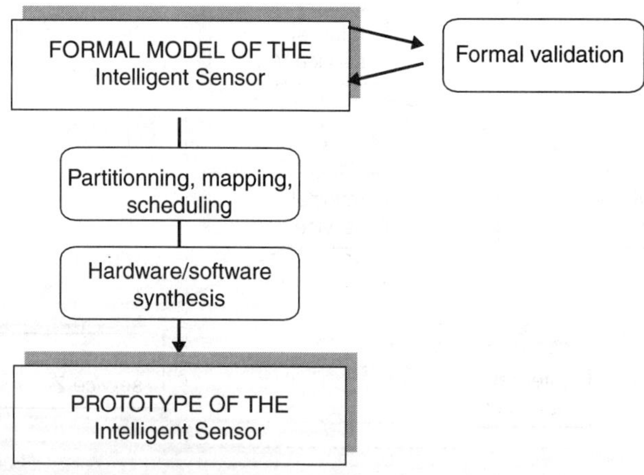

FIGURE 72.8   Design flow of an intelligent sensor.

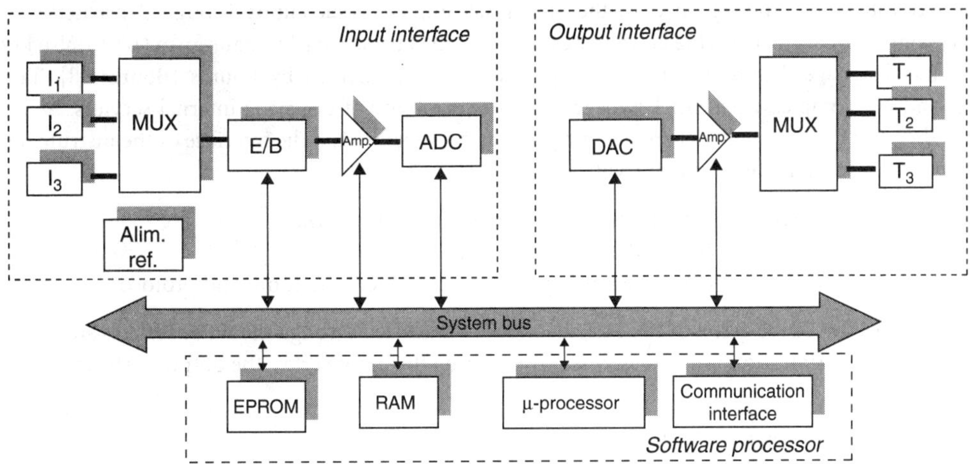

FIGURE 72.9　Typical model of an architecture for an intelligent sensor system.

of Quality of Service (QoS) of a message can be expressed using different means. For example, the response timing can be defined in terms of timeliness requirements (typically, deadlines) [Douglas-99].

Assigning an execution order to concurrent modules, and finding a sequence of instructions implementing a functional module are the primary challenges in software organization. These can be nontrivial issues to deal with, particularly when one must consider the performance, as well as the functional requirements, of the system. The software implementation can be facilitated by the use of real-time languages and their underlying executive kernel. Such languages provide a style of programming enabling the manipulation of events and/or state changes with constructions expressing the behavior through the parallelism, the synchronization, etc. Selecting a language to compute the model is not straightforward with basically several possibilities: developing a dedicated language or the use of an existing language, the use of a graphical or a textual input form, etc. The domain of applicability of a language must also be carefully studied with basically two approaches. The *synchronous* approach states that time is a sequence of instants between which nothing interesting occurs [Balarin-97]. In each instant, some events occur in the environment and a reaction is computed instantly by the modeled design. This means that computation and internal communication take no time. This hypothesis is very convenient, allowing modeling of the complete system as a single Finite State Machine with a totally predicable behavior. ESTEREL [Berry-92], [Esterel-www] and its graphical expression form, the SYNCCHARTS [Andre-96, Andre-01], are representatives of synchronous imperative programming languages. Like all language specialized to control-dominated systems programming, data manipulation cannot be done very naturally. In the asynchronous approach like in ELECTRE [Cassez-95], events are observed and processed immediately. This approach enhances the expression power of the language. Moreover, the design can be more efficiently implemented in heterogeneous hardware/software architectures. On the other hand, timing constraints are difficult to check.

72.3　The CAP Language

Description

The generic intelligent sensor model gave rise to a new language that can be akin to the category of asynchronous languages, the CAP language [Benoit-00]. Its ability of providing a rapid prototype of the intelligent sensor model on common microcontrollers available on the market has been one of the main reasons for developing this language. The developers have consequently limited the implementation to monoprocessors and to a sequential operation.

The CAP language is an incomplete language in the sense that it specifies only the interaction between computational modules (internal services), and not the computation performed by the modules. An interface with a host language specifies the behavioral contents of such units through C instructions. Like every conventional language, the grammar can be described using a backus naur form (BNF), also known as a metalanguage. Figure 72.10 shows a formal grammar defined by Bouras [Bouras-98]. Tailland [Tailland-00] complemented the language with the possibility of declaring internal services. As can be seen, the two parts of the model are normally successively expressed. The interface of the instrument follows the metavariable <instrument> with:

- A number referencing the instrument as a node in the network.
- Variables that can be exported or updated on the network.
- Communication links imported or exported to implement the command protocol.

The expression of the graph of the external modes is achieved through the set of vertices and the set of transitions. Each transition is then described by declaring the input vertex, the output vertex, and the link

```
EXTERNAL MODEL

<instrument> ::= <number_of_the_instrument>
                 <list_of_exported_variables> ::= {variables}
                 <list_of_imported_variables> ::= {variables}
                 <list_of_import_communication_links> ::= {name_of_the_link}
                 <list_of_export_communication_links> ::= {name_of_the_link}

<graph_of_the_USOM> ::=
        <list_of_the_external_modes> ::= {<name_of_the_mode>}
        <list_of_the_transitions> ::= {<transition>}

<service> ::=
        <name_of_the_service>
        <name_of_the_link of communication>
        <list_of_the_external_modes> ::= {<name_of_mode>}
        <list_of_the_internal_services> ::= {<iservice>}

<variable> ::=
        <name_of_variable>
        <type> :: <elementary type(C++ type)>

<transition> ::=
        <starting_mode> ::= <name_of_mode>
        <terminating_mode> :: <name_of_mode>
        <link_of_import_communication> ::= <name_of_the_link>

INTERNAL MODEL

<list_of_internal_events> ::= {<name_of_internal_event>}

<graph_of_the_internal_modes> ::=
        <list_of_internal_modes> ::= {<name_of_the_imode>}
        <list_of_the_transitions> ::= {<transition>}

<iservice> ::=
        <name_of_the_iservice>
        <condition_of_activation> ::= {<name_of_internal_event>}
        <list_of_internal_modes> ::= {<name_of_imode>}
        <treatments>

<itransition> ::=
        <starting_mode> ::= <name_of_imode>
        <terminating_mode> :: = <name_of_imode>
        <condition_of_fire> ::= {<name_of_internal_event>}
```

FIGURE 72.10   An overview of the formal grammar.

of communication source of the event of transition. Variables of the imported or exported lists are defined according to a type in C. Finally, the expression of a list of external modes in the definition of a service can be noted. These modes are the only ones in which the service can be launched. The description of the internal model is close to the external one and can be easily understood. As shown in Figure 72.11, the principle of the operation of the compiler is relatively straightforward: after the lexical and syntactical analyses, a set of lists reflecting the formal model is produced by the code generator. After being generated, the code is compiled and linked with the user-defined libraries and the processor-dependent startup code. The generated data structures are split into two sets of list: a first set of lists contains the names of the objects and is subdivided into eight lists: external modes, internal modes, external services, internal services, variables, export links, import links, and internal events. A second set of lists contains the detailed description of each transition. The conformity of the description to the formal model will be analyzed using these lists. They also contribute to providing a level of abstraction between the software synthesizer and the real machine.

Let us illustrate these principles in an intelligent instrument designed to process Ultrasonic sounds in order to produce a distance measurement value.

Illustration

To illustrate the approach, let us consider the following example [Tailland-00]: a measurement system is composed of one ultrasonic transmitter able to emit a signal toward a target in order to compute the distance between this target and the sensor with the help of two receivers. Figure 72.12 shows the basic configuration of the measurement system.

The intelligent instrument delivers the two measurements of distance d_1 and d_2, each of them with a validation degree d_{v1} and d_{v2}, respectively. As shown in Figure 72.13, the basic principle of the measurement is the following: the ultrasonic emitter sends a sinusoidal waveform linearly modulated in frequency to the $\Delta F = (f_{max} - f_{min})$ interval with a rate of variation of the frequency : $\alpha = \Delta F/T_r$. Then, the instantaneous frequency $F_{received}$ undergoes an offset relatively to the frequency $F_{emitted}$ of $\Delta f = | F_{emitted} - F_{received} |$. Mauris [Mauris93] has shown that the distance d can be determined by measuring the offset only inside the interval $[t_0, T_r]$, where its value is f_a. For this, the measurement is inhibited during the time t_a where the offset of the frequency has the value f_b.

The state space of the sensor is decomposed into two USOMs: the configuration mode (which is the default USOM) and the measurement mode (see Figure 72.14). The transition between each mode is triggered on the *cantcp_in* event.

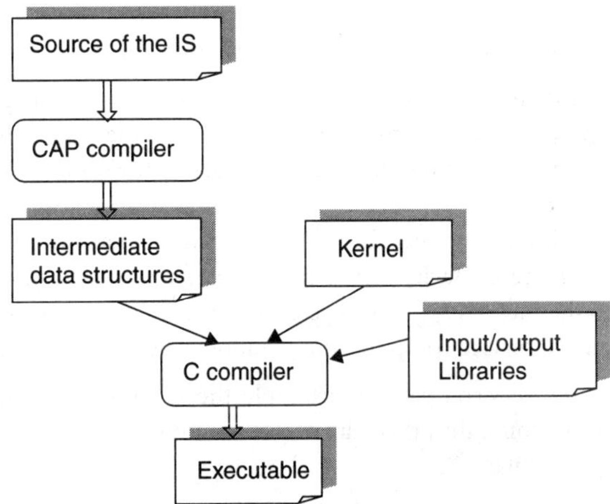

FIGURE 72.11   The CAP design flow.

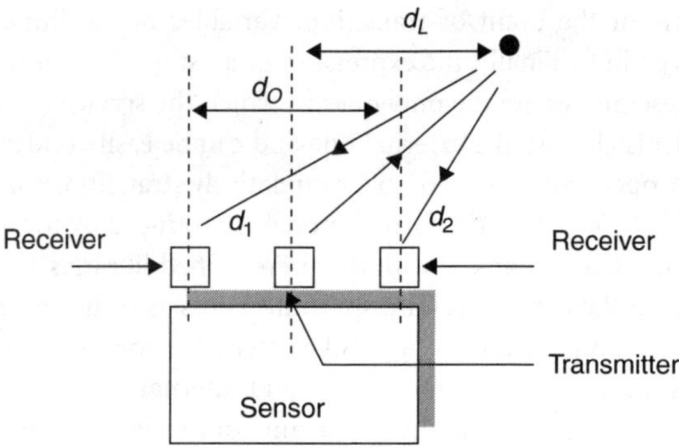

FIGURE 72.12   Configuration of the measurement system.

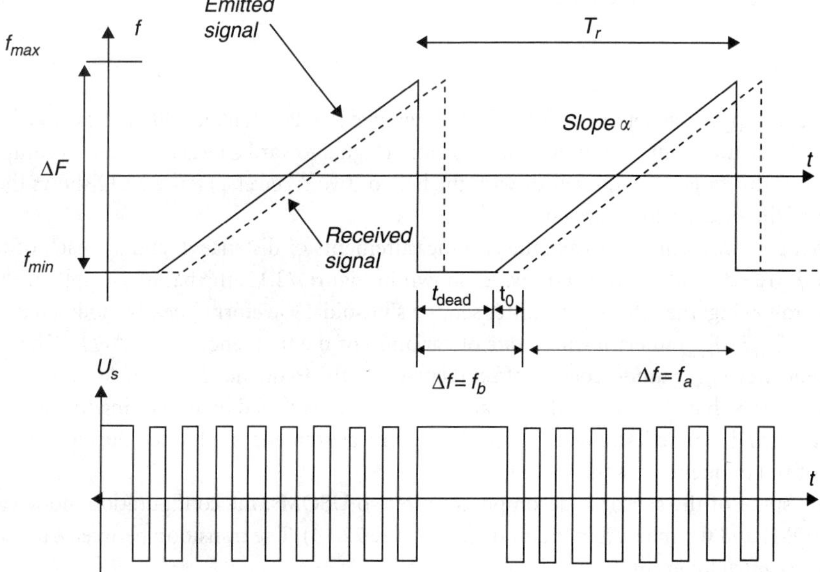

FIGURE 72.13   Principle for the measurement of f_a.

A number of internal services have been specifically developed for this application:

- Initialization and configuration services enable the modification of the parameter values used for the computation of d, duration of the slope T_r, slope p, time t_{dead}, voltage u_{min}. They take into account particular conditions of measurements (nature of the obstacle, environment, etc.).
- Internal services such as those depicted in Figure 72.15 contribute to the measurement elaboration; for example, the internal service for the *slope generation* has the responsibility of sending the signal in direction of the obstacle; *suppression of aberrant measurements* filters and keeps the impulsions close to the median f_{amoy} (average of the measured frequencies on T_r), while the others are discarded. The DPT service computes the pseudotriangular distribution of the possibilities.

Internal calculations are not very complex. For example, the *distance computation* service computes the distance on every f_a input. By doing this, there are N_{exp} measurements of d carried out during the interval T_r, using the following formula:

$$D = \frac{f_a}{2}\frac{T_r - t_{dead}}{\Delta F} * V \tag{72.1}$$

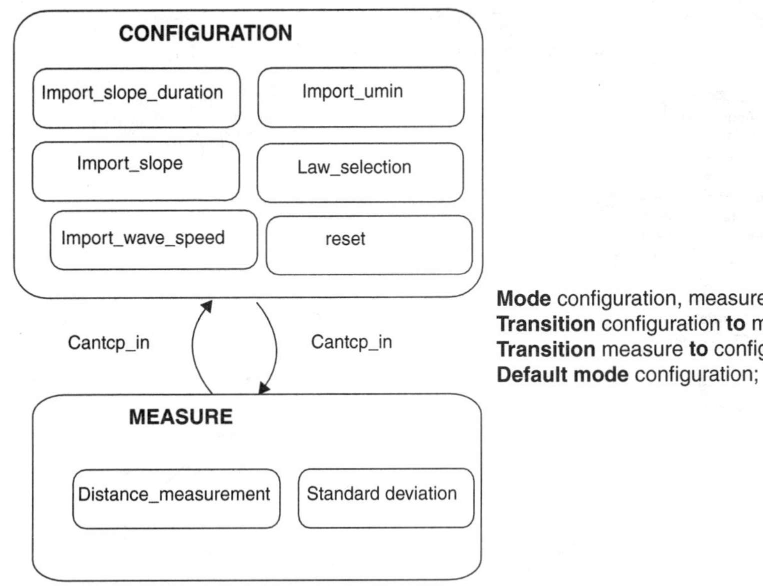

FIGURE 72.14   Graph of the USOMS and the corresponding CAP declarations.

The internal service named *validity estimation* computes a value in the [0,1] interval according to the following rule:

$$D_v = \frac{N_{exp}}{N_{th}} \qquad (72.2)$$

where N_{th} is the theoretical number of the periods of f_a that are supposed to be observed by the instrument.

As shown, the declaration of the *distance_measurement* external service includes the request (on *cantcp_in*) and the USOM (in *measure*) where it is available. The measurement functionality can be carried out according to four internal modes: *nominal, degraded1, degraded2,* and *critical.* For example, when the validity of the measurement d1 (see Equation [72.2]) decreases below a given threshold in the *nominal* mode, the instrument can automatically switch in the *degraded1* mode.

Part of the synthesized code (Figure 72.16) presents an excerpt of the synthesized code. As stated earlier, these symbolic data structures will be used by the verification tool to verify the properties of the formal model. The data structures (array of services, etc.) will be typically stored inside a volatile memory (RAM or SRAM) of the hardware architecture, while the automaton, not reachable to the user, will be set in a long-term memory (EPROM, FLASH, etc.). Depending on the operating mode, development, or exploitation mode, the user program can be stored inside the volatile memory or set in long-term memory.

Implementation

The execution of the various services is handled by an automaton that has the responsibility to interpret the formal model. As shown in Figure 72.17, the execution machine runs a cyclic program that processes the inputs, updating a FIFO storing the pending events. Then, depending on the nature of the event, a transition can be fired or a call to a procedure is achieved. Permanent functions are then executed. The loop ends with the emission of the output messages. For each step, the detailed behavior is as follows:

1. Reading a message on each communication link: if the intelligent sensor is concerned by the message, it will be processed according to its type:
   (a) In case of a *data* message, the corresponding variable is updated and the associated event is triggered.

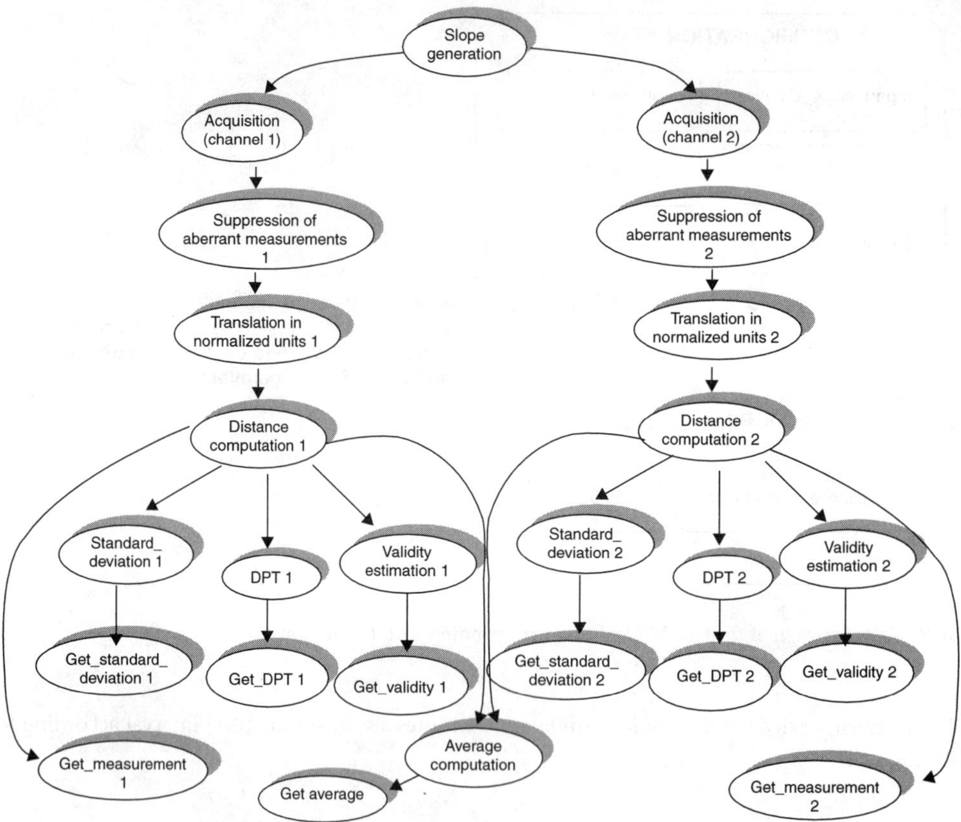

FIGURE 72.15   Functional decomposition of the measurement functionality.

    (b) In case of an *event* message, the corresponding event is inserted inside the FIFO event for an ulterior processing

    (c) In case of a *query* message, the current mode is exported on output links.

    (d) In case of a *variable* message, the variable is linked to import an external variable.

2. Event queue processing: three situations can arise and be analyzed in the following order:
    (a) The event is a request of a change of the external mode. Provided this change is authorized, the internal variable is updated and the new mode is exported on output links.
    (b) The same occurs for an event corresponding to a request of a change of the internal mode.
    (c) The event is an execution request of an external service. Provided the service(s) is (are) enabled in the current mode, these services are executed using the procedural entry point.
    (d) The same occurs for an event corresponding to a request for executing an internal service. If defined, an event noticing the end of execution is inserted into the FIFO event.

3. Execution of the EVER services: these are services that run permanently.
4. Diffusion of the messages on communication links (exported variables, change of mode, etc.).

We have to notice the reduced portability of the application, because of the tight link that exists between the application and the hardware. Open software platforms (EmbeddedJava, JavaCard, etc.

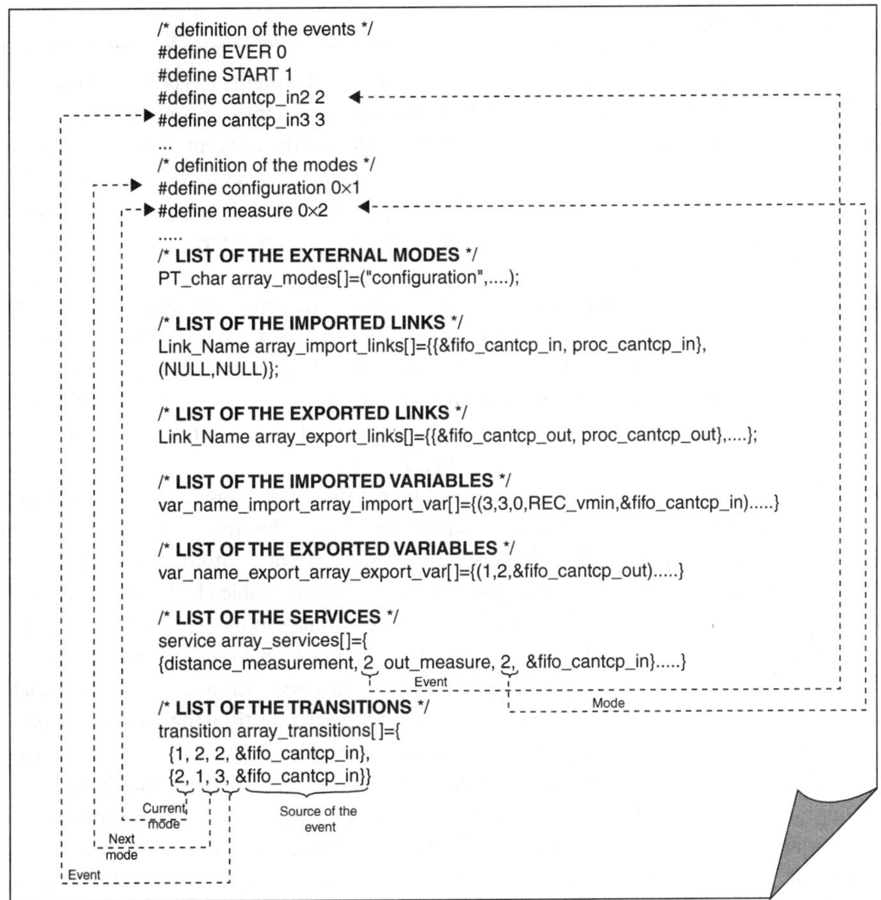

FIGURE 72.16 Part of the synthesized code.

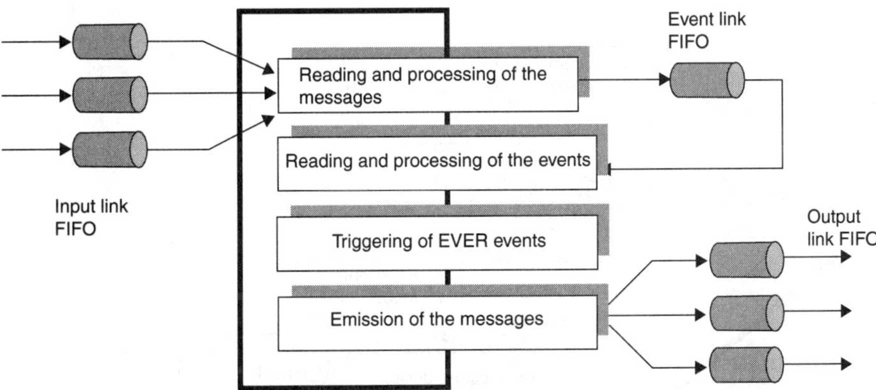

FIGURE 72.17 Principle of operation of the automaton.

[Sun-www]) can be considered as attractive solutions to this problem, by interpreting the applications via a virtual machine. But the designer has to be aware of the substantial performance penalty that could be paid by adopting this solution.

72.5 Conclusion

In this document, a generic model for the design of intelligent instruments has been discussed. In the formal model of the sensor, the specification of the external services is given from a user's point of view of the functionalities available in the sensor. The external model uses the concept of the USOMs to preserve the activation of services that are not available during the current external mode. Internal services define the basic units that will be assembled to describe the complex behavior of an external service, taking into account various temporal dependencies. Like the external services, the internal services can be gathered inside internal modes. Internal modes define the internal states of a service according to its possibilities of operating under various contextual situations. Some of the basic principles underlying the implementation of the model on hardware architectures have also been given. The advantages of the numeric instrumentation for the processing relative to the conventional instrumentation are well established today. We have finally discussed an implementation that can be generated automatically through the use of a language like CAP. The automatic generation of the implementation is conditioned to the formal verification of the properties underlying the generic model.

As stated earlier, the intelligent sensors to come will be composed of more complex and heterogeneous components. This trend will change the industrial landscape, making the trade and assembly of Intellectual Properties (IPs) embodied in layouts, RTL designs, and software programs indispensable [Balarin-97]. This aspect, not specific to the design of the intelligent sensors, is a considerable challenge. For example, hardware/software cosimulation is often performed with separated simulation models [Rowson-94]. This makes trade-off evaluation difficult because the models must be recompiled whenever a change in the architecture mapping is made. The *cofluent* system level design tool [CoFluent-www] introduces an intermediate level of abstraction, the functional level, between the specification and the architectural model of the sensor [Calvez-96]. As shown in Figure 72.18, this level defines the logical architecture of the system in terms of functional components (simply called *functions*) and the relations between them (*ports, shared variables, events* depending on the kind of relationship). Like in the Vulcan system [Gupta-94], the use of a control/data flow graph for the behavioral model facilitates the partitioning at the operation level. This environment has been used successfully for the design of an intelligent sensor for pattern recognition [Dekneuvel-01]. The functional model provides an environment for behavioral and performance analysis in a technology- and language-independent manner, which allows implementation of the same functionality on diverse physical architectures [Calvez-95]. Automatic synthesis can be achieved on hardware (VHDL descriptions) or RTOS primitives (VxWorks). A systemC simulation engine and code generator are also available for system-on-chip (SoC).

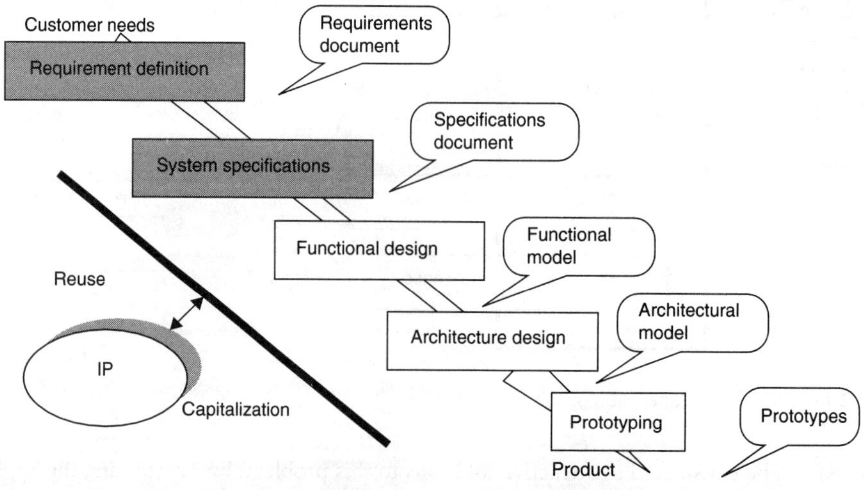

FIGURE 72.18 The MCSE design methodology for complex sensor hardware architectures.

References

[Andre-96] André, C., Representation and Analysis of Reactive Behaviors: A Synchronous Approach, in Proceedings of the CESA'96, Lille, France, 1996, pp. 19–29.

[Andre-01] André, C., F. Boulanger, and A. Girault, A Software Implementation of Synchronous Programs, in Proceedings of the Second International Conference on Application of Concurrency to System Design (ICACSD 2001), Newcastle upon Tyne, U.K., June 25–29, 2001, pp 133–142.

[Balarin-97] Balarin, F., M. Chiodo, P. Giusto, H. Hsieh, A. Jurecska, L. Lavagno, C. Passerone, A. Sangiovanni-Vincentelli, E. Sentovich, K. Suzuki, and B. Tabbara, *Hardware–Software Co-design of Embedded Systems*, Kluwer Academic Publishers, Dordrecht, 1997.

[Benoit-00] Benoit, E., J. Tailland, L. Foullooy, and G. Mauris, A Software Tool for Designing Intelligent Sensors, Proceedings of the IEEE Instrumentation & Measurement Technology IMTC/2000, Baltimore, Marynland, U.S.A., May 2000.

[Benoit-01] Benoit, E., R. Dapoigny, and L. Foulloy, Fuzzy-based Intelligent Sensors: Modelling, Design, Applications, in 8th IEEE International Conference on Emerging Technologies (ETFA'2001), Antibes, France, Oct. 2001.

[Berry-92] Berry, G. and G. Gonthier, The ESTEREL synchronous programming language: design, semantics, implementation, *Science of Computer programming*, 19:87–152, 1992.

[Bouras-98] Bouras, A. and M. Staroswiecki, Building Distributed Architectures by the Interconnection of Intelligent Instruments, IFAC INCOM'98, Nancy, June 1998.

[Calvez-93] Calvez, J.P., *Embedded Real-time Systems. A Specification and Design Methodology* Wiley and Sons Ed, New York, 1993.

[Calvez-95] Calvez, J.P. and O. Pasquier, Performance Assessment of Embedded Hw/Sw Systems, in International Conference on Computer Design, Austin, Texas, October 1995.

[Calvez-96] Calvez, J.P., A co-design case study with the MCSE Methodology, *Design Automation of Embedded Systems, Special Issue on Embedded Systems Case Studies*, 1, 183–211, 1996.

[Cassez-95] Cassez, F. and O. Roux, Compilation of the ELECTRE reactive language into Finite Transition Systems, *Theoretical Computer Science*, Vol. 146, 109–143, 1995.

[Dekneuvel-92] Dekneuvel, E., M. Ghallab, and J.P. Thibault, Hypotheses Management in a Multi-Sensory Perception Machine, in 10th European Conference on artificial Intelligence (ECAI), Vienna, Austria, Aug. 1992.

[Dekneuvel-01] Dekneuvel, E., F.muller, and T.Pitarque, Ultrasonic Smart Sensor Design for a Distributed Perception System, in 8th IEEE International Conference on Emerging Technologies and Factory Automation (ETFA), Antibes Juan le spins (France), 15–18 Oct. 2001.

[Douglas-99] Douglass, B.P., *Doing Hard Time. Developing Real-Time Systems with UML, Objects, Frameworks, and Patterns*, Addison-Wesley, Readings, MA, 1999.

[Edwards-97] Edwards, S., L. Lavagno, E.A. Lee, and A. Sangiovanni-Vincentelli, Design of embedded systems: formal models, validation and synthesis, *Proceedings of the IEE*, 85, 366–390, 1997.

[Fleury-94] Fleury, S., M. Herrb, and R. Chatila, Design of a Modular Architecture for Autonomous Robot, in IEEE International Conference on Robotics and Automation, San Diego, CA, U.S.A., 1994.

[Gupta-94] Gupta, R.K., C.N. Coelho, and G. De Micheli, Program implementation schemes for hardware/software systems, *IEEE Computer*, Vol. 27, no. 1. 48–55, 1994.

[Halbwachs-93] *Synchronous Programming of Reactive Systems*, Kluwer Academic Publishers, Dardrecht, 1993.

[Harel-90] Harel, D., H. Lachover, A. Naamad, A. Pnueli et al., STATEMATE: a working environment for the development of complex reactive systems, *IEEE Transactions on Software Engineering*, Vol. 4. 16, 1990.

[Kurshan-94] Kurshan, R.P., Automatic-Theoretic Verfication of Coordinating Processes, Princeton University Press, Princeton, NJ, 1994.

[Medvodovic-00] Medvodovic, N. and R.N Taylor, A classification and comparison framework for software architecture description language, *IEEE Transactions On Software Engineering*, 26, 70–93, 2000.

[Staroswiecki-96] Staroswiecki, M. and M. Bayart, Models and languages for the interoperability of smart instruments, *Automatica*, 32, 859–873, 1996.

[Staroswiecki-99] Staroswiecki, M., G. Hoblos, and A. Aitouche, Fault Tolerance Analysis of Sensor Systems, in 38th IEEE Conference on Decision and Control, Phoenix, U.S.A., 1999.

[Tailland-00] Tailland, J., L. Foulloy, and E. Benoit, Automatic Generation of Intelligent Instruments from Internal Model, Proceedings of International Conference SICICA'2000, Argentina, Sept. 2000.

[Mauris-93] Mauris, G., E. Benoit, and L. Foulloy, Ultrasonic Smart Sensors. *The Importance of the Measurement Principle*, Proceedings of the IEEE/SMC International conference on Systems Engineering in the service of Humans, Le touquet, France, Oct. 1993.

[Riviere-96] Riviere, J.M., M. Bayart, J.M. Thiriet, A. Bouras, and M. Robert, Intelligent instruments: some modeling approaches, *Measurement and Control*,179–166, 1996.

[Rowson-94] Hardware/Software Co-simulation, in Proceedings of the Design Automation Conference, 1994, pp. 439–440.

[Warrior-97] Smart Sensor Networks of the Future, *Sensor Magazine*, Mar. 1997, pp. 40–45.

Other references

[CoFluent-www] *CoFluent studio*, http://www.cofluentdesign.com.

[Sun-www] EmbeddedJava and Javacard, http://Java.sun.com.

[Esterel-www] Esterel Studio, http://www.esterel-technologies.com.

73

Robot Vision

Ray Jarvis
Intelligent Robotics Research Centre
Monash University

Seeing, in sighted humans, accounts for in excess of 85% of all sensory input data, the analysis of which supports our physical interactions with our everyday environment and our safe navigation through it as well as our appreciation of its richness as a quality-of-life component.

Not surprisingly, therefore, robot vision in its full realization would expand the scope of applicability of robotic systems, both manipulators and mobile vehicles, from the relative structure of classical industrial environments to the unstructuredness of natural and human-centered environments to embrace applications in mining, space exploration, undersea, service industries, healthcare, warehousing, mineral exploration, search and rescue, fire fighting, agriculture, forestry, recreation, catering, and domestic operations. Considerable research and development effort has gone into the theoretical as well as practical issues of robot vision (as a component of artificial intelligence). This effort continues today with the powerful support of rapidly improving and affordable sensor and computational technologies. This support promises an explosion of commercially viable vision-based robotic systems within the next decade.

Computer vision is usually distinguished from computer image processing by the explicit consideration of the three-dimensionality of the world of objects in the former case, even when the extraction of the depth dimension may be by the analysis of one or more two-dimensional images. However, this distinction is not always acknowledged and many image-processing procedures are claimed to be exercises in computer vision analysis.

Digital images are essentially two-dimensional arrays of values, each representing intensity. The dimensionality of the array specifies the spatial resolution of the image, while the number of data bits per cell specifies the intensity resolution. Each cell is referred to as a pixel (picture element). Three arrays of cells, one representing the red intensity, the second green, and the third blue, can collectively be regarded as a digital color image. Other color coding schemes such as hue, saturation, and intensity are also used. A typical digital color image may consist of three 512×512 element arrays with 8 bits per cell per array representing red, green, and blue values, respectively. The whole data structure is represented as $512\times512\times3$ bytes (8 bits) of information. A color pixel is the set of three values at a particular spatial location in the arrays.

When range data are used or derived with respect to a single viewpoint, it is usually also represented in a rectangular array of cells, each being referred to as a rangel (range element). This type of representation is not strictly three-dimensional, as it represents distances from the viewpoint to surfaces visible from that location. The surfaces of objects not visible because of self-obscuration or obstruction by other objects cannot be ranged to from that viewpoint and are thus not represented in the range array. Since, for a typical 3D scene, approximately half the surface points cannot be ranged to from a single specific viewpoint, this representation is often referred to as 2½D.

A full 3D representation might take the form of a three-dimensional volumetric array with solid or surface occupancy indicated. Such a representation is composed of volume elements (voxels). Simple occupancy can be supplemented by values indicating color components or surface normal directionality or other property values of interest. A full 3D volumetric representation including surface properties such as color, normal directionality, and perhaps other attributes can take up a considerable number of memory locations in a computer. For example, a $512 \times 5212 \times 512$ volume array with eight 8 bit property elements per cell would occupy 2^{30} bytes (≈ 8 billion bits of memory).

Robot vision can be interpreted as that part of computer vision that is ultimately intended to be used to guide the actions of robots. The quality of a robot vision system is judged in terms of it providing timely, reliable, and accurate information, extracted from visual or range data (or both), which can be used to direct correct action of a robot in carrying out a specified task and the extent to which its structure permits its use over a wider range of possible tasks. A superior system should also be affordable, physically robust, safe, and energy efficient.

While vision strictly refers to 3D perception derived from images, the analysis of directly acquired range data is usually accepted as part of this domain.

In the context of intelligent robotics (a cooperative interplay of perception, reasoning, and action; Figure 73.1), robot vision can be thought of as a powerful perceptual component. Since a robotic task can only be completed by also including reasoning (perhaps path planning and event sequencing) and action (perhaps following a planned path, avoiding obstacles, gripping an object, etc.), it is best to see robot vision as part of a whole system. In this way, its quality can be judged in relation to its contribution to the whole task.

Two robotic task scenarios are used to illustrate the functions of useful vision systems. While these examples are somewhat simplistic and traditional, they place vision unambiguously within a robot perception/reasoning/action loop to which humans can easily relate, thus improving the tutorial value of thus exercise.

The first example is a robot/camera "hand–eye" coordination task in which a robotic manipulator is used to pick up identified objects from a pile and sort them into groups of like objects (Figure 73.2).

A color camera and a rangefinder collect color intensity and range data in registered two-dimensional arrays, which collectively result in a 2½D representation of the 3D scene of the pile of objects to be dealt with. A database describing the types of objects likely to be encountered in terms of structure, geometry, and surface appearance is available.

Since for reasonably high-resolution color intensity/range, large amounts of data have to be processed, the usual first task of the vision system is to segment the scene into its major components without referring to the database of object descriptions. Some type of "semantic-free" clustering procedure can be used to group pixels/ranges together in clumps, which are homogeneous or continuous in some way. These initial clumps can be further refined, perhaps with some reference to broad distinguishing characteristics derived from the database. Once individual objects are more or less isolated, a search for a match with an object represented in the database can take place for each object in the scene. This is essentially a 2½D to 3D matching process and can be very difficult to carry out reliably, particularly for free-form objects. Some objects may be oriented so that they cannot be unambiguously matched, even if they are not severely obscured by other objects. Severely obscured objects cannot be identified, in general, until the objects obscuring them are removed. At the end of the first part of this process, a number of objects, mostly among those on top of the pile will be identified and their poses will be (placement geometry) determined. Even objects not yet identified have to be taken account of in a volumetric occupancy sense when determining possible grip sites with respect to identified objects and planning approach, grip, and withdraw trajectory components for the manipulator, all of which should preferably be collision-free and optimal in some sense (e.g., minimum path length, time, or energy).

Each identified object that can be reached for and withdrawn from the pile is sorted into groups by the robot once it is safely removed from the pile. This cycle of perception, reasoning (planning), and action is repeated until the entire pile has been sorted.

An interesting variation of the above cycle is to discover grip sites for object removal and to carry out trajectory planning and actuation prior to identifying the object. In this way, the object might be more easily identified once it is physically separated from the pole, as not only has segmentation been carried out physically, but the robot can present the object in various orientations to the vision system so that

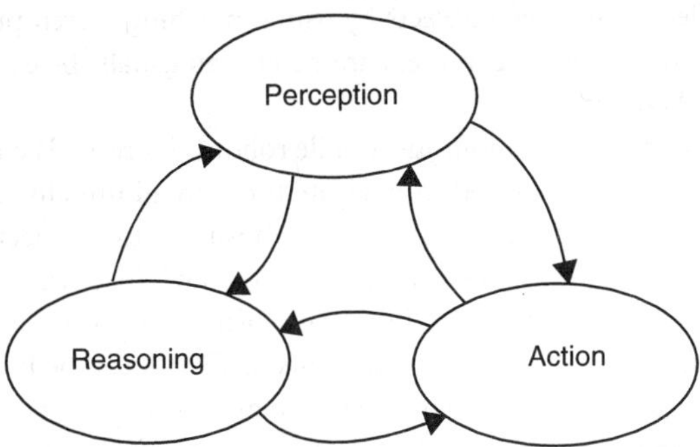

FIGURE 73.1   Intelligent robotics.

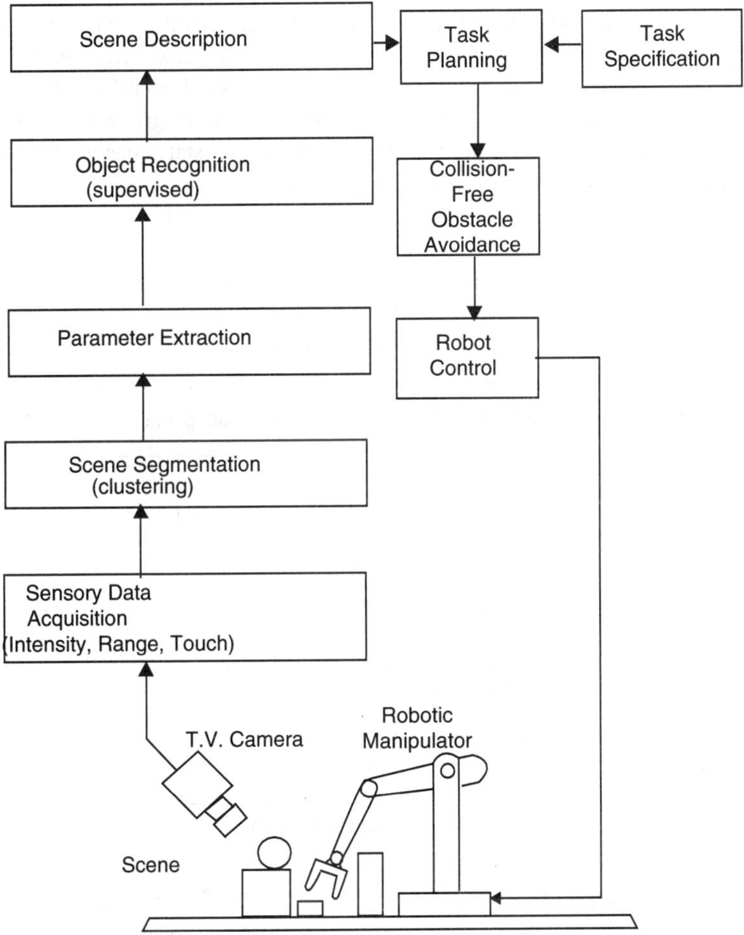

FIGURE 73.2   Hand/eye/touch coordination cycle.

unambiguous identification can be carried out before completing the sorting action. If all objects must be handled (as is the case for this example), this is a good approach; however, in a situation where only a particular object in a pile must be retrieved, this posthandling recognition approach will be cumbersome.

The type of vision system described above (both the pre- and posthandling recognition modes) is referred to as a model-reference system since an explicit database describing the various objects to be encountered is used for identifying scene objects by using matching search procedures. Most industrial robot vision systems are of this type since model parameters can usually be expected to be available from computer-aided design (CAD) data.

The second example concerns an autonomous mobile robot (Figure 73.3) equipped with beacon localization and range sensors, which is required to navigate through an initially unknown and slowly time-varying obstacle-strewn space from a specified start position to a specified goal position via a quasiminimal length collision-free path. Since a complete map of the environment is not initially available and any partial map generated incrementally using sensor data is subject to change, absolute global optimality of the path actually taken cannot be guaranteed; all that can be hoped is that what is known at any particular moment is taken fully into consideration at each stage of forward path planning.

At the start, the position and orientation of the robot are known; at any stage of its path toward its goal, position and orientation data (localization) can be determined using a bar code localizer, which determines the identity and angular placement of bar codes at known positions around the room (beacons).

The on-board rangefinder gathers position data from all surfaces visible all around the robot up to some maximum range measurable by the rangefinder. Obstacle identification is not required for the avoidance strategy. The floor projection of this range data is expanded in all directions by a distance equal to the radius of the robot (assumed circularly symmetric) plus a small collision tolerance amount that must accommodate the maximum range error. A high-resolution floor grid map of occupied and free space is built using the data available so far. Space whose occupancy status has not yet been determined is presumed empty until shown otherwise (the optimistic strategy).

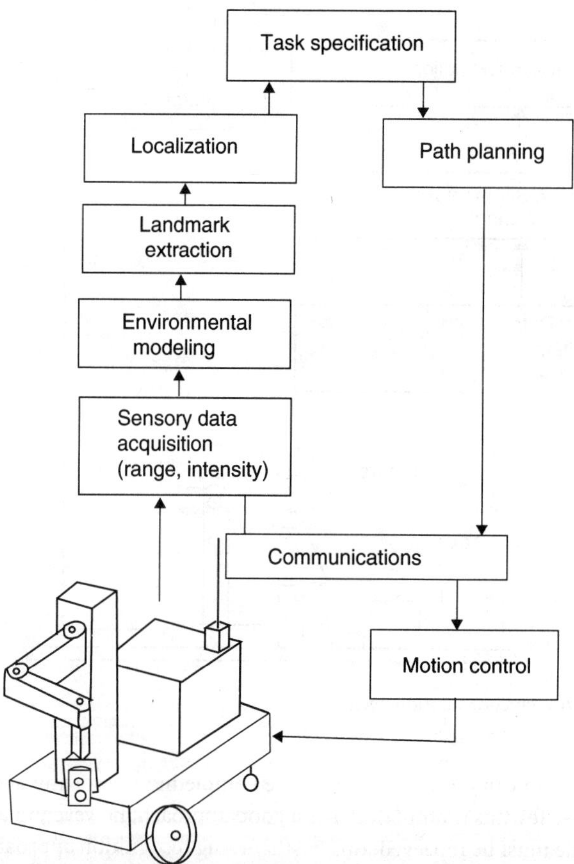

FIGURE 73.3 Sensory data driven mobile robot navigation.

A minimal length collision-free path is determined and the robot follows this path for a specified distance. Localization data collected during the move can be used to adjust the actual path to closely follow the planned path.

Range collection and incremental environmental modeling followed by path planning and partial actuation continues until the goal is reached. Change of occupancy status due to movement of obstacles must also be accommodated in the incremental map-building process.

A more ambitious task would be to carry out the navigation task without the use of beacon localization, relying only on natural landmarks for localization. One way of doing this is to match range data against the map so far derived without actually identifying particular landmarks. The continuity of the robot's movement can simplify this process by constraining the search space of the match task. A more elegant approach would be to identify appropriate landmarks and to recognize these same landmarks as the robot moves; this clearly requires more of the vision system. Note that the recognition of a landmark for localization purposes does not mean that the landmark must be recognized in the model reference sense as was the situation for the first example.

From the above two examples, it is easy to see the important role that vision can play in supporting robotic hand/eye coordination and autonomous robot navigation. In both cases, it was the 3D nature of the problem domain that made vision a particularly appropriate perceptive mechanism to apply, especially since noncontact analysis of the scenes involved was an important aspect of the tasks involved. Note also that vision may be applied both for volumetric modeling as well as shape recognition or both according to what is needed to solve the problem.

While instrumenting a system to collect or derive range data is only one part of robot vision, clearly the quality of such range data in terms of spatial density, accuracy, reliability, and timeliness is critical to the whole analysis process if it is to lead to robust plans for robots to follow. It is of interest to consider the variety of ways by which range might be extracted from a 3D scene.

Humans use a wide variety of depth queues, usually in selective combinations, to disambiguate many competing hypotheses about how images acquired by the retinas of two eyes might be explained in terms of 3D structures in the scene. Human depth cues include texture gradient, out-of-focus blur, aerial perspective, stereo disparity, binocular vergence, size constancy, and relative obscurance. Many robot vision ranging schemes have been inspired by human (and other animals) range extraction mechanisms. Some of these are better suited to robot vision than others and combinations can often be used to produce better results than can be obtained by using one approach alone. However, there are many ranging systems that are based on visual mechanisms not enjoyed by humans; these are also worth considering.

Three dichotomies define eight range-finding domains, not all of which are populated by feasible realizations. These dichotomies are (see Figure 73.4):

1. active or passive,
2. image-based or direct, and
3. single (monocular) or multiple viewpoint.

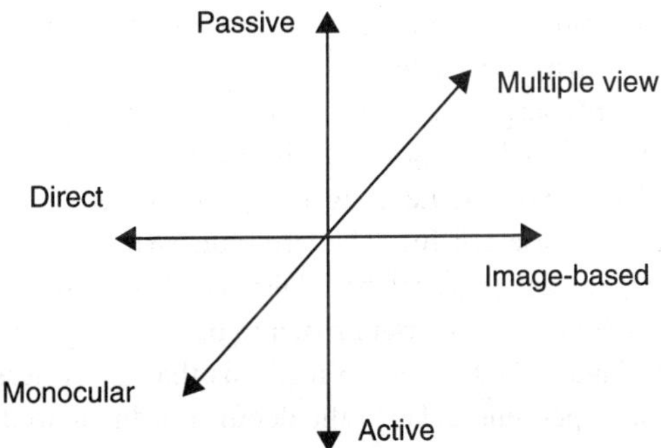

FIGURE 73.4 Range finder classification dichotomes.

A passive method uses only ambient lighting sources, while an active system probes the scene with an energy beam or a structured light pattern. Image-based methods analyze pixel arrays to derive range, while direct methods measure time-of-flight of known velocity light or ultrasonic energy beams to calculate range. Single-viewpoint methods use one fixed camera or one time-of-flight system while multiple-viewpoint methods are essentially based on triangulation. The more popular ranging modes within this categorizing structure will be briefly described below.

In indoor industrial settings, where the intrusion of contrived energy sources is not a severe disadvantage and where the contrived light source can be discriminated from the ambient light, the two most popular ranging methods are laser time-of-flight and striped lighting systems. Ultrasonic time-of-flight systems are also of interest but are not covered here.

Laser time-of-flight range instruments either measure the time it takes for a short pulse of laser light to beam to and be reflected back from a point on the surface of an object or the phase shift encountered by a modulated continuous laser beam during the round trip. Since light travels at approximately 30 cm per nanosecond (10^{-9} sec), subcentimeter accuracy range measurement required time to be measured with approximately ± 10 picosecond (10^{-12} sec) accuracy using the first approach. With both approaches, averaging over a number of cycles can be used to reduce range error but at the cost of increased measurement time.

A number of laser rangefinders are now readily available in the marketplace at modest cost ($2000–$7000). One supplier is Erwin Sick; another is Laser Optronix. Figure 73.5 shows the plot of range extracted using an Erwin Sick laser scanner that sweeps the laser beam in a single arc over 180°, capturing range to 50 m at several centimeter accuracy (improved models are now available). Figure 73.6 shows a large outdoor scene scanned in both pitch and pan dimensions, using a collimated color camera to associate each x, y, z point collected by two range finders with a particular surface color patch.

Striped lighting ranging methods are triangulation based and reply on analyzing the distortion in the image of a stripe of light projected on the scene obtained from a camera displaced from the contrived light source in a direction perpendicular to the stripe light plane (Figure 73.7). Using many stripes simultaneously can lead to ambiguity in identifying the individual lines in the image, while using only one line at a time and sweeping this across the scene can be slow for high-density ranging, particularly if a standard frame rate video camera is used. An elegant compromise consists of collecting a set of images for a stripe lighting pattern set, which allows each stripe to be binary coded among the images. For on stripes only ($\log_2 n$) + 1 images need to be collected; doubling the number of stripes requires only one extra image.

In terms of the categorization scheme introduced earlier, the laser time-of-flight ranging method would be regarded as in the direct/active/single-view class, while the striped lighting ranging method would be regarded as being in the image-based/active/multiple-view class. Other members of the active class are range from brightness (with contrived light sources) and range from attenuation (again with contrived light sources). Both time-of-flight and striped lighting systems have been used in industrial robotics for some years now.

Passive range methods have been actively researched over the last three decades, partly because their biological source of inspiration brings with it the confidence that these approaches have been tested through Darwinian natural selection over countless generations and partly because they promise a very wide application scope and are intrinsically safe. Computational and micro-electronic advances in recent times have brought fast and affordable general-purpose passive rangefinding to the marketplace for practical use in many industries and in the home with or without robotic implications.

Figure 73.8 shows a disparity map of a scene using a Digiclops stereo (actually three cameras are used) system available from Point Grey Research Inc. The larger disparities (separation of matching components in separate camera view) are shown as brighter. Disparity is inverse range related.

Passive stereopsis is clearly the most popular approach to passive rangefinding. There are two stereopsis modes. Lateral stereopsis (Figure 73.9) refers to methods that use at least two cameras in a known baseline configuration in a plane perpendicular to the depth coordinate to derive range, while temporal stereopsis refers to methods where time sequences of images from a camera moving through the 3D scene are used to derive range. In either case, there are two essential process components that must be dealt with

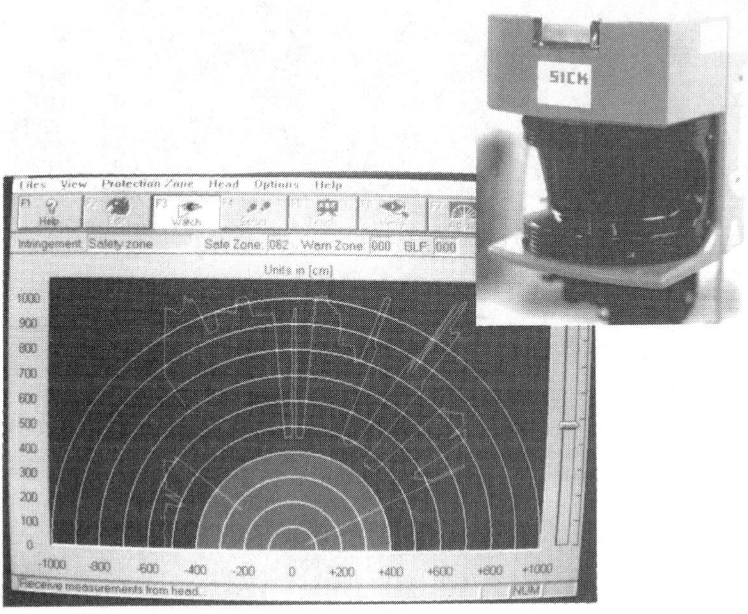

FIGURE 73.5 Erwin Sick scanning laser range finder scan and scene.

since both methods rely on matching elements among images, using the shifts of location (disparity) as inverse range measures (scaled appropriately):

1. The extraction or definition of the elements to be matched among images and
2. The search for correspondence between those selected elements among the images.

The image elements chosen for matching can be edge points, lines, corners, area patches, regions, object segments, or entire objects. More preparatory processing is required by choosing more complex elements, and, in the case of entire objects, the segmentation of these objects may be simplified by the use of range data, which is to be the result of the analysis, thus creating a potential deadlock situation. Thus, there is some advantage in matching the most primitive elements that can be discriminated clearly and matched unambiguously.

In recent times, mask correlation has been favored as a correspondence matching approach since the high computational complexity this entails now falls within the capacity of off-the-shelf computational resources.

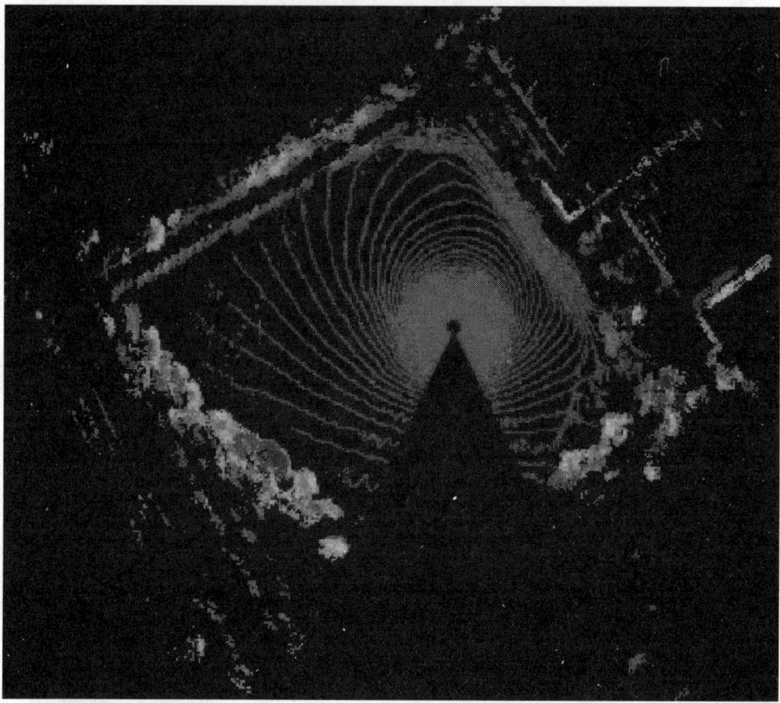

FIGURE 73.6   Large-scale laser range scan (using Laser Optronix device).

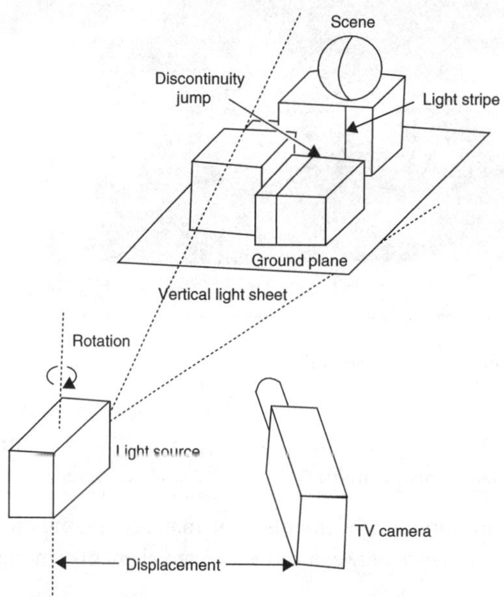

FIGURE 73.7   Striped light ranging principle.

For lateral stereopsis, the use of more than the minimum number of two cameras can improve both accuracy and reliability. Special hardware for very fast extraction of multiple camera stereopsis range data is under active research and development.

A variant of temporal stereopsis is known as optical flow; in this case, the vectors indicating the short-term movement of pixels are calculated from a video sequence with a single camera without explicitly

FIGURE 73.8 Point Gray Digiclops passive stereopsis.

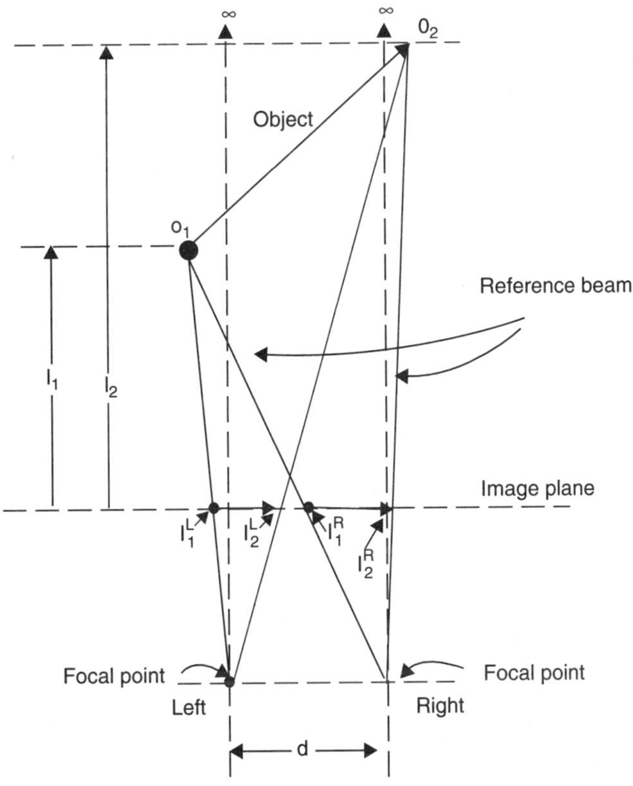

FIGURE 73.9 Geometry of passive stereopsis.

solving the correspondence problem. Depth can be calculated from this vector flow field as can the ego-motion of the camera through the scene. Analysis of moving objects in the scene can also be carried out using optical flow.

Once dense, accurate, and reliable range data have been extracted, the resulting 2½D representation of the 3D scene must be further processed. Full 3D scene representation can be in the form of an ordered or unordered set of 3D Cartesian coordinates with each point having a location in space as well as attributes such as surface color and normal vector data. Alternatively, a voxel-based representation may be preferred. Other data structures have also been proposed. The appropriateness of a data structure used to represent 3D scenes is measured in terms of compactness, lack of ambiguity, easy access, a modifiability, and its ability to support the application in mind.

The results of a complete scene analysis carried out by a robot vision system may take various forms, but would generally provide information on the identity, pose, and placement of individual items and their functional interrelationship (e.g., support, adjacency, occlusion, linkage, containment, etc.). In particular applications, perhaps a subset of this information would suffice; building vision systems to provide more than required is hardly a sensible thing to do for a specific task, but stretching vision methodologies toward generalization of applicability is also a worthwhile endeavor.

In summary, robot vision attempts to emulate human vision to the extent of correctly interpreting the makeup of 3D scenes, with the purpose of guiding robotic manipulators and autonomous mobile robots in fulfillment of useful tasks; many methods of range extraction support these ambitious goals and rapidly improving sensory device technologies and computational systems are likely to lead to an explosion of commercially available systems within the next decade.

74

Robot Tactile Sensing

R. Andrew Russell
Monash University

74.1 Introduction

As an aspect of robotics, tactile sensing covers any method of measurement involving physical contact between the sensor and sensed object. For this reason, it is also sometimes known as contact sensing. By its nature, tactile sensing is limited to contact and short-range situations. However, measurements gathered by tactile sensing usually require little processing to extract useful, unambiguous information. Mobile robots can use tactile sensing to warn of imminent collisions. The short-range but reliable information provided by touch is ideal as a last line of defense after obstacles have evaded longer-range sensors. Another aspect of tactile sensing is the measurement of contact forces. Many robotic tasks involve contact between the tool carried by the robot and external objects. By monitoring the resulting forces, the quality and safety of these operations can be improved. In robotics, there is considerable interest in producing touch sensor arrays that mimic the sensing abilities of the human skin. Such sensors can determine the regions of contact between robotic fingers and the objects they grip. Skin-like sensor arrays have also been developed to measure shear forces and the thermal properties of gripped objects. Individual, special-purpose sensors have also been designed to measure texture, slip, and incipient slip of a grasped object.

74.2 Whisker Sensors

A number of mobile research robots use whiskers as a short-range form of obstacle detection. The Stanford Research Institute's "Shakey" robot, developed between 1966 and 1972, was one of the first whisker-equipped mobile robots. This robot used its whiskers to detect obstacles and also the boxes that it pushed around its world. Whiskers have also been incorporated into legged robots, such as the four-legged robot Titan III built at the Tokyo Institute of Technology, to detect proximity between their feet and the ground (Hirose, et al., 1985). This allows the feet to decelerate before ground contact. Several researchers have mounted whisker sensors on robot grippers and have used them to locate small objects close to the gripper. Whisker sensors are simple and provide direct information about the close proximity of other objects. These attributes are also useful in mobile robotics.

A simple whisker sensor may be no more complicated than a short length of piano wire passing through a small hole at the end of a metal tube. Figure 74.1 shows a cross-section view of such a sensor. When the wire is deflected, it touches the metal tube, thus completing an electrical circuit to signal the contact.

Whisker sensors only provide information about a single contact point and this limits the rate at which they can gather information. For this reason, it can appear that they are unsuitable for all but very slow robotic manipulation tasks. Biological studies indicate that it should be possible to gather considerable useful information from whisker sensors and to do this rapidly. The domestic cat is one of the few animals where the sensing capabilities of whiskers have been studied in any detail. A cat captures and manipulates its prey using its mouth. Objects held in the mouth are too close for the cat's eyes to focus on and are also partially obscured from view. Whiskers allow the cat to determine the position and movements of prey animals close under its nose or held between its jaws. The upper lip of the cat is well supplied with whiskers (vibrissae). Each whisker is provided with about 200 sensory nerve endings, which respond to displacement and velocity stimuli. Muscles position the whiskers to sense objects close to the mouth of the cat or fold them back to keep them out of the way while eating. When springing onto its prey, or when carrying something, the whiskers are angled far forward to envelop the object. A blindfolded cat can locate a mouse. Within one tenth of a second of a mouse touching the cat's whiskers, the cat grasps the mouse with a fast and precise bite to the nape of the neck. There is also evidence that the cat uses its whiskers to find the direction of hairs on a prey animal's body, which is an aid to cutting up and eating the animal. In an attempt to emulate the sensory capabilities of biological whiskers, there has been increasing attention focused on augmenting the information provided by robotic whisker sensors. Determining the point along the length of a whisker where contact is made would be useful additional information that would help to localize a sensed object. Several techniques are currently being investigated to allow this information to be gathered. These methods use a combination of some sort of actuation together with measurements at the root of the otherwise insensitive whisker to infer the point of contact. Sensing principles that are being considered include:

- Whisker resonant frequency — a rotary actuator causes the whisker to strike the sensed object. The natural frequency of the whisker during contact is used to estimate the contact point along the whisker (Ueno and Kaneko, 1994).
- Whisker deflection resulting from movement of the robot while the whisker tip touches an external object. The whisker is free to rotate at its base (Russell, 1992).
- Starting with a flexible whisker in contact with an external object. If the whisker is rotated at its base, the contact force is increased. Measurements of the amount the whisker is rotated and the torque required to perform the rotation allow the contact point to be determined (Kaneko, 1994).
- Sweeping a whisker over the surface of an object by rotating it about its base. By measuring whisker rotational angle and forces and torques at the whisker base, an elastica model can be used to predict the shape of the bent whisker and the contact point with the sensed object (Scholz and Rahn, 2002).

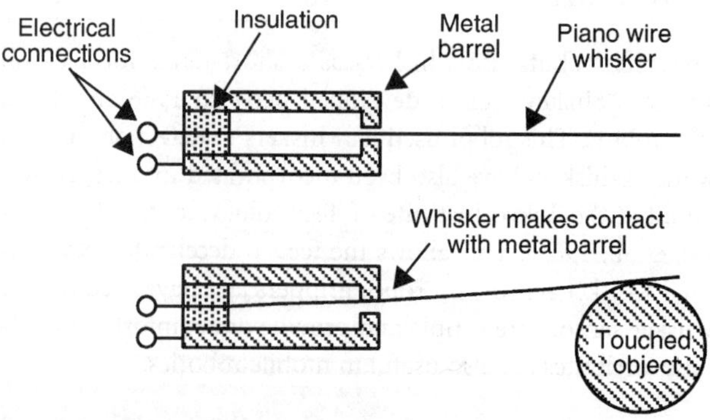

FIGURE 74.1   A simple whisker sensor.

74.3 Force/Torque Sensors

When two objects touch, the forces and torques generated by contact can be resolved into forces along three orthogonal axes and torques about these axes. In robotics, it is often important to measure these six force/torque component and special-purpose loadcells have been designed for this purpose. In practice, force/torque sensing load cells are commonly located between a robot manipulator arm and its tool or gripper. Here, they can monitor and help control contact forces between the tool or gripper and external objects. In recent years, miniature force/torque sensors have been made for mounting in the fingertips of a robot hand. In this location, they can measure forces and torques generated during grasping and manipulation tasks. Applications for force/torque sensors are usually centered around tasks requiring manipulation. By measuring applied force, a force/torque sensor can also help maintain optimum contact during robotic fettling, grinding, and burnishing operations. During assembly operations, jamming generates characteristic reaction forces and torques. If these forces and torques are detected, then appropriate actions can be taken to free the parts and complete the assembly. In principle, data from a force/torque sensor can also be used to find the point of contact between a tool or gripper mounted on the loadcell and the outside world.

The load cell structure consists of a number of interconnected metal beams usually machined from a single piece of metal. Strain gauges are attached to the beams so that each gauge is sensitive to different components of the applied forces and torques. Figure 74.2 shows a force/torque sensor machined from a single block of aluminum in the form of a spoked wheel. The outer rim of the wheel A is connected to the robot arm and the inner hub B to the gripper. There are four semiconductor strain gauges attached to each spoke of the wheel and they are wired in pairs as half-bridge circuits. The load cell produces eight output signals and these can be related to the three components of force and three components of torque acting on the load cell. It is difficult to make a sufficiently ideal mechanical structure that the relationship between the eight output signals and applied forces and torques can be calculated accurately. For this reason, a transfer matrix relating eight output voltages to six applied forces and torques is determined experimentally. A pseudoinverse of this calibration matrix allows the sensor output voltages to be interpreted in terms of applied forces and torques.

74.4 Skin-like Tactile Sensors

Artificial sensory skins have been made containing arrays of contact sensors. These can give an image of the area of contact between a robotic finger and a grasped object and may produce a graduated response that depends upon the degree of indentation into the artificial sensory skin. Essentially, skin-like contact sensors measure deflection of the active sensor surface. Although numerous different designs of skin-like sensors have been developed over many years, they have seen little or no use in practical robotic applications. However, similar sensing technology has several biomedical applications including sensing the distribution of foot pressure on the ground and measuring the bite (the way in which the upper and lower teeth meet).

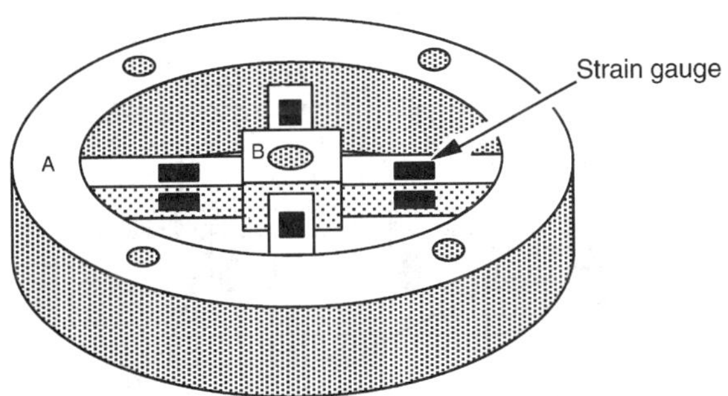

FIGURE 74.2 A 6-axis force/torque sensor.

Here is a list of desirable performance specifications for a robot tactile sensory skin. Some of the criteria were proposed in an early paper on tactile sensing by Harmon (Harmon, 1982):

- spatial resolution 1 to 2 mm on a finger tip. Tactile sensation on other parts of the human body such as the back is only of the order of 10 to 20 mm
- array size — human fingertip contains approximately 10 × 15 points
- threshold sensitivity about 0.5–10 g
- sampling rate 100 Hz to 1 kHz,
- rugged
- inexpensive
- conformable to adapt to the exterior shape of the robot and
- compact and easily integrated into the robot structure.

It is perhaps an indication of the lack of maturity in this area of robotic sensing that many different transducer mechanisms have been tried and still more are continually being proposed. Skin-like contact sensors have been designed that are based on simple switches, piezoresistivity, piezoelectricity, optical, opto-mechanical, fiber-optic, photoelasticity, magnetic, magnetoelastic, ultrasonic, capacitive, and electrochemical sensing methods. In such a brief outline, it is impossible to cover the full field. However, four sensors will be mentioned. They are examples of devices that have shown sufficient promise to be developed to the stage of a commercial product.

Opto-mechanical

One commercially manufactured opto-mechanical sensor array employs a rubber skin with an array of mushroom-shaped projections molded into its surface (Rebman and Morris, 1986). The head of the mushroom concentrates the applied force and the stalk acts as an optical shutter to control light transmission between a light-emitting diode and a photodetector. As normal force increases, light transmission is reduced. Construction of this sensor is quite labor-intensive. All sensor sites contain a photoemitter and photodetector and they must be individually matched and trimmed with an associated resistor to equalize their responses. This sensing technique was used in the first commercially available tactile sensor. Although each sensor site is quite complicated, sensor arrays were constructed with sensor spacings as small as 1.8 mm (Figure 74.3).

For the opto-mechanical tactile sensor, spatial resolution and sampling rate are adequate. A major drawback for this design is its complexity and this also leads to a relatively expensive solution. Modifying the design so that the sensor conformed to even a simple curve would be difficult. A further drawback of this design is that this relatively thick sensor would take up considerable space, particularly in a complex structure such as a dexterous robot hand.

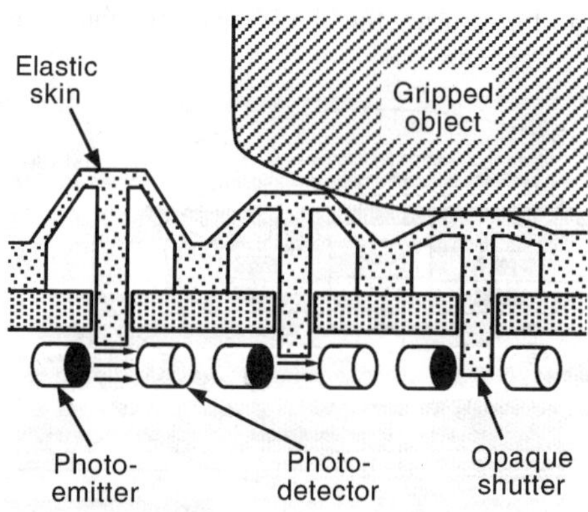

FIGURE 74.3 Opto-mechanical tactile sensor.

Frustrated Internal Reflection

This optical sensing technique is capable of producing very high-resolution tactile images (King and White, 1985). A major component of the sensor is a light guide, which may be a sheet of clear glass or plastic. Light introduced at one edge of the light guide propagates through the sheet by total internal reflection and emerges at the opposite edge (Figure 74.4). For total internal reflection to occur, light must strike the surface of the light guide at less than the critical angle. A sheet of reflective rubber material is suspended close to, but not touching the light guide. This is the active area of the sensor. When an object presses against the rubber sheet, the sheet deforms making contact with the light guide. Over the area of contact, the critical angle changes causing light to leave the light guide and effectively illuminate the region of contact. Diffusely reflected light from the area of contact passes back through the guide and is detected by a video camera or a similar optical sensor. Using a flat rubber sheet gives a binary image, which can only distinguish between contact or no contact. However, by embossing a pattern into the rubber surface, a certain amount of force information can also be distinguished.

This transducer technique produces very high-resolution tactile images of the order of 256 × 256 taxels over a 2 cm × 2 cm area. The sampling rate is limited by the scan rate of the optical sensor. However, the inherently serial nature of most camera signals greatly reduces the number of electrical connections to the sensor. Frustrated internal reflection sensors have been built with flat sensing surfaces and also in the form of a hemisphere for mounting on a fingertip. Even with the advent of miniature camera chips, the optical paths make this type of sensor inherently bulky. There is also a tendency for the rubber skin to stick to the light guide after the external object has been removed. This is another relatively expensive form of tactile sensor and probably would not be economically viable for areas of a robot where large area, low-density sensing is required.

Piezoresistive

Piezoresistive sensors are relatively simple to make and provide a large electrical output that requires little or no amplification. Figure 74.5 shows a cross-section view of a single sensor element. The sensor consists of two sheets of polyimide plastic film. On the inner surface of the lower sheet, two metallic electrodes separated by a small gap are deposited. The upper plastic sheet is coated with conductive ink. Such inks consist of semiconducting particles like molybdenum disulfide held together with a binder material such as acrylic resin. In this sensor, the piezoresistivity is a surface effect. Pressure reduces the contact resistance between the electrodes and the ink. Resistance between the electrodes depends upon compressive force applied to the sensor. This technology can be used to make simple and low-cost touch sensor arrays by what is essentially a printing process. The sensors are thin (<1 mm) and can be bent around simple curves. In many cases, the piezoresistive sensor arrays could be directly attached to the outer surface of the robot structure (Figure 74.5).

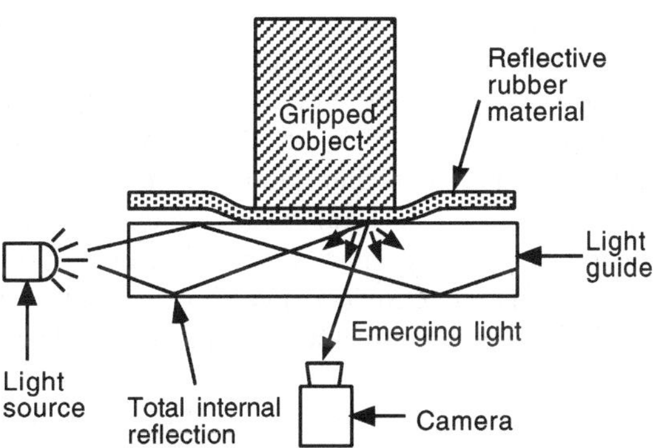

FIGURE 74.4   A tactile sensor using the principle of frustrated internal reflection.

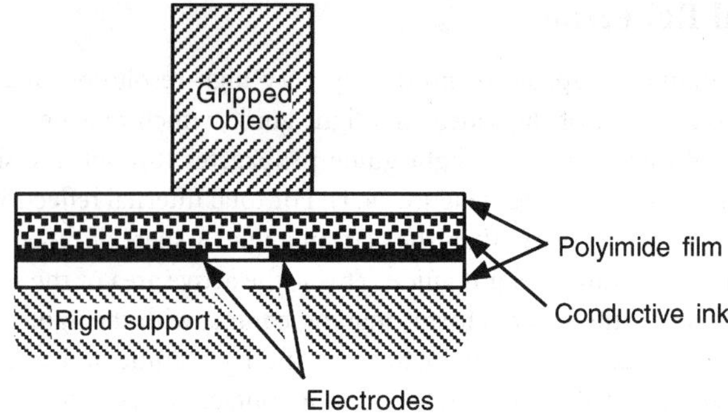

FIGURE 74.5   Cross-section through a single piezoresistive tactile sensor.

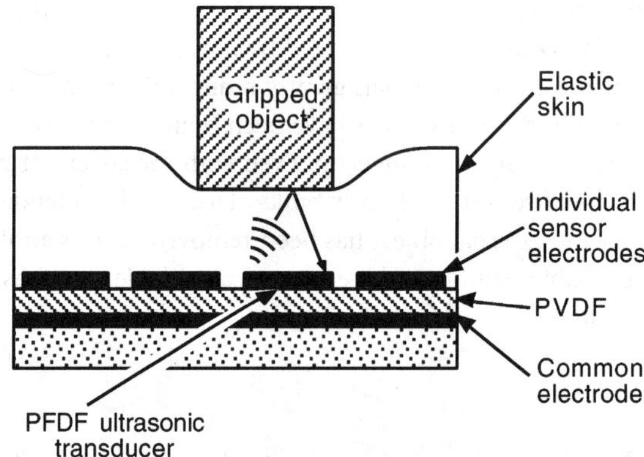

FIGURE 74.6   Cross-sectional view of a tactile sensor array using ultrasonic transduction.

Ultrasonic

Ultrasonic thickness gauges have been in use for many years to measure the thickness of paint layers, metal sheets, etc. An ultrasonic pulse is introduced into the sheet at one face, propagates through the thickness of the material, and is reflected from the opposite face. The returning echo has traveled twice through the thickness of the material. Knowing the speed of propagation of the ultrasound pulse and measuring the propagation delay allow the thickness to be determined. This principle has been used to construct a tactile sensor by using ultrasonic transducers to measure the thickness of a flexible elastomer layer at many closely spaced points (Grahn and Astle, 1984). The sensor is relatively simple consisting of a sheet of polyvinylidene fluoride (PVDF) piezoelectric plastic patterned with an array of electrodes. Overlying the sheet of PVDF is a layer of compliant elastomer, which changes shape as objects indent into it. The electronics associated with the sensor generate, detect, and time the ultrasonic pulses. Using this technology, a 16×16 element ultrasonic sensor array has been manufactured with 1.8 mm spacing between taxels (Figure 74.6).

The actual ultrasonic sensor array can be quite thin (most of the thickness is contributed by the elastic skin), and there seems to be no reason why the array could be molded around simple curves. Because measurements are based on time-of-flight, the readings of sensor indentation are very accurate and stable. A drawback is the large amount of support electronics required to generate high-voltage pulses to energize the ultrasonic transducers, amplify the returned echo, and measure transit time of the ultrasonic pulses.

Tactile Images

Figure 74.7 shows a wire-frame plot of a typical touch sensor image. The object pressed against the sensor, a slotted electrical lug, is small enough to fit entirely within the active area of the sensor. For most

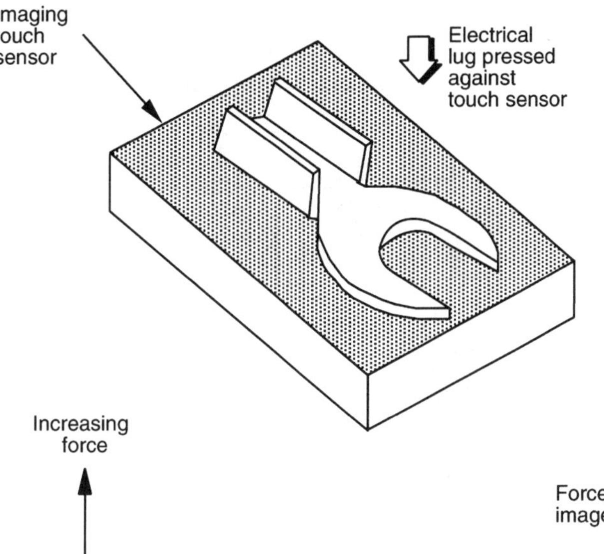

FIGURE 74.7 An example of a force image produced by pressing a slotted electrical lug against a 10 × 10 piezo-resistive array sensor.

objects, this would not be the case. Tactile information about a larger object would usually be gathered by a process of active search. This implies that the sensor is attached to a gripper and that a computer program directs the gripper to search for the required information.

Skin-like sensors have been developed to measure other quantities besides deflection in the region of contact. When we touch metal, it feels cold while wood and polystyrene foam feel warm, even though these materials are all at room temperature. The feelings of warmth or cold are caused by different materials cooling our fingers at different rates. This gives us additional information about the kinds of materials things are made of and robotic touch sensors have been designed using the same principle.

Skin-like Thermal Sensor

The human skin contains nerve endings that detect increases in temperature (Ruffini endings) and others that respond to lowering skin temperature (Kraus corpuscles). Acting together, these nerve endings provide information about the temperature and thermal characteristics of grasped objects. Knowledge of temperature provides obvious safety information particularly for things that will be grasped or ingested. Measurements of the thermal characteristics of objects at room temperature give information about the nature of their constituent materials. Steel feels cold, while glass is not quite as cold and cork appears warm to the touch.

For a robotic skin-like thermal sensor, a temperature-stabilized heat source warms the sensor in the same manner that the blood supply warms the skin. In Figure 74.8, thermistor TH2 provides temperature feedback for the temperature control circuit and the actual heating element is a power transistor. As shown in the figure, a layer of material of known thermal conductivity couples this heat source to the touched object. A thermistor on the outer surface of the coupling layer measures the contact point temperature. The thermistor replaces the thermally sensitive nerve endings in the human skin. When this sensor touches an object, the temperature change in the thermistor TH1 gives information about the material properties of the object structure. A skin-like array of such sensors can measure the extent of the temperature disturbance caused by contact with an object and hence give the outline of the touched object (Russell, 1988).

Figure 74.9 shows the temperature variation measured by a thermal sensor in contact with a number of different materials. Each material produces a characteristic response depending on its thermal capacity and thermal conductivity. A number of known materials can be discriminated by comparing the sensor response with values recorded for each material. Alternatively, there is a strong correlation between

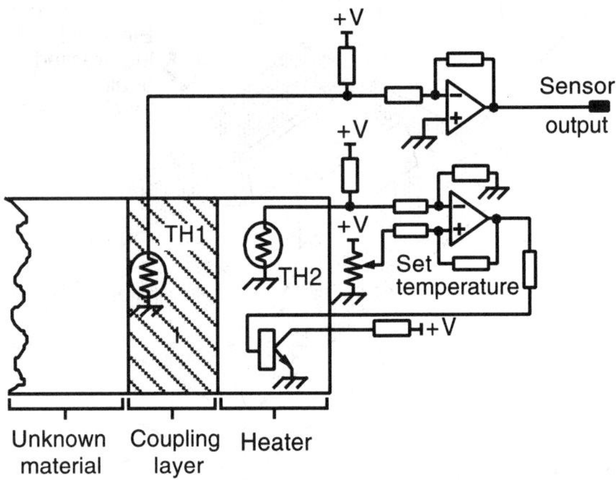

FIGURE 74.8   A schematic diagram of a thermal sensor.

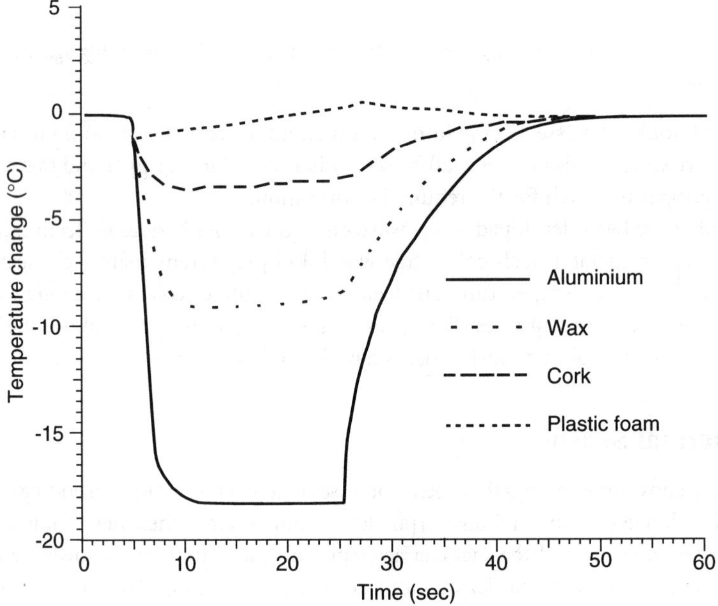

FIGURE 74.9   Thermal sensor output showing the response to contacting different materials for 20 sec.

the decline in thermal sensor temperature and specific mass of the material making up the touched object (Russell, 1990). This information about specific mass could provide information to help recognize an unknown object.

Slip Sensing

In addition to skin deflection due to contact and temperature change caused by the flow of heat between the skin and a touched object, the human skin can also gather more specialized information. Of particular interest during manipulation operations is the onset of slip that signals the imminent possibility of dropping a grasped object. Technological sensors have also been devised to give a robotic system information about the relative movement between a robot finger and the objects it is touching. As an example, Figure 74.10 shows a diagram of one form of special-purpose slip sensor (Ueda, et al., 1972). In this

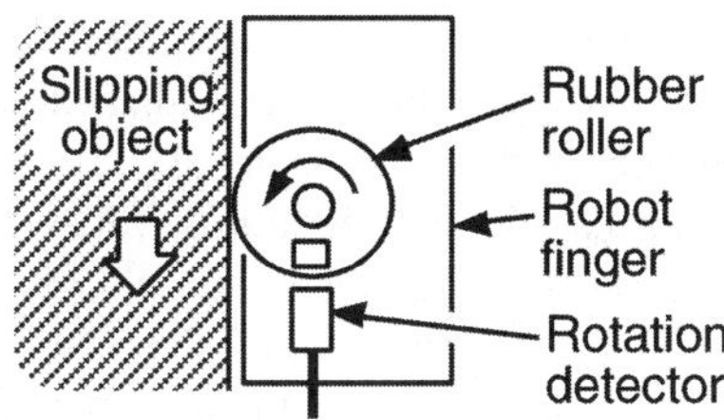

FIGURE 74.10 A sensor to determine slip.

example, a freely rotating rubber-coated wheel presses against the gripped object. As the object starts to slip, the wheel turns and a rotation sensor detects the movement.

Somewhat closer to the human slip sense is the incipient slip sensor described by Tremblay and Cutkosky (1993). The sensor is in the form of a highly compliant fingertip. A foam core provides compliance and this is covered by a rubber skin with an outer surface patterned with an array of raised bumps. Accelerometers attached to the inside surface of the skin measure skin vibration. Tangential forces acting on a gripped object tend to pull it from the grasp. As these forces increase, the skin distorts and some of the sensor bumps break free. The resulting skin vibration is detected by the accelerometers and can be used to identify the increased levels of tangential force that may lead to slip. It is thought that the fingerprint pattern on human fingers coupled with vibration-sensitive nerve endings provides a similar warning of incipient slip for humans.

74.5 Conclusion

Tactile sensing, like computer vision and many other aspects of robotics, all seem deceptively simple. As humans, we can use this sense with apparent ease. The reality is that it is extremely difficult to emulate the capabilities of the human sense of touch in a robotic system. However, we know from our own experience as to how useful touch sensing can be, especially when grasping and manipulating objects. For this reason, it seems certain that touch sensing will prove important for robotic systems also.

Many robotic tactile sensor designs have been proposed, particularly for skin-like touch sensing. Skin-like touch sensing is a very demanding application where a compact and sensitive sensor must be immune to severe thermal, chemical, impact, abrasion, and cutting challenges. Current designs fall well short of what is required. The large amounts of research and development necessary to bring an existing design to a suitable standard will only be committed when compelling applications appear. Particularly for robotic manipulation, the design of robotic grippers to carry the tactile sensors and the development of associated data processing and control programs are also important considerations. All of these elements will be required for successful tactile sensor-based robotic manipulation and object recognition. Perhaps less demanding but potentially very useful would be whole-body touch sensory skins for humanoid, mobile, and manipulator arm robots. These could warn of contact and collisions. Touching the tactile sensory skin of a robot could also provide an effective means of getting the robot's attention during human–robot interactions.

References

1. Grahn A.R. and Astle, L., Robotic Ultrasonic Force Sensor Arrays, Proceedings of the Robots 8 Conference, Detroit, Michigan, 1984, pp. 21.1–21.18.
2. Harmon, L.D., Automated tactile sensing, *International Journal of Robotics Research*, 1, 3–32, 1982.
3. Hirose, S. et al., Titan III: A Quadruped Walking Vehicle, Robotics Research The Second International Symposium, Hanafusa, Hideo and Inoue, Hirochika, Eds, The MIT Press, Cambridge, MA, 1985.
4. King, A.A. and White, R. M., Tactile Sensing Array Based on Forming and Detecting an Optical Image, *Sensors and Actuators*, 8, 49–63, 1985.

5. Rebman, J. and Morris, K.A., A Tactile Sensor with Electrooptical Transduction, *Robot Sensors, Vol. 2, Tactile and Non-Vision*, Pugh, A., Ed., IFS (Publications) Ltd., Bedford, U.K., 1986, pp. 145–155.
6. Russell, R.A., Thermal sensor for object shape and material constitution, *Robotica*, 6, 31–34, 1988.
7. Russell, R.A., *Robot Tactile Sensing*, Prentice Hall Australia, Sydney, 1990.
8. Russell, R.A., Using Tactile Whiskers to Measure Surface Contours, Proceedings of the IEEE International Conference on Robotics and Automation, Nice, 1992, pp. 1295–1299.
9. Scholz, G.R. and Rahn, C.D., Profile sensing with an actuated whisker, Proceedings of the International Engineering Conference and Exhibition, New Orleans, 2002, pp. 1–4.
10. Tremblay, M.R. and Cutkosky, M.R., Estimating Friction Using Incipient Sensing During a Manipulation Task, Proceedings of the IEEE International Conference on Robotics and Automation, Vol. 1, Atlanta, 1993, pp. 429–434.
11. Kaneko, M., Active Antenna, Proceedings of the IEEE International Conference on Robotics and Automation, San Diego, Vol. 2, 1994, pp. 2665–2671.
12. Ueno, N. and Kaneko, M., Dynamic Active Antenna, Proceedings of the IEEE International Conference on Robotics and Automation, Vol. 2, San Diego, 1994, pp. 1784–1790.
13. Ueda, M., *et al.*, Tactile Sensors for Industrial Robot to Detect Slip, Proceedings of the 2nd ISIR, IIT Research Institute, Chicago, 1972, pp. 63–76.

75

Giving Robots a Sense of Smell

R. Andrew Russell
Monash University

Currently, few robotic systems have been provided with access to the rich and complex dimension of chemical sensing. In biology, the response to chemical stimuli was the first sense to evolve and it is the most widespread sensory modality over the whole spectrum of living creatures. For the majority of plants and animals, chemical stimuli are essential to every aspect of their lives. Chemical senses are involved in finding food, avoiding danger, and locating a mate.

It has been estimated that the human sense of smell has been degenerating steadily over the past 5 million years. This is probably mirrored by a corresponding improvement of and reliance on visual sensing. However, in situations where odor is important, other animals, most notably dogs, are used to help substitute for the relatively poor human sense of smell. Among other things, sniffer dogs are used to detect:

- plant matter, drugs, and other materials important to customs officials,
- truffles,
- victims of avalanches and earthquakes,
- escaped prisoners,
- chemical leaks, and
- mines and unexploded bombs.

One of the drawbacks of sniffer dogs is that it is difficult or impossible to find out exactly which chemical species triggers their response. In many cases, it is not the target chemical but an impurity or by-product that is detected. Another problem is that sniffing is a natural action for dogs; hence, even if their sense of smell is impaired, this will be hard for a handler to determine. Such a condition could have serious consequences in some situations such as during mine clearance operations. Animals require extensive training, they must be cared for by an expert handler, their attention span is short, and they suffer from fatigue.

75-1

Robotic systems could perform tasks similar to those performed by sniffer dogs while offering a number of advantages including tireless operation, easy care, and supervision by relatively unskilled personnel. In addition, many new chemical sensing applications will be made possible by the varied forms that robots can take and their flexible capabilities. Robots of many different sizes can fly, swim, crawl, and burrow through the ground in search of chemical sources.

Among robotics researchers, interest in developing robots with a chemical sense is relatively recent. Larcombe and Helsall (1984) discussed techniques for utilizing chemically sensitive robots in a report to the European nuclear industry published in 1984. However, reports of the first practical experiments involving chemical sensing robots began to appear in the early 1990s (Russell, 1999a).

75.1 Application Scenarios

There are important economic, environmental, and humanitarian applications for systems that can find the source of a volatile chemical. Such sources could include pollutants being discharged by factories, leaking gas pipes, corroded underground chemical storage tanks, landmines, unexploded ordinance, and even victims of landslides and earthquakes.

As illustrated in Figure 75.1, a flying robot and a submersible are faced with a similar challenge of tracking a chemical plume in three dimensions in order to locate its source. Chemicals carried by the wind and by water currents are both subject to the effects of turbulence. Provided fluid velocity is high enough, then flow some distance away from the chemical source will be turbulent. When the plume of chemical enters this region, it will be torn into short segments. It then proceeds downwind/downstream as an irregular and expanding series of short chemical patches. Under these circumstances, interpreting the fluctuating chemical concentration to work out a path to the source is a significant challenge.

Terrestrial mobile robots are able to follow plumes of chemical released into an indoor environment or close to the ground. Such chemical plumes must be negatively buoyant if the robot is to detect them at a distance from their source. The terrestrial environment is often complicated by walls and other obstacles that deflect airflow and impede the progress of the robot. A chemical source locating robot must be able to deal with these complications. Mobile robots equipped to detect chemicals could patrol factories, chemical plants, and even homes searching for chemicals released into the atmosphere from sources such as leaks, spills, rotting food, and incipient fires (Figure 75.2).

Chemicals can be deposited on a solid surface in the form of a trail or percolate to the surface of the ground from an underground source. In both these situations, the location of the chemical source can be determined by detecting the traces of chemical released into the surrounding air (Figure 75.3).

Chemical markings on the ground could provide a trail for a robot to follow in much the same way that a worker ant follows an ant trail. Chemical percolating from the ground would signal the location of a buried

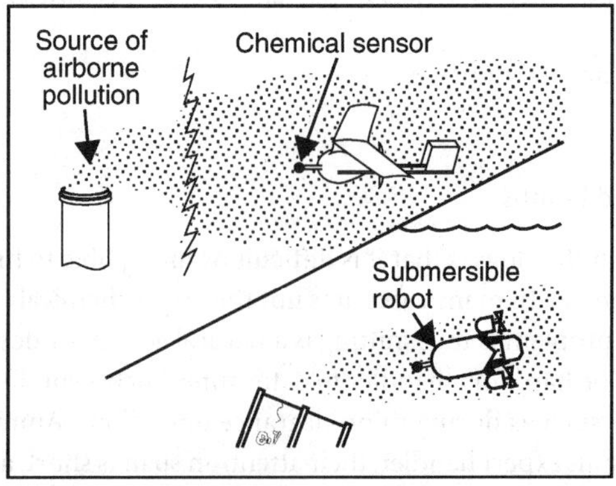

FIGURE 75.1 Both airborne and submersible robots track toward sources of an odor plume by maneuvering in three dimensions.

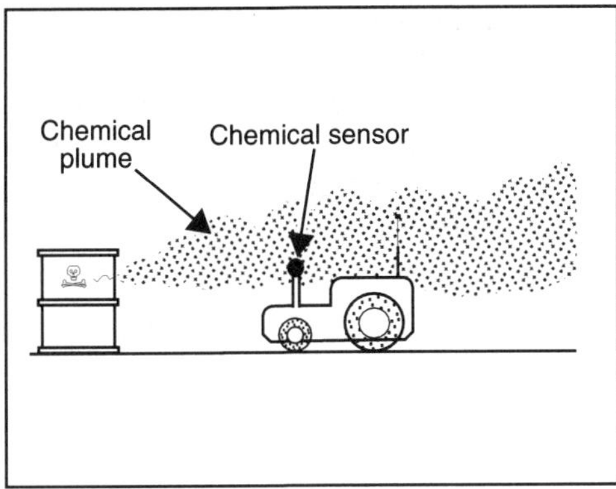

FIGURE 75.2   Chemical plumes released close to the ground can be traced by a terrestrial robot platform.

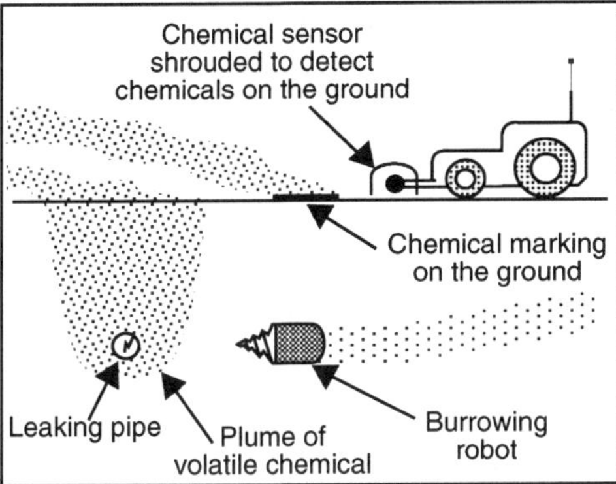

FIGURE 75.3   Locating the source of chemical markings on the ground or chemicals percolating out of the earth require specially designed sensor systems.

chemical source. In both these cases, careful control of airflow around the sensor is required to exclude chemicals released by distant sources and to ensure that chemicals from the ground are carried to the sensor.

The diffusion of volatile chemicals through the ground is much slower than through the air. Therefore, the belowground concentration of chemical from a buried source is much higher than the concentration in the air above the source. For this reason, it may be of benefit for the robot to burrow through the ground toward the source rather than trying to measure chemical concentration released into the air above.

As the scenarios presented in this Section have shown, there are many potential applications for chemical-sensing robots. An important first step in developing robots that can react to chemicals in their environment is the provision of suitable sensors.

75.2   Robotic Chemical-Sensing Technologies

There are many sensors available for detecting volatile chemicals. These range from the simple tin oxide sensors used by police to estimate alcohol consumption to mass spectrometers and gas chromatographs that analyze complex mixtures of chemicals. However, few of these sensors are suitable for robotics applications. Desirable attributes for robotic chemical sensors include:

- *High sensitivity* — increasing sensor sensitivity opens up many more application areas
- *Rapid response* — to provide real-time control of a robotic system, a response time of fractions of a second to a small number of seconds would usually be required

- *Low power consumption* — power for sensor systems will be drawn from the robot's power supply and must not impose an unreasonable power drain
- *Robustness* — robot sensor systems may be used in unconstrained environments and therefore should resist physical damage due to collisions and being poisoned or otherwise damaged by the chemicals it is likely encounter in its environment
- *Compact size* — space is usually limited; hence, a small size is preferable and would be related to the dimensions of the robot.

In addition, it would be of benefit if the sensor was inexpensive and commercially available or at least easily constructed from common materials. A number of sensors meet these requirements reasonably well and have been tested in robotic systems.

Tin Oxide Gas Sensors

The commercially available tin oxide sensors manufactured by companies such as Figaro Engineering Inc. have been used in the majority of robotic experiments. They are inexpensive, small in size, easy to interface, and the latest miniature devices made by thick-film technology have relatively low power consumption (the heater of the Figaro TGS2600 'air quality' sensor consumes 210 mW). The tin oxide sensor element is heated to about 300°C, where the presence of reducing gasses leads to a drop in sensor resistance. Sensitivities of the order of 1 ppm are available and sensors have been developed with responses optimized for solvents, halocarbons, and a range of combustible gasses. As an approximation sensor output resistance varies with gas concentration as follows:

$$\frac{R}{R_0} \cong KC^{\alpha} \tag{75.1}$$

where

R = sensor resistance in the presence of a reducing gas,
R_0 = sensor resistance in clean air,
C = gas concentration, and
α and K are constants relating to the particular sensor.

Quartz Crystal Microbalance

Quartz crystal microbalance sensors use a quartz crystal as a sensitive balance to weigh odor molecules. A chemical coating on the crystal is chosen to have a specific affinity for the target odorant molecules. When air containing molecules of this odor is drawn over the crystal, some of the molecules become temporarily attached to the coating. This increases the mass of the crystal and lowers its resonant frequency. A simple model proposed by Sauerbrey predicts the effect of small amounts of added mass:

$$\Delta f = -\frac{2f^2 \Delta m}{\rho v} \tag{75.2}$$

where

Δf = change in crystal resonant frequency,
f = crystal resonant frequency,
Δm = increase in the mass of the crystal per unit area,
ρ = density of crystal material, and
v = velocity of sound waves in the crystal material,

It is interesting to note that the change in crystal frequency resulting from added mass is proportional to the square of the crystal resonant frequency. Therefore, increasing the crystal resonant frequency has a very beneficial effect on its sensitivity as a chemical sensor.

Conductive Polymer

Conductive polymers can also be used to make chemical sensors for applications in robotics. An inexpensive and safe way of making polypyrrole has been reported, which uses electrochemical polymerization of pyrrole between a pair of electrodes. The resulting polypyrrole sensors have good sensitivity to common and relatively nontoxic chemicals such as ethanol and ammonia that can be used in robotic experiments. However, there are problems of consistency between sensors and of stability and aging.

If a robot is operating in a constrained environment and can be guaranteed to only encounter a single chemical, then the robot may be required to detect but not to recognize the chemical. In most other situations where the robot must respond to a specific chemical in the presence of other chemical species, the robot needs the ability to discriminate between different odors.

Odor Discrimination

In a number of commercial situations, assessing the quality of an odor is important. The aroma of foodstuffs such as bread, biscuits, coffee, beer, and wine are all important to the consumer. In the nonfood area, the success of perfume and scented products such as soap, washing powder, and air freshener all depend on their smell. Currently, quality control of these products is performed by humans. The human sense of smell is affected by age, health, and eating habits. There would be a number of commercial applications for an artificial odor discrimination system that could perform these quality-control tasks.

Odor discrimination systems are being developed, which have the structure shown in Figure 75.4.

The system comprises a number of odor sensors that are exposed to the odorant. Each sensor has a peak sensitivity to a different chemical species. However, their sensitivity is broad and overlaps between sensors. One specific odor will give a unique pattern of response from the odor sensors. This pattern is interpreted, in this case by a trained neural network, to identify the odor. The neural network may be trained using the back propagation algorithm by repeated exposures to the range of odors which the system will be required to discriminate. This kind of system is called an electronic nose, and several commercial versions are available on the market for assessing odors. The individual sensors used in an electronic nose would typically be tin oxide or quartz crystal microbalance chemical sensors. Electronic noses are used in robotic systems where an odor recognition capability is required.

Chemical sensors are not directional and do not form an image. They are only able to measure chemical concentration at a location in space. A robot requires additional information if it is to move in the direction of the chemical source. In some source location experiments, chemical gradient information has been used to determine the path that the robot must take. Because of the rapidly fluctuating nature of turbulent odor plumes, it is difficult to obtain reliable gradient information either by having widely spaced sensors on the robot or by taking chemical readings at different locations in the robot's environment. If the

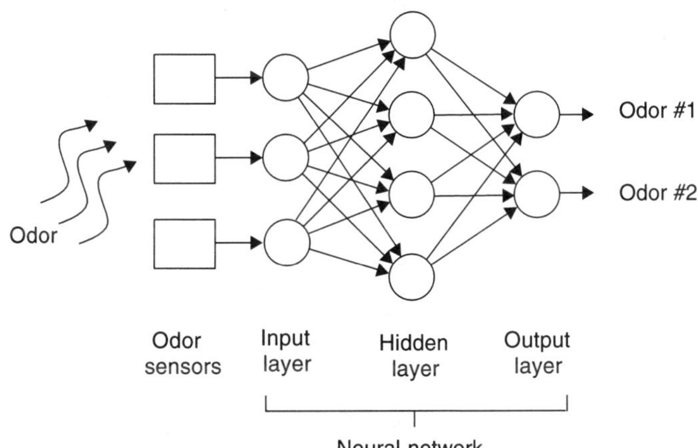

FIGURE 75.4   An odor discrimination system.

chemical plume is constrained, perhaps by passing down a narrow corridor, then provided the walls do not absorb the chemical, there will be no concentration gradient at all. For these reasons, it is often necessary for a plume tracking robot to have additional information about local airflow.

75.3 Airflow Monitoring

Since the early days of meteorology, wind direction has been commonly measured by wind vanes or wind socks. By contrast, many techniques are used for measuring wind velocity. These include turbine flowmeter, three-cup anemometer, pitotstatic tube, and hot-wire/film anemometer. In many applications, robots will be operating indoors involving wind velocities of well below 1 m/s. Using conventional techniques, measuring velocity and direction becomes difficult at these air speeds. A number of airflow measurement techniques have been developed especially for use by robots in an indoor environment.

Rotating Paddle

At Monash University, a novel active airflow velocity and direction sensor has been developed (Russell and Kennedy, 2000). This device works at low wind velocities and its simplicity, potential for miniaturization, and low power consumption make it suitable for mounting on small robots (10 cm diameter and smaller). The key feature of the sensor is a flat plate or paddle rotated by a precision dc motor (see Figure 75.5). In still air, the rotational speed of the paddle will be constant. However, when situated in an airflow, the paddle will slow down when moving upwind and speed up when moving downwind. The difference between the maximum and minimum paddle speed is proportional to wind velocity. Airflow direction can be inferred by noting the rotation angle corresponding to maximum or minimum paddle velocity and applying the appropriate 90° offset. A prototype version of this sensor was able to provide reliable airflow information with wind velocities below 0.1 m/sec.

Measuring Both Chemical Concentration and Airflow Direction

A group at Tokyo Institute of Technology has developed a number of innovative robotic sensor systems for use in locating the source of a chemical plume. These systems measure a combination of airflow direction and chemical concentration . Hiroshi Ishida and coworkers (1994) describe a system using four anemometric and four gas sensors mounted on a small mobile robot. As illustrated in Figure 75.6 the four anemometric sensors are sheltered by a square central pillar and for this reason the sensor that is located downwind records the lowest wind velocity. This effect can be used directly to determine the wind direction to within 90°. For a more accurate indication of airflow, patterns of anemometric sensor reading were normalized and recorded for wind directions rotated in increments of 45°. Wind direction could then be estimated to within 45° by finding the best match, based on minimum Euclidean distance,

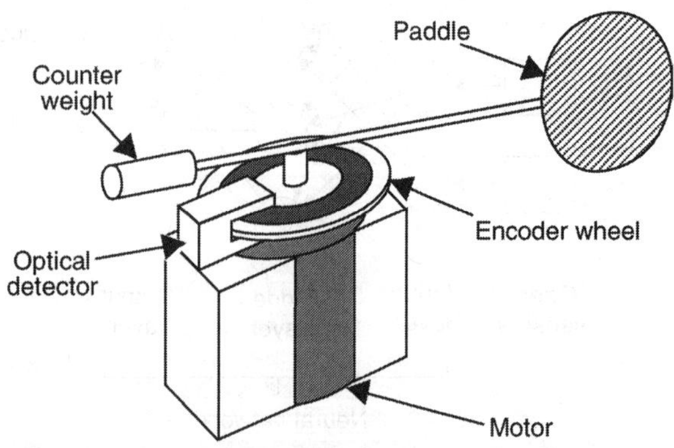

FIGURE 75.5   A diagram of the paddle airflow sensor.

between actual and stored anemometric readings. Once wind direction is established, then the chemical concentration gradient perpendicular to the flow direction can be estimated. This concentration gradient is found by subtracting readings from the two gas sensors that are at right angles to the measured wind direction. Airflow and concentration gradient information is then used to control the movements of the robot.

It has been observed that male silkworm moths flutter their wings when stimulated by the female pheromone Bombykol. Fluttering of the wings draws air over the moth's antennae and expels it behind the moth. It has been proposed that this flutter-induced airflow helps the moth to work out the direction to the pheromone. When the moth is facing toward the pheromone source, wing fluttering draws odor over the moth's antennae and an increase in pheromone is detected. If the moth is facing away from the source, then wing fluttering blows the pheromone away and less will be detected. Hiroshi Ishida and coworkers (1996) designed an active sensor based on this principle. In this sensor a small electric fan generates airflow and a TGS822 gas sensor detects the ethanol vapor used as the target odor. A diagram of the sensor is shown in Figure 75.7.

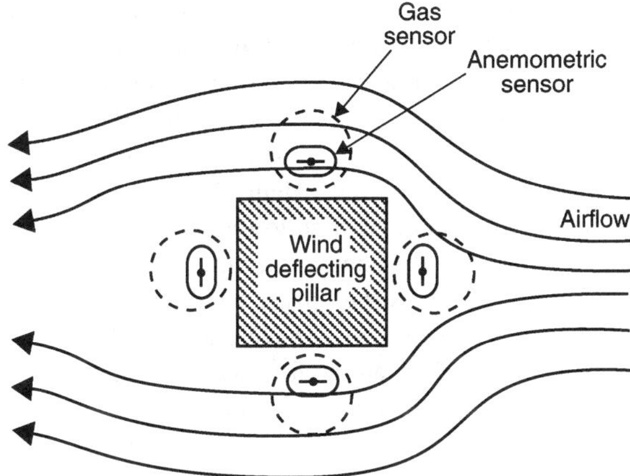

FIGURE 75.6 Plan view of a combination of gas and anemometric sensors used to estimate wind direction and chemical gradient.

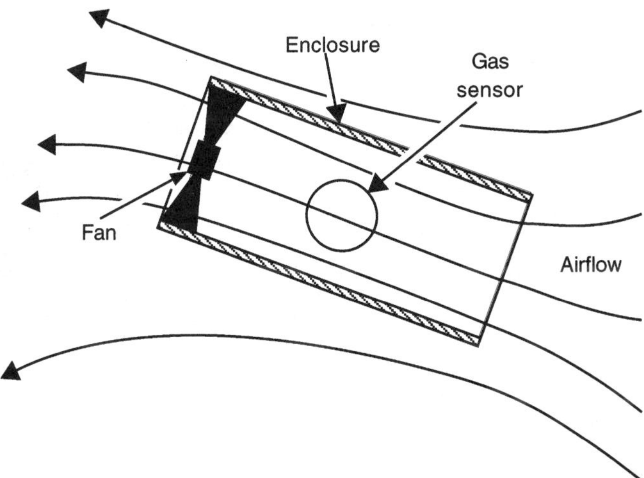

FIGURE 75.7 By rotating the active-sampling sensor the direction to the chemical source can be determined.

Whisker sensors to measure airflow

Humanoid robots will require a rich set of sensory inputs if they are to work effectively in a human environment. Chemical sensing in the form of an electronic nose would be a natural choice for part of a humanoid robot's sensor suite. In many cases, once a chemical has been detected and identified by the electronic nose, the task of the robot will be to locate the source. For this, the humanoid robot will need some means of determining airflow direction. It is a reasonable requirement that sensors carried by a humanoid robot blend in with its humanoid form and for this reason a wind vane is not appropriate. At Monash University, a sensor for determine airflow has been developed that is compatible with the humanoid form (Russell and Purnamadjaja, 2002). It consists of an array of sensory whiskers attached to a sweat band worn by the humanoid head (Figure 75.8). Turbulence in the airflow around the head causes the whiskers to vibrate and this is detected by an optical switch. Over a limited range, the frequency of whisker vibration is proportional to airflow velocity and the distribution of activation of the whisker sensors around the head provides information on the direction that airflow is incident on the head.

Sensing and Airflow

During experiments to develop robots that can follow chemical trails deposited on the ground, it has been found that unenclosed chemical sensors give very unreliable information. Sometime a sensor hardly responds at all when positioned immediately above the trail. In other situations, a strong but widely fluctuating response can be measured when the chemical marking is some distance away. These problems were found to be caused by air currents close to the floor. By careful control of the airflow around the sensor, these problems can be overcome. Figure 75.9 shows a cross-section view of a sensor that generates an air curtain to exclude odors carried to the sensor from some distance away (Russell, 1995). At the same time, air is drawn from the floor surface and over an odor-sensing quartz crystal. The airflow generated within and around the sensor is illustrated in Figure 75.9.

In addition to chemical and other sensors, a robot also requires control algorithms to allow it to make use of its sensory information. The insect world is an excellent source of simple algorithms for implementing chemical searching and tracking behaviors in robot systems, and some examples are given in the next section.

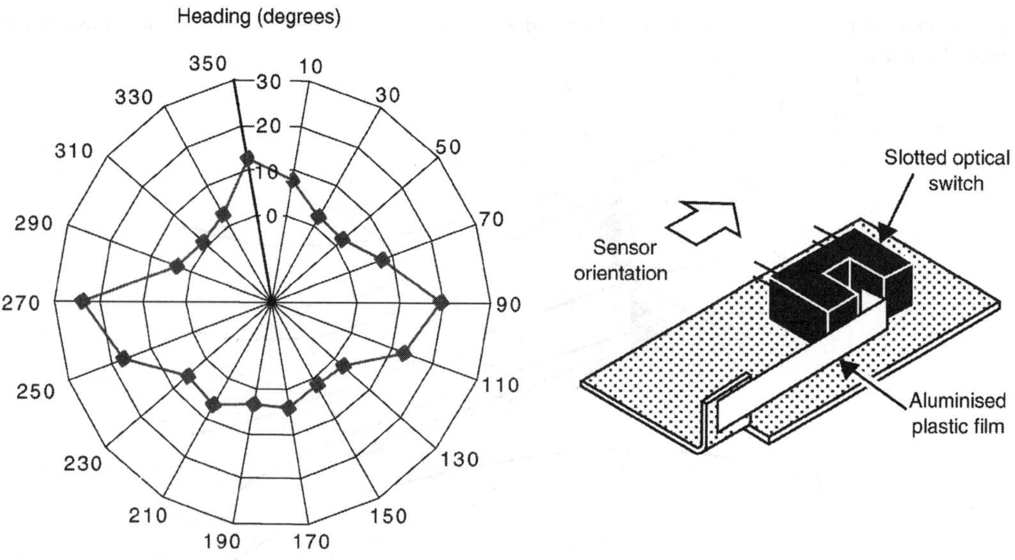

FIGURE 75.8   A graph of whisker airflow sensor readings taken at 20° intervals around the headband of a humanoid robot (robot facing an airflow) and a diagram of one whisker sensor.

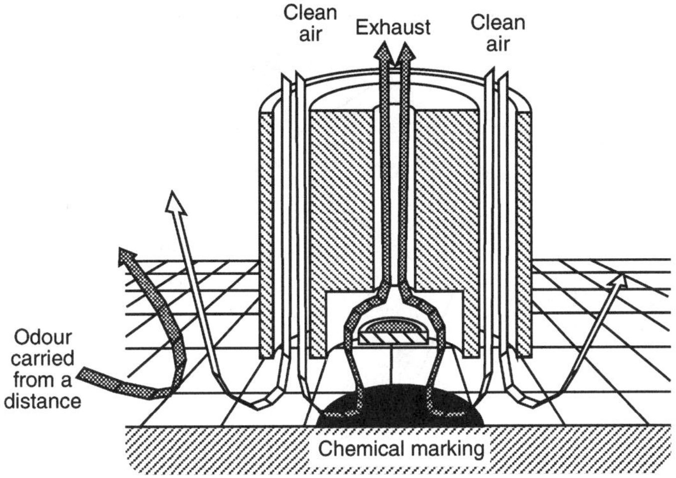

FIGURE 75.9 The air-curtain sensor.

75.4 Examples of Robotic Chemical Sensing

Finding the Source of a Chemical Plume Using the Dung Beetle Algorithm

The dung beetle *Geotrupes stercorarius* is reported to use a very simple strategy to follow the odor plume from a cow pat. Initially, the beetle flies across wind to give it the best chance of finding a plume of cow pat odor. Upon detecting this odor, it zigzags diagonally across the plume in an upwind direction turning back each time it leaves the edge of the plume. When the odor plume ends, as it does immediately above the cow pat, the beetle flies down and usually lands on or near its goal. In order to implement this algorithm, a robot requires information about the direction of airflow as well as chemical concentration. For the experiment illustrated in Figure 75.10, a small mobile robot was equipped with polypyrrole conducting polymer sensors that show a very strong response to ammonia (Russell, 1999b). The conducting polymer sensors had an initial resistance of about 25 kΩ and exhibited a maximum resistance change of almost 20% on exposure to concentrated ammonia vapor. In addition, the robot had a rotating paddle airflow sensor and tactile whiskers to detect contact with the odor source. The odor source consisted of a 1 l/min flow of air. The flow of air carried with it ammonia vapor resulting from bubbling through a flask of 5% ammonia solution. To establish an airflow in the laboratory, a cooling fan was run at a reduced voltage to give an airflow of 0.3 m/sec at a distance of 2 m from the fan.

In order to make contact with the ammonia plume, the robot starts by searching across wind. The dashed line in Figure 75.10 shows the path of the robot when the ammonia source was not present. This line is at right angles to the direction of airflow from the fan. The solid line marks the path of the robot as it detects the plume and then zigzags toward the source. The robot halts when it collides with the source.

Ant-like Pheromone Trail-Following

The trail-following algorithm used by the ant *Lasius fuliginousus* seems to be very simple. The normal mode of trail-following for the ant is to curve to the left until the right antenna detects the trail and then to curve to the right until the left antenna contacts the trail. In this way, the ant maintains the trail between its two antennae. An important feature of the ant's tracking algorithm involves turning back toward the trail after an extended period even if the appropriate antenna does not detect the pheromone. This prevents the ant walking in circles when it loses the trail. Instead, the ant moves forward in the general

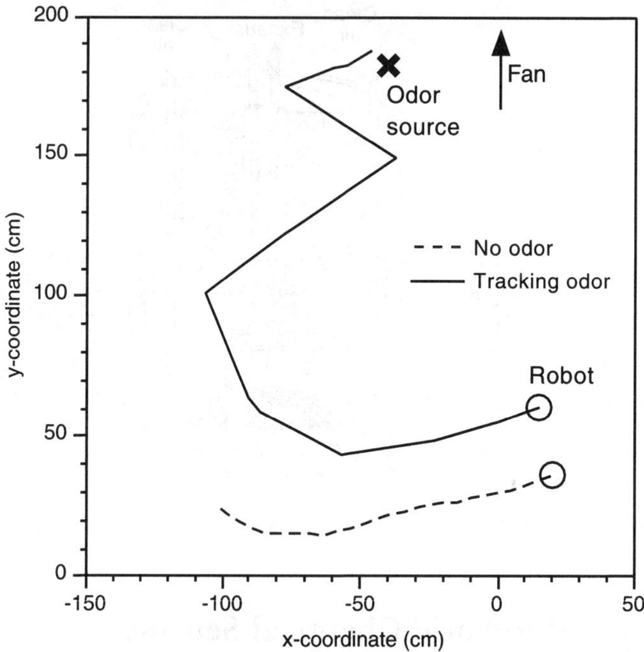

FIGURE 75.10   The path of a mobile robot as it locates the source of an ammonia plume using the dung beetle algorithm.

direction of the trail sweeping widely to left and right, which gives a good chance of picking up the trail again. The following algorithm was derived from this and implemented on a six-legged robot having two chemical sensors mounted ahead of the robot-simulating antennae (Russell, 1999c).

The ant trail-following algorithm:

- Take 8 paces bearing to the right until completed or until trail detected by the left antenna
- Take 8 paces bearing to the left until completed or until trail detected by the right antenna
- Repeat.

(If the left sensor detects the chemical trail, then the sequence of 8 paces to the left is started or restarted and stimulation of the right sensor starts or restarts the sequence of 8 paces to the right.)

Chemical trails were laid on the ground using a solution of camphor dissolved in ethanol. A few seconds after the trail was laid, the ethanol evaporated leaving a relatively long-lived layer of camphor. Chemical sensors were of the quartz crystal microbalance type consisting of an unencapsulated 10 MHz crystal coated with Silicone OV-17. The output of an oscillator incorporating the sensor crystal was subtracted from a reference oscillator to give a low-frequency difference signal that could be counted by a microcontroller.

One of the important advantages of the *Lasius fuliginousus* algorithm is the ability to accommodate large gaps in the chemical trail. This is not surprising because the pheromone trails produced by biological ants are made up of many discrete odor patches and probably have many gaps. To test the capabilities of this algorithm, a camphor trail was laid on the floor consisting of a 55 cm straight section, a gap of 40 cm, and then a further straight section finishing with a curve. The ant robot successfully followed the camphor trail in both right to left and left to right directions. Figure 75.11 shows the trajectory of the robot.

Locating the source of chemicals released underground

There are significant economic and humanitarian applications for methods of locating underground chemical sources. Such sources include land mines as well as chemicals leaking from gas pipelines and

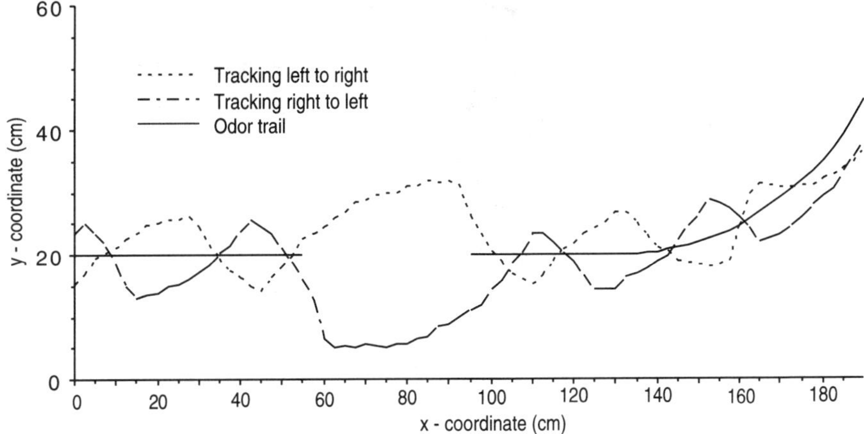

FIGURE 75.11   Practical results of robot chemical trail-following based on the ant *Lasius fulginousus*.

chemical storage tanks. If there is no pressure difference, chemicals are transported through the ground by diffusion. This provides a smooth distribution as opposed to fluctuating concentration produced by turbulence in air or water flows.

For the application of finding a chemical source, sensor readings must be taken in undisturbed soil and therefore the robot control strategy should take widely spaced readings. The following algorithm has been especially developed in the Intelligent Robotics Research Center at Monash University for this application.

The hex-path algorithm:

repeat

{

if (intensity at *n-2* > intensity at *n-1*) and (rotation direction at *n-1* was anticlockwise) or

(intensity at *n-2* < intensity at *n-1*) and (rotation direction at *n-1* was clockwise)

then rotate anticlockwise 60° and move forward *m*

else rotate clockwise 60° and move forward *m*

}

After burying the chemical source, a period of three hours was allowed to elapse for the chemical concentration to stabilize throughout the volume of sand. An airflow was established over the surface of the sand using a cooling fan to ensure that the surface concentration of ethanol was essentially zero. Without the fan, negatively buoyant ethanol vapor tends to concentrate above the surface and distort the chemical distribution in the sand.

Experiments were performed in a plastic container 60 cm long by 40 cm wide and filled to a depth of 10 cm with dry sand. The chemical source consisted of an open topped metal can 4 cm diameter and 2 cm deep half-filled with ethanol. A covering of cotton material over the open top of the can exclude sand while allowing the ethanol vapor to escape. This source was buried and covered by 4 cm of sand. To detect the ethanol vapor, a TGS2600 tin oxide sensor was used. In this experiment, the robot manipulator arm was used to guide a probe containing the sensor through the sand. Only the sensing tip of the probe penetrates the sand.

For the hex-path algorithm, two moves must be made before the algorithm can take over control of the robot. These two movements are shown by the dashed line in the plotted robot trajectory. These moves are deliberately chosen so that initially the robot is heading away from the source. Figure 75.12 shows a typical robot trajectory. When the probe is first inserted, a chemical reading is taken and then the robot proceeds by moving forward 3 cm, taking a chemical reading and then turning either clockwise 60° or anticlockwise 60°. The measured chemical sensor output is indicated by the size of the circle at each turning point. Recording stops when the sensor probe makes contact with the source.

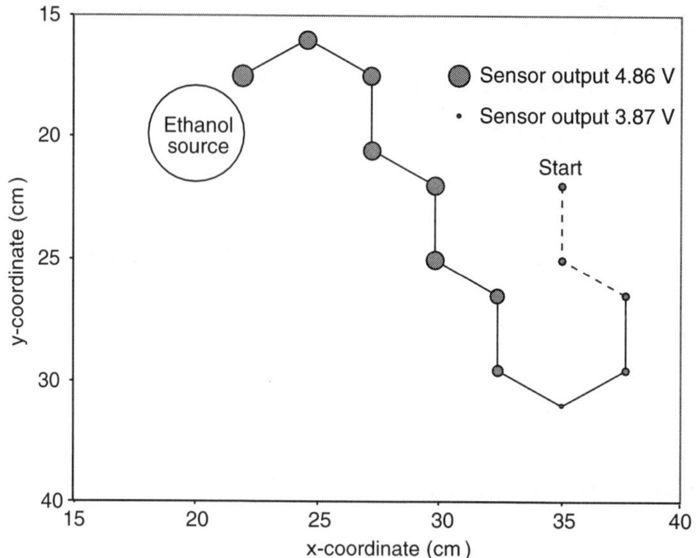

FIGURE 75.12   Experimental results showing sensor probe path as the robot locates the buried ethanol source.

75.5   Conclusion

Robots could use the ability to detect chemicals as part of an interrobot pheromone communication system. This would serve the same purpose as animal and insect pheromones to help organize large groups of robots and to pass information between them. Obvious applications include the laying of trails to aid navigation and the release of plumes of chemical to act as a distress call. In a manner similar to the use of sniffer dogs, chemical sensing robots could be used to enhance and project the human sense of smell.

The vital elements required to add chemical sensing capabilities onto robotic systems are appropriate sensors together with algorithms for processing the sensor data. It will be apparent from this survey that robotic odor sensing technology is still at an early stage of development. A small number of chemical sensors have been found suitable for use in robotics research. However, their sensitivity and selectivity are very poor compared with biological systems. The ability to follow chemical trails on the ground, to find an underground chemical source, and to locate the source of a chemical plume have been demonstrated experimentally. There are many important applications for this kind of capability but current systems have not been developed beyond an early experimental stage. The eventual availability of more sensitive, selective, and faster responding sensors will be one of the major factors governing the creation of chemically sensitive robots.

References

Ishida, H., Suetsugu, K., Nakamoto, T., and Moriizumi, T., Study of autonomous mobile sensing system for localization of odor source using gas sensors and anemometric sensors, *Sensors and Actuators A*, 45, 153–157,1994.

Ishida, H., Hayashi, K., Takakusaki, M., Nakomoto, T., Moriizumi, T., and Kanzaki, R., Odor-source localization system mimicking behaviour of silkworm moth, *Sensors and Actuators A*, 51, 225–230, 1996.

Larcombe, M.H.E. and Helsall, J.R., *Robotics in Nuclear Engineering*, Published by Graham and Trotman Ltd., for the Commission of the European Communities, 1984.

Russell, R.A., Laying and sensing odor markings as a strategy for assisting mobile robot navigation tasks, *IEEE Robotics and Automation Magazine*, Vol. 2, no. 3, September 1995, pp. 3–9.

Russell, R.A., *Odor Detection by Mobile Robots*, World Scientific, Singapore, 1999a.

Russell, R.A., The World of Odor: a Relatively Unexplored Sensory Dimension for Robots, Proceedings of the 1999 International Symposium on Robotics Research, Salt Lake City, Utah, 1996b, pp. 74–79.

Russell, R.A., Ant Trails — an Example for Robots to Follow? Proceedings IEEE International Conference on Robotics and Automation, Detroit, MI, 1999c, pp. 2698–2703.

Russell, R.A., and Kennedy, S., A novel airflow sensor for miniature mobile robots, *Mechatronics*, 10, 935–942, 2000.

Russell, R.A. and Purnamadjaja, A.H., Odour and Airflow: Complementary Senses for a Humanoid Robot, Proceedings of the IEEE International Conference on Robotics and Automation, Washington, DC, 2002, pp.1842–1847.

Ultrasonic Sensors in Robotics

Lindsay Kleeman
Monash University

76.1 Introduction

SOund NAvigation and Ranging (Sonar) sensing is a popular method for range measurement in robotics applications based on the active transmission of ultrasonic acoustic energy and the timed reception of the echo. The range to a reflector is computed from half the time of flight multiplied by the speed of sound, c. The speed of sound varies with temperature, humidity, and pressure. Detailed formulae for these effects can be found in [6]. The temperature dependence is most significant and is approximated by

$$c = 20.05 \sqrt{T_C + 273.16} \text{ msec}^{-1}$$

where T_C is the temperature in degrees Celsius. At 20°C, the speed of sound is approximately 343 msec^{-1}.

Sonar systems are popular in robotics applications due to their widespread availability, low cost, and ease of use. The most common form of sonar is the Polaroid 6500 Ranging Module (PRM) [12], which was originally designed for auto-focus cameras. The PRM uses pulse echo sonar whereby a short pulse of acoustic energy is transmitted and the same transducer acts as a receiver. The Polaroid ranging module reports the range to the nearest reflector based on a time-of-flight estimator using a fixed threshold on a received signal whose gain is increased in steps from the firing time. The position of the reflector is uncertain since the bearing angle to the reflector is only restricted to within the beamwidth of the transducer that can be as large as 30°. The PRM thus has poor angle measurement capability. Moreover, the echo amplitude is not directly measured either.

More recently, however, a technique has been developed by Roman Kuc [13], which can recover some echo amplitude information from the Polaroid ranging module by immediately resetting the thresholding circuitry when an echo is detected, thus allowing additional echo detections. The number of times the echo exceeds a threshold provides an estimate of the echo amplitude and this approach has been applied to differentiate specular targets (high amplitude) such as smooth walls from diffractive targets (low amplitude) such as edges. This approach is described in more detail in Section 76.2 of this chapter.

By processing the received signal using matched filtering on a Digital Signal Processor (DSP), significant improvements have been achieved in the estimation of time of flight [8]. Combining this time-of-flight estimation with the use of two transmitters and two receivers, an "advanced" sonar has been developed with the following enhancements. By using two receivers, the bearing angle to the reflector can be accurately estimated to an error standard deviation of approximately 0.1° in relatively still air conditions. Range error standard deviations are around 0.2 mm. Furthermore, by deploying two transmitters, reflectors can be classified into the geometric types of planes, concave corners and convex edges. One transmitter is fired a precisely controlled delay from the other (usually 150 to 500 μsec) to enable the echoes to be associated with the correct transmitter and also to enable interference from other sonar systems to be rejected on the basis of the arrival time separation. This advanced sonar system can classify reflectors, determine range and angle, and reject interference in a single transmit/receive cycle time. Processing can be achieved in near real time using DSP technology. Advanced sonar is described in more detail in Section 76.4 of this chapter.

An alternative approach to transmitting a short acoustic pulse is to use a Continuous Transmission Frequency Modulated (CTFM) approach. CTFM can allow greater energy to be transmitted compared to pulse echo techniques and hence the possibility of detecting weaker echoes exists. Arrival time is estimated from the frequency difference of the transmitted and received signals. Some researchers have also used this technique to classify plants for use as navigational beacons for mobile robots [3]. CTFM techniques are also described in [14].

A domestic vacuum cleaner robot is available from Electrolux [11], which incorporates a sonar system with a wide horizontal beamwidth of approximately 180° and a narrower vertical beamwidth of approximately 20° that reduces spurious reflections from floor surfaces. The transmitter is an electrostatic device 25 mm wide and covering 150° of the front of the robot. The transmitter operates at 60 kHz and echoes are detected with four microphones spaced around the front and recessed into the robot using tubes to further narrow the vertical beamwidth. More details are provided in [11].

Teruko Yata et al. [16] have developed a 360° ring of 30 transmitters and 30 receivers [15]. The transmitters are fired simultaneously to enable the surrounding environment to be sensed rapidly. Received signals are threshold with a decreasing voltage as a function of time from transmission, and receiver data are segmented to find the location of targets. The vertical beamwidth is narrowed using exponentially shaped wings. This system is described in more detail in Section 76.3.

76.2 PRM Enhancements

The PRM [12] provides a popular, low-cost, and simple to use sonar system. A single electrostatic transducer acts as a transmitter and receiver. A digital interface is provided whereby transmission of 16 periods of a 49.4 kHz square wave are emitted when the INIT digital input is asserted. On receiving of an echo, the digital output ECHO is asserted and the time of flight can be measured by timing the asserting of the INIT signal to the ECHO output. The echo signal is obtained from the same transducer used to transmit the ultrasonic pulse and is amplified with a time-varying gain. The gain is increased from the time of transmission in discrete steps. Rectification and lossy integration (for noise suppression) are used in the receiving electronics. When the result exceeds a fixed threshold, the ECHO output is asserted. Detailed modeling of the effect of beamwidth and thresholding is given in [12], which can assist in localizing reflectors given many PRM time-of-flight readings taken by scanning the direction of the transducer.

The PRM can be used to extract primitive echo amplitude information [13]. This works by applying a reset (using a pulse on BLNK input to the PRM) when the ECHO output of logic 1 is detected. Figure 76.1 taken from [13] illustrates the effect of this strategy. Notice that the number of ECHO pulses that appear are then related to the amplitude of the echo. Lines in Figure 76.2 represent ECHO pulses generated with this technique for various strength reflectors as the PRM is scanned across the scene.

76.3 Piezoelectric Sonar Ring

Piezoelectric transducers with wide beamwidth have been deployed in a ring of diameter 220 mm. Thirty transmitters and 30 receivers are alternatively positioned around the ring. The transmitters are fired simultaneously with a common drive voltage of 150 V that results in a receiver waveform lasting approximately

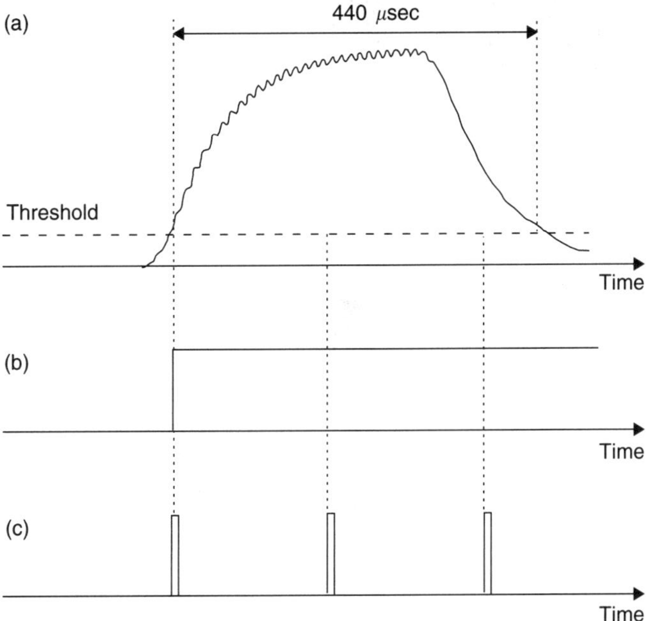

FIGURE 76.1   PRM operation modes: (a) processed echo waveform, (b) ECHO output produced in conventional time-of-flight mode, and (c) ECHO output produced in pseudoamplitude mode (taken from [13]).

Object		Arc extent (deg)
Plane		48.3
8.9 cm dia post		24.6
2.85 cm dia post		23.1
8 mm dia pole		21.9
1.5 mm dia wire		19.2
0.6 mm dia wire		10.5

FIGURE 76.2   Pseudoamplitude sonar maps of six objects located at 1 m range. The sonar is located below the figure (taken from [13]).

700 μsec or 28 cycles at 40 kHz. The operation of the ring is illustrated in Figure 76.3 taken from [15]. A variation in echo amplitude of 2:1 around the ring for the same reflector is reported. This variation can be attributed to sensitivity differences in the transmitters, receivers, and constructive and destructive interference between neighboring transmitters. The received signals are amplified and compared with a threshold that decreases with time. The binary result from the comparison is stored in memory every 1 μsec for each of the 30 channels. The binary echo results are processed to determine the leading edge of each echo and associated across different channels into groups, each of which corresponds to a reflector in the environment. By fitting geometric models to the echo arrival times in each group, the range and bearing to reflectors are determined. One practical problem that is addressed in the design is removing "cycle hopping" artifacts — that is, the arrival time estimation based on a threshold is susceptible to signals that just fall below the threshold in one cycle on one receiver yet achieve the threshold on an adjacent receiver. This may mean that errors of close to a full cycle can appear in the data and this is explicitly recognized, and discrete trials of one cycle error compensation are instigated to bring about better matching with the geometric models. Standard deviations of 0.4° in bearing angle errors for 1.5 m range to 45 mm diameter rods have

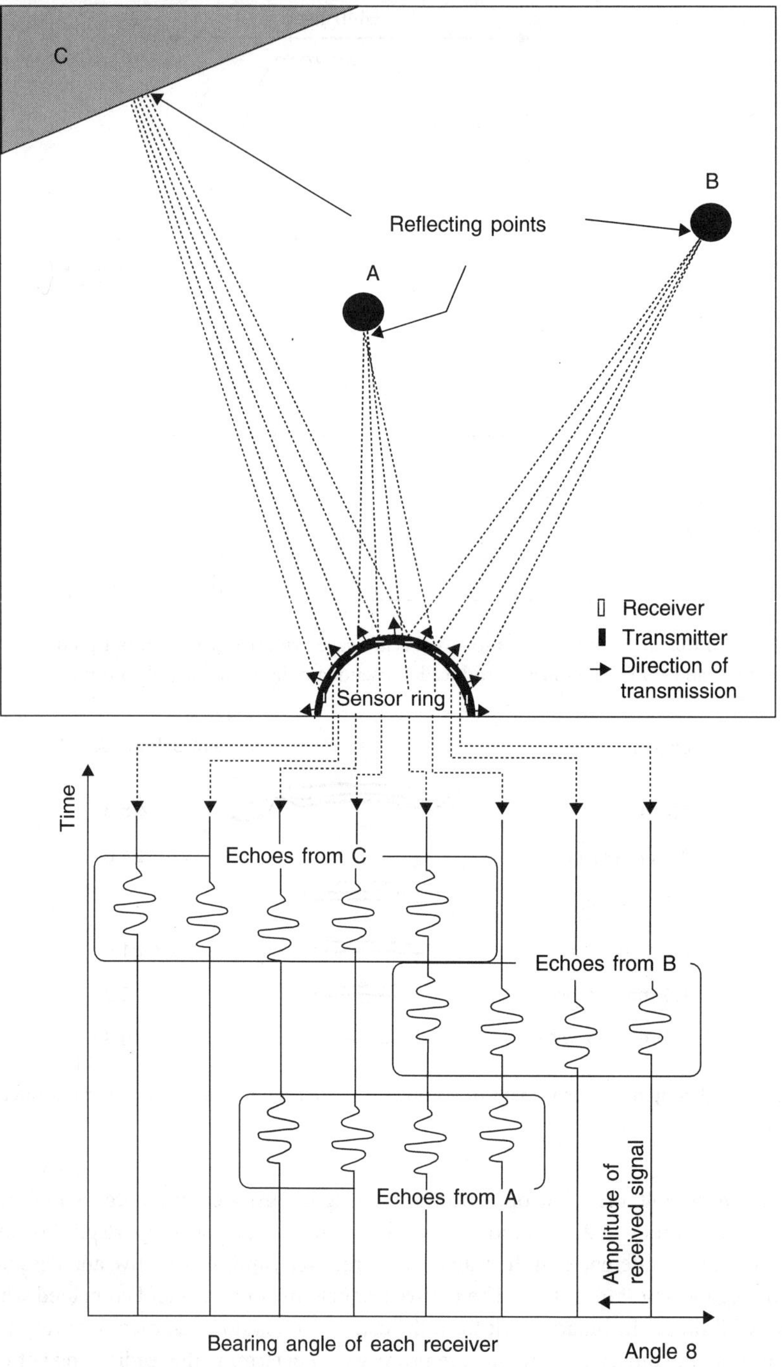

FIGURE 76.3　Representation of echo signals in a sonar ring due to three reflectors in the environment (taken from [15]).

been measured. Ranges beyond 2 m are not apparent in the experimental data. The system also has problems discriminating closely spaced targets, since the pulse duration of 700 μsec corresponds to approximately 240 mm in air. Thus, a second reflector closer than 120 mm in range would be masked by the first reflector due to pulse overlap.

76.4 Advanced Sonar Sensing

The term "advanced" sonar sensing refers to sensing that provides range, bearing, target classification, and interference rejection. This section describes how these features can be achieved and also the minimum sensor requirements. Range estimation can be achieved with a single transducer acting as a transmitter and a single receiver. Estimating the bearing angle in two dimensions of an echo can be achieved by two approaches.

The first approach uses just one receiver and estimates the angle based on the amplitude of the echo [16]. The echo amplitude at the receiver is a symmetrical function of the angle to the normal direction of the transducer, and consequently, only one side of the transducer beamwidth can be used to unambiguously measure the angle. By modeling the amplitude response of a circular transducer as a function of angle and range, an estimate can be achieved of the angle of reception from high reflectance targets. This approach is only practical for ranges less than 2 m due to the variation of absorption losses in air due to variations in humidity and temperature. Errors as low as 1° have been reported with this approach [16]. When the reflectance is low, it is difficult to differentiate the reduction in echo amplitude due to reflectance from angle reduction, unless an angular scan of the environment is taken and the amplitude model is fitted to the scan [7].

The second approach for bearing estimation uses two receivers and accurate time-of-flight estimation on each receiver. From the difference in arrival times of an echo, the angle of arrival can be accurately estimated. The accuracy is limited by the air conditions, such as turbulence and temperature gradients, and typical errors of 0.1° standard deviation have been reported in office-type environments [8]. The estimation of accurate arrival time is achieved by sampling the analog echo signal at 1 MHz with a 12-bit analog to digital converter and applying matched filtering digital signal processing. Matched filtering consists of calculating the correlation of the echo with an echo template for different positions of the template relative to the echo. The position of maximum correlation corresponds to the arrival time. The resolution of the arrival time can be improved over the signal sample time by fitting a parabola to the correlation as a function of time shift and using the peak of the parabola as the arrival time [8]. The template echo shape can be derived from a calibration procedure performed *a priori* that collects an echo from a plane reflector perpendicular to the transducer normal. This zero degree echo shape is distorted to account for different angles of arrival by double convolution with an elliptical impulse response with width proportional to the angle from normal [8]. A set of templates for different angles are generated and used in the matched filter of receiver data. The template shape also varies with range due to the dispersion of sound in air and this can be modeled to generate different template sets for each meter of range. When many echoes are received on the two receivers, the correspondence problem must be solved in order to extract bearing measurements. The correspondence problem is how to associate an echo on one receiver to an echo on the other. The problem can be alleviated by positioning the receivers as close together as the transducer dimensions allow, so that echoes on different receivers will be as close as possible in time, thus reducing the number of associations possible. Echo amplitude, template match angle, and other features can be used as attributes to assist in solving the correspondence problem. Nevertheless, certain reflector configurations can be constructed to produce ambiguous correspondences, but in practice, these occurrences are rarely found. A pragmatic approach is to flag any ambiguous correspondence and allow the resolution of the correspondence at a higher level in the data interpretation of even ignore these readings altogether.

Target classification with advanced sonar can be achieved with the following target types: plane, concave right angle corner, and point feature (often called an edge). A plane is analogous to a flat mirror in that a virtual sonar image is generated. Similarly, with a corner, two plane reflections occur and a reversed sonar image is formed. With a point feature, the echo appears to come from one position independent of the transmitter position and can correspond to high curvature objects such as table legs, or correspond to small concave corner reflectors where the dimensions of the corner are smaller than the transducer diameter. In order to classify targets into plane, corner, and point types, two transmitter positions and two receivers are required [8]. The classification process is analogous to observing one's image in a flat mirror, a corner mirror, and a shiny table leg. The image is reversed between the flat and corner mirrors, and the image is compressed into a small width in the table leg. In terms of sonar classification, one transmitter is fired and the angle of the echo is noted and then the other transmitter is fired

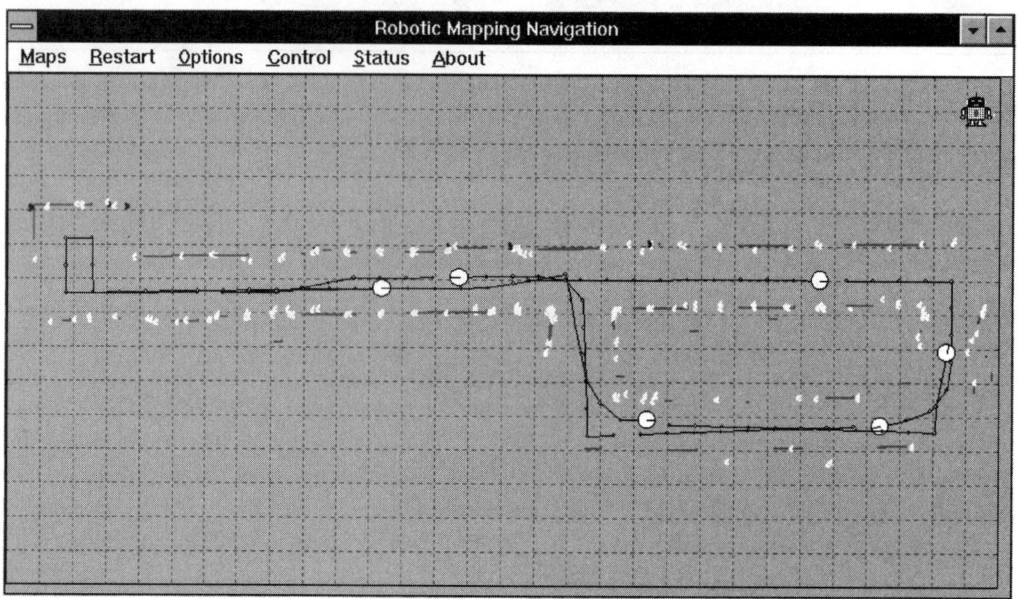

FIGURE 76.4   A large-scale sonar map generated by an advanced sonar sensor mounted on a mobile robot. Measurements are taken every 1 m of robot travel. A 1 m grid is marked with dashed lines and planes are shown as straight line segments, edges as "<" oriented marks, and corners by "<" oriented marks (taken from Figure 19 in [2]).

and the difference in angle is calculated. This angle difference can be used to determine the target classification. Large-scale simultaneous mapping and localization with a mobile robot have been achieved using this classifying sonar system [1, 2]. An example of a map constructed using advanced sonar is shown in Figure 76.4 taken from Figure 19 in [2].

Interference rejection allows multiple advanced sonar systems to operate in the same acoustic environment. The transmitter source of an echo needs to be identifiable. A double pulse coding approach is adopted in [10], whereby two pulses are transmitted separated in time by typically 150 to 500 μsec. Should two sonar systems interfere with each other, by examining the echo time-of-flight separations, the other sonar's transmissions can be rejected if they operate with different double pulse separations. One obvious problem is that echoes may overlap in time causing the arrival time estimation to fail. The echo energy has been concentrated into approximately 80 μsec or five or six cycles of 60 kHz to minimize this problem.

Classification and interference rejection have been combined into one sensing cycle [4], whereby two transmitters fire as the first and second pulses, respectively, of the double pulse code. In practice, the receiver arrival times can be associated with a transmitter by examining the relative timing. Moreover, double pulse spacing outside the expected range can be deemed to be interference and can be rejected. This approach has been implemented on a Digital Signal Processor (DSP) and runs in close to real time at 25 Hz repetition rate for ranges out to 5 m [9, 5].

References

[1] Chong, K.S. and L. Kleeman, Feature-based mapping in real, large scale environments using an ultrasonic array, *International Journal of Robotics Research*, 18, 3–19, 1999.

[2] Chong, K.S. and L. Kleeman, Mobile robot map building for an advanced sonar array and accurate odometry, *International Journal of Robotics Research*, 18, 20–36, 1999.

[3] Harper, N.L. and P.J. McKerrow, Recognition of Plants with CTFM Ultrasonic Range Data Using a Neural Network, Proceedings, 1997 IEEE International Conference on Robotics and Automation, Vol.4, April 20–25, 1997, pp. 3244–3249.

[4] Heale, A. and L. Kleeman, Fast Target Classification Using Sonar, IEEE/RSJ International Conference on Intelligent Robots and Systems, Hawaii, U.S.A., Oct. 2001, pp. 1446–1451.

[5] Heale, A. and L. Kleeman, A Real Time DSP Sonar Echo Processor, IEEE/RSJ International Conference on Intelligent Robots and Systems, Takamatsu, Japan, Oct. 2000, pp. 1261–1266.

[6] Kleeman, L., Ultrasonic sensors, *The Industrial Electronics Handbook*, CRC Press, IEEE Press, Boca Raton FL,1997, pp. 738–745.

[7] Kleeman, L., Scanned Monocular Sonar and the Doorway Problem, IEEE/RSJ International Conference on Intelligent Robots and Systems, Osaka, Nov. 1996, pp. 96–103.

[8] Kleeman, L., and R. Kuc, Mobile robot sonar for target localization and classification, *International Journal of Robotics Research*, 14, 295–318,1995.

[9] Kleeman, L., On-the-fly Classifying Sonar with Accurate Range and Bearing Estimation, IEEE/RSJ International Conference on Intelligent Robots and Systems, 2002, pp. 178–183.

[10] Kleeman, L., Fast and Accurate Sonar Trackers Using Double Pulse Coding, IEEE/RSJ International Conference on Intelligent Robots and Systems, Kyongju, Korea, Oct. 1999, pp. 1185–1190.

[11] Kleiner, M. and B. Riise, System and Device for a Self Orienting Device, U.S. Patent 5,935,179 August 10, 1999.

[12] Kuc, R., Forward model for sonar maps produced with the Polaroid ranging module *IEEE Transactions on Robotics and Automation*, 19, 358– 362,2003.

[13] Kuc, R., Pseudo-amplitude scan sonar maps, *IEEE Transactions on Robotics and Automation*, 17, 767–770, 2001.

[14] Politis, Z. and P.J. Probert, Target Localization and Identification using CTFM Sonar Imaging: the AURBIT Method, Proceedings, 1999 IEEE International Symposium on Computational Intelligence in Robotics and Automation, CIRA 99, 1999, pp. 256–261.

[15] Yata, T., A. Ohya, and S. Yuta, A Fast and Accurate Sonar-ring Sensor for a Mobile Robot, Proceedings, 1999 IEEE International Conference on Robotics and Automation, Vol.1, 1999, pp. 630–636.

[16] Yata, T., L. Kleeman, and S. Yuta, Fast bearing measurement with a single ultrasonic transducer, *International Journal of Robotics Research*, 17, 1202–1213, 1998.

77

Intelligent Space and Mobile Robots

Joo-Ho Lee
University of Tokyo &
Ritsumeikan University

Kazuyuki Morioka
University of Tokyo

Hideki Hashimoto
University of Tokyo

77.1 Introduction

Mobility is an essential function for serving people; thus, mobile robots are expected to be useful tools for this purpose. However, many problems remain to be solved to utilize mobile robots in this way. One of the biggest problems for a conventionally architectured mobile robot is that of localization. Unlike robot manipulators, a mobile robot does not have a fixed base. Thus, a localization method with boundary errors is required, and the robot should possess a variety of sensors to localize itself [1]. Unfortunately, however, most methods used so far are not efficient when encountering disturbances, for example, slipping of wheels, parameter changes, and collisions. Although GPS is considered a robust method for localization, only outdoor mobile robots are able to use this system. To navigate in an environment, a mobile robot needs to have sufficient information about the environment. However, to handle such information and sensors for localization, high computational power would be needed by each robot. Even if a robot adopts subsumption architecture [2], the computational complexity does not decrease much. Moreover, it is thought that mobile robots in subsumption architecture have not been developed sufficiently for practical use.

Another problem is in importing the mobile robot into the environment. If we want to utilize different types of mobile robots in a particular circumstance, we need to prepare different software for each robot, according to its resources. Thus, a mobile robot, tuned up for a particular circumstance, cannot be used easily and immediately in a different environment. If the robots use active sensors and the signals interfere with each other, countermeasures must be considered; in the case of mobile robots using laser or ultrasonic sensors, this can often occur. Static information in robots also causes problems; even though a robot contains a perfect map of its working area, the actual environment, especially when there are people around, varies dynamically and the static information becomes useless. Adding to these drawbacks, there are many others such as obstacle avoidance and path following. To solve these problems involving classical mobile robots, the Intelligent Space [3] concept is adopted.

77.2 Intelligent Space

What is an Intelligent Space?

In recent years, computers and computer networks have proliferated, becoming important parts of our daily lives. Research on computers and computer networks, as well as computer-aided research has rapidly proceeded in a variety of ways in the recent past. The concept of an intelligent environment is one example. Intelligent environments are able to monitor what is occurring in them, to build their own models, to communicate with their inhabitants, and to act on the basis of the decisions they make. Especially, the capability of the environment to act as a context-sensitive user interface (e.g., to respond to gestures) and react in certain situations (e.g., accidents, intruders) promises a range of application scenarios such as intelligent hospital rooms, offices, factories, and asylums, such as for the aged. We concluded that such a system is a natural outcome of technological progress, and have been developing the concept of an "Intelligent Space" since 1996.

The main purpose of an Intelligent Space is to be a human-centered system. Up to now, many intelligent systems have been developed; however, when humans want to use such systems, they first need to learn how to operate the system. Moreover, since most of such systems have no mobility capabilities, humans should move in front of them to operate. In an Intelligent Space, humans can express their will by intuitive actions or by unconscious usual actions so that they do not need to learn how to use the Intelligent Space. Since humans are in the Intelligent Space, they do not have to move around to interact with it. Even other intelligent systems can be operated through the Intelligent Space; they are its agents.

Classification of Space

According to its affordance [4], we can classify a space into three classes; Figure 77.1 describes these. In Figure 77.1(a), there is a road in the midst of a forest. Only local and simple information of the space is afforded to the client of the space. Most of our natural environment belongs to this type of space. We call it "Potential Information Space." This space includes some information; however, to acquire it from the space, a client needs to have high intelligence. A natural environment is a representative example. In Figure 77.1(a), a person is able to recognize that it is a road intuitively. However, it is ambiguous as to where this road goes to, and a person who wants to use the road should prepare map-like things to obtain additional information on this space.

From Figure 77.1(b) people can understand where the roads lead to by the road sign. Whether there is a road sign or not, it does not change the fact of where the road leads. A space with artificial signs is called "passive information space." In this space, artificial signs are included in the natural space. A client needs to have less intelligence than in a potential information space, to acquire the information from this space; but only a client who has enough intelligence to gain the information from the sign can know where this road goes. An example would be our general living environment. However, the sign affords only limited information, whether the client needs the information or not. Artificial signs help people to understand the affordance of the space that cannot be found in a potential information space. However, to comprehend the meaning of an artificial sign, the user should study it in advance.

In Figure 77.1(c), active devices are distributed in the space and the devices inform affordance of the space in the relevant format for clients. Therefore, the clients of the space do not need to study to understand the affordance of the space in advance. This is called "Active Information Space." In this space, clients are able to request information of the space and the space replies to those requests. Since the space itself has intelligence, a client needs the least intelligence, of the three examples, to gain information from the space. Information of the environment in potential information and passive information spaces is too difficult to be understood by robots. Although in a passive information space, artificial signs and symbols help to understand affordance more easily, such things are not designed for robots but for humans. However, in an active information space, information is transformed into an easy format for robots.

FIGURE 77.1   Classification of spaces: (a) Potential information space, (b) Passive information space, (c) Active information space.

The former two spaces are easily found in everyday life. However, the last space is hardly ever found around us. An Intelligent Space belongs to this active information type space. Due to the feature that the space adapts itself to clients, the Intelligent Space is a soft environment, not only for humans but also for artificial systems including robots.

Architecture Required for an Intelligent Space

The most important required property of an Intelligent Space is the scalability of the system. Since an Intelligent Space is a spatial system, it must be adaptable to any shape or size of space. An Intelligent Space should be coherently and easily applicable to the general environment that people live in. Therefore, large changes in environments will be undesirable for the application of an Intelligent Space. Incompatibility should also be considered, since the environment is not for the system, but for the people who inhabit the space. A designer of an Intelligent Space should make an effort to ensure that people do not feel an incompatibility with the system.

From the standpoint of Intelligent Space development, several points need to be satisfied. Since the functions of an Intelligent Space are continuously being developed, new or revised functions should be easily applicable to the Intelligent Space. If both hardware and software are made as modules, the Intelligent Space might be easily renewed. It should be simple to integrate existing intelligent components or services into a room. As many intelligent components are needed, the cost of a single component should be low. This requires the use of industry standard-type hardware. Setting up and maintaining an Intelligent Space should be possible with minimal effort. The space must be able to learn about itself (e.g., model building, autocalibration) and site-specific adaptation has to be easily achieved. In addition to the above properties, Intelligent Spaces should also be easy to administer.

Intelligent Space with Distributed Devices

To satisfy the required architecture described above, we devised the Intelligent Space based on Distributed Intelligent Network Device (DIND), which is composed of three basic elements [3]: Sensors, processors (computer), and communication devices. DIND is a small device based on three functions that the sensor watches: the dynamic environment, which contains people and robots; information, which is processed to be known easily by the clients by the processor; and the DIND communicates with other DINDs through networks. Figure 77.2 shows the basic structure of a DIND. By installing this DIND into the whole space, it can recognize objects inside the space easily, and it is enabled to perform informational exchanges and informational sharing by communicating through the network. Further, since such functions are obtained by just attaching DINDs, the space, where people live, is altered into an intelligent environment, without much effort.

Related Works

Integrated intelligent environments have their origin in ubiquitous computing. In 1991, Weiser introduced the area of ubiquitous computing where people could use computational resources everywhere in the environment [5]. Ubiquitous computing promises more than just infrastructure. It suggests new paradigms of interactions inspired by widespread access to information and computational capabilities. From the mid-1990s, research into intelligent environments has been undertaken. The recognition and awareness of people in environments are more focused in intelligent environment research than in other forms of ubiquitous computing research [6–9]. However, these studies have focused only on supporting humans.

Combined research on distributed sensor networks and network mobile robotics has recently appeared [10,11]. Horling et al. proposed Distributed Sensor Networks with early experimental results. The system consists of a set of reconfigurable sensors at fixed locations, each having local processing and low-bandwidth communication capabilities with other sensor nodes. The objective of the sensors is to track objects

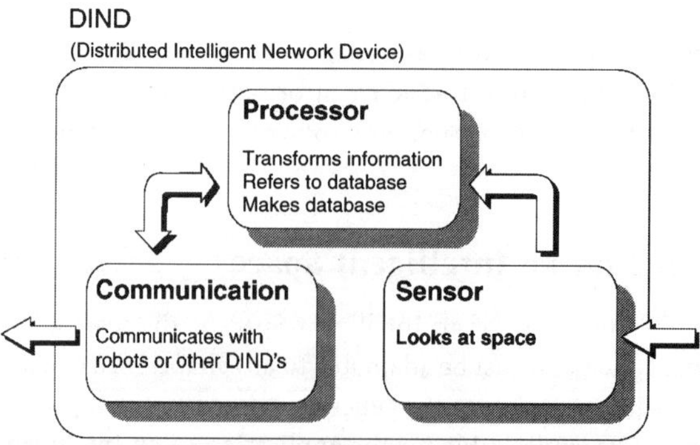

FIGURE 77.2   Architecture of DIND.

moving in the environment. Hoover and Olsen located video cameras in the environment to guide mobile robots to avoid collisions. This resembles some parts of the functions of an Intelligent Space. However, their system has some drawbacks in that the system needs several cameras to observe an area. Moreover, how to identify a mobile robot, which enters into the area within the system, is unclear. The main difference between their system and mobile robots in our Intelligent Space is the independence of the distributed devices. Their system is a centrally integrated system, where as each of the sensor devices in the Intelligent Space operates independently.

77.3 Intelligent Space for Mobile Robots

Among the functions of the Intelligent Space, in this chapter we focus on functions for mobile robots. As far as we know, the Intelligent Space is a unique system to guide mobile robots utilizing distributed intelligent sensor devices.

There are many situations where mobile robots are required. In homes, offices, hospitals and factories, robots are expected to play important roles, mostly supporting people. However, to be able to use a robot in such an environment requires many problems to be overcome (some reasons are described in Section 77.1). Among those situations, a factory is different from the others, since production lines and warehousing facilities in factories are well equipped with infrastructure and the center of the space is not focused on humans. Often, productivity and efficiency are the most important items in factories, so that the Intelligent Space, which needs to change the environment, is more easily adopted than in other environments.

Currently, most factories utilize AGVs for unmanned transportation. However, to use an AGV, some methods of guiding AGVs must be prepared. The representative method is one that uses special tapes as a guide-rail. In this case, it is difficult to cope with frequent structure changes in a factory. All the guide tapes must be restuck to the floor every time there is a structure change. To detect obstacles, which are mainly human, an AGV needs range sensors. If more than one robot is in the same area, some sensors such as an ultrasonic sensor or a laser rangefinder, must be used carefully. Since such sensors are easily interfered, some methods to overcome interference are required. Adding to these, there are many disadvantages in using AGVs in flexible and complex production lines.

In the Intelligent Space, all such kinds of problems are solved. Because there are no guide-rails in the Intelligent Space, the paths for robots are easily changed according to the change in the factory structure. The mobile robots do not require sensors for detecting obstacles since the Intelligent Space monitors the inside of the space. Obstacles, including humans, can be detected and information about them can be given to robots by the Intelligent Space. Apart from these points, there are many other positive reasons for adopting affordance. Figure 77.3 describes the differences between a conventional AGV and an AGV mobile robot in an Intelligent Space factory.

77.4 System Design of an Intelligent Space for Mobile Robots

We developed an Intelligent Space with mobile robots in a factory in cooperation with Toshiba IT & Control Systems Corporation [12]. Unlike other experimental systems in laboratories, this required a high degree of robustness for the components of the Intelligent Space. We developed the system based on the standards of the Japanese Standards Association (JSA) [13]. According to these standards, the maximum speed of mobile robots in factory is about 30 cm/sec and the minimum illumination is 200 lux.

Hardware Architecture

The factory is divided into several areas and each is monitored by a DIND. According to the placement of the DIND, important parameters, including size of the area covered and tracking precision, vary, since for the moment we adopt a vision camera as a sensor for a DIND. Therefore, the disposition of DINDs is one of the most important problems in composing an Intelligent Space. There is an on-going study to ascertain the optimal disposition of DINDs in any particular place [14].

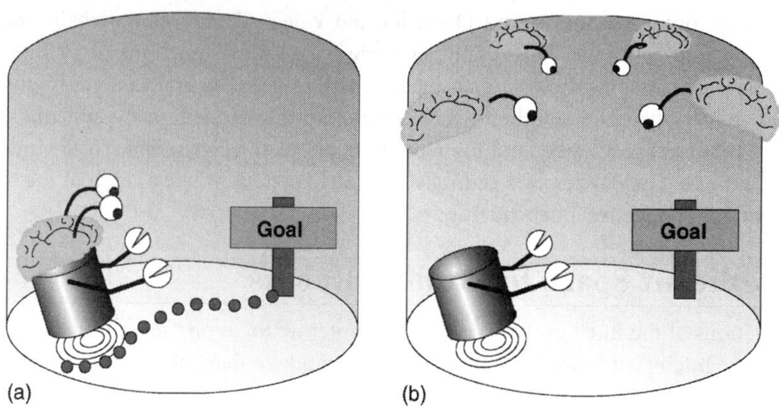

FIGURE 77.3   Mobile robot in environment: (a) conventional AGV and (b) AGV in Intelligent Space.

DIND

As explained in a previous section, Intelligent Space is composed of DINDs. In an actual system, a cheap CCD camera and a general bt848-based image capture board is adopted as a sensor. For the processor, an industrial standard Pentium III 500 MHz PC is used and a general 100baseT LAN card is used as the network device. With these components, one actual DIND can be assembled for less than 400 U.S. dollars.

Robots are able to use the resources of DINDs as if their own parts [24]. However, we could also consider that the robots in the Intelligent Space are a part of a DIND.

Color Bar-Codes

To find mobile robots in an Intelligent Space, four yellow targets are placed around the body of each mobile robot. These targets are found by DINDs, but since it is difficult to match the found blobs correctly, further clues are required to identify each target. For this, color bars, located under the targets, are utilized. A color bar consists of several colored fields. Each of the fields can have one of the eight predefined colors. With this technique, we are able to distinguish the mobile robots and localize that robot by considering the geometrical relations. There are some other methods to localize mobile robots with camera sensors, but the best way to localize the mobile robots is judged to be without adding any other artificial equipment to the robots. However, to achieve a high robustness for the system, we adopted color bars. A trade-off relationship exists between using additive equipment and robustness; but in this case, there were positive reasons, such as identification of mobile robots, and high precision positioning gained by using an energy-free passive piece of equipment.

Mobile Robots

Neither range sensors to detect obstacles around them nor positioning sensors to estimate their pose themselves are needed by mobile robots in the Intelligent Space. Robots only require passive color bars and a communication device to communicate with DINDs. Due to the communication between robots and DINDs, robots are able to freely use sensors in the DINDs. This is a kind of resource sharing, and this feature leads to lower cost robots, since expensive sensors, such as gyroscope sensors, and laser rangefinders are not needed. A simple Programmable Logic Controller (PLC) board, motors, and a wireless LAN (IEEE 802.11b) were installed in the robot that we used in this system.

Software Architecture

Linux was adopted as the operating system in the DINDs, and all software algorithms were written in C++ and the GUI parts were written in TCL/TK. In this application, the required functions of the Intelligent Space are summarized into two. One is to detect humans and unknown obstacles and the other is to handle mobile robots.

Localization procedures for humans are shown in Figure 77.4(a). A pair of DINDs are needed to generate the 3D position of humans or obstacles [15]. However, in this chapter, since we are focusing on functions for mobile robots, a detailed explanation is omitted. In the case of obstacles, whatever it is, background separation is used.

For the calibration of a DIND, we adopted Tsai's calibration algorithm [16] and made a GUI calibration module. The calibration module is an application module that is interactively launched by an operator. It uses a graphical interface to specify the points of the grid and their coordinates.

Localization of Mobile Robot

Unlike the case of detecting a human, to localize a mobile robot in the Intelligent Space, only one DIND is required. Because the height (z-axis) of the targets is known, only one camera is needed for a precise 3D construction. The procedure is shown in Figure 77.4(b). From the acquired image, a search for the color region and the background separation is performed simultaneously. In this system, since the color of the target is yellow, the yellow regions are searched. The background separation module compares the current image and the background image, and selects only that region where movement has occurred. We chose an adaptable background to make the background separation module work robustly. Changes in

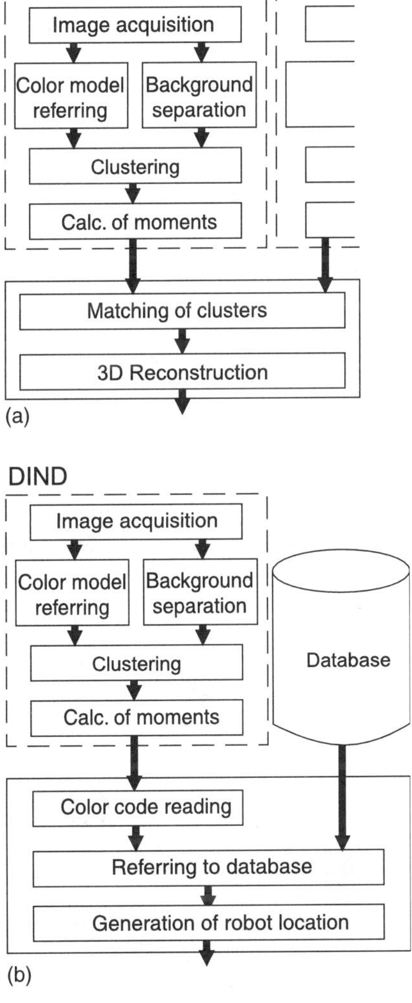

FIGURE 77.4 Localization module: (a) procedure for human or obstacles and (b) procedure for mobile robots.

lighting, moved objects, and camera signals will alter the image slightly and to cope with these, a continuously updated background model is needed. Since the captured background image gradually changes into the current image, the background image can always be retained in the current background. Outputs from color region searching and background separation are compared in the clustering process. Only common regions are conveyed for the calculation of the movement module, and 2D positions of the regions are calculated. Since a DIND has a database of the height of the color bar, one DIND is able to estimate the position of a colored target. The 2D positions of color targets in the image coordination are converted to 3D positions in the camera coordination. After the target detection procedure, to get rid of errors, the 2D euclidean distances between detected targets are checked. If the distance is too long or too short, the detected targets are neglected. Even after this, the result may still contain recognition errors. According to the DIND's database, each color code, placed below the yellow target, is checked. If color codes, which are not shown in the database, are detected, they are removed from the list. From this process, the DIND is able to recognize which robot it is detecting.

The geometrical relations, stored in the database, are compared between the color codes and it generates the pose of the robot. The pose of the robot is calculated from the pairs of color codes, with the same time stamp. The pose of a robot is estimated from the geometrical relation, if at least one pair is recognized. Below are equations to estimate positions and directions of robots. The parameters are described in Figure 77.5.

$$x_r = \left(\frac{x_1 + x_4}{2}\right) - \frac{l_2}{2} \cos \theta_r, \quad y_r = \left(\frac{y_1 + y_4}{2}\right) - \frac{l_2}{2} \sin \theta_r \tag{77.1}$$

$$x_r = \left(\frac{x_2 + x_3}{2}\right) + \frac{l_2}{2} \cos \theta_r, \quad y_r = \left(\frac{y_2 + y_3}{2}\right) + \frac{l_2}{2} \sin \theta_r \tag{77.2}$$

$$x_r = \left(\frac{x_1 + x_2}{2}\right) + \frac{l_1}{2} \sin \theta_r, \quad y_r = \left(\frac{y_1 + y_2}{2}\right) - \frac{l_1}{2} \cos \theta_r \tag{77.3}$$

$$x_r = \left(\frac{x_3 + x_4}{2}\right) - \frac{l_1}{2} \sin \theta_r, \quad y_r = \left(\frac{y_3 + y_4}{2}\right) + \frac{l_1}{2} \cos \theta_r \tag{77.4}$$

$$\theta_r = \tan^{-1}\left(\frac{y_1 - y_2}{x_1 - x_2}\right) = \tan^{-1}\left(\frac{y_4 - y_3}{x_4 - x_3}\right) \tag{77.5}$$

$$\theta_r = \tan^{-1}\left(\frac{y_4 - y_1}{x_4 - x_1}\right) + \frac{\pi}{2} = \tan^{-1}\left(\frac{y_3 - y_2}{x_3 - x_2}\right) + \frac{\pi}{2} \tag{77.6}$$

where (x_r, y_r, θ_r) is the pose of a robot and (x_n, y_n) is the position of a color bar ($n = 1,2,3,4$). Actually, there exist six possible pairs in four color bars. However, only four pairs are utilized in the system due to an empirical decision.

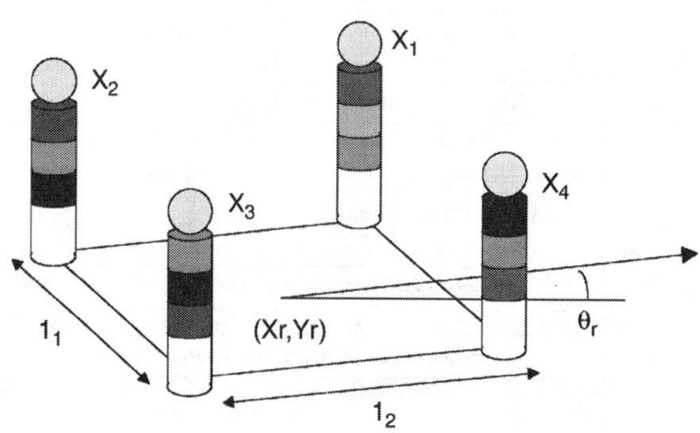

FIGURE 77.5   Color bar codes and parameters.

Mobile Robot Control with DIND

The mobile robot control module produces control input for the mobile robots. A DIND senses the mobile robots, and a vision server module estimates its pose. According to the desired path and the estimated pose, control inputs for mobile robots are generated and transferred to the mobile robots through the wireless LAN (IEEE802.11b). The flow of information is closed. Based on the current pose of the robot, the velocity generator module in the DIND generates both linear and angular velocity for the mobile robot to approach the target position. When the robot position is estimated clearly and the distance between the robot and the subgoal is larger than a defined value, the velocity generator produces maximum velocity and when robot detection is incomplete, it makes the robot stop. Therefore, when more than three color bars of a robot are occluded, the velocity generator module stops the robot.

Network Communication

DINDs and mobile robots organize a network and there are two kinds of communication within it. One is communication between DINDs; because each DIND only monitors its own area, other DINDs do not know that an event occurs at the other area. However, the real space is continuous to the clients of the DINDs, whether they are humans or robots. Thus, to supports themselves, DINDs should inform others of any event in their area that is apt to be shifted to adjacent areas. To guide mobile robots in the Intelligent Space, the dominant contents of communication between DINDs are mainly negotiations for control authority over mobile robots and information on the robots.

The other is communication between DINDs and the mobile robots. The mobile robot informs its internal state to the DINDs, and they mainly offer control signals.

Handing Over of Control Authority

A DIND is an independent device, and its functions, including localization and control of mobile robots, are performed completely within it. Thus, if a mobile robot is moving in the area that one DIND is monitoring, the robot is guided without any difficulty. However, to guide robots in wider areas, which a single DIND cannot cover, we defined the DIND that has control authority of a robot as the dominant DIND for that robot. When a robot moves from an area to a different area, the dominant DIND for the robot is changed automatically. We called this handing over of control authority. The dominant DIND has control authority of the robot and only one dominant DIND exists for a particular robot at one time. Therefore, the control authority should be smoothly handed over to the next DIND at the proper time and at the proper location. To solve this problem, a reliability rank is devised. High reliability stands for that DIND can guide a robot robustly and precisely. Generally, an area near the DIND and a center area of an image, which a DIND captures, have the highest reliability rank, and the boundary of the image and an area far from the DIND have the lowest reliability, since a vision camera is adopted as a sensor of a DIND. There are two cases for becoming a dominant DIND. One is that when a DIND finds a robot in its area and there is no dominant DIND for the robot. The other case occurs when there is a request to be the dominant DIND made to the robot from the current dominant DIND.

Figure 77.6 introduces how the reliability rank is determined. Figure 77.6(a) is an area monitored by a DIND and it is divided into four parts according to distance from the DIND. In Figure 77.6(b), the image is divided into three areas. Even though a good calibration algorithm is adopted, the outskirts of the image are greatly distorted since a vision sensor with a lens is used. Based on these two partition methods, actual reliability ranks in the monitored area are determined as shown in Figure 77.6(c).

Handing over of control authority on a robot is performed as follows. When DIND(n) is the dominant DIND for a robot, the dominant DIND(n) requests other DINDs about the reliability rank of the robot. The other DINDs reply with their reliability rank for the robot, and the current dominant DIND(n) compares those values with its own rank. If another DIND has a higher rank, then the current dominant DIND(n) transfers authority of control to that other DIND($n+1$) that has a higher rank for the robot. Then the new dominant DIND($n+1$) controls the robot. However, if authority of control is shifted to a

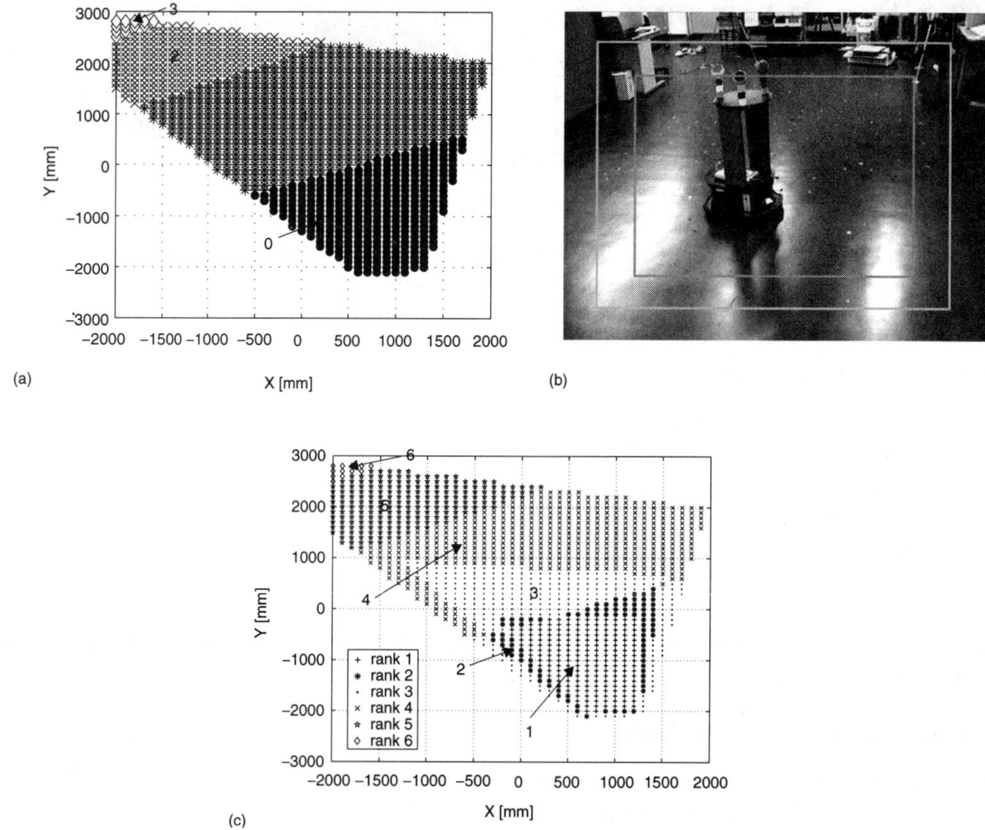

FIGURE 77.6   Reliability rank map: (a) based on distance from DIND, (b) based on capturing image, and (c) combined reliability map.

new DIND with the condition,

$$R_d > R_{other} \tag{77.7}$$

chattering may occur in handing over of control authority, where R_d and R_{other} are the reliability ranks of the dominant DIND and another DIND, respectively. Therefore, instead of (77.7), the following condition is used to avoid chattering:

$$R_d > R_{other} + \alpha \tag{77.8}$$

where α is the offset value to avoid chattering. Empirically, α is set to 1 in an actual system.

77.5   Demonstrations

Since the system that we describe in this chapter is obviously targeted at practical use, its operating characteristics need to be thoroughly examined and tested.

Position Estimation

When DINDs are distributed over an environment, the precision of 3D detection of robots should be considered. A visible area does not mean a serviceable area for a DIND. It is natural that the precision of an estimated position is very much related to the distance between robots and the DIND. The further away a target is from the DIND, the more the precision worsens. To divide areas in a space, and in order to allocate such areas to DINDs, we should measure estimated position errors all over the space. Based on the results, the areas are divided and their DINDs are determined.

We located two DINDs in our experimental space. Figure 77.7 shows the coordination and locations of those DINDs. To find a serviceable area for detecting robots, we located a robot at several points in the visible area of the DINDs and measured errors between real and estimated values. This process is performed over and over in the various directions of a robot. A robot was located at 15 different points and the two DINDs were located at $(-200, 4980, 2140)$ and $(8500, -180, 2050)$. Illumination conditions were dependent on time and position, and varied from 200 to 1000 lux. According to the measured results, directional error was less than $3.5°$ and the position error was less than 17 cm within the entire serviceable area.

Mobile Robot Control

A robot was controlled by a DIND to follow a path and the DIND is able to guide the mobile robot to reach a goal; but actually, goals are not always in a DIND serviceable area. The work area is divided into several serviceable areas of the DINDs. To enable a robot to reach a goal, which is not in the same serviceable area of a DIND, more than one DIND needs to cooperate. In this case, a DIND guides a mobile robot to reach an adjacent supervisory area of another DIND, and the control authority is handed over to the adjacent DIND. Thus, the robot can move freely and seamlessly go everywhere in the Intelligent Space. Figure 77.8 shows an example of this. Figure 77.8(a) and (b) show monitor windows of DIND A and DIND B, respectively (refer to Figure 77.7). Some recognition errors and failures of recognition are found in each monitor window. However, since the control authority is handed over according to the reliability rank value, the robot was properly supported by the two DINDs. In the monitor window, the line is the trajectory of the mobile robot. The circle and arrow indicate the mobile robot and its direction, respectively. The ladder shape is the expected progressing area of the mobile robot. The thick-lined square represents the workspace of the mobile robot.

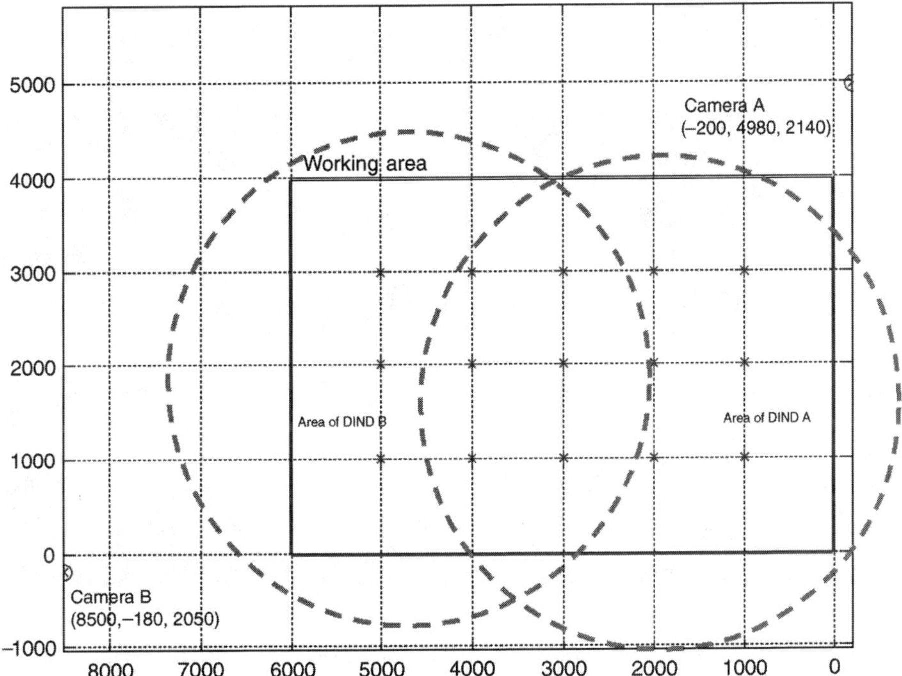

FIGURE 77.7   Coordination and position of DINDs.

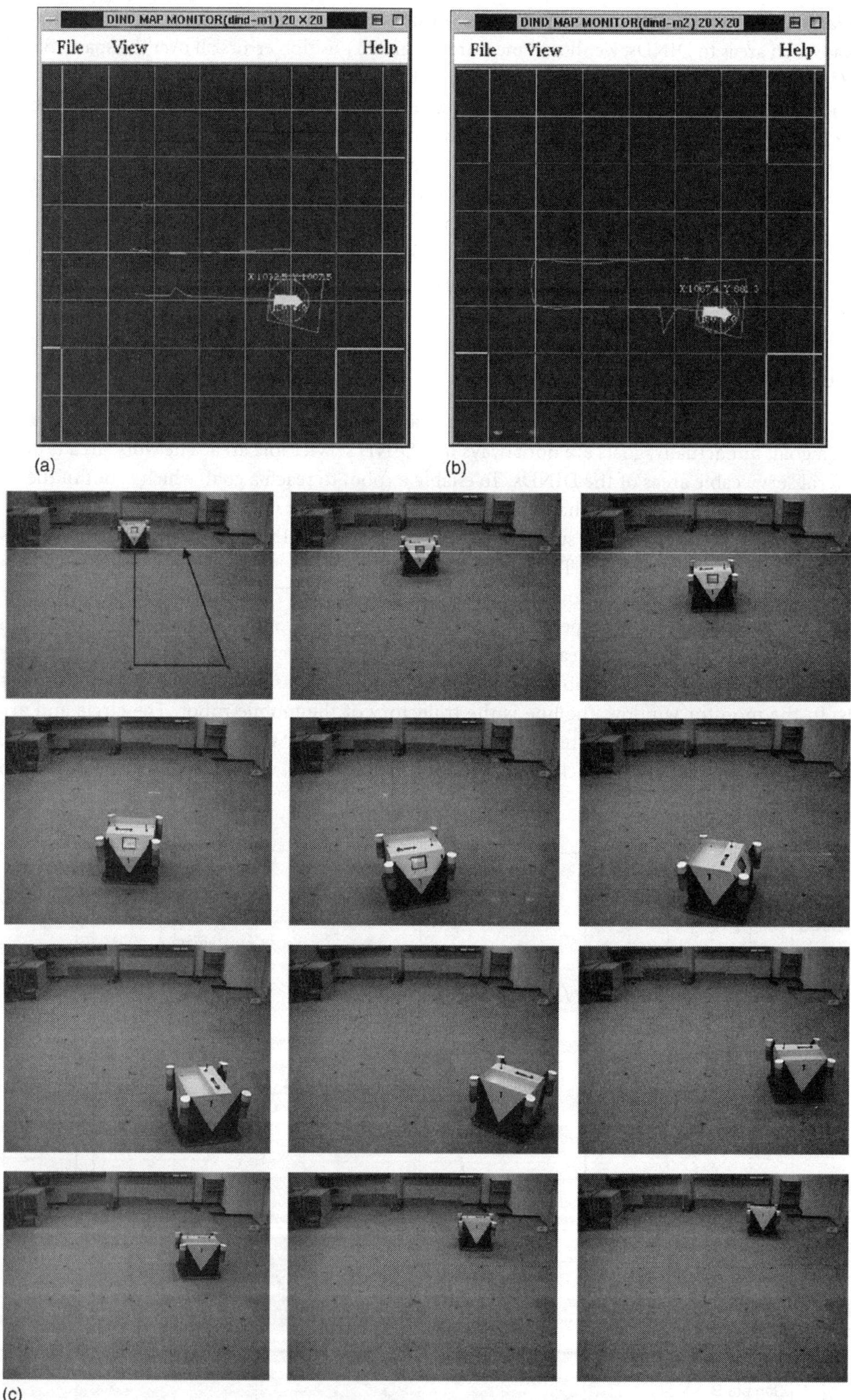

FIGURE 77.8   DIND's handing over of control authority: (a) monitor window of DIND A, (b) monitor window of DIND B, and (c) handover demonstration.

The length of the path in Figure 77.8 is about 8000 mm. The approximated serviceable areas of DIND A and DIND B are the right and left sides of the space, respectively, as shown in Figure 77.7. Figure 77.8(c) shows the actual movement of the robot. The photos in Figure 77.8(c) were captured from video stream with about a 3 sec interval. The robot was guided to follow a path that is shown in the first figure. The robot started from the area where DIND A monitors and moved to the area where DIND B monitors. The robot was controlled to change its direction and returned back to the area where DIND A monitors. The cycle of the control loop is 300 msec. This does not mean that the loop of the image processing is 300 msec, as it is not fixed, but it takes about 60 msec to find a robot from the camera image. It is possible to decrease the cycle to be faster than this. However, currently, since a real-time OS is not used, extra time is required to maintain a regular interval for the control cycle.

When DINDs guide more than one robot, a priority level among the robots exists, and it is contained in the database of the DINDs. According to this priority, one robot should give way when it blocks another's

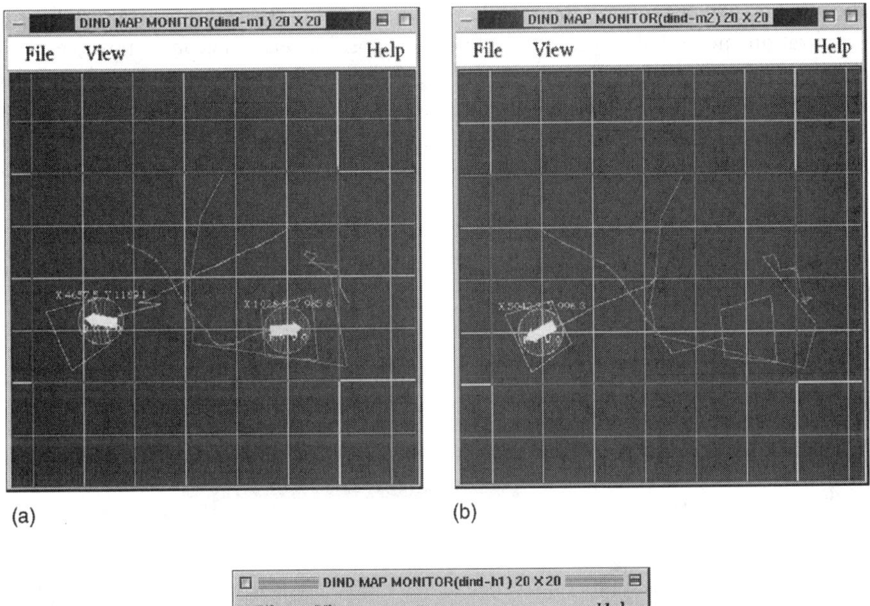

(a)   (b)

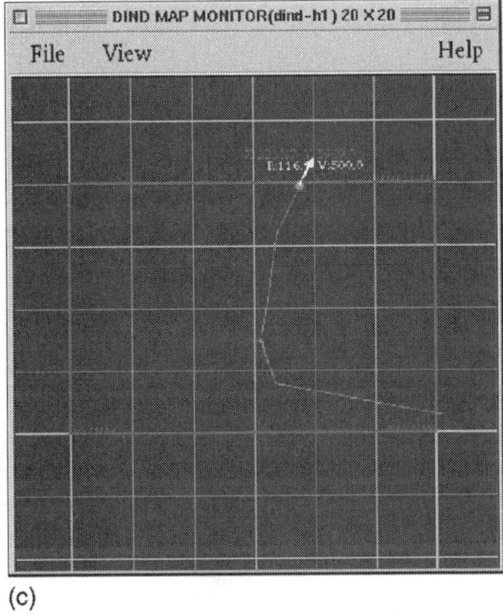

(c)

FIGURE 77.9 Trajectories of robots and a human: (a) monitor window of DIND A, (b) monitor window of DIND B, and (c) monitor window for a human.

path. A path for a robot with high priority is modified to avoid the other robots, which wait for the other robot to pass by. Similar situations occur when the ways of robots are blocked by dynamic obstacles, mainly humans. According to each situation, paths are modified or robots stop for the obstacles to pass by. We demonstrated this situation; first, a man blocked the way of a mobile robot, and then two mobile robots blocked each other's way. DINDs understood the situation and guided the mobile robots based on this. Figure 77.9 shows the monitor windows of DINDs for trajectories of both robots and the human.

77.6 Conclusion

In this chapter, an Intelligent Space for mobile robots was presented. The architecture of Intelligent Space, mobile robot localization, and control in an Intelligent Space were described and demonstrated. The research field of this study is well-known in computer engineering, but not in robotics. Even though an intelligent environment itself is not deeply related to robotics, robots coming under an Intelligent Space have many good features and people should focus on this.

Since the sensors, which monitor the mobile robots, are fixed at constant positions outside of the robot, a localization method that provides a bounded position error becomes possible. Due to the resource-sharing feature of the DINDs, sensors and high computational power are now not necessary elements for mobile robots. This feature leads to lower-cost mobile robots.

Since only camera sensors are used, interference problems do not occur. Even if there are many robots in a room, only a fixed number of DINDs are required. By experimentation and simulation, we concluded

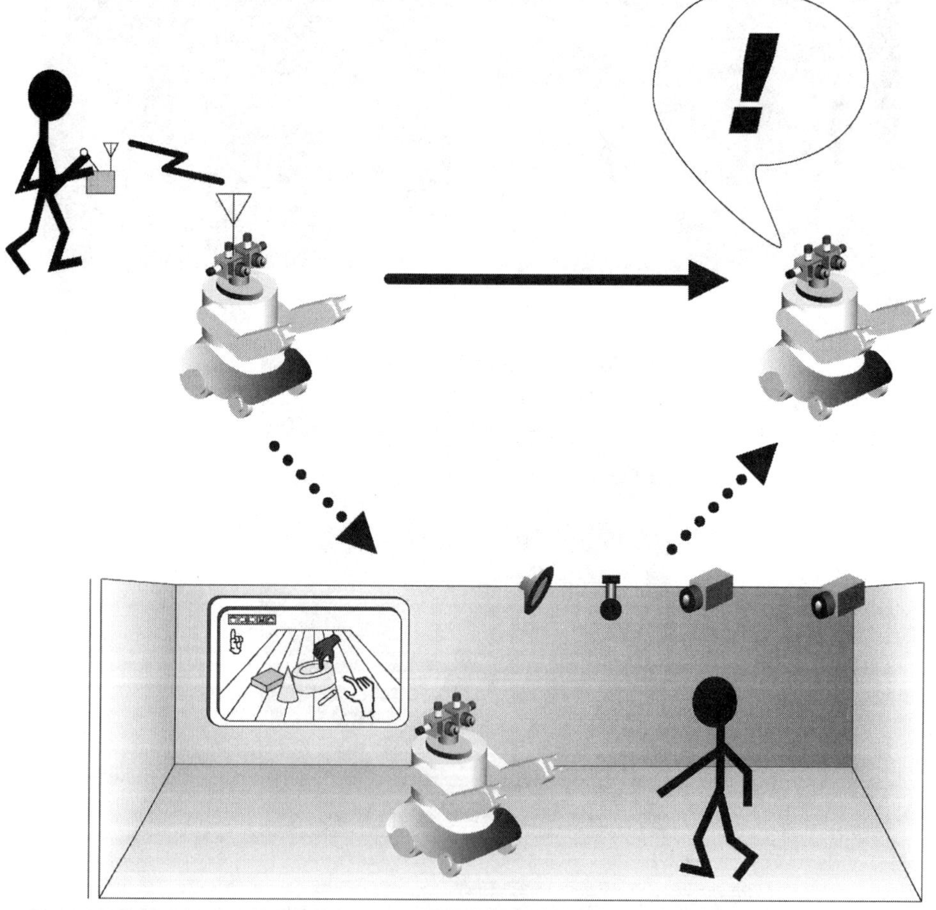

FIGURE 77.10   Flow of robot research.

that there are no limits on the number of robots, only on the performance of the system, which decreased according to the number of robots and was caused by occlusions and blocking of other robots' ways; but the system worked reliably. Especially, differing from an AGV, the routes are changed easily and no hardware changes are required.

This combination results in a proposal for modifying the robot research flow with the inclusion of an additional step before achieving a full autonomous intelligent robot. The new flow is expected to accelerate robot research and bring better results. As shown in Figure 77.10, several researchers are trying to create a full autonomous robot. However, this goal is far from being achievable with current technology. Currently, robots are utilized by borrowing human intelligence or by moving in very simple environments in which only low-level intelligence is needed to navigate. Since the goal of robot research is far ahead of the present stage of robot technology, robots in the Intelligent Space might be a valid and achievable subgoal. In this case, robots get intelligence/information that the robot lacks from the space. When a human operator wants to operate the robot, he communicates with the Intelligent Space and the space translates it to the robot. Especially, we want to emphasize that the intelligent space is an intelligent system to support robots as well as humans.

It can be considered that making such an environment may cost more than the savings achieved by reducing the numbers of sensors and computational power from the robots. However, the following two points should be considered. One is that the Intelligent Space is a system for supporting humans as well as robots. The other is the development of social surroundings. Computer networks are not unfamiliar concepts anymore in most people's daily lives. Currently, the target of the system described in this article is one where the infrastructure is well organized. However, as our living environment gets better infrastructure it will become easier to install an Intelligent Space in such an environment.

References

[1] Borenstein, J., H. Everett, and L. Feng, *Navigating Mobile Robots: Systems and Techniques*, A.K. Peters, 1996.

[2] Brooks, R.A., A robust layered control system for a mobile robot, *IEEE Transactions on Robotics and Automation*, 2, 14–23, 1986.

[3] Joo-Ho Lee and Hideki Hashimoto, Intelligent space — concept and contents, *Advanced Robotics*, 16, 265–280, 2002.

[4] Gibson, J., *The Ecological Approach to Visual Perception*, Houghton Mifflin, Boston, MA, 1979.

[5] Weiser, M., The computer for the twenty-first century, *Scientific American*, 265, 94–104, 1991.

[6] Streitz, N., J. GeiSler, T. Holmer et al., i-Land: An Interactive Landscape for Creativity and Innovation, Proceedings of ACM Conference CGI'99, 1999, pp. 120–127.

[7] http://oxygen.lcs.mit.edu/

[8] Pentland, A., R. Picard, and P. Maes, Smart rooms, desks, and clothes: toward seamlessly networked living, *British Telecommunications Engineering*, 15, 168–172, 1996.

[9] Brad Johanson, Armando Fox, and Terry Winograd, The Interactive Workspaces Project: Experiences with Ubiquitous Computing Rooms, *IEEE Pervasive Computing Magazine*, 1, no. 2, pp. 67–74, 2002.

[10] B. Horling, R. Vincent, R. Mailler et al., Distributed Sensor Network for Real Time Tracking, in Proceedings of the 5th International Conference on Autonomous Agents, 2001, pp. 417–424.

[11] Hoover, A. and B.D. Olsen, Sensor Network Perception for Mobile Robotics, IEEE International Conference on Robotics and Automation, 2000.

[12] http://www.toshiba-itc.com/en/

[13] http://www.jsa.or.jp/default_english.asp

[14] Joo-Ho Lee, Takashi Akiyama, and Hideki Hashimoto, Study on Optimal Camera Arrangement for Positioning People in Intelligent Space, International Conference on IROS, 2002, pp. 220–225.

[15] Joo-Ho Lee, Guido Appenzeller, and Hideki Hashimoto, Physical Agent for Sensed, Networked and Thinking Space, IEEE International Conference on Robotics and Automation, 1998, pp. 838–843.

[16] Tsai, R.Y., A versatile camera calibration technique for high-accuracy 3D machine vision metrology using off-the-shelf TV cameras and lenses, *IEEE Transactions on Robotics and Automation*, RA-3, 323–344, 1987.

78

A Survey on Self-Organizing Wireless Sensor Networks

Shashidhar Gandham
The University of Texas at Dallas

Ravi Musunuri
The University of Texas at Dallas

Praveen Rentala
The University of Texas at Dallas

Udit Saxena
Microsoft Corporation

78.1 Introduction

Due to advances in integrated circuits (ICs), fabrication technology, and Micro Electro Mechanical Systems (MEMS) [8, 11], it is now commercially feasible to manufacture ICs with sensing, signal processing, memory, and other relevant components built into them. Such ICs enabled with RF communication bring forth a new kind of networks, which are self-organizing and application-specific. These networks are referred to as wireless sensor networks.

A sensor network is a static *ad hoc* network consisting of hundreds of sensor nodes deployed on the fly for unattended operation. Each node consists of [1, 9, 10] sensors, processor, memory, radio, limited power battery, and software components such as operating system and protocols. The architecture of a sensor node is completely dependent on the purpose of the deployment. But we can generalize the architecture [11] as shown in the following Figure 78.1.

Sensor nodes are expected to monitor some surrounding environmental phenomena, process the data obtained, and forward these data toward a base station located on the periphery of the sensor network. Wireless sensor networks have numerous applications in fields like surveillance, security, environmental monitoring, habitat monitoring, smart spaces, precision agriculture, inventory tracking, and health care [9].

The main advantage of sensor networks is their ability to be deployed in almost any kind of remote terrain. Their unattended mode of operation makes them a preferable choice over ground-based radar

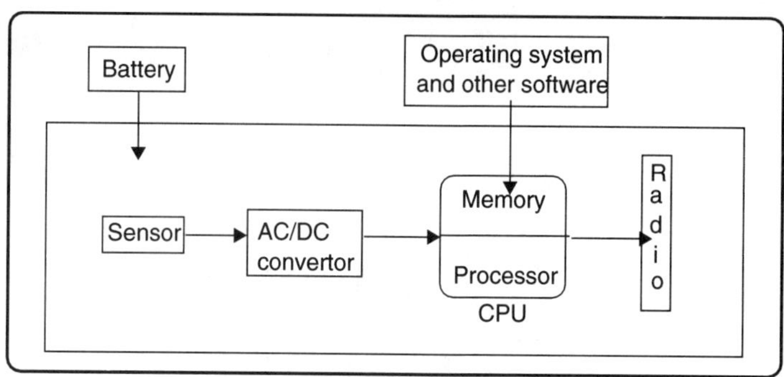

FIGURE 78.1   Sensor node architecture.

systems [10]. The spatial distribution of sensor nodes ensures greater Signal to Noise Ratio (SNR) by combining signals from various sensors. Furthermore, a higher level of redundancy allows greater fault tolerance. As sensor nodes are expected to be manufactured for a very less price, they can be deployed in large numbers. As a result, sensor networks can provide a large coverage area through the union of individual nodes' coverage area. Since sensor nodes are expected to be deployed close to the object of interest, obstruction of the line of sight for sensing activity is ruled out.

To illustrate the above-mentioned advantages, consider an example of seismic detection [9]. The earth generates seismic noise, which becomes attenuated and distorted with distance. Hence, to increase the probability of detection, it is advisable to have sensors closer to the source. To accomplish this, we need to detect the exact location and time of the seismic activity, which happens to be the goal of deploying sensors. If a distributed network of sensors was deployed across the entire geographical area of interest, then there would be no requirement to pinpoint the locations where sensors need to be deployed.

Sensor Networks Vs Mobile *Ad hoc* Networks

Wireless sensor networks are significantly different from *Mobile Ad hoc Networks (MANETS)* [7] due to the following reasons:

- *Mode of Communication*: In MANETS, potentially any node can send data to any other node. But in sensor networks, the mode of communication is restricted. In general, a base station will broadcast commands to all sensor nodes in its network and sensor nodes send back sensed data to the base station. Sometimes, sensor nodes may need to forward sensed data to other sensor nodes, if the base station is not reachable directly. Depending on the application, some sensor networks will use data aggregation at designated nodes to reduce the bandwidth usage. Most of the sensor network's messages are routed to base stations. Hence, sensor nodes need not maintain explicit routing tables.
- *Node Mobility*: In MANETS, every node can move. In general, sensor nodes are static and some architectures have mobile base stations [28].
- *Energy*: Nodes in MANETS have a rechargeable source of energy. Thus, energy conservation is of secondary importance. However, sensor networks consist several hundreds of nodes, which need to operate in remote terrain. Hence, battery replacement is not possible, due to which energy efficiency is critical for sensor networks.

Apart from the above-mentioned differences, sensor nodes have low computational power, and less cost as compared to *MANETS* nodes. Protocols designed for sensor networks should be more scalable, since they are expected to be deployed in hundreds.

The remaining part of this chapter is organized as follows. In Section 78.2, we describe system models used in the literature. Section 78.3 presents design issues in sensor networks. In Section 78.4, medium access layer issues and a few solutions proposed in literature are described. In Section 78.5, we move on to flat routing protocols and hierarchical routing protocols. We then describe other important issues like security, location determination, lifetime analysis, power management, and clock synchronization.

78.2   System Models

Various system models proposed in the literature can be classified based on the following factors:

1. Mobility of base stations.
2. Number of base stations.
3. Method of organization (Heirarchical/Flat).

System models considered by researchers until now consist of static sensor nodes randomly deployed in a geographical area of interest. This geographical area of interest is often referred to as a sensor field. Most of the models considered had a single, static base station [4, 7, 24, 22, 25]. In [30], the author evaluates the best position to locate a base station and proposes to split large sensor networks into small squares and move the base station to the center of each square to collect the data. In [28], the authors propose to deploy multiple, intermittently mobile base stations to increase the lifetime of the sensor networks.

Operational Model

Research on sensor networks done until now considered various operational models of the sensor nodes. These models can be broadly classified as below:

1. *Active*: In case of active sensor networks [4, 7, 22, 24, 25, 29], each sensor node would be sensing its environment continuously. Based on how frequently these sensed data are forwarded toward the base station, sensor networks can be further classified as
   (a) *Periodic*: Based on the application for which the sensor network is deployed, it might be required to gather the data from every sensor node periodically [4, 22].
   (b) *Event Driven*: Sensor networks that are deployed for monitoring specific events gather data only when an event of interest occurs [25, 27, 29]. For example, sensor nodes deployed to monitor seismic activity in a region need to route data only when they detect seismic currents in their proximity.

2. *Passive*: In the case of passive sensor networks, data forwarding would be triggered by a query from the base station. Passive sensor networks can be further classified as follows:
   (a) *Energized on Query*: In this operation mode, sensor nodes switch off their sensors most of the time. Only when a query is generated for the data would the sensor node switch on its sensor and record the data to be forwarded.
   (b) *Always Sensing*: This category of sensor nodes have their sensors running all the time. As soon as there is a query for data from it, a sensor node would generate the data packet based on the observations until now and forward the packet.

Radio Propagation Model

Most researchers assumed that energy spent in transmission over wireless medium is in accordance with the first-order radio model [4, 25]. In this model, the energy required to transmit a signal has a fixed part and a variable part. The variable part is directly proportional to the square of the distance. Some constant energy is required to receive a signal by a receiving antenna.

78.3 Design Issues in Sensor Networks

Most sensor networks encounter operational challenges [2] like *ad hoc* deployment, limited energy supply, dynamic environmental conditions, and unattended mode of operation. Any solution proposed for sensor networks should consider the following design issues:

1. Energy: Each sensor node is equipped with a limited battery-supplied energy. Sensor nodes spend more energy in comunication than local computations.[1] As sensor nodes are deployed in large numbers, it is not feasible to manually recharge the batteries. Thus, sensor nodes should conserve their energy by minimizing the number of messages that are to be transmitted. Based on the energy source, sensor nodes can be classified as follows:
   (a) *Rechargeable*: Sensor nodes equipped with solar cells can recharge their batteries when sunlight is available. For such sensor nodes, the main design criterion would be to maximize the number of nodes operational during times when no sunlight is available.
   (b) *Nonrechargeable*: Sensor nodes equipped with nonrechargeable batteries will cease to operate once they drain their energy. Thus, the main design issue in such sensor networks would be to maximize the operational time of every sensor node.
2. *Bandwidth*: Sensor nodes need to communicate over the Industrial, Scientific, and Medical (ISM) band. When many nodes make an attempt to use the same communication frequency, there might be a requirement to use the available bandwidth optimally.
3. *Limited Computation Power and Memory*: As the processing power at each sensor node is limited, proposed solutions for sensor networks should not expect sensor nodes to carry out computationally intensive tasks.
4. *Unpredictable Reliability, Failure Models*: Sensor networks are expected to be deployed in inaccessible and hostile environments. As a result, it is possible for sensor nodes to crash or malfunction due to external environmental factors. The proposed solutions should be based on failure models that account for such a possibility. Furthermore, failure of a few nodes should not bring down the network.
5. *Scalability*: Sensor nodes are expected to be deployed in thousands. As a result, scalability is a critical issue in the design of sensor networks. Any solution proposed should be scalable to large-sized sensor networks.
6. *Timeliness of Action (Latency)*: Latency is an important issue in sensor networks deployed for critical applications like security and surveillance. Hence, the time elapsed between the time an event is detected and the time the event is reported at the base station has to be minimized.

To address these design challenges, several strategies like *Cooperative signal processing, Exploiting redundancy, Adaptive signal processing, and Hierarchical architecture* are going to be key building blocks for sensor networks [1].

In the near future, we believe that sensor networks would find wide acceptance in day-to-day activities similar to computers. To attain such a wide-scale acceptance, sensor nodes should be affordable, easily available, easily configurable (plug and play) and deployable. To accomplish these objectives, we need to come up with suitable Medium Access Control (MAC) layer protocols, routing protocols, location discovery algorithms, power-management strategies, and solutions to other relevant problems. Some of these design problems are well studied by researchers. In the next section, we present a brief overview of existing solutions for each of these design problems.

78.4 MAC Layer Protocols

MAC layer provides topology information and channel allocation to the higher layers in the protocol stack. Channel allocation is critical for energy-efficient functioning of the link layer. Energy-efficiency

---

[1]To take an example of ground-to-ground communication [7], it takes 3 J of energy to transmit 1 kb of data a distance of 100 m. A general-purpose processor having a processing capability of 100 million instructions per second would execute 300 million instructions for the same amount of energy.

and scalability [56] are main issues in developing MAC protocols for sensor networks. Fairness, latency, and throughput are also important performance measures for channel allocation algorithms. A channel could be a time slot in Time Division Multiple Access (TDMA), a frequency band in Frequency Division Multiple Access (FDMA), or a code in Code Division Multiple Access (CDMA). Channel allocation algorithms should try to avoid energy wastage through:

- Collisions: When two or more nodes, which are in direct transmission range of each other, transmit packets in the same channel.
- Overhearing: Nodes receive data destined for other nodes.
- Idle listening: Unnecessarily listening to the channel when there are no packets to be received.
- Control packet: Bandwidth wastage due to exchange of too many control packets.

The existing solutions to channel allocation in *ad hoc* networks can be divided into two categories: contention-based and contention-free methods. In contention-based solutions, the sender continously senses the medium. *IEEE 802.11 Distributed Coordinated Function* (DCF), *MACAW* [58], and *PAMAS* [57] are examples of contention-based protocols. Contention-based schemes are not suitable for sensor networks because of energy wastage due to collisions and idle listening [31].

Sensor networks should use organized methods for channel allocation. The organized methods of channel allocation determine the network topology first and then assign the channels to the links. A channel assignment should avoid *cochannel* interference, which avoids two consecutive links being assigned to the same channel. Sensor networks channel allocation algorithm should be distributed because network-wide synchronization for calculation of a schedule would be an energy-intensive procedure. Another reason for distributed algorithms is that they scale well with an increase in network size and that they are robust to network partitions and node failures.

In [7], the authors proposed a Self-organizing MAC protocol (SMACS) for sensor networks. *SMACS* is a distributed protocol, which enables nodes to discover their neighbors and build a network topology for communication. SMACS builds a flat topology, that is, there are no clusters or cluster heads. In SMACS, each node allocates channels to links between itself and neighbors within a *TDMA* frame referred to as a *super frame*. In a given time slot, every node communicates with only one neighbor to avoid interference. Nodes communicate intermittently and hence they can power themselves off when they have no data to send. The *super frame* schedule is divided into two periods. In the first bootup period, nodes try to discover neighbors, and rebuild severed links. The second time period is reserved for communication between nodes. Authors in [7] also proposed an Eavesdrop and Register (EAR) protocol to handle channel allocation with moving base stations. In Piconet [59], authors used periodic sleep cycle to save energy. Here, if a node wishes to communicate with neighbors, then it has to wait until it receives a broadcast message from neighbors. Wei et al. [56] proposed an energy-efficient MAC protocol known as *S-MAC*, which saves energy by avoiding *collisions, overhearing,* and *idle listening,* and increases *latency.*

78.5　Routing

As stated earlier, each sensor node is expected to monitor some environmental phenomena and forward the corresponding data toward the base station. To forward the data packets, each node needs to have the routing information. Here, we would like to state that the flow of packets is mostly directed from sensor nodes toward the base station. As a result, each sensor node need not maintain explicit routing tables. Routing protocols can in general be divided into flat routing and cluster-based routing protocols.

Flat Routing Protocols

In flat routing protocols, the nodes in the network are considered to be homogeneous. Each node in the network participates in route discovery, maintainance, and forwarding of the data packets. Here, we describe a few existing flat routing protocols for the sensor networks.

Sequential Assignment Routing

Sequential Assignment Routing (SAR) [7] takes into consideration the energy and Quality of Service (QoS) for each path, and the priority level of each packet for making routing decisions. Every node maintains multiple paths to the sink to avoid the overhead of route recomputation due to the node or link failure. Multiple paths are built by building multiple trees rooted at the one hop sink neighbors. Each tree grows outward by successively adding more nodes to one hop sink neighbors, while avoiding nodes with lower QoS and energy reserves. Each sensor node can control which of the neighbors can be used for relaying a message. Each node associates two parameters, an additive QoS metric and energy measure, with every path. Energy is measured by estimating the maximum number of packets that can be routed without energy being depleted if this node is exclusively using the path. SAR then calculates a weighted QoS metric as the product of the additive QoS metric and a weighted coefficient associated with the priority level of the packet. The SAR algorithm attempts to minimize the average weighted QoS metric in the lifetime of the network. A periodic recomputation of paths is triggered by the sink to account for any changes in the topology. Failure recovery is done by a hand-shaking procedure between neighbors.

Directed Diffusion

Estrin et al. [3] proposed a diffusion-based scheme for routing queries from the base station to sensor nodes and forwarding corresponding replies. In directed diffusion, an attribute-based naming is used by the sensor nodes. Each sensor names data that it generates using one or more attributes. A sink may query for data by disseminating interests. Intermediate nodes propagate these interests. Interests establish gradients of data toward the sink that expressed that interest. For example, a seismic sensor may generate a data: (type = seismic, id = 12, location = NE, time stamp = 01.01.01, and footprint = vehicle/wheeled/over 40 t). A sink may send an interest of the form: (type = seismic, location = NE). The intermediate nodes may send an interest for data for vehicle data in the NE quadrant toward the approximate direction. The strength of the gradient may be different toward different neighbors, resulting in different amounts of information flow.

Minimum Cost Forwarding Algorithm for Large Sensor Networks

The minimum cost forwarding approach proposed by Ye et al. [24] exploits the fact that the data flow in sensor networks is in a single direction and is always toward the fixed base station. Their method neither requires sensor nodes to have a unique identity nor maintain routing tables to forward the messages. Each node maintains the least-cost estimate from itself to the base station. Each message to be forwarded is broadcasted by the node. On receiving a message, the node verifies if it is on the least cost path between the source sensor node and the base station. If so, it would forward the message by broadcasting.

In principle, the concept behind minimum cost forwarding is similar to the gravity field that drives waterfalls from top of mountain to the ground. At each point, water flows from a high post to a low post along the shortest path. For this algorithm to work, each node needs to have the least-cost estimate from itself to the base station. The base station broadcasts an advertisement message with the cost set to zero. Every node initially has the estimate set to infinity. On receiving an advertisement message, every node verifies if the estimate in the message plus the cost of link on which it is received is less than the current estimate. If so, the current estimate and the estimate in the advertisement message are updated. If the received advertisement message is updated with a new cost estimate, it is forwarded; else, it is purged. As a result of forwarding advertisement message immediately after updating, it was noticed by the authors that some nodes receive multiple updates and do multiple forwards as lesser cost estimates flow in. Furthermore, the nodes far away from the base station receive more updates than those close to the base station. To avoid this instability during the setup phase, a back-off algorithm was proposed. According to this back-off algorithm, on updating the current cost estimate, the advertisement message is not forwarded until $A * C_{node}$ units of time, where A is a constant determined through simulations and C_{node} is the cost of link on which the advertisement message was received.

Flow-Based Routing Protocol

In [28], the authors proposed to model the sensor network as a flow network and have proposed an Integer Linear Program (ILP)-based routing method. The objective of this ILP-based method is to minimize the maximum energy spent by any sensor node during a period of time. Through simulation results, the authors have shown that our ILP-based routing heuristic increases the lifetime of sensor network significantly.

In the above-mentioned study, the authors observed that the sensor nodes that are one hop away from a base station (sink) drain their energy much earlier than other nodes in the network. As a result, the base station is disconnected from the network. To address this problem, deployment of multiple, intermittently mobile base stations was proposed. The operation time of sensor network was split into equal periods of time referred to as rounds. Modifications were made to the flow network model and the ILP such that solving it gives locations of base stations in addition to the routing information. Apart from increasing the lifetime of the network, the authors argued that multiple, mobile base stations would decrease the average hop length taken by each packet and increase the robustness of the system.

Sensor Protocols for Information via Negotiation

Kaulik et al. [5] proposed a set of protocols to disseminate individual sensor information to all the sensor nodes. Sensor Protocols for Information via Negotiation (SPIN) overcomes information implosion and overlap by using negotiation and information descriptors (meta-data). Classic flooding suffers from the problem of implosion in that the information is sent to all nodes regardless of whether they have already seen that information or not. Another problem is that of overlap of information where two pieces of information might have some components in common; hence, it might be sufficient to just forward the information after removing the common part. SPIN uses three kinds of messages to communicate:

- ADV — When a node has data to send, it advertises this using this message.
- REQ — A node sends this message when it wishes to receive some data.
- DATA — Data message containing the data with a meta-data header.

The details of SPIN are as follows:

1. SPIN-PP: This protocol is designed for point-to-point communication, assuming that two nodes can communicate with each other without interfering with other nodes' communication. This protocol also assumes that energy is not a constraint and packets are never lost. This protocol works on a hop-by-hop basis. A node that has information to send advertises this by sending an ADV to its neighboring nodes. The nodes that are interested in receiving this information express their interest by sending an REQ. The originator of the ADV then sends the data to the nodes that sent an REQ. This process repeats itself.
2. SPIN-EC: This protocol adds an energy heuristic to the previous protocol. A node participates in the process only if it can complete all the stages in the protocol without going below a low energy threshold.
3. SPIN-BC: This protocol was proposed for broadcast channels. The advantage is that all the nodes within hearing range can hear a broadcast, while the disadvantage is that the nodes have to desist from transmitting if the channel is already in use. Another difference from the previous protocols is that nodes do not immediately send out REQ messages on hearing an ADV. Each node sets a random timer and on expiry of that timer sends out the REQ message. The other nodes whose timers have not yet expired cancel it on hearing the request thus preventing redundant copies of the request from being sent again.
4. SPIN-RL: This protocol was designed for lossy broadcast channels by incorporating two adjustments. First, each node keeps track of the advertisements it receives and rerequests data if a response from the requested node is not received within a specified time interval. Second, nodes limit the frequency with which they will resend data. Every node waits for a predetermined time period before servicing requests for the same piece of data again.

Multihop flat routing can also be subdivided according to the signal processing techniques. There are two types of cooperative signal processing techniques: noncoherent and coherent. For noncoherent processing, raw data are preprocessed at the node itself before forwarding it to the central node (CN) for further processing. For coherent processing, the data are forwarded after minimum processing to the central node. The processing at the node involves operations like time stamping. Thus, for energy efficiency, algorithmic techniques assume importance for coherent processing since the data traffic is low, while path optimality is important for coherent processing.

Cluster-Based Routing Protocols

In cluster-based routing protocols, special nodes, referred to as cluster heads, discover and maintain routes and non-cluster-head nodes join one of the clusters. All the data packets originating in the cluster are forwarded toward the cluster heads. Cluster heads in turn will forward these packets toward the destination using the routing information. Here, we describe some cluster-based routing protocols from the literature.

Low Energy-Adaptive Clustering Hierarchy

Chandrakasan et al. [17] proposed Low Energy-Adaptive Clustering Hierarchy (LEACH) as an energy-efficient communication protocol for wireless sensor networks. Authors of LEACH claim that this protocol will extend the life of wireless sensor networks by a factor of 8, when compared to protocols based on multihop routing and static clustering. In LEACH, self-elected cluster heads collect data from all the sensor nodes in their cluster, aggregate the collected data by data fusion methods and transmit the data directly to the base station. These self-elected cluster heads continue to be cluster heads for a period referred to as a round. At the beginning of each round, every node determines if it can be a cluster head during the current round. If it decides to be a cluster head for the current round, it announces its decision to its neighbors. Other nodes that choose not to be cluster heads opt to join one of the cluster heads on listening to these announcements, based on predetermined parameters like signal-to-noise ratio.

LEACH is proposed for routing data in wireless sensor networks, which have a fixed base station to which the recorded data need to be routed. All the sensor nodes are considered to be static, homogeneous, and energy constrained. The sensor nodes are expected to sense the environment continuously and thus have data to be sent at a fixed rate. Furthermore, radio channels are assumed to be symmetric. The term symmetric means that the energy required to transmit a particular message between two nodes is the same in either direction. A first-order radio model [4] is assumed to describe the transmission characteristics of the sensor nodes. Based on these assumptions, it is clear that having too many intermediate nodes to route the data might consume more energy, on a global perspective, when compared to direct transmission to the base station. The authors use this argument to support the decision to transmit the aggregated data directly from the cluster head to the base station.

The key features of LEACH are localized coordination for cluster setup and operation, randomized rotation of cluster heads, and local fusion of data to reduce global communication costs. LEACH is organized into rounds where each round starts with a setup phase followed by a longer steady-state data transfer phase. Here, we describe various subphases involved in both these phases.

1. *Advertisement phase*: A predetermined fraction of nodes, say p, elect themselves as cluster heads. The optimum value of p can be found from a plot between normalized energy dissipation and the percent of nodes acting as cluster heads. For a detailed description of this procedure, we refer the reader to [4]. The decision to be a cluster head is made by choosing a random number between 0 and 1. If the generated number is less than a threshold $T(n)$, then the node will be a cluster head for the current round. The threshold $T(n)$ is given by the expression $p/[1 - p(rmod(1/p))]$. This would ensure that every node would be a cluster head once in $1/p$ rounds. Once the decision is made, cluster heads advertise their id by using the CSMA MAC protocol.

2. *Cluster setup phase*: On listening to advertisements in the previous phase, non-cluster-head nodes determine which cluster head to join by comparing signal-to-noise ratios from various cluster

heads surrounding it. Each node informs the cluster head of the cluster which it decides to join by using the CSMA MAC protocol.

3. *Schedule creation*: On receiving all the messages, the cluster head creates a TDMA schedule and announces it to all the nodes in the cluster. In order to avoid interference between nodes in adjacent clusters, the cluster head determines the CDMA code to be used by all the nodes in its cluster. This CDMA code to be used in the current round is transmitted along with the TDMA schedule.

4. *Data transmission*: Once the schedule is known, each node will transmit the data during the time slot allocated to it. When the cluster head receives data from all the nodes in its cluster, it will run some data fusion algorithms to aggregate the data. The resulting data are transmitted directly to the base station.

Threshold-sensitive Energy-Efficient sensor Network protocol

In [25], the authors have classified sensor networks into proactive networks and reactive networks. Nodes in proactive networks continuously monitor the environment and thus have the data to be sent at a constant rate. LEACH suits such sensor networks in transmitting data efficiently to the base station. In case of the reactive sensor networks, nodes need to transmit the data only when an event of interest occurs. Hence, all the nodes in the network do not have equal amount of data to be transmitted. Manjeshwar et al. proposed Threshold-sensitive Energy-Efficient sensor Network protocol (TEEN) [25] for routing in reactive sensor networks.

TEEN uses the cluster formation strategy of LEACH, but adopts a different strategy in the data transmission phase. TEEN makes use of two user-defined parameters, hard threshold (Ht) and soft threshold (St), to determine if it needs to transmit the value it sensed currently. When the monitored value exceeds Ht for the first time, it is stored in a variable and is transmitted during the time slot of the node. Subsequently, if the monitored value exceeds the currently stored value by a magnitude of St, then the node will transmit the data. This transmitted value is stored for comparing in future.

Two-Level Clustering Algorithm

Estrin et al. [3] proposed a two-level clustering algorithm that can be extended to build a cluster hierarchy. In this algorithm, every sensor at a particular level is associated with a radius or the number of hops that its advertisements will reach. Sensors at a higher level are associated with higher radii. All sensors start with level 0. Each sensor sends out periodic advertisements to other nodes that are within its radius. The advertisements carry its current level, its parent's identity (if any), and the remaining energy. After transmitting the advertisement, each node will wait for a time proportional to its radius to receive advertisements from other nodes. At the end of the wait time, all level 0 nodes start a promotion timer that is proportional to its remaining energy reserves and the number of level 0 nodes whose advertisements it received. When the promotion timer expires, the node promotes itself to level 1 and starts sending out periodic advertisements. In these new advertisements, it lists its potential children, which are the level 0 nodes that it previously heard. A level 0 node then picks up its parent from one of the level 1 nodes, whose advertisements included its identity. Once a level 0 node picks up its parent, it cancels its promotion timer and drops out of the race. At the end, each level 1 node starts a wait timer and waits for its potential children's acknowledgements. If no level 0 node selected it as its parent or its energy dropped below a certain level, it demotes itself to a level 0 node. All level 0 and level 1 nodes periodically enter the wait stage to take into account any change in network conditions, and reclustering takes place.

78.6 Other Important Issues

In this Section, we will discuss other important issues such as security, location determination, lifetime analysis, power management, and clock synchronization. We describe why these issues are paramount for the functioning of sensor networks and some solutions proposed in the literature.

Security

Security is a very critical issue for the envisioned mass deployment of sensor networks. In particular, a strong security framework is a must in battlefield and border monitoring applications. The security framework in sensor networks should fulfill the following objectives:

- Authentication/nonrepudiation: Each sensor should be able to identify the sender of a message correctly and no node should deny its previous actions.
- Integrity: Messages sent over wireless medium should not be altered by unauthorized entities.
- Confidentiality: Messages should be kept secret from unauthorized entities.
- Freshness: Messages received by sensors should be current.

Sensor Networks Vs *Ad hoc* Networks: Security Perspective

Sensor networks share some similarities with *ad hoc* networks. But security in sensor networks is different from *ad hoc* networks due to the following reasons:

- *Node Power*: Sensor nodes have limited power supply and low computational capabilities as compared to *ad hoc* nodes. *Asymmetric Key* encryption [46] schemes require large computational power as compared to *Symmetric Key* encryption [46]. Thus, sensor networks can only use *Symmetric Key* encryption. To use *Symmetric Key* encryption mechanisms, we need to address the *Key Distribution* problem.
- *Mode of Communication*: As stated earlier, most of the communication in the sensor networks is from sensor nodes to the base station. At times, base station would issue commands to the sensor nodes. In this mode of communication, every node may not need to share keys with every other node in its network. Moreover, it is not practical to store keys, which are shared with every node, at every node.
- *Node Mobility*: In *ad hoc* networks, every node can move. In general, sensor nodes are static and in some architectures they have mobile base stations as in Reference [28].

The above differences make *ad hoc* network's or any other traditional network's security protocols impratical for sensor networks.

Proposed Security Protocols

Recently, there has been some work related to sensor network security. Perrig et al. [34] proposed *Security Protocols for Sensor Networks (SPINS)*. *The SPINS* framework consists of two protocols to satisfy the security objectives. *Secure Network Encryption Protocol (SNEP)* provides data integrity, two-party authentication, data freshness, and *micro Timed Efficient Streaming Loss-tolerent Authentication Protocol (µTESLA)* provides authenticated broadcast. In SNEP, each sensor node and base station share a unique key, which is bootstrapped. This shared key and an incremental message counter, maintained at both sensor nodes and base station, are used to derive new keys using the RC5 [46] algorithm. In µTESLA, a sender generates a chain of keys using one-way function MD5 [46]. The important property of the chain of keys is that if the sender authenticates the initial key, then other keys in the chain are self-authenticated. The sender divides time into equal intervals and each interval is assigned a key from the chain of keys. The sender and receiver agree upon the key disclosure schedule. The first key from the chain is authenticated using unicast authentication. After the first authentication, the receiver authenticates packets after receiving a symmetric key from the sender as per the disclosure shedule. µTESLA uses delayed disclosure of symmetric keys to authenticate packets after the first authentication of one key in the chain.

In [35], the authors proposed a security framework based on broadcast with end-to-end encryption of the data. This scheme avoids traffic analysis and also removes compromised and dead nodes from the network. Sasha et al. [36] divided the messages in sensor networks into three classes depending on the security required. Each class of messages is encrypted using different encryption keys. They showed that this multilevel scheme saves the resources at the nodes.

In general, the base stations will broadcast commands to all the sensor nodes. Hence, secure broadcast is a very important issue in the security framework. In µTESLA, the authentication of the first key in the chain is done using a unicast mechanism. This unicast authentication mechanism has a scalability problem. Authors in [37] replaced this unicast-based mechanism with a broadcast-based mechanism, which avoids *dental of service* [46] attacks. In [38], the authors proposed a routing-aware broadcast key distribution algorithm. Karlof et al. [39] described possible attacks on different routing protocols in the literature and suggested countermeasures.

The Asymmetric Key mechanism requires large computational power, bandwidth, and memory. Therefore, sensor networks use the Symmetric Key encryption to satisfy security objectives. A key distribution [40–44] in the Symmetric Key encryption mechanism is another important issue in sensor networks. Eschenauer et al. [40] proposed a probabilistic predistribution of keys scheme. In this scheme, every sensor node will be given a small set of m keys out of a large set of available keys such that every two sensor nodes will have one common key with the given probability p. This scheme dramatically reduces the number of keys stored in each sensor as compared to storing separate keys to every node in its network. In [41], the authors proposed three extentions to this basic key distribution scheme. In the first, *q-composite* keys extention, sensor nodes will share q common keys instead of one key with a given probability p. This extention improves the security against small-scale attacks such as eavesdroping on one link. The second extention, *multipath* extention, deals with setting up end-to-end path keys between two communicating nodes. In this extention, a path key is established between two nodes by sending random keys through every available path between them. A receiver uses all received random keys along all the paths to establish a path key. This improves the security against large-scale attacks such as eavesdropping many links. The third extention, *random pair-wise* keys scheme, provides the node-to-node authentication. In this scheme, unique node identities are generated randomly. Every node is randomly paired with m other nodes and m corresponding keys. Every node is aware of other node's ID in the pair and the corresponding key. This node ID information is used for node-to-node authentication.

Location Determination

Sensor nodes monitor surrounding phenomena like temperature, light, seismic currents, chemical leaks, radiation, and other parameters of interest. After detecting an event, sensor nodes forward the sensed data toward the nearest base station. In order to process any message reported by the sensor network, the base station is required to know the sender's location. For example, if the sensor network is deployed to detect forest fires, the base station should know the reporting sensor's location. Hence, the base station needs to be aware of the location of every sensor node deployed in the network. In this Section, we will explain different solutions proposed in the literature for location determination in sensor networks. Locationing algorithm's performance can be measured [47] by the following parameters:

- Resolution: The smallest distance between nodes that can be distinguished by the locationing system.
- Accuracy: Probability of locationing system finding the correct location.
- Robustness: Ability of the locationing system to find the correct location when subjected to node failures and link failures.

Global Positioning System (GPS) [48] has been used to locate outdoor nodes. Due to reflection and multipath fading, GPS is not a viable option for indoor locationing. Since sensor nodes can be deployed at indoor locations or on other planets, the GPS-based locationing system is not suggestable. Many non-GPS-based locationing solutions are proposed by the research community. Most of these solutions are either *proximity-* or *beacons*-based. In *proximity*-based solutions, some nodes will act as special nodes, whose locations are known. We can divide proximity-based solutions into two types. In the first type [50], beacons are sent by special nodes, from which other nodes can approximate their location. In the second type of proximity-based solutions [49], beacons are sent by nonspecial nodes, from which special nodes can approximate the locations of nonspecial nodes. *Cricket* [50] uses the difference in arrival times from known beacons as the basis for finding the location. In *RADAR* [51], authors used SNR as the basis for finding the location of nodes. The *SpotON* [52] system finds the location of nodes in the three-dimensional space. These solutions can be adopted for the location detection in sensor networks. In [23], the authors proposed a location detection scheme, which consists of *local positioning* and *global positioning*. In *local positioning,* nodes will approximate their relative location from anchor nodes, whose locations are assumed, by using *triangulation* method, *Global positioning* finds global location by using *cooperative ranging* approach, in which nodes iteratively converge to the global position by interacting with each other.

Saikat et al. [47] proposed a robust location detection algorithm for emergency applications. None of the above-explained solutions dealt with robustness. Robustness is an important issue in emergency scenarios such as building collapses. This location detection improves robustness by finding *identifying codes.* Estrin et al. [3] gave an interesting application of their clustering algorithm to pinpoint the location of an illegimate object. This algorithm is robust to link or node failures and the overhead is proportional to the local population density and a sublinear function of the total number of nodes.

Lifetime Analysis

Lifetime refers to the time period for which a sensor network is capable of sensing and transmitting the sensed data to the base station(s). In sensor networks, thousands of nodes are powered with very limited supply of battery power. As a result, lifetime analysis becomes an important tool to efficiently use the available energy. In sensor networks using rechargable energy such as solar energy, lifetime analysis helps the nodes to use the enery efficiently before recharging. Lifetime analysis may include an upper bound on the lifetime and factors influencing this upper bound.

Theoretical upper bound on the lifetime of a sensor network helps to understand the other protocol's efficiency. Bhardwaj et al. [27] proposed a theoretical upper bound on the lifetime of a sensor network deployed for tracking the movement of external objects. In [55], authors found the lifetime of a sensor network with *Hybrid Automata* modeling. Hybrid Automata is a mathematical method to analyze the systems with both discrete and continous behaviors. The authors used the trace data to analyze the power consumption and to estimate the lifetime of a sensor network.

Power Management

Sensor networks should operate with the minimum possible energy to increase the life of sensor nodes. This requires power-aware computation/communication component technology, low-energy signaling, and networking and power-aware software infrastructure.

Design challenges encountered in the building of wireless sensor networks can be broadly classified into hardware, wireless networking, and OS/applications. All three categories should minimize the power usage to increase the life of a sensor node. Hardware includes the design activities related to all hardware platforms that make up sensor networks. MEMS, digital circuit design, system integration, and RF are important categories in the design of hardware. The second aspect includes design of power-efficient algorithms and protocols. In the previous sections, we described a few energy-efficient protocols for MAC and routing. Next, we present a few OS/application-level strategies related to power management in sensor nodes.

Once the system is designed, additional power savings can be obtained by using the *Dynamic Power Management (DPM)* [13]. The basic idea behind DPM is to shut down (sleep mode) the devices when not needed and get them back when required. This requires an embedded operating system [14] that is able to support the DPM. The switching of a node from the sleep state to the active state takes some finite time and resource. Each sensor node could be equipped with multiple devices. The number of devices switched off determines the level of the sleep state. Each sleep state is characterized by the latency and the power consumption. The deeper the sleep state, the lesser the power consumption, and the more the latency. This requires a careful use of DPM to maximize the life of a sensor node. But in many cases, it is not known beforehand when a particular device is required. Hence, a stochastic analysis should be applied to predict the future events.

Energy can be conserved by using *Dynamic Voltage Scheduling (DVS)* [13, 19]. DVS minimizes the idle processor cycles by using a feedback control system. Energy savings can be obtained by optimizing the sensor nodes performance in active state. DVS is an effective tool to achieve this goal. The main idea behind DVS is to change the power supply to match the workload. This needs tuning of the processor to deliver the required throughput to avoid idle cycles. The crux of the problem lies in the fact that future workloads are nondeterministic. Thus, the efficiency depends on predicting the future workload.

Efficient Link-Layer strategies can be used to conserve energy at each sensor node. In [18], the authors propose to conserve energy by compromising on the quality of the link layer established. This is possible by maintaining the bit error rate (BER) just below the user requirements. Different error-controlling algorithms like Bose–Chaudhuri–Hocquen (BCH) coding, convolution coding, and turbo-coding can be used for error control. The algorithm with the lowest power consumption to support the predetermined BER and latency should be chosen.

Local computation and processing [14, 17] of sensor data in wireless networks can be made highly energy efficient. *Partitioning* the computation among multiple sensor nodes and performing the computation in parallel permit a greater control on latency and results in energy conservation through frequency scaling and voltage scaling.

Biomedical wireless sensor networks could use *Power-Efficient Topologies* [16] to save the energy spent in communication. Biomedical sensor nodes include monitors and implantable devices intended for long-term placement in the human body. Topology is predetermined in these sensor networks. Ayad et al. proposed *Directional Source-Aware routing Protocol* (DSAP) for this class of sensor networks. DSAP incorporates power considerations into routing tables. The authors explored various topologies to determine the most energy-efficient topology for biomedical sensor networks.

Clock Synchronization

Some of the communication algorithms for wireless sensor networks that are proposed in the literature make an inherent assumption that there exists some mechanism through which local clocks of all the sensor nodes are synchronized. Although this assumption is valid, we need to have an explicit way of synchronizing local clocks of all sensor nodes. Apart from the implementation of the communication algorithms, clock synchronization is required for accurate time stamps in cryptographic schemes, for recognizing duplicate detection of the same event from different sensor nodes, for data aggregation algorithms like beam forming, for ordering of logged events, and many other similar applications. In this Section, a *post facto* clock synchronization algorithm proposed by Elson and Estrin [26] is described.

The *post facto* clock synchronization algorithm discussed here is suitable for applications like beam forming, duplicate event detection, and other similar localized methods. This algorithm is expected to be implemented on systems similar to the Wireless Integrated Network Sensors (WINS) platform where a processor has various sleep modes and has the capability of powering down high-energy peripherals. Because of the capability of the sensor node processor to power down a device and power up when there is a requirement to sense data and transmit, existing clock synchronization methods for distributed systems are not applicable.

The basic idea behind *post facto* clock synchronization algorithm is that for certain applications like data fusion and beam forming, it is sufficient to order the events in a localized manner. In this scheme, node's clocks are normally unsynchronized. When a stimulus arrives (time to sense and transmit data), each node records the stimulus with respect to its local clock. Immediately following this event, a third party will broadcast a synchronization pulse. Every node receiving this pulse normalizes their stimulus time stamp with respect to the broadcasted synchronizing pulse. It is essential to note that the time elapsed between recording the stimulus and arrival of synchronized pulse needs to be measured accurately. For this reason, the algorithm is inappropriate for systems that need to communicate a time stamp over a long distance.

78.7   Conclusions

In this chapter, we made an attempt to present an overview of the wireless sensor networks and describe some design issues. We discussed various solutions proposed to prolong the lifetime of sensor networks.

Proposed solutions to issues like MAC layer, routing data from sensor node to the base station, power management, location determination, and clock synchronization were discussed.

References

1. Estrin, D., Girod L., Pottie G., and Srivastava M., Instrumenting the World with Wireless Sensor Networks. IEEE International Conference on Acoustics, Speech, and Signal Processing, 2001, pp. 2033–2036.
2. Elson, J. and Estrin, D., Time Synchronization for Wireless Sensor Networks, Proceedings 15th International Parallel and Distributed Processing Symposium, 2001, pp. 1965–1970.
3. Estrin, D., Govindan, R., Heidemann, J., and Kumar, S., Next Century Challenges: Scalable Coordination in Sensor Networks, Proceedings of the 5th Annual ACM/IEEE International Conference on Mobile Computing and Networking, 1999, pp. 263–270.
4. Heinzelman, W., Kulik, J., and Balakrishnan, H., Adaptive Protocols for Information Dissemination in Wireless Sensor Networks, Proceedings of the 5th Annual ACM/IEEE International Conference on Mobile Computing and Networking, 1999, pp. 174–185.
5. Heinzelman, W., Kulik, J., and Balakrishnan, H., Negotiation-Based Protocols for Disseminating Information in Wireless Sensor Networks, Proceedings of the 5th Annual ACM/IEEE International Conference on Mobile Computing and Networking, 1999.
6. Pottie, G.J. and Kaiser, W.J., Wireless integrated network sensors, *Communications of the ACM*, 43, 51–58, 2000.
7. Sohrabi, K., Gao, J., Ailawadhi, V., and Pottie, G.J., Protocols for self-organization of a wireless sensor network, *IEEE Personal Communications*, 7, 16–27, 2000.
8. Sohrabi, K. and Pottie, G.J., Performance of a Novel Self-organization Protocol for Wireless Ad-hoc Sensor Networks, IEEE Vehicular Technology Conference, Vol. 2, 1999, pp. 1222–1226.
9. Pottie, G.J., Wireless Sensor Networks, Information Theory Workshop, 1998, pp. 139–140.
10. Agre, J. and Clare, L., An integrated architecture for cooperative sensing networks, *Computer*, 33, *106–108*, 2000.
11. Min, R., Bhardwaj, M., Seong-Hwan Cho, Shih, E., Sinha, A., Wang, A., and Chandrakasan, A., Low-power Wireless Sensor Networks, Fourteenth International Conference on VLSI Design, 2001, pp 205–210.
12. Wang, A., Heinzelman, W.R., and Chandrakasan, A.P., Energy-scalable Protocols for Battery-operated Micro Sensor Networks, IEEE Workshop on Signal Processing Systems, 1999, pp. 483–490.
13. Sinha, A. and Chandrakasan, A., Dynamic power management in wireless sensor networks, *IEEE Design and Test of Computers, 18*, 62–74, 2001.
14. Wang, A. and Chandrakasan, A., Energy Efficient System Partitioning for Distributed Wireless Sensor Networks, IEEE International Conference on Acoustics, Speech, and Signal Processing, *Vol. 2*, 2001, pp. 905–908.
15. Slijepcevic, S. and Potkonjak, M., Power Efficient Organization of Wireless Sensor Networks, IEEE International Conference on Communications, Vol. 2, 2001, pp. 472–476.
16. Salhieh, A., Weinmann, J., Kochha, M., and Schwiebert, L., Power Efficient Topologies for Wireless Sensor Networks. International Conference on Parallel Processing, 2001, pp. 156–163.
17. Heinzelman, W.R., Chandrakasan, A., and Balakrishnan, H., Energy-efficient Communication Protocol for Wireless Micro Sensor Networks, Proceedings of the 33rd Annual Hawaii International Conference on System Sciences, 2000, pp. 3005–3014.
18. Shih, E., Calhoun, B.H., Seong Hwan Cho, and Chandrakasan, A.P., Energy-Efficient Link Layer for Wireless Micro Sensor Networks, Proceedings of IEEE Computer Society Workshop on VLSI, 2001, pp. 16–21.
19. lm, C, Huiseok Kim, and Soonhoi Ha, Dynamic Voltage Scheduling Technique for Low-Power Multimedia Applications Using Buffers, International Symposium on Low Power Electronics and Design, 2001, pp. 34–39.
20. Tsiatsis, V., Zimbeck, S., and Srivastava, M., Architecture Strategies for Energy-Efficient Packet Forwarding in Wireless Sensor Networks, Proceedings of the 2001 International Symposium on Low Power Electronics and Design, 2001, pp. 92–95.

21. Srisathapornphat, C., Jaikaeo, C., and Chien-Chung Shen., Sensor Information Networking Architecture, International Workshops on Parallel Processing, 2000, pp. 23–30.

22. Lindsey, S. and Raghavendra, C.S., PEGASIS: Power-Efficient Gathering in Sensor Information Systems, International Conference on Communications, 2001.

23. Savarese, C. and Rabaey, J., Locationing in Distributed Ad-Hoc Wireless Sensor Networks, IEEE proceedings on Acoustics, Speech, and Signal Processing, 2001, pp. 2037–2040.

24. Ye, F., Chen, A., Liu, S., and Zhang, L., A Scalable Solution to Minimum Cost Forwarding in Large Sensor Networks, Proceedings of Tenth International Conference on Computer Communications and Networks, 2001, pp. 304–309.

25. Manjeshwar, A. and Agrawal, D.P., TEEN: a Routing Protocol for Enhanced Efficiency in Wireless Sensor Networks, International Proceedings of 15th Parallel and Distributed Processing Symposium, 2001, pp. 2009–2015.

26. Elson, J. and Estrin, D., Time Synchronization for Wireless Sensor Networks, International Proceedings of 15th Parallel and Distributed Processing Symposium, 2001, pp. 1965–1970.

27. Bhardwaj, M., Chandrakasan, A., and Garnett, T., Upper Bounds on the Lifetime of Sensor Networks, IEEE International Conference on Communications, 2001, pp. 785–790.

28. Gandham S.R., Dawande M., Prakash R., and Venkatesan S., Energy Efficient Schemes for Wireless Sensor Networks with Multiple Mobile Stations. IEEE Globecom, 2003.

29. Youssef, M.A., Younis M.F., and Arisha K.A., A Constrained Shortest-Path Energy-Aware Routing Algorithm for Wireless Sensor Networks, Wireless Communications and Networking Conference, Vol. 2, 2002, pp. 794–799.

30. Gao, J., Analysis of Energy Consumption for Ad Hoc Wireless Sensor Networks Using the Watts-per-Meter Metric, IPN progress report, 2002, pp. 42–150.

31. Tanenbaum, A.S., *Computer Networks,* 3rd ed., Prentice-Hall Inc., Englewood Cliffs, NJ, 1996.

32. Nemhauser, G.L. and Wolsey, L.A., *Integer Programming and Combinatorial Optimization,* Wiley, New York, 1988.

33. Ahuja, R.K., Magnanti, T.L., and Orlin, J.B., *Network Flows,* Prentice-Hall, Englewood Cliffs, NJ, 1993.

34. Perrig, A., Szewczyk, R., Wen, V., Culler, D., and Tygar, J.D., SPINS: security protocols for sensor networks, *Wireless Networks Journal,* Vol. 8, Issue 5, pp. 521–534, Sept. 2002.

35. Undercoffer, J., Avancha, S., Joshi, A., and Pinkston, J., Security for Sensor Networks,. CADIP Research Symposium, 2002.

36. Sasha Slijepcevic, Miodrag Potkonjak, Vlasios Tsiatsis, Scott Zimbeck, and Srivastava, M.B., On Communication Security in Wireless Ad-Hoc Sensor Network, 11th IEEE International Workshops on Enabling Technologies: Infrastructure for Collaborative Enterprises, Pittsburgh, PA, U.S.A., June 10–12, 2002.

37. Donggang Liu and Peng Ning, Efficient Distribution of Key Chain Commitments for Broadcast Authentication in Distributed Sensor Networks, The 10th Annual Network and Distributed System Security Symposium, San Diego, CA, Feb. 2003.

38. Loukas Lazos and Radha Poovendran, Secure Broadcast in Energy-Aware Wireless Sensor Networks, IEEE International Symposium on Advances in Wireless Communications, Victoria, BC, Canada, Sept. 23–24, 2002.

39. Karlof, C. and Wagner, D., Secure Routing in Wireless Sensor Networks: Attacks and Countermeasures, First IEEE International Workshop on Sensor Network Protocols and Applications, May 2003.

40. Eschenauer, L. and Gligor, V.D., *A key-management scheme for distributed sensor networks,* Proceedings of the 9th ACM conference on Computer and communications security, Washington, DC, USA, pp. 41–47, 2002.

41. Haowen Chan, Adrian Perrig and Dawn Song, Random Key Predistribution Schemes for Sensor Networks, in 2003 IEEE Symposium on Research in Security and Privacy, 2003.

42. Carman, C.D.W., Matt, B.J., and Cirincione, G.H., Energy-efficient and Low-latency Key Management for Sensor Networks, Proceedings of 23rd Army Science Conference, Orlando, FL, Dec. 2–5, 2002.

43. Yee Wei Law, Corin, R., Etalle, S., and Hartel, P.H., Key Management with Group-Wise Pre-Deployed Keying and Secret Sharing Pre-Deployed Keying. Center for Telematics and

Information Technology, University of Twente, The Netherlands, Technical report (TR-CTIT-02-25), July 2002.

44. Yee Wei Law, Corin, R., Etalle, S., and Hartel, P.H., A Formally Verified Decentralized Key Management Architecture for Wireless Sensor Networks. Lecture Notes in Computer Science, PersonalWireless Communications, pp. 27–39, Oct. 2003.

45. Wood, A.D. and Stankovic, J.A., Denial of service in sensor networks, *IEEE Computer*, 35, 54–62, 2002.

46. Menezes, A.J., van Oorschot, P.C., and Vanstone, S.A., *Handbook of Applied Cryptography*, CRC Press, Boca Raton, Fl, October 1996.

47. Ray, S., Ungrangsi, R., De Pellegrini, F., Trachtenberg, A., and Starobinski, D., Robust Location Detection in Emergency Sensor Networks, Proceedings of INFOCOM, 2003.

48. Hofmann-Welleenhof, B., Lichtenegger, H., and Collins, J., *Global Positioning Sytem: Theory and Practice*, 4th ed., Springer-Verlag, *Berlin* 1997.

49. Want, R., Hopper, A., Falcao, V., and Gibbons, J., The active badge location system, *ACM Transactions on Information Sytems*, 10, 91–102, 1992.

50. Priyantha, N.B., Chakraborthy, A., and Balakrishnan, H., The Cricket Location-Support System, Proceedings of ACM MOBICOM Conference, Boston, MA, 2000.

51. Bahl, P. and Padmanabhan, V.N., RADAR: An In-Building RF- Based User Location and Tracking System, Proceedings of IEEE INFOCOM Conference, Tel Aviv, Israel, 2000.

52. Hightower, J., Borriello, G. and Want, R., SpotON: An Indoor 3D Location Sensing Technology Based on RF Signal Strength, Technical report, 2000-020-02, University of Washington, February 2000.

53. Castro, P., Patrick Chiu, Ted Kremenek, and Richard R. Muntz,. A Probalistic Room Location Service for Wireless Networked Environments, Proceedings of Ubicomp, Atlanta, GA, 2001.

54. Bulusu, N., Heidemann, J., and Estrin, D., GPS-Less Low Cost Outdoor Localization for Very Small Devices. Technical Report 00-729, USC/ISI, April, 2000.

55. Sinem Colerim Mustafa Ergen and T. John Koo, Lifetime Analysis of a Sensor Nework with Hybrid Automata Modelling, Proceedings of ACM WSNA Conference, Atlanta, GA, U.S.A., Sept. 2002.

56. Wei Ye, Heidemann, J., and Estrin, D., An Energy-Efficient MAC Protocol for Wireless Sensor Networks, Proceedings of IEEE INFOCOM, 2002.

57. Singh, S. and Ragavendra, C.S., PAMAS: power ware multi-access protocol with signalling for ad-hoc networks, *ACM Computer Communication Review*, 28, 5–26, 1998.

58. Bhargavan, V. Demers, A. Sheker, S., and Zhang, L., MACAW: A Media Access Protocol for Wireless LANS, Proceedings of ACM SIGCOMM Conference, 1994.

59. Bennett, F., Clarke, D., Evans, J.B., Hopper A., Jones, A., and Leask, D., Piconet: embedded mobile networking, *IEEE Personal Communications*, 4, 8–15 1997.

60. SMART DUST: Autonomous sensing and communication in a cubic millimeter, URL: http://www-bsac.eecs.berkeley.edu/ pister/SmartDust/

79

Software for Wireless Sensor Networks

Jan Blumenthal
University of Rostock

Frank Golatowski
University of Rostock

Marc Haase
University of Rostock

Matthias Handy
University of Rostock

79.1 Introduction

The increasing miniaturization of electronic components and advances in modern communication technologies enable the development of high-performance spontaneously networked and mobile systems. Wireless microsensor networks promise novel applications in several domains. Forest fire detection, battlefield surveillance, or telemonitoring of human physiological data are only in the vanguard of plenty of improvements encouraged by the deployment of microsensor networks. Hundreds or thousands of collaborating sensor nodes form a microsensor network. Sensor data is collected from the observed area, locally processed or aggregated, and transmitted to one or more base stations.

Sensor nodes can be spread out in dangerous or remote environments whereby new application fields can be opened. A sensor node combines the abilities to compute, communicate, and sense. Figure 79.1 shows the structure of a typical sensor node consisting of processing unit, communication module (radio interface), and sensing and/or actuator device.

Figure 79.2 shows a scenario taken from the environmental application domain: leakage detection of dykes. During floods, sandbags are used to reinforce dykes. Piled along hundreds of kilometers around lakes or rivers, sandbag dykes keep waters at bay and bring relief to residents. Sandbags are stacked against sluice gates and parts of broken dams to block off the tide. To find out spots of leakage each

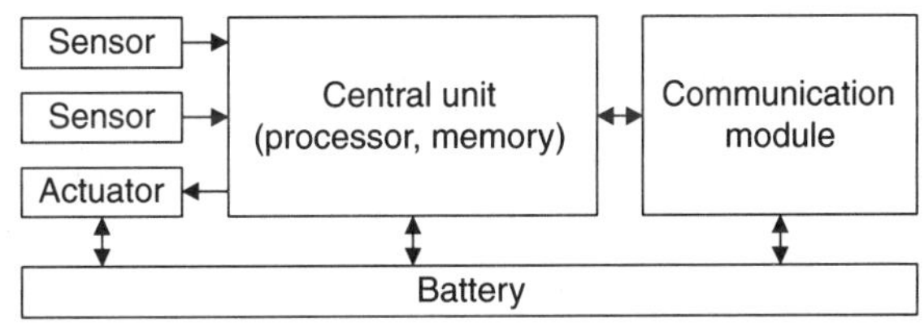

FIGURE 79.1 Structure of a sensor node.

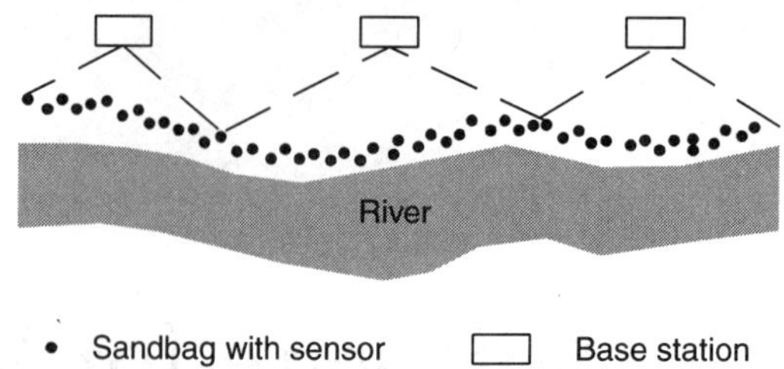

FIGURE 79.2   Example sensor network application: leakage detection.

sandbag is equipped with a moisture sensor and transmits sensor data to a base station next to the dyke. Thus, leakages can be detected earlier and reinforcement actions can be coordinated more efficiently.

Well-known research activities in the field of sensor networks are UCLA's WINS [3], Berkeley's Smart Dust [1], WEBS [2], and PicoRadio [4]. An example of European research activities is the EYES-Project [5]. Detailed surveys on sensor networks can be found in [6] and [7].

This chapter focuses on innovative architectures and basic concepts of current software development solutions for wireless sensor networks.

79.2 Preliminaries

Central unit of a sensor node is a low-power microcontroller that controls all functional parts of the node. Software for such a microcontroller has to be resource-aware on the one hand. On the other hand, several Quality-of-Service (QoS) aspects have to be met by sensor node software, such as latency, processing time for data fusion or compression, or flexibility regarding routing algorithms or MAC techniques.

Conventional software development for microcontrollers usually covers Hardware Abstraction Layer (HAL), operating system and protocols, and application layer. Often, software for microcontrollers is limited to an application-specific monolithic software block that is optimized for performance and resource usage. Abstracting layers such as HAL or operating system are often omitted due to resource constraints and low-power aspects.

Microcontrollers are often developed and programmed for a specific, well-defined task. This limitation of the application domain leads to high-performance embedded systems even with strict resource constraints. Development and programming of such systems is much effort. Furthermore, an application developed for one microcontroller is in most cases not portable to any other one, so that it has to be reimplemented from scratch. Microcontroller and application form an inseparable unit. If the application domain of an embedded system changes, often the whole microcontroller is replaced instead of writing and downloading a new program.

For sensor nodes, application-specific microcontrollers are preferred instead of general-purpose microprocessors. This is because of the small size and the low energy consumption of those controllers. However, requirements concerning a sensor node exceed the main characteristics of a conventional microcontroller and its software. The main reason for this is the dynamic character of a sensor node's task. Sensor nodes can adopt different tasks, such as sensor data acquisition, data forwarding, or information processing. The task assigned to a node with its deployment is not fixed until the end of its life cycle. Depending on, for instance, location, energy level, or neighborhood of a sensor node, a task change can become advantageous or even necessary.

Additionally, software for sensor nodes should be reusable. An application running on a certain sensor node should not be tied to a specific microcontroller but to some extent be portable onto different platforms to enhance interoperability of sensor nodes with different hardware platforms. Not limited to software development for wireless sensor networks is the general requirement for a straightforward programmability and, as a consequence, a short development time.

It is quite hard or even impossible to meet the requirements mentioned above with a monolithic application. Hence, at present there is much research effort in the areas of middleware and service architectures for wireless sensor networks. A middleware for wireless sensor networks should encapsulate required functionality in a layer between operating system and application. Incorporating a middleware layer has the advantage that applications get smaller and are not tied to a specific microcontroller. At the same time, development effort for sensor node applications reduces since a significant part of the functionality moves from application to middleware. Another research domain tends to service architectures for wireless sensor networks. A service layer is based on mechanisms of a middleware layer and makes its functionality more usable.

Architectural Layer Model

Like in other networking systems, the architecture of a sensor network can be divided into different layers (see Figure 79.3). The lower layers are hardware and HAL. The operating system layer and protocols are above the hardware-related layers. The operating system provides basic primitives like multithreading, resource management, and resource allocation that are needed by higher layers. Also, access to radio interface and input/output operations to sensing devices are supported by basic operating system primitives. Usually in node-level operating systems these primitives are rudimentary and there is no separation between user and kernel mode. On top of the operating system layer reside middleware, services, and application layer.

In recent years, much work has been done to develop sensor network node devices (e.g., Berkeley motes [9]), operating systems and algorithms, for example, for location awareness, power reduction, data aggregation, and routing. Today researchers are working on extended software solutions including middleware and service issues for sensor networks. The main focus of these activities is to simplify application development process and to support dynamic programming of sensor networks.

The overall development process of sensor node software usually ends with a manual download of an executable image over direct wired connections or over-the-air interface to target node.

After deployment of nodes, it is nearly impossible to improve or adapt new programs to the target nodes. But this feature is necessary in future wireless sensor networks to adapt the behavior of sensor networks dynamically by new injected programs. The task assigned to a node with its deployment is not fixed until the end of its life cycle. Depending on, for instance, location, energy level, or neighborhood of a sensor node, a task change can become advantageous or even necessary.

Middleware and Services for Sensor Networks

In sensor networks, design and development of solutions for higher-level middleware functionality and creation of service architectures are an open research issue. Middleware for sensor networks has two primary goals:

- Support of acceptable middleware Application Programming Interfaces (APIs), which abstract and simplify low-level APIs to ease application software development and to increase portability, and
- Distributed resource management and allocation.

Applications
Services
Middleware
Operating systems & protocols
Hardware abstraction layer
Hardware

FIGURE 79.3 Layered software model.

Besides the native network functions, such as routing and packet forwarding, future software architectures are required to enable location and utilization of services. A service is a program which can be accessed through standardized functions over a network. Services allow a cascading without previous knowledge of each other, and thus enable the solution of complex tasks.

A typical service used during the initialization of a node is the localization of a data sink for sensor data. Gateways or neighboring nodes can provide this service. To find services, nodes use a service discovery protocol.

Programming Aspect vs. Behavioral Aspect

Wireless sensor networks do not have to consist of homogeneous nodes. In reality, a network composed of several groups of different sensor nodes is imaginable. This fact changes the software development approach and points out new challenges as they are well-known from the distributed systems domain. In an inhomogeneous wireless sensor network, nodes contain different low-level system APIs, however, with similar functions. From a developers point of view it is hard to create programs, since APIs are mostly incompatible. To overcome the mentioned problems in heterogeneity and complexity, new software programming techniques are required. One attempt to accomplish this aspect is the definition of an additional API or an additional class library on top of each system API. But they are all limited by some means or other, for example, platform independency, flexibility, quantity, programming language. All approaches to achieve an identical API on different systems are covered by the *Programming Aspect* (Figure 79.4).

The programming aspect enables the developer to easily create programs on different hardware and software platforms. But an identical API on all platforms does not necessarily take the dynamics of the distributed system into account. Ideally, the application does not notice any dynamic system changes. This decoupling is termed as *Behavioral Aspect* and covers:

- Access to remote resources without previous knowledge, for example, Remote Procedure Calls (RPC) and discovered services
- Adaptations within the middleware layer to dynamic changes in the behavior of a distributed system, caused by incoming or leaving resources, mobility of nodes, or changes of the environment, and
- The ability of the network to evolve over time including modifications of the system's task, exchange or adaptation of running software parts, and mobile agents.

79.3　Current Software Solutions

This chapter presents five important software solutions for sensor networks. It starts with the most mature development TinyOS and depending software packages. It continues with SensorWare followed by two promising concepts, MiLAN and EnviroTrack. The Section finalizes with an introduction to SeNeTs.

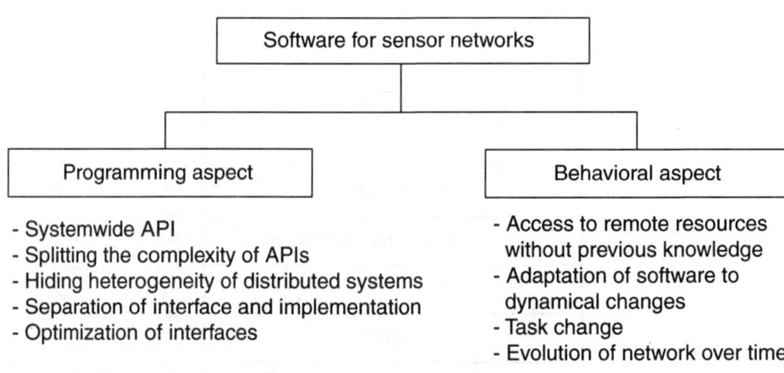

FIGURE 79.4　Two aspects of software for wireless sensor networks.

TinyOS

TinyOS is a component-based operating system for sensor networks developed at UC Berkeley. TinyOS can be seen as an advanced software framework [9] which has a large user community due to its open source character and its promising design. The framework contains numerous prebuilt sensor applications and algorithms, for example, multi-hop *ad hoc* routing and supports different sensor node platforms. Originally it was developed for Berkeley's Mica Motes. Programmers experienced with the C programming language can easily develop TinyOS applications written in a proprietary language called NesC [10].

The design of TinyOS is based on the specific sensor network characteristics: small physical size, low power consumption, concurrency-intensive operation, multiple flows, limited physical parallelism and controller hierarchy, diversity in design and usage, and robust operation to facilitate the development of reliable distributed applications. The main intention of the TinyOS developers was "retaining energy, computational and storage constraints of sensor nodes by managing the hardware capabilities effectively, while supporting concurrency-intensive operation in a manner that achieves efficient modularity and robustness"[16]. Therefore, TinyOS is optimized in terms of memory usage and energy efficiency. It provides defined interfaces between the components which reside in neighboring layers. A layered model is shown in Figure 79.5.

Elemental Properties

TinyOS utilizes an *event model* instead of a stack-based threaded approach, which would require more stack space and multitasking support for context switching, to handle high levels of concurrency in a very small amount of memory space. Event-based approaches are the favorite solution to achieve high performance in concurrency intensive applications. Additionally, the event-based approach uses CPU resources more efficiently and therefore takes care of the most precious resource, the energy.

An event is serviced by an event handler. More complex event handling can be done by a task. The event handler is responsible for posting the task to the task scheduler. Event and task scheduling is performed by a *two-level scheduling structure*. This kind of scheduling provides that *events*, associated with a small amount of processing, can be performed immediately, while longer running *tasks* can be interrupted by events. *Tasks* are handled rapidly, however no blocking or polling is permitted.

The TinyOS system is designed to *scale with the technology trends* supporting both, smaller designs and crossover of software components into hardware. The latter provides a straightforward integration of software components into hardware.

TinyOS Design

The architecture of a TinyOS system configuration is shown in Figure 79.6. It consists of the tiny scheduler and a graph of components. Components satisfy the demand for modular software architectures. Every component consists of four interrelated parts: a command handler, an event handler, an encapsulated fixed-size and statically allocated frame, and a bundle of simple tasks. The frame represents the internal state of the component. Tasks, commands and handlers execute in the context of the frame and operate on its state. In addition, the component declares the commands it uses and the events it signals. Through this declaration, modular component graphs can be composed. The composition process creates layers of components. Higher layer components issue commands to lower level components and these signal events to higher level components. To provide an abstract definition of the interaction of two components via commands and events, the bi-directional interface is introduced in TinyOS.

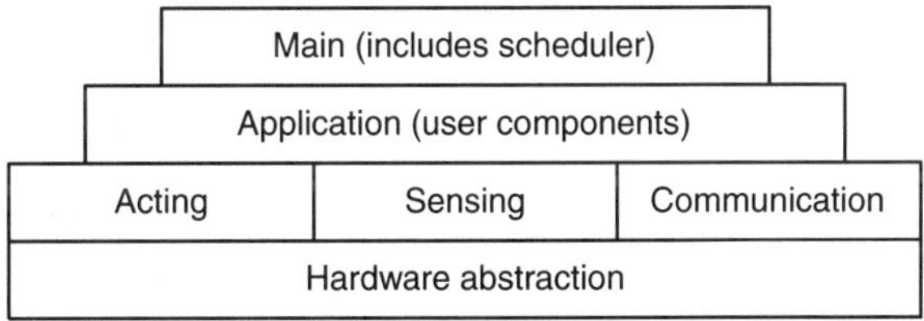

FIGURE 79.5 Software architecture of TinyOS.

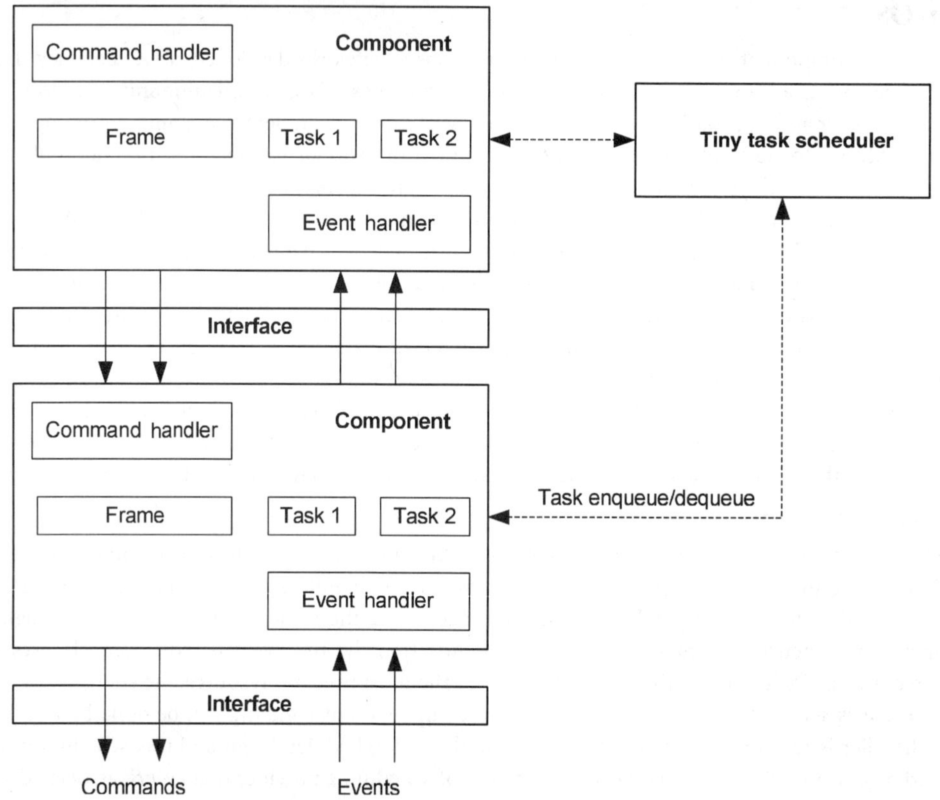

FIGURE 79.6 TinyOS architecture in detail.

Commands are nonblocking requests made to lower layer components. A command provides feedback to its caller by returning status information. Typically, the *command handler* puts the command parameters into the frame and posts a task into the task queue for execution. An event can signal the successful execution of a command.

Event handlers are invoked by *events* of lower layer components, or when directly connected to the hardware, by interrupts. Similar to commands, the frame will be modified and tasks are posted. Both, commands and tasks, perform a small fixed amount of work similar to interrupt service routines.

Tasks perform the primary work. They are atomic, run to completion, and can only be pre-empted by events. Tasks are queued in a FIFO *task scheduler* to perform an immediate return of event or command handling routines. Due to the FIFO scheduling, tasks are executed sequentially and should be short. Alternatively to the FIFO task scheduler, priority-based or deadline-based schedulers can be implemented into the TinyOS framework.

TinyOS distinguishes three categories of components. Hardware *abstraction components* map physical hardware into the component model. Mostly, these components export commands to the underlying hardware and handle hardware interrupts. *Synthetic hardware components* extend the functionality of hardware abstraction components by simulating the behavior of advanced hardware functions, for example, bit-to-byte transformation functions. For future hardware releases, these components can directly cast into hardware. *High-level software components* perform application specific tasks, for example, control, routing, data transmission, calculation on data, and data aggregation.

An interesting aspect of the TinyOS framework is the similarity of the component description to the description of hardware modules in hardware description languages, for example, VHDL or Verilog. A hardware module, for example, in VHDL is defined by an entity with input and output declarations, status registers to hold the internal state, and a finite state machine controlling the behavior of the module.

In comparison, a TinyOS component contains commands and events, the frame and a behavioral description. These similarities simplify the cast of TinyOS components to hardware modules. Future sensor nodes generations can benefit from this similarity in describing hardware and software components.

TinyOS Application

A TinyOS application consists of one or more components. These components are separated into modules and configurations. Modules implement application specific code, whereas configurations wire different components together. By using a top-level configuration, wired components can be compiled and linked to form an executable.

The interfaces between the components declare a set of commands and events which provide an abstract description of components. The application developer has to implement the appropriate handling routine into the component.

Figure 79.7 shows the component graph of a simple TinyOS application, that turns a LED on and off depending on the clock. The top-level configuration contains the application specific components (ClockC, LedsC, BlinkM) and a operating system specific component providing the tiny task scheduler and initialization functions. The Main component encapsulates the TinyOS specific components from the application. StdControl, Clock, and Leds are the interfaces used in this application. While BlinkM contains the application code, ClockC and LedsC are again configurations encapsulating further component graphs controlling the hardware clock and the LED's connected to the controller.

TinyOS provides a variety of additional extensions like the virtual machine MATÉ and the database TinyDB for cooperative data acquisition.

MATÉ

MATÉ [8] is a byte-code interpreter for TinyOS. It is a tiny communication-centric virtual machine designed as a component for the system architecture of TinyOS. MATÉ is located in the component graph on top of several system components, represented by sensor components, network component, timer component, and nonvolatile storage component.

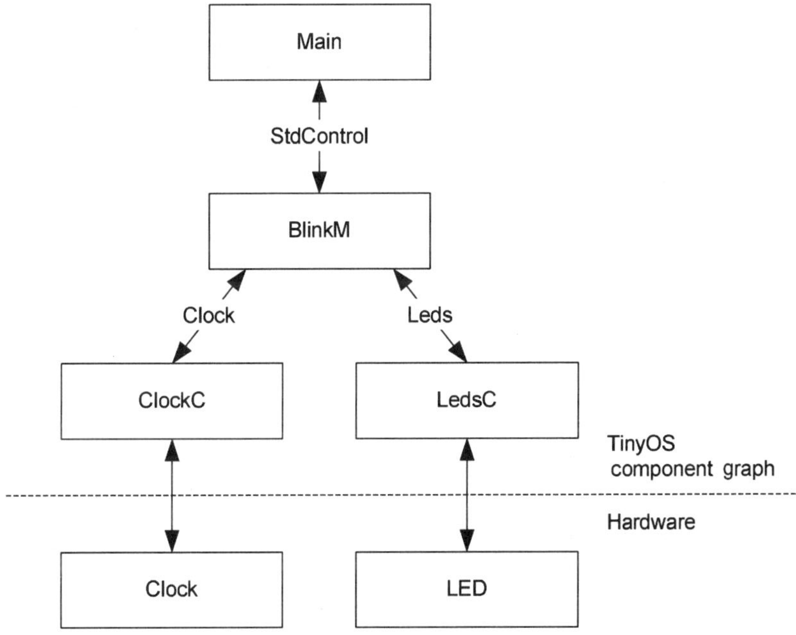

FIGURE 79.7   Simple TinyOS application.

The developer motivation for MATÉ was to solve novel problems in sensor network management and programming, in response to changing tasks, for example, exchange of the data aggregation function or routing algorithm. However, the associated inevitable reprogramming of hundreds or thousands of nodes is restricted to energy and storage resources of sensor nodes. Furthermore, the network is limited in bandwidth and network activity as a large energy draw. MATÉ attempts to overcome these problems, by propagating so-called code capsules through the sensor network.

The MATÉ virtual machine provides the possibility to compose a wide range of sensor network applications by the use of a small set of higher level primitives. In MATÉ, these primitives are one-byte instructions and they are stored into capsules of 24 instructions together with identifying and versioning information.

Architecture

MATÉ is a stack-based architecture which allows a concise instruction set. The use of instructions hides the asynchronous character of native TinyOS programming; because instructions are executed successively as several TinyOS tasks.

The MATÉ virtual machine shown in Figure 79.8 has three execution contexts: *Clock, Send,* and *Receive,* that can run concurrently at instruction granularity. *Clock* corresponds to timer events and *Receive* to message receive events, signaled from the underlying TinyOS components. *Send* can only be invoked from the *Clock* or *Receive* context. Each context holds an operand stack for handling data and a

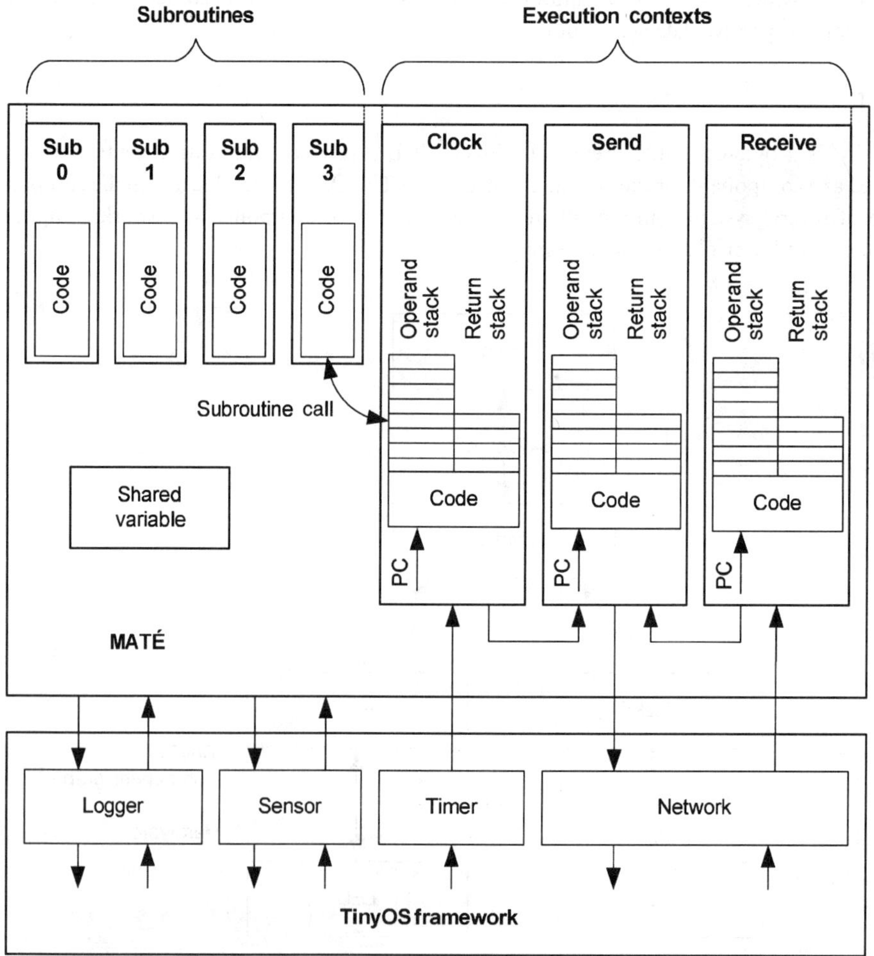

FIGURE 79.8 MATÉ architecture.

return stack for subroutines calls. Subroutines allow programs to be more complex as a single capsule can provide. Therefore, MATÉ has four spaces for subroutine code.

The code for the contexts and the subroutines is installed dynamically at runtime by code capsules. One capsule fits into the code space of a context or subroutine. The capsule installation process supports self-forwarding of capsules to reprogram a whole sensor network with new capsules. It is the task of the sensor network operator to inject code capsules in order to change the behavior of the network.

The program execution in MATÉ starts with a timer event or a packet receive event. The program counter jumps to the first instruction of the corresponding context (*Clock* or *Receive*) and executes until it reaches the *Halt* instruction. Each context can call subroutines for expanding functionality. The *Send* context is invoked from the other contexts to send a message in response to a sensor reading or to route an incoming message.

The MATÉ architecture provides separation of contexts. One context cannot access the state of another context. There is only one single shared variable among the three contexts that can be accessed by special instructions. The context separation qualifies MATÉ to fulfill the traditional role of an operating system. Compared to native TinyOS applications, the source code of MATÉ applications is much shorter.

TinyDB

TinyDB is a query processing system for extracting information from a network of TinyOS sensor nodes [11]. TinyDB provides a simple, SQL-like interface to specify the kind of data to be extracted from the network along with additional parameters, for example, the data refresh rate. The primary goal of TinyDB is to prevent the user from writing embedded C programs for sensor nodes or composing capsules of instructions regarding to MATÉ. The TinyDB framework allows data-driven applications to be developed and deployed much more quickly as developing, compiling, and deploying a TinyOS application.

Given a query specifying the data interests, TinyDB collects the data from sensor nodes in the environment, filters and aggregates the data, and routes it to the user autonomously. The network topology in TinyDB is a routing tree. Query messages flood down the tree and data messages flow back up the tree participating in more complex data query processing algorithms.

The TinyDB system is divided into two subsystems: sensor node software and a JAVA-based client interface on a PC. The sensor node software is the heart of TinyDB running on each sensor node. It consists of:

- sensor catalog and schema manager responsible for tracking the set of attributes, or types of readings and properties available on each sensor,
- query processor, utilizing the catalog to fetch the values of local attributes, to receive sensor readings from neighboring nodes, to combine and aggregate the values together, to filter, and to output the values to parents,
- small, handle-based dynamic memory manager,
- network topology manager to deal with the connectivity of nodes and to effectively route data and query subresults through the network.

The sensor node part of TinyDB is installed on top of TinyOS on each sensor node as an application. The JAVA-based client interface is used to access the network of TinyDB nodes from a PC physically connected to a bridging sensor node. It provides a simple graphical query-builder and a result display. The JAVA API simplifies writing PC applications that query and extract data from the network.

SensorWare

SensorWare is a software framework for wireless sensor networks that provides querying, dissemination, and fusion of sensor data as well as coordination of actuators [12]. A SensorWare platform has less stringent resource restrictions. The initial implementation runs on iPAQ handhelds (1 Mbyte ROM/128 kbyte RAM). The authors intended to develop a software framework regardless of present sensor node limitations.

SensorWare developed at the University of California, Los Angeles, aims at the programmability of an existing sensor network after its deployment. The functionality of sensor nodes can be dynamically modified through autonomous mobile agent scripts. SensorWare scripts can be injected into the network nodes as queries and tasks. After injection, scripts can replicate and migrate within the network.

Motivation for the SensorWare development was the observation that the distribution of updates and the download of complete images to sensor nodes is impractical for the following reasons. First, in a sensor network, a special sensor node may be not addressable because of missing node identifiers. Second, the distribution of complete images through a sensor network is highly energy consuming. Besides that, other nodes are affected by a download when multi-hop connections are necessary.

Updating complete images does not correspond to the low power requirements of sensor networks. As a consequence, it is more practicable to distribute only small scripts. In the following paragraphs, the basic architecture and concepts of SensorWare are described in detail.

Basic Architecture and Concepts

SensorWare consists of a scripting *language* and a *run-time environment*. The *language* contains various basic commands which control and execute specific tasks of sensor nodes. These tasks include, for example, communication with other nodes, collaboration of sensor data, sensor data filtering, or moving scripts to other nodes. The language comprises necessary constructs to generate appropriate control flows.

SensorWare utilizes Tcl as scripting language. However, SensorWare extends Tcl's core commands. These core extension commands are joined in several API groups, like Networking-API, Sensor-API, Mobility-API (see Figure 79.9).

SensorWare is event-based. Events are connected to special event handlers. If an event is signaled, an event handler serves the event according to its inherent state. Furthermore, an event handler is able to generate new events and to alter its current state by itself.

The *run-time environment* shown in Figure 79.10 contains fixed and platform-specific tasks. Fixed tasks are part of each SensorWare application. It is possible to add platform-specific tasks depending on specific application needs. The script manager task receives new scripts and forwards requests to the admission control task. The admission control task is responsible for script admission decisions and checks the overall energy consumption. Resource handlers manage different resources of the network.

Figure 79.11 shows the architecture of sensor nodes with included SensorWare software. The SensorWare layer uses operating system functions to provide the run-time environment and control scripts. Static node applications coexist with mobile scripts. To realize dynamic programmability of a deployed sensor network, a transient user can inject scripts into the network. After injection scripts are replicated within the network. The script code migrates between different nodes. SensorWare ensures that no script is loaded twice onto a node during the migration process.

MiLAN

Middleware Linking Applications and Networks (MiLAN) is a middleware concept introduced by Mark Perillo and Wendi B. Heinzelman from the University of Rochester [13, 14]. The main idea is to exploit the redundancy of information provided by sensor nodes. The performance of a cooperative algorithm in a distributed sensor network application depends on the number of involved nodes.

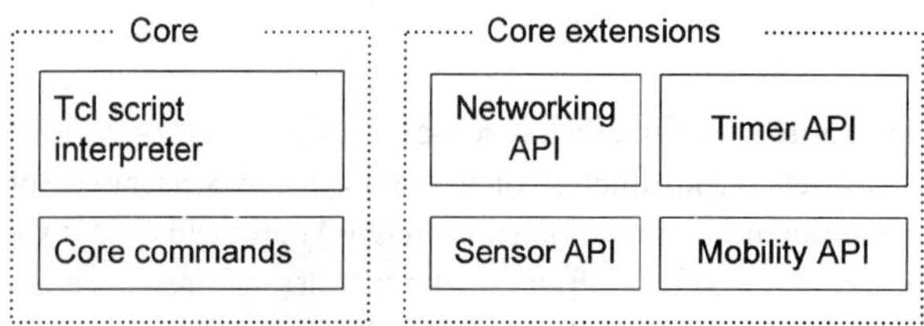

FIGURE 79.9   SensorWare scripting language.

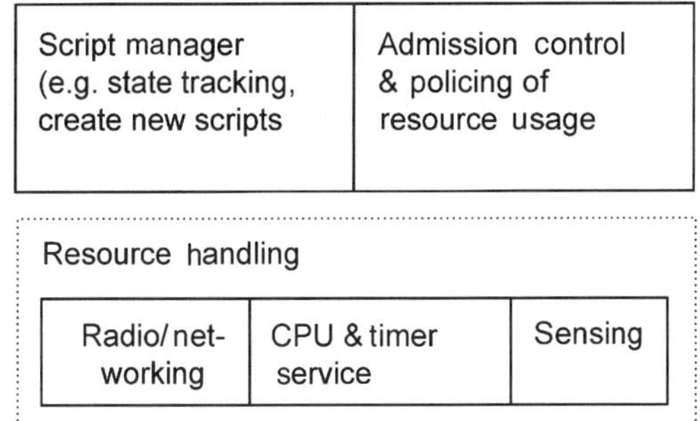

FIGURE 79.10 SensorWare run-time environment.

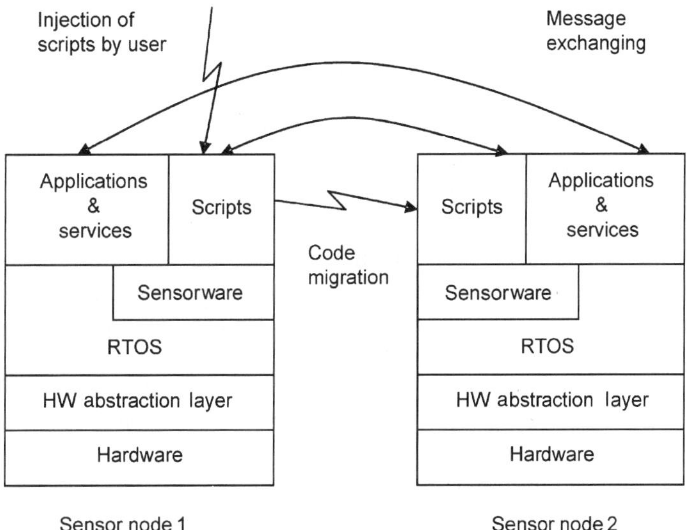

FIGURE 79.11 Sensor node architecture.

Because of the inherent redundancy of a sensor network where several sensor nodes provide similar or even equal information, evaluating all possible sensor nodes leads to high energy and network costs. Therefore, a sensor network application has to choose an appropriate set of sensor nodes to fulfill application demands.

Each application should have the ability to adopt its behavior in respect to the available set of components and bandwidth within the network. This can be achieved by a parameterized sensor node selection process with different cost values. These cost values are described by the following cost equations:

- Application performance: The minimum requirements for network performance are calculated from the needed reliability of monitored data. $F_R = \{S_i : \forall_j \in J \; R(S_i, j) \geq r_j\}$, where F_R stands for the *allowable set* of possible sensor node combinations, S_i represents the available sensor nodes, and $R(S_i)$ is their reliability.
- Network costs: Defines a subset of sensor nodes that meet the network n_o constraints. The *network feasible set* $F_N = \{S_i : N(S_i) \leq n_o\}$, where $N(S_i)$ represents the total cost and the maximal data rate the network can support.
- Application performance and network costs are combined to the *overall feasible set*: $F = F_R \cap F_N$.
- Energy: Describes the energy dissipation of the network. $C_P(S_i) \sum_{S_j \in S_i} = C_P(S_j)$, where $C_P(S_j)$ is the power cost to node S_j.

It is up to the application to decide how these equations are weighted. This decision-making process is completely hidden from the application, so the development process is simplified significantly. MiLAN uses two strategies to achieve the objective to balance QoS and energy costs:

- Turning off nodes with redundant information
- Use of energy efficient routing.

The MiLAN middleware is located between network and application layer. It can interface a great variety of underlying network protocols, such as Bluetooth and 802.11. MiLAN uses an API to abstract from network layer but gives the application access to low level network components. A set of commands identifies and configures the network layer.

EnviroTrack

EnviroTrack is a TinyOS-based application developed at the University of Virginia that solves a fundamental distributed computing problem, *environmental tracking of mobile entities* [15]. Therefore, EnviroTrack provides a convenient way to program sensor network applications that track activities in their physical environment. The programming model of EnviroTrack integrates objects living in physical time and space into the computational environment of the application through virtual objects, called *tracking objects*. A tracking object is represented by a group of sensor nodes in its vicinity and is addressed by *context labels*. If an object moves in the physical environment, then the corresponding virtual object moves too because it is not bound to a dedicated sensor node. Regarding the tracking of objects, EnviroTrack does not assume cooperation from the tracked entity.

Before a physical object or phenomenon can be tracked, the programmer has to specify its activities and corresponding actions. This specification enables the system to discover and tag those activities and to instantiate tracking objects. For example, to track an object warmer than 100°C, the programmer specifies a Boolean function, temperature > 100°C, a critical number or mass of sensor nodes, which fulfil the Boolean function within a certain time (a fact that is often referred to as *freshness* of information). These parameters of a tracking object are called aggregate state. All sensor nodes matching this aggregate state join a group. The network abstraction layer assigns a context label to this group. Using this label, different groups can be addressed independent of the set of nodes currently assigned to it. If the tracked object moves, nodes join or leave the group because of the changed aggregate state but the label resides persistent. This group management enables context-specific computation.

The EnviroTrack programming system consists of:

- *EnviroTrack compiler*: In EnviroTrack programs, a list of context declarations is defined. Each definition includes an activation statement, an aggregate state definition, and a list of objects attached to the definitions. The EnviroTrack compiler includes C program templates. The whole project is then built using the TinyOS development tools.
- *Group management protocol*: All sensors associated to a group are maintained by this protocol. A group leader is selected out of the group members when the critical mass of nodes and freshness of the approximate aggregate state is reached. The group management protocol ensures that only a single group leader per group exists. The leader sends a periodical heartbeat to inform its members that the leader is alive. Additionally, the heartbeat signal is used to synchronize the nodes and to inform nodes that are not part of the group, but fulfill the sensing condition.
- *Object naming and directory services*: These services maintain all active objects and their locations. The directory service provides a way to retrieve all objects of a given context type. It also assigns names to groups so they can be accessed easily. It handles dynamical joining and leaving of group members.
- *Communication and transport services*: The Migration Transport Protocol (MTP) is responsible for the transportation of data packets between nodes. All messages are routed via group leader nodes. Group leader nodes identify the context group of the target node and the position of its leader using the directory service. The packet is then forwarded to the leader of the destination group. All

leadership information provided by MTP packets is stored in the leaders on a least-recently-used-basis to keep the leader up-to-date and to reduce directory look-ups.

EnviroTrack enables the construction of an information infrastructure for tracking environmental conditions. It manages dynamic groups of redundant sensor nodes and attaches computation to external events in the environment. Furthermore, EnviroTrack implements noninterrupted communication between dynamically changing physical locales defined by environmental events.

SeNeTs

SeNeTs is a middleware architecture for wireless sensor networks. It is developed at the University of Rostock [17]. The SeNeTs middleware is primarily designed to support the developer of a wireless sensor network during the predeployment phase (programming aspect). SeNeTs supports the creation of small and energy-saving programs for heterogeneous networks. One of the key features of SeNeTs is the optimization of APIs. The required configuration, optimization, and compilation of software components is processed by a development environment. Besides the programming aspect, the middleware supports the behavioral aspect as well, such as task change or evolution over time (see Figure 79.4).

Architecture

SeNeTs is based on the software layer model introduced in Chapter 2. In order to increase flexibility and enhance scalability of sensor node software it separates small functional blocks as shown in Figure 79.12. In addition, the OS-layer is separated into a *Node-specific Operating System* and a *Driver Layer*, which contains at least one *Sensor Driver* and several *Hardware Drivers*, such as timer driver and RF driver. The *Node-specific Operating System* handles device-specific tasks, for example, boot-up, initialization of hardware, memory management, and process management as well as scheduling. *Host Middleware* is the superior software layer. Its main task is to organize the cooperation of distributed nodes in the network. *Middleware Core* handles four optional components, which can be implemented and exchanged according to the node's task. *Modules* are components that increase the functionality of the middleware. Typical modules are routing modules or security modules. *Algorithms* describe the behavior of modules. For example, the behavior of a security module can vary in the case the encryption algorithm changes. The *services* component contains the required software to perform local and cooperative services. This component usually cooperates with service components of other nodes to fulfill its task. *Virtual Machines (VM)* enable an execution of platform independent programs installed at run-time.

Figure 79.13 shows the expansion of the proposed architecture to a whole sensor network from the logical point of view. Nodes can only be contacted through services of the middleware layers. The *Distributed Middleware* coordinates the cooperation of services within the network. It is logically located

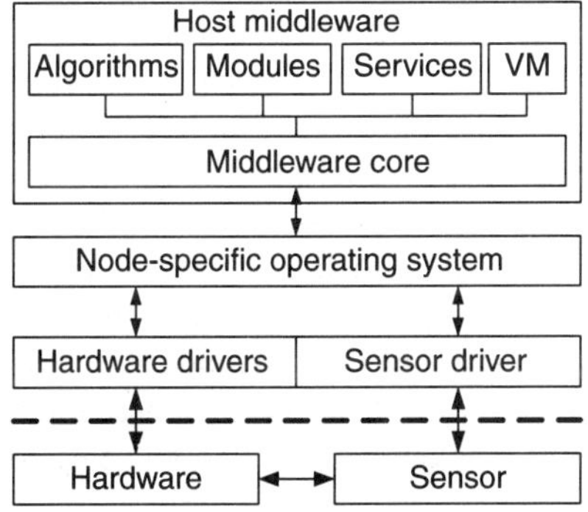

FIGURE 79.12 Structure of a node application.

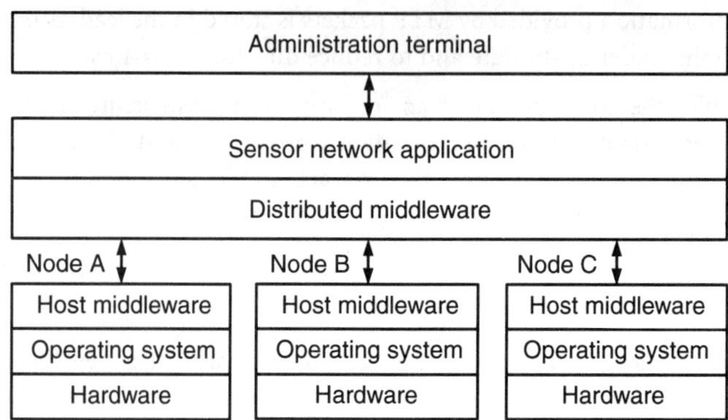

FIGURE 79.13   Structure of a sensor network.

in the network layer but physically exists in the nodes. All layers together in conjunction with their configuration compose the *sensor network application*. Thus, nodes do not perform any individual tasks.

The *Administration Terminal* is an external entity to configure the network and evaluate results. It can be connected to the network at any location.

All functional blocks of the described architecture are represented by *components* containing real source code and a description about dependencies, interfaces, and parameters in XML. One functional block can be rendered by alternative components. All components are predefined in libraries.

Interface Optimization

One of the key features in SeNeTs is interface optimization. Interfaces are the descriptions of functions between two software parts. As illustrated in Figure 79.14, higher level applications using services and middleware technologies require abstract software interfaces. The degree of hardware-dependent interfaces increases in lower software layers. Hardware-dependent interfaces are characterized by parameters to configure hardware components directly in contrast to abstract software interfaces whose parameters describe abstractions of the underlying system.

Software components require a static software interface to the application in order to minimize customization effort for other applications and to support compatibility. The use of identical components in different applications leads to a higher number of complex interfaces in these components. This is caused by component programming focused on supporting most possible use-cases of all possible applications whereby each application uses only a subpart of the functionality of a component. Reducing the remaining overhead is the objective of *generic software* and can be done by *interface optimization* during compile time.

Interface optimizations result in proprietary interfaces within a node (Figure 79.15). Parts of the software cannot be exchanged without sensible effort. In a sensor node, the software is mostly static except programs for virtual machines. Accordingly, static linking is preferred. Statically linked software in conjunction with interface optimization leads to in faster and smaller programs.

In SeNeTs, interfaces are customized to the application in contrast to common approaches used in desktop computer systems. These desktop systems are characterized by writing huge adaptation layers. The interface optimization can be propagated through all software layers and, therefore, saves resources. As an example of an optimization, a function OpenSocket(int name, int mode) identifies the network interface with its first parameter and the opening mode in the second parameter. However, a node that has only one interface opened with constant mode once or twice, does not need these parameters. Consequently, knowledge of this information at compile time can be used for optimizing, for example, by:

- Inlining the function
- Eliminating both parameters from the delivery process.

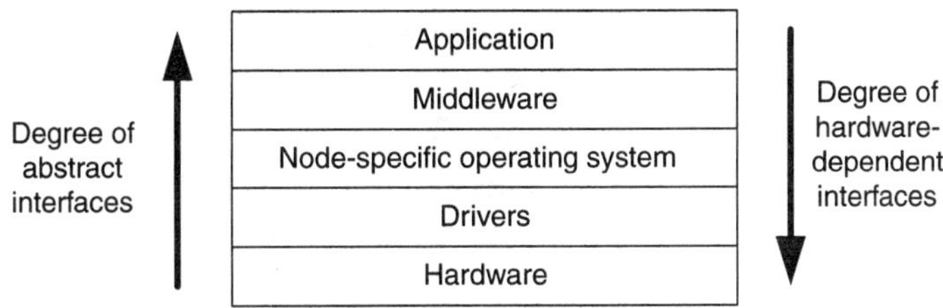

FIGURE 79.14　Interfaces within the software-layer model.

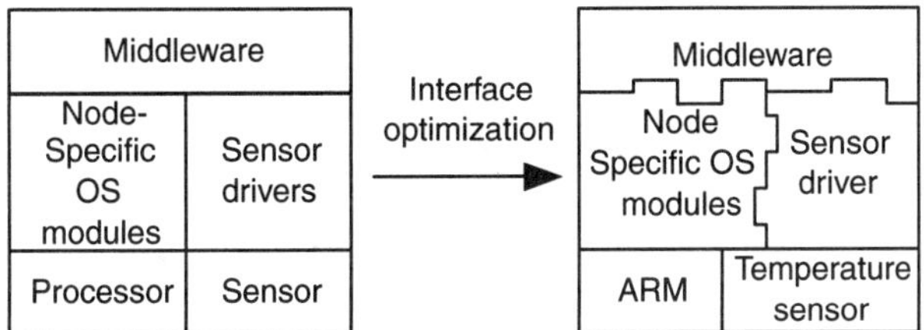

FIGURE 79.15　Interface optimization.

Another possibility is to change the semantics of data types. A potential use case is the definition of accuracy of addresses that results in changing data type's width. In SeNeTs, there are several types of interface optimizations proposed (Table 79.1).

Some optimizations such as "static parameters" are sometimes counterproductive in particular, if register-oriented parameter delivery is used. This is caused by the use of offset addresses once at parameter delivery instead of absolute addresses embedded in the "optimized" function. Consequently, the introduced optimizations strongly depend on:

- Processor and processor architecture
- Type of parameter delivery (stack or register-oriented)
- Memory management (small, huge, size of pointers)
- Objective of optimization (memory consumption, energy consumption, or compact code etc.), and
- Sensor network application.

Development Process

Figure 79.16 shows the development process of sensor node software in SeNeTs. First, for each functional block the components have to be identified and included into the project. During design phase, the chosen components are interconnected and configured depending on developer's settings. Then, interface as well as parameter optimization is performed. The final source codes are generated and logging

TABLE 79.1　Types of Interface Optimization

Optimization	Description
Parameter elimination	Parameters which are not used in one of the called subfunctions can be removed.
Static parameters	If a function is still called with same parameters, these parameters can be defined as constants or static variables in the global namespace. Thus, the parameter delivery to the function can be removed.
Parameter ordering	The sequence order of parameters is optimized in order to pass parameters through cascading functions with same or similar parameters. It is particular favorable in systems using processor registers instead of the system stack to deliver parameters to sub-functions.
Parameter aggregation	In embedded systems, many data types are not byte-aligned, for example, bits to configure hardware settings. If a function has several non byte-aligned parameters, these parameters may be combined.

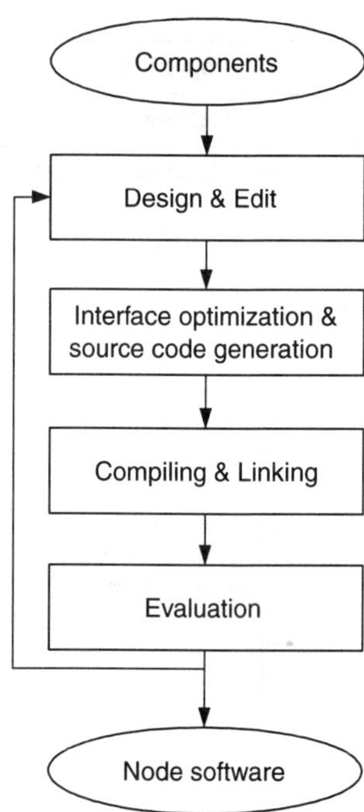

FIGURE 79.16   Development process of node software.

components can be included to monitor run-time behavior. The generated source codes are compiled and the executable is linked together. During evaluation phase, the created node application can be downloaded to the node and executed. Considering the monitoring results, a new design cycle can be started to improve project settings. As a result of the design flow, optimized node application software is generated. The node application now consists of special tailored parts only needed by the specific application of the node.

Optionally, software components in a node can be linked together statically or dynamically. Statically linking facilitates an optimization of interfaces between several components within a node. A dynamic link process is used for components exchanged during run-time, for example, algorithms downloaded from other nodes. This procedure results in system-wide interfaces with significant overhead and prevents interface optimization.

Currently, SeNeTs is in design state. Due to its initial design, the implementation of SeNeTs and proof of concept remain to be seen.

79.4   Summary

At the present time, TinyOS is the most mature operating system framework for sensor nodes. The component-based architecture of TinyOS allows an easy composition of sensor node applications. New components can be added easily to TinyOS to support novel sensing or transmission technologies or to support upcoming sensor node platforms.

MATÉ addresses the requirement to change a sensor node's behavior at run-time by introducing a VM on top of TinyOS. Via transmitting capsules containing high-level instructions, a wide range of sensor node applications can be installed dynamically into a deployed sensor network.

TinyDB was developed to simplify data querying from sensor networks. On top of TinyOS it provides an easy-to-use SQL interface to express data queries and addresses the group of users nonexperienced with writing embedded C code for sensor nodes.

EnviroTrack is an object-based programming model to develop sensor network applications for tracking activities in the physical environment. Its main feature is dynamical grouping of nodes depending on environmental changes described by predefined aggregate functions, critical mass, and freshness horizon.

SensorWare is a software framework for sensor networks employing lightweight and mobile control scripts that allow the dynamic deployment of distributed algorithms into a sensor network. In comparison to the MATÉ framework, the SensorWare run-time environment supports multiple applications to run concurrently on one SensorWare node.

The MiLAN middleware provides a framework to optimize network performance, needed sensing probability and energy costs based on equations. It is the programmer's decision to weight these equations.

SeNeTs is a new approach to optimize the interfaces of sensor network middleware. SeNeTs aims at the development of energy-saving applications and the resolving of component dependencies at compile-time.

References

[1] Kahn, J.M., R.H. Katz, and K.S.J. Pister, Next Century Challenges: Mobile Networking for Smart Dust, in Proceedings of the ACM MobiCom'99,Washington, U.S.A., 1999, pp. 271–278.

[2] Culler, D., E. Brewer, and D. Wagner, A Platform for WEbS (wireless embedded sensor actuator systems), Technical report, University of California, Berkeley, 2001.

[3] Pottie, G.J. and W.J. Kaiser, Wireless integrated network sensors, *CACM*, 43, 51–58, 2000.

[4] Rabaey, J. et.al., Picoradio supports ad hoc ultra-low power wireless networking, *IEEE Computer*, 42–48, Vol. 33, no. 7, July 2000.

[5] EYES- Energy efficient Energy-Efficient Sensor Networks, http.//eyes.eu.org

[6] Akyildiz, I.F., W. Su, Y. Sankarasubramaniam, and E. Cayirci, A survey on sensor networks, *IEEE Communications Magazine*, 102–114, Vol. 40, no. 8, Aug. 2002.

[7] Rentala, P., R. Musunuri, S. Gandham, and U. Saxena, Survey on Sensor Networks, Technical report UTDCS-10-03, University of Texas, http://www.utdallas.edu/~gshashi/survey.pdf

[8] Levis, P. and D. Culler, MATÉ: a Tiny Virtual Machine for Sensor Networks, in Proceedings of ACM Conference on Architecture Support for Programming Languages and Operating Systems (ASPLOS), Oct. 2002.

[9] Hill, J. et al., System Architecture Directions for Networked Sensors, in Proceedings of the Ninth International Conference on Architectural Support for Programming Languages and Operating Systems, Nov. 2000.

[10] Gay, D., P. Levis, R.V. Behren, M. Welsh, E. Brewer, and D. Culler, The nesC Language: A Holistic Approach to Networked Embedded Systems, Berkeley, 2002.

[11] Madden, S., J. Hellerstein, and W. Hong, TinyDB: In-Network Query Processing in TinyOS, Intel Research, IRB-TR-02-014, Oct. 1, 2002.

[12] Boulis, A. and M.B. Srivastava, A Framework for Efficient and Programmable Sensor Networks, The 5th IEEE Conference on Open Architectures and Network Programming (OPENARCH 2002), New York, June 2002.

[13] Murphy, A. and W. Heinzelman, MiLAN: Middleware Linking Applications and Networks, Technical report, Jan. 2003.

[14] Perillo, M. and W.B. Heinzelman, Providing Application QoS through Intelligent Sensor Management, 2002.

[15] Abdelzaher, T., B. Blum et al., EnviroTrack: An Environmental Programming Model for Tracking Applications in Distributed Sensor Networks, Technical report CS-2003-02, University of Virginia, 2003.

[16] Culler, D., TinyOS — a component-based OS for the networked sensor regime, URL: http://webs.cs.berkeley.edu/tos/, 2003.

[17] Blumenthal, J., M. Handy, F. Golatowski, M. Haase, and D. Timmermann, Wireless Sensor Networks — New Challenges in Software Engineering, 9th IEEE International Conference on Emerging Technologies and Factory Automation (ETFA), Lisbon, Portugal, Sept. 2003.

80

Introduction to Multi-sensor Data Fusion

Pascal Vasseur
*CREA (Centre de Robotique,
Electrotechnique et Automatique) -
UPJV*

El Mustapha Mouaddib
*CREA (Centre de Robotique,
Electrotechnique et Automatique) -
UPJV*

Claude Pegard
*CREA (Centre de Robotique,
Electrotechnique et Automatique) -
UPJV*

80.1 Introduction

Humans use multi-sensor data fusion largely and naturally in order to perceive the environment. We can give, for example, the testing of wine to estimate its vintage and to appreciate it (fusion of taste, smell, and vision). Another example is medical diagnosis where many parameters (temperature, context, tiredness, ...) must be taken into account. Artificial processes (robots, manufacturing process, vehicles, ...) work like natural ones, and they often need several parameters. By multiplying information origin, the goal is to increase accuracy, robustness, and reduce eventual conflicts and ambiguity. Wald, in [22], uses "increasing the quality" to define the main goal of data fusion.

B.V. Dasarathy, says: "*When you use information from one source, it's plagiarism; When you use information from many, it's information fusion*" to qualify fusion in [7]. There are other, more technical, definitions that emanated from working groups. Here is an example: "*Data fusion is a formal framework in which are expressed the means and tools for the alliance of data originating from different sources. Data fusion aims at obtaining information of greater quality; the exact definition of 'greater quality' well depend upon the application*" [22].

Here are two web sites where readers can find many references and some generalities about the subject: http://www.infofusion.org./isif/publications.html and http://www.data-fusion.org/. Multi-sensor data fusion has originally been used for military applications such as target recognition, battlefield surveillance, or guidance and control of autonomous vehicles. Current applications include monitoring of complex machinery or medical diagnosis [11].

Fusion in image processing could be used on one image to increase the potentialities of treatment and extraction of primitives [2,6,17,23]. But it can also be used to combine many homogenous or heterogeneous images, to increase the robustness and the accuracy.

For an autonomous system, it is essential to be able to sense its own enviornment. The sensor measurements usually contains noisy information with less than the expected accuracy. Thus, different types of sensors are often used to give a set of information allowing to extract parameters with a better accuracy. Robotics is a good application field for sensor fusion [4,5]. One can easily demonstrate the pertinence of using redundancy and complementarity of sensors [19]. Indeed, navigation in real enviornment requires a very large and complete perceptoin of the surroundings of the robot. For example, odometry and more generally the proprioceptives sensors are not sufficient because they systematically imply a bias in the estimation [14]. The exteroceptive perception (telemetry, vision) allows to solve this problem by performing absolute localization. However, sometimes, it is impossible to use these sensors. Hence, the fusion of odometry and exteroceptive sensors allows to estimate the state of the robot even if one is unable to give correct information.

In general, three types of multi-sensors data fusion are distinguished:

1. A complementary fusion: fusion of several disparate sensors that individually only give partial information of the environment. The multiplicity of ultrasonic or infrared range sensors can give multidirectional information.
2. A competitive fusion: in which information about the same parameter is collected through observations of several sensors, and the best estimate is evaluated. Stereovision and range sensors are often used in competitive fusion to estimate the distance from obstacles.
3. A cooperative fusion: fusion of different sensors of which one sensor relies on the observations of another to make its own observations.

A number of classifications have been proposed; most of them can be summarized in a three-level model:

Signal (or pixel)[1]-level fusion refers to the combination of the signals of a group of sensors in order to provide a signal that is usually of the same form as the original signals but of greater quality. The signals from sensors can be modeled as random variables corrupted by uncorrelated noise, and the fusion process can be considered as an estimation procedure. At this level, temporal or special concordance between signals to fuse is very important. When this concordance is not possible, a prediction process is used to synchronize all data. In this case, fusion reduces variance and increases image-processing tasks. The most common techniques for signal-level fusion consist in weighted averaging and Kalman filtering.

Feature extraction level: this stage increases the robustness of the features extracted from raw signals, or increases the variety of information provided about a target and the accuracy. Typical features are extracted from an image and are used for fusion include edges and regions of similar intensity.

Symbolic level or classification level: this level uses symbolic (linguistic, logical, semantical) techniques to infer more information on the state of the world or for object recognition tasks. Symbolic-level fusion can effectively integrate the information from multiple sensors at the highest level of abstraction. Symbol-level fusion is commonly used in the applications where multiple sensors are of different nature or refer to different regions of the environment. The symbols used for fusion can be derived from the processing of the individual sensory information, or through symbolic reasoning processes that may make use of prior knowledge from a world model or sources external to the system. This level increases the certitude of the decision.

In Section 80.2, we present some of the most relevant computational methods capable of performing the data fusion efficiently. With each one, we propose some examples of data fusion application to show how these methods can be implemented and their advantages. They have been chosen according to their relevance to the previously mentioned methods and because they correspond to very different experimental conditions. These applications are, respectively, mobile robotics and image processing. But the readers can find other applications, for example, in [12,15].

80.2 Methods of Fusion

These methods respectively belong to the probability theory, the evidence theory, and the possibility theory [10,24]. However, we can also mention other methods such as the AND operator, the weighted average, the voting, the information theory, or the geometric methods. Among this large set of methods, we will show that the choice can be made according to the learning stage, the mode of fusion, and the type of final decision.

---

[1] We use signal concept for signals and images, even if many authors treat these differently.

In order to provide the description of the methods in a homogeneous framework, we define the data fusion problem as follows: we first consider a centralized point of view in which all the information is available simultaneously. This restriction allows to avoid the problems linked to the synchronization or the datation of data, for example. We then suppose n heterogeneous sensors or sources S_j, which provide information about an observed phenomenon, and $D=\{D_1...,D_l\}$, the set of hypotheses or decision space in which each D_i represents a decision on an element e. Numerically, for each element e, each source S_j will provide for each decision D_i a value M_{ij} currently called a mesaure but that can take another particular signification according to the selected data fusion method. The final aim then consists in taking the most appropriate decision for each element by fusing the different measures.

Data Fusion in Probability Theory

The probabilistic framework is undoubtedly the most widely used in signal or image fusion. The probability theory was developed during the 18th century and allows to model some random phenomena and also to integrate the deviations between the true and the observed values. Probability theory is a huge domain and we will restrict our presentation to the Bayesian approach and the Kalman filter.

The Bayesian Approach

Bayesian data fusion has been the object of numerous works and is probably the most currently used method in probabilistic data fusion. This method defines the membership of a class for an element with the minimum error and estimates the risk of a decision. The combination of information is performed thanks to the generalization of the Bayes theorem. Consider an event E that may depend on N causes C_i that are different and mutually incompatible. The aim is to estimate the probability that C_i is the cause of E. We can deduce that $E = \bigcup_{i=1}^{N} E \cap C_i$ since $\{C_i\}$ is a complete system. From the total probability theorem, we obtain $P(E)= \sum_{i=1}^{N} P(E \cap C_i)$ and by the application of the conditional probability theorem, we have $P(C_i/E) = P(E)$ $P(C_i/E) = P(C_i) \, P(E/C_i)$ and then finally:

$$P(C_i/E) = \frac{P(C_i) \, P(E/C_i)}{\sum_{k=1}^{N} P(C_k) \, P(E/C_k)}$$

The expression $P(C_i/E)$ is called *the a posteriori probability*. $P(C_i)$ is the *a priori probability* of meeting this cause and $P(E/C_i)$ is the likelihood probability or the conditional probability. The denominator of the previous equation is a normalization factor (the sum of all a *posteriori probabilities* is equal to 1).

Fusion uses this theorem where E is a set of events. For example, we consider two events (E_1, E_2). In order to obtain more reliable information about the events, we can fuse by taking the product of the two probability density functions:

$$P(C_1/E_1, E_2) = \frac{P(C_1/E_2) \, P(C_1/E_1)}{P(C_1)}$$

Proof 1. Demonstration of the probabilities fusion.

Numerical example: We consider two doctors who respectively provide diagnosis M_1 and M_2 for the common decisions D_1 and D_2 about two sicknesses. $P(D_1/M_1)$ is the conditional probability density function and this corresponds to the likelihood of the disease D_1 after the diagnosis M_1.

In order to obtain more reliable information about the sickness, we can fuse the diagnosis of both doctors by taking the product of the two probability density functions:

$$P(D_1/M_1, M_2) = \frac{P(D_1/M_2) \, P(D_1/M_1)}{P(D_1)}$$

Decisions (likelihood that the patient has the sickness) of the two doctors are shown in Table 80.1.

To evaluate $P(D_1/M_1M_2)$, we need the *a priori probability* value $P(D_1)$. It could be estimated as the frequency of the apparition of the sickness D_1. In our case, we consider that $P(D_1)=P(D_2)=1$. Thus, we can only compute the numerator of the previous equation:

$$P(D_1/M_1,M_2)=0.6 \text{ and } P(D_2/M_1,M_2)=0.9$$

This means that the patient probably has the sickness D_2. To decide this, we choose the maximum *a posteriori* rule. They are many criteria to build decision rule: maximum likelihood, maximum entropy,… But the most used one is the maximum *a posteriori.*

The reader can find other implementations of Bayes rule with some examples in [12].

The Kalman Filter

The Kalman filter is usually used in automatic and systems control. It uses the system modeling by state variables and is particularly well adapted for the modelization of dynamic systems. From an observation, the aim consists in finding the value of the state vector according to the statistics of the additional noise of the system. An optimal estimation of the state vector is then obtained iteratively according to a least squares criterion. In the case of data fusion with Kalman filter, we consider measures from the different sensors as observations of the same state vector. As previously, the uncertainty of a measure can be modeled by the integration of the additional noise, contrary to the imprecision, which cannot be evaluated.

Assume that the state of the system obeys the linear dynamic model

$$x_k=Ax_{k-1}+Bu_k+w_{k-1}$$

and the measure is equal to

$$z_k=Hx_k+v_k$$

The random variables w_k and v_k, respectively, represent process noise and measurement noises. They are assumed to be independent, white, and with normal probabilty distributions

$$P(w)=N(0, Q) \text{ and } P(v)=N(0, R)$$

Q and R, respectively, represent the process and the measurement noise covariance matrices. The matrix A relates the state at the previous time step $k-1$ to the state at current step k. The matrix B relates the optional control input u to the state x. The matrix H relates the state x_k to the measurement z_k.

Figure 80.1 shows the different steps for the evaluation of the parameters of the system. There are two steps: the time update (prior or prediction) and the measurement update (posterior or correction). The system parameters are first estimated from the values of the previous step and are then corrected by the measures of the correction step.

This model is relatively general, and the solution gives the optimal way to utilize the measurements for reconstructing the system state. Still wider generality is offered by the *extended Kalman filter* (EKF): the model can be nonlinear (the solution in that case, however, is only approximative).

The interested readers can find more details in [9].

Example of application: In [1], the robot is equipped with an odometer, an omnidirectional laser rangefinder, and a camera in order to detect vertical lines. An EKF is used for the data fusion and the estimation of the robot position (*on-the-fly localization*). First, the position (state vector) is predicted from

TABLE 80.1   Likelihood Proposed by Two Doctors

$P(D/M)$	M_1	M_2
D_1	0.3	0.2
D_2	0.2	0.45

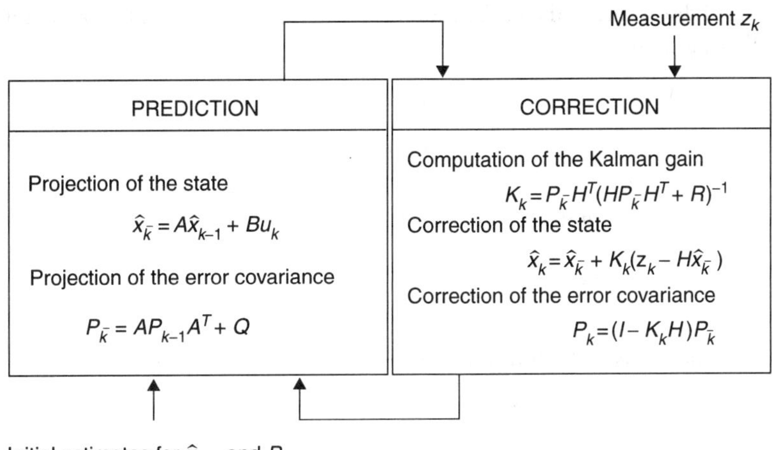

FIGURE 80.1 Scheme of fusion by using Kalman filter.

the odometry. Next, the vector of observations is created from the exteroceptives data: two parameters for the line segments (ρ, θ) and the azimuth angle for the vertical lines. After the prediction of the measures and their matching with the real measures, the next step consists in estimating the pose of the robot and updating the covariance matrices. The covariance matrices linked to the state vector and the observations represent, respectively, the uncertainties of the robot position and of the sensor measures. This robot has been successfully tested for several days and for more than 140,000 localizations.

Data Fusion in Evidence Theory

Evidence theory, initially introduced by Dempster [8], has been reformulated with a more complete formalism by Shafer [18]. This theory is described as a generalization of the Bayesian inference to the uncertainty treatment. In this way, it allows to manipulate unnecessary exclusive events. This capacity provides the advantage to explicitly represent the uncertainty for an event. Moreover, it takes into account the ignorance about an event and perfectly represents what is already known. The evidence theory allows to evaluate the best decision(s) according to the belief that describes the uncertainty for each decision. It is worth noting that the belief evaluation is performed in a subjective way in accordance with our knowledge.

Dempster–Shafer theory begins by the construction of a frame of discernment Θ, which is a set of mutually exclusive decisions. This frame can represent the possible values for any element. For example, if a medical diagnosis for a disease has to be performed, the frame of discernment will be composed of the possible diseases. We then define the mass function that represents the belief measure assigned to an element A of Θ. This function is also called basic probability assignment. Then, if Θ is a frame of discernment, a function $m : 2^{\Theta} \rightarrow [0,1]$ is called mass function or basic probability assignment if:

$$m(\phi) = 0$$

and

$$\sum_{A \subset \Theta} m(A) = 1.$$

In this way, the belief for the empty set must be null and a total belief for a decision is equal to 1. This quantity measures the belief that we exactly assign to A and not the total belief assigned to A. In order to obtain the total belief to A, it is necessary to add to $m(A)$ the quantities $m(B)$ for all proper subsets B of A:

$$Bel(A) = \sum_{B \subset A} m(B)$$

If Θ is a frame of discernment, then a function $Bel{:}2^{\Theta} \rightarrow [0,1]$ is a credibility function if and only if it satisfies the following conditions:

$$Bel(\phi)=0$$

$$Bel(\Theta)=1$$

$\forall n>0$ and for any collection $A_1, A_2,...,A_n$ of Θ:

$$Bel(A_1, \cup...\cup A_n) \geq \sum_{\substack{1 \subset \{1,...,n\} \\ 1 \neq \phi}} (-1)^{|I|+1} Bel\left(\bigcap_{i \in I} A_i\right)$$

In the same way, the plausibility function is defined as:

$$Pl(A) = \sum_{B \cap A = \phi} m(B) = 1 - Bel(\overline{A})$$

Thus, $1 - Pl(A)$ represents a measure of the doubt in A, noted $Dou(A) = Bel(\overline{A})$. A subset A of a frame Θ is called a focal element of a belief function over Θ if $m(A)>0$. If we suppose that $\Theta = \{D_1, D_2, D_3\}$ and $m(\{D_1\}) = 0.3$, $m(\{D_1, D_2\}) = 0.2$ and $m(\{D_1, D_2, D_3\}) = 0.5$ then we obtain the belief and the plausibility of Table 80.2.

Belief functions are well adapted to the representation of evidence because they admit a genuine combination rule. Given several belief functions over a common frame of discernment but based on different bodies of evidence, Dempster's combination rule permits the computation of their orthogonal sum, which is a new belief function based on the combined evidence. The combination rule is the following:

If $k = \sum_{\substack{B,C \\ B \cap C = \phi}} m_1(B)m_2(C) < 1$ then

$$(m_1 \oplus m_2)\,(A) = \frac{\sum_{B,C/A=B \cap C} m_1(B)m_2(C)}{1 - \sum_{B,C/B \cap C = \phi} m_1(B)m_2(C)}$$

k can obviously be used as a measure of conflict.

For example, if $\Theta = \{D, D'\}$, where D means "There is an obstacle in front of the robot" and D' means, "There is no obstacle in front of the robot." We suppose that m_1 is the mass function that assigns a belief that there is an obstacle from an acoustic sensor and m_2 the same measure from an infrared sensor with the values of Table 80.3, for example:

we obtain Table 80.4:

thus,

$$m_1 + m_2(\{D\}) = 0.72 + 0.08 + 0.18 = 0.98$$

TABLE 80.2 Results of the Belief and Plausibility Computation

$Bel(\{D_1\}) = 0.3$	$Pl(\{D_1\}) = 1.0$
$Bel(\{D_1\,D_2\}) = 0.5$	$Pl\{(D_1,D_2\}) = 1.0$
$Bel(\{D_1,D_3\}) = 0.3$	$Pl(\{D_1,D_3\}) = 1.0$
$Bel(\{D_1,D_2,D_3\}) = 1.0$	$Pl(\{D_1,D_2,D_3\}) = 1.0$

TABLE 80.3 Mass Distribution

$m_1(\{D\}) = 0.8$	$m_1(\{D'\}) = 0$	$m_1(\{D,D'\}) = 0.2$
$m_2(\{D\}) = 0.9$	$m_2(\{D'\}) = 0$	$m_2(\{D,D'\}) = 0.1$

Table 80.4 Results of the Fusion

m_1		m_2		
		0.9	0	0.1
		$\{D\}$	$\{D'\}$	$\{D,D'\}$
0.8	$\{D\}$	0.72	0	0.08
0	$\{D'\}$	0	0	0
0.2	$\{D,D'\}$	0.18	0	0.02

$$m_1 + m_2(\{D'\}) = 0$$

$$m_1 + m_2(\{D,D'\}) = 0.02$$

In order to combine a collection of belief functions, one can form the pairwise orthogonal sums:

$$((\ldots(m_1 \oplus m_2)\ldots) \oplus m_{n-1}) \oplus m_n$$

Finally, several criteria may be applied in order to make a decision among the frame of discernment [19]. Nonexhaustively, we can distinguish: maximum credibility, maximum plausibility, maximum credibility, and plausibility.

Application [21] proposes to use the perceptual organization combined with the Dempster–Shafer theory. The goal of the perceptual organization phase is to build global order in visual information with elementary information from one image. In our case, this means that we have to use some characteristics about straight lines to group the similar ones. The principal idea is to consider these characteristics as different sensory information and to perform data fusion on this information. The paper presents two parts, which, respectively, rectify the segmentation mistakes by restoring the coherence of the segments, and detects objects in the scene by forming groups of primitives. The authors show that without any prior knowledge and any threshold, their bottom-up algorithm efficiently detects the different objects even in a cluttered environment. Moreover, it is demonstrated that robustness and flexibility in indoor and outdoor scenes without any modification of parameters are more important.

Murphy [16] deals with mobile robot navigation by using the Dempster–Shafer theory.

Data Fusion in Possibility Theory

Another possibilistic approach, which received considerable attention, is fuzzy logic [25,26]. In possibility theory, the information available, relative to the value of parameter x, is represented by a possibility distribution π. A possibility distribution corresponds to an interval (or a set) representing imprecise information. Such a set is generally fuzzy. The union and the intersection of sets are both important ways, from others, to fuse the sets.

Let Ω be the referential. Let us then assume we have two sources of information, noted 1 and 2, which inform us about the value of parameter x as follows:

Source 1: $x \in E_1 \subseteq \Omega$
Source 2: $x \in E_2 \subseteq \Omega$

The decision about x then depends on whether sets E_1 and E_2 intersect more or less, or not at all, and whether the sources are reliable or not. In each case, the knowledge context of the situation will allow a conclusion to be reached. There is no method to fuse these two elements that is able to deal with all possible cases. But there exist different combinatory modes suiting each case.

It is possible to find a large range of schemes suiting each case. Among the ensemblist methods of data fusion, we can essentially remark:

The conjunctive fusion that corresponds to the fuzzy sets intersection. This approach assumes that the sources are totally reliable. The resulting set of this fusion, which concerns the value of the parameter x must largely tally with each other.

The disjunctive fusion that corresponds to the fuzzy sets union. This method is adequate if the sources are in conflict and if one of them is reliable contrary to the other. This approach allows to fuse sources that present nonintersecting sets.

The adaptive fusion that allows to take advantage of the conjunctive and the disjunctive fusions (after normalization, if necessary). For example, it is possible to apply the disjunctive fusion in order to complete information resulting from a conjunctive fusion in which a conflict appeared. If no agreement can be found between the sources, the conjunctive fusion is discarded in favor of the consideration of the disjunctive fusion.

A decision is usually taken from the largest of membership values after the combination step.

We propose a simple example in which two sensors $S1$ and $S2$ provide information about the shape of an object among the following propositions {cube, cylinder, sphere, prism}. We choose to use a conjunctive fusion as an intersection of the fuzzy sets. This kind of fusion assumes that the information sources are in full agreement and we particularly define the minimum function as:

$$\forall w \in \Omega,\ \pi_{conj}(w) = \min(\pi_1(w), \pi_2(w))$$

π_1 and π_2 are the possibility distributions that represent the opinion of the two sensors on the shape of the object x. Figure 80.2 describes the different cases that may be encountered according to the agreement of the sources.

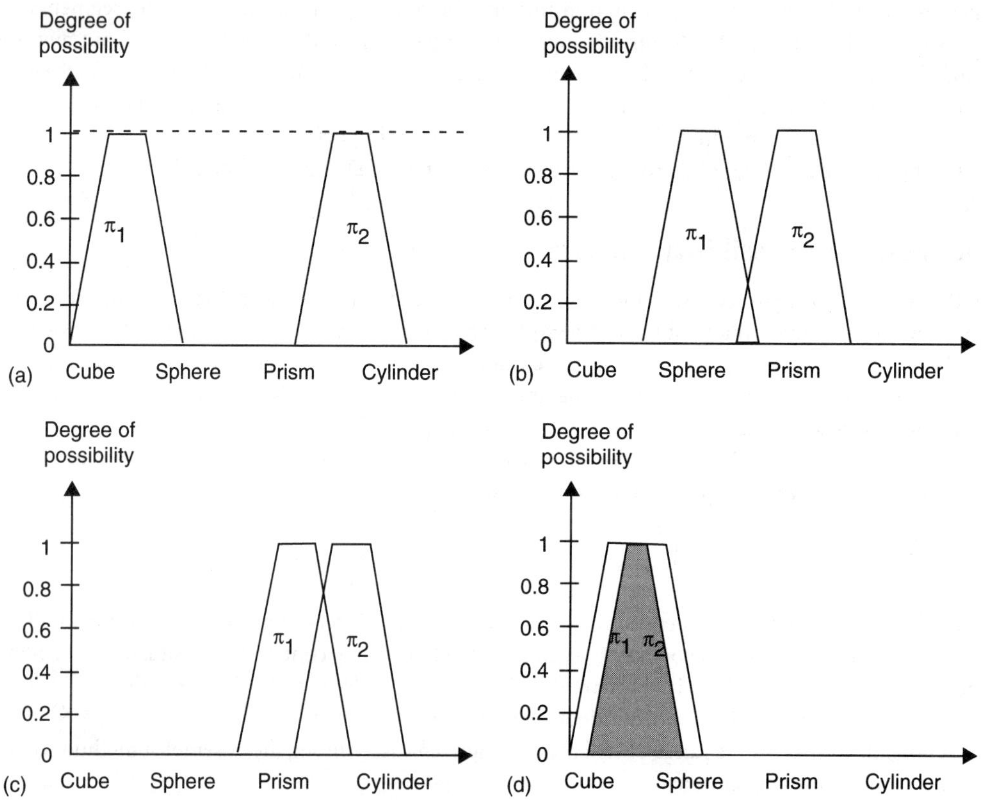

FIGURE 80.2   Illustration of the conjunctive fusion.

TABLE 80.5 Management of Imprecision and Uncertainty in Different Approaches

Approach	Imprecision	Uncertainty
Bayesian	Implicit	Explicit
Possibilistic	Explicit	Implicit
Evidential	Explicit	Explicit

We can see that the results of the fusion are, respectively, in total conflict (a), in strong conflict (b), in weak conflict (c), and in agreement (d). In fact, the last case is only appropriate for the conjunctive fusion and another kind of function must be used for the other situations.

Bloch [3] gives a very good statement and review of the fusion problem. Three approaches (Bayesian, Dempster–Shafer, and fuzzy set framework) to deal with imprecision and uncertainty in data fusion are described. These methods were applied and illustrated with medical images (dual echo MRI of the brain) in order to estimate the three classes (brain, ventricles, and pathology). Here, four stages are considered: modeling, estimation, combination, and decision. For each stage, a brief discussion about the three approaches is proposed.

Sossai et al. [20] presents one application of this approach to localize a mobile robot.

Discussion

Probabilistic approaches represent the category of the most used methods of fusion. This is not only due to its history but also and essentially due to its solid mathematical foundment. This is the large range of tools available for the modelization and the learning that make these methods successful. Moreover it is very easy to introduce information that can be represented as probabilities. However, some disadvantages can be noticeable. First, if the probabilistic methods allow to correctly represent the uncertainty, they do not provide the possibility to treat the inaccuracy. Moreover, the very severe constraints linked to the probabilistic methods can give rise to a very delicate learning phase. This remark is also valid for the combination rule, which often requires some simplifying hypotheses rarely verified.

The restrictions of the probabilistic approach also deal with the reasoning that can be done only on singletons in a closed world. Indeed, contrary to evidential and possibilistic theories, the probabilistic methods do not allow to treat the partial membership. In the same way, these theories allow the management of the ignorance and of the conflicts and offer powerful tools of modelization and management of uncertainty and inaccuracy. However, the decision rules are less formalized and efficient than the probabilistic methods.

We give, in Table 80.5, an abstract of the management of imprecision and uncertainty for the three approaches.

There is no real global solution in order to select a method of fusion rather than another one. Nevertheless, the nature of the data and of the sensors, the learning facility, and the system itself can direct the choice.

80.3 Conclusion

We have presented the multi-sensor data fusion problem, the principal classes of computational data fusion methods, and some examples of applications. Obviously, this presentation is not exhaustive and the methods have been proposed in a general framework. The aim was just to provide an overview of the different possibilities and to explain basically the functioning of these methods. Applications were chosen to show how these methods are implemented and their advantage; they give an idea about the variety of areas in which data fusion concepts can be applied. The contribution of data and sensor fusion was demonstrated through the increasing accuracy and robustness of developed systems. More generally, it is important to remark that the nonprobabilistic methods are becoming more and more popular and that they are better exploited.

References

[1] Arras Kai, O., Tomatis Nicola, Jensen Björn, T., and Siegwatt, R., Multisensor on-the-fly localization: precision and reliability for applications, *Robotics and Autonomous Systems*, 34, 131–143, 2001.

[2] Bloch, I., "Information combination operators for data fusion: a comparative review with classification", *SMC-A*, 26, 52–67, 1996.

[3] Bloch, I., "Some aspect of Dempster–Shafer evidence theory for classification of multi-modality medical images taking partial volume effect into account", *Pattern Recognition Letters*, 17, 905–916, 1996.

[4] Castellanos, Jose, A., and Tardos, Juan, D., *Mobile Robot Localization and Map Building — A Multisensor Fusion Approach*, Kluwer Academic Publishers Dordrecht, March 1, 2000.

[5] Chateau, T., Berducat, M., and Bonton, P., An Original Correlation and Data Fusion Based Approach to Detect a Reap Limit Into a Gray Level Image, IEEE/RSJ International Conference on Intelligent Robots and Systems, Grenoble, France, Sept. 7–11, 1997, pp. 1258–1263.

[6] Cremer, F., Schutte, K., Schavemaker, J.G.M., and Breejen E. den, A comparison of decision-level sensor-fusion methods for anti-personnel landmine detection, *Information Fusion*, 2, 187–208, 2001.

[7] Dasarathy, B.V., Information fusion — what, where, why, when, and how?, *Information Fusion*, 2, 75–76, 2001.

[8] Dempster, A.P., "Upper and lower probabilities induced by a multivalued mapping", *AMS*, 38, 325–339, 1967.

[9] Gao, J.B. and Harris, C.J., Some remarks on Kalman fitters for the multisensor fusion, *Information Fusion*, 3, 191–201, 2002.

[10] Hall, D.L., *Mathematical Techniques in Multi-Sensor Data Fusion*, Artech House, Inc., Norwood, MA, 1992, 301pp.

[11] Hall, D.L. and Linas, J., "*Handbook of Multisensor Data Fusion*", CRC Press, Boca Raton, FL, May 2001.

[12] Haralick, R.M. and Shapiro, L.G., *Computer and Robot Vision*, Vol. 1, Addison-Wesley, Reading, MA, 1992.

[13] lyengar, S. and Brooks, R.R., *Multi-Sensor Fusion: Fundamentals and Applications With Software*, Book and Disk edition, Prentice-Hall PTR, Englewood cliffs, NJ, 1998.

[14] Jetto, L., Longhi, S., and Vitali, D., Localization of a wheeled mobile robot by sensor data fusion based on a fuzzy logic adapted Kalman filter, *Control Engineering Practice*, 7, 763–771, 1999.

[15] Abidi, Mongi A. Ed., Rafael C. Gonzalez, and Ralph C. Gonzalez, Eds., *Data Fusion in Robotics and Machine Intelligence*, Academic Press, New York, October 1997.

[16] Murphy, R.R., "Dempster–Shafer theory for sensor fusion in autonomous mobile robots", *IEEE Transactions on Robotics and Automation*, Vol. 14, Issue 2, 197–206, April 1998.

[17] Pau, L.F., Sensor data fusion, *Journal of Intelligent and Robotic Systems*, 1, 103–116, 1988.

[18] Shafer, G., *A Mathematical Theory of Evidence*, Princeton University. Press, Princeton, NJ, 1976.

[19] Smets, Ph., The combination of evidence in the transferable belief model, *IEEE Transactions on Pattern Analysis and Machine Intelligence*, 12, 447–458, 1990.

[20] Sossai, C., Bison, P., Chemello G., and Trainito, G., Sensor fusion for localization using possibility theory, *Control Engineering Practice*, 7, 773–782, 1999.

[21] Vasseur, P., Pegard, C., Mouaddib, E., and Delahoche, L., Perceptual organization approach based on Dempster–Shafer theory, *Pattern Recognition*, 32, 1449–1462, 1999.

[22] Wald, L., A European proposal for terms of reference in data fusion, *International Archives of Photogrammetry and Remote Sensing*, XXXII, 651–654, 1998; Wald, L., Some terms of reference in data fusion, *IEEE Transactions on Geosciences and Remote Sensing*, 37, 1190–1193, 1999.

[23] Wald, L., << DATA FUSION: Definitions and Architectures — Fusion of images of different spatial resolutions >>, ISBN: 2-911762-38-X, In Presses de I'Ecole des mines de Paris, Paris, France, 2002.

[24] Waltz, E. and Llinas, J., *Multisensor Data Fusion*, Artech House, Norwood, MA, 1990.

[25] Zadeh, L., "Fuzzy sets", *Information and Control*, 8, 338–353, 1965.

[26] Zadeh, L., "Fuzzy sets as a basis for a theory of possibility", *Fuzzy Sets and Systems*, 1, 3–28, 1968.

Section 6: Real-Time Embedded Systems

81

Real-Time Systems

Hans Hansson
Mälardalen University

Mikael Nolin
Mälardalen University

Thomas Nolte
Mälardalen University

81.1 Introduction

Consider the *airbag* in the steering wheel of your car. It should, after the detection of a crash (and only then), inflate just in time to softly catch your head to prevent it from hitting the steering wheel; not too early — since this would make the airbag deflate before it can catch you, nor too late — since the exploding airbag could then injure you by blowing up in your face and/or catch you too late to prevent your head from banging into the steering wheel.

The computer-controlled airbag system is an example of a real-time system (RTS). But RTSs are of many different types, including vehicles, telecommunication systems, industrial automation systems, household appliances, etc.

There is no commonly agreed upon definition of what an RTS is, but the following characterization is (almost) universally accepted:

- RTSs are computer systems that physically interact with the real world.
- RTSs have requirements on the timing of these interactions.

Typically, the real-world interactions occur via sensors and actuators, rather than the keyboard and screen of your standard PC.

Real-time requirements typically express that an interaction should occur within a specified timing bound. It should be noted that this is quite different from requiring the interaction to be as fast as possible.

Essentially all RTSs are embedded in products, and the vast majority of *embedded computer systems* are RTSs. RTS is the dominating application of computer technology, as more than 99% of the manufactured processors (more than 8 billion in 2000 [Hal00]) are used in embedded systems.

Returning to the airbag system, we note that it in addition to being an RTS, it is a safety-critical system, that is, a system that, due to severe risks of damage, has strict *Quality of Service* (QoS) requirements, including requirements of functional behavior, robustness, reliability, and timeliness.

A typical strict timing property could be that a certain response to an interaction must always occur within some prescribed time, for example, that the charge in the airbag must detonate between 10 and 20 msec from the detection of a crash; violation of this must be avoided at any cost, since it would lead to something unacceptable, like you having to spend a couple of months in hospital. A system that is designed to meet strict timing requirements is often referred to as a *hard real-time system*. In contrast, systems for which occasional timing failures are acceptable — possibly because this will not lead to something terrible — are termed *soft real-time systems*.

An illustrative comparison between hard and soft RTSs, which highlights the difference between the extremes, is shown in Table 81.1. A typical hard RTS could in this context be an engine control system, which must operate with μs-precision, and that will severely damage the engine if timing requirements fail by more than a few *msec*. A typical soft RTS could be a banking system, for which timing is important, but where there are no strict deadlines and some variations in timing are acceptable.

Unfortunately, it is impossible to construct real systems that satisfy hard real-time requirements, since due to the imperfection of hardware (and designers), any system may break. The best that can be achieved is a system that with very high probability provides the intended behavior during a finite interval of time.

However, on the conceptual level, hard real-time makes sense, since it implies a certain amount of rigor in the way the system is designed, for example, it implies an obligation to prove that the strict timing requirements are met, at least under some simplifying, but realistic, assumptions.

Since the early 1980s, a substantial research effort has provided a sound theoretical foundation (e.g., [RTS, Klu]) and many practically useful results for the design of hard real-time systems. Most notably, hard RTS scheduling has evolved into a mature discipline, using abstract, but realistic, models of tasks executing on single CPU, multiprocessor, or distributed computer systems, together with associated methods for timing analysis. Such *schedulability analysis*, for example, the well-known rate-monotonic analysis [LL73, KRP+98, ABD+95], have also found significant use in some industrial segments.

However, hard real-time scheduling is not the cure for all RTSs. Its main weakness is that it is based on analysis of the worst possible scenario. For safety-critical systems, this is of course a must, but for other systems, where general customer satisfaction is the main criterion, it may be too costly to design the system for a worst-case scenario that may not occur during the system's lifetime.

If we look at the other end of the spectrum, we find the *best-effort approach*, which is still the dominating approach in industry. The essence of this approach is to implement the system using some best practice, and then use measurements, testing, and tuning to make sure that the system is of sufficient quality. Such a system will hopefully satisfy some soft real-time requirement, the weakness being that we do not know which. On the other hand, compared to the hard real-time approach, the system can be better

TABLE 81.1 Typical Characteristics of Hard and Soft RTSs [Kop03]

Characteristic	Hard Real-Time	Soft Real-Time
Timing requirements	Hard	Soft
Pacing	Environment	Computer
Peak-load performance	Predictable	Degraded
Error detection	System	User
Safety	Critical	Noncritical
Redundancy	Active	Standby
Time granularity	Millisecond	Second
Data files	Small	Large
Data integrity	Short term	Long term

optimized for the available resources. A further difference is that hard real-time system methods essentially are applicable to static configurations only, whereas it is less problematic to handle dynamic task creation, etc., in best-effort systems.

Having identified the weaknesses of the hard real-time and best-effort approaches, major efforts are now being put into more flexible techniques for soft RTSs. These techniques provide analyzability (like hard real-time), together with flexibility and resource efficiency (like best-effort). The bases for the flexible techniques are often quantified *Quality-of-Service* (QoS) characteristics. These are typically related to nonfunctional aspects, such as timeliness, robustness, dependability, and performance. To provide a specified QoS, some kind of resource management is needed. Such a QoS-management is either handled by the application, by the operating system, by some middleware, or by a mix of the above. The QoS-management is often a flexible on-line mechanism that dynamically adapts the resource allocation to balance between conflicting QoS-requirements.

81.2 Design of RTSs

The distinguishing factor for RTS design is the need to always have *timing predictability* in mind. Ignoring timing issues in the design phase, and hope to be able to understand, analyze, and modify the timing behavior of the completed system is a sure recipe for failure.

In this section, we describe the general structure of embedded RTSs and their major design structures. We also outline some commercial tools that are available for the design of RTSs.

Reference Architecture

A generic system architecture for an RTS is depicted in Figure 81.1. This architecture is a model of any computer-based system interacting with an external environment via sensors and actuators.

Since our focus is on the RTS, we will look more into different organizations of that part of the generic architecture in Figure 81.1. The simplest RTS is a single processor; but in many cases, the RTS is a distributed computer system consisting of a set of processors interconnected by a communications network. There could be several reasons for making an RTS distributed, including

- the physical distribution of the application,
- the computational requirements, which may not be conveniently provided by a single CPU,
- the need for redundancy to meet availability, reliability, or other safety requirements, or
- to reduce the cabling in the system.

Figure 81.2 shows an example of a distributed RTS. In a modern car, like the one depicted in the figure, there are some 20 to 100 computer nodes (which in the automotive industry are called Electronic Control Units; ECUs) interconnected with one or more communication networks. The initial

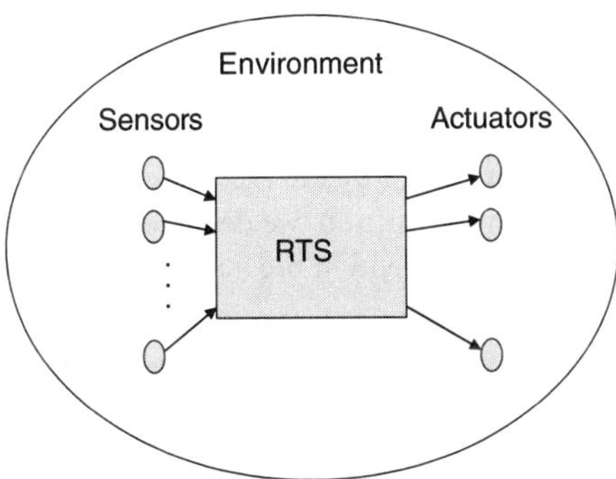

FIGURE 81.1   A generic RTS architecture.

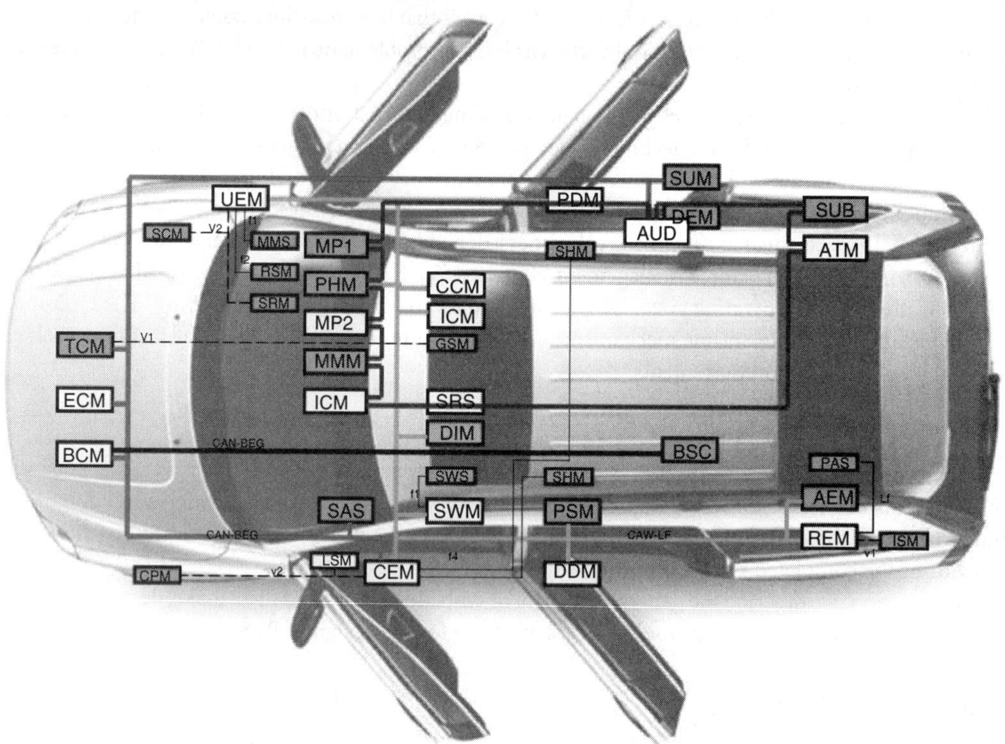

FIGURE 81.2 Network infrastructure of Volvo XC90.

motivation for this type of electronic architecture in cars was the need to reduce the amount of cabling. However, the electronic architecture has also led to other significant improvements, including substantial pollution reduction and new safety mechanisms, such as computer-controlled Electronic Stabilization Programs (ESPs). The current development is toward making the most safety-critical vehicle functions, like braking and steering, completely computer controlled. This is done by removing the mechanical connections (e.g., between steering wheel and front wheels, and between brake pedal and brakes), replacing them with computers and computer networks. Meeting the stringent safety requirements for such functions will require careful introduction of redundancy mechanisms in hardware and communication, as well as software, that is, a safety-critical system architecture is needed (an example of such an architecture is TTA [KB03]).

Models of Interaction

Above, we presented the physical organization of an RTS; but for an application programmer, this is not the most important aspect of the system architecture. Actually, from an application programmer's perspective, the system architecture is more given by the *execution paradigm* (execution strategy) and the *interaction model* used in the system. In this section, we describe what an interaction model is and how it affects the real-time properties of a system, and in the next section we discuss the execution strategies used in RTSs.

A model of interaction describes the rules by which components interact with each other (in this section, we will use the term component to denote any type of software unit, such as a task or a module). The interaction model can govern both control flow and data flow between system components. One of the most important design decisions, for all types of systems, is which interaction models to use (sadly, however, this decision is often implicit, and hidden in the system's architectural description).

When designing RTSs, attention should be paid to what are the timing properties of the interaction models chosen. Some models have a more predictable and robust behavior with respect to timing than other models. Examples of some of the more predictable models that are also commonly used in RTSs design are given below.

Pipes-and-filters. In this model, both data and control flow is specified using input and output ports of components. A component becomes eligible for execution when data have arrived on its input ports and when the component finishes execution it produces output on its output ports.

This model fits for many types of control programs very well, and control laws are easily mapped to this interaction model. Hence, it has gained widespread use in the real-time community. The real-time properties of this model are also quite good. Since both data and control flow unidirectionally through a series of components, the order of execution and end-to-end timing delay usually become highly predictable. The model also provides a high degree of decoupling in time; that is, components can often execute without having to worry about timing delays caused by other components. Hence, it is usually straightforward to specify the compound timing behavior of a set of components.

Publisher–subscriber. The publisher–subscriber model is similar to the pipes-and-filters model but it usually decouples data and control flow. That is, a subscriber can usually choose different forms for triggering its execution. If the subscriber chooses to be triggered on each new published value, the publisher–subscriber model takes on the form of the pipes-and-filters model. However, in addition, a subscriber could choose to ignore the timing of the published values and decide to use the latest published value. Also, for the publisher–subscriber model, the publisher is not necessarily aware of the identity, or even the existence, of its subscribers. This provides a higher degree of decoupling of components.

Similar to the pipes-and-filters model, the publisher–subscriber model provides good timing properties. However, a prerequisite for analysis of systems using this model is that subscriber components make explicit as to which values they will subscribe to (this is not mandated by the model itself). However, when using the publisher–subscriber model for embedded systems, it is the norm that subscription information is available (this information is used, for instance, to decide which values have to be published over a communications network, and to decide the receiving nodes of those values).

Blackboard. The blackboard model allows variables to be published on a globally available blackboard area. Thus, it resembles the use of global variables. The model allows any component to read or write values to variables in the blackboard. Hence, the software engineering qualities of the blackboard model are questionable. Nevertheless, it is a model that is commonly used, and in some situations it provides a pragmatic solution to problems that are difficult to address with more stringent interaction models.

Software engineering aspects aside, the blackboard model does not introduce any extra elements of unpredictable timing. On the other hand, the flexibility of the model does not help engineers to achieve predictable systems. Since the model does not address the control flow, components can execute relatively undisturbed and decoupled form other components.

On the other end of the spectrum of interaction models, there are models that increase the (timing) unpredictability of the system. These models should, if possible, be avoided when designing RTSs. The two most notable, and commonly used, are:

Client–server. In the client–server model, a client asynchronously invokes a service of a server. The service invocation passes the control flow (plus any input data) to the server, and control stays at the server until it has completed the service. When the server is done, the control flow (and any return data) is returned to the client, which in turn resumes execution.

The client–server model has an inherently unpredictable timing. Since services are invoked asynchronously, it is very difficult to a priori assess the load on the server for a certain service invocation. Thus, it is difficult to estimate the delay of the service invocation and, in turn, it is difficult to estimate the response-time of the client. This matter is furthermore complicated by the fact that most components often behave both as clients and servers (a server often uses other servers to implement its own services), leading to very complex and unanalyzable control flow paths.

Message-boxes. A component can have a set of message-boxes, and components communicate by posting messages in each other's message-boxes. Messages are typically handled in FIFO order, or in priority order (where the sender specifies a priority). Message passing does not change the flow of control for the sender. A component that tries to receive a message from an empty message-box, however, blocks on that message-box until a message arrives (often, the receiver can specify a timeout to prevent indefinite blocking).

From a sender's point of view, the message-box model has problems similar to the client–server model. The data sent by the sender (and the action that the sender expects the receiver to perform) may be delayed in an unpredictable way when the receiver is highly loaded. Also, the asynchronous nature of the message passing makes it difficult to foresee the load of a receiver at any particular moment.

Furthermore, from the receiver's point of view, the reading of message-boxes is unpredictable in the sense that the receiver may or may not block on the message-box. Also, since message boxes are often of limited size, a highly loaded receiver risks to lose some messages. Which messages that are lost is another source of unpredictability.

Execution Strategies

There are two main execution paradigms for RTSs: *time triggered* and *event triggered*. When using timed-triggered execution, activities occur at predefined instances of time, for example, a specific sensor value is read exactly every 10 msec and exactly 2 msec later the corresponding actuator receives an updated control parameter. In an event-triggered execution, on the other hand, actions are triggered by event occurrences, for example, when the level of toxic fluid in a tank reaches a certain level an alarm will go off. It should be noted that the same functionality typically can be implemented in both paradigms, for example, a time-triggered implementation of the above alarm would be to periodically read the level-measuring sensor and set off the alarm when the read level exceeds the maximum allowed. If alarms are rare, the time-triggered version will have a much higher computational overhead than the event-triggered one. On the other hand, the periodic sensor readings will facilitate detection of a malfunctioning sensor.

Time-triggered executions are used in many safety-critical systems with high dependability requirements (such as avionic control systems), whereas the majority of other systems are event triggered. Dependability can also be guaranteed in the event-triggered paradigm, but due to the observability provided by the exact timing of time-triggered executions, most experts argue for using time-triggered in ultradependable systems. The main argument against time-triggered systems is its lack of flexibility and the requirement of preruntime schedule generation (which is a nontrivial and possibly time-consuming task).

Time-triggered systems are mostly implemented by simple proprietary table-driven dispatchers [XP90] (see the section Off-line schedulers for a discussion on table-driven execution); but complete commercial systems including design tools are also available [TTT, KG94]. For the event-triggered paradigm, a large number of commercial tools and operating systems are available (examples are given in the section Commercial RTOSes). There are also examples of systems integrating the two execution paradigms, thereby aiming at getting the best of two worlds: time-triggered dependability and event-triggered flexibility. One example is the Basement system [HLS96] and its associated real-time kernel Rubus [Arc].

Since computations in time-triggered systems are statically allocated both in space (to a specific processor) and time, some kind of configuration tool is often used. This tool assumes that the computations are packaged into schedulable units (corresponding to tasks or threads in an event-triggered system). Typically, for example in Basement, computations are control-flow based, in the sense that they are defined by sequences of schedulable units, each unit performing a computation based on its inputs and producing outputs to the next unit in sequence. The system is configured by defining the sequences and their timing requirements. The configuration tool will then automatically (if possible[1]) generate a schedule, which guarantees that all timing requirements are met.

---

[1] This scheduling problem is theoretically intractable; thus, the configuration tool will have to rely on some heuristics that works well in practice, but that does not guarantee finding a solution in all cases when there is a solution.

Event-triggered systems typically have a richer and more complex APIs (Application Programmer Interfaces), defined by the used operating system and middleware, which will be elaborated on in Section 81.3.

Component-Based Design of Real-Time Systems

Component-Based Design (CBD) of software systems is an interesting idea for software engineering in general, and for engineering of RTSs in particular. In CBD, a *software component* is used to encapsulate some functionality. This functionality is only accessed through the *interface* of the component. A system is composed by assembling a set of components and connecting their interfaces.

The reason CBD could prove extra useful for RTSs is the possibility to extend components with *introspective interfaces*. An introspective interface does not provide any functionality *per se*, rather the interface can be used to retrieve information about *extrafunctional* properties of the component. Extrafunctional properties can include attributes such as memory consumption, execution-times, task periods, etc. For RTS, timing properties are of course of particular interest.

Unlike the functional interfaces of components, the introspective interfaces can be available off-line, that is, during the component assembly phase. This way, the timing attributes of the system components can be obtained at design time and tools to analyze the timing behavior of the system could be used. If the introspective interfaces are also available on-line they could be used in, for instance, admission control algorithms. An admission control could query new components for their timing behavior and resource consumption before deciding to accept a new component to the system.

Unfortunately, many industry standard software techniques are based on the client–server or the message-box models of interaction, which we deemed, in the section Models of Interaction, unfit for RTSs. This is especially true for the most commonly used component models. For instance, the Corba Component Model (CCM) [OMG02], Microsoft's COM [Mica] and .NET [Micb] models, and Java Beans [SUN] all have the client–server model as their core model. Also, none of these component technologies allow the specification of extrafunctional properties through introspective interfaces. Hence, from the real-time perspective, the biggest advantage of CBD is void for these technologies.

However, there are numerous research projects addressing CBD for real-time and embedded systems (e.g., [Sta01, vO02, MSZ02, SVK97]). These projects are addressing the issues left behind by the existing commercial technologies, such as timing predictability (using suitable computational models), support for off-line analysis of component assemblies, and better support for resource-constrained systems. Often, these projects strive to remove the considerable run-time flexibility provided by existing technologies. This run-time flexibility is judged to be the foremost contributor to unpredictability (the flexibility is also adding to the run-time complexity and prevents CBD for resource-constrained systems).

Tools for Design of RTSs

In industry, the term *RTS* is highly overloaded, and can mean anything from interactive systems to super-fast systems, or embedded systems. Consequently, it is not easy to judge what tools are suitable for developing RTSs (as we define real-time in this chapter).

For instance, UML [OMG03b] is commonly used for software design. However, UML's focus is mainly on client–server solutions, and it has proven inept for RTSs' design. As a consequence, UML-based tools that extend UML with constructs suitable for real-time programs have emerged. The two most well-known of such products are Rational's Rose RealTime [Rat] and i-Logix' Rhapsody [IL]. These tools provide UML-support with the extension *of real-time profiles*. While giving real-time engineers access to suitable abstractions and computational models, these tools do not provide means to describe timing properties or requirements in a formal way; thus, they do not allow automatic verification of timing requirements.

TeleLogic provides programming and design support using the language SDL [Tel]. SDL was originally developed as a specification language for the telecom industry, and is as such highly suitable for describing complex reactive systems. However, the fundamental model of computation is the message-box model, which has an inherently unpredictable timing behavior. However, for soft embedded RTSs, SDL can give very time- and space-efficient implementations.

For more resource-constrained hard RTSs, design tools are provided by, for example, Arcticus System [Arc], TTTech [TTT], and Vector [Veca]. These tools are instrumental during both system design and implementation, and also provide some timing analysis techniques that allow timing verification of the system (or parts of the system). However, these tools are based on proprietary formats and processes, and have as such reached a limited customer base (mainly within the automotive industry).

Within the near future, UML2 will become an adopted standard [OMG03c]. UML2 has support for computational models suitable for RTSs. This support comes mainly in the form of *ports* that can have *protocols* associated to them. Ports are either *provided* or *required,* hence allowing type-matching of connections between components. UML2 also includes much of the concepts from Rose RealTime, Rhapsody, and SDL. Other, future design techniques that are expected to have an impact on the design of RTSs include the EAST/EEA *Architecture Description Language* (EAST-ADL) [ITE]. The EAST-ADL is developed by the automotive industry and is a description language that will cover the complete development cycle of distributed, resource-constrained, safety-critical, RTSs. Tools to support development with EAST-ADL (which is a UML2-compliant language) are expected to be provided by automotive tool vendors such as ETAS [ETA], Vector [Vecb], and Siemens [Sie].

81.3 Real-Time Operating Systems

When building real-time computer systems, it is often beneficial to use a *Real-Time Operating System* (RTOS). An RTOS provides services for resource access and resource sharing, very much similar to a general-purpose operating system. The main reasons for not using a general-purpose operating system when developing RTSs are:

- High resource utilization, for example, large RAM and ROM footprints, and high internal CPU-demand.
- Difficult to access hardware and devices in a timely manner, for example, no application-level control over interrupts.
- Lack of services to allow timing-sensitive interactions between different processes.

Typical Properties of RTOSes

The state of practice in RTOS is reflected in [FAQ]. Not all operating systems are RTOSes. An RTOS is typically multithreaded and preemptible; there has to be a notion of thread priority, predictable thread synchronization has to be supported, priority inheritance should be supported, and the OS behavior should be known [Art03]. This means that the interrupt latency, worst-case execution time of system calls, and maximum time during which interrupts are masked must be known. A commercial RTOS is usually marketed as the run-time component of an embedded development platform.

As a general rule of thumb, one can say that RTOSes are:

- Being apt for the resource constrained environments where most RTSs operate.
  Most RTOSes can be configured preruntime (e.g., at compile time) to only include a subset of the total functionality. Thus, the application developer can choose to leave out unused portions of the RTOS in order to save resources.
  RTOSes typically store much of their configuration in ROM. This is done for mainly two purposes: (1) minimize the use of expensive RAM memory and (2) minimize the risk that critical data is overwritten by an erroneous application.
- Giving the application programmer easy access to hardware features such as interrupts and devices.
  Most often, the RTOSes give the application programmer means to install Interrupt Service Routines (ISRs) during compile time and/or during runtime. This means that the RTOS leaves all

interrupt handing to the application programmer, allowing fast, efficient, and predictable handling of interrupts.

In general-purpose operating systems, memory-mapped devices are usually protected from direct access using the Memory Management Unit (MMU) of the CPU, thus forcing all device accesses to go through the operating system. RTOSes typically do not protect such devices and allow the application to directly manipulate the devices. This gives faster and more efficient access to the devices. (However, this efficiency comes at the price of an increased risk of erroneous use of the device.)

- Providing services that allow implementation of timing-sensitive code.

An RTOS typically has many mechanisms to control the relative timing between different processes in the system. Most notably, the RTOS has a real-time process scheduler whose function is to make sure that the processes execute in the way the application programmer intended them to. We will elaborate more on the issues of scheduling in Section 81.4.

An RTOS also provides mechanisms to control the process's relative performance when accessing shared resources. This can, for instance, be done by priority queues, instead of plain FIFO-queues as is used in general-purpose operating systems. Typically an RTOS supports one or more real-time resource locking protocols, such as priority inheritance, or priority ceiling (the next section discusses resource locking protocols further).

- Being tailored to fit the development process that is typical when developing embedded systems (recall; most RTSs are also embedded systems).
- RTSs are usually constructed in a *host environment* that is different from the *target environment,* so-called *cross-platform* development. Also, it is typical that the whole memory image, including both RTOS and one or more applications, is created at the host platform and downloaded to the target platform. Hence, most RTOSes are delivered as source code modules or precompiled libraries that are statically linked with the applications at compile time.

Mechanisms for Real-Time

One of the most important functions of an RTOS is to arbitrate access to shared resources in such a way that the timing behavior of the system becomes predictable. The two most obvious resources that the RTOS manages to access are:

- The CPU — That is, the RTOS should allow processes to execute in a predictable manner.
- Shared memory areas — That is, the RTOS should resolve contention to shared memory in a way that gives predictable timing.

The CPU access is arbitrated with a *real-time scheduling policy.* Section 81.4 will describe real-time scheduling policies in more depth. Examples of scheduling policies that can be used in RTSs are priority scheduling, deadline scheduling, or rate scheduling. Some of these policies directly use timing attributes (like deadline) of the tasks to perform scheduling decisions, whereas other policies use scheduling parameters (like priority, rate, or bandwidth) that indirectly affect the timing of the tasks.

A special form of scheduling, which is also very useful for RTSs, is table-driven (static) scheduling. Table-driven scheduling is described further in the section Off-line schedulers on page **81**-11. To summarize, in table-driven scheduling all arbitration decisions have been made off-line and the RTOS scheduler just follows a simple table. This gives very good timing predictability, albeit at the expense of system flexibility.

The most important aspect of a real-time scheduling policy is that it should provide means to *a priori* analyze the timing behavior of the system, thus giving a predictable timing behavior of the system. Scheduling in general-purpose operating systems normally emphasizes properties like fairness, throughput, and guaranteed progress; these properties may all be adequate in their own respect; however, they are usually in conflict with the requirement that an RTOS should provide timing predictability.

Shared resources (such as memory areas, semaphores, and mutexes) are also arbitrated by the RTOS. When a task locks a shared resource, it will *block* all other tasks that subsequently tries to lock the resource. In order to achieve predictable blocking times, special real-time resource locking protocols have been proposed ([But97, BW96] provides more details about the protocols).

Priority Inheritance Protocol (PIP). PIP makes a low-priority task inherit the priority of any higher priority task that becomes blocked on a resource locked by the lower priority task.

This is a simple and straightforward method to lower the blocking time. However, it is computationally intractable to calculate the worst-case blocking (which may be infinite since the protocol does not prevent deadlocks). Hence, for hard RTSs or when timing performance needs to be calculated *a priori*, the PIP is not adequate.

Priority Ceiling inheritance Protocol (PCP). PCP associates to each resource a *ceiling value* that is equal to the highest priority of any task that may lock the resource. By clever use of the ceiling values of each resource, the RTOS scheduler will manipulate task priorities to avoid the problems of PIP.

PCP guarantees freedom from deadlocks, and the worst-case blocking is relatively easy to calculate. However, the computational complexity of keeping track of ceiling values and task priorities gives PCP high run-time overhead.

Immediate ceiling priority Inheritance Protocol (IIP). IIP also associates to each resource a *ceiling value*, which is equal to the highest priority of and task that may lock the resource. However, different from PCP, in IIP a task is immediately assigned the ceiling priority of the resource it is locking.

IIP has the same real-time properties as the PCP (including the same worst-case blocking time[2]). However, IIP is significantly more easy to implement. It is, in fact, for single node systems easier to implement than any other resource locking protocol (including non-real-time protocols). In IIP, no actual locks need to be implemented; it is enough for the RTOS to adjust the priority of the task that locks or releases a resource. IIP has other operational benefits, notably it paves the way for letting multiple tasks use the same stack area. Operating systems based on IIP can be used to build systems with footprints that are extremely small [Ast, Liv99].

Commercial RTOSes

There is an abundance of commercial RTOSes. Most of them provide adequate mechanisms to enable development of RTSs. Some examples are Tornado/VxWorks [Win], LYNX [LYN], OSE [Sys], QNX [QNX], RT-Linux [RTL], and ThreadX [Log]. However, the major problem with these tools is the rich set of primitives provided. These systems provide both primitives that are suitable for RTSs and primitives that are unfit for real-time systems (or that should be used with great care). For instance, they usually provide multiple resource locking protocols, some of which are suitable and some of which are not suitable for real-time.

This richness becomes a problem when these tools are used by inexperienced engineers and/or when projects are large and project management does not provide clear design guidelines/rules. In these situations, it is very easy to use primitives that will contribute to timing unpredictability of the developed system. Rather, an RTOS should help the engineers and project managers by providing only mechanisms that help in designing predictable systems. However, there is an obvious conflict between the desire/need of RTOS manufacturers to provide rich interfaces and stringency needed by designers of RTSs.

There is a smaller set of RTOSes that have been designed to resolve these problems, and at the same time also allow extreme lightweight implementations of predictable RTSs. The driving idea is to provide a small set of primitives that guide the engineers toward good design of their system. Typical examples are the research RTOS Asterix [Ast] and the commercial RTOS SSX5 [Liv99]. These systems provide a simplified task model, in which tasks cannot suspend themselves (e.g., there is no `sleep()` primitive) and tasks are restarted from their entry point on each invocation. The only resource locking protocol that is supported is IIP, and the scheduling policy is fixed priority scheduling. These limitations make it possible to build an RTOS that is able to run, for example, 10 tasks using less than 200 bytes of RAM, and at the same time giving predictable timing behavior [App98].Other commercial systems that follow a similar principle of reducing the degrees of freedom and hence promote stringent design of predictable RTSs include Arcticus Systems' Rubus OS [Arc].

---

[2]However, the average blocking time will be higher in IIP than in PCP.

Many of the commercial RTOSes provide standard APIs. The most important RTOS-standards are RT-POSIX [IEE98], OSEK [Gro], and APEX [Air96]. Here, we will only deal with POSIX since it is the most widely adopted RTOS standard, but those interested in automotive and avionic systems should take a closer look at OSEK and APEX, respectively.

The POSIX standard is based on Unix, and its goal is portability of applications at the source code level. The basic POSIX services include task and thread management, file system management, input and output, and event notification via signals. The POSIX real-time interface defines services facilitating concurrent programming and providing predictable timing behavior. Concurrent programming is supported by synchronization and communication mechanisms that allow predictability. Predictable timing behavior is supported by preemptive fixed priority scheduling, time management with high resolution, and virtual memory management. Several restricted subsets of the standard intended for different types of systems have been defined, as well as specific language bindings, for example, for Ada [ISO95].

81.4 Real-Time Scheduling

In this section, we will give an overview of the more common types of schedulers used in RTSs.

Traditionally, real-time schedulers are divided into *off-line* and *on-line* schedulers. Off-line schedulers are making all scheduling decisions before the system is executed. At runtime, a simple dispatcher is used to activate tasks according to off-line generated schedule. On-line schedulers, on the other hand, decide during execution, based on various parameters, which task should execute at any given time.

As there are several different schedulers developed in the research community, in this section we have focused on highlighting the main categories of schedulers that are readily available in existing RTOSes.

Introduction to Scheduling

An RTS consists of a set of real-time programs, which in turn consist of a set of *tasks*. These tasks are sequential pieces of code, executing on a platform with limited resources. The tasks have different timing properties, for example, *execution times, periods,* and *deadlines.* Several tasks can be allocated to a single processor. The scheduler decides, at each moment, which task to execute.

An RTS can be *preemptive* or *nonpreemptive.* In a preemptive system, tasks can preempt each other, letting the task with the highest priority execute. In a nonpreemptive system, a task that has been allowed to start will execute until its completion.

Tasks can be categorized into either being *periodic, sporadic,* or *aperiodic.* Periodic tasks are executing with a specified time (period) between task releases. Aperiodic tasks have no information stating when the task is to be released. Usually, aperiodics are triggered by interrupts. Similarly, sporadic tasks have no period, but in contrast with aperiodics, sporadic tasks have a known minimum time between releases. Typically, tasks that perform measurements are periodic, collecting some value(s) every nth time unit. A sporadic task is typically reacting to an event/interrupt that we know has a minimum inter arrival time, for example, an alarm or the emergency shutdown of a production robot. The minimum interarrival time can be constrained by physical laws, or it can be enforced by some hardware mechanism. If we do not know the minimum time between two consecutive events, we must classify the event-handling task to be aperiodic.

A real-time *scheduler* schedules the real-time tasks sharing the same resource (e.g., a CPU or a network link). The goal of the scheduler is to make sure that the timing requirements of these tasks are satisfied. The scheduler decides, based on the task timing properties, which task is to execute or to use the resource.

Off-line Schedulers

Off-line schedulers, or table-driven schedulers, work the following way: the schedulers create a schedule (the table) before the system is started (off-line). At runtime, a *dispatcher* follows the schedule, and makes sure that tasks are only executing at their predetermined time slots (according to the schedule). Off-line schedules are commonly used to implement the time-triggered execution paradigm (described in the section Execution strategies).

By creating a schedule off-line, complex timing constraints can be handled in a way that would be difficult to do on-line. The schedule that is created is the schedule that will be used at runtime. Therefore, the online behavior of table-driven schedulers is very deterministic. Because of this determinism, table-driven schedulers are the more commonly used schedulers in applications that have very high safety-critical demands. However, since the schedule is created off-line, the flexibility is very limited, in the sense that as soon as the system will change (due to, e.g., adding of functionality or change of hardware), a new schedule has to be created and given to the dispatcher. To create new schedules is nontrivial and sometimes very time consuming.

There also exist combinations of the predictable table-driven scheduling and the more flexible priority-based schedulers, and there exist methods to convert one policy to another [FLD03, Arc, MTS02].

On-line Schedulers

Scheduling policies that make their scheduling decisions during runtime are classified as *on-line schedulers*. These schedulers make their scheduling decisions based on some task properties, for example, task priority. Schedulers that make their scheduling decisions based on task priorities are called *priority-based schedulers*.

Using priority-based schedulers, the flexibility is increased (compared to table-driven schedulers), since the schedule is created on-line, based on the currently active task's constraints. Hence, priority-based schedulers can cope with changes in work-load and added functions, as long as the schedulability of the task-set is not violated. However, the exact behavior of priority-based schedulers is harder to predict. Therefore, these schedulers are not used as often in the most safety-critical applications.

Two common priority-based scheduling policies are *Fixed Priority Scheduling (FPS)* and *Earliest Deadline First (EDF)*. The difference between these scheduling policies is whether the priorities of the real-time tasks are fixed or if they can change during execution (i.e., they are dynamic).

In FPS, priorities are assigned to the tasks before execution (offline). The task with the highest priority among all tasks that are available for execution is scheduled for execution. It can be proved that some priority assignments are better than others. For instance, for a simple task model with strictly periodic noninterfering tasks with deadlines equal to the period of the task, a Rate Monotonic (RM) priority assignment has been shown by Liu and Layland [LL73] to be optimal. In RM, the priority is assigned based on the period of the task. The shorter the period, the higher the priority assigned to the task.

Using EDF, the task with the nearest (earliest) deadline among all available tasks is selected for execution. Therefore, the priority is not fixed; it changes with time. It has been shown that EDF is an optimal dynamic priority scheme [LL73].

In order for the priority-based schedulers to cope with aperiodic tasks, different service methods have been presented. The objective of these service methods is to give a good average response-time for aperiodic requests while preserving the timing properties of periodic and sporadic tasks. These services are implemented using special *server* tasks. In the scheduling literature, many types of servers are described. Using FPS, for instance, the Sporadic Server (SS) is presented by Sprunt et al. [SSL89]. SS has a fixed priority chosen according to the RM policy. Using EDF, Dynamic Sporadic Server (DSS) [SB94, SB96] extends SS. Other EDF-based schedulers are the Constant Bandwidth Server (CBS), presented by Abeni [AB98], and the Total Bandwidth Server (TBS) by Spuri and Buttazzo [SB94, SBS95]. Each server is characterized partly by its unique mechanism for assigning deadlines, and partly by a set of variables used to configure the server. Examples of such variables are bandwidth, period, and capacity.

In Section 81.6, we give examples of how timing properties of FPS can be calculated.

81.5 Real-Time Communications

As RTSs become distributed, data need to be transmitted between the system nodes. In many cases, this transmission must be timely and deterministic. Hence, there is a demand for real-time communication networks providing real-time guarantees. There are real-time communication networks of different types, ranging from small fieldbus-based control systems to large Ethernet/Internet-distributed applications. There is also a growing interest in wireless solutions.

In this section, we give an introduction to common techniques in communications in general and real-time communications in particular. We also provide an overview of the most popular real-time communication systems and protocols used today, both in industry and academia.

Communications Techniques

Common access mechanisms used in communication networks are CSMA/CD (Carrier Sense Multiple Access/Collision Detection), CSMA/CA (Carrier Sense Multiple Access/Collision Avoidance), TDMA (Time Division Multiple Access), Tokens, Central Master, and Mini Slotting. These techniques are all used both in real-time and non-real-time communication, and each of the techniques has different timing characteristics.

In CSMA/CD, collisions between messages are detected, causing the messages involved in the collision to be retransmitted. CSMA/CD is used, for example, in Ethernet. CSMA/CA, on the other hand, is avoiding collisions and is therefore more deterministic in its behavior compared to CSMA/CD. Hence, CSMA/CA is more suitable for hard real-time guarantees, whereas CSMA/CD can provide soft real-time guarantees. Examples of networks that implement CSMA/CA are CAN and ARINC 629.

TDMA is using time to achieve exclusive usage of the network. Messages are sent at predetermined instances in time. Hence, the behavior of TDMA-based networks is very deterministic, that is, very suitable to provide real-time guarantees. One example of a TDMA-based real-time network is TTP.

An alternative way of eliminating collisions on the network is to use tokens. In token-based networks, only the owner of the (unique within the network) token is allowed to send messages on the network. Once the token holder is done or has used its allotted time, the token is passed to another node. Tokens are used in, for example, Profibus.

It is also possible to eliminate collisions by letting one node in the network to be the master node. The master node is controlling the traffic on the network, and it decides which and when messages are allowed to be sent. This approach is used in, for example, LIN and TTP/A.

Finally, mini slotting can also be used to eliminate collisions. When using mini slotting, as soon as the network is idle and some node would like to transmit a message, the node has to wait for a unique (for each node) time before sending any messages. If there are several competing nodes wanting to send messages, the node with the longer waiting time will see that there is another node that already has started its transmission of a message. In such a situation, the node has to wait until the next time the network will become idle. Hence, collisions are avoided. Mini slotting can be found in, for example, FlexRay and ARINC 629.

Fieldbuses

Fieldbuses are a family of factory communication networks that have evolved as a response to the demand to reduce cabling costs in factory automation systems. By moving from a situation in which every controller has its own cables connecting the sensors to the controller (parallel interface), to a system with a set of controllers sharing a bus (serial interface), costs could be cut and flexibility could be increased. Pushing for this evolution of technology was both the fact that the number of cables in the system increased as the number of sensors and actuators grew, together with controllers moving from being specialized with their own microchip, to sharing a microprocessor with other controllers. Fieldbuses were soon ready to handle the most demanding applications on the factory floor.

Several fieldbus technologies, usually very specialized, were developed by different companies to meet the demands of their applications. Fieldbuses used in the automotive industry are, for example, CAN, TT-CAN, TTP, LIN, and FlexRay. In avionics, ARINC 629 is one of the more used communication standards. Profibus is widely used in automation and robotics, while in trains, TCN and WorldFIP are very popular communication technologies. We will now present each of these fieldbuses in some more detail, outlining key features and specific properties.

Controller Area Network (CAN). The CAN [CAN02] was standardized by the International Standardization Organization (ISO) [CAN92] in 1993. Today, CAN is a widely used fieldbus, mainly in automotive systems but also in other real-time applications, for example, medical equipment. CAN is an event-triggered broadcast bus designed to operate at speeds of up to 1 Mbps. CAN is using a fixed-priority-based arbitration mechanism that can provide timing guarantees using Fixed-Priority Scheduling (FPS) type of analysis (Tindell et al. [TBW95, THW94]). An example of this analysis will be provided in the section Example of analysis.

CAN is a collision-avoidance broadcast bus, using deterministic collision resolution to control access to the bus (so-called CSMA/CA). The basis for the access mechanism is the electrical characteristics of a CAN bus, allowing sending nodes to detect collisions in a nondestructive way. By monitoring the resulting bus value during message arbitration, a node detects if there are higher priority messages competing for access to the bus. If this is the case, the node will stop the message transmission, and try to retransmit the message as soon as the bus becomes idle again. Hence, the bus behaves like a priority-based queue.

Time-Triggered CAN. Time-Triggered communication on CAN (TT-CAN) [TTC] is a standardized session layer extension to the original CAN. In TT-CAN, the exchange of messages is controlled by the temporal progression of time, and all nodes are following a predefined static schedule. It is also possible to support original event-triggered CAN traffic together with the time-triggered traffic. This traffic is sent in dedicated arbitration windows, using the same arbitration mechanism as native CAN.

The static schedule is based on a time division (TDMA) scheme, where message exchanges may only occur during specific time slots or in time windows. Synchronization of the nodes is either done using a clock synchronization algorithm, or by periodic messages from a master node. In the latter case, all nodes in the system are synchronizing with this message, which gives a reference point in the temporal domain for the static schedule of the message transactions, that is, the master's view of time is referred to as the network's global time.

TT-CAN appends a set of new features to the original CAN, and being standardized, several semiconductor vendors are manufacturing TT-CAN-compliant devices.

Flexible Time-Triggered CAN. Flexible Time-Triggered communication on CAN (FTT-CAN) [AFF98, AFF99] provides a way to schedule CAN in a time-triggered manner with support for event-triggered traffic as well. In FTT-CAN, time is partitioned into Elementary Cycles (ECs), which are initiated by a special message, the Trigger Message (TM). This message triggers the start of the EC and contains the schedule for the time-triggered traffic that shall be sent within this EC. The schedule is calculated and sent by a master node. FTT-CAN supports both periodic and aperiodic traffic by dividing the EC into two parts. In the first part, the asynchronous window, the aperiodic messages are sent, and in the second part, the synchronous window, traffic is sent in a time-triggered manner according to the schedule delivered by the TM. FTT-CAN is still mainly an academic communication protocol.

Time-Triggered Protocol (TTP). TTP/C [TTT, TTT99], or the Time Triggered Protocol Class C, is a TDMA-based communication network intended for truly hard real-time communication. TTP/C is available for network speeds of up to 25 Mbps. TTP/C is part of the Time-Triggered Architecture (TTA) by Kopetz et al. [TTT, Kop98], which is designed for safety-critical applications. TTP/C has support for fault tolerance, clock synchronization, membership services, fast error detection, and consistency checks. Several major automotive companies are supporting this protocol, for example, Audi, VW, and Renault.

For the less hard RTSs (e.g., soft RTSs), there exists a scaled-down version of TTP/C called TTP/A [TTT].

Local Interconnect Network (LIN). LIN [LIN] was developed by the LIN Consortium (including Audi, BMW, Daimler Chrysler, Motorola, Volvo, and VW) as a low-cost alternative for small networks. LIN is cheaper than, for example, CAN. LIN is using the UART/SCI interface hardware, and transmission speeds are possible up to 20 kbps. Among the nodes in the network, one node is the master node, responsible for synchronization of the bus. The traffic is sent in a time-triggered manner.

FlexRay. FlexRay [BBE+02] was proposed in 1999 by several major automotive manufacturers, including Daimler-Chrystler and BMW, as a competitive next-generation fieldbus replacing CAN. FlexRay is a real-time communication network that provides both synchronous and asynchronous transmissions with network speeds up to 10 Mbps. For the synchronous traffic, FlexRay is using TDMA, providing deterministic data transmissions with a bounded delay. For the asynchronous traffic mini-slotting is used. Compared with CAN, FlexRay is more suitable for the dependable application domain, by including support for redundant transmission channels, bus guardians, and fast error detection and signaling.

ARINC 629. For avionic and aerospace communication systems, the ARINC 429 [ARI99] standard and its newer ARINC 629 [ARI99] successor are the most commonly used communication systems used today. ARINC 629 supports both periodic and sporadic communication. The bus is scheduled in bus cycles, which in turn are divided into two parts. In the first part, periodic traffic is sent, and in the second part the sporadic traffic is sent. The arbitration of messages is based on collision avoidance (i.e., CSMA/CA) using mini-slotting. Network speeds are as high as 2 Mbps.

Profibus. Profibus [PRO] is used in process automation and robotics. There are three different versions of Profibus: (1) Profibus-DP is optimized for speed and low cost. (2) Profibus-PA is designed for process automation. (3) Profibus-FMS is a general-purpose version of Profibus. Profibus provides master–slave communication together with token mechanisms. Profibus is available with data rates up to 12 Mbps.

Train Communication Network (TCN). The TCN [KZ01] is widely used in trains, and it is implementing the IEC 61275 standard as well as the IEEE 1473 standard. TCN is composed of two networks: the Wire Train Bus (WTB) and the Multifunction Vehicle Bus (MVB). The WTB is the network used to connect the whole train, that is, all vehicles of the train. Network data rate is up to 1 Mbps. The MVB is the network used within one vehicle. Here, the maximum data rate is 1.5 Mbps.

Both the WTB and the MVB are scheduled in cycles called basic periods. Each basic period consists of a periodic phase and a sporadic phase. Hence, there is a support for both periodic and sporadic type of traffic. The difference between the WTB and the MVB (apart from the data rate) is the length of the basic periods (1 or 2 msec for the MVB and 25 msec for the WTB).

WorldFIP. The WorldFIP [Wor] is a very popular communication network in train control systems. WorldFIP is based on the producer–distributor–consumers (PDC) communication model. Currently, network speeds are as high as 5 Mbps. The WorldFIP protocol defines an application layer that includes PDC- and messaging-services.

Ethernet for Real-Time Communication

In parallel with the search for the holy grail of real-time communication, Ethernet has established itself as the *de facto* standard for non-real-time communication. Comparing networking solutions for automation networks and office networks, fieldbuses were the choice for the former. At the same time, Ethernet developed as the standard for office automation, and due to its popularity, prices on networking solutions dropped. Ethernet is not originally developed for real-time communication since the original intention with Ethernet was to maximize throughput (bandwidth). However, nowadays, due to its popularity, a large effort is being made in order to provide real-time communication using Ethernet. The biggest challenge is to provide real-time guarantees using standard Ethernet components.

The reason why Ethernet is not very suitable for real-time communication is its handling of collisions on the network. Several proposals to minimize or eliminate the occurrence of collisions on Ethernet have been proposed. Below, we present some of these proposals.

TDMA. A simple solution would be to eliminate the occurrence of collisions on the network. This has been explored by, for example, Kopetz et al. [KDKM89], using a TDMA protocol on top of Ethernet.

Usage of tokens. Another solution to eliminate the occurrence of collisions is the usage of tokens. Token-based solutions [VC94, PMBL95] on the Ethernet also eliminate collisions, but is not compatible with standard hardware.

A token-based communication protocol is a way to provide real-time guarantees on most types of networks. This is because they are deterministic in their behavior, although a dedicated network is required. That is, all nodes sharing the network must obey the token protocol. Examples of token-based protocols are the Timed Token Protocol (TTP) [MZ94] and the IEEE 802.5 Token Ring Protocol.

Modified collision resolution algorithm. A different approach is to modify the collision resolution algorithm [RY94, Mol94]. Using standard Ethernet controllers, the modified collision resolution algorithm is nondeterministic. In order to make a deterministic modified collision resolution algorithm, a major modification of the Ethernet controllers is required [LR93].

Virtual time and window protocols. Another solution to real-time communication using Ethernet is the usage of the Virtual Time CSMA (VTCSMA) [MK85, ZR86, EDES90] protocol, where packets are delayed in a deterministic way in order to eliminate the occurrence of collisions. Moreover, Window

Protocols [ZSR90] are using a global window (synchronized time interval) that also remove collisions. The window protocol is more dynamic and somewhat more efficient in its behavior compared to the VTCSMA approach.

Master/slave. A fairly straightforward way of providing real-time traffic on Ethernet is by using a master/slave approach. As part of the Flexible Time-Triggered (FTT) framework [APF02], FTT-Ethernet [PAG02] is proposed as a master/multislave protocol. At the cost of some computational overhead at each node in the system, timely delivery of messages on Ethernet is provided.

Traffic smoothing. The most recent work, without modifications to the hardware or networking topology (infrastructure) is the usage of traffic smoothing. Traffic smoothing can be used to eliminate bursts of traffic [KSW00, CCLM02], which have a severe impact on the timely delivery of message packets on the Ethernet. By keeping the network load below a given threshold, a probabilistic guarantee of message delivery can be provided. Hence, traffic smoothing could be a solution for soft RTSs.

Black bursts. Black Burst [SK98] is using the fact that if a transmitting station is detecting a collision on the bus, the station will wait some time before retransmitting its message again (based on the collision resolution algorithm). However, suppose a station is jamming the network, causing stations to wait for retransmission. What the jamming station does is that it will send its message right after it has stopped the jamming. If all stations are using unique length jamming signals, then there will always be a unique winner. This is what the black burst approach is using.

Switches. Finally, a completely different approach to achieve real-time communication using Ethernet is by changing the infrastructure. One way of doing this is to construct the Ethernet using switches to separate collision domains. By using these switches, a collision-free network is provided. However, this requires new hardware supporting the IEEE 802. 1p standard. Therefore, it is not as attractive a solution for existing networks as, for example, traffic smoothing.

Wireless Communication

There are no commercially available wireless communication protocols providing real-time guarantees.[3] Two of the more common wireless protocols used today are the IEEE 802.11 (WLAN) and Bluetooth. However, these protocols are not providing the temporal guarantees needed for hard real-time communication. Today, a large effort is being made (as with Ethernet) to provide real-time guarantees for wireless communication, possibly by using either WLAN or Bluetooth.

81.6 Analysis of RTSs

Analysis techniques are techniques that are used to investigate what properties a component, or a system of components, exhibit. Of particular interest when dealing with RTSs is the issue of *temporal correctness*, that is, that actions are performed at the right time (i.e., we are not concerned with the actual values of an action, only the timing of the action).

Classical program-analysis techniques that focus on functional and structural correctness are well developed. However, to validate the correctness of an RTS, analysis techniques to validate the temporal correctness must be used *in addition to* other analysis techniques. In this section, we focus on techniques to analyze the temporal behavior of RTSs.

Timing Properties

Timing analysis is a complex problem. Not only are the used techniques sometimes complicated but also the problem itself is elusive; for instance, what is the meaning of the term "program execution-time?" Is it the average time to execute the program, or the worst possible time, or does it mean some form of "normal" Execution time? Under what conditions does a statement regarding program execution times

---

[3] Bluetooth provides real-time guarantees limited to streaming voice traffic.

apply? Is the program delayed by interrupts or higher priority tasks? Does the time include waiting for shared resources?, etc.

To straighten out some of these question marks, and to be able to study some existing techniques for timing analysis, we structure timing analysis into three major types. Each type has its own purpose, benefits, and limitations. The types are listed below.

Execution time. This refers to the execution time of a singe task (or program, or function, or any other unit of single threaded sequential code). The result of an execution-time analysis is the time (i.e., the number of clock cycles) the task takes to execute, when executing undisturbed on a single CPU, that is, the result should not account for interrupts, preemption, background DMA transfers, DRAM refresh delays, or any other types of interfering background activities.

At first glance, leaving out all types of interference from the execution-time analysis would give us unrealistic results. However, the purpose of the execution-time analysis is *not* to deliver estimates on "real-world" timing when executing the task. Instead, the role of execution-time analysis is to find out how much computing resources is needed to execute the task. (Hence, background activities that are not related to the task should not be accounted for.)

There are some different types of execution-times that can be of interest:

- Worst-Case Execution-Time (WCET). This is the worst possible execution time a task could exhibit, or equivalently, the maximum amount of computing resources required to execute the task. The WCET should include any possible atypical task execution such as exception handling or cleanup after abnormal task termination.
- Best-Case Execution-Time (BCET). During some types of real-time analysis, not only the WCET is used, but as we will describe later, having knowledge about the BCET of tasks is useful.
- Average Execution-Time (AET). The AET can be useful in calculating throughput figures for a system. However, for most RTS analysis, the AET is of less importance, simply since a reasonable approximation of the average case is easy to obtain during testing (where typically, the average system behavior is studied). Also, only knowing the average and not knowing any other statistical parameters such as standard deviation or distribution function makes statistical analysis difficult. For analysis purposes, a more pessimistic metric such as the 95%-quartile would be more useful. However, analytical techniques using statistical metrics of execution time are scarce and not very well developed.

Response time. The response time of a task is the time it takes from the *invocation* of the task to the *completion* of the task. In other words, the time from when the task first is placed in the operating-system's ready queue to the time when it is removed from the running state and placed in the idle or sleeping state.

Typically, for analysis purposes, it is assumed that a task does not voluntarily suspend itself during its execution. That is, the task may not call primitives such as `sleep()` or `delay()`. However, involuntarily suspension, such as blocking on shared resources, is allowed. That is, primitives such as `get_semaphore()` and `lock_database_tuple()` are allowed. When a program voluntarily suspends itself, then that program should be broken down into two (or more) analysis tasks.

The response-time is typically a *system level property*, in that it includes interference from other, unrelated, tasks and parts of the system. The response-time also includes delays caused by contention on shared resources. Hence, the response-time is only meaningful when considering a complete system, or in distributed systems, a complete node.

End-to-end delay. The above-described "execution time" and "response time" are useful concepts since they are relatively easy to understand and have well-defined scopes. However, when trying to establish the temporal correctness of a system, knowing the WCET and/or the response-times of tasks is often not enough. Typically, the correctness criterion is stated using *end-to-end* latency timing requirements, for instance, an upper bound on the delay between the input of a signal and the output of a response.

In a given implementation, there may be a chain of events taking place between the input of a signal and the output of a response. For instance, one task may be in charge of reading the input and another task for generating the response, and the two tasks may have to exchange messages on a communications link before the response can be generated. The end-to-end timing denotes the timing of externally visible events.

Jitter. The term *jitter* is used as a metric for variability in time. For instance, the jitter in execution time of a task is the difference between the task's BCET and WCET. Similarly, the response-time jitter of a task is the difference between its best-case response-time and its worst-case response-time. Often, control algorithms have requirements that the jitter of the output should be limited. Hence, the jitter is sometimes a metric equally important as the end-to-end delay.

Also, input to the system can have jitter. For instance, an interrupt that is expected to be periodic may have a jitter (due to some imperfection in the process generating the interrupt). In this case, the jitter-value is used as a bound on the maximum deviation from the ideal period of the interrupt. Figure 81.3 illustrates the relation between the period and the jitter for this example.

Note that jitter should not accumulate over time. For our example, even though two successive interrupts could arrive closer than one period, in the long run, the average interrupt interarrival time will be that of the period.

In the above list of types of time, we only mentioned time to execute programs. However, in many RTSs, other timing properties may also exist. Most typical are delays on a communications network, but also other resources such as hard disk drives may be causing delays and need to be analyzed. The above-introduced times can all be mapped to different types of resources, for instance, the WCET of a task corresponds the maximum size of a message to be transmitted, and the response time of message is defined analogous to the response time of a task.

Methods for Timing Analysis

When analyzing hard RTSs, it is essential that the estimates obtained during timing analysis are *safe*. An estimate is considered safe if it is guaranteed that it is not an underestimation of the actual worst-case time. It is also important that the estimate is *tight*, meaning that the estimated time is close to the actual worst-case time.

For the previously defined types of timings (section Timing properties), there are different methods available:

Execution-time estimation. For real-time tasks, the WCET is the most important execution time measure to obtain. Unfortunately, however, it is also often the most difficult measure to obtain.

Methods to obtain the WCET of a task can be divided into two categories: (1) static analysis and (2) dynamic analysis. Dynamic analysis is essentially equivalent to testing (i.e., executing the task on the target hardware) and has all the drawbacks/problems that testing exhibit (such as being tedious and error prone). One major problem with dynamic analysis is that it does not produce safe results. In fact, the result can never exceed the true WCET and it is very difficult to be sure that the estimated WCET is actually the true WCET.

Static analysis, on the other hand, can give guaranteed safe results. Static analysis is performed by analyzing the code (source and/or object code is used) and basically counting the number of clock cycles that the task may use to execute (in the worst possible case). Static analysis uses models of the hardware to predict execution times for each instruction. Hence, for modern hardware it may be very difficult to

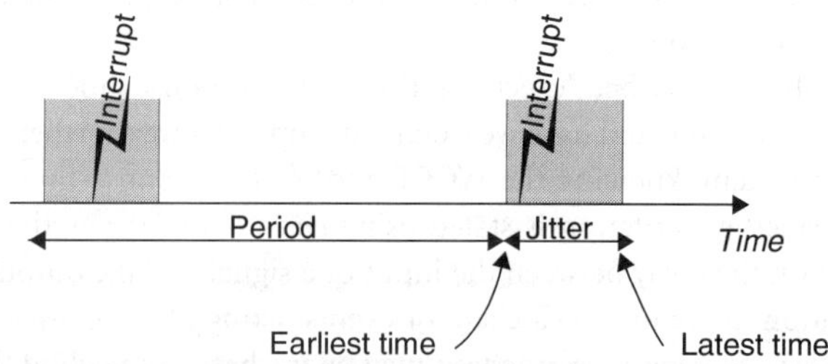

FIGURE 81.3   Jitter used as a bound on variability in periodicity.

produce static analyzers that give good results. One source of pessimism in the analysis (i.e., overestimation) is hardware caches; whenever an instruction or data-item cannot be guaranteed to reside in the cache, a static analyzer must assume a cache miss. And since modeling the exact state of caches (sometimes of multiple levels), branch predictors, etc., is very difficult and time consuming, few tools that give adequate results for advanced architectures exist. Also, to perform a program flow and data analysis that exactly calculates, for example, the number of times a loop iterates or the input parameters for procedures is difficult.

Methods for good hardware and software modeling do exist in the research community; however, combining these methods into good quality tools has proved to be tedious.

Schedulability analysis. The goal of schedulability analysis is to whether or not a system is *schedulable*. A system is deemed schedulable if it is guaranteed that all task deadlines will always be met. For statically scheduled (table driven) systems, calculation of response-times is trivially given from the static schedule. However, for dynamically scheduled systems (such as fixed priority or deadline scheduling), more advanced techniques have to be used.

There are two main classes of schedulability analysis techniques: (1) response-time analysis and (2) utilization analysis. As the name suggests, a response-time analysis calculates a (safe) estimate of the worst-case response-time of a task. That estimate can then be compared to the deadline of the task and if it does not exceed the deadline, then the task is schedulable. Utilization analysis, in contrast, does not directly derive the response-times for tasks; rather, they give a boolean result for each task, stating whether or not the task is schedulable. This result is based on the fraction of utilization of the CPU for a relevant subset of the tasks, hence the term utilization analysis.

Both types of analysis are based on similar types of task models. However, typically the task models used for analysis are not the task models provided by commercial RTOSes. This problem can be resolved by mapping one or more OS-tasks to one or more analysis tasks. However, this mapping has to be performed manually and requires an understanding of the limitations of the analysis task model and the analysis technique used.

End-to-end delay estimation. The typical way to obtain end-to-end delay estimations is to calculate the response time for each task/message in the end-to-end chain and to summarize these response times to obtain an end-to-end estimate. When using a utilization-based analysis technique (in which no response time is calculated), one has to resort to using the task/message deadlines as safe upper bounds on the response times.

However, when analyzing distributed RTSs, it may not be possible to calculate all response times in one pass. The reason for this is that delays on one node will lead to jitter on another node, and that this jitter may in turn affect the response times on that node. Since jitter can propagate in several steps between nodes, in both directions, there may not exist a right order to analyze the nodes. (If A sends a message to B, and B sends a message to A, which node should one analyze first?) Solutions to this type of problems are called *holistic* schedulability analysis methods (since they consider the whole system). The standard method for holistic response-time analysis is to repeatedly calculate response-times for each node (and update jitter values in the nodes affected by the node just analysed) until response-times do not change (i.e., a fix-point is reached).

Jitter estimation. To calculate the jitter, one need not only perform a worst-case analysis (of, for instance, response-time or end-to-end delay). It is also necessary to perform a best-case analysis.

However, even though best-case analysis techniques are often conceptually similar to worst-case analysis techniques, little attention has been paid to best-case analysis. One reason for not spending too much time on best-case analysis is that it is quite easy to make a conservative estimate of the best-case: the best-case time is never less than zero (0). Hence, in many tools it is simply assumed that the BCET (for instance) is zero, whereas great efforts can be spent in analyzing the WCET.

However, it is important to have tight estimates of the jitter, and to keep the jitter as low as possible. It has been shown that the number of *execution paths* a multitasking RTS can take dramatically increases if jitter increases [TH99]. Unless the number of possible execution paths is kept as low as possible, it becomes very difficult to achieve good coverage during testing.

Example of Analysis

In this section, we give simple examples of schedulability analysis. We show a very simple example of how a set of tasks running on a single CPU can be analyzed, and we also give an example of how the response-times for a set of messages sent on a CAN-bus can be calculated.

Analysis of tasks. This example is based on some 30-year-old task models and is intended to give the reader a feel for how these types of analysis work. Present-day methods allow for far richer and more realistic task models, with a resulting increase in the complexity of the equations used (hence, they are not suitable for use in our simple example).

In the first example, we will analyze a small task set described in Table 81.2, where T, C, and D denote the tasks' period, WCET and deadline, respectively. In this example, $T = D$ for all tasks and priorities have been assigned in *rate monotonic* order, that is, the highest rate gives the highest priority.

For the task set in Table 81.2 original analysis techniques of Liu and Layland [LL73] and Joseph and Pandya [JP86] are applicable, and we can perform both utilization-based and response-time-based schedulability analysis.

We start with the utilization-based analysis; for this task model, Liu and Layland's result is that a task set of n tasks is schedulable if its total utilization, U_{tot}, is bounded by the following equation:

$$U_{tot} \leq n(2^{1/n} - 1)$$

Table 81.3 shows the utilization calculations performed for the schedulability analysis. For our example task set $n = 3$ and the bound is approximately 0.78. However, the utilization ($U_{tot} = \sum_{i=1}^{n}(C_i/T_i)$) for our task set is 0.81, which exceeds the bound. Hence, the task set fails the *rate monotonic test* and cannot be deemed schedulable.

Joseph and Pandya's response-time analysis allows us to calculate worst-case response-time, R_i, for each task i in our example (Table 81.2). This is done using the following formula:

$$R_i = C_i + \sum_{j \in hp(i)} \left\lceil \frac{R_i}{T_j} \right\rceil C_j \tag{81.1}$$

where $hp(i)$ denotes the set of tasks with priority higher than i.

The observant reader may have noticed that Equation (81.1) is not in closed form, in that R_i is not isolated on the left-hand side of the equality. As a matter of fact, R_i cannot be isolated on the left-hand side of the equality; instead, Equation (81.1) has to be solved using *fix-point iteration*. This is done with the recursive formula in Equation (81.1), starting with $R_i^0 = 0$ and terminating when a fix-point has been reached (i.e., when $R_i^{m+1} = R_i^m$).

$$R_i^{m+1} = C_i + \sum_{j \in hp(i)} \left\lceil \frac{R_i^m}{T_j} \right\rceil C_j \tag{81.2}$$

For our example task-set, Table 81.4 shows the results of calculating Equation (81.1). From the table, we can conclude that no deadlines will be missed and that the system is schedulable.

Remarks. As we could see for our example task-set in Table 81.2, the utilization-based test could not deem the task-set as schedulable, whereas the response-time-based test could. This situation is symptomatic for the relation between utilization-based and response-time-based schedulability tests. That is, the response-time-based tests find more task-sets schedulable than does the utilization-based tests.

TABLE 81.2 Example Task-Set for Analysis

Task	T	C	D	Prio
X	301	0	30	High
Y	401	0	40	Medium
Z	521	0	52	Low

TABLE 81.3 Result of Rate Monotonic Test

Task	T	C	D	Prio	U
X	301	0	30	High	0.33
Y	401	0	40	Medium	0.25
Z	521	0	52	Low	0.23
				Total:	0.81
				Bound:	0.78

TABLE 81.4 Result of Response-Time Analysis for Tasks

Task	T	C	D	Prio	R	$R \leq D$
X	301	0	30	High	10	Yes
Y	401	0	40	Medium	20	Yes
Z	521	0	52	Low	52	Yes

However, also as shown by the example, the response-time-based test needs to perform more calculations than the utilization-based tests does. For this simple example, the extracomputational complexity of the response-time test is insignificant. However, when using modern task-models (which are capable of modelling realistic systems), the computational complexity of response-time-based tests is significant. Unfortunately, for these advanced models, utilization-based tests are not always available.

Analysis of messages. In our second example, we show how to calculate the worst-case response-times for a set of periodic messages sent over the CAN-bus (CAN is described in the section Fieldbuses). We use a response-time analysis technique similar to the one we used when we analyzed the task set in Table 81.2. In this example, our message set is given in Table 81.5, where T, S, D, and Id denote the messages' period, data size (in bytes), deadline, and CAN-identifier, respectively. (The time-unit used in this example is "bit-time," that is, the time it takes to send one bit. For a 1 Mbit CAN this means that 1 time-unit is 10^{-6} sec.)

Before we discuss the problem of calculating response-times, we extend Table 81.5 by two columns. First, we need the priority of each message; in CAN, this is given by the identifier; the lower the numerical value, the higher the priority. Second, we need to know the worst-case transmission time of each message. The transmission time is given partly by the message data-size but we also need to add time for the frame-header and for any *stuff bits*.[4] The formula to calculate the transmission time, C_i for a message i is given below:

$$C_i = S_i + 47 + \left\lceil \frac{34 + 8S_i + 1}{5} \right\rceil$$

In Table 81.6, the two columns Prio and C show the priority assignment and the transmission times for our example message set.

Now we have all the data needed to perform the response-time analysis. However, since CAN is a nonpreemptive resource, the structure of the equation is slightly different from Equation (81.1), which we used for analysis of tasks. The response-time equation for CAN is given as

$$R_i = w_i + C_i$$

$$w_i = 130 + \sum_{\forall j \in hp(i)} \left\lceil \frac{w_i + 1}{T_j} \right\rceil C_j \tag{81.3}$$

In Equation (81.3), $hp(i)$ denotes the set of messages with a priority higher than message i. Note that (similar to equation 1) w_i is not isolated on the left-hand side of the equation, and its value has to be calculated using fix-point iteration (compare to Equation [81.2]).

---

[4] CAN adds stuff bits, if necessary, to avoid the two reserved bit patterns 000000 and 111111. These stuff bits are never seen by the CAN-user but have to be accounted for in the timing analysis.

TABLE 81.5   Example CAN-Message Set

Message	T	S	D	Id
X	350	8	300	00010
Y	500	6	400	00100
Z	1000	5	800	00110

TABLE 81.6   Result of Response-Time Analysis for CAN

Message	T	S	D	Id	Prio	C	w	R	$R \leq D$
X	350	8	300	00010	High	130	130	260	Yes
Y	500	6	400	00100	Medium	111	260	371	Yes
Z	1000	5	800	00110	Low	102	612	714	Yes

Applying Equation (81.3), we can now calculate the worst-case response-time for our example messages. In Table 81.6, the two columns w and R show the the results of the calculations, and the final column shows the schedulablilty verdict for each message.

As we can see from Table 81.6, our example message set is schedulable, meaning that the messages will always be transmitted before their deadlines. Note that this analysis was carried out assuming that there will not be any retransmission of broken messages. CAN normally automatically retransmits any message that has been broken due to interference on the bus. To account for such automatic retransmissions, an *error model* needs to be adopted and the response-time equation adjusted accordingly, see, for example, [THW94].

Trends and Tools

As pointed our earlier, and also illustrated by our example in Table 81.2, there is a mismatch between the analytical task-models and the task-models provided by commonly used RTOSes. One of the basic problems is that there is no one-to-one mapping between analysis tasks and RTOS tasks. In fact, for many systems there is an N-to-N mapping between the task types. For instance, an interrupt handler may have to be modeled as several different analysis tasks (one analysis task for each type of interrupt it handles), and one OS task may have to be modeled as several analysis tasks (for instance, one analysis task per call to `sleep()` primitives).

Also, current schedulability analysis techniques cannot adequately model task synchronization other than locking/blocking on shared resources. Abstractions such as message queues are difficult to include in the schedulability analysis.[5] Furthermore, tools to estimate the WCET are also scarce. Currently, only two tools that give safe WCET estimates are commercially available [Abs, Bou].

These problems have led to low penetration of schedulability analysis in industrial software-development processes. However, in isolated domains, such real-time networks, some commercial tools that are based on real-time analysis do exist. For instance, Volcano [CRTM98, Vol] provides tools for the CAN bus that allow system designers to specify signals on and abstract level (giving signal attributes such as size, period, and deadline) and automatically derive a mapping of signals to CAN-messages where all deadlines are guaranteed to be met.

On the software side tools provided by, for instance, TimeSys [Timb], Arcticus Systems [Arc], and TTTech [TTT] can provide system development environments with timing analysis as an integrated part of the tool suite. However, these tools all require that the software development processes be under complete control of the respective tool. This requirement has limited the widespread use of these tools.

---

[5] Techniques to handle more advanced models include timed logic and model checking. However, the computational and conceptual complexity of these techniques has limited their industrial impact. There are, however, examples of commercial tools for this type of verification, for example, [Tima].

The widespread use of UML [OMG03b] in software design has led to some specialized UML-products for real-time engineering [Rat, IL]. However, these products, as of today, do support timing analysis of the designed systems. There is, however, recent work within the OMG that specifies a *profile* "Schedulability, Performance and Time" (SPT) [OMG03a], which allows specification of both timing properties and requirement in a standardized way. This will in turn lead to products that can analyze UML-models conforming to the SPT-profile.

The SPT-profile has, however, not been received without criticism. Critique has mainly come from researchers active in the timing analysis field, claiming both that the profile is not precise enough and that some important concepts are missing. For instance, the Universidad de Cantabria has instead developed the MAST-UML profile and an associated MAST-tool for analyzing MAST-UML models [MHD01, MAS]. MAST allows modeling of advanced timing properties and requirement, and the tool also provides state-of-the-art timing analysis techniques.

81.7 Testing and Debugging of RTSs

According to a recent study by NIST [U.S02], up to 80% of the life cycle cost for software is spent on testing and debugging. Despite the importance, there are few results on RTSs testing and debugging.

The main reason for this is that it is actually quite difficult to test and debug RTS. Remember that RTSs are timing critical and that they interact with the real world. Since testing and debugging typically involve some instrumentation of the code, the timing behavior of the system will be different when testing/debugging compared to when executing the deployed system. Hence, the test-cases that were passed during testing may very well lead to failures in the deployed system, and tests that failed may very well not cause any problem at all in the deployed system. For debugging, the situation is possibly even worse, since in addition to a similar effect when running the system in a debugger, entering a breakpoint will stop the execution for an unspecified time. The problem with this is that the controlled external process will continue to evolve (e.g., a car will not momentarily stop by stopping the execution of the controlling software). The result of this is that we obtain a behavior of the debugged system that will not be possible in the real system. Also, it is often the case that the external process cannot be completely controlled, which means that we cannot reproduce the observed behavior, indicating that it will be difficult to use (cyclic) debugging to track down an error that caused a failure.

The following are two possible solutions to the above-presented problems:

- To build a simulator that faithfully captures the functional as well as timing behavior of both the RTS and the environment that it is controlling. Since this is both time consuming and costly, this approach is only feasible in very special situations. Since such situations are rare, we will not further consider this alternative here.
- To record the RTS's behavior during testing or execution, and then if a failure is detected replay the execution in a controlled way. For this to work, it is essential that the timing behavior is the same during testing as in the deployed system. This can either be achieved by using nonintrusive hardware recorders, or by leaving the software used for instrumentation in the deployed system. The latter comes at a cost in memory space and execution time, but gives the additional benefit that it becomes possible to also debug the deployed system in case of a failure [RDBC+03].

An additional problem for most RTSs is that the system consists of several concurrently executing threads. This is also the case for the majority of non-real-time systems. This concurrency will *per se* lead to a problematic nondeterminism, since due to race conditions caused by slight variations in execution time, the exact preemtion points will vary, causing unpredictability, both in terms of the number of scenarios and in terms of being able to predict which scenario will actually be executed in a specific situation.

In conclusion, testing and debugging of RTSs are difficult and challenging tasks.

The following is a brief account of some of the results on testing of RTSs reported in the literature:

- Thane et al. [TH99] propose a method for deterministic testing of distributed RTSs. The key element here is to identify the different execution orderings (serializations of the concurrent system) and treat each of these orderings as a sequential program. The main weakness of this approach is the potentially exponential blowup of the number of execution orderings.
- For testing of temporal correctness, Tsai et al. [TFB90] provide a monitoring technique that records run-time information. This information is then used to analyze if temporal constraints are violated.
- Schütz [Sch94] have proposed testing a strategy for testing distributed RTSs. The strategy is tailored for the time-triggered MARS system [KDKM89].
- Zhu et al. [ZHM97] have proposed a framework for regression testing of real-time software in distributed systems. The framework is based on the Onomas [OTPS98] regression testing process.

When it comes to RTS debugging, the most promising approach is record/replay [CAN⁺01, MCL89, TCO91, TH00, ZN99] as mentioned above. Using record/replay, first, a reference execution of the system is executed and observed; second, a replay execution is performed based on the observations made during the reference execution. Observations are made by instrumenting the system, in order to extract information about the execution.

The industrial practice for testing and debugging of multitasking RTS is a time-consuming activity. At best, hardware emulators, for example, [Lau], are used to obtain some level of observability without interfering with the observed system. More often, it is an *ad hoc* activity, using intrusive instrumentations of the code to observe test results or try to track down intricate timing errors. However, some tools using the above record/replay method are now emerging on the market, for example, [Zea].

References

[AB98]   Abeni, L. and G. Buttazzo, Integrating Multimedia Applications in Hard Real-Time Systems, in Proceedings of the 19th IEEE Real-Time Systems Symposium (RTSS'98), Madrid, Spain, Dec. 1998, IEEE Computer Society, Silver Spring, MD, 1998, pp. 4–13.

[ABD⁺95] Audsley, N.C., A. Burns, R.I. Davis, K. Tindell, and A.J. Wellings, Fixed priority pre-emptive scheduling: an historical perspective, *Real-Time Systems*, 8, 129–154, 1995.

[Abs]   AbsInt, http://www.absint.com.

[AFF98]   Almeida, L., J.A. Fonseca, and P.F. Onseca, Flexible Time-Triggered Communication on a Controller Area Network, in Proceedings of the Work-In-Progress Session of the 19th IEEE Real-Time Systems Symposium (RTSS'98), Madrid, Spain, Dec. 1998, IEEE Computer Society, Silver Spring, MD, 1998.

[AFF99]   Almeida, L., J.A. Fonseca, and P. Fonseca, A Flexible Time-Triggered Communication System Based on the Controller Area Network: Experimental Results, in Proceedings of the IFAC International Conference on Filedbus Technology (FeT), 1999.

[Air96]   Airlines Electronic Engineering Committee (AEEC), ARINC 653: Avionics Application Software Standard Interface (Draft 15), June 1996.

[APF02]   Almeida, L., Pedreiras P., and Fonseca J.A.G, The FTT-CAN protocol: why and how, *IEEE Transactions on Industrial Electronics*, Vol. 49, Issue 6, pp. 1189–1201, Dec. 2002.

[App98]   Northern Real-Time Applications, Total time predictability, Whitepaper on SSX5, 1998.

[Arc]   Arcticus Systems, The Rubus Operating System, http://www.arcticus.se.

[ARI99]   ARINC/RTCA-SC-182/EUROCAE-WG-48, Minimal Operational Performance Standard for Avionics Computer Resources, 1999.

[Art03]   Roadmap — Adaptive Real-Time Systems for Quality of Service Management, ARTIST — Project IST-2001-34820, May 2003, http://www.artist-embedded. org/Roadmaps/.

[Ast]   The Asterix Real-Time Kernel, http://www.mrtc.mdh.se/projects/asterix/.

[BBE+02] Belschner, R., J. Berwanger, C. Ebner, H. Eisele, S. Fluhrer, T. Forest, T. Führer, F. Hartwich, B. Hedenetz, R. Hugel, A. Knapp, J. Krammer, A. Millsap, B. Müller, M. Peller, and A. Schedl, FlexRay — Requirements Specification, April 2002, http://www.flexray-group.com.

[Bou] Bound-T Execution Time Analyzer, http://www.bound-t.com.

[But97] Buttazzo, G.C., *Hard Real-Time Computing Systems*, Kluwer Academic Publishers, Dordrecht, 1997.

[BW96] Burns, A. and A. Wellings, *Real-Time Systems and Programming Languages, 2nd ed.*, Addison-Wesley, Reading, MA, 1996.

[CAN92] Road Vehicles — Interchange of Digital Information — Controller Area Network (CAN) for High Speed Communications, February 1992, ISO/DIS 11898.

[CAN+01] Choi, J.D., B. Alpern, T. Ngo, M. Sridharan, and J. Vlissides, A Pertrubation-Free Replay Platform for Cross-Optimized Multithreaded Applications, in Proceedings of the 15th International Parallel and Distributed Processing Symposium, IEEE Computer Society, Silver Spring, MD, April 2001.

[CAN02] CAN Specification 2.0, Part-A and Part-B, CAN in Automation (CiA), Am Weichselgarten 26, D-91058 Erlangen, 2002, http://www.can-cia.de.

[CCLM02] Carpenzano, A., R. Caponetto, L. LoBello, and O. Mirabella, Fuzzy Traffic Smoothing: an Approach for Real-Time Communication over Ethernet Networks, in Proceedings of the 4th IEEE International Workshop on Factory Communication Systems (WFCS'02), Västerås, Sweden, Aug. 2002, IEEE Industrial Electronics Society, pp. 241–248.

[CRTM98] Casparsson, L., A. Rajnak, K. Tindell, and P. Malmberg, Volcano — a revolution in on-board communications, *Volvo Technology Report*, 1, 9–19, 1998.

[EDES90] El-Derini, M. and M. El-Sakka, A Novel Protocol Under A Priority Time Constraint For Real-Time Communication Systems, in Proceedings of 2nd IEEE Workshop on Future Trends of Distributed Computing Systems (FTDCS'90), Cairo, Egypt, Sept. 1990, IEEE Computer Society, Silver Spring, MD, 1990, pp. 128–134.

[ETA] ETAS, http://en.etasgroup.com.

[FAQ] Comp.realtime FAQ, available at http://www.faqs.org/faqs/realtime-computing/faq/.

[FLD03] Fohler, G., T. Lennvall, and R. Dobrin, A component based real-time scheduling architecture, in *Architecting Dependable Systems*, Gacek, C. and Romanovsky, A., Ed., de Lemos, R., Lecture Notes in Computer Science, Vol. 2677, Springer-Verlag, Berlin, 2003.

[Gro] OSEK Group, OSEK/VDX Operating System Specification 2.2.1, http://www. osek-vdx.org/.

[Hal00] Halfhill, T.R., Embedded Markets Breaks New Ground, Microprocessor report, 17 January, 2000.

[HLS96] Hansson, H., H. Lawson, and M. Strömberg, BASEMENT a distributed real-time architecture for vehicle applications, *Real-Time Systems*, 3, 223–244, 1996.

[IEE98] IEEE, Standard for Information Technology — Standardized Application Environment Profile — POSIX Realtime Application Support (AEP), 1998, IEEE Standard P1003.13-1998.

[IL] I-Logix, Rhapsody, http://www.ilogix.com/products/rhapsody.

[ISO95] ISO, Ada95 Reference Manual, 1995, ISO/IEC 8652:1995(E).

[ITE] ITEA, EAST/EEA Project Site, http://www.east-eea.net.

[JP86] Joseph, M. and P. Pandya, Finding response times in a real-time System, *The Computer Journal*, 29, 390–395, 1986.

[KB03] Kopetz, H. and G. Bauer, The Time-Triggered Architecture, *Proceedings of the IEEE, Special Issue on Modeling and Design of Embedded Software*, 91, 112–126, 2003.

[KDKM89] Kopets, H., A. Damm, C. Koza, and M. Mullozzani, Distributed fault-tolerant real-time systems: the Mars approach, *Micro, IEEE*, Vol. 9, Issue 1, pp. 25–40, Feb. 1989.

[KG94] Kopetz, H. and G. Grünsteidl, TTP — a protocol for fault-tolerant real-time systems, *IEEE Computer*, Vol. 27, Issue 1, pp. 14–23, Jan. 1994.

[Klu]   Kluwer, Real-Time Systems (Journal), http://www.wkap.nl/kapis/CGI-BIN/WORLD/journal-home.htm?0922-6443.

[Kop98]   Kopetz, H., The Time-Triggered Model of Computation, in proceedings of the 19th IEEE Real-Time Systems Symposium (RTSS'98), Madrid, Spain, Dec. 1998, IEEE Computer Society, Silver Spring, MD, pp. 168–177.

[Kop03]   Kopetz, H., Introduction In Real-Time Systems: Introduction and Overview, Part XVIII of Lectures Notes from ESSES 2003 — European Summer School on Embedded Systems, Västerås, Sweden, Sept. 2003.

[KRP+98]   Klein, M.H., T. Ralya, B. Pollak, R Obenza, and M.G. Harbour, *A Practitioners Handbook for Rate-Monotonic Analysis,* Kluwer, Dordrecht, 1998.

[KSW00]   Kweon, S.K., K.G. Shin, and G. Workman, Achieving Real-Time Communication over Ethernet with Adaptive Traffic Smoothing, in Proceedings of the 6th IEEE Real-Time Technology and Applications Symposium (RTAS'00), Washington, DC, U.S.A., June 2000, IEEE Comuter Society, Silver Spring, MD, 2000, pp. 90–100.

[KZ01]   Kirrmann, H. and P.A. Zuber, The IEC/IEEE train communication network, *IEEE Micro,* 21, 81–92, 2001.

[Lau]   Lauterbach, http://www.laterbach.com.

[LIN]   LIN, Local Interconnect Network, http://www.lin-subbus.de.

[Liv99]   LiveDevices, Realogy Real-Time Architect, SSX5 Operating System, 1999, http://www. livedevices.com/realtime. shtml.

[LL73]   Liu, C. and J. Layland, Scheduling algorithms for multiprogramming in a hard-real-time environment, *Journal of the ACM,* 20, 46–61, 1973.

[Log]   Express Logic, Threadx, http://www.expresslogic.com.

[LR93]   Lann, G. and N. Riviere, Real-Time Communications over Broadcast Networks: the CSMA/DCR and the DOD-CSMA/CD Protocols, Technical report, TR 1863, INRIA, 1993.

[LYN]   Lynuxworks, http://www.lynuxworks.com.

[MAS]   MAST home-page, http://mast.unican.es/.

[MCL89]   Mellor-Crummey, J. and T. LeBlanc, A Software Instruction Counter, in Proceedings of the Third International Conference on Architectural Support for Programming Languages and Operating Systems, ACM, New York, April 1989, pp. 78–86.

[MHD01]   Medina, J.L., M. González Harbour, and J.M. Drake, MAST Real-Time View: A Graphic UML Tool for Modeling Object-Oriented Real-Time Systems, in Proceedings of the 22nd IEEE Real-Time Systems Symposium (RTSS), December 2001.

[Mica]   Microsoft, Microsoft COM Technologies, http://www.microsoft.com/com/.

[Micb]   Microsoft, .NET Home Page, http://www.microsoft.com/net/.

[MK85]   Molle, M. and L. Kleinrock, Virtual time CSMA: why two clocks are better than one. *IEEE Transactions on Communications,* 33, 919–933, 1985.

[Mol94]   Molle, M., A New Binary Logarithmic Arbitration Method for Ethernet, Technical report, TR CSRI-298, CRI, University of Toronto, Canada, 1994.

[MSZ02]   Müller, P.O., C.M. Stich, and C. Zeidler, Component based embedded systems, *Building Reliable Component-Based Software Systems,* Artech House Publisher, Norwood, MA, 2002, pp. 303–323.

[MTS02]   Mäki-Turja, J. and M. Sjödin, Combining Dynamic and Static Scheduling in Hard Real-Time Systems, Technical report MRTC no. 71, Mälardalen Real-Time Research Centre (MRTC), October 2002.

[MZ94]   Malcolm, N. and W. Zhao, The timed token protocol for real-time communication, *IEEE Computer,* 27, 35–41, 1994.

[OMG02]   OMG, CORBA Component Model 3.0, June 2002, http://www.omg.org/technology/documents/formal/components.htm.

[OMG03a]   OMG, UML Profile for Schedulability, Performance and Time Specification, September 2003, OMG document formal/2003-09-01.

[OMG03b] OMG, Unified Modeling Language (UML), Version 1.5, 2003, http://www.omg.org/technology/documents/formal/uml.htm.

[OMG03c] OMG, Unified Modeling Language (UML), Version 2.0 (draft), September 2003, OMG document ptc/03-09-15.

[OTPS98] Onoma, K., W.-T. Tsai, M. Poonawala, and H. Suganuma, Regression testing in an industrial Environment, *Communications of the ACM,* 41, 81–86, 1998.

[PAG02] Pedreiras, P., L. Almeida, and P. Gai, The FTT-Ethernet Protocol: Merging Flexibility, Timeliness and Efficiency, in Proceedings of the 14th Euromicro Conference on Real-Time Systems (ECRTS'02), Vienna, Austria, June 2002, IEEE Computer Society, Silver Spring, MD, 2002, pp. 152–160.

[PMBL95] Pritty, D.W., J.R. Malone, S.K. Banerjee, and N.L. Lawrie, A Real-Time Upgrade for Ethernet Based Factory Networking, in *Proceedings of IECON'95,* 1995, pp. 1631–1637.

[PRO] PROFIBUS, PROFIBUS International, http://www.profibus.com.

[QNX] QNX Software Systems, QNX realtime OS, http://www.qnx.com.

[Rat] Rational, Rational Rose RealTime, http://www.rational.com/products/rosert.

[RDBC⁺03] Ronsse, M., K. De Bosschere, M. Christiaens, J. Chassin de Kergommeaux, and D. Kranzlmüller, Record/replay for nondeterministic program executions, *Communications of the ACM,* 46, 62–67, 2003.

[RTL] List of real-time Linux variants, http://www.realtimelinuxfoundation.org/variants/variants.html.

[RTS] IEEE Computer Society, Technical Committee on Real-Time Systems Home Page, http://www.cs.bu.edu/pub/ieee-rts/.

[RY94] Ramakrishnan, K.K. and H. Yang, The Ethernet Capture Effect: Analysis and Solution, in Proceedings of 19th IEEE Local Computer Networks Conference (LCNC'94), Oct. 1994, pp. 228–240.

[SB94] Spuri, M. and G.C. Buttazzo, Efficient Aperiodic Service under Earliest Deadline Scheduling, in Proceedings of the 15th IEEE Real-Time Systems Symposium (RTSS'94), San Juan, Puerto Rico, Dec. 1994, IEEE Computer Society, Silver Spring, MD, 1994, pp. 2–11.

[SB96] Spuri, M. and G.C. Buttazzo, Scheduling aperiodic tasks in dynamic priority systems, *Real-Time Systems,* 10, 179–210, 1996.

[SBS95] Spuri, M., G.C. Buttazzo, and F. Sensini, Robust aperiodic scheduling under dynamic priority systems, in Proceedings of the 16th IEEE Real-Time Systems Symposium (RTSS'95), Pisa, Italy, Dec. 1995, IEEE Computer Society, Silver Spring, MD, 1995, pp. 210–219.

[Sch94] Schütz, W., Fundamental issues in testing distributed real-time systems, *Real-Time Systems,* 7, 129–157, 1994.

[Sie] Siemens, http://www.siemensvdo.com.

[SK98] Sobrinho, J.L. and A.S. Krishnakumar, EQuB-Ethernet Quality of Service using Black Bursts, in Proceedings of the 23rd IEEE Annual Conference on Local Computer Networks (LCN'98), Lowell, MA, U.S.A., Oct. 1998, IEEE Computer Society, Silver Spring, MD, 1998, pp. 286–296.

[SSL89] Sprunt, B., L. Sha, and J.P Lehoczky, Aperiodic task scheduling for hard real-time systems, *Real-Time Systems,* 1, 27–60, 1989.

[Sta01] Stankovic, J.A., VEST — a toolset for constructing and analyzing component based embedded systems, *Lecture Notes in Computer Science,* Vol. 2211/2001, pp. 390–402, 2001.

[SUN] SUN Microsystems, Introducing Java Beans. http://developer.java.sun.com/developer/online Training/Beans/Beans1/index.html.

[SVK97] Stewart, D.B., R.A. Volpe, and P.K. Khosla, Design of dynamically reconfigurable real-time software using port-based objects, *IEEE Transactions on Software Engineering,* Vol. 23, Issue 12, pp. 759–776, Dec. 1997.

[Sys] Enea OSE Systems, Ose, http://www.ose.com.

[TBW95] Tindell, K.W., A. Burns, and A.J. Wellings, Calculating controller area network (CAN) message response times, *Control Engineering Practice,* 3, 1163–1169, 1995.

[TCO91] Tai, K.C., R. Carver, and E. Obaid. Debugging concurrent ADA programs by deterministic execution, *IEEE Transactions on Software Engineering*, 17, 280–287, 1991.

[Tel] TeleLogic, Telelogic tau, http://www.telelogic.com/products/tau.

[TFB90] Tsai, J.J.P., K.Y. Fang, and Y.D. Bi, On Realtime Software Testing and Debugging, in Proceedings of Fourteenth Annual International Computer Software and Application Conference, 1990, pp. 512–518.

[TH99] Thane, H. and H. Hansson, Towards Systematic Testing of Distributed Real-Time Systems, in Proceedings of the 20th IEEE Real-Time Systems Symposium (RTSS), Dec. 1999, pp. 360–369.

[TH00] Thane, H. and H. Hansson, Using Deterministic Replay for Debugging of Distributed Real-Time Systems, in Proceedings of the 12th Euromicro Conference on Real-Time Systems, IEEE Computer Society, Silver Spring, MD, June 2000, pp. 265–272.

[THW94] Tindell, K., H. Hansson, and A. Wellings, Analysing Real-Time Communications: Controller Area Network (CAN), in Proceedings of the 15th IEEE Real-Time Systems Symposium (RTSS), IEEE Computer Society Press, Silver Spring, MD, Dec. 1994, pp. 259–263.

[Tima] The Times Tool, http://www.docs.uu.se/docs/rtmv/times.

[Timb] TimeSys, Timewiz — a modeling and simulation tool, http://www.timesys.com/.

[TTC] Road vehicles — Controller area network (CAN) — Part 4: Time triggered comunication, ISO/CD 11898-4.

[TTT] Time Triggered Technologies, http://www.tttech.com.

[TTT99] TTTech Computertechnik AG, Specification of the TTP/C Protocol v0.5, July 1999.

[U.S02] U.S. Department of Commerce, The Economic Impacts of Inadequate Infrastructure for Software Testing, NIST report, May 2002.

[VC94] Venkatramani, C. and T. Chiueh, Supporting Real-Time Traffic on Ethernet, in Proceedings of 15th IEEE Real-Time Systems Symposium (RTSS'94), San Juan, Puerto Rico, Dec. 1994, IEEE Computer Society, Silver Spring, MD, 1994, pp. 282–286.

[Veca] Vector, DaVinci Tool Suite, http://www.vector-informatik.de/.

[Vecb] Vector, http://www.vector-informatik.com.

[vO02] van Ommering, R., The koala component model, *Building Reliable Component-Based Software Systems*, Artech House Publishers, Norwood, MA, July 2002, pp. 223–236.

[Vol] Volcano automotive group, http://www.volcanoautomotive.com.

[Win] Wind River Systems Inc., VxWorks Programmer's Guide, http://www.windriver.com/.

[Wor] WorldFIP, WorldFIP Fieldbus. http://www.worldfip.org.

[XP90] Xu, J. and D.L. Parnas, Scheduling processes with release times, deadlines, precedence, and exclusion relations, *IEEE Transactions on Software Engineering*, 16, 360–369, 1990.

[Zea] ZealCore, ZealCore Embedded Solutions AB, http://www.zealcore.com.

[ZHM97] Zhu, H., P. Hall, and J. May, Software unit test coverage and adequacy, *ACM Computing Surveys (CSUR)*, Vol. 29, Issue 4, pp. 366–427, Dec. 1997.

[ZN99] Zambonelli, F. and R. Netzer, An Efficient Logging Algorithm for Incremental Replay of Message-Passing Applications, in Proceedings of the 13th International and 10th Symposium on Parallel and Distributed Processing, April 1999, IEEE, New York, 1999, pp. 392–398.

[ZR86] Zhao, W. and K. Ramamritham, A Virtual Time CSMA/CD Protocol for Hard Real-Time Communication, in Proceedings of 7th IEEE Real-Time Systems Symposium (RTSS'86), New Orleans, LA, U.S.A., Dec. 1986, IEEE Computer Society, Silver Spring, MD, 1986, pp. 120–127.

[ZSR90] Zhao, W., J. A. Stankovic, and K. Ramamritham, A window protocol for transmission of time-constrained messages, *IEEE Transactions on Computers*, 39, 1186–1203, 1990.

82

Design of Embedded Systems

Luciano Lavagno
Politecnico di Torino
Cadence Berkeley Labs

Claudio Passerone
Politecnico di Torino

82.1 The Embedded System Revolution

The world of electronics has witnessed a dramatic growth in its applications in the last few decades. From telecommunications to entertainment, from automotive to banking, almost any aspect of our everyday life uses some kind of electronic components. In most cases, these components are computer-based systems, which are not, however, used or perceived as computers. For instance, they often do not have a keyboard or a display to interact with the user, and they do not run standard operating systems and applications. Sometimes, these systems constitute a self-contained product themselves (e.g., a mobile phone), but they are frequently embedded inside another system, for which they provide better functionalities and performance (e.g., the engine control unit of a motor vehicle). We call these computer-based systems *embedded systems*.

The huge success of embedded electronics has several causes. The main one in our opinion is that embedded systems bring the advantages of Moore's Law into everyday life, that is, an exponential increase in performance and functionality at an ever decreasing cost. This is possible because of the capabilities of integrated circuit technology and manufacturing, which allow one to build more and more complex devices, and because of the development of new design methodologies, which allow one to efficiently and cleverly use those devices. Traditional steel-based mechanical development, on the other hand, reached a plateau near the middle of the 20th century, and thus it is not a significant source of innovation any longer, unless coupled to electronic manufacturing technologies (MEMS) or embedded systems, as argued above.

There are many examples of embedded systems in the real world. For instance, a modern car contains tens of electronic components (control units, sensors, and actuators) that perform very different tasks. The first embedded systems that appeared in a car were related to the control of mechanical aspects, such as the control of the engine, the antilock brake system, and the control of suspension and transmission. However, nowadays, cars also have a number of components that are not directly related to mechanical aspects, but are mostly related to the use of the car as a vehicle for moving around, or the communication needs of the passengers: navigation systems, digital audio and video players, and phones are just a few examples. Moreover, many of these embedded systems are connected together using a network, because they need to share information regarding the state of the car.

Other examples come from the communication industry: a cellular phone is an embedded system whose environment is the mobile network. These are very sophisticated computers, whose main task is to send and receive voice, but are also currently used as personal digital assistants, for games, to send and receive images and multimedia messages, and to wirelessly browse the Internet. They have been so successful and pervasive that in just a decade they became essential in our life. Other kinds of embedded systems significantly changed our life as well: for instance, ATM and Point-of-Sale (POS) machines modified the way we make payments, and multimedia digital players changed how we listen to music and watch videos.

We are just at the beginning of a revolution that will have an impact on every other industrial sector. Special-purpose embedded systems will proliferate and will be found in almost any object that we use. They will be optimized for the application and show a natural user interface. They will be flexible, in order to adapt to a changing environment. Most of them will also be wireless, in order to follow us wherever we go and keep us constantly connected with the information we need and the people we care for. Even the role of computers will have to be reconsidered, as many of the applications for which they are used today will be performed by specially designed embedded systems.

What are the consequences of this revolution in the industry? Modern car manufacturers today need to acquire a significant amount of skills in hardware and software design, in addition to the mechanical skills that they already had in-house, or they should outsource the requirements they have to an external supplier. In either case, a broad variety of skills need to be mastered, from the design of software architectures for implementing the functionality, to being able to model the performance, because real-time aspects are extremely important in embedded systems, especially those related to safety-critical applications. Embedded system designers must also be able to architect and analyze the performance of networks, as well as validate the functionality that has been implemented over a particular architecture and the communication protocols that are used.

A similar revolution has taken place or is about to take place to other industrial and socio-economical areas as well, such as entertainment, tourism, education, agriculture, government, and so on. It is therefore clear that new, more efficient, and easy-to-use embedded electronics design methodologies need to be developed, in order to enable the industry to make use of the available technology.

82.2 Design of Embedded Systems

Embedded systems are informally defined as a collection of programmable parts surrounded by Application-Specific Integrated Circuits (ASICs) and other standard components (Application-Specific Standard Parts, ASSPs), which interact continuously with an environment through sensors and actuators. The collection can be physically a set of chips on a board, or a set of modules on an integrated circuit. Software is used for features and flexibility, while dedicated hardware is used for increased performance and reduced power consumption. An example of an architecture of an embedded system is shown in Figure 82.1. The main programmable components are microprocessors and DSPs, which implement the software partition of the system. One can view reconfigurable components, especially if they can be reconfigured at runtime, as programmable components in this respect. They exhibit area, cost, performance, and power characteristics that are intermediate between dedicated hardware and processors. Custom and programmable hardware components, on the other hand, implement application-specific blocks and peripherals. All components are connected through standard and/or dedicated buses and networks, and data are stored on a set of memories. Often, several smaller subsystems are networked together to control, for example, an entire car, or to constitute a cellular or wireless network.

We can identify a set of typical characteristics that are commonly found in embedded systems. For instance, they are usually not very flexible and are designed to always perform the same task: if you buy an engine control embedded system, you cannot use it to control the brakes of your car, or to play games. A PC, on the other hand, is much more flexible because it can perform several very different tasks. An embedded system is often part of a larger controlled system. Moreover, cost, reliability, and safety are

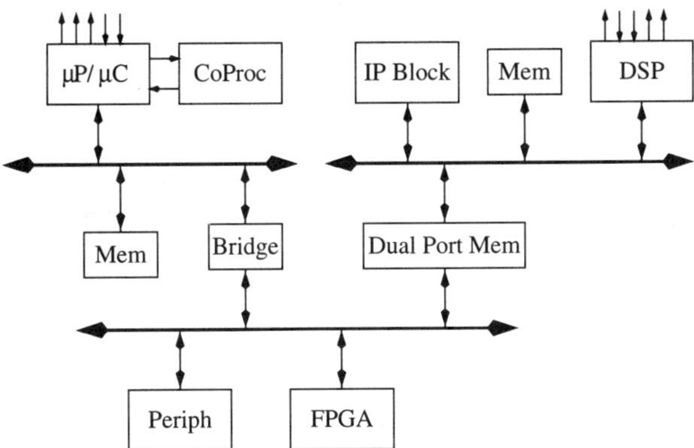

FIGURE 82.1 A reactive real-time embedded system architecture.

often more important criteria than performance, because the customer may not even be aware of the presence of the embedded system, and so he looks at other characteristics, such as the cost, the ease of use, or the lifetime of a product.

Another common characteristic of many embedded systems is that they need to be designed in an extremely short time to meet their time-to-market. Only a few months should elapse from conception of a consumer product to the first working prototypes. If these deadlines are not met, the result is a concurrent increase in design costs and decrease of the profits, because fewer items will be sold. Thus, delays in the design cycle may make a huge difference between a successful product and an unsuccessful one.

In the current state of the art, embedded systems are designed with an *ad hoc* approach that is heavily based on earlier experience with similar products and on manual design. Often, the design process requires several iterations to obtain convergence, because the system is not specified in a rigorous and unambiguous manner, and the level of abstraction, details, and design style in various parts are likely to be different. But as the complexity of embedded systems scales up, this approach is showing its limits, especially regarding design and verification time.

New methodologies are being developed to cope with the increased complexity and enhance designers' productivity. In the past, a sequence of two steps has always been used to reach this goal: *abstraction* and *clustering*. Abstraction means describing an object (i.e., a logic gate made of MOS transistors) using a model where some of the low-level details are ignored (i.e., the Boolean expression representing that logic gate). Clustering means connecting a set of models at the same level of abstraction, to obtain a new object, which usually shows new properties that are not part of the isolated models that constitute it. By successively applying these two steps, digital electronic design went from drawing layouts, to transistor schematics, to logic gate netlists, to register transfer level descriptions, as shown in Figure 82.2.

The notion of *platform* is key to the efficient use of abstraction and clustering. A platform is a single abstract model that hides the details of a set of different possible implementations as clusters of lower-level components. The platform, for example, a family of microprocessors, peripherals, and bus protocol, allows developers of designs at the higher level (generically called "applications" in the following) to operate without detailed knowledge of the implementation (e.g., the pipelining of the processor or the internal implementation of the UART). At the same time, it allows platform implementors to share design and fabrication costs among a broad range of potential users, broader than if each design was a one-of-a-kind type.

Today, we are witnessing the appearance of a new higher level of abstraction, as a response to the growing complexity of integrated circuits. Objects can be functional descriptions of complex behaviors, or architectural specifications of complete hardware platforms. They make use of formal high-level models that can be used to perform an early and fast validation of the final system implementation, although with reduced details with respect to a lower-level description.

The relationship between an "application" and elements of a platform is called a *mapping*. This exists, for example, between logic gates and geometric patterns of a layout, as well as between register transfer-level

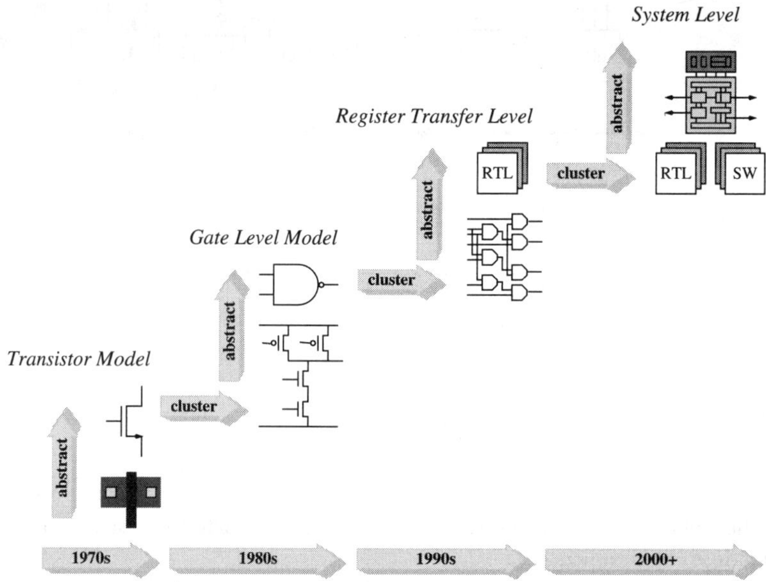

FIGURE 82.2   Abstraction and clustering levels in hardware design.

statements and gates. At the system level, the mapping is between functional objects with their communication links, and platform elements with their communication paths. Mapping at the system level means associating a functional behavior (e.g., an FFT or a filter) to an architectural element that can implement that behavior (e.g., a CPU or DSP or piece of dedicated hardware). It can also associate a communication link (e.g., an abstract FIFO) to some communication services available in the architecture (e.g., a driver, a bus, and some interfaces). The mapping step may also need to specify parameters for these associations (e.g., the priority of a software task or the size of a FIFO), in order to completely describe it. The object that we obtain after mapping shows properties that were not directly exposed in the separate descriptions, such as the performance of the selected system implementation. Performance is not just timing, but any other quantity that can be defined to characterize an embedded system, either physical (area, power consumption, etc.) or logical (quality of service, fault tolerance, etc.).

Since the system-level mapping operates on heterogeneous objects, it also allows one to elegantly separate different and orthogonal aspects such as:

1. Computation and communication. This separation is important because refinement of computation is generally done by hand, or by compilation and scheduling, while communication makes use of patterns.
2. Application and platform implementation (also called functionality and architecture, e.g., in [5]), because they are often defined and designed independently by different groups or companies.
3. Behavior and performance, which should be kept separate because performance information can either represent nonfunctional requirements (e.g., maximum response time of an embedded controller), or the result of an implementation choice (e.g., the worst-case execution time of a task). Nonfunctional constraint verification can be performed traditionally, by simulation and prototyping, or with static formal checks, such as schedulability analysis.

All these separations result in better *reuse,* because they decouple independent aspects, which would otherwise tie, for example, a given functional specification to low-level implementation details, by modeling it as assembler or Verilog code. This in turn allows one to reduce design time, by increasing the productivity and decreasing the time needed to verify the system.

A schematic representation of a methodology that can be derived from these abstraction and clustering steps is shown in Figure 82.3. At the functional level, a behavior for the system to be implemented is specified, designed, and analyzed, either through simulation or by proving that certain properties are satisfied (the algorithm always terminates, the computation performed satisfies a set of specifications, the complexity of the algorithm is polynomial, etc.). In parallel, a set of architectures are composed from a clustering of platform elements, and selected as candidates for the implementation of the behavior. These components may come from an existing library or may be specifications of components that will be designed later.

Now, functional operations are assigned to the various architecture components, and patterns provided by the architecture are selected for the defined communications. At this level, we are now able to verify the performance of the selected implementation, with much richer details than at the pure functional level. Different mappings to the same architecture, or mapping to different architectures, allow one to explore the design space to find the best solutions to important design challenges. These analyses allow the designer to identify and correct possible problems early in the design cycle, thus reducing drastically the time to explore the design space and weed out potentially catastrophic mistakes and bugs. At this stage, it is also very important to define the organization of the data storage units for the system. Various kinds of memories (e.g., ROM, SRAM, DRAM, Flash, etc.) have different performance and data persistency characteristics, and must be used judiciously to balance cost and performance. Mapping data structures to different memories, and even changing the organization and layout of arrays can have a dramatic impact on the satisfaction of a given latency in the execution of an algorithm, for example. In particular, a System-On-Chip designer can afford to do a very fine tuning of the number and sizes of embedded memories (especially SRAM, but now also Flash) to be connected to processors and dedicated hardware [11].

Finally, at the implementation level, the reverse transformation of abstraction and clustering occurs, that is, a lower-level specification of the embedded system is generated. This is obtained through a series of manual or automatic refinements and modifications that successively add more details, while checking their compliance with the higher-level requirements. This step does not need to generate directly a manufacturable final implementation, but rather produces a new description that in turn constitutes the input for another (recursive) application of the same overall methodology at a lower level of abstraction (e.g., synthesis, placement and routing for hardware, compilation, and linking for software). Moreover, the results obtained by these refinements can be back-annotated to the higher level, to perform a better and more accurate verification.

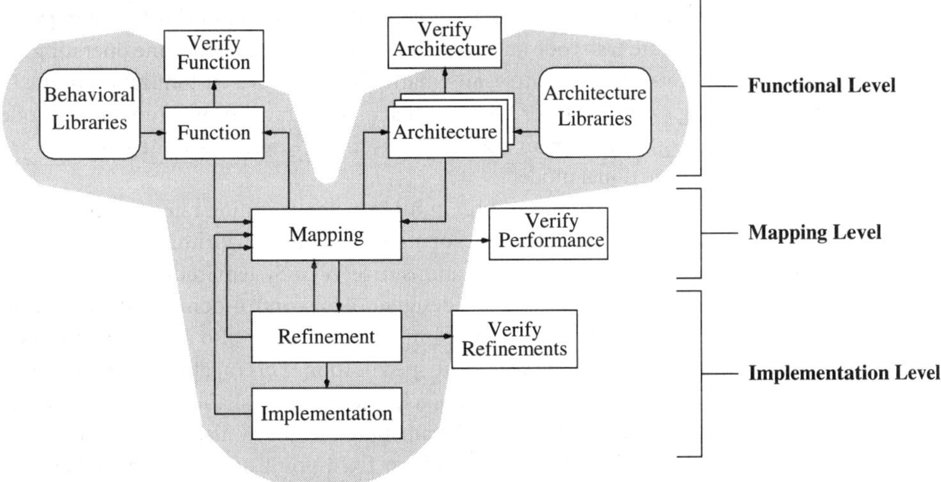

FIGURE 82.3   Design methodology for an embedded system.

82.3 Functional Design

As discussed in the previous section, system-level design of embedded electronics requires two distinct phases. In the first phase, functional and nonfunctional constraints are the key aspects. In the second phase, the available architectural platforms are taken into account, and detailed implementation can proceed after a mapping phase that defines the architectural component on which every functional model is implemented. This second phase requires a careful analysis of the trade-offs between algorithmic complexity, functional flexibility, and implementation costs.

In this section, we describe some of the tools that are used for requirements capture, focusing especially on those that permit *executable* specification. Such tools generally belong to two broad classes.

The first class is represented, for example, by Simulink [45], MATRIXx [33], Ascet-SD [2], SPW [54], SCADE [48], and SystemStudio [51]. It includes block-level editors and *libraries* using which the designer composes *data-dominated* digital signal processing and embedded control systems. The libraries include simple blocks, such as multiplication, addition, and multiplexing, as well as more complex ones, such as FIR filters, FFTs, and so on.

The second class is represented by tools such as Tau [52], StateMate [46], Esterel Studio [48], and StateFlow [45]. It is oriented to *control-dominated* embedded systems. In this case, the emphasis is placed on the decisions that must be taken by the embedded system in response to environment and user inputs, rather than on numerical computations. The notation is generally some form of Har'el's statecharts [24].

The Unified Modeling Language (UML), as standardized by the Object Management Group [53], is in a class by itself, since it focused historically more on general-purpose software (e.g., enterprise and commercial software), rather than on embedded real-time software. Only recently some embedded aspects such as performance and time have been incorporated in UML 2.0 [53,30] and emphasis has been placed on model-based software generation. However, tool support for UML 2.0 is still limited (Tau [52], Real Time Studio [47], and Rose RealTime [41] provide some), and UML-based hardware design is still in its infancy. Furthermore, the UML is a collection of notations, some of which (especially statecharts) are supported by several of the tools listed above in the control-dominated class.

Simulink and its related tools and toolboxes, both from Mathworks and from third parties such as dSPACE [16], are the workhorses of modern *model-based embedded system design*. In model-based design, a functional executable model is used for algorithm development. This is made easier in the case of Simulink by its tight integration with Matlab, the standard tool in DSP algorithm development. The same functional model, with added annotations such as bit widths and execution priorities, is then used for algorithmic refinements such as floating-point to fixed-point conversion and real-time task generation.

Then automated software generators such as Real-Time Workshop, Embedded Coder [45], and TargetLink [16] are used to generate task code and sometimes to customize a real-time operating system on which the tasks will run. Ascet-SD, for example, automatically generates a customization of the OSEK automotive real-time operating system [39] for the tasks that are generated from a functional model. In all these cases, a task is typically generated from a set of blocks that are executed at the same rate or triggered by the same event in the functional model.

Task formation algorithms can use either direct user input (e.g., the execution rate of each block in discrete time portions of a Simulink or Ascet-SD design) or static scheduling algorithms for dataflow models (e.g., based on relative block-to-block rate specifications in SPW or SystemStudio [31,7]).

Simulink is also tightly integrated with StateFlow, a design tool for control-dominated applications, in order to ease the integration of decision-making and computation code. It also allows one to smoothly generate both hardware and software from the very same specification. This capability, as well as the integration with some sort of statechart-based Finite State Machine editor, is available from most tools in the first class above. The difference in market share can be attributed to the availability of Simulink "toolboxes" for numerous embedded system design tasks (from fixed-point optimization to FPGA-based implementation) and its widespread adoption in undergraduate university courses, which makes it well known to most of today's engineers.

The second class of tools either plays an ancillary role in the design of embedded control systems (e.g., as StateFlow and EsterelStudio) or is devoted to inherently control-dominated application areas, such as telecommunication protocols. In the latter market, the clear dominator today is Tau. The underlying languages, SDL and Message Sequence Charts, are standardized by the International Telecommunication Union (ITU). They are commonly used to describe in a tool-independent way protocol standards; thus, modeling in SDL is quite natural in this application domain, since validation and refinement can proceed formally within a unified environment. Tau also has code generation capabilities for both application code and customization of real-time kernels on which the FSM-generated code will run. The use of Tau for embedded code generation (model-based design) significantly predates that of Simulink-based code generators, mostly due to the highly complex nature of telecom protocols and the less demanding memory and computing power constraints that switches and other networking equipment have.

Tau has links to the requirements capture tool Doors [52], also from Telelogic, which allows one to trace dependencies between multiple requirements written in English, and connect them to aspects of the embedded system design files that implement these requirements. The state of the art of such requirement tracing, however, is far from satisfactory, since there is no formal means in Doors to automatically check for violations. Similar capabilities are provided by Reqtify [42].

Techniques for automated functional constraint validation, starting from formal languages, are described in several books, for example, [35,29]. Deadline, latency, and throughput constraints are a special kind of nonfunctional requirements that have received extensive treatment in the Real-Time scheduling community. They are also covered in several books, for example, [9,22].

While model-based functional verification is quite attractive, due to its high abstraction level, it ignores cost and performance implications of algorithmic decisions. These are taken into account by the tools described in the next section.

82.4 Function/Architecture and Hardware/Software Codesign

In this section, we describe some of the tools that are available to help embedded system designers optimally architect the implementation of the system, and choose the best solution for each functional component. After these decisions have been made, detailed design can proceed using the languages, tools, and methods described in the following chapters in this book.

This step of the design process, whose general structure has been outlined in Section 82.2 by using the platform-based design paradigm, has received various names in the past. Early work [18,23] called it hardware/software codesign (or cosynthesis), because one of the key decisions at this level is what functionality has to be implemented in software and in dedicated hardware, and how the two partitions of the design interact together with minimum cost and maximum performance.

Later on, people came to realize that hardware/software was too coarse a granularity, and that more implementation choices had to be taken into account. For example, one could trade off single vs. multiple processors, general-purpose CPUs vs. specialized DSPs, and Application-Specific Instruction-set Processors (ASIPs), dedicated ASIC vs. ASSP (e.g., an MPEG coprocessor or an Ethernet Medium Access Controller), standard cells vs. FPGA. Thus, the term function/architecture codesign was coined [5], to refer to the more complex partitioning problem of a given functionality onto a heterogeneous architecture such as the one in Figure 82.1.

The term system-level design also had some popularity in the industry [54,36], to indicate "the level of design above Register Transfer, at which software and hardware interact." Other terms, such as timed functional model, have also been used [26].

The key problems that are tackled by tools acting as a bridge between the system-level application and the architectural platform are:

1. how to model the performance impact of making mapping decisions from a virtually "implementation-independent" functional specification to an architectural model, and

2. how to efficiently drive downstream code generation, synthesis, and validation tools to avoid redoing the modeling effort from scratch at the RTL, C, or assembly code levels, respectively. The notion of automated implementation generation from a high-level functional model is called "model-based design" in the software world.

In both cases, the notion of what is an "implementation-independent" functional specification, which can be retargeted indifferently to hardware and software implementations, must be carefully evaluated and considered. Taken in its most literal terms, this idea has often been taunted as a myth. However, current practice shows that it is already a reality, at least for some application domains (automotive electronics and telecommunication protocols). It is intuitively very appealing, since it can be considered as a high-level application of the platform-based design principle, by using a formal *system-level platform*. Such a platform, embodied in one of the several models of computation that are used in embedded system design, is a perfect candidate to maximize design reuse, and to optimally exploit different implementation options.

In particular, several of the tools that have been mentioned in the previous section (e.g., Simulink, TargetLink, StateFlow, SPW, System Studio, Tau, Ascet-SD, StateMate, Esterel Studio) have code generation capabilities that are considered good enough for *implementation* and not just for rapid prototyping and simulation acceleration. Moreover, several of them (e.g., Simulink, StateFlow, SPW, System Studio, StateMate, Esterel Studio) can generate indifferently C for software implementation, and synthesizable VHDL or Verilog for hardware implementation. Unfortunately, these code generation capabilities often require the laborious creation of implementation models for each target platform (e.g., software in C or assembler for a given DSP, synthesizable VHDL, or macro-block netlist for ASIC or FPGA, etc.). However, since these careful implementations are instances of the system-level platform mentioned above, their development cost can be shared among a multitude of designs performed using the tool.

Most block diagram-or statechart-based code generators work in a syntax-directed manner. A piece of C or synthesizable VHDL code is generated for each block and connection, or for each hierarchical state and transition. Thus, the designer has tight control over the complexity of the generated software or hardware. While this is a convenient means to bring manual optimization capabilities within the model-based design flow, it has a potentially significant disadvantage in terms of cost and performance (like disabling optimizations in the case of a C compiler). On the other hand, more recent tools like EsterelStudio and SystemStudio take a more radical approach to code generation, based on aggressive optimizations [6]. These optimizations, based on logic synthesis techniques also in the case of software implementation, destroy the original model structure, and thus make debugging and maintenance much harder. However, they can result in an order of magnitude improvement in terms of cost (memory size) and performance (execution speed) with respect to their syntax-directed counterparts [17].

Assuming that good automated code generation, or manual design, is available for each block in the functional model of the application, we are now facing the function-architecture codesign problem. This essentially means tuning the functional decomposition, as well as the algorithms used by the overall functional model and each block within it, to the available architecture, and vice versa.

Several design environments, for example:

- POLIS [5], COSYMA [18], Vulcan [23], COSMOS [28], and Roses [12] in the academic world, as well as
- VCC [54], Real Time Studio [47], Foresight [50], and CARDtools [10] in the commercial world,

help the designer in this task by somehow using the notion of independence between functional specification on the one hand, and hardware/software partitioning or architecture mapping choices on the other.

The step of performance evaluation is performed in an abstract, approximate manner by the tools listed above. Some of them use estimators to evaluate the cost and performance of mapping a functional block to an architectural block. Others (e.g., POLIS and VCC) rely on cycle-approximate simulation to perform the same task in a manner that better reflects real-life effects, such as burstiness of resource occupation and so on. Techniques for deriving both abstract static performance models (e.g., the Worst-Case Execution Time of a software task) and performance simulation models are discussed below.

In all cases, both the cost of computation and that of communication must be taken into account. This is because the best implementation, especially in the case of multimedia systems that manipulate large amounts of image and sound data, is often one that reduces the amount of transferred data between multiple memory locations, rather than one that finds the absolute best trade-off between software flexibility and hardware efficiency. In this area, the Atomium project at IMEC [11,3] has focused on finding the best memory architecture and schedule of memory transfers for data-dominated applications on mixed hardware/software platforms. By exploiting array access models based on polyhedra, they identify the best reorganization of inner loops of DSP kernels and the best embedded memory architecture. The goal is to reduce memory traffic due to register spills, and maximize the overall performance by accessing several memories in parallel (many DSPs offer this opportunity even in the embedded software domain). A very interesting aspect of Atomium, which distinguishes it from most other optimization tools for embedded systems, is the ability to return *a set of Pareto-optimal* solutions (i.e., solutions that are not strictly better than one another in at least one aspect of the cost function), rather than a single solution. This allows the designer to pick the best point based on the various aspects of cost and performance (e.g., silicon area vs. power and performance), rather than forcing him to "abstract" optimality into a single number.

Performance analysis can be based on simulation, as mentioned above, or rely on automatically constructed models that reflect the Worst-Case Execution Time (WCET) of pieces of software (e.g., RTOS tasks) running on an embedded processor. Such models, which must be both provably conservative and reasonably accurate, can be constructed using an execution model called *abstract interpretation* [15]. This technique traverses the software code, while building a symbolic model, often in the form of linear inequalities [32,1], which represents the requests that the software makes to the underlying hardware (e.g., code fetches, data loads and stores, code execution). A solution to those inequalities then represents the total "cost" of one execution of the given task. It can then be combined with processor, bus, cache, and main memory models that in turn compute the cost of each of these requests in terms of time (clock cycles) or energy. This finally results in a complete model for the cost of mapping that task to those architectural resources.

Another technique for software performance analysis, which does not require detailed models of the hardware, uses an approximate compilation step from the functional model to an executable model (rather than a set of inequalities as above) annotated with the same set of fetch, load, store, and execute requests. Then, simulation is used, in a more traditional setting, to analyze the cost of implementing that functionality on a given processor, bus, cache, and memory configuration. Simulation is more effective than WCET analysis in handling multiprocessor implementations, in which bus conflicts and cache pollution can be difficult, if not utterly impossible, to predict statically in a manner that is not too conservative. However, its success in identifying the true worst-case depends on the designer ability to provide the appropriate simulation scenarios. Coverage enhancement techniques from the hardware verification world [4,27] can be extended to also help in this case.

Similar abstract models can be constructed in the case of implementation as dedicated hardware, by using high-level synthesis techniques. Such techniques are not yet good enough to generate production-quality RTL code, but can be considered as a reasonable estimator of area, timing, and energy costs for both ASIC and FPGA implementations [49,55,14,51].

SystemC [26] and SpecC [19,20], on the other hand, are more traditional modeling and simulation languages, for which the design flow is based on successive refinement rather than codesign or mapping. Finally, OPNET [38] and NS [37] are simulators with a rich modeling library specialized for wireline and wireless networking applications. They help the designer in the more abstract task of generic performance analysis, without the notion of function/architecture separation and codesign.

Communication performance analysis, on the other hand, is generally not done using approximate compilation or WCET analysis techniques like those outlined above. Communication is generally implemented not by synthesis but by *refinement* using patterns and "recipes," such as interrupt-based, DMA-based, and so on. Thus, several design environments and languages at the function/architecture level, such as POLIS, COSMOS, Roses, VCC, SystemC and SpecC, as well as N2C [36], provide mechanisms to replace abstract communication, for example, FIFO-based or discrete event-based, with detailed protocol stacks using buses, interrupt controllers, memories, drivers, and so on. These refinements can then be estimated either using a library-based approach (they are generally part of a library of implementation choices anyway), or sometimes using the approaches described above for computation. Their cost and performance can thus be combined in an overall system-level performance analysis.

However, approximate performance analysis is often not good enough, and a more detailed simulation step is required. This can be achieved by using tools such as Seamless [44], CoMET [13], MaxSim [34], and N2C [36]. They work at a lower abstraction level, by cosimulating software running on Instruction Set Simulators and hardware running in a Verilog or VHDL simulator. While the simulation is often slower than with more abstract models, and dramatically slower than with static estimators, the precision can now be at the cycle level. Thus, it permits close investigation of detailed communication aspects, such as interrupt handling and cache behavior. These approaches are further discussed in the next section.

The key advantage of using the mapping-based approach over the traditional design–evaluate–redesign one is the speed with which design space exploration can be performed. This is done by setting up experiments that change either mapping choices or parameters of the architecture (e.g., cache size, processor speed, or bus bandwidth). Key decisions, such as the number of processors and the organization of the bus hierarchy, can thus be based on quantitative application-dependent data, rather than on past experience.

If mapping can then be used to drive synthesis, in addition to simulation and formal verification, advantages in terms of time-to-market and reduction of design effort are even more significant. Model-based code generation, as we mentioned in the previous section, is reasonably mature, especially for embedded software in application areas, such as avionics, automotive electronics, and telecommunications. In these areas, considerations other than absolute minimum memory footprint and execution time, for example, safety, sheer complexity, and time-to-market, dominate the design criteria.

At the very least, if some form of automated model-based synthesis is available, it can be used to rapidly generate FPGA - and processor-based prototypes of the embedded system. This significantly speeds up verification, with respect to workstation-based simulation. It permits even some hardware-in-the-loop validation for cases (e.g., the notion of "driveability" of a car) in which no formalization or simulation is possible, but a real physical experiment is required.

82.5   Hardware/Software Coverification and Hardware Simulation

Traditionally, the term "hardware/software codesign" has been identified with the ability to execute a simulation of the hardware and software at the same time. We prefer to use the term "hardware/software coverification" for this task, and leave codesign for the synthesis- and mapping-oriented approaches outlined in the previous section. In the form of simultaneously running an Instruction Set Simulator (ISS) and a Hardware Description Language (HDL) simulator, while keeping the timing of the two synchronized, the area is not new [43]. In recent years, however, we have seen a number of approaches to speeding up the task, in order to tackle platforms with several processors, and the need, for example, to boot an operating system in order to coverify a platform with a processor and its peripherals.

Recent techniques have been devoted to the three main ways in which cosimulation speed can be increased:

1. *Accelerate the hardware simulator.* Coverification generally works at the "clock cycle accurate" level, meaning that both the hardware simulator and the ISS view time as a sequence of discrete clock cycles, ignoring finer aspects of timing (sometimes clock phases are considered, e.g., for DSP systems, in which different memory banks are accessed in different phases of the same cycle). This allows one to speed up simulation with respect to traditional event-driven logic simulation, and yet retain enough precision to identify, for example, bottlenecks such as interrupt service latency or bus arbitration overhead.

   Native-code hardware simulation (e.g., NCSim[54]) and emulation (e.g., Quick-Turn [54] and Mentor Emulation [44]) can be used to further speed up hardware simulation, at the expense of longer compilation times and much higher costs, respectively.

2. *Accelerate the instruction set simulator.* Compiled-code simulation has been a popular topic in this area as well [56]. The technique compiles a piece of assembler or C code for a target processor into

object code that can be run on a host workstation. This code generally also contains annotations counting clock cycles by modeling the processor pipeline. The speedup that can be achieved with this technique over a traditional ISS, which fetches, decodes, and executes each target instruction individually, is significant (at least one order of magnitude). Unfortunately, this technique is not suitable for self-modifying code, such as that of a Real-Time Operating System (RTOS). This means that it is difficult to adapt to modern embedded software, which almost invariably runs under RTOS control, rather than on the bare CPU. However, hybrid techniques involving partial compilation on the fly are reportedly used by companies selling fast ISSs [13,34].

3. *Accelerate the interface between the two simulators.* This is the area where the earliest work has been performed. For example, Seamless [44] uses sophisticated filters to avoid sending requests for memory accesses over the CPU bus. This allows the bus to be used only for peripheral access, while memory data are provided to the processor directly by a "memory server," which is a simulation filter sitting in between the ISS and the HDL simulator. The filter reduces stimulation of the HDL simulator, and thus can result in speedups of one or more orders of magnitude, when most of the bus traffic consists of filtered memory accesses. Of course, also, precision of analysis drops, since, for example, it becomes harder to identify an overload in the processor bus due to a combination of memory and peripheral accesses, since no simulator component sees both.

In the HDL domain, as mentioned above, progress in the levels of performance has been achieved essentially by raising the level of abstraction. A "cycle-based" simulator, that is, one that ignores the timing information within a clock cycle, can be dramatically faster than one that requires the use of a timing queue to manage time-tagged events. This is due to mainly two reasons. The first one is that now most of the simulation can be executed always, at every simulation clock cycle. This means that it is much more parallelizable, while event-driven simulators do not fit well over a parallel machine due to the presence of the centralized timing queue. Of course, there is a penalty if most of the hardware is generally idle, since it has to be evaluated anyway, but clock gating techniques developed for low power consumption can obviously be applied here. The second one is that the overhead of managing the time queue, which often accounts for 50–90% of the event-driven simulation time, can now be completely eliminated.

Modern HDLs either are totally cycle-based (e.g., SystemC 1.0 [26]) or have a "synthesizable subset," which is fully synchronous and thus fully compilable to cycle-based simulation. The same synthesizable subset, by the way, is also supported by hardware emulation techniques, for obvious reasons.

Another interesting area of cosimulation in embedded system design is analog–digital cosimulation. This is because such systems quite often include analog components (amplifiers, filters, A/D and D/A converters, demodulators, oscillators, PLLs, etc.), and models of the environment quite often involve only continuous variables (distance, time, voltage, etc.). Simulink includes a component for simulating continuous-time models, using a variety of numerical integration methods, which can be freely mixed with discrete-time sampled-data subsystems. This is very useful when modeling and simulating, for example, a control algorithm for automotive electronics, in which the engine dynamics are modeled with differential equations, while the controller is described as a set of blocks implementing a sampled-time subsystem.

Simulink is still mostly used to drive software design, despite good toolkits implementing it in reconfigurable hardware [21,8]. Simulators in the hardware design domain, on the other hand, generally use HDLs as their input languages. Analog extensions of both VHDL [25] and Verilog [40] are available. In both cases, one can represent quantities that satisfy either of Kirchhoff's laws (i.e., conserved over cycles or nodes). Thus, one can easily build netlists of analog components interfacing with the digital portion, modeled using traditional Boolean or multivalued signals. The simulation environment will then take care of synchronizing the event-driven portion and the continuous time portion. A key problem here is to avoid causality errors, when an event that occurs later in "host workstation" time (because the simulator takes care of it later) has an effect on events that preceded it in "simulated time." In this case, one of the simulators has to "roll back" in time, undoing any potential changes in the state of the simulation, and restart with the new information that something has happened in the past (generally the analog simulator does it, since it is easier to reverse time in that case).

Also, in this case, as we have seen for hardware/software cosimulation, execution is much slower than in the pure event-driven or cycle-based case, due to the need to take small simulation steps in the analog part. There is only one case in which the performance of the interface between the two domains or of the continuous time simulator is not problematic. This occurs when the continuous time part is much slower in reality than the digital part. A classical example is automotive electronics, in which mechanical time constants are larger by several orders of magnitude than the clock period of a modern integrated circuit. Thus, the performance of continuous time electronics and mechanical cosimulation may not be the bottleneck, except in the case of extremely complex environment models with huge systems of differential equations (e.g., accurate combustion engine models). In that case, hardware emulation of the differential equation solver is the only option (see, e.g., [16]).

82.6 Conclusions

This chapter discussed several aspects of embedded system design, including both methodologies that allow one to perform judicious algorithmic and architectural decisions, and tools supporting various steps of these methodologies. One must not forget, however, that often embedded systems are complex compositions of parts that have been implemented by various parties, and thus the task of physical board or chip integration can be as difficult as, and much more expensive than, the initial architectural decisions.

In order to support the integration and system testing tasks, one must use formal models throughout the design process, and if possible perform early evaluation of the difficulties of integration, by virtual integration and rapid prototyping techniques. These allow one to find or avoid completely subtle bugs and inconsistencies earlier in the design cycle, and thus reduce overall design time and cost.

Thus, the flow and tools that we described in this chapter help not only with the initial design but also with the final integration. This is because they are based on executable specifications of the whole system (including models of its environment), early virtual integration, and systematic (often automated) refinement toward implementation.

References

1. AbsInt Worst-Case Execution Time Analyzers, http://www.absint.com.
2. ETAS Ascet-SD, http://www.etas.de.
3. IMEC ATOMIUM, http://www.imec.be/design/atomium/.
4. 0-In Design Automation, http://www.0-in.com/.
5. Balarin, F., E. Sentovich, M. Chiodo, P. Giusto, H. Hs ieh, B. Tabbara, A. Jurecska, L. Lavagno, C. Passerone, K. Suzuki, and A. Sangiovanni-Vincentelli, *Hardware–Software Co-design of Embedded Systems — The POLIS approach*, Kluwer Academic Publishers, Dordrecht, 1997.
6. Berry, G., The foundations of esterel, *in Proof, Language and Interaction: Essays in Honour of Robin Milner*, G. Plotkin, C. Stirling, M. Tofte, Eds., MIT Press, Cambridge, MA, 2000.
7. Buck, J. and R. Vaidyanathan, Heterogeneous Modeling and Simulation of Embedded Systems in El Greco, in *Proceedings of the International Conference on Hardware Software Codesign*, May 2000.
8. Altera DSP Builder, http://www.altera.com.
9. Buttazzo, G., *Hard Real-Time Computing Systems: Predictable Scheduling Algorithms and Applications*, Kluwer Academic Publishers, Dordrecht, 1997.
10. CARDtools, http://www.cardtools.com.
11. Catthoor, F., S. Wuytack, E. De Greef, F. Balasa, L. Nachtergaele, and A. Vandecapelle, *Custom Memory Management Methodology: Exploration of Memory Organisation for Embedded Multimedia System Design*, Kluwer Academic Publishers, Dordrecht, 1998.
12. Cesario, W., A. Baghdadi, L. Gauthier, D. Lyonnard, G. Nicolescu, Y. Paviot, S. Yoo, A.A. Jerraya, and M. Diaz-Nava, Component-Based Design Approach for Multicore Socs, in *Proceedings of the Design Automation Conference*, June 2002.
13. VAST Systems CoMET, http://www.vastsystems.com/.

14. Get2Chip Architectural Compiler, http://www.get2chip.com.
15. Cousot, P., and R. Cousot, Abstract Interpretation: a Unified Lattice Model for Static Analysis of Programs by Construction of Approximation of Fixpoints, in *Proceedings of the ACM Symposium on Principles of Programming Languages*, ACM Press, New York, 1977.
16. dSPACE TargetLink and Prototyper, http://www.dspace.de.
17. S.A. Edwards, Compiling Esterel into Sequential Code, in *International Workshop on Hardware/Software Codesign*, ACM Press, New York, May 1999.
18. Ernst, R., J. Henkel, and T. Benner, Hardware–software codesign for micro-controllers, *IEEE Design and Test of Computers*, 10, 64–75, 1993.
19. Gajski, D., J. Zhu, and R. Domer, *The SpecC Language*, Kluwer Academic Publishers, Dordrecht, 1997.
20. Gajski, D., J. Zhu, R. Domer, A. Gerstlauer, and S. Zhao, *SpecC: Specification Language and Methodology*, Kluwer Academic Publisher, Dordrecht, 2000.
21. Xilinx System Generator, http://www.xilinx.com.
22. Gomaa, H., *Software Design Methods for Concurrent and Real-Time Systems*, Addison-Wesley Publishing Company, Reading, MA,1993.
23. Gupta, R.K. and G. De Micheli, Hardware–software cosynthesis for digital systems, *IEEE Design and Test of Computers*, 10, 29–41, 1993.
24. Har'el D., H. Lachover, A. Naamad, A. Pnueli, M. Politi, R. Sherman, A. Shtull-Trauring, and M.B. Trakhtenbrot, STATEMATE: a working environment for the development of complex reactive systems, *IEEE Transactions on Software Engineering*, 16, pp. 403–414, 1990.
25. IEEE. Standard 1076.1, vhdl-ams, http://www.eda.org/vhdl-ams.
26. Open SystemC Initiative, http://www.systemc.org.
27. Norris Ip, C., Simulation Coverage Enhancement Using Test Stimulus Transformation, in Proceedings of the International Conference on Computer Aided Design, Nov. 2000.
28. Ismail, T.B., M. Abid, and A.A. Jerraya, COSMOS: a codesign approach for communicating systems, in International Workshop on Hardware/Software Codesign, ACM Press, New York, 1994.
29. Kurshan, R.P., *Automata-Theoretic Verification of Coordinating Processes*, Princeton University Press, Princeton, NJ,1994.
30. Lavagno, L., G. Martin, and B. Selic, Eds., UML for Real: Design of Embedded Real-Time Systems, Kluwer Academic Publishers, Dordrecht, 2003.
31. Lee, E.A. and D.G. Messerschmitt, Synchronous Data Flow, IEEE Proceedings, Sept. 1987.
32. Li, Y.T.S. and S. Malik, Performance Analysis of Embedded Software Using Implicit Path Enumeration, in Proceedings of the Design Automation Conference, June 1995.
33. National Instruments MATRIXx, http://www.ni.com/matrixx/.
34. Axys Design Automation MaxSim and MaxCore, http://www.axysdesign.com/.
35. McMillan K. *Symbolic Model Checking*, Kluwer Academic Publishers, Dordrecht, 1993.
36. CoWare N2C and LISATek, http://www.coware.com.
37. Network Simulator NS-2, http://www.isi.edu/nsnam/ns/.
38. OPNET, http://www.opnet.com.
39. OSEK/VDX, http://www.osek-vdx.org/.
40. OVI. Verilog-a standard, http://www.ovi.org.
41. IBM Rational Rose RealTime, http://www.rational.com/products/rosert/.
42. TNI Valiosys Reqtify, http://www.tni-valiosys.com.
43. Rowson, J., Hardware/Software Co-simulation, in Proceedings of the Design Automation Conference, 1994, pp. 439–440.
44. Mentor Graphics Seamless and Emulation, http://www.mentor.com.
45. The Mathworks Simulink and StateFlow, http://www.mathworks.com.
46. I-Logix Statemate and Rhapsody, http://www.ilogix.com.
47. Artisan Software Real Time Studio, http://www.artisansw.com/.
48. Esterel Technologies Esterel Studio, http://www.esterel-technologies.com.
49. Celoxica DK Design suite, http://www.celoxica.com.
50. Foresight Systems, http://www.foresight-systems.com.
51. Synopsys SystemStudio and Behavioral Compiler, http://www.synopsys.com.
52. Telelogic Tau and Doors, http://www.telelogic.com.

53. The Object Management Group UML, http://www.omg.org/uml/.
54. Cadence Design Systems Cierto VCC, SPW, and Quickturn, http://www.cadence.com.
55. Wakabayashi K., Cyber: High level synthesis system from software into ASIC, in *High Level VLSI Synthesis*, R. Camposano and W. Wolf, Eds., Kluwer Academic Publishers, Dordrecht, 1991.
56. Zivojnovic, V. and H. Meyr, Compiled HW/SW Co-simulation, *in Proceedings of the Design Automation Conference*, 1996.

83

Models of Computation for Embedded Systems

Luís Gomes
Universidade Nova de Lisboa
UNINOVA, Centro de Robótica
Inteligente

João Paulo Barros
UNINOVA, Centro de Robótica
Inteligente
Instituto Politécnico de Beja

83.1 Introduction

Embedded systems complexity has been continuously increasing and the embedded system requirements have been more and more demanding, trying to cope with different aspects, ranging from complexity management, heterogeneous platforms, design correctness, to costs, power consumption, time-to-market, and other issues.

The specification process is a key aspect in embedded system design. At that stage, a model of the system has to be built and it is described using some specific formalism. Formalisms should be selected to unambiguously describe design specifications. These formalisms are also called *models of computation*. A model of computation (MoC) is composed by a notation and by the rules for computation of the behavior. Instead of notation, we talk about the *syntax* of the model; the rules define the model *semantics*.

Different kinds of graphs are often used as an abstract syntax for MoC representation. Examples of graph-based MoC include Finite State Machines (with different "flavors" and expressiveness capabilities), Data flow, Petri nets, and networks of processes. With respect to semantics associated with a specific MoC, different notions of time and modes of concurrency and communication, within the model or among components, are commonly considered.

Depending on the selected MoC to represent system behavior, it will be possible to perform propriety verification through specific analysis methods.

There are many models of computation ready to be used by embedded system designers, and most of them use a graphical representation. The first task is to choose the right one either to be used for a specific system, or for describing subbehaviors of the system. This is a difficult task, as different embedded systems can emphasize different aspects. For instance, some models of computation for embedded systems are control-dominated (data processing and computation are minimal), while others emphasize data processing, containing complex data transformations, normally described by data flows. For example, in the first group, we found *reactive control systems*, and in the second one we found *digital signal processing applications*. As examples, we may say that for digital signal processing applications, it is common to select data flow models; on the other hand, finite state machine are usually used for control-intensive applications. The selection of the right formalism should match the specific system characteristics. Some MoCs were specifically defined for describing complex data processing functions, but completely inadequate for modeling complex control; others have complex control in mind, but lack good support for data transfer functions. Also, the need to handle different notions of time, or different modes of communication among components, may further complicate the search for the right model of computation.

The goal is to pick up the right formalism for the different subbehaviors specifications, and, afterwards, to be able to integrate all those models in a coherent way [22]. Several MoCs allow the modeler to describe the system as a collection of communicating modules. In this sense, modeling of behavior and communication among components often presents interdependencies. Yet, separating behavior and communication is a sound attitude as it allows handling system design complexity and reusability of components.

These aspects of time representation and communication support are addressed in the following sections. Later, several selected models of computation are presented.

83.2   Notions of Time

The notion of time is of major importance for many models of computation for embedded systems. Embedded systems are mostly regarded as real-time systems, where the time at which a computation takes place is typically critical.

Roughly speaking, we may identify three groups of attitudes:

- continuous time and differential equations;
- discrete time and difference equations; and
- discrete events.

The first group (see Figure 83.1(a)) uses continuous time functions, which are then expressed by differential equations. These are mostly used in embedded system design for modeling of some interface components. Application areas include modeling of analog circuits and physical systems in a broad sense.

In the second group (Figure 83.1(b)), time is made discrete, and differential equations become difference equations. A global clock (the tick) defines the specific points in time at which the signals have values. For some applications, involving heterogeneous components, it is useful to consider also multirate difference equations. Application areas include digital signal processing.

Finally, in the third group (see Figure 83.1(c)), a signal is seen as a sequence of events. Each event has associated value and time tags. The component will process events in a chronological order, based on a predefined precedence between events.

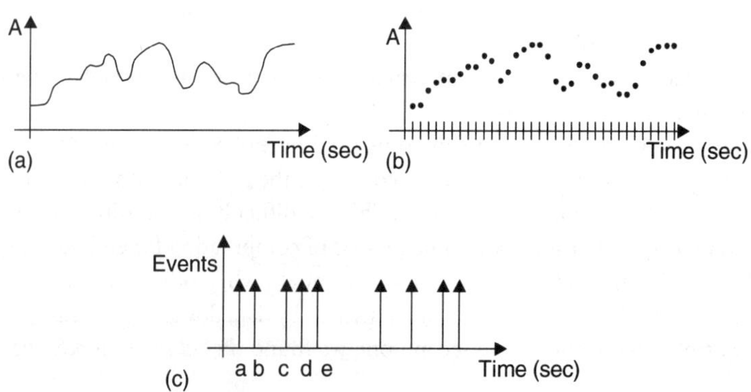

FIGURE 83.1   Time representations.

If the time tags are totally ordered [13], we are in the presence of a *timed system*. This means that, for any distinct t_1 and t_2, either $t_1 < t_2$ or $t_2 < t_1$ (this is called a *total order*). It is possible to define an associated metric, for instance, $f(t_1, t_2) = |t_1 - t_2|$.

A *continuous time system* is a metric time system where the metric is a continuum. A *discrete-event system* is a timed system where the time tags are order-isomorphic to a subset of the integers; hence, they are totally ordered.

Two events are synchronous if they have the same time tag attached (they occur simultaneously). Similarly, two signals are synchronous if for each event in one signal, there is a synchronous event in the other signal, and vice versa. A system is synchronous if every signal in the system is synchronous with every other signal in the system. In this sense, a discrete-time system is a synchronous discrete-event system.

Totally ordered events are used with digital hardware simulators, namely the ones associated with VHDL and Verilog hardware description languages.

Any two events are either simultaneous, which means that they have the same time tag, or any one of them can precede the other. Events can be considered partially ordered in the sense that the order does not include all the events in the system. When tags are partially ordered instead of totally ordered, the system is *untimed*. This means we can build several event sequences that do not contain all the system events. These missing events are included in other completely ordered event sequences. It is known [26] that totally order of the events cannot be maintained in distributed systems, where a partial order is sufficient to analyze system behavior. Partial orders have also been used to analyze Petri nets [37].

An asynchronous system is a system in which no two events can have the same tag [27]. If tags are totally ordered, the system is *asynchronous interleaved*, while if tags are partially ordered, the system is *asynchronous concurrent*.

As time is intrinsically continuous, real systems are asynchronous by nature. Yet, synchronicity is a very convenient abstraction, allowing efficient and robust implementations, through the use of a reference "clock" signal.

83.3 Communication Support

Due to complexity reasons, or reusability of already available components, it is common to decompose an embedded system into several components, which will interact with one another. Communication between processes is, in this sense, a key issue for embedded system design. It can be implicit or explicit [36]:

- Implicit communication generally requires totally ordered tag events, normally associated with physical time. In order to support this form of communication, it is necessary to have a physically shared signal (for instance, a clock signal), whose availability may be difficult or unfeasible in a large number of embedded system applications.
- Explicit communication imposes an order on the events, in the sense that the sender process will guarantee that the receiver process(es) is(are) informed about some part of its internal state.

Several modes of communication have been described in the literature, namely:

- *Handshake*, using a synchronization mechanism, where all parties are blocked, waiting for the conclusion;
- *Message passing*, using a send–receive scheme, where the receiver will wait for the message;
- *Shared variables*, where the blocking is decided by the control part of the memory where the shared variable is stored.

The referred communication modes are supported by a set of communication primitives (or by a combination of some of them), namely [36]:

Unsynchronized: producer and consumer(s) are not synchronized. There are no guarantees that the producer does not overwrite previously produced data, or that the consumer(s) will receive all produced data.

Read–modify–write: this is the common way to gain access to shared data structures from different processes in software. Access to the data structure is locked during a data access (either read–write or read–modify–write), which means that it is an atomic action (indivisible, and thus uninterruptible).

Unbounded FIFO buffered: the producer generates a sequence of data tokens and the consumer will read those tokens using a FIFO discipline.

Bounded FIFO buffered: also in this case, considering the simplest case of one producer and one consumer, the producer generates a sequence of data tokens and the consumer will also read those tokens using a FIFO discipline. Yet, here the buffer size is limited, so the difference between writings and readings will be bounded by some value. This means that writings can be blocked if the buffer is full.

Petri net places: producers generate sequences of data tokens and consumers will read those tokens.

Rendezvous: the writing and reading processes must simultaneously be at the point where the write and the read occur.

83.4 Common Models of Computation

In this section, we will present and compare some of the main MoCs that have been used for embedded system's design.

Dataflow Models

Dataflow models are graphs where nodes represent operations (also called *actors*) and arcs represent data paths (also called *channels*). These data paths contain totally ordered sequences of data (also called *tokens*). A Dataflow is a distributed MoC, in the sense that there is no single locus of control. Yet, the conventional dataflow model is not suitable for representing the control part of the system (only what is implied by the graph structure). With respect to time handling, dataflow systems are asynchronous concurrent.

Dataflow graphs have been extensively used in modeling data-dominated systems, namely digital signal processing applications, and applications dealing with streams of periodic/regular data samples. Computationally intensive systems, carrying complex data transformation, can be conveniently represented by a directed graph where the nodes represent computation and the arcs represent the order in which the computations are performed. Typically, signal-processing algorithms (encode/decode, filtering, convolution, compression, etc.) are expressed as block diagrams and coded in a specific programming language, like C, whose computation blocks match exactly the dataflow semantics.

Several dataflow graph-based MoCs have been referred in the literature, each one with its specific semantics, namely:

- Kahn Process Network [18];
- Dataflow Process Networks [11];
- Synchronous DataFlow Graphs [14].

In a Kahn Process Networks, and also in Dataflow Process Networks (which constitute a special case of Kahn Process Network), processes communicate by passing data tokens through unidirectional unbounded FIFO channels. A process can fire according to a set of firing rules. During each atomic firing, the actor consumes tokens from input channels, executes some behavior using a selected programming language, and produces tokens that are placed on output channels. Writing to the channel is nonblocking, although reading from the channel blocks the process until there is sufficient data in it. Kahn Process Networks have guaranteed deterministic behavior, in the sense that for a certain sequence of inputs, there is only one possible sequence of outputs. This is independent from the computation and communication durations, and also from the actors' firing order. The dataflow process analysis can be based only on graph inspection. Also, implementation can be very efficient, due to the freedom of choosing the actors' firing order. In fact, one common goal in many signal-processing applications is to schedule the actor firings, at compile time. This results in an interleaved implementation for the concurrent model, represented by a list of processes to be executed, allowing its implementation in single-processor architecture. Accordingly, for multiprocessor architectures, a list per processor is obtained. Kahn Process Networks cannot be schedule statically, as their firing rules do not allow us to build, at compile time, a firing sequence such that the system does not block under any circumstances. For this kind of MoC, one must use dynamic scheduling (with associated implementation overhead). Another disadvantage of Kahn Process Networks is associated with unbounded FIFO buffers, and the potentially growth of memory needs. In this sense, and without considering some limitations, Kahn Process Networks cannot be efficiently implemented.

Synchronous dataflow graphs are a kind of Kahn process networks with additional restrictions, namely:

- At each firing, a process consumes and produces a fixed amount of tokens on its incoming and outgoing channels, respectively.
- For a process to fire, it must have at least as many tokens on its input channels as it has to consume.
- Arcs are marked with the number of tokens produced or consumed.

Figure 83.2 shows a simple dataflow model and associated static scheduling list. This scheduling is a list of firings that can be repeated indefinitely. One cycle through the schedule should return the graph to its original state (by state, we mean the number of tokens in every arc).

This static schedule can be determined through the analysis of the graph in order to find the paths of firing sequences that satisfy, on each arc, the same amount of tokens to be produced and consumed. One balance equation is built per arc, considering that

$$\text{outgoing\_weight} * \text{origin\_node} = \text{incoming\_weight} * \text{destination node}$$

For the referred example, the following balance equations can be obtained (one per arc):

$$2a - 4b = 0$$

$$b - 2c = 0$$

$$2c - d = 0$$

$$2b - 2d = 0$$

$$2d - a = 0$$

The root of the set of presented equations is ($a = 4$, $b = 2$, $c = 1$, $d = 2$), which gives information about the firing vector associated with the static scheduling list. Yet, in other cases, it is possible that the mentioned balance equations do not have a root, implying that no static schedule is possible.

Many possible variants of data flow models have been explored. Specially interesting, among them, is an extension to Synchronous dataflow networks allowing *dynamic dataflow actors* [10]; dynamic actors are actors with at least one conditional input or output port. The canonical dynamic actors are "switch" and "selector," enabling the test of tokens on one specific incoming arc, and the consumption and production of tokens in a data-dependent way. A "Switch" process, for example, has one regular incoming arc, one control incoming

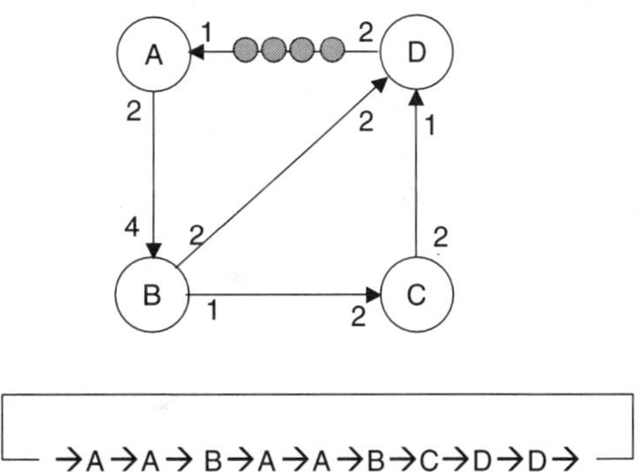

FIGURE 83.2 Simple dataflow model and associated static schedule list.

arc, and two outgoing arcs; whenever tokens are presented in the incoming arcs, the value carried by the token presented in the control arc will dynamically select the active outgoing arc: the process firing will produce one single token at the selected outgoing arc. Figure 83.3 illustrates dynamic dataflow actor behavior before and after firing. It has to be noted that control ports are never conditional ports. More complex dynamic actors can be used, namely dynamic actors with integer control (instead of Boolean control), giving rise to CASE and ENDCASE actors (generalizations of the SWITCH and SELECT actors, respectively).

Aggregation of the referred actors using specific patterns is commonly applied. For instance, integer-controlled dataflow graphs can result from cascading a CASE actor, a set of mutually exclusive processing nodes, and an ENDCASE actor; control ports of the CASE and ENDCASE actors receive the same value; only one of the processing nodes is executed at a time depending on the value presented at the control port.

Most simulation environments targeted for signal processing, like Matlab-Simulink or Khoros (for image processing), use dataflow models.

Discrete Event

In the discrete-event MoC, events carry a totally ordered time stamp, referring to the time at which the event occurs. The events are sorted by time stamp and they are analyzed in chronological order. A discrete-event

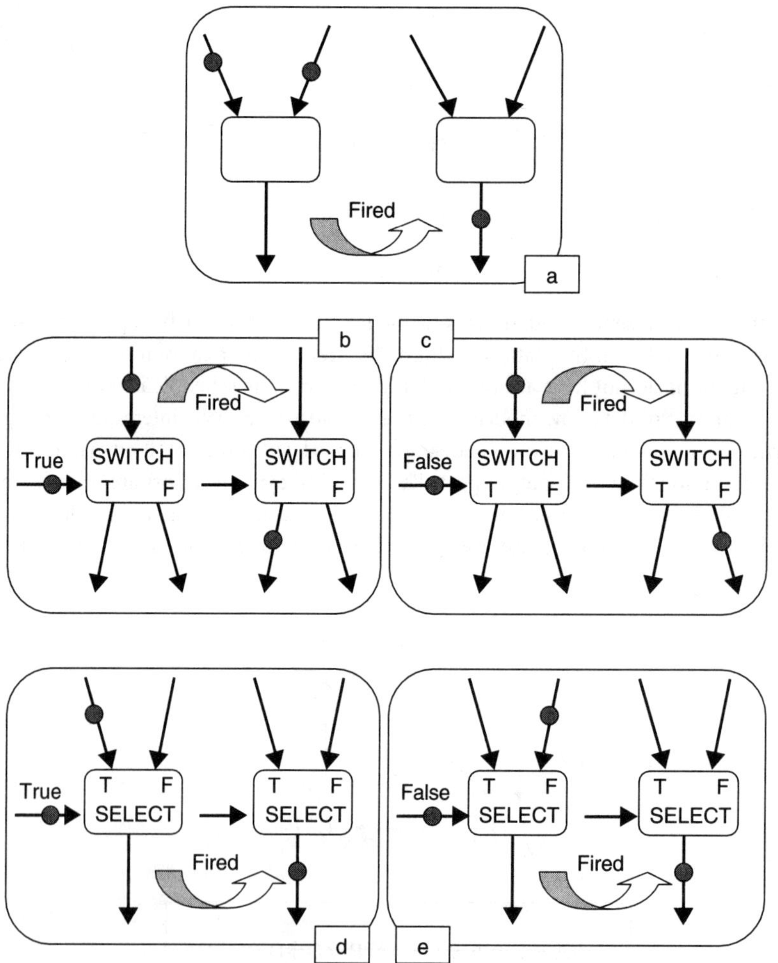

FIGURE 83.3 Dynamic dataflow actors and dataflow firings.

simulator usually maintains a global event queue. This can be computationally expensive, as it is necessary to maintain a sorted queue of all the occurred events, according to the respective time of occurrence.

The operational semantics of a discrete event system consists in the sequential firing of the processes according to the chronological time stamps of queued events. For each specific queued event, the processes, that have it as an input, will be fired and the event will be removed from the queue. New events will be produced due to the previous firing and should be added to the event queue, according to its time tag.

Hardware description languages, like VHDL and Verilog, as most digital hardware simulation engines, use a discrete-event based simulation approach.

Simultaneous events, especially those arising from zero-time delay, can cause some problems, due to the ambiguity resulting from equal time tags (they cannot be ordered). Figure 83.4 illustrates one simple situation.

Considering that X, Y and Z represent some processing modules/processes/components, and that module Y has zero-time delay, it will produce output events with the same time stamp as input events. If module X produces an event with time stamp t (Figure 83.4 (a)), both modules Y and Z will receive that event; there is ambiguity about whether Y or Z should be executed first. If Z is executed first, the event presented at its input will be consumed; afterwards, Y will be executed and will produce an event that will be consumed in turn by Z.

On the other hand, if Y is executed first, an event will be produced with the same time stamp, because module Y has zero-time delay (Figure 83.4 (b)). At this point, execution of Z can be accomplished in several ways: taking both events at the same invocation step, or taking one of the events first and the other afterwards (it is also not clear as to which should be processed first). In order to enable a coherent system simulation (producing the same result even if Y or Z are executed first), the concept of delta delay is introduced. Events with the same time stamp are ordered according to the delta value. Now, the event produced by module Y will have a time stamp of $t+\Delta$ (Figure 83.4 (c)), resulting in an execution of Z first consuming the event arising from X with time stamp t, and afterwards (in the next delta step) consuming the event arising from Y, with time stamp $t+\Delta$ (Figure 83.4 (d)). If a different zero-time delay process receives the event with time stamp $t+\Delta$, it will generate a further event with time stamp $t+2\Delta$. All events with the same (time tag, delta tag) will be consumed at the same time.

This means a two-level model of time was introduced: on top of the totally ordered events, each time instant can be further decomposed into (totally ordered) delta steps applicable to generated events. However, from the point of view of the simulated time reported to the user, no delta information is included. In this sense, delta steps do not relate to time as perceived by the observer. They are simply a mechanism to compute the behavior of the system at a point in time. Similar mechanisms can also be used in synchronous/reactive models. These are analyzed in the next section.

Finite State Machines

Finite State Machines (FSMs) are a common computational model used by system designers for decades. They are commonly represented in different ways. In this chapter, state diagrams are used. The modeling

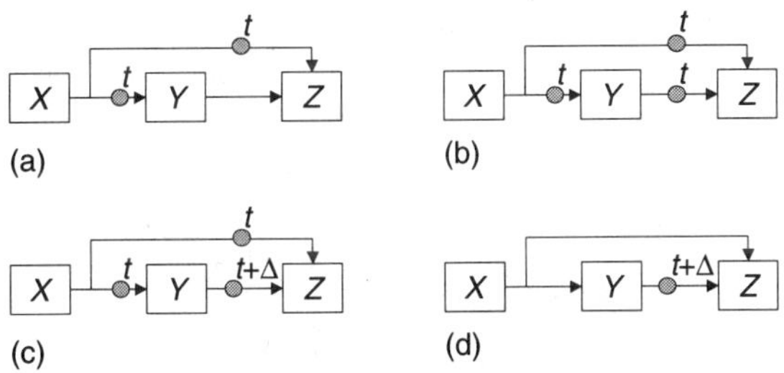

FIGURE 83.4 Handling of simultaneous events in a discrete-event system.

attitude is based on the characterization of the system in terms of the global states that it can exhibit, and also in terms of the conditions that can cause a change in those states. An FSM consists of a set of input signals, a set of output signals, a finite set of states (with a specified initial state), an output function, and a next-state function.

Figure 83.5 illustrates a basic notation for a state diagram. Circles or ellipses represent states; transitions between states use a directed arc. Each arc has an attached expression, potentially containing reference to the input event and/or to an external condition that will cause the change of state. Two basic models can be considered for output modeling: Moore-type machine [12], where outputs are associated with state activation, and Mealy-type machine [17], where outputs are associated with transition arcs. Both models have the same modeling capabilities.

They also have well-known strengths and weaknesses. Among the strengths, we should mention that they are simple and intuitive to understand. This is one of the reasons why designers have extensively used them in the past, and continue to use them. Unfortunately, several weaknesses prevent their usage for complex systems modeling. Namely, FSMs do not provide data processing capabilities, support for concurrency modeling, (practical) support for data memory, and hierarchical constructs.

FSMs are a control-dominated model of computation. We will use an FSM model to introduce a running example, inspired by the one in [30]. It has to do with the control of an electric car that should carry goods from one point to another, and come back. The order to start movement is specified by actuation on key "GO," while the command to return is specified through the actuation on the key "BACK." After receiving an order, the car motor is adequately activated, while the initial, or the final, position is not reached; there are two sensors available for this purpose. Figure 83.6(a) represents the external view of the controller in terms of inputs and outputs, and Figure 83.6(b) illustrates the layout of the plant.

Figure 83.6(c) and (d) present two possible (and equivalent) models for the control of the referred system. The first relies on the evaluation of external conditions (signal values are explicitly checked), while the second relies on events. It is clear that events usage will produce a lighter model, with less arcs and inscriptions (it is assumed in this representation that an event associated with a signal is generated when the signal changes its state from '0' to '1').

From the execution semantics point of view, we can find two attitudes (which correspond to different models of computation) [31]:

1. Synchronous FSMs;
2. Asynchronous FSMs.

In Synchronous FSMs, both computation and communication occur instantaneously at discrete time instants (under the control of clock ticks). One strong aspect in favor of Synchronous FSMs is its implementation robustness, especially when using synchronous hardware. However, when heterogeneous implementations are foreseen, some difficulties or inefficiencies may arise. Anyway, within a synchronous

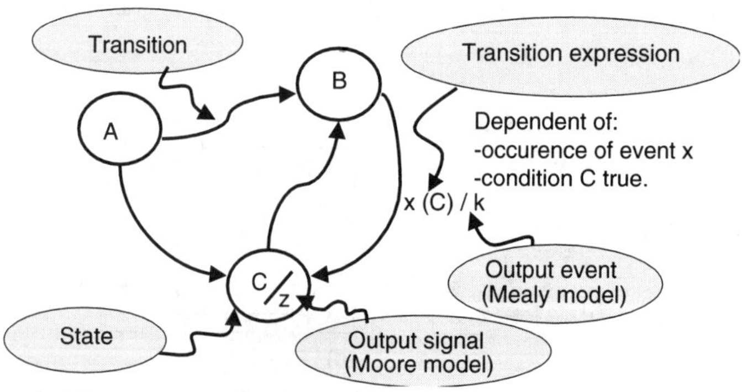

FIGURE 83.5   State diagram basic notation.

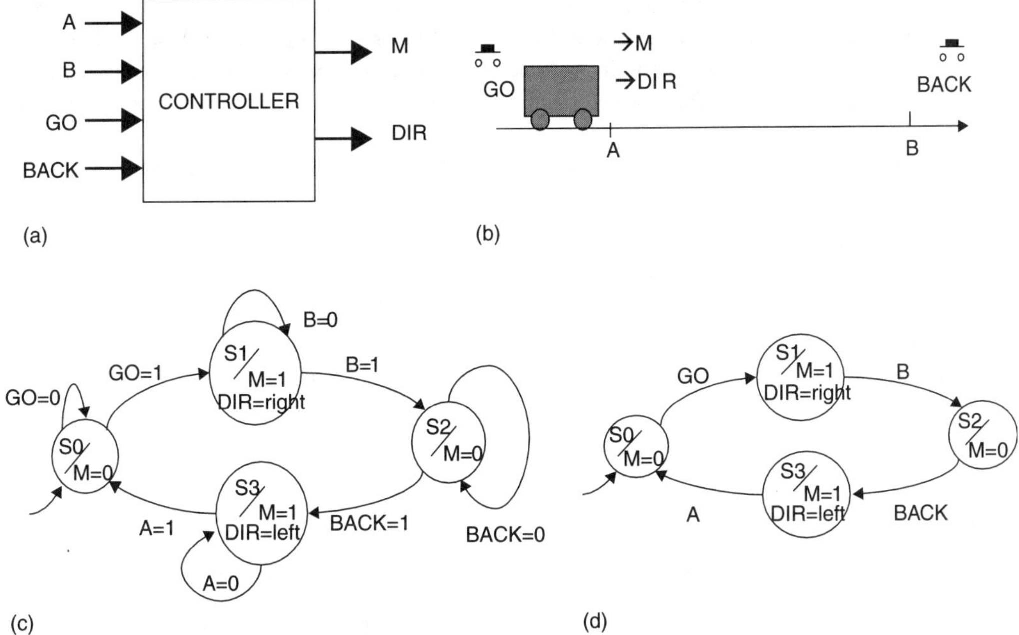

FIGURE 83.6   State diagram models of an electric car.

implementation island, it is possible to rely on robust compilation techniques, either to optimally map FSMs onto Boolean and sequential circuits (hardware) or to software code.

In Asynchronous FSMs, process behavior is similar to the one on synchronous FSMs, but without dependency on a clock tick. As already referred, an asynchronous system is a system in which two events cannot have the same time tag. In this sense, two asynchronous FSMs never execute a transition at the same time (asynchronous interleaving). For heterogeneous architectures or for multirate specifications, implementation can be easier than in the synchronous case, however, still somehow problematic. The difficulties come from the need to atomically synchronize communicating transitions, which can make a correct implementation of rendezvous on a distributed architecture very difficult to achieve.

Codesign Finite State Machines

Most MoCs are control-dominated or data-dominated. However, embedded systems are composed of a mixture of reactive behavior, control functions, and data processing, especially those targeted for networking and multimedia applications. Considering FSMs as the starting point, which proved to be adequate for low-to medium-complexity control-dominated system modeling, we can find in the literature numerous proposals extending FSMs in several directions. Each extension tries to overcome one or more intrinsic FSMs shortages, from concurrency modeling inability and the associated state space explosion problem, to data processing modeling and the absence of hierarchical structuring mechanisms (supporting specification at different levels of abstraction).

One line of thought is targeted to overcome the lack of modeling ability regarding data representation. For instance, to model an 8-bit variable with 256 possible values through an FSM, it is necessary to use 256 states, which is not manageable for the designer as the model loses its expressiveness. This problem can be eased through the use of FSM with data paths, where data-processing capabilities are added to FSM intrinsic characteristics: the system is decomposed into a control part and a data part. On the one hand, the control part is efficiently specified and synthesized using an FSM model, easier to understand if considered in synchronous terms. On the other hand, data computation can be efficiently specified and synthesized using data flow graphs (as usually found in digital signal processing and associated code generation tools).

Codesign Finite State Machines (CFSMs) are another MoC for embedded system design [29] [36] obtained in this line of thought. They extend *Finite State Machines* with support for data handling and asynchronous communication. Data handling is associated with transitions and these are associated with external instantaneous functions. This implies that data manipulation should typically exhibit a low algorithmic complexity.

In this sense, CFSMs supports the specification of embedded systems, involving both control and data flow aspects, and having the following key characteristics:

- Are composed by FSMs components.
- Support a data computation part.
- Use a Globally Asynchronous Locally Synchronous (GALS) reference model, where

  - Local behavior is synchronous (as seen inside each component), and
  - Global behavior is asynchronous (as seen by the rest of the system).

Any high-level language with precise semantics based on extended FSMs can be used to model individual CFSMs (currently, Esterel is directly supported).

Synchronous/reactive Models

Reactive systems are those that interact continuously with their environment. David Harel and Amir Pnueli gave the name in the early 1980s. In the synchronous model of computation, all events are synchronous. This means that all signals have events with equal tags to other events in other signals. The tags are totally ordered. In this computational model, the output responses are produced simultaneously with the input stimuli. Conceptually, Synchrony means that reactions take no time; thus, there is no observable time delay in the reaction (outputs become available as inputs become available). Unlike the discrete event model, all signals have events at all clock ticks, simplifying the simulator. Simulators that exploit this characterization are called cycle-based (as opposed to the so-called discrete-event simulators). A cycle is the processing of all events at a given clock tick. Cycle-based simulation is excellent for clocked synchronous circuits. It is also possible to consider, for some applications, a multirate cycle-based model, where every nth event in one signal aligns with the events in another signal.

Synchronous languages [34, 1], like Esterel [16], Lustre [33], and Signal [2], use the synchronous/reactive model of computation. Statecharts graphical formalism belongs to the same language family and is sometimes referred as quasi-synchronous; statecharts will be presented in a separated section.

Esterel is a language for describing a collection of interacting synchronous FSMs. Esterel supports the description of concurrent and sequential statements. *S1;S2* represents the sequential execution of S1 followed by S2, while *S1||S2* represents the parallel execution of S1 and S2; *[S1||S2]* executes until S1 and S2 terminate. The synchronized modules communicate through signals, using an instantaneous broadcast mechanism.

As an introductory example, we consider an FSM that implements the following behavior [19]: "Emit an output O as soon as two inputs A and B have occurred. Reset this behavior each time the input R occurs."

Figure 83.7 presents the associated state diagram. The associated Esterel representation is presented below:

```
module SMSM:
input A, B, R;
output O;
loop
  [ await A || await B ];
  emit O
each R
end module
```

The declaration part is trivial. Waiting for the occurrences of A and B is accomplished through "[await A || await B]." The "emit O" statement emits the signal O and terminates at the time it starts. Hence, O is

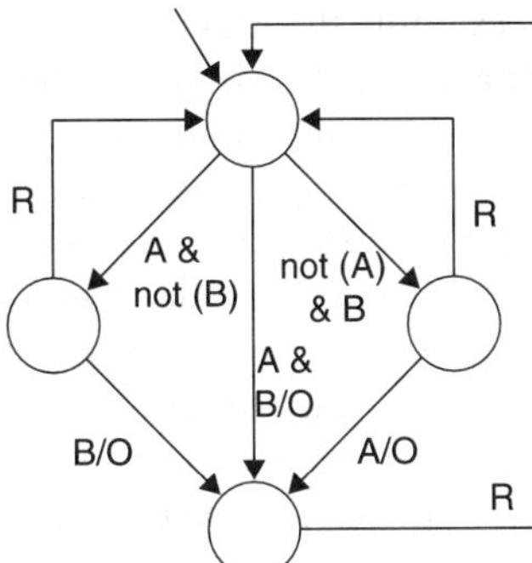

FIGURE 83.7   A simple Mealy state machine.

emitted exactly at the time where the last of *A* and *B* occurs. The handling of the reset condition is accomplished by the loop "loop p each R." When the loop starts, its body *p* runs freely until the occurrence of R. At that time, the body is aborted and it is immediately restarted. If the body terminates before the occurrence of R, one simply waits for R to restart the body.

It has to be noted that Esterel code grows linearly with the size of the specification (while FSM complexity grows exponentially). For instance, if we want to wait for the occurrence of three events, A, B, and C, the code change is minimal: "[await A || await B || await C]."

Esterel is a control flow-oriented language. On the other hand, Lustre is a language that adopts a data-oriented flavor. It follows a synchronous dataflow approach and supports multi-rate clocking mechanism.

In addition to the ability to translate these languages into finite-state descriptions, it is possible to compile these languages directly into hardware (execution in software of these languages is also achieved through the simulation of the generated circuit).

Statecharts

Statecharts [7, 6] provide a model of computation to specify complex reactive systems. They are based on state diagrams, plus the notions of hierarchy, parallelism, and communication between parallel subsystems. The main advantage of Statecharts over FSMs is the structuring of the specification, magnifying the legibility, and improving the system maintenance. These characteristics were key points that supported its adoption as one of the specification formalisms within the Unified Modeling Language (UML) [3, 4, 20].

Statecharts define three types of state instances: the set (implementing the AND refinement mechanism), the cluster (implementing the XOR refinement mechanism), and the simple state. The cluster supports the hierarchy concept, through encapsulation of state machines. The set supports the concurrency concept, through parallel execution of clusters. Figure 83.8(a) illustrates the usage of the cluster mechanism; the state diagram composed of states C, D, and associated arcs is encapsulated by the state A. This provides us with a top-level view of the model composed only of states A and B, but complemented by the inner level if one wishes to obtain further details regarding the system behavior. Figure 83.8(b) presents a simple model containing a set A composed of three AND components (B, C, and D); whenever A is activated/deactivated, the associated components B, C, and D will also be activated/deactivated, The reader should refer to [6, 7, 9] for further details regarding the Statecharts formalism.

Apart from the referred main characteristics, the Statecharts formalism presents some interesting features, such as:

- The default state that defines which state will take control in the case where a transition reaches a cluster state. In Figure 83.8(a), the system initial state is state B, and, after the occurrence of *w*, states A and C will become active.
- The notion of history, simple or deep, can be associated with cluster state instances. When the system enters a cluster with history property, the state that will be active upon entrance will be the

one that was active upon the last exit from that cluster. In the case of the first entrance in the cluster, the active state will be the default one. This is the case for cluster C of Figure 83.8(b), which holds the "*H*" attribute inside a circle. The history property, can also be deep, meaning that, all the clusters inside the cluster with the history property also have that property; this is the case for cluster B in Figure 83.8(b), which holds the "*H**" attribute inside a circle.

- The activation expressions present in transitions can also have special events, special conditions, and special actions. Special events can be the *entered(state)* event and the *exited(state)* event, which are generated whenever the system, respectively, enters or exits some state. One special condition is the *in(state)*, which indicates if the system is currently in the specified state. Examples of special actions are *clear history(state)* and *deep clear history(state)*. These initialize a cluster history state to the default state.

So far, we have only characterized syntactic issues for the Statecharts formalism. Its semantic characterization includes, among other things, the characterization of the *step algorithm* and the definition of the *criteria for solving conflicts* (which set of triggering transitions to choose among a set of conflicting transitions).

The step algorithm is of fundamental importance as it dictates the way the system evolves from one global state to another. The triggering of a finite set of transitions causes the evolution of the system. In [9], the trigger of this set of transitions is called microstep (conceptually it can be seen as equivalent to the delta concept, already introduced when presenting the discrete event MoC). Transitions fired in a microstep can generate internal events that, in turn, can fire other transitions. Thus, a step is defined as a finite sequence of microsteps. This means that, after all the microsteps have taken place, there will be no more active transitions and the system will remain in a stable state. External events are not considered during a step; hence, one must assume that a system can completely react to an external event before the occurrence of the next external event. According to many authors, the initial Statechart proposal [8] left their semantics partially open. As a consequence, different semantics associated with the handling of broadcasted event have been considered possible. Statechart semantics proposals are discussed in [32]. Yet, as pointed out by Harel in [8], that survey does not mention the semantics implemented in the STATEMATE tool (in the following citation, the original reference citations were replaced by the citation numbering used in this chapter): "the survey [32] does not mention the STATEMATE implementation of statecharts or the semantics adopted for it at all, although this semantics is different from the ones surveyed therein (and was developed earlier than all of them except for [8])" [8].

Even so, most common semantics consider that a generated event will only be valid during the next microstep (more used) or for the remaining microsteps within the same step (persistent events).

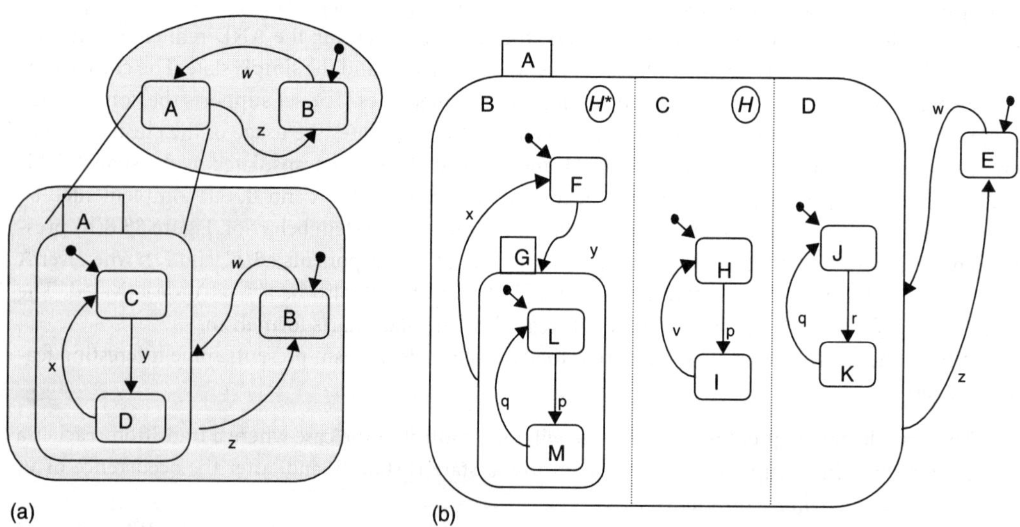

(a)   (b)

FIGURE 83.8   Statecharts basic models.

Coming back to the example of the electric car, if one considers controlling two or three of those cars using the same controller, and tries to use a state diagram to model the whole system, he/she will encounter the well-known problem of state explosion, due to the orthogonal composition of state diagrams: if we have two independent state diagrams with N1 and N2 states, the composed state diagram will contain N1 × N2, as for each state of machine 1, one may have machine 2 in any of its states. This illustrates the already presented weakness resulting from the lack of support for concurrency modeling of FSM.

Now consider that we want to keep the cars synchronized at both ends, which means that we will wait for all cars at each end, before launching them in the opposite direction. Figure 83.9 presents a statechart model enabling a compact modeling of the referred system. Those additional constraints associated with global start and global return are explicitly modeled through specific arc inscriptions; for instance, arc expression leaving state S01 is augmented with the condition [in(S02)], imposing that car 1 only starts its travel if car 2 is also in its original position (identical constraint for car 2).

Petri nets

Carl Adam Petri proposed Petri nets in 1962 [5]. They can be viewed as a generalization of a state machine where several states can be active simultaneously and transitions can start at a set of "states" and end in another set of "states." More formally, Petri nets are directed graphs with two distinct sets of nodes: places, drawn as circles or ellipses, are the passive entities, and transitions, drawn as bars, rectangles or squares, are the active entities. Arcs connect them in an alternate way: places can only be connected to transitions and transitions can only be connected to places. Places can contain tokens, also called marks. Small circles, inside places, specify these tokens. Figure 83.10 shows two nets, each with one transition (t_1), five places (p_1 to p_5), and five arcs. The nets have distinct markings: the net (b) corresponds to the net (a) marking after the transition firing. A transition can only fire if all its input places contain at least one token. From the transition point of view, one token is taken from each input place and one token is placed in each output place; this destruction and creation of tokens is an atomic operation, in the sense that it cannot be decomposed. Notice that a transition can have only input arcs (*sink transition*) or only output arcs (*source transition*).

More exactly, this Petri net model is a Place/Transition net [37, 24]. They are the most well-known and best-studied class of Petri nets. In fact, this class is sometimes just called Petri nets. Generalized Petri nets simply add the possibility of weights in arcs with the expected change in the semantics: each arc must now take or deposit the specified number of tokens.

It is important to note that, contrary to state machines, the number of places grows linearly with the size of the modeled system as not all the states have to be represented explicitly but can result from the combination of several already existent "states" (modeled by places). This implies that Petri nets models

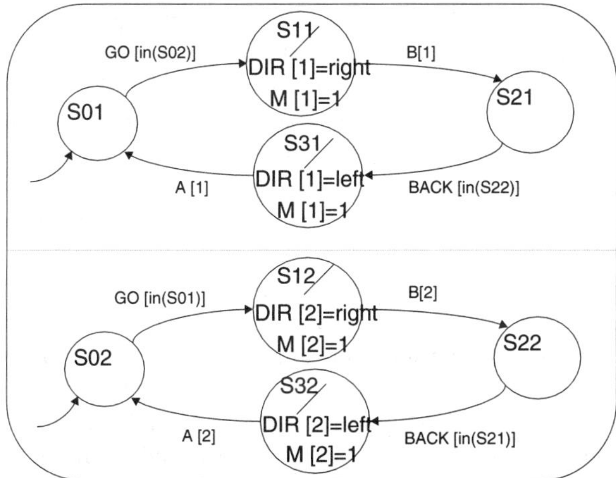

FIGURE 83.9 Simple statechart model for electric cars example.

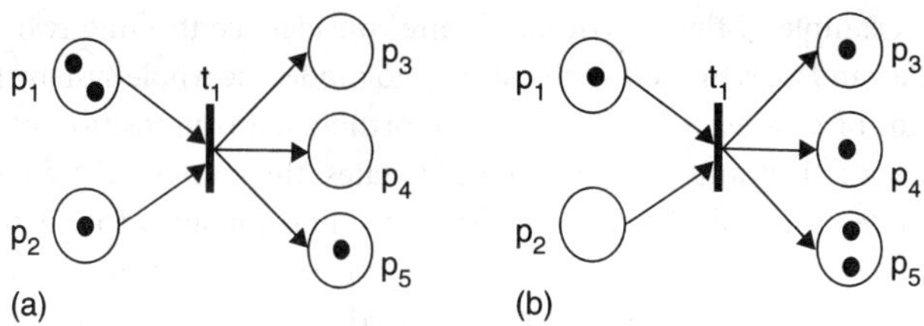

FIGURE 83.10   Transition t_1 firing: (a) before, and (b) after.

"scale better" that state machines (including statecharts). Even so, model growth is still a significant problem for Petri net models due to their graphical nature and low-level concepts. To minimize this, several abstraction and composition mechanisms exist.

Petri nets are well known for their inherent simplicity due to the small number of elementary concepts. Yet, Petri nets' simplicity is a two-edged sword [15]: on the one hand, it avoids the addition of further complications in the model resulting from the language complexity; on the other, it invariably implies that some desired feature is missing. As such, this basic type of Petri net has been modified in numerous ways to the point that *Petri nets* is actually a general name for a large group of different definitions and extensions, although all claim some close relation to the seminal work by Carl Petri [5]. A very readable, and informal, discussion about the existence of many different Petri nets classes can be found in the paper by Desel and Juhás [23]. This evidence has its origins in the fact, easily confirmed by experience, that conventional Petri nets, also known as low-level Petri nets, are sometimes difficult to use in practice not only due to the problem of rapid model growth, but also due to the lack of more specific constructions. Namely, time modeling, interface with the environment, and structuring mechanisms are frequent causes for extensions. While the former two are specific to some types of systems, structuring mechanisms are useful for even the simplest system model.

Coming back to our electric cars example, Figure 83.11(a) presents a Petri net model suitable for the modeling of a system with three electric cars, maintaining the constraint of synchronized starts and returns (in order to improve legibility, annotations associated with outputs are omitted, but can be associated with different places, or transitions, as well in state diagrams). It should be noted that transitions have the dynamics of the model associated with them, while places hold the current states of each car. Figure 83.11(b) uses a generalized Petri net model (arcs have weights associated with them), enabling a more compact representation of the system. Considering introducing more cars into our system (scability problem), it is clear that, considering this last model, it is only necessary to change the initial marking of the net: the left-most place should contain as many tokens as the number of electric cars in the system.

As even this simple example shows, Petri nets do not give special emphasis to states, as state diagrams, nor to actions, as data flows: states and actions are both "first-class citizens." In other words, Petri nets provide a *pure dual interpretation of the modeled world*: each entity is either passive or active. As already stated, passive entities (e.g., states, resources, buffers, channels, conditions) are modeled by places, and active entities (e.g., actions, events, signals, execution of procedures) are modeled as transitions. Another fundamental characteristic of Petri nets is their *locality of cause and effect*. More concretely, this means transitions are only red based on their vicinity, namely their input places and, possibly (for some classes of Petri nets), in their output places.

Two other fundamental aspects of Petri nets are their *concurrent and distributed execution*, and the *instantaneous firing* of transitions. The concurrent and distributed execution results from the locality of cause and effect: all transitions with no input or output places in common fire independently. This implies that the model is intrinsically distributed, as all transitions with disjoint locality do not depend on other transitions. The instantaneous firing means that each transition fires instantaneously: there are no intermediate states between token destruction in input places and token creation in output places.

In general terms, Petri nets structuring mechanisms can be classified into two groups: refinement/abstraction and composition mechanisms.

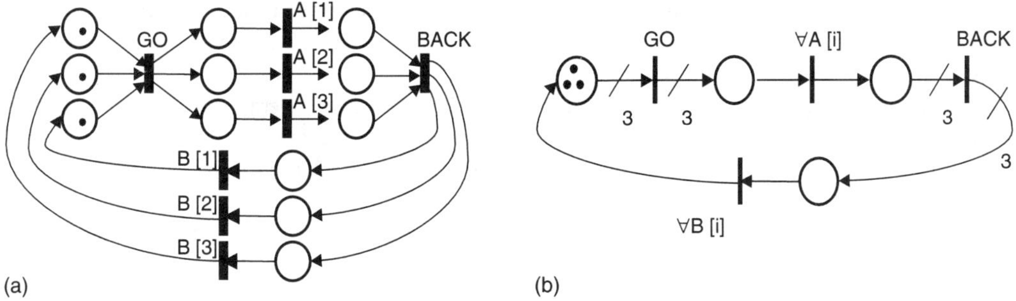

FIGURE 83.11 Petri net models for our electric cars example.

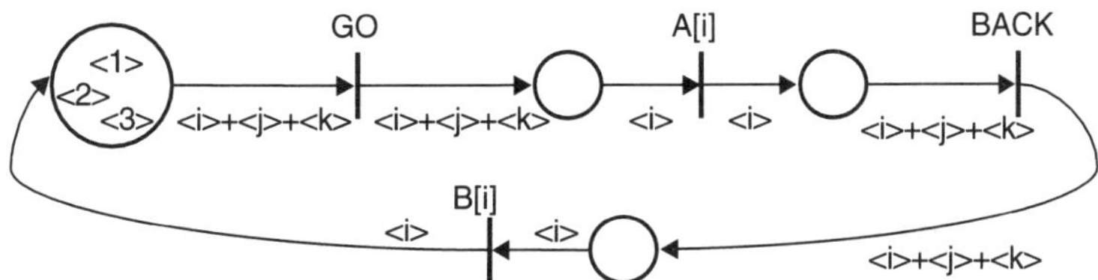

FIGURE 83.12 High-level Petri net models for our electric cars example.

Refinement/abstraction constructs correspond, roughly, to the programming languages concepts of macros or procedure invocation. Intuitively, macroexpansion is an "internal" composition where we get nets inside nets through the use of nodes (places or transitions). The use of places and transitions as encapsulated Petri nets goes back to the works of Carl Adam Petri [5] and was extensively presented in [30]. They are usually named macros (or macronodes) [28]. These constructs support hierarchical specification of systems, in the traditional sense of module decomposition/aggregation and "divide to conquer" attitude.

A common compositional mechanism is the *folding* of nets found in *high-level nets*. High-level nets can be seen as an internal composition made possible by structural symmetries inside a low-level net model. This is achieved by the use of high-level tokens. These tokens are no longer undistinguishable, but have data structures attached to them. Figure 83.12 shows the high-level Petri net model associated with our three electric cars system. Tokens presented at the left-most place (initial marking) now contain integer values, which enable the identification of a specific car status inside the model.

As tokens can carry data, it is necessary to add some inscriptions to the arcs, in order to properly select the right token to involve in a specific transition firing. For some classes of high-level nets, it is also possible to add within those arc inscriptions references to functions that support token attribute transformation. This is the case for Colored Petri nets [25]. In this way, data processing capabilities can also be embedded in the model.

It is important to note that many Petri net classes use several structuring mechanisms in an orthogonal way. For example, Hierarchical Colored Petri nets [25, 35] use abstraction in the form of macrotransitions, and composition by net folding and place fusion. Several proposals have even added object-oriented features to Petri nets (see [21] for an up-to-date detailed survey of some approaches).

The Tagged Signal Model

The Tagged-Signal Model is a denotational formalism that was defined as a meta-model for other MoCs [13, 36]. As such, it should not be seen as another MoC but as a very abstract framework within which different models of computation can be expressed and compared. The abstract nature of the Tagged-Signal Model also implies that the MoCs are expressed by imposing additional specific constraints on it.

The Tagged-Signal Model is also based on the notion of *event*. An event is a pair composed of a *value* and a *tag*. Tags are typically used to specify temporal behavior. A *signal* is a set of events. A functional or deterministic signal is defined as not containing two pairs with the same tag and a different value. The same is to say that a deterministic signal is a function from tags to values. The function can be partial, as not all tags have to be associated to values.

A process with *n* signals is a set of sequences of *n* signals. Processes operate concurrently. The synchronization among processes is defined by constraints imposed on their signal tags.

Functions are specified by the way events are related around a particular tag, or how inputs are used to compute outputs. Time is specified by the assignment of tags to events.

In [36], most significant models of computation are compared using the Tagged-Signal Model, where special attention is devoted to *Co-design Finite State Machines*.

References

1. Benveniste, A. and G. Berry, The synchronous approach to reactive and real-time systems, *Proceedings of the IEEE*, 79, 1270–1282, 1991.
2. Benveniste, A., and P. Le Guernic, Hybrid dynamical systems theory and the SIGNAL language, *IEEE Transactions on Automatic Control*, 35, 525–546, 1990.
3. Douglass, B.P., *Doing Hard Time — Developing Real-Time Systems with UML, Objects, Frameworks, and Patterns*, Object Technology Series, Addison-Wesley, Reading, MA, 1999.
4. Douglass, B.P., *Real-Time UML — Developing Efficient Objects for Embedded Systems*, Object Technology Series, Addison-Wesley, Reading, MA, 1998.
5. Petri, C.A., Kommunikation mit Automaten, Ph.D. thesis, University of Bonn, Bonn, West Germany, 1962.
6. Harel, D., On visual formalisms, *Communications of the ACM*, 31, 514–530. 1988.
7. Harel, D., Statecharts: a visual formalism for complex systems, *Science of Computer Programming*, 8, 231–274, 1987.
8. Harel, D., and Naamad, The statemate semantics of statecharts, *ACM Transactions on Software Engineering and Methodology (TOSEM)*, 5, 293–333, 1996.
9. Harel, D. and Politi, *Modeling Reactive Systems with Statecharts — The STATEMATE Approach*, McGraw-Hill, New York 1998.
10. Lee, E.A., Consistency in dataflow graphs, *IEEE Transactions on Parallel and Distributed Systems*, vol. 2, issue 2, pp. 223–235, April 1991.
11. Lee, E.A. and T. M. Parks, Dataflow process networks, *Proceedings of the IEEE*, vol. 83, no. 5, pp. 773–801, May 1995.
12. Moore, E.F., *Gedanken-Experiments on Sequential Machines*, Automata Studies, Princeton University Press, New Jersey, 1956
13. Lee, E.A. and A. Sangiovanni-Vincentelli, Comparing Models of Computation, Proceedings of ICCAD'1996.
14. Lee, E.A. and D.G. Messerschmitt, Synchronous data flow, *Proceedings of the IEEE*, vol. 75, no. 9, pp. 1235–1245, 1987.
15. Best, E., Design Methods Based on Nets: Esprit Basic Research Action., in *Advances in Petri Nets 1989*, Lecture Notes in Computer Science, Vol. 424, Rozenberg, G., Ed., Springer-Verlag, Berlin, Germany, 1990, pp. 487–506.
16. Boussinot, F. and R. De Simone, The ESTEREL language, *Proceedings of the IEEE*, 79, 1991.
17. Mealy, G.H.A., Method for synthesizing sequential circuits, *Bell System Technical Journal*, 34, 1045–1079, 1955.
18. Kahn, G., The Semantic of a Simple Language for Parallel Programming, in Proceedings of the IFIP Congress 74, North-Holland Publishing Co., Amsterdam, 1974.
19. Berry, G., The Esterel v5 Language Primer, 2000. Available from http://www.sop.inria.fr/esterel.org.
20. Booch, G., J. Rumbaugh, and I. Jacobson, *The Unified Modeling Language User Guide*, Object Technology Series, Addison-Wesley, Reading, MA, 1999.
21. Agha, G., F. de Cindio, and G. Rozenberg, Ed., *Concurrent object oriented programming and petri nets*, *Advances in Petri Nets*, Lecture Notes in Computer Science, Vol. 2001, Springer, Berlin, 2001.
22. Buck, J., S. Ha, E.A. Lee, and D.G. Messerschmitt, Ptolemy: a framework for simulating and prototyping heterogeneous systems, *Internatoinal Journal of Computer Simulation, Special Issue on Simulation Software Development*, vol. 4, pp. 155–182, April 1994.

23. Desel, J. and G. Juhas, What is a Petri net? Informal answers for the informed reader, in *Unifying Petri Nets*, Ehrig, H., Juhas, G., Padberg, J., and G. Rozenberg, Eds., Lecture Notes in Computer Science, Vol. 2128, Springer, Berlin, 2001, pp. 1–27.

24. Desel, J. and Reisig, W., Place/transition Petri nets, *Lectures on Petri Nets I: Basic Models*, Lecture Notes in Computer Science, Vol.1491, Springer-Verlag, Berlin, 1998, pp. 122–173.

25. Kurt Jensen, *Coloured Petri Nets. Basic Concepts, Analysis Methods and Practical Use,* Vols. 1–3, Monographs in Theoretical Computer Science, An EATCS Series, Springer-Verlag, Berlin, Germany, 1992–1997.

26. Lamport, L., Time, clocks, and the ordering of events in a distributed system, *Communications of the ACM*, 21,1978.

27. Lavagno, L., A. Sangiovanni-Vincentelli, and E. Sentovich, Models of Computation for Embedded System Design, NATO ASI Proceedings on System Synthesis, Il Ciocco, Italy, 1998.

28. Gomes, L., and João-Paulo Barros, On Structuring Mechanisms for Petri Nets Based System Design, ETFA'2003 — 2003 IEEE Conference on Emerging Technologies and Factory Automation Proceedings, Lisbon, Portugal, September 16–19, 2003.

29. Chiodo, M., P. Giusto, H. Hsieh, A. Jurecska, L. Lavagno, and A. Sangiovanni-Vicentelli, Hardware/software codesign, of embedded systems, *IEEE Micro*, 14, 26–36, 1994.

30. Silva, M., Las Redes de Petri: en la Automática y la Informática, Editorial AC, Madrid, 1985.

31. Sgroi, M., L. Lavagno, and A. Sangiovanni-Vincentelli, Formal models for embedded system design, *IEEE Design & Test of Computers*, vol. 17, no. 2, pp. 14–17, April–June 2000.

32. von der Beeck, M., A comparison of statecharts variants, *Formal Techniques in Real-Time and Fault-Tolerant Systems*, Lecture Notes in Computer Science, Vol. 863, Springer-Verlag, Berlin, 1994, pp. 128–148.

33. Halbwachs, N., P. Caspi, P. Raymond, and D. Pilaud, The synchronous data flow programming language LUSTRE, *Proceedings of the IEEE*, 79, 1305–1319, 1991.

34. Halbwachs, N., *Synchronous Programming for Reactive Systems*, Kluwer Academic Publishers, Dordrecht, 1993.

35. Huber, P., K. Jensen, and R.M. Shapiro, Hierarchies in Coloured Petri Nets, in Proceedings of the 10th International Conference on Application and Theory of Petri Nets, Bonn, Germany, 1989, pp. 192–209.

36. Edwards, S., L. Lavagno, E.A. Lee, and A. Sangiovanni-Vincentelli, Design of embedded systems: formal models, validation, and synthesis, *Proceedings of the IEEE*, 85, 366–390, 1997.

37. Reisig, W., *Petri Nets: An Introduction*, Springer-Verlag, New York, 1985.

84

Hardware-level Design Languages

Luís Gomes
Universidade Nova de Lisboa
UNINOVA, Centro de Robótica
Inteligente

Anikó Costa
Universidade Nova de Lisboa
UNINOVA, Centro de Robótica
Inteligente

84.1 Introduction

Digital circuit design has experienced a very rapid evolution during the last decades. Integrated circuits have become more and more complex. The number of gates in a single chip has been continuously increasing, from a few hundreds in Small Scale Integration (SSI) chips to thousands or millions in latest devices. At this point, the design process became very difficult to manage and Computer-Aided Design (CAD) tools became invaluable for chip designers. Some parts of the process began to be supported by computer tools.

With the advent of Very Large Scale Integration (VLSI) chips, with hundreds of thousands of transistors in a single chip, circuit complexity imposes a different role for CAD tools; it is no longer possible to have reliable test through breadboarding. Consequently, CAD tools became very popular, from logic simulators (supporting circuit functionality verification) to automatic placement and routing of circuit layout.

The role of logic simulation became more and more important, in direct correspondence with the circuit complexity (in terms of gate count). Designers felt the necessity to describe the circuit through the use of a neutral representation amenable to be used by them and by computer tools. This was the motivation leading to hardware description languages (HDLs).

To support circuit functionality description, designers have three major choices:

- Boolean equations.
- Schematics.
- Hardware description languages.

Boolean equations are interesting for small circuits. For complex circuits, only schematics and hardware description languages are widely used.

During the 1970s, the so-called Computer Hardware Description Languages were very popular enabling a compact representation of the large mainframe computers in use at the time. Designers used HDLs mostly for simulation and verification purposes and manually translated the circuit to schematics in order to produce the circuit.

Meanwhile, logic synthesis tools came into play and the process changed radically. Designers start with an HDL description, most of the time at the register transfer level (RTL), specifying how the data flow between registers and which transformations are required; afterwards, logic synthesis tools automatically generate the circuit description, in terms of a netlist (gates and interconnections).

84.2 What to Expect from Hardware Description Languages

Thus, by using HDLs, designers can produce a neutral representation of the hardware, amenable to represent the circuit structure (schematics) or the circuit behavior. The circuit representation should include the following facets:

- Project documentation, to support exchange of information among designers
- Modeling of the circuit, in order to support simulation
- Synthesis support
- Module representation, enabling reusage of predefined modules.

In the initial design attitude, schematics played a fundamental role in data exchange among designers, but, nowadays, automatic synthesis is fundamental to cope with system complexity. This means a shift from schematics to hardware description languages. In this sense, designers moved from describing the circuit from their structure perspective (gates, modules, and interconnections) to their behavior.

Emphasis on HDL-based synthesis tools received an additional significant push with the advent of *Programmable Logic Devices* (PLDs), from the "old" *Programmable Array Logic* (PALs), to *Complex Programmable Logic Devices* (CPLDs) and "recent" *Field Programmable Gate Arrays* (FPGAs). For these user-programmable platforms, the module concept plays a major role as it allows the reusability of predefined modules. Those modules range from user-defined modules to complex IPs (Intellectual Property), ready to implement functions for specific applications. These include signal processing (DCT, FFT), multimedia (MPEG, JPEG), and telecommunications (Viterbi), among others.

As already argued [1], compared to traditional schematic-based design, HDLs have many advantages:

- Designs can be described at several abstraction levels, independent of the technology to be used. HDLs support reusability, in the sense that logic synthesis tools can automatically convert the design to a specific implementation technology (thus, logic synthesis tools are responsible for the technology shifts).
- Functional verification of the design can be done early in the design cycle, using verification through simulation, minimizing risks associated with finding a bug in a later point in time at the gate-level netlist.
- Compared with gate-level schematics, which is difficult to use for complex designs, a textual description with comments is an easier way to develop and debug circuits.

Also, the usage of HDLs easily accommodates changes in the circuit. Consider a small example of a multiplexer of 3 inputs (A, B, C) and 1 output (Z), using two selection signals (X and Y). For a multiplexer implementation, we may have a circuit with a few OR gates, one three-input AND gate, and interconnecting nets; alternatively, we can describe this dependency using the associated Boolean equation:

$$Z = (A \text{ and } (X = 0 \text{ and } Y = 0)) \text{ or } (B \text{ and } (X = 0 \text{ and } Y = 1)) \text{ or } (C \text{ and } (X = 1 \text{ and } Y = 0)).$$

Consider, now, that we need to face a change in the specification, for instance, that we need to add an extra input (D) to the multiplexer. It is clear that we need to change significantly the schematic, even changing the supporting gate types (e.g., from three-input AND gate to four-input AND gate, if available, or to a composition of gates to support its implementation). On the other hand, textual description of the new multiplexer is close to the initial one:

$$Z = (A \text{ and } (X = 0 \text{ and } Y = 0)) \text{ or } (B \text{ and } (X = 0 \text{ and } Y = 1)) \text{ or}$$

$$(C \text{ and } (X = 1 \text{ and } Y = 0)) \text{ or } (D \text{ and } (X = 1 \text{ and } Y = 1))$$

An HDL description imposes a smaller change penalty.

On the one hand, a hardware description language can be characterized in a manner similar to a high-level programming language [2], and the main features should include:

- Describe the digital system in a precise way and as succinctly as possible.
- Support generation of associated documentation.
- Support hierarchical decomposition and multi-level simulation of the system.
- Shortening development time, accommodating time-to-market restrictions.
- Support earlier development of software parts of the system, as hardware interface and functionality can be described by blocks in advance.
- Reduce costs associated with changes.
- Detecting and reducing errors at earlier stages of the process, through support for system analysis at different complexity levels.
- Enabling comparison of different solutions for the same system, trying to reach specific goals in terms of power, performance, costs, or other features.
- Make available new technologies for integrated circuits to a broader range of persons, taking advantage of the fact that most of the technology-dependent tasks were transferred to the tools.

On the other hand, a hardware description language has to exhibit some hardware-intrinsic features in order to be a valuable tool supporting digital system development process, namely [3]:

- Describe the intrinsic parallel execution of hardware.
- Support separated description of architecture and control parts of the system, enabling different views of the system, namely structured- and behavior-dominated views.
- Support representation at several levels.
- Easy to learn and memorize.
- Support module representation, namely LSI and MSI modules commonly in use, and complex blocks, including IPs.
- Support for different approaches, namely top-down and bottom-up approaches, allowing the designer to define the adequate level of detail for the different components.

Several hardware description languages have been proposed during the last decades to reach these objectives. Yet, mainly due to the industry drive, it seems that "academic" or "experimental" languages do not seem to have a significant number of users. Currently, only VHDL and Verilog seem to survive in the hardware-level description languages arena. As such, and also due to time and space limitations, we will restrict our presentation to those two languages.

84.3　Levels for Digital System's Representation

The abstraction level in which a digital system can be described ranges from the physical level till the system level.

Figure 84.1 presents a common characterization of the reference abstraction levels. For each level, one can focus on two types of views. The first one emphasizes the structural view of the system: it is the "schematics" view of the system, with modules and interconnections between them. The second one emphasizes the behavioral view of the system: it is the "algorithm" view of the system, where, instead of specifying how the system is composed, as in the structural view, the designer specifies what should be implemented.

The "Physical" level gives the maximum system detail, containing information associated either with the layout of the circuit for discrete components-based implementation or with the geometric description associated with masks for integrated circuits production. This level is not covered by the techniques for digital circuit representation referred so far, namely schematics and hardware description languages.

At the "Circuits" level, we can find the traditional representation of electronic circuits, based on transistors and resistors interconnected by nets (wires). At this level, the building blocks (at the structural view) are transistors and resistors, which have an associated model amenable to be used for adequate simulators (within the behavioral view). The system model uses differential equations; SPICE is a common simulator for this description level. As the complexity of the circuit increases, simulation at this level starts becoming too expensive and time consuming. This abstraction level is also not associated with the digital circuits design.

The "Logic" level is the traditional representation level for digital circuits. From the structural point of view, the basic building blocks include gates and flip-flops. From the behavioral point of view, Boolean equations are used. Higher granularity blocks can also be included, like multiplexers and other modules. Each block has an associated model, normally expressed as a Boolean equation, and including time dependencies associated with delays between inputs and outputs. It is possible to simulate the circuit and determine some of its characteristics, namely typical delay times.

The next level, referred to as "Register transfer," can be seen as a generalization of the previous one. Here, the fundamental building block is the register, complemented by processing blocks, like arithmetic and logic units (ALUs), interconnected through buses. From the behavioral point of view, at this level, the circuit characterization is based on truth tables, state transition tables, and the way in which the operations on data are performed (namely how data are transferred among registers and which transformations on data are required). Also at this level, the notion of control part appears. It is a common level for usage of hardware description languages like VHDL and Verilog.

At the "Device" level, systems are decomposed into two interactive blocks: control and data parts. The data part is composed of register transfer architecture. The control part is responsible for the sequential configuration of the data part in order to accomplish a specific processing result or behavior; it is commonly characterized through a state machine like formalism. It is probably the most used abstraction level for digital circuit representation.

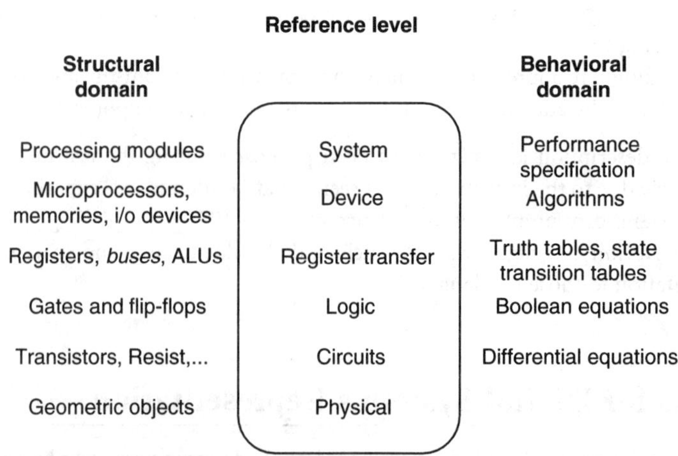

FIGURE 84.1   Levels of abstraction in digital systems.

The top abstraction level is the "System" level. Here, the system is decomposed into interconnected subsystems. Each subsystem is described from an external point of view and global reasoning about the behavior of the system is possible. Results on performance analysis and global proprieties can be obtained.

84.4 Hierarchical Design: A Key Concept

Complementing the reference abstraction levels, presented in the previous section, hierarchies are used for complexity management. It is a common way of structuring the system description, already started with the schematic representation. It is also a key concept in hardware description languages, like VHDL and Verilog.

Two main approaches can be followed for hierarchically structuring digital designs:

1. top-down design methodology;
2. bottom-up design methodology.

In the top-down approach, we start with the top-level system characterization. This block is decomposed in terms of the subblocks necessary to build the top-level block functionality. Each of the subblocks can be further decomposed until a representation amenable for implementation is reached.

In the bottom-up approach, we start by identifying the building blocks available for the task. Later, we build larger blocks using those available blocks. Further aggregations with other blocks will allow us to obtain the top-level block of the system.

Typically, in most designs we will use a mixed approach. Architectural design will define top-level goals that will be decomposed into blocks and subblocks according to functionality. At the same time, circuit designers will design or verify the viability to design-specific leaf-level blocks (starting by the critical blocks that will compromise overall performance or cost). In some way, analysis is top-down first, while circuit design is bottom-up first.

84.5 Verilog

The Verilog HDL was launched in early 1984 at Gateway Design Automation, and evolved to become a current industry standard. It started as a proprietary language supported by a simulation environment, which was the first environment supporting mixed-level design representations containing switches, gates, RTL, and a higher level of abstractions of digital circuits. In 1995, Verilog found a way to become an IEEE Standard, already updated to the IEEE Standard 1364-2001 [4]. Before, Verilog had been put in the public domain (as a response to VHDL also being an IEEE Standard).

Verilog offers many useful features for hardware design [1]:

- It is a general-purpose hardware description language that is easy to learn and use. It has a syntax similar to the C programming language. This allows an easy migration for designers with C programming experience.
- Allows different levels of abstractions to be mixed in the same model (from switches, gates, or RTL to behavioral code).
- Most logic synthesis tools support Verilog.
- Fabrication vendors provide Verilog libraries for postlogic synthesis simulation.
- Offers the Programming Language Interface (PLI) that allows the user to write custom C code and embedded it in the Verilog simulator, according to her/his needs.

The module is the basic building block in Verilog. A module can be an element or an aggregation of lower-level design blocks. It supports reusability of a predefined, already implemented, functionality. A module is described in upper levels by its interface ports (inputs and outputs), concealings internal implementation. This means that, from upper levels, it is seen as a black box. In this sense, it is possible to compare or test the usage of different implementations for a specific block without imposing changes at the upper levels. In Verilog, the keyword *module* declares a module. All interface ports have to be declared as *input* for input port, *output* for output port, or *inout* for bidirectional port. A module definition always ends with the *endmodule* keyword.

Verilog supports both behavioral and structural descriptions. Each model can be described at four levels of abstraction:

- *Behavioral or algorithmic level*: This is the highest level provided by Verilog. Hardware implementation details are omitted. Also, due to the C-like syntax of Verilog, descriptions at this level are similar to C programs.
- *Dataflow level*: At this level, emphasis is placed on how data flow between registers and how data are processed.
- *Gate level*: Design at this level is similar to describing a design in terms of "Logic" level, based on gates and interconnections between those gates.
- *Switch level*: This is the lowest level provided by Verilog, and requires that the designer master technological implementation details.

From the point of view of the circuit simulation, Verilog allows signals to have four logic values and eight strength levels. Possible logic values are: *0* (logic zero), *1* (logic one), *x* (unknown value), and *z* (high impedance). Verilog supports basic logic gates as predefined primitive modules. There are two classes of basic gates: and/or gates (composed by *and, nand, or, nor, xor,* and *xnor*) and buf/not gates (composed by *buf, not, bufif1, bufif0, notif1,* and *notif0*). For each basic gate, associated truth table is defined, considering the possible four logic values. Considering a 2-input AND gate as an example, the associated truth table is presented in Table 84.1.

In addition to logic values, the eight possible strength levels can be used for conflict resolution between drivers of different strengths. Possible strength levels are the following: *supply* (strongest), *strong, pull, large, weak, medium, small,* and *highz* (weakest). Considering the example of having two (or more) signals driving the same net with different values and strengths, the signal with stronger strength prevails; if several signals with incompatible values and with same strength are present, the result is *x* (unknown). For instance, if two signals with *strong1* and *weak0* are tied in the same net, the result will be *strong1* (stronger strength prevails); as a second example, if two signals with *strong1* and *strong0* are tied in the same wire, the result will be *strongx* (resulting unknown value).

84.6 VHDL

VHDL stands for VHSIC Hardware Description Language (and VHSIC stands for Very High Speed Integrated Circuit). The development of the language started in the beginning of 1980s by the United States Department of Defense with the purpose of obtaining a standardized method for describing electronic circuits. In 1987, the IEEE adopted VHDL as IEEE Standard 1076-1987. In 1993 and 2000, the standard was revised and established that:

"The VHSIC Hardware Description Language (VHDL) is a formal notation intended for use in all phases of the creation of electronic systems. Because it is both machine readable and human readable, it supports the development, verification, synthesis, and testing of hardware designs; the communication of hardware design data; and the maintenance, modification, and procurement of hardware" [5].

TABLE 84.1   AND-gate Truth Table

	and	0	*i1* 1	*x*	*z*
	0	0	0	0	0
i2	1	0	1	*x*	*x*
	x	0	*x*	*x*	*x*
	z	0	*x*	*x*	*x*

This means that VHDL can be applied throughout the design process: including capture, simulation, synthesis, and documentation.

Why use VHDL? The answer to this question should not be much different from "why use Verilog?" As a matter of fact, arguments are mostly interchangeable. In [6], the following arguments are presented as answers to the initial question:

- It is standardized; hence it is possible to move code among different development environments (although special attention should be paid to the synthesis of subset constructs supported by each tool).
- It supports top-down, bottom-up, and mixed methodologies.
- It is independent from the technology, as technology change support is transferred to automatic tools.
- It supports hierarchical structures, reusable components, error management, and verification.
- A component can be made in such a way that it can be adapted automatically to different sizes of elements (generic component). As an example, an adder can be defined with the bit size as an attribute; at instantiation time, the size attribute receives a value and the adder will be adjusted accordingly.
- It supports concurrent and sequential constructions.
- It uses the same language for simulation.

VHDL supports both behavioral and structural descriptions. Four abstraction levels are supported, according to [6]:

1. *Functional level*, where algorithms can be described.
2. *Behavioral level*: A behavioral module can be described as interconnected functional modules. The advantage of models at this level is that models for simulation can be built quickly.
3. *RTL (Register Transfer Level)*: At this level, it is possible to describe behaviors in several different ways, namely data flows, asynchronous and synchronous state machines, operations between registers, etc.
4. *Logic or gate levels* use descriptions based on Boolean algebra or associated netlist.

This means that lower levels of abstraction, like circuit level using transistor- and resistor-based descriptions, are not supported by VHDL.

Several criteria should be taken into account in the selection of the abstraction level to be used for the description of a specific circuit. For instance, if a shorter development time is required, then a higher abstraction level should be selected. Also, if the design has restrictive performance requirements, RTL should be selected, or even the gate level, depending on the set of synthesis tools to be used.

In VHDL, a signal has a type, which encapsulates a set of possible values that the signal can take. VHDL intrinsically supports a type known as *bit* (which allows signals to take values "0" and "1"), and also the *integer* type (using a 32-bit representation). Yet, the *bit* type is not a good support for simulation, as, for instance, high-impedance cannot be represented, and neither conflict can be solved. In this sense, it is common to use the *std_logic* type, defined in the IEEE 1164 standard. The *std_logic* type has nine values with the following meanings:

- U, uninitialized
- X, forcing (strong driven) unknown
- 0, forcing (strong driven) 0
- 1, forcing (strong driven) 1
- Z, high impedance
- W, weak (weakly driven) unknown
- L, weak (weakly driven) 0 (supports modeling of pull-downs)
- H, weak (weakly driven) 1 (supports modeling of pull-ups)
- -, don't care.

Circuit simulation is carried out in a similar way as the one described for Verilog.

84.7　From Boolean Algebra to Sequential Circuits, or How to Get VHDL into Action

In the current and following sections, VHDL code will be used to illustrate code implementations.

In VHDL (as in Verilog and in hardware design, in general), components are a central concept. Components can be used to build up component libraries, supporting reusability. In fact, VHDL is an object-based language, which means that the designer can create instances of the library's components. Additionally, generic components can be defined, enabling some modification before instantiation, for instance, the width of a bus.

A component representation contains two parts:

1. Entity and
2. Architecture.

Entity is responsible for the declaration of input and output ports; it is the black box view. Each entity port can have one of three modes associated with them: *in* for input signals, *out* for output signals, and *inout* for bidirectional signals. Note that the *inout* mode should only be used with bidirectional signals. If we need to read the status of an output signal, we should do it in an indirect way (to be referred later in the section). In this sense, the entity declaration defines the interface between the entity and the environment. The entity name is the same as the component name. Code Fragment 1 presents the definition of the entity interface ports for a component with three input signals (a, b, and c) and one output signal (z), all of type *std_logic*.

```
library IEEE;
use IEEE.STD_LOGIC_1164.ALL;
use IEEE.STD_LOGIC_ARITH.ALL;
use IEEE.STD_LOGIC_UNSIGNED.ALL;
entity logic is
    Port ( a  : in std_logic;
           b  : in std_logic;
           c  : in std_logic;
           z  : out std_logic);
end logic;
```

CODE FRAGMENT 1 — Entity representation.

Architecture contains the structural or behavioral description of the component. It defines the body for the component. The architecture name is different from the component name, although the architecture is tied to a specific entity.

VHDL offers several ways to describe the architecture, associated with the different ways in which one can describe a particular circuit. Here, we will use a multiplexer as an example, and several specifications will be produced.

Figure 84.2 summarizes different circuit modeling attitudes. Figure 84.2 (a) presents the black-box characterization of the component, which means an external view of the interface without knowing functionality. Figure 84.2(b) presents a schematic symbol commonly used to describe MUX functionality. The next step is related to the description of the component, either using a behavioral description or a structural view. Three views are shown: Figure 84.2(c) uses a Boolean expression to represent multiplexer dependencies, while in Figure 84.2(d) the multiplexer module is represented through the associated schematic, and in Figure 84.2(e) a linguistic description is used (algorithmic style).

Starting with the "old-fashioned" schematic style of Figure 84.2(d), we face a pure structural functionality description, based on the usage of the primitive components INV, OR2, and AND2. Each block has its own truth table, known by the simulator engine. The description of the structure is based on the instantiation of the primitive components, referring to how they will be interconnected with the other components. Associated architecture code is presented in Code Fragment 2.

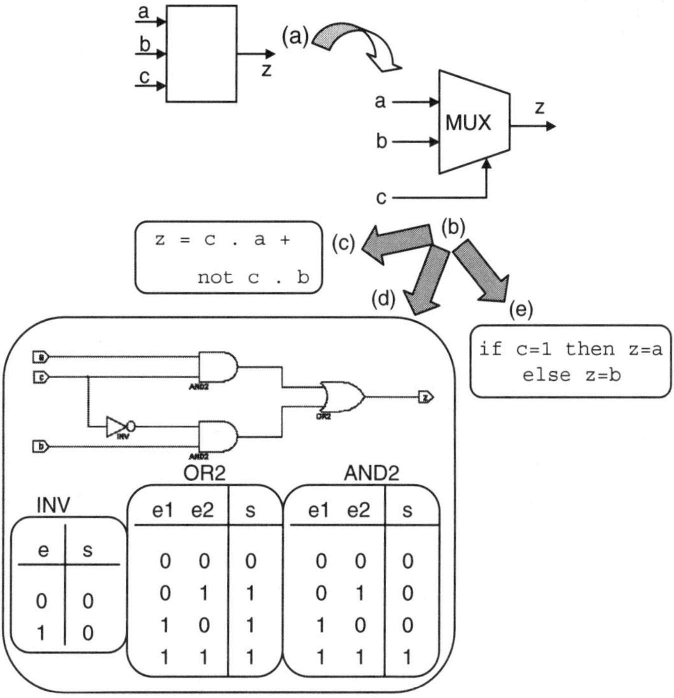

FIGURE 84.2   Equivalent multiplexer descriptions.

```
architecture Structural of logic is
signal node1, node2, node3 : std_logic;
begin
u1: and2 port map (I0    => a,
                   I1    => c,
                   O     => node1);
u2: inv port map (I      => c,
                   O     => node3);
u3: and2 port map (I0    => b,
                   I1    => node3,
                   O     => node2);
u4: or2 port map (I0     => node1,
                   I1    => node2,
                   O     => z);
end Structural;
```

CODE FRAGMENT 2 — Gate-based structure description of a multiplexer.

It has to be noted that internal nets are referred by signals that were declared at the beginning of the architecture body (signals *node1*, *node2*, and *node3* with *std_logic* type associated). This architecture has four modules concurrently running, obtained from the instantiation of primitive components.

An alternative method to describe the multiplexer is through the Boolean equation presented in Figure 84.2(c). In this case, the associated architecture has three statements running concurrently: the first and second ones are responsible for the generation of signals *node1* and *node2*, respectively, and the third one for producing the output. Associated code is presented in Code Fragment 3.

```
architecture Dataflow of logic is
signal node1, node2 : std_logic;
begin
  node1  <= c and a;
  node2  <= not c and b;
  z      <= node1 or node2;
end Dataflow;
```

CODE FRAGMENT 3 — Alternative description for a multiplexer.

It is important to note that in both previous alternative multiplexer representations, the hardware components were represented individually. Thus, as hardware is intrinsically concurrent, so are the statements that describe them. In this sense, the order of the different statements or component instantiation inside architecture is not relevant (we can change the order without consequences).

Coming to Figure 84.2(e), where a behavioral representation is used, it is useful to introduce an alternative language construct: the process. A process is a collection of sequential statements, in the sense that the statements will be evaluated in the order that they appear in the description; hence, for statements inside a process, the order is important. Associated architecture code for the multiplexer example is presented in Code Fragment 4. With this example, we introduce the *if..then..else* statement, which is a basic mechanism for building decision structures. The VHDL syntax is:

```
    IF condition THEN
    [{ sequence_of_statements }]
    [{ELSIF condition THEN}
    [{ sequence_of_statements }]
    [ELSE
    [{ sequence_of_statements }]
    END IF;
```

A minimum *if..then..else* statement is:

```
    IF condition THEN
    sequence_of_statements
    END IF;
```

```
architecture Behavioral of logic is
begin
process (a,b,c) begin
  if c = '1' then z  <= a;
            else z  <= b;
  end if;
end process;
end Behavioral;
```

CODE FRAGMENT 4 — Multiplexer behavioral description.

After the keyword *process*, a list of signals is referred. This is the sensitive list, and whenever an event occurs in one of these signals, the process will be executed. For the present combinatorial function example, whenever any input signal changes its value, it is necessary to recalculate z (which means running the process).

Several ways were already presented enabling the representation of combinatorial logic, using different constructs of the language. Before going into sequential circuits representation, an alternative representation for a multiplexer is presented in Code Fragment 5.

```
architecture Infered of logic is
begin
   z <= a when c='1' else b;
end Infered;
```

CODE FRAGMENT 5 — Multiplexer inferred architecture.

This time, the conditional construct will be recognized by the synthesis tool as amenable to be implemented by a multiplexer (similar to the one already used). This recognition technique is what is commonly known as inferred architecture.

An additional coding example on gate representation is presented in Code Fragment 6, where a tri-state buffer is represented in three alternative types (input signal is *d*, enable signal is *en*, and output signals are *x*, *y*, and *z*). The first representation uses the conditional statement, while the remaining two rely on process constructs, using different statements, namely the *if..then..else* and the *case* statements.

```
architecture Behavioral of tristate is
begin
   x <= d when en='1' else 'Z';
   process (d,en) begin
     if en = '1'
        then y <= d;
        else y <= 'Z';
     end if;
   end process;
   process (d,en) begin
     case en is
        when '1' =>
             z <= d;
        when others =>
             z <= 'Z';
     end case;
   end process;
end Behavioral;
```

CODE FRAGMENT 6 — Alternative representations for tri-state buffers.

Thus, this example is also used to introduce the *case* statement. It can be considered as a switch. An expression is evaluated and a sequence_of_statements associated with the matching choice is executed. The general syntax is:

```
CASE expression IS
WHEN choice => sequence_of_statements
[WHEN choice => sequence_of_statements]
END CASE;
```

The *case* statement (as well the *if..then..else* statement) will be extensively used to implement state machines, as presented later in this chapter.

A concluding coding example on gate-level representation uses the concept of bus. The goal is to connect 8-bit buses *A (7 downto 0)* and *B (7 downto 0)* to an 8-bit bus *Z (7 downto 0)*, through tri-state buffers controlled by the signal C (note that a vector was used to represent the group of signals under the same name). Figure 84.3 presents the associated schematics, and Code Fragment 7 illustrates one possible representation, introducing a new VHDL statement, *generate*.

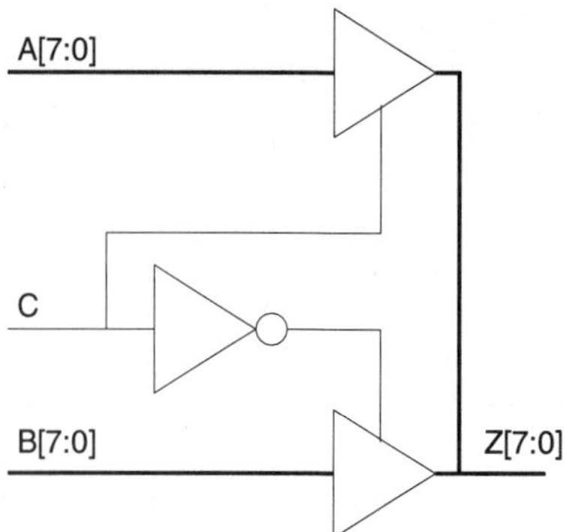

FIGURE 84.3   Bus access.

```
XPTO: for I in 0 to 7 generate
    Z(I) <= A(I) when C = '1' else 'Z';
    Z(I) <= B(I) when C = '0' else 'Z';
end generate;
```

CODE FRAGMENT 7 — Two drivers for one bus.

The *generate* statement generates logic by repeating a slice of logic. It is very useful in creating regular structures and can encapsulate concurrent statements. These concurrent statements can be either concurrent assignments, or component instances.

After presenting common possibilities to represent combinatorial logic functions, either structurally or behaviorally, we will focus on the representation of memory elements and sequential logic. There are two ways for describing memory elements in VHDL:

- Explicitly.
- Using implicit inference.

On the one hand, the explicit method relies on structural instantiation, where the component is already available in some library; its usage is similar to the instantiation of primitive components in Figure 84.2(d). In this situation, it is necessary to include at the beginning of the design the relevant library reference using the statement *use*, as it is not a primitive component.

On the other hand, implicit inference uses behavioral description based on the process construct.

Again, we can find alternative behaviorally equivalent descriptions for describing a memory element. The Code Fragment 8 presents the code for a register with eight D-type flip-flops, with asynchronous reset (*arst* signal) and synchronous reset (*srst* signal). The use of std_logic_vector-type signal allowing compact buses' representation has to be stressed.

```
entity reg is
    Port ( din : in std_logic_vector(7 downto 0);
           dout : out std_logic_vector(7 downto 0);
           arst, srst : in std_logic;
           clk : in std_logic);
end reg;
architecture Behavioral of reg is
begin
   reg : process (clk,arst)
     begin
       if (arst = '1') then
          dout <= "00000000";
```

```
      elsif (clk'event and clk='1') then
         if (srst = '1') then
            dout <= "00000000";
         else
            dout <= din;
         end if;
      end if;
   end process;
end Behavioral;
```

CODE FRAGMENT 8 — 8-bit D flip-flop register with synchronous and asynchronous resets.

```
entity cont1 is

   Port ( clk : in std_logic;

          rst : in std_logic;

          count : out std_logic_vector(7 downto 0));
end cont1;
architecture Behavioral of cont1 is
signal caux : std_logic_vector(7 downto 0);
begin
  process (clk,rst)
    begin
      if (rst = '1') then
         caux <= "00000000";
      elsif (clk'event and clk='1') then
         caux <= caux + 1;
      end if;
   end process;
   count <= caux;
end Behavioral;
```

CODE FRAGMENT 9 — Up-counter with asynchronous reset.

The sequential execution of the process statements starts whenever an event occurs on signals *clk* or *arst* (as the events on synchronous signals will not produce a change in the register status, they are not included in the sensitivity list). First, the asynchronous reset condition is evaluated; if it evaluates to true, register outputs are cleared. If not, a test on the rising edge of the signal *clk* is performed (by clk'event and clk="1"). In the presence of clock rising edge, a further test on synchronous reset condition is performed and the register is cleared or loaded with fresh data, accordingly. The statements, after the clock rising edge test, implement a register with two modes of operation under the control of one external signal (*srst*). It can be seen as a general template for multimode register implementation, for RTL design.

As an example, in Code Fragment 9 the implementation of an up-counter is presented.

```
LIBRARY ieee;
USE ieee.std_logic_1164.all;
ENTITY SM1 IS
      PORT (CLK,a,b,c,RESET: IN std_logic;
         y,z : OUT std_logic);
END;
```

CODE FRAGMENT 10 — Entity interface of the state machine represented in Figure 84.5.

Note that, due to output mode associated with *count*, it is not possible to involve *count* in the next counter state computation. In this sense, one internal signal (*caux*) is used and updated within the process. An additional concurrent statement assures that the value presented in *caux* is propagated to the output *count*.

84.8   State Machine Design

Finite state machines (FSMs) have been used for decades for control function implementation and state diagrams are, probably due to their graphical expressiveness, the most used notation representing FSMs.

Figure 84.4 presents an RTL block diagram associated with state machine implementation after synthesis from a state diagram, considering Moore and Mealy combinatorial outputs. Moore machine's outputs are a function of the current state, while Mealy machine' outputs are a function of the current state and all the inputs. It could also be of interest for some applications to consider registered outputs.

One common attitude to implement the block diagram of Figure 84.4 is to associate one process to each block. More specifically, we may have a process modeling the register that stores state variables, and another process computing the next state function and output functions.

The entity interface associated with the state machine of Figure 84.5 is presented in Code Fragment 10.

Code Fragment 11 implements the state diagram of Figure 84.5. Two processes are used: one for register modeling with reset capability (*sreg*) and another for combinatorial functions' modeling (next state function *next_sreg* and output functions *y* and *z*):

```
ARCHITECTURE BEHAVIOR OF SM1 IS
      SIGNAL sreg : std_logic_vector (1 DOWNTO 0);
      SIGNAL next_sreg : std_logic_vector (1 DOWNTO 0);
      CONSTANT STATE0 : std_logic_vector (1 DOWNTO 0) :="00";
      CONSTANT STATE1 : std_logic_vector (1 DOWNTO 0) :="01";
      CONSTANT STATE2 : std_logic_vector (1 DOWNTO 0) :="10";
BEGIN
      PROCESS (CLK, RESET)
      BEGIN
            IF ( RESET='1' ) THEN sreg <= STATE0;
            ELSIF CLK='1' AND CLK'event THEN
                  sreg <= next_sreg;
            END IF;
      END PROCESS;
      PROCESS (sreg,a,b,c)
      BEGIN
            y <= '0'; z <= '0';
            next_sreg<=STATE0;
            CASE sreg IS
                  WHEN STATE0 =>
                        z<='0'; y<='0';
                        IF ( a='1' ) THEN next_sreg<=STATE1;
                         ELSE next_sreg<=STATE0;
                        END IF;
                  WHEN STATE1 =>
                        y<='0'; z<='1';
```

```
                    IF ( b='1' ) THEN next_sreg<=STATE2;
                     ELSE next_sreg<=STATE1;
                    END IF;
              WHEN STATE2 =>
                    z<='0'; y<='1';
                    IF ( c='1' ) THEN next_sreg<=STATE0;
                     ELSE next_sreg<=STATE2;
                    END IF;
              WHEN OTHERS =>
          END CASE;
     END PROCESS;
END BEHAVIOR;
```

CODE FRAGMENT 11 — State machine representation.

One concept of key importance regarding state machine implementation is state encoding. Several techniques can be used for state encoding, namely sequential, Gray, one-hot, random, including outputs, *ad hoc,* among others. For the above-presented code, sequential coding is used. It is explicitly encoded through the use of the constants STATE0, STATE1, and STATE2.

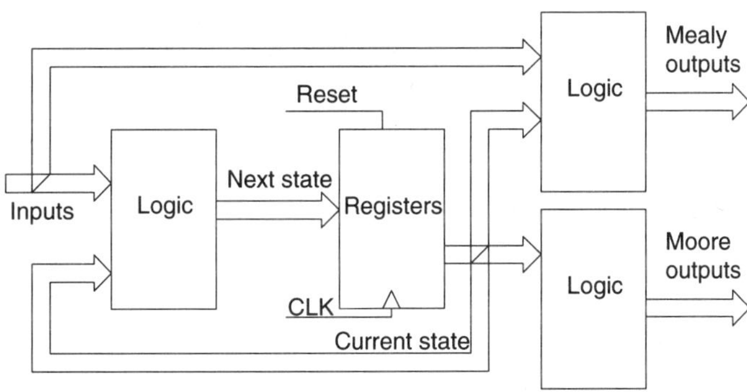

FIGURE 84.4   Block diagram for state machine implementation.

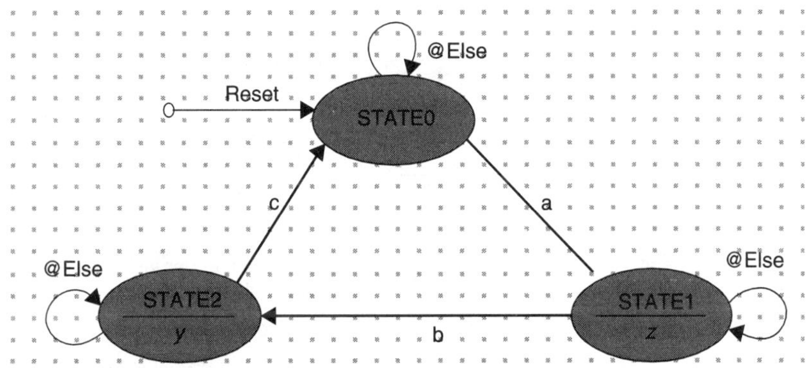

FIGURE 84.5   A simple state diagram.

Code Fragment 12 illustrates a particularly interesting "neutral" encoding scheme. Its is based on an enumerated type that will be defined later considering the synthesis tool active options.

```
ARCHITECTURE BEHAVIOR OF SM1 IS
      TYPE type_sreg IS (STATE0,STATE1,STATE2);
      SIGNAL sreg, next_sreg : type_sreg;
BEGIN ...
```

CODE FRAGMENT 12 — Symbolic state encoding scheme.

84.9 Simulation, Testing, and Design Verification

One of the initial goals of hardware description languages was to support simulation of the digital circuit design, and to verify its correctness against the specification document. Earlier design verification was accomplished through experimentation of the prototype, using signal generators, oscilloscopes, and data analyzers. But as circuit complexity increased, the importance of simulation has been proportional to it. VHDL also permits to build up a simulation environment, enabling test vectors generation, response capture, and analysis as well. Figure 84.6 presents the test-bench environment, where a VHDL top-level model represents the circuit under test.

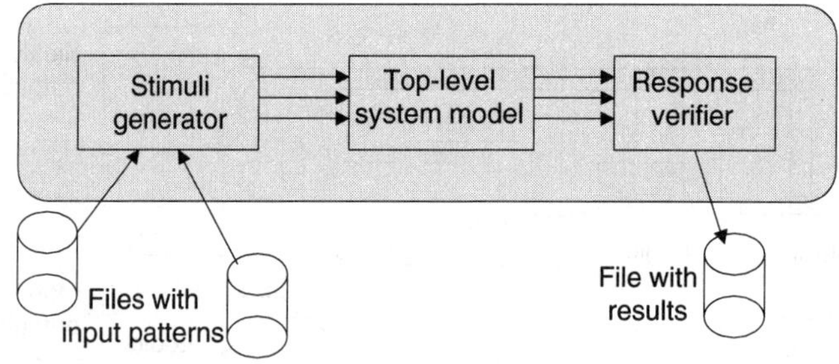

FIGURE 84.6   Test-bench environment.

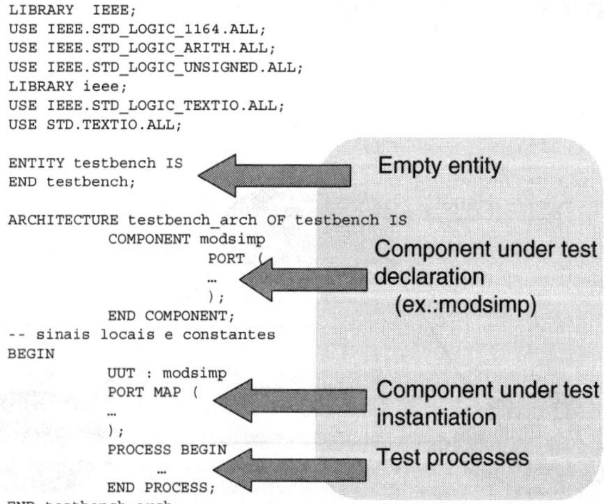

FIGURE 84.7   VHDL test-bench file structure.

Stimuli generator and response verifier are composed of pieces of VHDL code. Stimuli generator can explicitly generate adequate waveforms to exercise the model, or rely on a set of test vectors prestored in an external file. Similarly, verification of response can be accomplished by visual or automatic inspection of the results of the simulation, or stored in files for further analysis.

Anyway, considering that both stimuli generator and response verifier will be coded in VHDL, the test-bench code has no interface ports. Figure 84.7 presents a skeleton of the test-bench file.

The generation of test patterns is straightforward. It is possible to generate periodic or nonperiodic waveforms. Consider the statements contained in Code Fragment 13 as examples.

```
a <= '1', '0' after 150ns, '1' after 777ns, '0' after 2003ns;
databus <= "01010101", "11001100" after 245ns;
z <= x and y after 20ns;
```

CODE FRAGMENT 13 — Waveform generation.

The first one specifies the signal waveform identifying the time points where some change occurs; the second one does a similar job for a vector. The last statement models time delay associated with one *and* gate. It is possible to model inertial and transport delays [7].

```
PROCESS -- clock process for clk,
BEGIN
        clk <= '0';
        WAIT FOR 10 ns;
        clk <= '1';
        WAIT FOR 10 ns;
END PROCESS;
```

CODE FRAGMENT 14 — Clock generation.

Process constructs can also be used for producing periodic and nonperiodic signals. Code Fragment 14 illustrates the generation of a clock signal. Finally, Code Fragment 15 generates a nonperiodic signal, whose evolution depends on the status of some signals. Finally, the *wait* statement imposes waiting for the end of simulation.

```
PROCESS
BEGIN
        d <= '0';
        WAIT UNTIL reset='1';
        d <= '1';
        WAIT UNTIL clk='1';
        d <= '0' AFTER 10 ns;
        WAIT;
END PROCESS;
```

CODE FRAGMENT 15 — Conditional waveform generation.

References

[1] Palnitkar, S., *Verilog HDL — A Guide to Digital Design and Synthesis,* Prentice Hall, 1996, ISBN: 0134516753.
[2] McFarland, M.C., A.C. Parker, and R. Camposano, The high-level synthesis of digital systems, *Proceedings of the IEEE,* Vol. 78, no. 2, pp. 301–318, February 1990.

[3] Shiva, S.G., Computer hardware description languages — A Tutorial, *Proceedings of the IEEE,* Vol. 67, no. 12, pp. 1605–1615, December 1979.

[4] IEEE Std 1364-2001 (Revision of IEEE Std 1364-1995), IEEE Standard Verilog Hardware Description Language, 2001.

[5] IEEE Std 1076-2000 (Incorporates IEEE Std 1076-1993 and IEEE Std 1076a-2000), IEEE Standard VHDL Language Reference Manual, 2000.

[6] Sjoholm, S. and L. Lindh, *VHDL for Designers,* Prentice-Hall, Englewood Cliffs, NJ,1997.

[7] Perry, D.L., *VHDL*, McGraw-Hill Text, New York, 1998.

85

Languages for Embedded Systems[1]

Stephen A. Edwards
Columbia University

85.1 Introduction

An embedded system is a computer masquerading as a noncomputer that must perform a small set of tasks cheaply and efficiently. A typical system might have communication, signal processing, and user interface tasks to perform.

Because the tasks must solve diverse problems, a language general-purpose enough to solve them all would be difficult to write, analyze, and compile. Instead, a variety of languages have evolved, each best suited to a particular problem domain. The most obvious divide is between languages for software and hardware, but there are others. For example, a language for signal processing is often more convenient for a particular problem than, say, assembly, but might be poor for control-dominated behavior.

This chapter describes popular hardware, software, dataflow, and hybrid languages, each of which excels certain problems. Dataflow languages are good for signal processing, and hybrid languages combine ideas from the other three classes.

Due to space constraints, this chapter describes only the main features of each language. The author's book on the subject [9] provides many more details on all of these languages.

85.2 Software Languages

Software languages describe sequences of instructions for a processor to execute. As such, most consist of sequences of imperative instructions that communicate through memory: an array of numbers that hold their values until changed.

Each machine instruction typically does little more than, say, add two numbers, thus, high-level languages aim to specify many instructions concisely and intuitively. Arithmetic expressions are typical:

[1]Part of this chapter originally appeared in the Online Symposium for Electrical Engineers (OSEE).

TABLE 85.1 Software Language Features Compared

	C	C++	Java
Expressions	•	•	•
Control-flow	•	•	•
Recursive functions	•	•	•
Exceptions	○	•	•
Classes & Inheritance	•	•	
Templets		•	
Namespaces		•	•
Multiple inheritance		•	○
Threads & Locks			•
Garbage collection		○	•

•, full support;   ○, partial support.

coding an expression such as $ax^2 + bx + c$ in machine code is straightforward, tedious, and best done by a compiler. The C language provides such expressions, control-flow constructs such as loops and conditionals, and recursive functions (See Table 85.1). The C++ language adds classes as a way to build new data types, templates for polymorphic code, exceptions for error handling, and a standard library of common data structures. Java is a still higher-level language that provides automatic garbage collection, threads, and monitors to synchronize them.

Assembly Languages

An assembly language program (Figure 85.1) is a list of processor instructions written in a symbolic, human-readable form. Each instruction consists of an operation such as addition along with some operands. For example, `add r5, r2, r4` might add the contents of registers `r2` and `r4` and write the result to `r5`. Such arithmetic instructions are executed in order, but branch instructions can perform conditionals and loops by changing the processor's program counter — the address of the instruction being executed.

A processor's assembly language is defined by its opcodes, addressing modes, registers, and memories. The opcode distinguishes, say, addition from conditional branch, and an addressing mode defines how and where data are gathered and stored (e.g., from a register or from a particular memory location). Registers can be thought of as small, fast, easy-to-access pieces of memory.

There are roughly four categories of modern assembly languages (Table 85.2). The oldest are those for the so-called complex instruction set computers, or CISC. These are characterized by a rich set of instructions

```
        jmp   L2                    mov   %i0, %o1
    L1:                             b     .LL3
        movl  %ebx, %eax            mov   %i1, %i0
        movl  %ecx, %ebx            mov   %i0, %o1
    L2:                             b     .LL3
        xorl  %edx, %edx            mov   %i1, %i0
        divl  %ebx              .LL5:
        movl  %edx, %ecx            mov   %o0, %i0
        testl %ecx, %ecx        .LL3:
        jne   L1                    mov   %o1, %o0
                                    call  .rem, 0
                                    mov   %i0, %o1
                                    cmp   %o0, 0
                                    bne   .LL5
                                    mov   %i0, %o1

        (a)                             (b)
```

FIGURE 85.1 Euclid's algorithm (a) i386 assembly (CISC) and (b) SPARC assembly (RISC). SPARC has more registers and must call a routine to compute the remainder (the i386 has division instruction). The complex addressing modes of the i386 are not shown in this example.

TABLE 85.2 Typical Modern Processor Architectures

CISC	RISC	DSP	Microcontroller
x86	SPARC	TMS320	8051
68000	MIPS	DSP56000	PIC
	ARM	ASDSP-21xx	AVR

and addressing modes. For example, a single instruction in Intel's x86 family, a typical CISC processor, can add the contents of a register to a memory location whose address is the sum of two other registers and a constant offset. Such instruction sets are usually convenient for human programmers, who are generally fairly skilled at using a heterogeneous set of tools, and the code itself is usually quite compact. Figure 85.1(a) illustrates a small program in x86 assembly.

By contrast, reduced instruction set computers (RISC) tend to have fewer instructions and much simpler addressing modes. The philosophy is that while you generally need more RISC instructions to accomplish something, it is easier for a processor to execute them because it does not need to deal with the complex cases and it is easier for a compiler to produce them because they are simpler and more uniform. Figure 85.1(b) illustrates a small program in SPARC assembly.

The third category of assembly languages arises from more specialized processor architectures such as digital signal processors (DSPs) and very-long instruction word processors (VLIWs). The operations in these instruction sets are simple like those in RISC processors (e.g., add two registers); but they tend to be very irregular (only certain registers may be used with certain operations) and support a much higher degree of instruction-level parallelism. For example, Motorola's DSP56001 can, in a single instruction, multiply two registers, add the result to a third, load two registers from memory, and increase two circular buffer pointers. However, the instruction severely limits which registers (and even which memory) it may use. Figure 85.2(a) shows a filter implemented in 56001 assembly.

The fourth category includes instruction sets on small (4- and 8-bit) microcontrollers. In some sense, these combine the worst of all worlds: there are few instructions and each cannot do much, much like an RISC processor, and there are also significant restrictions on which registers can be used when, much like a CISC processor. The main advantage of such instruction sets is that they can be implemented very cheaply. Figure 85.2(b) shows a routine that writes to a parallel port in 8051 assembly.

```
move  #samples, r0                      START:
move  #coeffs, r4                          MOV   SP, #030H
move  #n-1, m0                             ACALL INITIALIZE
move  m0, m4                               ORL   P1,#0FFH
movep y:input, x:(r0)                      SETB  P3.5
clr   a                                  LOOP:
      x:(r0)+, x0 y:(r4)+, y0              CLR   P3.4
rep   #n-1                                 SETB  P3.3
mac   x0,y0,a                              SETB  P3.4
      x:(r0)+, x0 y:(r4)+, y0            WAIT:
macr  x0,y0,a  (r0)-                        JB    P3.5, WAIT
movep a, y:output
                                           CLR   P3.3
                                           MOV   A,P1
                                           ACALL SEND
                                           SETB  P3.3
                                           AJMP  LOOP
         (a)                                       (b)
```

FIGURE 85.2 (a) A finite impulse response filter in DSP56001 assembly. The *mac* instruction (multiply and accumulate) does most of the work, multiplying registers X0 and Y0, adding the result to accumulator A, fetching the next sample and coefficient from memory, and updating circular buffer pointers R0 and R4. The *rep* instruction repeats the *mac* instruction in a zero-overhead loop. (b) Writing to a parallel port in 8051 microcontroller assembly. This code takes advantage of the 8051's ability to operate on single bits.

The C Language

C is currently the most popular language for embedded system programming. C compilers exist for virtually every general-purpose processor, from the lowliest 4-bit microcontroller to the most powerful 64-bit processor for compute servers.

C was originally designed by Dennis Ritchie [23] as an implementation language for the Unix operating system being developed at Bell Labs for a 24K DEC PDP-11. Because the language was designed for systems programming, it provides very direct access to the processor through such constructs as untyped pointers and bit-manipulation operators, things appreciated today by embedded systems programmers. Unfortunately, the language also has many awkward aspects, such as the need to define everything before it is used, which are holdovers from the cramped execution environment in which it was first implemented.

A C program (Figure 85.3) contains functions built from arithmetic expressions structured with loops and conditionals. Instructions in a C program run sequentially, but control-flow constructs such as loops of conditionals can affect the order in which instructions execute. When control reaches a function call in an expression, control is passed to the called function, which runs until it produces a result, and control returns to continue evaluating the expression that called the function.

C derives its types from those that the processor manipulates directly: signed and unsigned integers ranging from bytes to words, floating point numbers, and pointers. These can be further aggregated into arrays and structures — groups of named fields.

C programs use three types of memory. Space for global data is allocated when the program is compiled, the stack stores automatic variables allocated and released when their function is called and returns, and the heap supplies arbitrarily sized regions of memory that can be deallocated in any order.

The C language is an ISO standard, but most people consult the book by Kernighan and Ritchie [16].

C succeeds because it can be compiled into very efficient code and because it allows the programmer almost arbitrarily low-level access to the processor when necessary. As a result, virtually every function can be written in C (exceptions include those that must manipulate specific processor registers) and can be expected to be fairly efficient. C's simple execution model also makes it fairly easy to estimate the efficiency of a piece of code and improve it if necessary.

While C compilers for workstation-class machines usually comply closely to ANSI/ISO standard C, C compilers for microcontrollers are often much less standard. For example, they often omit support for floating-point arithmetic and certain library functions. Many also provide language extensions that, while often very convenient for the hardware for which they were designed, can make porting the code to a different environment very difficult.

```c
#include <stdio.h>

int main(int argc, char *argv[])
{
  char *c;
  while (++argv, --argc > 0) {
    c = argv[0] + strlen(argv[0]);
    while (--c >= argv[0])
      putchar(*c);
    putchar('\n');
  }
  return 0;
}
```

FIGURE 85.3 A C program that prints each of its arguments backwards. The outermost while loop iterates through the arguments (count in argc, array of strings in argv), while the inner loop starts a pointer at the end of the current argument and walks it backwards, printing each character along the way. The $++$ and $--$ prefixes increment the variable they are attached to before returning its value.

C++

C++ (Figure 85.4) [24] extends C with structuring mechanisms for large programs: user-defined data types, a way to reuse code with different types, namespaces to group objects and avoid accidental name collisions when program pieces are assembled, and exceptions to handle errors. The C++ standard library includes a collection of efficient polymorphic data types such as arrays, trees, and strings for which the compiler generates custom implementations.

A class defines a new data type by specifying its representation and the operations that may access and modify it. Classes may be defined by inheritance, which extends and modifies existing classes. For example, a rectangle class might add length and width fields and an area method to a shape class.

A template is a function or class that can work with multiple types. The compiler generates custom code for each different use of the template. For example, the same *min* template could be used for both integers and floating-point numbers.

C++ also provides exceptions, a mechanism intended for error recovery. Normally, each method or function can only return directly to its immediate caller. Throwing an exception, however, allows control to return to an arbitrary caller, usually an error-handling mechanism in the *main* function or similar. Exceptions can be used, for example, to gracefully recover from out-of-memory conditions no matter where they occur, without the tedium of having to check whether every function encountered an out-of-memory condition.

Memory consumption is a disadvantage to C++'s exception mechanism. While most C++ compilers do not generate a slower code when exceptions are enabled, they do generate larger executables by including tables that record the location of the nearest exception handler. For this reason, many compilers, such as GNU's gcc, have a flag that completely disables exceptions.

C++ is being used more and more within embedded systems, but it is sometimes a less suitable choice than C for a number of reasons. First, C++ is a much more complicated language that demands a much larger compiler; hence, C++ has been ported to fewer architectures than C. Second, certain language features such as dynamic dispatch (virtual function calls) and exceptions can be too costly to implement in very small embedded systems. It is a more difficult language to learn and use properly, meaning there may be fewer qualified C++ programmers. Also, it is often more difficult to estimate the cost of a certain construct in C++ because the object-oriented programming style encourages many more function calls than the procedural style of C, and the cost of these is harder to estimate.

```cpp
class Cplx {
  double re, im;
public:
  Cplx(double v) : re(v), im(0) {}
  Cplx(double r, double i)
    : re(r), im(i) {}
  double abs() const {
    return sqrt(re*re + im*im);
  }
  void operator+= (const Cplx& a) {
    re += a.re; im += a.im;
  }
};

int main() {
  Cplx a(5), b(3,4);
  b += a;
  cout << b.abs() << '\n';
  return 0;
}
```

FIGURE 85.4   A C++ fragment illustrating a partial complex number type and how it can be used (the C++ library has a complete version). This class defines how to create a new complex number from either a scalar or by specifying the real and imaginary components, how to compute the absolute value of a complex number, and how to add a complex number to an existing one.

Java

Sun's Java language [1,11,19] resembles C++ but is not a superset. Like C++, Java is object-oriented, providing classes and inheritance. It is a higher-level language than C++ since it uses object references, arrays, and strings instead of pointers. Java's automatic garbage collection frees the programmer from memory management.

Java omits a number of C++'s more complicated features. Templates are absent, although there are plans to include them in a future release of the language because they make it possible to write type-safe container classes. Java also omits operator overloading, which can be a boon to readability (e.g., when performing operations on complex numbers) or a powerful obfuscating force. Java also does not support C++'s complex multiple inheritance mechanism completely. But it does provide the notion of an interface — a set of methods provided by a class — that is equivalent to one of the most common uses of multiple inheritance.

Java provides concurrent threads (Figure 85.5). Creating a thread involves extending the *Thread* class, creating instances of these objects, and calling their *start* methods to start a new thread of control that executes the objects' *run* methods.

Synchronizing a method or block uses a per-object lock to resolve contention when two or more threads attempt to access the same object simultaneously. A thread that attempts to gain a lock owned by another thread will block until the lock is released, which can be used to grant a thread exclusive access to a particular object.

For embedded systems, Java holds promise but also many caveats. On the positive side, it is a simple, powerful language that provides the programmer a convenient set of abstractions. For example, unlike C,

```java
import java.io.*;
class Counter {
    int value = 0;
    boolean present = false;
    public synchronized void count() {
        try { while (present) wait(); }
        catch (InterruptedException e) {}
        value++; present = true; notifyAll();
    }
    public synchronized int read() {
        try { while (!present) wait(); }
        catch (InterruptedException e) {}
        present = false; notifyAll();
        return value;
    }
}
class Count extends Thread {
    Counter cnt;
    public Count(Counter c) { cnt = c; start(); }
    public void run() { for (;;) cnt.count(); }
}
class Mod5 {
    public static void main(String args[]) {
        Counter c = new Counter();
        Count count = new Count(c);
        int v;
        for (;;) if ( (v = c.read()) % 5 == 0 )
            System.out.println(v);
    }
}
```

FIGURE 85.5 A contrived Java program that spawns a counting thread to print all numbers divisible by 5. The *main* method in the Mod5 class creates a new Counter, and then a new Count object. The Count class extends the Thread class and spawns a new thread in its constructor by executing *start*. This invokes its *run* method, which calls the method *count*. Both *count* and *read* are synchronized, meaning at most one may run on a particular Count object at once, here guaranteeing the counter is either counting or waiting for its value to be read.

Java provides true strings and variable-sized arrays. On the negative side, Java is a heavyweight language, even more so than C++. Its run-time system is large, consisting of either a bytecode interpreter, a just-in-time compiler, or perhaps both, and its libraries are absolutely vast. While work has been done on paring down these things, Java still requires a much larger footprint than C.

Unpredictable runtimes are a more serious problem for Java. For time-critical embedded systems, Java's automatic garbage collector, bytecode interpreter, or just-in-time compiler make runtimes both unpredictable and variable, making it difficult to assess efficiency both beforehand and in simulation.

The real-time Java specification [6] attempts to address many of these concerns. It introduces mechanisms for more precise control over the scheduling policy for concurrent threads (the standard Java specification is deliberately vague on this point to improve portability), memory regions for which automatic garbage collection can be disabled, synchronization mechanisms for avoiding priority inversion, and various other real-time features such as timers. It remains to be seen, however, whether this specification addresses enough real-time concerns and is sufficiently efficient to be practical. For example, a naive implementation of the memory management policies would be very inefficient.

Real-Time Operating Systems

Many embedded systems use a real-time operating system (RTOS) to simulate concurrency on a single processor. An RTOS manages multiple running processes, each written in sequential language such as C. The processes perform the system's computation and the RTOS schedules them — attempts to meet deadlines by deciding which process runs when. Labrosse [17] describes the implementation of a particular RTOS.

Most RTOSes uses fixed-priority preemptive scheduling in which each process is given a particular priority (a small integer) when the system is designed. At any time, the RTOS runs the highest-priority running process, which is expected to run for a short period of time before suspending itself to wait for more data. Priorities are usually assigned using rate-monotonic analysis [7] (due to Liu and Layland [20]), which assigns higher priorities to processes that must meet more frequent deadlines (Figure 85.6).

Priority inversion is a fundamental problem in fixed-priority preemptive scheduling that can lead to missed deadlines by enabling a lower-priority process to delay indefinitely the execution of a higher-priority one. Figure 85.7 illustrates the typical scenario: a low-priority process L runs and acquires a resource. Shortly thereafter, a high-priority process H preempts L, attempts to acquire the same resource, and blocks waiting for L to release it. This can cause H to miss its deadline even though it is at a higher priority than L. Even worse, if a process M with priority between L and now starts, it can delay the execution of H indefinitely. Process M does not allow L to run since M is at a higher priority; hence, L cannot execute and release the lock and H will continue to block.

Priority inversion is usually solved with priority inheritance. When a process L acquires a lock, its priority is temporarily raised to a level where it will not be preempted by any other process that will also attempt to acquire the lock. Many RTOSes provide a mechanism for doing this automatically.

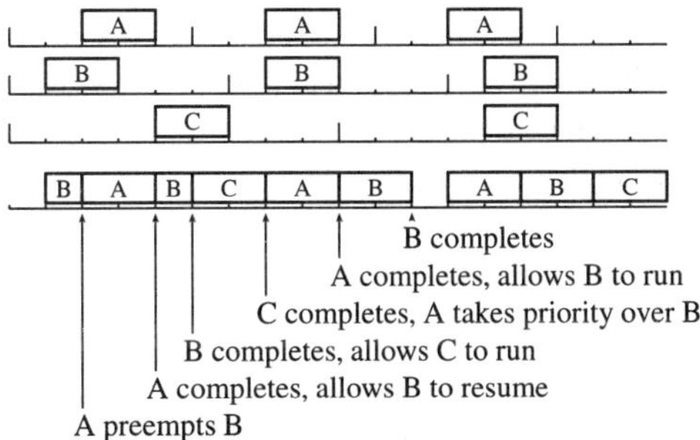

FIGURE 85.6   The behavior of an RTOS with fixed-priority preemptive scheduling. Rate-monotonic analysis gives process A the highest priority since it has the shortest period; C has the lowest.

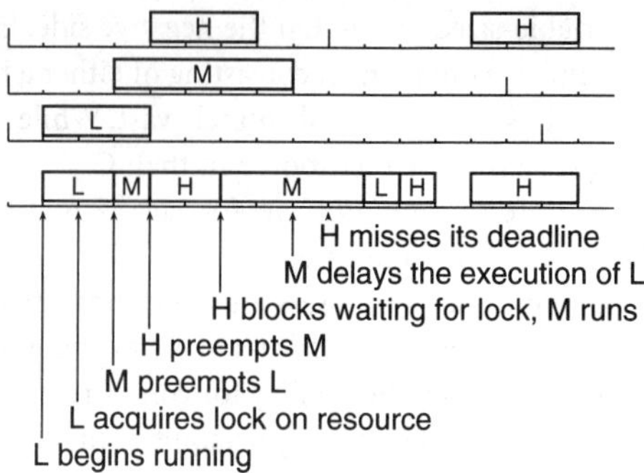

FIGURE 85.7 Priority inversion illustrated. When low-priority process L acquires a lock on a resource needed by process H, it effectively blocks process H, but then intermediate-priority process M preempts L, preventing it from running and releasing the resource needed by H. Priority inheritance, the common solution, temporarily raises the priority of L to that of H when H requests the resource held by L.

85.3 Hardware Languages

Concurrency and the notion of control is the fundamental difference between hardware and software. In hardware, every part of the "program" is always running, but in software, exactly one part of the program is running at any one time. Software languages naturally focus on sequential algorithms, while hardware languages enable concurrent function evaluation, speculation, and concurrency.

Ironically, efficient simulation *in software* is a main focus of the hardware languages presented here; thus, their discrete-event semantics are a compromise between what would be ideal for hardware and what simulates efficiently.

Verilog [13, 25] and VHDL [12, 22, 8, 2] are the most popular languages for hardware description and modeling (Figures 85.8 and 85.9). Both are model systems with discrete-event semantics that ignore idle portions of the design for efficient simulation. Both describe systems with structural hierarchy: a system consists of blocks that contain instances of primitives, other blocks, or concurrent processes. Connections are listed explicitly.

Verilog provides more primitives geared specifically toward hardware simulation. VHDL's primitive are assignments such as $a = b + c$ or procedural code. Verilog adds transistor and logic gate primitives, and allows new ones to be defined with truth tables.

Both languages allow concurrent processes to be described procedurally. Such processes sleep until awakened by an event that causes them to run, read, and write variables, and suspend. Processes may wait for a period of time (e.g., #10 in Verilog, `wait for 10 ns` in VHDL), a value change (`@(a or b)`, `wait on a,b`), or an event (`@(posedge clk)`, `wait on clk until clk='1'`).

VHDL communication is more disciplined and flexible. Verilog communicates through *wires* or *regs*: shared memory locations that can cause race conditions. VHDL's signals behave like wires but the resolution function may be user-defined. VHDL's variables are local to a single process unless declared shared.

Verilog's type system models hardware with four-valued bit vectors and arrays for modeling memory. VHDL does not include four-valued vectors, but its type system allows them to be added. Furthermore, composite types such as C *structs* can be defined.

Overall, Verilog is the leaner language more directly geared toward simulating digital integrated circuits. VHDL is a much larger, more verbose language capable of handing a wider class of simulation and modeling tasks.

Verilog

Verilog was first devised in 1984 as an input language for a discrete-event simulator for digital hardware design. It was one of the first hardware description languages able to specify both the circuit and a test bench in the same language, which remains one of its strengths.

Verilog has since been pressed into use as both a modeling language and a specification language. Although Verilog is still simulated frequently, it is also frequently fed to a logic synthesis system that translates it into an actual circuit. This is a technically challenging process and not all Verilog constructs can be translated into hardware since Verilog's semantics are nondeterministic and effectively defined by the behavior of an event-driven simulator.

Verilog provides both structural and behavioral modeling styles, and allows them to be combined at will. Consider the simple multiplexer circuit shown in Figure 85.8(a). It can be modeled in Verilog as a schematic

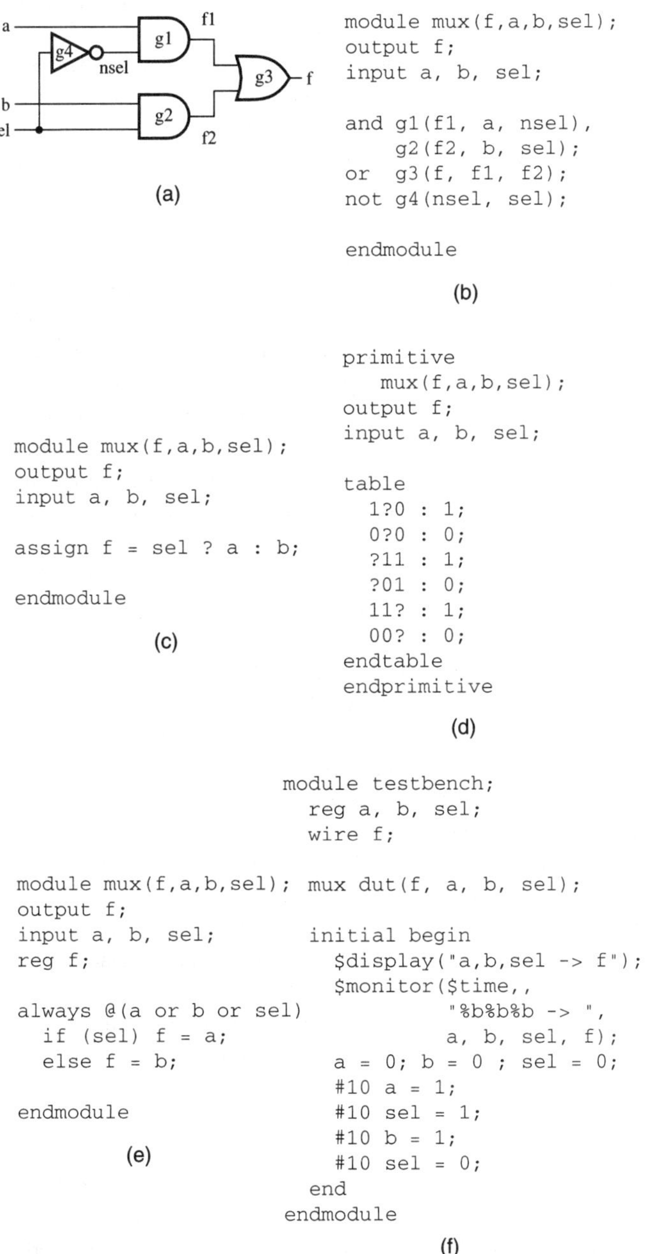

FIGURE 85.8   Verilog examples. (a) A multiplexer circuit. (b) The multiplexer described as a Verilog structural model. (c) The multiplexer described using a continuous assignment. (d) A user-defined primitive for the multiplexer. (e) The multiplexer described with imperative code. (f) A testbench for the multiplexer.

composed of logic gates (Figure 85.8[b]), with a continuous assignment statement that represents logic using an expression (Figure 85.8[c]), with a truth table as a "user-defined primitive" (Figure 85.8[d]), or with an imperative, event-driven code (Figure 85.8[e]).

The imperative modeling style is particularly useful for creating testbenches: models of an environment that stimulate a particular circuit and check its behavior. Figure 85.8(f) illustrates such a testbench, which instantiates a multiplexer (the instance is called "dut"— device under test) and starts a simple process (the `initial` block) to apply inputs and monitor outputs. Running Figure 85.8(f) in a Verilog simulator gives a partial truth table for the multiplexer.

As these examples illustrate, a Verilog program is composed of *modules.* Each module has an interface with named input and output ports and contains one or more instances of other modules, continuous assignments, and imperative code in *initial* and *always* blocks. Modules perform the same information hiding function as functions in imperative languages: a module's contents is not visible from outside and names for instances, wires, and whatnot inside a module do not have to differ from those in other modules.

Verilog programs manipulate four-valued bit vectors intended to model digital hardware. Each bit is 0, 1, X, representing unknown, or Z, used to represent an undriven tri-state bus. While such vectors are very convenient for modeling circuitry, one of Verilog's shortcomings is the lack of a more complicated type system. It does provide arrays of bit vectors but no other aggregate types.

The plumbing within a module comes in two varieties: one for structural modeling, and the other for behavioral. Structural components, such as instances of primitive logic gates and other modules, communicate through *wires,* each of which may be connected to drivers such as gates or continuous assignments. Conceptually, the value of a wire is computed constantly from whatever drives it. Practically, the simulator evaluates the expression in a continuous assignment whenever any of its inputs changes.

Behavioral components communicate through *regs,* which behave like memory in traditional programming languages. The value of a reg is set by an assignment statement executed within an initial or always block, and that value persists until the next time the reg is assigned. While a reg can be used to model a state-holding element such as a latch or flip-flop, it is important to remember that they are really just memory. Figure 85.8(e) illustrates this: a reg is used to store the output of the mux, even though it is not a state-holding element. This is because imperative code can only change the value of regs, not wires.

Verilog is a large language that contains many now-little-used features such as switch-level transistor models, pure event handling, and complicated delay specifications, all remnants of previous design methodologies. Today, switch-level modeling is rarely used because Verilog's precision is too low for circuits that take advantage of this behavior (a continuous simulator such as SPICE is preferred). Delays are rarely used because static timing analysis has replaced event-driven simulation as the timing analysis method of choice because of its speed and precision. Nevertheless, Verilog remains one of the most commonly used languages for hardware design.

System Verilog, a recently introduced standard (2002), is an extension to the Verilog language designed to aid in the creation of large specifications. It adds a richer set of datatypes, including C-like structures, unions, and multidimensional arrays, a richer set of processes (e.g., an `always_comb` block has an implied sensitivity to all variables it references), the concept of an interface to encapsulate communication and function between blocks, and many other features. Whether System Verilog supplants Verilog as a standard language for hardware specification remains to be seen, but it does have the advantage of being an obvious evolutionary improvement over previous versions of Verilog.

VHDL

The VHDL language (VHDL is a two-level acronym, standing for Very High Speed Integrated Circuit [VHSIC] Hardware Description Language) was designed to be a flexible modeling language for digital systems. It has fewer built-in features such as Verilog's four-valued bit vectors, gate, and transistor-level models. Instead, it has very flexible type and package systems that allow such things to be specified in the language.

Unlike Verilog, VHDL draws a strong distinction between the interface to a hierarchical object and its implementation. VHDL interfaces are called *entities* and their implementations are called *architectures.*

Figure 85.9 illustrates how these are used in a simple model: the entities are essentially named lists of ports and the architectures consist of named lists of component instances. While this increases the verbosity of the language, it makes it possible to use different implementations, perhaps at differing levels of abstraction.

```
entity NAND is
   port (a: in Bit; b: in Bit; y: out Bit);
end NAND;
```

(a)

```
architecture arch1 of mux2 is
   signal cc, ai, bi : Bit; -- internal signals

   component Inverter          -- component interface
      port (a:in Bit; y: out Bit);
   end component;

   component AndGate
      port (a1, a2:in Bit; y: out Bit);
   end component;

   component OrGate
      port (a1, a2:in Bit; y: out Bit);
   end component;

begin
   I1: Inverter port map(c => a, y => cc); -- by name
   A1: AndGate  port map(a, c, ai); -- by position
   A2: AndGate  port map(a1 => b, a2 => cc, y => bi);
   O1: OrGate   port map(a1 => ai, a2 => bi, y => d);
end;
```

(b)

```
architecture arch2 of mux2 is
   signal cc, ai, bi : Bit;
begin
   cc <= not c;
   ai <= a and c;
   bi <= b and cc;
   d  <= ai or bi;
end;
```

(c)

```
architecture arch3 of mux2 is
begin
   process(a, b, c)  -- sensitivity list
   begin
     if c = '1' then
        d <= a;
     else
        d <= b;
     end if;
   end process;
end;
```

(d)

FIGURE 85.9 VHDL examples. Compare with Figure 85.8. (a) The entity declaration for the multiplexer, which defines its interface. (b) A structural description of the multiplexer from Figure 85.8(a). (c) A dataflow description with one equation per gate. (d) An imperative behavioral description.

Like Verilog, VHDL supports structural, dataflow, and behavioral modeling styles, illustrated in Figure 85.9. As in Verilog, they can be mixed. In the three styles, an architecture is specified by listing components and their connections (structural), as a series of equations (dataflow, like Verilog's `assign` declarations), or as a sequence of imperative instructions (behavioral, like Verilog's `always` blocks).

In general, a process runs until it reaches a `wait` statement. This suspends the process until a particular event occurs, which may be an event on a signal, a condition on a signal, a timeout, or any combination of these. By itself, `wait` terminates a process. At the other extreme, `wait on Clk until Clk = "1" for 5 ns;` waits for the clock to rise or for 5 ns, whichever comes first.

Combinational processes, which always run in response to a change on any of their inputs, are common enough to warrant a shorthand. Thus, `process (A,B,C)` effectively executes a `wait on A,B,C` statement at the end.

VHDL's type system is much more elaborate than Verilog's. It provides integers, floating-point numbers, enumerations, and physical quantities. Integers and floating-point numbers include a range specification. For example, a 16-bit integer might be declared as

```
type address is range 16#0000# to 16#FFFF#;
```

Enumerated literals may be single characters or identifiers. Identifiers are useful for FSM states and single characters are useful for Boolean wire values. Typical declarations:

```
type Bit is ('0', '1');
type FourV is ('0', '1', 'X', 'Z');
type State is (Reset, Running, Halted);
```

Objects in VHDL, such as types, variables, and signals, have attributes such as size, base, and range. Such information can be useful for, say, iterating over all elements in an array. For example, if `type Index is range 31 downto 0`, then `Index'LOW` is 0. Access to information about signals can be used for collecting simulation statistics. For example, if `Count` is a signal, then `Count'EVENT` is true when there is an event on the signal.

VHDL has a powerful library and package facility for encapsulating and reusing definitions. For example, the standard logic library for VHDL includes types for representing wire states and standard functions such as AND and OR that operate on these types. Verilog has such facilities built in, but is not powerful enough to allow such functionality to be written as a library.

85.4 Dataflow Languages

The hardware and software languages described earlier have semantics very close to that of their implementations (e.g., as instructions on a sequential processor or as digital logic gates), which makes for efficient realizations, but some problems are better described using different models of computation.

Many embedded systems perform signal processing tasks such as reconstructing a compressed audio signal. While such tasks can be described and implemented using the hardware and software languages described earlier, signal-processing tasks are more conveniently represented with systems of processes that communicate through queues. Although clumsy for general applications, dataflow languages are a perfect fit for signal-processing algorithms, which use vast quantities of arithmetic derived from linear system theory to decode, compress, or filter data streams that represent periodic samples of continuously changing values such as sound or video. Dataflow semantics are natural for expressing the block diagrams typically used to describe signal-processing algorithms, and their regularity makes dataflow implementations very efficient because otherwise costly run-time scheduling decisions can be made at compile time, even in systems containing multiple sampling rates.

Kahn Process Networks

Kahn Process Networks [15] form a formal basis for dataflow computation. Kahn's systems consist of processes that communicate exclusively through unbounded point-to-point first-in, first-out queues

```
                process f(in int u, in int v, out int w)
                {
                  int i; bool b = true;
                  for (;;) {
                    i = b ? wait(u) : wait(w);
                    printf("%i\n", i);
                    send(i, w);
                    b = !b;
                  }
                }

                process g(in int u, out int v, out int w)
                {
                  for (;;) {
                    send(wait(u), v); send(wait(u), w);
                  }
                }

                process h(in int u, out int v, int init)
                {
                  send(v, init);
                  for(;;)
                    send(wait(u), v);
                }

                channel int X, Y, Z, T1, T2;
                f(Y, Z, X);
                g(X, T1, T2);
                h(T1, Y, 0);
                h(T2, Z, 1);
```

FIGURE 85.10   A Kahn Process Network written in a C-like dialect. Here, processes are functions that run continuously, may be attached to communication channels, and may call *wait* to wait for data on a particular port and *send* to write data to a particular port. The f process alternately copies from its u and v ports to its w port; the g process does the opposite, copying its u port to alternately v and w, and h simply copies its input to its output.

(Figure 85.10). Reading from a port makes a process wait until data are available, but writing to a port always completes immediately.

Deterministic behavior is the most unique aspect of Kahn's networks. Processes' blocking read behavior guarantees the overall system behavior (specifically, the sequence of data tokens that flow through each queue) is the same regardless of the relative execution rates of the processes, that is, regardless of the scheduling policy. This is generally a very desirable property because it provides a guarantee about the behavior of the system, ensures that simulation and reality will match, and greatly simplifies the design task since designers are not obligated to ensure this themselves.

Balancing processes' relative execution rates to avoid an unbounded accumulation of tokens is the challenge in scheduling a Kahn network. One general approach, proposed in Parks' thesis [21], places artificial limits on the size of each buffer. Any process that writes to a full buffer blocks until space is available, but if the system deadlocks because all buffers are full, the scheduler increases the capacity of the smallest buffer.

In practice, Kahn networks are rarely used in their pure form since they are fairly costly to schedule and their completely deterministic behavior is sometimes overly restrictive since they cannot easily handle sporadic events (e.g., an occasional change of volume level in a digital volume control) or server-like behavior where the environment may make requests in an unpredictable order. Nevertheless, Kahn's model still has useful properties and forms a starting point for other dataflow models.

Synchronous Dataflow

Lee and Messerschmitt's [18] Synchronous Dataflow (SDF) fixes the communication patterns of the blocks in a Kahn network (Figure 85.11 shows an example after Bhattacharyya et al. [5]). Each time a block runs, it consumes and produces a fixed number of data tokens on each of its ports. Although more

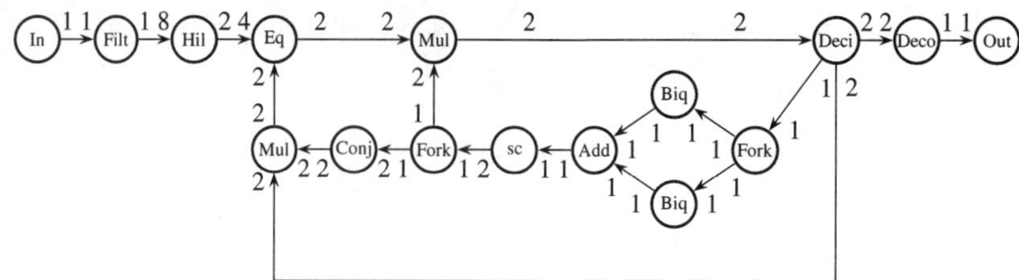

FIGURE 85.11 A modem in SDF. Each node represents a process. The labels on each arc indicate the number of tokens sent or received by a process each time it fires.

restrictive than Kahn networks, SDF's predictability allows it to be scheduled completely at compile time, producing a very efficient code.

Scheduling operates in two steps. First, the rate at which each block fires is established by considering the production and consumption rates of each block at the source and sink of each queue. For example, the arc between the Hil and Eq nodes in Figure 85.11 implies that Hil runs twice as frequently. Once the rates are established, any algorithm that simulates the execution of the network without buffer underflow will produce a correct schedule if one exists. However, more sophisticated techniques reduce generated code and buffer sizes by better ordering the execution of the blocks (see Bhattacharyya et al. [4]).

Synchronous dataflow specifications are built by assembling blocks typically written in an imperative language such as C. The SDF block interface is specific enough to make it easy to create libraries of general-purpose blocks such as adders, multipliers, and even FIR filters.

While SDF is often used as a simulation language, it is also well-suited to code generation. It enables a practical technique for generating code for digital signal processors, for which C compilers often cannot generate efficient code. Assembly code is handcrafted for each block in a library, and code synthesis consists of assembling these handwritten blocks, sometimes generating extra code that handles the interblock buffers. For large, specialized blocks such as fast Fourier transforms, this can be very effective because most of the generated code was carefully optimized by hand.

85.5 Hybrid Languages

The languages in this section use even more novel models of computation than the hardware, software, or dataflow languages presented earlier. While such languages are more restrictive than general-purpose ones, they are much better-suited for certain applications. Esterel excels at discrete control by blending software-like control flow with the synchrony and concurrency of hardware. Communication protocols are SDL's forte; it uses extended finite-state machines with single input queues. SystemC provides a very flexible discrete-event simulation environment built on C++.

Esterel

Intended for specifying control-dominated reactive systems, Esterel [3] combines the control constructs of an imperative software language with concurrency, preemption, and a synchronous model of time like that used in synchronous digital circuits. In each clock cycle, the program awakens, reads its inputs, produces outputs, and suspends (Table 85.3).

An Esterel program communicates through signals that are either present or absent in each cycle. In each cycle, each signal is absent unless an emit statement for the signal runs and makes the signal present for that cycle only. Esterel guarantees determinism by requiring each emitter of a signal to run before any statement that tests the signal.

Esterel is strongest at specifying hierarchical state machines. In addition to sequentially composing statements (separated by a semicolon), it has the ability to compose arbitrary blocks of code in parallel (the double vertical bars) and abort or suspend a block of code when a condition is true. For example, the *every-do* construct in Figure 85.12 effectively wraps a reset statement around two state machines running in parallel.

TABLE 85.3 Hybrid Language Features Compared

	Esterel	SDL	SystemC
Concurrency	•	•	•
Hierarchy	•	•	•
Preemption	•		o
Determinism	•	o	o
Synchronous communication	•		•
Buffered communication		•	•
FIFO communication		•	•
Procedural	•	o	•
Finite-state machines	•	•	
Dataflow		•	•
Multirate dataflow			o
Software implementation	•	•	
Hardware implementation	•		o

• full support   o partial support

SDL

SDL is a graphical specification language developed for describing telecommunication protocols defined by the ITU [14] (Ellsberger [10] is more readable). A system consists of concurrently running FSMs, each with a single input queue, connected by channels that define which messages they carry. Each FSM consumes the most recent message in its queue, reacts to it by changing internal state or sending messages to other FSMs, changes to its next state, and repeats the process. Each FSM is deterministic, but because messages from other FSMs may arrive in any order because of varying execution speed and communication delays, an SDL system may behave nondeterministically.

In addition to a fairly standard textual format, SDL has a formalized graphical notation. There are three types of diagrams. Flowcharts define the behavior of state machines at the lowest level (Figure 85.13). Block diagrams illustrating the communication among state machines local to a single processor are at the next level up. Each communication channel is labeled with the set of messages that it conveys. The top level is another block diagram that depicts the communication among processors. The communica-

```
module Example:

input  S, I;
output O;

signal R, A in
  every S do
    await I;
    weak abort
      sustain R
    when immediate A;
    emit O
  ||
    loop
      pause; pause;
      present R then emit A end;
    end
  end
end

end module
```

FIGURE 85.12   An Esterel program modeling a shared resource. This implements two parallel threads (separated by | |); one waits for an I signal, then asserts R until it received an A from the other thread and emits an O. Meanwhile, the second thread emits an R in response to an A in alternate cycles.

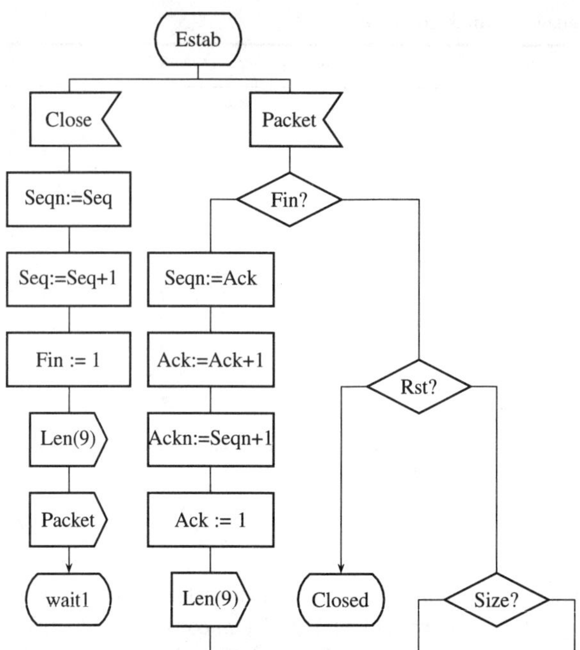

FIGURE 85.13   A fragment of an SDL flowchart specification for a TCP protocol. The rounded boxes denote states (Estab, wait1, and Closed). Immediately below Estab are inward-pointing boxes that receive signals (Close, Packet). The square and diamond boxes below these are actions and decisions. The outward-pointed boxes (e.g., Packet) emit signals.

tion channels in these diagrams are also labeled with the signals they convey, but are assumed to have significant delay, unlike the channels among FSMs in a single processor.

The behavior of an SDL state machine is straightforward. At the beginning of each cycle, it gets the next signal in its input queue and sees if there is a "receive" block for that signal off the current state. If there is, the associated code is executed, possibly emitting other signals and moving to a next state. Otherwise, the signal is simply discarded and the cycle repeats from the same state. By itself, such semantics have a hard time dealing with signals that arrive out-of-order, but SDL has an additional construct for handling this condition. The save construct is like the receive construct, a state appearing immediately and matching a signal, but when it matches it stores the signal in a buffer that holds it until another state has a matching rule.

SystemC

The SystemC language (Figure 85.14) is a C++ subset for system modeling. A SystemC specification is simulated by compiling it with a standard C++ compiler and linking in freely distributed class libraries from www.systemc.org.

The SystemC language builds systems from Verilog- and VHDL-like modules. Each has a collection of I/O ports and may contain instances of other modules or processes defined by a block of C++ code.

SystemC uses a discrete-event simulation model. The SystemC scheduler executes the code in a process in response to an event such as a clock signal, or a delay. This model resembles the one used in Verilog and VHDL, but has the flexibility of operating with a general-purpose programming language.

SystemC began life aiming to replace Verilog or VHDL as a hardware description language (it did not offer designers a sufficiently compelling reason to switch), but has since moved beyond that. Very often in system design, it is desirable to run simulations to estimate such high-level behavior as bus activity or memory accesses. Historically, designers had custom-written simulators in a general-purpose language such as C, but this was time-consuming because of the need to write a new simulation kernel (i.e., something that provided concurrency) for each new simulator.

```
                         #include "systemc.h"

                         struct complex_mult : sc_module {
                           sc_in<int>   a, b;
                           sc_in<int>   c, d;
                           sc_out<int>  x, y;
                           sc_in_clk    clock;

                           void do_mult() {
                             for (;;) {
                               x = a * c - b * d;
                               wait();
                               y = a * d + b * c;
                               wait();
                             }
                           }

                           SC_CTOR(complex_mult) {
                             SC_CTHREAD(do_mult, clock.pos());
                           }
                         };
```

FIGURE 85.14   A SystemC model for a complex multiplier.

SystemC is emerging as a standard for writing system-level simulations. While not perfect, it works well enough and makes it fairly easy to glue large pieces of existing software together. Although Verilog has a programming language interface (PLI) that allows arbitrary C/C++ code to be linked and run simultaneously with a simulation, the higher integration of the SystemC approach is more efficient.

SystemC supports transaction-level modeling, in which bus transactions, rather than being modeled on a per-cycle basis as would be done in a language such as Verilog, are modeled as function calls. For example, a burst-mode bus transfer would be modeled with a function that marks the bus as in use, advances simulation time according to the number of bytes to be transferred, actually copies the data in the simulator, and marks the bus as unused. Nowhere in the simulation would the actual sequence of signals and bits transferred over the bus appear.

85.6 Summary

Currently, most embedded systems are programmed using C for software and Verilog, or possibly VHDL, for hardware components such as FPGAs or ASICs, but this will probably change. The increased complexity of such designs makes a compelling case for different, higher-level languages. Years ago, designers made the jump from assembly to C, and the higher-level constructs of Java are growing more attractive despite its performance loss.

Domain-specific languages, especially for signal-processing problems, already have a significant beachhead, and will continue to make inroads. Most signal processing algorithms are already prototyped using a higher-level language (Matlab), but it remains to be seen whether synthesis from Matlab will ever be practical.

For hardware, the direction is less clear. While modeling languages such as SystemC will continue to grow in importance, there is currently no clear winner for the successor to VHDL and Verilog. Roughly a decade ago, a different, high-level subset of VHDL and Verilog was proposed as the new "behavioral" synthesis subset, but did not catch on because it was too limiting, largely because of restrictions placed on it by the synthesis algorithms. Additions such as System Verilog are incremental, if helpful, improvements, but will not provide the quantum leap forward that synthesis from the register-transfer level (RTL) subsets of Verilog and VHDL provided. Perhaps future hardware langauges may contain constructs such as Esterel's.

References

[1] Arnold, K., James Gosling, and David Holmes, *The Java Programming Language,* 3rd ed., Addison-Wesley, Reading, MA, 2000.

[2] Ashenden, P.J., *The Designer's Guide to VHDL,* Morgan Kaufmann, San Francisco, CA, 1996.

[3] Berry, G. and Georges Gonthier, The Esterel synchronous programming language: design, semantics, implementation, *Science of Computer Programming,* 19, 87–152, 1992.

[4] Bhattacharyya, S.S., Ranier Leupers, and Peter Marwedel, Software synthesis and code generation for signal processing systems, *IEEE Transactions on Circuits and Systems — II: Analog and Digital Signal Processing,* 47,849–875, 2000.

[5] Bhattacharyya, S.S., Praveen K. Murthy, and Edward A. Lee, Synthesis of embedded software from synchronous dataflow specifications, *Journal of VLSI Signal Processing Systems,* 21,151–166, 1999.

[6] Bollella, G., Ben Brosgol, Peter Dibble, Steve Furr, James Gosling, David Hardin, Mark Turnbull, Rudy Belliardi, Doug Locke, Scott Robbins, Pratik Solanki, and Dionisio de Niz, *The Real-Time Specification for Java,* Addison-Wesley, Reading, MA, 2000.

[7] Briand, L.P. and Daniel M. Roy, *Meeting Deadlines in Hard Real-Time Systems: The Rate Monotonic Approach,* IEEE Computer Society Press, New York, 1999.

[8] Cohen, B., *VHDL Coding Styles and Methodologies.* 2nd ed., Kluwer, Boston, MA, 1999.

[9] Edwards,. S.A., *Languages for Digital Embedded Systems,* Kluwer, Boston, MA, September 2000.

[10] Ellsberger, J., Dieter Hogrefe, and Amardeo Sarma, *SDL: Formal Object-Oriented Language for Communicating Systems,* 2nd ed., Prentice-Hall, Upper Saddle River, NJ, 1997.

[11] Gosling, J., Bill Joy, Guy Steele, and Gilad Bracha, *The Java Language Specification,* 2nd ed., Addison-Wesley, Reading, MA, 2000.

[12] IEEE Computer Society, 345 East 47th Street, New York, *IEEE Standard VHDL Language Reference Manual (1076-1993),* New York, 1994.

[13] IEEE Computer Society, *IEEE Standard Hardware Description Language Based on the Verilog Hardware Description Language (1364-1995),* New York, 1996.

[14] International Telecommunication Union, *ITU-T Recommendation Z.100: Specification and Description Language,* International Telecommunication Union, Geneva, Switzerland, 1999.

[15] Kahn, G., The Semantics of a Simple Language for Parallel Programming, in Information Processing 74: Proceedings of IFIP Congress 74, Stockholm, Sweden, Aug. 1974. North-Holland, Amsterdam, 1974, pp. 471–475.

[16] Kernighan, B.W. and Dennis M. Ritchie, *The C Programming Langage,* 2nd ed., Prentice-Hall, Upper Saddle River, NJ, 1988.

[17] Labrosse, J., *MicroC/OS-II,* CMP Books, Lawrence, KA, 1998.

[18] Lee, E.A. and David G. Messerschmitt, Synchronous Data Flow, *Proceedings of the IEEE,* 75, 1235–1245, 1987.

[19] Lindholm, T. and Frank Yellin, *The Java Virtual Machine Specification,* Addison-Wesley, Reading, MA, 1999.

[20] Liu, C.L. and James W. Layland, Scheduling algorithms for multiprogramming in a hard real-time environment, *Journal of the Association for Computing Machinery,* 20, 46–61, 1973.

[21] Parks, T.M., Bounded Scheduling of Process Networks, Ph.D. thesis, University of California, Berkeley, 1995. Available as UCB/ERL M95/105.

[22] Perry, D.L., *VHDL,* 3rd ed., McGraw-Hill, New York, 1998.

[23] Ritchie, D.M., The development of the C language, *History of Programming Languages,* Vol. II, Cambridge, MA, April 1993.

[24] Stroustrup, B., *The C++ Programming Language,* 3rd ed., Addison-Wesley, Reading, MA, 1997.

[25] Thomas, D.E. and Philip R. Moorby, *The Verilog Hardware Description Language,* 4th ed., Kluwer, Boston, MA, 1998.

86

Verification Languages

Aarti Gupta
NEC Laboratories America

Ali Alphan Bayazit
Princeton University

Yogesh Mahajan
Princeton University

86.1 Introduction

Verification is the process of checking whether a given design is correct with respect to its specification. The specification itself can be in different forms — it can be a nontangible entity such as a designer's intent, or a document written in a natural language, or expressions written in a formal language. In this chapter, we consider a *verification language* to be a language with a formal syntax, which supports expression of verification-related tasks. While it is useful for a verification language to have a precise unambiguous semantics, we do not regard it as a defining feature. Broadly speaking, we focus on the support it provides for dynamic verification based on simulation, as well as static verification based on formal techniques. Although we discuss some modeling languages also, we highlight their verification-related features. Other details on how they are used for representing higher-level models can be found in the accompanying chapters in this book [1], and a recent survey [2].

Verification is increasingly becoming a bottleneck in the design of embedded systems and system-on-chips (SoCs). The primary reason is the inability of design automation methodologies to keep up with growing design complexity, due to the increasing number of transistors achievable on a chip in accordance with the well-known Moore's Law. The emerging trend toward systematic reuse of design components and platforms can help amortize the verification costs across multiple designs. At the same time, it places an even greater responsibility on the verification methodology, because the correctness of reusable, parameterized components must be verified in potentially multiple contexts. The cost of detecting bugs late in the design cycle is very high, both in terms of design respins, and in terms of time lost to market.

A verification methodology for system-level design needs to address issues related to component-level verification, as well as system-level integration of these components. For embedded systems, this requires verifying the effects of distributed concurrent computation, not only for hardware but also for the embedded software. Additional complexity arises due to the heterogeneous nature of the components. For example, many embedded applications include digital controllers and analog sensors and actuators. This requires verifying the hybrid systems, where the dynamics consist of discrete as well as continuous behaviors. Another crucial problem for embedded systems is verifying the interfaces for parameterized components, and third-party IP components, mostly without access to the design details. Finally, real-time constraints play an important role in hardware–software partitioning and scheduling (typically implemented in an on-chip real-time operating system kernel). It is important to specify and verify these real-time assumptions and requirements.

Overview

Although there has been progress in verification of embedded systems, there has been relatively little effort at language standardization targeted at verification features described above. At the same time, much of the recent standardization activities for hardware and system-level verification languages (e.g., by Accellera [3], and IEEE [4]) are very much applicable for embedded systems. Therefore, in this chapter, we describe verification languages for hardware, software, and embedded systems. A comprehensive survey in any of these categories is beyond the scope of this chapter. Instead, we focus on representative languages, including language standards where available. Note that many of these languages and standards are still evolving, and are likely to undergo further changes from the details described in this chapter.

For hardware designs, the historic success of logic synthesis technology for register-transfer level (RTL) designs led to the popularity of hardware description languages (HDLs), such as Verilog and VHDL. This in turn has led to the emergence of *hardware verification languages* (HVLs). Examples of these languages, described in Section 86.3, include *e*, OpenVera, Sugar/PSL, and ForSpec. In the area of software verification related to embedded applications, described in Section 86.4, we focus on the use of standard programming languages such as C/C++, Java, and on software modeling languages such as UML, SDL, and Alloy. For embedded systems, described in Section 86.5, we focus on languages for system-level verification, such as SystemC, SpecC, and SystemVerilog. We also describe domain-specific verification efforts, such as those based on Esterel and hybrid systems.

86.2   Background

We start by providing some background for verification methods, in order to provide a sufficient context and terminology for our discussion on verification languages. More details can be found in the references cited.

Simulation-based Verification

Simulation has been, and continues to be, the primary method for functional verification of hardware and system level designs. It consists of providing input stimuli to the *design under verification* (DUV), and checking the correctness of the output response. In this sense, it is a dynamic verification method. The practical success of simulation has largely been due to automation through development of *testbenches*, which provide the verification context for a DUV. (Details of testbench development are described in a recent book [5].)

A typical testbench consists of a *generator* of testcases (input stimuli), a *checker* or *monitor* for checking the output response, and a *coverage analyzer* for reporting how much of the design functionality has been covered by the testcases. Depending on how much of the internal design state is observable and controllable by the generators and checkers, verification can be classified as black-box (none), white-box (full), or gray-box (partial, added to aid verification).

It is impossible to simulate all testcases for designs of modest size. Traditionally, *directed* testbenches are used to generate specific scenarios of interest, especially to nail down designer's intent in the early phase of the design cycle. In contrast, *constrained random* testbenches are used to generate random stimuli, subject to certain constraints added to the testbench. These are very useful for verifying unexpected scenarios, which help in finding bugs in later phases of the design cycle. In all cases, progress is evaluated in terms of *coverage metrics*, which have been used with varying degrees of success. These include code coverage metrics such as statement/branch/toggle/expression coverage, and functional coverage metrics such as state/transition coverage of the finite state machines (FSMs) in the design description.

As systems become increasingly complex, designs need to be specified at much higher levels of abstraction. A *transaction* is a sequence of lower-level tasks, which implement a logical operation at the higher level. For example, a read transaction by an agent on a shared bus typically consists of transferring the address to the bus, waiting for the data to be ready, and transferring the data from the bus. In complex systems, it is more natural to model intermodule communication at the transaction level, rather than at the signal level. This has led to the development of transaction-based testbenches for system-level verification. These testbenches use adaptors called transactors/bus functional models (BFM) to translate between the higher and lower levels.

Development of testbenches at RTL and system level can be quite tedious. It is useful to have language support for abstraction and object-oriented features, such as abstract data types, data encapsulation, dynamic object creation/destruction, inheritance, and polymorphism. The testbench language must also be able to specify and launch many tasks concurrently, and provide features for synchronization, monitoring multiple in-flight tasks, and reentrancy of tasks. Most of these features were absent from the HDLs, which motivated the emergence of testbench languages, HVLs, and system-level modeling languages, described in Sections 86.3 and 86.5, respectively.

Formal Verification

In contrast to simulation, formal verification methods do not rely on the dynamic response of a DUV to certain testcases. Rather, they perform static analysis on a formal mathematical model of the given DUV, to check its correctness with respect to a given specification under all possible input scenarios. In this section, we briefly describe some of these methods. Successful instances of industry applications have been described in a survey [6].

A popular formal method is *equivalence checking*, where a given DUV is checked for equivalence against a given reference design. Equivalence checking has been applied very successfully in automated hardware design, to check that no errors are introduced during logic synthesis flow from RTL to gate-level design. Since both the reference design and the DUV are adequately represented in standard HDLs, there has been little motivation to develop a special language for equivalence checking.

When reference models are unavailable, or where the correctness of the highest-level model needs to be checked, design requirements are expressed in terms of correctness properties. *Property checking* is used to check whether a given design satisfies a given correctness property, expressed as a formal specification. Most property specification languages are derived from formal logics, or automata theory (described in more detail in the Formal Specification of Properties section).

The two main techniques for property checking are *model checking* and *theorem-proving*. In model checking [7], the DUV is typically modeled as a finite state transition system, the property is specified as a temporal logic formula, and verification consists of checking whether the formula is true in that model.[1] In theorem-proving, both the DUV and the specification are modeled as logic formulas, and the satisfaction relation between them is proved as a theorem, using the deductive proof calculus of a theorem-prover. Model checking techniques have found better acceptance in the industry so far [8, 9], primarily because they can be easily automated for finite state systems, and they provide counterexamples that are useful for debugging. Although theorem-proving techniques [10–12] can handle more general problems, including infinite state systems, they are not as highly automated, and the learning curve for users is steeper.

---

[1] A related method is called *language containment*, where the specification and DUV are represented as automata, and verification consists of checking that the language of the DUV is contained in the language of the specification.

The practical application of model checking techniques is limited by the state explosion problem, that is, the state space to be searched grows exponentially with the number of state components. *Symbolic model checking* techniques [7, 8] use BDD-based methods [13] to symbolically manipulate sets of states without explicit enumeration. Although this improves scalability to some degree, these techniques can fail due to the memory explosion for BDDs. As an alternative, *Bounded model checking* techniques use SAT-based methods for finding bounded-length counterexamples [14]. These techniques scale better for finding bugs, but they need additional reasoning for obtaining conclusive proofs.

Assertion-Based Verification

Another emerging methodology is assertion-based verification. Although its basic elements have been part of industry practice for a while, it has gained attention more recently as a systematic means of enhancing the benefits of simulation and formal verification, and for combining them effectively. In particular, the Accellera organization [3] has been actively involved in developing and promoting language standards for top-down specification using a formal property language, as well as for bottom-up implementation assertions. (The chosen standards are described in the Property Specification Language, and the SystemVerilog and SystemVerilog Assertions sections. More details can be found in a recent book [15].)

The key ingredient of assertion-based verification is the formal specification of properties to capture designer intent at all levels of the design. Properties can be used as *assertions*, to check for violations of correct behavior or functionality. The checking can be done dynamically during simulation, statically using formal verification techniques, or by a combination of the two. Properties that specify interfaces between modules can also be used as *constraints*, that is, as assumptions on the input interface of a given module. For simulation, these constraints can be translated into stimulus generators, or added to the constraint-solvers for constrained random simulation. For formal verification, these constraints serve the role of an environment, which can be handled explicitly as an abstract module, or implicitly by the verification method.

Assertions and constraints are crucial in exploiting an *assume–guarantee* verification paradigm, whereby assertions (guarantees) on the output interface of a given module, serve as constraints (assumptions) on the input interface of the downstream module. (Care must be exercised in defining interfaces, in order to avoid a circular reasoning.) This allows verification of a large design to be handled modularly by verifying each of its components. Assertions and constraints also help to assess functional coverage achieved by simulation testbenches, for example how many assertions have been checked, what fraction of the constraint space has been covered, etc.

Formal Specification of Properties

Property specification languages are derived mostly from formal logics and automata theory. There are many related theoretical issues — expressiveness (i.e., what kinds of properties a logic can capture), complexity of the related property checking problem, etc. In general, the more expressive a logic, the higher the complexity of the property checking problem. There are also many practical issues — ease of use, available tool support, etc. In this section, we briefly describe some logics that form the basis of languages described in the rest of this chapter.

Propositional Logic: This is essentially the Boolean logic familiar to most hardware designers. Formulas consist of Boolean-valued variables connected using standard Boolean operators (not/and/or). The complexity of checking the satisfiability of a Boolean formula (SAT) is known to be NP-complete [16]. When both universal and existential quantifiers are added, the complexity of checking the satisfiability of a Quantified Boolean Formula (QBF) is PSPACE-complete [16]. The expressiveness of these logics is fairly limited, but they are well-suited for handling bit-level hardware designs.

First-order Predicate Logic: The expressiveness of Boolean logic is extended here by allowing variables to range over elements of an arbitrary set, not necessarily finite. This is very useful for handling higher-level systems and software, for example to reason directly about integers or real numbers. Although the problem of checking the validity of many interesting classes of formulas over these sets is undecidable, automated decision procedures exist for useful subsets of the logic [17, 18], which have been successfully utilized for verification of microprocessors, and word-level hardware designs. *Higher-order logics*, that is, where quantifiers are allowed to range over subsets of sets, have also been used, but mainly by specialists [10].

Temporal Logics: Temporal logics are used for specification of concurrent reactive systems, which are modeled as state transition systems labeled by propositions [19]. Here, the expressiveness of predicate logic is further extended, such that the interpretation of a proposition can change dynamically over the states. In addition to the standard Boolean operators, there are typically four temporal operators, with the intended semantics as shown below:

- $G\ p$: p is <u>G</u>lobally (Always) true
- $F\ p$: p is true some time in the <u>F</u>uture (Eventually)
- $X\ p$: p is true at the ne<u>X</u>t instant
- $p\ U\ q$: p is true <u>U</u>ntil q becomes true

Temporal logics are ideally suited for expressing qualitative temporal behavior. The correctness properties are typically categorized as follows:

- *Safety Properties*: nothing bad happens
  for example, $G\ (!p1\_critical + !p2\_critical)$
  This formula expresses the mutual exclusion property that processes p1 and p2 should never be in the critical section simultaneously.
- *Liveness Properties*: something good eventually happens
  for example, $G\ (request\ ->\ F\ grant)$
  This captures absence of starvation, that is, every request for a resource should be eventually granted.
- *Precedence Properties*: ordering of events
  for example, $G\ (!p2\_req\ U\ p1\_req\ ->\ !p2\_grant\ U\ p1\_grant)$
  This expresses an ordering requirement, that is, if process p1 requests a resource before process p2, then it should be granted the resource earlier.

Different kinds of temporal logics have been proposed, depending upon the view of time. Linear Temporal Logic (LTL) [19] takes a linear view, where formulas are interpreted over linear sequences of states. In contrast, Computation Tree Logic (CTL) [7] takes a branching view of time, where formulas are interpreted over a tree of possible computations starting from a state (with additional path quantifiers to denote all/some paths). In terms of expressiveness, LTL and CTL are incomparable, that is, each can express a property that cannot be expressed by the other. In terms of model checking complexity, LTL is linear in the size of the model, but exponential in the size of the specification, while CTL is linear in both. Despite the relative stability of these logics, there is an ongoing debate regarding their suitability for various verification applications [20].

Regular and ω-Regular Languages: Although temporal logics are quite useful, they cannot express regular (or ω-regular) properties of sequences. Such properties are easily expressed as languages recognized by finite state automata on finite (infinite) words [21]. Regular expressions are formed by using three basic operations — concatenation (" . "), choice ("|"), and bounded repetition ("*"). For example, a sequence with alternation of a *request* signal with *grant* or *error* signals can be expressed as *(request .(grant | error))**. An ω-regular expression consists of the form $U.V^{\omega}$, where U and V are *-free regular expressions, and V^{ω} denotes an infinite repetition of expression V. These are used for specifying properties of systems that do not terminate.

86.3 Languages for Hardware Verification

In this section, we focus on verification languages for hardware RTL designs (system-level hardware designs are covered in Section 86.5).

HDLs and Interfaces to Programming Languages

RTL designs are typically implemented in standard HDLs — Verilog [22], VHDL [23]. However, it is not practical to implement simulation testbenches in HDLs alone. Indeed, the testbench can be purely behavioral, because there are no synthesizability constraints. Furthermore, it can be implemented at levels higher than RTL. Historically, a popular testbench development approach has been to implement some of its parts in a software programming language such as C/C++ or Perl. These parts are integrated with HDL simulators through standard programming language interfaces. Unfortunately, this approach not only slows down the simulator but it also requires significant development effort, such as defining new data types (e.g., 128-bit bus), handling concurrency, dynamic memory objects, etc. [5].

For property specification, VHDL has some support for static assertions, with a variety of different severity levels. Although Verilog lacks such explicit constructs, it is straightforward to use the *if* and *$display* constructs to implement a similar effect.

Example 1: Suppose we need to check that two signals *A* and *B* cannot be high at the same time. The fragment below shows an assertion template and an instance in VHDL. The keyword *assert* specifies a property that must hold during simulation.

```
[label] assert expression
   [report message]
     [severity level]
assert (A NAND B)
   report "oops: A & B cannot both be 1"
     severity 0;
```

A similar effect can be achieved by using Verilog's if/$display combination, which specifies the undesired situation, as shown below:

```
always (A or B) begin
  if (A & B) begin
    $display("oops: A = B = 1");
    $finish; // end simulation
  end
end
```

Open Verification Library

Although simple assertions are quite useful, they do not provide a practical way for specifying properties. Temporal properties can be specified using checkers or monitors in HDLs, but this involves a significant development effort. Therefore, there has been a great deal of interest in developing a library of reusable monitors. An example is the Open Verification Library (OVL), available from Accellera [3]. It is not a stand-alone language, but a set of modules that can be used to check common temporal specifications within Verilog or VHDL design descriptions.

Example 2: As an example [15], consider the following PCI Local Bus Specification: *To prevent **AD**, **C/BE#**, and **PAR** signals from floating during reset, the central resource may drive these lines during reset (bus parking) but only to a logic low level; they may not be driven high.*

Suppose *ad, cbe_ , par, rst_* are the Verilog signal names corresponding to *AD, C/BE#, PAR,* and *RST#* respectively.[2] The given property can be specified within a Verilog implementation of OVL as follows:

assert_always master_reset (clk, !rst_, ! |(ad ,cbe_ ,par));

Here, *assert-always* is the name of the library monitor, and *master_reset* is an assertion instance. On every rising edge of *clk* signal, whenever *!rst_* is high, the monitor asserts that the last parameter should

---

[2] In PCI Local Bus Specification, a signal name ending with "#" indicates that signal is active low. In the examples with HDls, we will use "_" at the end of the signal name for the same purpose.

evaluate to true. Here, the last parameter is the negation (!) of the bitwise-or (|) over the given bits. Note that this is a simple safety property, which is expected to hold always during simulation.

Example 3: Consider another example [15] from the PCI Local Bus Specification: *The assertion of IRDY# and deassertion of FRAME# should occur as soon as possible after STOP# is asserted, preferably within one to three cycles.*

For simplicity, consider the *FRAME#* and *STOP#* signals only, that is, check whether *frame_* will be deasserted within 1 to 3 clock cycles after *stop_* is asserted, as shown below:

assert_frame # (1,1,3) check_frame_da (clk, **true**, !stop_, frame_)

Again, *check_frame_da* is an instance of the *assert_frame* module defined in the library. Three optional parameters #(1,1,3) are used, corresponding to the "severity level," "minimum number of clocks," and "maximum number of clocks," respectively. The monitor will check whether *frame_* goes high within 1 to 3 clock cycles after *!stop_*. Here, the severity level is set to 1 to continue the simulation even if the assertion is violated, and the reset parameter is set to **true**, as an example of where it is not needed.

OVL has many advantages. First, it can be used with any Verilog, VHDL or a mixed simulator, with no need for additional verification tools. Second, it is open, that is, the library can be modified easily, for example, for assessing functional coverage [15]. Another useful feature of OVL is that it does not slow down simulation, primarily because it is hard to specify very complex assertions. Unfortunately, OVL is used mainly for checking safety properties during simulation, and is not very useful for checking liveness or for formal verification. In some sense, OVL provides a transition from a traditional simulation-based methodology, to an assertion-based methodology.

Temporal *e*

Verisity's *e* language is an advanced verification language that is intended to cover many verification aspects. Like many high-level languages, it has constructs for Object-Oriented programming, such as class definitions, inheritance, and polymorphism [5]. It also provides elements of Aspect-Oriented Programming (AOP). AOP allows modifying the functionality of the environment without duplicating or modifying the original code, in a manner more advanced than simple inheritance. (See [5, 24] for more details.)

As a testbench language, *e* provides many constructs related to stimuli generation, such as specification of input constraints and facilities for data packing, as well as for assessing simulation coverage. It also provides support for property specification, and has been used widely both in simulation-based and formal verification. There is an ongoing effort to use *e* as the basis for IEEE's new verification language [4].

Example 4: As an example for stimuli generation, suppose we have a struct-type[3] *frame* for modeling an Ethernet frame, with one of the data fields defined as *%payload*. The type of payload can be defined in *e* as follows [24]:

```
struct payload {
    %id      : byte;
    %data    : list of byte;
    keep soft data.size() in [45..1499];
}
```

In this example, the "%" character in front of the field name means that the corresponding field is physical and represents data to be sent to the DUV. The **keep soft** keywords are used for bounding the values of the variable — the size of the *data* field in this case. It also allows specification of weighted ranges or constraints. In the example, the size will be varied automatically within the given range. (Using "!" character along with "%" would have indicated that the field would not be generated automatically.)

---

[3] A struct type basically corresponds to a class type in C++, that is, it allows method definitions along with data definitions. Since it is conceptually similar to other Object-Oriented languages, we omit the actual syntax.

Typically, a user-defined function is used for driving stimuli to the DUV. For example, suppose *my_frame* is an instance of the struct *frame*. The following *e* code can be used to input the frame data serially into the DUV:

Example 5:

```
var bitList: list of bit;
bitList = pack (packing.low, my_frame);
for each (b) in bitList{
     'testbench.duv.transmit_stream0'= b;
     wait cycle;
};
```

In this example, the keyword **pack** provides the mechanism to pack all data fields into a single list of bits, which is then fed serially to the Verilog signal *testbench.duv.transmit_stream0*. After each bit transfer, the function waits for one clock, denoted by the **wait cycle** keywords.

Support for specification of temporal properties is provided in *e* through the use of Temporal Expressions (TEs). A TE is defined as a combination of events and temporal operators. The language also supports the keyword **sync**, which is used as a point of synchronization for TEs.

Example 6: Returning back to PCI specifications, consider the following requirement, and its corresponding specification in *e*: *Once a master has asserted IRDY#, it cannot change IRDY# or FRAME# until the current data phase completes regardless of the state of TRDY#.*

```
expect @new_irdy => { [..] * ((not @irdy_rise) and (not change (frame_)));
   @ data_phase_complete } @sys.pci_clk ;
   else dut_error("Error, IRDY# or FRAME# changed before current data phase completed.") ;
```

Here, suppose that the events (shown as *@event*) have been defined already. The shown expression specifies that whenever IRDY# is asserted (*@new_irdy*), deassertion of IRDY# (*@irdy_rise*) or a change in FRAME# should not occur, until the data phase is complete (*@data_phase_complete*). The use of *@sys.pci_clk* denotes that the event *pci_clk* is used for sampling signals in evaluating the given TE. This feature is also useful for verifying multiple clocked designs.

OpenVera and OVA

OpenVera from Synopsis is another testbench language similar to *e* in terms of functionality and similar to C++ in terms of syntax. Since conceptually OpenVera is very similar to *e*, here we do not include testbench examples for OpenVera. It has similar constructs for coverage, random stimuli generation, data packing, etc.

OpenVera Assertions (OVA) is a stand-alone language, which is also part of the OpenVera suite [25]. OpenVera comes with a checker library (OVA IP) similar to OVL. OVA and OpenVera also have event definitions, repetition operators (*[..]), and sequencing, where different sequences can be combined to create more complex sequences using logical and repetition operators.

Example 7: The following example shows the OVA description of the PCI specification from Example 3:

```
clock posedge clk {
   event chk:
   if (negedge stop_) then
   #[1..3] posedge frame_;
}
assert frame_chk : check (chk)
```

Note the specification of the sampling clock, the event, and the corresponding action. The implication operation (**if–then**) is similar to conditionals in programming languages, and # is the cycle delay operator.

ForSpec

ForSpec is a specification and modeling language developed at Intel [26]. The underlying temporal logic in ForSpec is FTL, which is composed of regular language operators, and LTL-style temporal operators. ForSpec is aimed at an assume–guarantee verification paradigm, which is suitable for modular verification of large designs. The language also provides explicit support for multiple clocks, reset signals, and past-time temporal operators. Although none of these additional features increases the expressiveness of the basic language, they clearly ease specification of properties in practice. Example templates for asynchronous set/reset using FTL are shown below:

```
accept(boolean_expr & clock) in formula_expr
reject(boolean_expr & clock) in formula_expr
```

Recently, some constructs of ForSpec have also been added to OVA Version 2.3.[4] In particular, the concept of "temporal formulas" is added, which can be composed of applying temporal operators on sequences. The supported temporal operators are: **followed_by, triggers, until, wuntil, next, wnext** [25, 27]. Asynchronous set/reset have also been added.

Property Specification Language (PSL)

PSL is the language standard established by Accellera for formal property specification [3]. It originated from the language Sugar, developed at IBM. The first version of Sugar was based on CTL, and was aimed at easing expression of CTL properties by users for the RuleBase model checker [28]. PSL is based on Sugar Version 2.0, where the default temporal logic is based on LTL, called PSL/Sugar Foundation Language. The main advantage of LTL in assertion-based verification is that its semantics, that is, evaluation on a single execution path, is more natural for simulation-based methods where a single path is traced at a time. Although CTL is lower in model checking complexity, it is harder to support in simulation-based methods. For formal verification, CTL continues to be supported in PSL through an Optional Branching Extension (OBE). According to the Accellera standard, PSL assertions are declarative. Since many users prefer procedural assertions, nonstandard pragma-based assertions[5] are also supported.

PSL properties consist of four layers: Boolean layer, temporal layer, verification layer, and modeling layer. The bottom-most is the Boolean layer, which specifies the Boolean expressions that are combined using operators and constructs from the upper layers. The syntax of this layer depends on the hardware modeling language used to represent the design. Currently, PSL supports both Verilog and VHDL. (Sugar also supports the internal language EDL of RuleBase.)

The temporal layer specifies the temporal properties, through the use of temporal operators and regular expression operators. The temporal operators (**eventually, until, before, next**) and some regular expression operators have two different types. One is called the strong suffix, indicated with an "!" at the end (e.g., **eventually!**), which specifies that the operator must hold before the end of the simulation. The other one, called the weak suffix (without the "!"), denotes that it does not hold only when the chance of it being true disappears completely. PSL/Sugar also has some predefined temporal constructs regarded as syntactic sugar, targeted at easing usage without adding expressiveness.

The verification layer specifies the role of the property for the purpose of verification. This layer supports keywords — **assert, assume, assume–guarantee, restrict, restrict–guarantee, cover,** and **fairness.** The keyword **assert** indicates that the property should be checked as an assertion. The keywords **assume** and **assume–guarantee** denote that the property should be used as an assumption, rather than checked as an assertion. The keywords **restrict** and **restrict–guarantee** can be used to force the design into a specified state. The keyword **cover** is used for coverage analysis, and **fairness** to specify fairness constraints for the DUV. This layer also provides support for organizing properties into modular units.

---

[4] Not yet supported by Synopsys at the time of this writing.
[5] These assertions escape from simulators by defining them inside comments.

Finally, the modeling layer provides support for modeling the environment of the DUV, in terms of the behavior of the design inputs, and the auxiliary variables and signals required for verification.

Example 8: Going back to the same PCI specification, as shown in Examples 3 and 7, the property is written in PSL as follows:

assert always (!stop_-> **eventually!** frame_);

Note that the above specification is qualitative in terms of when the *frame_* signal should be deasserted, that is, the **eventually!** operator does not specify that it should do so within 1 to 3 cycles after *stop_* is asserted.

Example 9: Consider a property that states that a request–acknowledge sequence should be followed by exactly eight data transmissions, not necessarily consecutive. This can be expressed using a Sugar Extended Regular Expression (SERE) as follows:

always {req;ack} |=> {start_trans;data[=8];end_trans}

In addition to the number of data transmissions, the property also specifies that *start_trans* is asserted exactly one cycle after *ack* is asserted (|=>). The notation [=8] signifies a (not necessarily consecutive) repetition operator, with parameter 8. Another SERE that is equivalent to *data*[=8] is shown as follows:

{!data[*];data;!data[*]}[*8]

PSL/Sugar also allows specification of a sampling clock for each property, or to use the default clock (as shown in the previous examples). The **abort** operator can be used to specify reset-related properties, inspired by ForSpec. To support modularity and reusability, SEREs can be named and used in other SEREs or properties. (This feature is similar to events in *e* and OVA.)

Example 10: Another PCI specification states that the least significant address bit AD[0] should never be 1 during a data transfer. This can be expressed in PSL as follows [15]:

sequence SERE_MEM_ADDR_PHASE = {frame_; !frame_ && mem_cmd};
property PCI_VALID_MEM_BURST_ENCODING =
    always {SERE_MEM_ADDR_PHASE} |-> {!ad[0]} **abort** !rst_ @(**posedge** clk);
assert PCI_VALID_MEM_BURST_ENCODING;

In this example, "|->" is a weak suffix implication operator, which denotes that whenever the first sequence holds, we should expect to see the second sequence. Note the definition of the named SERE using keyword **sequence**, and its use within a property specification. The property is used as assertion, to be checked by the verification tool.

86.4   Languages for Software Verification

In this section, we focus on verification efforts for software relevant to embedded system applications. (For more general software systems and software analysis techniques, see [29, 30].) The correctness requirements are typically in the form of Floyd–Hoare style assertions [31] (invariants, pre/post conditions of actions represented in predicate logic) for functional verification, or temporal logic formulas for behavioral verification.

Programming Languages

Given the popularity of C/C++ and Java, verifying programs directly written in these languages is very attractive in principle. However, there are many challenging issues — handling of integers/floating point data variables, function/procedure calls, and object-oriented features such as classes, dynamic objects, and polymorphism.

Verification of C/C++ Programs

The VeriSoft project [32] at Lucent Technologies focused on guided search for deadlocks and violations of assertions expressed directly in C/C++, and has been used for verifying many communication protocols.

The SLAM project [33] has been successfully used at Microsoft for proving the correctness of device drivers written in C. Designers use a special language called SLIC [34] to provide correctness requirements. These requirements are translated automatically into special procedures, which essentially implement automata for checking safety properties. These procedures are added to the given program, such that a bug exists if a statement labeled "error" can be reached in the modified program. Verification is performed by obtaining a finite state abstract model, performing model checking, and refining the abstract model if needed. Other similar efforts [35, 36] have so far focused largely on the abstraction and refinement techniques. There has been relatively little effort in development of a separate verification language.

Verification of Java Programs

There have been many efforts for verification of Java programs also. The Extended Static Checker for Java (ESC/Java) [37] performs static checks that go beyond type checking. It automatically generates verifications conditions for catching common programming errors (such as null dereferences, array bounds errors, etc.), and synchronization errors (race conditions, deadlocks). These conditions are transparently checked by a backend theorem-prover. It also uses a simple annotation language for the user to provide object invariants and requirements, which aid the theorem-prover. The annotation language, called Java Modeling Language (JML), is part of a broader effort [38]. JML is used to specify the behavior and syntactic interfaces in Java programs, through pre/post conditions and invariants for classes and methods. Its syntax is quite close to that of Java, thereby making it easy to learn by programmers. JML is used by a host of verification tools, which span the spectrum from dynamic run-time assertion checkers and unit testers, to static theorem-provers and invariant generators. It has been applied successfully in smartcard applications implemented using a dialect of Java called Java Card [39].

Example 11: Consider the following code [37], which shows Java code for a class definition of integer multisets.

```
1     class Bag {
2     //@invariant 0<=size && size <=elements.length
3        int size;
4        int[] elements;
5        // @requires input != null
6        Bag(int[] input) {
7           ...
8        }
9        int extractMin( ) {
10          ...
11       }
12    }
```

As shown, the *bag* class provides two operations (details are not shown) — one for constructing a bag from an array of integers, and another for extracting the smallest element. The example also shows user annotations on Lines 2 and 5, to be used with ESC/Java. The annotation on Line 2 specifies an object invariant, which should hold for every initialized instance of the class. The annotation on Line 5 specifies a pre-condition, which is assumed to hold by the operation that follows.

Model checking techniques have also been used successfully for Java programs. Bandera [40] and Java PathFinder (JPF) [41] use various program analysis methods and abstractions to obtain a finite state model of the given program. This model is verified against temporal specifications provided by the user. In many cases, model checking is performed by translation of the (abstract) Java program into Promela, the input language of the model checker SPIN [9]. Promela focuses on system descriptions of asynchronous concurrent processes, and the correctness properties are specified in LTL. SPIN has been used successfully to verify the correctness of numerous protocols, distributed algorithms, controllers for reactive systems, etc.

The Bandera project has also focused on development of the Bandera Specification Language (BSL) [42]. BSL provides support for defining general assertions, pre/post conditions on methods, and atomic predicates for specifying temporal properties. There has also been related work [43] on developing a pattern-based approach for specifying complicated temporal properties, and translating them into input for model checking tools.

Software Modeling Languages

The Unified Modeling Language (UML) [44] is emerging as a popular standard for designing object-oriented software systems. There has been recent interest in its use for specification and modeling of real-time and embedded systems also [45, 46]. It provides different kinds of diagrams to specify the system structural and behavioral aspects, including Statecharts [47] and Message Sequence Charts [48, 49]. A language called Object Constraint Language (OCL) [50] has been developed as part of UML, for specifying requirements such as invariants, and pre/post conditions on operations in class diagrams. Tools for simulation of UML models, and for automatic code generation from UML into C/C++ programs, have become popular for embedded software development in automotive and avionics applications. However, formal verification efforts have been hindered by a lack of a precise formal semantics for UML and OCL, which are still under development.

The Specification and Description Language (SDL) [51] is another popular standard, especially targeted for communication systems. It provides both graphical and textual syntax, for specifying systems at different levels of hierarchy. Tool support for SDL is widely available, especially for automatic code generation to C, C++, and Java. Verification is supported by methods based on Message Sequence Charts, as well as model checking [52].

Another set of efforts is based on the use of declarative specification languages such as Z [53], and more recently Alloy [54]. Their main drawback is that there is little support for automatic extraction of such models from existing programs, or for automatic generation of programs from these models. Furthermore, Z is not very amenable to automatic analysis. However, Alloy does provide automated verification support, and has been used for some embedded applications. It uses a set-based syntax for expressing formulas in first-order relational logic. The analysis tool attempts to find a finite scope model that satisfies all given constraints, by translation to a Boolean SAT problem.

86.5   Languages for SoCs and Embedded Systems Verification

In this section, we describe verification languages for SoCs and embedded systems. We start by describing system-level modeling languages that have been heavily influenced by existing HDLs. Next, we describe verification support for domain-specific languages, such as Esterel, and efforts in the area of hybrid systems.

System-Level Modeling Languages

As systems become increasingly complex, it becomes useful to specify designs at higher levels of abstraction. This allows early design exploration for optimizing system requirements such as performance, power consumption, chip area, etc. Many system-level modeling languages focus on hardware–software partitioning issues, and development of system-level verification platforms.

SystemC and SystemC Verification Library

SystemC is a C++-based modeling platform, which supports hardware module abstractions at RTL, behavioral, and system levels, as well as software modules. Both the DUV and the testbench are written in the SystemC language. The platform provides a class library and a simulation kernel.

The SystemC Verification Library adds features for transaction-based testbench development. A typical example [55] is shown in Figure 86.1, where a transactor bridging the gap between the testbench and the DUV is shown in the setting of the SystemC class library. A technique called *data introspection*, which uses

C++ template specialization to attach a standard interface to arbitrary data types, is exploited to allow arbitrary data types to be used in constraints, assertions, transaction recording, and other high-level activities. Randomization of data generation for arbitrary data types is supported by defining a distribution through Boolean constraints and probability weights. The verification library also provides a callback mechanism to observe activities at transaction level during simulation, and a minimal set of HDL connection APIs, to permit interfacing with VHDL or Verilog. So far, there has not been much effort to add support for formal specification of properties within SystemC, although the standards can be potentially supported.

SpecC

SpecC is an executable modeling language based on C, targeted for hardware–software codesign [56]. The SpecC methodology advocates a clear separation between communication and computation in system-level designs. It provides support for creating an executable behavioral specification, design exploration including hardware–software partitioning, communication synthesis, and automatic generation of the software as well as hardware components. A significant effort has been made in providing a clean formal semantics, and in development of an associated simulation kernel. However, so far, there has not been much effort focused on its use in property specification or formal verification.

SystemVerilog and SystemVerilog Assertions

SystemVerilog is an extension of the Verilog hardware description language, with the latest version SystemVerilog 3.1 standardized by Accellera [57]. It adds many system-level modeling and verification capabilities to the older Verilog (IEEE Standard 1364). It also provides a Direct Programming Interface, which allows C functions to be called from Verilog code and vice versa.

For transaction-level testbench development, SystemVerilog has incorporated many features from C/C++ such as structures, pointers, classes, and inheritance. It also provides support for multiple threads, events, and semaphores for interprocess synchronization, mailboxes for interprocess communication, and random number classes to help generate random test vectors for simulation.

SystemVerilog also provides support for an assertion-based verification methodology, using the SystemVerilog Assertions (SVA) language, described below. The SystemVerilog standard has enhanced the scheduling scheme of Verilog, by defining new scheduling phases in which to sample stable signals, to evaluate the assertions, and to execute testbench code. This scheduling scheme avoids races between the DUV and testbench, guarantees that assertions and testbenches see stable values, and ensures common semantics across the simulation, verification, synthesis, and emulation tools.

SVA: SVA combines many ideas from the languages described in Section 86.3. Indeed, most examples described there have equivalent expressions in SVA. Properties in SVA are specified in terms of signal sequences, each of which is associated with a clock for sampling signals. Sequences are first-class objects

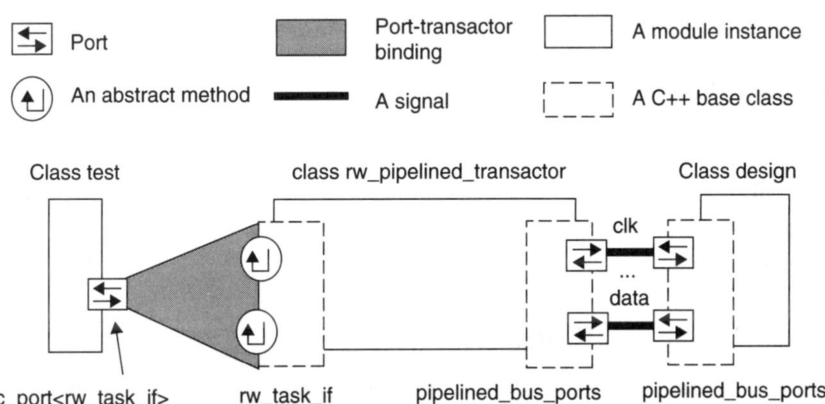

FIGURE 86.1 Transaction-level testbench in SystemC [55].

and can be declared, parameterized, or combined to build other sequences. It also provides limited support for combining sequences with multiple clocks.

Example 12: The following example defines a request–acknowledge sequence called *req_ack* [15]. It is defined in terms of parameter signals, and can be reused in multiple properties. The notation ##[1:3] indicates 1 to 3 cycles of the associated clock.

```
sequence req_ack (req, del, ack) ;
   // ack occurs within 1 to 3 cycles after req
   (req ##[1:3] ack);
endsequence
```

SVA also allows dynamic local variables to be declared within a sequence, which remain valid during the current invocation of the sequence. These can be used to check regular properties.

Example 13: Consider the following sequence which captures the I/O behavior of a pipeline register with depth 16 [15].

```
sequence pipe_operation;
   int x;
   (write_en, (x = data_in) ) |-> ##16 (data_out == x);
endsequence
```

Note that the input value is saved in the local variable *x* at the beginning of the sequence, which is compared with the value of *data_out* after 16 cycles. (Here, |-> denotes an implication operator that matches the end of the previous subsequence with the beginning of the next.)

Different directives act on properties to indicate their role — *assert* (check for violations), *cover* (observe and track when property is exercised), and *bind* (attaches an externally specified assertion to the code). Assertions are classified as immediate or concurrent. The immediate assertions are like assert statements in an imperative programming language such as C/C++. They are executed immediately when encountered in the SystemVerilog code, while following its event-based simulation semantics. On the other hand, concurrent assertions usually describe behavior that spans time. These are executed in the special scheduling phase using sampled values. In both cases, the action taken after evaluating an assertion may include system tasks to control severity like '$error,' '$fatal,' '$warning,' and '$info' [15]. Immediate assertions can be either declarative or procedural. Declarative assertions require enabling conditions to be explicitly provided by the user, which are monitored continuously on clock edges. In the case of procedural assertions, the enabling conditions are automatically inferred from the code context. This makes the procedural assertions easier to maintain when the code changes.

Example 14: Consider a simple paramaterized property declaration, and its use as a concurrent assertion [15]:

```
property mutex (clk, reset_n, a, b);
   @(posedge clk) disable iff (reset_n) (! (a & b ))
endproperty
assert_mutex: assert property(mutex (clk_a, master_reset_n, write_en, read_en));
```

The **disable iff** clause in the mutex property indicates that the property is disabled, that is, treated as if it holds, when the *reset_n* signal is active. The assertion checks that *write_en* and *read_en* cannot occur at the same time, provided *master_reset_n* is not active.

Domain-Specific System Languages

In this section, we describe some domain-specific languages for embedded system applications, and highlight the support they provide for verification. Unlike the system-level languages influenced by HDLs, there has been relatively little standardization effort aimed at their verification features.

Esterel

Esterel is a programming language used for the design of synchronous reactive systems [58]. The language provides features for describing the control aspects of parallel compositions of multiple processes, including elaborate clocking and exception-handling mechanisms. Typically, the control part of a reactive application is written in Esterel, which is combined with a functional part written in C. Esterel has a well-defined semantics, which is used by the associated compilers to automatically generate sequential C code for software modules, or Boolean gate-level netlists implementing finite state machines for hardware modules. Standard simulation or model checking can be performed on the synthesized finite state machines.

A central assumption in Esterel and other synchronous languages is the *synchrony hypothesis,* which assumes that a process can react infinitely fast to its environment. In practice, it is important to validate this assumption for the target machine. The tool TAXYS [59] checks this assumption by modeling the Esterel application and its environment as a real-time system. Two kinds of real-time constraints are specified as annotations in Esterel code — throughput constraints (which express the requirements that a system react fast enough for the given environment model), and deadline constraints (which express maximum delay between a given input and output of a system). These constraints are checked by Kronos [60], a model checker for real-time systems.

Example 15: An example of annotated application code [59] in Esterel with deadline constraints is shown below:

```
1     loop
2        await A; %{# Y = clock(last A) %}
3        call F( ); %{# Fmin < CPU < Fmax %}
4        %{# 0 < clock(last A) < d1 %}
5     end loop
6     ||
7     loop
8        await B;
9        call G( ); %{# Gmin < CPU < Gmax %}
10       %{# 0 < Y < d2 %}
11    end loop
```

In this example, it is required that the function F must terminate within $d1$ time units after arrival of event A, and also that the function G must terminate within $d2$ time units after the arrival of event A, which was consumed by F. These constraints are specified as shown in lines 4 and 10 (with line 2), respectively. Also note that the estimated runtimes for functions F and G are provided in lines 3 and 9, respectively.

Languages for Hybrid Systems

Many embedded system applications require an interaction of digital and analog components, for example automotive controllers, avionics systems, robotic systems, manufacturing plant controllers, etc. Since many such applications are also safety-critical, there has been a great deal of interest in the verification of such systems, called hybrid systems. The verification efforts can be broadly classified into those using classical control-theoretic methods, and others using automata-based methods.

A popular paradigm for control-theoretic methods is the MATLAB-based toolset, with the Simulink/Stateflow modeling language [61]. It provides standard control engineering components to model the continuous domain, and a Statechart-like language to model the discrete controller. MATLAB-based tools are used for analysis, optimization, and simulation of the continuous behavior specified in terms of differential equations. These tools have been used very successfully for applications dominated by continuous dynamics. However, complex interaction between the continuous and discrete control components has not received much attention. Furthermore, the simulation semantics has not been related to a formal semantics in any standard way.

The automata-based methods typically use discrete abstractions of the hybrid system in a way that preserves the properties of interest, typically expressed in temporal logic [62]. Many model checkers for handling these abstractions have been applied in industry settings, for example Hytech [63], Uppaal [64], Kronos [60], and Charon [65]. The details of the discrete abstractions, and the resulting automata models, are beyond the scope of this chapter. In most cases, the correctness requirement is to avoid a set of "bad" states, typically specified using the syntax of the modeling language itself. The model checkers perform an exact (or approximate) reachability analysis to (conservatively) check that all reachable states are safe. There has also been work on translating subsets of the popular Simulink/Stateflow-based models to the formal automata-based models, using abstraction and model checking techniques [66, 67].

Recently, there has been some standardization effort in this domain — the Hybrid Systems Interchange Format (HSIF) standard [68] is aimed at defining an interchange format for hybrid system models, which can be shared between modeling and analysis tools. So far, the focus has been on representation of the system dynamics, which include both continuous and discrete behaviors. Support for verification-related features will potentially follow its wider adoption.

86.6 Conclusions

We have presented a tutorial on verification languages in industry practice for verification of embedded systems and SoCs. We have described features that aid development of testbenches and specification of correctness properties to be used by simulation-based as well as formal verification methods. With verification becoming a critical activity in the design cycle of such systems today, these languages are receiving considerable attention, and there are several standardization efforts under way.

References

1. Edwards, S., Languages for embedded systems, in *The Industrial Information Technology Handbook*: CRC Press, 2004.
2. Bunker, A., G. Gopalakrishnan, and S. McKee, Formal hardware specification languages for protocol compliance verification, *ACM Transactions on Design Automation of Electronic Systems*, 9, 2004.
3. Accellera, http://www.accellera.org.
4. IEEE, Design Automation Standards Committee (DASC), http://www.dasc.org.
5. Bergeron, J., *Writing Testbenches, Functional Verification of HDL Models*, Kluwer Academic Publishers, Dordrecht, 2003.
6. Clarke, E.M., J. Wing et al., Formal methods: state of the art and future directions, *ACM Computing Surveys*, 28, 626–643, 1997.
7. Clarke, E.M., O. Grumberg, and D. Peled, *Model Checking*, MIT Press, Cambridge, MA, 1999.
8. McMillan, K.L., *Symbolic Model Checking: An Approach to the State Explosion Problem*, Kluwer Academic Publishers, Dordrecht, 1993.
9. Holzmann, G.J., The model checker SPIN, *IEEE Transactions of Software Engineering*, 23, 279–295, 1997.
10. Gordon, M.J.C., R. Milner, and C. P. Wadsworth, *Edinburgh LCF: A Mechanized Logic of Computation*, Vol. 78, Springer-Verlag, Berlin, 1979.
11. Boyer, R.S. and J.S. Moore, *A Computational Logic Handbook*, Academic Press, New York, 1988.
12. Owre, S., J.M. Rushby, and N. Shankar, PVS: A Prototype Verification System, in Proceedings of International Conference on Automatic Deduction (CADE), Lecture Notes in Computer Science, Vol. 607, Springer, Berlin, 1992.
13. Bryant, R.E., Graph-based algorithms for Boolean function manipulation, *IEEE Transactions on Computers*, C-35, 677–691, 1986.
14. Biere, A., A. Cimatti, E.M. Clarke, and Y. Zhu, Symbolic Model Checking without BDDs, in Proceedings of Workshop on Tools and Algorithms for Analysis and Construction of Systems (TACAS), Lecture Notes in Computer Science, Vol. 1579, Springer, Berlin, 1999.
15. Foster, H., A. Krolnik, and D. Lacey, *Assertion Based Design*, Kluwer Academic Publishers, Dordrecht, 2003.
16. Garey, M.R. and D.S. Johnson, *Computers and Intractability: A Guide to the Theory of NP-Completeness*, W. H. Freeman and Co., New York, 1979.

17. Barrett, C., D. Dill, and J. Levitt, Validity Checking for Combinations of Theories with Equality, in Proceedings of Formal Methods in CAD, Lecture Notes in Computer Science, Vol. 1166, Springer, 1996.

18. Bryant, R.E., S. Lahiri, and S. Seshia, Modeling and Verifying Systems using a Logic of Counter Arithmetic with Lambda Expressions and Uninterpreted Functions, in Proceedings of Conference on Computer Aided Verification, 2002.

19. Pnueli, A., The Temporal Logic of Programs, in Proceedings of the 18th IEEE Symposium on Foundation of Computer Science, 1977, pp. 46–57.

20. Vardi, M., Branching vs. Linear Time: Final Showdown, in Proceedings of Tools and Algorithms for Analysis and Construction of Systems (TACAS), 2003.

21. Thomas, W., Automata on infinite objects, *Handbook of Theoretical Computer Science*, Elsevier and MIT Press, Cambridge, MA, USA, Vol. B, 1990, pp. 133–191.

22. Thomas, D.E. and P.R. Moorby, *The Verilog Hardware Description Language*, Kluwer Academic Publishers, Norwell, MA, U.S.A., 1991.

23. Coelho, D.R., *The VHDL Handbook*, Kluwer Academic Publishers, Norwell, MA, U.S.A., 1989.

24. Kirshenbaum, Z., Understanding the "e" verification languages, in EE Design, http://www.eedesign.com, May 2003.

25. OpenVera Language Reference Manual: Assertions Version 2.3. http://www.openvera.org, 2003.

26. Armoni, R., L. Fix, A. Flaisher, R. Gerth, B. Ginsburg, T. Kanza, and A. Landver, The ForSpec Temporal Logic: A New Temporal Property-Specification Language, in Proceedings of Tools and Algorithms for Analysis and Construction of Systems (TACAS), Lecture Notes in Computer Science, Vol. 2280, Springer, Berlin, 2001.

27. Synopsis, OVA White Paper, www.openvera.org, 2003.

28. Beer, I., S. Ben-David, C. Eisner, D. Fisman, A. Gringauze, and Y. Rodeh, The Temporal Logic Sugar, in Proceedings of International Conference on Computer Aided Verification, 2001.

29. Craigen, D., S. Gerhart, and T. Ralston, Formal methods reality check: industrial usage, *IEEE Transactions of Software Engineering*, 21, 90–98, 1995.

30. Jackson, D. and M. Rinard, Software analysis: a roadmap, in *The Future of Software Engineering*, Finkelstein, A., Ed., ACM Press, New York, 2000.

31. Hoare, C.A.R., An axiomatic basis for computer programming, *Communications of the ACM*, 12, 576–580, 1969.

32. Godefroid, P., Model Checking for Programming Languages using VeriSoft, in Proceedings of ACM Symposium on Principles of Programming Languages, 1997.

33. Ball, T. and S. Rajamani, The SLAM Toolkit, in Proceedings of Conference on Computer Aided Verification, Lecture Notes in Computer Science, Vol. 2102, Springer, Berlin, 2001.

34. Ball, T. and S. Rajamani, SLIC: A Specification Language for Interface Checking (of C), Microsoft Research MSR-TR-2001-21, 2001.

35. Henzinger, T.A., R. Jhala, R. Majumdar, and G. Sutre, Software Verification with Blast, in Proceedings of the 10th SPIN Workshop on Model Checking, Lecture Notes in Computer Science, Vol. 2648, Springer-Verlag, Berlin, 2003.

36. Kroening, D., E.M. Clarke, and K. Yorav, Behavioral Consistency of C and Verilog Programs using Bounded Model Checking, in Proceedings of the Design Automation Conference, 2003.

37. Flanagan, C., K.R.M. Leino, M. Lillibridge, G. Nelson, J. Saxe, and R. Stata, Extended Static Checking for Java, in Proceedings of ACM Conference on Programming Language Design and Implementation (PLDI), 2002.

38. Leavens, G.T., K.R.M. Leino, E. Poll, C. Ruby, and B. Jacobs, JML: Notations and Tools Supporting Detailed Design in Java, in Proceedings of Conference on Object-Oriented Programming, Systems, Languages, and Applications, 2000.

39. Poll, E., J. van den Berg, and B. Jacobs, Specification of the JavaCard API in JML, in Proceedings of the Smart Card Research and Advanced Application Conference, 2000.

40. Corbett, J.C., M.B. Dwyer, J. Hatcliff, S. Laubach, C. S. Pasareanu, Robby, and H. Zheng, Bandera: Extracting Finite-State Models from Java Source Code, in Proceedings of the International Conference on Software Engineering, 2000.

41. Havelund, K. and T. Pressburger, Model checking Java programs using Java PathFinder, *International Journal on Software Tools for Technology Transfer (STTT)*, 2000.

42. Corbett, J.C., M.B. Dwyer, J. Hatcliff, and Robby, A Language Framework for Expressing Checkable Properties of Dynamic Software, in Proceedings of the SPIN Software Model Checking Workshop, 2000.

43. Dwyer, M.B., G. Avrunin, and J.C. Corbett, Patterns in Property Specifications for Finite-State Verification, in Proceedings of the International Conference on Software Engineering, 1999.

44. Rumbaugh, J., I. Jacobson, and G. Booch, The Unified Modeling Language User's Guide, Addison-Wesley, Reading, MA, 1999.

45. Martin, G., UML for Embedded Systems Specification and Design: Motivation and Overview, in Proceedings of Design Automation & Test Europe (DATE), 2002.

46. Selic, B., The Real-Time UML Standard: Definition and Application, in Proceedings of Design Automation & Test Europe (DATE), 2002.

47. Harel, D., StateCharts: A Visual Formalism for Complex Systems, *Science of Computer Programming*, Elsevier Science, 8, pp. 231–274, 1987.

48. International Telecommunication Union, Message Sequence Chart: ITU-T Recommendation, 1999.

49. Muscholl A. and D. Peled, From Finite State Communication Protocols to High-Level Message Sequence Charts, in Proceedings of International Symposium on Mathematical Foundations of Computer Science, Lecture Notes in Computer Science, Vol. 2076, Springer, Berlin, 2001.

50. Warmer J. and A. Kleppe, *The Object Constraint Language: Precise Modeling with UML,* Addison-Wesley, Reading, MA, 2000.

51. Ellsberger, J., D. Hogrefe, and A. Sarma, *SDL: Formal Object-Oriented Language for Communication Systems,* Prentice-Hall, Englewood Cliffs, NJ, 1997.

52. Levin V. and H. Yenigun, SDLcheck: a Model Checking Tool, in Proceedings of Conference on Computer Aided Verification, Lecture Notes in Computer Science, Vol. 2102, Springer, Berlin, 2001.

53. Spivey, J.M., *The Z Notation: A Reference Manual,* Prentice-Hall, Englewood Cliffs, NJ, 1992.

54. Jackson, D., Alloy: a lightweight object modeling notation, *ACM Transactions on Software Engineering and Methodology (TOSEM)*, 11, 256–290, 2002.

55. Ip C.N. and S. Swan, Using Transaction-based Verification in SystemC, http://www.systemc.org/, June 2002.

56. Gajski, D.D., J. Zhu, J. Doemer, A. Gerstlauer, and S. Zhao, *SpecC: Specification Language and Methodology,* Kluwer Academic Publishers, Dordrecht, 2000.

57. Accellera, SystemVerilog 3.1: Accellera's Extensions to Verilog, http://www.eda.org/sv-ec/SystemVerilog_3.1_final.pdf, 2003.

58. Berry G. and G. Gonthier, The Esterel synchronous programming language: design, semantics, implementation, *Science of Computer Programming*, 19, 87–152, 1992.

59. Closse, E., M. Poize, J. Pulou, J. Sifakis, P. Venier, D. Weill, and S. Yovine, TAXYS: A Tool for the Development and Verification of Real-Time Embedded Systems, in Proceedings of Conference on Computer Aided Verification (CAV), 2001.

60. Daws, C., A. Olivero, S. Tripakis, and S. Yovine, The tool Kronos, in Proceedings of Hybrid Systems III, Verification and Control, Lecture Notes in Computer Science, Vol. 1066, Springer, Berlin, 1996.

61. The Mathworks Inc., MATLAB/Simulink, http://www.mathworks.com.

62. Alur, R., T.A. Henzinger, G. Lafferriere, and G. Pappas, Discrete abstractions of hybrid systems, *Proceedings of the IEEE*, 88, 971–984, 2000.

63. Henzinger, T.A., P.-H. Ho, and H. Wong-Toi, HyTech: a model checker for hybrid systems, *Software Tools for Technology Transfer*, 1, 110–122, 1997.

64. Larsen, K., P. Pettersson, and W. Yi, UPPAAL in a nutshell, *International Journal of Software Tools for Technology Transfer*, Vol. 1, no. 1–2, pp. 134–152, Oct. 1997.

65. Alur, R., T. Dang, J. Esposito, R. Fierro, Y. Hur, F. Ivancic, V. Kumar, I. Lee, P. Mishra, G. Pappas, and O. Sokolsky, Hierarchical Hybrid Modeling of Embedded Systems, in Proceedings of EMSOFT '01: First Workshop on Embedded Software, Lecture Notes in Computer Science, Vol. 2211, Springer, Berlin, 2001.

66. Silva, B.I., K. Richeson, B.H. Krogh, and A. Chutinam, Modeling and Verification of Hybrid Dynamical System Using CheckMate, in Proceedings of Automation of Mixed Processes: Hybrid Dynamic Systems (ADPM), 2000.

67. Tiwari, A., N. Shankar, and J.M. Rushby, Invisible formal methods for embedded control systems, *Proceedings of the IEEE*, 91, 29–39, 2003.

68. HSIF: Hybrid System Interchange Format. http://micc.isis.vanderbilt.edu/HSIF.

87

Embedded Software in the SoC World. The Concept of HdS in View of the HW and SW Design Challenge

Frank Pospiech
Alcatel

87.1 Introduction

In modern complex Systems-on-Chip (SoC), software (SW) as an integral part of the SoC is gaining more and more importance. With increasing SoC complexity, SoC designers are facing an increasing complexity in the system's architecture other than the hardwre (HW) issues alone. The intricacy of the software needed to run on such devices is increasing tremendously as well.

Obviously, the challenges regarding design flow, design automation, and verification cannot be solved looking to the HW aspects of the design alone.

There is a large need for SoC designers to understand both their classical – the HW – world, and the world of Embedded Software as well. Particular attention needs to be paid on questions like

- How to deal with IP reuse that is composed of HW and SW.

- How does SW affect the SoC design flow.
- How to enable portability across HW platforms, different application domains, and so forth.

This chapter introduces Embedded Software for SoC and EDA designers, and tries to raise their awareness on SW issues they might be directly impacted by. However, both classical HW designers and Embedded Software designers have a lot to teach each other; therefore, mutual understanding is crucial for solving the SoC design challenges.

In the last part of this chapter, the concept of Hardware-dependent Software (HdS) is introduced, and how HdS can help to solve some of the issues mentioned before in this chapter.

This chapter is derived from a tutorial paper [1], which has been given by Steven Olsen from Mentor Graphics, and by the author during CICC 2003 in San Jose.

87.2 Embedded Software and its Challenge for ASIC/Custom/SoC Designers

Increasing SoC Complexity

As stated in [2], the semiconductor industry continues to follow Moore's Law, doubling the complexity of ICs every 18 months, to the point where they will soon be able to manufacture chips with 100 million gates. Entire systems can fit in always less area; hence, just as the hardware design industry has begun to accept the need for design reuse as a way to manage their exorbitantly increasing design costs, the problem is changing in them.

Now that whole systems can be put on a single piece of silicon, IC design is as much an SW design issue as it was an HW design problem before. The typical SoC design today has two or more processors, memory, dedicated subsystems specific to the application, and a complex, sometimes hierarchical communications system between them. As such, these chips contain multiprocessor real-time operating systems (RTOS), complete with I/O drivers, utilities, and diagnostic subsystems along with their specific SW applications. The exploding complexity of SW in these new SoC designs has already resulted in the SW design costs exceeding the HW design costs for some of the more complex designs.

Looking at this total complexity increase, the question now is how to manage the SoC design costs, not how to manage just the SoC HW design costs. Answering this question is virtually impossible without a deep understanding of what Embedded Software in the world of SoC actually means, what SoC SW design costs are, and how an Embedded Software design flow really works.

For the HW design and reuse challenge, in the past there have been several quite successful attempts to overcome Moore's law by standardization efforts. Thus, the Virtual Socket Interface Alliance (VSIA) [3] defined the concept of Virtual components (VCs) with rules on how to specify their interfaces and characteristics, in order to promote IP reuse on HW level. For Embedded Software, standardization efforts are just starting. There is an initiative in VSIA currently working on HdS standardization (see Section 87.4 of this chapter), and there have been some attempts to standardize driver interfaces on OS level by the Unified Driver Interface (UDI) project [4].

Changing the SoC Design Paradigm

What does the classical SoC design process look like?

To design complex chips, architects start with a high-level concept, which they translate into a high- level functional model. This model reflects the basic functions of their system but contains no distinction between the SW and HW. In the process of defining the appropriate HW architecture for the design, they must separate out which parts of the functions belong in HW and which parts belong in SW. This is typically done by reorganizing the functional model into its HW and SW components, along with a model of the communication between them. This has been done, and probably will continue to be done, by a few high-level architects.

The process of translating a cycle-accurate architectural model down to the actual silicon design is well understood and practiced, with sufficient design reuse and existing HW design techniques.

Likewise, there are no issues with translating the partitioned functional SW components into code for the finished working SoC.

What remains an issue, however, is the efficient HW/SW interface design and implementation. This is now becoming one of the most significant and growing efforts in SoC design.

It is clear that efficient HW/SW interface design is not possible without sufficient understanding of the other world's architectural principles, methods, and tools.

What does efficient interface design mean in this context?

First, those interfaces need to allow HW and SW designers to design their parts as independently as possible from each other, just to be able to parallelize their design processes.

Second, those interfaces need to be standardized. With the everincreasing number of HW architectures on the market, SW adaptations to any one of them become much too expensive. Ideally, some standard APIs could help to retain portability across different HW platforms.

Third, the way these interfaces are designed, as well as how the SoC overall design process is followed, needs to take into account, besides the "classical" HW design flow, the way in which modern embedded SW development works.

All this leads to a change in the SoC design paradigm that becomes much more SW centric than so far.

Validation and Verification Challenge

As we look at the overall SoC development costs, the SW/HW design ratio exceeds 2-to-1. While the development of the HW and the SW parts could easily be done in parallel, a greater effort must be spent to verify and validate the complete SoC. Meanwhile, verification and validation costs constitute up to 60% of overall SoC design costs.

Understanding how SW works and how it interacts with the HW can greatly reduce the overall verification complexity, and hence the cost.

Currently, verification and validation of the SoC largely rely on HW test benches to test the HW with the help of some deeply Embedded Software inside the SoC present to test the more complex internals of the SoC.

There needs to be a methodology that will allow system designers to model the core components irrespective of their HW or SW domain.

Some considerations based on SystemC as modeling language attempt to close this gap.

Another way to reduce complex SoC verification and validation costs is to reuse as much as possible prequalified IP blocks, and to define rules for their integration into the entire system. Here again, the right understanding of the HW–SW interface is key for lowering the overall system validation effort.

87.3   Embedded Software, Concepts, Methods, Tools

What is an Embedded System, What is Embedded SW?

For definitions used here, see [7].

Before talking about Embedded Software, it is necessary to understand what **embedded systems** are. Unlike general-purpose systems, like personal computers, embedded systems are **dedicated systems**. They serve a well-defined purpose, and usually all resources in the system are dimensioned according to their purpose.

Embedded systems typically interact with physical processes, be it via sensors, actuators, etc. Unlike general-purpose systems, critical properties of embedded systems tend to be nonfunctional: real-time, fault recovery, power, security, or robustness. They typically have to interact concurrently with multiple processes, and need to operate at the speed of their environment.

Simplistically speaking, **Embedded Software** is SW designed for and running on embedded systems. Since many of the properties mentioned before are actually implemented in HW, one of the most important properties of Embedded Software is its close link to HW. The close link is demonstrated not only by compiling for the direct HW dependence but also through intimate knowledge of the HW specifics (for instance, register and interrupt structures, bit orders, and so forth). Some of the properties mentioned above are increasingly implemented in SW, or directly influence the way Embedded Software is written. The classical example is real-time.

RTOS play an important role in Embedded Software. Besides serving as a layer on top of which parts of the Embedded Software can be implemented relatively independent of the actual HW platform, the RTOS also provides a framework to plug HW drivers.

Embedded Software itself can be structured into different SW layers:

- HdS (for details, see Section 87.4 of this chapter).
- HW-independent middleware or application software, which run on top of the HdS.
- The RTOS as a vertical "layer," which offers services that span both hardware-dependent and -independent parts that may be used by all the other Embedded Software layers. For details on real-time and RTOS, see [3.3.1.].

HW Related Trade-offs in Embedded SW

What are the specific challenges in designing Embedded Software — in particular HdS — in comparison with HW design on one hand, and classical application SW design on the other?

The HdS layer must expose HW internal features in a way that releases developers of higher layer SW from requiring too much processor-specific knowledge. Ideally, the SoC functionality should be exposed in a way, which completely hides the details of the underlying HW platform and its evolution. The best way to do so is via a functional API, the so-called HdS-API. For details, see Sectoin 87.4.

While application SW is reusable and portable per se via its large independence from actual HW platforms, Embedded Software (ESW) must be structured in such a way as to achieve portability or reuse. Highly optimized DSP algorithms, for instance, take full advantage of each HW feature. However, for this kind of SW, reuse methodologies are rather difficult to apply.

The specific properties of ESW impact its design methodology. Hence, the imminent HW dependency implies close coupling of ESW design with HW design. Early alignment on hardware features/interfaces is crucial for the design of the HdS parts of ESW. However, each HW change basically implies a modification in parts of the ESW (e.g., drivers or access functions). Conversely, stable HdS-APIs are crucial for starting application software design. ESW design principles have to deal with this problem.

Designing ESW means interfacing very different roles in the systems design process:

- *System designers*, for HW–SW partitioning.
- *HW engineers*: Concurrent development of drivers, load, and boot. Provision of HW test SW. Deep understanding of HW specifics.
- *SW engineers*: Provision of the HdS APIs. Sharing SW technology, methodology, tools. With respect to architecture, ESW needs to meet some specific requirements:
- *Portability*: HW-dependent parts are by definition not portable, but are intended to enable portability for higher layer SW.
- *Real-time*: Typical SW interprocess communication mechanisms like Common Object Request Broker Architecture (CORBA, an object-oriented communication mechanism that provides full location transparency) usually cannot be applied for real-time reasons. Often, ESW has to meet hard real-time requirements.

Parts of the ESW implement themselves RTOS-like services. Often, the implementation of ESW depends directly on the RTOS choice (message passing mechanism and memory management).

In summary, these are the most important differences between application SW and ESW, which are related to HW:

- Specific HW requirements: timing, (hard) real-time, performance of HW access functions, direct HW architecture dependence.

- Mixed top-down and bottom-up development of ESW.
- Complex synchronization in overall product development process.

ESW often needs to run with instable HW, or no target HW at all available.

For high-level SW or middleware in many application domains, there are a large number of standards or quasistandards defined. Examples are CORBA for interprocess communication, high-level specification languages like Unified Modeling Language (UML), Specification and Description Language (SDL), or programming languages like ADA, or telecommunication protocols standardized by ITU.

In the embedded world, standardization is only starting. Examples are Portable Operating System Interface (POSIX) for operating system interfaces, or some work that is currently done in VSIA.

Basic Concepts in ESW

RTOS

What is real-time? The following answers on this question are heavily influenced by a tutorial given by Krithi Ramamrithan on DAC 2002 [5].

Embedded systems usually have to deal with events in the environment they are designed for. The system has to follow the real speed of the environment. Otherwise, damages, even loss of human life, might be the consequence. Meeting deadlines that are determined by the environment, means "**real-time.**" Real-time does not necessarily mean "fast." However, real-time has to face the problem of the potentially changing "speed" of the environment.

Generally speaking, **real-time systems** are systems that react to events in the environment by performing predefined actions within specified time intervals.

Note that not all embedded systems are real-time.

The most important properties of real-time (RT) systems are:

- *Safety*: Nothing bad will happen
- *Liveness*: Something good will happen.
- *Timeliness*: Things will happen on time — by their deadlines, or periodically.

Different types of RT systems according to these questions:

- How *tight* are the deadlines?
- How *strict* are the deadlines?
- What are the *characteristics of the environment*; how static or dynamic must the system be?

According to these questions, the following types of RT systems can be distinguished:

- *Hard RT*: The results are useless or dangerous, if the deadline is not met.
- *Soft RT*: The results are of some lower value, if the deadline is exceeded.
- *Firm RT*: If the value drops to zero at the deadline.

Real-time behavior is mainly achieved by corresponding scheduling strategies. Examples are

- Strict table-driven approaches:
  - Perform static schedulability analysis by checking if a schedule is derivable.
  - The resulting scheduling table identifies the start times of each task.
  - This approach is applicable to periodic tasks.

- Static priority-driven preemptive approaches:
  - Tasks have, systematically assigned, static priorities.
  - Priorities take timing constraints into account.
  - Perform static schedulability analysis but no explicit schedule is constructed.
  - At runtime, tasks are executed highest-priority first, with preemptive-resume policy.

- Dynamic planning-based approaches:
  - Feasibility is checked at runtime — a dynamically arriving task is accepted only if it is feasible to meet its deadline.
  - One of the results of the feasibility analysis is a schedule or plan that determines start times.
- Dynamic best-effort approaches:
  - The system tries to do its best to meet the deadlines.
  - But since no guarantees are provided, a task may be aborted during its execution.
  - Until the deadline arrives, or until the task finishes, whichever comes first, one does not know if a timing constraint will be met.

RTOS support process management and synchronization, memory management, interprocess communications, and I/O.

There are several categories of RTOS:

- Small, proprietary kernels (e.g. VRTX32, QNX Neutrino, pSOS, VxWorks).
- Real-time extensions to commercial/noncommercial timesharing operating systems, for example, RT-Linux
- Research kernels (e.g., MARS, ARTS).

Many general-purpose operating systems like Windows NT can be seen as, or easily be transformed to, soft real-time OS.

ESW Development and Tools

Many of the development tools known from the general-purpose SW world are applied for ESW as well.

However, the intimate relation of ESW and the underlying HW platform ask for some specific tools, especially in the domains of

- Host and target compilation
- Target debugging
- Hardware testing and bringup.

Many processor platforms provide their dedicated development tools. An example is ARM's ADS (ARM Developer Suite). RTOS providers often offer development suites around their flagship OS. Examples are WindRiver's Tornado, or GreenHills' Multi2000.

The ESW development process has to take its specific dependencies on the HW and SW development cycles into account. The specification of APIs must be a condition for starting the application SW design. On the other hand, the ESW design cannot be finished before the final HW prototype availability. Many techniques to decouple HW and ESW development are currently in investigation, or available in first products. Most of these techniques try to provide early HW models, which are on an interface level identical to the later prototype.

In general, the ESW process follows classical SW development paradigms. Usually, all of the phases, like specification, top-level design, detailed design, implementation, unit test, integration and integration test, and system qualification and acceptance test can be found as well. However, integration and integration test here mean mainly integration with the HW platform.

By its nature, ESW design deals with HW specific tools, like instruction set simulators, hardware simulators and emulators, and distributed debuggers.

UML-Based Design and Development of ESW

For an excellent introduction into UML for embedded systems designers, see [6].

Formal SW design methods have been in place for many years in application SW. The object-oriented paradigm in analysis, design, and programming led to a broad adoption of UML as a quasistandard in recent years.

With the increasing use of the Object-orientation paradigm for ESW as well, UML-based formal design methods have also been applied in this domain.

What is **UML**? UML stands for "Unified Modeling Language." It intends to handle SW models. **Software models** in this sense are descriptions of the SW of a system that abstracts out irrelevant details, such as certain aspects of technology or implementation, so that we can more easily reason about the system itself.

Models allow for human communication, for formal analysis by mathematical methods, and for experimenting.

The usage of models in SW is steadily increasing. This way SW engineering gets closer to classical engineering disciplines. Based on executable models, automatic code generation becomes possible.

UML is based on the concept of the **object**. The idea is that all various features (procedures, data) of a logical unit are combined into one single package called an object. An SW system is defined as a structure of collaborating objects. More than one object can be constructed from the same definition — the **class**.

UML consists of:

- A basic set of (extensible) modeling concepts
- Formal rules of semantically meaningful composition (well-formedness)
- Graphical notation for modeling concepts.
- Different diagram types for different aspects (views) of the target system.

There are eight different diagram types (requirements, structure, behavior, and deployment). There are several **UML model views**:

- Requirements (use case diagrams)
- Static structure (class diagrams) → kind of objects and their relationships
- Object behavior (state machines) → possible life histories of an object
- Interobject behavior (activity, sequence, and collaboration diagrams) → flow and control among objects to achieve system-level behavior.

Physical implementation structures (component and deployment diagrams) → SW modules and deployment on physical nodes.

In application SW development, UML is meanwhile widely used for

- Requirements management
- Stepwise design refinement
- Code generation.

UML as a standard is a basis for a family of languages, for example Real-time UML, or UML for eCommerce.

It allows defining **UML profiles**. Profiles are packages of related specializations of general UML concepts that capture domain-specific variations and usage patterns. In this sense, UML profiles are domain-specific interpretations of UML.

For using UML in real-time systems, some standard models have been defined, which cover

- Physical time
- Timing specifications
- Timing services and mechanisms
- Concurrency and scheduling
- SW and HW infrastructure and their mapping.

A **Real-Time Profile** needs to

- Be able to specify quantitative information directly in UML models.
- Be flexible: users need to model their RT systems using modeling approaches and styles of their own choice.
- Facilitate the use of analysis methods. As much as possible, the generation of analysis models and of the analysis process itself needs to be automated.

RT profiles are included in UML 2.0, which has been standardized in 2003.

Using RT profiles in UML 2.0, SW performance engineering becomes possible. That is, "A systematic, quantitative approach to constructing software systems that meet performance objectives" (Connie Smith and Lloyd Williams).

87.4 Hardware dependent Software (HdS). Closing the Gap between HW and SW Design

As seen in the HW Related Trade-offs in Embedded SW section, an SoC can be regarded as an API to embedded SW developers. Hence, the SoC should have an abstraction layer, which exposes its functionality. We call it Hardware Abstraction Layer (HAL). The HAL might even enhance the SoC's capabilities. At the same time, it hides HW details that are irrelevant for upper-layer SW implementations. A typical example is an RTOS Board Support Package (BSP). It hides the processor internal timer and interrupt dispatching, but adds multitasking and timer functions.

Thus, it is useful to develop a standard abstraction for the SW that "sits directly on top" of the HW. This kind of SW is called **Hardware dependent Software (HdS)**.

What is HdS?

Under HdS, we understand that all SW that is directly dependent on the underlying HW platform. Examples are:

- HW drivers
- Boot strategy, load
- Built-in tests (basic level, off-line tests)
- HW-dependent parts of communication Stacks
- Algorithms implemented in SW on DSP.

The main purpose of HdS is to shield the HW for upper-layer application SW. Ideally, this SW communicates with the HW platform via stable APIs only. Another core motivation to define HdS is retaining portability across various simulation and target environments.

The introduction of HdS allows decoupling of HW and SW design. With HdS, a portability layer is defined, which:

- Offers stable API toward SW, enables platform transparency (prototype, emulator, simulator).
- Translates SW application and RTOS requirements to HW resource accesses.

HdS Seen from Different Perspectives

HdS can be regarded from different perspectives. For details, see the HdS taxonomy axes that are defined in VSIA's HdS taxonomy document [5], which are in the process of being published.

1. **Life Cycle View**
   As already discussed in the HW Related Trade-offs in Embedded SW section, the HdS design flow is tightly coupled to an HW (ASIC, FPGA, component) design flow. Examples are: OS simulators, ISS models of processors, HW emulation, and fast prototypes (FPGA based HW models).
   HdS is developed and used in
   - HW development, bringup, and test
   - HW production (manufacturing)
   - SoC in final product (drivers, boot&load, On-line and off-line diagnostics).
   HdS is one basis for SW development — it provides SW APIs.

2. **Run-time and Real-time View.**
   HdS implements services like Boot, Configure, Execute, Reload, Target monitor,…

With respect to real-time behavior, the HdS implementation takes scheduling models and timing characteristics into account.

Any HdS design has to incorporate communication mechanisms, for example address mapped, packet based, message based,… ones.

3. **HW architecture View**

   The HW platform defines a certain SW architecture. The HW architecture typically consists of a CPU subsystem (i.e., registers, address space, cache, interrupts), the memory subsystem, and the I/O subsystem.

   Multiprocessor architectures cause HdS to implement specific communication mechanisms. HdS cares for processor topology transparency toward application SW.

4. **SW Layering View**

   The HAL as part of HdS contains several hierarchical HW shielding levels:

   a. Access shielding
   b. Register shielding
   c. Functional shielding.

   HdS as an SW layer embeds into a general architectural model that is characterized by different abstraction levels, like Hardware layer, Primitive function layer, and Interface access layer, OS layer, and Application layer.

Standardization in SoC-based Embedded Systems

From an architectural point of view, ESW and HdS can be seen as SW layers on top of physical HW.

The following figure provides a view on HAL as part of HdS, and its relation to HW layer, OS layer, and application SW layer.

The layers introduced in the HdS seen from Different Perspectives section, can be mapped onto HW Space (the HW layer), Kernel space (the Primitive Function layer, the Interface Access Layer, and the OS Layer), and a User space (the application layer).

While the interface between Kernel space and User space is going to be standardized via the POSIX interface, VSIA has defined the OCB VCI interface for standardizing HW communication interfaces.

POSIX [8] is an RTOS interface standard toward middleware and application SW (IEEE standard P1003.1). It contains a minimal set of common OS commands, and was developed for UNIX. Currently, an extension toward real-time is under research. POSIX provides a portability interface toward SW.

VSIA's **On Chip Bus (OCB) VCI** was created to make HW IP more portable with a common bus interface. It includes a modeling language consisting of I/O operations at the transaction level (with read, write, open, close) to manage various types of multipacket transactions. OCB VCI provides a standardized IP block interface toward HW (Figure 87.1).

Inside the Kernel space, those parts that are directly dependent on HW register maps and HW interrupt structures can be separated from functional parts such as device driver logic, offline test segments, generic boot, and loads. This separation is done via a new HdS-API.

All parts below the HdS-API (the HAL, BSPs,) carry HW platform-specific information, and can partially be derived from the physical HW structure in terms of register maps.

The Kernel space parts above the HdS-API implement functional logic, and are independent of implementation details of the underlying HW platform.

Gaps in the SoC Design Flow. How A Standardized HdS-API could Help

Without any standard APIs between the OCB VCI and the POSIX, the entire kernel space SW (HdS + RTOS) depends on any HW modification. HW register structure changes need to be implemented in HdS, and in the RTOS' BSP, correspondingly.

After partitioning of high-level functional models into HW and SW IP blocks, they can be connected with a VCI-like interface on the HW side. Translation from the VCI transaction language to a specific VCI protocol is possible.

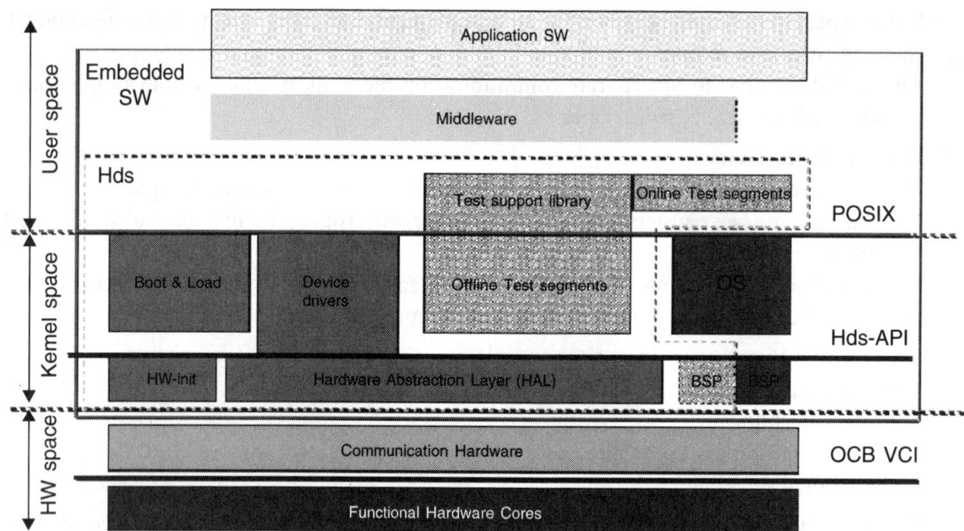

FIGURE 87.1   SoC Design needs a New Standard.

However, the entire kernel space depends on address-specific information. Hence, reuse and automation are difficult.

With a standard HdS-API, the SW above would be shielded from address-specific information.

It becomes possible to load initialization registers and device addresses from configuration databases. As a further step, this HdS-API allows for register-dependent interface code generation.

The HdS-API contains read, write, open, close, and control services — like standard OS or bus (i.e., VCI) interfaces. Drivers, diagnostics SW, and parts of the RTOS are built on top of this API. This way, this kernel space SW layer becomes portable over a large range of

- HW architectures, including processors, on the bottom
- SW applications, on the top.

With the potential for automating the generation, or at least reuse, of the kernel space SW below the HdS-API, SoC SW development effort can drastically be reduced.

Running Standardization Activities. The Role of VSIA's HdS-DWG

In 2002, VSIA investigated on SoC industry's needs to open for ESW. As a result, the HdS Design working group (DWG) was formed in 09/2001. In 2003, it was working with more than 25 members from 15 companies.

The DWG's mission consists of:

1. Improving company-internal and intercompany component reuse by defining or fixing an "HdS API."
2. Coming to industry specifications on HdS Virtual components.

An **HdS taxonomy document** [7] has been released by VSIA. This taxonomy

- Consolidates terms from the SoC and the SW world, which need to be agreed upon in the HdS context
- Defines HdS specific terms (HdS, HdS-API, kernel space, user space, driver, access shielding,…)
- Relates HdS concepts to different aspects, like life cycle, HW platform, SW platform, Real-time
- Determines HdS' architectural relation to HW, middleware, and application SW layers
- Provides a tutorial to train both HW and SW experts on HdS needs.

In 2003, the DWG was working on the definition of the **HdS-API**. The target is to determine the characteristics of a common API, which defines the way to provide SoC IP's functionality to upper-layer SW. HdS-APIs are defined for different application domains (e.g., telecommunications, automotive, multimedia). Certainly, this API is defined in relation to other driver standardization initiatives, like the Project UDI [4].

The API will be described using UML, and defines

- An HdS Framework in terms of an HAL architecture
- An API mechanism, which determines how to specify an HdS API
- Elementary Services: These are the basic services either to be offered directly to the SW, or of which services are they composed.

This HdS-API will be the basis for HdS-VC Reuse guidelines to be worked on in 2004.

87.5 Conclusion

Deep understanding of ESW issues is imperative in modern SoC design. Many of the methods, tools, and architectures known from SW development also need to be considered in designing SoCs as systems. Examples have been discussed in this paper. However, designing embedded real-time systems needs the special attention of both SW and HW engineers.

Achieving reuse and portability for systems containing complex SoCs needs reconsidering of classical HW and SW architecture approaches. HdS, and especially the HdS-API, are ways to overcome the current productivity crisis with huge efforts spent on integrating applications on top of SoCs, or on porting systems from one HW platform to another one.

VSIA's current work on HdS promises to advance this process by defining an industrywide HdS-API standard.

References

1. Embedded Software in the SoC World, How HdS Helps to Face the HW and SW Design Challenge, CICC 2003, San Jose.
2. Larry Cooke, API will bridge HW/SW design gap, EE Times, Nov. 21, 2003.
3. Virtual Socket Interface (VSI) Alliance, See http://www.vsi.org/
4. Unified Driver Interface (UDI) Project, See http://www.projectudi.org/
5. Ramaritham, K., System Support for Real-Time Embedded Systems, DAC 2002, New Orleans.
6. Selic, B., Embedded Software Development. Methodology and Tools, DAC 2002, New Orleans.
7. VSI Alliance, Hardware Dependent Software Taxonomy, Version 1.0, Hardware Dependent Software Development Working Group, 2003.
8. ISO/IEC 9945-x:2002, Information technology — Portable Operating System Interface (POSIX), ISO 2002.

88

Internal Architecture and Features of Real-time Embedded Operating Systems

Ivan Cibrario Bertolotti

IEIIT-CNR - Istituto di Elettronica e di Ingegneria dell'Informazione e delle Telecomunicazioni

88.1 Introduction

Informally speaking, a *real-time* computer system is a system where a computer senses events from the outside world and reacts to them; in such an environment, the *timely* availability of computation results is as important as their correctness.

Depending on the strictness of their timing requirements, real-time systems are often divided into three classes:

In *hard* real-time systems, the late availability of a computation makes its results completely useless and makes the application fail catastrophically; most safety-critical real-time applications, such as pacemakers, belong to this category;

In a *soft* real-time system, timing requirements are somewhat relaxed, so that the value of a computation decreases as its lateness increases, but the violation of a timing constraint, provided the probability of such an event is low, is acceptable. For example, if a video streaming application is unable to process all its input data on occasion, the video output will be degraded but will still be of acceptable quality;

In between are *firm* real-time systems. In these systems, a computation has no value at all if it is late, like in hard real-time; however, like in soft real-time, the system will not undergo an unrecoverable failure when a timing constraint is violated. Many financial applications fall in this category.

Instead, the exact definition of *embedded* system is somewhat less clear. In general, an embedded system is a special-purpose computer system built into a larger device and is usually not programmable by the end user.

The major areas of difference between general-purpose and embedded computer systems are cost, performance, and power consumption: often, embedded systems are mass-produced, thus reducing their unit cost is an important design goal; in addition, mobile battery-powered embedded systems like cellular phones have severe power budget constraints to enhance their battery life.

Both these constraints have a profound impact on system performance, because they entail simplifying the overall hardware architecture, reducing clock speed, and keeping memory requirements to a minimum.

Moreover, embedded computer systems often lack traditional peripheral devices, such as a disk drive, and interface with application-specific hardware instead.

Another common requirement for embedded computer systems is some kind of real-time behavior; the strictness of this requirement varies with the application, but it is so common that, for example, many operating system vendors often use the two terms interchangeably, and refer to their products either as "embedded operating systems" or "real-time operating systems for embedded applications."

In general, the term "embedded" is preferred when referring to smaller, uniprocessor computer systems, and "real-time" is generally used when referring to larger appliances, but the today's rapid increase of available computing power and hardware features in embedded systems contributes to shade this distinction.

An early example of a recognizable, modern real-time embedded system was the Apollo Guidance Computer (AGC) [28] and its custom operating system, whose design began in 1961 at the MIT Instrumentation Laboratory; its most important duty was to run the inertial guidance system of the Apollo Command Module and LEM.

More recent examples include many kinds of computer system, from large appliances like phone switches to mass-market consumer products such as printers and digital cameras.

Therefore, a real-time operating system must not only manage system resources and offer a well-defined set of services to application programs, like any other operating system does, but must also provide guarantees about the *timeliness* of such services and honor them, that is, its behavior must be *predictable*. Thus, for example, the maximum time the operating system will take to perform any service it offers must be known in advance.

This proves to be a tight constraint, and implies that real-time does *not* have the same meaning as "real fast," because it often conflicts with other operating system's goals, such as good resource utilization and coexistence of real-time and non-real-time jobs, and adds further complexity to the operating system duty.

Also, it is highly desirable that a real-time operating system optimize some operating system parameters, mainly context switch time and interrupt latency, which have a profound influence on the overall response time of the system to external events; moreover, in embedded systems, the operating system footprint, that is, its memory requirements, must be kept to a minimum to reduce costs.

Last, but not the least, due to the increasing importance of open system architectures in software design, the operating system services should be made available to the real-time application through a standard application programming interface. This approach promotes code reuse, interoperability, and portability, and reduces the software maintenance cost.

The chapter is organized as follows: Section 88.2 gives a brief refresher on the main design and architectural issues of operating systems and on how these concepts have been put in practice. Section 88.3 discusses the main set of international standards concerning real-time operating systems and their application programming interface; this section also includes some notes on mechanisms seldom mentioned in operating system theory but of considerable practical relevance, namely real-time signals and timers. Section 88.4 concludes the paper and contains a short survey on some commercial and open source real-time operating systems in use today.

88.2 Operating System Design and Architecture

The main goal of this section is to give a brief overview on the design and architectural issues of operating systems of interest to real-time application developers. See [27] for more general information, and [14] for an in-depth discussion about the design of the influential UNIX operating system.

Overall System Architecture

An operating system is a very complex piece of software; accordingly, its internal architecture can be built around several different designs. Four designs that have been tried in practice and are in common use are:

Monolithic Systems

This is the oldest design, but it is still popular for both real-time and general-purpose systems due to its simplicity and very low processor and memory overhead. For the same reason, it is also the design approach of choice for very small real-time executives for deeply embedded applications, such as [11], and for the real-time portion of more complex systems, like [6] and [19].

When using this approach, the operating system is written as a collection of procedures that are then bound together into a single executable file by the link editor. Each of them has a well-defined interface in terms of parameters, behavior, and results, and can freely call any other procedure when it needs to; hence, there is little or no information hiding.

In monolithic systems, the operating system as a whole runs in privileged mode, and the only internal structure is usually induced by the way operating system services are invoked: applications, running in user mode, request operating system services by placing service parameters in well-defined locations, for example, a set of registers or a stack frame, and then executing a special trap instruction, usually known as the *system call* instruction. This instruction brings the processor in privileged mode and transfers control to the system call dispatcher of the operating system. The system call dispatcher saves the user-mode registers into the appropriate *process control block*, determines which service must be carried out, and transfers control to the appropriate service procedure. Service procedures share a set of utility procedures, which implement generally useful functions on their behalf.

When the system call is finished, the operating system gives control back to the application, after restoring its user-mode registers; execution continues with the instruction following the system call. Figure 88.1 depicts the resulting structure.

The total *overhead* for the invocation of a nonblocking system service, that is, a service that the operating system is able to perform without any synchronization either among operating system modules or with an input/output (I/O) device, is therefore limited to the overhead of the system call instruction and return, plus the time needed to save and restore the user-mode registers of the invoking process. Besides, the interaction between operating system components is very fast, because they invoke each other by means of cheap function calls.

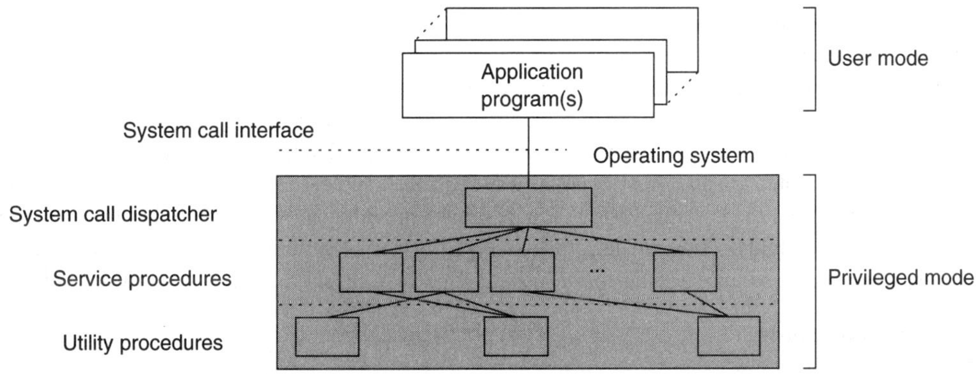

FIGURE 88.1   A monolithic operating system.

Interrupt handling is done directly in the kernel for the most part and interrupt handlers do not have a full processor control block allocated to them. As a consequence, the interrupt handling overhead is very small because there is no full task switching at interrupt arrival, but the interrupt handling code cannot invoke most system services, notably blocking synchronization primitives, and the processor performs all interrupt handling in interrupt service mode. During this time, the operating system scheduler is disabled, and only hardware prioritization of interrupt requests is in effect; hence, the interrupt handling code is implicitly executed at a priority higher than the priority of all application tasks in the system, without regard to the actual priority of the task that will eventually make use of the interrupt itself.

In addition, the only synchronization mechanism available between interrupt handlers and other operating system components is to temporarily disable interrupts. To keep interrupt response latencies bounded and small, it is therefore necessary to keep these critical code paths as short as possible, and to keep track of their actual length whenever they are updated. In turn, this constraint may introduce additional complexity in the operating system algorithms.

To further reduce processor overhead on small systems, it is also possible to run the application as a whole in supervisor mode. In this case, the application code is bound with the operating system at link time, and system calls become regular function calls. The interface between application code and operating system becomes much faster, because no user-mode state must be saved on system call invocation and no trap handling is needed. On the other hand, the overall control that the operating system can exercise on bad application behavior is greatly reduced and debugging may become harder.

In addition, it becomes impossible to upgrade individual software components, for example, an application module, without replacing the executable image as a whole and then rebooting the system. This constraint can be of concern in applications where software complexity implies frequent replacement of modules, and/or no system down time is allowed.

Layered Systems

A refinement and generalization of the monolithic system design consists of organizing the operating system as a hierarchy of layers at system design time. Each layer is built upon the services offered by the one below it and, in turn, offers a well-defined and usually richer set of services to the layer above it. Operating system interface and interrupt handling are implemented like in monolithic systems; hence, the corresponding overheads are very similar.

Better structure and modularity make maintenance easier, both because the operating system code is easier to read and understand, and because the inner structure of a layer can be changed at will without interfering with other layers, provided the interlayer interface does not change.

Moreover, the modular structure of the operating system enables the finegrained configuration of its capabilities, to tailor the operating system itself to its target platform and avoid wasting valuable memory space for operating system functions that are never used by the application. As a consequence, it is possible to enrich the operating system with many capabilities, for example, network support, without sacrificing its ability to run on very small platforms when these features are not needed. A number of operating systems in use today, like [1,2,9,21–23,32], evolved into this structure, often starting from a monolithic approach, and offer sophisticated build or link-time configuration tools.

Historically, this concept was taken to an extreme and further generalized by the MULTICS operating system [17]. MULTICS was organized as a set of concentric rings, of which the outer ones were less privileged than the inner ones. The cross-ring call mechanism was implemented in hardware much like a system call, and the associated protection and security policy was controlled and enforced at runtime by the hardware itself. Thus, in this case, the layering was not a mere design aid, but greatly contributed to the overall system security and reliability. Unfortunately, this concept has not been further pursued by contemporary operating systems, also because its efficient implementation requires special-purpose hardware assistance.

Microkernel Systems

This design, sketched in Figure 88.2, moves many operating system functions from the kernel up into operating system server processes running in user mode, leaving a minimal *microkernel* and reducing to

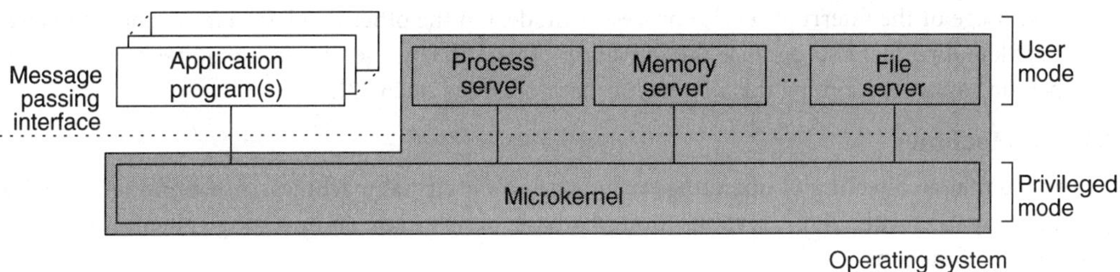

FIGURE 88.2 A microkernel-based operating system.

an absolute minimum the amount of privileged operating system code. Applications request operating system services by sending a message to the appropriate operating system server and waiting for a reply.

The main purpose of the microkernel is to handle the communication between applications and servers, to enforce an appropriate security policy on such communication, and to perform some critical operating system functions, such as accessing I/O device registers, that would be difficult, or inefficient, to do from user mode processes.

This approach has several advantages:

By splitting the operating system into smaller parts, each part becomes easier to manage and maintain. Also, the message-passing interface between user processes and operating system components encourages modularity and enforces a clear and well-understood structure on operating system components;

The reliability of the operating system is increased: since the operating system servers run in user mode, if one of them fails some operating system functions will no longer be available, but the system will not crash. Moreover, the failed component can be restarted and/or replaced without shutting down the whole system; the same mechanism can also be used to add and upgrade application modules without bringing the system down;

The model is easily extended to *distributed systems,* where operating system functions are split across a set of distinct machines connected by a communication network. Distributed systems are very promising in terms of performance, scalability, and fault tolerance, but many topics of interest for their design, such as distributed scheduling algorithms, resource sharing, and load balancing, are still an active research area. As a consequence, few operating system, for example [20,30], support them.

By contrast, making the message-passing communication mechanism efficient is usually a critical issue, as will be discussed further in the Interprocess Synchronization and Communication section. In addition, the system call invocation mechanism involves more overhead because, even for a nonblocking system call, the following steps must be performed:

- The application sends its request to the appropriate operating system server through the micro-kernel message-passing service, then blocks waiting for an answer.
- Assuming that the system server now has the highest priority, the scheduler wakes it up, and it starts executing the system service.
- When the system server finishes its job, it sends back a result message to the application, again through microkernel message-passing.
- At this point, the application can run again and the scheduler will eventually resume executing it.

Thus, the total overhead associated with the simplest kind of system call includes two microkernel service invocations, usually implemented by means of traps, two executions of the scheduler, and two full-fledged task switches.

Interrupts are handled in a similar way, that is, the interrupt handler proper runs in interrupt service mode and performs the minimum amount of work strictly required by the hardware, then wakes up an interrupt service task. In turn, the interrupt service task concludes interrupt handling running in user mode.

Being an ordinary task, the interrupt service task can, at least in principle, invoke the full range of operating system services, including blocking synchronization primitives, and must not concern itself with excessive usage of the interrupt service processor mode. On the other hand, system overheads related to interrupt handling increase, because the activation of the interrupt service task requires a task switch.

Operating systems adopting this design approach include [20,26,33].

Virtual Machines

The internal architecture of operating systems based on virtual machines, shown in Figure 88.3 and pioneered by IBM with CP/67, revolves around a basic observation: an operating system must perform two essential functions:

- multiprogramming and
- system services.

Accordingly, those operating systems fully separate these functions and implement them as two distinct operating system components:

- a *virtual machine monitor* runs in privileged mode; it does multiprogramming, and provides many virtual processors *identical* in all respects to the real processor it runs on. In addition, it implements basic synchronization and communication mechanisms between virtual processors, and partitions system resources between them;
- a *guest operating system,* which can even be monoprogrammed, runs on each virtual processor, and implements system services. Different virtual processors can run different operating systems, and they must not necessarily be aware of being run in a virtual machine.

In the oldest approach to virtual machine implementation, based on privileged instruction emulation, when an application program performs a system call instruction, the virtual machine monitor intercepts it and redirects it to the guest operating system running in its own virtual machine. The guest operating system itself is given the illusion of running in privileged mode, while the physical processor actually continues to operate in user mode; in this way, the virtual machine monitor is able to intercept all privileged instructions issued by the guest operating system, check them against the security policy of the system, and then perform them on behalf of the guest operating system, that is, *emulate* them.

Interrupt handling is implemented in a similar way: the virtual machine monitor catches all interrupt requests and then redirects them to the appropriate guest operating system handler, reverting to user mode in the process; thus, the virtual machine monitor can intercept all privileged instructions issued by the guest interrupt handler, and again emulate them as appropriate.

As a consequence, in this kind of virtual machine system the overhead experienced by system calls and interrupt handlers is strictly related to, and dominated by, the cost of emulation of privileged instructions issued by the guest operating system.

The full separation of roles described above has the advantage of making each system component smaller and easier to maintain. Moreover, the presence of a relatively small, centralized arbiter of all interactions between virtual machines makes the enforcement of security policies easier. The isolation of virtual machines

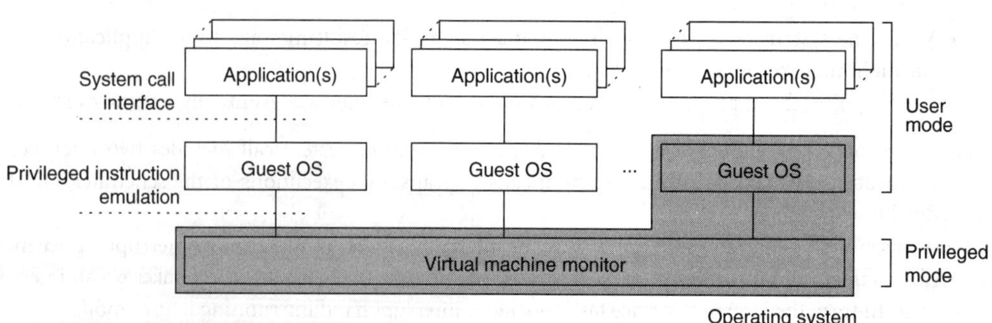

FIGURE 88.3   A virtual machine operating system.

from each other also enhances reliability because, even if one virtual machine fails, it does not bring down the system as a whole. In addition, it is possible to run a distinct operating system in each virtual machine thus supporting, for example, the coexistence between a real-time and a general-purpose operating system.

By contrast, the perfect implementation of virtual machines requires hardware assistance, both to make it feasible and to be able to emulate privileged instructions with a reasonable degree of efficiency. For example, for this approach to work, it is necessary that privileged instruction traps are nondestructive with respect to processor state, and that such a trap is actually taken on any attempt to access the privileged processor state from user mode, *including* read-only access.

Communication between virtual machines is usually implemented by means of shared virtual devices, like disks, or virtual network interfaces. In all cases, when virtual machines run software with different timing requirements, for example, one virtual machine is devoted to real-time code and another to the user interface, the communication channel must not introduce any undue synchronization between the two worlds. Lock-free, shared data structures are often used for this purpose.

A variant of this design adopts an interpretive approach, and allows the virtual machines to be *different* from the physical machine. For example, Java programs are compiled into byte-code instructions suitable for execution by an abstract Java virtual machine. On the target platform, an interpreter executes the byte-code on the physical processor, thus implementing the virtual machine.

More sophisticated approaches to virtual machine implementation are also possible, the most common one being on-the-fly code generation, also known as just-in-time compilation.

With this approach, virtual machine instructions are dynamically translated into sequences of native instructions that are then executed directly by the physical processor. After translation, instructions are executed much quicker, but the translation itself is a lengthy process and the translated code is often larger than the original one.

Therefore, an important design issue of just-in-time compilers is to appropriately determine when code must be translated and for how much time the translated code must be retained in view of its repeated execution, thus attempting to avoid both the translation of code that will be executed rarely and the waste of valuable storage area to hold translated code that will never be executed again.

Besides the now ubiquitous Java virtual machines, a real-world example of operating system based on virtual machines, and in which the virtual machine concept plays a central role, is [30].

Process and Thread Model

A convenient and easy to understand way to design real-time and non-real-time software applications is to organize them as a set of cooperating sequential *processes.* A process is an activity, namely the activity of executing a program, and encompasses the program being executed, the data it operates on, and the current processor state, including the program counter and registers. In particular, each process has its own address space.

Multiprogrammed operating systems give the user the illusion of executing multiple processes in parallel, usually by multiplexing the physical processors and switching them from one process to another as directed by an all-important part of the operating system, the *scheduler.*

During its life, a process transitions between several *states* along a process state diagram; the most basic process states are depicted in Figure 88.4:

Running: The process is currently running on a processor. From this state, a process can become *blocked* when it starts waiting for an external event to occur; also, the operating system can forcibly move it into the *runnable* state when the scheduling algorithm assigns the CPU to another process, an action known as *preemption.*

Runnable, or Ready: The process is ready to be run; the process will transition into the *running* state when the scheduler assigns an otherwise idle processor to it.

Blocked: The process is not ready to run, because it is waiting for some events, for example, an I/O operation, to occur. When the event eventually occurs, the process goes into the *runnable* state and starts competing for the CPU again.

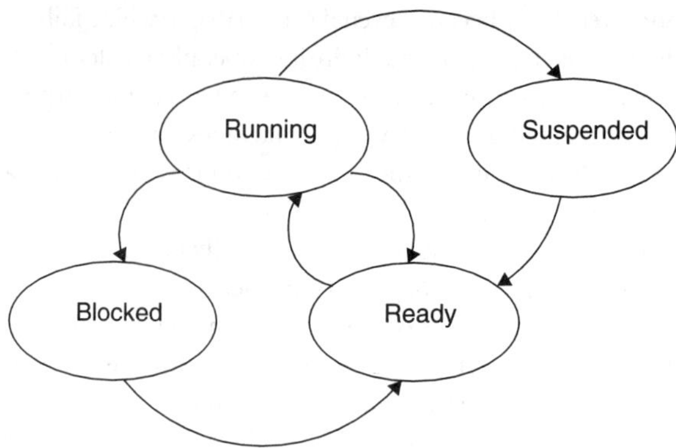

FIGURE 88.4   Basic process state diagram.

Suspended: The process is "passive," that is, it neither runs nor competes for processor usage nor is it waiting for an event to occur. Another process can activate it by means of a system service, and bring it into the *runnable* state. Also, a *running* process can voluntarily suspend itself, usually at the end of an execution cycle, thus entering the *suspended* state directly from the *running* state.

The process model just described is common to virtually all operating systems but; while general-purpose operating systems usually have one class of processes only and support the full state diagram outlined above for all of them, some real-time operating systems where process characteristics are well-known in advance, for example, those specified by [18], support a simplified state model for processes that cannot block, called *basic tasks* in [18], and remove the *blocked* state from their diagrams. The inherent advantage of basic tasks is that they require a lesser amount of run-time context to be stored in RAM, because it is never necessary to block them.

The efficient implementation of processes requires hardware assistance, namely the availability of a *Memory Management Unit* (MMU), to enforce separation between address spaces; also, the MMU must be reprogrammed to switch between address spaces while switching between processes. For this reason, in alternative or, in some cases, in addition to the process model described above, many real-time operating systems provide support for multiple *threads* of control within the same process. All threads in a process share the same address space; this gives both advantages and disadvantages:

Threads can be implemented for the most part in user mode, without the operating system's kernel intervention; moreover, when the processor is switched between threads, the address space remains the same and must not be switched. Both these facts make processor switching between threads very fast with respect to switching between processes;

Since all threads within a process share the same address space, there can be no protection among them; hence, for example, a thread is allowed to pollute by mistake another thread's data and the operating system has no way to detect errors of this kind.

Small operating systems for embedded applications, like [11,18,22], only support threads to keep overheads and hardware requirements to a minimum, while larger operating systems for more complex real-time applications offer the user a choice between a single or multiple process model to enhance the reliability of complex systems.

Another important operating system design issue is the choice between *static* and *dynamic* creation of processes and threads:

Some operating systems, usually oriented toward relatively simple embedded applications, for example [11,18], only support static tasks, that is, all tasks in the system are known in advance and it is not possible to create and destroy tasks while the system is running; thus, the total number of tasks in the system stays constant for all its life.

Other operating systems allow to create and destroy tasks at runtime, by means of a system call.

Dynamic task creation has the obvious advantage of making the application more flexible, but it increases the complexity of the operating system, because many operating system data structures, first of all the process table, must be allocated dynamically and their exact size cannot be known in advance. In addition, the application code requires a more sophisticated error-handling strategy, with its associated overheads, because it must be prepared to cope with the inability of the operating system to create a new task, due to lack of resources.

Processor Scheduling

The *scheduler* is one of the most important components of a real-time operating system, as it is responsible for deciding to which runnable threads the available processors must be assigned, and for how long. Schedulers can be divided into two broad categories:

Static

A static scheduler performs all scheduling computations *before* the application runs. Thus, there are no upper limits on the complexity of the scheduling algorithm to be used, because the time spent in those computations does not adversely affect the application at runtime.

Moreover, it is possible to develop exact schedulability tests for a given task set; these tests assure that the task set is schedulable under the worst possible conditions, a property most enjoyable in hard real-time systems.

However, the set of application tasks and their execution characteristics, such as period, deadline, and computation time, must all be known in advance, before scheduling decisions are made.

The output of a static scheduler is a set of tables, which are then linked with the operating system image and obeyed at runtime by a *dispatcher,* a software component much simpler, and more efficient, than the scheduler.

Dynamic

A dynamic scheduler performs scheduling computations at runtime, that is, *while* the application is running. Since in this case the computation time of the scheduler is seen as a system overhead by the application, scheduling algorithms must be simpler and often based on heuristics.

Also, in some cases it may be impossible to develop exact schedulability tests, and the application must be prepared to encounter and cope with an *overload* condition, where the scheduler is temporarily unable to satisfy all the processor requirements of the runnable tasks.

For these reasons, dynamic schedulers are best suited for firm and soft realtime systems, especially when the exact number and nature of application tasks is not fully known in advance and/or it is desirable to run together tasks with and without real-time requirements. A typical example of this situation are portable, wireless multimedia appliances that are gaining popularity nowadays.

Among dynamic schedulers, several algorithms are in use and offer different trade-offs between real-time predictability, implementation complexity, and overhead. Since the optimum compromise often depends on the application's characteristics, most real-time operating systems support multiple scheduling policies simultaneously and the responsibility of a correct choice falls on the application programmer. The most common scheduling algorithms supported by real-time operating systems and specified by international standards are:

First in, first out with priority classes: Under this algorithm, also known as fixed priority scheduling, there is a list of runnable threads for each priority level. When a processor is idle, the scheduler takes the runnable thread at the head of the highest-priority, nonempty thread list and runs it.

When the scheduler preempts a running thread, because a higher-priority task has become runnable, the preempted thread becomes the head of the thread list for its priority; when a blocked thread becomes runnable again, it becomes the tail of the thread list for its priority.

Thus, threads are placed into thread lists according to the time they have spent in the list without being executed: the thread that has been on the list the longest time will generally be at the head of the list, while the thread that has been on the list the shortest time will be at the tail.

The first in, first out scheduler never changes thread priorities at runtime; hence, the priority assignment is fully static; a well-known approach to static priority assignment for periodic tasks is the *rate monotonic* policy, in which task priorities are inversely proportional to their periods.

Since this is the simplest scheduling algorithm, virtually all real-time operating systems support it.

Round robin with priority classes: This algorithm extends the first in, first out algorithm just described with the additional condition that when a running thread has been executing for more than a maximum amount of time, the *quantum,* that thread is forcibly returned to the tail of its thread list and a new thread is selected for execution.

Similarly, when the scheduler resumes a thread after preemption, the thread is allowed to consume the unexpired portion of its *quantum.*

The effect of this extension is to ensure that when multiple, runnable threads share the same priority level, none of them can monopolize the processor; instead, the available processor time will be fairly divided among them.

Earliest deadline first: Unlike the scheduling policies discussed above, the earliest deadline first scheduler assigns thread priorities *dynamically.* In particular, this scheduler always executes the thread with the nearest deadline. It can be shown that this algorithm is optimal for uniprocessor systems, and supports full processor utilization in all situations.

However, its performance under overload is poor and dynamically updating thread priorities on the base of their deadlines may be computationally expensive, especially when this scheduler is layered on a fixed-priority lower-level scheduler.

Sporadic server: This scheduling algorithm was first introduced in [24], where a thorough description of the algorithm can be found.

The sporadic server algorithm is suitable for aperiodic event handling where, for timeliness, events must be handled at a certain, usually high, priority level, but lower priority threads with real-time requirements could suffer from excessive preemption if that priority level were maintained indefinitely. It acts on the base of two main scheduling parameters associated with each thread:

- the execution capacity and
- the replenishment period.

Informally, the execution capacity of a thread represents the maximum amount of processor time that the thread is allowed to consume at high priority in a replenishment period.

The execution capacity of a thread is preserved until an aperiodic request for that task occurs, thus making it runnable; then, thread execution depletes its execution capacity.

The sporadic server algorithm replenishes the thread's execution capacity after some or all of its capacity is consumed by thread execution; the schedule for replenishing the execution capacity is based on the thread's replenishment period.

Should the thread reach its processor usage upper limit, its execution capacity becomes zero and it is demoted to a lower priority level, thus avoiding excessive preemption against other threads. When replenishments have restored the execution capacity of the thread above a certain threshold level, the scheduler promotes the thread to its original priority again.

Most real-time operating systems implement at least the first in, first out and round-robin scheduling algorithms. More sophisticated policies are less common; for example, at the time of writing, the sporadic server algorithm was available in [16,20,33].

Interprocess Synchronization and Communication

An essential function of a multiprogrammed operating system is to allow processes to synchronize and exchange information; as a whole, these functions are known as IPC (InterProcess Communication). Many interprocess synchronization and communication mechanisms have been proposed and were objects of extensive theoretical study in the scientific literature. Among them, we recall:

Semaphores

A semaphore, first introduced by Dijkstra in 1965, is a synchronization device with an integer value and on which the following two primitive, *atomic* operations are defined:

The P operation, often called DOWN or WAIT, checks if the current value of the semaphore is greater than zero. If so, it decrements the value and returns to the caller; otherwise, the invoking process goes into the blocked state until another process performs a V on the same semaphore;

The V operation, also called UP, POST, or SIGNAL, checks whether there is any process currently blocked on the semaphore. In this case, it wakes exactly one of them up, allowing it to complete its P; otherwise, it increments the value of the semaphore. The V operation never blocks.

One common use of a semaphore with an initial value of 1 is to ensure the *mutual exclusion* between processes in the access to a critical region:

- in each process, the critical region code is bracketed by a P/V pair;
- when a process successfully enters its critical region, the semaphore value decrements to 0, denoting that the critical region is now "busy";
- any other process that subsequently attempts to enter its critical region will block in the P, ensuring mutual exclusion;
- when a process exits from its critical region and executes the V, another process, formerly blocked in the P, is allowed to enter the critical region; in this case, the semaphore value stays at 0, because the critical region is still "busy." If there are no processes waiting for their turn to enter the critical region, the semaphore value returns to 1, denoting that the critical region is now "free" again.

When used in this way, a semaphore can have only two meaningful values, 1 and 0, denoting that the critical region protected by the semaphore currently is "free" and "busy," respectively. Moreover, in any given thread, invocations of P and V always appear in pairs, and P always comes before V. Taking these peculiarities into account, most operating systems provide a specialized and optimized variety of semaphore, devoted to mutual exclusion only, and call it a *mutual exclusion device,* or *mutex* for short.

Semaphores are a very low-level IPC mechanism; therefore, they have the obvious advantage of being simple to implement, at least on uniprocessor systems, and of having a very low overhead. By contrast, they are difficult to use, especially in complex applications.

Also related to mutual exclusion with semaphores is the problem of *priority inversion.* Priority inversion occurs when a high-priority process is forced to wait for a lower-priority process to exit a critical region, a situation in contrast to the concept of relative task priorities. Most real-time operating systems take this kind of blocking into account and implement several protocols to bound it:

In the *priority inheritance* protocol, a lower-priority process is temporarily promoted to the priority of the higher-priority process it is blocking, and returns to its original priority when it leaves the critical region.

In the *priority ceiling* protocol, explicitly called for by [18], when a process acquires a critical region, its priority is temporarily raised to that of the highest priority process that will ever use the critical region. Again, the process returns to its original priority when it leaves the critical region.

An alternative approach, taken, for example, by [6,9] is to leave the burden of avoiding priority inversion to the application, thus saving the significant overhead required by the kernel to resolve it. Proponents of this approach show that it is often better and more efficient to attack the priority inversion issue at the application design level, instead of relying on universal, time-consuming run-time mechanisms. However, this topic is still the subject of considerable debate; see, for example, [12,34].

Monitors

To overcome the difficulties of semaphores, in 1974/1975, Hoare and Brinch Hansen introduced a higher level synchronization mechanism, the monitor.

A monitor is a set of data structures and procedures that operate on them; data structures are shared among all processes that can use the monitor, and it effectively hides the data structures it contains; hence, the only way to access the data associated with the monitor is through the procedures in the monitor itself.

In addition, all procedures in the same monitor are implicitly executed in mutual exclusion. Unlike semaphores, the responsibility of ensuring mutual exclusion falls on the compiler, not on the programmer, because monitors are a programming language construct, and the language compiler knows about them.

Inside a monitor, *condition variables* can be used to wait for events. Two *atomic* operations are defined on a condition variable:

The WAIT operation releases the monitor and blocks the invoking process until another process performs a SIGNAL on the same condition variable;

The SIGNAL operation unblocks exactly one process waiting on the condition variable. Then, to ensure that mutual exclusion is preserved, the invoking process is either forced to leave the monitor immediately (Brinch Hansen's approach), or is blocked until the monitor becomes free again (Hoare's approach).

Note that, unlike semaphores, condition variables have no memory; hence, a SIGNAL has no effect when performed on a condition variable on which no processes are currently blocked.

An alternative and simpler implementation of monitors, which does not require additional programming language constructs and can therefore be used with off-the-shelf compilers, consists of providing monitor enter and exit functions, and synchronization functions on condition variables, that must then be *explicitly* invoked by application code. However, in this case there is no information hiding and the mutual exclusion between monitor methods is not guaranteed, unless the application programmer bracketed them appropriately with the monitor enter and exit functions mentioned above.

It should also be noted that both semaphores and monitors merely convey synchronization signals among processes, and rely on the availability of a shared memory for data exchange.

Message Passing

Unlike all other IPC mechanisms described so far, message passing supports explicit *data* transfer between processes; hence, it does not require a shared memory and lends itself well to be extended to distributed systems. This IPC method provides for two primitives:

- the SEND primitive sends a message to a given destination;
- symmetrically, RECEIVE receives a message from a given source.

Many variants are possible on the exact semantics of these primitives; they mainly differ in the way messages are *addressed* and *buffered*.

A commonplace addressing scheme is to give to each process/thread in the system a unique address, and to send messages directly to processes. Otherwise, a message can be addressed to a *mailbox*, a message container whose maximum capacity is usually specified upon creation; in this case, message source and destination addresses are mailbox addresses.

When mailboxes are used, they also provide some amount of message buffering, that is, they hold messages that have been sent but have not been received yet. Moreover, a single task can own multiple mailboxes and can use them to classify messages depending on their source or priority.

A somewhat contrary approach to message buffering, simpler to implement but less flexible, is the *rendezvous* strategy: the system performs *no* buffering; hence, the sender and the receiver are forced to run in lockstep:

- the SEND does not complete until another process executes a matching RECEIVE;
- the RECEIVE waits until a matching SEND is executed.

Despite its simplicity and elegance, the message-passing mechanism suffers from several design issues, of which the most important is *performance*, because copying data between processes is always slower than using a shared memory and exchanging synchronization signals: considerable work has gone into making the mechanism efficient, and this still is an open research theme, especially for distributed systems.

For example, a common way to pose an upper bound on message passing time requirements, thus making the mechanism predictable, is to force a fixed message structure and/or limit the maximum message size.

Moreover, when a memory management unit is available, it is possible to leverage its capabilities to switch one or more physical pages from an address space to another one without copying their contents, and with a negligible and deterministic overhead.

Lock-free Data Structures

This kind of data structure allows two or more processes to exchange data without introducing any undue synchronization between them. This feature is most important when the communicating processes have different priority levels and timing requirements. An example are the real-time FIFOs used by [6,19] to exchange data between real-time and non-real-time processes.

Operating systems usually provide (at least) one IPC mechanism natively, and implement the other ones on top of it, for example, by means of utility libraries. Often, the choice of the basic, primary IPC mechanism of an operating system depends on its overall architecture: for example, a microkernel-based operating system like [20,26,33] will have extensive message-passing capabilties, whereas the natural choice for a monolithic or layered operating system like [9,16,22,23] is to adopt shared memory and semaphores as its primary IPC mechanism. Therefore, an in-depth understanding of the operating system architecture can help application designers to enhance the performance of their solutions.

Network Support

There are two basic approaches to implement network support in a real-time operating system and to offer it to applications:

The POSIX set of standards [7] specifies the *socket* paradigm for uniform access to any kind of network support and most real-time operating systems, like [2,3,13,16,19,20,26,33], provide them. Sockets, fully described in [14], were first introduced in the "Berkeley Unix" operating system and are now available on virtually all general-purpose operating systems; as a consequence, most programmers are likely to be proficient with them.

The main advantage of sockets is that they support in a *uniform* way any kind of communication network, protocol, naming conventions, hardware, and so on. Semantics of communication and naming are captured by *communication domains* and socket *types,* both specified upon socket creation. For example, communication domains are used to distinguish between IPv4 and X25 network environments, whereas the socket type determines whether communication will be stream-based or datagram-based and also implicitly selects which network protocol a socket will use.

Additional socket characteristics can be set up after creation through abstract *socket options;* for example, socket options provide a uniform, implementation-independent way to set the amount of receive buffer space associated with a socket.

Thus, sockets enhance software portability and hardware independence, at the expense of an abstract view of network characteristics that, in some cases, makes network-specific capabilities difficult to use efficiently.

Some operating systems, mostly focused on a specific class of embedded applications, offer network support through a less general, but more rich and efficient, application programming interface. For example, [18] is an operating system specification oriented to automotive applications; it specifies a communication environment (OSEK/VDX COM) less general than sockets and oriented to real-time message-passing networks, such as the Controller Area Network (CAN).

In this case, for example, the application programming interface allows applications to easily set message filters and perform out-of-order receives, thus enhancing their timing behavior; both these functions are supported with difficulty by sockets, because they do not fit well with the general socket paradigm.

In both cases, network device drivers are usually supplied by third-party hardware vendors and conform to a well-defined interface defined by the operating system vendor.

The network software itself, although it is often bundled with the operating system and provided by the same vendor, can be obtained from third-party software houses, too. For example, [4,10] are TCP/IP protocol stacks that run on a wide variety of hardware platforms and operating systems.

Often, these products come with source code; hence, it is also possible to port them to custom operating systems developed in-house, and can be extended and enhanced by the end user.

Hardware Requirements

Hardware platforms for embedded applications are much more varied than general-purpose platforms, such as personal computers. Accordingly, all real-time operating systems can run on more than one processor family and often accomplish this goal by interposing between the operating system kernel and the hardware a suitable, hardware-dependent abstraction layer (HAL). The main function of the HAL is to decouple abstract hardware functions that the kernel requires, for example, the "interrupt disable" operation, from the actual sequence of instructions required by the processor.

All popular 32-bit and many 16-bit processor families are currently supported by real-time operating systems, including, for example, *ARM, ColdFire, CPU32, MIPS, PowerPC, Sparc, SuperH, x86,* and many others. Some operating systems, like [20], are also starting to support some Digital Signal Processor families that in the past were confined to run either with no operating system at all, or with simple cyclic scheduling loops. Smaller, 8-bit processor families, such as the *8051,* are supported only by the simplest real-time executives, like [11].

To complement the HAL, operating system developers usually provide a set of platform-dependent device drivers, which enable operating system access to the peripheral devices the platform provides; in other cases, device drivers are provided by hardware developers directly.

The single most important hardware feature of interest to real-time operating systems for embedded applications is the availability of a Memory Management Unit.

This hardware component is crucial to the efficient implementation of multiple processes, each having its own, independent address space. By means of an MMU, the operating system can check every memory reference made by processes, and enforce memory protection, thus enhancing reliability because a process cannot pollute the address space of another process by effect of a software bug.

Also, the MMU makes memory management easier, because address spaces no longer have to be physically contiguous in memory, and the most sophisticated operating systems can use it to implement demand paging for non-time-critical tasks.

However, an MMU is a rather complex piece of hardware; hence, it is often not included in small microcontrollers, both for cost and power consumption reasons. When an MMU is not available, the operating system can still provide multiprogramming, but there will be a single process and address space in which multiple threads will run concurrently.

Another reason to avoid using the MMU, even when it is available, is that it must be reprogrammed on every context switch, thus adding a source of overhead; many modern MMUs attempt to reduce this overhead, for example, by tagging virtual addresses with per-process *Space Identifiers* (SI), but cannot fully eliminate it when the number of processes in the system grows beyond the available SI range.

Last, but not the least, MMUs often introduce a source of unpredictability in memory access timing; in fact, to accelerate address translation, they resort to a hardware component known as the *Translation Lookaside Buffer* (TLB). The TLB acts like a cache of the most recently performed address translations; however, since it has limited capacity and its entries are therefore subject to a replacement policy, its performance can be predicted only in statistical terms. The timing difference between a fast address translation performed through the TLB (TLB *hit)* and the much slower translation performed through a regular page table walk (TLB *miss)* can be of one or two orders of magnitude, especially when part of or all the page table walk must be performed in software, as is commonly done on modern RISC processors.

For this reason, in hard real-time applications it is important to be able to lock the most important TLB entries and thus avoid TLB misses for the most critical portions of code and data; this requires both hardware support and appropriate operating system interfaces.

An intermediate solution, adopted in some microprocessors and microcontrollers, is to implement a very simple memory management unit, often called *Protection Unit,* which does not perform address translation but still provides a selective protection mechanism for a physically contiguous area of memory and is fully deterministic with regard to timing.

Additional Functions

Even if real-time operating systems sometimes do not implement several major functions that are now commonplace in general-purpose operating systems, such as demand paging, swapping, and filesystem access, they must be concerned with other, less well-known functions that ensure or enhance system predictability, for example:

- asynchronous, real-time signals, and cancellation requests, to deal with unexpected events, such as software and hardware failures, and to gracefully degrade the system's performance should a processor overload occur;
- high-resolution clocks and timers, to give real-time processes an accurate notion of elapsed time; and
- asynchronous I/O operations, to decouple real-time processes from the inherent unpredictability of many I/O devices.

88.3 The POSIX Standard

The original version of the Portable Operating System Interface for Computing Environments, better known as the POSIX set of standards, was first published between 1988 and 1990, and defines a standard way for applications to interface with the operating system. The set now includes over 30 individual standards, and covers a wide range of topics, from the definition of basic operating system services such as process management, to specifications for testing the conformance of an operating system to the standard itself.

Among these, of particular interest is the System Interfaces (XSH) Volume of IEEE Std 1003.1-2001 [7], which defines a standard operating system interface and environment, including real-time extensions. The standard contains definitions for system service functions and subroutines, language-specific system services for the C programming language, and notes on portability, error handling, and error recovery.

This standard has been constantly evolving since it was first published in 1988; the latest developments have been crafted by a joint working group of members of the IEEE Portable Applications Standards Committee, members of The Open Group, and members of ISO/IEC Joint Technical Committee 1. The joint working group is known as the Austin Group, after the location of the inaugural meeting held at the IBM facility in Austin, Texas in September 1998.

The Austin group formally began its work in July 1999, after the subscription of a formal agreement between The Open Group and the IEEE, with the main goal of revising, updating, and combining into a single document the following standards: ISO/IEC 9945-1, ISO/IEC 9945-2, IEEE Std 1003.1, IEEE Std 1003.2, and the Base Specifications of The Open Group Single UNIX Specification Version 2.

For real-time systems, the latest version of IEEE Std 1003.1 [7] incorporates the real-time extension standards listed in Table 88.1. Since embedded systems can have strong resource limitations, the IEEE Std 1003.13-1998 [8] profile standard groups functions from the standards mentioned above into units of functionality. Implementation can then choose the profile most suited to their needs and to the computing resources of their target platforms.

Functions defined in the oldest standard, 1003.1b, are supported across a wider number of operating systems; for this reason, we concentrate only on them and on 1003.1c, which specifies support for multiple threads in the same process.

Table 88.2 summarizes the functional groups of IEEE Std 1003.1-2001 that will be discussed next.

TABLE 88.1 Real-time Extensions Incorporated into IEEE Std 1003.1-2001

Standard	Description
1003.1b	Basic real-time extensions; first published in 1993
1003.1c	Threads extensions; published in 1995
1003.1d	Additional real-time extensions; published in 1999
1003.1j	Advanced real-time extensions; published in 2000

TABLE 88.2 Basic Functional Groups of IEEE Std 1003.1-2001

Functional Group	Main Functions

Process Scheduling	`sched_get_priority_max sched_get_priority_min`
	`sched_getparam sched_getscheduler`
	`sched_rr_get_interval sched_setparam`
	`sched_setscheduler sched_yield`

Real-time Signals	`sigaction sigqueue sigtimedwait sigwait` `sigwaitinfo`

Interprocess
Synchronization
and Communication
	`mq_close mq_getattr mq_notify mq_open`
	`mq_receive mq_send mq_setattr mq_timedreceive`
	`mq_timedsend mq_unlink`
	`sem_close sem_destroy sem_getvalue sem_init`
	`sem_open sem_post sem_timedwait sem_trywait`
	`sem_unlink sem_wait shm_open shm_unlink`

Memory Management	`mlock mlockall mmap mprotect msync munlock` `munlockall munmap`

Asynchronous
and List-Directed I/O
	`aio_cancel aio_error aio_fsync aio_read`
	`aio_return aio_suspend aio_write`
	`lio_listio`

Clocks and Timers	`clock_getres clock_gettime clock_settime`
	`nanosleep`
	`timer_create timer_delete timer_gettime`
	`timer_getoverrun timer_settime`

Multiple Threads	`pthread_atfork pthread_cancel`
	`pthread_cleanup_pop pthread_cleanup_push`
	`pthread_cond_broadcast pthread_cond_destroy`
	`pthread_cond_init pthread_cond_signal`
	`pthread_cond_timedwait pthread_cond_wait`
	`pthread_create pthread_detach pthread_equal`
	`pthread_exit pthread_getschedparam`
	`pthread_getspecific pthread_join`
	`pthread_key_create pthread_key_delete`
	`pthread_kill pthread_mutex_destroy`
	`pthread_mutex_init pthread_mutex_lock`
	`pthread_mutex_trylock pthread_mutex_unlock`
	`pthread_once pthread_rwlock_destroy`
	`pthread_rwlock_init pthread_rwlock_rdlock`
	`pthread_rwlock_tryrdlock`
	`pthread_rwlock_trywrlock pthread_rwlock_unlock`
	`pthread_rwlock_wrlock pthread_self`
	`pthread_setcancelstate pthread_setcanceltype`
	`pthread_setschedparam pthread_setschedprio`
	`pthread_setspecific pthread_sigmask`
	`pthread_testcancel`

Process Scheduling

Functions in this group allow the application to select a specific policy that the operating system must follow to schedule a particular process, and to get and set the scheduling parameters associated with that process.

In order to support the orderly coexistence of multiple scheduling policies, the conceptual scheduling model defined by the standard assigns a global priority to all threads in the system and contains one ordered thread list for each priority; any runnable thread will be on the thread list for that thread's priority. When appropriate, the scheduler shall select the thread at the head of the highest priority nonempty thread list to become a running thread, regardless of its associated policy; this thread is then removed from its thread list. When a running thread yields the CPU, either voluntarily or by preemption, it is returned to the thread list it belongs to.

The purpose of a scheduling policy is then to determine how the operating system scheduler manages the thread lists, that is, how threads are moved between and within lists when they gain or lose access to the CPU. Associated with each scheduling policy is a priority range, which must span at least 32 distinct priority levels; all threads scheduled according to that policy must lie within that priority range, and priority ranges belonging to different policies can overlap in whole or in part.

The standard defines the following three scheduling policies, whose algorithms have been described in the Processor Scheduling section:

- first in, first out (SCHED_FIFO);
- round robin (SCHED_RR);
- optionally, a variant of the sporadic server scheduler (SCHED_SPORADIC).

A fourth scheduling policy, SCHED_OTHER, can be selected to denote that a thread no longer needs a specific real-time scheduling policy: general-purpose operating systems with real-time extensions usually revert to the default, non-real-time scheduler when this scheduling policy is selected.

Moreover, each implementation is free to redefine the exact meaning of the SCHED_OTHER policy and can provide additional scheduling policies besides those required by the standard, but any application using them will no longer be fully portable.

Real-time Signals

Signals are a mechanism to convey information to a process when it is not necessarily waiting for input; therefore, they allow an application to be notified asynchronously about the occurrence of an event.

In POSIX, there is a defined set of signals, and each member of the set has a unique signal number. Basically, the process can specify one of three possible actions to be taken when a signal arrives:

- ignore it completely,
- terminate, or
- execute a signal-handling procedure.

The real-time extensions of IEEE Std 1003.1-2001 add real-time features to the default signal-handling capabilities defined by the standard; in particular:

- they define a priority hierarchy between signals so that, when multiple signals are pending, the highest priority signal is serviced first; the default, non-real-time behavior is to process pending signals in an unspecified order; and
- they add support for signal queues, to keep track of multiple pending signals with the same signal number and to convey a limited amount of information with each signal request; under the default behavior, all signals but the first are lost and signal requests do not carry any additional information with them.

Interprocess Synchronization and Communication

The main interprocess synchronization and communication mechanisms offered by the standard are the semaphore and the message queue, both described in the Interprocess Synchronization and Communication section. The blocking synchronization primitives have a nonblocking and a timed counterpart, to enhance their real-time predictability. Moreover, asynchronous notification through a real-time signal is supported for message arrival. Areas of memory can be shared among multiple processes.

The multithreading extension, described in the Multiple Threads section, adds support for mutual exclusion devices, that is, binary semaphores, condition variables, and other synchronization mechanisms. The scope of these mechanisms can be limited to threads belonging to the same process to enhance their performance.

Memory Management

The standard allows processes to lock parts or all of their address space in main memory. The lock operation both forces the memory residence of the virtual memory pages involved, and prevents them from being paged out in the future. This is vital in operating systems that support demand paging, because the paging activity could introduce undue and highly unpredictable delays when a real-time process attempts to access a page that is currently not in the main memory and must therefore be retrieved from secondary storage.

Other memory management functions establish a mapping between the address space of the calling process and a memory object, possibly shared between multiple processes. The mapping facility is general enough; hence, it can also be used to map other kinds of object, such as files and devices, into the address space of a process, provided both the hardware and the operating system have this capability, which is not mandated by the standard. For example, once a file is mapped, a process can access it simply by reading and writing the data at the address range to which the file was mapped.

Finally, it is possible for a process to change the access protections of portions of its address space; in this case, it is assumed that protections will be enforced by the hardware. For example, to prevent inadvertent data corruption due to a software bug, one could protect critical data intended for read-only usage against write access.

Asynchronous and List-Directed Input and Output

Many operating systems carry out I/O operations synchronously with respect to the process requesting them. Thus, for example, if a process invokes a file read operation, it stays blocked until the operating system has finished it, either successfully or unsuccessfully. As a side effect, any process can have at most one pending I/O operation at any given time.

While this programming model is simple, intuitive, and perfectly adequate for general-purpose systems, it shows its limits in a real-time environment, namely:

- I/O device access timings can vary widely, especially when an error occurs; hence, it is not always wise to suspend the execution of a process until the operation completes, because this would introduce a source of unpredictability in the system;
- it is often desirable, for example, to enhance system performance by exploiting I/O hardware parallelism, to start more than one I/O operation simultaneously, under the control of a single process.

To satisfy these requirements, the standard defines a set of functions to start one or more I/O requests, to be carried out in parallel with process execution, and whose completion status can be retrieved asynchronously by the requesting process.

Clocks and Timers

Real-time applications very often rely on timing information to operate correctly; the standard specifies support for one or more timing bases, called *clocks,* of known resolution and whose value can be retrieved at will. In addition, applications can set one or more per-process *timers,* using a specified clock as a timing base. Each timer has a current value and, optionally, a reload value associated with it. The operating system decrements the current value of timers according to their clock and, when a timer expires, it notifies the owning process with an asynchronous timer expiration signal; the signal can either be handled directly, or trigger the awakening of a thread in the process. On timer expiration, the operating system also reloads the timer with its reload value, if it has been set, thus possibly realizing a repetitive timer.

Since, due to scheduling or processor load constraints, a process could lose one or more expiration signals, the standard also specifies a way for applications to retrieve the number of "missed" signals, that is, the number of extra timer expirations that occurred between the time at which a given timer expired, and when the signal associated with the expiration was eventually delivered to, or accepted by, the process.

Multiple Threads

The optional multithreading extension to the standard adds functions to populate a process with new threads and to cancel them in a carefully controlled way. Controlled cancellation is crucial, for example, to prevent the cancellation of a thread while it is inside a critical region; otherwise, the critical region would stay locked forever, likely inducing a deadlock in the system.

The multithreading extension also offers additional synchronization mechanisms, whose scope can be limited to threads belonging to the same process in order to achieve greater performance:

Mutex Devices: A *mutex* is a binary semaphore that can be used to ensure the mutual exclusion between multiple threads; it is simpler and more efficient than a full-fledged semaphore. Optionally, it is possible to associate to each mutex a protocol to deal with priority inversion.

Condition Variables: In concert with a *mutex*, condition variables can be used to implement a synchronization mechanism similar to the monitor.

Read/Write Locks: A read/write lock implements a specialized and widely used synchronization mechanism in which the access to the critical region is granted to multiple "readers," but to only one "writer" at a time. In addition, when a writer is in the critical region, no other process can enter it, not even readers.

Finally, there is support for multiple, thread-specific data areas. Each thread-specific data area is accessed by means of a data key; while the data key is shared between threads, each thread has got a corresponding, private data area. To make storage management easier, it is also possible to associate to each data key a destructor function, which the system will automatically invoke when a thread-specific data area must be freed, for example, because the owning thread is being canceled.

88.4 Real-time Operating Systems

Although in the general-purpose operating system camp a handful of products dominates the market, there are more than 30 real-time operating systems available for use today, both commercial and experimental, and new ones are still being developed. This is due both to the fact that there is much research in the latter area, and that real-time embedded applications are inherently less homogeneous than general-purpose applications, such as those found in office automation; hence, *ad hoc* operating system features are often needed. In addition, the computing power and the overall hardware architecture of embedded systems are much more varied than, for example, those of personal computers.

Therefore, due to lack of space, this section neither can nor pretends to be a thorough survey; instead, it will focus on some well-known operating systems, mainly to show the broad variety of design and implementation approaches tried in practice.

Commercial Operating Systems

Unless otherwise noted, operating systems in this category claim full conformance with the core services specified by IEEE Std 1003.1, real-time extensions 1003.1b, thread extensions 1003.1c, and possibly other additional extensions specified by the POSIX set of standards. However, although their application programming interface is uniform, they differ considerably in their internal architecture, namely:

AMX [9] from KADAK Products Ltd. is a layered multitasking kernel with a small memory footprint and low overhead, suitable for embedded applications. Of interest is the ability to reduce the kernel footprint down to about 30 kilobytes on most supported platforms. Additional kernel components, such as synchronization objects and memory managers, can be added at kernel configuration time, and consume more space. However, it does not handle the priority inversion problem directly, does not support multiple, independent address spaces, and does not claim conformance to the POSIX set of standards. A filesystem implementation and network support are available separately.

Inferno [30] from Vita Nuova Holdings Ltd. was originally developed by the Computing Science Research Center of Bell Labs with the main purpose of creating and supporting distributed services and has then evolved into a commercial product. It does not claim conformance with the POSIX set of standards.

The research work also included the specification of a new application programming language, Limbo. Limbo syntactically resembles C, but it has a number of additional features that make it easier to use, safer, and more suited to a networked environment.

The operating system design is based on the virtual machine concept, and all Limbo programs are compiled into instructions for a virtual machine called *Dis*.

Of particular interest is the notion of a unified *name space* for distributed systems. In Inferno, all components of a distributed system present their resources as "files" in a hierarchical name space. The objects appearing as files often represent actual files, but may very well be devices, interfaces to services, control points, and so on. This approach provides a uniform naming, structuring, and access control mechanism for all system resources. A network protocol, *Styx,* encodes file operations between client programs and the filesystem, and translates them into messages for transmission on a computer network.

LynxOS [13] from LynuxWorks Inc. is a layered system with a multithreaded, fully reentrant kernel. Of interest is an additional real-time scheduling policy, named *priority-based quantum;* it is similar to the round-robin policy, but it is possible to assign a different quantum length for each priority. In addition, it features full binary compatibility with Linux; hence, Linux program executables will run on it. Also supported are multiple address spaces and multiple processor privilege levels.

Nucleus [1] from Accelerated Technology is a highly modular and scalable operating system. Available modules range in complexity from an embedded electronic mail client, Nucleus EMAIL, to SSL support and can be configured at will. The kernel itself is seen as a replaceable module; several kernel "flavors" are available and are suitable for different application domains. For example, one of them, Nucleus OSEK, conforms with the OSEK/VDX specification [18], while another one, Nucleus MMU, incorporates support for memory protection when a hardware Memory Management Unit is available for use. Additional modules provide network support, the Controller Area Network in particular, filesystems, and graphics.

OS-9 [21] now from RadiSys Corporation, and formerly marketed by Microware, is based on a modular, layered kernel and supports a wide range of features, including networking, filesystems, and drivers for a variety of peripheral devices. Its footprint ranges from about 256 kilobytes for a base system, up to 8 megabytes for a full-fledged kernel including graphics and Java virtual machine support.

OSEK/VDX [18] is an open operating system specification developed as a joint project in the German automotive industry, later endorsed by many other partners. The main project goal is to support portability and reusability of automotive application software by means of a standard interface with the operating system and the communication network.

Thus, OSEK/VDX proposes itself as a less general, but more effective alternative to the POSIX set of standards in a well-defined application domain characterized by rather severe constraints on the available computing power, memory capacity, and network implementation.

The specification covers the operating system proper, communication, and network management; it has been implemented in many commercial products, for example [1,5,15,25,29,31]. Since the specification does not mandate any particular system architecture, implementors are free to choose the approach they prefer, provided the operating system interface conforms to the standard; even if most implementations adhere to the layered architecture, there is at least one microkernel-based implementation [25].

The OSEK/VDX specification does not support multiple address spaces, and the task set is defined statically at system generation time, thus removing the burden of dynamic task creation.

pSOSystem [32] is a modular, layered system with an extensive application programming interface, and conforms to POSIX multithreading and real-time signal standards. It includes filesystem and network support, and provides multiple, independent processes and memory protection on some platforms. Many kernel features are implemented as kernel components and libraries that can be freely included in the kernel configuration, or removed when they are not needed.

QNX [20] from QNX Software Systems Ltd. is based on a microkernel. It is a full-fledged, highly reliable operating system: thanks to its architecture, it is possible to cleanly terminate and restart any operating system module, including device drivers, upon software failure. In addition, the microkernel can be instrumented as a debugging aid in detecting timing problems and faults.

To aid the development of complex applications, it can also support symmetric multiprocessing (SMP) and distributed systems with little or no impact on application code, because the message passing paradigm implicit in microkernel-based operating systems can be extended transparently to handle those architectures. The native application programming interface fully conforms to the POSIX set of standard with many optional extensions implemented.

RTMX O/S [23] from RTMX Inc. is an extension to the freely available OpenBSD Unix-like operating system, which in turn was derived from the 4.4BSD system developed by the University of California at Berkeley and also informally known as "Berkeley Unix." Therefore, its design is heavily based on the traditional, monolithic design of the 4.4BSD operating system kernel.

RTMX further enhances the already rich set of OpenBSD application programming interfaces with real-time functions, but the basic OpenBSD kernel cannot be scaled down very much, and is not well suited for small, resource-limited embedded platforms.

VxWorks [33] from Wind River Systems, Inc. is based on a message-passing microkernel. With over 1800 application program interfaces, it is a full-fledged operating system and also supports networking and filesystems. In spite of this, it is highly scalable, thanks to its microkernel-based architecture that allows most operating system modules to be removed when not in use; hence, it can run on very small embedded systems as well. Moreover, VxWorks supports on-the-fly replacement of system modules without restarting the whole system, to enhance reliability and service-ability.

Open Source Operating Systems

Many open source real-time operating systems are available, whose source code can be used both to develop real-world applications, and for study and experimentation. They usually do not claim conformance with the POSIX set of standards; instead, they often implement the most advanced, state-of-the-art architectures and algorithms because researchers can play with them at will and their work can immediately be reflected in real applications. Among the open source operating systems that have found their way into commercial products, we recall:

ChorusOS [26] was developed in origin by Sun Microsystems Inc. for embedded telecommunication applications and then was open sourced. It is a highly scalable system based on a microkernel architecture and configurable components; real-time scheduling policies and synchronization objects are component themselves, and can be either included or not in the operating system configuration.

Conformance with the POSIX set of standard is available, and provided by a group of operating system components. ChorusOS also supports filesystems, networking (both TCP/IP and ISO/OSI), virtual address spaces, and demand paging.

CREEM [11] is among the smallest real-time executives for microcontrollers, as it has the ability to run on small 8-bit processors with as little as 128 bytes of RAM. It is based on an unconventional dataflow application design and message passing, and implements only three operating system services. The system scheduling policy and task synchronization are entirely based on messages waiting for delivery, and the bulk of the scheduling algorithm is performed statically off-line. By contrast, it can be difficult to use, because its programming paradigm is not well known among programmers.

eCos [22] development is coordinated by Red Hat Inc. and is based on a modular, layered real-time kernel. The most important innovation in eCos is its extensive configuration system that operates at kernel build time with a large number of configuration points placed at source code level, and allows a very fine-grained adaptation of the kernel itself to both application needs and hardware characteristics. The output of the configuration process is an operating system library that can then be linked with the application code.

Its application programming interface is compatible with the POSIX set of standards, but it does not support multiple processes with independent address spaces, even when a Memory Management Unit is available.

μClinux [2] is a stripped-down version of the well-known Linux operating system; its most interesting features are the ability to run on microcontrollers that lack a Memory Management Unit and its small size, compared with a standard Linux kernel. As it is, μClinux does *not* have any real-time capability, because it inherits its standard processor scheduler from Linux; however, both RT-Linux [6] and RTAI [19] real-time extensions are available for it.

RT-Linux [6] and *RTAI* [19] are hard real-time-capable extensions to the Linux operating system; they are similar, and their architecture was first outlined in 1997 [3]. The main design feature of both RT-Linux and RTAI is the clear separation between the real-time and non-real-time domains: in RT-Linux and RTAI, a small monolithic real-time kernel runs real-time tasks, and coexists with the Linux kernel.

As a consequence, non-real-time tasks running on the Linux kernel have the sophisticated services of a standard time-sharing operating system at their disposal, whereas real-time tasks operate in a protected, predictable, and low-latency environment.

The real-time kernel performs first-level real-time scheduling and interrupt handling, and runs the Linux kernel as its lowest priority task. In order to keep changes in the Linux kernel to an absolute minimum, the real-time kernel provides for an emulation of the interrupt control hardware. In particular, any interrupt disable/enable request issued by the Linux kernel is not passed to the hardware, but is emulated in the real-time kernel instead; thus, for example, when Linux disables interrupts, the hardware interrupts actually stay enabled and the real-time kernel queues and delays the delivery of any interrupt of interest to the Linux kernel. Real-time interrupts are not affected at all, and are handled as usual, without any performance penalty.

To handle communication between real-time and non-real-time tasks, RT-Linux and RTAI implement lock-free queues and shared memory. In this way, real-time applications can rely on Linux system services for non-time-critical operations, such as filesystem access and graphics user interface.

RTEMS [16] development is coordinated by On-Line Applications Research Corporation, which also offers paid technical support. Its application development environment is based on open-source GNU tools, and has a monolithic/layered architecture, commonly found in high-performance real-time executives.

RTEMS complies with the POSIX 1003.1b application programming interface and supports multiple threads of execution, but does not implement a multiprocess environment with independent application address spaces. It also supports networking and a filesystem.

References

[1] Accelerated Technology, Nucleus Product Overview, Available online, at `http://www.acceleratedtechnology.com/`.

[2] Arcturus Networks Inc., μClinux Documentation, Available online, at `http://www.uclinux.org/`.

[3] Barabanov, M. and Yodaiken, V. Real-time Linux, *Linux Journal*, February 1997, Also available online, at `http://www.rtlinux.org/`.

[4] DSPOS Inc., Fusion Net Overview, Available online, at `http://www.dspos.com/`.

[5] ETAS, ERCOS[EK] Information, Available online, at `http://www.etas.de/`.

[6] FSMLabs, Inc., RTLinuxPro Frequently Asked Questions, Available online, at `http://www.rtlinux.org/`.

[7] IEEE Std 1003.1-2001, The Open Group Base Specifications Issue 6, The IEEE and The Open Group, 2001, Also available online, at `http://www.opengroup.org/`.

[8] IEEE Std 1003.13-1998, Standardized application environment profile — POSIX Realtime Application Support (AEP), The IEEE, 1998.

[9] KADAK Products Ltd., AMX User's Guide and Programming Guide, Available online, at `http://www.kadak.com/`.

[10] KADAK Products Ltd., KwikNet TCP/IP Stack Reference Manual, Available online, at `http://www.kadak.com/`.

[11] Kauler, B., *Flow Design for Embedded Systems: A Radical New Unified Object-Oriented Methodology*, 2nd ed., CMP Books, Gilroy, CA, January 1999.

[12] Locke, D., Priority Inheritance: The Real Story, Available online, at `http://www.linuxdevices.com/`.

[13] LynuxWorks Inc., LynxOS Product Information, Available online, at `http://www.lynuxworks.com/`.

[14] McKusick, M.K., Bostic, K., Karels, M.J., and Quarterman, J.S., *The Design and Implementation of the 4.4BSD Operating System*, Addison-Wesley, Reading, MA, 1996.

[15] Metrowerks, OSEKturbo Real-Time Operating System brochure, Available online, at `http://www.metrowerks.com/`.

[16] On-Line Applications Research, RTEMS Documentation, Available online, at `http://www.rtems.com/`.

[17] Organick, E.I., *The Multics System: An Examination of its Structure*, MIT Press, Cambridge, MA, 1972.

[18] OSEK/VDX, OSEK/VDX Operating System Specification, Available online, at `http://www.osek-vdx.org/`.

[19] Politecnico di Milano, Dip. di Ingegneria Aerospaziale, The RTAI Manual, Available online, at `http://www.aero.polimi.it/~rtai/`.

[20] QNX Software Systems Ltd., QNX Neutrino RTOS Data Sheet, Available online, at `http://www.qnx.com/`.

[21] RadiSys Corporation, OS-9 Product Data Sheet, Available online, at `http://www.radisys.com/`.

[22] Red Hat Inc., eCos User Guide, Available online, at `http://sources.redhat.com/ecos/`.

[23] RTMX Inc., RTMX O/S Data Sheet, Available online, at `http://www.rtmx.com/`.

[24] Sprunt, B., Sha, L., and Lehoczky, J.P., Aperiodic task scheduling for hard real-time systems, *Real-Time Systems*, 1, 27–60, 1989.

[25] Kaiser, R., LEO/p4 — A μ-Kernel Based OSEK Implementation, SYSGO Real-Time Solutions GmbH, Available online, at `http://www.osek.de/`.

[26] Sun Microsystems Laboratories, ChorusOS Reference Manual Collection, Available online, at `http://docs.sun.com/?q=chorus+5.0`.

[27] Tanenbaum, A.S. and Woodhull, A.S., *Operating Systems — Design and Implementation*, Prentice-Hall, Englewood Cliffs, NJ,1997.

[28] Tomayko, J.E., Computers in Spaceflight — The NASA Experience, NASA's Center for Aerospace Information (CASI), 1998, Also available online, at `http://www.hq.nasa.gov/`.

[29] TRIALOG, OX-OSEK Datasheet, Available online, at `http://www.trialog.com/`.

[30] Vita Nuova Holdings Ltd., Inferno Overview, Available online, at `http://www.vitanuova.com/`.

[31] Wind River Systems, Inc., OSEKWorks Datasheet, Available online, at `http://www.windriver.com/`.

[32] Wind River Systems, Inc., pSOSystem Datasheet, Available online, at `http://www.windriver.com/`.

[33] Wind River Systems, Inc., VxWorks Datasheet, Available online, at `http://www.windriver.com/`.

[34] Yodaiken, V., Against Priority Inheritance, FSMLabs technical report, 2002, Also available online, at `http://www.rtlinux.org/`.

[19] [illegible] Investigación Aeroespacial. The RFM Manual. Available online at [illegible].

[20] QSR International website. [illegible] 9.0 [illegible] Tutorial. Available online at [illegible].

[21] Radio Components [illegible] CS 9.0 Product line [illegible]. Available [illegible].

[22] [illegible]

[23] Wine Barrel Systems the OSSERVOS Database. Available online at [illegible].

[illegible] Wine Barrel Systems the pROSystem Database available online at [illegible].

[illegible] Wind River Systems Inc. VxWorks Datasheet. Available online, it offers [illegible].

[24] Yodaiken V. A [illegible]. FSMLabs [illegible] abstract 2007. Also available online at [illegible].

89

Power Aware Embedded Computing

Margarida F. Jacome
University of Texas at Austin

Anand Ramachandran
University of Texas at Austin

89.1 Introduction

Embedded systems are pervasive in modern life. State-of-the-art embedded technology drives the ongoing revolution in consumer and communication electronics, and is the basis of substantial innovation in many other domains, including medical instrumentation, process control, etc. [MRW02]. The impact of embedded systems in well-established "traditional" industrial sectors, for example, automotive industry, is also increasing at a fast pace [MRW02, BGJ+97].

Unfortunately, as CMOS technology rapidly scales, enabling the fabrication of ever faster and denser integrated circuits (ICs), the challenges that must be overcome to deliver each new generation of electronic products multiply. In the last few years, power dissipation has emerged as a major concern. In fact, projections on power density increases due to CMOS scaling clearly indicate that this is one of the fundamental problems that will ultimately preclude further scaling [Bor99, ITR]. Although the power challenge is indeed considerable, much can be done to mitigate the deleterious effects of power dissipation, thus enabling performance and device density to be taken to truly unprecedented levels by the semiconductor industry throughout the next 10–15 years.

Power density has a direct impact on packaging and cooling costs, and can also affect system reliability, due to electro-migration and hot-electron degradation effects. Thus, the ability to decrease power density, while offering similar performance and functionality, critically enhances the competitiveness of a product. Moreover, for battery-operated portable systems, maximizing battery lifetime translates into maximizing the duration of service, an objective of paramount importance for this class of products. Power is thus a *primary figure of merit* in contemporaneous embedded system design.

Digital CMOS circuits have two main types of power dissipation: *dynamic* and *static*. Dynamic power is dissipated when the circuit performs the function(s) it was designed for, for example, logic and arithmetic operations (computation), data retrieval, storage, transport, etc. Ultimately, all these activities translate into switching of the logic states held on circuit nodes. Dynamic power dissipation is thus proportional to C, $V_{DD}^2 fr$, where C denotes the total circuit capacitance, V_{DD} and f denote the circuit supply voltage and clock frequency, respectively, and r denotes the fraction of transistors expected to switch at each clock cycle [PR02, GGH97]. In other words, dynamic power dissipation is impacted to first order by circuit size/complexity, speed/rate, and switching activity. In contrast, static power dissipation is associated with preserving the logic state of circuit nodes between such switching activity, and is caused by subthreshold leakage mechanisms. Unfortunately, as device sizes shrink, the severity of leakage power is increasing at an alarming pace [Bor99].

Clearly, the power problem must be addressed at all levels of the design hierarchy, from system to circuit, as well as through innovations on CMOS device technology [CSB92, PR02, JYWV03]. In this survey, we provide a snapshot on the state of the art on *system* and *architecture level* design techniques and methodologies aimed at reducing both static and dynamic power dissipation. Since such techniques focus on the highest level of the design hierarchy, their potential benefits are immense. In particular, at this high level of abstraction, the specifics of each particular class of embedded applications can be considered as a *whole* and, as it will be shown in our survey, such an ability is critical to designing power/energy efficient systems, that is, systems that spend energy strictly when and where it is needed. Broadly speaking, this requires a proper *design* and *allocation* of *system resources*, geared toward addressing critical performance bottlenecks in a power–efficient way. Substantial power/energy savings can also be achieved through the implementation of adequate *dynamic power management policies*, for example., tracking instantaneous workloads (or levels of resource utilization) and "shutting down" idling/unused resources, so as to reduce leakage power, or "slowing down" under utilized resources, so as to decrease dynamic power dissipation. These are clearly system level decisions/policies, in that their implementation typically impacts several architectural subsystems. Moreover, different decisions/policies may interfere or conflict with each other and, thus, assessing their overall effectiveness requires a system-level (i.e., *global)* view of the problem.

A typical embedded system architecture consists of a *processing subsystem* (including one or more processor cores, hardware accelerators, etc.), a *memory subsystem, peripherals,* and global and local *interconnect structures* (buses, bridges, crossbars, etc.). Figure 89.1 shows an abstract view of two such architecture instances. Broadly speaking, system-level design consists of defining the specific embedded system architecture to be used for a particular product, as well as defining how the target embedded application (implementing the required functionality/services) is to be mapped into that architecture.

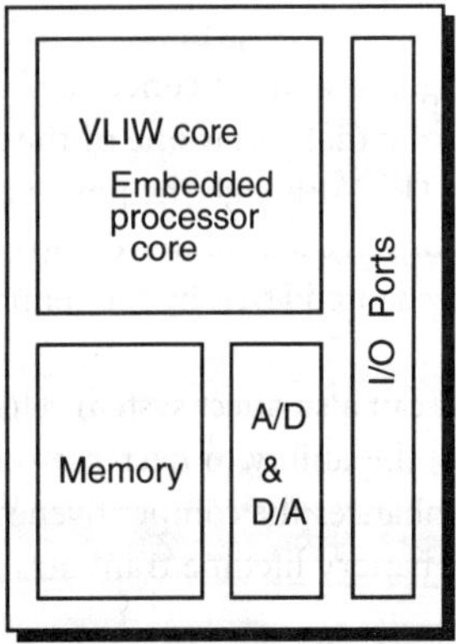

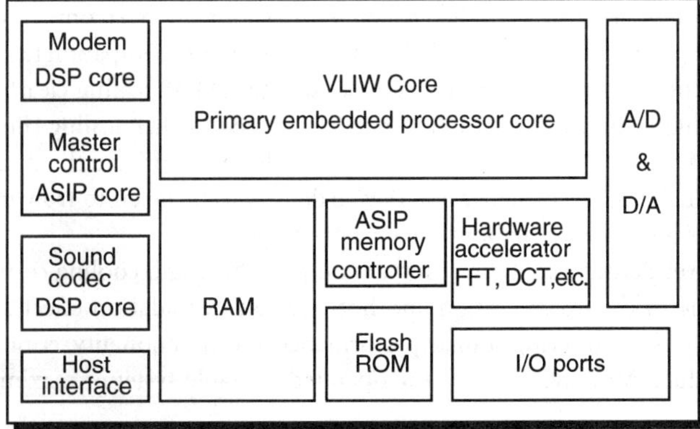

FIGURE 89.1   Illustrative examples of a simple and a more complex embedded system architecture.

Embedded systems come in many varieties and with many distinct design optimization goals and requirements. Even when two products provide the same basic functionality, say, video encoding/decoding, they may have fundamentally different characteristics, namely, different performance and quality-of-service requirements: one may be battery operated and the other not, etc. The implications of such product differentiation are of paramount importance when power/energy are considered. Clearly, the higher the system's required performance/speed (defined by metrics such as throughput, latency, bandwidth, response time, etc.), the higher its power dissipation. The key objective is thus to *minimize* the power dissipated to deliver the required level of performance [CB95a, PR02]. The trade-offs, techniques, and optimizations required to develop such *power-aware* or *power-efficient* designs vary widely across the vast spectrum of embedded systems available in today's market, encompassing many complex decisions driven by system requirements as well as intrinsic characteristics of the target applications [MSS+03].

Consider, for example, the design task of deciding on the number and type of processing elements to be instantiated on an embedded architecture, that is, defining its *processing subsystem.* Power, performance, cost, and time-to-market considerations dictate if one should rely entirely on readily available processors (i.e., off-the-shelf micro-controllers, DSPs, and/or general-purpose RISC cores), or one should also consider custom execution engines, namely, application-specific instruction set processors (ASIPs), possibly reconfigurable, and/or hardware accelerators (see Figure 89.1). *Hardware/software partitioning* is a critical step in this process [MRW02]. It consists of deciding which of an application's segments/functions should be implemented in *software* (i.e., run on a processor core) and which (if any) should be implemented in *hardware* (i.e., execute on high-performance, highly power-efficient custom hardware accelerators). Naturally, hardware/software partitioning decisions should reflect the power/performance criticality of each such segment/function. Clearly, this is a complex multiobjective optimization problem defined on a huge design space that encompasses, both *hardware-* and *software-*related decisions. To compound the problem, the performance and energy efficiency of an architecture's processing subsystem cannot be evaluated in isolation, since its effectiveness can be substantially impacted by the memory subsystem (i.e., the adopted memory hierarchy/organization) and the interconnect structures supporting communication/data-transfers between processing components and to/from the environment in which the system is embedded. Thus, decisions with respect to these other subsystems and components must be concurrently made and jointly assessed.

Targeting up front a specific *embedded system platform,* that is, an architectural "subspace" relevant to a particular class of products/applications, can considerably reduce the design effort [VM01, RV02]. Still, the design space remains (typically) so complex that a substantial *design space exploration* may be needed in order to identify power/energy efficient solutions for the specified performance levels. Since *time to market* is critical, methodologies to efficiently drive such an exploration, as well as fast simulators and low complexity (yet good fidelity) performance, power and energy estimation models, are critical to aggressively exploiting effective power/energy-driven optimizations, within a reasonable time frame.

Our survey starts by providing an overview on state-of-the-art models and tools used to evaluate the goodness of individual system design points. We then discuss power management techniques and optimizations aimed at aggressively improving the power/energy efficiency of the various subsystems of an embedded system.

89.2 Energy and Power Modeling

This section discusses *high-level* modeling and power estimation techniques aimed at assisting system and architecture-level design. It would be unrealistic to expect a high degree of accuracy on power estimates produced during such an early design phase, since accurate power modeling requires detailed physical-level information that may not yet be available. Moreover, highly accurate estimation tools (working with detailed circuit/layout-level information) would be too time consuming to allow for any reasonable degree of design space exploration [MRW02, PR02, BHLM94].

Thus, practically speaking, power estimation during early design space exploration should aim at ensuring a high degree of *fidelity* rather than necessarily *accuracy.* Specifically, the primary objective during this critical exploration phase is to assess the *relative* power efficiency of different candidate system architectures (populated with different hardware and/or software components), the *relative* effectiveness of alternative software implementations (of the same functionality), the *relative* effectiveness of different power management techniques, etc. Estimates that correctly expose such relative power trends across the design space region being explored provide the designer with the necessary information to guide the exploration process.

Instruction and Function Level Models

Instruction-level power models are used to assess the relative power/energy efficiency of different processors executing a given target embedded application, possibly with alternative memory subsystem configurations. Such models are thus instrumental during the definition of the main subsystems of an embedded architecture, as well as during hardware/software partitioning. Moreover, instruction-level power models can also be used to evaluate the relative effectiveness of different software implementations of the same embedded application, in the context of a specific embedded architecture/platform.

In their most basic form, instruction-level power models simply assign a power cost to each assembly instruction (or class of assembly instructions) of a programmable processor. The overall energy consumed by a program running on the target processor is estimated by summing up the instruction costs for a dynamic execution trace, which is representative of the application [TMW94, CP92, RJ98, BFSS00].

Instruction-level power models were first developed by experimentally measuring the current drawn by a processor while executing different instruction sequences [TMW94]. During this first modeling effort, it was observed that the power cost of an instruction may actually depend on previous instructions. Accordingly, the instruction-level power models developed in [TMW94] include several interinstruction effects. Later studies observed that, for certain processors, the power dissipation incurred by the hardware responsible for fetching, decoding, analyzing and issuing instructions, and then routing and reordering results, was so high that a simpler model that only differentiates between instructions that access *on-chip* resources and those that go *off-chip* would suffice for such processors [RJ98].

Unfortunately, power estimation based on instruction-level models can still be prohibitively time consuming during early design space exploration, since it requires collecting and analyzing large instruction traces and, for many processors, considering a quadratically large number of interinstruction effects. In order to "accelerate" estimation, processor-specific coarser *function-level* power models were later developed [QKUP00]. Such approaches are faster because they rely on the use of macromodels characterizing the *average energy consumption* of a library of functions/subroutines executing on a target processor [QKUP00]. The key challenge in this case is to devise macro-models that can properly quantify the power consumed by each subroutine of interest, as a function of "easily observable parameters." Thus, for example, a quadratic power model of the form $an^2 + bn + c$ could be first tentatively selected for a *insertion sort* routine, where n denotes the number of elements to be sorted. Actual power dissipation then needs to be measured for a large number of experiments, run with different values of n. Finally, the values of the macromodel's coefficients a, b, and c are derived, using regression analysis, and the overall accuracy of the resulting macro-model is assessed [QKUP00].

The "high-level" instruction and function-level power models discussed so far allow designers to "quickly" assess a large number of candidate system architectures and alternative software implementations, so as to narrow the design space to a few promising alternatives. Once this initial broad exploration is concluded, power models for each of the architecture's main subsystems and components are needed, in order to support the detailed architectural design phase that follows.

Micro-architectural Models

Micro-architectural power models are critical to evaluating the impact of different processing subsystem choices on power consumption, as well as the effectiveness of different (micro-architecture level) power management techniques implemented on the various subsystems.

In the late 1980s and early 1990s, cycle accurate (or more precisely, cycle-by-cycle) simulators, such as Simplescalar [BA97], were developed to study the effect of architectural choices on the performance of general-purpose processors. Such simulators are in general very flexible, allowing designers/architects to explore the complex design space of contemporaneous processors. Namely, they include built-in parameters that can be used to specify the number and mix of functional units to be instantiated in the processor's datapath, the issue width of the machine, the size and associativity of the L1 and L2 caches, etc. By varying such parameters, designers can study the performance of different machine configurations for representative applications/benchmarks. As power consumption became more important, simulators to estimate *dynamic* power dissipation (e.g., "Wattch [BTM00]," "Cai-Lim model [CL99]," and "Simplepower [YVKI00]") were later incorporated into these existing frameworks. Such an integration was performed seamlessly, by directly augmenting the "cycle-oriented" performance models for the various micro-architectural components with corresponding power models.

Naturally, the overall accuracy of these simulation-based power estimation techniques is determined by the level of detail of the power models used for the micro-architecture's constituent components. For out-of-order RISC cores, for example, the power consumed in finding independent instructions to issue is a function of the number of instructions currently in the instruction queue and of the actual dependencies between such instructions. Unfortunately, the use of detailed power models accurately capturing the impact of input and state data on the power dissipated by each component would prohibitively increase the already "long" micro-architectural simulation runtimes. Thus, most state-of-the-art simulators use very simple/straightforward empirical power models for datapath and control logic, and slightly more sophisticated models for regular structures such as caches [BTM00]. In their simplest form, such models capture "typical" or "average" power dissipation for each individual micro-architectural component. Specifically, each time a given component is accessed/used during a simulation run, it is assumed that it dissipates its corresponding "average" power. Slightly more sophisticated power macromodels for datapath components have been proposed in [JKSN99, BBM00, TGTS98, KE98, CRC00, MG01], and have shown to improve accuracy with a relatively small impact on simulation time.

So far, we have discussed power modeling of micro-architectural components; yet, a substantial percentage of the overall power budget of a processor is actually spent on the *global clock* (up to 40–45% [PR02]). Thus, global clock power models must also be incorporated in these frameworks. The power dissipated on global clock distribution is impacted to first order by the number of pipeline registers (and thus by a processor's pipeline depth) and by global and local wiring capacitances (and thus by a processor's core area) [PR02]. Accordingly, different processor cores and/or different configurations of the same core may dissipate substantially different clock distribution power. Power estimates incorporating such numbers are thus critical during processor core selection and configuration.

The component-level and clock distribution models discussed so far are used to estimate the dynamic power dissipation of a target micro-architecture. Yet, as mentioned above, static/leakage power dissipation is becoming a major concern, and thus, micro-architectural techniques aimed at reducing leakage power are increasingly relevant. Models to support early estimation of static power dissipation emerged along the same lines as those used for dynamic power dissipation. The "Butts–Sohi" model, which is one of the most influential static power models developed so far, quantifies static energy in CMOS circuits/components using a lumped parameter model that maps *technology* and *design* effects into corresponding characterizing parameters [BS00]. Specifically, static power dissipation is modeled as $V_{DD}Nk_{design}I_{leak}$, where V_{DD} is the supply voltage and N denotes the number of transistors in the circuit. k_{design} is the *design-dependent* parameter — it captures "circuit style"-related characteristics of a component, including average transistor aspect ratio, average number of transistors switched off during "normal/typical" component operation, etc. Finally, I_{leak} is the *technology-dependent* parameter. It accounts for the impact of threshold voltage, temperature, and other key parameters, on leakage current, for a specific fabrication process.

From a system designer's perspective, static power can be reduced by lowering supply voltage (V_{DD}), and/or by *power supply gating* or V_{DD}-gating (as opposed to clock-gating) unused/idling devices (N). Integrating models for estimating static power dissipation on cycle-by-cycle simulators thus enables embedded system designers to analyze critical static power versus performance tradeoffs enabled by "power-aware" features available in contemporaneous processors, such as dynamic voltage scaling and selective datapath (re)configuration. An improved version of the "Butts–Sohi" model, providing the ability to dynamically recalculate leakage currents (as temperature and voltage change due to operating conditions and/or dynamic voltage scaling), has been integrated into the Simplescalar simulation framework, called "HotLeakage," enabling such high-level trade-offs to be explored by embedded system designers [ZPS⁺03].

Memory and Bus Models

Storage elements, such as caches, register files, queues, buffers, and tables constitute a substantial part of the power budget of contemporaneous embedded systems [PAC⁺97]. Fortunately, the high regularity of some such memory structures (e.g., caches) permits the use of simple, yet reasonably accurate power estimation techniques, relying on automatically synthesized "structural designs" for such components.

The Cache Access and Cycle TIme (CACTI) framework implements this synthesis-driven power estimation paradigm. Specifically, given a specific cache hierarchy configuration (defined by parameters such as cache size, associativity, and line size), as well as information on the minimum feature size of the target technology [WJ96], it internally generates a coarse structural design for such cache configuration. It then derives delay and power estimates for that particular design, using parameterized built-in C models for the various constituent elements, namely, SRAM cells, row and column decoders, word and bit lines, precharge circuitry, etc. [KG97, RJ01].

CACTI's synthesis algorithms used to generate the structural design of the memory hierarchy (which include defining the aspect ratio of memory blocks, the number of instantiated subbanks, etc.) have been shown to consistently deliver reasonably "good" designs across a large range of cache hierarchy parameters [RJ01]. CACTI can thus be used to "quickly" generate power estimates (starting from high-level architectural parameters) exhibiting a reasonably good fidelity over a large region of the design space. During design space exploration, the designer may thus consider a number of alternative L1 and L2 cache configurations and use CACTI to obtain "access-based" power dissipation estimates for each such configuration with good fidelity. Naturally, the memory access traces used by CACTI should be generated by a micro-architecture simulator (e.g., Simplescalar) working with a memory simulator (e.g., Dinero [EH98]), so that they reflect the bandwidth requirements of the embedded application of interest.

Buses are also a significant contributor to dynamic power dissipation [PR02, CWD⁺98]. The dynamic power dissipation on a bus is proportional to $CVD_{DD}^2 fa$, where C denotes the total capacitance of the bus (including metal wires and buffers), V_{DD} denotes the supply voltage, and fa denotes the average switching frequency of the bus [CWD⁺98]. In this high-level model, the average switching frequency of the bus (fa) is defined by the product of two terms, namely, the average number of bus transitions per word, and the bus frequency (given in bus words per second). The average number of bus transitions per word can be estimated by simulating sample programs and collecting the corresponding transition traces. Although this model is coarse, it may suffice during the early design phases under consideration.

Battery Models

The capacity of a battery is a nonlinear function of the current drawn from it, that is, if one increases the average current drawn from a battery by a factor of two, the "remaining" *deliverable battery capacity,* and thus its lifetime, decreases by more than half. Peukert's formula models such nonlinear behavior by defining the capacity of a battery as k/I^α, where k is a constant depending on the battery design, I is the discharge current, and α quantifies the "nonideal" behavior of the battery [MS96].[1] More effective system-level trade-offs between quality/performance and duration of service can be implemented by taking such nonlinearity (also called *rate-capacity* effect) into consideration. However, in order to properly evaluate the effectiveness of such techniques/trade-offs during system-level design, adequate battery models and metrics are needed.

---

[1]For an "ideal" battery, that is, a battery whose capacity in independent of the way current is drawn, $\alpha = 0$, while for a real battery α may be as high as 0.7 [MS96].

Energy-delay product is a well-known metric used to assess the energy efficiency of a system. It basically quantifies a system's performance loss per unit gain in energy consumption. To emphasize the importance of accurately exploring key tradeoffs between *battery lifetime* and *system performance,* a new metric, viz., *battery-discharge delay product,* has recently been proposed [PW02]. Task scheduling strategies and dynamic voltage scaling policies adopted for battery-operated embedded systems should thus aim at minimizing battery-discharge delay product, rather than energy-delay product, since it captures the important rate-capacity effect alluded to above, whereas energy-delay product is insensitive to it. Yet, a metric such as battery-discharge delay product requires the use of precise/detailed battery models. One such detailed battery model has been recently proposed, which can predict the remaining capacity of a rechargeable lithium-ion battery in terms of several critical factors, viz., discharge-rate (current), battery output voltage, battery temperature, and cycle age (i.e., the number of times a battery has been charged and discharged) [RP03].

89.3 System/Application-Level Optimizations

When designing an embedded system, in particular, a battery-powered one, it may be useful to explore different *task implementations* exhibiting different power/energy vs. quality-of-service characteristics, so as to provide the system with the desired functionality while meeting cost, battery lifetime, and other critical requirements. Namely, one may be interested in trading off accuracy for energy savings on a handheld GPS system, or image quality for energy savings on an image decoder, etc. [SCB96, QP99, SBGM00]. Such application-level "tuning/optimizations" may be performed statically (i.e., one may use a single implementation for each task) or may be performed at runtime, under the control of a system-level power manager. For example, if the battery level drops below a certain threshold, the power manager may drop some services and/or swap some tasks to less power-hungry ("lower quality") software versions, so that the system can remain operational for a specified window of additional time. An interesting example of such dynamic power management was implemented for the Mars Pathfinder, an unmanned space robot that draws power from both a nonrechargeable battery and solar cells [LCBK01]. In this case, the power manager tracks the power available from the solar cells, ensuring that most of the robot's active work is done during daylight, since the solar energy cannot be stored for later use.

In addition to dropping and/or swapping tasks, a system's dynamic power manager may also shut down or slow down subsystems/modules that are idling or underutilized. The Advanced Configuration and Power Interface (ACPI) is a widely adopted standard that specifies an interface between an architecture's power managed subsystems/modules (e.g., display drivers, modems, hard-disk drivers, processors, etc.) and the system's dynamic power manager. It is assumed that the power-managed subsystems have at least two power states (ACTIVE and STANDBY) and that one can dynamically switch between them. Using ACPI, one can implement virtually any power management policy, including fixed time-out, predictive shutdown, predictive wake-up, etc. [ACP]. Since the transition from one state to another may take a substantial amount of time and consume a nonnegligible amount of energy, the selection of a proper power management policy for the various subsystems is critical to achieve good energy savings with a negligible impact on performance. The so-called time-out policy, widely used in modern computers, simply switches a subsystem to STANDBY mode when the elapsed time after the last utilization reaches a given fixed threshold. Predictive shutdown uses the previous history of the subsystem to predict the next expected idle time and, based on this, decides if it should or should not be shut down. More sophisticated policies, using stochastic methods [SCB96, QP99, SBGM00, SBA$^+$01], can also be implemented; yet, they are more complex, and thus the power associated with running the associated dynamic power management algorithms may render them inadequate or inefficient for certain classes of systems.

The system-level techniques discussed above act on each subsystem as a whole. Although the effectiveness of such techniques has been demonstrated across a wide variety of systems, finer-grained self-monitoring techniques, implemented at the subsystem level, can substantially add to these savings. Such techniques are discussed in the following sections.

89.4 Energy-Efficient Processing Subsystems

As mentioned earlier, hardware/software codesign methodologies partition the functionality of an embedded system into hardware and software components. Software components, executing on programmable micro-controllers or processors (either general-purpose or application-specific) are the preferred solution, since their use can substantially reduce design and manufacturing costs, as well as shorten time-to-market. Custom hardware components are typically used only when strictly necessary, namely, when an embedded system's power budget and/or performance constraints preclude the use of software. Accordingly, a large number of power-aware processor families are available in today's market, providing a vast gamut of alternatives suiting the requirements/needs of most embedded system [GH96, BB96]. In the sequel, we discuss power-aware features available in contemporaneous processors, as well as several power-related issues relevant to processor core selection.

Voltage and Frequency Scaling

Dynamic power consumption in a processor (be it general-purpose or application-specific) can be decreased by reducing two of its key contributors, viz., supply voltage and clock frequency. In fact, since the power dissipated in a CMOS circuit is proportional to the square of the supply voltage, the most effective way to reduce power is to scale down the supply voltage. Note, however, that the propagation delay across a CMOS transistor is proportional to $V_{DD}/(V_{DD} - V_T)^2$, where V_{DD} is the supply voltage and V_T is the threshold voltage. So, unfortunately, as the supply voltage decreases, the propagation delay increases as well, and so clock frequency (i.e., speed) may need to be decreased [IY98].

Accordingly, many contemporaneous processor families, such as Intel's XScale [INT], IBM's PowerPC 405LP [IBM], and Transmeta's Crusoe [TRA], offer *dynamic voltage and frequency scaling* features. For example, the Intel 80200 processor, which belongs to the XScale family of processors mentioned above, supports a software-programmable clock frequency. Specifically, the voltage can be varied from 1.0 to 1.5 V, in small increments, with the frequency varying correspondingly from 200 to 733 MHz, in steps of 33/66 MHz.

The simplest way to take advantage of the scaling features discussed above is by carefully identifying the "smallest" supply voltage (and corresponding operating frequency) that guarantee that the target embedded application meets its timing constraints, and run the processor for that fixed setting. If the workload is reasonably constant throughout execution, this simple scheme may suffice. However, if the workload varies substantially during execution, more sophisticated techniques that dynamically adapt the processor's voltage and frequency to the varying workload can deliver more substantial power savings. Naturally, when designing such techniques it is critical to consider the cost of transitioning from one setting to another, that is, the delay and power consumption overheads incurred by each transition. For example, for the Intel 80200 processor mentioned above, changing the processor frequency could take up to 20 μsec, while changing the voltage could take up to 1 msec [INT].

Most processors developed for the mobile/portable market already support some form of built-in mechanism for voltage/frequency scaling. Intel's SpeedStep technology, for example, detects if the system is currently plugged into a power outlet or running on a battery, and based on that, either runs the processor at the highest voltage/frequency or switches it to a less power-hungry mode. Transmeta's Crusoe processors offer a power manager called LongRun [TRA], which is implemented in the processor's firmware. LongRun relies on the historical utilization of the processor to guide clock rate selection: it increases the processor's clock frequency if the current utilization is high and decreases it if the utilization is low.

More sophisticated/aggressive dynamic scaling techniques should vary the core's voltage/frequency based on some predictive strategy, while carefully monitoring *performance*, so as to ensure that it does not drop beyond a certain threshold and/or that task deadlines are not consistently missed [WWDS94, GCW95, PBB98, PBB00] (see Figure 89.2). The use of such voltage/frequency scaling techniques must necessarily rely on adequate dynamic workload prediction and performance metrics, and thus requires the direct intervention of the operating system and/or of the applications themselves. Although more complex than the

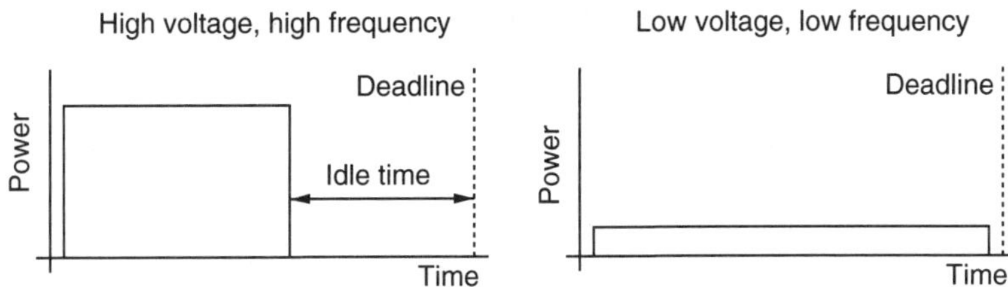

FIGURE 89.2 Power consumption without and with dynamic voltage and frequency scaling.

simple schemes discussed above, several such predictive techniques have been shown to deliver substantial gains for applications with well-defined task deadlines, for example, hard/soft real-time systems.

Simple *interval-based prediction schemes* consider the amount of idle time on a previous interval as a measure of the processor's utilization for the next time interval, and use that prediction to decide on the voltage/frequency settings to be used throughout its duration. Naturally, many critical issues must be factored when defining the duration of such an interval (or "prediction window"), including overhead costs associated with switching voltage/frequency settings. While a prediction scheme based on a single interval may deliver substantial power gains with marginal loss in performance [WWDS94], looking exclusively at a single interval may not suffice for many applications. Namely, the voltage/frequency settings may end up oscillating in an inefficient way between different settings [GCW95]. Simple smoothing techniques, for example, using an exponentially moving average of previous intervals, can be used to mitigate the problem [GCW95, FRM02].

Finally, note that, the benefits of dynamic voltage and frequency scaling are not limited to reducing dynamic power consumption. When the voltage/frequency of a battery-operated system is lowered, the instantaneous current drawn by the processor decreases accordingly, leading to a more effective utilization of the battery capacity, and thus to increased duration of service.

Dynamic Resource Scaling

Dynamic resource scaling refers to exploiting adaptive, fine-grained hardware resource reconfiguration techniques in order to improve power efficiency. Dynamic resource scaling requires enhancing the micro-architecture with the ability to selectively "disable" components, fully or partially, through either *clock gating* or V_{DD}-*gating*.[2] The effectiveness of dynamic resource scaling is predicated on the fact that many applications have variable workloads, that is, have execution phases with substantial instruction-level parallelism (ILP), and other phases with much less inherent parallelism. Thus, by dynamically "scaling down" micro-architecture components during such "low activity" periods, substantial power savings can potentially be achieved.

Techniques for reducing *static* power dissipation on processor cores can definitely exploit resource scaling features, once they become widely available on processors. Namely, underutilized or idling resources can be partially or fully V_{DD}-gated, thus reducing leakage power. Several utilization-driven techniques have already been proposed, which can selectively shut down functional units, segments of register files, and other datapath components, when such conditions arise [BM00, DKA+02, PKG01, BAB+02]. Naturally, the delay overhead incurred by power supply gating should be carefully factored into these techniques, so that the corresponding static energy savings are achieved with only a small degradation in performance.

Dynamic energy consumption can be also reduced by dynamically "scaling down" power-hungry micro-architecture components, for example, reducing the size of the issue window and/or reducing the effective width of the pipeline, during periods of low "activity" (say, when low ILP code is being executed). Moreover, if one is able to scale down these micro-architecture components that define the critical path delay of the machine's pipeline (typically located on the rename and window access stages), additional

---

[2]Note that, in contrast to clock gating, all state information is lost when a circuit is V_{DD}-gated

opportunities for voltage scaling, and thus dynamic power savings, can be created. This is thus a very promising area still undergoing intensive research [HSA01].

Processor Core Selection

The power aware techniques discussed thus far can be broadly applied to general-purpose as well as application-specific processors. Naturally, the selection of processor core(s) to be instantiated in the architecture's processing subsystem is likely to also have a substantial impact on overall power consumption, particularly when computation-intensive embedded systems are considered. A plethora of programmable processing elements, including micro-controllers, general-purpose processors, digital signal processors (DSPs), and application-specific instruction set processors (ASIPs), addressing the specific needs of virtually every segment of the embedded systems' market, is currently offered by vendors.

As alluded to above, for computation-intensive embedded systems with moderately high to stringent timing constraints, the selection of processor cores is particularly critical, since high-performance usually signifies high levels of power dissipation. For these systems, ASIPs and DSPs have the potential to be substantially more energy efficient than their general-purpose counterparts; yet, their "specialized" nature poses significant compilation challenges[3] [JdV00]. In contrast, general-purpose processors are easier to compile to, being typically shipped with good optimizing compilers, as well as debuggers and other development tools. Unfortunately, their "generality" incurs a substantial power overhead. In particular, high-performance general-purpose processors require the use of power-hungry hardware assists, including reservation stations, reorder buffers and rename logic, and complex branch prediction logic to alleviate control stalls. Still, their flexibility and high-quality development tools are very attractive for systems with stringent time-to-market constraints, making them definitively relevant for embedded systems.

IBM/Motorola's PowerPC family and the ARM family are examples of "general-purpose" processors enhanced with power-aware features that are widely used in modern embedded systems. A plethora of specialized/customizable processors is also offered by several vendors, including *specialized media cores* from Philips, Trimedia, MIPS, etc., *DSP cores* offered by Texas Instruments, StarCore and Motorola, and *customizable cores* from Hewlett-Packard-STMicroelectronics and Tensilica.

The ISAs and features offered on these specialized processors can vary substantially, since they are designed and optimized for different classes of applications. However, several of them, including TI's TMS320C6x family, HP-STS's Lx, the Starcore and Trimedia families, and Philip's Nexperia use a "Very Large Instruction Word" (VLIW) [CND+88, BYA93] or "Explicitly Parallel Instruction Computing" (EPIC) [SR00] paradigm. One of the key differences between VLIW and superscalar architectures is that VLIW machines rely on the compiler to extract instruction-level parallelism, and then schedule and bind instructions to functional units *statically*, while their high-performance superscalar counterparts use dedicated (power hungry) hardware to perform run-time dependence checking, instruction reordering, etc. Thus, in broad terms, VLIW machines eliminate power-hungry micro-architecture components by moving the corresponding functionality to the compiler.[4] Moreover, wide VLIW machines are generally organized as a set of small clusters with local register files.[5] Thus, in contrast to traditional superscalar machines, which rely on power-hungry multi-ported monolithic register files, multicluster VLIW machines scale better with increasing issue widths, for example, dissipate less dynamic power and can work at faster clock rates. Yet, they are harder to compile to [HHG+95, Ell85, DT93, DKK+99, JdVL00, LJdV02, PJ03]. In summary, the VLIW paradigm works very well for many classes of embedded applications, as attested to by the large number of VLIW processors currently available in the market. However, it poses substantial compilation challenges, some of which are still undergoing active research [MG95, Lie97, JYWV03]. Fortunately, many processing-intensive embedded applications have

---

[3]Several DSP/ASIP vendors provide preoptimized assembly libraries to mitigate this problem.

[4]Since functional unit binding decisions are made by the compiler, VLIW code is larger than RISC/superscalar code. We will discuss techniques to address this problem later in our survey.

[5]A cluster is a set of functional units connected to a local register file. Clusters communicate with each other through a dedicated interconnection network.

only a few time-critical loop kernels. Thus, only a very small percentage of the overall code needs to be actually subject to the complex, time-consuming compiler optimizations required by VLIW machines. In the case of media applications, for example, such loop kernels may represent as little as 3% of the overall program, and yet take up to 95% of the execution time [FWL99].

When the performance requirements of an embedded system are extremely high, using a dedicated coprocessor aimed at accelerating the execution of time-critical kernels (under the control of a host computer) may be the only feasible solution. Imagine, one such programmable coprocessor, was designed to accelerate the execution of kernels of streaming media applications [KDR+01]. It can deliver up to 20 GFLOPS at a relatively low power cost (2 GFLOPS/W); yet, such a power efficiency does not come without a cost [KDR+01]. As expected, Imagine has a complex programming paradigm, that requires extracting all time-critical kernels from the target application, and carefully reprogramming them using Imagine's "stream-oriented" coding style, so that both data and instructions can be efficiently routed from the host to the accelerator. Programming Imagine thus requires a substantial effort, yet its *power-efficiency* makes it very attractive for systems that demand such high levels of performance. Finally, at the highest end of the performance spectrum, one may need to consider using fully customized hardware accelerators. Comprehensive methodologies to design such accelerators are discussed in detail in [RMV+87].

89.5 Energy-Efficient Memory Subsystems

While processor speeds have been increasing at a very fast rate (about 60% a year), memory performance has increased at a comparatively modest rate (about 7% a year), leading to the well-known "processor-memory performance gap" [PAC+97]. In order to alleviate the memory access latency problem, modern processor designs use increasingly large on-chip caches, with up to 60% of the transistors dedicated to on-chip memory and support circuitry [PAC+97]. As a consequence, power dissipation in the memory subsystem contributes to a substantial fraction of the energy consumed by modern processors. A study targeting the StrongARM SA-110, a low-power processor widely used in embedded systems, revealed that more than 40% of the processor's power budget is taken up by on-chip data and instruction caches [MWA+96]. For high-performance general-purpose processors, this percentage is even higher, with up to 90% of the power budget consumed by memory elements and circuits aimed at alleviating the aforementioned memory bottleneck [PAC+97]. Power-aware memory designs have thus received considerable attention in recent years.

Cache Hierarchy Tuning

The energy cost of accessing data/instructions from off-chip memories can be as much as two orders of magnitude higher than that of an access to on-chip memory [HWO97]. By retaining instructions and data with high spatial and/or temporal locality on-chip, caches can substantially reduce the number of costly off-chip data transfers, thus leading to potentially quite substantial energy savings.

The "one-size-fits-all" nature of the general-purpose domain dictates that one should use large caches with a high degree of associativity, so as to try to ensure high hit rates (and thus low average memory access latencies) for as many applications as possible. Unfortunately, as one increases cache size and associativity, larger circuits and/or more circuits are activated on each access to the cache, leading to a corresponding increase in dynamic energy consumption.

Clearly, in the context of embedded systems, one can do much better. Specifically, by carefully tuning/scaling the configuration of the cache hierarchy, so that it more efficiently matches the bandwidth requirements and access patterns of the target embedded application, one can essentially achieve memory access latencies similar to those delivered by "larger" (general-purpose) memory subsystems, while substantially decreasing the average energy cost of such accesses.

Since several of the cache hierarchy parameters exhibit conflicting trends, an aggressive design space exploration over many candidate cache configurations is typically required in order to properly tune the cache hierarchy of an embedded system. Figure 89.3 summarizes the results of one such design space exploration performed for a media application. Namely, the graph in Figure 89.3 plots the energy-delay

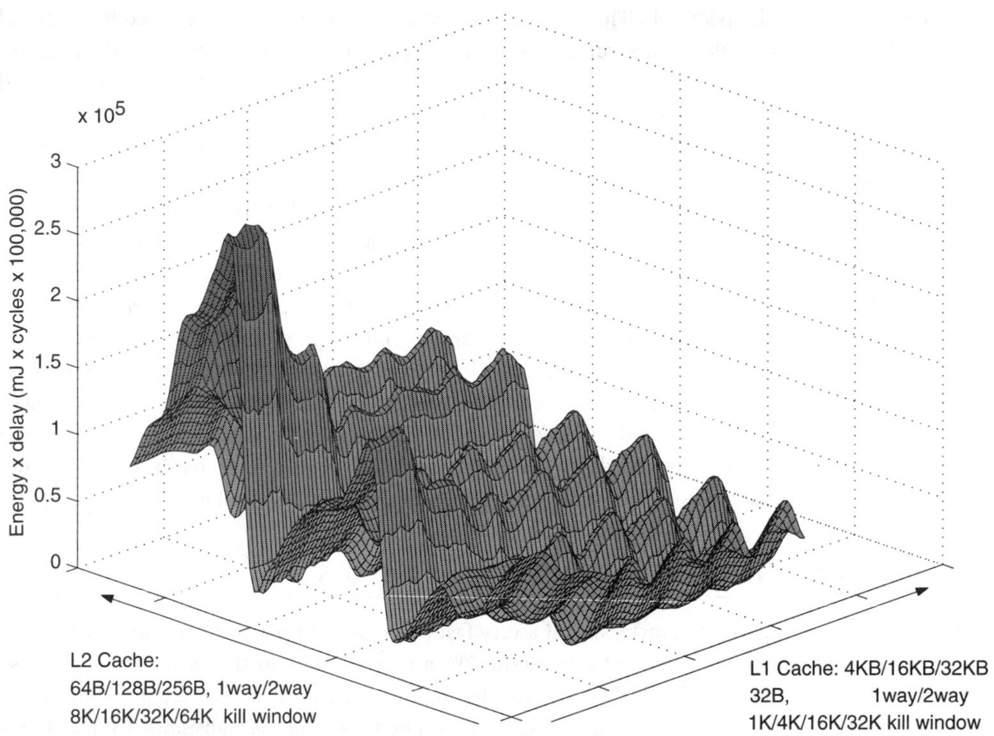

FIGURE 89.3　Design space exploration: energy-delay porduct for various L1 and 12 L2 D-cache configurations for a JPEG applicatoin running on an Xscale-like processor core.

product metric for a wide range of L1 and L2 on-chip D-cache configurations. The delay term is given by the number of cycles taken to process a representative data set from start to completion. The energy term accounts for the energy spent on data memory accesses for the particular execution trace.[6] As can be seen, the design space is very complex, reflecting the conflicting trends alluded to above. For this case study, the best set of cache hierarchy configurations exhibit an energy-delay product that is about one order of magnitude better (i.e., lower) than that of the worst configurations. Perhaps even more interesting, some of the worst memory subsystem configurations use quite large caches with a high degree of associativity, clearly indicating that no substantial performance gains would be achieved (for this particular media application) by using such an aggressively dimensioned memory subsystem.

For many embedded systems/applications, the power efficiency of the memory subsystems can be improved even more aggressively; yet, this requires the use of novel (nonstandard) memory system designs, as discussed in the sections that follow.

Novel Horizontal and Vertical Cache Partitioning Schemes

In recent years, several novel cache designs have been proposed to aggressively reduce the average *dynamic energy consumption* incurred by memory accesses. Energy efficiency is improved in these designs by taking direct advantage of specific characteristics of target classes of applications. The memory footprint of instructions and data in *media applications,* for example, tend to be very small, thus creating unique opportunities for energy savings [FWL99]. Since streaming media applications are pervasive in today's portable electronics market, they have been a preferred application domain for validating the effectiveness of such novel cache designs.

---

[6]Specifically, the energy term accounts for accesses to the on-chip L1 and L2 D-caches, and to main memory.

Vertical partition schemes [GK99, SD95, KGMS00, FTS95], as the name suggests, introduce additional small buffers/caches *before* the first level of the "traditional" memory hierarchy. For applications with "small" working sets, this strategy can lead to considerable dynamic power savings.

A concrete example of a vertical partition scheme is the *filter cache* [KGMS97], which is a very small cache placed in front of the standard L1 data cache. If the filter cache is properly dimensioned, dynamic energy consumption in the memory hierarchy can be substantially reduced, not only by accessing most of the data from the filter cache but also by powering down (clock-gating) the L1 D-cache to a STANDBY mode during periods of inactivity [KGMS97]. Although switching the L1 D-cache to STANDBY mode results in delay/energy penalties when there is a miss in the filter cache, it was observed that for media applications, the energy-delay product did improve quite significantly when the two techniques were combined.

Predecoded instruction buffers [BHK+97] and *loop buffers* [LMA99] are variants of the vertical partitioning scheme discussed above, yet applied to instruction caches (I-caches). The key idea of the first partitioning scheme mentioned above is to store recently used instructions on an instruction buffer, in a decoded form, so as to reduce the average dynamic power spent on fetching and decoding instructions. The second partitioning scheme allows one to hold time-critical loop bodies (identified a priori by the compiler or by the programmer) on small and thus energy-efficient dedicated loop buffers.

Horizontal partition schemes refer to the placement of additional (small) buffers or caches at the *same level* as the L1 cache. For each memory reference, the appropriate (level one) cache to be accessed is determined by dedicated decoding circuitry residing between the processor core and the memory hierarchy. Naturally, the method used to partition data across the set of first-level caches should ensure that the cache selection logic is simple, and thus cache access times are not significantly affected.

Region-based caches implement one such "horizontal partitioning" scheme, by adding two small 2 KB L1 D-caches to the first level of the memory hierarchy; one for stack and one for global data. This arrangement has also been shown to achieve substantial gains in dynamic energy consumption for streaming media applications with a negligible impact on performance [LT00].

Dynamic Scaling of Memory Elements

With an increasing number of on-chip transistors being devoted to storage elements in modern processors, of which only a very small set is active at any point in time, *static power dissipation* is expected to soon become a key contributor to a processor's power budget. State-of-the-art techniques to reduce static power consumption in on-chip memories are based on the simple observation that, in general, data or instructions fetched into a given cache line have an immediate flurry of accesses during a "small" interval of time, followed by a relatively "long" period of time where they are not used, before eventually being evicted to make way for new data/instructions [WHK91, BGK95]. If one can "guess" when that period starts, it is possible to "switch off" (i.e., V_{DD}-gate) the corresponding cache lines without introducing extra cache misses, thereby saving static energy consumption with no impact on performance [KHM01, ZTRC01].

Cache Decay was one of the earliest attempts to exploit such a "generational" memory usage behavior to decrease leakage power [KHM01]. The original Cache Decay implementation used a simple policy that turned off cache lines after a *fixed* number of cycles (decay-interval) since the last access. Note that if the selected decay interval happens to be too small, cache lines are switched off prematurely, causing extra cache misses, and if it is too large, opportunities for saving leakage energy are missed. Thus, when such a simple scheme is used, it is critical to tune the "fixed" decay-interval very carefully, so that it adequately matches the access patterns of the embedded application of interest. Adaptive strategies, varying the decay interval at runtime so as to dynamically adjust it to changing access patterns, have been proposed more recently, so as to enable the use of the cache decay principle across a wider range of applications [KHM01, ZTRC01]. Similar leakage energy reduction techniques have also been proposed for issue queues [BAB+02, FG01, PKG01] and branch prediction tables [HJS+02].

Naturally, leakage energy reduction techniques for instruction/program caches are also very critical [YPF+01]. A technique has been recently proposed that monitors the performance of the instruction cache over time, and dynamically scales (via V_{DD}-gating) its size, so as to closely match the size of the working set of the application [YPF+01].

Software-Controlled Memories, Scratch-Pad Memories

Most of novel designs and/or techniques discussed so far require an *application-driven* tuning of several architecturally visible parameters. However, similar to more "traditional" cache hierarchies, the memory subsystem interface implemented on these novel designs still exposes a flat view of the memory hierarchy to the compiler/software. That is, the underlying details of the memory subsystem architecture are essentially transparent to both.

Dynamic power dissipation incurred by accesses to basic memory modules occurs due to switching activity in bit-lines, word-lines, and input and output lines. Traditional caches have additional switching overheads, due to the circuitry (comparators, multiplexers, tags, etc.) needed to provide the "flat" memory interface alluded to above. Since the hardware assists necessary to support such a transparent view of the memory hierarchy are quite power hungry, additional energy-saving opportunities can be created by relying more on the compiler (and less on dedicated hardware) to manage the memory subsystem. The use of software-controlled (rather than hardware-controlled) memory components is thus becoming increasing prevalent in power-aware embedded system design.

Scratch-Pads are an example of such novel, software-controlled memories [PDA97, CJDR99, BMP00, BS+01, KRI+01, UWK+01]. Scratch-Pads are essentially on-chip partitions of main memory directly managed by the compiler. Namely, decisions concerning data/instruction placement in on-chip Scratch-Pads are made statically by the compiler, rather than dynamically, using dedicated hardware circuitry. Therefore, these memories are much less complex and thus less power hungry than traditional caches. As one would expect, the ability to aggressively improve energy-delay efficiency through the use of Scratch-Pads is predicated on the quality of the decisions made by the compiler on the subset of data/instructions that are to be assigned to that limited memory space [PDA97, PCD+01]. Several compiler-driven techniques have been proposed to identify the data/instructions that can be assigned to the Scratch-Pad more profitably, with the frequency of use being one of the key selection criteria [SFL98, IY00, SWLM02].

The *Cool Cache* architecture [UAK+03], also proposed for media applications, is a good example of a novel, power-aware memory subsystem that relies on the use of software-controlled memories. It uses a small Scratch-Pad and a "software controlled cache," each of which is implemented on a different on-chip SRAM. The program's scalars are mapped to the small (2 KB) Scratch-Pad [UWK+01].[7] Nonscalar data are mapped to the software-controlled cache, and the compiler is responsible for translating virtual addresses to SRAM lines, using a small register lookup area. Even though cache misses are handled in software, thereby incurring substantial latency/energy penalties, the overall architecture has been shown to yield substantial energy-delay product improvements for media applications, when compared to traditional cache hierarchies [UAK+03].

The effectiveness of techniques such as the above is so pronounced that several embedded processors currently offer a variety of software-controlled memory blocks, including configurable Scratch-Pads (TI's 320C6x [TI]), lockable caches (Intel's XScale [INT] and Trimedia [TRI]), and stream buffers (Intel's StrongARM [INT]).

Improving Access Patterns to Off-Chip Memory

During the last few decades, there has been a substantial effort in the compiler domain aimed at minimizing the number of off-chip memory accesses incurred by optimized code, as well as enabling the implementation of aggressive prefetching strategies. This includes devising compiler techniques to restructure, reorganize, and lay out data in off-chip memory, as well as techniques to properly reorder a program's memory access patters [WL91, CMT94, CM95, Wol95, KRCB01].

---

[7]This size was found to be sufficient for most media applications.

Prefetching techniques have received considerable attention lately, particularly in the domain of embedded streaming media applications. Instruction and data prefetching techniques can be hardware or software driven [Jou90, CB95b, FPJ92, PY96, CKP91, KL91, MLG92, ZLF00]. *Hardware-based* data prefetching techniques try to dynamically predict when a given piece of data will be needed, so as to load it into cache (or into some dedicated on-chip buffer), before it is actually referenced by the application (i.e., explicitly required by a demand access) [CB95b, FPJ92, PY96]. In contrast, *software-based* data prefetching techniques work by inserting prefetch instructions for selected data references at carefully chosen points in the program – such explicit prefetch instructions are executed by the processor, to move data into cache [CKP91, KL91, MLG92, ZLF00].

It has been extensively demonstrated that, when properly used, prefetching techniques can substantially improve average memory access latencies [Jou90, CB95b, FPJ92, PY96, CKP91, KL91, MLG92, ZLF00]. Moreover, techniques that prefetch substantial chunks of data (rather than, say, a single cache line), possibly to a dedicated buffer, can also simultaneously decrease dynamic power dissipation [CK03]. Namely, when data are brought from off-chip memory in large bursts, energy-efficient burst/page access modes can be more effectively exploited. Moreover, by prefetching large quantities of instructions/data, the average length of DRAM idle times is expected to increase, thus creating more profitable opportunities for the DRAM to be switched to a lower power mode. [FEL01, DSK$^+$02, RJ03]. Naturally, it is important to ensure that the overhead associated with the prefetching mechanism itself, as well as potential increases in static energy consumption due to additional storage requirements, do not outweigh the benefits achieved from enabling more energy-efficient off-chip accesses [RJ03].

Special-Purpose Memory Subsystems for Media Streaming

As alluded to before, streaming media applications have been a preferred application domain for validating the effectiveness of many novel, power-aware memory designs. Although the compiler is consistently given a more preeminent role in the management of these novel memory subsystems, they require no fundamental changes to the adopted programming paradigm. Additional opportunities for energy savings can be unlocked by adopting a programming paradigm that directly exposes those elements of an application that should be considered by an optimizing compiler, during performance versus power trade-off exploration. The two special-purpose memory subsystems discussed below do precisely that, in the context of streaming media applications.

Xtream-Fit is a special-purpose data memory subsystem targeted to generic uni-processor embedded system platforms executing media applications [RJ03]. Xtream-Fit's on-chip memory consists of a Scratch-Pad, to hold constants and scalars, and a novel software-controlled Streaming Memory, partitioned into regions, each of which holds one of the input or output streams used/produced by the target application. The use of software-controlled memories by Xtream-Fit ensures that dynamic energy consumption is low, while the region-based organization of the streaming memory enables the implementation of very simple and yet effective shutdown policies to turn off different memory regions, as the data they hold become "dead." Xtream-Fit's programming model is actually quite simple, requiring only a minor "reprogramming" effort. It simply requires organizing/partitioning the application code into a small set of processing and data transfer tasks. Data transfer tasks prefetch streaming media data (the amount required by the next set of processing tasks) into the streaming memory. The amount of prefetched data is explicitly exposed via a single customization parameter. By varying this single customization parameter, the compiler can thus aggressively minimize energy-delay product, by considering both dynamic and leakage power, dissipated in on-chip and off-chip memories [RJ03].

While Xtream-Fit provides sufficient memory bandwidth for generic uni-processor-embedded media architectures, it cannot support the very high bandwidth requirements of high-performance media accelerators. For example, imagine, the multicluster media accelerator alluded to previously, uses its own specialized memory hierarchy, consisting of a streaming memory, a 128 KB stream register file, and stream buffers and register files local to each of its eight clusters. Imagine's memory subsystem delivers a very high bandwidth (2.1 GB/sec) with very high energy efficiency; yet, it requires the use of a specialized programming paradigm. Namely, data transfers to/from the host are controlled by a stream controller, and between the stream register file and the functional units by a micro-controller, both of which have to be programmed separately, using Imagine's own "stream-oriented" programming style [Mat01].

Systems that demand still higher performance and/or energy efficiency may require memory architectures fully customized to the target application. Comprehensive methodologies for designing high-performance memory architectures for custom hardware accelerators are discussed in detail in [CWD$^+$98, GDN03].

Code Compression

Code size affects both program storage requirements and off-chip memory bandwidth requirements, and can thus have a first-order impact on the overall power consumed by an embedded system. Instruction compression schemes decrease both such requirements by storing in main memory (i.e., off-chip) frequently fetched/executed instruction sequences in an encoded/compressed form [WC92, LBCM97, LW99]. Naturally, when one such scheme is adopted, it is important to factor in the overhead incurred by the on-chip decoding circuitry, so that it does not outweigh the gains achieved on storage and interconnect elements. Furthermore, different approaches have considered storing such select instruction sequences *on-chip* in either *compressed* or *decompressed* forms. On-chip storage of instructions in compressed form saves on-chip storage; yet, instructions must be decoded every time they are executed, adding additional latency/power overheads.

Instruction subsetting is an alternative instruction compression scheme, where instructions not used commonly are discarded from the instruction set, thus enabling the "reduced" instruction set to be encoded using less bits [DPT98]. The Thumb instruction set is a classic example of a compressed instruction set, featuring the most commonly used 32-bit ARM instructions, compressed to 16-bit wide format. The Thumb instruction set is decompressed transparently to full 32-bit ARM instructions in real time, with no performance loss.

Interconnect Optimizations

Power dissipation in on- and off-chip interconnect structures is also a significant contributor to an embedded system's power budget [SK00]. A shared bus is a commonly used interconnect structure, as it offers a good trade-off between generality/simplicity and performance. Power consumption on the bus can be reduced by decreasing its supply voltage, capacitance, and/or switching activity. *Bus splitting,* for example, reduces bus capacitance by splitting long bus lines into smaller sections, with one section relaying the data to the next [HP00]. Power consumption in this approach is reduced at the expense of a small penalty in latency, incurred at each relay point. Bus switching activity, and thus dynamic power dissipation, can also be substantially reduced by using an appropriate bus encoding scheme [SB95, MOI96, BMM$^+$97, BMM$^+$98, PD99, CKC00]. *Bus-invert coding* [SB95], for example, is a simple, and yet widely used coding scheme. The first step of bus-invert coding is to compute the Hamming distance between the current bus value and the previous bus value. If this value is greater than half the number of total bits, then the data value is transmitted in inverted form, with an additional invert bit to interpret the data at the other end. Several other encoding schemes have been proposed, achieving lower switching activity at the expense of higher encoding and decoding complexity [MOI96, BMM$^+$97, BMM$^+$98, PD99, CKC00].

With the increasing adoption of system-on-chip (SOC) design methodologies for embedded systems, devising energy-delay efficient interconnect architectures for such large-scale systems is becoming increasingly critical and is still undergoing intensive research. [PR02].

89.6   Summary

Design methodologies for today's embedded systems must necessarily treat power consumption as a primary figure of merit. At the system and architecture levels of design abstraction, power-aware embedded system design requires the availability of high-fidelity power estimation and simulation frameworks. Such frameworks are essential to enabling designers to explore and evaluate, in reasonable time, the complex energy-delay trade-offs realized by different candidate architectures, subsystem realizations, and power management techniques, and thus quickly identify promising solutions for the target application of interest. The detailed system- and architecture-level design phases that follow should adequately combine coarse, system-level dynamic power management strategies, with fine-grained, self-monitoring techniques, exploiting voltage and frequency scaling, as well as advanced dynamic resource scaling and power-driven reconfiguration techniques.

References

[ACP] http://www.acpi.info/.

[BA97] D.C. Burger and T.M. Austin, The SimpleScalar tool set, version 2.0, *Computer Architecture News*, 25(3), pp. 13–25, 1997.

[BAB+02] A. Buyuktosunoglu, D. Albonesi, P. Bose, P. Cook, and S. Schuster, Tradeoffs in Power-Efficient Issue Queue Design, in Proceedings of the International Symposium on Low Power Electronics and Design, 2002.

[BB96] T.D. Burd and R.W. Brodersen, Processor design for portable systems. *Journal of VLSI Signal Processing*, 13(2, 3), pp. 203–221, 1996.

[BBM00] A. Bogliolo, L. Benini, and G.D. Micheli, Regression-based RTL power modeling, *ACM Transactions on Design Automation of Electronic Systems*, 5(3), pp. 337–372, 2000.

[BFSS00] C. Brandolese, W. Fornaciari, F. Salice, and D. Sciuto, An Instruction-level Functionality-based Energy Estimation Model for 32-bits Microprocessors, in Proceedings of the Design Automation Conference, 2000.

[BGJ+97] F. Balarin, P. Giusto, A. Jurecska, C. Passerone, E. Sentovich, B. Tabbara, M. Chiodo, H. Hsieh, L. Lavagno, A.L. Sangiovanni-Vincentelli, and K. Suzuki, *Hardware–Software Co-Design of Embedded Systems: The POLIS Approach*, Kluwer Academic Publishers, Dordrecht, 1997.

[BGK95] D.C. Burger, J.R. Goodman, and A. Kagi. The Declining Effectiveness of Dynamic Caching for General-Purpose Microprocessors, Technical report, University of Wisconsin-Madison Computer Sciences Technical Report 1261, 1995.

[BHK+97] R.S. Bajwa, M. Hiraki, H. Kojima, D.J. Gorny, K. Nitta, A. Shridhar, K. Seki, and K. Sasaki, Instruction buffering to reduce power in processors for signal processing, *IEEE Transactions on Very Large Scale Integration Systems*, 5(4), pp. 417–424, 1997.

[BHLM94] J.T. Buck, S. Ha, E.A. Lee, and D.G. Messerschmitt, Ptolemy: a framework for simulating and prototyping heterogeneous systems, *International Journal of Computer Simulation, (special issue on Simulation Software Development)*, 4(2), pp. 155–182, 1994.

[BM00] D. Brooks and M. Martonosi, Value-based clock gating and operation packing: dynamic strategies for improving processor power and performance,. *ACM Transactions on Computer Systems*, 18(2), pp. 89–126, 2000.

[BMM+97] L. Benini, G. De Micheli, E. Macii, M. Poncino, and S. Quez, System-level Power Optimization of Special Purpose Applications- The Beach Solution, in Proceedings of the International Symposium on Low Power Electronics and Design, 1997.

[BMM+98] L. Benini, G. De Micheli, E. Macii, D. Sciuto, and C. Silvano, Address Bus Encoding Techniques for System-Level Power Optimization, in Proceedings of the Design, Automation and Test in Europe, 1998.

BMP00] L. Benini, A. Macii, and M. Poncino, A Recursive Algorithm for Low-Power Memory Partitioning, in Proceedings of the International Symposium on Low Power Electronics and Design, 2000.

[Bor99] S. Borkar, Design challenges of technology scaling, *IEEE Micro*, 19(4), pp. 23–29, 1999.

[BS00] J.A. Butts and G.S. Sohi, A Static Power Model for Architects, in Proceedings of the International Symposium on Microarchitecture, 2000.

[BS[+]01] R. Banakar, S. Steinke, B.-S. Lee, M. Balakrishnan, and P. Marwedel, Scratchpad Memory: A Design Alternative for Cache On-chip memory in Embedded Systems, in Proceedings of the International Workshop on Hardware/Software Codesign, 2002.

[BTM00] D. Brooks, V. Tiwari, and M. Martonosi, Wattch: A Framework for Architectural Level Power Analysis and Optimizations, inProceedings of the International Symposium on Computer Architecture, 2000.

[BYA93] G.R. Beck, D.W.L. Yen, and T.L. Anderson. The Cydra 5: mini-supercomputer: architecture and implementation, *The Journal of Super computing*, 7(1, 2), pp. 143–180, 1993.

[CB95a] A.P. Chandrakasan and R.W. Brodersen, *Low Power Digital CMOS Design*, Kluwer Academic Publishers, Dordrecht, 1995.

[CB95b] T.F. Chen and J.L. Baer, Effective hardware-based data prefetching for high performance, processors, *IEEE Transactions on Computers*, 44(5), pp. 609–623, 1995.

[CJDR99] D. Chiou, P. Jain, S. Devadas, and L. Rudolph, Application-Specific Memory Management for Embedded Systems Using Software-Controlled Caches, in Proceedings of the Design Automation Conference, 2000.

[CK03] Y. Choi and T. Kim, Memory Layout Technique for Variables Utilizing Efficient DRAM Access Modes in Embedded System Design, in Proceedings of the Design Automation Conference, 2003.

[CKC00] N. Chang, K. Kim, and J. Cho, Bus Encoding for Low-Power High-Performance Memory Systems, in Proceedings of the Design Automation Conference, 2000.

[CKP91] D. Callahan, K. Kennedy, and A. Porterfield, Software Prefetching, in Proceedings of the International Conference on Architectural Support for Programming Languages and Operating Systems, 1991.

[CL99] G. Cai and C. H. Lim, Architectural Level Power/Performance Optimization and Dynamic Power Estimation, in Cool Chips Tutorial, International Symposium on Microarchitecture, 1999.

[CM95] S. Coleman and K.S. McKinley, Tile Size Selection Using Cache Organization and Data Layout, in Proceedings of the Conference on Programming Language Design and Implementation, 1995.

[CMT94] S. Carr, K.S. McKinley, and C. Tseng, Compiler Optimizations for Improving Data Locality, in Proceedings of the International Conference on Architectural Support for Programming Languages and Operating Systems, 1994.

[CND[+]88] R.P. Colwell, R.P. Nix, J.J.O. Donnell, D.B. Papworth, and P.K. Rodman, A VLIW architecture for a trace scheduling compiler, *IEEE Transactions on Computers*, 37(8), pp. 967–979, 1988.

[CP92] P.M. Chau and S.R. Powell, Power dissipation of VLSI array processing systems. *Journal of VLSI Signal Processing*, 4(2, 3), pp. 199–212, 1992.

[CRC00] Z. Chen, K. Roy, and E.K. Chong, Estimation of power dissipation using a novel power macro-modeling technique, *IEEE Transactions on Computer Aided Design of Integrated Circuits and Systems*, 19(11), pp. 1363–1369, 2000.

[CSB92] A.P. Chandrakasan, S. Sheng, and R.W. Brodersen, Low-power CMOS digital design, *IEEE Journal of Solid-State Circuits*, 27(4), pp. 473–484, 1992.

[CWD[+]98] F. Catthoor, S. Wuytack, E. DeGreef, F. Balasa, L. Nachtergaele, and A. Vandecappelle, *Custom Memory Management Methodology: Exploration of Memory Organization for Embedded Multimedia System Design*, Kluwer Academic Publishers, Dordrecht, 1998.

[DKA[+]02] S. Dropsho, V. Kursun, D.H. Albonesi, S. Dwarkadas, and E.G. Friedma, Managing Static Leakage Energy in Microprocessor Functional Units, in Proceedings of the International Symposium on Microarchitecture, 2002.

[DKK[+]99] C. Dulong, R. Krishnaiyer, D. Kulkarni, D. Lavery, W. Li, J. Ng, and D. Sehr, An overview of the Intel IA-64 compiler, *Intel Technology Journal*, Q4, 1999.

[DPT98] W.E. Dougherty, D.J. Pursley, and D.E. Thomas, Instruction Subsetting: Trading Power for Programmability, in Proceedings of the International Workshop on Hardware/Software Codesign, 1998.

[DSK[+]02] V. Delaluz, A. Sivasubramaniam, M. Kandemir, N. Vijaykrishnan, and M.J. Irwin, Scheduler-based DRAM Energy Management, in Proceedings of the Design Automation Conference, 2002.

[DT93] J. Dehnert and R. Towle, Compiling for the Cydra-5, *The Journal of Supercomputing*, 7(1, 2), pp. 181–227, 1993.

[EH98] J. Edler and M.D. Hill, Dinero IV Trace-Driven Uniprocessor Cache Simulator, 1998, http://www.cs.wisc.edu/~markhill/DineroIV/.

[Ell85] J.R. Ellis, *Bulldog: A Compiler for VLIW Architectures*, MIT Press, Cambridge, MA, 1985.

[FEL01] X. Fan, C.S. Ellis, and A.R. Lebeck, Memory Controller Policies for DRAM Power Management, in Proceedings of the International Symposium on Low Power Electronics and Design, 2001.

[FG01] D. Folegnani and A. Gonzalez, Energy-Effective Issue Logic, in Proceedings of the International Symposium on High-Performance Computer Architecture, 2001.

[FPJ92] J.W.C. Fu, J.H. Patel, and B.L. Janssens, Stride Directed Prefetching in Scalar Processor, in Proceedings of the International Symposium on Microarchitecture, 1992.

[FRM02] K. Flautner, S. Reinhardt, and T. Mudge, Automatic performance setting for dynamic voltage scaling, *ACM Journal of Wireless Networks*, 8(5), pp. 507–520, 2002.

[FTS95] A.H. Farrahi, G.E. Tellez, and M. Sarrafzadeh, Memory Segmentation to Exploit Sleep Mode Operation, in Proceedings of the Design Automation Conference, 1995.

[FWL99] J. Fritts, W. Wolf, and B. Liu, Understanding Multimedia Application Characteristics for Designing Programmable Media Processors, in SPIE Photonics West, Media Processors, 1999.

[GCW95] K. Govil, E. Chan, and H. Wasserman, Comparing Algorithm for Dynamic Speed-Setting of a Low-power CPU, in Proceedings of the International Conference on Mobile Computing and Networking, 1995.

[GDN03] P. Grun, N. Dutt, and A. Nicolau, *Memory Architecture Exploration for Programmable Embedded Systems*, Kluwer Academic Publishers, Dordrecht, 2003.

[GGH97] R. Gonzalez, B. Gordon, and M. Horowitz, Supply and threshold voltage scaling for low power CMOS, *IEEE Journal of Solid-State Circuits*, 32(8), pp. 1210–1216, 1997.

[GH96] R. Gonzalez and M. Horowitz, Energy dissipation in general purpose microprocessors, *IEEE Journal of Solid-State Circuits*, 31(9), pp. 1277–1284, 1996.

[GK99] K. Ghose and M.B. Kamble, Reducing Power in Superscalar Processor Caches Using Subbanking, Multiple Line Buffers and Bit-Line Segmentation, in Proceedings of the International Symposium on Low Power Electronics and Design, 1999.

[HHG$^+$95] W.W. Hwu, R.E. Hank, D.M. Gallagher, S.A. Mahlke, D.M. Lavery, G.E. Haab, J.C. Gyllenhaal, and D.I. August, Compiler technology for future microprocessors, *Proceedings of the IEEE*, 83(12), pp. 1625–1640, 1995.

[HJS$^+$02] Z. Hu, P. Juang, K. Skadron, D. Clark, and M. Martonosi,. Applying Decay Strategies to Branch Predictors for Leakage Energy Savings, in Proceedings of the International Conference on Computer Design, 2002.

[HP00] C.-T. Hsieh and M. Pedram, Architectural Power Optimization by Bus Splitting, in Proceedings of the Conference on Design, Automation and Test in Europe, 2000.

[HSA01] C.J. Hughes, J. Srinivasan, and S.V. Adve, Saving Energy with Architectural and Frequency Adaptations for Multimedia Applications, in Proceedings of the International Symposium on Microarchitecture, 2001.

[HWO97] P. Hicks, M. Walnock, and R.M. Owens, Analysis of Power Consumption in Memory Hierarchies, in Proceedings of the International Symposium on Low Power Electronics and Design, 1997.

[IBM] http://www.ibm.com/.

[INT] http://www.intel.com/.

[ITR] http://public.itrs.net/.

[IY98] T. Ishihara and H. Yasuura, Voltage Scheduling Problem for Dynamically Variable Voltage Processors, in Proceedings of the International Symposium on Low Power Electronics and Design, 1998.

[IY00] T. Ishihara and H. Yasuura, A Power Reduction Technique with Object Code Merging for Application Specific Embedded Processors, in Proceedings of the Design, Automation and Test in Europe, 2000.

[JdV00] M.F. Jacome and G. de Veciana, Design Challenges for New Application Specific Processors. *IEEE Design and Test of Computers, (special issue system Design of Embedded Systems)*, 12(2), pp. 40–50 2000.

[JdVL00] M.F. Jacome, G. de Veciana, and V. Lapinskii, Exploring Performance Tradeoffs for Clustered VLIW ASIPs, in Proceedings of the International Conference on Computer-Aided Design, 2000.

[JKSN99] G. Jochens, L. Kruse, E. Schmidt, and W. Nebel, A New Parameterizable Power Macro-Model for Datapath Components, in Proceedings of the Design Automation and Test in Europe, 1999.

[Jou90] N.P. Jouppi, Improving Direct-Mapped Cache Performance by the Addition of a Small Fully-Associative Cache and Prefetch Buffers, in Proceedings of the International Symposium on Computer Architecture, 1990.

[JYWV03] A.A. Jerraya, S. Yoo, N. Wehn, and D. Verkest, Eds., *Embedded Software for SoC*, Kluwer Academic Publishers, Dordrecht, 2003.

[KDR$^+$01] B. Khailany, W.J. Dally, S. Rixner, U.J. Kapasi, P. Mattson, J. Namkoong, J.D. Owens, B. Towles, and A. Chang, Imagine: media processing with streams, *IEEE Micro,* 21(2), pp. 35–46, 2001.

[KE98] M. Khellah and M.I. Elmasry, Effective Capacitance Macro-Modelling for Architectural-Level Power Estimation, in Proceedings of the Eighth Great Lakes Symposium on VLSI, 1998.

[KG97] M. Kamble and K. Ghose, Analytical Energy Dissipation Models For Low Power Caches, in Proceedings of the International Symposium on Low Power Electronics and Design, 1997.

[KGMS97] J. Kin, M. Gupta, and W.H. Mangione-Smith, The Filter Cache: An Energy Efficient Memory Structure, in Proceedings of the International Symposium on Microarchitecture, 1997.

[KGMS00] J. Kin, M. Gupta, and W.H. Mangione-Smith, Filtering memory references to increase energy efficiency,. *IEEE Transactions on Computers,* 49(1), pp. 1–15, 2000.

[KHM01] S. Kaxiras, Z. Hu, and M. Martonosi, Cache Decay: Exploiting Generational Behavior to Reduce Cache Leakage Power, in Proceedings of the International Symposium on Computer Architecture, 2001.

[KL91] A.C. Klaiber and H.M. Levy, An Architecture for Software Controlled Data Prefetching, in Proceedings of the International Symposium on Computer Architecture, 1991.

[KRCB01] M. Kandemir, J. Ramanujam, A. Choudhary, and P. Banerjee, A layout-conscious iteration space transformation technique, *IEEE Transactions on Computers,* 50(12), pp. 1321–1336, 2001.

[KRI$^+$01] M. Kandemir, J. Ramanujam, M. Irwin, N. Vijaykrishnan, I. Kadayif, and A. Parikh, Dynamic Management of Scratch-Pad Memory Space, in Proceedings of the Design Automation Conference, 2001.

[LBCM97] C. Lefurgy, P. Bird, I-C. Cheng, and T. Mudge, Improving Code Density using Compression Techniques, in Proceedings of the International Symposium on Microarchitecture, 1997.

[LCBK01] J. Liu, P. Chou, N. Bagherzadeh, and F. Kurdahi, A Constraint-based Application Model and Scheduling Techniques for Power-aware Systems, in Proceedings of the International Conference on Hardware/Software Codesign, 2001.

[Lie97] C. Liem, *Retargetable Compilers for Embedded Core Processors,* Kluwer Academic Publishers, Dordrecht 1997.

[LJdV02] V. Lapinskii, M.F. Jacome, and G. de Veciana. Application-specific clustered VLIW datapaths: early exploration on a parameterized design space, *IEEE Transactions on Computer Aided Design of Integrated Circuits and Systems,* 21(8), pp. 889–903, 2002.

[LMA99] L. Lee, B. Moyer, and J. Arends, Instruction Fetch Energy Reduction Using Loop Caches For Embedded Applications with Small Tight Loops, in Proceedings of the International Symposium on Low Power Electronics and Design, 1999.

[LT00] H.-H. Lee and G. Tyson, Region-Based Caching: An Energy-Delay Efficient Memory Architecture for Embedded Processors, in Proceedings of the International Conference on Compilers, Architectures and Synthesis for Embedded Systems, 2000.

[LW99] H. Lekatsas and W. Wolf, SAMC: a code compression algorithm for embedded processors, *IEEE Transactions on Computer Aided Design of Integrated Circuits and Systems,* 18(12), pp. 1689–1701, 1999.

[Mat01] P. Mattson, A Programming System for the Imagine Media Processor, Ph.D. thesis, Stanford University, 2001.

[MG95] P. Marwedel and G. Goosens, Eds., *Code Generation for Embedded Processors,* Kluwer Academic Publishers, Dordrecht,1995.

[MG01] R. Melhem and R. Graybill, Eds., Challenges for architectural level power modeling, *Power Aware Computing,* Kluwer Academic Publishers, Dordrecht, 2001.

[MLG92] T.C. Mowry, M.S. Lam, and A. Gupta, Design and Evaluation of a Compiler Algorithm for Prefetching, in Proceedings of the International Conference on Architectural Support for Programming Languages and Operating Systems, 1992.

[MOI96] H. Mehta, R.M. Owens, and M.J. Irwin, Some Issues in Gray Code Addressing, in Proceedings of the Sixth Great Lakes Symposium on VLSI, 1996.

[MRW02] G. Micheli, R. Ernst, and W. Wolf, Eds., *Readings in Hardware/Software Co-Design.,* Morgan Kaufman Publishers, Los dltos, CA, 2002.

[MS96] T. Martin and D. Siewiorek, A power metric for mobile systems, in International Symposium on Lower Power Electronics and Design, 1996.

[MSS$^+$03] T.L. Martin, D.P. Siewiorek, A. Smailagic, M. Bosworth, M. Ettus, and J. Warren, A case study of a system-level approach to power-aware computing, *ACM Transactions on Embedded Computing Systems, (special Issue on power-Aware Embedded Computing),* 2(3), pp. 255–276, 2003.

[MWA$^+$96] J. Montanaro, R.T. Witek, K. Anne, A.J. Black, E.M. Cooper, D.W. Dobberpuhl, P.M. Donahue, J. Eno, A. Farell, G.W. Hoeppner, D. Kruckemyer, T.H. Lee, P. Lin, L. Madden, D. Murray, M. Pearce, S. Santhanam, K.J. Snyder, R. Stephany, and S.C. Thierauf, A 160 MHz 32b 0.5 W CMOS RISC Microprocessor, Proceedings of the International Solid-State Circuits Conference, Digest of Technical Papers, 1996.

[PAC$^+$97] D. Patterson, T. Anderson, N. Cardwell, R. Fromm, K. Keeton, C. Kozyrakis, R. Thomas, and K. Yelick, A case for intelligent RAM, *IEEE Micro,* 17(2), pp. 34–44, 1997.

[PBB98] T. Pering, T. Burd, and R. Brodersen, The Simulation and Evaluation of Dynamic Voltage Scaling Algorithms, in Proceedings of the International Symposium on Low Power Electronics and Design, 1998.

[PBB00] T. Pering, T. Burd, and R. Brodersen, Voltage Scheduling in the lpARM Microprocessor System, in Proceedings of the International Symposium on Low Power Electronics and Design, 2000.

[PCD$^+$01] P.R. Panda, F. Catthoor, N.D. Dutt, K. Danckaert, E. Brockmeyer, C. Kulkarni, A. Van-dercappelle, and P.G Kjeldsberg, Data and memory optimization techniques for embedded systems, *ACM Transactions on Design Automation of Electronic Systems,* 6(2), pp. 149–206, 2001.

[PD99] P.R. Panda and N.D. Dutt, Low-Power Memory Mapping Through Reducing Address Bus Activity, *IEEE Transactions on Very Large Scale Integration Systems,* 7(3), pp. 309–320, 1999.

[PDA97] P.R. Panda, N.D. Dutt, and A. Nicolau, Efficient Utilization of Scratch-Pad Memory in Embedded Processor Applications, in Proceedings of the European Design and Test Conference, 1997.

[PJ03] S. Pillai and M.F. Jacome, Compiler-Directed ILP Extraction for Clustered VLIW/EPIC Machines: Predication, Speculation and Modulo Scheduling, in Proceedings of the Design Automation and Test in Europe, 2003.

[PKG01] D. Ponomarev, G. Kucuk, and K. Ghose, Reducing Power Requirements of Instruction Scheduling Through Dynamic Allocation of Multiple Datapath Resources, in Proceedings of the International Symposium on Microarchitecture, 2001.

[PR02] M. Pedram and J.M. Rabaey, *Power Aware Design Methodologies,* Kluwer Academic Publishers, Dordrecht, 2002.

[PW02] M. Pedram and Q. Wu, Battery-powered digital CMOS design, *IEEE Transactions on Very Large Scale Integration Systems,* 10(5), pp. 601–607, 2002.

[PY96] S.S. Pinter and A. Yoaz, A Hardware-based Data Prefetching Technique for Superscalar Processors, in Proceedings of the International Symposium on Computer Architecture, 1996.

[QKUP00] G. Qu, N. Kawabe, K. Usami, and M. Potkonjak, Function-Level Power Estimation Methodology for Microprocessors, in Proceedings of the Design Automation Conference, 2000.

[QP99] Q. Qiu and M. Pedram, Dynamic Power Management Based on Continuous-Time Markov Decision Processes, in Proceedings of the Design Automation Conference, 1999.

[RJ98] J. Russell and M. Jacome, Software Power Estimation and Optimization for High-Performance 32-bit Embedded Processors, in Proceedings of the International Conference on Computer Design, 1998.

[RJ01] G. Reinman and N.M. Jouppi, CACTI 2.0: An Integrated Cache Timing and Power Model, Technical report, Compaq Computer Corporation, Western Research Lab, 2001.

[RJ03] A. Ramachandran and M. Jacome, Xtream-Fit: An Energy-Delay Efficient Data Memory Subsystem for Embedded Media Processing, in Proceedings of the Design Automation Conference, 2003.

[RMV+87] J. Rabaey, H. De Man, J. Vanhoof, G. Goossens, and F. Catthoor, CATHEDRAL-II: a synthesis system for multiprocessor DSP systems, *Silicon Compilation*, Addison-Wesley, Reading, MA, 1987.

[RP03] P. Rong and M. Pedram, Remaining Battery Capacity Prediction for Lithium-Ion Batteries, in Proceedings of the Design Automation and Test in Europe, 2003.

[RV02] J.M. Rabaey and A.S. Vincentelli, System-on-a-Chip — A Platform Perspective, in *Keynote Presentation, Korean Semiconductor Conference*, 2002.

[SB95] M.R. Stan and W.P. Burleson, Bus-invert coding for low-power I/O, *IEEE Transactions on Very Large Scale Integration Systems*, 3(1), pp. 49–58, 1995.

[SBA+01] T. Simunic, L. Benini, A. Acquaviva, P. Glynn, and G. De Micheli, Dynamic Voltage Scaling for Portable Systems, in Proceedings of the Design Automation Conference, 2001.

[SBGM00] T. Simunic, L. Benini, P. Glynn, and G. De Micheli, Dynamic Power Management of Portable Systems, in Proceedings of the International Conference on Mobile Computing and Networking, 2000.

[SCB96] M. Srivastava, A. Chandrakasan, and R. Brodersen, Predictive system shutdown and other architectural techniques for energy efficient programmable computation, *IEEE Transactions on Very Large Scale Integration Systems*, 4(1), pp. 42–55, 1996.

[SD95] C.-L. Su and A.M. Despain, Cache Design Trade-offs for Power and Performance Optimization: A case study, in Proceedings of the International Symposium on Low Power Electronics and Design, 1995.

[SFL98] J. Sjodin, B. Froderberg, and T. Lindgren, Allocation of Global Data Objects in On-Chip RAM, in Proceedings of the Workshop on Compiler and Architectural Support for Embedded Computer Systems, 1998.

[SK00] D. Sylvester and K. Keutzer, A global wiring paradigm for deep submicron design, *IEEE Transactions on Computer Aided Design of Integrated Circuits and Systems*, 19(2), pp. 242–252, 2000.

[SR00] M.S. Schlansker and B.R. Rau, EPIC: An Architecture for Instruction-Level Parallel Processors, Hewlwtt Packard Laboratories, Technical report HPL-99-111, 2000.

[SWLM02] S. Steinke, L. Wehmeyer, B.-S. Lee, and P. Marwedel, Assigning Program and Data Objects to Scratchpad for Energy Reduction, in Proceedings of the Design Automation and Test in Europe, 2002.

[TGTS98] S.A. Theoharis, C.E. Goutis, G. Theodoridis, and D. Soudris, Accurate Data Path Models For RT-Level Power Estimation, in Proceedings of the International Workshop on Power and Timing Modeling, Optimization and Simulation, 1998.

[TI] http://www.ti.com/.

[TMW94] V. Tiwari, S. Malik, and A. Wolfe, Power analysis of embedded software: a first step towards software power minimization, *IEEE Transactions on Very Large Scale Integration Systems*, 2(4), pp. 437–445, 1994.

[TRA] http://www.transmeta.com/.

[TRI] http://www.trimedia.com/.

[UAK+03] O.S. Unsal, R. Ashok, I. Koren, C.M. Krishna, and C.A. Moritz, Cool cache: a compiler-enabled energy efficient data caching framework for embedded/multimedia processors, *ACM Transactions on Embedded Computing Systems, (special issue on power-Aware Embedded Computing)*, 2(3), pp. 373–392, 2003.

[UWK+01] O.S. Unsal, Z. Wang, I. Koren, C.M. Krishna, and C.A. Moritz. On Memory Behavior of Scalars in Embedded Multimedia Systems, in Proceedings of the Workshop on Memory Performance Issues, 2001.

[VM01] A.S. Vincentelli and G. Martin, A vision for embedded systems: platform-based design and software methodology, *IEEE Design and Test of Computers*, 18(6), pp. 23–33, 2001.

[WC92] A. Wolfe and A. Chanin, Executing Compressed Programs on an Embedded RISC Architecture, in Proceedings of the International Symposium on Microarchitecture, 1992.

[WHK91] D.A. Wood, M.D. Hill, and R.E. Kessler, A Model for Estimating Trace-Sample Miss Ratios, in Proceedings of the SIGMETRICS Conference on Measurement and Modeling of Computer Systems, 1991.

[WJ96] S.J.E. Wilton and N.M. Jouppi, CACTI: An Enhanced Cache Access and Cycle Time Model, Technical report, Digital Equipment Corporation, Western Research Lab, 1996.

[WL91] M.E. Wolf and M. Lam, A Data Locality Optimizing Algorithm, in Proceedings of the Conference on Programming Language Design and Implementation, 1991.

[Wol95] M.J. Wolfe, *High Performance Compilers for Parallel Computing*, Addison-Wesley Publishers, Reading MA, 1995.

[WWDS94] M. Weiser, B. Welch, A.J. Demers, and S. Shenker, Scheduling for Reduced CPU Energy, in Proceedings of the Symposium on Operating Systems Design and Implementation, 1994.

[YPF+01] S.-H. Yang, M.D. Powell, B. Falsafi, K. Roy, and T.N. Vijaykumar, An Integrated Circuit/Architecture Approach to Reducing Leakage in Deep-Submicron High-Performance I-Caches, in In Proceedings of the High-Performance Computer Architecture, 2001.

[YVKI00] W. Ye, N. Vijaykrishnan, M. Kandemir, and M.J. Irwin, The Design and Use of Sim-plepower: A Cycle-Accurate Energy Estimation Tool, in Proceedings of the Design Automation Conference, 2000.

[ZLF00] D.F. Zucker, R.B. Lee, and M.J. Flynn, Hardware and software cache prefetching techniques for MPEG benchmarks, *IEEE Transactions on Circuits and Systems for Video Technology*, 10(5), pp. 782–796, 2000.

[ZPS+03] Y. Zhang, D. Parikh, K. Sankaranarayanan, K. Skadron, and M. Stan, HotLeakage: A Temperature-Aware Model of Subthreshold and Gate Leakage for Architects, Technical report, Deptartment of Computer Science, University of Virginia, 2003.

[ZTRC01] H. Zhou, M.C. Toburen, E. Rotenberg, and T.M. Conte, Adaptive Mode Control: A Static-Power-Efficient Cache Design, in Proceedings of the International Conference on Parallel Architectures and Compilation Techniques, 2001.

90

HTTP Digest Authentication — Theory and Practice

Thomas P. von Hoff
ABB Switzerland Ltd.

Mario Crevatin
ABB Switzerland Ltd.

90.1 Introduction

Motivation

The application area of the Hypertext Transfer Protocol (HTTP) is becoming larger and larger. While it has been known as the protocol to transfer HTML files originally, it is being used more and more by other applications in order to pass data through a firewall. This is due to the fact that the port related to HTTP, is almost never blocked by a firewall. In other words, running an application on top of HTTP allows to by-pass network security elements like packet filters. Examples for such applications are web mail and Web-based Distributed Authoring and Versioning (WebDAV) [3,4]. Since these web services contain no security features in their standards, they depend on the security provided by HTTP and/or lower protocol layers. While most implementations of protocols below HTTP lack user authentication, this particular feature is provided by extensions to HTTP: Basic and Digest Access Authentication [6].

There is a tendency in industrial communication to replace proprietary communication protocols by the standardized TCP/IP protocol stack [8]. This involves an increase of connectivity in plant networks opening new opportunities to improve the efficiency of maintaining and running systems of automation devices. In the course of this development, the number of embedded web servers has increased rapidly. These web servers allow web-based configuring, control, and monitoring of devices and industrial processes. Due to the convergence of the communication networks of the different layers (control network, local area network, wide area network), establishing an access to smallest processors from any place in the plant or even the world becomes technically feasible. However, besides many opportunities, this technology involves many security challenges [1].

Usually embedded web servers are run on processors with limited resources, both in terms of memory and processor power. These restrictions favor the deployment of lightweight security mechanisms. Different vendors offer tailored versions of the comprehensive security protocol suites Secure Sockets Layer (SSL) and IP Security Protocol (IPSec). However, these versions are still not suitable for all types of processors and applications due to their consumption of memory and computational power. In the case where the range of applications is restricted to HTTP, Digest Access Authentication (DAA) is an alternative solution. This protocol extension to HTTP is economical with respect to memory and processor power requirements. Although designed for user authentication in particular, much more functionality has been foreseen in its original definition. In this chapter, we focus on the mechanisms and functionalities as well as the potential of HTTP Digest Authentication.

Security Objectives

We distinguish the following security objectives for communication systems:

- *Confidentiality*: Guarantee that information is shared only among authorized persons *or* organizations. Encryption of the transmitted data using cryptography makes unauthorized disclosure ineffective.
- *Integrity*: A system protects the integrity of data if it makes unauthorized modifications detectable. This can be achieved by adding a cryptographic check sum.
- *Authenticity:* Guarantee that a receiver of a message can ascertain its origin and that an intruder cannot masquerade as an authorized person. Authenticity is a prerequisite for access control.
- *Access control:* Guarantee that only authorized people or devices have access to specific information.
- *Availability*: Guarantee that a peer is always able to communicate.

In a business-oriented sense of security, auditibility, nonreputability, and third-party protection also belong to the set of security objectives. Note that the relevance of the individual security objectives varies from case to case and depends considerably on the specific application. For example, a business web application where potentially a monetary transaction is involved has security requirements different from an industrial application. While for the former application, confidentiality of the data transfer is a major issue, this is less sensitive in the latter case. In turn, other security objectives such as user authentication and integrity protection are much more critical in industrial communication. In other words, the protection of secure user authentication and of the integrity of the data has a higher priority than the confidentiality. This consideration becomes an issue in particular, when the embedded web server is not integrated in a well-protected network, but stands at a remote place. Such situations may occur in distributed applications.

Outline

First, an overview about the functionality of the security extensions in the TCP/IP protocol suite is given with a focus on SSL and IPSec. Starting with a brief review of the HTTP message exchange, the mechanisms of HTTP Basic and Digest Authentication are detailed and all its additional valuable options (integrity protection and mutual authentication) are discussed. Furthermore, the current implementation status of different (embedded) web servers (Apache 2.0.42, Allegro Rom Pager 4.05, GoAhead 2.1.2) and browsers (Mozilla 1.01, Internet Explorer 6.0.26, Opera 6.05) is investigated. The results of functionality and interoperability tests are presented.

90.2 Security Extensions in the TCP/IP Stack

Security services are provided at different layers in the TCP/IP communication protocol suite by protocol extensions [10]. An overview of these extensions is depicted in Figure 90.1. Due to the nature of the communication protocol stack, the security services are transparent to upper-layer protocols. Note that the security extensions on the Internet layer and the transport layer, IPSec and SSL, respectively, provide a large range of security services and, therefore, are widely implemented.

Link-Layer Security

As extensions to the Point-to-Point Protocol (PPP), the cryptographically weak Password Authentication Protocol (PAP) and the stronger Challenge Handshake Authentication Protocol (CHAP) provide authentication. To establish tunnels starting with a PPP connection into a LAN or a WAN, Point-to-Point Tunnel Protocol (PPTP) or the Layer 2 Tunnel Protocol (L2TP) can be used.

IPSec

This network-layer security protocol is particularly interesting if several network applications need to be secured. As protection is applied at the IP layer, IPSec provides a single means of protection for all the data exchanged (UDP and TCP applications). It is transparent to all upper layers. The security services provided by IPSec are:

- Access control (IP filtering).
- Data integrity.
- Encryption (optional).
- Data origin authentication (optional).

All these services are based on cryptographic mechanisms giving them a high security level when they are used with strong algorithms. However, it is a drawback of IPSec that for each target network a nontrivial, specific configuration is required. While IPSec provides machine-to-machine security, it cannot perform authentication of the user. Therefore, IPSec is mainly deployed to establish Virtual Private Networks (VPN).

A tested IPSec implementation on a Coldfire MCF5307 at 65 MHz showed a memory requirement of 64 kByte where the Internet Key Exchange (IKE) Protocol was not part of the implementation. Experiments consisting of ping requests between two Coldfires were executed. The time between a ping request and the reception of its reply was observed to become twice or even three times longer when IPSec was activated using the Authentication Header (AH) or the Encapsulation Security Payload (ESP) configuration, respectively.

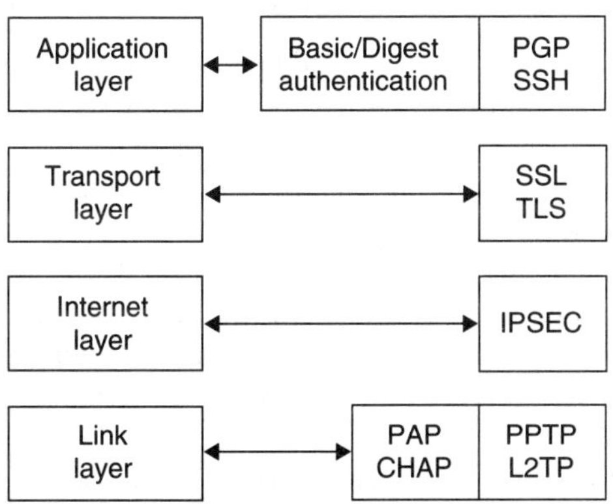

FIGURE 90.1 Network layers and respective security protocols.

SSL/TLS

The SSL is a protocol created by Netscape Communications Corporation. The standardized version is also known as Transport Layer Security (TLS). SSL is invisible to the end user and transparent to upper layers. It protects all applications running on top of TCP, but does not protect UDP applications. The "https" prefix on the URI and the lock icon indicate that the SSL protocol is in use. If the server's certificate is not signed by a certificate authority known to the client, the user has the choice to accept or to refuse the certificate. The security services provided by SSL/TLS are:

- Session key management and negotiation of cryptographic algorithms.
- Confidentiality by means of encryption.
- Strong server authentication using certificates.
- Data integrity protection.

Note that SSL is supposed to provide optional client authentication. However, client authentication is rarely performed by SSL in practice. Under the protection of the encryption provided by SSL, user authentication is often implemented at the application level. All in all, SSL provides a high level of security, but requires considerable memory and computation power, particularly in the dimensions of embedded web servers.

Application-Layer Security

With respect to authentication, the procedures mentioned in the previous sections work on lower layers and focus on the identity of machines.

On the application layer, the individual applications provide own security enhancements. Typical security tools are PGP/GnuPG to secure mail transfer and SSH (secure shell). With respect to HTTP, there exist the protocol extensions HTTP basic and digest authentication, which authenticate users to control their access to protected files. Authentication of the user stands in contrast to machine authentication provided by the tools detailed in the sections above. Since the focus is set on the protocol extensions of HTTP in the rest of the article, a brief review of HTTP is given in the Appendix.

90.3 Basic Access Authentication Scheme

The HTTP Basic Authentication scheme [6] is the simplest authentication scheme and provides little protection. This is due to the fact that username and password can be discovered when the message exchange is sniffed. The HTTP message exchange for basic authentication is depicted in Figure 90.2. On receipt of a "401 unauthorized" message, the browser prompts the user for its username and its password (see Figure 90.3). These are transmitted in the clear over the wire in the "Authorization" request-header-field for each accessed document within the same protection space.

1. The browser issues an HTTP GET command to the server, with the requested URI.
2. The server answers with a 401 unauthorized HTTP error code and requests the browser to send a valid username and password (credentials[1]) using basic authentication. The realm[2] (string) is also included in the challenge sent to the client. All this information can be found in the "WWW-authenticate" request-header-field.
3. The browser prompts the user for username and password. The realm (here "Basic Test Zone") is usually shown to the user. The credentials encoded in Base64 are sent with a new GET request. Decoding is trivial because Base64 is a simply invertible encoding scheme not based on any cryptography. The credentials are sent in the "Authorization" response-header-field.

---

[1]Credentials: Information that can be used to establish the identity of an entity. Credentials include such things as private keys, tickets, or simply a username and password pair. This is also known as the shared secret.

[2]Realm: Name identifying a protection space (zone) on a server. Usually shown to the user at password prompt.

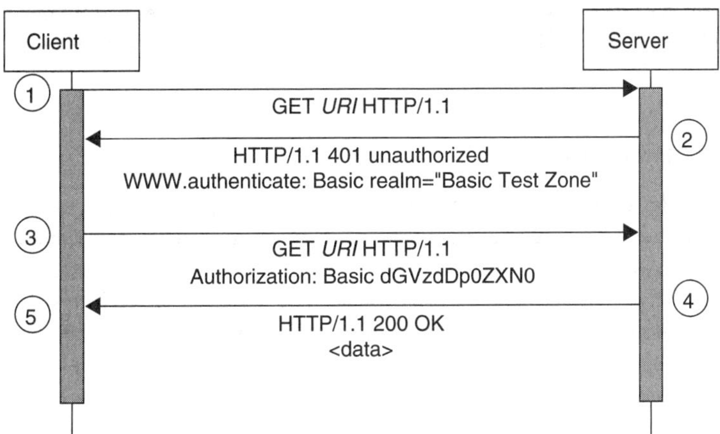

FIGURE 90.2   HTTP basic authentication negotiation.

4. After the server has verified and accepted the password, the page is sent back to the client with an HTTP 200 response.
5. The client displays the page, and automatically sends the same credentials for any subsequent request made under the same protection space. This means that the password is sent in clear with each request.

Note that unless confidentiality is provided by means of other security protocols on a lower layer (see Section 90.2), username and password are transmitted in an unprotected way.

90.4   Digest Access Authentication Scheme

Cryptographical Prerequisites

Unless public keys are used, authentication is based on a shared secret (credentials) between the authenticating and authenticated entity. Usually, these credentials consist of the relation between a username and its password. One possibility to authenticate a peer over the network is the submission of username and password as executed in the Basic Authentication scheme (see Section 90.3). However, since they are submitted in clear over the network, someone who has access to the network traffic can eavesdrop the credentials. The principle of challenge and response overcomes this drawback while avoiding sending the password in clear. Instead, the authenticating entity A sends x to the authenticated entity B. B calculates $z_B = f(x, y)$, where y is the shared secret between A and B. A calculates $z_A = f(x, y)$ as well and checks whether z_A coincides with z_B. If so, the identity of B is proven to A.

To make the procedure of challenge and response secure, there are two requirements. First, x needs to be unique so that z_B is of no value if someone intercepts it. Second, f has to be a one-way (hash) function. The properties of a hash function are:

- a finite-length output message (hash) is calculated from an arbitrary-length input message,
- it is easy to determine the output message,
- given an output message, it is hard to find a corresponding input message, and
- it is hard to find another input message with the same output message.

Functions with all the features mentioned above meet the requirements for a method to calculate cryptographic check sums. It is their quasiuniqueness (hard to find two input messages with the same output) that allows an owner B of a message z to show only the message's hash to another owner A and A can still trust that really z is owned by B. The same property is used to protect the integrity either in data storage or message transmission. A comparison of the hash of the received message or the stored data, respectively, with the original hash covers any change in the message or the data, provided the hash was stored or transmitted in a secure way. The most frequently used hash functions are MD5 and SHA-1 [9].

Digest Authentication

Although very similar to the basic authentication scheme, DAA [6] is much more secure (Figure 90.4). Instead of sending the username and the password in an unprotected way, a unique code is calculated from username, password, and a unique number received from the server. Figure 90.5 shows the HTTP DAA transactions between a web server and a browser. The "cnonce," "nonce," "responses," and "rspauth" parameter values have been shrunk for the sake of readibility of the chart.

1. The browser requests a document in the usual way, with an HTTP request message.
2. The server sends back a "401 unauthorized" challenge response message. The server generates a so-called nonce (number used once) and sends it to the client. Note that the nonce is different for every 401 message.
3. The browser prompts the user for its username and its password, and computes a 128-bit response using the MD5 algorithm [9] as a one-way (hash-) function:

   response=MD5[MD5(username:password):nonce:nc:cnonce:qop:MD5(method:uri)]

   This response is sent to the server along with the nonce received, the uri requested, its own generated cnonce (client nonce), and the username. Note that qop stands for Quality of Protection and indicates whether additional integrity protection is provided. Here, this is not the case; hence, qop="auth."
4. The server calculates an own response following the same scheme as given in Step 3, where the information sent to the client before and its own version of the username and password (or optionally a hashed form of them) are used. It compares the received version with the computed one and grants access to the resource (HTTP 200 OK response) if appropriate.

   If the authorization fails, a new 401 error message is sent. All the 401 error messages include an HTML error page to be displayed by the browser. Browsers usually reprompt the user for a new username and password three times before giving up and displaying the error page.

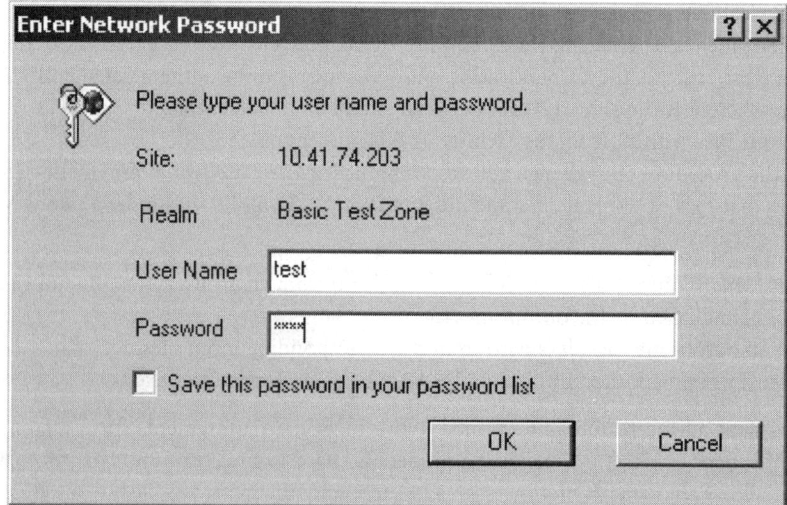

FIGURE 90.3   Internet Explorer's basic authentication prompt.

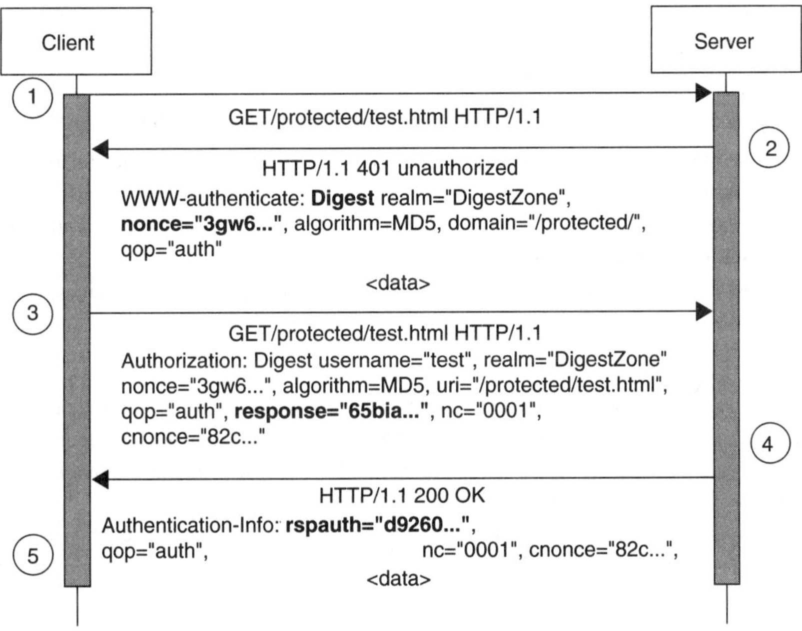

FIGURE 90.4 Internet Explorer's digest authentication prompt.

FIGURE 90.5 HTTP digest authentication: negotiation.

5. For any subsequent request, the client usually generates a different cnonce. A counter, nc, is incremented. This new cnonce and counter, along with the new uri is used to recompute a new valid response value. Usually, the browser stores the username and the password temporarily in the memory in order to enable a user to reaccess a given protection space without further typing.

Digest Authentication with Integrity Protection

The RFC for digest authentication [6] provides a way to include the hash of the entity (the payload, usually HTML code) in the computed MD5-hash response:

response = MD5[MD5(username:realm:password):nonce:nc:cnonce:qop:MD5(method:uri:MD5{entity})]

In that way, any modification of the transmitted information will result in a different MD5-hash response, easily detectable. While in a response message, the integrity of the served document is insured, the integrity of POST data is protected in a request message. To indicate that the server supports integrity protection, the argument Quality Of Protection is set to "auth-int."

Note: for GET requests with arguments, the integrity of the payload (the arguments) is already protected without the option qop=auth-int, because the URI with its arguments is included in the MD5-hash. For integrity protection of the response from the server the rspauth field must be present. See the next section on mutual authentication.

Digest Authentication with Mutual Authentication

We have already seen that digest authentication identifies the client. However, DAA also foresees the ability that the client checks whether the server is trustworthy, providing mutual authentication. The server already knows that the client is trustworthy, because the browser has sent the proof that the user knows their shared secret. This happened by sending the response to the server (see Step 3 in Figure 90.5). It is exactly the same mechanism used to authenticate the server. After receiving a correct credential along with the GET request, the server sends back a proof that it also knows the shared secret. This is done via the "rspauth" field sent in the "Authentication-Info" header of the HTTP 200 OK message, along with the document previously requested. The challenge initiating the server response is the cnonce from the client.

$$rspauth = MD5[MD5(username:realm:password):nonce:nc:cnonce:qop:MD5(:uri)]$$

The browser uses this information to authenticate the server. If integrity protection is activated, the hash for the entity is included in rspauth.

$$rspauth = MD5[MD5(username:realm:password):nonce:nc:cnonce:qop:MD5(:uri:MD5\{entity\})]^3$$

The nonce from the server is used to challenge the client with data that cannot be anticipated. In the same way, when server authentication is used, leading to mutual authentication with DAA, the cnonce from the client is used as a challenge that the server cannot anticipate. Therefore, it is not possible to precompute responses to those challenges.

This is known as a mutual challenge/response mechanism. Its concept is pictorially depicted in Figure 90.6.

Summary

DAA offers a secure, yet light authentication scheme, perfectly adapted to the embedded systems. In this environment, authentication and integrity protection is important, whereas confidentiality is often not an issue. Unfortunately, integrity protection and mutual authentication have not been supported by today's typical client/server implementations so far.

Table 90.1 compares DAA and SSL's mandatory and optional features as well as their memory requirements. Note that at the time of the tests, the *optional* features were supported by some implementations only.

90.5　Weaknesses and Attacks

Basic Authentication

If an attacker has access to the network link media, he can eavesdrop the HTTP transaction very easily and get the username and the password if there is no underlying encryption as provided by SSL or IPSec. In those cases, HTTP Basic Authentication should be replaced by the Digest scheme.

Replay Attacks

Security threats appear when data, for example, sensitive commands, are sent to the server. With POST and some other HTTP method, data are transmitted in the entity of the message. An attacker could replay

---

[3]The only difference from the client response is the missing field "method."

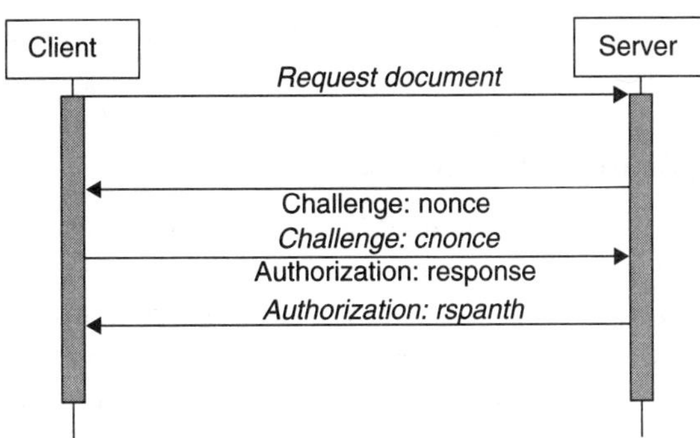

FIGURE 90.6　Mutual authentication mechanism.

TABLE 90.1　Comparison between DAA and SSL on Functionality and Footprint Size

	DAA	SSL
Mandatory Features		
Client authentication	✓	x
Server authentication	x	✓
Data integrity	x	✓
Data confidentiality	x	✓
Optional Features		
Client authentication	✓	✓
Server authentication	✓	✓
Data integrity	✓	✓
Data confidentiality	x	✓
Memory Requirements		
RAM	1648 Byte[a]	100kB 3-250kB[b]
ROM	6312 Byte[a]	200kB[c]

[a] According to own measurements on a Coldfire MCF5307.
[b] RAM required depending on the number of simultaneous secure connections (1 to 24) according to [12].
[c] According to information from Allegro.

valid credentials from a successfully intercepted POST request with modified form data (commands) thus taking control of the remotely controlled system. In contrast, the arguments sent in a GET request are part of the arguments to compute the digest. Thus, GET requests are safer and must be used in priority to send data to the server. The fact that the data are stored in the browser's cache using the GET method has no effect to the security, because of the uniqueness of the nonce.

Proper nonce generation together with a reliable check for uniqueness provides good protection against replay of previously used valid credentials. Although the definition of a nonce requires its uniqueness, implementors might be tempted to reuse a nonce. This must be avoided in any case. Within a session, a replay attack is prevented by the increase of the counter "nc," making the previously calculated hash value invalid. Note that the change of 1 bit in the argument of a hash function changes half of the bits of the hash value [9]. Furthermore, integrity protection would prevent from tampering information when POST is used.

On an embedded device, where usually only one legitimate user at a time is likely to access the system, locking the protection space (realm) to only one client at a time contributes to protection as well.

Man-in-the-Middle Attack

An attacker might be able to insert himself into the routing path between a client and a server. Capturing the packet containing the server's response with the challenge for digest authentication, he can replace the WWW-Authenticate field with that asking the client for Basic Authentication: Username and password can be gained because the client returns them in an unprotected form due to the faked WWW-Authenticate field.

Disabling of Basic Authentication in the browser and requiring mutual authentication could prevent such an attack.

Dictionary Attack/Brute Force Attack

This attack assumes that the user chooses a simple password. In the client's request message (Step 3), all information to calculate the response field apart from the password is available. Having such a message in his hands, an attacker might compute thousands of responses generated with a list of possible passwords from a dictionary and see if it coincides with the response sent by the browser. However, the success of a dictionary attack can be avoided by proper selection of the password, for example, avoid common words and include small and capital letters as well as special characters.

For a brute force attack the list of possible passwords is replaced by any combination of characters. However, the fact that a hash must be calculated for each password guess makes a brute force attack very cumbersome. Passwords with more than eight characters, small and capital letters, as well as special characters should be used. Use of common words should be avoided.

Buffer Overflow

In general, a buffer overflow attack occurs when an attacker tries to store too much information in a very small memory space [7]. With respect to embedded web servers, web pages often serve as a portal for CGI programs. Input data for such a program are transmitted in the HTTP message. If the size of such data is too large, it can be written in the memory over the executable code of the browser/server and can be executed by the operating system, unless a careful error check is performed. Mostly, this will crash the server, resulting in a Denial of Service. However, it can also allow a skilled attacker to gain access to everything the server has access to, for example, all confidential information and the code controlling the devices in the automation system.

Note that buffer overflow can be avoided when in all applications and in particular in those called by CGI requests careful data range validation checks are performed. Additionally, for integrated third-party code, it is highly recommended to update the code to the latest available version and to apply the security patches or the service packs as soon as they are released.

URI Check

Some web servers do not verify whether the "uri" field located in the "Authorization" header corresponds to the requested URI in the GET request. An attacker may replay a stored valid request from a client for a given "uri," but may modify the GET line to get some other protected documents he wants or, even worse, the server may accept modified parameters in a GET request. This might allow an attacker to send arbitrary commands to an embedded device. Details on uri check implementations are given in the next section.

90.6   Implementations

The available embedded web server implementations and browsers have been tested for their support of DAA. Investigations have shown that the implementations do not match the definition in every aspect. This section briefly outlines the results. A comparison of some server implementations is given in Table 90.2.

Servers

(1) *Apache 2.0.42*: Among the DAA implementations tested, Apache has the best one. While mutual authentication is in place and working, integrity protection is not yet implemented. The nonce lifetime is adjustable and the uri is checked. Apache is also the most robust server in terms of exploits, because a large community uses it daily. Note that DAA is compatible with the Opera browser only if the AuthDigestDomain directive is configured in the file htaccess. The Internet Explorer (IE) removes the

TABLE 90.2   Comparison between DAA Server and Implementation

| | Server | | |
Feature	Apache	RomPager	GoAhead
Basic Access Authentication	✓	✓	✓
Digest Access Authentication	✓	✓	✓
mutual authentication	✓	x	x
integrity protection	x	x	x
nonce check	✓	✓	x
uri check	✓	✓	x

arguments when copying the GET query in the uri field. Therefore, DAA does not work for GET requests with parameters between IE and Apache. The nonce is composed of a timestamp (in the clear), followed by the "=" sign, and the SHA-1 hash [9] of the previous timestamp, realm, and a server secret (private key):

$$nonce = timestamp:"=":SHA1(timestamp+secret)$$

(2) *Allegro RomPager 4.05*: In the RomPager web server [11], DAA is implemented without mutual authentication and integrity protection. It is known for its option "StrictDigestAndIp," that is, the validity of the unique nonce is time-limited and never more than one IP address are granted access simultaneously. This feature makes sense in embedded systems because it prevents replay attacks.

RomPager makes a full uri check. In addition, it is implemented to be able to cope with the less secure uri behavior of the IE.

RomPager assumes that requests received via HTTP 1.0 come from a browser not supporting digest. As a consequence, DAA does not work with IE and Opera when being connected via a proxy. Therefore, Opera needs to be configured without proxies. On IE, the "Use HTTP 1.1 through proxy connections" option can be set. The nonce is generated using the time, the server IP address, the previous nonce (if there was one), and the server name.

$$nonce = MD5(Time:Server-IP-Address: [previous nonce:] Server-Name)$$

(3) *GoAhead 2.1.2*: GoAhead is a free software and open source server, developed for embedded devices on a variety of platforms. As for RomPager, no mutual authentication and integrity protection is supported. A given nonce never expires and is never checked. Hence, there is no protection from replay attacks. The server removes the parameters of a GET query request in the digest "uri" field. This causes Mozilla not to work for those types of requests.

$$nonce = MD5(RANDOMKEY: timestamp:myrealm)$$

Browsers

It is difficult to select a browser among the three tested here. They all have their strengths and weaknesses. Internet Explorer is the only one using a different prompt for Basic and Digest Authentication (see Figures 90.3 and 90.4). Being aware of this feature, an attentive user can recognize a Man-in-the-Attack . However, the IE removes GET arguments in the uri. Mozilla is an open-source browser, and thus, very easy to modify. On the other hand, it is not very user-friendly at the time (slow and continually asking for the username and password). Opera is the pickier one in terms of security and the only one supporting mutual authentication. With respect to DAA, a combination of the three products mentioned above would probably meet the features expected from a perfect DAA implementation. These are as follows:

- Option to disable Basic Authentication.
- The user shall be notified by a visual clue that DAA is used, when he is prompted for username/password as well as during browsing.
- Support of DAA with mutual authentication with a visual clue that the server has been authenticated. Possibility to refuse pages served with DAA if the server has not been authenticated.

- Support of DAA with integrity protection, with a visual clue that it is used.
- Verification that the URI requested with DAA is in the protection space.

DAA Compatibility

Table 90.3 summarizes the compatibility of different clients vs. different servers. Note that the compatibility tests did not include server authentication and data integrity protection.

90.7 Conclusions

DAA is a lightweight, yet efficient way of providing user authentication. Applications running on top of HTTP can benefit from DAA's features. Typically, these are web services like WebDAV and HMI applications in automation systems migrating from proprietary communication protocols toward TCP/IP technology. Wherever Basic Authentication is still in use and not protected by a security protocol of a lower layer, it should be replaced by DAA. From the point of view of the embedded community, it is highly desirable that browser and web server vendors implement mutual authentication and integrity protection, because only together they achieve a high security level. In situations where encryption is not required, an implementation of DAA including all features proposed by the RFC [6], namely mutual authentication and integrity protection, would serve as a sufficient, yet lightweight, security mechanism for embedded systems.

Appendix: A Brief Review of the HTTP

HTTP is widely used to exchange text data across different platforms on a TCP/IP network. The definition of HTTP 1.1 is stated in [5]. HTTP is based on standard request/response messages transmitted between a client (browser) and a web server. An example of a typical HTTP handshake is depicted in Figure 90.7. The procedure is straightforward:

1. The browser sends a GET request to the server, indicating the requested resource.
2. The server responds with a 200 OK message along with the document requested.

Using Figure 90.7, the key aspects of HTTP are briefly explained. An HTTP message consists of a header and in most cases an entity. Note while all the data in the header are transferred as ASCII text, the entity might contain non-ASCII data, for example, .jpg-files.

Header
A Header is composed of a Start-Line and Header-Fields.

Start-Line
There are two types of start-lines:
A request start-line is of the form Method URI HTTP-Version, for example,
GET /simple.html HTTP/1.1.

TABLE 90.3 DAA: Client/Server Compatibility

	Clients		
Servers	Mozilla 1.01 Netscape 7.0	E 6.0.26	Opera 6.05
Apache 2.0.42 (win32)	✓	✓[a]	✓[b]
RomPager 4.05	✓	✓	✓
GoAhead 2.1.2	✓[b]	✓	x

[a] Not working for GET with parameters.
[b] Requires valid domain.

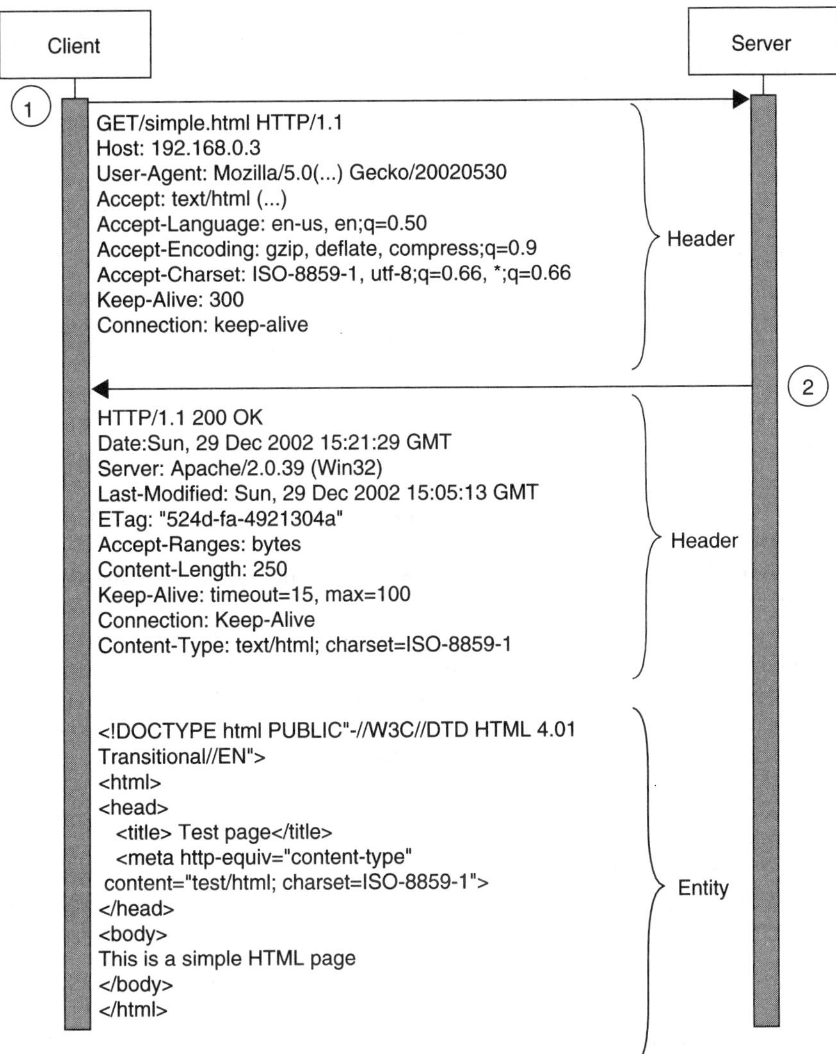

FIGURE 90.7 Example of an HTTP message exchange.

A status start-line has the form HTTP-Version Status-Code Phrase, for example, HTTP/1.1 200 OK.

Header-Fields

Header-Fields give various information, including date, language, and security information. In this document, the security-relevant header fields are discussed in the main part. Examples:

Host: 192.168.0.3

Date: Sun, 29 Dec 2002 15:21:29 GMT

WWW-authenticate: Digest realm="abc",...

Method

The most relevant methods are GET and POST.

GET: The most common method to request a document. The GET request can also include arguments in the URI. This is widely used to transmit commands from the client to the server.

POST: Method used to send information to the server, usually from a web page form.

URI
Uniform Resource Identifiers [2]. URIs in HTTP can be represented in absolute form or relative to some known base URI. Example:
http://www.test.ch/simple.html or /simple.html.

Entity
The rest of the message, for example, an HTML document.

Acknowledgment

The authors would like to thank Emanuel Corthay from the Swiss Federal institute of Technology Lausanne for his valuable contribution in the examination of the individual implementations.

References

[1] Anderson, R.J., *Security Engineering: A Guide to Building Dependable Distributed Systems*, Wiley, New York, 2001.

[2] Berners-Lee, T., R. Fielding, and L. Masinter, Uniform Resource Identifier (URI), RFC 2396, August 1998.

[3] Requirements for a Distributed Authoring and Versioning Protocol for the World Wide Web, RFC2291, 1998.

[4] HTTP Extensions for Distributed Authoring — WebDAV, RFC 2518, 1999.

[5] Fielding, R., J. Gettys, J. Mogul, H. Frystyk, L Masinter, P. Leach, and T. Berners-Lee, Hypertext Transfer Protocol — HTTP/1.1, RFC 2626, June 1999.

[6] Franks, J., P. Hallam-Baker, J. Hostetler, S. Lawrence, P. Leach, A. Luotonen, and L. Stewart, HTTP Authentication: Basic and Digest Access Authentication, RFC 2617, June 1999.

[7] Cole, E., *Hackers Beware*, New Riders, Indianapolis, 2002.

[8] Naedele, M., IT security for automation systems: motivations and mechanisms, *Automatisierungstechnische Praxis*, Vol. 45, no. 5, pp. 84–91, 2003.

[9] Schneier, B., *Applied Cryptography: Protocols, Algorithms, and Source Code in C*, John Wiley & Sons, New York, 1996.

[10] Stallings, W., *Network Security Essentials: Applications and Standards*, Prentice-Hall, Englewood Cliffs, NJ, 2000.

[11] Rom Pager Web Server Engine Porting & Configuration, Allegro Software Development Corporation, Boxborough, MA, U. S., 2000.

[12] Rom Pager Secure Programming Reference Version 4.20, Allegro Software Development Corporation, Boxborough, MA, U.S., 2002.

91

Security in Embedded Systems

Andreas Steffen
Zuercher Hochschule Winterthur

91.1   Introduction

In an increasing number of distributed embedded applications, the individual nodes must communicate with each other over insecure channels like, for example, the public Internet or via wireless communication links. In order to withstand malevolent attacks, the end-to-end communication channels must be secured using cryptographically strong encryption and authentication algorithms. Fortunately, powerful secure communication protocols exist today and most of them are readily available as OpenSource software implementations. The biggest obstacle to their widespread use in distributed embedded systems is the limited memory and the restricted processing power available from today's low-cost and mostly low-power microcontroller platforms. In this chapter, we are going to show some promising approaches regarding how security can be brought to embedded systems running time-critical applications.

This chapter on secure communications in embedded systems will be structured as follows:

- Section 91.2 of this chapter will give an overview of the secure communication solutions available for distributed embedded systems.
- Section 91.3 treats two of the key problems that have to be dealt with when implementing cryptagraphically strong security in embedded systems. One problem is the generation of random numbers used for keying material and the other problem is the efficient handling of the computationally expensive public key operations required in most of the modern security protocols.
- Section 91.4 gives a typical example of a secure communications solution for a low-end embedded platform. It is shown that an SSL/TLS-enabled Web server can be implemented with a small memory foot print.
- Section 91.5 concludes this chapter by summarizing the relevant facts.

91.2 Security in the OSI Communications Stack

Security can be implemented at different levels of the communications stack. The well-known OSI communications model defines a physical layer at the very bottom and goes up via the data link, network, and transport layers up to the topmost application layer (see Figure 91.1). The OSI presentation and session layers have been omitted since they are not relevant in our context. In the following paragraphs we will briefly describe how security can be applied at each layer.

Physical Layer Security

At the low end of the communications stack, the physical layer does not offer much protection in a strict cryptographical sense. Often, the binary payload bits are scrambled with a fixed pseudorandom code, primarily in order to avoid the occurrence of long runs of ones or zeroes and some wireless channels like, for example, Bluetooth use fast frequency hopping in order to increase the physical robustness of the link. Although these measures offer some protection from casual eavesdropping, they do not withstand any serious attempts at intercepting the communication traffic.

Data Link Layer Security

Most wireless communication systems like, for example, IEEE 802.11 WLAN, Bluetooth, or GSM use authentication and encryption at the data link layer in order to protect the data transmission over the vulnerable air interface. Unfortunately, the Wireline Equivalent Privacy (WEP) encryption protocol of the IEEE 802.11 Wireless LAN standard is known to possess several serious deficiencies that make transmission highly insecure [1, 2]. The same is true for the A5 encryption cipher used by the GSM global cellular phone system that can be cracked by a simple PC in real time [3]. Thus, if distributed embedded systems are communicating either over WLAN or GSM wireless links, no real security can be expected from the default layer 2 encryption and authentication protocols.

The most popular layer 2 protocol over wired links is Ethernet, which does not possess any inherent security mechanisms. Over asynchronous modem links or synchronous ISDN connections, the payload frames of the Point-to-Point Protocol (PPP) could optionally be encrypted using the PPP Encryption Control Protocol (ECP) [4], but two large vulnerabilities remain: first, PPP frames are not authenticated, making them prone to malicious changes and second, the PPP connection setup based on the auxiliary Link Control Protocol (LCP) is not secured at all, exposing it to man-in-the-middle attacks and session-hijackings.

Generally, it can be summarized that with the current layer 2 protocols in use, the data-link layer is not a suitable location for implementing strong communications security.

Communication layers	Security protocols
Application layer	ssh, S/MIME, PGP, http digest
Transport layer	SSL, TLS, WTLS
Network layer	IPsec
Data link layer	WEP (WLAN), A5 (GSM), PPP
Physical layer	Scrambling, frequency hopping

FIGURE 91.1   Security implemented at different layers of the OSI stack.

Network-Layer Security

The most flexible approach to security can be realized at the network layer since this is the lowest communications layer that allows the implementation of end-to-end security in routed, multihop networks. If we restrict ourselves to the ubiquitous IP network-layer protocol, then the corresponding IP security protocol (IPsec) [5, 6] is the preferred choice. The IPsec protocol comes in two flavors: the Authentication Header (AH) protocol [7] secures IP packets from unwanted modifications by adding a cryptographic check sum, thus guaranteeing data integrity of the payload and most of the IP headers including the source and destination addresses. The Encapsulating Security Payload (ESP) protocol [8] offers strong encryption of IP packets and is usually preferred over AH since optional authentication of the payload data is also available with ESP. Only in the special cases where encryption is not permitted by national law or where it would to put too much of a computational burden on the embedded platforms is AH used. Both ESP and AH are special IP protocols with protocol numbers 50 and 51, respectively. Because these protocols do not have port numbers, they cannot be masqueraded in the same way as UDP or TCP datagrams.

IPsec further differentiates between Transport Mode, which is mainly used for point-to-point connections between two single hosts and Tunnel Mode, which tunnels the traffic between two subnetworks that are protected by a security gateway. Since in embedded systems usually direct host-to-host connections predominate, we will restrict our discussion to the AH and ESP transport mode payloads only, which are shown in Figure 91.2.

The required overhead for ESP protection is shown in more detail in Figure 91.3. The ESP header consists of a 32-bit Security Parameters Index (SPI), which is used by the receiver to look up the session keys plus negotiated cryptographic parameters for the given IPsec connection. Each direction in a bidirectional IPsec connection has an SPI value of its own. The following 32-bit monotonically increasing Sequence Number prevents replay attacks. The ESP trailer is used to pad the payload data to an integer multiple of the block size required by the symmetric block cipher. The standard Triple DES (3DES) encryption algorithm works on 64-bit blocks of data, so that in the worst case, 7 bytes must be padded, whereas the Advanced Encryption Standard (AES), the official successor to 3DES, uses 128-bit data blocks, so that the overhead becomes 15 bytes at the worst. Add to this 2 bytes for the pad length and next header fields. The block encryption algorithms also need an initialization vector (IV) comprising the size of one data block. This means an additional 8 bytes of overhead for 3DES and 16 bytes for AES, respectively. Finally, the cryptographic checksum used for ESP authentication takes up another 12 bytes, resulting to a total overhead of 37 bytes for 3DES encryption and 53 bytes for AES. This is not much if large volumes of data are transferred as for, for example, during an ftp download, but it can have a large impact if many small IP packets are transmitted, for example, when running a telnet session.

The kernel part of an IPsec stack that implements the real-time ESP encapsulation of IP packets can be made quite compact. For distributed systems running an embedded Linux operating system like, for

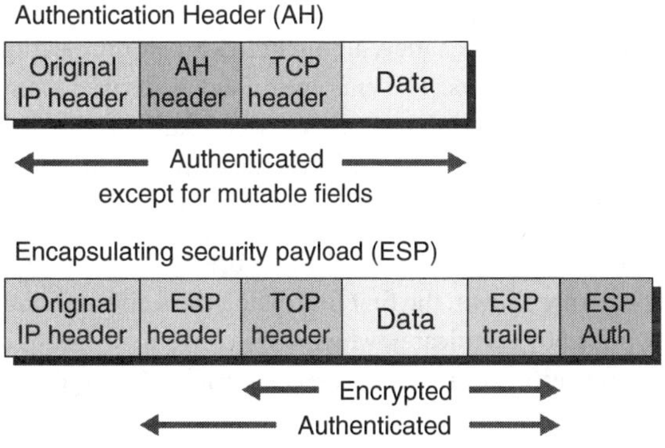

FIGURE 91.2 IPsec Transport Mode.

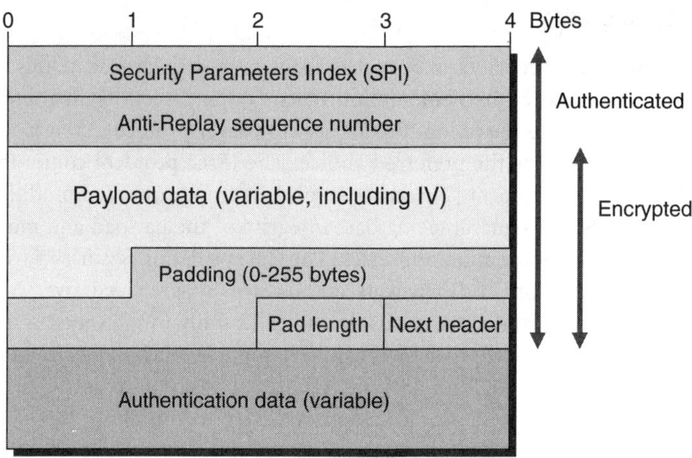

FIGURE 91.3   ESP payload structure.

example, μClinux [9], a free but nevertheless mature IPsec solution is available from the OpenSource Linux FreeS/WAN project [10]. Several commercial IPsec stacks for embedded platforms are also offered on the market. The RAM footprint is usually maintained between 200 and 300 kbytes.

The most serious drawback of IPsec is the complexity involved in setting up protected connections. Although, in principle, the configuration could be done manually by specifying two common SPI values for the two directions plus a set of shared authentication and encryption keys, this procedure would be too error-prone to be feasible in a practical application, especially if frequent rekeying is required. Therefore, an automatic connection setup is preferred using the standardized Internet Key Exchange (IKE) protocol [11], which uses the well-known UDP port 500. IKE is a rather complex protocol comprising a multitude of configuration options. It consists of a Phase 1 called Main Mode where six messages are exchanged in order to establish trust and to compute a common secret key between the two end points. Main Mode is followed by a Quick Mode or Phase 2 negotiation where another three messages are needed to set up the actual connection. A typical Main Mode exchange with mutual authentication based on preshared secrets is shown in Figure 91.4.

- The initiator sends a proposal, listing all supported encryption and authentication algorithms. The responder selects among them one set of common algorithms and sends the selection back to the initiator.
- The second message pair establishes a common secret with the help of a Diffie–Hellman (DH) Key Exchange. In order to prevent replay attacks, a pair of random values N_I and N_R, are also exchanged.
- Starting with the third message exchange, all further IKE communication is encrypted in order to prevent man-in-the-middle attacks. The encryption key is derived from the common DH secret, the random values N_I and N_R, and the preshared secret known only to the two end points. Over the secure channel, an ID string, which establishes the identity and a hash value authenticated with the pre-shared secret, is exchanged.

Another variant of the IKE Main Mode depicted in Figure 91.5 uses RSA public key authentication (only the last two IKE messages are shown, the first four being the same as in Figure 91.4). The use of RSA signatures in connection with X.509 certificates, which are a convenient means of distributing the needed RSA public keys, is the most powerful approach to IPsec, especially if a large number of connections must be maintained.

X.509 certificate support for embedded Linux IPsec solutions is available as a patch [12] developed by the Security Group of the Zurich University of Applied Sciences in Winterthur that can be applied to the

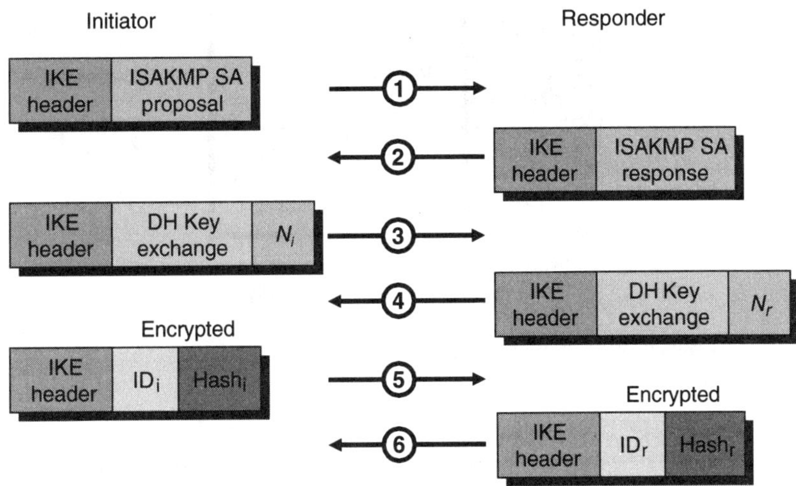

FIGURE 91.4 IKE Main Mode authentication based on preshared secrets.

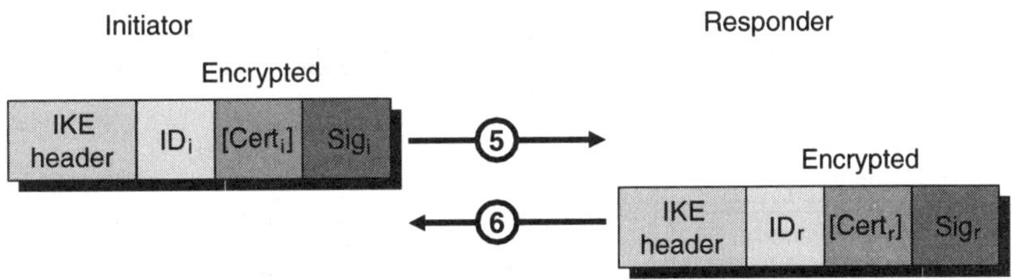

FIGURE 91.5 IKE Main Mode authentication based on X.509 certificates.

FreeS/WAN distribution [10]. Depending on the number of implemented features, an IKE daemon can become quite large. Footprints between 200 kbytes for a very lean client using preshared secrets only and more than 1 Mbytes for a full-fledged IKE implementation are not uncommon.

The complexity of the IKE protocol is the main reason that the use of IPsec is usually restricted to high-end 32-bit embedded platforms that possess the required memory and computational resources.

Transport-Layer Security

The most generic solution at the transport level is the well-known Secure Sockets Layer (SSL) [13, 15], as well as its official IETF successor called Transport Layer Security (TLS) [14, 15] and its lean variant WTLS for wireless applications. The SSL/TLS protocols are layered on top of the reliable TCP transport layer protocol and are typically used in client/server applications, for example, a central controller managing several distributed nodes, but SSL/TLS can also be used for peer-to-peer communication.

By taking a look at Figure 91.6, we see the fundamental difference between network and transport-layer security. Whereas IPsec requires large and potentially dangerous changes in the kernel of the given operating system, SSL/TLS leaves the built-in TCP/IP communications stack untouched because it uses regular TCP sockets. From the application's point of view, IPsec has the major advantage that the interface to the transport layer does not change at all, so that the application program is not even aware of the existence of an additional security layer and any transport protocol (TCP, UDP, ICMP, etc.) can be used. With SSL/TLS, special secure sockets must be opened by the application and TCP is the only transport layer supported. Fortunately, thanks to the widespread popularity of SSL/TLS, many applications now work with secure sockets, which in fact are very similar to normal TCP sockets.

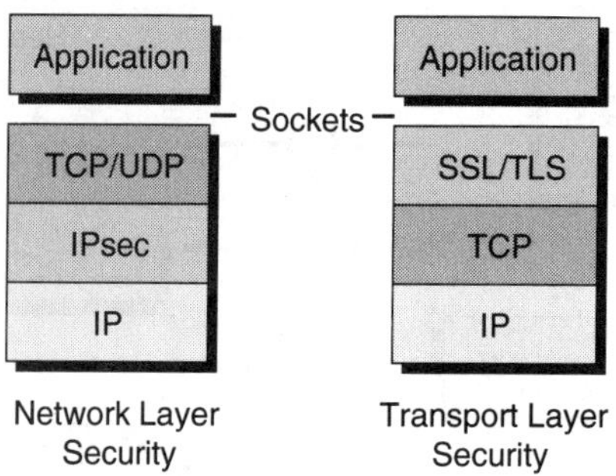

FIGURE 91.6   Comparison; network vs. transport-layer security.

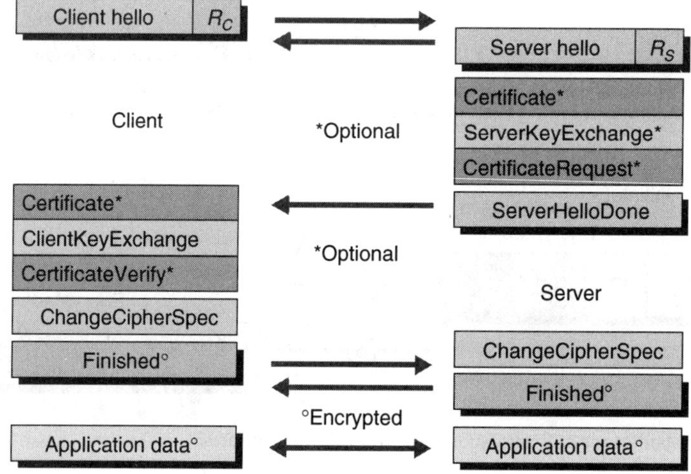

FIGURE 91.7   SSL/TLS handshake protocol.

SSL/TLS uses a record structure to encrypt and authenticate the application data and to embed them into a TCP/IP stream. Similar to IPsec and its companion IKE connection negotiation protocol, the SSL/TLS handshake protocol shown in Figure 91.7 is used to set up a secure SSL/TLS tunnel.

- The handshake is initiated by the client who sends a ClientHello message that lists all its suites of supported encryption, authentication, and compression protocols.
- The server answers by selecting one of the offered suites and includes the decision in a ServerHello message. Both client and server add random values R_C and R_S, respectively, to their hello messages in order to make each session unique, thereby preventing replay attacks.
- A server usually possesses a trusted server certificate, which it sends to the client right after the ServerHello message. By including an optional CertificateRequest, the client can be forced to authenticate itself in turn by requiring it to send a client certificate, too.
- The ServerHelloDone message terminates the first server message block and prompts the client to take over the protocol lead again.
- If requested to do so, the client sends its certificate together with a signature contained in the CertificateVerify message. As mentioned above, client-side authentication is optional and is not used very often.
- The client always generates a 48-byte random premaster secret, encrypts it with the public key extracted from the received server certificate, and puts the protected secret in the ClientKeyExchange message. Based on the premaster secret, the client then computes the master secret and derives all

required encryption and authentication session keys. It then activates the chosen cipher suite, signals this to the server by issuing a ChangeCipherSpec message, and adds an already encrypted Finished message that can be used by the peer side to test the successful establishment of the secure connection.
- The server in turn decrypts the received ClientKeyExchange message using its private key. With the extracted premaster secret, the server can then generate the same master secret and derived sessions keys as the client's. By sending the ChangeCipherSpec and Finished messages, the server signals its readiness to switch over to the secure channel, too.

From this moment on, all application data are exchanged in encrypted and authenticated form.

Every newly opened socket needs to be initialized first using the SSL/TLS handshake protocol. If, for example, the obsolete HTTP 1.0 protocol is used to access a web site, then each graphical element of an HTML page will be downloaded over a socket of its own. This would mean that every time a computationally expensive RSA public key authentication based on a 1024-bit key would have to be executed. In order to avoid a full handshake operation, SSL/TLS offers the possibility to resume or clone an already established session in the abbreviated manner detailed in Figure 91.8.

If the client wants to resume a session or open an additional socket, it sends the session ID of the existing session as part of the ClientHello message. If the server still has the corresponding session data including the premaster secret in its cache then it will reply by immediately sending the ChangeCipherSpec and Finished messages. Nevertheless, using a fresh set of the random values R_C and R_S, a set of new session keys are derived from the cached premaster secret.

A readily available OpenSource SSL/TLS implementation is the popular OpenSSL package [16]. Due to its innumerable options and supported algorithms, the required SSL and crypto libraries add up to a huge size of 1.6 Mbytes. OpenSSL in its direct form is therefore only suitable for high-end 32-bit platforms that usually run an embedded Linux operation system. In Section 91.4 we will demonstrate how a lightweight SSL/TLS stack can be derived from the original OpenSSL library.

Application-Layer Security

A typical example of an application-layer security application is the popular secure shell (ssh) [17], which allows a secure login for administration and monitoring purposes – functions that were classically done by system administrators using the insecure and now deprecated telnet protocol. The authentication mechanisms are similar to those used in the IKE and SSL/TLS handshake protocols. Either raw RSA keys or X.509 certificates are supported.

Another possible security approach at the application level would be to authenticate and encrypt the message bodies using the S/MIME [16] or PGP/GnuPG [18] packages known from secure e-mail. But be aware that additional precautions must be taken to thwart replay attacks using recorded past messages.

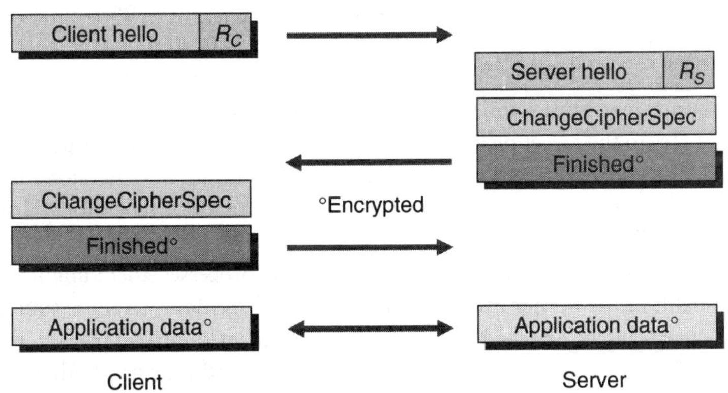

FIGURE 91.8   Resuming SSL/TLS session.

On many embedded systems, a Web server is running by default, allowing the configuration and monitoring of the embedded application via an interactive Web interface. A straightforward approach in this case would be to use the built-in security mechanisms of the HTTP protocol. Two authentication variants are available: Basic and Digest Access Authentication [19], whereas encryption is not supported and must be provided by other means. The Basic Access Authentication scheme defined by the HTTP 1.0 protocol is not secure at all and should not be used because the username/password pair of the client is transmitted in the clear and can easily be intercepted. The Digest Access Authentication introduced by the newer HTTP 1.1 protocol is much more secure because the secret password never crosses the communication channel. Instead, a random challenge string sent by the server is hashed together with the client's username and password to form an MD5 digest, which is sent back as a response to the HTTP server. This digest is vulnerable to brute-force dictionary attacks, though. If a weak password is chosen by the user, then an off-line cracking attempt will find the password in less than an hour if it has been chosen from a multilingual dictionary and has only been slightly modified, for example, by appending a number.

The major advantage of the HTTP Digest Access Authentication is the relative ease of its implementation and use, whereas the drawbacks are the lack of an encryption option and the potential vulnerability of the openly transmitted digest to dictionary attacks due to weakly chosen passwords.

91.3 Key Problems in Embedded Systems

In this section, we will examine two of the key problems that are typical for low-end embedded system platforms. The first topic is the generation of true random material in a low-entropy environment and the second item will concentrate on the computationally intensive operations required by the standard RSA public key cryptosystem as used by IPsec's Internet Key Exchange protocol, the SSL/TLS handshake protocol, and the SSH public key-based authentication.

Generation of True Random Numbers

The security of modern cryptographic algorithms is founded exclusively on the true randomness of the chosen session keys. The same is valid for the challenge/response protocols used for peer authentication that are based on the prerequisite that the challenge strings will never occur twice during the whole lifetime of the system. As a consequence, several random or pseudorandom material must be generated in order to satisfy the demands of a truly secure communication channel.

In a workstation environment, usually sufficient random information can be continuously collected from the timing of keyboard strokes and mouse movements. On unattended servers and firewalls, only disk access timing and arrival times of network packets remain as practical entropy sources, whereas the latter source might raise the concern of some paranoid crypto experts who postulate that the network traffic timing could potentially be manipulated by a malicious attacker. Fortunately, modern Intel-based platforms often possess a chip set [20] with a built-in random generator (RNG) that produces random numbers in copious quantities.

On low-end diskless embedded platforms, it becomes increasingly difficult to gather any random material at all. The usual makeshift solution consists of transferring an initialization file containing about 1024 true random bytes to the target platform during system configuration. This pool of random data is then used as a seed for a pseudorandom generator. After each access to the generator, the new state of the random pool is saved back to the file, thus guaranteeing that the random generator will always start from the last used state and will therefore never repeat itself.

Since embedded systems are often used to control real-time processes, there might be random information readily available at the numerous control inputs. As an alternative, a noise source or some other reliable random device could be attached to some of the unconnected inputs if available. Practical results undertaken at the Zurich University of Applied Sciences in Winterthur, using both a noisy diode and a free-running oscillator have shown that with little hardware effort, a low-cost, low-complexity random number generator can be built suited to low-end embedded platforms. Great care must be taken to deskew the generated stream of raw random bits, since it is very difficult to achieve a perfectly even distribution of 'zeros' and 'ones.' Either the simple von Neumann de-skewing algorithm [21] can be used or even better, all raw bits gathered from hardware random sources should first be hashed into a random pool using the MD5 or SHA-1 digest algorithms in order to improve the statistical properties of the derived key material.

Public Key Operations

The second obstacle in providing strong cryptography to embedded systems is the extremely time-consuming modular exponentiation operation

$$y = x^e \, mod \, n$$

that is the corner stone of both the RSA and Diffie–Hellman public key algorithms. Usually, the basis x, the exponent e, and the modulus n are large integers having a size k of 512..2048 bits. Using its RSA private key, an SSL/TLS server running on an embedded system must compute a singular modular exponentiation in order to decrypt the premaster secret that had been previously encrypted by the SSL/TLS client with the server's RSA public key. The processing time needed to execute this crucial RSA decryption operation was measured on the 20 MHz IPC@CHIP platform described in Section 91.4. The results are listed in Table 91.1.

The selected RSA key size k is equivalent to the size of the modulus n used in the modular exponentiation. The observed processing time t varies from a mere 8 sec for the insecure key size k of 512 bits, to 48 sec for the standard RSA key size of 1024 bits, and a horrendous 335 sec for the 2048-bit keys that might be required in a couple of years from now. These execution times listed above are typical values for machine code automatically generated by a standard C compiler. By manually coding the critical parts of the multiprecision multiplication in assembly code, an acceleration of nearly a factor of two can be achieved.

The processing time for the 16-bit IPC@CHIP platform can be well approximated by the formula

$$t \approx \frac{702(k/1024)^3 + 273(k/1024)^2}{f_c[\text{MHz}]} \, [\text{s}]$$

which is plotted together with our measurement data in Figure 91.9. For large key lengths k, the processing time increases with the third power of k. This asymptotic behavior can be easily explained by the fact that the execution time for a multi-precision multiplication of two k bit numbers is proportional to k^2 and that a modular exponentiation with a k bit exponent requires on the average about $k/2$ multi-precision multiplications. Besides the clock frequency, the processor's word size is very decisive for the execution speed. Halving the word size from 16 bits to 8 bits increases the time for a multiprecision multiplication by a factor of four, whereas doubling it to 32 bits decreases it by the same factor.

With the typical system clock of $f_c = 20$ MHz for a low-end embedded processor, an RSA decryption takes 48 sec to execute, whereas with today's 1000 MHz 32-bit Pentium processors, the same operation can be done within well below one second. Such an initial delay of nearly 1 min can be prohibitive in many real-time application, although as a workaround in the particular case of the SSL/TLS protocol, a session can be resumed after a break without the need to encrypt, transmit, and decrypt a new premaster secret.

TABLE 91.1 Processing Time for an RSA Decryption

RSA key Size k (bits)	Processing time t (s)
512	8
768	22
1024	48
1536	150
2048	335

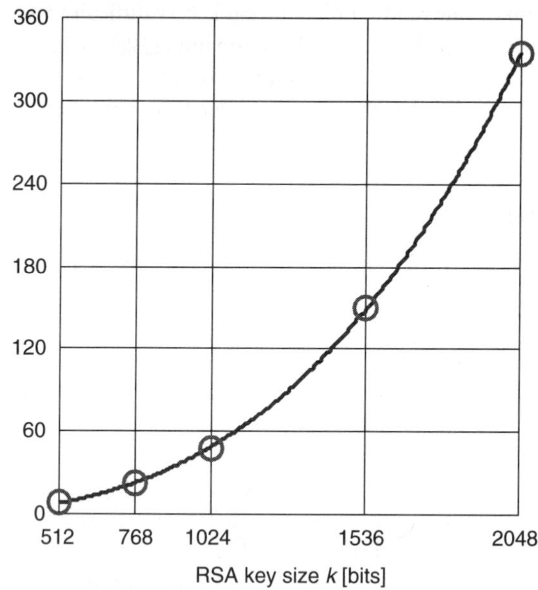

FIGURE 91.9   Processing time in [s] for RSA decryption.

TABLE 91.2   RSA/ECC Key Size Equivalent Strength Comparison (Data from [23])

Time to break in MIPS years	RSA key size (bits)	ECC key size (bits)	RSA/ECC key size ratio
10^4	512	106	5:1
10^8	768	132	6:1
10^{11}	1024	160	7:1
10^{20}	2048	210	10:1

The class of Elliptic Curve Public Key Cryptosystem (ECC) public key algorithms [22] can offer a significant improvement in processing delay, thus making many time-critical applications feasible. The main reason for this speed gain are the much shorter key lengths required (compared to the RSA cryptosystem) to achieve the same cryptographic strength. Table 91.2 shows that a 160-bit ECC key is equivalent in strength to a 1024-bit RSA key. In the years to come, 2048-bit RSA keys might be required to maintain adequate security, which doubles the current RSA key lengths, whereas the ECC key length will have to grow by 50 bits only.

Due to the much shorter key lengths, ECC promises acceleration factors of 10..100 for a 160-bit ECC key, compared to a 1024-bit RSA key [23], especially if the algorithm is realized in special hardware or on a smartcard crypto-coprocessor. ECC software implementations are still not widely deployed, with Certicom being the market leader. Fortunately, thanks to a software donation by Sun Microsystems, the [snapshot] version of the OpenSSL package [16] includes now full ECC support.

91.4   Example of an SSL/TLS-enabled Embedded Platform

In October 2000, the Security Group of the Zurich University of Applied Sciences in Winterthur started its activities in the field of lightweight security protocols by porting the OpenSSL library [16] to the tiny 16-bit 80186-based IPC@CHIP embedded Web server platform shown in Figure 91.10.

The IPC@CHIP is based on a 16-bit little-endian Intel-compatible 80186 processor running with a modest system clock of $f_c = 20$ MHz. The platform is equipped with 512 kbytes of RAM and 512 kbytes of nonvolatile flash memory and possesses an integrated 10Base-T Ethernet interface. Since about half of

FIGURE 91.10   SSL/TLS-enabled mini Web server.

the flash space is used to store the BIOS comprising a built-in TCP/IP stack, 256 kbytes of free memory are available for the SSL/TLS protocol implementation and the actual application running on the embedded system.

As part of their diploma thesis, the two ZHW students Bernhard Lenzlinger and Andreas Zingg succeeded in stripping the OpenSSL library from the original 1.6 Mbytes down to a mere 580 kbytes. Based on this shrunk 16-bit library, they were able to build an SSL/TLS-enabled Web server application. With its small size of 100 kbytes (after compression, but including the statically linked OpenSSL library), the SSL/TLS server fitted without problems into the available flash memory.

The features offered by the mini Web server depicted in Figure 91.10 are as follows:

- Support of the SSL 3.0 and TLS 1.0 protocols.
- File-based initialization of the random seed.
- MD5 and SHA-1 hash used for message authentication.
- RC4 stream cipher with 128-bit symmetric key used for encryption.
- X.509 certificate used for server authentication.
- 1024-bit RSA key embedded in the X.509 server certificate used for the encrypted transmission of the premaster secret from client to server.
- Secure client-side authentication based on HTTP passwords after SSL/TLS link has been successfully established.
- Single HTTP session supported, no multithreading.

Multiple sessions could be cached making session resuming possible, but only a single concurrent TCP/IP socket was supported.

Encouraged by the success of this lightweight SSL/TLS rapid prototype, the Security Group of the Zurich University of Applied Sciences in Winterthur decided to launch the project *"Secure Communications in Distributed Embedded Systems"* in July 2002, with the following extended objectives:

- Implementation of a modular and lightweight SSL 3.0 / TLS 1.0 protocol stack that can be quickly ported with little or no modifications to a large variety of 16- and 32-bit embedded platforms possessing limited memory resources and processing power.
- Support of the fast AES [24] with 128-bit symmetric keys, which is much more secure than the current default RC4 stream cipher.
- Support of the fast ECC [22] in addition to the traditional RSA Public Key Cryptosystem.
- Multithreading capability allowing multiple concurrent TCP/IP sockets.

The design of this extended and modular SSL/TLS stack will be tailored to the needs of a large percentage of the existing embedded system platforms, spanning the whole range from the low-end 16-bit microcontrollers with proprietary operating systems up to the powerful 32-bit processors running under embedded Linux.

91.5   Conclusions

In this chapter, we have shown that secure communications in distributed embedded systems can be implemented in nearly all layers of the communications stack, but we express a pronounced preference for the network and transport layers. A practical implementation of an SSL/TLS stack on a low-end 16-bit embedded platform has clearly demonstrated that lightweight security protocols are feasible. For high-end, embedded Linux-based platforms, IPsec might be the optimum choice due to the maximum flexibility that can achieved. With the advent of the fast AES and ECC algorithms, the speed handicap incurred by low-cost and low-power processors will be alleviated and time-critical applications will become feasible. In special cases, hardware acceleration using a crypto coprocessor might be required. Such devices are available, for example, in the form of smart card crypto coprocessors. For pure console-based applications, the ssh is certainly the right choice. And if your security requirements are not so stringent and encryption of the payload is not mandatory, then HTTP Digest Access Authentication might be the appropriate option.

References

[1] Borisov, N., I. Goldberg, and D. Wagner, Intercepting Mobile Communications: The Insecurity of 802.11, Proceedings of the Seventh Annual International Conference on Mobile Computing And Networking, July 16–21, 2001.

[2] Fluhrer, S., I. Mantin, and A. Shamir, Weaknesses in the Key Scheduling Algorithm of RC4, August 2001.

[3] Biryukow, A., A. Shamir, and D. Wagner, Real Time Cryptanalysis of A5/1 on a PC, Fast Software Encryption Workshop, Apr. 10–12, 2000.

[4] Meyer, G., The PPP Encryption Control Protocol (ECP), RFC 1968, June 1996.

[5] Kent, S. and R. Atkinson, Security Architecture for the Internet Protocol, RFC 2401, November 1998.

[6] Frankel, S., *Demystifying the IPsec Puzzle*, Artech House, Norwood, MA, 2001.

[7] Kent, S., IP Authentication Header (AH), RFC 2402, November 1998.

[8] Kent, S., IP Encapsulating Security Payload (ESP), RFC 2406, November 1998.

[9] µClinux Project, www.uclinux.org.

[10] Linux FreeS/WAN Project, www.freeswan.org.

[11] Harkins, D. and D. Carrel, The Internet Key Exchange (IKE), RFC 2409, November 1998.

[12] X.509 Certificate support for Linux FreeS/WAN, www.strongsec.com/freeswan.

[13] Freier, A., P. Karlton, and P. Kocher, The SSL Protocol Version 3.0, Internet draft, November 1996, http://wp.netscape.com/eng/ssl3/draft302.txt.

[14] Dierks, T. and C. Allen, The TLS Protocol Version 1.0, RFC 2246, January 1999.

[15] Rescorla, E., *SSL and TLS: Designing and Building Secure Systems*, Addison-Wesley, Reading, MA, 2001.

[16] OpenSSL Project, www.openssl.org.

[17] OpenSSH Project, www.openssh.org

[18] International PGP Home Page, www.pgpi.org.

[19] Franks, J. et al., HTTP Authentication: Basic and Digest Access Authentication, RFC 2617, June 1999.

[20] Intel Random Number Generator, www.intel.com/design/security/rng/rngbrf.htm.

[21] Eastlake, D., S. Crocker, and J. Schiller, Randomness Recommendations for Security, RFC 1750, December 1994.

[22] Menezes, A., *Elliptic Curve Public Key Cryptosystems*, Kluwer Academic Publishers, Dordresht, 1993.

[23] *The Elliptic Curve Cryptosystem for Smart Cards*, A Certicom White Paper, May 1998, www.certicom.com/resources/download/ECC_SC.pdf.

[24] Advanced Encryption Standard (AES), www.nist.gov/aes.

92

System-on-Chip and Network-on-Chip Design

Grant Martin
Tensilica, Inc.

92.1 Introduction

"System-on-Chip" is a phrase that has been much bandied about in recent years [1]. It is more than a design style, more than an approach to the design of application-specific integrated circuits (ASICs), more than a methodology. Rather, System-on-Chip (SoC) represents a major revolution in IC design — a revolution enabled by the advances in process technology allowing the integration of all or most of the major components and subsystems of an electronic product onto a single chip, or integrated chipset. [2] This revolution in design has been embraced by many designers of complex chips, as the performance, power consumption, cost, and size advantages of using the highest level of integration made available have proven to be extremely important for many designs. In fact, the design and use of SoCs is arguably one of the key problems in designing real-time embedded systems.

The move to SoC began sometime in the mid-1990s. At this point, the leading CMOS-based semiconductor process technologies of 0.35 and 0.25 microns were sufficiently capable of allowing the integration of many of the major components of a second-generation wireless handset or a digital set-top box onto a single chip. The digital baseband functions of a cellphone — a Digital Signal Processor (DSP), hardware support for voice encoding and decoding, and an RISC processor — could all be placed onto a single die. Although such a baseband SoC was far from the complete cellphone electronics — there were major components such as the RF transceiver, the analog power control, analog baseband, and passives that were not integrated — the evolutionary path with each new process generation, to integrate more and more onto a single die, was clear. Today's chipset would become tomorrow's chip. The problems of integrating hybrid technologies involved in making up a complete electronic system would be solved. Thus, eventually, SoC could encompass design components drawn from the standard and more adventurous domains of digital, analog, RF, reconfigurable logic, sensors, actuators, optical, chemical, microelectronic mechanical systems, and even biological and nanotechnology.

With this viewpoint, of continued process evolution leading to ever-increasing levels of integration into ever-more-complex SoC devices, the issue of an SoC being a single *chip* at any particular point in time is somewhat moot. Rather, the word "system" in SoC is more important than "chip." What is most important about an SoC whether packaged as a single chip, or integrated chipset, or System-in-Package (SiP) or System-on-Package (SoP) is that it is designed as an integrated *system*, making design tradeoffs across the processing domains and across the individual chip and package boundaries.

92.2 SoC

Let us define a SoC as a complex integrated circuit, or integrated chipset, which combines the major functional elements or subsystems of a complete end-product into a single entity. These days, all interesting SoC designs include at least one programmable processor, and very often a combination of at least one RISC control processor and one DSP. They also include on-chip communications structures — processor bus(es), peripheral bus(es), and perhaps a high-speed system bus. A hierarchy of on-chip memory units, and links to off-chip memory are important especially for SoC processors (cache, main memories, very often separate instruction, and data caches are included). For most signal-processing applications, some degree of hardware-based accelerating functional units are provided, offering higher performance and lower energy consumption. For interfacing to the external, real world, SoCs include a number of peripheral processing blocks, and due to the analog nature of the real world, this may include analog components as well as digital interfaces (e.g., to system buses at a higher packaging level). Although there is much interesting research in incorporating MEMS-based sensors and actuators, and in SoC applications incorporating chemical processing (lab-on-a-chip), these are, with rare exceptions, research topics only. However, future SoCs of a commercial nature may include such subsystems, as well as optical communications interfaces.

Figure 92.1 illustrates what a typical SoC might contain for consumer applications.

One key point about SoC that is often forgotten for those approaching them from a hardware-oriented perspective is that all interesting SoC designs encompass both hardware and software components, that is, programmable processors, real-time operating systems, and other aspects of hardware-dependent software such as peripheral device drivers, as well as middleware stacks for particular application domains, and possibly optimized assembly code for DSPs. Thus, the design and use of SoCs cannot remain a hardware-only concern — it involves aspects of system-level design and engineering, hardware-software tradeoff and partitioning decisions, and software architecture, design, and implementation.

92.3 System-on-a-Programmable-Chip

Recently, attention has increased in the SoC world from SoC implementations using custom, ASIC, or application-specific standard part (ASSP) design approaches, to include the design and use of complex reconfigurable logic parts with embedded processors and other application-oriented blocks of intellectual property. These complex Field-Programmable Gate Arrays (FPGAs) are offered by several vendors, including Xilinx (Virtex-II PRO Platform FPGA) and Altera (SOPC), but are referred to by several names: highly programmable SoCs, system-on-a-programmable-chip, and embedded FPGAs. The key idea behind this approach to SoC is to combine large amounts of reconfigurable logic with embedded RISC processors (either custom laid-out, "hardened" blocks, or synthesizable processor cores), in order to allow

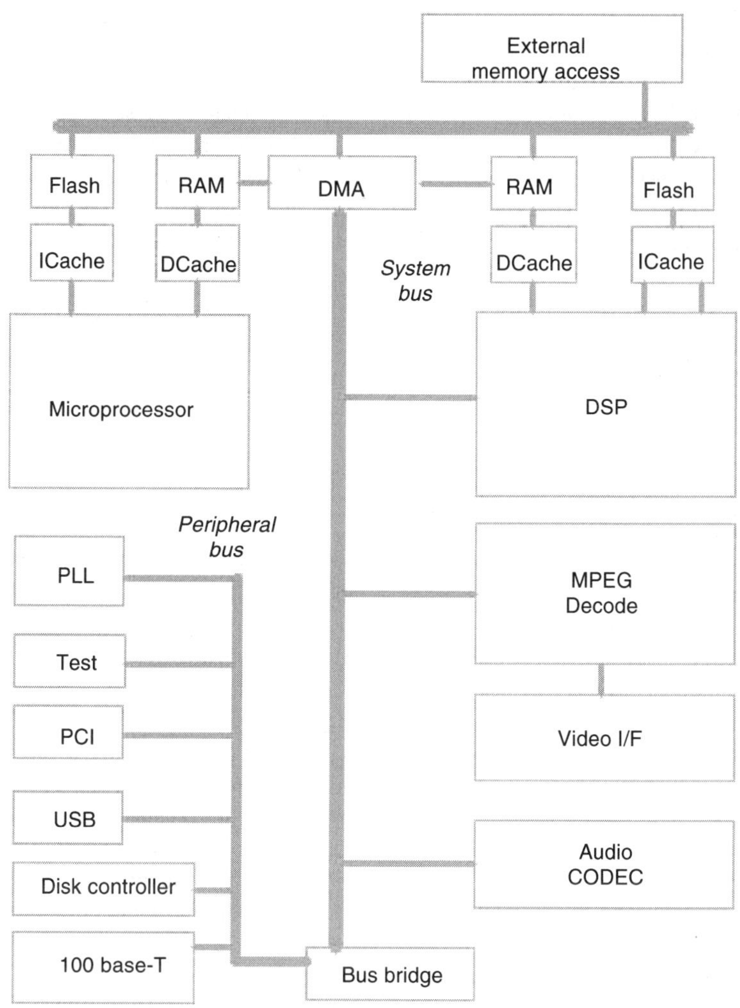

FIGURE 92.1 A typical SoC device for consumer applications.

very flexible and tailorable combinations of hardware and software processing to be applied to a particular design problem. Algorithms that consist of significant amounts of control logic, plus significant quantities of dataflow processing, can be partitioned into the control RISC processor (e.g., in Xilinx Virtex-II PRO, a PowerPC processor) and reconfigurable logic offering hardware acceleration. Although the resulting combination does not offer the highest performance, lowest energy consumption, or lowest cost, in comparison with custom IC or ASIC/ASSP implementations of the same functionality, it does offer tremendous flexibility in modifying the design in the field and avoiding expensive Non-Recurring Engineering (NRE) charges in the design. Thus, new applications, interfaces, and improved algorithms can be downloaded to products working in the field using this approach.

Products in this area also include other processing and interface cores, such as multiply-accumulate (MAC) blocks that are specifically aimed at DSP-type dataflow signal and image-processing applications, and high-speed serial interfaces for wired communications such as serializer/deserializer (SERDES) blocks. In this sense, system-on-a-programmable-chip SoCs are not exactly application-specific, but not completely generic either.

It remains to be seen whether system-on-a-programmable chip SoCs are going to be a successful way of delivering high-volume consumer applications, or will end up restricted to the two main applications for high-end FPGAs: rapid prototyping of designs that will be retargeted to ASIC or ASSP implementations, and used in high-end, relatively expensive parts of the communications infrastructure that require in-field flexibility and can tolerate the tradeoffs in cost, energy consumption, and performance. Certainly, the use of synthesizable processors on more moderate FPGAs to realize SoC style designs is one alternative to the cost issue. Intermediate forms, such as the use of metal-programmable gate-array style logic fabrics together with hard-core processor subsystems and other cores, such as is offered in the "Structured ASIC" offerings of LSI Logic (RapidChip) and NEC (Instant Silicon Solutions Platform) represents an intermediate form of SoC between the full-mask ASIC and ASSP approach and the field-programmable gate array approach. Here, the tradeoffs are a much slower design creation (a few weeks rather than a day or so), higher NRE than FPGA (but much lower than a full set of masks), and better cost, performance, and energy consumption than FPGA (perhaps 15–30% worse than an ASIC approach). Further interesting compromises or hybrid style approaches, such as ASIC/ASSP with on-chip FPGA regions, are also emerging to give design teams more choices.

92.4 IP Cores

The design of SoC would not be possible if every design started from scratch. In fact, the design of SoC depends heavily on the reuse of Intellectual Property blocks — what are called "IP Cores." IP reuse has emerged as a strong trend over the last 8–9 years [3] and has been one key element in closing what the International Technology Roadmap for Semiconductors [4] calls the "design productivity gap" — the difference between the rate of increase of complexity offered by advancing semiconductor process technology, and the rate of increase in designer productivity offered by advances in design tools and methodologies.

But reuse is not just important to offer ways of enhancing designer productivity — although it has dramatic impacts on this. It also provides a mechanism for design teams to create SoC products that span multiple design disciplines and domains. The availability of both hard (laid-out and characterized) and soft (synthesizable) processor cores from a number of processor IP vendors allows design teams who would not be able to design their own processor from scratch to drop them into their designs and thus add RISC control and DSP functionality to an integrated SoC without having to master the art of processor design within the team. In this sense, the advantages of IP reuse go beyond productivity — it offers both a large reduction in design risk, and also a way for SoC designs to be done that would otherwise be infeasible due to the length of time it would take to acquire expertise and design IP from scratch.

This ability when acquiring and reusing IP cores — to acquire, in a *prepackaged* form, design domain expertise outside one's own design team's set of core competencies, is a key requirement for the evolution of SoC design going forward. Up to this point, SoC has concentrated in large part on integrating digital components together, perhaps with some analog interface blocks that are treated as black boxes. The hybrid SoCs of the future, incorporating domains unfamiliar to the integration team, such as RF or MEMS, requires the concept of "drop-in" IP to be extended to these new domains. We are not yet at that state — considerable evolution in the IP business and the methodologies of IP creation, qualification, evaluation, integration, and verification are required before we will be able to easily specify and integrate truly heterogeneous sets of disparate IP blocks into a complete hybrid SoC.

However, the same issues existed at the beginning of the SoC revolution in the digital domain. They have been solved to a large extent, through the creation of standards for IP creation, evaluation, exchange, and integration — primarily for digital IP blocks but also extending to analog/mixed-signal (AMS) cores. Among the leading organizations in the identification and creation of such standards has been the Virtual Socket Interface Alliance (VSIA) [5], formed in 1996 and having at its peak membership more than 200 IP, systems, semiconductor, and Electronic Design Automation (EDA) corporate members. Although often criticized over the years for a lack of formal and acknowledged adoption of its IP standards, VSIA has had a more subtle influence on the electronics industry. Many companies instituting reuse programs internally, many IP, systems, and semiconductor companies engaging in IP creation and exchange, and many design groups have used VSIA IP standards as a key starting point for developing their own standards and methods for IP-based design. In this sense, use of VSIA outputs has enabled a kind of IP reuse in the IP business.

VSIA, for example, in its early architectural documents of 1996–1997, helped define the strong industry-adopted understanding of what it meant for an IP block to be considered to be in "hard" or "soft" form. Other important contributions to design included the widely read system-level design model taxonomy created by one of its working groups. Its standards, specifications, and documents thus represent a very useful resource for the industry.

Other important issues for the rise of IP-based design and the emergence of a third party industry in this area (which has taken much longer to emerge than originally hoped in the mid-1990s) are the business issues surrounding IP evaluation, purchase, delivery, and use. Organizations such as the Virtual Component Exchange (VCX) [6] emerged to look at these issues and provide solutions. Although still in existence, it is clear that the vast majority of IP business relationships between firms occurs within a more *ad hoc* supplier to customer business framework.

92.5   Virtual Components

The VSIA has had a strong influence on the nomenclature of the SoC- and IP-based design industry. The concept of the "virtual socket" — a description of all the design interfaces that an IP core must satisfy, and design models and integration information that must be provided with the IP core, required to allow it to be more easily integrated or "dropped into" an SoC design, comes from the concept of Printed Circuit Board (PCB) design where components are sourced and purchased in a prepackaged form and can be dropped into a board design in a standardized way.

The dual of the "virtual socket" then becomes the "virtual component." Specifically in the VSIA context, but also more generally in the interface, an IP core represents a design block that *might* be reusable. A virtual component represents a design block that is *intended* for reuse, and that has been *developed* and *qualified* to be highly reusable. The things that separate IP cores from virtual components are in general as follows:

- Virtual components conform in their development and verification processes to well-established design processes and quality standards.
- Virtual components come with design data, models, associated design files, scripts, characterization information, and other deliverables that conform to one or other well-accepted standards for IP reuse — for example, the VSIA deliverables, or another internal or external set of standards.
- Virtual components in general should have been fabricated at least once, and characterized post-fabrication to ensure that they have validated claims.
- Virtual components should have been reused at least once by an external design team, and usage reports and feedback should be available.
- Virtual components should have been rated for quality using an industry standard quality metric such as OpenMORE (created by Synopsys and Mentor Graphics) or the VSI Quality standard (which has OpenMORE as one of its inputs).

To a large extent, the developments over the last decade in IP reuse have been focused on defining the standards and processes to turn the *ad hoc* reuse of IP cores into a well-understood and reliable process for acquiring and reusing virtual components, thus enhancing the analogy with PCB design.

92.6   Platforms and Programmable Platforms

The emphasis in the preceding sections has been on IP (or virtual component) reuse on a somewhat *ad hoc* block-by-block basis in SoC design. Over the past several years, however, a more integrated approach to the design of complex SoCs and the reuse of virtual components has arisen — what has been called "Platform based design." This will be dealt with at a much greater length in another chapter in this book. Much more information is available in the references [7–10]. Suffice it here to define platform-based design in the SoC context from one perspective.

We can define platform-based design as a planned design methodology, which reduces the time and effort required, and risk involved, in designing and verifying a complex SoC. This is accomplished by extensive reuse of combinations of hardware and software IP. As an alternative to IP reuse in a block-by-block manner, platform-based design assembles groups of components into a reusable platform architecture. This reusable architecture, together with libraries of preverified and precharacterized, application-oriented HW and SW virtual components is an SoC integration platform.

There are several reasons for the growing popularity of the platform approach in industrial design. These include the increase in design productivity, the reduction in risk, the ability to utilize preintegrated virtual components from other design domains more easily, and the ability to reuse SoC architectures created by experts. Industrial platforms include full application platforms, reconfigurable platforms, and processor-centric platforms [11]. Full application platforms, such as Philips Nexperia and TI OMAP provide a complete implementation vehicle for specific product domains [12]. Processor-centric platforms, such as ARM PrimeXsys, concentrate on the processor, its required bus architecture and basic sets of peripherals, along with RTOS and basic software drivers. Reconfigurable or "highly programmable" platforms such as the Xilinx Platform FPGA and Altera's SOPC deliver hardcore processors plus reconfigurable logic along with associated IP libraries and design tool flows.

92.7 Integration Platforms and SoC Design

The use of SoC integration platforms changes the SoC design process in two fundamental ways:

1. The basic platform must be designed, using whatever *ad hoc* or formalized design process for SoC that the platform creators decide on. The next section outlines some of the basic steps required to build an SoC, whether building a platform or using a block-based more *ad hoc* integration process. However, when constructing an SoC platform for reuse in derivative design, it is important to remember that it may not be necessary to take the whole platform and its associated HW and SW component libraries through complete implementation. Enough implementation must be done to allow the platform and its constituent libraries to be fully characterized and modeled for reuse. It is also essential that the platform creation phase produce in an archivable and retrievable form all the design files required for the platform and its libraries to be reused in a derivative design process. This must also include the setup of the appropriate configuration programs or scripts to allow automatic creation of a configured platform during derivative design.

2. A design process must be created and qualified for all the *derivative* designs, which will be created based on the SoC integration platform. This must include processes for retrieving the platform from its archive, for entering the derivative design configuration into a platform configurator, the generation of the design files for the derivative, the generation of the appropriate verification environment(s) for the derivative, the ability for derivative design teams to select components from libraries, to modify these components and validate them within the overall platform context, and, to the extent supported by the platform, to create new components for their particular application.

Reconfigurable or highly programmable platforms introduce an interesting addition to the platform-based SoC design process [13]. Platform FPGAs and SOPC devices can be thought of as a "meta-platform": a platform for creating platforms. Design teams can obtain these devices from companies such as Xilinx and Altera, containing a basic set of more generic capabilities and IP — embedded processors, on-chip buses, special IP blocks such as MACs and SERDES, and a variety of other pre-qualified IP blocks. They can then customize the meta-platform to their own application space by adding application domain-specific IP libraries. Finally, the combined platform can be provided to derivative design teams, who can select the basic meta-platform and configure it within the scope intended by the intermediate platform creation team, selecting the IP blocks needed for their exact derivative application. More on platform-based design will be found in another chapter in this book.

92.8   Overview of the SoC Design Process

The most important thing to remember about SoC design is that it is a multidisciplinary design process, which needs to exercise design processes from across the spectrum of electronics. Design teams must gain some fluency with all these multiple disciplines, but the integrative and reuse nature of SoC design means that they may not need to become deep experts in all of them. Indeed, avoiding the need for designers to understand all methodologies, flows and domain-specific design techniques are one of the key reasons for reuse and enablers of productivity. Nevertheless, from DFT through digital and analog HW design, from verification through system-level design, from embedded SW through IP procurement and integration, from SoC architecture through IC analysis, a wide variety of knowledge is required by the team, if not every designer.

Figure 92.2 illustrates some of the basic constituents of the SoC design process.

We will now define each of these steps as illustrated:

SoC requirements analysis is the basic step of defining and specifying a complex SoC, based on the needs of the end product into which it will be integrated. The primary input into this step is the marketing definition of the end product and the resulting characteristics of what the SoC should be: both functional and nonfunctional (e.g., cost, size, energy consumption, performance: latency and throughput, package selection). This process of requirements analysis must ultimately answer the question: is the product feasible? Is the desired SoC feasible to design, and with what effort and in what timeframe? How much reuse will be possible? Is the SoC design based on legacy designs of previous generation products (or, in the case of platform-based design, to be built based on an existing platform offering)?

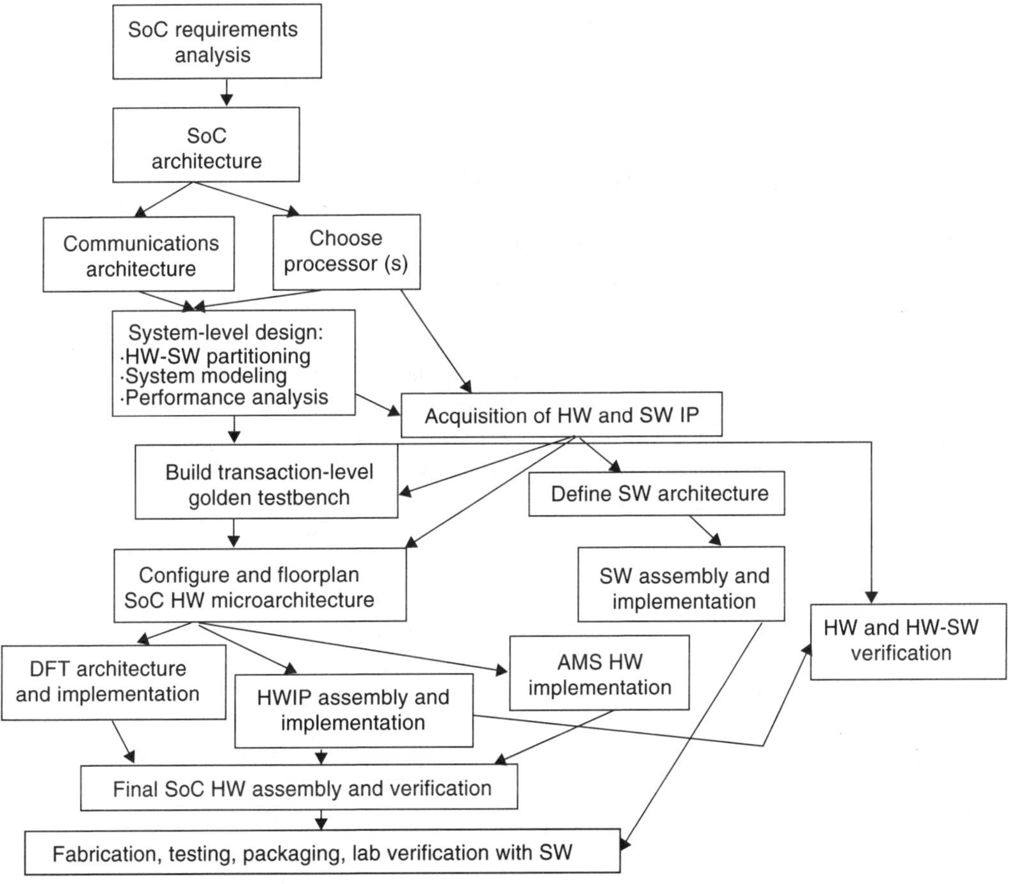

FIGURE 92.2   Steps in the SoC design process.

SoC architecture: In this phase, the basic structure of the desired SoC is defined. It is vitally important to decide on the *communications architecture* that will be used as the backbone of the SoC on-chip communications network. An inadequate communications architecture will cripple the SoC and have as large an impact as the use of an inappropriate processor subsystem. Of course, the choice of communications architecture is impossible to divorce from making the basic *processor(s) choice* — for example, do I use an RISC control processor? Do I have an on-board DSP? How many of each? What are the processing demands of my SoC application? Do I integrate the bare processor core, or use a whole processor subsystem provided by an IP company (most processor IP companies have moved from offering just processor cores, to whole processor subsystems including hierarchical bus fabrics tuned to their particular processor needs)? Do I have some ideas, based on legacy SoC design in this space, as to how SW and HW should be partitioned? What memory hierarchy is appropriate? What are the sizes, levels, performance requirements, and configurations of the embedded memories most appropriate to the application domain for the SoC?

System-level design is an important phase of the SoC process — but one that is often done in a relatively *ad hoc* way. The whiteboard and the spreadsheet are as much used by the SoC architects as more capable toolsets. However, *ad hoc* C/C++- based models have long been in use for the system design phase — to validate basic architectural choices. And designers of complex signal processing algorithms for voice and image processing have long adopted dataflow models and associated tools to define their algorithms, define optimal bit-widths, and validate performance whether destined for hardware or software implementation. A flurry of activity in the last few years on different C/C++ modeling standards for system architects has consolidated on SystemC [14]. The *system* nature of SoC demands a growing use of system-level design modeling and analysis as these devices grow more complex. The basic processes carried out in this phase include HW–SW partitioning (the allocation of functions to be implemented in dedicated HW blocks, in SW on processors (and the decision of RISC vs. DSP), or a combination of both, together with decisions on the communications mechanisms to be used to interface HW and SW, or HW–HW, and SW–SW). In addition, the construction of system-level models, and the analysis of correct functioning, performance, and other nonfunctional attributes of the intended SoC through simulation and other analytical tools, is necessary. Finally, all additional IP blocks required that can be sourced outside, or reused from the design group's legacy, must be identified — both HW and SW. The remaining new functions will need to be implemented as a part of the overall SoC design process.

After system-level design and the identification of the processors and communications architecture, and other HW or SW IP required for the design, the group must undertake an *IP acquisition* stage. To a large extent, this can be done at least in part in parallel with other work such as system-level design (assuming early identification of major external IP is made) or building golden transaction-level testbench models. Fortunate design groups will be working in companies with both a large legacy of existing well-crafted IP (rather, "virtual components") organized in easy to search databases, or those with access via supplier agreements to large external IP libraries, or at least those with experience at IP search, evaluation, purchase, and integration. For these lucky groups, the problems at this stage are greatly ameliorated. Others with less experience or infrastructure will need to explore these processes for the first time, hopefully making use of IP suppliers' experience with the legal and other processes required. Here, the external standards bodies such as VSIA and VCX have done much useful work that will smooth the path, at least a little. One key issue in IP acquisition is to conduct rigorous and thorough incoming inspection of IP to ensure its completeness and correctness to the greatest extent possible prior to use, and to resolve any problems with quality early with suppliers — long before SoC integration. Every hour spent on this at this stage will pay back in avoiding much longer schedule slip later. The IP quality guidelines discussed earlier are a foundation level for a quality process at this point.

Build a transaction-level golden testbench: The system model built up during the system-level design stage can form the basis for a more elaborated design model, using "transaction-level" abstractions [15], which represents the underlying HW–SW architecture and components in more detail — sufficient detail to act as a functional virtual prototype for the SoC design. This golden model can be used at this stage to verify the microarchitecture of the design and to verify detailed design models for HW IP at the Hardware Description Language (HDL) level within the overall system context. It thus can be reused all the way down the SoC design and implementation cycle.

Define the SoC SW architecture. SoC is, of course, not just about HW [16]. In addition to often defining the right on-chip communications architecture, the choice of processor(s) and the nature of the application domain have a very strong influence on the SW architecture. For example, Real-Time Operating System (RTOS) choice is limited by the processor ports that have been done and by the application domain (OSEK is an RTOS for automotive systems; Symbian OS for portable wireless devices; PalmOS for Personal Digital Assistants, etc.). In addition to the basic RTOS, every SoC peripheral device will need a device driver, hopefully based on reuse and configuration of templates. Various middleware application stacks (e.g., telephony, multimedia image processing) are important parts of the SW architecture; voice and image encoding and decoding on portable devices often is based on assembly code IP for DSPs. There is thus a strong need in defining the SoC to fully elaborate the SW architecture to allow reuse, easy customization, and effective verification of the overall HW–SW device.

Configure and floorplan SoC microarchitecture: At this point, we are beginning to deal with the SoC on a more physical and detailed logical basis. Of course, during high-level architecture and system-level design, the team has been looking at physical implementation issues (although our design process diagram shows everything as a waterfall kind of flow; in reality, SoC design like all electronics design is more of an iterative, incremental process — that is, more akin to the famous "spiral" model for SW). But before beginning detailed HW design and integration, it is important that there is agreement among the team on the basic physical floorplan, that all the IP blocks are properly and fully configured, that the basic microarchitectures (test, power, clocking, bus, timing) have been fully defined and configured, and that HW implementation can proceed. In addition, this process should also generate the downstream verification environments that will be used throughout the implementation processes — whether SW simulation based, emulation based, using rapid prototypes, or other hybrid verification approaches.

DFT architecture and implementation: The test architecture is only one of the key microarchitectures that must be implemented; it is complicated by IP legacy and the fact that it is often impossible to impose one Design-for-Test style (such as BIST or SCAN) on all IP blocks. Rather, wrappers or adaptations of standard test interfaces (such as JTAG ports) may be necessary to fit all IP blocks together into a coherent test architecture and plan.

AMS HW implementation: Most SoCs incorporating AMS blocks use them to interface to the external world. VSIA, among other groups, has done considerable work in defining how AMS IP blocks should be created to allow them to be more easily integrated into mainly digital SoCs (the "Big D/little a" SoC), and guidelines and rules for such integration. Experiences with these rules and guidelines and extra deliverables have been on the whole promising — but they have more impact between internal design groups today than on the industry as a whole. The "Big A/Big D" mixed-signal SoC is still relatively rare.

HW IP assembly and integration: This design step is in many ways the most traditional. Many design groups have experience in assembling design blocks done by various designers or subgroups in an incremental manner, into the agreed on architectures for communications, bussing, clocking, power, etc. The main difference with SoC is that many of the design blocks may be externally sourced IP. To avoid difficulties at this stage, the importance of rigorous qualification of incoming IP and the early definition of the SoC microarchitecture, to which all blocks must conform, cannot be overstated.

SW assembly and implementation: Just as with HW, the SW IP, together with new or modified SW tasks created for the particular SoC under design, must be assembled together and validated as to conformance to interfaces and expected operational quality. It is important to verify as much of the SW in its normal system operating context as possible (see below).

HW and HW–SW verification: Although represented as a single box on the diagram, this is perhaps one of the largest consumers of design time and effort and the major determinant of final SoC quality. Vital to effective verification is the setup of a targeted SoC verification environment, reusing the golden testbench models created at higher levels of the design process. In addition, highly capable, multi-language, mixed simulation environments are important (e.g., SystemC models and HDL implementation models need to be mixed in the verification process and effective links between them are crucial). There are a large number of different verification tools and techniques [17], ranging from SW-based simulation environments, to HW emulators, HW accelerators, and FPGA and bonded-core-based rapid prototyping approaches. In addition, formal techniques such as equivalence checking, and model/property checking have enjoyed some successful usage in verifying parts of SoC designs, or the design at multiple stages in the process.Mixed approaches to HW–SW verification range from incorporating Instruction Set Simulators (ISSs) of processors in SW-based simulation to linking HW emulation of the HW blocks (compiled from the HDL code) to SW running natively on a host workstation, linked in an *ad hoc* manner by design teams or using a commercial mixed verification environment. Alternatively, HDL models of new HW blocks running in an SW simulator can be linked to emulation of the rest of the system running in HW — a mix of emulation and use of bonded-out processor cores for executing SW. It is important that as much of the system SW be exercised in the context of the whole system as possible, using the most appropriate verification technology that can get the design team close to real-time execution speed (no more than 100X slower is the minimum to run significant amounts of software). The trend to transaction-based modeling of systems, where transactions range in abstraction from untimed functional communications via message calls, through abstract bus communications models, through cycle-accurate bus functional models, and finally to cycle and pin-accurate transformations of transactions to the fully detailed interfaces, allows verification to occur at several levels or with mixed levels of design description. Finally, a new trend in verification is assertion-based verification, using a variety of input languages (PSL/Sugar, e, Vera, or regular Verilog and VHDL) to model design properties, which can then be monitored during simulation, either to ensure that certain properties will be satisfied, or certain error conditions never occur. Combinations of formal property checking and simulation-based assertion checking have been created, viz. "semi-formal verification." The most important thing to remember about verification is that armed with a host of techniques and tools, it is essential for design teams to craft a well-ordered verification process, which allows them to definitively answer the question "how do we know that verification is done?" and thus allow the SoC to be fabricated.

Final SoC HW assembly and verification: Often done in parallel or overlapping "those final few simulation runs" in the verification stage, the final SoC HW assembly and verification phase includes final place and route of the chip, any hand-modifications required, and final physical verification (using design rule checking and layout-vs.-schematic (netlist) tools), as well as important analysis steps for issues that occur in advanced semiconductor processes such as IR drop, signal integrity, power network integrity, as well as satisfaction and design transformation for manufacturabilty (OPC, etc.).

Fabrication, testing, packaging, and lab verification: When an SoC has been shipped to fabrication, it would seem time for the design team to relax. Instead, this is an opportunity for additional verification to be carried out — especially more verification of system software running in the context of the hardware design — and for fixes, either of software, or of the SoC HW on hopefully no more than one expensive iteration of the design, to be determined and planned. When the tested packaged parts arrive back for verification in the lab, the ideal scenario is to load the software into the system and have the SoC and its system booted up and running software within a few hours. Interestingly, the most advanced SoC design teams, with well-ordered design methodologies and processes, are able to achieve this quite regularly.

92.9 System-Level Design

As we touched on earlier, when describing the overall SoC design flow, system-level design and SoC are essentially made for each other. A key aim of IP reuse and of SoC techniques such as platform-based design is to make the "back end" (RTL to GDS II) design implementation processes easier, fast and with low-risk, and to shift the major design phase for SoC up in time and in abstraction level to the system level. This also means that the back-end tools and flows for SoC designs do not necessarily differ from those used for complex ASIC, ASSP, and custom IC design — it is the methodology of how they are used, and how blocks are sourced and integrated, that overlays the underlying design tools and flows, which may differ for SoC. However, the fundamental nature of IP-based design of SoC has a stronger influence on the system level.

It is at the system level that the vital tasks of deciding on and validating the basic system architecture and choice of IP blocks are carried out. In general, this is known as "design space exploration." As part of this exploration, SoC platform customization for a particular derivative is carried out, should the SoC platform approach be used. Essentially, one can think of platform design space exploration (DSE) as being a task similar to general DSE, except that the scope and boundaries of the exploration are much more tightly constrained — the basic communications architecture and platform processor choices may be fixed, and the design team may be restricted to choosing certain customization parameters and choosing optional IP from a library. Other tasks include HW–SW partitioning, usually restricted to decisions about key processing tasks, which might be mapped into either HW or SW form and that have a large impact on system performance, energy consumption, on-chip communications bandwidth consumption, or other key attributes. Of course, in multi-processor systems, there are "SW–SW" partitioning or codesign issues as well, deciding on the assignment of SW tasks to various processor options. Again, perhaps 80–95% of these decisions can or are made *a priori*, especially if an SoC is either based on a platform or an evolution of an existing system; such codesign decisions are usually made on a small number of functions that have critical impact.

Because partitioning, codesign, and DSE tasks at the system level involve much more than HW–SW issues, a more appropriate term for this is "function-architecture codesign" [18, 19]. In this codesign model, systems are described on two equivalent levels:

- The functional intent of the system — for example, a network of applications, decomposed into individual sets of functional tasks, which may be modeled using a variety of models of computation such as discrete event, finite state machine, or dataflow.
- The architectural structure of the system — the communications architecture, major IP blocks such as processor(s), memorie(s), and HW blocks, captured or modeled, for example, using some kind of IP or platform configurator.

The methodology implied in this approach is then to build explicit mappings between the functional view of the system and the architectural view, which carry within them the implicit partitioning that is made for both computation and communications. This hybrid model can then be simulated, the results analyzed, and a variety of ancillary models (e.g., cost, power, performance, communications bandwidth consumption, etc.) can be utilized in order to examine the suitability of the system architecture as a vehicle for realizing or implementing the end-product functionality.

The function–architecture codesign approach has been implemented and used in both research and commercial tools [20] and forms the foundation of many system-level codesign approaches going forward. In addition, it has been found to be extremely suitable as the best system-level design approach for platform-based design of SoC [21].

92.10 Interconnection and Communication Architectures for Systems on Chip

This topic is dealt with in more detail in other chapters in this book. Suffice it to say here that current SoC architectures deal in fairly traditional hierarchies of standard on-chip buses: for example, processor-specific buses, high-speed system buses, and lower-speed peripheral buses, using standards such as ARM's AMBA and IBM's CoreConnect [12], and traditional master–slave bus approaches. Recently, there has been considerable interest in network-on-chip communications architectures, based on packet-sw, and a number of approaches have been reported in the literature but this remains primarily a research topic both in universities and industrial research labs. [22]

92.11 Computation and Memory Architectures for Systems on Chip

The primary processors used in SoC are embedded RISCs such as ARM processors, PowerPCs, MIPS architecture processors and some of the configurable processors, designed specifically for SoC such as Tensilica and ARC. In addition, embedded DSPs from traditional suppliers as TI, Motorola, ParthusCeva, and others are also quite common in many consumer applications, for embedded signal processing for voice and image data. Research groups have looked at compiling or synthesizing application-specific processors or coprocessors [23, 24] and these have interesting potential in future SoCs, which may incorporate networks of heterogeneous configurable processors collaborating to offer large amounts of computational parallelism. This is an especially interesting prospect given the wider use of reconfigurable logic that opens up the prospect of dynamic adaptation of SoC to application needs. However, most multi-processor SoCs today involve at most 2–4 processors of conventional design; the larger networks are more often found today in the industrial or university lab.

Although several years ago most embedded processors in early SoCs did not use cache memory-based hierarchies, this has changed significantly over the years, and most RISC and DSP processors now involve significant amounts of Level 1 Cache memory, as well as higher level memory units both on and off chip (off-chip flash memory is often used for embedded software tasks which may be only infrequently required). System design tasks and tools must consider the structure, size, and configuration of the memory hierarchy as one of the key SoC configuration decisions that must be made.

92.12 IP Integration Quality and Certification Methods and Standards

We have emphasized the design reuse aspects of SoC and the need for reuse of both internal and externally sourced IP blocks by design teams creating SoCs. In the discussion of the design process above, we mentioned issues such as IP quality standards and the need for incoming inspection and qualification of IP. The issue of IP quality remains one of the biggest impediments to the use of IP-based design for SoC [25]. The quality standards and metrics available from VSIA and OpenMORE, and their further enhancement help, but only to a limited extent. The industry could clearly use a formal certification body or lab for IP quality that would ensure conformance to IP transfer requirements and the integration quality of the blocks. Such a certification process would be of necessity quite complex due to the large number of configurations possible for many IP blocks and the almost infinite variety of SoC contexts into which they might be integrated. Certified IP would begin to deliver the "virtual components" of the VSIA vision.

In the absence of formal external certification (and such third-party labs seem a long way off, if they ever emerge), design groups must provide their own certification processes and real reuse quality metrics, based on their internal design experiences. Platform-based design methods help due to the advantages of pre-qualifying and characterizing groups of IP blocks and libraries of compatible domain-specific components. Short of independent evaluation and qualification, this is the best that design groups can do currently.

One key issue to remember is that IP not created for reuse, with all the deliverables created and validated according to a well-defined set of standards, is inherently not reusable. The effort required to make a reusable IP block has been estimated to be 50–200% more effort than that required to use it once; however, assuming the most conservative extra cost involved implies positive payback with three uses of the IP block. Planned and systematic IP reuse and investment in those blocks with greatest SoC use potential gives a high chance of achieving significant productivity soon after starting a reuse program. But *ad hoc* attempts to reuse existing design blocks not designed to reuse standards have failed in the past and are unlikely to provide the quality and productivity desired.

92.13 Summary

In this chapter, we have defined System-on-a-Chip and have surveyed a large number of the issues involved in its design. An outline of the important methods and processes involved in SoC design define a methodology that can be adopted by design groups and adapted to their specific requirements. Productivity in SoC design demands high levels of design reuse, and the existence of third-party and internal IP groups and the chance to create a library of reusable IP blocks (true virtual components) are all possible for most design groups today. The wide variety of design disciplines involved in SoC mean that unprecedented collaboration between designers of all backgrounds — from systems experts through embedded software designers through architects through HW designers — is required. But the rewards of SoC justify the effort required to succeed.

References

[1] Hunt, M. and J. Rowson, Blocking in a system on a chip, *IEEE Spectrum*, Vol. 33, no.11, pp. 35–41, Nov. 1996.

[2] Rajsuman, R., *System-on-a-Chip Design and Test*, Artech House, Norwood, MA, 2000.

[3] Keating, M. and P. Bricaud, *Reuse Methodology Manual for System-on-a-Chip Designs*, 1st ed., 1998 (2nd ed., 1999, 3rd ed., 2002), Kluwer Academic Publishers, Dordrecht.

[4] International Technology Roadmap for Semiconductors (ITRS), 2001 edition, URL: http://public.itrs.net/

[5] Virtual Socket Interface Alliance, on the web at URL: http://www.vsia.org. This includes access to its various public documents, including the original Reuse Architecture document of 1997, as well as more recent documents supporting IP reuse released to the public domain.

[6] The Virtual Component Exchange (VCX), Web URL: http://www.thevcx.com/

[7] Chang, H., L. Cooke, M. Hunt, G. Martin, A. McNelly, and L. Todd, *Surviving the SOC Revolution: A Guide to Platform-Based Design*, Kluwer Academic Publishers, Dordrecht, 1999.

[8] Keutzer, K., S. Malik, A.R. Newton, J. Rabaey, and A. Sangiovanni-Vincentelli, System-level design: orthogonalization of concerns and platform-based design, *IEEE Transactions on CAD of ICs and Systems*, 19, 1523, 2000.

[9] Sangiovanni-Vincentelli, A. and G. Martin, Platform-based design and software design methodology for embedded systems, *IEEE Design and Test of Computers*, 18, 23–33, 2001.

[10] *IEEE Design and Test, Special Issue on Platform-Based Design of SoCs*, Vol. 19, no. 6, pp. 4–63, Nov.-Dec. 2002.

[11] Martin, G. and F. Schirrmeister, A design chain for embedded systems, *IEEE Computer, Embedded Systems Column*, Vol. 35, Issue 3, pp. 100–103, March 2002.

[12] Martin, G. and H. Chang, Eds., *Winning the SOC Revolution: Experiences in Real Design*, Kluwer Academic Publishers, Dordrecht, May 2003.

[13] Lysaght, P., FPGAs as Meta-platforms for Embedded Systems, Proceedings of the IEEE Conference on Field Programmable Technology, Hong Kong, Dec., 2002.

[14] Groetker, T., S. Liao, G. Martin, and S. Swan, *System Design with SystemC*, Kluwer Academic Publishers, Dordrecht, May 2002.

[15] Bergeron, J., *Writing Testbenches*, 3rd ed., Kluwer Academic Publishers, Dordrecht, 2003.

[16] Martin, G. and C. Lennard, Invited CICC paper, Improving Embedded SW Design and Integration for SOCs, Custom Integrated Circuits Conference, May, 2000, pp. 101–108.

[17] Rashinkar, P., P. Paterson, and L. Singh, *System-on-a-Chip Verification : Methodology and Techniques*, Kluwer Academic Publishers, Dordrecht, 2001.

[18] Balarin, F., M. Chiodo, P. Giusto, H. Hsieh, A. Jurecska, L. Lavagno, C. Passerone, A. Sangiovanni-Vincentelli, E. Sentovich, K. Suzuki, and B. Tabbara, *Hardware–Software Co-Design of Embedded Systems: The POLIS Approach*, Kluwer Academic Publishers, Dordrecht, The Netherlands, 1997.

[19] Krolikoski, S., F. Schirrmeister, B. Salefski, J. Rowson and G. Martin, Methodology and Technology for Virtual Component Driven Hardware/Software Co-Design on the System Level, paper 94.1, ISCAS 99, Orlando, FL, May 30–June 2, 1999.

[20] Martin, G. and B. Salefski, System level design for SOC's: a progress report — two years on, in *System-on-Chip Methodologies and Design Languages*, Mermet, J., Ed., Kluwer Academic Publishers, Dordrecht, 2001, pp. 297–306, chap. 25.

[21] Martin, G., Productivity in VC reuse: linking SOC platforms to abstract systems design methodology, in *Virtual Component Design and Reuse*, Seepold, R. and Madrid, N.M., Eds., Kluwer Academic Publishers, Dordrecht, 2001, pp. 33–46, chap. 3.

[22] Jantsch, A. and H. Tenhunen, Ed., *Networks on Chip*, Kluwer Academic Publishers, Dordrecht, 2003.

[23] Kithail, V., S. Aditya, R. Schreiber, B. Ramakrishna Rau, D.C. Cronquist, and M. Sivaraman, PICO: automatically designing custom computers, *IEEE Computer*, 35, 39–47, 2002.

[24] Callahan, T.J., J.R. Hauser, and J. Wawrzynek, The Garp architecture and C compiler, *IEEE Computer*, 33, 62–69, 2000.

[25] DATE 2002 Proceedings, Session 1A: How to Choose Semiconductor IP?: Embedded Processors, Memory, Software, Hardware, Proceedings of DATE 2002, Paris, Mar., 2002, pp. 14–17.

93

Platform-Based and Derivative Design

Luca P. Carloni
University of California at Berkeley

Fernando De Bernardinis
University of California at Berkeley
Università di Pisa

Alberto L. Sangiovanni-Vincentelli
University of California at Berkeley

Marco Sgroi
University of California at Berkeley
DoCoMo Euro-labs

93.1 Introduction

The motivations behind *Platform-Based Design* [10] originate in the *disaggregation of the electronic industry*, a phenomenon that has been evolving over the past decade and whose ultimate result consists in the move from a vertically oriented model into a horizontally oriented one. In the past, electronic companies used to maintain full control of the production cycle from product definition to final manufacturing. Nowadays, the identification of a new market opportunity, the definition of the detailed system specifications, the development of the components, the assembly of these components, and the manufacturing of the final product are tasks that are mostly performed by distinct organizations. In fact, the complexity of electronic designs and the number of technologies that must be mastered to bring to market winning products have forced electronic companies to focus on their core competence. In this scenario, the integration of the design chain becomes a serious problem, whose most delicate aspects occur at the *hand-off points* from one company to another. These must be unambiguously defined to avoid the risk of costly redesigns. Furthermore, each organization that contributes a component to the final product naturally strives for a position that allows it to make continuous adjustments and accomodate last-minute engineering changes. A most dramatic consequence of this quest for flexibility is the rise of standardized chips at the expense of custom and ASIC solutions. In fact, due to the increasing cost of mask ownership, only high-volume ASICs remain economically feasible for those semiconductor companies that both design and manufacture silicon. Instead, standardized chips that are highly reconfigurable (in hardware) and programmable (in software) provide the advantage of spreading the multimillion-dollar cost of a design across a range of applications. This shift toward reconfigurable and programmable solutions is a defining characteristic of embedded system design.

Platform-based design addresses the above challenges by providing the following advantages:

It lays the foundation for developing economically feasible design flows because it is a structured methodology *that theoretically limits the space of exploration, yet still achieves superior results in the fixed time constraints of the design*. A platform is an abstraction layer in the design flow that facilitates a number of possible refinements into a subsequent abstraction layer in the design flow. The design process progresses through a series of platforms and each platform layer defines bounds for what is achievable by mapping into it, while offering a faster design and lower risk.

It provides a formal mechanism to identify the most critical hand-off points in the design chain: the hand-off point between system companies and IC design companies and the one between IC design companies and manufacturing companies represent the *articulation points* of the overall design process. For instance, semiconductor companies need to minimize risks when designing standardized chips. Hence, they need to have a fairly complete characterization of the application spaces that they wish to target together with the associated constraints in terms of affordable costs and performance levels. By the same token, system companies need to have an accurate characterization of the capabilities of the chips in terms of performance such as power consumption, size and timing, as well as *"Application Program Interfaces" (APIs)* that allow the mapping of their application into the chip at a fairly abstract level. APIs must then support a number of tools to ease the possibly automatic generation of the personalization of the programmable components of the chips.

It eliminates costly design iterations because it enables *derivative design*, that is, the technique of building an application-specific product by assembling and configuring platform components in a rapid and reliable manner. For instance, a hardware platform is a family of architectures satisfying a set of constraints imposed to allow reuse of hardware and software components. Then, an API platform can be developed to effectively extend the hardware platform toward the application software, thus enabling quick, reliable, and derivative design.

This paper is organized as follows. In Section 93.2, we define the main principles of platform-based design. Since the term "platform" has been used in several ways by the various segments of the electronic industry, our goal is to provide a precise reference that may be used as the basis for reaching a common understanding in the electronic system and circuit design community. In the following sections, we discuss the use of these principles in system design and in silicon implementation. First, we revisit the discussion on the platforms that define the articulation points between system definition and implementation (Section 93.3). Then, we discuss network platforms, which are an important example of application of *pure* system design (Section 93.4) and analog platforms, which are destined to have a key role in determining the overall design time and effort of silicon implementations (Section 93.5).

93.2 Platform-Based Design

The basic tenets of platform-based design are:

- The identification of design as a *meeting-in-the-middle process,* where successive refinements of specifications meet with abstractions of potential implementations.
- The identification of precisely defined layers where the refinement and abstraction processes take place. Each layer supports a design stage providing an opaque abstraction of lower layers that allows accurate performance estimations. This information is incorporated in appropriate parameters that annotate design choices at the present layer of abstraction. These layers of abstraction are called *platforms* to stress their role in the design process and their solidity.

In practice, a platform is a library of components that can be assembled to generate a design at that level of abstraction. This library not only contains *computational* blocks that carry out the appropriate computation but also *communication* components that are used to interconnect the functional components. Each element of the library has a characterization in terms of performance parameters, together with the functionality it can support. For every platform level, there is a set of methods used to map the upper layers of abstraction into the platform and a set of methods used to estimate performances of

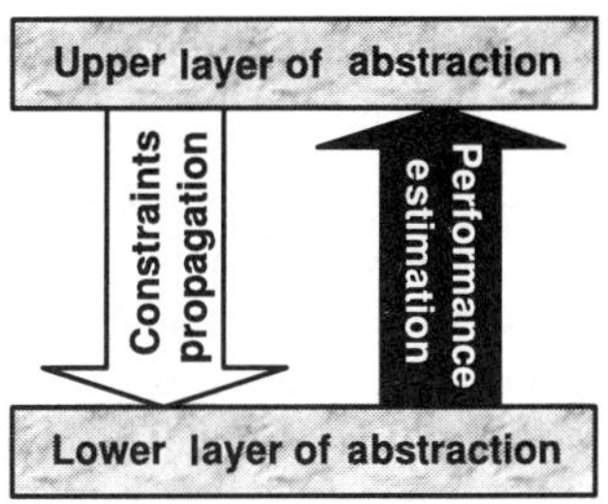

FIGURE 93.1   Interactions between abstraction layers.

lower level abstractions. As illustrated in Figure 93.1, the meeting-in-the-middle process is a combination of two efforts:

- top-down: map an instance of the top platform into an instance of the lower platform and propagate constraints;
- bottom-up: build a platform by defining the *library* that characterizes it and a performance abstraction (e.g., number of literals for tech. independent optimization, area and propagation delay for a cell in a standard cell library).

A *platform instance* is a set of architecture components that are selected from the library and whose parameters are set. Often, the combination of two consecutive layers and their "filling" can be interpreted as a unique abstraction layer with an "upper" view, the top abstraction layer, and a "lower" view, the bottom layer. A *platform stack* is a pair of platforms, along with the tools and methods that are used to map the upper layer of abstraction into the lower level. Note that we can allow a platform stack to include several substacks if we wish to span a large number of abstractions.

Platforms should be defined to eliminate large loop iterations for affordable designs: they should restrict design space via new forms of regularity and structure that surrender some design potential for lower cost and first-pass success. The library of function and communication components is the design space that we can explore at the appropriate level of abstraction. Establishing the number, location, and components of intermediate platforms is the essence of platform-based design. In fact, designs with different requirements and specification may use different intermediate platforms, hence different layers of regularity and design-space constraints. A critical step of the platform-based design process is the definition of intermediate platforms to support *predictability*, which enables the abstraction of implementation detail to facilitate higher-level optimization, and *verifiability*, that is, the ability to formally ensure correctness. On the other hand, when design-time and product volume permit it, it may be useful to skip intermediate platforms. This is equivalent to enlarge the design space and, therefore, can potentially produce a superior design. However, even if a "large-step-across-platform flow" can be adopted, there is still a benefit, from an evaluation standpoint, in having a lower bound on the optimality of the feasible design as the one that can be provided by a more constrained and predictable flow. Naturally, the larger the step across platforms, the more difficult it is to predict the performance, optimize at system level, and provide a tight lower bound. In fact, the design space for this approach may actually be smaller than the one obtained with smaller steps because it becomes harder to explore meaningful design alternatives and the restriction on search impedes complete design space exploration. Ultimately, predictions/abstractions may be so inaccurate that design optimizations are misguided and the lower bounds are incorrect.

It is important to emphasize that the Platform-Based Design paradigm applies to all levels of design. While it is rather easy to grasp the notion of a programmable hardware platform, the concept is completely general and should be exploited through the entire design flow to solve the design problem. In the following sections, we will show that platforms can be applied to low levels of abstraction such as analog components, where flexibility is minimal and performance is the main focus, as well as to very high levels of abstraction such as networks, where platforms have to provide connectivity and services. In the former case, platforms abstract hardware to provide (physical) implementation, while in the latter communication services abstract software layers (protocol) to provide global connectivity.

93.3   Platforms at the Articulation Points of the Design Process

As we mentioned above, the key to the application of the design principle is the careful definition of the platform layers. Platforms can be defined at several point of the design process. Some levels of abstraction are more important than others in the overall design trade-off space. In particular, the articulation point between system definition and implementation is a critical one for design quality and time. Indeed, the very notion of platform-based design originated at this point (see [1, 5–7]). In studying this articulation point (see [10] for full details), we have discovered that at this level there are indeed two distinct platforms that form the *system platform stack* that need to be defined together with the methods and tools necessary to link the two: an *architecture platform* and an API platform. The API platform allows system designers to use the *services* that a (micro-) architecture offers them. In the world of Personal Computers, this concept is well known and is the key to the development of application software on different hardware that share some commonalities allowing the definition of a unique API.

(Micro-) Architecture Platforms

Integrated circuits used for embedded systems will most likely be developed as an instance of a particular *(micro-) architecture platform.* That is, rather than being assembled from a collection of independently developed blocks of silicon functionalities, they will be derived from a specific *family of micro-architectures,* possibly oriented toward a particular class of problems, which can be extended or reduced by the system developer. The elements of this family are a kind of "hardware denominator" that could be shared across multiple applications. Hence, an architecture platform is a family of micro-architectures that share some commonality, the library of components that are used to define the micro-architecture. Every element of the family can be obtained quickly through the personalization of an appropriate set of parameters controlling the micro-architecture. Often,the family may have additional constraints on the components of the library that can or should be used. For example, a particular micro-architecture platform may be characterized by the same programmable processor and the same interconnection scheme, while the peripherals and the memories of a particular implementation may be selected from the predesigned library of components depending on the particular application. Depending on the implementation platform that is chosen, each element of the family may still need to go through the standard manufacturing process, including mask making. This approach then conjugates the need of saving design time with the optimization of the element of the family for the application at hand. Although it does not solve the mask cost issue directly, it should be noted that the mask cost problem is primarily due to generating multiple mask sets for multiple design spins, which is addressed by the architecture platform methodology.

The less constrained the platform, the more freedom a designer has in selecting an instance and the more potential there is for optimization, if time permits. However, more constraints mean stronger standards and easier addition of components to the library that defines the architecture platform (as with PC platforms). Note that the basic concept is similar to the cell-based design layout style, where regularity and the reuse of library elements allow faster design time at the expense of some optimality. The trade-off between design time and design "quality" needs to be kept in mind. The economics of the design problem must dictate the choice of design style. The higher the granularity of the library, the more leverage we have in shortening the design time. Given that the elements of the library are reused, there is a strong incentive to optimize them. In fact, we argue that the "macro-cells" should be designed with great care and attention to area and performance. It makes also sense to offer a variation of cells with the same functionality but with implementations that differ in performance, area and power dissipation. Architecture platforms are, in general, characterized by (but not limited to) the presence of programmable components. That means that each of the platform instances that can be derived from the architecture platform maintains enough flexibility to support an application space that guarantees the production volumes required for economically viable manufacturing.

The library that defines the architecture platform may also contain reconfigurable components. Reconfigurability comes in two flavors. With runtime reconfigurability, FPGA blocks can be customized by the user without the need of changing mask set, thus saving both design cost and fabrication cost. With design-time reconfigurability, where the silicon is still application-specific, only design time is reduced.

An *architecture platform instance* is derived from an architecture platform by choosing a set of components from the architecture platform library and/or by setting parameters of reconfigurable components of the library. The flexibility, or the capability of supporting different applications, of a platform instance is guaranteed by programmable components. Programmability will ultimately be of various forms. One is software programmability to indicate the presence of a microprocessor, DSP, or any other software-programmable component. Another is hardware programmability to indicate the presence of reconfigurable logic blocks such as FPGAs, whereby logic function can be changed by software tools without requiring a custom set of masks. Some of the new architecture and/or implementation platforms being offered on the market mix the two into a single chip. For example, Triscend, Altera, and Xilinx are offering FPGA fabrics with embedded hard processors. Software programmability yields a more flexible solution, since modifying software is, in general, faster and cheaper than modifying FPGA personalities. On the other hand, logic functions mapped on FPGAs execute orders of magnitude faster and with much less power than the corresponding implementation as a software program. Thus, here the trade-off is between flexibility and performance.

API Platform

The concept of architecture platform by itself is not enough to achieve the level of application software reuse we require. The architecture platform has to be abstracted at a level where the application software "sees" a high-level interface to the hardware that we call API or Programmers' Model. A software layer is used to perform this abstraction. This layer wraps the essential parts of the architecture platform:

- the programmable cores and the memory subsystem via a Real Time Operating System (RTOS),
- the I/O subsystem via the device drivers, and
- the network connection via the network communication subsystem.

In our framework, the API or Programmers' Model is a unique abstract representation of the architecture platform via the software layer. With an API thus defined, the application software can be reused for every platform instance. Indeed, the Programmers' Model (API) is a platform itself that we can call the API platform. Of course, the higher the abstraction level at which a platform is defined, the more instances it contains. For example, to share source code, we need to have the same operating system but not necessarily the same instruction set, while to share binary code, we need to add the architectural constraints that force us to use the same ISA, thus greatly restricting the range of architectural choices.

The RTOS is responsible for the scheduling of the available computing resources and of the communication between them and the memory subsystem. Note that in several embedded system applications, the available computing resources consist of a single microprocessor. In others, such as wireless handsets, the combination of an RISC microprocessor or controller and DSP has been used widely in 2G, now for 2.5G and 3G, and beyond. In set-top boxes, an RISC for control and a media processor has also been used. In general, we can imagine a multiple core architecture platform where the RTOS schedules software processes across different computing engines.

System Platform Stack

The basic idea of system platform stack is captured in Figure 93.2. The vertex of the two cones represents the combination of the API or Programmers' Model and the architecture platform. A system designer maps its application into the abstract representation, which "includes" a family of architectures that can be chosen to optimize cost, efficiency, energy consumption, and flexibility. The mapping of the application into the actual architecture in the family specified by the Programmers' Model or API can be carried out, at least in part, automatically if a set of appropriate software tools (e.g., software synthesis, RTOS synthesis, device-driver synthesis) is available. It is clear that the synthesis tools have to be aware of the architecture features as well as of the API. This set of tools makes use of the software layer to go from the API platform to the architecture platform. Note that the system platform effectively decouples the application development process (the upper triangle) from the architecture implementation process (the lower triangle). Note also

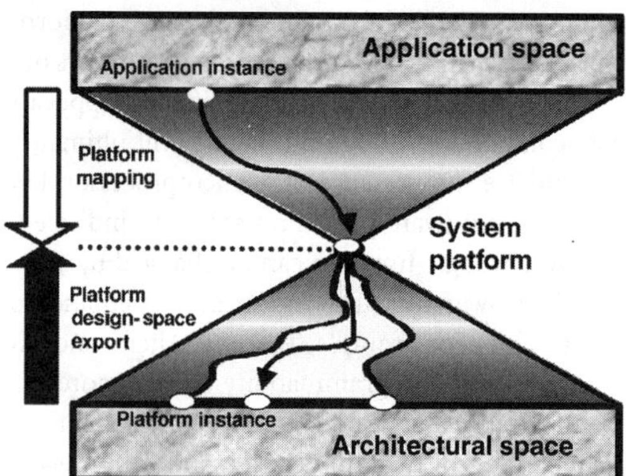

FIGURE 93.2   System platform stack.

that, once we use the abstract definition of "API" as described above, we may obtain extreme cases such as traditional PC platforms on the one hand and full hardware implementation on the other. Of course, the programmer model for a full custom hardware solution is trivial, since there is a one-to-one map between functions to be implemented and physical blocks that implement them. In this latter case, platform-based design amounts to adding to traditional design methodologies some higher level of abstractions.

93.4   Network Platforms

In distributed systems, the design of the communication protocols and channels that support the communication among the system components is a difficult task due to the tight constraints on performances and cost. To make the communication design problem more manageable, designers usually decompose the communication function into distinct protocol layers, and design each layer separately. According to this approach, of which the OSI Reference Model [12] is a particular instance, each protocol layer together with the lower layers defines a platform that provides communication services to the upper layers and to the application-level components. Identifying the most effective layered architecture for a given application requires one to solve a tradeoff between performance, which usually increases by minimizing the number of layers, and design manageability, which improves with the number of the intermediate steps. Present embedded system applications, due to their tight constraints, increasingly demand the codesign of protocol functions (e.g., MAC and routing protocols in sensor networks) that were traditionally assigned to different layers and designed separately. The definition of an optimal layered architecture, the design of the correct functionality for each protocol layer, and the design space exploration for the choice of the physical implementation must be supported by tools and methodologies that allow to evaluate the system performance and guarantee the satisfaction of the constraints after each step. For these reasons, we believe that the Platform-Based Design principles and methodology outlined above provide the right framework to design communication networks. In this section, first we formalize the concept of Network Platform, and then we outline a methodology for selecting, composing, and refining Network Platforms (NPS) [11].

Definitions

A *NP* is a library of resources that can be selected and composed together to form a Network Platform Instance (NPI) and support the interaction among a group of interacting components.

The structure of an NPI is defined by abstracting computation resources as nodes and communication resources as links. Ports interface nodes with links or with the environment of the NPI. The structure of an NPI is defined by a set of nodes and the links connecting them; the structure of a node or a link is defined by its input and output ports.

The behaviors and the performances of an NPI are defined in terms of the type and the quality of the communication services it offers. We formalize the behaviors of an NPI using the Tagged Signal Model [8]. NPI components are modeled as processes, and events model the instances of the send and receive actions of the processes. An event is associated with a message that has a type and a value and with tags that specify attributes of the corresponding action instance (e.g., when it occurs in time). The set of behaviors of an NPI is defined by the intersection of the behaviors of the component processes.

A NPI is defined as a tuple $NPI = (L, N, P, S)$, where

- $L = \{L_1, L_2, ..., L_{Nl}\}$ is a set of directed links.
- $N = \{N_1, N_2, ..., N_{Nn}\}$ is a set of nodes.
- $P = \{P_1, P_2, ..., P_{Np}\}$ is a set of ports. A port P_i is a triple (N_i, L_i, d), where $N_i \in N$ is a node, $L_i \in L \cup Env$ is a link or the NPI environment and $d = in$ if it is an input port, $d = out$ if it is an output port. The ports that interface the NPI with the environment define the sets $P^{in} \subseteq P$ (input ports), $P^{out} \subseteq P$ (output ports).
- $S = \cap_{Nn+Nl} R_i$ is the set of behaviors, where R_i indicates the set of behaviors of a resource that can be a link in L or a node in N.

The basic services provided by an NPI are called *Communication Services (CS)*. A CS consists of a sequence of message exchanges through the NPI from its input to its output ports. A CS can be accessed by NPI users through the invocation of send and receive *primitives*, whose instances are modeled as events. An *NPI API* consists of the set of primitives that can be invoked by the NPI users to access the CS. For the definition of an NPI API, it is essential to specify not only the service primitives but also the type of CS they provide access to (e.g., reliable send, out-of-order delivery, etc.). Formally, a CS is a tuple $(\bar{P}^{in}, \bar{P}^{out}, M, E, h, g, <^t)$, where $\bar{P}^{in} \subseteq P^{in}$ is a nonempty set of NPI input ports, $\bar{P}^{out} \subseteq P^{out}$ is a nonempty set of NPI output ports, M is a nonempty set of messages, E is a nonempty set of events, h is a mapping $h : E \to (\bar{P}^{in} \cup \bar{P}^{out})$ that associates each event with a port, g is a mapping $g : E \to M$ associating each event with a message, and $<^t$ is a total order on the events in E.

A CS is defined in terms of the number of ports, that determine, for example, if it is a unicast, multicast or broadcast CS, the set M of messages representing the exchanged information, the set E including the events that are associated with the messages in M and model the instances of the send and receive method invocations. The CS concept allows to express a *correlation* among events, and explicit, for example, if two events are from the same source or are associated with the same message.

Quality of Service

NPIs can be classified according to the type, the quality, and the cost of the CS they offer. Rather than in terms of event sequences, a CS is more conveniently described using *QoS parameters* like error rate, latency, throughput, jitter, and *cost parameters* like consumed power and manufacturing cost of the NPI components. QoS parameters can be defined using annotation functions that associate individual events with quantities, such as the time when an event occurs and the power consumed by an action. Then, one can compare the values of pairs of input and output events associated with the same message to quantify the error rate, or compare the timestamp of events observed at the same port to compute the delay or the jitter. The most relevant QoS parameters are defined below using a notation where $e^{i,j} \in e^{M,(\bar{P}^{in} \cup \bar{P}^{out})}$ indicates an event carrying the ith message and observed at the jth port, $v(e)$ and $t(e)$ represents, respectively, the value of the message carried by event e and the timestamp of the action modeled by event e.

delay: The communication delay of a message is given by the difference between the timestamps of the input and output events carrying that message. Assuming that the ith message is transferred from input port j_1 to output port j_2, the delay Δ_i of the ith message, the average delay Δ_{Av} and the peak delay Δ_{Peak} are defined, respectively, as $\Delta_i = t(e^{i,j^2}) - t(e^{i,j^1})$, $\Delta_{Av} = \sum_{i=1}^{|M|} t(e^{i,j^2}) - t(e^{i,j^1})/|M|$ *and* $\Delta_{peak} = max_i\{t(e^{i,j^2}) - t(e^{i,j^1})\}$.

Throughput: The throughput is given by the number of output events in an interval (t_0, t_1), that is, the cardinality of the set $\Theta = \{e_i \in E | h(e_i) \in \bar{P}^{out}, t(e_i) \in (t_0, t_1)\}$.

Error rate: The message error rate (MER) is given by the ratio between the number of lost or corrupted output events and the total number of input events. Given $LostM = \{e_i \in E | h(e_i) \in \bar{P}^{in}, \neg\, \exists e_j \in E \text{ s.t. } h(e_j) \in \bar{P}^{out}, g(e_j) = g(e_i)\}$, $CorrM = \{e_i \in E | h(e_i) \in \bar{P}^{in}, \exists e_j \in E \text{ s.t. } h(e_j) \in \bar{P}^{out}, g(e_j) = g(e_i), v(e_j) \neq v(e_i)\}$, and $InM = \{e_i \in E | h(e_i) \in \bar{P}^{in}\}$, the message error rate $MER = (\text{Lost } M| + |CorrM)/|InM|$. Using information on message encoding, MER can be converted to Packet and Bit Error Rate.

The number of CS that an NPI can offer is large; thus, the concept of Class of Communication Services (CCS) is introduced to simplify the description of an NPI. CCS define a new abstraction (and therefore another platform) that groups together CS of similar type and quality. For example, a CCS may include all the CS that transfer a periodic stream of messages with no errors, and another CCS all the CS that transfer a stream of input messages arriving at a bursty rate and with a 10^{-6} bit error rate. CCS can be identified based on the type of messages (e.g., packets, audio samples, video pixels, etc.), the input arrival pattern (e.g., periodic, bursty, etc.), and the range of QoS parameters. For each NPI supporting multiple CS, there are several ways to group them into CCS. It is task of the NPI designer to identify in each case the most useful CCS and provide the proper abstractions to facilitate the use of the NPI.

Design of Network Platforms

The design methodology we have defined derives an NPI implementation by successive refinement from the specification of the behaviors of the interacting components and the declaration of the constraints that the NPI implementation must satisfy. The most abstract NPI is defined by a set of end-to-end direct logical links connecting pairs of interacting components. Communication refinement of an NPI defines at each step a more detailed NPI' by replacing one or multiple links in the original NPI with a set of components or NPIs. During this process, another NPI can be used as a resource to build other NPIs. A correct refinement procedure generates an NPI' that provides CS equivalent to those offered by the original NPI with respect to the constraints defined at the upper level. A typical communication refinement step requires to define both the structure of the refined NPI', that is, its components and topology, and the behavior of these components, that is, the protocols deployed at each node. One or more NP components (or predefined NPIs) are selected from a library and composed to create CS of higher quality. Two types of compositions can be used. One type consists of choosing an NPI and extending it with a protocol layer to create CS at a higher level of abstraction (vertical composition). The other type is based on the concatenation of NPIs at the same abstraction level using an adapter (or gateway) that maps sequences of events between the ports being connected (horizontal composition). This design methodology has been applied to the design of Picoradio sensor networks as discussed in [11].

93.5   Analog Platforms

Emerging applications such as multimedia devices (video cell phones, digital cameras, wireless PDAs, to mention but a few) are driving the SoC market toward the integration of analog components in almost every system. Even when analog solutions could be functionally replaced by digital ones, power consumption or performance requirements may be so tight that the digital approach is out of the question. Moreover, we are experiencing a growing lag between the design of large digital blocks and the design of smaller (sometimes much smaller) analog components. The problem is that analog design has traditionally been the most difficult discipline of IC design. There are several reasons for this, ranging from the effects that physical implementations have on the functionality of analog circuits to the lack of abstraction levels and estimation mechanisms. In the digital case, functionality depends on discrete sequences of discrete (binary) signals. Not so in the analog case, where continuous sequences (waveforms) of continuous values encode the information that we need to manipulate and use. For this reason, any second-order physical effect may have a significant impact on the function and performance of an analog circuit. In order to overcome these issues, the design of analog components has been pivoted around low-level, nonsystematic "clever tricks" that involve transistor layout and parameter selection, thus making it virtually impossible to use higher levels of abstraction.

Today, system-level analog design is a design process dominated by heuristics. Given a set of specifications/requirements that describes the system to be realized, the selection of a feasible (let alone optimal) implementation architecture comes mainly out of experience. Usually, what is achieved is just a feasible point at the system level, while optimality is sought locally at the circuit level. The reason for this is the difficulty in the analog world of knowing whether something is realizable without actually attempting to design the circuit. The number of effects to consider and their complex interrelations make this problem approachable only through the experience of past designs. Platform-based design can provide the necessary insight to develop a methodology for analog components that takes into consideration system-level specifications and can choose among a set of possible solutions including digital approaches wherever it is feasible to do so. If the "productivity gap" between analog and digital components is not overcome, time-to-market and design quality of SoC will be seriously affected by the small analog sections required to interface with the real world. Moreover, SoC designs will expose system-level explorations that would be severely limited if the analog section is not provided with a proper abstraction level that allows system performance estimation in an efficient way and across the analog/digital boundary. Therefore, there is a strong need to develop more abstract design techniques that can encapsulate analog design into a methodology that could shorten design time without compromising the quality of the solutions, leading to a *hardware/software/analog* codesign paradigm for embedded systems.

Definitions

The term *Analog Platform* (AP) indicates an architectural resource to map analog functionalities during system-level exploration phases and constrain exploration to the feasible space of the considered implementation (or set of implementations). Therefore, an AP consists of a set of behavioral models and of matching performance models. *Behavioral models* are parameterized mathematical models of the analog functionality that introduce at the functional level a number of nonidealities attributable to the implementation and not intrinsic in the functionality itself, such as distortion, bandwidth limitations, and noise. Nonideal effects, as well as the principal performance figures of the functionality (e.g., gain, power consumption, and area) are embedded in the behavioral models in the form of parameters. *Performance models* constrain these parameters to satisfy a mathematical relation so that only feasible instances of the behavioral model (models with a complete set of parameter values) can be selected with respect to the considered architecture. Performance parameters for APs are derived bottom-up from simulation data. Different from common approaches that rely on regression schemes, relations on performance parameters are directly modeled by means of characteristic functions. Performance model can be formally defined with the following definitions:

1. *Input space* $\mathcal{I}$: Given a circuit C and m parameters controlling its instances, $\mathcal{I}_C \subseteq \mathbb{R}^m$ is the set of m-tuples (parameter values) over which we want to characterize C.
2. *Output space* $\mathcal{O}$: Given a circuit C and n performance figures completely characterizing its behavioral model, $\mathcal{O}_C \subseteq \mathbb{R}^n$ is the set of m-tuples (per formance values) that are achievable by C.
3. *Evaluation function* ϕ: Given a circuit C, $\mathcal{I}_C$, and $\mathcal{O}_C$, $\phi_C : \mathcal{I} \to \mathcal{O}$ allows translating a parameter m-tuple set into a performance n-tuple set.
4. *Performance relation* $\mathcal{P}$: Given a circuit C, $\mathcal{I}_C$, $\mathcal{O}_C$, and, ϕ, we define the performance relation of C given $\mathcal{I}_C$ and ϕ_C to be $\mathcal{P}_C$ on $\mathbb{R}^n$ that hold only for points $o \in \mathcal{O}_C$. With a little abuse in notation, we will denote both the performance relation characteristic function $\chi_\mathcal{P}(x) : \mathbb{R}^n \to \{0,1\}$ and the relation itself with $\mathcal{P}_C(x)$.

For example, in the Low Noise Amplifier (LNA) shown in Figure 93.3, we may define $\mathcal{I}$ as $\{W_{in}, L_{in}, I_{bias}\}$, where the first two quantities are related to the input transistor and *Ibias* is the bias current. A platform instance defines specific values for $\{W_{in}^*, L_{in}^*, I_{bias}^*\}$, that is, the actual circuit to be considered. Continuing with the LNA example, the output space $\mathcal{O}$ can be defined as $\{Gain, Power, Noise\}$. Finally, ϕ can be defined by circuit simulation and/or circuit equations. The LNA platform performances are then

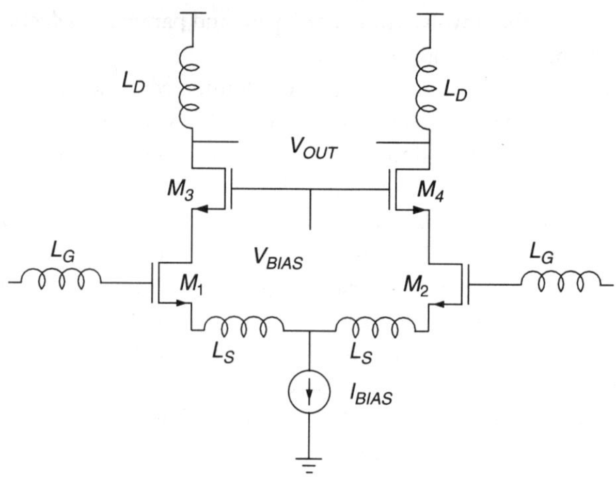

FIGURE 93.3   Schematics for the LNA used for generating the performance model $\mathcal{P}$ shown in Figure 93.4.

represented by $\mathcal{P}$ (*Gain, Power, Noise*) = 1, which are satisfied only by those n-tuples of performance figures that are actually obtainable. In this way, it is possible to capture all the interrelations among the different performance figures and annotate behavioral models consequently.

A key concept of APs is that only $\mathcal{O}$ has to be exported at the current level of abstraction (*platform opaqueness*), because the function $\phi(\cdot)$ and $\mathcal{I}$ are not necessary to evaluate the performances of mapped behaviors. Because of this, APs are also suitable candidates to protect sensitive data when exporting IPs.

In order to allow hierarchical exploitation of platform instances, the following operations have to be defined and implemented in terms of $\mathcal{P}$'s:

- Abstraction — given a platform relation $\mathcal{P}(\textit{Gain, Power, Noise, IP3, SR}) = 1$, we may want to derive an abstracted view of the platform (*virtual* platform) that only relates a subset of performance figures, for example, $\mathcal{P}'(\textit{Gain, Power, Noise}) = 1 \Leftrightarrow \exists IP3^*, SR^*$ s.t. $\mathcal{P}(\textit{Gain,Power, Noise, IP3}^*, SR^*)=1$. This operation is needed every time we use models at higher levels of abstraction (see Figure 93.5).
- Composition — given two platforms A and B with parameters ζ_a and ζ_b and some interface parameters λ (e.g., the output load for A/input load for B), we may want to derive $\mathcal{P}_{AoB}(\zeta_a,\zeta_b)=1 \Leftrightarrow \exists \lambda$ s.t. $\mathcal{P}_A(\zeta_a,\lambda)=1$ and $\mathcal{P}_B(\zeta_b,\lambda)=1$. $\mathcal{P}_{AoB}$ represents the set of compatible performances of the composition $A o B$.
- Merge — given n platforms for the same functionality, a super-set platform can be defined by or-ing the respective $\mathcal{P}$ after moving to a common level of abstraction. $\mathcal{P}_{merge} = 1 \Leftrightarrow \exists i$ s.t. $\mathcal{P}_i=1$.
- Closure — given a system implemented with a set of APs, closure generates a new platform at a different level of abstraction so as to implement system platform stacks.

The generation of platform performance models $\mathcal{P}s$ is in general accomplished through sampling $\phi(\cdot)$ over the input space $\mathcal{I}$. This is because in the general case $\phi(\cdot)$ is a complex nonlinear function, and no explicit representation for it might be available (in order to provide accurate characterizations, $\phi(\cdot)$ may be implicitly defined by circuit simulation). The goal is to obtain a smooth continuous approximation of the performance relation $\mathcal{P}$. This phase constitutes the bottom-up phase for building a platforms library and is the crucial step for achieving accurate performance annotations. The characterization of a platform over its input space is exponentially complex with the dimensionality of $\mathcal{I}$. Also, the characterization process generates large amounts of data that need to be effectively represented. The first problem requires some hints from the analog designer to limit the dimensionality of $\mathcal{I}$, that is, which "knobs" are most meaningful for defining the platform and which constraints on the value of each parameter can be exploited to prune the characterization space. Furthermore, the parameter space $\mathcal{I}$ of a circuit is usually much smaller than its bounding hypercube. For example, given n MOS sharing the same bias current $\mathcal{I}_{bias}$, their physical dimensions have to allow all of them to operate in saturation in order for the circuit to be functional. Therefore, the performance model is not affected if we limit $\mathcal{I}$ to some kind of "manifold" defined by a set of constraints over the variables in $\mathcal{I}$. Examples of constraints are correct biasing conditions, DC levels, or minimum gain/bandwidth required. The design of experiments techniques may also be used to achieve optimal accuracy/complexity tradeoffs. Even so, large numbers of multi-dimensional samples may be generated, on the order of thousands or tens of thousands. As Figure 93.4 illustrates, an effective way of representing the information provided by these simulations is to use machine learning techniques, in particular, Support Vector Machines (SVMs). As shown in [3], SVMs make it possible to store only a small portion of the original data (typically 10% or lower) while providing multidimensional approximation that enables the extrapolation of performance figures out of the available data based on some continuity assumptions on ϕ.

Analog Platform Stacks

An Analog Platform Stack contains several layers of abstraction with the appropriate representation at each level. In the analog world, finding a model for a platform is a complex matter. Behavioral models allow fast simulations, thus enabling more extensive explorations/optimizations. Furthermore, they are tailored for a specific platform to include specific idiosyncrasies of the platform at the behavioral level. By constraining models to reflect platform performances, that is, constraining model parameters to satisfy the platform $\mathcal{P}$, a mapping of functionality onto architecture is achieved.

Analog Platform Stacks provide hierarchies of behavioral models at different levels of abstraction to reflect the refinement process typical of top-down flows. In order to promote design space exploration, all the models are derived from root models characterized by functionality families (see Figure 93.5). For example, we can have LNA, mixer and PLL families, which in turn generate trees of models at different levels of abstraction. The nodes of these trees represent higher-level platforms and the leaves represent implementation platforms. By exploiting the abstraction and merge operations defined above, all

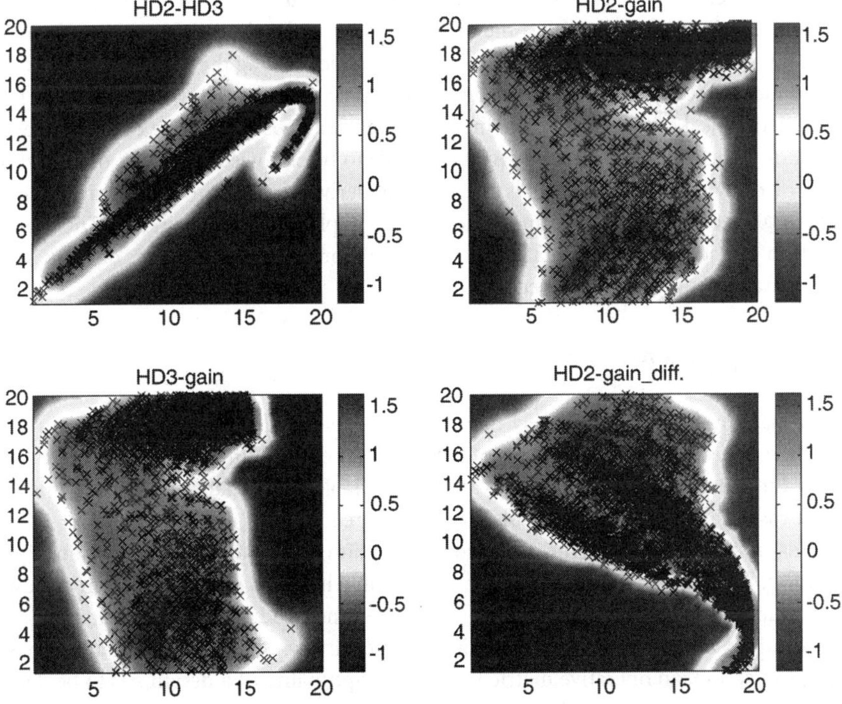

FIGURE 93.4   Projections of a seven-dimensional $\mathcal{P}$ for an LNA.

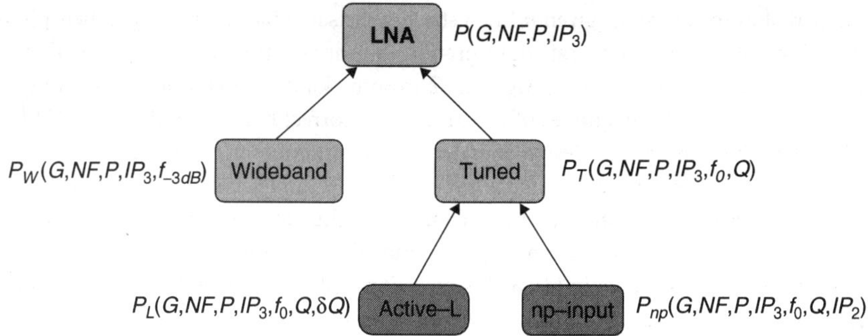

FIGURE 93.5 Sample model hierarchy for an LNA platform. The root node provides performance constraints for a generic LNA, which is then refined by more detailed $\mathcal{P}$ for specific classes of LNAs.

platforms in a given family can be included in a tree originating from the same root model (the most abstract platform). The refinement process proceeds by selecting branches in the tree and, therefore, more detailed platforms, transforming constraints from a more abstract platform to a more detailed platform (exactly as described in the digital case). The mapping process ends with the selection of a platform instance of a particular platform.

Platforms require to explicitly model communication. Communication is a main actor in APs, with even more emphasis than in the digital case. In fact, interactions between communication and computation in analog blocks are much more involved than for digital blocks. We can consider the communication between two blocks as well as the respective behaviors to be the fixed-point solution of the composition of the blocks. Because of this, orthogonalization becomes harder for detailed platforms and communication has to be modeled together with behavior. Therefore, the semantics of the component families requires to specify the allowed block interconnections. In this sense, the communication at the output of an LNA is modeled with respect to the mixer that is supposed to follow the LNA. At the most detailed level, this means that a given LNA can work with some mixers and not with others. Of course, different domains of application may require different compositional characteristics among the component families.

APs can be hierarchically composed to generate new APs at a higher level of abstraction. PLLs provide examples of hierarchical composition since their complexity allows them to be considered as systems *per se*. The result of the composition is a new AP that needs to be endowed with a set of behavioral models. The process of deriving new platforms is the same as deriving platform for circuits *(fractal nature of design)*, where $\phi\,(\cdot)$ now refers to the composition of APs in place of simple circuits. Therefore, hierarchical composition provides a more general way to generate more abstract platforms other than the abstraction/merge operations previously defined.

Design Flow with APs

The platform paradigm can be effectively used to drive exploration and eventually optimization at the system level. We call *exploration* the process of defining the set of optimal specifications compatible with the considered implementation architectures and the available performance models. In this sense, exploration can be cast as an optimization problem on the current level of abstraction.

The design process starts with the collection/generation of a suitable platform library for system implementation. The platform concept is extremely flexible in terms of encapsulation of design components. For example, schematics coming from previous designs, module generators, analog IPs, and eventually new solutions can be used to generate a platform library. Since the exploration phase leverages automatic optimization, an operative metric to measure optimality of a design has to be defined. Multi-objective optimization schemes may be adopted, for example, min max or weighted sums objective functions [4]. A number of test vectors (simulation scenarios) have to be defined in order to evaluate system performance under meaningful conditions (e.g., as dictated by a standard). A behavioral model of the system is then built, and exploration is performed having platform performance models as constraints to enforce feasibility. Since no mathematical properties are generally available for the optimization problem, global optimization schemes such as simulated annealing or genetic algorithms are exploited. Even if these schemes are potentially very expensive, the number of optimization parameters available at the system level is usually low (a few tens) and behavioral simulations are expected to be very efficient.

Since different implementation architectures may actually be merged at proper levels of abstraction (see Figure 93.5), platform-based analog design allows exploring a wide architectural space (depending on the platform library) while decomposing system specifications into block requirements. The selection of platform performances may have different implications on subsequent refinements: either the selected performance set is achievable by only one architecture (then the architecture has been selected), or by more than one architecture (which indicates that at the current abstraction level and with the current metrics, the selected platforms are equivalent). In the latter case, going to more detailed models or refining the metrics selects a proper architecture. Also, platforms may be locally refined to expand the design space or to explore interesting design corners with improved accuracy.

The result of the exploration phase is a set of specifications for each platform in the system. The exploration process can then be iterated over single subsystems. Eventually, the evaluation scheme used to estimate platform performances (e.g., simulation) can be inverted so that specifications get converted in to suitable configuration parameters. In practice, this task can be accomplished with some regression scheme on the performance data used to approximate the performance model. A vanilla implementation would simply return the configurations whose performances are closest to the requested specifications (nearest neighbors). This set of potential solutions could then be used as initial points of a local detailed optimization process, such as finalizing the circuit sizing and layout. From here on, it is up to the IP provider to generate an implementation with the required performances or up to designers to size the schematic and generate the layout. The novelty of the approach consists in the fact that the use of $\mathcal{P}$s guarantees by construction that the set of specifications generated is feasible, thus drastically reducing the number of iterations required to obtain feasible specifications and achieving optimal points over larger parameter spaces.

APs also separate the design of individual blocks from system design, thus allowing more effective explorations and optimizations at higher levels of abstraction where the design space is much larger. The introduction of the platform concept enables the integration of heterogeneous IPs, and does not put any constraint on analog designers for developing new circuit solutions for sensitive blocks. In this sense, APs may be exploited as a fast evaluation method for new solutions, generating coarse performance regions at first to evaluate the basic performances of a new solution at the system level, and, then, refining both the architecture and the functionality to take full advantage of the new solution (if convenient). This process can be pictorially represented by the classic Y chart that is essentially the same as in the digital case (Figure 93.6).

APs can also be used to model programmable fabrics. In the digital implementation platform domain, functionality with digital signal processings (FPGAs) provide a very intuitive example of platform, for example, including microprocessors on chip. The appearance of Field Programmable Analog Arrays [9] constitutes a new attempt to build reconfigurable AP. A platform stack can be built by exploiting the software tools that allow mapping complex functionalities (filters, amplifiers, triggers, and so on) directly on the array. The top-level platform, then, provides an API to map and configure analog functionalities, exposing analog hardware at the software level. By exploiting this abstraction, not only design exploration is greatly simplified, but new synergies between higher layers and analog components can be leveraged to further increase the flexibility/reconfigurability and optimize the system. From this abstraction level, implementing a functionality with digital signal processing (FPGA) or analog processing (FPAA) becomes subject to system-level optimization while exposing the same abstract interface. Moreover, very interesting tradeoffs can be explored exploiting different partitionings between analog and digital components and leveraging the reconfigurability of the FPAA. For example, limited analog performances can be mitigated by proper reconfiguration of the FPAA, so that a tight interaction between analog and digital subsystems can provide a new optimum from the system-level perspective.

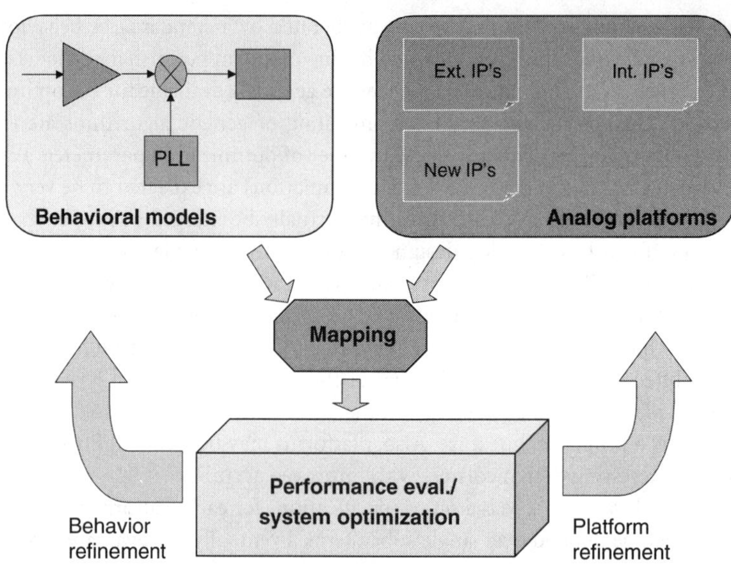

FIGURE 93.6   Analog design flow exploiting APs.

93.6   Concluding Remarks

We have defined platform-based design as an all-encompassing intellectual framework in which scientific research, design tool development, and design practices can be embedded and justified. A platform is an abstraction layer that hides the details of the several possible implementation refinements of the underlying layer. Platform-based design allows us to trade off various components of manufacturing, NRE, and design costs while sacrificing potential design performance as little as possible . We presented examples of these concepts at different key articulation points of the design process, including network platforms, implementation platforms, and analog platforms. The principles of platform-based have been applied in the development of the Metropolis design environment [2], a federation of integrated analysis, verification, and synthesis tools supported by a rigorous mathematical theory of metamodels and agents.

References

[1] Balarin, F., M. Chiodo, P. Giusto, H. Hsieh, A. Jurecska, L. Lavagno, C. Passerone, A. Sangiovanni-Vincentelli, E. Sentovich, K. Suzuki, and B. Tabbara, *Hardware–Software Co-Design of Embedded Systems: The POLIS Approach,* Kluwer Academic Publishers, Boston, 1997.

[2] Balarin, F., Y. Watanabe, H. Hsieh, L. Lavagno, C. Passerone, and A. Sangiovanni-Vincentelli, Metropolis: an integrated electronic system design environment, *IEEE Computer,* 36, 45–52, 2003.

[3] De Bernardinis, F., M.I. Jordan, and A.L. Sangiovanni-Vincentelli, Support Vector Machines for Analog Circuit Performance Representation, in Proceedings of the Design Automation Conference, June, 2003.

[4] Brayton, R.K., G.D. Hachtel, and A.L. Sangiovanni-Vincentelli, A survey of optimization techniques for integrated-circuit design, *Proceedings of the IEEE,* 69, 1334–1362, 1981.

[5] Chang, H., L. Cooke, M. Hunt, G. Martin, A. McNelly, and L. Todd, *Surviving the SOC Revolution: A Guide to Platform Based Design,* Kluwer Academic Publishers, Boston, 1999.

[6] Ferrari, A. and A.L. Sangiovanni-Vincentelli, System Design: Traditional Concepts and New Paradigms, in Proceedings of the Internatoinal Conference on Computer Design, Oct., 1999, pp. 1–12.

[7] Keutzer, K., S. Malik, A.R. Newton, J. Rabaey, and A. Sangiovanni Vincentelli, System level design: orthogonalization of concerns and platform-based design, *IEEE Transactions on Computer-Aided Design of Integrated Circuits and Systems,* 19, 2000.

[8] Lee, E.A., and A. Sangiovanni-Vincentelli, A framework for comparing models of computation, *IEEE Transactions on Computer-Aided Design of Integrated Circuits and Systems,* 17, 1217–1229, 1998.

[9] Macbeth, I., Programmable Analog Systems: the Missing Link, in EDA Vision, July 2001, www.edavision.com.

[10] Sangiovanni-Vincentelli, A.L., Defining Platform-Based Design, in EEDesign, February 2002, Available at www.eedesign.com/story/OEG20020204S0062.

[11] Sgroi, M, Platform-based Design Methodologies for Communication Networks, Ph.D. thesis, Electronics Research Laboratory, University of California, Berkeley, December 2002.

[12] Zimmerman, H., OSI reference model: the ISO model of architecture for open systems interconnection, *IEEE Transactions on Communications,* 28, 425–432, 1980.

94

Hardware/Software Interfaces Design for SoC

Wander O. Cesário
TIMA Laboratory

Flávio R. Wagner
Federal University of Rio Grande do Sul

A.A. Jerraya
TIMA Laboratory

94.1 Introduction

Modern system-on-chip (SoC) design shows a clear trend toward integration of multiple processor cores. The SoC System Driver section of the "International Technology Roadmap for Semiconductors [1]" predicts that the number of processor cores will increase fourfold per technology node in order to match the processing demands of the corresponding applications. Typical multiprocessor SoC (MPSoC) applications like network processors, multimedia hubs, and base-band telecom circuits have particularly tight time-to-market and performance constraints that require a very efficient design cycle.

Our conceptual model of the MPSoC platform is composed of four kinds of components: software tasks, processor and intellectual property (IP) cores, and a global on-chip interconnect IP (see Figure 94.1a). Moreover, to complete the MPSoC platform, we must also include hardware and software elements that adapt platform components to each other. MPSoC platforms are quite different from single-master processor SoCs (SMSoCs). For instance, their implementation of system communication is more complicated since heterogeneous processors may be involved and complex communication protocols and topologies may be used. The hardware adaptation layer must deal with some specific issues:

1. In SMSoC platforms, most peripherals (excluding DMA controllers) operate as slaves with respect to the shared communication interconnect. MPSoC platforms may use many different types of processor cores; in this case, sophisticated synchronization is needed to control shared communication between several heterogeneous masters.
2. While SMSoC platforms use simple master/slave shared-bus interconnections, MPSoC platforms often use several complex system buses or micro-networks as global interconnect. In MPSoC platforms, we can separate computation and communication design by using communication coprocessors and profiting from the multi-master architecture. Communication coprocessors/controllers (masters) implement high-level communication protocols in hardware and execute them in parallel with the computation executed on processor cores.

Application software is generally organized as a stack of layers that runs on each processor core (see Figure 94.1b). The lowest layer contains drivers and low-level routines to control/configure the platform. For the middle layer, we can use any commercial embedded OS and configure it according to the application. The upper layer is an application-programming interface (API), which provides some pre-defined routines to access the platform. All these layers correspond to the software adaptation layer in Figure 94.1a, coding application software can then be isolated from the design of the SoC platform (software coding is not the topic of this chapter and will be omitted). One of the main contributions of this work is to consider this layered approach also for the dedicated software (often-called firmware). Firmware is the software that controls the platform, and, in some cases, executes some nonperformance-critical application functions. In this case, it is not realistic to use a generic OS as the middle layer due to code size and performance reasons. A lightweight custom OS supporting an application- and platform-specific API is required.

Software and hardware adaptation layers isolate platform components enabling concurrent development as shown in Figure 94.1c. With this scheme, the software design team uses APIs for both application and dedicated software development. The hardware design team uses abstract interfaces provided by communication coprocessors/controllers. SoC design team can concentrate on implementing hardware and software abstraction layers for the selected communication interconnect IP. Designing these hardware/software (HW/SW) abstraction layers represents a major effort, and design tools are lacking. Established EDA tools are not well adapted to this new MPSoC design scenario, and consequently many challenges are emerging; some major issues are as follows:

1. *Higher abstraction level is needed*: the register-transfer level (RTL) is too time consuming to model and verify the interconnection between multiple processor cores.
2. *Higher-level programming is needed*: MPSoCs will include hundred thousands lines of dedicated software (firmware). This software cannot be programmed at the assembler level as today.

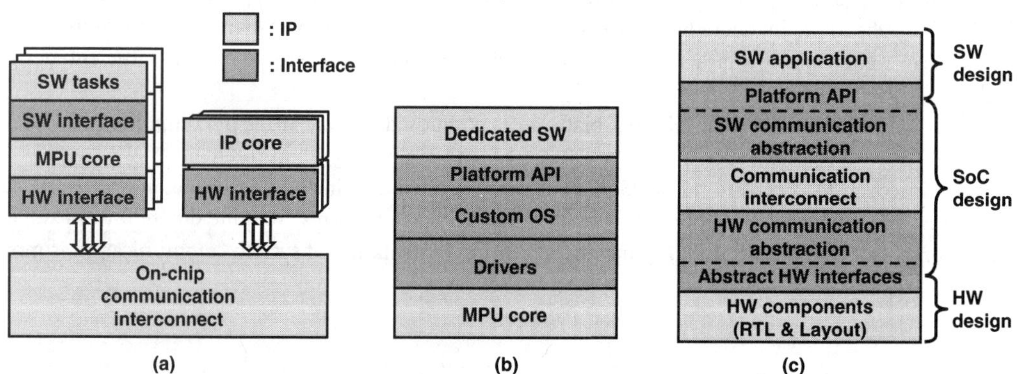

FIGURE 94.1 MPSoC platform (a), software stack (b), and concurrent development environment (c).

3. *Efficient HW/SW interfaces are required*: microprocessor interfaces, bank of registers, shared memories, software drivers, and operating systems must be optimized to each application.

This chapter presents a component-based design automation approach for MPSoC platforms. Section 94.2 introduces the basic concepts for MPSoC design and discusses some related platform- and component-based approaches. Section 94.3 details IP-based methodologies for hardware and software IP integration. Section 94.4 details our specification model and design flow. Section 94.5 presents the application of this flow for the design of a VDSL circuit and the analysis of the results.

94.2 System-on-Chip Design

System-Level Design Flow

This section gives an overview of current SoC design methodologies using a template design flow (see Figure 94.2). The basic theory behind this flow is the separation between communication and computation refinement for platform- and component-based design [2,3]; it has five main design steps:

1. *System specification*: system designers and the end-customer must agree on an informal model containing all application's functionality and requirements. Based on this model, system designers build a more formal specification that can be validated by the end-customer.
2. *Architecture exploration*: system designers build an executable model of the specification and iterate through a performance analysis loop to decide the HW/SW partitioning for the SoC architecture. This executable specification uses an abstract platform composed of abstract models for HW/SW components. For instance, an abstract software model can concentrate on I/O execution profiles, or most frequent use cases, or worst-case scheduling. Abstract hardware can be described using transaction-level models or behavioral models. This step produces the "golden" architecture model,

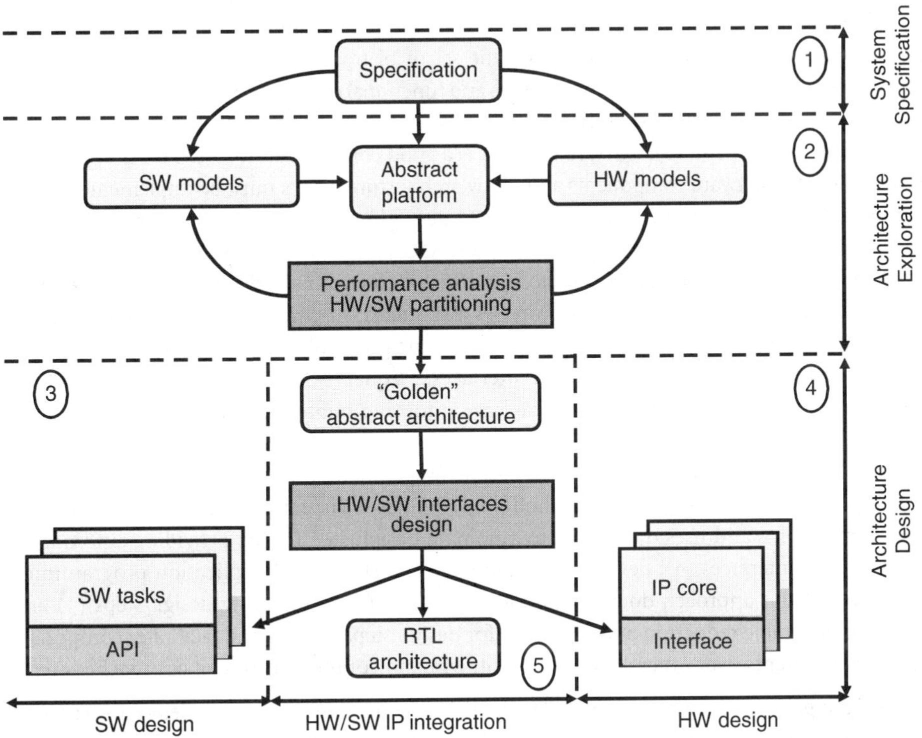

FIGURE 94.2 System-level design flow for SoC.

which is the customized SoC platform or a new architecture created by system designers after selecting processors, the global communication interconnect, and other IP components. Once HW/SW partitioning is decided, software and hardware development can be done concurrently.

3. *Software design*: since the final hardware platform will not be available during software development, some kind of hardware abstraction layer (HAL) or API must be provided to the software design team.
4. *Hardware design*: hardware IP designers implement the functionality described by the abstract hardware models at the RT level. Hardware IPs can use specific interfaces for a given platform or standard interface as defined by Virtual Socket Interface Alliance (VSIA [4]).
5. *HW/SW IP integration*: SoC designers create HW/SW interfaces to the global communication interconnect. The golden architecture model must specify performance constraints to assure a good HW/SW integration. Software and hardware communication interfaces are designed to conform to these constraints.

SoC Design Automation — An Overview

Many academic and industrial works propose tools for SoC design automation covering many, but not all, design steps presented before. Most approaches can be classified into three groups: system-level synthesis, platform-based design, and component-based design.

System-level synthesis methodologies are top-down approaches; the SoC architecture and software models are produced by synthesis algorithms from a system-level specification. COSY [5] proposes an HW/SW communication refinement process that starts with an extended Kahn Process Network model on design step (1), uses VCC [6] for step (2), callback signals over a standard RTOS for the API in (3), and VSIA interfaces for steps (4) and (5). SpecC [7] starts with an untimed functional specification model written in extended C on design step (1), uses performance estimation for a structural architecture model for step (2), HW/SW interface synthesis based on a timed bus-functional communication model for step (5), synthesized C code for step (3), and behavioral synthesis for step (4).

Platform-based design is a meet-in-the-middle approach that starts with a functional system specification and a predesigned SoC platform. Performance estimation models are used to try different mappings between the set of application's functional modules and the set of platform components. During these iterations, designers can try different platform customizations and functional optimizations. VCC [6] can produce a performance model using a functional description of the application and a structural description of the SoC platform for design steps (1) and (2). CoWare N2C [8] is a good complement for VCC for design steps (4) and (5). Still, the API for software components and many architecture details must be implemented manually.

Section 94.3 discusses hardware and software IP integration in the context of current IP-based design approaches. Most IP-based design approaches build SoC architectures from the bottom-up using predesigned components with standard interfaces and/or a standard bus. For instance, IBM defined a standard bus called CoreConnect [9], Sonics proposes a standard on-chip network called Silicon Backplane μNetwork [10], and VSIA defined a standard component protocol called VCI. When needed, wrappers adapt incompatible buses and/or component interfaces. Frequently, internally developed components are tied to inhouse (nonpublic) standards; in this case, adopting public standards implies a large effort to redesign interfaces or wrappers for old components.

Section 94.4 introduces a higher-level IP-based design methodology for HW/SW interface design called component-based design. This methodology defines a virtual architecture model composed of HW/SW components and uses this model to automate design step (5), by providing automatic generation of hardware interfaces (4), device drivers, operating systems, and application programming interfaces (3). Even if this approach does not provide much help in automating design steps (1) and (2), it provides a considerable reduction of design time for design steps (3)–(5) and facilitates component reuse. The key improvements over other state-of-art platform and component-design approaches are:

1. *Strong support for software design and integration*: the generated API completely abstracts the hardware platform and OS services. Software development can be concurrent to and independent of platform customization.

2. *Higher-level abstractions*: the use of a virtual architecture model allows designers to deal with HW/SW interfaces at a high abstraction level. Behavior and communication are separated in the system specification; thus, they can be refined independently.
3. *Flexible HW/SW communication*: automatic HW/SW interfaces generation is based on the composition of library elements. It can be used with a variety of IP interconnect components by adding the necessary supporting library.

94.3 HW/SW IP Integration

There are two major approaches for the integration of HW/SW IP components into a given design. In the first one, component interfaces follow a given standard (such as a bus or core interface, for hardware components, or a set of high-level communication primitives, for software components) and can be thus directly connected to each other. In the second approach, components are heterogeneous in nature and their integration requires the generation of hardware and/or software wrappers. In both cases, an RTOS must be used to provide services that are needed in order that the application software fits into the SoC architecture. This section describes different solutions to the integration of HW/SW IP components.

Introduction to IP Integration

The design of an embedded SoC starts with a high-level functional specification, which can be validated. This specification must already follow a clear separation between computation and communication [11], in order to allow their concurrent evolution and design. An abstract architecture is then used to evaluate this functionality based on a mapping that assigns functional blocks to architectural ones. This high-level architectural model abstracts away all low-level implementation details. A performance evaluation of the system is then performed, by using estimates of the computation and communication costs. Communication refinement is now possible, with a selection of particular communication mechanisms and a more precise performance evaluation.

According to the platform-based design approach [2], the abstract architecture follows an architectural template that is usually domain-specific. This template includes both a hardware platform, consisting of a given communication structure and given types of components (processors, memories, hardware blocks), and a software platform, in the form of a high-level API. The target-embedded SoC will be designed as a derivative of this template, where the communication structure, the components, and the software platform are all tailored to fit the particular application needs.

The IP-based design approach follows the idea that the architectural template may be implemented by assembling reusable HW/SW IP components, maybe even delivered by third-party companies.

The IP integration step comprises a set of tasks that are needed to assemble predesigned components in order to fulfill system requirements. As shown in Figure 94.3, it takes as inputs the abstract architecture and a set of HW/SW IP components that have been selected to implement the architectural blocks. Its output is a micro-architecture where hardware components are described at the RT level with all cycle-and-pin accurate details that are needed for a further automatic synthesis. Software components are described in an appropriate programming language, such as C, and can be directly compiled to the target processors of the architecture.

In an ideal situation, IP components would fit directly together (or to the communication structure) and exactly match the desired SoC functionality. In a more general situation, the designer may need to adapt each component's functionality (a step called IP derivation) and synthesize HW/SW wrappers to interconnect them. For programmable components, although adaptation may be easily performed by programming the desired functionality, the designer may still need to develop software wrappers (usually device and/or bus drivers) to match the application software to the communication infrastructure. The generation of HW/SW wrappers is usually known as interface or communication synthesis. Besides these, application software may also need to be retargeted to the processors and OS of the chosen architecture.

In the following subsections, different approaches to IP integration are introduced and their impact on the possible integration subtasks is analyzed.

Bus- and Core-Based Approaches

In the bus-based design approach [9,12,13], IP components communicate through one or more buses (interconnected by bus bridges). Since the bus specification can be standardized, libraries of components whose interfaces directly match this specification can be developed. Even if components follow the bus standard, very simple bus interface adapters may still be needed [14]. For components that do not directly match the specification, wrappers have to be built. Companies offer very rich component libraries and specialized development and simulation environments for designing systems around their buses.

A somewhat different approach is the core-based design, as proposed by the VSIA VCI standard [4] and by the OCP-IP organization [15]. In this case, IP components are compliant to a bus-independent and standardized interface and can thus be directly connected to each other. Although the standard may support a wide range of functionalities, each component may have an interface containing only the functions that are relevant for it. These components may also be interconnected through a bus, in which case standard wrappers can adapt the component interface to the bus. Sonics [13] follows this approach, proposing wrappers to adapt the bus-independent OCP socket to the MicroNetwork bus.

For particular needs, the SoC may be built around a sophisticated and dedicated network-on-chip (NoC) [16] that may deliver a very high performance for connecting a large number of components. Even in this case, a bus- or core-based approach may be adopted to connect the components to the network.

Bus- and core-based design methodologies are integration approaches that depend on standardized component or bus interfaces. They allow the integration of homogeneous IP components that follow these standards and can be directly connected to each other, without requiring the development of complex wrappers. The problem we face is that many *de facto* standards exist, coming from different companies or organizations, thus preventing a real interchange of libraries of IP components developed for different substandards.

Integrating Software IP

Programmable components are important in a reusable architectural platform, since it is very cost-effective to tailor a platform to different applications by simply adapting the low-level software and maybe only configuring certain hardware parameters, such as memory sizes and peripherals.

As illustrated in Figure 94.3, the software view of an embedded system shows three different layers:

1. The bottom layer is composed of services directly provided by hardware components (processor and peripherals) such as instruction sets, memory and peripheral accesses, and timers.

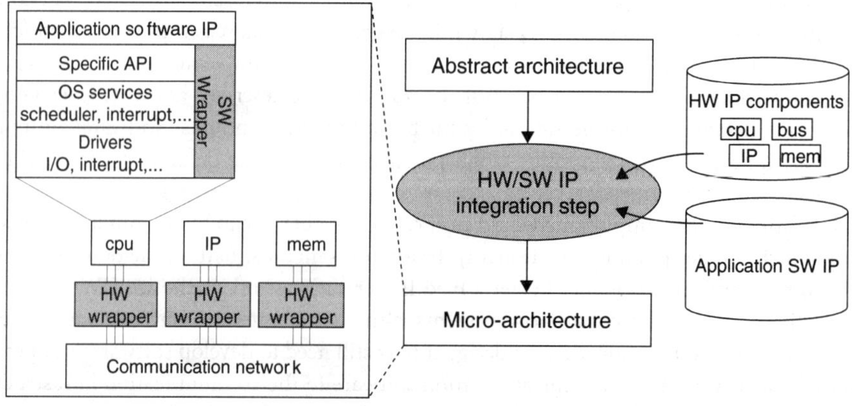

FIGURE 94.3   The HW/SW IP integration design step.

2. The top layer is the application software, which should remain completely independent of the underlying hardware platform.
3. The middle layer is composed of three different sub-layers, as seen from bottom to top:
    (a) hardware-dependent software (HdS), consisting, for instance, of device drivers, boot code, parts of an RTOS (such as context switching code and configuration code to access the MMU), and even some domain-oriented algorithms that directly interact with the hardware;
    (b) hardware-independent software, typically high-level RTOS services, such as task scheduling and high-level communication primitives; and
    (c) the API , which defines a system platform that isolates the application software from the hardware platform and from all basic software layers, and enables their concurrent design.

The standardization of this API, which can be seen as a collection of services usually offered by an operating system, is essential for software reuse above and below it. At the application software level, libraries of reusable software IP components can implement a large number of functions that are necessary for developing systems for given application domains. If, however, one tries to develop a system by integrating application software components that do not directly match a given API, software retargeting to the new platform will be necessary. This can be a very tedious and error-prone manual process, which is a candidate for an automatic software synthesis technique.

Nevertheless, reuse can also be obtained below the API. Software components implementing the hardware-independent parts of the RTOS can be more easily reused, especially if the interface between this layer and the HdS layer is standardized. Although the development of reusable HdS may be harder to accomplish, because of the diversity of hardware platforms, it can be at least obtained for platforms aimed at specific application domains.

There are many academic and industrial alternatives providing RTOS services. The problem with most approaches, however, is that they do not consider specific requirements for SoC, such as minimizing memory usage and power consumption. Recent research efforts propose the development of application-specific RTOS containing only the minimal set of functions needed for a given application [17,18] or including dynamic power management techniques [19]. IP integration methodologies should thus consider the generation of application-specific RTOS that are compliant to a standard API and optimized for given system requirements.

In recent years, many standardization efforts aimed at hardware IP reuse have been developed. Similar efforts for software IP reuse are now needed. VSIA [4] has recently created working groups to deal with HdS- and platform-based design.

Communication Synthesis

Solutions for the automatic synthesis of communication wrappers to connect hardware IP components that have incompatible interfaces have already been proposed. In the PIG tool [20], component interfaces are specified as protocols described as regular expressions, and an FSM interface for connecting two arbitrary protocols is automatically generated. The Polaris tool [21] generates adapters based on state machines for converting component protocols into a standard internal protocol, together with send and receive buffers and an arbiter.

These approaches, however, do not address the integration of software IP components. The TEReCS tool [18] synthesizes communication software to connect software IP components, given a specification of the communication architecture and a binding of IP components to processors. In the IPChinook environment [22], abstract communication protocols are synthesized into low-level bus protocols according to the target architecture. While the IPChinook environment also generates a scheduler for a given partitioning of processes into processors, the TEReCS approach is associated to the automatic synthesis of a minimal OS, assembled from a general-purpose library of reusable objects that are configured according to application demands and the underlying hardware.

Recent solutions uniformly handle HW/SW interfaces between IP components. In the COSY approach [5], design is performed by an explicit separation between function and architecture. Functions are then mapped to architectural components. Interactions between functions are modeled by high-level transactions and then mapped to hardware–software communication schemes. A library provides a fixed set of wrapper IPs, containing hardware and/or software implementations for given communication schemes.

IP Derivation

Hardware IP components may come in several forms [23]. They may be hard, when all gates and interconnects are placed and routed, soft, with only an RTL representation, or firm, with an RTL description together with some physical floorplanning or placement. The integration of hard IP components cannot be performed by adapting their internal behavior and/or structure. If they have the advantage of a more predictable performance, in turn they are less flexible and therefore less reusable than adaptable components.

Several approaches for enhancing reusability are based on adaptable components. Although one can think of very simple component configurations (for instance, by selecting a bit width), a higher degree of reusability can be achieved by components whose behavior can be more freely modified. Object-orientation is a natural vehicle for high-level modeling and adaptation of reusable components [24,25]. This approach, which can be better classified as IP derivation, is adequate for not only firm and soft hardware IP components but also for software IP [26]. Although component reusability is enhanced by this approach, the system integrator has a greater design effort, and it becomes more difficult to predict IP performance.

IP derivation and communication synthesis are different approaches to solve the same problem of integration between heterogeneous IP components, which do not follow standards (or the same substandards). IP derivation is a solution usually based on object-oriented concepts coming from the software community. It can be applied to the integration of application software components and for hardware soft and firm components, but it cannot be used for hard IP components. Communication synthesis, on the other hand, follows the path of the hardware community on automatic logic and high-level synthesis. It is the only solution to the integration of heterogeneous hard IP components, although it can also be used for integrating software IP and soft and firm hardware IP. While IP derivation is essentially a user-guided manual process, communication synthesis is an automatic process, with no user intervention.

94.4   Component-Based SoC Design

This section introduces the component-based design methodology, a high-level IP-based methodology aimed at the integration of heterogeneous HW/SW IP components. It follows an automatic communication synthesis approach, generating both HW/SW wrappers. It also generates a minimal and dedicated OS for programmable components. It uses a high-level API, which isolates the application software from the implementation of a hardware–software solution for the system platform, such that software retargeting is not necessary.

This approach enables the automatic integration of heterogeneous (components that do not follow a given bus or core standard) and hard IP components (whose internal behavior or structure is not known). However, the approach is also very well suited to the integration of homogeneous and soft IP components. The methodology has been conceived to fit to any communication structure, such as an NoC [16] or a bus.

The component-based methodology is based on a clear definition of three abstraction levels that are also adopted by other current approaches: system (pure functional), macro-architecture, and micro-architecture (RTL). These levels constitute clear "interfaces" between design steps, promoting reuse of both components and tools for design tasks at each of these levels.

Design Methodology Principles

The design flow starts with a virtual architecture model that corresponds to the "golden" architecture in Figure 94.2, and allows automatic generation of wrappers, device drivers, operating systems, and application programming interfaces. The goal is to produce a synthesizable RTL model of the MPSoC platform that is composed of processor cores, IP cores, the communication interconnect IP, and HW/SW wrappers.

The latter are automatically generated from the interfaces of virtual components (as indicated by the arrows in Figure 94.4). Software written for the virtual architecture specification runs without modification on the implementation because the same APIs are provided by the generated custom OSs.

The input abstract architecture (see Figure 94.4a) is composed of virtual modules (VM), corresponding to processing and memory IPs, connected by any communication structure, also encapsulated within a VM. This abstract architecture model clearly separates computation from communication, allowing independent and concurrent implementation paths for components and for communication. Virtual modules that correspond to processors may be hierarchically decomposed into submodules containing software tasks assigned to this processor. Virtual modules communicate through virtual ports, which are sets of hierarchical internal and external ports through which services are requested and provided. The separation between internal and external ports makes possible the connection of modules described at different abstraction levels.

Virtual Architecture

The virtual architecture represents a system as an abstract netlist of virtual components (see Figure 94.4a). It is described in VADeL, a SystemC [27] extension that includes a platform-independent API offering high-level communication primitives. This API abstracts the underlying hardware platform, thus enhancing the free development of reusable components. In the abstract architecture model, the interfaces of software tasks are the same for SW–SW and SW–HW connections, even if the software tasks are executed by different

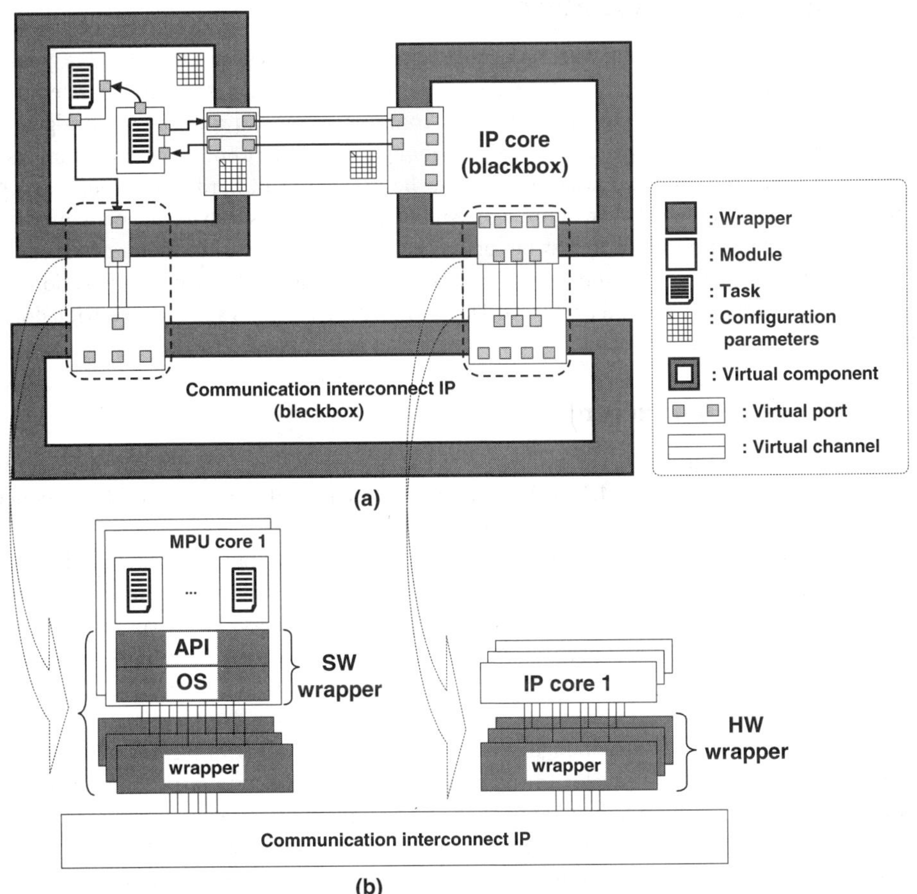

FIGURE 94.4 MPSoC design flow: virtual architecture (a) and target MPSoC platform (b).

processors. Different hardware/software realizations of this API are possible. Architectural design space exploration can thus be achieved without influencing the functional description of the application.

Virtual components use wrappers to adapt accesses from the internal component (a set of software tasks or a hardware function) to the external channels. The wrapper is modeled as a set of virtual ports that contain internal and external ports that can be different in terms of: (1) communication protocol, (2) abstraction level, and/or (3) specification language. This model is not directly synthesizable or executable because wrapper's behavior is not described. These wrappers can be generated automatically, in order to produce a detailed architecture that can be both synthesized and simulated.

The required SystemC extensions implemented in VADeL are:

1. virtual module: consists of a module and its wrapper;
2. virtual port: groups some internal and external ports that have a conversion relationship. The wrapper is the set of virtual ports for a given virtual module;
3. virtual channel: groups several channels having a logical relationship (e.g., multiple channels belonging to the same communication protocol); and
4. parameters: used to customize hardware interfaces (e.g., buffer size and physical addresses of ports), operating systems, and drivers.

In VADeL, there are also predefined ports with special semantics called Service Access Ports (SAP). They can be used to access some services that are implemented by hardware or software wrapper components. For instance, the *timer* SAP can be used to request an interrupt from a hardware timer after a given delay.

Target MPSoC Architecture Model

We use a generic MPSoC architecture where processors and other IP cores are connected to a global communication interconnect IP via wrappers (see Figure 94.4b). In fact, processors are separated from the physical communication IP by wrappers that act as communication coprocessors or bridges, freeing processors from communication management and enabling parallel execution of computation tasks and communication protocols. Software tasks also need to be isolated from hardware through an OS that plays the role of an SW wrapper. When defining this model, our goal was to have a generic model where both computation and communication may be customized to fit the specific needs of the application. For computation, we may change the number and kind of components and, for communication, we can select specific communication IP and protocols. This architecture model is suitable to a wide domain of applications; more details can be found in [31].

HW/SW Wrapper Architecture

Wrappers are automatically generated as point-to-point adapters between each VM and the communication structure, as shown in Figure 94.4b [31]. This approach allows the connection of components to standard buses as well as point-to-point connections between cores.

Wrappers may have HW/SW parts. The internal architecture of a wrapper on the hardware side is shown in Figure 94.5b. It consists of a processor adapter, one or more channel adapters, and an internal bus. The number of channel adapters depends on the number of channels that are connected to the corresponding virtual module. This architecture allows the easy generation of multi-point, multi-protocol wrappers. The wrapper dissociates communication from computation, since it can be considered as a communication coprocessor that operates concurrently with other processing functions.

On the software side [17], as shown in Figure 94.5a, wrappers provide the implementation of the high-level communication primitives (available through the API) used in the system specification and drivers to control the hardware. If required, the wrapper will also provide sophisticated OS services such as task scheduling and interrupt management minimally tailored for the particular application.

The synthesis of wrappers is based on libraries of basic modules from which hardware wrappers and dedicated OSs are assembled. These libraries may be easily extended with modules that are needed to build wrappers for processors, memories, and other components that follow various bus and core standards.

Design Tools

Figure 94.6 shows an overall view of our design environment, which is called *ROSES*. The input model may be imported from a specification analysis tools (e.g., [6]) or manually coded using our extended SystemC library. All design tools use a unified design model that contains an abstract HW/SW netlist annotated with parameters (Colif [28]). Hardware wrapper generation [31] transforms the input model into a synthesizable architecture. The software wrapper generator [17] produces a custom OS for each processor on the target platform. For validation, we use the cosimulation wrapper generator [30] to produce simulation models. Details about these tools can be found in the references; only their principles will be discussed here.

Hardware wrapper generation assembles library components using the architecture template presented before (Figure 94.5b) to produce the RTL architecture. This library contains generalized descriptions of hardware components in a macro language (**m4** like); it has two parts: the processor library and the protocol library. The former contains local template architectures for processors with four types of

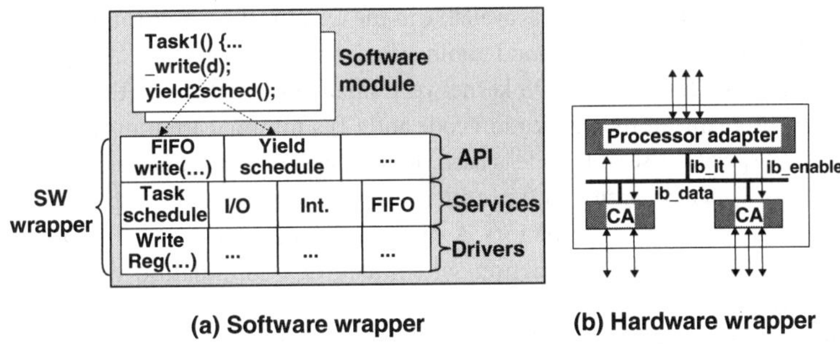

FIGURE 94.5   HW/SW wrapper architecture.

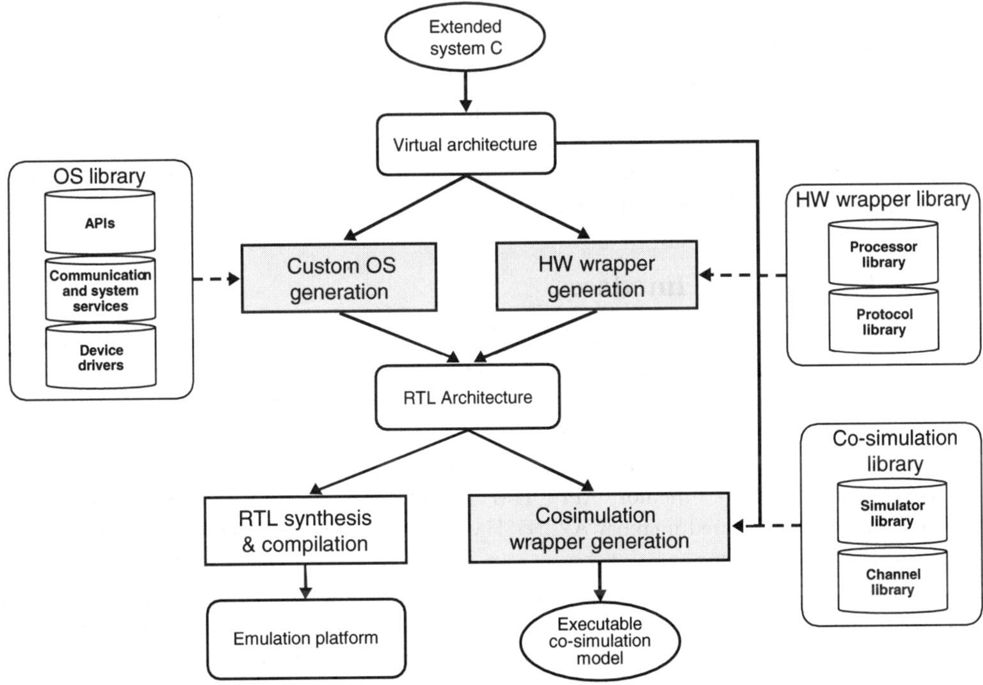

FIGURE 94.6   Design automation tools for MPSoC.

elements: processor cores, local buses, local IP components (e.g., local memory, address decoder, coprocessors, etc.), and processor adapters. The latter consists of a list of channel adapters. Each channel adapter has simulation, estimation, and synthesis models that are parameterized (by the channel parameters, e.g., direction, storage size, and data type) as the elements in the processor library.

The software wrapper generator produces operating systems streamlined and preconfigured for the software module(s) that run(s) on each target processor. It uses a library organized in three parts: APIs, communication/system services, and device drivers. Each part contains elements that will be used in a given software layer in the generated OS. The generated OS provides services: communication services (e.g., FIFO communication), I/O services (e.g., AMBA bus drivers), memory services (e.g., cache or virtual memory usage), etc. Services have dependency between them, for instance, communication services are dependent on I/O services. Elements of the OS library also have dependency information. This mechanism is used to keep the size of the generated OS at a minimum; the elements that provide unnecessary services are not included.

There are two types of service code: re-usable (or existing) code and expandable code. As an example of existing code, AMBA bus-master service code can exist in the OS library in the form of C language. As an example of expandable code, OS kernel functions can exist in the OS library in the form of macrocode (**m4** like). There are several preemptive schedulers available in the OS library such as round-robin scheduler, priority-based scheduler, etc. In the case of round-robin scheduler, time-slicing (i.e., assigning different CPU load to tasks) is supported. To make the OS kernel very small and flexible, (1) the task scheduler can be selected from the requirements of the application code and (2) a minimal amount (less than 10% of kernel code size) of processor-specific assembly code is used (for context switching and interrupt service routines).

The cosimulation wrapper generator [30] produces an executable model composed of a SystemC simulator that acts as a master for other simulators. A variety of simulators can participate in this cosimulation: SystemC, VHDL, Verilog, and Instruction-set simulators. Cosimulation wrappers have the same structure as hardware wrappers (see Figure 94.5b), with simulation adapters in the place of processor adapters and simulation models of channel adapters. In the cosimulation wrapper library, there are simulation adapters for the different simulators supported and channel adapters that implement all supported protocols in different languages.

In terms of functionality, the cosimulation wrapper transforms channel access(es) via internal port(s) to channel access(es) via external port(s) using the following functional chain: channel interface, channel resolution, data conversion, and module communication behavior. Internal ports use channel functions (e.g., FIFO available, FIFO write) to exchange data. Channel interface provides the implementation of these channel functions. Channel resolution maps N-to-M correspondence between internal and external ports. Data conversion is required since different abstraction levels can use different data types to represent the same data. Module communication behavior is required to exchange data via external port(s), that is, to call port functions of external ports.

Defining IP-Component Interfaces

Hardware and software component interfaces must be composed using basic elements of the hardware wrapper and software wrapper generators libraries (respectively). Table 94.1 lists some API functions available for different kinds of software task interfaces and some services provided by channel adapters available to be used in hardware component interfaces.

Software tasks must communicate through API functions provided by the software-wrapper generator library. For instance, the shared-memory API (SHM) provides read/write functions for intertask communication. The guarded-shared memory API (GSHM) adds semaphores services to the SHM API by providing lock/unlock functions.

Hardware IP components must communicate through communication primitives provided by the channel adapters of the hardware-wrapper generator library. For instance, FIFO channel adapters (sender and receiver) implement a buffered two-phase handshake protocol (Put/Get) and provide Full/Empty functions for accessing the state of the buffer. ASFIFO channel adapters use instead a single-phase handshake protocol and can generate an interrupt for signaling the full and empty state of the buffer.

A recurrent problem in library-based approaches is library size explosion. In ROSES, this problem is minimized by the use of layered library structures where a service is factorized so that its implementation uses elements of different layers. This scheme increases reuse of library elements since the elements of the upper layers must use the services provided by the elements in the immediate lower layer.

Designers are able to extend ROSES libraries since they are implemented in an open format. This is an important feature since it enables the support of different standards while reusing most of the basic elements in the libraries.

Table 94.2 shows some of the existing hardware and software components in the current ROSES IP library and gives the type of communication they use in their interfaces.

Figure 94.7 a shows the "stream" software IP and part of its code to demonstrate the utilization of the communication APIs. Its interface is composed of four ports: two for the FIFO API (P3 and P4), one for the signal API (P2), and one for the GSHM API (P1). In line 7 of Figure 94.4a, the stream IP uses P1 to lock the access to the shared memory that contains the data that will be streamed. P2 is used to suspend the task that fills up the shared memory (line 8). Then, some header information is obtained from the input FIFO using P3 (line 11) and streamed to the output FIFO using P4 (line 12). When streaming is finished, P1 is used to unlock the access to the shared memory (line 14).

TABLE 94.1   HW/SW communication APIs

Basic component interfaces	API functions
SW	
Register	Put/Get
Signal	Sleep/Wakeup
FIFO	Put/Get
SHM	Read/Write
GSHM	Lock/Unlock/Read/Write
HW	
Register	Put/Get
FIFO	Put/Get/Full/Empty
ASFIFO	Put/Get/IT(Full/Empty)
Buffer	BPut/BGet
Event	Send/IT(receiver)
AHB master/slave	Read/Write
Timer	Set/Wait

TABLE 94.2   Sample IP library

IP	Description	Interfaces
SW		
Host-if	Host PC interface	Register/signal
Rand	Random number generator	Signal/FIFO
mult-tx	Multipoint FIFO data transmission	FIFO
reg-config	Register configuration	Register/FIFO/SHM
shm-sync	Shared memory synchronization	SHM/signal
Stream	FIFO data streaming	GSHM/FIFO/signal
HW		
ARM7	Processor core	ARM7 pins
TX_Framer	Data package framing	17 registers, 1 FIFO

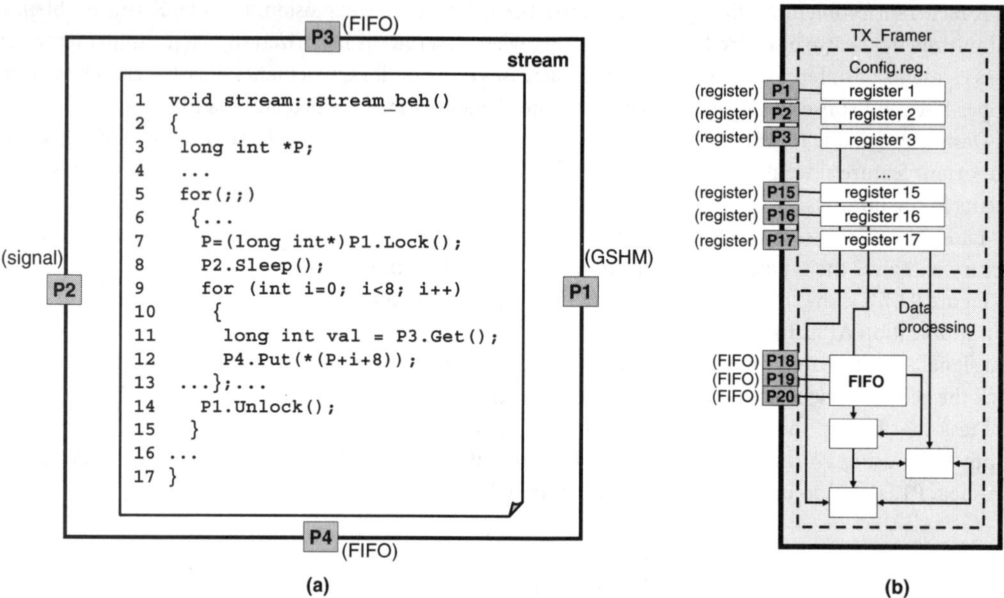

FIGURE 94.7   The stream software IP (a) and the TX_Framer hardware IP (b).

Figure 94.7b shows the "TX_Framer" hardware IP, which is part of a VDSL modem and responsible for packaging data into ATM-network-compatible frames. Its interface is composed of 17 configuration registers (P1–P17) and one single-handshake input FIFO (P18–P20). The registers are used to configure the IP functionality and have bit sizes varying from 2 to 11, while the FIFO is used to store data packets that will be inserted into specific places in the output ATM frames. These ports are driven directly by the compatible outputs from register and ASFIFO channel adapters that are generated by the hardware wrapper generator.

94.5   Component-Based Design of a VDSL Application

Specification

The design presented in this section illustrates the IP integration capabilities of ROSES. We redesigned part of a VDSL modem that was prototyped by Diaz-Nava and Okvist [29] using discrete components (the shaded part in Figure 94.8a). The block diagram for the modem subset used in the rest of this paper is shown in Figure 94.8b. It corresponds to a deframing/framing unit (DFU), composed of two ARM7 processors and the TX_Framer. The TX_Framer is part of the VDSL Protocol Processor. In this experiment, it is used as a hard IP component described at the RT level. The partition of processors/tasks was suggested by the design team of the VDSL-modem prototype.

Processors exchange data using three asynchronous FIFO buffers. The TX_Framer IP has some configuration registers and input a data stream through a synchronous FIFO buffer. Tasks use a variety of control and data-transmission protocols to communicate. For instance, a task can block/unblock the execution of other tasks by sending them an OS signal. Tasks use for data transmission: a FIFO memory buffer, two shared memories (with or without semaphores), and direct register access. Despite representing only a subset of the VDSL modem, the design of the DFU remains quite challenging. In fact, it uses two processors executing parallel tasks. The control over the three modules of the specification is fully distributed. All three modules act as masters when interacting with their environment. Additionally, the application includes multipoint communication channels requiring sophisticated OS services.

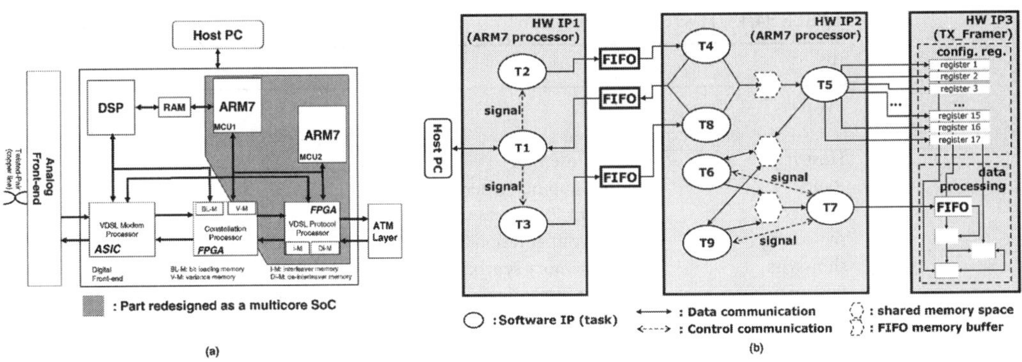

FIGURE 94.8 VDSL modem prototype (a) and DFU block diagram (b).

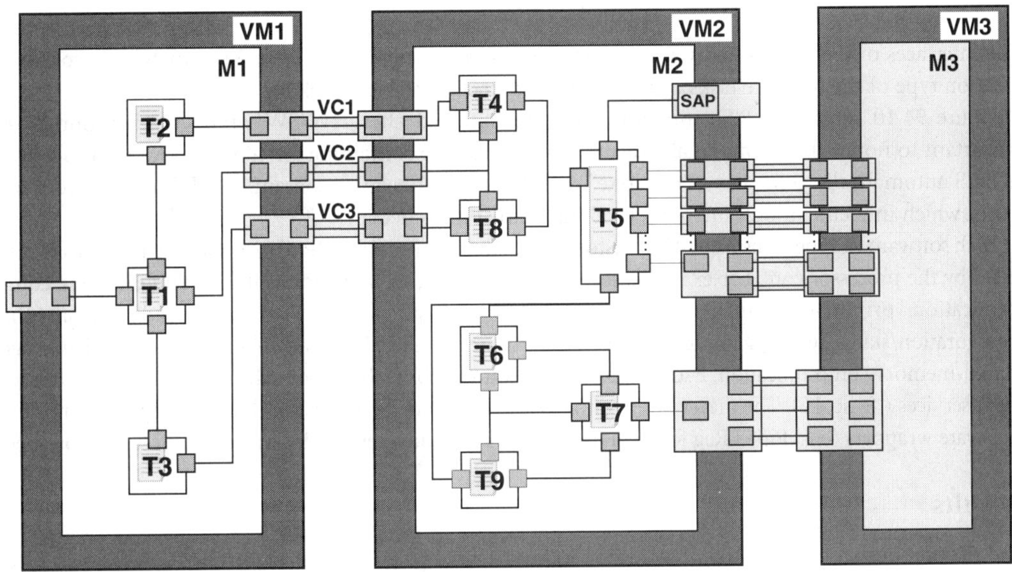

FIGURE 94.9 DFU abstract architecture specification.

DFU Abstract Architecture

Figure 94.9 shows the abstract architecture model that captures the DFU specification with point-to-point communications between the three main IP cores. *VM1* and *VM2* are two virtual processors, and *VM3* corresponds to the TX_Framer function. VM1 and VM2 include several submodules corresponding to software tasks *T1–T9* assigned to these processors. This abstract model can be mapped onto different concrete micro-architectures depending on the selected IP components and on desired performance, area, and power constraints. For instance, the three point-to-point connections (*VC1–VC3*) between VM1 and VM2 can be mapped onto a bus or onto a shared memory.

MPSoC RTL Architecture

For the implementation of the DFU virtual architecture, two hardware IP cores have been selected: an ARM7 processor and the TX_Framer. The application software has been built by reusing several available software IP components for implementing tasks T1–T9. Table 94.3 lists the selected IP components and indicates their correspondence to the virtual modules and submodules in the DFU virtual architecture.

TABLE 94.3 HW/SW IP utilization

IP	Description	Use
SW		
Host-if	Host PC interface	T1
Rand	Random number generator	T2, T3
mult-tx	Multipoint FIFO data transmission	T4, T8
reg-config	Register configuration	T5
shm-sync	Shared memory synchronization	T6, T9
Stream	FIFO data streaming	T7
HW		
ARM7	Processor core	VM1, VM2
TX_Framer	Data package framing	VM3

The interfaces of the selected software IP components in Table 94.3 (see Table 94.2) match the communication type of the software tasks of the virtual architecture in Figure 94.8b.

Figure 94.10 shows the RTL micro-architecture obtained after HW/SW wrapper generation. It is important to notice that, from an abstract target architecture containing an "abstract" ARM7 processor, ROSES automatically generates a "concrete" ARM7 local architecture containing additional IP components, which implement local memory, local bus, and address decoder.

Each software wrapper (custom OS) is customized to the set of software IPs corresponding to the tasks executed by the processor core. For example, software IPs running on VM2 access the custom OS using communication primitives available through the API: Register is used to write/read to/from the configuration/status registers inside the TX_Framer block, while SHM and GSHM are used to manage shared-memory communication. Each OS contains a round-robin scheduler (Sched) and resource management services (Sync, IT). The driver layer contains low-level code to access the channel adapters within the hardware wrappers (e.g., Pipe LReg for the HNDSHK channel adapter), and some low-level kernel routines.

Results

The manual design of a full VDSL modem requires several person-years; the presented DFU was estimated as a more than five persons-year effort. When using the ROSES IP integration capabilities, the overall experiment took only one person during four months, including all validation and verification time (but not counting the effort to develop library components and to debug design tools). This corresponds to a 15-times reduction in design effort (a more detailed presentation can be found in [32]).

Application code and generated OS are compiled and linked together to be executed on each ARM7 processor. The HW wrapper can be synthesized using RTL synthesis. As can be seen in Table 94.4, most OS code is generated in C; only a small part of it is in assembly and includes some low-level routines (e.g., context switching and processor boot) that are specific to each processor. If we compare the numbers presented in Table 94.4 with commercial embedded operating systems, the results are still very good. The minimum size for such operating systems is around 4 kbytes; but with this size, few of them could provide the required functionality. Table 94.5 shows the numbers obtained after RTL synthesis of the HW wrappers using a CMOS 0.35 μm technology. These are good results because wrappers account for less than 5% of the ARM7s core surface and have a critical path that corresponds to less than 15% of the clock cycle for the 25 MHz ARM7 processors used in this case study.

Evaluation

Results show that the component-based approach can generate HW/SW interfaces and operating systems that are as efficient as the manually coded/configured ones. The HW/SW frontier in wrapper

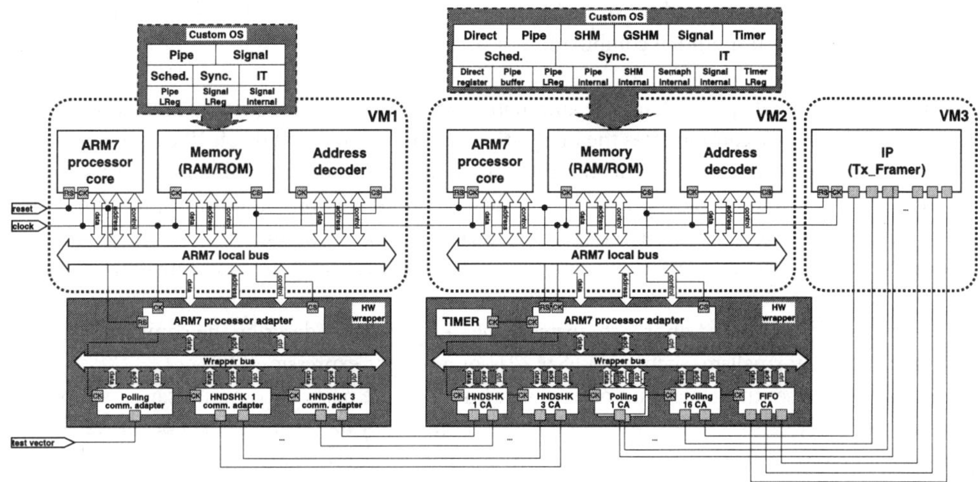

FIGURE 94.10 Generated MPSoC architecture.

TABLE 94.4 Results for OS generation

OS results	No. of lines in C	# of lines in Assembly	Code size (bytes)	Data size (bytes)
VM1	968	281	3829	500
VM2	1872	281	6684	1020

Context switch (cycles)	36
Latency for interrupt treatment (cycles)	59(OS) + 28(ARM7)
System call latency (cycles)	50
Resume of task execution (cycles)	26

TABLE 94.5 Results for hardware wrapper generation

HW interfaces	# of Gates	Critical path delay (ns)	Max. freq. (MHz)
VM1	3284	5.95	168
VM2	3795	6.16	162

Latency for read operation (clock cycles)	6
Latency for write operation (clock cycles)	2
Number of code lines (RTL VHDL)	2168

implementation can be easily displaced by changing some library components. This choice is transparent to the final user since everything that implements the interconnect API is generated automatically (the API does not change only its implementation does). Furthermore, correctness and coherence can be verified inside tools and libraries against the API semantics without having to impose fixed boundaries to the HW/SW frontier (in contrast to standardized component interfaces and/or buses).

The utilization of layered library components provides considerable flexibility; the design environment can be easily adapted to accommodate different languages to describe system behavior, different task scheduling and resource management policies, different global communication interconnect topologies and protocols, a diversity of processor cores and IP cores, and different memory architectures. In most cases, inserting a new design element into this environment only requires to add the appropriate library components. Layered library components are at the roots of the methodology; the principle followed is to contain a unique functionality, and to respect well-defined interfaces that enable easy composition. This layered structure prevents library size explosion since composition is used to implement complex functionality and to increase component reutilization.

As explained in this chapter and illustrated by the design case study, ROSES use a component-based methodology that presents a unique combination of features:

It implements a general approach for the automatic integration of heterogeneous and hard IP components, although it easily accommodates the integration of homogeneous and soft IP components.

It offers an architectural-independent API, integrated into SystemC, containing high-level communication primitives and enhancing the free development of reusable components. Application software accessing this API does not need to be retargeted to each system implementation.

It adopts a library-based approach to wrapper generation. As long as components communicate through a known protocol, communication synthesis can be done automatically, without any additional design effort. In a formal-based approach, instead, the designer must describe the component interface by some appropriate formalism, such as an FSM [21] or a regular expression [20].

It uniformly addresses the generation of hardware and software parts of communication wrappers for programmable components. While some approaches consider only hardware [20, 21] or software [18, 22] wrappers, others also consider hardware and software parts but are restricted to predefined wrapper libraries for given communication schemes [5]. The library-based approach of ROSES, in turn, allows the synthesis of software interfaces for various communication schemes.

It can be used with any architectural template and communication structure, such as a bus, an NoC, or point-to-point connections between components. It is also configurable to synthesize wrappers for any bus or core standard.

94.6 Conclusions

Reuse of IP components is a major requirement for the design of complex embedded SoCs. However, reuse is a complex process that involves many steps, requires support from specialized tools and methodologies, and influences current design practices. The integration of IP components into a particular design is may be the most complex step of the reuse process. Many design approaches, such as bus-, core-, and platform-based designs, are aimed at an easier IP integration. Nevertheless, many problems are still open, in particular, the automatic synthesis of hardware and software wrappers between heterogeneous and hard IP components.

The chapter has shown that the component-based design methodology provides a complete, generic, and efficient solution to the hardware/software interfaces design problem. Starting from a high-level function specification and an abstract architecture, design tools can automatically generate hardware and software wrappers that are necessary to integrate heterogeneous IP components that have been selected to implement the application. Designers do not need to design any low-level interfacing details manually. The chapter has also shown how HW/SW component interfaces can be decomposed and easily adapted to different communication structures and bus and core standards.

References

[1] ITRS, available at http://public.itrs.net/

[2] Keutzer, K., A.R. Newton, J.M. Rabaey, and A. Sangiovanni-Vincentelli, System-level design: orthogonalization of concerns and platform-based design, *IEEE Transactions on Computer-Aided Design of Integrated Circuits and Systems*, Vol. 19, no. 12. pp. 1523–1543, Dec. 2000.

[3] Sgroi, M., M. Sheets, A. Mihal, K. Keutzer, S. Malik, J. Rabaey, and A. Sangiovanni-Vincentelli, Addressing the System-on-Chip Interconnect woes through Communication-Based Design, Proceedings of 38th DAC, Las Vegas, June, 2001.

[4] VSIA, http://www.vsi.org

[5] Brunel, J.-Y., W.M. Kruijtzer, H.J.H.N. Kenter, F. Pétrot, L. Pasquier, E.A. de Kock, and W.J.M. Smits, COSY Communication IPs," Proceedings of 37th DAC, Los Angeles, June, 2000.

[6] Cadence Design Systems, Inc., Virtual Component Co-design, http://www.cadence.com/products/vcc.html

[7] Gajski, D., J. Zhu, R. Domer, A. Gerslauer, and S. Zhao, *SpecC Specification Language and Methodology*, Kluwer Academic Publishers, Dordrecht, 2000.

[8] CoWare, http://www.coware.com

[9] IBM CoreConnect Bus Architecture, http://www.chips.ibm.com/bluelogic

[10] Wingard, D., MicroNetwork-Based Integration for SOCs, Proceedings of 38th DAC, Las Vegas, June, 2001.

[11] Rowson, J. and A.Sangiovanni-Vincentelli, Interface-based Design, DAC'1997.

[12] ARM AMBA, http://www.arm.com

[13] Sonics SiliconBackplane MicroNetwork, http://www.sonicsinc.com

[14] Bergamaschi, R.A. and W.R. Lee. Designing Systems-on-Chip Using Cores, DAC'2000.

[15] Open Core Protocol. http://www.ocpip.org

[16] de Micheli, G. and L. Benini, Networks-on-Chip: A New Paradigm for Systems-on-Chip Design, DATE'2002.

[17] Gauthier, L., S. Yoo, and A.A. Jerraya, Automatic generation and targeting of application specific operating systems and embedded systems software, *IEEE TCAD*, 20, 2001.

[18] Böke, C., Combining Two Customization Approaches: Extending the Customization Tool TEReCS for Software Synthesis of Real-Time Execution Platforms, Workshop on Architectures of Embedded Systems (AES2000), Jan., 2000.

[19] Benini, L., A. Bogliolo, and G. De Micheli, Dynamic Power Management of Electronic Systems, ICCAD 1998.

[20] Passerone, R., J.A. Rowson, and A. Sangiovanni-Vincentelli, Automatic Synthesis of Interfaces between Incompatible Protocols, DAC'1998.

[21] Smith, J. and G. deMicheli, Automated Composition of Hardware Components, DAC'1998.

[22] Chou, P. et al., IPChinook: An Integrated IP-Based Design Framework for Distributed Embedded Systems," DAC'1999.

[23] Birnbaum, M. and H. Sachs, How VSIA answers the SoC dilemma, *IEEE Computer*, Vol. 32, no. 6, pp. 42–50, June 1999.

[24] Barna, C. and W. Rosenstiel, Object-Oriented Reuse Methodology for VHDL, DATE'1999.

[25] Schaumont, P. et al., Hardware Reuse at the Behavioral Level, DAC'1999.

[26] Rammig, F.J., Web-based System Design with Components off the Shelf, FDL'2000.

[27] SystemC, http://www.systemc.org

[28] Cesário, W.O., G. Nicolescu, L. Gauthier, D. Lyonnard, and A.A. Jerraya, Colif: A design representation for application-specific multiprocessor SOCs, *IEEE Design & Test of Computers*, 18, 2001.

[29] Diaz-Nava, M. and G.S. Okvist, The Zipper Prototype: A Complete and Flexible VDSL Multicarrier Solution, *IEEE Communications Magazine*, Vol. 40, no. 12, pp. 92–105, Dec. 2002.

[30] Yoo, S., G. Nicolescu, D. Lyonnard, A. Baghdadi, and A. A. Jerraya, A Generic Wrapper Architecture for Multi-Processor SoC Cosimulation and Design, International Symposium on HW/SW Codesign (CODES), 2001.

[31] Lyonnard, D., S. Yoo, A. Baghdadi, and A. A. Jerraya, Automatic Generation of Application-Specific Architectures for Heterogeneous Multiprocessor System-on-Chip, Proceedings of 38th DAC, Las Vegas, June, 2001.

[32] Cesário, W., A. Baghdadi, L. Gauthier, D. Lyonnard, G. Nicolescu, Y. Paviot, S. Yoo, M. Diaz-Nava, and A.A. Jerraya, Component-Based Design Approach for Multicore SoCs, Proceedings of 39th DAC, New Orleans, June, 2002.

95

Network-on-Chip Design for Gigascale Systems-on-Chip

Davide Bertozzi
University of Bologna

Luca Benini
University of Bologna

Giovanni De Micheli
Stanford University

95.1 Introduction

The increasing integration densities made available by shrinking device geometries will have to be exploited to meet the computational requirements of parallel applications such as multimedia processing, automotive, multi-window TV, ambient intelligence, etc.

As an example, systems designed for ambient intelligence will be based on high-speed digital signal processing with computational loads ranging from 10 MOPS for lightweight audio processing, 3 GOPS for video processing, 20 GOPS for multilingual conversation interfaces, and up to 1 TOPS for synthetic video generation. This computational challenge will have to be addressed at manageable power levels and affordable costs [1].

Such a performance cannot be provided by a single processor, but requires a heterogeneous on-chip multi-processor system containing a mix of general-purpose programmable cores, application-specific processors, and dedicated hardware accelerators.

In this context, performance of gigascale *Systems-on-Chip* (SoCs) will be communication dominated, and only an interconnect-centric system architecture will be able to cope with this problem. Current on-chip interconnects consist of low-cost shared arbitrated buses, based on the serialization of bus access requests; only one master at a time can be granted access to the bus. The main drawback of this solution is its lack of scalability, which will result in unacceptable performance degradation for complex SoCs (more than a dozen of integrated cores). Moreover, the connection of new blocks to a shared bus increases its associated load capacitance, resulting in more energy-consuming bus transactions.

A scalable communication infrastructure that better supports the trend of SoC integration consists of an on-chip micro-network of interconnects, generally known as *Network-on-Chip* (NoC) architecture [2,6,7]. The basic idea is borrowed from the wide-area networks domain, and envisions router (or switch)-based networks on which on-chip packetized communication takes place, as depicted in Figure 95.1. Cores access the network by means of proper interfaces, and have their packets forwarded to destination through a certain number of hops.

The scalable and modular nature of NoCs and their support for efficient on-chip communication potentially lead to NoC-based multi-processor systems characterized by high structural complexity and functional diversity. On the one hand, these features need to be properly addressed by means of new design methodologies [3], while on the other more efforts have to be devoted to modeling on-chip communication architectures and integrating them into a single modeling and simulation environment combining both processing elements and communication infrastructures [4,9,10]. These efforts are needed to include on-chip communication architecture in any quantitative evaluation of system design during design space exploration [8,11], to enable the assessment of the impact of the interconnect on achieving a target system performance.

An important design decision for NoCs regards the choice of topology. Several researchers [7,12,3,13] envision NoCs as regular tile-based topologies (such as mesh networks and fat trees), which are suitable for interconnecting homogeneous cores in a chip multiprocessor. However, SoC component specialization (used by designers to optimize performance at low power consumption and competitive cost) leads to the on-chip integration of heterogeneous cores having varied functionality, size, and communication requirements. If a regular interconnect is designed to match the requirements of a few communication-hungry components, it is bound to be largely overdesigned with respect to the needs of the remaining components. This is the main reason why most current SoCs use irregular topologies like bridged buses and/or dedicated point-to-point links [14].

This chapter introduces the basic principles and guidelines for the NoC design. At first, the motivation for the design paradigm shift of SoC communication architectures from shared buses to NoCs is examined. Then, the chapter goes into the details of NoC building blocks (switch, network interface, and switch-to-switch links), discussing the design guidelines and presenting a case study where some of the most advanced concepts in NoC design have been applied to a real NoC architecture (called *Xpipes* and developed at University of Bologna [15]).

Finally, the challenging issue of heterogeneous NoC design will be addressed, and the effects of mapping the communication requirements of an application onto a domain-specific NoC, instead of a network with regular topology, will be detailed by means of an illustrative example.

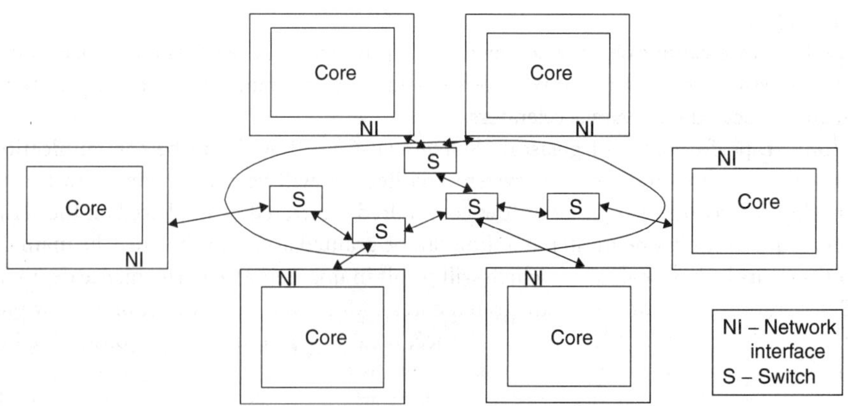

FIGURE 95.1   Example of network-on-chip architecture.

95.2 Design Challenges for On-Chip Communication Architectures

SoC design challenges that are driving the evolution of traditional bus architectures toward NoCs can be outlined as follows:

Technology issues: While gate delays scale down with technology, global wire delays typically increase or remain constant as repeaters are inserted. It is estimated that in 50 nm technology, at a clock frequency of 10 GHz, a global wire delay can be up to 6–10 clock cycles [2]. Therefore, limiting the on-chip distance traveled by critical signals will be key to guarantee the performance of the overall system, and will be a common design guideline for all kinds of system interconnects. On the contrary, other challenges posed by deep sub-micron technologies are leading to a paradigm shift in the design of SoC communication architectures. For instance, global synchronization of cores on future SoCs will be unfeasible due to deep submicron effects (clock skew, power associated with clock distribution tree, etc.), and an alternative scenario consists of self-synchronous cores that communicate with one another through a network-centric architecture [16]. Finally, signal integrity issues (cross-talk, power supply noise, soft errors, etc.) will lead to more transient and permanent failures of signals, logic values, devices, and interconnects, thus raising the reliability concern for on-chip communication [17]. In many cases, on-chip networks can be designed as regular structures, allowing electrical parameters of wires to be optimized and well controlled. This leads to lower communication failure probabilities, thus enabling the use of low swing signaling techniques [18], and to the capability of exploiting performance optimization techniques such as wavefront pipelining [19].

Performance issues: In traditional buses, all communication actors share the same bandwidth. As a consequence, performance does not scale with the level of system integration, but degrades significantly, although once the bus is granted to a master, access occurs with no additional delay. On the contrary, NoCs can provide a much better performance scalability. No delays are experienced for accessing the communication infrastructure, since multiple outstanding transactions originating from multiple cores can be handled at the same time, resulting in a more efficient network resources utilization. However, given a certain network dimension (e.g., the number of instantiated switches), large latency fluctuations for packet delivery could be experienced as a consequence of network congestion. This is unacceptable when hard real-time constraints of an application have to be met, and two solutions are viable: network overdimensioning (for NoCs designed to support best-effort traffic only) or implementation of dedicated mechanisms to provide guarantees for timing-constrained traffic (e.g., loss-less data transport, minimal bandwidth, bounded latency, minimal throughput, etc.) [20].

Design productivity issues: It is well known that synthesis and compiler technology development do not keep up with IC manufacturing technology development [21]. Moreover, times-to-market need to be kept as low as possible. Reuse of complex preverified design blocks is an efficient means to increase productivity, and considers both computation resources and the communication infrastructure [22]. It would be highly desirable to have processing elements that could be used in different platforms by means of a plug-and-play design style. To this end, a scalable and modular on-chip network represents a more efficient communication infrastructure compared with shared bus-based architectures. However, the reuse of processing elements is facilitated by the definition of standard network interfaces, which also make the modularity property of the NoC effective. The *Virtual Socket Interface Alliance* (VSIA) has attempted to set the characteristics of this interface industrywide [23]. OCP [24] is another example of standard interface sockets for cores. It is worth remarking that such network interfaces also decouple the development of new cores from the evolution of new communication architectures. The core developer will not have to make assumptions about the system, when the core will be plugged into. Similarly, designers of new on-chip interconnects will not be constrained by the knowledge of detailed interfacing requirements for particular legacy SoC components. Finally, let us observe that NoC components (e.g., switches or interfaces) can be instantiated multiple times in the same design as opposed to the arbiter of traditional shared buses (which is instance-specific) and reused in a large number of products targeting a specific application domain.

The developments of NoC architectures and protocols are fueled by the aforementioned arguments, in spite of the challenges represented by the need for new design methodologies and an increased complexity of system design.

95.3 Related Work

The need to progressively replace on-chip buses with micro-networks was extensively discussed in [2,7]. A number of NoC architectures have been proposed in the literature so far.

Sonics MicroNetwork [25] is an on-chip network making use of communication architecture-independent interface sockets. The *MicroNetwork* is an example of evolutionary solutions [26], which move from a physical implementation as a shared bus, and propose generalizations to support a higher bandwidth (such as partial and full crossbars).

STBUS interconnect from STMicroelectronics is another example of evolutionary architecture, which provides designers with the capability to instantiate both shared bus or partial or full crossbar interconnect configurations.

Even though these architectures provide a bandwidth higher than simple buses, addressing the wiring delay and scalability challenge in the long term requires more radical solutions.

One of the earliest contributions in this area is the *Maia* heterogeneous signal processing architecture, proposed by Zhang et al., [27] based on a hierarchical mesh network. Unfortunately, *Maia*'s interconnect is fully instance-specific. Furthermore, routing is static at configuration time: network switches are programmed once and for all for a given application (as in an FPGA). Thus, communication is based on circuit switching, as opposed to packet switching.

In this direction, Dally and Lacy sketch the architecture of a VLSI multi-computer using 2009 technology [35]. A chip with 64 processor-memory tiles is envisioned. Communication is based on packet switching. This seminal work draws upon past experiences in designing parallel computers and reconfigurable architectures (FPGAs and their evolutions) [28–30].

Most proposed NoC platforms are packet switched and exhibit a regular structure. An example is a mesh interconnection, which can rely on a simple layout and the switch independence on the network size. The *NOSTRUM* network described in [3] adopts this approach: the platform includes both a mesh architecture and the design methodology. The *Scalable Programmable Integrated Network (SPIN)* described in [31] is another regular, fat-tree-based network architecture. It adopts cut-through switching to minimize message latency and storage requirements in the design of network switches. The *Linkoeping SoCBUS* [32] is a two-dimensional mesh network, which uses a *packet connected circuit* (PCC) to set up routes through the network: a packet is switched through the network locking the circuit as it goes. This notion of virtual circuit leads to deterministic communication behavior but restricts routing flexibility for the rest of the communication traffic.

The need to map communication requirements of heterogeneous cores may lead to the adoption of irregular topologies. The motivation for such architectures lies in the fact that each block can be optimized for a specific application (e.g., video or audio processing), and link characteristics can be adapted to the communication requirements of the interconnected cores. Supporting heterogeneous architectures requires a major design effort and leads to coarser-granularity control of physical parameters. Many recent heterogeneous SoC implementations are still based on shared buses (such as the single chip MPEG-2 codec reported in [33]), but the growing complexity of customizable media-embedded processor architectures for digital media processing will soon require NoC-based communication architectures and proper hardware/software development tools. The *Aethereal* NoC design framework presented in [34] aims at providing a complete infrastructure for developing heterogeneous NoC with end-to-end quality of service guarantees. The network supports *guaranteed throughput* (GT) for real-time applications and *best-effort* (BE) traffic for timing unconstrained applications.

Support for heterogeneous architectures requires highly configurable network building blocks, customizable at instantiation time for a specific application domain. For instance, the *Proteo* NoC [36] consists of a small library of predefined, parameterized components that allow the implementation of a large range of different topologies, protocols, and configurations.

Xpipes interconnect [15] and its synthesizer *XpipesCompiler* [41] push this approach to the limit, by instantiating an application-specific NoC from a library of composable soft macros (network interface, link, and switch). The components are highly parameterizable and provide a reliable and latency-insensitive operation.

95.4   Network-on-Chip Architecture

Most of the terminology for on-chip packet switched communication is adapted from computer network and multiprocessor domain. Messages that have to be transmitted across the network are usually partitioned into fixed-length packets. Packets in turn are often broken into message flow control units called *flits*. In the presence of channel width constraints, multiple physical channel cycles can be used to transfer a single flit. A *phit* is the unit of information that can be transferred across a physical channel in a single step. Flits represent logical units of information, as opposed to phits that correspond to physical quantities. In many implementations, a flit is set to be equal to a phit. The basic building blocks for packet-switched communication across NoCs are:

  I. network link
 II. switch
III. network interface

and will be described hereafter.

Network Link

The performance of interconnect is a major concern in scaled technologies. As geometries shrink, gate delay improves much faster than the delay in long wires. Therefore, the long wires increasingly determine the maximum clock rate, and hence performance, of the entire design. The problem becomes particularly serious for domain-specific heterogeneous SoCs, where the wire structure is highly irregular and may include both short and extremely long switch-to-switch links. Moreover, it has been estimated that only a fraction of the chip area (between 0.4 and 1.4%) will be reachable in one clock cycle [42].

A solution to overcome the interconnect-delay problem consists of pipelining interconnects [37,38]. Wires can be partitioned into segments (or relay stations, which have a function similar to the one of latches on a pipelined data path), whose length satisfies predefined timing requirements (e.g., desired clock speed of the design). In this way, link delay is changed into latency, but data introduction rate is not bounded by the link delay any more. Now, the latency of a channel connecting two modules may end up being more than one clock cycle. Therefore, if the functionality of the design is based on the sequencing of the signals and not on their exact timing, then link pipelining does not change the functional correctness of the design. This requires the system to be composed of modules whose behavior does not depend on the latency of the communication channels (latency-insensitive operation). As a consequence, the use of interconnect pipelining can be seen as a part of a new and more general methodology for *deep submicron* (DSM) designs, which can be envisioned as synchronous distributed systems composed of functional modules that exchange data on communication channels according to a latency-insensitive protocol. This protocol ensures that functionally correct modules behave correctly independent of the channel latencies [37]. The effectiveness of the latency-insensitive design methodology is strongly related to the ability of maintaining a sufficient communication throughput in the presence of increased channel latencies.

The *International Technology Roadmap for Semiconductors* (ITRS) 2001 [16] assumes that interconnect pipelining is the strategy of choice in its estimates of achievable clock speeds for MPUs. Some industrial designs already make use of interconnect pipelining. For instance, the NETBURST micro-architecture of Pentium 4 contains instances of a stage dedicated exclusively to handle wire delays: in fact, a so-called drive stage is used only to move signals across the chip without performing any computation and, therefore, can be seen as a physical implementation of a relay station [39].

Xpipes interconnect makes use of pipelined links and of a latency-insensitive operation in the implementation of its building blocks. Switch-to-switch links are subdivided into basic segments whose length guarantees that the desired clock frequency (i.e., the maximum speed provided by a certain technology) can be used. In this way, the system operating frequency is not bound by the delay of long links. According to the link length, a certain number of clock cycles are needed by a flit to cross the interconnect. If network switches are designed in such a way that their functional correctness depends on the flit arriving order and not on their timing, input links of the switches can be different and of any length. These design choices are the bases of latency-insensitive operation of the NoC and allow the construction of an arbitrary network topology and hence support for heterogeneous architectures.

Figure 95.2 illustrates the link model, which is equivalent to a pipelined shift register. Pipelining has been used both for data and control lines. The figure also illustrates how pipelined links are used to support latency-insensitive link-level error control, ensuring robustness against communication errors. The retransmission of a corrupted flit between two successive switches is represented. Multiple outstanding flits propagate across the link during the same clock cycle. When flits are correctly received at the destination switch, an ACK is propagated back to the source. After at least $2N$ clock cycles since transmission (where N is the length of the link expressed in number of repeater stages), the flit will be discarded from the buffer of the source

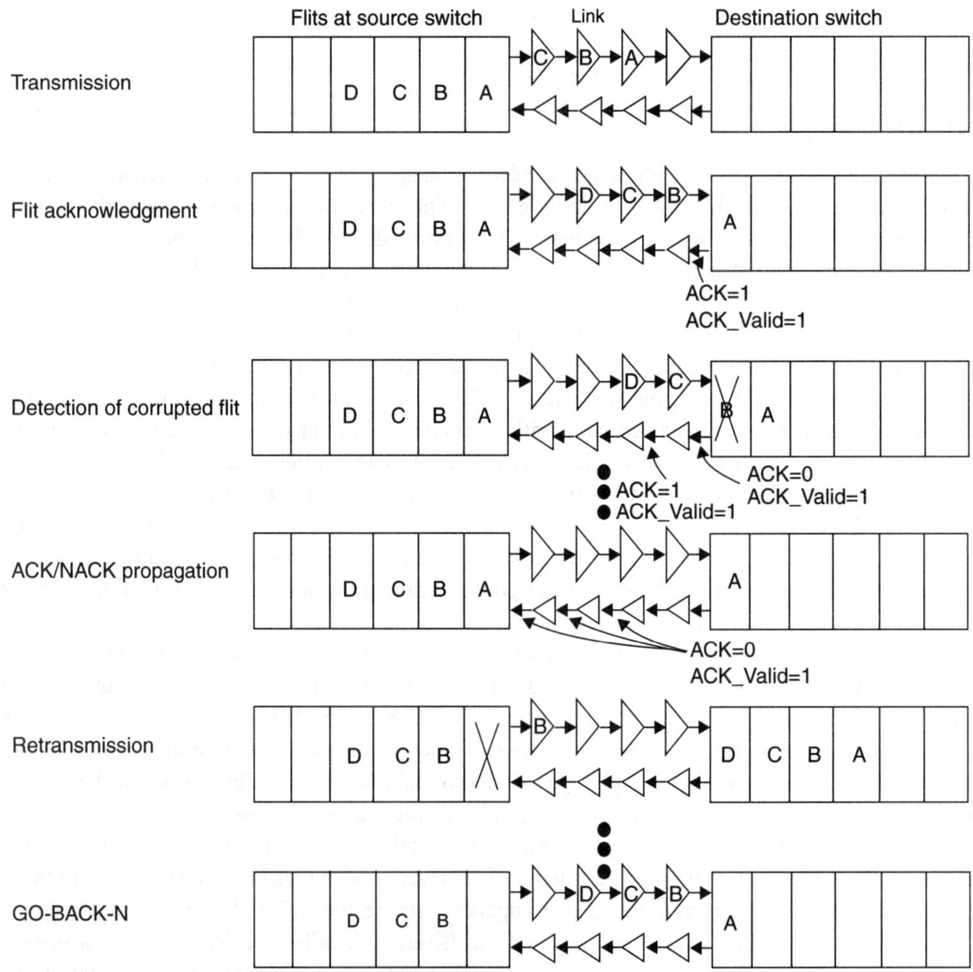

FIGURE 95.2   Pipelined link model and latency-insensitive link-level error control.

switch. On the contrary, a corrupted flit is NACKed and will be retransmitted in due time. The implemented retransmission policy is GO-BACK-N, to keep the switch complexity as low as possible.

Switch

The task of the switch is to carry packets injected into the network to their final destination, following a statically defined or dynamically determined routing path. The switch transfers packets from one of its input ports to one or more of its output ports.

Switch design is usually characterized by a power-performance trade-off: power-hungry switch memory resources can be required by the need to support high-performance on-chip communication. A specific design of a switch may include both input and output buffers or only one type of buffers. Input queuing uses fewer buffers, but suffers from head-of-line blocking. Virtual output queuing has a higher performance, but at the cost of more buffers.

Network flow control (or routing mode) specifically addresses the limited amount of buffering resources in switches. Three policies are feasible in this context [5].

In *store-and-forward routing*, an entire packet is received and entirely stored before being forwarded to the next switch. This is the most demanding approach in terms of memory requirements and switch latency. Also, *virtual cut-through routing* requires buffer space for an entire packet, but allows lower latency communication, in that a packet is forwarded as soon as the next switch guarantees that the complete packet will be accepted. If this is not the case, the current router must be able to store the whole packet.

Finally, a *wormhole routing* scheme can be used to reduce switch memory requirements and to permit low latency communication. The first flit of a packet contains routing information, and header flit decoding enables the switches to establish the path and subsequent flits simply follow this path in a pipelined manner by means of switch output port reservation. A flit is passed to the next switch as soon as enough space is available to store it, even though there is not enough space to store the whole packet. If a certain flit faces a busy channel, subsequent flits have to wait at their current locations and are therefore spread over multiple switches, thus blocking the intermediate links. This scheme avoids buffering the full packet at one switch and keeps end-to-end latency low, although it is more sensitive to deadlock and may result in low link utilization.

Guaranteeing quality of service in switch operation is another important design issue, which needs to be addressed when time-constrained (hard or soft real time) traffic is to be supported. Throughput guarantees or latency bounds are examples of time-related guarantees.

Contention-related delays are responsible for large fluctuations of performance metrics, and a fully predictable system can be obtained only by means of contention-free routing schemes. With *circuit switching*, a connection is set up over which all subsequent data are transported. Therefore, contention resolution takes place at setup at the granularity of connections, and time-related guarantees during data transport can be given. In *time division circuit switching* (see [25] for an example), bandwidth is shared by time division multiplexing connections over circuits.

In packet switching, contention is unavoidable since packet arrival cannot be predicted. Therefore, arbitration mechanisms and buffering resources must be implemented at each switch, thus delaying data in an unpredictable manner and making it difficult to provide guarantees. Best-effort NoC architectures can mainly rely on network overdimensioning to bound fluctuations of performance metrics.

The *Aethereal* NoC architecture makes use of a router that tries to combine GT and BE services [34]. The GT router subsystem is based on a time-division multiplexed circuit switching approach. A router uses a *slot table* to (i) avoid contention on a link, (ii) divide up bandwidth per link between connections, and (iii) switch data to the correct output. Every slot table T has S time slots (rows), and N router outputs (columns). There is a logical notion of synchronicity: all routers in the network are in the same fixed-duration slot. In a slot s, at most one *block* of data can be read/written per input/output port. In the next slot, the read blocks are written to their appropriate output ports. Blocks thus propagate in a store and forward manner. The latency a block incurs per router is equal to the duration of a slot and bandwidth is guaranteed in multiples of block size per S slots. The BE router uses packet switching, and it has been showed that both input queuing with wormhole routing or virtual cut-through routing and virtual output queuing with wormhole routing are feasible in terms of buffering cost. The BE and GT router subsystems are combined in the *Aethereal* router architecture of Figure 95.3. The GT router offers a fixed end-to-end latency for its traffic, which is given the highest priority by the arbiter. The BE router uses all the bandwidth (slots) that has not been reserved or used by GT traffic. GT router slot tables are programmed by means of BE packets (see the arrow "program" in Figure 95.3). Negotiations, resulting in slot allocation, can be done at compile time, and be configured deterministically at runtime. Alternatively, negotiations can be done at runtime.

A different perspective has been taken in the design of the switch for the best-effort *Xpipes* NoC. Figure 95.4 shows an example configuration with four inputs, four outputs and two virtual channels multiplexed across the same physical output link. A physical link is assigned to different virtual channels on a flit-by-flit basis, thereby improving network throughput. Switch operation is latency insensitive, in that correct operation is guaranteed for arbitrary link pipeline depth. In fact, as explained above, network links in *Xpipes* interconnect are pipelined with a flexible number of stages, thereby decoupling link data introduction rate from its physical length.

For a latency-insensitive operation, the switch has virtual channel registers to store $2N + M$ flits, where N is the link length (expressed as the number of basic repeater stages) and M is a switch architecture-related contribution (12 cycles in this design). The reason is that each transmitted flit has to be acknowledged before being discarded from the buffer. Before an ACK is received, the flit has to travel across the link (N cycles), an ACK/NACK decision has to be taken at the destination switch (a portion of M cycles), and the ACK/NACK signal has to be propagated back (N cycles) and recognized by the source switch (remaining portion of M cycles). During this time, other $2N + M$ flits are transmitted but not yet ACKed.

Output buffering was chosen for *Xpipes* switches, and the resulting architecture is reported in Figure 95.5. It consists of a replication of the same output module, accepting all input ports as its own inputs. Flow control signals generated by each output block are directed to a centralized module, which takes care of generating proper ACKs or NACKs for the incoming flits from the different input ports.

Each output module is deeply pipelined (seven pipeline stages) to maximize the operating clock frequency of the switch. Architectural details on the pipelined output module are illustrated in Figure 95.6. Forward flow control is used and a flit is transmitted to the next switch only when adequate storage is available. The CRC decoders for error detection work in parallel with the switch operation, thereby concealing their impact on switch latency.

The first pipeline stage checks the header of incoming packets on different input ports to determine whether those packets have to be routed through the output port under consideration. Only matching packets are forwarded to the second stage, which resolves contention based on a round-robin policy. Arbitration is carried out against receipt of the tail flits of preceding packets, so that all other flits of a

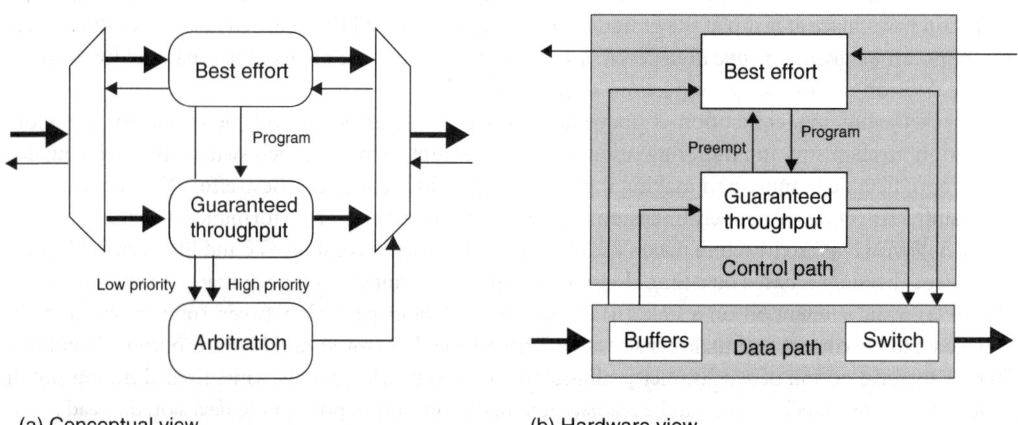

FIGURE 95.3 A combined GT–BE router.

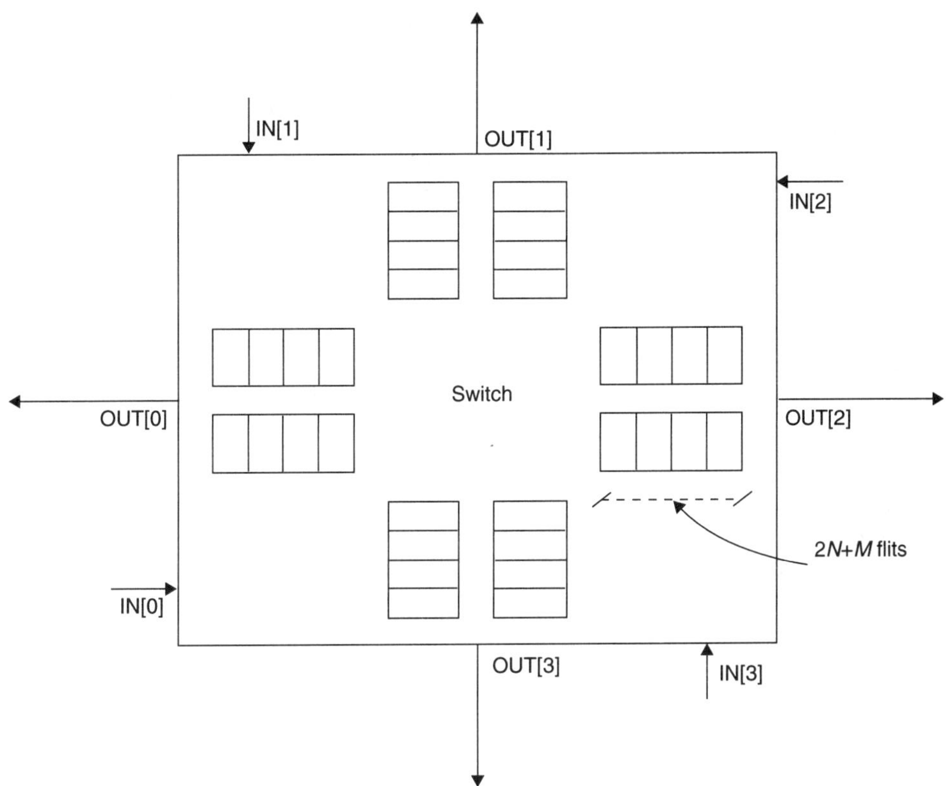

FIGURE 95.4 Example of switch configuration with two virtual channels.

packet can be propagated without contention resolution at this stage. A NACK for flits of nonselected packets is generated. The third stage is just a multiplexer, which selects the prioritized input port. The following arbitration stage retains the status of virtual channel registers and decides whether the flits can be stored in the registers or not. A header flit is sent to the register with more free locations, followed by successive flits of the same packet. The fifth stage is the actual buffering stage, and the ACK/NACK message at this stage indicates whether a flit has been successfully stored or not. The following stage takes care of forward flow control and finally a last arbitration stage multiplexes the virtual channels on the physical output link.

Finally, the switch is highly parameterizable. Design parameters are: number of I/O ports, flit width, number of virtual channels, length of switch-to-switch links, and size of output registers.

Network Interface

The most relevant tasks of the *network interface* (NI) are: (i) concealing the details about the network communication protocol to the cores, so that they can be developed independent of the communication infrastructure, (ii) communication protocol conversion (from end-to-end to network protocol), and (iii) data packetization (packet assembly, delivery, and disassembly).

The former objective can be achieved by means of standard interfaces. For instance, the VSIA vision [23] is to specify open standards and specifications that facilitate the integration of software and hardware virtual components from multiple sources. Different complexity interfaces are described in the standard, from Peripheral *Virtual Complexity Interfaces* (VCI) to Basic VCI and Advanced VCI.

Another example of standard socket to interface cores to networks is represented by *Open Core Protocol* (OCP) [24]. Its main characteristics are a high degree of configurability to adapt to the core's

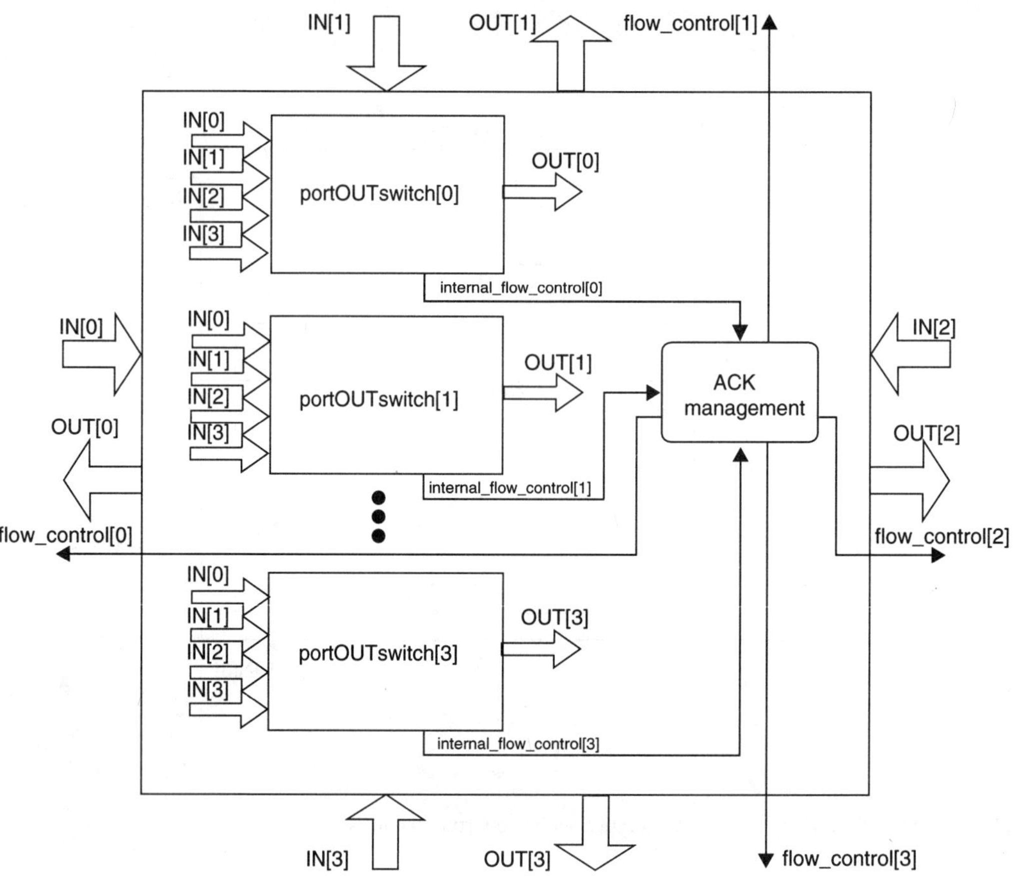

FIGURE 95.5   Architecture of output-buffered *Xpipes* switch.

functionality and the independence of request and response phases, thus supporting multiple outstanding requests and pipelining of transfers.

Data packetization is a critical task for the network interface, and has an impact on the communication latency, besides the latency of the communication channel. The packet preparation process consists of building packet header, payload, and packet tail. The header contains the necessary routing and network control information (e.g., source and destination address). When source routing is used, the destination address is ignored and replaced with a route field that specifies the route to the destination. This overhead in terms of packet header is counterbalanced by the simpler routing logic at the network switches: they simply have to look at the route field and route the packet over the specified switch output port. The packet tail indicates the end of a packet and usually contains parity bits for error-detecting or error-correcting codes.

An insight into the *Xpipes* network interface implementation will provide an example of these concepts. It provides a standardized OCP-based interface to network nodes. The NI for cores that initiate communication (initiators) needs to turn OCP-compliant transactions into packets to be transmitted across the network. It represents the slave side of an OCP end-to-end connection, and it is therefore referred to as *network interface slave* (NIS). Its architecture is shown in Figure 95.7.

The NIS has to build the packet header, which has to be spread over a variable number of flits depending on the length of the path to the destination node. In fact, *Xpipes* relies on a static routing algorithm called *street sign routing*. Routes are derived by the network interface by accessing a lookup table based on the destination address. Such information consists of direction bits read by each switch and indicating the output port of the switch that flits belonging to a certain packet have to be directed to.

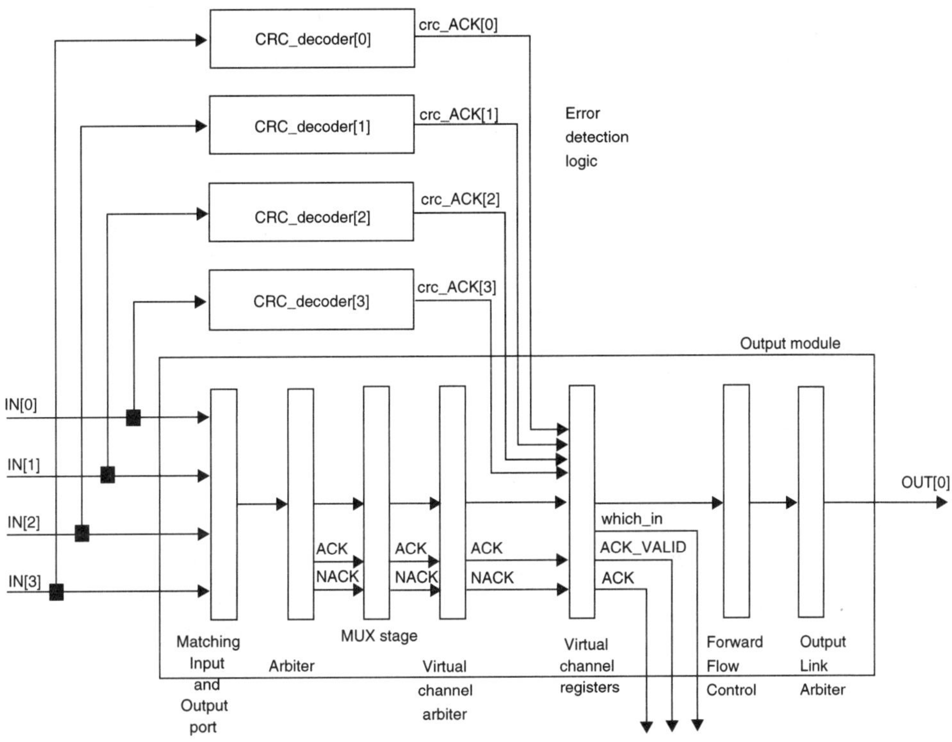

FIGURE 95.6 Architecture of an *Xpipes* switch output module.

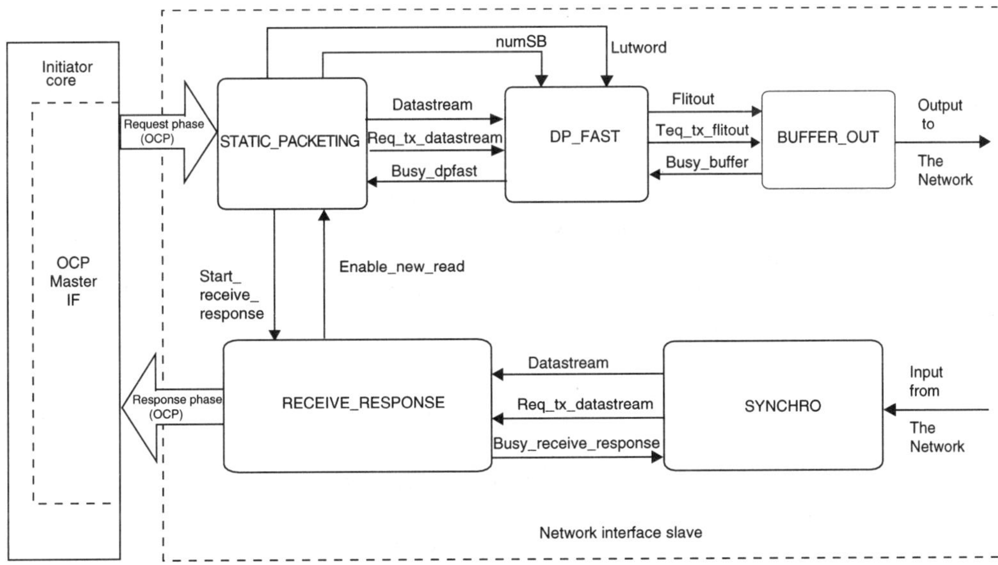

FIGURE 95.7 Architecture of the *Xpipes* network interface slave.

The lookup table is accessed by the *STATIC_PACKETING* block, a finite state machine that forwards the routing information *numSB* (number of hops to destination) and *lutword* (word read from the lookup table) as well as the request-related information *datastream* from the initiator core to the *DP_FAST* block, provided the enable signal *busy_dpfast* is not asserted.

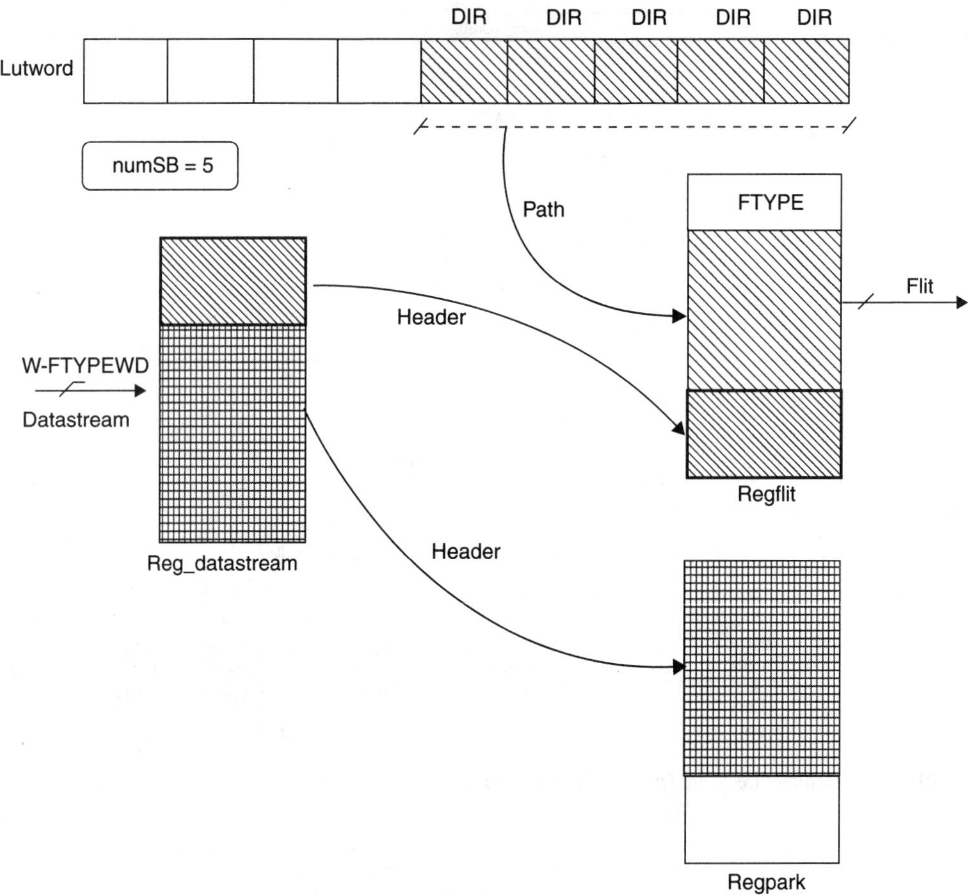

FIGURE 95.8   Mechanism for building header flits.

Based on the input data, module *DP_FAST* has the task of building the flits to be transmitted via the output buffer *BUFFER_OUT*, according to the mechanism illustrated in Figure 95.8. Let us assume that a packet requires *numSB* = 5 hops to get to the destination, and that the direction to be taken at each switch is expressed by *DIR*. Module *DP_FAST* builds the first flit by concatenating the flit-type field with path information. If there is some space left in the flit, it is filled with header information derived from the input *datastream*. The unused part of the *datastream* is stored in a *regpark* register, so that a new datastream can be read from the *STATIC_PACKETING* block. The following header and/or payload flits will be formed by combining data stored in *regpark* and *reg_datastream*. No partially filled flits are transmitted to make transmission more efficient.

Finally, module *BUFFER_OUT* stores flits to be sent across the network, and allows the NIS to keep preparing successive flits when the network is congested. The size of this buffer is a design parameter.

The response phase is carried out by means of two modules. *SYNCHRO* receives incoming flits and reads out only useful information (e.g., it discards route fields). At the same time, it contains buffering resources to synchronize the network's requests to transmit remaining packet flits with the core consuming rate. The *RECEIVE_RESPONSE* module translates useful header and payload information into OCP-compliant response fields.

When a read transaction is initiated by the master core, the *STATIC_PACKETING* block asserts a *start_receive_response* signal that triggers the waiting phase of the *RECEIVE_RESPONSE* module for the requested data. As a consequence, the NIS supports only one outstanding read operation to keep interface

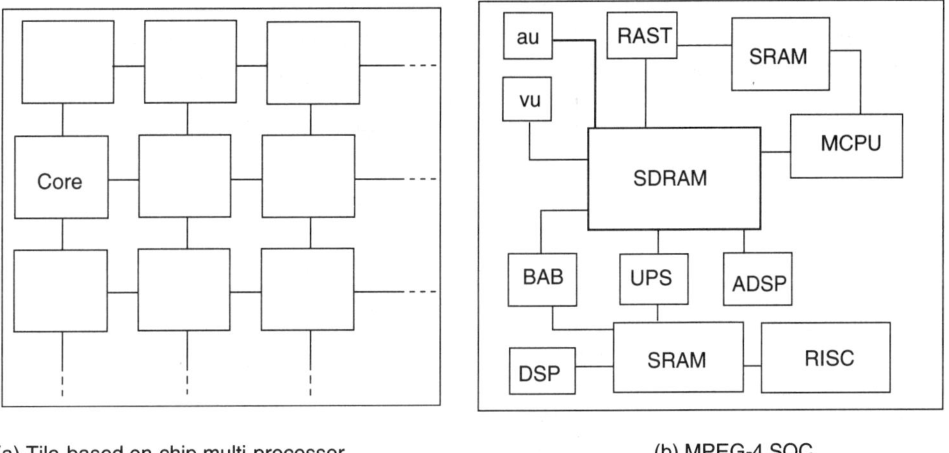

(a) Tile-based on-chip multi-processor   (b) MPEG-4 SOC

FIGURE 95.9   Homogeneous versus heterogeneous architectural template.

complexity low. Although no read after read transactions can be initiated unless the previous one has been completed, an indefinite number of write transactions can be carried out after an outstanding read has been initiated.

The architecture of a network interface master is similar to the one just described, and is not reported here due to lack of space. At instantiation time, the main network interface-related parameters to be set are: total number of core blocks, flit width, and maximum number of hops across the network.

95.5   Network-on-Chip Topology

The individual components of SoCs are inherently heterogeneous with widely varying functionality and communication requirements. The communication infrastructure should optimally match communication patterns among these components accounting for the individual component needs.

As an example, consider the implementation of an MPEG4 decoder [40], depicted in Figure 95.9 (b), where blocks are drawn roughly to scale and links represent interblock communication. First, the embedded memory (SDRAM) is much larger than all other cores and it is a critical communication bottleneck. Block sizes are highly nonuniform and the floorplan does not match the regular, tile-based floorplan shown in Figure 95.9 (a). Second, the total communication bandwidth to/from the embedded SDRAM is much larger than that required for communication among the other cores. Third, many neighboring blocks do not need to communicate. Even though it may be possible to implement MPEG4 onto a homogeneous fabric, there is a significant risk of either underutilizing many tiles and links, or, at the opposite extreme, of achieving poor performance because of localized congestion. These factors motivate the use of an application-specific on-chip network [27].

With an application-specific network, the designer is faced with the additional task of designing network components (e.g., switches) with different configurations (e.g., different I/Os, virtual channels, buffers) and interconnecting them with links of uneven length. These steps require significant design time and the need to verify network components and their communications for every design.

The library-based nature of network building blocks seems the more appropriate solution to support domain-specific custom NoCs. Two relevant examples have been reported in the open literature: *Proteo* and *Xpipes Interconnects. Proteo* consists of a fully reusable and scalable component library where the components can be used to implement networks from very simple bus emulation structures to complex packet networks. It uses a standardized VCI interface between the functional cores and the communication network. *Proteo* is described using synthesizable VHDL and relies on an interconnect node architecture that targets flexible on-chip communication. It is used as a testing platform when the efficiency of network topologies and routing schemes are investigated for on-chip environments. The node is constructed from a collection of parameterized and reusable hardware blocks, including components such as FIFO buffers, routing controllers, and standardized interface wrappers. A node can be tuned to fulfill the desired characteristics of communication by properly selecting the internal architecture of the node itself.

Xpipes NoC adopts a similar approach. As described throughout this chapter, its network building blocks have been designed as highly configurable and design-time composable soft macros described in SystemC at the cycle-accurate level.

An optimal system solution will also require an efficient mapping of high-level abstractions onto the underlying platform. This mapping procedure involves optimizations and trade-offs between many complex constraints, including the quality of service, real-time response, power consumption, area, etc. Tools are urgently needed to explore this mapping process, and assist and automate optimization where possible.

The first challenge for these tools is to bridge the gap in building custom NoCs that optimally match the communication requirements of the system. The network components they build should be highly optimized for that particular NoC design, providing large savings in area, power, and latency with respect to standard NoCs based on regular structures.

In the following section, an example of design methodology for heterogeneous SoCs is briefly illustrated. It is relative to Xpipes interconnect and relies on a tool that automatically instantiates an application-specific NoC for heterogeneous on-chip multi-processors (called *XpipesCompiler* [41]).

Domain Specific Network-on-Chip Synthesis Flow

The complete *XpipesCompiler*-based NoC design flow is depicted in Figure 95.10. From the specification of an application, the designer (or a high-level analysis and exploration tool) creates a high-level view of the SoC floorplan, including nodes (with their network interfaces), links and switches. Based on clock speed target and link routing, the number of pipeline stages for each link is also specified. The information on the network architecture is specified in an input file for the *XpipesCompiler*. Routing tables for the network interfaces are also specified. The tool takes as additional input the SystemC library of soft network components. The output is a SystemC hierarchical description, which includes all switches, links,

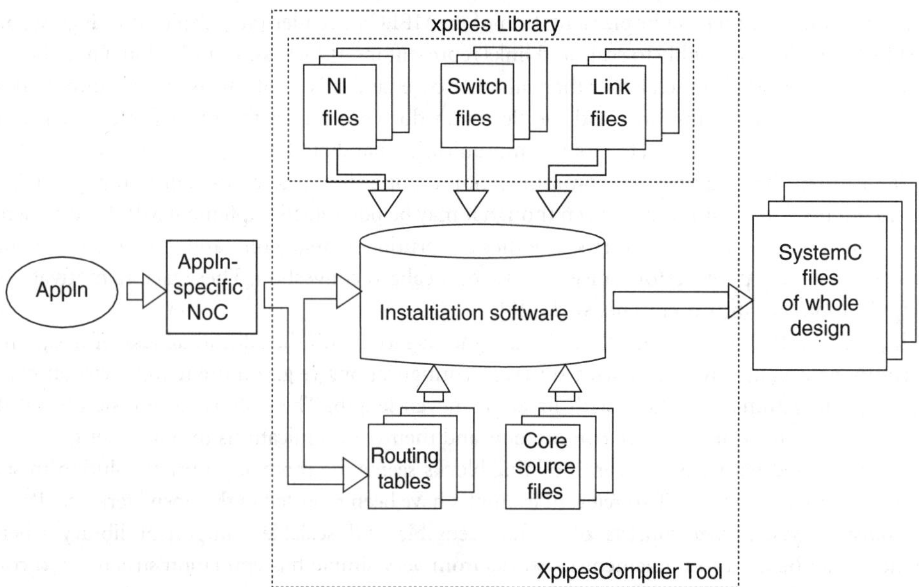

FIGURE.95.10 NoC synthesis flow with *XpipesCompiler*.

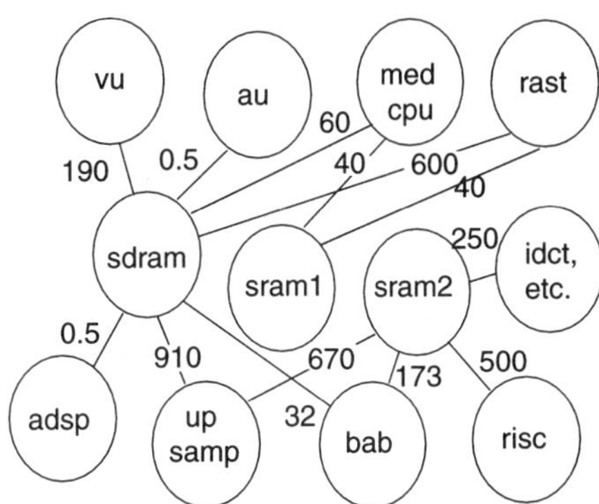

FIGURE 95.11   Core graph representation of an MPEG4 design with annotated average communication requirements.

TABLE 95.1   Area and power estimates for the MPEG4-related NoC configurations

NoC Configuration	Area (mm²)	Ratio mesh/cust	Power (mW)	Ratio mesh/cust
Mesh	1.31		114.36	
Custom 1	0.86	1.52	110.66	1.03
Custom 2	0.71	1.85	93.66	1.22

network nodes and interfaces, and specifies their topological connectivity. The final description can then be compiled and simulated at the cycle-accurate and signal-accurate level. At this point, the description can be fed to back-end RTL synthesis tools for silicon implementation.

In a nutshell, the *XpipesCompiler* generates a set of network component instances that are custom-tailored to the specification contained in its input network description file. This tool allows a very instructive comparison of the effects (in terms of area, power, and performance) of mapping applications on customized domain-specific NoCs and regular mesh NoCs.

Let us focus on the MPEG4 decoder already introduced in this chapter. Its core graph representation together with its communication requirements are reported in Figure 95.11. The edges are annotated with the average bandwidth requirements of the cores in MB/s. Customized application-specific NoCs that closely match the application's communication characteristics have been manually developed and compared to a regular mesh topology. The different NoC configurations are reported in Figure 95.12. In the MPEG4 design considered, many of the cores communicate with each other through the shared SDRAM. Thus, a large switch is used for connecting the SDRAM with other cores (Figure 95.12 (b)), while smaller switches are connected to less communication-intensive cores. An alternate custom NoC is also considered (Figure 95.12 (c)): it is an optimized mesh network, with superfluous switches and switch I/Os removed.

Area (in 0.1 μm technology) and power estimates for the different NoC configurations are reported in Table 95.1. Since all cores communicate with many other cores, many switches are needed and therefore area savings are not extremely significant for custom NoCs.

Based on the average traffic through each network component, the power dissipation for each NoC design has been calculated. Power savings for the custom solutions are not very significant, as most of the traffic traverses the larger switches connected to the memories. As power dissipation on a switch increases nonlinearly with an increase in switch size, there is more power dissipation in the switches of custom NoC1 (that has an 8×8 switch) than the mesh NoC. However, most of the traffic traverses short links in this custom NoC, thereby giving marginal power savings for the whole design.

Figure 95.13 reports the variation of average packet latency (for 64B packets, 32-bit flits) with link bandwidth. Custom NoCs, as synthesized by *XpipesCompiler*, have lower packet latencies as the average number of switches and link traversals is lower. At the minimum plotted bandwidth value, almost 10% latency saving is achieved. Moreover, the latency increases more rapidly with the mesh NoC as the link

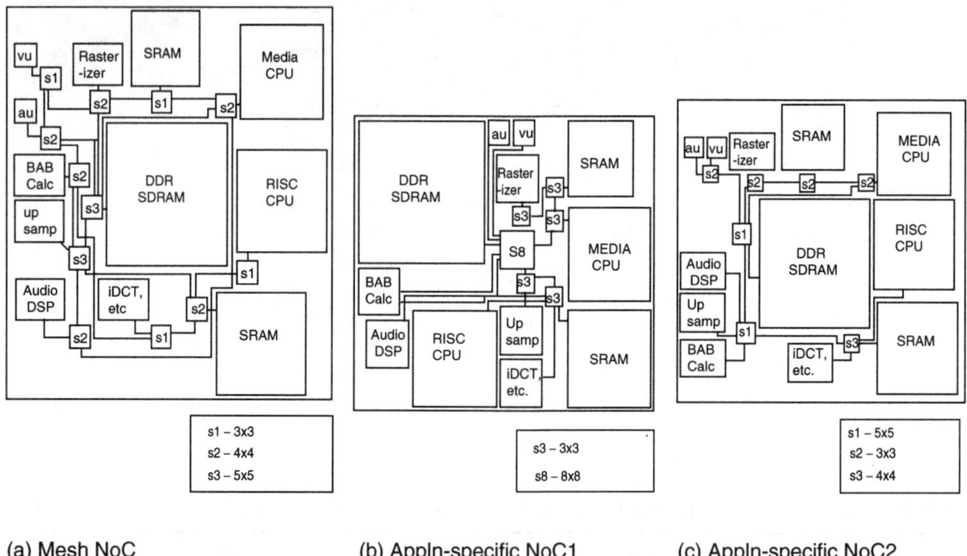

(a) Mesh NoC   (b) Appln-specific NoC1   (c) Appln-specific NoC2

FIGURE 95.12 NoC configurations for MPEG4 decoder.

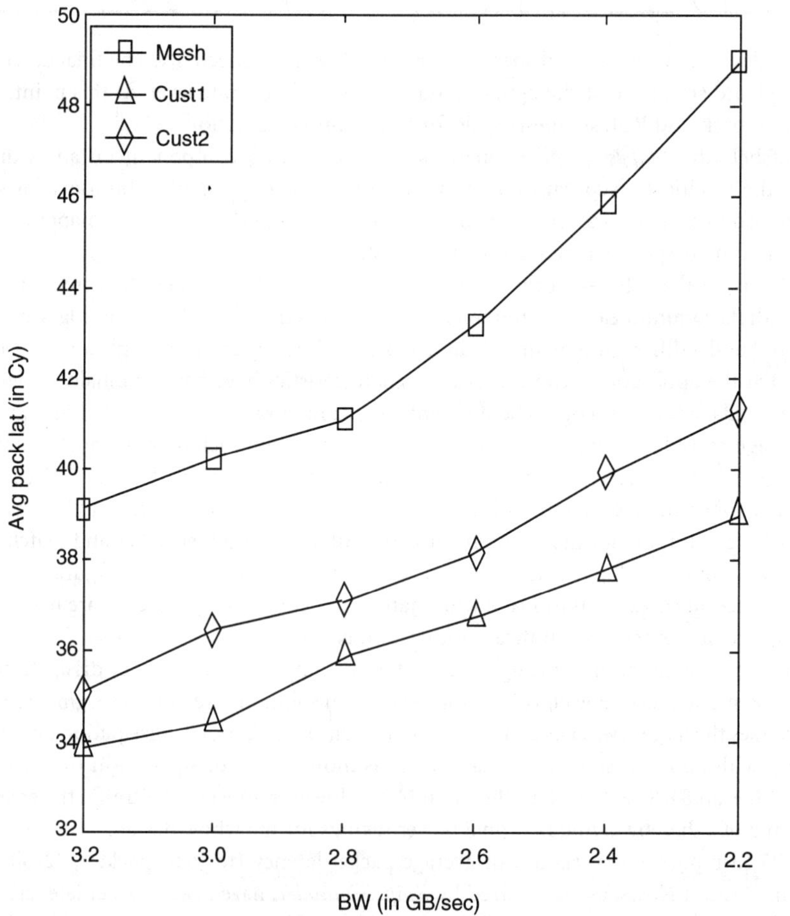

FIGURE 95.13 Average packet latency as a function of the link bandwidth.

bandwidth decreases. Also, custom NoCs have better link utilization: around 1.5 times the link utilization of a mesh topology.

Area, power, and performance optimizations by means of custom NoCs turn out to be more difficult for MPEG4 than other applications such as Video Object Plane Decoders and Multi-Window Displayer [41].

95.6   Conclusions

This chapter describes the motivation for packet-switched networks as communication paradigm for deep submicron SoCs. After an overview of NoC proposals from the open literature, this chapter goes into the details of NoC architectural components (switch, network interface and point-to-point links), introducing the *Xpipes* library of composable soft macros as a case study. Finally, the challenging issue of heterogeneous NoC design is addressed, showing an example NoC synthesis flow and detailing area, power, and performance metrics of customized application-specific NoC architectures with respect to regular mesh topologies. The chapter aims at highlighting the main guiding principles and open issues for NoC design on gigascale multi-processor SoCs.

Acknowledgement

This work was supported in part by MARCO/DARPA Gigascale Silicon Research Center.

References

[1]   Boekhorst, F., *Ambient Intelligence, the Next Paradigm for Consumer Electronics: How will it Affect Silicon?*, ISSCC 2002, Vol. 1, Feb. 2002, pp. 28–31.

[2]   Benini, L. and G. De Micheli, *Networks On Chips: a New SoC Paradigm*, IEEE Computer, Vol. 35, Issue 1, pp. 70–78, 2002.

[3]   Kumar, S., A. Jantsch, J.P. Soininen, M. Forsell, M. Millberg, J. Oeberg, K. Tiensyrja, and A.Hemani, *A Network on Chip Architecture and Design Methodology*, IEEE Symposium on VLSI ISVLSI02, April 2002, pp. 105–112.

[4]   Benini, L., D. Bertozzi, D. Bruni, N. Drago, F. Fummi, and M. Poncino, *SystemC Cosimulation and Emulation of Multiprocessor SoC Designs*, IEEE Computer, Vol. 36, Issue 4, pp. 53–59, 2003.

[5]   Duato, J., S. Yalamanchili, and L. Ni, *Interconnection Networks: an Engineering Approach*, IEEE Computer Society Press, Silver Spring, MD, 1997.

[6]   Wielage, P. and K. Goossens, *Networks on Silicon: Blessing or Nightmare?*, Proceedings of the Euromicro Symposium on Digital System Design DSD02, Sept. 2002, pp. 196–200.

[7]   Dally, W.J. and B. Towles, *Route Packets, not Wires: On-Chip Interconnection Networks*, Design and Automation Conference DAC01, June 2001, pp. 684–689.

[8]   Blume, H., H. Huebert, H.T. Feldkaemper, and T.G. Noll, *Model-based Exploration of the Design Space for Heterogeneous Systems on Chip*, IEEE Conference On Application-Specific Systems, Architectures and Processors ASAP02, pp. 29–40, 2002.

[9]   Nugent, S., D.S. Wills, and J.D. Meindl, *A Hierarchical Block-Based Modeling Methodology for SoC in GENESYS*, IEEE ASIC/SOC Conference, Sept. 2002, pp. 239–243.

[10]   Gerin, P., S. Yoo, G. Nicolescu, and A.A. Jerraya, *Scalable and Flexible Cosimulation of SoC Designs with Heterogeneous Multi-Processor Target Architecture*, Proceedings of the ASP-DAC 2001, Jan./Feb. 2001, pp. 63–68.

[11]   Paulin, P.G., C. Pilkington, and E. Bensoudane, *StepNP: a System-level Exploration Platform for Network Processors*, IEEE Design and Test of Computers, Vol. 19, Issue 6, pp. 17–26, Nov–Dec. 2002.

[12]   Guerrier, P. and A. Greiner, *A Generic Architecture for On-Chip Packet Switched Interconnections*, Design, Automation and Testing in Europe DATE00, March 2000, pp. 250–256.

[13]   Lee, S.J. et al., *An 800 MHz Star-Connected On-Chip Network for Application to Systems on a Chip*, ISSCC03, Vol. 1, pp. 468–469, Feb. 2003.

[14]   Yamauchi, H. et al., *A 0.8 W HDTV Video Processor with Simultaneous Decoding of Two MPEG2 MP@HL Streams and Capable of 30 Frames/s Reverse Playback*, ISSCC02, Vol.1, Feb. 2002, pp. 473–474.

[15] Dall'Osso, M., G. Biccari, L. Giovannini, D. Bertozzi, and L. Benini, *Xpipes: a Latency Insensitive Parameterized Network-on-Chip Architecture for Multi-Processor SoCs*, ICCD03, pp. 536–539. Oct. 2003.

[16] ITRS 2001, Http://public.itrs.net/Files/2001ITRS/Home.htm

[17] Bertozzi, D., L. Benini, and G. De Micheli, *Energy-Reliability Trade-off for NoCs*, in *Networks on Chip*, Jantsch, A. and Tenhunen, H. Eds., Kluwer, Dordrecht, 2003, pp. 107–129.

[18] Zhang, H., V. George, and J.M. Rabaey, *Low-Swing On-Chip Signaling Techniques: Effectiveness and Robustness*, IEEE Transactions on VLSI Systems, Vol. 8, Issue 3, pp. 264–272, 2000.

[19] Xu, J. and W. Wolf, *Wave Pipelining for Application-Specific Networks-on-Chips*, CASES02, Oct. 2002, pp. 198–201.

[20] Goossens, K., J. Dielissen, J. van Meerbergen, P. Poplavko, A. Radulescu, E. Rijpkema, E. Waterlander, and P. Wielage, *Guaranteeing the Quality of Services in Networks on Chip*, in *Networks on Chip*, Jantsch, A. and Tenhunen, H. Eds., Kluwer, Dordrecht, 2003, pp. 61–82.

[21] ITRS 1999, Http://public.itrs.net/files/1999_SIA_Roadmap/

[22] Jantsch, A. and H. Tenhunen, *Will Networks on Chip Close the Productivity Gap?*, in *Networks on Chip*, Jantsch, A. and Tenhunen, H. Eds., Kluwer, Dordrecht, 2003, pp. 3–18.

[23] VSI Alliance, Virtual Component Interface Standard, 2000.

[24] OCP International Partnership, Open Core Protocol Specification, 2001.

[25] Wingard, D., *MicroNetwork-based Integration for SoCs*, Design Automation Conference DAC01, June, 2001, pp. 673–677.

[26] Flynn, D., *AMBA: Enabling Reusable on-chip Designs*, IEEE Micro, Vol. 17, Issue 4, pp. 20–27, 1997.

[27] Zhang, H. et al., *A 1V Heterogeneous Reconfigurable DSP IC for Wireless Baseband Digital Signal Processing*, IEEE Journal of Solid-State Circuits, Vol. 35, Issue 11, pp. 1697–1704, 2000.

[28] Culler, D., J.P. Singh, and A. Gupta, *Parallel Computer Architecture, a Hardware/Software Approach*, Morgan Kaufmann, Los Altos, CA, 1999.

[29] Compton, K. and S. Hauck, Reconfigurable Computing: a Survery of System and Software, ACM Computing Surveys, Vol. 34, Issue 2, pp. 171–210, 2002.

[30] Tessier, R. and W. Burleson, *Reconfigurable Computing and Digital Signal Processing: a survey*, Journal of VLSI Signal Processing, Vol. 28, Issue 3, pp. 7–27, 2001.

[31] Walrand, J. and P. Varaja, *High Performance Communication Networks*, Morgan Kaufmann, San Francisco, 2000.

[32] Liu, D. et al., *SoCBUS: The Solution of High Communication Bandwidth on Chip and Short TTM*, invited paper in Real Time and Embedded Computing Conference, Sept. 2002.

[33] Ishiwata, S. et al., *A Single Chip MPEG-2 Codec Based on Customizable Media Embedded Processor*, IEEE Journal of Solid-State Circuits, Vol. 38, Issue 3, pp. 530–540, 2003.

[34] Rijpkema, E., K. Goossens, A. Radulescu, J. van Meerbergen, P. Wielage, and E. Waterlander, *Trade Offs in the Design of a Router with both Guaranteed and Best-Effort Services for Networks on Chip*, Design Automation and Test in Europe DATE03, March 2003, pp. 350–355.

[35] Dally, W.J. and S. Lacy, *VLSI Architecture: Past, Present and Future*, Conference on Advanced Research in VLSI, 1999, pp. 232–241.

[36] Saastamoinen, I., D. Siguenza-Tortosa, and J. Nurmi, *Interconnect IP Node for Future Systems-on-Chip Designs*, IEEE Workshop on Electronic Design, Test and Applications, Jan., 2002, pp. 116–120.

[37] Carloni, L.P., K.L. McMillan, and A.L. Sangiovanni-Vincentelli, *Theory of Latency-Insensitive Design*, IEEE Transactions on CAD of ICs and Systems, Vol. 20, Issue 9, pp. 1059–1076, 2001.

[38] Scheffer, L., *Methodologies and Tools for Pipelined On-Chip Interconnects*, International Conference On Computer Design, 2002, pp. 152–157.

[39] Glaskowsky, P., *Pentium 4 (partially) previewed*, Microprocessor Report, Vol. 14, no. 8, pp. 10–13, 2000.

[40] Van der Tol, E.B. and E.G.T. Jaspers, *Mapping of MPEG4 Decoding on a Flexible Architecture Platform*, SPIE 2002, January 2002, pp. 1–13.

[41] Jalabert, A., Murali, S., Benini, L., De Micheli, G., *XpipesCompiler: a Tool for Instantiating Application Specific Networks on Chip*, DATE 2004, pp. 884–889, 2004.

[42] Agarwal, V., M.S. Hrishikesh, S.W. Keckler, and D. Burger, *Clock Rate versus IPC: The End of the Road for Conventional Microarchitectures*, in Proceedings 27th Annual International Symposium on Computer Architecture, June 2000, pp. 248–250.

96

An Introductory Survey of Networked Embedded Systems

Hiren D. Patel
Virginia Tech

Sumit Gupta
University of California at Irvine

Sandeep K. Shukla
Virginia Tech

Rajesh Gupta
University of California

96.1 Introduction

Rapid advances in microelectronic technology coupled with integration of microelectronic radios on the same board or even on the same chip has been a powerful driver of growth and revolutionary advances in the telecommunications industry over the last decade. Recently, a new breed of *Networked Embedded Systems* (NES) has emerged, which are multiple distributed computing devices with wireline and/or wireless communication interfaces embedded in myriad products such as automobiles, medical components, sensor networks, consumer products, and personal mobile devices. These systems have been variously referred to as Embedded Network Systems (EmNets), Networked Embedded System Technology (NEST), and NES [1–3].

NES are often distributed embedded systems that must interact not only with the environment and the user but also with each other to coordinate computing and communication. And yet, these devices must often operate in very constrained environments related to their size, energy availability, network connectivity, etc. The challenges posed by the design and deployment of NES have captured the imagination of a large number of researchers and galvanized whole new communities into action. The design of NES requires multi-disciplinary, multi-level cooperation and development to address the diverse hardware (processor cores, radios, security cores) and software (applications, middleware, operating systems, networking protocols) needs. In this paper, we briefly highlight some of the design concerns and challenges in deploying NES.

Some examples of NES are wireless data acquisition systems such as habitat [4–7], agriculture and weather monitoring [8], disaster management and civil monitoring, Cooperative Engagement Capability (CEC) [9] for military use [10], and fabric e-textile [11, 12]. Common to all these systems/applications is their ability to provide an interaction between the environment and humans through a medium of devices such as sensors for data collection, computation processors to perform data computation, and remote storage devices to preserve and collate the information.

A good exposition of the characteristics, parameters, examples, and design challenges of networked embedded systems is presented in [1]. We draw heavily on this book for material and examples. Similar surveys and expositions of challenges in the applications and design and implementation of sensor networks are given in [13–17].

96.2   Characteristics of NES

The realm of possibilities where NES applications can be implemented makes characterizing these systems an inherently difficult task. However, we make an attempt at characterizing the basic functionality and constraints, distributed nature and usability, dependability, and availability of such systems. Then, we describe NES through some examples.

Functionality and Constraints

Networked embedded systems are typically designed to interact with and react to the environment and people that are around them. Thus, often, NES have sensors that measure temperature, moisture, movement, light, and so on. By definition, NES have a communication mechanism — either a wireline connection or a wireless radio. Also, they typically have computation engines that can do at least a minimal amount of computing on the data they acquire.

The environment and user needs place constraints on NES such as small size, low weight, harsh working conditions, safety and reliability concerns, low cost, and poor resource availability in terms of low computational ability and low energy availability (limited battery) [18]. NES devices have to be small in size so that their deployment does not interfere with the environment; that is, they must function almost invisibly to the environment. For example, animals must not be aware of the habitat-monitoring sensors that are embedded on them or around them. This example also demonstrates the need for these systems to be low weight and be able to work under harsh working conditions, that is, be tolerant of temperature changes, physical abuse, vibration, shock, and corrosion. Since NES are frequently deployed in the field with little or no access to renewable energy sources, they have to live off a limited energy source or battery. Due to real-time and mission-critical requirements, NES have to frequently meet safety and reliability constraints. For example, the cruise control, antilock braking, and airbag systems in automobiles have to respond within given real-time constraints to meet safety requirements. The small form factor and wide distribution of NES also place a constraint on cost; price fluctuations of even a few cents on each device have a large impact as the volume of devices deployed increases.

Figure 96.1 shows the Berkeley NES device that consists of a MICA processor and a radio board [19]. As technology advances, these devices are becoming smaller. In fact, the latest Berkeley "mote" is as small as a coin. This has led to the notion of *smart dust* or a massively distributed sensor network that is self-contained, networked, and provides multiple sensor and coordinated computational capabilities [20, 21, 7, 22].

Distributed Nature

The application spectrum of NES often means that these systems are physically distributed. In fact, the distributed nature of NES extends to distributed functionality and communication as well. *Distributed in function* refers to NES components that perform specific roles and work together with other NES components to complete the system. Automotive electronics is a good example where many different function-specific components work in unison, such as the power control modules, the engine, airbag

FIGURE 96.1 U. C. Berkeley NES device — the MICA processor and radio board [19].

deployment, cruise control, suspension, etc. Figure 96.2 shows the components from Kyocera that are widely used in automotive electronics. Similarly, *distributed communication* refers to local and global communication between the embedded systems distributed throughout the system. For example, automotive systems have local wires from actuators/sensors to Electronic Control Units (ECUs) and global wires/buses between ECUs [23].

Usability, Dependability, and Availability

NES are becoming an increasingly dominant part of a number of devices and systems in all aspects of our daily lives from entertainment, transportation, personal communications to biomedical devices.

The pervasiveness of these systems, however, raises concerns about dependability and availability. *Availability* generally means access to the system. In scenarios where some of the components of the NES fail, there must be a mechanism through which the users can interface and interact with the components to investigate and rectify the problems. Mediums of access can be through Personal Digital Assistants (PDAs), wired serial access points, infra-red technology, etc. Another dimension of availability is the *long life* expected from NES components. Often, NES do not have access to a renewable energy source and sometimes it is not possible to change the battery source either. For example, sensors deployed to measure traffic on roads or sensor tags placed on animals are inaccessible or difficult to reach after deployment.

Dependability or *reliability* is also a major concern that goes hand in hand with the availability requirement. The system must guarantee a certain level of service that the user can depend on. For example, temperature sensors and smoke detectors are critical to the fire safety requirements of any building.

Availability and dependability characteristics are especially crucial to safety-critical systems such as avionics and biomedical applications. The sensors used in an airplane to monitor cabin pressure, oxygen levels, elevation, relative speed, etc, are all important to maintain safety. For example, the release of oxygen into the oxygen masks in airplanes is controlled via sensors that monitor the cabin oxygen levels. For biomedical applications such as electronic heart pacemakers, the need for dependability is obvious since malfunctioning components can be life threatening. Military personnel monitoring is another example for biomedical applications with NES where devices are used to transmit the location and vital statistics of the personnel.

Since NES consist of diverse hardware and software components that interact with each other, component *interoperability* becomes another important concern [1]. Today, cars have tens or sometimes even a hundred embedded computing systems that are designed and manufactured by different contractors. Such complex distributed, interacting embedded systems raise difficult challenges in system integration, component interoperability, and system testing and validation.

These networked embedded systems characteristics present new challenges and constraints for systems engineers, which have not been fully addressed in the design of past networking and distributed systems. Software and hardware tools and techniques are required to satisfy the need for low cost, low power, QoS guarantees, and fast time-to-market. Formalization of methodologies to ensure functional correctness of

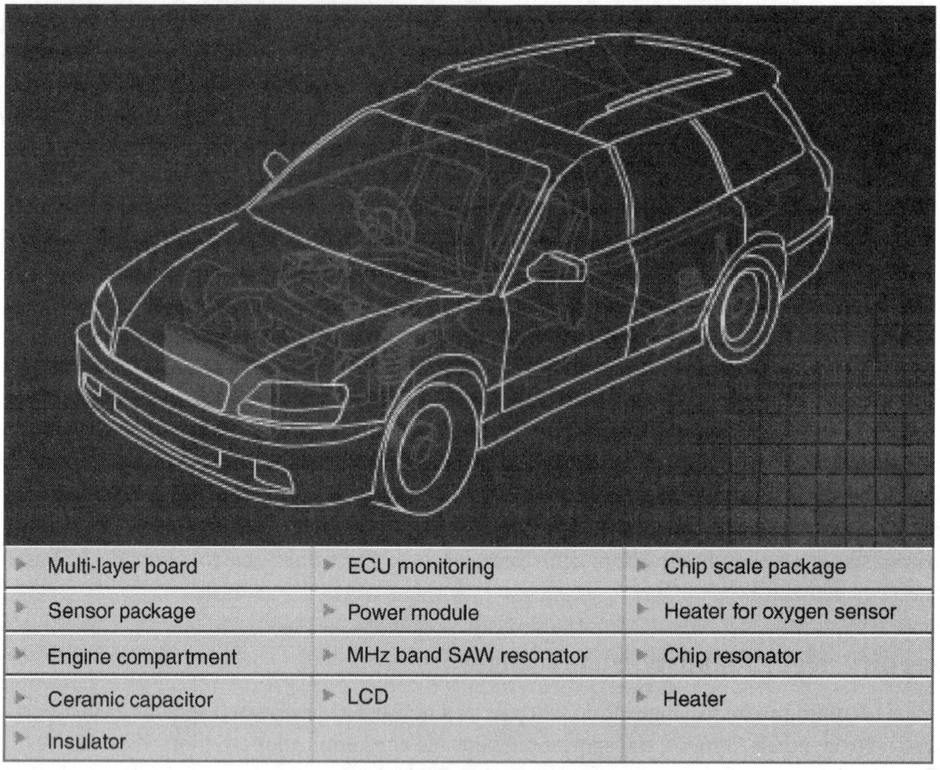

(a)

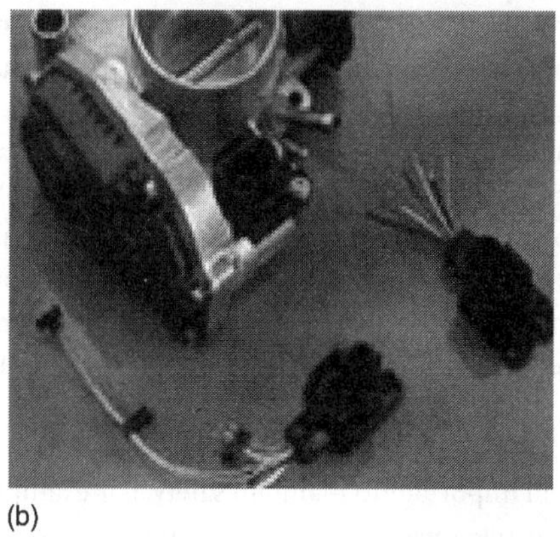

(b)

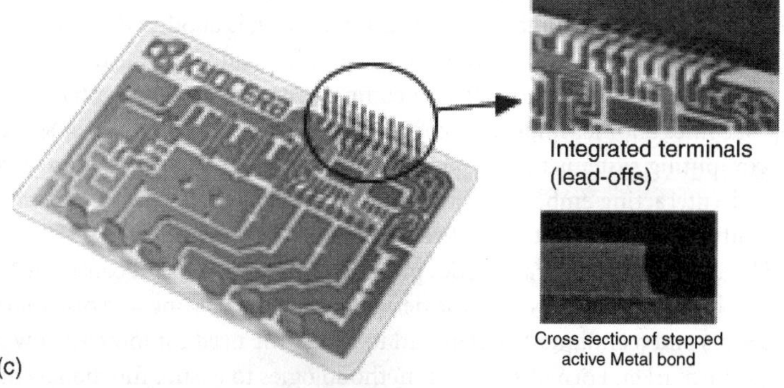

(c)

FIGURE 96.2 Automotive electronic components by Kyocera [24].

design and efficiency of design is paramount in meeting the time-to-market requirements by reducing the iterations in the design process.

In the following sections, we discuss some of the interesting applications of NES and then describe NES tools and methodologies such as programming languages, simulation environments, and performance measurement tools designed specifically to address the design challenges posed by NES.

96.3   Examples of NES

To demonstrate the characteristics, constraints, and design challenges of NES, we present several examples from current and future NES. These examples are representative of the diverse application domains where NES can be found. We start with three examples from [1].

Automobile: Safety-Critical Vs. Telematics

Cars today typically have tens to a hundred microprocessors controlling all aspects of the automobile from entertainment systems to the emergency airbag release mechanisms. Figure 96.3 shows some of the telematic components in a Mitsubishi car [25]. The microprocessors in charge of the functionality frequently communicate and interact with other processors. For example, the stereo volume is automatically reduced when the driver receives (or answers) a call on his or her cell phone.

Thus, a range of devices that perform different tasks are beginning to be organized in sophisticated networks as distributed systems. Broadly speaking, there are two such distributed systems: *safety-critical processing systems* and *telematics* systems. Clearly, the safety-critical aspects cannot be sacrificed or compromised in any way. These two systems are an integral part of the design and construction of the automobile and dictate several design parameters. Since automobiles have times-to-market that can span up to five years from concept to final product, frequently the technology used for the telematics and safety-critical components in the automobile is already outdated. This is a rising concern, especially for the safety-critical components, because upgrading or switching out components is generally not performed or usually not even feasible. Note that systems that cannot be upgraded or altered after final production are considered to be *closed systems*.

Conversely, open systems allow plugging in newer components with more capabilities and features similar to a plug-and-play environment. Thus, to make automobiles open systems, we have to develop technologies that enable automobile designers to construct the safety-critical and telematics systems in an abstracted manner such that components with a standardized communication protocol and interface

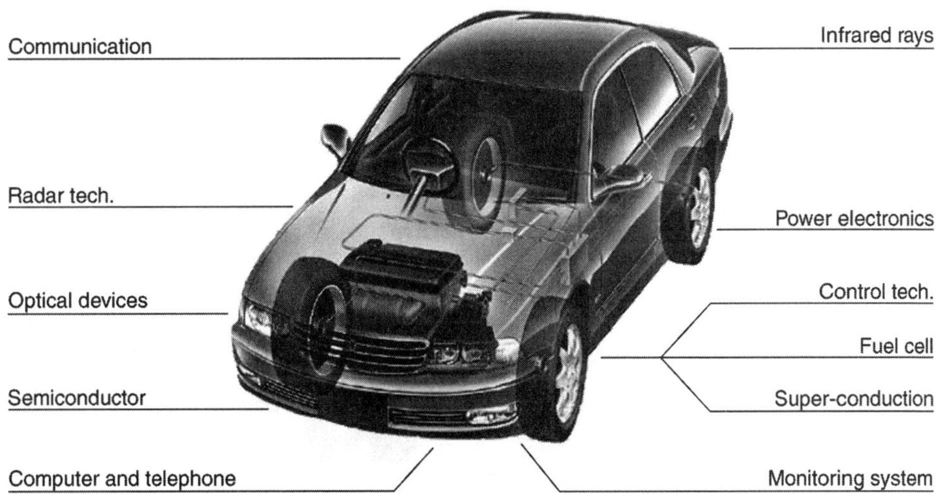

FIGURE 96.3   Telematics components in an automobile [25].

can simply be plugged into the final product. This resolves the disparity between the long design cycles for automobiles and the rapid advances in NES components used in them.

The increasing popularity of wireless technology has spawned interesting applications for automobiles. For example, the OnStar system from General Motors can monitor the location of a car and customer service staff can remotely unlock the car, detect when airbags have been deployed, and so on. Wireless communication opens up infinite possibilities such as automobile service requests and data collection for automobile users, dealers, and manufacturers.

Data Acquisition: Precision Agriculture and Habitat Monitoring

The use of sensor nodes for data acquisition is becoming a useful tool for agricultural and habitat monitoring. In [7], the authors present a study in which they used wireless sensor networks in a real-world habitat monitoring project. The small footprint and weight of modern sensor nodes make them attractive for habitat and agricultural monitoring since they cause minimal disturbance to the animals, plant

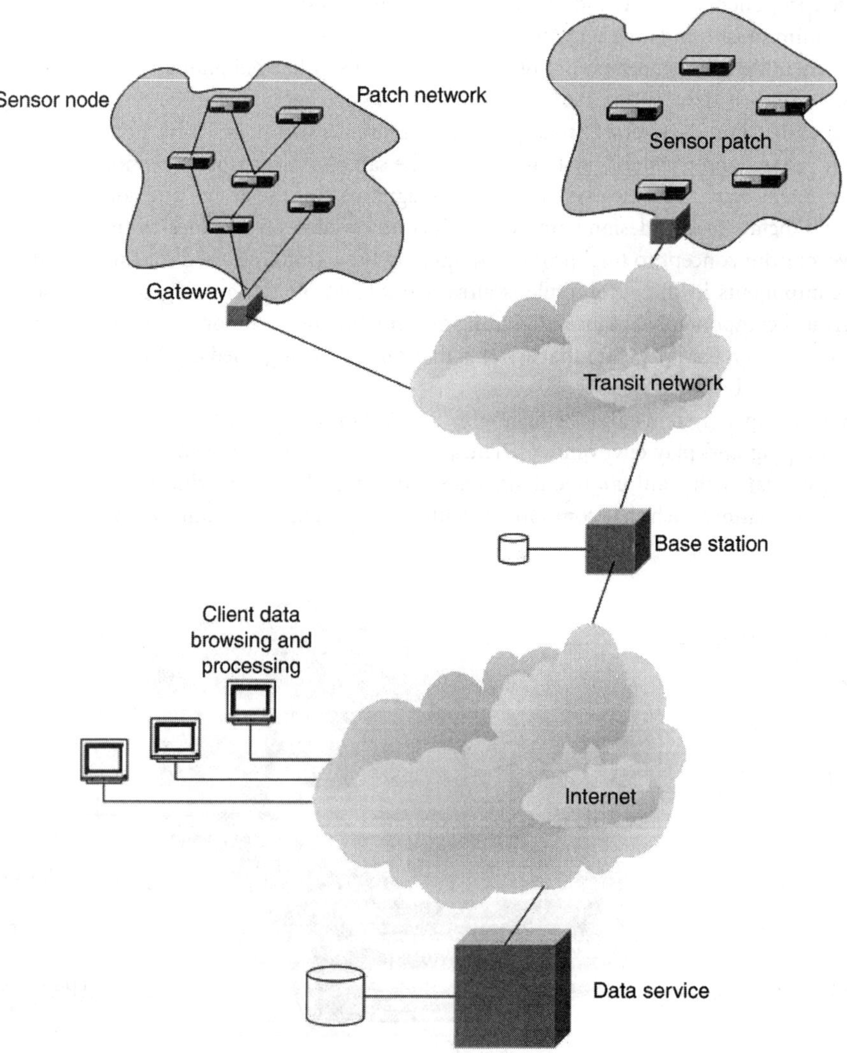

FIGURE 96.4   System Architecture of NES for common data acquisition scenarios [7].

population, and other natural elements of the habitats being monitored. This solution to monitoring automates some of menial tasks such as data collection for researchers (Figure 96.4).

Precision agriculture is an important area of research, where NES technology is likely to have a large impact [1, 26, 27]. Precision agriculture envisages sensor deployment to monitor and manage crop productivity, quality, and growth. Besides increasing productivity, better crop quality and crop management are also key aspects of using NES in precision agriculture.

Crop management provides monitoring and adjusting the level of fertilizer, pesticides, water for particular areas, resulting in better yields at less pollution, emissions, and definitely lower costs. The automation of these functions requires an adaptive behavior to the changing surroundings such as water levels if it rains, or pesticides for a particular season when bug problems are more common. This adaptation is an integral aspect of precision agriculture. While there exist models that dictate the necessary amount of fertilizer, water, nutrient combinations, these models are not always accurate for the specific locale. Thus, NES can also perform on-the-side data acquisition functions for purposes of reconstructing appropriate models and recalibrating or reconfiguring the sensor metrics accordingly to better suit the specific climate and locale.

Feedback into such systems is crucial to develop the notion of true automated precision agriculture. Fine-grained tuning to crop management can be done automatically based on these regularly updated models that can also be monitored by researchers. However, manual adjustments of some kind would require appropriate interfaces between the deployed NES and the end-user attempting to make the change. Once again, wireless interfaces can be used for such manual fine-tunes. Configuration and management of the network can be handled remotely via hand-held devices or even desktops. A practical example of such a deployment is shown in [27] and wireless sensor networks for precision agriculture in [26]. A similar two-tiered network that couples wireless and wired networks together has been proposed for structural monitoring of civil structures that can be affected by natural disasters such as earthquakes [28]. The small size and wireless nature of NES sensor nodes enable field researchers to deploy these sensors in small and sensitive locations.

Defense Applications: Battle-Space Surveillance

NES are projected to become crucial for future defense applications, particularly in battle-space surveillance and condition and location monitoring of vehicles, equipment, and personnel. A military application called CEC, developed by Raytheon Systems [9, 29], acts as a force multiplier for naval air and missile defense systems by distributing sensor and weapons data to multiple CEC ships and airborne units. Data from each unit are distributed to other CEC units, after which the data are filtered and combined via an identical algorithm in each unit to construct a common radar-like aerial picture for missile engagements.

DARPA has funded research in several areas of defense systems under the aegis of their future combat systems program [30]. Manipulating battle environments and critical threat-identification are projected uses of NES in such systems. Manipulating battle environments refers to controlling opposition by detecting their presence and either altering their route or constraining their advance. Threat-identification involves identifying a threat early for force protection. A force-protection scenario involves deployment of sensors around a perimeter that requires protection so that forced entry can be identified and automated responses such as alarms can be triggered based on a certain event.

System deployment used to be of concern in the past because sensors were bulky and large in size and required manual deployment. However, with the advances in technology, entire sensor networks can now be deployed by airdrop, personnel, or even via artillery. The small sizes have enabled NES to be deployed for monitoring vehicles in a manner similar to the automobile example we discussed earlier.

A relatively new technology called *e-textiles* has emerged whereby sensors or other computation devices are integrated into wearable material [11]. Nakad et al. [11] are investigating the communication requirements between sensing nodes of e-textile and the computing elements embedded with them. One key and obvious application of e-textiles is for data acquisition for human monitoring where sensor nodes can be used to track the location and vital statistics of military personnel.

Biomedical Applications

A "civilian" application for e-textiles is for monitoring the health of people — particularly, old people. A sensor node embedded in an e-textile worn by patients with heart problems can automatically alert doctors or emergency services when the patient suffers from heart failure. We have already seen the value of heart pacemakers for helping millions of people around the world maintain a regular heart beat.

Work is in progress to make sensors small enough and body-friendly enough that they can either be surgically inserted or swallowed for temporary monitoring. These devices can be used to monitor, diagnose, and even correct anomalies in the health of a patient. Of course, surgical insertion or ingestion of microelectronic devices raises several concerns about safety, and the ability of the body to adapt to the foreign bodies. These are active areas of research.

Disaster Management

Scenarios that involve disaster management can be seen as data acquisition applications where certain information is gathered, based on which a response is computed and performed. A good example of an implemented scenario is provided in [31], where remote villages are monitored by four sensors measuring the seismic activity and water levels for earthquakes and floods, respectively. Through wireless media, these sensors are connected to the nearest emergency rescue stations signaling emergency events when the thresholds for maximum water levels and seismic activity are crossed. As mentioned earlier, Kottapalli et al. [28] propose a sensor network for structural monitoring of civil structures in case of natural disasters such as earthquakes. Other applications of disaster management systems include severe cold (or heat) monitoring, fire monitoring (smoke detectors, heat sensors), volcano monitoring, etc.

96.4 Design Considerations for NES

The examples presented in the last section given an idea of the breadth of the application domains in which NES can be deployed. By studying these examples, we understand the various requirements, constraints, issues, and concerns involved in developing these kind of systems. Furthermore, as NES proliferate, the true potential of these systems will be realized when they are deployed at a massive scale, in the order of thousands or more components. Such a large-scale deployment, however, raises some problems [17, 13, 15, 16]:

Deployment: Deployment refers to the physical distribution of the nodes in the NES. The first concerns for deployment are safety, durability, and sturdiness; if devices are dropped from the air, they should not do damage to other objects (people, animals, plants, or material) while landing, and should not be damaged themselves either. This is clearly important for defense applications where surveillance sensors may be airdropped in the battle-space.

Several deployment strategies are available that can be classified into either random or strategic deployment. As the name suggests, *random deployment* refers to deploying NES nodes in an arbitrary manner in the field. Random deployment is useful when the region being monitored is not accessible for precise placement of sensors [32]. The problem then becomes of determining region coverage and also the possible redeployment or movement of nodes to improve coverage. *Strategic deployment* refers to placing NES nodes at well-planned points so that the coverage is maximized or to place nodes strategically in a small field of concentration such that these motes are not easily subjected to natural damages (e.g., for habitat monitoring).

The number of NES nodes deployed must be considered in the cost and performance/quality of monitoring tradeoff because some nodes are generally bound to be destroyed by some means so that there should be sufficient reserves or fault tolerance in the network to continue the monitoring.

Environment-interaction: NES components often need to interact with the environment without human interaction. Thus, a requirement of NES is an ability to work on their own and perhaps also have a feedback loop so that nodes can adapt to changes (failure of nodes, movement of objects) in the environment and continue functioning correctly. Systems such as those used in precision agriculture, chemical and hazardous gas monitoring, and so on are designed to interact and react to changes in the system. For example, in agriculture, the release of water can be tied to the moisture content in the air.

Life-expectancy of nodes: As discussed earlier, an essential requirement for nodes in an NES is a long life expectancy. This is because once deployed, it is very difficult to access and refurbish the batteries in the nodes. Also, these nodes must sustain environmental challenges such as inclement weather and unexpected loss of nodes to animal interaction or due to component failure. Thus, a whole body of work has gone into identifying node failure and subsequent reconfiguration of the network to have some amount of fault tolerance [33].

Communication protocol between devices: A combination of wired and wireless links can be used to establish a NES. Furthermore, the nodes in the network may be stationary or mobile. Mobile nodes bring in a whole range of issues related to dynamic route and neighbor discovery, dynamic routing, etc. The NES should also be able to reconfigure and adjust to tolerate loss of nodes from a communication point of view. That is, if a node that is a relay point fails or dies, then the network should be able to use other nodes for relaying instead.

Reconfigurability: In many scenarios, it is not possible to physically reach nodes. However, NES frequently require nodes to reconfigured after deployment. This may be to add, remove, or change functionality or to adjust parameters of the functionality. For example, hand-held devices or even desktops may be used to reconfigure nodes to fine tune certain aspects of the system. For example, the water level can be increased in precision agriculture when the weather report suggests a sudden heat wave for the following few days [26, 27].

Security: NES — particularly those that use wireless communication — are prone to malicious attack [34]. This is most evident in military equipment where communication has to be secure from enemy eavesdropping. Security in hand-held devices is becoming an increasing concern with their widespread use in office environments for everything from checking email to exchanging sensitive documents and data. Running security protocols is computationally expensive and hence, power hungry, and several researchers are proposing ways to reduce these power requirements for sensor networks and hand-held devices [35–37].

Energy-constrained: The small form factor, low weight, and the deployment of NES nodes in inaccessible and remote regions imply that these nodes have access to a limited nonrenewable energy source. Thus, one major focus of the research community is to develop networking protocols, applications, operating systems, etc (besides devices), that are energy efficient and utilize robust, high throughput but low-power communication schemes [17, 13].

Operating System: There is a need for special or optimized operating systems due to the stringent hardware constraints (small form factor, limited energy source, limited memory space) and strict application requirements (real-time constraints, adaptability). Several real-time operating systems have been proposed for embedded devices such as eCos [38], LynxOS from LynuxWorks [39], QNX RTOS [40], etc.

Adequate design methodologies: Standard design methodologies and design flows have to be modified or new ones have to be created to address the special needs of NES. For example, there is a need for design methodologies for low-power system-on-a-chip implementations to enable integration of the large number of diverse components that form an NES device [41].

96.5 System Engineering and Engineering Tradeoffs in NES

The design considerations presented in the previous section raise opportunities for interesting tradeoffs between the hardware and software components in NES. Whereas area, power, and weight constraints limit the amount of hardware that can be put in an NES node, integration, debugging, and complexity issues are hindering, increased dependence on software.

Hardware

Rapid advances in silicon technology are ushering in an era where we will see the widespread use of smart dust or very small sensor nodes with reasonably complex computational and communication abilities

[20–22]. Besides a small size, these nodes have low power and have a variety of actuators and sensors, along with radio/wireless communication devices, and processors for computation. This enables these nodes to move from beyond being just data acquisition sensors that send their data to a central server. They can now also act as computation points that first collate and process the data before sending it to a server or even coordinate computation among themselves independent of a central server.

The power and area constraints on NES nodes mean that general-purpose microprocessors cannot be used in them. However, low-power application-specific instruction processors (ASIPs) augmented with application-specific integrated circuits (ASICs) will provide the necessary computational ability at a relatively low power. Whereas the ASIPs are easily programmable, the ASICs can be used for executing computationally expensive and/or time-sensitive portions of applications. For example, target identification in defense systems or airbag release mechanisms in cars require ASICs to meet their timing and computational needs.

In fact, Henkel and Li [42] and Brodersen et al. [1] have shown that less power is consumed by custom-made processors than by general-purpose processors. The reason for this is that with custom chips, parallelism can be effectively exploited to gain better power consumption. Also, hardwiring the execution of each function eliminates the need for instruction storage and decoding, thus reducing power as well.

On the other hand, applications such as habitat monitoring and precision agriculture do not have high timing or computational requirements; hence, generic microprocessors or ASIPs can be used. The compromise is speed and computational ability versus programmability. ASICs have a high design and manufacturing cost and are inflexible when compared to programmable processors. A change in applications or protocols leads to a large redesign effort. Programmable processors, on the other hand, can be reused for several generations of an application (provided computational requirements do not increase).

Reconfigurable hardware such as field programmable gate arrays (FPGA) provides a middle path between programmable processors and hardwired ASICs. As the name suggests, FPGAs can be reprogrammed after being deployed in the field and hence provide the flexibility of microprocessors while providing the hardwired speed of ASICs. In fact, FPGAs can be configured at runtime as suggested by Nitsch and Kebschull [43]. They propose storing the functional behavior and structure of applications in an XML format. When a client wishes to execute an application, the XML is analyzed and the appropriate mapping is done to the FPGA. The drawbacks of FPGAs are that they require large chip area and have a low clock frequency.

Software

Small memory storage devices and low computational processors limit the size and complexity of the software that can run on NES nodes. Porting commonly used operating systems and applications to NES is difficult because of the limitations posed by the hardware. Hence, software development for NES nodes is another challenge that embedded systems designers have to overcome.

The Tiny microthreading operating system (TinyOS) [44] has been proposed to address the unique characteristics of NES nodes. TinyOS is a component-based, highly configurable embedded operating system with a small footprint. TinyOS has a highly efficient multithreading engine and maintains a two-level FIFO scheduler. TinyOS consists of a set of interconnected modular components. Each component has tasks, events, and command handlers associated with it. Tasks are the processing units of components. Components can signal event, issue commands, and execute other tasks. They are allocated a static area of memory to hold the state information of the thread associated with the component. TinyOS does not provide the functionality of dynamic memory allocation due to the restrictions imposed by the hardware. The component-based structure allows TinyOS to be a highly application-specific operating system that can be configured by altering configuration files (.comp and .desc files) (Figure 96.5).

Volgyesi and Ledeczi [2] provide a model-based approach to the development of applications based on TinyOS. They presented a graphical environment, called *GRATIS*, through which the application and operating system components are automatically glued together to produce an application. GRATIS provides automatic code generation capability and a graphical user interface to construct the .comp and .desc

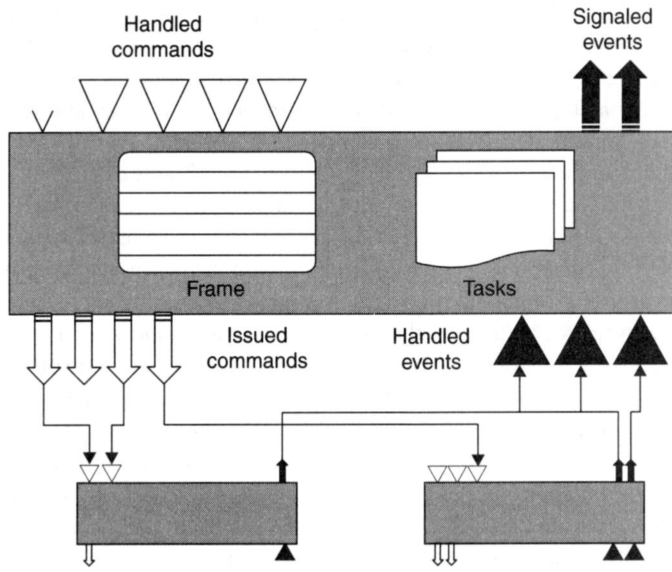

FIGURE 96.5   Events and commands in TinyOS.

files automatically, thus simplifying the task of component description and wiring for building TinyOS-based applications [2]. This increases design productivity and reconfigurability.

Another design effort to provide functionality to program operating system in sensor nodes is the development of a programming language framework called *nesC* [41]. nesC provides a programming paradigm based on an event-driven execution, flexible concurrency, and component-based design. TinyOS is an example where this language has been used to develop a commonly used Linux-based operating system for sensor networks. This programming language manages to successfully integrate concurrency, reactivity to the environment, and communication.

The distributed nature of NES means that these systems are inherently concurrent. For example, data processing and event arrival are two processes that need to be concurrently executed on the NES node. Concurrency management then has to ensure that race conditions do not occur. For example, in the emergency airbag release system, the sensor needs to be able to sense the impact as well as react to it based on processing the collected data. These type of real-time demands along with the small size and low cost of NES nodes make concurrency management a challenging task.

nesC addresses these issues by drawing upon several existing language concepts. Three of the main contributions that nesC provides are: definition of a component model, expressive concurrency model, and program analysis to improve the reliability and reduce code volume. The component model supports sensor node-like event-targeted systems with bidirectional channel interfaces to ease event communication. It also provides flexible hardware and software boundaries and avoids dynamic component instantiation and the use of virtual functions. Expressive concurrency model is tied in with compile time analysis yielding data race detection at compile time to allow for comprehensive concurrent behaviors in NES nodes. Reduction of code size and improvement of reliability are natural goals for any programming language.

TinyOS influenced the design of nesC due to the specific features of the operating system. Firstly, TinyOS provides a collection of reusable system components best suited for component-based architectures. The channel interface connecting components is called the wiring specification, which is independent of the specific implementation of the component. Tasks and events are inherent in TinyOS where tasks are regarded as nonpreemptive computation mechanisms and events are similar to tasks except that they can preempt another task or event. The event-task-based concurrency scheme in TinyOS makes event-driven and expressive concurrency closely related to the implementation of nesC.

Components in nesC are of either modules or configuration types, where the former consists of application code and the latter provides interfaces for communicating between components. Modules are written in C-style code and a top-level configuration is used to wire components together. This resembles the VHDL-like component-architecture scheme where components are defined and the architecture is the top-level model that connects signals between components. The component-based architecture brings flexibility to application implementations and allows users to write highly concurrent programs for a very small-scale platform with limited physical resources. Fortunately, with the aid of nesC and graphics, configuration tools such as GRATIS construction of dedicated operating systems based on TinyOS is gradually becoming easier [2]. These tools allow a designer to build their own operating system with relative ease, but application-specific functionality still requires implementation at a programming level.

Another area of software for NES nodes that has received considerable attention is network protocols [14, 17]. Power and energy constraints in NES nodes necessitate efficient network protocols for transmission of sensed data and intermediary communication. Two broad classifications of sensor networks are proactive and reactive. *Proactive*, as the word suggests, periodically sends the sensed attribute to the data collection location or base station. The period is known a priori, allowing the sensors to migrate to their idle, sleep, or off modes to conserve energy. Applications that require periodic monitoring are best suited for this type of sensor network. A protocol called low-energy adaptive clustering hierarchy (LEACH) [45] is one of the many proposed proactive protocols.

Reactive networks, on the other hand, continuously sense the environment and transmit their data to the base station only upon sensing that the attribute has exceeded a specified threshold. This type of network is useful for time-critical data so that the user or base station receives the sensitive information immediately. One such time-sensitive protocol for reactive systems that has been proposed recently is the threshold-sensitive energy-efficient-sensor network (TEEN) protocol [46].

Hybrid networks constitute a third type of sensor networks that are a combination of proactive and reactive networks and are an attempt to overcome the drawbacks experienced by the other two network systems [47]. In hybrid networks, sensor nodes send sensed data at periodic intervals and also when sensed data exceed the set threshold. The periodic interval is generally longer than the ones found in proactive networks such that the functionality of the two network types can be incorporated within one. Furthermore, hybrid systems can be made to work in only proactive or only reactive modes as well.

96.6   Design Methodologies and Tools

Deploying large-scale distributed networked embedded systems is inherently a complex and error-prone task. Designers of such systems rely on system-level modeling and simulation tools during the initial architectural definition phase for design space exploration, to come up with possible architectures that will satisfy the constraints and requirements and then to verify the functionality of the system.

Design verification at the highest levels of abstraction down to final implementation is an important concern with any complex system. With distributed systems, this need becomes even more acute due to the inability of a system designer to foresee all possible events and sequences of events that may occur.

Several tools have been developed for simulating NES at the highest level of abstraction as communicating network models composed of basic network models. Network Simulator 2 (NS-2), OPNET, SensorSim, and NESLsim are popular network simulation tools widely used in the community [48–51]. SensorSim and NESLsim are simulation frameworks specifically designed for sensor networks.

SensorSim closely ties in sensor networks with their power considerations by constructing a two-pronged approach to creating the models. The first prong involves creating a sensor functional model that represents the software functions of the sensor consisting of the network protocol stack, middleware, user applications, and the sensor protocol stack. The second prong is the power model that simulates the hardware abstracts such as CPU, and radio module to provide the sensor functional model to execute. The architecture of the SensorSim simulator is shown in Figure 96.6.

This two-pronged model implies that the sensor functional model dictates the execution of the tasks to the power model and these two models work in parallel with each other. An added feature is the sensor channel that allows sensing devices to detect events. In this way, the sensor channel exposes external signals to the sensor modules such as microphones, infra-red detectors, etc. The signals that are transmitted through this channel can be of any available form such as infra-red light and sound waves for microphones. Every type of signal has different characteristics based on the medium through which they travel and this is the primary goal of the sensor channel — to simulate these characteristics accurately and to detect and monitor the events in a sensor network.

The use of a power model in SensorSim follows from the importance placed in designing low-power NES devices [52]. Efficient power control is a basic requirement for the longevity of these devices. The basis behind the power model is that there is a single power supplier, the battery, and all other components or models are energy consumers (as shown in Figure 96.6). The consumers such as the CPU model and radio model drain energy from the battery through events.

An attractive feature of SensorSim is its capability to perform hybrid simulations. This refers to the ability of SensorSim to behave as a network emulator and interact with real external components such as network nodes and user applications. However, network emulation for sensor networks differs from traditional network emulation. The large number and speed of input/output events in sensor networks mandate readjusting the real-time delays for the events and reordering the events, making the implementation of an emulator for such networks a much more difficult task.

SensorSim enables reprogramming the sensor channel to monitor external inputs, thus using real inputs instead of models for these channels. For example, instead of modeling waves traveling through a wired (e.g., coaxial) cable, a microphone can be connected using a sensor channel to send waveforms through a wired link to the simulator.

NESLsim is another modeling framework for sensor networks. It is based on the Parallel Simulation Environment for Complex Systems (PARSEC) simulator [51]. NESLsim abstracts a sensor node into two entities: the *node entity* and the *radio entity*. The node entity is responsible for computation tasks such as scheduling, traffic monitoring, and congestion control, whereas the radio entity undertakes the responsibility to maintain the communication between the sensor nodes in the NES. A third entity that is not part of the sensor node is the *channel entity*. This models the wireless medium through which communication is performed.

NS-2 is a discrete event simulator based on an open-source C++ framework that implements a discrete event simulator developed by Virtual Inter Network Test bed (VINT) collaborative research project at University of Southern California and the University of California, Berkeley. They provide substantial support for simulation of the routing and multicast protocols in the TCP/IP networking stack (IP, TCP, UDP, etc) over both wired and wireless channels. *OPNET* also performs similar tasks but is a proprietary software developed by Opnet Technologies.

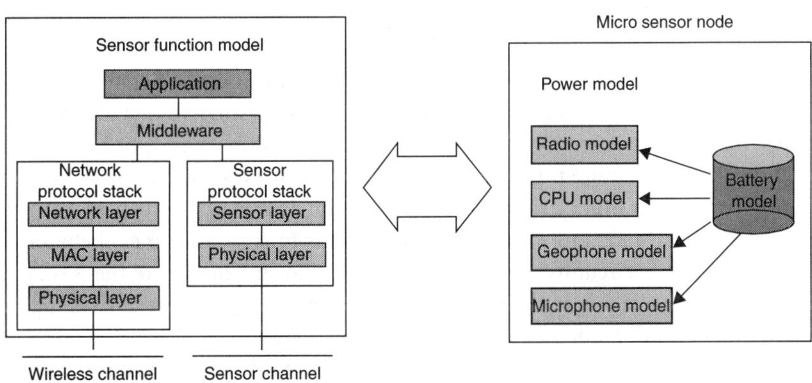

FIGURE 96.6   SensorSim architecture [52].

SystemC [53, 54] is a system-level description language developed to model both the hardware and software of a behavioral specification. Drago et al. [55] developed a methodology to combine the simulation environments of NS-2 and SystemC to simulate and test the functionality of NES. They promote the use of NS-2 for modeling the network topology and communication infrastructure and SystemC for representing and simulating the hardware/software components of the embedded system. Using NS-2 relieves the designer of writing detailed high-level network protocols that already exist and are available for simulation in NS, whereas SystemC allows modeling and simulation of implementations of embedded systems. Together, these simulation frameworks preserve simulation integrity and reduce the modeling effort with an admissible degradation in simulation performance.

Integration of simulators is regarded as a valuable resource for systems designers. However, to be able to perform such a link between simulators, the underlying development platform must be similar. For example, NS-2 and SystemC both have a C++ underlying framework on which these simulators were built. Also, the basic simulation paradigm in NS-2 and SystemC is similar. NS-2 is a discrete event-driven simulator where the scheduler runs by selecting the next event, executing it to completion, and looping back to execute the next event. Similarly, the SystemC simulator also has a discrete-event-based kernel where processes are executed and signals are updated at clocked transitions, working on the evaluate–update paradigm. Drago et al. [55] use a shared memory queue to pass tokens and packets for communication between the NS-2 kernel and SystemC kernel.

System-level design methodologies for embedded systems are based on a hardware–software codesign approach [56–59]. Hardware–software codesign is a methodology where the hardware and software are designed and developed concurrently and in collaboration. This leads to a more efficient and optimized implementation of applications.

Ramanathan et al. [3,60] present a timing-driven design methodology for NES and explore the need for temporal correctness while designing these systems. Determining temporal correctness of system models is difficult because these models are not cycle accurate and usually have no notion of hardware and/or software implementation. Determining the correctness of timing constraints after the hardware has been manufactured naturally leads to costly redesign iterations for both the hardware and software subsystems. Instead, the authors propose a solution whereby they specify, explore, and exploit temporal information in high-level network models and bridge the gap between requirement analysis and system design. At each stage of the design refinement, timing information modeled at the higher-level network models is trickled down to finer and lower level models. The authors used NS-2 as their modeling framework.

This timing-driven design methodology uses high-level network models generated using a rate derivation technique called *RADHA-RATAN* [60, 61]. RADHA-RATAN works on generalized task graphs that have nodes that represent functionality or tasks and edges that are asynchronous, unidirectional communication channels between producers and consumers. RADHA-RATAN is a collation of algorithms that generate timing budgets for the task graph nodes based on their preset execution (firing) and data production and consumption rates.

Along with RADHA-RATAN, network-level models at the highest level are used to represent functionalities such as routing, congestion, and quality of service. The designer can specify distributions, protocols, and such settings to generate the network graph in NS-2. The nodes of the network graph simulate the transfer of packets and tokens among themselves. This enables testing of the functionality of the protocols. However, further refinement by an experienced designer of this high-level network graph results in network subsystems that capture timing requirements. In a process known as *timing-driven task structuring*, the designer can then mutate the task graph until the desired timing behavior is achieved, after which partitioning of hardware and software can be performed. The disconnect between requirement analysis and system design is circumvented by this mutation, allowing the mix of the NS-2 modeling paradigm and formal timing analysis techniques to provide a methodology whereby timing requirements seep from high-level network models to low-level synthesis models.

Simulation and timing analysis is one part of the puzzle in hardware–software codesign. Automated hardware synthesis and software synthesis and compilation techniques are the next step in generating implementations from the system-level models. To this end, Gupta et al. [62] have proposed the *SPARK* parallelizing high-level synthesis framework that performs automated synthesis of behavioral descriptions specified in C to synthesizable register-transfer level VHDL. This framework can then be used in a system level codesign methodology to implement an application on a core-based platform target [63].

Such hardware–software codesign methodologies are crucial for the design and development of networked embedded systems. Automated or semiautomated methodologies are less error-prone, lead to faster times-to-market, and can help realize hardware–software tradeoffs and design exploration that may not be obvious in large systems.

96.7 Conclusions

The increasing interest in networked embedded systems is quite timely as evident from the several important application areas where such systems can be used. The development process for these applications and systems remains an *ad hoc* process. We need methodologies and tools to provide system designers with the flexibility and capability to quickly construct efficient and optimized system designs. In this paper, we have examined the range of design and verification challenges faced by the NES designers. Among existing solutions, we have presented techniques that promise to increase the efficiency of the design process by raising the level of design abstraction and by enhancing the scope of system models. These include design tools such as GRATIS and nesC for operating system configuration and NS-2, SystemC, NESLsim, and SensorSim for simulation and codesign of embedded systems. There are several open research problems. Reducing device size and power and increasing device speed further remain important objectives. There is a need for distributed applications, along with middleware and operating system support and support for network protocols for distributed coordinated collaboration. Continued progress in these technologies will fulfill the promise of networked embedded systems as ubiquitous computing systems.

References

[1] National Research Council, *Embedded Everywhere*, National Academy Press, Washington DC, 2001.
[2] Volgyesi, P., and A. Ledeczi, Component-Based Development of Networked Embedded Applications, in Proceedings of EuroMicro, 2002.
[3] Ramanathan, D., R. Jejurikar, and R. Gupta, Timing Driven Co-design of Networked Embedded Systems, in Proceedings of ASPDAC, 2000, pp. 117–122.
[4] H. Wang, J. Elson, L. Girod, D. Estrin, and K Yao, Target Classification and Localization in Habitat Monitoring, in Proceedings of IEEE International Conference on Acoustics, Speech, and Signal Processing (ICASSP 2003), 2003.
[5] S. Simic and S Sastry, Distributed Environmental Monitoring Using Random Sensor Networks, in Second International Workshop, IPSN 2003, 2003.
[6] B. West, P. Flikkema, T. Sisk, and G. Koch, Wireless Sensor Networks for Dense Spatio-temporal Monitoring of the Environment: A Case for Integrated Circuit, System and Network Design, in IEEE CAS Workshop on Wireless Communications and Networking, 2001.
[7] A. Mainwaring, J. Polsatre, R. Szewczyk, D. Culler, and J. Anderson, Wireless Sensor Networks for Habitat Monitoring, in Proceedings of WSNA '02, 2002.
[8] P. Flikkema and B. West, Wireless Sensor Networks: From the Laboratory to the Field, in National Conference for Digital Government Research, 2002.
[9] Cooperative engagement capability, `http://www.fas.org/man/dod-101/sys/ship/weaps/cec.htm`.
[10] Tian He, B.M. Blum, John A. Stankovic, and Tarek F. Abdelzaher, Aida: adaptive application independent data aggregation in wireless sensor Networks, *ACM Transactions on Embedded Computing System, Special issue on Dynamically Adaptable Embedded Systems*, 2003, pp. 426–457.
[11] Z. Nakad, M. Jones, and T. Martin, Communications in Electronic Textile Systems, in International Conference on Communications in Computing (CIC), 2003.
[12] D. Meoli and T.M.-Plumlee, Interactive electronic textile, *Journal of Textile and Apparel, Technology and Management*, 2, 2002.

[13] D. Estrin, R. Govindan, J. Heidemann, and Satish Kumar, Next Century Challenges: Scalable Coordination in Sensor Networks, in International Conference on Mobile Computing and Networking (MobiCom), 1999.

[14] D. Estrin, A. Sayeed, and M. Srivastava, Wireless Sensor Networks, in International Conference on Mobile Computing and Networking (MobiCom), 2002.

[15] J. Kahn, R. Katz, and K. Pister, Emerging challenges: mobile networking for 'smart dust', *Journal of Communication Networks*, Vol. 2, pp. 188–196, 2000.

[16] I.F. Akyildiz, W. Su, Y. Sankarasubramaniam, and E. Cayirci, A survey of wireless sensor networks, *IEEE Communications Magazine*, Vol. 38, No.4, pp. 393–422, March 2002.

[17] C.E. Jones, K.M. Sivalingam, P. Agrawal, and J.C. Chen, A survey of energy efficient network protocols for wireless networks, *Wireless Networks*, 7, pp. 343–358, 2001.

[18] P. Koopman, Embedded System Design Issues — The Rest of the Story, in International Conference on Computer Design, 1996.

[19] M. Horton, D. Culler, K. Pister, J. Hill, R. Szewczyk, and Alec Woo, Mica: the commercialization of microsensor motes, *Sensors Magazine*, Vol. 19, No. 4, pp. 40–48, April 2002.

[20] K.S.J. Pister, J.M. Kahn, and B.E. Boser, Smart Dust: Wireless Networks of Millimeter-scale Sensor Nodes, Technical report, Highlight Article in 1999 Electronics Research Laboratory Research Summary, 1999.

[21] J.M. Kahn, R.H. Katz, and K.S.J. Pister, Mobile Networking for Smart Dust, in ACM/IEEE International. Conference. on Mobile Computing and Networking (MobiCom), 1999.

[22] Crossbow: Smarter sensors in silicon, `http://www.xbow.com/`.

[23] Gabriel Leen and Donal Heffernan, Expanding automotive electronic systems, *IEEE Computer*, 35, pp. 88–93, 2002.

[24] Kyocera, `http://global.kyocera.com/application/automotive/auto_elec/`.

[25] Mitsubishi electric, `http://www.mitsubishielectric.ca/automotive/`.

[26] Y. Li and R. Wang, Precision Agriculture: Smart Farm Stations, IEEE 802 Plenary Meeting Tutorials.

[27] Board on Agriculture and Natural Resources, *Precision Agriculture in the 21st Century: Geospatial and Information Technologies in Crop Management*, National Academy Press, Washington DC, 1998.

[28] V. Kottapalli, A. Kiremidjian, J. Lynch, E. Carryer, T. Kenny, K. Law, and Y. Lei, Two-tiered Wireless Sensor Network Architecture for Structural Health Monitoring, in SPIE's 10th Annual International Symposium on Smart Structures and Materials, 2003.

[29] Raytheon systems co., `http://www.raytheon.com/`.

[30] Darpa, `http://www.darpa.mil/fcs/index.html`.

[31] N. Sarwabhotla and S. Seetharamaiah, Intelligent Disaster Management System for Remote Villages in India, in *Development by Design, Bangalore, India*, 2002.

[32] T. Clouqueur, V. Phipatanasuphorn, P. Ramanathan, and K. Saluja, Sensor Deployment Strategy for Target Detection, in Proceedings of WSNA 02, 2002.

[33] F. Koushanfar, M. Potkonjak, and A. Sangiovanni-Vincentelli, Fault Tolerance in Wireless Ad hoc Sensor Networks, in IEEE International Conference on Sensors, 2002.

[34] C. Karlof and D. Wagner, Secure Routing in Wireless Sensor Networks: Attacks and Countermeasures, in IEEE International Workshop on Sensor Network Protocols and Applications, 2003.

[35] A. Perrig, R. Szewczyk, J.D. Tygar, V. Wen, and D.E. Culler, Spins: security protocols for sensor networks, *Wireless Networks*, 8, pp. 521–534, 2002.

[36] H. Cam, S. Ozdemir, D. Muthuaavinashiappan, and Prashant Nair, Energy-efficient Security Protocol for Wireless Sensor Networks, in IEEE VTC Fall 2003 Conference, 2003.

[37] N.R. Potlapally, S. Ravi, A. Raghunathan, and N.K. Jha, Analyzing the Energy Consumption of Security Protocols, in International Symposium on Low Power Electronics and Design, 2003.

[38] Ecos open-source real-time operating system for embedded systems, `http://sources.redhat.com/ecos/`.

[39] Lynxos real-time operating system for embedded systems, `http://www.lynuxworks.com/`.

[40] Qnx real-time operating system for embedded systems, `http://www.qnx.com/`.

[41] D. Gay, P. Levis, R. Behren, M. Welsh, E. Brewer, and D. Culler, The NESC Language: A Holistic Approach to Networked Embedded Systems, in Proceedings of the ACM SIGPLAN 2003 Conference on Programming Language Design and Implementation, 2002.

[42] J. Henkel and Y. Li, Energy-conscious HW/SW-partitioning of Embedded Systems: A Case Study on an Mpeg-2 Encoder, In International Workshop on Hardware/Software Codesign, 1998.

[43] C. Nitsch and U. Kebschull, The Use of Runtime configuration Capabilities for Network Embedded Systems, in Proceedings of Design, Automation and Test in Europe Conference and Exhibition, 2002. pp. 1093–2002.

[44] J. Hill, R. Szewczyk, A. Woo, S. Hollar, D. Culler, and K. Pister, System Architecture Directions for Networked Sensors, in international Conference on Architectural Support for Programming Languages and Operating Systems, 2000.

[45] W. Ye, J. Heidemann, and D. Estrin, An Energy Efficient Mac Protocol for Wireless Sensor Networks, in INFOCOM 2002: 21st Annual Joint Conference of IEEE, 2002.

[46] A. Manjeshwar and D.P. Agrawal, Teen: A Protocol for Enhanced Efficiency in Wireless Sensor Networks, in International Workshop on Parallel and Distributed Computing Issues in Wireless Networks and Mobile Computing, 2001.

[47] A. Manjeshwar and D.P. Agrawal, Apteen: A Hybrid Protocol for Efficient Routing and Comprehensive Information Retrieval in Wireless Sensor Networks, in International Workshop on Parallel and Distributed Computing Issues in Wireless Networks and Mobile Computing, 2002.

[48] Opnet, http://www.opnet.com/.

[49] Network Simulator 2, http://www.isi.edu/nsnam/ns/.

[50] SensorSim, http://nesl.ee.ucla.edu/projects/sensorsim/.

[51] NESLsim, http://www.ee.ucla.edu/~saurabh/NESLsim/.

[52] S. Park, A. Savvides, and M. Srivastava, Sensorsim: A Simulation Framework for Sensor Networks, in Proceedings of MSWiM 2000, 2000.

[53] R. K. Gupta and S. Y. Liao, Using a programming language for digital system design, *IEEE Design and Test of Computers*, Vol. 14, Issue 2, pp. 72–80, April-June 1997.

[54] SystemC, http://www.systemc.org.

[55] N. Drago, F. Fummio, and M. Poncino, Modeling Network Embedded Systems with NS-2 and SystemC, in Proceedings of ICCSC: Circuits and Systems for Communication, 2002, pp. 240–245.

[56] R.K. Gupta and G. De Micheli, Hardware–Software Cosynthesis for digital systems, *IEEE Design and Test of Computers*, Vol. 10, Issue 3, pp. 29–41, Sept. 1993.

[57] G. Micheli and R. Gupta, Hardware/Software Co-design, *Proceedings of IEEE*, 85, 349–365, 1997.

[58] R. Ernst and J. Henkel, Hardware-Software Codesign of Embedded Controllers Based on Hardware Extraction, in International Workshop on Hardware/Software Codesign, 1992.

[59] J. Henkel and R. Ernst, A Hardware-Software Partitioner Using a Dynamically Determined Granularity, in Proceedings of the Design Automation Conference, 1997.

[60] A. Dasdan, D. Ramanathan, and R.K. Gupta, A timing-driven design and validation methodology for embedded real-time systems, *ACM Transactions on Design Automation of Electronic Systems*, 3:533–553, 1998.

[61] A. Dasdan, D. Ramanathan, and R. K. Gupta, Rate Derivation and its Applications to Reactive, Real-time Embedded Systems, in Design Automation Conference, 1998.

[62] S. Gupta, N.D. Dutt, R.K. Gupta, and A. Nicolau, SPARK: A High-level Synthesis Framework for Applying Parallelizing Compiler Transformations, in International Conference on VLSI Design, 2003.

[63] M. Luthra, S. Gupta, N.D. Dutt, R.K. Gupta, and A. Nicolau, Interface Synthesis Using Memory Mapping for an FPGA Platform, in International Conference on Computer Design, Oct. 2003.

Section 7: Integration Technologies

97

Introduction to e-Manufacturing

Muammer Koç
University of Michigan

Jun Ni
University of Michigan

Jay Lee
University of Wisconsin-Milwaukee

Pulak Bandyopadhyay
GM R&D Center

97.1 Introduction

For the past decade, the impact of web-based technologies has added *"velocity"* to the design, manufacturing, and aftermarket service of a product. Today's competition in manufacturing industry depends not just on lean manufacturing but also on the ability to provide customers with total solutions and life-cycle costs for sustainable value. Manufacturers are now under tremendous pressure to improve their responsiveness and efficiency in terms of product development, operations, and resource utilization with a transparent visibility of production and quality control. Lead times must be cut short to their extreme extent to meet the changing demands of customers in different regions of the world. Products are required to be made-to-order with no or minimum inventory, requiring (a) an efficient information flow between customers, manufacturing, and product development (i.e., plant floor, suppliers, and designers), (b) a tight control between customers and manufacturing, and (c) near-zero downtime of the plant floor assets. Figure 97.1 summarizes the trends in manufacturing and function of predictive intelligence as an enabling tool to meet the needs [1–4].

With emerging applications of Internet and tether-free communication technologies, the impact of e-intelligence is forcing companies to shift their manufacturing operations from the traditional factory integration philosophy to an e-factory and e-supply chain philosophy. It transforms companies from a local factory automation to a global enterprise and business automation. The technological advances for achieving this highly collaborative design and manufacturing environment are based on multimedia-type information-based engineering tools and a highly reliable communication system for enabling distributed procedures in concurrent engineering design, remote operation of manufacturing processes, and operation of distributed production systems. As shown in Figure 97.2, e-manufacturing fills the gaps existing in the traditional manufacturing systems. The gaps between product development and supply chain consist of lack of life-cycle information and lack of information about supplier capabilities. Hence, designers, unless with years of experience, work in a vacuum, design the product according to the specification given, and wait for the next step. Most of the time, the design made according to specifications is realized to be infeasible for manufacturing with suppliers' machinery. As a result, lead times become longer. Similarly, for instance, because of the lack of information and synchronization between suppliers

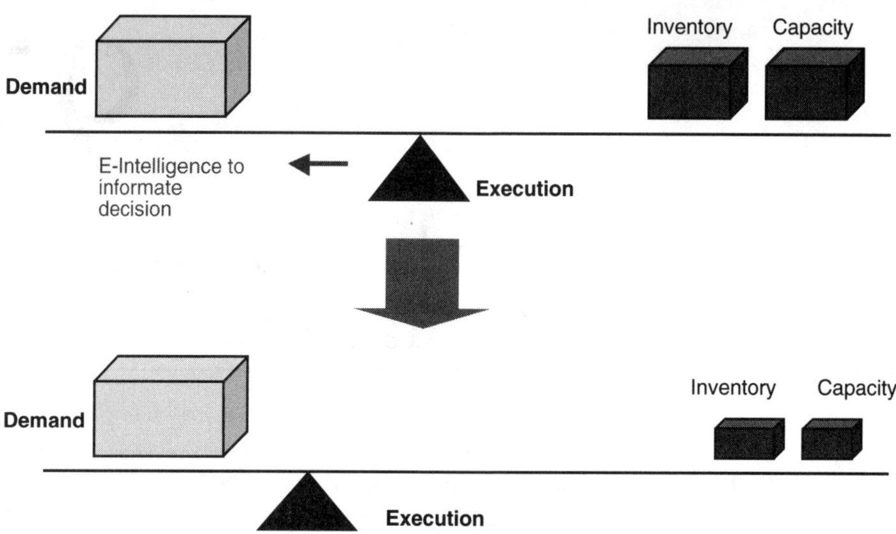

FIGURE 97.1 The transformation of e-Manufacturing for unmet needs.

and assembly plants, just-in-time manufacturing and on-time shipment become possible only with a substantial amount of inventory whereas with e-manufacturing, real-time information regarding reliability and status of supplier's equipment will also be available as a part of the product quality information. With these information and synchronization capabilities, less and less inventory will be necessary contributing to the profitability of the enterprise.

97.2 e-Manufacturing: Rationale and Definitions

e-Manufacturing is a transformation system that enables the manufacturing operations to achieve predictive near-zero-downtime performance as well as to synchronize with the business systems through the use of web-enabled and tether-free (i.e., wireless, web, etc.) infotronics technologies. It integrated information and decision-making among data flow (of machine/process level), information flow (of factory and supply system level), and cash flow (of business system level) [5–7]. e-Manufacturing is a business strategy as well as a core competency for companies to compete in today's e-business environment. It is aimed to complete integration of all the elements of a business *including suppliers, customer service network, manufacturing enterprise, and plant floor assets* with connectivity and intelligence brought by the web-enabled and tether-free technologies and intelligent computing to meet the demands of e-business/e-commerce practices that gained great acceptance and momentum over the last decade. e-Manufacturing is a transformation system that enables e-Business systems to meet the increasing demands through tightly coupled supply chain management (SCM), enterprise resource planning (ERP), and customer relation management (CRM) systems as well as environmental and labor regulations and awareness, (Figure 97.3) [4–7].

e-Manufacturing includes the ability to monitor the plant floor assets, predict the variation of product quality and performance loss of any equipment for dynamic rescheduling of production and maintenance operations, and synchronize with related business services to achieve a seamless integration between manufacturing and higher level enterprise systems. Dynamically updated information and knowledge about the capabilities, limits, and variation of manufacturing assets for various suppliers guarantee the best decisions for outsourcing at the early stages of design. In addition, it enables customer orders autonomously across the supply chain, bringing unprecedented levels of speed, flexibility, and visibility to the production process reducing inventory, excess capacity, and uncertainties.

The intrinsic value of an e-Manufacturing system is to enable real-time decision making among product designers, process capabilities, and suppliers as illustrated in Figure 97.4. It provides tools to access

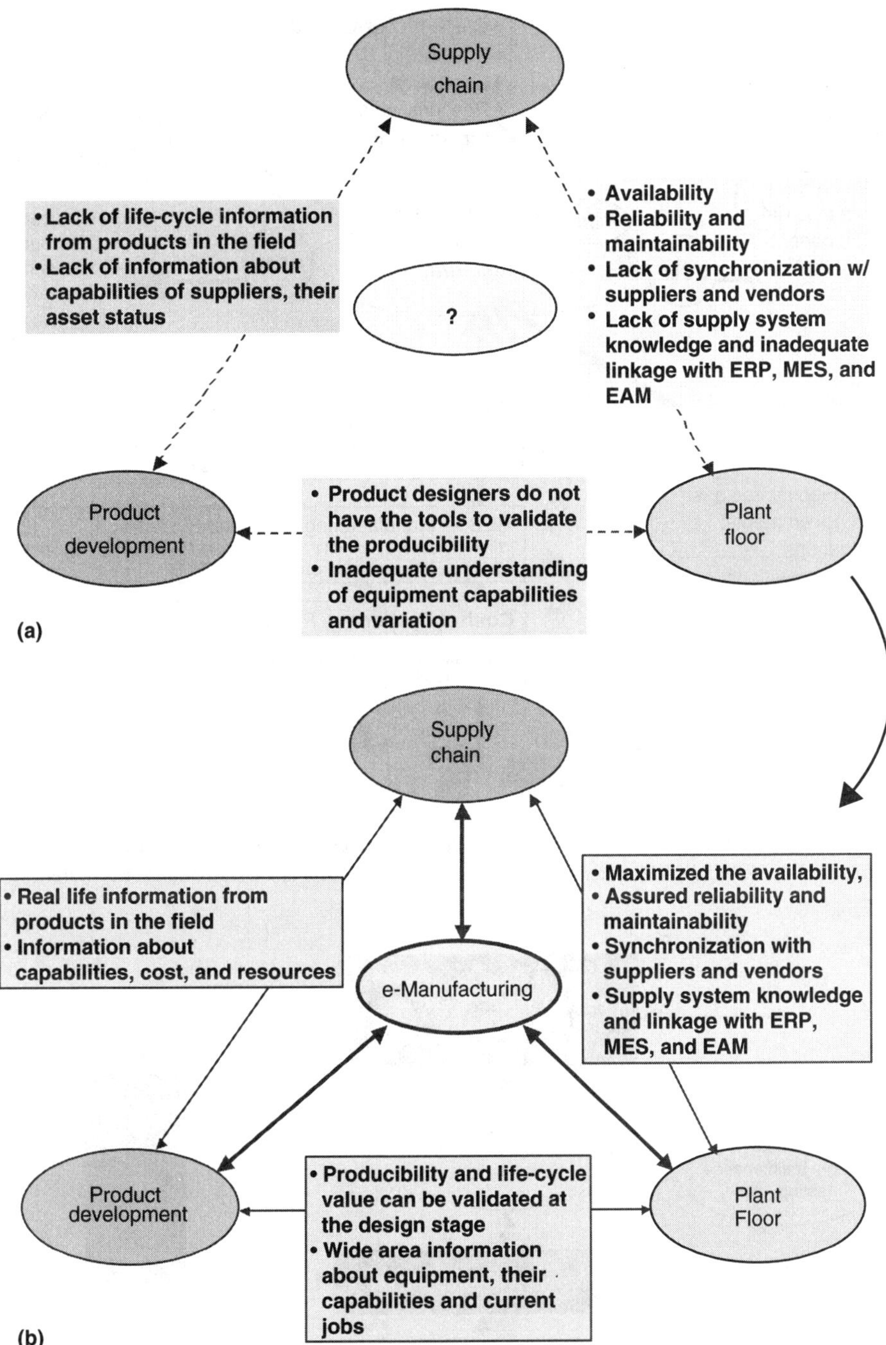

FIGURE 97.2 The transformation of e-Manufacturing for unmet needs.

life-cycle information of a product or equipment for continuous design improvement. Traditionally, product design or changes take weeks or months to be validated with suppliers. With the e-Manufacturing system platform, designers can validate product attributes within hours using the actual process characteristics and machine capabilities. It also provides efficient configurable information exchanges and synchronization with various e-business systems.

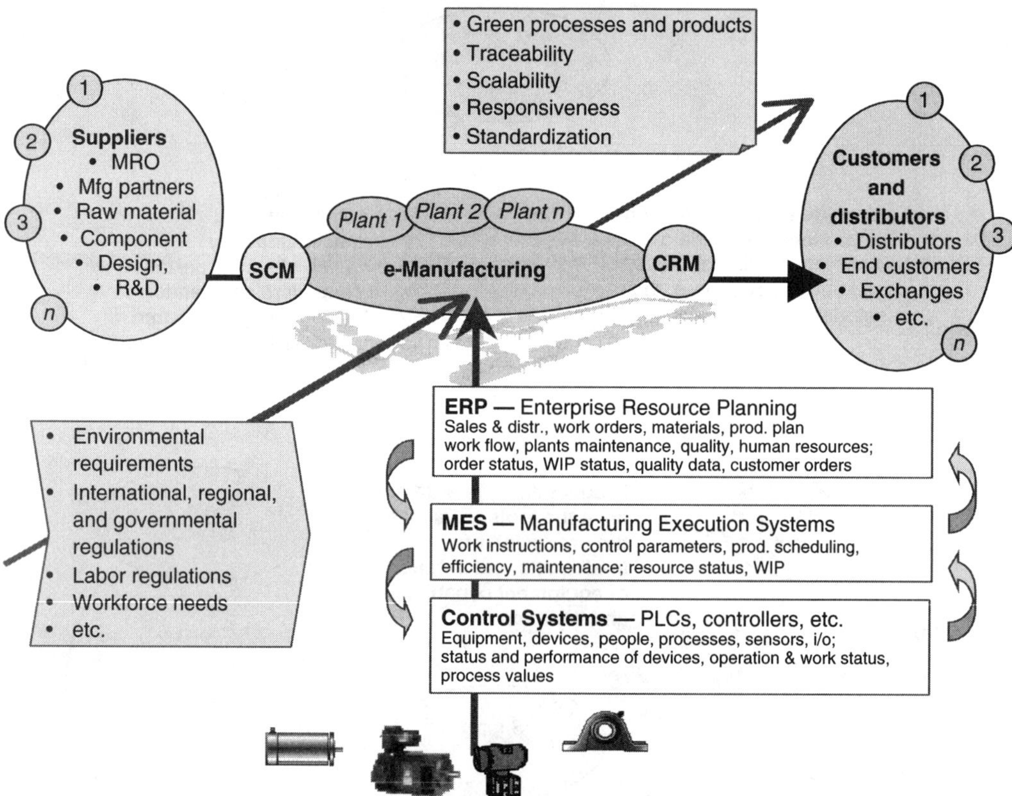

FIGURE 97.3   Integration of e-Manufacturing into e-Business systems to meet the increasing demands through tightly coupled SCM, ERP, and CRM systems as well as environmental and labor regulations and awareness.

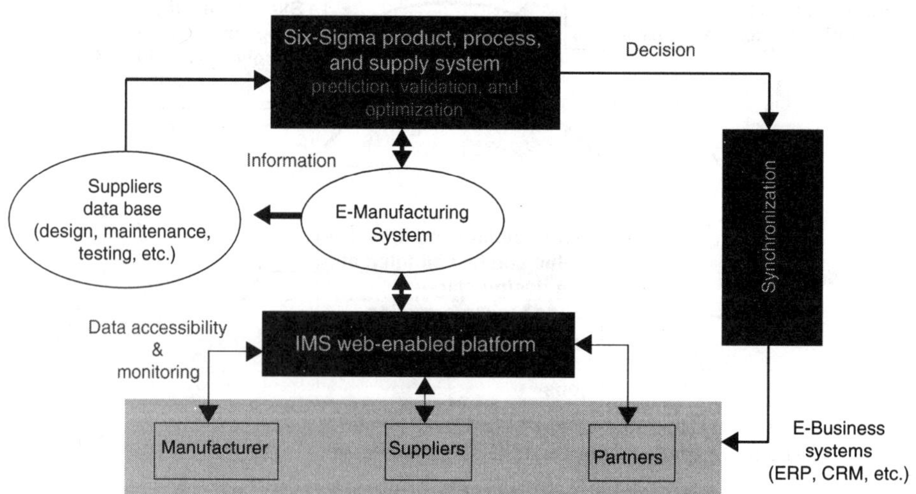

FIGURE 97.4   Using e-Manufacturing for product deign validation.

97.3   e-Manufacturing: Architecture

Currently, Manufacturing Execution Systems (MES) enable the data flow among design, process, and manufacturing systems. The ERP systems serve as an engine for driving the operations and the supply chain systems. However, the existing structure of the ERP and MES cannot *informate* (i.e., communicate the information in real-time) the decision across the supply chain systems. The major functions and objectives of e-Manufacturing are: (1) enable an only handle information once (OHIO) environment; (2) predict and optimize total asset utilization in the plant floor; (3) synchronize asset information with supply chain network; and (4) automate business and customer service processes. The proposed e-manufacturing architecture in this position paper addresses the above needs.

To address these needs, an e-Manufacturing system should offer comprehensive solutions by addressing the following requirements: (1) development of intelligent agents for continuous, real time, remote, and distributed monitoring of devices, machinery, and systems to predict machine's performance status (health condition) and to enable capabilities of producing quality parts; (2) development of infotronics platform that is scalable and reconfigurable for data transformation, prognostics, performance optimization, and synchronization; and (3) development of virtual design platform for collaborative design and manufacturing among suppliers, design, and process engineers as well as customers for fast validation and decision making. Figure 97.5 illustrates the proposed e-Manufacturing architecture and its elements [5–7].

Data gathering and transformation: This has already been done at various levels. However, massive raw data are not useful unless it is reduced and transformed into useful information format (i.e., XML) for responsive actions. Hence, data reconfiguration and mining tools for data reduction, representation for plant floor data

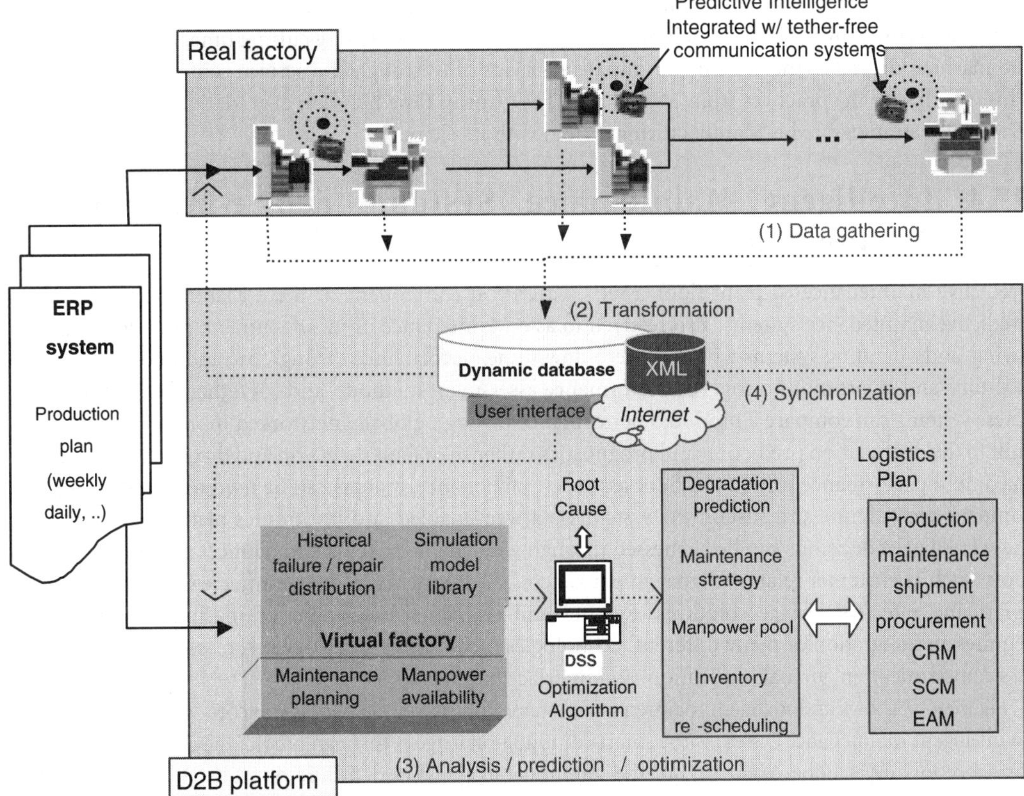

FIGURE 97.5   An e-Manufacturing architecture that comprises (1) and (2) data gathering and transformation, (3) prediction and optimization, and (4) synchronization [5].

need to be developed. An infotronics platform, namely, Device-to-Business (D2B™) has been developed by the Intelligent Maintenance Systems (IMS) Center. To make pervasive impacts to different industrial applications, existing industrial standards should be used (i.e., IEEE 802.xx standard committees, MIMOSA, etc.)

Prediction and optimization: Advanced prediction methods and tools need to be developed in order to measure degradation, performance loss, or implications of failure, etc. For *prediction* of degradation on components/machinery, computational and statistical tools should be developed to measure and predict the degradation using intelligent computational tools.

Synchronization: Tools and agent technologies are needed to enable autonomous business automation among factory floor, suppliers, and business systems. Embedded intelligent machine infotronics agent that links between the devices/machinery and business systems and enables products, machinery, and systems to (1) learn about their status and environment, (2) predict degradation of performance, (3) reconfigure itself to sustain functional performance, and (4) informate business decisions directly from the device itself [1–7].

Under this architecture, many web-enabled applications can be performed. For example, we can perform remote machine calibration and experts from machine tool manufacturers can assist users to analyze machine calibration data and perform prognostics for preventive maintenance. Users from different factories or locations can also share this information through these web tools. This will enable users to exchange high-quality communications since they are all sharing the same set of data formats without any language barriers.

Moreover, by knowing the degradation of machines in the production floor, the operation supervisor can estimate their impacts to the materials flow and volume and synchronize it with the ERP systems. The revised inventory needs and materials delivery can also be synchronized with other business tools such as CRM system. When cutting tools wear out on a machining center, the information can be directly channeled to the tool providers and update the tool needs for tool performance management. In this case, the cutting tool company is no longer selling cutting tools, but instead, selling cutting time. In addition, when the machine degrades, the system can initiate a service call through the service center for prognostics. This will change the practices from MTTR to MTBD (mean time between degradation) [10–13]. Figure 97.6 shows an integrated e-Manufacturing system with its elements.

97.4 Intelligent Maintenance Systems and e-Maintenance Architecture

Predictive maintenance of plant floor assets is a critical component of the e-Manufacturing concept. Predictive maintenance systems, also referred to as e-Maintenance in this document, provides manufacturing, and operating systems with near-zero downtime performance through use and integration of (a) real-time and smart monitoring, (b) performance assessment methods, and (c) tether-free technologies. These systems can compare a product's performance through globally networked monitoring systems to shift to the degradation prediction and prognostics rather than fault detection and diagnostics. To achieve maximum performance from plant floor assets, e-maintenance systems can be used to monitor, analyze, compare, reconfigure, and sustain the system via a web-enabled and infotronics platform. In addition, these intelligent decisions can be harnessed through web-enabled agents and connect them to e-business tools (such as customer relation management systems, ERP systems, and e-commerce systems) to achieve smart and effective service solutions. Remote and real-time assessment of machine's performance requires an integration of many different technologies including sensory devices, reasoning agents, wireless communication, virtual integration, and interface platforms [14–17].

Figure 97.7 shows an intelligent maintenance system with its key elements. The core-enabling element of an intelligent maintenance system is the smart computational agent that can predict the degradation or performance loss (Watchdog Agent™), not the traditional diagnostics of failure or faults. A complete understanding and interpretation of states of degradation is necessary to accurately predict and prevent failure of a component or a machine once it has been identified as a critical element to the overall production system. The degradation is assessed through the performance assessment methods explained in the previous sections.

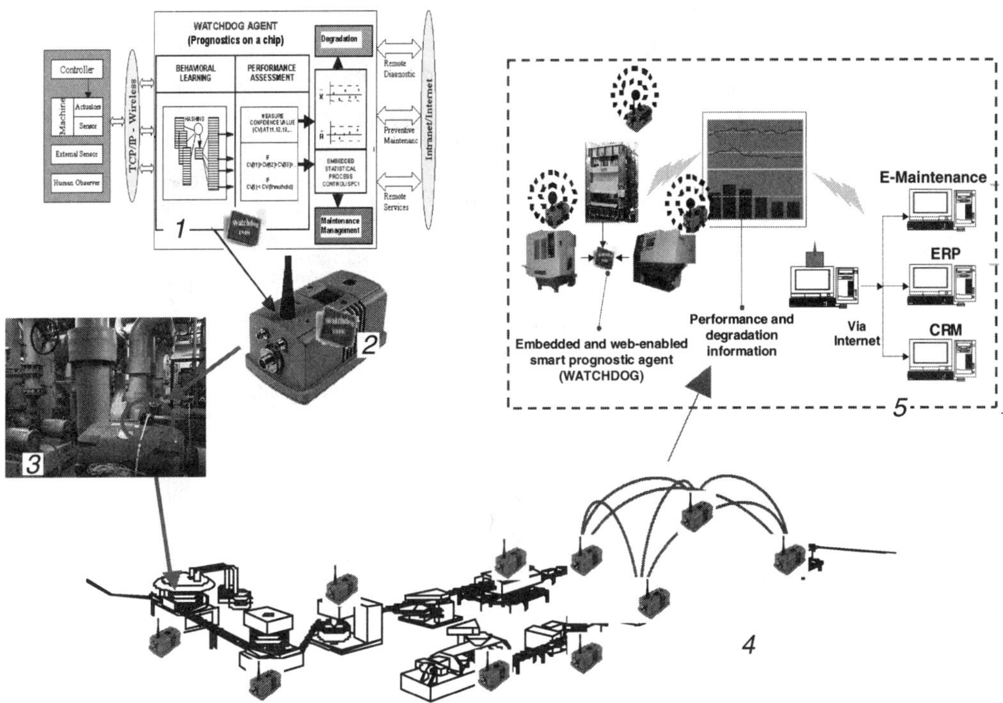

FIGURE 97.6 Various elements of an e-Manufacturing System: (1) data gathering and predictive intelligence-D2B™ platform and Watchdog Agent™, (2)–(4) tether-free communication technologies, and (5) optimization and synchronization tools for business automation.

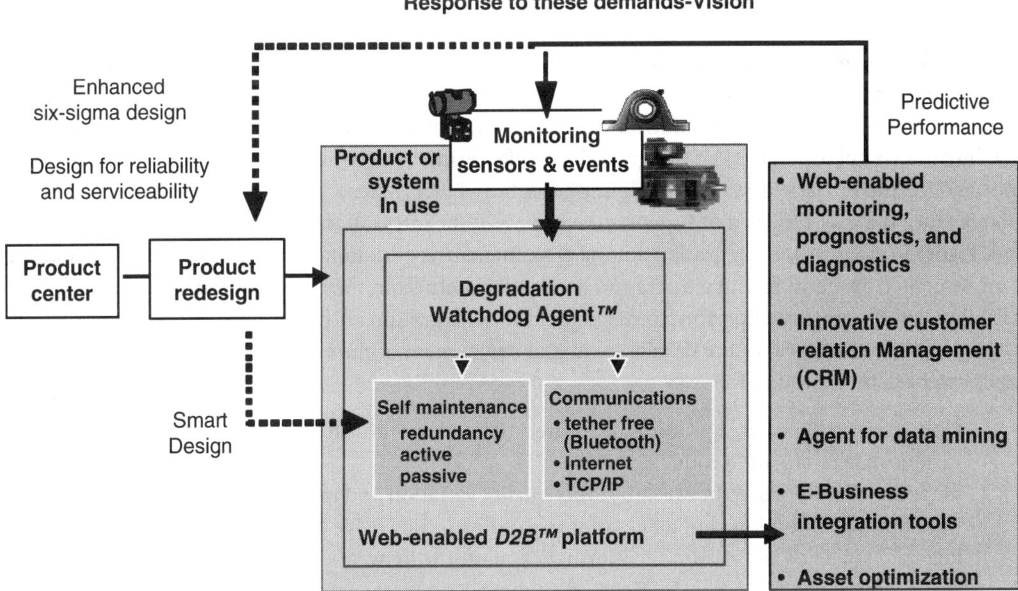

FIGURE 97.7 An Intelligent e-Maintenance System.

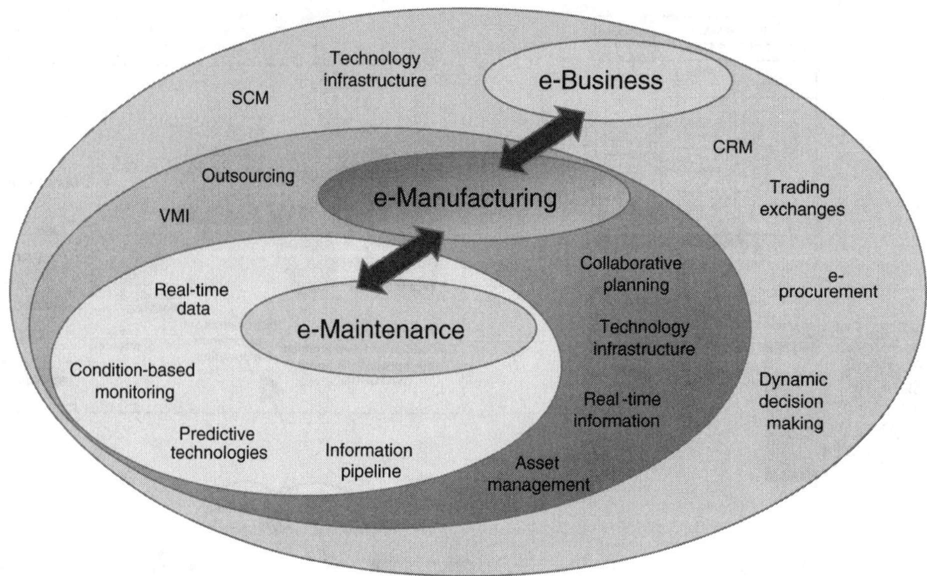

FIGURE 97.8   e-Manufacturing and Its Integrations with e-Maintenance and e-Business.

A product's performance degradation behavior is often associated with multi-symptom-domain information cluster, which consists of degradation behavior of functional components in a chain of actions. The acquisition of specific sensory information may contain multiple behavior information such as nonlinear vibration, thermal or materials surface degradation, and misalignment. All of the information should be correlated for product behavior assessment and prognostics.

97.5   Conclusions and Future Work

This position paper introduced an e-Manufacturing architecture, and outlined its fundamental requirements and elements as well as expected impact to achieve high-velocity and high-impact manufacturing performance. Web-enabled and infotronics technologies play indispensable roles in supporting and enabling the complex practices of design and manufacturing by providing the mechanisms to facilitate and manage the integrated system discipline with the higher system levels such as SCM and ERP. E-Maintenance is a major pillar that supports the success of the integration of e-Manufacturing and e-business. Figure 97.8 shows the integration among e-Maintenance, e-Manufacturing, and e-Business systems. If implemented properly, manufacturers and users will benefit from the increased equipment and process reliability with optimal asset performance and seamless integration with suppliers and customers.

In order to further advance the development and deployment of the e-Manufacturing system, research needs can be summarized as follows:

1. Predictive intelligence (algorithms, software, and agents) with a focus on degradation detection on various machinery and products.
2. Mapping of relationship between product quality variation and machine and process degradation.
3. Data mining, reduction, and data-to-information-to-knowledge conversion tools.
4. Reliable, scalable, and common informatics platform between devices and business, including implementation of wireless, Internet, and Ethernet networks in the manufacturing environment to achieve flexible and low-cost installations and commissioning.
5. Data/information security and vulnerability issues at the machine/product level.
6. Distributed and web-based computing and optimization and synchronization systems for dynamic decision making.

7. Education and training of technicians, engineers, and leaders to make them capable of pacing with the speed of information flow and understanding the overall structure.
8. Develop a new enterprise culture that resonates the spirit of e-manufacturing.

References

[1] Zipkin, P., Seminar on the Limits of Mass Customization, Center for Innovative Manufacturing and Operations Management (CIMOM), Apr., 22, 2002.

[2] Waurzyniak, P., Moving towards e-factory, *SME Manufacturing Magazine*, 127(5), November 2001, http://www.sme.org/gmn/mag/2001/01nom042/01nom042.pdf.

[3] Waurzyniak, P., 2001, Web tools catch on, *SME Manufacturing Magazine*, v. 127(4), Oct. 2001, www.sme.org.

[4] Rockwell Automation e-Manufacturing Industry Road Map, http://www.rockwellautomation.com.

[5] Koç, M. and J. Lee, e-Manufacturing and e-Maintenance — Applications and Benefits, International Conference on Responsive Manufacturing (ICRM) 2002, Gaziantep, Turkey, June 26–29, 2002.

[6] Koç, M. and J. Lee, A System Framework for Next-Generation E-Maintenance System, EcoDesign 2001, Second International Symposium on Environmentally Conscious Design and Inverse Manufacturing, Tokyo Big Sight, Tokyo, Japan, Dec. 11–15, 2001.

[7] Lee, J., Ahad Ali, and M. Koç, e-Manufacturing — Its Elements and Impact, Proceedings of the Annual Institute of Industrial Engineering (IIE) Conference, Advances in Production Session, Dallas, TX, U.S.A., May 21–23, 2001.

[8] Albus, J.S., A new approach to manipulator control: the CMAC, *Journal of Dynamic Systems and Control, Transactions of ASME*, Series, G., 97, 220, 1975.

[9] Lee, J., Measurement of machine performance degradation using a neural network model, Journal of Computers in Industry, 30, 193, 1996.

[10] Wong, Y. and S. Athanasios, Learning convergence in the CMAC, *IEEE Transactions on Neural Networks*, 3, 115, 1992.

[11] Lee, J. and B. Wang, *Computer-aided Maintenance: Methodologies and Practices*, Kluwer Academic Publishing, Dordrecht, 1999.

[12] Lee, J., Machine Performance Assessment Methodology and Advanced Service Technologies, Report of Fourth Annual Symposium on Frontiers of Engineering, National Academy Press, Washington, DC, 1999, pp. 75–83.

[13] Lee, J. and B.M. Kramer, Analysis of machine degradation using a neural networks based pattern discrimination model, *Journal of Manufacturing Systems*, 12, 379–387, 1992.

[14] Maintenance is not as mundane as it sounds *Manufacturing News*, 8(21), Nov. 30, 2001. http://www.manufacturingnews.com/news/01/1130/art1.html.

[15] Society of Manufacturing Engineers (SME), Less factory downtime with 'predictive intelligence', *Manufacturing Engineering Journal*, Feb. 2002, www.sme.org.

[16] Lee, J., E-Intelligence heads quality transformation, *Quality in Manufacturing Magazine*, March/April 2001, www.manufacturingcenter.com.

[17] How the Machine Will Fix Itself in Tomorrow's World, *Tooling and Productions Magazine*, Nov. 2000, www.manufacturingcenter.com.

98

XML for the Exchange of Automation Project Information

Alexander Fay
ABB Corporate Research Center

98.1 Introduction: The Need for Information Exchange in the Engineering of Automation Systems

The engineering of industrial automation systems, that is, the design and implementation of manufacturing and production plants and the related control and automation systems, comprises a huge number of tasks: in the manufacturing industry, these are mainly the mechanical engineering, electrical engineering, and control engineering, and in the production industry, these are process engineering, plant engineering (including construction), and process automation engineering. Furthermore, usually a variety of companies are involved, which partner in different ways to provide a running plant. For each of these tasks, specialized software tools, for example, CAD and CAE tools, have been designed and implemented and used already for many years. However, the information exchange between the different involved parties has mostly been based on paper, firstly to fulfill the legal documentation requirements, secondly because the software tools have been designed individually and independently and specially for different tasks, without spending much attention to the need for electronic information exchange with tools from different vendors, built for different purposes. With the rising need for electronic information exchange, export and import interfaces have been built to support a variety of data formats, most of them proprietary and tool vendor-specific. This has led to a huge number of data formats and interfaces that have to be maintained and supported, and still considerable information has to be reentered manually, which is tedious and error-prone.

Today, with the need for quick and seamless information exchange to shorten the engineering time, the lacks regarding electronic information exchange become obvious, and urgent activities are needed to overcome today's deficiencies.

The most straightforward solution is to agree among all involved companies in a project on one common project database and related tool suite. However, this approach usually fails for several reasons:

- very few tool suites exist that cover all phases of the engineering life cycle, from process design to commissioning & test and beyond,
- not all the tools in these tool suites are regarded optimal for the individual tasks by all the respective companies involved (which are each responsible for their task and their results),
- instead, each company has a preferred tool that they are familiar with and, for which they have solution libraries, and using a different tool would increase training and engineering effort for them,
- at the beginning of an automation project, not all companies are known which will become partners in the project (e.g., the providers of electrical and control system equipment will be chosen later in the project based on requirements and tendering).

Therefore, the approach must be to exchange information electronically via a commonly accepted data format with commonly agreed syntax and semantics.

Extended markup language (XML) [1] has gained considerable momentum during the last few years as a meta-language for the file-based exchange of information. During the first hype, the impression was widely common that XML would overcome all the problems of data exchange of the past. Realistically seen, XML is a very valuable means because it defines (in device type documents (DTD) or extensible schema description (XSD)) how information can be ordered in a file in a hierarchical tree, with the possibility to define tags that describe the information items in between them. XML has the advantage that it especially defines the syntax of the files, and the elements of the metalanguage allow to impose some restrictions regarding the "well-formedness" of XML files, for example, which file entries are mandatory and which are not, and the number of elements of a certain kind allowed. Thus, XML provides a means to bridge the syntactical gaps for information exchange.

However, XML does not overcome the intrinsic problem of communication: the partners involved do not only have to agree on the syntax of information exchange but also on the semantics.

To give an example: Company A might use

```
< SetPoint>0.8</SetPoint>
```

as a means to internally store the set point for a control loop in their control system. However, another Company B might have defined

```
<controller_setpoint_value>0.8</controller_setpoint_value>
```

in their XML schema, and the XML parser of their controller configuration tool will — unlike humans — not understand the tags used by Company A. Further complications might arise in the interpretation of the value itself — is it in engineering units, on a scale between 0 and 1, or on a different standard scale like 4–20 mA?

The simple example given above already illustrates the continuing need to agree upon the semantics of the information to be exchanged. Once the semantics and the structuring of the information have been agreed upon, either as a company standard, or as an agreement between project partners, or even beyond as a national or international standard, one can straightforwardly create an XML schema to define the XML files for information exchange based on this standard, and commercially available XML parsers can be used to read such an XML file and to transfer it into a tool-internal object model.

Recently, various attempts to use XML as an information exchange vehicle have emerged. In the following, three examples will be described to provide an idea of the possibilities, the difficulties, and the chances of XML for the exchange of industrial automation project information.

98.2 XML for the Description of Control System Hardware Components

An automation system for a manufacturing or production plant is usually composed of several controllers being connected by a communication network, each of the controllers comprising several units with ports for (a) receiving data from the sensors and/or (b) sending data to the actuators. According to the digital or analog nature of the signals, the ports are called digital input (DI), analog input (AI), digital output (DO), and analog output (AO). Usually, the units combine 4 or 8 or 16 or 32 ports of the same type, thus being named DI boards, AI boards, DO boards, and AO boards, respectively. The boards further differ in the type of physical signal level (0–10V, 4–20mA, etc.) and additional features like signal quality checking etc. Therefore, usually a control system vendor offers dozens of types of input/output boards.

In a large automation project, for example, a power plant or a large chemical plant or rolling mill, thousands of input and output signals have to be connected to suitable input and output ports, and therefore hundreds of input/output boards have to be installed in the plant. The engineers have to select the input/output boards from product catalogs, matching them with the requirements and at the same time keeping the actual number of boards and the costs of the control system low.

To support the engineer, a tool has been designed [2] that, based on a specification of the required input/output signals, selects the needed input/output boards from a product catalog, while optimizing the selection and configuration according to predefined optimization criteria, such as costs, redundancy, ease of extension, rack and cubicle space, etc. Figure 98.1 shows the basic concept.

XML plays a twofold crucial role in this concept:

(a) XML is used to describe the requirements on the input/output signals, and
(b) XML is used to describe the technical features of the available products in the product database.

Thus, the selection and optimization algorithm can operate on a common format, easily matching requirements with available products. The special advantages of XML, which have been the reasons for choosing XML for this particular project, are that XML allows for a description that is independent of the control system, the application domain, the operating system, and even the vendor of the automation system components. The format of the XML files for requirements and for product descriptions have been defined jointly and specified in a DTD. During application, the XML parser reads the requirements specification document and compares against the DTD to automatically detect inconsistencies with the required format. This reduces the need for manually programming consistency check functions.

The approach described above has been specific for one control system vendor. However, other vendors follow similar approaches, and the effort to join the approaches (provided there is a will) is rather low, thanks to extensible stylesheet language translation (XSLT), which is a language for transforming XML documents into other XML documents [3].

98.3 XML for the Description of Control Programs According to IEC 61131-3

With the general tendency toward *open* industrial automation systems, which allow to combine components (field devices, remote I/O, fieldbus connectors, etc.) from various vendors, there are also increasing

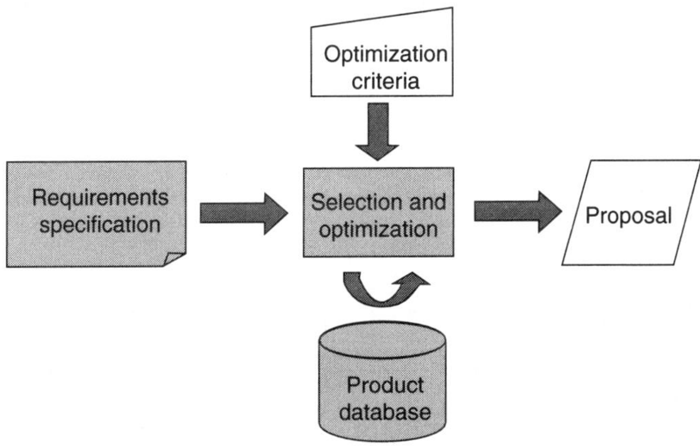

FIGURE 98.1   Basic concept of the automation system configuration optimizer.

user demands regarding the interoperability and interchangeability of the control units of the automation systems, that is, the components that execute the programs that control the process. In the past, these programs have been programmed and debugged with engineering tools provided by the vendor of the respective control unit. The introduction of the international standard IEC 61131-3, which defines syntax and semantics of five industrial control programming languages (Function Block Diagram, Sequential Function Chart, Structured Text, Instruction List and Ladder Diagram), has led to some standardization among the vendors of industrial control systems. However, there are still many proprietary features among the programming languages used and the implementation of the standard.

The users, that is, those who implement and operate industrial automation systems, especially long for a possibility to exchange the source code of control programs between control systems of different vendors. One use case, as an example, is a globally active company with highly automated plants on several continents. Due to local preferences, automation systems from different vendors are installed at the different production sites. This company would like to create and maintain the source code for their automation solutions only once in a central lab and then distribute it to the production sites and compile and download it into the different control systems.

Therefore, there is an increasing need for a generic, vendor-independent description of control programs written in the above-mentioned languages.

Figure 98.2 shows an example of a (very much simplified) control program written as a Function Block Diagram, with one instance "FB1" of the Function Block "MyType" and two "AND" Functions.

A standardized description of such Function Block Diagram should not only cover the logics of the signal flow and control functions (such as connections of variables with functions and *vice versa*) but also the graphical information, that is, the positioning and size of the elements (e.g., Function Blocks, Functions, and connections).

The organization "PLCopen" [4] is a vendor- and product-independent worldwide association supporting the standard IEC 61131-3, for the benefit of all the users of industrial automation systems. PLCopen has recently started to seize this issue [5]. XML has been chosen as a basis for an exchange format for graphical and textual languages of IEC 61131-3, due to its high profile and the wide range of available tools (viewers, parsers, etc.). The goal of PLCopen is to provide in XML format the logic information with optional explicit graphics for the languages Function Block Diagram, Sequential Function Chart, and Instruction List and Ladder Diagram. Proposals have been provided already by Schneider Automation, Rockwell Automation, and ABB. Further companies participating in this joint approach are, among others, Siemens, Matsushita, and Triconex. An example of a proposed XML description of the Function Block Diagram shown in Figure 98.2 is shown partially below in Figure 98.3. The *x* and *y* coordinates refer to a predefined grid. The object positions given for connections denote the positions of corners of the connecting line:

The expectation is that a first draft of a commonly agreed XML description format will be elaborated and presented in 2003, with subsequent refinements and possible standardization in the following years.

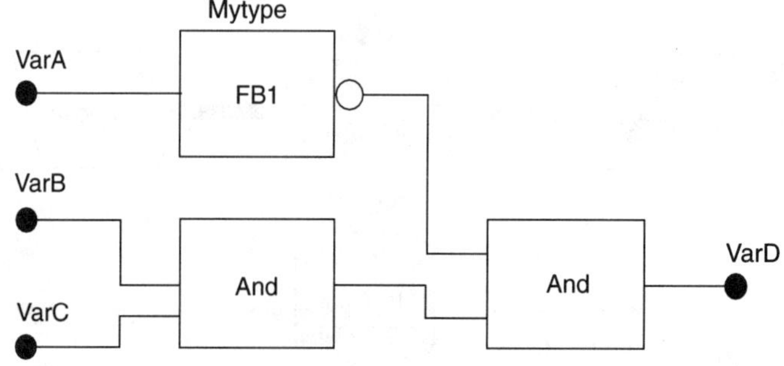

FIGURE 98.2 Example of a Function Block Diagram control program.

```
<FBDCodeBlock xmlns:xsi="http://wwww.w3.org/2001/XIVILSchema-instance"
xsi:noNamespaceSchemaLocation="FBD.xsd">
   <FBDVariable name="VarA">
      <objPosition posX="5" posY="20"/>
   </FBDVariable>
...
   <FBDFunction functionName="and" executionId="2">
      <objPosition posX="30" posY="40"/>
      <inputVariables>
         <inputVariable formalParameter="in1">
            <connection name="VarB">
               <objPosition posX="20" posY="50"/>
               <objPosition posX="20" posY="40"/>
            </connection>
         </inputVariable>
         <inputVariable formalParameter="in2">
            <connection name="VarC">|
               <objPosition posX="20" posY="55"/>
               <objPosition posX="20" posY="60"/>
            </connection>
         </inputVariable>
      </inputVariables>
      <outputVariables>
         <outputVariable formalParameter="out1" invertedPin="false"/>
      </outputVariables>
   </FBDFunction>
...
</FBDCodeBlock>
```

FIGURE 98.3   Proposed XML representation of a Function Block Diagram (draft).

98.4   XML for the Exchange of Plant Engineering Information

At a certain point in the engineering workflow of a process plant (like a power, chemical, or pharmaceutical plant), the process engineer hands over to the control engineer. At this time, the process engineer has designed the so-called "Pipe & Instrumentation Diagrams" (P&IDs), which document most of the process engineer's work (Figure 98.4). This P&ID is supplemented by tables ("Instrument Lists") which specify for each control item the setpoint, the measurement range, etc. On this basis, the control engineer designs and implements the continuous control loops, the interlocks, and the sequence control.

In the past, the handover of these documents has always been via printouts, and the control engineer had to manually reenter a huge amount of data, which is time-consuming and error-prone. The main reasons, among others, are that process engineering and control engineering are usually carried out by different companies, and that the paper format was prescribed for documentation purposes. The latter can be accomplished with similar effect by electronic document management systems today, and the former is no more accepted as an excuse for inefficient work processes. Therefore, the electronic transfer of process engineering information to the control engineer is required today. With respect to the Instrument Lists, this can be achieved with little effort, since the table headlines are few and easy to agree upon. With respect to the P&IDs, however, there is little progress in electronic exchange, due to the reasons described in Section 98.1.

To bridge the gap between process engineering and control engineering, a new attempt has been made recently by automation and engineering tool manufacturers and an institute of the RWTH Aachen (Aachen University) to define a meta model that allows to exchange information about a plant (especially about the process and its control system) in a format independent of a proprietary engineering or control tool. This information can be exchanged based on XML.

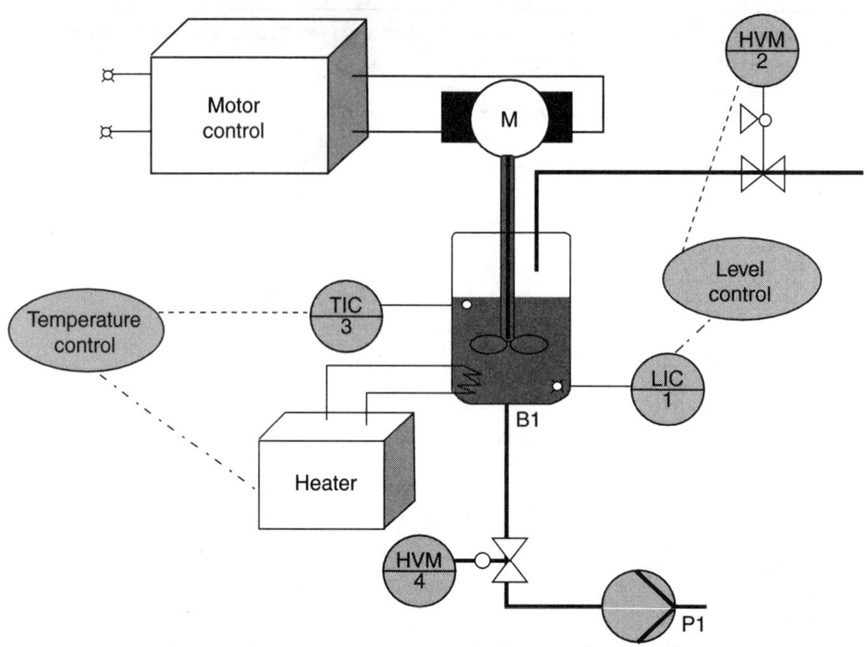

FIGURE 98.4 Pipe & Instrumentation Diagram example of a Technical Unit.

The requirements for this attempt have been: The proposed neutral XML description should

1. be able to store each state of the plant design workflow (to document defined steps and progress),
2. be prepared to support several workflow types (e.g., subcontracting of various tasks, bottom-up versus top-down, draft & refine iterations),
3. offer object-oriented concepts (which become more and more state of the art in engineering)
4. support hierarchical design of plants (e.g., plants consisting of components),
5. include not only the process and control objects but also the links in between (process material flow and information flow),
6. support user libraries and vendor libraries (which are always required in addition to agreed basic standards),
7. support several methods of reusing (e.g., reuse by "cut&paste" of objects from former projects).

Instead of presenting the complete XML schema, the following discussion concentrates on the key requirements and how they can be implemented with XML in general. Further information about the XML schema and its use can be found in [6].

To ensure consistency of the information given in one XML file, it is not sufficient to provide only the data about the object instances in the file — furthermore, the definition of the object type where this instance is derived from has to be provided as well. To support any type of libraries, both well-known public standards and proprietary project libraries, the information about the object type can be provided

- by a link to an entry in a public library or
- by attaching a library to the XML file or
- by directly including parts of a library.

In case the object type is inherited from other object types, these also have to be referenced or included.

To allow for different types of workflow and to also store intermediate planning results in a consistent manner, the concept allows to create instances from generic object types first (such as "pump") and to assign a more detailed specification as a refinement later ("pump for corrosive liquids, capacity 10 m^3/hour," "Vendor x, purchase order no. 341-45467").

```xml
<PlantComplex xmlns="http://www.acplt.de/xmlschema/caeclassmodel" xmlns:om=
"http://www.acplt.de/xmlschema/basicobjectmodel"
xmlns:xsi="http://www.w3.org/2001/ XMLSchema-instance"
xsi:schemaLocation="http://www.acplt.de/Xmlschema/caeclassmodel
D:\Daten\Projekte\PROCES~1\XML\NEUTRA~1\EPPLEV~1\VERSIO~1.0\CaeClassModel.xsd"
TechnicalUnitName="DKE Beispielanlage">
  <om: Version>1.0</om:Version>
  <PlantObjectInstance TechnicalUnitName="Anlage Süd">
    <om:InternalPlantObjectInstance TechnicalUnitName="BA1"refSystemUnitClass=
    ""refElementClass=""/>
    <om:InternalPlantObjectInstance TechnicalUnitName="BA2"refSystemUnitClass=
    ""refElementClass=""/>
  </PlantObjectInstance>
  <PlantObjectInstance TechnicalUnitName="Heizwerk" refSystemUnitClass="
  SU_Heizwerk">
    <om:InternalPlantObjectInstance TechnicalUnitName="Heizeinrichtung"
    refSystemUnitClass="SU_Heizeinrichtung">
      <om:InternalPlantObjectInstance TechnicalUnitName=
      "B1"refElementClass="EC_Tank2Stutzen"/>
      <om:InternalPlantObjectInstance TechnicalUnitName=
      "S1"refElementClass="EC_GenericStirer"/>
      <om:InternalPlantObjectInstance TechnicalUnitName="MotorControl"
      refElementClass="C_M"/>
      <om:InternalPlantObjectInstance TechnicalUnitName="TIC3"
      refElementClass="C_T"/>
      <om:InternalPlantObjectInstance TechnicalUnitName="TemperatureControl"
      refElementClass="C_K"/>
      <om:InternalPlantObjectInstance TechnicalUnitName="Heater"
      refElementClass="EC_GenericHeater"/>
      <om:InternalPlantObjectInstance TechnicalUnitName="V1"
      refSystemUnitClass="SU_A3711-3A7" refElementClass="EC_GenericValvePLT"/>
      <om:InternalPlantObjectInstance TechnicalUnitName="HMV2"
      refElementClass="C_Y"/>
      <om:InternalPlantObjectInstance TechnicalUnitName="V2"
      refSystemUnitClass="SU_A3711-3 A8" refElementClass=
      "EC_GenericValvePLT"/>
      <om:InternalPlantObjectInstance TechnicalUnitName="HMV4"
      refElementClass="C_Y"/>
      <om:InternalPlantObjectInstance TechnicalUnitName="Pipe1"
      refElementClass="EC_Pipe"/>
      <om:InternalPlantObjectInstance TechnicalUnitName="Pipe2"
      refElementClass="EC_Pipe"/>
      <om:InternalPlantObjectInstance TechnicalUnitName="Pipe3"
      refElementClass="EC_Pipe"/>
      <om:InternalPlantObjectInstance TechnicalUnitName="Pipe4"
      refElementClass="EC_Pipe"/>
      <om:InternalPlantObjectInstance TechnicalUnitName="Pipe5"
      refElementClass="EC_Pipe"/>
    </om:InternalPlantObjectInstance>
  </PlantOBjectInstance>
</PlantComplex>
```

FIGURE 98.5   XML representation of a Technical Unit (draft).

Figure 98.5 shows the XML representation of the Technical Unit of Figure 98.4 within a simplified plant hierarchy. Note the references to libraries in the header of the file, as well as that most PlantObjectInstances refer to a reference ElementClass (which is a generic object type) except for the two valves V1 and V2 to which already a specified reference SystemUnitClass is assigned.

This approach can serve the automation engineer to gain consistent information about the plant that should be automated. Since the XML definition allows to mix different levels of detail in the specification of components, the information exchange can already take place early during the workflow, which gives the automation engineer the possibility for iterative and concurrent engineering and the chance to deliver optimal results with respect to the quality of the automation solution.

98.5 Conclusion

Information technologies do not only have an impact on the automation system itself but also on the methods and tools used to engineer the automation system. The wealth of special tools for dedicated tasks, together with the need for efficient work, requires quick and seamless information exchange. Many description means and exchange formats for engineering information have been developed in the past, and few have gained widespread use. The key to success is to find a good compromise between simplicity (imagine to interpret comma-separated-value files like "2, 300, 1, 10.0, ..." without explanation) and expressiveness (where the definition of all mandatory information items might already be outdated when published). XML does not guarantee a solution, but has the capability to, if used carefully, form the basis for a valuable information exchange, as shown for example, by the examples given above.

References

[1] http://www.w3.org/XML/
[2] Kirmair, S. and Fay, A., Nutzung von XML zur Beschreibung von Prozessleittechnik-Hardware-Komponenten (XML as a description means for process control hardware components), *Automatisierungstechnische Praxis*, 43, 55–58, 2001, *(in German)*.
[3] http://www.w3.org/TR/xslt
[4] http://www.plcopen.org/
[5] http://www.plcopen.org/plcopen/TC6_XML_kick_off.htm
[6] Fedai, M., Epple, U., Drath, R., and Fay, A., A Metamodel for generic data exchange between various CAE systems, MathMod 2003, Vienna, January 5–7, 2003, pp. 1247–1255.

99

Enterprise-Manufacturing Data Exchange Using XML

David Emerson

Yokogawa Corporation of America

99.1 Introduction

The integration of enterprise-level business systems with manufacturing systems is increasingly an important factor driving productivity increases and making businesses more responsive to supply chain demands. As a result, more and more businesses are making integration a priority and are searching for standards and tools to make integration projects easier. The World Batch Forum's (WBF) Business To Manufacturing Markup Language (B2MML) is an Extensible Markup Language (XML) vocabulary based upon the ANSI-ISA 95 (ISA-95) standards and the international equivalent IEC/ISO 62264-1 standard.

99.2 Integration Challenges

While there are many software tools that provide varying levels of assistance for integrating systems, integration projects typically require extensive labor in order to overcome differences between terminology, data formats, interfaces, and communications options in the systems to be integrated.

A significant difference in the software tools commonly used within the enterprise and manufacturing domains exists. For example, many enterprises use middleware products that provide robust communications between systems from the same and different vendors. This type of middleware is not commonly seen in manufacturing systems, primarily due to the cost of the software and the technical expertise required by the middleware. Higher level manufacturing systems, such as Plant Information

Management Systems that collect and aggregate data from manufacturing systems, are sometimes used with enterprise middleware systems, although often the interfaces with enterprise systems are a custom-developed or single vendor solution.

At this point in time, manufacturing systems are predominantly based upon Microsoft Windows. This is in large part a result of the constant drive to reduce manufacturing costs, the mature nature of the manufacturing system marketplace, and the reluctance to replace manufacturing systems with newer versions or operating systems. These factors often preclude the use of enterprise middleware solutions and has fostered a *de facto* industry standard for communication called OPC.

OPC is a set of communication protocols developed by the OPC Foundation for the exchange of manufacturing data using Microsoft's Distributed Common Object Model (DCOM) technology. The OPC Foundation is a nonprofit organization, primarily funded by manufacturing system vendors, that develops and maintains the OPC protocols. While limited to Microsoft platforms, the OPC Foundation is starting to develop web service implementations of their protocols to enable cross-platform connectivity. While OPC is a common tool for interoperability in the manufacturing domain, it is infrequently seen in the enterprise domain and is not a match for high-end middleware products.

The use of the World Wide Consortium's (W3C) XML is a common trait in many recent integration projects. As a mainstream technology, XML offers universal support by software vendors, a large number of tools for developing and using it, and the promise of a common language that will work with disparate systems. However, while XML provides interoperability on the protocol level, there remains an application-level integration issue of what structure the XML being exchanged should have, the elements/attributes to use, and the organization of the data.

When integration projects settle on communication formats and protocols, there is still the need to identify the data to be exchanged followed by a data mapping exercise. When cross-functional project teams assemble, there is usually a learning curve as team members from different parts of the organization learn the type of data available and needed in other parts of the organization.

99.3 Solutions

While the communication issues involving the physical and transport layers should be resolved as appropriate for each project, taking into account corporate and local requirements and infrastructure it is assumed that the resulting architecture will utilize XML for the protocol layer. XML provides the following benefits as the protocol for integration projects:

- It is a mainstream technology supported by all major operating system and application software vendors.
- Numerous tools are available for manipulating XML making the task of data mapping/conversion simpler.
- As a mainstream technology, it has a better chance of providing a longer lived technology than proprietary and older technologies. This is an important consideration in determining the total cost of ownership of a solution.

With XML as the common protocol for an integration project, the issue of standardizing the XML vocabulary for the project becomes critical. B2MML provides a solution to this issue. Also, since B2MML is based on the ISA-95 and IEC/ISO 62264-1 standards. Coupled together, B2MML and ISA-95 permit designers to define the data mapping using a standardized, common terminology and models that can be carried over to the B2MML XML vocabulary.

If custom interface development is required to integrate a computer system, there is a long-term benefit to using ISA-95 and B2MML. By interfacing individual systems to B2MML, a single format is used for all data received by a system, the number of interfaces is reduced, programmers may more easily move between interfaces, and the same terminology used for designing the data mapping is used in the interfaces. These factors will reduce software maintenance costs, make the integrated system easier to upgrade, and integrate new systems. Figure 99.1 shows a comparison of the number of interfaces required when

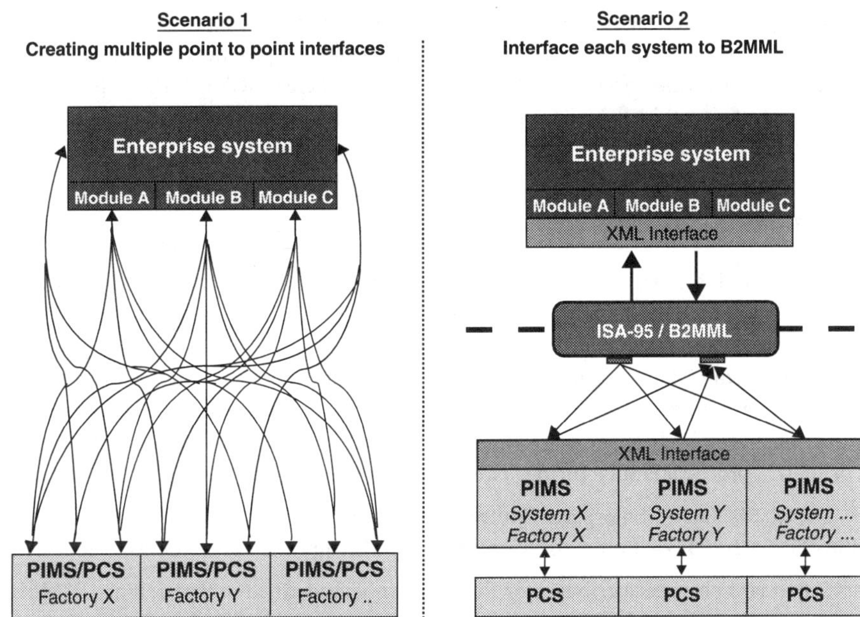

FIGURE 99.1   Point to point vs. common interfaces.

point-to-point interfaces are used as in scenario 1 vs. fewer interfaces required when a common format such as B2MML is used as in scenario 2.

99.4 B2MML

The B2MML is a set of XML schemas that are based on the ISA-95 Enterprise-Control System Integration Standards. The XML schemas comply with the W3C's XML schema format and define a vocabulary using the terminology and models in the ISA-95 standard. XML documents based on B2MML may be used to exchange data between the business/enterprise and manufacturing systems.

B2MML was developed by a group of volunteers working for the World Batch Forum (WBF), a nonprofit educational professional organization. While the WBF is the owner of B2MML, the licensing terms make the schemas available royalty free for any use. B2MML was created with the intention of fostering the use of the ISA-95 standards by providing XML schemas that could be used, and modified as necessary, for integration projects.

The existence of a core set of ISA-95 XML schemas is critical since without it each company, or even work group, would have to develop their own definitions of XML elements and types based on the ISA-95 standards. This would inevitably lead to numerous variations with enough structural and nomenclature differences to make the exchange of data using XML more difficult than expected. The creation of B2MML will not make XML-based data exchange easy, but it should make it easier. Even when the B2MML schemas are used to derive proprietary schemas that extend and constrain the originals if the B2MML element names and type definitions are used, there will be a common footing that can be used to establish a data mapping between applications.

B2MML has advantages over proprietary interfaces in that it is independent of any one vendor, based on an international standard and that representatives from the manufacturing domain, both vendors and end users, have been very active in its development.

As a vendor-independent and standards-based XML vocabulary B2MML can be used to implement ISA-95-based designs using most XML enabled middleware and application interfaces. This provides the ability for project teams to use a vendor-independent framework during analysis and design and the ability to carry it directly to the implementation phase.

While other organizations, such as the Open Applications Group (OAG), provide standard interfaces for enterprise applications, they do not provide the level of detail and completeness required for full functioned interfaces with manufacturing systems that B2MML provide. Where OAG's OAGIS XML schemas provide interfaces primarily for within the enterprise domain, B2MML's interfaces are totally focused on the exchange of data between the enterprise and manufacturing domains.

99.5 ISA-95 Standard

In order to understand B2MML, one must have a basic understanding of the ISA-95 standards. A complete explanation of the ISA-95 standards is beyond the scope of this paper; however, a brief overview of the standards is provided.

ISA is a nonprofit educational organization that serves instrumentation, systems, and automation professionals in manufacturing industries. ISA is an accredited standards body under agreement with the American National Standards Institute (ANSI). ISA develops standards relating to the manufacturing industry, primarily, process manufacturing. The ISA-95 standards that B2MML is based upon are:

ANSI/ISA-95.00.01-2000 — Enterprise-Control System Integration Part 1: Models and Terminology and
ANSI/ISA-95.00.02-2001 — Enterprise-Control System Integration Part 2: Object Model Attributes.

The Part 2 standard provides attributes for the object models defined in Part 1. Since B2MML uses the models and terminology defined in Part 1 and the attributes defined in Part 2, it is said to be based upon the ISA-95 standards.

After ISA-95 was accepted as a U.S. standard by ANSI, it was submitted to the IEC and ISO for acceptance as international standards. While slight modifications were made to the standards, the international versions are substantially the same as the ISA version. IEC and ISO agree to release the international standard as a dual logo standard; therefore, it is available from either organization. The international version of Part 1 is called IEC/ISO 62264-1. At the time of writing, the Part 2 version of the international standard was progressing through the joint IEC/ISO working group; hence, it is not yet a released international standard.

ISA-95 builds upon existing work, its models are based upon "The Purdue Reference Model for CIM" developed in the 1990s by a group of chemical company representatives under the leadership of Dr. Theodore Williams at Purdue University (Purdue Model); the MESA International Functional Model as defined in "MES Functionality and MRP to MES Data Flow Possibilities — White Paper Number 2 (1994); and IEC 61512-1 batch control — Part 1: models and terminology (ANSI/ISA-88)." The value of ISA-95 is in providing a more comprehensive and detailed definition of the data exchange between the enterprise and manufacturing domains than the previous works.

The terms enterprise and manufacturing domains are defined in ISA-95 in order to put terms to the reality of the different business issues, needs, and drivers at the different levels of a business. ISA-95 uses the levels defined in the Purdue Model as shown in Figure 99.2.

Levels 0–2 represent process control and supervisory functions and are not addressed in the standard. Level 3, manufacturing operations and control, is considered the manufacturing domain and represents the highest level of manufacturing functions. Level 4, business planning and logistics, encompasses all enterprise- or business-level functions that interact with manufacturing and is referred to as the enterprise domain. The focus of the standards is on the interfaces between levels 3 and 4.

Is it important to note that the ISA-95 enterprise and manufacturing domains refer to functions, not organizations, individuals, or computer systems. Any one organization, person, or computer system may perform functions in both domains.

In practice, there is no single boundary between domains that applies to all industries, companies, divisions, and manufacturing plants. The standards draw an arbitrary line based upon commonly accepted practices. In recognition of the flexible boundaries between domains, the ISA is currently working on further parts of the standards that will define the functions an data flows inside level 3 so that when different boundaries exist, the standard may be used to identify the data flows that cross the specific level 3–4 boundary in use.

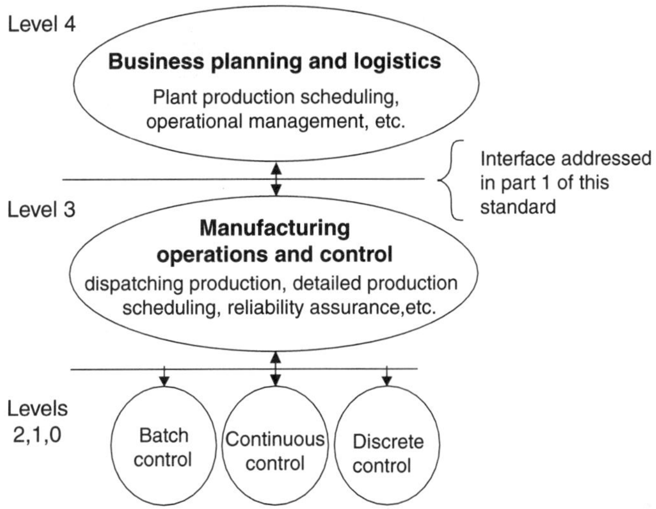

FIGURE 99.2   Levels in a manufacturing enterprise.

Figure 99.3 illustrates the concept of the flexible boundary between the enterprise and manufacturing domains. When boundary #1 between the enterprise and manufacturing domains is used, functions 1 and 3 in the enterprise domain must interface with functions 4 and 5 in the manufacturing domain. Functions 2 and 6 do not interface with functions in the other domains and therefore would not be the focus of an integration project. However, when boundary #2 is used, all the functions except for 1 and 2 would be the focus of an integration project since they interface with functions in the other domain.

Drawing heavily upon the Purdue Model, ISA-95 defines data flows that may cross between the enterprise and manufacturing domains. These data flows are grouped into categories of information, which are the foundations of the ISA-95 models and the B2MML schemas. The key information categories are listed in Figure 99.4.

Figure 99.5 illustrates the overlap of information in the enterprise and manufacturing domains and how the three key information categories provide a conduit for the flow of information between domains.

In addition to the three categories of information, three types of resources used by each category are identified in the standard as:

Each category of information may include information about some or all of the resource types and may include information about multiple instances of each resource type (Figure 99.6).

99.6 ISA-95 Models

ISA-95 defines nine object models defining the structure of the categories of information and resources. Each object model defines the data associated with the category of information or resource. The models are listed in Figure 99.7.

Communicating actual production results from the manufacturing domain to the enterprise domain is one of the most common and important goals of integration projects. The production performance model, shown in Figure 99.8, addresses this function.

This model is typical of the category of information models in that it defines a hierarchy built upon resources.

The hierarchy starts with the production performance object, which is made up of one or more production responses. This permits manufacturing requests from the enterprise domain to be split into multiple elements, for example, if the request was for more than to be manufactured at one time. In this case, each production response would report the results of an element of the manufacturing request, with the sum of all the production responses making up the production performance associated with the manufacturing request.

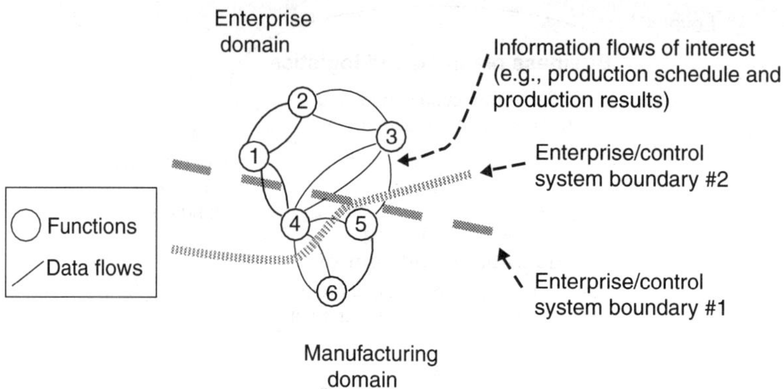

FIGURE 99.3   Flexible domain boundary.

Production capability	Information describing the manufacturing capability for a period of time. Total capability is the sum of committed, available and unattainable capabilities. This information is used to inform enterprise systems of a manufacturing area's ability to produce which is required to develop accurate plans and schedules.
Product definition	Information describing how a product is produced. When product definitions are maintained at the enterprise level this information must be sent to the manufacturing domain when product modifications are made or new products introduced.
Production information	Information instructing the manufacturing domain what to make and when in the form of a schedule and the report by the manufacturing domain up to the enterprise domain of actual production accomplishments including material usage, units of labor and equipment used for a product.

FIGURE 99.4   Key information categories between the enterprise and manufacturing domains.

Moving down the hierarchy each production response is made up of one or more segment responses. A segment response is the "production response for a specific segment of production." The production capability and process segment models are used to define segments for each application and would map into production performance at this level. The objects that each segment response consists of are listed in Figure 99.9.

Taken together, the production performance object defines the actual performance of the process that is reported by the manufacturing domain to the enterprise domain, most likely in response to a production request.

Each of the category of information models is constructed in a similar manner. The resource models for personnel, equipment, and material are themselves similar. The material model, shown in Figure 99.10, is a good example of this.

Reading the material model from left to right shows a hierarchy of material information.

Material classes (e.g., oils) define a grouping of material definitions (e.g., peanut oil) that are used to define material lots that may be made up of material sublots. Sublots may themselves by made up of multiple sublots. Material classes, definitions, and lots are further defined by lists of properties. Sublots do not have properties since each sublot must have the same properties of the parent lot. The QA objects provide a means to document test specifications and results for each property.

When used with the category of information models, any of the four levels of material may be referenced as appropriate. For example, production performance typically references specific lots and sublots used in production, production schedule may reference a material definition, or for tracking purposes a material lot or sublot. Production capability and product definition would probably reference material classes and definitions since they deal with more abstract information.

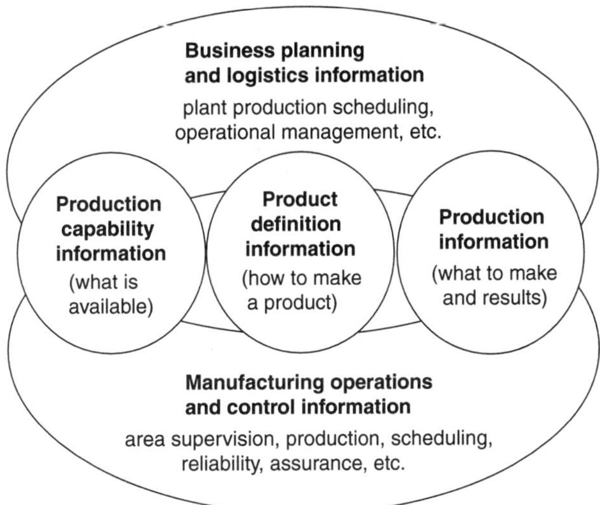

FIGURE 99.5   Categories of information.

Personnel	Individuals, or classes of people, with certain qualifications may be identified as a capability, required as part of a product definition, scheduled or reported as units of labor for production performance.
Equipment	Pieces of equipment, or classes of equipment, with certain characteristics may be identified as a capability, required as part of a product definition, scheduled or reported as utilized as part of production performance.
Material	Material sublots, lots, material definitions or material classes with certain properties may be identified as a capability, required as part of a product definition, scheduled or reported as consumed or produced as part of production performance.

Note: The standards considers energy to be a material.

FIGURE 99.6   Resources use in the three categories of information.

Production capability model

Process segment capability model

Process segment model

Product definition model

Production schedule model

Production performance model

Personnel model

Equipment model

Material model

FIGURE 99.7   List of ISA-95 object models.

While each of the models may be used by itself when used together, they are able to provide an integrated set of data exchanges. The interrelationships of the nine models are shown in Figure 99.11. Below each model title is a summary of the model's purpose. The horizontal dashed lines indicate how from the right side each model builds upon the model to the left. Note that the process segment capability model and the process seg-

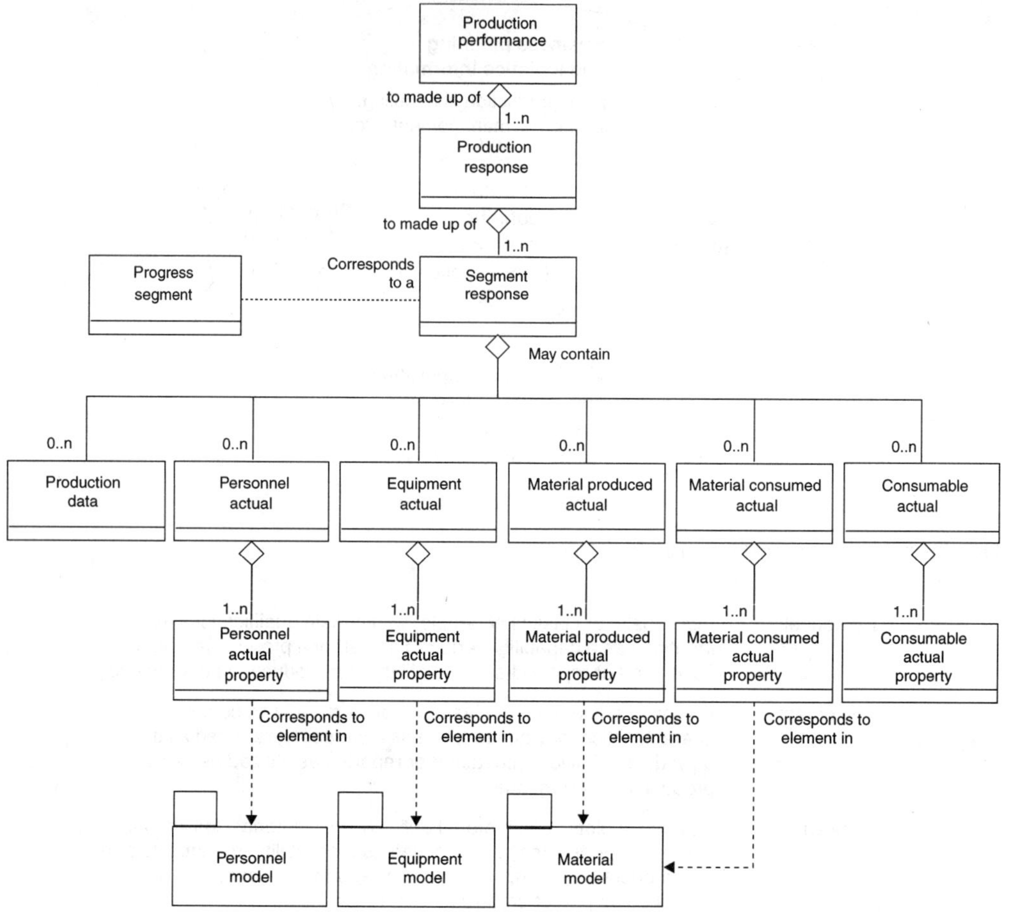

FIGURE 99.8   ISA-95 production performance model.

ment model have been combined under Process Capability. In the standard, these two models were shown separately in order to make them clearer but they both define the capabilities of the manufacturing process.

99.7   B2MML Architecture

B2MML is a collection of XML schemas organized to align with the ISA-95 standard's object models. The basis for each of the schemas, including the mapping to the standard's data models, is listed in Figure 99.12.

There is a separate schema for each model, with the exception of the equipment model, which has two schemas, one for the equipment objects, the other for the maintenance objects. This was done to provide the flexibility of using equipment and maintenance objects separately.

The separate schemas permit applications to only reference the schemas required thereby eliminating unused elements from populating application namespaces.

The common schema, B2MML-V02-Common.xsd, does not directly relate to an ISA-95 model, rather it is used to contain type definitions, which are referenced by more than 1 schema.

The internal structure of the model-related schemas follows the ISA-95 standard's object model structures. The root element in a schema is named after each data model's root element and each object in an object model is generally represented as an XML element.

Production data	Data associated with the products being produced, the process segment or waste material but not directly identified as a resource.
Personnel actual	Units of labor for the personnel classes or persons related with the process segment.
Equipment actual	Equipment or classes of equipment used by the segment.
Material produced actual	Material produced by the segment. This may include one or multiple products or intermediate materials as well as byproducts and waste products. Material may be identified by sublot, lot, material definition or material class.
Material consumed actual	Material consumed by the segment. Material may be identified by sublot, lot, material definition or material class.
Consumable actual	Material not tracked by lots, not included in bills of material, or not individually tracked that have been consumed by the segment.

FIGURE 99.9   Objects that make up the segment response object.

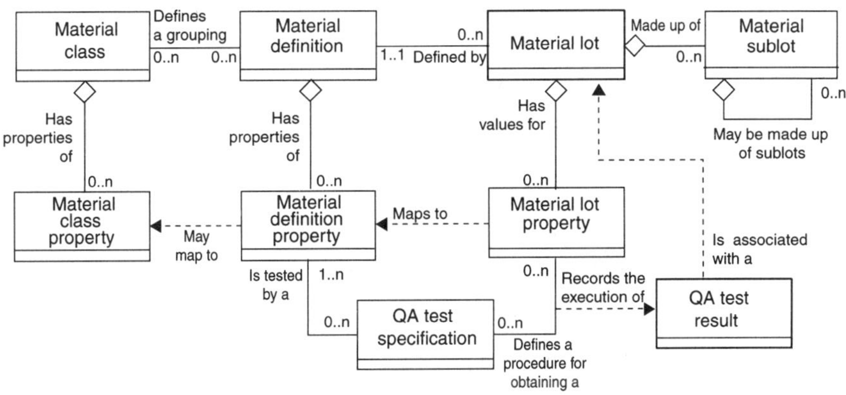

FIGURE 99.10   ISA-95 material model.

The standard's technique of using application-specific properties is implemented in the schemas using property types, which may be used to list any number of application-specific properties in an XML document.

All elements in the schemas are declared using simple and complex types. The common schema is included by each of the other schemas that use the types as needed. Any type that is used in only one schema is defined in that schema.

In B2MML, only a few elements are declared globally, meaning they may be used in other schemas or XML documents. Generally, the objects in the ISA-95 standard that represent data to be exchanged between systems are implemented as global elements with the addition of a few container elements for the equipment, personnel, and material models. The other objects, which are generally part of the exchanged objects, are defined as local elements. The global elements are listed in Figure 99.13.

Most of the elements in the schemas are optional. This enables XML documents based upon them to only contain the elements applicable to the application, resulting in more concise XML documents.

The B2MML schemas permit most XML types to be expanded with additional elements. This is accomplished by placing an element called "Any" as the last element in a type's definition. The "Any" type is defined using the "AnyType," which is based upon the XML schema wildcard component ##any. The XML schema wildcard component permits any element to be included inside the "Any" element that is at the end of the type's list of elements.

The use of this wildcard is a compromise between maintaining the ability to rigorously validate XML documents against the schema and the pragmatic recognition that diverse integration projects have unique requirements that can best be served by permitting application-specific elements to be used to extend B2MML types.

While the application-specific addition of elements can hurt interoperability, this can be limited by having XML processors expect to find either nothing or some unknown (from the B2MML viewpoint) element after the last standard B2MML element in each type. This technique will make XML processors more robust and ensure that the standard B2MML data can be processed.

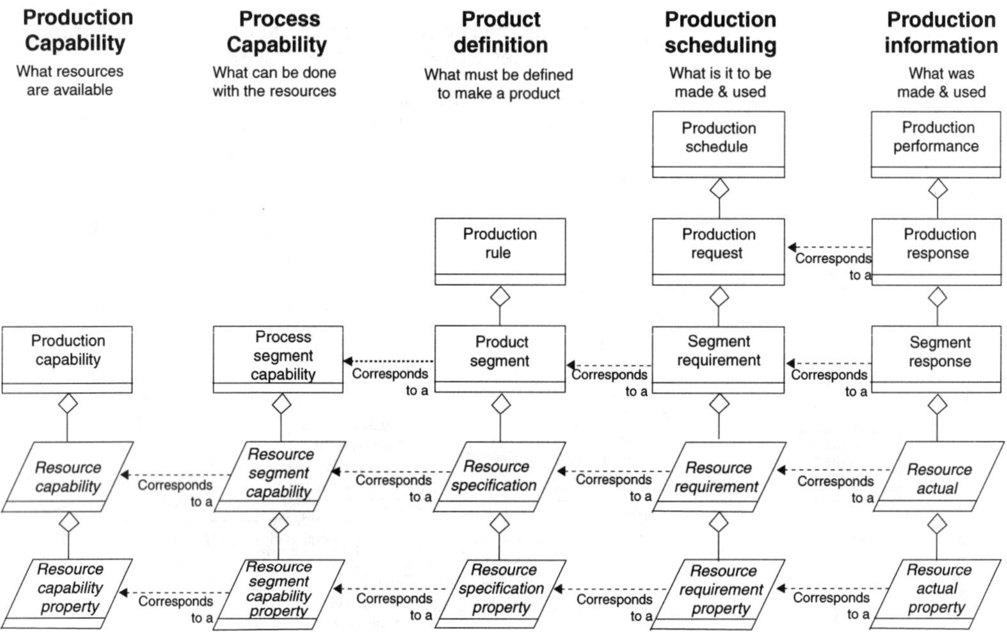

FIGURE 99.11   Interrelationships of models.

B2MML schema	Schema basis
B2MML-V02-Common.xsd	All elements and types used in more than 1 other schema are defined here
B2MML- V02-Personnel.xsd	ISA-95 Personnel Model
B2MML- V02-Equipment.xsd	ISA-95 Equipment Model (except for the maintenance objects)
B2MML- V02-Maintenance.xsd	ISA-95 Equipment Model (maintenance objects only)
B2MML- V02-Material.xsd	ISA-95 Material Model
B2MML- V02-ProcessCapability.xsd	ISA-95 Production Capability Model
	ISA-95 Process Segment Capability Model
B2MML- V02-ProcessSegment.xsd	ISA-95 Process Segment Model
B2MML- V02-ProductDefinition.xsd	ISA-95 Product Definition Model
B2MML- V02-ProductionSchedule.xsd	ISA-95 Production Schedule Model
B2MML- V02-ProductionPerformance.xsd	ISA-95 Production Performance Model

FIGURE 99.12   B2MML schemas and corresponding to ISA-95 models.

99.8   Using the B2MML Schemas in XML Documents

The root element in a B2MML XML document must be a globally defined element. For example, a production performance document may use either the ProductionPerformance or ProductionResponse element while a material document may use one of the MaterialClass, MaterialDefinition, Material Information, MaterialLot, or MaterialSubLot elements.

Individual XML documents may reference one or more of the model-based, resource, or common schemas as required. This is done by placing namespace references attributes in the root element as shown in Figure 99.14.

In Figure 99.14, the xmlns attribute declares the namespace for the document. While the string http://www.wbf.org/xml/b2mml-v02 has the form of a URL, it is merely a unique string used to identify the version of B2MML used by the document.

A simple B2MML document is shown in Figure 99.15.

This document uses MaterialInformation as the root element, references the B2MML namespace and schema location, as well as the standard XML W3C namespace. The data contents of the file provide information about the material lot with an ID of M-1215. Note that many optional elements in MaterialLot are not in this file; this is an example of how unneeded optional elements can be omitted.

Many elements are based upon types whose content has been restricted to be an enumerated list. This means that the value of the element must be one of the values listed in the schema. For example, EquipmentElementLevel is based upon EquipmentElementLevelType, which in turn is based upon EquipmentElementLevel1Type. These two types are shown in Figure 99.16.

Whenever there is an enumerated list in B2MML, a simple type's content is restricted to the values in the enumerated list and a companion complex type is declared, which extends the simple type by adding an attribute named OtherValue. This is required in order to provide XML document authors the ability to extend the list. The enumerated list may be extended by an XML document author by giving the EquipmentElementLevel element a content of "Other" and an attribute of "OtherValue" whose content is

Equipment	PersonnelClass
EquipmentCapabilityTestSpecification	PersonnelInformation
EquipmentClass	ProcessSegment
EquipmentInformation	ProcessSegmentInformation
MaintenanceInformation	ProductInformation
MaintenanceRequest	ProductionCapability
MaintenanceResponse	ProductionPerformance
MaintenanceWorkOrder	ProductionRequest
MaterialClass	ProductionResponse
MaterialDefinition	ProductionSchedule
MaterialInformation	ProductDefinition
MaterialLot	QAMaterialTestSpecification
MaterialSubLot	QualificationTestSpecification
Person	

FIGURE 99.13   B2MML global elements.

```
<MaterialInformation
    xmlns              = "http://www.wbf.org/xml/b2mml-v02"
    xmlns:xsi          = "http://www.w3.org/2001/XMLSchema-instance"
    xsi:schemaLocation = "http://www.wbf.org/xml/b2mml-v02
                          B2MML-V02-Material.xsd">
```

FIGURE 99.14   Sample XML root element with namespace references.

```xml
<?xml version="1.0" encoding="UTF-8"?>
<MaterialInformation
  xmlns="http://www.wbf.org/xml/b2mml-v02"
  xmlns:xsi="http://www.w3.org/2001/XMLSchema-instance"
  xsi:schemaLocation="http://www.wbf.org/xml/b2mml-v02
                      B2MML-V02-Material.xsd">

  <MaterialLot>
    <ID>M-1215</ID>
    <Description>Sample Lot</Description>
    <MaterialDefinitionID>M42</MaterialDefinitionID>
    <Status>Available</Status>
    <StorageLocation>T-942</StorageLocation>
  </MaterialLot>
</MaterialInformation>
```

FIGURE 99.15   Simple B2MML material information XML document.

```xml
<xsd:simpleType name = "EquipmentElementLevel1Type">
  <xsd:restriction base = "xsd:string">
    <xsd:enumeration value = "Enterprise" />
    <xsd:enumeration value = "Site" />
    <xsd:enumeration value = "Area" />
    <xsd:enumeration value = "ProcessCell" />
    <xsd:enumeration value = "Unit" />
    <xsd:enumeration value = "ProductionLine" />
    <xsd:enumeration value = "WorkCell" />
    <xsd:enumeration value = "ProductionUnit" />
    <xsd:enumeration value = "Other" />
  </xsd:restriction>
</xsd:simpleType>

<xsd:complexType name = "EquipmentElementLevelType">
  <xsd:simpleContent>
    <xsd:extension base = "EquipmentElementLevel1Type">
      <xsd:attribute name = "OtherValue" type = "xsd:string"/>
    </xsd:extension>
  </xsd:simpleContent>
</xsd:complexType>
```

FIGURE 99.16   Use of enumerated lists in type declarations.

the extended value. Figure 99.17 contains a sample B2MML document that demonstrates the use of the enumeration list extension method.

In Figure 99.17, the bold text `other` is one of the permitted enumerated values for Equipment Element Level. When this value is used, the XML processor must look for the attribute OtherValue, which

```
<?xml version="1.0" encoding="UTF-8"?>
<MaterialInformation
   xmlns="http://www.wbf.org/xml/b2mml-v02"
   xmlns:xsi="http://www.w3.org/2001/XMLSchema-instance"
   xsi:schemaLocation="http://www.wbf.org/xml/b2mml-v02
                       B2MML-V02-Material.xsd">

   <MaterialLot>
     <ID>M-1215</ID>
     <Description>Sample Lot</Description>
     <MaterialDefinitionID>M42</MaterialDefinitionID>
     <Status>Available</Status>
     <Location>
       <EquipmentID>T-942</EquipmentID>
       <EquipmentElementLevel OtherValue="WorkCenter">
         Other
       </EquipmentElementLevel>
     </Location>
     <StorageLocation>T-942</StorageLocation>
   </MaterialLot>
</MaterialInformation>
```

FIGURE 99.17   Example of extending an enumerated list in an XML document.

in this case has the value of `Work Center` and then uses the attributes value as the value of the element. This technique may be used on any of the enumerated lists.

The B2MML schemas have been designed to permit most XML types to be expanded with additional elements. The element "Any" that appears as the last element in most complex types serves as a container for any other elements the XML document author wants to insert into an element. The Any element is based upon the AnyType complex type that is defined in the B2MML common schema. The Any element and AnyType ComplexType declarations are shown in Figure 99.18.

The string "##any" seen in Figure 99.18 is a W3C XML schema wildcard component that permits any other element to be added to the end of the type's list of elements. The use of this wildcard is a compromise between maintaining the ability to rigorously validate XML documents against the schema and the pragmatic recognition that diverse integration projects have unique requirements that can best be served by permitting application-specific elements to be used to extend B2MML types.

When elements are added to an existing B2MML element, they must be added within the Any element; otherwise, the XML documents will not be valid. While it is good practice for all added elements to use a prefix that identifies the XML schema, they are defined in this reference as optional and not required by B2MML.

While the application-specific addition of elements can hurt interoperability, this has been limited by having XML processors expect to find either nothing or some unknown (from the B2MML viewpoint) element after the last standard B2MML element in each type. This technique will make XML processors more robust and ensure that the standard B2MML data can be processed.

Figure 99.19 contains an example of adding elements not in B2MML to a B2MML element.

In this case, three elements not defined in B2MML are included as part of the MaterialLot element by placing them inside the Any element. The extended elements have a prefix of "ext:" which is defined in the namespace declarations at the top of the document. There the "ext:" prefix is defined to point to the ExtensionExample.xsd XML schema. The figure contains this schema, which has been used to declare the

```
Any element declaration:
<xsd:element name="Any" type="AnyType"
                        minOccurs="0"
                        maxOccurs="unbounded" />



Any type complex type declaration:
<xsd:complexType name="AnyType" >
    <xsd:sequence>
        <xsd:any namespace="##any" processContents="skip"
                                   minOccurs = "0"
                                   maxOccurs="unbounded"/>
    </xsd:sequence>
</xsd:complexType>
```

FIGURE 99.18   Any element and AnyType complextype declarations.

simple types used in the XML document. Of note is the fact that while XML processors will check for well-formed XML, they will not validate the content within the Any element since the AnyType has been defined with the attribute processContents= "skip."

99.9 Usage Scenario

The following scenario provides an example of using B2MML's Production Performance schema to report production results from a manufacturing system to an enterprise system. Figure 99.20 lists the manufacturing data to be reported.

Figure 99.21 contains a production performance XML document containing these results. The document has been broken into parts for clarity and for reference in the description below. If the XML in the each box were concatenated, it would create one production performance document.

Header: The header information in figure 21 includes an XML declaration, the start of the document's root element, ProductionPerformance, and attributes declaring XML namespaces, identification of the XML schema the document is based upon, and a suggestion as to the schema's location. To further understand the XML syntax, refer to the W3C's XML and XML schema recommendations.

Production performance and response information: The production performance and production response elements provide information to the receiving system regarding where this information fits into the overall production performance data. There may be one or many production performance XML documents per lot of product. Therefore, sufficient information must be included in the document to permit the receiving system to know where to store or send each piece of data.

In this case, the overall production performance ID of MT593 is a batch ID and the production response, MT593-1, is a subdivision of the batch operating on one unit.

Segment response information: The segment response information identifies the product or process segment within the production response. In this case, the segment maps to a product segment since that element is used and the process segment element is not. The actual start and end times provide potentially important information that can be used by the enterprise system for costing or utilization purposes.

Production data: This section contains four production data elements from the Production Data table above. Each measurement has been placed in its own ProductionData element with a unique ID and containing its value, data type, and units of measure.

<u>XML Document with extensions:</u>

```
<?xml version="1.0" encoding="UTF-8"?>
<MaterialInformation
    xmlns="http://www.wbf.org/xml/b2mml-v02"
    xmlns:xsi="http://www.w3.org/2001/XMLSchema-instance"
    xmlns:ext="ExtensionExample"
    xsi:schemaLocation="http://www.wbf.org/xml/b2mml-v02 B2MML-V02-Material.xsd">

    <MaterialLot>
        <ID>M-1215</ID>
        <Description>Sample Lot</Description>
        <MaterialDefinitionID>M42</MaterialDefinitionID>
        <Status>Available</Status>
        <MaterialLotProperty>
            <ID>Purity</ID>
            <Description>Measurement of purity</Description>
            <Value>
                <ValueString>99.4</ValueString>
                <DataType>float</DataType>
                <UnitOfMeasure>Percent</UnitOfMeasure>
            </Value>
        </MaterialLotProperty>
        <Location>
            <EquipmentID>T-942</EquipmentID>
            <EquipmentElementLevel>Unit</EquipmentElementLevel>
        </Location>
        <StorageLocation>T-942</StorageLocation>
        <Quantity>
            <QuantityString>200</QuantityString>
            <DataType>float</DataType>
            <UnitOfMeasure>Kg</UnitOfMeasure>
        </Quantity>
        <Any>
            <ext:ExtendedElement1>sample content</ext:ExtendedElement1>
            <ext:ExtendedElement1>sample content</ext:ExtendedElement1>
            <ext:ExtendedElement2>472.5</ext:ExtendedElement2>
        </Any>
    </MaterialLot>
</MaterialInformation>


Custom schema (extensionexample.xsd) containing definitions of extended elements

<?xml version="1.0"?>
<xsd:schema targetNamespace="http://Extensions"
            xmlns="http://Extensions"
            xmlns:xsd=http://www.w3.org/2001/XMLSchema
            elementFormDefault="qualified"
            attributeFormDefault="unqualified">

    <xsd:simpleType name="ExtendedElement1">
        <xsd:restriction base="xsd:string"/>
    </xsd:simpleType>

    <xsd:simpleType name="ExtendedElement2">
        <xsd:restriction base="xsd:float"/>
    </xsd:simpleType>

</xsd:schema>
```

FIGURE 99.19 B2MML document with extended elements.

Production data:

Date	Time	Event	Temperature 1 (Deg C)	Temperature 2 (Deg C)
2003-08-05	14:34:03	Start Charging Milk	25.0	26.4
2003-08-05	14:39:29	End Charging Milk	19.6	21.3

Material used

Date	Time	Material	Target (Kg)	Actual Quantity (Kg)
2003-08-05	14:34:03	Milk	400	402.4
2003-08-05	14:59:43	Flour	750	750.3

FIGURE 99.20 Production data to be reported to an enterprise system.

Material consumed — milk: The material consumed — milk section is used to transmit the amount of milk actually added to the process, the target (i.e., amount of milk that was supposed to be added), and the time the milk was added.

The MaterialConsumedActual element contains identifying information about the material, the location the material was added from, the amount added, and properties of the material consumed. The properties have been used to convey the time the milk was consumed and the target amount. In any integration project, the sending and receiving systems must be programmed to use the same property IDs as part of the data-mapping exercise.

This is an example of how an element's properties can be used to provide extended information without using the Any element. This type of extension should be easier for receiving systems since properties will be expected.

Material consumed — flour: This section is similar to the milk material consumed section; only, it refers to the addition of flour. This is an example of how each material consumed may be documented.

End of elements: These three lines indicate the end of each of the elements opened in the earlier sections. </ProductionPerformance> indicates the end of the XML document.

Many optional elements have been omitted from this example, as will often be the case in actual implementations. When empty elements are shown above, it is because they are required by the B2MML schemas.

While elements such as MaterialProducedActual, PersonnelActual, and EquipmentActual have not been shown, their usage closely follows the above example.

99.10 Schema Customization

While the ISA-95 standards provide a firm basis for many integration projects, they cannot satisfy every requirement. If the addition of elements using the "Any" type is insufficient, the schemas may be used to derive custom corporate or application-specific schemas. While the derivation of new schemas may seem contradictory to the use of a standard, it is a pragmatic recognition that companies have requirements beyond the core functionality of the standards and B2MML.

B2MML types and elements may be referenced or included in other schemas. This may be done to build new types that are extensions or restrictions of B2MML types or to include B2MML elements inside corporate or project-specific schemas.

Since the B2MML schemas are freely distributed with no restrictions placed on their use, each user is free to change their contents or include them in other work. It is strongly recommended that if modifications are

Header

```
<?xml version="1.0" encoding="UTF-8"?>
<ProductionPerformance
    xmlns="http://www.wbf.org/xml/b2mml-v02"
    xmlns:xsi="http://www.w3.org/2001/XMLSchema-instance"
    xsi:schemaLocation="http://www.wbf.org/xml/b2mml-v02 B2MML-V02-ProductionPerformance.xsd">
```

Production performance and response information

```
<ID>B-1</ID>
<PublishedDate>2003-08-05T15:12:34-05:00</PublishedDate>
<ProductionScheduleID>MT593</ProductionScheduleID>

<ProductionResponse>
    <ID>UR1</ID>
    <ProductionRequestID>MT593-1</ProductionRequestID>
```

Segment response information

```
<SegmentResponse>
    <ID>SR1</ID>
    <ProductSegmentID>UR1-Charge Milk</ProductSegmentID>
    <ActualStartTime>2003-08-05T14:34:03-05:00</ActualStartTime>
    <ActualEndTime>2003-08-05T14:39:29-05:00</ActualEndTime>
```

Production data

```
<ProductionData>
    <ID>Charge Start Temp 1</ID>
    <Value>
        <ValueString>25.0</ValueString>
        <DataType>float</DataType>
        <UnitOfMeasure>Deg C</UnitOfMeasure>
    </Value>
</ProductionData>

<ProductionData>
    <ID>Charge Start Temp 2</ID>
    <Value>
        <ValueString>26.4</ValueString>
        <DataType>float</DataType>
        <UnitOfMeasure>Deg C</UnitOfMeasure>
    </Value>
</ProductionData>

<ProductionData>
    <ID>Charge End Temp 1</ID>
    <Value>
        <ValueString>19.6</ValueString>
        <DataType>float</DataType>
        <UnitOfMeasure>Deg C</UnitOfMeasure>
    </Value>
</ProductionData>

<ProductionData>
    <ID>Charge End Temp 2</ID>
    <Value>
        <ValueString>21.3</ValueString>
        <DataType>float</DataType>
        <UnitOfMeasure>Deg C</UnitOfMeasure>
    </Value>
</ProductionData>
```

Material consumed - milk

```
<MaterialConsumedActual>
    <MaterialClassID>Milk</MaterialClassID>
    <MaterialDefinitionID>Milk Low-Fat</MaterialDefinitionID>
    <MaterialLotID>MLF-3948</MaterialLotID>
    <Location>
        <EquipmentID>T-19</EquipmentID>
        <EquipmentElementLevel OtherValue="EquipmentModule">
            Other
        </EquipmentElementLevel>
    </Location>
```

FIGURE 99.21 B2MML production performance document.

```
                        <Quantity>
                            <QuantityString>402.4</QuantityString>
                            <DataType>float</DataType>
                            <UnitOfMeasure>Kg</UnitOfMeasure>
                        </Quantity>
                        <MaterialConsumedActualProperty>
                            <ID>Time Consumed</ID>
                            <Value>
                                <ValueString>2003-08-05T14:39:29-05:00</ValueString>
                                <DataType>time</DataType>
                                <UnitOfMeasure></UnitOfMeasure>
                            </Value>
                        </MaterialConsumedActualProperty>

                        <MaterialConsumedActualProperty>
                            <ID>Target</ID>
                            <Value>
                                <ValueString>400</ValueString>
                                <DataType>float</DataType>
                                <UnitOfMeasure>Kg</UnitOfMeasure>
                            </Value>
                        </MaterialConsumedActualProperty>
                    </MaterialConsumedActual>
```

Material consumed - flour

```
                    <MaterialConsumedActual>
                        <MaterialClassID>Flour</MaterialClassID>
                        <MaterialDefinitionID>Enriched Flour</MaterialDefinitionID>
                        <MaterialLotID>EF-382</MaterialLotID>
                        <Location>
                            <EquipmentID>R-43</EquipmentID>
                            <EquipmentElementLevel>Unit</EquipmentElementLevel>
                        </Location>
                        <Quantity>
                            <QuantityString>750.3</QuantityString>
                            <DataType>float</DataType>
                            <UnitOfMeasure>Kg</UnitOfMeasure>
                        </Quantity>
                        <MaterialConsumedActualProperty>
                            <ID>Time Consumed</ID>
                            <Value>
                                <ValueString>2003-08-05T14:59:43-05:00</ValueString>
                                <DataType>time</DataType>
                                <UnitOfMeasure></UnitOfMeasure>
                            </Value>
                        </MaterialConsumedActualProperty>
                        <MaterialConsumedActualProperty>
                            <ID>Target</ID>
                            <Value>
                                <ValueString>750</ValueString>
                                <DataType>float</DataType>
                                <UnitOfMeasure>Kg</UnitOfMeasure>
                            </Value>
                        </MaterialConsumedActualProperty>
                    </MaterialConsumedActual>
```

End of elements

```
            </SegmentResponse>
        </ProductionResponse>
    </ProductionPerformance>
```

FIGURE 99.21   (continued)

made to the B2MML types, they be done as part of another schema using a different namespace and filename.
If a B2MML schema file has its contents changed without the namespace and filename being changed, there
is an increased risk of errors in the future from incompatible versions of the same file being mixed up.

99.11 Conclusion

B2MML, the Business To Manufacturing Markup Language, is an XML-based implementation of the
ISA-95 standard. This industry markup language will enable the use of mainstream information technology with a standards-based approach to integrating enterprise and manufacturing systems.

References

ISA, www.isa.org

OPC Foundation, www.opcfoundation.org

Open Applications Group, www.oag.org

Using XML with S88.02, by David Emerson, presented at the World Batch Forum 2000 European Conference, Brussels, Belgium, Oct., 2000.

World Batch Forum, www.wbf.org

XML Schema Part 0: Primer; W3C Recommendation, 2 May 2001, http://www.w3.org/TR/2001/REC-xmlschema-0-20010502/

XML Schema Part 1: Structures; W3C Recommendation 2 May 2001, http://www.w3.org/TR/2001/REC-xmlschema-1-20010502/

XML Schema Part 2: Datatypes, W3C Recommendation 02 May 2001, http://www.w3.org/TR/2001/REC-xmlschema-2-20010502/

100

Web Services for Integrated Automation Systems — Challenges, Solutions and Future

Zaijun Hu
ABB Corporate Research Center

Eckhard Kruse
ABB Corporate Research Center

100.1 Introduction

Integrated automation systems are gaining more and more momentum in the automation industry. They address not only the vertical integration that covers layers from devices via manufacturing execution systems to business applications but also the horizontal integration ranging from design, engineering, and operation to maintenance and support. The emerging Web Services technology with growing acceptance in industry are a good way to create an open, flexible, and platform-neutral integrated system. In this chapter, we attempt to analyze and describe the main challenges in using Web Services for integrated automation systems. We believe that performance, client compatibility, client addressability, object designation, and ability to deal with multiple structures are important and essential issues for deploying Web Services in automation systems. We present some solution concepts including architecture, mechanisms and methods such as the structure cursor, Web Services bundling, the event service, the object designator, and so forth. Finally, we discuss the future of using Web Services, where Ontology will play an important role for efficient system engineering and assembling.

100.2   Background

Integration is a strong trend in the current development of the automation technology. Integrated control systems, integrated factories, or integrated manufacturing are some examples. A large integrated automation system covers not only the whole production life cycle including purchase, design, engineering, operation, and maintenance but it also involves the different control levels ranging from the field device layer to Enterprise Resource Planning (ERP) layer [1–3]. The creation of such systems thus poses the challenge of addressing various requirements from different areas at the same time, of assembling heterogeneous applications, integrating data models, and binding the applications to the data models. Typically, the diverse applications developed for handling issues of different business areas such as purchase, design, and engineering are distributed via network and unstructured. It is difficult for an engineer to find a suitable application for his specific purpose. There is no common and structured way of organizing or describing the applications in the automation area. Another challenge for large integrated automation system is the heterogeneity of platforms on which applications are developed. On the one hand, Microsoft's COM technology is widely used to create applications for traditional automation systems such as human–machine interface (HMI) or Supervisory Control And Data Acquisition (SCADA). OLE for Process Control (OPC), originally based on the COM technology, provides a standard specification for data access. It greatly facilitates interoperability for access to control instruments and devices. On the other hand, many applications in other areas such as ERP or Supply Chain Management (SCM) are based on CORBA or EJB. Interoperability between heterogeneous platforms is always a headache for integration — a uniform base would extremely reduce development costs. Appropriate data models are another challenge when building large integrated automation system. A unified description method, easy transformation and mapping, and efficient navigation mechanisms are natural requirements on data modeling. Last but not the least, the binding of applications to data models of a large automation system is crucial. The engineering costs for finding appropriate applications for specific data is quite high. An efficient way to reduce costs will greatly influence the development direction of automation technology. Web Services will play an ever-more important role in addressing the challenges in integrated automation systems due to their open, flexible, standard,- and service-oriented architecture.

100.3   ABB Industrial IT Platform

To address the integration challenges in automation systems, ABB has created an integration platform for integrated automation systems called Aspect Integrator Platform (AIP), which follows the paradigm of decoupling the data model from its computational model. It is subjected to IEC 61346 [13]. The basic elements in the model are Aspect Object, Aspect, and Structure. An *Aspect Object* in AIP is a container that holds different parts of an object in an automation system. Such an object might be, for example, a reactor, a pump, or a node (computer). The Aspect Object covers data modeling, including data type, relationship among data, and its structure. The *Aspect* represents operations that are associated with an object. It can contain its own data. Examples of aspects are signal flow diagram, CAD drawing, analysis program, simulation, trend display, and so on. The Aspect focuses on the operational aspect. Figure 100.1 shows an AIP example.

To create an automation system based on the AIP platform, a data model should usually be built at first. Then, the engineer chooses suitable applications in the form of aspects and binds them to the data model. For an integrated automation system, which covers the whole production life cycle and all control levels, a large number of Aspects and Aspect Objects result. Thus, binding suitable Aspects to a certain Aspect Object requires significant engineering effort. Web Services could help to simplify this process. By creating an additional layer to cover Aspects of AIP, they could be searched and accessed in a unified way, using the standard Web Service discovery and description mechanisms. Besides, it should be noted that an Aspect itself can also be a Web Service.

A *Structure* — another element defined in AIP architecture and conforming to IEC 61346 — represents the semantic relationship of a data model. In IEC 61346 it is separated from the objects and expressed through an additional aspect such that an object can be organized in different structures at the same time. IEC 61346 presents three examples of information structures that are important for design, engineering, operation, and maintenance: function-oriented, location-oriented, and product-oriented structures. A structure is determined through a defined hierarchy, which describes the semantic relationships between Aspect Objects from a certain point of view. For example, the function-oriented structure organizes objects based on their purpose or function in the system, while the location-oriented structure results from the spatial constitution relationship, for example, ground area, building, floor, room, and so on. IEC 61346 provides the structure concept to address the semantics of a data model, but it does not define mechanisms to describe the semantics in different structures. Figure 100.2 shows three structures regarding function, location, and maintenance. The maintenance structure presented in the figure is useful for a maintenance engineer.

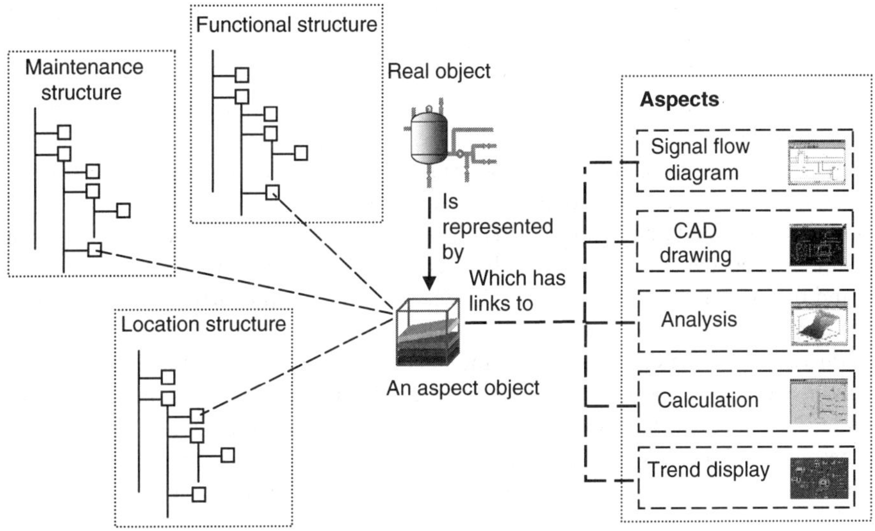

FIGURE 100.1 An example showing AIP architecture concept.

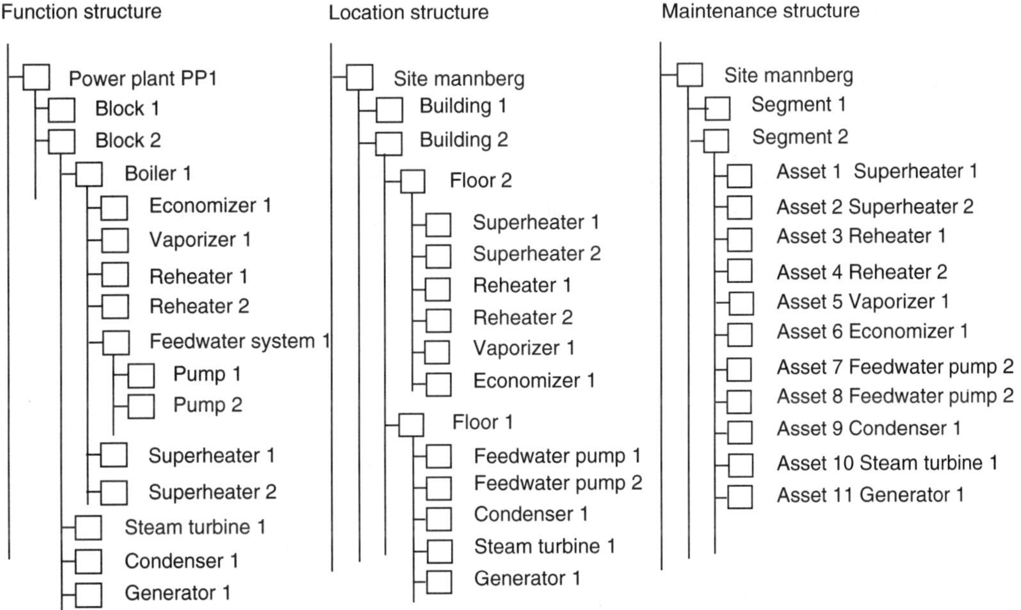

FIGURE 100.2 Example for functional, location, and maintenance structure.

100.4 Web Services

Definition

Web Services can be defined in different ways [5]. From a business point of view, Web Services present a common service-oriented architecture for companies and organizations to provide their key businesses in the form of services. From an application point of view, Web Services create a platform-independent and programming language-neutral middleware for interoperable interaction among applications. In this chapter, we concentrate on the technical aspect of Web Services and use the definition from the W3C [4]. "A Web service is a software system identified by a URI [RFC 2396], whose public interfaces and bindings are defined and described using XML. Its definition can be discovered by other software systems. These systems may then interact with the Web Service in a manner prescribed by its definition, using XML based messages conveyed by Internet protocols." Web Services have the following key features:

They can be described according to their nonoperational service information and operational information. The nonoperational information includes service category, service description, and expiration date, as well as business information about the service provider (e.g., company name, address, and contact information). The typical description language used for the nonoperational information is Universal Description, Discovery, and Integration (UDDI). The operational information describes the behaviors of Web Services. It covers dynamical aspects such as service interface, implementation binding, interaction protocol, and the invoking endpoint (URL). Web Service Description Language (WSDL) is usually used to describe the operational information.

Web Services have repositories for storing the nonoperational and operational information of Web Services. By means of the repositories, Web Services can be published, located, or discovered anywhere and anytime. They can also be invoked over a network such as the World Wide Web. SOAP is used to describe messages for Web Services. HTTP, TCP/IP, etc., can be used as communication protocols.

Web Services are standard-based, platform, and programming language-independent. They use standards for the description of services.

In comparison with the traditional middleware and component-based technologies, the differentiating features of Web Services are description and discovery mechanisms based on standards that enable the platform- and programming language-neutrality. Web Services provide a way for integrating applications developed on different platforms; it is thus a natural choice to use them within integrated automation systems.

Architecture

Basic Components

The Web Service architecture consists of a set of building blocks, which represent different roles. The key components are Service Provider, Service Requester, and Service Broker. Their relationship is illustrated in Figure 100.3.

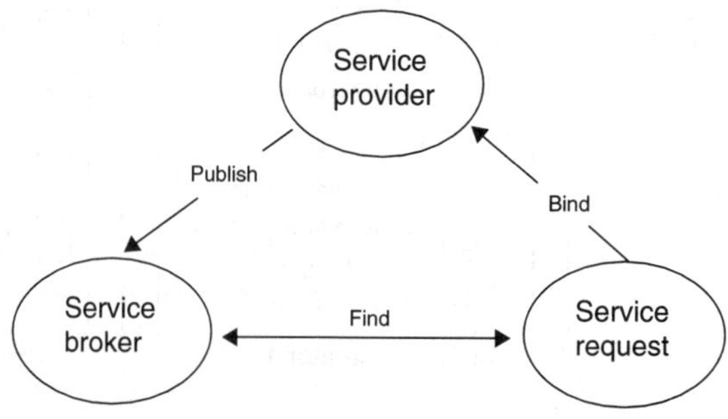

FIGURE 100.3 Basic components of Web Services.

The Service Provider deploys and publishes services by registering them with the Server Broker. It provides an environment for running Web Services so that consumers can use them. The Service Broker has a repository to register and manage the service description including nonoperational and operational information. It can also provide some mechanism for efficiently organizing and structuring Web Services. The Service Requester finds required services using the Service Broker, binds them to the Service Provider, and then uses them.

Technology Stacks

The Web Service concept comprises different aspects such as description, discovery, composition, management, interaction, and communication. These are addressed through different layered and interrelated technologies. Figure 100.4 gives an overview of their relationship.

The figure shows how Web Services include the basic technology stacks and communications. The security and management of Web Services are also important for the development of a Web Service system. The process stack in the basic technology stacks is responsible for the discovery, aggregation, and choreography of Web Services, while the description stack defines how to describe a Web Service. The messages stack is related to the method of exchanging information between Web Services.

Web Service Style

There are two Web Service styles: remote-procedure-call (RPC)-style and message-style.

RPC-style: A remote procedure call (RPC)-style Web Service is like a remote object for a client application. When the client application invokes a Web Service, it sends parameter values to the Web Service, which executes the required methods and then sends back the return values. Because of this back and forth conversation between the client and the Web Service, RPC-style Web Services are tightly coupled and resemble traditional distributed object paradigms, such as RMI or DCOM. RPC-style Web Services are synchronous, meaning that when a client sends a request, it waits for a response before doing anything else.

Message-style: Message-style Web Services are loosely coupled and document-driven rather than being associated with a service-specific interface. When a client invokes a message-style Web Service, the client typically sends an entire document, for example, a purchase order, rather than a discrete set of parameters. The Web Service accepts the entire document, processes it, and may or may not return a result message. Because there is no tightly coupled request–response between the client and the Web Service, message-style Web Services provide a looser coupling between the client and the server. Message-style Web Services are usually asynchronous, meaning that a client that invokes a Web Service does not wait for a response before it does something else. The response from the Web Service, if any, can appear hours

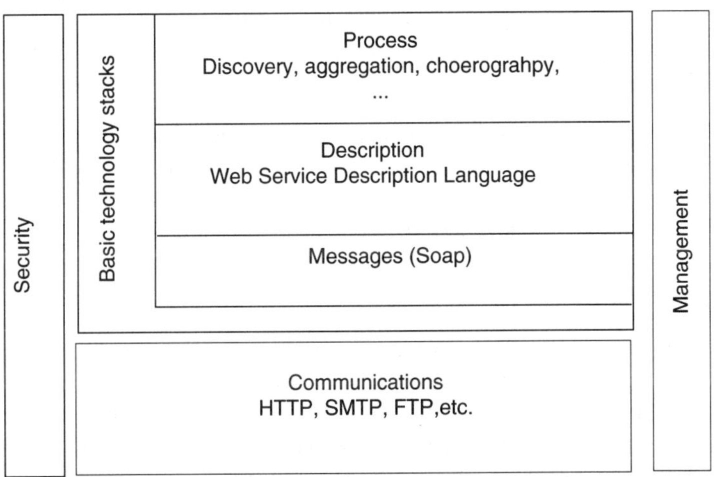

FIGURE 100.4　Technology stacks.

or days later, making interactions more efficient. Asynchronous Web Services may be a requirement for enterprise-class Web Services.

100.5 Challenges of Using Web Services for Integrated Automation Systems

Web Services provide many attractive features for integrated automation systems, but there are also several challenges.

Multiple Structures

For automation systems, it is crucial to have close integration of the information, which is carefully selected for specific purposes. Multiple structures of industrial information systems characterize the current trend in information modeling in the automation industry. They vertically cover different business layers from the process, sensor/actuator, field bus, HMI to manufacturing execution system (MES), and ERP. Horizontally, they adress different life-cycle phases ranging from ordering, design, engineering, to operation, optimization, and maintenance.

The information required for each business layer or life-cycle phase is different. It has to be structured accordingly, taking into account the specific properties of the layers and phases, and it has to be provided in a consistent way for the integrated solution. In this context, a multiple structural representation is inevitable. Even for the same business layer or the same life-cycle phase, multi-structure or multiple views of information are sometimes desired to provide an insight into the system from different points of view. The challenges in handling multiple structures can be characterized as follows:

- Information structures and hierarchies are closely related and interactive, that is, they are subject to the global goal of the information system.
- Connection points are clearly defined. The connection points determine how the multiple structures are associated and interrelated. For example, for a plant-centric automation architecture, the connection points are plant objects such as valves, pumps, etc.
- It is possible to navigate between structures.
- Each structure has a clearly defined semantic and serves one purpose. For example, for the engineering process, the product information structure is used to organize the product information.
- Designation is required to uniquely identify an information entity or an object of an information system.
- A multiple hierarchies-based information model should enable the integration of the computation model that provides the operation model for information manipulation and utilization. Additionally, it supports the integration of external computation applications that can use or process the information.

Web Services cover the dynamic aspect of an application, namely functions in the form of interfaces described by WSDL. They promote the separation of operations from the data model. For data presented in the form of multiple structures, it is necessary to provide the corresponding services for navigation, identification of data entities, and moving from one structure to another.

Client Compatibility

One benefit of Web Services is that they allow to integrate heterogeneous applications and to "webalize" legacy applications. A legacy automation system usually has client–server and peer-to-peer architecture using a defined protocol for the communication such as a socket or COM/DCOM. Web Services do not destroy this client–server architecture style. They only change the way of message and interface description and make it independent of the platform and the programming language. One possible scenario is the migration of a legacy automation system to Web service-based architecture. Web Services are usually used to provide mediator-like interfaces to the clients of the legacy automation system. One requirement could be that the clients who are using the legacy systems should not be required to do any adaptation, at least in the earlier phase of the migration. For example, process graphics displaying process data from the process server machines should not be changed if the data server of an automation system is just wrapped in a web service. This client compatibility guarantees low development cost and incremental evolution of an automation system. Interfaces between clients and servers do not have to be changed, including data types, data models, and invocation methods. For example, if a COM-based application provides an automation model to its clients, the client compatibility requires that the clients can use Web Services in the same way as if nothing has been changed.

Performance

An automation system is a real-time system with a large amount of process data, which changes over time. Thus, data transfer capacity and speed are two essential quality attributes.

Selective Data Access and Presentation

Web Services usually use Simple Object Access Protocol (SOAP) to describe messages exchanged between service requestors and service providers. Different communication protocols (Figure 100.4) can be used. When Web Services are invoked via the Internet or Intranet, the time for communication may be considerably longer than the time for data access, processing, and presentation. For monitoring and controlling an automation system, for example, a SCADA system, data access and presentation are typical functions. Here, it is not necessary to constantly obtain all data from the data server that is connected to instrumentation and devices. Efficient data access and presentation are required. However, in the Web Service environment due to SOAP, much data overhead is introduced. Time-costly roundtrips may occur frequently if no optimization is applied.

Additionally, Web Services communicate with the external world by sending XML messages, which have the advantage of being a platform-independent textual representation of information. Consequently, for the communication between the service provider and the service requestor, it is necessary to package the message, to transfer it to the service provider, and to unpack or parse it. Again, this might take a considerable amount of time and is opposed to the high-performance requirements of a real-time automation system.

An intelligent caching mechanism can help to tackle these problems by creating efficient data access and presentation, and by reducing the data overhead and round-trip time.

"Chatty" Interfaces

Multiple sequential calls between an interface and a business logic layer are acceptable in a stand-alone application for an automation system, but they cause a large performance loss when it comes to Web Services. In an automation system such as SCADA or an HMI system, the amount of process data exchanged between clients and servers is large. Transferring these data via Internet or Intranet through sequential calls will lead to large performance loss. This is a challenge for Web Services applications. Avoiding repetitive data transmission and reducing the number of interactions between the service requestor and the service provider are important issues to be solved.

Object Designation

Web Services are usually stateless, meaning that after the invocation of a Web Service, all state-related data created during the Web Service call are deleted and thus not available anymore. One way to address this problem is to use session management. Each Web Service requestor is allocated with a session on the server side that manages all client-specific state-related data such as intermediate variables, global variables for the client, and so on. But creation of a sessions on the server side for a client is always a burden for the server, impairs the scalability of the system, and thus should be avoided whenever possible. An integrated automation system usually has a structured data repository containing asset- and process-related data, which are organized in different structures for satisfying a variety of requirements. The ABB Industrial IT platform is such an example. The repository is a kind of data pool that is connected to processes over OPC or other communication channels. Careful design of methods for designating a data entity on the server is essential for efficient data browsing, navigation, and access. Uniqueness and multiple-structure characteristics of data should be taken into account. Well-designed designation methods are a condition for the use of stateless Web Services.

Client Addressability

In a client–server application, the client usually initiates the communication. It sends requests to the server, and the server responds and gives the requested data back to the client. In an automation system, sometimes it is required that the server triggers the interaction between the client and the server. An alarm and event server is such an example, which informs a client that a process parameter such as pressure or temperature has exceeded an upper or lower limit value. For this purpose, it supplies condition-related events.

There are also simple and tracking-related events. For example, a message about the failure of a unit can be represented by a simple event. The information about intervention in a process (corrective action on site) can be represented by a tracking-related event.

Events are organized in the event space. There are a variety of methods by which the client can influence the behavior of the server. Condition-related events, for example, can be enabled, disabled, and acknowledged. Web Services uses SOAP for description of messages and usually HTTP as a communication protocol. But HTTP is not good at delivering event notifications to clients or supporting long-lived message exchanges.

Security

Security is a very important aspect, especially in automation systems. Exposing a Web Service entails that the location and execution mechanism of the code changes, and this change requires a revision of the security policies mechanism. All data sent and received by a Web Service are formatted using SOAP on top of an XML specification. SOAP messages are easily readable; thus, it is necessary to encrypt certain data such as passwords. In this chapter, security is not the focus and thus is not discussed in detail.

100.6 Solution Concepts

We have listed some challenges for using Web Services with integrated automation systems. In this section, we are proposing concepts and solutions to address these challenges.

Overall Architecture of Web Services for an Automation System

Web Services are usually implemented based on the client-server architecture. They require client-side proxies and server-side implementations of the Web Service interfaces. For an integrated automation system, some special Web Services such as the structure navigation service, the service request unpacking service, the event service, and so on are needed on the service provider side. For each Web Service, there exists a service proxy on the service requestor side. Figure 100.5 illustrates the overall architecture of such a system.

Figure 100.5 also contains typical components in a traditional automation system, HMI client, communication channels over control bus (FDT, Profibus) and OPC, and devices. Usually in a client–server architecture, the connection between the HMI client and the automation system server is established through certain programming interfaces, as shown in Figure 100.5. For an automation system with Web Service support, the connections between the client and the server can be realized through Web Services. The façade pattern [8] on the service provider side controls the communication and request handling to simplify the implementation. The client- and server-side caches improve the performance of Web Services. The adapter pattern on the service requestor side solves the problem of client compatibility.

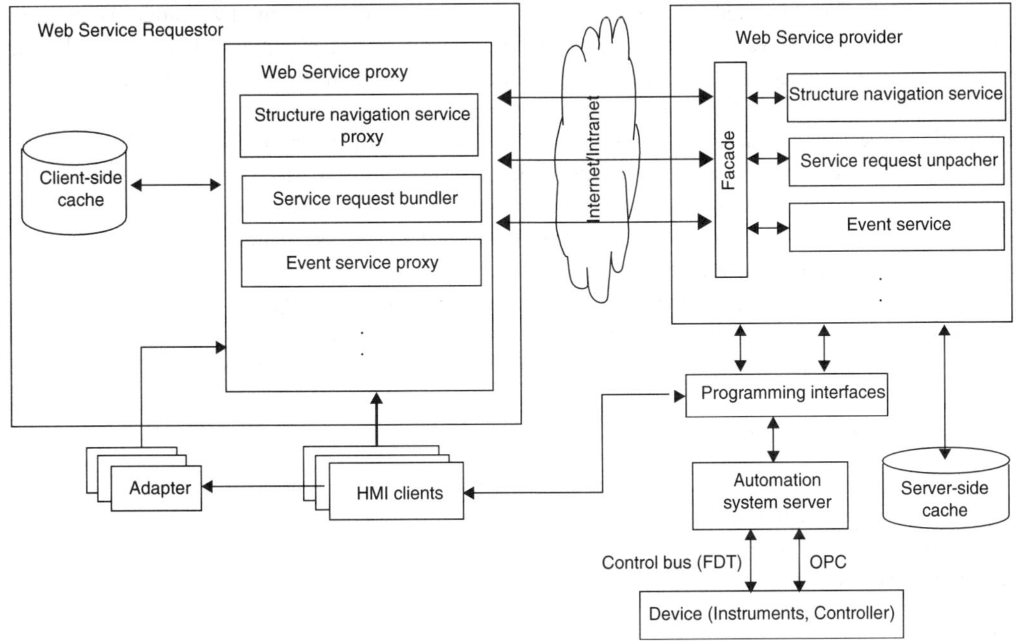

FIGURE 100.5 The overall architecture of an automation system with Web Service Implementation.

The main purpose of this architecture is to migrate an integrated automation system to the Web Service platform to improve interoperability of systems. The service broker (one of the basic components of Web Services) and the process stack including service discovery (one of basic technology stacks of Web Services) are not exploited here. Both must not necessarily be deployed for interoperability improvement.

Structure Cursor

The structure cursor is a concept for implementing the structure navigation service. As mentioned before, multiple structures are a characteristic of an integrated automation system. The structure cursor is used to navigate in a certain structure or to move from one structure to another. It enables access to all information in multiple structures such as product information, real-time information, asset information, location information, and so on. The structure cursor has two parts: the structure navigation service on the service provider side and the structure navigation service proxy on the service requestor side, as illustrated in Figure 99.5. There could be different structure cursors for different purposes [9] so that the structure-specific semantics can be taken into account in navigation.

Client Compatibility

An adapter [8] is an effective way to address client compatibility. Figure 100.5. shows the role of adapters. Their original purpose is to "convert the interface of a class into another interface clients expect." An adapter lets classes work together that otherwise could not because of incompatible interfaces. For the migration of legacy automation systems to the Web Service platform, the implementation usually begins with the server side of automation systems, which provides the processing functionalities. To avoid the forced change on the client-side, the adapter pattern can be used. It addresses incompatibility in interface, type, access logic, and processing logic. For example, if an automation system provides a COM automation object model for accessing information in the automation system, the adapter pattern can be used to solve incompatibility problems when the COM interfaces are converted to the Web Service interfaces. The same also applies for type. The adapter pattern does not add any new functionality; it just plays a role of conversion or transformation.

Another kind of incompatibility in exposing an automation system to its clients by means of Web Services is the possibly different access logic. We could consider COM automation object model as an example again. A COM automation object model is a way to expose the functionalities of an application to its environment so that clients can exploit the functionalities or that the application can be controlled from outside. A COM automation object model usually contains a set of classes, which implement a set of COM interfaces. It is object-oriented, that is, it can be used to navigate the whole object tree to access the information of a concrete object. Identification of objects is accomplished by object names. In contrast, Web Services use URLs to identify themselves, and do not automatically provide mechanisms for identifying an object. Therefore, two kinds of potential incompatibility may occur: identification of objects and navigation in the object tree. One way to work around the potential problems is to use one-to-one mapping and introduce an object designator for each function defined in Web Services. The one-to-one mapping means that each interface implemented in the automation object model is exposed as a Web Service identified through a URL.

Design for Performance

To address the performance challenge when using Web Services for an automation system, it is essential to design a proper mechanism for handling roundtrips and the amount of data transferred between the service requestor and the service provider. The following recommendations should be considered when designing a Web Service.

Caching

Caching is an effective mechanism for increasing performance. Web Services performance in an integrated automation system can be maximized by carefully studying the data characteristics and correctly using data caching. There are three major choices to use caches: near the service client or consumer (client-oriented), near the service provider (provider-oriented), or at strategic points in the network [7]. In this chapter, only client- and server-oriented caches are considered (Figure 100.5).

A client-oriented cache intercepts the requests from a client, and if it finds the requested objects in the cache it returns them to the client. The content to be cached fully depends on the client's needs. Data requested by the client can also be prefetched and stored in the cache if necessary. Typical client-oriented caching is the proxy caching, transparent caching, and so forth [7].

The provider-oriented cache is located on the server side. It is content-dependent, meaning that if many clients require the same data (or the same data are required repeatedly), these data can be put into the cache for sharing. Examples for the provider-oriented cache are the reverse proxy cache and the push cache [7]. The provider-oriented cache is useful if it is very time-consuming to make data available on the provider-side for transfer to the client. An example of such a case is large simulation programs, which require intensive computation and thus a long processing time. In this case, the provider-oriented cache can avoid unnecessary redundant computations and thus reduce the waiting time.

To properly cache data, it is necessary to take the following issues into account:

What kind of data can be cached? You should consider using caching in a Web Service when the service's requested information is primarily read only. An integrated automation system contains not only the real-time process data such as temperature, pressure, and flow rate but also other static data that includes information of equipment or components (name, size, location, etc.), data on the producer of the equipment, data on the features of the equipment, price information, and so on. For a client, it is not necessary to update the static data constantly. Such a kind of data can be cached on the client side. For the provider-oriented cache, it is necessary to identify which data can be shared by many clients, or which are required repeatedly. The key criterion is how long it will take to make data available for the client.

Data marking: Data marking is a mechanism for identifying data entities to be cached. Therefore, object designation plays an important role. This task becomes difficult if the data are organized in multiple structures. Caching can be used for a single property of a data entity, for the whole data entity, or for a structure, such as the functional structure, location structure, or maintenance structure, which contains a group of data entities for a certain purpose.

Time window: It should be possible to define a time window for caching. The time window defines a range in which data should not be obtained from the Web Services server. Only after the defined time window the system refreshes the data. The time window is similar to an aging mechanism. It is necessary to have an update or cleanup mechanism to force the refreshing of the cache. The time window should also allow to differently deal with slowly changing and quickly changing data, as typically both types coexist in automation systems. For example, the temperature of a boiler changes at a comparably slower rate than the pressure in response to the disturbance. Process data changing at a slower rate can have a relatively longer time window.

Data model for cache: As mentioned above, Web Services are usually stateless. Web Services represent a set of functions that can be invoked by the service requestor. Caches only deal with data. To associate functions represented by Web Services and the data to be accessed, the object identification is needed, meaning that each function should contain object identification to specify which objects are treated in the function. Different data models for caching can be used, for example, hierarchical data structures (trees) or hash tables. Today, many libraries are available for the implementation of such data models.

Granularity of data: To avoid unnecessary roundtrips in the client–server communication, it is important to find an optimal granularity of the data handled by the Web Services and transferred between the service requestor and the service provider. While fine-grained data entities lead to smaller sets of data, coarse-grained entities create relatively large data chunks. For Web Services using SOAP as a protocol, each invocation of a Web service needs to parse the XML document request and construct the XML response. Fine grained granularity may cause more roundtrips and more efforts for parsing and constructing the XML data. A tradeoff between the fine- and coarse-grained strategy will help to increase performance. For the coarse-grained strategy, the service provider may provide more information than the client needs for a particular request. However, if the client issues similar requests, caching the data may improve response time. This is especially true for clients making synchronous requests, since they must consider the time to construct the response in addition to the time to transfer the data.

For an automation system, the proper granularity can be found based on an analysis of data, regarding which data are logically related and typically used together. These data entities can be put together and transferred as a chunk. For example, if for monitoring and controlling some process variables such as temperature and pressure shall be displayed together, they can be grouped into a data chunk. For manufacturing execution systems, all information on the work order can build in a single data set.

Bundling Web Services

Another way to improve performance and to reduce the number of roundtrips is to bundle Web Service calls. As already mentioned, each web service invocation requires dealing with XML data, including constructing the XML request and response as well as XML document parsing. This may become critical if a great number of sequential Web Service calls are involved to fulfill a task. A potential solution to improve performance is to bundle the sequential Web service calls to create one single call. The Service Request Bundler on the client side and the Service Request Unpacker on the server side (Figure 100.5) can be used to implement this mechanism.

Serialization

Complex objects and data structures must be serialized to be transmitted, causing an overhead for serialization and deserialization and the volume of serialized data. There are two kinds of serialization:

XML Serialization: This is the default serialization model. When a Web Service returns a complex data structure, it is serialized to XML, producing a significant overhead in the size of the data being transmitted. XML, or SOAP serialization, is platform-independent. This kind of serialization can be used for static data such as asset information, plant structure, and so on.

Binary serialization: Objects are serialized into a sequence of bytes, and transmitted inside a SOAP envelope. This reduces the overhead introduced by XML serialization, but platform independence is lost. Also, it is necessary to introduce some code to manage the serialization and deserialization processes. This kind of serialization is especially suited for real-time data.

Object Designator

The object designator [9] identifies objects and their properties, which need to be processed by Web Services, both on the service requestor and the service provider side. As mentioned earlier, all functions in a Web service should have an object designator as one of their parameters. There are two methods to identify an object in an information system: direct and indirect.

The direct identification method uses a globally unique ID (GUID) to reference an object. The prerequisite is that all information objects or entities are assigned such GUID when they are created. The information system also has to provide the structure to access objects by using GUIDs. Direct identification methods are a very easy way to identify objects, because a client can obtain an object by just supplying a GUID without any complex or complicated navigation. Another advantage is that the server running the information system can be switched to a backup system without affecting the current clients if the same GUIDs are used in both systems. A drawback, however, is the consumption of additional memory and hard disk capacity to manage the potentially large number of GUIDs.

The indirect identification method uses relationships such as aggregation, composition, etc. among objects to identify an object. The indirect identification method usually needs less memory and hard disk capacity, but the reference may be much more complicated.

The object designator can also be used to identify the position of an object in the structures. Obviously, the designator depends on the structure it is addressing, that is, it is structure-specific.

Client Addressability

In an automation system, client addressability regards

- how the service provider finds the suitable service requestors and
- how the service provider informs its service requestors of what has happened on the server side.

An alarm and event server is an example where client addressability is important. Two basic mechanisms are necessary to have client addressability in an automation system with Web Service support. The first mechanism allows for subscribing and unsubscribing events so that the service requestor can be notified about messages coming from the service provider. The second one regards cyclically querying (polling) the service provider to check if any events or messages have occurred. The basic components are the event service proxy on the client-side and the event service on the server-side (Figure 100.5). The event service proxy on the client-side deals with registering, polling, and managing event handlers for the service requestor. The event service on the provider-side is responsible for event queue management and functionalities like registering and managing event handlers. The Event Service Proxy and the Event Service in Figure 100.6 are the basis for the implementation of event-handling mechanisms.

100.7 Future

We have discussed various challenges and solution concepts to address interoperability — a key issue in integrated automation systems. Platform and programming language neutrality is the important feature of Web Services to improve interoperability among various applications for automation tasks such as simulation, data processing, presentation, and management. With increasing complexity and size of integrated automation systems, especially when more and more applications from business management, MES, and different phases of product lifecycle are involved, efficient system engineering and assembling may emerge as a new challenge. Data modeling, computation modeling, association of the computation elements (software applications, components, process modules) to data models [10], efficient composition of systems with existing applications and modules that are implemented in the form of Web Services are just a few examples for the new challenges.

The further development of Web Services technology — automatic Web Service discovery, automatic Web Service execution, and automatic Web Service composition and interoperation [11] — will help to address these issues.

Ontology as an explicit specification of conceptualization [12] will likely play a more important role in the development of Web Service and integration technology. The traditional software (applications, modules, components), especially component-based software, uses the interface description language (IDL) to describe functionalities. It provides a way to describe the semantics of applications or components on a very low level. The description is platform- and programming language-dependent and can only be used for certain platforms such as COM or CORBA. Web Services use WSDL that is based on XML and thus independent from platform and programming language. From that point of view, it is better than IDL, but it still cannot address the description of semantics on a high level such as relationships among Web Services, domain knowledge, concepts, and so forth. Web Ontology [13] is a natural next step of technology development to address this problem. It uses controlled vocabularies or terms to encode classes and subclasses of concepts and relations. It can be used as an additional semantic layer that may sit on top of the data model and computation model including software applications, modules, or components, as illustrated in Figure 100.7. In this way, data model and computation model may share the same ontologies, or the ontologies used for data and computation model can be mapped or transformed in a simple way. The essential issue for successful use of ontologies is efficient ontology engineering. It includes creation of unified and standards-based ontologies, ontology management, ontology mapping and transformation, ontology matching, and so forth. Obviously, ontology engineering, which is aimed at creation of unified and widely accepted ontologies, is not an easy work. It is a long-term process and needs cooperation from different related stakeholders. Creation of ontologies based

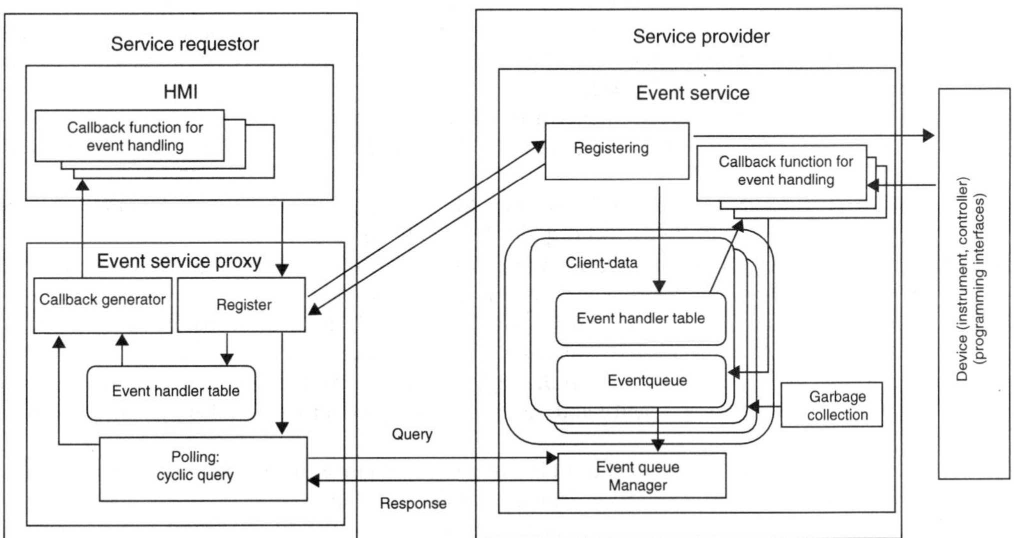

FIGURE 100.6 Event Service architecture.

Web ontology		
Model		Web service discovery
Data		Web service description
		Component and module
		Specification
		Implementation

FIGURE 100.7 Data and computation model with ontology layer.

on existing standards could be an effective way. For the integration in automation systems, different ontologies such as domain ontologies and computation ontologies are needed.

For engineering, the full use of other Web Services' features, namely the service broker and service discovery, aggregation, and choreography will facilitate searching for suitable components and aggregating applications to build an integrated automation system.

100.8　Conclusion

In this chapter, we have discussed the use of Web Services for implementing integrated automation systems. We believe that performance, client compatibility, client addressability, object designation, structure navigation, and security are very important for integrated automation systems and pose major challenges. The solution concepts based on Web Services have been presented, such as the structure cursor, the client- and server-oriented cache, Web Services bundling, the object designator, and the event service. More efficient system engineering and assembling will definitely benefit from the further development of Web Services technology such as automatic discovery, execution, composition, and interoperation.

Reference

[1] Ragaller, K., An Inside Look at Industrial IT Commitment, ABB Technology Day, 14 November, 2001.

[2] Krantz, L., Industrial IT — The next way of thinking, *ABB Review* 1, pp. 4–10, 2000.

[3] Bratthall, L.G., R. van der Geest, H. Hofmann, E. Jellum, Z. Korendo, R. Martinez, M. Orkisz, C. Zeidler, and J.S. Andersson, Integrating Hundred's of Products through One Architecture — The Industrial IT Architecture, ICSE 2002.

[4] Web Services Architecture Requirements, http://www.w3.org/TR/wsa-reqs#id2604831

[5] Thompson, M., Defining Web Services, TECH/CPS 1004, Butler Direct Limited by Addax Media Limited, December 2001.

[6] Booth, D., H. Haas, F. McCabe, E. Newcomer, M. Champion, C. Ferris, and D. Orchard, Web Services Architecture, http://www.w3.org/TR/2003/WD-ws-arch-20030808/

[7] Barish, G. and K. Obraczka, World Wide Web Caching: Trends and Techniques, *IEEE Communications Magazine*, Internet Technology Series, May 2000.

[8] Gamma, E., R. Helm, R. Johnson, and J. Vlissides, *Design Patterns Elements of Reusable Object-Oriented Software*, Addison-Wesley Publishing Company, Reading, MA, 1995

[9] Hu, Z., A Web Service Model for the Industrial Information System with Multi-Structures, The International Association of Science and Technology for Development, Tokyo, Japan, September 25–27, 2002

[10] Hu, Z., E. Kruse, and L. Draws, Intelligent binding in the engineering of automation systems using ontology and web services, *IEEE SMC Transactions Part C*, 33, pp. 403–412, August 2003.

[11] Mcilraith, S.A., T.C. Son, and H. Zeng, Semantic Web Services, *IEEE Intelligent Systems*, Vol. 16, Issue 2, 46–53. March/April 2001.

[12] Gruber, T.R., A translation approach to portable ontology specifications, *Knowledge Acquisition*, 5, 199–220, 1993.

[13] International Electrical Commission (IEC), IEC 1346-1, "Industrial Systems, Installations and Equipment and Industrial Products — Structuring Principles and Reference Designations, 1st ed., 1996.

101

Integration Between Production and Business Systems

Claus Vetter
ABB Switzerland Ltd.

Thomas Werner
ABB Switzerland Ltd.

101.1 Introduction

More efficient use of production equipment, best possible scheduling, and optimized production processes are the challenges in today's process and manufacturing industries. These trends become apparent for a wide range of industries such as electric utilities [1] and production facilities in batch and chemical operations [2,3].

With numerous software systems already in place in most production facilities, the challenges ahead lie in seamless integration of the IT landscape, with one of the focus points in manufacturing and process industries being the connectivity between plant floor execution and business systems. Effective and accurate exchange of information is the means to meet those challenges. Existing IT integration projects focused on delivering data from one system to the other, but usually the integration methodology has not concentrated on reusable solutions.

The goal is to shift efforts from point-to-point solutions, which come with a high coding effort when connecting single data sets toward reusable components, which form building blocks of an enterprise — plant floor integration architecture. The challenge is to leverage existing mainstream technologies such as enterprise application integration (EAI) [4], web services [5] or XML [6], and emerging industry standards (e.g., ISA 95 [7] or CIM [8]) and apply them in the context of the usage scenarios found in today's production environments.

This chapter presents the concepts of an integration study between production and business systems. A software architecture combines a set of functional components for implementing interfaces between manufacturing execution systems and business systems, containing all the elements necessary to develop, execute, and operate the integration in all likely scenarios, including real-time data exchange from shop floor to board room, near-real-time message and event oriented, and periodic data exchange that extracts and imports data in bulk.

The benefits of the integration concept include:

- Reduced overall effort by developing a single architecture for all connectivity scenarios.
- Faster time-to-market for new functionality, since the focus lies on developing this add-on components instead of developing "infrastructure."
- Reduced interface development effort by utilizing tools and templates to jumpstart development.
- Reduced maintenance costs since each interface is implemented using the same technology concepts.

The study presents thoughts on features to be considered in the architecture and technologies that should be used to implement the functional building blocks. The concepts presented are based upon the experiences of the authors and research of recently available technologies. Key elements of the concept are confirmed in proof-of-concept and prototyping activities.

Objectives and Scope

The objectives of the document are to provide the reader with an understanding of the following:

- Initial functional and technical requirements driving the development of the architecture.
- Technology guiding principles used as the basis for making decisions on the selected technology.
- Available integration technologies and approaches.
- High-level design for the integration components.

This chapter is technical in nature and is intended to be read by (technical) project managers and software architects. It assumes basic knowledge regarding functionality performed by the key applications in the integration context — Manufacturing Execution Systems (MES) and Enterprise Resource Planning (ERP).

Document Organization

The chapter is organized as follows: Section 101.2 presents example functional scenarios that provide fundamental requirements for the overall software architecture. Section 101.3 presents integration types, features, and the technology guiding principles that are required to fulfill the base requirements outlined. Integration types define the different technical integration scenarios (view, functional, and data). Required features address the characteristics of architectures common to each integration type. The guiding principles provide high-level criteria that are used in selecting between the various technologies available in constructing the architectures. Additionally, the main components of the presented prototypes — an ABB production system based on the Industrial^{IT} framework and SAP's R/3 system — are outlined with a focus on its main interfacing points and structuring concepts relevant for integration. Sections 101.4–101.6 outline the available technologies being considered for the integration architectures. Each of them addresses an exemplary use case, the architectural concept, and the prototype design of a specific integration type as defined in Section 101.3.

101.2 Integration Scenarios in a Production Environment

Functional Interaction

In a production environment, three generic scenarios (Figure 101.1) can be distinguished:

- *Make* (*intraenterprise integration*): This scenario includes data exchange between business and execution systems within one plant (or company) in order to automate data synchronization, and to optimize production and maintenance schedules.
- *Make/buy* (*business-to-business*): Information from the supplier side is included for optimization reasons, as, for example, scheduling the production triggers placement of orders to a supplier, or production scheduling depending on the availability of raw material from a supplier.
- *Make/sell* (*business-to-consumer*): Production information is available toward customers, for example, availability or production capacity information. Production can be optimized toward certain customers (urgency), or customers can track the production progress of the placed orders.

In the following sections, the focus lies on the intraenterprise integration between MES and business systems. It has broad applicability across several industries such as chemical, food, primary pharmaceutical, and to some extent discrete manufacturing. It is also of importance in the context of process industries and utilities, where the trend of interaction between production and back-office becomes more and more apparent.

Interenterprise Integration Scenarios

Table 101.1 summarizes the transactional "data" interfaces between production planning and materials management modules of an ERP system and the MES systems that typically result from the above-outlined

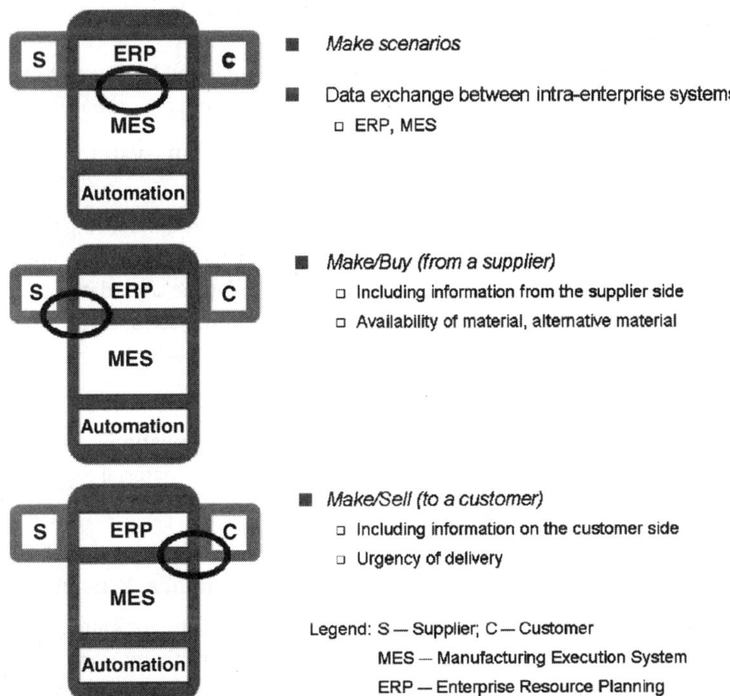

FIGURE 101.1 Integration usage: three main scenarios.

TABLE 101.1 Sample Functional Integration Requirements for the "Make" Process

From	To	Description
Production planning	MES	Released production orders — Production orders created in production planning or scheduling are released and then sent to MES.
Production planning	MES	Production order changes (dates, quantity, released bill-of-materials [BOMs], recipes and production parameters, routings, and cancellations) — Production order changes are sent from ERP to MES.
Production planning	MES	Master data (BOMs, recipes, routings) and changes are sent to MES.
MES	Production planning	Production order start/stop (typically included in confirmation transactions) — MES users input command: START/STOP and data are sent to ERP system.
MES	Production planning	Production confirmations (in-process and final confirmations) — MES sends production confirmations to ERP.
MES	Production planning	Production order and schedule changes (quantities, dates, cuts, adds) — MES input changes and data are sent to ERP.
MES	Production planning	Material and process deviations (changes to released BOM, recipe and routings) — MES users input command: CHANGES and data is sent to ERP.
MES	Production planning	Real capacity information — Data sent from MES to ERP for improving scheduling.
MES	Production planning	Execution events (batch history) — Data are sent from MES to ERP.
MES	Material management	Inventory transactions (moves, issues, receipts) — Data from MES is sent to ERP.
Plant maintenance	MES	Maintenance time is synchronized with production plan and spare parts.
Plant maintenance	MES	Maintenance schedule of a device in order to chose alternate production routings.
MES	Plant maintenance	Detection that an asset is performing poorly. Send a work order notification to plant maintenance.

application scenario of ERP and MES functionality, showing the many interfacing points for optimizing and automating data exchange.

Out of the above, the most generic patterns of the functional interaction between production and business systems can be extracted and are outlined in the following sections.

Download of Production-Relevant Information to an Execution System

This step typically involves scheduled production orders, which are released from the business system (production planning) and are transferred to a production execution system. During this process, a number of data mappings take place: the order recipe is mapped to a production recipe and recipe parameters are resolved (e.g., quantity, production line, equipment, and time and date). If unplanned, late changes occur, information must be communicated to the production system, as, for example, order size or order quality since such changes may impact the current production schedule.

Upload of Status Information from a Production System

Upload of status information from the production system includes deviations of the produced order from the actual production status, by, for example, order size, quality, material locations, and material usage. Furthermore, if the status information is continuously sent to the business system, the data can be used for detailed tracking of the production progress.

Data Exchange Between a Production System and a Maintenance Management System

Automating the process of creating work orders due to equipment malfunctions and performing rescheduling of production due to lines in service can be achieved by synchronizing events from maintenance and production planning modules. Maintenance planning can be taken into account in order to calculate actual and planned capacity of production lines. This allows timely shifting production to alternate lines if maintenance actions are necessary in parts of the production facilities. Users can access equipment maintenance reports to identify causes for degrading production quality and schedule an order on alternate production lines already on the planning level.

Data Exchange between a Production and Scheduling System

Using history information from production (e.g., planned vs. actual material used or time needed to perform a production recipe in planning systems) allows more accurate forecasts and optimized production schedules. A planning system allocates all equipment resources for a specific recipe during the execution of the order. If the production system additionally provides detailed information on the equipment usage (e.g., when and how long an equipment is needed), capacity forecasts and scheduling can be optimized.

Summarizing, the integration scenarios range from simple data exchange to complex data mapping and routing functionality. To each of these, one or several appropriate technical integration solutions can be mapped.

101.3 Technical Integration

Integration Options

Depending on the functional integration requirements of data exchange, synchronization, and user access, four technical integration options (Figure 101.2) can be distinguished. They differ with respect to the number of involved systems and therefore the level of complexity, which in turn has a direct effect on the customization effort and adaptations that have to be made. While less complex systems with fewer numbers of interfacing points can be productized quite easily, larger systems with a high number of inter-systems communications are usually tailored toward customer requirements.

View integration allows access to a target system, for example, ERP through its Graphical User Interface (GUI), which is embedded in the calling system. This can either be realized by linking the application GUI by simple command calls or through web-based navigation, which effectively depends on the display capabilities of the target system. In other words, for example, a user has access to transaction screens in an ERP system from his MES workplace and therefore can update order changes or material consumptions directly. View integration becomes more powerful if context-based information is shared between the source and target system, as, for example, a device or batch identifier is shared between the systems, which in turn eases the usage for users since it minimizes navigation effort to the requested information. View integration is a real-time user interaction with another system — accessing ERP transaction from

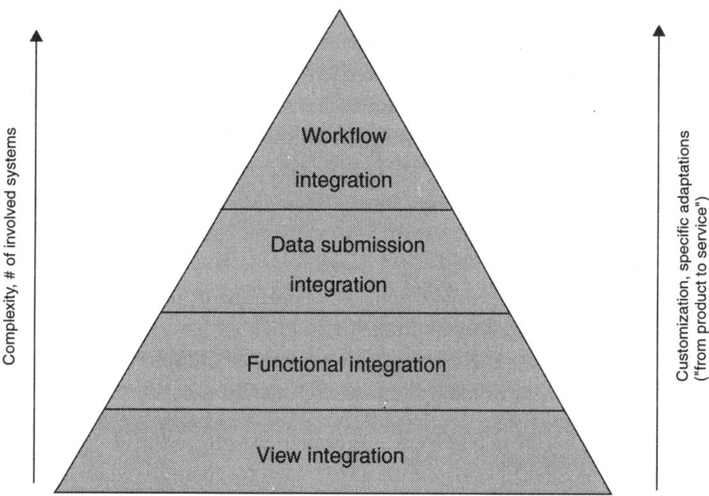

FIGURE 101.2   Technical integration options.

within MES environment — and by thus blocking other activities while a transaction is in process. Also, changes to data do not become visible until the transaction has completed.

The *functional integration* option realizes interaction between plant-floor and business systems through Application Programming Interfaces (APIs). Data are exchanged through programmable services at two sides, the production and the business system, by thus separating user input (GUI) and data exchange points. View access to the data can be fit to specific requirements of certain user groups such as operators or production engineers. Similar to the view integration, the functional integration also requires context information to be shared between the systems.

An Enterprise Application Integration (EAI) component provides the integration layer for the *data submission* scenario. The component provides data transformation and message-based routing as key functionalities, since both involved end-points (business and production systems) do not track the published information, such as releases of production orders (business system) or the status of the production itself (production system). Data submission requires APIs on all involved systems.

Data submission functionality is either triggered from *events,* or scheduled in the form of *bulk data transfer*. Since EAI components, in most cases, work through message queues, both the business and the production system must not be connected all the time — the information is stored in queues and delivered as soon as the message channel is available.

- *Event-based* functionality is primarily used to monitor for events on one system, such as a data change, that immediately initiates a transaction on the other system. A context (message identification) is required to decide how to handle the information received.
- *Bulk data transfer* is used for interfaces that are either time dependent or involve large volumes of data. It allows the scheduled extract of multiple records from one system an intermediate transformation and the scheduled import into the other system. The architecture is developed to address interfaces originating in either ERP or MES.

Workflow processing components manage their own data sources and computation states and update this information while processing data flows. A simple workflow (e.g., based on the content of the data) can create, modify, or delete order entries in the production or business system. Examples for advanced workflow processing are advanced scheduling and balancing algorithms or order management, which might not be available as functional modules in either business or production systems themselves.

All components — view and functional integration, as well as data submission — are needed for workflows as the functionality is a mix of user-driven and event-based business process steps.

In general, two concurrent concepts can be distinguished in terms of workflow definition, administration and execution.

- A *centralized* solution using standard EAI tools. They provide means to define workflows through an orchestration engine, which supports the users in graphically defining the steps involved in a business workflow, the specification of data exchange maps for the single steps of the workflow, and the integration of the components from, in our case, MES and ERP system.
- A *decentralized* solution where each participating component administers its own workflow part (function, engine, data translation) but does not know of the overall business scenario. The entry point of the function is responsible for orchestrating the different subfunctions, which are registered in a lookup service (e.g., UDDI) on a domain basis (e.g., each participating system registers a function supporting "outage management"). A component manager on both systems supports the translation from the service lookup to the individual services and functions on the systems that are included in the workflow. Each participating function will only know of its subworkflow and performs the individual services lookup for the following step.

However, both concepts still have to be tested and evaluated with respect to both their applicability in a production environment and their performance for the general characteristics such as reliability, transaction safety, or scalability, which are of utmost importance for a guaranteed execution of workflows. Therefore, they will not be covered in detail in this chapter.

Guiding Principles

The following principles shall be used to guide the development of the technical architecture and product selection decisions throughout the design process:

- utilize standards to ensure compatibility with current and future IT developments and encapsulate existing APIs of MES and ERP;
- design for large-scale, complex enterprise environments and provide an infrastructure for secure, reliable, and predictable execution;
- facilitate the rapid generation of new functional interfaces by leveraging "off-the-shelf" technology and tools from known vendors; and
- create an extensible integration architecture that can eventually be used to integrate MES with other applications in addition to ERP.

The following characteristics were taken into account for the design of the prototype architectures:

- *Reliability*: Guaranteed transaction processing, logging, and audit trails.
- *Scalability*: Support for multiple processor machines and server farms; distribution across multiple tiers.
- *Interoperability*: Support for multiple platforms and message transport protocols; support for web service technologies.
- *Adaptability*: Highly configurable allowing for fast adoptions; support the ability to adopt new client user interfaces, and modular to isolate changes within a tier.
- *Availability*: No single points of failure; support for clustering, load balancing, backup, and restore; and monitoring services.
- *Maintainability*: Tools that assist in interface development, usage of widely available technologies, and support for templates and programming standard guidelines.

Integration Approach and Typical Use Cases

The integration concept supports large-scale, high-volume integration scenarios typically encountered in enterprise environments.

Table 101.2 describes real-life examples in a business context of different types of integration that are usually found across a production environment. These include a subset of the user interface and "data"-oriented interfaces. Subsequent sections will refer back to this table.

In the following chapters, the main technical concepts for the integration scenarios are described covering in detail the *view, functional,* and *data submission (event-based and bulk data)* approach.

For each scenario, a typical use case is presented from both a manufacturing- and process industry-related scenario or from an energy production and delivery scenario of electric utilities. The main technical concept behind the integration is presented and a prototype integration that uses specific technologies is outlined.

Prototype Components

MES

Today's MES systems are of a monolithic architecture exposing a system-level interface only to exchange production-relevant information, such as product information (recipes, parameters) or production status information. Scheduling is most often provided from the ERP or advanced planning level; however, the MES application will often perform the detailed scheduling of operations that are highly dependent on the manufacturing process (e.g., creating a detailed schedule to consume a roll of paper to produce an end product, where the production scheduling optimization depends on the quality of the paper as it relates to the location on the roll) [9].

TABLE 101.2 Example Scenarios

Description	Real-Life Example	Illustrated Approach
Production orders and confirmations — Production orders from ERP are downloaded into MES. These orders include recipes stored in ERP.	Production orders created in ERP production planning or scheduling are released and then sent to MES. Recipes are copied at the time of a production order release and sent to MES. The MES creates line-specific versions of the recipes, prepares, and initiates production.	Data submission (event)
	Under control of the control system, a box is filled with a product. After packaging, one unit of production is recorded in the MES system. Once an hour, material production confirmations for a production order are summarized, by material, and sent to ERP as a single production confirmation.	Data submission (bulk)
	Alternatively, the single unit of production is sent immediately from the MES to ERP.	Data submission (event)
Detailed status and schedule changes — Dates, quantities, and cancellations are input by a scheduler into ERP or by a shop floor supervisor into MES.	A master scheduler receives a phone call from customer service with an urgent request to increase a released production order by 10%. After ensuring that the order has not started and that there are sufficient components, the scheduler increases the order by 10%. This change is sent to the MES for execution.	Data submission (event)
	Alternatively, at the beginning of a production order the supervisor discovers that a key component is 50% short and that additional components cannot be expedited into the shop. The supervisor navigates to the ERP production order screen and changes the quantity expected by 50% giving visibility to customer service and production scheduling.	View or functional (UI)
Materials — A shop floor supervisor may make a material supply inquiry or a material demand inquiry.	The material pick list in the MES system is calling for more material than is physically present. The shop floor supervisor navigates to the material supply screen in the ERP system to investigate alternate locations for the required materials.	View or functional (UI)
	Additionally, a shop floor supervisor must decide which of two different high-priority component orders to run. From the MES system, the supervisor navigates to the material demand function in the ERP system to view the "pegged" demand for that component.	View or functional (UI)

Throughout this chapter, the MES referenced to is ABB's batch production system, based on ABB's Industrial[IT] architecture. Production equipment and recipes can be configured and managed from the framework. Users can view progress of production orders or details on the execution.

The following paragraphs summarize the architectural building blocks as realized in ABB's Aspect Integrator Platform (AIP) being the central component of ABB Industrial IT architecture (Figure 101.3) AIP constitutes the glue between all components of the control system and the MES system.

Aspect Objects and Aspect Object Types. Concepts, actors, and entities that are relevant within the enterprise and plant context, for example, motors, robots, productions cells, valves, pumps, products, processes, customers, or locations are represented in the control and production system as aspect objects. In the following, we will refer to it shortly as objects.

Each object in the system is an instance of an object type. Object types, which are related by inheritance hierarchies, determine which aspects and aspect systems (see below) are associated with each object instance.

Instances of objects are organized hierarchically in so-called structures — means to define groupings from a user perspective, such as a location or functional structure. Structures also describe the dependencies between real objects in a certain navigation context. An object can exist in multiple structures (Figure 101.4), for example, a motor object in both a functional and location unit structure. Each structure represents one meaningful context and notion of relatedness between objects. For example, an

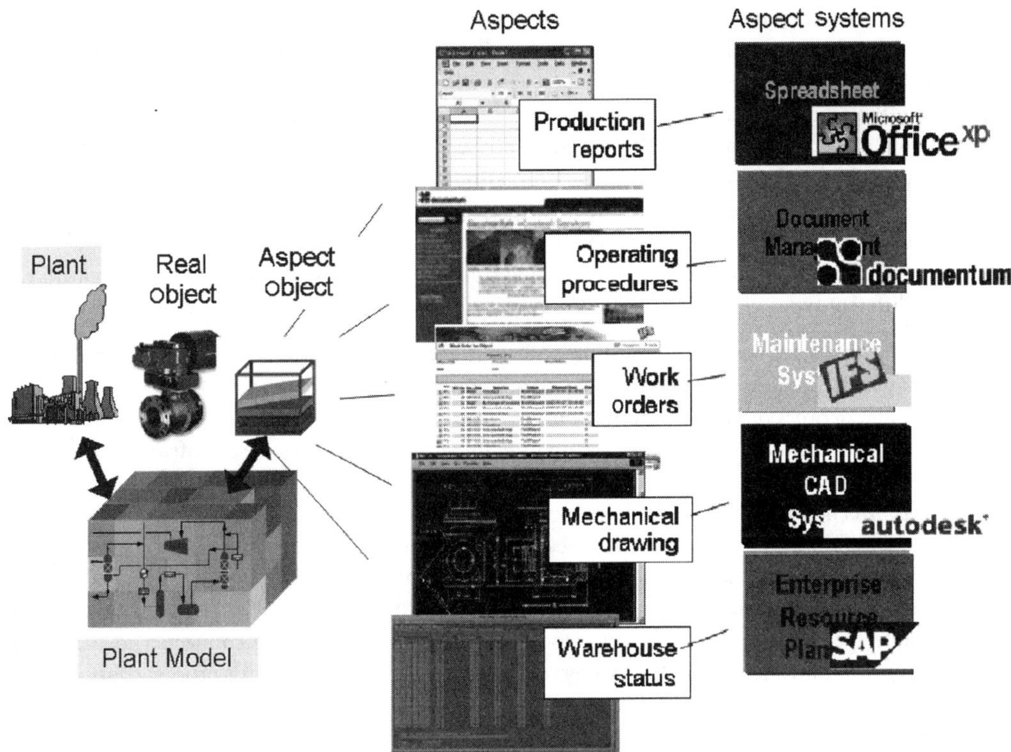

FIGURE 101.3 Aspect Systems and Aspect Objects.

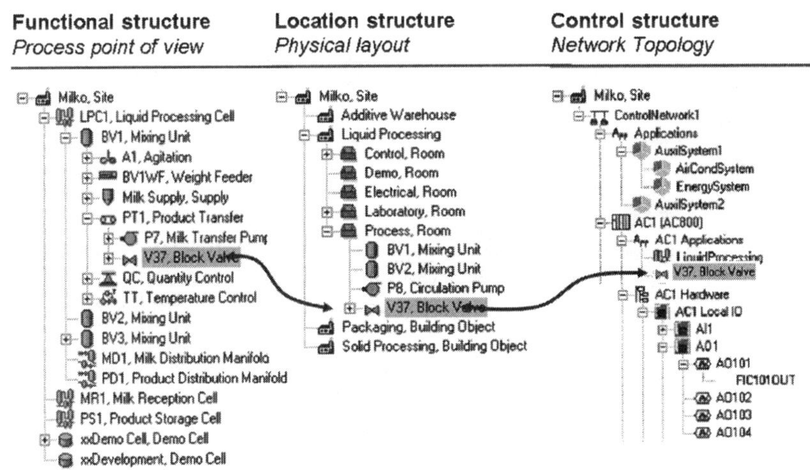

FIGURE 101.4 Structuring concept of Aspect Object.

asset is (a) a technical equipment, (b) a member of an organizational unit such as a production cell, (c) located at a certain place, and (d) participates in a production process.

Aspects and Aspect Systems. Each object carries a number of so-called aspects. An aspect is a component encapsulating a subset of data (attributes) and corresponding methods, associated with the object and relating to a common context or purpose (e.g., all maintenance attributes of the equipment objects of a production cell object are part of the consolidated maintenance data aspect of the, e.g., pump object).

Aspects also serve as ports to external data sources. Data are exchanged via protocols like OPC, HTTP, or specific connectors to various ERP systems. Aspect systems are the information access applications with or without user interface, which are used to view, edit, maintain, store, and process the information contained in the aspects of an object.

Within AIP, aspect systems can also transparently access any other aspects that allow navigating between aspects in the same context. From the process-related information displayed as process graphics, the user can thus navigate seamlessly to the maintenance related information (e.g., a work order) without losing the context of the object (e.g., object name, object identifier).

Its Aspect Object architecture assumes a system of computers and devices that communicate with each other over different types of communication networks. This layered network (Figure 101.5) consists of

- The *Intranet,* user for communication with thin clients such as mobile devices or browser-only work panels, but also access to third-party nonproduction systems, such as ERP.
- The *plant network,* used for communication between *servers,* and between servers and *workplaces.* Servers run software that provides system functionality, and workplaces run software that provides various forms of user interaction, such as process graphics, alarming, or trending.
- The *control network,* a Local Area Network (LAN) that is optimized for high performance and reliable communication with predictable response times in real time. It is used to connect *controllers* to the servers. Controllers are nodes that run control software.
- *Fieldbuses,* used to interconnect field devices, such as I/O modules, smart sensors and actuators, variable speed drives, or small single loop devices. These devices are connected to the system, either via a controller or directly to a server through, for example, OLE for process control (OPC).

ERP System

SAP's Production Planning and Plant Maintenance modules act as the respective ERP components in the outlined prototype scenarios.

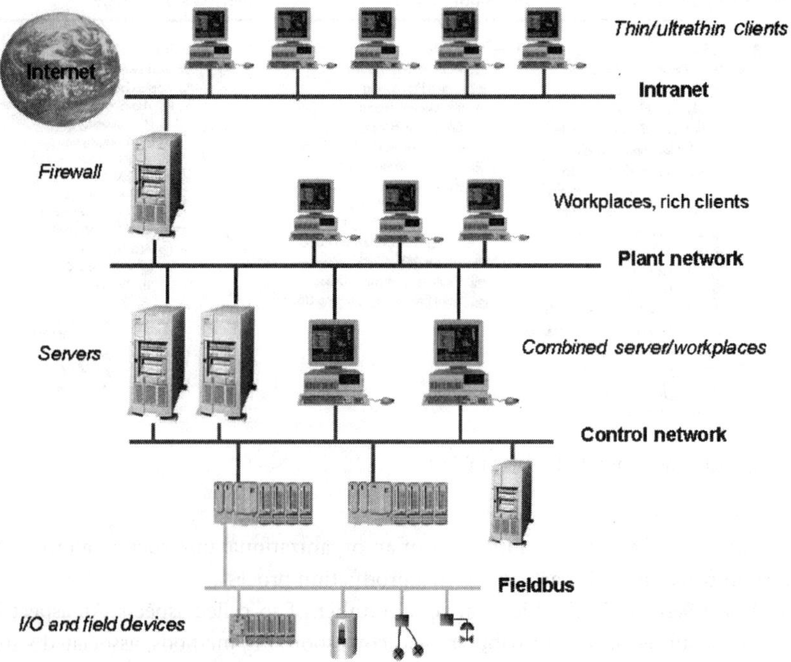

FIGURE 101.5   Industrial IT network architecture.

Calling specific functionality in SAP is available through different mechanisms. In this scope, the Business Application Programming Interfaces (BAPI) and Intermediate Documents (IDoc) mechanisms are covered. While the first one is a functional interface, IDocs utilize message queues.

An object-oriented approach for access to SAP systems has been introduced by SAP providing business processes and their relevant data as business objects. External applications can access the SAP business objects via standard interfaces called BAPIs. BAPIs offer an abstract, object-oriented view and keep implementation details well concealed. As a consequence, BAPIs represent the building blocks for the construction of interacting components, with the outside world being accessible through a variety of technology connectors (e.g., JAVA, COM, .NET).

IDocs are used in an SAP context to exchange messages with other SAP systems or third-party systems. Messages are sent asynchronously, possibly even as batches. A message carries data, which typically are split into multiple semantically meaningful segments. These segments may be structured hierarchically. Together with the data, a message exchange is accompanied by a control record, which specifies the source and routing of the message, and a status record, which tracks the life cycle of the message.

The IDoc and partner definition process consists of the following sequence of activities:

- IDoc definition (segment definition and message type definition).
- Linkage of IDoc to message type.
- Definition of ports and RFC destination.
- Definition of partner profile.
- Linkage of IDoc and message type to application object.

Figure 101.6 graphically describes an inbound IDoc message exchange. Typically, two partners (two connected software systems) exchange a message, which in this case is a request for a list of active work orders. The message is called GetActiveWorkOrders. Both the external partner, BIZTALKCH, and the message type, GetActiveWorkOrders, have to be registered in the SAP system and configured with appropriate settings. As many partners as needed may be configured in an SAP system.

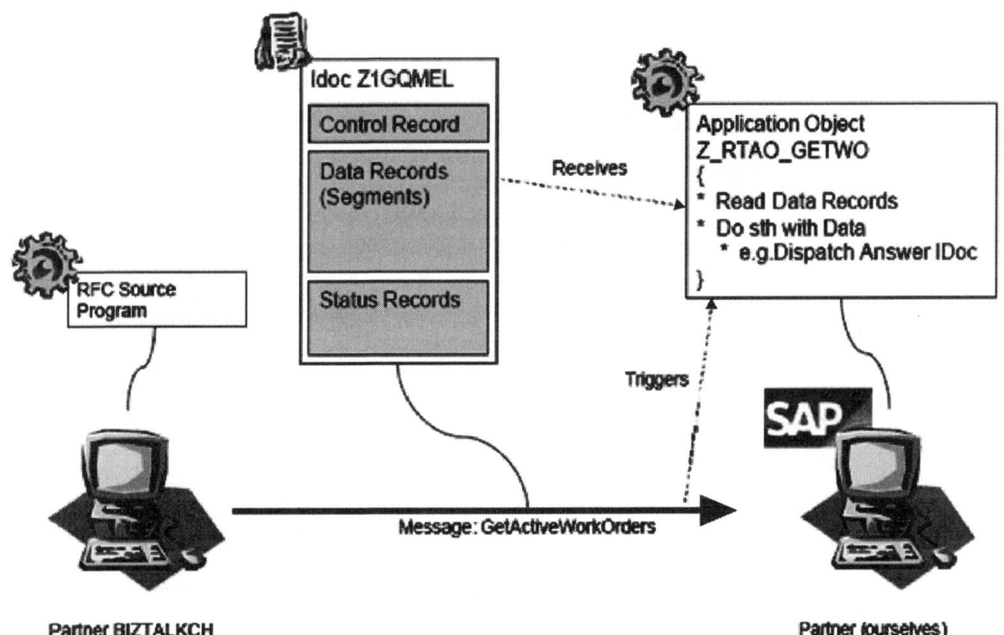

FIGURE 101.6   IDoc exchange between business partners.

101.4   View Integration

The view integration architecture is intended to provide a solution to user interface access to all business system functionality from the MES user environment. The architecture allows a production system user direct access to all user-relevant ERP transactions with context information and allows them to view, update, create, or delete data. For example, a user would be able to go directly to the list of material associated with a specific production order.

Use Case

The responsible engineer of a paint shop production line notices that the output quality of a production cell is poor; in fact, looking at color quality trends, he notices that the numbers of repainting have increased steadily over the last 24 h. With integrated ERP-CMMS (Computerized Maintenance Management System) connectivity into MES workplaces, he can access the maintenance history for the paint shop cell and its components in order to identify possible causes for the poor quality. Selecting the paint shop cell on the workplace screen, he can invoke "show maintenance history" as an action.

Last planned maintenance on the cell has been done 3 weeks ago, and the report summary indicates that a "normal" maintenance service has been carried out. Also, the performance history of the paint shop cell components does not reveal any unusual behavior.

Navigating on a GUI, the workplace, the user can select a piece of equipment and invoke through interaction with the UI one of the available CMMS functions for accessing maintenance information, which is stored in the CMMS system for the selected equipment. The CMMS data are displayed through the proprietary CMMS user interface (Figure 101.7).

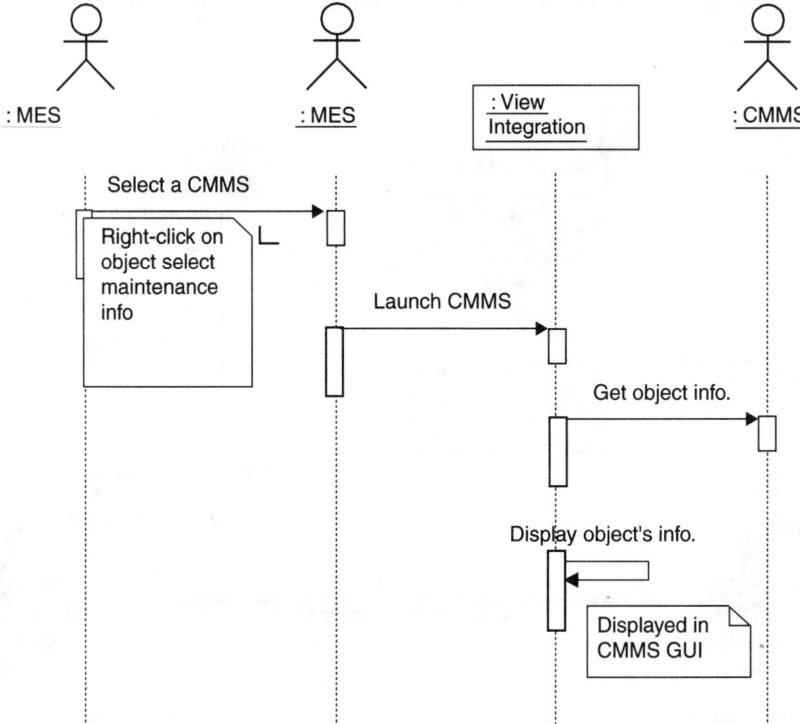

FIGURE 101.7   Accessing maintenance information.

Technical Concept

The user is given the possibility to invoke an ERP transaction screen from an MES workstation. While this is a relatively easy solution to implement, it does not provide navigation from the ERP screen to other production system applications nor does it provide for data translation between the systems.

The architecture utilizes the "wrapper" capabilities of available software components such as Internet Explorer or other to encapsulate the client's transactions (ERP GUI). Through these containers, a variety of target components in the form of ActiveX controls, HTML pages, or Java applets can be accessed. To alleviate the manual configuration required, "smart wrappers" can be developed for specific ERP transactions, which automatically collect relevant context data associated with the entity when the function is initiated and embed it into the ERP GUI invocation. The equipment identifier and ERP transaction number are examples of context data that are passed. The concept of embedded context information allows navigating directly to the required transaction screens and avoiding that the user has to step through a number of menus. The architecture also provides a single sign-on capability to enable access to ERP transactions without having to provide credentials for each transaction. The primary components and message flow required for this architecture are illustrated in Figure 101.8.

Prototype Realization

The prototype builds upon the concept of representing a real-world object through the AIP Aspect Object container. By attaching relevant context information, such as an equipment identifier from the different systems (MES, ERP) as aspects, this concept allows to map and share this information when invoking the view functionality.

The prototype architecture with specific technologies is illustrated in Figure 101.9. The focus of the implementation builds on the functionality of the SAP Internet Transaction Server (ITS), which allows accessing any SAP transaction through a URL as an HTML page. Therefore, a distribution of client software to each MES workstation is not required, which in turn increases the likelihood of supporting direct access to specific transactions. Simple HTML wrappers are used to provide application access to SAP transactions with static or hard-coded context data, whereas the "smart" wrappers are able to collect the required context data to navigate to a specific instance of a transaction.

The user is able to enter the original SAP screen and to navigate in the transaction context with all SAP functionality being available (Figure 101.10). The main disadvantage of this concept is that the screens typically are not customized toward their specific needs by, for example, limiting the information presented, but present the ERP transaction screen that was usually built on requirements from other user groups.

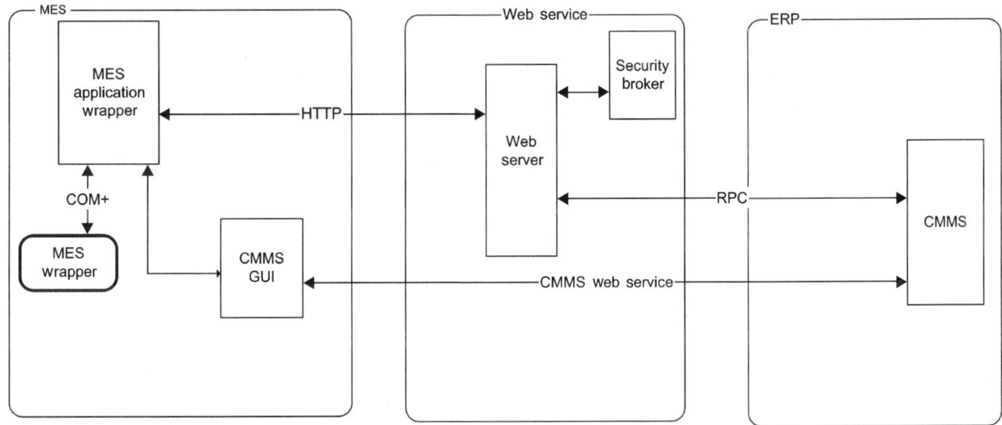

FIGURE 101.8 View integration concept.

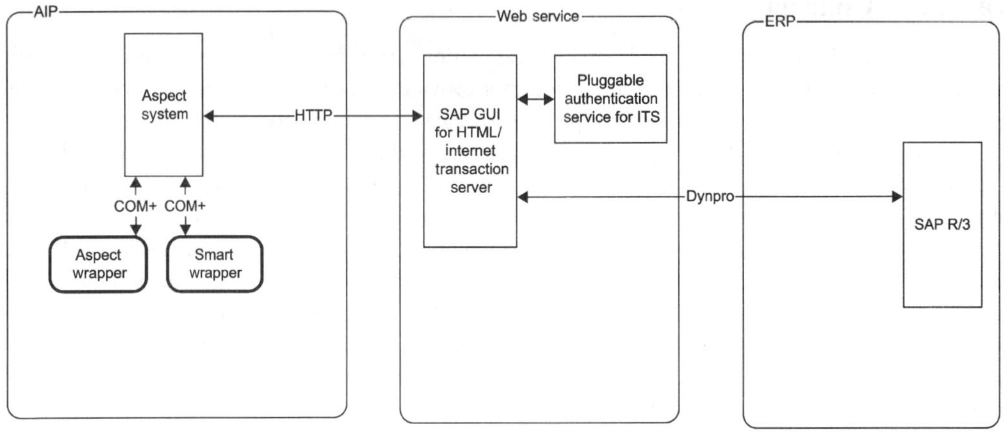

FIGURE 101.9   Prototype: view integration.

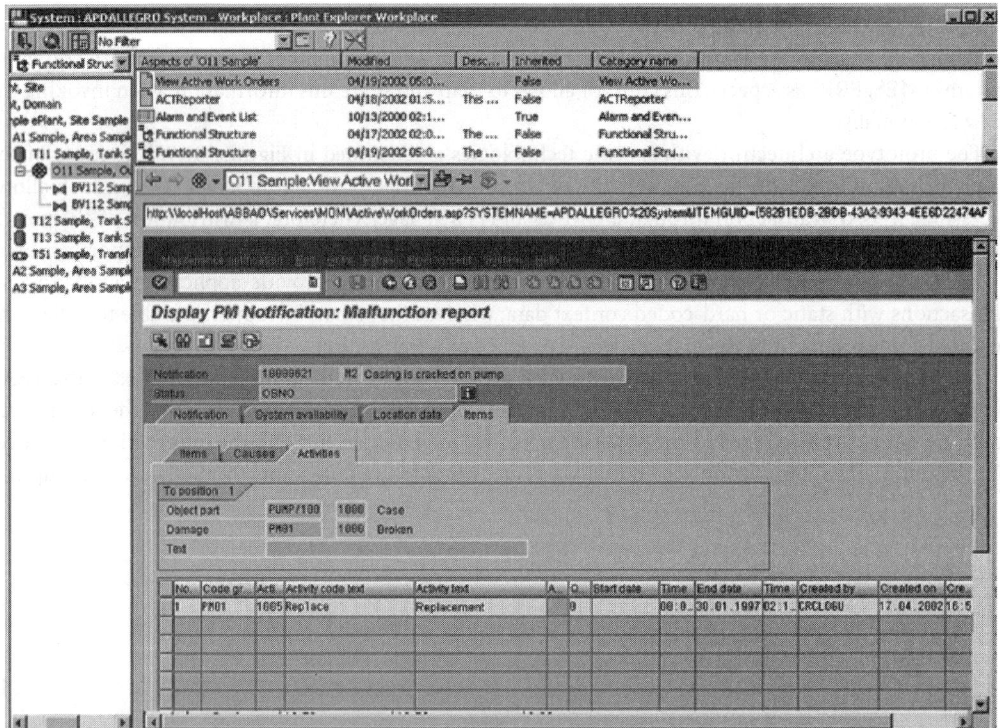

FIGURE 101.10   View integration — example screenshot.

101.5   Functional Integration

This architecture provides interface integration from MES to specific transactions in ERP and vice versa through APIs. This is enabled by defining a set of interfaces exposed toward either MES or ERP. Data exchange through interfaces and graphical representation of the data is separated, allowing developing custom GUI depending on user requirements while reducing the interface development efforts. Through definition of domain interfaces, for example, for maintenance or production, only the adapters to the various ERP systems have to be developed, while the exposure of interfaces toward the integration system remains consistent. Thus, adapting to other systems is eased, since only the adapter needs to be exchanged.

Use Case

In the production scenario from the paint shop production line, the user still is not satisfied with the quality of the cell. The next planned maintenance shutdown is scheduled in 1 week. However, since the engineer could not identify the root cause of the quality degradation, he decides to issue a maintenance request for the paint shop cell with severity "high."

Creating a maintenance request with integrated CMMS connectivity is simply done by selecting the paint shop cell on his workplace screen, and invoking "maintenance request" as action. A data screen is shown to the engineer, which already has data entered specific to the selected cell, such as cell identifier, date and time, name of the user, etc. The engineer selects priority "high," and attaches links to the quality trend diagrams. As soon as the work order is submitted, a new notification is generated within the CMMS system and the corresponding data sets are stored. Now, the maintenance planning department can analyze the attached information and decide if an unplanned shutdown is necessary (Figure 101.11).

In addition to retaining a common context as already described in the previous section on view integration, the data semantics between the systems have to be defined, ideally through a common application data model that abstracts specific application functions calls into domain APIs, such as managing maintenance requests or scheduling orders. Only such domain interfaces guarantee interoperability between systems from various vendors.

Technical Concept

Functional integration offers the ability to separate data representation and the communication between involved systems through defined APIs. In this architecture, the MES initiates a call to a web service that

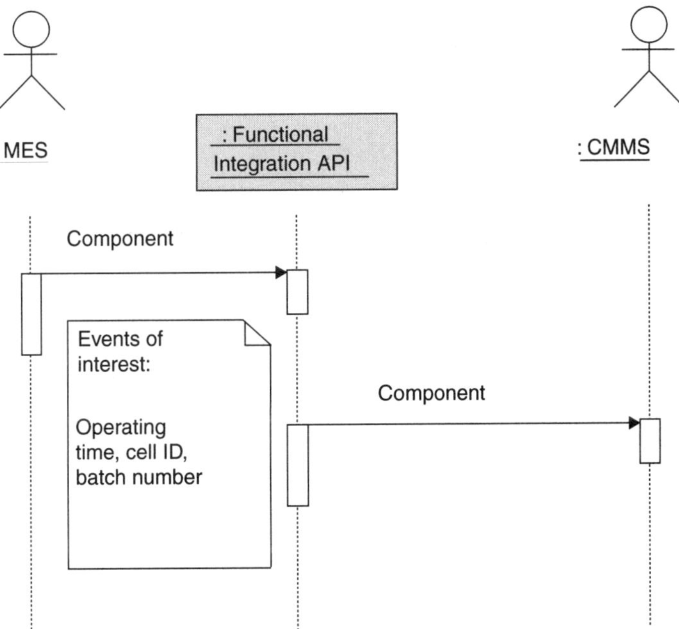

FIGURE 101.11   Transfer of quality information from MES to CMMS.

exposes domain interfaces (e.g., for maintenance) and in turn uses specific ERP APIs to implement the functionality. The web service brokers access either directly to ERP through an ERP proxy component or through an EAI component and works across firewalls. The latter provides functionality such as context-based data mapping and translation. User interface components use the web services to request or post data; the transaction itself is executed with the call to a specific method on the web service itself.

The primary components and message flow required for this architecture are illustrated in Figure 101.12.

As with the view integration, this architecture could be utilized for integration points as described in Table 101.2.

Prototype Realization

The prototype architecture with specific technologies is illustrated in Figure 101.13. The user interface component gathers the required information necessary to request or process information in SAP and package it into a Simple Object Access Protocol (SOAP) packet. The SOAP packet is then submitted via HTTP to a web service, where the information from within the SOAP message is used to identify the

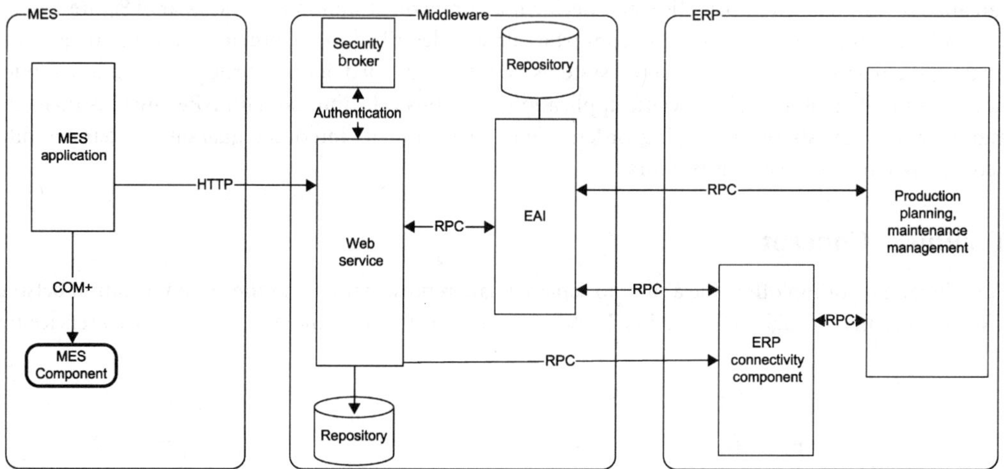

FIGURE 101.12 Functional integration concept.

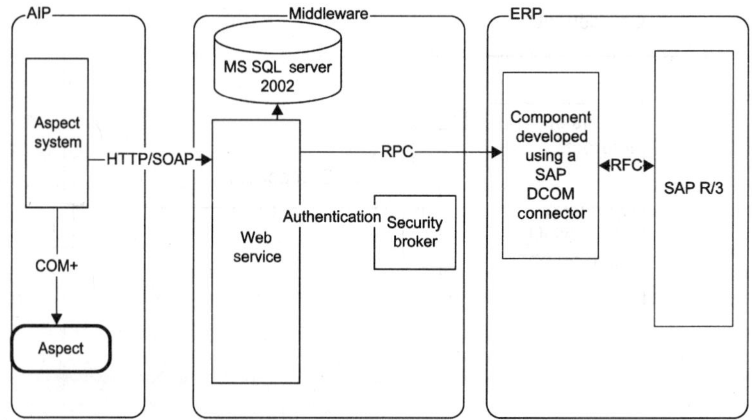

FIGURE 101.13 Prototype: functional integration.

required request for SAP. As a next step, the web service queries the repository for the necessary information needed to call the SAP Microsoft Distributed Component Model (DCOM) generated proxy, instantiate the proxy, and pass the necessary information gathered from the SOAP message.

In the functional integration architecture, GUIs are typically customized (Figure 101.14) and only context-relevant information is shown in order to perform a specific task. Changing requirements result in coding efforts to adapt data mapping and API calls. Information from the ERP on the selected object is retrieved and displayed to the user through a customized screen that is tailored to his needs and omitting information which is more specific for, for example, maintenance planning. For example, a user would be able to go directly to the fault notification site associated with a specific piece of plant equipment. However, the interface definitions and the ERP specific code do not change.

101.6 Data Submission

Data submission requires functional interfaces as described in the preceding section on both business and plant-floor systems. An EAI component provides the integration layer for the data submission scenario. This component provides data transformation and message-based routing as key functionalities, since both involved end-points (business and production systems) do not track the published information, such as releases of production orders (business system) or the status of the production itself (production system). The EAI component uses functions, which query or listen for information from one system and convert data from the format delivered from the source to the format required by the target system. Since EAI components, in most cases, work through message queues, both the business and the production systems must not be connected all the time — the information is stored in queues and delivered as soon as the message channel is available. A context (message identification) is required to decide how to handle the information received through the queues.

Event-Based Data Submission

Use Case

As a precondition to the described scenario, a customer order is received by a dispatcher, who enters the order into the planning system as specified by the customer: product type, quantity, and delivery date. The manufacturing plant is manually chosen by the dispatcher. After the order has been entered into the

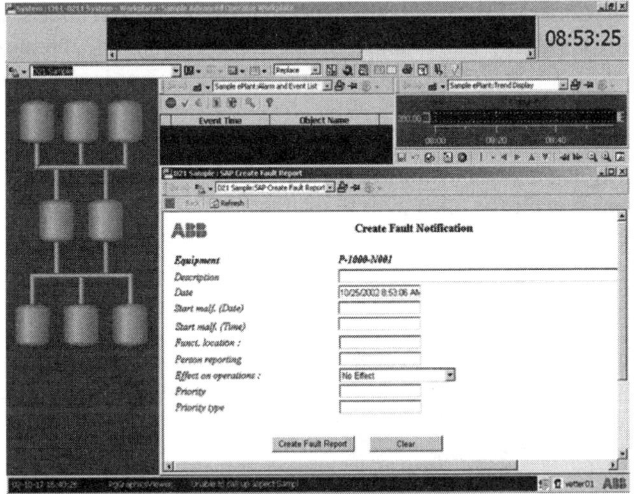

FIGURE 101.14 Functional integration — example screenshot.

planning system, verification is carried out against available material in stock. The order can be released to the production system, and the relevant data are transferred.

As soon as the order is completed, status information (e.g., material used) is transferred back to the production planning system, which triggers an update of the material and inventory stocks (planned vs. actual). The dispatcher can look up the inventories and check if the order has been executed against the given customer order. As soon as the information is updated in the inventory, the customer can be notified and the delivery can be initiated.

Technical Concept

The presented architecture provides a loosely coupled integration between the MES and a production planning system. This will be beneficial in the cases where the transaction has the potential to "block" the user's workstation while they wait for a response from ERP. Instead, a user can make the asynchronous request and perform other tasks while waiting for a message to return indicating that the transaction has been processed. Requests are submitted automatically through system-generated events originating in either ERP or production system.

The implementation involves an EAI component, which interacts both with the ERP system, as well as with the production system. It is the central component of the integration architecture. Many of the other integration architecture components are determined by the EAI application selected. Utilizing an off-the-shelf EAI application greatly reduces the effort involved in developing and maintaining the integration architecture between production systems and ERP. EAI applications typically provide the following functionality:

- *Process flow integration and management*: GUI tools for process integration and management, workflow, and state management across applications and enterprise boundaries.
- *Development tools*: Including configuration management, source control, debugging tools, and general coding environment.
- *Technology architecture*: Reliability, scalability, availability, adaptability, as well as operations support.
- *Transformation and formatting*: Transformation, translation, mapping, and formatting for integration purposes (i.e., to reconcile the differences between data from multiple systems and data sources).
- *Business-to-business capabilities*: Integration with trading partners, partner management, and Internet standards support — XML, HTTP, SMTP, FTP, SSL.

Utilizing an EAI application to develop interfaces helps to reduce the long-term maintenance effort by providing a central repository for data mapping between data sources and targets. The repository allows developers to reuse mappings and translations consistently across multiple interfaces. EAI also provides customers with a platform to integrate the MES with other applications in addition to production planning.

Interaction with the production system is realized by two custom services (Figure 101.15), which hook up to the proprietary APIs provided by the production system. One service listens to production orders from the EAI component, while the other polls the status interface of the production system.

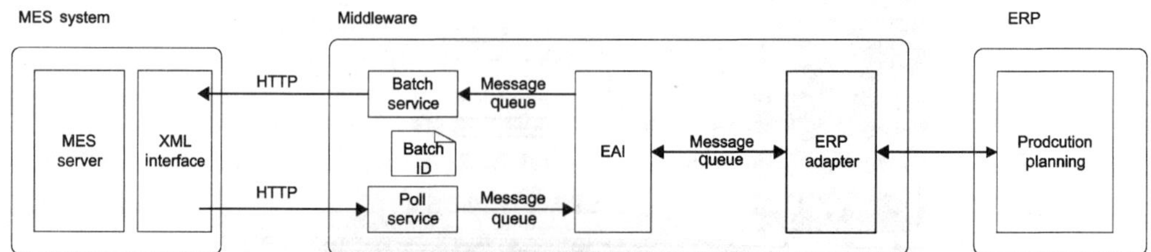

FIGURE 101.15   Event-based data submission concept.

Production orders are sent from ERP to the EAI component as ERP messages and are placed in an EAI inbound message queue. The information contained in the message is read and transformed according to the defined mapping and placed in an outbound message queue. The custom service listens to outbound messages, reads the message, and calls the appropriate CreateOrder function on the production system and passes the corresponding parameters. At the same time, a log file is created containing the batch identifier number that allows the second service to correlate the status to active orders.

The second service continuously polls the status interface of the production system and checks the status of the order number that is logged. As soon as the production status changes from "running" to "completed," the service reads the status information (actual material used) from the production system, composes a message, and places the message to an EAI inbound message queue. Again, the message content is read and mapped according to the PollStatus mapping definition, and the resulting message is placed in an outbound message queue. Finally, the EAI architecture for ERP reads the message, composes an ERP message, and routes it into the ERP system.

Prototype Realization

Microsoft BizTalk Server 2002 has been chosen as an EAI component, providing the server, tools, and plug-ins needed to integrate and automate the business between the production system and ERP. A key benefit of BizTalk Server is its ability to integrate XML web services and to supply a central repository for mappings and transformations, which are stored natively in XML. BizTalk Server comes with Software Development Kits (SDKs) available for transports, document types, and application architectures. Custom BizTalk mapping functions (functoids) can be developed and reused for multiple interfaces to accommodate the specific transformations of batch data to ERP data and vice versa.

Development is performed using wizard-based design tools such as BizTalk Mapper, Orchestration Designer, and Message Manager. If needed, C# and scripting languages (as, e.g., VBScript) supporting COM or the .NET framework can be used as well.

Biztalk supports Remote Function Call (RFC), BAPI, and iDOC integration with SAP through plug-ins. There are a number of different connectors available both from Microsoft and independent system vendors that provide ERP-specific integration capabilities for BizTalk. The chosen Microsoft connector provides the following:

- retrieves iDoc structure, and generates XML schemas for iDoc automatically;
- defines routes for documents within BizTalk Server environment;
- guarantees successful delivery of an iDoc both into and out of ERP; and
- supports both BizTalk Orchestration Services using a COM component and the BizTalk Messaging Manager using a BizTalk Application Integration Component (AIC).

The architecture makes extensive use of the data mapping and transformation capabilities of EAI. Figure 101.16 depicts how BizTalk Server as one example of an EAI integration product can be utilized to define data mappings between batch and ERP, while highlighting BizTalk Server's Message Mapper tool.

Two mappings have been defined for the architecture: CreateOrder and PollStatus. The first one defines the relationship between ERP's message format and a custom message format in order to deliver the necessary information with corresponding parameters to schedule and initiate a production order. The second mapping contains status information definition for updating ERP's inventories.

The prototype architecture provides two means of submitting asynchronous requests. Requests can be submitted either through an MES user interface or automatically through system-generated events originating in either SAP or MES.

Custom UI control that collect the required input data from the user and submit or receive requests are developed through the web services features of the .NET framework. The user will receive a message indicating that the transaction has been processed as the response corresponding to the initial transaction.

An MES service monitor receives system generated events from the production system. When an event is received, the MES service formats the request into a SOAP packet and submits the packet into an

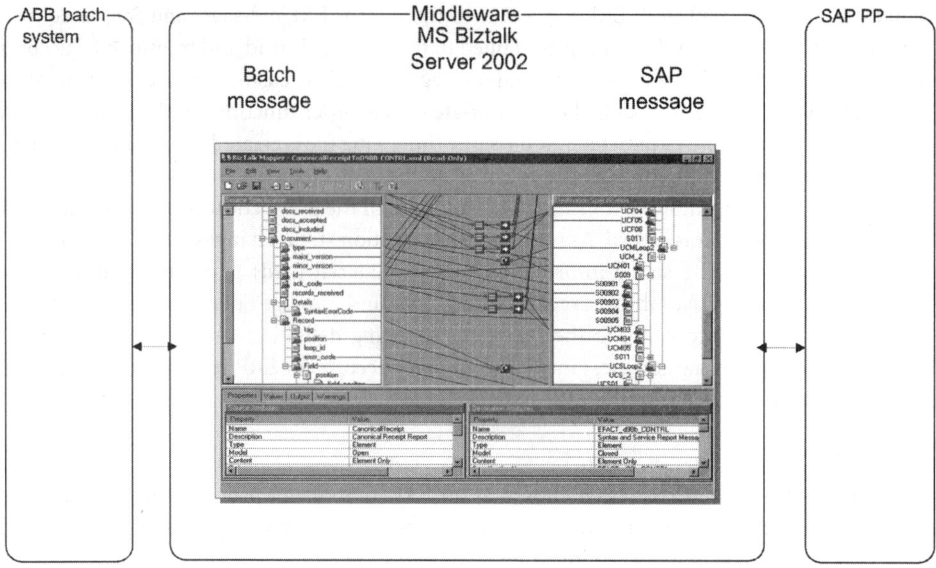

FIGURE 101.16   Message mapping.

outbound message queue. The architecture utilizes Microsoft Message Queue (MSMQ) as the transport mechanism between MES and BizTalk Server to enable HTTP as the transport protocol. The data transported using MSMQ are received by BizTalk Server, transforms, or maps the XML stream received from the MSMQ into an IDoc format to be processed by SAP. BizTalk Server then invokes BizTalk Adapter for SAP connector to submit the IDoc into SAP. BizTalk Adapter utilizes the DCOM Connector to initiate the receipt of the IDoc from SAP.

Requests from SAP to MES travel a similar route. SAP initiates the Remote Function Call (RFC) Server, COM4ABAP, through a transactional RFC (tRFC) call. COM4ABAP deposits the IDoc into an MSMQ. Once the IDoc is in MSMQ, BizTalk Server can initiate any transformations needed for the MES integration. The transformed message is dropped on an outgoing MSMQ to be delivered to an MES inbound MSMQ. The MES service reads the MSMQ, and based on the information supplied within the message, the appropriate production system API is called.

The prototype architecture with specific technologies is illustrated in Figure 101.17.

Data Submission using Bulk Data Transfer

Bulk data integration architecture provides the ability to extract data from either MES or ERP, apply a custom data transformation, and import it into the other system. The architecture is used for moving large volumes of data and for transactions that are time dependent. An example would be the periodic update of selected master data records from ERP to MES. Another example would be the scheduled summarizing of material production confirmations.

Use Case

Production schedules are released to the MES according to the incoming orders from the customer. While the MES performs a detailed scheduling of the production orders and distributes the schedules among the available production lines, the ERP system has to be notified of finished production in order to initiate delivery to the customer and fill up material stocks for remaining production. Therefore, finished production batches are collected at the MES and sent in bulk to the ERP. Additionally, every 24 h, the material consumption or the various production lines is recorded at the MES and sent to the ERP for updating stock lists and initiating further material orders (Figure 101.18).

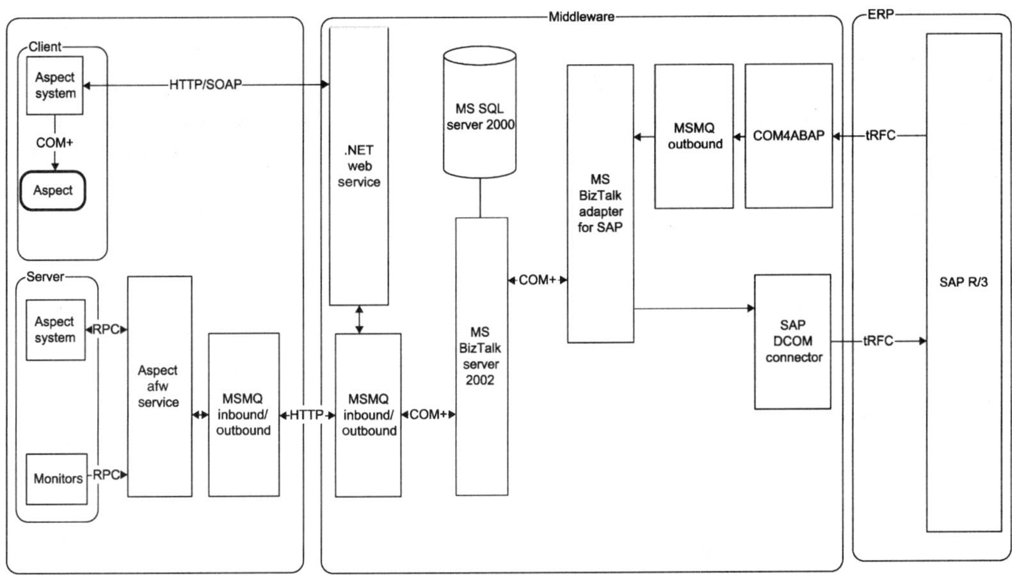

FIGURE 101.17 Prototype event-based data submission.

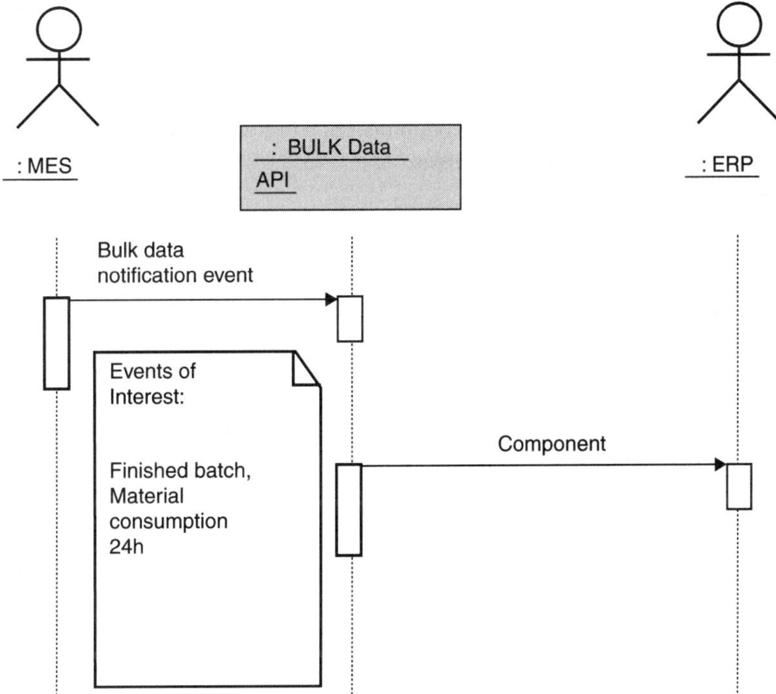

FIGURE 101.18 Transfer of quality information from MES to ERP.

Technical Concept

Bulk data integration architecture is driven by an external scheduling system that initiates jobs on each system through a scheduling agent.

A scheduling agent on the MES system triggers export and import jobs through a custom data manager, which determines the appropriate import and export routines to execute, log audit data, and initiate outbound data transfers. The scheduling agent on the EAI server initiates the data transformation routines and outbound data transfers. On the ERP system side, the scheduling agent initiates export and import jobs through a data manager that determines the appropriate import and export objects to execute, log audit data, and initiate outbound data transfers.

The primary components and message flow required for this architecture are illustrated in Figure 101.19.

The architecture relies on the availability of an enterprise job scheduling application that can initiate jobs across multiple systems and platforms. The scheduling agents are provided by the enterprise scheduler, required on the MES and ERP servers where extracts and imports are running. The agent is also required on the EAI server. The agents initiate jobs on the local servers and report status back to the enterprise scheduler.

Providing the correct containers for moving large data volumes is of equal importance as the scheduling architecture. The following protocols are being considered for transporting files from server to server.

- *File Transfer Protocol (FTP)* — FTP is a commonly used client–server protocol that allows a user on one computer to transfer files to and from another computer over a TCP/IP network. FTP is often used as a reliable data transfer method between dissimilar systems. Utilities to ensure the complete transfer of files could easily be developed or purchased at a nominal cost.
- *File System* — The files could be transferred by copying files between file system shares. This would most likely be supported in a Windows-based environment, but would require additional utilities if the ERP server was on a platform other than Windows.
- *HTTP* — Utilizing HTTP would ensure access through firewalls; however, HTTP has been shown to have poor performance in transferring larger files.
- *Message Queues* — MQ support the guaranteed delivery of messages. Using message queues for this architecture would require the development of a utility to break large files into smaller pieces depending on the maximum message size.

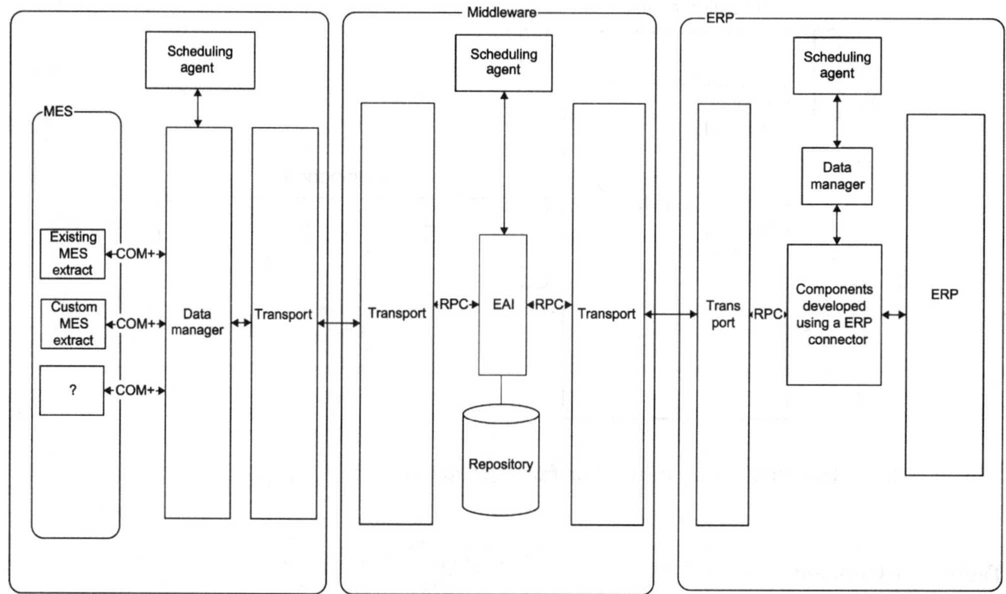

FIGURE 101.19 Bulk data submission concept.

Prototype Realization

The prototype architecture utilizes the customer's enterprise scheduling system and agents to control the execution of the bulk data integration interfaces. The scheduler provides agents for the MES, BizTalk Server, and the SAP system.

The scheduling agent on the MES calls the import and export programs directly. Upon successful completion of an export, the scheduling agent initiates an FTP transfer of the export file to BizTalk Server. The scheduling agent then calls a component on the BizTalk Server to audit the file, confirms a successful transfer, and initiates an XLANG schedule in BizTalk Server. XLANG is an extension of the Web Service Definition Language (WSDL) and provides both the model of an orchestration of services as well as collaboration contracts between orchestrations. The XLANG schedule performs any required transformations and mappings, adds new audit information to the file, and transfers the file to the SAP R/3 server. BizTalk Server utilizes a connector component for FTP to send the file to the SAP server. On the SAP side, the scheduling agent calls the file transfer audit program to confirm the successful file transfer, and then initiates the data import and log the results. A similar process is implemented for interfaces from SAP to MES.

The realized architecture with specific technologies is illustrated in Figure 101.20.

101.7 Conclusions and Outlook

We have presented integration scenarios in an intraenterprise production environment, and have outlined technical integration options as building blocks for reusable integration architectures. Prototype concepts were outlined for each of the use cases presented.

The following conclusions can be drawn regarding integration between plant-floor and business systems:

- point-to-point data exchange solutions restrict integration flexibility;
- an integration solution is composed of different technical components;
- for a reusable solution, domain interfaces have to be standardized and developed, which allow connecting different systems without the need to redevelop the whole architecture;
- systems with fewer numbers of interfacing points can be productized quite easily; larger systems with a high number of intersystems communications are usually tailored toward customer requirements;

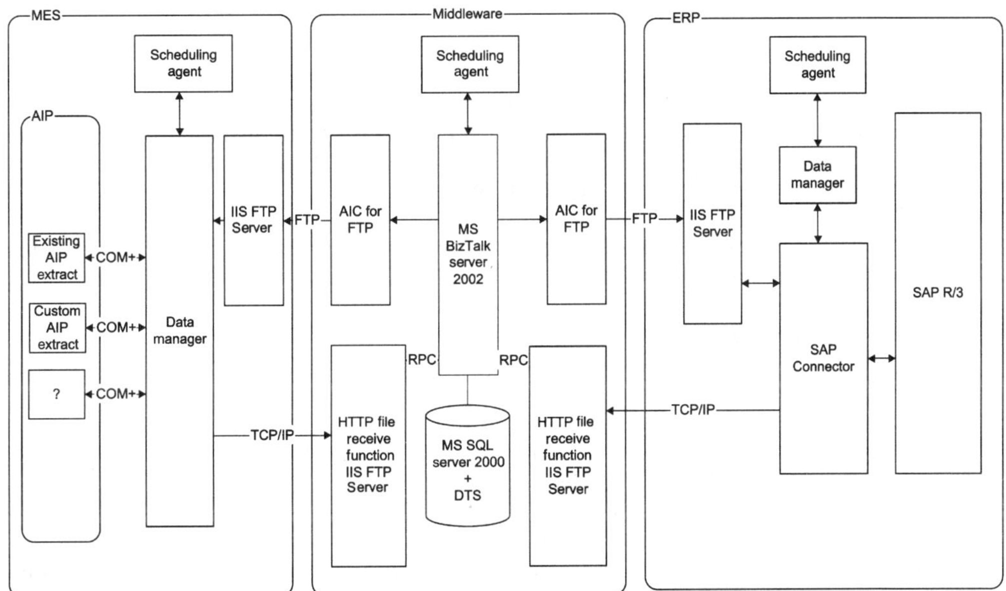

FIGURE 101.20 Prototype bulk data submission.

- standards serving cross-system integration, such as ISA S95, will further strengthen the role of EAI as a bridging layer between automation and business systems through their data mapping and orchestration capabilities; and
- using standard technologies such as XML and web services eases the task of application integration across vendors and platforms.

However, the full solution potential can only be achieved if the following integration issues are carefully taken into account:

- *Data Consistency and Engineering*: Data entry must be facilitated and promoted across the various subsystems and must, by all means, be kept synchronized across the various subsystems; this implies detecting changes (insert, modify, delete) in systems on objects (e.g., equipment, batches) and their attributes (e.g., status data) and replicate changes to "connected" systems according to replication rules and in a defined sequence.
- *Data Exchange*: Common sets of functionality for various subsystems (such as production planning), high-level APIs to facilitate the access to these systems, and common data description to overcome semantic differences of the involved systems must be developed. A uniform data access for users and applications, hiding the origin of data from different systems, can be achieved by combing functional and data integration capabilities as described in this document. To the outside (user or application), a uniform access interface allows requesting information of a specified "category" — for example, operational data, maintenance-related information, performance-related information, etc, which are composed of attributes of object instances from different systems.

In order to achieve this functionality, additional concepts, such as data and attribute views (i.e., typed concepts) should be introduced with the assignment of object instances to types, and the ability to define relations between the content of types.

Transparent access to object attribute information will be the functionality that enables applications and users to access information according to defined "data views" independent of the information sources. The origin of the information (source systems) is hidden. Benefits of this functionality include a better access management to applications, simplified aggregation of data across systems, and the synchronization of objects and their attributes between systems.

References

1. Deliang, Z., L. Jizhen, and L. Yanquan, Control-Decision and Integrated Information System of Power Plant, in Proceedings of Powercon '98, Beijing, China, Vol. 2, 1998, pp. 1241–1245.
2. Canton, J., D. Ruiz, C. Benqlilou, J.M Nougues, and L. Puigjaner, Integrated Information System for Monitoring, Scheduling and Control Applied to Batch Chemical Processes, in Proceedings of IEEE International Conference on Emerging Technologies and Factory Automation, Vol.1, 1999, pp. 211–217.
3. Herzog, U., R. Hantikainen, and M. Buysse, A Real World Innovative Concept for Plant Information Integration, in Proceedings of Cement Industry Technical Conference, IEEE-IAS/PCA 44th, 2002, pp. 323–334.
4. Wells, D. et al., Enterprise Application Integration, Report, Ovum Ltd., 2002.
5. Axtor, C. et al., Web Services for the Enterprise, Opportunities and Challenges, Report, Ovum Ltd., 2002.
6. XML resources, http://www.w3c.org/xml.
7. Enterprise-Control System Integration, ANSI/ISA-95.00.01-2000, Part 1: Models and Terminology, ANSI/ISA-95.00.02-2001, Part 2: Object Model Attributes.
8. De Vos, A., S. Widergren, and J. Zhu, XML for CIM Model Exchange, *in Proceedings of IEEE Conference for Power Industry Computer Applications, PICA 2001*, Sydney, Australia, 2001, pp. 31–37.
9. McClellan, Michael, *Applying Manufacturing Execution Systems*, APICS Series on Resource Management, St. Lucie Press, Boca Raton, FL, USA, 1997.

102

Principles and Features of PROFINET

Manfred Popp
Automation & Drives

Joachim Feld
Automation & Drives

Ralph Büsgen
Automation & Drives

102.1 Introduction

Automation technology is undergoing continuous change due to the ever-shorter innovation cycles for new products. The use of field bus technology in recent years has represented a significant innovation. It has enabled the migration of automation systems from centralized to decentralized systems. In this regard, PROFIBUS has set the standard as the market leader for more than 15 years.

Moreover, in today's automation technology, information technology (IT) with established standards, such as TCP/IP and XML, is increasingly dictating changes. Integration of information technology into automation is opening up significant advances in communication options between automation systems, far-reaching configuration and diagnostic options, and networkwide service functions. These functions have been a fixed component of PROFINET from the start.

PROFINET is the open standard for industrial automation based on Industrial Ethernet. PROFINET enables problem-free realization of distributed automation, integration of existing field devices, and operation of demanding, time-critical applications (such as motion control).

In addition to utilization of IT technology, protection of investment also plays an important role with PROFINET. PROFINET enables existing field bus systems such as PROFIBUS to be integrated without modifications with existing devices. This protects the investments of plant operators, machinery/plant construction firms, and device manufacturers.

Automation technology requirements are thoroughly covered by PROFINET. It was possible to transfer the many years of experience in the PROFIBUS sphere to PROFINET standardization. The use of open standards, simple handling ability, and integration into existing plant units have defined PROFINET from the start. PROFINET is currently integrated in IEC 61158.

A long-term perspective is offered to users through continuous advancements in PROFINET.

Costs incurred by plant or mechanical system engineers for installation, engineering, and startup are minimized through the use of PROFINET. For plant operators, PROFINET enables plants to be easily expanded and achieve a high level of availability through independently operating plant units.

Establishment of certification by the PROFIBUS User Organization (PNO) guarantees a high standard of quality for PROFINET products.

This chapter describes in detail how experience gained with IT standards in the PROFIBUS sphere has been converted to PROFINET.

102.2   PROFINET at a Glance

The motivation to create PROFINET comes from the user requirements outlined in Section 102.1 and the anticipated cost reduction resulting from manufacturer-independent, plantwide engineering. With PROFINET, a modern automation concept has emerged that is based on Ethernet and enables simple integration of existing field bus systems (in particular, PROFIBUS). This represents an important aspect for satisfying uniformity of requirements from the corporate management level to the field level.

Decentralized Field Devices (PROFINET IO)

Simple field devices are integrated in PROFINET using PROFINET IO and are described by the familiar IO view in PROFIBUS DP. Decentralized peripherus for PROFINET are also integrated using this approach. The essential feature of this integration is the use of decentralized field devices with their input and output data, which are processed in the PLC user program (Figure 102.1).

PROFINET IO describes a device model that differentiates slots and channels in a similar way as the model for PROFIBUS DP. The device properties are described by an XML-based description file (GSD).

PROFINET IO devices are engineered using the same approach that has long been familiar to system integrators of PROFIBUS DP. This includes assignment of the decentralized field devices to a controller during configuration. Productive data are then exchanged between the controller and the assigned field devices.

Distributed Automation (Component Model)

Distributed automation systems typically consist of several subunits that act autonomously for the most part and coordinate with each other using signals for synchronization, sequence control, and information exchange.

The PROFINET component model refers to these subunits as *technological modules*. The technological modules form an intelligent functional unit. Through the use of the component technology proven in the IT sphere, the overall functionality of a technological module is encapsulated in an associated software

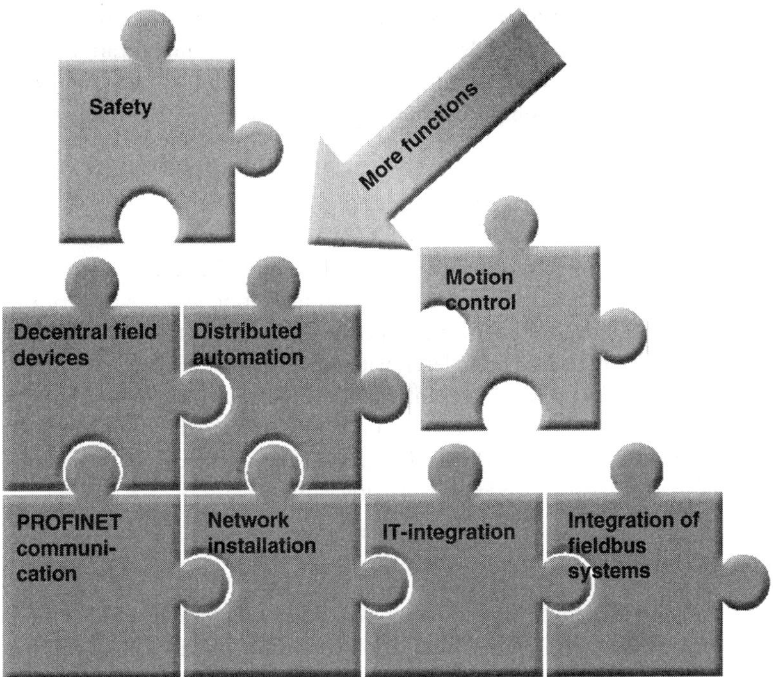

FIGURE 102.1 Architecture of PROFINET IO.

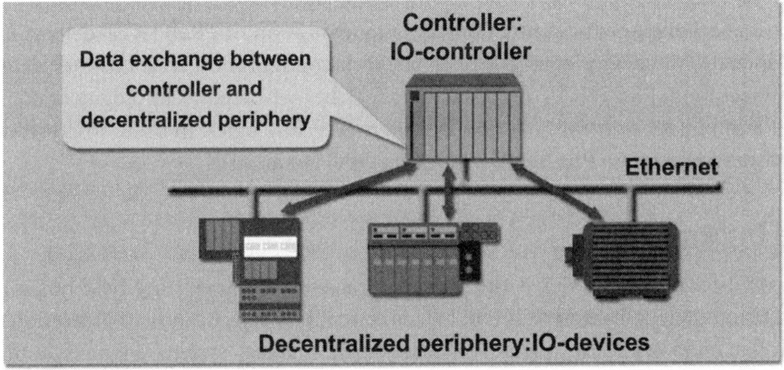

FIGURE 102.2 Mechanical, electrical/electronic, and software aspects are combined in technological modules.

component. Such a component is modeled as an object and regarded as a black box. An outside technological component interface is defined in order for the component to communicate with other components within the distributed system (Figure 102.2).

A distributed automation system designed in this way constitutes the prerequisite for modularization of plants and machinery and hence for reuse of plant and machine parts. This significantly reduces engineering costs.

A technological module is described in PROFINET within the component model using PROFINET Component Description (PCD). PCD is XML-based and produced either by a component generator of a manufacturer-specific configuration tool or the PROFINET Component Editor.

A manufacturer-neutral *engineering concept* is available for user-friendly configuration of a PROFINET system.

Engineering of distributed automation systems distinguishes between programming of control logic for individual technological modules (manufacturer-specific configuration tools) and the technological configuration of the entire system (interconnection editor). A systemwide application is formed in the three steps: create components, interconnect components, and download interconnection information.

Communication

Different performance levels are available for PROFINET communication.

Parameters, configuration data, and interconnection information that are not critical with respect to time are transferred in PROFINET via the standard channel based on TCP/UDP and IP. This satisfies the prerequisites for interfacing the automation levels with other networks (MES, ERP). For transfer of time-critical process data within the production plant, the real-time channel known as Soft Realtime (SRT) is available. This channel is implemented as software on the basis of existing controllers.

For isochronous applications, the Isochrone Realtime Communication (IRT) is available that enables clock pulse rates of less than 1 msec and a jitter accuracy of 1 μsec.

Network Installation

PROFINET network installations are oriented toward specific requirements for Ethernet networks in industrial environments. The "PROFINET Installation Guideline" provides plant construction engineers and plant operators with simple rules for installing Ethernet networks and associated cabling. This guideline provides device manufacturers with clear specifications for device interfaces.

IT Integration

The network management includes functions for administration of PROFINET devices in Ethernet networks. This includes the device configuration, network configuration, and network diagnostics. In the case of web integration, PROFINET makes use of the Ethernet-based technologies and enables access to a PROFINET component by means of standard Internet technologies. In order to preserve an open connection to other system types, PROFINET supports OPC DA and DX.

Field Bus Integration

An important aspect of PROFINET is the seamless transition from existing field bus solutions such as PROFIBUS DP to Ethernet-based PROFINET. This contributes significantly to protection of investments by the device manufacturer, the plant construction/mechanical system engineer, and the end user (Figure 102.3).

PROFINET offers two alternatives for integrating field bus systems:

- *Integration of field bus devices by means of proxies*: The proxy is the representative for the lower-level field devices on the Ethernet. Through the proxy principle, PROFINET offers a completely transparent transition from existing to newly installed plant units.
- *Integration of whole field bus applications*: A field bus segment represents a self-contained component. The representative for this component is the PROFINET device that operates a field bus such as PROFIBUS DP at a lower level. The entire functionality of a loswer-level field bus is thereby implemented in the form of a component in the proxy, which is available on the Ethernet.

102.3 Decentralized Field Devices (PROFINET IO)

Decentralized field devices are integrated directly on the Ethernet using PROFINET IO. To accomplish this, the master–slave system familiar in PROFIBUS DP is transferred over to a provider–consumer

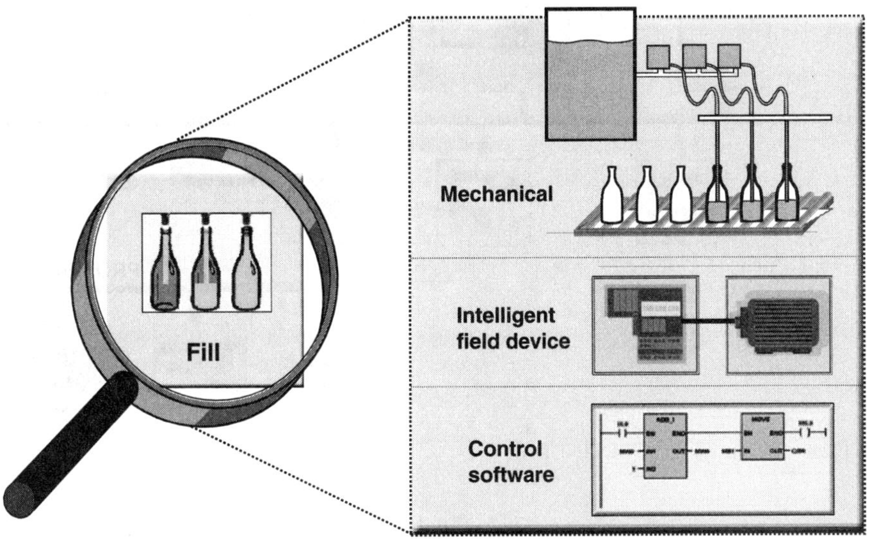

FIGURE 102.3   PROFIBUS systems, for example, can be integrated in PROFINET using a proxy.

model. Although all devices on the Ethernet have equal communication rights, the configuration specifies which field devices are assigned to a centralized controller. In this way, the familiar user view in PROFIBUS is transferred to PROFINET IO. I/O signals are read in, processed by the PLC, and resent to the outputs.

Functional Scope

PROFINET IO distinguishes between three device types: PN-IO controller, PN-IO device, and PN-IO supervisor (Figure 102.4):

- *PN-IO controller*: A PLC on which the automation program runs.
- *PN-IO device*: Decentralized field device that is assigned to a PN-IO controller (such as remote IO, valve terminals, frequency converters).
- *PN-IO supervisor*: Programming device or PC with commissioning and diagnostic functions.

Data can be transferred between the IO controller and IO device by means of the following channels:

- Cyclic user data via the real-time channel.
- Event-triggered interrupts (diagnostics) via real-time channel.
- Parameter assignment and configuration as well as reading of diagnostic information via the standard channel based on UDP/IP.

At the start, a communication relationship called Application Relation (IO-AR) is established between the IO controller and the IO device based on the acyclic UDP/IP channel. Then, the IO controller transfers the configuration data for the IO device by means of this established channel. Based on the configuration data: (1) the correct operating mode is determined, for example, and the IO device is uniquely identified, (2) high-speed, cyclic useful data exchange via the real-time channel (IO-CR) is started. If a diagnostic event occurs (such as a wire break), an interrupt is sent to the IO controller via the high-speed, acyclic real-time channel (interrupt CR) for processing in the PLC program located on the IO controller (Figure 102.5).

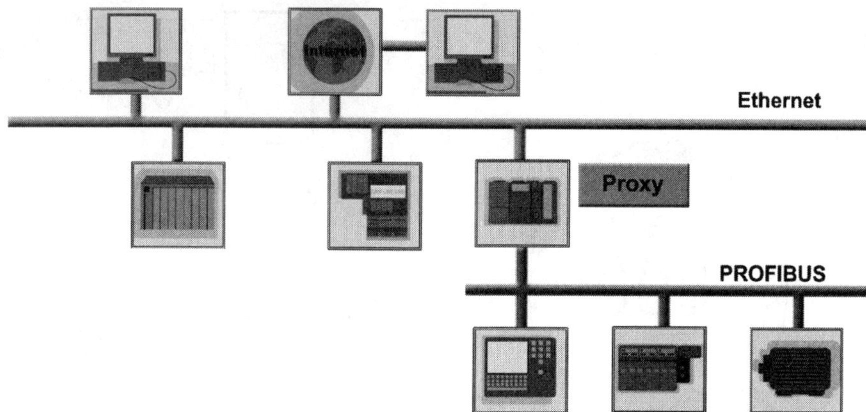

FIGURE 102.4   Device types in PROFINET IO.

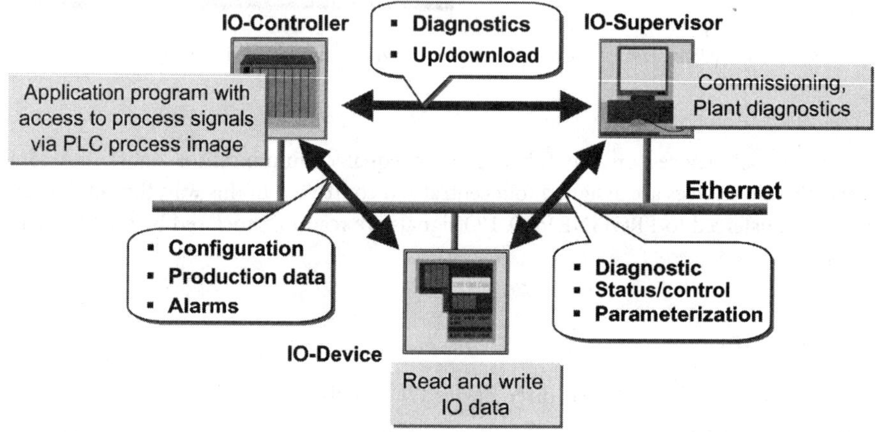

FIGURE 102.5   Communication relationships in PROFINET IO governed by the consumer/provider model.

Device Model

A uniform device model has been specified for the PROFINET IO device. This model enables modular and compact field devices to be modeled. This model is oriented toward the main features in PROFIBUS DP and extends the advantages of these features into the future (Figure 102.6).

An IO device with a modular configuration consists of slots in which modules are inserted. The modules contain channels over which process signals are read in or read out. The representative of the IO device is the interface module, which receives data from the IO controller and forwards it to the modules via the backplane bus. Conversely, it receives the process and diagnostic information from the modules via the backplane bus and forwards this information to the IO controller. Each IO device receives a global device identification that is uniquely assigned within the framework of PROFINET IO. This 32-bit device ID number is divided into a 16-bit manufacturer identifier and a 16-bit device identifier. This device ID is assigned by the PROFIBUS user organization (PNO).

Device Description (GSD)

As in PROFIBUS, a device description is used to integrate a PROFINET device into the configuration tool of an IO controller. The properties of an IO device are described in the form of General Station

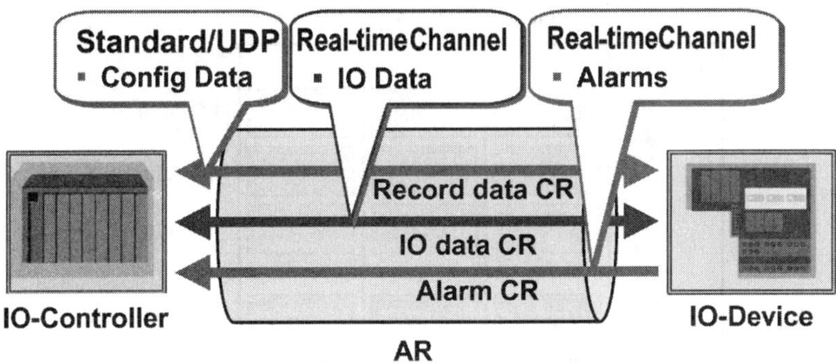

FIGURE 102.6　PROFINET IO device model is similar to that of PROFIBUS DP.

Description (GSD), which contains all necessary information:

- Properties of the IO device (e.g., communication parameters).
- Insertable modules (number of type).
- Configuration data for individual modules (e.g., 4 to 20 mA analog input).
- Parameters of modules.
- Error texts for diagnostics (e.g., wire break, short circuit).

XML is the description basis for the GSD of PROFINET IO devices. Because XML is an open, widespread, and accepted standard for describing data, appropriate tools and derived properties are automatically available, including:

- creation and validation through a standard tool,
- foreign language integration, and
- hierarchical structuring.

The GSD structure corresponds to ISO 15745 and consists of a header, the device description in the application layer (e.g., configuration data and module parameters), and the communication properties description in the transport layer.

Configuration and Data Exchange

The description files of the IO devices are imported into the configuration tool. IO addresses are assigned to the individual IO channels of the field devices. The IO input addresses contain the received process values. The user program evaluates and processes these values. The user program forms the IO output values and outputs them to the process via the IO output addresses. In addition, parameters are assigned to the individual IO modules or channels in the configuration tool, for example, 4 to 20 mA current range for an analog channel.

After conclusion of the configuration, the configuration data are downloaded to the IO controller. The IO devices are assigned and configured automatically by the IO controller and then enters into the cyclic data exchange (Figure 102.7).

Diagnostics

PROFINET IO supports a multilevel diagnostic concept that enables efficient fault localization and correction. When a fault occurs, the faulty IO device generates a diagnostic alarm to the IO controller. This alarm triggers a call in the PLC program to the appropriate program routine in order to be able to

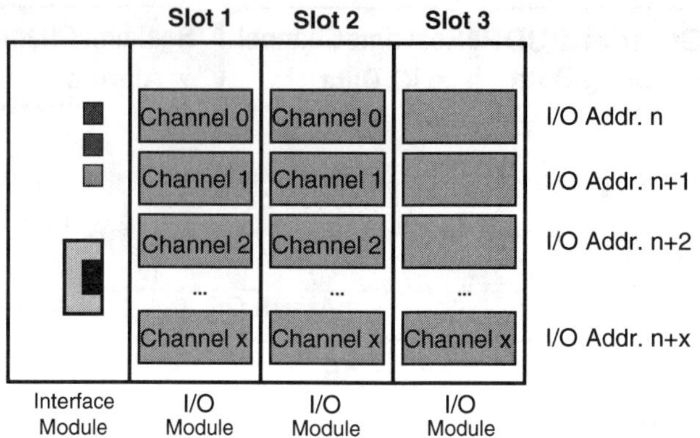

FIGURE 102.7　Configuration path for data exchange.

respond to the fault. If a device or module defect requires a complete replacement of the device or module, the IO controller automatically performs a parameter assignment and configuration of the new device or module.

The diagnostic information is structured hierarchically:

- Slot number (module).
- Channel number.
- Channel type (input/output).
- Coded fault cause (e.g., wire break, short circuit).
- Additional manufacturer-specific information.

When an error occurs in a channel, the IO device generates a diagnostic alarm to the IO controller. This alarm triggers a call in the control program to the appropriate fault routine. After processing of the fault routine, the IO controller acknowledges the fault to the IO device. This acknowledgment mechanism ensures that a sequential fault processing is possible in the IO controller.

102.4　Distributed Automation

Automation development has given rise to modular plants and machinery. This structuring has triggered further development in automation to produce distributed automation systems. PROFINET also offers a solution in this area. The PROFINET solution involves separation of plant units into technological modules.

Technological Modules

The function of an automated plant or machine — for a goods manufacturing process — is produced by a defined interaction of mechanical, electrical/electronic, and control logic/software aspects. According to this principle, PROFINET defines the mechanical, electrical/electronic, and control logic/software aspects for a *technological module* (see Figure 102.2). A technological module is in turn modeled by a software component, that is, the PROFINET component.

PROFINET Components

The representative of a technological module during plant engineering is the so-called PROFINET component. Each PROFINET component has an interface that contains the technological variables to be exchanged with other components.

The PROFINET components are modeled using the standardized COM technology. COM is an advanced object orientation that enables applications to be developed on the basis of preassembled components. The components are characterized by the formation of complete units that can be in relationship to other components. Like blocks, the components can be combined flexibly and easily reused, irrespective of how they are implemented internally. Access mechanisms to the component interfaces are uniformly defined in PROFINET.

Granularity of Technological Modules

When the granularity of modules is being specified, the ability to reuse the modules in different plants must be examined with cost and availability in mind. The objective is to be able to merge individual components into an overall plant as flexibly as possible according to the building block principle. On the one hand, with too fine a granularity the technological view of the plant can become more complex, resulting in higher engineering costs. On the other hand, with too coarse a granularity, the degree of reusability is reduced. This, in turn, results in higher implementation costs.

The machine or system manufacturer creates the software component. The *component design* has a major influence on reduction of engineering and hardware costs and the time response of the automation system. When a component is being designed, the granularity (i.e., size of machine/plant or the specified machine parts/plant units) can extend from an individual device to a complete machine containing a number of devices.

PROFINET Engineering

A manufacturer-neutral engineering concept has been created for user-friendly configuration of a PROFINET system. For one thing, this engineering concept enables development of configuration tools that can be used for components of different manufacturers. Also, it permits manufacturer-specific and application-specific function expansions.

The engineering model distinguishes between programming of control logic for individual technological modules and the technological configuration of the overall plant. A plantwide application is generated in three stages.

Component Creation

Components are created as an image of the technological modules by the machinery or plant construction firms. Devices are programmed and configured as before with the respective manufacturer-specific tools. That way, available user programs and the know-how of programmers and service personnel can continue to be used. Then the user software is encapsulated in the form of a PROFINET component. In so doing, a component description (PCD) in the form of an XML file is created. The contents of the

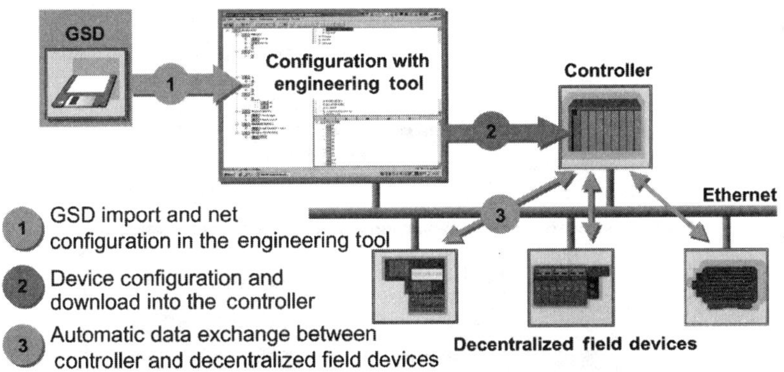

FIGURE 102.8　Component creation is standardized in PROFINET.

component description are specified in PROFINET. These component descriptions are imported into the library of the interconnection editor (Figure 102.8).

Component Interconnection

The created PROFINET components are moved from a library to an application and interconnected at the click of a mouse using the PROFINET *interconnection editor* (Figure 102.9).

Interconnection replaces the previous labor-intensive programming of communication relationships with simple graphical configuration. During programming, detailed knowledge about the integration and sequence of communication functions in the device is required. At the time programming is performed, the following must already have been specified: which devices will communicate with one another, when the communication will occur, and which bus system will be used for the communication. By contrast, knowledge of the communication functions is not required during configuration because these functions run automatically in the devices. The interconnection editor consolidates the individual distributed applications on a plantwide basis. Operation of the interconnection editor is manufacturer-neutral, that is, the editor interconnects any manufacturer's PROFINET components.

Downloading

Following component interconnection, the interconnection information as well as the code and configuration data for the components are downloaded to the PROFINET devices at the click of a mouse. As a

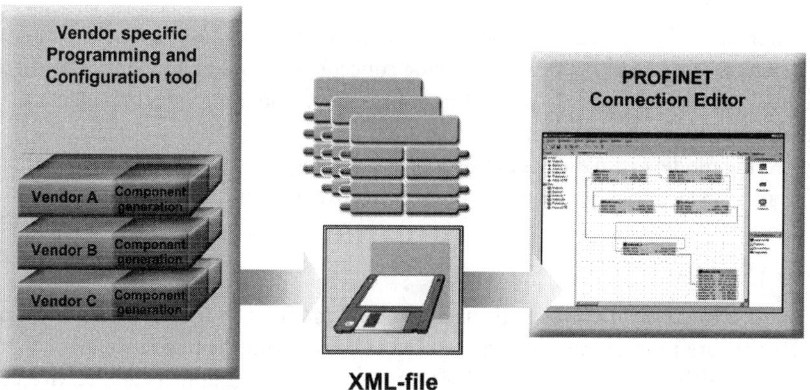

FIGURE 102.9   Interconnected components.

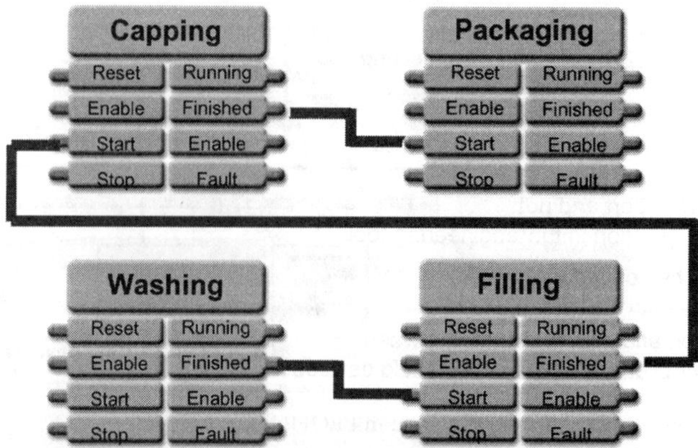

FIGURE 102.10   Downloading the interconnection information for PROFINET devices.

result, each device knows all of its communication peers, communication relationships, and information to be exchanged. The distributed application can be executed afterward (Figure 102.10).

Component Description (PCD)

The PCD is an XML file, which is created by manufacturer-specific tools. This assumes that these tools have a component generator. Alternatively, a manufacturer-neutral "PROFINET Component Editor," available for download on the PROFIBUS website, can be used to create the PCD file. The PCD file contains information on the functions and objects of the PROFINET component. Specifically, this information includes:

- Description of components as library elements: component ID, component name.
- *Hardware description*: IP address storage, diagnostic data access, interconnection download.
- *Description of software functionality*: software <-> hardware assignment, component interface, properties of variables such as technological name, data type, and direction (input or output).
- Storage location of component project.

Component libraries are generated to support reusability.

Interconnection Editor

In general, an interconnection editor has two views: plant view and network view.

In the plant view, required components are imported from the library and placed on the screen, and the individual interconnections are established. This yields a *technological* structure and its local relationships within a plant. By contrast, the *topological* structure of the automation system is created in the network view. Here, the field devices and automation devices are assigned to a communication system or bus system, and the device addresses are specified according to the rules for the underlying bus system (Figures 102.11 and 102.12).

PROFINET Runtime

The PROFINET run-time model defines functions and utilities that require cooperating automation components to accomplish an automation task. This model establishes and monitors the interconnec-

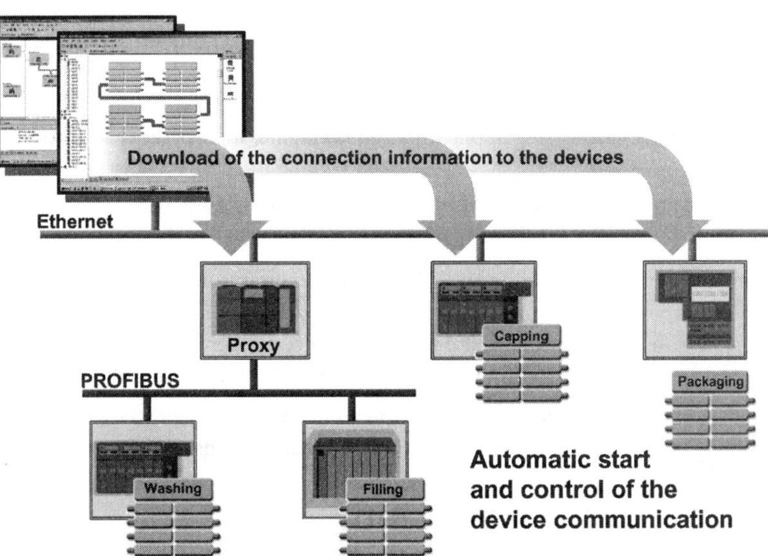

FIGURE 102.11   Plant view in the interconnection editor shows the interconnected components.

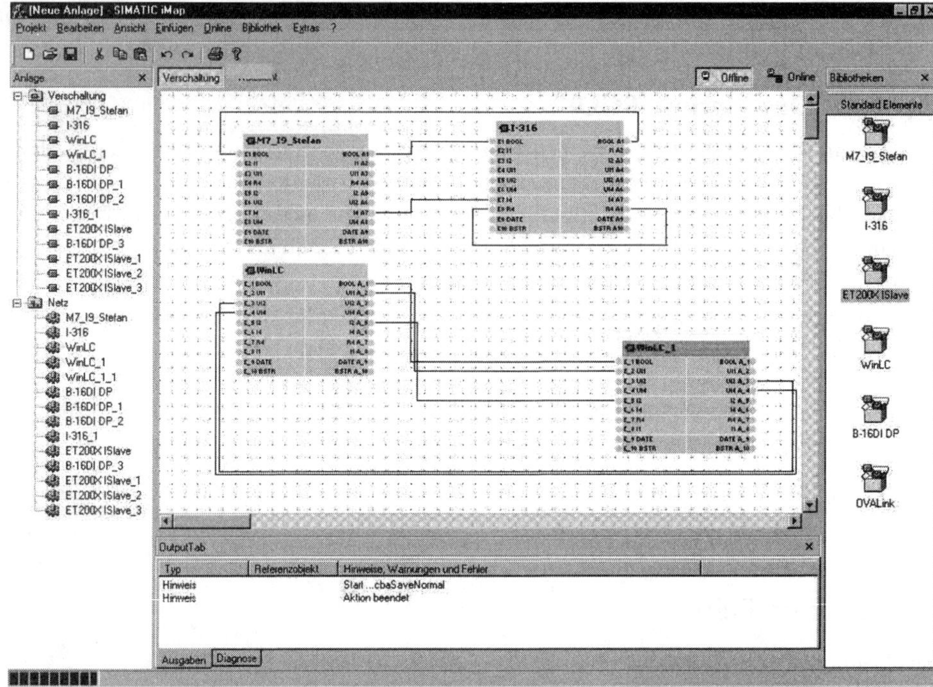

FIGURE 102.12   Network view in the interconnection editor shows the connected field devices.

tions between PROFINET components that have been configured by engineering. The PROFINET runtime model sets up a provider–consumer model in which the provider creates and sends data and the consumer receives and processes data.

102.5   PROFINET Communication

Ethernet-based communication is scalable in PROFINET. The following three performance levels are differentiated:

1. TCP/UDP and IP for data that are not time critical, such as parameter assignment and configuration data.
2. SRT for time-critical process data in factory automation.
3. IRT for particularly challenging application requirements, such as those for motion control.

These three PROFINET communication performance levels cover the spectrum of automation applications in their overall diversity. The PROFINET communication standard is characterized by the following in particular:

- Coexisting utilization of real-time and TCP-based IT communication on one line.
- Uniform real-time protocol for all applications both for communication between components in distributed systems and for communication between the controller and the distributed field devices.
- Scalable real-time communication from performant to high-performant, and isochronous mode.

Scalability and a uniform communication basis represent two of the major strengths of PROFINET. They guarantee continuity to the corporate management level and fast response times in the automation process.

Standard Communication with TCP/UDP

PROFINET uses Ethernet and TCP/UDP with IP as the basis for communication. TCP/UDP with IP is a *de facto* standard with respect to communication protocols in the IT landscape. However, for interoperability (i.e., the interaction between applications), establishment of a common communication channel over the field devices, based on TCP/UDP (layer 4), is insufficient. TCP or UDP represent only the foundation on which Ethernet devices can exchange data via a transport channel in local and distributed networks. Therefore, additional specifications and protocols beyond TCP/UDP — the so-called application protocols — are required. Interoperability is only guaranteed if the same application protocol is used on the devices. Typical application protocols are SMTP (for e-mail), FTP (for file transfer), and HTTP (used on the Internet).

Real-Time Communication

Real-time applications in manufacturing automation require update, or response times, in the range of 5 to 10 msec. The update time refers to the time that elapses when a variable is generated by an application in a device, then sent to a peer device via the communication system, and then received updated by the application. For devices, the implementation of a real-time communication causes only a small load on the processor so that execution of the user program continues to take precedence. From experience, in the case of Fast Ethernet (100 Mb/sec Ethernet), the transmission rate on the line in proportion to the execution in the devices can be disregarded. Most of the time gets lost in the application. The time it takes to provide data to the application of the provider is not influenced by the communication. This also applies to processing of data received in the consumer. As a result, noteworthy improvements in update rates and thus the real-time performance can be obtained primarily through proper optimization of the communication stack in the provider and consumer.

RT

In order to satisfy real-time demands in automation, PROFINET uses an optimized real-time communication channel (Figure 102.13).

This channel is based on Ethernet (layer 2). The solution minimizes the throughput time in the communication stack considerably and increases performance in terms of the process data update rate. By doing away with several protocol layers, the message frame length is reduced. By not including these

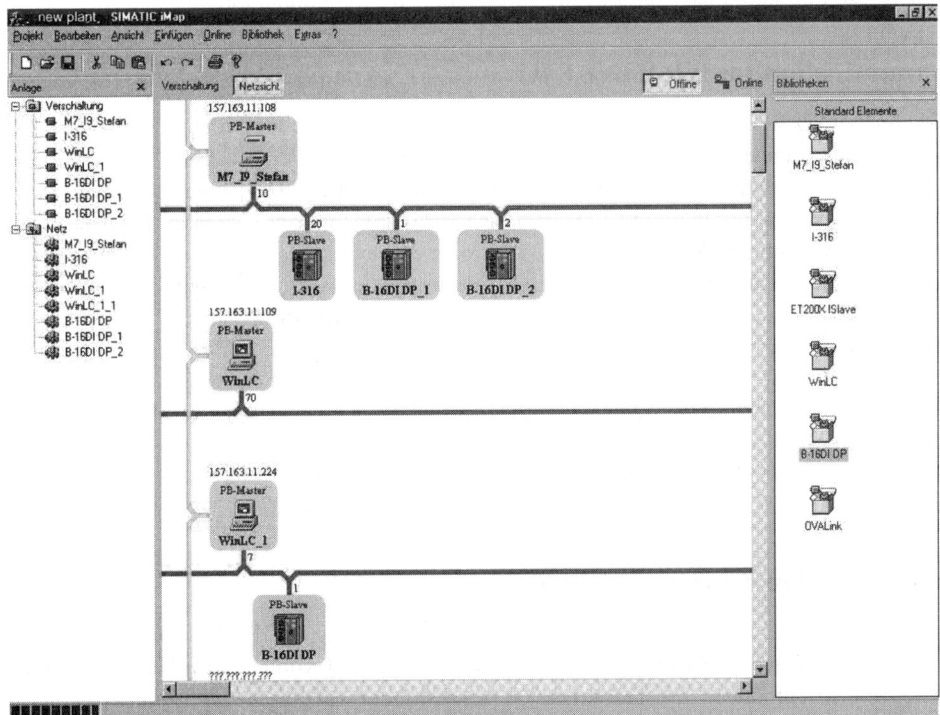

FIGURE 102.13　Communication channels in PROFINET.

additional layers, the data to be transmitted from the provider can be sent sooner, or are available earlier to the application on the consumer side. At the same time, this significantly reduces the processor power required for communication in the device.

Optimized Data Transmission through Prioritization.
In addition to the minimized communication stack in the automation devices, the transmission of data in the network is also optimized in PROFINET. In order to achieve an optimal result, the packets are prioritized in PROFINET according to IEEE 802.1Q. The network components use this priority to control the dataflow between the devices. The standard priority for real-time data is based on Prio 6. Thus, priority handling over other applications, such as Internet telephony, is guaranteed because they are using Prio 5.

IRT

The presented solutions are not sufficient for motion control applications, in particular. These applications require update rates in the range of 1 msec along with a jitter for consecutive update cycles of 1 μsec for cases involving up to 100 stations. To satisfy these requirements, PROFINET defines the IRT time slot-controlled transmission process on layer 2 for Fast Ethernet. This means every device knows exactly in which time slot it is allowed to send data over the bus. Through synchronization of the devices involved (network components and PROFINET devices) with the accuracy indicated above, a time slot can be specified during which data critical for the automation task are transferred. The communication cycle is split into a deterministic part and an open part. The cyclic real-time message frames are dispatched in the deterministic channel, while the TCP/IP message frames are transported in the open channel. The process is comparable to the traffic on a highway where the left lane is reserved for time-critical traffic (real-time traffic), and the remaining traffic elements (TCP/IP traffic) are prevented from switching to this lane. Even if there is a traffic jam in the right lane, the time-critical traffic is not affected.

Isochronous data transmission is realized based on hardware, for example, it is burned in an ASIC. Such an ASIC covers the cycle synchronization and time slot reservation functionality for the real-time data. Realization in hardware ensures the accuracy requirements to be achieved within the order of magnitude required. Furthermore, the processor in the PROFINET device is relieved from communication tasks. The resulting additional runtime can be made available for automation tasks.

Communication for PROFINET IO

For PROFINET IO, RPC based on UDP/IP is used in the startup phase for initiation of the data exchange between devices, parameter assignment of distributed field devices, and diagnostics. Through the open and standardized RPC protocol, HMI stations or engineering systems (IO supervisor) can also access PROFINET IO devices. The PROFINET real-time channel is used for transmission of user data and alarms.

In a typical IO configuration, there is an IO controller that exchanges the user data cyclically by means of communication relationships that are established during the startup phase with multiple distributed field devices (IO devices). In each cycle, the input data are sent from the assigned field devices to the IO controller and, in return, the output data are sent back to the appropriate field devices. The communication relationship is monitored by keeping track of cyclic messages. If, for example, cyclic messages fail in three cycles, the IO controller recognizes that the corresponding IO device has failed.

The data transmission layer of PROFINET is defined in IEEE 802.3, which describes the protocol design and malfunction monitoring. A user data message frame consists of a minimum of 64 bytes and a maximum of 1500 bytes. The overall protocol overhead for real-time data is 28 bytes.

Communication between Technological Modules

Distributed Component Object Model (DCOM) is specified in the PROFINET component view as the common TCP/IP-based application protocol between PROFINET components. DCOM represents an enhancement to Component Object Model (COM) for distribution of objects and their interaction in a network. DCOM is based on the standardized RPC protocol. DCOM is used for loading of interconnections,

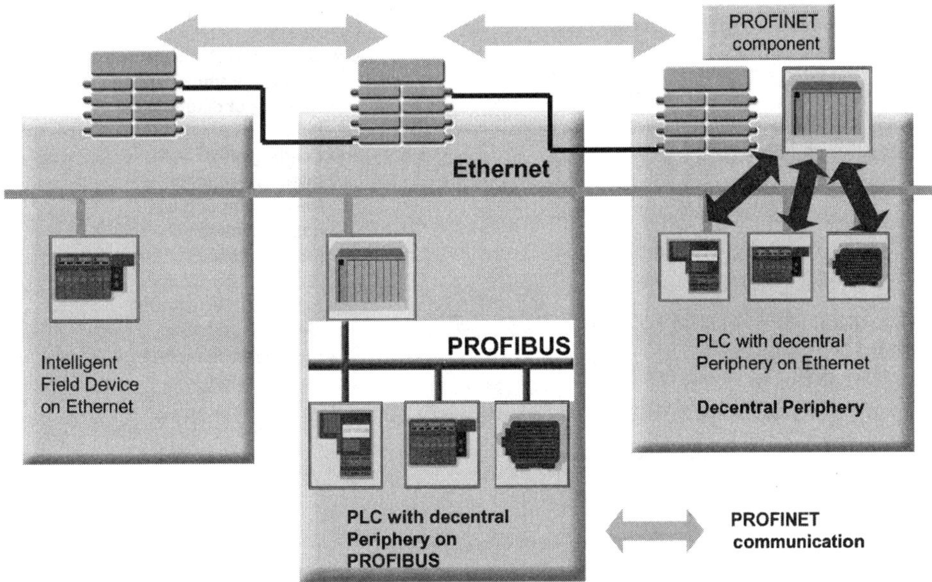

FIGURE 102.14 PROFINET communication between PROFINET components and PROFINET IO devices.

reading of diagnostic data, and device parameter assignment and configuration, establishing interconnections and, to some extent, for exchanging user data between components in PROFINET. However, DCOM does not have to be used for exchange of user data between PROFINET components. Whether user data exchange is to take place via DCOM, or the real-time channel is configured by the user in the engineering system, devices can then negotiate the use of a real-time-capable protocol. This is because communication between such plant or machinery modules can require real-time conditions that cannot be satisfied through TCP/IP and UDP.

TCP/IP and DCOM form the common "language" that can be used to start communication between the devices in all cases. The PROFINET real-time channel is then used for real-time communication between individual stations in time-critical applications. In the configuration tool, the user can select the update time, the so-called quality of service, to determine whether the values are to be transferred between components cyclically during operation or when a change is made. A cyclic transfer is more advantageous for high frequently update times since checking for changes and acknowledging them imposes less processor load than cyclic sending (Figure 102.14).

102.6 Installation Technology for PROFINET

The international standard ISO/IEC 11801 and its European equivalent EN 50173 define an application-neutral, information-oriented standard networking for a complex of buildings. The content of the two standards is essentially identical. Both standards assume that the buildings are similar to an office environment and assert a claim to being application-neutral. The following specific requirements for Ethernet networks in the industrial environment are not taken into consideration in the two standards:

- Plant-specific cable routing.
- Individual degree of networking for each machine/plant.
- Linear network structures.
- Robust industrial cable and connectors with special requirements for EMC, temperature, moisture, dust, and vibration.

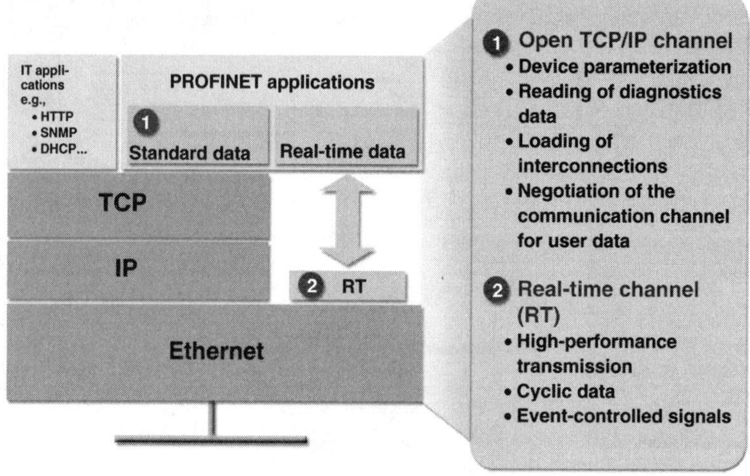

FIGURE 102.15   Structure of Ethernet networks in office systems.

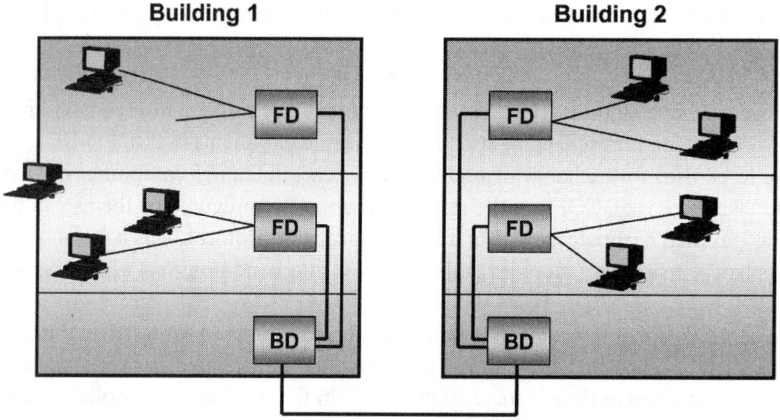

Office Setting	Manufacturing and Field Setting
Fixed basic installation in buildings	Robust plant-independent cabling
Cabling in false floors	Plant-specific cable routing
Variable device connection at workspace	Connection points seldom changed
Precut/assembled device connection cable	Device connections can be field assembled
Tree-shaped network structures	Linear network structures and (redundant) ring structures are frequently used
Large data packets (e.g., pictures)	Small data packets (measured values)
Average network availability	Very high network availability
Moderate temperatures	Extreme temperatures
No humidity	Humidity possible (IP65)
Minimal vibration	Vibrating machinery
Low EMC-load	High EMC-Load
Low mechanical hazard	Mechanical damage hazard
Minimal chemical hazard	Chemical hazard through oil or corrosive atmosphere

FIGURE 102.16   Differences in cable installation in office systems and automation systems.

For this reason, the "PROFINET Transmission Technology and Cabeling" guideline defines an industrial cable installation for the "Fast Ethernet" application based on the fundamental requirements in IEC 11801 (Figures 102.15 and 102.16).

PROFINET Cable Installation

PROFINET Cable Installation with Symmetrical Copper Cable

Signals are transmitted over symmetrical copper cable (twisted pair) in accordance with 100BASE-TX at a transmission rate of 100 Mbits/sec (Fast Ethernet). The transmission medium is a two-pair, twisted-conductor, shielded copper cable (twisted pair or star quad) with a characteristic impedance of 100 Ω.

Only shielded cable and connection elements are permitted. The individual components must satisfy Category 5 requirements according to IEC 11801. The overall transmission line must satisfy the Class D requirements according to IEC 11801. Removable connections are produced using an RJ45 or M12 connector system. Sockets are used as device connections. Connecting cables (device connection cables, marshaling cables) are provided with suitable plugs at both ends. An active network component is used to connect all devices. To ensure that installation is as simple as possible, the transmission cable has been defined to be identical at both ends. This connecting cable satisfies the function of a cable assembly with two identical ends (Figure 102.17).

The maximum segment length is 100 m.

PROFINET Cable Installation with Optical Fibers

PROFINET can be operated with multimode or single-mode fiber optic cables. Signals are transferred over two-strand optical fibers in accordance with 100BASE-FX at a transmission rate of 100 Mbits/sec. The optical interfaces comply with the ISO/IEC 9314-3 (multimode) or ISO/IEC 9314-4 (single-mode) specifications. For applications external to the control cabinet, the outer sheath must satisfy the applicable requirements at the point of use (mechanical, chemical, thermal). The maximum segment length is 2 km for multimode and 14 km for single-mode specifications.

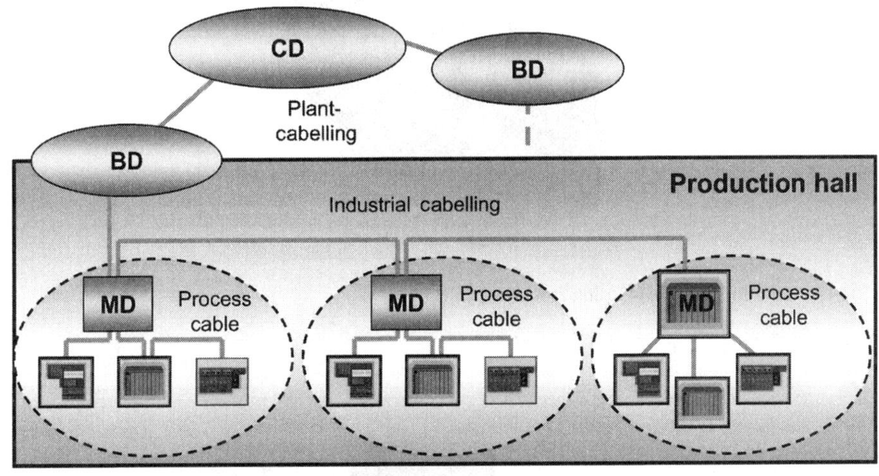

FIGURE 102.17   Ethernet networks in an industrial environment.

Plug Connectors

Plug connectors for M12 and for RJ45 are available in PROFINET. They can be easily cut to length and terminated on site (Figure 102.18).

RJ45 type is used in control cabinet settings in PROFINET. It is compatible to an office connector. Plug connectors external to the control cabinet must notably accommodate industrial requirements. The RJ45 types in IP65 or IP67 or the M12 type are utilized in this case (Figure 102.19).

Fiberoptic connections are implemented according to ISO/IEC 11801, preferably with a duplex SC plug connector system. This system is described in IEC 60874-14. Devices are equipped with the socket, and the connecting cable is equipped with the plug. Alternatively, the fiberoptic plug connector BFOC/2,5 in accordance with IEC 60874-10 can be used.

The hybrid plug connector is used in situations where decentralized field modules are connected using a combined plug connector containing data and power supply. A plug connector that is completely protected against accidental contact enables the use of plug connectors that are identical at both ends because a pin-socket reversal is not required due to the integrated accidental contact protection (Figure 102.20).

FIGURE 102.18   Example of an RJ45 plug connector in IP20.

FIGURE 102.19   Example of an RJ45 plug connector in IP67.

FIGURE 102.20   Example of a hybrid plug connector with RJ45 in IP67.

This example involves an RJ45 in IP67 with a two-pair, shielded data cable for communication and four copper conductors for voltage supply.

Switches as Network Components

PROFINET uses general switches as a network component. Switches are devices that are situated in the transmission path between the end devices and regenerate received signals and forward them selectively. They are used to structure networks. ISO/IEC 15802-3 contains the basic specifications. Switches suitable for PROFINET are sized for Fast Ethernet (100 Mbits/sec, IEEE 802.3u) and full duplex transmission. During full duplex operation, a switch receives and sends data simultaneously at the same port. No collisions occur when switches are utilized. As a result, bandwidth is not lost due to the Ethernet collision process. Network configuration is simplified significantly because route length checking is not required within a collision domain. To ensure compatibility with old plants or individual, older end devices or hubs, 10BASE-TX (10 Mbits/sec, CSMA/CD) is supported.

Moreover, a PROFINET switch supports prioritized message frames in accordance with IEEE 802.1Q, standardized diagnostic paths, and auto polarity exchange, autonegotiation mode, and auto crossover function. Port mirroring for diagnostic purposes is optional.

As a general rule, office-type switches cannot be used, even if the functionality described above is satisfied. Special switches are applied for industrial use. For one thing, these are designed for harsh industrial use based on their mechanical (IP degree of protection, etc.) and electrical (24 V power supply, etc.) properties. For another thing, they must fulfill the EMC requirements for industrial machine applications to enable safe operation.

102.7 IT Integration

In addition to the automation functionalities described, the use of Ethernet as a communication medium in the PROFINET context enables IT functions to be integrated in PROFINET as well. As is the case in the field bus world, Ethernet places additional network management requirements in connection with TCP/IP. In order to manage all technical aspects of integration of PROFINET devices in such networks, a concept for network management has been specified in PROFINET. The topics of network infrastructure, IP management, network diagnostics, and time-of-day synchronization are the primary elements of the concept. Network management simplifies Ethernet administration and management through the use of standard protocols from the IT world. The use of Internet technologies in automation systems represents an additional aspect. PROFINET has specified a concept under the scope of web integration that enables access to a PROFINET component. This access is achieved by using web utilities based on standard Internet technologies, such as http, XML, and HTML.

Network Management

Network management comprises all functions for administering the network, such as configuration (assignment of IP addresses), fault monitoring (diagnostics), and performance optimization.

IP Management

The use of TCP/IP in the PROFINET context means that an IP address has to be assigned to the network stations (i.e., PROFINET devices):

- *Address assignment with manufacturer-specific configuration systems:* This alternative is required because a network management system is not always available. In PROFINET, the DCP (Discovery and Basic Configuration) protocol is specified, enabling IP parameters to be assigned using manufacturing-specific configuration/programming tools or during plantwide engineering (e.g., in the PROFINET interconnection editor). The use of DCP is mandatory for PROFINET devices. In this way, a uniform behavior of PROFINET devices is ensured.

- *Automatic address assignment with DHCP*: The Dynamic Host Configuration Protocol (DHCP) has been established for assigning and managing IP addresses in office networks with network management systems. PROFINET provides for the use of this standard and describes how DHCP can be applied in a useful way in the PROFINET environment. Implementation of DHCP in PROFINET devices is optional.

Diagnostics Management

The reliability of network operation has a very high priority in network management. The *Simple Network Management Protocol (SNMP)* has been established in existing networks as the *de facto* standard for maintaining and monitoring network components and their functions. In order for PROFINET devices to be monitored with established management systems, it is useful to implement SNMP. SNMP provides for both read access from (monitoring, diagnostics) and write access to (administration) a device.

To begin with, only read access of device parameters was specified in PROFINET. Like the IP management functions, SNMP is optional. When SNMP is implemented in components, only the usual standard information for SNMP is accessed (MIB 2).

A specific diagnostic for PROFINET components is possible by means of the mechanisms described in the PROFINET specification. In this context, SNMP will not open an additional diagnostic path. Rather, SNMP enables integration in network management systems that do not normally process PROFINET-specific information.

Web Utilities

In addition to the use of modern Ethernet-based technologies in PROFINET, it is also possible to access a PROFINET component using web clients based on standard Internet technologies, such as http, XML, HTML, or scripting.

Data are transmitted in a standardized format (HTML, XML), and through standardized frontends (browsers such as Netscape, MS Internet Explorer, Opera, etc.). This enables integration of information from PROFINET components into modern, multimedia-supported information systems. The advantages of web integration in the IT sphere, such as utilization of browsers as uniform user interfaces, access to information on any number of clients from any location, platform independence of clients, and reduced effort for installing and servicing software on the client side, are thus also made available for PROFINET components.

Functional Properties

Web integration of PROFINET has been designed with an emphasis on commissioning and diagnostics. Web-based concepts can be used very effectively within these application areas.

- No special tools are required to access components. Established standard tools can be used.
- Global accessibility means that user support can be easily obtained when commissioning is performed by the component manufacturer.
- Auto description of components enables access with a standard tool without configuration information.

Possible scenarios for PROFINET web integration in the areas of commissioning and maintenance include: testing and commissioning, overview of device database, device diagnostics, and plant and device documentation.

The information provided should be represented both in a human-readable format (e.g., via a browser) and a machine-readable format (e.g., via a XML file). Both variants can be provided in integral form using the PROFINET web integration. For certain information, the PROFINET web integration also provides standardized XML schemes.

Technical Properties

The basic component of web integration is the web server. The web server forms the interface between the PROFINET object model and the basic technologies for web integration.

The PROFINET web integration can be scaled according to the performance and properties of the web server. This means that, in addition to a PROFINET device with an "MS Internet Information Server" or "Apache Webserver," small PROFINET devices equipped with only an "embedded web server" can also participate in the web integration with equal rights.

The web integration for PROFINET has been created in such a way that it can be made available optionally for each device. Certain functions are optional and can be used for the device depending on its load capacity. This enables implementation of scalable solutions that are adapted to the respective use case to the maximum extent possible. The PROFINET-specific elements can be integrated seamlessly in an existing component web implementation. Based on uniform interfaces and access mechanisms, the creator of a technological component can provide his technology-related data via the web. Through the namespace specified in the PROFINET web integration and the addressing concept, elements of the PROFINET component object model (technological variables) can be referenced by the web server. In this way, dynamic web sites configured with current data from the PROFINET component can be created.

Scope

The basic architecture model for the web integration concept is shown in Figure 102.21. Web integration is optional for PROFINET.

With respect to the system architecture of an automation system with PROFINET, all architectural forms are supported by web integration. In particular, the use of proxies for interfacing to any field bus is supported. The specification contains appropriate models that describe the relationships among the PROFINET components, the existing web components, and the PROFINET web integration elements.

Security

The PROFINET web integration specification has been created in such a way that access to the PROFINET devices is the same whether it occurs from the Internet or an Intranet. As a result, all advantages of web integration can be reaped even if the device itself is not connected to the Internet. In the case of local access, there is little risk for an unauthorized access (comparable to today's HMI systems).

For networking within a larger factory or over the Internet, the PROFINET web integration is based on a graded security concept. PROFINET web integration recommends a security concept that has been optimized for the specific use case and includes one or more upstream security zones. No structural restrictions are imposed in the web integration concept because the security measures are always arranged around PROFINET devices. As a result, PROFINET devices are not burdened, and the security concept can be optimized to the changing security requirements for a continuous automation solution.

FIGURE 102.21   Web integration enables web access to PROFINET components

The best-practice recommendations for PROFINET web integration contain scenarios and exam-ples of how requirement-dependent security measures can be implemented around PROFINET devices.

In this way, for example, security mechanisms can be used in the transport protocols (TCP/IP and HTTP). In addition, encoding, authentication, and access management can be scaled in the utilized web servers. Advanced security elements such as application gateways can be added for web services, if required.

OPC

The PROFINET component model and OPC have the same technological basis, that is, DCOM. This yields user-friendly options for data interchange between different plant units. OPC is a widely used interface for data exchange between applications in automation systems. OPC enables a flexible selection of stations of different manufacturers and data exchange between the stations without programming. OPC is not object-oriented like PROFINET, rather, it is tag-oriented. That is, the automation objects do not exist as COM objects but rather as names (tag).

OPC Data Access

OPC Data Access (DA) is an industry standard that defines a uniform user interface to access process data. As a result of this standard, the following are harmonized: access to data of process and control devices, locating of OPC servers, and simple browsing in the namespaces of the OPC servers.

OPC Data Exchange

OPC Data Exchange (DX) defines a communication standard for the higher-level exchange of non-time-critical user data on the system level between controllers of different manufacturers and types (e.g., between PROFINET and Ethernet/IP. However, OPC DX does not permit any direct access to the field level of another system. OPC DX represents an expansion of the OPC DA specification and defines interfaces for interoperable data exchange and server-to-server communication in Ethernet networks.

OPC DX is commonly used by the following:

- User and system integrators who integrate devices, controllers, and software of different manufac-turers and want to enable access to jointly used data in multivendor systems.
- Manufacturers who want to offer products that are based on an open industry standard for inter-operability and data exchange.

OPC DX and PROFINET

OPC DX was developed with the objective of having a minimum degree of interoperability between the different field bus systems and Ethernet-based communication protocols. In order to maintain an open connection to other systems, OPC DX was integrated in PROFINET. Integration is accomplished as follows:

- Each PROFINET node can be referenced as an OPC server because the basic capacity already exists in the form of the PROFINET Runtime implementation.
- Each OPC server can be operated as a PROFINET node by means of a standard adapter. This is accomplished by the OPC objectizer, an SW component that implements a PROFINET device on the basis of an OPC server in a PC. This SW component need only be implemented once and can then be used for all OPC servers.

The functionality and performance of PROFINET are significantly greater compared to OPC. Moreover, PROFINET offers the required real-time capability for automation solutions. On the other hand, OPC exhibits a higher degree of interoperability (Figure 102.22).

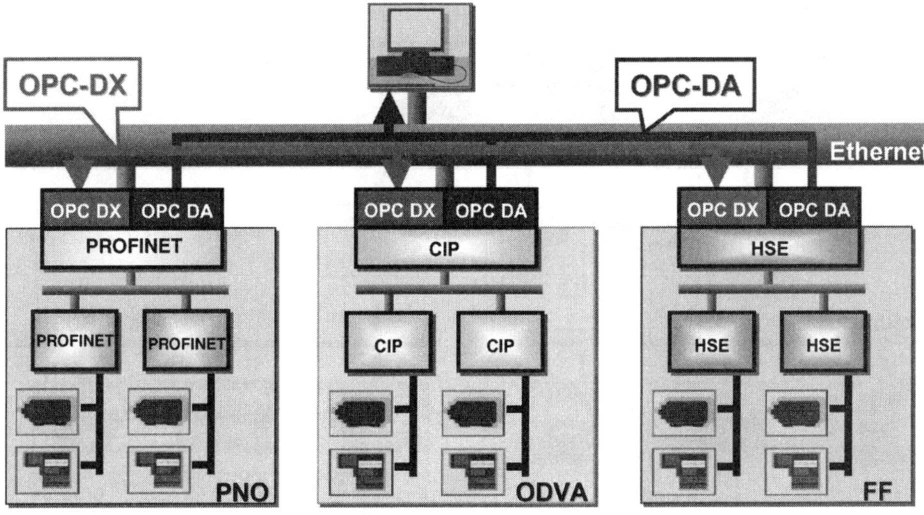

FIGURE 102.22   Cross-system DX with OPC DA and OPC DX

102.8   Integration of Field Bus Systems

PROFINET offers a model for integrating the existing PROFIBUS and other field bus systems. This enables any combination of field bus and Ethernet-based subsystems to be established. In this way, a continuous technology transfer from field bus-based systems to PROFINET is possible.

Migration Strategies

In view of the large number of existing PROFIBUS systems, it is essential for purposes of protection of investment that these systems be able to be easily integrated into PROFINET (migrated) without any changes. The following cases are distinguished:

- The *plant operator* would like to be able to easily integrate his existing installations into a new PROFINET automation concept that is to be installed.
- The *plant construction engineer* would like to be able to use his proven and documented range of devices for PROFINET automation projects without any changes.
- The *device manufacturer* or OEM would like to be able to integrate his existing field devices in PROFINET systems without expending any effort on changes.

There are two ways available in PROFINET for connection of field bus systems:

- Integration of field bus devices by means of proxies and
- integration of field bus applications (Figure 102.23).

Integration by Means of Proxies

The proxy concept in PROFINET enables simple, highly transparent integration of existing field bus systems. The proxy is the representative on the Ethernet for one or more field bus devices (e.g., on the PROFIBUS). This representative ensures transparent communication (no protocol tunneling) between the networks. For example, the representative forwards cyclical data to the field bus devices transparently. In the case of PROFIBUS DP, the proxy is, on the one hand, the PROFIBUS master that coordinates data exchange among the PROFIBUS stations and, on the other, an Ethernet station with

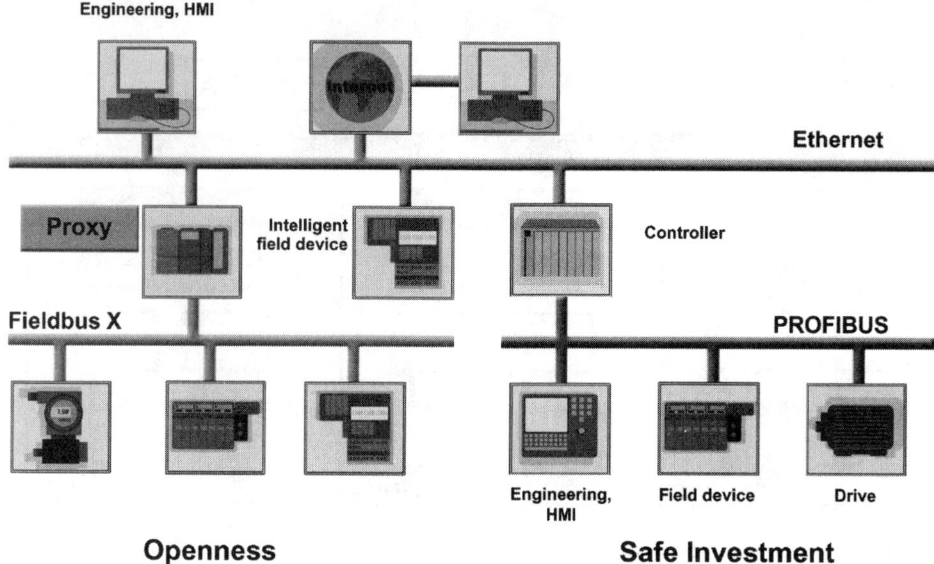

FIGURE 102.23 PROFINET offers openness and protection of investment to device manufacturers and end customers.

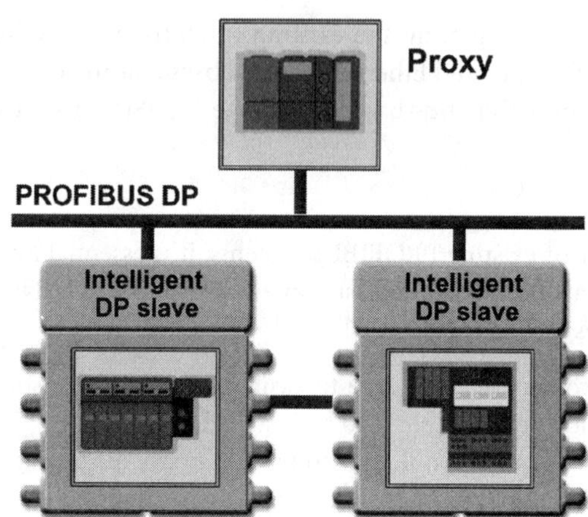

FIGURE 102.24 Principles of integration of individual field bus devices using a proxy: PROFIBUS example.

PROFINET communication. A proxy can be implemented, for example, as a PLC- or PC-based control or purely as a gateway (Figure 102.24).

The DP slaves on the PROFIBUS are handled as IO devices in the PROFINET IO context. In the component view, the intelligent DP slaves are used as stand-alone PROFINET components. Within the PROFINET interconnection editor, such PROFIBUS components cannot be distinguished from the components on the Ethernet. The use of proxies enables transparent communication between devices on different bus systems.

Integration of Field Bus Applications

An entire field bus application can be copied as a PROFINET component within the framework of the component model. This is always important if an existing operating plant is to be expanded using PROFINET. Which field bus was used to automate the plant unit plays no role in this case (Figure 102.25).

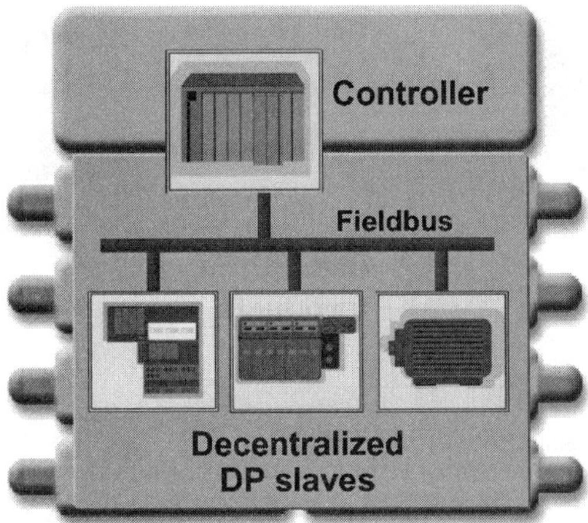

FIGURE 102.25   Principles of integration of field bus applications: PROFIBUS example.

To communicate with the existing plant using PROFINET, the field bus master in the PROFINET component must be PROFINET-capable. Consequently, the existing field bus mechanisms (e.g., PROFIBUS DP) are used within the component and the PROFINET mechanisms are used outside the component.

This migration options ensures that the investment of the user (plant operators and plant construction engineers) in existing plants and cabling is protected. In addition, existing know-how in the user programs is protected. This enables a seamless transition to new plant units with PROFINET.

PROFINET and Other Field Bus Systems

The proxy concept can be used to integrate other field bus systems besides PROFIBUS in PROFINET (e.g., Foundation Fieldbus, DeviceNet, CC-LInk, etc.) A bus-specific image of the component interfaces must be defined and stored in the proxy for all possible data transmissions on each bus. This enables any field bus to be integrated in PROFINET with a manageable amount of effort.

102.9   PNO Offer

Optimal support by the PROFIBUS User Organization (PNO) is important for rapid dissemination of PROFINET in the market. In order to guarantee this, a powerful offer of services and products has been established (Figure 102.26).

Technology Development

PROFINET IO

A specification is available for PROFINET IO, which provides a detailed description of the device model and the behavior of a field device in the form of protocols and communication sequences (so-called state machines). This type of description has already been proven with PROFIBUS DP. The level of detail in the PROFINET IO specification permits software creation of a standard stack from different stack suppliers.

It must therefore be anticipated that different implementations by a few firms will be offered. For example, Siemens offers implementation in the form of a development package.

Component Model

Like PROFINET IO, the PROFINET component technology is available as a detailed specification. The specification includes all aspects including communication, device model, engineering, network

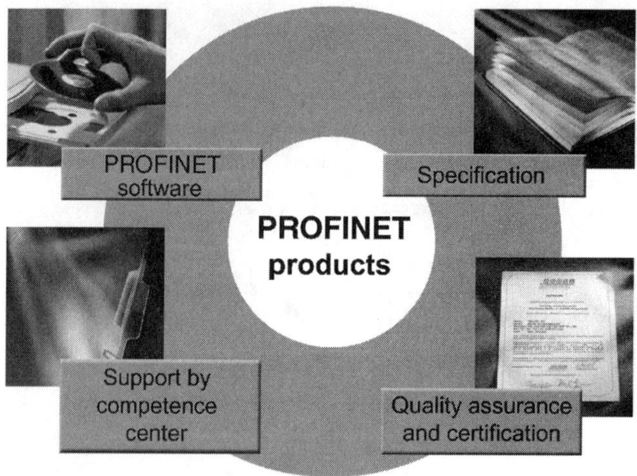

FIGURE 102.26 PROFIBUS User Organization (PNO) offer.

management, web integration, and field bus connection. In addition to the specification, the PROFIBUS User Organization (PNO) is offering PROFINET software (in the form of source code) for the component technology. The PROFINET software includes the entire run-time communication. The combination of a specification and operating system-independent software (as source code) has created the opportunity for easy, time-saving integration of PROFINET into a wide range of device operating system environments.

The PROFINET run-time software is structured in such a way that it supports a simple integration of existing application software in the run-time object model. Sample portings for Win32, Linux, VxWorks, and WinCE are already available for PROFINET. The PROFINET run-time software has a modular design and consists of various layers that must be adapted to each system environment. These adaptations are confined to the porting interfaces for the different functional parts of the environment, the operating system (e.g., WinCE), and the device application (e.g., PLC). Instructions for porting are available, enabling the device developer to easily understand the individual porting steps.

Quality Measures

From the start, the PROFIBUS User Organization (PNO) has developed PROFINET with the assurance that the entire life cycle from the PROFINET specification to plant engineering is supported by measures that guarantee a high level of quality in each phase.

QA for the Specification and Implementation Processes

The PROFINET specification and software are created in a working group represented by multiple companies ("PROFINET Core Team"), whose mission is to ensure that the entire development process ranging from the initial formulation of requirements to the release of the PROFINET software is conducted under the auspices of quality management (QM). Quality measures are governed by a QA manual adapted to the boundary conditions of the multicompany development team. This ensures that the source code corresponds to the quality management rules in effect at that time. The QA manual describes the process model to be applied. It defines the terms, methods, and tools that must be applied in the QA measures. It also specifies the responsibilities in the overall QA process. Defect management represents a major component. Defect management includes a unique error classification and understandable error message communication.

Testing and Certification

To ensure the interaction of all PROFINET devices and a high level of product quality from the start, a certification system has been established. Proven models for PROFIBUS products in accordance with the certification system have been established. Certification tests by accredited test laboratories authorized by the PROFIBUS User Organization (PNO) form the core of the process. The test for obtaining a test certificate by these competent test laboratories ensures that the products offered conform to specifications and are free of errors.

Defect Database

In order for defects and requests from end customers and device manufacturers to be systematically addressed in the run-time software, defect databases have been set up by the PROFIBUS user organization (PNO). A defect database containing a record of all defects and their status is available. The entries in the database comply with the QA process rules.

Technical Support

An important factor for the success of PROFINET is that a sufficient number of PROFINET products from different manufacturers are available on the market within a short time.

Competence Center

The PROFINET Competence Center has been established for support of the product development process. This ensures that porting to the different operation systems and adaptation to product-specific boundary conditions occurs in an optimal manner, particularly in the initial phase when companies lack experience. The Competence Center builds up know-how in interested companies so that development departments can develop additional products skillfully without additional support. The services of the PROFINET Competence Center also include a telephone hotline and customized workshops.

Tools

A tool for device manufacturers is required for creation of a component description of Ethernet devices in the form of an XML file. The PROFIBUS User Organization offers the PROFINET Component Editor — similar to the GSD Editor in PROFIBUS DP — for download on its website (Figure 102.27).

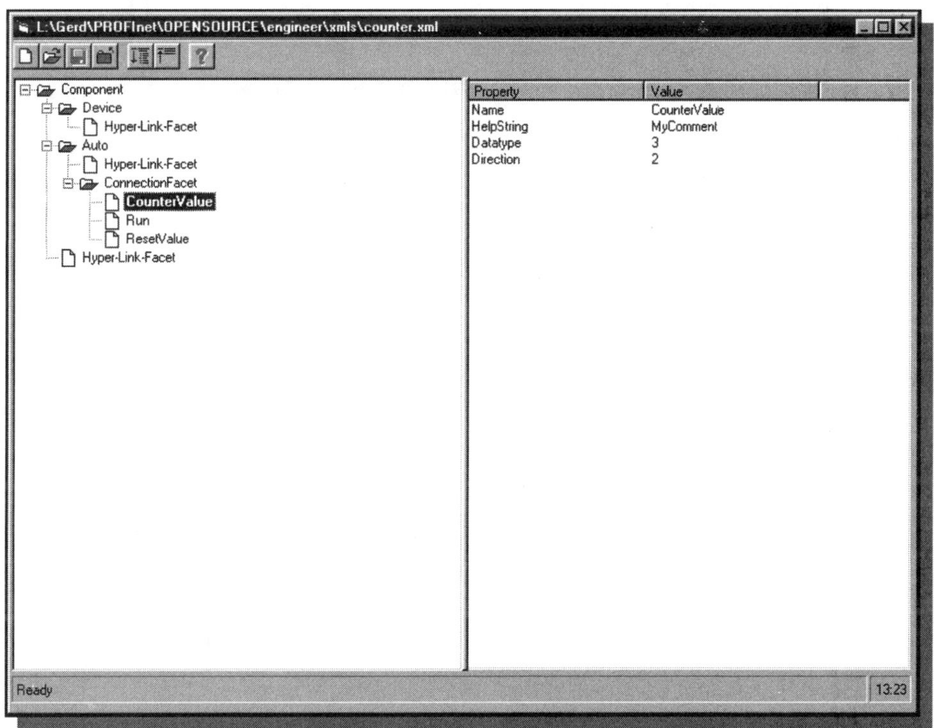

FIGURE 102.27   PROFINET component editor.

In order to prepare newly developed products for certification, the PROFIBUS User Organization (PNO) offers a PROFINET testing tool for download on its website. The PROFINET testing tool enables the device manufacturer to perform static tests prior to certification (Figure 102.28).

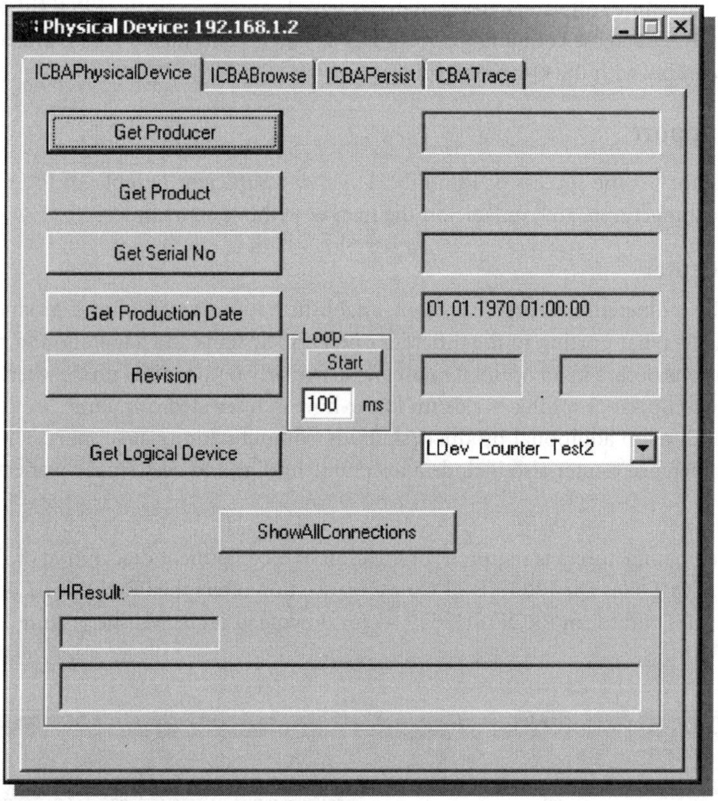

FIGURE 102.28   PROFINET test tool.

103

The IDA Standard

Martin Buchwitz
Jetter AG, Germany

103.1 Introduction

IDA stands for "Interface for Distributed Automation." It is an open standard for automation technology, based on distributed intelligence, aiming to achieve the interplay of tools and devices in a nonhierarchical network, in which each participant is able to communicate with any other participant in realtime.

IDA is based on Ethernet, containing IT and Web technologies, besides safety technology. This goes to show that IDA is far more than just Ethernet technology or simply a protocol. The specifications presented in this chapter illustrate significant elements of the standard.

The Modbus-IDA is an association with the aim to develop a comprehensive standard for distributed intelligence in automation technology, containing software, hardware (i.e., devices as well as their description), and communication. IDA includes all automation devices, even the company's electronic data processing. Whenever standards can be used, IDA integrates them into its standard, thus, for example, the Internet protocols FTP, http, and others, as well as OPC from the automation world. IDA consists of the following functions and technologies:

- Ethernet TCP/IP and web technologies.
- All communication services and interfaces to devices and software.
- Interoperability of devices by differing manufacturers.
- Horizontal integration, that is, communication without interfaces or extra programming.
- Vertical integration, that is, the connection between production and data processing, including the Internet.
- Safety on Ethernet is integrated in the concept.

The aim of all this is to be able to use software and devices by differing manufacturers in a common network with distributed intelligence, which can then be integrated into the overall network by plug and play. What we are talking about are devices such as PLCs, Soft-PLCs, drive controllers, remote I/Os, as well as user interfaces. Any Modbus-IDA compatible tool may be chosen for programming, (Figure 103.1).

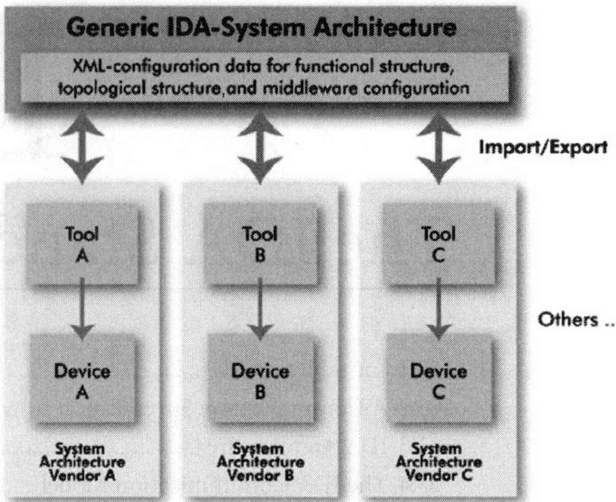

FIGURE 103.1   IDA is a multivendor standard that gives priority to interoperability of devices.

103.2   A Comprehensive Specification is Available

The specification available goes to show as to what is at the core of IDA technology: it is subdivided into the areas architecture, communication, web technologies, and safety, just like the respective work-groups. The specification is supplemented by chapter XDDML (Extensible Device Description Markup Language). XDDML is a system and network-independent description language for devices. This language may contain descriptive elements in various languages in an XML file.

The IDA specification is available and it is open for anyone, ready for download from the IDA website http://www2.modbus-ida.org/idagroup/. The main features of the specification will be outlined in this chapter (Figure 103.2).

103.3   IDA Architecture

The IDA system architecture model is based on the following technological pillars:

1. The application model describes the structure and basic elements of a modular application, based on the reference architecture defined in IEC 61499. Thus, the foundation is laid for subdividing an automation application into any structure of hierarchy, which makes for an optimum detailed presentation of the individual functions required for a specific task.
2. The engineering model contains the plant and machine model, illustrating the relationship between the application model and the engineering framework components. It describes how a concrete automation application is set up, which conditions have to be met, as well as the relationship between the individual elements.
3. IDA defines the logical layer of the programs within the application model, independent of the physical layer of the devices concerned. The process model describes the image of application model elements for their execution, thus forming a bridge between logic and physics, which now merely needs to be configured and will have no influence on the program structure.
4. The presentation model includes the description of the external behavior of the application model elements. To this end, all model elements supply ports of a defined structure through which individual elements may be connected to form an overall network. For an IDA block, these are the Input and Output ports of the event and data connections through which the IDA block cooperates with other IDA blocks. Within an active network of co-operating IDA blocks, the presentation model projects this structure onto the communication aspects of the real-time middleware.

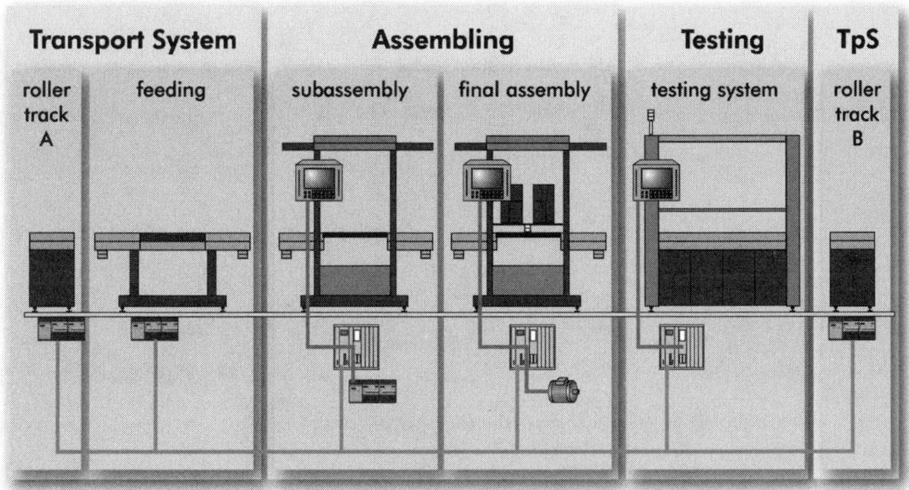

FIGURE 103.2 Especially with modular arrangement of machines, the advantages of distributed intelligence with Ethernet, such as transparent communication structures, become noticeable.

5. Within a distributed automation system, the static connection between visualization and control is difficult to handle, as logic and physics are principally deconnected. As a program can only be connected to the executing resource comparatively late and may even have to be reconfigured, visualization is connected to the logic in the HMI model independent of the physics. In order to have access from any visualization system or user interface simultaneously and independent of their location, the HMI model is based on Web technologies and supplies standardized access through a browser-based user interface for supervising and diagnosing an automation solution.

Thus, the IDA system model addresses all architectonic requirements that have to be met with the design and configuration of hierarchically structured distributed automation systems. As the development of the function block-oriented standard IEC 61499 is based on equivalent requirements, the model defined in this standard has been adopted as a reference architecture for distributed systems in the IDA system model. Proprietary manufacturer-bound technologies have been avoided (Figure 103.3).

103.4 The IDA Communication Model

Communication with IDA is based on an integrated approach containing the modeling of communication aspects as well as the network view of functionality. IDA device communication is based on existing Ethernet communication standards and protocols (such as IP, UDP, TCP, HTTP, FTP, SNMP, DHCP, NTP, and SMTP). Important is the fact that the IDA communication system offers real-time as well as non-real-time communication services. The non-real-time services are mainly based on the Ethernet communication protocols mentioned above, while real-time communication services (such as data distribution, Remote Method Invocation, and Event Notification) are working with the RTPS (Real-Time Publish/Subscribe) protocol, which is based on the UDP protocol. Publish/Subscribe means that information is sent directly from the information source (e.g., the sensor) to the user of the information (e.g., PLC). The source takes the initiative for sending the data, in contrast to the request/demand method in master/slave solutions. The publisher makes data available under a name in the network, which the subscriber will use to identify the data it wants to receive.

In order to realize the configuration and execution of real-time communication services, IDA specifies an object-oriented model building a hierarchy of communication objects, which can be accessed through a Application Program Interface (API). In the process, specific requirements of safety-relevant applications

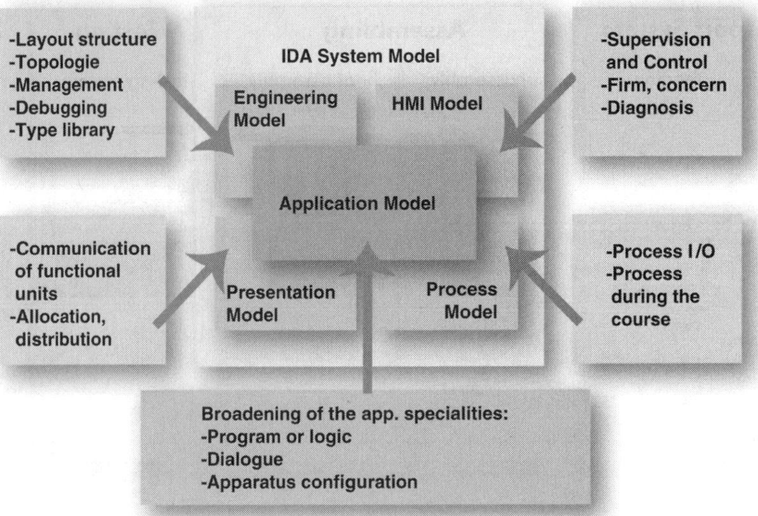

FIGURE 103.3 The IDA system model takes into account the whole cycle of an automation project, from configuration to runtime, and maintenance.

are supported through a special safety API. The illustration of IDA functions is based on a hierarchical view of objects describing the physical and logical structure of a devices with its communication performance and the application data connected to the network.

Real-time communication with IDA is based on the use of the RTPS protocol. In general, real-time services have top priority among all IDA communication services. Depending on the requirements of the application, various kinds of real-time communication relations and resulting network traffic will appear: preconfigured or dynamic, cyclical or on-demand, point-to-point or group oriented.

103.5 IDA and Modbus TCP/IP

Worldwide, most devices communicate via industrial Ethernet using Modbus TCP/IP. The IDA Group has decided to closely cooperate with the Modbus user group, and is adopting Modbus into their IDA Standard as a quasistandard for Ethernet communication in automation technology. Modbus TCP/IP has been presented to the IETF (Internet Engineering Task Force) in order to officially establish the protocol as an Internet standard. Thus, Modbus TCP/IP will be implemented as a standard in all common operating systems, as well as FTP or http. Apart from this, Modbus occupies one of the first 1000 defined ports, namely port no. 502. The fact that this port is no longer available for other applications is another factor that helps to establish Modbus as a standard (Figure 103.4).

Modbus TCP/IP is a part of the IDA standard, which leads to a situation where a large installed basis and several hundred manufacturers will be available for it. As Modbus TCP/IP is completely transparent, it perfectly fits into the IDA strategy of open standards. IDA will actively support the activities of the Modbus organization in their work groups, participating in the development of tool kits. Apart from this, there may be an even closer cooperation in the future.

Strategically, two things are achieved by integrating the Modbus TCP stacks into the IDA stack: firstly, the largest actually existing Ethernet standard for devices meets the lowest IDA conformity level, and has the option for other upward compatible services based on the Client–Server and the Publish–Subscribe model. Secondly, in this way, IDA is making itself available for being implemented in very low-performing devices, due to an especially lean stack.

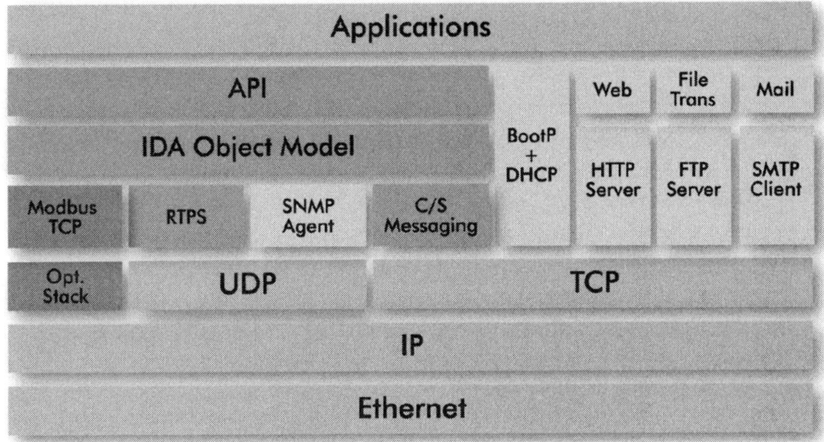

FIGURE 103.4 The IDA standard now also contains the Modbus protocol, thus addressing the major part of the Industrial Ethernet market. Modbus is in the process of becoming an Internet standard, which means that in the future, it will be available as standard in all relevant operating systems.

103.6 Using Web Technologies

The IDA standard defines varying web-based device services:

- access to device-specific data required for the definition of user-defined web-sites,
- display of device characteristics for device-specific web-sites,
- modification of the device configuration for the illustration of the state of the device, and of parameters,
- device diagnostics for local or remote maintenance,
- remote maintenance with save/restore function for data download by means of FTP,
- monitoring and user input,
- creating and saving of user-defined web-sites,
- device documentation, and
- Web HMI services for alarm functions, data history, and mathematical functions in the form of Java code.

The IDA standard defines four differing types of web servers: depending on the device and its performance and characteristics, there will be a scalable solution. The fact that CPU and memory resources for integrated systems are limited must be taken into consideration. In the case of IDA devices, these resources are mainly reserved for real-time functions, such as real-time communication.

The web server as such may contain various functions, as well as an installation capacity ranging from 7 to more than 50 kB, with a complete server. According to the large range of device characteristics, the server concept has been subdivided into four categories.

Each category describes a web server fitting into a certain type of device and meeting certain IDA Web requirements. The hardware and firmware characteristics as well as requirements regarding the server are decisive for the choice of a web server. The following classes have been defined (Figure 103.5):

- Minimum Server.
- Device Configuration Server.
- Configurable Server.
- Web Active Server.

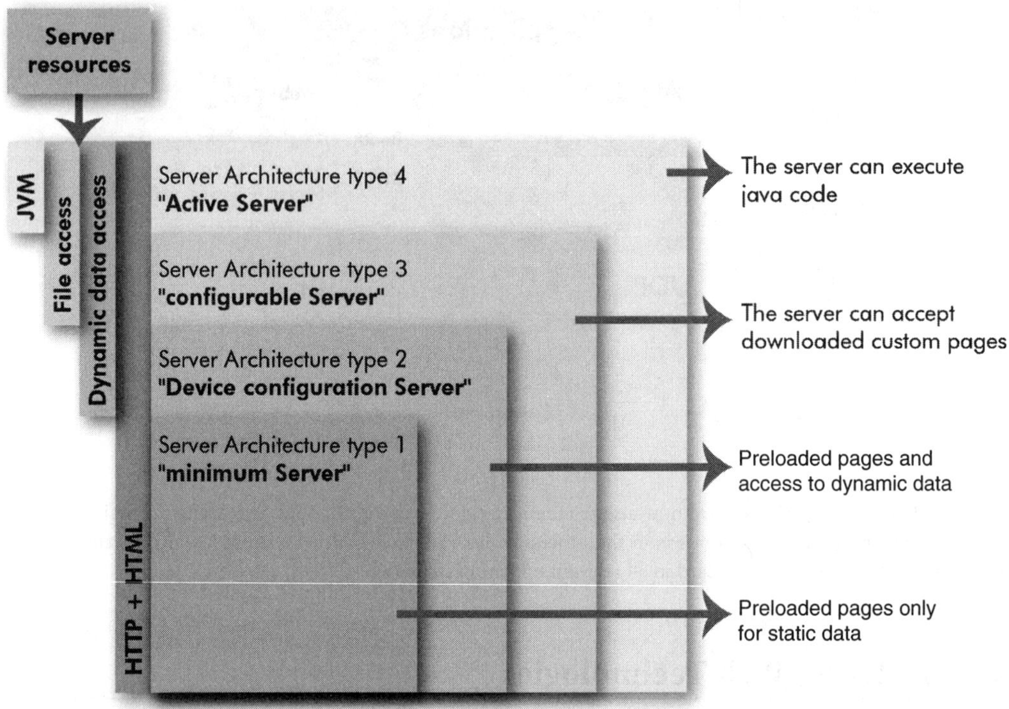

FIGURE 103.5 An important requirement for the universal use of Internet technologies is the availability of devices. To accelerate this process, the Modbus-IDA has developed a four-stage Web-server model for devices of differing performance.

103.7 Safety Technology Integrated

Basically, the current Ethernet transmission methods and formats do not meet the safety-oriented automation standards. Ethernet-specific safety measures are able to detect simple failures; effective algorithms for identifying and handling repetitive statistical or systematic faults are missing, however. Beyond this, the Ethernet standard is only relevant for data transmission, as well as their formats and services. Demands on a safety-oriented Ethernet, however, are much higher: in this case, the failure model must not only refer to the bus system and data traffic as such but also include application, interface technology, power supply, as well as the transmission medium.

The communication model chosen for Ethernet contains a specific safety layer (IDA Safety API), supplying all safety data with formats and services that are able to meet all common faults related to data transmission. As subordinate layers are not altered, the entire communication is fully compatible with standard Ethernet.

103.8 Summary

IDA is more than a simple protocol on Ethernet, and more, too, than just an Ethernet technology. IDA is a standard for modern structures of distributed intelligence, using the most recent software, hardware, and network technologies. IDA eliminates traditional barriers within automation technology (horizontal integration) as well as barriers between production and data processing (vertical integration). Whenever possible, already existing standards such as Ethernet TCP/IP, XML, http, OPC, and others are used and integrated into the entire IDA concept. IDA comprises the area of engineering using its tools, as well as the entire communication sector with its demands on realtime and safety.

104

Open System Architecture for Controls within Automation Systems (OSACA)

Michael Seyfarth
ISW-University of Stuttgart

Andreas Kahmen
Werkzeugmaschinenlabor

104.1 About OSACA

Open System Architecture for Controls within Automation Systems (OSACA) is a European initiative to define a vendor-neutral, open controller architecture in order to improve the competitiveness and flexibility of suppliers and users of control systems: machine tool builders, control vendors, and end users.

To cope with these aims, the requirements for a new generation of control systems were analyzed. Out of this preparatory work, an application programming interface (API) for control applications and an appropriate infrastructure (the so-called system platform) were specified and implemented.

104.2 OSACA Technical Overview

The control functionality was divided into several so-called Architecture Objects (AOs), which use the API and the underlying communication system to exchange data. Using this concept, the OSACA architecture supports the characteristics of an open control system: interoperability, portability, scaleability, and reusability of applications Figure 104.1.

Pilot applications and control prototypes were implemented on different operating systems in order to verify and improve the specifications.

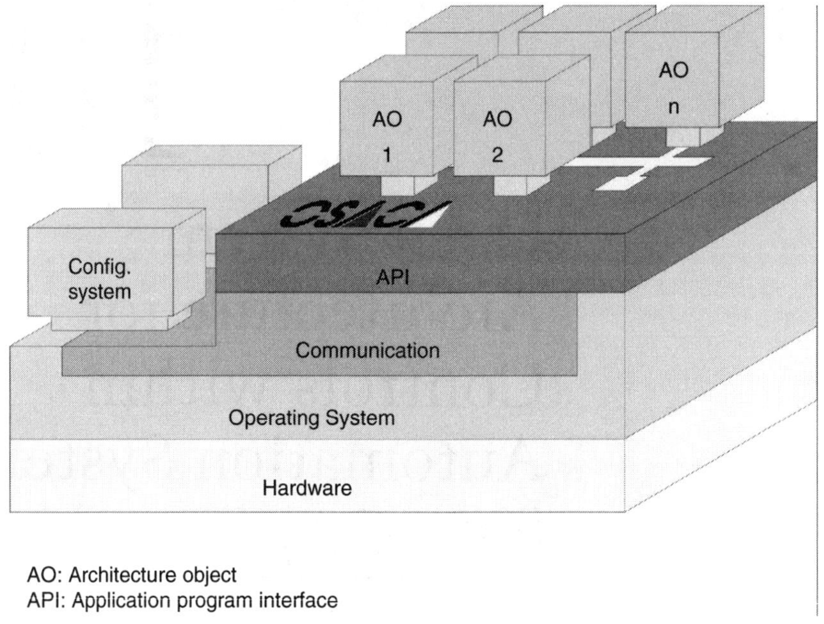

FIGURE 104.1   OSACA system architecture.

Three main features of OSACA include:

Communication Platform

This defines a hardware- and system-software-independent interface to exchange information between different application modules of a controller. The interface hides any specific details of the underlying transport system and therefore enables the combination of modules distributed on one or several processor-boards.

Reference Architecture

This determines the functional units of a controller, such as, for example, Motion Control (MC) and Logic Control (LC) and specifies their external interfaces. This makes it possible to use and integrate functionality of external units by interacting with internal data in a well-defined manner.

Configuration System

This enables the dynamic configuration of a controller by combining different application modules at bootup time. This allows not only the setup of a specific topology for a given functionality, but also the synchronization between distributed processes. A graphical configuration tool allows the simple projection of control applications.

Communication Platform

The communication system is the only means of the system platform for the exchange of data between application modules, so-called AOs. It must support both, the exchange of information between tightly coupled AOs, for example, located on the same processor-board as well as between AOs distributed over different hardware units connected through a bus-system. A standardized protocol is defined to ensure uniform data formats and a fixed set of messages. The protocol architecture is derived from the OSI base reference model. However, it comprises only two layers: a Message Transport System (MTS) equivalent to the OSI layers 1 to 4 and an Application Services System (ASS) equivalent to layers 5 to 7 (Figure 104.2).

The *MTS* offers connection-oriented services for a transparent transport of arbitrary messages between AOs. It can be adapted to use any kind of existing mechanisms for information exchange, for example, operating system services like message queues and shared memory, as well as LAN-protocols like TCP/IP.

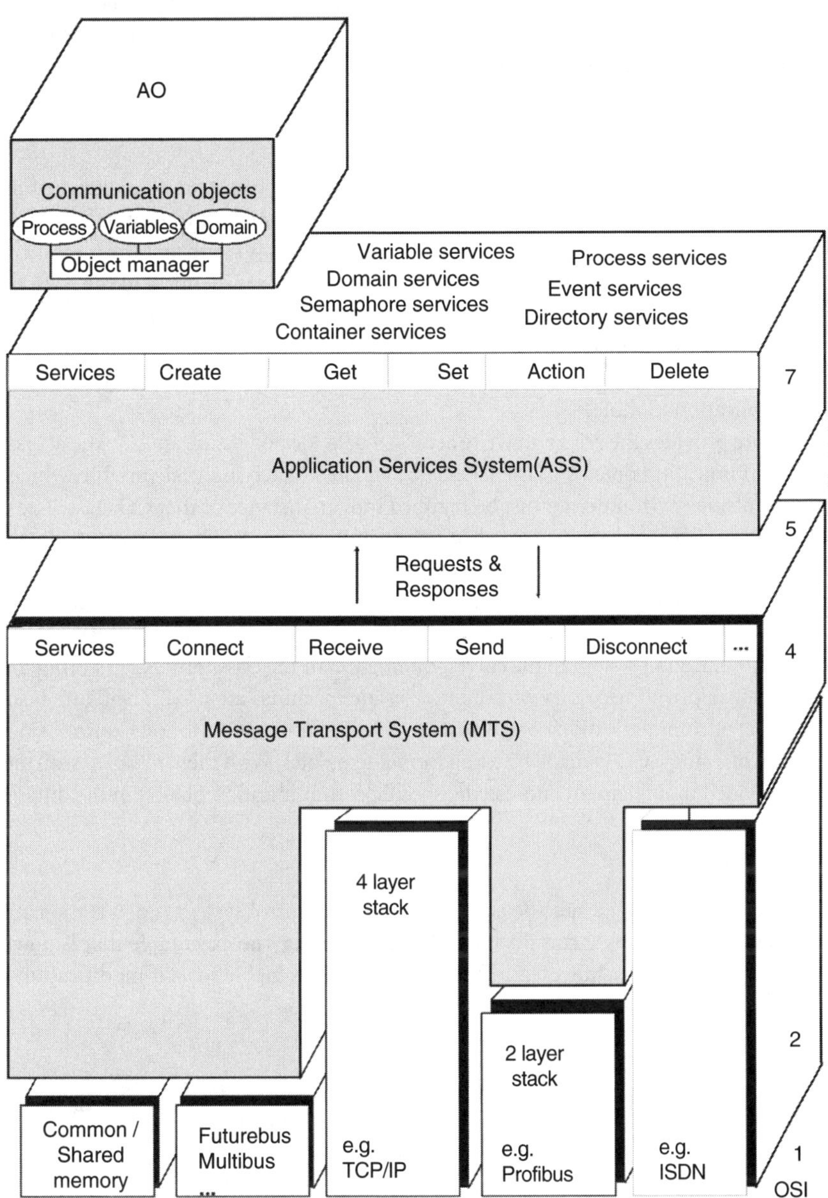

FIGURE 104.2   OSACA communication platform.

The *ASS* has the task to handle the application protocol. This includes connection management, the assembly/disassembly of messages, and data conversion between different data representations like Little and Big Endian. Additional functionality covers synchronous and asynchronous services as well as the segmentation of large data into smaller message units.

The OSACA application protocol is realized on a client/server basis using object-oriented principles. With the help of so-called communication objects, a server AO provides access to its data and functionality. From a client's point of view, a server includes a set of communication objects that can be accessed by sending and receiving messages using the communication system.

There is a fixed set of classes for Communication Objects (Cos), which are managed by the so-called Communication Object Manager (COM). Through the separation of application and communication services, the application programmer can fully focus on solving his control-specific problem (Figure 104.3).

Reference Architecture

As described above, the OSACA communication system allows for a transparent exchange of information between client and server applications. In order to enable an unequivocal interpretation and use of the exchanged data, additional definitions are required. These definitions cover *first* the logical classification of functionality into so-called functional units, which in most cases are identical to the AOs. Examples of functional units are: Man Machine Control (MMC), LC, MC, and Axis Control (AC).

Second, for each of the identified units, the functionality and behavior of external modules have to be described. This is done by using communication objects that form the data interface and the process interface of an application module.

The *data interface* provides the read and write access to data located inside an AO. The access is enabled by the use of communication objects of the CO class "variable." Each internal variable, which should be accessible by the outside environment, must be mapped into an instance of that CO class. The tuple comprising the internal variable and the CO is called an attribute of the AO. From the outside, these attributes can be accessed by calling appropriate services of the communication system.

The *process interface* defines the dynamic behavior of an application module. It allows to trigger and control the functionality that is provided by an AO. The call to an internal function is initiated by an "action" service via the communication system to the corresponding CO of the class "process." To bring the functions into a logical relationship with necessary preconditions state machines are used. The figure below gives an overview on a subset of communication objects that are defined for the AO "Motion Control" (Figure 104.4).

The description of a single CO is made by using formal templates. With these templates, all the necessary information is provided to implement and use the specific communication objects in the different AOs.

Configuration System

Typically, the process of realizing a specific configuration of a control system is of a static nature. Before runtime, the software is completely compiled and linked, forming one executable that is downloaded to the control level. This solution is, however, inflexible and requires high efforts if modifications and additions are necessary after the delivery of the control level.

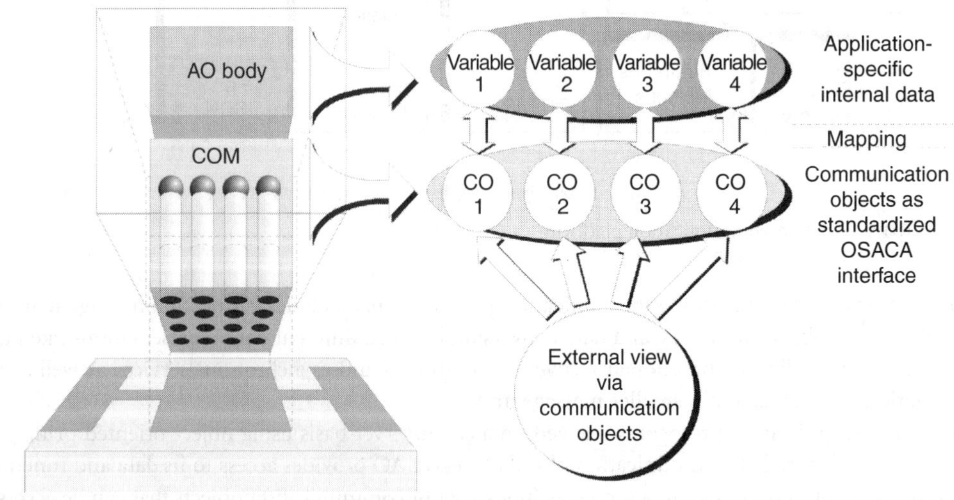

FIGURE 104.3   Data exchange via Cos.

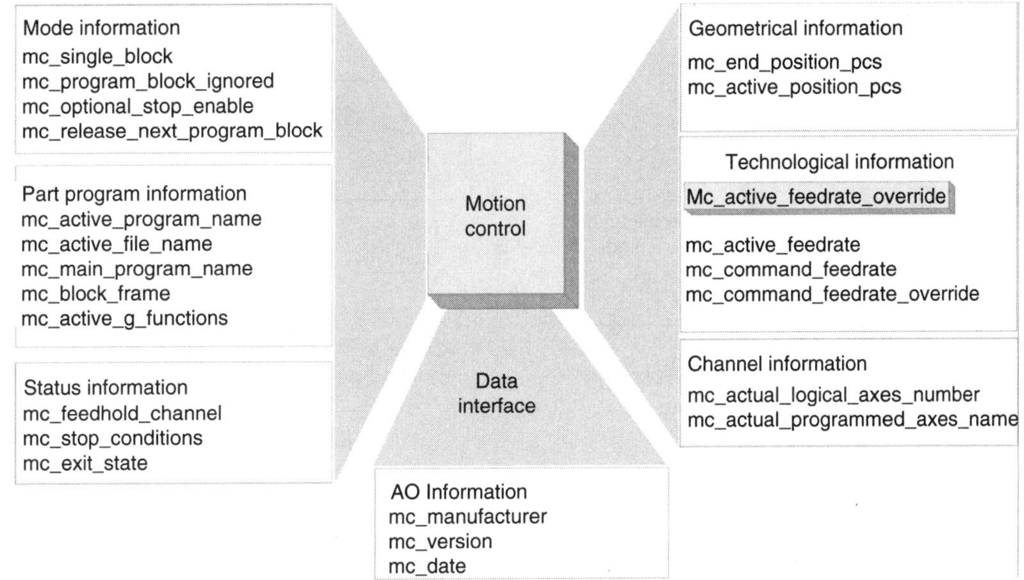

FIGURE 104.4   A subset of Cos.

To overcome this situation, a fully modular system is needed where the actual topology of the software is generated at the bootup of the system. For this, the system platform has to contain a configuration system that is able to handle a library of AO classes. At bootup, AOs of the different classes are instantiated and communication connections between AOs are established.

The actual topology of the control system is described in an externally generated ASCII configuration list, which is interpreted by the Configuration Run-Time System (CRS). A graphical configuration editor can be used to define the configuration order in the manner as CAD-systems are used to define the layout of a processor-board.

The configuration order contains a list of all AO instances that are necessary for a specific control system satisfying the requirements of a control application. It is possible that there are several instances of the same AO class, for example, class AC. For every AO, the client/server relationships are defined. Every relationship between a client and a server AO consists of a list of communication objects of the server, which will be used under specific names in the client (Figure 104.5).

104.3   Applications

Demonstrators

OSACA-Demonstrator at ISW

In order to demonstrate the benefits and capabilities of OSACA and to show the feasibility of a control based on its architecture, a demonstrator machine was set up at the Institute for Control Engineering of Machine Tools and Manufacturing Units (ISW), at the University of Stuttgart in 1994. It is a portable three-axis milling machine with conventional interfaces to the drives, the analog, and the digital I/Os. Figure 104.6 shows the front side of this demonstrator.

The control system runs on different hardware platforms, using Windows95 for operator-related functionality and VxWorks for machine-related control functionality. The NC-kernel is a compilation of software control units written in C-language. These units belong to a modular control construction system that is commercially available. For accessing the functionality of the NC-kernel, an OSACA compliant gateway process was implemented in the real-time part.

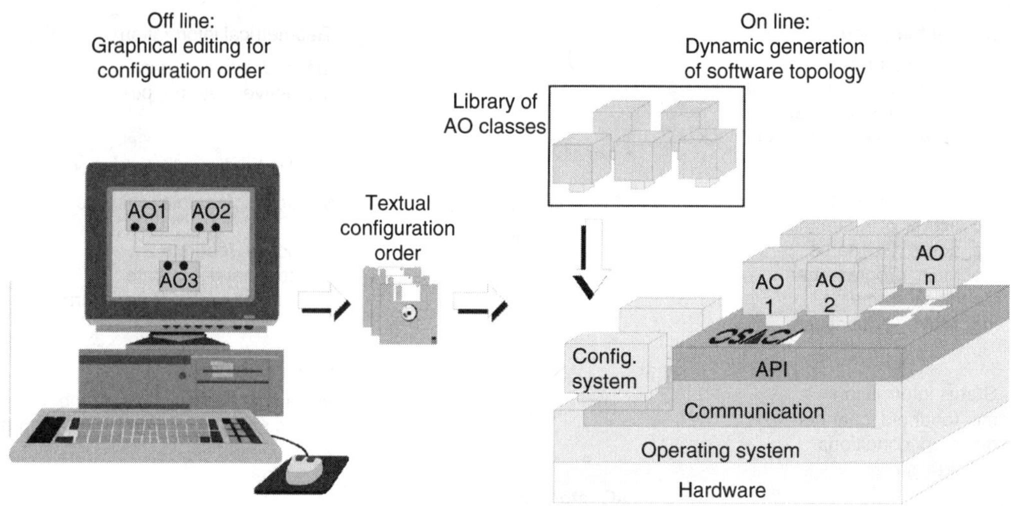

FIGURE 104.5   System configuration in OSACA.

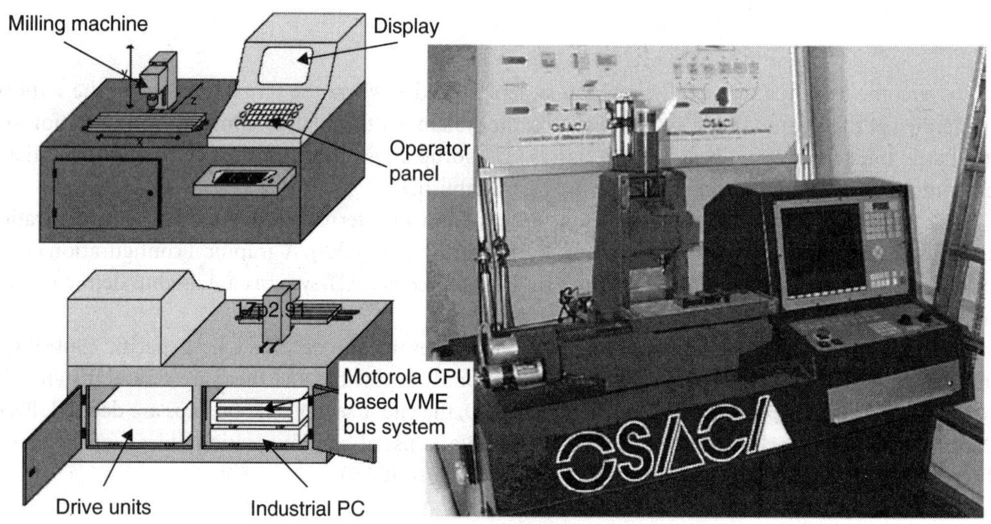

FIGURE 104.6   The OSACA demonstrator (three-axis milling machine).

The demonstration does not only show the communication between clients (e.g., HMC) and servers (e.g., NC-kernel) but also proves that existing control applications can easily be equipped with an OSACA-compliant access.

OSACA-Demonstrator at WZL

In the beginning of 1995, the development of a new NC-controller named "WZL-NC" commenced at the WZL (Laboratory for Machine Tools and Production Engineering). This was the time when the first OSACA software was available. It was decided to use OSACA as the platform basis for that controller to gain the benefits of OSACA (modular, configurable, extendible platform, independent from hardware and operating system). A second reason for choosing OSACA was to develop a test-bed for OSACA to gain experiences in an industrial environment and to improve the platform software.

The WZL-NC is connected to a five-axis milling machine as shown in Figure 104.7. The NC offers fully spline-based algorithms for milling sculptured surfaces.

Pilot Applications

At the end of the HÜMNOS project, in 1998, one unit of a transfer-line at Daimler Benz was equipped with an OSACA-control System (CNC and SPS). A newly developed MMI Control Panel was connected via the OSACA interface. Another Pilot Application was realized at the BMW engine factory in Munich, also in 1998. Here, an assembling machine unit was driven by a SIEMENS S5-PLC with an OSACA architecture.

OSACA CNC-kernels

ISG-NC

The Industrielle Steuerungstechnik GmbH (ISG, www.isg-stuttgart.de) has realized an OSACA interface for their modular CNC-software-kit. The OSACA interface is running on the real-time operating system (VxWorks or OS-9) together with the CNC-kernel.

WZL-NC

At the Laboratory for Machine Tools and Production Engineering, a five-axis numerical control for milling was developed. This control serves WZL as a development platform and test environment for most diverse research tasks within the range of the numerical control technology.

The WZL-NC was realized as an OSACA control. Its software modules and its structure correspond to the OSACA reference architecture. Over the thus available open control interfaces, external developments can be integrated and tested simply in the control environment.

At present, the control runs on a serial 5-Achs MAHO milling machine and a Dyna-M kinetics.

HÜMNOS Applications

The HÜMNOS project (1995–1998) was based on the OSACA results. Within HÜMNOS, applications for the OSACA architecture were elaborated in different areas, such as CNC, PLC, HMI, Tool Management, Pallet Management, Maintenance and Diagnosis. Also, tools like a data monitor or a configuration system were developed (Figure 104.8).

FIGURE 104.7 Five-axis milling machine at WZL.

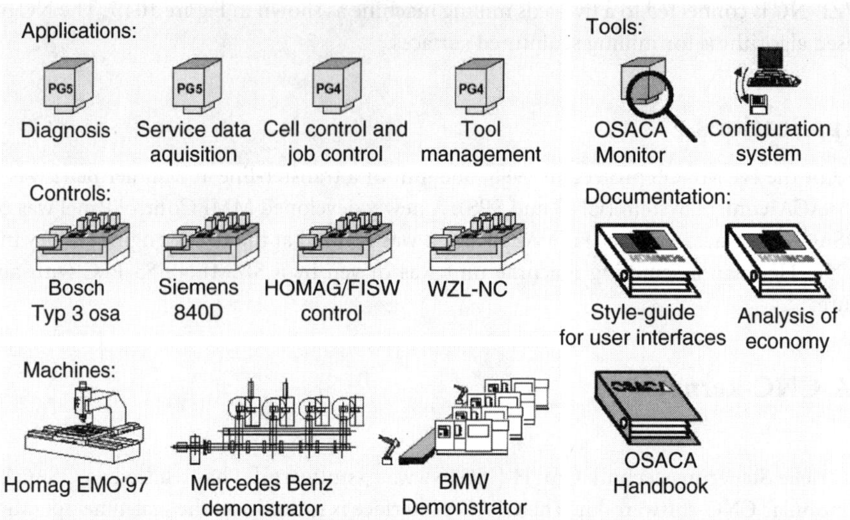

FIGURE 104.8   Application and tools within OSACA.

Selected applications include:

- Windows-based Tool Management for OSACA controllers.
- Windows-based Diagnosis Application for OSACA controllers.
- Windows-based MDA-Application for OSACA controllers.
- Windows-based Application for Order Administration and Processing Management on OSACA controllers.

Tools

Configuration Tool for OSACA Controls

New possibilities for flexibly generating control software of high quality result from using open control architectures; control systems can be configured uniformly and individually from libraries, using modular concepts in accordance to the demanded function range and efficiency. With the right tool, one can thus save cost and time. Therefore, a configuration tool for open controls was developed at the ISW (www.isw.uni-stuttgart.de).

The configuration process

Control software elements are deposited in a module library. They are encapsulated in the independent units, which contain both structural and functional information. This causes a high reusability of the modules.

Based on this library, new configurations are generated by dragging and dropping the modules into the project and by graphically connecting the communication interfaces. Due to this intuitive operation, the tool can be used by individuals after relatively short theoretical and practical training courses. With the help of automatic test mechanisms, configuration errors can be avoided.

After the configuration process has terminated, the target code for the control is generated.

Characteristics of the configuration tool

- Support of the model-based configuration: In a kind of "building plan," the project-neutral interactions of the control software elements that have been deposited in the library are described.
- Intuitive operation: Drags and drops elements from the library into the project, and provides an easy creation of communication relations by graphically connecting two pins (= communication interfaces).

- Parameter setting of control software elements: The functionality of a module can be adjusted easily using comfortable editors.
- Automatic testing mechanisms: The generation process of a configuration is checked dynamically by integrated modules. In this way, faults can be avoided to the maximum.
- Generation of different control target codes: By using post processors that can be exchanged easily, the configuration result can be adapted to any arbitrary control.

Summary

Using the configuration tool, the time for creating a control configuration by using libraries can be reduced. Software can thus be produced at a lower cost. At the same time, the quality of software improves as a result of automated testing, and reuse of models.

OSACA Development Environment

For an efficient and user-friendly compilation of open control applications, a development tool has been realized by the Laboratory for Machine Tools and Production Engineering (WZL).

The OSACA Development Environment (ODE) enables the user to integrate process-specific know-how comfortably into controls without knowing details on how to program OSACA applications. The ODE continuously supports the software development all the way from the idea up to the finished application. It comprises the following functionalities:

Specification Editor

The Specification Editor supports the description of the modules' interfaces and the test of the system design already in early development phases.

Code Generator

The Code Generator provides a software framework for a module. Through this, the particularities of the communication system are completely hidden from the application programmer. In this way, he can concentrate on filling the framework with his specific functionality.

Data Monitor

The Data Monitor enables to analyze the interfaces of modular, distributed systems and to manipulate their contents. The optional logging and evaluation of exchanged data telegrams supports debugging on the lowest communication level.

OSACA-OPC Server

The OSACA-OPC (OLE for Process Control) Server is a Windows-based application for OSACA-compliant control systems to exchange data with OPC client applications, for example, Wonderware, Visual Basic, etc. The server application translates the OSACA protocol into the OPC protocol (Figure 104.9).

Projects Based on OSACA

SFB 368 "Autonomous Production Cell"

At Aachen University of Technology, a group of about 20 scientists from 10 different research institutes is currently developing the new concept of manufacturing systems called the autonomous production cell (APC) (http://sfb368.rwth-aachen.de/).

In this project, a "manufacturing cell" consists of a single machine including necessary handling systems and a worker. The goal is to develop the necessary technical means to enable such manufacturing cells to undertake complex machining tasks largely independent of other units within the company. To assure the failure-safety of the system, several process monitoring and process control modules are being incorporated into the control of the APC.

To realize the integration of the process monitoring and control functionality into the control, an open control system is needed. While first tests were undertaken using a commercially available control system offering limited openness, further implementations required a completely open control. The Laboratory

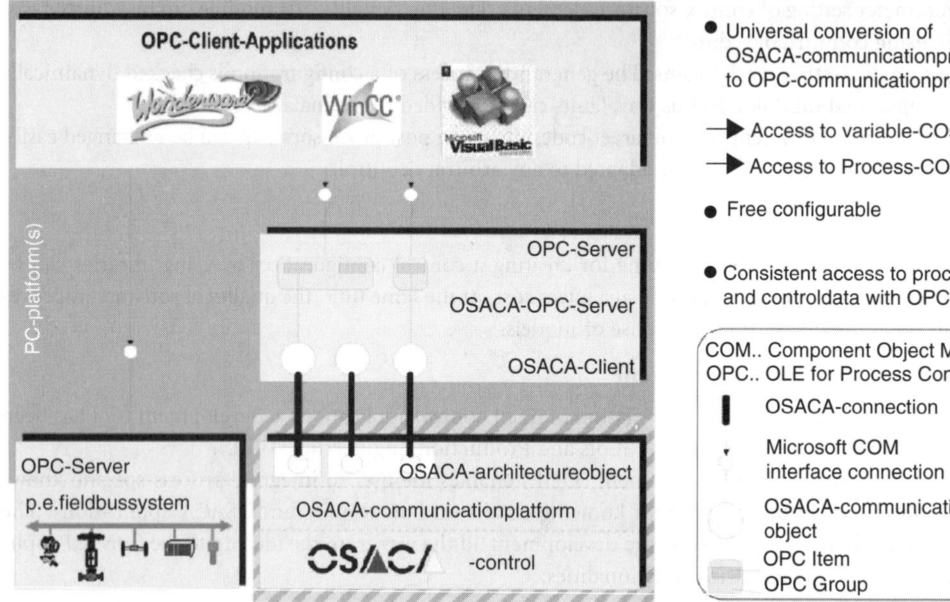

FIGURE 104.9   Structure of the OSACA-OPC Server.

for Machine Tools and Production Engineering developed a fully OSACA-compatible NC, which is the basis of the APC control. This OSACA control offers the necessary preconditions for the realization of the cell's autonomous functions.

SFB467 "Wandlungsfähige Unternehmensstrukturen für die variantenreiche Serienproduktion" (Transformable Business Structures for Multiple-Variant Series Production)

At the University of Stuttgart, a group of about 25 scientists from 9 different research institutes are currently working on the development of new concpts for changeable company structures demanded by high frequently changes driven by global competition.

One task is the realization of reconfigurable machining systems, including the control system. To include the new mechanisms of reconfiguration, there is a need for an fully open, modular, and configurable control system. The system is based on the principles of OSACA and OCEAN and allows the implementation of new modules like automatical reconfiguration.

OSACA Project Background and Phases

In Europe in 1992, the research project OSACA was started within the ESPRIT program of the European Commission. Within this project, the basics for a vendor neutral control architecture were defined. The focus was on the specification of a reference architecture for the system platform, the definition of a neutral application layer, and the development of communication mechanisms. This resulted in a strict division of hardware, system software, and application software. Furthermore, the OSACA architecture is characterized by the availability of its unique programming interface as well as the modular structure of its application software.

The OSACA project was carried out in the following phases:
OSACA Phase I (EP 6379); duration: May 1992 until October 1994.
OSACA Phase II (EP 9115); duration: November 1994 until April 1996.
OSACA - IDAS (EP 22168); duration: January 1997 until June 1998.
HÜMNOS; duration: January 1996 until December 1997.

In August 2002, the research project OCEAN, http://www.fidia.it/english/research_ocean_fr.htm (Open Controller Enabled by an Advanced Real-Time Network), funded by the European Commission, commenced. The OCEAN (IST-2001-37394) is envisaged to run from August 2002 until July 2005. This project aims at the development of a real-time capable platform for distributed control. This platform will enable a dynamic integration of control components that are based on an open specification. Through this approach, a flexible and application-specific configuration of the control system will be possible. The OCEAN project draws from OSACA by adopting solutions and incorporating newly available technologies such as Real-Time Linux and Real-Time CORBA.

105

Open Controller Enabled by an Advanced Real-Time Network (OCEAN)

Fabrizio Meo
FIDIA S.p.A.

105.1 OCEAN Background

Open Controller Enabled by an Advanced real-time Network (OCEAN) is a project funded by the European Commission under the Information Society Technologies (IST) priority. The activities are conducted through a consortium of ten partners, as of the end of 2003. The project started in August 2002 and will be completed by July 2005.

OCEAN is realizing a real-time-capable platform for distributed control applications. This platform will enable a dynamic integration of control components that are based on an open specification. Through this, a flexible and application-specific configuration of control systems will be possible.

The basic idea is to continue the approach of OSACA by adopting its benefits and realizing a fundamental improvement on the basis of newly available technologies. This strategy promises a broad industrial acceptance of open control technologies based on the DCRF.

105.2 Objectives

The two main objectives of the project are:

1. The development of a "Distributed Control System Real-Time Framework" (DCRF) for numerical controls based on standardized communication systems and delivered as an open source. The DCRF is envisaged to host control components in distributed open platforms and provide a real-time communication Application Programming Interface (API). This framework is to be coupled with standardized interfaces enabling the integration of external real-time-critical and non-real-time-critical control components. The DCRF is based on open source components (e.g., RT-Linux as operating system, RT-Common Object Request Broker Architecture (CORBA) for standardized communication mechanisms) and will be available, as an open source, at the end of the project. Figure 105.1 shows the structure of the DCRF and motion control components.

   The communication within a complex, dynamic environment requires a maximum flexibility for the data transfer. Hence, the deployment of transparent communication standards is essential. Nowadays, reliable open communication systems like CORBA are available, which follow the object-oriented concept for data exchange. This property guarantees a transparent handling of data objects and furthermore makes use of the network transparency, which hides from the software developer the problems caused by the communication between distributed components. Embracing this philosophy means bringing a component-based view of the software to a network distributed environment. Applications use object interfaces without the need to know whether they use a local or a remote object. In terms of communication, this approach is the backbone of the DCRF that closes the gap between the distributed control components. With respect to communications, a main focus of the OCEAN project is on the real-time capability of the DCRF. Real-time communication is crucial for distributed control systems because it is a prerequisite for a productive, high-quality, and fail-safe production. The DCRF will be real-time capable in contrast to former approaches like OSACA, OPC, etc. This capability is crucial to distributed control systems.

   The DCRF will enable a flexible composition and reconfiguration of control systems with manufacturing task-specific functionality on the basis of a common communication platform. Suitable

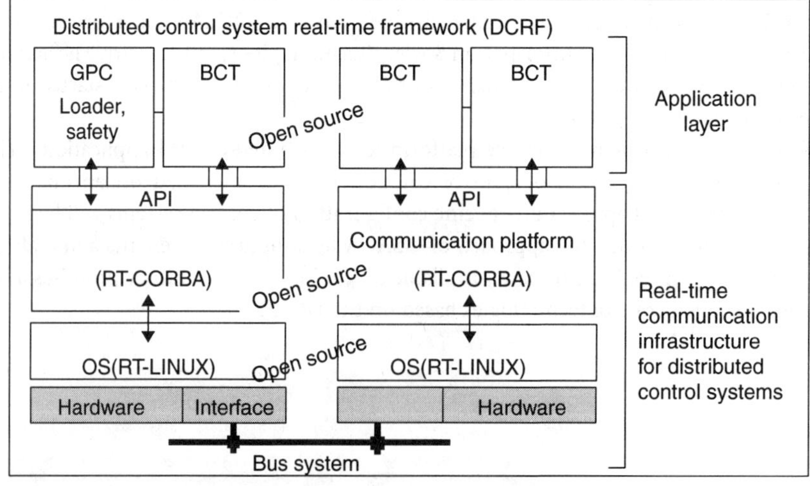

FIGURE 105.1 Distributed Control System Real-Time Framework (DCRF).

lower-level layers (hardware, bus system, RTOS) and the communication platform (RT-CORBA) have to be extended and integrated, so that they can serve as a distributed real-time framework for the whole control system with all its single components.

General-purpose components (GPC in Figure 105.1) for configuration, loader, and safety aspects are also a very important part of the DCRF. In particular, safety functions will be designed and implemented, which can assure a fail-safe operation of component-based control systems in industrial production.

2. The definition and realization of a component-based open numerical control reference architecture for machine tool; in order to take advantage of the open control systems, it is necessary to extend the existing reference architecture and to decompose the monolithic blocks into components with clearly defined interfaces, described in a standardized format. The open control reference architecture will not be delivered as an open source but with publicly available new standardized interfaces for motion control components of machine tools, which will make use of the DCRF. These interfaces will be published for further comments and implementations by users in the field of control techniques (the Base Component Templates (BCT) visible in Figure 105.1). Thus, it will become possible to integrate additional functionality and third-party software by just using the standardized interface description without a need to adapt interfaces. The component-based reference architecture, which will be developed in the OCEAN project, has to be seen as a starting point for the specification of further control system applications in the future. The first step is the design and implementation of motion control base components that are mandatory to operate a machine such as HMI, motion control kernel, PLC, and kinematics components (Figure 105.2).

The second step aims at the design and implementation of motion control extension components that are not yet covered by any reference architecture, such as filter and process control components. Particular focus will be put on a new safety component that is specifically control related and includes mechanisms like real-time comparators for a multichannel environment. The component-based concept will be proved with tests and simulations of the single components as well as of the complete control system on different machine tool demonstrators.

Clearly, the OCEAN project has a potential to substantially influence the practices of major vendors by demonstrating that the adoption of the OCEAN open source-based architecture (Linux, CORBA, etc.) combined with standards like POSIX, IEC 61131, and IEC 61508 will stimulate the development of competitive products. In particular, the OCEAN project will allow to integrate additional functionality and

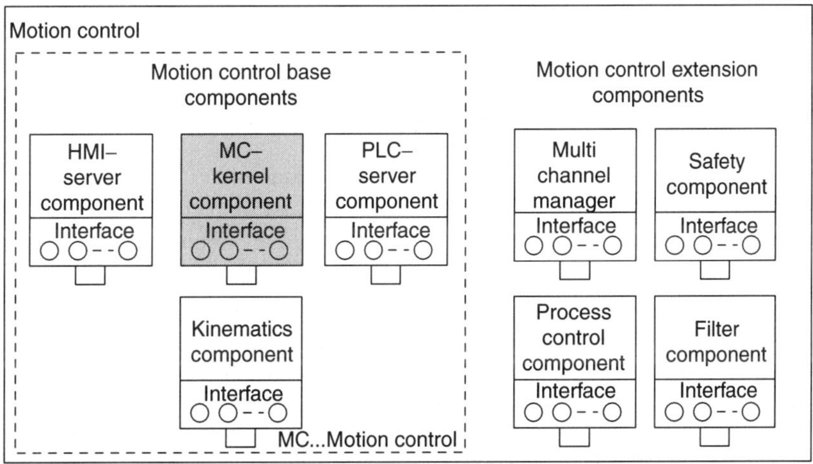

FIGURE 105.2 Subdivision of complex motion control structure based on components.

third-party software by just using the standardized interface description, and to merge components provided by different CNC manufacturers. Two examples will clarify the importance of the results; it will become possible and easy for the machine tool builder to implement his own kinematics component or filter component and even reuse it for different CNCs from different control manufacturers. Moreover, the possibility to substitute the user interface leaving the rest of the numerical control unchanged (hardware and software) will overcome one of the major problems that machine tool builders have to face today, namely the obvious desire by end users to have uniform user interfaces for all machine tools in the same plant.

105.3   Status of the Project

Analysis of Communication Systems, Platforms, and Tools

The first part of the project dealt with the analysis of Communication Systems, Platforms, and Tools. The main objective was to find an appropriate communication infrastructure for a "DCRF," which would suit the technical and cost aspects, and could be provided to the end user as an open source.

CORBA implementations, as well as available Linux real-time extensions, were analyzed and compared. The two main criteria for this analysis were the real-time capability and the Open Source availability of the investigated software. As a result, TAO was found to be the most suitable CORBA Object Request Broker (ORB), and RTAI the preferable Linux real-time extension.

TAO has been developed at the Washington University of St. Louis with a special focus on its real-time capability. TAO offers a variety of CORBA services as well as the quality-of-service (QoS) functionality that can be applied to the supervision of modular software systems. Additionally, an implementation of the CORBA Component Model (CCM) is currently being developed for TAO. The CCM allows for a better software reuse and a dynamic configuration of CORBA applications. RTAI is preferred as the Linux real-time extension rather than RTLinux because RTAI offers the best real-time support, which is crucial for the realization of the DCRF. RTAI allows to run hard real-time processes in kernel space simultaneously with soft real-time processes in user space, for instance. Furthermore, RTAI offers the best C++ support and a more complete feature set. An additional criteria on is the availability of LXRT for RTAI that enables the operation of hard real-time tasks in user space. This allows the use of various standard operating system functionalities in real-time applications.

The consortium has performed analysis of the benefits of existing open controller approaches that can be considered for the design of the DCRF. Due to the fact that DCRF will be based on CORBA, the main focus of the analysis was on the comparison between CORBA as a middleware and the communication infrastructure of existing open controller approaches. The following open controller solutions were investigated:

- OSACA (Europe),
- OMAC (U.S.A.), and
- OSEC and FAOP (Japan).

With respect to the communication infrastructure and the communication system APIs, all benefits of the different open controller approaches are fulfilled by CORBA. Compared with OSACA, the total functional range of CORBA Services will even have to be reduced to the minimum functionality that is necessary for a control platform. This will result in a smaller footprint of DCRF. The modularization of control functionality achieved by the existing open controller approaches, as well as the configuration and startup of the software modules, can be mapped into the respective functions of the CORBA Component Model. Finally the OSACA reference architecture that specifies the software interfaces of control modules can be mapped into a reference architecture based on the Object Management Group Component Interface Definition Language (OMG CIDL).

The requirements for the infrastructure and the general-purpose components of DCRF were compiled. These requirements address on the one hand the data exchange between different real-time components and on the other the communication between hard real-time and less time-critical software components. Additionally, requirements for safety functions in DCRF were addressed.

Requirements were compiled for the following categories:

- Functional requirements,
- Latency requirements,
- Task Scheduling Jitter,
- Clock synchronization,
- Task interruption, and
- Software safety aspects.

The first part of the project was concluded once all information and data needed for designing the DCRF were collected. As a result of the investigations, the TAO–RTAI solution has been selected as the best combination to fulfill the requirements of DCRF. It allows a distribution and cross-platform deployment of components in a real-time environment. The benefits of existing open controller approaches have been identified and can be mapped into the CORBA-based development of DCRF.

Design, Implementation, and Validation of a DCRF

The second phase of the project, currently in progress, is the design, implementation, and validation of a DCRF. The first step for this implementation is porting of TAO to RTAI, done by the consortium and OCI. After this phase, the consortium will verify if the services of the TAO implementation satisfy the real-time approach. In this case, it will adapt the result to be the basis for the DCRF. One of the main aims of the DCRF is to have CNC components from different vendors interoperable with the DCRF framework, which means they have to be hardware independent or at least easily adaptable to the hardware of different control. To solve this problem, hardware abstractions have to be investigated and evaluated so that an exchange of components will be easy and feasible.

Moreover, it is mandatory for the real time components that their communication channels meet their time constraints (QoS). Therefore, suitable transport mechanisms have to be chosen. The lower level CORBA layer, which is called the "pluggable protocol framework" in TAO, has to be extended to meet the needs of DCRF.

Another parallel activity is the design and implementation of general-purpose components (for loading, configuration, and safety).

Design, Implementation, and Validation of Motion Control Base Components for an Open Numerical Control System

As of 2003, the detailed functionality of the components is another focus area. The aim here is to achieve a definition of a reference architecture for the motion control components of the open control systems by extending existing architectures (OSACA, OMAC, JOP) and new proposals. The goal is to provide standardized and published interface descriptions for components that are up to now part of monolithic control blocks and that will be extracted from there; this way, machine tool builders or end users will be able to integrate additional functionalities independent of control manufacturers. The components and their interfaces will be defined with formal methods like IDL or XML.

105.4　Foundations of the OCEAN Project, and Results

OSACA

The first goal of the OCEAN project is the development of DCRF based on Open Source Software, which provides a communication infrastructure for distributed real-time control applications. OCEAN draws considerably from the results and experience of the OSACA project.

The OSACA system platform is composed of three main parts: reference architecture, configuration system, and communication system. The reference architecture specifies a data model for numerical controls. The configuration system coordinates the setup of control systems. The communication system handles data exchange.

In the OSACA reference architecture, five main functional units of numerical control have been specified: Motion Control, Motion Control Manager, Axis Control, Spindle Control, and Logic Control. The reference architecture that was developed in OSACA mainly deals with non-real-time control tasks. It was intended to be a specification for uniform interfaces between the NC kernel and the Human–Machine Interface (HMI).

The OCEAN project aims at the development of real-time capable control tasks. Therefore, data models for components within the NC kernel have to be defined and implemented. The development is split into two steps. In the first step, the basic components that are mandatory to operate an NC are designed and implemented, followed by the development of extension components that will enhance the functionality of numerical control. The base components are Motion Control Kernel Component, Kinematics Component, PLC-Server Component, and HMIServer Component. The benefits of the OSACA Motion Control, Axis Control, and Spindle Control units are taken into account for specifying the OCEAN Motion Control Kernel and HMI-Server Components. Furthermore, the OSACA Logic Control unit, which is very basic and only offers three data items, is taken into consideration for the OCEAN PLC-Server Component. The specification for the OCEAN Kinematics Component was developed from scratch. With respect to the Extension Components, all specifications were developed from scratch, except for the Multi-Channel Manager Component. This component can partly be based on the OSACA Motion Control Manager unit. The experiences from the OSACA configuration system are used for the development of the General-Purpose Components. In OCEAN, these are components for the configuration of distributed control systems, for supervision concerning safety, and for dynamically loading additional control components. For this purpose, the OSACA configuration system is extended by the CCM specification if applicable. The scope of the CCM and its impact in "resource management," "QoS," and usability has to be considered and adapted for the use in DCRF.

The OSACA communication system was completely replaced by RT CORBA. In this context, migration from OSACA to RT CORBA of the three different types of communication objects developed in OSACA (Variable Objects for data exchange, Process Objects for commands, and Event Objects for notifications about data updates) was required. This exchange of the communication systems also affects the development of the General-Purpose Components. For instance, the OSACA configuration system had to be further redeveloped to allow for the use of RT-CORBA.

LINUX

Linux is an operating system created by Linus Torvalds, at the University of Helsinki in Finland. The first version, 1.0, of the Linux Kernel was released in 1994. At present, Linux is developed under the GNU General Public License and its source code is freely available to everyone. This however, does not mean that Linux and its assorted distributions are free. Companies and developers may charge money for it as long as the source code remains available. Linux may be used for a wide variety of purposes including networking, software development, and as an end-user platform. Linux is often considered an excellent, low-cost alternative to other more expensive operating systems.

Due to the very nature of Linux's functionality and availability, it has become quite popular worldwide and a vast number of software programmers have taken Linux's source code and adapted it to meet their individual needs.

Linux Real-Time Extensions

For real-time extensions, a distinction between Soft Real-Time Extensions and Hard Real-Time Extensions can be made. Linux Real-Time Extensions can be defined as Hard if they guarantee deterministic worst-case latencies of interrupt servicing and scheduling. They can be defined as Soft if a fast response in terms of the Hard real-time requirements cannot be guaranteed. In the case of Numerical Control, a Hard Real-Time Extension that guarantees fast response is mandatory.

Available Implementations.
The two available implementations of real-time Linux are RTAI and RTLinux. RTAI and RTLinux are both patches (additions, but patch is the real technical term) to the standard Linux kernel. The multi-threaded real-time kernel runs standard Linux as the lowest priority thread that is always preemptible. This makes hard real-time functionality as well as most features of the non-real-time Linux kernel available to programmers. RTLinux is a small POSIX 1003.13/PSE51 compatible hard real-time operating system. RTAI is a variant of RTLinux that has gone off on its own path. It provides compatibility with POSIX 1003.1c and 1003.1b (Threads and Queues).

Some characteristics of RTL are:

- Majority of APIs are not portable.
- Dynamic thread creation is dangerous (uses kmalloc non-atomic).
- Support for Mutexes and Condition Variables.
- POSIX calls are part of the scheduler core.

Characteristics of RTAI:

- Modeled after standard LinuxThreads (by Xavier Leroy).
- Supports dynamic thread creation (from preallocated pool).
- Supports mutexes with priority inheritance.
- Supports condition variables.
- POSIX API provided by an optional module.
- Good basis for extending to meet the EL/IX specification.
- Supports user mode real-time tasks.
- Provided through a GPL License.

RTAI now has two separate approaches toward user space real-time: one extends the microkernel with system calls from the user space (via the trap mechanism); the other approach extends the Linux task API. The former is meant to work with ADEOS, as a completely Linux-independent real-time operating system, if desired. The latter can never work without Linux. The development in this issue is ongoing and not yet finished.

Based on the comparison, it can be judged that RTAI provides better RT support, meaning that:

- The modification required to install the patch on the Linux Kernel is smaller.
- It allows soft real time in user space along with hard real time in kernel space.
- It provides excellent performance in terms of low jitter and low latency.
- It provides better C++ support and a more complete feature set than RTLinux.
- RTAI has the better open source approach with lots of feedback from developers.

Furthermore, RTAI has evolved by gaining a good support for real-time access to system peripherals (e.g., real-time drivers for serial ports are now available).

RTAI now offers development of interrupt handlers in user mode and an enhanced scheduling policy.

License Agreements

The basic mechanism that stands at the basis of RTLinux is protected under U.S. Patent 5,995,745 and licensed without a fee to RTLinux users for commercial and noncommercial purposes. This means that RTLinux can be used royalty-free under two conditions:

- Under the terms of the GPL-Licence. It can be used and even modified, if it is used with any software that also runs under the terms of the GPL-License.
- Together with software that uses "Open RTLinux Execution Environment." In this case, any software can be run together with RTLinux even if the software is not open or not complying with the

terms of GPL (e.g., LGPL software). "Open RTLinux Execution Environment" means a computer hardware system where the interrupt control hardware of processors and system boards is under the direct control of an unmodified Open RTLinux Software in binary form. A configuration done by using the options of a configuration tool is not considered a modification.

RTAI is a real-time extension that has partly been changed from LGPL conditions to GPL conditions. It is covered by the terms of Version 2 of the Open RTLinux Patent License, which says: "The Patented Process may be used, without any payment of a royalty."

CORBA

CORBA is a standard architecture for distributed object systems (Figure 105.3). It allows a distributed, heterogeneous collection of objects to interoperate. The definition of CORBA is managed by the Object Management Group (OMG). The OMG comprises over 700 companies and organizations, including almost all the major vendors and developers of distributed object technology, such as platform, database, and application vendors as well as software tool and corporate developers. An extensive introduction to Corba can be found in the Computer Software and Web Technologies section of this book.

Comparison of Available CORBA Implementations

For the comparison of available CORBA implementations, an Internet research was performed. In the scope of this research, open source ORBs as well as commercial solutions were investigated.

The following ORBs were included in the investigation:

- TAO, Distributed Object Group at Washington University, St. Louis, U.S.A.,
- ROFES, Aachen University of Technology, Germany,
- MICO, University of Frankfurt, Germany,
- ORBacus, IONA, Dublin, Ireland,
- ORBExpress, Object Interface Systems, Herndon, VA, U.S.A.,
- e*ORB, Vertel, Woodland Hills, CA, U.S.A.,
- ORBit, GNOME Project,
- omniORB2, AT&T Laboratories, Cambridge, U.K.
- ORBIX E2A, IONA, Dublin, Ireland,
- VisiBroker, Borland, Scotts Valley, CA, U.S.A., and
- JacORB, Software Engineering and Software Group, FU Berlin.

The research was based on the following criteria:

- Open Source availability,
- CORBA version implemented,
- real-time capability and implemented RT CORBA features,

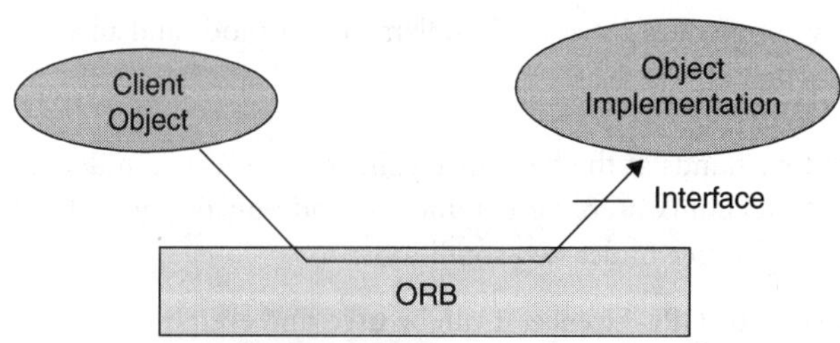

FIGURE 105.3   Basic Corba concept.

- available language bindings,
- implemented CORBA services,
- supported multithreading concept,
- supported protocols, and
- tested Linux and compiler support.

Except for the real-time version of VisiBroker and JacORB, all investigated ORBs have been ported into and tested on Linux platforms. All ORBs (except the Java-based JacORB) offer language bindings for C++. Real-time capability is provided by the Open Source ORBs TAO and ROFES. Further real-time versions of ORBExpress, e*ORB, and VisiBroker are available but only in commercial distributions. The range of CORBA services that is implemented for each ORB is quite different. The basic services like Naming, Event, and Trading are provided by almost all ORBs. Additional CORBA Services like Notification, Time, Scheduling, etc., vary among implementations. The largest range of implemented CORBA Services offer TAO and ORBit.

Interoperability between Different ORBs

The interoperability between different ORBs is an essential feature for the realization of distributed control systems using CORBA as middleware. The OMG specification covers the aspect of ORB interoperability since CORBA 2.0. This revision of the specification introduced a general ORB interoperability architecture, which is called General Inter-ORB Protocol (GIOP). GIOP is an abstract protocol that specifies the transfer syntax and a standard set of message formats to allow independently developed ORBs to communicate over any connection-oriented transport. The Internet Inter-ORB Protocol (IIOP) specifies how GIOP is implemented over TCP/IP (Figure 105.4).

ORB interoperability also requires standardized object reference formats. Object references are opaque to applications but they contain information that ORBs need in order to establish communications between clients and target objects. The standard object reference format, called the Interoperable Object reference (IOR), is flexible enough to store information for almost any inter-ORB protocol imaginable. An IOR identifies one or more supported protocols and contains information specific to that protocol. For IIOP, an IOR contains a host name, a TCP/IP port number, and an object key that identifies the target object at the given host name and port combination.

All investigated ORBs were implemented in conformance with the CORBA 2.0 specification. Therefore, all of them support IIOP, which is the basis for interoperability. Statements for successful interoperability test were found for the combination of the three ORBs: TAO, MICO, and JacORB.

To ensure and to test the interoperability between different ORBs, research projects have been launched for the development of test suits. One of them is the EC-funded project CORVAL2 (Enhanced Techniques for CORBA Validation; IST-1999-11131) and a second one is the open source projects COST (CORBA Open Source Testing). The main objective of these projects is to investigate new mechanisms and tools for validating the implementation of Object-Oriented technology. The resulting test suits can be applied by users and providers of CORBA technology to test products against the CORBA specifications, and hence to ensure product interoperability and application portability.

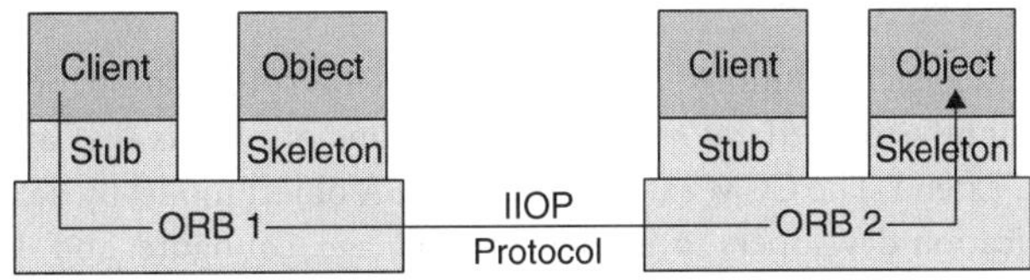

FIGURE 105.4   CORBA interoperability concept.

Investigation of the CCM

Introduction to the CCM

To provide higher-level reusable components, the OMG specifies a set of CORBA Object Services that define standard interfaces to access common distribution services, such as naming, trading, and event notification. By using CORBA and its Object Services, system developers can integrate and assemble large, complex distributed applications, and systems using features and services from different providers.

Unfortunately, the traditional CORBA object model, as defined by CORBA 2.4, has the following limitations.

1. *No standard way to deploy object implementations.* The earlier CORBA specification did not define a standard for deployment of object implementations in server processes. Deployment involves distributing object implementations, installing those implementations in their execution contexts, and activating the implementations in an ORB. Thus, system designers developed *ad hoc* strategies to instantiate all objects in a system. Moreover, since objects may depend on one another, the deployment and instantiation of objects in a large-scale distributed system are complicated and nonportable.

2. *Limited standard support for common CORBA server programming patterns.* The CORBA family of specifications provides a rich set of features to implement servers. For example, the CORBA 2.2 specification introduced *Portable Object Adapter* (POA), which is the ORB mechanism that forwards client requests to concrete object implementations. The POA specification provides standard APIs to register object implementations with the ORB, to deactivate those objects, and to activate object implementations on demand. The POA is flexible and provides numerous policies to configure its behavior. In many application domains, however, only a limited subset of these features is ever used repeatedly; yet, server developers face a steep learning curve to understand how to confugure POA policies *selectively* to obtain their desired behavior.

3. *Limited extension of object functionality.* In the traditional CORBA object model, objects can be extended only via inheritance. To support new interfaces, therefore, application developers must: (1) use CORBA's Interface Definition Language (IDL) to define a new interface that inherits from all the required interfaces; (2) implement the new interface; and (3) deploy the new implementation across all their servers. Multiple inheritance in CORBA IDL is fragile, because overloading is not supported in CORBA; therefore, multiple inheritance has limited applicability. Moreover, applications may need to expose the same IDL interface multiple times to allow developers to either provide multiple implementations or multiple instances of the service through a single access point. Unfortunately, multiple inheritance cannot expose the same interface more than once, nor can it alone determine which interface should be exported to clients.

4. *Availability of* CORBA Object *Services not defined in advance.* The CORBA specification does not mandate which Object Services are available at runtime. Thus, object developers used *ad hoc* strategies to configure and activate these services when deploying a system.

5. *No standard object life cycle management.* Although the CORBA Object Service defines a Life Cycle Service, its use is not mandated. Therefore, clients often manage the life cycle of an object explicitly in *ad hoc* ways. Moreover, the developers of CORBA objects controlled through the life cycle service must define auxiliary interfaces to control the object life cycle. Defining these interfaces is tedious and should be automated when possible, but earlier CORBA specifications lacked the capabilities required to implement such automation.

In summary, the inadequacies, outlined above, of the CORBA specification, prior to and including version 2.4, often yield tightly coupled, *ad hoc* implementations of objects that are hard to design, reuse, deploy, maintain, and extend. The CCM extends the CORBA object model by defining features and services that enable application developers to implement, manage, configure, and deploy components that integrate commonly used CORBA services — such as transaction, security, persistent state, and event notification services — in a standard environment. In addition, the CCM standard allows greater software reuse for servers and provides greater flexibility for dynamic configuration of CORBA applications.

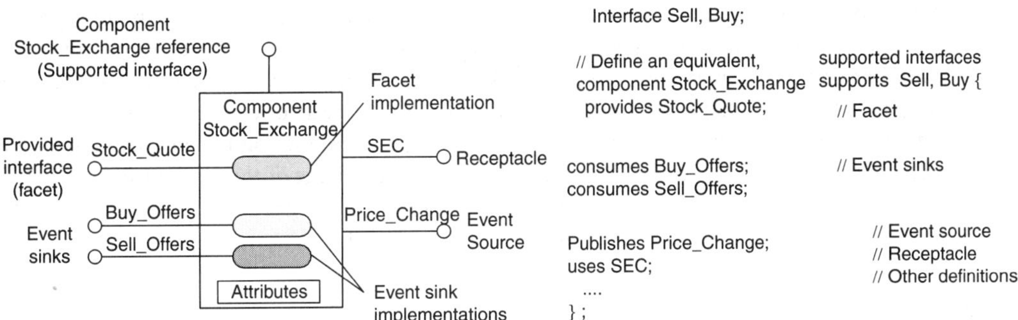

FIGURE 105.5 Example CCM component.

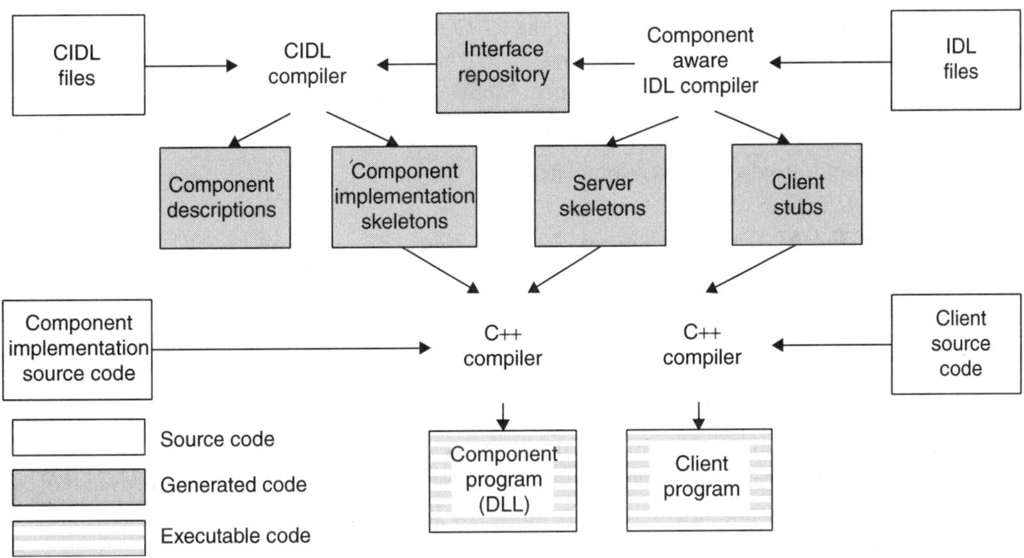

FIGURE 105.6 Tool chain using the Component Implementation Definition Language.

Figure 105.5 shows an example CCM component with IDL specification and Figure 105.6 shows the tool chain using the Component Implementation Definition Language.

Conclusion

The CORBA object model is increasingly gaining acceptance as the industry standard, cross-platform, cross-language distributed object computing model. The recent addition of the CCM integrates a successful component programming model from EJB, while maintaining the interoperability and language-neutrality of CORBA. The CCM programming model is thus suitable for leveraging proven technologies and existing services to develop the next -generation of highly scalable distributed applications. However, the CCM specification is large and complex. Therefore, ORB providers have only started implementing the specification recently. As with first-generation CORBA implementations several years ago, it is still hard to evaluate the quality and performance of CCM implementations. Moreover, the interoperability of components and containers from different providers is not yet well understood.

CCM providers are now implementing the complete specification, as well as support value-added enhancements to their implementations, just as operating system and ORB providers have done historically. In particular, containers provided by the CCM component model implementation provide QoS capabilities for CCM components, and can be extended to provide more services to components to relieve components from implementing these functionalities in an *ad hoc* way. These container QoS extensions provide services that can monitor and control certain aspects of components behaviors that crosscut different programming layers or require close interaction among components, containers, and operating systems. As CORBA and the CCM evolve, we expect that some of these enhancements will be incorporated into the CCM specification. CIAO will be the first implementation available with QoS RT-CCM enhancements and it is worth taking it into account for DCRF.

Selection of suitable ORB

For the development of DCRF, TAO is favored as ORB. This selection is based on the following arguments:

- TAO is available in open source,
- offers the majority of features defined in the RT CORBA 1.0 specification,
- comprises the main CORBA services, and
- a release including the CORBA Component Model is planned for 2003.

The only drawback of using TAO is its size, which should not be a problem in most cases, since a large amount of memory is needed to host a complete CNC.

105.5 OCEAN Consortium Members

As of 2003, the consortium involves:

1. **FIDIA S.p.A.**, coordinator: Manufacturer of Numerical Controls, High speed Milling Machines and CAM products (I);
2. **USTUTT:** The Numerical Controls Institute of the University of Stuttgart (D);
3. **RWTH:** The Machine Tool Institute of the Aachen University (D);
4. **KUL:** The Department of Mechanical Engineering of the Katholieke Universiteit Leuven (B);
5. **HOMAG:** Manufacturer of Numerical controls and Wood Working Machines (D);
6. **FATRONIK:** Research Institute for Machine Tool and Production Techniques (E);
7. **FAGOR:** Numerical Controls Manufacturer (E);
8. **OSAI:** Numerical Controls Manufacturer (I);
9. **GORATU:** Machine-tool (lathes, milling and High speed milling machine) Manufacturer, (E); and
10. **ITIA:** The Industrial Technologies and Automation Institute, of the Italian National Research Council (I).

Holonic Manufacturing Systems: A Technical Overview

Robert W. Brennan
University of Calgary

James H. Christensen
Rockwell Automation

William A. Gruver
Simon Fraser University

Dilip B. Kotak
*National Research Council Canada
Institute for Fuel Cell Innovation*

Douglas H. Norrie
University of Calgary

Edwin H. van Leeuwen
*BHP Billiton, Exploration and Mining
Technologies*

106.1 Introduction

Holonic manufacturing systems (HMS) have their roots in using multiple interacting decision-makers in place of a single centralized decision-maker to solve complex manufacturing problems. In order to illustrate this, the inherently difficult task of shopfloor scheduling provides a good example of why a distributed approach is more promising than a centralized approach for computer control. The traditional approach to shopfloor scheduling uses a single, centralized scheduler. As Parunak (1993) notes, "scheduling and control take place sequentially" and are analogous to a large sequential computer program when this approach is used. If there are no disturbances on the shop floor (e.g., rush orders, missing tools or material, machine processing time variability, or failures), this approach is very effective and can generate solutions to the scheduling problem.

Real manufacturing systems, however, are not deterministic. In addition, they consist of physically distributed materials and resources that behave in a highly concurrent and dynamic manner. As a result, centralized schedules quickly become obsolete and must be either ignored or frequently updated, although this is not always possible if the schedule generation and dissemination times are large.

An alternative approach is to match the scheduling model more closely with the physical system. In other words, the problem can be decomposed such that decision-making is performed by many simple, autonomous, and cooperative entities. Rather than following a predetermined plan, the schedule emerges from the interaction of these intelligent entities. This approach leads to a scheduling system that is "decentralised rather than centralised, emergent rather than planned, and concurrent rather than sequential" (Parunak, 1996).

The disadvantages of this approach are that global optima cannot be guaranteed and predictions of the system's behavior can only be made at the aggregate level (Parunak, 1996). However, even with conventional approaches, global optima often cannot be achieved in practice and the detailed predictions provided by these systems are frequently invalidated by unanticipated events. Alternatively, distributed intelligent control provides the advantages of adaptability, ease of upgradeability and maintenance, and emergent behavior.

In the next section, we provide some background on distributed problem solving in the manufacturing domain. Next, we introduce the concept of HMS and provide a brief history of HMS research. We follow this with a more detailed description of holonic systems and focus on the various architectures proposed by researchers in HMS. Finally, we provide a brief summary of this work as well as our views on the future of HMS research.

106.2 Background

An early contribution in this area in the manufacturing domain was based on the negotiation metaphor that stemmed from the research of Davis and Smith (1983) on distributed sensor systems. Their distributed problem-solving approach requires the "co-operative solution of problems by a decentralised, loosely coupled collection of problem solvers" (Davis, 1983) and differs from distributed computing in that it is concerned with a "single task envisioned for the system;" the goal is to create an "environment for co-operative behaviour."

Parunak (1987) extended the concept of the contract net from the information domain to the manufacturing domain with the development of the Yet Another Manufacturing System (YAMS) software. This system uses Davis' and Smith's negotiation metaphor to achieve coordination of a network of distributed computing nodes. The YAMS system uses an "open system model" that allows for the addition and removal of problem solvers and separates knowledge from the control structure through information hiding.

Duffie and Prabhu (1994) have investigated this distributed approach with a "heterarchical," or nonhierarchical opportunistic scheduling approach for machine cell control. One characteristic of this approach that sets it apart from traditional hierarchical control structures is that part flow is achieved by part-oriented requests as opposed to machine-oriented requests. This requires the parts to be "intelligent enough to know when they are done at one machine and need another" (Duffie and Piper, 1986). Similar work in the area of bidding-based control of manufacturing cells has also been investigated by other researchers (e.g., Veeramani [1992]). Work in this area has served as a catalyst for research on extending concepts from the emerging field of Multi-Agent Systems (MAS) to the manufacturing domain. This next phase in the transition from FMS to HMS is discussed next.

MAS can be thought of as a "general software technology that was motivated by fundamental research questions" (Bussmann, 1998) such as how groups of software agents can work together to solve a common problem (Moulin and Chaib-draa, 1996). Huhns and Singh (1998) define agents as "active, persistent (software) components that perceive, reason, act, and communicate." This definition follows from the notion of a software object where the focus is on data abstraction, encapsulation, modularity, and inheritance (Booch, 1994).

A major advantage of agent-based software over conventional software is this autonomous nature of individual agents. As Parunak (1993) notes, agents "decide locally not only how to act (as subroutines do), and what actions to take (as objects do), but also when to initiate their own activity." This, in combination with the cooperative nature of agents in MAS, is particularly well suited to distributed intelligent control problems as noted previously.

Not surprisingly, there has already been considerable utilization of the agent concept in the manufacturing domain. In particular, autonomous agents or MAS are an attractive software engineering tool for the development of systems in which "data, control, expertise, or resources are distributed; agents provide a natural metaphor for delivering system functionality; or a number of legacy systems must be made to interwork" (Wooldridge and Jennings, 1999). Manufacturing applications are characteristically "modular, decentralised, changeable, ill-structured, and complex" (Parunak, 1999), and as a result present a host of problems that are well-suited to agent technology. Despite the traditional conservatism of the manufacturing domain, it is "one of the oldest and strongest areas of agent research" (Aparicio, 1999).

Research into the application of MAS and distributed artificial intelligence in the manufacturing domain has been steadily growing over the last 10 years and has focused on all areas of the manufacturing enterprise ranging from product design to real-time control. For a comprehensive overview of agent-based systems in manufacturing, see Shen and Norrie (1999) and for MAS in concurrent design and manufacturing, Shen et al. (2000) can be consulted.

One motivation for the use of the agent model for the representation of manufacturing entities is that manufacturing entities, like agents, typically have well-defined state variables that are distinct from those of their environment (Parunak, 1999). As a result, a physical decomposition of the manufacturing system is obvious: agents are used to represent entities in the physical world, such as workers, machines, tools, fixtures, products, parts, features, operations, etc. Evidence from natural systems suggests that it may be more effective to assign agents to physical entities in the system (Weiss, 1999). In the manufacturing domain, critical information is usually organized by physical entities and the agents that represent these entities are the natural locus for maintaining the information. This approach naturally defines distinct sets of state variables that can be managed efficiently by individual agents with limited interactions; however, it requires a large number of resource-related agents. This leads to other problems such as communication overheads and complex agent management. A common example of this approach is the use of part and machine agents for manufacturing planning and scheduling (Duffie and Prabhu, 1994; Lin and Soldberg, 1992; Parunak, 1987).

An alternative approach to problem decomposition that has been used in the manufacturing domain is functional decomposition. With this approach, agents are used to encapsulate functionality such as order acquisition, planning, scheduling, material handling, transportation management, and product distribution, and have no explicit relationship with physical entities. This approach tends to exhibit high coupling (Booch, 1994), that is, the sharing of many state variables across different functions, which leads to problems of consistency and unintended interactions. Examples of this approach include the use of agents to encapsulate special functionality, such as facilitator agents (Cutkosky et al., 1993), broker agents (Peng et al., 1998), and mediator agents (Maturana and Norrie, 1996). A second set of examples is provided by systems such as ARCHON (Cockburn and Jennings, 1996) and EXPORT (Monceyron and Barthes, 1992), which utilize agents to resolve problems in the integration of legacy systems.

As noted previously, MAS can be thought of as a general software technology; manufacturing, however, is fundamentally concerned with "ironware" (physical equipment and products). As a result, when these two worlds merge, it becomes useful to start thinking about a type of agent different from those that populate classic agent-based systems. In particular, it becomes useful to think about "physical agents" that possess both classical software agent capabilities as well as the ability to interface with physical equipment on the shop floor. In the next section, we provide an overview of the international collaborative effort to develop systems of these unique manufacturing-specific agents.

106.3 Holonic Concepts

The concepts of HMS can be considered as constituting a distributed intelligent control paradigm for manufacturing. Unlike MAS, which is a broader software approach that can be also used for distributed intelligent control, an HMS is, by definition, a manufacturing-specific approach to distributed intelligent control.

Suda (1989) first introduced the concept of HMS in the 1990s. This work was motivated by the need to address these shortcomings as well as by new pressures faced by manufacturers in the 1990s such as increasingly stringent customer requirements for high-quality, customizable, low-cost products that can be delivered quickly, as well as the increasing levels of system complexity due to the distributed, concurrent, and stochastic nature of manufacturing systems.

The research in distributed control described in the previous section led many to the realization that an "autonomous, distributed and co-operative" (Valckenaers and van Brussel, 1994) approach was required to address these issues. This new control software and hardware approach appeared to hold the most promise of realizing manufacturing systems that are both flexible (capable of reconfiguration) and responsive (capable of recovering from disturbances). However, past experiences with "green field" approaches like Flexible

Manufacturing Systems (FMS) left a bad taste in the mouths of many manufacturers. Clearly, an incremental approach was required if new techniques of flexible automation were to be accepted by industry.

HMS is one of the Intelligent Manufacturing Systems (IMS) program's six major projects resulting from a 2-yr feasibility study started in February 1992 (Christensen, 1994). In very general terms, holonics can be thought of as a marriage between a general philosophy (of living organisms and social organizations) and an emerging software approach (distributed artificial intelligence) that is intended to ultimately lead to a better understanding of how next-generation manufacturing systems should be designed. The objective of the HMS Consortium is to "attain in manufacturing the benefits that holonic organisation provides to living organisms and societies, e.g., stability in the face of disturbances, adaptability and flexibility in the face of change, and efficient use of available resources" (HMS, 2003).

The term "holon" was coined by Arthur Koestler (Koestler, 1967). He observed a dichotomy of wholeness and partness in living organisms and social organizations, and summarized this with the statement: "wholes and parts in the absolute sense do not exist anywhere." To understand these systems, Koestler used the "Janus Effect" as a metaphor for this dichotomy of wholeness and partness observed in many such systems: that is, "like the Roman god Janus, members of a hierarchy have two faces looking in opposite directions." In other words, these members can be thought of as self-contained wholes looking toward the subordinate level and/or dependent parts looking upward. To explain this concept, Koestler suggested a new term to describe the members of these systems: "holon," from the Greek *holos* meaning "whole" and the suffix *on* implying particle as in "proton" or "neutron."

In order to achieve the benefits of Koestler's holonic organizations, HMS "consist of autonomous, self-reliant manufacturing units, called holons" (Bussmann, 1998) that cooperate to achieve the overall manufacturing system objectives. A holon is defined by the Holonic Manufacturing Systems Consortium (HMS, 2003) as "an autonomous and co-operative building block of a manufacturing system for transforming, transporting, storing and/or validating information and physical objects."

Another other important characteristic of holons that is particularly relevant to the manufacturing domain is that they consist of an information processing part and often a physical processing part. This distinction really sets HMS apart from the mainstream MAS community, although another organization formed in 1996, the Foundation for Intelligent Physical Agents (FIPA, 2003), also recognized this use of agent technology quite early. FIPA however, did not pursue the notion of *physical agents* closely until recently (Ulieru, 2001).

Another important holonic concept that is also relevant to the manufacturing domain is the notion of functional decomposition. Simon's (1996) observation that "complex systems evolve from simple systems much more rapidly if there are stable intermediate forms than if there are not" is at the heart of this. In other words, the complexity of manufacturing systems can be dealt with by decomposing the system into smaller parts. A consequence of this is the idea that holons can contain other holons (i.e., they are recursive). In addition, problem solving is achieved by holarchies, or groups of autonomous and cooperative basic holons and/or recursive holons that are themselves holarchies.

As noted previously, HMS began in 1992 as a feasibility study. At the time, it was felt that an international effort would be required if manufacturing systems based on the concepts described above were to be developed in a reasonable period of time. It was not clear, however, whether a collaborative effort among competitive industries and across various cultures would work in practice (Valckenaers and Van Brussel, 1994). Despite these early fears, the results of this feasibility study were very promising (Kriz, 1995), leading to the launch of the full-scale program in 1995.

During this first phase of the HMS Project (1995–2000), work was conducted around three main areas: generic technologies, benchmarking, and organization (e.g., project management, technology transfer, dissemination, and exploitation). In order to provide a clear focus for the project, this work was organized into the seven work packages summarized below (Gruver et al., 2003):

- Systems Architecture and Engineering (WP1): Generation of reference models, information models, standards, and support tools for holons, holonic systems, and the holonic systems engineering process.
- Systems Operation (WP2): Generic operational aspects of HMS such as communication, negotiation, and coordination, planning, resource allocation, fault management, and reconfiguration.

- Holonic Resource Management (WP3): Development of strategies, techniques, and architectures for resource management in HMS.
- Holonic Manufacturing Unit (WP4): Machining, finishing, inspection, and tooling.
- Holonic Fixturing (WP5): Fixturing in the machining and assembly process.
- Holonic Handling Systems (WP6): Robots, end-effectors, fixtures, feeders, mobile handling systems, and sensors.
- Holomobiles (WP7): Mobile systems for transport, maintenance, and supervision.

As can be seen from these work package descriptions, this first phase of the HMS project focused on the development of a generic holonic systems technology (WP1–WP3) as well as the demonstration of this approach in specific application areas (WP4–WP7).

A significant contribution was made during the first phase of the HMS project by van Brussel's group at Katholieke Universiteit Leuven. Van Brussel et al. (1998) proposed a reference architecture for HMS called PROSA that specifies four main holon types: Product, Resource, Order, and StAff. The generic nature of this architecture has resulted in it serving as the foundation for many of the proposed holonic architectures that followed during the first and second phases of the HMS project (Langer, 1999), (Suessmann et al., 2002; Weichhart et al., 2002; Monch et al., 2003). PROSA will be discussed in further detail in the next section.

Another significant advance during this phase of the project involved the progress toward the definition of standards for holonic systems that were primarily based on the International Electrotechnical Commission (IEC) 61499 function block standard (IEC, 2000). For example, Fletcher et al. (2000) describe a suitable architecture for holonic systems that uses a layered framework of IEC 61499 function blocks called a holonic kernel. In order to enable holonic cooperation and autonomy, this kernel consists of four specialized services (each implemented by an IEC 61499 service interface function block): a Function Block Manager, a Coordination Manager, a Cooperation Domain Interface, and a Data/Knowledge Base Interface (this will be described in more detail in the next section).

The primary motivation for the use of the function block model was its suitability for event-driven control. As Marik and Pechoucek (2001) noted, the scan-based approach to control used by traditional programmable logic controllers "does not enable the required flexibility in scheduling of events, especially in the case of hard real-time control." Alternatively, by explicitly representing events, the function block model focuses on process abstraction and synchronization; this makes this approach particularly suitable for control of an environment that is concurrent, asynchronous, and distributed.

However, the work during this phase of the HMS Consortium showed that this model does have some limitations in its current form. For example, as pointed out by Christensen (1994), the model must support mechanisms for dynamic reconfiguration if the agility requirements of FMS are to be met. In addition, although the function block is particularly suited for low-level, real-time process/machine control and lends itself well to various hardware implementations such as (Fletcher et al., 2001), it is not well suited for higher-level reasoning (Brennan and Norrie, 2001).

As a result, it was recognized that the best approach is to encapsulate the function block solution into a higher-level software when and where it is required to enable more sophisticated reasoning and a richer knowledge representation than function blocks alone. These resulting components have been referred to as "holonic agents" (Marik and Pechoucek, 2001) because of their integration of a holonic part (for hard real-time) and software agent part (responsible for higher-level, soft real-time or non-real-time intelligent decision-making). This concept is illustrated in Figure 106.1, which shows a multilayer architecture consisting of four temporally decomposed layers of agents (i.e., the ovals in Figure 106.1) and devices: execution control (EC), control execution (CE), execution (E), and hardware (Brennan and Norrie, 2002). As one moves down the layers, time scales become shorter and real-time constraints change from soft to hard real time; also, the degree of agency decreases in higher-level agents are more sophisticated but slower, while lower-level agents are fast and light-weight.

Having explored holonic systems technologies since 1995, the HMS Consortium entered its second phase in 2001, focusing on the integration and application of these technologies. In order to provide for the entire scope of the manufacturing enterprise, the Consortium took the conventional view of manufacturing

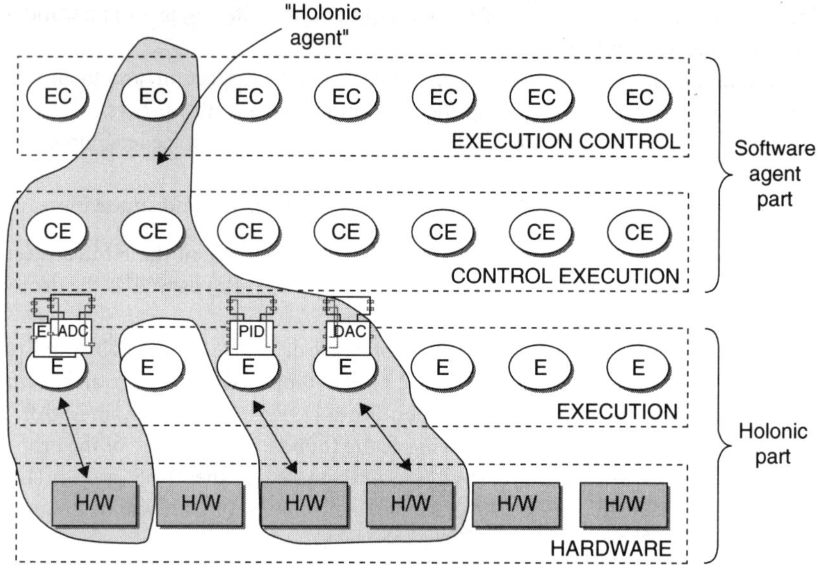

FIGURE 106.1　Software agents and holons.

planning and control systems as being a hierarchy of decision levels that range from long-term, strategic decisions at the highest level to short-term, detailed execution and control at the lowest level. As a result, the focus of HMS is now on three fundamental levels of the manufacturing enterprise (Gruver et al., 2003):

- Holonic Production Execution Systems: The factory and supply chain level.
- Holonic Production Sites and Physical Equipment: The work cell level (which consists of a number of holonic control devices).
- Holonic Control Devices: The physical, real-time control level.

This decomposition can be thought of as a simplification of the widely accepted National Bureau of Standards (NBS) (Jones and McLean, 1986) five-level model and the International Standards Organisation (ISO) (ISO, 1986) six-level model. In addition, the three HMS levels now have a close parallel with the recently proposed FIPA "Holonic Enterprise" model (Ulieru, 2001). Figure 106.2 provides an overview of these models.

In the next section, we look at holonics in more detail. In particular, we focus on the main architectures proposed to implement the holonic concepts described above in the manufacturing domain.

106.4　Holonic Architectures

It was noted previously that an important distinguishing characteristic of holons is that they consist of an information processing part and often a physical processing part. This is reflected in Bussmann's (1998) basic architecture of a holon illustrated in Figure 106.3.

Given this basic building block, the HMS community next looked at how holons could be implemented to solve basic manufacturing problems. As noted previously, Van Brussel et al. (1998) proposed a reference architecture to address this issue, which consists of four fundamental holon types:

- Resource Holon: The resource holon is an abstraction of the production processes (factory, shop, machine tools, conveyors, robots, and other intelligent devices) in a manufacturing system that provides production capacity and functionality to other holons.
- Product Holon: The product holon contains process and product knowledge in order to ensure that products are made correctly and in sufficient quantity. As a result, the product holon is

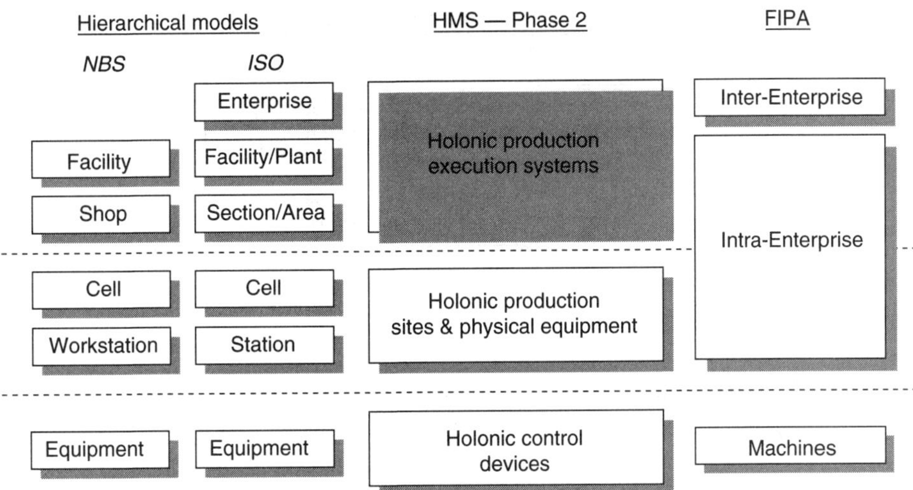

FIGURE 106.2 HMS, FIPA, and the hierarchical models of the manufacturing enterprise.

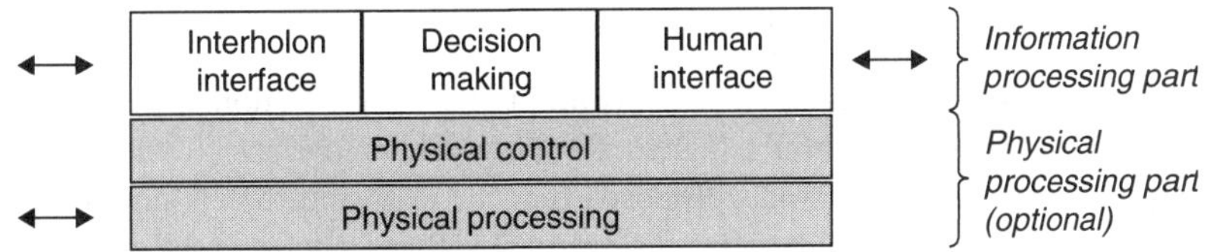

FIGURE 106.3 Architecture of a basic holon (from Bussmann, 1998).

responsible for the functionalities that are traditionally covered by product design, process planning, and quality assurance.

- Order Holon: The order holon manages the set of physical products being produced. As a result, it is concerned with logistical information such as customer orders and make-to-stock orders.
- Staff Holon: The staff holon is an optional, centralized element of the PROSA architecture that is used to assist the basic holons (Resource, Product, and Order) in performing their tasks. For example, the staff holon may play the role of a mediator as in the Metamorph Architecture of Maturana and Norrie (1996).

The basic building blocks of the PROSA architecture are illustrated in Figure 106.4, where each of the three main types of holons are intended to be responsible for manufacturing control. The staff holon can be combined with these basic holons to provide expert knowledge for longer-term, strategic decision-making, fault-diagnosis, maintenance, and to facilitate the incorporation of legacy systems.

Another important property of holons, not illustrated in Figure 106.4, is that they can contain other holons (i.e., they are recursive). Problem solving is achieved by holarchies, groups of autonomous, and co-operative basic holons and/or recursive holons that are themselves holarchies. In order to achieve this form of cooperation, the notion of a "Cooperation Domain" (CD) was introduced in holonic systems, which, as Fletcher et al. (2000) observed, "is considered a logical space in which holons communicate and operate, that provides the context where holons may locate, contact, and

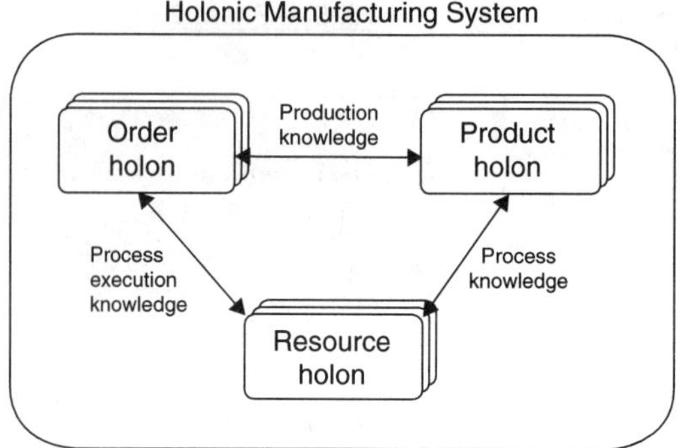

FIGURE 106.4   The PROSA architecture (from Van Brussel et al., 1998).

interact with each other." Within this framework, two types of cooperation occur among holons (Fletcher et al., 2000):

- Simple cooperation: A holon is committed to answer queries from another holon, even if the response is noncooperative.
- Complex cooperation: Holons achieve a joint goal (e.g., agreeing upon a mutual plan of executing tasks for solving a distributed problem).

This second form of cooperation is based on the concept of cooperation domains, and requires an integration of the physical device level with the higher reasoning/deliberative level of the manufacturing system. For example, Christensen (2003) has proposed that the architecture of a holonic manufacturing system is split into two distinct levels: Low-Level Control (LLC) and High-Level Control (HLC).

LLC refers to normal, nonholonic control and automation functions where response times range from 10 μsec to 100 msec (Christensen, 2003). At this level, the IEC 61499 standard for the use of function blocks in distributed automation is particularly relevant (IEC, 2000; Lewis, 2001). As a result, the LLC architecture shown in Figure 106.5 addresses the functions associated with real-time control.

Unlike the LLC, the HLC sits more firmly in the IT world of software agents and, as such, is concerned with higher-level reasoning. More specifically, the HLC architecture, shown in Figure 106.6, "addresses the functions associated with the domain of inter-holon cooperation, including negotiation and coordination of mutually agreed tasks and mutual action to recover from operational faults" (Christensen, 2003). Response times in this domain are much slower (100 msec to 10 sec), and rather than the control-oriented function block, the main entity is the agent as described by the FIPA architecture standard for software agent systems (FIPA, 2003).

As can be seen in Figure 106.6, each holon is composed of three basic building blocks, which together form the "holonic kernel" (Fletcher et al., 2000): Cooperation Domain Interface (CDI), Coordination Manager (CM), and Function Block Manager (FBM). The CDI provides an interface from holons to one or more cooperation domains to transfer knowledge relating to tasks and ontologies. It also serves as "yellow pages," allowing services to be found and offered. The CM controls interactions within the holon and among holons. For example, the CM is responsible for task decomposition, planning, conflict resolution, and the aggregation of results. It is also responsible for maintaining knowledge bases in the holon's local repository. Finally, the FBM serves as the interface to the LLC layer. It is responsible for the management of function block configurations upon a resource (i.e., physical device). For example, the FBM manages the creation, configuration, running, and deletion of function block applications.

Figures 106.5 and 106.6 illustrate the internal structure of holons at the LLC and HLC levels, and also how these holons interface with the cooperation domain. When combined, the LLC and HLC architectures provide the link between the information and physical worlds that is unique to holonic systems. For example, higher-level reasoning tasks such as planning, scheduling, fault monitoring, and fault recovery

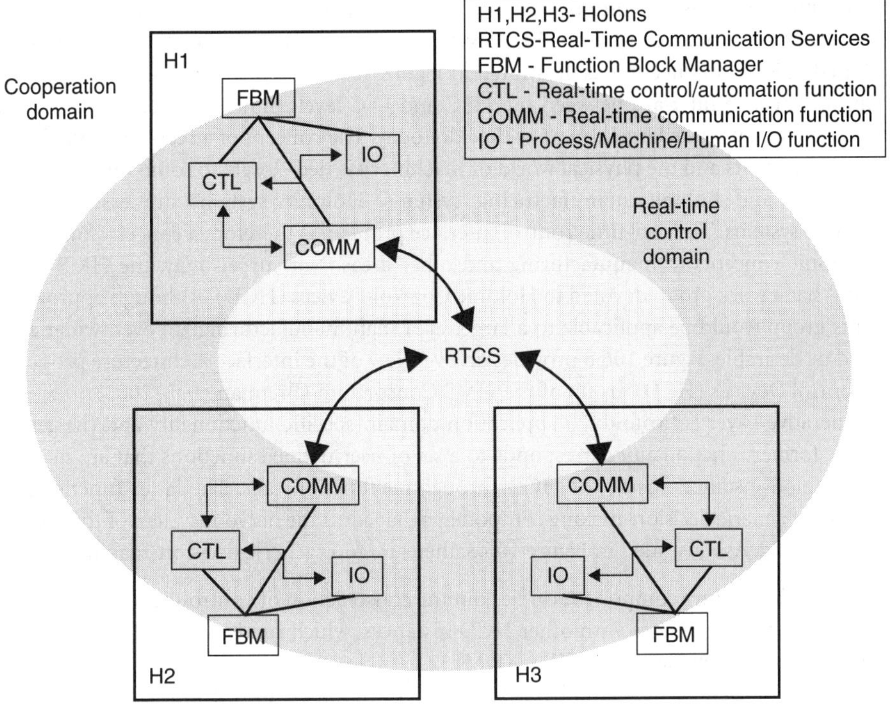

FIGURE 106.5 LLC architecture (from Christensen, 2003).

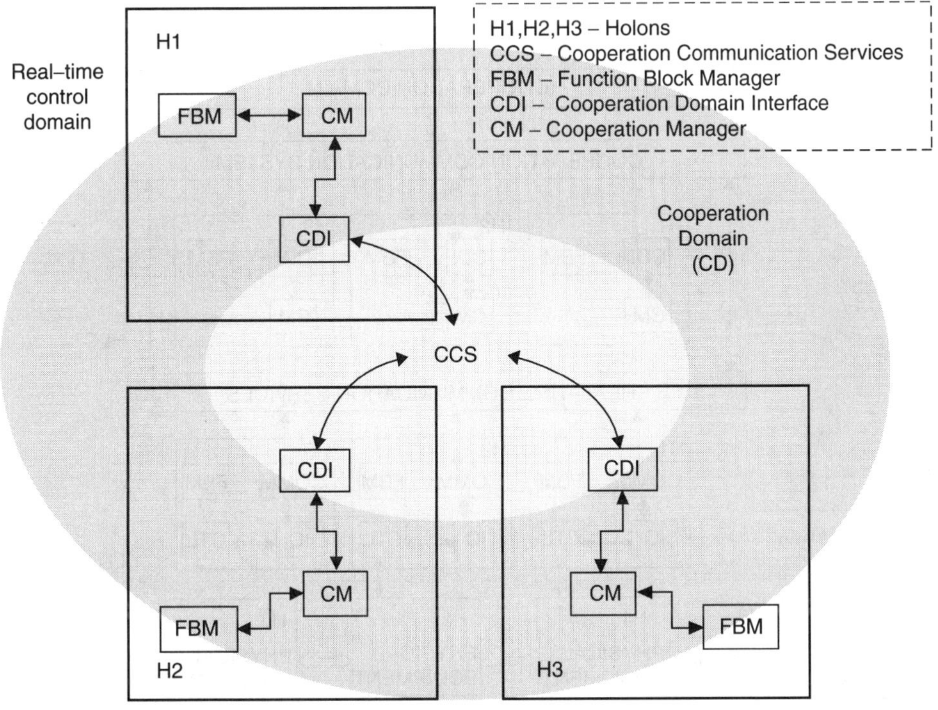

FIGURE 106.6 HLC architecture (from Christensen, 2003).

can be accomplished by task decomposition and virtual clustering at the HLC level. Plans can be then realized by the physical machinery at the LLC level. Neligwa and Fletcher (2003) provide an example of this combined holonic architecture as illustrated in Figure 106.7.

Without an effective interface between the HLC and LLC levels, however, agents and machines will continue to exist and operate largely apart as they do today. The concept of a real-time interface between the soft world of agents and the physical world of machinery is tied closely to some fundamental ideas in holonic systems and holonic manufacturing systems. Holonic systems are essentially adaptive agent–machine systems. The real-time control interface problem is therefore a concern for those wishing to apply holonic concepts to manufacturing and other areas. Not surprisingly, the HMS Consortium (HMS, 2003) has a work group devoted to Holonic Control Devices (HCD). Although approaches developed by this group would be applicable to a larger area than manufacturing, an even wider application focus would be desirable. Figure 106.8 provides an overview of the interface architecture proposed by the Holonic Control Devices (HCD) group of the HMS Consortium (Brennan et al., 2003).

The Deliberative Layer is twofold: (i) application domain-specific functionality and (ii) generic functionality. The former functionality corresponds to a set of user-defined functions that are made available to the agent/holon instances inside the HCD throughout function calls. The latter functionality corresponds to a set of generic decision-making components that act as the nervous system of the HCD instance for decision-making. As illustrated in Figure 106.8, there are four generic decision-making modules:

- Planner: The planner component carries out the construction of control-action steps for the HCD instance. It also negotiates with other HCD instances, which can be local or remote, via an agent communication language (e.g., FIPA, KQML).
- Process Model: The process model, which carries out a local model simulation (which can be a mathematical expression that quantifies performance metrics) of input data to quantify the current control step performance, thereby evaluates present execution states or forecasting near-future states.
- Execution Control: The execution control component coordinates the execution of control action steps, thereby carrying out the HCD plans, which are built and committed by the planner during the

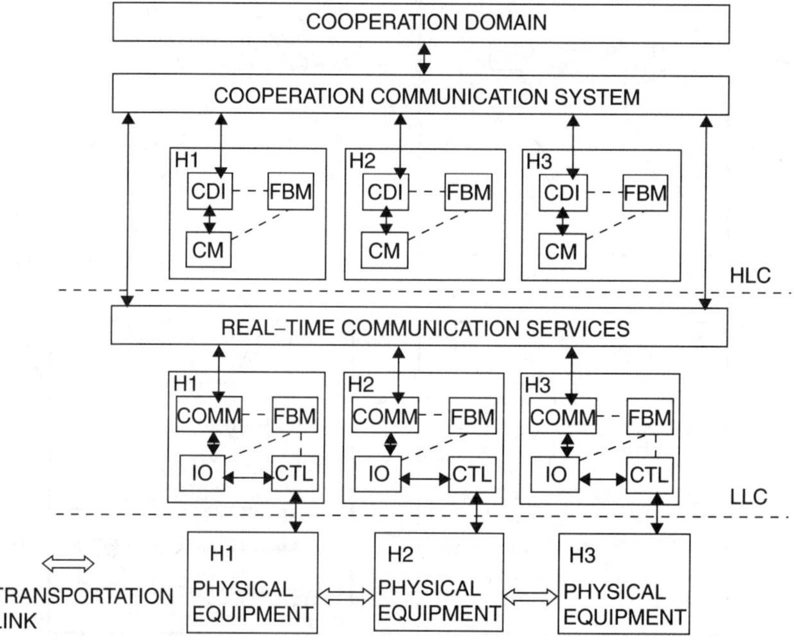

FIGURE 106.7 Integration of the HLC and LLC architectures (from Neligwa and Fletcher, 2003).

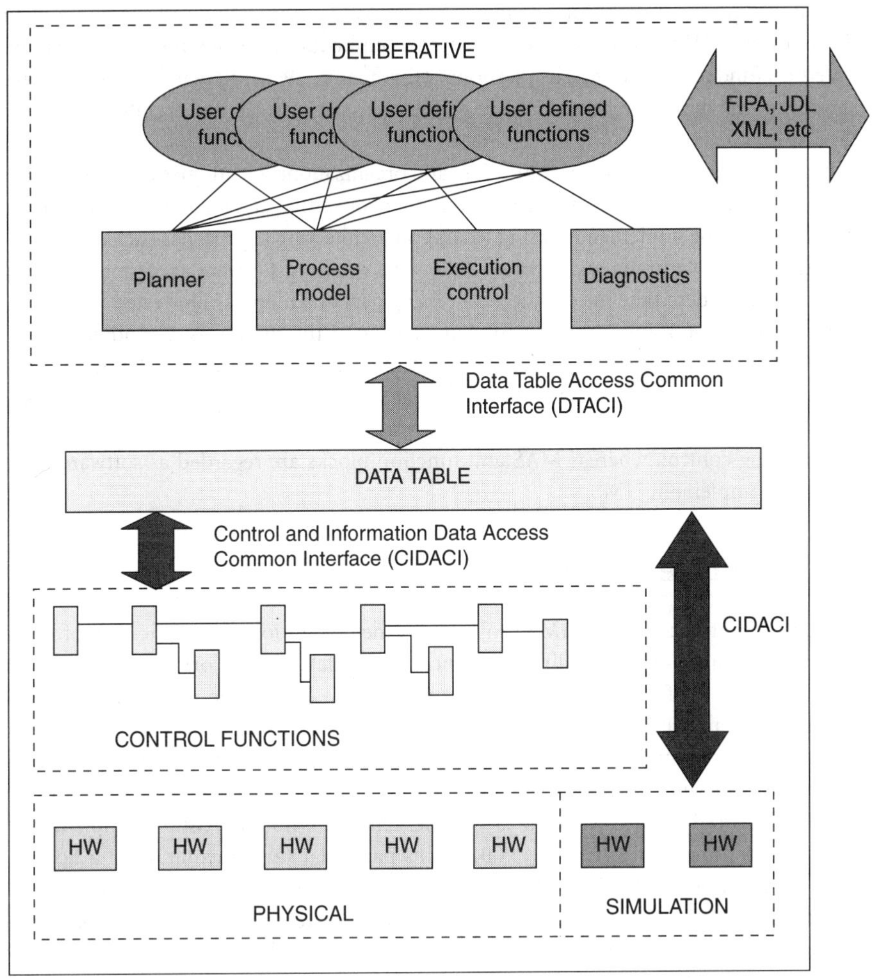

FIGURE 106.8 An interface architecture for holonic control devices.

inter-HCD instance negotiation. This is a further refinement of the planner's functionality, which results in the higher-level goals being translated into a form that can be used at the control level.

- Diagnostics: The diagnostics component carries out a local self-assessment activity to verify the correct operation of hardware attached to the HCD and/or HCD instances.

The *Deliberative Layer* communicates with other layers throughout the device's data table via a Data Table Access Common Interface (DTACI). The *Data Table Layer* is the state data repository for the HCD instances. The *Control Functions Layer* is the user-defined application logic that controls the activities of the physical hardware or process that is attached to the HCD. Although Figure 106.8 shows IEC 61131-3 function blocks, any of the IEC 61131-3 languages for programmable logic controllers (i.e., IL, ST, LD, SFC, or LD) could be used to specify the control logic. In addition, distributed intelligent devices at the physical layer can also be supported at this layer by the IEC 61499 model.

106.5 Holonic Systems and Information Technology

In this chapter, we have provided an overview of the evolution in manufacturing systems control from distributed decision-making to holonic systems. A common thread that runs throughout this work is the close link between distributed artificial intelligence and multiagent systems and holonic systems.

Like HMS, multiagent manufacturing systems also consist of cooperative and autonomous manufacturing units, but unlike HMS, MAS can be considered as embedding a "general software technology that was motivated by fundamental research questions" (Bussmann, 1998). Research in MAS, however, has played a key role in the development of HMS (e.g., Bussman's agent-oriented architecture for a holon as shown in Figure 106.3).

As a result, various cooperation, communication, and organizational techniques from the MAS literature (e.g., Weiss, 1999) can be used to implement autonomous, cooperative, and recursive agents ("holons"). For example, holarchies can be implemented using federation architecture approaches such as facilitators, brokers, or mediators (e.g., Maturana and Norrie, 1996). The resulting software agent may be considered a holon, or it may be considered the information processing part of a holon as illustrated in Figure 106.3.

As shown in Figures 106.7 and 106.8, the implementation of the "holon" is dependent on the role the holon plays and its position in the manufacturing hierarchy. In other words, it is the stance that is taken toward the manufacturing systems control problem that will influence whether agents, holons, or function blocks are used. More specifically, HMS can be considered a general paradigm for distributed intelligent manufacturing control, whereas MAS and function blocks are regarded as software technologies that can be used to implement HMS.

106.6   The Future of HMS

The focus of the current phase of the HMS project is on the integration and application of holonic concepts. This is a particularly important direction, since, to this date, the majority of the developments in HMS have been primarily of an exploratory nature with only a few proven industrial implementations (McFarlane and Bussman, 2003). If HMS are to gain wider industrial acceptance, evidence will have to be shown that holonic manufacturing systems are economically viable and that they do indeed perform well in the presence of disturbances.

One notable industrial implementation is DaimlerChrysler's holonic control implementation for engine assembly (Bussman and Sieverding, 2001). This particular implementation of a holonic system has now been in operation for 3 years and has demonstrated that many of the claimed technical advantages are achievable using a holonic systems approach. Unfortunately, it is difficult to weigh the potential economic advantages of this approach (e.g., as a result of increased flexibility) against the up-front cost associated with the implementation of this technology. This illustrates the need to not only demonstrate the technological advantages of holonic systems but also the economic benefits.

Despite this concern, there is interest in the holonic systems community for extending the scope for the holonic approach. Rather than just the relatively narrow focus on manufacturing systems control during the first phase of the HMS project, there is now considerable activity in applying holonic concepts to a variety of nonmanufacturing domains. For example, Maturana et al. (2003) have recently applied holonic and agent-based techniques to the design of reconfigurable ship-based chilled-water systems. Fleetwood et al. (2003) have been investigating the integration of the holonic system architecture with the evolving hydrogen infrastructure for fueling hydrogen to fuel cell-powered vehicles. A considerable amount of interest has also been shown recently in the application of holonic systems concepts to supply chain management (e.g., Weichhart et al., 2003).

As the HMS Consortium (HMS, 2003) draws to a close in 2004, the momentum is building for applications to a variety of industrial problems. As McFarlane and Bussman (2003) observed, a number of open issues must still be in the areas of holonic control system design and implementation; however, promise of this approach appears to be strong enough to warrant further work in these areas in the years to come.

References

Aparicio, M., IV, Internet-scale network intelligence, *IEEE Internet Computing*, Vol. 3, Issue 5, 38–40, 1999.
Booch, G., *Object-Oriented Analysis and Design with Applications*, 2nd ed., Addison-Wesley, Reading, MA, 1994.

Brennan, R., Hall, K., Marik, V., Maturana, F., and Norrie, D., A real-time interface for holonic control devices, in *Holonic and Multiagent Systems for Manufacturing*, Marik, V., McFarlane, D., and Valckenaers, P., Eds. Lecture Notes in Artificial Intelligence, Vol. 2744, Springer, Berlin, 2003, pp. 25–34.

Brennan, R. and Norrie, D. Managing Fault Monitoring and Recovery in Distributed Real-time Control Systems, Proceedings. of the 5th IEEE/IFIP International Conference on Information Technology for Balanced Automation Systems in Manufacturing and Services, Cancun, Mexico, 2002, pp. 247–254.

Brennan, R. and Norrie, D., Agents, holons and function blocks: distributed intelligent control in manufacturing, *Journal of Applied Systems Studies, Special Issue on Industrial Applications of Multi-Agent and Holonic Systems*, 2, 1–19, 2001.

Bussmann, S., An Agent-oriented Architecture for Holonic Manufacturing Control, Proceedings of the First Open Workshop on Intelligent Manufacturing Systems Europe, Hannover, Germany, 1998, pp. 1–12.

Bussmann, S. and Sieverding, J., Holonic Control of an Engine Assembly Plant — an Industrial Evaluation, Proceedings of the 2001 IEEE International Conference on Systems, Man, and Cybernetics, Tuscon, AZ, U.S.A, Oct. 2001.

Christensen, J., HMS/FB architecture and its implementation, in *Agent-Based Manufacturing: Advances in the Holonic Approach*, Deen, M., Ed., 2003, pp. 53–88.

Christensen, J., HMS: Initial Architecture and Standards Directions, Proceedings of the 1st European Conference on Holonic Manufacturing Systems, Hannover, Germany, 1994, pp. 1–20.

Cockburn, D. and Jennings, N., ARCHON: a DAI system for industrial applications, in *Foundations of Distributed Artificial Intelligence*, O'Hare, G. and Jennings, N., Eds., John Wiley & Sons, New York, 1996, pp. 319–344.

Cutkosky, M., Englemor, R., Fikes, R., Gruber, T., Genesereth, M., Mark, W., Tenenbaum, J., and Weber, J., PACT: an experiment in integrating concurrent engineering systems, *IEEE Computer*, 26, 28–37, 1993.

Davis, R. and Smith, R., Negotiation as a metaphor for distributed problem solving, *Artificial Intelligence*, 20, 63–109, 1983.

Duffie, N. and Piper, R., Nonhierarchical control of manufacturing systems, *Journal of Manufacturing Systems*, 5, 137–139, 1986.

Duffie, N. and Prabhu, V., Real-time distributed scheduling of heterarchical manufacturing systems, *Journal of Manufacturing Systems*, 13, 94–107, 1994.

Fleetwood, M., Kotak, D.B., Wu, S., and Tamoto, H., Holonic System Architecture for Scalable Infrastructures, Proceedings of the 2003 IEEE International Conference on Systems, Man, and Cybernetics, Washington, DC, U.S.A., 2003.

Fletcher, M., Brennan, R.W., and Norrie, D.H., Design and Evaluation of Real-time Distributed Manufacturing Control Systems Using UML Capsules, Proceedings of the7th International Conference on Object-oriented Information Systems, Springer-Verlag, Berlin, 2001 pp. 382–386.

Fletcher, M., Garcia-Herreros, E., Christensen, J., Deen, S., and Mittmann, R., An Open Architecture for Holonic Cooperation and Autonomy, Proceedings of the 11th International Workshop on Database and Expert Systems Applications, 2000 pp. 224–230.

Foundation for Intelligent Physical Agents, Web Site, http://www.fipa.org/., 2001.

Gruver, W., Kotak, D., van Leeuwen, E., and Norrie, D., Holonic manufacturing systems: Phase I, in *Holonic and Multiagent Systems for Manufacturing*, Marik, V., McFarlane, D., and Valckenaers, P., Eds., Lecture Notes in Artificial Intelligence, Vol. 2744, Springer, Berlin, 2003, pp. 1–14.

Holonic Manufacturing Systems Consortium, Holonic manufacturing systems overview, Holonic Manufacturing Systems Consortium Web Site, http://hms.ifw.uni-hannover.de/, 2001.

Huhns, M., and Singh, M., *Readings in Agents*, Morgan Kaufmann, Los AeEos, CA, 1998.

IEC TC65/WG6, Voting Draft — Publicly Available Specification — Function Blocks for Industrial Process-measurement and Control Systems, Part 1 — Architecture, International Electrotechnical Commission, 2000.

ISO, The Ottawa Report on Reference Models for Automated Manufacturing Systems, Geneva, Switzerland, 1986.

Jones, A. and McLean, C., A proposed hierarchical control model for automated manufacturing systems, *Journal of Manufacturing Systems*, 5, pp. 15–25, 1986.

Koestler, A., *The Ghost in the Machine*, Arkana Books, London, 1967.

Kriz, D., Holonic manufacturing systems: case study of an IMS Consortium, Holonic Manufacturing Systems Consortium Web Site, http://hms.ifw.uni-hannover.de/, 1995.

Langer, G., HoMuCS: A Methodology and Architecture for Holonic Multi-Cell Control Systems, Ph.D. thesis, Technical University of Denmark, 1999.

Lewis, R., Modelling Control Systems Using IEC 61499: Applying Function Blocks to Distributed Systems, The Institution of Electrical Engineers, London, 2001.

Lin, G. and Solberg, J., Integrated shop floor control using autonomous agents, *IIE Transactions: Design and Manufacturing*, 24, 57–71, 1992.

Marik, V. and Pechoucek, M. Holons and Agents: Recent Developments and Mutual Impacts, Proceedings of the Twelfth International Workshop on Database and Expert Systems Applications, IEEE Computer Society, Silver Spring, MD,2001, pp. 605–607.

Maturana, F., Tichy, P., Slechta, P. Staron, R., Discenzo, F., Hall, K., and Marik, V., Cost-based dynamic reconfiguration system for evolving holarchies, in *Holonic and Multiagent Systems for Manufacturing*, Marik, V., McFarlane, D. and Valckenaers, P., Eds., Lecture Notes in Artificial Intelligence,Vol. 2744, Springer-Verlag,Berlin, 2003.

Maturana, F. and Norrie, D., Multi-agent mediator architecture for distributed manufacturing, *Journal of Intelligent Manufacturing*, 7, 257–270, 1996.

McFarlane, D.C. and Bussman, S., Holonic manufacturing control: rationales, developments and open issues, in *Agent-Based Manufacturing: Advances in the Holonic Approach*, Deen, M., Ed., 2003, pp. 303–326.

Monceyron, E., and Barthes, J., Architecture for ICAD systems: an example from harbor design, *Revue Sciences et Techniques de la Conception*, 1, 49–68, 1992.

Monch, L. Stehli, M., and Zimmermann, J., FABMAS: an agent-based system for production control of semiconductor manufacturing processes, in *Holonic and Multiagent Systems for Manufacturing*, Marik, V., McFarlane, D., and Valckenaers, P., Eds.,Lecture Notes in Artificial Intelligence, Vol. 2744, Springer, Berlin, 2003, pp.258–267.

Moulin, B. and Chaib-draa, B., (1996): An overview of distributed artificial intelligence, in *Foundations of Distributed Artificial Intelligence*, O'Hare, G. and Jennings, N., Eds., John Wiley & Sons, New York, 1996, pp. 3–56.

Neligwa, T. and Fletcher, M., An HMS operational model, in *Agent-Based Manufacturing: Advances in the Holonic Approach*, Deen, M., Ed., Springer-Verlag, Berlin, 2003, pp. 163–192.

Parunak, H., Industrial and practical applications of DAI, in *Multiagent Systems: A Modern Approach to Distributed Artificial Intelligence*, Weiss, G., Ed., MIT Press, Cambridge, MA, 1999, pp. 377–424.

Parunak, H., Applications of distributed artificial intelligence in industry, in *Foundations of Distributed Artificial Intelligence*, O'Hare, G. and Jennings, N., Eds., John Wiley & Sons, New York, 1996, pp. 139–164.

Parunak, H., Autonomous Agent Architectures: A Non-Technical Introduction, Industrial Technology Institute Report, 1993.

Parunak, H., Manufacturing experience with the contract net, *Distributed Artificial Intelligence*, Vol. 1, Pitman, London, 1987.

Peng, Y., Finin, T., Labrou, Y., Chu, B., Long, J., Tolone, W., and Boughannam, A., A Multi-agent System for Enterprise Integration, Proceedings. of PAAM'98, 1998, pp. 533–548.

Suda, Future factory automation system formulated in Japan, *Techno Japan*, 22, 15–25, 1989.

Suessmann, B., Colombo, A., and Neubert, R., An Agent-Based Approach Towards the Design of Industrial Holonic Control Systems, Proceedings of the 5th IEEE/IFIP International Conference on Information Technology for Balanced Automation Systems in Manufacturing and Services, Cancun, Mexico, 2002, pp. 255–270.

Shen, W. and Norrie, D., Agent-based systems for intelligent manufacturing: a state-of-the-art survey, *Knowledge and Information Systems*, 1, 129–156, 1999.

Shen, W., Norrie, D., and Barthes, J., *MultiAgent Systems for Concurrent Design and Manufacturing*, Taylor & Francis, London, 2000.

Simon, H., *The Sciences of the Artificial*, MIT Press, Cambriidge, MA, 1996.

Ulieru, M., FIPA technology overview: holonic enterprise, *FIPA Inform!*, 2, 3–4, 2001.

Valckenaers, P. and Van Brussel, H., IMS TC5: holonic manufacturing systems technical overview, Holonic Manufacturing Systems Consortium Web Site, http://hms.ifw.uni-hannover.de/, 1994.

Van Brussel, H., Wyns, J., Valckenaers, P., Bongaerts, L., and Peeters, P., Reference architecture for holonic manufacturing systems: PROSA, *Computers in Industry*, 37, 255–274, 1998.

Veeramani, D., Task and Resource Allocation Via Auctioning, Proceedings of the 1992 Winter Simulation Conference, 1992, pp. 945–954.

Weichhart, G., Hammerle, A., and Fessl, K., Service-oriented Concept of a Holonic Enterprise-enabling Adaptive Networks Along the Value Chain, Proceedings of the 5th IEEE/IFIP International Conference on Information Technology for Balanced Automation Systems in Manufacturing and Services, Cancun, Mexico, 2002, pp. 289–304.

Weiss, G., *Multi-Agent Systems*, MIT Press, Cambridge, MA, 1999.

Wooldridge, M. and Jennings, N., Software engineering with agents: pitfalls and pratfalls, *IEEE Internet Computing*, Vol. 3, Issue 3, 20–27, 1999.

107

From Holonic Control to Virtual Enterprises: The Multi-Agent Approach

Pavel Vrba
Rockwell Automation Research Center

Vladimir Marik
Czech Technical University &
Rockwell Automation Research Center

107.1 Introduction

Both the complexity of manufacturing environments as well as the complexity of tasks to be solved are growing continuously. In many manufacturing scenarios, traditional centralized and hierarchical approaches applied to production control, planning and scheduling, supply chain management, and manufacturing and business solutions in general are not adequate and can fail because of the insufficient means to cope with the high degree of complexity and practical requirements for *generality* and *reconfigurability*.

These issues naturally lead to a development of new manufacturing architectures and solutions based on the consideration of highly distributed, autonomous, and efficiently cooperating units integrated by the plug-and-play approach. This trend of application of multi-agent systems (MAS) techniques is clearly visible at all levels of manufacturing and businesses. On the lowest, real-time level, where these units are tightly linked with the physical manufacturing hardware, we refere to them as *holons* or *holonic agents* [1]. Intelligent agents are also used in solving production planning and scheduling tasks both on the work-shop and factory levels. More generic visions of intensive cooperation among enterprises connected via communication network have led to the ideas of *virtual enterprises*:

A virtual enterprise is a temporary alliance of enterprises that come together to share skills or core competencies and resources in order to better respond to business opportunities, and whose cooperation is supported by computer networks [2].

The philosophical background of all these highly distributed solutions is the same: the community of autonomous, intelligent, and goal-oriented units efficiently cooperating and coordinating their behavior in order to reach the global level goals. The decision-making knowledge stored and exploited locally in the agents/holons invokes the global behavior of the system that is not deterministic, but rather emergent — such a behavior cannot be precisely predicted at the design time of the community. The experimental test-ing of the global behavior with the physical manufacturing/control environment being involved is not only extremely expensive, but nonrealistic as well. The only possible solution is the simulation, both of the con-trolled process of the manufacturing facility as well as the simulation of the interagent interactions.

107.2   Technology Overview

The architecture of an agent usually consists of the *agent's body* and the *agent's wrapper*. We can also say that the body, the functional core of an agent, is encapsulated by the wrapper to create an agent [3]. The wrapper accounts for the interagent communication and real-time reactivity. The body is an agent's rea-soning component, responsible for carrying out the main functionality of the agent. It is usually not aware of the other members of the community, their capabilities, duties, etc. This is the wrapper, which is responsible for communicating with the other agents, for collecting information about the intents, goals, capabilities, load, reliability, etc., of the other units in the agents' community.

From the implementation point of view, there are two types of agents: (i) *custom-tailored agents*, which are implemented in order to provide a specific service to the community (e.g., service brokering) and (ii) *integrated agents*, which encapsulate a preexisting, "inherited," or "legacy" piece of software/hardware by the agent's wrapper into the appropriate agent structure. In this case, the wrapper provides a standard-ized communication interface enabling to plug the legacy system into the corresponding agent commu-nity. From the outside, such a wrapped software system cannot be distinguished from the *custom-tailored agents* as it communicates in a standard way with the others and understands the predefined language used for the interagent communication (e.g., Agent Communication Language — ACL defined by Foundation for Intelligent Physical Agents [FIPA]). These *agentification processes* provide an elegant mechanism for *system integration* — a technique supporting the *technology migration* from the central-ized systems toward the distributed agent-based architectures.

We distinguish three different concepts of agency that we need to explain in more detail.

Mobile Agents

Mobile agents are generally pieces of code traveling freely inside a certain communication network (usu-ally inside the Internet). Such agents, being usually developed and studied in the area of computer sci-ence, are stand-alone, executable software modules, which are being sent to different host computers/servers to carry out a specific computational tasks (usually upon the locally stored data). They are expected to report back the results of the computation process. The mobile agents can travel across the network, can be cloned or destroyed by their own decision when fulfilling their specific task, etc.

MAS

MAS can be described as goal-oriented communities of cooperative/self-interested agents in a certain inter-action environment. They have been developed within the domain of the distributed artificial intelligence and they explore the principles of artificial intelligence for reasoning, communication, and cooperation. The important attribute of each agent is its *autonomy* — the agent resides on a computer platform where it autonomously carries out a particular task/functionality. The agent owns only a part of the global information about the goals of the community that is sufficient for its local decision-making and behavior. However, in some situations, for example, when the agent is not capable of fulfilling the requested operation alone, the agent is capable of *cooperation* with other agents (asking them for help) usually via message sending. The agents asked for cooperation are still autonomous in their decisions, that is, can either agree on cooperation or can even refuse to cooperate (e.g., due to lack of their own resources). Another important attribute of MAS is the *plug-and-play* approach — the agents, which can be grouped into different types of communities (such as teams, coalitions, platforms, etc., see Section 107.3), can freely join and leave the communities. Usually, the community offers a yellow-pages-like mechanism that is used by agents to offer their services to the other agents as well as to find out suitable agents for possible cooperation. This allows to dynamically change the overall capabilities and goals of the MAS (according to user requests) by adding new agents with desired functionalities.

The cooperation among the agents supported by their *social behavior* is the dominant feature of the activities of the agents in the community. The term social behavior means that the agents are able to communicate, to understand the goals, states, capabilities, etc., of the others and to respect the general rules and constraints of behavior valid for each of the community members. Communication (exchange of information usually in the form of messages) plays the crucial and decisive role in the agents' communities with social behavior of either a cooperative or competitive nature.

Holons

Holons are agents dedicated to real-time manufacturing tasks [4]. These are tightly physically coupled, really "hard-wired" with the manufacturing hardware (devices, machines, workshop cells), thus with a low degree of freedom in their mutual communication. The holons operate in the "hard" real time, and their patterns of behavior are strictly preprogrammed. This means that their reactions in certain situations are predictable and emergent behavior is not highly appreciated (usually not allowed). The Holonic Manufacturing Systems (HMS) are designed mainly to enable a fast and efficient system *reconfiguration* in the case of any machine failure. One of the pioneering features of holonic systems is a complete separation of the data flow from the control instructions. The communication principles, supporting the holons' interoperability, are already standardized (the IEC 61499 standard; see Section 107.6).

107.3 Cooperation and Coordination Models

As we have already mentioned, communication among the agents is an important enabler of their social behavior. The agents usually explore specific communication language with standardized types of messages. The set of messages is chosen so that it represents the most typical communicative acts, often called *performatives* (according to the speech act theory of Searle [5]), used by agents in a particular domain. Examples of such performatives can be `request` used by an agent to request the other one to perform a particular operation, `agree` used to confirm the willingness to cooperate, or, for example `refuse` to deny to perform requested action. Along with the performative, the message bears the information about the sender and receiver, the content of the message, and the identification of the language used for the message's content. Additional attributes can be included in the message, for example, the reference to the appropriate knowledge ontology describing semantics of the message (see Section 107.5) or the brief description of the negotiation strategy used (the structure of replies expected).

The communication among the agents is usually not just a random exchange of messages, but the message flow is managed by a set of standard *communication protocols*. These protocols range from a simple "question–reply," via "subscribe–inform" through to more complex negotiation protocols like "Contract-Net-Protocol" (CNP), different versions of auctions (Dutch, sealed, Vickery, etc.). The communication protocols and communication traffic in general can be represented graphically by means of interaction diagrams. Figure 107.1 shows the *Request* and *Contract-net* protocols from FIPA specifications captured in AUML — Agent-based extensions to the standard UML.

The more knowledge is available locally, which means it is "owned" by individual agents, the smaller the communication traffic needed to achieve cooperative social behavior. One of the crucial questions is how to store and use the knowledge locally. For this purpose, different *acquaintance models* are located in the wrappers of individual agents. These are used to organize, maintain, and explore knowledge about the other agents (about their addresses, capabilities, load, reliability, etc.). This kind of knowledge, which strongly supports collaboration activities among the agents, is called *social knowledge* [3]. These models can be used to organize both the long-term as well as temporary or semipermanent knowledge/data concerning cooperation partners. To keep the temporary and semipermanent knowledge fresh, several *knowledge maintenance techniques* have been developed, namely (i) *periodic knowledge revisions* — the knowledge is updated periodically by regular "question–answer" processes, (ii) *subscription-based update* — the knowledge update is preordered by a specific subscription mechanism. The field is strongly influenced and motivated by Rao and Georgeff's BDI (Beliefs-Desires-Intentions) model [11] used to express and model agents' beliefs, desires and intentions, and other generic aspects of MAS.

The MAS research community provides various techniques and components for creating the architecture of an agent community. The crucial categories of agents, according to their intra- and inter-community functionalities (e.g., resource agents, order or customer agents, and information agents) have been identified. Services have been defined for specific categories of agents (e.g., white pages, yellow pages, brokerage, etc). In some of the architectures, the agents do not communicate directly among themselves. They send messages via facilitators, which play the role of communication interfaces among collaborating agents [6]. Other architectures are based on utilization of matchmakers, which proactively try to find the best possible collaborator, brokers that act on behalf of the agent [7], or mediators that coordinate the agents by suggesting and promoting new cooperation patterns among them [8]. Increasing attention has recently been paid to the concept of the meta-agent, which independently observes the interagent communication and suggests possible operational improvements [9].

The techniques for organizing long-term *alliances* and short-term *coalitions* as well as techniques for planning of their activities (*team action planning*) have been developed recently [10]. These algorithms can help to solve certain types of tasks more efficiently and to allocate the load among the agents in an

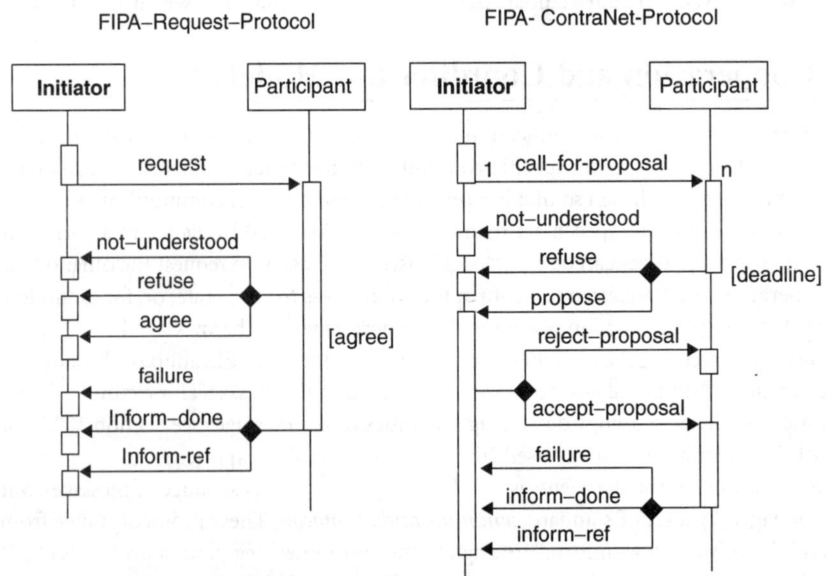

FIGURE 107.1   FIPA's *Request* and *Contract-net* communication protocols.

optimal way. They can be explored with an advantage for automated creation and dissolving of virtual organizations as well as for an optimal load distribution among the individual bodies in a virtual organization. Alliance and coalition formation techniques can be linked with methodologies and techniques for administration and maintenance of *private, semiprivate,* and *public knowledge.* This seems to be an important issue to tackle, especially in virtual organizations of temporary nature where units competing in one business project are also contracted to cooperate in the other one. It is also possible to classify and measure the necessary leakage of private/semiprivate knowledge to reflect this fact in the future contracts.

107.4 Agents Interoperability Standardization — FIPA

The FIPA is a nonprofit association registered in Geneva, Switzerland, founded in December 1995. The main goal of FIPA is to maximize interoperability across agent-based applications, services, and equipment. This is done through FIPA specifications.

FIPA provides specifications of basic agent technologies that can be integrated by agent systems developers to make complex systems with a high degree of interoperability. FIPA specifies the set of interfaces, which the agent uses for interaction with various components in the agent's environment, that is, humans, other agents, nonagent software, and the physical world. It focuses on specifying *external communication among agents* rather than the internal processing of the communication at the receiver.

The FIPA Abstract Architecture defines a high-level organizational model for agent communication and core support for it. It is neutral with respect to any particular network protocol for message transport or any service implementation. This abstract architecture cannot be directly implemented; it should be viewed as a basis or specification framework for the development of particular architectural specifications.

The FIPA Abstract Architecture contains agent system specifications in the form of both descriptive and formal models. It covers three important areas: (i) *agent communication,* (ii) *agent management,* and (iii) *agent message transport.*

Agent Communication and Agent Communication Language

FIPA provides standards for agent communication languages. The messages exchanged among the agents must comply with a FIPA-ACL specification. FIPA-ACL is based on speech act theory [5] and resembles Knowledge Query Manipulation Language (KQML). Each message is labeled by a *performative,* denoting a corresponding *communicative act* (see previous section), such as `inform` or `request`. Along with the performative, each message contains information about its sender and receiver, the content of the message, a content language specification, and an ontology identifier. Other important attributes of an message are the information about the conversation protocol that applies to the current message (e.g., FIPA-Request) and the *conversation ID* that uniquely identifies to which "conversation thread" this message belongs (since there can be more conversations between two agents following the same protocol at the same time).

The core of the FIPA ACL message is its content, which is encoded in language denoted in the *language* slot of the message. FIPA offers a Semantic Language (*FIPA-SL*) as a general-purpose knowledge representation formalism for different agent domains. This formalism maps each agent message type (performative) to an SL formula that defines constraints (called feasibility conditions) that the sender has to satisfy, and to another formula that defines the rational effect of the corresponding action. Nevertheless, any other existing or user-defined language can be used as a content language of FIPA messages. One of the most popular languages used today to express the message syntax is the XML.

Bellow is an example of a FIPA message, where the agent "the-sender" requests the agent "the-receiver" to deliver a specific box to a specific location, using the XML content language and the FIPA-Request conversation protocol.

```
(request
    :sender (agent-identifier :name the-sender)
    :receiver (set (agent-identifier :name the-receiver))
    :content
```

```
        <deliver item="box017" locX="12" locY="19" />
   :protocol fipa-request
   :language XML
   :conversation-id order567
 )
```

Agent Management

The FIPA Agent Management Specification provides a specification of how the members of the multi-agent community shall be registered, organized, and managed. According to the FIPA philosophy, the agents are grouped into high-level organizational structures called *agent platforms* (APs). Members of each platform are usually geographically "close" — for example, they may be run on one computer or be located in a local area network. Each platform must provide its agents with the following two mandatory services: the Agent Management System (AMS) and the Directory Facilitator (DF).

The AMS administers a list of agents registered with the platform. This component implements the creation and deletion of the running agents and provides the agents with a "white-page-list" type of service, for example, a list of all agents accommodated on the given agent platform and their addresses. Unlike the AMS, the DF supplies the community with a "yellow-page-list" type of service. Agents register their services with the DF and can query the DF to find out what services are offered by other agents, or to find all agents that provide a particular service. Thus, agents can find the addresses of others that can assist in accomplishing the desired goals.

Message Transport Service

The Message Transport Service (MTS) is a third mandatory component (besides AMS and DF) that the agent platform has to offer to the agents. The MTS is provided by so-called Agent Communication Channel (ACC), which is responsible for the physical transportation of messages among agents local to a single AP as well as among agents hosted by different APs. For the former case, FIPA does not mandate to use a specific communication protocol or interface — different protocols are being used today in agent platform implementations, such as the TCP/IP sockets, the UDP protocol, or, particularly in JAVA implementations, the JAVA Remote Method Invocation (RMI). In the latter case, the FIPA defines a *message transport protocol* (MTP) that ensures the interoperability between agents from different agent platforms. For this purpose, the ACC must implement the MTP for at least one of the following communication protocols specified by FIPA: the IIOP (Internet Inter-Orb Protocol), WAP, or HTTP.

107.5 Ontologies

Ontologies play a significant role not only in the interagent communication, where the content of messages exchanged among agents must conform to some ontology in order to be understood, but also in knowledge capturing, sharing, and reuse. One of the main reasons why ontologies are being used is the *semantic interoperability* enabling among others:

- to share knowledge — by sharing the understanding of the structure of information exchanged among software agents and people,
- to reuse knowledge — ontology can be reused for other systems operating on a similar domain, and
- to make assumptions about a domain explicit — for example, for easier communication.

Basically, ontology can be referred to as a vocabulary providing the agents with the semantics of symbols, terms, or keywords used in messages [12]. Thus, if an agent sends a message to another agent using particular ontology, it can be sure that the other agent (of course, if it shares the same ontology) will understand the message.

Ontologies for MAS

Two examples can be selected to illustrate multi-agent-oriented ontological efforts in the area of manufacturing. The first one, FIPA Ontology Service Recommendation, is a part of a set of practical recommendations on how to implement agents in a standardized way. The other one, Process Specification Language (PSL) project [13], tries to develop general ontology for representing manufacturing processes. Its aim is to serve as interlingua for translating between process ontologies. The transformation between ontologies for translation using PSL is expected to be defined by humans.

The former approach seems to be much more general and applicable. FIPA uses Open Knowledge Base Connectivity, (OKBC) [14], as a base for expressing ontologies. OKBC is an API for accessing and modifying multiple, heterogeneous knowledge bases. Its knowledge model defines a meta-ontology for expressing ontologies in an object-oriented frame-based manner. OKBC can be mapped to the object-oriented languages, so that classes in programming languages can be built on the underlying ontology and be used for exchanging information. Semantics of OKBC constructs is defined in KIF as a description of what the constructs intuitively mean. However, no reasoning engine that would enable to use this information for, for example, ontology integration, is provided. Moreover, it could be difficult to provide a reasoning support for some of the constructs.

Ontologies in FIPA proposals and related ontologies for practical applications [15] are motivated mainly by the need to have something that would work immediately, because currently more attention is paid to the functional behavior of agents. There is nothing wrong with this approach, if we want to have a working solution in a short time where we do not care about the possibility of reasoning about the ontologies and further interoperability. However, the need for reasoning about ontologies can easily arise, for example, when requiring interoperability in open multiagent systems, that is, systems where new agents with possibly other ontologies can join the community.

107.6 HMS

Over the past 10 years, researches attempted to apply the agent technology to various manufacturing areas such as supply chain management, manufacturing planning, scheduling, and execution control. This effort resulted in the development of a new concept, the *HMS*, based on the ideas of *holons* presented by Koestler [17] and strongly influenced by the requirements of industrial control.

Holons are autonomous, cooperative units that can be considered as elementary building blocks of manufacturing systems with decentralized control [16]. They can be organized in hierarchical or heterarchical structures. Holons, especially those for real-time control, are usually directly linked to the physical hardware of the manufacturing facility and are able to physically influence the real world (e.g., they may be linked to a device, tool or other manufacturing unit, or to a transportation belt or a storage manipulator).

Holons for real-time control are expected to provide reactive behavior rather than being capable of deliberative behavior based on complex "mental states" and strongly proactive strategies. They are expected mainly to react to changes in the manufacturing environment (e.g., when a device failure or a change in the global plan occurs). Under "stable circumstances," during routine operation, they are not required to change the environment proactively. The reason for the prevalence of reactive behavior of real-time control holons is that each of them is linked to a physical manufacturing facility/environment, changes to which are not very simple, cheap, or desirable in a comparatively "stable" manufacturing facility. The physical linkage to physical equipment seems to be a strong limiting factor of the holons' freedom in decision-making.

The more generic holonic ideas and considerations have led to the vision of a *holonic factory* [18]. Here, all the operations (starting from product ordering, planning, scheduling, and manufacturing, to invoicing the customer) are based entirely on holonic principles. A holonic factory contains a group of principal system components (holons) that represent physical manufacturing entities such as machines or products as well as virtual entities like orders or invoices. The holons work autonomously and cooperate together in order to achieve the global goals of the factory. Thus, the factory can be managed toward global goals by the activities of individual autonomous holons operating locally. The community of researchers trying to implement the vision of the holonic factory is well organized around the international HMS consortium.

The vision of the holonic factory covers several levels of information processing for manufacturing. We can distinguish at least three separate levels, namely,

- *real-time control*, which is tightly linked with the physical manufacturing equipment;
- *production planning and scheduling*, both on the workshop and on the factory level; and
- *supply chain management*, integrating a particular plant with external entities (suppliers, customers, cooperators, sales network, etc.).

At the lowest RT-control level, the main characteristics of holons is their linkage to the physical manufacturing devices — these holons read data from sensors and send control signals to actuators. Within the HMS activities, the standard IEC-61499 known as *function blocks* has been developed for these RT-control purposes. It is based on function blocks part of the well-known IEC-1131-3 standard for languages in Programmable Logical Controllers (PLCs). The major advantage is the separation between the data flow and the event flow among various function blocks. Multiple function blocks can be logically grouped together, across multiple devices into an application, to perform some process control.

Since the IEC-61499 fits well these RT-control purposes, it does not address the higher level aspects of holons acting as cooperative entities capable of communication, negotiation, and high-level decision making. It is obvious that this is the field where the techniques of multi-agent systems have to be applied. Thus, a general architecture combining function-bocks with agents was presented in [1]. As shown in Figure 107.2, a software agent and function block control application (connected to the physical layer) are encapsulated into a single structure.

In such a holon equipped with a higher-level software component, three communication channels should be considered:

- *Intraholon* communication between the function block part and the software agent component.
- *Interholon* communication that is aimed at communication among the agent-based parts of multiple holons — FIPA standards are used more and more often for this purpose.
- A *direct* communication channel between function block parts of neighboring holons. If we are prepared to break the autonomy of an independent holon, then this communication is standardized by IEC 61499 already; otherwise, a new type of real-time coordination technology is needed to ensure real-time coordination.

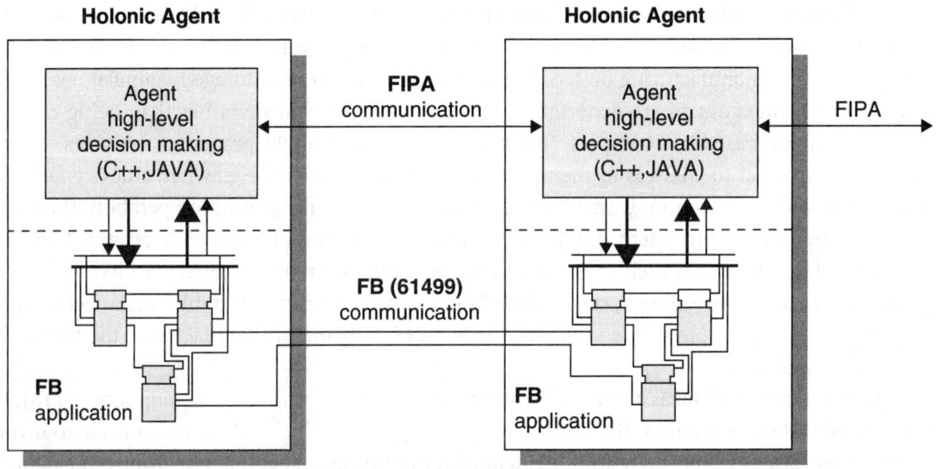

FIGURE 107.2   Holonic agent: combination of function block application and software agent.

As a matter of fact, the holons defined in this way behave — on the level of interholon communication — like standard software agents. They can communicate widely among themselves, carry out complex negotiations, cooperate, develop manufacturing scenarios, etc. We can call them holonic agents (or agentified holons), as they consist of both a holonic part connected with the physical layer of the manufacturing system (operating in hard real time) and a software agent for higher-level, soft real-time or non-real-time intelligent decision making. It has already been mentioned that the interholon communication is usually standardized by the FIPA approach, and direct communication could be achieved by IEC standards (not necessarily IEC 61499). Let us stress that the FIPA standards are not applicable for the low-level real-time control purposes as they do not take account of the real-time control aspects.

The attention of system developers is currently directed mainly at the intraholon communication, which is usually both application- and company-specific and is usually connected with the solution of the "migration problem" (the problem of exploring the classical real-time control hardware for holonic control). McFarlane et al. [19] introduced a blackboard system for accomplishing the intra-holon communication, while others [20] proposed using a special management service interface function block.

It is expected that the communication among holonic agents will be standardized for many reasons. One reason is that these holonic agents should be involved in global communities of company agents, where they can directly participate in supply chain management negotiations or contribute to virtual enterprise simulation games, etc. The FIPA communication standards are considered preferable for implementing the inter-holon communication. To develop these standards, the HMS community must declare messages, define their semantics, and develop the appropriate *knowledge ontologies* (see Section 107.5). This seems to be quite a demanding task, as manufacturing, material-handling, production planning, and supply chain management requirements differ significantly between different industries and between different types of production.

From a wider perspective, for the FIPA standards to be applicable to holonic manufacturing, they should take account of the preexistence and coexistence of other standards. In manufacturing industry, there is STEP (Standard for the Exchange of Product Model Data), which is a comprehensive ISO standard (ISO 10303) that is used for representation and exchange of engineering product data and specifies the EXPRESS language for product data representation in any kind of industry. Integration of these widely accepted concepts with the HMS and FIPA effort seems to be of high importance. On the level of physical interoperability, there are various standards, such as the TCP/IP and UDP protocols, Common Request Broker Architecture (CORBA), Distributed Common Object Model (DCOM), and others. Similarly, the use of higher-level cooperation standards in the area of multi-agent systems such as KQML, FIPA, and JINI is inevitable for dynamic, flexible, and reconfigurable manufacturing enterprises.

107.7 Agent Platforms

A complex nature of the agents (high-level decision-making units capable of mutual collaboration) requires using a high-level programming language such as C++ or JAVA for their implementation. As mentioned in the previous section, for manufacturing purposes the software agents' parts of holonic agents have to be able to interact with the low-level control layer. In the majority of current holonic testbed implementations, the low-level holonic control (connected to the physical layer) is usually carried out by IEC 1131-3 (mainly ladder logic) or IEC 61499 function block programs that run on industrial PLC-based automation controllers. However, the software agent parts, implemented in C++ or JAVA, are running separately on a standard PC and communicate, for example, via a *blackboard system* (part of the data storage area in a controller allocated for each holonic agent and shared by the agent- and holonic-subsystem of the holonic agent).

It is obvious that for real industrial deployment, particularly where a high degree of robustness is required, the use of PC(s) for running agent-components of holonic agents is not safe and is also not possibly feasible for certain types of control systems. The only acceptable solution is to *run holonic agents as wholes directly within PLC-based controllers*. One controller can host one or more holonic agents, but not all of them — they have to be distributed in reasonable groups over several controllers and allowed to communicate with each other either within a single controller or among different controllers.

The major issue of such a solution is to extend the current architecture of a PLC in such a way that it is able to run software agents written in a high-level programming language in parallel with the low-level control code and also provide the interface for interactions between these two layers. The programming language in which the software agents should be implemented can either be C++ or JAVA. However, there are many reasons to prefer the JAVA language to be the target one. One of its advantages is the portability of JAVA programs, which the user develops independent of hardware platforms or operating systems — the same application can run either on a PC with Microsoft Windows or Unix/Linux or on a small device like Personal Digital Assistant (PDA) or a mobile phone with Windows CE, Symbian, or other operating systems with JAVA support. Another reason to choose JAVA is that currently there are a large number of JAVA-based agent development tools available, either as commercial products or open-source projects, that simplify the development of agent systems. Moreover, some of them are fully compliant with the FIPA specifications, which insures the desired interoperability.

Agent Development Tools Characteristics

Basically, the agent development tool, often called an *agent platform*, provides the user with a set of JAVA libraries for specification of user agent classes with specific attributes and behaviors. A kind of a *run-time environment* that is provided by the agent platform is then used to actually run the agent application. This run-time environment, implemented in JAVA as well, particularly ensures transport of messages among agents, registration, and deregistration of agents in the community (white pages services) and also registration and lookup for services provided by agents themselves (yellow pages services). Some other optional tools can also be a part of the agent platform runtime, for instance, a graphical viewer of messages sent among agents, etc.

The implementation of JAVA-based agents in the automation controllers obviously requires such an agent platform runtime to be embedded into the controller architecture. Since it is used as a background for the *real-time* holonic agents, there are specific requirements on the properties of the agent platform, such as speed, memory footprint, reliability, etc. The evaluation of available JAVA agent platforms, presented in the following paragraphs, has been conducted [21] in order to find out to what extent they fulfill these criteria and therefore which ones are best suitable for the purposes of manufacturing control.

FIPA Compliancy

Compliance with the FIPA standards has been recognized as a crucial property ensuring the interoperability of holonic agents not only at the lowest real-time control level (allowing, e.g., communication of different kinds of holonic agents hosted by PLC controllers from different vendors) but also the interoperability between holonic agents and other agents at higher levels of information processing within the company, for example, data-mining agents, ERP agents, supply chain management agents, and so on.

The FIPA specification of the message transport protocol (see the Section Message transport service) defines how the messages should be delivered among agents within the same agent community and particularly between different communities. For the latter case, the protocol based on IIOP or HTTP ensures the full interoperability between different agent platform implementations. It means that the agent running, for example, on the JADE agent platform can easily communicate with the agent hosted by the FIPA-OS platform, etc.

Costs and Maintainability of the Source Code

From the cost point of view, the agent platforms that are currently available can basically be divided into two categories: free and commercial ones. Majority of the free agent platforms are distributed under a kind of an open source license (e.g., GNU Lesser General Public License), which means that you are provided with the source codes and allowed to modify them. This is an important characteristic since the integration of the agent platform into the PLC-based controllers certainly requires some modifications to be made, for example, due to different versions of JAVA virtual machine supported by the controller, the specifics of the TCP/IP communication support, or other possible issues and limitations.

On the other hand, in the case of commercial products the cost in order of thousands USD per each installation, for example, can considerably increase the total cost of the agent-based control solution where a large number of PLC controllers, PCs, and possibly other devices running agents are expected to be deployed. Moreover, the source codes are not available, so that all modifications of the platform that need to be made in order to port it to another device has to be committed to the company developing the agent platform.

Memory Requirements

An issue that has to be taken into account is usually a limited memory available for user applications on the controller. Within the RAM memory of the controller, which can, for example, be about 4 to 8 MB, the agent platform run-time environment, the agents themselves and also the low-level control code (ladder logic or function blocks) have to fit inside. There are also smaller PLC-like devices that can have only 256 KB of memory available, which would be a strong limitation factor for integrating the run-time part of the agent platform. Fortunately, the agent platform developers, especially in the telecommunication area, are seriously interested in deploying agents on small devices like mobile phones or PDAs, that is, on devices with similar memory limitations. Due to this fact, for some of the agent platforms, their lightweight versions have been developed, usually implemented in Java2 Micro Edition (CLDC/MIDP) [22]. It has been documented [23] that the memory footprint of such an agent platform runtime can be less than 100KB, that is, small enough to fit well within the memory capacity limits of majority of small mobile devices and thus the PLC-based automation controllers as well.

Message Sending Speed

The last factor considered in this evaluation is the speed of the message sending between the agents. It has already been argued that the holonic agents are expected to be used for real-time control applications where a fast reaction can be a vital characteristic. In Figure 107.2 (Chapter 6), a direct communication channel between RT control subsystems of neighboring holons is conceded but it obviously breaks the autonomy of holonic agents. If we are not willing to accept such a violation, communication at the agent level is the only allowable way of interaction among holonic agents. Thus, the agent platform runtime, carrying out such interactions, should be fast enough to ensure reasonable message delivery times (i.e., in the order of milliseconds or tens of milliseconds).

We have conducted a series of tests to compare the message-sending speed of different agent platforms. Detailed information about the benchmarking testbed configuration and the speed measuring results can be found in the subsequent section.

Agent Platforms Overview

Table 107.1 gives an overview of majority of currently available agent development tools with respect to the properties discussed in previous paragraphs. A security attribute has been added as a property of the agent platform ensuring secure communication (usually via SSL), authorization, authentication, permissions, etc. The ✓ sign indicates that an agent platform has a particular property; meanwhile, the ✗ sign indicates that such a property is missing. If a ? sign is used, there is no reference to such a property in available sources and it can be assumed that the platform does not have it.

The Java Agent Services (JAS) have also been included in Table 107.1. However, this project is aimed at the development of the standard JAVA APIs (under the `javax.agent` namespace), that is, a set of classes and interfaces for the development of your own FIPA-compliant agent-based systems. From this perspective, JAS cannot be considered as a classical agent platform, since it does not provide any run-time environment that could be used to run your agents (either on a PC or possibly on an automation controller).

The JINI technology [24] has not been considered in this evaluation either. Similar to JAS, JINI is a set of APIs and network protocols (based on JAVA Remote Method Invocation) that can help you to build and deploy distributed systems. It is based on the idea of services providing useful functions on the network and the lookup service that helps clients to locate these services. Although JINI provides a solid framework for various agent implementations (see, e.g., [25]), it cannot itself be regarded as an agent platform.

TABLE 107.1 Agent Platforms Overview

JAVA-Based Agent Development Toolkits/Platforms — Overview

Agent Platform	Developer	FIPA Compatibility			Open-source	J2ME version (Lightweight)	Security (authent., SSL, …)
		Agent Communication	Agent Management	Inter-platform Messaging (MTP)			
JADE (Java Agent DEvelopment Framework)	CSELT http://jadecselt.it	√	√	√	√	√	√
FIPA-OS	Emorphia http://fipa-os.sourceforge.net	√	√	√	√	√	√
ZEUS	British Telecom www.labs.bt.com/projects/agents/zeus	√	√	√	√	×	√
JACK (Jack Intelligent Agents)	Agent Oriented Software http://www.agent-software.com	√	√	√	×	×	√
GRASSHOPPER 2	IKV ++ Technologies AG http://www.grasshopper.de	√	√	√	×	√	√
ADK (Agent Development Kit)	Tryllian http://www.tryllian.com	√	×	×	×	?	√
JAS (Java Agent Services API)	Fujitsu, HP, IBM, SUN, … http://java-agent.org	√	√	√	√	×	?
AgentBuilder	IntelliOne Technologies http://www.agerbuilder.com	×	×	×	×	×	?
MadKit (Multi-Agent Development Kit)	MadKit Team http://www.madkit.org	×	×	×	√	×	?
Comtec Agent Platform	Communication Technologies http://cas.comtec.co.jp/ap	√	√	√	√	×	×
Bee-gent	Toshiba http://www2.toshiba.co.jp/ beeagent/index.htm	×	×	×	×	×	√
Aglets	IBM Japan http://Iri.ibm.com/aglets	×	×	×	√	×	√

Message-Sending Speed Benchmarks

It has been discussed earlier that a speed at which the messages are exchanged among agents can be a crucial factor in agent-based real-time manufacturing applications. Thus, we have put selected agent platforms through a series of tests where the message delivery times have been observed under different conditions.

In each test, the so-called average round-trip time (avgRTT) is measured. This is the time period needed for a pair of agents (let say A and B) to send a message (from A to B) and receive a reply (from B to A). We use a JAVA `System.currentTimeMillis()` method, which returns the current time as the number of milliseconds since midnight, January 1, 1970. The round-trip time is computed by the A-agent when a reply from B is received as a difference between the receive time and the send time. An issue is that a millisecond precision cannot be mostly reached; the time grain is mostly 10 or 15 msec (depending on the hardware configuration and the operating system). However, it can easily be solved by repeating a message exchange several times (1000 times in our testing) and computing the average from all the trials.

As can be seen in Table 107.2, three different numbers of agent pairs have been considered: 1 agent pair (A–B) with 1000 messages exchanged, ten agent pairs (A1–B1, A2–B2, …, A10–B10) with 100 messages exchanged within each pair, and finally 100 agent pairs (A1–B1, A2–B2, …, A100–B100) with ten messages per pair. Moreover, for each of these configurations, two different ways of executing the tests are applied. In the serial test, the A agent from each pair sends one message to its B counterpart and when a reply is received, the round-trip time for this trial is computed. It is repeated in the same manner N-times (N is 1000/100/10 according to number of agents) and after the Nth round-trip is finished, the average response time is computed from all the trials. The parallel test differs in such a way that the A agent from each pair sends all N messages to B at once and then waits until all N replies from B are received. In both the cases, when all the agent pairs are finished, from their results the total average round-trip time is computed.

As the agent-based systems are distributed in their nature, all the agent platforms provide the possibility to distribute agents on several computers (hosts) as well as run agents on several agent platforms (or parts of the same platform) within one computer. Thus, for each platform, three different configurations have been considered: (i) all agents running on one host within one agent platform, (ii) agents running on one host but within two agent platforms (i.e., within two Java Virtual Machines — JVM), and (iii) agents distributed on two hosts. The distribution in the last two cases was obviously done by separation of the A–B agent pairs.

The overall benchmark results are presented in Table 107.2. Recall that the results for serial tests are in milliseconds (msec) while for parallel testing, seconds (sec) have been used. Different protocols used by agent platforms for the interplatform communication are also mentioned: Java RMI for JADE and FIPA-OS, TCP/IP for ZEUS, and UDP for JACK. To give some technical details, two Pentium II processor-based computers running on 600 MHz (256 MB memory) with Windows 2000 and Java2 SDK v1.4.1_01 were used. Some of the tests, especially in the case of 100 agents, were not successfully completed mainly because of communication errors or errors connected with the creation of agents. These cases (particularly for FIPA-OS and ZEUS platforms) are marked by a ✕ symbol.

Platforms — Conclusion

On the basis of the results of this study, the *JADE* agent platform seems to be the most suitable open-source candidate for the development tool and the run-time environment for agent-based manufacturing solutions. In comparison with its main competitor, *FIPA-OS*, the JADE platform offers approximately twice the speed in message sending and, above all, a much more stable environment, especially in the case of larger number of agents deployed.

Among the commercial agent platforms, to date, only *JACK* can offer full FIPA compliancy and also cross-platform interoperability through a special plug-in — JACK agents can send messages, for example, to JADE or FIPA-OS agents (and vice versa) via the FIPA message transport protocol based on HTTP. In the intraplatform communication, based on the UDP protocol, JACK (unlike other platforms) keeps

TABLE 107.2 Message Delivery Time Results for Selected Agent Platforms

JAVA-based Agent Development Toolkits/Platforms — Benchmark Results

Agent Platform	Message sending - average roundtrip time (RTT)					
	agents: 1 pair messages: $1{,}000 \times \rightleftarrows$		agents: 10 pairs messages: $100 \times \rightleftarrows$		agents: 100 pairs messages: $10 \times \rightleftarrows$	
	serial [ms]	parallel [s]	serial [s]	parallel [s]	serial [s]	parallel [s]
JADE v2.5	0.4	0.36	4.4	0.22	57.8	0.21
JADE v2.5	8.8	4.30	85.7	4.34	1426.5	4.82
1 host, 2 JVM, RMI						
JADE v2.5	6.1	3.16	56.2	3.60	939.7	3.93
2 hosts, RMI						
FIPA-OS v2.1.0	28.6	14.30	607.1	30.52	2533.9	19.50
FIPA-OS v2.1.0	20.3	39.51	205.2	12.50	×	×
1 host, 2 JVM, RMI						
FIPA-OS v2.1.0	12.2	5.14	96.2	5.36	×	×
2 hosts, RMI						
ZEUS v1.04	101.0	50.67	224.8	13.28	×	×
ZEUS v1.04	101.7	51.80	227.9	×	×	×
1 host, 2 JVM, ?						
ZEUS v1.04	101.1	50.35	107.6	8.75	×	×
2 hosts, TCP/IP						
JACK v3.51	2.1	1.33	21.7	1.60	221.9	1.60
JACK v3.51	3.7	2.64	31.4	3.65	185.2	2.24
1 host, 2 JVM, UDP						
JACK v3.51	2.5	1.46	17.6	1.28	165.0	1.28
2 hosts, UDP						
NONAME	141.3	1.98	2209.3	0.47	×	×
NONAME	N/A	N/A	N/A	N/A	N/A	N/A
1 host, 2 JVM, ?						
NONAME	158.9	×	×	×	×	×
2 hosts, IIOP						

pace with JADE in case of one host and even surpasses it in other cases, being approximately 2–3 times faster. Considering the full implementation of the Belief–Desire–Intention (BDI) model, JACK can be regarded as a good alternative to both the open-source JADE and FIPA-OS platforms.

107.8 Role of Agent-Based Simulation

The process of developing and implementing a holonic system relies on several phases and widely explores the simulation principles. The simulation process and its fast development using efficient simulation tools represent the key tasks for any implementation of a real-life holonic/agent-based system. The following stages of the design process based on simulation can be gathered like this:

1. *Identification of holons/agents*: The design of each holonic system starts from a thorough analysis of:

   (a) The system to be controlled or manufacturing facility to be deployed.
   (b) The control/manufacturing requirements, constraints, and hardware/software available.

   The result of this analysis is the first specification of *holon/agent classes* (types) to be introduced. This specification is based on the application and its ontology knowledge. The obvious design principle is that each device, or each segment of the transportation path or each workcell is represented by a holon.

2. *Implementation/Instantiation* of holon classes from the holon/agent type-library. The holon/agent-type library is either developed (step 1) or reused (if already available). Particular holons/agents

are created as instances of the holonic definitions in the holon/agent-type library. Furthermore, the implementation of communication links among these holon/agent instances is established within the framework of initialization from these generic holon/agent classes (for instance, holons are given the names of their partners for cooperation).

3. *Simulation*: The behavior of a holonic system is not deterministic, but rather emergent — the decision-making knowledge stored locally in the agents/holons invokes the global behavior of the system in a way that cannot be precisely predicted. Yet, the direct experimental testing of the global behavior with the physical manufacturing/control environment being involved is not only extremely expensive, but nonrealistic as well. Simulation is the only way out. For this purpose, it is necessary to have:

    (a) A suitable tool used to model and simulate the physical processes in the manufacturing facility. Standard simulation tools like, for example, Arena, Grasp, Silk, or Matlab can be used for these purposes.

    (b) A suitable agent run-time environment for modeling the interactions of holonic agent parts. On the basis of the results of agent platforms comparison (Section. 107.7), the JADE platform as open source or JACK as a commercial tool can be recommended.

    (c) A good simulation environment to model the real-time parts of multiple holons. There are function block emulation tools from Rockwell Automation (Holobloc [26]) and the modified 4-control platform from Softing [27]. Yet, these tools do not adequately generate and handle real-time control problems. A more sophisticated real-time solution would be to use embedded firmware systems like JBED, with its time-based scheduler, to run JAVA objects (which simulate function blocks) and manage events in a realistic manner.

    (d) Human–Machine Interfaces (HMI) for all the phases of the system design and simulation.

4. *Implementation of the target control/manufacturing system*: In this stage, the target holonic control or manufacturing system is reimplemented into the (real-time) running code. This implementation usually relies on ladder logic, structured text, or function blocks at the lowest level of control. However, some parts of the targeted manufacturing systems (such as resource or operation planning subsystems) are often reused as in the phase 3. For example, in the eXPlanTech production planning MAS [28], there was 70% of the real code reused from the simulation prototype. Therefore, the choice of the multi-agent platform in the phases 3 and 4 is critical (it has been advised to operate with one platform only).

107.9 Conclusions

Why does the agent technology seem to be so important for the area of manufacturing? What are the reasons and advantages of applying them? What do they really bring? Let us try to summarize the current experience shared by both the holonic and multiagent communities. The main advantages can be summed up as follows:

1. *Robustness and flexibility of the control/diagnostic systems*:
    (a) Robustness is achieved mainly due to the fact that there is *no central element*, no centralized decision making. Any loss of any subsystem cannot cause a fatal failure of any other subsystem.
    (b) The agent technology enables to handle the problems of *production technology failures* in a very efficient way. Optimal reconfiguration of the available equipment (which remains in operation) can be carried out in a very fast way. Thus, sustainable continuation of the production task or operation or safety stopping of the manufacturing process can be achieved. (Similarly, accomplishment of the life-critical part of a mission — after an important part of the equipment has been destroyed — can be achieved in the military environment).
    (c) *Changes in the production facility* (adding a machine, deleting a transportation path, etc.) can be handled on the fly, without any need to reprogram the software system as a whole. Just a couple of messages are exchanged, and the agents are aware of the change and behave accordingly.

(d) *Changes in production plan or schedule* can be handled easily, without the need for stopping the process or bringing it back to some of the initial states. The changes in the production plan or schedule can be handled in parallel to solving the tasks connected with changes in the facility equipment and/or failures.

2. The *plug-and-play approach* is strongly supported. This enables to change/add/delete the hardware equipment as well as software modules on the fly. The migration process from the old to the new technology can be carried out smoothly, on a permanent basis, without any need to stop the operation. This also makes the system maintenance costs significantly cheaper.

3. Control and diagnostics are carried out as *near to the physical processes* as possible; control and diagnostic subsystems can cooperate on the lowest level (and in a much faster way). Control and diagnostics can be really fully integrated. This fact improves the behavior of control/diagnostic systems in the hard real-time control/diagnostic tasks. Moreover, it is possible to change the principles of behavior centrally, just by changes in the rules or policies known to each of the agents.

4. *The same agent-based philosophy can be used on different levels,* in different subsystems of the manufacturing facility and company. The same agent-oriented principles and techniques can be, for example, applied on the hard real-time level (holonic control), soft real-time control, strategic decision making for control tasks, for integrated diagnostics or diagnostics running as a separate process aside of control, for production planning and scheduling, for higher-level decision making on the company level, for supply-chain management, as well as for the purposes of virtual enterprises (viewed as coalitions of cooperating companies). Despite the same communication standards and negotiation scenarios being used across all the tasks mentioned above, a very high efficiency resulting from automatic communication and negotiation between units on different levels and located in different subsystems can be obtained.

Besides the advantages of the agent-based solutions, several disadvantages can also be easily identified:

1. The *investments* needed to implement the agents-based manufacturing system *are higher.* Unfortunately, the available flexibility, which is the payoff for these expenses, is usually so enormous that the manufacturing process can leverage just a very small portion of it.

2. As there is no central control element present (in an ideal agent-based factory), in the society of mutually communicating agents, unpredictable, *emergent behavior* can be expected. This causes several obstacles for the agent-based solutions to be easily accepted by the company management. The only way out seems to be a very thorough *simulation of the agent-community behavior.* From the authors' own experience, the simulation detects just a limited number of patterns of emergent behavior. The "dangerous" patterns can be avoided by introducing appropriate policies across the system. Thus, the system simulation helps to understand the patterns of emergent behavior and their nature and to find protective measures, if necessary.

3. The current control systems offered by all the important vendors support the centralized control solutions only. The *migration toward autonomous, independent controllers,* communicating asynchronously (when needed) among themselves in the peer-to-peer way, seems to be the necessary technology enabler for a wider application of the agent-based solutions. Rockwell Automation, as a pioneering company in solving the migration process, is currently extending the classical PLC controller architecture to enable to run JAVA agents as well as a JAVA-based agent run-time environment directly within existing PLCs in parallel with the classical real-time scan-based control. Thus, the concept of holonic agents presented in Section 107.6 shifts from mainly academic considerations to the actual implementation.

4. Nearly the entire community *of control engineers has been educated to design,* run, and maintain strictly *centralized solutions.* This is quite a serious obstacle, as the engineers with the "classical" centralization-oriented approach (stressed in the last three decades under the CIM label) are really not ready and able to support the agent-based solutions. Much more educational efforts will be needed to overcome this serious hurdle.

5. Not all the tasks can be solved by the agent-based approach (the estimates talk about 30% of the control tasks and 60% of the diagnostic tasks to be suitable for application of the agent-based techniques). But certain areas with a higher degree of applicability of the agent-based technology have been identified already (see below). In general, *applying the agent-based technology in inappropriate tasks can lead to frustration.*

The areas suitable for application of the agent-based techniques are as follows:

1. *Transportation of material/material handling.* The transportation paths (conveyors, pipelines, AGVs, etc.) and their sensing and switching elements (diverters, crossings, storages, valves, tag readers, pressure sensors, etc.) can be easily represented by agents; their mutual communication can be defined and organized in a quite natural way. Interesting pioneering testbeds have been built. They document the viability and efficiency of this approach in the given category of tasks.
2. *Intelligent control of highly distributed systems*, namely in the *chemical industry* and in the area of *utility distribution* control (electrical energy, gas, waste water treatment, etc.). Many decisions can be made locally, in a very fast way; the communication among the autonomous unit is carried out only if really needed.
3. *Flexible manufacturing in automotive industry.* For this industry (aimed at mass production of individually customized products), very variable customization requirements, changes in the plans and schedules, changes in technology, as well as equipment failures seem to be quite obvious features of everyday operation. All these requirements and emergency situations can be easily handled by the agent technology.
4. *Complex military systems* (like aircraft and their groups, ships, army troops in the battlefield) can be *modeled and managed as groups of agents.* For instance, very high flexibility of the technical equipment on board a ship enables to accomplish at least a part of its mission if certain subsystems are destroyed or permanently out of operation.

The research in the field of agent-based control and diagnostic systems for manufacturing has been concentrated namely around the HMS consortium within the frame of the international initiative Intelligent Manufacturing Systems. Currently, there can be recognized several leading academic and industrial centers active in this field and bringing important results.

Let us mention the following academic sites: University of Cambridge, Center for Distributed Automation and Control (CDAC), Cambridge, U.K.; University of Calgary, Department of Mechanical Engineering, Canada; Katholieke University of Leuwen, Department of PMA, Belgium; Vienna University of Technology, INFA Institute, Vienna, Austria; University of Hannover, IPA, Hannover, Germany; Czech Technical University, Gerstner Lab, Prague, Czech Republic.

Among the industrial leaders in agent-based control and diagnostics, following companies should be mentioned: Rockwell Automation, Milwaukee, WI; Rockwell Scientific comp., Thousand Oaks, CA; Daimler-Chrysler, Central Research Institute, Stuttgart, Germany; Toshiba + Fanuc, Japan; ProFactor, Steyr, Austria; SoftIng, Munich, Germany; CertiCon, a.s., Prague, Czech Republic; CSIRO, Melbourne, Australia.

The agent-based technology for manufacturing is developing in a very fast way. This development trend strictly follows the current trend in MAS research in the field of Artificial Intelligence as well as all the recommendations of the FIPA standardization consortium. But a long way remains in front of us: it is necessary, for example, (i) to change the way of thinking of industrial designers and engineers of control systems, (ii) to document the reliability and manageability of the emergent behavior of the agent-based systems (for this purpose, much more robust simulation tools should be developed), (iii) to support the migration processes from the centralized to agent-based control, which concern both the hardware and software, (iv) to solve the technical problems of interoperability, communication, and negotiation among the agents, and (v) to work toward widely acceptable ontology structures and languages.

References

[1] Marik, V., Pechoucek, M., Vrba, P., and Hrdonka, V., FIPA standards and holonic manufacturing, in *Agent Based Manufacturing: Advances in the Holonic Approach*, Deen, S. M., Ed., Springer-Verlag, Berlin, 2003, pp. 89–121.

[2] Camarinha-Matos, L.M. and Afsarmanesh, H., Eds., Interim Green Report on New Collaborative Forms and Their Needs, September 2002.

[3] Marik, V., Pechoucek, M., and Stepankova, O., Social knowledge in multi-agent systems, in *Multi-Agent Systems and Applications*, Lecture Notes in Artificial Intelligence, Michael Luck, Vladimir Marik, Olga Stepankova, Robert Trappl, Eds., Vol. 2086, Springer-Verlag, Heidelberg, 2001, pp. 211–245.

[4] Deen, S.M., Ed., *Agent Based Manufacturing: Advances in the Holonic Approach*, Springer-Verlag, Berlin, 2003.

[5] Searle, J.R., *Speech Acts*, Cambridge University Press, Cambridge, 1969.

[6] McGuire, J., Kuokka, D., Weber, J., Tenebaum, J., Gruber, T., and Olsen, G., SHADE: technology for knowledge-based collaborative engineering, *Concurrent Engineering: Research and Applications*, 1, 137–146, 1993.

[7] Decker, K., Sycara, K., and Williamson, M., Middle Agents for Internet, in Proceedings of the International Joint Conference on Artificial Intelligence 97, Nagoya, Vol. 1, 1997, pp. 578–583.

[8] Shen, W., Norrie, D.H., and Barthes, J.A., *Multi-Agent Systems for Concurrent Intelligent Design and Manufacturing*, Taylor & Francis, London, 2001.

[9] Marik, V., Pechoucek, M., Stepankova, O., and Lazansky, J., ProPlanT: multi-agent system for production planning, *Applied Artificial Intelligence Journal*, 14, 727–762, 2000.

[10] Pechoucek, M., Marik, V., and Barta, J., A knowledge-based approach to coalition formation, *IEEE Intelligent Systems*, 17, 17–25, 2002.

[11] Rao, A.S. and Georgeff, M.P., An Abstract Architecture for Rational Agents, in Proceedings of Knowledge Representation and Reasoning KR&R-92, 1992, pp. 439–449.

[12] Obitko, M. and Marik, V., Mapping between ontologies in agent communication, in *Multi-agent Systems and Applications III*, Lecture Notes in Artificial Intelligence, Vol. 2619, Springer-Verlag, Heidelberg, 2003, pp. 177–188.

[13] Knutilla, A., Schlenoff, C., and Ivester, R., A Robust Ontology for Manufacturing Systems Integration, in Proceedings of the 2nd International Conference on Engineering Design and Automation, 1998.

[14] Chaudhri, A.F., Fikes, R., Karp, P., and Rice, J., OKBC: A Programmatic Foundation for Knowledge Base Interoperability, in Proceedings of AAAI-98, 1998.

[15] Vrba, P. and Hrdonka, V., Material Handling Problem: FIPA compliant agent implementation, in *Multi-Agent Systems and Applications II*, Lecture Notes in Artificial Intelligence, Vladimir Marik, Olga Stepankova, Hana Krautwurmova, Michael Luck, Eds., Vol. 2322, Springer-Verlag, Berlin, 2002.

[16] Van Leeuwen, E.H. and Norrie, D., Intelligent manufacturing: holons and holarchies, *Manufacturing Engineer*, 76, 86–88, 1997.

[17] Koestler, A., *The Ghost in the Machine*, Arkana Books, London, 1967.

[18] Chirn, J.L. and McFarlane, D.C., Building holonic systems in today's factories: a migration strategy, *Journal of Applied Systems Studies*, 2, 82–105, 2001.

[19] McFarlane, D.C., Kollingbaum, M., Matson, J., and Valckenaers, P., Development of Algorithms for Agent-based Control of Manufacturing Flow Shops, in Proceedings of the IEEE International Conference on Systems, Man and Cybernetics, Tucson, 2001.

[20] Fletcher, M. and Brennan, R.W., Designing a Holonic Control System with IEC 61499 Function Blocks, in Proceedings of the International Conference on Intelligent Modeling and Control, Las Vegas, 2001.

[21] Vrba, P., Agent platforms evaluation, in *Holonic and Multi-Agent Systems for Manufacturing*, Lecture Notes in Artificial Intelligence, Vladimir Marik, Duncan McFarlane, Paul Valckenaers, Eds., Vol. 2744, Springer-Verlag, Heidelberg, 2003, pp. 47–58.

[22] Java2 Platform, Micro Edition: website http://java.sun.com/j2me.

[23] Berger, M., Rusitschka, S., Toropov, D., Watzke, M., and Schlichte, M., Porting Distributed Agent-Middleware to Small Mobile Devices, in Proceedings of the Workshop on Ubiquitous Agents on Embedded, Wearable, and Mobile Devices, Bologna, Italy, 2002.

[24] JINI technology: website http://www.jini.org.

[25] Ashri, R. and Luck, M., Paradigma: Agent Implementation through Jini, in Proceedings of the Eleventh International Workshop on Database and Expert Systems Applications, Tjoa, A.M., Wagner, R. R., and Al-Zobaidie, A., Eds., IEEE Computer Society, Silver spring, MD, 2000, pp. 53–457.

[26] Holobloc, Rockwell Automation, http://www.holobloc.com.

[27] 4-Control, Softing AG, http://www.softing.com.

[28] Riha, A., Pechoucek, M., Krautwurmova, H., Charvat, P. and Koumpis, A., Adoption of an Agent-Based Production Planning Technology in the Manufacturing Industry, in Proceedings of the 12th International Workshop on Database and Expert Systems Applications, Munich, Germany, 3–7 Sept. 2001, pp. 640–646.

108

Multiagent-based Architecture for Plant Automation

Axel Klostermeyer
Otto-von-Guericke-University

Eckehardt Klemm
Phoenix Contact GmbH & Co. KG

108.1 Motivation and Introduction

The evolution during the last few years in the field of production sciences and production practice has induced major changes. Many companies have found that formerly successful actions, methods, and strategies — which were based on the paradigm of a stable production environment with predictable processes and a minimum of external disturbances — were no longer suitable to produce good results [KühM00]. This challenge to production science and practice is mainly driven by a turbulent industrial environment, which is characterized by an aggressive economic competition on a global scale, more educated and demanding customers, and a rapid pace of change in process technology. The overall reaction can be outlined by:

- an increasing frequency in the introduction of new products;
- smaller lot sizes and a rising amount of variants;
- changes in parts of existing products;
- large fluctuations in product demand and mix;
- changes in governmental regulations (e.g., safety and environment); and
- changes in process technology [KoUa99].

Consequently, the markets are characterized by an individualization of the customers' wishes. To be able to cope with these wishes — and by this, simultaneously, to be able to cope with the competitors — enterprises answer with increasingly more (customer) specific products and services. The number of variants that have to be produced in the corresponding plants are also increasing. This dynamics concerning the alteration of processes is accompanied by a decreasing possibility to plan the production program and the amount of sold products [Förs99].

Summarizing, it has to be stated that one basic thinking model is replaced by another one, which partially overthrows rules and regularity or even leads to contradictions. The point is a basic paradigm change toward highly flexible and reconfigurable production systems, which are able to reflect the new balance among economy, technology, and society.

Unfortunately, due to an insufficient technological infrastructure, implementations of such concepts in industrial plants were for a long time unworkable until relatively recently. Nowadays, with rapidly declining computer hardware cost, a viable infrastructure is in place [RePM01]. Furthermore, the latest developments in information technology and science support the required reconfigurability and flexibility of modern production systems and simultaneously allow for a dislocation and decentralization of production to open production networks. Under the catchwords "Extended Enterprise," "eProduction," and "B2B," such open production networks come more and more into the focus of production sciences and production practice [Klm02]. As the corresponding "enabling technologies" for information and communication in such production systems, the Internet with connections down to single field devices, the Extensible Markup Language (XML) as integrating technology, and so-called active technologies such as Java and .Net increasingly come into the focus of vendors and end users of automation systems [Klm02, Pesc02]. Against the background of Java's platform independence — which is displayed in the "write once, run anywhere" concept of this language — and Java's advantages as an object-oriented high-level language, (Java based) mobile software agents seem to be worth taking a deeper look inside. Mobile software agents are a relatively new concept to realize distributed (software) systems. Distributed systems — also referred to as distributed control systems or decentralized intelligent automation technology — are a number of autonomous, subnet-connected processors and repositories that communicate with each other to achieve a common goal. By means of communication via the network, the respective distributed processes autonomously coordinate their activities and exchange information [Ensl78]. While traditional distributed systems are based on stationary software programs, exchanging their data via a network, mobile agents are designed as software entities that autonomously migrate from one node in the network to the next one. This means that the algorithms are going to be mobile, not only the data which have to be calculated [McOS98, MiHM99].

Particularly, the above-mentioned technologies are the basis for the PABADIS (Plant Automation based on Distributed Systems, an IMS project under reference IST-1999-60016) project, which aims at providing a flexible plant automation solution based on the use of information technologies as mentioned above and described in several other chapters of this book (see [Spec03] for more details).

108.2 PABADIS Architecture for Reconfigurable Manufacturing Systems

General Overview

In this chapter, a control system is described, which enforces the use of distributed intelligence and therefore enables flexibility in plant, product, and device management. Agents are responsible for scheduling and resource allocation of their respective products and manufacturing steps. Most important is that the agents are able to use less complex algorithms than those used in centralized systems and can react much more flexibly than those.

The above-mentioned reconfigurability can be seen in two ways: first, in a horizontal way and second, in a vertical way. Reconfigurability in a horizontal way refers to the plug-and-play capabilities of the used hardware and means the possibility to simply plug in new production equipment, and use it without major changes within the production system. By an appropriate network technology, simple mechanisms are provided that enable machines to plug together to form an impromptu community — a community put together without any planning, installation, or human intervention. Each machine provides services that other entities in the community may use. These machines provide their own interfaces, which ensures reliability and compatibility. Such mechanisms already do exist for the embedded systems market; it is one objective of PABADIS to make this technology also applicable in factory automation.

In a vertical way, a dynamic architecture means that especially the scheduling of the produced articles has to be very dynamic. This allows to react on urgent jobs, failures of machinery, and other possible disturbances, which are due to the turbulence of the production's environment. Scheduling within a company means to consider the office level with its ERP system as well as the field level where programmable logic controllers and machine controls are located. In fact, the aim is to dissolve the MES layer and divide its functionality into a centralized part that can be attached to the ERP system and a decentralized part that can be implemented by partially mobile software agents; hence, the baseline vision of the project is that every work piece has an agent "attached to it" carrying the necessary product information and moving through the plant along with the work piece. This requires the introduction of an information-oriented "alter ego" for each product. Software agents are a suitable approach for this as it is very difficult to design a conventional network that adjusts well to either changing usage patterns over short time scales or to evolving needs and circumstances over the long haul. In contrast, a mobile agent-based system can be reconfigured on the fly, in response to new situations or demands, and with or without human intervention [MiHM99].

The transition from restrictive and rigid communication and information in central production systems to flexible and reconfigurable information streams in reconfigurable, decentralized units is outlined in Figure 108.1.

From the communication point of view, this approach both necessitates and supports the current trend in industrial automation to flatten the network hierarchy and to use IP-based networks down to the control level [DieS00]. It can be stated that in the near future, industrial automation will change significantly in all three aspects of hardware, software, and communication systems. At the communication level, Ethernet more and more pervades the shop floor and platform organizations such as the IAONA e.V. (Industrial Automation Open Network Alliance) foster this trend and smooth Ethernet's way down into the plant. With respect to the software, first applications of Java™ and .Net™ in industrial control can be found, and first products already do exist. For these technologies, PC platforms are needed. Fortunately, from a hardware point of view, industrial PCs (iPCs) have been conquering industrial plants for some time, and, generally speaking, there are no major challenges to cope with in the first step.

The following sections are dedicated to first, an overview of the architecture, second, the Multi-Agent System (MAS), third, the corresponding Agency, and fourth, the connection to the machinery by a generic interface layer, the so-called Cooperative Manufacturing Unit (CMU).

Overview of Architecture and Processes

Contrary to the conventional centralized job control approach, the Multiagent Architecture described in this chapter focuses on job control using autonomous production units. The objective is to dissolve the Manufacturing Execution System (MES) layer and divide its functionality into a centralized part that can be attached to the Enterprise Resource Planning (ERP) system and a decentralized part that can be implemented by (partially) mobile software agents. This effectively comes down to reducing the hierarchy described in Figure 108.1 into two layers.

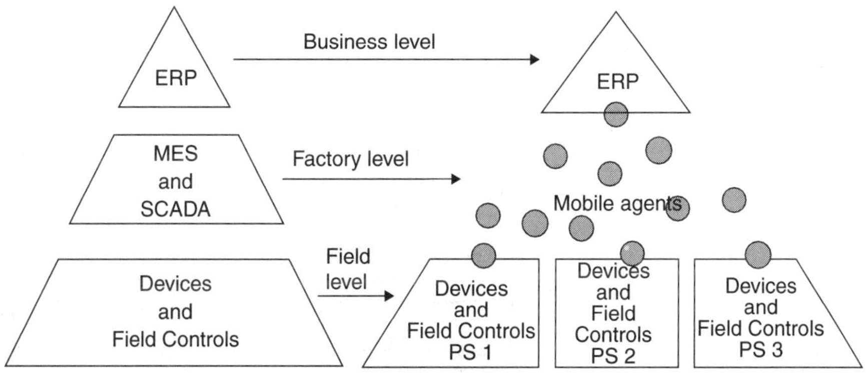

FIGURE 108.1 Transition of information and communication from rigid to reconfigurable systems [Klm02].

As outlined above, the baseline vision of the architecture is that every work piece has a mobile agent "assigned to it," carrying the necessary product information and moving through the plant along with the work piece. In fact, every production system needs two main inputs: the actual physical work piece and information. If we consider a single piece production system, most of this information is tightly connected to the individual product, such as

- production sequence and schedule;
- machine-related production data;
- status of the processing; and
- general administrative information about the order.

In addition, there is information associated with the entire production system, such as

- overall resource use;
- overall production schedule;
- machine status information; and
- quality control information.

The overall production scheduling and resource planning data should, of course, be consistent with the product-specific data sets; hence, they can be compiled or deduced from each other. Traditionally, these data are generated by the ERP system in a strictly centralized manner. Detailed planning, adjustments to the overall scheduling, and job control are subsequently done by the MES. With a view to the information distribution sketched above, it seems reasonable to largely remove the detailed planning and job control functionality from the ERP and distribute it on the level below among the "products" that can independently keep track of their processing needs and status. As mentioned above, this requires the introduction of an information-oriented "alter ego" for each product, the so-called *Product Agent* (PA). Figure 108.2 gives an overview of the corresponding architecture.

The production means are modeled as abstract entities with a generic interface between PABADIS and the actual production equipment. Such an entity is called CMU. It provides available functions to the manufacturing process. This definition is deliberately abstract and makes no assumptions about the physical realization. In fact, we can distinguish two different types of CMUs:

- *Manufacturing CMUs* are used for the physical processing of products and involve some kind of manipulation of the work pieces, for example, welding machines.

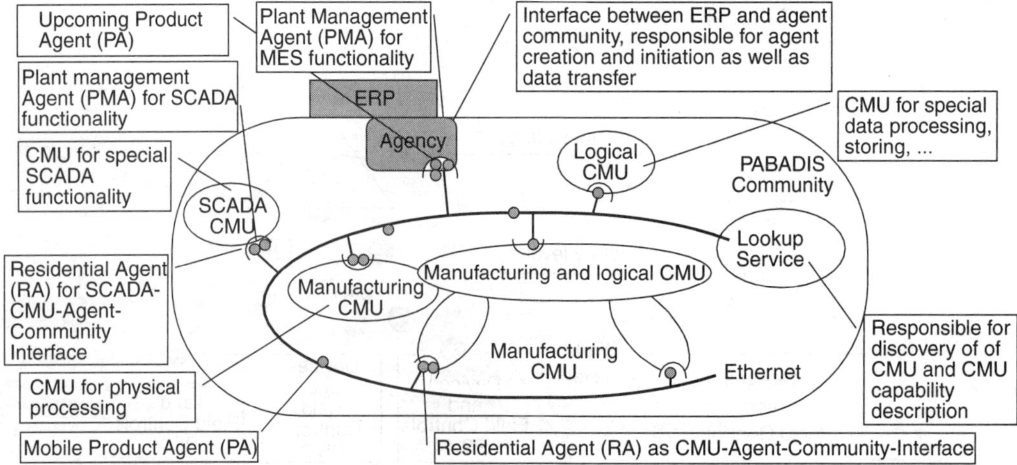

FIGURE 108.2   Agent-based architecture for job control in turbulent industrial environments.

- *Logical CMUs* provide computational services and are consulted by the agents for special tasks like complex scheduling algorithms, database search, or similar things.

A detailed description of the CMU concept can be seen in a further section.

The manufacturing process according to the PABADIS idea starts with the generation of a conventional manufacturing order by the ERP system to process a product. This order comprises the sequence of required processing steps together with the appropriate parameters for product processing. The manufacturing order is passed to the *Agency*, where it is translated into one or several work orders. A mobile Product Agent will be responsible for each work order, which is generated and initialized at the *Agent Fabricator* as part of the Agency and joins the MAS.

Step by step, the PA executes its production plan in the course of which it always follows the same basic procedure: it consults a *Lookup Service*, which is available in the network to find a CMU (see Figure 108.2) that can provide the needed manufacturing service. The Lookup Service acts as a central service broker and informs the PA about the access path and the abilities of possibly useable CMUs. Subsequently, the PA contacts the CMU and asks for additional information that is necessary to decide which service provider should be chosen. Based on these data, the agent selects the CMU that it considers optimal.

Throughout the manufacturing process, the PA guides the work piece. It ensures the processing of the work piece by the CMU in an appropriate way in cooperation with the *Residential Agent* of each CMU.

Upon completion, the PA returns to the Agency to deliver all product-dependent information that was collected during processing at the respective CMUs to the data collector and terminates. The Agency then generates a report to the ERP system using the data the PA collected on its way through the production (if so desired by the ERP system in the production order). The whole product processing flow is depicted in Figure 108.3.

One main characteristic of this system is the correlation of work piece and product data by Product Agents. Nevertheless, the association between PA, product data, and work pieces is not only a 1:1:1 association. Of course, a set of unique products with the same product information and therefore a set of work pieces can be handled by one PA. The only prerequisite is the necessity of control over the work pieces and their production. If the work pieces belonging to one PA are always transported together, for example, on the same pallet, this control is ensured. Hence, the PABADIS architecture can be used in a broader context than mentioned here.

In this context, let us have a look at the system behavior resulting from the occurrence of a manufacturing order with a lot size of χ products. If η products can be transported on/in a common transport unit (pallet/container), the *Agency* needs to create $\lceil \chi/\eta \rceil$ PAs guiding $\lceil \chi/\eta \rceil$ transportation units with a maximum amount of $\lceil \chi/\eta \rceil$ times η products. The total amount of transported products is equal to or

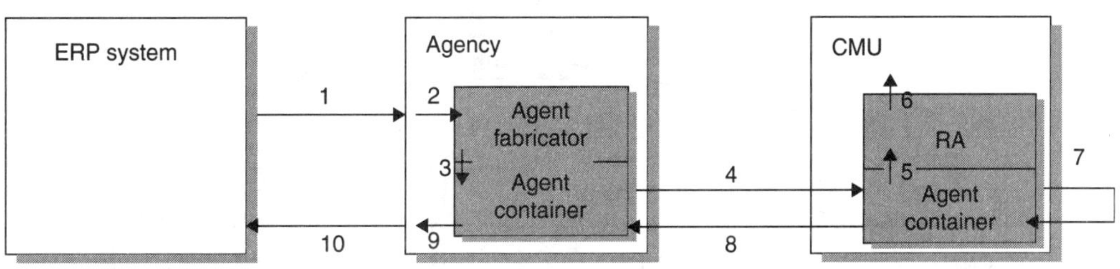

1. Manufacturing order from ERP to Agency	6. Access from the RA to the machine functions
2. Distribution of the manufacturing order to one or more Product Agents (PAs)	7. PA migration to the next CMU
3. PA initialization and start	8. PA migration back to the Agency
4. Migration to a Cooperative Manufacturing Unit (CMU)	9. Aggregation of the product data from each PA
5. PA contacts Residential Agent (RA) at the respective CMU	10. Manufacturing report to the ERP system

FIGURE 108.3 Product processing steps in a PABADIS system.

less than $\lceil \chi/\eta \rceil$ times η products, due to the fact that at least one transport unit needs not necessarily to be full. These PAs (or PA sets) can behave as described above.

In addition to the PAs, *Plant Management Agents* are created by the Agency to fulfill specific control or supervision tasks that are not related to individual products as, for example, maintenance- and quality-related tasks.

The MAS

The MAS has been designed with respect to decentralized control in plant automation and distributed intelligence. In the context at hand, it is a set of independently acting intelligent software entities that are working autonomously and communicate with each other in order to perform their own respective tasks. An important achievement of this combination of independent process execution and agent communication is the realization of two objectives of the plant automation control system:

- product manufacturing control and
- an efficient machine usage.

As shown in Figures 108.2 and 108.4, there are three types of agents defined in a PABADIS agent community: PAs, Residential Agents (RAs), and Plant Management Agents (PMAs). According to the general objectives of job control in a manufacturing plant, the PA community implements the main job control tasks by performing the relevant work orders decentralized, independently, and by their own "intelligence."

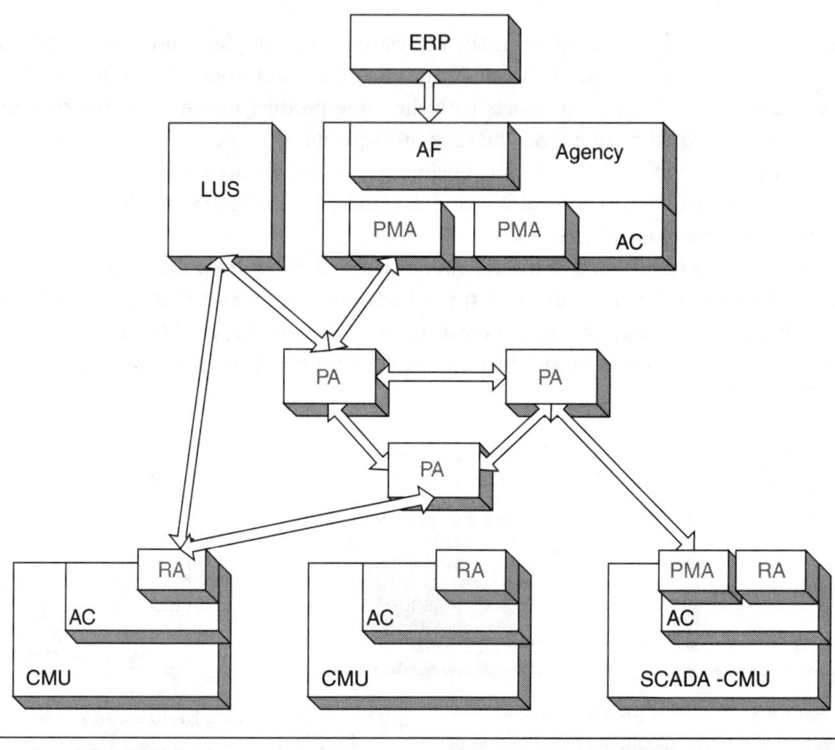

AC	Agent Container		PA	Product Agent
AF	Automation Function		PMA	Plant Management Agent
CMU	Cooperative Manufacturing Unit		RA	Residential Agent
ERP	Enterprise Resource Planning		SCADA	Supervisory Control and Data Acquisition
LUS	Lookup Service			

FIGURE 108.4   MAS and the respective communication relations.

This independent order execution includes the necessary communication with the appropriate entities of the whole system. Against this background, it is important to note that all PAs work in concurrence with other agents (either other PAs or PMAs) and perform the "egoistic" target of producing their respective work order.

An RA provides the functionality of a CMU to the PABADIS system and its entities. Contrary to PAs, RAs do not communicate with each other via the network. If at all, they communicate locally on a shared device. This is due to the fact that there might be the necessity for negotiations between RAs — providing parallel accessible functions of a CMU and the respective RAs providing alternative functions that always depend on the parallel functions provided by the other RAs residing on the same machine. This prohibited communication between RAs over the network is a principal difference between the PABADIS concept and many other agent-based manufacturing concepts like the Holonic Manufacturing Systems (HMS) (see [FGCD00]). All "intelligence" of the PABADIS system is concentrated in PAs, while in HMS, the intelligent modules are "machine"-dependent.

The duty of a PMA is to provide an interface between the MAS and the ERP system. It also takes part in the communication between a SCADA system and the MAS, but the main reason for introducing PMAs in the PABADIS concept is the necessity of having special units for MES functions that cannot be implemented by PAs or RAs, which means MES functions independent of a definite work piece and a definite CMU.

The following three subsections shall provide a deeper insight into the reason and structure of the above-mentioned types of agents.

The PA

The community of PAs is seen as a key component of the PABADIS MAS, because it provides the main "intelligence" of the system. As mentioned above, the main task of a PA is to perform a work order provided by the Agency. Due to this objective, a life cycle of a PA can be described by the following steps:

Initialization
During this step, a PA receives a work order, setup variables, recognizes its place in the community, and prepares a work order to be analyzed.

Work order parsing
As the name suggests, work order parsing means that a PA parses a work order in order to construct an execution plan. To do this, the PA analyzes all branches of the work order, which are represented in a graph view, and creates a string array with a sequence of actions, which will be used by the PA for resource allocation.

Scheduling
Scheduling means that a PA reserves resources for its tasks, representing the processing steps. In order to do this, it communicates with the RAs of the needed CMUs. Depending on the depth of the scheduling (provided by the Agency with a work order), a PA can allocate all resources in advance, for a first set of tasks, or only for the very next task (step-by-step scheduling). At the end of this step, a PA knows where it wants to migrate to in order to perform the next task of the work order.

Migration
During this step, a PA migrates to a CMU, checks its location, and waits for an execution slot. A PA initiates a migration, but does not control the process itself — this is done by the agent containers of the concerned CMUs (current location of an agent and destination CMU).

Task performing
As long as a special manufacturing step is performed at the respective CMU, the PA just waits for the end of this execution process. The CMU's RA, correspondingly, informs the waiting PA about the results of the workpiece processing; hence a PA is just a passive element during a task's performance. For critical situations, the possibility is foreseen to interrupt the task-performing process.

Termination
As the final step, after finishing its work order processing, the termination of a PA is foreseen. Termination either occurs when the work order is finished or, due to disturbances in the system or other failures, cannot

be finished. In both cases, a PA starts a termination procedure: it migrates to the Agency, creates a report to the Agency/ERP, and terminates itself. This means that a PA removes all its files from the system and, accordingly, cannot be restarted; only a creation of a new agent using the old PA's report is possible.

During all these steps, a PA bases its behavior on the work order provided by the Agency and the status of the available CMUs. Correspondingly, if it should become necessary to react on a sudden change within a work order — for example, the customer decides *during* the manufacturing process that he would prefer a blue car instead of the ordered yellow one — by a change of the respective work order the job can be changed "on the fly." This work order is an XML file that is accessible via an XML parser; this XML -parser converts XML data into a view that is understandable for an agent as well.

From a logical point of view, a PA consists of several modules that execute specific parts of the agent functionality.

PAServiceModule — provides the main functionality of the PA, including initialization, agent logic, calculation, and termination. This module includes the starting method of the PA and does most of the functionality related to the work order.

PAMigrationModule — a specific module for the PA, which implements the mobility of the PA, including pre- and postmigration methods. The Migration Module is closely connected to the agent platform. In fact, it is based on the methods provided by the agent platform. In order to make the whole system more flexible, the usage of the mechanisms of the used agent platform is hidden.

PACommunicationModule — this module includes all mechanisms that are used by a PA for communicating with other components of the system, such as an RA, a PMA, and other PAs. Communications are represented by a set of interfaces that a PA uses in order to organize the communication.

PADataModule — all agent data and methods for data manipulation are contained within the data module. This module includes an XML parser, which is used for work order interpretation. All other data related to a PA, such as history data, status data, and other MES data, are also represented by the data module.

The RA

As stated above, the RA is an interface between the agent community and the CMU community. The main functionality of an RA is to provide a set of interfaces and methods in order to provide a communication mechanism between two different entities. Logically, an RA is a part of the MAS and a CMU. In contrast to PAs, an RA does not have intelligence in the sense of task decisions based on its own calculations. Nevertheless, this does not mean that an RA cannot control processes. Specifically, an RA can be an active element in communications or negotiations. All decisions of an RA are based on inputs from other components of the PABADIS system (CMU and PA), but mostly an RA just reacts to the events that occur in the system.

Logically, an RA consists of the same set of modules as a PA does, except for the migration module. However, the functionality of these modules is slightly different:

RAServiceModule — In contrast to the PA concept, this module is an additional component, which does not have logical importance. The main method is still implemented in this module, but there is no life cycle for the RA to be implemented. A corresponding event handler is also implemented here because an RA most of the time just reacts on certain events.

RACommunicationModule — the most important part of an RA. It contains all logic of this agent type and is some kind of analog to the PA's service module. Specifically, the functionality of the agent is implemented there.

RADataModule — all agent data and methods for data manipulation are realized in the data module. This module deals with a CMU's XML parser. All other data related to an RA, such as history data, status data, and other MES data, are represented by the data module, but stored in the Common Feature Module (CFM), which is part of the CMU (see The CMU concept). The RA uses this database to perform its tasks.

The PMA

The purpose of a PMA is to provide work piece-independent MES functions. Some of these functions cannot be implemented completely without human control. Hence, PMAs become necessary that are responsible for requesting appropriate information from the system needed by, for example, ERP or SCADA system and for giving the responses back to the operators, who initialized these requests. The "human control," nevertheless, is optional and might sometimes be replaced by an automatically initialized request.

Logically, a PMA can be seen as a combination of a PA and an RA. On the one hand, a PMA is a stationary agent that is located in the Agency. On the other, a PMA is created in order to perform a special work order that is based on an ERP-based manufacturing order. As mentioned above, a PMA is responsible for several tasks, which will be listed in the following in more detail:

Some *special MES functions* need a direct communication between the ERP and the Agent Community. In this case, the PMA is a permanent and stationary agent residing at the Agency, which behaves like an arbitrary RA. Furthermore, this PMA type is stable and will never be removed from the system. Otherwise the system would lose some of the MES functionality provided by the appropriate PMA. For this reason, an agent life cycle like that of a PA can be stated.

For *temporary functions*. In this case, the ERP orders the Agency to perform specific tasks; the Agent Fabricator creates a PMA, which executes an appropriate sequence of actions related to the given order. As a result, the PMA sends the report back to the ERP system and terminates in the Agency. Hence, it shows an agent life cycle like a PA, but is stationary.

A third type of PMAs is the *mobile PMA for special tasks*. It is absolutely equal to a PA with the exception of the work order definition. In contrast to a PA, the work order of this PMA type is to execute some special tasks, mostly to collect appropriate information as, for example, particular quality data. Hence, the PMA does not have a work order of the type "make a product," but logically the aforementioned tasks of the mobile PMA are conceptually equal to the PAs work orders.

The Agency

The Agency is a program that interfaces the ERP level with the MAS. The Agency processes ERP demands, creates agents, and transmits process and status information about production operations between ERP and agent community in both directions. Therefore, the ERP system produces a manufacturing order, which has to be decomposed by the Agency into elementary work orders to allow parallel processing. For each work order, a specific PA is generated.

In case products are made with concurrent operations like assembly, disassembly, scrap, and recycling, the Agency creates as many PAs as necessary and combines the different issues. Against this background, the Agency provides PAs with the execution plan of the whole work order, so that synchronization and coordination can be managed by the agents themselves. This plan may have the form of a Gantt chart as shown in Figure 108.5.

At completion of the elementary order (work order), the PA reports to the Agency and is then terminated; after completion of the entire manufacturing order, the Agency finally sends a report to the ERP system. This report accounts for the whole manufacturing process and has to be archived to enable traceability.

As mentioned before, the elaboration of work orders occurs consecutively to a manufacturing order issued by the ERP system. Manufacturing orders start a production process with all data from product and process databases. For flexibility reasons, in PABADIS the allocation of resources is no longer an operation planned beforehand, but is now an on-line and dynamic operation performed by PAs and CMUs.

Before presenting the structure of the Agency, first it shall be explained how work orders are generated. A manufacturing order can be decomposed into elementary work orders following different ways. Figure 108.6 shows a simple example.

Hours	0	1	2	3	4	5	6
PA1	Task 1		Task 2		Task 4		
PA2			Task 3				
PA3						Task 5	

FIGURE 108.5 Example of a manufacturing order.

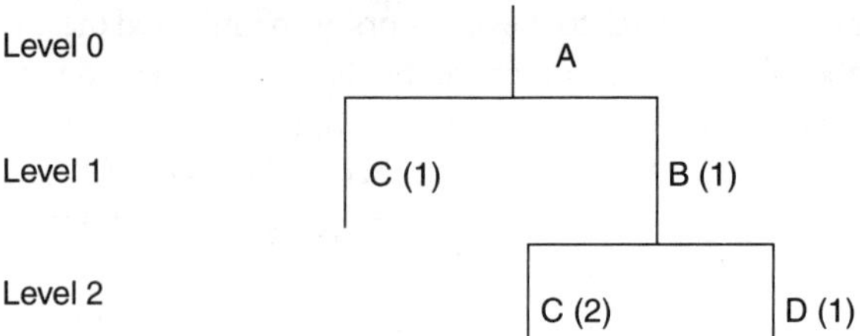

FIGURE 108.6   Example of a bill of material.

In this example, the end product A is made up of C plus B, and B in turn is made up of two Cs plus D.

Level code allows for building the structure of the product by providing the tree decomposition or parent/child relationships. As a first assumption, it shall be supposed that all endings of the tree (here C and D) are purchased parts, components, or raw materials that are available in the inventory at the start of the manufacturing order. This means that the ERP system has correctly released all previous orders (buying, supply, or production orders) and that the stocks are sufficient.

In this example, we then have two possibilities of defining work orders (WOs) (Figure 108.7):

- One WO: this WO consists in making product B with C and D, and then making product A with B and C.
- Two WOs: in this case, we define one WO for product B, and one WO for product A.

One of these alternatives is predefined in the manufacturing order or — to prevent complex situations that could arise depending on the nature of manufacturing processes involved — the second one is chosen automatically, which means one WO per manufactured product. In a general case, there are distinguished products or parts in two categories: Manufactured (M) or Purchased (P). More details can be seen in [Spec03]

With each work order, a routing is associated (from a routing file, accessed by ERP) and a sequence of tasks (processing steps) necessary to achieve the work order. This routing has to be expressed as a list of services, and each service will be matched with CMU functions. Routings are associated to products by means of a correspondence table (part and service correspondence table).

Figure 108.8 shows the technical composition of the Agency. Modules are differentiated between static and dynamic elements.

Static elements:

- Interface to ERP: This module in fact represents the communication link with ERP on the Agency side. It is limited to the access methods regarding the Agency database.
- MOManager, IRManager, PSObserver have exactly the same behavior: each time a new request is delivered, the corresponding module creates a specific PMA, gets its report, and finally deletes it.
- Agentfabricator is a module that offers agent creation methods and, thus, in the narrow sense is the entity that produces the agents.
- AgencyRA is the Agency Residential Agent, created at the start of the Agency.

Dynamic elements:

PMAs are dynamically created in the Agency: an MOPMA for a Manufacturing Order, an IRPMA for an Information Request, and a PSPMA for a Plant Structure request. In addition, an MOPMA will create the PAs that will be in charge of the Work Orders.

1. An MOPMA is created with a list of WOs:
   - in turn it creates a PA for each WO;
   - it maintains a PA reference table (unique agent identifiers), a WO table, and a WO report table;
   - when MO is completed, MOPMA is deleted by MOManager; and
   - in addition, MOPMA disposes of agent management methods (from AgentManager) for the control of the PAs.

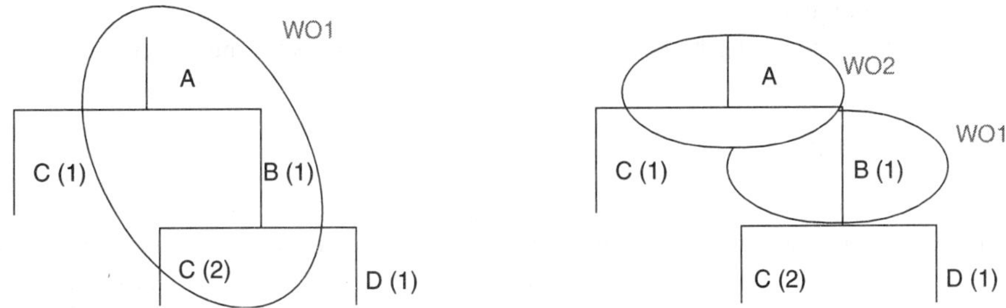

FIGURE 108.7 Decomposition in WOs: (a) one WO, and (b) two WOs.

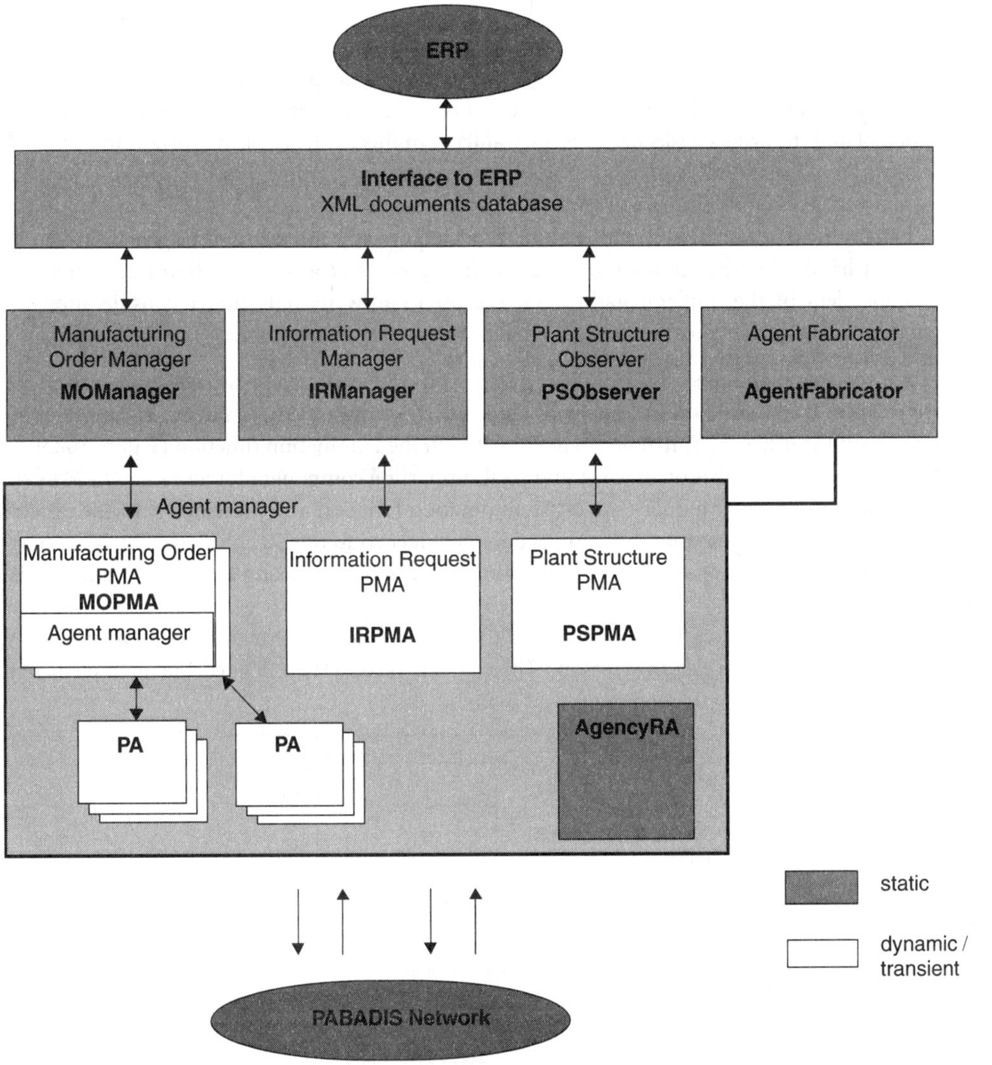

FIGURE 108.8 Structure of the Agency.

2. IRPMA sends requests to CMUs and waits for the reports.
3. PSPMA sends a request to the Look Up Service (LUS) and waits for the report.

The CMU Concept

In the agent-based architecture at hand, manufacturing operations as such (drilling, welding, handling, etc.) and the corresponding automation functions are represented by autonomous units, the so-called CMU. Each CMU contains an agent run-time environment and an RA as an interface between PA and the corresponding automation function. To enable a broad applicability of the described architecture, some basic requirements have to be met:

- Different control systems must be allowed to be integrated into the concept. This means that the CMU architecture has to be independent of the used control hardware.
- A broad spectrum of different, partially unknown automation functions shall be covered and allow the usage of the whole architecture in various manufacturing branches.
- At runtime, the real-time capabilities of the automation functions of the manufacturing system must not be influenced by chance. To achieve this, a strict separation of the automation functions' control and the agents' tasks has to be guaranteed. In this context, it is important to note that the agents are slower for several magnitudes than conventional controls and with regard to job control normally do not require real-time in a narrow sense.
- Changes in an automation function should not require any changes in the agent-based architecture at hand. The description and the evaluation process of any automation function shoud be independent of the implementation and the used hardware, only under consideration of the manipulation (or the possibility to manipulate) the work piece at hand.

The requirements mentioned above are best met by the loose coupling of two components independent from each other; these are on the one hand the PABADIS software environment (as described in this chapter) and on the other the run-time environment with the automation function (Figure 108.9).

In principle, a CMU consists of an RA, a universal supporting component for all PAs and PMAs — the so-called Common Feature Module — and the automation function as representative of the processing of the work piece. Each automation function is definitely related to one RA; thus, the RA is in charge of "his" automation function. Nevertheless, a fix relation between hardware and automation function is not necessarily given.

Between RA and automation function, the generic interface Residential Agent Function Interface (RAFI) is located for control and data exchange reasons. This interface is completely defined, but an implementation

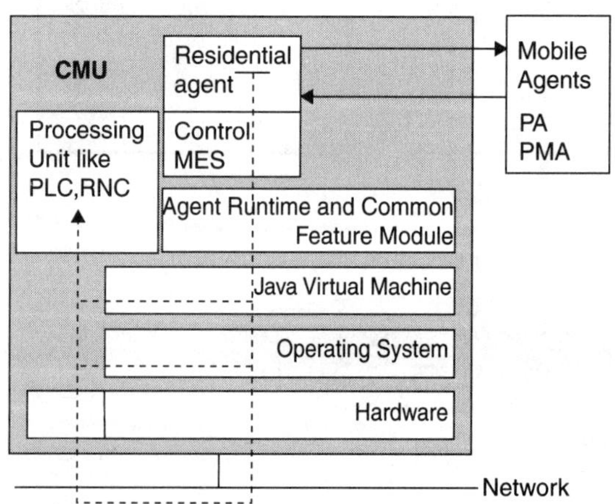

FIGURE 108.9   CMU hardware/software-structure.

is not prescribed. In this way, different manners of coupling the PABADIS software architecture to the automation function are possible; a Java-based Programmable Logic Controller (PLC), for example, under the same hardware could run a second Java Virtual Machine (Java linked CMU), or an already existing control system can be used, which is coupled via the communication network (COM-linked CMU). Several other variants between these two extremes can be realized. *The following subchapters give an overview of the main components of the CMU concept. A more detailed description of the CMU concept can also be found in [B1DeK102].*

The Residential Agent Function Interface

Residential Agent Function Interface (RAFI) offers a universal view on a CMU's automation function and its connection (also universal) to the MAS (Figure 108.10).

At the moment, although not necessary, it seems most appropriate to implement the agents and the other parts of the PABADIS architecture in the Java programming language. The access to the data of the automation function should be realized in the corresponding language of the control system used. Since the RAFI methods are specified in Java and are also called in Java, within the RAFI also the translation from Java into the target language is performed (Java Native Interface).

As Java is independent of the respective automation function, only one RAFI implementation per control system (e.g., PLC type) is necessary, which has to be developed by the producer of the respective control system.

The Function Control Module

The Function Control Module (FCM) offers the generic access path to the automation function, whereas the real-time environment of the automation function and the MAS are decoupled from each other.

Transitions from one state to the other are triggered by the automation function as well as by the RA. To avoid couplings and additional synchronizations, the state machine (Figure 108.11) is defined such that each time only one of these two components determines the access. If a transition is performed, the other (passive) component is informed.

This distribution of competencies as well as the mutual information result in definite messages between RA and automation function; thus, the FCM state machine can be differentiated into two independent state machines and the respective messages during implementation. This property enables the required flexibility when coupling the MAS to existing control systems. Hence, the FCM can also be seen as a communication protocol between RA and automation function.

Parameter Interface and Information Request Interface

Before an automation function can be started to process a certain manufacturing step, normally several parameters have to be set. The function-specific parameter sets are described within the CD (Capability

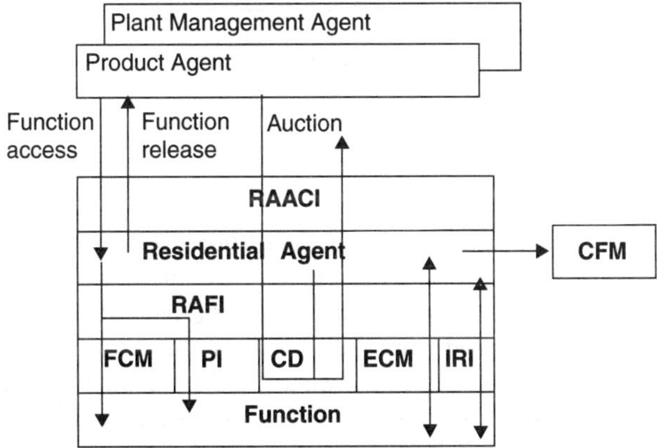

FIGURE 108.10   Access path to the function.

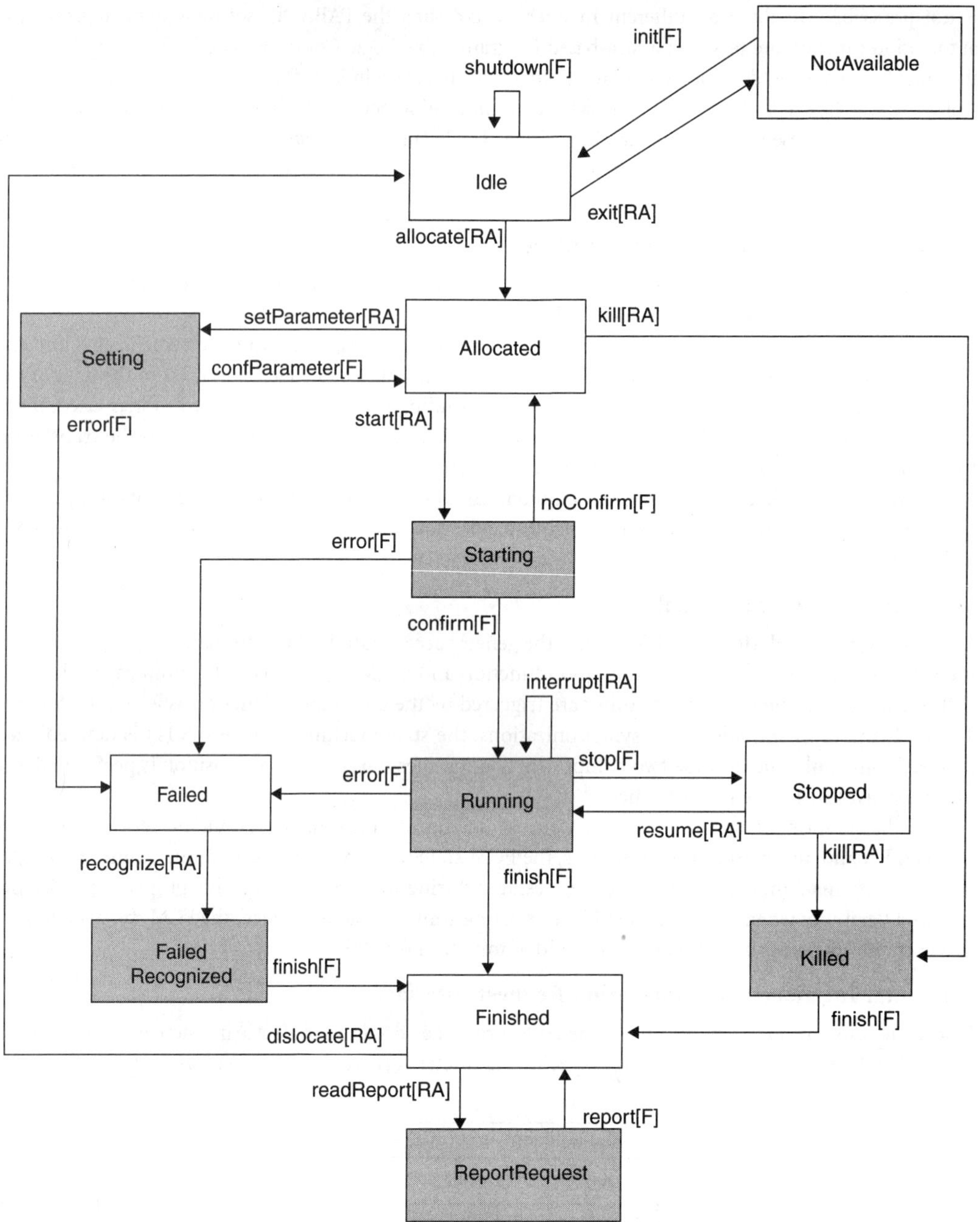

FIGURE 108.11 Function Control Machine.

Description, see below), and the PA contains the special manufacturing step-relevant parameter values. The setting of these parameters occurs via a generic interface, the Parameter Interface (PI), which is embedded into the FCM. As can be seen from the state machine (Figure 108.11), setting of parameter values is only possible in the state "allocated" with the message "setParameter." By means of a call, the data of a parameter set are transmitted; within the PI, it will be confirmed whether all mandatory parameters are set by the PA and whether all parameters that shall be set by the PA are extant. When arrays of parameters are set, these are also verified; if in the course of verification no mistakes are detected, the set of parameters is passed to the automation function.

As during or after the processing of an automation function, possibly a PMA needs data from this automation function, the reading of the data of any automation function is possible via the Information Request Interface (IRI). Nevertheless, data that are read at a certain point of time, depending on the automation function, are not always consistent over time. For this reason, additionally so-called reports are defined, which can only be read after an automation function has finished. The handling of reports within the IRI occurs analogously to the parameter sets within the PI. Reading of reports is controlled via the FCM and is only possible in the state "finished" — otherwise, an error message is produced.

The Event Control Module

The Event Control Module (ECM) serves for the asynchronous information exchange between the automation function and the RA. To be able to support different control platforms, similar to the FCM concept, a state machine is defined, which loosely couples the two systems to each other. To have a bidirectional asynchronous course of events, two instances for the state machine exist. The events, which are supported by the automation function, and their parameters are defined within the CD; the processing of the parameter data occurs in a manner similar to PI and IRI and, thus, are not explained in more detail at this point.

The Capability Description

The Capability Description (CD) is an XML-based description [REC00], which, in a clear, standardized form contains the complete description of all data objects needed for the application of a CMU in MAS. Against this background, a data structure was defined and translated into a set of XML-Schema [REC01a, b]; hence, a CD instance is an XML file based on these schemes. Each data object contains a clear identification (<xxx.ID>) and an extensive description (<xxx.Description>) as well as several other elements, depending on the respective object. Per CMU, there exists exactly one CD that has to be built by the developer of the respective automation function. During the definition phase of the CD former works in the area "description of control components within a network" as the FDCML [ISO03] and methods for process description as the PSL [PSL03] were evaluated and considered.

The CD data structure consists of five main elements, see Figure 108.12 below, under which the respective data elements in a tree structure are positioned. <Identity> contains information for the definite identification of the respective CMU (e.g., name, version). <System Data> contains data with respect to the used technology and system topology. Further objects — which are independent from the state of the automation function — are summarized under the term <data management>. These contain information about events concerning the CMU, reports that the CMU has to deliver, and product-independent data that the CMU is able to offer. In contrast, <function support> contains information with direct relation to the automation function. Here, information that refers to tools, consumables, and the necessary supervisory control, which are required for the respective CMU's execution of its automation function, is available.

This structure is resolved in very much more detail [Spec03], but will not be further described in this context.

108.3 Benefits of the Agent-Based Architecture for Job Control in Turbulent Industrial Environments

The economic relevance and the corresponding benefits for the end user of the proposed agent-based architecture can be seen in three main fields: the improved application of industrial production systems by enhancing the system flexibility, the simplification of system design by introducing new system design strategies, and the refinement of the supply chain management by increasing system clarity and system openness.

The *enhancement of flexibility* of a plant by a PABADIS system is based on two main effects. The distribution of MES functions and the use of mobile product agents, which are correlated to the work pieces, ensure an improvement of vertical flexibility. New products can be easily integrated into the production plan. This also applies to products that were unknown at the moment of plant design. The use of a suitable plug-and-play system (which is not described in more detail in this chapter) enables a horizontal flexibility. New machines can be integrated and used in the system without any changes in the configuration

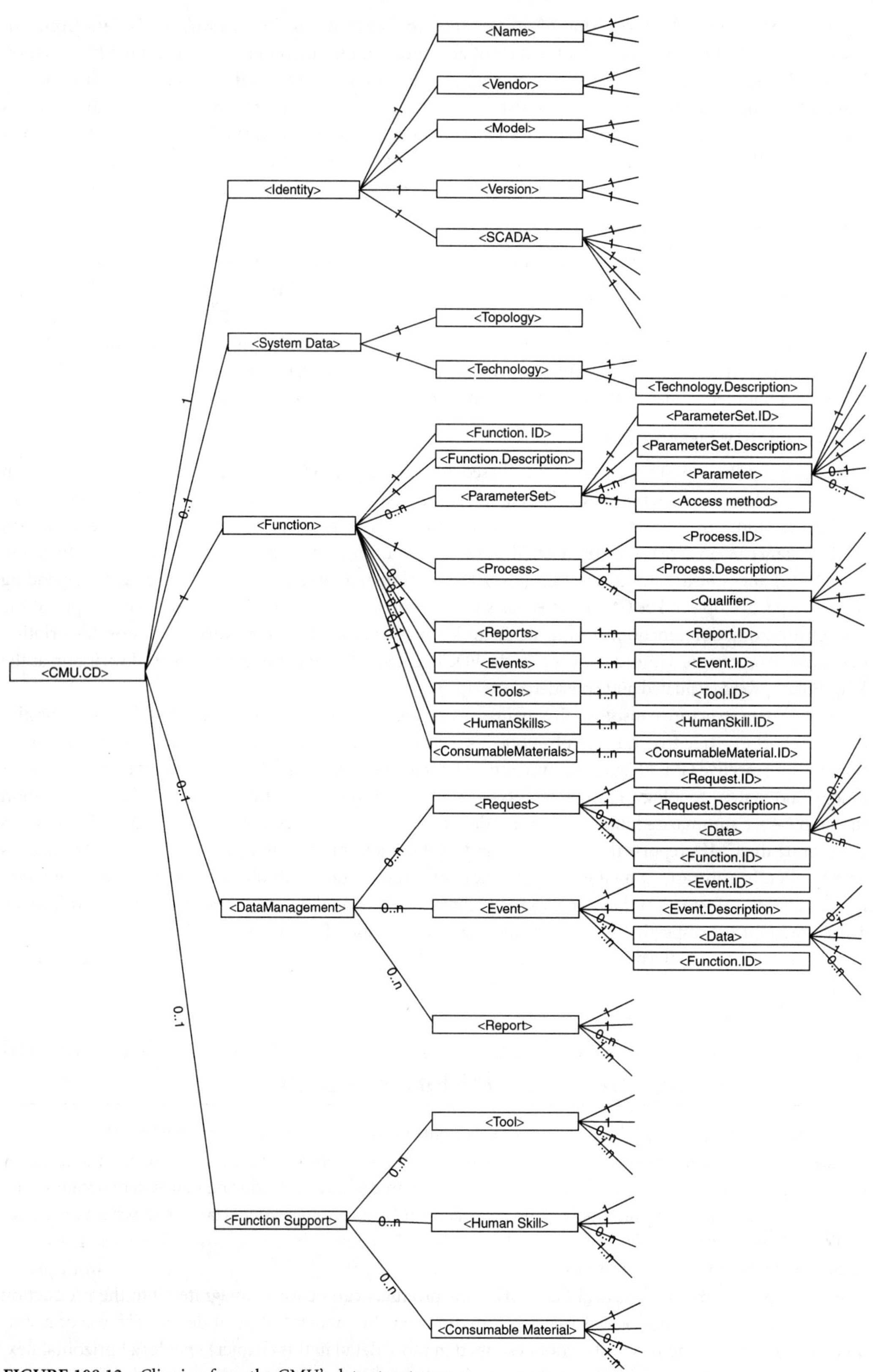

FIGURE 108.12 Clipping from the CMU's data structure.

of other machines. Moreover, a system redesign is easily possible. Further horizontal flexibility arises from the use of Product Agents. These agents react spontaneously to a system change (by the introduction of new machines as well as by failure of existing machines). In both cases, a rescheduling and a new resource allocation might be initiated.

The *system design* is normally based on a system design method that uses a design specification and a design tool. The use of a PABADIS system for plant design enables the plant designer to reduce the complexity of the machine control design by reducing the control to production process control. In contrast to this, in a conventional control system communication with ERP and MES as well as logistical behavior are implemented. In a plant as described here, these functions are implemented by the pregiven and preimplemented RA of the machine containing CMU. Thereby, the reuse of software is also improved. The plug-and-play technology enables an immediate availability of the systems/ CMUs joining the CMU community. The HTTP-server of each CMU, required for the plug-and-play system, can also be used for system redesign by control program uploading or, more generally speaking, for Web Integrated Manufacturing and Web-Based Management.

The *supply chain management is improved* by the use of mobile PAs correlated to work pieces. Data mining with respect to the product-dependent data as well as the machine-dependent data becomes much more easy since direct data acquisition based on a PMA is an essential part of the described system structure. As a consequence, product tracing becomes possible. In addition, the direct relation between product data and the PA can be used for the improvement of the supply chain management. Particularly, it is conceivable to extend the system to the whole logistical supply chain: this means that the agents (PAs) may move from one company to another company of the supply chain.

Finally, it can be stated that the proposed PABADIS concept makes it possible to retrofit a conventional controlled plant, while maintaining the majority of investments already made. These investments can be saved by integrating existing ERP, SCADA, and control systems with slight changes. This prevents the PABADIS system from a main drawback in the field of practical application by enabling a soft change from centralized to decentralized plant control.

108.4   Summary and Conclusions

In this chapter, an agent-based architecture for job control in turbulent industrial environments was presented. It was shown that the presented structure could be applied to nearly all product processing systems, but it is most fitted for small and medium lot sizes and works very well even for lot size one. Correspondingly, the architecture presented is a technical solution to several of the concepts, which, under the term "Mass Customization" try to manage a synthesis of two competing management systems: mass production systems versus those that provide products and services individually. The mass customization paradigm is based on flexibility and quick responsiveness in developing, producing, marketing, and delivering products that can satisfy as wide a range of customers as possible without substantially increasing costs. [Knol01].

The practical use and, of course, the success of a new control structure within plant automation requires a strategy to integrate existing solutions. Otherwise, investments already made cannot be preserved and existing knowledge is lost. The integration of existing control structures into a system as described in this chapter depends on the level of these structures in the control pyramid shown in Figure 108.1.

The integration of existing ERP systems is uncomplicated. An ERP system is connected to the MAS via the Agency. Here, manufacturing orders and data depending on these orders as well as general plant relevant data are exchanged. To integrate an arbitrary ERP system, the interface between Agency and ERP was also designed (although not part of this chapter). This interface is based on an XML service module so that the only requirement for the ERP system is an XML interface, which may be given in advance or needs to be designed, based, for example, on SQL database queries.

At the field control level, the integration of existing solutions is assured by the use of the generic CMU interfaces (see The CMU Concept). Hence, the corresponding controls need not be changed, only adapted. The type of integration also depends on the hardware and software environment of the respective control's device.

As mentioned, one main advantage of the described structure is the distribution and decentralization of MES functions with the aim to be able to concentrate more on the special requirements of the single product. Here, mobile as well as stationary agents are used to fulfill the corresponding jobs, whereas each MES functionality can be translated into corresponding agent functions. The integration of existing MES systems, which are based on a centralized system view, is thus not reasonable.

The integration of SCADA systems is possible by the introduction of a so-called SCADA-CMU. In this case, the existing systems are extended in the same way as a normal CMU, with the SCADA functions used in the sense of a service. If the SCADA functions act as a client, the SCADA system is extended by a residential PMA, acting as an interface to the agent system. Hence, any conventional SCADA system can be integrated if these extensions are suitable.

As an upshot, the authors would like to state that the technical possibility exists and the economic reason dictates the use of office information technologies also in plant environments. In conjunction with intelligent job control architectures like PABADIS, this offers chances to production science and practice, which will eventually provide the end user with the flexibility needed to compete and will finally open new markets and jobs for component suppliers and plant erectors.

References

[BlDeKl02] Blume, R., Deter, S., and Klemm, E., Generic Machine Representation in the PABADIS Community, in *Challenges and Achievements in E-business and E-work,* Vol II, Stanford-Smith, Brian, Chiozza, Enrica, and Edin, Mirelle, Eds., IOS Press, Amsterdam, The Netherlands, 2002, pp. 1134–1140.

[DieS00] Dietrich, D. and Sauter, T., Evolution Potentials of Fieldbus Systems, IEEE Workshop on Factory Communication Systems, Proceedings, Porto, Sept. 6–8 2000, pp. 343–350.

[Ensl78] Enslow, P.H., What is a distributed data processing system, *Computer,* Vol. 11, No. 1, Jan. 1978, pp. 13–21

[FGCD00] Fletcher, M., Garcia-Herreros, E., Christensen, J.H., Deen, S.M., and Mittmann, R., An Open Architecture for Holonic Cooperation and Autonomy, DEXA Conference, London, Sept. 2000.

[Förs99] Förster, T., Entwicklung einer Methode zur Bewertung der Wandlungsfähigkeit von Produktionsbereichen für die variantenreiche Serienfertigung, Ph.D. Thesis, University of Magdeburg, 1999.

[ISO03] ISO 15745-3, Industrial Automation Systems and Integration — Open Systems Application Integration Framework — Part 3, Reference Description for EN 50170 and EN 50254 Based Control Systems, ISO TC 184/ SC 5/ WG 5, ISO/ FDIS 15745-3:2003(E).

[Klm02] Klostermeyer, A., Agentengestützte Navigation wandlungsfähiger Produktionssysteme, Ph.D. thesis, University of Magdeburg, 2002.

[Knol01] Knolmayer, G.F., On the Optimal Extent of Mass Customization, in Proceedings of The International NAISO Congress on Information Science Innovations ISI' 2001, Fares Sebaaly, M., Ed., Mar. 17–21, 2001, pp. 122–128.

[KoUa99] Koren, Y., Heisel, U., Jovane, F., Moriwaki, T., Pritschow, G., Ulsoy, G., and Van Brussel, H., Reconfigurable manufacturing systems, *Annals of the CIRP,* 48, S. 527–540, 1999.

[KühM00] Kühnle, H., and Martinetz, J., Paradigmatic Options for Future Production, in IMS/ GNOSIS Consortium: GNOSIS Open Day Conference – Proceedings, Tokyo, Mar. 2000.

[McOS98] McEleney, B., O'Hare, G. M. P., and Sampson, J., An Agent Based System to reducing Changeover Delays in a Job-Shop Factory Environment, in Proceedings of PAAM'98, London, 1998, pp. 591–613.

[MiHM99] Minar, N., Hultmann Kramer, K., and Maes, P., Cooperating mobile agents for dynamic network routing, in *Software Agents for Future Communications Systems,* Hayzelden, A., Ed., Springer, Berlin, 1999, chap. 12.

[Pesc02] Peschke, J., Internet technologies in automation, in *Praxis Profiline, Industrial Ethernet – Products and Applications,* Vogel, Würzburg, 2002, pp. 21–24.

[PSL03]   The Process Specification Language Project, National Institute of Standards and Technologies, http://ats.nist.gov/psl/.

[REC00]   REC-XML-20001006 — Extensible Markup Language (XML) 1.0., 2nd Ed., W3C Recommendation, October 6, 2000.

[REC01a]   REC-XMLschema-1-20010502, XML Schema Part 1: Structures, W3C — Recommendation, May 2, 2001.

[REC01b]   REC-XMLschema-2-20010502, XML Schema Part 1: Datatypes, W3C — Recommendation, May 2, 2001.

[RePM01]   Reichwald, R., Piller, F.T., and Möslein, K., Mass Customization Concepts for the E-Conomy — Four Strategies to Create Competitive Advantage with Mass Customized Goods and Services on the Internet, in Proceedings of the International NAISO Congress on Information Science Innovations ISI' 2001, Fares Sebaaly, M., Ed., Mar. 17–21, 2001, pp. 129–135.

[Spec03]   The PABADIS Specification — Structure and Behaviour, The PABADIS Consortium, Magdeburg, 2003.

109

Collaborative (Agent-Based) Factory Automation

Armando Walter Colombo
Schneider Electric GmbH (Germany)

Ronald Schoop
Schneider Automation SA (France)

Ralf Neubert
Schneider Electric GmbH (Germany)

109.1 Introduction

From CIM to Heter-archical Control and Production Management

New revolutionary manufacturing concepts and emerging technologies, which take advantage of the newest mechatronics, information, and communication technologies and paradigms, and address many of the fundamental problems described above, have been researched and developed since the last decade of the 20th century (see, e.g., [9, 39], and the references therein).

The process of globalization has forced the development and operation of traditional manufacturing systems to evolve into inherently multidisciplinary tasks. Manufacturing paradigms, such as mass customization, instigated rapid changes in economic, technical, and organizational manufacturing environments. Considering all this, the manufacturing system of the 21st century [22] constitutes a complicated mixture of people, software systems, processes, and equipment (hardware). The management and control of such complex systems is, as a consequence, a multidisciplinary task based on knowledge of manufacturing strategies, planning, operations, and the integration of communication, information, and control functions, and this across the entire enterprise[8,27,43,45].

The Computer Integrated Manufacturing (CIM) concept has been promoted as a solution that can somehow deal with all the above-addressed challenges, for example, providing more flexibility in product spectrum and processes, improving the agility of the production system, better response, and integration of hardware and software components [1,28]. Nevertheless, this centralized and sequential manufacturing planning, scheduling, and control mechanism is increasingly being found to be insufficient in flexibility and agility when responding to changing production styles and highly dynamic variations in product requirements. Moreover, its construction always entails the risk of huge investment, long development schedules, and the generation of very rigid systems due to increased size and centralization. For this reason, before the CIM vision even made its way into practice [31], the original approach changed from a basically centralized model to a decentralized one.

One promising architecture, in this respect, is to have a conglomerate of distributed, autonomous, intelligent, fault-tolerant, reusable manufacturing units, which operate as a set of cooperating entities. Each entity is capable of dynamically interacting with the others to achieve both local and global manufacturing objectives, from the physical/machine control level on the shop floor to the higher levels of factory management systems [12,18]. Due to the tremendous amount of interaction between the different components and the variety of performed functions, the control of such manufacturing systems is currently based on a hierarchical and distributed structure, that is, heter-archical structure, as shown in Figure 109.1 [18,35].

Collaborative Factory Automation. A Result of the Integration of Emerging Technologies and Paradigms: Agent Technology, Holonic Control Systems, and Mechatronics

In the context of today's markets, many industrial companies are looking for a flexible, network-shaped, but sometimes temporally restricted, virtual cooperation of decentral and distributed production competencies. This decentralization originates within a single company (spread over several different production sites) or through the association of several different companies within a single supply chain (virtual enterprise). Collaborative automation approaches are required for such scenarios, that is, a shared but remote supervision is necessary. Autonomous automation units with local supervision functionality installed in each production site interact or cooperate, providing a global (networkwide) supervision (control, monitoring, diagnosis, HMI, maintenance).

A software agent approach seems well suited in relation to the control and supervision of each mechatronics component in an intelligent manufacturing system. Agent-based software systems are becoming a key control software technology for manufacturing control systems. A multiagent-based software platform can offer distributed intelligent control functions with communication, cooperation, and synchronization capabilities, and also provide for the behavior specifications of the mechatronics components and the production specifications to be fulfilled by the manufacturing system (see [2–7,11,15,25,26, 36]).

This trend is accompanied by the fact that modular design of machines and processes, combined with the aggressive augmentation of IT-SW technologies, has pushed automation architectures from a central concentrated system, to a Collaborative Automation, requiring new services via Ethernet TCP/IP, to synchronize distributed applications, new engineering tools, distributed web HMI, and integrate mechatronics specifications.

The answer from Schneider Electric IA to these requirements is a new industrial collaborative automation architecture. It is a global application consisting of several applications deployed as differ-

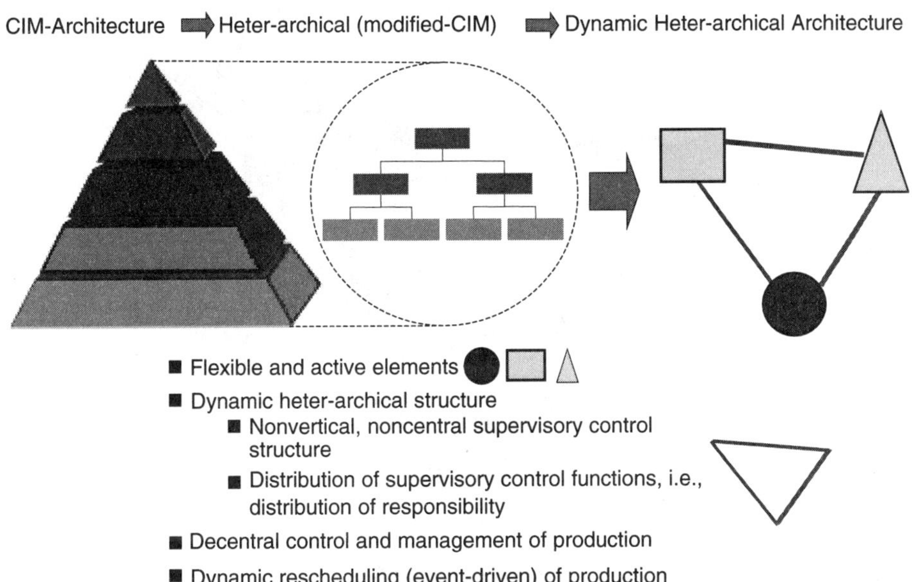

FIGURE 109.1 Heter-archical control and management of production.

ent control units. The interconnection of the control units is ensured using Modbus TCP. An Ethernet TCP/IP application layer controls messaging between devices and assures automatic data refreshing in web browsers. High-performance data sharing for PLC synchronization via Ethernet is now available thanks to Global Data services, providing for automatic and powerful multicast exchanges and improving determinism.

In this chapter, the underlying principles on which the "collaborative factory automation" concept is based (see Figure 109.2) are described. Following this description, this chapter proposes an approach that supports the design and the implementation of a collaborative control platform for industrial manufacturing automation systems based on a multiagent architecture with functions closely related to real-time "shop-floor distributed control and dynamic scheduling." The system has embedded knowledge about the layout and current behavior of mechatronics components that allows it to perform real-time decision-making processes and to run in a synchronized manner with the real production world. It is well suited to accommodating heterogeneous hardware and software components in both their manufacturing control and information systems, in integrating existing/legacy systems, and in adapting new production plans and/or schedules to changing environments (Dynamic System Reconfiguration).

109.2 From Holonic Manufacturing Systems and Agent Technology to Collaborative Factory Automation

The Technical Challenge of Holonic Manufacturing Systems and its Relation to Collaborative Automation

Approximately 35 years ago, Arthur Koestler proposed the word "Holon" [17] to describe the hybrid nature of "wholes/parts." It is an amalgamation from the Greek word "holo = whole," and the suffix "on," which suggests a particle or part, as in "electron, neutron, or proton." Hence, Holon, meaning "whole and part." Holons are self-contained wholes to their subordinated parts but are simultaneously dependent parts themselves when viewed from another angle. Transferring Koestler's model of evolution and life to the shop floor means the installation of intelligent components (hardware and software) and a lateral

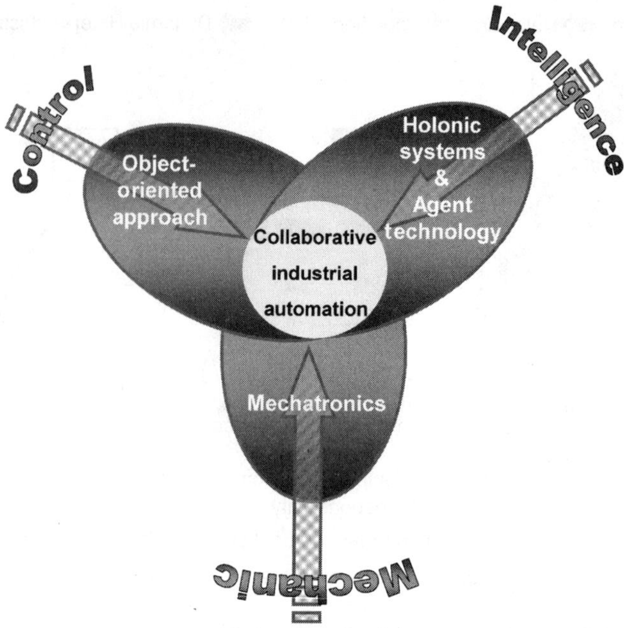

FIGURE 109.2 Collaborative industrial automation paradigm.

means of communication between the production entities and individual machines (viewed as Holons) to enable autonomous decisions to be made within a predefined decision range.

Collaborative Control Units/Holons are responsible for interfacing the system with humans, other Holons for the information system, and peripherals, such as sensors and actuators.

The system entails:

- Multiagent-based Automation Software Technology (physical agent collaborative control component).
- Real-time control functions that implement and supervise the required sequences of operations as well as detecting and diagnosing malfunctions.
- Physical interfaces between the control functions and the sensors and actuators of the physical processing equipment.

The range of applications for collaborative manufacturing automation transcends the traditional views of restricting autonomous operations to machines, manufacturing cells, or other geographically distributed units and/or humans. In fact, a Collaborative Automation Unit can be purely informational (e.g., a process plan or diagnosis module) or may consist of physical objects that are endowed with additional information processing capabilities (e.g., machines with an informational shell or intelligent wrapper), as shown in Figure 109.3.

To summarize, an Intelligent Control System, the humans, and the physical processing system can all constitute separate "Collaborative Automation Units," or be a "Collaborative Automation Unit" in itself.

Multiagent-based Automation Software Technology

The components of a system based on "collaborative principles" naturally suggests a distributed software implementation with autonomously executing cooperative entities as modules, that is, Collaborative Automation Units. Two well-known software engineering technologies are well suited to implement an abstraction of a problem [40]: Multi-Agent Systems (MAS) because of their distributed nature and Object-Oriented Software Engineering, given its recursive structure.

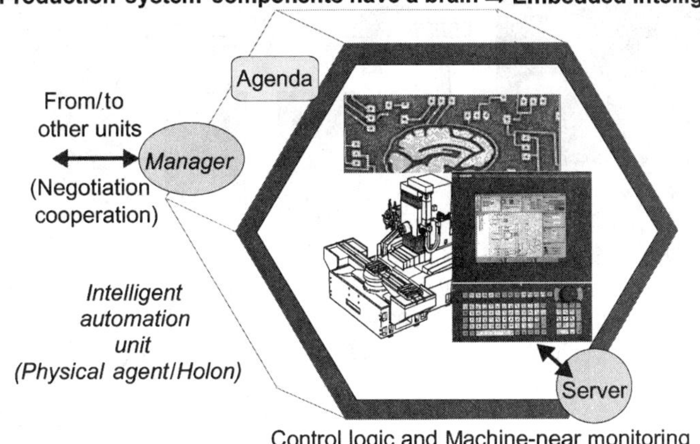

FIGURE 109.3   Collaborative Automation Unit.

The agent technology and the MAS paradigms have been considered over more than a decade as an important approach for developing and implementing the software components of intelligent manufacturing systems. Researchers and practitioners alike have been applying agent technology to manufacturing systems planning, scheduling, execution control, materials handling, and new manufacturing paradigms (see [36] and references therein).

A brief analysis of the reported results shows that the following deficiencies have not yet been solved [16]:

- An agreed definition: Agents built by different teams have different capabilities.
- Duplication of effort: There has been little or no reuse of agent architectures, designs, or components.
- An inability to satisfy industrial strength requirements: Agents must integrate with existing hardware, software, and computer infrastructure, that is, integration of legacy systems.

Agent and Collaborative Automation Units

An agent-based representation of a manufacturing system organization at the physical machine level allows for the conception of an intelligent control component being part of the physical transformation, the transportation, and the storage resources. In fact, each manufacturing resource, including work-pieces and tools, is mapped into an agent, which abstracts all those parameters of the resource needed for the control and supervision of the physical processing equipment. Moreover, the communication- and information-processing capabilities that are endowed in the agent-based control software of the manufacturing resources, transforms it into a self-reconfiguring, intelligent element, that is, a Collaborative Automation Unit. The result is a distributed intelligent control system associated with the lowest layer of a manufacturing holarchy (as depicted in Figure 109.4), with the following main functions:

- Interunit communication allowing the intelligent manufacturing resource to negotiate and coordinate the execution of manufacturing plans (sequences of manufacturing operations) and recovery from abnormal operations.
- Real-time supervisory control functions, that is, control, monitoring, as well as detection and diagnosis of malfunctions.
- Physical interfaces between the control software and the sensor/actuator interface of the physical manufacturing resources.

In the following, this multiagent-based control system is called a "Collaborative Automation System."

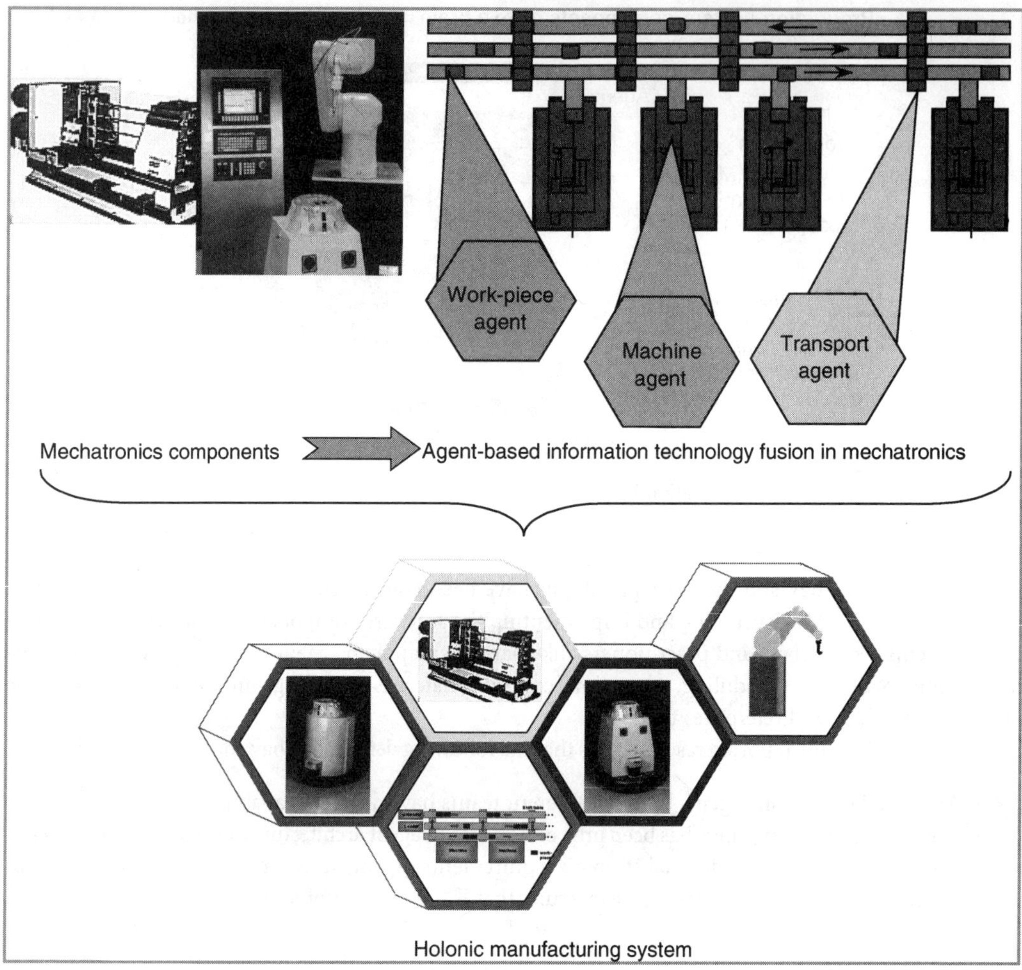

FIGURE 109.4   Principle of an agent-based Collaborative Automation System.

109.3 A Collaborative (Agent-Based) Factory Automation Platform

Motivation and General Characteristics

The application of new strategies and technologies to attain an agile production, standardized machines, flexible material flow systems, and hardware redundancy (e.g., each processing operation is provided by multiple processing units) generates a new set of problems with their corresponding requirements for the control and management of operations:

- Either all possible error cases and/or abnormal states/situations are preprogrammed or more intelligence is distributed to the hardware components.
- The basic components of the manufacturing system are composed of mechatronical components (mechanic plus electrical units), which must be treated as a unique element.
- The components become autonomous and have more "responsibility."

Taking into account the above factors, and based on our previous research results and the prototypes and agent framework specifications presented in [6,13,41], this work summarizes the main characteristics of the heterogeneous agent-oriented collaborative control system *FactoryBroker*™ that has been developed and implemented by Schneider Electric GmbH (Industrial Automation), in cooperation with DaimlerChrysler AG, Research and Technology, Berlin, Germany.

FactoryBroker has been developed as an automation tool to support the implementation, at industrial level, of the "Collaborative Manufacturing Automation System" paradigm. It is particularly relevant in an industrial process control system to control widely distributed, heterogeneous (i.e., different hardware manufacturers) devices in environments that are prone to disruptions [37] and where stiff real-time constraints must be met to achieve safe system operation. The system embodies the idea of the "agile manufacturing" paradigm, where "reprogrammable, reconfigurable, continuously changeable production systems, are integrated into an information intensive manufacturing system, making the lot size of an order irrelevant" [22].

Design Specifications

The "collaborative (agent-based) components" of FactoryBroker are basically formed on a functional modularization of a shop floor. It is complemented by all the essential attributes that are necessary in a holarchy: cooperation, autonomy, intelligence, and transparency.

The following two key constraints have been applied to the design of this Collaborative Control System, which reduce the problems stated above:

- The interfaces of the agents are specified in such a way that different (heterogeneous) agents may cooperate with each other. This supports the interoperability of the Collaborative Control System with other MAS-platforms.
- Agents are used to model the intelligent software part of Collaborative Automation Units, which are software and hardware entities. The scope of responsibility of the designed agents is then derived from the physical system — machines have machine agents, transport systems have transport agents, and work-pieces have work-piece agents.

This differs from functionally derived agents (e.g., prescheduling agent, evaluation agent, simulation agents, etc., as in [2]).

To reduce the effort, existing platforms were used in the design and prototype phase. However, limitations are encountered when the agents have to be embedded into PLC and CNC systems.

The key aspect for applications at the shop floor level is indeed the integration with existing machine control systems, that is, legacy systems [10,33]. As depicted in Figure 109.5, this includes embedding into conventional control equipment (e.g., Programmable Logic Controllers [PLC] and Computer Numerical Control [CNC]), the use of specific operating systems at the shop floor (e.g., Windows NT), and interfacing with established communication interfaces (e.g., DCOM, Ethernet).

Summarizing, a FactoryBroker-Agent is:

- Intelligent software, which has a defined set of goals related to production specifications:
  - to Harmonize overall transport,
  - to Reduce down-time: optimizing load of machines, and
  - to allow different types and degrees of flexibility, that is, allowing variants in work-pieces.
- Intelligent software, which extends existing platforms
  - IEC 61131 function blocks for PLCs, and
  - Windows NT.
- Intelligent software, which is acting autonomously and communicating interactively with other components, using different communication methods
  - COM/DCOM [32] and
  - TCP/IP (XWAY- Schneider Electric protocol).

Architecture and Communication Interface

The architecture of FactoryBroker provides a framework for an unambiguous specification of the structure and relationships between the functional units of the system. During the design phase, this distributed

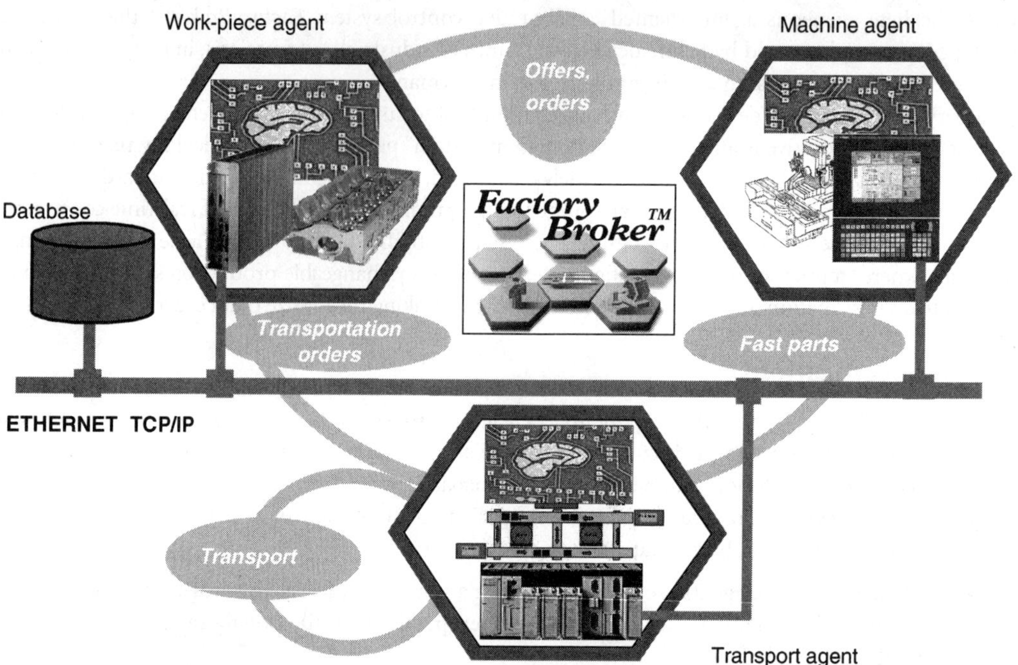

FIGURE 109.5 Collaborative Automation Architecture, FactoryBroker principle.

control system passed through several incremental steps of simplification. The starting point was a mobile agent platform based on JAVA with a simplified agent architecture, that is, the capabilities for mobile agents had been removed. Corresponding to the static arrangement of the production system, the system was redesigned with static, nonmobile agents, as shown in Figure 109.6.

The derived architecture is characterized by:

- A communication framework as a Windows NT-Service that provides standardized agent-orientated interfaces, a user-defined level of fault-detection, and no blocking mechanism of internal functions if communication fails (this was reached by specific implemented additions to Windows NT).
- Encapsulation of the agent type-specific algorithm (COM DLL).
- Definition of critical (wait for a response) and noncritical (continuing after timeout if no response is received) transactions.

The Collaborative Control System depicted in Figure 109.7 covers both conventional real-time machine control and multi-agent-based control of production. These control components are assigned to the distributed manufacturing components in a 1-to-1 relationship:

1. Each machine is controlled by a CNC and PLC, both implemented on an Industrial PC (IPC). Each CNC includes a machine agent that acts as an "Operator" of the machine.
2. Each shift table of a transport system is controlled by a PLC that includes a transport agent that is able to choose transport jobs.
3. Each work-piece is identified by a unique name in the system. A work-piece agent manages a real work-piece and is responsible for organizing all needed manufacturing tasks for the specific real component.

In the following, the different agents constituting the Collaborative Control System architecture are briefly described.

Work-piece Agent: This PC-based type of agent manages planning and processing of one or more work-pieces.

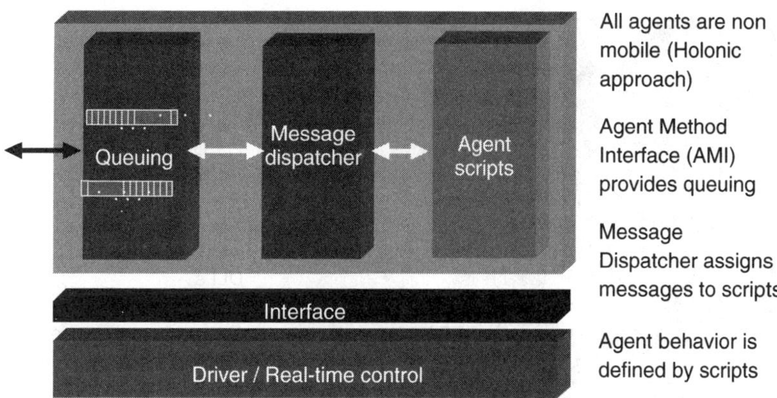

FIGURE 109.6   General architecture of a FactoryBroker agent.

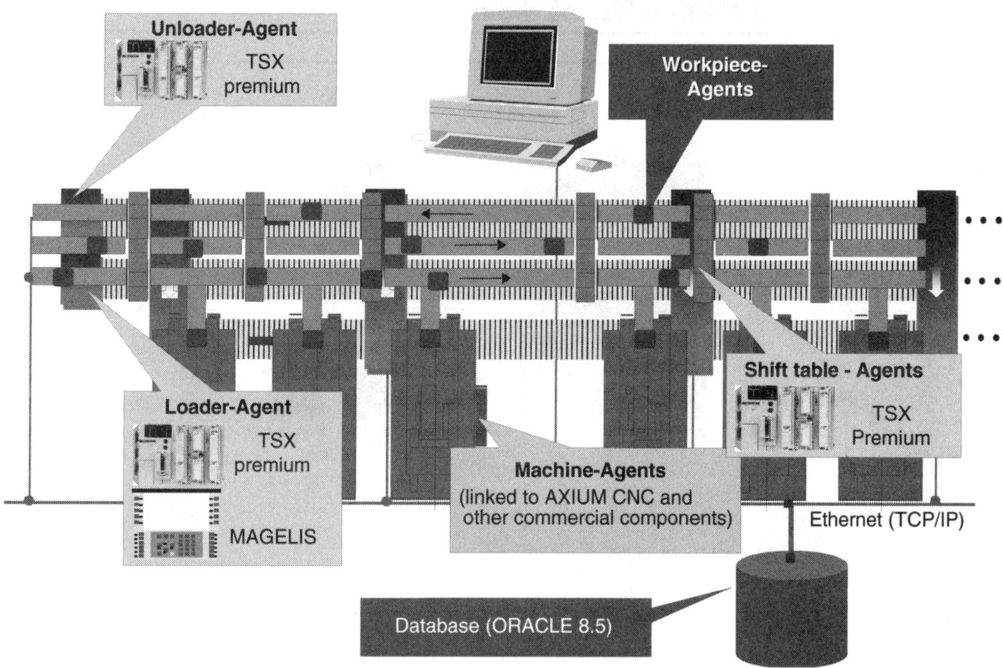

FIGURE 109.7   Architecture of the FactoryBroker Collaborative Manufacturing Control System.

In order to build a redundant system, several workpiece agents are active and share the load of the production system simultaneously. If one agent fails, other work-piece agents take over its work-pieces. At present, running interactions are switched to the new agent responsible. Work-piece agents are implemented as a Windows-NT service and act as a supervisor (identifying the workpiece, checking its required operations, looking for available machines, and initiating the transport).

Machine Agent: A machine agent not only knows the possible tools required for drilling or milling, etc. but also the state and load of its corresponding machine. It acts like a virtual operator, who would be responsible to fill the machine with workpieces and to generate job requests for the machine. Like the workpiece agent, the machine agent is implemented as a Windows-NT service, as depicted in Figure 109.8, and interconnected with the collaborative control platform using, for example, the Internet communication platform.

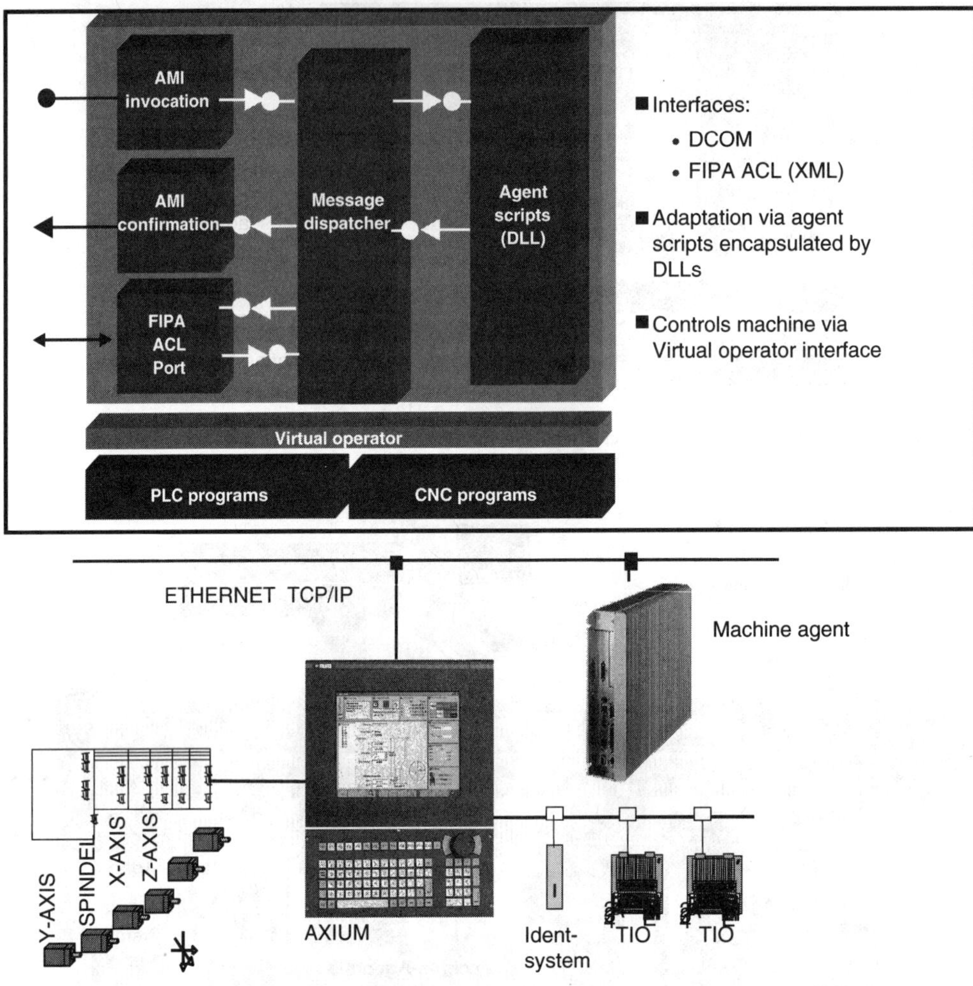

FIGURE 109.8   Structure of a FactoryBroker machine agent.

Transport Agent: There are different types of transport agents but all of them participate in cooperation control/scheduling activity. A *Loader Agent* organizes the loading of new work-pieces into the system. A *Shift Table Agent* is responsible for moving work-pieces between different conveyor belts. This second type knows the destination and decides autonomously which path must be selected to attain transportation to the target machine. Work-pieces that are finished are handled by an *Unloader Agent*, which releases these workpieces to the external environment of the production system.

Note: All types of transportation agents are parameterized corresponding to the topology and design of the flexible material flow system. All these agents are PLC-based and they are implemented as an IEC 61131 DFB inside a PLC application [14].

Communication Interface: The interface of each agent is message-orientated designed as usual. Based on the desired DCOM interface, the messages are mapped to method calls. Therefore, the *FactoryBroker* agent interface is described by the methods.

The subset of interactions were described as a set of methods offered by various Distributed Component Object Model (DCOM)-based classes, which provide a remote Agent Method Interface (AMI) for these interactions for the PC-based agents. The design of PLC-based agent function blocks has been adapted to dedicated network protocols based on TCP/IP.

With respect to the limited set of interactions within the fixed arrangement of the manufacturing system and the restricted degree of freedom of its components, a modification of the behavior can be achieved using a pre-compiled and exchangeable set of script objects. The script objects run as function blocks inside a PLC application as well as Component Object Model (COM) objects inside a DCOM-based framework implemented for PC-based agents such as a work-piece agent [34].

109.4 Negotiation Mechanisms and Interactions between Production Agents for Factory Automation

Holonic Manufacturing Systems proposed that cooperation among "Holons" occurs within logical spaces called *Cooperation Domains*. Following this concept, which maps directly to the FIPA definition of *domains*, the agents in FactoryBroker communicate and operate within cooperation domains defined by the layout and production specifications of the shopfloor components to be controlled.

An agent has to cover the behavior of the component for which it is responsible and has to interact with other agents to accomplish a specific task.

The interactions between agents within FactoryBroker can also be defined as "conversations" [19], and the overall system behavior is formalized by auction-based interactions between the different agent types [42] (see Figure 109.9).

All steps of processing the manufacturing of a work-piece can be described in a precedence graph (see Figure 109.10). In order to simplify the precedence graph (A), several manufacturing steps are summarized in a resulting manufacturing graph (C) with a partial order formed by these steps. As presented below, in the actual implementation of the system, the workpiece agent drives most of the interactions between the agents of the MAS based on the information given in this manufacturing graph.

The principles of some of the conversation specifications that have been implemented are:

- *Offering/Ordering*: Just after loading a work-piece into the system, the responsible work-piece agent determines the type of work-piece and the manufacturing process to be performed. Based on this, an offering process will be initiated. Machine agents, which are applicable for the necessary manufacturing steps, are requested to send an offer for the work-piece (see Figure 109.9).

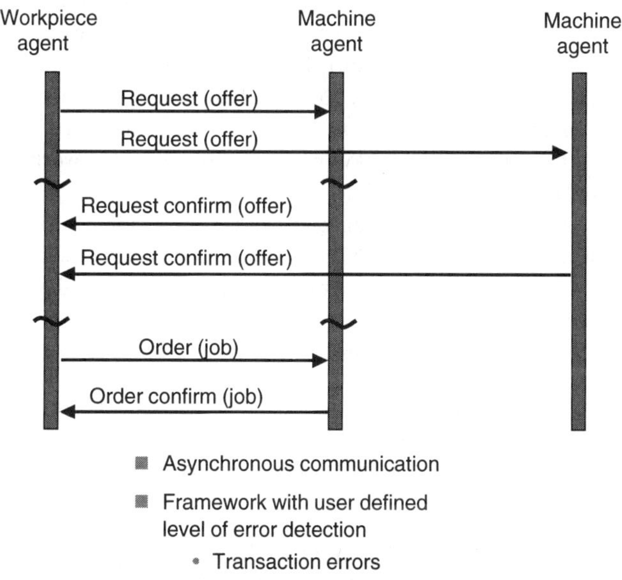

FIGURE 109.9 Interaction mechanisms between agents.

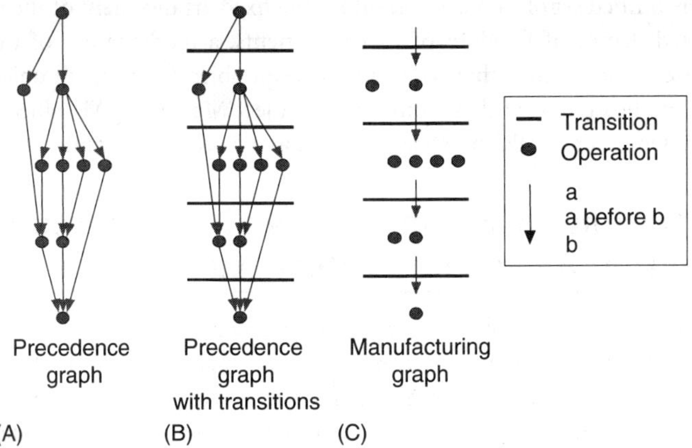

FIGURE 109.10   Manufacturing graph.

The machine agent is responsible for managing in and out buffers for work-pieces, for bidding, and also for controlling the bottleneck capacity of the system. For these purposes, a machine agent applies the algorithm reported by Bussmann and Schild in [6], which was completed with a fairness approach, counting the lifetimes of the work-pieces to be manufactured. All responses are validated and prioritized by the work-piece agent. The best offer will be selected first and the appropriate machine agent is contracted for the job, which must be confirmed. If no confirmation is received, the next offer within the remaining list of best validations is activated. In the case that no responses or confirmations are received, the whole offering process is restarted.

Note: The failure or the partial failure of a machine has an impact on the offering process. Since other machines, however, may provide the required manufacturing steps, the continuous operation of the manufacturing system is guaranteed.

- *Transportation*: In the case that a destination must be defined for a work-piece and it is successfully stated by the offering process, shift table agents are responsible for the selection of the right path within the distributed manufacturing environment. If a shift table fails, a default route is used as an alternate path (e.g., forward direction). The next shift table must redirect the work-piece if necessary. If all machines are occupied or the next destination cannot be precisely defined, the material flow system acts as a dynamic buffer for these work-pieces. In this case, the shift table agents, between the forward and the backward conveyors, rotate the work-piece, until a destination machine is available for the work-piece.
- *Processing*: Usually, a machine agent manages a work-piece as long as it is processed inside the machine. It acts as a virtual operator, which generates and starts a dedicated Numeric Control (NC) program for the manufacture of the workpiece. The progress in the manufacturing process is monitored within a database updated by the work-piece agent each time the machine agent confirms the state of the manufacturing process.

Note: The work-piece is released into the material flow system again after the machine has processed it. In the case that some steps of the manufacturing work still have to be completed, the offering process is restarted. This sequence of negotiations continues as long as necessary until the manufacturing process of the work-piece is finalized.

- *Unloading*: If all manufacturing steps are complete, the work-piece is transported by the shift table agents to the "un-loader agent," which is selected by the work-piece agent. The un-loader agent acts in a manner very similar to the machine agent, except that it does not manufacture the workpiece of course.

Note: The un-loader agent releases the work-piece if, and only if, all data related to this work-piece have been archived in the database by the agent. After completing this step, the work-piece unloads the flexible production system.

109.5 Industrial Applications. Implemented Solutions

Collaborative Automation of an Industrial Production System

Applying the "hardware-in-the-loop" principle, the collaborative control platform is implemented at the industrial level and set into operation. A flexible transfer "line" situated at the DaimlerChrysler cylinder-head manufacturing department, Germany, is currently completely controlled by the developed collaborative (agent-based) controller and operates as the first commercial industrial Collaborative Manufacturing System.

Figure 109.11 shows the shop floor together with its layout. The installation consists of three conveyor belts (two in forward and one in backward motion) and machines that are connected to the material flow system by the lowest conveyor. Each machine has its own set of resources, such as buffers, tools, component trays, etc. Shift tables between the machines connect the different conveyors, providing flexible routing of pallets between the different work places, that is, machines. Each pallet is responsible for carrying a work-piece, that is, a cylinder head. All steps of processing the manufacturing of a work-piece (e.g., drilling 8 (mm)) are dedicated to its specific type (e.g., 4, 6, or 8-cylinder head) and this information is extracted from the manufacturing graph.

Results at the ShopFloor Level

The industrial Collaborative Manufacturing System is operated and fully supervised by the agent-based intelligent control system both during normal shop floor manufacturing as well as performance tests. Throughout the analysis of the architectural and behavioral specifications of the production system under real-time operating conditions, it is possible to evaluate many important "Collaborative" characteristics of the whole system, that is, production equipment with an integrated collaborative (agent-based) control system.

Evaluation of the system's flexibility: As stated earlier in this chapter, one of the main goals for implementing the multiagent-based intelligent controller was to maintain and, when possible, to extend the hardware flexibility with a high degree of control software flexibility. The machine-, routing-, material handling system-, process-, volume-, and expansion-flexibility that is offered by the mechatronics components of the shop floor (see [38]) have all been enhanced by the agent-based software components associated with each of them. Figure 109.12 shows the reaction of the manufacturing system during a working day when different types of workpieces, for example, 2, 3, 4, 5, and 6-cylinder heads are simultaneously manufactured (chaotic manufacturing paradigm).

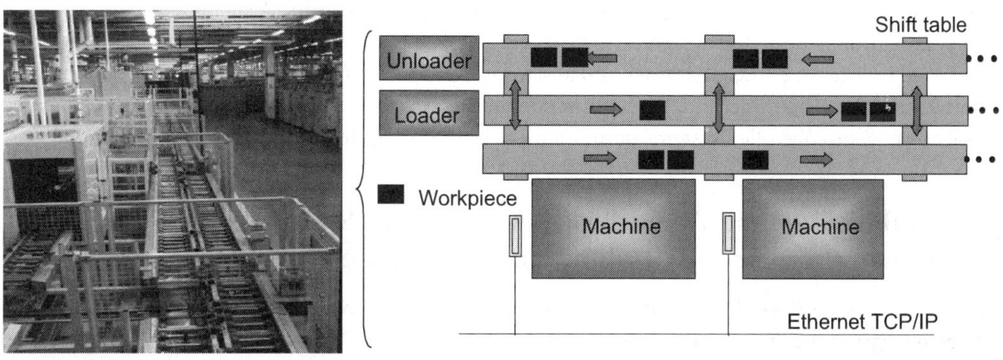

FIGURE 109.11   Industrial shopfloor installation principle.

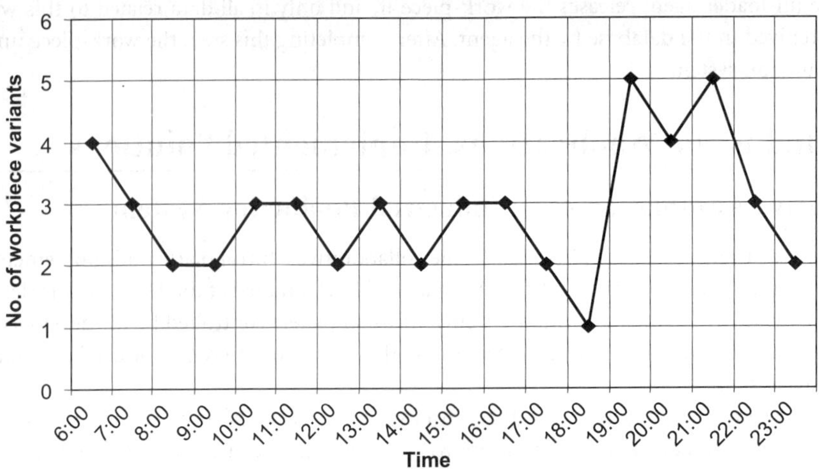

FIGURE 109.12   Flexible behavior in relation to the type of parts manufactured in the system.

Evaluation of the system's agility and reliability: The ability of the intelligent control system to adapt quickly and profitably to continuously changing and unexpected work conditions in the production environment, that is, agility, was tested with respect to the handling of subsystem failures. In particular, a redundant behavior with high system robustness was validated under different working conditions. A self-reconfiguration capability and modular expandability, both from hardware and software points of view, allow the collaborative control system to perform dynamic (re)scheduling. It focuses on the reservation and dynamic allocation of collaborative automation units to manufacturing tasks, aimed at a good overall system performance, both under normal working conditions and in reaction to unexpected situations.

As shown in Figure 109.13, the system is able to maintain its operability and consequently to offer a minimum degree of performance, even when one "collaborative unit," or more, breaks down. This behavior proves a high degree of robustness (see Figure 109.14). In the example, the "proactive" behavior of the system is described as follows: *FactoryBroker* detects an unexpected situation, that is, machine No. 2 fails (e.g., due to a broken tool) at midday. In this case, the workpiece agent, which recognizes the situation, switches its running negotiations from the machine agent of the failed machine to other machine agents (machine No. 3 and later machine No. 1). These agents will take over and share the current jobs subsequently if they can provide the correct handling and machining scope.

Evaluation of system availability: As shown in Figure 109.13, the agent-orientated manufacturing system overcomes the tested failure-situation with a high degree of system availability. The relation of the real availability to the expected availability of a system is defined here as the Availability-factor *A*.

The system attains an *A*-factor 0.89, which represents a substantial improvement in comparison to conventional transfer lines (i.e., *A*-factor 0.66) and a lower production rate remains guaranteed compared to a conventional transfer line. Indeed, both the better behavior in terms of *A*-factor and the increased availability of agent-controlled systems permit a more precise planning of the production output.

Interoperability among Agent-Based Automation Systems: An Approach on how to Implement a Holonic Intraenterprise Platform

Once production objectives, that is, work-plans, have been deployed from the Interenterprise layer to each collaborative partner, each enterprise (now considered as a "Collaborative Unit") has to organize its own resources to fulfill its commitments [40]. The practical experiences of the authors show that two production sections of a holonic enterprise, each controlled by its own MA-based control system (such as the

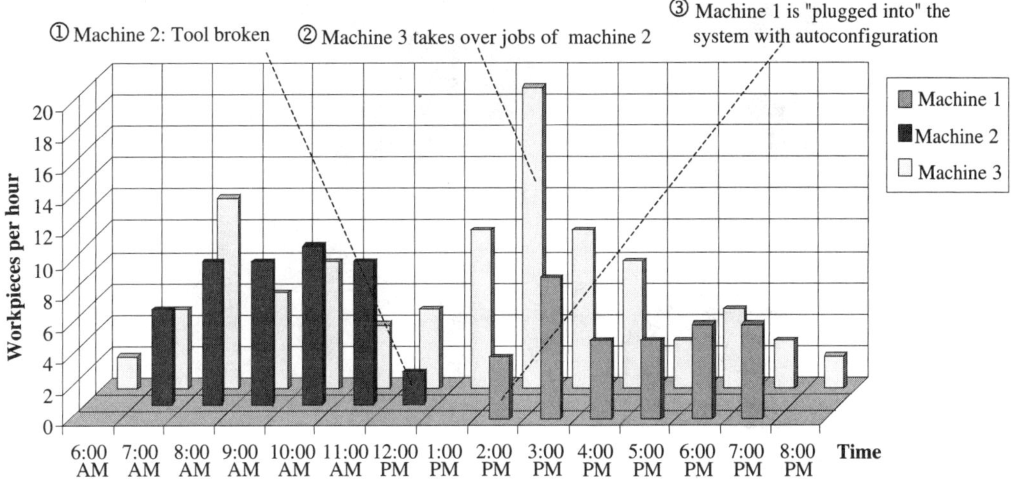

FIGURE 109.13 Manufacturing system's agile evolution when supervised by the multiagent-based control system.

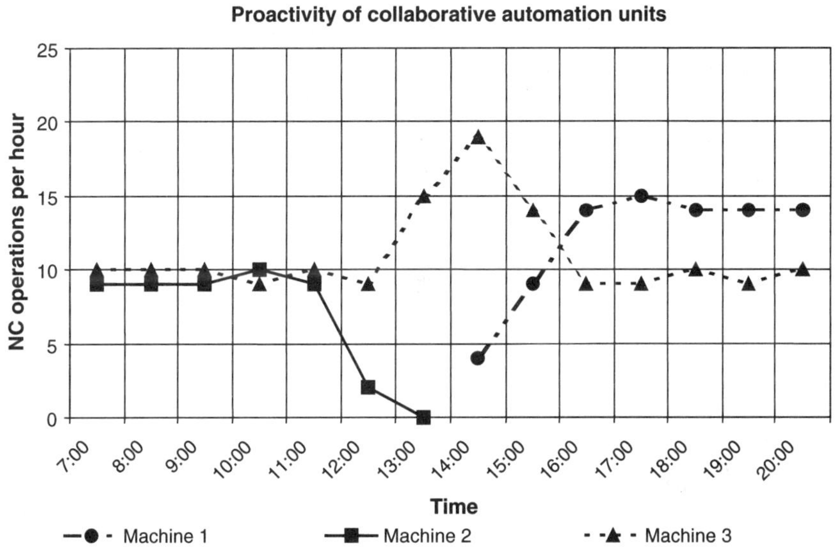

FIGURE 109.14 Evaluation of system's robustness.

collaborative manufacturing cell described above, and a flexible transport system composed of AGVs (see [29] and [30])) can be seen as two collaborative sections within the intraenterprise level (see [20] and Figure 109.15).

There is a set of forces governing the collaborative work of such a constellation of production resources, which are mainly related to optimization of resources and rapid (re)configuration of schedules to accommodate new orders in a timely manner. Typical driving forces at this level are both production specifications:

- on-line ordering and negotiation,
- on-line and flexible production (re)configuration, fault-detection, and recovery,
- on-line and dynamic (re)scheduling of orders,

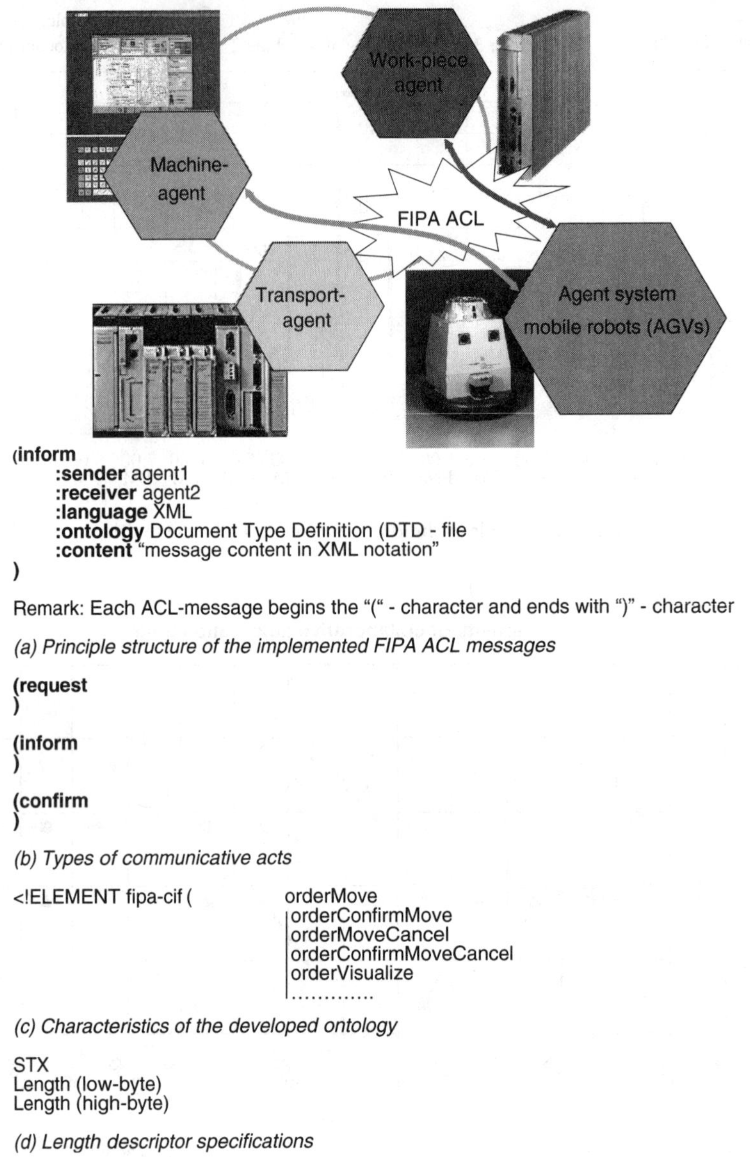


```
(inform
      :sender agent1
      :receiver agent2
      :language XML
      :ontology Document Type Definition (DTD - file
      :content "message content in XML notation"
)
```

Remark: Each ACL-message begins the "(" - character and ends with ")" - character

(a) Principle structure of the implemented FIPA ACL messages

```
(request
)

(inform
)

(confirm
)
```

(b) Types of communicative acts

```
<!ELEMENT fipa-cif (       orderMove
                           |orderConfirmMove
                           orderMoveCancel
                           orderConfirmMoveCancel
                           orderVisualize
                           .............
```

(c) Characteristics of the developed ontology

```
STX
Length (low-byte)
Length (high-byte)
```

(d) Length descriptor specifications

FIGURE 109.15   Intraenterprise Holonic/Collaborative Architecture.

and from an infrastructure perspective:

- registration of agents in the system,
- resources for finding information (including other agents from the other MA-platform), and
- message routing, security management, and error handling.

All of these capabilities can only be implemented if a reliable and safe communication and the corresponding "interoperability" of both MA-platform, is guaranteed, both in their functional and technical aspects.

Integration Approach between MA-Platforms

The main functional specifications of the integration of the two implemented MA-platforms, using the Agent Communication Language (ACL) released by the Foundation for Intelligent Physical Agents (FIPA) as a communication tool [13,44], are depicted in Figure 109.15, and will be briefly summarized here.

The functional communication between the MA-systems is based on the speech–act theory and the technical communication is based on a socket technology.

Each host, the work-piece agent of *FactoryBroker*, as well as the agentified mobile robot system, all implement and set up a socket-server and a socket-client. The workpiece agent receives all the messages from the robot system on its socket-server and sends all the messages to the robot system via the socket-lient. The mobile robot system performs the same function. Once established, the socket connection continues to exist until the computer is shut down or the connection is broken. (The socket communication is done using the port of a communication channel.)

Note: As shown in Figures 109.15(a)–(d), in the socket layer, each FIPA-ACL message is enhanced with a message length descriptor. To specify the communication interface, a Document Type Definition "DTD-file" is used to describe the grammar of the XML-content included in each FIPA ACL message.

109.6 Industrial Trends and Future Requirements

Many manufacturing industry sectors and particularly that of automation technology are developing roadmaps defining the future path for their particular branch. Both national and international organizations, responsible for necessary supportive R&D activities, are observing and supporting this activity. The main objective is to identify technology requirements and make the advances that will help in reducing costs, increase profitability, improve quality, shorten time-to-market, respond to regulatory standards, and to better serve their customers and other stakeholders [23,24].

While a tremendous volume of R&D resources is being expended on developing and implementing new manufacturing technologies, it is clear that:

1. There is much redundant effort being focused in a few key areas.
2. Many manufacturing infrastructure issues that affect the entire industry are receiving insufficient attention.
3. Huge investments in proprietary solutions are either not delivering the promised solutions or are being rendered obsolete by new technologies or by unpredicted changes in the business environment [23].

Taking into account these conclusions and the experience of the authors of this report (see [35]), this section can be summarized as below:

1. There is a set of major challenges for future intelligent automation systems.
2. A set of Key Enabling Technologies (KETs) have to be developed, optimized, and/or adapted to meet these challenges.
3. A set of Technology Research Initiatives (TRIs) is needed to support the development of the KETs.
4. Develop a set of Technology Standardization Initiatives (TSIs) associated with the KETs.

Challenges

Today's manufacturing automation technology requires a growing number of functions, increased flexibility, agility, reliability, and a life-cycle approach encompassing the adaptability to changing operations and conditions in the manufacturing environment. The reason for this development is the varying nature and growing number of industrial processes having, among other things, the following main characteristics:

1. Concurrency in all operations.
2. Rapid reconfiguration of structure in response to changing needs and opportunities, that is, customized functionality.
3. Instantaneous transformation of the information gathered from a vast array of sources into useful knowledge for effective decision making.

4. Integration of human and technical resources to enhance workforce performance and satisfaction (understanding the effects of the agent-based technology on the intelligent manufacturing workforces, working environment, and the surrounding community).

KETs

Automation industry is responding to these challenges with the development of intelligent automation systems, that is, systems with embedded intelligent functions. A new generation of agent-based intelligent automation systems based on multifunctional hardware-platforms for the flexible integration of hardware and software functionality (Collaborative Automation Units) is required. The main characteristic of this kind of system is that it must be focused on the system behavior instead of on the process itself, as in conventional automation technology.

Of course, such intelligent automation systems have not only embedded agent-based technology as one of the main KETs but also other principal information and communication technologies that cover a broad spectrum, such as:

- Web-based Automation.
- Security technologies (secure remote access, peer-to-peer).
- HMI paradigms, Virtual Manufacturing, Transparent/Digital Factory.
- Pervasive Automation.

TRIs

The development of the KETs addressed above is only possible if research effort is put into practice in the areas listed below. This is only possible when a close relation of collaborative R&D, supporting the critical requirements of the automation industry, and research resources is reached.

1. Validation of large, distributed control systems (hundreds of intelligent nodes):
   (a) Self-validation in Collaborative Automation Systems
   (b) Automatic Testing Approaches
2. Distributed planning, dynamic rescheduling, and intelligent supervision:
   (a) Planning and rescheduling capabilities included in an automation island
   (b) Monitoring, Diagnosis, Recovery [decision-making processes]
   (c) Remote monitoring, diagnosis, and maintenance
3. Interoperability between Local Control, Coordination Control Level, and Intraenterprise Level
   (a) Functional, Logic, and Physical System Architecture; Agent-based MES systems
4. Automation Design Environment
   (a) Support the optimization of costs during the life cycle of Automation Solutions
   (b) Integration of Discrete-Event Control Simulation with Process- and Component-Simulation.

TSIs

Proposed improvements in standardization should take into account the new possibilities offered by the KETs above. Standards in intelligent manufacturing automation should consider the functional integration on multipurpose platforms. The seamless exchange and the integration of functions across products require a new type of standard that manages the interchange of information within a configurable network of functions. There is an obvious requirement for a new standardization, supporting the global exchange of automation software functions and the reusability of functional automation software on a variety of platforms [21]. Taking into account these concepts and the new Information–Communication Technologies (ICT), it is obvious that new concepts, such as hardware-independent intracontrol communication, are required.

Finally, it is important to remember that any standardization initiatives should involve the different key players, in particular the equipment vendors, in order to obtain global support and a critical authority for the resulting standardization activities [24].

From the point of view of the authors, intelligent automation and, in particular, agent-based automation systems, should be closely related to any process standardization (plant modularity/flexibility).

We strongly suggest the following areas as candidates for standardization and TSI:

1. Intelligent devices on Ethernet/web (Transparent Ready™):
    (a) Hard real time on Ethernet for automation at the factory floor
    (b) Web pages for intelligent devices (IAONA, IDA)
    (c) Device description for intelligent nodes
2. Agent interfaces, communication, and architectures:
    (a) Foundation for Intelligent Physical Agents (FIPA), including
    (b) Product Design and Manufacturing Group (PDM-WG)
3. Control languages and system models:
    (a) Functional description: flow chart, state graph languages
    (b) Object description: object-oriented frameworks and libraries for application objects+views
    (c) Model description: IEC 61499 (Function Blocks)
4. Enterprise-Control System Integration:
    (a) ISO/IEC 62264-1.

109.7 Conclusions

This chapter briefly summarizes our latest results concerning the industrial application of the emerging "collaborative (agent-based) factory automation technology."

This report indicates that multiagent-technology is an appropriate aid in making the application/implementation of the collaborative manufacturing automation paradigm a reality. Further to discussing the fundamentals of the Multiagent-based "Collaborative" manufacturing control technology, this chapter presents the main specifications of a new collaborative automation and control platform for industrial intelligent manufacturing systems. The developed and implemented collaborative factory automation system, (FactoryBroker), merges the latest research results obtained in the field of Holonic Manufacturing Systems with the latest developments offered by the MAS technology. Besides discussing the suitability of the system for making the intelligent manufacturing system paradigm a reality, this chapter emphasizes the compatibility of the solution with currently implemented industrial production systems. Moreover, this paper presents the results of an application and integration of the mechatronics levels of a collaborative manufacturing automation system at shop floor level with the upper management- and control-layer of an intraenterprise architecture.

As a result of different industrial automation roadmaps, and taking into consideration the experience reported by the authors and other producers of automation systems, a set of challenges and resolutions for future intelligent automation systems is presented. Since these challenges can only be met if KETs are developed, optimized, and/or adapted, some of these KETs are identified and discussed and lead to a summarized set of TRIs that are required to support the development of the addressed KETs.

Finally, a group of TSIs associated with the KETs is proposed, which complete, from the point of view of the authors, *a set of future requirements* on the main subject of this report, that is, Collaborative Factory Automation.

Note: The results presented in this chapter are a summary of our latest reports that can be found in the proceedings of the IECON'01, ETFA'01, ITM'01, IECON'02, and in the IEEE Transaction on Industrial Electronics.

References

[1] Asai, K. and S. Takashima, *Manufacturing Automation Systems and CIM Factories*, Chapman & Hall, London, 1994.

[2] Bohnenberger, T., K. Fischer, and C. Gerber, Agents in Manufacturing: Online Scheduling and Production Plant Configuration, Proceedings of 1st International Symposium on Agent Systems and Applications, 1998.

[3] Bongaerts, L., Integration of Scheduling and Control in Holonic Manufacturing Systems, Ph.D. thesis, PMA/K.U. Leuven, Belgium, 1998.

[4] Brennan, R.W., M. Fletcher, and D.H. Norrie, An agent-based approach to reconfiguration of real-time distributed control Systems, *IEEE Transactions on Robotics and Automation*, 18, 444–451, 2002.

[5] Brückner, S., S. Bussmann, K. Schild, and H. Windisch, Grobspezifikation des Agentensystems im Prototypen, Technical report, DaimlerChrysler AG, Berlin/Germany, 1998.

[6] Bussmann, S. and K. Schild, Self-Organizing Manufacturing Control: An Industrial Application of Agent Technology, Proceedings of the 4th International Conference on Multi-Agent Systems (ICMAS '2000), Boston, 2000.

[7] Bussmann, S. and K. Schild, An Agent-Based Approach to The Control of Flexible Production Systems, Proceedings of the 8th IEEE International Conference on Emerging Technologies and Factory Automation (ETFA'01), Sophia/Nice, France, 2001.

[8] Camarinha-Matos, L.M. and H. Afsarmanesh, Infrastructures for Virtual Enterprises: A Summary of Achievements, Proceedings of IFIP Inernational. Conference on Infrastructures for Virtual Enterprises (PRO-VE'99), Kluwer Academic Publishers, Dordrecht, 1999.

[9] Colombo, A.W., Integration of High-Level Petri Net-based Formal Methods for the Supervision of Flexible Production Systems, Tutorial Lecture at the 1st Online Symposium for Electronics Engineers, U.S.A., Feb. 20, 2001.

[10] Colombo, A.W., R. Neubert, and R. Schoop, A Solution to Holonic Control Systems, Proceedings of the 8th IEEE International Conference on Emerging Technologies and Factory Automation (ETFA'01), Sophia/Nice, France, 2001.

[11] Colombo, A.W., R. Neubert, and B. Süssmann, A Colored Petri Net-based Approach Towards a Formal Specification of Agent-Controlled Production Systems, Proceedings of the 2002IEEE International Conference on Systems, Man and Cybernetics (SMC'02), Hammamet, Tunisia, 2002.

[12] Fanti, M., B. Maione, G. Piscitelli, and B. Turchiano, System Approach to Design Generic Software for Real-Time Control of Flexible Manufacturing Systems, IEEE Transactions as Systems, Man and Cybernetics SMC–A 26, 190–202, 1996.

[13] FIPA. FIPA 97 specification part 2: Agent communication language, FIPA — Foundation for Intelligent Physical Agents, 1997.

[14] IEC CDV 61499-1, Function Blocks for Industrial-Process Measurement and Control Systems Part 1 — Architecture, 1999.

[15] Jennings, N.R. and M.J. Wooldridge, Applications of intelligent agents, in *Agent Technology: Foundations, Applications, and Markets, Nicholas R. Jennings and Michael J. Wooldridge, Eds.,* Springer, Berlin, 1998, pp. 3–28.

[16] Kendall, E.A., Patterns of Intelligent and Mobile Agents, Proceedings of "Autonomous Agents", Minneapolis, U.S.A., 1998.

[17] Koestler, A., The Ghost in the Machine, Arcan Books, London, 1967.

[18] Lauzon, S., A. Ma, J. Mills, and B. Benhabib, Application of Discrete-Event-System Theory to Flexible Manufacturing, *IEEE Control Systems Magazine*, 6, 41–48, 1996.

[19] Lin, F. and D. H. Norrie, "Schema-based Conversation Modelling for Intelligent Agent-oriented Manufacturing Systems", IMS'99, Belgium, 1999.

[20] Neubert, R., A.W. Colombo, and R. Schoop, An Approach to Integrate a Multiagent-based Production Controller into a Holonic Enterprise Platform, Proceedings of the IEEE International Conference on Information Technology in Mechatronics (ITM'01), Istanbul, Turkey, 2001.

[21] Neugebauer, J., and M. Hoepf, The Role of Automation and Control in the Information Society, Report of the Fraunhofer Institut for Production and Automation (IPA), European Commission, DGIS, Fraunhofer IRB Verlag, 1999.

[22] N.N., Visionary Manufacturing Challenges for 2020, Committee on Visionary Manufacturing Challenges, Commission on Engineering and Technical Systems, National Research Council, National Academy Press, Washington, DC., 1998.

[23] The National Institute of Standards and Technology (NIST), U.S. Department of Energy (DOE), National Science Foundation (NSF), and Defense Advanced Research Projects Agency (DARPA), Integrated Manufacturing Technology Initiative, Integrated Manufacturing Technology Road mapping (IMTR) Initiative, NIST, DOE, NSF and DARPA – Executive Summary, July 2000.

[24] European Commission, Dietlind Jering, Paolo Garello — European IMS Secretariat, Christophe Lesniak, Jyrki Suominen — DG Research, GROWTH programme, Erastos Filos, Florent Frederix — DG Information Society, IST programme, Minutes of the IMS Workshop on Standardization Issues for Manufacturing — European Commission, Intelligent Manufacturing System Program, Brussels, 27 Feb., 2002.

[25] Parunak, V.D., A. Baker, and S. Clark, The AARIA Agent Architecture: From Manufacturing Requirements to Agent-Based System Design, Working Notes of the Agent-Based Manufacturing Workshop, Minneapolis, MN, 1998.

[26] Rannanjärvi, L. and T. Heikkila, Software development for holonic manufacturing systems, *Computer in Industry,* 37, 233–253, 1998.

[27] Rathwell, G., Design of Plant Control and Information Systems within an Enterprise Architecture, PERA-Standard documentation, 2001.

[28] Rembold, U. and B.O. Nnaji, *Computer Integrated Manufacturing and Engineering,* Addison-Wesley, Reading, MA, 1993.

[29] Ritter, A., J. Neugebauer, R.D. Schraft, and E. Westkämper, Holonic Task Generation for Mobile Robots, Proceedings of the 30th International Symposium on Robotics, Tokyo, Japan, Oct. 27–29, 1999, pp. 553–560.

[30] Ritter, A., A. Braatz, and W. Baum Transport Agents in Manufacturing, Tagungsband/SPS IPC DRIVES Nürnberg 2000, 12th Fachmesse und Kongress, Hüthig, Heidelberg, Nov. 2001.

[31] Scheer, A.W., *CIM Computer-Integrated-Manufacturing Towards the Factory of the Future,* Springer-Verlg, Berlin, 1993.

[32] Schoop, R. and R. Neubert, Agent-Oriented Material Flow Control System Based on DCOM, Proceedings of the 3rd International Symposium on Object-Oriented Real-Time Distributed Computing (ISORC 2000), Newport Beach, U.S.A., 2000.

[33] Schoop, R., R. Neubert, and B. Süssmann, Flexible Manufacturing Control with PLC, CNC and Software-Agents, Proceedings of the 5th International Symposium on Autonomous Decentralized Systems, Texas, U.S.A., 2001, pp. 365–371.

[34] Schoop, R., R. Neubert, and A.W. Colombo, A Multiagent-based Distributed Control Platform for Industrial Flexible Production Systems, Proceedings of the IEEE IECON′2001, Denver, U.S.A., 2001.

[35] Schoop, R., A.W. Colombo, B. Süssmann, and R. Neubert, Industrial Experiences, Trends and Future Requirements on Agent-based Intelligent Automation, Proceedings of the IEEE IECON′2002, Sevilla, Spain, 2002.

[36] Shen, W. and D. Norrie, Agent-based systems for intelligent manufacturing: a state-of-the-art survey, *International Journal on Knowledge and Information Systems,* 1, 129–156, 1999.

[37] Siegel, K., Final Report on Project Production 2000+, Technical report, DaimlerChrysler AG, Germany, 1999.

[38] Tsourveloudis, N. and Y. Phillis, Manufacturing Flexibility Measurement: A Fuzzy Logic Framework. *IEEE Transactions on Robotics and Automation,* 14, 513–524,1998.

[39] Tzafestas, S., *Modern Manufacturing Systems: An Information Technology Perspective,* Advanced manufacturing Series — Computer-Assisted Management and Control of Manufacturing Systems, Springer Verlag, London, 1997, pp. 1–56.

[40] Ulieru, M., S. Walker, and R. Brennan, Holonic Enterprise as a Collaborative Information Ecosystem, Proceedings of Workshop: Holons, Autonomous and Cooperative Agents for the Industry, Montreal, Canada, May 20, 2001 (AA 2001).

[41] Weber, J. and T. Finin, Specification of the Kqml Agent Communication Language (draft), Technical report, The DARPA Knowledge Sharing Initiative, 1993.

[42] Wellmann, M., W. Walsh, P. Wurman, and J. MacKie-Mason, Auction Protocols for Decentralized Scheduling, revised and extended version of "Some Economics of Market-Based Distributed Scheduling", Proceedings of the 18th International Conference On Distributed Computing Systems, Amsterdam, 1998, June 1999.

[43] Williams, T. and H. Li, PERA and GERAM — Enterprise Reference Architectures in Enterprise Integration, Information Infrastructure Systems for Manufacturing, Vol. II, IFIP, Kluwer Academic Publishers, Dordrecht, 1998.

[44] http://www.fipa.org/docs/wps/f-wp-00009/f-wp-00009.html

[45] http://www.pera.net/

110

Smart Power Systems Rely on Standards for Information Models and Messaging — IEC 61850

Karlheinz Schwarz

Schwarz Consulting Company (SCC)

110.1 Introduction

Automation and information systems in power systems are widely used. Usually, they are based on an immense number of proprietary specifications or (*de facto*) standards. To reduce this variety and to meet future requirements, the standard IEC 61850 (Communication networks and systems in substations) has been developed by users, vendors, and system integrators.

It is not sufficient to develop systems that only produce, transmit, or distribute power. Fully automated — remotely supervised — systems that require little or no human intervention seem to be ideal. Technologies bundled into the power system, therefore, have to include protection and control equipment, as well as interfaces to supervisory control and data acquisition (SCADA) of control centers. Additional applications that are necessary may include: metering, fault diagnosis, automated dispatch and control, data retrieval, asset management, as well as condition monitoring and diagnosis.

Globally, utility deregulation is expanding and requiring demands to integrate, consolidate, and disseminate real-time information quickly and accurately within all kinds of utility systems — from power plants to customer interfaces. Utilities and vendors spend an everincreasing amount for real-time information exchange; expenses for data integration and maintenance are exploding.

Vendors of power systems have — because of the fast-growing market and market deregulation — very limited resources to implement and apply hundreds of proprietary communication and information systems.

The standard IEC 61850 provides the basis for many application domains. One of the biggest success stories in power systems is the deployment of wind power — now the world's fastest-growing energy source. Utilities and operators of wind turbines are struggling with the fast growing diversity of communication systems for wind turbines. Utilities have initiated the standardization of a communication system in 2001: IEC 61400-25 (Communications for monitoring and control of wind power plants). IEC 61400-25 is based on IEC 61850.

Experts of further application domains like hydro power or distributed energy resources (DER) are specifying information and communication systems based on IEC 61850.

The standards chosen in IEC 61850, for example, Ethernet, TCP/IP, MMS (ISO 9506), and XML make use of advanced IT solutions, the reduced bandwidth costs, and increased processor capabilities in the end devices. IEC 61850 focuses on process data and meta-data. More than 3000 standardized objects with concrete names and type information have been defined. The standard overcomes the huge point lists to be created for each installation in case traditional communication systems are applied.

These definitions can be (re)used by applications for on-line verification of the integration and configuration of databases throughout the utility. The self-description of the device's database significantly reduces the cost of data engineering, management, commissioning, operation, monitoring, diagnostics, and maintenance. It reduces system down times due to configuration errors. An example of an analog measurement data is the value "magnitude." Meta-data are, for example, "SI unit," "offset," "scale," "dead band for reporting," and "description."

The objective of IEC 61850 is to provide for seamless information integration across the utility enterprise using off-the-shelf products implementing these international standards.

The Standard IEC 61850 is developed in international cooperation with broad vendor and utility participation. The primary target is the electrical substation automation (switchyards and transformers in the medium- and high-voltage transport and distribution). Most major utilities (e.g., AEP, EdF, E.ON, ENEL, RWE, etc.) and system suppliers (ABB, Alstom, General Electric, SAT, Siemens, etc.) have contributed to the development of the standard.

110.2 Primary Application of the Standard IEC 61850

Figure 110.1 shows an example of a typical hierarchical automation system (substation). At process level, there are the process interfaces hard-wired in the past and serially linked by the process bus in the future. The protection and control devices are connected by the interbay/station bus. At the station level, there is very often a station computer with human–machine interface (HMI) and a gateway to the control center at the higher network level. The focus of IEC 61850 is the support of substation automation functions by the communication of (numbers in brackets refer to the figure):

- sampled value exchange for current and voltage sensors (1),
- fast exchange of I/O data for protection and control (2),
- control and trip signals (3),
- engineering and configuration (4),
- monitoring and supervision (5),
- control-center communication (6),
- time-synchronization, etc.

The standard IEC 61850 is a multipart standard with 14 parts as depicted in Table 110.1.

The information exchange mechanisms access well-defined information models. These information models and the modeling methods are crucial for IEC 61850. The IEC 61850 uses the approach of modeling the common information found in real devices as shown in Figure 110.2. All common information made available to be exchanged with other devices is defined. The standard applies an advanced name space concept. The name space allows to define private extensions for information that is required by an application but not yet standardized.

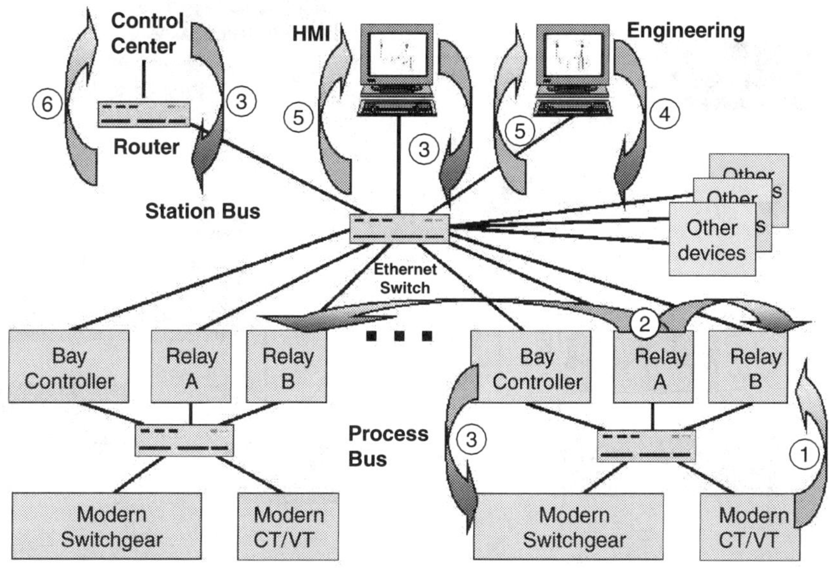

FIGURE 110.1 Modern substation topology.

TABLE 110.1 Parts of IEC 61850

System Aspects	
Part 1	Introduction and overview
Part 2	Glossary
Part 3	General requirements
Part 4	System and project management
Part 5	Communication requirements for functions and devices models
Configuration	
Part 6	Configuration description language for communication in electrical substations related to IEDs
Information Modeling Basics and Information Exchange Model	
Part 7-1	Basic communication structure for substation and feeder equipment — Principles and models
Part 7-2	Basic communication structure for substation and feeder equipment — Abstract communication service interface (ACSI)
Information models	
Part 7-3	Basic communication structure for substation and feeder equipment — Common data classes
Part 7-4	Basic communication structure for substation and feeder equipment — Compatible logical node classes and data classes
Specific mappings to existing protocols	
Part 8-1	Specific communication service mapping (SCSM) — Mappings to MMS (ISO/IEC 9506-1 and ISO/IEC 9506-2) and to ISO/IEC 8802-3
Part 9-1	Specific communication service mapping (SCSM) — Sampled values over serial unidirectional multidrop point to point link
Part 9-2	Specific communication service mapping (SCSM) — Sampled values over ISO/IEC 8802-3
Testing	
Part 10	Conformance testing

IEC 61850 defines the information and information exchange independent of a concrete implementation (i.e., it uses abstract models). The standard also uses the concept of virtualization. Virtualization provides a view of those aspects of a real device that are of interest for the information exchange with other devices. Only those details that are required to provide interoperability of devices are defined in IEC 61850.

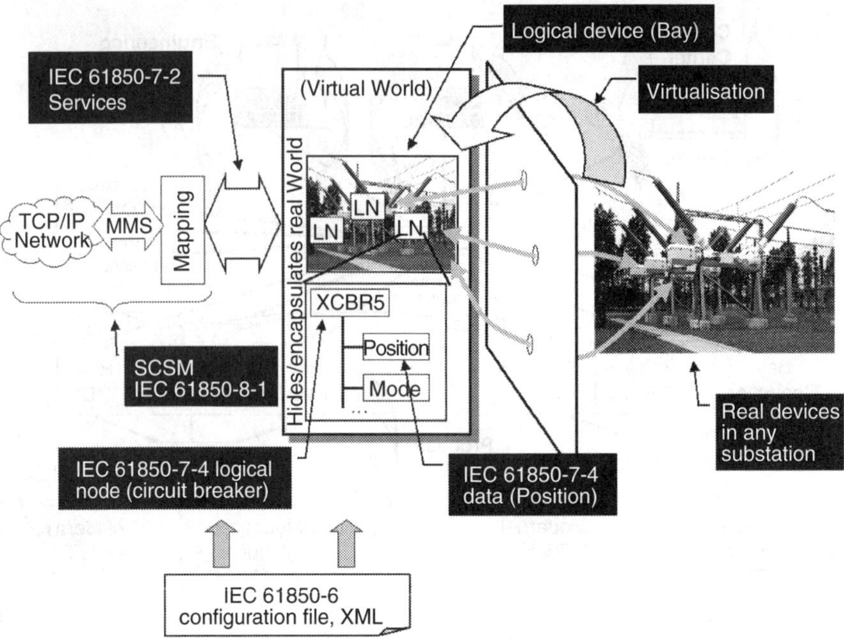

FIGURE 110.2   Conceptual modeling approach.

The approach of the standard is to decompose the application functions into the smallest entities, which are used to exchange information. The granularity is given by a reasonable distributed allocation of these entities to dedicated devices. These entities are called logical nodes (e.g., a virtual representation of a circuit breaker class, with the standardized class name XCBR). Several logical nodes build a logical device (e.g., a representation of a Bay unit). A logical device is always implemented in one Intelligent Electronic Device (IED).

Real devices on the right-hand side of Figure 110.2 are modeled as a virtual model in the middle of the figure. The logical nodes defined in the logical device (e.g., bay) correspond to well-known functions in the real devices. In this example, the logical node XCBR5 represents a specific circuit breaker of the bay to the right.

Based on their functionality, a logical node contains a list of data (e.g., position) with dedicated data attributes. The data have a structure and a well-defined semantic (meaning in the context of substation automation systems). The semantic is partly very common (status or measurement). Other information is substation-specific (directional mode information of a distance protection function). The information represented by the data and their attributes are accessed by various services. The services are implemented by specific and concrete communication means (e.g., using MMS, TCP/IP, and Ethernet among others). The logical nodes and the data contained in the logical nodes are crucial for the description and information exchange of substation automation systems to reach interoperability.

The logical devices, the logical nodes, and the data they contain need to be configured. The main reason for the configuration is to select the appropriate logical nodes and data from the standard and to assign the instance-specific values, for example, concrete references between instances of the logical nodes (their data) and the exchange mechanisms, and initial values for process data.

The four main building blocks (see Figure 110.3) are

Substation automation system information models

IEC 61850-7-4 defines *substation-specific* information models for substation automation functions (e.g., breaker with the status of a breaker position, settings for a protection function, etc.) — *what* is modeled and could be accessed for information exchange. Includes an electronic data sheet embedded in the IED for self-identification and self-description.

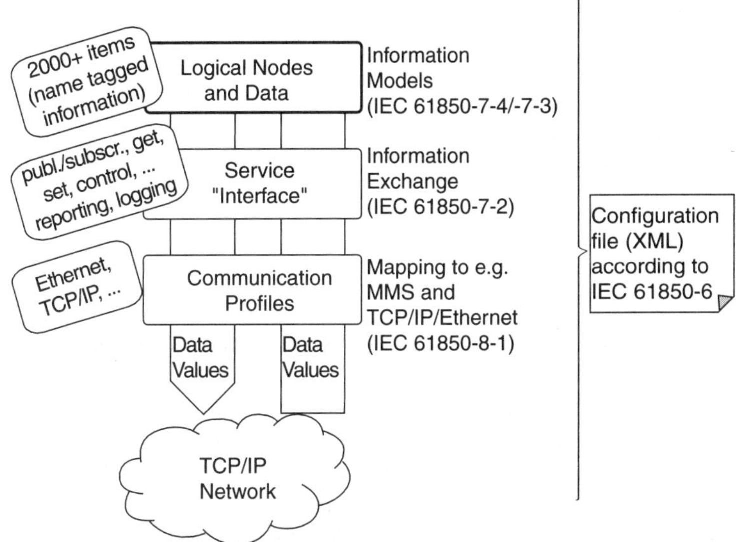

FIGURE 110.3   Levels of the IEC 61850 model.

IEC 61850-7-3 defines a list of *commonly used* information templates (e.g., double-point control, three-phase measurand value, etc.) — *what* is the common basic information.

Information exchange methods

IEC 61850-7-2 provides the services to exchange information for the different kinds of functions (e.g., control, report, log, get and set, etc.) — *how* to exchange information. The information to be accessed also comprises the IED's self-identification and self-description.

Mapping to concrete communication protocols

IEC 61850-8-1 defines the concrete means to communicate the information between IEDs (e.g., the application layer, the encoding, etc.) — *how* to serialize the information in corresponding messages.

Configuration of a substation IED (SCL — substation configuration language)

IEC 61850-6 offers the formal configuration description of a substation IED including the description of the relations with other IEDs and with the power process (single line diagram) — *how* to describe the configuration. This off-line data sheet completely describes the IED. The SCL file (XML document) eliminates the need to manually input this data when configuring an IED. This not only reduces system configuration time but also increases the general integrity and reliability of systems by reducing human error.

These four building blocks are independent of each other to a high degree. The information models can easily be extended by definition of new logical nodes and new data according to specific and flexible rules — as required by another application domains (e.g., wind power plants; see below). In the same way, communication stacks may be exchanged following the state of the art in communication technology.

The information is separated from the presentation and from the information exchange services. The information exchange services are separated from the concrete communication profiles.

110.3 Information Models

Some 100 logical nodes covering the most common applications of substation and feeder equipment are defined. The applications are:

- System information
- Protection functions
- Protection-related functions

- Supervisory control
- Generic references
- Interfacing and archiving
- Automatic control
- Metering and measurement
- Sensors and monitoring
- Switchgear
- Instrument transformer
- Power transformer
- Further power system equipment.

Most logical nodes provide information that can be categorized as depicted in Figure 110.4. The semantic of a logical node is represented by data and data attributes. Logical nodes may provide a few or up to 30 data. Data may contain a few or even more than 20 data attributes. Logical nodes may contain more than 100 individual information (points) organized in a hierarchical structure.

The mean number of specific data provided by logical nodes defined in IEC 61850-7-4 is approximately 20. Each of the data (e.g., position of a circuit breaker) comprises several details (the data attributes). The position (named "Pos") of a circuit breaker is defined in the logical node "XCBR" (see Figure 110.5). The position is defined as data.

The position "Pos" is more than just a simple "point" in the sense of traditional RTU protocols. It is made up of several data attributes. The data attributes are categorized as follows:

- control (status, measured/metered values, or settings),
- substitution, and
- configuration, description, and extension.

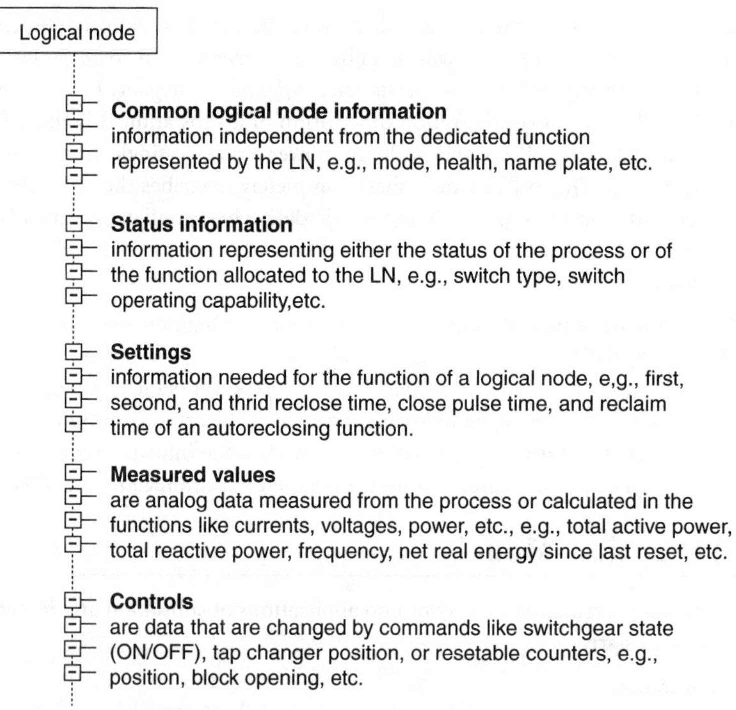

FIGURE 110.4   Logical node information categories.

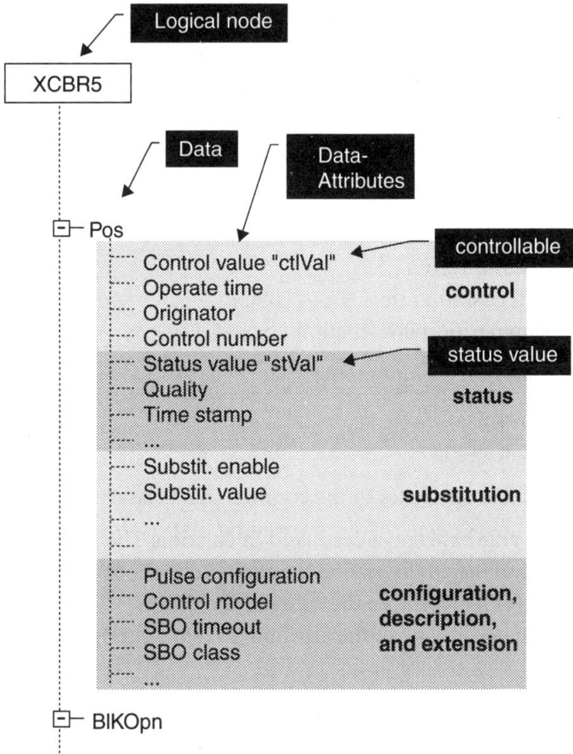

FIGURE 110.5 Position information depicted as a tree (conceptual).

The data example "Pos" has approximately 20 data attributes. The data attribute "Pos.ctlVal" represents the controllable information (can be set to ON or OFF). The data attribute "Pos.stVal" represents the position of the real breaker (could be in intermediate-state, off, on, or bad-state).

The position also has information about when to process the control command (Operate time), the originator that issued the command, and the control number (given by the originator in the request). The quality and time stamp information indicates the current validity of the status value and the time of the last change of the status value.

The current values for "stVal," "q" (quality), and "t" (time stamp) can be read, reported, or logged in a buffer of the IED.

The values for "stVal" and "q" can be remotely substituted. The substituted values take effect immediately after enabling substitution.

Common data classes (templates) are used to define data attributes of data. The crucial common data classes defined in IEC 61850-7-3 are:

Common data classes for status information

- Single point status (SPS)
- Double point status (DPS)
- Integer status (INS)
- Binary counter reading (BCR)

Common data classes for measurand information

- Measured value (MV)
- Complex measured value (CMV)
- Sampled value (SAV)
- Phase to ground-related measured values of a three-phase system (WYE)

- Phase to phase-related measured values of a three-phase system (DEL)
- Harmonic Value (HMV)
- Harmonic value for WYE (HWYE)
- Harmonic value for DEL (HDEL)

Common data classes for controllable status information

- Controllable single point (SPC)
- Controllable double point (DPC)
- Controllable integer status (INC)
- Binary-controlled step position information (BSC)
- Integer-controlled step position information (ISC)

Common data classes for description information

- Device name plate (DPL)
- Logical node name plate (LPL)

The minimum subsets of data attributes of the common data class MV are:

mag: Deadbanded value. Value based on a deadband calculation. The value of mag is updated when the value has changed according to the configuration parameter db (deadband — on-line changeable).

q: Quality of the attribute(s) representing the value of the data.

t: Timestamp of the last change in one of the attribute(s) representing the value of the data or in the q attribute.

These three data attributes can be used for immediate reading, reporting, or logging. Almost all RTU protocols like IEC 60870-5 or DNP3 communicate these three data attributes. The complete set of data attributes of the common data class MV is listed in Table 110.2.

Changes in the data attributes "mag," "range," or "q" can be used to issue a spontaneous report to registered clients indicating the change. Or the change event can be used to log the change in a log of the IED.

The logical nodes, data, and data attributes are defined mainly to specify the information required to perform an application, and for the exchange of information between IEDs. The information exchange is defined by means of services.

TABLE 110.2   Common Data Class MV

Attribute Category	Data Attribute	Explanation
Measured attributes	instMag	Instantaneous value
	mag	Deadbanded value (according to db in per cent)
	range	Normal|high|low|high-high|low-low|...
	q	Quality information
	t	Timestamp
Substitution	subEna	Enable the subMag instead of the process value
	subMag	Value to be used for mag
	subQ	Value to be used for q
	subID	Substitution ID — who substituted the value?
Configuration, description and extension	units	Engineering units (SI units)
	db	Deadband value (in per cent)
	zeroDb	mag is Zero as long as value is less than zeroDb
	sVC	Scale and offset for INTEGER representation of mag
	rangeC	Values for hhLim, hLim, lLim, llLim, min, max
	smpRate	Sampling rate used to determine the analogue value
	d	Textual description
	dU	Textual description (based on UNICODE)
	cdcNs	Common data class name space
	cdcName	Common data class name
	dataNs	Data name space

110.4 Information Exchange Models

An excerpt of the information exchange services is displayed in Figure 110.6. The circles with the numbers refer to the bulleted list below.

The operate service manipulates the control-specific data attributes of a circuit breaker position (open or close the breaker). The report services spontaneously inform another device that the position of the circuit breaker has been changed. The substitute service forces a specific data attribute to be set to a value independent of the process.

The categories of services (defined in IEC 61850-7-2) are as follows:

- control devices (operate service or by multicast trip signals) (1),
- fast and reliable peer-to-peer exchange of status information (tripping or blocking of functions or devices) (2),
- reporting of any set of data (data attributes), SoE — cyclic and event triggered (3),
- logging and retrieving of any set of data (data attributes) — cyclic and event triggered (4),
- substitution (5),
- handling and setting of parameter setting groups,
- transmission of sampled values from sensors,
- time synchronization,
- file transfer,
- on-line configuration (6), and
- retrieving the self-description of a device (7).

The information exchange models are as listed in Table 110.3.

The information exchange supports various methods to access process data (see Table 110.4).

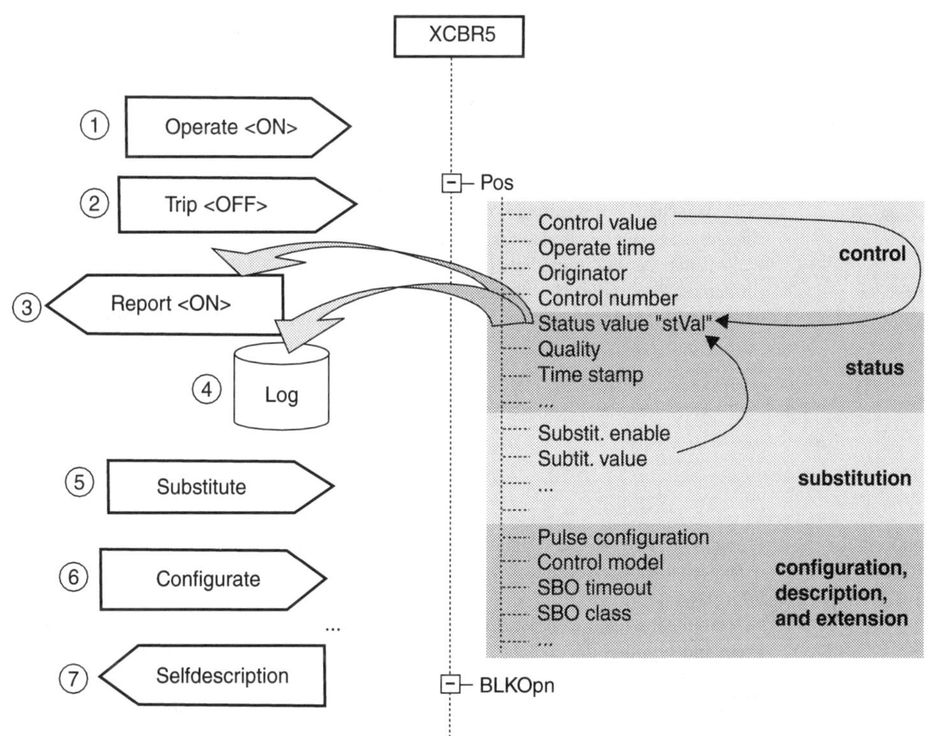

FIGURE 110.6 Service excerpt.

TABLE 110.3 Information Exchange Models

Information Exchange Model	Description	Services
Server	Represents the external visible behaviour of a device. All other ACSI models are part of the server.	ServerDirectory
Application association	Provision of how two or more devices can be connected. Provides different views to a device: restricted access to the server's information and functions.	Associate Abort Release
Logical device	Represents a group of functions; each function is defined as a logical node.	LogicalDeviceDirectory GetAllDataValues
Logical node	Represents a specific function of the substation system, for example, overvoltage protection.	LogicalNodeDirectory
Data	Provides a means to specify typed information, for example, position of a switch with quality information, and timestamp.	GetDataValues SetDataValues GetDataDefinition GetDataDirectory
Data set	Allow to group various data together.	GetDataSetValue SetDataSetValue CreateDataSet DeleteDataSet GetDataSetDirectory
Substitution	The client can request the server to replace a process value by a value set by the client, for example, in the case of an invalid measurement value.	SetDataValues
Setting group control	Defines how to switch from one set of setting values to another one and how to edit setting groups.	SelectActiveSG SelectEditSG SetSGValues ConfirmEditSGValues GetSGValues GetSGCBValues

Reporting and logging	Describes the conditions for generating reports and logs based on parameters set by the client. Reports may be triggered by changes of process data values (e.g., state change or deadband) or by quality changes. Logs can be queried for later retrieval. Reports may be sent immediately or deferred (buffered). Reports provide change-of-state and sequence-of-events information exchange.	*Buffered RCB*
		Report
		  GetBRCBValues
		  SetBRCBValues
		Unbuffered RCB:
		Report
		  GetURCBValues
		  SetURCBValues
		Log CB
		  GetLCBValues
		  SetLCBValues
		Log
		  QueryLogByTime
		  QueryLogAfter
		  GetLogStatusValues

Generic substation events (GSE)	Provides fast and reliable system-wide distribution of data; peer-to-peer exchange of IED binary status information.
	GOOSE means Generic Object Oriented Substation Event and supports the exchange of a wide range of possible common data organised by a DATA-SET
	GSSE means Generic Substation State Event and provides the capability to convey state change information (bit pairs).
		GOOSE CB
		  SendGOOSEMessage
		  GetGoReference
		  GetGOOSEElementNumber
		  GetGoCBValues
		  SetGoCBValues
		GSSE CB
		  SendGSSEMessage
		  GetGsReference
		  GetGSSEElementNumber

TABLE 110.3   (*Continued*)

Information Exchange Model	Description	Services
		GetGsCBValues SetGsCBValues
Transmission of sampled values	Fast and cyclic transfer of samples, for example, of instrument transformers.	*Multicast SVC* SendMSVMessage GetMSVCBValues SetMSVCBValues *Unicast SVC* SendUSVMessage GetUSVCBValues SetUSVCBValues
Control	Describes the services to control, for example, devices or parameter setting groups.	Select SelectWithValue Cancel Operate CommandTermination TimeActivatedOperate
Time and time synchronization	Provides the time base for the device and system.	Services in SCSM
File transfer	Defines the exchange of huge data blocks such as programs.	GetFile SetFile DeleteFile GetFileAttributeValues

TABLE 110.4   Retrieval Methods

Retrieval Method	Time-Critical Information Exchange	Can Loose Changes (of Sequence)	Multiple Clients to Receive Information
Polling (GetDataValues)	NO	YES	YES
Unbuffered Reporting	YES	YES	NO
Buffered Reporting	YES	NO	NO
Log (used for SOE logging)	NO	NO	YES

The polling is the most common method. The use is restricted to applications that do not rely on sequence-of-events (SoE). Data values may be lost because the value may have been overwritten by the application between two GetDataValue requests.

The buffered and unbuffered reporting starts with the configuration of the report control blocks. The basic buffered reporting mechanism is shown in Figure 110.7. The reporting starts with setting the enable buffer attribute to TRUE; setting to FALSE stops the reporting.

The specific characteristic of the buffered report control block is that it continues buffering the event data as they occur according to the enabled trigger options in case of, for example, a communication loss. The reporting process continues as soon as the communication is available again. The buffered report control block guarantees the SoEs up to some practical limits (e.g., buffer size and maximum interruption time).

Figure 110.8 shows an example of a log and three log control blocks. The first step is to configure and enable log control blocks. After enabling, the association with that server may be closed. The log entries are stored in the log as they arrive for inclusion into the log. The logs are stored in time sequence order. This allows retrieval of a SoEs list.

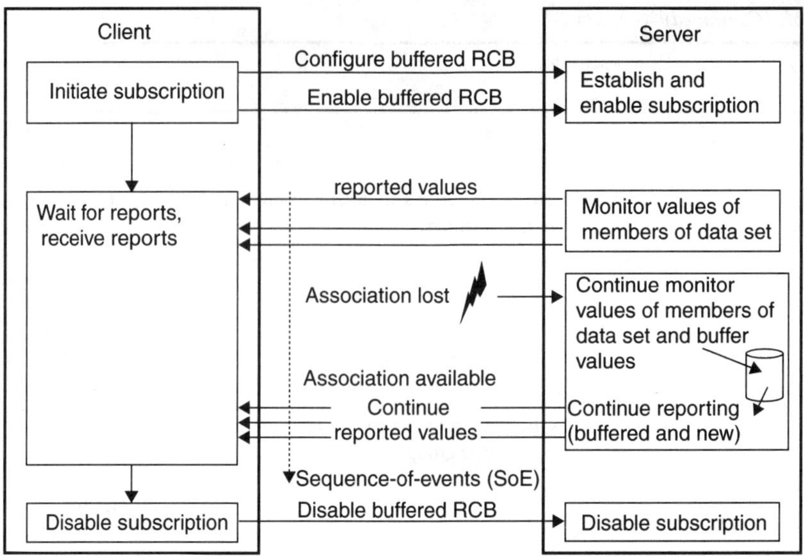

FIGURE 110.7 Buffered reporting (conceptual).

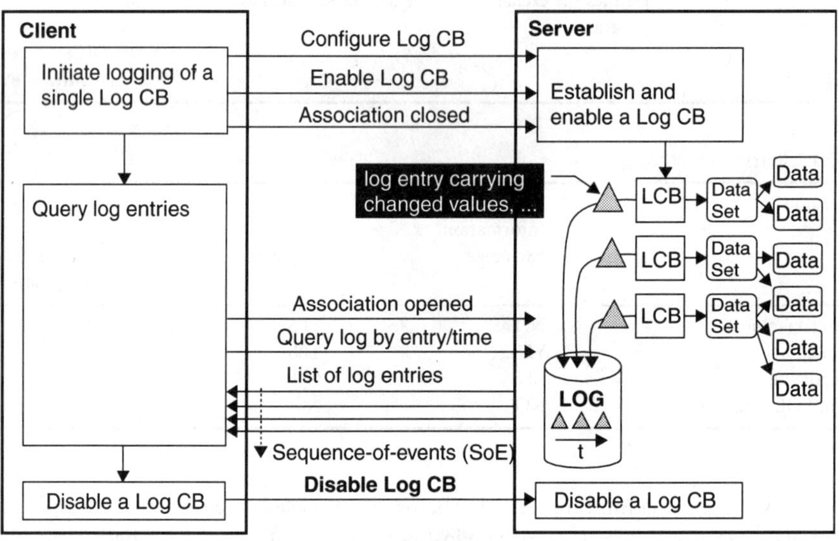

FIGURE 110.8 Log control block (conceptual).

The different log control blocks allow storage of information from different data sets into a log instance. Each log control block is independent of the other control blocks.

Two models are dedicated to the transmission of information with high priority:

GOOSE (Generic Object Oriented System Event) is used to model the transmission of high-priority information like trip commands or interlocking information. The model is based on cyclic and high-priority transmission of status information. Information like a trip command is transmitted spontaneously and then cyclically at increasing intervals.

SV (Sampled Value) is used to model the exchange of the sampled measured values from current and voltage transducers to any IED that needs the samples. The model is based on an unconfirmed transmission of a set of sampled values. A counter is added to time correlate samples from different sources and to detect the loss of a set of samples.

GOOSE and SV service models are defined in IEC 61850-7-2. The mapping of the SMV model to a concrete communication system is specified in IEC 61850-9-1.

IEC 61850-9-1 defines a unidirectional serial communication interface connecting current/voltage transducers with digital output to electrical metering and protection devices. The goal of the standard is to support interoperability between such devices from different manufacturers. With devices supporting this standard, the customer has the possibility to select a current/voltage transducer of one manufacturer and connect it to a protection device or a meter of another manufacturer. Being convinced that this is of real benefit for the customer, ABB and SIEMENS decided to support this standard. In order to demonstrate the feasibility, the real-time exchange of sampled measured values between ABB and SIEMENS devices was shown at the UCA user group and utility initiative meeting in Dana Point, CA (U.S.A.) in January 2002.

Both ABB and SIEMENS each developed a device called Merging Unit converting their own proprietary signals from the current/voltage transducers (CT/VT) to messages according to IEC 61850-9-1 transmitted over Ethernet [1]. Each message contains sampled values of currents and voltages for the three phases and neutral.

On the data sink side, ABB and SIEMENS each developed a distance protection relay, supporting the IEC 61850-9-1 messages as input signals. In addition, SIEMENS developed a meter with the same interface.

An overview of the five devices is given in Figure 110.9.

Each merging unit transmits synchronized samples with a transmission rate of 1000 messages/sec. Two sample rates — 1000 and 4000 samples/sec — are supported. In the case of 4000 samples/sec, four sets of samples are transmitted in one message.

110.5 IEC 61400-25 communication for wind power plants

The IEC Technical Committee 88 has set up a new project to develop a communication standard for distributed generation (primary scope per TC 88: wind power plants) in 2001:

IEC 61400 Part 25: Communications for monitoring and control of wind power plants

This standard defines like IEC 61850 several levels:

- wind power plant-specific information,
- information description methods (reference to IEC 61850),
- substation configuration method (reference to IEC 61850),
- information exchange for monitoring and control systems for wind power plants (reference to IEC 61850), and
- communications profiles (reference to IEC 61850 and definition of additional mappings).

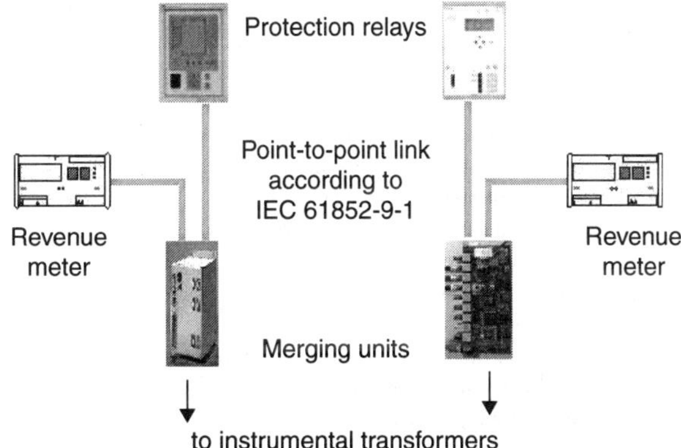

FIGURE 110.9 Devices with IEC 61850-9-1 interfaces.

The information defined in this standard comprises mainly wind power plant-specific information like status, counters, measurands, and control information of various parts of a wind power plant, for example, turbine, generator, gear, rotor, and grid.

The standard IEC 61400-25 mainly defines additional information models and common data classes. The following wind power plant-specific logical nodes (Table 110.5) comprise a total of some 150 data.

The data are more comprehensive than those defined in IEC 61850. An analog value, for example, comprises the Instantaneous value, Deadbanded value, Average value, Root-mean-square value (effective), as well as the characteristics and historical information listed in Table 110.6.

An analog value is very comprehensive (not complex!). Most data attributes defined in these templates are optional. That means if they are needed then the standardized names can be used.

Many status and measurement information can optionally be stored in various logs for later retrieval. Advanced turbine controller may comprise several thousand information points to be reported and logged, respectively. IEC 61400-25 defines many common logs, for example, turbine commands, turbine status, high urgent alarms, low urgent alarms, counting information, and event timing information.

Figure 110.10 shows the current situation. Among other challenges, equipment having different proprietary protocols cannot be integrated without the installation of Remote Terminal Units (RTUs) to perform the translation work (Gateway), which is both expensive and labor-intensive. Usually raw data are exchanged on a cyclic basis.

The move to international standards provides the basis for the migration of common SCADA processing functions, for example, logging of historical and statistical information in the turbine controller. Self-identification and self-description may also be stored in the IED's database.

Today's advanced controllers already provide most of the historical and statistical information. In this regard, the standard IEC 61400-25 follows the market — not vice versa.

The object-oriented information description methods allow precise and complete specification of the information.

The information exchange provides:

- real-time data access and retrieval,
- controlling devices,
- event/alarm reporting and logging,
- self-description of devices,
- data typing and discovery of data types, and
- file transfer.

IEC 61400-25 has been developed in order to provide a uniform communications basis for the monitoring and control of wind power plants. It defines wind power plant-specific information, the mecha-

TABLE 110.5 Wind Power Plant-Specific Logical Nodes

Logical Node	Description
WTUR	Wind turbine general information
WROT	Wind turbine rotor information
WTRM	Wind turbine transmission information
WGEN	Wind turbine generator information
WCNV	Wind turbine converter information
WGDC	Wind turbine grid connection information
WNAC	Wind turbine nacelle information
WYAW	Wind turbine yawing information
WTOW	Wind turbine tower information
WMET	Wind power plant meteorological information
WALM	Wind turbine alarm information
WSLG	Wind turbine state log information
WALG	Wind turbine analog log information
WREP	Wind turbine report information

TABLE 110.6 Additional Information

Attribute Category	Data Attribute	Explanation
Characteristics infor-	maxVal	Maximum value
mation	minVal	Minimum value of data
	totAvgVal	Total average value of data
	sdvVal	Standard deviation of data
	opRs	Operator identifier of last reset
	tRs	Timestamp at last reset
	q	Quality
Characteristics control	rsMan	Manual forced reset
Information	rsPer	Time periodical reset: hly | dly | wly | mly | manual
Historical information	hlyMax	hourly max value of 25 hours
	dlyMax	daily max value of 32 days
	mlyMax	monthly max value of 13 months
	ylyMax	yearly max value of 2 years
	hlyMin	hourly min value of 25 hours
	dlyMin	daily min value of 32 days
	mlyMin	monthly min value of 13 months
	ylyMin	yearly min value of 2 years
	hlyAvg	hourly average of 25 hours
	dlyAvg	daily average of 32 days
	mlyAvg	monthly average of 13 months
	ylyAvg	yearly average of 2 years
	tRs	Operator identifier of last reset
	opRs	Timestamp of last reset
Historical control and	rsHis	reset log hly | dly | mly | yly | all
setpoint information	d	textual description
	dU	textual description (based on UNICODE)
	cdcNs	common data class name space
	cdcName	common data class name
	dataNs	data name space

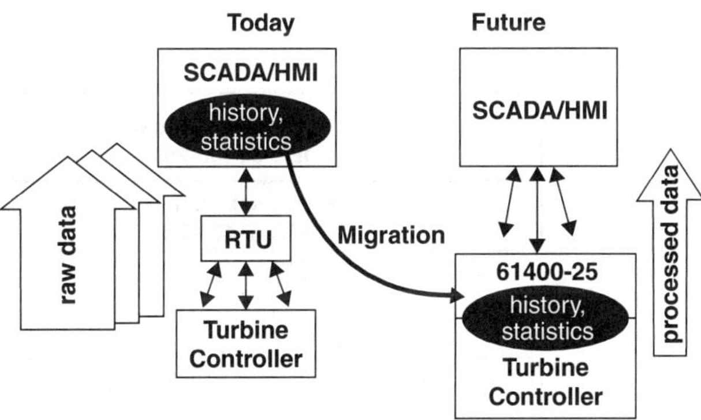

FIGURE 110.10 Migration of processing to IEDs.

nisms for information exchange and the mapping to communication protocols. In this regard, the standard defines all details required to connect wind power plant components in a multivendor environment and to exchange the information made available by a component. This is done by definitions given in this chapter or by reference to other commonly used standards.

The wind power plant-specific information models the crucial and common process data and metadata of a wind power plant. Process information is hierarchically structured and covers, for example, common process information found in the rotor, generator, converter, grid connection, and the like. The data may be simple (value, timestamp, and quality) or more comprehensive (adding more meta-data, e.g., engineering unit, scale, description, short hand reference, statistical, and historical information of the process value). All information of a wind power plant defined in this standard is name tagged — it defines a comprehensive name space. A concise meaning of each signal is given. The standardized wind power plant information can be easily extended by means of a name space extension rule.

All process and meta-data can be exchanged by corresponding services like get, set, publish–subscribe (report), logging, and control. Access to the meta-data (including configuration information with regard to the wind power plant information model and services and communication stacks) provides the so-called self-description of a device. The self-description could also be contained in an XML-based configuration file. The references include commonly applied standards like XML, ISO 9506 (MMS), SOAP, OPC XML-DA, IEC 60870-5-104, DNP3, and TCP/IP (see Figure 110.11).

This standard allows SCADA systems to communicate with wind turbine controllers from multiple vendors. The standardized self-description (contained either in an XML file or retrieved on-line from a device) can be used to configure SCADA applications. Standardization of SCADA applications is excluded in IEC 61400-25 but standardized common wind turbine information provides means for reuse of applications and operator screens for wind turbines from different vendors. From a utility perspective, unified definitions of common data minimize conversion and recalculation of data values for evaluation and comparison of all their wind power plants.

The standard can be applied to any wind power plant operation concept, that is, both in individual and integrated operations. The application area of IEC 61400-25 covers all components required for the operation of wind power plants, that is, not only the wind turbine but also the meteorological system (reference wind mast), the electrical system, and the wind power plant management system. The wind power plant-specific information in IEC 61400-25 excludes information associated with feeders and substations. Substation and feeder communication is covered by IEC 61850.

110.6 Implementation

An easy software architecture applicable for getting started is shown in Figure 110.12.

The DLL approach is an easy way of integration of the server into an existing software environment. The server "serves" one or more clients for real-time information exchange implemented in the IEC 61850 server.

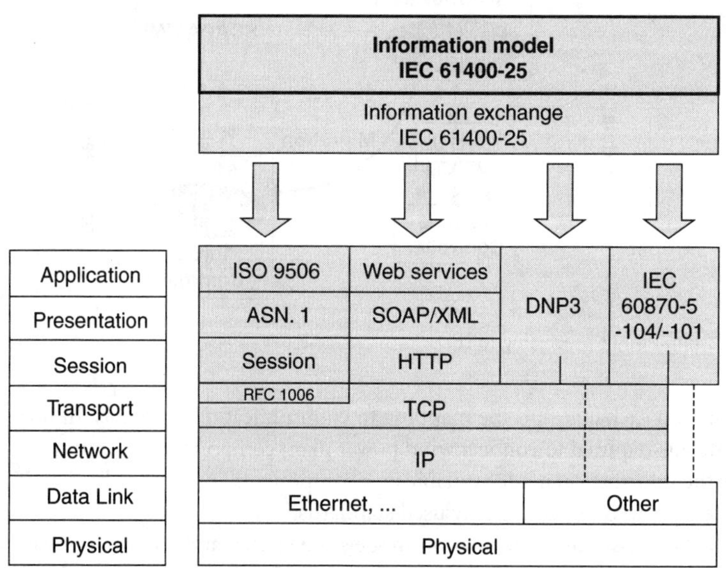

FIGURE 110.11 Communication profiles.

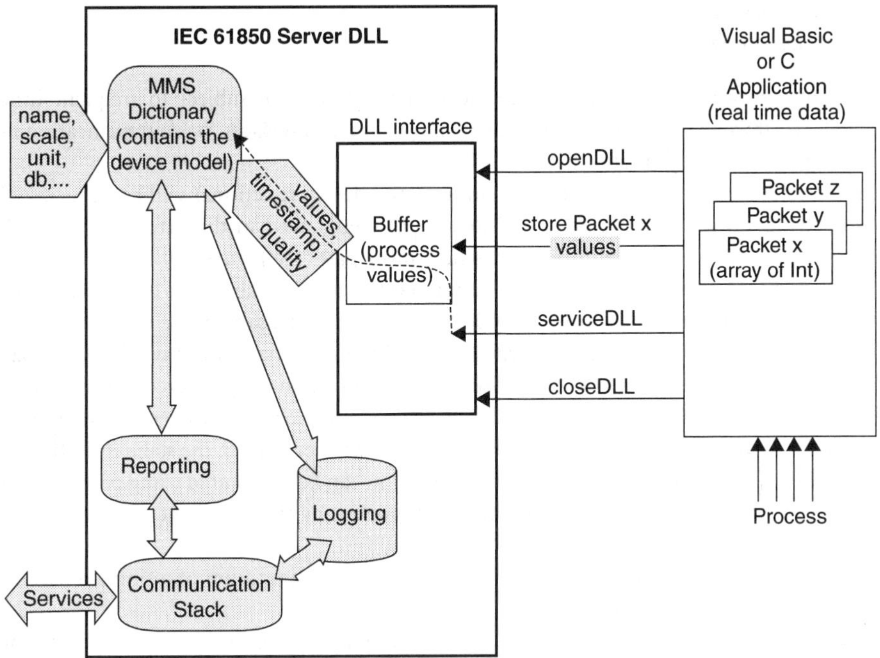

FIGURE 110.12 IEC 61850 DLL software.

The server provides an additional software for the HTTP access (exchanging HTML and XML coded pages) providing values in a non-real-time manner.

The interface between the Visual Basic application and the server DLL is defined by four local calls: openDLL, storePacket x, serviceDLL, and coseDLL. The application controls the DLL. The first action is to open the DLL (openDLL). The process data values are stored in the server applying the "storePacketx" calls. When all data values are store, the application calls the "serviceDLL" call. This call starts the processing of the server DLL.

The information model is stored in the DLL. All services like reporting (including the monitoring process if the data value has changed more than specified by the configuration attribute "db" since the last report) and logging are autonomously processed by the DLL.

This DLL approach is the easiest way to get started with IEC 61400-25 and IEC 61850. The interface between the communication software and the application may require an approach other than a DLL. The DLL runs only when it is called by the application. As a consequence of this DLL-typical behavior, the server DLL processes incoming and outgoing messages as often as the application calls the server DLL.

110.7 Reusability and device modeling

Describing device functionality by specifying the data (syntax and semantic) and the dynamic behavior (state machines) of devices (as seem from remote) is one of the fundamental challenges in the standardization. Many standardization groups have started defining different views of domain-specific device types. The views are, for example:

- Engineering (in the context of a plant),
- Commissioning,
- Configuration,
- Operation,
- Asset management,

- Maintenance, and
- Decommissioning.

Hardware and software, as well as communication networks, are subject to frequent innovation. Therefore, it is worthwhile to standardize independent (abstract) interfaces for communication networks and the access to the application objects.

The abstract objects (objects define the semantic of the device functions) will continuously be used (with minor changes only). The object definitions will be enhanced in the future to meet additional requirements, that is, reusing the definitions specified in the past (see Figure 110.13).

The most important objective of the device description is to define reusable parts to be used for specifying the data models and behavior of various types of industrial devices. Reusability has two aspects. First, reuse of a given functionality in many devices throughout an application domain (we may call this: horizontal reuse). Second, reuse of a given function in the definition of an enhanced or specialized function (we may call this vertical reuse). The reusability is a crucial factor in reducing the costs of the overall system design, engineering, operation, and maintenance. Support of reusability is the key issue in the standardization!

The reusable parts describe, for example, how a substation can be configured using part 6 (SCL). A diagram as shown in Figure 110.14 maps to an XML file representing the use of the classes defined in part IEC 61850-7-4.

The application of the SCL allows to describe the complete configuration of a single device or a substation. An excerpt is shown in Figure 110.15.

The real benefit of device modeling is the reuse of (common) definitions made in the past. This is our daily practice! We are using common terms at work (key board, laser printer, office, etc.) or at home (kitchen, chair, wheel chair, bath room, etc.). Just misunderstandings are the result if terms are not understood uniquely on both sides (sender and receiver). It is not only a matter to define something completely — it is more important to understand it uniquely. All technical specifications in the area of distributed systems have to follow distinct rules for defining, exchanging, and unique interpreting exchanged information.

Interpretation is quite easy if we can reuse common terms learned in the past. In our daily life, we reuse (instantiate) the term "laser printer" (more precisely, we reuse the class definition that is associated with term "laser printer") for a laser printer next to you "laser printer in room 23" or we may reuse the term for a special type of a laser printer: A4 laser printer ("A4 laser printer in room 23").

Distributed systems should operate in the way they have been told to do. If they do not? This may have many reasons. A major issue is that independently developed devices may follow the specification of their

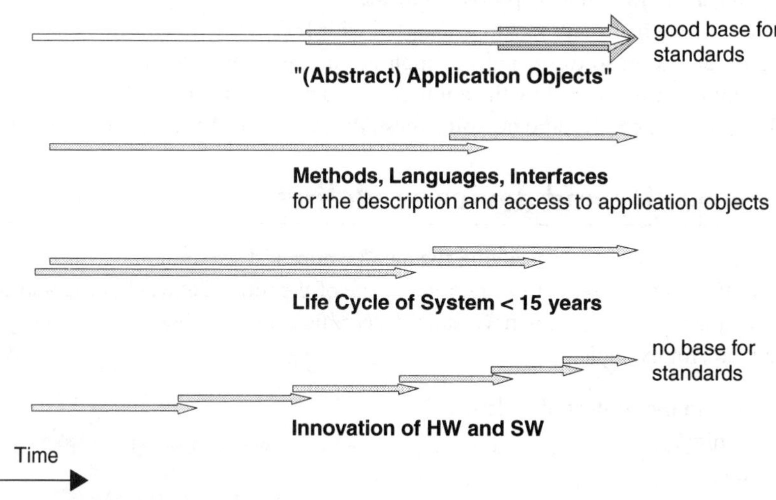

FIGURE 110.13   What is important to be standardized?

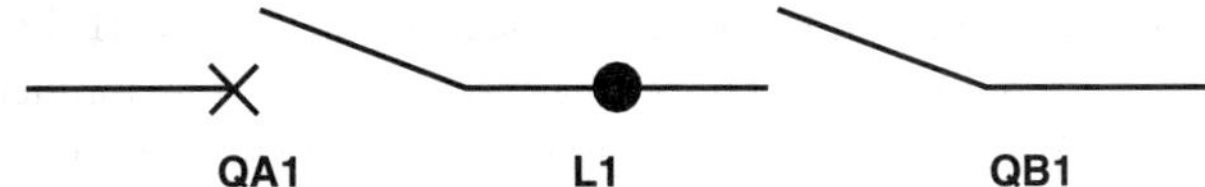

FIGURE 110.14   Application diagram.

```
<Substation Ref= " " >
   <VoltageLevel Ref= "E1" >
      <Bay Ref="Q1">
          <Device Ref= "QA1" Type= "CBR">
             <LNode Ref="1" LNClass= "XCBR" />
             <Connection TNodeRef= "L1" />
          </Device>
          <Device Ref= "QB1" Type= "DIS">
             <LNode Ref= "2" LNClass= "XSWI"/>
             <Connection TNodeRef= "L1" />
          </Device>
      </Bay>
   </VoltageLevel>
</Substation>
```

FIGURE 110.15   Application of SCL (excerpt).

implementers but the implementers may have different interpretations of the specification that describes the cooperation of the devices!

Devices will not operate in the way they should do, if the human beings (the implementers) do not understand each other!

Device models are collections of terms with associated semantics and a description of the dynamical behavior.

Usually, models are abstract in the sense that they do describe only those aspects that are visible to the remote user of a device. It is sufficient to know the external visible data and behavior of the device (the WHAT). The concrete realization of the device, its internal interfaces, and programming language or operating system (the HOW) are not of interest for the view from outside.

IEC 61400-25 reuses IEC 61850 instead of developing this new standard from scratch.

110.8   Resume

Deregulation will place greater demands for information on utilities than they have experienced before. IEC 61850, IEC 61400-25, and IEEE's UCA "UCA" is a Trademark of EPRI, Palo Alto, CA, USA provide a timely, cost-effective, and standardized solution to allow advanced IED functions and distributed systems to form the foundation for "next-generation" electric utility systems.

The benefactors of the results of open device data integration span the entire industry and include all of the stakeholders in this industry. With the standard IEC 61850, intelligent protection relays and other real-time devices are becoming more common. Utilities could take advantage of these new developments, and make the power systems safer than before — taking into account that all critical information (status and measurements) is available (at any time and anywhere) when making control decisions.

The customers are in a position to save large sums of money and time. The vendors who provide solutions that meet or exceed expectations will become very successful. This is an exciting time in the industry with an inexorable move toward practical software components.

The most important issues are the models of the real device data and the rules (service interface) how to access these data. On the other hand, it is obvious that an appropriate transport mechanism (communication profiles), for example, the TCP/IP or a point-to-point link, must be used to exchange the messages between devices.

By providing a common communications protocol stack, IEC 61850 and IEC 61400-25 support the "plug-and-play" equipment from different vendors. The specification of the uniquely tagged semantic of the most important device model data leads to a tremendous cost reduction during engineering, commissioning, operation, asset management, and maintenance. The solution provides plant- and enterprisewide seamless integration.

IEC 61850 is based mainly on UCA. The UCA 2.0 specification will be harmonized with the final IEC 61850 standard.

An evaluation CD ROM is available, which includes original software of several vendors and documentation helping to get started applying the IEC 61850, IEC 61400-25, UCA, and MMS approach (also including the IEEE TR 1550). For details, see:

www.Nettedautomation.com/solutions/uca/evalkit/index.html

A comprehensive free Demo Software and Tutorial for IEC 61850/IEC 61400-25/UCA™/MMS (with Web/XML support) executable on PCs (Win 95, 98, NT, 2000, XP) could be downloaded from the following URL:

http://www.nettedautomation.com/solutions/demo/20020114/index.html

References

1. Christoph Brunner (ABB) and Holger Schubert, (SIEMENS), The ABB – SIEMENS IEC 61850 interoperability projects, January 2002, http://www.nettedautomation.com/solutions/uca/products/9-1/index.html
2. IEEE Technical Report 1550, Utility Communications Architecture, UCA, 1999, http://www.nettedautomation.com/standardization/IEEE_SCC36_UCA
3. IEC 60870-6-TASE.2:2003, Telecontrol application service element 2.
4. Standards and committee drafts IEC 61850, Communication networks and systems in substations, http://www.scc-online.de/std/61850.
5. Working Draft IEC 61400-25, Communications for monitoring and control of wind power plants, 2001, http://www.scc-online.de/std/61400.
6. Becker, Gerhard, Gärtner, W., Kimpel, T., Link, V., März, W., Schmitz, W., and Schwarz, K., Open Communication Platforms for Telecontrol Applications — Benefits from the New Standard IEC 60870-6 TASE.2 (ICCP), etz-Report 32, VDE-Verlag Berlin, 1999, www.Nettedautomation.com/standardization/IEC_TC57/WG07/etz_report.html.
7. Comparison of IEC 60870-5-101 (-103, 104), DNP3, IEC 60870-6-TASE.2 with the new standard IEC 61850, http://www.nettedautomation.com/news/n_51.html.
8. http://www.nettedautomation.com/news/n_66.html.
9. Smart power systems thanks to IEC 61850, http://www.iec.ch/online_news/etech/arch_2003/etech_0603/prodserv.htm#iec61850
10. IEC 61850-5:2003, Communication Networks and Systems in Substations, Part 5: Communication Requirements for Functions and Device Models.
11. IEC 61850-7-3:2003, Communication Networks and Systems in Substations, Part 7-2: Basic communication structure for substations and feeder equipment — Abstract communication service interface (ACSI).
12. IEC 61850-7-3, Communication Networks and Systems in Substations, Part 7-3: Basic communication structure for substations and feeder equipment — Common data classes.
13. IEC 61850-7-4, Communication Networks and Systems in Substations, Part 7-4: Basic communication structure for substations and feeder equipment — Compatible logical node classes and data classes.
14. Object Models for Power Quality Monitoring in UCA2.0 and IEC 61850, A. Apostolov, DistribuTech 2003, Las Vegas, NV, February 4–6, 2003.
15. Object Modeling of Metering Functions in IEC 61850 Based IEDs, A. Apostolov, Distributed Generation and Advanced Metering Conference, Clemson, SC, March 12–14, 2003.
16. Andersson, L., Brunner, Ch., and Engler, F., Substation Automation based on IEC 61850 with new process-close Technologies, IEEE Powertech 2003, Bologna, Italy, June 23–26, 2003.

17. Elforsk Rapport Number 2:14, Wind Power Communication — Verification report and recommendation, Stockholm, April 2002.
18. Elforsk Rapport Number 2:16, Wind Power Communication — Design & Implementation of Test Environment for IEC61850/UCA2, Stockholm, April 2002.

Useful web pages that provide additional information:
http://www.livedata.com
http://www.sisconet.com
http://www.tamarack.com
http://scc-online.de
http://nettedautomation.com

111

The JEVIS Service Platform — Distributed Energy Data Acquisition and Management

Peter Palensky

Vienna University of Technology

111.1 Introduction

The deregulation and liberalization of large parts of the European energy market had a number of essential implications. Besides the wanted and expected consequences like increased competition — and therefore decreased prices — the energy customers became more and more aware of energy as being a part of their daily financial balance. Having the opportunity to choose between different suppliers that offer in deed different "products" resulted in the need to know what and how much energy is consumed where and by whom.

The types of customer that is currently taking advantage of the liberalized energy market are typically small and medium enterprises (SMEs), multi-site customers (e.g., supermarket chains), and medium industrial customers. Large customers and large industrial sites like metal or glass industry are — as a part of their customer retention efforts — looked after by the respective electric utilities' key account managers anyway, while the medium customers do not enjoy such attention.

These customers might change the supplying utility company but have no basis for this decision. It is generally not possible to estimate if tariff A is more suitable for some customer than tariff B unless you know the exact consumption behavior.

This demand for information was the starting point of the JEVis project (Java Envidatec Visualization). The Institute of Computer Technology (ICT) at the Vienna University of Technology (Austria) and Envidatec GmbH, a service provider in the domain of energy in Hamburg (Germany) created a concept of a new and modern data acquisition system that is capable of processing millions of new records per day and that offers high availability and scalability called the JEVis system (see also www.my-jevis.com).

Customers that join (i.e., that are connected to) this JEVis system can browse through and analyze the consumption of every process within their production site or shopping mall. This transparency enables them to optimize their business and to choose the optimal energy tariff. Some of the customers are subsequently equipped with an automatic load scheduling system, which helps to avoid expensive consumption peaks and can directly determine their savings via the JEVis system.

Modern energy bills for SMEs are based on the so-called load chart, a consumption chart with 96 samples per day (15-min consumption values). A typical tariff in Mid-Europe is that not only the consumed energy (counted by an energy counter, in kWh) but also the three largest 15-min consumption values (the consumption "peaks") are taken into account. Additionally, the power peaks constitute quite a large portion of the bill; thus, avoiding or decreasing these peaks can significantly reduce the energy bill. The JEVis system stores the energy consumption charts and offers the recorded data via a web-portal.

The next step after having a detailed load chart of the consumption is to find out what processes or what parts of the customer installation are responsible for a particular power peak or the load shape in general. As already mentioned, sometimes it is even possible by simply looking at the chart while knowing the customer's processes. Usually, it is, however, not that easy. Large buildings and industrial sites with a large number of electrical consumers sometimes cannot be overlooked in terms of power consumption. It is thus necessary to apply additional sensors, besides the existing energy meter, to increase the granularity of the information.

The JEVis system is, for instance, used for "benchmarking" (e.g., comparing) multiple branches of one business to identify the most efficient and the least efficient ones. This massively depends on how much information you "get out" of the building or the process and into your calculation and optimization algorithms. The sources of this information are a large number of sensors and actuators that are queried by a database that is capable of storing these large amounts of data. The main challenges in the JEVis project are the highly distributed nature of the customer installations. Long-range communication is entirely based on the Internet or other IP-based communication infrastructure like virtual private networks (VPNs). Topics like reliability and security are vital for the JEVis system, as well as how IP can be used for flat peer-to-peer networking as it is needed for "global" automation applications.

111.2 JEVis Architecture

The customer-side part of the JEVis system interconnects all necessary sensors and actuators at the customer's site via a local control network (also known as field area network or "fieldbus" [1]) that can be accessed via the Internet. See later in this chapter for a more detailed description of the respective communication infrastructure.

Having this infrastructure, the JEVis system can not only record and analyze energy consumption data but virtually any kind of measurement data. The usage of machinery, temperatures, the number of persons entering a supermarket, the status of the air conditioning equipment, and many other things are fed into the JEVis database for further processing. The customer can then not only see the load chart but also its correlation to other measurement charts and thus identify those parts of his facilities that are responsible for power peaks. The combination of process data and consumption data enables the customer to evaluate the performance of different branches of his enterprise and to optimize his overall business. The JEVis system offers an all-in-one solution to acquire, process, and use these data.

The IGUANA project (1999–2001, [2]) aimed toward generating the customer-side infrastructure for sensor networks and data acquisition. IGUANA is the direct predecessor to the JEVis project and resulted in a flexible Internet/fieldbus gateway that meanwhile runs on various different hardware platforms. The JEVis system uses the IGUANA gateway software on an embedded rail-mounted industrial PC called "VIDA." The key properties of this VIDA are

- Intel-architecture industrial PC
- Linux operating system with web-server, SSL (secure socket layer), SSH (secure shell), etc.

- Robust transaction-based nonvolatile flash memory, no moving parts
- Digital input channels for energy meters (electricity, gas, water, etc.)
- Digital outputs channels for relays
- LonWorks control network interface for additional sensors and actuators
- Ethernet or analog, ISDN (integrated services digital network) or GSM (global system for mobile communication) modem for IP-connectivity.

The VIDA hardware is designed for robustness and usage in a rough and EMC-critical (electromagnetic compatibility) environment. Therefore, all inputs and outputs (I/Os) are galvanically insulated. It has no moving or maintainable parts and can be installed in closed switchboards. When powered up, the IGUANA gateway software scans the attached LonWorks network for sensors and actuators. After the discovery procedure, each node that was found is registered in an on-board database and can thereafter be used and operated by the VIDA software.

Being originally intended to serve as a remote configuration node for the Envidatec Load Management System, it turned out that the customers were much more interested in an "auxiliary" feature of the VIDA: data logging. Meanwhile, providing historical measurement values and generating alarms depending on the system state and some rules is the majority of VIDA's usage.

The VIDA was initially used as an autonomous system, playing the role of an "agent" in terms of network management (= a server that simply responds to queries and simultaneously an alarm notificator that can initiate messages as well), using its on-board http-server and other server- and alarm-features. Now, the features and services of the VIDA are massively "grafted" and extended by including it in the JEVis system.

An alarm message is not only an E-Mail that is sent out by the VIDA anymore, it is part of a whole alarm management concept: the JEVis system takes alarm messages from VIDAs and creates FAX messages, GSM SMS (short message service), telephone calls, and other ways to reach the intended recipient of the alarm message. If the receiving of the alarm is not confirmed within a defined time, further alarm messages are sent to other responsible persons until the problem is solved. Managing all this is too much for embedded nodes like the VIDA, but is no big deal for sophisticated platforms like the JEVis system.

These extended services were the driving factors for JEVis. The result is a flexible platform that can collect, calculate, distribute, and manage data that are typically acquired via some measurement system. The increased granularity of knowledge about a building or an industrial process makes it more easy to identify problems and to optimize the processes. Devices and facilities that were previously controlled and run independently can now be analyzed and viewed by using a global and correlating view. The shared energy supply always "links" the individual processes of a plant when it comes to energy consumption, although they are not controlled collectively. JEVis now enables the customer to obtain an overview and to identify and overcome logistical and organizational problems.

A typical situation that can be found at customer sites is sketched in Figure 111.1 A number of energy-consuming customer processes (manufacturing lanes, compressed air production, etc.) are controlled individually but supplied collectively. Since the behavior of the individual processes is not coordinated, unnecessary and unwanted consumption peaks may very likely occur.

The customer-side part of the JEVis system interconnects the individual devices and their controlling processes with a control network and schedules them with an energy management algorithm. Such energy management systems (EMS) coordinate processes in a more or less sophisticated way. Sometimes, like in the case of the Envidatec EMS, even a short-time prognosis of the individual consumers is used to optimize the consumption shape.

Interprocess coordination is, however, just one side of the medal. Even within the processes consumption, coordination is usually missing. Equipping the consuming appliances with the respective JEVis sensors that measure consumption and operation enables the customers to find out what parts of his system are responsible for consumption peaks. The detailed knowledge about the processes is then the basis for the configuration of the energy management.

Some of the JEVis customers are not consumers of electrical energy but rather producers. A typical example is a wind power station, which is usually not allowed to feed energy into the grid whenever possible. The respective policy of energy supply and production determines how much a particular wind power station might produce, depending on the time, the situation on the grid, and so forth. The optimization process here is no consumption optimization and load shaping but production and injection optimization. The algorithms are similar and the hardware equipment is almost identical. A control network measures the power production, the wind, etc. and feeds energy into the grid or not. Having logs about the production is essential, since produced energy is a valuable thing, in addition to the consumed energy.

The JEVis service provider side consists basically of a database that stores and processes the collected measurement data. The results are subsequently offered as Internet-content via web-interfaces to the customer. Hence, the JEVis system consists of three parts (Figure 111.2).

- The customer-side installation: the customer processes, equipped with an Internet-gateway (the VIDA, for instance) and probably some control network connected to this VIDA for additional sensors and actuators.
- The JEVis server, an Internet-able database.
- The customer as a client to the JEVis server, using a standard web-browser.

The VIDA gateway is, as already mentioned, an embedded rail-mounted personal computer, which is in charge of offering a remote control channel to a local control network, of evaluating alarm conditions, and of measuring and logging data like water consumption, temperatures, and so forth.

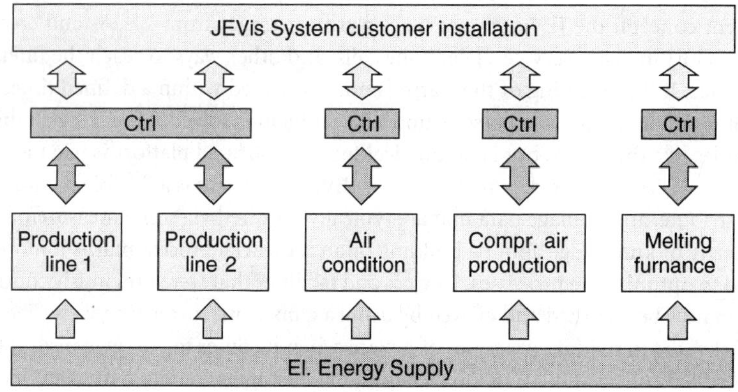

FIGURE 111.1   Separate processes linked via the JEVis system.

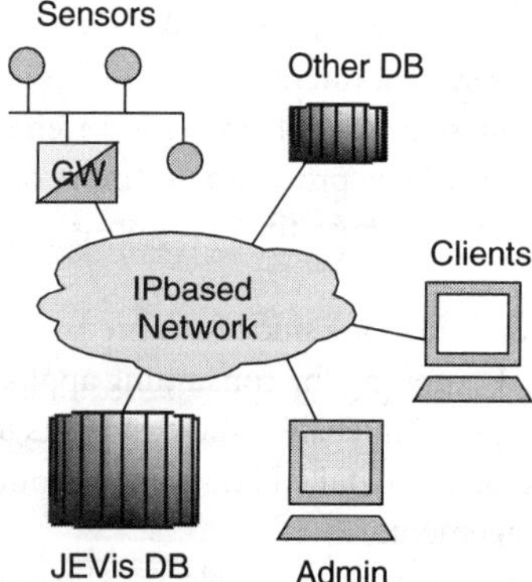

FIGURE 111.2   The JEVis database with its sinks and sources of data.

The main innovation of the VIDA is its network capabilities. Equipped with a flexible Linux-based set of software and some modem(s), it is able to

- pick up incoming calls and establish IP connectivity,
- to dial into a number of ISPs on-demand,
- to be permanently connected to an IP network like digital subscriber line (DSL), local area network (LAN), etc., and
- to be "woken up" by a telephone signal (e.g., a "ring") and to call back for IP connectivity.

These flexible networking options were the key to success. The VIDA can be installed virtually everywhere, in corporate networks, "in the field," etc. The customer's sites are subsequently equipped with VIDAs and the respective sensors. If only typical consumption data like electrical energy or heat consumption are needed, there is no need for a local control network, if temperatures, air pressure, and other values are needed as well, external sensor nodes must be attached. The protocols of the VIDA were carefully selected to safely pass firewalls and proxies while offering strong data security. Since conventional proxies often only allow http (hypertext transfer protocol) traffic, the gateway protocols must be embedded in http. The security-relevant layers must therefore be either embedded in http as well or be based on widely accepted standard protocols like secure socket layer (SSL). The JEVis system does both: it has internal authentication and wraps the traffic in SSL, when public networks like the Internet are used for connectivity. Key distribution for the cryptographic parts of the security layers is based on electronic chip cards, which is beyond the scope of this chapter. See [3] for a discussion on security and fieldbus/Internet connectivity.

One of the main challenges of the system was availability and reliability. The domain of energy supply is dominated by the high quality of energy supply in the last decades. Electrical energy was always available and had a constant quality. This is what the customers in this domain expect from anything that has to do with energy. When it comes to electronics, communication systems, servers, and the Internet, we find a totally different level of quality. These systems are, due to their complexity, much more instable and endangered by hardware and software problems than the proved components of energy supply. Therefore, all software parts of the JEVis system are monitored by watchdog-like programs for proper functionality and valid states. Two communicating software components (a client and a server), for instance, mutually monitor each other for timeouts, stability, correct behavior, valid status, and other things. Only these measures made it possible to offer the required availability. Figure 111.3 depicts two replicated instances of the JEVis system that exchange data for replication purposes and additionally monitor the system behavior of the peer besides two processes inside a system that have a client/server relation.

The monitoring occurs in a variety of ways. Typically, customer queries are issued (emulated) and the result is analyzed. This guarantees that at least the customer requests are processed correctly. Additionally, there are system-internal channels for supervision, which provide a deeper insight into the current state of the observed system. Key aspects like memory usage, system load, and other operating system-level attributes can be transported via these channels as well as state information of the individual processes of the systems. Figure 111.3 shows two processes on one and the same or even (or say "even better") on individual machines of the JEVis network. The processes are usually part of a client/server relation. Operational data are transmitted via these relations as well as monitoring data. In this way, one process A can be a watchdog for another process B. Since all parts of the JEVis system use or can use secured interprocess communication channels, the individual processes might be located on geographically

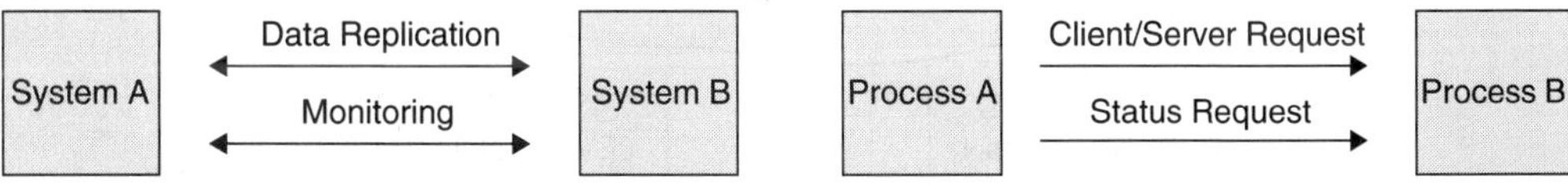

FIGURE 111.3 Systems and processes monitor each other.

distributed servers. This fundamental design decision makes it possible to install the system in a distributed and fault-tolerant way. One important rule of mutual supervision is that the final decision chain of "who resets who" (in the case of a detected faulty state) must be strictly hierarchical in order to prevent loops and meshes in the decision process. The propagation of rollback or reset actions must be absolutely deterministic. The internal client/server relations for the status requests and reset commands of all components must therefore follow a strict hierarchy.

111.3 Distributed Data Acquisition, Storage, and Access

Data are acquired via a number of different channels since the gateways can establish outgoing connections as well as incoming ones as already mentioned. If, for instance, one ISP is not reachable for a gateway, it tries to contact other ones and probably even via another channel (a GSM modem as a backup solution for the ordinary ISDN connection).

Let us take the incoming modem connection ("pick up" by the gateway) as an example. A connectivity component of the JEVis database calls the gateways via modems and transfers data via the subsequently established IP point-to-point connection. The JEVis network in Figure 111.4 has two replicated instances of its database and all its components (JEVis System 1 and JEVis System 2).

It serves two customer sites (customer A and customer B); one of the services might be to read out the data loggers of the gateways once a day. Let us suppose that the customer's sites are in different countries (customer A and JEVis System 1 in the same country as well as customer B and JEVis System 2) and the connectivity costs money, depending on the distance (public telephone network). The JEVis system therefore chooses the instance with the lowest on-line costs for each particular data retrieval job so that customer A will be queried by JEVis system 1 while customer B is read out by JEVis system 2.

The databases in the JEVis system instances synchronize with and replicate each other — some parts permanently, some parts at defined times or system states. The problems arise when parts of the JEVis instances are considered to fail. If one session that retrieves some thousands of data samples breaks because of hardware reasons or something similar, another data retrieval component has to finish the job. Thus, it might happen that JEVis system 2 continues a job that was initially started by JEVis system 1. System 2 therefore checks the status of the retrieval queue of system 1 and vice versa. If a job hangs or fails, the partner system tries to continue it. A transaction-oriented retrieval, storage, and replication policy ensures that the databases remain consistent.

The database has a number of storages and registries that are beyond the scope of this chapter, but two important data storages shall be explained as examples. The first one is the "samples storage." Databases for SCADA (supervisory control and data acquisition) like the JEVis system have to store "engineering values," that is, retrieved measurement values that are slightly more than plain numbers (See the ISO 16484 standard for an impression of the variety of datapoints that can be found in a modern building). A "sample" is an "extended" physical engineering value with a timestamp. The "extended" means that a single sample might

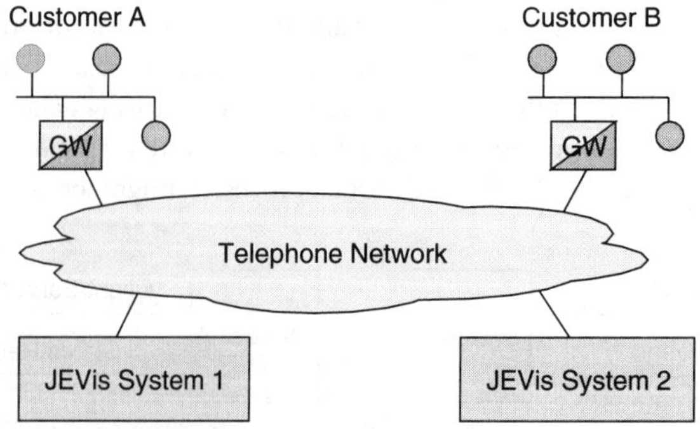

FIGURE 111.4 Geographically distributed customers and JEVis systems.

be a measurement value, a setpoint, a digital photo image, the status of a machine, a statistical, or organizational value typed in by some user and so on. Such samples typically consist of a set of attributes like

- relative accuracy of the value (in percent),
- absolute accuracy of the value (in the respective unit),
- physical unit in SI units,
- absolute accuracy of the timestamp (in sec),
- an absolute validity (from timestamp A to timestamp B),
- a relative validity: the interval in seconds around the timestamp (e.g., $[0, 900[$ or $]-450, 450]$),
- the value itself, etc.

All JEVis services use these samples. Statistics are calculated, alarm conditions are evaluated, prognoses are estimated, etc., and all of this is stored in the samples storage. All entries in the samples storage provide additional information about the retrieval procedure so that it is possible to determine how a sample got into the sample storage. Samples are never deleted; they can only be updated by a newer sample with a more recent timestamp.

Data samples can be assigned to data rows like measurement rows (samples from one data source) or other rows like consumption samples from holidays or the temperatures of the 15th of March of the last 10 years.

Another important registry in the JEVis database is the "organizational hierarchy registry." Customers might see their sites and branches in different views. One example is that a supermarket chain sorts its branches by

- geographical location or
- financial clusters,

since a troubleshooting team for the freezers needs the first "view" of the sites, while the accounting department would use the latter one. Although both views finally contain the same sites, the views are different. Users that authenticate themselves to the JEVis portal can choose their preferred views and work with them.

The database replicates its tables and data structures depending on their contents and their role in the system. Configuration data like "data retrieval schedules," for instance, are replicated synchronously and immediately. Data storages are replicated on demand and asynchronously since they are unique in the system anyway. If two samples represent a value from the same data source, logged at the same time, a deterministic collision and conflict resolution algorithm cleans the samples storage.

Data are put into and read out of the database via components that tightly attach themselves to the database via standard interfaces like ODBC or JDBC (open database connectivity, Java database connectivity). These components typically reside on the same system and machine as the database itself. These components either have a client- or a server-interface to the rest of the system or to the outside. Data input components (DIs) and data output components (DOs) furthermore might be of an interactive nature or of a batch nature. The first one usually interfaces to humans, which results in the requirement of low latency and quick responses.

Figure 111.5 shows typical data input components and data output components. The HTTP Server DO, for instance, would be an interactive DO because it is typically queried by human users using a web browser. The SMTP Client DO is noninteractive since the simple mail transfer protocol (SMTP) does not define any qualities of service regarding interactivity.

Server components are always queried by some client while client components might contact other servers. The server DIs and DOs are therefore the front-ends of the system, while the clients are back-ends that retrieve data or send data somewhere.

The simple object access protocol (SOAP) client DI in Figure 111.5 is a classical DI. It queries JEVis gateways for measurement data and other information from the customer site via SOAP. This client DI is used by the database and abstracts or encapsulates all gateway-relevant technology like special protocols, authentication mechanisms, datatype abstraction, and so forth. The HTTP Server DO is, as already

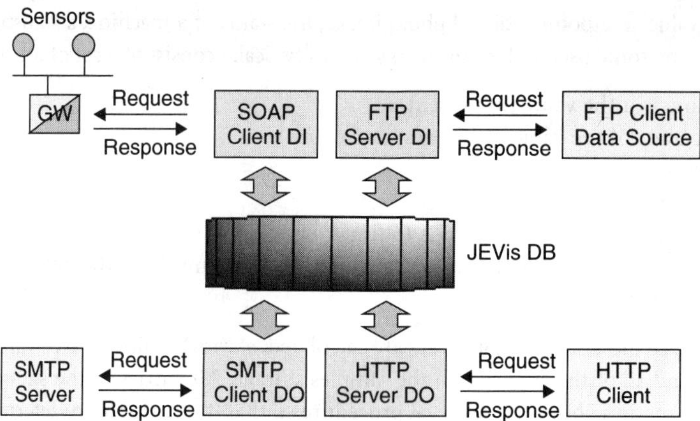

FIGURE 111.5   Paths of data into and out of the database.

mentioned, an interactive DO. Usability, response time, user interfaces (typically graphical user interfaces, GUIs), and other aspects are the main concerns for developing these components, since they are the direct "face" to the customer that wants to be satisfied. Interactive DIs and DOs are therefore typically web-enabled forms and GUIs, based on Java servlet/applet technology.

The distributed nature of the system requires some considerations to protect data and other aspects of security. The data replication channels, for instance, should be protected from unauthorized access or manipulation. Therefore, JEVis implements the following methods:

- secure data transport between all communication entities and components of the system,
- secure data storage,
- powerful and flexible access right methods,
- usage of smart card security, and
- secure customer front-ends.

See [3] for further details on data security for control networks and smart card technology.

111.4   Global Energy Management

Considering influence on the consumption behavior is generally called energy management or more precisely demand side management (DSM). DSM usually consists of three parts, namely

- retrofitting (changing insulation, devices, etc.),
- logistical optimization, and
- load scheduling.

In our context, energy management means load scheduling, the active influence on the operation of equipment. The customer-side part of the JEVis system contains an energy management subsystem that actively schedules electrical consumers. The rules for this scheduling are based on the specific energy tariff. After having analyzed the customer processes with the JEVis system, and after having identified the reasons for power peaks and other unwanted behavior of the system, it is easy to configure and tune the energy management system in order to avoid this behavior.

Global energy management (GEM) goes one step further. The deregulated energy market in Europe moves toward a direction where customers are allowed to group their geographically distributed sites ("multi site customers") onto one energy bill. The overall consumption load is then used as a basis for the energy bill. The energy meters might record the individual load charts, which are then summed up, but the individual local energy management systems usually do not cooperate globally.

The problem of "global" energy management is basically the same as local energy management: a number of resource-consuming entities are to be organized and scheduled in order to impose some overall behavior. The difference to local energy management lies in the communication channels. The problems that a GEM application faces in the world of the Internet are perfectly transferable to other "global" automation problems. The weak quality of services, the limited ways of establishing network connectivity, and the restricted transport of information in the Internet are problems that any distributed automation application faces so that global energy management can be seen as an example.

Local EM interconnects the energy consumers or say the controllers of the energy consumers via some kind of local network — a control network or a LAN. The availability of these network resources is generally very high and not connected to any "on-line costs." The local EM peers can count on the fact that the connectivity is a permanent one, and every member of the EM system can reach any other member whenever needed.

GEM faces a different network infrastructure. Long-distance connectivity is costly; the nodes might be connected to the Internet via GSM channels, telephone lines, or other ways of "going on-line." The Internet is virtually the only affordable network that offers long-range peer-to-peer connectivity at reasonable prices, but it still costs. Therefore, the members of the GEM system will not stay on-line 24 h a day but only at certain times or on demand. Not only the frequency of data exchange but also the amount of exchanged data is relevant since it directly influences communication costs.

As a consequence, the JEVis GEM system estimates the future behavior of its members and uses this prognosis to calculate the schedule in an as-off-line-as-possible way. The algorithms of this optimization process will not be discussed here but another important aspect of the system: the need for flexible communication. The GEM system consists of a number of nodes that represent an energy-consuming or-producing entity like

- a wind power station,
- a private home with a fuel cell system,
- a bakery,
- an industrial site, or
- a supermarket with a number of freezers.

Some of these nodes might "store virtual energy" like the freezers or the bakery; others can produce energy like the fuel cell and the wind power stations, and the majority simply consumes energy. Thus, the system consists of a geographically distributed number of energy consumers, storages, and producers with restricting rules of consumption, storage, and production. A wind power station, for instance, cannot be influenced in order to produce more energy and a bakery may not be switched off anytime but only under certain circumstances. Thus, the optimization process tries to take advantage of the degrees of freedom that the system offers in order to find an acceptable schedule for all members. Finding the right schedule that satisfies the "global goal" (some overall load chart, some special consumption peak avoidance in a region, etc.) strongly depends on the amount of information that the algorithm has about the situation. The degrees of freedom of every member sum up to a very complex optimization problem, which results in large amounts of data to be exchanged. Independent of the type of algorithm (negotiation-based, search for solution, etc.), the members actually need "flat" peer-to-peer connectivity in order to exchange data like tables, statistics, states, commands, etc.

The Internet usually offers peer-to-peer IP networking once the nodes are on-line. Unfortunately, depending on the Internet service provider, some of the dial-up nodes are hidden behind an IP proxy so that they cannot offer any services (open and "listening" IP) to the outside network. In this case, methods used by Internet Relay Chat (IRC) must be applied, where a publicly available server routes the requests of the GEM members that appear as IP clients. Figure 111.6, for instance, shows a wind power station and a gateway GW1 that want to communicate. Since the wind power station is behind a proxy, they use the JEVis relay to communicate instead of using native IP peer-to-peer communication. The same would occur if a GW goes on-line via standard GPRS (general-purpose radio switching, a packet-oriented GSM service), since a GPRS node gets an "internal" IP address behind a masquerading firewall

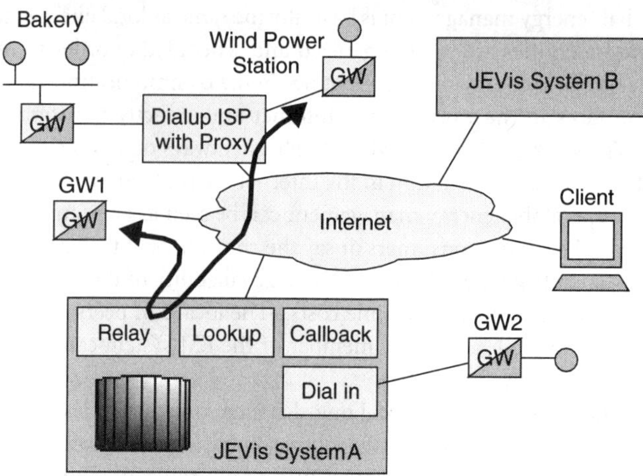

FIGURE 111.6   Connectivity services of the JEVis system.

and is not accessible from outside. Such firewalls, proxies, and other phenomena on the Internet restrict the communication massively and therefore the system must use technologies that are widely accepted and therefore supported and not inhibited by the Internet service providers. Additionally, the nodes might be organized in clusters and groups where not every node knows who and where its peers are. As a consequence, there must be some lookup tables where nodes can register in order to be found by other nodes.

The JEVis system therefore does not only host a database but also a number of other useful services that make a global automation application like global energy management possible:

- managing new dial-up connections (call back) on behalf of other nodes,
- dial-in service,
- lookup service for nodes to find peers, and
- communication relay for nodes with restricted transport (behind proxies, firewalls, etc.).

Another problem is the quality of services (QoS) like latency, availability, real-time aspects, and so forth. The IP protocol usually does not offer any guarantees. If the algorithm relies on the assumption that an offer (as a part of a negotiation procedure) is transmitted for sure within 5 sec, the algorithm is a bad choice for an Internet-based solution. Additionally, the Internet protocol itself is not the only unreliable part of the system. The lower-layer transports like dial-up telephone lines or GSM networks have a nondeterministic availability. It is not guaranteed that the node gets a free telephone line when it is needed. Therefore, the algorithm must be designed in such a way that it copes with unavailable information and unreliable and nondeterministic communication. The nodes must have as much "intelligence" as necessary to work fine even when communication is not possible for hours or days. This "local intelligence" results in a very robust and fault-tolerant system, which is one of the typical goals of distributed systems. The drawback of such a distributed architecture is that distributed network management is much more complicated. Therefore, the central JEVis system must be the location for network management-related things like node-configuration, diagnosis, and the like.

The distributed and robust nature of the GEM network would be destroyed if the JEVis system would not be redundant and distributed as well. If only one central server offers the communication relay and the other services, this server represents a single point of failure. The replicated and distributed nature of the JEVis system is therefore a good contribution to the fault tolerance of the GEM system.

111.5 Outlook

The JEVis system is, although very mature and massively in use, still subject to intensive research and development. The current research and development activities are system availability, alternative database systems, and, as already described, the infrastructure for GEM, which is probably the biggest challenge.

The availability of the overall system is still more limited than necessary due to insufficiencies of the Internet technology itself. By using dynamic host resolving, automatic query relays, and other dynamic methods, we hope to further increase the fault tolerance and availability of the system.

The JEVis database is currently based on an Oracle 9i® system, and proved stability and high performance. The philosophy of the JEVis services, however, would actually suggest an object-oriented design and not a relational database. Object-oriented databases (OODBs) are expected to be more flexible when it comes to different types of data that have to be stored. As an example, the configuration of all equipment used in the JEVis system (IP gateways, sensors, fieldbus nodes, etc.) is stored in and maintained by the JEVis database. A new type of device can result in a change of the data structures because it demands new, specialized data fields to be stored. An OODB has more flexibility in that aspect and can abstract devices and encapsulate data structures.

Another interesting branch of database technology is active databases (ADBMS) [4]. Active databases and especially active real-time databases are preferably used for job shop scheduling, work flow management, and production processes. Reactive behavior, triggered actions, and the action scheduling features of ADBMS are actually ideal for the JEVis services — the JEVis system also reacts to various changes in its environment. However, Oracle triggers are currently sufficient for the reactivity of the JEVis system.

Extensions like new services for customers, new devices, and components like alternative gateways and protocols are further fields of development.

References

[1] Loy, D., Dietrich, D., and Schweinzer, H.-J., Eds., *Open Control Networks, LonWorks/EIA 709 Technology*, Kluwer Academic Publishers, Dordrecht, 2001.

[2] Lobachov, M. and P. Palensky, Bringing Energy-related Services to Reality, Proceedings of the International Conference on Energy Economics (IEWT01), Vienna, Austria, 2001.

[3] Sauter, T. and P. Palensky, Security Considerations for FAN-Internet Connections, 3rd IEEE Workshop on Factory Communication Systems, Barcelona, 2000.

[4] Buchmann, A.P., Architecture of Active Database Systems, in *Active Rules in Database Systems*, Paton, Norman, Ed., Springer-Verlag, New York, 1998.

Industrial IT-Based Network Management

Yauheni Veryha
ABB Corporate Research

Peter Bort
ABB Corporate Research

112.1 Introduction

Today, companies involved in power distribution and transmission require network control systems not only with advanced graphical user interface, like Windows look & feel, panning, zooming, etc., but also with advanced integration and migration opportunities. The integration opportunities cover the Supervisory Control and Data Acquisition (SCADA) application itself and other systems including commercial applications like, for example, Enterprise Asset Management and technical systems like Geographical Information System. All of them can be integrated with a reasonable effort using the Industrial IT Aspect Integrator Platform (AIP) [2,5]. Additionally, one will always have an option of migrating to other SCADA systems by simply replacing the SCADA servers without losing system integration work and developed Human–Machine Interface (HMI). This means that the complete system will not have to be discarded, as was the case earlier.

In today's scenario, SCADA users often carry out most of the data engineering and picture building work themselves using an efficient engineering concept for both data engineering and picture building. For example, it should be possible to copy and paste the engineering data of one station (e.g., complete electrical or gas substation), like in Windows scenario, with the possibility of later editing it to create a new station. This requirement can be simply reused from ABB's Industrial IT AIP by saving time on engineering and concentrating mainly on system configuration, customization, and optimization.

112.2 System Architecture

Industrial control applications for power distribution and transformation usually perform different functions such as HMI, data logging, advanced control, input/output operations, enterprise connectivity, etc. [1,3,4]. Personal computer (PC) systems running Windows-based operating systems, like Windows 2000 and Windows XP, are increasingly being used in industrial applications due to their open PC architecture, Windows look & feel, availability of common Office Tools, a wide variety of off-the-shelf software and hardware products, which provide a range of data acquisition, analysis, presentation, management tools, and relatively easy connection to various computer systems within a network. The practical question here is how to use all these applications effectively, namely, perform data exchange, navigation, substitution of components, and system maintenance in a fast and easy way [6, 7]. Another important aspect in power distribution and transformation applications is the possibility of migration from one SCADA system to another or the integration of more than one SCADA system behind one graphical user interface. The industrial IT concept presents a unique opportunity here, namely, the iterative migration of components on the Industrial IT basis. In Figure 112.1, we show one of the solutions, namely, the exemplary architecture of the network control system based on the migration of SCADA system to Industrial IT AIP.

In the system architecture (see Figure 112.1), process data are transferred between SCADA server and AIP server using process data cache with standard OPC (OLE for Process Control) interface inherited from the AIP that has its own OPC server and OPC client with advanced caching features. This provides a high system performance due to the fact that process data cache has a short access time for retrieving cached data. An intelligent supervisory manager, associated with the process data cache on the client machine, controls storage of the process data in the cache and selectively transfers process data from the server computer system to data cache of client node using data subscription mechanism.

The system components of a given SCADA system based on AIP can be defined for each particular implementation separately, depending on the given requirements. In Figure 112.1, various box types show that components from different systems have been easily integrated into the AIP. The conceptual scheme of network management system for power transformation and distribution applications includes

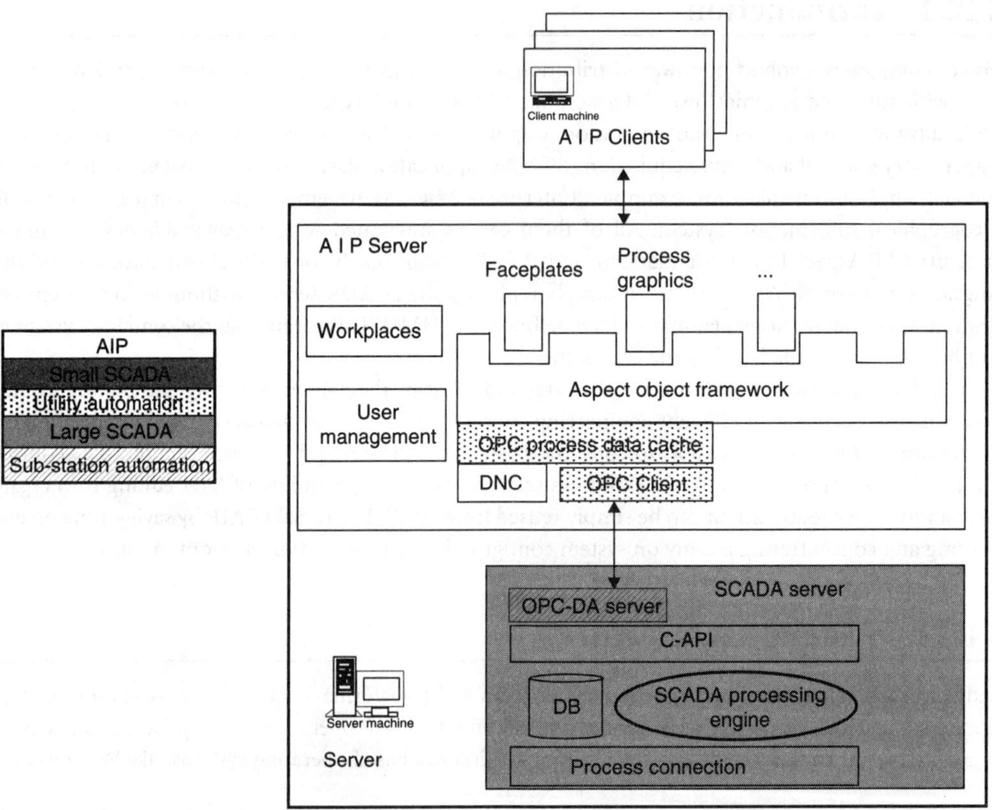

FIGURE 112.1 Exemplary architecture of network control system based on Industrial IT platform.

the following main exchangeable components and systems integrated on the Industrial IT basis (other components, like Alarm & Event Lists, Trend Displays, etc. are reused from the AIP):

- Process Graphics (WS 500, DataViews, GraphWorX, etc.);
- SCADA server;
- Client OPC Process Data Cache;
- OPC–DA server (Data Access) of SCADA server; and
- DNC (Dynamic Network Coloring application).

SCADA server application includes real-time database (responsible for the handling of a subset of the process variables related to data acquisition, alarm handling, and archiving), process connection module that interfaces controllers of I/O devices, processing module, and OPC server. The process connection module is used to poll the controllers of I/O devices at a user-defined polling rate. The polling rate may be different for different parameters. Time stamping of the process parameters is typically performed in the controllers and this time stamp is taken over by the OPC server that receives process data at the end to make them available for subscription to various OPC clients located on client machines.

In the given AIP-based implementation of the OPC, the OPC server exposes an OPC–DA interface that has functions to read and write variable values. These functions are marshaled across the network using stable CSLib (ABB's proprietary protocol) so that clients are able to call them. After an OPC item on the server has been subscribed to from the client, any data value change in the server application automatically updates the associated value in the client application. The CSLib-based implementation of the OPC has an advantage that DCOM (raising a number of stability problems) is not used for implementing OPC–DA.

The AIP-based process data cache contains a set of classes that implement all mandatory OPC DA interfaces, as shown in Figure 112.2. OPC Items contain process data in the form of the attributes that are listed in Table 112.1.

Supervisory manager of AIP-based process data cache offers features that allow customizing how items are cached and how long they are cached. It is possible to instruct supervisory manager to give certain items priority over other items when the supervisory manager performs removal of seldom used or unimportant items to free some memory.

The AIP-based system architecture provides a high flexibility to define customized SCADA solutions and data flows, in particular, on the common OPC basis with advanced integration and iterative migration opportunities. The components coming from different control systems can be quite easily integrated using Industrial IT Aspect Integrator Platform using standard OPC and COM interfaces inherited from the AIP.

112.3 Real Case

Generally, it is not very difficult to integrate modern, component-based applications supporting well-known integration standards like CORBA, COM, etc. [3, 6] and running on the same platforms at which the integration framework is targeted. However, integrating monolithic applications with poor or not clearly defined application programming interfaces is not only difficult but sometimes is even highly questionable due to the very high cost of integration work. Unfortunately, the latter kinds of legacy applications are often used in power distribution and transmission systems. In this case, some easy and flexible integration approaches are required.

The implementation of the integration concept based on ABB's Industrial IT AIP for power distribution and transformation companies was done for public municipalities in Germany by ABB Utility Automation. The conceptual scheme of the Industrial IT-based system for network management on the Industrial IT basis is shown in Figure 112.3. Some of the public municipalities in Germany requested new workstations with advanced Windows look & feel graphical user interface, advanced navigation

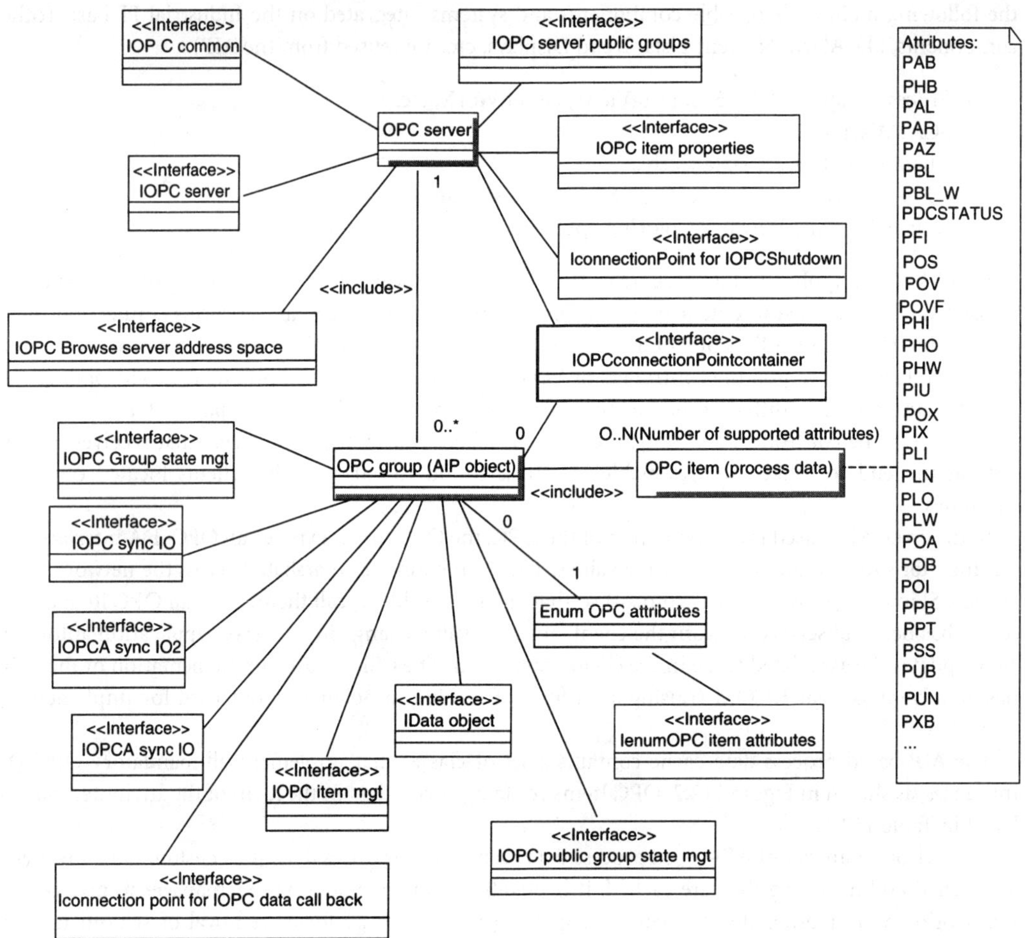

FIGURE 112.2   Class diagram for OPC implementation in AIP.

possibilities within system components (Trends, Alarm & Event Lists, Faceplates, Workplaces, and Geographical Information System), but with the use of its previous SCADA server and unchanged process connections. User management in the Industrial IT-based control system was implemented based on Windows accounts. This means that every authenticated user has a certain role within the control system and certain permissions and rights to available objects, aspects, and aspect systems in the AIP. Standard technologies, like ActiveX, COM, DCOM, and OPC [3], have been used to implement ABB's concept of Aspects and Objects.

To minimize development costs in integration projects, it is a good approach to concentrate on reuse and, thus, on Industrial IT with AIP.

112.4   Implementation Approaches

One of the typical examples of AIP use for network management applications is the new way of implementing Interlocking and Sequential Control functionality. The goal has been to design Interlocking and Sequential Control components in such a way that they will easily work with different SCADA servers. The purpose of the Interlocking component is to prevent prohibited commands, set point values, and manual entries of inappropriate data. Normally, for a two-state device, the operator may define interlock

TABLE 112.1   Selected Attributes of SCADA Process Object Presented as OPC Items in AIP

Attribute	Description	Possible Values	Data Type
AB	Alarm Blocking	0 = no blocking 1 = blocking	Long
AL	Alarm	0 = no alarm 1 = alarm is prevailing	Long
AR	Alarm Receipt	0 = alarm is not acknowledged 1 = alarm is acknowledged	Long
AZ	Alarm Zone	0 = normal state 1 = low alarm 2 = high alarm 3 = low warning 4 = high warning	Long
BL	Blocked	0 = no blocking 1 = blocking	Long
HB	History Blocking	0 = no blocking 1 = blocking	Long
IU	In Use	0 = not in use 1 = in use	Long
IX	Index	Integer	Long
LI	Lower Input	Float	Float
LN	Logical Name	Max 255 characters	BSTR
LO	Lower Output	Float	Float
LW	Lower Warning	Float	Float
OA	Object Address	Integer	Long
OB	Object Bit address	Integer	Long
OI	Object Identifier	Max 255 characters	BSTR
OV	Object Value	Integer or float (depending on Object Type)	Long or Float
…	…	…	…

conditions that must be valid to allow switching of a device. Control requests that do not meet these conditions will be rejected but can be bypassed in emergency and test situations. The Interlocking component allows the user to define an interlock condition, or a sequence of conditions, to be checked each time a control command is given or a new status of an indication is manually entered.

The Sequential Control (automatic switching order) function provides for the control of a number of devices by means of predefined sequences of control commands, which can also include safety checks, delays, etc. Sequential Control can be used for:

1. connection/disconnection of lines to bus-bars with a sequence of control commands for isolators and breakers in a bay,
2. reconfiguration of bus-bars,
3. multiple On/Off commands to breakers for load shedding and load restoration,
4. regulation of objects, and
5. supplying the operators with work instructions, etc.

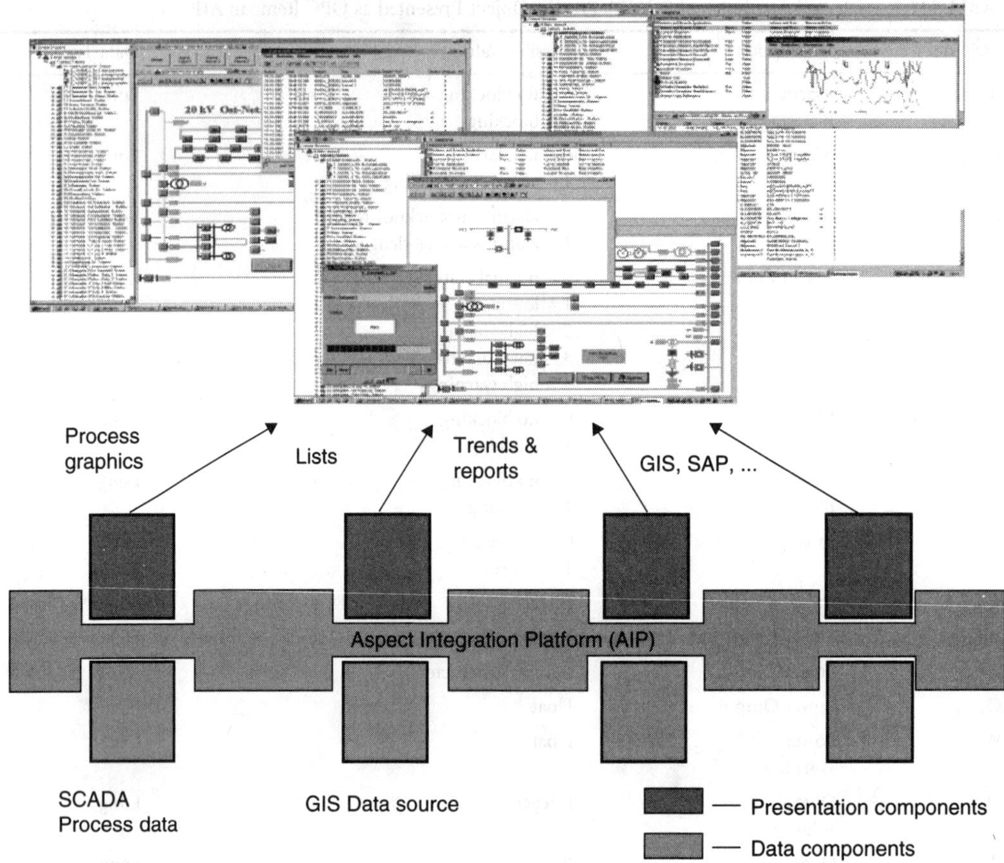

FIGURE 112.3 Industrial IT-based network management for power distribution and transformation applications.

Possible technical solutions for Interlocking and Sequential Control are:

- SCADA server-based solution (old approach) and
- AIP-based solution (new approach).

The AIP-based solution suggests that Interlocking and Sequential Control components are implemented as Aspect Systems in the AIP and, thus, can be easily reused. The schematic presentation of the given solution is shown in Figure 112.4(a). In this case, one can easily use Interlocking and Sequential Control aspects in Faceplates and distribute them between different objects using Aspect Object Model of AIP.

As an alternative to AIP-based solution, the native Interlocking and Sequential Control functionality of the given SCADA server can be used in the given network control application. In this case, the common API (Application Programming Interface) of the given SCADA can be used to access the Interlocking functions from external applications like AIP or others. The schematic presentation of the SCADA server-based solution is shown in Figure 112.4(b). The comparison analysis of AIP-based and SCADA server-based solutions for Interlocking and Sequential Control is presented in Table 112.2.

The main benefit of the SCADA server-based solution is that high robustness can be provided due to the fact that the Interlocking and Sequential Control modules run on the SCADA server (closer to the process in comparison to the AIP-based solution). The main drawback of this solution is that it can be used only with the given SCADA server. Additionally, some SCADA servers have a limited Interlocking and Sequential Control functionality, which is not easy to configure or reuse in other SCADA applications.

The main benefit of AIP-based Interlocking and Sequential Control solution is that it can be used with almost any SCADA server. Additionally, the implementation of the Interlocking and Sequential Control

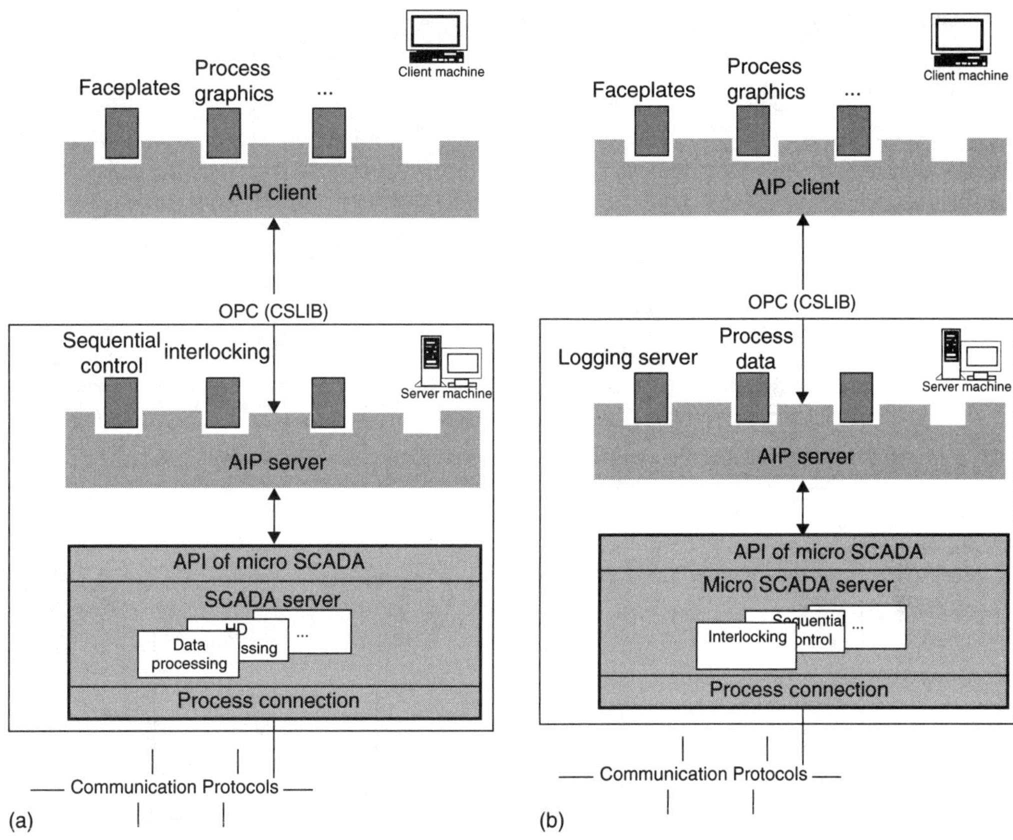

FIGURE 112.4 Interlocking and Sequential Control: (a) AIP-based solution and (b) SCADA server-based solution.

TABLE 112.2 Comparison Analysis of Possible Solutions for Interlocking and Sequential Control

Criterion/Solution	AIP-Based Solution	SCADA Server-Based Solution
Robustness and Stability	Middle	High
Reusability	High	Low
Easy to implement and maintain	Yes	No
Flexibility (easy to extend)	Yes	No

functionality based on the AIP is relatively easy due to the availability of many reuse opportunities of components that provide similar functionality. As an example, the use of the standard AIP's Property Translation Aspect allows implementing Interlocking using Logical Expressions and without any additional programming. Figure 112.5 presents schematically using AIP's Plant Explorer how the Logical Expression can be defined for a given valve in the gas network.

Typically, multifunctional AIP's Property Translation Aspect can be used to create text strings presenting the result of an evaluation or calculation of one or several AIP's object properties. The result is available as an OPC String Property. All Aspect Systems in the AIP can access this information as an ordinary OPC Item. The data collection can be done from any public aspect object property. This means that object properties in controllers and SCADA systems are accessible as data sources. Object properties are usually made available through OPC servers connected with Generic OPC Solution of AIP. The usage of the Property Translation Aspect is quite easy. One has to add this aspect to the object that has to contain

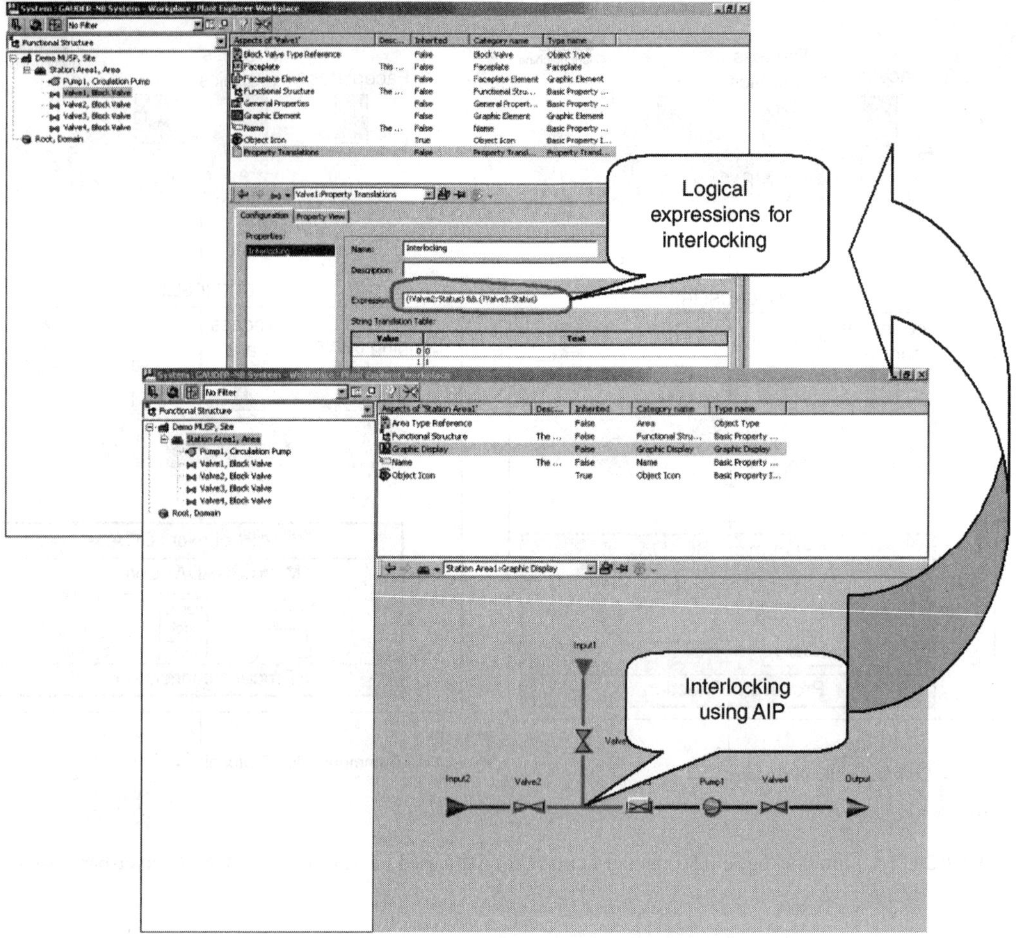

FIGURE 112.5   Interlocking using AIP's Property Translation Aspect.

an Interlocking rule and configure this aspect by setting Interlocking rule Name, Description, Values, and the logical expression that define the Interlocking rule.

112.5   Benefits

The use of ABB's Industrial IT AIP provided the following key benefits to the power distribution and transmission applications:

- component-based network control system architecture;
- companywide migration opportunities with a low investment;
- reduced effort for data engineering (40–50%);
- reuse (configuration is required) of Industrial IT AIP components like User Management, Faceplates, Workplaces, Navigation, Data Structuring, etc.;
- easy integration and reuse of common commercial and technical systems, like Enterprise Asset Management, Geographical Information System, and others on Industrial IT basis; and
- windows look & feel based on standard Industrial IT graphical user interface.

In this case, the integration and reuse based on Industrial IT allow locating appropriate and related information timely with minimum modifications in various existing legacy systems.

112.6 Summary

We have shown the use of ABB's Industrial IT AIP in power distribution and transmission applications, for instance, in SCADA systems. Some of the key customer benefits provided by Industrial IT include component-based network control system architecture and companywide migration opportunities with a low investment. Additionally, many reuse options became available for SCADA users. It is especially important for companies that wish to minimize their development costs in integration projects. ABB's Industrial IT Aspect Integrator Platform provided not only a run-time and design environment but also a number of standard ABB's and non-ABB's reusable components on the Industrial IT basis as well as a concept of how to easily build a common GUI of network control systems using Industrial IT.

References

1. Boyer, S., *SCADA: Supervisory Control and Data Acquisition*, ISA — The Instrumentation, Systems, and Automation Society, Triangle Park, NC, U.S.A., 1999.
2. Industrial IT — the next way of thinking, URL: http://www.abb.com.
3. Iwanitz, F. and Lange, J., *OLE for Process Control*, Huethig, Heidelberg, Germany, 2001.
4. Patrick, D. and Fardo, S., *Industrial Process Control Systems*, Delmar Publishing, Albany, NY, 1997.
5. Rytoft, C. and Normark, B., Industrial[IT] and the utility industry, *ABB Review: Focus for the Utility Industry*, 3, 23–28, 2002.
6. Sharma, R., Stearns, B., and Ng, T., *J2EE(TM) Connector Architecture and Enterprise Application Integration*, Pearson Education, Harlow, U.S.A., 2001.
7. Szyperski, C., *Component Software — Beyond Object-Oriented Programming*, Addison-Wesley, Harlow, U.S.A., 1998.

Indexes

Author Index